CROSS-REFERENCE INDEX OF HAZARDOUS CHEMICALS, SYNONYMS, AND CAS REGISTRY NUMBERS

Chemical and Environmental Safety and Health in Schools and Colleges Series

Compact School and College Administrator's Guide for Compliance with Federal and State Right-to-Know Regulations

Written Hazard Communication Program for Schools and Colleges

Concise Manuals of Chemical and Environmental Safety in Schools and Colleges

Volume 1 *Basic Principles*
Volume 2 *Hazardous Chemical Classes*
Volume 3 *Chemical Interactions*
Volume 4 *Safe Chemical Storage*
Volume 5 *Safe Chemical Disposal*

Pocket Guides to Chemical and Environmental Safety in Schools and Colleges (Five condensed, portable versions of the *Concise Manuals*, with the same volume numbers and titles as above)

Handbook of Chemical and Environmental Safety in Schools and Colleges

Compendium of Hazardous Chemicals in Schools and Colleges

List of Lists of Worldwide Hazardous Chemicals and Pollutants

Cross-Reference Index of Hazardous Chemicals, Synonyms, and CAS Registry Numbers

Index of Hazardous Contents of Commercial Products in Schools and Colleges

CROSS-REFERENCE INDEX OF HAZARDOUS CHEMICALS, SYNONYMS, AND CAS REGISTRY NUMBERS

The Forum for Scientific Excellence, Inc.

J. B. Lippincott Company
Philadelphia
Grand Rapids New York St. Louis San Francisco
London Sydney Tokyo

 Printed in the United States of America. For information write J. B. Lippincott Company, East Washington Square, Philadelphia, Pennsylvania 19105.

6 5 4 3 2 1

Library of Congress Cataloging-in-Publication Data

Cross-reference index of hazardous chemicals, synonyms, and CAS registry numbers.

p. cm. — (Chemical and environmental safety and health in schools and colleges series)

Includes bibliographical references.

ISBN 0-397-53021-8

1. Hazardous substances.—Handbooks, manuals, etc. I. Forum for Scientific Excellence, Inc. II. Series.

T55.3.H3C76 1990

604.7′.0216—dc20 89-28851

CIP

Foreword to the Series

The Forum for Scientific Excellence, Inc. (FSE) has spent many years developing an integrated hazardous chemical management program for schools and colleges. The elements incorporated into this program are the result of careful consideration of all existing regulatory requirements at both the federal and state levels, practical experience and awareness of the actual needs within the educational environment, and the best ideas from employee committees and management teams. As our on-the-job experience in schools and colleges has revealed new informational needs, we have expanded our data base and services to meet these needs.

This comprehensive school and college series has been consciously and carefully designed to benefit administrators and employees alike. In order to effectively modify poor employee work habits, often established over the course of many years, employees must be provided with hazardous chemical information in a variety of complementary ways that will remind them that their health and safety on the job depend on their own knowledge and work practices. Simultaneously, management needs to become more aware that job safety and health are good business. Administrators should implement not only "bare minimum" procedures, but also a responsible hazardous chemical management program that fulfills all three of the following objectives:

1. *Public Responsibility*: What must be done to properly protect employees, students, and contractors who work in an educational facility, and maintain public confidence.
2. *Tort Liability*: What can be done to minimize the potential for lawsuits.
3. *Regulatory Compliance*: What the laws require.

We have found that educational institutions often have neither enough time nor money to do the job right the first time, but always have enough time *and* money to do it over—after there has been an accident, lawsuit, or fine. Similarly, employees often "know a better way," even if it is *not* safer, until there is an accident on the job. Then they often retort: "*You* should have made me listen!"

Two books in this series have been specifically designed to increase the knowledge and awareness of school and college administrators:

Compact School and College Administrator's Guide for Compliance with Federal and State Right-to-Know Regulations

Written Hazard Communication Program for Schools and Colleges

Eleven books have been designed to progressively increase employee knowledge and awareness of hazardous chemicals:

Concise Manual of Chemical and Environmental Safety in Schools and Colleges

Volume 1 *Basic Principles*
Volume 2 *Hazardous Chemical Classes*
Volume 3 *Chemical Interactions*
Volume 4 *Safe Chemical Storage*
Volume 5 *Safe Chemical Disposal*

Pocket Guide to Chemical and Environmental Safety in Schools and Colleges (condensed editions of the five volumes listed above)

Handbook of Chemical and Environmental Safety in Schools and Colleges

Four reference books have been developed to give both administrators and employees access to complete hazard information for chemicals and products:

Compendium of Hazardous Chemicals in Schools and Colleges

List of Lists of Worldwide Hazardous Chemicals and Pollutants

Cross-Reference Index of Hazardous Chemicals, Synonyms, and CAS Registry Numbers

Index of Hazardous Contents of Commercial Products in Schools and Colleges

Since considerable experience and expertise have gone into the development and integration of this hazardous chemical management system, customized adjustments within your school or college should be carefully implemented to assure that all factors have been considered. Certainly, specific state or institutional regulations, or priorities, may warrant some modifications in this system, but we have given considerable thought and effort to accommodating as many contingencies as we could envision.

The breadth and depth of the series of publications will certainly leave considerable opportunity for improving areas of each book. We invite comments and suggestions that will improve these publications for future editions.

Dr. George R. Thompson, CEO
The Forum for Scientific Excellence, Inc.

CONTENTS

LIST OF ABBREVIATIONS

DOT	U.S. Department of Transportation
ACGIH	American Conference of Governmental Industrial Hygienists
IARC	International Agency for Research on Cancer
NTP	National Toxicology Program
NCI	National Cancer Institute
RCRA	Resource Conservation and Recovery Act
TSCA	Toxic Substance Control Act
ECRA	Environmental Cleanup and Responsibility Act
EPA	Environmental Protection Agency
OSHA	Occupational Safety and Health Act

CHAPTER 1

INTRODUCTION

The *Cross Reference Index of Hazardous Chemicals, Synonyms, and CAS Registry Numbers* provides a means for identifying hazardous chemical substances based on the substance name commonly used in chemistry and by industry. The need for such a dictionary became apparent when surveys for federal and state hazard communication were undertaken. A hazardous substance indicated on a product label by the manufacturer quite often is not the name that appears on various federal, state, and local regulatory lists of hazardous substances. Although numerous common names and trade names are included, this dictionary does not contain all of the available substance names for the many chemical substances listed.

This dictionary provides information to allow individuals to be able to identify hazardous substances in their workplace and to obtain a Chemical Abstract Service (CAS) registry number and/or an index name for these substances. The CAS registry number is used because of its universal use in literature. Once a substance is identified as hazardous, other sources, such as the *Compendium of Hazardous chemicals in Schools and Colleges* should be consulted to determine safe and proper handling of the substance and what regulations cover the substance's use, storage, and shipping.

The dictionary consists of two parts: the name section (Chapter 2) and the number section (Chapter 3). The name section lists, in alphabetic order, common names and associated CAS registry numbers. The number section lists, in CAS registry numerical order, all the synonyms from the name section associated with each CAS registry number and indicates the index name used in the *Compendium of Hazardous Chemicals in Schools and Colleges*.

NAME SECTION

The name section lists alphabetically common names, trade names, trivial names, U.S. Department of Transportation (DOT) number, Resource Conservation and Recovery Act (RCRA) waste number, etc. Complex names in excess of 100 characters have been truncated to 100 characters. Each substance name is associated with a CAS registry number. An extension appears with some substance names to indicate list index names, source of synonym, or give more specific detail.

Key to the Name Section

50-00-00	Formaldehyde	(DOT)
(1)	(2)	(3)

1. The *CAS registry number*.
2. The *substance name*, appearing in computerized alphabetic order. The many organic chemistry prefixes are *not* disregarded in this alphabetizing. These include *ortho-*, *meta-*, *para-*, *o-*, *m-*, *p-*, *sec-*, *tert-*, *cis-*, *trans-*, etc. as well as all numerals denoting chemical structure, numerals preceding characters, blanks, and brackets.
3. The *extension*, when it appears after the substance name, giving more information about the substance name. This may indicate a foreign spelling, such as Dutch or Italian, a source index name, such as DOT, ACGIH, 9CI, type of substance, such as pesticide, pharmaceutical, or chemical formula (CCl_4).

Uses of the Name Section

The name section can be used to identify a substance by a CAS registry number based on a commonly used name. This section lists over 35,000 synonyms for over 3,000 hazardous substances.

Many common names have become associated with more than one CAS registry number. In such cases, the user of this dictionary must determine which is the proper number for the particular substance being researched. Consulting the number section, where all names associated with the CAS registry numbers are grouped together, may provide further information to determine the correct CAS registry number. In compiling the dictionary, great effort was taken to try to provide accurate names and associated CAS registry numbers. In most cases discrepancies that may appear in spelling and CAS registry numbers are from the original sources. We encourage you to write to us concerning any errors observed and if possible provide corrected information for use in future editions and products.

NUMBER SECTION

The number section groups together all the names from the name section associated with a specific CAS registry number. The entries in this section appear in increasing CAS registry numerical order.

Key to the Number Section

50-00-00	*	Formaldehyde	(ACGIH)
(1)	(2)	(3)	(4)

1. The *CAS registry number*, appearing in increasing number order.
2. The *index name indicator*, indicates the index name used in the *compendium of Hazardous Chemicals for Schools and Colleges* in the series.
3. The *substance name,* appearing in computer alphabetical order for each CAS registry number.
4. The *extension*, when it appears indicating source index name, language, substance type, or chemical formula.

USES OF THE NUMBER SECTION

The number section can be used to identify synonymous common names for the 3,000 hazardous substances listed herein.

This section is useful for researchers looking for the various names of a given chemical for a literature search. This section is also useful to determine if a substance is regulated, by whom, and under what name. For example:

MOTH BALLS (DOT)
NAFTALEN (POLISH)
NAPHTHALENE (ACGIH, DOT)
UN 1334 (DOT)
UN 2304 (DOT)
RCRA Waste Number U165
NCI-C52904

A DOT extension indicates regulation by the U.S. Department of Transportation, giving both the number and index name. A RCRA Waste Number synonym indicates the substance falls under the Resource Conservation and Recovery Act. An NCI-C Number indicates the substance has been assigned a number by the National Cancer Institute. The 9CI and 8CI extensions indicate the index name used in the *Chemical Abstracts* subject indexes.

CHAPTER 2

CHEMICAL BY NAME

CAS No.	Chemical Name
7723-14-0	1 PALLET 120UF
2074-50-2	1',1'-DIMETHYL-4,4'-DIPYRIDINIUM DI(METHYL SULFATE)
16543-55-8	1'-NITROSONORNICOTINE
76-44-8	1(3A),4,5,6,7,8,8-HEPTACHLORO-3A(1),4,7,7A-TETRAHYDRO-4,7-METHANOINDENE
25251-51-8	1(3H)-ISOBENZOFURANONE, 3-[(1,1-DIMETHYLETHYL)DIOXY]-3-PHENYL-
122-20-3	1,1',1''-NITRILOTRI-2-PROPANOL
122-20-3	1,1',1''-NITRILOTRIS(2-PROPANOL)
52-24-4	1,1',1''-PHOSPHINOTHIOYLIDYNETRISAZIRIDINE
545-55-1	1,1',1''-PHOSPHINYLIDYNETRISAZIRIDINE
122-51-0	1,1',1'-(METHYLIDYNETRIS(OXY))TRIS(ETHANE)
72-43-5	1,1'-(2,2,2-TRICHLOROETHYLIDENE)BIS(4-METHOXYBENZENE)
85-00-7	1,1'-AETHYLEN-2,2'-BIPYRIDINIUM-DIBROMID (GERMAN)
1464-53-5	1,1'-BI(ETHYLENE OXIDE)
92-52-4	1,1'-BIPHENYL
827-52-1	1,1'-BIPHENYL, 1,2,3,4,5,6-HEXAHYDRO-
643-58-3	1,1'-BIPHENYL, 2-METHYL-
92-66-0	1,1'-BIPHENYL, 4-BROMO-
92-93-3	1,1'-BIPHENYL, 4-NITRO-
1336-36-3	1,1'-BIPHENYL, CHLORO DERIVS.
8004-13-5	1,1'-BIPHENYL, MIXT WITH 1,1'-OXYBIS(BENZENE)
28984-85-2	1,1'-BIPHENYL, NITRO-
2407-94-5	1,1'-DIHYDROXYDICYCLOHEXYL PEROXIDE
4685-14-7	1,1'-DIMETHYL-4,4'-BIPYRIDINIUM
2074-50-2	1,1'-DIMETHYL-4,4'-BIPYRIDINIUM BIS(METHYL SULFATE)
4685-14-7	1,1'-DIMETHYL-4,4'-BIPYRIDINIUM CATION
1910-42-5	1,1'-DIMETHYL-4,4'-BIPYRIDINIUM DICHLORIDE
2074-50-2	1,1'-DIMETHYL-4,4'-BIPYRIDINIUM DIMETHOSULFATE
2074-50-2	1,1'-DIMETHYL-4,4'-BIPYRIDYLIUM DIMETHYLSULFATE
2074-50-2	1,1'-DIMETHYL-4,4'-BIPYRIDYNIUM DIMETHYLSULFATE
2074-50-2	1,1'-DIMETHYL-4,4'-DIPYRIDINIUM DIMETHOSULFATE
1910-42-5	1,1'-DIMETHYL-4,4'-DIPYRIDINIUM-DICHLORID (GERMAN)
1910-42-5	1,1'-DIMETHYL-4,4'-DIPYRIDYLIUM CHLORIDE
1910-42-5	1,1'-DIMETHYL-4,4'-DIPYRIDYLIUM DICHLORIDE
92-52-4	1,1'-DIPHENYL
93-96-9	1,1'-DIPHENYLDIETHYL ETHER
2764-72-9	1,1'-ETHYLENE-2,2'-BIPYRIDINIUM
85-00-7	1,1'-ETHYLENE-2,2'-BIPYRIDINIUM DIBROMIDE
85-00-7	1,1'-ETHYLENE-2,2'-BIPYRIDYLIUM DIBROMIDE
85-00-7	1,1'-ETHYLENE-2,2'-DIPYRIDYLIUM DIBROMIDE
110-97-4	1,1'-IMINOBIS[2-PROPANOL]
110-97-4	1,1'-IMINODI-2-PROPANOL
101-68-8	1,1'-METHYLENEBIS[4-ISOCYANATOBENZENE]
111-44-4	1,1'-OXYBIS(2-CHLORO)ETHANE
142-96-1	1,1'-OXYBISBUTANE
109-93-3	1,1'-OXYBISETHENE
111-43-3	1,1'-OXYBIS[PROPANE]
25265-71-8	1,1'-OXYDI-2-PROPANOL
2407-94-5	1,1'-PEROXYDICYCLOHEXANOL
505-60-2	1,1'-THIOBIS(2-CHLOROETHANE)
67-72-1	1,1,1,2,2,2-HEXACHLOROETHANE
76-11-9	1,1,1,2-TETRACHLORO-2,2-DIFLUOROETHANE
116-16-5	1,1,1,3,3,3-HEXACHLORO-2-PROPANONE
116-16-5	1,1,1,3,3,3-HEXACHLOROPROPANONE
684-16-2	1,1,1,3,3,3-HEXAFLUORO-2-PROPANONE
999-97-3	1,1,1,3,3,3-HEXAMETHYLDISILAZANE
71-55-6	1,1,1-TCE
71-55-6	1,1,1-TRICHLOORETHAAN (DUTCH)
115-32-2	1,1,1-TRICHLOR-2,2-BIS(4-CHLORPHENYL)-AETHANOL (GERMAN)
72-43-5	1,1,1-TRICHLOR-2,2-BIS(4-METHOXY-PHENYL)-AETHAN (GERMAN)
71-55-6	1,1,1-TRICHLORAETHAN (GERMAN)
71-55-6	1,1,1-TRICHLORETHANE
50-29-3	1,1,1-TRICHLORO-2,2-BIS(4,4'-DICHLORODIPHENYL)ETHANE
72-43-5	1,1,1-TRICHLORO-2,2-BIS(4-METHOXYPHENYL)ETHANE
72-43-5	1,1,1-TRICHLORO-2,2-BIS(P-ANISYL)ETHANE
50-29-3	1,1,1-TRICHLORO-2,2-BIS(P-CHLOROPHENYL)ETHANE
72-43-5	1,1,1-TRICHLORO-2,2-BIS(P-METHOXYPHENOL)ETHANOL
72-43-5	1,1,1-TRICHLORO-2,2-BIS(P-METHOXYPHENYL)ETHANE
72-43-5	1,1,1-TRICHLORO-2,2-DI(4-METHOXYPHENYL)ETHANE
71-55-6	1,1,1-TRICHLOROETHANE
71-55-6	1,1,1-TRICHLOROETHANE (DOT)
71-55-6	1,1,1-TRICLOROETANO (ITALIAN)
463-82-1	1,1,1-TRIMETHYLETHANE
115-77-5	1,1,1-TRIS(HYDROXYMETHYL)ETHANOL
79-27-6	1,1,2,2-TETRABROMAETHAN (GERMAN)
79-27-6	1,1,2,2-TETRABROMOETANO (ITALIAN)
79-27-6	1,1,2,2-TETRABROMOETHANE
79-27-6	1,1,2,2-TETRABROMOETHYLENE
79-27-6	1,1,2,2-TETRABROOMETHAAN (DUTCH)
79-34-5	1,1,2,2-TETRACHLORETHANE
76-12-0	1,1,2,2-TETRACHLORO-1,2-DIFLUOROETHANE
79-34-5	1,1,2,2-TETRACHLOROETHANE
127-18-4	1,1,2,2-TETRACHLOROETHENE
127-18-4	1,1,2,2-TETRACHLOROETHYLENE
116-14-3	1,1,2,2-TETRAFLUOROETHYLENE
79-29-8	1,1,2,2-TETRAMETHYLETHANE
563-79-1	1,1,2,2-TETRAMETHYLETHYLENE
116-15-4	1,1,2,3,3,3-HEXAFLUORO-1-PROPENE
87-68-3	1,1,2,3,4,4-HEXACHLORO-1,3-BUTADIENE
76-13-1	1,1,2-TRICHLORO-1,2,2-TRIFLUOROETHANE
79-00-5	1,1,2-TRICHLOROETHANE
79-01-6	1,1,2-TRICHLOROETHYLENE
76-13-1	1,1,2-TRICHLOROTRIFLUOROETHANE
76-13-1	1,1,2-TRIFLUORO-1,2,2-TRICHLOROETHANE
79-38-9	1,1,2-TRIFLUORO-2-CHLOROETHYLENE
76-13-1	1,1,2-TRIFLUOROTRICHLOROETHANE
78-78-4	1,1,2-TRIMETHYLETHANE
513-35-9	1,1,2-TRIMETHYLETHYLENE
22288-43-3	1,1,3,3-TETRAMETHYL BUTYL PEROXY-2-ETHYL HEXANOATE, * TECHNICALLY PURE
107-45-9	1,1,3,3-TETRAMETHYLBUTANAMINE
5809-08-5	1,1,3,3-TETRAMETHYLBUTYL HYDROPEROXIDE
5809-08-5	1,1,3,3-TETRAMETHYLBUTYL HYDROPEROXIDE, TECHNICALLY PURE
5809-08-5	1,1,3,3-TETRAMETHYLBUTYL HYDROPEROXIDE, TECHNICALLY PURE (DOT)
107-45-9	1,1,3,3-TETRAMETHYLBUTYLAMINE
107-41-5	1,1,3-TRIMETHYLTRIMETHYLENEDIOL
3025-88-5	1,1,4,4-TETRAMETHYLTETRAMETHYLENE DIHYDROPEROXIDE
3006-86-8	1,1- DI-(TERT-BUTYLPEROXY)CYCLOHEXANE, 47% WITH INERT INORGANIC SOLID, WITH ≥ 13% PHLEGMATIZER
92-52-4	1,1-BIPHENYL
50-29-3	1,1-BIS(4-CHLOROPHENYL)-2,2,2-TRICHLOROETHANE
115-32-2	1,1-BIS(4-CHLOROPHENYL)-2,2,2-TRICHLOROETHANOL
72-54-8	1,1-BIS(4-CHLOROPHENYL)-2,2-DICHLOROETHANE
115-32-2	1,1-BIS(CHLOROPHENYL)-2,2,2-TRICHLOROETHANOL
50-29-3	1,1-BIS(P-CHLOROPHENYL)-2,2,2-TRICHLOROETHANE
115-32-2	1,1-BIS(P-CHLOROPHENYL)-2,2,2-TRICHLOROETHANOL
72-54-8	1,1-BIS(P-CHLOROPHENYL)-2,2-DICHLOROETHANE
72-55-9	1,1-BIS(P-CHLOROPHENYL)-2,2-DICHLOROETHYLENE
72-43-5	1,1-BIS(P-METHOXYPHENYL)-2,2,2-TRICHLOROETHANE
6731-36-8	1,1-BIS(TERT-BUTYLDIOXY)-3,3,5-TRIMETHYLCYCLOHEXANE
3006-86-8	1,1-BIS(TERT-BUTYLDIOXY)CYCLOHEXANE
6731-36-8	1,1-BIS(TERT-BUTYLPEROXY)-3,3,5-TRIMETHYLCYCLOHEXANE
3006-86-8	1,1-BIS(TERT-BUTYLPEROXY)CYCLOHEXANE
3006-86-8	1,1-DI(TERT—BUTYLPEROXY)CYCLOHEXANE
6731-36-8	1,1-DI-(tert-BUTYLPEROXY)-3,3,5-TRIMETHYL CYCLOHEXANE
6731-36-8	1,1-DI-(TERT-BUTYLPEROXY)-3,3,5-TRIMETHYL CYCLOHEXANE, 57% IN SOLUTION
6731-36-8	1,1-DI-(TERT-BUTYLPEROXY)-3,3,5-TRIMETHYLCYCLOHEXANE (DOT)
3006-86-8	1,1-DI-(TERT-BUTYLPEROXY)CYCLOHEXANE (DOT)
3006-86-8	1,1-DI-(TERT-BUTYLPEROXY)CYCLOHEXANE, NOT MORE THAN 77% IN SOLUTION (DOT)
3006-86-8	1,1-DI-(TERT-BUTYLPEROXY)CYCLOHEXANE, TECHNICALLY PURE
105-57-7	1,1-DIAETHOXY-AETHAN (GERMAN)
75-34-3	1,1-DICHLOOREETHAAN (DUTCH)
75-34-3	1,1-DICHLORAETHAN (GERMAN)
75-34-3	1,1-DICHLORETHANE
594-72-9	1,1-DICHLORO-1-NITROETHANE
595-44-8	1,1-DICHLORO-1-NITROPROPANE
72-54-8	1,1-DICHLORO-2,2-BIS(4-CHLOROPHENYL)ETHANE
72-54-8	1,1-DICHLORO-2,2-BIS(P-CHLOROPHENYL)ETHANE
72-55-9	1,1-DICHLORO-2,2-BIS(P-CHLOROPHENYL)ETHYLENE
72-55-9	1,1-DICHLORO-2,2-DI(P-CHLOROPHENYL)ETHYLENE
79-01-6	1,1-DICHLORO-2-CHLOROETHYLENE
622-44-6	1,1-DICHLORO-N-PHENYLMETHANIMINE
75-34-3	1,1-DICHLOROETHANE
75-34-3	1,1-DICHLOROETHANE (ACGIH,DOT)
75-35-4	1,1-DICHLOROETHENE
75-35-4	1,1-DICHLOROETHYLENE
78-99-9	1,1-DICHLOROPROPANE
75-34-3	1,1-DICLOROETANO (ITALIAN)
3054-95-3	1,1-DIETHOXY-2-PROPENE
105-57-7	1,1-DIETHOXY-ETHAAN (DUTCH)
105-57-7	1,1-DIETHOXYETHANE

CAS No.	Chemical Name
462-95-3	1,1-DIETHOXYMETHANE
760-21-4	1,1-DIETHYLETHENE
105-57-7	1,1-DIETOSSIETANO (ITALIAN)
75-68-3	1,1-DIFLUORO-1-CHLOROETHANE
25497-29-4	1,1-DIFLUORO-1-CHLOROETHANE
75-37-6	1,1-DIFLUOROETHANE
25497-28-3	1,1-DIFLUOROETHANE
75-37-6	1,1-DIFLUOROETHANE (DOT)
25497-28-3	1,1-DIFLUOROETHANE (DOT)
75-38-7	1,1-DIFLUOROETHENE
75-38-7	1,1-DIFLUOROETHYLENE
75-38-7	1,1-DIFLUOROETHYLENE (DOT)
76-11-9	1,1-DIFLUOROPERCHLOROETHANE
97-97-2	1,1-DIMETHOXY-2-CHLOROETHANE
534-15-6	1,1-DIMETHOXYETHANE
57-14-7	1,1-DIMETHYL HYDRAZINE
75-85-4	1,1-DIMETHYL-1-PROPANOL
4461-48-7	1,1-DIMETHYL-2-BUTENE
124-68-5	1,1-DIMETHYL-2-HYDROXYETHYLAMINE
115-19-5	1,1-DIMETHYL-2-PROPYN-1-OL
115-19-5	1,1-DIMETHYL-2-PROPYNOL
330-54-1	1,1-DIMETHYL-3-(3,4-DICHLOROPHENYL)UREA
926-56-7	1,1-DIMETHYLBUTADIENE
75-28-5	1,1-DIMETHYLETHANE
75-66-1	1,1-DIMETHYLETHANETHIOL
75-65-0	1,1-DIMETHYLETHANOL
540-88-5	1,1-DIMETHYLETHYL ACETATE
507-19-7	1,1-DIMETHYLETHYL BROMIDE
507-20-0	1,1-DIMETHYLETHYL CHLORIDE
75-91-2	1,1-DIMETHYLETHYL HYDROPEROXIDE
75-64-9	1,1-DIMETHYLETHYLAMINE
115-11-7	1,1-DIMETHYLETHYLENE
57-14-7	1,1-DIMETHYLHYDRAZIN (GERMAN)
57-14-7	1,1-DIMETHYLHYDRAZINE (ACGIH)
115-19-5	1,1-DIMETHYLPROPARGYL ALCOHOL
625-16-1	1,1-DIMETHYLPROPYL ACETATE
594-36-5	1,1-DIMETHYLPROPYL CHLORIDE
115-19-5	1,1-DIMETHYLPROPYNOL
126-33-0	1,1-DIOXIDETETRAHYDROTHIOFURAN
126-33-0	1,1-DIOXIDETETRAHYDROTHIOPHENE
126-33-0	1,1-DIOXOTHIOLAN
85-00-7	1,1-ETHYLENE 2,2-DIPYRIDYLIUM DIBROMIDE
75-34-3	1,1-ETHYLIDENE DICHLORIDE
101-68-8	1,1-METHYLENEBIS(4-ISOCYANATOBENZENE)
193-39-5	1,10-(1,2-PHENYLENE)PYRENE
193-39-5	1,10-(O-PHENYLENE)PYRENE
112-57-2	1,11-DIAMINO-3,6,9-TRIAZAUNDECANE
191-24-2	1,12-BENZOPERYLENE
191-24-2	1,12-BENZPERYLENE
2385-85-5	1,1A,2,2,3,3A,4,5,5,5A,5B,6-DODECACHLOROOCTAHYDRO-1,3,4-METHENO-1H-CYCLOBUTA(CD)PENTALENE
143-50-0	1,1A,3,3A,4,5,5,5A,5B,6-DECACHLOROOCTAHYDRO-1,3,4-METHENO-2H-CYCLOBUTA(CD)PENTALEN-2-ONE
79-00-5	1,2,2-TRICHLOROETHANE
79-01-6	1,2,2-TRICHLOROETHYLENE
76-13-1	1,2,2-TRICHLOROTRIFLUOROETHANE
30900-23-3	1,2,3,4,10,10-HEXACHLORO-1,4,4A,5,8,8A-HEXAHYDRO-EXO-1,4-ENDO-5,8-DIMETHANONAPHTHALENE
30900-23-3	1,2,3,4,10,10-HEXACHLORO-1,4,4A,5,8,8A-HEXAHYDRO-1,4,5,8-DIMETHANONAPHTHALENE
30900-23-3	1,2,3,4,10,10-HEXACHLORO-1,4,4A,5,8,8A-HEXAHYDRO-1,4-ENDO-EXO-5,8-DIMETHANONAPHTHALENE
77-47-4	1,2,3,4,5,5-HEXACHLORO-1,3-CYCLOPENTADIENE
2385-85-5	1,2,3,4,5,5-HEXACHLORO-1,3-CYCLOPENTADIENE DIMER
2234-13-1	1,2,3,4,5,6,7,8-OCTACHLORONAPHTHALENE
58-89-9	1,2,3,4,5,6-HEXACHLOROCYCLOHEXANE
608-73-1	1,2,3,4,5,6-HEXACHLOROCYCLOHEXANE
58-89-9	1,2,3,4,5,6-HEXACHLOROCYCLOHEXANE, GAMMA-ISOMER
115-29-7	1,2,3,4,7,7-HEXACHLOROBICYCLO(221)HEPTEN-5,6-BIOXYMETHYLENESULFITE
299-75-2	1,2,3,4-BUTANETETROL, 1,4-DIMETHANESULFONATE, [S-(R*,R*)]-
1464-53-5	1,2,3,4-DIEPOXYBUTANE
771-29-9	1,2,3,4-TETRAHYDRO-1-NAPHTHYL HYDROPEROXIDE
110-83-8	1,2,3,4-TETRAHYDROBENZENE
119-64-2	1,2,3,4-TETRAHYDRONAPHTHALENE
143-50-0	1,2,3,5,6,7,8,9,10,10-DECACHLORO(5210(SUP 2,6)0(SUP 3,9)0(SUP 5,8))DECANO-4-ONE
133-06-2	1,2,3,6-TETRAHYDRO-N-(TRICHLOROMETHYLTHIO)PHTHALIMIDE
100-50-5	1,2,3,6-TETRAHYDROBENZALDEHYDE
1321-16-0	1,2,3,6-TETRAHYDROBENZALDEHYDE
100-50-5	1,2,3,6-TETRAHYDROBENZALDEHYDE (DOT)
85-43-8	1,2,3,6-TETRAHYDROPHTHALIC ACID ANHYDRIDE
85-43-8	1,2,3,6-TETRAHYDROPHTHALIC ANHYDRIDE
694-05-3	1,2,3,6-TETRAHYDROPYRIDINE
61215-72-3	1,2,3,6-TETRAHYDROPYRIDINE
61215-72-3	1,2,3,6-TETRAHYDROPYRIDINE (DOT)
19408-74-3	1,2,3,7,8,9-HEXACHLORODIBENZO-P-DIOXIN
87-66-1	1,2,3-BENZENETRIOL
86-50-0	1,2,3-BENZOTRIAZIN-4(3H)-ONE, 3-(MERCAPTOMETHYL)-, O,O-DIMETHYL PHOSPHORODITHIOATE
86-50-0	1,2,3-BENZOTRIAZINE, PHOSPHORODITHIOIC ACID DERIV
2642-71-9	1,2,3-BENZOTRIAZINE, PHOSPHORODITHIOIC ACID DERIV
1185-57-5	1,2,3-PROPANETRICARBOXYLIC ACID, 2-HYDROXY-, AMMONIUM IRON(3+) SALT
1185-57-5	1,2,3-PROPANETRICARBOXYLIC ACID, 2-HYDROXY-, AMMONIUM IRON(III) SALT
3012-65-5	1,2,3-PROPANETRICARBOXYLIC ACID, 2-HYDROXY-, DIAMMONIUM SALT
56-81-5	1,2,3-PROPANETRIOL
1301-70-8	1,2,3-PROPANETRIOL, MONO(DIHYDROGEN PHOSPHATE), IRON(3+) SALT (3:2)
55-63-0	1,2,3-PROPANETRIOL, TRINITRATE
55-63-0	1,2,3-PROPANETRIYL NITRATE
96-19-5	1,2,3-TRICHLORO-1-PROPENE
87-61-6	1,2,3-TRICHLOROBENZENE
96-18-4	1,2,3-TRICHLOROPROPANE
96-19-5	1,2,3-TRICHLOROPROPENE
87-66-1	1,2,3-TRIHYDROXYBENZEN (CZECH)
87-66-1	1,2,3-TRIHYDROXYBENZENE
56-81-5	1,2,3-TRIHYDROXYPROPANE
12789-03-6	1,2,4,5,6,7,10,10-OCTACHLORO-4,7,8,9-TETRAHYDRO-4,7-METHYLENEINDANE
12789-03-6	1,2,4,5,6,7,8,8-OCTACHLOR-2,3,3A,4,7,7A-HEXAHYDRO-4,7-METHANOINDANE
12789-03-6	1,2,4,5,6,7,8,8-OCTACHLORO-2,3,3A,4,7,7A-HEXAHYDRO-4,7-METHANO-1H-INDENE
12789-03-6	1,2,4,5,6,7,8,8-OCTACHLORO-2,3,3A,4,7,7A-HEXAHYDRO-4,7-METHANOINDENE
12789-03-6	1,2,4,5,6,7,8,8-OCTACHLORO-3A,4,7,7A-HEXAHYDRO-4,7-METHYLENE INDANE
12789-03-6	1,2,4,5,6,7,8,8-OCTACHLORO-3A,4,7,7A-TETRAHYDRO-4,7-METHANOINDAN
12789-03-6	1,2,4,5,6,7,8,8-OCTACHLORO-3A,4,7,7A-TETRAHYDRO-4,7-METHANOINDANE
12789-03-6	1,2,4,5,6,7,8,8-OCTACHLORO-4,7-METHANO-3A,4,7,7A-TETRAHYDROINDANE
552-30-7	1,2,4-BENZENETRICARBOXYLIC ACID, ANHYDRIDE
552-30-7	1,2,4-BENZENETRICARBOXYLIC ACID, CYCLIC 1,2-ANHYDRIDE
552-30-7	1,2,4-BENZENETRICARBOXYLIC ANHYDRIDE
7421-93-4	1,2,4-METHENOCYCLOPENTA[CD]PENTALENE-5-CARBOXALDEHYDE, 2,2A,3,3,4,7-HEXACHLORODECAHYDRO-, (1ALPHA,
21087-64-9	1,2,4-TRIAZIN-5(4H)-ONE, 4-AMINO-6-(1,1-DIMETHYLETHYL)-3-(METHYLTHIO)-
61-82-5	1,2,4-TRIAZOLE-3-AMINE
120-82-1	1,2,4-TRICHLOROBENZENE
120-82-1	1,2,4-TRICHLOROBENZOL
95-63-6	1,2,4-TRIMETHYL BENZENE
226-36-8	1,2,5,6-DIBENZACRIDINE
226-36-8	1,2,5,6-DIBENZOACRIDINE
100-50-5	1,2,5,6-TETRAHYDROBENZALDEHYDE
61215-72-3	1,2,5,6-TETRAHYDROPYRIDINE
120-82-1	1,2,5-TRICHLOROBENZENE
95-63-6	1,2,5-TRIMETHYLBENZENE
224-42-0	1,2,7,8-DIBENZACRIDINE
206-44-0	1,2-(1,8-NAPHTHYLENE)BENZENE
120-58-1	1,2-(METHYLENEDIOXY)-4-PROPENYLBENZENE
94-58-6	1,2-(METHYLENEDIOXY)-4-PROPYLBENZENE
56-55-3	1,2-BENZANTHRACENE
56-55-3	1,2-BENZANTHRENE
95-83-0	1,2-BENZENEDIAMINE, 4-CHLORO-
85-44-9	1,2-BENZENEDICARBOXYLIC ACID ANHYDRIDE
84-72-0	1,2-BENZENEDICARBOXYLIC ACID, 2-ETHOXY-2-OXOETHYL ETHYL ESTER
85-71-2	1,2-BENZENEDICARBOXYLIC ACID, 2-ETHOXY-2-OXOETHYL METHYL ESTER
117-81-7	1,2-BENZENEDICARBOXYLIC ACID, BIS(2-ETHYLHEXYL) ESTER
85-68-7	1,2-BENZENEDICARBOXYLIC ACID, BUTYL PHENYLMETHYL ESTER

CAS No.	Chemical Name
131-17-9	1,2-BENZENEDICARBOXYLIC ACID, DI-2-PROPENYL ESTER
84-74-2	1,2-BENZENEDICARBOXYLIC ACID, DIBUTYL ESTER
84-66-2	1,2-BENZENEDICARBOXYLIC ACID, DIETHYL ESTER
3648-21-3	1,2-BENZENEDICARBOXYLIC ACID, DIHEPTYL ESTER
26761-40-0	1,2-BENZENEDICARBOXYLIC ACID, DIISODECYL ESTER
26761-40-0	1,2-BENZENEDICARBOXYLIC ACID, DIISODECYL ESTER (9CI)
131-11-3	1,2-BENZENEDICARBOXYLIC ACID, DIMETHYL ESTER
117-84-0	1,2-BENZENEDICARBOXYLIC ACID, DIOCTYL ESTER
85-44-9	1,2-BENZENEDICARBOXYLIC ANHYDRIDE
120-80-9	1,2-BENZENEDIOL
98-29-3	1,2-BENZENEDIOL, 4-(1,1-DIMETHYLETHYL)-
81-07-2	1,2-BENZISOTHIAZOL-3(2H)-ONE, 1,1-DIOXIDE
128-44-9	1,2-BENZISOTHIAZOL-3(2H)-ONE, 1,1-DIOXIDE, SODIUM SALT
81-07-2	1,2-BENZISOTHIAZOLIN-3-ONE, 1,1-DIOXIDE
128-44-9	1,2-BENZISOTHIAZOLIN-3-ONE, 1,1-DIOXIDE, SODIUM DERIV.
128-44-9	1,2-BENZISOTHIAZOLIN-3-ONE, 1,1-DIOXIDE, SODIUM SALT
56-55-3	1,2-BENZOANTHRACENE
218-01-9	1,2-BENZOPHENANTHRENE
128-44-9	1,2-BENZOTHIAZOL-3(2H)-ONE 1,1-DIOXIDE SODIUM SALT
218-01-9	1,2-BENZPHENANTHRENE
56-55-3	1,2-BENZ[A]ANTHRACENE
107-06-2	1,2-BICHLOROETHANE
1954-28-5	1,2-BIS(2-(2,3-EPOXYPROPOXY)ETHOXY)ETHANE
112-26-5	1,2-BIS(2-CHLOROETHOXY)ETHANE
112-27-6	1,2-BIS(2-HYDROXYETHOXY)ETHANE
112-26-5	1,2-BIS(CHLOROETHOXY)ETHANE
110-18-9	1,2-BIS(DIMETHYLAMINO)ETHANE
110-18-9	1,2-BIS(DIMETHYLAMINO)ETHANE (DOT)
106-88-7	1,2-BUTENE OXIDE
106-88-7	1,2-BUTYLENE EPOXIDE
106-88-7	1,2-BUTYLENE OXIDE
156-59-2	1,2-CIS-DICHLOROETHYLENE
107-06-2	1,2-DCE
110-18-9	1,2-DI(DIMETHYLAMINO)ETHANE
110-18-9	1,2-DI(DIMETHYLAMINO)ETHANE (DOT)
121-75-5	1,2-DI(ETHOXYCARBONYL)ETHYL O,O-DIMETHYL PHOSPHORO-DITHIOATE
111-55-7	1,2-DIACETOXYETHANE
629-14-1	1,2-DIAETHOXY-AETHEN (GERMAN)
95-83-0	1,2-DIAMINO-4-CHLOROBENZENE
107-15-3	1,2-DIAMINOETHANE
78-90-0	1,2-DIAMINOPROPANE
96-12-8	1,2-DIBROM-3-CHLOR-PROPAN (GERMAN)
106-93-4	1,2-DIBROMAETHAN (GERMAN)
300-76-5	1,2-DIBROMO-2,2-DICHLOROETHYL DIMETHYL PHOSPHATE
96-12-8	1,2-DIBROMO-3-CHLOROPROPANE
96-12-8	1,2-DIBROMO-3-CHLOROPROPANE (DOT)
96-12-8	1,2-DIBROMO-3-CLORO-PROPANO (ITALIAN)
106-93-4	1,2-DIBROMOETANO (ITALIAN)
106-93-4	1,2-DIBROMOETHANE
106-93-4	1,2-DIBROMOETHANE (DOT)
96-12-8	1,2-DIBROOM-3-CHLOORPROPAAN (DUTCH)
106-93-4	1,2-DIBROOMETHAAN (DUTCH)
107-06-2	1,2-DICHLOORETHAAN (DUTCH)
107-06-2	1,2-DICHLOR-AETHAN (GERMAN)
540-59-0	1,2-DICHLOR-AETHEN (GERMAN)
107-06-2	1,2-DICHLORETHANE
95-50-1	1,2-DICHLOROBENZENE
616-21-7	1,2-DICHLOROBUTANE
107-06-2	1,2-DICHLOROETHANE
540-59-0	1,2-DICHLOROETHENE
10140-87-1	1,2-DICHLOROETHYL ACETATE
540-59-0	1,2-DICHLOROETHYLENE
540-59-0	1,2-DICHLOROETHYLENE (ACGIH)
78-87-5	1,2-DICHLOROPROPANE
8003-19-8	1,2-DICHLOROPROPANE-1,3-DICHLOROPROPENE MIXT.
78-87-5	1,2-DICHLOROPROPYLENE
107-06-2	1,2-DICLOROETANO (ITALIAN)
629-14-1	1,2-DIETHOXY ETHENE
629-14-1	1,2-DIETHOXY ETHYLENE
629-14-1	1,2-DIETHOXYETHANE
1615-80-1	1,2-DIETHYL HYDRAZINE
135-01-3	1,2-DIETHYLBENZENE
76-12-0	1,2-DIFLUORO-1,1,2,2-TETRACHLOROETHANE
147-47-7	1,2-DIHYDRO-2,2,4-TRIMETHYLQUINOLINE
504-29-0	1,2-DIHYDRO-2-IMINOPYRIDINE
81-07-2	1,2-DIHYDRO-2-KETOBENZISOSULFONAZOLE
83-32-9	1,2-DIHYDROACENAPHTHYLENE
96-24-2	1,2-DIHYDROXY-3-CHLOROPROPANE
98-29-3	1,2-DIHYDROXY-4-TERT-BUTYLBENZENE
120-80-9	1,2-DIHYDROXYBENZENE
107-21-1	1,2-DIHYDROXYETHANE
57-55-6	1,2-DIHYDROXYPROPANE
110-71-4	1,2-DIMETHOXYETHANE
110-71-4	1,2-DIMETHOXYETHANE (DOT)
56-25-7	1,2-DIMETHYL-3,6-EPOXYPERHYDROPHTHALIC ANHYDRIDE
95-47-6	1,2-DIMETHYLBENZENE
540-73-8	1,2-DIMETHYLHYDRAZIN (GERMAN)
540-73-8	1,2-DIMETHYLHYDRAZINE
598-75-4	1,2-DIMETHYLPROPANOL
528-29-0	1,2-DINITROBENZENE
528-29-0	1,2-DINITROBENZOL
122-66-7	1,2-DIPHENYLHYDRAZINE
26447-14-3	1,2-EPOXY-3-(TOLYLOXY)PROPANE
106-92-3	1,2-EPOXY-3-ALLYLOXYPROPANE
3132-64-7	1,2-EPOXY-3-BROMOPROPANE
930-22-3	1,2-EPOXY-3-BUTENE
930-22-3	1,2-EPOXY-3-BUTYLENE
106-89-8	1,2-EPOXY-3-CHLOROPROPANE
4016-11-9	1,2-EPOXY-3-ETHOXYPROPANE
4016-11-9	1,2-EPOXY-3-ETHYLOXY PROPANE (DOT)
4016-11-9	1,2-EPOXY-3-ETHYLOXYPROPANE
556-52-5	1,2-EPOXY-3-HYDROXYPROPANE
4016-14-2	1,2-EPOXY-3-ISOPROPOXYPROPANE
122-60-1	1,2-EPOXY-3-PHENOXYPROPANE
106-87-6	1,2-EPOXY-4-(EPOXYETHYL)CYCLOHEXANE
106-88-7	1,2-EPOXYBUTANE
75-21-8	1,2-EPOXYETHANE
96-09-3	1,2-EPOXYETHYLBENZENE
75-56-9	1,2-EPOXYPROPANE
25322-69-4	1,2-EPOXYPROPANE POLYMER
107-15-3	1,2-ETHANEDIAMINE
39990-99-3	1,2-ETHANEDIAMINE, COMPD WITH LITHIUM ACETYLIDE (Li(C_2H))
112-24-3	1,2-ETHANEDIAMINE, N,N'-BIS(2-AMINOETHYL)-
110-18-9	1,2-ETHANEDIAMINE, N,N,N',N'-TETRAMETHYL-
110-18-9	1,2-ETHANEDIAMINE, N,N,N',N'-TETRAMETHYL- (9CI)
100-36-7	1,2-ETHANEDIAMINE, N,N-DIETHYL-
91-80-5	1,2-ETHANEDIAMINE, N,N-DIMETHYL-N'-2-PYRIDINYL-N'-(2-THIENYLMETHYL)-
111-40-0	1,2-ETHANEDIAMINE, N-(2-AMINOETHYL)-
112-57-2	1,2-ETHANEDIAMINE, N-(2-AMINOETHYL)-N'-(2-((2-AMINOETHYL) AMINO)ETHYL)-
107-21-1	1,2-ETHANEDIOL
111-55-7	1,2-ETHANEDIOL DIACETATE
110-71-4	1,2-ETHANEDIOL, DIMETHYL ETHER
628-96-6	1,2-ETHANEDIOL, DINITRATE
123-73-9	1,2-ETHANEDIOL, DIPROPANOATE (9CI)
4170-30-3	1,2-ETHANEDIOL, DIPROPANOATE (9CI)
107-22-2	1,2-ETHANEDIONE
947-02-4	1,2-ETHANEDITHIOL, CYCLIC ESTER WITH P,P-DIETHYL PHOS-PHONODITHIOIMIDOCARBONATE
947-02-4	1,2-ETHANEDITHIOL, CYCLIC S,S-ESTER WITH PHOSPHONO-DITHIOIMIDOCARBONIC ACID P,P-DIETHYL ESTER
142-59-6	1,2-ETHANEDIYLBISCARBAMODITHIOIC ACID DISODIUM SALT
110-17-8	1,2-ETHENEDICARBOXYLIC ACID, TRANS-
106-93-4	1,2-ETHYLENE DIBROMIDE
107-06-2	1,2-ETHYLENE DICHLORIDE
107-15-3	1,2-ETHYLENEDIAMINE
110-17-8	1,2-ETHYLENEDICARBOXYLIC ACID, (E)
107-06-2	1,2-ETHYLIDENE DICHLORIDE
107-41-5	1,2-HEXANEDIOL
100-79-8	1,2-ISOPROPYLIDENEGLYCERIN
646-06-0	1,2-ISOPROPYLIDENEGLYCERIN
100-79-8	1,2-ISOPROPYLIDENEGLYCEROL
646-06-0	1,2-ISOPROPYLIDENEGLYCEROL
94-58-6	1,2-METHYLENEDIOXY-4-PROPYL-BENZENE
100-79-8	1,2-O,O-ISOPROPYLIDENEGLYCERIN
646-06-0	1,2-O,O-ISOPROPYLIDENEGLYCERIN
100-79-8	1,2-O-ISOPROPYLIDENEGLYCEROL
646-06-0	1,2-O-ISOPROPYLIDENEGLYCEROL
1120-71-4	1,2-OXATHIOLANE 2,2-DIOXIDE
930-22-3	1,2-OXIDO-3-BUTENE
463-49-0	1,2-PROPADIENE
463-49-0	1,2-PROPADIENE (9CI)
78-90-0	1,2-PROPANEDIAMINE
56-18-8	1,2-PROPANEDIAMINE, N-(AMINOMETHYLETHYL)-
57-55-6	1,2-PROPANEDIOL

CAS No.	Chemical Name
999-61-1	1,2-PROPANEDIOL, 1-ACRYLATE
96-24-2	1,2-PROPANEDIOL, 3-CHLORO-
6423-43-4	1,2-PROPANEDIOL, DINITRATE
25584-83-2	1,2-PROPANEDIOL, MONOACRYLATE
27813-02-1	1,2-PROPANEDIOL, MONOMETHACRYLATE
950-10-7	1,2-PROPANEDITHIOL, CYCLIC ESTER WITH P,P-DIETHYL PHOSPHONODITHIOIMIDOCARBONATE
57-55-6	1,2-PROPYLENE GLYCOL
6423-43-4	1,2-PROPYLENE GLYCOL DINITRATE
75-56-9	1,2-PROPYLENE OXIDE
78-90-0	1,2-PROPYLENEDIAMINE
75-55-8	1,2-PROPYLENIMINE
156-60-5	1,2-TRANS-DICHLOROETHENE
156-60-5	1,2-TRANS-DICHLOROETHYLENE
95-47-6	1,2-XYLENE
1954-28-5	1,2:15,16-DIEPOXY-4,7,10,13-TETRAOXAHEXADECANE
192-65-4	1,2:4,5-DIBENZOPYRENE
53-70-3	1,2:5,6-BENZANTHRACENE
53-70-3	1,2:5,6-DIBENZANTHRACENE
53-70-3	1,2:5,6-DIBENZOANTHRACENE
53-70-3	1,2:5,6-DIBENZ[A]ANTHRACENE
189-55-9	1,2:7,8-DIBENZPYRENE
297-78-9	1,3,4,5,6,7,10,10-OCTACHLORO-4,7-ENDO-METHYLENE-4,7,8,9-TETRAHYDROPHTHALAN
297-78-9	1,3,4,5,6,7,8,8-OCTACHLORO-2-OXA-3A,4,7,7A-TETRAHYDRO-4,7-METHANOINDENE
297-78-9	1,3,4,5,6,8,8-OCTACHLORO-1,3,3A,4,7,7A-HEXAHYDRO-4,7-METHANOISOBENZOFURAN
2385-85-5	1,3,4-METHENO-1H-CYCLOBUTA(CD)PENTALENE, 1,1A,2,2,3,3A,4,5,5,5A,5B,6-DODECACHLOROOCTAHYDRO-
2385-85-5	1,3,4-METHENO-1H-CYCLOBUTA(CD)PENTALENE, DODECACHLOROOCTAHYDRO-
143-50-0	1,3,4-METHENO-2H-CYCLOBUTA(CD)PENTALEN-2-ONE, 1,1A,3,3A,4,5,5,5A,5B,6-DECACHLOROCTAHYDRO-
712-68-5	1,3,4-THIADIAZOL-2-AMINE, 5-(5-NITRO-2-FURANYL)-
712-68-5	1,3,4-THIADIAZOLE, 2-AMINO-5-(5-NITRO-2-FURYL)-
120-82-1	1,3,4-TRICHLOROBENZENE
95-63-6	1,3,4-TRIMETHYLBENZENE
629-20-9	1,3,5,7-CYCLOOCTATETRAENE
100-97-0	1,3,5,7-TETRAAZAADAMANTANE
101-25-7	1,3,5,7-TETRAAZABICYCLO(331)NONANE, 3,7-DINITROSO-
100-97-0	1,3,5,7-TETRAAZATRICYCLO[33113,7]DECANE
131-73-7	1,3,5,7-TETRAAZATRICYCLO[33113,7]DECANE
108-62-3	1,3,5,7-TETROXOCANE, 2,4,6,8-TETRAMETHYL-
37273-91-9	1,3,5,7-TETROXOCANE, 2,4,6,8-TETRAMETHYL-
9002-91-9	1,3,5,7-TETROXOCANE, 2,4,6,8-TETRAMETHYL-
3058-38-6	1,3,5-TRIAMINO-2,4,6-TRINITROBENZENE
121-82-4	1,3,5-TRIAZA-1,3,5-TRINITROCYCLOHEXANE
101-05-3	1,3,5-TRIAZIN-2-AMINE, 4,6-DICHLORO-N-(2-CHLOROPHENYL)-
12654-97-6	1,3,5-TRIAZIN-2-AMINE, 4,6-DICHLORO-N-(2-CHLOROPHENYL)-
108-77-0	1,3,5-TRIAZINE, 2,4,6-TRICHLORO-
675-14-9	1,3,5-TRIAZINE, 2,4,6-TRIFLUORO-
121-82-4	1,3,5-TRIAZINE, HEXAHYDRO-1,3,5-TRINITRO-
87-90-1	1,3,5-TRIAZINE-2,4,6(1H,3H,5H)-TRIONE, 1,3,5-TRICHLORO-
2782-57-2	1,3,5-TRIAZINE-2,4,6(1H,3H,5H)-TRIONE, 1,3-DICHLORO-
2244-21-5	1,3,5-TRIAZINE-2,4,6(1H,3H,5H)-TRIONE, 1,3-DICHLORO-, POTASSIUM SALT
2893-78-9	1,3,5-TRIAZINE-2,4,6(1H,3H,5H)-TRIONE, 1,3-DICHLORO-, SODIUM SALT
51580-86-0	1,3,5-TRIAZINE-2,4,6(1H,3H,5H)-TRIONE, 1,3-DICHLORO-, SODIUM SALT, DIHYDRATE
13057-78-8	1,3,5-TRIAZINE-2,4,6(1H,3H,5H)-TRIONE, 1-CHLORO-
1912-24-9	1,3,5-TRIAZINE-2,4-DIAMINE, 6-CHLORO-N-ETHYL-N'-(1-METHYLETHYL)-
1912-24-9	1,3,5-TRIAZINE-2,4-DIAMINE, 6-CHLORO-N-ETHYL-N'-(1-METHYLETHYL)- (9CI)
87-90-1	1,3,5-TRICHLORO-2,4,6-TRIOXOHEXAHYDRO-S-TRIAZINE
87-90-1	1,3,5-TRICHLORO-S-TRIAZINE-2,4,6(1H,3H,5H)-TRIONE
87-90-1	1,3,5-TRICHLOROISOCYANURIC ACID
108-77-0	1,3,5-TRICHLOROTRIAZINE
123-63-7	1,3,5-TRIMETHYL-2,4,6-TRIOXANE
108-67-8	1,3,5-TRIMETHYLBENZENE
108-67-8	1,3,5-TRIMETHYLBENZENE (DOT)
121-82-4	1,3,5-TRINITRO-1,3,5-TRIAZACYCLOHEXANE
99-35-4	1,3,5-TRINITROBENZENE
121-82-4	1,3,5-TRINITROHEXAHYDRO-1,3,5-TRIAZINE
121-82-4	1,3,5-TRINITROHEXAHYDRO-S-TRIAZINE
121-82-4	1,3,5-TRINITROPERHYDRO-1,3,5-TRIAZINE
110-88-3	1,3,5-TRIOXANE

CAS No.	Chemical Name
123-63-7	1,3,5-TRIOXANE, 2,4,6-TRIMETHYL-
9002-81-7	1,3,5-TRIOXANE, HOMOPOLYMER
108-45-2	1,3-BENZENEDIAMINE
39156-41-7	1,3-BENZENEDIAMINE, 4-METHOXY-, SULFATE (1:1)
95-80-7	1,3-BENZENEDIAMINE, 4-METHYL-
626-17-5	1,3-BENZENEDICARBONITRILE
121-91-5	1,3-BENZENEDICARBOXYLIC ACID
1477-55-0	1,3-BENZENEDIMETHANAMINE
108-46-3	1,3-BENZENEDIOL
519-44-8	1,3-BENZENEDIOL, 2,4-DINITRO-
35860-51-6	1,3-BENZENEDIOL, 2,4-DINITRO-
519-44-8	1,3-BENZENEDIOL, DINITRO-
35860-51-6	1,3-BENZENEDIOL, DINITRO-
626-17-5	1,3-BENZODINITRILE
120-58-1	1,3-BENZODIOXOLE, 5-(1-PROPENYL)-
94-59-7	1,3-BENZODIOXOLE, 5-(2-PROPENYL)-
94-58-6	1,3-BENZODIOXOLE, 5-PROPYL-
1477-55-0	1,3-BIS(AMINOMETHYL)BENZENE
106-99-0	1,3-BUTADIENE
25339-57-5	1,3-BUTADIENE
106-99-0	1,3-BUTADIENE (ACGIH)
25339-57-5	1,3-BUTADIENE (ACGIH)
1464-53-5	1,3-BUTADIENE DIEPOXIDE
930-22-3	1,3-BUTADIENE MONOEPOXIDE
930-22-3	1,3-BUTADIENE OXIDE
87-68-3	1,3-BUTADIENE, 1,1,2,3,4,4-HEXACHLORO-
1653-19-6	1,3-BUTADIENE, 2,3-DICHLORO-
126-99-8	1,3-BUTADIENE, 2-CHLORO-
78-79-5	1,3-BUTADIENE, 2-METHYL-
87-68-3	1,3-BUTADIENE, HEXACHLORO-
590-88-5	1,3-BUTANEDIAMINE
32280-46-9	1,3-BUTANEDIAMINE, N,N'-DIETHYL-
109-70-6	1,3-CHBP
542-92-7	1,3-CYCLOPENTADIENE
77-47-4	1,3-CYCLOPENTADIENE, 1,2,3,4,5,5-HEXACHLORO-
2385-85-5	1,3-CYCLOPENTADIENE, 1,2,3,4,5,5-HEXACHLORO-, DIMER
77-73-6	1,3-CYCLOPENTADIENE, DIMER
542-75-6	1,3-D
616-29-5	1,3-DIAMINO-2-HYDROXYPROPANE
616-29-5	1,3-DIAMINO-2-PROPANOL
2231-57-4	1,3-DIAMINO-2-THIOUREA
95-80-7	1,3-DIAMINO-4-METHYLBENZENE
108-45-2	1,3-DIAMINOBENZENE
590-88-5	1,3-DIAMINOBUTANE
109-76-2	1,3-DIAMINOPROPANE
542-75-6	1,3-DICHLORO-1-PROPENE
926-57-8	1,3-DICHLORO-2-BUTENE
7415-31-8	1,3-DICHLORO-2-BUTENE
96-23-1	1,3-DICHLORO-2-PROPANOL
534-07-6	1,3-DICHLORO-2-PROPANONE
542-75-6	1,3-DICHLORO-2-PROPENE
118-52-5	1,3-DICHLORO-5,5-DIMETHYL-2,4-IMIDAZOLIDINEDIONE
118-52-5	1,3-DICHLORO-5,5-DIMETHYLHYDANTOIN
2244-21-5	1,3-DICHLORO-S-TRIAZINE-2,4,6(1H,3H,5H)TRIONE POTASSIUM SALT
2782-57-2	1,3-DICHLORO-S-TRIAZINE-2,4,6-TRIONE
534-07-6	1,3-DICHLOROACETONE
534-07-6	1,3-DICHLOROACETONE (DOT)
541-73-1	1,3-DICHLOROBENZENE
926-57-8	1,3-DICHLOROBUTENE-2
142-28-9	1,3-DICHLOROPROPANE
96-23-1	1,3-DICHLOROPROPANOL
96-23-1	1,3-DICHLOROPROPANOL-2
542-75-6	1,3-DICHLOROPROPENE
542-75-6	1,3-DICHLOROPROPYLENE
626-17-5	1,3-DICYANOBENZENE
141-93-5	1,3-DIETHYLBENZENE
108-46-3	1,3-DIHYDROXYBENZENE
105-30-6	1,3-DIMETHYL BUTANOL
54972-97-3	1,3-DIMETHYL BUTANOL
591-21-9	1,3-DIMETHYL CYCLOHEXANE
1118-58-7	1,3-DIMETHYL-1,3-BUTADIENE
105-30-6	1,3-DIMETHYL-1-BUTANOL
54972-97-3	1,3-DIMETHYL-1-BUTANOL
108-38-3	1,3-DIMETHYLBENZENE
108-09-8	1,3-DIMETHYLBUTANAMINE
108-84-9	1,3-DIMETHYLBUTYL ACETATE
108-09-8	1,3-DIMETHYLBUTYLAMINE
108-09-8	1,3-DIMETHYLBUTYLAMINE (DOT)

CAS No.	*Chemical Name*
97-00-7	1,3-DINITRO-4-CHLOROBENZENE
25567-67-3	1,3-DINITRO-4-CHLOROBENZENE
99-65-0	1,3-DINITROBENZENE
99-65-0	1,3-DINITROBENZOL
100-79-8	1,3-DIOXACYCLOPENTANE
646-06-0	1,3-DIOXACYCLOPENTANE
552-30-7	1,3-DIOXO-5-PHTHALANCARBOXYLIC ACID
100-79-8	1,3-DIOXOLAN
646-06-0	1,3-DIOXOLAN
96-49-1	1,3-DIOXOLAN-2-ONE
100-79-8	1,3-DIOXOLANE
646-06-0	1,3-DIOXOLANE
126-39-6	1,3-DIOXOLANE, 2-ETHYL-2-METHYL-
100-79-8	1,3-DIOXOLANE-4-METHANOL, 2,2-DIMETHYL-
646-06-0	1,3-DIOXOLANE-4-METHANOL, 2,2-DIMETHYL-
100-79-8	1,3-DIOXOLE, DIHYDRO-
646-06-0	1,3-DIOXOLE, DIHYDRO-
85-44-9	1,3-DIOXOPHTHALAN
21548-32-3	1,3-DITHIETHANE, PHOSPHORAMIDIC ACID DERIV
950-10-7	1,3-DITHIOLANE, 2-(DIETHOXYPHOSPHINYLIMINO)-4-METHYL-
108-57-6	1,3-DIVINYLBENZENE
96-45-7	1,3-ETHYLENETHIOUREA
87-68-3	1,3-HEXACHLOROBUTADIENE
3691-35-8	1,3-INDANDIONE, 2-((P-CHLOROPHENYL)PHENYLACETYL)-
82-66-6	1,3-INDANDIONE, 2-(DIPHENYLACETYL)-
83-26-1	1,3-INDANDIONE, 2-PIVALOYL-
83-26-1	1,3-INDANDIONE, 2-PIVALYL-
85-44-9	1,3-ISOBENZOFURANDIONE
85-43-8	1,3-ISOBENZOFURANDIONE, 3A,4,7,7A-TETRAHYDRO-
504-60-9	1,3-PENTADIENE
504-60-9	1,3-PENTADIENE(CIS & TRANS-MIXED)
1118-58-7	1,3-PENTADIENE, 2-METHYL-
926-56-7	1,3-PENTADIENE, 4-METHYL-
149-31-5	1,3-PENTANEDIOL, 2-METHYL-
108-45-2	1,3-PHENYLENEDIAMINE
85-44-9	1,3-PHTHALANDIONE
109-76-2	1,3-PROPANEDIAMINE
104-78-9	1,3-PROPANEDIAMINE, N,N-DIETHYL-
109-55-7	1,3-PROPANEDIAMINE, N,N-DIMETHYL-
56-18-8	1,3-PROPANEDIAMINE, N-(3-AMINOPROPYL)-
3312-60-5	1,3-PROPANEDIAMINE, N-CYCLOHEXYL-
115-77-5	1,3-PROPANEDIOL, 2,2-BIS(HYDROXYMETHYL)-
115-76-4	1,3-PROPANEDIOL, 2,2-DIETHYL-
115-84-4	1,3-PROPANEDIOL, 2-BUTYL-2-ETHYL-
1120-71-4	1,3-PROPANESULTONE
156-87-6	1,3-PROPANOLAMINE
57-57-8	1,3-PROPIOLACTONE
109-76-2	1,3-PROPYLENEDIAMINE
109-76-2	1,3-TRIMETHYLENEDIAMINE
108-38-3	1,3-XYLENE
1477-55-0	1,3-XYLYLENEDIAMINE
76-44-8	1,4,5,6,7,10,10-HEPTACHLORO-4,7,8,9-TETRAHYDRO-4,7-ENDO-METHYLENEINDENE
76-44-8	1,4,5,6,7,10,10-HEPTACHLORO-4,7,8,9-TETRAHYDRO-4,7-METHYL-ENEINDENE
115-29-7	1,4,5,6,7,7-HEXACHLORO-5-NORBORNENE-2,3-DIMETHANOL CYCLIC SULFITE
76-44-8	1,4,5,6,7,8,8-EPTACLORO-3A,4,7,7A-TETRAIDRO-4,7-ENDO-METANO-INDENE (ITALIAN)
76-44-8	1,4,5,6,7,8,8-HEPTACHLOOR-3A,4,7,7A-TETRAHYDRO-4,7-ENDO-METHANO-INDEEN (DUTCH)
76-44-8	1,4,5,6,7,8,8-HEPTACHLOR-3A,4,7,7,7A-TETRAHYDRO-4,7-ENDO-METHANO-INDEN (GERMAN)
76-44-8	1,4,5,6,7,8,8-HEPTACHLORO-3A,4,7,7,7A-TETRAHYDRO-4,7-METHYLENE INDENE
76-44-8	1,4,5,6,7,8,8-HEPTACHLORO-3A,4,7,7A-TETRAHYDRO-4,7-ENDO-METHANOINDENE
76-44-8	1,4,5,6,7,8,8-HEPTACHLORO-3A,4,7,7A-TETRAHYDRO-4,7-METHANOINDENE
76-44-8	1,4,5,6,7,8,8-HEPTACHLORO-3A,4,7,7A-TETRAHYDRO-4,7-METHANOL-1H-INDENE
76-44-8	1,4,5,6,7,8,8A-HEPTACHLORO-3A,4,7,7A-TETRAHYDRO-4,7-METHANOINDANE
112-57-2	1,4,7,10,13-PENTAAZATRIDECANE
112-24-3	1,4,7,10-TETRAAZADECANE
111-40-0	1,4,7-TRIAZAHEPTANE
106-50-3	1,4-BENZENEDIAMINE
3081-14-9	1,4-BENZENEDIAMINE, N,N'-BIS(1,4-DIMETHYLPENTYL)-
101-96-2	1,4-BENZENEDIAMINE, N,N'-BIS(1-METHYLPROPYL)-

CAS No.	*Chemical Name*
93-05-0	1,4-BENZENEDIAMINE, N,N-DIETHYL-
99-98-9	1,4-BENZENEDIAMINE, N,N-DIMETHYL-
99-98-9	1,4-BENZENEDIAMINE, N,N-DIMETHYL- (9CI)
100-20-9	1,4-BENZENEDICARBONYL CHLORIDE
100-20-9	1,4-BENZENEDICARBONYL DICHLORIDE
120-61-6	1,4-BENZENEDICARBOXYLIC ACID, DIMETHYL ESTER
120-61-6	1,4-BENZENEDICARBOXYLIC ACID, DIMETHYL ESTER (9CI)
123-31-9	1,4-BENZENEDIOL
106-51-4	1,4-BENZOQUINE
106-51-4	1,4-BENZOQUINONE
577-11-7	1,4-BIS(2-ETHYLHEXYL) SODIUM SULFOSUCCINATE
111-50-2	1,4-BIS(CHLOROCARBONYL)BUTANE
55-98-1	1,4-BIS(METHANESULFONYLOXY)BUTANE
55-98-1	1,4-BIS[METHANESULFONOXY]BUTANE
124-04-9	1,4-BUTANEDICARBOXYLIC ACID
110-63-4	1,4-BUTANEDIOL
55-98-1	1,4-BUTANEDIOL DIMESYLATE
55-98-1	1,4-BUTANEDIOL DIMETHANESULPHONATE
55-98-1	1,4-BUTANEDIOL DIMETHYLSULFONATE
55-98-1	1,4-BUTANEDIOL, DIMETHANESULFONATE
110-63-4	1,4-BUTYLENE GLYCOL
110-65-6	1,4-BUTYNEDIOL
10070-67-0	1,4-BUTYNEDIOL
110-65-6	1,4-BUTYNEDIOL (DOT)
10070-67-0	1,4-BUTYNEDIOL (DOT)
106-51-4	1,4-CYCLOHEXADIENE DIOXIDE
106-51-4	1,4-CYCLOHEXADIENEDIONE
106-50-3	1,4-DIAMINOBENZENE
106-50-3	1,4-DIAMINOBENZOL
280-57-9	1,4-DIAZABICYCLO(2,2,2)OCTANE
280-57-9	1,4-DIAZABICYCLOOCTANE
110-85-0	1,4-DIAZACYCLOHEXANE
106-46-7	1,4-DICHLOORBENZEEN (DUTCH)
106-46-7	1,4-DICHLOR-BENZOL (GERMAN)
106-46-7	1,4-DICHLOROBENZENE
110-56-5	1,4-DICHLOROBUTANE
106-46-7	1,4-DICLOROBENZENE (ITALIAN)
111-69-3	1,4-DICYANOBUTANE
105-05-5	1,4-DIETHYLBENZENE
123-91-1	1,4-DIETHYLENE DIOXIDE
110-85-0	1,4-DIETHYLENEDIAMINE
110-65-6	1,4-DIHYDROXY-2-BUTYNE
10070-67-0	1,4-DIHYDROXY-2-BUTYNE
123-31-9	1,4-DIHYDROXY-BENZEEN (DUTCH)
123-31-9	1,4-DIHYDROXY-BENZOL (GERMAN)
123-31-9	1,4-DIHYDROXYBENZEN (CZECH)
123-31-9	1,4-DIHYDROXYBENZENE
110-63-4	1,4-DIHYDROXYBUTANE
123-31-9	1,4-DIIDROBENZENE (ITALIAN)
55-98-1	1,4-DIMETHANESULFONOXYBUTANE
106-42-3	1,4-DIMETHYLBENZENE
589-90-2	1,4-DIMETHYLCYCLOHEXANE
589-90-2	1,4-DIMETHYLCYCLOHEXANE (DOT)
624-29-3	1,4-DIMETHYLCYCLOHEXANE-CIS
2207-04-7	1,4-DIMETHYLCYCLOHEXANE-TRANS
55-98-1	1,4-DIMETHYLSULFONYLOXYBUTANE
100-25-4	1,4-DINITROBENZENE
78-34-2	1,4-DIOSSAN-2,3-DIYL-BIS(O,O-DIETIL-DITIOFOSFATO) (ITALIAN)
106-51-4	1,4-DIOSSIBENZENE (ITALIAN)
78-34-2	1,4-DIOXAAN-2,3-DIYL-BIS(O,O-DIETHYL-DITHIOFOSFAAT) (DUTCH)
123-91-1	1,4-DIOXACYCLOHEXANE
123-91-1	1,4-DIOXAN
78-34-2	1,4-DIOXAN-2,3-DIYL BIS(O,O-DIETHYL PHOSPHOROTHIOLOTHIO-NATE)
78-34-2	1,4-DIOXAN-2,3-DIYL BIS(O,O-DIETHYLPHOSPHOROTHIOLOTHIO-NATE)
78-34-2	1,4-DIOXAN-2,3-DIYL O,O,O',O'-TETRAETHYL DI(PHOS-PHOROMITHIOATE)
78-34-2	1,4-DIOXAN-2,3-DIYL-BIS(O,O-DIAETHYL-DITHIOPHOSPHAT) (GERMAN)
123-91-1	1,4-DIOXANE
25136-55-4	1,4-DIOXANE, DIMETHYL-
123-91-1	1,4-DIOXIN, TETRAHYDRO-
106-51-4	1,4-DIOXY-BENZOL (GERMAN)
106-51-4	1,4-DIOXYBENZENE
280-57-9	1,4-ETHYLENEPIPERAZINE
592-45-0	1,4-HEXADIENE
117-80-6	1,4-NAPHTHALENEDIONE, 2,3-DICHLORO-

CAS No.	*Chemical Name*
84-80-0	1,4-NAPHTHALENEDIONE, 2-METHYL-3-(3,7,11,15-TETRAMETHYL-2-HEXADECENYL)-
84-80-0	1,4-NAPHTHALENEDIONE, 2-METHYL-3-(3,7,11,15-TETRAMETHYL-2-HEXADECENYL)-, [R-[R*,R*-(E)]]-
117-80-6	1,4-NAPHTHOQUINONE, 2,3-DICHLORO-
84-80-0	1,4-NAPHTHOQUINONE, 2-METHYL-3-PHYTYL-
15980-15-1	1,4-OXATHIANE
15980-15-1	1,4-OXATHIIN, 2,3,5,6-TETRAHYDRO-
140-80-7	1,4-PENTANEDIAMINE, N(SUP 1),N(SUP 1)-DIETHYL-
140-80-7	1,4-PENTANEDIAMINE, N1,N1-DIETHYL-
106-50-3	1,4-PHENYLENEDIAMINE
110-85-0	1,4-PIPERAZINE
15980-15-1	1,4-THIOXANE
106-42-3	1,4-XYLENE
309-00-2	1,4:5,8-DIMETHANONAPHTHALENE, 1,2,3,4,10,10-HEXACHLORO-1,4,4 A,5,8,8A-HEXAHYDRO-, (1ALPHA,4ALPHA,4AB
30900-23-3	1,4:5,8-DIMETHANONAPHTHALENE, 1,2,3,4,10,10-HEXACHLORO-1,4,4 A,5,8,8A-HEXAHYDRO-, ENDO, EXO-
309-00-2	1,4:5,8-DIMETHANONAPHTHALENE, 1,4,4A,5,8,8A-HEXAHYDRO-1,2, 3,4,10,10-HEXACHLORO-, ENDO, EXOMIXTURE (
78-59-1	1,5,5-TRIMETHYL-3-OXOCYCLOHEXENE
4904-61-4	1,5,9-CYCLODODECATRIENE
56-18-8	1,5,9-TRIAZANONANE
111-78-4	1,5-CYCLOOCTADIENE
1552-12-1	1,5-CYCLOOCTADIENE
111-40-0	1,5-DIAMINO-3-AZAPENTANE
55-86-7	1,5-DICHLORO-3-METHYL-3-AZAPENTANE HYDROCHLORIDE
111-44-4	1,5-DICHLORO-3-OXAPENTANE
628-76-2	1,5-DICHLOROPENTANE
101-25-7	1,5-ENDOMETHYLENE-3,7-DINITROSO-1,3,5,7-TETRAAZACY-CLOOCTANE
821-08-9	1,5-HEXADIEN-3-YNE
101-25-7	1,5-METHYLENE-3,7-DINITROSO-1,3,5,7-TETRAAZACYCLOOCTANE
111-30-8	1,5-PENTANEDIAL
111-30-8	1,5-PENTANEDIONE
124-09-4	1,6-DIAMINO-N-HEXANE
124-09-4	1,6-DIAMINOHEXANE
822-06-0	1,6-DIISOCYANATOHEXANE
822-06-0	1,6-HEXAMETHYLENE DIISOCYANATE
124-09-4	1,6-HEXAMETHYLENEDIAMINE
124-09-4	1,6-HEXANEDIAMINE
4835-11-4	1,6-HEXANEDIAMINE, N,N'-DIBUTYL-
25620-58-0	1,6-HEXANEDIAMINE, TRIMETHYL-
124-04-9	1,6-HEXANEDIOIC ACID
822-06-0	1,6-HEXANEDIOL DIISOCYANATE
105-60-2	1,6-HEXOLACTAM
822-06-0	1,6-HEXYLENE DIISOCYANATE
124-09-4	1,6-HEXYLENEDIAMINE
4835-11-4	1,6-N,N'-DIBUTYLHEXANEDIAMINE
2001-95-8	1,7,13,19,25,31-HEXAOXA-4,10,16,22,28,34-HEXAAZACYCLOHEXA-TRIACONTANE, CYCLIC PEPTIDE DERIV
76-22-2	1,7,7-TRIMETHYLBICYCLO(221)-2-HEPTANONE
76-22-2	1,7,7-TRIMETHYLNORCAMPHOR
56-18-8	1,7-DIAMINO-4-AZAHEPTANE
138-86-3	1,8(9)-P-MENTHADIENE
112-24-3	1,8-DIAMINO-3,6-DIAZAOCTANE
112-26-5	1,8-DICHLORO-3,6-DIOXAOCTANE
83-32-9	1,8-ETHYLENENAPHTHALENE
138-86-3	1,8-P-MENTHADIENE
86-88-4	1-(1-NAPHTHYL)-2-THIOUREA
30553-04-9	1-(1-NAPHTHYL)-2-THIOUREA
86-88-4	1-(1-NAPHTHYL)THIOUREA
6358-53-8	1-(2,5-DIMETHOXYPHENYLAZO)-2-NAPHTHOL
140-31-8	1-(2-AMINOETHYL)PIPERAZINE
124-16-3	1-(2-BUTOXYETHOXY)-2-PROPANOL
13010-47-4	1-(2-CHLOROETHYL)-3-CYCLOHEXYL-1-NITROSOUREA
13010-47-4	1-(2-CHLOROETHYL)-3-CYCLOHEXYLNITROSOUREA
5344-82-1	1-(2-CHLOROPHENYL)-2-THIOUREA
5344-82-1	1-(2-CHLOROPHENYL)THIOUREA
111-41-1	1-(2-HYDROXYETHYLAMINO)-2-AMINOETHANE
330-54-1	1-(3,4-DICHLOROPHENYL)-3,3-DIMETHYLUREA
330-54-1	1-(3,4-DICHLOROPHENYL)-3,3-DIMETHYLUREE (FRENCH)
5836-73-7	1-(3,4-DICHLOROPHENYL)-3-TRIAZENETHIO-CARBOXAMIDE
53558-25-1	1-(3-PYRIDYLMETHYL)-3-(4-NITROPHENYL)UREA
81-81-2	1-(4'-HYDROXY-3'-COUMARINYL)-1-PHENYL-3-BUTANONE
1982-47-4	1-(4-(4-CHLORO-PHENOXY)PHENYL)-3,3-D'METHYLUREE (FRENCH)
3691-35-8	1-(4-CHLORPHENYL)-1-PHENYL-ACETYL-INDAN-1,3-DION (GER-MAN)
555-84-0	1-(5-NITROFURFURYLIDENE)AMINO)-2-IMIDAZOLIDINONE
555-84-0	1-(5-NITROFURFURYLIDENEAMINO)IMIDAZOLIDIN-2-ONE
124-16-3	1-(BUTOXYETHOXY)-2-PROPANOL
100-14-1	1-(CHLOROMETHYL)-4-NITROBENZENE
3312-60-5	1-(CYCLOHEXYLAMINO)-3-AMINOPROPANE
140-80-7	1-(DIETHYLAMINO)-4-AMINOPENTANE
108-16-7	1-(DIMETHYLAMINO)-2-PROPANOL
138-89-6	1-(DIMETHYLAMINO)-4-NITROSOBENZENE
106-87-6	1-(EPOXYETHYL)-3,4-EPOXYCYCLOHEXANE
96-20-8	1-(HYDROXYMETHYL)PROPYLAMINE
22224-92-6	1-(METHYLETHYL)-ETHYL 3-METHYL-4-(METHYLTHIO)PHENYL PHOSPHORAMIDATE
5344-82-1	1-(O-CHLOROPHENYL)THIOUREA
53-86-1	1-(P-CHLOROBENZOYL)-2-METHYL-5-METHOXY-3-INDOLE-ACETIC ACID
53-86-1	1-(P-CHLOROBENZOYL)-2-METHYL-5-METHOXYINDOLE-3-ACETIC ACID
53-86-1	1-(P-CHLOROBENZOYL)-5-METHOXY-2-METHYLINDOLE-3-ACETIC ACID
102-01-2	1-(PHENYLCARBAMOYL)-2-PROPANONE
98-07-7	1-(TRICHLOROMETHYL)BENZENE
2243-62-1	1-5-NAPHTHALENEDIAMINE
103-89-9	1-ACETAMIDO-4-METHYLBENZENE
97-36-9	1-ACETOACETYLAMINO-2,4-DIMETHYLBENZENE
92-15-9	1-ACETOACETYLAMINO-2-METHOXYBENZENE
108-22-5	1-ACETOXY-1-METHYLETHYLENE
542-58-5	1-ACETOXY-2-CHLOROETHANE
111-15-9	1-ACETOXY-2-ETHOXYETHANE
108-05-4	1-ACETOXYETHYLENE
109-60-4	1-ACETOXYPROPANE
28314-03-6	1-ACETYLAMINOFLUORENE
106-92-3	1-ALLILOSSI-2,3 EPOSSIPROPANO (ITALIAN)
94-59-7	1-ALLYL-3,4-METHYLENEDIOXYBENZENE
106-92-3	1-ALLYLOXY-2,3-EPOXY-PROPAAN (DUTCH)
106-92-3	1-ALLYLOXY-2,3-EPOXYPROPAN (GERMAN)
106-92-3	1-ALLYLOXY-2,3-EPOXYPROPANE
58-89-9	1-ALPHA,2-ALPHA,3-BETA,4-ALPHA,5-ALPHA,6-BETA-HEXA-CHLOROCYCLOHEXANE
98-84-0	1-AMINO-1-PHENYLETHANE
88-05-1	1-AMINO-2,4,6-TRIMETHYL BENZENE
88-05-1	1-AMINO-2,4,6-TRIMETHYLBENZEN (CZECH)
97-02-9	1-AMINO-2,4-DINITROBENZENE
102-56-7	1-AMINO-2,5-DIMETHOXYBENZENE
87-62-7	1-AMINO-2,6-DIMETHYLBENZENE
929-06-6	1-AMINO-2-(2-HYDROXYETHOXY)ETHANE
100-36-7	1-AMINO-2-DIETHYLAMINOETHANE
104-75-6	1-AMINO-2-ETHYLHEXAN (CZECH)
104-75-6	1-AMINO-2-ETHYLHEXANE
141-43-5	1-AMINO-2-HYDROXYETHANE
78-96-6	1-AMINO-2-HYDROXYPROPANE
120-71-8	1-AMINO-2-METHOXY-5-METHYLBENZENE
99-59-2	1-AMINO-2-METHOXY-5-NITROBENZENE
90-04-0	1-AMINO-2-METHOXYBENZENE
29191-52-4	1-AMINO-2-METHOXYBENZENE
82-28-0	1-AMINO-2-METHYLANTHRAQUINONE
95-53-4	1-AMINO-2-METHYLBENZENE
78-81-9	1-AMINO-2-METHYLPROPANE
2038-03-1	1-AMINO-2-MORPHOLINOETHANE
89-62-3	1-AMINO-2-NITRO-4-METHYLBENZENE
88-74-4	1-AMINO-2-NITROBENZENE
78-96-6	1-AMINO-2-PROPANOL
79-19-6	1-AMINO-2-THIOUREA
104-78-9	1-AMINO-3-(DIETHYLAMINO)PROPANE
109-55-7	1-AMINO-3-(DIMETHYLAMINO)PROPANE
98-16-8	1-AMINO-3-(TRIFLUOROMETHYL)BENZENE
156-87-6	1-AMINO-3-HYDROXYPROPANE
5332-73-0	1-AMINO-3-METHOXYPROPANE
108-44-1	1-AMINO-3-METHYLBENZENE
123-00-2	1-AMINO-3-MORPHOLINOPROPANE
119-32-4	1-AMINO-3-NITRO-4-METHYLBENZENE
156-87-6	1-AMINO-3-PROPANOL
99-98-9	1-AMINO-4-(DIMETHYLAMINO)BENZENE
27134-26-5	1-AMINO-4-CHLORO BENZENE
371-40-4	1-AMINO-4-FLUOROBENZENE
123-30-8	1-AMINO-4-HYDROXYBENZENE
104-94-9	1-AMINO-4-METHOXYBENZENE
106-49-0	1-AMINO-4-METHYLBENZENE
100-01-6	1-AMINO-4-NITROBENZENE
29757-24-2	1-AMINO-4-NITROBENZENE

CAS No.	Chemical Name
109-73-9	1-AMINO-BUTAAN (DUTCH)
109-73-9	1-AMINOBUTAN (GERMAN)
109-73-9	1-AMINOBUTANE
108-91-8	1-AMINOCYCLOHEXANE
2016-57-1	1-AMINODECANE
75-04-7	1-AMINOETHANE
75-39-8	1-AMINOETHANOL
140-31-8	1-AMINOETHYLPIPERAZINE
111-68-2	1-AMINOHEPTANE
111-26-2	1-AMINOHEXANE
134-32-7	1-AMINONAPHTHALENE
111-86-4	1-AMINOOCTANE
2570-26-5	1-AMINOPENTADECANE
110-58-7	1-AMINOPENTANE
107-11-9	1-AMINOPROP-2-ENE
78-96-6	1-AMINOPROPAN-2-OL
107-10-8	1-AMINOPROPANE
79-19-6	1-AMINOTHIOUREA
109-97-7	1-AZA-2,4-CYCLOPENTADIENE
105-60-2	1-AZA-2-CYCLOHEPTANONE
111-49-9	1-AZACYCLOHEPTANE
91-22-5	1-AZANAPHTHALENE
545-55-1	1-AZIRIDINYL PHOSPHINE OXIDE (TRIS)
545-55-1	1-AZIRIDINYL PHOSPHINE OXIDE (TRIS) (DOT)
91-22-5	1-BENZAZINE
91-22-5	1-BENZINE
598-73-2	1-BROMO-1,2,2-TRIFLUOROETHYLENE
3132-64-7	1-BROMO-2,3-EPOXYPROPANE
4784-77-4	1-BROMO-2-BUTENE
592-55-2	1-BROMO-2-ETHOXYETHANE
592-55-2	1-BROMO-2-ETHOXYETHYLENE
95-46-5	1-BROMO-2-METHYLBENZENE
78-77-3	1-BROMO-2-METHYLPROPANE
598-31-2	1-BROMO-2-PROPANONE
106-95-6	1-BROMO-2-PROPENE
106-96-7	1-BROMO-2-PROPYNE
109-70-6	1-BROMO-3-CHLOROPROPANE
107-82-4	1-BROMO-3-METHYLBUTANE
107-82-4	1-BROMO-3-METHYLBUTANE (DOT)
106-38-7	1-BROMO-4-METHYLBENZENE
586-78-7	1-BROMO-4-NITROBENZENE
101-55-3	1-BROMO-4-PHENOXYBENZENE
109-65-9	1-BROMOBUTANE
110-53-2	1-BROMOPENTANE
106-94-5	1-BROMOPROPANE
106-94-5	1-BROMOPROPANE (DOT)
590-86-3	1-BUTANAL, 3-METHYL-
109-73-9	1-BUTANAMINE
26094-13-3	1-BUTANAMINE, (Z)-9-OCTADECENOATE
3037-72-7	1-BUTANAMINE, 4-(DIETHOXYMETHYLSILYL)-
102-82-9	1-BUTANAMINE, N,N-DIBUTYL-
111-92-2	1-BUTANAMINE, N-BUTYL-
924-16-3	1-BUTANAMINE, N-BUTYL-N-NITROSO-
13360-63-9	1-BUTANAMINE, N-ETHYL-
110-68-9	1-BUTANAMINE, N-METHYL-
109-52-4	1-BUTANECARBOXYLIC ACID
109-79-5	1-BUTANETHIOL
71-36-3	1-BUTANOL
96-20-8	1-BUTANOL, 2-AMINO-
97-95-0	1-BUTANOL, 2-ETHYL-
137-32-6	1-BUTANOL, 2-METHYL-
123-51-3	1-BUTANOL, 3-METHYL-
123-92-2	1-BUTANOL, 3-METHYL-, ACETATE
5593-70-4	1-BUTANOL, TITANIUM(4+) SALT
78-94-4	1-BUTEN-3-ONE
689-97-4	1-BUTEN-3-YNE
78-80-8	1-BUTEN-3-YNE, 2-METHYL-
106-98-9	1-BUTENE
106-88-7	1-BUTENE OXIDE
594-56-9	1-BUTENE, 2,3,3-TRIMETHYL-
563-78-0	1-BUTENE, 2,3-DIMETHYL-
760-21-4	1-BUTENE, 2-ETHYL-
563-46-2	1-BUTENE, 2-METHYL-
64037-54-3	1-BUTENE, 3,4-DICHLORO-, (.+-.)-
930-22-3	1-BUTENE, 3,4-EPOXY-
563-45-1	1-BUTENE, 3-METHYL-
689-97-4	1-BUTENYNE
2426-08-6	1-BUTOXY-2,3-EPOXYPROPANE
142-96-1	1-BUTOXYBUTANE
123-86-4	1-BUTYL ACETATE
71-36-3	1-BUTYL ALCOHOL
109-21-7	1-BUTYL BUTYRATE
109-79-5	1-BUTYL MERCAPTAN
109-73-9	1-BUTYLAMINE
104-51-8	1-BUTYLBENZENE
106-98-9	1-BUTYLENE
106-88-7	1-BUTYLENE OXIDE
689-97-4	1-BUTYN-3-ENE
107-00-6	1-BUTYNE
106-89-8	1-CHLOOR-2,3-EPOXY-PROPAAN (DUTCH)
97-00-7	1-CHLOOR-2,4-DINITROBENZEEN (DUTCH)
25567-67-3	1-CHLOOR-2,4-DINITROBENZEEN (DUTCH)
100-00-5	1-CHLOOR-4-NITROBENZEEN (DUTCH)
106-89-8	1-CHLOR-2,3-EPOXY-PROPAN (GERMAN)
97-00-7	1-CHLOR-2,4-DINITROBENZENE
25567-67-3	1-CHLOR-2,4-DINITROBENZENE
100-00-5	1-CHLOR-4-NITROBENZOL (GERMAN)
76-15-3	1-CHLORO-1,1,2,2,2-PENTAFLUOROETHANE
75-68-3	1-CHLORO-1,1-DIFLUOROETHANE
25497-29-4	1-CHLORO-1,1-DIFLUOROETHANE
25497-29-4	1-CHLORO-1,1-DIFLUOROETHANE (DOT)
79-38-9	1-CHLORO-1,2,2-TRIFLUOROETHENE
79-38-9	1-CHLORO-1,2,2-TRIFLUOROETHYLENE
96-24-2	1-CHLORO-1-DEOXYGLYCEROL
598-92-5	1-CHLORO-1-NITROETHANE
600-25-9	1-CHLORO-1-NITROPROPANE
590-21-6	1-CHLORO-1-PROPENE
79-01-6	1-CHLORO-2,2-DICHLOROETHYLENE
97-97-2	1-CHLORO-2,2-DIMETHOXYETHANE
96-12-8	1-CHLORO-2,3-DIBROMOPROPANE
96-24-2	1-CHLORO-2,3-DIHYDROXYPROPANE
106-89-8	1-CHLORO-2,3-EPOXYPROPANE
96-24-2	1-CHLORO-2,3-PROPANEDIOL
97-00-7	1-CHLORO-2,4-DINITROBENZENE
97-00-7	1-CHLORO-2,4-DINITROBENZENE 4-CHLORO-1,3-DINITROBENZENE
25567-67-3	1-CHLORO-2,4-DINITROBENZENE 4-CHLORO-1,3-DINITROBENZENE
97-00-7	1-CHLORO-2,4-DINITROBENZOL (GERMAN)
25567-67-3	1-CHLORO-2,4-DINITROBENZOL (GERMAN)
2100-42-7	1-CHLORO-2,5-DIMETHOXYBENZENE
111-44-4	1-CHLORO-2-(BETA-CHLOROETHOXY)ETHANE
505-60-2	1-CHLORO-2-(BETA-CHLOROETHYLTHIO)ETHANE
88-16-4	1-CHLORO-2-(TRIFLUOROMETHYL)BENZENE
591-97-9	1-CHLORO-2-BUTENE
542-76-7	1-CHLORO-2-CYANOETHANE
123-04-6	1-CHLORO-2-ETHYLHEXANE
127-00-4	1-CHLORO-2-HYDROXYPROPANE
59-50-7	1-CHLORO-2-METHYL-4-HYDROXYBENZENE
1321-10-4	1-CHLORO-2-METHYL-4-HYDROXYBENZENE
95-49-8	1-CHLORO-2-METHYLBENZENE
513-36-0	1-CHLORO-2-METHYLPROPANE
88-73-3	1-CHLORO-2-NITROBENZENE
127-00-4	1-CHLORO-2-PROPANOL
78-95-5	1-CHLORO-2-PROPANONE
107-05-1	1-CHLORO-2-PROPENE
107-84-6	1-CHLORO-3,3-DIMETHYLPROPANE
98-15-7	1-CHLORO-3-(TRIFLUOROMETHYL)BENZENE
109-70-6	1-CHLORO-3-BROMOPROPANE
109-70-6	1-CHLORO-3-BROMOPROPANE (DOT)
1912-24-9	1-CHLORO-3-ETHYLAMINO-5-ISOPROPYLAMINO-2,4,6-TRIAZINE
1912-24-9	1-CHLORO-3-ETHYLAMINO-5-ISOPROPYLAMINO-S-TRIAZINE
627-30-5	1-CHLORO-3-HYDROXYPROPANE
107-84-6	1-CHLORO-3-METHYLBUTANE
121-73-3	1-CHLORO-3-NITROBENZENE
627-30-5	1-CHLORO-3-PROPANOL
98-56-6	1-CHLORO-4-(TRIFLUOROMETHYL)BENZENE
104-83-6	1-CHLORO-4-CHLOROMETHYLBENZENE
106-43-4	1-CHLORO-4-METHYLBENZENE
777-37-7	1-CHLORO-4-NITRO-2-(TRIFLUOROMETHYL)BENZENE
100-00-5	1-CHLORO-4-NITROBENZENE
7005-72-3	1-CHLORO-4-PHENOXYBENZENE
78-95-5	1-CHLOROACETONE
532-27-4	1-CHLOROACETOPHENONE
109-69-3	1-CHLOROBUTANE
25154-42-1	1-CHLOROBUTANE
109-69-3	1-CHLOROBUTANE (DOT)
544-10-5	1-CHLOROHEXANE
127-00-4	1-CHLOROISOPROPYL ALCOHOL
598-92-5	1-CHLORONITROETHANE

CAS No.	Chemical Name
543-59-9	1-CHLOROPENTANE
540-54-5	1-CHLOROPROPANE
96-24-2	1-CHLOROPROPANE-2,3-DIOL
107-05-1	1-CHLOROPROPENE-2
590-21-6	1-CHLOROPROPENE
590-21-6	1-CHLOROPROPYLENE
106-89-8	1-CLORO-2,3-EPOSSIPROPANO (ITALIAN)
97-00-7	1-CLORO-2,4-DINITROBENZENE (ITALIAN)
25567-67-3	1-CLORO-2,4-DINITROBENZENE (ITALIAN)
100-00-5	1-CLORO-4-NITROBENZENE (ITALIAN)
4784-77-4	1-CROTYL BROMIDE
591-97-9	1-CROTYL CHLORIDE
78-82-0	1-CYANO-1-METHYLETHANE
110-67-8	1-CYANO-2-METHOXYETHANE
109-74-0	1-CYANOPROPANE
108-93-0	1-CYCLOHEXANOL
100-50-5	1-CYCLOHEXENE-4-CARBOXALDEHYDE
2016-57-1	1-DECANAMINE
143-10-2	1-DECANETHIOL
112-30-1	1-DECANOL
872-05-9	1-DECENE
2016-57-1	1-DECYLAMINE
1642-54-2	1-DIETHYLCARBAMOYL-4-METHYLPIPERAZINE DIHYDROGEN CITRATE
109-55-7	1-DIMETHYLAMINO-3-AMINOPROPANE
644-64-4	1-DIMETHYLCARBAMOYL-5-METHYL-3-PYRAZOLYL DIMETHYL-CARBAMATE
112-55-0	1-DODECANETHIOL
112-53-8	1-DODECANOL
112-41-4	1-DODECENE
112-53-8	1-DODECYL ALCOHOL
112-55-0	1-DODECYL MERCAPTAN
60-24-2	1-ETHANOL-2-THIOL
103-44-6	1-ETHENOXY-2-ETHYLHEXANE
928-55-2	1-ETHOXY-1-PROPENE
592-55-2	1-ETHOXY-2-BROMOETHANE
628-81-9	1-ETHOXYBUTANE
141-97-9	1-ETHOXYBUTANE-1,3-DIONE
109-92-2	1-ETHOXYETHENE
109-92-2	1-ETHOXYETHYLENE
628-32-0	1-ETHOXYPROPANE
928-55-2	1-ETHOXYPROPENE
766-09-6	1-ETHYL PIPERIDINE
766-09-6	1-ETHYL PIPERIDINE (DOT)
759-73-9	1-ETHYL-1-NITROSOUREA
584-02-1	1-ETHYL-1-PROPANOL
123-07-9	1-ETHYL-4-HYDROXYBENZENE
100-50-5	1-FORMYL-3-CYCLOHEXENE
109-27-3	1-GUANYL-4-NITROAMINOGUANYL-1-TETRAZENE
30207-98-8	1-HENDECANOL
57-11-4	1-HEPTADECANECARBOXYLIC ACID
111-68-2	1-HEPTANAMINE
1639-09-4	1-HEPTANETHIOL
111-70-6	1-HEPTANOL
592-76-7	1-HEPTENE
111-68-2	1-HEPTYLAMINE
112-02-7	1-HEXADEÇANAMINIUM, N,N,N-TRIMETHYL-, CHLORIDE
16919-58-7	1-HEXADECANAMINIUM, N-ETHYL-N,N-DIMETHYL-, BROMIDE (9CI)
2917-26-2	1-HEXADECANETHIOL
66-25-1	1-HEXANAL
111-26-2	1-HEXANAMINE
104-75-6	1-HEXANAMINE, 2-ETHYL-
106-20-7	1-HEXANAMINE, 2-ETHYL-N-(2-ETHYLHEXYL)-
143-16-8	1-HEXANAMINE, N-HEXYL-
111-31-9	1-HEXANETHIOL
142-62-1	1-HEXANOIC ACID
111-27-3	1-HEXANOL
104-76-7	1-HEXANOL, 2-ETHYL-
103-09-3	1-HEXANOL, 2-ETHYL-, ACETATE
103-11-7	1-HEXANOL, 2-ETHYL-, ACRYLATE
298-07-7	1-HEXANOL, 2-ETHYL-, HYDROGEN PHOSPHATE
3452-97-9	1-HEXANOL, 3,5,5-TRIMETHYL-
592-41-6	1-HEXENE
592-41-6	1-HEXENE (DOT)
142-92-7	1-HEXYL ACETATE
111-27-3	1-HEXYL ALCOHOL
544-10-5	1-HEXYL CHLORIDE
111-26-2	1-HEXYLAMINE

CAS No.	Chemical Name
771-29-9	1-HYDROPEROXYTETRALIN
52-68-6	1-HYDROXY-2,2,2-TRICHLORO-ETHYLE PHOSPHONATE DE DI-METHYLE (FRENCH)
52-68-6	1-HYDROXY-2,2,2-TRICHLOROETHYLPHOSPHONIC ACID DI-METHYL ESTER
556-52-5	1-HYDROXY-2,3-EPOXYPROPANE
105-67-9	1-HYDROXY-2,4-DIMETHYLBENZENE
51-28-5	1-HYDROXY-2,4-DINITROBENZENE
6117-91-5	1-HYDROXY-2-BUTENE
96-20-8	1-HYDROXY-2-BUTYLAMINE
78-89-7	1-HYDROXY-2-CHLOROPROPANE
60-24-2	1-HYDROXY-2-MERCAPTOETHANE
109-86-4	1-HYDROXY-2-METHOXYETHANE
95-48-7	1-HYDROXY-2-METHYLBENZENE
107-19-7	1-HYDROXY-2-PROPYNE
111-90-0	1-HYDROXY-3,6-DIOXAOCTANE
108-39-4	1-HYDROXY-3-METHYLBENZENE
554-84-7	1-HYDROXY-3-NITROBENZENE
150-76-5	1-HYDROXY-4-METHOXYBENZENE
106-44-5	1-HYDROXY-4-METHYLBENZENE
100-02-7	1-HYDROXY-4-NITROBENZENE
96-20-8	1-HYDROXY-SEC-BUTYLAMINE
71-36-3	1-HYDROXYBUTANE
112-53-8	1-HYDROXYDODECANE
50-21-5	1-HYDROXYETHANECARBOXYLIC ACID
513-86-0	1-HYDROXYETHYL METHYL KETONE
111-70-6	1-HYDROXYHEPTANE
111-27-3	1-HYDROXYHEXANE
78-83-1	1-HYDROXYMETHYLPROPANE
111-87-5	1-HYDROXYOCTANE
87-86-5	1-HYDROXYPENTACHLOROBENZENE
71-23-8	1-HYDROXYPROPANE
513-38-2	1-IODO-2-METHYLPROPANE
563-46-2	1-ISOAMYLENE
141-79-7	1-ISOBUTENYL METHYL KETONE
4098-71-9	1-ISOCYANATO-3,3,5-TRIMETHYL-5-ISOCYANATOMETHYLCYCLO-HEXANE
4098-71-9	1-ISOCYANATO-3-ISOCYANATOMETHYL-3,5,5-TRIMETHYLCYCLO-HEXANE
4098-71-9	1-ISOCYANATO-5-(ISOCYANATOMETHYL)-3,3,5-TRIMETHYLCYCLO-HEXANE
111-36-4	1-ISOCYANATOBUTANE
110-78-1	1-ISOCYANATOPROPANE
119-38-0	1-ISOPROPYL-3-METHYL-5-PYRAZOLYL DIMETHYLCARBAMATE
119-38-0	1-ISOPROPYL-3-METHYLPYRAZOLYL-(5)-DIMETHYLCARBAMATE
99-87-6	1-ISOPROPYL-4-METHYLBENZENE
60-24-2	1-MERCAPTO-2-HYDROXYETHANE
112-55-0	1-MERCAPTODODECANE
111-88-6	1-MERCAPTOOCTANE
110-66-7	1-MERCAPTOPENTANE
1663-35-0	1-METHOXY-2-(VINYLOXY)ETHANE
99-59-2	1-METHOXY-2-AMINO-4-NITROBENZENE
107-98-2	1-METHOXY-2-HYDROXYPROPANE
107-98-2	1-METHOXY-2-PROPANOL
100-17-4	1-METHOXY-4-NITROBENZENE
105-45-3	1-METHOXYBUTANE-1,3-DIONE
7786-34-7	1-METHOXYCARBONYL-1-PROPEN-2-YL DIMETHYL PHOSPHATE
107-25-5	1-METHOXYETHYLENE
557-17-5	1-METHOXYPROPANE
70-25-7	1-METHYL 3-NITRO 1-NITROSO-GUANIDINE
109-01-3	1-METHYL PIPERAZINE
78-92-2	1-METHYL PROPANOL
504-60-9	1-METHYL-1,3-BUTADIENE
6032-29-7	1-METHYL-1-BUTANOL
70-25-7	1-METHYL-1-NITROSO-3-NITROGUANIDINE
684-93-5	1-METHYL-1-NITROSOUREA
80-15-9	1-METHYL-1-PHENYLETHYL HYDROPEROXIDE
98-83-9	1-METHYL-1-PHENYLETHYLENE
513-53-1	1-METHYL-1-PROPANETHIOL
78-92-2	1-METHYL-1-PROPANOL
594-56-9	1-METHYL-1-TERT-BUTYLETHYLENE
118-96-7	1-METHYL-2,4,6-TRINITROBENZENE
121-14-2	1-METHYL-2,4-DINITROBENZENE
606-20-2	1-METHYL-2,6-DINITROBENZENE
54-11-5	1-METHYL-2-(3-PYRIDYL)PYRROLIDINE
29790-52-1	1-METHYL-2-(3-PYRIDYL)PYRROLIDINE SALICYLATE
95-53-4	1-METHYL-2-AMINOBENZENE
78-96-6	1-METHYL-2-AMINOETHANOL
95-49-8	1-METHYL-2-CHLOROBENZENE

CAS No.	Chemical Name
88-72-2	1-METHYL-2-NITROBENZENE
366-70-1	1-METHYL-2-P-(ISOPROPYLCARBAMOYL)BENZYLHYDRAZINE HYDROCHLORIDE
764-35-2	1-METHYL-2-PROPYLACETYLENE
872-50-4	1-METHYL-2-PYRROLIDINONE
872-50-4	1-METHYL-2-PYRROLIDONE
671-16-9	1-METHYL-2-[P-(ISOPROPYLCARBAMOYL)BENZYL]HYDRAZINE
610-39-9	1-METHYL-3,4-DINITROBENZENE
62450-07-1	1-METHYL-3-AMINO-5H-PYRIDO[4,3-B]INDOLE
99-08-1	1-METHYL-3-NITROBENZENE
138-86-3	1-METHYL-4-(1-METHYLETHENYL)CYCLOHEXENE
99-87-6	1-METHYL-4-(1-METHYLETHYL)BENZENE
140-80-7	1-METHYL-4-(DIETHYLAMINO)BUTYLAMINE
1642-54-2	1-METHYL-4-DIETHYLCARBAMOYLPIPERAZINE CITRATE
106-44-5	1-METHYL-4-HYDROXYBENZENE
138-86-3	1-METHYL-4-ISOPROPENYL-1-CYCLOHEXENE
138-86-3	1-METHYL-4-ISOPROPENYLCYCLOHEXENE
99-87-6	1-METHYL-4-ISOPROPYLBENZENE
99-99-0	1-METHYL-4-NITROBENZENE
98-51-1	1-METHYL-4-TERT-BUTYLBENZENE
622-97-9	1-METHYL-4-VINYLBENZENE
872-50-4	1-METHYL-5-PYRROLIDINONE
625-30-9	1-METHYL-N-BUTYLAMINE
872-50-4	1-METHYLAZACYCLOPENTAN-2-ONE
504-60-9	1-METHYLBUTADIENE
6032-29-7	1-METHYLBUTANOL
626-38-0	1-METHYLBUTYL ACETATE
53496-15-4	1-METHYLBUTYL ACETATE
625-30-9	1-METHYLBUTYLAMINE
108-87-2	1-METHYLCYCLOHEXANE
75-33-2	1-METHYLETHANETHIOL
108-21-4	1-METHYLETHYL ACETATE
638-11-9	1-METHYLETHYL BUTANOATE
105-48-6	1-METHYLETHYL MONOCHLOROACETATE
75-31-0	1-METHYLETHYLAMINE
110-43-0	1-METHYLHEXANAL
60-34-4	1-METHYLHYDRAZINE
109-02-4	1-METHYLMORPHOLINE
90-12-0	1-METHYLNAPHTHALENE
626-67-5	1-METHYLPIPERIDINE
626-67-5	1-METHYLPIPERIDINE (DOT)
13952-84-6	1-METHYLPROPANAMINE
105-46-4	1-METHYLPROPYL ACETATE
78-92-2	1-METHYLPROPYL ALCOHOL
78-86-4	1-METHYLPROPYL CHLORIDE
13952-84-6	1-METHYLPROPYLAMINE
120-94-5	1-METHYLPYRROLIDINE
872-50-4	1-METHYLPYRROLIDINONE
872-50-4	1-METHYLPYRROLIDONE
108-22-5	1-METHYLVINYL ACETATE
872-05-9	1-N-DECENE
592-76-7	1-N-HEPTENE
592-41-6	1-N-HEXENE
124-11-8	1-N-NONENE
30553-04-9	1-NAFTIL-TIOUREA (ITALIAN)
30553-04-9	1-NAFTYLTHIOUREUM (DUTCH)
134-32-7	1-NAPHTHALAMINE
134-32-7	1-NAPHTHALENAMINE
63-25-2	1-NAPHTHALENOL, METHYLCARBAMATE
30553-04-9	1-NAPHTHALENYLTHIOUREA
63-25-2	1-NAPHTHOL N-METHYLCARBAMATE
63-25-2	1-NAPHTHYL METHYLCARBAMATE
63-25-2	1-NAPHTHYL N-METHYLCARBAMATE
63-25-2	1-NAPHTHYL-N-METHYL-KARBAMAT (GERMAN)
30553-04-9	1-NAPHTHYL-THIOHARNSTOFF (GERMAN)
30553-04-9	1-NAPHTHYL-THIOUREE (FRENCH)
134-32-7	1-NAPHTHYLAMINE
134-32-7	1-NAPHTHYLAMINE, TECHNICAL GRADE
86-88-4	1-NAPHTHYLTHIOUREA
88-74-4	1-NITRO-2-AMINOBENZENE
88-73-3	1-NITRO-2-CHLOROBENZENE
129-15-7	1-NITRO-2-METHYLANTHRAQUINONE
402-54-0	1-NITRO-4-(TRIFLUOROMETHYL)BENZENE
100-00-5	1-NITRO-4-CHLOROBENZENE
92-93-3	1-NITRO-4-PHENYLBENZENE
556-88-7	1-NITROGUANIDINE
463-04-7	1-NITROPENTANE
108-03-2	1-NITROPROPANE
684-93-5	1-NITROSO-1-METHYLUREA
70-25-7	1-NITROSO-3-NITRO-1-METHYLGUANIDINE
100-75-4	1-NITROSOPIPERIDINE
930-55-2	1-NITROSOPYRROLIDINE
1455-21-6	1-NONANETHIOL
143-08-8	1-NONANOL
124-11-8	1-NONENE
2646-17-5	1-O-TOLUENEAZO-2-NAPHTHOL
614-78-8	1-O-TOLYL-2-THIOUREA
2646-17-5	1-O-TOLYLAZO-2-NAPHTHOL
2885-00-9	1-OCTADECANETHIOL
124-13-0	1-OCTANAL
111-86-4	1-OCTANAMINE
1120-48-5	1-OCTANAMINE, N-OCTYL-
111-66-0	1-OCTANE
111-88-6	1-OCTANETHIOL
111-87-5	1-OCTANOL
112-14-1	1-OCTANOL ACETATE
111-66-0	1-OCTENE
112-14-1	1-OCTYL ACETATE
111-88-6	1-OCTYL MERCAPTAN
111-88-6	1-OCTYL THIOL
111-86-4	1-OCTYLAMINE
110-91-8	1-OXA-4-AZACYCLOHEXANE
15980-15-1	1-OXA-4-THIACYCLOHEXANE
53-86-1	1-P-CLORO-BENZOIL-5-METOXI-2-METILINDOL-3-ACIDO ACETICO (SPANISH)
622-97-9	1-P-TOLYLETHENE
2570-26-5	1-PENTADECANAMINE
629-76-5	1-PENTADECANOL
2570-26-5	1-PENTADECYLAMINE
110-58-7	1-PENTANAMINE
621-77-2	1-PENTANAMINE, N,N-DIPENTYL-
2050-92-2	1-PENTANAMINE, N-PENTYL-
142-62-1	1-PENTANECARBOXYLIC ACID
110-66-7	1-PENTANETHIOL
71-41-0	1-PENTANOL
628-63-7	1-PENTANOL ACETATE
105-30-6	1-PENTANOL, 2-METHYL-
54972-97-3	1-PENTANOL, 2-METHYL-
108-11-2	1-PENTANOL, METHYL-
54972-97-3	1-PENTANOL, METHYL-
109-67-1	1-PENTENE
565-76-4	1-PENTENE, 2,3,4-TRIMETHYL-
107-39-1	1-PENTENE, 2,4,4-TRIMETHYL-
763-29-1	1-PENTENE, 2-METHYL-
691-37-2	1-PENTENE, 4-METHYL-
12772-47-3	1-PENTOL
12772-47-3	1-PENTOL (DOT)
628-63-7	1-PENTYL ACETATE
71-41-0	1-PENTYL ALCOHOL
110-53-2	1-PENTYL BROMIDE
540-18-1	1-PENTYL BUTYRATE
1002-16-0	1-PENTYL NITRATE
463-04-7	1-PENTYL NITRITE
110-58-7	1-PENTYLAMINE
627-19-0	1-PENTYNE
98-85-1	1-PHENETHYL ALCOHOL
127-82-2	1-PHENOL-4-SULFONIC ACID ZINC SALT
122-60-1	1-PHENOXY-2,3-EPOXYPROPANE
24017-47-8	1-PHENYL-1,2,4-TRIAZOLYL-3-(O,O-DIETHYLTHIONOPHOSPHATE)
96-09-3	1-PHENYL-1,2-EPOXYETHANE
98-84-0	1-PHENYL-1-ETHANAMINE
98-86-2	1-PHENYL-1-ETHANONE
98-85-1	1-PHENYL-1-HYDROXYETHANE
98-83-9	1-PHENYL-1-METHYLETHYLENE
7568-93-6	1-PHENYL-2-AMINOETHANOL
103-85-5	1-PHENYL-2-THIOUREA
24017-47-8	1-PHENYL-3-(O,O-DIETHYL-THIONOPHOSPHORYL)-1,2,4-TRIAZOLE
538-68-1	1-PHENYL-N-PENTANE
104-51-8	1-PHENYLBUTANE
104-72-3	1-PHENYLDECANE
123-01-3	1-PHENYLDODECANE
98-84-0	1-PHENYLETHANAMINE
98-85-1	1-PHENYLETHANOL
98-86-2	1-PHENYLETHANONE
98-85-1	1-PHENYLETHYL ALCOHOL
98-84-0	1-PHENYLETHYLAMINE
538-68-1	1-PHENYLPENTANE
103-65-1	1-PHENYLPROPANE

CAS No.	Chemical Name
2097-19-0	1-PHENYLSILATRANE
1459-10-5	1-PHENYLTETRADECANE
103-85-5	1-PHENYLTHIOUREA
64-10-8	1-PHENYLUREA
1642-54-2	1-PIPERAZINECARBOXAMIDE, N,N-DIETHYL-4-METHYL-, 2-HYDROXY-1,2,3-PROPANETRICARBOXYLATE (1:1)
1642-54-2	1-PIPERAZINECARBOXAMIDE, N,N-DIETHYL-4-METHYL-, CITRATE (1:1)
140-31-8	1-PIPERAZINEETHANAMINE
140-31-8	1-PIPERAZINEETHYLAMINE
123-38-6	1-PROPANAL
107-10-8	1-PROPANAMINE
78-81-9	1-PROPANAMINE, 2-METHYL-
110-96-3	1-PROPANAMINE, 2-METHYL-N-(2-METHYLPROPYL)-
56-18-8	1-PROPANAMINE, 3,3'-IMINOBIS-
5332-73-0	1-PROPANAMINE, 3-METHOXY-
926-63-6	1-PROPANAMINE, N,N-DIMETHYL-
102-69-2	1-PROPANAMINE, N,N-DIPROPYL-
621-64-7	1-PROPANAMINE, N-NITROSO-N-PROPYL-
142-84-7	1-PROPANAMINE, N-PROPYL-
107-92-6	1-PROPANECARBOXYLIC ACID
1120-71-4	1-PROPANESULFONIC ACID, 3-HYDROXY-, .GAMMA.-SULTONE
107-03-9	1-PROPANETHIOL
107-03-9	1-PROPANETHIOL (DOT)
79869-58-2	1-PROPANETHIOL (DOT)
71-23-8	1-PROPANOL
75-84-3	1-PROPANOL, 2,2-DIMETHYL-
126-72-7	1-PROPANOL, 2,3-DIBROMO-, PHOSPHATE (3:1)
556-52-5	1-PROPANOL, 2,3-EPOXY-
106-91-2	1-PROPANOL, 2,3-EPOXY-, METHACRYLATE
124-68-5	1-PROPANOL, 2-AMINO-2-METHYL-
78-89-7	1-PROPANOL, 2-CHLORO-
78-83-1	1-PROPANOL, 2-METHYL-
156-87-6	1-PROPANOL, 3-AMINO-
627-30-5	1-PROPANOL, 3-CHLORO-
3087-37-4	1-PROPANOL, TITANIUM(4+) SALT
3087-37-4	1-PROPANOL, TITANIUM(4+) SALT (9CI)
123-38-6	1-PROPANONE
10118-72-6	1-PROPEN-2-CHLORO-1,3-DIOL-DIACETATE
108-22-5	1-PROPEN-2-OL, ACETATE
108-22-5	1-PROPEN-2-YL ACETATE
107-18-6	1-PROPEN-3-OL
115-07-1	1-PROPENE
115-07-1	1-PROPENE (9CI)
116-15-4	1-PROPENE, 1,1,2,3,3,3-HEXAFLUORO-
96-19-5	1-PROPENE, 1,2,3-TRICHLORO-
542-75-6	1-PROPENE, 1,3-DICHLORO-
8003-19-8	1-PROPENE, 1,3-DICHLORO-, MIXT. WITH 1,2-DICHLOROPROPANE
590-21-6	1-PROPENE, 1-CHLORO-
928-55-2	1-PROPENE, 1-ETHOXY-
78-88-6	1-PROPENE, 2,3-DICHLORO-
557-98-2	1-PROPENE, 2-CHLORO-
115-11-7	1-PROPENE, 2-METHYL-
7756-94-7	1-PROPENE, 2-METHYL-, TRIMER
7756-94-7	1-PROPENE, 2-METHYL-, TRIMER (9CI)
98-83-9	1-PROPENE, 2-PHENYL-
557-40-4	1-PROPENE, 3,3'-OXYBIS-
3054-95-3	1-PROPENE, 3,3-DIETHOXY-
3917-15-5	1-PROPENE, 3-(ETHENYLOXY)-
106-95-6	1-PROPENE, 3-BROMO-
107-05-1	1-PROPENE, 3-CHLORO-
563-47-3	1-PROPENE, 3-CHLORO-2-METHYL-
557-31-3	1-PROPENE, 3-ETHOXY-
557-31-3	1-PROPENE, 3-ETHOXY- (9CI)
556-56-9	1-PROPENE, 3-IODO-
556-56-9	1-PROPENE, 3-IODO- (9CI)
57-06-7	1-PROPENE, 3-ISOTHIOCYANATO-
26952-23-8	1-PROPENE, DICHLORO-
6842-15-5	1-PROPENE, TETRAMER
6842-15-5	1-PROPENE, TETRAMER (9CI)
13987-01-4	1-PROPENE, TRIMER
107-18-6	1-PROPENOL-3
109-60-4	1-PROPYL ACETATE
71-23-8	1-PROPYL ALCOHOL
105-66-8	1-PROPYL BUTYRATE
110-78-1	1-PROPYL ISOCYANATE
107-03-9	1-PROPYL MERCAPTAN
107-10-8	1-PROPYLAMINE
103-65-1	1-PROPYLBENZENE
115-07-1	1-PROPYLENE
107-19-7	1-PROPYN-3-OL
107-19-7	1-PROPYN-3-YL ALCOHOL
74-99-7	1-PROPYNE
106-96-7	1-PROPYNE, 3-BROMO-
107-19-7	1-PROPYNE-3-OL
98-51-1	1-TERT-BUTYL-4-METHYLBENZENE
112-72-1	1-TETRADECANOL
27196-00-5	1-TETRADECANOL
1120-36-1	1-TETRADECENE
41083-11-8	1-TRI(CYCLOHEXYL) STANNYL-1H-1,2,4-TRIAZOLE
2437-56-1	1-TRIDECENE
5332-52-5	1-UNDECANETHIOL
821-95-4	1-UNDECENE
100-40-3	1-VINYL-3-CYCLOHEXENE
106-87-6	1-VINYL-3-CYCLOHEXENE DIOXIDE
2646-17-5	1-[(2-METHYLPHENYL)AZO]-2-NAPHTHOL
205-82-3	10,11-BENZOFLUORANTHENE
357-57-3	10,11-DIMETHYSTRYCHNINE
260-94-6	10-AZAANTHRACENE
578-94-9	10-CHLORO-5,10-DIHYDROARSACRIDINE
578-94-9	10-CHLORO-5,10-DIHYDROPHENARSAZINE
92-84-2	10H-PHENOTHIAZINE
207-08-9	11,12-BENZOFLUORANTHENE
1694-09-3	11386 VIOLET
1300-73-8	11460 BROWN
633-03-4	12415 GREEN
20830-75-5	12BETA-HYDROXYDIGITOXIN
23214-92-8	14-HYDROXYDAUNOMYCIN
8030-30-6	160 DEGREE BENZOL
3761-53-3	1695 RED
57-63-6	17-ETHINYL-3,17-ESTRADIOL
57-63-6	17-ETHINYLESTRADIOL
57-63-6	17-ETHYNYLESTRADIOL
72-33-3	17-ETHYNYLESTRADIOL 3-METHYL ETHER
57-63-6	17ALPHA-ETHINYL-17BETA-ESTRADIOL
68-22-4	17ALPHA-ETHINYL-19-NORTESTOSTERONE
57-63-6	17ALPHA-ETHINYLESTRADIOL
72-33-3	17ALPHA-ETHINYLESTRADIOL 3-METHYL ETHER
68-22-4	17ALPHA-ETHYNYL-19-NORTESTOSTERONE
57-63-6	17ALPHA-ETHYNYLESTRADIOL
72-33-3	17ALPHA-ETHYNYLESTRADIOL 3-METHYL ETHER
72-33-3	17ALPHA-ETHYNYLESTRADIOL METHYL ETHER
50-28-2	17BETA-ESTRADIOL
50-28-2	17BETA-OESTRADIOL
7440-50-8	1721 GOLD
68-22-4	19-NOR-17ALPHA-ETHYNYLTESTOSTERONE
68-22-4	19-NOR-17ALPHA-PREGN-4-EN-20-YN-3-ONE, 17-HYDROXY-
72-33-3	19-NOR-17ALPHA-PREGNA-1,3,5(10)-TRIEN-20-YN-17-OL, 3-METHOXY-
57-63-6	19-NOR-17ALPHA-PREGNA-1,3,5(10)-TRIEN-20-YNE-3,17-DIOL
68-22-4	19-NORPREGN-4-EN-20-YN-3-ONE, 17-HYDROXY-, (17ALPHA)-
72-33-3	19-NORPREGNA-1,3,5(10)-TRIEN-20-YN-17-OL, 3-METHOXY-, (17ALPHA)-
57-63-6	19-NORPREGNA-1,3,5(10)-TRIEN-20-YNE-3,17-DIOL, (17ALPHA)-
68-22-4	19-NORTESTOSTERONE, 17-ETHYNYL-
61-82-5	1H-1,2,4-TRIAZOL-3-AMINE
24017-47-8	1H-1,2,4-TRIAZOL-3-OL, 1-PHENYL-, O-ESTER WITH O,O-DIETHYL PHOSPHOROTHIOATE
111-49-9	1H-AZEPINE, HEXAHYDRO-
151-56-4	1H-AZIRINE, DIHYDRO-
3878-19-1	1H-BENZIMIDAZOLE, 2-(2-FURANYL)-
95-14-7	1H-BENZOTRIAZOLE
4316-42-1	1H-IMIDAZOLE, 1-BUTYL- (9CI)
4316-42-1	1H-IMIDAZOLE, BUTYL-
443-48-1	1H-IMIDAZOLE-1-ETHANOL, 2-METHYL-5-NITRO-
4342-03-4	1H-IMIDAZOLE-4-CARBOXAMIDE, 5-(3,3-DIMETHYL-1-TRIAZENYL)-
95-13-6	1H-INDENE
83-26-1	1H-INDENE-1,3(2H)-DIONE, 2-(2,2-DIMETHYL-1-OXOPROPYL)-
82-66-6	1H-INDENE-1,3(2H)-DIONE, 2-(DIPHENYLACETYL)-
3691-35-8	1H-INDENE-1,3(2H)-DIONE, 2-[(4-CHLOROPHENYL)PHENYLACETYL]-
53-86-1	1H-INDOLE-3-ACETIC ACID, 1-(4-CHLOROBENZOYL)-5-METHOXY-2-METHYL-
133-06-2	1H-ISOINDOLE-1,3(2H)-DIONE, 3A,4,7,7A-TETRAHYDRO-2-((TRICHLOROMETHYL)THIO)-
2425-06-1	1H-ISOINDOLE-1,3(2H)-DIONE, 3A,4,7,7A-TETRAHYDRO-2-[(1,1,2,2-TETRACHLOROETHYL)THIO]-

CAS No.	Chemical Name
446-86-6	1H-PURINE, 6-[(1-METHYL-4-NITRO-1H-IMIDAZOL-5-YL)THIO]-
109-97-7	1H-PYRROLE
27417-39-6	1H-PYRROLE, METHYL-
50-76-0	1H-PYRROLO[2,1-I][1,4,7,10,13]OXATETRAAZACYCLOHEXA-DECINE, CYCLIC PEPTIDE DERIV.
64-00-6	1PC
1912-24-9	2 -AETHYLAMINO-4-ISOPROPYLAMINO-6-CHLOR-1,3,5-TRIAZIN (GERMAN)
94-58-6	2',3'-DIHYDROSAFROLE
84-80-0	2',3'-TRANS-VITAMIN K1
97-56-3	2',3-DIMETHYL-4-AMINOAZOBENZENE
97-36-9	2',4'-ACETOACETOXYLIDIDE
97-36-9	2',4'-DIMETHYLACETOACETANILIDE
1420-04-8	2',5-DICHLORO-4'-NITROSALICYLANILIDE, 2-AMINOETHANOL SALT
92-15-9	2'-METHOXYACETOACETANILIDE
3691-35-8	2(2-(4-CHLOOR-FENYL-2-FENYL)-ACETYL)-INDAAN-1,3-DION (DUTCH)
3691-35-8	2(2-(4-CHLOR-PHENYL-2-PHENYL)ACETYL)INDAN-1,3-DION (GERMAN)
3691-35-8	2(2-(4-CHLOROPHENYL)-2-PHENYLACETYL)INDAN-1,3-DIONE
3691-35-8	2(2-(4-CLORO-FENIL-2FENIL)-ACETIL)INDAN-1,3-DIONE (ITALIAN)
3570-75-0	2(2-FORMYLHYDRAZINO)4-(5-NITRO-2-FURYL)THIAZOLE
111-40-0	2, 2'-DIAMINODIETHYLAMINE
102-71-6	2,2',2''-NITRILOTRIETHANOL
102-71-6	2,2',2''-NITRILOTRIS[ETHANOL]
555-77-1	2,2',2''-TRICHLOROTRIETHYLAMINE
70-30-4	2,2',3,3',5,5'-HEXACHLORO-6,6'-DIHYDROXYDIPHENYLMETHANE
2217-06-3	2,2',4,4',6,6'-HEXANITRODIPHENYL SULFIDE
100-97-0	2,2',4,4',6,6'-HEXANITRODIPHENYLAMINE
131-73-7	2,2',4,4',6,6'-HEXANITRODIPHENYLAMINE
20062-22-0	2,2',4,4',6,6'-HEXANITROSTILBENE
112-27-6	2,2'-(1,2-ETHANEDIYLBIS(OXY))BISETHANOL
1954-28-5	2,2'-(2,5,8,11-TETRAOXA-1,12-DODECANEDIYL)BISOXIRANE
102-79-4	2,2'-(BUTYLIMINO)DIETHANOL
139-87-7	2,2'-(ETHYLIMINO)DIETHANOL
91-99-6	2,2'-(M-TOLYLIMINO)DIETHANOL
105-59-9	2,2'-(METHYLIMINO)DIETHANOL
2160-93-2	2,2'-(TERT-BUTYLIMINO)DIETHANOL
78-67-1	2,2'-AZOBIS(2-CYANOBUTANE)
78-67-1	2,2'-AZOBIS(2-METHYLPROPIONITRILE)
78-67-1	2,2'-AZOBIS(ISOBUTYRONITRILE)
78-67-1	2,2'-AZOBIS[2-METHYLPROPANENITRILE]
15545-97-8	2,2'-AZODI-(2,4-DIMETHYL-4-METHOXYVALERONITRILE)
28604-91-3	2,2'-AZODI-(2,4-DIMETHYLVALERONITRILE)
78-67-1	2,2'-AZODIISOBUTYRONITRILE
1464-53-5	2,2'-BIOXIRANE
2764-72-9	2,2'-BIPYRIDINIUM, 1,1'-(1,2-ETHANEDIYL)-
96-69-5	2,2'-DI-TERT-BUTYL-5,5'-DIMETHYL-4,4'-THIODIPHENOL
111-40-0	2,2'-DIAMINODIETHYLAMINE
111-44-4	2,2'-DICHLOORETHYLETHER (DUTCH)
111-44-4	2,2'-DICHLOR-DIAETHYLAETHER (GERMAN)
111-44-4	2,2'-DICHLORETHYL ETHER
51-75-2	2,2'-DICHLORO-N-METHYLDIETHYLAMINE
55-86-7	2,2'-DICHLORO-N-METHYLDIETHYLAMINE HYDROCHLORIDE
302-70-5	2,2'-DICHLORO-N-METHYLDIETHYLAMINE N-OXIDE HYDRO-CHLORIDE
111-44-4	2,2'-DICHLORODIETHYL ETHER
505-60-2	2,2'-DICHLORODIETHYL SULFIDE
111-44-4	2,2'-DICHLOROETHYL ETHER
505-60-2	2,2'-DICHLOROETHYL SULFIDE
63283-80-7	2,2'-DICHLOROISOPROPYL ETHER
538-07-8	2,2'-DICHLOROTRIETHYLAMINE
111-44-4	2,2'-DICLOROETILETERE (ITALIAN)
78-67-1	2,2'-DICYANO-2,2'-AZOPROPANE
106-20-7	2,2'-DIETHYLDIHEXYLAMINE
70-30-4	2,2'-DIHYDROXY-3,3',5,5',6,6'-HEXACHLORODIPHENYLMETHANE
70-30-4	2,2'-DIHYDROXY-3,5,6,3',5',6'-HEXACHLORODIPHENYLMETHANE
111-42-2	2,2'-DIHYDROXYDIETHYLAMINE
25265-71-8	2,2'-DIHYDROXYDIPROPYL ETHER
111-46-6	2,2'-DIHYDROXYETHYL ETHER
25265-71-8	2,2'-DIHYDROXYISOPROPYL ETHER
78-67-1	2,2'-DIMETHYL-2,2'-AZODIPROPIONITRILE
838-88-0	2,2'-DIMETHYL-4,4'-METHYLENEDIANILINE
56-18-8	2,2'-DIMETHYLBIS(2-AMINOETHYL)AMINE
112-27-6	2,2'-ETHYLENEDIOXYBIS(ETHANOL)
112-27-6	2,2'-ETHYLENEDIOXYDIETHANOL
112-27-6	2,2'-ETHYLENEDIOXYETHANOL
111-40-0	2,2'-IMINOBIS(ETHANAMINE)
111-42-2	2,2'-IMINOBISETHANOL
111-42-2	2,2'-IMINODI-1-ETHANOL
111-42-2	2,2'-IMINODIETHANOL
86-73-7	2,2'-METHYLENEBIPHENYL
70-30-4	2,2'-METHYLENEBIS[3,4,6-TRICHLOROPHENOL]
111-46-6	2,2'-OXYBISETHANOL
111-46-6	2,2'-OXYDIETHANOL
111-46-6	2,2'-OXYETHANOL
111-48-8	2,2'-THIOBISETHANOL
111-48-8	2,2'-THIODIETHANOL
111-48-8	2,2'-THIODIGLYCOL
4985-85-7	2,2'-[(3-AMINOPROPYL)IMINO]DIETHANOL
112-60-7	2,2'-[OXYBIS(2,1-ETHANEDIYLOXY)]BISETHANOL
115-32-2	2,2,2-TRICHLOOR-1,1-BIS(4-CHLOORFENYL)-ETHANOL (DUTCH)
115-32-2	2,2,2-TRICHLOR-1,1-BIS(4-CHLOR-PHENYL)-AETHANOL (GERMAN)
115-32-2	2,2,2-TRICHLORO-1,1-BIS(4-CHLOROPHENYL)-ETHANOL (FRENCH)
50-29-3	2,2,2-TRICHLORO-1,1-BIS(4-CHLOROPHENYL)ETHANE
115-32-2	2,2,2-TRICHLORO-1,1-BIS(4-CHLOROPHENYL)ETHANOL
115-32-2	2,2,2-TRICHLORO-1,1-BIS(4-CLORO-FENIL)-ETANOLO (ITALIAN)
72-43-5	2,2,2-TRICHLORO-1,1-BIS(4-METHOXYPHENYL)ETHANE
115-32-2	2,2,2-TRICHLORO-1,1-BIS(P-CHLOROPHENYL)ETHANOL
115-32-2	2,2,2-TRICHLORO-1,1-DI(4-CHLOROPHENYL)ETHANOL
52-68-6	2,2,2-TRICHLORO-1-HYDROXYETHYL-PHOSPHONATE, DIMETHYL ESTER
75-87-6	2,2,2-TRICHLOROACETALDEHYDE
75-87-6	2,2,2-TRICHLOROETHANAL
76-05-1	2,2,2-TRIFLUOROACETIC ACID
7154-79-2	2,2,3,3-TETRAMETHYL PENTANE
1186-53-4	2,2,3,4-TETRAMETHYL PENTANE
428-59-1	2,2,3-TRIFLUORO-3-(TRIFLUOROMETHYL)OXIRANE
464-06-2	2,2,3-TRIMETHYLBUTANE
564-02-3	2,2,3-TRIMETHYLPENTANE
107-40-4	2,2,4-TRIMETHYL-3-PENTENE
107-39-1	2,2,4-TRIMETHYL-4-PENTENE
540-84-1	2,2,4-TRIMETHYLPENTANE
3522-94-9	2,2,5-TRIMETHYLHEXANE
72-54-8	2,2-BIS(4-CHLOROPHENYL)-1,1-DICHLOROETHANE
72-55-9	2,2-BIS(4-CHLOROPHENYL)-1,1-DICHLOROETHYLENE
115-77-5	2,2-BIS(HYDROXYMETHYL)-1,3-PROPANEDIOL
72-43-5	2,2-BIS(P-ANISYL)-1,1,1-TRICHLOROETHANE
50-29-3	2,2-BIS(P-CHLOROPHENYL)-1,1,1-TRICHLOROETHANE
72-54-8	2,2-BIS(P-CHLOROPHENYL)-1,1-DICHLOROETHANE
72-43-5	2,2-BIS(P-METHOXYPHENYL)-1,1,1-TRICHLOROETHANE
2167-23-9	2,2-BIS(TERT-BUTYLDIOXY)BUTANE
2167-23-9	2,2-BIS(TERT-BUTYLPEROXY)BUTANE
2167-23-9	2,2-BIS-(TERT-BUTYLPEROXY)-BUTANE (CONCENTRATION >70%)
2167-23-9	2,2-DI(TERT-BUTYLPEROXY)BUTANE
72-43-5	2,2-DI-(P-METHOXYPHENYL)-1,1,1-TRICHLOROETHANE
2167-23-9	2,2-DI-(TERT-BUTYLPEROXY)BUTANE, NOT MORE THAN 55% IN SOLUTION (DOT)
72-43-5	2,2-DI-P-ANISYL-1,1,1-TRICHLOROETHANE
108-60-1	2,2-DICHLORO ISOPROPYL ETHER
72-55-9	2,2-DICHLORO-1,1-BIS(4-CHLOROPHENYL)ETHYLENE
79-43-6	2,2-DICHLOROACETIC ACID
79-36-7	2,2-DICHLOROACETYL CHLORIDE
62-73-7	2,2-DICHLOROETHENYL DIMETHYL PHOSPHATE
2636-26-2	2,2-DICHLOROETHENYL DIMETHYL PHOSPHATE
2636-26-2	2,2-DICHLOROETHENYL PHOSPHORIC ACID DIMETHYL ESTER
75-99-0	2,2-DICHLOROPROPANOIC ACID
75-99-0	2,2-DICHLOROPROPIONIC ACID
75-99-0	2,2-DICHLOROPROPIONIC ACID (ACGIH,DOT)
62-73-7	2,2-DICHLOROVINYL DIMETHYL PHOSPHATE
2636-26-2	2,2-DICHLOROVINYL DIMETHYL PHOSPHATE
2636-26-2	2,2-DICHLOROVINYL DIMETHYL PHOSPHORIC ACID ESTER
115-76-4	2,2-DIETHYL-1,3-PROPANEDIOL
76-11-9	2,2-DIFLUORO-1,1,1,2-TETRACHLOROETHANE
2614-76-8	2,2-DIHYDROPEROXY PROPANE
2614-76-8	2,2-DIHYDROPEROXYPROPANE (CONCENTRATION ≥ 30%)
100-79-8	2,2-DIMETHYL-1,3-DIOXOLAN-4-YLMETHANOL
646-06-0	2,2-DIMETHYL-1,3-DIOXOLAN-4-YLMETHANOL
100-79-8	2,2-DIMETHYL-1,3-DIOXOLANE-4-METHANOL
646-06-0	2,2-DIMETHYL-1,3-DIOXOLANE-4-METHANOL
75-84-3	2,2-DIMETHYL-1-PROPANOL
1563-66-2	2,2-DIMETHYL-2,3-DIHYDRO-7-BENZOFURANYL N-METHYLCAR-BAMATE
79-92-5	2,2-DIMETHYL-3-METHYLENEBICYCLO[221]HEPTANE
79-92-5	2,2-DIMETHYL-3-METHYLENENORBORNANE
100-79-8	2,2-DIMETHYL-4-HYDROXYMETHYLDIOXOLANE
646-06-0	2,2-DIMETHYL-4-HYDROXYMETHYLDIOXOLANE

CAS No.	Chemical Name
100-79-8	2,2-DIMETHYL-4-OXYMETHYL-1,3-DIOXOLANE
646-06-0	2,2-DIMETHYL-4-OXYMETHYL-1,3-DIOXOLANE
100-79-8	2,2-DIMETHYL-5-HYDROXYMETHYL-1,3-DIOXOLANE
646-06-0	2,2-DIMETHYL-5-HYDROXYMETHYL-1,3-DIOXOLANE
1563-66-2	2,2-DIMETHYL-7-COUMARANYL N-METHYLCARBAMATE
75-83-2	2,2-DIMETHYLBUTANE
463-82-1	2,2-DIMETHYLPROPANE
463-82-1	2,2-DIMETHYLPROPANE (DOT)
75-98-9	2,2-DIMETHYLPROPANOIC ACID
3282-30-2	2,2-DIMETHYLPROPANOYL CHLORIDE
75-98-9	2,2-DIMETHYLPROPIONIC ACID
3282-30-2	2,2-DIMETHYLPROPIONIC ACID CHLORIDE
3282-30-2	2,2-DIMETHYLPROPIONYL CHLORIDE
75-84-3	2,2-DIMETHYLPROPYL ALCOHOL
75-99-0	2,2-DPA
111-42-2	2,2-IMINODIETHANOL
677-71-4	2,2-PROPANEDIOL, 1,1,1,3,3,3-HEXAFLUORO-
677-71-4	2,2-PROPANEDIOL, HEXAFLUORO-
97-18-7	2,2-THIOBIS(4,6-DICHLORO-PHENOL
4418-66-0	2,2-THIOBIS(4-CHLORO-6-METHYL)PHENOL
207-08-9	2,3,1',8'-BINAPHTHYLENE
594-56-9	2,3,3-TRIMETHYL-1-BUTENE
594-56-9	2,3,3-TRIMETHYLBUTENE
560-21-4	2,3,3-TRIMETHYLPENTANE
87-86-5	2,3,4,5,6-PENTACHLOROPHENOL
126-33-0	2,3,4,5-TETRAHYDROTHIOPHENE-1,1-DIOXIDE
565-76-4	2,3,4-TRIMETHYL-1-PENTENE
68-76-8	2,3,5-ETHYLENIMINE-1,4-BENZOQUINONE
933-78-8	2,3,5-TRICHLOROPHENOL
68-76-8	2,3,5-TRIETHYLENEIMINO-P-BENZOQUINONE
68-76-8	2,3,5-TRIS(AZIRIDINYL)-1,4-BENZOQUINONE
68-76-8	2,3,5-TRIS(ETHYLENIMINO)-1,4-BENZOQUINONE
68-76-8	2,3,5-TRIS(ETHYLENIMINO)-P-BENZOQUINONE
68-76-8	2,3,5-TRISETHYLENEIMINOBENZOQUINONE
50-31-7	2,3,6-TBA
933-75-5	2,3,6-TRICHLOROPHENOL
1746-01-6	2,3,7,8-TCDD
1746-01-6	2,3,7,8-TETRACHLORODIBENZO-1,4-DIOXIN
1746-01-6	2,3,7,8-TETRACHLORODIBENZO-P-DIOXIN
100-79-8	2,3-(ISOPROPYLIDENEDIOXY)PROPANOL
205-99-2	2,3-BENZFLUORANTHENE
56-55-3	2,3-BENZOPHENANTHRENE
260-94-6	2,3-BENZOQUINOLINE
78-34-2	2,3-BIS(DIETHOXYPHOSPHINOTHIOYLTHIO)-1,4-DIOXANE
431-03-8	2,3-BUTADIONE
431-03-8	2,3-BUTANEDIONE
513-86-0	2,3-BUTANOLONE
2303-16-4	2,3-DCDT
1653-19-6	2,3-DICHLORO-1,3-BUTADIENE
117-80-6	2,3-DICHLORO-1,4-NAPHTHALENEDIONE
117-80-6	2,3-DICHLORO-1,4-NAPHTHOQUINONE
78-88-6	2,3-DICHLORO-1-PROPENE
96-19-5	2,3-DICHLOROALLYL CHLORIDE
2303-16-4	2,3-DICHLOROALLYL N,N-DIISOPROPYLTHIOLCARBAMATE
1653-19-6	2,3-DICHLOROBUTA-1,3-DIYNE
1653-19-6	2,3-DICHLOROBUTADIENE
1653-19-6	2,3-DICHLOROBUTADIENE-1,3
7581-97-7	2,3-DICHLOROBUTANE
117-80-6	2,3-DICHLORONAPHTHOQUINONE-1,4
78-88-6	2,3-DICHLOROPROPENE
78-88-6	2,3-DICHLOROPROPYLENE
3333-52-6	2,3-DICYANO-2,3-DIMETHYLBUTANE
1563-66-2	2,3-DIHYDRO-2,2-DIMETHYL-7-BENZOFURANYL METHYLCARBAMATE
1563-66-2	2,3-DIHYDRO-2,2-DIMETHYLBENZOFURANYL-7-N-METHYLCARBAMATE
81-07-2	2,3-DIHYDRO-3-OXOBENZISOSULFONAZOLE
110-87-2	2,3-DIHYDRO-4H-PYRAN
3174-74-1	2,3-DIHYDRO-4H-PYRAN
110-87-2	2,3-DIHYDROPYRAN
87-69-4	2,3-DIHYDROSUCCINIC ACID
3164-29-2	2,3-DIHYDROXY-BUTANEDIOIC ACID, DIAMMONIUM SALT (9CI)
87-66-1	2,3-DIHYDROXYPHENOL
96-24-2	2,3-DIHYDROXYPROPYL CHLORIDE
431-03-8	2,3-DIKETOBUTANE
357-57-3	2,3-DIMETHOXYSTRYCHNINE
32749-94-3	2,3-DIMETHYL PENTALDEHYDE
563-78-0	2,3-DIMETHYL-1-BUTENE
563-79-1	2,3-DIMETHYL-2-BUTENE
56-25-7	2,3-DIMETHYL-7-OXABICYCLO(221)HEPTANE-2,3-DICARBOXYLIC ANHYDRIDE
87-59-2	2,3-DIMETHYLANILINE
87-59-2	2,3-DIMETHYLBENZENAMINE
79-29-8	2,3-DIMETHYLBUTANE
79-29-8	2,3-DIMETHYLBUTANE (DOT)
42195-92-6	2,3-DIMETHYLCYCLOHEXYLAMINE
584-94-1	2,3-DIMETHYLHEXANE
32749-94-3	2,3-DIMETHYLPENTANAL
565-59-3	2,3-DIMETHYLPENTANE
87-59-2	2,3-DIMETHYLPHENYLAMINE
32749-94-3	2,3-DIMETHYLVALERALDEHYDE
431-03-8	2,3-DIOXOBUTANE
624-92-0	2,3-DITHIABUTANE
122-60-1	2,3-EPOXY-1-PHENOXYPROPANE
765-34-4	2,3-EPOXY-1-PROPANAL
556-52-5	2,3-EPOXY-1-PROPANOL
765-34-4	2,3-EPOXYPROPANAL
75-56-9	2,3-EPOXYPROPANE
765-34-4	2,3-EPOXYPROPIONALDEHYDE
122-60-1	2,3-EPOXYPROPOXYBENZENE
2426-08-6	2,3-EPOXYPROPYL BUTYL ETHER
106-89-8	2,3-EPOXYPROPYL CHLORIDE
106-91-2	2,3-EPOXYPROPYL METHACRYLATE
122-60-1	2,3-EPOXYPROPYL PHENYL ETHER
100-79-8	2,3-ISOPROPYLIDENEGLYCEROL
646-06-0	2,3-ISOPROPYLIDENEGLYCEROL
100-79-8	2,3-O-ISOPROPYLIDENEGLYCEROL
646-06-0	2,3-O-ISOPROPYLIDENEGLYCEROL
78-34-2	2,3-P-DIOXAN-S,S'-BIS(O,O-DIAETHYLDITHIOPHOSPHAT)(GERMAN)
78-34-2	2,3-P-DIOXANDITHIOL S,S-BIS(O,O-DIETHYL PHOSPHORODITHIOATE)
78-34-2	2,3-P-DIOXANE S,S-BIS(O,O-DIETHYLPHOSPHOROITHIOATE)
78-34-2	2,3-P-DIOXANEDITHIOL S,S-BIS(O,O-DIETHYL PHOSPHORODITHIOATE)
57-55-6	2,3-PROPANEDIOL
87-59-2	2,3-XYLIDINE
87-59-2	2,3-XYLYLAMINE
314-40-9	2,4(1H,3H)-PYRIMIDINEDIONE, 5-BROMO-6-METHYL-3-(1-METHYLPROPYL)-
51-21-8	2,4(1H,3H)-PYRIMIDINEDIONE, 5-FLUORO-
66-75-1	2,4(1H,3H)-PYRIMIDINEDIONE, 5-[BIS(2-CHLOROETHYL)AMINO]-
25167-70-8	2,4,4-TRIMETHYL PENTENE
107-39-1	2,4,4-TRIMETHYL-1-PENTENE
5809-08-5	2,4,4-TRIMETHYL-2-HYDROPEROXYPENTANE
141-59-3	2,4,4-TRIMETHYL-2-PENTANETHIOL
107-40-4	2,4,4-TRIMETHYL-2-PENTENE
5809-08-5	2,4,4-TRIMETHYL-2-PENTYL HYDROPEROXIDE
107-45-9	2,4,4-TRIMETHYL-2-PENTYLAMINE
25167-70-8	2,4,4-TRIMETHYLPENTENE
93-76-5	2,4,5-T
93-76-5	2,4,5-T (ACGIH,DOT)
42589-07-1	2,4,5-T AMINE
93-79-8	2,4,5-T BUTOXYETHANOL ESTER
2545-59-7	2,4,5-T BUTOXYETHANOL ESTER
93-79-8	2,4,5-T BUTOXYETHYL ESTER
2545-59-7	2,4,5-T BUTOXYETHYL ESTER
93-79-8	2,4,5-T BUTYL ESTER
2545-59-7	2,4,5-T BUTYL ESTER
6369-97-7	2,4,5-T DIMETHYLAMINE SALT
93-79-8	2,4,5-T ESTER [BUTYL ESTER]
25168-15-4	2,4,5-T [ISOOCTYL ESTER
93-78-7	2,4,5-T ESTER [ISOPROPYL ESTER]
25168-15-4	2,4,5-T [ISOOCTYL ESTER
93-79-8	2,4,5-T N-BUTYL ESTER
2545-59-7	2,4,5-T N-BUTYL ESTER
13560-99-1	2,4,5-T SODIUM SALT
3813-14-7	2,4,5-T TRIETHANOLAMINE SALT
6369-96-6	2,4,5-T TRIMETHYLAMINE SALT
93-72-1	2,4,5-TP
93-72-1	2,4,5-TP (DOT)
95-95-4	2,4,5-TRICHLOROPHENOL
933-78-4	2,4,5-TRICHLOROPHENOL
93-76-5	2,4,5-TRICHLOROPHENOXY ACETIC ACID
93-76-5	2,4,5-TRICHLOROPHENOXYACETIC ACID (DOT)
93-79-8	2,4,5-TRICHLOROPHENOXYACETIC ACID AMINE, ESTER, OR SALT
2545-59-7	2,4,5-TRICHLOROPHENOXYACETIC ACID AMINE, ESTER, OR SALT
2545-59-7	2,4,5-TRICHLOROPHENOXYACETIC ACID BUTOXYETHYL ESTER

CAS No.	Chemical Name
6369-97-7	2,4,5-TRICHLOROPHENOXYACETIC ACID DIMETHYL AMINE SALT
3813-14-7	2,4,5-TRICHLOROPHENOXYACETIC ACID TRIETHANOLAMINE SALT
93-79-8	2,4,5-TRICHLOROPHENOXYACETIC ACID, BUTYL ESTER
2545-59-7	2,4,5-TRICHLOROPHENOXYACETIC ACID, BUTYL ESTER
1319-72-8	2,4,5-TRICHLOROPHENOXYACETIC ACID, ISOPROPANOL AMINE
13560-99-1	2,4,5-TRICHLOROPHENOXYACETIC ACID, SODIUM SALT
93-72-1	2,4,5-TRICHLOROPHENOXYPROPIONIC ACID
93-72-1	2,4,5-TRICHLOROPHENOXYPROPIONIC ACID (DOT)
32534-95-5	2,4,5-TRICHLOROPHENOXYPROPIONIC ACID ESTER
137-17-7	2,4,5-TRIMETHYLANILINE
2217-06-3	2,4,6,2',4',6'-HEXANITRODIPHENYL SULFIDE
100-97-0	2,4,6,2',4',6'-HEXANITRODIPHENYLAMINE
131-73-7	2,4,6,2',4',6'-HEXANITRODIPHENYLAMINE
479-45-8	2,4,6-TETRYL
933-78-4	2,4,6-TRICHLORFENOL (CZECH)
108-77-0	2,4,6-TRICHLORO-1,3,5-TRIAZINE
108-77-0	2,4,6-TRICHLORO-S-TRIAZINE
108-77-0	2,4,6-TRICHLORO-SYM-TRIAZINE
88-06-2	2,4,6-TRICHLOROPHENOL
933-78-4	2,4,6-TRICHLOROPHENOL
108-77-0	2,4,6-TRICHLOROTRIAZINE
675-14-9	2,4,6-TRIFLUORO-1,3,5-TRIAZINE
675-14-9	2,4,6-TRIFLUORO-S-TRIAZINE
675-14-9	2,4,6-TRIFLUOROTRIAZINE
88-05-1	2,4,6-TRIMETHYL ANILINE
123-63-7	2,4,6-TRIMETHYL-1,3,5-TRIOXACYCLOHEXANE
123-63-7	2,4,6-TRIMETHYL-1,3,5-TRIOXANE
123-63-7	2,4,6-TRIMETHYL-S-TRIOXANE
638-17-5	2,4,6-TRIMETHYLDIHYDRO-1,3,5-DITHIAZINE
638-17-5	2,4,6-TRIMETHYLPERHYDRO-1,3-DITHIAZINE
88-05-1	2,4,6-TRIMETHYLPHENYLAMINE
81-15-2	2,4,6-TRINITRO-1,3-DIMETHYL-5-TERT-BUTYLBENZENE
81-15-2	2,4,6-TRINITRO-3,5-DIMETHYL-TERT-BUTYLBENZENE
100-97-0	2,4,6-TRINITRO-N-(2,4,6-TRINITROPHENYL)BENZENAMINE
131-73-7	2,4,6-TRINITRO-N-(2,4,6-TRINITROPHENYL)BENZENAMINE
606-35-9	2,4,6-TRINITROANISOLE
129-66-8	2,4,6-TRINITROBENZOIC ACID
4732-14-3	2,4,6-TRINITROPHENETOLE
88-89-1	2,4,6-TRINITROPHENOL
131-74-8	2,4,6-TRINITROPHENOL AMMONIUM SALT
479-45-8	2,4,6-TRINITROPHENYL METHYLNITRAMINE
2217-06-3	2,4,6-TRINITROPHENYL SULFIDE
479-45-8	2,4,6-TRINITROPHENYL-N-METHYLNITRAMINE
479-45-8	2,4,6-TRINITROPHENYLMETHYLNITROAMINE
82-71-3	2,4,6-TRINITRORESORCINOL
118-96-7	2,4,6-TRINITROTOLUEEN (DUTCH)
118-96-7	2,4,6-TRINITROTOLUENE
118-96-7	2,4,6-TRINITROTOLUENE (ACGIH)
118-96-7	2,4,6-TRINITROTOLUENE (TNT)
118-96-7	2,4,6-TRINITROTOLUOL (GERMAN)
97-36-9	2,4-ACETOACETOXYLIDIDE
94-75-7	2,4-D
94-75-7	2,4-D (ACGIH,DOT)
1929-73-3	2,4-D 2-BUTOXYETHYL ESTER
94-75-7	2,4-D ACID
1929-73-3	2,4-D BUTOXYETHANOL ESTER
1929-73-3	2,4-D BUTOXYETHYL ESTER
94-80-4	2,4-D BUTYL ESTER
2971-38-2	2,4-D CHLOROCROTYL ESTER
25168-26-7	2,4-D ISOOCTYL ESTER
94-11-1	2,4-D ISOPROPYL ESTER
1928-38-7	2,4-D METHYL ESTER
1320-18-9	2,4-D PROPYLENE GLYCOL BUTYL ETHER ESTER
94-80-4	2,4-DBE
1929-73-3	2,4-DBEE
136-78-7	2,4-DES SODIUM
138-00-1	2,4-DI-N-PENTYLPHENOL
95-80-7	2,4-DIAMINO-1-METHYLBENZENE
615-05-4	2,4-DIAMINOANISOLE
39156-41-7	2,4-DIAMINOANISOLE SULFATE
39156-41-7	2,4-DIAMINOANISOLE SULPHATE
95-80-7	2,4-DIAMINOTOLUENE
138-00-1	2,4-DIAMYLPHENOL
1836-75-5	2,4-DICHLORO-4'-NITRODIPHENYL ETHER
101-03-3	2,4-DICHLORO-6-(2-CHLOROANILINO)-1,3,5-TRIAZINE
12654-97-6	2,4-DICHLORO-6-(2-CHLOROANILINO)-1,3,5-TRIAZINE
101-05-3	2,4-DICHLORO-6-(O-CHLOROANILINO)-S-TRIAZINE
101-05-3	2,4-DICHLORO-6-O-CHLORANILINO-S-TRIAZINE
12654-97-6	2,4-DICHLORO-6-O-CHLORANILINO-S-TRIAZINE
470-90-6	2,4-DICHLORO-ALPHA-(CHLOROMETHYLENE)BENZYLDIETHYL PHOSPHATE
133-14-2	2,4-DICHLOROBENZOYL PEROXIDE
133-14-2	2,4-DICHLOROBENZOYL PEROXIDE, NOT MORE THAN 52% AS A PASTE (DOT)
133-14-2	2,4-DICHLOROBENZOYL PEROXIDE, NOT MORE THAN 52% IN SOLUTION (DOT)
120-83-2	2,4-DICHLOROPHENOL
94-75-7	2,4-DICHLOROPHENOXYACETIC ACID
94-75-7	2,4-DICHLOROPHENOXYACETIC ACID (DOT)
94-75-7	2,4-DICHLOROPHENOXYACETIC ACID (HIGH VOLATILE ESTERS)
1929-73-3	2,4-DICHLOROPHENOXYACETIC ACID BUTOXYETHANOL ESTER
1929-73-3	2,4-DICHLOROPHENOXYACETIC ACID BUTOXYETHYL ESTER
94-80-4	2,4-DICHLOROPHENOXYACETIC ACID BUTYL ESTER
25168-26-7	2,4-DICHLOROPHENOXYACETIC ACID ISOOCTYL ESTER
94-11-1	2,4-DICHLOROPHENOXYACETIC ACID ISOPROPYL ESTER
1928-38-7	2,4-DICHLOROPHENOXYACETIC ACID METHYL ESTER
94-80-4	2,4-DICHLOROPHENOXYACETIC ACID N-BUTYL ESTER
1320-18-9	2,4-DICHLOROPHENOXYACETIC ACID, PROPYLENE GLYCOL BUTYL ETHER ESTER
94-75-7	2,4-DICHLOROPHENOXYETHANOIC ACID
1836-75-5	2,4-DICHLOROPHENYL 4-NITROPHENYL ETHER
1836-75-5	2,4-DICHLOROPHENYL P-NITROPHENYL ETHER
94-75-7	2,4-DICHLORPHENOXYACETIC ACID
1464-53-5	2,4-DIEPOXYBUTANE
107-41-5	2,4-DIHYDROXY-2-METHYLPENTANE
584-84-9	2,4-DIISOCYANATO-1-METHYLBENZENE
26471-62-5	2,4-DIISOCYANATO-1-METHYLBENZENE
26471-62-5	2,4-DIISOCYANATO-1-METHYLBENZENE (9CI)
584-84-9	2,4-DIISOCYANATOTOLUENE
26471-62-5	2,4-DIISOCYANATOTOLUENE
26419-73-8	2,4-DIMETHYL-1,3-DITHIOLANE-2-CARBOXALDEHYDE O-METHYLCARBAMOYLOXIME
1068-87-7	2,4-DIMETHYL-3-ETHYLPENTANE
589-43-5	2,4-DIMETHYLHEXANE
108-08-7	2,4-DIMETHYLPENTANE
105-67-9	2,4-DIMETHYLPHENOL
97-02-9	2,4-DINITRANILINE
97-00-7	2,4-DINITRO-1-CHLOROBENZENE
25567-67-3	2,4-DINITRO-1-CHLOROBENZENE
14797-65-0	2,4-DINITRO-1-THIOCYANOBENZENE
88-85-7	2,4-DINITRO-6-(1-METHYL-PROPYL)PHENOL (FRENCH)
88-85-7	2,4-DINITRO-6-(1-METHYLPROPYL)PHENOL
131-89-5	2,4-DINITRO-6-CYCLOHEXYLPHENOL
1335-85-9	2,4-DINITRO-6-METHYLPHENOL
534-52-1	2,4-DINITRO-6-METHYLPHENOL
88-85-7	2,4-DINITRO-6-SEC-BUTYLPHENOL
1420-07-1	2,4-DINITRO-6-TERT-BUTYLPHENOL
1335-85-9	2,4-DINITRO-O-CRESOL
14797-65-0	2,4-DINITRO-RHODANBENZOL (GERMAN)
97-02-9	2,4-DINITROANILIN (GERMAN)
97-02-9	2,4-DINITROANILINA (ITALIAN)
97-02-9	2,4-DINITROANILINE
97-00-7	2,4-DINITROCHLOROBENZENE
25567-67-3	2,4-DINITROCHLOROBENZENE
51-28-5	2,4-DINITROFENOL (DUTCH)
51-28-5	2,4-DINITROPHENOL
97-00-7	2,4-DINITROPHENYL CHLORIDE
25567-67-3	2,4-DINITROPHENYL CHLORIDE
14797-65-0	2,4-DINITROPHENYL THIOCYANATE
519-44-8	2,4-DINITRORESORCINOL
35860-51-6	2,4-DINITRORESORCINOL
14797-65-0	2,4-DINITROTHIOCYANATOBENZENE
14797-65-0	2,4-DINITROTHIOCYANOBENZENE
121-14-2	2,4-DINITROTOLUENE
121-14-2	2,4-DINITROTOLUOL
109-87-5	2,4-DIOXAPENTANE
123-54-6	2,4-DIOXOPENTANE
138-00-1	2,4-DIPENTYLPHENOL
541-53-7	2,4-DITHIOBIURET
51-28-5	2,4-DNP
121-14-2	2,4-DNT
94-75-7	2,4-DWUCHLOROFENOKSYOCTOWY KWAS (POLISH)
142-83-6	2,4-HEXADIENAL
80466-34-8	2,4-HEXADIENAL
118-52-5	2,4-IMIDAZOLIDINEDIONE, 1,3-DICHLORO-5,5-DIMETHYL-
57-41-0	2,4-IMIDAZOLIDINEDIONE, 5,5-DIPHENYL-
94-75-7	2,4-PA

CAS No.	Chemical Name
123-54-6	2,4-PENTADIONE
107-41-5	2,4-PENTANEDIOL, 2-METHYL-
123-54-6	2,4-PENTANEDIONE
123-54-6	2,4-PENTANEDIONE (DOT)
37187-22-7	2,4-PENTANEDIONE PEROXIDE, WITH NO MORE THAN 9% BY WEIGHT ACTIVE OXYGEN
37187-22-7	2,4-PENTANEDIONE, PEROXIDE
584-84-9	2,4-TDI
26471-62-5	2,4-TDI
584-84-9	2,4-TOLUENE DIISOCYANATE
26471-62-5	2,4-TOLUENE DIISOCYANATE
95-80-7	2,4-TOLUENEDIAMINE
25376-45-8	2,4-TOLUYLENEDIAMINE
584-84-9	2,4-TOLYLENE DIISOCYANATE
26471-62-5	2,4-TOLYLENE DIISOCYANATE
95-80-7	2,4-TOLYLENEDIAMINE
105-67-9	2,4-XYLENOL
1300-71-6	2,4-XYLENOL
111-96-6	2,5,8-TRIOXANONANE
2618-77-1	2,5-BIS(BENZOYLDIOXY)-2,5-DIMETHYLHEXANE
2618-77-1	2,5-BIS(BENZOYLPEROXY)-2,5-DIMETHYLHEXANE
3025-88-5	2,5-BIS(HYDROPEROXY)-2,5-DIMETHYLHEXANE
78-63-7	2,5-BIS(TERT-BUTYLDIOXY)-2,5-DIMETHYLHEXANE
1068-27-5	2,5-BIS(TERT-BUTYLPEROXY)-2,5-DIMETHYL-3-HEXYNE
78-63-7	2,5-BIS(TERT-BUTYLPEROXY)-2,5-DIMETHYLHEXANE
106-51-4	2,5-CYCLOHEXADIENE-1,4-DIONE
68-76-8	2,5-CYCLOHEXADIENE-1,4-DIONE, 2,3,5-TRIS(1-AZIRIDINYL)-
2618-77-1	2,5-DIBENZOYLPEROXY-2,5-DIMETHYLHEXANE
1918-00-9	2,5-DICHLORO-6-METHOXYBENZOIC ACID
2275-14-1	2,5-DICHLOROPHENYLTHIOMETHYL O,O-DIETHYL PHOSPHORO-DITHIOATE
3025-88-5	2,5-DIHYDROPEROXY-2,5-DIMETHYLHEXANE
2935-44-6	2,5-DIHYDROXYHEXANE
102-56-7	2,5-DIMETHOXYANILINE
102-56-7	2,5-DIMETHOXYBENZENAMINE
6358-53-8	2,5-DIMETHOXYBENZENEAZO-.BETA.-NAPHTHOL
2100-42-7	2,5-DIMETHOXYCHLOROBENZENE
2618-77-1	2,5-DIMETHYL-2,5-BIS(BENZOYLPEROXY)HEXANE
3025-88-5	2,5-DIMETHYL-2,5-BIS(HYDROPEROXY)HEXANE
78-63-7	2,5-DIMETHYL-2,5-BIS(TERT-BUTYLDIOXY)HEXANE
78-63-7	2,5-DIMETHYL-2,5-BIS(TERT-BUTYLPEROXY)HEXANE
78-63-7	2,5-DIMETHYL-2,5-BIS-(TERT-BUTYLPEROXY)HEXANE, MAX CONCENTRATION 52% WITH INERT SOLID (DOT)
2618-77-1	2,5-DIMETHYL-2,5-DI(BENZOYLPEROXY)HEXANE
78-63-7	2,5-DIMETHYL-2,5-DI(T-BUTYLPEROXY)HEXANE
1068-27-5	2,5-DIMETHYL-2,5-DI(T-BUTYLPEROXY)HEXYNE-3
1068-27-5	2,5-DIMETHYL-2,5-DI(TERT-BUTYLPEROXY)-3-HEXYNE
78-63-7	2,5-DIMETHYL-2,5-DI(TERT-BUTYLPEROXY)HEXANE
2618-77-1	2,5-DIMETHYL-2,5-DI-(BENZOYLPEROXY)HEXANE TECHNICALLY PURE (DOT)
2618-77-1	2,5-DIMETHYL-2,5-DI-(BENZOYLPEROXY)HEXANE, NOT MORE THAN 82% WITH INERT SOLID (DOT)
2618-77-1	2,5-DIMETHYL-2,5-DI-(BENZOYLPEROXY)HEXANE, TECHNICALLY PURE
78-63-7	2,5-DIMETHYL-2,5-DI-(TERT-BUTYLPEROXY)HEXANE, TECHNI-CALLY PURE
78-63-7	2,5-DIMETHYL-2,5-DI-(TERT-BUTYLPEROXY)HEXANE, TECHNI-CALLY PURE (DOT)
1068-27-5	2,5-DIMETHYL-2,5-DI-(tert-BUTYLPEROXY)HEXYNE-3, MAXIMUM 52% PEROXIDE IN INERT SOLID
1068-27-5	2,5-DIMETHYL-2,5-DI-(TERT-BUTYLPEROXY)HEXYNE-3, TECHNI-CALLY PURE
1068-27-5	2,5-DIMETHYL-2,5-DI-(TERT-BUTYLPEROXY)HEXYNE-3, TECHNI-CALLY PURE (DOT)
110-18-9	2,5-DIMETHYL-2,5-DIAZAHEXANE
3025-88-5	2,5-DIMETHYL-2,5-DIHYDROPEROXYHEXANE
2618-77-1	2,5-DIMETHYL-2,5-HEXANEDIHYDROPEROXIDE DIBENZOATE
2618-77-1	2,5-DIMETHYL-2,5-HEXANEDIOL BIS(PEROXYBENZOATE)
2618-77-1	2,5-DIMETHYL-2,5-HEXANEDIOL DIPEROXYBENZOATE
2618-77-1	2,5-DIMETHYL-2,5-HEXANEDIYL BIS(PEROXYBENZOATE)
2618-77-1	2,5-DIMETHYL-2,5-HEXANEDIYL PEROXYBENZOATE
3025-88-5	2,5-DIMETHYL-2,5-HEXYLENE DIHYDROPEROXIDE
625-86-5	2,5-DIMETHYLFURAN
3025-88-5	2,5-DIMETHYLHEXANE 2,5-BIS(HYDROPEROXIDE)
3025-88-5	2,5-DIMETHYLHEXANE-2,5-DIHYDROPEROXIDE
2618-77-1	2,5-DIMETHYLHEXANE-2,5-DIPEROXYBENZOATE
2618-77-1	2,5-DIMETHYLHEXANE-2,5-DIYL DIPERBENZOATE
3025-88-5	2,5-DIMETHYLHEXYL-2,5-DIHYDROPEROXIDE
329-71-5	2,5-DINITROPHENOL

CAS No.	Chemical Name
110-71-4	2,5-DIOXAHEXANE
142-46-1	2,5-DITHIOBIUREA
329-71-5	2,5-DNP
108-31-6	2,5-FURANDIONE
504-20-1	2,5-HEPTADIEN-4-ONE, 2,6-DIMETHYL-
2935-44-6	2,5-HEXANEDIOL
2618-77-1	2,5-HEXANEDIOL, 2,5-DIMETHYL-, BIS(PEROXYBENZOATE)
1024-57-3	2,5-METHANO-2H-INDENO[1,2-B]OXIRENE, 2,3,4,5,6,7,7-HEPTA-CHLORO-1A,1B,5,5A,6,6A-HEXAHYDRO-, (1A ALPHA
121-46-0	2,5-NORBORNADIENE
28324-52-9	2,6,6-TRIMETHYL NORPINANYL HYDROPEROXIDE, TECHNICALLY PURE (DOT)
80-56-8	2,6,6-TRIMETHYLBICYCLO[3.1.1]HEPT-2-ENE
123-17-1	2,6,8-TRIMETHYL-4-NONANOL
123-18-2	2,6,8-TRIMETHYL-4-NONANONE
128-37-0	2,6-BIS(1,1-DIMETHYLETHYL)-4-METHYLPHENOL
128-37-0	2,6-DI-TERT-BUTYL-4-HYDROXYTOLUENE
128-37-0	2,6-DI-TERT-BUTYL-4-METHYLPHENOL
128-37-0	2,6-DI-TERT-BUTYL-P-CRESOL
128-37-0	2,6-DI-TERT-BUTYL-P-METHYLPHENOL
128-37-0	2,6-DI-TERT-BUTYLCRESOL
128-37-0	2,6-DI-TERT-BUTYLMETHYLPHENOL
94-78-0	2,6-DIAMINO-3-PHENYLAZOPYRIDINE
136-40-3	2,6-DIAMINO-3-PHENYLAZOPYRIDINE HYDROCHLORIDE
136-40-3	2,6-DIAMINO-3-PHENYLAZOPYRIDINE MONOHYDROCHLORIDE
1194-65-6	2,6-DICHLOROBENZONITRILE
1194-65-6	2,6-DICHLOROCYANOBENZENE
66-75-1	2,6-DIHYDROXY-5-BIS[2-CHLOROETHYL]AMINOPYRIMIDINE
91-08-7	2,6-DIISOCYANATO-1-METHYLBENZENE
91-08-7	2,6-DIISOCYANATOTOLUENE
108-82-7	2,6-DIMETHYL HEPTANOL-4
2971-90-6	2,6-DIMETHYL-3,5-DICHLORO-4-PYRIDINOL
108-82-7	2,6-DIMETHYL-4-HEPTANOL
108-83-8	2,6-DIMETHYL-4-HEPTANONE
108-83-8	2,6-DIMETHYL-HEPTAN-4-ON (DUTCH, GERMAN)
87-62-7	2,6-DIMETHYLANILINE
87-62-7	2,6-DIMETHYLBENZENAMINE
108-83-8	2,6-DIMETHYLHEPTAN-4-ONE
108-83-8	2,6-DIMETHYLHEPTANONE
141-91-3	2,6-DIMETHYLMORPHOLINE
108-83-8	2,6-DIMETIL-EPTAN-4-ONE (ITALIAN)
4726-14-1	2,6-DINITRO-4-METHYLSULFONYL-N,N-DIPROPYLANILINE
1582-09-8	2,6-DINITRO-4-TRIFLUORMETHYL-N,N-DIPROPYLANILIN (GER-MAN)
1582-09-8	2,6-DINITRO-N,N-DI-N-PROPYL-ALPHA,ALPHA,ALPHA-TRI-FLUORO-P-TOLUIDINE
1582-09-8	2,6-DINITRO-N,N-DIPROPYL-4-(TRIFLUOROMETHYL)BEN-ZENAMINE
573-56-8	2,6-DINITROPHENOL
606-20-2	2,6-DINITROTOLUENE
606-20-2	2,6-DNT
66-81-9	2,6-PIPERIDINEDIONE, 4-[2-(3,5-DIMETHYL-2-OXOCYCLOHEXYL)-2-HYDROXYETHYL]-, [1S-[1 ALPHA (S*),3 ALPHA
94-78-0	2,6-PYRIDINEDIAMINE, 3-(PHENYLAZO)-
136-40-3	2,6-PYRIDINEDIAMINE, 3-(PHENYLAZO)-, MONOHYDROCHLORIDE
91-08-7	2,6-TDI
91-08-7	2,6-TOLUENE DIISOCYANATE
87-62-7	2,6-XYLIDENE
87-62-7	2,6-XYLIDINE
87-62-7	2,6-XYLYLAMINE
33857-26-0	2,7-DICHLORODIBENZO-P-DIOXIN
2602-46-2	2,7-NAPHTHALENEDISULFONIC ACID, 3,3'-[[1,1'-BIPHENYL]-4,4'-DIYLBIS(AZO)]BISP5-AMINO-4-HYDROXY-, TETR
3564-09-8	2,7-NAPHTHALENEDISULFONIC ACID, 3-HYDROXY-4-[(2,4,5-TRIMETHYLPHENYL)AZO]-, DISODIUM SALT
1937-37-7	2,7-NAPHTHALENEDISULFONIC ACID, 4-AMINO-3-[[4'-[(2,4-DIAMINOPHENYL)AZO][1,1'-BIPHENYL]-4-YL]AZO]-5-H
3761-53-3	2,7-NAPHTHALENEDISULFONIC ACID, 4-[(2,4-DIMETHYLPHENYL) AZO]-3-HYDROXY-, DISODIUM SALT
2097-19-0	2,8,9-TRIOXA-5-AZA-1-SILABICYCLO(333)UNDECANE, PHENYL-
1120-23-6	2,BETA-BUTOXYETHOXYETHYL CHLORIDE
3691-35-8	2-((4-CHLOROPHENYL)PHENYLACETYL)-1H-INDENE-1,3(2H)-DIONE
51-83-2	2-((AMINOCARBONYL)OXY)-N,N,N-TRIMETHYLETHANAMINIUM CHLORIDE
3691-35-8	2-((P-CHLOROPHENYL)PHENYLACETYL)-1,3-INDANDIONE
1420-07-1	2-(1,1-DIMETHYLETHYL)-4,6-DINITROPHENOL
114-26-1	2-(1-METHYLETHOXY)PHENYL N-METHYLCARBAMATE
88-85-7	2-(1-METHYLPROPYL)-4,6-DINITROPHENOL

CAS No.	Chemical Name
89-72-5	2-(1-METHYLPROPYL)PHENOL
3878-19-1	2-(2'-FURYL)BENZIMIDAZOLE
83-26-1	2-(2,2-DIMETHYL-1-OXOPROPYL)-1H-INDENE-1,3(2H)-DIONE
93-72-1	2-(2,4,5-TRICHLOROPHENOXY)PROPIONIC ACID
112-50-5	2-(2-(2-ETHOXYETHOXY)ETHOXY)ETHANOL
24800-44-0	2-(2-(2-HYDROXYPROPOXY)PROPOXY-1-PROPANOL
929-06-6	2-(2-AMINOETHOXY)ETHANOL
929-06-6	2-(2-AMINOETHOXY)ETHANOL (DOT)
112-34-5	2-(2-BUTOXYETHOXY) ETHANOL
124-17-4	2-(2-BUTOXYETHOXY)ETHANOL ACETATE
124-17-4	2-(2-BUTOXYETHOXY)ETHYL ACETATE
101-05-3	2-(2-CHLORANILIN)-4,6-DICHLOR-1,3,5-TRIAZIN (GERMAN)
12654-97-6	2-(2-CHLORANILIN)-4,6-DICHLOR-1,3,5-TRIAZIN (GERMAN)
112-26-5	2-(2-CHLOROETHOXY)ETHYL 2'-CHLOROETHYL ETHER
111-90-0	2-(2-ETHOXYETHOXY)ETHANOL
3878-19-1	2-(2-FURYL)BENZIMIDAZOLE
111-46-6	2-(2-HYDROXYETHOXY)ETHANOL
929-06-6	2-(2-HYDROXYETHOXY)ETHYLAMINE
111-41-1	2-(2-HYDROXYETHYLAMINO)ETHYLAMINE
111-96-6	2-(2-METHOXYETHOXY)-1-METHOXYETHANE
111-77-3	2-(2-METHOXYETHOXY)ETHANOL
3691-35-8	2-(2-PHENYL-2-(4-CHLOROPHENYL)ACETYL)-1,3-INDANDIONE
2312-35-8	2-(4-(1,1-DIMETHYLETHYL)PHENOXY)CYCLOHEXYL 2-PROPYNYL SULFITE
622-40-2	2-(4-MORPHOLINYL)ETHANOL
2038-03-1	2-(4-MORPHOLINYL)ETHYLAMINE
50-78-2	2-(ACETYLOXY)BENZOIC ACID
818-61-1	2-(ACRYLOYLOXY)ETHANOL
3691-35-8	2-(ALPHA-P-CHLOROPHENYLACETYL)INDANE-1,3-DIONE
617-89-0	2-(AMINOMETHYL)FURAN
622-08-2	2-(BENZYLOXY)ETHANOL
3132-64-7	2-(BROMOMETHYL)OXIRANE
106-89-8	2-(CHLOROMETHYL)OXIRANE
2842-38-8	2-(CYCLOHEXYLAMINO)ETHANOL
102-81-8	2-(DIBUTYLAMINO)ETHANOL
947-02-4	2-(DIETHOXYPHOSPHINYLIMINO)-1,3-DITHIOLANE
950-10-7	2-(DIETHOXYPHOSPHINYLIMINO)-4-METHYL-1,3-DITHIOLANE
100-37-8	2-(DIETHYLAMINO)ETHANOL
2426-54-2	2-(DIETHYLAMINO)ETHYL ACRYLATE
100-37-8	2-(DIETHYLAMINO)ETHYL ALCOHOL
100-36-7	2-(DIETHYLAMINO)ETHYLAMINE
96-80-0	2-(DIISOPROPYLAMINO)ETHYL ALCOHOL
108-01-0	2-(DIMETHYLAMINO)-1-ETHANOL
23135-22-0	2-(DIMETHYLAMINO)-N-(((METHYLAMINO)CARBONYL)OXY)-2-OXOETHANIMIDOTHIOIC ACID METHYL ESTER
82-66-6	2-(DIPHENYLACETYL)-1H-INDENE-1,3(2H)-DIONE
82-66-6	2-(DIPHENYLACETYL)INDAN-1,3-DIONE
92-50-2	2-(ETHYLPHENYLAMINO)ETHANOL
929-06-6	2-(HYDROXYETHOXY)ETHYLAMINE
556-52-5	2-(HYDROXYMETHYL)OXIRANE
434-07-1	2-(HYDROXYMETHYLENE)-17-METHYLDIHYDROTESTOSTERONE
119-36-8	2-(METHOXYCARBONYL)PHENOL
109-83-1	2-(METHYLAMINO)ETHANOL
93-90-3	2-(METHYLPHENYLAMINO)ETHANOL
100-37-8	2-(N,N-DIETHYLAMINO)ETHANOL
100-36-7	2-(N,N-DIETHYLAMINO)ETHYLAMINE
108-01-0	2-(N,N-DIMETHYLAMINO)ETHANOL
2867-47-2	2-(N,N-DIMETHYLAMINO)ETHYL METHACRYLATE
644-64-4	2-(N,N-DIMETHYLCARBAMYL)-3-METHYLPYRAZOLYL-5 N,N-DIMETHYLCARBAMATE
92-50-2	2-(N-ETHYL-N-PHENYLAMINO)ETHANOL
92-50-2	2-(N-ETHYLANILINO)ETHANOL
93-90-3	2-(N-METHYL-N-PHENYLAMINO)ETHANOL
109-83-1	2-(N-METHYLAMINO)ETHANOL
93-90-3	2-(N-METHYLANILINO)ETHANOL
2312-35-8	2-(P-T-BUTYLPHENOXY)CYCLOHEXYL PROPARGYL SULFITE
2312-35-8	2-(P-TERT-BUTYLPHENOXY)CYCLOHEXYL 2-PROPYNYL SULFITE
2312-35-8	2-(P-TERT-BUTYLPHENOXY)CYCLOHEXYL PROPARGYL SULFITE
140-57-8	2-(P-TERT-BUTYLPHENOXY)ISOPROPYL 2-CHLOROETHYL SULFITE
122-98-5	2-(PHENYLAMINO)ETHANOL
135-88-6	2-(PHENYLAMINO)NAPHTHALENE
3775-90-4	2-(TERT-BUTYLAMINO)ETHYL METHACRYLATE
21564-17-0	2-(THIOCYANOMETHYLTHIO)BENZOTHIAZOLE
21564-17-0	2-(THIOCYANOMETHYLTHIO)BENZOTHIAZOLE, 60%
777-37-7	2-(TRIFLUOROMETHYL)-4-NITROCHLOROBENZENE
88-17-5	2-(TRIFLUOROMETHYL)ANILINE
88-17-5	2-(TRIFLUOROMETHYL)BENZENAMINE
83-26-1	2-(TRIMETHYLACETYL)-1,3-INDANDIONE
83-26-1	2-(TRIMETIL-ACETIL)-INDAN-1,3-DIONE (ITALIAN)
75-83-2	2-2-DIMETHYLBUTANE
53-96-3	2-AAF
513-49-5	2-AB
13952-84-6	2-AB
53-96-3	2-ACETAMIDOFLUORENE
92-15-9	2-ACETOACETYLAMINOANISOLE
542-58-5	2-ACETOXY-1-CHLOROETHANE
50-78-2	2-ACETOXYBENZOIC ACID
105-46-4	2-ACETOXYBUTANE
542-58-5	2-ACETOXYETHYL CHLORIDE
626-38-0	2-ACETOXYPENTANE
53496-15-4	2-ACETOXYPENTANE
108-21-4	2-ACETOXYPROPANE
108-22-5	2-ACETOXYPROPENE
108-22-5	2-ACETOXYPROPYLENE
53-96-3	2-ACETYLAMINOFLUORENE
563-80-4	2-ACETYLPROPANE
111-76-2	2-AETHOXY-AETHYLACETAT (GERMAN)
1912-24-9	2-AETHYLAMINO-4-CHLOR-6-ISOPROPYLAMINO-1,3,5-TRIAZIN (GERMAN)
104-76-7	2-AETHYLHEXANOL (GERMAN)
61-82-5	2-AMINO-1,3,4-TRIAZOLE
88-05-1	2-AMINO-1,3,5-TRIMETHYLBENZENE
87-62-7	2-AMINO-1,3-DIMETHYLBENZENE
87-62-7	2-AMINO-1,3-XYLENE
96-20-8	2-AMINO-1-BUTANOL
141-43-5	2-AMINO-1-ETHANOL
124-68-5	2-AMINO-1-HYDROXY-2-METHYLPROPANE
96-20-8	2-AMINO-1-HYDROXYBUTANE
99-59-2	2-AMINO-1-METHOXY-4-NITROBENZENE
95-53-4	2-AMINO-1-METHYLBENZENE
78-96-6	2-AMINO-1-METHYLETHANOL
7568-93-6	2-AMINO-1-PHENYLETHANOL
929-06-6	2-AMINO-2'-HYDROXYDIETHYL ETHER
124-68-5	2-AMINO-2,2-DIMETHYLETHANOL
124-68-5	2-AMINO-2-METHYL-1-PROPANOL
75-64-9	2-AMINO-2-METHYLPROPANE
124-68-5	2-AMINO-2-METHYLPROPANOL
95-83-0	2-AMINO-4-CHLOROANILINE
95-85-2	2-AMINO-4-CHLOROPHENOL
95-85-2	2-AMINO-4-CHLOROPHENOL (DOT)
95-79-4	2-AMINO-4-CHLOROTOLUENE
108-09-8	2-AMINO-4-METHYLPENTANE
99-59-2	2-AMINO-4-NITROANISOLE
99-55-8	2-AMINO-4-NITROTOLUENE
712-68-5	2-AMINO-5(5-NITRO-2-FURYL)-1,3,4-THIADIZOLE
95-69-2	2-AMINO-5-CHLOROTOLUENE
3165-93-3	2-AMINO-5-CHLOROTOLUENE HYDROCHLORIDE
140-80-7	2-AMINO-5-DIETHYLAMINOPENTANE
140-80-7	2-AMINO-5-DIETHYLAMINOPENTANE (DOT)
121-66-4	2-AMINO-5-NITROTHIAZOLE
117-79-3	2-AMINO-9,10-ANTHRAQUINONE
87-62-7	2-AMINO-M-XYLENE
75-31-0	2-AMINO-PROPAAN (DUTCH)
75-31-0	2-AMINO-PROPANO (ITALIAN)
90-04-0	2-AMINOANISOLE
29191-52-4	2-AMINOANISOLE
117-79-3	2-AMINOANTHRAQUINONE
88-17-5	2-AMINOBENZOTRIFLUORIDE
90-41-5	2-AMINOBIPHENYL
513-49-5	2-AMINOBUTANE
13952-84-6	2-AMINOBUTANE
513-49-5	2-AMINOBUTANE BASE
96-20-8	2-AMINOBUTYL ALCOHOL
90-41-5	2-AMINODIPHENYL
141-43-5	2-AMINOETHANOL
929-06-6	2-AMINOETHOXYETHANOL
929-06-6	2-AMINOETHYL 2-HYDROXYETHYL ETHER
75-64-9	2-AMINOISOBUTANE
124-68-5	2-AMINOISOBUTANOL
88-05-1	2-AMINOMESITYLENE
4795-29-3	2-AMINOMETHYLTETRAHYDROFURAN
91-59-8	2-AMINONAPHTHALENE
88-74-4	2-AMINONITROBENZENE
625-30-9	2-AMINOPENTANE
94-70-2	2-AMINOPHENETOLE
75-31-0	2-AMINOPROPAN (GERMAN)
75-31-0	2-AMINOPROPANE
504-29-0	2-AMINOPYRIDINE

CAS No.	Chemical Name
95-53-4	2-AMINOTOLUENE
61-82-5	2-AMINOTRIAZOLE
122-98-5	2-ANILINOETHANOL
135-88-6	2-ANILINONAPHTHALENE
135-02-4	2-ANISALDEHYDE
90-04-0	2-ANISIDINE
29191-52-4	2-ANISIDINE
105-60-2	2-AZACYCLOHEPTANONE
17804-35-2	2-BENZIMIDAZOLECARBAMIC ACID, 1-(BUTYLCARBAMOYL)-, METHYL ESTER
149-30-4	2-BENZOTHIAZOLETHIOL
140-82-9	2-BETA-DIETHYLAMINOETHOXYETHANOL
90-41-5	2-BIPHENYLAMINE
76-22-2	2-BORNANONE
70-11-1	2-BROMO-1-PHENYLETHANONE
507-19-7	2-BROMO-2-METHYLPROPANE
507-19-7	2-BROMO-2-METHYLPROPANE (DOT)
70-11-1	2-BROMOACETOPHENONE
598-21-0	2-BROMOACETYL BROMIDE
78-76-2	2-BROMOBUTANE
78-76-2	2-BROMOBUTANE (DOT)
592-55-2	2-BROMOETHOXYETHANE
592-55-2	2-BROMOETHYL ETHYL ETHER
592-55-2	2-BROMOETHYL ETHYL ETHER (DOT)
507-19-7	2-BROMOISOBUTANE
29756-38-5	2-BROMOPENTANE
29756-38-5	2-BROMOPENTANE (DOT)
95-46-5	2-BROMOTOLUENE
513-49-5	2-BUTANAMINE
13952-84-6	2-BUTANAMINE
513-49-5	2-BUTANAMINE, (S)-
626-23-3	2-BUTANAMINE, N-(1-METHYLPROPYL)-
513-53-1	2-BUTANETHIOL
2084-18-6	2-BUTANETHIOL, 3-METHYL-
78-92-2	2-BUTANOL
105-46-4	2-BUTANOL ACETATE
75-85-4	2-BUTANOL, 2-METHYL-
625-16-1	2-BUTANOL, 2-METHYL-, ACETATE
598-75-4	2-BUTANOL, 3-METHYL-
513-86-0	2-BUTANOL-3-ONE
78-93-3	2-BUTANONE
126-39-6	2-BUTANONE ETHYLENE KETAL
1338-23-4	2-BUTANONE PEROXIDE, MAXIMUM CONCENTRATION 60%
1338-23-4	2-BUTANONE PEROXIDE, WITH NOT MORE THAN 9% BY WEIGHT ACTIVE OXYGEN
39196-18-4	2-BUTANONE, 3,3-DIMETHYL-1-(METHYLTHIO)-, O-((METHYL-AMINO)CARBONYL)OXIME
513-86-0	2-BUTANONE, 3-HYDROXY-
563-80-4	2-BUTANONE, 3-METHYL-
126-39-6	2-BUTANONE, CYCLIC 1,2-ETHANEDIYL ACETAL
1338-23-4	2-BUTANONE, PEROXIDE
6117-91-5	2-BUTEN-1-OL
123-73-9	2-BUTENAL
4170-30-3	2-BUTENAL
123-73-9	2-BUTENAL (9CI)
4170-30-3	2-BUTENAL (9CI)
123-73-9	2-BUTENAL, (E)-
4170-30-3	2-BUTENAL, (E)-
107-01-7	2-BUTENE
624-64-6	2-BUTENE, (E)-
590-18-1	2-BUTENE, (Z)-
41070-66-9	2-BUTENE, 1,1,1,2,3,4,4,4-OCTAFLUORO-
926-57-8	2-BUTENE, 1,3-DICHLORO-
4784-77-4	2-BUTENE, 1-BROMO-
591-97-9	2-BUTENE, 1-CHLORO-
563-79-1	2-BUTENE, 2,3-DIMETHYL-
4461-41-0	2-BUTENE, 2-CHLORO-
513-35-9	2-BUTENE, 2-METHYL-
26760-64-5	2-BUTENE, 2-METHYL-
590-18-1	2-BUTENE-CIS
624-64-6	2-BUTENE-TRANS
110-17-8	2-BUTENEDIOIC ACID, (E)-
110-16-7	2-BUTENEDIOIC ACID, (Z)-
627-63-4	2-BUTENEDIOYL DICHLORIDE, (E)-
3724-65-0	2-BUTENOIC ACID
3724-65-0	2-BUTENOIC ACID (9CI)
303-34-4	2-BUTENOIC ACID, 2-METHYL-, 7-[[2,3-DIHYDROXY-2-(1-METHOXYETHYL)-3-METHYL-1-OXYBUTOXY]METHYL]-2,3,5,
7786-34-7	2-BUTENOIC ACID, 3-((DIMETHOXYPHOSPHINYL)OXY)-, METHYL ESTER (9CI)
7786-34-7	2-BUTENOIC ACID, 3-[(DIMETHOXYPHOSPHINYL)OXY]-, METHYL ESTER
14861-06-4	2-BUTENOIC ACID, ETHENYL ESTER
10544-63-5	2-BUTENOIC ACID, ETHYL ESTER
10544-63-5	2-BUTENOIC ACID, ETHYL ESTER, (E)- (9CI)
78-94-4	2-BUTENONE
6117-91-5	2-BUTENYL ALCOHOL
4784-77-4	2-BUTENYL BROMIDE
591-97-9	2-BUTENYL CHLORIDE
111-76-2	2-BUTOXY ETHANOL
124-16-3	2-BUTOXY-1-(2-HYDROXYPROPOXY)ETHANE
111-76-2	2-BUTOXY-1-ETHANOL
1929-73-3	2-BUTOXYETHYL 2,4-DICHLOROPHENOXYACETATE
105-46-4	2-BUTYL ACETATE
78-92-2	2-BUTYL ALCOHOL
78-76-2	2-BUTYL BROMIDE
513-53-1	2-BUTYL MERCAPTAN
13952-84-6	2-BUTYLAMINE
149-57-5	2-BUTYLBUTANOIC ACID
503-17-3	2-BUTYNE
503-17-3	2-BUTYNE (8CI,9CI)
110-65-6	2-BUTYNE-1,4-DIOL
10070-67-0	2-BUTYNE-1,4-DIOL
110-65-6	2-BUTYNEDIOL
10070-67-0	2-BUTYNEDIOL
76-22-2	2-CAMPHANONE
7786-34-7	2-CARBOMETHOXY-1-METHYLVINYL DIMETHYL PHOSPHATE
7786-34-7	2-CARBOMETHOXY-1-PROPEN-2-YL DIMETHYL PHOSPHATE
69-72-7	2-CARBOXYPHENOL
50-78-2	2-CARBOXYPHENYL ACETATE
535-89-7	2-CHLOOR-4-DIMETHYLAMINO-6-METHYL-PYRIMIDINE (DUTCH)
107-07-3	2-CHLOORETHANOL (DUTCH)
535-89-7	2-CHLOR-4-DIMETHYLAMINO-6-METHYLPYRIMIDIN (GERMAN)
107-07-3	2-CHLORAETHANOL (GERMAN)
999-81-5	2-CHLORAETHYL-TRIMETHYLAMMONIUMCHLORID (GERMAN)
95-06-7	2-CHLORALLYL DIETHYLDITHIOCARBAMATE
107-07-3	2-CHLORETHANOL (GERMAN)
107-14-2	2-CHLORO ACETONITRILE
88-16-4	2-CHLORO(TRIFLUOROMETHYL)BENZENE
777-37-7	2-CHLORO- ALPHA, ALPHA, ALPHA -TRIFLUORO-5-NITROTOL-UENE
88-16-4	2-CHLORO- ALPHA, ALPHA, ALPHA -TRIFLUOROTOLUENE
79-38-9	2-CHLORO-1,1,2-TRIFLUOROETHYLENE
97-97-2	2-CHLORO-1,1-DIMETHOXYETHANE
126-99-8	2-CHLORO-1,3-BUTADIENE
2100-42-7	2-CHLORO-1,4-DIMETHOXYBENZENE
470-90-6	2-CHLORO-1-(2,4-DICHLOROPHENYL)VINYL DIETHYL PHOSPHATE
107-20-0	2-CHLORO-1-ETHANAL
107-07-3	2-CHLORO-1-ETHANOL
95-49-8	2-CHLORO-1-METHYLBENZENE
88-73-3	2-CHLORO-1-NITROBENZENE
532-27-4	2-CHLORO-1-PHENYLETHANONE
78-89-7	2-CHLORO-1-PROPANOL
557-98-2	2-CHLORO-1-PROPENE
4461-41-0	2-CHLORO-2-BUTENE
594-36-5	2-CHLORO-2-METHYLBUTANE
507-20-0	2-CHLORO-2-METHYLPROPANE
594-71-8	2-CHLORO-2-NITROPROPANE
29027-17-6	2-CHLORO-3-METHYL ANILINE (CHLOROTOLUIDINE)
42350-99-2	2-CHLORO-4,6-DI-TERT-AMYL-PHENOL
1912-24-9	2-CHLORO-4-ETHYLAMINEISOPROPYLAMINE-S-TRIAZINE
1912-24-9	2-CHLORO-4-(2-PROPYLAMINO)-6-ETHYLAMINO-S-TRIAZINE
535-89-7	2-CHLORO-4-(DIMETHYLAMINO)-6-METHYLPRIMONE
1912-24-9	2-CHLORO-4-(ETHYLAMINO)-6-(ISOPROPYLAMINO)TRIAZINE
1912-24-9	2-CHLORO-4-ETHYLAMINO-6-ISOPROPYLAMINO-S-TRIAZINE
1912-24-9	2-CHLORO-4-ETHYLAMINO-6-ISOPROPYLAMINO-1,3,5-TRIAZINE
615-65-6	2-CHLORO-4-METHYL ANILINE (CHLOROTOLUIDINE)
535-89-7	2-CHLORO-4-METHYL-6-(DIMETHYLAMINO)PYRIMIDINE
92-04-6	2-CHLORO-4-PHENYLPHENOL
98-28-2	2-CHLORO-4-TERT-BUTYLPHENOL
121-17-5	2-CHLORO-5-(TRIFLUOROMETHYL)NITROBENZENE
59-50-7	2-CHLORO-5-HYDROXYTOLUENE
1321-10-4	2-CHLORO-5-HYDROXYTOLUENE
95-81-8	2-CHLORO-5-METHYL ANILINE (CHLOROTOLUIDINE)
777-37-7	2-CHLORO-5-NITRO-1-TRIFLUOROMETHYLBENZENE
777-37-7	2-CHLORO-5-NITROBENZOTRIFLUORIDE
1929-82-4	2-CHLORO-6-(TRICHLOROMETHYL)PYRIDINE
59-50-7	2-CHLORO-HYDROXYTOLUENE

CAS No.	*Chemical Name*
1321-10-4	2-CHLORO-HYDROXYTOLUENE
999-81-5	2-CHLORO-N,N,N-TRIMETHYLETHANAMINIUM CHLORIDE
535-89-7	2-CHLORO-N,N-6-TRIMETHYL-4-PYRIMIDINAMINE
61702-44-1	2-CHLORO-P-PHENYLENEDIAMINE SULFATE
92-04-6	2-CHLORO-P-PHENYLPHENOL
107-20-0	2-CHLOROACETALDEHYDE
97-97-2	2-CHLOROACETALDEHYDE DIMETHYL ACETAL
532-27-4	2-CHLOROACETOPHENONE
80-63-7	2-CHLOROACRYLIC ACID METHYL ESTER
78-88-6	2-CHLOROALLYL CHLORIDE
95-06-7	2-CHLOROALLYL DIETHYLDITHIOCARBAMATE
95-06-7	2-CHLOROALLYL N,N-DIETHYLDITHIOCARBAMATE
2698-41-1	2-CHLOROBENZALMALONONITRILE
88-16-4	2-CHLOROBENZOTRIFLUORIDE
2698-41-1	2-CHLOROBENZYLIDENEMALONINITRILE
2698-41-1	2-CHLOROBENZYLIDENEMALONONITRILE
126-99-8	2-CHLOROBUTADIENE
78-86-4	2-CHLOROBUTANE
4461-41-0	2-CHLOROBUTENE-2
107-20-0	2-CHLOROETHANAL
107-07-3	2-CHLOROETHANOL
542-58-5	2-CHLOROETHYL ACETATE
107-07-3	2-CHLOROETHYL ALCOHOL
627-11-2	2-CHLOROETHYL CHLOROCARBONATE
627-11-2	2-CHLOROETHYL CHLOROFORMATE
111-44-4	2-CHLOROETHYL ETHER
110-75-8	2-CHLOROETHYL VINYL ETHER
507-20-0	2-CHLOROISOBUTANE
91-58-7	2-CHLORONAPHTHALENE
88-73-3	2-CHLORONITROBENZENE
95-57-8	2-CHLOROPHENOL
5344-82-1	2-CHLOROPHENYL THIOUREA
75-29-6	2-CHLOROPROPANE
17639-93-9	2-CHLOROPROPANOIC ACID METHYL ESTER
78-89-7	2-CHLOROPROPANOL
557-98-2	2-CHLOROPROPENE
557-98-2	2-CHLOROPROPENE (DOT)
79435-04-4	2-CHLOROPROPIONIC ACID ISOPROPYL ESTER
17639-93-9	2-CHLOROPROPIONIC ACID METHYL ESTER
78-89-7	2-CHLOROPROPYL ALCOHOL
557-98-2	2-CHLOROPROPYLENE
109-09-1	2-CHLOROPYRIDINE
29154-12-9	2-CHLOROPYRIDINE
29154-12-9	2-CHLOROPYRIDINE (DOT)
2039-87-4	2-CHLOROSTYRENE
95-49-8	2-CHLOROTOLUENE
541-25-3	2-CHLOROVINYLDICHLOROARSINE
535-89-7	2-CLORO-4-DIMETILAMINO-6-METIL-PIRIMIDINA (ITALIAN)
107-07-3	2-CLOROETANOLO (ITALIAN)
95-48-7	2-CRESOL
109-78-4	2-CYANO-1-ETHANOL
126-98-7	2-CYANO-1-PROPENE
75-86-5	2-CYANO-2-HYDROXYPROPANE
75-86-5	2-CYANO-2-PROPANOL
372-09-8	2-CYANOACETIC ACID
137-05-3	2-CYANOACRYLIC ACID METHYL ESTER
109-78-4	2-CYANOETHANOL
106-71-8	2-CYANOETHYL ACRYLATE
109-78-4	2-CYANOETHYL ALCOHOL
110-67-8	2-CYANOETHYL METHYL ETHER
106-71-8	2-CYANOETHYL PROPENOATE
75-86-5	2-CYANOPROPAN-2-OL
78-82-0	2-CYANOPROPANE
126-98-7	2-CYANOPROPENE
126-98-7	2-CYANOPROPENE-1
2244-16-8	2-CYCLOHEXEN-1-ONE, 2-METHYL-5-(1-METHYLETHENYL)-, (S)-
78-59-1	2-CYCLOHEXEN-1-ONE, 3,5,5-TRIMETHYL-
131-89-5	2-CYCLOHEXYL-4,6-DINITROFENOL (DUTCH)
131-89-5	2-CYCLOHEXYL-4,6-DINITROPHENOL
119-42-6	2-CYCLOHEXYLPHENOL
584-79-2	2-CYCLOPENTEN-1-ONE, 2-ALLYL-4-HYDROXY-3-METHYL-, 2,2-DIMETHYL-3-(2-METHYLPROPENYL)CYCLOPROPANECARBO
23505-41-1	2-DIETHYLAMINO-6-METHYLPYRIMIDIN-4-YL DIETHYLPHOS-PHOROTHIONATE
100-37-8	2-DIETHYLAMINOETHANOL (ACGIH)
96-80-0	2-DIISOPROPYLAMINOETHANOL
141-66-2	2-DIMETHYL CIS-2-DIMETHYL-CARBAMOYL-1-METHYLVINYL PHOSPHATE
23135-22-0	2-DIMETHYLAMINO-1-(METHYLTHIO)GLYOXAL O-METHYLCAR-BAMOYLMONOXIME
926-64-7	2-DIMETHYLAMINOACETO-NITRILE
926-64-7	2-DIMETHYLAMINOACETONITRILE (DOT)
108-01-0	2-DIMETHYLAMINOETHANOL
2867-47-2	2-DIMETHYLAMINOETHYL METHACRYLATE
644-64-4	2-DIMETHYLCARBAMOYL-3-METHYL-5-PYRAZOLYL DIMETHYL-CARBAMATE
644-64-4	2-DIMETHYLCARBAMOYL-3-METHYLPYRAZOLYL-(5)-N,N-DI-METHYLCARBAMAT (GERMAN)
82-66-6	2-DIPHENYLACETYL-1,3-DIKETOHYDRINDENE
82-66-6	2-DIPHENYLACETYL-1,3-INDANDIONE
56-18-8	2-DIPROPYLENETRIAMINE
141-43-5	2-ETHANOLAMINE
110-90-0	2-ETHOXY ETHANOL
103-75-3	2-ETHOXY-2,3-DIHYDRO-4H-PYRAN
103-75-3	2-ETHOXY-2,3-DIHYDRO-GAMMA-PYRAN
103-75-3	2-ETHOXY-3,4-DIHYDRO-1,2-PYRAN
103-75-3	2-ETHOXY-3,4-DIHYDRO-2H-PYRAN
103-75-3	2-ETHOXY-3,4-DIHYDROPYRAN
103-95-3	2-ETHOXY-3-4-DIHYDRO-2-PYRAN
111-76-2	2-ETHOXY-ETHYLACETAAT (DUTCH)
94-70-2	2-ETHOXYANILINE
103-75-3	2-ETHOXYDIHYDROPYRAN, IN PREGNANCY DIAGNOSIS
110-80-5	2-ETHOXYETHANOL
111-15-9	2-ETHOXYETHANOL ACETATE
111-76-2	2-ETHOXYETHANOL, ESTER WITH ACETIC ACID
111-15-9	2-ETHOXYETHYL ACETATE
111-76-2	2-ETHOXYETHYL ACETATE
111-76-2	2-ETHOXYETHYL ACETATE (ACGIH)
110-80-5	2-ETHOXYETHYL ALCOHOL
592-55-2	2-ETHOXYETHYL BROMIDE
629-14-1	2-ETHOXYETHYL ETHYL ETHER
111-76-2	2-ETHOXYETHYLE, ACETATE DE (FRENCH)
578-54-1	2-ETHYL BENZENAMINE
104-76-7	2-ETHYL HEXANOL
104-75-6	2-ETHYL HEXYLAMINE
625-27-4	2-ETHYL-1,1-DIMETHYLETHYLENE
97-95-0	2-ETHYL-1-BUTANOL
760-21-4	2-ETHYL-1-BUTENE
104-75-6	2-ETHYL-1-HEXANAMINE
149-57-5	2-ETHYL-1-HEXANOIC ACID
104-76-7	2-ETHYL-1-HEXANOL
103-09-3	2-ETHYL-1-HEXANOL ACETATE
298-07-7	2-ETHYL-1-HEXANOL HYDROGEN PHOSPHATE
103-09-3	2-ETHYL-1-HEXYL ACETATE
104-75-6	2-ETHYL-1-HEXYLAMINE
115-84-4	2-ETHYL-2-BUTYL-1-3 PROPANEDIOL
115-84-4	2-ETHYL-2-BUTYLPROPANEDIOL-1,3
645-62-5	2-ETHYL-2-HEXENAL
5309-52-4	2-ETHYL-2-HEXENOIC ACID
126-39-6	2-ETHYL-2-METHYL-1,3-DIOXOLANE
126-39-6	2-ETHYL-2-METHYLDIOXOLANE
75-85-4	2-ETHYL-2-PROPANOL
496-03-7	2-ETHYL-3-HYDROXYHEXANAL
645-62-5	2-ETHYL-3-PROPYLACROLEIN
5309-52-4	2-ETHYL-3-PROPYLACRYLIC ACID
88-09-5	2-ETHYL-N-BUTYRIC ACID
110-73-6	2-ETHYLAMINO-1-ETHANOL
110-73-6	2-ETHYLAMINOETHANOL
578-54-1	2-ETHYLANILINE
578-54-1	2-ETHYLANILINE (DOT)
97-96-1	2-ETHYLBUTANAL
88-09-5	2-ETHYLBUTANOIC ACID
97-95-0	2-ETHYLBUTANOL
10031-87-5	2-ETHYLBUTYL ACETATE
3953-10-4	2-ETHYLBUTYL ACRYLATE
97-95-0	2-ETHYLBUTYL ALCOHOL
97-96-1	2-ETHYLBUTYRALDEHYDE
97-96-1	2-ETHYLBUTYRALDEHYDE (DOT)
88-09-5	2-ETHYLBUTYRIC ACID
97-96-1	2-ETHYLBUTYRIC ALDEHYDE
149-57-5	2-ETHYLCAPROIC ACID
123-05-7	2-ETHYLHEXALDEHYDE
123-05-7	2-ETHYLHEXANAL
104-75-6	2-ETHYLHEXANAMINE
149-57-5	2-ETHYLHEXANOIC ACID
94-04-2	2-ETHYLHEXANOIC ACID, VINYL ESTER
103-09-3	2-ETHYLHEXANYL ACETATE
645-62-5	2-ETHYLHEXENAL

CAS No.	Chemical Name
149-57-5	2-ETHYLHEXOIC ACID
103-11-7	2-ETHYLHEXYL 2-PROPENOATE
103-09-3	2-ETHYLHEXYL ACETATE
103-11-7	2-ETHYLHEXYL ACRYLATE
104-76-7	2-ETHYLHEXYL ALCOHOL
123-04-6	2-ETHYLHEXYL CHLORIDE
24468-13-1	2-ETHYLHEXYL CHLOROFORMATE
24468-13-1	2-ETHYLHEXYL CHLOROFORMATE (DOT)
103-09-3	2-ETHYLHEXYL ETHANOATE
117-81-7	2-ETHYLHEXYL PHTHALATE
577-11-7	2-ETHYLHEXYL SULFOSUCCINATE SODIUM
103-44-6	2-ETHYLHEXYL VINYL ETHER
123-05-7	2-ETHYLHEXYLALDEHYDE
104-75-6	2-ETHYLHEXYLAMINE
104-75-6	2-ETHYLHEXYLAMINE (DOT)
106-88-7	2-ETHYLOXIRANE
589-34-4	2-ETHYLPENTANE
686-31-7	2-ETHYLPEROXYHEXANOIC ACID TERT-PENTYL ESTER
578-54-1	2-ETHYLPHENYLAMINE
640-15-3	2-ETHYLTHIOETHYL O,O-DIMETHYL PHOSPHORODITHIOATE
111-76-2	2-ETOS SIETIL-ACETATO (ITALIAN)
15271-41-7	2-EXO-CHLORO-6-ENDO-CYANO-2-NORBORNANONE-O-(METHYL-CARBAMOYL)OXIME
53-96-3	2-FAA
98-82-8	2-FENILPROPANO (ITALIAN)
98-82-8	2-FENYL-PROPAAN (DUTCH)
640-19-7	2-FLUOROACETAMIDE
144-49-0	2-FLUOROACETIC ACID
371-62-0	2-FLUOROETHANOL
100-73-2	2-FORMYL-3,4-DIHYDRO-2H-PYRAN
96-17-3	2-FORMYLBUTANE
98-01-1	2-FORMYLFURAN
123-15-9	2-FORMYLPENTANE
90-02-8	2-FORMYLPHENOL
98-01-1	2-FURALDEHYDE
3688-53-7	2-FURANACETAMIDE, ALPHA-[(5-NITRO-2-FURANYL)METHYL-ENE]-
3688-53-7	2-FURANACRYLAMIDE, ALPHA-2-FURYL-5-NITRO-
98-01-1	2-FURANALDEHYDE
98-00-0	2-FURANCARBINOL
98-01-1	2-FURANCARBONAL
98-01-1	2-FURANCARBOXALDEHYDE
617-89-0	2-FURANMETHANAMINE
4795-29-3	2-FURANMETHANAMINE, TETRAHYDRO-
98-00-0	2-FURANMETHANOL
97-99-4	2-FURANMETHANOL, TETRAHYDRO-
617-89-0	2-FURANMETHYLAMINE
98-00-0	2-FURANYLMETHANOL
617-89-0	2-FURANYLMETHYLAMINE
98-01-1	2-FURFURAL
98-01-1	2-FURFURALDEHYDE
98-00-0	2-FURFUROL
98-00-0	2-FURFURYL ALCOHOL
98-00-0	2-FURFURYLALKOHOL (CZECH)
617-89-0	2-FURFURYLAMINE
98-01-1	2-FURIL-METANALE (ITALIAN)
98-01-1	2-FURYL-METHANAL
98-01-1	2-FURYLALDEHYDE
98-00-0	2-FURYLCARBINOL
98-01-1	2-FURYLCARBOXALDEHYDE
98-00-0	2-FURYLMETHANOL
617-89-0	2-FURYLMETHYLAMINE
110-43-0	2-HEPTANONE
14686-13-6	2-HEPTENE, (E)-
591-78-6	2-HEXANONE
110-12-3	2-HEXANONE, 5-METHYL-
645-62-5	2-HEXENAL, 2-ETHYL-
592-43-8	2-HEXENE(MIXED CIS & TRANS)
7688-21-3	2-HEXENE-CIS
5309-52-4	2-HEXENOIC ACID, 2-ETHYL-
764-35-2	2-HEXYNE
75-91-2	2-HYDROPEROXY-2-METHYLPROPANE
616-29-5	2-HYDROXY-1,3-DIAMINOPROPANE
616-29-5	2-HYDROXY-1,3-PROPANEDIAMINE
60-24-2	2-HYDROXY-1-ETHANETHIOL
78-96-6	2-HYDROXY-1-PROPANAMINE
78-96-6	2-HYDROXY-1-PROPYLAMINE
75-86-5	2-HYDROXY-2-CYANOPROPANE
115-19-5	2-HYDROXY-2-METHYL-3-BUTYNE
75-86-5	2-HYDROXY-2-METHYLPROPANENITRILE
75-86-5	2-HYDROXY-2-METHYLPROPIONITRILE
7568-93-6	2-HYDROXY-2-PHENYLETHYLAMINE
513-86-0	2-HYDROXY-3-BUTANONE
95-85-2	2-HYDROXY-5-CHLOROANILINE
109-83-1	2-HYDROXY-N-METHYLETHYLAMINE
107-16-4	2-HYDROXYACETONITRILE
90-02-8	2-HYDROXYBENZALDEHYDE
69-72-7	2-HYDROXYBENZENECARBOXYLIC ACID
69-72-7	2-HYDROXYBENZOIC ACID
119-36-8	2-HYDROXYBENZOIC ACID METHYL ESTER
78-92-2	2-HYDROXYBUTANE
95-57-8	2-HYDROXYCHLOROBENZENE
109-78-4	2-HYDROXYCYANOETHANE
141-43-5	2-HYDROXYETHANAMINE
60-24-2	2-HYDROXYETHANETHIOL
107-21-1	2-HYDROXYETHANOL
818-61-1	2-HYDROXYETHYL ACRYLATE
107-07-3	2-HYDROXYETHYL CHLORIDE
109-78-4	2-HYDROXYETHYL CYANIDE
60-24-2	2-HYDROXYETHYL MERCAPTAN
109-83-1	2-HYDROXYETHYL-N-METHYLAMINE
141-43-5	2-HYDROXYETHYLAMINE
929-06-6	2-HYDROXYETHYLOXYETHYLAMINE
75-86-5	2-HYDROXYISOBUTYRONITRILE
124-68-5	2-HYDROXYMETHYL-2-PROPYLAMINE
98-00-0	2-HYDROXYMETHYLFURAN
88-75-5	2-HYDROXYNITROBENZENE
120-80-9	2-HYDROXYPHENOL
78-96-6	2-HYDROXYPROPANAMINE
50-21-5	2-HYDROXYPROPANOIC ACID
138-22-7	2-HYDROXYPROPANOIC ACID BUTYL ESTER
57-55-6	2-HYDROXYPROPANOL
50-21-5	2-HYDROXYPROPIONIC ACID
78-97-7	2-HYDROXYPROPIONITRILE
999-61-1	2-HYDROXYPROPYL ACRYLATE
78-96-6	2-HYDROXYPROPYLAMINE
95-48-7	2-HYDROXYTOLUENE
100-37-8	2-HYDROXYTRIETHYLAMINE
96-45-7	2-IMIDAZOLIDINETHIONE
61-57-4	2-IMIDAZOLIDINONE, 1-(5-NITRO-2-THIAZOLYL)-
555-84-0	2-IMIDAZOLIDINONE, 1-[(5-NITROFURFURYLIDENE)AMINO]-
555-84-0	2-IMIDAZOLIDINONE, 1-[[(5-NITRO-2-FURANYL)METHYLENE] AMINO]-
96-45-7	2-IMIDAZOLINE-2-THIOL
513-48-4	2-IODOBUTANE
25267-27-0	2-IODOBUTANE
25267-27-0	2-IODOBUTANE (DOT)
75-66-1	2-ISOBUTANETHIOL
4439-24-1	2-ISOBUTOXYETHANOL
30674-80-7	2-ISOCYANATOETHYL METHACRYLATE
1795-48-8	2-ISOCYANATOPROPANE
109-59-1	2-ISOPROPOXYETHANOL
114-26-1	2-ISOPROPOXYPHENYL METHYLCARBAMATE
114-26-1	2-ISOPROPOXYPHENYL N-METHYLCARBAMATE
108-20-3	2-ISOPROPOXYPROPANE
109-56-8	2-ISOPROPYLAMINOETHANOL
76-22-2	2-KETO-1,7,7-TRIMETHYLNORCAMPHANE
105-60-2	2-KETOHEXAMETHYLENEIMINE
105-60-2	2-KETOHEXAMETHYLENIMINE
60-24-2	2-ME
60-24-2	2-MERCAPTO-1-ETHANOL
96-45-7	2-MERCAPTO-2-IMIDAZOLINE
68-11-1	2-MERCAPTOACETIC ACID
513-53-1	2-MERCAPTOBUTANE
68-11-1	2-MERCAPTOETHANOIC ACID
60-24-2	2-MERCAPTOETHANOL
60-24-2	2-MERCAPTOETHYL ALCOHOL
96-45-7	2-MERCAPTOIMIDAZOLINE
75-33-2	2-MERCAPTOPROPANE
79-42-5	2-MERCAPTOPROPANOIC ACID
79-42-5	2-MERCAPTOPROPIONIC ACID
513-42-8	2-METHALLYL ALCOHOL
563-47-3	2-METHALLYL CHLORIDE
90-04-0	2-METHOXY-1-AMINOBENZENE
29191-52-4	2-METHOXY-1-AMINOBENZENE
109-86-4	2-METHOXY-1-ETHANOL
107-98-2	2-METHOXY-1-METHYLETHANOL
1918-00-9	2-METHOXY-3,6-DICHLOROBENZOIC ACID

CAS No.	Chemical Name
120-71-8	2-METHOXY-5-METHYLANILINE
99-59-2	2-METHOXY-5-NITROANILINE
99-59-2	2-METHOXY-5-NITROBENZENAMINE
109-86-4	2-METHOXY-AETHANOL (GERMAN)
110-49-6	2-METHOXY-ETHYL ACETAAT (DUTCH)
92-15-9	2-METHOXYACETOACETANILIDE
110-49-6	2-METHOXYAETHYLACETAT (GERMAN)
90-04-0	2-METHOXYANILINE
29191-52-4	2-METHOXYANILINE
134-29-2	2-METHOXYANILINE HYDROCHLORIDE
135-02-4	2-METHOXYBENZALDEHYDE
90-04-0	2-METHOXYBENZENAMINE
29191-52-4	2-METHOXYBENZENAMINE
135-02-4	2-METHOXYBENZENECARBOXALDEHYDE
7786-34-7	2-METHOXYCARBONYL-1-METHYLVINYL DIMETHYL PHOSPHATE
109-86-4	2-METHOXYETHANOL
109-86-4	2-METHOXYETHANOL (ACGIH)
110-49-6	2-METHOXYETHANOL ACETATE
110-49-6	2-METHOXYETHYL ACETATE
110-49-6	2-METHOXYETHYL ACETATE (ACGIH)
109-86-4	2-METHOXYETHYL ALCOHOL
1663-35-0	2-METHOXYETHYL ETHENYL ETHER
1663-35-0	2-METHOXYETHYL VINYL ETHER
110-49-6	2-METHOXYETHYLE, ACETATE DE (FRENCH)
110-66-7	2-METHYL 2-BUTANETHIOL
75-85-4	2-METHYL BUTANOL-2
75-86-5	2-METHYL LACTONITRILE
107-41-5	2-METHYL PENTANE-2,4-DIOL
78-83-1	2-METHYL PROPANOL
110-19-0	2-METHYL PROPYL ACETATE
643-58-3	2-METHYL-1,1'-BIPHENYL
118-96-7	2-METHYL-1,3,5-TRINITROBENZENE
78-79-5	2-METHYL-1,3-BUTADIENE
78-79-5	2-METHYL-1,3-BUTADIENE (DOT)
606-20-2	2-METHYL-1,3-DINITROBENZENE
1118-58-7	2-METHYL-1,3-PENTADIENE
149-31-5	2-METHYL-1,3-PENTANEDIOL
109-08-0	2-METHYL-1,4-DIAZINE
95-53-4	2-METHYL-1-AMINOBENZENE
82-28-0	2-METHYL-1-ANTHRAQUINONYLAMINE
137-32-6	2-METHYL-1-BUTANOL
814-78-8	2-METHYL-1-BUTEN-3-ONE
78-80-8	2-METHYL-1-BUTEN-3-YNE
563-46-2	2-METHYL-1-BUTENE
563-46-2	2-METHYL-1-BUTENE (TECHNICAL)
78-80-8	2-METHYL-1-BUTENYNE
513-36-0	2-METHYL-1-CHLOROPROPANE
583-60-8	2-METHYL-1-CYCLOHEXANONE
129-15-7	2-METHYL-1-NITROANTHRAQUINONE
88-72-2	2-METHYL-1-NITROBENZENE
105-30-6	2-METHYL-1-PENTANOL
54972-97-3	2-METHYL-1-PENTANOL
763-29-1	2-METHYL-1-PENTENE
538-93-2	2-METHYL-1-PHENYLPROPANE
78-84-2	2-METHYL-1-PROPANAL
78-81-9	2-METHYL-1-PROPANAMINE
78-83-1	2-METHYL-1-PROPANOL
115-11-7	2-METHYL-1-PROPENE
110-19-0	2-METHYL-1-PROPYL ACETATE
107-41-5	2-METHYL-2,4-PENTANDIOL
107-41-5	2-METHYL-2,4-PENTANEDIOL
116-06-3	2-METHYL-2-(METHYLTHIO)PROPANAL, O-((METHYLAMINO)CARBONYL)OXIME
116-06-3	2-METHYL-2-(METHYLTHIO)PROPIONALDEHYDE O-(METHYLCARBAMOYL)OXIME
124-68-5	2-METHYL-2-AMINO-1-PROPANOL
75-64-9	2-METHYL-2-AMINOPROPANE
124-68-5	2-METHYL-2-AMINOPROPANOL
507-19-7	2-METHYL-2-BROMOPROPANE
75-85-4	2-METHYL-2-BUTANOL
513-35-9	2-METHYL-2-BUTENE
26760-64-5	2-METHYL-2-BUTENE
513-35-9	2-METHYL-2-BUTENE (DOT)
26760-64-5	2-METHYL-2-BUTENE (DOT)
625-16-1	2-METHYL-2-BUTYL ACETATE
115-19-5	2-METHYL-2-BUTYNOL
594-36-5	2-METHYL-2-CHLOROBUTANE
507-20-0	2-METHYL-2-CHLOROPROPANE
126-39-6	2-METHYL-2-ETHYL-1,3-DIOXOLANE
126-39-6	2-METHYL-2-ETHYLDIOXOLANE
115-19-5	2-METHYL-2-HYDROXY-3-BUTYNE
75-85-4	2-METHYL-2-HYDROXYBUTANE
116-06-3	2-METHYL-2-METHYLTHIO-PROPIONALDEHYD-O-(N-METHYL-CARBAMOYL)-OXIM (GERMAN)
123-42-2	2-METHYL-2-PENTANOL-4-ONE
141-79-7	2-METHYL-2-PENTEN-4-ONE
625-27-4	2-METHYL-2-PENTENE
141-79-7	2-METHYL-2-PENTENONE-4
98-06-6	2-METHYL-2-PHENYLPROPANE
75-64-9	2-METHYL-2-PROPANAMINE
75-66-1	2-METHYL-2-PROPANETHIOL
75-65-0	2-METHYL-2-PROPANOL
513-42-8	2-METHYL-2-PROPEN-1-OL
78-85-3	2-METHYL-2-PROPENAL
126-98-7	2-METHYL-2-PROPENENITRILE
79-41-4	2-METHYL-2-PROPENOIC ACID
760-93-0	2-METHYL-2-PROPENOIC ACID ANHYDRIDE
80-62-6	2-METHYL-2-PROPENOIC ACID METHYL ESTER
513-42-8	2-METHYL-2-PROPENOL
563-47-3	2-METHYL-2-PROPENYL CHLORIDE
507-20-0	2-METHYL-2-PROPYL CHLORIDE
105-30-6	2-METHYL-2-PROPYLETHANOL
54972-97-3	2-METHYL-2-PROPYLETHANOL
148-01-6	2-METHYL-3,5-DINITROBENZAMIDE
84-80-0	2-METHYL-3-(3,7,11,15-TETRAMETHYL-2-HEXADECENYL)-1,4-NAPHTHALENEDIONE
598-75-4	2-METHYL-3-BUTANOL
563-45-1	2-METHYL-3-BUTENE
115-19-5	2-METHYL-3-BUTYN-2-OL
563-47-3	2-METHYL-3-CHLOROPROPENE
609-26-7	2-METHYL-3-ETHYLPENTANE
4461-48-7	2-METHYL-3-PENTENE
84-80-0	2-METHYL-3-PHYTYL-1,4-NAPHTHOCHINON (GERMAN)
534-52-1	2-METHYL-4,6-DINITROPHENOL
123-51-3	2-METHYL-4-BUTANOL
95-69-2	2-METHYL-4-CHLOROANILINE
95-79-4	2-METHYL-4-CHLOROANILINE
3165-93-3	2-METHYL-4-CHLOROANILINE HYDROCHLORIDE
3074-75-7	2-METHYL-4-ETHYLHEXANE
141-79-7	2-METHYL-4-OXO-2-PENTENE
108-11-2	2-METHYL-4-PENTANOL
54972-97-3	2-METHYL-4-PENTANOL
108-10-1	2-METHYL-4-PENTANONE
98-27-1	2-METHYL-4-TERT-BUTYLPHENOL
19594-40-2	2-METHYL-4-UNDECANONE
97-56-3	2-METHYL-4-[(O-TOLYL)AZO]ANILINE
95-79-4	2-METHYL-5-CHLOROANILINE
104-89-2	2-METHYL-5-ETHYLPIPERIDINE
104-90-5	2-METHYL-5-ETHYLPYRIDINE
104-90-5	2-METHYL-5-ETHYLPYRIDINE (DOT)
110-12-3	2-METHYL-5-HEXANONE
99-55-8	2-METHYL-5-NITROANILINE
140-76-1	2-METHYL-5-VINYLPYRIDINE
563-47-3	2-METHYL-ALLYLCHLORID (GERMAN)
97-88-1	2-METHYL-BUTYLACRYLAAT (DUTCH)
97-88-1	2-METHYL-BUTYLACRYLAT (GERMAN)
97-88-1	2-METHYL-BUTYLACRYLATE
91-08-7	2-METHYL-M-PHENYLENE ISOCYANATE
91-08-7	2-METHYL-META-PHENYLENE DIISOCYANATE
91-08-7	2-METHYL-META-PHENYLENE ISOCYANATE
137-32-6	2-METHYL-N-BUTANOL
763-29-1	2-METHYL-PENTENE-1
78-85-3	2-METHYLACROLEIN
79-41-4	2-METHYLACRYLIC ACID
126-98-7	2-METHYLACRYLONITRILE
513-42-8	2-METHYLALLYL ALCOHOL
563-47-3	2-METHYLALLYL CHLORIDE
109-83-1	2-METHYLAMINO-1-ETHANOL
109-83-1	2-METHYLAMINOETHANOL
95-53-4	2-METHYLANILINE
636-21-5	2-METHYLANILINE HYDROCHLORIDE
75-55-8	2-METHYLAZIRIDINE
95-53-4	2-METHYLBENZENAMINE
643-58-3	2-METHYLBIPHENYL
95-46-5	2-METHYLBROMOBENZENE
78-79-5	2-METHYLBUTADIENE
563-80-4	2-METHYLBUTAN-3-ONE
96-17-3	2-METHYLBUTANAL

CAS No.	Chemical Name
78-78-4	2-METHYLBUTANE
26760-64-5	2-METHYLBUTENE
78-80-8	2-METHYLBUTENYNE
137-32-6	2-METHYLBUTYL ALCOHOL
96-17-3	2-METHYLBUTYRALDEHYDE
96-17-3	2-METHYLBUTYRIC ALDEHYDE
95-49-8	2-METHYLCHLOROBENZENE
583-60-8	2-METHYLCYCLOHEXANONE
12108-13-3	2-METHYLCYCLOPENTADIENYLMANGANESE TRICARBONYL
78-85-3	2-METHYLENEPROPANAL
75-55-8	2-METHYLETHYLENIMINE
534-22-5	2-METHYLFURAN
27137-41-3	2-METHYLFURAN (CZECH)
27137-41-3	2-METHYLFURAN (DOT)
591-76-4	2-METHYLHEXANE
88-72-2	2-METHYLNITROBENZENE
3221-61-2	2-METHYLOCTANE
123-15-9	2-METHYLPENTALDEHYDE
123-15-9	2-METHYLPENTANAL
107-83-5	2-METHYLPENTANE
105-30-6	2-METHYLPENTANOL-1
54972-97-3	2-METHYLPENTANOL-1
763-29-1	2-METHYLPENTENE
95-48-7	2-METHYLPHENOL
95-53-4	2-METHYLPHENYLAMINE
614-78-8	2-METHYLPHENYLTHIOUREA
513-42-8	2-METHYLPROP-1-EN-3-OL
78-83-1	2-METHYLPROPAN-1-OL
78-84-2	2-METHYLPROPANAL
75-28-5	2-METHYLPROPANE
78-82-0	2-METHYLPROPANENITRILE
79-31-2	2-METHYLPROPANOIC ACID
79-30-1	2-METHYLPROPANOYL CHLORIDE
78-85-3	2-METHYLPROPENAL
78-85-3	2-METHYLPROPENAL (CZECH)
115-11-7	2-METHYLPROPENE
126-98-7	2-METHYLPROPENENITRILE
79-41-4	2-METHYLPROPENOIC ACID
78-84-2	2-METHYLPROPIONALDEHYDE
79-31-2	2-METHYLPROPIONIC ACID
97-85-8	2-METHYLPROPIONIC ACID ISOBUTYL ESTER
78-82-0	2-METHYLPROPIONITRILE
79-30-1	2-METHYLPROPIONYL CHLORIDE
97-85-8	2-METHYLPROPYL 2-METHYLPROPANOATE
97-85-8	2-METHYLPROPYL 2-METHYLPROPIONATE
106-63-8	2-METHYLPROPYL ACRYLATE
78-83-1	2-METHYLPROPYL ALCOHOL
513-36-0	2-METHYLPROPYL CHLORIDE
542-55-2	2-METHYLPROPYL FORMATE
97-85-8	2-METHYLPROPYL ISOBUTYRATE
97-86-9	2-METHYLPROPYL METHACRYLATE
108-10-1	2-METHYLPROPYL METHYL KETONE
540-42-1	2-METHYLPROPYL PROPANOATE
540-42-1	2-METHYLPROPYL PROPIONATE
78-81-9	2-METHYLPROPYLAMINE
109-08-0	2-METHYLPYRAZINE
109-06-8	2-METHYLPYRIDINE
91-63-4	2-METHYLQUINOLINE
96-47-9	2-METHYLTETRAHYDROFURAN
123-15-9	2-METHYLVALERALDEHYDE
116-06-3	2-METIL-2-TIOMETIL-PROPIONALDEID-O-(N-METIL-CARBAMOIL)-OSSIMA (ITALIAN)
109-86-4	2-METOSSIETANOLO (ITALIAN)
110-49-6	2-METOSSIETILACETATO (ITALIAN)
107-07-3	2-MONOCHLOROETHANOL
2038-03-1	2-MORPHOLINOETHANAMINE
622-40-2	2-MORPHOLINOETHANOL
2038-03-1	2-MORPHOLINOETHYLAMINE
102-81-8	2-N-DIBUTYLAMINOETHANOL
100-37-8	2-N-DIETHYLAMINOETHANOL
110-73-6	2-N-MONOETHYLAMINOETHANOL
109-83-1	2-N-MONOMETHYLAMINOETHANOL
91-59-8	2-NAPHTHALENAMINE
494-03-1	2-NAPHTHALENAMINE, N,N-BIS(2-CHLOROETHYL)-
135-88-6	2-NAPHTHALENAMINE, N-PHENYL-
18811-72-8	2-NAPHTHALENEDIAZONIUM, 1-HYDROXY-5-SULFO-, HYDROXIDE, INNER SALT, SODIUM SALT
6358-53-8	2-NAPHTHALENOL, 1-[(2,5-DIMETHOXYPHENYL)AZO]-
2646-17-5	2-NAPHTHALENOL, 1-[(2-METHYLPHENYL)AZO]-
91-59-8	2-NAPHTHYLAMIN (GERMAN)
91-59-8	2-NAPHTHYLAMINE
91-59-8	2-NAPHTHYLAMINE MUSTARD
494-03-1	2-NAPHTHYLAMINE, N,N-BIS(2-CHLOROETHYL)-
135-88-6	2-NAPHTHYLAMINE, N-PHENYL-
135-88-6	2-NAPHTHYLPHENYLAMINE
5809-08-5	2-NEOPENTYL-2-PROPYL HYDROPEROXIDE
121-17-5	2-NITRO-4-(TRIFLUOROMETHYL)-1-CHLOROBENZENE
119-32-4	2-NITRO-4-AMINOTOLUENE
89-62-3	2-NITRO-4-METHYLANILINE
121-17-5	2-NITRO-4-TRIFLUOROMETHYLCHLOROBENZENE
5307-14-2	2-NITRO-P-PHENYLENEDIAMINE
89-62-3	2-NITRO-P-TOLUIDINE
88-74-4	2-NITROANILINE
88-74-4	2-NITROBENZENAMINE
88-73-3	2-NITROCHLOROBENZENE
88-75-5	2-NITROPHENOL
79-46-9	2-NITROPROPANE
25322-01-4	2-NITROPROPANE
79-46-9	2-NITROPROPANE (ACGIH,DOT)
25322-01-4	2-NITROPROPANE (ACGIH,DOT)
88-72-2	2-NITROTOLUENE
15271-41-7	2-NORBORNANECARBONITRILE, 5-CHLORO-6-OXO-, O-(METHYLCARBAMOYL) OXIME, (E)-ENDO-2,EXO-5-
15271-41-7	2-NORBORNANONE, ENDO-3-CHLORO-EXO-6-CYANO-, O-(METHYLCARBAMOYL)OXIME
16219-75-3	2-NORBORNENE, 5-ETHYLIDENE-
79-46-9	2-NP
25322-01-4	2-NP
139-91-3	2-OXAZOLIDINONE, 5-(4-MORPHOLINYLMETHYL)-3-[[(5-NITRO-2-FURANYL)METHYLENE]AMINO]-
139-91-3	2-OXAZOLIDINONE, 5-(MORPHOLINOMETHYL)-3-[(5-NITROFURFURYLIDENE)AMINO]-
57-57-8	2-OXETANONE
3068-88-0	2-OXETANONE, 4-METHYL-
674-82-8	2-OXETANONE, 4-METHYLENE-
625-33-2	2-OXO-3-PENTENE
105-60-2	2-OXOHEXAMETHYLENEIMINE
105-60-2	2-OXOHEXAMETHYLENIMINE
591-78-6	2-OXOHEXANE
616-45-5	2-OXOPYRROLIDINE
99-87-6	2-P-TOLYLPROPANE
625-30-9	2-PENTANAMINE
107-45-9	2-PENTANAMINE, 2,4,4-TRIMETHYL-
108-09-8	2-PENTANAMINE, 4-METHYL-
141-59-3	2-PENTANETHIOL, 2,4,4-TRIMETHYL-
6032-29-7	2-PENTANOL
108-11-2	2-PENTANOL, 4-METHYL-
54972-97-3	2-PENTANOL, 4-METHYL-
108-84-9	2-PENTANOL, 4-METHYL-, ACETATE
626-38-0	2-PENTANOL, ACETATE
53496-15-4	2-PENTANOL, ACETATE (8CI, 9CI)
107-87-9	2-PENTANONE
54693-46-8	2-PENTANONE, 4-HYDROXY-4-METHYL-, PEROXIDE, MORE THAN 57% IN SOLUTION
123-42-2	2-PENTANONE, 4-HYDROXY-4-METHYL-
54693-46-8	2-PENTANONE, 4-HYDROXY-4-METHYL-, PEROXIDE
54693-46-8	2-PENTANONE, 4-HYDROXY-4-METHYL-, PEROXIDE, NOT MORE THAN 57% IN SOLUTION
108-10-1	2-PENTANONE, 4-METHYL-
37206-20-5	2-PENTANONE, 4-METHYL-, PEROXIDE
37206-20-5	2-PENTANONE, 4-METHYL-, PEROXIDE, WITH NOT MORE THAN 9% BY WEIGHT ACTIVE OXYGEN
625-33-2	2-PENTEN-4-ONE
12772-47-3	2-PENTEN-4-YN-1-OL, 3-METHYL-
646-04-8	2-PENTENE, (E)-
627-20-3	2-PENTENE, (Z)-
107-40-4	2-PENTENE, 2,4,4-TRIMETHYL-
625-27-4	2-PENTENE, 2-METHYL-
598-96-9	2-PENTENE, 3,4,4-TRIMETHYL-
4461-48-7	2-PENTENE, 4-METHYL-
626-38-0	2-PENTYL ACETATE
53496-15-4	2-PENTYL ACETATE
6032-29-7	2-PENTYL ALCOHOL
625-30-9	2-PENTYLAMINE
136-81-2	2-PENTYLPHENOL
105-60-2	2-PERHYDROAZEPINONE
98-83-9	2-PHENYL-1-PROPENE
7568-93-6	2-PHENYL-2-HYDROXYETHYLAMINE

CAS No.	Chemical Name
98-83-9	2-PHENYL-2-PROPENE
80-15-9	2-PHENYL-2-PROPYL HYDROPEROXIDE
140-29-4	2-PHENYLACETONITRILE
90-41-5	2-PHENYLANILINE
90-41-5	2-PHENYLBENZENAMINE
135-98-8	2-PHENYLBUTANE
96-09-3	2-PHENYLOXIRANE
90-43-7	2-PHENYLPHENOL
98-82-8	2-PHENYLPROPANE
98-83-9	2-PHENYLPROPENE
98-83-9	2-PHENYLPROPYLENE
109-06-8	2-PICOLINE
104-90-5	2-PICOLINE, 5-ETHYL-
140-76-1	2-PICOLINE, 5-VINYL-
80-56-8	2-PINENE
140-31-8	2-PIPERAZINYLETHYLAMINE
83-26-1	2-PIVALOYL-1,3-INDANDIONE
83-26-1	2-PIVALOYL-INDAAN-1,3-DION (DUTCH)
83-26-1	2-PIVALOYL-INDAN-1,3-DION (GERMAN)
83-26-1	2-PIVALOYLINDANE-1,3-DIONE
83-26-1	2-PIVALYL-1,3-INDANDIONE
75-31-0	2-PROPANAMINE
75-64-9	2-PROPANAMINE, 2-METHYL-
26264-05-1	2-PROPANAMINE, DODECYLBENZENESULFONATE
108-18-9	2-PROPANAMINE, N-(1-METHYLETHYL)-
75-33-2	2-PROPANETHIOL
75-33-2	2-PROPANETHIOL (DOT)
75-66-1	2-PROPANETHIOL, 2-METHYL-
67-63-0	2-PROPANOL
122-20-3	2-PROPANOL, 1,1',1"-NITRILOTRI-
122-20-3	2-PROPANOL, 1,1',1"-NITRILOTRIS-
110-97-4	2-PROPANOL, 1,1'-IMINOBIS-
110-97-4	2-PROPANOL, 1,1'-IMINOBIS- (9CI)
110-97-4	2-PROPANOL, 1,1'-IMINODI-
25265-71-8	2-PROPANOL, 1,1'-OXYDI-
616-29-5	2-PROPANOL, 1,3-DIAMINO-
124-16-3	2-PROPANOL, 1-(2-BUTOXYETHOXY)-
2109-64-0	2-PROPANOL, 1-(DIBUTYLAMINO)-
108-16-7	2-PROPANOL, 1-(DIMETHYLAMINO)-
78-96-6	2-PROPANOL, 1-AMINO-
1319-72-8	2-PROPANOL, 1-AMINO-, (2,4,5-TRICHLOROPHENOXY)ACETATE (SALT)
54590-52-2	2-PROPANOL, 1-AMINO-, 4-DODECYLBENZENESULFONATE (SALT)
1319-72-8	2-PROPANOL, 1-AMINO-, COMPD. WITH (2,4,5-TRICHLORO-PHENOXY)ACETIC ACID (1:1)
127-00-4	2-PROPANOL, 1-CHLORO-
107-98-2	2-PROPANOL, 1-METHOXY-
75-65-0	2-PROPANOL, 2-METHYL-
67-64-1	2-PROPANONE
116-16-5	2-PROPANONE, 1,1,1,3,3,3-HEXACHLORO-
684-16-2	2-PROPANONE, 1,1,1,3,3,3-HEXAFLUORO-
534-07-6	2-PROPANONE, 1,3-DICHLORO-
598-31-2	2-PROPANONE, 1-BROMO-
598-31-2	2-PROPANONE, 1-BROMO- (9CI)
78-95-5	2-PROPANONE, 1-CHLORO-
123-54-6	2-PROPANONE, ACETYL-
598-31-2	2-PROPANONE, BROMO-
75-86-5	2-PROPANONE, CYANOHYDRIN
116-16-5	2-PROPANONE, HEXACHLORO-
684-16-2	2-PROPANONE, HEXAFLUORO-
677-71-4	2-PROPANONE, HEXAFLUORO-, HYDRATE
107-11-9	2-PROPEN-1-AMINE
107-11-9	2-PROPEN-1-AMINE (9CI)
102-70-5	2-PROPEN-1-AMINE, N,N-DI-2-PROPENYL-
102-70-5	2-PROPEN-1-AMINE, N-N-DI-2-PROPENYL- (9CI)
107-18-6	2-PROPEN-1-OL
513-42-8	2-PROPEN-1-OL, 2-METHYL-
107-02-8	2-PROPEN-1-ONE
107-02-8	2-PROPENAL
78-85-3	2-PROPENAL, 2-METHYL-
100-73-2	2-PROPENAL, DIMER
869-29-4	2-PROPENAL, MONOHYDRATE, DIACETATE
79-06-1	2-PROPENAMIDE
79-06-1	2-PROPENAMIDE (9CI)
107-11-9	2-PROPENAMINE
869-29-4	2-PROPENE-1,1-DIOL, DIACETATE
107-18-6	2-PROPENE-1-OL
107-13-1	2-PROPENENITRILE
126-98-7	2-PROPENENITRILE, 2-METHYL-

CAS No.	Chemical Name
1931-62-0	2-PROPENEPEROXOIC ACID, 3-CARBOXY-, 1-(1,1-DIMETHYL-ETHYL) ESTER, (Z)-
79-10-7	2-PROPENOIC ACID
79-10-7	2-PROPENOIC ACID (9CI)
141-32-2	2-PROPENOIC ACID BUTYL ESTER
140-88-5	2-PROPENOIC ACID ETHYL ESTER
96-33-3	2-PROPENOIC ACID METHYL ESTER
2426-54-2	2-PROPENOIC ACID, 2-(DIETHYLAMINO)ETHYL ESTER
80-63-7	2-PROPENOIC ACID, 2-CHLORO-, METHYL ESTER
80-63-7	2-PROPENOIC ACID, 2-CHLORO-, METHYL ESTER (9CI)
137-05-3	2-PROPENOIC ACID, 2-CYANO-, METHYL ESTER
106-71-8	2-PROPENOIC ACID, 2-CYANOETHYL ESTER
3953-10-4	2-PROPENOIC ACID, 2-ETHYLBUTYL ESTER
103-11-7	2-PROPENOIC ACID, 2-ETHYLHEXYL ESTER
103-11-7	2-PROPENOIC ACID, 2-ETHYLHEXYL ESTER (9CI)
818-61-1	2-PROPENOIC ACID, 2-HYDROXYETHYL ESTER
818-61-1	2-PROPENOIC ACID, 2-HYDROXYETHYL ESTER (9CI)
999-61-1	2-PROPENOIC ACID, 2-HYDROXYPROPYL ESTER
79-41-4	2-PROPENOIC ACID, 2-METHYL-
2867-47-2	2-PROPENOIC ACID, 2-METHYL-, 2-(DIMETHYLAMINO)ETHYL ESTER
30674-80-7	2-PROPENOIC ACID, 2-METHYL-, 2-ISOCYANATOETHYL ESTER
97-86-9	2-PROPENOIC ACID, 2-METHYL-, 2-METHYLPROPYL ESTER
3775-90-4	2-PROPENOIC ACID, 2-METHYL-, 2-[(1,1-DIMETHYLETHYL)AMI-NO]ETHYL ESTER
760-93-0	2-PROPENOIC ACID, 2-METHYL-, ANHYDRIDE
760-93-0	2-PROPENOIC ACID, 2-METHYL-, ANHYDRIDE (9CI)
97-88-1	2-PROPENOIC ACID, 2-METHYL-, BUTYL ESTER
97-63-2	2-PROPENOIC ACID, 2-METHYL-, ETHYL ESTER
80-62-6	2-PROPENOIC ACID, 2-METHYL-, METHYL ESTER
27813-02-1	2-PROPENOIC ACID, 2-METHYL-, MONOESTER WITH 1,2-PROPANE-DIOL
106-91-2	2-PROPENOIC ACID, 2-METHYL-, OXIRANYLMETHYL ESTER
106-63-8	2-PROPENOIC ACID, 2-METHYLPROPYL ESTER
106-63-8	2-PROPENOIC ACID, 2-METHYLPROPYL ESTER (9CI)
2156-96-9	2-PROPENOIC ACID, DECYL ESTER
1330-61-6	2-PROPENOIC ACID, ISODECYL ESTER
1330-61-6	2-PROPENOIC ACID, ISODECYL ESTER (9CI)
96-33-3	2-PROPENOIC ACID, METHYL ESTER (9CI)
25584-83-2	2-PROPENOIC ACID, MONOESTER WITH 1,2-PROPANEDIOL
2499-59-4	2-PROPENOIC ACID, OCTYL ESTER
107-18-6	2-PROPENOL
814-68-6	2-PROPENOYL CHLORIDE
814-68-6	2-PROPENOYL CHLORIDE (9CI)
107-18-6	2-PROPENYL ALCOHOL
106-95-6	2-PROPENYL BROMIDE
107-05-1	2-PROPENYL CHLORIDE
57-06-7	2-PROPENYL ISOTHIOCYANATE
107-11-9	2-PROPENYLAMINE
108-21-4	2-PROPYL ACETATE
75-29-6	2-PROPYL CHLORIDE
75-31-0	2-PROPYLAMINE
75-33-2	2-PROPYLMERCAPTAN
107-19-7	2-PROPYN-1-OL
115-19-5	2-PROPYN-1-OL, 1,1-DIMETHYL-
107-19-7	2-PROPYNOL
107-19-7	2-PROPYNYL ALCOHOL
106-96-7	2-PROPYNYL BROMIDE
504-29-0	2-PYRIDINAMINE
1918-02-1	2-PYRIDINECARBOXYLIC ACID, 4-AMINO-3,5,6-TRICHLORO-
2921-88-2	2-PYRIDINOL, 3,5,6-TRICHLORO-, O-ESTER WITH O,O-DIETHYL PHOSPHOROTHIOATE
504-29-0	2-PYRIDYLAMINE
616-45-5	2-PYROL
616-45-5	2-PYRROLIDINONE
872-50-4	2-PYRROLIDINONE, 1-METHYL-
616-45-5	2-PYRROLIDONE
88-85-7	2-SEC-BUTYL-4,6-DINITROPHENOL
89-72-5	2-SEC-BUTYLPHENOL
1420-07-1	2-TERT-BUTYL-4,6-DINITROPHENOL
564-02-3	2-TERT-BUTYLBUTANE
75-18-3	2-THIAPROPANE
56-04-2	2-THIO-6-METHYLURACIL
60-24-2	2-THIOETHANOL
68-11-1	2-THIOGLYCOLIC ACID
96-45-7	2-THIOIMIDAZOLIDINE
79-42-5	2-THIOLACTIC ACID
96-45-7	2-THIONOIMIDAZOLIDINE
75-18-3	2-THIOPROPANE

CAS No.	Chemical Name
79-19-6	2-THIOSEMICARBAZIDE
62-56-6	2-THIOUREA
95-53-4	2-TOLUIDINE
95-46-5	2-TOLYL BROMIDE
624-64-6	2-TRANS-BUTENE
646-04-8	2-TRANS-PENTENE
88-17-5	2-TRIFLUOROMETHYL ANILINE
88-17-5	2-TRIFLUOROMETHYL ANILINE (DOT)
5408-74-2	2-VINYL-5-ETHYLPYRIDINE
110-75-8	2-VINYLOXYETHYL CHLORIDE
111-41-1	2-[(2-AMINOETHYL)AMINO]ETHANOL
111-42-2	2-[(2-HYDROXYETHYL)AMINO]ETHANOL
21564-17-0·	2-[(THIOCYANATOMETHYL)THIO]BENZOTHIAZOLE
140-82-9	2-[2-(DIETHYLAMINO)ETHOXY]ETHANOL
366-70-1	2-[P-(ISOPROPYLCARBAMOYL)BENZYL]-1-METHYLHYDRAZINE HYDROCHLORIDE
91-80-5	2-[[2-(DIMETHYLAMINO)ETHYL]-2-THENYLAMINO]PYRIDINE
56-49-5	20-MC
56-49-5	20-METHYLCHOLANTHRENE
315-22-0	20-NORCROTALANAN-11,15-DIONE, 14,19-DIHYDRO-12,13-DIHYDROXY-, (13 ALPHA,14 ALPHA)-
50-18-0	2H-1,3,2-OXAZAPHOSPHORIN-2-AMINE, N,N-BIS(2-CHLOROETHYL) TETRAHYDRO-, 2-OXIDE
50-18-0	2H-1,3,2-OXAZAPHOSPHORINE, 2-[BIS(2-CHLOROETHYL)AMINO] TETRAHYDRO-, 2-OXIDE
110-91-8	2H-1,4-OXAZINE, TETRAHYDRO-
117-52-2	2H-1-BENZOPYRAN-2-ONE, 3-(1-(2-FURANYL)-3-OXOBUTYL)-4-HYDROXY- (9CI)
117-52-2	2H-1-BENZOPYRAN-2-ONE, 3-[1-(2-FURANYL)-3-OXOBUTYL]-4-HYDROXY-
28772-56-7	2H-1-BENZOPYRAN-2-ONE, 3-[3-(4'-BROMO[1,1'-BIPHENYL]-4-YL)-3-HYDROXY-1-PHENYLPROPYL]-4-HYDROXY-
5836-29-3	2H-1-BENZOPYRAN-2-ONE, 4-HYDROXY-3-(1,2,3,4-TETRAHYDRO-1-NAPHTHALENYL)-
5836-29-3	2H-1-BENZOPYRAN-2-ONE, 4-HYDROXY-3-(1,2,3,4-TETRAHYDRO-1-NAPHTHALENYL)- (9CI)
81-81-2	2H-1-BENZOPYRAN-2-ONE, 4-HYDROXY-3-(3-OXO-1-PHENYLBUTYL)-
81-81-2	2H-1-BENZOPYRAN-2-ONE, 4-HYDROXY-3-(3-OXO-1-PHENYLBUTYL)- (9CI)
129-06-6	2H-1-BENZOPYRAN-2-ONE, 4-HYDROXY-3-(3-OXO-1-PHENYLBUTYL)-, SODIUM SALT
129-06-6	2H-1-BENZOPYRAN-2-ONE, 4-HYDROXY-3-(3-OXO-1-PHENYLBUTYL)-, SODIUM SALT (9CI)
120-80-9	2H-1-BENZOPYRAN-3,5,7-TRIOL, 2-(3,4-DIHYDROXYPHENYL)-3,4-DIHYDRO-, (2R-TRANS)-
154-23-4	2H-1-BENZOPYRAN-3,5,7-TRIOL, 2-(3,4-DIHYDROXYPHENYL)-3,4-DIHYDRO-, (2R-TRANS)-
110-87-2	2H-3,4-DIHYDROPYRAN
105-60-2	2H-AZEPIN-2-ONE, HEXAHYDRO-
105-60-2	2H-AZEPIN-7-ONE, HEXAHYDRO-
316-42-7	2H-BENZO[A]QUINOLIZINE, EMETAN DERIV
103-75-3	2H-PYRAN, 2-ETHOXY-3,4-DIHYDRO-
110-87-2	2H-PYRAN, 3,4-DIHYDRO-
3174-74-1	2H-PYRAN, 3,6-DIHYDRO-
110-87-2	2H-PYRAN, DIHYDRO-
25512-65-6	2H-PYRAN, DIHYDRO-
3174-74-1	2H-PYRAN, DIHYDRO-
142-68-7	2H-PYRAN, TETRAHYDRO-
100-73-2	2H-PYRAN-2-CARBOXALDEHYDE, 3,4-DIHYDRO-
315-22-0	2H-[1,6]DIOXACYCLOUNDECINO[2,3,4-GH]PYRROLIZINE, 20-NORCROTALANAN-11,15-DIONE DERIV.
2763-96-4	3(2H)-ISOXAZOLONE, 5-(AMINOMETHYL)-
50-28-2	3,17-EPIDIHYDROXYESTRATRIENE
50-28-2	3,17 BETA-DIHYDROXYESTRA-1,3,5(10)-TRIENE
50-28-2	3,17 BETA-ESTRADIOL
56-18-8	3,3'-DIAMINODIPROPYLAMINE
91-94-1	3,3'-DICHLORO-4,4'-DIAMINOBIPHENYL
91-94-1	3,3'-DICHLORO-4,4'-DIAMINODIPHENYL
28434-86-8	3,3'-DICHLORO-4,4'-DIAMINODIPHENYL ETHER
101-14-4	3,3'-DICHLORO-4,4'-DIAMINODIPHENYLMETHANE
91-94-1	3,3'-DICHLOROBENZIDINE
91-94-1	3,3'-DICHLOROBIPHENYL-4,4'-DIAMINE
514-73-8	3,3'-DIETHYL-2,2'-THIADICARBOCYANINE IODIDE
514-73-8	3,3'-DIETHYLDITHIACARBODICYANINE IODIDE
514-73-8	3,3'-DIETHYLPENTAMETHINETHIACYANINE IODIDE
514-73-8	3,3'-DIETHYLTHIADICARBOCYANINE IODIDE
119-90-4	3,3'-DIMETHOXY-4,4'-DIAMINODIPHENYL
119-90-4	3,3'-DIMETHOXYBENZIDINE
119-93-7	3,3'-DIMETHYL-4,4'-BIPHENYLDIAMINE
119-93-7	3,3'-DIMETHYL-4,4'-DIAMINOBIPHENYL
838-88-0	3,3'-DIMETHYL-4,4'-DIAMINODIPHENYLMETHANE
119-93-7	3,3'-DIMETHYLBENZIDINE
56-18-8	3,3'-IMINOBIS(PROPYLAMINE)
56-18-8	3,3'-IMINOBISPROPYLAMINE (DOT)
56-18-8	3,3'-IMINODIPROPYLAMINE
119-93-7	3,3'-TOLIDINE
6731-36-8	3,3,5-TRIMETHYL-1,1-BIS(TERT-BUTYLPEROXY)CYCLOHEXANE
116-02-9	3,3,5-TRIMETHYL-1-CYCLOHEXANOL
4098-71-9	3,3,5-TRIMETHYL-5-(ISOCYANATOMETHYL)CYCLOHEXYL ISOCYANATE
116-02-9	3,3,5-TRIMETHYLCYCLOHEXANOL
22397-33-7	3,3,6,6,9,9-HEXAMETHYL-1,2,4,5-TETRAOXOCYCLONONANE, TECHNICAL PURE
22397-33-7	3,3,6,6,9,9-HEXAMETHYL-1,2,4,5-TETROXACYCLONONANE (CONC. ≥75%)
115-84-4	3,3-BIS(HYDROXYMETHYL)HEPTANE
56-18-8	3,3-DIAMINODIPROPYLAMINE
91-94-1	3,3-DICHLOROBENZIDENE
3054-95-3	3,3-DIETHOXY-1-PROPENE
3054-95-3	3,3-DIETHOXYPROPENE
3054-95-3	3,3-DIETHOXYPROPENE (DOT)
1067-20-5	3,3-DIETHYLPENTANE
119-90-4	3,3-DIMETHOXYBENZIDINE
91-93-0	3,3-DIMETHOXYBENZIDINE-4,4-DIISOCYANATE
39196-18-4	3,3-DIMETHYL-1-(METHYLTHIO)-2-BUTANONE-O-((METHYLAMINO)CARBONYL)OXIME
79-92-5	3,3-DIMETHYL-2-METHYLENENORBORNANE
79-92-5	3,3-DIMETHYL-2-METHYLENENORCAMPHANE
119-93-7	3,3-DIMETHYLBENZIDINE
4032-86-4	3,3-DIMETHYLHEPTANE
598-96-9	3,4,4-TRIMETHYL-2-PENTENE
76-44-8	3,4,5,6,7,8,8-HEPTACHLORODICYCLOPENTADIENE
76-44-8	3,4,5,6,7,8,8A-HEPTACHLORODICYCLOPENTADIENE
60-57-1	3,4,5,6,9,9-HEXACHLORO-1A,2,2A,3,6,6A,7,7A-OCTAHYDRO-2,7:3,6-DIMETHANONAPHTH(2,3-B)OXIRENE
72-20-8	3,4,5,6,9,9-HEXACHLORO-1A,2,2A,3,6,6A,7,7A-OCTAHYDRO-2,7:3,6-DIMETHANONAPHTH(2,3-B)OXIRENE
194-59-2	3,4,5,6-DIBENZOCARBAZOLE
609-19-8	3,4,5-TRICHLOROPHENOL
149-91-7	3,4,5-TRIHYDROXYBENZOIC ACID
50-55-5	3,4,5-TRIMETHOXYBENZOYL METHYL RESERPATE
2167-23-9	3,4,6,7-TETRAOXANONANE, 5-ETHYL-2,2,5,8,8-PENTAMETHYL-
1068-27-5	3,4,9,10-TETRAOXADODEC-6-YNE, 2,2,5,5,8,8,11,11-OCTAMETHYL-
120-58-1	3,4-(METHYLENEDIOXY)-1-PROPENYLBENZENE
205-99-2	3,4-BENZFLUORANTHENE
205-99-2	3,4-BENZOFLUORANTHENE
50-32-8	3,4-BENZOPYRENE
50-32-8	3,4-BENZPYRENE
50-32-8	3,4-BENZ[A]PYRENE
205-99-2	3,4-BENZ[E]ACEPHENANTHRYLENE
95-76-1	3,4-DCA
101-25-7	3,4-DI-N-NITROSOPENTAMETHYLENETETRAMINE
95-83-0	3,4-DIAMINO-1-CHLOROBENZENE
95-83-0	3,4-DIAMINOCHLOROBENZENE
95-76-1	3,4-DICHLORANILINE
95-76-1	3,4-DICHLOROANILINE
95-76-1	3,4-DICHLOROBENZENAMINE
616-21-7	3,4-DICHLOROBUTANE
64037-54-3	3,4-DICHLOROBUTENE-1
102-36-3	3,4-DICHLOROPHENYL ISOCYANATE
103-75-3	3,4-DIHYDRO-2-ETHOXY-2H-PYRAN
103-75-3	3,4-DIHYDRO-2-ETHOXYPYRAN
100-73-2	3,4-DIHYDRO-2-FORMYL-2H-PYRAN
110-87-2	3,4-DIHYDRO-2H-PYRAN
100-73-2	3,4-DIHYDRO-2H-PYRAN-2-CARBOXALDEHYDE
2642-71-9	3,4-DIHYDRO-4-OXO-3-BENZOTRIAZINYLMETHYL O,O-DIETHYL PHOSPHORODITHIOATE
110-87-2	3,4-DIHYDROPYRAN
1401-55-4	3,4-DIHYDROXY-5-((3,4,5-TRIHYDROXYBENZOYL)OXY)BENZOIC ACID
565-59-3	3,4-DIMETHYLPENTANE
610-39-9	3,4-DINITROTOLUENE
610-39-9	3,4-DNT
930-22-3	3,4-EPOXY-1-BUTENE
930-22-3	3,4-EPOXYBUTENE
94-59-7	3,4-METHYLENEDIOXY-ALLYLBENZENE
189-64-0	3,4:8,9-DIBENZOPYRENE

CAS No.	Chemical Name
189-55-9	3,4:9,10-DIBENZOPYRENE
3452-97-9	3,5,5-TRIMETHYL-1-HEXANOL
78-59-1	3,5,5-TRIMETHYL-2-CYCLOHEXENONE
3452-97-9	3,5,5-TRIMETHYLHEXANOL
3452-97-9	3,5,5-TRIMETHYLHEXYL ALCOHOL
1918-02-1	3,5,6-TRICHLORO-4-AMINOPICOLINIC ACID
128-37-0	3,5-DI-TERT-BUTYL-4-HYDROXYTOLUENE
2971-90-6	3,5-DICHLORO-2,6-DIMETHYL-4-HYDROXYPYRIDINE
2971-90-6	3,5-DICHLORO-2,6-DIMETHYL-4-PYRIDINOL
2971-90-6	3,5-DICHLORO-2,6-DIMETHYLPYRIDINOL
2971-90-6	3,5-DICHLORO-4-HYDROXY-2,6-DIMETHYLPYRIDINE
23950-58-5	3,5-DICHLORO-N-(1,1-DIMETHYL-2-PROPYNYL)BENZAMIDE
23950-58-5	3,5-DICHLORO-N-(1,1-DIMETHYLPROPYNYL)BENZAMIDE
2032-65-7	3,5-DIMETHYL-4-(METHYLTHIO)PHENOL METHYLCARBAMATE
2032-65-7	3,5-DIMETHYL-4-(METHYLTHIO)PHENYL METHYLCARBAMATE
2032-65-7	3,5-DIMETHYL-4-METHYL-THIOPHENYL-N-CARBAMAT (GERMAN)
2032-65-7	3,5-DIMETHYL-4-METHYLTHIOPHENYL N-METHYLCARBAMATE
108-67-8	3,5-DIMETHYLTOLUENE
534-52-1	3,5-DINITRO-2-HYDROXYTOLUENE
1335-85-9	3,5-DINITRO-2-HYDROXYTOLUENE
148-01-6	3,5-DINITRO-2-METHYLBENZAMIDE
148-01-6	3,5-DINITRO-O-TOLUAMIDE
2032-65-7	3,5-XYLENOL, 4-(METHYLTHIO)-, METHYLCARBAMATE
112-57-2	3,6,9-TRIAZAUNDECANE-1,11-DIAMINE
112-50-5	3,6,9-TRIOXAUNDECAN-1-OL
112-60-7	3,6,9-TRIOXAUNDECANE-1,11-DIOL
112-24-3	3,6-DIAZAOCTANE-1,8-DIAMINE
60-00-4	3,6-DIAZAOCTANEDIOIC ACID, 3,6-BIS(CARBOXYMETHYL)-
1918-00-9	3,6-DICHLOOR-2-METHOXY-BENZOEIZUUR (DUTCH)
1918-00-9	3,6-DICHLOR-3-METHOXY-BENZOESAEURE (GERMAN)
1918-00-9	3,6-DICHLORO-2-METHOXYBENZOIC ACID
1918-00-9	3,6-DICHLORO-O-ANISIC ACID
3174-74-1	3,6-DIHYDRO-2H-PYRAN
112-34-5	3,6-DIOXA-1-DECANOL
111-77-3	3,6-DIOXA-1-HEPTANOL
629-14-1	3,6-DIOXAOCTANE
112-27-6	3,6-DIOXAOCTANE-1,8-DIOL
121-46-0	3,6-METHANO-1,4-CYCLOHEXADIENE
498-15-7	3,7,7-TRIMETHYLBICYCLO(410)-3-HEPTENE
101-25-7	3,7-DI-N-NITROSOPENTAMETHYLENETETRAMINE
101-25-7	3,7-DINITROSO-1,3,5,7-TETRAAZABICYCLO-(3,3,1)-NONANE
61-82-5	3,A-T
7786-34-7	3-((DIMETHOXYPHOSPHINYL)OXY)-2-BUTENOIC ACID METHYL ESTER
81-81-2	3-(ALPHA-ACETONYLBENZYL)-4-HYDROXYCOUMARIN
81-81-2	3-(ALPHA-PHENYL-BETA-ACETYLETHYL)-4-HYDROXYCOUMARIN
81-81-2	3-(1'-PHENYL-2'-ACETYLETHYL)-4-HYDROXYCOUMARIN
5836-29-3	3-(1,2,3,4-TETRAHYDRO-1-NAPHTHYL)-4-HYDROXYCUMARIN (GERMAN)
5836-29-3	3-(1,2,3,4-TETRAHYDRO-1-NAPHTYL)-4-HYDROXYCOUMARINE (FRENCH)
106-87-6	3-(1,2-EPOXYETHYL)-7-OXABICYCLO[4.1.0]HEPTANE
117-52-2	3-(1-FURYL-3-ACETYLETHYL)-4-HYDROXYCOUMARIN
10213-74-8	3-(2-ETHYLBUTOXY)PROPIONIC ACID
330-54-1	3-(3,4-DICHLOOR-FENYL)-1,1-DIMETHYLUREUM (DUTCH)
330-54-1	3-(3,4-DICHLOR-PHENYL)-1,1-DIMETHYL-HARNSTOFF (GERMAN)
330-54-1	3-(3,4-DICHLOROPHENOL)-1,1-DIMETHYLUREA
330-54-1	3-(3,4-DICHLOROPHENYL)-1,1-DIMETHYLUREA
330-54-1	3-(3,4-DICLORO-FENYL)-1,1-DIMETIL-UREA (ITALIAN)
94-59-7	3-(3,4-METHYLENEDIOXYPHENYL)PROP-1-ENE
28772-56-7	3-(3-(4'-BROMO(1,1'-BIPHENYL)-4-YL)3-HYDROXY-1-PHENYLPROPYL)-4-HYDROXY-2H-1-BENZOPYRAN-2-ONE
1982-47-4	3-(4-(4-CHLOOR-FENOXY)-FENOXY)-FENYL)-1,1-DIMETHYLUREUM (DUTCH)
1982-47-4	3-(4-(4-CHLOR-PHENOXY)-PHENYL)-1,1-DIMETHYLHARNSTOFF (GERMAN)
1982-47-4	3-(4-(4-CHLORO-FENOSSIL)-1,1-DIMETIL-UREA (ITALIAN)
123-00-2	3-(4-MORPHOLINYL)PROPYLAMINE
81-81-2	3-(ACETONYLBENZYL)-4-HYDROXYCOUMARIN
28772-56-7	3-(ALPHA-(P-(P-BROMOPHENYL)-BETA-HYDROXYPHENETHYL) BENZYL)-4-HYDROXYCOUMARIN
129-06-6	3-(ALPHA-ACETONYLBENZYL)-4-HYDROXY-COUMARIN SODIUM SALT
81-81-2	3-(ALPHA-ACETONYLBENZYL)-4-HYDROXYCOUMARIN
129-06-6	3-(ALPHA-ACETONYLBENZYL)-4-HYDROXYCOUMARIN SODIUM
117-52-2	3-(ALPHA-ACETONYLFURFURYL)-4-HYDROXYCOUMARIN
117-52-2	3-(ALPHA-FURYL-BETA-ACETYLAETHYL)-4-HYDROXYCUMARIN (GERMAN)
81-81-2	3-(ALPHA-PHENYL-BETA-ACETYLAETHYL)-4-HYDROXYCUMARIN (GERMAN)
81-81-2	3-(ALPHA-PHENYL-BETA-ACETYLETHYL)-4-HYDROXYCOUMARIN
5836-29-3	3-(ALPHA-TETRAL)-4-OXYCOUMARIN
5836-29-3	3-(ALPHA-TETRALINYL)-4-HYDROXYCOUMARIN
5836-29-3	3-(ALPHA-TETRALYL)-4-HYDROXYCOUMARIN
1477-55-0	3-(AMINOMETHYL)BENZYLAMINE
4985-85-7	3-(AMINOPROPYL)DIETHANOLAMINE
123-04-6	3-(CHLOROMETHYL)HEPTANE
6959-48-4	3-(CHLOROMETHYL)PYRIDINE HYDROCHLORIDE
3312-60-5	3-(CYCLOHEXYLAMINO)-1-PROPYLAMINE
702-03-4	3-(CYCLOHEXYLAMINO)PROPIONITRILE
5836-29-3	3-(D-TETRALYL)-4-HYDROXYCOUMARIN
104-78-9	3-(DIETHYLAMINO)-N-PROPYLAMINE
104-78-9	3-(DIETHYLAMINO)PROPYLAMINE (DOT)
141-66-2	3-(DIMETHOXYPHOSPHINYLOXY)-N,N DIMETHYLISOCROTONAMIDE
141-66-2	3-(DIMETHOXYPHOSPHINYLOXY)-N,N-DIMETHYL-CIS-CROTONAMIDE
6923-22-4	3-(DIMETHOXYPHOSPHINYLOXY)N-METHYL-CIS-CROTONAMIDE
109-55-7	3-(DIMETHYLAMINO)-1-AMINOPROPANE
141-66-2	3-(DIMETHYLAMINO)-1-METHYL-3-OXO-1-PROPENYL DIMETHYL PHOSPHATE
109-55-7	3-(DIMETHYLAMINO)-1-PROPANAMINE
109-55-7	3-(DIMETHYLAMINO)-1-PROPYLAMINE
109-55-7	3-(DIMETHYLAMINO)PROPANAMINE
109-55-7	3-(DIMETHYLAMINO)PROPYLAMINE
106-87-6	3-(EPOXYETHYL)-7-OXABICYCLO[4.1.0]HEPTANE
86-50-0	3-(MERCAPTOMETHYL)-1,2,3-BENZOTRIAZIN-4(3H)-ONE O,O-DIMETHYL PHOSPHORODITHIOATE S-ESTER
3268-49-3	3-(METHYLMERCAPTO)PROPIONALDEHYDE
3268-49-3	3-(METHYLTHIO)PROPANAL
3268-49-3	3-(METHYLTHIO)PROPIONALDEHYDE
104-78-9	3-(N,N-DIETHYLAMINO)-1-PROPYLAMINE
54-11-5	3-(N-METHYLPYRROLIDINO)PYRIDINE
123-00-2	3-(N-MORPHOLINO)-1-AMINOPROPANE
1982-47-4	3-(P-(P-CHLOROPHENOXY)PHENYL)-1,1-DIMETHYLUREA
122-60-1	3-(PHENYLOXY)-1,2-EPOXYPROPANE
777-37-7	3-(TRIFLUOROMETHYL)-4-CHLORONITROBENZENE
98-16-8	3-(TRIFLUOROMETHYL)BENZENAMINE
928-96-1	3-(Z)-HEXENOL
591-87-7	3-ACETOXY-1-PROPENE
591-87-7	3-ACETOXYPROPENE
584-79-2	3-ALLYL-2-METHYL-4-OXO-2-CYCLOPENTEN-1-YL CHRYSANTHEMATE
584-79-2	3-ALLYL-4-KETO-2-METHYLCYCLOPENTENYL CHRYSANTHEMUM-MONOCARBOXYLATE
462-08-8	3-AMINO PYRIDINE
2454-37-7	3-AMINO-ALPHA-METHYLBENZYL ALCOHOL
61-82-5	3-AMINO-1,2,4-TRIAZOLE
62450-06-0	3-AMINO-1,4-DIMETHYL-5H-PYRIDO[4,3-B]INDOLE
3312-60-5	3-AMINO-1-(CYCLOHEXYLAMINO)PROPANE
109-55-7	3-AMINO-1-(DIMETHYLAMINO)PROPANE
62450-07-1	3-AMINO-1-METHYL-5H-PYRIDO[4,3-B]INDOLE
108-44-1	3-AMINO-1-METHYLBENZENE
156-87-6	3-AMINO-1-PROPANOL
107-11-9	3-AMINO-1-PROPENE
61-82-5	3-AMINO-1H-1,2,4-TRIAZOLE
17026-81-2	3-AMINO-4-ETHOXYACETANILIDE
120-71-8	3-AMINO-4-METHOXYTOLUENE
132-32-1	3-AMINO-9-ETHYLCARBAZOLE, HYDROCHLORIDE
61-82-5	3-AMINO-A-TRIAZOLE
87-59-2	3-AMINO-O-XYLENE
61-82-5	3-AMINO-S-TRIAZOLE
108-45-2	3-AMINOANILINE
98-16-8	3-AMINOBENZOTRIFLUORIDE
108-44-1	3-AMINOPHENYLMETHANE
156-87-6	3-AMINOPROPANOL
107-11-9	3-AMINOPROPENE
156-87-6	3-AMINOPROPYL ALCOHOL
5332-73-0	3-AMINOPROPYL METHYL ETHER
107-11-9	3-AMINOPROPYLENE
108-44-1	3-AMINOTOLUEN (CZECH)
108-44-1	3-AMINOTOLUENE
61-82-5	3-AMINOTRIAZOLE
111-40-0	3-AZAPENTANE-1,5-DIAMINE
81-07-2	3-BENZISOTHIAZOLINONE 1,1-DIOXIDE
3132-64-7	3-BROMO-1,2-EPOXYPROPANE
109-70-6	3-BROMO-1-CHLOROPROPANE
106-95-6	3-BROMO-1-PROPENE

CAS No.	Chemical Name
106-96-7	3-BROMO-1-PROPYNE
106-95-6	3-BROMOPROPENE
109-70-6	3-BROMOPROPYL CHLORIDE
106-95-6	3-BROMOPROPYLENE
106-96-7	3-BROMOPROPYNE
106-96-7	3-BROMOPROPYNE (DOT)
107-89-1	3-BUTANOLAL
78-93-3	3-BUTANONE
689-97-4	3-BUTEN-1-YNE
78-94-4	3-BUTEN-2-ONE
814-78-8	3-BUTEN-2-ONE, 3-METHYL-
78-94-4	3-BUTENE-2-ONE
674-82-8	3-BUTENO-BETA-LACTONE
674-82-8	3-BUTENOIC ACID, 3-HYDROXY-, BETA-LACTONE
2426-08-6	3-BUTOXY-1,2-EPOXYPROPANE
115-19-5	3-BUTYN-2-OL, 2-METHYL-
498-15-7	3-CARENE
498-15-7	3-CARENE, (1S,6R)-(+)-
563-47-3	3-CHLOR-2-METHYL-PROP-1-EN (GERMAN)
98-15-7	3-CHLORO(TRIFLUOROMETHYL)BENZENE
96-12-8	3-CHLORO-1,2-DIBROMOPROPANE
96-24-2	3-CHLORO-1,2-DIHYDROXYPROPANE
106-89-8	3-CHLORO-1,2-EPOXYPROPANE
96-24-2	3-CHLORO-1,2-PROPANEDIOL
96-24-2	3-CHLORO-1,2-PROPYLENE GLYCOL
106-89-8	3-CHLORO-1,2-PROPYLENE OXIDE
109-70-6	3-CHLORO-1-BROMOPROPANE
627-30-5	3-CHLORO-1-HYDROXYPROPANE
121-73-3	3-CHLORO-1-NITROBENZENE
627-30-5	3-CHLORO-1-PROPANOL
107-05-1	3-CHLORO-1-PROPENE
107-05-1	3-CHLORO-1-PROPYLENE
4461-41-0	3-CHLORO-2-BUTENE
87-60-5	3-CHLORO-2-METHYL ANILINE (CHLOROTOLUIDINE)
563-47-3	3-CHLORO-2-METHYL-1-PROPENE
563-47-3	3-CHLORO-2-METHYLPROPENE
92-04-6	3-CHLORO-4-BIPHENYLOL
95-74-9	3-CHLORO-4-METHYL ANILINE (CHLOROTOLUIDINE)
56-72-4	3-CHLORO-4-METHYL-7-COUMARINYL DIETHYL PHOSPHOROTHIOATE
56-72-4	3-CHLORO-4-METHYL-7-HYDROXYCOUMARIN DIETHYL THIOPHOSPHORIC ACID ESTER
28479-22-3	3-CHLORO-4-METHYLPHENYL ISOCYANATE
28479-22-3	3-CHLORO-4-METHYLPHENYL ISOCYANATE (DOT)
56-72-4	3-CHLORO-4-METHYLUMBELLIFERONE O-ESTER WITH O,O-DIETHYL PHOSPHOROTHIOATE
15271-41-7	3-CHLORO-6-CYANO-2-NORBORNANONE-O-(METHYLCARBAMOYL) OXIME
15271-41-7	3-CHLORO-6-CYANONORBORNANONE-2 OXIME O,N-METHYLCARBAMATE
56-72-4	3-CHLORO-7-HYDROXY-4-METHYL-COUMARIN O,O-DIETHYL PHOSPHOROTHIOATE
56-72-4	3-CHLORO-7-HYDROXY-4-METHYL-COUMARIN O-ESTER WITH O,O-DIETHYL PHOSPHOROTHIOATE
98-15-7	3-CHLORO-ALPHA,ALPHA,ALPHA-TRIFLUOROTOLUENE
28479-22-3	3-CHLORO-P-TOLYL ISOCYANATE
542-75-6	3-CHLOROALLYL CHLORIDE
98-15-7	3-CHLOROBENZOTRIFLUORIDE
76-44-8	3-CHLOROCHLORDENE
121-73-3	3-CHLORONITROBENZENE
937-14-4	3-CHLOROPERBENZOIC ACID
937-14-4	3-CHLOROPEROXYBENZOIC ACID
937-14-4	3-CHLOROPEROXYBENZOIC ACID, MAXIMUM CONCENTRATION 86% (DOT)
108-43-0	3-CHLOROPHENOL (CHLOROPHENATE)
107-05-1	3-CHLOROPRENE
96-24-2	3-CHLOROPROPANE-1,2-DIOL
96-24-2	3-CHLOROPROPANEDIOL
542-76-7	3-CHLOROPROPANENITRILE
627-30-5	3-CHLOROPROPANOL
627-30-5	3-CHLOROPROPANOL-1
627-30-5	3-CHLOROPROPANOL-1 (DOT)
542-76-7	3-CHLOROPROPANONITRILE
107-05-1	3-CHLOROPROPENE
107-05-1	3-CHLOROPROPENE-1
106-89-8	3-CHLOROPROPENE-1,2-OXIDE
542-75-6	3-CHLOROPROPENYL CHLORIDE
542-76-7	3-CHLOROPROPIONITRILE
109-70-6	3-CHLOROPROPYL BROMIDE
107-05-1	3-CHLOROPROPYLENE
96-24-2	3-CHLOROPROPYLENE GLYCOL
106-89-8	3-CHLOROPROPYLENE OXIDE
627-30-5	3-CHLORPROPAN-1-OL (GERMAN)
107-05-1	3-CHLORPROPEN (GERMAN)
563-47-3	3-CLORO-2-METIL-PROP-1-ENE (ITALIAN)
108-39-4	3-CRESOL
154-23-4	3-CYANIDANOL, (+)-
626-17-5	3-CYANOBENZONITRILE
100-50-5	3-CYCLOHEXEN-1-ALDEHYDE
100-50-5	3-CYCLOHEXENE-1-CARBOXALDEHYDE
104-78-9	3-DIETHYLAMINO-1-PROPYLAMINE
4016-11-9	3-ETHOXY-1,2-EPOXYPROPANE
557-31-3	3-ETHOXY-1-PROPENE
2806-85-1	3-ETHOXYPROPANAL
4324-38-3	3-ETHOXYPROPANOIC ACID
2806-85-1	3-ETHOXYPROPIONALDEHYDE
1331-11-9	3-ETHOXYPROPIONIC ACID
4324-38-3	3-ETHOXYPROPIONIC ACID
1068-87-7	3-ETHYL-2,4-DIMETHYLPENTANE
514-73-8	3-ETHYL-2-(5-(3-ETHYL-2-BENZOTHIAZOLINYLIDENE)-1,3-PENTADIENYL)BENZOTHIAZOLIUM IODIDE
609-26-7	3-ETHYL-2-METHYLPENTANE
3074-77-9	3-ETHYL-4-METHYLHEXANE
104-90-5	3-ETHYL-6-METHYLPYRIDINE
645-62-5	3-FORMYLHEPT-3-ENE
123-05-7	3-FORMYLHEPTANE
97-96-1	3-FORMYLPENTANE
1401-55-4	3-GALLOYL GALLIC ACID
149-57-5	3-HEPTANECARBOXYLIC ACID
106-35-4	3-HEPTANONE
541-85-5	3-HEPTANONE, 5-METHYL-
589-38-8	3-HEXANONE
928-96-1	3-HEXEN-1-OL, (Z)-
1068-27-5	3-HEXYNE, 2,5-DIMETHYL-2,5-DI(T-BUTYLPEROXY)-
1068-27-5	3-HEXYNE, 2,5-DIMETHYL-2,5-DI(T-BUTYLPEROXY)-, MAXIMUM CONCENTRATION 52% WITH INERT SOLID
556-52-5	3-HYDROXY-1,2-EPOXYPROPANE
1120-71-4	3-HYDROXY-1-PROPANESULFONIC ACID SULTONE
107-18-6	3-HYDROXY-1-PROPENE
156-87-6	3-HYDROXY-1-PROPYLAMINE
107-19-7	3-HYDROXY-1-PROPYNE
519-44-8	3-HYDROXY-2,4-DINITROPHENOL
35860-51-6	3-HYDROXY-2,4-DINITROPHENOL
513-86-0	3-HYDROXY-2-BUTANONE
496-03-7	3-HYDROXY-2-ETHYLHEXANAL
513-42-8	3-HYDROXY-2-METHYLPROPENE
115-19-5	3-HYDROXY-3-METHYL-1-BUTYNE
2763-96-4	3-HYDROXY-5-AMINOMETHYLISOXAZOLE
2763-96-4	3-HYDROXY-5-AMINOMETHYLISOXAZOLE-AGARIN
141-66-2	3-HYDROXY-N,N-DIMETHYL-CIS-CROTONAMIDE DIMETHYL PHOSPHATE
6923-22-4	3-HYDROXY-N-METHYL-CIS-CROTONAMIDE DIMETHYL PHOSPHATE
6923-22-4	3-HYDROXY-N-METHYLCROTONAMIDE DIMETHYL PHOSPHATE
81-07-2	3-HYDROXYBENZISOTHIAZOLE-S,S-DIOXIDE
107-89-1	3-HYDROXYBUTANAL
107-89-1	3-HYDROXYBUTYRALDEHYDE
3068-88-0	3-HYDROXYBUTYRIC ACID LACTONE
7786-34-7	3-HYDROXYCROTONIC ACID METHYL ESTER DIMETHYL PHOSPHATE
108-46-3	3-HYDROXYCYCLOHEXADIEN-1-ONE
141-66-2	3-HYDROXYDIMETHYL CROTONAMIDE DIMETHYL PHOSPHATE
554-84-7	3-HYDROXYNITROBENZENE
108-46-3	3-HYDROXYPHENOL
109-78-4	3-HYDROXYPROPANENITRILE
107-18-6	3-HYDROXYPROPENE
57-57-8	3-HYDROXYPROPIONIC ACID LACTONE
109-78-4	3-HYDROXYPROPIONITRILE
156-87-6	3-HYDROXYPROPYLAMINE
556-52-5	3-HYDROXYPROPYLENE OXIDE
108-39-4	3-HYDROXYTOLUENE
556-56-9	3-IODO-1-PROPENE
556-56-9	3-IODOPROPENE
556-56-9	3-IODOPROPYLENE
4098-71-9	3-ISOCYANATOMETHYL-3,5,5-TRIMETHYLCYCLOHEXYLISOCYANATE
4016-14-2	3-ISOPROPOXY-1,2-EPOXYPROPANE
2631-37-0	3-ISOPROPYL-5-METHYLPHENYL METHYLCARBAMATE

CAS No.	Chemical Name
2631-37-0	3-ISOPROPYL-5-METHYLPHENYL N-METHYLCARBAMATE
4016-14-2	3-ISOPROPYLOXYPROPYLENE OXIDE
64-00-6	3-ISOPROPYLPHENYL METHYLCARBAMATE
64-00-6	3-ISOPROPYLPHENYL N-METHYLCARBAMATE
57-06-7	3-ISOTHIOCYANATO-1-PROPENE
2763-96-4	3-ISOXAZOLOL, 5-(AMINOMETHYL)-
56-49-5	3-MC
107-03-9	3-MERCAPTOPROPANOL
79869-58-2	3-MERCAPTOPROPANOL
5332-73-0	3-METHOXY-1-PROPANAMINE
72-33-3	3-METHOXY-17 ALPHA-ETHINYLESTRADIOL
72-33-3	3-METHOXY-17 ALPHA-ETHYNYLESTRADIOL
5332-73-0	3-METHOXY-N-PROPYLAMINE
72-33-3	3-METHOXYETHYNYLESTRADIOL
110-67-8	3-METHOXYPROPANENITRILE
110-67-8	3-METHOXYPROPIONITRILE
5332-73-0	3-METHOXYPROPYLAMINE
110-67-8	3-METHOXYPROPYLNITRILE
563-80-4	3-METHYL BUTAN-2-ONE (DOT)
123-51-3	3-METHYL BUTANOL
115-19-5	3-METHYL BUTYNOL
78-79-5	3-METHYL-1,3-BUTADIENE
107-82-4	3-METHYL-1-BROMOBUTANE
590-86-3	3-METHYL-1-BUTANAL
123-51-3	3-METHYL-1-BUTANOL
123-51-3	3-METHYL-1-BUTANOL (CZECH)
563-45-1	3-METHYL-1-BUTENE
123-92-2	3-METHYL-1-BUTYL ACETATE
115-19-5	3-METHYL-1-BUTYN-3-OL
2084-18-6	3-METHYL-2-BUTANETHIOL
598-75-4	3-METHYL-2-BUTANOL
563-80-4	3-METHYL-2-BUTANONE
513-35-9	3-METHYL-2-BUTENE
12772-47-3	3-METHYL-2-PENTEN-4-YN-1-OL
78-81-9	3-METHYL-2-PROPYLAMINE
78-80-8	3-METHYL-3-BUTEN-1-YNE
814-78-8	3-METHYL-3-BUTEN-2-ON (GERMAN)
59-50-7	3-METHYL-4-CHLOROPHENOL
1321-10-4	3-METHYL-4-CHLOROPHENOL
3074-77-9	3-METHYL-4-ETHYLHEXANE
2631-37-0	3-METHYL-5-(1-METHYLETHYL)PHENOLMETHYLCARBAMATE
541-85-5	3-METHYL-5-HEPTANONE
2631-37-0	3-METHYL-5-ISOPROPYL N-METHYLCARBAMATE
2631-37-0	3-METHYL-5-ISOPROPYLPHENYL METHYLCARBAMATE
2631-37-0	3-METHYL-5-ISOPROPYLPHENYL-N-METHYLCARBAMATE
115-19-5	3-METHYL-BUTIN-(1)-OL-(3) (GERMAN)
102-27-2	3-METHYL-N-ETHYLANILINE
3724-65-0	3-METHYLACRYLIC ACID
6117-91-5	3-METHYLALLYL ALCOHOL
108-44-1	3-METHYLANILINE
108-44-1	3-METHYLBENZENAMINE
123-51-3	3-METHYLBUTAN-1-OL
75-85-4	3-METHYLBUTAN-3-OL
590-86-3	3-METHYLBUTANAL
563-80-4	3-METHYLBUTANONE
123-92-2	3-METHYLBUTYL ACETATE
107-82-4	3-METHYLBUTYL BROMIDE
107-84-6	3-METHYLBUTYL CHLORIDE
123-92-2	3-METHYLBUTYL ETHANOATE
115-19-5	3-METHYLBUTYN-3-OL
590-86-3	3-METHYLBUTYRALDEHYDE
56-49-5	3-METHYLCHOLANTHRENE
814-78-8	3-METHYLENE-2-BUTANONE
760-21-4	3-METHYLENEPENTANE
72-33-3	3-METHYLETHYNYLESTRADIOL
589-34-4	3-METHYLHEXANE
99-08-1	3-METHYLNITROBENZENE
2216-33-3	3-METHYLOCTANE
97-95-0	3-METHYLOLPENTANE
96-14-0	3-METHYLPENTANE
108-39-4	3-METHYLPHENOL
1129-41-5	3-METHYLPHENYL METHYLCARBAMATE
1129-41-5	3-METHYLPHENYL N-METHYLCARBAMATE
108-44-1	3-METHYLPHENYLAMINE
123-51-3	3-METIL-BUTANOLO (ITALIAN)
108-11-2	3-MIC
123-00-2	3-MORPHOLINO-1-PROPYLAMINE
123-00-2	3-MORPHOLINOPROPANAMINE
123-00-2	3-MORPHOLINOPROPYLAMINE

CAS No.	Chemical Name
89-62-3	3-NITRO-4-AMINOTOLUENE
121-17-5	3-NITRO-4-CHLORO-ALPHA,ALPHA,ALPHA-TRIFLUOROTOLUENE
121-17-5	3-NITRO-4-CHLOROBENZOTRIFLUORIDE (DOT)
121-17-5	3-NITRO-4-CHLOROTRIFLUOROMETHYLBENZENE
119-32-4	3-NITRO-4-METHYLANILINE
119-32-4	3-NITRO-4-TOLUIDIN (CZECH)
119-32-4	3-NITRO-4-TOLUIDINE
99-59-2	3-NITRO-6-METHOXYANILINE
1777-84-0	3-NITRO-P-ACETOPHENETIDE
119-32-4	3-NITRO-P-TOLUIDINE
98-46-4	3-NITROBENZOTRIFLUORIDE
121-73-3	3-NITROCHLOROBENZENE
554-84-7	3-NITROPHENOL
504-88-1	3-NITROPROPIONIC ACID
99-08-1	3-NITROTOLUENE
99-08-1	3-NITROTOLUOL
498-15-7	3-NORCARENE, 3,7,7-TRIMETHYL-
72-33-3	3-O-METHYLETHYNYLESTRADIOL
106-68-3	3-OCTANONE
111-46-6	3-OXA-1,5-PENTANEDIOL
111-76-2	3-OXA-1-HEPTANOL
111-46-6	3-OXAPENTAMETHYLENE-1,5-DIOL
60-29-7	3-OXAPENTANE
111-46-6	3-OXAPENTANE-1,5-DIOL
102-01-2	3-OXO-N-PHENYLBUTANAMIDE
141-97-9	3-OXOBUTANOIC ACID ETHYL ESTER
105-45-3	3-OXOBUTANOIC ACID METHYL ESTER
78-94-4	3-OXOBUTENE
88-09-5	3-PENTANECARBOXYLIC ACID
584-02-1	3-PENTANOL
96-22-0	3-PENTANONE
625-33-2	3-PENTEN-2-ONE
141-79-7	3-PENTEN-2-ONE, 4-METHYL-
584-02-1	3-PENTYL ALCOHOL
122-60-1	3-PHENOXY-1,2-EPOXYPROPANE
122-60-1	3-PHENOXY-1,2-PROPYLENE OXIDE
25251-51-8	3-PHENYL-3-(TERT-BUTYLPEROXY)PHTHALIDE
1031-47-6	3-PHENYL-5-AMINO-1,2,4-TRIAZOLYL-(1)-(N,N'-TETRAMETHYL) DIAMIDOPHOSPHONATE
156-87-6	3-PROPANOLAMINE
57-57-8	3-PROPANOLIDE
57-57-8	3-PROPIOLACTONE
107-19-7	3-PROPYNOL
25251-51-8	3-TERT-BUTYL PEROXY-3-PHENYLPHTHALIDE, TECHNICALLY PURE
25251-51-8	3-TERT-BUTYLPEROXY-3-PHENYLPHTHALIDE
25251-51-8	3-TERT-BUTYLPEROXY-3-PHENYLPHTHALIDE, TECHNICAL PURE (DOT)
108-44-1	3-TOLUIDINE
1129-41-5	3-TOLYL METHYLCARBAMATE
1129-41-5	3-TOLYL-N-METHYLCARBAMATE
98-16-8	3-TRIFLUOROMETHYL ANILINE
98-16-8	3-TRIFLUOROMETHYL ANILINE (DOT)
928-96-1	3-Z-HEXEN-1-OL
22259-30-9	3-[(DIMETHYLAMINO)METHYLENIMINO]PHENYL N-METHYLCARBAMATE
4985-85-7	3-[BIS(2-HYDROXYETHYL)AMINO]PROPYLAMINE
148-82-3	3025 C. B.
77-73-6	3A,4,7,7A-TETRAHYDRO-4,7-METHANOINDENE
133-06-2	3A,4,7,7A-TETRAHYDRO-N-(TRICHLOROMETHANESULPHENYL) PHTHALIMIDE
50-76-0	3H-PHENOXAZINE, ACTINOMYCIN D DERIV.
126-72-7	3PBR
613-35-4	4',4'''-BIACETANILIDE
62-44-2	4'-ETHOXYACETANILIDE
122-82-7	4'-ETHOXYACETOACETANILIDE
103-89-9	4'-METHYLACETANILIDE
1836-75-5	4'-NITRO-2,4-DICHLORODIPHENYL ETHER
56-04-2	4(1H)-PYRIMIDINONE, 2,3-DIHYDRO-6-METHYL-2-THIOXO-
51-52-5	4(1H)-PYRIMIDINONE, 2,3-DIHYDRO-6-PROPYL-2-THIOXO-
622-40-2	4(2-HYDROXYETHYL)MORPHOLINE
72-43-5	4,4'-(2,2,2-TRICHLOROETHYLIDENE)DIANISOLE
119-90-4	4,4'-BI-O-ANISIDINE
92-87-5	4,4'-BIANILINE
92-87-5	4,4'-BIPHENYLDIAMINE
4685-14-7	4,4'-BIPYRIDINIUM, 1,1'-DIMETHYL-
2074-50-2	4,4'-BIPYRIDINIUM, 1,1'-DIMETHYL-, BIS(METHYL SULFATE)
1910-42-5	4,4'-BIPYRIDINIUM, 1,1'-DIMETHYL-, DICHLORIDE
90-94-8	4,4'-BIS(DIMETHYLAMINO)BENZOPHENONE

CAS No.	Chemical Name
101-61-1	4,4'-BIS(DIMETHYLAMINO)DIPHENYLMETHANE
101-61-1	4,4'-BIS(DIMETHYLAMINOPHENYL)METHANE
80-51-3	4,4'-BIS(HYDRAZINOSULFONYL)DIPHENYL ETHER
72-54-8	4,4'-DDD
72-55-9	4,4'-DDE
50-29-3	4,4'-DDT
613-35-4	4,4'-DIACETAMIDOBIPHENYL
613-35-4	4,4'-DIACETYLAMINOBIPHENYL
613-35-4	4,4'-DIACETYLBENZIDINE
92-87-5	4,4'-DIAMINO-1,1'-BIPHENYL
81-11-8	4,4'-DIAMINO-2,2'-STILBENEDISULFONIC ACID
91-94-1	4,4'-DIAMINO-3,3'-DICHLOROBIPHENYL
91-94-1	4,4'-DIAMINO-3,3'-DICHLORODIPHENYL
119-90-4	4,4'-DIAMINO-3,3'-DIMETHOXYBIPHENYL
119-90-4	4,4'-DIAMINO-3,3'-DIMETHOXYDIPHENYL
119-93-7	4,4'-DIAMINO-3,3'-DIMETHYLBIPHENYL
838-88-0	4,4'-DIAMINO-3,3'-DIMETHYLDIPHENYLMETHANE
92-87-5	4,4'-DIAMINOBIPHENYL
92-87-5	4,4'-DIAMINODIPHENYL
101-80-4	4,4'-DIAMINODIPHENYL ETHER
101-80-4	4,4'-DIAMINODIPHENYL OXIDE
139-65-1	4,4'-DIAMINODIPHENYL SULFIDE
101-77-9	4,4'-DIAMINODIPHENYLMETHAN (GERMAN)
101-77-9	4,4'-DIAMINODIPHENYLMETHANE
101-77-9	4,4'-DIAMINODIPHENYLMETHANE (DOT)
101-80-4	4,4'-DIAMINOPHENYL ETHER
139-65-1	4,4'-DIAMINOPHENYL SULFIDE
115-32-2	4,4'-DICHLORO-ALPHA-(TRICHLOROMETHYL)BENZHYDROL
510-15-6	4,4'-DICHLOROBENZILIC ACID ETHYL ESTER
50-29-3	4,4'-DICHLORODIPHENYLTRICHLOROETHANE
56-53-1	4,4'-DIHYDROXY-ALPHA, BETA-DIETHYLSTILBENE
56-53-1	4,4'-DIHYDROXYDIETHYLSTILBENE
5124-30-1	4,4'-DIISOCYANATODICYCLOHEXYLMETHANE
101-68-8	4,4'-DIISOCYANATODIPHENYLMETHANE
1910-2-5	4,4'-DIMETHYLDIPYRIDYL DICHLORIDE
92-87-5	4,4'-DIPHENYLENEDIAMINE
101-68-8	4,4'-DIPHENYLMETHANE DIISOCYANATE
101-68-8	4,4'-DIPHENYLMETHANE ISOCYANATE
101-77-9	4,4'-DIPHENYLMETHANEDIAMINE
13552-44-8	4,4'-METHYLENDIANILINE DIHYDROCHLORIDE
101-77-9	4,4'-METHYLENE DIANILINE
101-14-4	4,4'-METHYLENE-BIS(2-CHLOROANILINE)
101-14-4	4,4'-METHYLENEBIS(2-CHLORANILINE)
101-77-9	4,4'-METHYLENEBIS(ANILINE)
101-77-9	4,4'-METHYLENEBIS(BENZENEAMINE)
5124-30-1	4,4'-METHYLENEBIS(CYCLOHEXYL ISOCYANATE)
101-68-8	4,4'-METHYLENEBIS(ISOCYANATOBENZENE)
101-14-4	4,4'-METHYLENEBIS(O-CHLOROANILINE)
838-88-0	4,4'-METHYLENEBIS(O-TOLUIDINE)
101-68-8	4,4'-METHYLENEBIS(PHENYL ISOCYANATE)
101-14-4	4,4'-METHYLENEBIS-(2-CHLOROANILINE)
101-77-9	4,4'-METHYLENEBISANILINE
838-88-0	4,4'-METHYLENEBIS[2-METHYLANILINE]
101-61-1	4,4'-METHYLENEBIS[N,N-DIMETHYLANILINE]
838-88-0	4,4'-METHYLENEDI-O-TOLUIDINE
101-68-8	4,4'-METHYLENEDI-P-PHENYLENE DIISOCYANATE
101-77-9	4,4'-METHYLENEDIANILINE
101-77-9	4,4'-METHYLENEDIBENZENAMINE
101-68-8	4,4'-METHYLENEDIPHENYL DIISOCYANATE
101-68-8	4,4'-METHYLENEDIPHENYL ISOCYANATE
101-68-8	4,4'-METHYLENEDIPHENYLENE ISOCYANATE
28434-86-8	4,4'-OXYBIS(2-CHLOROANILINE)
101-80-4	4,4'-OXYBIS(ANILINE)
80-51-3	4,4'-OXYBIS(BENZENESULFONIC ACID HYDRAZIDE)
80-51-3	4,4'-OXYBIS(BENZENESULFONIC ACID) DIHYDRAZIDE
80-51-3	4,4'-OXYBIS(BENZENESULFONYL HYDRAZIDE)
80-51-3	4,4'-OXYDI(BENZENESULFONIC ACID HYDRAZIDE)
101-80-4	4,4'-OXYDIANILINE
80-51-3	4,4'-OXYDIBENZENESULFONIC ACID DIHYDRAZIDE
101-80-4	4,4'-OXYDIPHENYLAMINE
56-53-1	4,4'-STILBENEDIOL, ALPHA, ALPHA'-DIETHYL-, (E)-
101-61-1	4,4'-TETRAMETHYLDIAMINODIPHENYLMETHANE
96-69-5	4,4'-THIOBIS(3-METHYL-6-TERT-BUTYLPHENOL)
96-69-5	4,4'-THIOBIS[2-TERT-BUTYL-5-METHYLPHENOL]
96-69-5	4,4'-THIOBIS[6-TERT-BUTYL-3-METHYLPHENOL]
96-69-5	4,4'-THIOBIS[6-TERT-BUTYL-M-CRESOL]
139-65-1	4,4'-THIOBIS[ANILINE]
139-65-1	4,4'-THIODIANILINE
101-68-8	4,4-DIISOCYANATE DIPHENYL METHANE
80-05-7	4,4-ISOPROPYLIDENDIPHENOL
80-05-7	4,4-ISOPROPYLIDENEDIPHENOL
101-77-9	4,4-METHYLENE DIANILINE
101-61-1	4,4-METHYLENEBIS(N-N-DIMETHYL)BENZENEAMINE
101-77-9	4,4-METHYLENEDIANILINE (ACGIH)
139-65-1	4,4-THIODIANILINE
50-32-8	4,5-BENZPYRENE
95-76-1	4,5-DICHLOROANILINE
96-45-7	4,5-DIHYDRO-2-MERCAPTOIMIDAZOLE
2179-59-1	4,5-DITHIA-1-OCTENE
101-05-3	4,6-DICHLORO-N-(2-CHLOROPHENYL)-1,3,5-TRIAZIN-2-AMINE
12654-97-6	4,6-DICHLORO-N-(2-CHLOROPHENYL)-1,3,5-TRIAZIN-2-AMINE
120-83-2	4,6-DICHLOROPHENOL
105-67-9	4,6-DIMETHYLPHENOL
88-85-7	4,6-DINITRO-2-(1-METHYL-N-PROPYL)PHENOL
534-52-1	4,6-DINITRO-2-METHYLPHENOL
88-85-7	4,6-DINITRO-2-SEC-BUTYLPHENOL
88-85-7	4,6-DINITRO-2-SECBUTYLFENOL (CZECH)
1335-85-9	4,6-DINITRO-O-CRESOL
534-52-1	4,6-DINITRO-O-CRESOL
1335-85-9	4,6-DINITRO-O-CRESOLO (ITALIAN)
131-89-5	4,6-DINITRO-O-CYCLOHEXYLPHENOL
1335-85-9	4,6-DINITRO-O-KRESOL (CZECH)
88-85-7	4,6-DINITRO-O-SEC-BUTYLPHENOL
1335-85-9	4,6-DINITROKRESOL (DUTCH)
57-24-9	4,6-METHANO-6H,14H-INDOLO[3,2,1-IJ]OXEPINO[2,3,4-DE]PYRROLO[2,3-H]QUINOLINE, STRYCHNIDIN-1
357-57-3	4,6-METHANO-6H,14H-INDOLO[3,2,1-IJ]OXEPINO[2,3,4-DE]PYRROLO[2,3-H]QUINOLINE, STRYCHNIDIN-10-ONE DERI
1954-28-5	4,7,10,13-TETRAOXAHEXADECANE, 1,2:15,16-DIEPOXY-
1954-28-5	4,7,10,13-TETRAOXAHEXADECANE, 1,2:15,16-DIEPOXY- (8CI)
498-15-7	4,7,7-TRIMETHYL-3-NORCARENE
124-16-3	4,7-DIOXAUNDECAN-2-OL
56-25-7	4,7-EPOXYISOBENZOFURAN-1,3-DIONE, HEXAHYDRO-3A,7A-DIMETHYL-, (3AALPHA,4BETA,7BETA,7AALPHA)-
57-74-9	4,7-METHANO-1H-INDENE, 1,2,4,5,6,7,8,8-OCTACHLORO-2,3,3A,4,7,7A-HEXAHYDRO-
12789-03-6	4,7-METHANO-1H-INDENE, 1,2,4,5,6,7,8,8-OCTACHLORO-2,3,3A,4,7,7A-HEXAHYDRO-
76-44-8	4,7-METHANO-1H-INDENE, 1,4,5,6,7,8,8-HEPTACHLORO-3A,4,7,7A-TETRAHYDRO-
77-73-6	4,7-METHANO-1H-INDENE, 3A,4,7,7A-TETRAHYDRO-
991-42-4	4,7-METHANO-1H-ISOINDOLE-1,3(2H)-DIONE, 3A,4,7,7A-TETRAHYDRO-5-(HYDROXYPHENYL-2-PYRIDINYLMETHYL)-8-(
57-74-9	4,7-METHANOINDAN, 1,2,4,5,6,7,8,8-OCTACHLORO-3A,4,7,7A-TETRAHYDRO-
12789-03-6	4,7-METHANOINDAN, 1,2,4,5,6,7,8,8-OCTACHLORO-3A,4,7,7A-TETRAHYDRO-
1024-57-3	4,7-METHANOINDAN, 1,4,5,6,7,8,8-HEPTACHLORO-2,3-EPOXY-3A,4,7,7A-TETRAHYDRO-
12789-03-6	4,7-METHANOINDEN, 1,2,4,5,6,7,8,9-OCTACHLORO-3A,4,7,7A-TETRAHYDRO-
76-44-8	4,7-METHANOINDENE, 1,4,5,6,7,8,8-HEPTACHLORO-3A,4,7,7A-TETRAHYDRO-
77-73-6	4,7-METHANOINDENE, 3A,4,7,7A-TETRAHYDRO-
297-78-9	4,7-METHANOISOBENZOFURAN, 1,3,4,5,6,7,8,8-OCTACHLORO-1,3,3A,4,7,7A-HEXAHYDRO-
297-78-9	4,7-METHANOISOBENZOFURAN, 1,3,4,5,6,7,8,8-OCTACHLORO-3A,4,7,7A-TETRAHYDRO-
93-05-0	4- AMINO-N,N-DIETHYLANILINE
98-29-3	4-(1,1-DIMETHYLETHYL)BENZENE-1,2-DIOL
98-29-3	4-(1,1-DIMETHYLETHYL)CATECHOL
80-46-6	4-(1,1-DIMETHYLPROPYL)PHENOL
1836-75-5	4-(2,4-DICHLOROPHENOXY)NITROBENZENE
2038-03-1	4-(2-AMINOETHYL)MORPHOLINE
123-00-2	4-(3-AMINOPROPYL)MORPHOLINE
101-77-9	4-(4-AMINOBENZYL)ANILINE
103-89-9	4-(ACETYLAMINO)TOLUENE
15159-40-7	4-(CHLOROFORMYL) MORPHOLINE
100-14-1	4-(CHLOROMETHYL)NITROBENZENE
1582-09-8	4-(DI-N-PROPYLAMINO)-3,5-DINITRO-1-TRIFLUOROMETHYLBENZENE
3037-72-7	4-(DIETHOXYMETHYLSILYL)BUTYLAMINE
140-80-7	4-(DIETHYLAMINO)-1-METHYLBUTYLAMINE
93-05-0	4-(DIETHYLAMINO)ANILINE
2032-59-9	4-(DIMETHYLAMINO)-M-TOLYL METHYLCARBAMATE
99-98-9	4-(DIMETHYLAMINO)ANILINE
99-98-9	4-(DIMETHYLAMINO)BENZENAMINE
138-89-6	4-(DIMETHYLAMINO)NITROSOBENZENE

CAS No.	Chemical Name
53324-05-3	4-(DIMETHYLAMINO)NITROSOBENZENE
4342-03-4	4-(DIMETHYLTRIAZENO)IMIDAZOLE-5-CARBOXAMIDE
123-00-2	4-(GAMMA-AMINOPROPYL)MORPHOLINE
4726-14-1	4-(METHYLSULFONYL)-2,6-DINITRO-N,N-DIPROPYLANILINE
4726-14-1	4-(METHYLSULFONYL)-2,6-DINITRO-N,N-DIPROPYLBENZENE-AMINE
2032-65-7	4-(METHYLTHIO)-3,5-XYLYL METHYLCARBAMATE
60-11-7	4-(N,N-DIMETHYLAMINO)AZOBENZENE
97-56-3	4-(O-TOLYLAZO)-O-TOLUIDINE
1126-78-9	4-(PHENYLAMINO)BUTANE
60-11-7	4-(PHENYLAZO)-N,N-DIMETHYLANILINE
98-56-6	4-(TRIFLUOROMETHYL)CHLOROBENZENE
402-54-0	4-(TRIFLUOROMETHYL)NITROBENZENE
101-68-8	4-4'-DIISOCYANATE DE DIPHENYLMETHANE (FRENCH)
103-89-9	4-ACETOTOLUIDE
1696-20-4	4-ACETYLMORPHOLINE
94-59-7	4-ALLYL-1,2-(METHYLENEDIOXY)BENZENE
140-80-7	4-AMINO-1-(DIETHYLAMINO)PENTANE
123-30-8	4-AMINO-1-HYDROXYBENZENE
27598-85-2	4-AMINO-1-HYDROXYBENZENE
106-49-0	4-AMINO-1-METHYLBENZENE
97-56-3	4-AMINO-2',3-DIMETHYLAZOBENZENE
119-34-6	4-AMINO-2-NITROPHENOL
119-32-4	4-AMINO-2-NITROTOLUENE
1918-02-1	4-AMINO-3,5,6-TRICHLOROPICOLINIC ACID
89-62-3	4-AMINO-3-NITROTOLUENE
54-62-6	4-AMINO-4-DEOXYPTEROYLGLUTAMATE
99-98-9	4-AMINO-N,N-DIMETHYLANILINE
54-62-6	4-AMINO-PGA
106-50-3	4-AMINOANILINE
104-94-9	4-AMINOANISOLE
60-09-3	4-AMINOAZOBENZENE
92-67-1	4-AMINOBIPHENYL
3037-72-7	4-AMINOBUTYL(METHYL)DIETHOXYSILANE
616-45-5	4-AMINOBUTYRIC ACID LACTAM
92-67-1	4-AMINODIPHENYL
54-62-6	4-AMINOFOLIC ACID
100-01-6	4-AMINONITROBENZENE
1321-31-9	4-AMINOPHENETOLE
156-43-4	4-AMINOPHENETOLE
123-30-8	4-AMINOPHENOL
27598-85-2	4-AMINOPHENOL
101-80-4	4-AMINOPHENYL ETHER
123-00-2	4-AMINOPROPYLMORPHOLINE
54-62-6	4-AMINOPTEROYLGLUTAMIC ACID
504-24-5	4-AMINOPYRIDIN
504-24-5	4-AMINOPYRIDINE
504-24-5	4-AMINOPYRIDINE (DOT)
106-49-0	4-AMINOTOLUEN (CZECH)
106-49-0	4-AMINOTOLUENE
1918-02-1	4-AMINOTRICHLOROPICOLINIC ACID
104-94-9	4-ANISIDINE
100-07-2	4-ANISOYL CHLORIDE
504-24-5	4-AP
56-18-8	4-AZAHEPTAMETHYLENEDIAMINE
56-18-8	4-AZAHEPTANE-1,7-DIAMINE
92-66-0	4-BIPHENYL BROMIDE
4301-50-2	4-BIPHENYLACETIC ACID, 2-FLUOROETHYL ESTER
92-67-1	4-BIPHENYLAMINE
92-04-6	4-BIPHENYLOL, 3-CHLORO-
92-66-0	4-BROMO-1,1'-BIPHENYL
106-38-7	4-BROMO-1-METHYLBENZENE
107-82-4	4-BROMO-2-METHYLBUTANE
92-66-0	4-BROMOBIPHENYL
92-66-0	4-BROMODIPHENYL
101-55-3	4-BROMODIPHENYL ETHER
586-78-7	4-BROMONITROBENZENE
101-55-3	4-BROMOPHENOXYBENZENE
101-55-3	4-BROMOPHENYL PHENYL ETHER
106-38-7	4-BROMOTOLUENE
552-30-7	4-CARBOXYPHTHALIC ANHYDRIDE
27134-26-5	4-CHLORANILIN (CZECH)
95-83-0	4-CHLORO-1,2-BENZENEDIAMINE
95-83-0	4-CHLORO-1,2-DIAMINOBENZENE
95-83-0	4-CHLORO-1,2-PHENYLENEDIAMINE
97-00-7	4-CHLORO-1,3-DINITROBENZENE
106-43-4	4-CHLORO-1-METHYLBENZENE
100-00-5	4-CHLORO-1-NITROBENZENE
95-85-2	4-CHLORO-2-AMINOPHENOL
95-79-4	4-CHLORO-2-AMINOTOLUENE
95-69-2	4-CHLORO-2-METHYLANILINE
3165-93-3	4-CHLORO-2-METHYLANILINE HYDROCHLORIDE
95-69-2	4-CHLORO-2-METHYLBENZENAMINE
3165-93-3	4-CHLORO-2-METHYLBENZENAMINE HYDROCHLORIDE
95-79-4	4-CHLORO-2-METHYLBENZENEAMINE
107-84-6	4-CHLORO-2-METHYLBUTANE
95-69-2	4-CHLORO-2-TOLUIDINE
3165-93-3	4-CHLORO-2-TOLUIDINE HYDROCHLORIDE
777-37-7	4-CHLORO-3-(TRIFLUOROMETHYL)NITROBENZENE
59-50-7	4-CHLORO-3-CRESOL
1321-10-4	4-CHLORO-3-CRESOL
7149-75-9	4-CHLORO-3-METHYLANILINE (CHLOROTOLUIDINE)
59-50-7	4-CHLORO-3-METHYLPHENOL
1321-10-4	4-CHLORO-3-METHYLPHENOL
121-17-5	4-CHLORO-3-NITRO-1-(TRIFLUOROMETHYL)BENZENE
121-17-5	4-CHLORO-3-NITRO-ALPHA,ALPHA,ALPHA-TRIFLUOROTOLUENE
121-17-5	4-CHLORO-3-NITROBENZOTRIFLUORIDE
121-17-5	4-CHLORO-3-NITROBENZYLIDYNE FLUORIDE
59-50-7	4-CHLORO-5-METHYLPHENOL
1321-10-4	4-CHLORO-5-METHYLPHENOL
95-69-2	4-CHLORO-6-METHYLANILINE
3165-93-3	4-CHLORO-6-METHYLANILINE HYDROCHLORIDE
121-17-5	4-CHLORO-ALPHA,ALPHA,ALPHA-TRIFLUORO-3-NITROTOLUENE
98-56-6	4-CHLORO-ALPHA,ALPHA,ALPHA-TRIFLUOROTOLUENE
115-32-2	4-CHLORO-ALPHA-(4-CHLOROPHENYL)-ALPHA-(TRICHLORO-METHYL)BENZENEMETHANOL
59-50-7	4-CHLORO-M-CRESOL
1321-10-4	4-CHLORO-M-CRESOL
5131-60-2	4-CHLORO-M-PHENYLENEDIAMINE
95-83-0	4-CHLORO-O-PHENYLENEDIAMINE
95-69-2	4-CHLORO-O-TOLUIDINE
3165-93-3	4-CHLORO-O-TOLUIDINE HYDROCHLORIDE
3165-93-3	4-CHLORO-O-TOLUIDINE HYDROCHLORIDE (DOT)
27134-26-5	4-CHLOROANILINE
104-88-1	4-CHLOROBENZALDEHYDE
27134-26-5	4-CHLOROBENZENAMINE
98-56-6	4-CHLOROBENZOTRIFLUORIDE
94-17-7	4-CHLOROBENZOYL PEROXIDE
104-83-6	4-CHLOROBENZYL CHLORIDE
628-20-6	4-CHLOROBUTANENITRILE
628-20-6	4-CHLOROBUTYRONITRILE
7005-72-3	4-CHLORODIPHENYL ETHER
100-00-5	4-CHLORONITROBENZENE
106-48-9	4-CHLOROPHENOL
25167-80-0	4-CHLOROPHENOL
106-48-9	4-CHLOROPHENOL (CHLOROPHENATE)
7005-72-3	4-CHLOROPHENYL PHENYL ETHER
27134-26-5	4-CHLOROPHENYLAMINE
106-43-4	4-CHLOROTOLUENE
106-44-5	4-CRESOL
133-06-2	4-CYCLOHEXENE-1,2-DICARBOXIMIDE, N-(TRICHLOROMETHYL) THIO-
2425-06-1	4-CYCLOHEXENE-1,2-DICARBOXIMIDE, N-[(1,1,2,2-TETRACHLORO-ETHYL)THIO]-
85-43-8	4-CYCLOHEXENE-1,2-DICARBOXYLIC ACID ANHYDRIDE
85-43-8	4-CYCLOHEXENE-1,2-DICARBOXYLIC ANHYDRIDE
100-50-5	4-CYCLOHEXENE-1-CARBOXALDEHYDE
827-52-1	4-CYCLOHEXYLBENZENE
315-18-4	4-DIMETHYLAMINO-3,5-DIMETHYLPHENYL N-METHYLCARBA-MATE
315-18-4	4-DIMETHYLAMINO-3,5-XYLYL METHYLCARBAMATE
315-18-4	4-DIMETHYLAMINO-3,5-XYLYL N-METHYLCARBAMATE
60-11-7	4-DIMETHYLAMINOAZOBENZENE
622-97-9	4-ETHENYLMETHYLBENZENE
62-44-2	4-ETHOXYACETANILIDE
156-43-4	4-ETHOXYANILINE
1321-31-9	4-ETHOXYANILINE
156-43-4	4-ETHOXYBENZENAMINE
3074-75-7	4-ETHYL-2-METHYLHEXANE
3074-77-9	4-ETHYL-3-METHYLHEXANE
100-74-3	4-ETHYLMORPHOLINE
123-07-9	4-ETHYLPHENOL
371-40-4	4-FLUORANILIN (CZECH)
371-40-4	4-FLUOROANILINE
371-40-4	4-FLUOROANILINE (DOT)
371-40-4	4-FLUOROBENZENAMINE
462-23-7	4-FLUOROBUTYRIC ACID
37759-72-1	4-FLUOROCROTONIC ACID

CAS No.	Chemical Name
100-50-5	4-FORMYLCYCLOHEXENE
108-82-7	4-HEPTANOL, 2,6-DIMETHYL-
123-19-3	4-HEPTANONE
108-83-8	4-HEPTANONE, 2,6-DIMETHYL-
105-67-9	4-HYDROXY-1,3-DIMETHYLBENZENE
123-42-2	4-HYDROXY-2-KETO-4-METHYLPENTANE
128-37-0	4-HYDROXY-3,5-DI-TERT-BUTYLTOLUENE
5836-29-3	4-HYDROXY-3-(1,2,3,4-TETRAHYDRO-1-NAFTYL)-CUMARINE (DUTCH)
5836-29-3	4-HYDROXY-3-(1,2,3,4-TETRAHYDRO-1-NAPHTHYL)-CUMARIN
81-81-2	4-HYDROXY-3-(3-OXO-1-FENYL-BUTYL) CUMARINE (DUTCH)
81-81-2	4-HYDROXY-3-(3-OXO-1-PHENYL-BUTYL)-CUMARIN (GERMAN)
81-81-2	4-HYDROXY-3-(3-OXO-1-PHENYLBUTYL)-2H-1-BENZOPYRAN-2-ONE
92-04-6	4-HYDROXY-3-CHLOROBIPHENYL
123-42-2	4-HYDROXY-4-METHYL PENTAN-2-ONE
123-42-2	4-HYDROXY-4-METHYL-2-PENTANONE
123-42-2	4-HYDROXY-4-METHYL-PENTAN-2-ON (GERMAN, DUTCH)
123-42-2	4-HYDROXY-4-METHYLPENTANONE-2
123-30-8	4-HYDROXYANILINE
27598-85-2	4-HYDROXYANILINE
150-76-5	4-HYDROXYANISOLE
123-30-8	4-HYDROXYBENZENAMINE
106-48-9	4-HYDROXYCHLOROBENZENE
100-79-8	4-HYDROXYMETHYL-2,2-DIMETHYL-1,3-DIOXOLANE
100-02-7	4-HYDROXYNITROBENZENE
123-31-9	4-HYDROXYPHENOL
123-07-9	4-HYDROXYPHENYLETHANE
106-44-5	4-HYDROXYTOLUENE
5836-29-3	4-IDROSSI-3-(1,2,3,4-TETRAIDRO-1-NAFTIL)-CUMARINA (ITALIAN)
81-81-2	4-IDROSSI-3-(3-OXO-)-FENIL-BUTIL)-CUMARINE (ITALIAN)
123-42-2	4-IDROSSI-4-METIL-PENTAN-2-ONE (ITALIAN)
138-86-3	4-ISOPROPENYL-1-METHYL-1-CYCLOHEXENE
138-86-3	4-ISOPROPENYL-1-METHYLCYCLOHEXENE
99-87-6	4-ISOPROPYL-1-METHYLBENZENE
586-62-9	4-ISOPROPYLIDENE-1-METHYLCYCLOHEXENE
99-87-6	4-ISOPROPYLTOLUENE
2763-96-4	4-ISOXAZOLIN-3-ONE, 5-(AMINOMETHYL)-
100-17-4	4-METHOXY-1-NITROBENZENE
107-70-0	4-METHOXY-4-METHYLPENTAN-2-ONE
39156-41-7	4-METHOXY-M-PHENYLENEDIAMINE SULFATE
104-94-9	4-METHOXYANILINE
104-94-9	4-METHOXYBENZENAMINE
100-07-2	4-METHOXYBENZOIC ACID CHLORIDE
100-07-2	4-METHOXYBENZOYL CHLORIDE
100-17-4	4-METHOXYNITROBENZENE
150-76-5	4-METHOXYPHENOL
610-39-9	4-METHYL-1,2-DINITROBENZENE
95-80-7	4-METHYL-1,3-BENZENEDIAMINE
926-56-7	4-METHYL-1,3-PENTADIENE
95-80-7	4-METHYL-1,3-PHENYLENEDIAMINE
106-38-7	4-METHYL-1-BROMOBENZENE
591-47-9	4-METHYL-1-CYCLOHEXENE
691-37-2	4-METHYL-1-PENTENE
107-41-5	4-METHYL-2,4-PENTANEDIOL
128-37-0	4-METHYL-2,6-DI-TERT-BUTYLPHENOL
120-71-8	4-METHYL-2-AMINOANISOLE
108-09-8	4-METHYL-2-AMINOPENTANE
89-62-3	4-METHYL-2-NITROANILINE
89-62-3	4-METHYL-2-NITROBENZENAMINE
3068-88-0	4-METHYL-2-OXETANONE
108-10-1	4-METHYL-2-OXOPENTANE
108-11-2	4-METHYL-2-PENTANOL
108-84-9	4-METHYL-2-PENTANOL, ACETATE
108-10-1	4-METHYL-2-PENTANONE
4461-48-7	4-METHYL-2-PENTENE
108-84-9	4-METHYL-2-PENTYL ACETATE
108-11-2	4-METHYL-2-PENTYL ALCOHOL
56-04-2	4-METHYL-2-THIOURACIL
119-32-4	4-METHYL-3-NITROANILINE
119-32-4	4-METHYL-3-NITROBENZENAMINE
141-79-7	4-METHYL-3-PENTEN-2-ON (DUTCH, GERMAN)
141-79-7	4-METHYL-3-PENTEN-2-ONE
625-27-4	4-METHYL-3-PENTENE
141-79-7	4-METHYL-3-PENTENE-2-ONE
123-42-2	4-METHYL-4-HYDROXY-2-PENTANONE
763-29-1	4-METHYL-4-PENTENE
89-62-3	4-METHYL-6-NITROANILINE
584-84-9	4-METHYL-M-PHENYLENE DIISOCYANATE
26471-62-5	4-METHYL-M-PHENYLENE DIISOCYANATE

CAS No.	Chemical Name
584-84-9	4-METHYL-M-PHENYLENE ISOCYANATE
26471-62-5	4-METHYL-M-PHENYLENE ISOCYANATE
95-80-7	4-METHYL-M-PHENYLENEDIAMINE
89-62-3	4-METHYL-O-NITROANILINE
26471-62-5	4-METHYL-PHENYLENE DIISOCYANATE
26471-62-5	4-METHYL-PHENYLENE ISOCYANATE
98-51-1	4-METHYL-TERT-BUTYLBENZENE
103-89-9	4-METHYLACETANILIDE
106-49-0	4-METHYLANILINE
106-49-0	4-METHYLBENZENAMINE
104-15-4	4-METHYLBENZENESULFONIC ACID
25231-46-3	4-METHYLBENZENESULFONIC ACID
106-38-7	4-METHYLBROMOBENZENE
591-47-9	4-METHYLCYCLOHEXENE
674-82-8	4-METHYLENE-2-OXETANONE
99-87-6	4-METHYLISOPROPYLBENZENE
2032-65-7	4-METHYLMERCAPTO-3,5-DIMETHYLPHENYL N-METHYLCARBAMATE
2032-65-7	4-METHYLMERCAPTO-3,5-XYLYL METHYLCARBAMATE
109-02-4	4-METHYLMORPHOLINE
99-99-0	4-METHYLNITROBENZENE
2216-34-4	4-METHYLOCTANE
108-84-9	4-METHYLPENT-2-YL ETHANOATE
108-11-2	4-METHYLPENTANOL-2
106-44-5	4-METHYLPHENOL
106-38-7	4-METHYLPHENYL BROMIDE
106-49-0	4-METHYLPHENYLAMINE
108-89-4	4-METHYLPYRIDINE
622-97-9	4-METHYLSTYRENE
2032-65-7	4-METHYLTHIO-3,5-DIMETHYLPHENYL METHYLCARBAMATE
3254-63-5	4-METHYLTHIOPHENYLDIMETHYL PHOSPHATE
106-42-3	4-METHYLTOLUENE
141-79-7	4-METIL-3-PENTEN-2-ONE (ITALIAN)
108-11-2	4-METILPENTAN-2-OLO (ITALIAN)
2038-03-1	4-MORPHOLINEETHANAMINE
622-40-2	4-MORPHOLINEETHANOL
123-00-2	4-MORPHOLINEPROPANAMINE
123-00-2	4-MORPHOLINEPROPYLAMINE
28805-86-9	4-N-BUTYLPHENOL
100-01-6	4-NITRANILINE
29757-24-2	4-NITRANILINE
100-00-5	4-NITRO-1-CHLOROBENZENE
1836-75-5	4-NITRO-2',4'-DICHLOROPHENYL ETHER
99-55-8	4-NITRO-2-AMINOTOLUENE
100-01-6	4-NITROANILINE
29757-24-2	4-NITROANILINE
100-17-4	4-NITROANISOLE
100-01-6	4-NITROBENZENAMINE
29757-24-2	4-NITROBENZENAMINE
100-14-1	4-NITROBENZYLCHLORIDE
92-93-3	4-NITROBIPHENYL
586-78-7	4-NITROBROMOBENZENE
100-00-5	4-NITROCHLOROBENZENE
92-93-3	4-NITRODIPHENYL
100-02-7	4-NITROFENOL (DUTCH)
100-02-7	4-NITROPHENOL
586-78-7	4-NITROPHENYL BROMIDE
100-17-4	4-NITROPHENYL METHYL ETHER
1124-33-0	4-NITROPYRIDINE 1-OXIDE
1124-33-0	4-NITROPYRIDINE N-OXIDE
1124-33-0	4-NITROPYRIDINE OXIDE
56-57-5	4-NITROQUINOLINE 1-OXIDE
56-57-5	4-NITROQUINOLINE N-OXIDE
56-57-5	4-NITROQUINOLINE OXIDE
138-89-6	4-NITROSO-N,N-DIMETHYLANILINE
156-10-5	4-NITROSO-N-PHENYLANILINE
138-89-6	4-NITROSODIMETHYLANILINE
53324-05-3	4-NITROSODIMETHYLANILINE
156-10-5	4-NITROSODIPHENYLAMINE
59-89-2	4-NITROSOMORPHOLINE
99-99-0	4-NITROTOLUENE
99-99-0	4-NITROTOLUOL
123-17-1	4-NONANOL, 2,6,8-TRIMETHYL-
123-18-2	4-NONANONE, 2,6,8-TRIMETHYL-
56-57-5	4-NQO
111-43-3	4-OXAHEPTANE
156-43-4	4-PHENETIDINE
92-04-6	4-PHENYL-2-CHLOROPHENOL
92-67-1	4-PHENYLANILINE

CAS No.	Chemical Name
92-53-5	4-PHENYLMORPHOLINE
108-89-4	4-PICOLINE
504-24-5	4-PYRIDINAMINE
2971-90-6	4-PYRIDINOL, 3,5-DICHLORO-2,6-DIMETHYL-
504-24-5	4-PYRIDYLAMINE
535-89-7	4-PYRIMIDINAMINE, 2-CHLORO-N,N,6-TRIMETHYL-
333-41-5	4-PYRIMIDINOL, 2-ISOPROPYL-6-METHYL-, O-ESTER WITH O,O-DIETHYL PHOSPHOROTHIOATE
80-46-6	4-T-AMYLPHENOL
98-29-3	4-T-BUTYLCATECHOL
98-29-3	4-TBC
2049-92-5	4-TERT-AMYLANILINE
80-46-6	4-TERT-AMYLPHENOL
98-29-3	4-TERT-BUTYL CATECHOL
98-29-3	4-TERT-BUTYL-1,2-BENZENEDIOL
98-29-3	4-TERT-BUTYL-1,2-DIHYDROXYBENZENE
98-51-1	4-TERT-BUTYL-1-METHYLBENZENE
98-28-2	4-TERT-BUTYL-2-CHLOROPHENOL
299-86-5	4-TERT-BUTYL-2-CHLOROPHENYL METHYL METHYLPHOSPHORAMIDATE
98-27-1	4-TERT-BUTYL-2-METHYLPHENOL
98-27-1	4-TERT-BUTYL-O-CRESOL
98-29-3	4-TERT-BUTYLCATECHIN
98-29-3	4-TERT-BUTYLCATECHOL
98-29-3	4-TERT-BUTYLPYROCATECHOL
98-51-1	4-TERT-BUTYLTOLUENE
80-46-6	4-TERT-PENTYLPHENOL
3268-49-3	4-THIAPENTANAL
3268-49-3	4-THIAPENTANAL (DOT)
104-15-4	4-TOLUENESULFONIC ACID
25231-46-3	4-TOLUENESULFONIC ACID
106-49-0	4-TOLUIDINE
19594-40-2	4-UNDECANONE, 2-METHYL-
106-87-6	4-VINYL-1,2-CYCLOHEXENE DIEPOXIDE
106-87-6	4-VINYL-1-CYCLOHEXENE DIEPOXIDE
106-87-6	4-VINYL-1-CYCLOHEXENE DIOXIDE
100-40-3	4-VINYLCYCLOHEXENE
106-87-6	4-VINYLCYCLOHEXENE DIEPOXIDE
106-87-6	4-VINYLCYCLOHEXENE DIOXIDE
622-97-9	4-VINYLTOLUENE
671-16-9	4-[(2-METHYLHYDRAZINO)METHYL]-N-ISOPROPYLBENZAMIDE
305-03-3	4-[BIS(2-CHLOROETHYL)AMINO]PHENYLBUTYRIC ACID
305-03-3	4-[P-[BIS(2-CHLOROETHYL)AMINO]PHENYL]BUTYRIC ACID
638-17-5	4H-1,3,5-DITHIAZINE, DIHYDRO-2,4,6-TRIMETHYL-, (2 ALPHA,4 ALPHA,6 ALPHA)-
110-91-8	4H-1,4-OXAZINE, TETRAHYDRO-
91681-63-9	4H-1-BENZOPYRAN-4-ONE, 6-(2,3-DIHYDROXY-3-METHYLBUTYL)-3-(2,4-DIHYDROXYPHENYL)-7-HYDROXY-
83-79-4	5'BETA-ROTENONE
25316-40-9	5,12-NAPHTHACENEDIONE, 10-[(3-AMINO-2,3,6-TRIDEOXY-ALPHA-L-LYXO-HEXOPYRANOSYL)OXY]-7,8,9,10-TETRAH
23214-92-8	5,12-NAPHTHACENEDIONE, 10-[(3-AMINO-2,3,6-TRIDEOXY-ALPHA-L-LYXO-HEXOPYRANOSYL)OXY]-7,8,9,10-TETRAH
20830-81-3	5,12-NAPHTHACENEDIONE, 8-ACETYL-10-[(3-AMINO-2,3,6-TRIDEOXY-ALPHA-L-LYXO-HEXOPYRANOSYL)OXY]-7,8,9,(+)DAUNOMYCIN
57-41-0	5,5-DIPHENYL-2,4-IMIDAZOLIDINEDIONE
57-41-0	5,5-DIPHENYLHYDANTOIN
85-00-7	5,6-DIHYDRO-DIPYRIDO(1,2A;2,1C)PYRAZINIUM DIBROMIDE
51787-44-1	5,7,8,9-TETRAMETHYL-3,4-BENZOACRIDINE
112-98-1	5,8,11,14,17-PENTAOXAHENEICOSANE
4342-03-4	5-(3,3-DIMETHYLTRIAZENO)IMIDAZOLE-4-CARBOXAMIDE
2763-96-4	5-(AMINOMETHYL)-3(2H)-ISOXAZOLONE
2763-96-4	5-(AMINOMETHYL)-3-ISOXAZOLOL
66-75-1	5-(BIS(2-CHLOROETHYL)AMINO)URACIL 2,4(1H,3H) PYRIMIDINEDIONE, 5-BIS(2-CHLOROETHYL)AMINO
140-80-7	5-(DIETHYLAMINO)-2-PENTYLAMINE
4342-03-4	5-(DIMETHYLTRIAZENO)IMIDAZOLE-4-CARBOXAMIDE
94-59-7	5-ALLYL-1,3-BENZODIOXOLE
61-82-5	5-AMINO-1,2,4-TRIAZOLE
1031-47-6	5-AMINO-1-(BIS(DIMETHYLAMINO)PHOSPHINYL)-3-PHENYL-1,2,4-TRIAZOLE
1031-47-6	5-AMINO-1-BIS(DIMETHYLAMIDE)PHOSPHORYL-3-PHENYL-1,2,4-TRIAZOLE
1031-47-6	5-AMINO-1-BIS(DIMETHYLAMIDO)PHOSPHORYL-3-PHENYL-1,2,4-TRIAZOLE
61-82-5	5-AMINO-1H-1,2,4-TRIAZOLE
1031-47-6	5-AMINO-3-FENIL-1-BIS(-DIMETILAMINO)-FOSFORIL-1,2,4-TRIAZOLO (ITALIAN)
1031-47-6	5-AMINO-3-FENYL-1-BIS(DIMETHYL-AMINO)-FOSFORYL-1,2,4-TRIAZOOL (DUTCH)
1031-47-6	5-AMINO-3-PHENYL-1,2,4-TRIAZOLE-1-YL-N,N,N',N'-TETRAMETHYLPHOSPHODIAMIDE
1031-47-6	5-AMINO-3-PHENYL-1,2,4-TRIAZOLYL-1-BIS(DIMETHYLAMIDO)PHEOSPHATE
1031-47-6	5-AMINO-3-PHENYL-1,2,4-TRIAZOLYL-NNN'N'-TETRAMETHYLPHOSPHONAMIDE
1031-47-6	5-AMINO-3-PHENYL-1,2,4-TRIAZOLYLBIS(DIMETHYLAMINO)-PHOSPHINOXID (GERMAN)
1031-47-6	5-AMINO-3-PHENYL-1-BIS (DIMETHYL-AMINO)-PHOSPHORYLE-1,2,4-TRIAZOLE (FRENCH)
1031-47-6	5-AMINO-3-PHENYL-1-BIS(DIMETHYLAMINO)-PHOSPHORYL-1H-1,2,4-TRIAZOL (GERMAN)
2763-96-4	5-AMINOMETHYL-3-HYDROXYISOXAZOLE
2763-96-4	5-AMINOMETHYL-3-ISOXYZOLE
578-94-9	5-AZA-10-ARSENAANTHRACENE CHLORIDE
320-67-2	5-AZACYTIDINE
95-79-4	5-CHLORO-2-AMINOTOLUENE
3165-93-3	5-CHLORO-2-AMINOTOLUENE HYDROCHLORIDE
95-85-2	5-CHLORO-2-HYDROXYANILINE
95-79-4	5-CHLORO-2-METHYLANILINE
95-79-4	5-CHLORO-2-METHYLBENZENAMINE
95-79-4	5-CHLORO-2-TOLUIDINE
578-94-9	5-CHLORO-5,10-DIHYDROPHENARSAZINE
15271-41-7	5-CHLORO-6-((((METHYLAMINO)CARBONYL)OXY)IMINO)BICYCLO-(221)HEPTANE-2-CARBONITRILE
95-79-4	5-CHLORO-O-TOLUIDINE
104-90-5	5-ETHYL-ALPHA-PICOLINE
104-89-2	5-ETHYL-2-METHYLPIPERIDINE
104-90-5	5-ETHYL-2-METHYLPYRIDINE
104-90-5	5-ETHYL-2-PICOLINE
104-89-2	5-ETHYL-2-PIPECOLINE
5408-74-2	5-ETHYL-2-VINYLPYRIDINE
104-90-5	5-ETHYL-ALPHA-PICOLINE
16219-75-3	5-ETHYLIDENE-2-NORBORNENE
16219-75-3	5-ETHYLIDENEBICYCLO(221)HEPT-2-ENE
51-21-8	5-FLUORACIL (GERMAN)
51-21-8	5-FLUORO-2,4(1H,3H)-PYRIMIDINEDIONE
51-21-8	5-FLUORO-2,4-PYRIMIDINEDIONE
51-21-8	5-FLUOROPYRIMIDINE-2,4-DIONE
51-21-8	5-FLUOROURACIL
51-21-8	5-FLUOROURACIL 2,4(1H,3H)-PYRIMIDINEDIONE 5-FLUORO
51-21-8	5-FLUORURACIL (GERMAN)
51-21-8	5-FU
100-73-2	5-HEXENAL, 2,6-EPOXY-
98-00-0	5-HYDROXYMETHYLFURAN
481-39-0	5-HYDROXYNAPHTHALENE-1,4-DIONE
552-30-7	5-ISOBENZOFURANCARBOXYLIC ACID, 1,3-DIHYDRO-1,3-DIOXO-
4098-71-9	5-ISOCYANATO-1-ISOCYANATOMETHYL-1,3,3-TRIMETHYLCYCLOHEXANE
2778-04-3	5-METHOXY-2-(DIMETHOXYPHOSPHINYLTHIOMETHYL)PYRONE-4
106-68-3	5-METHYL 3-HEPTATONE
644-64-4	5-METHYL-1H-PYRAZOL-3-YL DIMETHYLCARBAMATE
110-12-3	5-METHYL-2-HEXANONE
119-38-0	5-METHYL-2-ISOPROPYL-3-PYRAZOLYL DIMETHYLCARBAMATE
541-85-5	5-METHYL-3-HEPTANONE
120-71-8	5-METHYL-O-ANISIDINE
3697-24-3	5-METHYLCHRYSENE
534-22-5	5-METHYLFURAN
110-12-3	5-METHYLHEXAN-2-ONE
110-12-3	5-METHYLHEXAN-2-ONE (DOT)
139-91-3	5-MORPHOLINOMETHYL-3-(5-NITROFURFURYLIDENEAMINO)OXAZOLIDONE
66-75-1	5-N,N-BIS(2-CHLOROETHYL)AMINOURACIL
99-59-2	5-NITRO-2-METHOXYANILINE
99-55-8	5-NITRO-2-METHYLANILINE
99-55-8	5-NITRO-2-TOLUIDINE
119-32-4	5-NITRO-4-TOLUIDINE
99-59-2	5-NITRO-O-ANISIDINE
99-55-8	5-NITRO-O-TOLUIDINE
602-87-9	5-NITROACENAPHTHENE
2338-12-7	5-NITROBENZOTRIAZOL
991-42-4	5-NORBORNENE-2,3-DICARBOXIMIDE, 5-(ALPHA-HYDROXY-ALPHA-2-PYRIDYLBENZYL)-7-(ALPHA-2-PYRIDYLBENZYLIDEN
1031-07-8	5-NORBORNENE-2,3-DIMETHANOL, 1,4,5,6,7,7-HEXACHLORO-, CYCLIC SULFATE
115-29-7	5-NORBORNENE-2,3-DIMETHANOL, 1,4,5,6,7,7-HEXACHLORO-, CYCLIC SULFITE

CAS No.	Chemical Name
959-98-8	5-NORBORNENE-2,3-DIMETHANOL, 1,4,5,6,7,7-HEXACHLORO-, CYCLIC SULFITE, ENDO-
33213-65-9	5-NORBORNENE-2,3-DIMETHANOL, 1,4,5,6,7,7-HEXACHLORO-, CYCLIC SULFITE, EXO-
94-58-6	5-PROPYL-1,3-BENZODIOXOLE
140-76-1	5-VINYL-2-PICOLINE
66-75-1	5-[BIS(2-CHLOROETHYL)AMINO]URACIL
66-75-1	5-[DI(BETA-CHLOROETHYL)AMINO]URACIL
434-07-1	5 ALPHA-ANDROSTAN-3-ONE, 17 BETA-HYDROXY-2-(HYDROXY-METHYLENE)-17-METHYL-
81-07-2	550 SACCHARINE
62450-06-0	5H-PYRIDO[4,3-B]INDOL-3-AMINE, 1,4-DIMETHYL-
62450-07-1	5H-PYRIDO[4,3-B]INDOL-3-AMINE, 1-METHYL-
115-29-7	6,7,8,9,10,10-HEXACHLORO-1,5,5A,6,9,9A-HEXAHYDRO-6,9-METHANO-2,4,3-BENZODIOXATHIEPIN-3-OXIDE
50-32-8	6,7-BENZOPYRENE
85-00-7	6,7-DIHYDROPYRIDO(1,2-A;2',1'-C)PYRAZINEDIUM DIBROMIDE
2244-16-8	6,8(9)-P-MENTHADIEN-2-ONE
1031-07-8	6,9-METHANO-2,4,3-BENZODIOXATHIEPIN, 6,7,8,9,10,10-HEXA-CHLORO-1,5,5A,6,9,9A-HEXAHYDRO-, 3,3-DIOXIDE
959-98-8	6,9-METHANO-2,4,3-BENZODIOXATHIEPIN, 6,7,8,9,10,10-HEXA-CHLORO-1,5,5A,6,9,9A-HEXAHYDRO-, 3-OXIDE, (3.
33213-65-9	6,9-METHANO-2,4,3-BENZODIOXATHIEPIN, 6,7,8,9,10,10-HEXA-CHLORO-1,5,5A,6,9,9A-HEXAHYDRO-, 3-OXIDE, (3.
446-86-6	6-(1-METHYL-4-NITROIMIDAZOL-5-YL)THIOPURINE
88-85-7	6-(1-METHYL-PROPYL)-2,4-DINITROFENOL (DUTCH)
88-85-7	6-(1-METIL-PROPIL)-2,4-DINITRO-FENOLO (ITALIAN)
120-58-1	6-(1-PROPENYL)-1,3-BENZODIOXOLE
551-16-6	6-AMINO-PENICILLANIC ACID
105-60-2	6-AMINOCAPROIC ACID LACTAM
105-60-2	6-AMINOHEXANOIC ACID CYCLIC LACTAM
105-60-2	6-CAPROLACTAM
25567-67-3	6-CHLORO-1,3-DINITROBENZENE
97-00-7	6-CHLORO-1,3-DINITROBENZENE
1929-82-4	6-CHLORO-2-(TRICHLOROMETHYL)PYRIDINE
87-63-8	6-CHLORO-2-METHYL ANILINE
59-50-7	6-CHLORO-3-HYDROXYTOLUENE
1321-10-4	6-CHLORO-3-HYDROXYTOLUENE
59-50-7	6-CHLORO-M-CRESOL
1321-10-4	6-CHLORO-M-CRESOL
1912-24-9	6-CHLORO-N-ETHYL-N'-(1-METHYLETHYL)-1,3,5-TRIAZINE-2,4-DIAMINE
131-89-5	6-CICLOESIL-2,4-DINITR-FENOLO (ITALIAN)
131-89-5	6-CYCLOHEXYL-2,4-DINITROPHENOL
105-60-2	6-HEXANELACTAM
135-02-4	6-METHOXYBENZALDEHYDE
56-04-2	6-METHYL 2-THIOURACIL
534-52-1	6-METHYL-2,4-DINITROPHENOL
56-04-2	6-METHYL-2-MERCAPTOURACIL
56-04-2	6-METHYL-2-THIOURACIL
104-90-5	6-METHYL-3-ETHYLPYRIDINE
99-55-8	6-METHYL-3-NITROANILINE
51-52-5	6-N-PROPYL-2-THIOURACIL
51-52-5	6-N-PROPYLTHIOURACIL
91-59-8	6-NAPHTHYLAMINE
94-52-0	6-NITROBENZIMIDAZOLE
51-52-5	6-PROPYL-2-THIO-2,4(1H,3H)PYRIMIDINEDIONE
51-52-5	6-PROPYL-2-THIOURACIL
51-52-5	6-PROPYLTHIOURACIL
88-85-7	6-SEC-BUTYL-2,4-DINITROPHENOL
872-10-6	6-THIAUNDECANE
56-04-2	6-THIO-4-METHYLURACIL
57-97-6	7,12-DIMETHYL-1,2-BENZANTHRACENE
57-97-6	7,12-DIMETHYLBENZ(A)ANTHRACENE
57-97-6	7,12-DIMETHYLBENZANTHRACENE
57-97-6	7,12-DIMETHYLBENZO[A]ANTHRACENE
57-97-6	7,12-DMBA
205-82-3	7,8-BENZFLUORANTHENE
3457-61-2	7-(TERT-BUTYLPEROXY)CUMENE
30580-75-7	7-(TERT-BUTYLPEROXY)CUMENE
12001-29-5	7-45 ASBESTOS
64-86-8	7-ALPHA-H-COLCHICINE
50-07-7	7-AMINO-9-ALPHA-METHOXYMITOSANE
194-59-2	7-AZA-7H-DIBENZO[C,G]FLUORENE
226-36-8	7-AZADIBENZ[A,H]ANTHRACENE
224-42-0	7-AZADIBENZ[A,J]ANTHRACENE
1563-66-2	7-BENZOFURANOL, 2,3-DIHYDRO-2,2-DIMETHYL-, METHYLCAR-BAMATE
80-15-9	7-CUMYL HYDROPEROXIDE
56-25-7	7-OXABICYCLO(221)HEPTANE-2,3-DICARBOXYLIC ANHYDRIDE, 2,3-DIMETHYL-
106-87-6	7-OXABICYCLO[4.1.0]HEPTANE, 3-(EPOXYETHYL)-
106-87-6	7-OXABICYCLO[4.1.0]HEPTANE, 3-OXIRANYL-
194-59-2	7H-DIBENZO[C,G]CARBAZOLE
10048-13-2	7H-FURO[3',2':4,5]FURO[2,3-C]XANTHEN-7-ONE, 3A,12C-DIHYDRO-8-HYDROXY-6-METHOXY-, (3AR-CIS)-
1332-21-4	7N05
1332-21-4	7RF10
207-08-9	8,9-BENZFLUORANTHENE
207-08-9	8,9-BENZOFLUORANTHENE
3085-26-5	8-METHYLNONANAL
60-51-5	8014 BIS HC
72-33-3	8027 C. B.
298-00-0	8056HC
122-14-5	8057HC
121-75-5	8059HC
82-28-0	9,10-ANTHRACENEDIONE, 1-AMINO-2-METHYL-
117-79-3	9,10-ANTHRACENEDIONE, 2-AMINO-
129-15-7	9,10-ANTHRACENEDIONE, 2-METHYL-1-NITRO-
85-00-7	9,10-DIHYDRO-8A,10,-DIAZONIAPHENANTHRENE DIBROMIDE
2764-72-9	9,10-DIHYDRO-8A,10A-DIAZONIAPHENANTHRENE
85-00-7	9,10-DIHYDRO-8A,10A-DIAZONIAPHENANTHRENE DIBROMIDE
85-00-7	9,10-DIHYDRO-8A,10A-DIAZONIAPHENANTHRENE(1,1'-ETHYLENE-2,2'-BIPYRIDYLIUM)DIBROMIDE
57-97-6	9,10-DIMETHYL-1,2-BENZANTHRACENE
57-97-6	9,10-DIMETHYLBENZ[A]ANTHRACENE
112-80-1	9,10-OCTADECENOIC ACID
50-14-6	9,10-SECOERGOSTA-5,7,10(19),22-TETRAEN-3-OL, (3BETA-,5Z,7E,22E)-
260-94-6	9-AZAANTHRACENE
112-80-1	9-CIS-OCTADECENOIC ACID
112-80-1	9-OCTADECENOIC ACID (Z)-
544-60-5	9-OCTADECENOIC ACID (Z)-, AMMONIUM SALT
26094-13-3	9-OCTADECENOIC ACID (Z)-, COMPD. WITH 1-BUTANAMINE (1:1)
12772-47-3	9-OCTADECENOIC ACID (Z)-, ESTER WITH 2,2-BIS(HYDROXY-METHYL)-1,3-PROPANEDIOL
1191-80-6	9-OCTADECENOIC ACID (Z)-, MERCURY(2+) SALT
143-18-0	9-OCTADECENOIC ACID (Z)-, POTASSIUM SALT
143-19-1	9-OCTADECENOIC ACID (Z)-, SODIUM SALT
140-04-5	9-OCTADECENOIC ACID, 12-(ACETYLOXY)-, BUTYL ESTER, [R-(Z)]-
112-80-1	9-OCTADECENOIC ACID, CIS-
13387-02-0	9-PHOSPHABICYCLONONANE [...3,3,1-NONANE]
13396-80-0	9-PHOSPHABICYCLONONANE [...4,2,1-NONANE]
56-81-5	90 TECHNICAL GLYCERINE
86-73-7	9H-FLUORENE
80-62-6	'MONOCITE' METHACRYLATE MONOMER
52-68-6	((2,2,2-TRICHLORO-1-HYDROXYETHYL) DIMETHYLPHOSPHO-NATE)
3691-35-8	((4-CHLORPHENYL)-1-PHENYL)-ACETYL-1,3-INDANDION (GER-MAN)
121-75-5	((DIMETHOXYPHOSPHINOTHIOYL)THIO)BUTANEDIOIC ACID DIETHYL ESTER
124-65-2	((DIMETHYLARSINO)OXY)SODIUM-AS-OXIDE
26447-14-3	((METHYLPHENOXY)METHYL)OXIRANE
513-49-5	(+)-2-BUTYLAMINE
154-23-4	(+)-3',4',5,7-TETRAHYDROXY-2,3-TRANS-FLAVAN-3-OL
498-15-7	(+)-3-CARENE
584-79-2	(+)-ALLELRETHONYL (+)-CIS,TRANS-CHRYSANTHEMATE
498-15-7	(+)-CARENE-3
2244-16-8	(+)-CARVONE
154-23-4	(+)-CATECHIN
154-23-4	(+)-CATECHOL
154-23-4	(+)-CYANIDAN-3-OL
154-23-4	(+)-CYANIDANOL-3
121-29-9	(+)-PYRETHRONYL (+)-PYRETHRATE
121-21-1	(+)-PYRETHRONYL (+)-TRANS-CHRYSANTHEMATE
513-49-5	(+)-SEC-BUTYLAMINE
54-11-5	(-)-3-(1-METHYL-2-PYRROLIDYL)PYRIDINE
54-11-5	(-)-3-(N-METHYLPYRROLIDINO)PYRIDINE
357-57-3	(-)-BRUCINE
316-42-7	(-)-EMETINE DIHYDROCHLORIDE
54-11-5	(-)-NICOTINE
65-31-6	(-)-NICOTINE BITARTRATE
65-31-6	(-)-NICOTINE HYDROGEN TARTRATE
57-47-6	(-)-PHYSOSTIGMINE
83-79-4	(-)-ROTENONE
111-41-1	(BETA-HYDROXYETHYL)ETHYLENEDIAMINE
4301-50-2	(1,1'-BIPHENYL)-4-ACETIC ACID, 2-FLUOROETHYL ESTER

CAS No.	Chemical Name
78-63-7	(1,1,4,4-TETRAMETHYLTETRAMETHYLENE)BIS(TERT-BUTYL PEROXIDE)
98-84-0	(1-AMINOETHYL)BENZENE
98-85-1	(1-HYDROXYETHYL)BENZENE
119-38-0	(1-ISOPROPIL-3-METIL-1H-PIRAZOL-5-IL)-N,N-DIMETIL-CARBAMMATO (ITALIAN)
119-38-0	(1-ISOPROPYL-3-METHYL-1H-PYRAZOL-5-YL)-N,N-DIMETHYL-CARBAMAT (GERMAN)
119-38-0	(1-ISOPROPYL-3-METHYL-1H-PYRAZOL-5-YL)-N,N-DIMETHYLCARBAMAAT (DUTCH)
98-83-9	(1-METHYLETHENYL)BENZENE
98-82-8	(1-METHYLETHYL)BENZENE
135-98-8	(1-METHYLPROPYL)BENZENE
53-86-1	(1-P-CHLOROBENZOYL-5-METHOXY-2-METHYLINDOL-3-YL)ACETIC ACID
52-68-6	(2,2,2-TRICHLORO-1-HYDROXYETHYL)PHOSPHONIC ACID DIMETHYL ESTER
2636-26-2	(2,2-DICHLOOR-VINYL)-DIMETHYL-FOSFAAT (DUTCH)
2636-26-2	(2,2-DICHLOR-VINYL)-DIMETHYL-PHOSPHAT (GERMAN)
2636-26-2	(2,2-DICHLORO-VINIL)DIMETIL-FOSFATO (ITALIAN)
93-76-5	(2,4,5-TRICHLOOR-FENOXY)-AZIJNZUUR (DUTCH)
93-76-5	(2,4,5-TRICHLOR-PHENOXY)-ESSIGSAEURE (GERMAN)
93-79-8	(2,4,5-TRICHLOROPHENOXY)ACETIC ACID 2-BUTOXYETHYL ESTER
2545-59-7	(2,4,5-TRICHLOROPHENOXY)ACETIC ACID 2-BUTOXYETHYL ESTER
2545-59-7	(2,4,5-TRICHLOROPHENOXY)ACETIC ACID BUTOXYETHANOL ESTER
2545-59-7	(2,4,5-TRICHLOROPHENOXY)ACETIC ACID BUTYL ESTER
93-79-8	(2,4,5-TRICHLOROPHENOXY)ACETIC ACID BUTYL ESTER
94-75-7	(2,4-DICHLOOR-FENOXY)-AZIJNZUUR (DUTCH)
94-75-7	(2,4-DICHLOR-PHENOXY)-ESSIGSAEURE (GERMAN)
94-75-7	(2,4-DICHLOROPHENOXY)ACETIC ACID
94-79-1	(2,4-DICHLOROPHENOXY)ACETIC ACID, SEC-BUTYL ESTER
111-41-1	(2-AMINOETHYL)ETHANOLAMINE
51-83-2	(2-CARBAMOYLOXYETHYL)TRIMETHYLAMMONIUM CHLORIDE
63283-80-7	(2-CHLORO-1-METHYLETHYL) ETHER
627-11-2	(2-CHLOROETHOXY)CARBONYL CHLORIDE
110-75-8	(2-CHLOROETHOXY)ETHENE
999-81-5	(2-CHLOROETHYL)TRIMETHYLAMMONIUM CHLORIDE
78-53-5	(2-DIETHYLAMINO)ETHYLPHOSPHOROTHIOIC ACID O,O-DIETHYL ESTER
100-37-8	(2-HYDROXYETHYL)DIETHYLAMINE
108-01-0	(2-HYDROXYETHYL)DIMETHYLAMINE
110-73-6	(2-HYDROXYETHYL)ETHYLAMINE
111-41-1	(2-HYDROXYETHYL)ETHYLENEDIAMINE
109-83-1	(2-HYDROXYETHYL)METHYLAMINE
51-83-2	(2-HYDROXYETHYL)TRIMETHYL AMMONIUM CHLORIDE CARBAMATE
7786-34-7	(2-METHOXYCARBONYL-1-METHYL-VINYL)-DIMETHYL-FOSFAAT (DUTCH)
7786-34-7	(2-METHOXYCARBONYL-1-METHYL-VINYL)-DIMETHYL-PHOSPHAT (GERMAN)
111-96-6	(2-METHOXYETHYL) ETHER
538-93-2	(2-METHYLPROPYL)BENZENE
7786-34-7	(2-METOSSICARBONIL-1-METIL-VINIL)-DIMETIL-FOSFATO (ITALIAN)
299-75-2	(2S,3S)-THREITOL 1,4-BISMETHANESULFONATE
109-55-7	(3-AMINOPROPYL)DIMETHYLAMINE
2631-37-0	(3-METHYL-5-ISOPROPYLPHENYL)-N-METHYLCARBAMAT (GERMAN)
121-17-5	(3-NITRO-4-CHLOROPHENYL)TRIFLUOROMETHANE
3037-72-7	(4-AMINOBUTYL)DIETHOXYMETHYLSILANE
127-85-5	(4-AMINOPHENYL)ARSONIC ACID SODIUM SALT
104-83-6	(4-CHLOROPHENYL)METHYL CHLORIDE
629-20-9	(8)ANNULENE
62-38-4	(ACETATO)PHENYLMERCURY
12002-03-8	(ACETATO-O)(TRIMETAARSENITO)DICOPPER
62-38-4	(ACETATO-O)PHENYLMERCURY
62-38-4	(ACETOXYMERCURI)BENZENE
100-46-9	(AMINOMETHYL)BENZENE
999-81-5	(BETA-CHLOROETHYL)TRIMETHYLAMMONIUM CHLORIDE
111-41-1	(BETA-HYDROXYETHYL)ETHYLENEDIAMINE
100-39-0	(BROMOMETHYL)BENZENE
3132-64-7	(BROMOMETHYL)ETHYLENE OXIDE
3132-64-7	(BROMOMETHYL)OXIRANE
2426-08-6	(BUTOXYMETHYL)OXIRANE
10108-56-2	(BUTYLAMINO)CYCLOHEXANE
106-89-8	(CHLOROMETHYL)ETHYLENE OXIDE
1558-25-4	(CHLOROMETHYL)TRICHLOROSILANE
502-39-6	(CYANOGUANIDINO)METHYLMERCURY
140-29-4	(CYANOMETHYL)BENZENE
3037-72-7	(DELTA-AMINOBUTYL)METHYLDIETHOXYSILANE
98-87-3	(DICHLOROMETHYL)BENZENE
947-02-4	(DIETHOXYPHOSPHINYL)DITHIOIMIDOCARBONIC ACID CYCLIC ETHYLENE ESTER
950-10-7	(DIETHOXYPHOSPHINYL)DITHIOIMIDOCARBONIC ACID CYCLIC PROPYLENE ESTER
21548-32-3	(DIETHOXYPHOSPHINYLIMINO)-1,3-DITHIETANE
91-66-7	(DIETHYLAMINO)BENZENE
121-44-8	(DIETHYLAMINO)ETHANE
100-37-8	(DIETHYLAMINO)ETHANOL
121-69-7	(DIMETHYLAMINO)BENZENE
79-44-7	(DIMETHYLAMINO)CARBONYL CHLORIDE
108-01-0	(DIMETHYLAMINO)ETHANOL
4342-03-4	(DIMETHYLTRIAZENO)IMIDAZOLECARBOXAMIDE
106-89-8	(DL)-ALPHA-EPICHLOROHYDRIN
156-60-5	(E)-1,2-DICHLOROETHENE
156-60-5	(E)-1,2-DICHLOROETHYLENE
624-64-6	(E)-2-BUTENE
14686-13-6	(E)-2-HEPTENE
646-04-8	(E)-2-PENTENE
6923-22-4	(E)-DIMETHYL 1-METHYL-3-(METHYLAMINO)-3-OXO-1-PROPENYL PHOSPHATE
96-09-3	(EPOXYETHYL)BENZENE
4016-11-9	(ETHOXYMETHYL)OXIRANE
517-16-8	(ETHYLMERCURIC)-p-TOLUENESULPHONANILIDE
28772-56-7	(HYDROXY-4 COUMARINYL 3)-3 PHENYL-3 (BROMO-4 BIPHENYLYL-4)-1 PROPANOL-1 (FRENCH)
109-83-1	(HYDROXYETHYL)METHYLAMINE
100-51-6	(HYDROXYMETHYL)BENZENE
620-05-3	(IODOMETHYL)BENZENE
4016-14-2	(ISOPROPOXYMETHYL)OXIRANE
2454-37-7	(M-AMINOPHENYL)METHYL CARBINOL
100-53-8	(MERCAPTOMETHYL)BENZENE
29191-52-4	(METHOXYPHENYL)AMINE
100-61-8	(METHYLAMINO)BENZENE
12108-13-3	(METHYLCYCLOPENTADIENYL)TRICARBONYLMANGANESE
624-92-0	(METHYLDITHIO)METHANE
96-80-0	(N,N-DIISOPROPYLAMINO)ETHANOL
109-56-8	(N-HYDROXYETHYL)ISOPROPYLAMINE
9016-45-9	(NONYLPHENOXY)POLYETHYLENE OXIDE
950-37-8	(O,O-DIMETHYL)-S-(-2-METHOXY-DELTA(SUP 2)-1,3,4-THIADIAZOLIN-5-ON-4-YLMETHYL)DITHIOPHOSPHATE
732-11-6	(O,O-DIMETHYL-PHTHALIMIDIOMETHYL-DITHIOPHOSPHATE)
60-51-5	(O,O-DIMETHYL-S-(N-METHYL-CARBAMOYL-METHYL)-DITHIOPHOSPHAT) (GERMAN)
12654-97-6	(O-CHLOROANILINO)DICHLOROTRIAZINE
101-05-3	(O-CHLOROANILINO)DICHLOROTRIAZINE
2698-41-1	(O-CHLOROBENZAL)MALONONITRILE
2698-41-1	(O-CHLOROBENZYLIDENE)MALONONITRILE
5344-82-1	(O-CHLOROPHENYL)THIOUREA
140-89-6	(O-ETHYL DITHIOCARBONATO)POTASSIUM
98-56-6	(P-CHLOROPHENYL)TRIFLUOROMETHANE
122-60-1	(PHENOXYMETHYL)OXIRANE
81-81-2	(PHENYL-1 ACETYL-2 ETHYL) 3-HYDROXY-4 COUMARINE (FRENCH)
100-46-9	(PHENYLMETHYL)AMINE
2244-16-8	(S)-(+)-CARVONE
513-49-5	(S)-2-BUTANAMINE
6505-86-8	(S)-3-(1-METHYL-2-PYRROLIDINYL)PYRIDINE SULFATE (2:1)
65-30-5	(S)-3-(1-METHYL-2-PYRROLIDINYL)PYRIDINE SULFATE (2:1)
2244-16-8	(S)-CARVONE
54-11-5	(S)-NICOTINE
10380-29-7	(TETRAAMMINE)COPPER SULFATE HYDRATE
110-05-4	(TRIBUTYL)PEROXIDE
98-07-7	(TRICHLOROMETHYL)BENZENE
98-08-8	(TRIFLUOROMETHYL)BENZENE
428-59-1	(TRIFLUOROMETHYL)TRIFLUOROOXIRANE
156-59-2	(Z)-1,2-DICHLOROETHENE
156-59-2	(Z)-1,2-DICHLOROETHYLENE
590-18-1	(Z)-2-BUTENE
627-20-3	(Z)-PENTENE
928-96-1	(Z)-HEX-3-EN-1-OL
1477-55-0	ALPHA, ALPHA'-DIAMINO-M-XYLENE
542-88-1	ALPHA, ALPHA'-DICHLORODIMETHYL ETHER
56-53-1	ALPHA, ALPHA'-DIETHYL-4,4'-STILBENEDIOL
56-53-1	ALPHA, ALPHA'-DIETHYLSTILBENEDIOL

CAS No.	Chemical Name
1477-55-0	ALPHA, ALPHA'-M-XYLENEDIAMINE
75-99-0	ALPHA, ALPHA-DICHLOROPROPANOIC ACID
6607-45-0	ALPHA, BETA-DICHLOROSTYRENE
96-09-3	ALPHA, BETA-EPOXYSTYRENE
542-75-6	ALPHA, GAMMA-DICHLOROPROPYLENE
2449-49-2	ALPHA,N,N-TRIMETHYLBENZYLAMINE
80-56-8	ALPHA-(+)-PINENE
32534-95-5	ALPHA-(2,4,5-TRICHLOROPHENOXY)PROPIONIC ACID ISOOCTYL ESTER
7568-93-6	ALPHA-(AMINOMETHYL)BENZYL ALCOHOL
319-84-6	ALPHA-1,2,3,4,5,6-HEXACHLORCYCLOHEXANE
319-84-6	ALPHA-1,2,3,4,5,6-HEXACHLOROCYCLOHEXANE
28314-03-6	ALPHA-ACETAMIDOFLUORENE
102-01-2	ALPHA-ACETYL-N-PHENYLACETAMIDE
102-01-2	ALPHA-ACETYLACETANILIDE
1344-28-1	ALPHA-ALUMINA
1344-28-1	ALPHA-ALUMINUM OXIDE
134-32-7	ALPHA-AMINONAPHTHALENE
504-29-0	ALPHA-AMINOPYRIDINE
319-84-6	ALPHA-BENZENE HEXACHLORIDE
959-98-8	ALPHA-BENZOEPIN
319-84-6	ALPHA-BHC
106-98-9	ALPHA-BUTENE
106-98-9	ALPHA-BUTYLENE
106-88-7	ALPHA-BUTYLENE OXIDE
9004-34-6	ALPHA-CELLULOSE
591-97-9	ALPHA-CHLORO-BETA-BUTYLENE
14464-46-1	ALPHA-CRISTOBALITE
14464-46-1	ALPHA-CRYSTOBALITE
137-05-3	ALPHA-CYANOACRYLIC ACID METHYL ESTER
57-50-1	ALPHA-D-GLUCOPYRANOSIDE, BETA-D-FRUCTOFURANOSYL
959-98-8	ALPHA-ENDOSULFAN
645-62-5	ALPHA-ETHYL-2-HEXENAL
88-09-5	ALPHA-ETHYLBUTYRIC ACID
1309-37-1	ALPHA-FERRIC OXIDE
319-84-6	ALPHA-HEXACHLORAN
319-84-6	ALPHA-HEXACHLORANE
319-84-6	ALPHA-HEXACHLORCYCLOHEXANE
1309-37-1	ALPHA-IRON OXIDE
563-45-1	ALPHA-ISOAMYLENE
78-59-1	ALPHA-ISOPHORON
78-59-1	ALPHA-ISOPHORONE
319-84-6	ALPHA LINDANE
100-53-8	ALPHA-MERCAPTOTOLUENE
96-17-3	ALPHA-METHYL-N-BUTANAL
96-17-3	ALPHA-METHYLBUTANAL
96-17-3	ALPHA-METHYLBUTYRALDEHYDE
96-17-3	ALPHA-METHYLBUTYRIC ALDEHYDE
583-60-8	ALPHA-METHYLCYCLOHEXANONE
534-22-5	ALPHA-METHYLFURAN
90-12-0	ALPHA-METHYLNAPHTHALENE
78-82-0	ALPHA-METHYLPROPANENITRILE
79-31-2	ALPHA-METHYLPROPANOIC ACID
78-84-2	ALPHA-METHYLPROPIONALDEHYDE
79-31-2	ALPHA-METHYLPROPIONIC ACID
134-32-7	ALPHA-NAPHTHYLAMINE
98-84-0	ALPHA-PHENETHYLAMINE
98-84-0	ALPHA-PHENYLETHYLAMINE
9002-81-7	ALPHA-POLY(OXYMETHYLENE)
127-00-4	ALPHA-PROPYLENE CHLOROHYDRIN
504-29-0	ALPHA-PYRIDINAMINE
504-29-0	ALPHA-PYRIDYLAMINE
616-45-5	ALPHA-PYRROLIDINONE
616-45-5	ALPHA-PYRROLIDONE
14808-60-7	ALPHA-QUARTZ
9005-25-8	ALPHA-STARCH
959-98-8	ALPHA-THIODAN
33213-65-9	ALPHA-THIONEX
100-53-8	ALPHA-TOLUENETHIOL
100-53-8	ALPHA-TOLUOLTHIOL
100-53-8	ALPHA-TOLYL MERCAPTAN
15468-32-3	ALPHA-TRIDYMITE
111-48-8	BETA, BETA'-DIHYDROXYDIETHYL SULFIDE
111-48-8	BETA, BETA'-DIHYDROXYETHYL SULFIDE
2698-41-1	BETA, BETA-DICYANO-O-CHLOROSTYRENE
2426-54-2	BETA-(DIETHYLAMINO)ETHYL ACRYLATE
319-85-7	BETA-1,2,3,4,5,6-HEXACHLOROCYCLOHEXANE
156-87-6	BETA-ALANINOL
1344-28-1	BETA-ALUMINUM OXIDE
117-79-3	BETA-AMINOANTHRAQUINONE
141-43-5	BETA-AMINOETHANOL
141-43-5	BETA-AMINOETHYL ALCOHOL
124-68-5	BETA-AMINOISOBUTANOL
9004-34-6	BETA-AMYLOSE
122-98-5	BETA-ANILINOETHANOL
319-85-7	BETA-BENZENE HEXACHLORIDE
33213-65-9	BETA-BENZOEPIN
319-85-7	BETA-BHC
107-01-7	BETA-BUTENE
107-01-7	BETA-BUTYLENE
542-58-5	BETA-CHLOROETHYL ACETATE
110-75-8	BETA-CHLOROETHYL VINYL ETHER
91-58-7	BETA-CHLORONAPHTHALENE
590-18-1	BETA-CIS-BUTYLENE
109-78-4	BETA-CYANOETHANOL
57-50-1	BETA-D-FRUCTOFURANOSYL .ALPHA.-D-GLUCOPYRANOSIDE
14901-08-7	BETA-D-GLUCOPYRANOSIDE, (METHYL-ONN-AZOXY)METHYL
108-01-0	BETA-DIMETHYLAMINOETHYL ALCOHOL
33213-65-9	BETA-ENDOSULFAN
141-43-5	BETA-ETHANOLAMINE
110-80-5	BETA-ETHOXYETHANOL
2806-85-1	BETA-ETHOXYPROPIONALDEHYDE
92-50-2	BETA-ETHYLANILINOETHYL ALCOHOL
1024-57-3	BETA-HEPTACHLOREPOXIDE
319-85-7	BETA-HEXACHLORAN
319-85-7	BETA-HEXACHLOROBENZENE
7568-93-6	BETA-HYDROXY-BETA-PHENYLETHYLAMINE
109-59-1	BETA-HYDROXYETHYL ISOPROPYL ETHER
111-48-8	BETA-HYDROXYETHYL SULFIDE
141-43-5	BETA-HYDROXYETHYLAMINE
7568-93-6	BETA-HYDROXYPHENETHYLAMINE
7568-93-6	BETA-HYDROXYPHENYLETHYLAMINE
999-61-1	BETA-HYDROXYPROPYL ACRYLATE
102-01-2	BETA-KETOBUTYRANILIDE
319-85-7	BETA-LINDANE
110-67-8	BETA-METHOXYPROPIONITRILE
3068-88-0	BETA-METHYL-BETA-PROPIOLACTONE
110-67-8	BETA-METHYOXYPROPIONITRILE
622-40-2	BETA-MORPHOLINOETHANOL
2038-03-1	BETA-MORPHOLINOETHYLAMINE
494-03-1	BETA-NAPHTHYLBIS(BETA-CHLOROETHYL)AMINE
494-03-1	BETA-NAPHTHYLDI(2-CHLOROETHYL)AMINE
135-88-6	BETA-NAPHTHYLPHENYLAMINE
7568-93-6	BETA-PHENETHANOLAMINE
136-40-3	BETA-PHENYLAZO-ALPHA,ALPHA'-DIAMINOPYRIDINE HYDROCHLORIDE
7568-93-6	BETA-PHENYLETHANOLAMINE
14808-60-7	BETA-QUARTZ
79-00-5	BETA-T
33213-65-9	BETA-THIODAN
111-48-8	BETA-THIODIGLYCOL
959-98-8	BETA-THIONEX
624-64-6	BETA-TRANS-BUTYLENE
79-00-5	BETA-TRICHLOROETHANE
319-86-8	DELTA-1,2,3,4,5,6-HEXACHLOROCYCLOHEXANE
1344-28-1	DELTA-ALUMINUM OXIDE
319-86-8	DELTA-BENZENE HEXACHLORIDE
319-86-8	DELTA-BHC
319-86-8	DELTA-HCH
319-86-8	DELTA-HEXACHLOROCYCLOHEXANE
319-86-8	DELTA-LINDANE
72-33-3	DELTA-MVE
58-22-0	DELTA4-ANDROSTEN-17.BETA-OL-3-ONE
57-83-0	DELTA4-PREGNENE-3,20-DIONE
1344-28-1	GAMMA-ALUMINA
1344-28-1	GAMMA-ALUMINUM OXIDE
616-45-5	GAMMA-AMINOBUTYRIC ACID LACTAM
616-45-5	GAMMA-AMINOBUTYRIC LACTAM
616-45-5	GAMMA-AMINOBUTYROLACTAM
156-87-6	GAMMA-AMINOPROPANOL
616-45-5	GAMMA-BUTYROLACTAM
542-75-6	GAMMA-CHLOROALLYL CHLORIDE
109-55-7	GAMMA-DIMETHYLAMINOPROPYLAMINE
1309-37-1	GAMMA-FERRIC OXIDE
156-87-6	GAMMA-HYDROXY-1-PROPYLAMINE
1309-37-1	GAMMA-IRON OXIDE (FE2O3)
563-46-2	GAMMA-ISOAMYLENE
591-97-9	GAMMA-METHALLYL CHLORIDE

CAS No.	Chemical Name
5332-73-0	GAMMA-METHOXYPROPYLAMINE
591-97-9	GAMMA-METHYLALLYL CHLORIDE
108-89-4	GAMMA0-METHYLPYRIDINE
1309-37-1	GAMMA-MYD
122-60-1	GAMMA-PHENOXYPROPYLENE OXIDE
108-89-4	GAMMA-PICOLINE
1120-71-4	GAMMA-PROPANE SULTONE
305-03-3	GAMMA-[P-DI(2-CHLOROETHYL)AMINOPHENYL]BUTYRIC ACID
12079-65-1	PI-CYCLOPENTADIENYLMANGANESE TRICARBONYL
12108-13-3	PI-METHYLCYCLOPENTADIENYLMANGANESE TRICARBONYL
122-14-5	009
102-01-2	[(ACETOACETYL)AMINO]BENZENE
106-92-3	[(ALLYLOXY)METHYL]OXIRANE
121-75-5	[(DIMETHOXYPHOSPHINOTHIOYL)THIO]BUTANEDIOIC ACID DIETHYL ESTER
90-41-5	[1,1'-BIPHENYL]-2-AMINE
92-87-5	[1,1'-BIPHENYL]-4,4'-DIAMINE
91-94-1	[1,1'-BIPHENYL]-4,4'-DIAMINE, 3,3'-DICHLORO-
119-90-4	[1,1'-BIPHENYL]-4,4'-DIAMINE, 3,3'-DIMETHOXY-
119-93-7	[1,1'-BIPHENYL]-4,4'-DIAMINE, 3,3'-DIMETHYL-
92-67-1	[1,1'-BIPHENYL]-4-AMINE
92-04-6	[1,1'-BIPHENYL]-4-OL, 3-CHLORO-
514-73-8	[2-BIS(3-ETHYLBENZOTHIAZOLYL)] PENTAMETHINE CYANINE IODIDE
71-43-2	[6]ANNULENE
1344-28-1	ETA-ALUMINA
9000-07-1	KAPPALAMBDA-CARRAGEENAN
154-23-4	(+)-(2R:3S)-5,7,3',4'-TETRAHYDROXYFLAVAN-3-OL
56-53-1	(E)-3,4-BIS(4-HYDROXYPHENYL)-3-HEXENE
7429-90-5	A 00
1344-28-1	A 1
1344-28-1	A 1 (SORBENT)
25155-30-0	A 1-1575
9004-66-4	A 100
9004-66-4	A 100 (PHARMACEUTICAL)
67-68-5	A 10846
7429-90-5	A 1200P
75-94-5	A 150
75-94-5	A 150 (SILANE)
1309-64-4	A 1530
1309-64-4	A 1582
1309-64-4	A 1588LP
12001-26-2	A 21 (MINERAL)
75-28-5	A 31
75-28-5	A 31 (HYDROCARBON)
91-80-5	A 3322
1912-24-9	A 361
12001-26-2	A 41 (MINERAL)
9004-70-0	A 5021
545-55-1	A 6366
9016-45-9	A 730
9016-45-9	A 730 (SURFACTANT)
7429-90-5	A 97
7429-90-5	A 99
7429-90-5	A 99 (METAL)
7429-90-5	A 99N
13463-67-7	A-FIL CREAM
298-00-0	A-GRO
91-59-8	A-NAPHTHYLAMIN (GERMAN)
54-62-6	A-NINOPTERIN
2646-17-5	A.F. ORANGE NO. 2
1694-09-3	A.F. VIOLET NO. 1
50-78-2	A.S.A. EMPIRIN
105-60-2	A1030
105-60-2	A1030N0
7429-90-5	A999
7429-90-5	A999V
107-18-6	AA
7429-90-5	AA 1099
7429-90-5	AA 1193
7429-90-5	AA 1199
25155-30-0	AA-10
25155-30-0	AA-9
133-06-2	AACAPTAN
106-93-4	AADIBROOM
60-51-5	AADIMETHOAL
53-96-3	AAF
14484-64-1	AAFERTIS
76-44-8	AAHEPTA

CAS No.	Chemical Name
58-89-9	AALINDAN
137-26-8	AAPIROL
8020-83-5	AAR 1
56-38-2	AAT
137-26-8	AATACK
137-26-8	AATIRAM
88-85-7	AATOX
56-38-2	AATP
1912-24-9	AATREX
1912-24-9	AATREX 4L
1912-24-9	AATREX 80W
1912-24-9	AATREX NINE-O
50-29-3	AAVERO-EXTRA
115-19-5	AB 32
3383-96-8	ABATE
3383-96-8	ABATHION
25155-30-0	ABESON NAM
12001-26-2	ABHRAK
9004-34-6	ABICEL
514-73-8	ABMINTHIC
8020-83-5	ABOLIUM
64-17-5	ABSOLUTE ETHANOL
97-77-8	ABSTENSIL
97-77-8	ABSTINIL
97-77-8	ABSTINYL
297-97-2	AC 18133
2275-18-5	AC 18682
8052-42-4	AC 20
563-12-2	AC 3422
999-81-5	AC 38555
298-02-2	AC 3911
947-02-4	AC 47031
950-10-7	AC 47470
3383-96-8	AC 52160
50-78-2	AC 5230
78-34-2	AC 528
64-00-6	AC 5727
21548-32-3	AC 64475
8052-42-4	AC 8 (ASPHALT)
13071-79-9	AC 92100
60-51-5	AC-12880
2778-04-3	AC-18,737
60-51-5	AC-18682
122-14-5	AC-47300
123-54-6	ACAC
9000-01-5	ACACIA
510-15-6	ACAR
510-15-6	ACARABEN
115-32-2	ACARIN
786-19-6	ACARITHION
297-97-2	ACC 18133
97-77-8	ACCEL TET
137-26-8	ACCEL TMT
137-26-8	ACCELERANT T
5459-93-8	ACCELERATOR HX
137-26-8	ACCELERATOR T
137-26-8	ACCELERATOR THIURAM
138-89-6	ACCELERINE
53324-05-3	ACCELERINE
95-79-4	ACCO FAST RED KB BASE
21548-32-3	ACCONEM
122-14-5	ACCOTHION
75-86-5	ACETONCYAANHYDRINE (DUTCH)
124-17-4	ACEATATE
1319-77-3	ACEDE CRESYLIQUE (FRENCH)
71-63-6	ACEDOXIN
83-32-9	ACENAPHTHENE
602-87-9	ACENAPHTHENE, 5-NITRO-
208-96-8	ACENAPHTHYLENE
83-32-9	ACENAPHTHYLENE, 1,2-DIHYDRO-
602-87-9	ACENAPHTHYLENE, 1,2-DIHYDRO-5-NITRO-
50-78-2	ACENTERINE
122-14-5	ACEOTHION
10265-92-6	ACEPHATE-MET
10118-76-0	ACERDOL
50-78-2	ACESAL
105-57-7	ACETAAL (DUTCH)
105-57-7	ACETAL
105-57-7	ACETAL (DOT)

CAS No.	Chemical Name
105-57-7	ACETAL DIETHYLIQUE (FRENCH)
75-07-0	ACETALDEHYD (GERMAN)
75-07-0	ACETALDEHYDE
75-07-0	ACETALDEHYDE (ACGIH,DOT)
75-39-8	ACETALDEHYDE AMMONIA
75-39-8	ACETALDEHYDE AMMONIA (DOT)
105-57-7	ACETALDEHYDE ETHYL ACETAL
16568-02-8	ACETALDEHYDE METHYLFORMYLHYDRAZONE
16568-02-8	ACETALDEHYDE N-FORMYL-N-METHYLHYDRAZONE
107-29-9	ACETALDEHYDE OXIME
107-29-9	ACETALDEHYDE OXIME (DOT)
37273-91-9	ACETALDEHYDE POLYMER
75-39-8	ACETALDEHYDE, AMINE SALT
107-20-0	ACETALDEHYDE, CHLORO-
97-97-2	ACETALDEHYDE, CHLORO-, DIMETHYL ACETAL
78-97-7	ACETALDEHYDE, CYANOHYDRIN
105-57-7	ACETALDEHYDE, DIETHYL ACETAL
108-62-3	ACETALDEHYDE, HOMOPOLYMER
9002-91-9	ACETALDEHYDE, HOMOPOLYMER
37273-91-9	ACETALDEHYDE, HOMOPOLYMER
108-62-3	ACETALDEHYDE, POLYMERS
9002-91-9	ACETALDEHYDE, POLYMERS
108-62-3	ACETALDEHYDE, TETRAMER
9002-91-9	ACETALDEHYDE, TETRAMER
37273-91-9	ACETALDEHYDE, TETRAMER
75-87-6	ACETALDEHYDE, TRICHLORO-
75-87-6	ACETALDEHYDE, TRICHLORO- (9CI)
123-63-7	ACETALDEHYDE, TRIMER
16568-02-8	ACETALDEHYDE-N-METHYL-N-FORMYLHYDRAZONE
107-89-1	ACETALDOL
107-29-9	ACETALDOXIME
105-57-7	ACETALE (ITALIAN)
60-35-5	ACETAMIDE
56-75-7	ACETAMIDE, 2,2-DICHLORO-N-[2-HYDROXY-1-(HYDROXY-METHYL)-2-(4-NITROPHENYL)ETHYL]-, [R-(R*,R*)]-
640-19-7	ACETAMIDE, 2-FLUORO-
613-35-4	ACETAMIDE, N,N'-[1,1'-BIPHENYL]-4,4'-DIYLBIS-
127-19-5	ACETAMIDE, N,N-DIMETHYL-
62-44-2	ACETAMIDE, N-(4-ETHOXYPHENYL)-
103-89-9	ACETAMIDE, N-(4-METHYLPHENYL)-
64-86-8	ACETAMIDE, N-(5,6,7,9-TETRAHYDRO-1,2,3,10-TETRAMETHOXY-9-OXOBENZO(ALPHA)HEPTALEN-7-YL)-
64-86-8	ACETAMIDE, N-(5,6,7,9-TETRAHYDRO-1,2,3,10-TETRAMETHOXY-9-OXOBENZO[A]HEPTALEN-7-YL)-, (S)-
28314-03-6	ACETAMIDE, N-9H-FLUOREN-1-YL-
53-96-3	ACETAMIDE, N-9H-FLUOREN-2-YL-
1119-49-9	ACETAMIDE, N-BUTYL-
91-49-6	ACETAMIDE, N-BUTYL-N-PHENYL-
28314-03-6	ACETAMIDE, N-FLUOREN-1-YL-
53-96-3	ACETAMIDE, N-FLUOREN-2-YL-
2540-82-1	ACETAMIDE, N-FORMYL-2-MERCAPTO-N-METHYL-, S-ESTER WITH O,O-DIMETHYL PHOSPHORODITHIOATE
2275-18-5	ACETAMIDE, N-ISOPROPYL-2-MERCAPTO-, S-ESTER WITH O,O-DIETHYL PHOSPHORODITHIOATE
103-84-4	ACETAMIDE, N-PHENYL-
531-82-8	ACETAMIDE, N-[4-(5-NITRO-2-FURANYL)-2-THIAZOLYL]-
531-82-8	ACETAMIDE, N-[4-(5-NITRO-2-FURYL)-2-THIAZOLYL]-
62-55-5	ACETAMIDE, THIO-
103-84-4	ACETAMIDOBENZENE
103-84-4	ACETANIL
103-84-4	ACETANILIDE
91-49-6	ACETANILIDE, N-BUTYL-
50-78-2	ACETARD
112-14-1	ACETATE C-8
628-63-7	ACETATE D'AMYLE (FRENCH)
111-76-2	ACETATE D'ETHYLGLYCOL (FRENCH)
110-19-0	ACETATE D'ISOBUTYLE (FRENCH)
108-21-4	ACETATE D'ISOPROPYLE (FRENCH)
123-86-4	ACETATE DE BUTYLE (FRENCH)
105-46-4	ACETATE DE BUTYLE SECONDAIRE (FRENCH)
111-76-2	ACETATE DE CELLOSOLVE (FRENCH)
142-71-2	ACETATE DE CUIVRE (FRENCH)
111-76-2	ACETATE DE L'ETHER MONOETHYLIQUE DE L'ETHYLENE-GLYCOL (FRENCH)
110-49-6	ACETATE DE L'ETHER MONOMETHYLIQUE DE L'ETHYLEN E-GLYCOL (FRENCH)
110-49-6	ACETATE DE METHYLE GLYCOL (FRENCH)
109-60-4	ACETATE DE PROPYLE NORMAL (FRENCH)
108-05-4	ACETATE DE VINYLE (FRENCH)
82-28-0	ACETATE FAST ORANGE R
62-38-4	ACETATE PHENYLMERCURIQUE (FRENCH)
111-76-2	ACETATO DI CELLOSOLVE (ITALIAN)
110-49-6	ACETATO DI METIL CELLOSOLVE (ITALIAN)
62-38-4	ACETATOPHENYLMERCURY
127-19-5	ACETDIMETHYLAMIDE
74-85-1	ACETENE
7558-79-4	ACETEST
64-19-7	ACETIC ACID
64-19-7	ACETIC ACID (ACGIH)
64-19-7	ACETIC ACID (AQUEOUS SOLUTION)
64-19-7	ACETIC ACID (AQUEOUS SOLUTION) (DOT)
123-92-2	ACETIC ACID 3-METHYLBUTYL ESTER
103-09-3	ACETIC ACID ALPHA-ETHYLHEXYL ESTER
142-71-2	ACETIC ACID CUPRIC SALT
108-05-4	ACETIC ACID ETHENYL ESTER
141-78-6	ACETIC ACID ETHYL ESTER
108-22-5	ACETIC ACID ISOPROPENYL ESTER
301-04-2	ACETIC ACID LEAD SALT (2:1)
123-86-4	ACETIC ACID N-BUTYL ESTER
109-60-4	ACETIC ACID N-PROPYL ESTER
64-19-7	ACETIC ACID SOLUTION, MORE THAN 25% BUT NOT MORE THAN 80% ACID, BY WEIGHT (DOT)
108-05-4	ACETIC ACID VINYL ESTER
93-76-5	ACETIC ACID, (2,4,5-TRICHLOROPHENOXY)-
93-79-8	ACETIC ACID, (2,4,5-TRICHLOROPHENOXY)-, 2-BUTOXYETHYL ESTER
2545-59-7	ACETIC ACID, (2,4,5-TRICHLOROPHENOXY)-, 2-BUTOXYETHYL ESTER
93-79-8	ACETIC ACID, (2,4,5-TRICHLOROPHENOXY)-, BUTYL ESTER
2545-59-7	ACETIC ACID, (2,4,5-TRICHLOROPHENOXY)-, BUTYL ESTER
1319-72-8	ACETIC ACID, (2,4,5-TRICHLOROPHENOXY)-, COMPD. WITH 1-AMINO-2-PROPANOL (1:1)
3813-14-7	ACETIC ACID, (2,4,5-TRICHLOROPHENOXY)-, COMPD. WITH 2,2',2''-NITRILOTRIETHANOL (1:1)
3813-14-7	ACETIC ACID, (2,4,5-TRICHLOROPHENOXY)-, COMPD. WITH 2,2',2''-NITRILOTRIS[ETHANOL] (1:1)
6369-97-7	ACETIC ACID, (2,4,5-TRICHLOROPHENOXY)-, COMPD. WITH DIMETHYLAMINE (1:1)
6369-96-6	ACETIC ACID, (2,4,5-TRICHLOROPHENOXY)-, COMPD. WITH N,N-DIMETHYLMETHANAMINE (1:1)
6369-97-7	ACETIC ACID, (2,4,5-TRICHLOROPHENOXY)-, COMPD. WITH N-METHYLMETHANAMINE (1:1)
6369-96-6	ACETIC ACID, (2,4,5-TRICHLOROPHENOXY)-, COMPD. WITH TRIMETHYLAMINE (1:1)
13560-99-1	ACETIC ACID, (2,4,5-TRICHLOROPHENOXY)-, SODIUM SALT
94-75-7	ACETIC ACID, (2,4-DICHLOROPHENOXY)-
94-11-1	ACETIC ACID, (2,4-DICHLOROPHENOXY)-, 1-METHYLETHYL ESTER
94-79-1	ACETIC ACID, (2,4-DICHLOROPHENOXY)-, 1-METHYLPROPYL ESTER
1929-73-3	ACETIC ACID, (2,4-DICHLOROPHENOXY)-, 2-BUTOXYETHYL ESTER
1320-18-9	ACETIC ACID, (2,4-DICHLOROPHENOXY)-, 2-BUTOXYMETHYL-ETHYL ESTER
2971-38-2	ACETIC ACID, (2,4-DICHLOROPHENOXY)-, 4-CHLORO-2-BUTENYL ESTER
1320-18-9	ACETIC ACID, (2,4-DICHLOROPHENOXY)-, BUTOXY PROPYLENE DERIV
94-80-4	ACETIC ACID, (2,4-DICHLOROPHENOXY)-, BUTYL ESTER
25168-26-7	ACETIC ACID, (2,4-DICHLOROPHENOXY)-, ISOOCTYL ESTER
94-11-1	ACETIC ACID, (2,4-DICHLOROPHENOXY)-, ISOPROPYL ESTER
1928-38-7	ACETIC ACID, (2,4-DICHLOROPHENOXY)-, METHYL ESTER
1928-61-6	ACETIC ACID, (2,4-DICHLOROPHENOXY)-, PROPYL ESTER
94-79-1	ACETIC ACID, (2,4-DICHLOROPHENOXY)-, SEC-BUTYL ESTER
60-00-4	ACETIC ACID, (ETHYLENEDINITRILO)TETRA-
123-86-4	ACETIC ACID, (P-METHOXYPHENOXY), BUTYL ESTER
540-88-5	ACETIC ACID, 1,1-DIMETHYLETHYL ESTER
540-88-5	ACETIC ACID, 1,1-DIMETHYLETHYL ESTER (9CI)
108-84-9	ACETIC ACID, 1,3-DIMETHYLBUTYL ESTER
108-21-4	ACETIC ACID, 1-METHYLETHYL ESTER
108-21-4	ACETIC ACID, 1-METHYLETHYL ESTER (9CI)
105-46-4	ACETIC ACID, 1-METHYLPROPYL ESTER
105-46-4	ACETIC ACID, 1-METHYLPROPYL ESTER (9CI)
60-00-4	ACETIC ACID, 2,2',2'',2'''-(1,2-ETHANEDIYLDINITRILO)TETRAKIS-
25168-15-4	ACETIC ACID, 2,4,5-TRICHLOROPHENOXY-, ISOOCTYL ESTER
105-46-4	ACETIC ACID, 2-BUTOXY ESTER
111-76-2	ACETIC ACID, 2-ETHOXYETHYL ESTER
10031-87-5	ACETIC ACID, 2-ETHYLBUTYL ESTER

CAS No.	Chemical Name
103-09-3	ACETIC ACID, 2-ETHYLHEXYL ESTER
110-19-0	ACETIC ACID, 2-METHYLPROPYL ESTER
53496-15-4	ACETIC ACID, 2-PENTYL ESTER
591-87-7	ACETIC ACID, 2-PROPENYL ESTER
591-87-7	ACETIC ACID, 2-PROPENYL ESTER (9CI)
4301-50-2	ACETIC ACID, 4-BIPHENYLYL-, 2-FLUOROETHYL ESTER
591-87-7	ACETIC ACID, ALLYL ESTER
631-61-8	ACETIC ACID, AMMONIUM SALT
628-63-7	ACETIC ACID, AMYL ESTER
108-24-7	ACETIC ACID, ANHYDRIDE
108-24-7	ACETIC ACID, ANHYDRIDE (9CI)
103-09-3	ACETIC ACID, BIS(2-ETHYLHEXYL) ESTER
105-36-2	ACETIC ACID, BROMO-, ETHYL ESTER
96-32-2	ACETIC ACID, BROMO-, METHYL ESTER
5798-79-8	ACETIC ACID, BROMOPHENYL-, NITRILE
123-86-4	ACETIC ACID, BUTYL ESTER
543-90-8	ACETIC ACID, CADMIUM SALT
75-36-5	ACETIC ACID, CHLORIDE
79-11-8	ACETIC ACID, CHLORO-
105-48-6	ACETIC ACID, CHLORO-, 1-METHYLETHYL ESTER
2549-51-1	ACETIC ACID, CHLORO-, ETHENYL ESTER
105-39-5	ACETIC ACID, CHLORO-, ETHYL ESTER
105-48-6	ACETIC ACID, CHLORO-, ISOPROPYL ESTER
96-34-4	ACETIC ACID, CHLORO-, METHYL ESTER
3926-62-3	ACETIC ACID, CHLORO-, SODIUM SALT
2549-51-1	ACETIC ACID, CHLORO-, VINYL ESTER
1066-30-4	ACETIC ACID, CHROMIUM(3+) SALT
71-48-7	ACETIC ACID, COBALT(2+) SALT
142-71-2	ACETIC ACID, COPPER(2+) SALT
372-09-8	ACETIC ACID, CYANO-
105-56-6	ACETIC ACID, CYANO-, ETHYL ESTER
79-43-6	ACETIC ACID, DICHLORO-
88-09-5	ACETIC ACID, DIETHYL-
127-19-5	ACETIC ACID, DIMETHYLAMIDE
108-05-4	ACETIC ACID, ETHYLENE ETHER
144-49-0	ACETIC ACID, FLUORO-
23745-86-0	ACETIC ACID, FLUORO-, POTASSIUM SALT
62-74-8	ACETIC ACID, FLUORO-, SODIUM SALT
64-19-7	ACETIC ACID, GLACIAL
64-19-7	ACETIC ACID, GLACIAL (DOT)
64-19-7	ACETIC ACID, GLACIAL OR ACETIC ACID SOLUTION, MORE THAN 80% ACID
142-92-7	ACETIC ACID, HEXYL ESTER
110-19-0	ACETIC ACID, ISOBUTYL ESTER
123-92-2	ACETIC ACID, ISOPENTYL ESTER
108-21-4	ACETIC ACID, ISOPROPYL ESTER
301-04-2	ACETIC ACID, LEAD(2+) SALT
546-67-8	ACETIC ACID, LEAD(4+) SALT
68-11-1	ACETIC ACID, MERCAPTO-
123-81-9	ACETIC ACID, MERCAPTO-, 1,2-ETHANEDIYL ESTER
123-81-9	ACETIC ACID, MERCAPTO-, ETHYLENE ESTER
1600-27-7	ACETIC ACID, MERCURIDI-
631-60-7	ACETIC ACID, MERCURY(1+) SALT
1600-27-7	ACETIC ACID, MERCURY(2+) SALT
79-20-9	ACETIC ACID, METHYL ESTER
64-19-7	ACETIC ACID, MORE THAN 25% BUT NOT MORE THAN 80% ACID, BY WEIGHT
373-02-4	ACETIC ACID, NICKEL(2+) SALT
139-13-9	ACETIC ACID, NITRILOTRI-
60-51-5	ACETIC ACID, O,O-DIMETHYLDITHIOPHOSPHORYL-, N-MONO-METHYLAMIDE SALT
112-14-1	ACETIC ACID, OCTYL ESTER
628-63-7	ACETIC ACID, PENTYL ESTER
62-38-4	ACETIC ACID, PHENYLMERCURY DERIV.
109-60-4	ACETIC ACID, PROPYL ESTER
105-46-4	ACETIC ACID, SEC-BUTYL ESTER
53496-15-4	ACETIC ACID, SEC-PENTYL ESTER
563-63-3	ACETIC ACID, SILVER(1+) SALT
540-88-5	ACETIC ACID, TERT-BUTYL ESTER
507-09-5	ACETIC ACID, THIO-
76-03-9	ACETIC ACID, TRICHLORO-
598-99-2	ACETIC ACID, TRICHLORO-, METHYL ESTER
76-05-1	ACETIC ACID, TRIFLUORO-
75-98-9	ACETIC ACID, TRIMETHYL-
557-34-6	ACETIC ACID, ZINC SALT
557-34-6	ACETIC ACID, ZINC SALT (8CI,9CI)
557-34-6	ACETIC ACID, ZINC(II) SALT
5153-24-2	ACETIC ACID, ZIRCONIUM SALT
4229-34-9	ACETIC ACID, ZIRCONIUM SALT
3227-63-2	ACETIC ACID, ZIRCONIUM SALT
3227-63-2	ACETIC ACID, ZIRCONIUM(2+) SALT
4229-34-9	ACETIC ACID, ZIRCONIUM(4+) SALT
75-07-0	ACETIC ALDEHYDE
108-24-7	ACETIC ANHYDRIDE
108-24-7	ACETIC ANHYDRIDE (ACGIH,DOT)
75-36-5	ACETIC CHLORIDE
141-78-6	ACETIC ETHER
108-24-7	ACETIC OXIDE
79-21-0	ACETIC PEROXIDE
50-78-2	ACETICYL
141-78-6	ACETIDIN
50-78-2	ACETILUM ACIDULATUM
62-55-5	ACETIMIDIC ACID, THIO-
50-78-2	ACETISAL
78-67-1	ACETO AZIB
101-25-7	ACETO DNPT 100
101-25-7	ACETO DNPT 40
101-25-7	ACETO DNPT 80
100-97-0	ACETO HMT
131-73-7	ACETO HMT
135-88-6	ACETO PBN
137-26-8	ACETO TETD
62-44-2	ACETO-4-PHENETIDINE
76-03-9	ACETO-CAUSTIN
93-68-5	ACETOACET-O-TOLUIDIDE
122-82-7	ACETOACET-P-PHENETIDIDE
102-01-2	ACETOACETAMIDOBENZENE
102-01-2	ACETOACETANILIDE
102-01-2	ACETOACETIC ACID ANILIDE
97-36-9	ACETOACETIC ACID M-XYLIDIDE
92-15-9	ACETOACETIC ACID O-ANISIDIDE
122-82-7	ACETOACETIC ACID P-PHENETIDIDE
141-97-9	ACETOACETIC ACID, ETHYL ESTER
105-45-3	ACETOACETIC ACID, METHYL ESTER
102-01-2	ACETOACETIC ANILIDE
141-97-9	ACETOACETIC ESTER
105-45-3	ACETOACETIC METHYL ESTER
97-36-9	ACETOACETO-M-XYLIDIDE
123-54-6	ACETOACETONE
97-36-9	ACETOACETYL-M-XYLIDIDE
92-15-9	ACETOACETYL-O-ANISIDIDE
92-15-9	ACETOACETYL-O-ANISIDINE
102-01-2	ACETOACETYLANILINE
103-84-4	ACETOANILIDE
12002-03-8	ACETOARSENITE DE CUIVRE (FRENCH)
78-97-7	ACETOCYANOHYDRIN
968-81-0	ACETOHEXAMIDE
513-86-0	ACETOIN
50-78-2	ACETOL
105-57-7	ACETOL
67-64-1	ACETON (GERMAN, DUTCH, POLISH)
75-86-5	ACETONCIANHIDRINEI (ROMANIAN)
75-86-5	ACETONCIANIDRINA (ITALIAN)
75-86-5	ACETONCYANHYDRIN (GERMAN)
67-64-1	ACETONE
67-64-1	ACETONE (ACGIH,DOT)
75-86-5	ACETONE CYANOHYDRIN
75-86-5	ACETONE CYANOHYDRIN (DOT)
75-86-5	ACETONE CYANOHYDRIN, STABILIZED (DOT)
100-79-8	ACETONE GLYCERIN KETAL
100-79-8	ACETONE MONOGLYCEROL KETAL
1752-30-3	ACETONE THIOSEMICARBAZIDE
123-54-6	ACETONE, ACETYL-
78-95-5	ACETONE, CHLORO-
100-79-8	ACETONE, CYCLIC (HYDROXYMETHYL)ETHYLENE ACETAL
116-16-5	ACETONE, HEXACHLORO-
684-16-2	ACETONE, HEXAFLUORO-
677-71-4	ACETONE, HEXAFLUORO-, HYDRATE
141-79-7	ACETONE, ISOPROPYLIDENE-
78-93-3	ACETONE, METHYL-
78-94-4	ACETONE, METHYLENE-
75-86-5	ACETONECYANHYDRINE (FRENCH)
50-21-5	ACETONIC ACID
75-05-8	ACETONITRIL (GERMAN, DUTCH)
75-05-8	ACETONITRILE
75-05-8	ACETONITRILE (ACGIH,DOT)
926-64-7	ACETONITRILE, (DIMETHYLAMINO)-
5798-79-8	ACETONITRILE, BROMOPHENYL-

CAS No.	Chemical Name
107-14-2	ACETONITRILE, CHLORO-
140-29-4	ACETONITRILE, PHENYL-
75-86-5	ACETONKYANHYDRIN (CZECH)
598-31-2	ACETONYL BROMIDE
78-95-5	ACETONYL CHLORIDE
123-42-2	ACETONYLDIMETHYLCARBINOL
50-78-2	ACETOPHEN
62-44-2	ACETOPHENETIDIN
62-44-2	ACETOPHENETIDINE
62-44-2	ACETOPHENETIN
98-86-2	ACETOPHENON
98-86-2	ACETOPHENONE
70-11-1	ACETOPHENONE, 2-BROMO-
532-27-4	ACETOPHENONE, 2-CHLORO-
1341-24-8	ACETOPHENONE, CHLORO-
82-28-0	ACETOQUINONE LIGHT ORANGE JL
50-78-2	ACETOSAL
50-78-2	ACETOSALIC ACID
50-78-2	ACETOSALIN
62-55-5	ACETOTHIOAMIDE
141-78-6	ACETOXYETHANE
108-05-4	ACETOXYETHYLENE
94-36-0	ACETOXYL
62-38-4	ACETOXYPHENYLMERCURY
644-31-5	ACETOZONE
108-24-7	ACETYL ACETATE
37187-22-7	ACETYL ACETONE PEROXIDE
37187-22-7	ACETYL ACETONE PEROXIDE SOLUTION WITH ≤ 9% BY WEIGHT ACTIVE OXYGEN
108-24-7	ACETYL ANHYDRIDE
644-31-5	ACETYL BENZOYL PEROXIDE
644-31-5	ACETYL BENZOYL PEROXIDE SOLUTION, NOT OVER 40% PEROXIDE (DOT)
506-96-7	ACETYL BROMIDE
506-96-7	ACETYL BROMIDE (DOT)
598-21-0	ACETYL BROMIDE, BROMO-
75-36-5	ACETYL CHLORIDE
75-36-5	ACETYL CHLORIDE (DOT)
79-04-9	ACETYL CHLORIDE, CHLORO-
79-36-7	ACETYL CHLORIDE, DICHLORO-
359-06-8	ACETYL CHLORIDE, FLUORO-
103-80-0	ACETYL CHLORIDE, PHENYL-
76-02-8	ACETYL CHLORIDE, TRICHLORO-
3282-30-2	ACETYL CHLORIDE, TRIMETHYL-
3179-56-4	ACETYL CYCLOHEXANE SULFONYL PEROXIDE
3179-56-4	ACETYL CYCLOHEXANE SULPHONYL PEROXIDE :MAXIMUM CONCENTRATION 32%
3179-56-4	ACETYL CYCLOHEXANE SULPHONYL PEROXIDE :MAXIMUM CONCENTRATION 82%
3179-56-4	ACETYL CYCLOHEXANEPERSULFONATE
3179-56-4	ACETYL CYCLOHEXANESULFONYL PEROXIDE
3179-56-4	ACETYL CYCLOHEXANESULFONYL PEROXIDE (DOT)
3179-56-4	ACETYL CYCLOHEXANESULFONYL PEROXIDE, ≤ 82%, WETTED WITH ≥ 12% WATER
3179-56-4	ACETYL CYCLOHEXANESULFONYL PEROXIDE, NOT MORE THAN 32% IN SOLUTION (DOT)
3179-56-4	ACETYL CYCLOHEXYLSULFONYL PEROXIDE
108-24-7	ACETYL ETHER
78-94-4	ACETYL ETHYLENE
79-21-0	ACETYL HYDROPEROXIDE
507-02-8	ACETYL IODIDE
507-02-8	ACETYL IODIDE (DOT)
507-09-5	ACETYL MERCAPTAN
598-31-2	ACETYL METHYL BROMIDE
513-86-0	ACETYL METHYL CARBINOL
110-49-6	ACETYL METHYL CELLOSOLVE
513-86-0	ACETYL METHYLCARBINOL (DOT)
108-24-7	ACETYL OXIDE
110-22-5	ACETYL PEROXIDE
110-22-5	ACETYL PEROXIDE (SOLUTION)
110-22-5	ACETYL PEROXIDE SOLUTION, NOT OVER 25% PEROXIDE (DOT)
107-71-1	ACETYL TERT-BUTYL PEROXIDE
103-89-9	ACETYL-P-TOLUIDINE
123-54-6	ACETYLACETONE
37187-22-7	ACETYLACETONE PEROXIDE
37187-22-7	ACETYLACETONE PEROXIDE (DOT)
20830-81-3	ACETYLADRIAMYCIN
75-07-0	ACETYLALDEHYDE
103-84-4	ACETYLANILINE
98-86-2	ACETYLBENZENE
74-86-2	ACETYLEN
74-86-2	ACETYLENE
74-86-2	ACETYLENE (DOT)
74-86-2	ACETYLENE (LIQUID)
1333-86-4	ACETYLENE BLACK
540-59-0	ACETYLENE DICHLORIDE
79-27-6	ACETYLENE TETRABROMIDE
79-27-6	ACETYLENE TETRABROMIDE (ACGIH,DOT)
79-34-5	ACETYLENE TETRACHLORIDE
79-01-6	ACETYLENE TRICHLORIDE
7572-29-4	ACETYLENE, DICHLORO-
74-86-2	ACETYLENE, DISSOLVED (DOT)
75-20-7	ACETYLENOGEN
50-78-2	ACETYLIN
50-78-2	ACETYLSAL
50-78-2	ACETYLSALICYLIC ACID
50-78-2	ACETYSAL
1066-33-7	ACID AMMONIUM CARBONATE
1341-49-7	ACID AMMONIUM FLUORIDE
7803-63-6	ACID AMMONIUM SULFATE
12788-93-1	ACID BUTYL PHOSPHATE
12788-93-1	ACID BUTYL PHOSPHATE (DOT)
10290-12-7	ACID COPPER ARSENITE
1694-09-3	ACID FAST VIOLET 5BN
7784-40-9	ACID LEAD ARSENATE
1300-73-8	ACID LEATHER BROWN 2G
3761-53-3	ACID LEATHER RED KPR
3761-53-3	ACID LEATHER RED P 2R
3761-53-3	ACID LEATHER SCARLET IRW
1300-73-8	ACID ORANGE 24
3761-53-3	ACID PONCEAU 2RL
3761-53-3	ACID PONCEAU R
3761-53-3	ACID PONCEAU SPECIAL
7646-93-7	ACID POTASSIUM SULFATE
3761-53-3	ACID RED 26
3761-53-3	ACID SCARLET
3761-53-3	ACID SCARLET 2B
3761-53-3	ACID SCARLET 2R
3761-53-3	ACID SCARLET 2RL
64742-24-1	ACID SLUDGE
1694-09-3	ACID VIOLET
1694-09-3	ACID VIOLET 49
1694-09-3	ACID VIOLET 4BNP
1694-09-3	ACID VIOLET 4BNS
1694-09-3	ACID VIOLET 5B
1694-09-3	ACID VIOLET 5BN
1694-09-3	ACID VIOLET 6B
1694-09-3	ACID VIOLET S
7789-75-5	ACID-SPAR
3761-53-3	ACIDAL PONCEAU G
93-76-5	ACIDE 2,4,5-TRICHLORO PHENOXYACETIQUE (FRENCH)
94-75-7	ACIDE 2,4-DICHLORO PHENOXYACETIQUE (FRENCH)
64-19-7	ACIDE ACETIQUE (FRENCH)
1327-53-3	ACIDE ARSENIEUX (FRENCH)
7778-39-4	ACIDE ARSENIQUE LIQUIDE (FRENCH)
10035-10-6	ACIDE BROMHYDRIQUE (FRENCH)
75-60-5	ACIDE CACODYLIQUE (FRENCH)
79-11-8	ACIDE CHLORACETIQUE (FRENCH)
7647-01-0	ACIDE CHLORHYDRIQUE (FRENCH)
372-09-8	ACIDE CYANACETIQUE (FRENCH)
74-90-8	ACIDE CYANHYDRIQUE (FRENCH)
75-60-5	ACIDE DIMETHYLARSINIQUE (FRENCH)
60-00-4	ACIDE ETHYLENEDIAMINETETRACETIQUE (FRENCH)
7664-39-3	ACIDE FLUORHYDRIQUE (FRENCH)
16961-83-4	ACIDE FLUOROSILICIQUE (FRENCH)
16961-83-4	ACIDE FLUOSILICIQUE (FRENCH)
64-18-6	ACIDE FORMIQUE (FRENCH)
121-91-5	ACIDE ISOPHTALIQUE (FRENCH)
79-11-8	ACIDE MONOCHLORACETIQUE (FRENCH)
7697-37-2	ACIDE NITRIQUE (FRENCH)
144-62-7	ACIDE OXALIQUE (FRENCH)
7664-38-2	ACIDE PHOSPHORIQUE (FRENCH)
79-09-4	ACIDE PROPIONIQUE (FRENCH)
7783-06-4	ACIDE SULFHYDRIQUE (FRENCH)
7664-93-9	ACIDE SULFURIQUE (FRENCH)
1401-55-4	ACIDE TANNIQUE (FRENCH)
68-11-1	ACIDE THIOGLYCOLIQUE (FRENCH)
76-03-9	ACIDE TRICHLORACETIQUE (FRENCH)

CAS No.	Chemical Name
144-49-0	ACIDE-MONOFLUORACETIQUE (FRENCH)
93-76-5	ACIDO (2,4,5-TRICLORO-FENOSSI)-ACETICO (ITALIAN)
1918-00-9	ACIDO (3,6-DICLORO-2-METOSSI)-BENZOICO (ITALIAN)
64-19-7	ACIDO ACETICO (ITALIAN)
10035-10-6	ACIDO BROMIDRICO (ITALIAN)
74-90-8	ACIDO CIANIDRICO (ITALIAN)
7647-01-0	ACIDO CLORIDRICO (ITALIAN)
7664-39-3	ACIDO FLUORIDRICO (ITALIAN)
16961-83-4	ACIDO FLUOSILICICO (ITALIAN)
64-18-6	ACIDO FORMICO (ITALIAN)
7664-38-2	ACIDO FOSFORICO (ITALIAN)
144-49-0	ACIDO MONOFLUOROACETIO (ITALIAN)
7697-37-2	ACIDO NITRICO (ITALIAN)
144-62-7	ACIDO OSSALICO (ITALIAN)
69-72-7	ACIDO SALICILICO (ITALIAN)
7664-93-9	ACIDO SOLFORICO (ITALIAN)
76-03-9	ACIDO TRICLOROACETICO (ITALIAN)
94-75-7	ACIDO(2,4-DICLORO-FENOSSI)-ACETICO (ITALIAN)
124-47-0	ACIDOGEN NITRATE
1338-24-5	ACIDOL (PETROLEUM BY-PRODUCT)
79-11-8	ACIDOMONOCLOROACETICO (ITALIAN)
1338-24-5	ACIDS, CARBOXYLIC
1338-24-5	ACIDS, NAPHTHENIC
50-78-2	ACIDUM ACETYLSALICYLICUM
124-04-9	ACIFLOCTIN
70-30-4	ACIGENA
1694-09-3	ACILAN VIOLET S 4BN
124-04-9	ACINETTEN
138-86-3	ACINTENE DP
138-86-3	ACINTENE DP DIPENTENE
50-78-2	ACISAL
2244-21-5	ACL 59
2893-78-9	ACL 60
2782-57-2	ACL 70
87-90-1	ACL 85
94-36-0	ACNEGEL
630-60-4	ACOCANTHERIN
76-06-2	ACQUINITE
107-02-8	ACQUINITE
107-02-8	ACRALDEHYDE
107-02-8	ACRALDEHYDEACROLEINA (ITALIAN)
260-94-6	ACRIDINE
260-94-6	ACRIDINE (DOT)
79-10-7	ACROLEIC ACID
107-02-8	ACROLEIN
107-02-8	ACROLEIN (ACGIH)
3054-95-3	ACROLEIN ACETAL
869-29-4	ACROLEIN DIACETATE
100-73-2	ACROLEIN DIMER
100-73-2	ACROLEIN DIMER, STABILIZED
100-73-2	ACROLEIN DIMER, STABILIZED (DOT)
645-62-5	ACROLEIN, 2-ETHYL-3-PROPYL-
78-85-3	ACROLEIN, 2-METHYL-
3054-95-3	ACROLEIN, DIETHYL ACETAL
107-02-8	ACROLEIN, INHIBITED
107-02-8	ACROLEIN, INHIBITED (DOT)
107-02-8	ACROLEINA (ITALIAN)
107-02-8	ACROLEINE (DUTCH, FRENCH)
7008-42-6	ACRONYCINE
107-02-8	ACRYLALDEHYD (GERMAN)
107-02-8	ACRYLALDEHYDE
3054-95-3	ACRYLALDEHYDE DIETHYL ACETAL
79-06-1	ACRYLAMIDE
79-06-1	ACRYLAMIDE (ACGIH,DOT)
79-06-1	ACRYLAMIDE SOLUTION [COMBUSTIBLE LIQUID LABEL]
79-06-1	ACRYLAMIDE SOLUTION [FLAMMABLE LIQUID LABEL]
140-88-5	ACRYLATE D'ETHYLE (FRENCH)
96-33-3	ACRYLATE DE METHYLE (FRENCH)
79-10-7	ACRYLIC ACID
79-10-7	ACRYLIC ACID (ACGIH,DOT)
79-10-7	ACRYLIC ACID (GLACIAL)
141-32-2	ACRYLIC ACID BUTYL ESTER
814-68-6	ACRYLIC ACID CHLORIDE
140-88-5	ACRYLIC ACID ETHYL ESTER
96-33-3	ACRYLIC ACID METHYL ESTER
141-32-2	ACRYLIC ACID N-BUTYL ESTER
2426-54-2	ACRYLIC ACID, 2-(DIETHYLAMINO)ETHYL ESTER
80-63-7	ACRYLIC ACID, 2-CHLORO-, METHYL ESTER
137-05-3	ACRYLIC ACID, 2-CYANO-, METHYL ESTER
3953-10-4	ACRYLIC ACID, 2-ETHYLBUTYL ESTER
103-11-7	ACRYLIC ACID, 2-ETHYLHEXYL ESTER
818-61-1	ACRYLIC ACID, 2-HYDROXYETHYL ESTER
999-61-1	ACRYLIC ACID, 2-HYDROXYPROPYL ESTER
79-41-4	ACRYLIC ACID, 2-METHYL-
80-62-6	ACRYLIC ACID, 2-METHYL-, METHYL ESTER
3724-65-0	ACRYLIC ACID, 3-METHYL-
2156-96-9	ACRYLIC ACID, DECYL ESTER
106-71-8	ACRYLIC ACID, ESTER WITH HYDRACRYLONITRILE
79-10-7	ACRYLIC ACID, INHIBITED (DOT)
106-63-8	ACRYLIC ACID, ISOBUTYL ESTER
1330-61-6	ACRYLIC ACID, ISODECYL ESTER
25584-83-2	ACRYLIC ACID, MONOESTER WITH 1,2-PROPANEDIOL
2499-59-4	ACRYLIC ACID, OCTYL ESTER
107-02-8	ACRYLIC ALDEHYDE
79-06-1	ACRYLIC AMIDE
107-13-1	ACRYLON
107-13-1	ACRYLONITRILE
814-68-6	ACRYLOYL CHLORIDE
140-88-5	ACRYLSAEUREAETHYLESTER (GERMAN)
96-33-3	ACRYLSAEUREM ETHYLESTER (GERMAN)
814-68-6	ACRYLYL CHLORIDE
7681-49-4	ACT
66-81-9	ACTI-AID
66-81-9	ACTI-DIONE
66-81-9	ACTI-DIONE BR
66-81-9	ACTI-DIONE PM
66-81-9	ACTI-DIONE TGF
1333-86-4	ACTICARBON AC 35
7631-86-9	ACTICEL
66-81-9	ACTIDION
1912-24-9	ACTINITE PK
50-76-0	ACTINOMYCIN AIV
50-76-0	ACTINOMYCIN C1
50-76-0	ACTINOMYCIN D
50-76-0	ACTINOMYCIN I1
50-76-0	ACTINOMYCIN IV
50-76-0	ACTINOMYCINDIOIC D ACID, DILACTONE
8020-83-5	ACTIPRON
141-97-9	ACTIVE ACETYL ACETATE
137-32-6	ACTIVE AMYL ALCOHOL
80-43-3	ACTIVE DICUMYL PEROXIDE
137-32-6	ACTIVE PRIMARY AMYL ALCOHOL
123-30-8	ACTIVOL
27598-85-2	ACTIVOL
25322-69-4	ACTOCOL 51-530
7553-56-2	ACTOMAR
1314-13-2	ACTOX 14
1314-13-2	ACTOX 16
1314-13-2	ACTOX 216
97-64-3	ACTYLOL
50-78-2	ACYLPYRIN
97-64-3	ACYTOL
50-76-0	AD
1333-86-4	AD 200
75-56-9	AD 6
75-56-9	AD 6 (SUSPENDING AGENT)
112-41-4	ADACENE 12
578-94-9	ADAMSIT
578-94-9	ADAMSITE
2465-27-2	ADC AURAMINE O
633-03-4	ADC BRILLIANT GREEN CRYSTALS
64-18-6	ADD-F
25322-69-4	ADEKA P 1000
25322-69-4	ADEKA P 3000
25322-69-4	ADEKA P 700
9016-45-9	ADEKATOL NP 700
7631-86-9	ADELITE 30
7631-86-9	ADELITE A
8006-54-0	ADEPS LANE
137-05-3	ADHERE
124-04-9	ADILACTETTEN
123-79-5	ADIMOLL DO
124-04-9	ADIPIC ACID
111-50-2	ADIPIC ACID DICHLORIDE
111-69-3	ADIPIC ACID DINITRILE
111-69-3	ADIPIC ACID NITRILE
103-23-1	ADIPIC ACID, BIS(2-ETHYLHEXYL) ESTER
123-79-5	ADIPIC ACID, BIS(2-ETHYLHEXYL) ESTER

CAS No.	Chemical Name
123-79-5	ADIPIC ACID, DIOCTYL ESTER
111-50-2	ADIPIC DICHLORIDE
120-92-3	ADIPIC KETONE
120-92-3	ADIPIN KETON
111-69-3	ADIPODINITRILE
103-23-1	ADIPOL 2EH
111-69-3	ADIPONITRILE
111-50-2	ADIPOYL CHLORIDE
111-50-2	ADIPOYL DICHLORIDE
111-50-2	ADIPYL CHLORIDE
112-02-7	ADOGEN 444
23214-92-8	ADRIAMYCIN
25316-40-9	ADRIAMYCIN, HYDROCHLORIDE
23214-92-8	ADRIBLASTIN
108-93-0	ADRONAL
108-93-0	ADRONOL
434-07-1	ADROYD
51-21-8	ADRUCIL
128-37-0	ADVASTAB 401
8002-74-2	ADVAWAX 165
7429-90-5	AE
1344-28-1	AERO 100
156-62-7	AERO CYANAMID GRANULAR
156-62-7	AERO CYANAMID SPECIAL GRADE
74-90-8	AERO LIQUID HCN
156-62-7	AERO-CYANAMID
7782-42-5	AERODAG G
7631-86-9	AEROGEL 200
52-68-6	AEROL 1
52-68-6	AEROL 1 (PESTICIDE)
7631-86-9	AEROSIL
7631-86-9	AEROSIL 130
7631-86-9	AEROSIL 130V
7631-86-9	AEROSIL 175
7631-86-9	AEROSIL 200
7631-86-9	AEROSIL 200V
7631-86-9	AEROSIL 300
7631-86-9	AEROSIL 308
7631-86-9	AEROSIL 380
7631-86-9	AEROSIL A 130
7631-86-9	AEROSIL A 175
7631-86-9	AEROSIL A 300
7631-86-9	AEROSIL A 380
7631-86-9	AEROSIL BS 50
7631-86-9	AEROSIL E 300
7631-86-9	AEROSIL K 7
7631-86-9	AEROSIL M 300
7631-86-9	AEROSIL OX 50
7631-86-9	AEROSIL PST
7631-86-9	AEROSIL TT 600
7631-86-9	AEROSIL-DEGUSSA
577-11-7	AEROSOL GPG
577-11-7	AEROSOL OT
577-11-7	AEROSOL OT 100
577-11-7	AEROSOL OT 75
577-11-7	AEROSOL OT-B
577-11-7	AEROSOL OT-S
7779-88-6	AEROTEX ACCELERATOR NUMBER 5
75-09-2	AEROTHENE MM
71-55-6	AEROTHENE TT
75-08-1	AETHANETHIOL (GERMAN)
64-17-5	AETHANOL (GERMAN)
60-29-7	AETHER
122-51-0	AETHON
141-78-6	AETHYLACETAT (GERMAN)
140-88-5	AETHYLACRYLAT (GERMAN)
64-17-5	AETHYLALKOHOL (GERMAN)
75-04-7	AETHYLAMINE (GERMAN)
103-69-5	AETHYLANILIN (GERMAN)
100-41-4	AETHYLBENZOL (GERMAN)
75-00-3	AETHYLCHLORID (GERMAN)
106-93-4	AETHYLENBROMID (GERMAN)
107-06-2	AETHYLENCHLORID (GERMAN)
107-07-3	AETHYLENECHLORHYDRIN (GERMAN)
109-86-4	AETHYLENGLYKOL-MONOMETHYLAETHER (GERMAN)
111-76-2	AETHYLENGLYKOLA ETHERACETAT (GERMAN)
110-49-6	AETHYLENGLYKOLMETHYLAETHERACETAT (GERMAN)
151-56-4	AETHYLENIMIN (GERMAN)
109-94-4	AETHYLFORMIAT (GERMAN)
75-34-3	AETHYLIDENCHLORID (GERMAN)
75-00-3	AETHYLIS
75-00-3	AETHYLIS CHLORIDUM
75-08-1	AETHYLMERCAPTAN (GERMAN)
78-93-3	AETHYLMETHYLKETON (GERMAN)
542-90-5	AETHYLRHODANID (GERMAN)
1310-73-2	AETZNATRON
330-54-1	AF 101
3688-53-7	AF 2
53894-28-3	AF 2
51004-61-6	AF 2
91681-63-9	AF 2
53894-28-3	AF 2 (CROSS-LINKING AGENT)
51004-61-6	AF 2 (FOAMING AGENT)
3688-53-7	AF 2 (PRESERVATIVE)
9002-84-0	AFG 80VS
58-89-9	AFICIDE
640-19-7	AFL 1081
7220-81-7	AFLATOXIN B2
1165-39-5	AFLATOXIN G1
1402-68-2	AFLATOXINS
2540-82-1	AFLIX
9002-84-0	AFLON G 8
9002-84-0	AFLON G 80
3691-35-8	AFNOR
7782-42-5	AG 1500
14807-96-6	AGALITE
2763-96-4	AGARIN
2763-96-4	AGARINE
106-92-3	AGE
1330-61-6	AGEFLEX FA-10
2867-47-2	AGEFLEX FM-1
8004-09-9	AGEL TG 37
8004-09-9	AGEL TG 67
1338-24-5	AGENAP
112-30-1	AGENT 504
75-60-5	AGENT BLUE
135-88-6	AGERITE POWDER
128-37-0	AGIDOL
128-37-0	AGIDOL 1
8006-54-0	AGNOLIN
8006-54-0	AGNOLIN NO. 1
57-83-0	AGOLUTIN
56-53-1	AGOSTILBEN
9016-45-9	AGRAL
9016-45-9	AGRAL 90
9016-45-9	AGRAL LN
9016-45-9	AGRAL R
1582-09-8	AGREFLAN
122-14-5	AGRIA 1050
8001-35-2	AGRICIDE MAGGOT KILLER (F)
1317-65-3	AGRICULTURAL LIMESTONE
115-90-2	AGRICUR
56-72-4	AGRIDIP
1582-09-8	AGRIFLAN 24
37273-91-9	AGRIMORT
327-98-0	AGRISIL
58-89-9	AGRISOL G-20
50-29-3	AGRITAN
327-98-0	AGRITOX
122-14-5	AGRIYA 1050
76-44-8	AGROCERES
58-89-9	AGROCIDE
58-89-9	AGROCIDE 2
58-89-9	AGROCIDE 6G
58-89-9	AGROCIDE 7
58-89-9	AGROCIDE III
58-89-9	AGROCIDE WP
17804-35-2	AGROCIT
52-68-6	AGROFOROTOX
58-89-9	AGRONEXIT
62-38-4	AGROSAN
62-38-4	AGROSAN D
62-38-4	AGROSAN GN 5
502-39-6	AGROSOL
133-06-2	AGROSOL S
94-75-7	AGROTECT
122-14-5	AGROTHION
133-06-2	AGROX 2-WAY AND 3-WAY

CAS No.	Chemical Name
1317-65-3	AGSTONE
1910-42-5	AH 501
91-80-5	AH-42
1937-37-7	AHCO DIRECT BLACK GX
3761-53-3	AHCOCID FAST SCARLET R
52-24-4	AI 3-24916
78-67-1	AIBN
20859-73-8	AIP
10290-12-7	AIR-FLO GREEN
1937-37-7	AIREDALE BLACK ED
2602-46-2	AIREDALE BLUE 2BD
1332-58-7	AIRFLO V 8
57-06-7	AITC
57-06-7	AITK
78-67-1	AIVN
1694-09-3	AIZEN ACID VIOLET 5BH
2465-27-2	AIZEN AURAMINE
2465-27-2	AIZEN AURAMINE CONC.SFA
2465-27-2	AIZEN AURAMINE OH
633-03-4	AIZEN DIAMOND GREEN GH
2602-46-2	AIZEN DIRECT BLUE 2BH
1937-37-7	AIZEN DIRECT DEEP BLACK EH
1937-37-7	AIZEN DIRECT DEEP BLACK GH
1937-37-7	AIZEN DIRECT DEEP BLACK RH
633-03-4	AIZEN MALACHITE GREEN GH
3761-53-3	AIZEN PONCEAU RH
16071-86-6	AIZEN PRIMULA BROWN BRLH
16071-86-6	AIZEN PRIMULA BROWN PLH
510-15-6	AKAR
510-15-6	AKAR 338
786-19-6	AKARITHION
7783-28-0	AKOUSTAN A
107-02-8	AKROLEIN (CZECH)
107-02-8	AKROLEINA (POLISH)
79-06-1	AKRYLAMID (CZECH)
140-88-5	AKRYLANEM ETYLU (POLISH)
1912-24-9	AKTICON
1912-24-9	AKTIKON
1912-24-9	AKTIKON PK
1912-24-9	AKTINIT A
1912-24-9	AKTINIT PK
144-62-7	AKTISAL
105-60-2	AKULON
105-60-2	AKULON M 2W
9016-45-9	AKYPO NP 70
9016-45-9	AKYPOROX NP 105
9016-45-9	AKYPOROX NP 95
2235-54-3	AKYPOSAL ALS 33
151-21-3	AKYPOSAL SDS
139-96-8	AKYPOSAL TLS
1344-28-1	AL 13
1344-28-1	AL 13 (OXIDE)
20859-73-8	AL-PHOS
2223-93-0	ALAIXOL 11
7784-21-6	ALANE
75-24-1	ALANE, TRIMETHYL-
531-76-0	ALANINE, 3-[P-[BIS(2-CHLOROETHYL)AMINO]PHENYL]-, DL-
148-82-3	ALANINE, 3-[P-[BIS(2-CHLOROETHYL)AMINO]PHENYL]-, L-
8006-54-0	ALAPURIN
75-99-0	ALATEX
91-20-3	ALBOCARBON
7722-84-1	ALBONE
7722-84-1	ALBONE 35
7722-84-1	ALBONE 35CG
7722-84-1	ALBONE 50
7722-84-1	ALBONE 50CG
7722-84-1	ALBONE 70
7722-84-1	ALBONE 70CG
7722-84-1	ALBONE DS
64-17-5	ALCARE HAND DEGERMER
10049-04-4	ALCIDE
1344-28-1	ALCOA F 1
7681-49-4	ALCOA SODIUM FLUORIDE
64-17-5	ALCOHOL
64-17-5	ALCOHOL (ETHYL)
64-17-5	ALCOHOL ANHYDROUS
112-30-1	ALCOHOL C-10
30207-98-8	ALCOHOL C-11
112-53-8	ALCOHOL C-12

CAS No.	Chemical Name
111-87-5	ALCOHOL C-8
143-08-8	ALCOHOL C-9
64-17-5	ALCOHOL DEHYDRATED
8013-75-0	ALCOHOLS, FUSEL
67-63-0	ALCOJEL
107-18-6	ALCOOL ALLILCO (ITALIAN)
107-18-6	ALCOOL ALLYLIQUE (FRENCH)
123-51-3	ALCOOL AMILICO (ITALIAN)
71-41-0	ALCOOL AMYLIQUE (FRENCH)
71-36-3	ALCOOL BUTYLIQUE (FRENCH)
78-92-2	ALCOOL BUTYLIQUE SECONDAIRE (FRENCH)
75-65-0	ALCOOL BUTYLIQUE TERTIAIRE (FRENCH)
64-17-5	ALCOOL ETHYLIQUE (FRENCH)
64-17-5	ALCOOL ETILICO (ITALIAN)
123-51-3	ALCOOL ISOAMYLIQUE (FRENCH)
78-83-1	ALCOOL ISOBUTYLIQUE (FRENCH)
108-11-2	ALCOOL METHYL AMYLIQUE (FRENCH)
54972-97-3	ALCOOL METHYL AMYLIQUE (FRENCH)
67-56-1	ALCOOL METHYLIQUE (FRENCH)
67-56-1	ALCOOL METILICO (ITALIAN)
71-23-8	ALCOOL PROPILICO (ITALIAN)
71-23-8	ALCOOL PROPYLIQUE (FRENCH)
97-77-8	ALCOPHOBIN
577-11-7	ALCOPOL O
67-63-0	ALCOSOLVE 2
9002-89-5	ALCOTEX 17F-H
9002-89-5	ALCOTEX 725L
9002-89-5	ALCOTEX 75L
9002-89-5	ALCOTEX 88/05
9002-89-5	ALCOTEX 88/10
9002-89-5	ALCOTEX 99/10
30525-89-4	ALDACIDE
116-06-3	ALDECARB
97-96-1	ALDEHYDE 2-ETHYLBUTYRIQUE (FRENCH)
75-07-0	ALDEHYDE ACETIQUE (FRENCH)
107-02-8	ALDEHYDE ACRYLIQUE (FRENCH)
75-39-8	ALDEHYDE AMMONIA
123-72-8	ALDEHYDE BUTYRIQUE (FRENCH)
66-25-1	ALDEHYDE C-6
124-13-0	ALDEHYDE C-8
123-73-9	ALDEHYDE CROTONIQUE (FRENCH)
4170-30-3	ALDEHYDE CROTONIQUE (FRENCH)
50-00-0	ALDEHYDE FORMIQUE (FRENCH)
123-38-6	ALDEHYDE PROPIONIQUE (FRENCH)
104-90-5	ALDEHYDECOLLIDINE
104-90-5	ALDEHYDINE
75-07-0	ALDEIDE ACETICA (ITALIAN)
107-02-8	ALDEIDE ACRILICA (ITALIAN)
123-72-8	ALDEIDE BUTIRRICA (ITALIAN)
50-00-0	ALDEIDE FORMICA (ITALIAN)
111-30-8	ALDESAN
116-06-3	ALDICARB
116-06-3	ALDICARBE (FRENCH)
51-28-5	ALDIFEN
309-00-2	ALDOCIT
107-89-1	ALDOL
107-89-1	ALDOL (DOT)
555-30-6	ALDOMET
107-29-9	ALDOXIME
309-23-3	ALDREX
309-23-3	ALDREX 30
309-00-2	ALDREX 40
309-00-2	ALDRIN
309-23-3	ALDRIN (ACGIH,DOT)
60-57-1	ALDRIN EPOXIDE
309-00-2	ALDRIN MIXTURE, DRY (WITH 65% OR LESS ALDRIN) (DOT)
309-00-2	ALDRIN MIXTURE, DRY (WITH MORE THAN 65% ALDRIN) (DOT)
309-00-2	ALDRIN MIXTURE, LIQUID (WITH 60% OR LESS ALDRIN) (DOT)
309-00-2	ALDRIN MIXTURE, LIQUID (WITH MORE THAN 60% ALDRIN) (DOT)
309-23-3	ALDRIN, CAST SOLID (DOT)
309-23-3	ALDRINE (FRENCH)
309-23-3	ALDRITE
309-23-3	ALDROSOL
21564-17-0	ALENTISAN
494-03-1	ALEUKON
57-41-0	ALEVIATIN
333-41-5	ALFA-TOX
9016-45-9	ALFENOL

CAS No.	Chemical Name
9016-45-9	ALFENOL 10
9016-45-9	ALFENOL 18
9016-45-9	ALFENOL 22
9016-45-9	ALFENOL 28
9016-45-9	ALFENOL 710
9016-45-9	ALFENOL 8
56-75-7	ALFICETYN
2425-06-1	ALFLOC 7020
2425-06-1	ALFLOC 7046
112-30-1	ALFOL 10
112-53-8	ALFOL 12
112-72-1	ALFOL 14
27196-00-5	ALFOL 14
111-87-5	ALFOL 8
123-23-9	ALFOZONO
7440-22-4	ALGAEDYN
75-45-6	ALGEON 22
62-38-4	ALGIMYCIN
62-38-4	ALGIMYCIN 200
117-80-6	ALGISTAT
9002-84-0	ALGOFLON
9002-84-0	ALGOFLON SV
75-45-6	ALGOFRENE 22
75-45-6	ALGOFRENE 6
75-69-4	ALGOFRENE TYPE 1
75-71-8	ALGOFRENE TYPE 2
75-43-4	ALGOFRENE TYPE 5
75-45-6	ALGOFRENE TYPE 6
75-37-6	ALGOFRENE TYPE 67
64-17-5	ALGRAIN
79-01-6	ALGYLEN
112-02-7	ALIQUAT 6
105-60-2	ALKAMID
25155-30-0	ALKANATE DC
75-75-2	ALKANE SULFONIC ACID
102-71-6	ALKANOLAMINE 244
25322-69-4	ALKAPOL PPG-1200
25322-69-4	ALKAPOL PPG-2000
25322-69-4	ALKAPOL PPG-4000
124-65-2	ALKARSODYL
9016-45-9	ALKASURF NP
9016-45-9	ALKASURF NP 11
9016-45-9	ALKASURF NP 8
148-82-3	ALKERAN
56-04-2	ALKIRON
128-37-0	ALKOFEN BP
64-17-5	ALKOHOL (GERMAN)
64-17-5	ALKOHOLU ETYLOWEGO (POLISH)
9002-89-5	ALKOTEX
56-38-2	ALKRON
123-01-3	ALKYLATE P 1
7440-50-8	ALLBRI NATURAL COPPER
463-49-0	ALLENE
56-38-2	ALLERON
584-79-2	ALLETHRIN
584-79-2	ALLETHRIN (DOT)
584-79-2	ALLETHRIN I
584-79-2	ALLETHRINE
3254-63-5	ALLIED GC 6506
106-92-3	ALLIL-GLICIDIL-ETERE (ITALIAN)
107-05-1	ALLILE (CLORURO DI) (ITALIAN)
107-18-6	ALLILOWY ALKOHOL (POLISH)
110-17-8	ALLOMALEIC ACID
541-53-7	ALLOPHANIMIDIC ACID, DITHIO-
8001-35-2	ALLTEX
8001-35-2	ALLTOX
7446-70-0	ALLUMINIO(CLORURO DI) (ITALIAN)
106-92-3	ALLYL 2,3-EPOXYPROPYL ETHER
591-87-7	ALLYL ACETATE
591-87-7	ALLYL ACETATE (DOT)
107-18-6	ALLYL AL
107-18-6	ALLYL ALCOHOL
107-18-6	ALLYL ALCOHOL (ACGIH,DOT)
556-52-5	ALLYL ALCOHOL OXIDE
107-02-8	ALLYL ALDEHYDE
1693-71-6	ALLYL BORATE
1693-71-6	ALLYL BORATE $((C_3H_5O)_3B)$
106-95-6	ALLYL BROMIDE
106-95-6	ALLYL BROMIDE (DOT)
107-05-1	ALLYL CHLORIDE
107-05-1	ALLYL CHLORIDE (ACGIH,DOT)
2937-50-0	ALLYL CHLOROCARBONATE
2937-50-0	ALLYL CHLOROCARBONATE (DOT)
2937-50-0	ALLYL CHLOROFORMATE
2937-50-0	ALLYL CHLOROFORMATE (DOT)
584-79-2	ALLYL CINERIN
584-79-2	ALLYL CINERIN I
557-40-4	ALLYL ETHER
557-31-3	ALLYL ETHYL ETHER
557-31-3	ALLYL ETHYL ETHER (DOT)
1838-59-1	ALLYL FORMATE
1838-59-1	ALLYL FORMATE (DOT)
106-92-3	ALLYL GLYCIDYL ETHER
106-92-3	ALLYL GLYCIDYL ETHER (ACGIH,DOT)
584-79-2	ALLYL HOMOLOG OF CINERIN I
556-56-9	ALLYL IODIDE
556-56-9	ALLYL IODIDE (DOT)
57-06-7	ALLYL ISORHODANIDE
57-06-7	ALLYL ISOSULFOCYANATE
57-06-7	ALLYL ISOSULPHOCYANATE
57-06-7	ALLYL ISOTHIOCYANATE
57-06-7	ALLYL ISOTHIOCYANATE, INHIBITED
57-06-7	ALLYL ISOTHIOCYANATE, STABILIZED (DOT)
57-06-7	ALLYL MUSTARD OIL
131-17-9	ALLYL PHTHALATE
2179-59-1	ALLYL PROPYL DISULFIDE
57-06-7	ALLYL SEVENOLUM
57-06-7	ALLYL THIOCARBONIMIDE
96-18-4	ALLYL TRICHLORIDE
107-37-9	ALLYL TRICHLOROSILANE
107-37-9	ALLYL TRICHLOROSILANE (DOT)
107-37-9	ALLYL TRICHLOROSILANE, STABILIZED (DOT)
3917-15-5	ALLYL VINYL ETHER
107-18-6	ALLYLALKOHOL (GERMAN)
107-11-9	ALLYLAMINE
107-11-9	ALLYLAMINE (DOT)
94-59-7	ALLYLCATECHOL METHYLENE ETHER
107-05-1	ALLYLCHLORID (GERMAN)
94-59-7	ALLYLDIOXYBENZENE METHYLENE ETHER
107-05-1	ALLYLE (CHLORURE D') (FRENCH)
74-99-7	ALLYLENE
106-92-3	ALLYLGLYCIDAETHER (GERMAN)
107-18-6	ALLYLIC ALCOHOL
869-29-4	ALLYLIDENE ACETATE
869-29-4	ALLYLIDENE DIACETATE
2937-50-0	ALLYLOXYCARBONYL CHLORIDE
94-59-7	ALLYLPYROCATECHOL METHYLENE ETHER
584-79-2	ALLYLRETHRONYL DL-CIS-TRANS-CHRYSANTHEMATE
57-06-7	ALLYLSENEVOL
57-06-7	ALLYLSENFOEL
57-06-7	ALLYLSENFOEL (GERMAN)
57-06-7	ALLYLSEVENOLUM
70-30-4	ALMEDERM
1344-28-1	ALMITE
100-52-7	ALMOND ARTIFICIAL ESSENTIAL OIL
1344-28-1	ALON
1344-28-1	ALON C
1344-28-1	ALOXITE
7429-90-5	ALPASTE 0230T
7429-90-5	ALPASTE 1500MA
7429-90-5	ALPASTE 240T
105-74-8	ALPEROX C
105-74-8	ALPEROX F
6607-45-0	ALPHA, BETA-DICHLOROSTYRENE
104-83-6	ALPHA,4-DICHLOROTOLUENE
139-13-9	ALPHA,ALPHA',ALPHA''-TRIMETHYLAMINETRICARBOXYLIC ACID
78-67-1	ALPHA,ALPHA'-AZOBIS(ISOBUTYRONITRILE)
78-67-1	ALPHA,ALPHA'-AZOBISISOBUTYLONITRILE
78-67-1	ALPHA,ALPHA'-AZODIISOBUTYRIC ACID DINITRILE
78-67-1	ALPHA,ALPHA'-AZODIISOBUTYRONITRILE
534-07-6	ALPHA,ALPHA'-DICHLOROACETONE
28347-13-9	ALPHA,ALPHA'-DICHLOROXYLENE
80-43-3	ALPHA,ALPHA'-DICUMYL PEROXIDE
137-26-8	ALPHA,ALPHA'-DITHIOBIS(DIMETHYLTHIO)FORMAMIDE
107-41-5	ALPHA,ALPHA,ALPHA'-TRIMETHYLTRIMETHYLENE GLYCOL
98-07-7	ALPHA,ALPHA,ALPHA-TRICHLOROTOLUENE
1582-09-8	ALPHA,ALPHA,ALPHA-TRIFLUORO-2,6-DINITRO-N,N-DIPROPYL-P-TOLUIDINE

CAS No.	Chemical Name
98-56-6	ALPHA,ALPHA,ALPHA-TRIFLUORO-4-CHLOROTOLUENE
98-16-8	ALPHA,ALPHA,ALPHA-TRIFLUORO-M-TOLUIDINE
88-17-5	ALPHA,ALPHA,ALPHA-TRIFLUORO-O-TOLUIDINE
402-54-0	ALPHA,ALPHA,ALPHA-TRIFLUORO-P-NITROTOLUENE
98-08-8	ALPHA,ALPHA,ALPHA-TRIFLUOROTOLUENE
50-29-3	ALPHA,ALPHA-BIS(P-CHLOROPHENYL)-BETA,BETA,BETA-TRICHLORETHANE
79-36-7	ALPHA,ALPHA-DICHLOROACETYL CHLORIDE
107-30-2	ALPHA,ALPHA-DICHLORODIMETHYL ETHER
75-99-0	ALPHA,ALPHA-DICHLOROPROPIONIC ACID
98-87-3	ALPHA,ALPHA-DICHLOROTOLUENE
80-15-9	ALPHA,ALPHA-DIMETHYLBENZYL HYDROPEROXIDE
80-43-3	ALPHA,ALPHA-DIMETHYLBENZYL PEROXIDE
115-19-5	ALPHA,ALPHA-DIMETHYLPROPARGYL ALCOHOL
75-98-9	ALPHA,ALPHA-DIMETHYLPROPIONIC ACID
115-29-7	ALPHA,BETA-1,2,3,4,7,7-HEXACHLOROBICYCLO(221)-2-HEPTENE-5,6-BISOXYMETHYLENE SULFITE
106-93-4	ALPHA,BETA-DIBROMOETHANE
107-06-2	ALPHA,BETA-DICHLOROETHANE
110-71-4	ALPHA,BETA-DIMETHOXYETHANE
646-06-0	ALPHA,BETA-ISOPROPYLIDENEGLYCEROL
106-99-0	ALPHA,GAMMA-BUTADIENE
25339-57-5	ALPHA,GAMMA-BUTADIENE
534-07-6	ALPHA,GAMMA-DICHLOROACETONE
104-83-6	ALPHA,P-DICHLOROTOLUENE
53-86-1	ALPHA-(1-(P-CHLOROBENZOYL)-2-METHYL-5-METHOXY-3-INDOLYL) ACETIC ACID
93-72-1	ALPHA-(2,4,5-TRICHLOROPHENOXY)PROPIONIC ACID
9016-45-9	ALPHA-(NONYLPHENYL)-OMEGA-HYDROXYPOLY(OXY-1,2-ETHANEDIYL)
9016-45-9	ALPHA-(NONYLPHENYL)-OMEGA-HYDROXYPOLYOXYETHYLENE
7786-34-7	ALPHA-2-CARBOMETHOXY-1-METHYLVINYL DIMETHYL PHOSPHATE
7784-21-6	ALPHA-ALUMINUM TRIHYDRIDE
75-39-8	ALPHA-AMINOETHYL ALCOHOL
78-96-6	ALPHA-AMINOISOPROPYL ALCOHOL
100-46-9	ALPHA-AMINOTOLUENE
5798-79-8	ALPHA-BROMO-ALPHA-TOLUNITRILE
598-31-2	ALPHA-BROMOACETONE
70-11-1	ALPHA-BROMOACETOPHENONE
5798-79-8	ALPHA-BROMOBENZENEACETONITRILE
5798-79-8	ALPHA-BROMOBENZYL CYANIDE
5798-79-8	ALPHA-BROMOBENZYLNITRILE
776-74-9	ALPHA-BROMODIPHENYLMETHANE
790-83-0	ALPHA-BROMOETHANOIC ACID
5798-79-8	ALPHA-BROMOPHENYLACETONITRILE
598-31-2	ALPHA-BROMOPROPANONE
3724-65-0	ALPHA-BUTENOIC ACID
96-24-2	ALPHA-CHLORHYDRIN
100-14-1	ALPHA-CHLORO-4-NITROTOLUENE
100-14-1	ALPHA-CHLORO-P-NITROTOLUENE
107-20-0	ALPHA-CHLOROACETALDEHYDE
79-11-8	ALPHA-CHLOROACETIC ACID
78-95-5	ALPHA-CHLOROACETONE
107-14-2	ALPHA-CHLOROACETONITRILE
532-27-4	ALPHA-CHLOROACETOPHENONE
532-27-4	ALPHA-CHLOROACETOPHENONE (ACGIH)
98-88-4	ALPHA-CHLOROBENZALDEHYDE
25497-29-4	ALPHA-CHLOROETHYLIDENE FLUORIDE
96-24-2	ALPHA-CHLOROHYDRIN
17639-93-9	ALPHA-CHLOROPROPIONIC ACID METHYL ESTER
107-05-1	ALPHA-CHLOROPROPYLENE
29154-12-9	ALPHA-CHLOROPYRIDINE
100-44-7	ALPHA-CHLOROTOLUENE
100-44-7	ALPHA-CHLORTOLUOL (GERMAN)
3724-65-0	ALPHA-CROTONIC ACID
10544-63-5	ALPHA-CROTONIC ACID ETHYL ESTER
80-15-9	ALPHA-CUMENE HYDROPEROXIDE
80-15-9	ALPHA-CUMYL HYDROPEROXIDE
80-43-3	ALPHA-CUMYL PEROXIDE
26748-47-0	ALPHA-CUMYL PEROXYNEODECANOATE
140-29-4	ALPHA-CYANOTOLUENE
872-05-9	ALPHA-DECENE
75-99-0	ALPHA-DICHLOROPROPIONIC ACID
51-28-5	ALPHA-DINITROPHENOL
112-41-4	ALPHA-DODECENE
106-89-8	ALPHA-EPICHLOROHYDRIN
645-62-5	ALPHA-ETHYL-2-HEXENAL
645-62-5	ALPHA-ETHYL-BETA-N-PROPYLACROLEIN
645-62-5	ALPHA-ETHYL-BETA-PROPYLACROLEIN
97-96-1	ALPHA-ETHYLBUTYRALDEHYDE
123-05-7	ALPHA-ETHYLCAPROALDEHYDE
149-57-5	ALPHA-ETHYLCAPROIC ACID
123-05-7	ALPHA-ETHYLHEXANAL
149-57-5	ALPHA-ETHYLHEXANOIC ACID
144-49-0	ALPHA-FLUOROACETIC ACID
98-00-0	ALPHA-FURFURYL ALCOHOL
617-89-0	ALPHA-FURFURYLAMINE
98-01-1	ALPHA-FUROLE
98-00-0	ALPHA-FURYLCARBINOL
319-84-6	ALPHA-HCH
319-84-6	ALPHA-HEXACHLOROCYCLOHEXANE
25322-69-4	ALPHA-HYDRO-OMEGA-HYDROXYPOLY(OXYPROPYLENE)
123-31-9	ALPHA-HYDROQUINONE
98-85-1	ALPHA-HYDROXYETHYLBENZENE
75-86-5	ALPHA-HYDROXYISOBUTYRONITRILE
50-21-5	ALPHA-HYDROXYPROPANOIC ACID
50-21-5	ALPHA-HYDROXYPROPIONIC ACID
78-97-7	ALPHA-HYDROXYPROPIONITRILE
100-51-6	ALPHA-HYDROXYTOLUENE
620-05-3	ALPHA-IODOTOLUENE
138-86-3	ALPHA-LIMONENE
68-11-1	ALPHA-MERCAPTOACETIC ACID
79-42-5	ALPHA-MERCAPTOPROPANOIC ACID
79-42-5	ALPHA-MERCAPTOPROPIONIC ACID
78-85-3	ALPHA-METHACROLEIN
79-41-4	ALPHA-METHACRYLIC ACID
126-98-7	ALPHA-METHACRYLONITRILE
557-17-5	ALPHA-METHOXYPROPANE
98-83-9	ALPHA-METHYL STYRENE
98-83-9	ALPHA-METHYL STYRENE (ACGIH)
123-15-9	ALPHA-METHYL VALERALDEHYDE
123-15-9	ALPHA-METHYL VALERALDEHYDE (DOT)
98-83-9	ALPHA-METHYL-STYROL (GERMAN)
78-85-3	ALPHA-METHYLACROLEIN
78-85-3	ALPHA-METHYLACRYLALDEHYDE
79-41-4	ALPHA-METHYLACRYLIC ACID
126-98-7	ALPHA-METHYLACRYLONITRILE
98-85-1	ALPHA-METHYLBENZENEMETHANOL
98-85-1	ALPHA-METHYLBENZYL ALCOHOL
98-85-1	ALPHA-METHYLBENZYL ALCOHOL (DOT)
2449-49-2	ALPHA-METHYLBENZYL DIMETHYL AMINE
93-96-9	ALPHA-METHYLBENZYL ETHER
98-84-0	ALPHA-METHYLBENZYLAMINE
555-30-6	ALPHA-METHYLDOPA
123-15-9	ALPHA-METHYLPENTENAL
79-30-1	ALPHA-METHYLPROPIONYL CHLORIDE
109-06-8	ALPHA-METHYLPYRIDINE
98-83-9	ALPHA-METHYLSTYREEN (DUTCH)
98-83-9	ALPHA-METHYLSTYROL
100-41-4	ALPHA-METHYLTOLUENE
98-83-9	ALPHA-METIL-STIROLO (ITALIAN)
96-24-2	ALPHA-MONOCHLOROHYDRIN
109-67-1	ALPHA-N-AMYLENE
63-25-2	ALPHA-NAFTYL-N-METHYLKARBAMAT (CZECH)
63-25-2	ALPHA-NAPHTHALENYL METHYLCARBAMATE
30553-04-9	ALPHA-NAPHTHALTHIOHARNSTOFF (GERMAN)
30553-04-9	ALPHA-NAPHTHOTHIOUREA
63-25-2	ALPHA-NAPHTHYL METHYLCARBAMATE
63-25-2	ALPHA-NAPHTHYL-N-METHYLCARBAMATE
86-88-4	ALPHA-NAPHTHYLTHIOCARBAMIDE
30553-04-9	ALPHA-NAPHTHYLTHIOCARBAMIDE
86-88-4	ALPHA-NAPHTHYLTHIOUREA
30553-04-9	ALPHA-NAPHTHYLTHIOUREA
30553-04-9	ALPHA-NAPHTHYLTHIOUREA (DOT)
86-88-4	ALPHA-NAPTHYL THIOUREA (ANTU)
556-88-7	ALPHA-NITROGUANIDINE
124-11-8	ALPHA-NONENE
111-66-0	ALPHA-OCTENE
111-66-0	ALPHA-OCTYLENE
119-61-9	ALPHA-OXODIPHENYLMETHANE
119-61-9	ALPHA-OXODITANE
98-85-1	ALPHA-PHENETHYL ALCOHOL
103-80-0	ALPHA-PHENYLACETYL CHLORIDE
98-85-1	ALPHA-PHENYLETHANOL
98-85-1	ALPHA-PHENYLETHYL ALCOHOL
103-85-5	ALPHA-PHENYLTHIOUREA
84-80-0	ALPHA-PHYLLOQUINONE

CAS No.	*Chemical Name*
109-06-8	ALPHA-PICOLINE
80-56-8	ALPHA-PINENE
57-55-6	ALPHA-PROPYLENE GLYCOL
71-55-6	ALPHA-T
1120-36-1	ALPHA-TETRADECENE
771-29-9	ALPHA-TETRALIN HYDROPEROXIDE
118-96-7	ALPHA-TNT
100-51-6	ALPHA-TOLUENOL
140-29-4	ALPHA-TOLUNITRILE
71-55-6	ALPHA-TRICHLOROETHANE
2437-56-1	ALPHA-TRIDECENE
821-95-4	ALPHA-UNDECENE
7440-32-6	ALPHA-VT 1-0
1332-58-7	ALPHACOTE
30553-04-9	ALPHANAPHTYL THIOUREE (FRENCH)
577-11-7	ALPHASOL OT
123-23-9	ALPHOZONE
134-32-7	ALPHA-NAPHTHYLAMINE
28981-97-7	ALPRAZOLAM
86-88-4	ALRATO
30553-04-9	ALRATO
139-91-3	ALTABACTINA
139-91-3	ALTAFUR
30900-23-3	ALTOX
50-28-2	ALTRAD
25167-93-5	ALTRITAN
10043-01-3	ALUM
1344-28-1	ALUMINA
1344-28-1	ALUMINA C GAMMA
1344-28-1	ALUMINASOL 100
16853-85-3	ALUMINATE (1-), TETRAHYDRO-, LITHIUM
13770-96-2	ALUMINATE (1-), TETRAHYDRO-, SODIUM
16853-85-3	ALUMINATE(1-), TETRAHYDRO-, LITHIUM, (T-4)-
16853-85-3	ALUMINATE(1-), TETRAHYDRO-, LITHIUM, (T-4)- (9CI)
13770-96-2	ALUMINATE(1-), TETRAHYDRO-, SODIUM, (T-4)-
13770-96-2	ALUMINATE(1-), TETRAHYDRO-, SODIUM, (T-4)- (9CI)
1344-28-1	ALUMINITE 37
7429-90-5	ALUMINIUM
7446-70-0	ALUMINIUM CHLORIDE
7429-90-5	ALUMINIUM FLAKE
20859-73-8	ALUMINIUM FOSFIDE (DUTCH)
20859-73-8	ALUMINIUM PHOSPHIDE
20859-73-8	ALUMINIUM PHOSPHIDE (ALP)
7446-70-0	ALUMINIUMCHLORID (GERMAN)
7784-30-7	ALUMINOPHOSPHORIC ACID
7429-90-5	ALUMINUM
7429-90-5	ALUMINUM (FUME OR DUST)
7429-90-5	ALUMINUM A 00
7784-30-7	ALUMINUM ACID PHOSPHATE
10043-01-3	ALUMINUM ALUM
16962-07-5	ALUMINUM BOROHYDRIDE
16962-07-5	ALUMINUM BOROHYDRIDE (DOT)
7727-15-3	ALUMINUM BROMIDE
7727-15-3	ALUMINUM BROMIDE (AlBr3)
7727-15-3	ALUMINUM BROMIDE (ANHYDROUS)
7727-15-3	ALUMINUM BROMIDE SOLUTION
7727-15-3	ALUMINUM BROMIDE SOLUTION (DOT)
7727-15-3	ALUMINUM BROMIDE, ANHYDROUS
7727-15-3	ALUMINUM BROMIDE, ANHYDROUS (DOT)
1299-86-1	ALUMINUM CARBIDE
12656-43-8	ALUMINUM CARBIDE
1299-86-1	ALUMINUM CARBIDE (Al4C3)
12656-43-8	ALUMINUM CARBIDE (Al4C3)
1299-86-1	ALUMINUM CARBIDE (DOT)
12656-43-8	ALUMINUM CARBIDE (DOT)
7446-70-0	ALUMINUM CHLORIDE
7446-70-0	ALUMINUM CHLORIDE (1:3)
7446-70-0	ALUMINUM CHLORIDE (AlCl3)
7446-70-0	ALUMINUM CHLORIDE SOLUTION (DOT)
7446-70-0	ALUMINUM CHLORIDE, ANHYDROUS
7446-70-0	ALUMINUM CHLORIDE, ANHYDROUS (DOT)
7446-70-0	ALUMINUM CHLORIDE, SOLUTION
7429-90-5	ALUMINUM DEHYDRATED
12003-41-7	ALUMINUM FERROSILICON
7784-18-1	ALUMINUM FLUORIDE
7784-21-6	ALUMINUM HYDRIDE
7784-21-6	ALUMINUM HYDRIDE (AlH3)
7784-21-6	ALUMINUM HYDRIDE (DOT)
16962-07-5	ALUMINUM HYDROBORATE
16962-07-5	ALUMINUM HYDROBORATE (Al[BH4]3)
16853-85-3	ALUMINUM LITHIUM HYDRIDE
16853-85-3	ALUMINUM LITHIUM HYDRIDE (LiAlH4)
16853-85-3	ALUMINUM LITHIUM TETRAHYDRIDE
7784-30-7	ALUMINUM MONOPHOSPHATE
20859-73-8	ALUMINUM MONOPHOSPHIDE
13473-90-0	ALUMINUM NITRATE
7784-30-7	ALUMINUM ORTHOPHOSPHATE
1344-28-1	ALUMINUM OXIDE
1344-28-1	ALUMINUM OXIDE (Al2O3)
1344-28-1	ALUMINUM OXIDE (BROCKMANN)
7784-30-7	ALUMINUM PHOSPHATE
7784-30-7	ALUMINUM PHOSPHATE (1:1)
7784-30-7	ALUMINUM PHOSPHATE (Al(PO4))
7784-30-7	ALUMINUM PHOSPHATE SOLUTION
7784-30-7	ALUMINUM PHOSPHATE SOLUTION (DOT)
20859-73-8	ALUMINUM PHOSPHIDE
20859-73-8	ALUMINUM PHOSPHIDE (AlP)
20859-73-8	ALUMINUM PHOSPHIDE (DOT)
7429-90-5	ALUMINUM POWDER
61789-65-9	ALUMINUM RESINATE
1344-28-1	ALUMINUM SESQUIOXIDE
10043-01-3	ALUMINUM SESQUISULFATE
12042-55-6	ALUMINUM SILICON (AlSi)
57485-31-1	ALUMINUM SILICON (AlSi2)
50810-25-8	ALUMINUM SILICON (AlSi5)
13770-96-2	ALUMINUM SODIUM HYDRIDE
13770-96-2	ALUMINUM SODIUM HYDRIDE (AlNaH4)
11138-49-1	ALUMINUM SODIUM OXIDE
10043-01-3	ALUMINUM SULFATE
10043-01-3	ALUMINUM SULFATE (2:3)
10043-01-3	ALUMINUM SULFATE (Al2(SO4)3)
10043-01-3	ALUMINUM SULPHATE
16962-07-5	ALUMINUM TETRAHYDROBORATE
7727-15-3	ALUMINUM TRIBROMIDE
7727-15-3	ALUMINUM TRIBROMIDE (AlBr3)
1116-70-7	ALUMINUM TRIBUTYL
7446-70-0	ALUMINUM TRICHLORIDE
7446-70-0	ALUMINUM TRICHLORIDE (AlCl3)
7784-21-6	ALUMINUM TRIHYDRIDE
75-24-1	ALUMINUM TRIMETHYL
75-24-1	ALUMINUM TRIMETHYL (DOT)
1344-28-1	ALUMINUM TRIOXIDE
10043-01-3	ALUMINUM TRISULFATE
10043-01-3	ALUMINUM(III) SULFATE
96-10-6	ALUMINUM, CHLORODIETHYL-
563-43-9	ALUMINUM, DICHLOROETHYL-
96-10-6	ALUMINUM, DICHLOROTETRAETHYLDI-
871-27-2	ALUMINUM, DIETHYLHYDRO-
1191-15-7	ALUMINUM, HYDROBIS(2-METHYLPROPYL)-
1191-15-7	ALUMINUM, HYDRODIISOBUTYL-
2036-15-9	ALUMINUM, HYDRODIPROPYL-
7429-90-5	ALUMINUM, METAL AND OXIDE AND WELDING FUMES
12263-85-3	ALUMINUM, TRIBROMOTRIMETHYLDI-
12075-68-2	ALUMINUM, TRICHLOROTRIETHYLDI-
12542-85-7	ALUMINUM, TRICHLOROTRIMETHYLDI-
97-93-8	ALUMINUM, TRIETHYL-
100-99-2	ALUMINUM, TRIISOBUTYL-
102-67-0	ALUMINUM, TRIPROPYL-
100-99-2	ALUMINUM, TRIS(2-METHYLPROPYL)-
100-99-2	ALUMINUM, TRIS(2-METHYLPROPYL)- (9CI)
7429-90-5	ALUMINUM-27
1344-28-1	ALUMITE
1344-28-1	ALUMITE (OXIDE)
1344-28-1	ALUMOGEL A 1
1344-28-1	ALUNDUM
1344-28-1	ALUNDUM 600
7784-30-7	ALUPHOS
60-57-1	ALVIT
60-57-1	ALVIT 55
300-76-5	ALVORA
9002-89-5	ALVYL
156-62-7	ALZODEF
420-04-2	ALZOGUR
1333-86-4	AM BLACK
7664-41-7	AM-FOL
119-90-4	AMACEL DEVELOPED NAVY SD
3761-53-3	AMACID LAKE SCARLET 2R
9005-25-8	AMAIZO 310

CAS No.	Chemical Name
9005-25-8	AMAIZO W 13
1314-13-2	AMALOX
1937-37-7	AMANIL BLACK GL
1937-37-7	AMANIL BLACK WD
2602-46-2	AMANIL BLUE 2BX
16071-86-6	AMANIL FAST BROWN BRL
72-57-1	AMANIL SKY BLUE R
16071-86-6	AMANIL SUPRA BROWN LBL
89-62-3	AMARTHOL FAST RED GL BASE
89-62-3	AMARTHOL FAST RED GL SALT
95-79-4	AMARTHOL FAST RED TR BASE
3165-93-3	AMARTHOL FAST RED TR BASE
3165-93-3	AMARTHOL FAST RED TR SALT
99-55-8	AMARTHOL FAST SCARLET G BASE
99-55-8	AMARTHOL FAST SCARLET G SALT
118-74-1	AMATIN
1332-58-7	AMAZON 88
8006-54-0	AMBER LANOLIN
61-57-4	AMBILHAR
305-03-3	AMBOCHLORIN
305-03-3	AMBOCLORIN
116-06-3	AMBUSH
76-03-9	AMCHEM GRASS KILLER
12125-02-9	AMCHLOR
7773-06-0	AMCIDE
316-42-7	AMEBICIDE
58-89-9	AMEISENATOD
58-89-9	AMEISENMITTEL MERCK
64-18-6	AMEISENSAEURE (GERMAN)
58-89-9	AMEISENTOD
57-63-6	AMENORON
133-06-2	AMERCIDE
57-50-1	AMERFOND
60-51-5	AMERICAN CYANAMID 12880
297-97-2	AMERICAN CYANAMID 18133
2275-18-5	AMERICAN CYANAMID 18682
298-02-2	AMERICAN CYANAMID 3,911
56-38-2	AMERICAN CYANAMID 3422
121-75-5	AMERICAN CYANAMID 4,049
947-02-4	AMERICAN CYANAMID 47031
947-02-4	AMERICAN CYANAMID AC 47,031
122-14-5	AMERICAN CYANAMID CL-47,300
947-02-4	AMERICAN CYANAMID CL-47031
950-10-7	AMERICAN CYANAMID CL-47470
50-07-7	AMETYCIN
50-07-7	AMETYCINE
7773-06-0	AMICIDE
9005-25-8	AMICOL 1B
9005-25-8	AMICOL C
420-04-2	AMIDOCYANOGEN
299-86-5	AMIDOFOS
10102-43-9	AMIDOGEN, OXO-
299-86-5	AMIDOPHOS
5329-14-6	AMIDOSULFONIC ACID
5329-14-6	AMIDOSULFURIC ACID
563-41-7	AMIDOUREA HYDROCHLORIDE
94-75-7	AMIDOX
109-83-1	AMIETOL M 11
108-01-0	AMIETOL M 21
9005-25-8	AMIGEL 12014
9005-25-8	AMIGEL 30076
105-60-2	AMILAN CM 1001
105-60-2	AMILAN CM 1001C
105-60-2	AMILAN CM 1001G
105-60-2	AMILAN CM 1011
80-46-6	AMILFENOL
93-76-5	AMINE 2,4,5-T FOR RICE
5329-14-6	AMINESULFONIC ACID
64-18-6	AMINIC ACID
504-24-5	AMINO-4 PYRIDINE
62-53-3	AMINOBENZENE
56-18-8	AMINOBIS(PROPYLAMINE)
105-60-2	AMINOCAPROIC LACTAM
95-85-2	AMINOCHLOROPHENOL
108-91-8	AMINOCYCLOHEXANE
101-83-7	AMINODICYCLOHEXANE
1300-73-8	AMINODIMETHYLBENZENE
75-04-7	AMINOETHANE
141-43-5	AMINOETHANOL

CAS No.	Chemical Name
111-41-1	AMINOETHYL ETHANOLAMINE
593-67-9	AMINOETHYLENE
111-40-0	AMINOETHYLETHANDIAMINE
140-31-8	AMINOETHYLPIPERAZINE
100-97-0	AMINOFORM
100-97-0	AMINOFORMALDEHYDE
108-91-8	AMINOHEXAHYDROBENZENE
102-56-7	AMINOHYDROQUINONE DIMETHYL ETHER
10124-48-8	AMINOMERCURIC CHLORIDE
88-05-1	AMINOMESITYLENE
74-89-5	AMINOMETHANE
29757-24-2	AMINONITROBENZENE
62-53-3	AMINOPHEN
27598-85-2	AMINOPHENOL
27598-85-2	AMINOPHENOLS
27598-85-2	AMINOPHENOLS (O-, M-, P-)
4985-85-7	AMINOPROPYLDIETHANOLAMINE
4985-85-7	AMINOPROPYLDIETHANOLAMINE (DOT)
123-00-2	AMINOPROPYLMORPHOLINE
54-62-6	AMINOPTERIDINE
54-62-6	AMINOPTERIN
54-62-6	AMINOPTERINE
26445-05-6	AMINOPYRIDINE
26445-05-6	AMINOPYRINE
5329-14-6	AMINOSULFONIC ACID
5329-14-6	AMINOSULFURIC ACID
139-13-9	AMINOTRIACETIC ACID
61-82-5	AMINOTRIAZOLE
61-82-5	AMINOTRIAZOLE (PLANT REGULATOR)
66-75-1	AMINOURACIL MUSTARD
563-41-7	AMINOUREA HYDROCHLORIDE
9002-84-0	AMIP 15M
25321-22-6	AMISIA-MOTTENSCHUTZ
78-53-5	AMITON
3734-97-2	AMITON OXALATE
61-82-5	AMITROL
61-82-5	AMITROL 90
61-82-5	AMITROLE
61-82-5	AMIZOL
7773-06-0	AMMAT
7773-06-0	AMMATE
7773-06-0	AMMATE X
1336-21-6	AMMMONIUM HYDROXIDE ((NH4)(OH))
100-97-0	AMMOFORM
12125-02-9	AMMONERIC
7664-41-7	AMMONIA
7664-41-7	AMMONIA (ACGIH)
7664-41-7	AMMONIA ANHYDROUS
1336-21-6	AMMONIA AQUEOUS
7664-41-7	AMMONIA GAS
9080-17-5	AMMONIA POLYSULFIDE
544-60-5	AMMONIA SOAP
1336-21-6	AMMONIA SOLUTION, CONTAINING 44% OR LESS AMMONIA (DOT)
7664-41-7	AMMONIA SOLUTION, CONTAINING MORE THAN 44% AMMONIA (DOT)
7664-41-7	AMMONIA SOLUTION, CONTAINING MORE THEN 50% AMMONIA
1336-21-6	AMMONIA WATER
1336-21-6	AMMONIA WATER 29%
7664-41-7	AMMONIA, ANHYDROUS
7664-41-7	AMMONIA, ANHYDROUS (DOT)
1336-21-6	AMMONIA, AQUA
1336-21-6	AMMONIA, MONOHYDRATE
7664-41-7	AMMONIA-14N
7664-41-7	AMMONIAC (FRENCH)
7664-41-7	AMMONIACA (ITALIAN)
7664-41-7	AMMONIAK (GERMAN)
10380-29-7	AMMONIATED CUPRIC SULFATE MONOHYDRATE
10124-48-8	AMMONIATED MERCURIC CHLORIDE
10124-48-8	AMMONIATED MERCURY
7789-09-5	AMMONIO (BICROMATO DI) (ITALIAN)
7789-09-5	AMMONIO (DICROMATO DI) (ITALIAN)
100-97-0	AMMONIOFORMALDEHYDE
7789-09-5	AMMONIUM (DICHROMATE D') (FRENCH)
631-61-8	AMMONIUM ACETATE
631-61-8	AMMONIUM ACETATE (DOT)
7784-44-3	AMMONIUM ACID ARSENATE
7803-63-6	AMMONIUM ACID SULFATE
10192-30-0	AMMONIUM ACID SULFITE

CAS No.	Chemical Name
7664-41-7	AMMONIUM AMIDE
7773-06-0	AMMONIUM AMIDOSULFATE
7773-06-0	AMMONIUM AMIDOSULFONATE
7773-06-0	AMMONIUM AMIDOSULPHATE
1111-78-0	AMMONIUM AMINOFORMATE
7773-06-0	AMMONIUM AMINOSULFONATE
7784-44-3	AMMONIUM ARSENATE
7784-44-3	AMMONIUM ARSENATE ((NH4)2HAsO4)
7784-44-3	AMMONIUM ARSENATE (DOT)
7784-44-3	AMMONIUM ARSENATE, SOLID
7784-44-3	AMMONIUM ARSENATE, SOLID (DOT)
1863-63-4	AMMONIUM BENZOATE
1863-63-4	AMMONIUM BENZOATE (DOT)
1066-33-7	AMMONIUM BICARBONATE
1066-33-7	AMMONIUM BICARBONATE (1:1)
1066-33-7	AMMONIUM BICARBONATE (DOT)
7789-09-5	AMMONIUM BICHROMATE
7789-09-5	AMMONIUM BICHROMATE (DOT)
1341-49-7	AMMONIUM BIFLUORIDE
1341-49-7	AMMONIUM BIFLUORIDE (NH4HF2)
1341-49-7	AMMONIUM BIFLUORIDE (NH5F2)
1341-49-7	AMMONIUM BIFLUORIDE, SOLID (DOT)
1341-49-7	AMMONIUM BIFLUORIDE, SOLUTION (DOT)
5972-73-6	AMMONIUM BINOXALATE MONOHYDRATE
7803-63-6	AMMONIUM BISULFATE
10192-30-0	AMMONIUM BISULFITE
10192-30-0	AMMONIUM BISULFITE (NH4HSO3)
10192-30-0	AMMONIUM BISULFITE SOLUTION
10192-30-0	AMMONIUM BISULFITE, SOLID
10192-30-0	AMMONIUM BISULFITE, SOLID (DOT)
10192-30-0	AMMONIUM BISULFITE, SOLUTION (DOT)
12007-89-5	AMMONIUM BORATE (NH4B5O8)
13826-83-0	AMMONIUM BOROFLUORIDE
13826-83-0	AMMONIUM BOROFLUORIDE (NH4BF4)
12007-89-5	AMMONIUM BORON OXIDE ((NH4)B5O8)
12124-97-9	AMMONIUM BROMIDE
12124-97-9	AMMONIUM BROMIDE ((NH4)Br)
1111-78-0	AMMONIUM CARBAMATE
1111-78-0	AMMONIUM CARBAMATE (DOT)
131-74-8	AMMONIUM CARBAZOATE
506-87-6	AMMONIUM CARBONATE
1066-33-7	AMMONIUM CARBONATE
10361-29-2	AMMONIUM CARBONATE
506-87-6	AMMONIUM CARBONATE ((NH4)2CO3)
506-87-6	AMMONIUM CARBONATE (DOT)
12125-02-9	AMMONIUM CHLORIDE
12125-02-9	AMMONIUM CHLORIDE ((NH4)Cl)
12125-02-9	AMMONIUM CHLORIDE (ACGIH,DOT)
12125-02-9	AMMONIUM CHLORIDE FUME
16919-58-7	AMMONIUM CHLOROPLATINATE
7788-98-9	AMMONIUM CHROMATE
7788-98-9	AMMONIUM CHROMATE ((NH4) 2CrO4)
7789-09-5	AMMONIUM CHROMATE ((NH4)2Cr2O7)
7788-98-9	AMMONIUM CHROMATE (DOT)
7788-98-9	AMMONIUM CHROMATE(VI)
3012-65-5	AMMONIUM CITRATE
3012-65-5	AMMONIUM CITRATE DIBASIC
3012-65-5	AMMONIUM CITRATE DIBASIC (DOT)
3164-29-2	AMMONIUM D-TARTRATE
7783-28-0	AMMONIUM DIBASIC PHOSPHATE
7789-09-5	AMMONIUM DICHROMATE
7789-09-5	AMMONIUM DICHROMATE (DOT)
7789-09-5	AMMONIUM DICHROMATE(VI)
1341-49-7	AMMONIUM DIFLUORIDE
1341-49-7	AMMONIUM DIFLUORIDE (NH4HF2)
29595-25-3	AMMONIUM DINITRO-O-CRESOLATE
29595-25-3	AMMONIUM DINITRO-O-CRESOLATE (DOT)
15699-18-0	AMMONIUM DISULFATONICKELATE(II)
2235-54-3	AMMONIUM DODECYL SULFATE
1185-57-5	AMMONIUM FERRIC CITRATE
14221-47-7	AMMONIUM FERRIC OXALATE
14221-47-7	AMMONIUM FERRIOXALATE
10045-89-3	AMMONIUM FERROUS SULFATE
13826-83-0	AMMONIUM FLUOBORATE
13826-83-0	AMMONIUM FLUOBORATE (DOT)
12125-01-8	AMMONIUM FLUORIDE
1341-49-7	AMMONIUM FLUORIDE ((NH4)(HF2))
12125-01-8	AMMONIUM FLUORIDE (DOT)
12125-01-8	AMMONIUM FLUORIDE (NH4F)

CAS No.	Chemical Name
1341-49-7	AMMONIUM FLUORIDE COMPD. WITH HYDROGEN FLUORIDE (1:1)
13826-83-0	AMMONIUM FLUOROBORATE
13826-83-0	AMMONIUM FLUOROBORATE (NH4BF4)
16919-19-0	AMMONIUM FLUOROSILICATE
16919-19-0	AMMONIUM FLUOROSILICATE ((NH4)2SiF6)
12125-01-8	AMMONIUM FLUORURE (FRENCH)
1309-32-6	AMMONIUM FLUOSILICATE
16919-19-0	AMMONIUM FLUOSILICATE
540-69-2	AMMONIUM FORMATE
19222-41-4	AMMONIUM GLUCONATE
13106-76-8	AMMONIUM HEPTAMOLYBDATE
16919-58-7	AMMONIUM HEXACHLOROPLATINATE
16919-58-7	AMMONIUM HEXACHLOROPLATINATE(IV)
16919-19-0	AMMONIUM HEXAFLUOROSILICATE
1341-49-7	AMMONIUM HYDROFLUORIDE
1341-49-7	AMMONIUM HYDROFLUORIDE (NH4HF2)
1341-49-7	AMMONIUM HYDROFLUORIDE (NH5F2)
1341-49-7	AMMONIUM HYDROGEN BIFLUORIDE
1066-33-7	AMMONIUM HYDROGEN CARBONATE
1341-49-7	AMMONIUM HYDROGEN DIFLUORIDE
1341-49-7	AMMONIUM HYDROGEN DIFLUORIDE (NH4(HF2))
1341-49-7	AMMONIUM HYDROGEN FLUORIDE
1341-49-7	AMMONIUM HYDROGEN FLUORIDE (NH4HF2)
1341-49-7	AMMONIUM HYDROGEN FLUORIDE SOLUTION
1341-49-7	AMMONIUM HYDROGEN FLUORIDE SOLUTION (DOT)
1341-49-7	AMMONIUM HYDROGEN FLUORIDE, SOLID
1341-49-7	AMMONIUM HYDROGEN FLUORIDE, SOLID (DOT)
7783-28-0	AMMONIUM HYDROGEN PHOSPHATE
7783-28-0	AMMONIUM HYDROGEN PHOSPHATE ((NH4)2HPO4)
7803-63-6	AMMONIUM HYDROGEN SULFATE
7803-63-6	AMMONIUM HYDROGEN SULFATE (DOT)
7803-63-6	AMMONIUM HYDROGEN SULFATE (NH4HSO4)
10192-30-0	AMMONIUM HYDROGEN SULFITE
12124-99-1	AMMONIUM HYDROSULFIDE SOLUTION
1336-21-6	AMMONIUM HYDROXIDE
1336-21-6	AMMONIUM HYDROXIDE ((NH4)(OH))
1336-21-6	AMMONIUM HYDROXIDE (DOT)
1336-21-6	AMMONIUM HYDROXIDE, CONTAINING LESS THAN 12% AMMONIA
1336-21-6	AMMONIUM HYDROXIDE, CONTAINING LESS THAN 12% AMMONIA (DOT)
1336-21-6	AMMONIUM HYDROXIDE, CONTAINING NOT LESS THAN 12% BUT NOT MORE THAN 44% AMMONIA (DOT)
1336-21-6	AMMONIUM HYDROXIDE, WITH 12% AMMONIA
7803-65-8	AMMONIUM HYPOPHOSPHITE
7783-18-8	AMMONIUM HYPOSULFITE
12027-06-4	AMMONIUM IODIDE
12027-06-4	AMMONIUM IODIDE ((NH4)I)
10045-89-3	AMMONIUM IRON SULFATE
10045-89-3	AMMONIUM IRON SULFATE (2:2:1)
1185-57-5	AMMONIUM IRON(III) CITRATE
1762-95-4	AMMONIUM ISOTHIOCYANATE
515-98-0	AMMONIUM LACTATE
2235-54-3	AMMONIUM LAURYL SULFATE
7803-55-6	AMMONIUM METAVANADATE
7803-55-6	AMMONIUM METAVANADATE (DOT)
7803-55-6	AMMONIUM METAVANADATE (NH4VO3)
13106-76-8	AMMONIUM MOLYBDATE
13106-76-8	AMMONIUM MOLYBDATE [(NH4)2MoO4]
3012-65-5	AMMONIUM MONOHYDROGEN CITRATE
1341-49-7	AMMONIUM MONOHYDROGEN DIFLUORIDE
7783-28-0	AMMONIUM MONOHYDROGEN ORTHOPHOSPHATE
7803-63-6	AMMONIUM MONOHYDROGEN SULFATE
12135-76-1	AMMONIUM MONOSULFIDE
10192-30-0	AMMONIUM MONOSULFITE
7803-55-6	AMMONIUM MONOVANADATE
12125-02-9	AMMONIUM MURIATE
2235-54-3	AMMONIUM N-DODECYL SULFATE
15699-18-0	AMMONIUM NICKEL SULFATE
15699-18-0	AMMONIUM NICKEL SULFATE ((NH4)2NI(SO4)2)
6484-52-2	AMMONIUM NITRATE
6484-52-2	AMMONIUM NITRATE (DOT)
6484-52-2	AMMONIUM NITRATE (NO ORGANIC COATING)
6484-52-2	AMMONIUM NITRATE (ORGANIC COATING)
6484-52-2	AMMONIUM NITRATE (SOLUTION)
57608-40-9	AMMONIUM NITRATE MIXED WITH AMMONIUM PHOSPHATE
57608-40-9	AMMONIUM NITRATE PHOSPHATE
6484-52-2	AMMONIUM NITRATE, SOLUTION (CONTAINING NOT LESS THAN

CAS No.	Chemical Name
	15% WATER) (DOT)
6484-52-2	AMMONIUM NITRATE, WITH MORE THAN 0.2% COMBUSTIBLE SUBSTANCES (DOT)
6484-52-2	AMMONIUM NITRATE, WITH NOT MORE THAN 0.2% COMBUSTIBLE SUBSTANCES (DOT)
57608-40-9	AMMONIUM NITRATE-PHOSPHATE (DOT)
544-60-5	AMMONIUM OLEATE
1113-38-8	AMMONIUM OXALATE
1113-38-8	AMMONIUM OXALATE ((NH4)2C2O4) MONOHYDRATE
6009-70-7	AMMONIUM OXALATE ((NH4)2C2O4) MONOHYDRATE
1113-38-8	AMMONIUM OXALATE (DOT)
1113-38-8	AMMONIUM OXALATE MONOHYDRATE
6009-70-7	AMMONIUM OXALATE MONOHYDRATE
5972-73-6	AMMONIUM OXALATE, NH4C2HO4, HYDRATE
13106-76-8	AMMONIUM PARAMOLYBDATE
12007-89-5	AMMONIUM PENTABORATE
12007-89-5	AMMONIUM PENTABORATE ((NH4)B5O8)
14639-98-6	AMMONIUM PENTACHLOROZINCATE
7790-98-9	AMMONIUM PERCHLORATE
7790-98-9	AMMONIUM PERCHLORATE (DOT)
7790-98-9	AMMONIUM PERCHLORATE (NH4ClO4)
7790-98-9	AMMONIUM PERCHLORATE, AVERAGE PARTICLE SIZE LESS THAN 45 MICRONS (DOT)
13446-10-1	AMMONIUM PERMANGANATE
13446-10-1	AMMONIUM PERMANGANATE (DOT)
7727-54-0	AMMONIUM PEROXIDODISULFATE
7727-54-0	AMMONIUM PEROXYDISULFATE
7727-54-0	AMMONIUM PEROXYDISULFATE ((NH4)2S2O8)
7727-54-0	AMMONIUM PEROXYSULFATE
7727-54-0	AMMONIUM PERSULFATE
7727-54-0	AMMONIUM PERSULFATE (ACGIH,DOT)
7783-28-0	AMMONIUM PHOSPHATE
7783-28-0	AMMONIUM PHOSPHATE ((NH4)2HPO4)
7783-28-0	AMMONIUM PHOSPHATE DIBASIC
57608-40-9	AMMONIUM PHOSPHATE, MIXED WITH AMMONIUM NITRATE
131-74-8	AMMONIUM PICRATE
131-74-8	AMMONIUM PICRATE (WET)
131-74-8	AMMONIUM PICRATE, DRY
131-74-8	AMMONIUM PICRATE, DRY OR CONTAINING, BY WEIGHT, LESS THAN 10% WATER (DOT)
131-74-8	AMMONIUM PICRATE, WET WITH 10% OR MORE WATER (DOT)
131-74-8	AMMONIUM PICRONITRATE
16919-58-7	AMMONIUM PLATINIC CHLORIDE
9080-17-5	AMMONIUM POLYSULFIDE
12259-92-6	AMMONIUM POLYSULFIDE
9080-17-5	AMMONIUM POLYSULFIDE SOLUTION
9080-17-5	AMMONIUM POLYSULFIDE SOLUTION (DOT)
1762-95-4	AMMONIUM RHODANATE
1762-95-4	AMMONIUM RHODANIDE
16919-19-0	AMMONIUM SILICOFLUORIDE
16919-19-0	AMMONIUM SILICOFLUORIDE (DOT)
16919-19-0	AMMONIUM SILICON FLUORIDE ((NH4)2SiF6)
1002-89-7	AMMONIUM STEARATE
7773-06-0	AMMONIUM SULFAMATE
7773-06-0	AMMONIUM SULFAMATE (ACGIH,DOT)
7783-20-2	AMMONIUM SULFATE
7783-20-2	AMMONIUM SULFATE ((NH4)2SO4)
7803-63-6	AMMONIUM SULFATE ((NH4)HSO4)
7783-20-2	AMMONIUM SULFATE (SOLUTION)
12135-76-1	AMMONIUM SULFIDE
9080-17-5	AMMONIUM SULFIDE ((NH4)2(SX))
9080-17-5	AMMONIUM SULFIDE (POLY-)
12135-76-1	AMMONIUM SULFIDE SOLUTION
12135-76-1	AMMONIUM SULFIDE SOLUTION (DOT)
9080-17-5	AMMONIUM SULFIDE SOLUTION, RED
10196-04-0	AMMONIUM SULFITE
10196-04-0	AMMONIUM SULFITE (DOT)
10192-30-0	AMMONIUM SULFITE (NH4HSO3)
1762-95-4	AMMONIUM SULFOCYANATE
1762-95-4	AMMONIUM SULFOCYANIDE
7773-06-0	AMMONIUM SULPHAMATE
7783-20-2	AMMONIUM SULPHATE
12135-76-1	AMMONIUM SULPHIDE, SOLUTION (DOT)
3164-29-2	AMMONIUM TARTRATE
3164-29-2	AMMONIUM TARTRATE (DOT)
13820-41-2	AMMONIUM TETRACHLOROPLATINATE
14639-97-5	AMMONIUM TETRACHLOROZINCATE
13826-83-0	AMMONIUM TETRAFLUOROBORATE
13826-83-0	AMMONIUM TETRAFLUOROBORATE (NH4BF4)
13826-83-0	AMMONIUM TETRAFLUOROBORATE(1-)
1762-95-4	AMMONIUM THIOCYANATE
1762-95-4	AMMONIUM THIOCYANATE (DOT)
7783-18-8	AMMONIUM THIOSULFATE
7783-18-8	AMMONIUM THIOSULFATE ((NH4)2S2O3)
14221-47-7	AMMONIUM TRIOXALATOFERRATE(III)
14221-47-7	AMMONIUM TRIS(OXALATO)FERRATE(III)
9080-17-5	AMMONIUM TRISULFIDE
7803-55-6	AMMONIUM VANADATE
7803-55-6	AMMONIUM VANADATE ((NH4)VO3)
7803-55-6	AMMONIUM VANADATE(V) ((NH4)VO3)
7803-55-6	AMMONIUM VANADIUM OXIDE (NH4VO3)
7803-55-6	AMMONIUM VANADIUM TRIOXIDE
52628-25-8	AMMONIUM ZINC CHLORIDE
14639-98-6	AMMONIUM ZINC CHLORIDE (NH4)3ZnCl5)
14639-97-5	AMMONIUM ZINC CHLORIDE (2NH4ClZnCl2)
6484-52-2	AMMONIUM(I) NITRATE (1:1)
999-81-5	AMMONIUM, (2-CHLOROETHYL)TRIMETHYL-, CHLORIDE
633-03-4	AMMONIUM, (4-(P-(DIETHYLAMINO)-ALPHA-PHENYLBENZYLIDENE)-2,5-CYCLOHEXADIEN-1-YLIDENE)-DIETHYL-, SULFA
56-93-9	AMMONIUM, BENZYLTRIMETHYL-, CHLORIDE
112-02-7	AMMONIUM, HEXADECYLTRIMETHYL-, CHLORIDE
75-59-2	AMMONIUM, TETRAMETHYL-, HYDROXIDE
7789-09-5	AMMONIUMBICHROMAAT (DUTCH)
506-87-6	AMMONIUMCARBONAT (GERMAN)
12125-02-9	AMMONIUMCHLORID (GERMAN)
7789-09-5	AMMONIUMDICHROMAAT (DUTCH)
7789-09-5	AMMONIUMDICHROMAT (GERMAN)
7773-06-0	AMMONIUMSALZ DER AMIDOSULFONSAURE (GERMAN)
7664-41-7	AMONIAK (POLISH)
12001-28-4	AMORPHOUS CROCIDOLITE ASBESTOS
7631-86-9	AMORPHOUS SILICA
1332-21-4	AMOSITE
12172-73-5	AMOSITE
12172-73-5	AMOSITE ASBESTOS
12172-73-5	AMOSITE, GRUNERITE
94-75-7	AMOXONE
124-68-5	AMP 95
124-68-5	AMP REGULAR
56-75-7	AMPHENICOL
300-62-9	AMPHETAMINE
56-75-7	AMPHICOL
69-53-4	AMPICILLIN
137-88-2	AMPROLIUM HYDROCHLORIDE
7773-06-0	AMS
7773-06-0	AMS (SALT)
8030-30-6	AMSCO H-J
8030-30-6	AMSCO H-SB
8020-83-5	AMSCO OMS
6842-15-5	AMSCO TETRAMER
56-75-7	AMSECLOR
1762-95-4	AMTHIO
53-86-1	AMUNO
628-63-7	AMYL ACETATE
628-63-7	AMYL ACETATE (DOT)
628-63-7	AMYL ACETIC ESTER
628-63-7	AMYL ACETIC ETHER
12789-46-7	AMYL ACID PHOSPHATE
71-41-0	AMYL ALCOHOL
71-41-0	AMYL ALCOHOL, NORMAL
110-62-3	AMYL ALDEHYDE
110-58-7	AMYL AMINE
628-63-7	AMYL AZETAT (GERMAN)
110-53-2	AMYL BROMIDE
540-18-1	AMYL BUTYRATE
543-59-9	AMYL CHLORIDE
543-59-9	AMYL CHLORIDE (DOT)
106-68-3	AMYL ETHYL KETONE
638-49-3	AMYL FORMATE
638-49-3	AMYL FORMATE (DOT)
109-66-0	AMYL HYDRIDE (DOT)
110-66-7	AMYL HYDROSULFIDE
5350-03-8	AMYL LAURATE
110-66-7	AMYL MERCAPTAN
110-66-7	AMYL MERCAPTAN (DOT)
105-30-6	AMYL METHYL ALCOHOL
108-11-2	AMYL METHYL ALCOHOL
54972-97-3	AMYL METHYL ALCOHOL
110-43-0	AMYL METHYL KETONE

CAS No.	Chemical Name
110-43-0	AMYL METHYL KETONE (DOT)
1002-16-0	AMYL NITRATE
1002-16-0	AMYL NITRATE (DOT)
110-46-3	AMYL NITRITE
463-04-7	AMYL NITRITE
463-04-7	AMYL NITRITE (DOT)
1322-06-1	AMYL PHENOL
110-66-7	AMYL SULFHYDRATE
872-10-6	AMYL SULFIDE
110-66-7	AMYL THIOALCOHOL
1320-01-0	AMYL TOLUENE
107-72-2	AMYL TRICHLOROSILANE
107-72-2	AMYL TRICHLOROSILANE (DOT)
1320-21-4	AMYL XYLYL ETHER
110-43-0	AMYL-METHYL-CETONE (FRENCH)
110-58-7	AMYLAMINE (DOT)
110-58-7	AMYLAMINE (MIXED ISOMERS)
538-68-1	AMYLBENZENE
111-27-3	AMYLCARBINOL
109-67-1	AMYLENE
513-35-9	AMYLENE
26760-64-5	AMYLENE
513-35-9	AMYLENE (DOT)
26760-64-5	AMYLENE (DOT)
75-85-4	AMYLENE HYDRATE
513-35-9	AMYLENE, NORMAL
109-67-1	AMYLENE, NORMAL (DOT)
513-35-9	AMYLENE, NORMAL (DOT)
1322-06-1	AMYLHYDROXYBENZENE
1320-27-0	AMYLNAPHTHALENE
71-41-0	AMYLOL
9005-25-8	AMYLOMAIZE VII
9005-25-8	AMYLOSE, MIXT. WITH AMYLOPECTIN
123-51-3	AMYLOWY ALKOHOL (POLISH)
9005-25-8	AMYLUM
9005-25-8	AMYSIL K
494-52-0	ANABASINE
7440-50-8	ANAC 110
434-07-1	ANADROL
60-29-7	ANAESTHETIC ETHER
443-48-1	ANAGIARDIL
119-36-8	ANALGIT
79-01-6	ANAMENTH
434-07-1	ANAPOLON
434-07-1	ANASTERON
434-07-1	ANASTERONAL
434-07-1	ANASTERONE
8001-35-2	ANATOX
7637-07-2	ANCA 1040
101-77-9	ANCAMINE TL
58-22-0	ANDROLIN
58-22-0	ANDRONAQ
58-22-0	ANDROST-4-EN-3-ONE, 17-HYDROXY-, (17 BETA)-
434-07-1	ANDROSTAN-3-ONE, 17-HYDROXY-2-(HYDROXYMETHYLENE)-17-METHYL-, (5 ALPHA,17 BETA)-
58-22-0	ANDRUSOL
514-73-8	ANELMID
109-87-5	ANESTHENYL
60-29-7	ANESTHESIA ETHER
60-29-7	ANESTHETIC ETHER
126-72-7	ANFRAM 3PB
55-63-0	ANGIBID
55-63-0	ANGININE
55-63-0	ANGIOLINGUAL
7446-14-2	ANGLISLITE
15739-80-7	ANGLISLITE
55-63-0	ANGORIN
514-73-8	ANGUIFUGAN
85-43-8	ANHYDRID KYSELINY TETRAHYDROFTALOVE (CZECH)
108-24-7	ANHYDRIDE ACETIQUE (FRENCH)
1327-53-3	ANHYDRIDE ARSENIEUX (FRENCH)
1303-28-2	ANHYDRIDE ARSENIQUE (FRENCH)
124-38-9	ANHYDRIDE CARBONIQUE (FRENCH)
8070-50-6	ANHYDRIDE CARBONIQUE ET OXYDE D'ETHYLENE MELANGES (FRENCH)
13530-68-2	ANHYDRIDE CHROMIQUE (FRENCH)
85-44-9	ANHYDRIDE PHTALIQUE (FRENCH)
1314-62-1	ANHYDRIDE VANADIQUE (FRENCH)
81-07-2	ANHYDRO-O-SULFAMINEBENZOIC ACID

CAS No.	Chemical Name
64-17-5	ANHYDROL
10034-81-8	ANHYDRONE
552-30-7	ANHYDROTRIMELLITIC ACID
7664-41-7	ANHYDROUS AMMONIA (DOT)
1330-43-4	ANHYDROUS BORAX
75-87-6	ANHYDROUS CHLORAL
302-01-2	ANHYDROUS HYDRAZINE (DOT)
10034-85-2	ANHYDROUS HYDRIODIC ACID
10035-10-6	ANHYDROUS HYDROBROMIC ACID
7647-01-0	ANHYDROUS HYDROCHLORIC ACID
7664-39-3	ANHYDROUS HYDROFLUORIC ACID
7664-39-3	ANHYDROUS HYDROFLUORIC ACID (DOT)
8006-54-0	ANHYDROUS LANOLIN
8006-54-0	ANHYDROUS LANUM
10034-81-8	ANHYDROUS MAGNESIUM PERCHLORATE
7558-79-4	ANHYDROUS SODIUM ACID PHOSPHATE
127-85-5	ANHYDROUS SODIUM ARSANILATE
7722-88-5	ANHYDROUS TETRASODIUM PYROPHOSPHATE
108-24-7	ANIDRIDE ACETICA (ITALIAN)
13530-68-2	ANIDRIDE CROMICA (ITALIAN)
85-44-9	ANIDRIDE FTALICA (ITALIAN)
101-05-3	ANILAZIN
12654-97-6	ANILAZIN
101-05-3	ANILAZINE
12654-97-6	ANILAZINE
62-53-3	ANILIN (CZECH)
62-53-3	ANILINA (ITALIAN, POLISH)
62-53-3	ANILINE
62-53-3	ANILINE (ACGIH,DOT)
142-04-1	ANILINE CHLORIDE
633-03-4	ANILINE GREEN
142-04-1	ANILINE HYDROCHLORIDE
142-04-1	ANILINE HYDROCHLORIDE (DOT)
62-53-3	ANILINE OIL
62-53-3	ANILINE OIL, LIQUID
62-53-3	ANILINE OIL, LIQUID (DOT)
88-05-1	ANILINE, 2,4,6-TRIMETHYL-
97-02-9	ANILINE, 2,4-DINITRO-
102-56-7	ANILINE, 2,5-DIMETHOXY-
4726-14-1	ANILINE, 2,6-DINITRO-N,N-DIPROPYL-4-(METHYLSULFONYL)-
578-54-1	ANILINE, 2-ETHYL-
95-53-4	ANILINE, 2-METHYL-
95-76-1	ANILINE, 3,4-DICHLORO-
108-44-1	ANILINE, 3-METHYL-
101-14-4	ANILINE, 4,4'-METHYLENEBIS[2-CHLORO-
101-61-1	ANILINE, 4,4'-METHYLENEBIS[N,N-DIMETHYL-
101-77-9	ANILINE, 4,4'-METHYLENEDI-
28434-86-8	ANILINE, 4,4'-OXYBIS[2-CHLORO-
101-80-4	ANILINE, 4,4'-OXYDI-
139-65-1	ANILINE, 4,4'-THIODI-
4726-14-1	ANILINE, 4-(METHYLSULFONYL)-2,6-DINITRO-N,N-DIPROPYL-
27134-26-5	ANILINE, 4-CHLORO-
371-40-4	ANILINE, 4-FLUORO-
100-01-6	ANILINE, 4-NITRO-
29757-24-2	ANILINE, 4-NITRO-
27134-26-5	ANILINE, CHLORO-
41587-36-4	ANILINE, CHLORONITRO-
27134-27-6	ANILINE, DICHLORO-
27134-27-6	ANILINE, DICHLORO- (MIXED ISOMERS)
26471-56-7	ANILINE, DINITRO-
26471-56-7	ANILINE, DINITRO- (MIXED ISOMERS)
613-29-6	ANILINE, N,N-DIBUTYL-
91-66-7	ANILINE, N,N-DIETHYL-
121-69-7	ANILINE, N,N-DIMETHYL-
138-89-6	ANILINE, N,N-DIMETHYL-P-NITROSO-
53324-05-3	ANILINE, N,N-DIMETHYL-P-NITROSO-
53324-05-3	ANILINE, N,N-DIMETHYLNITROSO-
1126-78-9	ANILINE, N-BUTYL-
103-69-5	ANILINE, N-ETHYL-
768-52-5	ANILINE, N-ISOPROPYL-
100-61-8	ANILINE, N-METHYL-
479-45-8	ANILINE, N-METHYL-N,2,4,6-TETRANITRO-
122-39-4	ANILINE, N-PHENYL-
29757-24-2	ANILINE, NITRO-
578-54-1	ANILINE, O-ETHYL-
578-54-1	ANILINE, O-ETHYL- (8CI)
88-74-4	ANILINE, O-NITRO-
27134-26-5	ANILINE, P-CHLORO-
1321-31-9	ANILINE, P-ETHOXY-

CAS No.	*Chemical Name*
371-40-4	ANILINE, P-FLUORO-
106-49-0	ANILINE, P-METHYL-
100-01-6	ANILINE, P-NITRO-
29757-24-2	ANILINE, P-NITRO-
122-39-4	ANILINOBENZENE
103-69-5	ANILINOETHANE
100-61-8	ANILINOMETHANE
1309-48-4	ANIMAG
29191-52-4	ANISIDINE
29191-52-4	ANISIDINE, ISOMERS
100-66-3	ANISOL
100-66-3	ANISOLE
100-66-3	ANISOLE (DOT)
100-17-4	ANISOLE, P-NITRO-
100-07-2	ANISOYL CHLORIDE
1300-64-7	ANISOYL CHLORIDE
100-07-2	ANISOYL CHLORIDE (DOT)
101-05-3	ANIYALINE
12654-97-6	ANIYALINE
127-18-4	ANKILOSTIN
137-26-8	ANLES
75-00-3	ANODYNON
108-93-0	ANOL
108-94-1	ANON
108-94-1	ANONE
68-22-4	ANOVULE
84-66-2	ANOZOL
50-55-5	ANQUIL
75-60-5	ANSAR
75-60-5	ANSAR 138
124-65-2	ANSAR 160
2163-80-6	ANSAR 170
2163-80-6	ANSAR 170 HC
2163-80-6	ANSAR 170L
2163-80-6	ANSAR 529
2163-80-6	ANSAR 529 HC
124-65-2	ANSAR 560
1309-48-4	ANSCOR P
95-79-4	ANSIBASE RED KB
97-77-8	ANTABUS
97-77-8	ANTABUSE
97-77-8	ANTADIX
97-77-8	ANTAETHYL
112-30-1	ANTAK
97-77-8	ANTALCOL
9016-45-9	ANTAROX CO
9016-45-9	ANTAROX CO 430
9016-45-9	ANTAROX CO 530
9016-45-9	ANTAROX CO 630
9016-45-9	ANTAROX CO 730
9016-45-9	ANTAROX CO 850
9016-45-9	ANTAROX CO 880
9016-45-9	ANTAROX CO 970
97-77-8	ANTETHAN
97-77-8	ANTETIL
2540-82-1	ANTHIO
2540-82-1	ANTHIO 25
7727-21-1	ANTHION
10049-04-4	ANTHIUM DIOXCIDE
52-68-6	ANTHON
1332-21-4	ANTHOPHYLLITE
120-12-7	ANTHRACEN (GERMAN)
120-12-7	ANTHRACENE
120-12-7	ANTHRACIN
119-36-8	ANTHRAPOLE ND
84-65-1	ANTHRAQUINONE
82-28-0	ANTHRAQUINONE, 1-AMINO-2-METHYL-
117-79-3	ANTHRAQUINONE, 2-AMINO-
129-15-7	ANTHRAQUINONE, 2-METHYL-1-NITRO-
7632-00-0	ANTI-RUST
56-04-2	ANTIBASON
2001-95-8	ANTIBIOTIC N-329 B
7681-49-4	ANTIBULIT
118-74-1	ANTICARIE
97-77-8	ANTICOL
62-38-4	ANTICON
97-77-8	ANTIETANOL
97-77-8	ANTIETIL
103-84-4	ANTIFEBRIN

CAS No.	*Chemical Name*
124-13-0	ANTIFOAM LF
7681-52-9	ANTIFORMIN
333-41-5	ANTIGAL
56-53-1	ANTIGESTIL
84-80-0	ANTIHEMORRHAGIC VITAMIN
100-97-0	ANTIHYDRAL
97-77-8	ANTIKOL
108-62-3	ANTIMILACE
55-86-7	ANTIMIT
7647-18-9	ANTIMOINE (PENTACHLORURE D') (FRENCH)
10025-91-9	ANTIMOINE (TRICHLORURE D') (FRENCH)
7783-56-4	ANTIMOINE FLUORURE (FRENCH)
28300-74-5	ANTIMONATE(1-), AQUA[TARTRATO(4-)]-, POTASSIUM, HEMIHYDRATE, DIMER
28300-74-5	ANTIMONATE(1-), OXO(TARTRATO)-, POTASSIUM HEMIHYDRATE, DIMER
28300-74-5	ANTIMONATE(2-), BIS(MU-TARTRATO(4-))DI-, DIPOTASSIUM, TRIHYDRATE
12627-52-0	ANTIMONIAL GLASS
1315-04-4	ANTIMONIAL SAFFRON
1315-04-4	ANTIMONIC SULFIDE
7647-18-9	ANTIMONIO (PENTACLORURO DI) (ITALIAN)
10025-91-9	ANTIMONIO (TRICLORURO DI) (ITALIAN)
1309-64-4	ANTIMONIOUS OXIDE
10025-91-9	ANTIMONOUS CHLORIDE
10025-91-9	ANTIMONOUS CHLORIDE (DOT)
7783-56-4	ANTIMONOUS FLUORIDE
12627-52-0	ANTIMONOUS SULFIDE
7647-18-9	ANTIMONPENTACHLORID (GERMAN)
10025-91-9	ANTIMONTRICHLORID (GERMAN)
7803-52-3	ANTIMONWASSERSTOFFES (GERMAN)
7440-36-0	ANTIMONY
7440-36-0	ANTIMONY (ACGIH)
10025-91-9	ANTIMONY (III) CHLORIDE
7647-18-9	ANTIMONY (V) CHLORIDE
7783-70-2	ANTIMONY (V) FLUORIDE
7440-36-0	ANTIMONY BLACK
7789-61-9	ANTIMONY BROMIDE
7789-61-9	ANTIMONY BROMIDE (SbBr3)
10025-91-9	ANTIMONY BUTTER
10025-91-9	ANTIMONY CHLORIDE
10025-91-9	ANTIMONY CHLORIDE (DOT)
10025-91-9	ANTIMONY CHLORIDE (Sb2Cl6)
10025-91-9	ANTIMONY CHLORIDE (SbCl3)
7647-18-9	ANTIMONY CHLORIDE (SbCl5)
7440-36-0	ANTIMONY ELEMENT
7783-70-2	ANTIMONY FLUORIDE
7783-56-4	ANTIMONY FLUORIDE (SbF3)
7783-70-2	ANTIMONY FLUORIDE (SbF5)
12627-52-0	ANTIMONY GLANCE
7803-52-3	ANTIMONY HYDRIDE
7803-52-3	ANTIMONY HYDRIDE (SbH3)
58164-88-8	ANTIMONY LACTATE
12627-52-0	ANTIMONY ORANGE
1309-64-4	ANTIMONY OXIDE
1309-64-4	ANTIMONY OXIDE (Sb2O3)
7647-18-9	ANTIMONY PENTACHLORIDE
7647-18-9	ANTIMONY PENTACHLORIDE (DOT)
7647-18-9	ANTIMONY PENTACHLORIDE (SbCl5)
7647-18-9	ANTIMONY PENTACHLORIDE SOLUTION
7647-18-9	ANTIMONY PENTACHLORIDE SOLUTION (DOT)
7647-18-9	ANTIMONY PENTACHLORIDE, LIQUID (DOT)
7783-70-2	ANTIMONY PENTAFLUORIDE
7783-70-2	ANTIMONY PENTAFLUORIDE (DOT)
1315-04-4	ANTIMONY PENTASULFIDE
7647-18-9	ANTIMONY PERCHLORIDE
1309-64-4	ANTIMONY PEROXIDE
28300-74-5	ANTIMONY POTASSIUM TARTRATE
28300-74-5	ANTIMONY POTASSIUM TARTRATE (DOT)
28300-74-5	ANTIMONY POTASSIUM TARTRATE SOLID (DOT)
28300-74-5	ANTIMONY POTASSIUM TARTRATE, SOLID
7440-36-0	ANTIMONY POWDER
7440-36-0	ANTIMONY POWDER (DOT)
1315-04-4	ANTIMONY RED
7440-36-0	ANTIMONY, REGULUS
1309-64-4	ANTIMONY SESQUIOXIDE
12627-52-0	ANTIMONY SULFIDE
1315-04-4	ANTIMONY SULFIDE (Sb2S5)
1315-04-4	ANTIMONY SULFIDE GOLDEN

CAS No.	Chemical Name
12627-52-0	ANTIMONY SULFIDE, SOLID
12627-52-0	ANTIMONY SULFIDE, SOLID (DOT)
7789-61-9	ANTIMONY TRIBROMIDE
7789-61-9	ANTIMONY TRIBROMIDE SOLUTION
7789-61-9	ANTIMONY TRIBROMIDE, SOLID
7789-61-9	ANTIMONY TRIBROMIDE, SOLID (DOT)
7789-61-9	ANTIMONY TRIBROMIDE, SOLUTION (DOT)
10025-91-9	ANTIMONY TRICHLORIDE
10025-91-9	ANTIMONY TRICHLORIDE SOLUTION
10025-91-9	ANTIMONY TRICHLORIDE SOLUTION (DOT)
10025-91-9	ANTIMONY TRICHLORIDE, LIQUID (DOT)
10025-91-9	ANTIMONY TRICHLORIDE, SOLID
10025-91-9	ANTIMONY TRICHLORIDE, SOLID (DOT)
7783-56-4	ANTIMONY TRIFLUORIDE
7783-56-4	ANTIMONY TRIFLUORIDE SOLUTION
7783-56-4	ANTIMONY TRIFLUORIDE, SOLID
7783-56-4	ANTIMONY TRIFLUORIDE, SOLID (DOT)
7803-52-3	ANTIMONY TRIHYDRIDE
1309-64-4	ANTIMONY TRIOXIDE
1309-64-4	ANTIMONY TRIOXIDE (ACGIH,DOT)
1309-64-4	ANTIMONY TRIOXIDE (Sb2O3)
12627-52-0	ANTIMONY TRISULFIDE
12627-52-0	ANTIMONY TRISULFIDE COLLOID
1309-64-4	ANTIMONY WHITE
1309-64-4	ANTIMONY(3+) OXIDE
10025-91-9	ANTIMONY(III) CHLORIDE
7783-56-4	ANTIMONY(III) FLUORIDE (1:3)
7647-18-9	ANTIMONY(V) CHLORIDE
7783-70-2	ANTIMONY(V) FLUORIDE
7783-70-2	ANTIMONY(V) PENTAFLUORIDE
28300-74-5	ANTIMONYL POTASSIUM TARTRATE
28300-74-5	ANTIMONYL POTASSIUM TARTRATE, HEMIHYDRATE
7647-18-9	ANTIMOONPENTACHLORIDE (DUTCH)
10025-91-9	ANTIMOONTRICHLRIDE (DUTCH)
62-38-4	ANTIMUCIN WBR
62-38-4	ANTIMUCIN WDR
1397-94-0	ANTIMYCIN
1397-94-0	ANTIMYCIN A
534-52-1	ANTINONIN
1335-85-9	ANTINONIN
534-52-1	ANTINONNIN
1335-85-9	ANTINONNIN
2540-82-1	ANTIO
135-88-6	ANTIOXIDANT 116
101-96-2	ANTIOXIDANT 22
128-37-0	ANTIOXIDANT 264
128-37-0	ANTIOXIDANT 29
128-37-0	ANTIOXIDANT 30
128-37-0	ANTIOXIDANT 4
3081-14-9	ANTIOXIDANT 4030
128-37-0	ANTIOXIDANT 4K
135-88-6	ANTIOXIDANT D
128-37-0	ANTIOXIDANT DBPC
128-37-0	ANTIOXIDANT KB
135-88-6	ANTIOXIDANT PBN
1397-94-0	ANTIPIRICULLIN
110-85-0	ANTIREN
127-18-4	ANTISAL 1
108-88-3	ANTISAL 1A
7664-39-3	ANTISAL 2B
7601-54-9	ANTISAL 4
79-09-4	ANTISCHIM B
127-18-4	ANTISOL 1
60-41-3	ANTIVAMPIRE
97-77-8	ANTIVITIUM
105-36-2	ANTOL
1309-64-4	ANTOX
86-88-4	ANTU
30553-04-9	ANTU
30553-04-9	ANTU (ACGIH)
86-88-4	ANTURAT
30553-04-9	ANTURAT
7440-36-0	ANTYMON (POLISH)
7803-52-3	ANTYMONOWODOR (POLISH)
999-81-5	ANTYWYLEGACZ
62-53-3	ANYVIM
128-37-0	AO 29
96-69-5	AO 4
128-37-0	AO 4
128-37-0	AO 4K
577-11-7	AOT
577-11-7	AOT I
128-37-0	AOX 4
128-37-0	AOX 4K
1309-64-4	AP 50
9080-17-5	AP-S
470-90-6	APACHLOR
6923-22-4	APADRIN
58-89-9	APARASIN
7631-86-9	APASIL
2636-26-2	APAVAP
7786-34-7	APAVINPHOS
108-45-2	APCO 2330
126-72-7	APEX 462-5
54-62-6	APGA
56-38-2	APHAMITE
545-55-1	APHOXIDE
58-89-9	APHTIRIA
8002-74-2	APIEZON M
8002-74-2	APIEZON N
8002-74-2	APIEZON W
58-89-9	APLIDAL
545-55-1	APO
1937-37-7	APOMINE BLACK GX
7568-93-6	APOPHEDRIN
50-55-5	APOPLON
732-11-6	APPA
106-48-9	APPLIED 3-78
1332-58-7	APSILEX
59-50-7	APTAL
1321-10-4	APTAL
111-90-0	APV
1344-28-1	AQ 10
1344-28-1	AQ 25
1344-28-1	AQ 50
1336-21-6	AQUA AMMONIA
1336-21-6	AQUA AMMONIA, SOLUTION (DOT)
7697-37-2	AQUA FORTIS
8007-56-5	AQUA REGIA
94-75-7	AQUA-KLEEN
7440-48-4	AQUACAT
85-00-7	AQUACIDE
53-16-7	AQUACRINE
7782-42-5	AQUADAG
50-28-2	AQUADIOL
7631-86-9	AQUAFIL
107-02-8	AQUALIN
107-02-8	AQUALINE
84-80-0	AQUAMEPHYTON
151-21-3	AQUAREX ME
151-21-3	AQUAREX METHYL
639-58-7	AQUATIN
144-62-7	AQUISAL
7429-90-5	AR 2
30030-25-2	AR-(CHLOROMETHYL)STYRENE
57-64-7	AR-44
25168-05-2	AR-CHLOROTOLUENE
104-15-4	AR-TOLUENESULFONIC ACID
25231-46-3	AR-TOLUENESULFONIC ACID
1319-77-3	AR-TOLUENOL
30030-25-2	AR-VINYLBENZYL CHLORIDE
81-81-2	ARAB RAT DETH
9002-89-5	ARACET APV
9002-89-5	ARACET APV 50/88
8002-03-7	ARACHIS OIL
101-77-9	ARALDITE HARDENER 972
112-24-3	ARALDITE HARDENER HY 951
85-44-9	ARALDITE HT 901
112-24-3	ARALDITE HY 951
56-38-2	ARALO
140-57-8	ARAMIT
140-57-8	ARAMITE
137-26-8	ARASAN
137-26-8	ARASAN 42S
137-26-8	ARASAN 50 RED
137-26-8	ARASAN 70
137-26-8	ARASAN 70-S RED
137-26-8	ARASAN 75

CAS No.	Chemical Name
137-26-8	ARASAN M
137-26-8	ARASAN-SF
137-26-8	ARASAN-SF-X
76-44-8	ARBINEX 30TN
58-89-9	ARBITEX
9004-34-6	ARBOCEL
9004-34-6	ARBOCEL BC 200
9004-34-6	ARBOCELL B 600/30
122-14-5	ARBOGAL
93-79-8	ARBORICID
534-52-1	ARBOROL
2425-06-1	ARBORSEAL
9016-45-9	ARCOPAL N 100
8002-74-2	ARCOWAX 1150G
8002-74-2	ARCOWAX 2143G
8002-74-2	ARCOWAX 4154G
8002-74-2	ARCOWAX 4158G
75-46-7	ARCTON
75-73-0	ARCTON 0
75-46-7	ARCTON 1
75-69-4	ARCTON 11
75-71-8	ARCTON 12
75-45-6	ARCTON 22
75-72-9	ARCTON 3
75-45-6	ARCTON 4
1330-45-6	ARCTON 50
75-71-8	ARCTON 6
76-13-1	ARCTON 63
75-43-4	ARCTON 7
75-69-4	ARCTON 9
123-31-9	ARCTUVIN
109-94-4	AREGINAL
88-85-7	ARETIT
7783-95-1	ARGENT FLUORURE (FRENCH)
506-61-6	ARGENTATE(1-), BIS(CYANO-C)-, POTASSIUM
506-61-6	ARGENTATE(1-), DICYANO-, POTASSIUM
7783-95-1	ARGENTIC FLUORIDE
20667-12-3	ARGENTOUS OXIDE
7440-22-4	ARGENTUM
1912-24-9	ARGEZIN
1332-58-7	ARGIFLEX
1332-58-7	ARGILLA
1332-58-7	ARGIREC KN 15
1332-58-7	ARGIREK B 22
7440-37-1	ARGON
7440-37-1	ARGON, ISOTOPE OF MASS 40
7440-37-1	ARGON, LIQUID PRESSURIZED (DOT)
7440-37-1	ARGON, REFRIGERATED LIQUID
7440-37-1	ARGON, REFRIGERATED LIQUID (DOT)
7440-37-1	ARGON-40
8006-54-0	ARGOWAX
63-25-2	ARILAT
63-25-2	ARILATE
108-62-3	ARIOTOX
9002-91-9	ARIOTOX
37273-91-9	ARIOTOX
8002-74-2	ARISTOWAX
8002-09-3	ARIZOLE
76-13-1	ARKLONE P
9016-45-9	ARKOPAL N
9016-45-9	ARKOPAL N 040
9016-45-9	ARKOPAL N 060
9016-45-9	ARKOPAL N 080
9016-45-9	ARKOPAL N 090
9016-45-9	ARKOPAL N 100
9016-45-9	ARKOPAL N 110
9016-45-9	ARKOPAL N 150
9016-45-9	ARKOPAL N 300
50-29-3	ARKOTINE
9002-84-0	ARMALON XT 2663
111-86-4	ARMEEN 8
111-86-4	ARMEEN 8D
7758-29-4	ARMOFOS
101-61-1	ARNOLD'S BASE
11104-28-2	AROCHLOR 1221
11097-69-1	AROCHLOR 1254
11096-82-5	AROCHLOR 1260
12767-79-2	AROCLOR
12674-11-2	AROCLOR 1016

CAS No.	Chemical Name
11104-28-2	AROCLOR 1221
11141-16-5	AROCLOR 1232
53469-21-9	AROCLOR 1242
12672-29-6	AROCLOR 1248
11097-69-1	AROCLOR 1254
11096-82-5	AROCLOR 1260
37324-23-5	AROCLOR 1262
11100-14-4	AROCLOR 1268
37324-24-6	AROCLOR 2565
11120-29-9	AROCLOR 4465
12642-23-8	AROCLOR 5442
7631-86-9	AROGEN 500
8020-83-5	AROMAX 3
114-26-1	ARPROCARB
112-02-7	ARQUAD 16-29
112-02-7	ARQUAD 16-50
1314-84-7	ARREX E
9005-25-8	ARROWROOT STARCH
127-85-5	ARSAMIN
75-60-5	ARSAN
127-85-5	ARSANILIC ACID SODIUM SALT
127-85-5	ARSANILIC ACID, MONOSODIUM SALT
124-65-2	ARSECODILE
7440-38-2	ARSEN (GERMAN,POLISH)
7778-39-4	ARSENATE
10102-49-5	ARSENATE OF IRON, FERRIC
10102-50-8	ARSENATE OF IRON, FERROUS
28838-01-9	ARSENENIC ACID, ZINC SALT
13464-44-3	ARSENENIC ACID, ZINC SALT
13464-33-0	ARSENENIC ACID, ZINC SALT
10031-13-7	ARSENENOUS ACID, LEAD(2+) SALT
10124-50-2	ARSENENOUS ACID, POTASSIUM SALT
7784-46-5	ARSENENOUS ACID, SODIUM SALT
7784-46-5	ARSENENOUS ACID, SODIUM SALT (9CI)
10326-24-6	ARSENENOUS ACID, ZINC SALT
10326-24-6	ARSENENOUS ACID, ZINC SALT (9CI)
7778-44-1	ARSENIATE DE CALCIUM (FRENCH)
10103-50-1	ARSENIATE DE MAGNESIUM (FRENCH)
7440-38-2	ARSENIC
7440-38-2	ARSENIC (ACGIH)
1327-53-3	ARSENIC (III) OXIDE
7778-39-4	ARSENIC (V) ACID
1303-28-2	ARSENIC (V) OXIDE
7778-39-4	ARSENIC ACID
7778-39-4	ARSENIC ACID (H3AsO4)
13464-44-3	ARSENIC ACID (H3AsO4) ZINC SALT
13464-33-0	ARSENIC ACID (H3AsO4) ZINC SALT
7778-44-1	ARSENIC ACID (H3AsO4), CALCIUM SALT (2:3)
7784-44-3	ARSENIC ACID (H3AsO4), DIAMMONIUM SALT
10102-50-8	ARSENIC ACID (H3AsO4), IRON(2+) SALT (2:3)
10102-49-5	ARSENIC ACID (H3AsO4), IRON(3+) SALT (1:1)
7645-25-2	ARSENIC ACID (H3AsO4), LEAD SALT
7784-40-9	ARSENIC ACID (H3AsO4), LEAD(2+) SALT (1:1)
10102-48-4	ARSENIC ACID (H3AsO4), LEAD(4+) SALT (3:2)
10103-50-1	ARSENIC ACID (H3AsO4), MAGNESIUM SALT
7784-37-4	ARSENIC ACID (H3AsO4), MERCURY(2+) SALT (1:1)
7784-41-0	ARSENIC ACID (H3AsO4), MONOPOTASSIUM SALT
7631-89-2	ARSENIC ACID (H3AsO4), SODIUM SALT
13464-33-0	ARSENIC ACID (H3AsO4), ZINC SALT (1:1)
28838-01-9	ARSENIC ACID (H3AsO4), ZINC SALT (2:3)
13464-44-3	ARSENIC ACID (H3AsO4), ZINC SALT (2:3)
28838-02-0	ARSENIC ACID (H4As2O7), ZINC SALT (1:2)
28838-01-9	ARSENIC ACID (HAsO3), ZINC SALT
1303-28-2	ARSENIC ACID ANHYDRIDE
7784-40-9	ARSENIC ACID LEAD SALT
10103-50-1	ARSENIC ACID MAGNESIUM SALT
7778-39-4	ARSENIC ACID SOLUTION
7778-39-4	ARSENIC ACID SOLUTION (DOT)
7778-39-4	ARSENIC ACID, (SOLUTION)
7778-44-1	ARSENIC ACID, CALCIUM SALT(2:3)
7784-44-3	ARSENIC ACID, DIAMMONIUM SALT
7778-39-4	ARSENIC ACID, LIQUID
7778-39-4	ARSENIC ACID, LIQUID (DOT)
7784-41-0	ARSENIC ACID, MONOPOTASSIUM SALT
7778-39-4	ARSENIC ACID, SOLID
7778-39-4	ARSENIC ACID, SOLID (DOT)
7631-89-2	ARSENIC ACID, TRISODIUM SALT
13464-33-0	ARSENIC ACID, ZINC SALT
13464-44-3	ARSENIC ACID, ZINC SALT

CAS No.	Chemical Name
28838-01-9	ARSENIC ACID, ZINC SALT
1303-28-2	ARSENIC ANHYDRIDE
7440-38-2	ARSENIC BLACK
1327-53-3	ARSENIC BLANC (FRENCH)
7784-33-0	ARSENIC BROMIDE
64973-06-4	ARSENIC BROMIDE
64973-06-4	ARSENIC BROMIDE (DOT)
64973-06-4	ARSENIC BROMIDE, SOLID
64973-06-4	ARSENIC BROMIDE, SOLID (DOT)
7784-34-1	ARSENIC BUTTER
7784-34-1	ARSENIC CHLORIDE
37226-49-6	ARSENIC CHLORIDE
7784-34-1	ARSENIC CHLORIDE (AsCl3)
7784-34-1	ARSENIC CHLORIDE (DOT)
7784-34-1	ARSENIC CHLORIDE, LIQUID (DOT)
56320-22-0	ARSENIC DISULFIDE
7440-38-2	ARSENIC ELEMENT
7784-42-1	ARSENIC HYDRID
7784-42-1	ARSENIC HYDRIDE
7784-42-1	ARSENIC HYDRIDE (AsH3)
7784-45-4	ARSENIC IODIDE
7784-45-4	ARSENIC IODIDE (AsI3)
7784-45-4	ARSENIC IODIDE, SOLID
7784-45-4	ARSENIC IODIDE, SOLID (DOT)
12044-79-0	ARSENIC MONOSULFIDE
1303-28-2	ARSENIC OXIDE
1327-53-3	ARSENIC OXIDE
1327-53-3	ARSENIC OXIDE (As2O3)
1303-28-2	ARSENIC OXIDE (As2O5)
1303-28-2	ARSENIC PENTAOXIDE
1303-28-2	ARSENIC PENTOXIDE
1303-28-2	ARSENIC PENTOXIDE (DOT)
1303-28-2	ARSENIC PENTOXIDE, SOLID
1303-28-2	ARSENIC PENTOXIDE, SOLID (DOT)
1327-53-3	ARSENIC SESQUIOXIDE
1303-33-9	ARSENIC SESQUISULFIDE
1303-33-9	ARSENIC SESQUISULFIDE (As2S3)
1303-33-9	ARSENIC SESQUISULPHIDE
1303-33-9	ARSENIC SULFIDE
12044-79-0	ARSENIC SULFIDE (As2S2)
1303-33-9	ARSENIC SULFIDE (As2S3)
12044-79-0	ARSENIC SULFIDE (As3S3)
12044-79-0	ARSENIC SULFIDE (AsS)
12044-79-0	ARSENIC SULFIDE RED
1303-33-9	ARSENIC SULFIDE YELLOW
1303-33-9	ARSENIC SULFIDE, SOLID (DOT)
1303-33-9	ARSENIC SULPHIDE
1303-33-9	ARSENIC TERSULPHIDE
64973-06-4	ARSENIC TRIBROMIDE
7784-34-1	ARSENIC TRICHLORIDE
60646-36-8	ARSENIC TRICHLORIDE (ARSENIC)
7784-34-1	ARSENIC TRICHLORIDE (ARSENOUS)
7784-34-1	ARSENIC TRICHLORIDE (DOT)
7784-34-1	ARSENIC TRICHLORIDE, LIQUID
7784-34-1	ARSENIC TRICHLORIDE, LIQUID (DOT)
7784-42-1	ARSENIC TRIHYDRIDE
7784-45-4	ARSENIC TRIIODIDE
8012-54-2	ARSENIC TRIIODIDE MIXED WITH MERCURIC IODIDE
1327-53-3	ARSENIC TRIOXIDE
1327-53-3	ARSENIC TRIOXIDE (ACGIH)
1327-53-3	ARSENIC TRIOXIDE, SOLID
1327-53-3	ARSENIC TRIOXIDE, SOLID (DOT)
1303-33-9	ARSENIC TRISULFIDE
1303-33-9	ARSENIC TRISULFIDE (As2S3)
1303-33-9	ARSENIC TRISULFIDE (DOT)
1303-33-9	ARSENIC YELLOW
64973-06-4	ARSENIC(II) BROMIDE
7784-34-1	ARSENIC(III) CHLORIDE
1303-33-9	ARSENIC(III) SULFIDE
7440-38-2	ARSENIC, METALLIC (DOT)
7440-38-2	ARSENIC, SOLID
7440-38-2	ARSENIC, SOLID (DOT)
1327-53-3	ARSENIC, WHITE, SOLID (DOT)
7440-38-2	ARSENIC-75
8028-73-7	ARSENICAL DUST
8028-73-7	ARSENICAL DUST (DOT)
8028-73-7	ARSENICAL FLUE DUST
7440-38-2	ARSENICALS
1327-53-3	ARSENICUM ALBUM

CAS No.	Chemical Name
1327-53-3	ARSENIGEN SAURE (GERMAN)
1327-53-3	ARSENIOUS ACID
10290-12-7	ARSENIOUS ACID (H3AsO3), COPPER(2+) SALT (1:1)
10124-50-2	ARSENIOUS ACID (H3AsO3), POTASSIUM SALT
7784-08-9	ARSENIOUS ACID (H3AsO3), TRISILVER(1+) SALT
13464-37-4	ARSENIOUS ACID (H3AsO3), TRISODIUM SALT
10031-13-7	ARSENIOUS ACID (HAsO2), LEAD(2+) SALT
10326-24-6	ARSENIOUS ACID (HAsO2), ZINC SALT
13464-37-4	ARSENIOUS ACID SODIUM SALT (NA3AsO3)
52740-16-6	ARSENIOUS ACID, CALCIUM SALT
10290-12-7	ARSENIOUS ACID, COPPER(II) SALT (1:1)
7784-46-5	ARSENIOUS ACID, MONOSODIUM SALT
10124-50-2	ARSENIOUS ACID, POTASSIUM SALT
7784-46-5	ARSENIOUS ACID, SODIUM SALT
1327-53-3	ARSENIOUS ACID, SOLID (DOT)
91724-16-2	ARSENIOUS ACID, STRONTIUM SALT
7784-08-9	ARSENIOUS ACID, TRISILVER(1+) SALT
10326-24-6	ARSENIOUS ACID, ZINC SALT
8012-54-2	ARSENIOUS AND MERCURIC IODIDE SOLUTION
8012-54-2	ARSENIOUS AND MERCURIC IODIDE SOLUTION (DOT)
7784-34-1	ARSENIOUS CHLORIDE
1327-53-3	ARSENIOUS OXIDE
1303-33-9	ARSENIOUS SULFIDE
1303-33-9	ARSENIOUS SULPHIDE
1327-53-3	ARSENIOUS TRIOXIDE
1327-53-3	ARSENITE
10124-50-2	ARSENITE DE POTASSIUM (FRENCH)
7784-46-5	ARSENITE DE SODIUM (FRENCH)
7784-42-1	ARSENIURETTED HYDROGEN
1327-53-3	ARSENOLITE
1327-53-3	ARSENOUS ACID
1327-53-3	ARSENOUS ACID ANHYDRIDE
52740-16-6	ARSENOUS ACID, CALCIUM SALT
10124-50-2	ARSENOUS ACID, POTASSIUM SALT
7784-08-9	ARSENOUS ACID, TRISILVER(1+) SALT
7784-08-9	ARSENOUS ACID, TRISILVER(1+) SALT (9CI)
13464-37-4	ARSENOUS ACID, TRISODIUM SALT
1327-53-3	ARSENOUS ANHYDRIDE
64973-06-4	ARSENOUS BROMIDE
7784-34-1	ARSENOUS CHLORIDE
7784-42-1	ARSENOUS HYDRIDE
7784-45-4	ARSENOUS IODIDE
1327-53-3	ARSENOUS OXIDE
1327-53-3	ARSENOUS OXIDE ANHYDRIDE
1303-33-9	ARSENOUS SULFIDE
64973-06-4	ARSENOUS TRIBROMIDE
7784-34-1	ARSENOUS TRICHLORIDE
7784-34-1	ARSENOUS TRICHLORIDE (9CI)
7784-45-4	ARSENOUS TRIIODIDE
7784-45-4	ARSENOUS TRIIODIDE (9CI)
7784-42-1	ARSENOWODOR (POLISH)
1327-53-3	ARSENTRIOXIDE
7784-42-1	ARSENWASSERSTOFF (GERMAN)
124-65-2	ARSICODILE
7631-86-9	ARSIL
7784-42-1	ARSINE
7784-42-1	ARSINE (ACGIH,DOT)
75-60-5	ARSINE OXIDE, DIMETHYLHYDROXY-
124-65-2	ARSINE OXIDE, DIMETHYLHYDROXY-, SODIUM SALT
75-60-5	ARSINE OXIDE, HYDROXYDIMETHYL-
124-65-2	ARSINE OXIDE, HYDROXYDIMETHYL-, SODIUM SALT
541-25-3	ARSINE, (2-CHLOROVINYL)DICHLORO-
712-48-1	ARSINE, CHLORODIPHENYL-
541-25-3	ARSINE, DICHLORO(2-CHLOROVINYL)-
696-28-6	ARSINE, DICHLOROPHENYL-
75-60-5	ARSINIC ACID, DIMETHYL-
75-60-5	ARSINIC ACID, DIMETHYL- (9CI)
124-65-2	ARSINIC ACID, DIMETHYL-, SODIUM SALT
124-65-2	ARSINIC ACID, DIMETHYL-, SODIUM SALT (9CI)
12044-79-0	ARSINO, THIOXO-
127-85-5	ARSINOSOLVIN
712-48-1	ARSINOUS CHLORIDE, DIPHENYL-
712-48-1	ARSINOUS CHLORIDE, DIPHENYL- (9CI)
1327-53-3	ARSODENT
2163-80-6	ARSONATE LIQUID
127-85-5	ARSONIC ACID, (4-AMINOPHENYL)-, MONOSODIUM SALT
127-85-5	ARSONIC ACID, (4-AMINOPHENYL)-, MONOSODIUM SALT (9CI)
52740-16-6	ARSONIC ACID, CALCIUM SALT (1:1)
10290-12-7	ARSONIC ACID, COPPER(2+) SALT (1:1)

CAS No.	Chemical Name
10290-12-7	ARSONIC ACID, COPPER(2+) SALT (1:1) (9CI)
2163-80-6	ARSONIC ACID, METHYL-, MONOSODIUM SALT
98-05-5	ARSONIC ACID, PHENYL-
10124-50-2	ARSONIC ACID, POTASSIUM SALT
541-25-3	ARSONOUS DICHLORIDE, (2-CHLOROETHENYL)-
541-25-3	ARSONOUS DICHLORIDE, (2-CHLOROETHENYL)- (9CI)
598-14-1	ARSONOUS DICHLORIDE, ETHYL- (9CI)
593-89-5	ARSONOUS DICHLORIDE, METHYL- (9CI)
696-28-6	ARSONOUS DICHLORIDE, PHENYL-
696-28-6	ARSONOUS DICHLORIDE, PHENYL- (9CI)
124-65-2	ARSYCODILE
74-87-3	ARTIC
100-52-7	ARTIFICIAL ALMOND OIL
98-01-1	ARTIFICIAL ANT OIL
100-52-7	ARTIFICIAL ESSENTIAL OIL OF ALMOND
57-06-7	ARTIFICIAL MUSTARD OIL
98-01-1	ARTIFICIAL OIL OF ANTS
57-06-7	ARTIFICIAL OIL OF MUSTARD
82-28-0	ARTISIL ORANGE 3RP
8020-83-5	ARTOL 10
53-86-1	ARTRACIN
53-86-1	ARTRINOVO
53-86-1	ARTRIVIA
51-21-8	ARUMEL
7440-50-8	ARWOOD COPPER
63-25-2	ARYLAM
26264-05-1	ARYLAN PWS
25155-30-0	ARYLAN SBC
9004-70-0	AS
7782-42-5	AS 1
3570-75-0	AS 17665
8020-83-5	AS 6
57-14-7	AS-DIMETHYLHYDRAZINE
98-83-9	AS-METHYLPHENYLETHYLENE
21087-64-9	AS-TRIAZIN-5(4H)-ONE, 4-AMINO-6-TERT-BUTYL-3-(METHYLTHIO)-
95-63-6	AS-TRIMETHYLBENZENE
50-78-2	ASA
50-78-2	ASAGRAN
1333-86-4	ASAHITHERMAL
7440-66-6	ASARCO L 15
2163-80-6	ASAZOL
14807-96-6	ASBESTINE
1332-21-4	ASBESTOS
12001-29-5	ASBESTOS (ACGIH)
12001-28-4	ASBESTOS (ACGIH)
1332-21-4	ASBESTOS (FRIABLE)
1332-21-4	ASBESTOS DUST
12172-73-5	ASBESTOS, AMOSITE
1332-21-4	ASBESTOS, AMPHIBOLE
77536-67-5	ASBESTOS, ANTHOPHYLLITE
12001-28-4	ASBESTOS, BLUE
12001-29-5	ASBESTOS, CHRYSOTILE
12001-28-4	ASBESTOS, CROCIDOLITE
12172-73-5	ASBESTOS, FERROGEDRITE
1332-21-4	ASBESTOS, FIBERS
12172-73-5	ASBESTOS, GRUNERITE
12001-29-5	ASBESTOS, WHITE
12001-29-5	ASBESTOS, WHITE (DOT)
1310-73-2	ASCARITE
7775-09-9	ASEX
7681-11-0	ASMOFUG E
1332-58-7	ASP 170
3689-24-5	ASP 47
9015-68-3	ASPARAGINASE
8052-42-4	ASPHALT
8052-42-4	ASPHALT CUTBACK
8052-42-4	ASPHALT FUMES
8052-42-4	ASPHALT(LIQUID RAPID-CURING)
8052-42-4	ASPHALTUM
50-78-2	ASPIRIN
12789-03-6	ASPON-CHLORDANE
50-78-2	ASPRO
50-18-0	ASTA B 518
50-78-2	ASTERIC
71-63-6	ASTHENTHILO
7440-32-6	ASTM B348 GR 2
633-03-4	ASTRA DIAMOND GREEN GX
633-03-4	ASTRAZON GREEN D
630-60-4	ASTROBAIN
2636-26-2	ASTROBOT
7778-74-7	ASTRUMAL
7704-34-9	ASULFA-SUPRA
56-72-4	ASUNTHOL
56-72-4	ASUNTOL
57-14-7	ASYMMETRIC DIMETHYLHYDRAZINE
95-63-6	ASYMMETRICAL TRIMETHYLBENZENE
61-82-5	AT
7782-42-5	AT 20
70-30-4	AT 7
1332-21-4	AT 7-1
61-82-5	ATA
9002-89-5	ATACTIC POLY(VINYL ALCOHOL)
1912-24-9	ATAZINAX
1918-02-1	ATCP
1333-86-4	ATG 60
1333-86-4	ATG 70
62-73-7	ATGARD
62-73-7	ATGARD V
129-06-6	ATHROMBIN
81-81-2	ATHROMBIN-K
81-81-2	ATHROMBINE-K
2642-71-9	ATHYL-GUSATHION
74-85-1	ATHYLEN (GERMAN)
7440-32-6	ATI 24
137-26-8	ATIRAM
7782-42-5	ATJ-S
7782-42-5	ATJ-S GRAPHITE
7775-09-9	ATLACIDE
1694-09-3	ATLANTIC ACID VIOLET 4BNS
1937-37-7	ATLANTIC BLACK BD
1937-37-7	ATLANTIC BLACK C
1937-37-7	ATLANTIC BLACK E
1937-37-7	ATLANTIC BLACK EA
1937-37-7	ATLANTIC BLACK GAC
1937-37-7	ATLANTIC BLACK GG
1937-37-7	ATLANTIC BLACK GXCW
1937-37-7	ATLANTIC BLACK GXOO
1937-37-7	ATLANTIC BLACK SD
2602-46-2	ATLANTIC BLUE 2B
16071-86-6	ATLANTIC FAST BROWN BRL
16071-86-6	ATLANTIC RESIN FAST BROWN BRL
7784-46-5	ATLAS 'A'
26264-05-1	ATLAS G 3300
26264-05-1	ATLAS G 711
105-60-2	ATM 2(NYLON)
7440-70-2	ATOMIC CALCIUM
7440-59-7	ATOMIC HELIUM
7704-34-9	ATOMIC SULFUR
1309-64-4	ATOX S
63-25-2	ATOXAN
127-85-5	ATOXYL
1912-24-9	ATRANEX
1912-24-9	ATRASINE
1912-24-9	ATRATAF
7775-09-9	ATRATOL
1912-24-9	ATRATOL A
1912-24-9	ATRAZIN
1912-24-9	ATRAZINE
1912-24-9	ATRAZINE (ACGIH)
1912-24-9	ATRED
1912-24-9	ATREX
8001-35-2	ATTAC 4-2
8001-35-2	ATTAC 4-4
8001-35-2	ATTAC 6
8001-35-2	ATTAC 6-3
8001-35-2	ATTAC 8
1937-37-7	ATUL DIRECT BLACK E
2602-46-2	ATUL DIRECT BLUE 2B
1912-24-9	ATZ
9000-07-1	AUBYGUM X 2
137-26-8	AULES
2465-27-2	AURAMIN
492-80-8	AURAMINE
2465-27-2	AURAMINE
2465-27-2	AURAMINE 0-100
2465-27-2	AURAMINE A1
2465-27-2	AURAMINE CHLORIDE
2465-27-2	AURAMINE EXTRA

CAS No.	Chemical Name
2465-27-2	AURAMINE EXTRA CONC. A
2465-27-2	AURAMINE FA
2465-27-2	AURAMINE FWA
2465-27-2	AURAMINE HYDROCHLORIDE
2465-27-2	AURAMINE II
2465-27-2	AURAMINE LAKE YELLOW O
2465-27-2	AURAMINE N
2465-27-2	AURAMINE O
2465-27-2	AURAMINE O EXTRA CONC. A EXPORT
2465-27-2	AURAMINE ON
2465-27-2	AURAMINE OO
2465-27-2	AURAMINE OOO
2465-27-2	AURAMINE OS
2465-27-2	AURAMINE PURE
2465-27-2	AURAMINE SP
2465-27-2	AURAMINE YELLOW
2465-27-2	AURAMINE (TECHNICAL GRADE)
1303-33-9	AURIPIGMENT
1333-86-4	AUSTIN BLACK
13463-67-7	AUSTIOX R-CR 3
56-75-7	AUSTRACOL
13717-00-5	AUSTRALIAN MAGNESITE
18454-12-1	AUSTRIAN CINNABAR
7428-48-0	AUSTROSTAB 110E
115-77-5	AUXINUTRIL
9016-45-9	AUXIPON NP
7429-90-5	AV 000
7429-90-5	AV00
2303-16-4	AVADEX
67-63-0	AVANTIN
67-63-0	AVANTINE
9002-84-0	AVCOAT 8029-1
9002-84-0	AVCOAT 8029-2
9002-84-0	AVCOAT 8029-3
14808-60-7	AVENTURINE
14808-60-7	AVENTURINE (QUARTZ)
97-77-8	AVERSAN
97-77-8	AVERZAN
12001-29-5	AVIBEST C
9004-34-6	AVICEL
9004-34-6	AVICEL 101
9004-34-6	AVICEL 102
9004-34-6	AVICEL PH 101
9004-34-6	AVICEL PH 105
82-68-8	AVICOL (PESTICIDE)
8020-83-5	AVIOL
151-21-3	AVIROL 101
151-21-3	AVIROL 118 CONC
2235-54-3	AVIROL 200
504-24-5	AVITROL
67-72-1	AVLOTHANE
131-11-3	AVOLIN
633-03-4	AVON GREEN A 4379
8020-83-5	AVTOL 10
8020-83-5	AW 409
8020-83-5	AWK 1
8001-58-9	AWPA #1
1333-86-4	AX 3023
1954-28-5	AYERST 62013
1405-87-4	AYFIVIN
8052-42-4	AZ-IP 90
110-86-1	AZABENZENE
111-49-9	AZACYCLOHEPTANE
110-89-4	AZACYCLOHEXANE
123-75-1	AZACYCLOPENTANE
151-56-4	AZACYCLOPROPANE
3165-93-3	AZANIL RED SALT TRD
61-82-5	AZAPLANT
115-02-6	AZASERIN
115-02-6	AZASERINE
446-86-6	AZATHIOPRIN
446-86-6	AZATHIOPRINE
78-67-1	AZDH
78-67-1	AZDN
1941-79-3	AZELAIC DIPERACID
26628-22-8	AZIDE
633-03-4	AZIEN MALACHITE GREEN GH
64-19-7	AZIJNZUUR (DUTCH)
108-24-7	AZIJNZUURANHYDRIDE (DUTCH)
334-88-3	AZIMETHYLENE
110-86-1	AZINE
1937-37-7	AZINE DEEP BLACK EW
2642-71-9	AZINFOS-ETHYL (DUTCH)
86-50-0	AZINFOS-METHYL (DUTCH)
86-50-0	AZINOPHOS-METHYL
2642-71-9	AZINOS
86-50-0	AZINPHOS
2642-71-9	AZINPHOS ETHYL
86-50-0	AZINPHOS METHYL, LIQUID
86-50-0	AZINPHOS METHYL, LIQUID (DOT)
2642-71-9	AZINPHOS-AETHYL (GERMAN)
2642-71-9	AZINPHOS-ETILE (ITALIAN)
86-50-0	AZINPHOS-METHYL (ACGIH,DOT)
86-50-0	AZINPHOS-METILE (ITALIAN)
151-56-4	AZIRAN
151-56-4	AZIRANE
151-56-4	AZIRIDIN (GERMAN)
151-56-4	AZIRIDINE
52-24-4	AZIRIDINE, 1,1',1"-PHOSPHINOTHIOYLIDYNETRIS-
545-55-1	AZIRIDINE, 1,1',1"-PHOSPHINYLIDYNETRIS-
75-55-8	AZIRIDINE, 2-METHYL-
26628-22-8	AZIUM
1314-13-2	AZO 22
89-62-3	AZOAMINE RED A
100-01-6	AZOAMINE RED ZH
29757-24-2	AZOAMINE RED ZH
99-59-2	AZOAMINE SCARLET K
89-62-3	AZOBASE NAT
103-33-3	AZOBENZENE
78-67-1	AZOBISISOBUTYLONITRILE
78-67-1	AZOBISISOBUTYRONITRILE
1937-37-7	AZOCARD BLACK EW
2602-46-2	AZOCARD BLUE 2B
25551-14-8	AZODI-(1,1'-HEXAHYDROBENZONITRILE)
78-67-1	AZODI-ISOBUTYRONITRILE
78-67-1	AZODIISOBUTYRONITRILE (DOT)
1314-13-2	AZODOX
6923-22-4	AZODRIN
6923-22-4	AZODRIN INSECTICIDE
136-40-3	AZODYNE
88-74-4	AZOENE FAST ORANGE GR BASE
88-74-4	AZOENE FAST ORANGE GR SALT
95-79-4	AZOENE FAST RED KB BASE
89-62-3	AZOENE FAST RED RED GL SALT
95-79-4	AZOENE FAST RED TR BASE
3165-93-3	AZOENE FAST RED TR SALT
99-55-8	AZOENE FAST SCARLET GC BASE
99-55-8	AZOENE FAST SCARLET GC SALT
88-74-4	AZOFIX ORANGE GR SALT
100-01-6	AZOFIX RED GG SALT
29757-24-2	AZOFIX RED GG SALT
89-62-3	AZOFIX RED GL SALT
99-55-8	AZOFIX SCARLET G SALT
298-00-0	AZOFOS
119-90-4	AZOGENE FAST BLUE B
88-74-4	AZOGENE FAST ORANGE GR
95-79-4	AZOGENE FAST RED TR
3165-93-3	AZOGENE FAST RED TR
99-55-8	AZOGENE FAST SCARLET G
3165-93-3	AZOIC DIAZO COMPONENT 11 BASE
95-79-4	AZOIC DIAZO COMPONENT 11, BASE
100-01-6	AZOIC DIAZO COMPONENT 37
29757-24-2	AZOIC DIAZO COMPONENT 37
88-74-4	AZOIC DIAZO COMPONENT 6
89-62-3	AZOIC DIAZO COMPONENT 8
123-30-8	AZOL
27598-85-2	AZOL
109-97-7	AZOLE
123-75-1	AZOLIDINE
1937-37-7	AZOMINE BLACK EWO
2602-46-2	AZOMINE BLUE 2B
298-00-0	AZOPHOS
10102-44-0	AZOTE (FRENCH)
446-86-6	AZOTHIOPRINE
7697-37-2	AZOTIC ACID
10102-44-0	AZOTO (ITALIAN)
7697-37-2	AZOTOWY KWAS (POLISH)
50-29-3	AZOTOX M-33

CAS No.	Chemical Name
26628-22-8	AZOTURE DE SODIUM (FRENCH)
94-36-0	AZTEC BPO
56-72-4	AZUNTHOL
5593-70-4	B 1
5593-70-4	B 1 (TITANATE)
115-90-2	B 25141
70-30-4	B 32
9016-45-9	B 350
2032-65-7	B 37344
91-22-5	B 500
50-18-0	B 518
22224-92-6	B 68138
9004-66-4	B 75
3878-19-1	B-33172
101-05-3	B-622
12654-97-6	B-622
7775-09-9	B-HERBATOX
57-13-6	B-I-K
7681-52-9	B-K LIQUID
7778-54-3	B-K POWDER
94-75-7	B-SELEKTONON
789-61-7	B-TGDR
6834-92-0	B-W
92-93-3	BA 2794
61-57-4	BA 32644
108-89-4	BA 35846
9004-70-0	BA 85
91-49-6	BAA
1405-87-4	BACI-JEL
1405-87-4	BACIGUENT
1405-87-4	BACILIQU IN
1319-77-3	BACILLOL
1405-87-4	BACITEK OINTMENT
1405-87-4	BACITRACIN
140-04-5	BAKERS P 6
59-50-7	BAKTOL
1321-10-4	BAKTOL
59-50-7	BAKTOLAN
1321-10-4	BAKTOLAN
123-92-2	BANANA OIL
50-55-5	BANASIL
1918-00-9	BANEX
133-06-2	BANGTON
1918-00-9	BANLEN
1642-54-2	BANOCIDE
86-88-4	BANTU
30553-04-9	BANTU
1918-00-9	BANVEL
1918-00-9	BANVEL 4S
1918-00-9	BANVEL 4WS
1918-00-9	BANVEL CST
1918-00-9	BANVEL D
1918-00-9	BANVEL HERBICIDE
1918-00-9	BANVEL II HERBICIDE
12788-93-1	BAP
12788-93-1	BAP (ESTER)
50-28-2	BARDIOL
8002-74-2	BARECO 170/175
1304-29-6	BARIO (PEROSSIDO DI) (ITALIAN)
7440-39-3	BARIUM
7440-39-3	BARIUM (ACGIH)
7440-39-3	BARIUM METAL, NON-PYROPHORIC
7440-39-3	BARIUM ALLOY
7440-39-3	BARIUM ALLOY, PYROPHORIC
18810-58-7	BARIUM AZIDE
18810-58-7	BARIUM AZIDE (Ba(N3)2)
18810-58-7	BARIUM AZIDE, (WET)
18810-58-7	BARIUM AZIDE, DRY OR CONTAINING, BY WEIGHT, LESS THAN 50% WATER
18810-58-7	BARIUM AZIDE, WET, 50% OR MORE WATER (DOT)
1304-29-6	BARIUM BINOXIDE
13967-90-3	BARIUM BROMATE
13967-90-3	BARIUM BROMATE (DOT)
513-77-9	BARIUM CARBONATE
513-77-9	BARIUM CARBONATE (1:1)
513-77-9	BARIUM CARBONATE (BaCO3)
13477-00-4	BARIUM CHLORATE
13477-00-4	BARIUM CHLORATE (Ba(ClO3)2)
13477-00-4	BARIUM CHLORATE (DOT)

CAS No.	Chemical Name
542-62-1	BARIUM CYANIDE
542-62-1	BARIUM CYANIDE (Ba(CN)2)
542-62-1	BARIUM DICYANIDE
10022-31-8	BARIUM DINITRATE
1304-29-6	BARIUM DIOXIDE
13465-95-7	BARIUM DIPERCHLORATE
7440-39-3	BARIUM ELEMENT
22326-55-2	BARIUM HYDROXIDE (Ba(OH)2), MONOHYDRATE
22326-55-2	BARIUM HYDROXIDE MONOHYDRATE
13477-10-6	BARIUM HYPOCHLORITE
13477-10-6	BARIUM HYPOCHLORITE, CONTAINING MORE THAN 22% AVAILABLE CHLORINE (DOT)
513-77-9	BARIUM MONOCARBONATE
1304-28-5	BARIUM MONOOXIDE
1304-28-5	BARIUM MONOXIDE
10022-31-8	BARIUM NITRATE
10022-31-8	BARIUM NITRATE (Ba(NO3)2)
10022-31-8	BARIUM NITRATE (DOT)
1304-28-5	BARIUM OXIDE
1304-28-5	BARIUM OXIDE (BaO)
1304-29-6	BARIUM OXIDE (BaO2)
1304-28-5	BARIUM OXIDE (DOT)
13465-95-7	BARIUM PERCHLORATE
13465-95-7	BARIUM PERCHLORATE (DOT)
7787-36-2	BARIUM PERMANGANATE
7787-36-2	BARIUM PERMANGANATE (DOT)
1304-29-6	BARIUM PEROXIDE
1304-29-6	BARIUM PEROXIDE (Ba(O2))
1304-29-6	BARIUM PEROXIDE (DOT)
1304-28-5	BARIUM PROTOXIDE
7787-41-9	BARIUM SELENATE
13718-59-7	BARIUM SELENITE
1304-29-6	BARIUM SUPEROXIDE
10022-31-8	BARIUM(II) NITRATE (1:2)
7440-39-3	BARIUM, ALLOYS, NON-PYROPHORIC (DOT)
7440-39-3	BARIUM, ALLOYS, PYROPHORIC (DOT)
7440-39-3	BARIUM, METAL, NON-PYROPHORIC (DOT)
1304-29-6	BARIUMPEROXID (GERMAN)
1304-29-6	BARIUMPEROXYDE (DUTCH)
999-81-5	BARLEYQUAT B
57-11-4	BAROLUB FTA
112-02-7	BARQUAT CT 29
140-04-5	BARYL
1304-28-5	BARYTA
88-85-7	BASANITE
56-04-2	BASECIL
110-91-8	BASF 238
106-50-3	BASF URSOL D
123-30-8	BASF URSOL P BASE
27598-85-2	BASF URSOL P BASE
75-99-0	BASFAPON
75-99-0	BASFAPON B
75-99-0	BASFAPON/BASFAPON N
633-03-4	BASIC BRIGHT GREEN
633-03-4	BASIC BRIGHT GREEN SULFATE
633-03-4	BASIC BRILLIANT GREEN
18454-12-1	BASIC CHROMIUM LEAD OXIDE (CRPB2O5)
633-03-4	BASIC GREEN 1
633-03-4	BASIC GREEN V
11119-70-3	BASIC LEAD CHROMATE
1312-03-4	BASIC MERCURIC SULFATE
2465-27-2	BASIC YELLOW 2
7699-43-6	BASIC ZIRCONIUM CHLORIDE
75-99-0	BASINEX
75-99-0	BASINEX P
12002-03-8	BASLE GREEN
333-41-5	BASSADINON
333-41-5	BASUDIN
333-41-5	BASUDIN 10G
333-41-5	BASUDIN 5G
82-68-8	BATRILEX
1309-37-1	BAUXITE RESIDUE
8065-48-3	BAY 10756
298-00-0	BAY 11405
52-68-6	BAY 15922
2642-71-9	BAY 16255
86-50-0	BAY 17147
919-86-8	BAY 18436
298-04-4	BAY 19639

CAS No.	Chemical Name
56-72-4	BAY 21/199
640-15-3	BAY 23129
2497-07-6	BAY 23323
115-90-2	BAY 25141
5836-29-3	BAY 25634
55-38-9	BAY 29493
68-76-8	BAY 3231
3878-19-1	BAY 33172
2636-26-2	BAY 34727
327-98-0	BAY 37289
2032-65-7	BAY 37344
4104-14-7	BAY 38819
114-26-1	BAY 39007
122-14-5	BAY 41831
2032-65-7	BAY 5024
114-26-1	BAY 5122
1420-04-8	BAY 6076
21087-64-9	BAY 6159
21087-64-9	BAY 61597
21087-64-9	BAY 6159H
22224-92-6	BAY 68138
1563-66-2	BAY 70143
10265-92-6	BAY 71628
1420-04-8	BAY 73
2032-65-7	BAY 9026
86-50-0	BAY 9027
21087-64-9	BAY 94337
298-00-0	BAY E-601
5836-29-3	BAY ENE 11183 B
122-14-5	BAY S 5660
2636-26-2	BAY-19149
3689-24-5	BAY-E-393
35400-43-2	BAY-NTN 9306
10101-53-8	BAYCHROM A
10101-53-8	BAYCHROM F
55-38-9	BAYCID
52-68-6	BAYER 15922
2642-71-9	BAYER 16259
86-50-0	BAYER 17147
298-04-4	BAYER 19639
8022-00-2	BAYER 21/116
56-72-4	BAYER 21/199
5836-29-3	BAYER 25/634
919-86-8	BAYER 25/154
3878-19-1	BAYER 33172
327-98-0	BAYER 37289
2032-65-7	BAYER 37344
4104-14-7	BAYER 38819
114-26-1	BAYER 39007
122-14-5	BAYER 41831
327-98-0	BAYER 5081
443-48-1	BAYER 5360
10265-92-6	BAYER 5546
1420-04-8	BAYER 6076
21087-64-9	BAYER 6159
22224-92-6	BAYER 68138
10265-92-6	BAYER 71628
1420-04-8	BAYER 73
8065-48-3	BAYER 8169
55-38-9	BAYER 9007
86-50-0	BAYER 9027
114-26-1	BAYER B 5122
56-38-2	BAYER E-605
52-68-6	BAYER L 13/59
35400-43-2	BAYER NTN 9306
327-98-0	BAYER S 4400
122-14-5	BAYER S 5660
1309-37-1	BAYER S11
3689-24-5	BAYER-E 393
13463-67-7	BAYERTITAN A
13463-67-7	BAYERTITAN R-FD 1
13463-67-7	BAYERTITAN R-U-F
1309-37-1	BAYFERROX 110M
1309-37-1	BAYFERROX 130M
114-26-1	BAYGON
1420-04-8	BAYLUCIT
1420-04-8	BAYLUSCIDE
56-72-4	BAYMIX
56-72-4	BAYMIX 50

CAS No.	Chemical Name
55-38-9	BAYTEX
333-41-5	BAZINON
333-41-5	BAZUDEN
5798-79-8	BBC
17804-35-2	BBC
96-12-8	BBC 12
58-89-9	BBH
5798-79-8	BBN
85-68-7	BBP
25168-05-2	BC
91681-63-9	BC 3
17804-35-2	BC 6597
353-59-3	BCF
93-76-5	BCF-BUSHKILLER
3457-61-2	BCP
30580-75-7	BCP
7758-98-7	BCS COPPER FUNGICIDE
9002-84-0	BDH 29-801
133-06-2	BEAN SEED PROTECTANT
14807-96-6	BEAVER WHITE 200
434-07-1	BECOREL
57-50-1	BEET SUGAR
103-23-1	BEHA
117-84-0	BEHP
1937-37-7	BELAMINE BLACK GX
2602-46-2	BELAMINE BLUE 2B
16071-86-6	BELAMINE FAST BROWN BRLL
1305-62-0	BELL MINE
12789-03-6	BELT
13010-47-4	BELUSTINE
58-89-9	BEN-HEX
50-78-2	BENASPIR
1937-37-7	BENCIDAL BLACK E
2602-46-2	BENCIDAL BLUE 2B
72-57-1	BENCIDAL BLUE 3B
22781-23-3	BENDIOCARB
2636-26-2	BENFOS
58-89-9	BENHEXOL
17804-35-2	BENLATE
17804-35-2	BENLATE 50
17804-35-2	BENLATE 50W
17804-35-2	BENOMYL
17804-35-2	BENOMYL-IMEX
94-36-0	BENOXYL
7704-34-9	BENSULFOID
59-96-1	BENSYLYT
21564-17-0	BENTHIAZOLE
58-89-9	BENTOX 10
57-13-6	BENURAL 70
56-55-3	BENZ(A)ANTHRACENE
57-97-6	BENZ(A)ANTHRACENE, 7,12-DIMETHYL-
94-36-0	BENZAC
94-36-0	BENZAKNEW
100-51-6	BENZAL ALCOHOL
98-87-3	BENZAL CHLORIDE
100-52-7	BENZALDEHYDE
100-52-7	BENZALDEHYDE (DOT)
100-52-7	BENZALDEHYDE FFC
633-03-4	BENZALDEHYDE GREEN
90-02-8	BENZALDEHYDE, 2-HYDROXY-
135-02-4	BENZALDEHYDE, 2-METHOXY-
104-88-1	BENZALDEHYDE, 4-CHLORO-
98-88-4	BENZALDEHYDE, ALPHA-CHLORO-
90-02-8	BENZALDEHYDE, O-HYDROXY-
104-88-1	BENZALDEHYDE, P-CHLORO-
55-21-0	BENZAMIDE
148-01-6	BENZAMIDE, 2-METHYL-3,5-DINITRO-
23950-58-5	BENZAMIDE, 3,5-DICHLORO-N-(1,1-DIMETHYL-2-PROPYNYL)-
1420-04-8	BENZAMIDE, 5-CHLORO-N-(2-CHLORO-4-NITROPHENYL)-2-HYDROXY-, COMPD. WITH 2-AMINOETHANOL (1:1)
671-16-9	BENZAMIDE, N-(1-METHYLETHYL)-4-[(2-METHYLHYDRAZINO)-METHYL]-
366-70-1	BENZAMIDE, N-(1-METHYLETHYL)-4-[(2-METHYLHYDRAZINO)-METHYL]-, MONOHYDROCHLORIDE
72-57-1	BENZAMINE BLUE
1937-37-7	BENZANIL BLACK E
2602-46-2	BENZANIL BLUE 2B
72-57-1	BENZANIL BLUE 3BN
16071-86-6	BENZANIL SUPRA BROWN BRLL

CAS No.	Chemical Name
16071-86-6	BENZANIL SUPRA BROWN BRLN
56-55-3	BENZANTHRACENE
56-55-3	BENZANTHRENE
62-53-3	BENZENAMINE
87-59-2	BENZENAMINE, 2,3-DIMETHYL-
88-05-1	BENZENAMINE, 2,4,6-TRIMETHYL-
88-05-1	BENZENAMINE, 2,4,6-TRIMETHYL- (9CI)
100-97-0	BENZENAMINE, 2,4,6-TRINITRO-N-(2,4,6-TRINITROPHENYL)-
131-73-7	BENZENAMINE, 2,4,6-TRINITRO-N-(2,4,6-TRINITROPHENYL)-
97-02-9	BENZENAMINE, 2,4-DINITRO-
97-02-9	BENZENAMINE, 2,4-DINITRO- (9CI)
102-56-7	BENZENAMINE, 2,5-DIMETHOXY-
87-62-7	BENZENAMINE, 2,6-DIMETHYL-
1582-09-8	BENZENAMINE, 2,6-DINITRO-N,N-DIPROPYL-4-(TRIFLUOROMETHYL)-
1582-09-8	BENZENAMINE, 2,6-DINITRO-N,N-DIPROPYL-4-(TRIFLUOROMETHYL)- (9CI)
88-17-5	BENZENAMINE, 2-(TRIFLUOROMETHYL)-
88-17-5	BENZENAMINE, 2-(TRIFLUOROMETHYL)- (9CI)
94-70-2	BENZENAMINE, 2-ETHOXY-
578-54-1	BENZENAMINE, 2-ETHYL-
578-54-1	BENZENAMINE, 2-ETHYL- (9CI)
90-04-0	BENZENAMINE, 2-METHOXY-
90-04-01	BENZENAMINE, 2-METHOXY- (9CI)
134-29-2	BENZENAMINE, 2-METHOXY-, HYDROCHLORIDE
120-71-8	BENZENAMINE, 2-METHOXY-5-METHYL-
99-59-2	BENZENAMINE, 2-METHOXY-5-NITRO-
95-53-4	BENZENAMINE, 2-METHYL-
95-53-4	BENZENAMINE, 2-METHYL- (9CI)
636-21-5	BENZENAMINE, 2-METHYL-, HYDROCHLORIDE
97-56-3	BENZENAMINE, 2-METHYL-4-[(2-METHYLPHENYL)AZO]-
99-55-8	BENZENAMINE, 2-METHYL-5-NITRO-
88-74-4	BENZENAMINE, 2-NITRO-
95-76-1	BENZENAMINE, 3,4-DICHLORO-
98-16-8	BENZENAMINE, 3-(TRIFLUOROMETHYL)-
108-44-1	BENZENAMINE, 3-METHYL-
2465-27-2	BENZENAMINE, 4,4'-CARBONIMIDOYLBIS[N,N-DIMETHYL-, MONOHYDROCHLORIDE
101-77-9	BENZENAMINE, 4,4'-METHYLENEBIS-
101-14-4	BENZENAMINE, 4,4'-METHYLENEBIS[2-CHLORO-
838-88-0	BENZENAMINE, 4,4'-METHYLENEBIS[2-METHYL-
101-61-1	BENZENAMINE, 4,4'-METHYLENEBIS[N,N-DIMETHYL-
101-80-4	BENZENAMINE, 4,4'-OXYBIS-
28434-86-8	BENZENAMINE, 4,4'-OXYBIS[2-CHLORO-
139-65-1	BENZENAMINE, 4,4'-THIOBIS-
2049-92-5	BENZENAMINE, 4-(1,1-DIMETHYLPROPYL)-
4726-14-1	BENZENAMINE, 4-(METHYLSULFONYL)-2,6-DINITRO-N,N-DIPROPYL-
4726-14-1	BENZENAMINE, 4-(METHYLSULFONYL)-2,6-DINITRO-N,N-DIPROPYL- (9CI)
95-69-2	BENZENAMINE, 4-CHLORO-2-METHYL-
3165-93-3	BENZENAMINE, 4-CHLORO-2-METHYL-, HYDROCHLORIDE
156-43-4	BENZENAMINE, 4-ETHOXY-
1321-31-9	BENZENAMINE, 4-ETHOXY- (9CI)
13410-72-5	BENZENAMINE, 4-ETHOXY-N-[(5-NITRO-2-FURANYL)METHYLENE]-
371-40-4	BENZENAMINE, 4-FLUORO-
104-94-9	BENZENAMINE, 4-METHOXY-
106-49-0	BENZENAMINE, 4-METHYL-
89-62-3	BENZENAMINE, 4-METHYL-2-NITRO-
119-32-4	BENZENAMINE, 4-METHYL-3-NITRO-
100-01-6	BENZENAMINE, 4-NITRO-
100-01-6	BENZENAMINE, 4-NITRO- (9CI)
29757-24-2	BENZENAMINE, 4-NITRO- (9CI)
156-10-5	BENZENAMINE, 4-NITROSO-N-PHENYL-
95-79-4	BENZENAMINE, 5-CHLORO-2-METHYL-
27134-27-6	BENZENAMINE, AR,AR-DICHLORO-
27134-27-6	BENZENAMINE, AR,AR-DICHLORO- (9CI)
1300-73-8	BENZENAMINE, AR,AR-DIMETHYL-
26471-56-7	BENZENAMINE, AR,AR-DINITRO-
26471-56-7	BENZENAMINE, AR,AR-DINITRO- (9CI)
1321-31-9	BENZENAMINE, AR-ETHOXY-
29191-52-4	BENZENAMINE, AR-METHOXY-
29757-24-2	BENZENAMINE, AR-NITRO-
27134-26-5	BENZENAMINE, CHLORO-
41587-36-4	BENZENAMINE, CHLORONITRO-
108-91-8	BENZENAMINE, HEXAHYDRO-
142-04-1	BENZENAMINE, HYDROCHLORIDE
121-69-7	BENZENAMINE, N,N,-DIMETHYL- (9CI)
613-29-6	BENZENAMINE, N,N-DIBUTYL-
91-66-7	BENZENAMINE, N,N-DIETHYL-
91-66-7	BENZENAMINE, N,N-DIETHYL- (9CI)
121-69-7	BENZENAMINE, N,N-DIMETHYL-
60-11-7	BENZENAMINE, N,N-DIMETHYL-4-(PHENYLAZO)-
138-89-6	BENZENAMINE, N,N-DIMETHYL-4-NITROSO-
138-89-6	BENZENAMINE, N,N-DIMETHYL-4-NITROSO- (9CI)
53324-05-3	BENZENAMINE, N,N-DIMETHYL-4-NITROSO- (9CI)
53324-05-3	BENZENAMINE, N,N-DIMETHYLNITROSO-
603-34-9	BENZENAMINE, N,N-DIPHENYL-
768-52-5	BENZENAMINE, N-(1-METHYLETHYL)-
10137-80-1	BENZENAMINE, N-(2-ETHYLHEXYL)-
122-98-5	BENZENAMINE, N-(2-HYDROXYETHYL)-
622-44-6	BENZENAMINE, N-(DICHLOROMETHYLENE)-
103-84-4	BENZENAMINE, N-ACETYL-
1126-78-9	BENZENAMINE, N-BUTYL-
1126-78-9	BENZENAMINE, N-BUTYL- (9CI)
103-69-5	BENZENAMINE, N-ETHYL-
103-69-5	BENZENAMINE, N-ETHYL- (9CI)
102-27-2	BENZENAMINE, N-ETHYL-3-METHYL-
102-27-2	BENZENAMINE, N-ETHYL-3-METHYL- (9CI)
622-57-1	BENZENAMINE, N-ETHYL-4-METHYL-
135-20-6	BENZENAMINE, N-HYDROXY-N-NITROSO-, AMMONIUM SALT
100-61-8	BENZENAMINE, N-METHYL-
479-45-8	BENZENAMINE, N-METHYL-N,2,4,6-TETRANITRO-
479-45-8	BENZENAMINE, N-METHYL-N,2,4,6-TETRANITRO- (9CI)
86-30-6	BENZENAMINE, N-NITROSO-N-PHENYL-
122-39-4	BENZENAMINE, N-PHENYL-
122-39-4	BENZENAMINE, N-PHENYL- (9CI)
71-43-2	BENZENE
100-52-7	BENZENE CARBALDEHYDE
108-90-7	BENZENE CHLORIDE
58-89-9	BENZENE HEXACHLORIDE
58-89-9	BENZENE HEXACHLORIDE-GAMMA-ISOMER
644-97-3	BENZENE PHOSPHOROUS DICHLORIDE
14684-25-4	BENZENE PHOSPHOROUS THIODICHLORIDE
644-97-3	BENZENE PHOSPHORUS DICHLORIDE
644-97-3	BENZENE PHOSPHORUS DICHLORIDE (DOT)
14684-25-4	BENZENE PHOSPHORUS THIODICHLORIDE
14684-25-4	BENZENE PHOSPHORUS THIODICHLORIDE (DOT)
98-09-9	BENZENE SULFOCHLORIDE
98-09-9	BENZENE SULFONYL CHLORIDE
110-83-8	BENZENE TETRAHYDRIDE
98-06-6	BENZENE, (1,1-DIMETHYLETHYL)-
6607-45-0	BENZENE, (1,2-DICHLOROETHENYL)-
98-83-9	BENZENE, (1-METHYLETHENYL)-
98-82-8	BENZENE, (1-METHYLETHYL)-
98-82-8	BENZENE, (1-METHYLETHYL)- (9CI)
135-98-8	BENZENE, (1-METHYLPROPYL)-
538-93-2	BENZENE, (2-METHYLPROPYL)-
62-38-4	BENZENE, (ACETOXYMERCURI)-
62-38-4	BENZENE, (ACETOXYMERCURIO)-
100-39-0	BENZENE, (BROMOMETHYL)-
100-44-7	BENZENE, (CHLOROMETHYL)-
30030-25-2	BENZENE, (CHLOROMETHYL)ETHENYL-
98-87-3	BENZENE, (DICHLOROMETHYL)-
96-09-3	BENZENE, (EPOXYETHYL)-
620-05-3	BENZENE, (IODOMETHYL)-
620-05-3	BENZENE, (IODOMETHYL)- (9CI)
122-39-4	BENZENE, (PHENYLAMINO)-
98-07-7	BENZENE, (TRICHLOROMETHYL)-
98-08-8	BENZENE, (TRIFLUOROMETHYL)-
50-29-3	BENZENE, 1,1'-(2,2,2-TRICHLOROETHYLIDENE)BIS[4-CHLORO-
72-43-5	BENZENE, 1,1'-(2,2,2-TRICHLOROETHYLIDENE)BIS[4-METHOXY-
72-54-8	BENZENE, 1,1'-(2,2-DICHLOROETHYLIDENE)BIS[4-CHLORO-
776-74-9	BENZENE, 1,1'-(BROMOMETHYLENE)BIS-
72-55-9	BENZENE, 1,1'-(DICHLOROETHENYLIDENE)BIS[4-CHLORO-
93-96-9	BENZENE, 1,1'-(OXYDIETHYLIDENE)BIS-
122-66-7	BENZENE, 1,1'-HYDRAZOBIS-
101-68-8	BENZENE, 1,1'-METHYLENEBIS(4-ISOCYANATO- (9CI)
101-68-8	BENZENE, 1,1'-METHYLENEBIS[4-ISOCYANATO-
101-84-8	BENZENE, 1,1'-OXYBIS-
8004-13-5	BENZENE, 1,1'-OXYBIS-, MIXT CONTG
2217-06-3	BENZENE, 1,1'-THIOBIS[2,4,6-TRINITRO-
87-66-1	BENZENE, 1,2,3-TRIHYDROXY-
120-82-1	BENZENE, 1,2,4-TRICHLORO-
95-63-6	BENZENE, 1,2,4-TRIMETHYL-
95-63-6	BENZENE, 1,2,5-TRIMETHYL-
206-44-0	BENZENE, 1,2-(1,8-NAPHTHALENEDIYL)-
120-58-1	BENZENE, 1,2-(METHYLENEDIOXY)-4-PROPENYL-

CAS No.	*Chemical Name*
94-58-6	BENZENE, 1,2-(METHYLENEDIOXY)-4-PROPYL-
95-50-1	BENZENE, 1,2-DICHLORO-
102-36-3	BENZENE, 1,2-DICHLORO-4-ISOCYANATO-
135-01-3	BENZENE, 1,2-DIETHYL-
95-47-6	BENZENE, 1,2-DIMETHYL-
528-29-0	BENZENE, 1,2-DINITRO-
108-67-8	BENZENE, 1,3,5-TRIMETHYL-
99-35-4	BENZENE, 1,3,5-TRINITRO-
541-73-1	BENZENE, 1,3-DICHLORO-
541-73-1	BENZENE, 1,3-DICHLORO- (9CI)
108-57-6	BENZENE, 1,3-DIETHENYL-
141-93-5	BENZENE, 1,3-DIETHYL-
91-08-7	BENZENE, 1,3-DIISOCYANATO-2-METHYL-
108-38-3	BENZENE, 1,3-DIMETHYL-
99-65-0	BENZENE, 1,3-DINITRO-
106-46-7	BENZENE, 1,4-DICHLORO-
105-05-5	BENZENE, 1,4-DIETHYL-
106-42-3	BENZENE, 1,4-DIMETHYL-
100-25-4	BENZENE, 1,4-DINITRO-
81-15-2	BENZENE, 1-(1,1-DIMETHYLETHYL)-3,5-DIMETHYL-2,4,6-TRI-NITRO-
98-51-1	BENZENE, 1-(1,1-DIMETHYLETHYL)-4-METHYL-
100-14-1	BENZENE, 1-(CHLOROMETHYL)-4-NITRO-
95-46-5	BENZENE, 1-BROMO-2-METHYL-
106-38-7	BENZENE, 1-BROMO-4-METHYL-
586-78-7	BENZENE, 1-BROMO-4-NITRO-
101-55-3	BENZENE, 1-BROMO-4-PHENOXY-
27458-20-4	BENZENE, 1-BUTYL-2-METHYL- (9CI)
97-00-7	BENZENE, 1-CHLORO-2,4-DINITRO-
25567-67-3	BENZENE, 1-CHLORO-2,4-DINITRO-
88-16-4	BENZENE, 1-CHLORO-2-(TRIFLUOROMETHYL)-
2039-87-4	BENZENE, 1-CHLORO-2-ETHENYL-
95-49-8	BENZENE, 1-CHLORO-2-METHYL-
88-73-3	BENZENE, 1-CHLORO-2-NITRO-
121-17-5	BENZENE, 1-CHLORO-2-NITRO-4-(TRIFLUOROMETHYL)-
98-15-7	BENZENE, 1-CHLORO-3-(TRIFLUOROMETHYL)-
98-15-7	BENZENE, 1-CHLORO-3-(TRIFLUOROMETHYL)- (9CI)
121-73-3	BENZENE, 1-CHLORO-3-NITRO-
104-83-6	BENZENE, 1-CHLORO-4-(CHLOROMETHYL)-
98-56-6	BENZENE, 1-CHLORO-4-(TRIFLUOROMETHYL)-
106-43-4	BENZENE, 1-CHLORO-4-METHYL-
100-00-5	BENZENE, 1-CHLORO-4-NITRO-
777-37-7	BENZENE, 1-CHLORO-4-NITRO-2-(TRIFLUOROMETHYL)-
7005-72-3	BENZENE, 1-CHLORO-4-PHENOXY-
622-97-9	BENZENE, 1-ETHENYL-4-METHYL-
622-97-9	BENZENE, 1-ETHENYL-4-METHYL- (9CI)
100-17-4	BENZENE, 1-METHOXY-4-NITRO-
100-17-4	BENZENE, 1-METHOXY-4-NITRO- (9CI)
121-14-2	BENZENE, 1-METHYL-2,4-DINITRO-
88-72-2	BENZENE, 1-METHYL-2-NITRO-
99-08-1	BENZENE, 1-METHYL-3-NITRO-
99-87-6	BENZENE, 1-METHYL-4-(1-METHYLETHYL)-
99-99-0	BENZENE, 1-METHYL-4-NITRO-
402-54-0	BENZENE, 1-NITRO-4-(TRIFLUOROMETHYL)-
402-54-0	BENZENE, 1-NITRO-4-(TRIFLUOROMETHYL)- (9CI)
81-15-2	BENZENE, 1-TERT-BUTYL-3,5-DIMETHYL-2,4,6-TRINITRO-
1836-75-5	BENZENE, 2,4-DICHLORO-1-(4-NITROPHENOXY)-
584-84-9	BENZENE, 2,4-DIISOCYANATO-1-METHYL-
26471-62-5	BENZENE, 2,4-DIISOCYANATO-1-METHYL-
91-08-7	BENZENE, 2,6-DIISOCYANATO-1-METHYL-
2100-42-7	BENZENE, 2-CHLORO-1,4-DIMETHOXY-
28479-22-3	BENZENE, 2-CHLORO-4-ISOCYANATO-1-METHYL-
118-96-7	BENZENE, 2-METHYL-1,3,5-TRINITRO-
606-20-2	BENZENE, 2-METHYL-1,3-DINITRO-
606-20-2	BENZENE, 2-METHYL-1,3-DINITRO- (9CI)
94-59-7	BENZENE, 4-ALLYL-1,2-(METHYLENEDIOXY)-
118-83-2	BENZENE, 4-CHLORO, 1-NITRO, 4-(TRIFLUOROMETHYL)
610-39-9	BENZENE, 4-METHYL-1,2-DINITRO-
610-39-9	BENZENE, 4-METHYL-1,2-DINITRO- (9CI)
98-86-2	BENZENE, ACETYL-
62-53-3	BENZENE, AMINO
122-39-4	BENZENE, ANILINO-
119-61-9	BENZENE, BENZOYL-
28347-13-9	BENZENE, BIS(CHLOROMETHYL)-
28347-13-9	BENZENE, BIS(CHLOROMETHYL)- (9CI)
108-86-1	BENZENE, BROMO-
35884-77-6	BENZENE, BROMODIMETHYL-
104-51-8	BENZENE, BUTYL-
27458-20-4	BENZENE, BUTYLMETHYL-
108-90-7	BENZENE, CHLORO-
97-00-7	BENZENE, CHLORODINITRO-
25567-67-3	BENZENE, CHLORODINITRO-
97-00-7	BENZENE, CHLORODINITRO- (MIXED ISOMERS)
25567-67-3	BENZENE, CHLORODINITRO- (MIXED ISOMERS)
25168-05-2	BENZENE, CHLOROMETHYL-
61878-61-3	BENZENE, CHLOROMETHYLNITRO-
25567-68-4	BENZENE, CHLOROMETHYLNITRO-
25167-93-5	BENZENE, CHLORONITRO-
100-47-0	BENZENE, CYANO-
827-52-1	BENZENE, CYCLOHEXYL-
104-72-3	BENZENE, DECYL-
26249-12-7	BENZENE, DIBROMO-
25321-22-6	BENZENE, DICHLORO-
1321-74-0	BENZENE, DIETHENYL-
25340-17-4	BENZENE, DIETHYL-
1320-21-4	BENZENE, DIMETHYL(PENTYLOXY)-
1330-20-7	BENZENE, DIMETHYL-
25168-04-1	BENZENE, DIMETHYLNITRO-
25154-54-5	BENZENE, DINITRO-
1321-74-0	BENZENE, DIVINYL-
123-01-3	BENZENE, DODECYL-
100-42-5	BENZENE, ETHENYL-
1319-73-9	BENZENE, ETHENYL-, MONOMETHYL DERIV.
25013-15-4	BENZENE, ETHENYLMETHYL-
100-41-4	BENZENE, ETHYL-
462-06-6	BENZENE, FLUORO-
25496-08-6	BENZENE, FLUOROMETHYL-
118-74-1	BENZENE, HEXACHLORO-
110-82-7	BENZENE, HEXAHYDRO-
538-93-2	BENZENE, ISOBUTYL-
103-71-9	BENZENE, ISOCYANATO-
98-83-9	BENZENE, ISOPROPENYL-
98-82-8	BENZENE, ISOPROPYL
541-73-1	BENZENE, M-DICHLORO-
141-93-5	BENZENE, M-DIETHYL-
108-46-3	BENZENE, M-DIHYDROXY-
99-65-0	BENZENE, M-DINITRO-
108-57-6	BENZENE, M-DIVINYL-
100-66-3	BENZENE, METHOXY-
108-88-3	BENZENE, METHYL-
30143-13-6	BENZENE, METHYL-, DIAMINO DERIV
25376-45-8	BENZENE, METHYL-, DIAMINO DERIV
25567-68-4	BENZENE, METHYL-, MONOCHLORO MONONITRO DERIV
61878-61-3	BENZENE, METHYL-, MONOCHLORO MONONITRO DERIV
25321-14-6	BENZENE, METHYLDINITRO-
1321-12-6	BENZENE, METHYLNITRO-
1320-01-0	BENZENE, METHYLPENTYL-
28729-54-6	BENZENE, METHYLPROPYL-
98-95-3	BENZENE, NITRO-
95-50-1	BENZENE, O-DICHLORO-
135-01-3	BENZENE, O-DIETHYL-
528-29-0	BENZENE, O-DINITRO-
106-46-7	BENZENE, P-DICHLORO-
105-05-5	BENZENE, P-DIETHYL-
123-31-9	BENZENE, P-DIHYDROXY-
100-25-4	BENZENE, P-DINITRO-
82-68-8	BENZENE, PENTACHLORONITRO-
538-68-1	BENZENE, PENTYL-
101-84-8	BENZENE, PHENOXY-
103-65-1	BENZENE, PROPYL-
135-98-8	BENZENE, SEC-BUTYL-
98-06-6	BENZENE, TERT-BUTYL-
1459-10-5	BENZENE, TETRADECYL-
110-83-8	BENZENE, TETRAHYDRO-
12002-48-1	BENZENE, TRICHLORO-
25340-18-5	BENZENE, TRIETHYL-
25340-18-5	BENZENE, TRIETHYL- (MIXED ISOMERS)
25551-13-7	BENZENE, TRIMETHYL-
25551-13-7	BENZENE, TRIMETHYL- (MIXED ISOMERS)
99-35-4	BENZENE, TRINITRO- (10% TO 30% WATER)
6742-54-7	BENZENE, UNDECYL-
100-42-5	BENZENE, VINYL-
121-91-5	BENZENE-1,3-DICARBOXYLIC ACID
84-74-2	BENZENE-O-DICARBOXYLIC ACID DI-N-BUTYL ESTER
510-15-6	BENZENEACETIC ACID, 4-CHLORO-ALPHA-(4-CHLOROPHENYL)-ALPHA-HYDROXY-, ETHYL ESTER
140-29-4	BENZENEACETONITRILE
140-29-4	BENZENEACETONITRILE (9CI)

CAS No.	Chemical Name
5798-79-8	BENZENEACETONITRILE, ALPHA-BROMO-
5798-79-8	BENZENEACETONITRILE, ALPHA-BROMO- (9CI)
103-80-0	BENZENEACETYL CHLORIDE
103-80-0	BENZENEACETYL CHLORIDE (9CI)
27134-26-5	BENZENEAMINE, 4-CHLORO
98-05-5	BENZENEARSONIC ACID
305-03-3	BENZENEBUTANOIC ACID, 4-[BIS(2-CHLOROETHYL)AMINO]-
100-51-6	BENZENECARBINOL
100-52-7	BENZENECARBONAL
98-88-4	BENZENECARBONYL CHLORIDE
2618-77-1	BENZENECARBOPEROXOIC ACID, 1,1,4,4-TETRAMETHYL-1,4-BUTANEDIYL ESTER
614-45-9	BENZENECARBOPEROXOIC ACID, 1,1-DIMETHYLETHYL ESTER
4511-39-1	BENZENECARBOPEROXOIC ACID, 1,1-DIMETHYLPROPYL ESTER
937-14-4	BENZENECARBOPEROXOIC ACID, 3-CHLORO-
100-52-7	BENZENECARBOXALDEHYDE
65-85-0	BENZENECARBOXYLIC ACID
140-29-4	BENZENEDIACETONITRILE
30143-13-6	BENZENEDIAMINE, AR-METHYL-
25376-45-8	BENZENEDIAMINE, AR-METHYL-
12385-08-9	BENZENEDIOL
7568-93-6	BENZENEETHANAMINE, BETA-HYDROXY-
65-85-0	BENZENEFORMIC ACID
100-46-9	BENZENEMETHANAMINE
98-84-0	BENZENEMETHANAMINE, ALPHA-METHYL-
2449-49-2	BENZENEMETHANAMINE, N,N,ALPHA-TRIMETHYL-
772-54-3	BENZENEMETHANAMINE, N,N-DIETHYL-
59-96-1	BENZENEMETHANAMINE, N-(2-CHLOROETHYL)-N-(1-METHYL-2-PHENOXYETHYL)-
63-92-3	BENZENEMETHANAMINE, N-(2-CHLOROETHYL)-N-(1-METHYL-2-PHENOXYETHYL)-, HYDROCHLORIDE
92-59-1	BENZENEMETHANAMINE, N-ETHYL-N-PHENYL-
56-93-9	BENZENEMETHANAMINIUM, N,N,N-TRIMETHYL-, CHLORIDE
56-93-9	BENZENEMETHANAMINIUM, N,N,N-TRIMETHYL-, CHLORIDE (9CI)
100-53-8	BENZENEMETHANETHIOL
65-85-0	BENZENEMETHANOIC ACID
100-51-6	BENZENEMETHANOL
7568-93-6	BENZENEMETHANOL, ALPHA-(AMINOMETHYL)-
2454-37-7	BENZENEMETHANOL, 3-AMINO-ALPHA-METHYL-
115-32-2	BENZENEMETHANOL, 4-CHLORO-ALPHA-(4-CHLOROPHENYL)-ALPHA-(TRICHLOROMETHYL)-
98-85-1	BENZENEMETHANOL, ALPHA-METHYL-
100-61-8	BENZENENAMINE, N-METHYL- (9CI)
100-47-0	BENZENENITRILE
2104-64-5	BENZENEPHOSPHONIC ACID, THIONO-, ETHYL-P-NITROPHENYL ESTER
80-17-1	BENZENESULFOHYDRAZIDE
80-51-3	BENZENESULFONIC ACID, 4,4'-OXYBIS-, DIHYDRAZIDE
80-51-3	BENZENESULFONIC ACID, 4,4'-OXYDI-, DIHYDRAZIDE
54590-52-2	BENZENESULFONIC ACID, 4-DODECYL-, COMPD WITH 1-AMINO-2-PROPANOL (1:1)
127-82-2	BENZENESULFONIC ACID, 4-HYDROXY-, ZINC SALT (2:1)
104-15-4	BENZENESULFONIC ACID, 4-METHYL-
25231-46-3	BENZENESULFONIC ACID, 4-METHYL-
80-48-8	BENZENESULFONIC ACID, 4-METHYL-, METHYL ESTER
27176-87-0	BENZENESULFONIC ACID, DODECYL-
26264-06-2	BENZENESULFONIC ACID, DODECYL-, CALCIUM SALT
27323-41-7	BENZENESULFONIC ACID, DODECYL-, COMPD WITH 2,2',2"-NITRILOTRIETHANOL (1:1)
27323-41-7	BENZENESULFONIC ACID, DODECYL-, COMPD WITH 2,2',2"-NITRILOTRIS[ETHANOL] (1:1)
26264-05-1	BENZENESULFONIC ACID, DODECYL-, COMPD WITH 2-PROPANAMINE (1:1)
26264-05-1	BENZENESULFONIC ACID, DODECYL-, COMPD WITH ISOPROPYLAMINE
26264-05-1	BENZENESULFONIC ACID, DODECYL-, COMPD WITH ISOPROPYLAMINE (1:1)
25155-30-0	BENZENESULFONIC ACID, DODECYL-, SODIUM SALT
80-17-1	BENZENESULFONIC ACID, HYDRAZIDE
1333-39-7	BENZENESULFONIC ACID, HYDROXY-
104-15-4	BENZENESULFONIC ACID, METHYL-
25231-46-3	BENZENESULFONIC ACID, METHYL-
31212-28-9	BENZENESULFONIC ACID, NITRO-
80-51-3	BENZENESULFONIC ACID, OXYBIS-, DIHYDRAZIDE (9CI)
127-82-2	BENZENESULFONIC ACID, P-HYDROXY-, ZINC SALT (2:1)
80-17-1	BENZENESULFONIC HYDRAZIDE
80-17-1	BENZENESULFONOHYDRAZIDE
80-17-1	BENZENESULFONYL HYDRAZIDE
108-98-5	BENZENETHIOL
108-98-5	BENZENETHIOL (DOT)
108-95-2	BENZENOL
98-07-7	BENZENYL CHLORIDE
98-08-8	BENZENYL FLUORIDE
98-07-7	BENZENYL TRICHLORIDE
115-32-2	BENZHYDROL, 4,4'-DICHLORO-ALPHA-(TRICHLOROMETHYL)-
776-74-9	BENZHYDRYL BROMIDE
62-53-3	BENZIDAM
92-87-5	BENZIDINE
91-94-1	BENZIDINE, 3,3'-DICHLORO-
119-90-4	BENZIDINE, 3,3'-DIMETHOXY-
119-93-7	BENZIDINE, 3,3'-DIMETHYL-
510-15-6	BENZILAN
100-44-7	BENZILE (CLORURO DI) (ITALIAN)
510-15-6	BENZILIC ACID, 4,4'-DICHLORO-, ETHYL ESTER
3878-19-1	BENZIMIDAZOLE, 2-(2-FURYL)-
8030-30-6	BENZIN
8030-30-6	BENZIN B70
8030-30-6	BENZINE
8032-32-4	BENZINE (LIGHT PETROLEUM DISTILLATE)
86290-81-5	BENZINE (MOTOR FUEL)
56-23-5	BENZINOFORM
79-01-6	BENZINOL
72-57-1	BENZO BLUE
2602-46-2	BENZO BLUE 2B
72-57-1	BENZO BLUE 3BS
2602-46-2	BENZO BLUE BBA-CF
2602-46-2	BENZO BLUE BBN-CF
2602-46-2	BENZO BLUE GS
1937-37-7	BENZO DEEP BLACK E
1937-37-7	BENZO LEATHER BLACK E
64-86-8	BENZO(A)HEPTALEN-9(5H)-ONE, 7-ACETAMIDO-6,7-DIHYDRO-1,2,3,10-TETRAMETHOXY-
50-32-8	BENZO(A)PYRENE
205-99-2	BENZO(B)FLUORANTHENE
91-22-5	BENZO(B)PYRIDINE
260-94-6	BENZO(B)QUINOLINE
129-00-0	BENZO(DEF)PHENANTHRENE
205-82-3	BENZO-12,13-FLUORANTHENE
106-51-4	BENZO-CHINON (GERMAN)
56-55-3	BENZOANTHRACENE
119-64-2	BENZOCYCLOHEXANE
115-29-7	BENZOEPIN
1031-07-8	BENZOEPIN SULFATE
1937-37-7	BENZOFORM BLACK BCN-CF
106-50-3	BENZOFUR D
123-30-8	BENZOFUR P
27598-85-2	BENZOFUR P
123-31-9	BENZOHYDROQUINONE
65-85-0	BENZOIC ACID
100-47-0	BENZOIC ACID NITRILE
129-66-8	BENZOIC ACID, 2,4,6-TRINITRO-
50-78-2	BENZOIC ACID, 2-(ACETYLOXY)-
69-72-7	BENZOIC ACID, 2-HYDROXY-
29790-52-1	BENZOIC ACID, 2-HYDROXY-, COMPD WITH (S)-3-(1-METHYL-2-PYRROLIDINYL)PYRIDINE (1:1)
119-36-8	BENZOIC ACID, 2-HYDROXY-, METHYL ESTER
149-91-7	BENZOIC ACID, 3,4,5-TRIHYDROXY-
1401-55-4	BENZOIC ACID, 3,4-DIHYDROXY-5-((3,4,5-TRIHYDROXYBENZOYL)-OXY)- (9CI)
1918-00-9	BENZOIC ACID, 3,6-DICHLORO-2-METHOXY-
1863-63-4	BENZOIC ACID, AMMONIUM SALT
136-60-7	BENZOIC ACID, BUTYL ESTER
98-88-4	BENZOIC ACID, CHLORIDE
93-89-0	BENZOIC ACID, ETHYL ESTER
583-15-3	BENZOIC ACID, MERCURY(2+) SALT
93-58-3	BENZOIC ACID, METHYL ESTER
94-36-0	BENZOIC ACID, PEROXIDE
129-66-8	BENZOIC ACID, TRINITRO-
129-66-8	BENZOIC ACID, TRINITRO- (10% TO 30% WATER)
100-52-7	BENZOIC ALDEHYDE
81-07-2	BENZOIC SULFIMIDE
81-07-2	BENZOIC SULPHINIDE
98-07-7	BENZOIC TRICHLORIDE
119-53-9	BENZOIN
71-43-2	BENZOL
71-43-2	BENZOLE
8032-32-4	BENZOLINE
100-47-0	BENZONITRIL

CAS No.	Chemical Name
100-47-0	BENZONITRILE
100-47-0	BENZONITRILE (DOT)
1194-65-6	BENZONITRILE, 2,6-DICHLORO-
94-36-0	BENZOPEROXIDE
119-61-9	BENZOPHENONE
90-94-8	BENZOPHENONE, 4,4'-BIS(DIMETHYLAMINO)-
50-32-8	BENZOPYRENE
91-22-5	BENZOPYRIDINE
123-31-9	BENZOQUINOL
106-51-4	BENZOQUINONE
106-51-4	BENZOQUINONE (DOT)
81-07-2	BENZOSULFINIDE
21564-17-0	BENZOTHIAZOLE, 2-[(THIOCYANATOMETHYL)THIO]-
514-73-8	BENZOTHIAZOLIUM, 3-ETHYL-2-(5-(3-ETHYL-2-BENZOTHIAZOLINYLIDENE)-1,3-PENTADIENYL)-, IODIDE
514-73-8	BENZOTHIAZOLIUM, 3-ETHYL-2-[5-(3-ETHYL-2(3H)-BENZOTHIAZOLYLIDENE)-1,3-PENTADIENYL]-, IODIDE
86-50-0	BENZOTRIAZINE DERIVATIVE OF A METHY L DITHIOPHOSPHATE
2642-71-9	BENZOTRIAZINE DERIVATIVE OF AN ETHYL DITHIOPHOSPHATE
86-50-0	BENZOTRIAZINEDITHIOPHOSPHORIC ACID DIMETHOXY ESTER
98-07-7	BENZOTRICHLORIDE
98-07-7	BENZOTRICHLORIDE (DOT)
98-08-8	BENZOTRIFLUORIDE
98-08-8	BENZOTRIFLUORIDE (DOT)
121-17-5	BENZOTRIFLUORIDE, 4-CHLORO-3-NITRO-
94-36-0	BENZOYL
644-31-5	BENZOYL ACETYL PEROXIDE
100-51-6	BENZOYL ALCOHOL
98-88-4	BENZOYL CHLORIDE
98-88-4	BENZOYL CHLORIDE (DOT)
100-07-2	BENZOYL CHLORIDE, 4-METHOXY-
100-07-2	BENZOYL CHLORIDE, METHOXY-
100-07-2	BENZOYL CHLORIDE, METHOXY- (9CI)
93-89-0	BENZOYL ETHYL ETHER
98-86-2	BENZOYL METHIDE
94-36-0	BENZOYL PEROXIDE
94-36-0	BENZOYL PEROXIDE (ACGIH,DOT)
94-36-0	BENZOYL PEROXIDE (DOT)
94-36-0	BENZOYL PEROXIDE, MORE THAN 72% BUT LESS THAN 95%
94-36-0	BENZOYL PEROXIDE, MORE THAN 72% BUT LESS THAN 95% AS A PASTE (DOT)
94-36-0	BENZOYL PEROXIDE, MORE THAN 77% BUT LESS THAN 95% WITH WATER (DOT)
94-36-0	BENZOYL PEROXIDE, NOT LESS THAN 30% BUT NOT MORE THAN 52% WITH INERT SOLID
94-36-0	BENZOYL PEROXIDE, TECHNICAL PURE (DOT)
94-36-0	BENZOYL SUPEROXIDE
614-45-9	BENZOYL TERT-BUTYL PEROXIDE
94-36-0	BENZOYL-PEROXIDE, MORE THAN 52% WITH INERT SOLID (DOT)
119-61-9	BENZOYLBENZENE
70-11-1	BENZOYLMETHYL BROMIDE
94-36-0	BENZOYLPEROXID (GERMAN)
94-36-0	BENZOYLPEROXYDE (DUTCH)
644-31-5	BENZOZONE
56-55-3	BENZO[A]ANTHRACENE
64-86-8	BENZO[A]HEPTALENE, ACETAMIDE DERIV
218-01-9	BENZO[A]PHENANTHRENE
56-55-3	BENZO[B]PHENANTHRENE
50-32-8	BENZO[D,E,F]CHRYSENE
205-99-2	BENZO[E]FLUORANTHENE
191-24-2	BENZO[GHI]PERYLENE
206-44-0	BENZO[JK]FLUORENE
207-08-9	BENZO[K]FLUORANTHENE
205-82-3	BENZO[L]FLUORANTHENE
189-55-9	BENZO[RST]PENTAPHENE
100-51-6	BENZYL ALCOHOL
7568-93-6	BENZYL ALCOHOL, ALPHA-(AMINOMETHYL)-
470-90-6	BENZYL ALCOHOL, 2,4-DICHLORO-ALPHA-(CHLOROMETHYLENE)-, DIETHYL PHOSPHATE
26748-47-0	BENZYL ALCOHOL, ALPHA,ALPHA-DIMETHYL-, PEROXYNEODECANOATE
98-85-1	BENZYL ALCOHOL, ALPHA-METHYL-
2454-37-7	BENZYL ALCOHOL, M-AMINO-ALPHA-METHYL-
100-39-0	BENZYL BROMIDE
85-68-7	BENZYL BUTYL ESTER PHTHALIC ACID
85-68-7	BENZYL BUTYL PHTHALATE
100-44-7	BENZYL CHLORIDE
100-44-7	BENZYL CHLORIDE (ACGIH,DOT)
501-53-1	BENZYL CHLOROFORMATE

CAS No.	Chemical Name
140-29-4	BENZYL CYANIDE
98-87-3	BENZYL DICHLORIDE
103-83-3	BENZYL DIMETHYLAMINE
620-05-3	BENZYL IODIDE
620-05-3	BENZYL IODIDE (DOT)
100-53-8	BENZYL MERCAPTAN
85-68-7	BENZYL N-BUTYL PHTHALATE
140-29-4	BENZYL NITRILE
98-07-7	BENZYL TRICHLORIDE
1694-09-3	BENZYL VIOLET
1694-09-3	BENZYL VIOLET 3B
1694-09-3	BENZYL VIOLET 4B
100-46-9	BENZYLAMINE
98-84-0	BENZYLAMINE, ALPHA-METHYL-
2449-49-2	BENZYLAMINE, N,N,ALPHA-TRIMETHYL-
772-54-3	BENZYLAMINE, N,N-DIETHYL-
59-96-1	BENZYLAMINE, N-(2-CHLOROETHYL)-N-(1-METHYL-2-PHENOXYETHYL)-
63-92-3	BENZYLAMINE, N-(2-CHLOROETHYL)-N-(1-METHYL-2-PHENOXYETHYL)-, HYDROCHLORIDE
92-59-1	BENZYLAMINE, N-ETHYL-N-PHENYL-
100-44-7	BENZYLCHLORID (GERMAN)
772-54-3	BENZYLDIETHYLAMINE
100-44-7	BENZYLE (CHLORURE DE) (FRENCH)
98-87-3	BENZYLENE CHLORIDE
92-59-1	BENZYLETHYLPHENYLAMINE
98-87-3	BENZYLIDENE CHLORIDE
98-87-3	BENZYLIDENE CHLORIDE (DOT)
98-07-7	BENZYLIDYNE CHLORIDE
98-08-8	BENZYLIDYNE FLUORIDE
61-33-6	BENZYLPENICILLIN
100-53-8	BENZYLTHIOL
56-93-9	BENZYLTRIMETHYLAMMONIUM CHLORIDE
59-96-1	BENZYLYT
56-55-3	BENZ[A]ANTHRACENE
57-97-6	BENZ[A]ANTHRACENE, 7,12-DIMETHYL-
50-32-8	BENZ[A]PYRENE
51787-44-1	BENZ[C]ACRIDINE, 7,8,9,11-TETRAMETHYL-
205-99-2	BENZ[E]ACEPHENANTHRYLENE
50-55-5	BENZ[G]INDOLO[2,3-A]QUINOLIZINE, YOHIMBAN-16-CARBOXYLIC ACID DERIV.
56-49-5	BENZ[J]ACEANTHRYLENE, 1,2-DIHYDRO-3-METHYL-
115-29-7	BEOSIT
115-84-4	BEP
14484-64-1	BERCEMA FERTAM 50
63-25-2	BERCEMA NMC50
9016-45-9	BEROL 02
9016-45-9	BEROL 09
9016-45-9	BEROL 259
9016-45-9	BEROL 26
9016-45-9	BEROL 267
9016-45-9	BEROL 296
151-21-3	BEROL 452
577-11-7	BEROL 478
7782-50-5	BERTHOLITE
3811-04-9	BERTHOLLET'S SALT
1302-52-9	BERYL
1302-52-9	BERYL ($Al_2Be_3(SiO_3)_6$)
1302-52-9	BERYL ORE
1304-56-9	BERYLLIA
7440-41-7	BERYLLIUM
7440-41-7	BERYLLIUM (ACGIH)
13106-47-3	BERYLLIUM CARBONATE
13106-47-3	BERYLLIUM CARBONATE ($BeCO_3$)
7787-47-5	BERYLLIUM CHLORIDE
7787-47-5	BERYLLIUM CHLORIDE ($BeCl_2$)
7787-47-5	BERYLLIUM CHLORIDE (DOT)
7787-47-5	BERYLLIUM DICHLORIDE
7787-49-7	BERYLLIUM DIFLUORIDE
13327-32-7	BERYLLIUM DIHYDROXIDE
13597-99-4	BERYLLIUM DINITRATE
7440-41-7	BERYLLIUM ELEMENT
7787-49-7	BERYLLIUM FLUORIDE
7787-49-7	BERYLLIUM FLUORIDE (BeF_2)
7787-49-7	BERYLLIUM FLUORIDE (DOT)
13327-32-7	BERYLLIUM HYDRATE
13327-32-7	BERYLLIUM HYDROXIDE
13327-32-7	BERYLLIUM HYDROXIDE ($Be(OH)_2$)
1304-56-9	BERYLLIUM MONOXIDE

CAS No.	Chemical Name
13597-99-4	BERYLLIUM NITRATE
13597-99-4	BERYLLIUM NITRATE (Be(NO3)2)
13597-99-4	BERYLLIUM NITRATE (DOT)
7787-55-5	BERYLLIUM NITRATE TRIHYDRATE
15191-85-2	BERYLLIUM ORTHOSILICATE
15191-85-2	BERYLLIUM ORTHOSILICATE (Be2SiO4)
1304-56-9	BERYLLIUM OXIDE
1304-56-9	BERYLLIUM OXIDE (BeO)
35089-00-0	BERYLLIUM PHOSPHATE
13598-15-7	BERYLLIUM PHOSPHATE
13598-26-0	BERYLLIUM PHOSPHATE
13598-15-7	BERYLLIUM PHOSPHATE (BeHPO4)
58500-38-2	BERYLLIUM SILICATE
15191-85-2	BERYLLIUM SILICATE
15191-85-2	BERYLLIUM SILICATE (Be2SiO4)
15191-85-2	BERYLLIUM SILICON OXIDE (Be2SiO4)
13510-49-1	BERYLLIUM SULFATE
13510-49-1	BERYLLIUM SULFATE (1:1)
13510-49-1	BERYLLIUM SULFATE (BeSO4)
13510-49-1	BERYLLIUM SULPHATE
7440-41-7	BERYLLIUM, METAL POWDER (DOT)
7440-41-7	BERYLLIUM, POWDER
7440-41-7	BERYLLIUM-9
319-85-7	BETA -HCH
11138-49-1	BETA"-ALUMINA
111-44-4	BETA,BETA'-DICHLORODIETHYL ETHER
505-60-2	BETA,BETA'-DICHLORODIETHYL SULFIDE
51-75-2	BETA,BETA'-DICHLORODIETHYL-N-METHYLAMINE
111-44-4	BETA,BETA'-DICHLOROETHYL ETHER
505-60-2	BETA,BETA'-DICHLOROETHYL SULFIDE
111-46-6	BETA,BETA'-DIHYDROXYDIETHYL ETHER
96-24-2	BETA,BETA'-DIHYDROXYISOPROPYL CHLORIDE
505-60-2	BETA,BETA-DICHLOR-ETHYL-SULPHIDE
111-44-4	BETA,BETA-DICHLORODIETHYL ETHER
929-06-6	BETA-(BETA-HYDROXYETHOXY)ETHYLAMINE
108-01-0	BETA-(DIMETHYLAMINO)ETHANOL
109-83-1	BETA-(METHYLAMINO)ETHANOL
3268-49-3	BETA-(METHYLMERCAPTO)PROPIONALDEHYDE
3268-49-3	BETA-(METHYLTHIO)PROPIONALDEHYDE
2867-47-2	BETA-(N,N-DIMETHYLAMINO)ETHYL METHACRYLATE
470-90-6	BETA-2-CHLORO-1-(2',4'-DICHLOROPHENYL) VINYL DIETHYL-PHOSPHATE
11138-49-1	BETA-ALUMINA
107-15-3	BETA-AMINOETHYLAMINE
78-96-6	BETA-AMINOISOPROPANOL
627-20-3	BETA-AMYLENE-CIS
646-04-8	BETA-AMYLENE-TRANS
111-76-2	BETA-BUTOXYETHANOL
3068-88-0	BETA-BUTYROLACTONE
107-07-3	BETA-CHLOROETHANOL
107-07-3	BETA-CHLOROETHYL ALCOHOL
627-11-2	BETA-CHLOROETHYL CHLOROFORMATE
999-81-5	BETA-CHLOROETHYLTRIMETHYLAMMONIUM CHLORIDE
126-99-8	BETA-CHLOROPRENE
557-98-2	BETA-CHLOROPROPENE
542-76-7	BETA-CHLOROPROPIONITRILE
557-98-2	BETA-CHLOROPROPYLENE
541-25-3	BETA-CHLOROVINYLBICHLOROARSINE
109-78-4	BETA-CYANOETHANOL
100-37-8	BETA-DIETHYLAMINOETHANOL
100-37-8	BETA-DIETHYLAMINOETHYL ALCOHOL
108-01-0	BETA-DIMETHYLAMINOETHANOL
108-01-0	BETA-DIMETHYLAMINOETHYL ALCOHOL
2867-47-2	BETA-DIMETHYLAMINOETHYL METHACRYLATE
1738-25-6	BETA-DIMETHYLAMINOPROPIONITRILE
573-56-8	BETA-DINITROPHENOL
50-28-2	BETA-ESTRADIOL
111-76-2	BETA-ETHOXYETHYL ACETATE
103-09-3	BETA-ETHYLHEXYL ACETATE
104-75-6	BETA-ETHYLHEXYLAMINE
371-62-0	BETA-FLUOROETHANOL
4301-50-2	BETA-FLUOROETHYL 4-BIPHENYLACETATE
319-85-7	BETA-HEXACHLOROCYCLOHEXANE
109-78-4	BETA-HPN
929-06-6	BETA-HYDROXY-BETA'-AMINOETHYL ETHER
107-89-1	BETA-HYDROXYBUTYRALDEHYDE
60-24-2	BETA-HYDROXYETHANETHIOL
818-61-1	BETA-HYDROXYETHYL ACRYLATE
107-07-3	BETA-HYDROXYETHYL CHLORIDE
108-01-0	BETA-HYDROXYETHYLDIMETHYLAMINE
60-24-2	BETA-HYDROXYETHYLMERCAPTAN
109-78-4	BETA-HYDROXYPROPIONITRILE
26760-64-5	BETA-ISO-AMYLENE
513-35-9	BETA-ISOAMYLENE
30674-80-7	BETA-ISOCYANATOETHYL METHACRYLATE
67-64-1	BETA-KETOPROPANE
60-24-2	BETA-MERCAPTOETHANOL
583-15-3	BETA-MERCURIBENZOATE
1344-48-5	BETA-MERCURIC SULFIDE
513-42-8	BETA-METHALLYL ALCOHOL
563-47-3	BETA-METHALLYL CHLORIDE
111-77-3	BETA-METHOXY-BETA'-HYDROXYDIETHYL ETHER
109-86-4	BETA-METHOXYETHANOL
110-49-6	BETA-METHOXYETHYL ACETATE
4170-30-3	BETA-METHYL ACROLEIN
123-73-9	BETA-METHYL ACROLEIN (DOT)
4170-30-3	BETA-METHYL ACROLEIN (DOT)
123-73-9	BETA-METHYLACROLEIN
3724-65-0	BETA-METHYLACRYLIC ACID
513-42-8	BETA-METHYLALLYL ALCOHOL
563-47-3	BETA-METHYLALLYL CHLORIDE
78-79-5	BETA-METHYLBIVINYL
590-86-3	BETA-METHYLBUTANAL
110-19-0	BETA-METHYLPROPYL ETHANOATE
91-59-8	BETA-NAPHTHYLAMINE
91-59-8	BETA-NAPHTHYLAMINE (ACGIH,DOT)
556-88-7	BETA-NITROGUANIDINE
79-46-9	BETA-NITROPROPANE
25322-01-4	BETA-NITROPROPANE
98-83-9	BETA-PHENYLPROPENE
98-83-9	BETA-PHENYLPROPYLENE
57-57-8	BETA-PROPIOLACTONE
57-57-8	BETA-PROPIOLACTONE (ACGIH)
57-57-8	BETA-PROPIONOLACTONE
123-05-7	BETA-PROPYL-ALPHA-ETHYLACROLEIN
129-00-0	BETA-PYRENE
54-11-5	BETA-PYRIDYL-ALPHA-N-METHYLPYRROLIDINE
123-31-9	BETA-QUINOL
62-56-6	BETA-THIOPSEUDOUREA
57-57-8	BETAPRONE
137-26-8	BETOXIN
58-89-9	BEXOL
115-26-4	BFP
115-26-4	BFPO
50-00-0	BFV
94-75-7	BH 2,4-D
75-99-0	BH DALAPON
58-89-9	BHC
128-37-0	BHT
60-51-5	BI 58
60-51-5	BI 58 EC
431-03-8	BIACETYL
50-78-2	BIALPIRINA
92-52-4	BIBENZENE
62-73-7	BIBESOL
2636-26-2	BIBESOL
74-85-1	BICARBURRETTED HYDROGEN
79-43-6	BICHLORACETIC ACID
2385-85-5	BICHLORENDO
7487-94-7	BICHLORIDE OF MERCURY
107-06-2	BICHLORURE D'ETHYLENE (FRENCH)
7487-94-7	BICHLORURE DE MERCURE (FRENCH)
7789-09-5	BICHROMATE D'AMMONIUM (FRENCH)
10588-01-9	BICHROMATE DE SODIUM (FRENCH)
7778-50-9	BICHROMATE OF POTASH
10588-01-9	BICHROMATE OF SODA
280-57-9	BICYCLO(2,2,2)-1,4-DIAZAOCTANE
76-22-2	BICYCLO(221)HEPTAN-2-ONE, 1,7,7-TRIMETHYL-
1330-16-1	BICYCLO(311)HEPTANE, 2,6,6-TRIMETHYL-, DIDEHYDRO DERIV
498-15-7	BICYCLO(410)HEPT-3-ENE, 3,7,7-TRIMETHYL- (9CI)
15271-41-7	BICYCLOHEPTANE-2-CARBONITRILE, 5-CHLORO-6-((((METHYL-AMINO)CARBONYL)OXY)IMINO)-
77-73-6	BICYCLOPENTADIENE
121-46-0	BICYCLO[2.2.1]HEPTA-2,5-DIENE
121-46-0	BICYCLO[2.2.1]HEPTADIENE
507-70-0	BICYCLO[2.2.1]HEPTAN-2-OL, 1,7,7-TRIMETHYL-, ENDO-
16219-75-3	BICYCLO[221]HEPT-2-ENE, 5-ETHYLIDENE-
79-92-5	BICYCLO[221]HEPTANE, 2,2-DIMETHYL-3-METHYLENE-

CAS No.	Chemical Name
15271-41-7	BICYCLO[221]HEPTANE-2-CARBONITRILE, 5-CHLORO-6-[[[(METHYLAMINO)CARBONYL]OXY]IMINO]-, [1S-(1ALPHA,2BE
80-56-8	BICYCLO[3.1.1]HEPT-2-ENE, 2,6,6-TRIMETHYL-
91-17-8	BICYCLO[4.4.0]DECANE
498-15-7	BICYCLO[410]HEPT-3-ENE, 3,7,7-TRIMETHYL-, (1S)-
102-79-4	BIDE
141-66-2	BIDRIN
106-99-0	BIETHYLENE
25339-57-5	BIETHYLENE
7681-38-1	BIF
18130-74-0	BIFLUORIDE
18130-74-0	BIFLUORIDE (HF2 -)
18130-74-0	BIFLUORIDE ANION (HF2 1-)
18130-74-0	BIFLUORIDE, N.O.S.
7789-29-9	BIFLUORURE DE POTASSIUM (FRENCH)
107-22-2	BIFORMAL
107-22-2	BIFORMYL
122-39-4	BIG DIPPER
79-29-8	BIISOPROPYL
52-68-6	BILARCIL
6923-22-4	BILOBORN
6923-22-4	BILOBRAN
64-18-6	BILORIN
108-01-0	BIMANOL
74-84-0	BIMETHYL
584-79-2	BINAMIN FORTE
99-65-0	BINITROBENZENE 2,4-DINITROBENZENE
115-29-7	BIO 5,462
56-53-1	BIO-DES
57-83-0	BIO-LUTON
25155-30-0	BIO-SOFT D 40
25155-30-0	BIO-SOFT D-35X
25155-30-0	BIO-SOFT D-40
58253-04-6	BIO-SOFT D-60
25155-30-0	BIO-SOFT D-62
27176-87-0	BIO-SOFT S 100
584-79-2	BIOALLETHRIN
1305-62-0	BIOCALC
4342-03-4	BIOCARBAZINE R
107-02-8	BIOCIDE
15663-27-1	BIOCISPLATINUM
74-82-8	BIOGAS
50-55-5	BIOSERPINE
3383-96-8	BIOTHION
1464-53-5	BIOXIRAN
1464-53-5	BIOXIRANE
10102-43-9	BIOXYDE D'AZOTE (FRENCH)
1309-60-0	BIOXYDE DE PLOMB (FRENCH)
8020-83-5	BIPHAGITTOL
92-52-4	BIPHENYL
92-52-4	BIPHENYL (ACGIH)
101-84-8	BIPHENYL OXIDE
643-58-3	BIPHENYL, 2-METHYL-
92-66-0	BIPHENYL, 4-BROMO-
92-93-3	BIPHENYL, 4-NITRO-
1336-36-3	BIPHENYL, CHLORINATED
8004-13-5	BIPHENYL, MIXED WITH BIPHENYL OXIDE (3:7)
28984-85-2	BIPHENYL, NITRO-
1336-36-3	BIPHENYL, POLYCHLORO-
8004-13-5	BIPHENYL-DIPHENYL ETHER MIXTURE
8004-13-5	BIPHENYL-PHENYL ETHER MIXTURE
7789-00-6	BIPOTASSIUM CHROMATE
470-90-6	BIRLAN
470-90-6	BIRLANE
470-90-6	BIRLANE 10G
470-90-6	BIRLANE 24
628-63-7	BIRNENOEL
137-26-8	BIS((DIMETHYLAMINO)CARBONOTHIOYL) DISULPHIDE
93-96-9	BIS(ALPHA-METHYLBENZYL) ETHER
93-96-9	BIS(ALPHA-PHENYLETHYL) ETHER
111-91-1	BIS(BETA-CHLOROETHYL) FORMAL
111-48-8	BIS(BETA-HYDROXYETHYL) SULFIDE
102-79-4	BIS(BETA-HYDROXYETHYL)BUTYLAMINE
102-54-5	BIS(ETA-CYCLOPENTADIENYL)IRON
1271-28-9	BIS(ETA 5-2,4-CYCLOPENTADIEN-1-YL)NICKEL
110-05-4	BIS(1,1-DIMETHYLETHYL) PEROXIDE
101-68-8	BIS(1,4-ISOCYANATOPHENYL)METHANE
112-58-3	BIS(1-HEXYL) ETHER
143-16-8	BIS(1-HEXYL)AMINE
2407-94-5	BIS(1-HYDROXYCYCLOHEXYL)PEROXIDE
80-43-3	BIS(1-METHYL-1-PHENYLETHYL) PEROXIDE
626-23-3	BIS(1-METHYLPROPYL)AMINE
2238-07-5	BIS(2,3-EPOXYPROPYL)ETHER
100-97-0	BIS(2,4,6-TRINITROPHENYL)AMINE
131-73-7	BIS(2,4,6-TRINITROPHENYL)AMINE
2217-06-3	BIS(2,4,6-TRINITROPHENYL)SULFIDE
133-14-2	BIS(2,4-DICHLOROBENZOYL) PEROXIDE
638-56-2	BIS(2-(2-CHLOROETHOXY)ETHYL)ETHER
111-40-0	BIS(2-AMINOETHYL)AMINE
56-18-8	BIS(2-AMINOPROPYL)AMINE
63283-80-7	BIS(2-CHLORO-1-METHYLETHYL) ETHER
108-60-1	BIS(2-CHLORO-1-METHYLETHYL)ETHER
112-26-5	BIS(2-CHLOROETHOXY)ETHANE
111-91-1	BIS(2-CHLOROETHOXY)METHANE
111-44-4	BIS(2-CHLOROETHYL) ETHER
111-91-1	BIS(2-CHLOROETHYL) FORMAL
538-07-8	BIS(2-CHLOROETHYL)ETHYLAMINE
51-75-2	BIS(2-CHLOROETHYL)METHYLAMINE
55-86-7	BIS(2-CHLOROETHYL)METHYLAMINE HYDROCHLORIDE
50-18-0	BIS(2-CHLOROETHYL)PHOSPHORAMIDE CYCLIC PROPANOLAMIDE ESTER
505-60-2	BIS(2-CHLOROETHYL)SULFIDE
505-60-2	BIS(2-CHLOROETHYL)SULPHIDE
63283-80-7	BIS(2-CHLOROISOPROPYL) ETHER
39638-32-9	BIS(2-CHLOROISOPROPYL) ETHER
542-88-1	BIS(2-CHLOROMETHYL)ETHER
117-81-7	BIS(2-ETHYLHEXYL) 1,2-BENZENEDICARBOXYLATE
123-79-5	BIS(2-ETHYLHEXYL) ADIPATE
117-81-7	BIS(2-ETHYLHEXYL) O-PHTHALATE
16111-62-9	BIS(2-ETHYLHEXYL) PERDICARBONATE
16111-62-9	BIS(2-ETHYLHEXYL) PEROXYDICARBONATE
577-11-7	BIS(2-ETHYLHEXYL) S-SODIUM SULFOSUCCINATE
577-11-7	BIS(2-ETHYLHEXYL) SODIOSULFOSUCCINATE
577-11-7	BIS(2-ETHYLHEXYL) SULFOSUCCINATE SODIUM SALT
117-84-0	BIS(2-ETHYLHEXYL)-1,2-BENZENEDICARBOXYLATE
103-23-1	BIS(2-ETHYLHEXYL)ADIPATE
106-20-7	BIS(2-ETHYLHEXYL)AMINE
298-07-7	BIS(2-ETHYLHEXYL)HYDROGEN PHOSPHATE
298-07-7	BIS(2-ETHYLHEXYL)ORTHOPHOSPHORIC ACID
298-07-7	BIS(2-ETHYLHEXYL)PHOSPHATE
298-07-7	BIS(2-ETHYLHEXYL)PHOSPHORIC ACID
117-81-7	BIS(2-ETHYLHEXYL)PHTHALATE
117-84-0	BIS(2-ETHYLHEXYL)PHTHALATE
577-11-7	BIS(2-ETHYLHEXYL)SODIUM SULFOSUCCINATE
70-30-4	BIS(2-HYDROXY-3,5,6-TRICHLOROPHENYL)METHANE
111-46-6	BIS(2-HYDROXYETHYL) ETHER
105-59-9	BIS(2-HYDROXYETHYL) METHYL AMINE
111-48-8	BIS(2-HYDROXYETHYL) SULFIDE
111-48-8	BIS(2-HYDROXYETHYL) THIOETHER
111-42-2	BIS(2-HYDROXYETHYL)AMINE
143-00-0	BIS(2-HYDROXYETHYL)AMMONIUM LAURYL SULFATE
110-97-4	BIS(2-HYDROXYPROPYL)AMINE
111-96-6	BIS(2-METHOXYETHYL)ETHER
96-69-5	BIS(2-METHYL-4-HYDROXY-5-TERT-BUTYLPHENYL) SULFIDE
96-69-5	BIS(2-METHYL-5-TERT-BUTYL-4-HYDROXYPHENYL) SULFIDE
80-43-3	BIS(2-PHENYL-2-PROPYL) PEROXIDE
110-97-4	BIS(2-PROPANOL)AMINE
70-30-4	BIS(3,5,6-TRICHLORO-2-HYDROXYPHENYL)METHANE
56-18-8	BIS(3-AMINOPROPYL)AMINE
123-23-9	BIS(3-CARBOXYPROPIONYL) PEROXIDE
101-14-4	BIS(3-CHLORO-4-AMINOPHENYL)METHANE
62207-76-5	BIS(3-FLUOROSALICYLAL)ETHYLENEDIAMINECOBALT(II)
838-88-0	BIS(3-METHYL-4-AMINOPHENYL)METHANE
101-14-4	BIS(4-AMINO-3-CHLOROPHENYL)METHANE
101-80-4	BIS(4-AMINOPHENYL) ETHER
139-65-1	BIS(4-AMINOPHENYL) SULFIDE
101-77-9	BIS(4-AMINOPHENYL)METHANE
94-17-7	BIS(4-CHLOROBENZOYL) PEROXIDE
90-94-8	BIS(4-DIMETHYLAMINOPHENYL) KETONE
96-69-5	BIS(4-HYDROXY-5-TERT-BUTYL-2-METHYLPHENYL) SULFIDE
5124-30-1	BIS(4-ISOCYANATOCYCLOHEXYL)METHANE
101-68-8	BIS(4-ISOCYANATOPHENYL)METHANE
15520-11-3	BIS(4-TERT-BUTYLCYCLOHEXYL) PEROXYDICARBONATE
71-48-7	BIS(ACETATO)COBALT
1335-32-6	BIS(ACETATO)DIHYDROXYTRILEAD
541-09-3	BIS(ACETATO)DIOXOURANIUM
4229-34-9	BIS(ACETATO)OXOZIRCONIUM
3227-63-2	BIS(ACETATO)OXOZIRCONIUM

CAS No.	Chemical Name
5153-24-2	BIS(ACETATO)OXOZIRCONIUM
543-90-8	BIS(ACETOXY)CADMIUM
1600-27-7	BIS(ACETYLOXY)MERCURY
80-43-3	BIS(ALPHA,ALPHA-DIMETHYLBENZYL)PEROXIDE
13479-54-4	BIS(AMINOACETATO)COPPER
101-77-9	BIS(AMINOPHENYL)METHANE
56-18-8	BIS(AMINOPROPYL)AMINE
7209-38-3	BIS(AMINOPROPYL)PIPERAZINE
506-87-6	BIS(AMMONIUM) CARBONATE
111-40-0	BIS(BETA-AMINOETHYL)AMINE
111-44-4	BIS(BETA-CHLOROETHYL) ETHER
51-75-2	BIS(BETA-CHLOROETHYL)METHYLAMINE
505-60-2	BIS(BETA-CHLOROETHYL)SULFIDE
111-46-6	BIS(BETA-HYDROXYETHYL) ETHER
152-16-9	BIS(BISDIMETHYLAMINOPHOSPHONOUS)ANHYDRIDE
111-44-4	BIS(CHLORO-2-ETHYL) OXIDE
28347-13-9	BIS(CHLOROMETHYL)BENZENE
542-88-1	BIS(CHLOROMETHYL)ETHER
534-07-6	BIS(CHLOROMETHYL)KETONE
102-54-5	BIS(CYCLOPENTADIENYL)IRON
97-77-8	BIS(DIETHYLTHIOCARBAMOYL) DISULFIDE
137-26-8	BIS(DIMETHYL-THIOCARBAMOYL)-DISULFID (GERMAN)
115-26-4	BIS(DIMETHYLAMIDO)FLUOROPHOSPHATE
115-26-4	BIS(DIMETHYLAMIDO)FLUOROPHOSPHINE OXIDE
115-26-4	BIS(DIMETHYLAMIDO)PHOSPHORYL FLUORIDE
1031-47-6	BIS(DIMETHYLAMINO)-3-AMINO-5-PHENYLTRIAZOLYL PHOSPHINE OXIDE
115-26-4	BIS(DIMETHYLAMINO)FLUOROPHOSPHATE
152-16-9	BIS(DIMETHYLAMINO)PHOSPHONOUS ANHYDRIDE
152-16-9	BIS(DIMETHYLAMINO)PHOSPHORIC ANHYDRIDE
137-26-8	BIS(DIMETHYLTHIOCARBAMOYL) DISULFIDE
137-26-8	BIS(DIMETHYLTHIOCARBAMOYL) DISULPHIDE
137-26-8	BIS(DIMETHYLTHIOCARBAMYL) DISULFIDE
78-34-2	BIS(DITHIOPHOSPHATE DE O,O-DIETHYLE) DE S,S'-(1,4-DIOXANNE-2,3-DIYLE) (FRENCH)
13426-91-0	BIS(ETHYLENEDIAMINE)COPPER ION
13426-91-0	BIS(ETHYLENEDIAMINE)COPPER(2+)
13426-91-0	BIS(ETHYLENEDIAMINE)COPPER(2+) ION
13426-91-0	BIS(ETHYLENEDIAMINE)COPPER(II)
577-11-7	BIS(ETHYLHEXYL) ESTER OF SODIUM SULFOSUCCINIC ACID
16111-62-9	BIS(ETHYLHEXYL) PEROXYDICARBONATE
2235-25-8	BIS(ETHYLMERCURI)PHOSPHATE
13479-54-4	BIS(GLYCINATO)COPPER
13479-54-4	BIS(GLYCINATO)COPPER(II)
111-42-2	BIS(HYDROXYETHYL)AMINE
10039-54-0	BIS(HYDROXYLAMINE) SULFATE
110-65-6	BIS(HYDROXYMETHYL)ACETYLENE
10070-67-0	BIS(HYDROXYMETHYL)ACETYLENE
794-93-4	BIS(HYDROXYMETHYL)FURATRIZINE
1191-15-7	BIS(ISO-BUTYL)ALUMINUM HYDRIDE
1191-15-7	BIS(ISOBUTYL)HYDROALUMINUM
26761-40-0	BIS(ISODECYL)PHTHALATE
108-20-3	BIS(ISOPROPYL) ETHER
371-86-8	BIS(ISOPROPYLAMIDO) FLUOROPHOSPHATE
371-86-8	BIS(MONOISOPROPYLAMINO)FLUOROPHOSPHATE
97-77-8	BIS(N,N-DIETHYLTHIOCARBAMOYL) DISULFIDE
101-80-4	BIS(P-AMINOPHENYL) ETHER
101-77-9	BIS(P-AMINOPHENYL)METHANE
94-17-7	BIS(P-CHLOROBENZOYL) PEROXIDE
101-68-8	BIS(P-ISOCYANATOPHENYL)METHANE
563-12-2	BIS(S-(DIETHOXYPHOSPHINOTHIOYL)MERCAPTO)METHANE
110-05-4	BIS(TERT-BUTYL) PEROXIDE
116-16-5	BIS(TRICHLOROMETHYL) KETONE
999-97-3	BIS(TRIMETHYLSILYL)AMINE
2407-94-5	BIS-(1-HYDROXYCYCLOHEXYL)PEROXIDE, TECHNICAL PURE (DOT)
152-16-9	BIS-N,N,N',N'-TETRAMETHYLPHOSPHORODIAMIDIC ANHYDRIDE
84-74-2	BIS-N-BUTYL PHTHALATE
107-49-3	BIS-O,O-DIETHYLPHOSPHORIC ANHYDRIDE
3689-24-5	BIS-O,O-DIETHYLPHOSPHOROTHIONIC ANHYDRIDE
101-77-9	BIS-P-AMINOFENYLMETHAN (CZECH)
154-93-8	BISCHLOROETHYL NITROSOUREA
77-73-6	BISCYCLOPENTADIENE
115-26-4	BISDIMETHYLAMINOFLUOROPHOSPHINE OXIDE
371-86-8	BISISOPROPYLAMINOFLUOROPHOSPHINE OXIDE
1304-82-1	BISMUTH SESQUITELLURIDE
1304-82-1	BISMUTH TELLURIDE
37293-14-4	BISMUTH TELLURIDE
1304-82-1	BISMUTH TELLURIDE (Bi2Te3)
1304-82-1	BISMUTH(3+) TELLURIDE
7568-93-6	BISNOREPHEDRINE
117-84-0	BISOFLEX 81
117-81-7	BISOFLEX 81
103-23-1	BISOFLEX DOA
123-79-5	BISOFLEX DOA
117-81-7	BISOFLEX DOP
117-84-0	BISOFLEX DOP
818-61-1	BISOMER 2HEA
136-40-3	BISTERIL
7681-38-1	BISULFATE OF SODA
7446-09-5	BISULFITE
7631-90-5	BISULFITE DE SODIUM (FRENCH)
101-61-1	BIS[4-(DIMETHYLAMINO)PHENYL]METHANE
101-61-1	BIS[4-(N,N-DIMETHYLAMINO)PHENYL]METHANE
101-61-1	BIS[P-(DIMETHYLAMINO)PHENYL]METHANE
90-94-8	BIS[P-(N,N-DIMETHYLAMINO)PHENYL] KETONE
101-61-1	BIS[P-(N,N-DIMETHYLAMINO)PHENYL]METHANE
4044-65-9	BITOSCANATE
8052-42-4	BITUMEN
8052-42-4	BITUMENS, ASPHALT
8052-42-4	BITUMINOUS MATERIALS, ASPHALT
8052-42-4	BITUSIZE B
541-53-7	BIURET, 2,4-DITHIO-
541-53-7	BIURET, DITHIO-
106-99-0	BIVINYL
25339-57-5	BIVINYL
9004-70-0	BK2-W
9004-70-0	BK2-Z
1937-37-7	BLACK 2EMBL
1937-37-7	BLACK 4EMBL
123-31-9	BLACK AND WHITE BLEACHING CREAM
1333-86-4	BLACK FW
7782-42-5	BLACK LEAD
65-30-5	BLACK LEAF 40
1333-86-4	BLACK PEARLS
79-01-6	BLACOSOLV
3689-24-5	BLADAFUM
3689-24-5	BLADAFUME
107-49-3	BLADAN
757-58-4	BLADAN
757-58-4	BLADAN BASE
56-38-2	BLADAN F
298-00-0	BLADAN-M
21725-46-2	BLADEX
1929-73-3	BLADEX B
93-79-8	BLADEX H
2545-59-7	BLADEX H
79-01-6	BLANCOSOLV
8020-83-5	BLANDOL WHITE MINERAL OIL
7775-14-6	BLANKIT
7775-14-6	BLANKIT IN
65996-68-1	BLAST FURNACE GAS (FERROUS METAL)
55-63-0	BLASTING GELATIN
55-63-0	BLASTING GELATIN (DOT)
55-63-0	BLASTING OIL
114-26-1	BLATTANEX
928-96-1	BLATTERALKOHOL
114-26-1	BLATTOSEP
74-90-8	BLAUSAEURE (GERMAN)
74-90-8	BLAUWZUUR (DUTCH)
7778-54-3	BLEACHING POWDER
7428-48-0	BLEISTEARAT (GERMAN)
7446-14-2	BLEISULFAT (GERMAN)
15739-80-7	BLEISULFAT (GERMAN)
299-84-3	BLITEX
9016-45-9	BLM
7440-70-2	BLOOD-COAGULATION FACTOR IV
8006-20-0	BLOW GAS
2602-46-2	BLUE 2B
2602-46-2	BLUE 2B SALT
72-57-1	BLUE 3B
12001-28-4	BLUE ASBESTOS
12001-28-4	BLUE ASBESTOS (DOT)
119-90-4	BLUE BASE IRGA B
119-90-4	BLUE BASE NB
119-90-4	BLUE BN BASE
7758-98-7	BLUE COPPER
712-48-1	BLUE CROSS

CAS No.	Chemical Name
72-57-1	BLUE EMB
8021-92-9	BLUE GAS
62-53-3	BLUE OIL
7440-66-6	BLUE POWDER
7758-98-7	BLUE STONE
7758-98-7	BLUE VITRIOL
1314-84-7	BLUE-OX
108-31-6	.BM 10
88-85-7	BNP 20
88-85-7	BNP 30
9016-45-9	BO
8001-86-3	BOLAKO OIL
8001-86-3	BOLEKO OIL
110-17-8	BOLETIC ACID
114-26-1	BOLFO
59164-68-0	BOLLS-EYE
124-65-2	BOLLS-EYE
35400-43-2	BOLSTAR
1332-58-7	BOLUS ALBA
105-60-2	BONAMID
7733-02-0	BONAZEN
126-33-0	BONDELANE A
126-33-0	BONDOLANE A
7723-14-0	BONIDE BLUE DEATH RAT KILLER
79-34-5	BONOFORM
10043-35-3	BORACIC ACID
19287-45-7	BORANE (B2H6)
10294-33-4	BORANE, TRIBROMO-
97-94-9	BORANE, TRIETHYL-
7637-07-2	BORANE, TRIFLUORO-
353-42-4	BORANE, TRIFLUORO-, COMPD WITH OXYBIS[METHANE] (1:1)
109-63-7	BORANE, TRIFLUORO-, COMPD. WITH 1,1'-OXYBIS[ETHANE] (1:1)
13319-75-0	BORANE, TRIFLUORO-, DIHYDRATE
1303-96-4	BORASCU
13826-83-0	BORATE(1-), TETRAFLUORO-, AMMONIUM
14486-19-2	BORATE(1-), TETRAFLUORO-, CADMIUM
14486-19-2	BORATE(1-), TETRAFLUORO-, CADMIUM (2:1)
13814-96-5	BORATE(1-), TETRAFLUORO-, LEAD (2+)
13814-96-5	BORATE(1-), TETRAFLUORO-, LEAD(2+) (2:1)
14708-14-6	BORATE(1-), TETRAFLUORO-, NICKEL(2+)
14708-14-6	BORATE(1-), TETRAFLUORO-, NICKEL(2+) (2:1)
14708-14-6	BORATE(1-), TETRAFLUORO-, NICKEL(2+) (2:1) (9CI)
13826-88-5	BORATE(1-), TETRAFLUORO-, ZINC
13826-88-5	BORATE(1-), TETRAFLUORO-, ZINC (2:1)
16962-07-5	BORATE(1-), TETRAHYDRO-, ALUMINUM
16962-07-5	BORATE(1-), TETRAHYDRO-, ALUMINUM (3:1)
16962-07-5	BORATE(1-), TETRAHYDRO-, ALUMINUM (3:1) (9CI)
16962-07-5	BORATE(1-), TETRAHYDRO-, ALUMINUM SALT
16949-15-8	BORATE(1-), TETRAHYDRO-, LITHIUM
13762-51-1	BORATE(1-), TETRAHYDRO-, POTASSIUM
13762-51-1	BORATE(1-), TETRAHYDRO-, POTASSIUM (8CI,9CI)
16940-66-2	BORATE(1-), TETRAHYDRO-, SODIUM
1303-96-4	BORATE, TETRASODIUM SALT
1303-96-4	BORAX
1303-96-4	BORAX (B4Na2O7.10H2O)
1332-07-6	BORAX 2335
1303-96-4	BORAX DECAHYDRATE
1330-43-4	BORAX GLASS
107-06-2	BORER SOL
688-74-4	BORESTER 2
121-43-7	BORESTER O
10043-35-3	BORIC ACID
1330-43-4	BORIC ACID (H2B4O7), DISODIUM SALT
1303-96-4	BORIC ACID (H2B4O7), DISODIUM SALT, DECAHYDRATE
10043-35-3	BORIC ACID (H3BO3)
1693-71-6	BORIC ACID (H3BO3), TRI-2-PROPENYL ESTER
1693-71-6	BORIC ACID (H3BO3), TRIALLYL ESTER
688-74-4	BORIC ACID (H3BO3), TRIBUTYL ESTER
13195-76-1	BORIC ACID (H3BO3), TRIISOBUTYL ESTER
121-43-7	BORIC ACID (H3BO3), TRIMETHYL ESTER
13195-76-1	BORIC ACID (H3BO3), TRIS(2-METHYLPROPYL) ESTER
12007-89-5	BORIC ACID (HB5O8), AMMONIUM SALT
1303-86-2	BORIC ACID (HBO2), ANHYDRIDE
121-43-7	BORIC ACID TRIMETHYL ESTER
51845-86-4	BORIC ACID, ETHYL ESTER
1693-71-6	BORIC ACID, TRI-2-PROPENYL ESTER
1693-71-6	BORIC ACID, TRIALLYL ESTER
5419-55-6	BORIC ACID, TRIISOPROPYL ESTER
5419-55-6	BORIC ACID, TRIS(1-METHYLETHYL) ESTER

CAS No.	Chemical Name
1332-07-6	BORIC ACID, ZINC SALT
1303-86-2	BORIC ANHYDRIDE
1303-86-2	BORIC OXIDE
1303-86-2	BORIC OXIDE (B2O3)
1303-96-4	BORICIN
76-22-2	BORNANE, 2-OXO-
507-70-0	BORNEOL
28324-52-9	BORNYL CHLORIDE
19287-45-7	BOROETHANE
10043-35-3	BOROFAX
13762-51-1	BOROHYDRURE DE POTASSIUM (FRENCH)
16940-66-2	BOROHYDRURE DE SODIUM (FRENCH)
16940-66-2	BOROL
10294-33-4	BORON BROMIDE
10294-33-4	BORON BROMIDE (BBr3)
7637-07-2	BORON FLUORIDE
7637-07-2	BORON FLUORIDE (BF3)
13319-75-0	BORON FLUORIDE (BF3) DIHYDRATE
353-42-4	BORON FLUORIDE (BF3), COMPD WITH METHYL ETHER (1:1)
109-63-7	BORON FLUORIDE (BF3), COMPD. WITH ETHYL ETHER (1:1)
353-42-4	BORON FLUORIDE COMPLEX WITH DIMETHYL ETHER
109-63-7	BORON FLUORIDE DIETHYL ETHER COMPLEX
109-63-7	BORON FLUORIDE DIETHYL ETHERATE
13319-75-0	BORON FLUORIDE DIHYDRATE
109-63-7	BORON FLUORIDE ETHERATE
109-63-7	BORON FLUORIDE MONOETHERATE
109-63-7	BORON FLUORIDE-DIETHYL ETHER COMPOUND
109-63-7	BORON FLUORIDE-ETHYL ETHER COMPLEX
109-63-7	BORON FLUORIDE-ETHYL ETHERATE
19287-45-7	BORON HYDRIDE
17702-41-9	BORON HYDRIDE (B10H14)
19287-45-7	BORON HYDRIDE (B2H6)
5419-55-6	BORON ISOPROPOXIDE
1303-86-2	BORON OXIDE
54566-73-3	BORON OXIDE
1303-86-2	BORON OXIDE (B2O3)
1303-86-2	BORON SESQUIOXIDE
1330-43-4	BORON SODIUM OXIDE (B4Na2O7)
1303-96-4	BORON SODIUM OXIDE (B4Na2O7), DECAHYDRATE
10294-33-4	BORON TRIBROMIDE
688-74-4	BORON TRIBUTOXIDE
10294-34-5	BORON TRICHLORIDE
7637-07-2	BORON TRIFLUORIDE
753-53-7	BORON TRIFLUORIDE ACETIC ACID COMPLEX
353-42-4	BORON TRIFLUORIDE COMPD WITH METHYL ETHER
353-42-4	BORON TRIFLUORIDE COMPOUND WITH METHYL ETHER (1:1)
109-63-7	BORON TRIFLUORIDE DIETHYL ETHERATE
13319-75-0	BORON TRIFLUORIDE DIHYDRATE
13319-75-0	BORON TRIFLUORIDE DIHYDRATE (DOT)
353-42-4	BORON TRIFLUORIDE DIMETHYL ETHERATE
353-42-4	BORON TRIFLUORIDE DIMETHYL ETHERATE (DOT)
109-63-7	BORON TRIFLUORIDE ETHERATE
109-63-7	BORON TRIFLUORIDE ETHYL ETHERATE (1:1)
109-63-7	BORON TRIFLUORIDE-DIETHYL ETHER 1:1 COMPLEX
109-63-7	BORON TRIFLUORIDE-DIETHYL ETHER COMPLEX
109-63-7	BORON TRIFLUORIDE-DIETHYL ETHER COMPLEX (1:1)
353-42-4	BORON TRIFLUORIDE-DIMETHYL ETHER
353-42-4	BORON TRIFLUORIDE-DIMETHYL ETHER COMPLEX
109-63-7	BORON TRIFLUORIDE-ETHER COMPLEX
109-63-7	BORON TRIFLUORIDE-ETHYL ETHER
109-63-7	BORON TRIFLUORIDE-ETHYL ETHER (1:1)
109-63-7	BORON TRIFLUORIDE-ETHYL ETHER COMPLEX
109-63-7	BORON TRIFLUORIDE-ETHYL ETHERATE
10043-35-3	BORON TRIHYDROXIDE
5419-55-6	BORON TRIISOPROPOXIDE
1303-86-2	BORON TRIOXIDE
109-63-7	BORON, TRIFLUORO[1,1'-OXYBIS[ETHANE]]-, (T-4)-
353-42-4	BORON, TRIFLUORO[OXYBIS[METHANE]]-, (T-4)-
10043-35-3	BORSAURE (GERMAN)
101-05-3	BORTRYSAN
12654-97-6	BORTRYSAN
114-26-1	BORUHO
114-26-1	BORUHO 50
50-29-3	BOSAN SUPRA
82-68-8	BOTRILEX
7664-93-9	BOV
50-29-3	BOVIDERMOL
52-68-6	BOVINOX
9002-89-5	BOVLON

CAS No.	Chemical Name
9004-70-0	BOX TOE GUM
50-32-8	BP
57-57-8	BPL
2312-35-8	BPPS
1937-37-7	BRASILAMINA BLACK GN
2602-46-2	BRASILAMINA BLUE 2B
72-57-1	BRASILAMINA BLUE 3B
97-56-3	BRASILAZINA OIL YELLOW R
82-68-8	BRASSICOL
82-68-8	BRASSICOL 75
82-68-8	BRASSICOL SUPER
111-46-6	BRECOLANE NDG
88-74-4	BRENTAMINE FAST ORANGE GR BASE
88-74-4	BRENTAMINE FAST ORANGE GR SALT
95-79-4	BRENTAMINE FAST RED TR BASE
3165-93-3	BRENTAMINE FAST RED TR SALT
639-58-7	BRESTANOL
62-73-7	BREVINYL
62-73-7	BREVINYL E 50
8001-58-9	BRICK OIL
633-03-4	BRILLANT-GRUN (GERMAN)
1937-37-7	BRILLIANT CHROME LEATHER BLACK H
60-11-7	BRILLIANT FAST OIL YELLOW
60-11-7	BRILLIANT FAST SPIRIT YELLOW
633-03-4	BRILLIANT GREEN
633-03-4	BRILLIANT GREEN ASEPTIC
633-03-4	BRILLIANT GREEN B
633-03-4	BRILLIANT GREEN BP
633-03-4	BRILLIANT GREEN BP CRYSTALS
633-03-4	BRILLIANT GREEN BPC
633-03-4	BRILLIANT GREEN CRYSTALS
633-03-4	BRILLIANT GREEN CRYSTALS H
633-03-4	BRILLIANT GREEN DSC
633-03-4	BRILLIANT GREEN G
633-03-4	BRILLIANT GREEN GX
633-03-4	BRILLIANT GREEN LAKE
633-03-4	BRILLIANT GREEN P
633-03-4	BRILLIANT GREEN SPECIAL
633-03-4	BRILLIANT GREEN SULFATE
633-03-4	BRILLIANT GREEN WP CRYSTALS
633-03-4	BRILLIANT GREEN Y
633-03-4	BRILLIANT GREEN YN
633-03-4	BRILLIANT GREEN YNS
633-03-4	BRILLIANT LAKE GREEN Y
60-11-7	BRILLIANT OIL YELLOW
633-03-4	BRILLIANT TUNGSTATE GREEN TONER GT 288
7704-34-9	BRIMSTONE
52-68-6	BRITON
52-68-6	BRITTEN
107-06-2	BROCIDE
2921-88-2	BRODAN
7726-95-6	BROM (GERMAN)
8004-09-9	BROM-O-GAS
314-40-9	BROMACIL
28772-56-7	BROMADIALONE
28772-56-7	BROMADIOLON
28772-56-7	BROMADIOLONE
106-95-6	BROMALLYLENE
7789-38-0	BROMATE DE SODIUM (FRENCH)
314-40-9	BROMAZIL
5798-79-8	BROMBENZYL CYANIDE
300-76-5	BROMCHLOPHOS
7726-95-6	BROME (FRENCH)
300-76-5	BROMEX
300-76-5	BROMEX (INSECTICIDE)
300-76-5	BROMEX 50
13967-90-3	BROMIC ACID, BARIUM SALT
14519-17-6	BROMIC ACID, MAGNESIUM SALT
7758-01-2	BROMIC ACID, POTASSIUM SALT
7789-38-0	BROMIC ACID, SODIUM SALT
14519-07-4	BROMIC ACID, ZINC SALT
74-96-4	BROMIC ETHER
7726-95-6	BROMINE
7726-95-6	BROMINE (ACGIH,DOT)
13863-41-7	BROMINE CHLORIDE
506-68-3	BROMINE CYANIDE
506-68-3	BROMINE CYANIDE (BrCN)
7726-95-6	BROMINE ELEMENT
7787-71-5	BROMINE FLUORIDE (BrF3)
7789-30-2	BROMINE FLUORIDE (BrF5)
7726-95-6	BROMINE MOLECULE (br2)
506-68-3	BROMINE MONOCYANIDE
7789-30-2	BROMINE PENTAFLUORIDE
7726-95-6	BROMINE SOLUTION
7726-95-6	BROMINE SOLUTION (DOT)
7787-71-5	BROMINE TRIFLUORIDE
126-72-7	BROMKAL P 67-6HP
7726-95-6	BROMO (ITALIAN)
598-31-2	BROMO-2-PROPANONE
790-83-0	BROMOACETIC ACID, SOLUTION (DOT)
105-36-2	BROMOACETIC ACID ETHYL ESTER
96-32-2	BROMOACETIC ACID METHYL ESTER
790-83-0	BROMOACETIC ACID SOLUTION
79-08-3	BROMOACETIC ACID, SOLID OR SOLUTION
598-31-2	BROMOACETONE
598-31-2	BROMOACETONE (DOT)
598-31-2	BROMOACETONE, LIQUID
598-31-2	BROMOACETONE, LIQUID (DOT)
598-21-0	BROMOACETYL BROMIDE
598-21-0	BROMOACETYL BROMIDE (DOT)
108-86-1	BROMOBENZENE
108-86-1	BROMOBENZENE (DOT)
5798-79-8	BROMOBENZYL CYANIDE
5798-79-8	BROMOBENZYL CYANIDE (DOT)
5798-79-8	BROMOBENZYLNITRILE
353-59-3	BROMOCHLORODIFLUOROMETHANE
74-97-5	BROMOCHLOROMETHANE
74-97-5	BROMOCHLOROMETHANE (DOT)
506-68-3	BROMOCYAN
506-68-3	BROMOCYANIDE
506-68-3	BROMOCYANIDE (BrCN)
506-68-3	BROMOCYANOGEN
75-27-4	BROMODICHLOROMETHANE
776-74-9	BROMODIPHENYLMETHANE
74-96-4	BROMOETHANE
593-60-2	BROMOETHENE
593-60-2	BROMOETHENE (9CI)
592-55-2	BROMOETHYL ETHYL ETHER
593-60-2	BROMOETHYLENE
75-63-8	BROMOFLUOROFORM
75-25-2	BROMOFORM
75-25-2	BROMOFORM (ACGIH,DOT)
75-25-2	BROMOFORME (FRENCH)
75-25-2	BROMOFORMIO (ITALIAN)
106-93-4	BROMOFUME
3132-64-7	BROMOHYDRIN
74-83-9	BROMOMETHANE
598-31-2	BROMOMETHYL METHYL KETONE
70-11-1	BROMOMETHYL PHENYL KETONE
107-82-4	BROMOMETHYLBUTANE
507-19-7	BROMOMETHYLPROPANE
28772-56-7	BROMONE
29756-38-5	BROMOPENTANE
100-39-0	BROMOPHENYLMETHANE
106-94-5	BROMOPROPANE
26446-77-5	BROMOPROPANE
106-96-7	BROMOPROPYNE
598-73-2	BROMOTRIFLUOROETHENE
598-73-2	BROMOTRIFLUOROETHYLENE
598-73-2	BROMOTRIFLUOROETHYLENE (DOT)
75-63-8	BROMOTRIFLUOROMETHANE
75-63-8	BROMOTRIFLUOROMETHANE (DOT)
507-19-7	BROMOTRIMETHYLMETHANE
10035-10-6	BROMOWODOR (POLISH)
74-96-4	BROMURE D'ETHYLE
506-68-3	BROMURE DE CYANOGEN (FRENCH)
593-60-2	BROMURE DE VINYLE (FRENCH)
35884-77-6	BROMURE DE XYLYLE (FRENCH)
106-93-4	BROMURO DI ETILE (ITALIAN)
10035-10-6	BROMWASSERSTOFF (GERMAN)
7726-95-6	BROOM (DUTCH)
10035-10-6	BROOMWATERSTOF (DUTCH)
16071-86-6	BROWN 4EMBL
8004-09-9	BROZONE
300-76-5	BRP
357-57-3	BRUCIN (GERMAN)
357-57-3	BRUCINA (ITALIAN)
357-57-3	BRUCINE

CAS No.	Chemical Name
357-57-3	BRUCINE (DOT)
357-57-3	BRUCINE ALKALOID
357-57-3	BRUCINE, SOLID
357-57-3	BRUCINE, SOLID (DOT)
81-81-2	BRUMOLIN
1918-00-9	BRUSH BUSTER
1929-73-3	BRUSH KILLER 64
93-76-5	BRUSH RHAP
93-76-5	BRUSH-OFF 445 LOW VOLATILE BRUSH KILLER
94-75-7	BRUSH-RHAP
93-76-5	BRUSHTOX
114-26-1	BRYGOU
7440-32-6	BS 2TA6
7631-86-9	BS 30
7631-86-9	BS 30 (FILLER)
7631-86-9	BS 50
7631-86-9	BS 50 (SILICA)
56-93-9	BTM
112-34-5	BUCB
50-14-6	BUCO-D
2163-80-6	BUENO
2163-80-6	BUENO 6
62-38-4	BUFEN
62-38-4	BUFEN 30
56-53-1	BUFON
7733-02-0	BUFOPTO ZINC SULFATE
8003-34-7	BUHACH
8020-83-5	BUKOMKLEEN
128-37-0	BUKS
118-74-1	BUNT-CURE
118-74-1	BUNT-NO-MORE
1303-96-4	BURA
7775-14-6	BURMOL
1305-78-8	BURNT LIME
106-97-8	BURSHANE
68476-85-7	BURSHANE
9016-45-9	BURTEMUL N
21564-17-0	BUSAN
21564-17-0	BUSAN 15
21564-17-0	BUSAN 30
21564-17-0	BUSAN 30-1
21564-17-0	BUSAN 30A
21564-17-0	BUSAN 30I
21564-17-0	BUSAN 70
21564-17-0	BUSAN 71
21564-17-0	BUSAN 72
21564-17-0	BUSAN 72A
55-98-1	BUSULFAN
55-98-1	BUSULPHAN
106-99-0	BUTA-1,3-DIEEN (DUTCH)
25339-57-5	BUTA-1,3-DIEEN (DUTCH)
106-99-0	BUTA-1,3-DIEN (GERMAN)
25339-57-5	BUTA-1,3-DIEN (GERMAN)
106-99-0	BUTA-1,3-DIENE
25339-57-5	BUTA-1,3-DIENE
106-99-0	BUTADIEEN (DUTCH)
25339-57-5	BUTADIEEN (DUTCH)
106-99-0	BUTADIEN (POLISH)
25339-57-5	BUTADIEN (POLISH)
1464-53-5	BUTADIENDIOXYD (GERMAN)
106-99-0	BUTADIENE
25339-57-5	BUTADIENE
1464-53-5	BUTADIENE DIEPOXIDE
1464-53-5	BUTADIENE DIOXIDE
930-22-3	BUTADIENE EPOXIDE
930-22-3	BUTADIENE MONOEPOXIDE
930-22-3	BUTADIENE MONOOXIDE
930-22-3	BUTADIENE MONOXIDE
930-22-3	BUTADIENE OXIDE
87-68-3	BUTADIENE, HEXACHLORO-
106-99-0	BUTADIENE, INHIBITED
25339-57-5	BUTADIENE, INHIBITED
106-99-0	BUTADIENE, INHIBITED (DOT)
25339-57-5	BUTADIENE, INHIBITED (DOT)
85-43-8	BUTADIENE-MALEIC ANHYDRIDE ADDUCT
112-34-5	BUTADIGOL
513-49-5	BUTAFUME
13952-84-6	BUTAFUME
123-72-8	BUTAL

CAS No.	Chemical Name
123-72-8	BUTALDEHYDE
123-72-8	BUTALYDE
71-36-3	BUTAN-1-OL
78-92-2	BUTAN-2-OL
123-72-8	BUTANAL
110-69-0	BUTANAL OXIME
97-96-1	BUTANAL, 2-ETHYL-
96-17-3	BUTANAL, 2-METHYL-
107-89-1	BUTANAL, 3-HYDROXY-
590-86-3	BUTANAL, 3-METHYL-
123-72-8	BUTANALDEHYDE
102-01-2	BUTANAMIDE, 3-OXO-N-PHENYL-
97-36-9	BUTANAMIDE, N-(2,4-DIMETHYLPHENYL)-3-OXO-
92-15-9	BUTANAMIDE, N-(2-METHOXYPHENYL)-3-OXO-
122-82-7	BUTANAMIDE, N-(4-ETHOXYPHENYL)-3-OXO-
106-97-8	BUTANE
106-97-8	BUTANE (ACGIH,DOT)
109-99-9	BUTANE ALPHA,DELTA-OXIDE
1464-53-5	BUTANE DIEPOXIDE
142-96-1	BUTANE, 1,1'-OXYBIS-
616-21-7	BUTANE, 1,2-DICHLORO-
106-88-7	BUTANE, 1,2-EPOXY-
1464-53-5	BUTANE, 1,2:3,4-DIEPOXY-
110-56-5	BUTANE, 1,4-DICHLORO-
109-99-9	BUTANE, 1,4-EPOXY-
111-34-2	BUTANE, 1-(ETHENYLOXY)-
109-65-9	BUTANE, 1-BROMO-
107-82-4	BUTANE, 1-BROMO-3-METHYL-
109-69-3	BUTANE, 1-CHLORO-
107-84-6	BUTANE, 1-CHLORO-3-METHYL-
628-81-9	BUTANE, 1-ETHOXY-
111-36-4	BUTANE, 1-ISOCYANATO-
464-06-2	BUTANE, 2,2,3-TRIMETHYL-
75-83-2	BUTANE, 2,2-DIMETHYL-
7581-97-7	BUTANE, 2,3-DICHLORO-
79-29-8	BUTANE, 2,3-DIMETHYL-
78-76-2	BUTANE, 2-BROMO-
78-86-4	BUTANE, 2-CHLORO-
594-36-5	BUTANE, 2-CHLORO-2-METHYL-
25267-27-0	BUTANE, 2-IODO-
78-78-4	BUTANE, 2-METHYL-
25267-27-0	BUTANE, IODO-
109-52-4	BUTANECARBOXYLIC ACID
3333-52-6	BUTANEDINITRILE, TETRAMETHYL-
28300-74-5	BUTANEDIOIC ACID, 2,3-DIHYDROXY- [R-(R*,R*)]-, ANTIMONY COMPLEX
3164-29-2	BUTANEDIOIC ACID, 2,3-DIHYDROXY- [R-(R*,R*)]-, DIAMMONIUM SALT
815-82-7	BUTANEDIOIC ACID, 2,3-DIHYDROXY-, (R-(R*,R*)-, COPPER (2+) SALT
87-69-4	BUTANEDIOIC ACID, 2,3-DIHYROXY-
577-11-7	BUTANEDIOIC ACID, SULFO-, 1,4-BIS(2-ETHYLHEXYL) ESTER, SODIUM SALT
577-11-7	BUTANEDIOIC ACID, SULFO-, 1,4-BIS(2-ETHYLHEXYL) ESTER, SODIUM SALT (9CI)
121-75-5	BUTANEDIOIC ACID, [(DIMETHOXYPHOSPHINOTHIOYL)THIO]-, DIETHYL ESTER
431-03-8	BUTANEDIONE
106-97-8	BUTANEN (DUTCH)
109-74-0	BUTANENITRILE
628-20-6	BUTANENITRILE, 4-CHLORO-
628-20-6	BUTANENITRILE, 4-CHLORO- (9CI)
109-79-5	BUTANETHIOL
106-97-8	BUTANI (ITALIAN)
107-92-6	BUTANIC ACID
107-92-6	BUTANOIC ACID
105-54-4	BUTANOIC ACID ETHYL ESTER
540-18-1	BUTANOIC ACID PENTYL ESTER
638-11-9	BUTANOIC ACID, 1-METHYLETHYL ESTER
88-09-5	BUTANOIC ACID, 2-ETHYL-
3068-88-0	BUTANOIC ACID, 3-HYDROXY-, BETA-LACTONE
556-24-1	BUTANOIC ACID, 3-METHYL-, METHYL ESTER
556-24-1	BUTANOIC ACID, 3-METHYL-, METHYL ESTER (9CI)
141-97-9	BUTANOIC ACID, 3-OXO-, ETHYL ESTER
105-45-3	BUTANOIC ACID, 3-OXO-, METHYL ESTER
105-45-3	BUTANOIC ACID, 3-OXO-, METHYL ESTER (9CI)
123-23-9	BUTANOIC ACID, 4,4'-DIOXYBIS[4-OXO-
616-45-5	BUTANOIC ACID, 4-AMINO-, LACTAM
106-31-0	BUTANOIC ACID, ANHYDRIDE

CAS No.	Chemical Name
106-31-0	BUTANOIC ACID, ANHYDRIDE (9CI)
109-21-7	BUTANOIC ACID, BUTYL ESTER
123-20-6	BUTANOIC ACID, ETHENYL ESTER
623-42-7	BUTANOIC ACID, METHYL ESTER
105-66-8	BUTANOIC ACID, PROPYL ESTER
106-31-0	BUTANOIC ANHYDRIDE
71-36-3	BUTANOL
71-36-3	BUTANOL (DOT)
71-36-3	BUTANOL (FRENCH)
78-92-2	BUTANOL SECONDAIRE (FRENCH)
75-65-0	BUTANOL TERTIAIRE (FRENCH)
78-92-2	BUTANOL-2
71-36-3	BUTANOLEN (DUTCH)
71-36-3	BUTANOLO (ITALIAN)
78-93-3	BUTANONE
78-93-3	BUTANONE 2 (FRENCH)
1338-23-4	BUTANOX LPT
1338-23-4	BUTANOX M 105
1338-23-4	BUTANOX M 50
141-75-3	BUTANOYL CHLORIDE
88-85-7	BUTAPHEN
88-85-7	BUTAPHENE
94-80-4	BUTAPON
26760-64-5	BUTENE, 2-METHYL-
41070-66-9	BUTENE, OCTAFLUORO-
51023-22-4	BUTENE, TRICHLORO-
106-98-9	BUTENE-1
110-17-8	BUTENEDIOIC ACID, (E)-
78-94-4	BUTENONE
689-97-4	BUTENYNE
97-88-1	BUTIL METACRILATO (ITALIAN)
123-86-4	BUTILE (ACETATI DI) (ITALIAN)
592-34-7	BUTOXYCARBONYL CHLORIDE
112-34-5	BUTOXYDIETHYLENE GLYCOL
112-34-5	BUTOXYDIGLYCOL
1929-73-3	BUTOXYETHANOL ESTER OF 2,4-D
111-34-2	BUTOXYETHENE
112-34-5	BUTOXYETHOXYETHANOL
124-17-4	BUTOXYETHOXYETHYL ACETATE
1929-73-3	BUTOXYETHYL (2,4-DICHLOROPHENOXY)ACETATE
93-79-8	BUTOXYETHYL 2,4,5-T
2545-59-7	BUTOXYETHYL 2,4,5-T
2545-59-7	BUTOXYETHYL 2,4,5-TRICHLOROPHENOXYACETATE
111-34-2	BUTOXYETHYLENE
4435-53-4	BUTOXYL
107-92-6	BUTRIC ACID
10025-91-9	BUTTER OF ANTIMONY
7647-18-9	BUTTER OF ANTIMONY
7784-34-1	BUTTER OF ARSENIC
7646-85-7	BUTTER OF ZINC
60-11-7	BUTTER YELLOW
13530-65-9	BUTTERCUP YELLOW
37224-57-0	BUTTERCUP YELLOW
107-92-6	BUTTERSAEURE (GERMAN)
2545-59-7	BUTYL (2,4,5-TRICHLOROPHENOXY)ACETATE
94-80-4	BUTYL (2,4-DICHLOROPHENOXY)ACETATE
93-79-8	BUTYL 2,4,5-T
2545-59-7	BUTYL 2,4,5-T
93-79-8	BUTYL 2,4,5-TRICHLOROPHENOXYACETATE
2545-59-7	BUTYL 2,4,5-TRICHLOROPHENOXYACETATE
94-80-4	BUTYL 2,4-D
97-88-1	BUTYL 2-METHACRYLATE
97-88-1	BUTYL 2-METHYL-2-PROPENOATE
141-32-2	BUTYL 2-PROPENOATE
123-86-4	BUTYL ACETATE
123-86-4	BUTYL ACETATE (DOT)
140-04-5	BUTYL ACETYL RICINOLEATE
12788-93-1	BUTYL ACID PHOSPHATE
141-32-2	BUTYL ACRYLATE
71-36-3	BUTYL ALCOHOL
71-36-3	BUTYL ALCOHOL (DOT)
109-47-7	BUTYL ALCOHOL, HYDROGEN PHOSPHITE
1809-19-4	BUTYL ALCOHOL, HYDROGEN PHOSPHITE
5593-70-4	BUTYL ALCOHOL, TITANIUM(4+) SALT
123-72-8	BUTYL ALDEHYDE
110-69-0	BUTYL ALDEHYDE, OXIME
104-51-8	BUTYL BENZENE
136-60-7	BUTYL BENZOATE
85-68-7	BUTYL BENZYL PHTHALATE
688-74-4	BUTYL BORATE
109-65-9	BUTYL BROMIDE
109-65-9	BUTYL BROMIDE (DOT)
109-65-9	BUTYL BROMIDE, NORMAL
109-65-9	BUTYL BROMIDE, NORMAL (DOT)
109-21-7	BUTYL BUTANOATE
109-21-7	BUTYL BUTYLATE
109-21-7	BUTYL BUTYRATE
112-34-5	BUTYL CARBITOL
124-17-4	BUTYL CARBITOL ACETATE
111-76-2	BUTYL CELLOSOLVE
111-76-2	BUTYL CELLU-SOL
109-69-3	BUTYL CHLORIDE
109-69-3	BUTYL CHLORIDE (DOT)
592-34-7	BUTYL CHLOROCARBONATE
94-80-4	BUTYL DICHLOROPHENOXYACETATE
124-17-4	BUTYL DIETHYLENE GLYCOL ACETATE
112-34-5	BUTYL DIGLYCOL
112-34-5	BUTYL DIGOL
112-34-5	BUTYL DIOXITOL
123-86-4	BUTYL ETHANOATE
142-96-1	BUTYL ETHER
142-96-1	BUTYL ETHER (DOT)
123-05-7	BUTYL ETHYL ACETALDEHYDE
628-81-9	BUTYL ETHYL ETHER
106-35-4	BUTYL ETHYL KETONE
110-62-3	BUTYL FORMAL
592-84-7	BUTYL FORMATE
2426-08-6	BUTYL GLYCIDYL ETHER
111-76-2	BUTYL GLYCOL
12788-93-1	BUTYL HYDROGEN PHOSPHATE
5809-08-5	BUTYL HYDROPEROXIDE, 1,1,3,3-TETRAMETHYL-
71-36-3	BUTYL HYDROXIDE
4316-42-1	BUTYL IMIDAZOLE
111-36-4	BUTYL ISOCYANATE
109-19-2	BUTYL ISOVALERATE
138-22-7	BUTYL LACTATE
109-79-5	BUTYL MERCAPTAN
97-88-1	BUTYL METHACRYLATE
591-78-6	BUTYL METHYL KETONE
111-76-2	BUTYL MONOETHER GLYCOL
928-45-0	BUTYL NITRATE
544-16-1	BUTYL NITRITE
5593-70-4	BUTYL ORTHOTITANATE
142-96-1	BUTYL OXIDE
111-76-2	BUTYL OXITOL
112-34-5	BUTYL OXITOL GLYCOL ETHER
614-45-9	BUTYL PEROXYBENZOATE
23474-91-1	BUTYL PEROXYCROTONATE
16215-49-9	BUTYL PEROXYDICARBONATE
28805-86-9	BUTYL PHENOL
28805-86-9	BUTYL PHENOL, LIQUID
85-68-7	BUTYL PHENYLMETHYL 1,2-BENZENEDICARBOXYLATE
126-73-8	BUTYL PHOSPHATE
1809-19-4	BUTYL PHOSPHITE
1809-19-4	BUTYL PHOSPHONATE ((BUO)2HPO)
12788-93-1	BUTYL PHOSPHORIC ACID
12788-93-1	BUTYL PHOSPHORIC ACID (DOT)
84-74-2	BUTYL PHTHALATE
590-01-2	BUTYL PROPANOATE
590-01-2	BUTYL PROPIONATE
590-01-2	BUTYL PROPIONATE (DOT)
5593-70-4	BUTYL TITANATE
5593-70-4	BUTYL TITANATE(IV)
5593-70-4	BUTYL TITANATE(IV) ((BUO)4Ti)
27458-20-4	BUTYL TOLUENE
7521-80-4	BUTYL TRICHLOROSILANE
7521-80-4	BUTYL TRICHLOROSILANE (DOT)
111-34-2	BUTYL VINYL ETHER
111-34-2	BUTYL VINYL ETHER, INHIBITED (DOT)
123-86-4	BUTYLACETA TEN (DUTCH)
1119-49-9	BUTYLACETAMIDE
91-49-6	BUTYLACETANILIDE
123-86-4	BUTYLACETAT (GERMAN)
142-62-1	BUTYLACETIC ACID
110-43-0	BUTYLACETONE
141-32-2	BUTYLACRYLATE, INHIBITED (DOT)
110-69-0	BUTYLALDOXIME
138-22-7	BUTYLALPHA-HYDROXYPROPIONATE

CAS No.	Chemical Name
109-73-9	BUTYLAMINE
109-73-9	BUTYLAMINE (DOT)
26094-13-3	BUTYLAMINE OLEATE
107-45-9	BUTYLAMINE, 1,1,3,3-TETRAMETHYL-
108-09-8	BUTYLAMINE, 1,3-DIMETHYL-
625-30-9	BUTYLAMINE, 1-METHYL-
3037-72-7	BUTYLAMINE, 4-(DIETHOXYMETHYLSILYL)-
13360-63-9	BUTYLAMINE, N-ETHYL-
110-68-9	BUTYLAMINE, N-METHYL-
75-64-9	BUTYLAMINE, TERTIARY
26094-13-3	BUTYLAMMONIUM OLEATE
1126-78-9	BUTYLANILINE
93-79-8	BUTYLATE 2,4,5-T
2545-59-7	BUTYLATE 2,4,5-T
128-37-0	BUTYLATED HYDROXYTOLUENE
102-79-4	BUTYLBIS(2-HYDROXYETHYL)AMINE
592-34-7	BUTYLCHLOROFORMATE
70042-58-9	BUTYLCYCLOHEXYCHLOROFORMATE
10108-56-2	BUTYLCYCLOHEXYLAMINE
102-79-4	BUTYLDIETHANOLAMINE
123-86-4	BUTYLE (ACETATE DE) (FRENCH)
25167-67-3	BUTYLENE
78-92-2	BUTYLENE HYDRATE
106-88-7	BUTYLENE OXIDE
123-75-1	BUTYLENIMINE
149-57-5	BUTYLETHYLACETIC ACID
13360-63-9	BUTYLETHYLAMINE
97-88-1	BUTYLMETHACRYLAAT (DUTCH)
110-68-9	BUTYLMETHYLAMINE
71-36-3	BUTYLOWY ALKOHOL (POLISH)
111-34-2	BUTYLOXYETHENE
106-99-0	BUTYNE
25339-57-5	BUTYNE
110-65-6	BUTYNEDIOL
10070-67-0	BUTYNEDIOL
123-72-8	BUTYRAL
123-72-8	BUTYRALDEHYDE
123-72-8	BUTYRALDEHYDE (DOT)
97-96-1	BUTYRALDEHYDE, 2-ETHYL-
96-17-3	BUTYRALDEHYDE, 2-METHYL-
107-89-1	BUTYRALDEHYDE, 3-HYDROXY-
590-86-3	BUTYRALDEHYDE, 3-METHYL-
110-69-0	BUTYRALDEHYDE, OXIME
496-03-7	BUTYRALDOL
110-69-0	BUTYRALDOXIME
110-69-0	BUTYRALDOXIME (DOT)
107-92-6	BUTYRATE
107-92-6	BUTYRIC ACID
107-92-6	BUTYRIC ACID (DOT)
106-31-0	BUTYRIC ACID ANHYDRIDE
141-75-3	BUTYRIC ACID CHLORIDE
109-74-0	BUTYRIC ACID NITRILE
88-09-5	BUTYRIC ACID, 2-ETHYL-
305-03-3	BUTYRIC ACID, 4-[P-[BIS(2-CHLOROETHYL)AMINO]PHENYL]-
109-21-7	BUTYRIC ACID, BUTYL ESTER
105-54-4	BUTYRIC ACID, ETHYL ESTER
638-11-9	BUTYRIC ACID, ISOPROPYL ESTER
623-42-7	BUTYRIC ACID, METHYL ESTER
540-18-1	BUTYRIC ACID, PENTYL ESTER
105-66-8	BUTYRIC ACID, PROPYL ESTER
123-20-6	BUTYRIC ACID, VINYL ESTER
123-72-8	BUTYRIC ALDEHYDE
106-31-0	BUTYRIC ANHYDRIDE
106-31-0	BUTYRIC ANHYDRIDE (DOT)
141-75-3	BUTYRIC CHLORIDE
105-54-4	BUTYRIC ETHER
71-36-3	BUTYRIC OR NORMAL PRIMARY BUTYL ALCOHOL
616-45-5	BUTYROLACTAM
123-19-3	BUTYRONE
123-19-3	BUTYRONE (DOT)
109-74-0	BUTYRONITRILE
109-74-0	BUTYRONITRILE (DOT)
628-20-6	BUTYRONITRILE, 4-CHLORO-
141-75-3	BUTYRYL CHLORIDE
141-75-3	BUTYRYL CHLORIDE (DOT)
106-31-0	BUTYRYL OXIDE
123-72-8	BUTYRYLALDEHYDE
109-74-0	BUTYRYLONITRILE
111-34-2	BVE

CAS No.	Chemical Name
446-86-6	BW 57-322
94-36-0	BZF-60
112-30-1	C 10 ALCOHOL
12001-26-2	C 1000
7440-50-8	C 10200
11099-02-8	C 11-2S
11099-02-8	C 11-9
7440-50-8	C 12200
3771-19-5	C 13437SU
6923-22-4	C 1414
9002-89-5	C 17
1982-47-4	C 1983
9004-70-0	C 2018
87-68-3	C 46
25155-30-0	C 550
141-66-2	C 709
3165-93-3	C I 37085
470-90-6	C-10015
124-13-0	C-8 ALDEHYDE
1306-23-6	C. P. GOLDEN YELLOW 55
88-89-1	C.I. 10305
60-11-7	C.I. 11020
97-56-3	C.I. 11160
97-56-3	C.I. 11160B
2646-17-5	C.I. 12100
6358-53-8	C.I. 12156
3761-53-3	C.I. 16150
3564-09-8	C.I. 16155
2602-46-2	C.I. 22610
91-94-1	C.I. 23060
72-57-1	C.I. 23850
16071-86-6	C.I. 30145
1937-37-7	C.I. 30235
115-02-6	C.I. 337
102-56-7	C.I. 35811
100-01-6	C.I. 37035
99-55-8	C.I. 37105
106-49-0	C.I. 37107
89-62-3	C.I. 37110
134-29-2	C.I. 37115
119-93-7	C.I. 37230
134-32-7	C.I. 37265
91-59-8	C.I. 37270
434-07-1	C.I. 406
2465-27-2	C.I. 41000
1694-09-3	C.I. 42640
82-28-0	C.I. 60700
62-53-3	C.I. 76000
142-04-1	C.I. 76001
95-83-0	C.I. 76015
39156-41-7	C.I. 76051
106-50-3	C.I. 76060
108-46-3	C.I. 76505
7429-90-5	C.I. 77000
7440-36-0	C.I. 77050
1309-64-4	C.I. 77052
10025-91-9	C.I. 77056
1315-04-4	C.I. 77061
12044-79-0	C.I. 77085
1303-33-9	C.I. 77086
7440-43-9	C.I. 77180
543-90-8	C.I. 77185
7782-42-5	C.I. 77265
1333-86-4	C.I. 77266
10101-53-8	C.I. 77305
7440-48-4	C.I. 77320
7440-50-8	C.I. 77400
12002-03-8	C.I. 77410
1309-37-1	C.I. 77491
7439-92-1	C.I. 77575
10101-63-0	C.I. 77613
7446-27-7	C.I. 77622
7446-14-2	C.I. 77630
1314-87-0	C.I. 77640
7722-64-7	C.I. 77755
7440-02-0	C.I. 77775
7440-06-4	C.I. 77795
7782-49-2	C.I. 77805
7440-22-4	C.I. 77820

CAS No.	Chemical Name
7440-31-5	C.I. 77860
13463-67-7	C.I. 77891
1314-62-1	C.I. 77938
27774-13-6	C.I. 77940
3486-35-9	C.I. 77950
2650-18-2	C.I. ACID BLUE 9, DIAMMONIUM SALT
3844-45-9	C.I. ACID BLUE 9, DISODIUM SALT
4680-78-8	C.I. ACID GREEN 3
2429-80-3	C.I. ACID ORANGE 45
6459-94-5	C.I. ACID RED 114, DISODIUM SALT
3761-53-3	C.I. ACID RED 26
3761-53-3	C.I. ACID RED 26, DISODIUM SALT
3567-65-5	C.I. ACID RED 85
1694-09-3	C.I. ACID VIOLET 49
1694-09-3	C.I. ACID VIOLET 49, SODIUM SALT
91-92-9	C.I. AZOIC COUPLING COMPONENT
91-96-3	C.I. AZOIC COUPLING COMPONENT 5
106-49-0	C.I. AZOIC COUPLING COMPONENT 107
92-87-5	C.I. AZOIC DIAZO COMPONENT 112
119-93-7	C.I. AZOIC DIAZO COMPONENT 113
134-32-7	C.I. AZOIC DIAZO COMPONENT 114
99-55-8	C.I. AZOIC DIAZO COMPONENT 12
100-01-6	C.I. AZOIC DIAZO COMPONENT 37
89-62-3	C.I. AZOIC DIAZO COMPONENT 8
633-03-4	C.I. BASIC GREEN 1
569-64-2	C.I. BASIC GREEN 4
12768-82-0	C.I. BASIC ORANGE 15
989-38-8	C.I. BASIC RED 1
2465-27-2	C.I. BASIC YELLOW 2
2465-27-2	C.I. BASIC YELLOW 2, MONOHYDROCHLORIDE
106-50-3	C.I. DEVELOPER 13
100-01-6	C.I. DEVELOPER 17
108-46-3	C.I. DEVELOPER 4
61703-05-7	C.I. DIRECT BLACK 114
1937-37-7	C.I. DIRECT BLACK 38
1937-37-7	C.I. DIRECT BLACK 38, DISODIUM SALT
25156-49-4	C.I. DIRECT BLACK 4
2429-83-6	C.I. DIRECT BLACK 4, DISODIUM SALT
6739-62-4	C.I. DIRECT BLACK 91, TRISODIUM SALT
2610-05-1	C.I. DIRECT BLUE 1
72-57-1	C.I. DIRECT BLUE 14
72-57-1	C.I. DIRECT BLUE 14, TETRASODIUM SALT
2429-74-5	C.I. DIRECT BLUE 15
25180-19-2	C.I. DIRECT BLUE 2
2429-73-4	C.I. DIRECT BLUE 2, TRISODIUM SALT
2586-57-4	C.I. DIRECT BLUE 22, DISODIUM SALT
25180-27-2	C.I. DIRECT BLUE 25
2150-54-1	C.I. DIRECT BLUE 25, TETRASODIUM SALT
2602-46-2	C.I. DIRECT BLUE 6
2602-46-2	C.I. DIRECT BLUE 6, TETRASODIUM SALT
2429-71-2	C.I. DIRECT BLUE 8, DISODIUM SALT
3811-71-0	C.I. DIRECT BROWN 1
12222-20-7	C.I. DIRECT BROWN 111
6360-54-9	C.I. DIRECT BROWN 154
25255-06-5	C.I. DIRECT BROWN 2
2429-82-5	C.I. DIRECT BROWN 2, DISODIUM SALT
25180-41-0	C.I. DIRECT BROWN 31
2429-81-4	C.I. DIRECT BROWN 31, TETRASODIUM SALT
6247-51-4	C.I. DIRECT BROWN 59
3476-90-2	C.I. DIRECT BROWN 59, DISODIUM SALT
25180-39-6	C.I. DIRECT BROWN 6
2893-80-3	C.I. DIRECT BROWN 6, DISODIUM SALT
8014-91-3	C.I. DIRECT BROWN 74
16071-86-6	C.I. DIRECT BROWN 95
3626-28-6	C.I. DIRECT GREEN 1
3626-28-6	C.I. DIRECT GREEN 1, DISODIUM SALT
25180-46-5	C.I. DIRECT GREEN 6
25180-46-5	C.I. DIRECT GREEN 6, DISODIUM SALT
25180-47-6	C.I. DIRECT GREEN 8
5422-17-3	C.I. DIRECT GREEN 8, TRISODIUM SALT
54579-28-1	C.I. DIRECT ORANGE 1
6637-88-3	C.I. DIRECT ORANGE 6, DISODIUM SALT
64083-59-6	C.I. DIRECT ORANGE 8
25188-24-3	C.I. DIRECT RED 1
2429-84-7	C.I. DIRECT RED 1, DISODIUM SALT
25188-29-8	C.I. DIRECT RED 10
2429-70-1	C.I. DIRECT RED 10, DISODIUM SALT
25188-30-1	C.I. DIRECT RED 13
1937-35-5	C.I. DIRECT RED 13, DISODIUM SALT
992-59-6	C.I. DIRECT RED 2, DISODIUM SALT
573-58-0	C.I. DIRECT RED 28
3530-19-6	C.I. DIRECT RED 37
6358-29-8	C.I. DIRECT RED 39, DISODIUM SALT
25188-44-7	C.I. DIRECT VIOLET 1
2586-60-9	C.I. DIRECT VIOLET 1, DISODIUM SALT
25329-82-2	C.I. DIRECT VIOLET 22
6426-67-1	C.I. DIRECT VIOLET 22, TRISODIUM SALT
6426-62-6	C.I. DIRECT YELLOW 20
119-90-4	C.I. DISPERSE BLACK 6
82-28-0	C.I. DISPERSE ORANGE 11
2832-40-8	C.I. DISPERSE YELLOW 3
81-88-9	C.I. FOOD RED 15
3761-53-3	C.I. FOOD RED 5
1694-09-3	C.I. FOOD VIOLET 2
95-80-7	C.I. OXIDATION BASE
106-50-3	C.I. OXIDATION BASE 10
108-46-3	C.I. OXIDATION BASE 31
95-80-7	C.I. OXIDATION BASE 35
7782-42-5	C.I. PIGMENT BLACK 10
1333-86-4	C.I. PIGMENT BLACK 7
12002-03-8	C.I. PIGMENT GREEN 21
7440-50-8	C.I. PIGMENT METAL 2
7439-92-1	C.I. PIGMENT METAL 4
7440-31-5	C.I. PIGMENT METAL 5
1309-37-1	C.I. PIGMENT RED 101
1309-64-4	C.I. PIGMENT WHITE 11
7446-14-2	C.I. PIGMENT WHITE 3
13463-67-7	C.I. PIGMENT WHITE 6
7789-06-2	C.I. PIGMENT YELLOW 32
1303-33-9	C.I. PIGMENT YELLOW 39
12044-79-0	C.I. PIGMENT YELLOW 39
2646-17-5	C.I. SOLVENT ORANGE 2
82-28-0	C.I. SOLVENT ORANGE 35
3118-97-6	C.I. SOLVENT ORANGE 7
6358-53-8	C.I. SOLVENT RED 80
60-09-3	C.I. SOLVENT YELLOW 1
842-07-9	C.I. SOLVENT YELLOW 14
60-11-7	C.I. SOLVENT YELLOW 2
97-56-3	C.I. SOLVENT YELLOW 3
128-66-5	C.I. VAT YELLOW 4
470-90-6	C8949
5798-79-8	CA
7440-50-8	CA 122
9004-70-0	CA 80
9004-70-0	CA 80-15
372-09-8	CAA
1068-27-5	CAB-O-CURE 2P
1344-28-1	CAB-O-GRIP
7631-86-9	CAB-O-SIL
7631-86-9	CAB-O-SIL H 5
7631-86-9	CAB-O-SIL L 5
7631-86-9	CAB-O-SIL M 5
7631-86-9	CAB-O-SIL MS 7
13463-67-7	CAB-O-TI
7631-86-9	CABOSIL N 5
7631-86-9	CABOSIL ST 1
1333-86-4	CABOT 607
112-53-8	CACHALOT L-50
112-53-8	CACHALOT L-90
124-65-2	CACODYLATE DE SODIUM (FRENCH)
75-60-5	CACODYLIC ACID
75-60-5	CACODYLIC ACID (DOT)
124-65-2	CACODYLIC ACID SODIUM SALT
10108-64-2	CADDY
94-36-0	CADET
7440-43-9	CADMIUM
543-90-8	CADMIUM ACETATE
543-90-8	CADMIUM ACETATE (DOT)
7789-42-6	CADMIUM BROMIDE
7789-42-6	CADMIUM BROMIDE (CdBr2)
7789-42-6	CADMIUM BROMIDE (DOT)
10108-64-2	CADMIUM CHLORIDE
10108-64-2	CADMIUM CHLORIDE (CdCl2)
10108-64-2	CADMIUM CHLORIDE (DOT)
543-90-8	CADMIUM DIACETATE
7789-42-6	CADMIUM DIBROMIDE
10108-64-2	CADMIUM DICHLORIDE
10325-94-7	CADMIUM DINITRATE

CAS No.	Chemical Name
2223-93-0	CADMIUM DISTEARATE
14486-19-2	CADMIUM FLUOBORATE
14486-19-2	CADMIUM FLUOROBORATE
1306-23-6	CADMIUM GOLDEN 366
1306-23-6	CADMIUM LEMON YELLOW 527
10124-36-4	CADMIUM MONOSULFATE
1306-23-6	CADMIUM MONOSULFIDE
1306-19-0	CADMIUM MONOXIDE
10325-94-7	CADMIUM NITRATE
2223-93-0	CADMIUM OCTADECANOATE
1306-19-0	CADMIUM OXIDE
1306-19-0	CADMIUM OXIDE (ACGIH)
1306-19-0	CADMIUM OXIDE (CdO)
1306-23-6	CADMIUM PRIMROSE 819
2223-93-0	CADMIUM STEARATE
10124-36-4	CADMIUM SULFATE
10124-36-4	CADMIUM SULFATE (1:1)
1306-23-6	CADMIUM SULFIDE
1306-23-6	CADMIUM SULFIDE (CdS)
1306-23-6	CADMIUM SULFIDE YELLOW
10124-36-4	CADMIUM SULPHATE
1306-23-6	CADMIUM SULPHIDE
14486-19-2	CADMIUM TETRAFLUOROBORATE
1306-23-6	CADMIUM YELLOW
1306-23-6	CADMIUM YELLOW 000
1306-23-6	CADMIUM YELLOW 10G CONC
1306-23-6	CADMIUM YELLOW 892
1306-23-6	CADMIUM YELLOW CONC. DEEP
1306-23-6	CADMIUM YELLOW CONC. GOLDEN
1306-23-6	CADMIUM YELLOW CONC. LEMON
1306-23-6	CADMIUM YELLOW CONC. PRIMROSE
1306-23-6	CADMIUM YELLOW OZ DARK
1306-23-6	CADMIUM YELLOW PRIMROSE 47-4100
543-90-8	CADMIUM(II) ACETATE
1306-23-6	CADMOPUR GOLDEN YELLOW N
1306-23-6	CADMOPUR YELLOW
110-05-4	CADOX
94-36-0	CADOX
94-36-0	CADOX 40E
94-36-0	CADOX B 50 P
94-36-0	CADOX BS
94-17-7	CADOX PS
75-91-2	CADOX TBH
110-05-4	CADOX TBP
133-14-2	CADOX TDP
133-14-2	CADOX TS
133-14-2	CADOX TS 40,50
1314-13-2	CADOX XX 78
56-75-7	CAF
532-27-4	CAF
56-75-7	CAF (PHARMACEUTICAL)
3691-35-8	CAID
7722-64-7	CAIROX
138-86-3	CAJEPUTEN
138-86-3	CAJEPUTENE
10043-01-3	CAKE ALUM
1305-78-8	CALCIA
7440-70-2	CALCICAT
592-01-8	CALCID
50-14-6	CALCIFEROL
1304-28-5	CALCINED BARYTA
1309-48-4	CALCINED MAGNESIA
12401-86-4	CALCINED SODA
7440-70-2	CALCIUM
75-20-7	CALCIUM ACETYLIDE
75-20-7	CALCIUM ACETYLIDE (Ca(C2))
7778-44-1	CALCIUM ARSENATE
7778-44-1	CALCIUM ARSENATE (Ca3(AsO4)2)
7778-44-1	CALCIUM ARSENATE (DOT)
7778-44-1	CALCIUM ARSENATE, SOLID
7778-44-1	CALCIUM ARSENATE, SOLID (DOT)
52740-16-6	CALCIUM ARSENITE
52740-16-6	CALCIUM ARSENITE (CaHAsO3)
52740-16-6	CALCIUM ARSENITE, SOLID
52740-16-6	CALCIUM ARSENITE, SOLID (DOT)
7440-70-2	CALCIUM ATOM
14307-33-6	CALCIUM BICHROMATE
13780-03-5	CALCIUM BISULFIDE
15512-36-4	CALCIUM BISULFITE
15512-36-4	CALCIUM BISULFITE SOLUTION (DOT)
75-20-7	CALCIUM CARBIDE
75-20-7	CALCIUM CARBIDE (CaC2)
75-20-7	CALCIUM CARBIDE (DOT)
156-62-7	CALCIUM CARBIMIDE
10137-74-3	CALCIUM CHLORATE
10137-74-3	CALCIUM CHLORATE (DOT)
10137-74-3	CALCIUM CHLORATE :AQUEOUS: SOLUTION
10137-74-3	CALCIUM CHLORATE SOLUTION
10137-74-3	CALCIUM CHLORATE, AQUEOUS SOLUTION (DOT)
10043-52-4	CALCIUM CHLORIDE
10043-52-4	CALCIUM CHLORIDE (CaCl2)
14674-72-7	CALCIUM CHLORITE
14674-72-7	CALCIUM CHLORITE (DOT)
13765-19-0	CALCIUM CHROMATE
14307-33-6	CALCIUM CHROMATE (CaCr2O7)
13765-19-0	CALCIUM CHROMATE (CaCrO4)
13765-19-0	CALCIUM CHROMATE (DOT)
13765-19-0	CALCIUM CHROMATE (VI)
13765-19-0	CALCIUM CHROME YELLOW
13765-19-0	CALCIUM CHROMIUM OXIDE (CaCrO4)
156-62-7	CALCIUM CYANAMID
156-62-7	CALCIUM CYANAMIDE
156-62-7	CALCIUM CYANAMIDE (ACGIH)
156-62-7	CALCIUM CYANAMIDE (CaCN2)
156-62-7	CALCIUM CYANAMIDE, NOT HYDRATED (CONTAINING MORE THAN 0.1% CALCIUM CARBIDE) (DOT)
592-01-8	CALCIUM CYANIDE
592-01-8	CALCIUM CYANIDE (Ca(CN)2)
592-01-8	CALCIUM CYANIDE, SOLID
592-01-8	CALCIUM CYANIDE, SOLID (DOT)
75-20-7	CALCIUM DICARBIDE
10043-52-4	CALCIUM DICHLORIDE
14307-33-6	CALCIUM DICHROMATE
14307-33-6	CALCIUM DICHROMATE (CaCr2O7)
7789-75-5	CALCIUM DIFLUORIDE
7789-78-8	CALCIUM DIHYDRIDE
57308-10-8	CALCIUM DIHYDRIDE
1305-62-0	CALCIUM DIHYDROXIDE
10124-37-5	CALCIUM DINITRATE
1305-79-9	CALCIUM DIOXIDE
13477-36-6	CALCIUM DIPERCHLORATE
15512-36-4	CALCIUM DITHIONITE
26264-06-2	CALCIUM DODECYLBENZENE SULFONATE
26264-06-2	CALCIUM DODECYLBENZENESULFONATE (DOT)
7440-70-2	CALCIUM ELEMENT
7789-75-5	CALCIUM FLUORIDE
7789-75-5	CALCIUM FLUORIDE (CaF2)
1305-62-0	CALCIUM HYDRATE
7789-78-8	CALCIUM HYDRIDE
57308-10-8	CALCIUM HYDRIDE
7789-78-8	CALCIUM HYDRIDE (CaH2)
57308-10-8	CALCIUM HYDRIDE (CaH2)
7789-78-8	CALCIUM HYDRIDE (DOT)
57308-10-8	CALCIUM HYDRIDE (DOT)
13780-03-5	CALCIUM HYDROGEN SULFITE
15512-36-4	CALCIUM HYDROGEN SULFITE SOLUTION
15512-36-4	CALCIUM HYDROGEN SULFITE SOLUTION (DOT)
1344-95-2	CALCIUM HYDROSILICATE
15512-36-4	CALCIUM HYDROSULFITE
15512-36-4	CALCIUM HYDROSULPHITE (DOT)
1305-62-0	CALCIUM HYDROXIDE
1305-62-0	CALCIUM HYDROXIDE (ACGIH)
1305-62-0	CALCIUM HYDROXIDE (Ca(OH)2)
7778-54-3	CALCIUM HYPOCHLORITE
7778-54-3	CALCIUM HYPOCHLORITE MIXTURE, DRY (CONTAINING MORE THAN 39% AVAILABLE CHLORINE) (DOT)
7778-54-3	CALCIUM HYPOCHLORITE MIXTURE, DRY, WITH > 10% BUT ≤ 39% AVAILABLE CHLORINE
7778-54-3	CALCIUM HYPOCHLORITE MIXTURE, DRY, WITH > 39% AVAILABLE CHLORINE
7778-54-3	CALCIUM HYPOCHLORITE, HYDRATED (≥ 55% BUT < 10% WATER, AND CONTAINING > 39% AVAILABLE CHLORINE)
13765-19-0	CALCIUM MONOCHROMATE
1344-95-2	CALCIUM MONOSILICATE
1305-78-8	CALCIUM MONOXIDE
26264-06-2	CALCIUM N-DODECYLBENZENESULFONATE
10124-37-5	CALCIUM NITRATE
10124-37-5	CALCIUM NITRATE (Ca(NO3)2)

CAS No.	Chemical Name
10124-37-5	CALCIUM NITRATE (DOT)
7778-44-1	CALCIUM ORTHOARSENATE
1305-78-8	CALCIUM OXIDE
1305-78-8	CALCIUM OXIDE (ACGIH,DOT)
1305-78-8	CALCIUM OXIDE (CaO)
1305-79-9	CALCIUM OXIDE (CaO2)
7778-54-3	CALCIUM OXYCHLORIDE
13477-36-6	CALCIUM PERCHLORATE
13477-36-6	CALCIUM PERCHLORATE (DOT)
10118-76-0	CALCIUM PERMANGANATE
10118-76-0	CALCIUM PERMANGANATE (DOT)
10118-76-0	CALCIUM PERMANGANATE, CA(MNO4)2
1305-79-9	CALCIUM PEROXIDE
1305-79-9	CALCIUM PEROXIDE (Ca(O2))
1305-79-9	CALCIUM PEROXIDE (DOT)
10103-46-5	CALCIUM PHOSPHATE
1305-99-3	CALCIUM PHOSPHIDE
1305-99-3	CALCIUM PHOSPHIDE (Ca3P2)
1305-99-3	CALCIUM PHOSPHIDE (DOT)
1305-99-3	CALCIUM PHOTOPHOR
1344-95-2	CALCIUM POLYSILICATE
9007-13-0	CALCIUM RESINATE
9007-13-0	CALCIUM RESINATE (DOT)
9007-13-0	CALCIUM RESINATE, FUSED
9007-13-0	CALCIUM RESINATE, FUSED (DOT)
9007-13-0	CALCIUM RESINATE, TECHNICALLY PURE (DOT)
14019-91-1	CALCIUM SELENATE
14019-91-1	CALCIUM SELENATE (CaSeO4)
1344-95-2	CALCIUM SILICATE
12737-18-7	CALCIUM SILICIDE
12737-18-7	CALCIUM SILICIDE (DOT)
12013-56-8	CALCIUM SILICON
12737-18-7	CALCIUM SILICON (DOT)
12737-18-7	CALCIUM SILICON (POWDER)
12737-18-7	CALCIUM SILICON, POWDER (DOT)
10124-37-5	CALCIUM(II) NITRATE (1:2)
7440-70-2	CALCIUM, METAL
7440-70-2	CALCIUM, METAL (DOT)
7440-70-2	CALCIUM, METAL, CRYSTALLINE
7440-70-2	CALCIUM, METAL, CRYSTALLINE (DOT)
7440-70-2	CALCIUM, METAL, PYROPHORIC
7440-70-2	CALCIUM, NON-PYROPHORIC (DOT)
7440-70-2	CALCIUM, PYROPHORIC (DOT)
7778-44-1	CALCIUMARSENAT
3761-53-3	CALCOCID SCARLET 2R
16071-86-6	CALCODUR BROWN BRL
3761-53-3	CALCOLAKE SCARLET 2R
1937-37-7	CALCOMINE BLACK
1937-37-7	CALCOMINE BLACK EXL
2602-46-2	CALCOMINE BLUE 2B
10043-52-4	CALCOSAN
633-03-4	CALCOZINE BRILLIANT GREEN G
2465-27-2	CALCOZINE YELLOW OX
592-01-8	CALCYAN
592-01-8	CALCYANIDE
56-18-8	CALDINE
88-85-7	CALDON
1344-95-2	CALFLO E
10124-56-8	CALGON
10124-56-8	CALGON (OLD)
10124-56-8	CALGON S
1332-21-4	CALIDREA HPP
1332-21-4	CALIDRIA R-G 244
12001-29-5	CALIDRIA RG 100
12001-29-5	CALIDRIA RG 144
12001-29-5	CALIDRIA RG 600
121-75-5	CALMATHION
1317-65-3	CALMOT AD
7487-94-7	CALOCHLOR
7546-30-7	CALOGREEN
7546-30-7	CALOMEL
7546-30-7	CALOMELANO (ITALIAN)
9005-25-8	CALOREEN
7546-30-7	CALOSAN
1305-78-8	CALOXOL CP 2
1305-78-8	CALOXOL W 3
1305-79-9	CALPER
1305-79-9	CALPER G
10043-52-4	CALPLUS
1344-95-2	CALSIL
25155-30-0	CALSOFT F 90
25155-30-0	CALSOFT L 40
25155-30-0	CALSOFT L-60
27176-87-0	CALSOFT LAS 99
27323-41-7	CALSOFT T 60
10043-52-4	CALTAC
1305-62-0	CALVIT
1305-62-0	CALVITAL
1305-78-8	CALX
1305-78-8	CALXYL
56-75-7	CAM
5798-79-8	CAMITE
8001-35-2	CAMPHECHLOR
79-92-5	CAMPHENE
79-92-5	CAMPHENE (DOT)
8001-35-2	CAMPHOCHLOR
8001-35-2	CAMPHOCLOR
8001-35-2	CAMPHOFENE HUILEUX
99-87-6	CAMPHOGEN
507-70-0	CAMPHOL
76-22-2	CAMPHOR
8008-51-3	CAMPHOR OIL
8008-51-3	CAMPHOR OIL (DOT)
8008-15-3	CAMPHOR OIL (LIGHT)
8008-51-3	CAMPHOR OIL WHITE
8008-51-3	CAMPHOR OIL YELLOW
8008-51-3	CAMPHOR OIL, RECTIFIED
91-20-3	CAMPHOR TAR
76-22-2	CAMPHOR, SYNTHETIC
76-22-2	CAMPHOR, SYNTHETIC (ACGIH,DOT)
76-22-2	CAMPHOR—NATURAL
506-68-3	CAMPILIT
56-25-7	CAN
28772-56-7	CANADIEN 2000
8032-32-4	CANADOL
59-50-7	CANDASEPTIC
1321-10-4	CANDASEPTIC
1912-24-9	CANDEX
57-50-1	CANE SUGAR
8001-35-2	CANFECLOR
7782-42-5	CANLUB
62-73-7	CANOGARD
56-25-7	CANTHARIDES CAMPHOR
56-25-7	CANTHARIDIN
56-25-7	CANTHARIDINE
56-25-7	CANTHARONE
128-37-0	CAO 1
128-37-0	CAO 3
56-75-7	CAP
532-27-4	CAP
7778-54-3	CAPORIT
105-60-2	CAPRAN 77C
105-60-2	CAPRAN 80
112-30-1	CAPRIC ALCOHOL
50-78-2	CAPRIN
112-30-1	CAPRINIC ALCOHOL
66-25-1	CAPROALDEHYDE
142-62-1	CAPROIC ACID
66-25-1	CAPROIC ALDEHYDE
105-60-2	CAPROLACTAM
105-60-2	CAPROLACTAM MONOMER
105-60-2	CAPROLATTAME (FRENCH)
63-25-2	CAPROLIN
105-60-2	CAPROLON B
105-60-2	CAPROLON V
762-16-3	CAPROLYL PEROXIDE
105-60-2	CAPRON
105-60-2	CAPRON 8250
105-60-2	CAPRON 8252
105-60-2	CAPRON 8253
105-60-2	CAPRON 8256
105-60-2	CAPRON 8257
105-60-2	CAPRON B
105-60-2	CAPRON GR 8256
105-60-2	CAPRON GR 8258
105-60-2	CAPRON PK4
66-25-1	CAPRONALDEHYDE
142-62-1	CAPRONIC ACID

CAS No.	Chemical Name
111-27-3	CAPROYL ALCOHOL
762-16-3	CAPRYL PEROXIDE
124-13-0	CAPRYLALDEHYDE
111-86-4	CAPRYLAMINE
111-66-0	CAPRYLENE
111-64-8	CAPRYLIC ACID CHLORIDE
111-87-5	CAPRYLIC ALCOHOL
124-13-0	CAPRYLIC ALDEHYDE
111-64-8	CAPRYLOYL CHLORIDE
762-16-3	CAPRYLOYL PEROXIDE (DOT)
112-14-1	CAPRYLYL ACETATE
111-64-8	CAPRYLYL CHLORIDE
762-16-3	CAPRYLYL PEROXIDE
7530-07-6	CAPRYLYL PEROXIDE (N-OCTANOYL PEROXIDE)
762-16-3	CAPRYLYL PEROXIDE SOLUTION
762-16-3	CAPRYLYL PEROXIDE SOLUTION (DOT)
111-86-4	CAPRYLYLAMINE
1306-23-6	CAPSEBON
1335-85-9	CAPSINE
133-06-2	CAPTAF
133-06-2	CAPTAF 85W
2425-06-1	CAPTAFOL
133-06-2	CAPTAN
133-06-2	CAPTAN (ACGIH,DOT)
133-06-2	CAPTAN 50W
133-06-2	CAPTAN-STREPTOMYCIN 75-01 POTATO SEED PIECE PROTECTANT
133-06-2	CAPTANCAPTENEET 26,538
133-06-2	CAPTANE
133-06-2	CAPTEX
1309-37-1	CAPUT MORTUUM
1309-37-1	CAPUT MORTUUM LIGHT
101-68-8	CARADATE 30
51-83-2	CARBACHOL
51-83-2	CARBACHOL CHLORIDE
51-83-2	CARBACHOLIN
51-83-2	CARBACHOLINE
51-83-2	CARBACHOLINE CHLORIDE
51-83-2	CARBACOLINA
107-13-1	CARBACRYL
1333-86-4	CARBALAC 2
75-12-7	CARBAMALDEHYDE
1111-78-0	CARBAMIC ACID, AMMONIUM SALT
95-06-7	CARBAMIC ACID, DIETHYLDITHIO-, 2-CHLOROALLYL ESTER
4384-82-1	CARBAMIC ACID, DIETHYLDITHIO-, SODIUM SALT
2303-16-4	CARBAMIC ACID, DIISOPROPYLTHIO-, S-(2,3-DICHLOROALLYL) ESTER
644-64-4	CARBAMIC ACID, DIMETHYL-, 1-((DIMETHYLAMINO)CARBONYL)-5-METHYL-1H-PYRAZOL-3-YL ESTER
644-64-4	CARBAMIC ACID, DIMETHYL-, 1-DIMETHYLCARBAMOYL-5-METHYLPYRAZOL-3-YL ESTER
119-38-0	CARBAMIC ACID, DIMETHYL-, 1-ISOPROPYL-3-METHYLPYRAZOL-5-YL ESTER
644-64-4	CARBAMIC ACID, DIMETHYL-, 1-[(DIMETHYLAMINO)CARBONYL]-5-METHYL-1H-PYRAZOL-3-YL ESTER
119-38-0	CARBAMIC ACID, DIMETHYL-, 3-METHYL-1-(1-METHYLETHYL)-1H-PYRAZOL-5-YL ESTER
644-64-4	CARBAMIC ACID, DIMETHYL-, 5-METHYL-1H-PYRAZOL-3-YL ESTER
644-64-4	CARBAMIC ACID, DIMETHYL-, ESTER WITH 3-HYDROXY-N,N,5-TRIMETHYLPYRAZOLE-1-CARBOXAMIDE
4384-82-1	CARBAMIC ACID, DITHIO-, ION(1-)
51-83-2	CARBAMIC ACID, ESTER WITH CHOLINE CHLORIDE
51-79-6	CARBAMIC ACID, ETHYL ESTER
142-59-6	CARBAMIC ACID, ETHYLENEBIS(DITHIO-, DISODIUM SALT
116-06-3	CARBAMIC ACID, METHYL-, 0-((2-METHYL-2-(METHYLTHIO)PROPYLIDENE)AMINO) DERIV
63-25-2	CARBAMIC ACID, METHYL-, 1-NAPHTHYL ESTER
1563-66-2	CARBAMIC ACID, METHYL-, 2,2-DIMETHYL-2,3-DIHYDROBENZOFURAN-7-YL ESTER
1563-66-2	CARBAMIC ACID, METHYL-, 2,3-DIHYDRO-2,2-DIMETHYL-7-BENZOFURANYL ESTER
2032-65-7	CARBAMIC ACID, METHYL-, 3,5-DIMETHYL-4-(METHYLTHIO)PHENYL ESTER
64-00-6	CARBAMIC ACID, METHYL-, 3-(1-METHYLETHYL)PHENYL ESTER
2631-37-0	CARBAMIC ACID, METHYL-, 3-METHYL-5-(1-METHYLETHYL)PHENYL ESTER
1129-41-5	CARBAMIC ACID, METHYL-, 3-METHYLPHENYL ESTER
1129-41-5	CARBAMIC ACID, METHYL-, 3-TOLYL ESTER
315-18-4	CARBAMIC ACID, METHYL-, 4-(DIMETHYLAMINO)-3,5-XYLYL ESTER
2032-65-7	CARBAMIC ACID, METHYL-, 4-(METHYLTHIO)-3,5-XYLYL ESTER
57-47-6	CARBAMIC ACID, METHYL-, ESTER WITH ESEROLINE
22259-30-9	CARBAMIC ACID, METHYL-, ESTER WITH N'-(M-HYDROXYPHENYL)-N,N-DIMETHYLFORMAMIDINE
23422-53-9	CARBAMIC ACID, METHYL-, ESTER WITH N'-(M-HYDROXYPHENYL)-N,N-DIMETHYLFORMAMIDINE, MONOHYDROCHLORIDE
22259-30-9	CARBAMIC ACID, METHYL-, M-(((DIMETHYLAMINO)METHYLENE)AMINO)PHENYL ESTER
64-00-6	CARBAMIC ACID, METHYL-, M-CUMENYL ESTER
2631-37-0	CARBAMIC ACID, METHYL-, M-CYM-5-YL ESTER
1129-41-5	CARBAMIC ACID, METHYL-, M-TOLYL ESTER
114-26-1	CARBAMIC ACID, METHYL-, O-ISOPROPOXYPHENYL ESTER
23135-22-0	CARBAMIC ACID, METHYL-, O-[[[(DIMETHYLCARBAMOYL)-METHYLTHIO]METHYLENE]AMINO] DERIV
615-53-2	CARBAMIC ACID, METHYLNITROSO-, ETHYL ESTER
1111-78-0	CARBAMIC ACID, MONOAMMONIUM SALT
64-00-6	CARBAMIC ACID, N-METHYL-, 3-ISOPROPYLPHENYL ESTER
2631-37-0	CARBAMIC ACID, N-METHYL-, 3-METHYL-5-ISOPROPYLPHENYL ESTER
2032-65-7	CARBAMIC ACID, N-METHYL-, 4-(METHYLTHIO)-3,5-XYLYL ESTER
17804-35-2	CARBAMIC ACID, [1-[(BUTYLAMINO)CARBONYL]-1H-BENZIMIDAZOL-2-YL]-, METHYL ESTER
88-10-8	CARBAMIC CHLORIDE, DIETHYL-
79-44-7	CARBAMIC CHLORIDE, DIMETHYL-
57-13-6	CARBAMIDE
124-43-6	CARBAMIDE PEROXIDE
57-13-6	CARBAMIDE RESIN
57-13-6	CARBAMIMIDIC ACID
63-25-2	CARBAMINE
51-83-2	CARBAMINOCHOLINE CHLORIDE
51-83-2	CARBAMINOYLCHOLINE CHLORIDE
51-83-2	CARBAMIOTIN
34731-32-3	CARBAMODITHIOIC ACID, 1,2-ETHANEDIYL ESTER
142-59-6	CARBAMODITHIOIC ACID, 1,2-ETHANEDIYLBIS-, DISODIUM SALT
95-06-7	CARBAMODITHIOIC ACID, DIETHYL-, 2-CHLORO-2-PROPENYL ESTER
4384-82-1	CARBAMODITHIOIC ACID, DIETHYL-, SODIUM SALT (9CI)
14484-64-1	CARBAMODITHIOIC ACID, DIMETHYL-, IRON COMPLEX
4384-82-1	CARBAMODITHIOIC ACID, ION(1-)
420-04-2	CARBAMONITRILE
2303-16-4	CARBAMOTHIOIC ACID, BIS(1-METHYLETHYL)-, S-(2,3-DICHLORO-2-PROPENYL) ESTER
88-10-8	CARBAMOYL CHLORIDE, DIETHYL-
79-44-7	CARBAMOYL CHLORIDE, DIMETHYL-
51-83-2	CARBAMOYLCHOLINE CHLORIDE
51-83-2	CARBAMOYLCHOLINE-HYDROCHLORIDE
2631-37-0	CARBAMULT
79-44-7	CARBAMYL CHLORIDE, N,N-DIMETHYL-
51-83-2	CARBAMYLCHOLINE CHLORIDE
563-41-7	CARBAMYLHYDRAZINE HYDROCHLORIDE
103-71-9	CARBANIL
116-06-3	CARBANOLATE
63-25-2	CARBARIL (ITALIAN)
63-25-2	CARBARYL
63-25-2	CARBARYL (ACGIH,DOT)
63-25-2	CARBATOX
63-25-2	CARBATOX 60
63-25-2	CARBATOX 75
63-25-2	CARBAVUR
115-32-2	CARBAX
88-89-1	CARBAZOTIC ACID
121-75-5	CARBETHOXY MALATHION
84-72-0	CARBETHOXYMETHYL ETHYL PHTHALATE
121-75-5	CARBETOVUR
121-75-5	CARBETOX
141-66-2	CARBICRON
1937-37-7	CARBIDE BLACK E
420-04-2	CARBIMIDE
74-89-5	CARBINAMINE
67-56-1	CARBINOL
111-46-6	CARBITOL
111-90-0	CARBITOL
111-90-0	CARBITOL CELLOSOLVE
111-90-0	CARBITOL SOLVENT
501-53-1	CARBOBENZOXY CHLORIDE
51-83-2	CARBOCHOL
51-83-2	CARBOCHOLIN

CAS No.	Chemical Name
51-83-2	CARBOCHOLINE
786-19-6	CARBOFENOTHION
786-19-6	CARBOFENOTHION (DUTCH)
786-19-6	CARBOFENTHION
121-75-5	CARBOFOS
409-21-2	CARBOFRAX M
1563-66-2	CARBOFURAN
1563-66-2	CARBOFURAN (ACGIH,DOT)
1563-66-2	CARBOFURAN MIXTURE, LIQUID
8063-77-2	CARBOGEN
8063-77-2	CARBOGEN (8CI)
2231-57-4	CARBOHYDRAZIDE, THIO-
1333-86-4	CARBOLAC 1
108-95-2	CARBOLIC ACID
63-25-2	CARBOMATE
463-51-4	CARBOMETHENE
141-66-2	CARBOMICRON
12604-58-9	CARBON ALLOY, NONBASE, V,C,Fe (FERROVANADIUM)
127-18-4	CARBON BICHLORIDE
75-15-0	CARBON BISULFIDE
75-15-0	CARBON BISULFIDE (DOT)
75-15-0	CARBON BISULPHIDE
1333-86-4	CARBON BLACK
1333-86-4	CARBON BLACK MONARCH 81
558-13-4	CARBON BROMIDE
558-13-4	CARBON BROMIDE (CBr4)
56-23-5	CARBON CHLORIDE
56-23-5	CARBON CHLORIDE (CCl4)
463-71-8	CARBON CHLOROSULFIDE
142-59-6	CARBON D
127-18-4	CARBON DICHLORIDE
75-44-5	CARBON DICHLORIDE OXIDE
353-50-4	CARBON DIFLUORIDE OXIDE
124-38-9	CARBON DIOXIDE
124-38-9	CARBON DIOXIDE (ACGIH,DOT)
8070-50-6	CARBON DIOXIDE AND ETHYLENE OXIDE MIXTURES, WITH MORE THAN 6% ETHYLENE OXIDE (DOT)
124-38-9	CARBON DIOXIDE, LIQUEFIED (DOT)
53569-62-3	CARBON DIOXIDE, MIXTURE WITH NITROGEN OXIDE (N2O) (9CI)
124-38-9	CARBON DIOXIDE, REFRIGERATED LIQUID
124-38-9	CARBON DIOXIDE, REFRIGERATED LIQUID (DOT)
124-38-9	CARBON DIOXIDE, SOLID
124-38-9	CARBON DIOXIDE, SOLID (DOT)
8070-50-6	CARBON DIOXIDE-ETHYLENE OXIDE MIXTURE, WITH ≥ 6% ETHYLENE OXIDE
8070-50-6	CARBON DIOXIDE-ETHYLENE OXIDE MIXTURE, WITH NOT MORE THAN 6% ETHYLENE OXIDE
53569-62-3	CARBON DIOXIDE-NITROGEN OXIDE (N2O) MIXTURE
53569-62-3	CARBON DIOXIDE-NITROUS OXIDE MIXTURE
53569-62-3	CARBON DIOXIDE-NITROUS OXIDE MIXTURE (DOT)
8063-77-2	CARBON DIOXIDE-OXYGEN MIXTURE
8063-77-2	CARBON DIOXIDE-OXYGEN MIXTURE (DOT)
75-15-0	CARBON DISULFIDE
75-15-0	CARBON DISULFIDE (ACGIH,DOT)
75-15-0	CARBON DISULPHIDE
75-73-0	CARBON FLUORIDE
75-73-0	CARBON FLUORIDE (CF4)
353-50-4	CARBON FLUORIDE OXIDE
353-50-4	CARBON FLUORIDE OXIDE (COF2)
67-72-1	CARBON HEXACHLORIDE
74-90-8	CARBON HYDRIDE NITRIDE (CHN)
124-38-9	CARBON ICE (DOT)
75-63-8	CARBON MONOBROMIDE TRIFLUORIDE
630-08-0	CARBON MONOXIDE
630-08-0	CARBÒN MONOXIDE (ACGIH,DOT)
463-58-1	CARBON MONOXIDE MONOSULFIDE
630-08-0	CARBON MONOXIDE, CRYOGENIC
630-08-0	CARBON MONOXIDE, CRYOGENIC LIQUID (DOT)
460-19-5	CARBON NITRIDE
460-19-5	CARBON NITRIDE (C2N2)
57-12-5	CARBON NITRIDE ION (CN(SUP 1-))
57-12-5	CARBON NITRIDE ION (CN1-)
630-08-0	CARBON OXIDE (CO)
124-38-9	CARBON OXIDE (CO2)
463-58-1	CARBON OXIDE SULFIDE
463-58-1	CARBON OXIDE SULFIDE (9CI)
463-58-1	CARBON OXIDE SULFIDE (COS)
75-44-5	CARBON OXYCHLORIDE
353-50-4	CARBON OXYFLUORIDE
353-50-4	CARBON OXYFLUORIDE (COF2)
463-58-1	CARBON OXYSULFIDE
463-58-1	CARBON OXYSULFIDE (COS)
409-21-2	CARBON SILICIDE
75-15-0	CARBON SULFIDE
75-15-0	CARBON SULFIDE (CS2)
75-15-0	CARBON SULPHIDE (DOT)
56-23-5	CARBON TET
558-13-4	CARBON TETRABROMIDE
558-13-4	CARBON TETRABROMIDE (ACGIH,DOT)
56-23-5	CARBON TETRACHLORIDE
56-23-5	CARBON TETRACHLORIDE (ACGIH,DOT)
75-73-0	CARBON TETRAFLUORIDE
75-46-7	CARBON TRIFLUORIDE
75-47-8	CARBON TRIIODIDE
56-23-5	CARBONA
75-44-5	CARBONE (OXYCHLORURE DE) (FRENCH)
630-08-0	CARBONE (OXYDE DE) (FRENCH)
75-15-0	CARBONE (SUFURE DE) (FRENCH)
4452-58-8	CARBONIC ACID DISODIUM SALT, COMPD WITH HYDROGEN PEROXIDE (2:3)
15630-89-4	CARBONIC ACID DISODIUM SALT, COMPD WITH HYDROGEN PEROXIDE (2:3)
4452-58-8	CARBONIC ACID DISODIUM SALT, COMPD WITH HYDROGEN PEROXIDE (H2O2) (2:3)
3313-92-6	CARBONIC ACID DISODIUM SALT, COMPD WITH HYDROGEN PEROXIDE (H2O2) (2:3)
15630-89-4	CARBONIC ACID DISODIUM SALT, COMPD WITH HYDROGEN PEROXIDE (H2O2) (2:3)
124-38-9	CARBONIC ACID GAS
10361-29-2	CARBONIC ACID, AMMONIUM SALT
513-77-9	CARBONIC ACID, BARIUM SALT (1:1)
13106-47-3	CARBONIC ACID, BERYLLIUM SALT (1:1)
30714-78-4	CARBONIC ACID, BUTYL ETHYL ESTER
96-49-1	CARBONIC ACID, CYCLIC ETHYLENE ESTER
506-87-6	CARBONIC ACID, DIAMMONIUM SALT
506-87-6	CARBONIC ACID, DIAMMONIUM SALT (8CI,9CI)
105-58-8	CARBONIC ACID, DIETHYL ESTER
616-38-6	CARBONIC ACID, DIMETHYL ESTER
534-16-7	CARBONIC ACID, DISILVER(1+) SALT
6533-73-9	CARBONIC ACID, DITHALLIUM(1+) SALT
140-89-6	CARBONIC ACID, DITHIO-, O-ETHYL ESTER, POTASSIUM SALT
1066-33-7	CARBONIC ACID, MONOAMMONIUM SALT
16337-84-1	CARBONIC ACID, NICKEL SALT
3333-67-3	CARBONIC ACID, NICKEL(2+) SALT (1:1)
17237-93-3	CARBONIC ACID, NICKEL(2+) SALT (2:1)
2941-64-2	CARBONIC ACID, THIO-, ANHYDROSULFIDE WITH THIOHYPO-CHLOROUS ACID, ETHYL ESTER
26555-35-1	CARBONIC ACID, THIO-, ANHYDROSULFIDE WITH THIOHYPO-CHLOROUS ACID, ETHYL ESTER
3486-35-9	CARBONIC ACID, ZINC SALT (1:1)
124-38-9	CARBONIC ANHYDRIDE
75-44-5	CARBONIC DICHLORIDE
622-44-6	CARBONIC DICHLORIDE, (PHENYLIMINO)-
463-71-8	CARBONIC DICHLORIDE, THIO-
353-50-4	CARBONIC DIFLUORIDE
353-50-4	CARBONIC DIFLUORIDE (9CI)
630-08-0	CARBONIC OXIDE
622-44-6	CARBONIMIDIC DICHLORIDE, PHENYL-
622-44-6	CARBONIMIDIC DICHLORIDE, PHENYL- (9CI)
75-44-5	CARBONIO (OSSICLORURO DI) (ITALIAN)
630-08-0	CARBONIO (OSSIDO DI) (ITALIAN)
75-15-0	CARBONIO (SOLFURO DI) (ITALIAN)
108-23-6	CARBONOCHLORIDE ACID, 1-METHYLETHYL ESTER
70042-58-9	CARBONOCHLORIDIC ACID, (1,1-DIMETHYLETHYL)CYCLOHEXYL ESTER
108-23-6	CARBONOCHLORIDIC ACID, 1-METHYLETHYL ESTER
627-11-2	CARBONOCHLORIDIC ACID, 2-CHLOROETHYL ESTER
24468-13-1	CARBONOCHLORIDIC ACID, 2-ETHYLHEXYL ESTER
2937-50-0	CARBONOCHLORIDIC ACID, 2-PROPENYL ESTER
592-34-7	CARBONOCHLORIDIC ACID, BUTYL ESTER
592-34-7	CARBONOCHLORIDIC ACID, BUTYL ESTER (9CI)
22128-62-7	CARBONOCHLORIDIC ACID, CHLOROMETHYL ESTER
22128-62-7	CARBONOCHLORIDIC ACID, CHLOROMETHYL ESTER (9CI)
81228-87-7	CARBONOCHLORIDIC ACID, CYCLOBUTYL ESTER
541-41-3	CARBONOCHLORIDIC ACID, ETHYL ESTER
79-22-1	CARBONOCHLORIDIC ACID, METHYL ESTER
1885-14-9	CARBONOCHLORIDIC ACID, PHENYL ESTER
109-61-5	CARBONOCHLORIDIC ACID, PROPYL ESTER

CAS No.	Chemical Name
2941-64-2	CARBONOCHLORIDOTHIOIC ACID, S-ETHYL ESTER
13889-92-4	CARBONOCHLORIDOTHIOIC ACID, S-PROPYL ESTER
140-89-6	CARBONODITHIOIC ACID, O-ETHYL ESTER, POTASSIUM SALT
4452-58-8	CARBONOPEROXOIC ACID, DISODIUM SALT
3313-92-6	CARBONOPEROXOIC ACID, DISODIUM SALT
15630-89-4	CARBONOPEROXOIC ACID, DISODIUM SALT
3313-92-6	CARBONOPEROXOIC ACID, DISODIUM SALT (9CI)
4452-58-8	CARBONOPEROXOIC ACID, DISODIUM SALT (9CI)
15630-89-4	CARBONOPEROXOIC ACID, DISODIUM SALT (9CI)
2372-21-6	CARBONOPEROXOIC ACID, OO-(1,1-DIMETHYLETHYL) O-(1-METHYLETHYL) ESTER
2941-64-2	CARBONOTHIOIC ACID, ANHYDROSULFIDE WITH THIOHYPOCHLOROUS ACID, ETHYL ESTER
26555-35-1	CARBONOTHIOIC ACID, ANHYDROSULFIDE WITH THIOHYPOCHLOROUS ACID, ETHYL ESTER
463-71-8	CARBONOTHIOIC DICHLORIDE
463-71-8	CARBONOTHIOIC DICHLORIDE (9CI)
2231-57-4	CARBONOTHIOIC DIHYDRAZIDE
2231-57-4	CARBONOTHIOIC DIHYDRAZIDE (9CI)
75-44-5	CARBONYL CHLORIDE
75-44-5	CARBONYL CHLORIDE (DOT)
463-71-8	CARBONYL CHLORIDE, THIO-
57-13-6	CARBONYL DIAMIDE
75-44-5	CARBONYL DICHLORIDE
353-50-4	CARBONYL DIFLUORIDE
353-50-4	CARBONYL DIFLUORIDE (COF2)
353-50-4	CARBONYL FLUORIDE
353-50-4	CARBONYL FLUORIDE (ACGIH,DOT)
353-50-4	CARBONYL FLUORIDE (COF2)
463-58-1	CARBONYL SULFIDE
463-58-1	CARBONYL SULFIDE (COS)
463-58-1	CARBONYL SULFIDE (DOT)
463-71-8	CARBONYL SULFIDE DICHLORIDE
463-58-1	CARBONYL SULFIDE-(SUP 32)S
463-58-1	CARBONYL SULFIDE-32S
75-44-5	CARBONYLCHLORID (GERMAN)
57-13-6	CARBONYLDIAMINE
786-19-6	CARBOPHENOTHION
121-75-5	CARBOPHOS
409-21-2	CARBORUNDUM
140-88-5	CARBOSET 511
57-06-7	CARBOSPOL
8070-50-6	CARBOXID
8070-50-6	CARBOXIDE
8070-50-6	CARBOXIDE (PESTICIDE)
65-85-0	CARBOXYBENZENE
79-09-4	CARBOXYETHANE
1338-24-5	CARBOXYLIC ACIDS, NAPHTHENIC
51-83-2	CARBYL
51-83-2	CARCHOLIN
630-60-4	CARD-20(22)-ENOLIDE, 3-[(6-DEOXY-ALPHA-L-MANNOPYRANOSYL)OXY]-1,5,11,14,19-PENTAHYDROXY-, (1BETA,3BET
55-63-0	CARDAMIST
71-63-6	CARDIDIGIN
71-63-6	CARDIGIN
71-63-6	CARDITOXIN
498-15-7	CARENE
86-50-0	CARFENE
1642-54-2	CARICIDE
1642-54-2	CARITROL
154-93-8	CARMUSTINE
8020-83-5	CARNEA 21
8020-83-5	CARNEA OIL 31
92-52-4	CAROLIDAL
50-55-5	CARPACIL
7631-86-9	CARPLEX
7631-86-9	CARPLEX 1120
7631-86-9	CARPLEX 30
7631-86-9	CARPLEX 67
7631-86-9	CARPLEX 80
7631-86-9	CARPLEX FPS 1
7631-86-9	CARPLEX FPS 3
63-25-2	CARPOLIN
9000-07-1	CARRAGEENAN
9000-07-1	CARRAGEENAN GH
9000-07-1	CARRAGEENAN GUM
9000-07-1	CARRAGEENIN
9000-07-1	CARRAGHEEN
9000-07-1	CARRAGHEENAN
7681-52-9	CARREL-DAKIN SOLUTION
151-21-3	CARSONOL SLS
151-21-3	CARSONOL SLS PASTE B
151-21-3	CARSONOL SLS SPECIAL
8070-50-6	CARTOX
2244-16-8	CARVOL
2244-16-8	CARVONE
2244-16-8	CARVONE, (+)-
63-25-2	CARYLDERM
51-75-2	CARYOLYSIN
55-86-7	CARYOLYSINE
23422-53-9	CARZOL
23422-53-9	CARZOL SP
51-21-8	CARZONAL
1194-65-6	CASORON
1194-65-6	CASORON 133
12001-29-5	CASSIAR AK
535-89-7	CASTRIX
26264-06-2	CASUL 70HF
128-37-0	CATALIN CAO-3
7631-86-9	CATALOID
7631-86-9	CATALOID HS 40
7631-86-9	CATALOID S 30H
7631-86-9	CATALOID S 30L
7631-86-9	CATALOID SI 350
1344-28-1	CATAPAL S
154-23-4	CATECHIN
120-80-9	CATECHIN
154-23-4	CATECHIN (FLAVAN)
120-80-9	CATECHIN (PHENOL)
120-80-9	CATECHINIC ACID
154-23-4	CATECHINIC ACID
154-23-4	CATECHOL
120-80-9	CATECHOL
154-23-4	CATECHOL (FLAVAN)
120-80-9	CATECHOL (PHENOL)
154-23-4	CATECHUIC ACID
154-23-4	CATERGEN
541-41-3	CATHYL CHLORIDE
56-75-7	CATILAN
112-02-7	CATION PB 40
1309-48-4	CAUSTIC MAGNESITE
1310-58-3	CAUSTIC POTASH
1310-58-3	CAUSTIC POTASH, DRY, SOLID, FLAKE, BEAD, OR GRANULAR (DOT)
1310-58-3	CAUSTIC POTASH, LIQUID OR SOLUTION (DOT)
1310-73-2	CAUSTIC SODA
7681-49-4	CAVI-TROL
494-03-1	CB 1048
305-03-3	CB 1348
55-98-1	CB 2041
148-82-3	CB 3025
671-16-9	CB 400-497
7782-42-5	CB 50
545-55-1	CBC 906288
87-90-1	CBD 90
558-13-4	CBR4
1333-86-4	CC 40-220
156-62-7	CCC
999-81-5	CCC
9016-45-9	CCC JELLY
999-81-5	CCC PLANT GROWTH REGULANT
13010-47-4	CCNU
71-36-3	CCS 203
78-92-2	CCS 301
57-74-9	CD 68
7440-50-8	CDA 101
7440-50-8	CDA 102
7440-50-8	CDA 110
7440-50-8	CDA 122
2782-57-2	CDB 60
2893-78-9	CDB 63
87-90-1	CDB 90
2893-78-9	CDB CLEARON
95-06-7	CDEC
97-00-7	CDNB
25567-67-3	CDNB
479-45-8	CE

CAS No.	Chemical Name
7440-50-8	CE 1110
999-81-5	CE CE CE
79-01-6	CECOLENE
330-54-1	CEKIURON
63-25-2	CEKUBARYL
115-32-2	CEKUDIFOL
52-68-6	CEKUFON
108-62-3	CEKUMETA
9002-91-9	CEKUMETA
37273-91-9	CEKUMETA
298-00-0	CEKUMETHION
2636-26-2	CEKUSAN
62-38-4	CEKUSIL
60-51-5	CEKUTHOATE
122-14-5	CEKUTROTHION
1912-24-9	CEKUZINA-T
21923-23-9	CELA S 2957
21923-23-9	CELAMERCK S 2957
58-89-9	CELANEX
577-11-7	CELANOL DOS 65
577-11-7	CELANOL DOS 75
21923-23-9	CELATHION
21923-23-9	CELATHION (PESTICIDE)
9004-70-0	CELEX
7631-86-9	CELITE SUPERFLOSS
9004-34-6	CELLEX MX
9004-70-0	CELLINE 200
119-90-4	CELLITAZOL B
82-28-0	CELLITON ORANGE R
101-25-7	CELLMIC A
101-25-7	CELLMIC A 80
80-51-3	CELLMIC S
9004-70-0	CELLOIDIN
79-34-5	CELLON
110-80-5	CELLOSOLVE
111-15-9	CELLOSOLVE ACETATE
111-76-2	CELLOSOLVE ACETATE (DOT)
7779-88-6	CELLOXAN
1330-78-5	CELLUFLEX 179C
84-74-2	CELLUFLEX DPB
115-86-6	CELLUFLEX TPP
9004-34-6	CELLULOSE
9004-34-6	CELLULOSE 248
9004-34-6	CELLULOSE CRYSTALLINE
9004-70-0	CELLULOSE NITRATE
9004-70-0	CELLULOSE NITRATE (CONTAINING NOT MORE THAN 12.6% NITROGEN)
9004-70-0	CELLULOSE TETRANITRATE
9004-70-0	CELLULOSE, NITRATE (9CI)
126-73-8	CELLUPHOS 4
62-38-4	CELMER
106-93-4	CELMIDE
9004-70-0	CELNOVA BTH 1/2
80-17-1	CELOGEN BSH
80-51-3	CELOGEN OT
60-00-4	CELON A
60-00-4	CELON ATH
20859-73-8	CELPHIDE
20859-73-8	CELPHINE
20859-73-8	CELPHOS
121-75-5	CELTHION
7440-58-6	CELTIUM
9004-34-6	CELUFI
137-05-3	CEMEDINE 3000
137-05-3	CEMEDINE 3000 TYPE-II
65997-15-1	CEMENT, PORTLAND, CHEMICALS
9016-45-9	CEMULSOL NP 10
9016-45-9	CEMULSOL NP 8
9016-45-9	CEMULSOL NP 9
9016-45-9	CEMULSOL NP-EO 6
80-51-3	CENITRON OB
57-11-4	CENTURY 1210
57-11-4	CENTURY 1220
57-11-4	CENTURY 1230
57-11-4	CENTURY 1240
112-80-1	CENTURY CD FATTY ACID
9004-34-6	CEPO
9004-34-6	CEPO CFM
9004-34-6	CEPO S 20
9004-34-6	CEPO S 40
60-11-7	CERASINE YELLOW GG
8002-74-2	CERATAK
62-38-4	CERESAN
2235-25-8	CERESAN NI
62-38-4	CERESAN UNIVERSAL
62-38-4	CERESOL
8002-74-2	CERETAL 165
7440-45-1	CERIUM
8049-18-1	CERIUM ALLOY, BASE, (FERROCERIUM)
69523-06-4	CERIUM ALLOY, BASE, (FERROCERIUM)
60475-66-3	CERIUM ALLOY, BASE, (FERROCERIUM)
8049-20-5	CERIUM MISCH METAL
7440-45-1	CERIUM, CRUDE
7440-45-1	CERIUM, CRUDE, POWDER (DOT)
7440-45-1	CERIUM, CRUDE, SLABS OR INGOTS (DOT)
3761-53-3	CERTICOL PONCEAU MXS
123-30-8	CERTINAL
27598-85-2	CERTINAL
57-24-9	CERTOX
8020-83-5	CERTREX 39
20830-81-3	CERUBIDIN
1309-37-1	CERVEN H
140-57-8	CES
7440-46-2	CESIUM
21351-79-1	CESIUM HYDRATE
21351-79-1	CESIUM HYDROXIDE
21351-79-1	CESIUM HYDROXIDE (ACGIH)
21351-79-1	CESIUM HYDROXIDE (Cs(OH))
21351-79-1	CESIUM HYDROXIDE DIMER
21351-79-1	CESIUM HYDROXIDE SOLUTION
21351-79-1	CESIUM HYDROXIDE, SOLID (DOT)
21351-79-1	CESIUM HYDROXIDE, SOLUTION (DOT)
7440-46-2	CESIUM METAL
7440-46-2	CESIUM METAL (DOT)
7789-18-6	CESIUM NITRATE
7789-18-6	CESIUM NITRATE (DOT)
7789-18-6	CESIUM(I) NITRATE (1:1)
7440-46-2	CESIUM, POWDERED (DOT)
7440-46-2	CESIUM-133
1912-24-9	CET
112-02-7	CETAC
112-02-7	CETRIMONIUM CHLORIDE
112-02-7	CETYLTRIMETHYLAMMONIUM CHLORIDE
7782-42-5	CEYLON BLACK LEAD
75-71-8	CF 12
71-55-6	CF 2
75-69-4	CFC 11
75-71-8	CFC 12
75-43-4	CFC 21
75-45-6	CFC 22
7446-18-6	CFS
10031-59-1	CFS
470-90-6	CFV
75-44-5	CG
2385-85-5	CG-1283
470-90-6	CGA 26351
563-41-7	CH
3771-19-5	CH 13-437
37187-22-7	CHALOXYD AAP-NA 1
1338-23-4	CHALOXYD MEKP-HA 1
1338-23-4	CHALOXYD MEKP-LA 1
2167-23-9	CHALOXYD P 1200AL
3006-86-8	CHALOXYD P 1250AL
2167-23-9	CHALOXYD P 1293AL
3006-82-4	CHALOXYD P 1310
3006-82-4	CHALOXYD P 1327
614-45-9	CHALOXYD TBPB
7722-64-7	CHAMELEON MINERAL
1333-86-4	CHANNEL BLACK
57-63-6	CHEE-O-GEN
57-63-6	CHEE-O-GENF
60-00-4	CHEELOX
60-00-4	CHEELOX BF ACID
139-13-9	CHEL 300
75-00-3	CHELEN
7784-46-5	CHEM PELS C
50-21-5	CHEM-CAST
8001-35-2	CHEM-PHENE

CAS No.	Chemical Name
98-00-0	CHEM-REZ 200
7784-46-5	CHEM-SEN 56
87-86-5	CHEM-TOL
115-90-2	CHEMAGRO 25141
327-98-0	CHEMAGRO 37289
124-65-2	CHEMAID
128-37-0	CHEMANOX 11
121-75-5	CHEMATHION
142-59-6	CHEMBAM
60-00-4	CHEMCOLOX 340
136-60-7	CHEMCRYL C 101N
101-84-8	CHEMCRYL JK-EB
1309-64-4	CHEMETRON FIRE SHIELD
72-43-5	CHEMFORM
10124-56-8	CHEMI-CHARL
86-88-4	CHEMICAL 109
30553-04-9	CHEMICAL 109
532-27-4	CHEMICAL MACE
56-75-7	CHEMICETIN
56-75-7	CHEMICETINA
7778-54-3	CHEMICHLON G
7681-49-4	CHEMIFLUOR
8020-83-5	CHEMKLEEN
105-60-2	CHEMLON
88-85-7	CHEMOX GENERAL
88-85-7	CHEMOX PE
101-25-7	CHEMPOR N 90
101-25-7	CHEMPOR PC 65
83-26-1	CHEMRAT
1335-85-9	CHEMSECT DNOC
9002-89-5	CHEMTREND 39
1333-86-4	CHESACARB K 2
10265-92-6	CHEVRON 9006
67-64-1	CHEVRON ACETONE
10265-92-6	CHEVRON ORTHO 9006
75-60-5	CHEXMATE
7631-99-4	CHILE SALTPETER
111-76-2	CHIMEC NR
9016-45-9	CHIMIPAL WN 6
1332-58-7	CHINA CLAY
91-63-4	CHINALDINE
1344-48-5	CHINESE VERMILION
9004-66-4	CHINOFER
91-22-5	CHINOLEINE
91-22-5	CHINOLIN (CZECH)
91-22-5	CHINOLINE
106-51-4	CHINON (DUTCH, GERMAN)
106-51-4	CHINONE
82-68-8	CHINOZAN
1563-66-2	CHINUFUR
7778-44-1	CHIP-CAL
7778-44-1	CHIP-CAL GRANULAR
137-26-8	CHIPCO THIRAM 75
3734-97-2	CHIPMAN 6199
78-53-5	CHIPMAN 6200
3734-97-2	CHIPMAN R-6,199
106-87-6	CHISSONOX 206 MONOMER
101-25-7	CHKHZ 18
78-67-1	CHKHZ 57
80-17-1	CHKHZ 9
56-75-7	CHLOMYCOL
87-86-5	CHLON
7782-50-5	CHLOOR (DUTCH)
74-87-3	CHLOOR-METHAAN (DUTCH)
108-90-7	CHLOORBENZEEN (DUTCH)
12789-03-6	CHLOORDAAN (DUTCH)
75-00-3	CHLOORETHAAN (DUTCH)
3691-35-8	CHLOORFACINON (DUTCH)
76-06-2	CHLOORPIKRINE (DUTCH)
7647-01-0	CHLOORWATERSTOF (DUTCH)
1336-36-3	CHLOPHEN
7782-50-5	CHLOR (GERMAN)
12789-03-6	CHLOR KIL
74-87-3	CHLOR-METHAN (GERMAN)
76-06-2	CHLOR-O-PIC
79-11-8	CHLORACETIC ACID
78-95-5	CHLORACETONE (FRENCH)
107-14-2	CHLORACETONITRILE
532-27-4	CHLORACETOPHENONE

CAS No.	Chemical Name
79-04-9	CHLORACETYL CHLORIDE
52-68-6	CHLORAK
75-87-6	CHLORAL
75-87-6	CHLORAL, ANHYDROUS, INHIBITED
75-87-6	CHLORAL, ANHYDROUS, INHIBITED (DOT)
95-06-7	CHLORALLYL DIETHYLDITHIOCARBAMATE
107-05-1	CHLORALLYLENE
133-90-4	CHLORAMBEN
305-03-3	CHLORAMBUCIL
79-22-1	CHLORAMEISENSAEURE METHYLESTER (GERMAN)
541-41-3	CHLORAMEISENSAEUREAETHYLESTER (GERMAN)
56-75-7	CHLORAMEX
55-86-7	CHLORAMIN
55-86-7	CHLORAMINE
10599-90-3	CHLORAMINE
1937-37-7	CHLORAMINE BLACK C
1937-37-7	CHLORAMINE BLACK EC
1937-37-7	CHLORAMINE BLACK ERT
1937-37-7	CHLORAMINE BLACK EX
1937-37-7	CHLORAMINE BLACK EXR
1937-37-7	CHLORAMINE BLACK XO
2602-46-2	CHLORAMINE BLUE 2B
72-57-1	CHLORAMINE BLUE 3B
1937-37-7	CHLORAMINE CARBON BLACK S
1937-37-7	CHLORAMINE CARBON BLACK SJ
1937-37-7	CHLORAMINE CARBON BLACK SN
16071-86-6	CHLORAMINE FAST BROWN BRL
16071-86-6	CHLORAMINE FAST CUTCH BROWN PL
305-03-3	CHLORAMINOPHENE
56-75-7	CHLORAMPHENICOL
56-75-7	CHLORAMSAAR
16071-86-6	CHLORANTINE FAST BROWN BRLL
14866-68-3	CHLORATE
10137-74-3	CHLORATE DE CALCIUM (FRENCH)
3811-04-9	CHLORATE DE POTASSIUM (FRENCH)
14866-68-3	CHLORATE ION
14866-68-3	CHLORATE ION (ClO3-)
3811-04-9	CHLORATE OF POTASH
3811-04-9	CHLORATE OF POTASH (DOT)
7775-09-9	CHLORATE OF SODA
7775-09-9	CHLORATE OF SODA (DOT)
10326-21-3	CHLORATE SALT OF MAGNESIUM
7775-09-9	CHLORATE SALT OF SODIUM
14866-68-3	CHLORATE(1-)
7775-09-9	CHLORAX
1937-37-7	CHLORAZOL BLACK E
1937-37-7	CHLORAZOL BLACK EA
1937-37-7	CHLORAZOL BLACK EN
2602-46-2	CHLORAZOL BLUE B
2602-46-2	CHLORAZOL BLUE BP
1937-37-7	CHLORAZOL BURL BLACK E
1937-37-7	CHLORAZOL LEATHER BLACK ENP
1937-37-7	CHLORAZOL SILK BLACK G
108-90-7	CHLORBENZENE
510-15-6	CHLORBENZILAT
108-90-7	CHLORBENZOL
510-15-6	CHLORBENZYLATE
305-03-3	CHLORBUTIN
999-81-5	CHLORCHOLINCHLORID (CZECH,GERMAN)
999-81-5	CHLORCHOLINE CHLORIDE
506-77-4	CHLORCYAN
12789-03-6	CHLORDAN
12789-03-6	CHLORDANE
12789-03-6	CHLORDANE (ACGIH)
12789-03-6	CHLORDANE [FLAMMABLE LIQUID LABEL]
12789-03-6	CHLORDANE, TECHNICAL
12789-03-6	CHLORDANE,LIQUID (DOT)
143-50-0	CHLORDECONE
58-25-3	CHLORDIAZEPOXIDE
438-41-5	CHLORDIAZEPOXIDE HYDROCHLORIDE
107-30-2	CHLORDIMETHYLETHER (CZECH)
7782-50-5	CHLORE (FRENCH)
87-90-1	CHLOREAL
75-00-3	CHLORENE
58-89-9	CHLORESENE
66-75-1	CHLORETHAMINACIL
51-75-2	CHLORETHAZINE
75-01-4	CHLORETHENE
75-00-3	CHLORETHYL

CAS No.	Chemical Name
75-01-4	CHLORETHYLENE
111-44-4	CHLOREX
1336-36-3	CHLOREXTOL
3691-35-8	CHLORFACINON (GERMAN)
470-90-6	CHLORFENVINFOS
470-90-6	CHLORFENVINPHOS
470-90-6	CHLORFENWINFOSEM (POLISH)
52-68-6	CHLORFOS
142-04-1	CHLORHYDRATE D'ANILINE (FRENCH)
3165-93-3	CHLORHYDRATE DE 4-CHLOROORTHOTOLUIDINE (FRENCH)
2820-51-1	CHLORHYDRATE DE NICOTINE (FRENCH)
96-24-2	CHLORHYDRIN
7790-93-4	CHLORIC ACID
7790-93-4	CHLORIC ACID (DOT)
7790-93-4	CHLORIC ACID SOLUTION, CONTAINING NOT MORE THAN 10% ACID (DOT)
13477-00-4	CHLORIC ACID, BARIUM SALT
10137-74-3	CHLORIC ACID, CALCIUM SALT
26506-47-8	CHLORIC ACID, COPPER SALT
14866-68-3	CHLORIC ACID, ION(1-)
10326-21-3	CHLORIC ACID, MAGNESIUM SALT
3811-04-9	CHLORIC ACID, POTASSIUM SALT
7775-09-9	CHLORIC ACID, SODIUM SALT
7791-10-8	CHLORIC ACID, STRONTIUM SALT
13453-30-0	CHLORIC ACID, THALLIUM(1+) SALT
10361-95-2	CHLORIC ACID, ZINC SALT
12125-02-9	CHLORID AMONNY (CZECH)
142-04-1	CHLORID ANILINU (CZECH)
10025-91-9	CHLORID ANTIMONITY (CZECH)
10026-04-7	CHLORID KREMICITY (CZECH)
7447-39-4	CHLORID MEDNY (CZECH)
7487-94-7	CHLORID RTUTNATY (CZECH)
7778-54-3	CHLORIDE OF LIME
7719-12-2	CHLORIDE OF PHOSPHORUS (DOT)
10545-99-0	CHLORIDE OF SULFUR (DOT)
10025-67-9	CHLORIDE OF SULFUR (DOT)
75-00-3	CHLORIDUM
79-01-6	CHLORILEN
1336-36-3	CHLORINATED BIPHENYL
8001-35-2	CHLORINATED CAMPHENE
8001-35-2	CHLORINATED CAMPHENE 60% (ACGIH)
1336-36-3	CHLORINATED DIPHENYL
55720-99-5	CHLORINATED DIPHENYL OXIDE
51289-10-2	CHLORINATED DIPHENYL OXIDE
1336-36-3	CHLORINATED DIPHENYLENE
76-13-1	CHLORINATED FLUOROCARBON
75-34-3	CHLORINATED HYDROCHLORIC ETHER
7778-54-3	CHLORINATED LIME
57-74-9	CHLORINDAN
7782-50-5	CHLORINE
7782-50-5	CHLORINE (ACGIH,DOT)
506-77-4	CHLORINE CYANIDE
506-77-4	CHLORINE CYANIDE (ClCN)
10049-04-4	CHLORINE DIOXIDE
7790-93-4	CHLORINE DIOXIDE HYDRATE, FROZEN
1318-59-8	CHLORINE DIOXIDE ION(1-)
14998-27-7	CHLORINE DIOXIDE ION(1-)
7790-91-2	CHLORINE FLUORIDE
7790-91-2	CHLORINE FLUORIDE (Cl2F6)
7790-91-2	CHLORINE FLUORIDE (ClF3)
13637-63-3	CHLORINE FLUORIDE (ClF5)
7616-94-6	CHLORINE FLUORIDE OXIDE (ClO3F)
7790-99-0	CHLORINE IODIDE
7790-99-0	CHLORINE IODIDE (CiI)
7782-50-5	CHLORINE MOLECULE (Cl2)
7790-99-0	CHLORINE MONOIODIDE
7791-21-1	CHLORINE MONOOXIDE
7791-21-1	CHLORINE MONOXIDE
7791-21-1	CHLORINE MONOXIDE (Cl2O)
7791-21-1	CHLORINE OXIDE
10049-04-4	CHLORINE OXIDE
7791-21-1	CHLORINE OXIDE (Cl2O)
14380-61-1	CHLORINE OXIDE (ClO1-)
10049-04-4	CHLORINE OXIDE (ClO2)
14866-68-3	CHLORINE OXIDE (ClO31-)
7616-94-6	CHLORINE OXYFLUORIDE (ClO3F)
13637-63-3	CHLORINE PENTAFLUORIDE
13637-63-3	CHLORINE PENTAFLUORIDE (DOT)
10049-04-4	CHLORINE PEROXIDE
10545-99-0	CHLORINE SULFIDE
10545-99-0	CHLORINE SULFIDE (Cl2S)
7790-91-2	CHLORINE TRIFLUORIDE
7790-91-2	CHLORINE TRIFLUORIDE (ACGIH,DOT)
7790-91-2	CHLORINE TRIFLUORIDE (ClF3)
10049-04-4	CHLORINE(IV) OXIDE
1318-59-8	CHLORITE
14998-27-7	CHLORITE
1318-59-8	CHLORITE (MINERAL CLASS)
14998-27-7	CHLORITE (MINERAL CLASS)
1318-59-8	CHLORITE ION
14998-27-7	CHLORITE ION
1318-59-8	CHLORITE, INORGANIC
14998-27-7	CHLORITE, INORGANIC
1318-59-8	CHLORITE-GROUP MINERALS
14998-27-7	CHLORITE-GROUP MINERALS
1318-59-8	CHLORITE-TYPE MINERALS
14998-27-7	CHLORITE-TYPE MINERALS
24934-91-6	CHLORMEPHOS
999-81-5	CHLORMEQUAT
999-81-5	CHLORMEQUAT CHLORIDE
51-75-2	CHLORMETHINE
55-86-7	CHLORMETHINUM
24934-91-6	CHLORMETHYLFOS
494-03-1	CHLORNAPHAZINE
1336-36-3	CHLORO 1,1-BIPHENYL
1336-36-3	CHLORO BIPHENYL
7778-54-3	CHLORO LIME CHEMICAL
79-38-9	CHLORO TRIFLUOROETHYLENE (DOT)
2100-42-7	CHLORO-1,4-DIMETHOXYBENZENE
78-95-5	CHLORO-2-PROPANONE
73090-69-4	CHLORO-4-TERT-AMYLPHENOL
121-73-3	CHLORO-M-NITROBENZENE
88-73-3	CHLORO-O-NITROBENZENE
61878-61-3	CHLORO-O-NITROTOLUENE (DOT)
25567-68-4	CHLORO-O-NITROTOLUENE (DOT)
107-20-0	CHLOROACETALDEHYDE
107-20-0	CHLOROACETALDEHYDE (ACGIH,DOT)
97-97-2	CHLOROACETALDEHYDE DIMETHYL ACETAL
107-20-0	CHLOROACETALDEHYDE MONOMER
79-11-8	CHLOROACETIC ACID
79-04-9	CHLOROACETIC ACID CHLORIDE
105-39-5	CHLOROACETIC ACID ETHYL ESTER
105-48-6	CHLOROACETIC ACID ISOPROPYL ESTER
3926-62-3	CHLOROACETIC ACID SODIUM SALT
79-11-8	CHLOROACETIC ACID, LIQUID
79-11-8	CHLOROACETIC ACID, LIQUID (DOT)
79-11-8	CHLOROACETIC ACID, SOLID
79-11-8	CHLOROACETIC ACID, SOLID (DOT)
79-11-8	CHLOROACETIC ACID, SOLUTION (DOT)
79-04-9	CHLOROACETIC CHLORIDE
78-95-5	CHLOROACETONE
78-95-5	CHLOROACETONE, STABILIZED (DOT)
107-14-2	CHLOROACETONITRILE
107-14-2	CHLOROACETONITRILE (DOT)
532-27-4	CHLOROACETOPHENONE
1341-24-8	CHLOROACETOPHENONE
532-27-4	CHLOROACETOPHENONE (DOT)
532-27-4	CHLOROACETOPHENONE, GAS, LIQUID, OR SOLID (DOT)
79-04-9	CHLOROACETYL CHLORIDE
79-04-9	CHLOROACETYL CHLORIDE (ACGIH,DOT)
75-00-3	CHLOROAETHAN (GERMAN)
107-05-1	CHLOROALLYLENE
27134-26-5	CHLOROANILINE
27134-26-5	CHLOROANILINE, LIQUID
27134-26-5	CHLOROANILINE, SOLID
8007-56-5	CHLOROAZOTIC ACID
95-50-1	CHLOROBEN
98-87-3	CHLOROBENZAL
108-90-7	CHLOROBENZEN (POLISH)
108-90-7	CHLOROBENZENE
108-90-7	CHLOROBENZENE (ACGIH,DOT)
108-90-7	CHLOROBENZENU (CZECH)
510-15-6	CHLOROBENZILATE
108-90-7	CHLOROBENZOL (DOT)
98-15-7	CHLOROBENZOTRIFLUORIDE
52181-51-8	CHLOROBENZOTRIFLUORIDE
104-83-6	CHLOROBENZYL CHLORIDE
1332-21-4	CHLOROBESTOS 25

CAS No.	Chemical Name
353-59-3	CHLOROBROMODIFLUOROMETHANE
74-97-5	CHLOROBROMOMETHANE
74-97-5	CHLOROBROMOMETHANE (ACGIH)
109-70-6	CHLOROBROMOPROPANE
109-69-3	CHLOROBUTANE
25154-42-1	CHLOROBUTANE
305-03-3	CHLOROBUTINE
8001-35-2	CHLOROCAMPHENE
56-75-7	CHLOROCAPS
541-41-3	CHLOROCARBONATE D'ETHYLE (FRENCH)
79-22-1	CHLOROCARBONATE DE METHYLE (FRENCH)
541-41-3	CHLOROCARBONIC ACID ETHYL ESTER
79-22-1	CHLOROCARBONIC ACID METHYL ESTER
999-81-5	CHLOROCHOLINE CHLORIDE
56-75-7	CHLOROCID
56-75-7	CHLOROCIDE
56-75-7	CHLOROCIDIN C
56-75-7	CHLOROCIDIN C TETRAN
59-50-7	CHLOROCRESOL
1321-10-4	CHLOROCRESOLS
3926-62-3	CHLOROCTAN SODNY (CZECH)
506-77-4	CHLOROCYAN
506-77-4	CHLOROCYANIDE
506-77-4	CHLOROCYANIDE (ClCN)
506-77-4	CHLOROCYANOGEN
542-18-7	CHLOROCYCLOHEXANE
12789-03-6	CHLORODANE
95-50-1	CHLORODEN
96-24-2	CHLORODEOXYGLYCEROL
124-48-1	CHLORODIBROMOMETHANE
814-49-3	CHLORODIETHOXYPHOSPHINE OXIDE
96-10-6	CHLORODIETHYLALUMINUM
1609-19-4	CHLORODIETHYLSILANE
353-59-3	CHLORODIFLUOROBROMOMETHANE
353-59-3	CHLORODIFLUOROBROMOMETHANE (DOT)
75-68-3	CHLORODIFLUOROETHANE
25497-29-4	CHLORODIFLUOROETHANE
25497-29-4	CHLORODIFLUOROETHANE (DOT)
25497-29-4	CHLORODIFLUOROETHANES
75-45-6	CHLORODIFLUOROMETHANE
75-45-6	CHLORODIFLUOROMETHANE (ACGIH,DOT)
353-59-3	CHLORODIFLUOROMONOBROMOMETHANE
2524-03-0	CHLORODIMETHOXYPHOSPHINE SULFIDE
107-30-2	CHLORODIMETHYL ETHER
75-29-6	CHLORODIMETHYLMETHANE
97-00-7	CHLORODINITROBENZENE
25567-67-3	CHLORODINITROBENZENE
97-00-7	CHLORODINITROBENZENE (DOT)
25567-67-3	CHLORODINITROBENZENE (DOT)
3691-35-8	CHLORODIPHACINONE
53449-21-9	CHLORODIPHENYL (42% CHLORINE)
11097-69-1	CHLORODIPHENYL (54% CHLORINE)
55720-99-5	CHLORODIPHENYL OXIDE
712-48-1	CHLORODIPHENYLARSINE
75-00-3	CHLOROETHANE
79-11-8	CHLOROETHANOIC ACID
107-07-3	CHLOROETHANOL
71-55-6	CHLOROETHENE
75-01-4	CHLOROETHENE
71-55-6	CHLOROETHENE NU
627-11-2	CHLOROETHYL CHLOROFORMATE
111-44-4	CHLOROETHYL ETHER
13010-47-4	CHLOROETHYLCYCLOHEXYLNITROSOUREA
75-01-4	CHLOROETHYLENE
25497-29-4	CHLOROETHYLIDENE FLUORIDE
107-07-3	CHLOROETHYLOWY ALKOHOL (POLISH)
470-90-6	CHLOROFENVINPHOS
75-71-8	CHLOROFLUOROCARBON 12
67-66-3	CHLOROFORM
67-66-3	CHLOROFORM (ACGIH,DOT)
71-55-6	CHLOROFORM, METHYL-
76-06-2	CHLOROFORM, NITRO-
67-66-3	CHLOROFORME (FRENCH)
79-22-1	CHLOROFORMIATE DE METHYLE (FRENCH)
24468-13-1	CHLOROFORMIC ACID 2-ETHYLHEXYL ESTER
592-34-7	CHLOROFORMIC ACID BUTYL ESTER
79-44-7	CHLOROFORMIC ACID DIMETHYLAMIDE
541-41-3	CHLOROFORMIC ACID ETHYL ESTER
108-23-6	CHLOROFORMIC ACID ISOPROPYL ESTER
79-22-1	CHLOROFORMIC ACID METHYL ESTER
1885-14-9	CHLOROFORMIC ACID PHENYL ESTER
627-11-2	CHLOROFORMIC ACID, 2-CHLOROETHYL ESTER
20830-75-5	CHLOROFORMIC DIGITALIN
75-44-5	CHLOROFORMYL CHLORIDE
52-68-6	CHLOROFOS
52-68-6	CHLOROFTALM
544-10-5	CHLOROHEXANE
25495-90-3	CHLOROHEXANE
7647-01-0	CHLOROHYDRIC ACID
96-24-2	CHLOROHYDRIN
106-89-8	CHLOROHYDRINS
2100-42-7	CHLOROHYDROQUINONE DIMETHYL ETHER
13057-78-8	CHLOROISOCYANURIC ACID
74-87-3	CHLOROMETHANE
22128-62-7	CHLOROMETHOXYCARBONYL CHLORIDE
3188-13-4	CHLOROMETHOXYETHANE
107-30-2	CHLOROMETHOXYMETHANE
6806-86-6	CHLOROMETHYL
22128-62-7	CHLOROMETHYL CHLOROFORMATE
107-14-2	CHLOROMETHYL CYANIDE
542-88-1	CHLOROMETHYL ETHER
3188-13-4	CHLOROMETHYL ETHYL ETHER
3188-13-4	CHLOROMETHYL ETHYL ETHER (DOT)
107-30-2	CHLOROMETHYL METHYL ETHER
107-30-2	CHLOROMETHYL METHYL ETHER (TECHNICAL GRADE)
78-95-5	CHLOROMETHYL METHYL KETONE
532-27-4	CHLOROMETHYL PHENYL KETONE
100-44-7	CHLOROMETHYLBENZENE
25168-05-2	CHLOROMETHYLBENZENE
22128-62-7	CHLOROMETHYLCHLOROFORMATE (DOT)
106-89-8	CHLOROMETHYLOXIRANE
28479-22-3	CHLOROMETHYLPHENYLISOCYANATE
993-00-0	CHLOROMETHYLSILANE
56-75-7	CHLOROMYCETIN
494-03-1	CHLORONAPHTHINA
494-03-1	CHLORONAPHTHINE
56-75-7	CHLORONITRIN
41587-36-4	CHLORONITROANILINE
41587-36-4	CHLORONITROANILINES (DOT)
25167-93-5	CHLORONITROBENZENE
25167-93-5	CHLORONITROBENZENES
25567-68-4	CHLORONITROTOLUENE
61878-61-3	CHLORONITROTOLUENE
8007-56-5	CHLORONITROUS ACID
76-15-3	CHLOROPENTAFLUOROETHANE
76-15-3	CHLOROPENTAFLUOROETHANE (ACGIH,DOT)
76-15-3	CHLOROPERFLUOROETHANE
10049-04-4	CHLOROPEROXYL
3691-35-8	CHLOROPHACINON
3691-35-8	CHLOROPHACINONE
3691-35-8	CHLOROPHACINONE (ROZOL)
87-86-5	CHLOROPHEN
25167-80-0	CHLOROPHENOL
25167-80-0	CHLOROPHENOL, LIQUID
106-48-9	CHLOROPHENOL, SOLID
50-29-3	CHLOROPHENOTHANE
26571-79-9	CHLOROPHENYL TRICHLOROSILANE
100-44-7	CHLOROPHENYLMETHANE
26571-79-9	CHLOROPHENYLTRICHLOROSILANE (DOT)
52-68-6	CHLOROPHOS
52-68-6	CHLOROPHOSE
814-49-3	CHLOROPHOSPHORIC ACID, DIETHYL ESTER
52-68-6	CHLOROPHTHALM
76-06-2	CHLOROPICRIN
76-06-2	CHLOROPICRIN (ACGIH,DOT)
8004-09-9	CHLOROPICRIN AND METHYL BROMIDE, MIXTURE (DOT)
76-06-2	CHLOROPICRIN MIXTURE (DOT)
76-06-2	CHLOROPICRIN MIXTURE, FLAMMABLE (DOT)
76-06-2	CHLOROPICRIN MIXTURE, FLAMMABLE (PRESSURE = 147 PSIA; FLASH POINT 100 DEG F)
76-06-2	CHLOROPICRIN MIXTURE, WITH NO COMPRESSED GAS OR POISON A LIQUID
76-06-2	CHLOROPICRIN, ABSORBED
76-06-2	CHLOROPICRIN, ABSORBED (DOT)
76-06-2	CHLOROPICRIN, LIQUID
76-06-2	CHLOROPICRIN, LIQUID (DOT)
76-06-2	CHLOROPICRINE (FRENCH)
16941-12-1	CHLOROPLATINIC ACID

CAS No.	Chemical Name
16941-12-1	CHLOROPLATINIC(IV) ACID
16941-12-1	CHLOROPLATINIC(IV) ACID (H2PtCl6)
126-99-8	CHLOROPRENE
96-24-2	CHLOROPROPANEDIOL
78-95-5	CHLOROPROPANONE
28554-00-9	CHLOROPROPIONIC ACID
28554-00-9	CHLOROPROPIONIC ACID (DOT)
106-89-8	CHLOROPROPYLENE OXIDE
56-75-7	CHLOROPTIC
29154-12-9	CHLOROPYRIDINE
2921-88-2	CHLOROPYRIFOS
2921-88-2	CHLOROPYRIPHOS
54-05-7	CHLOROQUINE
7681-52-9	CHLOROS
13465-78-6	CHLOROSILANE
13465-78-6	CHLOROSILANE (ClShI3)
4109-96-0	CHLOROSILANE (SiH2Cl2)
13465-78-6	CHLOROSILANE, [CORROSIVE LABEL]
13465-78-6	CHLOROSILANE, [EMITS FLAMMABLE GAS WHEN WET,CORROSIVE LABELS]
13465-78-6	CHLOROSILANE, [FLAMMABLE, CORROSIVE LABELS]
10025-67-9	CHLOROSULFANE
7790-94-5	CHLOROSULFONIC ACID
7790-94-5	CHLOROSULFONIC ACID (DOT)
7791-27-7	CHLOROSULFONIC ANHYDRIDE
7790-94-5	CHLOROSULFURIC ACID
7790-94-5	CHLOROSULPHONIC ACID
71-55-6	CHLOROTENE
63938-10-3	CHLOROTETRAFLUOROETHANE
63938-10-3	CHLOROTETRAFLUOROETHANE (DOT)
1897-45-6	CHLOROTHALONIL
71-55-6	CHLOROTHANE NU
71-55-6	CHLOROTHENE
71-55-6	CHLOROTHENE NU
71-55-6	CHLOROTHENE SM
71-55-6	CHLOROTHENE VG
71-55-6	CHLOROTHENE(INHIBITED)
106-43-4	CHLOROTOLUENE
100-44-7	CHLOROTOLUENE
25168-05-2	CHLOROTOLUENE
108-77-0	CHLOROTRIAZINE
7790-91-2	CHLOROTRIFLUORIDE
1330-45-6	CHLOROTRIFLUOROETHANE
1330-45-6	CHLOROTRIFLUOROETHANE (DOT)
79-38-9	CHLOROTRIFLUOROETHENE
79-38-9	CHLOROTRIFLUOROETHYLENE
75-72-9	CHLOROTRIFLUOROMETHANE
75-72-9	CHLOROTRIFLUOROMETHANE (DOT)
507-20-0	CHLOROTRIMETHYLMETHANE
75-77-4	CHLOROTRIMETHYLSILANE
1066-45-1	CHLOROTRIMETHYLSTANNANE
1066-45-1	CHLOROTRIMETHYLTIN
28260-61-9	CHLOROTRINITROBENZENE
639-58-7	CHLOROTRIPHENYLSTANNANE
639-58-7	CHLOROTRIPHENYLTIN
14674-72-7	CHLOROUS ACID, CALCIUM SALT
1318-59-8	CHLOROUS ACID, ION(1-)
14998-27-7	CHLOROUS ACID, ION(1-)
7758-19-2	CHLOROUS ACID, SODIUM SALT
541-25-3	CHLOROVINYLARSINE DICHLORIDE
7647-01-0	CHLOROWODOR (POLISH)
1982-47-4	CHLOROXIFENIDIM
94-75-7	CHLOROXONE
1982-47-4	CHLOROXURON
52-68-6	CHLOROXYPHOS
7699-43-6	CHLOROZIRCONYL
3691-35-8	CHLORPHACINON (ITALIAN)
3691-35-8	CHLORPHACINONE
3691-35-8	CHLORPHENACONE
50-29-3	CHLORPHENOTHAN
50-29-3	CHLORPHENOTOXUM
470-90-6	CHLORPHENVINFOS
470-90-6	CHLORPHENVINPHOS
76-06-2	CHLORPIKRIN (GERMAN)
2921-88-2	CHLORPYRIFOS
2921-88-2	CHLORPYRIFOS (ACGIH,DOT)
2921-88-2	CHLORPYRIFOS-ETHYL
2921-88-2	CHLORPYRIPHOS
7775-09-9	CHLORSAURE (GERMAN)
71-55-6	CHLORTEN
115-29-7	CHLORTHIEPIN
21923-23-9	CHLORTHIOPHOS
21923-23-9	CHLORTHIOPHOS I
12789-03-6	CHLORTOX
79-38-9	CHLORTRIFLUORAETHYLEN (GERMAN)
10025-91-9	CHLORURE ANTIMONIEUX (FRENCH)
7784-34-1	CHLORURE ARSENIEUX (FRENCH)
7446-70-0	CHLORURE D'ALUMINIUM (FRENCH)
7784-34-1	CHLORURE D'ARSENIC (FRENCH)
75-00-3	CHLORURE D'ETHYLE (FRENCH)
107-06-2	CHLORURE D'ETHYLENE (FRENCH)
75-34-3	CHLORURE D'ETHYLIDENE (FRENCH)
98-07-7	CHLORURE DE BENZENYLE (FRENCH)
100-44-7	CHLORURE DE BENZYLE (FRENCH)
98-87-3	CHLORURE DE BENZYLIDENE (FRENCH)
109-69-3	CHLORURE DE BUTYLE (FRENCH)
79-04-9	CHLORURE DE CHLORACETYLE (FRENCH)
14977-61-8	CHLORURE DE CHROMYLE (FRENCH)
506-77-4	CHLORURE DE CYANOGENE (FRENCH)
79-36-7	CHLORURE DE DICHLORACETYLE (FRENCH)
627-63-4	CHLORURE DE FUMARYLE (FRENCH)
563-47-3	CHLORURE DE METHALLYLE (FRENCH)
74-87-3	CHLORURE DE METHYLE (FRENCH)
75-09-2	CHLORURE DE METHYLENE (FRENCH)
75-01-4	CHLORURE DE VINYLE (FRENCH)
7646-85-7	CHLORURE DE ZINC (FRENCH)
7546-30-7	CHLORURE MERCUREUX (FRENCH)
7487-94-7	CHLORURE MERCURIQUE (FRENCH)
7705-08-0	CHLORURE PERRIQUE (FRENCH)
62-73-7	CHLORVINPHOS
7647-01-0	CHLORWASSERSTOFF (GERMAN)
75-00-3	CHLORYL
75-00-3	CHLORYL ANESTHETIC
10049-04-4	CHLORYL RADICAL
79-01-6	CHLORYLEN
56-49-5	CHOLANTHRENE, 3-METHYL-
50-70-4	CHOLAXINE
51-83-2	CHOLINE CARBAMATE CHLORIDE
51-83-2	CHOLINE CHLORIDE, CARBAMOYL-
51-83-2	CHOLINE CHLORINE CARBAMATE
999-81-5	CHOLINE DICHLORIDE
51-83-2	CHOLINE, CHLORIDE CARBAMATE(ESTER)
51-83-2	CHOLINE, CHLORIDE, CARBAMATE
79-01-6	CHORYLEN
1561-49-5	CHPC
15468-32-3	CHRISTENSENITE
106-42-3	CHROMAR
7789-00-6	CHROMATE OF POTASSIUM
7775-11-3	CHROMATE OF SODA
37224-57-0	CHROMATE(1-), HYDROXYOCTAOXODIZINCATEDI-, POTASSIUM
7440-47-3	CHROME
13530-68-2	CHROME (TRIOXYDE DE) (FRENCH)
7788-97-8	CHROME FLUORURE (FRENCH)
1937-37-7	CHROME LEATHER BLACK E
1937-37-7	CHROME LEATHER BLACK EC
1937-37-7	CHROME LEATHER BLACK EM
1937-37-7	CHROME LEATHER BLACK G
2602-46-2	CHROME LEATHER BLUE 2B
72-57-1	CHROME LEATHER BLUE 3B
1937-37-7	CHROME LEATHER BRILLIANT BLACK ER
16071-86-6	CHROME LEATHER BROWN BRLL
16071-86-6	CHROME LEATHER BROWN BRSL
11119-70-3	CHROME ORANGE
9004-34-6	CHROMEDIA CC 31
9004-34-6	CHROMEDIA CF 11
13530-68-2	CHROMIC (VI) ACID
1066-30-4	CHROMIC ACETATE
1066-30-4	CHROMIC ACETATE (DOT)
1066-30-4	CHROMIC ACETATE (III)
7738-94-5	CHROMIC ACID
13530-68-2	CHROMIC ACID
13530-68-2	CHROMIC ACID (H2Cr2O7)
14307-33-6	CHROMIC ACID (H2Cr2O7), CALCIUM SALT (1:1)
7789-09-5	CHROMIC ACID (H2Cr2O7), DIAMMONIUM SALT
7778-50-9	CHROMIC ACID (H2Cr2O7), DIPOTASSIUM SALT
10588-01-9	CHROMIC ACID (H2Cr2O7), DISODIUM SALT
14018-95-2	CHROMIC ACID (H2Cr2O7), ZINC SALT (1:1)
1189-85-1	CHROMIC ACID (H2CrO4), BIS(1,1-DIMETHYLETHYL) ESTER

CAS No.	Chemical Name
13765-19-0	CHROMIC ACID (H2CrO4), CALCIUM SALT (1:1)
1189-85-1	CHROMIC ACID (H2CrO4), DI-TERT-BUTYL ESTER
7788-98-9	CHROMIC ACID (H2CrO4), DIAMMONIUM SALT
14307-35-8	CHROMIC ACID (H2CrO4), DILITHIUM SALT
7789-00-6	CHROMIC ACID (H2CrO4), DIPOTASSIUM SALT
7775-11-3	CHROMIC ACID (H2CrO4), DISODIUM SALT
7758-97-6	CHROMIC ACID (H2CrO4), LEAD(2+) SALT (1:1)
13423-61-5	CHROMIC ACID (H2CrO4), MAGNESIUM SALT (1:1)
7789-06-2	CHROMIC ACID (H2CrO4), STRONTIUM SALT (1:1)
13530-65-9	CHROMIC ACID (H2CrO4), ZINC SALT (1:1)
18454-12-1	CHROMIC ACID (H4CrO5), LEAD(2+) SALT (1:2)
13530-68-2	CHROMIC ACID (MIXTURE)
7788-98-9	CHROMIC ACID AMMONIUM SALT
13530-68-2	CHROMIC ACID MIXTURE, DRY
13530-68-2	CHROMIC ACID MIXTURE, DRY (DOT)
13530-68-2	CHROMIC ACID SOLUTION
1189-85-1	CHROMIC ACID, BIS(1,1-DIMETHYLETHYL) ESTER
13765-19-0	CHROMIC ACID, CALCIUM SALT (1:1)
7788-98-9	CHROMIC ACID, DIAMMONIUM SALT
7778-50-9	CHROMIC ACID, DIPOTASSIUM SALT
7789-00-6	CHROMIC ACID, DIPOTASSIUM SALT
7775-11-3	CHROMIC ACID, DISODIUM SALT
10588-01-9	CHROMIC ACID, DISODIUM SALT
11119-70-3	CHROMIC ACID, LEAD SALT
11119-70-3	CHROMIC ACID, LEAD SALT, BASIC
37224-57-0	CHROMIC ACID, POTASSIUM ZINC SALT (2:2:1)
12680-48-7	CHROMIC ACID, SODIUM SALT
13530-68-2	CHROMIC ACID, SOLID
13530-68-2	CHROMIC ACID, SOLID (DOT)
7738-94-5	CHROMIC ACID, SOLUTION
13530-68-2	CHROMIC ACID, SOLUTION (DOT)
7789-06-2	CHROMIC ACID, STRONTIUM SALT (1:1)
13530-65-9	CHROMIC ACID, ZINC SALT (ZnCrO4)
1333-82-0	CHROMIC ANHYDRIDE
13530-68-2	CHROMIC ANHYDRIDE
13530-68-2	CHROMIC ANHYDRIDE (DOT)
10025-73-7	CHROMIC CHLORIDE
10025-73-7	CHROMIC CHLORIDE (CrCl3)
7788-97-8	CHROMIC FLUORIDE
7788-97-8	CHROMIC FLUORIDE SOLUTION
7788-97-8	CHROMIC FLUORIDE, SOLID
7788-97-8	CHROMIC FLUORIDE, SOLID (DOT)
7788-97-8	CHROMIC FLUORIDE, SOLUTION (DOT)
16065-83-1	CHROMIC ION
13548-38-4	CHROMIC NITRATE
14977-61-8	CHROMIC OXYCHLORIDE
10101-53-8	CHROMIC SULFATE
10101-53-8	CHROMIC SULFATE (Cr2(SO4)3)
10101-53-8	CHROMIC SULFATE (DOT)
10101-53-8	CHROMIC SULPHATE
7788-97-8	CHROMIC TRIFLUORIDE
1333-82-0	CHROMIC TRIOXIDE
13530-68-2	CHROMIC TRIOXIDE (DOT)
10101-53-8	CHROMITAN B
10101-53-8	CHROMITAN MS
10101-53-8	CHROMITAN NA
7440-47-3	CHROMIUM
16065-83-1	CHROMIUM (Cr3+)
18540-29-9	CHROMIUM (Cr6+)
22541-79-3	CHROMIUM (II)
10101-53-8	CHROMIUM (III) SULFATE (2:3)
7440-47-3	CHROMIUM (METAL)
14977-61-8	CHROMIUM (VI) DIOXYCHLORIDE
13530-68-2	CHROMIUM (VI) OXIDE
1066-30-4	CHROMIUM ACETATE
1333-82-0	CHROMIUM ANHYDRIDE
10049-05-5	CHROMIUM CHLORIDE
10025-73-7	CHROMIUM CHLORIDE
10049-05-5	CHROMIUM CHLORIDE (CrCl2)
10025-73-7	CHROMIUM CHLORIDE (CrCl3)
14977-61-8	CHROMIUM CHLORIDE OXIDE
14977-61-8	CHROMIUM CHLORIDE OXIDE (CrCl2O2)
10025-73-7	CHROMIUM CHLORIDE, ANHYDROUS
10049-05-5	CHROMIUM DICHLORIDE
14977-61-8	CHROMIUM DICHLORIDE DIOXIDE
18454-12-1	CHROMIUM DILEAD PENTAOXIDE
14977-61-8	CHROMIUM DIOXIDE DICHLORIDE
7775-11-3	CHROMIUM DISODIUM OXIDE
7788-97-8	CHROMIUM FLUORIDE (CrF3)
18540-29-9	CHROMIUM HEXAVALENT ION
10101-53-8	CHROMIUM III SULFATE
16065-83-1	CHROMIUM ION (Cr3+)
11119-70-3	CHROMIUM LEAD OXIDE
18454-12-1	CHROMIUM LEAD OXIDE (CrPb2O5)
14307-35-8	CHROMIUM LITHIUM OXIDE (CrLi2O4)
13423-61-5	CHROMIUM MAGNESIUM OXIDE (MgCrO4)
13548-38-4	CHROMIUM NITRATE
13548-38-4	CHROMIUM NITRATE (Cr(NO3)3)
13530-68-2	CHROMIUM OXIDE
1333-82-0	CHROMIUM OXIDE (Cr4O12)
1333-82-0	CHROMIUM OXIDE (CrO3)
14977-61-8	CHROMIUM OXYCHLORIDE
14977-61-8	CHROMIUM OXYCHLORIDE (CrO2Cl2)
14977-61-8	CHROMIUM OXYCHLORIDE (DOT)
37224-57-0	CHROMIUM POTASSIUM ZINC OXIDE
7775-11-3	CHROMIUM SODIUM OXIDE
10588-01-9	CHROMIUM SODIUM OXIDE
12680-48-7	CHROMIUM SODIUM OXIDE
10588-01-9	CHROMIUM SODIUM OXIDE (Cr3Na2O7)
7775-11-3	CHROMIUM SODIUM OXIDE (CrNa2O4)
14489-25-9	CHROMIUM SULFATE
10101-53-8	CHROMIUM SULFATE
10101-53-8	CHROMIUM SULFATE (2:3)
10101-53-8	CHROMIUM SULFATE (Cr2(SO4)3)
10101-53-8	CHROMIUM SULPHATE
10101-53-8	CHROMIUM SULPHATE (2:3)
1066-30-4	CHROMIUM TRIACETATE
10025-73-7	CHROMIUM TRICHLORIDE
10025-73-7	CHROMIUM TRICHLORIDE (CrCl3)
7788-97-8	CHROMIUM TRIFLUORIDE
13548-38-4	CHROMIUM TRINITRATE
1333-82-0	CHROMIUM TRIOXIDE
13530-68-2	CHROMIUM TRIOXIDE
13530-68-2	CHROMIUM TRIOXIDE, ANHYDROUS (DOT)
13530-65-9	CHROMIUM ZINC OXIDE (ZnCrO4)
22541-79-3	CHROMIUM(2+) ION
16065-83-1	CHROMIUM(3+)
13548-38-4	CHROMIUM(3+) NITRATE
18540-29-9	CHROMIUM(6+) ION
13530-68-2	CHROMIUM(6+) TRIOXIDE
22541-79-3	CHROMIUM(II)
10049-05-5	CHROMIUM(II) CHLORIDE
10049-05-5	CHROMIUM(II) CHLORIDE (1:2)
22541-79-3	CHROMIUM(II) ION
16065-83-1	CHROMIUM(III)
1066-30-4	CHROMIUM(III) ACETATE
16065-83-1	CHROMIUM(III) CATION
10025-73-7	CHROMIUM(III) CHLORIDE
10025-73-7	CHROMIUM(III) CHLORIDE (1:3)
7788-97-8	CHROMIUM(III) FLUORIDE
16065-83-1	CHROMIUM(III) ION
13548-38-4	CHROMIUM(III) NITRATE
1308-38-9	CHROMIUM(III) OXIDE(2:3)
18540-29-9	CHROMIUM(VI)
1333-82-0	CHROMIUM(VI) OXIDE
13530-68-2	CHROMIUM(VI) OXIDE (1:3)
1333-82-0	CHROMIUM(VI) OXIDE(1:3)
14977-61-8	CHROMIUM, DICHLORODIOXO-
14977-61-8	CHROMIUM, DICHLORODIOXO-, (T-4)-
22541-79-3	CHROMIUM, ION (CR2+)
16065-83-1	CHROMIUM, ION (CR3+)
18540-29-9	CHROMIUM, ION (CR6+)
9002-84-0	CHROMOSORB T
14489-25-9	CHROMOSULFURIC ACID
14489-25-9	CHROMOSULPHURIC ACID (DOT)
10049-05-5	CHROMOUS CHLORIDE
10049-05-5	CHROMOUS CHLORIDE (DOT)
22541-79-3	CHROMOUS ION
14977-61-8	CHROMOXYCHLORID (GERMAN)
1912-24-9	CHROMOZIN
13530-68-2	CHROMSAEUREANHYDRID (GERMAN)
13530-68-2	CHROMTRIOXID (GERMAN)
14977-61-8	CHROMYL CHLORIDE
14977-61-8	CHROMYL CHLORIDE (ACGIH,DOT)
14977-61-8	CHROMYL CHLORIDE (CrO2Cl2)
14977-61-8	CHROMYLCHLORID (GERMAN)
14489-25-9	CHRONISULFAT (GERMAN)
14977-61-8	CHROOMOXYCHLORIDE (DUTCH)

CAS No.	Chemical Name
13530-68-2	CHROOMTRIOXYDE (DUTCH)
13530-68-2	CHROOMZUURANHYDRIDE (DUTCH)
8003-34-7	CHRYSANTHEMUM CINERAREAEFOLIUM
121-21-1	CHRYSANTHEMUMMONOCARBOXYLIC ACID PYRETHROLONE ESTER
218-01-9	CHRYSENE
3697-24-3	CHRYSENE, 5-METHYL-
12001-29-5	CHRYSOTILE
75364-04-4	CHRYSOTILE
12001-29-5	CHRYSOTILE (DOT)
12001-29-5	CHRYSOTILE (H4Mg3(Si2O9))
12001-29-5	CHRYSOTILE (Mg3H2(SiO4)2.H2O)
12001-29-5	CHRYSOTILE ASBESTOS
12001-29-5	CHRYSOTILE DUST
111-40-0	CHS-P 1
122-39-4	CI 10355
88-74-4	CI 37025
29757-24-2	CI 37035
106-49-0	CI 37077
95-53-4	CI 37077
108-44-1	CI 37077
95-53-4	CI 37107
108-44-1	CI 37107
106-49-0	CI 37107
633-03-4	CI 42040
142-04-1	CI 76001
108-45-2	CI 76025
99-98-9	CI 76075
120-80-9	CI 76500
87-66-1	CI 76515
95-85-2	CI 76525
123-30-8	CI 76550
12627-52-0	CI 77060
513-77-9	CI 77099
13765-19-0	CI 77223
10025-73-7	CI 77295
1317-36-8	CI 77577
1309-60-0	CI 77580
592-05-2	CI 77610
1344-40-7	CI 77620
15739-80-7	CI 77630
21908-53-2	CI 77760
7784-37-4	CI 77762
7546-30-7	CI 77764
1344-48-5	CI 77766
7782-49-2	CI 77805
7772-99-8	CI 77864
7440-66-6	CI 77945
3486-35-9	CI 77950
106-49-0	CI AZOIC COUPLING COMPONENT 107
3165-93-3	CI AZOIC DIAZO COMPONENT 11
29757-24-2	CI AZOIC DIAZO COMPONENT 37
88-74-4	CI AZOIC DIAZO COMPONENT 6
633-03-4	CI BASIC GREEN 1, SULFATE (1:1)
108-45-2	CI DEVELOPER 11
100-01-6	CI DEVELOPER 17
29757-24-2	CI DEVELOPER 17
62-53-3	CI OXIDATION BASE 1
95-85-2	CI OXIDATION BASE 18
120-80-9	CI OXIDATION BASE 26
87-66-1	CI OXIDATION BASE 32
123-30-8	CI OXIDATION BASE 6
123-30-8	CI OXIDATION BASE 6A
27598-85-2	CI OXIDATION BASE 6A
7440-66-6	CI PIGMENT BLACK 16
12002-03-8	CI PIGMENT GREEN 21 (9CI)
7440-66-6	CI PIGMENT METAL 6
1344-48-5	CI PIGMENT RED 106
12627-52-0	CI PIGMENT RED 107
513-77-9	CI PIGMENT WHITE 10
15739-80-7	CI PIGMENT WHITE 3
1303-33-9	CI PIGMENT YELLOW
13765-19-0	CI PIGMENT YELLOW 33
1303-33-9	CI PIGMENT YELLOW 39
1317-36-8	CI PIGMENT YELLOW 46
592-05-2	CI PIGMENT YELLOW 48
947-02-4	CI-47031
2636-26-2	CIAFOS
154-23-4	CIANIDANOL

CAS No.	Chemical Name
592-04-1	CIANURINA
143-33-9	CIANURO DI SODIO (ITALIAN)
333-41-5	CIAZINON
3771-19-5	CIBA 13437SU
6923-22-4	CIBA 1414
1982-47-4	CIBA 1983
61-57-4	CIBA 32644
61-57-4	CIBA 32644-BA
141-66-2	CIBA 709
143-50-0	CIBA 8514
950-37-8	CIBA-GEIGY GS 13005
119-90-4	CIBACETE DIAZO NAVY BLUE 2B
52-68-6	CICLOSOM
111-30-8	CIDEX 7
82-28-0	CILLA ORANGE R
121-75-5	CIMEXAN
138-86-3	CINEN
138-86-3	CINENE
584-79-2	CINERIN I ALLYL HOMOLOG
8003-34-7	CINERIN I OR II
100-42-5	CINNAMENE
100-42-5	CINNAMENOL
100-42-5	CINNAMOL
56-75-7	CIPLAMYCETIN
9002-89-5	CIPOVIOL W 72
79-01-6	CIRCOSOLV
156-59-2	CIS-1,2-DICHLORETHYLENE
156-59-2	CIS-1,2-DICHLOROETHENE
156-59-2	CIS-1,2-DICHLOROETHYLENE
590-18-1	CIS-1,2-DIMETHYLETHYLENE
110-16-7	CIS-1,2-ETHYLENEDICARBOXYLIC ACID
10061-01-5	CIS-1,3-DICHLOROPROPENE
624-29-3	CIS-1,4-DIMETHYLCYCLOHEXANE
6923-22-4	CIS-1-METHYL-2-METHYL CARBAMOYL VINYL PHOSPHATE
590-18-1	CIS-2-BUTENE
590-18-1	CIS-2-BUTYLENE
141-66-2	CIS-2-DIMETHYLCARBAMOYL-1-METHYLVINYL DIMETHYLPHOSPHATE
627-20-3	CIS-2-PENTENE
928-96-1	CIS-3-HEXEN-1-OL
928-96-1	CIS-3-HEXENOL
112-80-1	CIS-9-OCTADECENOIC ACID
627-20-3	CIS-BETA-AMYLENE
590-18-1	CIS-BUTENE
590-18-1	CIS-BUTENE-2
110-16-7	CIS-BUTENEDIOIC ACID
108-31-6	CIS-BUTENEDIOIC ANHYDRIDE
590-18-1	CIS-BUTYLENE
15663-27-1	CIS-DDP
112-80-1	CIS-DELTA(SUP 9)-OCTADECENOIC ACID
112-80-1	CIS-DELTA9-OCTADECENOIC ACID
15663-27-1	CIS-DIAMINODICHLOROPLATINUM(II)
15663-27-1	CIS-DIAMMINEDICHLOROPLATINUM
15663-27-1	CIS-DIAMMINEDICHLOROPLATINUM(II)
15663-27-1	CIS-DICHLORODIAMMINEPLATINUM
15663-27-1	CIS-DICHLORODIAMMINEPLATINUM(II)
156-59-2	CIS-DICHLOROETHYLENE
540-59-0	CIS-DICHLOROETHYLENE
15663-27-1	CIS-DPP
112-80-1	CIS-OCTADEC-9-ENOIC ACID
112-80-1	CIS-OLEIC ACID
627-20-3	CIS-PENTENE
15663-27-1	CIS-PLATINE
15663-27-1	CIS-PLATINOUS DIAMINODICHLORIDE
15663-27-1	CIS-PLATINUM
15663-27-1	CIS-PLATINUM DIAMINODICHLORIDE
15663-27-1	CIS-PLATINUM II
15663-27-1	CIS-PLATINUM(II) DIAMINODICHLORIDE
15663-27-1	CIS-PLATINUMDIAMINE DICHLORIDE
15663-27-1	CISPLATIN
15663-27-1	CISPLATINUM
123-30-8	CITOL
27598-85-2	CITOL
8020-83-5	CITOL OIL
50-29-3	CITOX
78-53-5	CITRAM
3734-97-2	CITRAM
1185-57-5	CITRIC ACID, AMMONIUM IRON(3+) SALT
3012-65-5	CITRIC ACID, DIAMMONIUM SALT

CAS No.	Chemical Name
10045-94-0	CITRINE OINTMENT
37224-57-0	CITRON YELLOW
6358-53-8	CITRUS RED 2
6358-53-8	CITRUS RED NO. 2
1333-86-4	CK 4
1333-86-4	CK 4 (CARBON BLACK)
1344-28-1	CKA
10265-92-6	CKB 1220
60-51-5	CL 12880
947-02-4	CL 47031
122-14-5	CL 47300
950-10-7	CL 47470
21548-32-3	CL 64475
50-18-0	CLAFEN
50-18-0	CLAPHENE
712-48-1	CLARK I
9005-25-8	CLARO 5591
1327-53-3	CLAUDELITE
1327-53-3	CLAUDETITE
1332-58-7	CLAY 347
3251-23-8	CLAYCOP
10421-48-4	CLAYFEN
1332-58-7	CLAYS, CHINA
94-36-0	CLEARASIL BENZOYL PEROXIDE LOTION
94-36-0	CLEARASIL BP ACNE TREATMENT
9005-25-8	CLEARJEL
2893-78-9	CLEARON
577-11-7	CLESTOL
60-00-4	CLEWAT TAA
50-29-3	CLOFENOTAN
50-29-3	CLOFENOTANE
470-90-6	CLOFENVINFOS
50-41-9	CLOMIPHENE CITRATE
1420-04-8	CLONITRALID
1420-04-8	CLONITRALIDE
443-48-1	CLONT
1336-36-3	CLOPHEN
2971-90-6	CLOPIDOL
8001-35-2	CLOR CHEM T-590
75-87-6	CLORALIO (ITALIAN)
56-75-7	CLORAMFICIN
56-75-7	CLORAMICOL
51-75-2	CLORAMIN
57109-90-7	CLORAZEPATE DIPOTASSIUM
12789-03-6	CLORDAN (ITALIAN)
143-50-0	CLORDECONE
75-00-3	CLORETILO
111-44-4	CLOREX
93-58-3	CLORIUS
7782-50-5	CLORO (ITALIAN)
95-50-1	CLOROBEN
108-90-7	CLOROBENZENE (ITALIAN)
56-75-7	CLOROCYN
75-00-3	CLOROETANO (ITALIAN)
67-66-3	CLOROFORMIO (ITALIAN)
52-68-6	CLOROFOS (RUSSIAN)
74-87-3	CLOROMETANO (ITALIAN)
56-75-7	CLOROMISAN
494-03-1	CLORONAFTINA
76-06-2	CLOROPICRINA (ITALIAN)
7681-52-9	CLOROX
107-06-2	CLORURO DI ETHENE (ITALIAN)
75-00-3	CLORURO DI ETILE (ITALIAN)
75-34-3	CLORURO DI ETILIDENE (ITALIAN)
7546-30-7	CLORURO DI MERCURIO (ITALIAN)
7487-94-7	CLORURO DI MERCURIO (ITALIAN)
563-47-3	CLORURO DI METALLILE (ITALIAN)
74-87-3	CLORURO DI METILE (ITALIAN)
75-01-4	CLORURO DI VINILE (ITALIAN)
7546-30-7	CLORURO MERCUROSO (ITALIAN)
105-60-2	CM 1001
105-60-2	CM 1011
105-60-2	CM 1031
105-60-2	CM 1041
21923-23-9	CM S 2957
7786-34-7	CMDP
107-30-2	CMME
2275-14-1	CMP
2275-14-1	CMP (PESTICIDE)

CAS No.	Chemical Name
532-27-4	CN
115-02-6	CN 15757
9004-70-0	CN 85
9004-70-0	CN 88
112-53-8	CO 12
9016-45-9	CO 630
112-53-8	CO-1214
112-53-8	CO-1214N
112-53-8	CO-1214S
118-74-1	CO-OP HEXA
56-72-4	CO-RAL
81-81-2	CO-RAX
8001-58-9	COAL CREOSOTE
71-43-2	COAL NAPHTHA
8002-05-9	COAL OIL
8007-45-2	COAL TAR
8007-45-2	COAL TAR (COAL TAR PITCH)
8001-58-9	COAL TAR CREOSOTE
65996-92-1	COAL TAR DISTILLATE, COMBUSTIBLE, LIQUID
65996-92-1	COAL TAR DISTILLATE, FLAMMABLE, LIQUID
8007-45-2	COAL TAR EXTRACT
65996-91-0	COAL TAR LIGHT OIL
65996-79-4	COAL TAR NAPHTHA
8030-30-6	COAL TAR NAPHTHA (DOT)
8001-58-9	COAL TAR OIL
8001-58-9	COAL TAR OIL (DOT)
8007-45-2	COAL TAR OINTMENT
8007-45-2	COAL TAR PITCH VOLATILES
65996-91-0	COAL TAR UPPER DISTILLATE
7783-20-2	COALTROL LPA 40
7783-28-0	COALTROL LPA 445
137-05-3	COAPT.
7440-48-4	COBALT
10124-43-3	COBALT (2+) SULFATE
7440-48-4	COBALT (ACGIH)
10124-43-3	COBALT (II) SULFATE (1:1)
71-48-7	COBALT ACETATE
71-48-7	COBALT ACETATE (Co(OAC)2)
10141-05-6	COBALT BIS(NITRATE)
7789-43-7	COBALT BROMIDE (CoBr2)
10210-68-1	COBALT CARBONYL
37264-96-3	COBALT CARBONYL
10210-68-1	COBALT CARBONYL (ACGIH)
10210-68-1	COBALT CARBONYL (Co2(CO)8)
7646-79-9	COBALT CHLORIDE
7646-79-9	COBALT CHLORIDE (CoCl2)
71-48-7	COBALT DIACETATE
7789-43-7	COBALT DIBROMIDE
7646-79-9	COBALT DICHLORIDE
7646-79-9	COBALT DICHLORIDE (CoCl2)
11113-57-8	COBALT DIFLUORIDE
10026-17-2	COBALT DIFLUORIDE
544-18-3	COBALT DIFORMATE
10141-05-6	COBALT DINITRATE
7440-48-4	COBALT ELEMENT
10026-17-2	COBALT FLUORIDE
11113-57-8	COBALT FLUORIDE
10026-17-2	COBALT FLUORIDE (CoF2)
11113-57-8	COBALT FLUORIDE (CoF2)
544-18-3	COBALT FORMATE
544-18-3	COBALT FORMATE (Co(O2CH)2)
16842-03-8	COBALT HYDROCARBONYL
16842-03-8	COBALT HYDROCARBONYL [CoH(CO)4]
7440-48-4	COBALT METAL, DUST, AND FUME
7646-79-9	COBALT MURIATE
61789-51-3	COBALT NAPHTHENATE
61789-51-3	COBALT NAPHTHENATE, POWDER
61789-51-3	COBALT NAPHTHENATE, POWDER (DOT)
61789-51-3	COBALT NAPHTHENATES
10141-05-6	COBALT NITRATE
10141-05-6	COBALT NITRATE (Co(NO3)2)
10210-68-1	COBALT OCTACARBONYL
68956-82-1	COBALT RESINATE, PRECIPITATED
68956-82-1	COBALT RESINATE, PRECIPITATED (DOT)
14017-41-5	COBALT SULFAMATE
10124-43-3	COBALT SULFATE
10124-43-3	COBALT SULFATE (1:1)
10124-43-3	COBALT SULFATE (CoSO4)
10210-68-1	COBALT TETRACARBONYL

CAS No.	*Chemical Name*
10210-68-1	COBALT TETRACARBONYL DIMER
16842-03-8	COBALT TETRACARBONYL HYDRIDE
71-48-7	COBALT(2+) ACETATE
7646-79-9	COBALT(2+) CHLORIDE
544-18-3	COBALT(2+) FORMATE
10141-05-6	COBALT(2+) NITRATE
10124-43-3	COBALT(2+) SULFATE
71-48-7	COBALT(II) ACETATE
7789-43-7	COBALT(II) BROMIDE
7646-79-9	COBALT(II) CHLORIDE
11113-57-8	COBALT(II) FLUORIDE
10026-17-2	COBALT(II) FLUORIDE
10141-05-6	COBALT(II) NITRATE
10141-05-6	COBALT(II) NITRATE (1:2)
10124-43-3	COBALT(II) SULFATE
10124-43-3	COBALT(II) SULPHATE
62207-76-5	COBALT(II), N,N'-ETHYLENEBIS(3-FLUOROSALICYLIDENE-IMINATO)-
62207-76-5	COBALT, ((2,2'-(1,2-ETHANEDIYLBIS(NITRILOMETHYLIDENE)BIS-(PHENALOTO))(2-)-N,N',O,O')-
62207-76-5	COBALT, BIS(3-FLUOROSALICYLALDEHYDE)-ETHYLENEDIIMINE-
10210-68-1	COBALT, DI-MU-CARBONYLHEXACARBONYLDI-, (CO-CO)
16842-03-8	COBALT, TETRACARBONYLHYDRO-
62207-76-5	COBALT, [[2,2'-[1,2-ETHANEDIYLBIS(NITRILOMETHYLIDYNE)]-BIS[6-FLUOROPHENOLATO]](2-)-N,N',O,O']-, (SP-4
7440-48-4	COBALT-59
71-48-7	COBALTOUS ACETATE
7789-43-7	COBALTOUS BROMIDE
7789-43-7	COBALTOUS BROMIDE (DOT)
7646-79-9	COBALTOUS CHLORIDE
71-48-7	COBALTOUS DIACETATE
7646-79-9	COBALTOUS DICHLORIDE
11113-57-8	COBALTOUS FLUORIDE
10026-17-2	COBALTOUS FLUORIDE
544-18-3	COBALTOUS FORMATE
544-18-3	COBALTOUS FORMATE (DOT)
10141-05-6	COBALTOUS NITRATE
14017-41-5	COBALTOUS SULFAMATE
14017-41-5	COBALTOUS SULFAMATE (DOT)
10124-43-3	COBALTOUS SULFATE
148-01-6	COCCIDINE A
2971-90-6	COCCIDIOSTAT C
148-01-6	COCCIDOT
124-87-8	COCCULIN
124-87-8	COCCULUS
124-87-8	COCCULUS, SOLID
124-87-8	COCCULUS, SOLID (FISHBERRY) (DOT)
8001-69-2	COD LIVER OIL DISTILLATE
8001-69-2	COD OIL
8001-69-2	COD-LIVER OIL
58-89-9	CODECHINE
1937-37-7	COIR DEEP BLACK C
577-11-7	COLACE
3761-53-3	COLACID PONCEAU SPECIAL
141-43-5	COLAMINE
64-86-8	COLCHICIN (GERMAN)
64-86-8	COLCHICINA (ITALIAN)
64-86-8	COLCHICINE
64-86-8	COLCHINEOS
64-86-8	COLCHISOL
64-86-8	COLCIN
1309-37-1	COLCOTHAR
2665-30-7	COLEP
51-83-2	COLETYL
50-78-2	COLFARIT
104-90-5	COLLIDINE, ALDEHYDECOLLIDINE
1309-37-1	COLLIRON
7757-79-1	COLLO-BO
64-18-6	COLLO-BUEGLATT
64-18-6	COLLO-DIDAX
1310-73-2	COLLO-GRILLREIN
1310-73-2	COLLO-TAPETTA
9004-70-0	COLLODION
9004-70-0	COLLODION (DOT)
9004-70-0	COLLODION COTTON
9004-70-0	COLLODION WOOL
7440-38-2	COLLOIDAL ARSENIC
1309-37-1	COLLOIDAL FERRIC OXIDE
7439-96-5	COLLOIDAL MANGANESE

CAS No.	*Chemical Name*
7439-97-6	COLLOIDAL MERCURY
7631-86-9	COLLOIDAL SILICA
7631-86-9	COLLOIDAL SILICON DIOXIDE
7704-34-9	COLLOIDAL SULFUR
7704-34-9	COLLOIDAL-S
7704-34-9	COLLOKIT
9004-70-0	COLLOXYLIN
9004-70-0	COLLOXYLIN VNV
64-17-5	COLOGNE SPIRITS
64-17-5	COLOGNE SPIRITS (ALCOHOL) (DOT)
67-56-1	COLONIAL SPIRIT
93-72-1	COLOR-SET
64-86-8	COLSALOID
7704-34-9	COLSUL
1937-37-7	COLUMBIA BLACK EP
1333-86-4	COLUMBIA CARBON
67-56-1	COLUMBIAN SPIRITS
67-56-1	COLUMBIAN SPIRITS (DOT)
67-63-0	COMBI-SCHUTZ
84-80-0	COMBINAL K1
52-68-6	COMBOT
52-68-6	COMBOT EQUINE
56-53-1	COMESTROL
2312-35-8	COMITE
7723-14-0	COMMON SENSE COCKROACH AND RAT PREPARATIONS
1344-28-1	COMPALOX
577-11-7	COMPLEMIX
139-13-9	COMPLEXON I
60-00-4	COMPLEXON II
514-73-8	COMPOUND 01748
640-19-7	COMPOUND 1081
64-00-6	COMPOUND 10854
309-00-2	COMPOUND 118
143-50-0	COMPOUND 1189
7786-34-7	COMPOUND 2046
72-20-8	COMPOUND 269
8001-35-2	COMPOUND 3956
121-75-5	COMPOUND 4049
470-90-6	COMPOUND 4072
81-81-2	COMPOUND 42
60-57-1	COMPOUND 497
117-80-6	COMPOUND 604
63-25-2	COMPOUND 7744
117-81-7	COMPOUND 889
1918-00-9	COMPOUND B DICAMBA
640-15-3	COMPOUND M-81
62-74-8	COMPOUND NO 1080
991-42-4	COMPOUND S-6,999
15271-41-7	COMPOUND UC-20047 A
58-89-9	COMPOUND-666
50-28-2	COMPUDOSE
25155-30-0	CONCO AAS 35
25155-30-0	CONCO AAS 35H
25155-30-0	CONCO AAS 45S
27323-41-7	CONCO AAS SPECIAL
54590-52-2	CONCO AAS SPECIAL 3
25155-30-0	CONCO AAS-40
25155-30-0	CONCO AAS-65
25155-30-0	CONCO AAS-90
9016-45-9	CONCO NI
9016-45-9	CONCO NI 190
2235-54-3	CONCO SULFATE A
151-21-3	CONCO SULFATE WA
151-21-3	CONCO SULFATE WA-12 45
151-21-3	CONCO SULFATE WA-1200
151-21-3	CONCO SULFATE WAG
151-21-3	CONCO SULFATE WAN
151-21-3	CONCO SULFATE WAS
151-21-3	CONCO SULFATE WN
143-00-0	CONDANOL DLS
50-14-6	CONDOC
1333-86-4	CONDUCTEX 40-220
1333-86-4	CONDUCTEX 900
1333-86-4	CONDUCTEX 950
1333-86-4	CONDUCTEX 975
1333-86-4	CONDUCTEX CC 40-220
1333-86-4	CONDUCTEX N 472
1333-86-4	CONDUCTEX SC
7722-64-7	CONDY'S CRYSTALS

CAS No.	Chemical Name
64-86-8	CONDYLON
57-50-1	CONFECTIONER'S SUGAR
53-86-1	CONFORTID
72-57-1	CONGO BLUE
68-22-4	CONLUDAF
25155-30-0	CONOCO C 550
25155-30-0	CONOCO C-50
25155-30-0	CONOCO C-60
25155-30-0	CONOCO SD 40
1344-28-1	CONOPAL
577-11-7	CONSTONATE
8002-05-9	CONTAINERS, OIL TANKS
92-84-2	CONTAVERM
7440-32-6	CONTIMET 30
7440-32-6	CONTIMET 55
1333-86-4	CONTINEX N 356
62-38-4	CONTRA CREME
28772-56-7	CONTRAC
97-77-8	CONTRALIN
1694-09-3	COOMASSIE VIOLET
104-89-2	COPELLIDINE
7440-50-8	COPPER
3251-23-8	COPPER (II) NITRATE
7758-98-7	COPPER (II) SULFATE (1:1)
142-71-2	COPPER ACETATE
142-71-2	COPPER ACETATE (Cu(C2H3O2)2)
12002-03-8	COPPER ACETATE ARSENITE
12002-03-8	COPPER ACETOARSENITE
12002-03-8	COPPER ACETOARSENITE (DOT)
12002-03-8	COPPER ACETOARSENITE, SOLID
12002-03-8	COPPER ACETOARSENITE, SOLID (DOT)
13479-54-4	COPPER AMINOACETATE
10290-12-7	COPPER ARSENITE
33382-64-8	COPPER ARSENITE
10290-12-7	COPPER ARSENITE (CuHASO3)
10290-12-7	COPPER ARSENITE (DOT)
10290-12-7	COPPER ARSENITE, SOLID
10290-12-7	COPPER ARSENITE, SOLID (DOT)
7447-39-4	COPPER BICHLORIDE
11129-27-4	COPPER BROMIDE
7787-70-4	COPPER BROMIDE (Cu4Br4)
7787-70-4	COPPER BROMIDE (CuBr)
7787-70-4	COPPER BROMIDE (OUS)
26506-47-8	COPPER CHLORATE
26506-47-8	COPPER CHLORATE (DOT)
1344-67-8	COPPER CHLORIDE
7447-39-4	COPPER CHLORIDE
7447-39-4	COPPER CHLORIDE (CuCl2)
544-92-3	COPPER CYANIDE
14763-77-0	COPPER CYANIDE
544-92-3	COPPER CYANIDE (Cu(CN))
544-92-3	COPPER CYANIDE (DOT)
142-71-2	COPPER DIACETATE
7447-39-4	COPPER DICHLORIDE
544-19-4	COPPER DIFORMATE
13479-54-4	COPPER DIGLYCINATE
3251-23-8	COPPER DINITRATE
544-19-4	COPPER FORMATE
13479-54-4	COPPER GLYCINATE
1335-23-5	COPPER IODIDE
16039-52-4	COPPER LACTATE
7440-50-8	COPPER M 1
7787-70-4	COPPER MONOBROMIDE
7758-98-7	COPPER MONOSULFATE
1338-02-9	COPPER NAPHTHANATE SOLUTION, ≤8% COPPER NAPHTHANATE
1338-02-9	COPPER NAPHTHENATE
3251-23-8	COPPER NITRATE (Cu(NO3)2)
10290-12-7	COPPER ORTHOARSENITE
814-91-5	COPPER OXALATE
814-91-5	COPPER OXALATE (CuC2O4)
7440-50-8	COPPER POWDER
9007-39-0	COPPER RESINATE
70027-50-8	COPPER SELENATE
10214-40-1	COPPER SELENITE
51325-42-9	COPPER SELENITE
10214-40-1	COPPER SELENITE (Cu(SeO3))
51325-42-9	COPPER SELENITE (Cu(SeO3))
14264-31-4	COPPER SODIUM CYANIDE
7758-98-7	COPPER SULFATE
7758-98-7	COPPER SULFATE (1:1)
7758-98-7	COPPER SULFATE (CuSO4)
7758-98-7	COPPER SULFATE BASIC
10380-29-7	COPPER TETRAAMINE SULFATE MONOHYDRATE
1338-02-9	COPPER UVERSOL
7787-70-4	COPPER(1+) BROMIDE
7787-70-4	COPPER(1+) BROMIDE TETRAMER
142-71-2	COPPER(2+) ACETATE
7447-39-4	COPPER(2+) CHLORIDE
142-71-2	COPPER(2+) DIACETATE
544-19-4	COPPER(2+) FORMATE
3251-23-8	COPPER(2+) NITRATE
814-91-5	COPPER(2+) OXALATE
7758-98-7	COPPER(2+) SULFATE
7758-98-7	COPPER(2+) SULFATE (1:1)
13426-91-0	COPPER(2+), BIS(1,2-ETHANEDIAMINE-N,N')-
13426-91-0	COPPER(2+), BIS(ETHYLENEDIAMINE)-, ION
10380-29-7	COPPER(2+), TETRAAMMINE-, SULFATE (1:1), MONOHYDRATE
13479-54-4	COPPER(GLYCINE)2
7787-70-4	COPPER(I) BROMIDE
544-92-3	COPPER(I) CYANIDE
13682-73-0	COPPER(I) POTASSIUM CYANIDE
142-71-2	COPPER(II) ACETATE
7447-39-4	COPPER(II) CHLORIDE
544-19-4	COPPER(II) FORMATE
13479-54-4	COPPER(II) GLYCINATE
3251-23-8	COPPER(II) NITRATE
814-91-5	COPPER(II) OXALATE
10214-40-1	COPPER(II) SELENITE
51325-42-9	COPPER(II) SELENITE
7758-98-7	COPPER(II) SULFATE
16039-52-4	COPPER, BIS(2-HYDROXYPROPANOATO-O1,O2)-
12002-03-8	COPPER, BIS(ACETATO)HEXAMETAARSENITOTETRA-
13479-54-4	COPPER, BIS(GLYCINATO)-
13479-54-4	COPPER, BIS(GLYCINATO-N,O)-
16039-52-4	COPPER, BIS(LACTATO)-
13426-91-0	COPPER-ETHYLENEDIAMINE COMPLEX
7720-78-7	COPPERAS
8001-31-8	COPRA
577-11-7	COPROL
124-87-8	COQUES DU LEVANT (FRENCH)
1937-37-7	CORANIL DIRECT BLACK F
7631-86-9	CORASIL II
1333-86-4	CORAX 3HS
1333-86-4	CORAX A
1333-86-4	CORAX L
1333-86-4	CORAX L 6
1333-86-4	CORAX P
20830-75-5	CORDIOXIL
9004-70-0	CORIAL EM FINISH F
57-83-0	CORLUTIN
57-83-0	CORLUTINA
57-83-0	CORLUVITE
8001-30-7	CORN OIL
9005-25-8	CORN STARCH
2921-88-2	COROBAN
12789-03-6	CORODANE
9016-87-9	CORONATE MR 200
7704-34-9	COROSUL D AND S
50-28-2	CORPAGEN
57-83-0	CORPORIN
57-83-0	CORPUS LUTEUM HORMONE
7487-94-7	CORROSIVE MERCURY CHLORIDE
7487-94-7	CORROSIVE SUBLIMATE
57-74-9	CORTILAN-NEU
12789-03-6	CORTILAN-NEU
7704-34-9	COSAN
7704-34-9	COSAN 80
50-76-0	COSMEGEN
8006-54-0	COSMELAN
86-50-0	COTNEON
86-50-0	COTNION
86-50-0	COTNION METHYL
2642-71-9	COTNION-ETHYL
70-30-4	COTOFILM
8001-29-4	COTTONSEED OIL
81-81-2	COUMADIN
129-06-6	COUMADIN
129-06-6	COUMADIN SODIUM

CAS No.	Chemical Name
81-81-2	COUMAFEN
81-81-2	COUMAFENE
129-06-6	COUMAFENE SODIUM
56-72-4	COUMAFOS
117-52-2	COUMAFURYL
81-81-2	COUMAPHEN
56-72-4	COUMAPHOS
56-72-4	COUMAPHOS (DOT)
81-81-2	COUMARIN
28772-56-7	COUMARIN, 3-(3-(4'-BROMO-1,1'-BIPHENYL-4-YL)-3-HYDROXY-1-PHENYLPROPYL)-4-HYDROXY-
28772-56-7	COUMARIN, 3-(ALPHA-(P-(P-BROMOPHENYL)-BETA-HYDROXY-PHENETHYL)BENZYL)-4-HYDROXY- (8CI)
81-81-2	COUMARIN, 3-(ALPHA-ACETONYLBENZYL)-4-HYDROXY-
129-06-6	COUMARIN, 3-(ALPHA-ACETONYLBENZYL)-4-HYDROXY-, SODIUM SALT
117-52-2	COUMARIN, 3-(ALPHA-ACETONYLFURFURYL)-4-HYDROXY-
56-72-4	COUMARIN, 3-CHLORO-7-HYDROXY-4-METHYL-, O-ESTER WITH O,O-DIETHYL PHOSPHOROTHIOATE
5836-29-3	COUMARIN, 4-HYDROXY-3-(1,2,3,4-TETRAHYDRO-1-NAPHTHYL)-
5836-29-3	COUMATETRALYL
81-81-2	COUMEFENE
13071-79-9	COUNTER
13071-79-9	COUNTER 15G
13071-79-9	COUNTER 15G SOIL INSECTICIDE
13071-79-9	COUNTER 15G SOIL INSECTICIDE-NEMATICIDE
81-81-2	COV-R-TOX
9002-89-5	COVAL 9700
9002-89-5	COVOL
9002-89-5	COVOL 971
2971-90-6	COYDEN
2971-90-6	COYDEN 25
1929-82-4	CP
14807-96-6	CP 10-40
106-91-2	CP 105
297-78-9	CP 14,957
2303-16-4	CP 15336
108-88-3	CP 25
108-90-7	CP 27
110-86-1	CP 32
110-02-1	CP 34
14807-96-6	CP 38-33
95-06-7	CP 4,742
2665-30-7	CP 40294
122-14-5	CP 47114
2540-82-1	CP 53926
7758-98-7	CP BASIC SULFATE
7440-32-6	CP TITANIUM
7782-42-5	CPB 5000
9005-25-8	CPC 3005
9005-25-8	CPC 6448
115-32-2	CPCA
15663-27-1	CPDD
56-75-7	CPH
115-26-4	CR 409
136-78-7	CRAG HERBICIDE
136-78-7	CRAG HERBICIDE 1
136-78-7	CRAG SESONE
63-25-2	CRAG SEVIN
79-01-6	CRAWHASPOL
7487-94-7	CRC
7681-49-4	CREDO
8001-58-9	CREOSOTE
8001-58-9	CREOSOTE OIL
8001-58-9	CREOSOTE P1
8001-58-9	CREOSOTE, COAL TAR
8001-58-9	CREOSOTE, FROM COAL TAR
8001-58-9	CREOSOTUM
120-71-8	CRESIDINE
1314-77-3	CRESOL
1319-77-3	CRESOL (ACGIH,DOT)
1319-77-3	CRESOL (MIXED ISOMERS)
26447-14-3	CRESOL GLYCIDYL ETHER
59-50-7	CRESOL, CHLORO-
1321-10-4	CRESOL, CHLORO-
12167-20-3	CRESOL, NITRO-
1319-77-3	CRESOLI (ITALIAN)
26471-62-5	CRESORCINOL DIISOCYANATE
8001-58-9	CRESOTE
8001-58-9	CRESOTE OIL
2602-46-2	CRESOTINE BLUE 2B
72-57-1	CRESOTINE BLUE 3B
8001-35-2	CRESTOXO
26447-14-3	CRESYL GLYCIDYL ETHER
1330-78-5	CRESYL PHOSPHATE
26447-14-3	CRESYLGLYCIDE ETHER
1319-77-3	CRESYLIC ACID
1319-77-3	CRESYLIC ACID (DOT)
8001-58-9	CRESYLIC CREOSOTE
535-89-7	CRIMIDIN (GERMAN)
535-89-7	CRIMIDINA (ITALIAN)
535-89-7	CRIMIDINE
535-89-7	CRIMITOX
12627-52-0	CRIMSON ANTIMONY
53-16-7	CRINOVARYL
1582-09-8	CRISALIN
75-99-0	CRISAPON
1912-24-9	CRISATRINA
1912-24-9	CRISAZINE
6923-22-4	CRISODIN
6923-22-4	CRISODRIN
128-44-9	CRISTALLOSE
53-16-7	CRISTALLOVAR
71-63-6	CRISTAPURAT
58-22-0	CRISTERONA T
14464-46-1	CRISTOBALITE
14464-46-1	CRISTOBALITE (SiO_2)
14464-46-1	CRISTOBALITE DUST
8001-35-2	CRISTOXO
8001-35-2	CRISTOXO 90
115-29-7	CRISULFAN
330-54-1	CRISURON
12001-28-4	CROCIDOLITE
61105-31-5	CROCIDOLITE
53799-46-5	CROCIDOLITE
12001-28-4	CROCIDOLITE (DOT)
61105-31-5	CROCIDOLITE ($Fe_2Mg_3Na_2(SiO_3)_8$)
53799-46-5	CROCIDOLITE ($Fe_5Na_2(SiO_3)_8$)
12001-28-4	CROCIDOLITE ASBESTOS
12001-28-4	CROCIDOLITE DUST
1309-37-1	CROCUS (IRON OXIDE)
8006-54-0	CRODAPUR
107-02-8	CROLEAN
8002-74-2	CROLENE LC
14977-61-8	CROMILE, CLORURO DI (ITALIAN)
13530-68-2	CROMO(TRIOSSIDO DI) (ITALIAN)
14977-61-8	CROMO, OSSICLORURO DI (ITALIAN)
97-77-8	CRONETAL
94-75-7	CROP RIDER
2971-38-2	CROTILIN
2971-38-2	CROTILINE
123-73-9	CROTONAL
4170-30-3	CROTONAL
123-73-9	CROTONALDEHYDE
4170-30-3	CROTONALDEHYDE
123-73-9	CROTONALDEHYDE (ACGIH)
4170-30-3	CROTONALDEHYDE (ACGIH)
123-73-9	CROTONALDEHYDE (DOT)
4170-30-3	CROTONALDEHYDE (DOT)
123-73-9	CROTONALDEHYDE, (E)-
123-73-9	CROTONALDEHYDE, INHIBITED (DOT)
4170-30-3	CROTONALDEHYDE, INHIBITED (DOT)
141-66-2	CROTONAMIDE, 3-HYDROXY-N,N-DIMETHYL-, CIS-, DIMETHYL PHOSPHATE
6923-22-4	CROTONAMIDE, 3-HYDROXY-N-METHYL-, DIMETHYLPHOSPHATE, (E)-
6923-22-4	CROTONAMIDE, 3-HYDROXY-N-METHYL-, DIMETHYLPHOSPHATE, CIS-
141-66-2	CROTONAMIDE, 3-HYDROXY-N-N-DIMETHYL-, DIMETHYL PHOSPHATE, (E)-
141-66-2	CROTONAMIDE, 3-HYDROXY-N-N-DIMETHYL-, DIMETHYL PHOSPHATE, CIS-
10544-63-5	CROTONATE D'ETHYLE (FRENCH)
3724-65-0	CROTONIC ACID
3724-65-0	CROTONIC ACID (DOT)
7786-34-7	CROTONIC ACID, 3-HYDROXY-, METHYL ESTER, DIMETHYL PHOSPHATE
10544-63-5	CROTONIC ACID, ETHYL ESTER

CAS No.	Chemical Name
10544-63-5	CROTONIC ACID, ETHYL ESTER, (E)-
14861-06-4	CROTONIC ACID, VINYL ESTER
123-73-9	CROTONIC ALDEHYDE
4170-30-3	CROTONIC ALDEHYDE
6117-91-5	CROTONOL
6117-91-5	CROTONYL ALCOHOL
4784-77-4	CROTONYL BROMIDE
503-17-3	CROTONYLENE
503-17-3	CROTONYLENE (DOT)
6117-91-5	CROTYL ALCOHOL
4784-77-4	CROTYL BROMIDE
591-97-9	CROTYL CHLORIDE
123-73-9	CROTYLALDEHYDE
4170-30-3	CROTYLALDEHYDE
2971-38-2	CROTYLIN
13463-67-7	CRS 31
1327-53-3	CRUDE ARSENIC
8002-05-9	CRUDE OIL
299-86-5	CRUFOMAT
299-86-5	CRUFOMATE
15096-52-3	CRYOLITE
87-86-5	CRYPTOGIL OL
16919-19-0	CRYPTOHALITE
7631-86-9	CRYSTALITE A 1
14808-60-7	CRYSTALITE AA
14807-96-6	CRYSTALITE CRS 6002
71-63-6	CRYSTALLINE
71-63-6	CRYSTALLINE DIGITALIN
142-71-2	CRYSTALLIZED VERDIGRIS
128-44-9	CRYSTALLOSE
142-71-2	CRYSTALS OF VENUS
6834-92-0	CRYSTAMET
7704-34-9	CRYSTEX
2642-71-9	CRYSTHION
86-50-0	CRYSTHION 2L
86-50-0	CRYSTHYON
14464-46-1	CRYSTOBALITE
71-63-6	CRYSTODIGIN
53-16-7	CRYSTOGEN
409-21-2	CRYSTOLON B
50-55-5	CRYSTOSERPINE
2698-41-1	CS
2698-41-1	CS (LACRIMATOR)
2425-06-1	CS 5623
57-47-6	CS 58525
1344-95-2	CS LAFARGE
111-76-2	CSAC
7446-18-6	CSF-GIFTWEIZEN
10031-59-1	CSF-GIFTWEIZEN
1333-86-4	CSX 147
1333-86-4	CSX 150A2
1333-86-4	CSX 174
1333-86-4	CSX 200A
112-02-7	CTAC
79-38-9	CTFE
112-02-7	CTMA
7440-50-8	CU M2
7440-50-8	CU M3
83-79-4	CUBE-PULVER
7631-99-4	CUBIC NITER
7778-44-1	CUCUMBER DUST
7440-50-8	CuEP
7440-50-8	CuEPP
129-06-6	CUMADIN
56-72-4	CUMAFOS (DUTCH)
117-52-2	CUMAFURYL (GERMAN)
5836-29-3	CUMATETRALYL (GERMAN, DUTCH)
98-82-8	CUMEEN (DUTCH)
80-15-9	CUMEENHYDROPEROXYDE (DUTCH)
98-82-8	CUMENE
98-82-8	CUMENE (ACGIH)
80-15-9	CUMENE HYDROPEROXIDE
80-15-9	CUMENE HYDROPEROXIDE (DOT)
80-15-9	CUMENE HYDROPEROXIDE, TECHNICALLY PURE (DOT)
80-43-3	CUMENE PEROXIDE
80-15-9	CUMENYL HYDROPEROXIDE
98-82-8	CUMOL
80-15-9	CUMOLHYDROPEROXID (GERMAN)
80-15-9	CUMYL HYDROPEROXIDE
80-15-9	CUMYL HYDROPEROXIDE, TECHNICAL PURE (DOT)
26748-47-0	CUMYL PERNEODECANOATE
80-43-3	CUMYL PEROXIDE
26748-47-0	CUMYL PEROXYNEODECANOATE
26748-47-0	CUMYL PEROXYNEODECANOATE, MAXIMUM CONCENTRATION 77% IN SOLUTION (DOT)
3457-61-2	CUMYL TERT-BUTYL PEROXIDE
30580-75-7	CUMYL TERT-BUTYL PEROXIDE
135-20-6	CUPFERRON
4384-82-1	CUPRAL
13682-73-0	CUPRATE(1-), BIS(CYANO-C)-, POTASSIUM
13682-73-0	CUPRATE(1-), DICYANO-, POTASSIUM
14264-31-4	CUPRATE(2-), TRIS(CYANO-C)-, DISODIUM
142-71-2	CUPRIC ACETATE (DOT)
142-71-2	CUPRIC ACETATE
12002-03-8	CUPRIC ACETOARSENITE
13479-54-4	CUPRIC AMINOACETATE
10290-12-7	CUPRIC ARSENITE
26506-47-8	CUPRIC CHLORATE
1344-67-8	CUPRIC CHLORIDE
7447-39-4	CUPRIC CHLORIDE
142-71-2	CUPRIC DIACETATE
7447-39-4	CUPRIC DICHLORIDE
544-19-4	CUPRIC DIFORMATE
3251-23-8	CUPRIC DINITRATE
544-19-4	CUPRIC FORMATE
13479-54-4	CUPRIC GLYCINATE
10290-12-7	CUPRIC GREEN
3251-23-8	CUPRIC NITRATE
3251-23-8	CUPRIC NITRATE (DOT)
3251-23-8	CUPRIC NITRATE SOLUTION
814-91-5	CUPRIC OXALATE
814-91-5	CUPRIC OXALATE (1:1)
814-91-5	CUPRIC OXALATE (DOT)
10214-40-1	CUPRIC SELENITE
51325-42-9	CUPRIC SELENITE
7758-98-7	CUPRIC SULFATE
7758-98-7	CUPRIC SULFATE (DOT)
7758-98-7	CUPRIC SULFATE ANHYDROUS
10380-29-7	CUPRIC SULFATE, AMMONIATED
10380-29-7	CUPRIC SULFATE, AMMONIATED (DOT)
7758-98-7	CUPRIC SULPHATE
815-82-7	CUPRIC TARTRATE
815-82-7	CUPRIC TARTRATE (DOT)
9004-34-6	CUPRICELLULOSE
544-92-3	CUPRICIN
13426-91-0	CUPRIETHYLENE DIAMINE
13426-91-0	CUPRIETHYLENE-DIAMINE SOLUTION
13426-91-0	CUPRIETHYLENE-DIAMINE SOLUTION (DOT)
1338-02-9	CUPRINOL
16071-86-6	CUPROFIX BROWN GL
7787-70-4	CUPROUS BROMIDE
7787-70-4	CUPROUS BROMIDE (CuBr)
7447-39-4	CUPROUS CHLORIDE
544-92-3	CUPROUS CYANIDE
7447-39-4	CUPROUS DICHLORIDE
13682-73-0	CUPROUS POTASSIUM CYANIDE
74-83-9	CURAFUME
1563-66-2	CURATERR
101-77-9	CURITHANE
96-33-3	CURITHANE 103
91-94-1	CURITHANE C 126
7439-96-5	CUTAVAL
94-36-0	CUTICURA ACNE CREAM
470-90-6	CVP
470-90-6	CVP (PESTICIDE)
156-62-7	CY-L 500
62-53-3	CYANOL
74-90-8	CYAANWATERSTOF (DUTCH)
105-56-6	CYANACETATE ETHYLE (GERMAN)
156-62-7	CYANAMID
156-62-7	CYANAMID GRANULAR
156-62-7	CYANAMID SPECIAL GRADE
156-62-7	CYANAMIDE
156-62-7	CYANAMIDE CALCIQUE (FRENCH)
156-62-7	CYANAMIDE, CALCIUM SALT (1:1)
1467-79-4	CYANAMIDE, DIMETHYL-
372-09-8	CYANESSIGSAEURE (GERMAN)
75-86-5	CYANHYDRINE D'ACETONE (FRENCH)

CAS No.	Chemical Name
154-23-4	CYANIDANOL
57-12-5	CYANIDE
57-12-5	CYANIDE (CN(SUP 1-))
57-12-5	CYANIDE (CN 1-)
57-12-5	CYANIDE ANION
57-12-5	CYANIDE COMPOUNDS
57-12-5	CYANIDE ION
151-50-8	CYANIDE OF POTASSIUM
143-33-9	CYANIDE OF SODIUM
57-12-5	CYANIDE OR CYANIDE MIXTURE, DRY (DOT)
2074-87-5	CYANIDE RADICAL
57-12-5	CYANIDE SOLUTION
57-12-5	CYANIDE, DRY
57-12-5	CYANIDE, SOLUTION (DOT)
57-12-5	CYANIDES
2074-87-5	CYANO RADICAL
2074-87-5	CYANO RADICAL (CN)
502-39-6	CYANO(METHYLMERCURI)GUANIDINE
372-09-8	CYANOACETIC ACID
105-56-6	CYANOACETIC ACID ETHYL ESTER
105-56-6	CYANOACETIC ESTER
420-04-2	CYANOAMINE
100-47-0	CYANOBENZENE
137-05-3	CYANOBOND SS
143-33-9	CYANOBRIK
506-68-3	CYANOBROMIDE
506-77-4	CYANOCHLORIDE (CNCl)
1467-79-4	CYANODIMETHYLAMINE
107-12-0	CYANOETHANE
107-13-1	CYANOETHENE
107-13-1	CYANOETHYLENE
592-01-8	CYANOGAS
460-19-5	CYANOGEN
2074-87-5	CYANOGEN
460-19-5	CYANOGEN (ACGIH,DOT)
460-19-5	CYANOGEN (C2N2)
2074-87-5	CYANOGEN (CN)
506-68-3	CYANOGEN BROMIDE
506-68-3	CYANOGEN BROMIDE ((CN)Br)
506-68-3	CYANOGEN BROMIDE (BrCN)
506-68-3	CYANOGEN BROMIDE (DOT)
506-77-4	CYANOGEN CHLORIDE
506-77-4	CYANOGEN CHLORIDE ((CN)Cl)
506-77-4	CYANOGEN CHLORIDE (ACGIH)
506-77-4	CYANOGEN CHLORIDE (ClCN)
506-77-4	CYANOGEN CHLORIDE, CONTAINING LESS THAN 0.9% WATER (DOT)
506-77-4	CYANOGEN CHLORIDE, INHIBITED (DOT)
506-77-4	CYANOGEN CHLORIDE, WITH 0.9% WATER
460-19-5	CYANOGEN GAS
460-19-5	CYANOGEN GAS (DOT)
506-78-5	CYANOGEN IODIDE
506-68-3	CYANOGEN MONOBROMIDE
506-78-5	CYANOGEN MONOIODIDE
420-04-2	CYANOGEN NITRIDE
2074-87-5	CYANOGEN RADICAL
420-04-2	CYANOGENAMIDE
460-19-5	CYANOGENE (FRENCH)
143-33-9	CYANOGRAN
137-05-3	CYANOLIT
75-05-8	CYANOMETHANE
107-16-4	CYANOMETHANOL
2636-26-2	CYANOPHOS
2636-26-2	CYANOX
3734-95-0	CYANTHOATE
108-77-0	CYANUR CHLORIDE
108-77-0	CYANURCHLORIDE
57-12-5	CYANURE (FRENCH)
506-64-9	CYANURE D'ARGENT (FRENCH)
592-01-8	CYANURE DE CALCIUM (FRENCH)
592-04-1	CYANURE DE MERCURE (FRENCH)
75-05-8	CYANURE DE METHYL (FRENCH)
592-05-2	CYANURE DE PLOMB (FRENCH)
151-50-8	CYANURE DE POTASSIUM (FRENCH)
143-33-9	CYANURE DE SODIUM (FRENCH)
557-21-1	CYANURE DE ZINC (FRENCH)
108-77-0	CYANURIC ACID CHLORIDE
108-77-0	CYANURIC CHLORIDE
108-77-0	CYANURIC CHLORIDE (DOT)
675-14-9	CYANURIC FLUORIDE
108-77-0	CYANURIC TRICHLORIDE
108-77-0	CYANURIC TRICHLORIDE (DOT)
675-14-9	CYANURIC TRIFLUORIDE
108-77-0	CYANURYL CHLORIDE
74-90-8	CYANWASSERSTOFF (GERMAN)
2636-26-2	CYAP
1912-24-9	CYAZIN
14901-08-7	CYCASIN
96-49-1	CYCLIC ETHYLENE CARBONATE
947-02-4	CYCLIC ETHYLENE P,P-DIETHYL PHOSPHONODITHIOIMIDOCARBONATE
947-02-4	CYCLIC ETHYLENE(DIETHOXYPHOSPHINOTHIOYL)-DITHIOIMIDOCARBONATE
950-10-7	CYCLIC PROPYLENE (DIETHOXYPHOSPHINYL)DITHIOIMIDOCARBONATE
7785-84-4	CYCLIC SODIUM TRIMETAPHOSPHATE
126-33-0	CYCLIC TETRAMETHYLENE SULFONE
287-23-0	CYCLOBUTANE
287-23-0	CYCLOBUTANE (DOT)
4806-61-5	CYCLOBUTANE, ETHYL-
115-25-3	CYCLOBUTANE, OCTAFLUORO-
81228-87-7	CYCLOBUTYL CHLOROFORMATE
81228-87-7	CYCLOBUTYLCHLOROFORMATE (DOT)
999-81-5	CYCLOCEL
115-29-7	CYCLODAN
291-64-5	CYCLOHEPTANE
544-25-2	CYCLOHEPTATRIENE
628-92-2	CYCLOHEPTENE
110-83-8	CYCLOHEX-1-ENE
106-51-4	CYCLOHEXADIENEDIONE
111-49-9	CYCLOHEXAMETHYLENIMINE
108-91-8	CYCLOHEXANAMINE
42195-92-6	CYCLOHEXANAMINE, 2,3-DIMETHYL-
1195-42-2	CYCLOHEXANAMINE, N-(1-METHYLETHYL)-
5432-61-1	CYCLOHEXANAMINE, N-(2-ETHYLHEXYL)-
10108-56-2	CYCLOHEXANAMINE, N-BUTYL-
101-83-7	CYCLOHEXANAMINE, N-CYCLOHEXYL-
3129-91-7	CYCLOHEXANAMINE, N-CYCLOHEXYL-, NITRITE
5459-93-8	CYCLOHEXANAMINE, N-ETHYL-
100-60-7	CYCLOHEXANAMINE, N-METHYL-
100-60-7	CYCLOHEXANAMINE, N-METHYL- (9CI)
110-82-7	CYCLOHEXANE
5124-30-1	CYCLOHEXANE, 1,1'-METHYLENEBIS[4-ISOCYANATO-
608-73-1	CYCLOHEXANE, 1,2,3,4,5,6-HEXACHLORO-
319-86-8	CYCLOHEXANE, 1,2,3,4,5,6-HEXACHLORO-, (1ALPHA,2ALPHA,3ALPHA,4BETA,5ALPHA,6BETA
319-84-6	CYCLOHEXANE,1,2,3,4,5,6-HEXACHLORO-, (1ALPHA,2ALPHA,-3BETA,4ALPHA,5BETA,6BETA)-
319-85-7	CYCLOHEXANE, 1,2,3,4,5,6-HEXACHLORO-, (1ALPHA,2BETA,3ALPHA,4BETA,5ALPHA,6BETA)-
58-89-9	CYCLOHEXANE, 1,2,3,4,5,6-HEXACHLORO-, (1ALPHA,2ALPHA,3BETA,4ALPHA,5ALPHA,6BETA)-
6108-11-8	CYCLOHEXANE, 1,2,3,4,5,6-HEXACHLORO-, ALPHA-ISOMER
319-85-7	CYCLOHEXANE, 1,2,3,4,5,6-HEXACHLORO-, BETA-ISOMER
58-89-9	CYCLOHEXANE, 1,2,3,4,5,6-HEXACHLORO-, GAMMA-ISOMER
591-21-9	CYCLOHEXANE, 1,3-DIMETHYL-
589-90-2	CYCLOHEXANE, 1,4-DIMETHYL-
624-29-3	CYCLOHEXANE, 1,4-DIMETHYL-, CIS-
2207-04-7	CYCLOHEXANE, 1,4-DIMETHYL-, TRANS-
4098-71-9	CYCLOHEXANE, 5-ISOCYANATO-1-(ISOCYANATOMETHYL)-1,3,3-TRIMETHYL-
4098-71-9	CYCLOHEXANE, 5-ISOCYANATO-1-(ISOCYANATOMETHYL)-1,3,3-TRIMETHYL- (9CI)
542-18-7	CYCLOHEXANE, CHLORO-
1331-43-7	CYCLOHEXANE, DIETHYL-
1678-91-7	CYCLOHEXANE, ETHYL-
108-87-2	CYCLOHEXANE, METHYL-
1122-60-7	CYCLOHEXANE, NITRO-
827-52-1	CYCLOHEXANE, PHENYL-
1569-69-3	CYCLOHEXANETHIOL
108-93-0	CYCLOHEXANOL
2407-94-5	CYCLOHEXANOL, 1,1'-DIOXYBIS-
2407-94-5	CYCLOHEXANOL, 1,1'-DIOXYBIS- (9CI)
2407-94-5	CYCLOHEXANOL, 1,1'-DIOXYDI-
116-02-9	CYCLOHEXANOL, 3,3,5-TRIMETHYL-
15520-11-3	CYCLOHEXANOL, 4-TERT-BUTYL-, PEROXYDICARBONATE (2:1)
25639-42-3	CYCLOHEXANOL, METHYL-
1321-60-4	CYCLOHEXANOL, TRIMETHYL-

CAS No.	Chemical Name
108-94-1	CYCLOHEXANONE
105-60-2	CYCLOHEXANONE ISO-OXIME
12262-58-7	CYCLOHEXANONE PEROXIDE
583-60-8	CYCLOHEXANONE, 2-METHYL-
1331-22-2	CYCLOHEXANONE, METHYL-
11118-65-3	CYCLOHEXANONE, METHYL-, PEROXIDE
71-43-2	CYCLOHEXATRIENE
110-83-8	CYCLOHEXENE
138-86-3	CYCLOHEXENE, 1-METHYL-4-(1-METHYLETHENYL)-
586-62-9	CYCLOHEXENE, 1-METHYL-4-(1-METHYLETHYLIDENE)-
100-40-3	CYCLOHEXENE, 4-ETHENYL-
591-47-9	CYCLOHEXENE, 4-METHYL-
100-40-3	CYCLOHEXENE, 4-VINYL-
100-50-5	CYCLOHEXENE-4-CARBOXALDEHYDE
1321-16-0	CYCLOHEXENECARBOXALDEHYDE
108-94-1	CYCLOHEXENONE
10137-69-6	CYCLOHEXENYL TRICHLOROSILANE
66-81-9	CYCLOHEXIMIDE
622-45-7	CYCLOHEXYL ACETATE
108-93-0	CYCLOHEXYL ALCOHOL
542-18-7	CYCLOHEXYL CHLORIDE
3173-53-3	CYCLOHEXYL ISOCYANATE
98-12-4	CYCLOHEXYL TRICHLOROSILANE
3312-60-5	CYCLOHEXYL-1,3-PROPANEDIAMINE
10108-56-2	CYCLOHEXYL-N-BUTYLAMINE
108-91-8	CYCLOHEXYLAMINE
42195-92-6	CYCLOHEXYLAMINE, 2,3-DIMETHYL-
702-03-4	CYCLOHEXYLAMINE, CYANOETHYL-
10108-56-2	CYCLOHEXYLAMINE, N-BUTYL-
5459-93-8	CYCLOHEXYLAMINE, N-ETHYL-
1195-42-2	CYCLOHEXYLAMINE, N-ISOPROPYL-
100-60-7	CYCLOHEXYLAMINE, N-METHYL-
34216-34-7	CYCLOHEXYLAMINE, TRIMETHYL-
827-52-1	CYCLOHEXYLBENZENE
1195-42-2	CYCLOHEXYLISOPROPYLAMINE
108-87-2	CYCLOHEXYLMETHANE
100-60-7	CYCLOHEXYLMETHYLAMINE
74-90-8	CYCLON
74-90-8	CYCLONE B
121-82-4	CYCLONITE
116-02-9	CYCLONOL
29965-97-7	CYCLOOCTADIENE
29965-97-7	CYCLOOCTADIENES (DOT)
115-25-3	CYCLOOCTAFLUOROBUTANE
629-20-9	CYCLOOCTATETRAENE
629-20-9	CYCLOOCTATETRAENE (DOT)
542-92-7	CYCLOPENTADIENE
77-73-6	CYCLOPENTADIENE DIMER
2385-85-5	CYCLOPENTADIENE, HEXACHLORO-, DIMER
12079-65-1	CYCLOPENTADIENYLMANGANESE TRICARBONYL
12079-65-1	CYCLOPENTADIENYLTRICARBONYLMANGANESE
287-92-3	CYCLOPENTANE
287-92-3	CYCLOPENTANE (ACGIH,DOT)
1640-89-7	CYCLOPENTANE, ETHYL-
96-37-7	CYCLOPENTANE, METHYL-
96-41-3	CYCLOPENTANOL
120-92-3	CYCLOPENTANONE
208-96-8	CYCLOPENTA[DE]NAPHTHALENE
142-29-0	CYCLOPENTENE
110-89-4	CYCLOPENTIMINE
50-18-0	CYCLOPHOSPHAMID
50-18-0	CYCLOPHOSPHAMIDE
50-18-0	CYCLOPHOSPHAN
50-18-0	CYCLOPHOSPHANE
75-19-4	CYCLOPROPANE
584-79-2	CYCLOPROPANECARBOXYLIC ACID, 2,2-DIMETHYL-3-(2-METHYL-1-PROPENYL)-,2-METHYL-4-OXO-3-(2-PROPENYL)-2-C
584-79-2	CYCLOPROPANECARBOXYLIC ACID, 2,2-DIMETHYL-3-(2-METHYL-PROPENYL)-, ESTER WITH 2-ALLYL-4-HYDROXY-3-METH
151-21-3	CYCLORYL 21
151-21-3	CYCLORYL 31
151-21-3	CYCLORYL 580
151-21-3	CYCLORYL 585N
139-96-8	CYCLORYL TAWF
139-96-8	CYCLORYL WAT
7546-30-7	CYCLOSAN
2312-35-8	CYCLOSULFYNE
109-99-9	CYCLOTETRAMETHYLENE OXIDE
126-33-0	CYCLOTETRAMETHYLENE SULFONE

CAS No.	Chemical Name
2691-41-0	CYCLOTETRAMETHYLENETETRANITRAMINE
121-82-4	CYCLOTRIMETHYLENENITRAMINE
121-82-4	CYCLOTRIMETHYLENETRINITRAMINE
999-81-5	CYCOCEL
999-81-5	CYCOCEL-EXTRA
999-81-5	CYCOGAN
999-81-5	CYCOGAN EXTRA
122-14-5	CYFEN
60-51-5	CYGON
60-51-5	CYGON 2E
60-51-5	CYGON 4E
60-51-5	CYGON INSECTICIDE
13121-70-5	CYHEXATIN
74-90-8	CYJANOWODOR (POLISH)
947-02-4	CYLAN
56-75-7	CYLPHENICOL
143-33-9	CYMAG
12079-65-1	CYMANTRENE
25155-15-1	CYMENE
144-49-0	CYMONIC ACID
297-97-2	CYNEM
297-97-2	CYNOPHOS
999-81-5	CYOCEL
947-02-4	CYOLANE
947-02-4	CYOLANE CYLAN
947-02-4	CYOLANE INSECTICIDE
110-89-4	CYPENTIL
2636-26-2	CYPONA
56-53-1	CYREN
56-53-1	CYREN A
100-97-0	CYSTAMIN
100-97-0	CYSTOGEN
147-94-4	CYSTOSINE ARABINOSIDE
122-14-5	CYTEL
122-14-5	CYTEN
121-75-5	CYTHION
50-18-0	CYTOPHOSPHAN
50-18-0	CYTOXAN
61-82-5	CYTROL
950-10-7	CYTROLANE
61-82-5	CYTROLE
137-26-8	CYURAM DS
104-15-4	CYZAC 4040
25231-46-3	CYZAC 4040
56-23-5	CZTEROCHLOREK WEGLA (POLISH)
127-18-4	CZTEROCHLOROETYLEN (POLISH)
78-00-2	CZTEROETYLEK OLOWIU (POLISH)
2312-35-8	D 014
1563-66-2	D 1221
298-07-7	D 2EHPA
280-57-9	D 33LV
94-75-7	D 50
3564-09-8	D AND C RED 15
3761-53-3	D AND C RED NO. 5
1694-09-3	D AND C VIOLET NO. 1
584-79-2	D,L-2-ALLYL-4-HYDROXY-3-METHYL-2-CYCLOPENTEN-1-ONE-D,L-CHRYSANTHEMUM MONOCARBOXYLATE
2244-16-8	D-(+)-CARVONE
120-80-9	D-(+)-CATECHIN
154-23-4	D-(+)-CATECHIN
57-50-1	D-(+)-SACCHAROSE
57-50-1	D-(+)-SUCROSE
56-75-7	D-(-)-CHLORAMPHENICOL
50-70-4	D-(-)-SORBITOL
56-75-7	D-(-)-THREO-CHLORAMPHENICOL
23135-22-0	D-1410
50-28-2	D-3,17BETA-ESTRADIOL
584-79-2	D-ALLETHRIN
50-14-6	D-ARTHIN
2244-16-8	D-CARVONE
120-80-9	D-CATECHIN
154-23-4	D-CATECHIN
120-80-9	D-CATECHOL
154-23-4	D-CATECHOL
56-75-7	D-CHLORAMPHENICOL
81-81-2	D-CON
8003-19-8	D-D
8003-19-8	D-D (PESTICIDE)
8003-19-8	D-D SOIL FUMIGANT

CAS No.	Chemical Name
50-70-4	D-GLUCITOL
63937-14-4	D-GLUCONIC ACID, MERCURY COMPLEX
19222-41-4	D-GLUCONIC ACID, MONOAMMONIUM SALT
18883-66-4	D-GLUCOSE, 2-DEOXY-2-[[(METHYLNITROSOAMINO)CARBONYL]-AMINO]-
7775-14-6	D-OX
577-11-7	D-S-S
50-70-4	D-SORBITOL
50-70-4	D-SORBOL
57-50-1	D-SUCROSE
56-75-7	D-THREO-CHLORAMPHENICOL
712-48-1	DA
60-11-7	DAB
60-11-7	DAB (CARCINOGEN)
280-57-9	DABCO
280-57-9	DABCO 33LV
280-57-9	DABCO CRYSTAL
280-57-9	DABCO EG
280-57-9	DABCO R-8020
280-57-9	DABCO S-25
93-76-5	DACAMINE
39196-18-4	DACAMOX
4342-03-4	DACARBAZINE
8052-42-4	DACHOLEUM
2163-80-6	DACONATE
2163-80-6	DACONATE 6
115-90-2	DACONIT
118-52-5	DACTIN
83-79-4	DACTINOL
50-76-0	DACTINOMYCIN
50-76-0	DACTINOMYCIN D
333-41-5	DACUTOX
101-77-9	DADPM
117-84-0	DAF 68
786-19-6	DAGADIP
9004-70-0	DAICEL RS 1
9004-70-0	DAICEL RS 1/2H
9004-70-0	DAICEL RS 7
79-38-9	DAIFLON
75-69-4	DAIFLON 11
75-45-6	DAIFLON 22
75-69-4	DAIFLON S 1
76-13-1	DAIFLON S 3
330-54-1	DAILON
99-55-8	DAINICHI FAST SCARLET G BASE
95-69-2	DAITO RED BASE TR
95-79-4	DAITO RED BASE TR
3165-93-3	DAITO RED SALT TR
99-55-8	DAITO SCARLET BASE G
7681-52-9	DAKIN'S SOLUTION
118-52-5	DAKTIN
2163-80-6	DAL-E-RAD
2163-80-6	DAL-E-RAD 120
75-99-0	DALAPON
75-99-0	DALAPON 85
298-00-0	DALF
114-26-1	DALF DUST
8003-34-7	DALMATION INSECT FLOWERS
128-37-0	DALPAC
102-71-6	DALTOGEN
105-60-2	DANAMID
52-68-6	DANEX
92-84-2	DANIKOROPA
118-52-5	DANTOIN
869-29-4	DAP
869-29-4	DAP (PESTICIDE)
60-51-5	DAPHENE
101-77-9	DAPM
131-17-9	DAPON R
131-17-9	DAPPU
80-08-0	DAPSONE
57-11-4	DAR-CHEM 14
12125-02-9	DARAMMON
115-90-2	DASANIT
333-41-5	DASSITOX
2303-16-4	DATC
20830-81-3	DAUNOMYCIN
20830-81-3	DAUNORUBICIN
20830-81-3	DAUNORUBICINE
12001-26-2	DAVENITE P 12
7631-86-9	DAVISON 951
20830-75-5	DAVOXIN
56-53-1	DAWE'S DESTROL
333-41-5	DAZZEL
53-70-3	DBA
96-12-8	DBCP
86-50-0	DBD
106-93-4	DBE
58-89-9	DBH
4835-11-4	DBHMD
1194-65-6	DBN
1194-65-6	DBN (PESTICIDE)
88-85-7	DBNF
84-74-2	DBP
84-74-2	DBP (ESTER)
128-37-0	DBPC
7782-42-5	DC 2
79-43-6	DCA
95-76-1	DCA
79-43-6	DCA (ACID)
95-50-1	DCB
25321-22-6	DCB
133-14-2	DCBP
111-44-4	DCEE
75-09-2	DCM
330-54-1	DCMU
120-83-2	DCP
2032-65-7	DCR 736
7440-50-8	DCUP1
8003-19-8	DD NEMATOCIDE
79-44-7	DDC
72-54-8	DDD
72-55-9	DDE
101-77-9	DDM
50-29-3	DDT
2636-26-2	DDVF
62-73-7	DDVP
114-26-1	DDVP
2636-26-2	DDVP
62-73-7	DDVP (INSECTICIDE)
114-26-1	DDVP (PROPOXUR)
60-51-5	DE-FEND
7775-09-9	DE-FOL-ATE
10326-21-3	DE-FOL-ATE
91-66-7	DEA
111-42-2	DEA
109-89-7	DEA
143-00-0	DEA-LAURYL SULFATE
111-46-6	DEACTIVATOR E
111-46-6	DEACTIVATOR H
100-37-8	DEAE
96-10-6	DEAK
108-01-0	DEANOL
1309-37-1	DEANOX
56-53-1	DEB
1464-53-5	DEB
94-75-7	DEBROUSSAILLANT 600
93-76-5	DEBROUSSAILLANT CONCENTRE
93-76-5	DEBROUSSAILLANT SUPER CONCENTRE
94-36-0	DEBROXIDE
105-58-8	DEC
91-17-8	DEC
17702-41-9	DECABORANE
17702-41-9	DECABORANE (B10H14)
17702-41-9	DECABORANE(14)
1163-19-5	DECABROMODIPHENYL ETHER
143-50-0	DECACHLORO-1,3,4-METHENO-2H-CYCLOBUTA(CD)PENTALEN-2-ONE
143-50-0	DECACHLOROKETONE
143-50-0	DECACHLOROOCTAHYDRO-1,3,4-METHENO-2H-CYCLOBUTA(CD)-PENTALEN-2-ONE
143-50-0	DECACHLOROPENTACYCLO(5210(SUP 2,6)0(SUP 3,9)0(SUP 5,8))DE-CAN-4-ONE
143-50-0	DECACHLOROPENTACYCLO(5300(SUP 2,6)0(SUP 4,10)0(SUP 5,9)DE-CAN-3-ONE
143-50-0	DECACHLOROPENTACYCLO[52102,603,905,8]DECAN-4-ONE
143-50-0	DECACHLOROTETRACYCLODECANONE
143-50-0	DECACHLOROTETRAHYDRO-4,7-METHANOINDENEONE

CAS No.	Chemical Name
91-17-8	DECAHYDRONAPHTHALENE
91-17-8	DECALIN
94-75-7	DECAMINE
93-76-5	DECAMINE 4T
2016-57-1	DECANAMINE
124-18-5	DECANE
104-72-3.	DECANE, 1-PHENYL-
2385-85-5	DECANE,PERCHLOROPENTACYCLO-
112-30-1	DECANOL
762-12-9	DECANOX
762-12-9	DECANOX F
762-12-9	DECANOYL PEROXIDE
762-12-9	DECANOYL PEROXIDE (DOT)
762-12-9	DECANOYL PEROXIDE, TECHNICALLY PURE
13952-84-6	DECCOTANE
732-11-6	DECEMTHION
732-11-6	DECEMTHION P-6
3129-91-7	DECHAN
2385-85-5	DECHLORANE
2385-85-5	DECHLORANE 4070
1309-64-4	DECHLORANE A-O
115-32-2	DECOFOL
27323-41-7	DECOL T 70
7664-38-2	DECON 4512
2156-96-9	DECYL ACRYLATE
112-30-1	DECYL ALCOHOL
2016-57-1	DECYLAMINE
104-72-3	DECYLBENZENE
112-30-1	DECYLIC ALCOHOL
75-99-0	DED-WEED
94-75-7	DED-WEED
93-76-5	DED-WEED BRUSH KILLER
93-76-5	DED-WEED LV-6 BRUSH KIL AND T-5 BRUSH KIL
94-75-7	DED-WEED LV-69
4384-82-1	DEDC
62-73-7	DEDEVAP
2636-26-2	DEDEVAP
4384-82-1	DEDK
128-37-0	DEENAX
7789-06-2	DEEP LEMON YELLOW
577-11-7	DEFILIN
443-48-1	DEFLAMON
7440-21-3	DEFOAMER S-10
111-46-6	DEG
534-52-1	DEGRASSAN
1335-85-9	DEGRASSAN
8001-22-7	DEGUMMED SOYBEAN OIL
1333-86-4	DEGUSSA BLACK FW
111-40-0	DEH 20
112-24-3	DEH 24
112-57-2	DEH 26
103-23-1	DEHA
123-79-5	DEHA
3710-84-7	DEHA
117-81-7	DEHP
298-07-7	DEHPA EXTRACTANT
9016-45-9	DEHSCOXID 781
151-21-3	DEHYDAG SULFATE GL EMULSION
151-21-3	DEHYDAG SULPHATE GL EMULSION
10034-81-8	DEHYDRITE
112-02-7	DEHYQUART A
85-00-7	DEIQUAT
514-73-8	DEJO
96-22-0	DEK
91-17-8	DEKALIN
534-52-1	DEKRYSIL
1335-85-9	DEKRYSIL
78-34-2	DELANOV
20859-73-8	DELICIA
20859-73-8	DELICIA GASTOXIN
78-34-2	DELNATEX
78-34-2	DELNAV
3691-35-8	DELTA
110-87-2	DELTA(SUP 2)-DIHYDROPYRAN
3174-74-1	DELTA(SUP 2)-DIHYDROPYRAN
25512-65-6	DELTA(SUP 2)-DIHYDROPYRAN
78-94-4	DELTA(SUP 3)-2-BUTENONE
498-15-7	DELTA(SUP 3)-CARENE
61215-72-3	DELTA(SUP 3)-PIPERIDINE
85-43-8	DELTA(SUP 4)-TETRAHYDROPHTHALIC ANHYDRIDE
119-64-2	DELTA(SUP 5,7,9)-NAPHTHANTRIENE
2244-16-8	DELTA(SUP 6,8)-(9)-TERPADIENONE-2
140-80-7	DELTA-(DIETHYLAMINO)-ALPHA-METHYLBUTYLAMINE
138-86-3	DELTA-1,8-TERPODIENE
2244-16-8	DELTA-1-METHYL-4-ISOPROPENYL-6-CYCLOHEXEN-2-ONE
3037-72-7	DELTA-AMINOBUTYLMETHYLDIETHOXYSILANE
107-07-3	DELTA-CHLOROETHANOL
110-87-2	DELTA2-DIHYDROPYRAN
3174-74-1	DELTA2-DIHYDROPYRAN
25512-65-6	DELTA2-DIHYDROPYRAN
85-43-8	DELTA4-TETRAHYDROPHTHALIC ANHYDRIDE
112-80-1	DELTA9-CIS-OCTADECENOIC ACID
112-80-1	DELTA9-CIS-OLEIC ACID
67-68-5	DELTAN
1314-84-7	DELUSAL
514-73-8	DELVEX
67-68-5	DEMASORB
67-68-5	DEMAVET
919-86-8	DEMENTON-S-METHYL
67-68-5	DEMESO
66-75-1	DEMETHYLDOPAN
298-00-0	DEMETHYLFENITROTHION
8065-48-3	DEMETON
8065-48-3	DEMETON (ACGIH)
8022-00-2	DEMETON METHYL
8065-48-3	DEMETON-O + DEMETON-S
919-86-8	DEMETON-S-METHYL
919-86-8	DEMETON-S-METILE (ITALIAN)
60-51-5	DEMOS-L40
8065-48-3	DEMOX
67-68-5	DEMSODROX
55-18-5	DEN
55-18-5	DENA
63-25-2	DENAPON
64-17-5	DENATURED ALCOHOL
9002-89-5	DENKA POVAL B 17
9002-89-5	DENKA POVAL G 05
1333-86-4	DENKABLACK
79-01-6	DENSINFLUAT
57-41-0	DENYL
8020-83-5	DEOBASE
633-03-4	DEORLENE GREEN JJO
7681-52-9	DEOSAN
50-29-3	DEOVAL
52-68-6	DEP
115-76-4	DEP
52-68-6	DEP (PESTICIDE)
2497-07-6	DEPD
52-68-6	DEPTHON
2636-26-2	DERIBAN
83-79-4	DERIL
16071-86-6	DERMA FAST BROWN W-GL
16071-86-6	DERMAFIX BROWN PL
299-84-3	DERMAFOS
557-05-1	DERMARONE
67-68-5	DERMASORB
470-90-6	DERMATON
1333-86-4	DERMMAPOL BLACK G
2636-26-2	DERRIBANTE
83-79-4	DERRIN
83-79-4	DERRIS
83-79-4	DERRIS (INSECTICIDE)
56-53-1	DES
62-73-7	DES
64-67-5	DES
56-53-1	DES (SYNTHETIC ESTROGEN)
333-41-5	DESAPON
14807-96-6	DESERTALC 57
1305-78-8	DESICAL P
7778-39-4	DESICCANT L-10
88-85-7	DESICOIL
66-75-1	DESMETHYLDOPAN
101-68-8	DESMODUR 44
9016-87-9	DESMODUR PU 1520A20
26471-62-5	DESMODUR T100
26471-62-5	DESMODUR T80
25322-69-4	DESMOPHEN 360C
7775-09-9	DESOLET

CAS No.	Chemical Name
94-75-7	DESORMONE
79-21-0	DESOXON 1
53-16-7	DESTRONE
16893-85-9	DESTRUXOL APPLEX
107-06-2	DESTRUXOL BORER-SOL
54-11-5	DESTRUXOL ORCHID SPRAY
111-40-0	DETA
534-52-1	DETAL
151-21-3	DETERGENT 66
123-01-3	DETERGENT ALKYLATE
123-01-3	DETERGENT ALKYLATE NO 2
25155-30-0	DETERGENT HD-90
25155-30-0	DETERLON
52-68-6	DETF
81-81-2	DETHMOR
81-81-2	DETHNEL
20859-73-8	DETIA
20859-73-8	DETIA GAS EX-B
20859-73-8	DETIA-EX-B
121-75-5	DETMOL MA
121-75-5	DETMOL MA 96%
2921-88-2	DETMOL UA
58-89-9	DETMOL-EXTRAKT
50-29-3	DETOX
58-89-9	DETOX 25
50-29-3	DETOXAN
56-75-7	DETREOMYCIN
7782-39-0	DEUTERIUM
7782-39-0	DEUTERIUM (D2)
7782-39-0	DEUTERIUM (DOT)
7782-39-0	DEUTERIUM MOLECULE
95-79-4	DEVAL RED K
95-79-4	DEVAL RED TR
108-45-2	DEVELOPER 11
108-45-2	DEVELOPER C
108-45-2	DEVELOPER H
108-45-2	DEVELOPER M
108-46-3	DEVELOPER O
100-01-6	DEVELOPER P
29757-24-2	DEVELOPER P
106-50-3	DEVELOPER PF
108-46-3	DEVELOPER R
108-46-3	DEVELOPER RS
63-25-2	DEVICARB
60-51-5	DEVIGON
2636-26-2	DEVIKOL
75-99-0	DEVIPON
115-29-7	DEVISULPHAN
7704-34-9	DEVISULPHUR
298-00-0	DEVITHION
72-33-3	DEVOCIN
88-74-4	DEVOL ORANGE B
88-74-4	DEVOL ORANGE SALT B
89-62-3	DEVOL RED G
100-01-6	DEVOL RED GG
29757-24-2	DEVOL RED GG
3165-93-3	DEVOL RED K
89-62-3	DEVOL RED SALT G
3165-93-3	DEVOL RED TR SALT
3165-93-3	DEVOL RED TR
99-55-8	DEVOL SCARLET B
99-55-8	DEVOL SCARLET G SALT
1332-58-7	DEVOLITE
58-89-9	DEVORAN
79-20-9	DEVOTON
9004-66-4	DEXTROFER 100
9004-66-4	DEXTROFER 75
85-00-7	DEXTRONE
91-20-3	DEZODORATOR
9002-84-0	DF 100
122-39-4	DFA
1333-86-4	DG 100
2238-07-5	DGE
2238-07-5	DGE (DIGLYCIDYL ETHER)
9004-70-0	DHX 30/50
1758-61-8	DI(1-HYDROXY CYCLOHEXYL)PEROXIDE
2238-07-5	DI(2,3-EPOXY)PROPYL ETHER
111-91-1	DI(2-CHLOROETHOXY)METHANE
111-44-4	DI(2-CHLOROETHYL) ETHER

CAS No.	Chemical Name
494-03-1	DI(2-CHLOROETHYL)-BETA-NAPHTHYLAMINE
51-75-2	DI(2-CHLOROETHYL)METHYLAMINE
16111-62-9	DI(2-ETHYLHEXYL) PEROXYDICARBONATE
106-20-7	DI(2-ETHYLHEXYL)AMINE
298-07-7	DI(2-ETHYLHEXYL)ORTHOPHOSPHORIC ACID
117-84-0	DI(2-ETHYLHEXYL)ORTHOPHTHALATE
16111-62-9	DI(2-ETHYLHEXYL)PEROXYDICARBONATE
16111-62-9	DI(2-ETHYLHEXYL)PEROXYDICARBONATE, 77% IN SOLUTION (DOT)
298-07-7	DI(2-ETHYLHEXYL)PHOSPHATE
298-07-7	DI(2-ETHYLHEXYL)PHOSPHORIC ACID
298-07-7	DI(2-ETHYLHEXYL)PHOSPHORIC ACID (DOT)
117-84-0	DI(2-ETHYLHEXYL)PHTHALATE
117-81-7	DI(2-ETHYLHEXYL)PHTHALATE
577-11-7	DI(2-ETHYLHEXYL)SULFOSUCCINATE SODIUM SALT
111-48-8	DI(2-HYDROXYETHYL) SULFIDE
111-42-2	DI(2-HYDROXYETHYL)AMINE
3034-79-5	DI(2-METHYLBENZOYL)PEROXIDE
101-68-8	DI(4-ISOCYANATOPHENYL)METHANE
15520-11-3	DI(4-TERT-BUTYLCYCLOHEXYL) PEROXYDICARBONATE
111-44-4	DI(BETA-CHLOROETHYL) ETHER
117-81-7	DI(ETHYLHEXYL) PHTHALATE
84-74-2	DI(N-BUTYL) 1,2-BENZENEDICARBOXYLATE
139-65-1	DI(P-AMINOPHENYL) SULFIDE
72-43-5	DI(P-METHOXYPHENYL)-TRICHLOROMETHYL METHANE
3689-24-5	DI(THIOPHOSPHORIC) ACID, TETRAETHYL ESTER
2407-94-5	DI-(1-HYDROXYCYCLOHEXYL) PEROXIDE, TECHNICALLY PURE
2407-94-5	DI-(1-HYDROXYCYCLOHEXYL)PEROXIDE, TECHNICAL PURE (DOT)
16111-62-9	DI-(2-ETHYLHEXYL) PEROXYDICARBONATE, TECHNICALLY PURE
577-11-7	DI-(2-ETHYLHEXYL) SODIUM SULFOSUCCINATE
16111-62-9	DI-(2-ETHYLHEXYL)PEROXYDICARBONATE, MAXIMUM CONCENTRATION 32% (DOT)
16111-62-9	DI-(2-ETHYLHEXYL)PEROXYDICARBONATE, TECHNICAL PURE (DOT)
101-77-9	DI-(4-AMINOPHENYL)METHANE
94-17-7	DI-(4-CHLOROBENZOYL) PEROXIDE, NOT MORE THAN 75% WITH WATER (DOT)
15520-11-3	DI-(4-TERT-BUTYLCYCLOHEXYL)PEROXYDICARBONATE, NOT MORE THAN 42% IN WATER (DOT)
15520-11-3	DI-(4-TERT-BUTYLCYCLOHEXYL)PEROXYDICARBONATE, TECHNICAL PURE (DOT)
15520-11-3	DI-(4-TERT-BUTYLCYCLOHEXYL)PEROXYDICARBONATE, TECHNICALLY PURE
115-32-2	DI-(P-CHLOROPHENYL)TRICHLOROMETHYLCARBINOL
102-54-5	DI-PI-CYCLOPENTADIENYL IRON
1271-28-9	DI-PI-CYCLOPENTADIENYLNICKEL
56-18-8	DI-1,2-PROPANETRIAMINE
102-54-5	DI-2,4-CYCLOPENTADIEN-1-YLIRON
133-14-2	DI-2,4-DICHLOROBENZOYL PEROXIDE, MAXIMUM CONCENTRATION 52% AS A PASTE OR IN SOLUTION (DOT)
626-23-3	DI-2-BUTYLAMINE
111-91-1	DI-2-CHLOROETHYL FORMAL
505-60-2	DI-2-CHLOROETHYL SULFIDE
103-23-1	DI-2-ETHYLHEXYL ADIPATE
123-79-5	DI-2-ETHYLHEXYL ADIPATE
298-07-7	DI-2-ETHYLHEXYL HYDROGEN PHOSPHATE
16111-62-9	DI-2-ETHYLHEXYL PEROXYDICARBONATE
80-43-3	DI-ALPHA-CUMYL PEROXIDE
112-27-6	DI-BETA-HYDROXYETHOXYETHANE
106-46-7	DI-CHLORICIDE
80-43-3	DI-CUP
80-43-3	DI-CUP 40C
80-43-3	DI-CUP 40HAF
80-43-3	DI-CUP 40KE
80-43-3	DI-CUP R
80-43-3	DI-CUP T
56-53-1	DI-ESTRYL
57-41-0	DI-HYDAN
26471-62-5	DI-ISO-CYANATOLUENE
108-83-8	DI-ISOBUTYLCETONE (FRENCH)
25167-70-8	DI-ISOBUTYLENE
26471-62-5	DI-ISOCYANATE DE TOLUYLENE (FRENCH)
57-41-0	DI-LAN
10210-68-1	DI-MU-CARBONYLHEXACARBONYLDICOBALT
2050-92-2	DI-N-AMYLAMINE
2050-92-2	DI-N-AMYLAMINE (DOT)
142-96-1	DI-N-BUTYL ETHER
142-96-1	DI-N-BUTYL ETHER (DOT)

CAS No.	*Chemical Name*
1809-19-4	DI-N-BUTYL HYDROGEN PHOSPHITE
107-66-4	DI-N-BUTYL PHOSPHATE
84-74-2	DI-N-BUTYL PHTHALATE
111-92-2	DI-N-BUTYLAMINE
2109-64-0	DI-N-BUTYLAMINO-2-PROPANOL
3648-21-3	DI-N-HEPTYL PHTHALATE
112-58-3	DI-N-HEXYL ETHER
143-16-8	DI-N-HEXYLAMINE
101-25-7	DI-N-NITROSOPENTAMETHYLENETETRAMINE
762-13-0	DI-N-NONANOYL PEROXIDE
762-13-0	DI-N-NONANOYL PEROXIDE, TECHNICALLY PURE (DOT)
762-16-3	DI-N-OCTANOYL PEROXIDE
762-16-3	DI-N-OCTANOYL PEROXIDE, TECHNICAL PURE (DOT)
103-23-1	DI-N-OCTYL ADIPATE
123-79-5	DI-N-OCTYL ADIPATE
117-84-0	DI-N-OCTYL PHTHALATE
1120-48-5	DI-N-OCTYLAMINE
872-10-6	DI-N-PENTYL SULFIDE
2050-92-2	DI-N-PENTYLAMINE
111-43-3	DI-N-PROPYL ETHER
123-19-3	DI-N-PROPYL KETONE
16066-38-9	DI-N-PROPYL PEROXYDICARBONATE
16066-38-9	DI-N-PROPYL PEROXYDICARBONATE (CONCENTRATION <80%)
16066-38-9	DI-N-PROPYL PEROXYDICARBONATE, TECHNICALLY PURE
16066-38-9	DI-N-PROPYL PEROXYDICARBONATE, TECHNICALLY PURE (DOT)
142-84-7	DI-N-PROPYLAMINE
142-59-6	DI-NATRIUM-AETHYLENBISDITHIOCARBAMAT (GERMAN)
330-54-1	DI-ON
19910-65-7	DI-SEC-BUTYL PEROXYDICARBONATE
19910-65-7	DI-SEC-BUTYL PEROXYDICARBONATE (CONCENTRATION <80%)
19910-65-7	DI-SEC-BUTYL PEROXYDICARBONATE :MAXIMUM CONCENTRATION 52% IN SOLUTION
626-23-3	DI-SEC-BUTYLAMINE
117-84-0	DI-SEC-OCTYL PHTHALATE
117-84-0	DI-SEC-OCTYL PHTHALATE (ACGIH)
298-04-4	DI-SYSTON
298-04-4	DI-SYSTON G
110-05-4	DI-T-BUTYL PEROXIDE
1189-85-1	DI-TERT-BUTOXYCHROMYL
1189-85-1	DI-TERT-BUTYL CHROMATE
110-05-4	DI-TERT-BUTYL PEROXIDE
110-05-4	DI-TERT-BUTYL PEROXIDE, TECHNICALLY PURE
110-05-4	DI-TERT-BUTYL PEROXIDE, TECHNICALLY PURE (DOT)
110-05-4	DI-TERT-BUTYL PEROXYDE (DUTCH)
128-37-0	DI-TERT-BUTYL-4-METHYLPHENOL
128-37-0	DI-TERT-BUTYL-P-CRESOL
128-37-0	DI-TERT-BUTYLCRESOL
15520-11-3	DI-TERT-BUTYLDICYCLOHEXYL PEROXYDICARBONATE
110-05-4	DI-TERT-BUTYLPEROXID (GERMAN)
2155-71-7	DI-TERT-BUTYLPEROXYPHTHALATE, MAXIMUM 55% IN PASTE
2155-71-7	DI-TERT-BUTYLPEROXYPHTHALATE, MAXIMUM 55% IN SOLUTION
2155-71-7	DI-TERT-BUTYLPEROXYPHTHALATE, TECHNICAL PURE
99-55-8	DIABASE SCARLET G
1333-86-4	DIABLACK G
119-90-4	DIACEL NAVY DC
141-97-9	DIACETIC ETHER
123-42-2	DIACETONALCOHOL (DUTCH)
123-42-2	DIACETONALCOOL (ITALIAN)
123-42-2	DIACETONALKOHOL (GERMAN)
123-42-2	DIACETONE
123-42-2	DIACETONE ALCOHOL
123-42-2	DIACETONE ALCOHOL (ACGIH,DOT)
54693-46-8	DIACETONE ALCOHOL PEROXIDE
54693-46-8	DIACETONE ALCOHOL PEROXIDE, >57% IN SOLUTION WITH >9% HYDROGEN PEROXIDE, <26% DIACETONE ALCOHO
54693-46-8	DIACETONE ALCOHOL PEROXIDE, MORE THAN 57% IN SOLUTION (DOT)
54693-46-8	DIACETONE ALCOHOL PEROXIDE, NOT MORE THAN 57% IN SOLUTION (DOT)
123-42-2	DIACETONE ALCOHOL [COMBUSTIBLE LIQUID LABEL]
123-42-2	DIACETONE ALCOHOL [FLAMMABLE LIQUID LABEL]
123-42-2	DIACETONE-ALCOHOL (FRENCH)
1600-27-7	DIACETOXYMERCURY
431-03-8	DIACETYL
110-22-5	DIACETYL PEROXIDE
110-22-5	DIACETYL PEROXIDE (SOLUTION)
613-35-4	DIACETYLBENZIDINE
123-54-6	DIACETYLMETHANE

CAS No.	*Chemical Name*
2602-46-2	DIACOTTON BLUE BB
1937-37-7	DIACOTTON DEEP BLACK
1937-37-7	DIACOTTON DEEP BLACK RX
111-42-2	DIAETHANOLAMIN (GERMAN)
105-57-7	DIAETHYLACETAL (GERMAN)
60-29-7	DIAETHYLAETHER (GERMAN)
109-89-7	DIAETHYLAMIN (GERMAN)
100-37-8	DIAETHYLAMINOAETHANOL (GERMAN)
91-66-7	DIAETHYLANILIN (GERMAN)
105-58-8	DIAETHYLCARBONAT (GERMAN)
123-31-9	DIAK 5
50-70-4	DIAKARMON
80-62-6	DIAKON
10311-84-9	DIALIFOS
2303-16-4	DIALLATE
557-40-4	DIALLYL ETHER
131-17-9	DIALLYL PHTHALATE
124-02-7	DIALLYLAMINE
16071-86-6	DIALUMINOUS BROWN BRS
10043-01-3	DIALUMINUM SULFATE
1344-28-1	DIALUMINUM TRIOXIDE
10043-01-3	DIALUMINUM TRISULFATE
101-14-4	DIAMET KH
302-01-2	DIAMIDE
302-01-2	DIAMINE
2602-46-2	DIAMINE BLUE 2B
72-57-1	DIAMINE BLUE 3B
2602-46-2	DIAMINE BLUE BB
1937-37-7	DIAMINE DEEP BLACK EC
1937-37-7	DIAMINE DIRECT BLACK E
72-57-1	DIAMINEBLUE
101-80-4	DIAMINODIPHENYL ETHER
101-77-9	DIAMINODIPHENYL METHANE
107-15-3	DIAMINOETHANE
25376-45-8	DIAMINOTOLUENE
30143-13-6	DIAMINOTOLUENE
25376-45-8	DIAMINOTOLUENE (MIXED ISOMERS)
1596-84-5	DIAMINOZIDE
7783-28-0	DIAMMONIUM ACID PHOSPHATE
7784-44-3	DIAMMONIUM ARSENATE
506-87-6	DIAMMONIUM CARBONATE
7788-98-9	DIAMMONIUM CHROMATE
7788-98-9	DIAMMONIUM CHROMATE ((NH4)2CrO4)
3012-65-5	DIAMMONIUM CITRATE
3164-29-2	DIAMMONIUM D-TARTRATE
7789-09-5	DIAMMONIUM DICHROMATE
10045-89-3	DIAMMONIUM FERROUS DISULFATE
16919-19-0	DIAMMONIUM FLUOSILICATE ((NH4)2SiF6)
16919-58-7	DIAMMONIUM HEXACHLOROPLATINATE
16919-58-7	DIAMMONIUM HEXACHLOROPLATINATE(2-)
16919-19-0	DIAMMONIUM HEXAFLUOROSILICATE
16919-19-0	DIAMMONIUM HEXAFLUOROSILICATE(2-)
7784-44-3	DIAMMONIUM HYDROGEN ARSENATE
3012-65-5	DIAMMONIUM HYDROGEN CITRATE
7783-28-0	DIAMMONIUM HYDROGEN ORTHOPHOSPHATE
7783-28-0	DIAMMONIUM HYDROGEN PHOSPHATE
10045-89-3	DIAMMONIUM IRON DISULFATE
3164-29-2	DIAMMONIUM L-(+)-TARTRATE
13106-76-8	DIAMMONIUM MOLYBDATE
13106-76-8	DIAMMONIUM MOLYBDATE ((NH4)2MoO4)
7784-44-3	DIAMMONIUM MONOHYDROGEN ARSENATE
7783-28-0	DIAMMONIUM MONOHYDROGEN PHOSPHATE
7783-28-0	DIAMMONIUM ORTHOPHOSPHATE
1113-38-8	DIAMMONIUM OXALATE
1113-38-8	DIAMMONIUM OXALATE MONOHYDRATE
6009-70-7	DIAMMONIUM OXALATE MONOHYDRATE
7727-54-0	DIAMMONIUM PEROXYDISULFATE
7727-54-0	DIAMMONIUM PEROXYDISULPHATE
7727-54-0	DIAMMONIUM PERSULFATE
7783-28-0	DIAMMONIUM PHOSPHATE
16919-58-7	DIAMMONIUM PLATINUM HEXACHLORIDE
9080-17-5	DIAMMONIUM POLYSULFIDE
16919-19-0	DIAMMONIUM SILICON HEXAFLUORIDE
7783-20-2	DIAMMONIUM SULFATE
12135-76-1	DIAMMONIUM SULFIDE
10196-04-0	DIAMMONIUM SULFITE
7783-20-2	DIAMMONIUM SULPHATE
3164-29-2	DIAMMONIUM TARTRATE
14639-97-5	DIAMMONIUM TETRACHLOROZINCATE

CAS No.	Chemical Name
14639-97-5	DIAMMONIUM TETRACHLOROZINCATE(2-)
7783-18-8	DIAMMONIUM THIOSULFATE
9080-17-5	DIAMMONIUM TRISULFIDE
633-03-4	DIAMOND GREEN G
39196-18-4	DIAMOND SHAMROCK DS-15647
2050-92-2	DIAMYL AMINE
872-10-6	DIAMYL SULFIDE
3129-91-7	DIANA
1918-00-9	DIANAT
1918-00-9	DIANAT (RUSSIAN)
1918-00-9	DIANATE
101-77-9	DIANILINEMETHANE
101-77-9	DIANILINO METHANE
119-90-4	DIANISIDINE
72-43-5	DIANISYLTRICHLORETHANE
333-41-5	DIANON
1309-64-4	DIANTIMONY TRIOXIDE
141-66-2	DIAPADRIN
1937-37-7	DIAPHTAMINE BLACK V
2602-46-2	DIAPHTAMINE BLUE BB
72-57-1	DIAPHTAMINE BLUE TH
16071-86-6	DIAPHTAMINE LIGHT BROWN BRLL
100-42-5	DIAREX HF 77
13464-44-3	DIARSENIC ACID, ZINC SALT (1:2)
1303-28-2	DIARSENIC PENTOXIDE
1327-53-3	DIARSENIC TRIOXIDE
1303-33-9	DIARSENIC TRISULFIDE
1303-33-9	DIARSENIC TRISULPHIDE
1303-28-2	DIARSONIC PENTOXIDE
9000-92-4	DIASTASE
330-54-1	DIATER
333-41-5	DIATERR-FOS
105-58-8	DIATOL
61790-53-2	DIATOMACEOUS EARTH
7726-95-6	DIATOMIC BROMINE
7782-50-5	DIATOMIC CHLORINE
7727-37-9	DIATOMIC NITROGEN
280-57-9	DIAZABICYCLOOCTANE
333-41-5	DIAZAJET
333-41-5	DIAZATOL
439-14-5	DIAZEPAM
333-41-5	DIAZIDE
1937-37-7	DIAZINE BLACK E
2602-46-2	DIAZINE BLUE 2B
72-57-1	DIAZINE BLUE 3B
1937-37-7	DIAZINE DIRECT BLACK E
1937-37-7	DIAZINE DIRECT BLACK G
16071-86-6	DIAZINE FAST BROWN RSL
333-41-5	DIAZINON
333-41-5	DIAZINON (ACGIH,DOT)
333-41-5	DIAZINON AG 500
333-41-5	DIAZINONE
334-88-3	DIAZIRINE
333-41-5	DIAZITOL
88-74-4	DIAZO FAST ORANGE GR
100-01-6	DIAZO FAST RED GG
29757-24-2	DIAZO FAST RED GG
89-62-3	DIAZO FAST RED GL
3165-93-3	DIAZO FAST RED TR
95-79-4	DIAZO FAST RED TRA
3165-93-3	DIAZO FAST RED TRA
99-55-8	DIAZO FAST SCARLET G
87-31-0	DIAZODINITROPHENOL
333-41-5	DIAZOL
1937-37-7	DIAZOL BLACK 2V
2602-46-2	DIAZOL BLUE 2B
72-57-1	DIAZOL BLUE 3B
16071-86-6	DIAZOL LIGHT BROWN BRN
334-88-3	DIAZOMETHANE
334-88-3	DIAZONIUM METHYLIDE
7632-00-0	DIAZOTIZING SALTS
1191-15-7	DIBAL
7784-44-3	DIBASIC AMMONIUM ARSENATE
3012-65-5	DIBASIC AMMONIUM CITRATE
7783-28-0	DIBASIC AMMONIUM PHOSPHATE
301-04-2	DIBASIC LEAD ACETATE
1344-40-7	DIBASIC LEAD METAPHOSPHATE
1344-40-7	DIBASIC LEAD PHOSPHITE
56189-09-4	DIBASIC LEAD STEARATE

CAS No.	Chemical Name
7558-79-4	DIBASIC SODIUM PHOSPHATE
557-05-1	DIBASIC ZINC STEARATE
108-83-8	DIBC
59-96-1	DIBENYLIN
59-96-1	DIBENYLINE
226-36-8	DIBENZ(A,H)ACRIDINE
53-70-3	DIBENZ(A,H)ANTHRACENE
224-42-0	DIBENZ(A,J)ACRIDINE
192-65-4	DIBENZO(A,E)PYRENE
189-64-0	DIBENZO(A,H)PYRENE
189-55-9	DIBENZO(A,I)PYRENE
260-94-6	DIBENZO(B,E)PYRIDINE
92-84-2	DIBENZO-1,4-THIAZINE
1746-01-6	DIBENZO-P-DIOXIN, 2,3,7,8-TETRACHLORO-
132-64-9	DIBENZOFURAN
92-84-2	DIBENZOTHIAZINE
94-36-0	DIBENZOYL PEROXIDE
94-36-0	DIBENZOYLPEROXYDE (DUTCH)
94-36-0	DIBENZOYLPEROXID (GERMAN)
53-70-3	DIBENZO[A,H]ANTHRACENE
205-82-3	DIBENZO[A,JK]FLUORENE
189-64-0	DIBENZO[B,DEF]CHRYSENE
1746-01-6	DIBENZO[B,E][1,4]DIOXIN, 2,3,7,8-TETRACHLORO-
189-55-9	DIBENZO[B,H]PYRENE
207-08-9	DIBENZO[B,JK]FLUORENE
2144-45-8	DIBENZYL PEROXYDICARBONATE
2144-45-8	DIBENZYL PEROXYDICARBONATE (CONCENTRATION ≤90%)
2144-45-8	DIBENZYL PEROXYDICARBONATE (DOT)
2144-45-8	DIBENZYL PEROXYDICARBONATE, MAXIMUM CONCENTRATION 87% WITH WATER (DOT)
18414-36-3	DIBENZYLDICHLOROSILANE
59-96-1	DIBENZYLINE
63-92-3	DIBENZYLINE CHLORIDE
63-92-3	DIBENZYLINE HYDROCHLORIDE
226-36-8	DIBENZ[A,D]ACRIDINE
224-42-0	DIBENZ[A,F]ACRIDINE
15191-85-2	DIBERYLLIUM MONOSILICATE
56-53-1	DIBESTROL 2 PREMIX
1304-82-1	DIBISMUTH TRITELLURIDE
108-83-8	DIBK
19287-45-7	DIBORANE
19287-45-7	DIBORANE (ACGIH,DOT)
19287-45-7	DIBORANE (B2H6)
19287-45-7	DIBORANE(6)
19287-45-7	DIBORON HEXAHYDRIDE
1303-86-2	DIBORON TRIOXIDE
50-29-3	DIBOVIN
300-76-5	DIBROM
96-12-8	DIBROMCHLORPROPAN (GERMAN)
300-76-5	DIBROMFOS
7726-95-6	DIBROMINE
26249-12-7	DIBROMOBENZENE
26249-12-7	DIBROMOBENZENE (DOT)
3479-86-5	DIBROMOBUTANONE
124-48-1	DIBROMOCHLOROMETHANE
96-12-8	DIBROMOCHLOROPROPANE
75-61-6	DIBROMODIFLUOROMETHANE
106-93-4	DIBROMOETHANE
7789-47-1	DIBROMOMERCURY
124-48-1	DIBROMOMONOCHLOROMETHANE
106-93-4	DIBROMURE D' ETHYLENE (FRENCH)
128-37-0	DIBUNOL
88-85-7	DIBUTOX
112-98-1	DIBUTOXY TETRAGLYCOL
1809-19-4	DIBUTOXYPHOSPHINE OXIDE
84-74-2	DIBUTYL 1,2-BENZENEDICARBOXYLATE
107-66-4	DIBUTYL ACID PHOSPHATE
84-74-2	DIBUTYL ESTER PHTHALIC ACID
142-96-1	DIBUTYL ETHER
107-66-4	DIBUTYL HYDROGEN PHOSPHATE
109-47-7	DIBUTYL HYDROGEN PHOSPHITE
1809-19-4	DIBUTYL HYDROGEN PHOSPHITE
84-74-2	DIBUTYL O-PHTHALATE
142-96-1	DIBUTYL OXIDE
107-66-4	DIBUTYL PHOSPHATE
109-47-7	DIBUTYL PHOSPHITE
1809-19-4	DIBUTYL PHOSPHITE
1809-19-4	DIBUTYL PHOSPHONATE
84-74-2	DIBUTYL PHTHALATE

CAS No.	Chemical Name
84-74-2	DIBUTYL PHTHALATE (ACGIH)
111-92-2	DIBUTYLAMINE
924-16-3	DIBUTYLAMINE, N-NITROSO-
102-81-8	DIBUTYLAMINOETHANOL
613-29-6	DIBUTYLANILINE
128-37-0	DIBUTYLATED HYDROXYTOLUENE
102-81-8	DIBUTYLETHANOLAMINE
4835-11-4	DIBUTYLHEXAMETHYLENEDIAMINE
2109-64-0	DIBUTYLISOPROPANOLAMINE
924-16-3	DIBUTYLNITROSAMINE
1067-33-0	DIBUTYLTIN DIACETATE
1918-00-9	DICAMBA
1918-00-9	DICAMBA (DOT)
762-16-3	DICAPRYLYL PEROXIDE
63-25-2	DICARBAM
121-75-5	DICARBOETHOXYETHYL O,O-DIMETHYL PHOSPHORODITHIOATE
557-34-6	DICARBOMETHOXYZINC
506-87-6	DICARBONIC ACID, DIAMMONIUM SALT
1642-54-2	DICAROCIDE
26322-14-5	DICETYL PEROXYDICARBONATE
3129-91-7	DICHAN
3129-91-7	DICHAN (CZECH)
23745-86-0	DICHAPETULUM CYMOSUM (HOOK) ENGL
1194-65-6	DICHLOBENIL
62-73-7	DICHLOFOS
117-80-6	DICHLONE
2636-26-2	DICHLOORVO (DUTCH)
78-87-5	DICHLOR
107-06-2	DICHLOR-MULSION
2244-21-5	DICHLOR-S-TRIAZIN-2,4,6(1H,3H,5H)TRIONE POTASSIUM
79-43-6	DICHLORACETIC ACID
79-36-7	DICHLORACETYL CHLORIDE
51-75-2	DICHLORAMINE
118-52-5	DICHLORANTIN
107-06-2	DICHLOREMULSION
55-86-7	DICHLOREN
51-75-2	DICHLOREN (GERMAN)
79-43-6	DICHLORETHANOIC ACID
330-54-1	DICHLORFENIDIM
2636-26-2	DICHLORFOS (POLISH)
627-63-4	DICHLORID KYSELINY FUMAROVE (CZECH)
7782-50-5	DICHLORINE
7791-21-1	DICHLORINE MONOXIDE
7791-21-1	DICHLORINE OXIDE
62-73-7	DICHLORMAN
2636-26-2	DICHLORMAN
79-36-7	DICHLORO ACETYL CHLORIDE (DOT)
50-29-3	DICHLORO DIPHENYL TRICHLOROETHANE
541-25-3	DICHLORO(2-CHLOROVINYL)ARSINE
622-44-6	DICHLORO(PHENYLIMINO)METHANE
14684-25-4	DICHLORO(PHENYLTHIO)PHOSPHINE
107-06-2	DICHLORO-1,2-ETHANE (FRENCH)
540-59-0	DICHLORO-1,2-ETHYLENE (FRENCH)
2244-21-5	DICHLORO-S-TRIAZINE-2,4,6(1H,3H,5H)-TRIONE POTASSIUM
2244-21-5	DICHLORO-S-TRIAZINE-2,4,6(1H,3H,5H)-TRIONE POTASSIUM DERIV
2782-57-2	DICHLORO-S-TRIAZINETRIONE
79-43-6	DICHLOROACETIC ACID
79-43-6	DICHLOROACETIC ACID (DOT)
79-36-7	DICHLOROACETIC ACID CHLORIDE
79-36-7	DICHLOROACETYL CHLORIDE
7572-29-4	DICHLOROACETYLENE
27134-27-6	DICHLOROANILINE
27134-27-6	DICHLOROANILINES (DOT)
8023-53-8	DICHLOROBENZALKONIUM CHLORIDE
25321-22-6	DICHLOROBENZENE
25321-22-6	DICHLOROBENZENE (MIXED ISOMERS)
95-50-1	DICHLOROBENZENE, ORTHO, LIQUID
95-50-1	DICHLOROBENZENE, ORTHO, LIQUID (DOT)
106-46-7	DICHLOROBENZENE, PARA, SOLID
106-46-7	DICHLOROBENZENE, PARA, SOLID (DOT)
91-94-1	DICHLOROBENZIDINE
75-27-4	DICHLOROBROMOMETHANE
1653-19-6	DICHLOROBUTADIENE
11069-19-5	DICHLOROBUTENE
10108-64-2	DICHLOROCADMIUM
12789-03-6	DICHLOROCHLORDENE
2782-57-2	DICHLOROCYANURIC ACID
1719-53-5	DICHLORODIETHYLSILANE
27156-03-2	DICHLORODIFLUORO-ETHYLENE

CAS No.	Chemical Name
75-71-8	DICHLORODIFLUOROMETHANE
75-71-8	DICHLORODIFLUOROMETHANE (ACGIH,DOT)
63283-80-7	DICHLORODIISOPROPYL ETHER
542-88-1	DICHLORODIMETHYL ETHER
75-78-5	DICHLORODIMETHYLSILANE
75-78-5	DICHLORODIMETHYLSILICON
14977-61-8	DICHLORODIOXOCHROMIUM
72-54-8	DICHLORODIPHENYL DICHLOROETHANE
50-29-3	DICHLORODIPHENYL TRICHLOROETHANE
80-10-4	DICHLORODIPHENYLSILANE
10025-67-9	DICHLORODISULFANE
79-43-6	DICHLOROETHANOIC ACID
79-36-7	DICHLOROETHANOYL CHLORIDE
111-44-4	DICHLOROETHER
111-44-4	DICHLOROETHYL ETHER
111-44-4	DICHLOROETHYL ETHER (ACGIH,DOT)
111-44-4	DICHLOROETHYL OXIDE
563-43-9	DICHLOROETHYLALUMINUM
107-06-2	DICHLOROETHYLENE
156-59-2	DICHLOROETHYLENE-CIS
156-60-5	DICHLOROETHYLENE-TRANS
1789-58-8	DICHLOROETHYLSILANE
7572-29-4	DICHLOROETHYNE
75-43-4	DICHLOROFLUOROMETHANE
75-43-4	DICHLOROFLUOROMETHANE (ACGIH)
75-54-7	DICHLOROHYDRIDOMETHYLSILICON
2782-57-2	DICHLOROISOCYANURATE
2782-57-2	DICHLOROISOCYANURIC ACID
2244-21-5	DICHLOROISOCYANURIC ACID POTASSIUM SALT
2244-21-5	DICHLOROISOCYANURIC ACID POTASSIUM SALT (DOT)
2893-78-9	DICHLOROISOCYANURIC ACID SODIUM SALT
2782-57-2	DICHLOROISOCYANURIC ACID, DRY
2782-57-2	DICHLOROISOCYANURIC ACID, DRY (DOT)
63283-80-7	DICHLOROISOPROPYL ETHER
63283-80-7	DICHLOROISOPROPYL ETHER (DOT)
115-32-2	DICHLOROKELTHANE
7487-94-7	DICHLOROMERCURY
75-09-2	DICHLOROMETHANE
75-09-2	DICHLOROMETHANE (DOT)
542-88-1	DICHLOROMETHYL ETHER
75-34-3	DICHLOROMETHYLMETHANE
149-74-6	DICHLOROMETHYLPHENYLSILANE
676-83-5	DICHLOROMETHYLPHOSPHINE
676-97-1	DICHLOROMETHYLPHOSPHINE OXIDE
676-98-2	DICHLOROMETHYLPHOSPHINE SULFIDE
75-54-7	DICHLOROMETHYLSILANE
75-27-4	DICHLOROMONOBROMOMETHANE
563-43-9	DICHLOROMONOETHYLALUMINUM
75-43-4	DICHLOROMONOFLUOROMETHANE
75-43-4	DICHLOROMONOFLUOROMETHANE (DOT)
7791-21-1	DICHLOROMONOXIDE
7699-43-6	DICHLOROOXOZIRCONIUM
30586-10-8	DICHLOROPENTANE
94-75-7	DICHLOROPHENOXYACETIC ACID
27137-85-5	DICHLOROPHENYL TRI-CHLOROSILANE
102-36-3	DICHLOROPHENYL-ISOCYANATE (1,2-DICHLORO-4-ISOCYANATO-BENZENE)
39920-37-1	DICHLOROPHENYL-ISOCYANATE (1,3-DICHLORO-2-ISOCYANATO-BENZENE)
34893-92-0	DICHLOROPHENYL-ISOCYANATE (1,3-DICHLORO-5-ISOCYANATO-BENZENE)
5392-82-5	DICHLOROPHENYL-ISOCYANATE (1,4-DICHLORO-2-ISOCYANATO-BENZENE)
2612-57-9	DICHLOROPHENYL-ISOCYANATE (2,4-DICHLORO-1-ISOCYANATO-BENZENE)
696-28-6	DICHLOROPHENYLARSINE
98-87-3	DICHLOROPHENYLMETHANE
644-97-3	DICHLOROPHENYLPHOSPHINE
2636-26-2	DICHLOROPHOS
1498-51-7	DICHLOROPHOSPHORIC ACID, ETHYL ESTER
534-07-6	DICHLOROPROPANE
542-75-6	DICHLOROPROPENE
26952-23-8	DICHLOROPROPENE
26952-23-8	DICHLOROPROPYLENE
4109-96-0	DICHLOROSILANE
4109-96-0	DICHLOROSILANE (DOT)
10545-99-0	DICHLOROSULFANE
76-14-2	DICHLOROTETRAFLUOROETHANE
1320-37-2	DICHLOROTETRAFLUOROETHANE

CAS No.	Chemical Name
1320-37-2	DICHLOROTETRAFLUOROETHANE (DOT)
463-71-8	DICHLOROTHIOCARBONYL
463-71-8	DICHLOROTHIOFORMALDEHYDE
7772-99-8	DICHLOROTIN
62-73-7	DICHLOROVOS
7791-21-1	DICHLOROXIDE
28347-13-9	DICHLOROXYLYLENE
6341-97-5	DICHLORPHENOXYACETIC ACID ESTER (2,4-DICHLOROACETATE PHENOL)
28165-71-1	DICHLORPHENOXYACETIC ACID ESTER (2,6-DICHLOROACETATE PHENOL)
2636-26-2	DICHLORPHOS
78-87-5	DICHLORPROPEN-GEMISCH (GERMAN)
62-73-7	DICHLORVOS
62-73-7	DICHLORVOS MIXTURE, DRY
13530-68-2	DICHROMIC ACID
13530-68-2	DICHROMIC ACID (H2Cr2O7)
14307-33-6	DICHROMIC ACID (H2Cr2O7), CALCIUM SALT (1:1)
7789-09-5	DICHROMIC ACID (H2Cr2O7), DIAMMONIUM SALT
7778-50-9	DICHROMIC ACID (H2Cr2O7), DIPOTASSIUM SALT
10588-01-9	DICHROMIC ACID (H2Cr2O7), DISODIUM SALT
14018-95-2	DICHROMIC ACID (H2Cr2O7), ZINC SALT (1:1)
7789-09-5	DICHROMIC ACID, DIAMMONIUM SALT
10588-01-9	DICHROMIC ACID, DISODIUM SALT
14018-95-2	DICHROMIC ACID, ZINC SALT (1:1)
13530-68-2	DICHROMIC(VI) ACID
14307-33-6	DICHROMIC(VI) ACID, CALCIUM SALT (1:1)
10101-53-8	DICHROMIUM SULFATE
10101-53-8	DICHROMIUM SULPHATE
10101-53-8	DICHROMIUM TRIS(SULFATE)
10101-53-8	DICHROMIUM TRISULFATE
10101-53-8	DICHROMIUM TRISULPHATE
333-41-5	DICID
598-14-1	DICK (GERMAN)
117-80-6	DICLONE
94-75-7	DICLORDON
10210-68-1	DICOBALT CARBONYL
10210-68-1	DICOBALT CARBONYL (Co2(CO)8)
10210-68-1	DICOBALT OCTACARBONYL
115-32-2	DICOFOL
111-46-6	DICOL
50-29-3	DICOPHANE
94-75-7	DICOPUR
16071-86-6	DICOREL BROWN LMR
94-75-7	DICOTOX
1129-41-5	DICRESYL
141-66-2	DICROTOFOS (DUTCH)
141-66-2	DICROTOPHOS
141-66-2	DICROTOPHOS (ACGIH)
80-43-3	DICUMENE HYDROPEROXIDE
80-43-3	DICUMENYL PEROXIDE
80-43-3	DICUMYL PEROXIDE
80-43-3	DICUMYL PEROXIDE, DRY
80-43-3	DICUMYL PEROXIDE, DRY (DOT)
80-43-3	DICUMYL PEROXIDE, TECHNICAL PURE OR WITH INERT SOLID (DOT)
80-43-3	DICUP 40
97-77-8	DICUPRAL
460-19-5	DICYAN
460-19-5	DICYANOGEN
592-04-1	DICYANOMERCURY
557-19-7	DICYANONICKEL
1561-49-5	DICYCLOHEXYL PEROXIDE CARBONATE
1561-49-5	DICYCLOHEXYL PEROXYDICARBONATE
1561-49-5	DICYCLOHEXYL PEROXYDICARBONATE (DOT)
1561-49-5	DICYCLOHEXYL PEROXYDICARBONATE, NOT MORE THAN 91% WITH WATER (DOT)
1561-49-5	DICYCLOHEXYL PEROXYDICARBONATE, TECHNICALLY PURE
101-83-7	DICYCLOHEXYLAMINE
3129-91-7	DICYCLOHEXYLAMINE, NITRITE
3129-91-7	DICYCLOHEXYLAMINONITRITE
3882-06-2	DICYCLOHEXYLAMMONIUM NITRATE
3129-91-7	DICYCLOHEXYLAMMONIUM NITRITE
5124-30-1	DICYCLOHEXYLMETHANE 4,4'-DIISOCYANATE
77-73-6	DICYCLOPENTADIENE
77-73-6	DICYCLOPENTADIENE (ACGIH,DOT)
76-44-8	DICYCLOPENTADIENE, 3,4,5,6,7,8,8A-HEPTACHLORO-
102-54-5	DICYCLOPENTADIENYL IRON
1271-28-9	DICYCLOPENTADIENYLNICKEL

CAS No.	Chemical Name
3129-91-7	DICYKLOHEXYLAMIN NITRIT (CZECH)
77-73-6	DICYKLOPENTADIEN (CZECH)
3129-91-7	DICYNIT (CZECH)
127-18-4	DIDAKENE
82-66-6	DIDANDIN
762-12-9	DIDECANOYL PEROXIDE
762-12-9	DIDECANOYL PEROXIDE, TECHNICAL PURE
762-12-9	DIDECANOYL PEROXIDE, TECHNICAL PURE (DOT)
82-66-6	DIDION
105-74-8	DIDODECANOYL PEROXIDE
26761-40-0	DIDP
26761-40-0	DIDP (PLASTICIZER)
60-57-1	DIELDREX
60-57-1	DIELDRIN
60-57-1	DIELDRIN (ACGIH,DOT)
60-57-1	DIELDRINE (FRENCH)
60-57-1	DIELDRITE
25322-69-4	DIELECTROL VI
60-57-1	DIELMOTH
84-17-3	DIENOESTROL
1464-53-5	DIEPOXYBUTANE
7440-70-2	DIETARY CALCIUM
109-89-7	DIETHAMINE
111-48-8	DIETHANOL SULFIDE
91-99-6	DIETHANOL-M-TOLUIDINE
111-42-2	DIETHANOLAMIN (CZECH)
111-42-2	DIETHANOLAMINE
111-42-2	DIETHANOLAMINE (ACGIH)
143-00-0	DIETHANOLAMINE LAURYL SULFATE
139-87-7	DIETHANOLETHYLAMINE
105-59-9	DIETHANOLMETHYLAMINE
1116-54-7	DIETHANOLNITROSAMINE
1321-74-0	DIETHENYLBENZENE
563-12-2	DIETHION
8065-48-3	DIETHOXY THIOPHOSPHORIC ACID ESTER OF 2-ETHYLMERCAP-TOETHANOL
78-62-6	DIETHOXYDIMETHYL SILANE
629-14-1	DIETHOXYLETHANE
462-95-3	DIETHOXYMETHANE
462-95-3	DIETHOXYMETHANE (DOT)
21548-32-3	DIETHOXYPHOSPHINYLIMINO-2 DITHIETANNE-1,3 (FRENCH)
814-49-3	DIETHOXYPHOSPHORUS OXYCHLORIDE
814-49-3	DIETHOXYPHOSPHORYL CHLORIDE
3054-95-3	DIETHOXYPROPENE
106-97-8	DIETHYL
950-10-7	DIETHYL (4-METHYL-1,3-DITHIOLAN-2-YLIDENE)PHOSPHORO-AMIDATE
121-75-5	DIETHYL (DIMETHOXYPHOSPHINOTHIOYLTHIO) BUTANEDIOATE
121-75-5	DIETHYL (DIMETHOXYPHOSPHINOTHIOYLTHIO)SUCCINATE
84-66-2	DIETHYL 1,2-BENZENEDICARBOXYLATE
947-02-4	DIETHYL 1,3-DITHIOLAN-2-YLIDENEPHOSPHORAMIDATE
470-90-6	DIETHYL 1-(2,4-DICHLOROPHENYL)-2-CHLOROVINYL PHOSPHATE
470-90-6	DIETHYL 2-CHLORO-1-(2,4-DICHLOROPHENYL)VINYL PHOSPHATE
333-41-5	DIETHYL 2-ISOPROPYL-4-METHYL-6-PYRIMIDINYL PHOS-PHOROTHIONATE
56-72-4	DIETHYL 3-CHLORO-4-METHYLUMBELLIFERYL THIONOPHOS-PHATE
333-41-5	DIETHYL 4-(2-ISOPROPYL-6-METHYLPYRIMIDINYL)PHOS-PHOROTHIONATE
311-45-5	DIETHYL 4-NITROPHENYL PHOSPHATE
56-38-2	DIETHYL 4-NITROPHENYL PHOSPHOROTHIONATE
105-57-7	DIETHYL ACETAL
97-96-1	DIETHYL ACETALDEHYDE
91-66-7	DIETHYL ANILINE
25340-17-4	DIETHYL BENZENE
88-10-8	DIETHYL CARBAMYL CHLORIDE
105-58-8	DIETHYL CARBONATE
105-58-8	DIETHYL CARBONATE (DOT)
629-14-1	DIETHYL CELLOSOLVE
814-49-3	DIETHYL CHLOROPHOSPHATE
814-49-3	DIETHYL CHLOROPHOSPHONATE
1719-53-5	DIETHYL DICHLOROSILANE
628-37-5	DIETHYL DIOXIDE
95-92-1	DIETHYL ETHANEDIOATE
60-29-7	DIETHYL ETHER
60-29-7	DIETHYL ETHER (DOT)
109-63-7	DIETHYL ETHER COMPOUND WITH BORON TRIFLUORIDE
109-63-7	DIETHYL ETHER TRIFLUOROBORANE COMPLEX
111-96-6	DIETHYL GLYCOL DIMETHYL ETHER

CAS No.	Chemical Name
96-22-0	DIETHYL KETONE
96-22-0	DIETHYL KETONE (ACGIH,DOT)
121-75-5	DIETHYL MERCAPTOSUCCINATE S-ESTER WITH O,O-DIMETHYL PHOSPHORODITHIOATE
121-75-5	DIETHYL MERCAPTOSUCCINATE, O,O-DIMETHYL DITHIOPHOSPHATE, S-ESTER
121-75-5	DIETHYL MERCAPTOSUCCINATE, O,O-DIMETHYL PHOSPHORODITHIOATE
121-75-5	DIETHYL MERCAPTOSUCCINATE, O,O-DIMETHYL THIOPHOSPHATE
121-75-5	DIETHYL MERCAPTOSUCCINIC ACID O,O-DIMETHYL PHOSPHORODITHIOATE
297-97-2	DIETHYL O-2-PYRAZINYL PHOSPHOROTHIONATE
84-72-0	DIETHYL O-CARBOXYBENZOYLOXYACETATE
95-92-1	DIETHYL OXALATE
60-29-7	DIETHYL OXIDE
56-38-2	DIETHYL P-NITROPHENYL PHOSPHOROTHIONATE
56-38-2	DIETHYL P-NITROPHENYL THIONOPHOSPHATE
56-38-2	DIETHYL PARATHION
628-37-5	DIETHYL PEROXIDE
14666-78-5	DIETHYL PEROXYDICARBONATE
14666-78-5	DIETHYL PEROXYDICARBONATE (CONCENTRATION ≤30%)
14666-78-5	DIETHYL PEROXYDICARBONATE, NOT MORE THAN 27% IN SOLUTION (DOT)
14666-78-5	DIETHYL PEROXYDIFORMATE
814-49-3	DIETHYL PHOSPHOROCHLORIDATE
814-49-3	DIETHYL PHOSPHOROCHLORIDE
84-66-2	DIETHYL PHTHALATE
78-53-5	DIETHYL S-2-DIETHYLAMINOETHYL PHOSPHOROTHIOATE
627-53-2	DIETHYL SELENIDE
627-53-2	DIETHYL SELENIUM
4384-82-1	DIETHYL SODIUM DITHIOCARBAMATE
64-67-5	DIETHYL SULFATE
540-82-9	DIETHYL SULFATE
352-93-2	DIETHYL SULFIDE
64-67-5	DIETHYL SULPHATE
540-82-9	DIETHYL SULPHATE
56-72-4	DIETHYL THIOPHOSPHORIC ACID ESTER OF 3-CHLORO-4-METHYL-7-HYDROXYCOUMARIN
2524-04-1	DIETHYL THIOPHOSPHORIC CHLORIDE
2524-04-1	DIETHYL THIOPHOSPHORYL CHLORIDE
557-20-0	DIETHYL ZINC
557-20-0	DIETHYL ZINC (DOT)
100-37-8	DIETHYL(2-HYDROXYETHYL)AMINE
120-22-9	DIETHYL-P-NITROSOANILINE
93-05-0	DIETHYL-P-PHENYLENEDIAMINE
93-05-0	DIETHYL-PARA-PHENYLENEDIAMINE
88-09-5	DIETHYLACETIC ACID
871-27-2	DIETHYLALANE
96-10-6	DIETHYLALUMINIUM CHLORIDE
96-10-6	DIETHYLALUMINIUM CHLORIDE (DOT)
96-10-6	DIETHYLALUMINUM CHLORIDE
871-27-2	DIETHYLALUMINUM HYDRIDE
96-10-6	DIETHYLALUMINUM MONOCHLORIDE
871-27-2	DIETHYLALUMINUM MONOHYDRIDE
109-89-7	DIETHYLAMINE
109-89-7	DIETHYLAMINE (ACGIH,DOT)
111-40-0	DIETHYLAMINE, 2,2'-DIAMINO-
51-75-2	DIETHYLAMINE, 2,2'-DICHLORO-N-METHYL-
55-86-7	DIETHYLAMINE, 2,2'-DICHLORO-N-METHYL-, HYDROCHLORIDE
302-70-5	DIETHYLAMINE, 2,2'-DICHLORO-N-METHYL-, N-OXIDE, COMPD. WITH HYDROCHLORIC ACID
302-70-5	DIETHYLAMINE, 2,2'-DICHLORO-N-METHYL-, N-OXIDE, HYDROCHLORIDE
111-42-2	DIETHYLAMINE, 2,2'-DIHYDROXY-
55-18-5	DIETHYLAMINE, N-NITROSO-
100-37-8	DIETHYLAMINOETHANOL
100-37-8	DIETHYLAMINOETHANOL (DOT)
140-82-9	DIETHYLAMINOETHOXYETHANOL
140-82-9	DIETHYLAMINOETHOXYETHANOL [COMBUSTIBLE LIQUID LABEL]
140-82-9	DIETHYLAMINOETHOXYETHANOL [FLAMMABLE LIQUID LABEL]
2426-54-2	DIETHYLAMINOETHYL ACRYLATE
104-78-9	DIETHYLAMINOPROPYLAMINE
104-78-9	DIETHYLAMINOTRIMETHYLENAMINE
25340-17-4	DIETHYLBENZENE (DOT)
772-54-3	DIETHYLBENZYLAMINE
1642-54-2	DIETHYLCARBAMAZANE CITRATE
1642-54-2	DIETHYLCARBAMAZINE ACID CITRATE
1642-54-2	DIETHYLCARBAMAZINE CITRATE
1642-54-2	DIETHYLCARBAMAZINE HYDROGEN CITRATE
4384-82-1	DIETHYLCARBAMODITHIOIC ACID, SODIUM SALT
88-10-8	DIETHYLCARBAMOYL CHLORIDE
584-02-1	DIETHYLCARBINOL
584-02-1	DIETHYLCARBINOL (DOT)
96-22-0	DIETHYLCETONE (FRENCH)
96-10-6	DIETHYLCHLOROALUMINUM
1609-19-4	DIETHYLCHLOROSILANE
2524-04-1	DIETHYLCHLOROTHIOPHOSPHATE
2524-04-1	DIETHYLCHLORTHIOFOSFAT (CZECH)
1331-43-7	DIETHYLCYCLOHEXANE
1719-53-5	DIETHYLDICHLOROSILANE (DOT)
1719-53-5	DIETHYLDICHLOROSILICON
4384-82-1	DIETHYLDITHIOCARBAMATE SODIUM
95-06-7	DIETHYLDITHIOCARBAMIC ACID 2-CHLOROALLYL ESTER
4384-82-1	DIETHYLDITHIOCARBAMIC ACID SODIUM
4384-82-1	DIETHYLDITHIOCARBAMIC ACID, SODIUM SALT
123-91-1	DIETHYLENE DIOXIDE
123-91-1	DIETHYLENE ETHER
111-46-6	DIETHYLENE GLYCOL
929-06-6	DIETHYLENE GLYCOL AMINE
112-34-5	DIETHYLENE GLYCOL BUTYL ETHER
124-17-4	DIETHYLENE GLYCOL BUTYL ETHER ACETATE
111-44-4	DIETHYLENE GLYCOL DICHLORIDE
111-96-6	DIETHYLENE GLYCOL DIMETHYL ETHER
693-21-0	DIETHYLENE GLYCOL DINITRATE
111-90-0	DIETHYLENE GLYCOL ETHYL ETHER
111-77-3	DIETHYLENE GLYCOL METHYL ETHER
112-34-5	DIETHYLENE GLYCOL MONO-N-BUTYL ETHER
929-06-6	DIETHYLENE GLYCOL MONOAMINE
112-34-5	DIETHYLENE GLYCOL MONOBUTYL ETHER
124-17-4	DIETHYLENE GLYCOL MONOBUTYL ETHER ACETATE
111-90-0	DIETHYLENE GLYCOL MONOETHYL ESTER
111-90-0	DIETHYLENE GLYCOL MONOETHYL ETHER
111-77-3	DIETHYLENE GLYCOL MONOMETHYL ETHER
112-34-5	DIETHYLENE GLYCOL N-BUTYL ETHER
110-91-8	DIETHYLENE IMIDOXIDE
109-99-9	DIETHYLENE OXIDE
123-91-1	DIETHYLENE OXIDE
110-91-8	DIETHYLENE OXIMIDE
110-85-0	DIETHYLENEDIAMINE
142-64-3	DIETHYLENEDIAMINE DIHYDROCHLORIDE
110-91-8	DIETHYLENEIMIDE OXIDE
110-85-0	DIETHYLENEIMINE
111-40-0	DIETHYLENETRIAMINE
111-40-0	DIETHYLENETRIAMINE (ACGIH,DOT)
56-18-8	DIETHYLENETRIAMINE, 2,6-DIMETHYL-
56-18-8	DIETHYLENETRIAMINE, DIMETHYL-
110-91-8	DIETHYLENIMIDE OXIDE
105-58-8	DIETHYLESTER CARBONIC ACID
84-66-2	DIETHYLESTER PHTHALIC ACID
100-37-8	DIETHYLETHANOLAMINE
100-36-7	DIETHYLETHYLENE DIAMINE
462-95-3	DIETHYLFORMAL
103-23-1	DIETHYLHEXYL ADIPATE
123-79-5	DIETHYLHEXYL ADIPATE
871-27-2	DIETHYLHYDROALUMINUM
3710-84-7	DIETHYLHYDROXYLAMINE
557-18-6	DIETHYLMAGNESIUM
557-18-6	DIETHYLMAGNESIUM (DOT)
100-37-8	DIETHYLMONOETHANOLAMINE
627-53-2	DIETHYLMONOSELENIDE
55-18-5	DIETHYLNITROSAMIDE
55-18-5	DIETHYLNITROSAMINE
55-18-5	DIETHYLNITROSOAMINE
111-42-2	DIETHYLOLAMINE
91-66-7	DIETHYLPHENYLAMINE
814-49-3	DIETHYLPHOSPHORIC ACID CHLORIDE
56-53-1	DIETHYLSTILBESTROL
514-73-8	DIETHYLTHIADICARBOCYANINE IODIDE
2524-04-1	DIETHYLTHIOPHOSPHORYL CHLORIDE
2524-04-1	DIETHYLTHIOPHOSPHORYL CHLORIDE (DOT)
109-89-7	DIETILAMINA (ITALIAN)
101-68-8	DIFENIL-METAN-DIISOCIANATO (ITALIAN)
3383-96-8	DIFENPHOS
101-68-8	DIFENYLMETHAAN-DISSOCYANAAT (DUTCH)
7783-41-7	DIFLUORINE MONOOXIDE
7783-41-7	DIFLUORINE MONOXIDE

CAS No.	Chemical Name
7783-41-7	DIFLUORINE OXIDE
75-68-3	DIFLUORO-1-CHLOROETHANE
353-59-3	DIFLUOROCHLOROBROMOMETHANE
75-45-6	DIFLUOROCHLOROMETHANE
75-61-6	DIFLUORODIBROMOMETHANE
75-71-8	DIFLUORODICHLOROMETHANE
75-37-6	DIFLUOROETHANE
25497-28-3	DIFLUOROETHANE
353-50-4	DIFLUOROFORMALDEHYDE
25497-29-4	DIFLUOROMONOCHLOROETHANE
25497-29-4	DIFLUOROMONOCHLOROETHANE (DOT)
75-45-6	DIFLUOROMONOCHLOROMETHANE
353-50-4	DIFLUOROOXOMETHANE
353-50-4	DIFLUOROPHOSGENE
13779-41-4	DIFLUOROPHOSPHORIC ACID
13779-41-4	DIFLUOROPHOSPHORIC ACID, ANHYDROUS
13779-41-4	DIFLUOROPHOSPHORIC ACID, ANHYDROUS (DOT)
115-26-4	DIFO
2425-06-1	DIFOLATAN
2425-06-1	DIFOLATAN 4F
2425-06-1	DIFOLATAN 4F1
2425-06-1	DIFOLATAN BOW
944-22-9	DIFONATE
944-22-9	DIFONATUL
107-22-2	DIFORMAL
107-22-2	DIFORMYL
3383-96-8	DIFOS
20830-75-5	DIGACIN
1401-55-4	DIGALLIC ACID
111-46-6	DIGENOS
1582-09-8	DIGERMIN
71-63-6	DIGILONG
71-63-6	DIGIMED
71-63-6	DIGIMERCK
71-63-6	DIGISIDIN
71-63-6	DIGITALIN
71-63-6	DIGITALINE (FRENCH)
71-63-6	DIGITALINE CRISTALLISEE
71-63-6	DIGITALINE NATIVELLE
71-63-6	DIGITALINUM VERUM
20830-75-5	DIGITALIS GLYCOSIDE
71-63-6	DIGITOPHYLLIN
71-63-6	DIGITOXIGENIN-TRIDIGITOXOSID (GERMAN)
71-63-6	DIGITOXIGENIN TRIDIGITOXOSIDE
71-63-6	DIGITOXIN
71-63-6	DIGITOXOSIDE
71-63-6	DIGITRIN
2238-07-5	DIGLYCIDYL ETHER
2238-07-5	DIGLYCIDYL ETHER (ACGIH)
1675-54-3	DIGLYCIDYL ETHER OF BISPHENOL A
1954-28-5	DIGLYCIDYLTRIETHYLENE GLYCOL
111-46-6	DIGLYCOL
112-34-5	DIGLYCOL MONOBUTYL ETHER
124-17-4	DIGLYCOL MONOBUTYL ETHER ACETATE
111-90-0	DIGLYCOL MONOETHYL ETHER
111-77-3	DIGLYCOL MONOMETHYL ETHER
929-06-6	DIGLYCOLAMINE
111-96-6	DIGLYME
111-46-6	DIGOL
20830-75-5	DIGOSIN
20830-75-5	DIGOXIGENIN-TRIDIGITOXOSID (GERMAN)
20830-75-5	DIGOXIN
20830-75-5	DIGOXINE
3648-21-3	DIHEPTYL PHTHALATE
112-58-3	DIHEXYL ETHER
143-16-8	DIHEXYLAMINE
106-20-7	DIHEXYLAMINE, 2,2'-DIETHYL-
57-41-0	DIHYCON
80-51-3	DIHYDRAZIDE SDO
142-64-3	DIHYDRO PIP WORMER
151-56-4	DIHYDRO-1H-AZIRINE
108-31-6	DIHYDRO-2,5-DIOXOFURAN
151-56-4	DIHYDROAZIRENE
126-33-0	DIHYDROBUTADIENE SULPHONE
50-28-2	DIHYDROFOLLICULAR HORMONE
50-28-2	DIHYDROFOLLICULIN
1333-74-0	DIHYDROGEN
7722-84-1	DIHYDROGEN DIOXIDE
16941-12-1	DIHYDROGEN HEXACHLOROPLATINATE
16941-12-1	DIHYDROGEN HEXACHLOROPLATINATE(2-)
16961-83-4	DIHYDROGEN HEXAFLUOROSILCATE(2-)
16961-83-4	DIHYDROGEN HEXAFLUOROSILICATE
1623-24-1	DIHYDROGEN ISOPROPYL PHOSPHATE
7783-06-4	DIHYDROGEN MONOSULFIDE
7783-07-5	DIHYDROGEN SELENIDE
7664-93-9	DIHYDROGEN SULFATE
7783-06-4	DIHYDROGEN SULFIDE
50-28-2	DIHYDROMENFORMON
75-21-8	DIHYDROOXIRENE
110-87-2	DIHYDROPYRAN
3174-74-1	DIHYDROPYRAN
25512-65-6	DIHYDROPYRAN
110-87-2	DIHYDROPYRAN (DOT)
3174-74-1	DIHYDROPYRAN (DOT)
25512-65-6	DIHYDROPYRAN (DOT)
123-31-9	DIHYDROQUINONE
94-58-6	DIHYDROSAFROLE
128-46-1	DIHYDROSTREPTOMYCIN
50-28-2	DIHYDROTHEELIN
123-31-9	DIHYDROXYBENZENE
12385-08-9	DIHYDROXYBENZENE
111-46-6	DIHYDROXYDIETHYL ETHER
50-28-2	DIHYDROXYESTRIN
794-93-4	DIHYDROXYMETHYLFURATRIZINE
7553-56-2	DIIODINE
7774-29-0	DIIODOMERCURY
75-11-6	DIIODOMETHANE
1309-37-1	DIIRON TRIOXIDE
10028-22-5	DIIRON TRISULFATE
25167-70-8	DIISOBUTENE
504-20-1	DIISOBUTENYL KETONE
108-83-8	DIISOBUTILCHETONE (ITALIAN)
108-82-7	DIISOBUTYL CARBINOL
108-83-8	DIISOBUTYL KETONE
108-83-8	DIISOBUTYL KETONE (ACGIH,DOT)
1191-15-7	DIISOBUTYLALUMINUM
1191-15-7	DIISOBUTYLALUMINUM HYDRIDE
110-96-3	DIISOBUTYLAMINE
110-96-3	DIISOBUTYLAMINE (DOT)
25167-70-8	DIISOBUTYLENE
25167-70-8	DIISOBUTYLENE (DOT)
1191-15-7	DIISOBUTYLHYDROALUMINUM
108-83-8	DIISOBUTYLKETON (DUTCH, GERMAN)
3437-84-1	DIISOBUTYRYL PEROXIDE
3437-84-1	DIISOBUTYRYL PEROXIDE (DOT)
3437-84-1	DIISOBUTYRYL PEROXIDE, ≤52% IN SOLUTION
26471-62-5	DIISOCYANAT-TOLUOL (GERMAN)
26471-62-5	DIISOCYANATOMETHYLBENZENE
26471-62-5	DIISOCYANATOTOLUENE
26761-40-0	DIISODECYL PHTHALATE
27215-10-7	DIISOOCTYL ACID PHOSPHATE
27215-10-7	DIISOOCTYL ACID PHOSPHATE (DOT)
27215-10-7	DIISOOCTYL PHOSPHATE
2935-44-6	DIISOPROPANOL
110-97-4	DIISOPROPANOLAMINE
79-29-8	DIISOPROPYL
96-80-0	DIISOPROPYL ETHANOLAMINE
108-20-3	DIISOPROPYL ETHER
108-20-3	DIISOPROPYL OXIDE
105-64-6	DIISOPROPYL PERCARBONATE
105-64-6	DIISOPROPYL PERDICARBONATE
105-64-6	DIISOPROPYL PEROXYDICARBONATE
105-64-6	DIISOPROPYL PEROXYDICARBONATE, MAXIMUM CONCENTRATION 52% IN SOLUTION (DOT)
105-64-6	DIISOPROPYL PEROXYDICARBONATE, TECHNICAL PURE (DOT)
105-64-6	DIISOPROPYL PEROXYDIFORMATE
108-18-9	DIISOPROPYLAMINE
108-18-9	DIISOPROPYLAMINE (ACGIH,DOT)
26762-93-6	DIISOPROPYLBENZENE HYDROPEROXIDE
26762-93-6	DIISOPROPYLBENZENE HYDROPEROXIDE SOLUTION , NOT OVER 72% PEROXIDE (DOT)
80-43-3	DIISOPROPYLBENZENE PEROXIDE
108-20-3	DIISOPROPYLETHER (DOT)
504-20-1	DIISOPROPYLIDENE ACETONE
10103-46-5	DIKAL 21
674-82-8	DIKETENE
674-82-8	DIKETENE, INHIBITED (DOT)
123-42-2	DIKETONE ALCOHOL

CAS No.	Chemical Name
2893-78-9	DIKONIT
57-41-0	DILABID
1330-20-7	DILAN
20830-75-5	DILANACIN
95-50-1	DILATIN DB
127-18-4	DILATIN PT
105-74-8	DILAUROYL PEROXIDE
105-74-8	DILAUROYL PEROXIDE, TECHNICAL PURE (DOT)
105-74-8	DILAURYL PEROXIDE
72-54-8	DILENE
75-60-5	DILIC
14307-35-8	DILITHIUM CHROMATE (Li2CrO4)
12031-80-0	DILITHIUM PEROXIDE
534-52-1	DILLEX
514-73-8	DILOMBRIN
514-73-8	DILOMBRINE
7647-01-0	DILUTE HYDROCHLORIC ACID
60-51-5	DIMATE 267
298-04-4	DIMAZ
57-14-7	DIMAZIN
57-14-7	DIMAZINE
13171-21-6	DIMECRON
13171-21-6	DIMECRON 100
13171-21-6	DIMECRON 50
13171-21-6	DIMECRON-20
115-26-4	DIMEFOX
50-28-2	DIMENFORMON
77-73-6	DIMER CYKLOPENTADIENU (CZECH)
15385-58-7	DIMERCURY DIBROMIDE
15385-57-6	DIMERCURY DIIODIDE
60-51-5	DIMETATE
121-69-7	DIMETHLYANILINE (ACGIH)
60-51-5	DIMETHOAAT (DUTCH)
60-51-5	DIMETHOAT (GERMAN)
60-51-5	DIMETHOAT TECHNISCH 95%
60-51-5	DIMETHOATE
60-51-5	DIMETHOATE-267
60-51-5	DIMETHOGEN
357-57-3	DIMETHOXY STRYCHNINE
357-57-3	DIMETHOXY STRYCHNINE (DOT)
2524-03-0	DIMETHOXY THIOPHOSPHONYL CHLORIDE
52-68-6	DIMETHOXY-2,2,2-TRICHLORO-1-HYDROXY-ETHYL-PHOSPHINE OXIDE
72-43-5	DIMETHOXY-DDT
110-71-4	DIMETHOXYETHANE
127-19-5	DIMETHOXYL ACETAMIDE
79-44-7	DIMETHOXYL CARBAMYL CHLORIDE
109-87-5	DIMETHOXYMETHANE
74-84-0	DIMETHYL
7786-34-7	DIMETHYL (1-METHOXYCARBOXYPROPEN-2-YL)PHOSPHATE
52-68-6	DIMETHYL (2,2,2-TRICHLORO-1-HYDROXYETHYL)PHOSPHONATE
131-11-3	DIMETHYL 1,2-BENZENEDICARBOXYLATE
300-76-5	DIMETHYL 1,2-DIBROMO-2,2-DICHLOROETHYL PHOSPHATE
300-76-5	DIMETHYL 1,2-DIBROMO-2,2-DICHLOROETHYL PHOSPHATE (NALED)
120-61-6	DIMETHYL 1,4-BENZENEDICARBOXYLATE
52-68-6	DIMETHYL 1-HYDROXY-2,2,2-TRICHLOROETHYL PHOSPHONATE
6923-22-4	DIMETHYL 1-METHYL-2-(METHYLCARBAMOYL)VINYL PHOSPHATE, CIS
62-73-7	DIMETHYL 2,2-DICHLOROETHENYL PHOSPHATE
62-73-7	DIMETHYL 2,2-DICHLOROVINYL PHOSPHATE
644-64-4	DIMETHYL 2-CARBAMYL-3-METHYLPYRAZOLYLDIMETHYLCARBAMATE
141-66-2	DIMETHYL 2-DIMETHYLCARBAMOYL-1-METHYLVINYL PHOSPHATE
7786-34-7	DIMETHYL 2-METHOXYCARBONYL-1-METHYLVINYL PHOSPHATE
122-10-1	DIMETHYL 3-HYDROXYGLUTACONATE DIMETHYL PHOSPHATE
122-14-5	DIMETHYL 3-METHYL-4-NITROPHENYL PHOSPHOROTHIONATE
122-14-5	DIMETHYL 4-NITRO-M-TOLYL PHOSPHOROTHIONATE
298-00-0	DIMETHYL 4-NITROPHENYL PHOSPHOROTHIONATE
127-19-5	DIMETHYL ACETAMIDE
127-19-5	DIMETHYL ACETAMIDE (ACGIH)
131-11-3	DIMETHYL BENZENEORTHODICARBOXYLATE
79-44-7	DIMETHYL CARBAMOYL CHLORIDE (ACGIH)
111-96-6	DIMETHYL CARBITOL
616-38-6	DIMETHYL CARBONATE
616-38-6	DIMETHYL CARBONATE (DOT)
110-71-4	DIMETHYL CELLOSOLVE
97-97-2	DIMETHYL CHLORACETAL
2524-03-0	DIMETHYL CHLOROTHIONOPHOSPHATE
2524-03-0	DIMETHYL CHLOROTHIOPHOSPHATE
2524-03-0	DIMETHYL CHLOROTHIOPHOSPHATE (DOT)
62-73-7	DIMETHYL DICHLOROVINYL PHOSPHATE
431-03-8	DIMETHYL DIKETONE
25136-55-4	DIMETHYL DIOXANE
624-92-0	DIMETHYL DISULFIDE
624-92-0	DIMETHYL DISULFIDE (DOT)
624-92-0	DIMETHYL DISULPHIDE
131-11-3	DIMETHYL ESTER PHTHALIC ACID
120-61-6	DIMETHYL ESTER TERAPHTHALIC ACID
115-10-6	DIMETHYL ETHER
115-10-6	DIMETHYL ETHER (DOT)
109-87-5	DIMETHYL FORMAL
68-12-2	DIMETHYL FORMAMIDE
67-64-1	DIMETHYL KETONE
593-74-8	DIMETHYL MERCURY
7786-34-7	DIMETHYL METHOXYCARBONYLPROPENYL PHOSPHATE
77-78-1	DIMETHYL MONOSULFATE
75-18-3	DIMETHYL MONOSULFIDE
131-11-3	DIMETHYL O-PHTHALATE
115-10-6	DIMETHYL OXIDE
3254-63-5	DIMETHYL P-(METHYLTHIO)PHENYL PHOSPHATE
120-61-6	DIMETHYL P-BENZENEDICARBOXYLATE
298-00-0	DIMETHYL P-NITROPHENYL MONOTHIOPHOSPHATE
298-00-0	DIMETHYL P-NITROPHENYL PHOSPHOROTHIONATE
298-00-0	DIMETHYL P-NITROPHENYL THIOPHOSPHATE
120-61-6	DIMETHYL P-PHTHALATE
298-00-0	DIMETHYL PARATHION
6923-22-4	DIMETHYL PHOSPHATE ESTER OF 3-HYDROXY-N-METHYL-CIS-CROTONAMIDE
6923-22-4	DIMETHYL PHOSPHATE ESTER WITH (E)-3-HYDROXY-N-METHYL-CROTONAMIDE
141-66-2	DIMETHYL PHOSPHATE ESTER WITH 3-HYDROXY-N,N-DIMETHYL-CIS-CROTONAMIDE
6923-22-4	DIMETHYL PHOSPHATE OF 3-HYDROXY-N-METHYL-CIS-CROTONAMINE
63917-41-9	DIMETHYL PHOSPHORAMIDOCYANIDIC ACID
2524-03-0	DIMETHYL PHOSPHOROCHLORIDOTHIOATE
2524-03-0	DIMETHYL PHOSPHOROCHLORIDOTHIOATE (DOT)
131-11-3	DIMETHYL PHTHALATE
131-11-3	DIMETHYL PHTHALATE (ACGIH)
919-86-8	DIMETHYL S-(2-ETHTHIOETHYL)THIOPHOSPHATE
77-78-1	DIMETHYL SULFATE
77-78-1	DIMETHYL SULFATE (ACGIH,DOT)
75-18-3	DIMETHYL SULFIDE
75-18-3	DIMETHYL SULFIDE (DOT)
67-68-5	DIMETHYL SULFOXIDE
77-78-1	DIMETHYL SULPHATE
75-18-3	DIMETHYL SULPHIDE
67-68-5	DIMETHYL SULPHOXIDE
120-61-6	DIMETHYL TEREPHTHALATE
75-18-3	DIMETHYL THIOETHER
2524-03-0	DIMETHYL THIONOPHOSPHOROCHLORIDATE
2524-03-0	DIMETHYL THIOPHOSPHOROCHLORIDATE
2524-03-0	DIMETHYL THIOPHOSPHORYL CHLORIDE
4685-14-7	DIMETHYL VIOLOGEN
1910-42-5	DIMETHYL VIOLOGEN CHLORIDE
60-11-7	DIMETHYL YELLOW
108-01-0	DIMETHYL(2-HYDROXYETHYL)AMINE
108-16-7	DIMETHYL(2-HYDROXYPROPYL)AMINE
108-01-0	DIMETHYL(HYDROXYETHYL)AMINE
138-89-6	DIMETHYL(P-NITROSOPHENYL)AMINE
53324-05-3	DIMETHYL(P-NITROSOPHENYL)AMINE
300-76-5	DIMETHYL-1,2-DIBROMO-2-DICHLOROETHYL PHOSPHATE
7786-34-7	DIMETHYL-1-CARBOMETHOXY-1-PROPEN-2-YL PHOSPHATE
119-38-0	DIMETHYL-5-(L-ISOPROPYL-3-METHYL-PYRAZOLYL)-CARBAMATE
75-78-5	DIMETHYL-DICHLORSILAN (CZECH)
78-62-6	DIMETHYL-DIETHOXYSILAN (CZECH)
926-63-6	DIMETHYL-N-PROPYLAMINE
926-63-6	DIMETHYL-N-PROPYLAMINE (DOT)
62-73-7	DIMETHYL-O-O-DICHLOROVINYL-2-2-PHOSPHATE (TECHNICAL)
25136-55-4	DIMETHYL-P-DIOXANE (DOT)
298-00-0	DIMETHYL-P-NITROPHENYL THIONPHOSPHATE
138-89-6	DIMETHYL-P-NITROSOANILINE
138-89-6	DIMETHYL-P-NITROSOANILINE (DOT)
53324-05-3	DIMETHYL-P-NITROSOANILINE (DOT)
99-98-9	DIMETHYL-P-PHENYLENEDIAMINE
534-15-6	DIMETHYLACETAL

CAS No.	Chemical Name
79-31-2	DIMETHYLACETIC ACID
96-22-0	DIMETHYLACETONE
127-19-5	DIMETHYLACETONE AMIDE
78-82-0	DIMETHYLACETONITRILE
503-17-3	DIMETHYLACETYLENE
115-19-5	DIMETHYLACETYLENECARBINOL
115-19-5	DIMETHYLACETYLENYLCARBINOL
108-01-0	DIMETHYLAETHANOLAMIN (GERMAN)
127-19-5	DIMETHYLAMIDE ACETATE
77-81-6	DIMETHYLAMIDOETHOXYPHOSPHORYL CYANIDE
124-40-3	DIMETHYLAMINE
124-40-3	DIMETHYLAMINE (ACGIH)
6369-97-7	DIMETHYLAMINE, (2,4,5-TRICHLOROPHENOXY)ACETATE
124-40-3	DIMETHYLAMINE, ANHYDROUS
124-40-3	DIMETHYLAMINE, ANHYDROUS (DOT)
124-40-3	DIMETHYLAMINE, AQUEOUS SOLUTION
124-40-3	DIMETHYLAMINE, AQUEOUS SOLUTION (DOT)
62-75-9	DIMETHYLAMINE, N-NITROSO-
124-40-3	DIMETHYLAMINE, SOLUTION (DOT)
926-64-7	DIMETHYLAMINOACETONITRILE
108-01-0	DIMETHYLAMINOAETHANOL (GERMAN)
1300-73-8	DIMETHYLAMINOBENZENE
77-81-6	DIMETHYLAMINOCYANPHOSPHORSAEUREAETHYLESTER (GERMAN)
108-01-0	DIMETHYLAMINOETHANOL
2867-47-2	DIMETHYLAMINOETHYL METHACRYLATE
2867-47-2	DIMETHYLAMINOETHYL METHACRYLATE (DOT)
121-69-7	DIMETHYLANILINE
75-60-5	DIMETHYLARSENIC ACID
75-60-5	DIMETHYLARSINIC ACID
57-97-6	DIMETHYLBENZANTHRACENE
1330-20-7	DIMETHYLBENZENE
57-97-6	DIMETHYLBENZ[A]ANTHRACENE
119-38-0	DIMETHYLCARBAMATE D'L-ISOPROPYL 3-METHYL 5-PYRAZOLYLE (FRENCH)
644-64-4	DIMETHYLCARBAMIC ACID 1-((DIMETHYLAMINO)CARBONYL)-5-METHYL-1H-PYRAZOL-3-YL ESTER
119-38-0	DIMETHYLCARBAMIC ACID 3-METHYL-1-(1-METHYLETHYL)-1H-PYRAZOL-5-YL ESTER
79-44-7	DIMETHYLCARBAMIC ACID CHLORIDE
644-64-4	DIMETHYLCARBAMIC ACID ESTER WITH 3-HYDROXY-N,N,5-TRIMETHYLPYRAZOLE-1-CARBOXAMIDE
79-44-7	DIMETHYLCARBAMIC CHLORIDE
79-44-7	DIMETHYLCARBAMIDOYL CHLORIDE
79-44-7	DIMETHYLCARBAMOYL CHLORIDE
79-44-7	DIMETHYLCARBAMYL CHLORIDE
67-63-0	DIMETHYLCARBINOL
107-30-2	DIMETHYLCHLOROETHER
79-44-7	DIMETHYLCHLOROFORMAMIDE
2524-03-0	DIMETHYLCHLORTHIOFOSFAT (CZECH)
1467-79-4	DIMETHYLCYANAMIDE
926-64-7	DIMETHYLCYANOMETHYLAMINE
98-94-2	DIMETHYLCYCLOHEXYL AMINE
75-78-5	DIMETHYLDICHLOROSILANE
75-78-5	DIMETHYLDICHLOROSILANE (DOT)
78-62-6	DIMETHYLDIETHOXYSILANE
78-62-6	DIMETHYLDIETHOXYSILANE (DOT)
25136-55-4	DIMETHYLDIOXANE
14484-64-1	DIMETHYLDITHIOCARBAMIC ACID IRON(3+) SALT
86-50-0	DIMETHYLDITHIOPHOSPHORIC ACID N-METHYLBENZAZIMIDE ESTER
75-21-8	DIMETHYLENE OXIDE
107-15-3	DIMETHYLENEDIAMINE
151-56-4	DIMETHYLENEIMINE
463-49-0	DIMETHYLENEMETHANE
151-56-4	DIMETHYLENIMINE
77-78-1	DIMETHYLESTER KYSELINY SIROVE (CZECH)
120-61-6	DIMETHYLESTER KYSELINY TEREFTALOVE (CZECH)
108-01-0	DIMETHYLETHANOLAMINE
108-01-0	DIMETHYLETHANOLAMINE (DOT)
98-06-6	DIMETHYLETHYLBENZENE
75-85-4	DIMETHYLETHYLCARBINOL
115-19-5	DIMETHYLETHYNYLCARBINOL
115-19-5	DIMETHYLETHYNYLMETHANOL
67-64-1	DIMETHYLFORMALDEHYDE
68-12-2	DIMETHYLFORMAMID (GERMAN)
68-12-2	DIMETHYLFORMAMIDE (ACGIH)
431-03-8	DIMETHYLGLYOXAL
3025-88-5	DIMETHYLHEXANE DIHYDROPEROXIDE, (WITH 18% OR MORE WATER) (DOT)
3025-88-5	DIMETHYLHEXANE DIHYDROPEROXIDE, DRY (DOT)
57-14-7	DIMETHYLHYDRAZINE
57-14-7	DIMETHYLHYDRAZINE UNSYMMETRICAL (DOT)
540-73-8	DIMETHYLHYDRAZINE, SYMMETRICAL
540-73-8	DIMETHYLHYDRAZINE, SYMMETRICAL (DOT)
57-14-7	DIMETHYLHYDRAZINE, UNSYMMETRICAL
108-16-7	DIMETHYLISOPROPANOLAMINE
67-64-1	DIMETHYLKETAL
513-86-0	DIMETHYLKETOL
2999-74-8	DIMETHYLMAGNESIUM
2999-74-8	DIMETHYLMAGNESIUM (DOT)
74-98-6	DIMETHYLMETHANE
108-01-0	DIMETHYLMONOETHANOLAMINE
79-46-9	DIMETHYLNITROMETHANE
25322-01-4	DIMETHYLNITROMETHANE
62-75-9	DIMETHYLNITROSAMIN (GERMAN)
62-75-9	DIMETHYLNITROSAMINE
62-75-9	DIMETHYLNITROSOAMINE
1300-71-6	DIMETHYLPHENOL
121-69-7	DIMETHYLPHENYLAMINE
1300-73-8	DIMETHYLPHENYLAMINE
77-81-6	DIMETHYLPHOSPHORAMIDOCYANIDIC ACID, ETHYL ESTER
463-82-1	DIMETHYLPROPANE
926-63-6	DIMETHYLPROPYLAMINE
77-78-1	DIMETHYLSULFAAT (DUTCH)
77-78-1	DIMETHYLSULFAT (CZECH)
75-18-3	DIMETHYLSULFID (CZECH)
52-68-6	DIMETHYLTRICHLOROHYDROXYETHYL PHOSPHONATE
544-97-8	DIMETHYLZINC
544-97-8	DIMETHYLZINC (DOT)
110-18-9	DIMETHYL[2-(DIMETHYLAMINO)ETHYL]AMINE
644-64-4	DIMETILAN
644-64-4	DIMETILANE
68-12-2	DIMETILFORMAMIDE (ITALIAN)
77-78-1	DIMETILSOLFATO (ITALIAN)
60-51-5	DIMETON
52-68-6	DIMETOX
68-12-2	DIMETYLFORMAMIDU (CZECH)
60-51-5	DIMEVUR
67-68-5	DIMEXIDE
55-86-7	DIMITAN
12001-26-2	DIMONITE
12001-26-2	DIMONITE DM(NA-TS)
333-41-5	DIMPYLAT
333-41-5	DIMPYLATE
53220-22-7	DIMYRISTYL PEROXYDICARBONATE
53220-22-7	DIMYRISTYL PEROXYDICARBONATE, ≤42%, IN WATER
53220-22-7	DIMYRISTYL PEROXYDICARBONATE, NOT MORE THAN 22% IN WATER (DOT)
53220-22-7	DIMYRISTYL PEROXYDICARBONATE, TECHNICALLY PURE (DOT)
142-59-6	DINATRIUM-(N,N'-AETHYLEN-BIS(DITHIOCARBAMAT)) (GERMAN)
142-59-6	DINATRIUM-(N,N'-ETHYLEEN-BIS(DITHIOCARBAMAAT)) (DUTCH)
131-89-5	DINEX
1314-06-3	DINICKEL TRIOXIDE
8004-13-5	DINIL
148-01-6	DINITOLMIDE
10102-06-4	DINITRATODIOXOURANIUM
88-85-7	DINITRO
534-52-1	DINITRO
88-85-7	DINITRO-3
534-52-1	DINITRO-O-CRESOL
1335-85-9	DINITRO-O-CRESOL
1335-85-9	DINITRO-O-CRESOL (ACGIH)
131-89-5	DINITRO-O-CYCLOHEXYLPHENOL
26471-56-7	DINITROANILINE
26471-56-7	DINITROANILINES
26471-56-7	DINITROANILINES (DOT)
528-29-0	DINITROBENZENE
25154-54-5	DINITROBENZENE
25154-54-5	DINITROBENZENE (ACGIH)
25154-54-5	DINITROBENZENE SOLID (DOT)
25154-54-5	DINITROBENZENE SOLUTION
25154-54-5	DINITROBENZENE, SOLID
25154-54-5	DINITROBENZENE, SOLUTION (DOT)
25154-54-5	DINITROBENZOL SOLID (DOT)
88-85-7	DINITROBUTYLPHENOL
97-00-7	DINITROCHLOROBENZENE
25567-67-3	DINITROCHLOROBENZENE

CAS No.	Chemical Name
97-00-7	DINITROCHLOROBENZENE (DOT)
25567-67-3	DINITROCHLOROBENZENE (DOT)
97-00-7	DINITROCHLOROBENZOL
25567-67-3	DINITROCHLOROBENZOL
97-00-7	DINITROCHLOROBENZOL (DOT)
25567-67-3	DINITROCHLOROBENZOL (DOT)
534-52-1	DINITROCRESOL
1335-85-9	DINITROCRESOL
131-89-5	DINITROCYCLOHEXYLPHENOL
131-89-5	DINITROCYCLOHEXYLPHENOL (DOT)
534-52-1	DINITRODENDTROXAL
1335-85-9	DINITRODENDTROXAL
51-28-5	DINITROFENOLO (ITALIAN)
7727-37-9	DINITROGEN
10024-97-2	DINITROGEN MONOXIDE
10024-97-2	DINITROGEN OXIDE
10036-47-2	DINITROGEN TETRAFLUORIDE
10544-72-6	DINITROGEN TETRAOXIDE
10544-72-6	DINITROGEN TETROXIDE
10544-72-6	DINITROGEN TETROXIDE (DOT)
10544-73-7	DINITROGEN TRIOXIDE
534-52-1	DINITROL
1335-85-9	DINITROL
1335-85-9	DINITROMETHYL CYCLOHEXYLTRIENOL
25550-58-7	DINITROPHENOL
25550-58-7	DINITROPHENOL SOLUTION
25550-58-7	DINITROPHENOL SOLUTION (DOT)
25550-58-7	DINITROPHENOL, DRY OR CONTAINING, BY WEIGHT, LESS THAN 15% WATER (DOT)
25550-58-7	DINITROPHENOL, WETTED WITH, BY WEIGHT, AT LEAST 15% WATER (DOT)
25321-14-6	DINITROPHENYLMETHANE
519-44-8	DINITRORESORCINOL
35860-51-6	DINITRORESORCINOL
519-44-8	DINITRORESORCINOL, WETTED WITH, BY WEIGHT, AT LEAST 15% WATER (DOT)
35860-51-6	DINITRORESORCINOL, WETTED WITH, BY WEIGHT, AT LEAST 15% WATER (DOT)
25550-55-4	DINITROSOBENZENE
101-25-7	DINITROSOPENTAMETHENETETRAMINE
101-25-7	DINITROSOPENTAMETHYLENE TETRAMINE
121-14-2	DINITROTOLUENE
25321-14-6	DINITROTOLUENE
121-14-2	DINITROTOLUENE (ACGIH)
25321-14-6	DINITROTOLUENE, LIQUID
25321-14-6	DINITROTOLUENE, LIQUID (DOT)
25321-14-6	DINITROTOLUENE, MOLTEN (DOT)
25321-14-6	DINITROTOLUENE, SOLID
25321-14-6	DINITROTOLUENE, SOLID (DOT)
1332-58-7	DINKIE A
534-52-1	DINOC
1335-85-9	DINOC
51-28-5	DINOFAN
117-84-0	DINOPOL NOP
88-85-7	DINOSEB
88-85-7	DINOSEBE (FRENCH)
1420-07-1	DINOTERB
1420-07-1	DINOTERBE
94-75-7	DINOXOL
93-76-5	DINOXOL
57-41-0	DINTOINA
534-52-1	DINURANIA
1335-85-9	DINURANIA
8004-13-5	DINYL
762-16-3	DIOCTANOYL PEROXIDE
577-11-7	DIOCTLYN
103-23-1	DIOCTYL ADIPATE
123-79-5	DIOCTYL ADIPATE
577-11-7	DIOCTYL ESTER OF SODIUM SULFOSUCCINATE
577-11-7	DIOCTYL ESTER OF SODIUM SULFOSUCCINIC ACID
117-84-0	DIOCTYL O-PHTHALATE
117-84-0	DIOCTYL PHTHALATE
117-81-7	DIOCTYL PHTHALATE
577-11-7	DIOCTYL SODIUM SULFOSUCCINATE
577-11-7	DIOCTYL SULFOSUCCINATE SODIUM
577-11-7	DIOCTYL SULFOSUCCINATE SODIUM SALT
577-11-7	DIOCTYL-MEDO FORTE
577-11-7	DIOCTYLAL
1120-48-5	DIOCTYLAMINE

CAS No.	Chemical Name
540-59-0	DIOFORM
50-28-2	DIOGYN
57-63-6	DIOGYN-E
50-28-2	DIOGYNETS
123-91-1	DIOKAN
123-91-1	DIOKSAN (POLISH)
25322-69-4	DIOL 1000
110-63-4	DIOL 14B
25322-69-4	DIOL 2000
25322-69-4	DIOL 400
111-42-2	DIOLAMINE
107-41-5	DIOLANE
56-72-4	DIOLICE
577-11-7	DIOMEDICONE
123-91-1	DIOSSANO-1,4 (ITALIAN)
577-11-7	DIOSUCCIN
577-11-7	DIOTILAN
577-11-7	DIOVAC
577-11-7	DIOX
123-91-1	DIOXAAN-1,4 (DUTCH)
123-91-1	DIOXAN
123-91-1	DIOXAN-1,4 (GERMAN)
123-91-1	DIOXANE
123-91-1	DIOXANE (ACGIH,DOT)
123-91-1	DIOXANE, TECH. GRADE
123-91-1	DIOXANE-1,4
123-91-1	DIOXANNE (FRENCH)
52-68-6	DIOXAPHOS
78-34-2	DIOXATHION
78-34-2	DIOXATHION (ACGIH)
78-34-2	DIOXATION
1746-01-6	DIOXIN
1746-01-6	DIOXIN (HERBICIDE CONTAMINANT)
111-90-0	DIOXITOL
14977-61-8	DIOXODICHLOROCHROMIUM
100-79-8	DIOXOLAN
646-06-0	DIOXOLAN
100-79-8	DIOXOLANE
646-06-0	DIOXOLANE
100-79-8	DIOXOLANE (DOT)
646-06-0	DIOXOLANE (DOT)
126-33-0	DIOXOTHIOLAN
78-34-2	DIOXOTHION
1312-03-4	DIOXOTRIMERCURY SULFATE
1464-53-5	DIOXYBUTADIENE
1304-29-6	DIOXYDE DE BARYUM (FRENCH)
123-91-1	DIOXYETHYLENE ETHER
7782-44-7	DIOXYGEN
14915-07-2	DIOXYGEN ION(2-)
110-97-4	DIPA
108-18-9	DIPA
138-86-3	DIPANOL
82-66-6	DIPAXIN
762-13-0	DIPELARGONYL PEROXIDE
101-25-7	DIPENTAX
138-86-3	DIPENTEN
138-86-3	DIPENTENE
138-86-3	DIPENTENE (DOT)
872-10-6	DIPENTYL SULFIDE
2050-92-2	DIPENTYLAMINE
1941-79-3	DIPERAZELAIC ACID
1941-79-3	DIPEROXYAZELAIC ACID
1941-79-3	DIPEROXYAZELAIC ACID, MAXIMUM CONCENTRATION 27% (DOT)
82-66-6	DIPHACIN
82-66-6	DIPHACINON
82-66-6	DIPHACINONE
57-41-0	DIPHANTOIN
57-41-0	DIPHEDAN
82-66-6	DIPHENACIN
82-66-6	DIPHENADION
82-66-6	DIPHENADIONE
92-52-4	DIPHENYL
2602-46-2	DIPHENYL BLUE 2B
72-57-1	DIPHENYL BLUE 3B
2602-46-2	DIPHENYL BLUE KF
2602-46-2	DIPHENYL BLUE M 2B
1937-37-7	DIPHENYL DEEP BLACK G
80-10-4	DIPHENYL DICHLOROSILANE
80-10-4	DIPHENYL DICHLOROSILANE (DOT)

CAS No.	Chemical Name
101-84-8	DIPHENYL ETHER
80-51-3	DIPHENYL ETHER 4,4'-DISULFOHYDRAZIDE
16071-86-6	DIPHENYL FAST BROWN BRL
119-61-9	DIPHENYL KETONE
101-68-8	DIPHENYL METHANE DIISOCYANATE
776-74-9	DIPHENYL METHYL BROMIDE SOLUTION
776-74-9	DIPHENYL METHYL BROMIDE, SOLID
776-74-9	DIPHENYL METHYL BROMIDE, SOLID (DOT)
776-74-9	DIPHENYL METHYL BROMIDE, SOLUTION (DOT)
8004-13-5	DIPHENYL MIXED WITH DIPHENYL OXIDE
101-84-8	DIPHENYL OXIDE
80-51-3	DIPHENYL OXIDE 4,4'-DISULFOHYDRAZIDE
1336-36-3	DIPHENYL, CHLORINATED
122-39-4	DIPHENYLAMINE
122-39-4	DIPHENYLAMINE (ACGIH)
100-97-0	DIPHENYLAMINE, 2,2',4,4',6,6'-HEXANITRO-
131-73-7	DIPHENYLAMINE, 2,2',4,4',6,6'-HEXANITRO-
156-10-5	DIPHENYLAMINE, 4-NITROSO-
86-30-6	DIPHENYLAMINE, N-NITROSO-
578-94-9	DIPHENYLAMINECHLORARSINE
578-94-9	DIPHENYLAMINECHLOROARSINE
578-94-9	DIPHENYLAMINECHLOROARSINE (DOT)
776-74-9	DIPHENYLBROMOMETHANE
712-48-1	DIPHENYLCHLOORARSINE (DUTCH)
712-48-1	DIPHENYLCHLOROARSINE
712-48-1	DIPHENYLCHLOROARSINE (DOT)
86-73-7	DIPHENYLENEMETHANE
94-36-0	DIPHENYLGLYOXAL PEROXIDE
57-41-0	DIPHENYLHYDANTOIN
555-54-4	DIPHENYLMAGNESIUM
101-68-8	DIPHENYLMETHAN-4,4'-DIISOCYANAT (GERMAN)
101-68-8	DIPHENYLMETHANE 4,4'-DIISOCYANATE
101-68-8	DIPHENYLMETHANE DIISOCYANATE
101-68-8	DIPHENYLMETHANE-4,4'-DIISOCYANATE (DOT)
119-61-9	DIPHENYLMETHANONE
776-74-9	DIPHENYLMETHYL BROMIDE
776-74-9	DIPHENYLMETHYL BROMIDE (DOT)
86-30-6	DIPHENYLNITROSAMINE
80-10-4	DIPHENYLSILICON DICHLORIDE
80-10-4	DIPHENYLSILYL DICHLORIDE
1314-80-3	DIPHOPSHORUS PENTASULFIDE
3383-96-8	DIPHOS
3383-96-8	DIPHOS (PESTICIDE)
152-16-9	DIPHOSPHORAMIDE, OCTAMETHYL-
152-16-9	DIPHOSPHORAMIDE, OCTAMETHYL- (9CI)
107-49-3	DIPHOSPHORIC ACID, TETRAETHYL ESTER
7722-88-5	DIPHOSPHORIC ACID, TETRASODIUM SALT
1314-56-3	DIPHOSPHORUS PENTAOXIDE
1314-80-3	DIPHOSPHORUS PENTASULFIDE
1314-56-3	DIPHOSPHORUS PENTOXIDE
1314-24-5	DIPHOSPHORUS TRIOXIDE
12165-69-4	DIPHOSPHORUS TRISULFIDE
8004-13-5	DIPHYL
2217-06-3	DIPICRYL SULFIDE
2217-06-3	DIPICRYL SULPHIDE, WETTED WITH, BY WEIGHT, AT LEAST 10% WATER (DOT)
100-97-0	DIPICRYLAMINE
131-73-7	DIPICRYLAMINE
67-68-5	DIPIRARTRIL-TROPICO
7782-39-0	DIPLOGEN
333-41-5	DIPOFENE
7778-50-9	DIPOTASSIUM BICHROMATE
7778-50-9	DIPOTASSIUM BICHROMATE (K2Cr2O7)
7789-00-6	DIPOTASSIUM CHROMATE
7789-00-6	DIPOTASSIUM CHROMATE (K2CrO4)
7778-50-9	DIPOTASSIUM DICHROMATE
4429-42-9	DIPOTASSIUM DISULFITE
16731-55-8	DIPOTASSIUM DISULFITE
16871-90-2	DIPOTASSIUM HEXAFLUOROSILICATE
16871-90-2	DIPOTASSIUM HEXAFLUOROSILICATE(2-)
16923-95-8	DIPOTASSIUM HEXAFLUOROZIRCONATE
16923-95-8	DIPOTASSIUM HEXAFLUOROZIRCONATE(2-)
4429-42-9	DIPOTASSIUM METABISULFITE
16731-55-8	DIPOTASSIUM METABISULFITE
7789-00-6	DIPOTASSIUM MONOCHROMATE
1312-73-8	DIPOTASSIUM MONOSULFIDE
1312-73-8	DIPOTASSIUM MONOSULFIDE (K2S)
583-52-8	DIPOTASSIUM OXALATE
17014-71-0	DIPOTASSIUM PEROXIDE
7727-21-1	DIPOTASSIUM PEROXODISULFATE
7727-21-1	DIPOTASSIUM PEROXYDISULFATE
7727-21-1	DIPOTASSIUM PERSULFATE
4429-42-9	DIPOTASSIUM PYROSULFITE
16731-55-8	DIPOTASSIUM PYROSULFITE
1312-73-8	DIPOTASSIUM SULFIDE
12136-49-1	DIPOTASSIUM TETRASULFIDE
16923-95-8	DIPOTASSIUM ZIRCONIUM HEXAFLUORIDE
7664-93-9	DIPPING ACID
3248-28-0	DIPROPIONYL PEROXIDE
3248-28-0	DIPROPIONYL PEROXIDE (DOT)
111-43-3	DIPROPYL ETHER
111-43-3	DIPROPYL ETHER (DOT)
123-19-3	DIPROPYL KETONE
123-19-3	DIPROPYL KETONE (ACGIH,DOT)
142-82-5	DIPROPYL METHANE
111-43-3	DIPROPYL OXIDE
16066-38-9	DIPROPYL PEROXYDICARBONATE
110-97-4	DIPROPYL-2,2'-DIHYDROXY-AMINE
2036-15-9	DIPROPYLALUMINUM HYDRIDE
142-84-7	DIPROPYLAMINE
142-84-7	DIPROPYLAMINE (DOT)
56-18-8	DIPROPYLAMINE, 3,3'-DIAMINO-
621-64-7	DIPROPYLAMINE, N-NITROSO-
25265-71-8	DIPROPYLENE GLYCOL
12002-25-4	DIPROPYLENE GLYCOL METHYL ETHER
34590-94-8	DIPROPYLENE GLYCOL METHYL ETHER
34590-94-8	DIPROPYLENE GLYCOL MONOMETHYL ETHER
56-18-8	DIPROPYLENE TRIAMINE
56-18-8	DIPROPYLENTRIAMIN (GERMAN)
621-64-7	DIPROPYLNITROSAMINE
52-68-6	DIPTEREX
52-68-6	DIPTEREX 50
52-68-6	DIPTEVUR
85-00-7	DIPYRIDO(1,2-A;2',1'-C)PYRAZINEDIIUM, 6,7-D[HYDRO-, DIBROMIDE
2764-72-9	DIPYRIDO[1,2-A:2',1'-C]PYRAZINEDIIUM, 6,7-DIHYDRO-
85-00-7	DIQUAT
2764-72-9	DIQUAT
85-00-7	DIQUAT (ACGIH,DOT)
85-00-7	DIQUAT DIBROMIDE
2764-72-9	DIQUAT DICATION
86-88-4	DIRAX
30553-04-9	DIRAX
1937-37-7	DIRECT BLACK 38
1937-37-7	DIRECT BLACK 38(TECHNICAL) GRADE
1937-37-7	DIRECT BLACK A
1937-37-7	DIRECT BLACK BRN
1937-37-7	DIRECT BLACK CX
1937-37-7	DIRECT BLACK CXR
1937-37-7	DIRECT BLACK E
1937-37-7	DIRECT BLACK EW
1937-37-7	DIRECT BLACK EX
1937-37-7	DIRECT BLACK FR
1937-37-7	DIRECT BLACK GAC
1937-37-7	DIRECT BLACK GW
1937-37-7	DIRECT BLACK GX
1937-37-7	DIRECT BLACK GXR
1937-37-7	DIRECT BLACK JET
1937-37-7	DIRECT BLACK META
1937-37-7	DIRECT BLACK METHYL
1937-37-7	DIRECT BLACK N
1937-37-7	DIRECT BLACK RX
1937-37-7	DIRECT BLACK SD
1937-37-7	DIRECT BLACK WS
1937-37-7	DIRECT BLACK Z
1937-37-7	DIRECT BLACK ZSH
2602-46-2	DIRECT BLUE 6
72-57-1	DIRECT BLUE 14
2602-46-2	DIRECT BLUE 2B
72-57-1	DIRECT BLUE 3B
2602-46-2	DIRECT BLUE 6
2602-46-2	DIRECT BLUE 6 (TECHNICAL GRADE)
2602-46-2	DIRECT BLUE A
2602-46-2	DIRECT BLUE BB
2602-46-2	DIRECT BLUE GS
2602-46-2	DIRECT BLUE K
2602-46-2	DIRECT BLUE M 2B
72-57-1	DIRECT BLUE M3B

CAS No.	Chemical Name
16071-86-6	DIRECT BROWN 95
16071-86-6	DIRECT BROWN 95 (TECHNICAL GRADE)
108-45-2	DIRECT BROWN BR
16071-86-6	DIRECT BROWN BRL
108-45-2	DIRECT BROWN GG
1937-37-7	DIRECT DEEP BLACK E
1937-37-7	DIRECT DEEP BLACK E EXTRA
1937-37-7	DIRECT DEEP BLACK E-EX
1937-37-7	DIRECT DEEP BLACK EA-CF
1937-37-7	DIRECT DEEP BLACK EAC
1937-37-7	DIRECT DEEP BLACK EW
1937-37-7	DIRECT DEEP BLACK EX
1937-37-7	DIRECT DEEP BLACK WX
16071-86-6	DIRECT FAST BROWN BRL
16071-86-6	DIRECT FAST BROWN LMR
16071-86-6	DIRECT LIGHT BROWN BRS
16071-86-6	DIRECT LIGHTFAST BROWN M
16071-86-6	DIRECT SUPRA LIGHT BROWN ML
330-54-1	DIREX 4L
101-05-3	DIREZ
12654-97-6	DIREZ
136-40-3	DIRIDONE
1642-54-2	DIROCIDE
54497-43-7	DISELENIOUS ACID, ZINC SALT (1:1)
126-73-8	DISFLAMOLL TB
1330-78-5	DISFLAMOLL TKP
115-86-6	DISFLAMOLL TP
999-97-3	DISILAZANE, 1,1,1,3,3,3-HEXAMETHYL-
2157-42-8	DISILOXANE, HEXAETHOXY-
534-16-7	DISILVER CARBONATE
10294-26-5	DISILVER MONOSULFATE
20667-12-3	DISILVER MONOXIDE
20667-12-3	DISILVER OXIDE
10294-26-5	DISILVER SULFATE
10294-26-5	DISILVER(1+) SULFATE
7558-79-4	DISODIUM ACID ORTHOPHOSPHATE
7558-79-4	DISODIUM ACID PHOSPHATE
4452-58-8	DISODIUM CARBONATE COMPOUND WITH HYDROGEN PEROXIDE (2:3)
15630-89-4	DISODIUM CARBONATE COMPOUND WITH HYDROGEN PEROXIDE (2:3)
7775-11-3	DISODIUM CHROMATE
7775-11-3	DISODIUM CHROMATE (Na2CrO4)
10588-01-9	DISODIUM DICHROMATE
7681-49-4	DISODIUM DIFLUORIDE
1313-60-6	DISODIUM DIOXIDE
7681-57-4	DISODIUM DISULFITE
7757-74-6	DISODIUM DISULFITE
7775-14-6	DISODIUM DITHIONITE
142-59-6	DISODIUM ETHYLENE-1,2-BISDITHIOCARBAMATE
142-59-6	DISODIUM ETHYLENEBIS(DITHIOCARBAMATE)
16893-85-9	DISODIUM HEXAFLUOROSILICATE
16893-85-9	DISODIUM HEXAFLUOROSILICATE (2-)
16893-85-9	DISODIUM HEXAFLUOROSILICATE (Na2SIF6)
10039-32-4	DISODIUM HYDROGEN ORTHOPHOSPHATE DODECAHYDRATE
7558-79-4	DISODIUM HYDROGEN PHOSPHATE
10039-32-4	DISODIUM HYDROGEN PHOSPHATE DODECAHYDRATE
7558-79-4	DISODIUM HYDROPHOSPHATE
7775-14-6	DISODIUM HYDROSULFITE
6834-92-0	DISODIUM METASILICATE
7558-79-4	DISODIUM MONOHYDROGEN PHOSPHATE
10039-32-4	DISODIUM MONOHYDROGEN PHOSPHATE DODECAHYDRATE
6834-92-0	DISODIUM MONOSILICATE
1313-82-2	DISODIUM MONOSULFIDE
12401-86-4	DISODIUM MONOXIDE
7558-79-4	DISODIUM ORTHOPHOSPHATE
10039-32-4	DISODIUM ORTHOPHOSPHATE DODECAHYDRATE
62-76-0	DISODIUM OXALATE
12401-86-4	DISODIUM OXIDE
1313-60-6	DISODIUM PEROXIDE
7775-27-1	DISODIUM PEROXODISULFATE
4452-58-8	DISODIUM PEROXYDICARBONATE
3313-92-6	DISODIUM PEROXYDICARBONATE
15630-89-4	DISODIUM PEROXYDICARBONATE
7775-27-1	DISODIUM PEROXYDISULFATE
7775-27-1	DISODIUM PERSULFATE
28831-12-1	DISODIUM PERSULFATE
15593-29-0	DISODIUM PERSULFATE
7558-79-4	DISODIUM PHOSPHATE
7558-79-4	DISODIUM PHOSPHATE (Na2HPO4)
10039-32-4	DISODIUM PHOSPHATE (Na2HPO4) DODECAHYDRATE
10039-32-4	DISODIUM PHOSPHATE (Na2HPO4.12H2O)
10039-32-4	DISODIUM PHOSPHATE DODECAHYDRATE
10039-32-4	DISODIUM PHOSPHATE DODECAHYDRATE (Na2HPO4.12H2O)
7558-79-4	DISODIUM PHOSPHORIC ACID
7757-74-6	DISODIUM PYROSULFITE
7681-57-4	DISODIUM PYROSULFITE
13410-01-0	DISODIUM SELENATE
10102-18-8	DISODIUM SELENITE
10102-18-8	DISODIUM SELENIUM TRIOXIDE
6834-92-0	DISODIUM SILICATE
16893-85-9	DISODIUM SILICOFLUORIDE
1313-82-2	DISODIUM SULFIDE
10102-20-2	DISODIUM TELLURITE
1330-43-4	DISODIUM TETRABORATE
1303-96-4	DISODIUM TETRABORATE DECAHYDRATE
137-26-8	DISOLFURO DI TETRAMETILTIOURAME (ITALIAN)
577-11-7	DISONATE
333-41-5	DISONEX
1344-28-1	DISPAL
1344-28-1	DISPAL M
9016-45-9	DISPERGATOR BO
110-85-0	DISPERMINE
96-69-5	DISPERSE MB 61
82-28-0	DISPERSE ORANGE
82-28-0	DISPERSE ORANGE (ANTHRAQUINONE DYE)
111-46-6	DISSOLVANT APV
52326-66-6	DISTEARYL PEROXYDICARBONATE
52326-66-6	DISTEARYL PEROXYDICARBONATE, ≤85% WITH STEARYL ALCOHOL
52326-66-6	DISTEARYL PEROXYDICARBONATE, NOT MORE THAN 85% WITH STEARYL ALCOHOL (DOT)
56-53-1	DISTILBENE
65996-91-0	DISTILLATES (COAL TAR), UPPER
505-60-2	DISTILLED MUSTARD
70-30-4	DISTODIN
67-72-1	DISTOKAL
67-72-1	DISTOPAN
67-72-1	DISTOPIN
123-23-9	DISUCCINIC ACID PEROXIDE
123-23-9	DISUCCINIC ACID PEROXIDE, MAXIMUM CONCENTRATION 72% (DOT)
123-23-9	DISUCCINIC ACID PEROXIDE, TECHNICAL PURE (DOT)
123-23-9	DISUCCINOYL PEROXIDE
136-78-7	DISUL-SODIUM
298-04-4	DISULFATON
14644-61-2	DISULFATOZIRCONIC ACID
2179-59-1	DISULFIDE, 2-PROPENYL PROPYL
2179-59-1	DISULFIDE, ALLYL PROPYL
97-77-8	DISULFIDE, BIS(DIETHYLTHIOCARBAMOYL)
137-26-8	DISULFIDE, BIS(DIMETHYLTHIOCARBAMOYL)
624-92-0	DISULFIDE, DIMETHYL
97-77-8	DISULFIRAM
298-04-4	DISULFOTON
298-04-4	DISULFOTON (ACGIH,DOT)
2497-07-6	DISULFOTON DISULIDE
298-04-4	DISULFOTON MIXTURE, DRY
298-04-4	DISULFOTON MIXTURE, LIQUID
2497-07-6	DISULFOTON SULFOXIDE
10025-67-9	DISULFUR DICHLORIDE
7791-27-7	DISULFUR PENTOXYDICHLORIDE
97-77-8	DISULFURAM
137-26-8	DISULFURE DE TETRAMETHYLTHIOURAME (FRENCH)
7783-05-3	DISULFURIC ACID
4429-42-9	DISULFUROUS ACID, DIPOTASSIUM SALT
16731-55-8	DISULFUROUS ACID, DIPOTASSIUM SALT
7681-57-4	DISULFUROUS ACID, DISODIUM SALT
7757-74-6	DISULFUROUS ACID, DISODIUM SALT
7791-27-7	DISULFURYL CHLORIDE
8014-95-7	DISULPHURIC ACID
7783-05-3	DISULPHURIC ACID
53-16-7	DISYNFORMON
298-04-4	DISYSTON
2497-07-6	DISYSTON S
2497-07-6	DISYSTON SULFOXIDE
2497-07-6	DISYSTON SULPHOXIDE
298-04-4	DISYSTOX
71-63-6	DITAVEN

CAS No.	Chemical Name
53220-22-7	DITETRADECYL PEROXYDICARBONATE
6533-73-9	DITHALLIUM CARBONATE
7446-18-6	DITHALLIUM SULFATE
10031-59-1	DITHALLIUM SULFATE
1314-32-5	DITHALLIUM TRIOXIDE
7446-18-6	DITHALLIUM(1+) SULFATE
10031-59-1	DITHALLIUM(1+) SULFATE
142-59-6	DITHANE A 40
100-25-4	DITHANE A-4
142-59-6	DITHANE D 14
514-73-8	DITHIAZANIN IODIDE
514-73-8	DITHIAZANINE IODIDE
514-73-8	DITHIAZINE
514-73-8	DITHIAZINE (DYE)
514-73-8	DITHIAZININE
541-53-7	DITHIOBIURET
4384-82-1	DITHIOCARB
4384-82-1	DITHIOCARBAMATE
4384-82-1	DITHIOCARBAMATE ANION
4384-82-1	DITHIOCARBAMATE PESTICIDE, LIQUID [FLAMMABLE LIQUID LABEL]
4384-82-1	DITHIOCARBAMATE PESTICIDE, LIQUID [POISON B LABEL]
4384-82-1	DITHIOCARBAMATE PESTICIDE, SOLID
75-15-0	DITHIOCARBONIC ANHYDRIDE
298-04-4	DITHIODEMETON
3689-24-5	DITHIODIPHOSPHORIC ACID, TETRAETHYL ESTER
3689-24-5	DITHIOFOS
640-15-3	DITHIOMETHON
640-15-3	DITHIOMETON (FRENCH)
3689-24-5	DITHION
3689-24-5	DITHIONE
7783-05-3	DITHIONIC ACID
8014-95-7	DITHIONIC ACID
15512-36-4	DITHIONOUS ACID, CALCIUM SALT (1:1)
14293-73-3	DITHIONOUS ACID, DIPOTASSIUM SALT
7775-14-6	DITHIONOUS ACID, DISODIUM SALT
7779-86-4	DITHIONOUS ACID, ZINC SALT (1:1)
3689-24-5	DITHIOPHOS
298-02-2	DITHIOPHOSPHATE DE O,O-DIETHYLE ET D'ETHYLTHIOMETHYLE (FRENCH)
786-19-6	DITHIOPHOSPHATE DE O,O-DIETHYLE ET DE (4-CHLORO-PHENYL) THIOMETHYLE (FRENCH)
2275-14-1	DITHIOPHOSPHATE DE O,O-DIETHYLE ET DE S(2,5-DICHLORO-PHENYL) THIOMETHYLE (FRENCH)
298-04-4	DITHIOPHOSPHATE DE O,O-DIETHYLE ET DE S-(2-ETHYLTHIO-ETHYLE) (FRENCH)
60-51-5	DITHIOPHOSPHATE DE O,O-DIMETHYLE ET DE S(-N-METHYLCAR-BAMOYL-METHYLE) (FRENCH)
121-75-5	DITHIOPHOSPHATE DE O,O-DIMETHYLE ET DE S-(1,2-DICARBO-ETHOXYETHYLE) (FRENCH)
640-15-3	DITHIOPHOSPHATE DE O,O-DIMETHYLE ET DE S-(2-ETHYLTHIO-ETHYLE) (FRENCH)
3689-24-5	DITHIOPYROPHOSPHATE DE TETRAETHYLE (FRENCH)
298-04-4	DITHIOSYSTOX
3689-24-5	DITHIOTEP
1642-54-2	DITRAZIN
1642-54-2	DITRAZIN CITRATE
1642-54-2	DITRAZINE
1642-54-2	DITRAZINE CITRATE
52-68-6	DITRIFON
534-52-1	DITROSOL
1335-85-9	DITROSOL
330-54-1	DIUREX
330-54-1	DIUROL
330-54-1	DIURON
330-54-1	DIURON (ACGIH,DOT)
330-54-1	DIURON 4L
1314-62-1	DIVANADIUM PENTAOXIDE
1314-62-1	DIVANADIUM PENTOXIDE
1314-34-7	DIVANADIUM TRIOXIDE
106-99-0	DIVINYL
25339-57-5	DIVINYL
821-08-9	DIVINYL ACETYLENE
108-57-6	DIVINYL BENZENE
1321-74-0	DIVINYL BENZENE
109-93-3	DIVINYL ETHER
109-93-3	DIVINYL ETHER (DOT)
109-93-3	DIVINYL ETHER, INHIBITED (DOT)
109-93-3	DIVINYL OXIDE

CAS No.	Chemical Name
110-00-9	DIVINYLENE OXIDE
110-02-1	DIVINYLENE SULFIDE
109-97-7	DIVINYLENIMINE
62-73-7	DIVIPAN
2636-26-2	DIVIPAN
109-93-3	DIVYNYL OXIDE
20830-75-5	DIXINA
9002-84-0	DIXON 164
95-50-1	DIZENE
333-41-5	DIZINON
79-43-6	DKHUK
584-79-2	DL-3-ALLYL-2-METHYL-4-OXOCYCLOPENT-2-ENYL DL-CIS TRANS CHRYSANTHEMATE
50-21-5	DL-LACTIC ACID
138-86-3	DL-LIMONENE
531-76-0	DL-PHENYLALANINE MUSTARD
531-76-0	DL-PHENYLALANINE, 4-[BIS(2-CHLOROETHYL)AMINO]-
531-76-0	DL-SARCOLYSIN
531-76-0	DL-SARCOLYSINE
53558-25-1	DLP 787
53558-25-1	DLP-87
9002-84-0	DLX 6000
9002-84-0	DLX 7000
578-94-9	DM
578-94-9	DM (ARSENIC COMPOUND)
127-19-5	DMA
124-40-3	DMA
94-75-7	DMA-4
75-60-5	DMAA
60-11-7	DMAB
127-19-5	DMAC
108-01-0	DMAE
57-97-6	DMBA
79-44-7	DMCC
72-43-5	DMDT
110-71-4	DME
9016-45-9	DME
68-12-2	DMF
115-26-4	DMF
68-12-2	DMF (AMIDE)
131-11-3	DMF (INSECT REPELLENT)
68-12-2	DMFA
1333-86-4	DMG 105A
57-14-7	DMH
540-73-8	DMH
62-75-9	DMN
62-75-9	DMNA
131-11-3	DMP
99-98-9	DMPD
75-18-3	DMS
77-78-1	DMS
67-68-5	DMS 70
67-68-5	DMS 90
67-68-5	DMSO
115-90-2	DMSP
120-61-6	DMT
950-37-8	DMTP
950-37-8	DMTP(JAPAN)
330-54-1	DMU
131-89-5	DN
1335-85-9	DN
131-89-5	DN (PESTICIDE)
131-89-5	DN 1
88-85-7	DN 289
131-89-5	DN DRY MIX NO 1
131-89-5	DN DUST NO 12
1335-85-9	DN-DRY MIX NO 2
97-02-9	DNA
88-85-7	DNBP
1335-85-9	DNC
97-00-7	DNCB
25567-67-3	DNCB
534-52-1	DNOC
1335-85-9	DNOC
131-89-5	DNOCHP
1335-85-9	DNOK (CZECH)
117-84-0	DNOP
88-85-7	DNOSBP
101-25-7	DNPMT

CAS No.	Chemical Name
101-25-7	DNPT
14797-65-0	DNRB
88-85-7	DNSBP
121-14-2	DNT
25321-14-6	DNT
14797-65-0	DNTB
.1420-07-1	DNTBP
56-38-2	DNTP
103-23-1	DOA
123-79-5	DOA
27176-87-0	DOBANIC ACID 83
27176-87-0	DOBANIC ACID JN
577-11-7	DOCUSATE SODIUM
50-29-3	DODAT
2385-85-5	DODECACHLOROOCTAHYDRO-1,3,4-METHENO-2H-CYCLOBUTA-(C,D)PENTALENE
2385-85-5	DODECACHLOROPENTACYCLO(3220(SUP 2,6),0(SUP 3,9),0(SUP 5,10))DECANE
2385-85-5	DODECACHLOROPENTACYCLODECANE
101-83-7	DODECAHYDRODIPHENYLAMINE
3129-91-7	DODECAHYDROPHENYLAMINE NITRITE
123-01-3	DODECANE, 1-PHENYL-
5350-03-8	DODECANOIC ACID, PENTYL ESTER
112-53-8	DODECANOL
105-74-8	DODECANOYL PEROXIDE
6842-15-5	DODECENE
112-53-8	DODECYL ALCOHOL
151-21-3	DODECYL ALCOHOL, HYDROGEN SULFATE, SODIUM SALT
2235-54-3	DODECYL AMMONIUM SULFATE
25155-30-0	DODECYL BENZENE SODIUM SULFONATE
151-21-3	DODECYL HYDROGEN SULFATE, SODIUM SALT
112-55-0	DODECYL MERCAPTAN
151-21-3	DODECYL SODIUM SULFATE
143-00-0	DODECYL SULFATE DIETHANOLAMINE SALT
151-21-3	DODECYL SULFATE SODIUM
151-21-3	DODECYL SULFATE SODIUM SALT
139-96-8	DODECYL SULFATE TRIETHANOLAMINE SALT
143-00-0	DODECYL SULFATE, COMPD WITH 2,2'-IMINODIETHANOL
4484-72-4	DODECYL TRICHLOROSILANE
4484-72-4	DODECYL TRICHLOROSILANE (DOT)
123-01-3	DODECYLBENZENE
26264-05-1	DODECYLBENZENESULFONATE ISOPROPYLAMINE SALT
27176-87-0	DODECYLBENZENESULFONIC ACID
27176-87-0	DODECYLBENZENESULFONIC ACID (DOT)
26264-06-2	DODECYLBENZENESULFONIC ACID CALCIUM SALT
26264-05-1	DODECYLBENZENESULFONIC ACID ISOPROPYLAMINE SALT
25155-30-0	DODECYLBENZENESULFONIC ACID SODIUM SALT
27323-41-7	DODECYLBENZENESULFONIC ACID TRIETHANOLAMINE SALT
26264-05-1	DODECYLBENZENESULFONIC ACID, ISOPROPYLAMINE SALT
25155-30-0	DODECYLBENZENESULPHONATE, SODIUM SALT
27176-87-0	DODECYLBENZENESULPHONIC ACID
25155-30-0	DODECYLBENZENSULFONAN SODNY (CZECH)
6842-15-5	DODECYLENE
112-02-7	DODIGEN 1383
76-06-2	DOJYOPICRIN
58-89-9	DOL GRANULE
7783-20-2	DOLAMIN
57-24-9	DOLCO MOUSE CEREAL
99-87-6	DOLCYMENE
50-78-2	DOLEAN PH 8
87-68-3	DOLEN-PUR
67-68-5	DOLICUR
67-68-5	DOLIGUR
2646-17-5	DOLKWAL ORANGE SS
3564-09-8	DOLKWAL PONCEAU 3R
76-06-2	DOLOCHLOR
53-86-1	DOLOVIN
56-53-1	DOMESTROL
67-68-5	DOMOSO
8012-54-2	DONOVAN'S SOLUTION
117-84-0	DOP
117-81-7	DOP
94-75-7	DORMONE
51-83-2	DORYL
51-83-2	DORYL (PHARMACEUTICAL)
60-57-1	DORYTOX
577-11-7	DOSS
577-11-7	DOSS 70
24934-91-6	DOTAN

CAS No.	Chemical Name
1344-28-1	DOTMENT 324
1344-28-1	DOTMENT 358
3926-62-3	DOW DEFOLIANT
131-52-2	DOW DORMANT FUNGICIDE
299-84-3	DOW ET 14
299-84-3	DOW ET 57
88-85-7	DOW GENERAL
88-85-7	DOW GENERAL WEED KILLER
87-86-5	DOW PENTACHLOROPHENOL DP-2 ANTIMICROBIAL
88-85-7	DOW SELECTIVE WEED KILLER
76-03-9	DOW SODIUM TCA INHIBITED
127-18-4	DOW-PER
79-01-6	DOW-TRI
109-59-1	DOWANAL EIPAT
111-90-0	DOWANOL
112-34-5	DOWANOL DB
111-90-0	DOWANOL DE
111-77-3	DOWANOL DM
34590-94-8	DOWANOL DPM
111-76-2	DOWANOL EB
110-80-5	DOWANOL EE
109-86-4	DOWANOL EM
112-50-5	DOWANOL TE
107-98-2	DOWANOL-33B
57-74-9	DOWCHLOR
12789-03-6	DOWCHLOR
299-86-5	DOWCO 132
315-18-4	DOWCO 139
2921-88-2	DOWCO 179
13121-70-5	DOWCO 213
9016-45-9	DOWFAX 9N9
10043-52-4	DOWFLAKE
57-55-6	DOWFROST
37286-64-9	DOWFROTH 250
106-93-4	DOWFUME 40
106-93-4	DOWFUME EDB
8004-09-9	DOWFUME MC 2
8004-09-9	DOWFUME MC 33
8003-19-8	DOWFUME N
106-93-4	DOWFUME W-100
106-93-4	DOWFUME W-8
106-93-4	DOWFUME W-85
106-93-4	DOWFUME W-90
95-95-4	DOWICIDE 2
933-78-4	DOWICIDE 2
15950-66-0	DOWICIDE 2
25167-82-2	DOWICIDE 2
88-06-2	DOWICIDE 2S
933-78-4	DOWICIDE 2S
25167-82-2	DOWICIDE 2S
15950-66-0	DOWICIDE 2S
92-04-6	DOWICIDE 4
87-86-5	DOWICIDE 7
87-86-5	DOWICIDE EC-7
87-86-5	DOWICIDE G
131-52-2	DOWICIDE G
131-52-2	DOWICIDE G-ST
75-99-0	DOWPON
75-99-0	DOWPON M
131-89-5	DOWSPRAY 17
8004-13-5	DOWTHERM
107-98-2	DOWTHERM 209
8004-13-5	DOWTHERM A
95-50-1	DOWTHERM E
107-21-1	DOWTHERM SR 1
142-64-3	DOWZENE DHC
10049-04-4	DOXCIDE 50
577-11-7	DOXINATE
577-11-7	DOXOL
23214-92-8	DOXORUBICIN
25316-40-9	DOXORUBICIN HYDROCHLORIDE
8020-83-5	DP 11
8002-74-2	DP 652
298-07-7	DP 8R
75-99-0	DPA
122-39-4	DPA
93-05-0	DPD
57-41-0	DPH
56-38-2	DPP

CAS No.	Chemical Name
23135-22-0	DPX 1410
23135-22-0	DPX 1410L
14808-60-7	DQ 12
12001-26-2	DR 1
65-85-0	DRACYLIC ACID
8020-83-5	DRAKEOL
8020-83-5	DRAKEOL 10
8020-83-5	DRAKEOL 13
8020-83-5	DRAKEOL 15
8020-83-5	DRAKEOL 19
8020-83-5	DRAKEOL 21
8020-83-5	DRAKEOL 32
8020-83-5	DRAKEOL 33
8020-83-5	DRAKEOL 9
3691-35-8	DRAT
2032-65-7	DRAZA
14797-65-0	DRB
1129-41-5	DRC 3341
4104-14-7	DRC 714
151-21-3	DREFT
139-96-8	DRENE
110-91-8	DREWAMINE
577-11-7	DREWFAX 007
330-54-1	DREXEL
7775-09-9	DREXEL DEFOL
330-54-1	DREXEL DIURON 4L
298-00-0	DREXEL METHYL PARATHION 4E
56-38-2	DREXEL PARATHION 8E
7631-86-9	DRI-DIE
83-79-4	DRI-KIL
7601-54-9	DRI-TRI
58-89-9	DRILL TOX-SPEZIAL AGLUKON
76-44-8	DRINOX
30900-23-3	DRINOX
76-44-8	DRINOX H-34
67-68-5	DROMISOL
7775-09-9	DROP LEAF
94-36-0	DRY AND CLEAR
124-38-9	DRY ICE
124-38-9	DRY ICE (DOT)
131-89-5	DRY MIX NO 1
39196-18-4	DS 15647
9016-45-9	DS 3195
25155-30-0	DS 60
78-53-5	DSDP
142-59-6	DSE
7558-79-4	DSP
577-11-7	DSS
541-53-7	DTB
110-05-4	DTBP
514-73-8	DTDC
4342-03-4	DTIC
115-32-2	DTMC
16752-77-5	DU PONT 1179
23135-22-0	DU PONT 1410
17804-35-2	DU PONT 1991
101-96-2	DU PONT GASOLINE ANTIOXIDANT NO. 22
16752-77-5	DU PONT INSECTICIDE 1179
75-00-3	DUBLOFIX
100-97-0	DUIREXOL
131-73-7	DUIREXOL
79-01-6	DUKERON
105-60-2	DULL 704
577-11-7	DULSIVAC
120-92-3	DUMASIN
2636-26-2	DUO-KILL
577-11-7	DUOSOL
151-21-3	DUPONAL
151-21-3	DUPONAL WAQE
151-21-3	DUPONOL
151-21-3	DUPONOL C
151-21-3	DUPONOL ME
151-21-3	DUPONOL METHYL
151-21-3	DUPONOL QC
151-21-3	DUPONOL QX
151-21-3	DUPONOL WA
151-21-3	DUPONOL WA DRY
151-21-3	DUPONOL WAQ
151-21-3	DUPONOL WAQA
151-21-3	DUPONOL WAQE
151-21-3	DUPONOL WAQM
87-86-5	DURA TREET II
1330-78-5	DURAD
106-50-3	DURAFUR BLACK R
123-30-8	DURAFUR BROWN RB
27598-85-2	DURAFUR BROWN RB
120-80-9	DURAFUR DEVELOPER C
108-46-3	DURAFUR DEVELOPER G
50-78-2	DURAMAX
330-54-1	DURAN
82-28-0	DURANOL ORANGE G
7681-49-4	DURAPHAT
7786-34-7	DURAPHOS
67-68-5	DURASORB
919-86-8	DURATOX
2636-26-2	DURAVOS
16071-86-6	DURAZOL BROWN BR
105-60-2	DURETHAN BK
105-60-2	DURETHAN BK 30S
105-60-2	DURETHAN BKV 30H
105-60-2	DURETHAN BKV 55H
7720-78-7	DURETTER
1333-86-4	DUREX O
16071-86-6	DUROFAST BROWN BRL
7720-78-7	DUROFERON
9002-84-0	DUROID 5650
9002-84-0	DUROID 5650M
9002-84-0	DUROID 5813
9002-84-0	DUROID 5870
9002-84-0	DUROID 5870M
9002-84-0	DUROID X
9002-84-0	DUROID X 026
87-86-5	DUROTOX
8002-74-2	DUROWAX FT 300
2921-88-2	DURSBAN
2921-88-2	DURSBAN 10CR
2921-88-2	DURSBAN 4E
2921-88-2	DURSBAN F
10022-31-8	DUSICNAN BARNATY (CZECH)
13746-89-9	DUSICNAN ZIRKONICITY (CZECH)
3129-91-7	DUSITAN DICYKLOHEXYLAMINU (CZECH)
7632-00-0	DUSITAN SODNY (CZECH)
107-06-2	DUTCH LIQUID
107-06-2	DUTCH OIL
124-65-2	DUTCH-TREAT
8020-83-5	DUTEREX
298-04-4	DUTION
106-93-4	DWUBROMOETAN (POLISH)
1320-37-2	DWUCHLOROCZTEROFLUOROETAN (POLISH)
111-44-4	DWUCHLORODWUETYLOWY ETER (POLISH)
75-71-8	DWUCHLORODWUFLUOROMETAN (POLISH)
75-43-4	DWUCHLOROFLUOROMETAN (POLISH)
109-89-7	DWUETYLOAMINA (POLISH)
60-29-7	DWUETYLOWY ETER (POLISH)
121-69-7	DWUMETYLOANILINA (POLISH)
68-12-2	DWUMETYLOFORMAMID (POLISH)
77-78-1	DWUMETYLOWY SIARCZAN (POLISH)
1335-85-9	DWUNITRO-O-KREZOL (POLISH)
99-65-0	DWUNITROBENZEN (POLISH)
62-38-4	DYANACIDE
7789-38-0	DYETONE
944-22-9	DYFONAT
944-22-9	DYFONATE
944-22-9	DYFONATE 10G
1336-36-3	DYKANOL
50-29-3	DYKOL
57-63-6	DYLOFORM
52-68-6	DYLOX
52-68-6	DYLOX-METASYSTOX-R
75-68-3	DYMEL 142
75-37-6	DYMEL 152
25497-28-3	DYMEL 152
75-45-6	DYMEL 22
115-10-6	DYMEL A
98-86-2	DYMEX
63-25-2	DYNA-CARBYL
10103-46-5	DYNAFOS
78-10-4	DYNASIL A

CAS No.	Chemical Name
330-54-1	DYNEX
105-74-8	DYP-97F
944-22-9	DYPHONATE
101-05-3	DYRENE
12654-97-6	DYRENE
101-05-3	DYRENE 50W
12654-97-6	DYRENE 50W
52-68-6	DYREX
112-53-8	DYTOL J-68
111-87-5	DYTOL M-83
112-72-1	DYTOL R-52
27196-00-5	DYTOL R-52
112-30-1	DYTOL S-91
88-85-7	DYTOP
52-68-6	DYVON
333-41-5	DYZOL
8065-48-3	E 1059
7440-50-8	E 115
7440-50-8	E 115 (METAL)
9004-70-0	E 1440
7440-22-4	E 20
76-44-8	E 3314
9004-70-0	E 375
9016-87-9	E 534
298-00-0	E 601
56-38-2	E 605
56-38-2	E 605 F
56-38-2	E 605 FORTE
27176-87-0	E 7256
7440-50-8	E-COPPER
7440-50-8	E-CU57
106-93-4	E-D-BEE
105-60-2	E-KAPROLAKTAM (CZECH)
6923-22-4	E-MONOCROTOPHOS
10326-21-3	E-Z-OFF
3689-24-5	E393
9016-45-9	EA 120
77-81-6	EA 1205
107-44-8	EA 1208
9016-45-9	EA 80
141-97-9	EAA
106-68-3	EAK
92-84-2	EARLY BIRD WORMER
8002-03-7	EARTHNUT OIL
81-81-2	EASTERN STATES DUOCIDE
514-73-8	EASTMAN 7663
137-05-3	EASTMAN 910
680-31-9	EASTMAN INHIBITOR HPT
3081-14-9	EASTOZONE 33
100-41-4	EB
56-38-2	ECATOX
510-15-6	ECB
7631-86-9	ECCOSPHERES SI
7446-18-6	ECCOTHAL
10031-59-1	ECCOTHAL
541-41-3	ECF
106-89-8	ECH
305-03-3	ECLORIL
50-78-2	ECM
3383-96-8	ECOPRO 1707
50-78-2	ECOTRIN
299-84-3	ECTORAL
598-14-1	ED
60-00-4	EDATHAMIL
106-93-4	EDB
106-93-4	EDB-85
107-06-2	EDC
60-00-4	EDETIC
60-00-4	EDETIC ACID
3761-53-3	EDICOL SUPRA PONCEAU R
1333-86-4	EDO
1333-86-4	EDO (CARBON BLACK)
60-00-4	EDTA
60-00-4	EDTA (CHELATING AGENT)
60-00-4	EDTA (DOT)
60-00-4	EDTA ACID
72-33-3	EE 3ME
103-23-1	EFFEMOLL DOA
123-79-5	EFFEMOLL DOA
51-21-8	EFFLUDERM (FREE BASE)
103-23-1	EFFOMOLL DOA
123-79-5	EFFOMOLL DOA
534-52-1	EFFUSAN
1335-85-9	EFFUSAN
534-52-1	EFFUSAN 3436
1335-85-9	EFFUSAN 3436
51-21-8	EFUDEX
51-21-8	EFUDIX
7782-42-5	EG 0
9002-89-5	EG 40
110-71-4	EGDME
628-96-6	EGDN
67-72-1	EGITOL
109-86-4	EGM
109-86-4	EGME
151-56-4	EI
999-81-5	EI 38555
947-02-4	EI 47031
122-14-5	EI 47300
950-10-7	EI 47470
3383-96-8	EI 52160
60-51-5	EI-12880
298-02-2	EI3911
9002-84-0	EK 1108GY-A
97-77-8	EKAGOM DTET
100-97-0	EKAGOM H
137-26-8	EKAGOM TB
97-77-8	EKAGOM TEDS
97-77-8	EKAGOM TETDS
640-15-3	EKATIN
640-15-3	EKATIN AEROSOL
298-04-4	EKATIN TD
640-15-3	EKATIN ULV
640-15-3	EKATIN WF & WF ULV
640-15-3	EKATINE-25
56-38-2	EKATOX
112-34-5	EKTASOLVE DB
141-66-2	EKTAFOS
124-17-4	EKTASOLVE DB ACETATE
111-90-0	EKTASOLVE DE
111-77-3	EKTASOLVE DM
111-76-2	EKTASOLVE EB
110-80-5	EKTASOLVE EE
111-76-2	EKTASOLVE EE ACETATE SOLVENT
109-86-4	EKTASOLVE EM
121-75-5	EL 4049
123-63-7	ELALDEHYDE
1582-09-8	ELANCOLAN
84-74-2	ELAOL
1309-48-4	ELASTOMAG 100
1309-48-4	ELASTOMAG 170
74-85-1	ELAYL
9004-34-6	ELCEMA F 150
9004-34-6	ELCEMA G 250
9004-34-6	ELCEMA P 050
9004-34-6	ELCEMA P 100
123-31-9	ELDOPAQUE
123-31-9	ELDOQUIN
75-69-4	ELECTRO CF 11
75-71-8	ELECTRO-CF 12
75-45-6	ELECTRO-CF 22
7782-42-5	ELECTROGRAPHITE
7783-07-5	ELECTRONIC E-2
1332-58-7	ELECTROS
1314-13-2	ELECTROX 2500
2551-62-4	ELEGAS
1333-86-4	ELF 78
1333-86-4	ELF-O
139-96-8	ELFAN 4240 T
25155-30-0	ELFAN WA 35
25155-30-0	ELFAN WA 50
25155-30-0	ELFAN WA POWDER
27176-87-0	ELFAN WA SULPHONIC ACID
27323-41-7	ELFAN WAT
9016-45-9	ELFAPUR N 70
1333-86-4	ELFTEX 150
1333-86-4	ELFTEX 5
1333-86-4	ELFTEX 8

CAS No.	Chemical Name
88-85-7	ELGETOL
534-52-1	ELGETOL
1335-85-9	ELGETOL
534-52-1	ELGETOL 30
1335-85-9	ELGETOL 30
88-85-7	ELGETOL 318
16071-86-6	ELIAMINA LIGHT BROWN BRL
534-52-1	ELIPOL
1335-85-9	ELIPOL
1401-55-4	ELLAGI- TANNIN ANALYSIS
7704-34-9	ELOSAL
50-55-5	ELSERPINE
9002-89-5	ELVANOL
9002-89-5	ELVANOL 50-42
9002-89-5	ELVANOL 51-05G
9002-89-5	ELVANOL 5105
9002-89-5	ELVANOL 52-22
9002-89-5	ELVANOL 52-22G
9002-89-5	ELVANOL 522-22
9002-89-5	ELVANOL 70-05
9002-89-5	ELVANOL 71-30
9002-89-5	ELVANOL 73125G
9002-89-5	ELVANOL 90-50
9002-89-5	ELVANOL T 25
151-21-3	EMAL 0
151-21-3	EMAL 10
139-96-8	EMAL 20T
2235-54-3	EMAL A
2235-54-3	EMAL AD
151-21-3	EMAL O
139-96-8	EMAL T
139-96-8	EMAL TD
9016-45-9	EMALEX NP 15
7440-66-6	EMANAY ZINC DUST
1314-13-2	EMAR
56-75-7	EMBACETIN
74-83-9	EMBAFUME
563-12-2	EMBATHION
55-86-7	EMBICHIN
51-75-2	EMBICHIN
633-03-4	EMERALD GREEN
12002-03-8	EMERALD GREEN
151-21-3	EMERSAL 6400
139-96-8	EMERSAL 6434
57-11-4	EMERSOL 120
57-11-4	EMERSOL 132
57-11-4	EMERSOL 150
57-11-4	EMERSOL 153
112-80-1	EMERSOL 210
112-80-1	EMERSOL 211
112-80-1	EMERSOL 213
112-80-1	EMERSOL 220 WHITE OLEIC ACID
112-80-1	EMERSOL 221
112-80-1	EMERSOL 221 LOW TITER WHITE OLEIC ACID
112-80-1	EMERSOL 233LL
112-80-1	EMERSOL 6321
57-11-4	EMERSOL 6349
2235-54-3	EMERSOL 6430
12415-34-8	EMERY
60-24-2	EMERY 5791
316-42-7	EMETAN, 6',7',10,11-TETRAMETHOXY-, DIHYDROCHLORIDE
316-42-7	EMETINE HYDROCHLORIDE
316-42-7	EMETINE, DIHYDROCHLORIDE
28300-74-5	EMETIQUE (FRENCH)
7487-94-7	EMISAN 6
9005-25-8	EMJEL 200
9005-25-8	EMJEL 300
110-80-5	EMKANOL
25322-69-4	EMKAPYL
121-75-5	EMMATOS
121-75-5	EMMATOS EXTRA
9016-45-9	EMMON 15332
7440-32-6	EMO 140
54-11-5	EMO-NIK
2235-25-8	EMP
151-21-3	EMPICOL LPZ
151-21-3	EMPICOL LS 30
151-21-3	EMPICOL LX 28
9016-45-9	EMPILAN NP 9
50-78-2	EMPIRIN
62-50-0	EMS
14807-96-6	EMTAL 500
14807-96-6	EMTAL 549
14807-96-6	EMTAL 596
14807-96-6	EMTAL 599
9016-45-9	EMU 02
9016-45-9	EMU 09
9016-45-9	EMULGATOR NP 10
9016-45-9	EMULGEN 900
9016-45-9	EMULGEN 903
9016-45-9	EMULGEN 905
9016-45-9	EMULGEN 906
9016-45-9	EMULGEN 909
9016-45-9	EMULGEN 910
9016-45-9	EMULGEN 911
9016-45-9	EMULGEN 913
9016-45-9	EMULGEN 920
9016-45-9	EMULGEN 930
9016-45-9	EMULGEN 931
9016-45-9	EMULGEN 935
9016-45-9	EMULGEN 950
9016-45-9	EMULGEN 985
9016-45-9	EMULGEN PI 20T
25155-30-0	EMULIN B 22
94-75-7	EMULSAMINE BK
94-75-7	EMULSAMINE E-3
151-21-3	EMULSIFIER NO 10 4
7601-54-9	EMULSIPHOS 440/660
9016-45-9	EMULSON 20B
9016-45-9	EMULSON 9B
297-97-2	EN 18133
72-20-8	EN 57
111-70-6	ENANTHIC ALCOHOL
9002-89-5	ENBRA OV
507-70-0	ENDO-2-HYDROXY-1,7,7-TRIMETHYLNORBORNANE
15271-41-7	ENDO-3-CHLORO-EXO-6-CYANO-2-NORBORNANONE O-(METHYL-CARBAMOYL)OXIME
507-70-0	ENDO-BORNEOL
115-29-7	ENDOCEL
2778-04-3	ENDOCID
2778-04-3	ENDOCIDE
53-16-7	ENDOFOLLICULINA
115-29-7	ENDOSOL
50-78-2	ENDOSPRIN
115-29-7	ENDOSULFAN
115-29-7	ENDOSULFAN (ACGIH,DOT)
959-98-8	ENDOSULFAN 1
33213-65-9	ENDOSULFAN 2
115-29-7	ENDOSULFAN 35EC
959-98-8	ENDOSULFAN A
33213-65-9	ENDOSULFAN B
959-98-8	ENDOSULFAN I
33213-65-9	ENDOSULFAN II
1031-07-8	ENDOSULFAN SULFATE
115-29-7	ENDOSULPHAN
129-67-9	ENDOTHALL
2778-04-3	ENDOTHION
5836-29-3	ENDOX
50-18-0	ENDOXAN
60-00-4	ENDRATE
72-20-8	ENDREX
72-20-8	ENDRICOL
72-20-8	ENDRIN
72-20-8	ENDRIN (ACGIH,DOT)
7421-93-4	ENDRIN ALDEHYDE
72-20-8	ENDRINE (FRENCH)
5836-29-3	ENDROCID
5836-29-3	ENDROCIDE
50-78-2	ENDYDOL
786-19-6	ENDYL
5836-29-3	ENE 11183 B
13838-16-9	ENFLURANE
1309-37-1	ENGLISH IRON OXIDE RED
60-11-7	ENIAL YELLOW 2G
1937-37-7	ENIANIL BLACK CN
2602-46-2	ENIANIL BLUE 2BN
16071-86-6	ENIANIL LIGHT BROWN BRL
56-75-7	ENICOL

CAS No.	Chemical Name
470-90-6	ENOLOFOS
16893-85-9	ENS-ZEM WEEVIL BAIT
115-29-7	ENSURE
88-85-7	ENT 1,122
16893-85-9	ENT 1,501
50-29-3	ENT 1,506
107-06-2	ENT 1,656
72-43-5	ENT 1,716
127-18-4	ENT 1,860
56-38-2	ENT 15,108
76-44-8	ENT 15,152
106-93-4	ENT 15,349
309-00-2	ENT 15,949
30900-23-3	ENT 15,949
1335-85-9	ENT 154
131-89-5	ENT 157
60-57-1	ENT 16,225
3689-24-5	ENT 16,273
143-50-0	ENT 16,391
121-75-5	ENT 17,034
72-20-8	ENT 17,251
152-16-9	ENT 17,291
298-00-0	ENT 17,292
8065-48-3	ENT 17,295
584-79-2	ENT 17,510
2104-64-5	ENT 17,798
56-72-4	ENT 17,957
119-38-0	ENT 19,060
115-26-4	ENT 19,109
333-41-5	ENT 19,507
52-68-6	ENT 19,763
65-30-5	ENT 2,435
6505-86-8	ENT 2,435
62-73-7	ENT 20,738
2636-26-2	ENT 20,738
3734-97-2	ENT 20,993
2642-71-9	ENT 22,014
7786-34-7	ENT 22,374
78-34-2	ENT 22,897
86-50-0	ENT 23,233
299-84-3	ENT 23,284
298-04-4	ENT 23,437
115-32-2	ENT 23,648
786-19-6	ENT 23,708
63-25-2	ENT 23,969
115-29-7	ENT 23,979
298-02-2	ENT 24,042
563-12-2	ENT 24,105
141-66-2	ENT 24,482
60-51-5	ENT 24,650
2275-18-5	ENT 24,652
2778-04-3	ENT 24,653
545-55-1	ENT 24,915
115-90-2	ENT 24,945
470-90-6	ENT 24,969
300-76-5	ENT 24,988
78-53-5	ENT 24,980-X
61-82-5	ENT 25,445
64-00-6	ENT 25,500
55-38-9	ENT 25,540
64-00-6	ENT 25,543
297-78-9	ENT 25,545
297-78-9	ENT 25,545-X
12789-03-6	ENT 25,552-X
3735-23-7	ENT 25,554-X
297-97-2	ENT 25,580
1024-57-3	ENT 25,584
2275-14-1	ENT 25,585
644-64-4	ENT 25,595-X
114-26-1	ENT 25,671
732-11-6	ENT 25,705
327-98-0	ENT 25,712
122-14-5	ENT 25,715
2385-85-5	ENT 25,719
2032-65-7	ENT 25,726
3254-63-5	ENT 25,734
315-18-4	ENT 25,766
2665-30-7	ENT 25,787
944-22-9	ENT 25,796
947-02-4	ENT 25,830
644-64-4	ENT 25,922
15271-41-7	ENT 25,962
950-10-7	ENT-25,991
51-75-2	ENT-25,294
131-11-3	ENT 262
101-05-3	ENT 26,058
12654-97-6	ENT 26,058
54-62-6	ENT-26,079
133-06-2	ENT 26,538
1464-53-5	ENT-26,592
116-06-3	ENT 27,093
6923-22-4	ENT 27,129
1563-66-2	ENT 27,164
2312-35-8	ENT 27,226
2540-82-1	ENT 27,257
2631-37-0	ENT 27,300
2631-37-0	ENT 27,300-A
2921-88-2	ENT 27,311
13194-48-4	ENT 27,318
10265-92-6	ENT 27,396
22224-92-6	ENT 27,572
21923-23-9	ENT 27,635
39196-18-4	ENT 27,851
54-11-5	ENT 3,424
117-80-6	ENT 3,776
92-84-2	ENT 38
2499-59-4	ENT 3827
111-44-4	ENT 4,504
56-23-5	ENT 4,705
50-55-5	ENT 50,146
28300-74-5	ENT 50,434
66-75-1	ENT 50,439
680-31-9	ENT 50,882
991-42-4	ENT 51,762
58-89-9	ENT 7,796
56-72-4	ENT 7,957
94-75-7	ENT 8,538
58-89-9	ENT 8,601
12002-03-8	ENT 884
8001-35-2	ENT 9,735
12789-03-6	ENT 9,932
57-74-9	ENT 9,932
56-75-7	ENTEROMYCETIN
50-78-2	ENTEROSAREIN
50-78-2	ENTEROSARINE
55-38-9	ENTEX
443-48-1	ENTIZOL
58-89-9	ENTOMOXAN
50-78-2	ENTROPHEN
118-96-7	ENTSUFON
759-73-9	ENU
94-75-7	ENVERT 171
94-75-7	ENVERT DT
93-76-5	ENVERT-T
9002-89-5	EP 160
25322-69-4	EP 240
87-86-5	EP 30
2631-37-0	EP 316
23422-53-9	EP 332
556-61-6	EP-161E
112-30-1	EPAL 10
112-53-8	EPAL 12
111-27-3	EPAL 6
111-87-5	EPAL 8
1333-86-4	EPC
1333-86-4	EPC (CARBON BLACK)
94-36-0	EPI-CLEAR
96-24-2	EPIBLOC
3132-64-7	EPIBROMHYDRIN
3132-64-7	EPIBROMOHYDRIN
3132-64-7	EPIBROMOHYDRIN (DOT)
3132-64-7	EPIBROMOHYDRINE
106-89-8	EPICHLOORHYDRINE (DUTCH)
106-89-8	EPICHLORHYDRIN (GERMAN)
106-89-8	EPICHLORHYDRINE (FRENCH)
106-89-8	EPICHLOROHYDRIN
106-89-8	EPICHLOROHYDRIN (ACGIH,DOT)
106-89-8	EPICHLOROHYDRYNA (POLISH)

CAS No.	Chemical Name
106-89-8	EPICLORIDRINA (ITALIAN)
101-77-9	EPICURE DDM
556-52-5	EPIHYDRIN ALCOHOL
765-34-4	EPIHYDRINALDEHYDE
765-34-4	EPIHYDRINE ALDEHYDE
101-77-9	EPIKURE DDM
2104-64-5	EPN
2104-64-5	EPN (ACGIH)
2104-64-5	EPN 300
1954-28-5	EPODYL
4016-11-9	EPOXY ETHYLOXY PROPANE
106-88-7	EPOXYBUTANE
75-21-8	EPOXYETHANE
1024-57-3	EPOXYHEPTACHLOR
765-34-4	EPOXYPROPANAL
75-56-9	EPOXYPROPANE
96-09-3	EPOXYSTYRENE
105-60-2	EPSILON-CAPROLACTAM
105-60-2	EPSILON-CAPROLACTAM (ACGIH)
105-60-2	EPSYLON KAPROLAKTAM (POLISH)
76-44-8	EPTACLORO (ITALIAN)
142-82-5	EPTANI (ITALIAN)
151-21-3	EQUEX S
62-73-7	EQUIGARD
2636-26-2	EQUIGARD
62-73-7	EQUIGEL
2636-26-2	EQUIGEL
52-68-6	EQUINO-ACID
52-68-6	EQUINO-AID
2921-88-2	ERADEX
7553-56-2	ERANOL
75-60-5	ERASE
55-86-7	ERASOL
110-85-0	ERAVERM
50-14-6	ERGOCALCIFEROL
103-23-1	ERGOPLAST ADDO
123-79-5	ERGOPLAST ADDO
84-74-2	ERGOPLAST FDB
117-81-7	ERGOPLAST FDO
117-84-0	ERGOPLAST FDO
117-81-7	ERGOPLAST FDO-S
379-79-3	ERGOTAMINE TARTRATE
1937-37-7	ERIE BLACK B
1937-37-7	ERIE BLACK BF
1937-37-7	ERIE BLACK GAC
1937-37-7	ERIE BLACK GXOO
1937-37-7	ERIE BLACK JET
1937-37-7	ERIE BLACK NUG
1937-37-7	ERIE BLACK RXOO
1937-37-7	ERIE BRILLIANT BLACK S
1937-37-7	ERIE FIBRE BLACK VP
7632-00-0	ERINITRIT
2426-08-6	ERL 0810
57-47-6	ERSERINE
105-60-2	ERTALON 6SA
8020-83-5	ERVOL
8020-83-5	ERVOL WHITE MINERAL OIL
494-03-1	ERYSAN
106-99-0	ERYTHRENE
25339-57-5	ERYTHRENE
431-03-8	ERYTHRITOL ANHYDRIDE
1464-53-5	ERYTHRITOL ANHYDRIDE
78-10-4	ES 100
78-10-4	ES 28
78-10-4	ES 28 (ESTER)
11099-06-2	ES 32
11099-06-2	ES 40
11099-06-2	ES 40 (SILICATE)
126-72-7	ES 685
118-74-1	ESACLOROBENZENE (ITALIAN)
100-97-0	ESAMETILENTETRAMINA (ITALIAN)
131-73-7	ESAMETILENTETRAMINA (ITALIAN)
110-54-3	ESANI (ITALIAN)
50-70-4	ESASORB
298-07-7	ESCAID 100
85-44-9	ESEN
57-47-6	ESERINE
57-64-7	ESERINE SALICYLATE
57-47-6	ESEROLEIN, METHYLCARBAMATE (ESTER)

CAS No.	Chemical Name
50-55-5	ESERPINE
1910-2-5	ESGRAM
8002-74-2	ESKAR R 25
50-55-5	ESKASERP
75-69-4	ESKIMON 11
75-71-8	ESKIMON 12
75-45-6	ESKIMON 22
97-77-8	ESPENAL
97-77-8	ESPERAL
16111-62-9	ESPERCARB 840M
614-45-9	ESPEROX 10
109-13-7	ESPEROX 24M
3006-82-4	ESPEROX 28
927-07-1	ESPEROX 31M
26748-41-4	ESPEROX 33M
29240-17-3	ESPEROX 551M
26748-47-0	ESPEROX 939M
57-47-6	ESROMIOTIN
98-95-3	ESSENCE OF MIRBANE
98-95-3	ESSENCE OF MYRBANE
93-58-3	ESSENCE OF NIOBE
141-78-6	ESSIGESTER (GERMAN)
64-19-7	ESSIGSAEURE (GERMAN)
108-24-7	ESSIGSAEUREANHYDRID (GERMAN)
8002-74-2	ESSO 3150
133-06-2	ESSO FUNGICIDE 406
94-80-4	ESSO HERBICIDE 10
8007-45-2	ESTAR
8007-45-2	ESTAR (SKIN TREATMENT)
57-63-6	ESTEED
93-76-5	ESTERCIDE T-2 AND T-245
93-76-5	ESTERON
94-75-7	ESTERON
93-76-5	ESTERON 245
93-76-5	ESTERON 245 BE
94-11-1	ESTERON 44
94-75-7	ESTERON 44 WEED KILLER
94-75-7	ESTERON 76 BE
94-75-7	ESTERON 99
94-75-7	ESTERON 99 CONCENTRATE
94-75-7	ESTERON BRUSH KILLER
93-76-5	ESTERON BRUSH KILLER
94-75-7	ESTERONE FOUR
57-63-6	ESTIGYN
56-53-1	ESTILBIN MCO
57-63-6	ESTINYL
57-63-6	ESTON-E
50-29-3	ESTONATE
94-75-7	ESTONE
8001-35-2	ESTONOX
57-63-6	ESTORAL
57-63-6	ESTORALS
79-21-0	ESTOSTERIL
53-16-7	ESTRA-1,3,5(10)-TRIEN-17-ONE, 3-HYDROXY-
50-28-2	ESTRA-1,3,5(10)-TRIENE-3,17-DIOL (17BETA)-
50-28-2	ESTRA-1,3,5(10)-TRIENE-3,17BETA-DIOL
50-28-2	ESTRACE
50-28-2	ESTRADIOL
22966-79-6	ESTRADIOL MUSTARD
57-63-6	ESTRADIOL, 17-ETHYNYL-
50-28-2	ESTRALDINE
56-53-1	ESTROBENE
50-28-2	ESTROGENS (NOT CONJUGATED): ESTRADIOL 17BETA
53-16-7	ESTROGENS (NOT CONJUGATED): ESTRONE
57-63-6	ESTROGENS (NOT CONJUGATED): ETHINYLOESTRADIOL
72-33-3	ESTROGENS (NOT CONJUGATED): MESTRANOL
56-53-1	ESTROMENIN
53-16-7	ESTRON
53-16-7	ESTRONE
62-73-7	ESTROSEL
62-73-7	ESTROSOL
56-53-1	ESTROSYN
53-16-7	ESTROVARIN
50-28-2	ESTROVITE
53-16-7	ESTRUGENONE
53-16-7	ESTRUSOL
299-84-3	ET 14
299-84-3	ET 57
97-77-8	ETABUS

CAS No.	Chemical Name
7646-78-8	ETAIN (TETRACHLORURE D') (FRENCH)
64-17-5	ETANOLO (ITALIAN)
75-08-1	ETANTIOLO (ITALIAN)
60-29-7	ETERE ETILICO (ITALIAN)
75-08-1	ETHAANTHIOL (DUTCH)
71-55-6	ETHANA NU
75-07-0	ETHANAL
107-29-9	ETHANAL OXIME
75-04-7	ETHANAMINE
75-04-7	ETHANAMINE, (AQUEOUS SOLUTION)
98-84-0	ETHANAMINE, 1-PHENYL-
555-77-1	ETHANAMINE, 2-CHLORO-N,N-BIS(2-CHLOROETHYL)-
538-07-8	ETHANAMINE, 2-CHLORO-N-(2-CHLOROETHYL)-N-ETHYL-
51-75-2	ETHANAMINE, 2-CHLORO-N-(2-CHLOROETHYL)-N-METHYL-
55-86-7	ETHANAMINE, 2-CHLORO-N-(2-CHLOROETHYL)-N-METHYL-, HYDROCHLORIDE
302-70-5	ETHANAMINE, 2-CHLORO-N-(2-CHLOROETHYL)-N-METHYL-, N-OXIDE, HYDROCHLORIDE
121-44-8	ETHANAMINE, N,N-DIETHYL-
109-89-7	ETHANAMINE, N-ETHYL-
3710-84-7	ETHANAMINE, N-ETHYL-N-HYDROXY-
55-18-5	ETHANAMINE, N-ETHYL-N-NITROSO-
10595-95-6	ETHANAMINE, N-METHYL-N-NITROSO-
51-83-2	ETHANAMINIUM, 2-(AMINOCARBONYL)OXY-N,N,N-TRIMETHYL-, CHLORIDE
999-81-5	ETHANAMINIUM, 2-CHLORO-N,N,N-TRIMETHYL-, CHLORIDE
999-81-5	ETHANAMINIUM, 2-CHLORO-N,N,N-TRIMETHYL-, CHLORIDE (9CI)
633-03-4	ETHANAMINIUM, N-[4-[[4-(DIETHYLAMINO)PHENYL]PHENYL-METHYLENE]-2,5-CYCLOHEXADIEN-1-YLIDENE]-N-ETHYL-,
107-22-2	ETHANDIAL
74-84-0	ETHANE
107-06-2	ETHANE DICHLORIDE
144-62-7	ETHANE DIONIC ACID
67-72-1	ETHANE HEXACHLORIDE
76-01-7	ETHANE PENTACHLORIDE
79-21-0	ETHANE PEROXOIC ACID
3188-13-4	ETHANE, (CHLOROMETHOXY)-
3188-13-4	ETHANE, (CHLOROMETHOXY)- (9CI)
122-51-0	ETHANE, 1,1',1"-[METHYLIDYNETRIS(OXY)]TRIS-
111-44-4	ETHANE, 1,1'-OXYBIS(2-CHLORO-
111-96-6	ETHANE, 1,1'-OXYBIS(2-METHOXY- (9CI)
60-29-7	ETHANE, 1,1'-OXYBIS-
109-63-7	ETHANE, 1,1'-OXYBIS-, COMPD. WITH TRIFLUOROBORANE (1:1)
111-96-6	ETHANE, 1,1'-OXYBIS[2-METHOXY-
627-53-2	ETHANE, 1,1'-SELENOBIS-
505-60-2	ETHANE, 1,1'-THIOBIS[2-CHLORO-
462-95-3	ETHANE, 1,1'-[METHYLENEBIS(OXY)]BIS-
111-91-1	ETHANE, 1,1'-[METHYLENEBIS(OXY)]BIS[2-CHLORO-
76-11-9	ETHANE, 1,1,1,2-TETRACHLORO-2,2-DIFLUORO-
71-55-6	ETHANE, 1,1,1-TRICHLORO-
50-29-3	ETHANE, 1,1,1-TRICHLORO-2,2-BIS(4-CHLOROPHENYL)-
50-29-3	ETHANE, 1,1,1-TRICHLORO-2,2-BIS(P-CHLOROPHENYL)-
72-43-5	ETHANE, 1,1,1-TRICHLORO-2,2-BIS(P-METHOXYPHENYL)-
79-27-6	ETHANE, 1,1,2,2-TETRABROMO-
79-34-5	ETHANE, 1,1,2,2-TETRACHLORO-
76-12-0	ETHANE, 1,1,2,2-TETRACHLORO-1,2-DIFLUORO-
79-00-5	ETHANE, 1,1,2-TRICHLORO-
76-13-1	ETHANE, 1,1,2-TRICHLORO-1,2,2-TRIFLUORO-
75-34-3	ETHANE, 1,1-DICHLORO-
594-72-9	ETHANE, 1,1-DICHLORO-1-NITRO-
72-54-8	ETHANE, 1,1-DICHLORO-2,2-BIS(P-CHLOROPHENYL)-
105-57-7	ETHANE, 1,1-DIETHOXY-
75-37-6	ETHANE, 1,1-DIFLUORO-
25497-28-3	ETHANE, 1,1-DIFLUORO-
112-26-5	ETHANE, 1,2-BIS(2-CHLOROETHOXY)-
1954-28-5	ETHANE, 1,2-BIS[2-(2,3-EPOXYPROPOXY)ETHOXY]-
13426-91-0	ETHANE, 1,2-DIAMINO-, COPPER COMPLEX
106-93-4	ETHANE, 1,2-DIBROMO-
107-06-2	ETHANE, 1,2-DICHLORO-
629-14-1	ETHANE, 1,2-DIETHOXY-
110-71-4	ETHANE, 1,2-DIMETHOXY-
96-09-3	ETHANE, 1,2-EPOXY-1-PHENYL-
592-55-2	ETHANE, 1-BROMO-2-ETHOXY-
592-55-2	ETHANE, 1-BROMO-2-ETHOXY- (9CI)
75-68-3	ETHANE, 1-CHLORO-1,1-DIFLUORO-
25497-29-4	ETHANE, 1-CHLORO-1,1-DIFLUORO-
598-92-5	ETHANE, 1-CHLORO-1-NITRO-
1663-35-0	ETHANE, 1-METHOXY-2-(VINYLOXY)-
72-43-5	ETHANE, 2,2-BIS(P-ANISYL)-1,1,1-TRICHLORO-

CAS No.	Chemical Name
97-97-2	ETHANE, 2-CHLORO-1,1-DIMETHOXY-
74-96-4	ETHANE, BROMO-
75-00-3	ETHANE, CHLORO-
25497-29-4	ETHANE, CHLORODIFLUORO-
76-15-3	ETHANE, CHLOROPENTAFLUORO-
63938-10-3	ETHANE, CHLOROTETRAFLUORO-
1330-45-6	ETHANE, CHLOROTRIFLUORO-
1320-37-2	ETHANE, DICHLOROTETRAFLUORO-
75-37-6	ETHANE, DIFLUORO-
25497-28-3	ETHANE, DIFLUORO-
353-36-6	ETHANE, FLUORO-
67-72-1	ETHANE, HEXACHLORO-
76-16-4	ETHANE, HEXAFLUORO-
109-90-0	ETHANE, ISOCYANATO-
540-67-0	ETHANE, METHOXY-
79-24-3	ETHANE, NITRO-
76-01-7	ETHANE, PENTACHLORO-
74-84-0	ETHANE, REFRIGERATED
74-84-0	ETHANE, REFRIGERATED LIQUID (DOT)
25322-20-7	ETHANE, TETRACHLORO-
542-90-5	ETHANE, THIOCYANATO-
27987-06-0	ETHANE, TRIFLUORO-
79-09-4	ETHANECARBOXYLIC ACID
107-22-2	ETHANEDIAL
460-19-5	ETHANEDINITRILE
144-62-7	ETHANEDIOIC ACID
1113-38-8	ETHANEDIOIC ACID DIAMMONIUM SALT
2944-67-4	ETHANEDIOIC ACID, AMMONIUM IRON(3+) SALT (3:3:1)
14258-49-2	ETHANEDIOIC ACID, AMMONIUM SALT
814-91-5	ETHANEDIOIC ACID, COPPER(2+) SALT (1:1)
1113-38-8	ETHANEDIOIC ACID, DIAMMONIUM SALT, MONOHYDRATE
6009-70-7	ETHANEDIOIC ACID, DIAMMONIUM SALT, MONOHYDRATE
95-92-1	ETHANEDIOIC ACID, DIETHYL ESTER
6153-56-6	ETHANEDIOIC ACID, DIHYDRATE
583-52-8	ETHANEDIOIC ACID, DIPOTASSIUM SALT
10043-22-8	ETHANEDIOIC ACID, DIPOTASSIUM SALT
583-52-8	ETHANEDIOIC ACID, DIPOTASSIUM SALT (9CI)
10043-22-8	ETHANEDIOIC ACID, DIPOTASSIUM SALT (9CI)
62-76-0	ETHANEDIOIC ACID, DISODIUM SALT
516-03-0	ETHANEDIOIC ACID, IRON(2+) SALT (1:1)
5972-73-6	ETHANEDIOIC ACID, MONOAMMONIUM SALT, MONOHYDRATE
127-95-7	ETHANEDIOIC ACID, MONOPOTASSIUM SALT
583-52-8	ETHANEDIOIC ACID, POTASSIUM SALT
10043-22-8	ETHANEDIOIC ACID, POTASSIUM SALT
111-55-7	ETHANEDIOL DIACETATE
107-22-2	ETHANEDIONE
75-05-8	ETHANENITRILE
107-71-1	ETHANEPEROXOIC ACID, 1,1-DIMETHYLETHYL ESTER
1622-32-8	ETHANESULFONYL CHLORIDE, 2-CHLORO
62-55-5	ETHANETHIOAMIDE
507-09-5	ETHANETHIOIC ACID
75-08-1	ETHANETHIOL
640-15-3	ETHANETHIOL, 2-(ETHYLTHIO)-, S-ESTER WITH O,O-DIMETHYL PHOSPHORODITHIOATE
919-86-8	ETHANETHIOL, 2-(ETHYLTHIO)-, S-ESTER WITH O,O-DIMETHYL PHOSPHOROTHIOATE
507-09-5	ETHANETHIOLIC ACID
23135-22-0	ETHANIMIDOTHIOIC ACID, 2-(DIMETHYLAMINO)-N-[[(METHYL-AMINO)CARBONYL]OXY]-2-OXO-, METHYL ESTER
16752-77-5	ETHANIMIDOTHIOIC ACID, N-[[(METHYLAMINO)CARBONYL]OXY]-,METHYL ESTER
64-19-7	ETHANOIC ACID
108-05-4	ETHANOIC ACID, ETHENYL ESTER
108-24-7	ETHANOIC ANHYDRATE
108-24-7	ETHANOIC ANHYDRIDE
64-17-5	ETHANOL
64-17-5	ETHANOL (DOT)
64-17-5	ETHANOL 200 PROOF
64-17-5	ETHANOL SOLUTION (DOT)
10140-87-1	ETHANOL, 1,2-DICHLORO- , ACETATE
75-39-8	ETHANOL, 1-AMINO-
75-39-8	ETHANOL, 1-AMINO- (8CI,9CI)
98-85-1	ETHANOL, 1-PHENYL-
102-71-6	ETHANOL, 2,2',2"-NITRILOTRI-
3813-14-7	ETHANOL, 2,2',2"-NITRILOTRI-, (2,4,5-TRICHLOROPHENOXY)-ACETATE (SALT)
27323-41-7	ETHANOL, 2,2',2"-NITRILOTRI-, DODECYLBENZENESULFONATE (SALT)
102-71-6	ETHANOL, 2,2',2"-NITRILOTRIS-

CAS No.	Chemical Name
3813-14-7	ETHANOL, 2,2',2"-NITRILOTRIS-, (2,4,5-TRICHLOROPHENOXY)ACETATE (1:1) (SALT)
139-96-8	ETHANOL, 2,2',2"-NITRILOTRIS-, DODECYL SULFATE (SALT)
27323-41-7	ETHANOL, 2,2',2"-NITRILOTRIS-, DODECYLBENZENESULFONATE (SALT)
4985-85-7	ETHANOL, 2,2'-(AMINOPROPYLIMINO)-
102-79-4	ETHANOL, 2,2'-(BUTYLIMINO)BIS-
102-79-4	ETHANOL, 2,2'-(BUTYLIMINO)DI-
112-27-6	ETHANOL, 2,2'-(ETHYLENEDIOXY)DI-
139-87-7	ETHANOL, 2,2'-(ETHYLIMINO)BIS-
139-87-7	ETHANOL, 2,2'-(ETHYLIMINO)DI-
91-99-6	ETHANOL, 2,2'-(M-TOLYLIMINO)DI-
105-59-9	ETHANOL, 2,2'-(METHYLIMINO)BIS-
105-59-9	ETHANOL, 2,2'-(METHYLIMINO)DI-
1116-54-7	ETHANOL, 2,2'-(NITROSOIMINO)BIS-
112-60-7	ETHANOL, 2,2'-(OXYBIS(ETHYLENEOXY))DI-
2160-93-2	ETHANOL, 2,2'-(TERT-BUTYLIMINO)DI-
111-42-2	ETHANOL, 2,2'-IMINOBIS-
143-00-0	ETHANOL, 2,2'-IMINOBIS-, DODECYL SULFATE (SALT)
143-00-0	ETHANOL, 2,2'-IMINODI-, DODECYL SULFATE (SALT)
111-42-2	ETHANOL, 2,2'-IMINODI-
143-00-0	ETHANOL, 2,2'-IMINODI-, COMPD WITH DODECYL SULFATE
1116-54-7	ETHANOL, 2,2'-NITROSIMINODI-
111-46-6	ETHANOL, 2,2'-OXYBIS-
111-96-6	ETHANOL, 2,2'-OXYBIS-, DIMETHYL ETHER
112-34-5	ETHANOL, 2,2'-OXYBIS-, MONOBUTYL ETHER
111-90-0	ETHANOL, 2,2'-OXYBIS-, MONOETHYL ETHER
111-77-3	ETHANOL, 2,2'-OXYBIS-, MONOMETHYL ETHER
111-46-6	ETHANOL, 2,2'-OXYDI-
111-48-8	ETHANOL, 2,2'-THIOBIS-
111-48-8	ETHANOL, 2,2'-THIODI-
2160-93-2	ETHANOL, 2,2'-[(1,1-DIMETHYLETHYL)IMINO]BIS-
4985-85-7	ETHANOL, 2,2'-[(3-AMINOPROPYL)IMINO]BIS-
4985-85-7	ETHANOL, 2,2'-[(3-AMINOPROPYL)IMINO]DI-
91-99-6	ETHANOL, 2,2'-[(3-METHYLPHENYL)IMINO]BIS-
112-27-6	ETHANOL, 2,2'-[1,2-ETHANEDIYLBIS(OXY)]BIS-
112-60-7	ETHANOL, 2,2'-[OXYBIS(2,1-ETHANEDIYLOXY)]BIS-
115-32-2	ETHANOL, 2,2,2-TRICHLORO-1,1-BIS(4-CHLOROPHENYL)-
115-32-2	ETHANOL, 2,2,2-TRICHLORO-1,1-BIS(P-CHLOROPHENYL)-
109-56-8	ETHANOL, 2-((1-METHYLETHYL)AMINO)- (9CI)
111-41-1	ETHANOL, 2-((2-AMINOETHYL)AMINO)-
109-59-1	ETHANOL, 2-(1-METHYLETHOXY)-
136-78-7	ETHANOL, 2-(2,4-DICHLOROPHENOXY)-, HYDROGEN SULFATE SODIUM SALT
112-50-5	ETHANOL, 2-(2-(2-ETHOXYETHOXY)ETHOXY)-
929-06-6	ETHANOL, 2-(2-AMINOETHOXY)-
112-34-5	ETHANOL, 2-(2-BUTOXYETHOXY)-
124-17-4	ETHANOL, 2-(2-BUTOXYETHOXY)-, ACETATE
111-90-0	ETHANOL, 2-(2-ETHOXYETHOXY)-
111-77-3	ETHANOL, 2-(2-METHOXYETHOXY)-
4439-24-1	ETHANOL, 2-(2-METHYLPROPOXY)-
622-08-2	ETHANOL, 2-(BENZYLOXY)-
2842-38-8	ETHANOL, 2-(CYCLOHEXYLAMINO)-
102-81-8	ETHANOL, 2-(DIBUTYLAMINO)-
100-37-8	ETHANOL, 2-(DIETHYLAMINO)-
96-80-0	ETHANOL, 2-(DIISOPROPYLAMINO)-
2867-47-2	ETHANOL, 2-(DIMETHYLAMINO)-, METHACRYLATE
110-73-6	ETHANOL, 2-(ETHYLAMINO)-
92-50-2	ETHANOL, 2-(ETHYLPHENYLAMINO)-
109-56-8	ETHANOL, 2-(ISOPROPYLAMINO)-
109-83-1	ETHANOL, 2-(METHYLAMINO)-
93-90-3	ETHANOL, 2-(METHYLPHENYLAMINO)-
92-50-2	ETHANOL, 2-(N-ETHYLANILINO)-
93-90-3	ETHANOL, 2-(N-METHYLANILINO)-
122-98-5	ETHANOL, 2-(PHENYLAMINO)-
622-08-2	ETHANOL, 2-(PHENYLMETHOXY)-
3775-90-4	ETHANOL, 2-(TERT-BUTYLAMINO)-, METHACRYLATE (ESTER)
141-43-5	ETHANOL, 2-AMINO-
1420-04-8	ETHANOL, 2-AMINO-, COMPD. WITH 2',5-DICHLORO-4'-NITROSALICYLANILIDE (1:1)
122-98-5	ETHANOL, 2-ANILINO-
111-76-2	ETHANOL, 2-BUTOXY-
2545-59-7	ETHANOL, 2-BUTOXY-, (2,4,5-TRICHLOROPHENOXY)ACETATE
1929-73-3	ETHANOL, 2-BUTOXY-, (2,4-DICHLOROPHENOXY)ACETATE
107-07-3	ETHANOL, 2-CHLORO-
140-57-8	ETHANOL, 2-CHLORO-, 2-(P-TERT-BUTYLPHENOXY)-1-METHYLETHYL SULFITE
542-58-5	ETHANOL, 2-CHLORO-, ACETATE
108-01-0	ETHANOL, 2-DIMETHYLAMINO-

CAS No.	Chemical Name
110-80-5	ETHANOL, 2-ETHOXY-
111-15-9	ETHANOL, 2-ETHOXY-, ACETATE
371-62-0	ETHANOL, 2-FLUORO-
4301-50-2	ETHANOL, 2-FLUORO-, 4-BIPHENYLACETATE
4439-24-1	ETHANOL, 2-ISOBUTOXY-
109-59-1	ETHANOL, 2-ISOPROPOXY-
60-24-2	ETHANOL, 2-MERCAPTO-
109-86-4	ETHANOL, 2-METHOXY-
110-49-6	ETHANOL, 2-METHOXY-, ACETATE
109-56-8	ETHANOL, 2-[(1-METHYLETHYL)AMINO]-
10138-74-6	ETHANOL, 2-[(2-AMINO-1-METHYLETHYL)AMINO]-
140-82-9	ETHANOL, 2-[2-(DIETHYLAMINO)ETHOXY]-
96-80-0	ETHANOL, 2-[BIS(1-METHYLETHYL)AMINO]-
141-43-5	ETHANOLAMINE
111-41-1	ETHANOLETHYLENE DIAMINE
109-56-8	ETHANOLISOPROPYLAMINE
98-86-2	ETHANONE, 1-PHENYL-
98-86-2	ETHANONE, 1-PHENYL- (9CI)
1341-24-8	ETHANONE, 1-PHENYL-, MONOCHLORO DERIV.
70-11-1	ETHANONE, 2-BROMO-1-PHENYL-
70-11-1	ETHANONE, 2-BROMO-1-PHENYL- (9CI)
532-27-4	ETHANONE, 2-CHLORO-1-PHENYL-
563-12-2	ETHANOX
75-36-5	ETHANOYL CHLORIDE
593-67-9	ETHENAMINE
593-67-9	ETHENAMINE (9CI)
4549-40-0	ETHENAMINE, N-METHYL-N-NITROSO-
74-85-1	ETHENE
107-07-3	ETHENE CHLOROHYDRIN
75-21-8	ETHENE OXIDE
110-75-8	ETHENE, (2-CHLOROETHOXY)-
1663-35-0	ETHENE, (2-METHOXYETHOXY)-
109-93-3	ETHENE, 1,1'-OXYBIS-
75-35-4	ETHENE, 1,1-DICHLORO-
75-38-7	ETHENE, 1,1-DIFLUORO-
540-59-0	ETHENE, 1,2-DICHLORO-
156-60-5	ETHENE, 1,2-DICHLORO-, (E)-
156-59-2	ETHENE, 1,2-DICHLORO-, (Z)-
593-60-2	ETHENE, BROMO-
598-73-2	ETHENE, BROMOTRIFLUORO-
598-73-2	ETHENE, BROMOTRIFLUORO- (9CI)
75-01-4	ETHENE, CHLORO-
79-38-9	ETHENE, CHLOROTRIFLUORO-
109-92-2	ETHENE, ETHOXY
689-97-4	ETHENE, ETHYNYL-
75-02-5	ETHENE, FLUORO-
107-25-5	ETHENE, METHOXY-
127-18-4	ETHENE, TETRACHLORO-
116-14-3	ETHENE, TETRAFLUORO-
9002-84-0	ETHENE, TETRAFLUORO-, HOMOPOLYMER
79-01-6	ETHENE, TRICHLORO-
9002-89-5	ETHENOL HOMOPOLYMER (9CI)
62-73-7	ETHENOL, 2,2-DICHLORO-, DIMETHYL PHOSPHATE
2636-26-2	ETHENOL, 2,2-DICHLORO-, DIMETHYL PHOSPHATE
9002-89-5	ETHENOL, HOMOPOLYMER
463-51-4	ETHENONE
674-82-8	ETHENONE, DIMER
108-05-4	ETHENYL ACETATE
108-05-4	ETHENYL ETHANOATE
100-42-5	ETHENYLBENZENE
930-22-3	ETHENYLOXIRANE
109-93-3	ETHENYLOXYETHENE
60-29-7	ETHER
9016-45-9	ETHER
142-96-1	ETHER BUTYLIQUE (FRENCH)
75-00-3	ETHER CHLORATUS
107-12-0	ETHER CYANATUS
111-44-4	ETHER DICHLORE (FRENCH)
629-14-1	ETHER DIETHYLIQUE DE L'ETHYLENE-GLYCOL (FRENCH)
97-95-0	ETHER ETHYLBUTYLIQUE (FRENCH)
60-29-7	ETHER ETHYLIQUE (FRENCH)
75-00-3	ETHER HYDROCHLORIC
108-20-3	ETHER ISOPROPYLIQUE (FRENCH)
107-30-2	ETHER METHYLIQUE MONOCHLORE (FRENCH)
109-86-4	ETHER MONOMETHYLIQUE DE L'ETHYLENE-GLYCOL (FRENCH)
75-00-3	ETHER MURIATIC
1836-75-5	ETHER, 2,4-DICHLOROPHENYL P-NITROPHENYL
592-55-2	ETHER, 2-BROMOETHYL ETHYL
110-75-8	ETHER, 2-CHLOROETHYL VINYL

CAS No.	Chemical Name
103-44-6	ETHER, 2-ETHYLHEXYL VINYL
106-92-3	ETHER, ALLYL 2,3-EPOXYPROPYL
557-31-3	ETHER, ALLYL ETHYL
3917-15-5	ETHER, ALLYL VINYL
93-96-9	ETHER, BIS(ALPHA-METHYLBENZYL)
2238-07-5	ETHER, BIS(2,3-EPOXYPROPYL)
63283-80-7	ETHER, BIS(2-CHLORO-1- METHYLETHYL)
111-44-4	ETHER, BIS(2-CHLOROETHYL)
111-96-6	ETHER, BIS(2-METHOXYETHYL)
111-44-4	ETHER, BIS(CHLOROETHYL)
542-88-1	ETHER, BIS(CHLOROMETHYL)
112-98-1	ETHER, BIS[2-(2-BUTOXYETHOXY)ETHYL]
628-81-9	ETHER, BUTYL ETHYL
111-34-2	ETHER, BUTYL VINYL
3188-13-4	ETHER, CHLOROMETHYL ETHYL
107-30-2	ETHER, CHLOROMETHYL METHYL
111-43-3	ETHER, DI-N-PROPYL-
2238-07-5	ETHER, DIGLYCIDYL
115-10-6	ETHER, DIMETHYL
107-30-2	ETHER, DIMETHYL CHLORO
101-84-8	ETHER, DIPHENYL
109-93-3	ETHER, DIVINYL
927-80-0	ETHER, ETHYL ETHYNYL
540-67-0	ETHER, ETHYL METHYL
928-55-2	ETHER, ETHYL PROPENYL
628-32-0	ETHER, ETHYL PROPYL
109-92-2	ETHER, ETHYL VINYL
109-53-5	ETHER, ISOBUTYL VINYL
108-20-3	ETHER, ISOPROPYL
926-65-8	ETHER, ISOPROPYL VINYL
115-10-6	ETHER, METHYL
100-66-3	ETHER, METHYL PHENYL
557-17-5	ETHER, METHYL PROPYL
107-25-5	ETHER, METHYL VINYL
101-55-3	ETHER, P-BROMOPHENYL PHENYL
7005-72-3	ETHER, P-CHLOROPHENYL PHENYL
1320-21-4	ETHER, PENTYL XYLYL
109-92-2	ETHER, VINYL ETHYL
9002-84-0	ETHICON PTFE
594-72-9	ETHIDE
57-63-6	ETHIDOL
74-86-2	ETHINE
57-63-6	ETHINORAL
79-01-6	ETHINYL TRICHLORIDE
57-63-6	ETHINYLESTRADIOL
72-33-3	ETHINYLESTRADIOL 3-METHYL ETHER
57-63-6	ETHINYLESTRIOL
68-22-4	ETHINYLNORTESTOSTERONE
57-63-6	ETHINYLOESTRADIOL
563-12-2	ETHIOL
563-12-2	ETHIOL 100
121-75-5	ETHIOLACAR
563-12-2	ETHION
563-12-2	ETHION (ACGIH,DOT)
2921-88-2	ETHION, DRY
536-33-4	ETHIONAMIDE
1344-48-5	ETHIOPS MINERAL
56-38-2	ETHLON
563-12-2	ETHODAN
1642-54-2	ETHODRYL CITRATE
1954-28-5	ETHOGLUCID
1954-28-5	ETHOGLUCIDE
122-51-0	ETHONE
563-12-2	ETHOPAZ
13194-48-4	ETHOPROP
13194-48-4	ETHOPROPHOS
111-76-2	ETHOXY ACETATE
111-90-0	ETHOXY DIGLYCOL
112-50-5	ETHOXY TRIGLYCOL
2104-64-5	ETHOXY-4-NITROPHENOXYPHENYLPHOSPHINE SULFIDE
927-80-0	ETHOXYACETYLENE
1321-31-9	ETHOXYANILINE
541-41-3	ETHOXYCARBONYL CHLORIDE
140-88-5	ETHOXYCARBONYLETHYLENE
105-36-2	ETHOXYCARBONYLMETHYL BROMIDE
84-72-0	ETHOXYCARBONYLMETHYL ETHYL PHTHALATE
85-71-2	ETHOXYCARBONYLMETHYL METHYL PHTHALATE
2941-64-2	ETHOXYCARBONYLSULFENYL CHLORIDE
26555-35-1	ETHOXYCARBONYLSULFENYL CHLORIDE
3188-13-4	ETHOXYCHLOROMETHANE
103-75-3	ETHOXYDIHYDROPYRAN
60-29-7	ETHOXYETHANE
109-92-2	ETHOXYETHENE
111-76-2	ETHOXYETHYL ACETATE
109-92-2	ETHOXYETHYLENE
927-80-0	ETHOXYETHYNE
105-58-8	ETHOXYFORMIC ANHYDRIDE
9016-45-9	ETHOXYLATED NONYLPHENOL
3188-13-4	ETHOXYMETHYL CHLORIDE
462-95-3	ETHOXYMETHYL ETHYL ETHER
112-50-5	ETHOXYTRIETHYLENE GLYCOL
563-43-9	ETHYALUMINUM DICHLORIDE
2025-56-1	ETHYL
2941-64-2	ETHYL (CHLOROSULFENYL)FORMATE
26555-35-1	ETHYL (CHLOROSULFENYL)FORMATE
26555-35-1	ETHYL (CHLOROTHIO)FORMATE
10544-63-5	ETHYL (E)-CROTONATE
928-55-2	ETHYL 1-PROPENYL ETHER
105-36-2	ETHYL 2-BROMOACETATE
10544-63-5	ETHYL 2-BUTENOATE
105-39-5	ETHYL 2-CHLOROACETATE
105-56-6	ETHYL 2-CYANOACETATE
97-64-3	ETHYL 2-HYDROXYPROPANOATE
97-64-3	ETHYL 2-HYDROXYPROPIONATE
97-63-2	ETHYL 2-METHYL-2-PROPENOATE
97-63-2	ETHYL 2-METHYLACRYLATE
541-85-5	ETHYL 2-METHYLBUTYL KETONE
97-62-1	ETHYL 2-METHYLPROPANOATE
97-62-1	ETHYL 2-METHYLPROPIONATE
140-88-5	ETHYL 2-PROPENOATE
557-31-3	ETHYL 2-PROPENYL ETHER
22224-92-6	ETHYL 3-METHYL-4-(METHYLTHIO)PHENYL (1-METHYLETHYL)-PHOSPHORAMIDATE
141-97-9	ETHYL 3-OXOBUTANOATE
141-97-9	ETHYL 3-OXOBUTYRATE
510-15-6	ETHYL 4,4'-DICHLOROBENZILATE
22224-92-6	ETHYL 4-(METHYLTHIO)-M-TOLYL ISOPROPYLPHOSPHORAMI-DATE
141-78-6	ETHYL ACETATE
141-78-6	ETHYL ACETATE (ACGIH,DOT)
141-78-6	ETHYL ACETIC ESTER
141-97-9	ETHYL ACETOACETATE
107-87-9	ETHYL ACETONE
141-97-9	ETHYL ACETONECARBOXYLATE
141-97-9	ETHYL ACETYL ACETATE
141-97-9	ETHYL ACETYLACETONATE
107-00-6	ETHYL ACETYLENE
107-00-6	ETHYL ACETYLENE, INHIBITED
107-00-6	ETHYL ACETYLENE, INHIBITED (DOT)
140-88-5	ETHYL ACRYLATE
140-88-5	ETHYL ACRYLATE (ACGIH)
140-88-5	ETHYL ACRYLATE, INHIBITED
140-88-5	ETHYL ACRYLATE, INHIBITED (DOT)
64-17-5	ETHYL ALCOHOL
64-17-5	ETHYL ALCOHOL (ACGIH,DOT)
64-17-5	ETHYL ALCOHOL ANHYDROUS
75-07-0	ETHYL ALDEHYDE
75-07-0	ETHYL ALDEHYDE (DOT)
557-31-3	ETHYL ALLYL ETHER
105-36-2	ETHYL ALPHA-BROMOACETATE
105-39-5	ETHYL ALPHA-CHLOROACETATE
535-13-7	ETHYL ALPHA-CHLOROPROPIONATE
97-64-3	ETHYL ALPHA-HYDROXYPROPIONATE
97-63-2	ETHYL ALPHA-METHYL ACRYLATE
563-43-9	ETHYL ALUMINUM DICHLORIDE
563-43-9	ETHYL ALUMINUM DICHLORIDE (DOT)
12075-68-2	ETHYL ALUMINUM SESQUICHLORIDE
12075-68-2	ETHYL ALUMINUM SESQUICHLORIDE (DOT)
106-68-3	ETHYL AMYL KETONE
541-85-5	ETHYL AMYL KETONE
106-68-3	ETHYL AMYL KETONE (DOT)
2642-71-9	ETHYL AZINPHOS
100-41-4	ETHYL BENZENE
100-41-4	ETHYL BENZENE (ACGIH,DOT)
93-89-0	ETHYL BENZENECARBOXYLATE
93-89-0	ETHYL BENZOATE
34099-73-5	ETHYL BORATE
51845-86-4	ETHYL BORATE

CAS No.	Chemical Name
51845-86-4	ETHYL BORATE (DOT)
105-36-2	ETHYL BROMACETATE
74-96-4	ETHYL BROMIDE
74-96-4	ETHYL BROMIDE (ACGIH,DOT)
105-36-2	ETHYL BROMOACETATE
105-36-2	ETHYL BROMOACETATE (DOT)
105-54-4	ETHYL BUTANOATE
30714-78-4	ETHYL BUTYL CARBONATE
628-81-9	ETHYL BUTYL ETHER
97-95-0	ETHYL BUTYL ETHER (DOT)
106-35-4	ETHYL BUTYL KETONE
97-96-1	ETHYL BUTYRALDEHYDE
97-96-1	ETHYL BUTYRALDEHYDE (DOT)
105-54-4	ETHYL BUTYRATE
105-54-4	ETHYL BUTYRATE (DOT)
123-66-0	ETHYL CAPROATE
106-32-1	ETHYL CAPRYLATE
51-79-6	ETHYL CARBAMATE
84-72-0	ETHYL CARBETHOXYMETHYL PHTHALATE
71-23-8	ETHYL CARBINOL
111-90-0	ETHYL CARBITOL
105-58-8	ETHYL CARBONATE
105-58-8	ETHYL CARBONATE ((ETO)2CO)
541-41-3	ETHYL CARBONOCHLORIDATE
786-19-6	ETHYL CARBOPHENOTHION
110-80-5	ETHYL CELLOSOLVE
111-76-2	ETHYL CELLOSOLVE ACETAAT (DUTCH)
111-15-9	ETHYL CELLOSOLVE ACETATE
105-39-5	ETHYL CHLORACETATE
75-00-3	ETHYL CHLORIDE
75-00-3	ETHYL CHLORIDE (ACGIH,DOT)
105-39-5	ETHYL CHLOROACETATE
105-39-5	ETHYL CHLOROACETATE (DOT)
541-41-3	ETHYL CHLOROCARBONATE
541-41-3	ETHYL CHLOROCARBONATE (DOT)
105-39-5	ETHYL CHLOROETHANOATE
541-41-3	ETHYL CHLOROFORMATE
541-41-3	ETHYL CHLOROFORMATE (DOT)
3188-13-4	ETHYL CHLOROMETHYL ETHER
535-13-7	ETHYL CHLOROPROPIONATE
2941-64-2	ETHYL CHLOROTHIOFORMATE
26555-35-1	ETHYL CHLOROTHIOFORMATE
2812-73-9	ETHYL CHLOROTHIOFORMATE (CARBONOCHLORIDOTHIOC ACID, S-ETHYL ESTE
2941-64-2	ETHYL CHLOROTHIOLFORMATE
26555-35-1	ETHYL CHLOROTHIOLFORMATE
2941-64-2	ETHYL CHLOROTHIOLOFORMATE
26555-35-1	ETHYL CHLOROTHIOLOFORMATE
623-70-1	ETHYL CROTONATE
10544-63-5	ETHYL CROTONATE
10544-63-5	ETHYL CROTONATE (DOT)
105-56-6	ETHYL CYANACETATE
107-12-0	ETHYL CYANIDE
105-56-6	ETHYL CYANOACETATE
105-56-6	ETHYL CYANOACETATE (DOT)
105-56-6	ETHYL CYANOETHANOATE
4806-61-5	ETHYL CYCLOBUTANE
1678-91-7	ETHYL CYCLOHEXANE
1640-89-7	ETHYL CYCLOPENTANE
1498-51-7	ETHYL DICHLOROPHOSPHATE
1789-58-8	ETHYL DICHLOROSILANE
1789-58-8	ETHYL DICHLOROSILANE (DOT)
111-90-0	ETHYL DIETHYLENE GLYCOL
111-90-0	ETHYL DIGOL
77-81-6	ETHYL DIMETHYLAMIDOCYANOPHOSPHATE
77-81-6	ETHYL DIMETHYLPHOSPHORAMIDOCYANIDATE
140-88-5	ETHYL ESTER ACRYLIC ACID
141-78-6	ETHYL ETHANOATE
60-29-7	ETHYL ETHER
60-29-7	ETHYL ETHER (ACGIH,DOT)
109-63-7	ETHYL ETHER, COMPD. WITH BORON FLUORIDE (BF3) (1:1)
109-63-7	ETHYL ETHER-BORON TRIFLUORIDE COMPLEX
927-80-0	ETHYL ETHYNYL ETHER
2025-56-1	ETHYL FLUID
353-36-6	ETHYL FLUORIDE
353-36-6	ETHYL FLUORIDE (DOT)
109-94-4	ETHYL FORMATE
109-94-4	ETHYL FORMATE (ACGIH,DOT)
122-51-0	ETHYL FORMATE (ORTHO)
109-94-4	ETHYL FORMIC ESTER
4016-11-9	ETHYL GLYCIDYL ETHER
110-80-5	ETHYL GLYCOL
111-76-2	ETHYL GLYCOL ACETATE
633-03-4	ETHYL GREEN
2642-71-9	ETHYL GUSATHION
2642-71-9	ETHYL GUTHION
123-05-7	ETHYL HEXALDEHYDE (2-ETHYL HEXANAL)
123-66-0	ETHYL HEXANOATE
104-75-6	ETHYL HEXYLAMINE
24468-13-1	ETHYL HEXYLCHLOROFORMATE
64-17-5	ETHYL HYDRATE
74-84-0	ETHYL HYDRIDE
540-82-9	ETHYL HYDROGEN SULFATE
75-08-1	ETHYL HYDROSULFIDE
64-17-5	ETHYL HYDROXIDE
97-62-1	ETHYL ISOBUTANOATE
97-62-1	ETHYL ISOBUTYRATE
109-90-0	ETHYL ISOCYANATE
109-90-0	ETHYL ISOCYANATE (DOT)
96-22-0	ETHYL KETONE
97-64-3	ETHYL LACTATE
97-64-3	ETHYL LACTATE (DOT)
75-08-1	ETHYL MERCAPTAN
75-08-1	ETHYL MERCAPTAN (ACGIH,DOT)
2235-25-8	ETHYL MERCURY PHOSPHATE
62-50-0	ETHYL MESYLATE
97-63-2	ETHYL METHACRYLATE
97-63-2	ETHYL METHACRYLATE, INHIBITED (DOT)
62-50-0	ETHYL METHANESULFONATE
62-50-0	ETHYL METHANESULPHONATE
109-94-4	ETHYL METHANOATE
78-93-3	ETHYL METHYL CETONE (FRENCH)
540-67-0	ETHYL METHYL ETHER
540-67-0	ETHYL METHYL ETHER (DOT)
78-93-3	ETHYL METHYL KETONE
78-93-3	ETHYL METHYL KETONE (DOT)
1338-23-4	ETHYL METHYL KETONE PEROXIDE
1338-23-4	ETHYL METHYL KETONE PEROXIDE, MAXIMUM CONCENTRATION 50% (DOT)
1338-23-4	ETHYL METHYL KETONE PEROXIDE, MAXIMUM CONCENTRATION 60% (DOT)
563-12-2	ETHYL METHYLENE PHOSPHORODITHIOATE
563-12-2	ETHYL METHYLENE PHOSPHORODITHIOATE ([(ETO)2P(S)S]2CH2)
105-36-2	ETHYL MONOBROMOACETATE
105-39-5	ETHYL MONOCHLORACETATE
105-39-5	ETHYL MONOCHLOROACETATE
104-90-5	ETHYL MORPHINE
77-81-6	ETHYL N,N-DIMETHYLAMINO CYANOPHOSPHATE
77-81-6	ETHYL N,N-DIMETHYLPHOSPHORAMIDOCYANIDATE
106-68-3	ETHYL N-AMYL KETONE
628-81-9	ETHYL N-BUTYL ETHER
105-54-4	ETHYL N-BUTYRATE
615-53-2	ETHYL N-METHYLNITROSOCARBAMATE
106-32-1	ETHYL N-OCTANOATE
628-32-0	ETHYL N-PROPYL ETHER
625-58-1	ETHYL NITRATE
625-58-1	ETHYL NITRATE (DOT)
75-05-8	ETHYL NITRILE
109-95-5	ETHYL NITRITE
109-95-5	ETHYL NITRITE (DOT)
109-95-5	ETHYL NITRITE, SOLUTION (DOT)
85-71-2	ETHYL O-[O-(METHOXYCARBONYL)BENZOYL]GLYCOLATE
106-32-1	ETHYL OCTANOATE
106-32-1	ETHYL OCTOATE
122-51-0	ETHYL ORTHOFORMATE
122-51-0	ETHYL ORTHOFORMATE (DOT)
78-10-4	ETHYL ORTHOSILICATE
95-92-1	ETHYL OXALATE
95-92-1	ETHYL OXALATE (DOT)
2104-64-5	ETHYL P-NITROPHENYL BENZENETHIONOPHOSPHONATE
2104-64-5	ETHYL P-NITROPHENYL BENZENETHIOPHOSPHATE
2104-64-5	ETHYL P-NITROPHENYL BENZENETHIOPHOSPHONATE
2104-64-5	ETHYL P-NITROPHENYL PHENYLPHOSPHONOTHIOATE
2104-64-5	ETHYL P-NITROPHENYL THIONOBENZENEPHOSPHATE
2104-64-5	ETHYL P-NITROPHENYL THIONOBENZENEPHOSPHONATE
56-38-2	ETHYL PARATHION
106-68-3	ETHYL PENTYL KETONE
628-37-5	ETHYL PEROXIDE

CAS No.	Chemical Name
14666-78-5	ETHYL PEROXYCARBONATE
1125-27-5	ETHYL PHENYL DICHLORO-SILANE
1498-40-4	ETHYL PHOSPHONOUS DICHLORIDE
814-49-3	ETHYL PHOSPHOROCHLORIDATE (Cl(ETO)2PO)
814-49-3	ETHYL PHOSPHOROCHLORIDATE (ETO)2ClPO
1498-51-7	ETHYL PHOSPHORODICHLORIDATE
1498-51-7	ETHYL PHOSPHORODICHLORIDATE (DOT)
84-66-2	ETHYL PHTHALATE
84-72-0	ETHYL PHTHALYL ETHYL GLYCOLATE
766-09-6	ETHYL PIPERIDINE
23505-41-1	ETHYL PIRIMIPHOS
11099-06-2	ETHYL POLYSILICATE
140-89-6	ETHYL POTASSIUM XANTHATE
140-89-6	ETHYL POTASSIUM XANTHOGENATE
105-37-3	ETHYL PROPANOATE
140-88-5	ETHYL PROPENOATE
928-55-2	ETHYL PROPENYL ETHER
105-37-3	ETHYL PROPIONATE
105-37-3	ETHYL PROPIONATE (DOT)
628-32-0	ETHYL PROPYL ETHER
628-32-0	ETHYL PROPYL ETHER (DOT)
589-38-8	ETHYL PROPYL KETONE
13194-48-4	ETHYL PROPYL PHOSPHORODITHIOATE ((ETO)(PRS)2PO)
297-97-2	ETHYL PYRAZINYL PHOSPHOROTHIOATE
107-49-3	ETHYL PYROPHOSPHATE (ET4P2O7)
2025-56-1	ETHYL RADICAL
542-90-5	ETHYL RHODANATE
541-85-5	ETHYL SEC-AMYL KETONE
627-53-2	ETHYL SELENIDE
78-10-4	ETHYL SILICATE
2157-42-8	ETHYL SILICATE
4521-94-2	ETHYL SILICATE
11099-06-2	ETHYL SILICATE
78-10-4	ETHYL SILICATE ((ETO)4Si)
78-10-4	ETHYL SILICATE (ACGIH,DOT)
11099-06-2	ETHYL SILICATE 32
11099-06-2	ETHYL SILICATE 40
11099-06-2	ETHYL SILICATE 50
115-21-9	ETHYL SILICON TRICHLORIDE
64-67-5	ETHYL SULFATE
540-82-9	ETHYL SULFATE
75-08-1	ETHYL SULFHYDRATE
352-93-2	ETHYL SULFIDE
542-90-5	ETHYL SULFOCYANATE
8065-48-3	ETHYL SYSTOX
30145-38-1	ETHYL TELLURAC
757-58-4	ETHYL TETRAPHOSPHATE
757-58-4	ETHYL TETRAPHOSPHATE, HEXA-
75-08-1	ETHYL THIOALCOHOL
2941-64-2	ETHYL THIOCHLOROFORMATE
26555-35-1	ETHYL THIOCHLOROFORMATE
542-90-5	ETHYL THIOCYANATE
298-04-4	ETHYL THIOMETON
3689-24-5	ETHYL THIOPYROPHOSPHATE
3689-24-5	ETHYL THIOPYROPHOSPHATE ([(ETO)2PS]2O)
97-77-8	ETHYL THIRAM
97-77-8	ETHYL THIURAD
622-57-1	ETHYL TOLUIDINE
10544-63-5	ETHYL TRANS-CROTONATE
327-98-0	ETHYL TRICHLOROPHENYLETHYLPHOSPHONOTHIOATE
115-21-9	ETHYL TRICHLOROSILANE
115-21-9	ETHYL TRICHLOROSILANE (DOT)
97-77-8	ETHYL TUADS
97-77-8	ETHYL TUEX
51-79-6	ETHYL URETHANE
109-92-2	ETHYL VINYL ETHER
92-50-2	ETHYL(.BETA.-HYDROXYETHYL)ANILINE
535-13-7	ETHYL-2-CHLOROPROPIONATE
535-13-7	ETHYL-2-CHLOROPROPIONATE (DOT)
5459-93-8	ETHYL-N-CYCLOHEXYLAMINE
538-07-8	ETHYL-S
141-78-6	ETHYLACETAAT (DUTCH)
107-92-6	ETHYLACETIC ACID
140-88-5	ETHYLACRYLAAT (DUTCH)
140-88-5	ETHYLAKRYLAT (CZECH)
462-95-3	ETHYLAL
64-17-5	ETHYLALCOHOL (DUTCH)
563-43-9	ETHYLALUMINIUM DICHLORIDE
75-04-7	ETHYLAMINE
75-04-7	ETHYLAMINE (ACGIH,DOT)
75-04-7	ETHYLAMINE SOLUTION
75-04-7	ETHYLAMINE SOLUTION, IN WATER, CONCENTRATIONS UP TO 70% (DOT)
111-40-0	ETHYLAMINE, 2,2'-IMINOBIS-
32280-46-9	ETHYLAMINE, N,N'-(1-METHYLTRIMETHYLENE)BIS-
10595-95-6	ETHYLAMINE, N-METHYL-N-NITROSO-
1642-54-2	ETHYLAMINOAZINE CITRATE
110-73-6	ETHYLAMINOETHANOL
110-73-6	ETHYLAMINOETHANOL [COMBUSTIBLE LIQUID LABEL]
110-73-6	ETHYLAMINOETHANOL [FLAMMABLE LIQUID LABEL]
9016-45-9	ETHYLAN 20
9016-45-9	ETHYLAN 44
9016-45-9	ETHYLAN 55
9016-45-9	ETHYLAN BCP
9016-45-9	ETHYLAN HA
9016-45-9	ETHYLAN N
9016-45-9	ETHYLAN N 55
9016-45-9	ETHYLAN TU
103-69-5	ETHYLANILINE
100-41-4	ETHYLBENZEEN (DUTCH)
100-41-4	ETHYLBENZOL
92-59-1	ETHYLBENZYLANILINE
119-94-8	ETHYLBENZYLTOLUIDINE
538-07-8	ETHYLBIS(2-CHLOROETHYL)AMINE
139-87-7	ETHYLBIS(2-HYDROXYETHYL)AMINE
538-07-8	ETHYLBIS(BETA-CHLOROETHYL)AMINE
97-95-0	ETHYLBUTANOL
40780-64-1	ETHYLBUTYL ACETATE
123-05-7	ETHYLBUTYLACETALDEHYDE
13360-63-9	ETHYLBUTYLAMINE
541-41-3	ETHYLCHLOORFORMIAAT (DUTCH)
107-07-3	ETHYLCHLOROHYDRIN
598-14-1	ETHYLDICHLORARSINE
563-43-9	ETHYLDICHLOROALUMINUM
598-14-1	ETHYLDICHLOROARSINE (DOT)
139-87-7	ETHYLDIETHANOLAMINE
75-85-4	ETHYLDIMETHYLCARBINOL
141-78-6	ETHYLE (ACETATE D') (FRENCH)
109-94-4	ETHYLE (FORMIATE D') (FRENCH)
541-41-3	ETHYLE, CHLOROFORMIAT D' (FRENCH)
107-07-3	ETHYLEEN-CHLOORHYDRINE (DUTCH)
107-06-2	ETHYLEENDICHLORIDE (DUTCH)
151-56-4	ETHYLEENIMINE (DUTCH)
593-67-9	ETHYLENAMINE
74-85-1	ETHYLENE
74-85-1	ETHYLENE (DOT)
111-55-7	ETHYLENE ACETATE
107-21-1	ETHYLENE ALCOHOL
107-02-8	ETHYLENE ALDEHYDE
593-67-9	ETHYLENE AMINES [COMBUSTIBLE LIQUID LABEL]
593-67-9	ETHYLENE AMINES [CORROSIVE LABEL]
593-67-9	ETHYLENE AMINES [FLAMMABLE LIQUID LABEL]
593-67-9	ETHYLENE AMINES [FLAMMABLE LIQUID, CORROSIVE LABELS]
142-59-6	ETHYLENE BIS DITHIOCARBAMATE
123-81-9	ETHYLENE BIS(MERCAPTOACETATE)
34731-32-3	ETHYLENE BISDITHIOCARBAMATE
106-93-4	ETHYLENE BROMIDE
96-49-1	ETHYLENE CARBONATE
107-07-3	ETHYLENE CHLORHYDRIN
107-06-2	ETHYLENE CHLORIDE
107-07-3	ETHYLENE CHLOROHYDRIN
107-07-3	ETHYLENE CHLOROHYDRIN (ACGIH,DOT)
109-78-4	ETHYLENE CYANOHYDRIN
111-55-7	ETHYLENE DIACETATE
106-93-4	ETHYLENE DIBROMIDE
106-93-4	ETHYLENE DIBROMIDE (ACGIH,DOT)
107-06-2	ETHYLENE DICHLORIDE
107-06-2	ETHYLENE DICHLORIDE (ACGIH,DOT)
111-55-7	ETHYLENE DIETHANOATE
111-46-6	ETHYLENE DIGLYCOL
111-90-0	ETHYLENE DIGLYCOL MONOETHYL ETHER
111-77-3	ETHYLENE DIGLYCOL MONOMETHYL ETHER
107-21-1	ETHYLENE DIHYDRATE
110-71-4	ETHYLENE DIMETHYL ETHER
628-96-6	ETHYLENE DINITRATE
123-73-9	ETHYLENE DIPROPIONATE
4170-30-3	ETHYLENE DIPROPIONATE
85-00-7	ETHYLENE DIPYRIDYLIUM DIBROMIDE

CAS No.	Chemical Name
75-37-6	ETHYLENE FLUORIDE
25497-28-3	ETHYLENE FLUORIDE
371-62-0	ETHYLENE FLUOROHYDRIN
107-21-1	ETHYLENE GLYCOL
111-55-7	ETHYLENE GLYCOL ACETATE
110-49-6	ETHYLENE GLYCOL ACETATE MONOMETHYL ETHER
123-81-9	ETHYLENE GLYCOL BIS(MERCAPTOACETATE)
123-81-9	ETHYLENE GLYCOL BIS(THIOGLYCOLATE)
123-81-9	ETHYLENE GLYCOL BIS(THIOGLYCOLIC ESTER)
111-76-2	ETHYLENE GLYCOL BUTYL ETHER
96-49-1	ETHYLENE GLYCOL CARBONATE
111-55-7	ETHYLENE GLYCOL DIACETATE
629-14-1	ETHYLENE GLYCOL DIETHYL ETHER
112-27-6	ETHYLENE GLYCOL DIHYDROXYDIETHYL ETHER
110-71-4	ETHYLENE GLYCOL DIMETHYL ETHER
628-96-6	ETHYLENE GLYCOL DINITRATE
110-80-5	ETHYLENE GLYCOL ETHYL ETHER
111-15-9	ETHYLENE GLYCOL ETHYL ETHER ACETATE
100-79-8	ETHYLENE GLYCOL FORMAL
646-06-0	ETHYLENE GLYCOL FORMAL
109-59-1	ETHYLENE GLYCOL ISOPROPYL ETHER
109-86-4	ETHYLENE GLYCOL METHYL ETHER
110-49-6	ETHYLENE GLYCOL METHYL ETHER ACETATE
109-59-1	ETHYLENE GLYCOL MONISOPROPYL ETHER
111-76-2	ETHYLENE GLYCOL MONO-N-BUTYL ETHER
818-61-1	ETHYLENE GLYCOL MONOACRYLATE
622-08-2	ETHYLENE GLYCOL MONOBENZYL ETHER
111-76-2	ETHYLENE GLYCOL MONOBUTYL ETHER
110-80-5	ETHYLENE GLYCOL MONOETHYL ETHER
111-15-9	ETHYLENE GLYCOL MONOETHYL ETHER ACETATE
4439-24-1	ETHYLENE GLYCOL MONOISOBUTYL ETHER
109-59-1	ETHYLENE GLYCOL MONOISOPROPYL ETHER
109-86-4	ETHYLENE GLYCOL MONOMETHYL ETHER
109-86-4	ETHYLENE GLYCOL MONOMETHYL ETHER (DOT)
110-49-6	ETHYLENE GLYCOL MONOMETHYL ETHER ACETATE
110-49-6	ETHYLENE GLYCOL MONOMETHYL ETHER ACETATE (DOT)
111-76-2	ETHYLENE GLYCOL N-BUTYL ETHER
107-21-1	ETHYLENE GLYCOL PARTICULATE AND VAPOR
818-61-1	ETHYLENE GLYCOL, ACRYLATE
107-07-3	ETHYLENE GLYCOL, CHLOROHYDRIN
123-73-9	ETHYLENE GLYCOL, DIPROPIONATE (8CI)
4170-30-3	ETHYLENE GLYCOL, DIPROPIONATE (8CI)
109-86-4	ETHYLENE GLYCOL, MONOMETHYL ETHER
60-24-2	ETHYLENE GLYCOL, MONOTHIO-
112-27-6	ETHYLENE GLYCOL-BIS-(2-HYDROXYETHYL ETHER)
67-72-1	ETHYLENE HEXACHLORIDE
151-56-4	ETHYLENE IMINE , INHIBITED
151-56-4	ETHYLENE IMINE, INHIBITED (DOT)
123-81-9	ETHYLENE MERCAPTOACETATE
75-01-4	ETHYLENE MONOCHLORIDE
628-96-6	ETHYLENE NITRATE
75-21-8	ETHYLENE OXIDE
8070-50-6	ETHYLENE OXIDE AND CARBON DIOXIDE MIXTURES (DOT)
106-88-7	ETHYLENE OXIDE, ETHYL-
75-56-9	ETHYLENE OXIDE, METHYL-
8070-50-6	ETHYLENE OXIDE, MIXED WITH CARBON DIOXIDE
8070-50-6	ETHYLENE OXIDE-CARBON DIOXIDE MIXTURE
9016-45-9	ETHYLENE OXIDE-NONYLPHENOL CONDENSATE
9016-45-9	ETHYLENE OXIDE-NONYLPHENOL POLYMER
123-73-9	ETHYLENE PROPIONATE
4170-30-3	ETHYLENE PROPIONATE
127-18-4	ETHYLENE TETRACHLORIDE
116-14-3	ETHYLENE TETRAFLUORIDE
96-45-7	ETHYLENE THIOUREA
79-01-6	ETHYLENE TRICHLORIDE
75-35-4	ETHYLENE, 1,1-DICHLORO-
75-35-4	ETHYLENE, 1,1-DICHLORO- (8CI)
75-35-4	ETHYLENE, 1,1-DICHLORO-, INHIBITED
72-55-9	ETHYLENE, 1,1-DICHLORO-2,2-BIS(P-CHLOROPHENYL)-
75-38-7	ETHYLENE, 1,1-DIFLUORO-
540-59-0	ETHYLENE, 1,2-DICHLORO-
156-60-5	ETHYLENE, 1,2-DICHLORO-, (E)-
156-59-2	ETHYLENE, 1,2-DICHLORO-, (Z)-
629-14-1	ETHYLENE, 1,2-DIETHOXY-
593-60-2	ETHYLENE, BROMO-
598-73-2	ETHYLENE, BROMOTRIFLUORO-
75-01-4	ETHYLENE, CHLORO-
79-38-9	ETHYLENE, CHLOROTRIFLUORO-
74-85-1	ETHYLENE, COMPRESSED (DOT)
75-02-5	ETHYLENE, FLUORO-
75-02-5	ETHYLENE, FLUORO- (8CI)
100-42-5	ETHYLENE, PHENYL-
74-85-1	ETHYLENE, REFRIGERATED LIQUID
74-85-1	ETHYLENE, REFRIGERATED LIQUID (DOT)
127-18-4	ETHYLENE, TETRACHLORO-
116-14-3	ETHYLENE, TETRAFLUORO-
9002-84-0	ETHYLENE, TETRAFLUORO-, POLYMERS
79-01-6	ETHYLENE, TRICHLORO-
79-38-9	ETHYLENE, TRIFLUOROCHLORO-
513-35-9	ETHYLENE, TRIMETHYL-
26760-64-5	ETHYLENE, TRIMETHYL-
593-67-9	ETHYLENEAMINE
142-59-6	ETHYLENEBIS(DITHIOCARBAMATE), DISODIUM SALT
142-59-6	ETHYLENEBIS(DITHIOCARBAMIC ACID) DISODIUM SALT
123-81-9	ETHYLENEBIS(THIOGLYCOLATE)
79-06-1	ETHYLENECARBOXAMIDE
79-10-7	ETHYLENECARBOXYLIC ACID
107-15-3	ETHYLENEDIAMINE
60-00-4	ETHYLENEDIAMINE TETRA-ACETIC ACID
112-24-3	ETHYLENEDIAMINE, N,N'-BIS(2-AMINOETHYL)-
110-18-9	ETHYLENEDIAMINE, N,N,N',N'-TETRAMETHYL-
100-36-7	ETHYLENEDIAMINE, N,N-DIETHYL-
111-40-0	ETHYLENEDIAMINE, N-(2-AMINOETHYL)-
60-00-4	ETHYLENEDIAMINE-N,N,N',N'-TETRAACETIC ACID
60-00-4	ETHYLENEDIAMINETETRAACETATE
60-00-4	ETHYLENEDIAMINETETRAACETIC ACID
60-00-4	ETHYLENEDIAMINETETRAACETIC ACID (DOT)
60-00-4	ETHYLENEDIAMINOTETRAACETIC ACID
60-00-4	ETHYLENEDINITRILOTETRAACETIC ACID
151-56-4	ETHYLENEIMINE
151-56-4	ETHYLENEIMINE (ACGIH)
151-56-4	ETHYLENIMINE
122-51-0	ETHYLESTER KYSELINY ORTHOMRAVENCI (CZECH)
110-73-6	ETHYLETHANOLAMINE
106-98-9	ETHYLETHYLENE
106-88-7	ETHYLETHYLENE OXIDE
107-00-6	ETHYLETHYNE
109-94-4	ETHYLFORMIAAT (DUTCH)
79-09-4	ETHYLFORMIC ACID
111-76-2	ETHYLGLYCOL ACETATE
111-76-2	ETHYLGLYKOLACETAT (GERMAN)
123-05-7	ETHYLHEXALDEHYDE
123-05-7	ETHYLHEXALDEHYDE (DOT)
149-57-5	ETHYLHEXANOIC ACID
104-76-7	ETHYLHEXANOL
149-57-5	ETHYLHEXOIC ACID
103-09-3	ETHYLHEXYL ACETATE
117-81-7	ETHYLHEXYL PHTHALATE
117-84-0	ETHYLHEXYL PHTHALATE
64-19-7	ETHYLIC ACID
75-34-3	ETHYLIDENE CHLORIDE
75-34-3	ETHYLIDENE DICHLORIDE
105-57-7	ETHYLIDENE DIETHYL ETHER
75-37-6	ETHYLIDENE DIFLUORIDE
25497-28-3	ETHYLIDENE DIFLUORIDE
75-37-6	ETHYLIDENE FLUORIDE
25497-28-3	ETHYLIDENE FLUORIDE
16568-02-8	ETHYLIDENE GYROMITRIN
16219-75-3	ETHYLIDENE NORBORNENE
16219-75-3	ETHYLIDENE NORBORNENE (ACGIH)
625-33-2	ETHYLIDENEACETONE
107-29-9	ETHYLIDENEHYDROXYLAMINE
50-21-5	ETHYLIDENELACTIC ACID
151-56-4	ETHYLIMINE
97-62-1	ETHYLISOBUTYRATE (DOT)
75-08-1	ETHYLMERCAPTAAN (DUTCH)
107-27-7	ETHYLMERCURIC CHLORIDE
2235-25-8	ETHYLMERCURIC PHOSPHATE
75-08-1	ETHYLMERKAPTAN (CZECH)
78-92-2	ETHYLMETHYL CARBINOL
78-93-3	ETHYLMETHYLKETON (DUTCH)
10595-95-6	ETHYLMETHYLNITROSAMINE
100-74-3	ETHYLMORPHOLINE
141-43-5	ETHYLOLAMINE
106-88-7	ETHYLOXIRANE
109-92-2	ETHYLOXYETHENE
103-69-5	ETHYLPHENYLAMINE
92-50-2	ETHYLPHENYLETHANOLAMINE

CAS No.	Chemical Name
1498-51-7	ETHYLPHOSPHORIC ACID DICHLORIDE
540-82-9	ETHYLSULFURIC ACID
540-82-9	ETHYLSULPHURIC ACID (DOT)
298-04-4	ETHYLTHIOMETON B
2497-07-6	ETHYLTHIOMETON SULFOXIDE
112-50-5	ETHYLTRIGLYCOL
140-89-6	ETHYLXANTHIC ACID POTASSIUM SALT
74-86-2	ETHYNE
75-20-7	ETHYNE, CALCIUM DERIV
7572-29-4	ETHYNE, DICHLORO-
927-80-0	ETHYNE, ETHOXY-
107-19-7	ETHYNYLCARBINOL
115-19-5	ETHYNYLDIMETHYLCARBINOL
57-63-6	ETHYNYLESTRADIOL
72-33-3	ETHYNYLESTRADIOL 3-METHYL ETHER
72-33-3	ETHYNYLESTRADIOL METHYL ETHER
68-22-4	ETHYNYLNORTESTOSTERONE
57-63-6	ETICYCLIN
57-63-6	ETICYCLOL
140-88-5	ETIL ACRILATO (ITALIAN)
541-41-3	ETIL CLOROCARBONATO (ITALIAN)
541-41-3	ETIL CLOROFORMIATO (ITALIAN)
140-88-5	ETILACRILATULUI (ROMANIAN)
75-04-7	ETILAMINA (ITALIAN)
100-41-4	ETILBENZENE (ITALIAN)
141-78-6	ETILE (ACETATO DI) (ITALIAN)
109-94-4	ETILE (FORMIATO DI) (ITALIAN)
151-56-4	ETILENIMINA (ITALIAN)
75-08-1	ETILMERCAPTANO (ITALIAN)
56-38-2	ETILON
57-63-6	ETINESTROL
57-63-6	ETINESTRYL
57-63-6	ETINOESTRYL
121-75-5	ETIOL
57-63-6	ETISTRADIOL
75-21-8	ETO
1954-28-5	ETOGLUCID
1954-28-5	ETOGLUCIDE
9016-45-9	ETOLAT 914
8070-50-6	ETOX
299-84-3	ETROLENE
11099-06-2	ETS 32
11099-06-2	ETS 40
96-45-7	ETU
75-04-7	ETYLOAMINA (POLISH)
100-41-4	ETYLOBENZEN (POLISH)
64-17-5	ETYLOWY ALKOHOL (POLISH)
56-38-2	ETYLPARATION (CZECH)
74-96-4	ETYLU BROMEK (POLISH)
75-00-3	ETYLU CHLOREK (POLISH)
78-10-4	ETYLU KRZEMIAN (POLISH)
95-80-7	EUCANINE GB
29790-52-1	EUDERMOL
105-58-8	EUFIN
138-86-3	EULIMEN
143-19-1	EUNATROL
8020-83-5	EUPHYTAN EXTRA
7778-54-3	EUSOL BPC
7775-09-9	EVAU-SUPER
109-92-2	EVE
74-90-8	EVERCYN
117-81-7	EVIPLAST 80
117-84-0	EVIPLAST 80
117-84-0	EVIPLAST 81
117-81-7	EVIPLAST 81
7664-38-2	EVITS
106-46-7	EVOLA
8002-74-2	EVORAL PL
8002-74-2	EVORAL SP
14807-96-6	EX-II
75-18-3	EXACT-S
58-89-9	EXAGAMA
119-36-8	EXAGIEN
25322-69-4	EXCENOL 1020
630-08-0	EXHAUST GAS
97-77-8	EXHORRAN
1309-64-4	EXITELITE
56-25-7	EXO-1,2-CIS-DIMETHYL-3,6-EPOXYHEXAHYDROPHTHALIC ANHYDRIDE

CAS No.	Chemical Name
15271-41-7	EXO-5-CHLORO-6-OXO-ENDO-2-NORBORNANECARBONITRILE O-(METHYLCARBAMOYL)OXIME
60-57-1	EXO-DIELDRIN
333-41-5	EXODIN
70-30-4	EXOFENE
7723-14-0	EXOLIT 405
7723-14-0	EXOLIT LPKN 275
7723-14-0	EXOLIT VPK-N 361
7723-14-0	EXOLITE 405
1344-28-1	EXOLON XW 60
2778-04-3	EXOTHION
1333-86-4	EXP
1333-86-4	EXP (CARBON BLACK)
1333-86-4	EXP 1
1333-86-4	EXP 2
7782-42-5	EXP-F
136-78-7	EXPERIMENTAL HERBICIDE 1
60-51-5	EXPERIMENTAL INSECTICIDE 12,880
72-20-8	EXPERIMENTAL INSECTICIDE 269
298-02-2	EXPERIMENTAL INSECTICIDE 3911
121-75-5	EXPERIMENTAL INSECTICIDE 4049
3383-96-8	EXPERIMENTAL INSECTICIDE 52,160
63-25-2	EXPERIMENTAL INSECTICIDE 7744
297-97-2	EXPERIMENTAL NEMATOCIDE 18,133
131-74-8	EXPLOSIVE D
7488-56-4	EXSEL
7720-78-7	EXSICCATED FERROUS SULFATE
7720-78-7	EXSICCATED FERROUS SULPHATE
7558-79-4	EXSICCATED SODIUM PHOSPHATE
2646-17-5	EXT D AND C ORANGE NO. 4
3564-09-8	EXT D AND C RED NO. 15
121-75-5	EXTERMATHION
584-79-2	EXTHRIN
534-52-1	EXTRAR
1335-85-9	EXTRAR
78-10-4	EXTREMA
1309-64-4	EXTREMA
7789-60-8	EXTREMA
10026-04-7	EXTREMA
105-60-2	EXTROM 6N
7631-86-9	EXTRUSIL
57-47-6	EZERIN
9002-84-0	F 103
75-69-4	F 11
75-69-4	F 11 (HALOCARBON)
76-12-0	F 112
76-13-1	F 113
76-15-3	F 115
76-16-4	F 116
75-71-8	F 12
1333-86-4	F 122
75-72-9	F 13
75-63-8	F 13B1
75-73-0	F 14
139-91-3	F 150
75-43-4	F 21
75-45-6	F 22
7631-86-9	F 307
1344-28-1	F 360
1344-28-1	F 360 (ALUMINA)
9002-84-0	F 4DP
9002-84-0	F 4K20
9002-84-0	F 4ZH20
25155-30-0	F 90
23214-92-8	F.I. 106
7681-49-4	F1-TABS
50-00-0	FA
53-96-3	FAA
640-19-7	FAA
144-49-0	FAA
1185-57-5	FAC
2275-18-5	FAC
2275-18-5	FAC 20
10024-97-2	FACTITIOUS AIR
11104-93-1	FACTITIOUS AIR
2275-18-5	FAK-40
137-26-8	FALITIRAM
67-72-1	FALKITOL
7775-09-9	FALL

CAS No.	Chemical Name
52-85-7	FAMPHUR
7681-38-1	FANAL
50-00-0	FANNOFORM
1332-21-4	FAPM 410-120
1333-86-4	FARBRUSS FW 1
1333-86-4	FARBRUSS S 160
9005-25-8	FARINEX 100
9005-25-8	FARINEX TSC
94-75-7	FARMCO
1912-24-9	FARMCO ATRAZINE
330-54-1	FARMCO DIURON
93-76-5	FARMCO FENCE RIDER
6369-97-7	FARMCO TA 20
2971-90-6	FARMCOCCID
56-75-7	FARMICETINA
82-68-8	FARTOX
67-72-1	FASCIOLIN
81-81-2	FASCO FASCRAT POWDER
8001-35-2	FASCO-TERPENE
1344-28-1	FASERTON
1344-28-1	FASERTONERDE
1694-09-3	FAST ACID VIOLET 5BN
119-90-4	FAST BLUE B BASE
119-90-4	FAST BLUE BASE B
119-90-4	FAST BLUE DSC BASE
92-87-5	FAST CORINTH BASE B
119-93-7	FAST DARK BLUE BASE R
134-32-7	FAST GARNET BASE B
633-03-4	FAST GREEN J
633-03-4	FAST GREEN JJO
60-11-7	FAST OIL YELLOW B
88-74-4	FAST ORANGE BASE GR
88-74-4	FAST ORANGE BASE JR
88-74-4	FAST ORANGE GR BASE
88-74-4	FAST ORANGE GR SALT
88-74-4	FAST ORANGE O BASE
88-74-4	FAST ORANGE O SALT
88-74-4	FAST ORANGE SALT GR
88-74-4	FAST ORANGE SALT JR
100-01-6	FAST RED 2G BASE
29757-24-2	FAST RED 2G BASE
100-01-6	FAST RED 2G SALT
29757-24-2	FAST RED 2G SALT
89-62-3	FAST RED 3NT BASE
89-62-3	FAST RED 3NT SALT
95-79-4	FAST RED 5CT BASE
95-69-2	FAST RED 5CT BASE
3165-93-3	FAST RED 5CT SALT
100-01-6	FAST RED BASE 2J
29757-24-2	FAST RED BASE 2J
100-01-6	FAST RED BASE GG
29757-24-2	FAST RED BASE GG
89-62-3	FAST RED BASE GL
89-62-3	FAST RED BASE JL
95-69-2	FAST RED BASE TR
95-79-4	FAST RED BASE TR
134-29-2	FAST RED BB BASE
89-62-3	FAST RED G BASE
100-01-6	FAST RED GG BASE
29757-24-2	FAST RED GG BASE
100-01-6	FAST RED GG SALT
29757-24-2	FAST RED GG SALT
89-62-3	FAST RED GL
89-62-3	FAST RED GL BASE
95-79-4	FAST RED KB AMINE
95-79-4	FAST RED KB BASE
95-79-4	FAST RED KB SALT
95-79-4	FAST RED KB SALT SUPRA
95-79-4	FAST RED KB-T BASE
95-79-4	FAST RED KBS SALT
89-62-3	FAST RED MGL BASE
100-01-6	FAST RED MP BASE
29757-24-2	FAST RED MP BASE
100-01-6	FAST RED P BASE
29757-24-2	FAST RED P BASE
100-01-6	FAST RED P SALT
29757-24-2	FAST RED P SALT
100-01-6	FAST RED SALT 2J
29757-24-2	FAST RED SALT 2J
100-01-6	FAST RED SALT GG
29757-24-2	FAST RED SALT GG
3165-93-3	FAST RED SALT TR
3165-93-3	FAST RED SALT TRA
3165-93-3	FAST RED SALT TRN
99-55-8	FAST RED SG BASE
95-79-4	FAST RED TR
95-69-2	FAST RED TR BASE
95-79-4	FAST RED TR BASE
3165-93-3	FAST RED TR SALT
95-69-2	FAST RED TR-T BASE
95-79-4	FAST RED TR11
95-69-2	FAST RED TRO BASE
95-79-4	FAST RED TRO BASE
91-59-8	FAST SCARLET BASE B
99-55-8	FAST SCARLET BASE J
99-55-8	FAST SCARLET G
99-55-8	FAST SCARLET G BASE
99-55-8	FAST SCARLET G SALT
99-55-8	FAST SCARLET GC BASE
99-55-8	FAST SCARLET J SALT
99-55-8	FAST SCARLET M 4NT BASE
99-55-8	FAST SCARLET T BASE
7446-14-2	FAST WHITE
15739-80-7	FAST WHITE
97-56-3	FAST YELLOW AT
16071-86-6	FASTOLITE BROWN BRL
16071-86-6	FASTUSOL BROWN LBRSA
16071-86-6	FASTUSOL BROWN LBRSN
2646-17-5	FAT ORANGE II
2646-17-5	FAT ORANGE RR
60-11-7	FAT YELLOW
60-11-7	FAT YELLOW A
60-11-7	FAT YELLOW AD OO
97-56-3	FAT YELLOW B
60-11-7	FAT YELLOW ES
60-11-7	FAT YELLOW ES EXTRA
60-11-7	FAT YELLOW EXTRA CONC
60-11-7	FAT YELLOW R
8006-54-0	FATS, LANOLIN
8006-54-0	FATS, WOOL
7784-30-7	FB 67
3313-92-6	FB SODIUM PERCARBONATE
4452-58-8	FB SODIUM PERCARBONATE
15630-89-4	FB SODIUM PERCARBONATE
85-00-7	FB/2
9002-84-0	FBF 74D
75-69-4	FC 11
75-69-4	FC 11 (HALOCARBON)
76-12-0	FC 112
76-13-1	FC 113
76-15-3	FC 115
75-71-8	FC 12
75-72-9	FC 13
75-63-8	FC 13B1
75-73-0	FC 14
75-68-3	FC 142B
75-37-6	FC 152A
25497-28-3	FC 152A
75-43-4	FC 21
75-45-6	FC 22
41070-66-9	FC-1318
115-25-3	FC-C 318
25497-29-4	FC142B
2646-17-5	FD AND C ORANGE NO. 2
3564-09-8	FD AND C RED NO. 1
1694-09-3	FD AND C VIOLET 1
696-28-6	FDA
7681-49-4	FDA 0101
584-79-2	FDA 1446
3564-09-8	FDC RED 1
80-51-3	FE 9
1912-24-9	FE NATROL
9004-66-4	FE-DEXTRAN
62-73-7	FECAMA
2636-26-2	FECAMA
127-18-4	FEDAL-UN
92-84-2	FEENO
3811-04-9	FEKABIT

CAS No.	Chemical Name
62-73-7	FEKAMA
1309-37-1	FELAC
124-41-4	FELDALAT NM
7773-06-0	FELIDERM K
57-06-7	FEMA NO 2034
50-28-2	FEMESTRAL
53-16-7	FEMESTRONE INJECTION
53-16-7	FEMIDYN
62-38-4	FEMMA
50-28-2	FEMOGEN
16071-86-6	FENALUZ BROWN BRL
1912-24-9	FENAMIN
1937-37-7	FENAMIN BLACK E
2602-46-2	FENAMIN BLUE 2B
1912-24-9	FENAMINE
140-56-7	FENAMINOSULF (LESAN)
22224-92-6	FENAMIPHOS
22224-92-6	FENAMIPHOS (ACGIH)
3761-53-3	FENAZO SCARLET 2R
7778-44-1	FENCAL
93-76-5	FENCE RIDER
299-84-3	FENCHLORFOS
299-84-3	FENCHLORPHOS
299-84-3	FENCLOFOS
1336-36-3	FENCLOR
299-84-3	FENCLORVUR
56-75-7	FENICOL
62-44-2	FENIDINA
696-28-6	FENILDICLOROARSINA (ITALIAN)
100-63-0	FENILIDRAZINA (ITALIAN)
62-44-2	FENINA
122-14-5	FENITION
122-14-5	FENITROTHION
2275-14-1	FENKAPTON (DUTCH)
58-89-9	FENOFORM FORTE
9016-45-9	FENOPAL
327-98-0	FENOPHOSPHON
93-72-1	FENOPROP
93-72-1	FENORMONE
92-84-2	FENOVERM
51-28-5	FENOXYL CARBON N
115-90-2	FENSULFOTHION
55-38-9	FENTHION
55-38-9	FENTHION-METHYL
639-58-7	FENTIN CHLORIDE
1885-14-9	FENYLESTER KYSELINY CHLORMRAVENCI (CZECH)
100-63-0	FENYLHYDRAZINE (DUTCH)
62-38-4	FENYLMERCURIACETAT (CZECH)
2097-19-0	FENYLSILATRAN (CZECH)
7720-78-7	FEOSOL
7720-78-7	FEOSPAN
13463-40-6	FER PENTACARBONYLE (FRENCH)
7720-78-7	FER-IN-SOL
14484-64-1	FERBAM
14484-64-1	FERBAM 50
14484-64-1	FERBERK
9004-66-4	FERDEX 100
60-51-5	FERKETHION
14484-64-1	FERMATE
7446-09-5	FERMENICIDE LIQUID
7446-09-5	FERMENICIDE POWDER
64-17-5	FERMENTATION ALCOHOL
123-51-3	FERMENTATION AMYL ALCOHOL
78-83-1	FERMENTATION BUTYL ALCOHOL
7446-09-5	FERMENTICIDE LIQUID
137-26-8	FERMIDE
131-11-3	FERMINE
137-26-8	FERNACOL
137-26-8	FERNASAN
137-26-8	FERNASAN A
94-80-4	FERNESTA
23505-41-1	FERNEX
137-26-8	FERNIDE
94-75-7	FERNIMINE
94-75-7	FERNOXONE
7720-78-7	FERO-GRADUMET
1332-21-4	FERODO C3C
14484-64-1	FERRADOW
7720-78-7	FERRALYN
14221-47-7	FERRATE(3-), TRIS(ETHANEDIOATO(2-)-O,O')-, TRIAMMONIUM, (OC-6-11)- (9CI)
14221-47-7	FERRATE(3-), TRIS(OXALATO)-, TRIAMMONIUM
13601-19-9	FERRATE(4-), HEXACYANO-, TETRASODIUM
13601-19-9	FERRATE(4-), HEXAKIS(CYANO-C)-, TETRASODIUM, (OC-6-11)-
2385-85-5	FERRIAMICIDE
1185-57-5	FERRIC AMMONIUM CITRATE
1185-57-5	FERRIC AMMONIUM CITRATE (DOT)
14221-47-7	FERRIC AMMONIUM OXALATE
14221-47-7	FERRIC AMMONIUM OXALATE (DOT)
10102-49-5	FERRIC ARSENATE
10102-49-5	FERRIC ARSENATE, SOLID
10102-49-5	FERRIC ARSENATE, SOLID (DOT)
63989-69-5	FERRIC ARSENITE
63989-69-5	FERRIC ARSENITE, BASIC
63989-69-5	FERRIC ARSENITE, SOLID
63989-69-5	FERRIC ARSENITE, SOLID (DOT)
7705-08-0	FERRIC CHLORIDE
7705-08-0	FERRIC CHLORIDE SOLUTION
7705-08-0	FERRIC CHLORIDE, SOLID (DOT)
7705-08-0	FERRIC CHLORIDE, SOLID, ANHYDROUS
7705-08-0	FERRIC CHLORIDE, SOLID, ANHYDROUS (DOT)
7705-08-0	FERRIC CHLORIDE, SOLUTION (DOT)
9004-66-4	FERRIC DEXTRAN
14484-64-1	FERRIC DIMETHYLDITHIOCARBAMATE
7783-50-8	FERRIC FLUORIDE
7783-50-8	FERRIC FLUORIDE (DOT)
7783-50-8	FERRIC FLUORIDE (FeF3)
1301-70-8	FERRIC GLYCEROPHOSPHATE
14484-64-1	FERRIC N,N-DIMETHYLDITHIOCARBAMATE
10421-48-4	FERRIC NITRATE
10421-48-4	FERRIC NITRATE (DOT)
1309-37-1	FERRIC OXIDE
10028-22-5	FERRIC SULFATE
10028-22-5	FERRIC SULFATE (DOT)
7705-08-0	FERRIC TRICHLORIDE
7783-50-8	FERRIC TRIFLUORIDE
9004-66-4	FERRIDEXTRAN
7758-94-3	FERRO 66
1306-23-6	FERRO LEMON YELLOW
1306-23-6	FERRO ORANGE YELLOW
1306-23-6	FERRO YELLOW
7720-78-7	FERRO-GRADUMET
7720-78-7	FERRO-THERON
102-54-5	FERROCENE
69523-06-4	FERROCERIUM
69523-06-4	FERROCERIUM (DOT)
9004-66-4	FERRODEXTRAN
12172-73-5	FERROGEDRITE ASBESTOS
9004-66-4	FERROGLUCIN
9004-66-4	FERROGLUKIN
9004-66-4	FERROGLUKIN 75
12604-53-4	FERROMANGANESE
8049-17-0	FERROSILICON
8049-17-0	FERROSILICON (DOT)
8049-17-0	FERROSILICON, CONTAINING MORE THAN 30% BUT LESS THAN 90% SILICON (DOT)
8049-17-0	FERROSILICON, WITH ≥30% BUT ≤70% SILICON
7720-78-7	FERROSULFAT (GERMAN)
7720-78-7	FERROSULFATE
102-54-5	FERROTSEN
10045-89-3	FERROUS AMMONIUM SULFATE
10045-89-3	FERROUS AMMONIUM SULFATE (DOT)
10045-89-3	FERROUS AMMONIUM SULFATE (Fe(NH4)2(SO4)2)
10102-50-8	FERROUS ARSENATE
10102-50-8	FERROUS ARSENATE (DOT)
10102-50-8	FERROUS ARSENATE, SOLID
10102-50-8	FERROUS ARSENATE, SOLID (DOT)
7758-94-3	FERROUS CHLORIDE
7758-94-3	FERROUS CHLORIDE, SOLID
7758-94-3	FERROUS CHLORIDE, SOLID (DOT)
7758-94-3	FERROUS CHLORIDE, SOLUTION
7758-94-3	FERROUS CHLORIDE, SOLUTION (DOT)
10045-89-3	FERROUS DIAMMONIUM DISULFATE
7758-94-3	FERROUS DICHLORIDE
592-87-0	FERROUS ISOTHIOCYANATE
516-03-0	FERROUS OXALATE
516-03-0	FERROUS OXALATE (1:1)
7720-78-7	FERROUS SULFATE

CAS No.	Chemical Name
7720-78-7	FERROUS SULFATE (1:1)
7720-78-7	FERROUS SULFATE (DOT)
7720-78-7	FERROUS SULPHATE
12604-58-9	FERROVANADIUM
12604-58-9	FERROVANADIUM DUST
516-03-0	FERROX
1309-37-1	FERRUGO
7720-78-7	FERSOLATE
7757-74-6	FERTISILO
7681-57-4	FERTISILO
94-75-7	FERXONE
70-30-4	FESIA-SIN
534-16-7	FETIZON'S REAGENT
7784-30-7	FFB 32
9002-84-0	FG 15
2893-78-9	FI CLOR 60S
2893-78-9	FI CLOR CLEARON
1937-37-7	FIBRE BLACK VF
14807-96-6	FIBRENE C 400
12001-28-4	FIBROUS CROCIDOLITE ASBESTOS
87-90-1	FICHLOR 91
2782-57-2	FICLOR 71
1642-54-2	FILAZINE
10265-92-6	FILITOX
9004-70-0	FILM (DOT)
7632-00-0	FILMERINE
151-21-3	FINASOL OSR(SUB 2)
151-21-3	FINASOL OSR2
7631-86-9	FINESIL B
13463-67-7	FINNTITAN
13463-67-7	FINNTITAN RF 2
60-51-5	FIP
74-82-8	FIRE DAMP
59536-65-1	FIREMASTER BP 6
126-72-7	FIREMASTER LV-T 23P
126-72-7	FIREMASTER T 23
126-72-7	FIREMASTER T 23P
8003-34-7	FIRMOTOX
8002-74-2	FISCHER-TROPSCH-WAX
124-87-8	FISH BERRY
8001-69-2	FISH-LIVER OILS
327-98-0	FITOSOL
1937-37-7	FIXANOL BLACK E
2602-46-2	FIXANOL BLUE 2B
7631-86-9	FK 160
16961-83-4	FKS
75-69-4	FKW 11
7778-44-1	FLAC
126-72-7	FLACAVON R
443-48-1	FLAGESOL
443-48-1	FLAGIL
443-48-1	FLAGYL
13463-67-7	FLAMENCO
131-74-8	FLAMMABLE SOLID
126-72-7	FLAMMEX AP
126-72-7	FLAMMEX LV-T 23P
126-72-7	FLAMMEX T 23P
1341-49-7	FLAMMON
1333-86-4	FLAMMRUSS 101
1402-68-2	FLAVATOXINS
8020-83-5	FLAVEX 937
534-52-1	FLAVIN-SANDOZ
57-83-0	FLAVOLUTAN
8001-26-1	FLAXSEED OIL
79-01-6	FLECK-FLIP
108-67-8	FLEET-X
9004-70-0	FLEXIBLE COLLODION
117-84-0	FLEXIMEL
117-81-7	FLEXIMEL
103-23-1	FLEXOL A 26
123-79-5	FLEXOL A 26
117-84-0	FLEXOL DOP
117-81-7	FLEXOL DOP
103-23-1	FLEXOL PLASTICIZER 10-A
123-79-5	FLEXOL PLASTICIZER 10-A
123-79-5	FLEXOL PLASTICIZER A-26
103-23-1	FLEXOL PLASTICIZER A-26
117-84-0	FLEXOL PLASTICIZER DOP
1330-78-5	FLEXOL PLASTICIZER TCP

CAS No.	Chemical Name
8020-83-5	FLEXON 791
8002-74-2	FLEXOWAX C
140-04-5	FLEXRICIN P 6
3081-14-9	FLEXZONE 4L
52-68-6	FLIBOL E
52-68-6	FLIEGENTELLER
133-06-2	FLIT 406
72-43-5	FLO PRO MCSEED PROTECTANT
137-26-8	FLO PRO T SEED PROTECTANT
30525-89-4	FLO-MOR
79-01-6	FLOCK FLIP
1313-60-6	FLOCOOL 180
1344-95-2	FLOLITE R
7631-86-9	FLOLITE S 700
93-79-8	FLOMORE
75-45-6	FLON 22
7705-08-0	FLORES MARTIS
7681-49-4	FLORIDINE
7681-49-4	FLOROCID
9002-84-0	FLOROLON 4M
7803-62-5	FLOTS 100SCO
7704-34-9	FLOUR SULPHUR
1309-64-4	FLOWERS OF ANTIMONY
7704-34-9	FLOWERS OF SULFUR (DOT)
7704-34-9	FLOWERS OF SULPHUR
1314-13-2	FLOWERS OF ZINC
7681-49-4	FLOZENGES
16871-71-9	FLUAT-VOGEL
79-01-6	FLUATE
119-36-8	FLUCARMIT
7681-49-4	FLUDENT
8028-73-7	FLUE DUST, ARSENIC-CONTAINING
67711-90-4	FLUE DUST, POISONOUS
630-08-0	FLUE GAS
65996-68-1	FLUE GASES, FERROUS METAL, BLAST FURNACE
4301-50-2	FLUENETHYL
4301-50-2	FLUENETIL
4301-50-2	FLUENYL
75-45-6	FLUGENE 22
353-59-3	FLUGEX 12B1
75-63-8	FLUGEX 13B1
56-23-5	FLUKOIDS
9002-84-0	FLUO-KEM
16872-11-0	FLUOBORIC ACID
74-90-8	FLUOHYDRIC ACID GAS
2164-17-2	FLUOMETURON
62207-76-5	FLUOMIN
62207-76-5	FLUOMINE
62207-76-5	FLUOMINE DUST
9002-84-0	FLUON
9002-84-0	FLUON 169
9002-84-0	FLUON AD 704
9002-84-0	FLUON CD 023
9002-84-0	FLUON CD 042
9002-84-0	FLUON CD 1
9002-84-0	FLUON G 163
9002-84-0	FLUON G 201
9002-84-0	FLUON G 308
9002-84-0	FLUON G 4
9002-84-0	FLUON GPI
9002-84-0	FLUON L 169
9002-84-0	FLUON L 169A
9002-84-0	FLUON L 169B
9002-84-0	FLUON L 170
9002-84-0	FLUON L 171
9002-84-0	FLUON VP 25
9002-84-0	FLUON VX 2
9002-84-0	FLUON VXI
353-50-4	FLUOPHOSGENE
13779-41-4	FLUOPHOSPHORIC ACID (F2(HO)PO)
115-26-4	FLUOPHOSPHORIC ACID DI(DIMETHYLAMIDE)
7681-49-4	FLUOR-O-KOTE
7681-49-4	FLUORADAY
640-19-7	FLUORAKIL 100
7681-49-4	FLUORAL
206-44-0	FLUORANTHENE
86-73-7	FLUORENE
62-74-8	FLUORESSIGAEURE (GERMAN)
7664-39-3	FLUORHYDRIC ACID

CAS No.	Chemical Name
7681-49-4	FLUORID SODNY (CZECH)
16984-48-8	FLUORIDE
18130-74-0	FLUORIDE (HF2-)
16984-48-8	FLUORIDE ION
16984-48-8	FLUORIDE ION (F-)
16984-48-8	FLUORIDE(1-)
7681-49-4	FLUORIDE, SODIUM
7681-49-4	FLUORIDENT
7681-49-4	FLUORIGARD
7782-41-4	FLUORINE
16984-48-8	FLUORINE ION(1-)
16984-48-8	FLUORINE ION(F1-)
7783-41-7	FLUORINE MONOXIDE
7783-41-7	FLUORINE MONOXIDE (F2O)
7783-41-7	FLUORINE OXIDE
7783-41-7	FLUORINE OXIDE (F2O)
7782-41-4	FLUORINE, CRYOGENIC LIQUID
16984-48-8	FLUORINE, ION
7782-41-4	FLUORINE-19
7681-49-4	FLUORINEED
7681-49-4	FLUORINSE
7783-47-3	FLUORISTAN
7681-49-4	FLUORITAB
7789-75-5	FLUORITE (9CI)
12001-26-2	FLUORMICA
144-49-0	FLUORO ACETIC ACID (DOT)
640-19-7	FLUOROACETAMIDE
640-19-7	FLUOROACETAMIDE/1081
144-49-0	FLUOROACETATE
144-49-0	FLUOROACETIC ACID
640-19-7	FLUOROACETIC ACID AMIDE
62-74-8	FLUOROACETIC ACID SODIUM SALT
359-06-8	FLUOROACETYL CHLORIDE
371-40-4	FLUOROANILINE
462-06-6	FLUOROBENZENE
462-06-6	FLUOROBENZENE (DOT)
371-86-8	FLUOROBISISOPROPYLAMINOPHOSPHINE OXIDE
51-21-8	FLUOROBLASTIN
74-97-5	FLUOROCARBON 1011
75-69-4	FLUOROCARBON 11
76-12-0	FLUOROCARBON 112
76-11-9	FLUOROCARBON 112A
76-13-1	FLUOROCARBON 113
76-14-2	FLUOROCARBON 114
76-15-3	FLUOROCARBON 115
75-71-8	FLUOROCARBON 12
353-59-3	FLUOROCARBON 1211
75-63-8	FLUOROCARBON 1301
75-45-6	FLUOROCARBON 22
25497-29-4	FLUOROCARBON FC142B
75-69-4	FLUOROCARBON NO 11
75-69-4	FLUOROCHLOROFORM
7681-49-4	FLUOROCID
75-43-4	FLUORODICHLOROMETHANE
353-36-6	FLUOROETHANE
144-49-0	FLUOROETHANOIC ACID
75-02-5	FLUOROETHENE
75-02-5	FLUOROETHYLENE
9002-84-0	FLUOROFLEX
75-46-7	FLUOROFORM
353-50-4	FLUOROFORMYL FLUORIDE
7681-49-4	FLUOROL
9002-84-0	FLUOROLON 4
9002-84-0	FLUOROLON 4D
593-53-3	FLUOROMETHANE
593-53-3	FLUOROMETHANE (CH3F)
558-25-8	FLUOROMETHYL SULFONE
9002-84-0	FLUORON AD 2
9002-84-0	FLUOROPAK 80
353-50-4	FLUOROPHOSGENE
13537-32-1	FLUOROPHOSPHONIC ACID (F(HO)2PO)
13537-32-1	FLUOROPHOSPHORIC ACID
13537-32-1	FLUOROPHOSPHORIC ACID, ANHYDROUS
13537-32-1	FLUOROPHOSPHORIC ACID, ANHYDROUS (DOT)
79-38-9	FLUOROPLAST 3
116-14-3	FLUOROPLAST 4
9002-84-0	FLUOROPLAST 4
9002-84-0	FLUOROPLAST 4B
9002-84-0	FLUOROPLAST 4D

CAS No.	Chemical Name
9002-84-0	FLUOROPLAST 4M
51-21-8	FLUOROPLEX
9002-84-0	FLUOROPORE F 045
9002-84-0	FLUOROPORE FP 022
9002-84-0	FLUOROPORE FP 045
9002-84-0	FLUOROPORE FP 120
9002-84-0	FLUOROPORE FP 200
9002-84-0	FLUOROPORE FP 500
17084-08-1	FLUOROSILICATE
17084-08-1	FLUOROSILICATE (SiF62-)
16961-83-4	FLUOROSILICIC ACID
16961-83-4	FLUOROSILICIC ACID (DOT)
16961-83-4	FLUOROSILICIC ACID (H2SiF6)
7789-21-1	FLUOROSULFONIC ACID
7789-21-1	FLUOROSULFONIC ACID (DOT)
7789-21-1	FLUOROSULFURIC ACID
7789-21-1	FLUOROSULFURIC ACID (HSO3F)
7789-21-1	FLUOROSULPHONIC ACID
25496-08-6	FLUOROTOLUENE
75-69-4	FLUOROTRICHLOROMETHANE
75-69-4	FLUOROTROJCHLOROMETAN (POLISH)
51-21-8	FLUOROURACIL
7664-39-3	FLUOROWODOR (POLISH)
7789-75-5	FLUORSPAR
371-86-8	FLUORURE DE N,N'-DIISOPROPYLE PHOSPHORODIAMIDE (FRENCH)
115-26-4	FLUORURE DE N,N,N',N'-TETRAMETHYLE PHOSPHORO-DIAMIDE (FRENCH)
7789-23-3	FLUORURE DE POTASSIUM (FRENCH)
7681-49-4	FLUORURE DE SODIUM (FRENCH)
2699-79-8	FLUORURE DE SULFURYLE (FRENCH)
7664-39-3	FLUORWASSERSTOFF (GERMAN)
7664-39-3	FLUORWATERSTOF (DUTCH)
75-46-7	FLUORYL
17084-08-1	FLUOSILICATE
16919-19-0	FLUOSILICATE DE AMMONIUM (FRENCH)
16893-85-9	FLUOSILICATE DE SODIUM
16961-83-4	FLUOSILICIC ACID
16961-83-4	FLUOSILICIC ACID (DOT)
7789-21-1	FLUOSULFONIC ACID
7789-21-1	FLUOSULFONIC ACID (DOT)
7789-21-1	FLUOSULFURIC ACID
7681-49-4	FLURA
7681-49-4	FLURA DROPS
7681-49-4	FLURA-GEL
7681-49-4	FLURA-LOZ
51-21-8	FLURACIL
17617-23-1	FLURAZEPAM
7681-49-4	FLURCARE
51-21-8	FLURI
51-21-8	FLURIL
7681-49-4	FLURSOL
640-19-7	FLUTRITEX 1
371-62-0	FLUTRITEX 2
54-11-5	FLUX MAAG
2636-26-2	FLY FIGHTER
2636-26-2	FLY-DIE
333-41-5	FLYTROL
7440-02-0	FM 1208
8020-83-5	FM 5.6AP
9004-70-0	FM-NTS
62-38-4	FMA
1563-66-2	FMC 10242
584-79-2	FMC 249
115-29-7	FMC 5462
8001-29-4	FMC 710
563-12-2	FMC-1240
9002-84-0	FN 3
3570-75-0	FNT
121-75-5	FOG 3
510-15-6	FOLBEX
2425-06-1	FOLCID
56-38-2	FOLI
54-62-6	FOLIC ACID, 4-AMINO-
56-38-2	FOLIDOL
56-38-2	FOLIDOL E
56-38-2	FOLIDOL E & E 605
56-38-2	FOLIDOL E605
298-00-0	FOLIDOL M

CAS No.	*Chemical Name*
298-00-0	FOLIDOL M 40
56-38-2	FOLIDOL OIL
298-00-0	FOLIDOL-80
53-16-7	FOLIKRIN
53-16-7	FOLIPEX
53-16-7	FOLISAN
122-14-5	FOLITHION
122-14-5	FOLITHION EC 50
53-16-7	FOLLESTRINE
53-16-7	FOLLESTROL
57-63-6	FOLLICORAL
53-16-7	FOLLICULAR HORMONE
53-16-7	FOLLICULIN
53-16-7	FOLLICUNODIS
50-28-2	FOLLICYCLIN
53-16-7	FOLLIDRIN
57-83-0	FOLOGENON
82-68-8	FOLOSAN
82-68-8	FOMAC 2
56-53-1	FONATOL
944-22-9	FONOFOS
944-22-9	FONOFOS (ACGIH)
944-22-9	FONOPHOS
3761-53-3	FOOD RED 5
1694-09-3	FOOD VIOLET 2
298-02-2	FORAAT (DUTCH)
76-13-1	FORANE 113
75-71-8	FORANE 12
75-45-6	FORANE 22
94-75-7	FOREDEX 75
58-89-9	FORLIN
109-87-5	FORMAL
100-79-8	FORMAL GLYCOL
646-06-0	FORMAL GLYCOL
50-00-0	FORMALDEHYD (CZECH, POLISH)
50-00-0	FORMALDEHYDE
50-00-0	FORMALDEHYDE (ACGIH)
50-00-0	FORMALDEHYDE (CONCENTRATION ≤90%)
50-00-0	FORMALDEHYDE (GAS)
50-00-0	FORMALDEHYDE (SOLUTION)
111-91-1	FORMALDEHYDE BIS(BETA-CHLOROETHYL) ACETAL
111-91-1	FORMALDEHYDE BIS(2-CHLOROETHYL) ACETAL
107-16-4	FORMALDEHYDE CYANOHYDRIN
462-95-3	FORMALDEHYDE DIETHYL ACETAL
109-87-5	FORMALDEHYDE DIMETHYL ACETAL
109-87-5	FORMALDEHYDE METHYL KETAL
9002-81-7	FORMALDEHYDE POLYMER
50-00-0	FORMALDEHYDE SOLUTION
50-00-0	FORMALDEHYDE SOLUTION (DOT)
50-00-0	FORMALDEHYDE SOLUTION (FLASH POINT >141 DEG F; IN CONTAINERS >110 GALLONS
50-00-0	FORMALDEHYDE, AS FORMALIN SOLUTION (DOT)
50-00-0	FORMALDEHYDE, GAS
9002-81-7	FORMALDEHYDE, HOMOPOLYMER
110-88-3	FORMALDEHYDE, TRIMER
50-00-0	FORMALIN
50-00-0	FORMALIN (DOT)
50-00-0	FORMALIN 40
50-00-0	FORMALIN-LOESUNGEN (GERMAN)
50-00-0	FORMALINA (ITALIAN)
50-00-0	FORMALINE (GERMAN)
1937-37-7	FORMALINE BLACK C
50-00-0	FORMALITH
137-26-8	FORMALSOL
75-12-7	FORMAMIDE
137-26-8	FORMAMIDE, 1,1'-DITHIOBIS(N,N-DIMETHYLTHIO-
68-12-2	FORMAMIDE, N,N-DIMETHYL-
22259-30-9	FORMAMIDINE, N'-(M-HYDROXYPHENYL)-N,N-DIMETHYL-, METHYLCARBAMATE (ESTER)
100-97-0	FORMAMINE
22259-30-9	FORMETANAT
22259-30-9	FORMETANATE
23422-53-9	FORMETANATE HYDROCHLORIDE
23422-53-9	FORMETANATE MONOHYDROCHLORIDE
107-31-3	FORMIATE DE METHYLE (FRENCH)
110-74-7	FORMIATE DE PROPYLE (FRENCH)
64-18-6	FORMIC ACID
64-18-6	FORMIC ACID (ACGIH,DOT)
540-69-2	FORMIC ACID AMMONIUM SALT

CAS No.	*Chemical Name*
64-18-6	FORMIC ACID SOLUTION
625-55-8	FORMIC ACID, 1-METHYLETHYL ESTER
542-55-2	FORMIC ACID, 2-METHYLPROPYL ESTER
1838-59-1	FORMIC ACID, 2-PROPENYL ESTER
3570-75-0	FORMIC ACID, 2-[4-(5-NITRO-2-FURYL)-2-THIAZOLYL]HYDRAZIDE
1838-59-1	FORMIC ACID, ALLYL ESTER
592-84-7	FORMIC ACID, BUTYL ESTER
627-11-2	FORMIC ACID, CHLORO-, 2-CHLOROETHYL ESTER
24468-13-1	FORMIC ACID, CHLORO-, 2-ETHYLHEXYL ESTER
2937-50-0	FORMIC ACID, CHLORO-, ALLYL ESTER
592-34-7	FORMIC ACID, CHLORO-, BUTYL ESTER
22128-62-7	FORMIC ACID, CHLORO-, CHLOROMETHYL ESTER
541-41-3	FORMIC ACID, CHLORO-, ETHYL ESTER
108-23-6	FORMIC ACID, CHLORO-, ISOPROPYL ESTER
79-22-1	FORMIC ACID, CHLORO-, METHYL ESTER
1885-14-9	FORMIC ACID, CHLORO-, PHENYL ESTER
109-61-5	FORMIC ACID, CHLORO-, PROPYL ESTER
2941-64-2	FORMIC ACID, CHLOROTHIO-, ETHYL ESTER
26555-35-1	FORMIC ACID, CHLOROTHIO-, ETHYL ESTER
2941-64-2	FORMIC ACID, CHLOROTHIO-, S-ETHYL ESTER
26555-35-1	FORMIC ACID, CHLOROTHIO-, S-ETHYL ESTER
13889-92-4	FORMIC ACID, CHLOROTHIO-, S-PROPYL ESTER
544-18-3	FORMIC ACID, COBALT(2+) SALT
544-19-4	FORMIC ACID, COPPER(2+) SALT
544-19-4	FORMIC ACID, COPPER(2+) SALT (1:1)
109-94-4	FORMIC ACID, ETHYL ESTER
16568-02-8	FORMIC ACID, ETHYLIDENEMETHYLHYDRAZIDE
542-55-2	FORMIC ACID, ISOBUTYL ESTER
625-55-8	FORMIC ACID, ISOPROPYL ESTER
107-31-3	FORMIC ACID, METHYL ESTER
3349-06-2	FORMIC ACID, NICKEL(2+) SALT
638-49-3	FORMIC ACID, PENTYL ESTER
110-74-7	FORMIC ACID, PROPYL ESTER
64-18-6	FORMIC ACID, SOLUTION (DOT)
557-41-5	FORMIC ACID, ZINC SALT
50-00-0	FORMIC ALDEHYDE
74-90-8	FORMIC ANAMMONIDE
1937-37-7	FORMIC BLACK C
1937-37-7	FORMIC BLACK CW
1937-37-7	FORMIC BLACK EA
1937-37-7	FORMIC BLACK MTG
1937-37-7	FORMIC BLACK TG
109-94-4	FORMIC ETHER
75-12-7	FORMIMIDIC ACID
100-97-0	FORMIN
100-97-0	FORMIN (HETEROCYCLE)
64-18-6	FORMIRA
64-18-6	FORMISOTON
50-00-0	FORMOL
74-90-8	FORMONITRILE
76-22-2	FORMOSA CAMPHOR
8008-51-3	FORMOSA CAMPHOR OIL
8008-51-3	FORMOSE OIL OF CAMPHOR
2540-82-1	FORMOTHION
17702-57-7	FORMPARANATE
57-11-4	FORMULA 300
94-75-7	FORMULA 40
67-66-3	FORMYL TRICHLORIDE
64-18-6	FORMYLIC ACID
504-20-1	FORON
52-68-6	FOROTOX
93-76-5	FORRON
93-76-5	FORST U 46
58-89-9	FORST-NEXEN
7782-42-5	FORTAFIL 5Y
93-76-5	FORTEX
121-75-5	FORTHION
60-51-5	FORTION NM
1405-87-4	FORTRACIN
300-76-5	FOSBROM
52-68-6	FOSCHLOR
52-68-6	FOSCHLOR 25
52-68-6	FOSCHLOR R
52-68-6	FOSCHLOR R-50
52-68-6	FOSCHLOREM (POLISH)
7786-34-7	FOSDRIN
60-51-5	FOSFAMID
563-12-2	FOSFATOX E
60-51-5	FOSFATOX R

CAS No.	Chemical Name
56-38-2	FOSFERNO
298-00-0	FOSFERNO M 50
56-38-2	FOSFEX
56-38-2	FOSFIVE
563-12-2	FOSFONO 50
7723-14-0	FOSFORO BIANCO (ITALIAN)
10026-13-8	FOSFORO(PENTACLORURO DI) (ITALIAN)
7719-12-2	FOSFORO(TRICLORURO DI) (ITALIAN)
10026-13-8	FOSFORPENTACHLORIDE (DUTCH)
7719-12-2	FOSFORTRICHLORIDE (DUTCH)
122-52-1	FOSFORYN TROJETYLOWY (CZECH)
121-45-9	FOSFORYN TROJMETYLOWY (CZECH)
7664-38-2	FOSFORZUUROPLOSSINGEN (DUTCH)
121-75-5	FOSFOTHION
121-75-5	FOSFOTION
60-51-5	FOSFOTOX
60-51-5	FOSFOTOX R
60-51-5	FOSFOTOX R 35
20859-73-8	FOSFURI DI ALLUMINIO (ITALIAN)
75-44-5	FOSGEEN (DUTCH)
75-44-5	FOSGEN (POLISH)
75-44-5	FOSGENE (ITALIAN)
56-38-2	FOSOVA
7631-86-9	FOSSIL FLOUR
56-38-2	FOSTERN
94-36-0	FOSTEX
21548-32-3	FOSTHIETAN
2275-18-5	FOSTION
60-51-5	FOSTION MM
56-38-2	FOSTOX
70-30-4	FOSTRIL
117-52-2	FOUMARIN
87-66-1	FOURAMINE BROWN AP
106-50-3	FOURAMINE D
95-80-7	FOURAMINE J
123-30-8	FOURAMINE P
27598-85-2	FOURAMINE P
120-80-9	FOURAMINE PCH
95-85-2	FOURAMINE PY
108-46-3	FOURAMINE RS
106-50-3	FOURRINE 1
120-80-9	FOURRINE 68
108-46-3	FOURRINE 79
123-30-8	FOURRINE 84
27598-85-2	FOURRINE 84
87-66-1	FOURRINE 85
95-80-7	FOURRINE 94
106-50-3	FOURRINE D
108-46-3	FOURRINE EW
95-80-7	FOURRINE M
123-30-8	FOURRINE P BASE
27598-85-2	FOURRINE P BASE
87-66-1	FOURRINE PG
9002-84-0	FP 4
143-18-0	FR 14
1338-23-4	FR 222
1330-43-4	FR 28
7631-90-5	FR-62
25322-69-4	FRA 1173
1642-54-2	FRANOCIDE
1642-54-2	FRANOZAN
7631-86-9	FRANSIL 251
62-74-8	FRATOL
8002-74-2	FREEMAN 155/160
7446-14-2	FREEMANS WHITE LEAD
12002-03-8	FRENCH GREEN
75-45-6	FREON
75-69-4	FREON 11
127-18-4	FREON 1110
76-12-0	FREON 112
76-13-1	FREON 113
76-13-1	FREON 113 TR-T
76-15-3	FREON 115
76-16-4	FREON 116
75-69-4	FREON 11A
75-69-4	FREON 11B
75-71-8	FREON 12
75-61-6	FREON 1282
353-59-3	FREON 12B1
75-61-6	FREON 12B2
75-72-9	FREON 13
75-63-8	FREON 13B1
75-73-0	FREON 14
25497-29-4	FREON 142
25497-29-4	FREON 142B
75-37-6	FREON 152
25497-28-3	FREON 152
75-43-4	FREON 21
76-19-7	FREON 218
75-45-6	FREON 22
75-46-7	FREON 23
75-09-2	FREON 30
593-53-3	FREON 41
115-25-3	FREON C 318
75-43-4	FREON F 21
76-13-1	FREON F-113
75-71-8	FREON F-12
75-46-7	FREON F-23
75-69-4	FREON HE
75-69-4	FREON MF
76-12-0	FREON R 112
76-13-1	FREON TF
9004-34-6	FRESENIUS D 6
107-21-1	FRIDEX
75-45-6	FRIGEN
75-69-4	FRIGEN 11
76-13-1	FRIGEN 113
76-13-1	FRIGEN 113A
76-13-1	FRIGEN 113TR
76-13-1	FRIGEN 113TR-N
76-13-1	FRIGEN 113TR-T
75-69-4	FRIGEN 11A
75-71-8	FRIGEN 12
75-72-9	FRIGEN 13
75-45-6	FRIGEN 22
75-69-4	FRIGEN S 11
513-49-5	FRUCOTE
93-76-5	FRUITONE A
93-72-1	FRUITONE T
298-04-4	FRUMIN
298-04-4	FRUMIN AL
298-04-4	FRUMIN G
8002-74-2	FT 150
8002-74-2	FT 300
9002-84-0	FT 4
85-44-9	FTAALZUURANHYDRIDE (DUTCH)
732-11-6	FTALOPHOS
85-44-9	FTALOWY BEZWODNIK (POLISH)
9002-84-0	FTORLON 4
9002-84-0	FTORLON 4D
9002-84-0	FTORLON 4DP
9002-84-0	FTORLON 4K20
9002-84-0	FTORLON 4M
9002-84-0	FTOROLON 4
9002-84-0	FTOROPLAST 4
9002-84-0	FTOROPLAST 4B
9002-84-0	FTOROPLAST 4D
9002-84-0	FTOROPLAST 4DP
9002-84-0	FTOROPLAST 4G10
9002-84-0	FTOROPLAST 4K15M5L-EA
9002-84-0	FTOROPLAST 4K20
9002-84-0	FTOROPLAST 4M
9002-84-0	FTOROPLAST AMIP 15M
9002-84-0	FTOROPLAST F 4
9002-84-0	FTOROPLAST FBF 74D
9002-84-0	FTOROPLAST FP 4D
51-21-8	FU
3878-19-1	FUBERIDATOL
3878-19-1	FUBERIDAZOL
3878-19-1	FUBERIDAZOLE
3878-19-1	FUBERISAZOL
3878-19-1	FUBRIDAZOLE
68476-26-6	FUEL GAS, GENERATOR
68476-26-6	FUEL GASES
68476-26-6	FUEL GASES, GENERATOR
68476-26-6	FUEL GASES, MANUFD.
8006-20-0	FUEL GASES, PRODUCER GAS
8021-92-9	FUEL GASES, WATER GAS

CAS No.	Chemical Name
8008-20-6	FUEL OIL #1
8006-14-2	FUELS, GAS
68476-26-6	FUELS, GAS
86290-81-5	FUELS, GASOLINE
106-97-8	FUELS, LIQUEFIED PETROLEUM GAS
68476-85-7	FUELS, LIQUEFIED PETROLEUM GAS
14484-64-1	FUKLASIN ULTRA
14484-64-1	FUKLAZIN
96-12-8	FUMAGON
110-17-8	FUMARIC ACID
110-17-8	FUMARIC ACID (DOT)
627-63-4	FUMARIC ACID CHLORIDE
627-63-4	FUMARIC ACID DICHLORIDE
627-63-4	FUMARIC DICHLORIDE
117-52-2	FUMARIN
627-63-4	FUMAROYL CHLORIDE
627-63-4	FUMAROYL DICHLORIDE
627-63-4	FUMARYL CHLORIDE
627-63-4	FUMARYL CHLORIDE (DOT)
627-63-4	FUMARYLCHLORID (CZECH)
117-52-2	FUMASOL
96-12-8	FUMAZONE
96-12-8	FUMAZONE 86
96-12-8	FUMAZONE 86E
54-11-5	FUMETOBAC
558-25-8	FUMETTE
8070-50-6	FUMIGEN 10
107-13-1	FUMIGRAIN
7784-34-1	FUMING LIQUID ARSENIC
8014-95-7	FUMING SULFURIC ACID
8014-95-7	FUMING SULFURIC ACID (DOT)
20859-73-8	FUMITOXIN
106-93-4	FUMO-GAS
17804-35-2	FUNDAZOL
7487-94-7	FUNGCHEX
17804-35-2	FUNGICIDE D-1991
62-38-4	FUNGICIDE R
87-86-5	FUNGIFEN
62-38-4	FUNGITOX OR
17804-35-2	FUNGOCHROM
16871-71-9	FUNGOL
7681-49-4	FUNGOL B
16871-71-9	FUNGONIT GF 2
133-06-2	FUNGUS BAN TYPE II
106-50-3	FUR BLACK 41867
106-50-3	FUR BROWN 41866
106-50-3	FUR YELLOW
1563-66-2	FURADAN
1563-66-2	FURADAN 3G
1563-66-2	FURADAN 75 WP
98-01-1	FURAL
98-01-1	FURALDEHYDE
98-01-1	FURALE
139-91-3	FURALTADONE
110-00-9	FURAN
110-00-9	FURAN (DOT)
625-86-5	FURAN, 2,5-DIMETHYL-
534-22-5	FURAN, 2-METHYL-
27137-41-3	FURAN, 2-METHYL-
27137-41-3	FURAN, METHYL-
109-99-9	FURAN, TETRAHYDRO-
96-47-9	FURAN, TETRAHYDRO-2-METHYL-
25265-68-3	FURAN, TETRAHYDROMETHYL-
110-02-1	FURAN, THIO-
98-01-1	FURANCARBONAL
109-99-9	FURANIDINE
531-82-8	FURATHIAZOLE
62-74-8	FURATOL
794-93-4	FURATONE
794-93-4	FURATRIZINE, BIS(HYDROXYMETHYL)-
139-91-3	FURAZOLIN
139-91-3	FURAZOLINE
98-01-1	FURFURAL
98-01-1	FURFURAL (ACGIH,DOT)
98-00-0	FURFURAL ALCOHOL
98-00-0	FURFURALCOHOL
98-01-1	FURFURALDEHYDE
98-01-1	FURFURALE (ITALIAN)
98-01-1	FURFURALU (POLISH)
110-00-9	FURFURAN
98-01-1	FURFUROL
98-01-1	FURFUROLE
98-00-0	FURFURYL ALCOHOL
98-00-0	FURFURYL ALCOHOL (ACGIH,DOT)
97-99-4	FURFURYL ALCOHOL, TETRAHYDRO-
98-01-1	FURFURYLALDEHYDE
617-89-0	FURFURYLAMINE
617-89-0	FURFURYLAMINE (DOT)
4795-29-3	FURFURYLAMINE, TETRAHYDRO-
3878-19-1	FURIDAZOL
3878-19-1	FURIDAZOLE
712-68-5	FURIDIAZINA
531-82-8	FURIUM
117-52-2	FURMARIN
139-91-3	FURMETHANOL
139-91-3	FURMETHONOL
139-91-3	FURMETONOL
1333-86-4	FURNACE BLACK
1333-86-4	FURNAL 500
1333-86-4	FURNEX
1333-86-4	FURNEX N 765
1563-66-2	FURODAN
98-01-1	FUROLE
531-82-8	FUROTHIAZOLE
106-50-3	FURRO D
123-30-8	FURRO P BASE
27598-85-2	FURRO P BASE
98-00-0	FURYL ALCOHOL
3688-53-7	FURYLAMIDE
98-00-0	FURYLCARBINOL
3688-53-7	FURYLFURAMIDE
2235-25-8	FUSARIOL UNIVERSAL
1330-43-4	FUSED BORAX
1303-86-2	FUSED BORIC ACID
8013-75-0	FUSEL OIL
8013-75-0	FUSEL OIL (DOT)
8013-75-0	FUSELOEL (GERMAN)
640-19-7	FUSSOL
106-50-3	FUTRAMINE D
1333-86-4	FW 200
1333-86-4	FW 200 (CARBON)
115-32-2	FW 293
1836-75-5	FW 925
14807-96-6	FW-XO
50-00-0	FYDE
121-75-5	FYFANON
7773-06-0	FYRAN 206K
126-72-7	FYROL HB 32
1330-78-5	FYRQUEL 150
111-49-9	G 0
1344-28-1	G 0
111-49-9	G 0 (AMINE)
1344-28-1	G 0 (OXIDE)
70-30-4	G 11
9002-84-0	G 163
1333-86-4	G 2
1344-28-1	G 2
1333-86-4	G 2 (CARBON BLACK)
1344-28-1	G 2 (OXIDE)
94-36-0	G 20
644-64-4	G 22870
119-38-0	G 23611
510-15-6	G 23992
333-41-5	G 24480
76-06-2	G 25
2275-14-1	G 28029
1912-24-9	G 30027
333-41-5	G 301
3735-23-7	G 30494
26264-05-1	G 3300
510-15-6	G 338
26264-05-1	G 711
630-60-4	G-STROPHANTHIN
630-60-4	G-STROPHICOR
77-81-6	GA
77-81-6	GA (CHEMICAL WARFARE AGENT)
8001-69-2	GADISTOL
8001-69-2	GADUOL

CAS No.	Chemical Name
9016-45-9	GAFAC CO 99 0
111-76-2	GAFCOL EB
1314-87-0	GALENA
333-41-5	GALESAN
9005-90-7	GALIPOT
149-91-7	GALLIC ACID
1401-55-4	GALLIC ACID 3-MONOGALLATE
1401-55-4	GALLIC ACID, 3-GALLATE
7440-55-3	GALLIUM
13450-90-3	GALLIUM CHLORIDE
13450-90-3	GALLIUM CHLORIDE (GaCl3)
7440-55-3	GALLIUM ELEMENT
7440-55-3	GALLIUM METAL, LIQUID
7440-55-3	GALLIUM METAL, LIQUID (DOT)
7440-55-3	GALLIUM METAL, SOLID
7440-55-3	GALLIUM METAL, SOLID (DOT)
13450-90-3	GALLIUM TRICHLORIDE
13450-90-3	GALLIUM(3+) CHLORIDE
1401-55-4	GALLO-TANNIN ANALYSIS
58-89-9	GALLOGAMA
1401-55-4	GALLOTANNIC ACIDS
1401-55-4	GALLOTANNINS
62-38-4	GALLOTOX
9002-89-5	GALVATOL 1-60
58-89-9	GAMACARBATOX
58-89-9	GAMACID
58-89-9	GAMAPHEX
67-68-5	GAMASOL 90
58-89-9	GAMENE
58-89-9	GAMISO
58-89-9	GAMMA BENZENE HEXACHLORIDE
58-89-9	GAMMA ISOMER OF BENZENE HEXACHLORIDE
58-89-9	GAMMA-1,2,3,4,5,6-HEXACHLOROCYCLOHEXANE
58-89-9	GAMMA-666
504-24-5	GAMMA-AMINOPYRIDINE
58-89-9	GAMMA-BENZENE HEXACHLORIDE
58-89-9	GAMMA-BENZOHEXACHLORIDE
58-89-9	GAMMA-BHC
106-96-7	GAMMA-BROMOALLYLENE
115-11-7	GAMMA-BUTYLENE
51-83-2	GAMMA-CARBAMOYL CHOLINE CHLORIDE
12789-03-6	GAMMA-CHLORDAN
628-20-6	GAMMA-CHLOROBUTYRONITRILE
563-47-3	GAMMA-CHLOROISOBUTYLENE
106-89-8	GAMMA-CHLOROPROPYLENE OXIDE
58-89-9	GAMMA-COL
104-78-9	GAMMA-DIETHYLAMINOPROPYLAMINE
329-71-5	GAMMA-DINITROPHENOL
58-89-9	GAMMA-HCH
58-89-9	GAMMA-HEXACHLOR
58-89-9	GAMMA-HEXACHLORAN
58-89-9	GAMMA-HEXACHLORANE
58-89-9	GAMMA-HEXACHLOROBENZENE
58-89-9	GAMMA-HEXACHLOROCYCLOHEXANE
513-86-0	GAMMA-HYDROXY-BETA-OXOBUTANE
58-89-9	GAMMA-LINDANE
123-00-2	GAMMA-MORPHOLINOPROPYLAMINE
78-94-4	GAMMA-OXO-ALPHA-BUTYLENE
58-89-9	GAMMAHEXA
58-89-9	GAMMAHEXANE
58-89-9	GAMMALIN
58-89-9	GAMMALIN 20
58-89-9	GAMMATERR
58-89-9	GAMMEX
58-89-9	GAMMEXANE
58-89-9	GAMMOPAZ
63-25-2	GAMONIL
70-30-4	GAMOPHEN
70-30-4	GAMOPHENE
81-07-2	GARANTOSE
333-41-5	GARDEN TOX
7631-86-9	GAROSIL GB
7631-86-9	GAROSIL N
94-36-0	GAROX
786-19-6	GARRATHION
1333-86-4	GAS BLACK
65996-68-1	GAS BLAST FURNACE
8006-14-2	GAS NATURAL
64741-44-2	GAS OIL

CAS No.	Chemical Name
8006-20-0	GAS PRODUCER
68476-26-6	GASES, FUEL
8006-14-2	GASES, NATURAL
8006-61-9	GASOLINE
86290-81-5	GASOLINE
68425-31-0	GASOLINE (NATURAL GAS), NATURAL
86290-81-5	GASOLINE, SYNTHETIC
20859-73-8	GASTION
94-78-0	GASTRACID
8002-74-2	GATCH
107-44-8	GB
123-19-3	GBL
7681-38-1	GBS
143-50-0	GC 1189
2385-85-5	GC 1283
82-68-8	GC 3944-3-4
470-90-6	GC 4072
3254-63-5	GC 6506
684-16-2	GC 7887
639-58-7	GC 8993
116-16-5	GC-1106
7440-50-8	GE 1110
56-38-2	GEARPHOS
88-85-7	GEBUTOX
950-37-8	GEIGY 13005
644-64-4	GEIGY 22870
333-41-5	GEIGY 24480
2275-14-1	GEIGY 28029
1912-24-9	GEIGY 30027
3735-23-7	GEIGY 30494
510-15-6	GEIGY 338
950-37-8	GEIGY GS 13005
644-64-4	GEIGY GS 13332
119-38-0	GEIGY G 23611
2275-14-1	GEIGY G 28029
3735-23-7	GEIGY G 30494
7681-49-4	GEL II
7783-47-3	GEL-TIN
77-81-6	GELAN I
7723-14-0	GELBER PHOSPHOR (GERMAN)
13765-19-0	GELBIN
9000-07-1	GELCARIN HWG
9000-07-1	GELLOID J
9000-07-1	GELOZONE
7681-49-4	GELUTION
9002-89-5	GELUTOL
9002-89-5	GELVATOL
9002-89-5	GELVATOL 1-30
9002-89-5	GELVATOL 1-60
9002-89-5	GELVATOL 1-90
9002-89-5	GELVATOL 20-30
9002-89-5	GELVATOL 2060
9002-89-5	GELVATOL 209
9002-89-5	GELVATOL 3-60
9002-89-5	GELVATOL 3-91
6731-36-8	GEM-BIS(TERT-BUTYLPEROXY)-3,3,5-TRIMETHYLCYCLOHEXANE
6731-36-8	GEM-DI-TERT-BUTYLPEROXY-3,3,5-TRIMETHYLCYCLOHEXANE
79-01-6	GEMALGENE
112-02-7	GENAMIN CTAC
95-79-4	GENAZO RED KB SOLN
143-50-0	GENERAL CHEMICALS 1189
639-58-7	GENERAL CHEMICALS 8993
33213-65-9	GENERAL WEED KILLER
68476-26-6	GENERATOR GAS
75-37-6	GENETRON 100
25497-28-3	GENETRON 100
75-68-3	GENETRON 101
25497-29-4	GENETRON 101
75-69-4	GENETRON 11
79-38-9	GENETRON 1113
76-12-0	GENETRON 112
76-13-1	GENETRON 113
75-38-7	GENETRON 1132A
76-15-3	GENETRON 115
75-71-8	GENETRON 12
75-72-9	GENETRON 13
75-68-3	GENETRON 142B
25497-29-4	GENETRON 142B
75-37-6	GENETRON 152A

CAS No.	Chemical Name
25497-28-3	GENETRON 152A
75-43-4	GENETRON 21
76-19-7	GENETRON 218
75-45-6	GENETRON 22
75-46-7	GENETRON 23
8001-35-2	GENIPHENE
56-38-2	GENITHION
78-67-1	GENITRON
78-67-1	GENITRON AZDN
78-67-1	GENITRON AZDN-FF
80-17-1	GENITRON BSH
80-51-3	GENITRON OB
58-22-0	GENO-CRISTAUX GREMY
84-74-2	GENOPLAST B
50-18-0	GENOXAL
111-70-6	GENTANOL
25497-29-4	GENTRON 142B
9000-07-1	GENUGOL RLV
12002-03-8	GENUINE PARIS GREEN
9000-07-1	GENUVISCO J
58-89-9	GEOBILAN
21548-32-3	GEOFOS
58-89-9	GEOLIN G 3
2163-80-6	GEPIRON
101-84-8	GERANIUM CRYSTALS
70-30-4	GERMA-MEDICA
63-25-2	GERMAIN'S
79-01-6	GERMALGENE
7782-65-2	GERMANE
7782-65-2	GERMANE (DOT)
7782-65-2	GERMANIUM HYDRIDE
7782-65-2	GERMANIUM HYDRIDE (GeH4)
7782-65-2	GERMANIUM TETRAHYDRIDE
7782-65-2	GERMANIUM TETRAHYDRIDE (ACGIH)
1303-96-4	GERSTLEY BORATE
50-29-3	GESAFID
1912-24-9	GESAPRIM
1912-24-9	GESAPRIM 50
1912-24-9	GESAPRIM 500
50-29-3	GESAROL
299-84-3	GESEKTIN K
7786-34-7	GESFID
1912-24-9	GESOPRIM
57-83-0	GESTEROL
68-22-4	GESTEST
7786-34-7	GESTID
57-83-0	GESTONE
57-83-0	GESTORMONE
57-83-0	GESTRON
110-54-3	GETTYSOLVE-B
142-82-5	GETTYSOLVE-C
58-89-9	GEXANE
9002-89-5	GH 20
1314-13-2	GIAP 10
100-79-8	GIE
646-06-0	GIE
23745-86-0	GIFBLAAR
144-49-0	GIFBLAAR POISON
10124-56-8	GILTEX
55-63-0	GILUCOR NITRO
443-48-1	GINEFLAVIR
57-63-6	GINESTRENE
50-28-2	GINOSEDOL
13397-24-5	GIPS
52-24-4	GIROSTAN
1344-28-1	GK
1344-28-1	GK (OXIDE)
7782-42-5	GK 2
7782-42-5	GK 3
9002-89-5	GL 02
9002-89-5	GL 03
9002-89-5	GL 05
64-19-7	GLACIAL ACETIC ACID
53-16-7	GLANDUBOLIN
57-83-0	GLANDUCORPIN
9002-84-0	GLASROCK POREX P 1000
87-86-5	GLAZD PENTA
298-04-4	GLEBOFOS
7681-49-4	GLEEM

CAS No.	Chemical Name
107-07-3	GLICOL MONOCLORIDRINA (ITALIAN)
9002-89-5	GLO 5
56-75-7	GLOBENICOL
50-78-2	GLOBENTYL
50-78-2	GLOBOID
55-63-0	GLONOIN
8020-83-5	GLORIA
8020-83-5	GLORIA WHITE MINERAL OIL
56-75-7	GLOVETICOL
19222-41-4	GLUCAMONIX
81-07-2	GLUCID
7440-41-7	GLUCINIUM
7440-41-7	GLUCINUM
50-70-4	GLUCITOL
50-70-4	GLUCITOL, D-
71-63-6	GLUCODIGIN
19222-41-4	GLUCONIC ACID, MONOAMMONIUM SALT, D-
18883-66-4	GLUCOPYRANOSE, 2-DEOXY-2-(3-METHYL-3-NITROSOUREIDO)-, D-
9001-37-0	GLUCOSE OXIDASE
81-07-2	GLUSIDE
111-30-8	GLUTACLEAN
54-62-6	GLUTAMIC ACID, N-(P-(((2,4-DIAMINO-6-PTERIDINYL)METHYL)-AMINO)BENZOYL)-, L-
111-30-8	GLUTARAL
111-30-8	GLUTARALDEHYDE
111-30-8	GLUTARALDEHYDE SOLUTION
111-30-8	GLUTARDIALDEHYDE
111-30-8	GLUTAREX 28
111-30-8	GLUTARIC ACID DIALDEHYDE
111-30-8	GLUTARIC DIALDEHYDE
66-81-9	GLUTARIMIDE, 3-[2-(3,5-DIMETHYL-2-OXOCYCLOHEXYL)-2-HYDROXYETHYL]-
124-43-6	GLY-OXIDE
56-81-5	GLYCERIN
56-81-5	GLYCERIN (ACGIH)
96-24-2	GLYCERIN ALPHA-MONOCHLORHYDRIN
96-24-2	GLYCERIN EPICHLOROHYDRIN
100-79-8	GLYCERIN ISOPROPYLIDENE ETHER
646-06-0	GLYCERIN ISOPROPYLIDENE ETHER
55-63-0	GLYCERIN TRINITRATE
56-81-5	GLYCERIN, ANHYDROUS
56-81-5	GLYCERIN, SYNTHETIC
56-81-5	GLYCERINE
56-81-5	GLYCERINE, CRUDE, CONCENTRATED
55-63-0	GLYCERINTRINITRATE (CZECH)
56-81-5	GLYCERITOL
56-81-5	GLYCEROL
96-24-2	GLYCEROL 3-CHLOROHYDRIN
100-79-8	GLYCEROL ACETONIDE
646-06-0	GLYCEROL ACETONIDE
100-79-8	GLYCEROL ALPHA,BETA-ISOPROPYLIDENE ETHER
646-06-0	GLYCEROL ALPHA,BETA-ISOPROPYLIDENE ETHER
96-24-2	GLYCEROL ALPHA-CHLOROHYDRIN
96-24-2	GLYCEROL ALPHA-MONOCHLOROHYDRIN
96-24-2	GLYCEROL CHLOROHYDRIN
100-79-8	GLYCEROL DIMETHYLKETAL
646-06-0	GLYCEROL DIMETHYLKETAL
106-89-8	GLYCEROL EPICHLORHYDRIN
106-89-8	GLYCEROL EPICHLOROHYDRIN
96-18-4	GLYCEROL TRICHLOROHYDRIN
55-63-0	GLYCEROL TRINITRATE
55-63-0	GLYCEROL(TRINITRATE DE) (FRENCH)
100-79-8	GLYCEROL, 1,2-O-ISOPROPYLIDENE
646-06-0	GLYCEROL, 1,2-O-ISOPROPYLIDENE
1301-70-8	GLYCEROL, MONO(DIHYDROGEN PHOSPHATE), IRON(3+) SALT (3:2)
55-63-0	GLYCEROL, NITRIC ACID TRIESTER
96-24-2	GLYCEROL-ALPHA-MONOCHLOROHYDRIN
96-24-2	GLYCEROL-ALPHA-MONOCHLOROHYDRIN (DOT)
100-79-8	GLYCEROLACETONE
646-06-0	GLYCEROLACETONE
55-63-0	GLYCEROLTRINITRAAT (DUTCH)
96-24-2	GLYCERYL CHLORIDE
55-63-0	GLYCERYL NITRATE
96-18-4	GLYCERYL TRICHLOROHYDRIN
55-63-0	GLYCERYL TRINITRATE
55-63-0	GLYCERYL TRINITRATE SOLUTION
96-24-2	GLYCERYL-ALPHA-CHLOROHYDRIN
765-34-4	GLYCIDAL

CAS No.	*Chemical Name*
765-34-4	GLYCIDALDEHYDE
765-34-4	GLYCIDALDEHYDE (DOT)
556-52-5	GLYCIDE
556-52-5	GLYCIDOL
106-91-2	GLYCIDOL METHACRYLATE
122-60-1	GLYCIDOL PHENYL ETHER
106-92-3	GLYCIDYL 2-PROPENYL ETHER
556-52-5	GLYCIDYL ALCOHOL
106-92-3	GLYCIDYL ALLYL ETHER
106-91-2	GLYCIDYL ALPHA-METHYL ACRYLATE
2426-08-6	GLYCIDYL BUTYL ETHER
106-89-8	GLYCIDYL CHLORIDE
2238-07-5	GLYCIDYL ETHER
4016-14-2	GLYCIDYL ISOPROPYL ETHER
106-91-2	GLYCIDYL METHACRYLATE
26447-14-3	GLYCIDYL METHYLPHENYL ETHER
2426-08-6	GLYCIDYL N-BUTYL ETHER
122-60-1	GLYCIDYL PHENYL ETHER
26447-14-3	GLYCIDYL TOLYL ETHER
765-34-4	GLYCIDYLALDEHYDE
13479-54-4	GLYCINE, COPPER(2+) SALT (2:1)
60-00-4	GLYCINE, N,N'-1,2-ETHANEDIYLBIS(N-(CARBOXYMETHYL)- (9CI)
60-00-4	GLYCINE, N,N'-1,2-ETHANEDIYLBIS[N-(CARBOXYMETHYL)-
139-13-9	GLYCINE, N,N-BIS(CARBOXYMETHYL)-
139-13-9	GLYCINE, N,N-BIS(CARBOXYMETHYL)- (9CI)
13256-22-9	GLYCINE, N-METHYL-N-NITROSO-
141-43-5	GLYCINOL
926-64-7	GLYCINONITRILE, N,N-DIMETHYL-
107-21-1	GLYCOL
107-21-1	GLYCOL ALCOHOL
112-27-6	GLYCOL BIS(HYDROXYETHYL) ETHER
123-81-9	GLYCOL BIS(MERCAPTOACETATE)
106-93-4	GLYCOL BROMIDE
111-76-2	GLYCOL BUTYL ETHER
96-49-1	GLYCOL CARBONATE
107-07-3	GLYCOL CHLOROHYDRIN
109-78-4	GLYCOL CYANOHYDRIN
111-55-7	GLYCOL DIACETATE
106-93-4	GLYCOL DIBROMIDE
107-06-2	GLYCOL DICHLORIDE
629-14-1	GLYCOL DIETHYL ETHER
123-81-9	GLYCOL DIMERCAPTOACETATE
110-71-4	GLYCOL DIMETHYL ETHER
628-96-6	GLYCOL DINITRATE
111-46-6	GLYCOL ETHER
112-34-5	GLYCOL ETHER DB
111-76-2	GLYCOL ETHER EE ACETATE
109-86-4	GLYCOL ETHER EM
110-49-6	GLYCOL ETHER EM ACETATE
111-46-6	GLYCOL ETHYL ETHER
123-91-1	GLYCOL ETHYLENE ETHER
100-79-8	GLYCOL FORMAL
646-06-0	GLYCOL FORMAL
109-86-4	GLYCOL METHYL ETHER
622-08-2	GLYCOL MONOBENZYL ETHER
111-76-2	GLYCOL MONOBUTYL ETHER
111-76-2	GLYCOL MONOBUTYL ETHER ACETATE
107-07-3	GLYCOL MONOCHLOROHYDRIN
110-80-5	GLYCOL MONOETHYL ETHER
111-15-9	GLYCOL MONOETHYL ETHER ACETATE
109-86-4	GLYCOL MONOMETHYL ETHER
110-49-6	GLYCOL MONOMETHYL ETHER ACETATE
68-11-1	GLYCOLIC ACID, 2-THIO-
84-72-0	GLYCOLIC ACID, ETHYL ESTER, ETHYL PHTHALATE
85-71-2	GLYCOLIC ACID, ETHYL ESTER, METHYL PHTHALATE
68-11-1	GLYCOLIC ACID, THIO-
107-16-4	GLYCOLIC NITRILE
107-07-3	GLYCOLMONOCHLOORHYDRINE (DUTCH)
107-16-4	GLYCOLONITRILE
107-16-4	GLYCOLONITRILE (8CI)
9016-45-9	GLYCOLS, POLYETHYLENE, MONO(NONYLPHENYL) ETHER
25322-69-4	GLYCOLS, POLYPROPYLENE
107-07-3	GLYCOMONOCHLORHYDRIN
57-11-4	GLYCON DP
112-80-1	GLYCON RO
57-11-4	GLYCON S-70
57-11-4	GLYCON S-80
57-11-4	GLYCON S-90
57-11-4	GLYCON TP

CAS No.	*Chemical Name*
112-80-1	GLYCON WO
107-16-4	GLYCONITRILE
56-81-5	GLYCYL ALCOHOL
110-71-4	GLYME
111-96-6	GLYME-2
133-06-2	GLYODEX 3722
107-22-2	GLYOXAL
107-22-2	GLYOXAL ALDEHYDE
56-81-5	GLYROL
56-81-5	GLYSANIN
9002-89-5	GM 14
9002-89-5	GOHSEFIMER L 7514
9002-89-5	GOHSENOL
9002-89-5	GOHSENOL AH 22
9002-89-5	GOHSENOL EG 40
9002-89-5	GOHSENOL GH
9002-89-5	GOHSENOL GH 17
9002-89-5	GOHSENOL GH 20
9002-89-5	GOHSENOL GH 23
9002-89-5	GOHSENOL GL 02
9002-89-5	GOHSENOL GL 03
9002-89-5	GOHSENOL GL 05
9002-89-5	GOHSENOL GL 08
9002-89-5	GOHSENOL GM 14
9002-89-5	GOHSENOL GM 14L
9002-89-5	GOHSENOL GM 94
9002-89-5	GOHSENOL KH 17
9002-89-5	GOHSENOL KH 20
9002-89-5	GOHSENOL KL 05
9002-89-5	GOHSENOL KP 06
9002-89-5	GOHSENOL KP 08
9002-89-5	GOHSENOL L 5307
9002-89-5	GOHSENOL MG 14
9002-89-5	GOHSENOL N 300
9002-89-5	GOHSENOL NH 05
9002-89-5	GOHSENOL NH 14
9002-89-5	GOHSENOL NH 17
9002-89-5	GOHSENOL NH 18
9002-89-5	GOHSENOL NH 20
9002-89-5	GOHSENOL NH 26
9002-89-5	GOHSENOL NK 114
9002-89-5	GOHSENOL NL 05
9002-89-5	GOHSENOL NM 11
9002-89-5	GOHSENOL NM 114
9002-89-5	GOHSENOL NM 14
9002-89-5	GOHSENOL NM 300
9002-89-5	GOHSENOL T
9002-89-5	GOHSENOL T 330
107-22-2	GOHSEZAL P
1315-04-4	GOLDEN ANTIMONY SULFIDE
117-84-0	GOOD-RITE GP 264
117-81-7	GOOD-RITE GP 264
4104-14-7	GOPHACIDE
9002-84-0	GORE-TEX
86-50-0	GOTHNION
7782-42-5	GP 60
7782-42-5	GP 60S
7782-42-5	GP 63
7631-86-9	GP 71
87-68-3	GP-40-66:120
76-44-8	GPKH
1024-57-3	GPKH EPOXIDE
131-52-2	GR 48-11PS
131-52-2	GR 48-32S
56-53-1	GRAFESTROL
7782-42-5	GRAFOIL
7782-42-5	GRAFOIL GTA
64-17-5	GRAIN ALCOHOL
7775-09-9	GRAIN SORGHUM HARVEST-AID
75-99-0	GRAMEVIN
1910-42-5	GRAMIXEL
1910-2-5	GRAMONOL
1910-2-5	GRAMOXON
1910-42-5	GRAMOXONE
1910-42-5	GRAMOXONE D
1910-42-5	GRAMOXONE DICHLORIDE
2074-50-2	GRAMOXONE METHYL SULFATE
1910-42-5	GRAMOXONE S
1910-42-5	GRAMOXONE W

CAS No.	*Chemical Name*
1910-2-5	GRAMURON
7775-09-9	GRANEX O
2235-25-8	GRANOSAN M
118-74-1	GRANOX NM
133-06-2	GRANOX PFM
7440-66-6	GRANULAR ZINC
57-50-1	GRANULATED SUGAR
298-02-2	GRANUTOX
7782-42-5	GRAPHITE
7782-42-5	GRAPHITE (NATURAL) DUST
77-47-4	GRAPHLOX
7782-42-5	GRAPHNOL N 3M
1333-86-4	GRAPHON C
1333-86-4	GRAPHTOL BLACK BLN
60-11-7	GRASAL BRILLIANT YELLOW
75-87-6	GRASEX
630-60-4	GRATIBAIN
630-60-4	GRATUS STROPHANTHIN
633-03-4	GREEN EN
120-12-7	GREEN OIL
1314-13-2	GREEN SEAL 8
7720-78-7	GREEN VITRIOL
1318-59-8	GREENITE
14998-27-7	GREENITE
7440-38-2	GREY ARSENIC
1912-24-9	GRIFFEX
7758-98-7	GRIFFIN SUPER CU
105-60-2	GRILON
112-80-1	GROCO 2
112-80-1	GROCO 4
57-11-4	GROCO 54
57-11-4	GROCO 55
57-11-4	GROCO 55L
57-11-4	GROCO 58
57-11-4	GROCO 59
112-80-1	GROCO 5L
112-80-1	GROCO 6
56-81-5	GROCOLENE
7704-34-9	GROUND VOCLE SULPHUR
8002-03-7	GROUNDNUT OIL
87-86-5	GRUNDIER ARBEZOL
12172-73-5	GRUNERITE ASBESTOS
14797-65-0	GRYZBOL
950-37-8	GS 13005
7782-42-5	GS 2
644-64-4	GS-13332
55-98-1	GT 41
55-63-0	GTN
107-70-0	GUAIACOL
506-93-4	GUANIDINE NITRATE
506-93-4	GUANIDINE NITRATE (1:1)
506-93-4	GUANIDINE NITRATE (DOT)
70-25-7	GUANIDINE, 1-METHYL-3-NITRO-1-NITROSO-
502-39-6	GUANIDINE, CYANO(METHYLMERCURIO)-
502-39-6	GUANIDINE, CYANO-, MERCURY COMPLEX
502-39-6	GUANIDINE, CYANO-, METHYLMERCURY DERIV
506-93-4	GUANIDINE, MONONITRATE
70-25-7	GUANIDINE, N-METHYL-N'-NITRO-N-NITROSO-
506-93-4	GUANIDINE, NITRATE
556-88-7	GUANIDINE, NITRO-
506-93-4	GUANIDINIUM NITRATE
872-05-9	GULFTENE 10
111-66-0	GULFTENE 8
8002-74-2	GULFWAX
50-70-4	GULITOL
9000-01-5	GUM ARABIC
76-22-2	GUM CAMPHOR
9000-07-1	GUM CARRAGEENAN
9000-07-1	GUM CHON 2
9000-07-1	GUM CHOND
9005-90-7	GUM THUS
9005-90-7	GUM TURPENTINE
9004-70-0	GUNCOTTON
9004-70-0	GUNCOTTON (DOT)
86-50-0	GUSATHION
86-50-0	GUSATHION 25
2642-71-9	GUSATHION A
2642-71-9	GUSATHION ETHYL
2642-71-9	GUSATHION H AND K

CAS No.	*Chemical Name*
86-50-0	GUSATHION K
86-50-0	GUSATHION M
86-50-0	GUSATHION METHYL
86-50-0	GUSATHION-20
133-06-2	GUSTAFSON CAPTAN 30-DD
86-50-0	GUTHION
86-50-0	GUTHION (DOT)
2642-71-9	GUTHION ETHYL
86-50-0	GUTHION, LIQUID (DOT)
9000-32-2	GUTTA PERCHA, SOLUTION (DOT)
9000-32-2	GUTTA-PERCHA
9000-32-2	GUTTA-PERCHA SOLUTION
7782-42-5	GY 70
8001-35-2	GY-PHENE
50-28-2	GYNERGON
57-83-0	GYNLUTIN
50-28-2	GYNOESTRYL
57-83-0	GYNOLUTONE
13397-24-5	GYPSITE
13397-24-5	GYPSUM
13397-24-5	GYPSUM (Ca(SO4).2H2O)
26499-65-0	GYPSUM HEMIHYDRATE
16568-02-8	GYROMITRIN
9004-70-0	H 1/2
1194-65-6	H 133
8002-74-2	H 1N3
2032-65-7	H 321
7782-42-5	H 451
64-00-6	H 5727
64-00-6	H 8757
538-07-8	H N1
76-44-8	H-34
65-85-0	HA 1
3254-63-5	HA-1200
7440-58-6	HAFNIUM
7440-58-6	HAFNIUM (ACGIH)
7440-58-6	HAFNIUM ELEMENT
7440-58-6	HAFNIUM METAL, DRY
7440-58-6	HAFNIUM METAL, DRY (DOT)
7440-58-6	HAFNIUM METAL, WET
7440-58-6	HAFNIUM METAL, WET (DOT)
7440-58-6	HAFNIUM, WET WITH NOT LESS THAN 25% WATER (DOT)
10124-56-8	HAGAN PHOSPHATE
540-72-7	HAIMASED
2425-06-1	HAIPEN 50
7428-48-0	HAL-LUB-N
118-52-5	HALANE
23092-17-3	HALAZEPAM
999-81-5	HALLOWEEN
75-69-4	HALOCARBON 11
75-38-7	HALOCARBON 1132A
76-15-3	HALOCARBON 115
75-72-9	HALOCARBON 13/UCON 13
75-73-0	HALOCARBON 14
75-37-6	HALOCARBON 152A
25497-28-3	HALOCARBON 152A
75-46-7	HALOCARBON 23
115-25-3	HALOCARBON C-138
75-71-8	HALON
74-88-4	HALON 10001
74-83-9	HALON 1001
74-97-5	HALON 1011
75-69-4	HALON 11
75-61-6	HALON 1202
353-59-3	HALON 1211
75-63-8	HALON 1301
75-73-0	HALON 14
74-96-4	HALON 2001
9002-84-0	HALON G 183
9002-84-0	HALON G 700
9002-84-0	HALON G 80
9002-84-0	HALON TFE
9002-84-0	HALON TFEG 180
151-67-7	HALOTHANE
1321-65-9	HALOWAX
1335-87-1	HALOWAX 1014
2234-13-1	HALOWAX 1051
74-83-9	HALTOX
75-45-6	HALTRON 22

CAS No.	Chemical Name
10265-92-6	HAMIDOP
60-00-4	HAMP-ENE ACID
139-13-9	HAMPSHIRE NTA ACID
115-26-4	HANANE
470-90-6	HAPTASOL
8002-74-2	HARD PARAFFIN
8002-74-2	HAROWAX L 1
8002-74-2	HAROWAX L 2
67-63-0	HARTOSOL
7775-09-9	HARVEST-AID
117-84-0	HATCOL DOP
60-00-4	HAVIDOTE
6923-22-4	HAZODRIN
116-16-5	HCA
7440-02-0	HCA 1
118-74-1	HCB
87-68-3	HCBD
58-89-9	HCC
58-89-9	HCCH
77-47-4	HCCPD
1024-57-3	HCE
58-89-9	HCH
7647-01-0	HCL
74-90-8	HCN
57-74-9	HCS 3260
12789-03-6	HCS 3260
505-60-2	HD
89-62-3	HD FAST RED GL BASE
298-07-7	HDEHP
822-06-0	HDI
7631-86-9	HDK-N 20
7631-86-9	HDK-S 15
7631-86-9	HDK-V 15
9004-70-0	HE 2000
123-31-9	HE 5
8001-58-9	HEAVY OIL
58-89-9	HECLOTOX
1335-85-9	HEDOLIT
534-52-1	HEDOLIT
534-52-1	HEDOLITE
1335-85-9	HEDOLITE
94-75-7	HEDONAL
94-75-7	HEDONAL (HERBICIDE)
110-54-3	HEKSAN (POLISH)
88-85-7	HEL-FIRE
50-78-2	HELICON
16071-86-6	HELION BROWN BRSL
7440-59-7	HELIUM
7440-59-7	HELIUM (DOT)
7440-59-7	HELIUM, COMPRESSED (DOT)
7440-59-7	HELIUM, REFRIGERATED LIQUID
7440-59-7	HELIUM, REFRIGERATED LIQUID (DOT)
7440-59-7	HELIUM-4
58933-55-4	HELIUM-OXYGEN (MIXTURE)
58933-55-4	HELIUM-OXYGEN MIXTURE (DOT)
35400-43-2	HELOTHION
26499-65-0	HEMIHYDRATE GYPSUM
71-36-3	HEMOSTYP
680-31-9	HEMPA
1120-21-4	HENDECANE
30207-98-8	HENDECANOIC ALCOHOL
30207-98-8	HENDECYL ALCOHOL
60-57-1	HEOD
76-44-8	HEPTA
76-44-8	HEPTACHLOOR (DUTCH)
76-44-8	HEPTACHLOR
76-44-8	HEPTACHLOR (ACGIH,DOT)
1024-57-3	HEPTACHLOR CIS-OXIDE
1024-57-3	HEPTACHLOR EPOXIDE
76-44-8	HEPTACHLORANE
76-44-8	HEPTACHLORE(FRENCH)
76-44-8	HEPTAGRAN
76-44-8	HEPTAMUL
142-82-5	HEPTAN (POLISH)
123-19-3	HEPTAN-4-ONE
142-82-5	HEPTANE
142-82-5	HEPTANE (ACGIH,DOT)
142-82-5	HEPTANE :AND ITS ISOMERS:
142-82-5	HEPTANE(N-HEPTANE)

CAS No.	Chemical Name
4032-86-4	HEPTANE, 3,3-DIMETHYL-
123-04-6	HEPTANE, 3-(CHLOROMETHYL)-
103-44-6	HEPTANE, 3-[(ETHENYLOXY)METHYL]-
30586-18-6	HEPTANE, PENTAMETHYL-
142-82-5	HEPTANEN (DUTCH)
111-70-6	HEPTANOL
592-76-7	HEPTENE
25339-56-4	HEPTENE
81624-04-6	HEPTENE
111-70-6	HEPTYL ALCOHOL
111-87-5	HEPTYL CARBINOL
142-82-5	HEPTYL HYDRIDE
19594-40-2	HEPTYL ISOBUTYL KETONE
3648-21-3	HEPTYL PHTHALATE
111-68-2	HEPTYLAMINE
25339-56-4	HEPTYLENE
14686-13-6	HEPTYLENE-2-TRANS
64-00-6	HER 5727
100-97-0	HERAX UTS
2163-80-6	HERB-ALL
2163-80-6	HERBAN M
330-54-1	HERBATOX
1912-24-9	HERBATOXOL
1910-2-5	HERBAXON
314-40-9	HERBICIDE 976
86290-81-5	HERBICIDE ES
94-75-7	HERBIDAL
61-82-5	HERBIDAL TOTAL
1420-07-1	HERBOGIL
1910-2-5	HERBOXONE
6484-52-2	HERCO PRILLS
117-84-0	HERCOFLEX 260
8001-35-2	HERCULES 3956
78-34-2	HERCULES 528
64-00-6	HERCULES 5727
64-00-6	HERCULES AC 5727
78-34-2	HERCULES AC528
115-77-5	HERCULES P6
8001-35-2	HERCULES TOXAPHENE
2636-26-2	HERKAL
2636-26-2	HERKOL
137-26-8	HERMAL
137-26-8	HERMAT TMT
137-26-8	HERYL
757-58-4	HET
100-97-0	HETERIN
757-58-4	HETP
1642-54-2	HETRAZAN
9004-34-6	HEWETEN 10
592-41-6	HEX-1-ENE
592-41-6	HEX-1-ENE (DOT)
58-89-9	HEXA
100-97-0	HEXA
131-73-7	HEXA
100-97-0	HEXA (VULCANIZATION ACCELERATOR)
131-73-7	HEXA (VULCANIZATION ACCELERATOR)
118-74-1	HEXA CB
100-97-0	HEXA-FLO-PULVER
131-73-7	HEXA-FLO-PULVER
70-30-4	HEXABALM
133-06-2	HEXACAP
58-89-9	HEXACHLOR
87-68-3	HEXACHLOR-1,3-BUTADIEN (CZECH)
67-72-1	HEXACHLOR-AETHAN (GERMAN)
58-89-9	HEXACHLORAN
58-89-9	HEXACHLORANE
118-74-1	HEXACHLORBENZOL (GERMAN)
87-68-3	HEXACHLORBUTADIENE
77-47-4	HEXACHLORCYKLOPENTADIEN (CZECH)
67-72-1	HEXACHLORETHANE
55720-99-5	HEXACHLORO DIPHENYL OXIDE
87-68-3	HEXACHLORO-1,3-BUTADIENE
77-47-4	HEXACHLORO-1,3-CYCLOPENTADIENE
116-16-5	HEXACHLORO-2-PROPANONE
116-16-5	HEXACHLOROACETONE
116-16-5	HEXACHLOROACETONE (DOT)
118-74-1	HEXACHLOROBENZENE
118-74-1	HEXACHLOROBENZENE (DOT)
87-68-3	HEXACHLOROBUTADIENE

CAS No.	Chemical Name
87-68-3	HEXACHLOROBUTADIENE (ACGIH,DOT)
58-89-9	HEXACHLOROCYCLOHEXANE
58-89-9	HEXACHLOROCYCLOHEXANE, GAMMA-ISOMER
77-47-4	HEXACHLOROCYCLOPENTADIENE
77-47-4	HEXACHLOROCYCLOPENTADIENE (ACGIH,DOT)
2385-85-5	HEXACHLOROCYCLOPENTADIENE DIMER
72-20-8	HEXACHLOROEPOXYOCTAHYDRO-ENDO,ENDO-DIMETHANO-NAPHTHALENE
60-57-1	HEXACHLOROEPOXYOCTAHYDRO-ENDO,EXO-DIMETHANONAPH-THALENE
67-72-1	HEXACHLOROETHANE
67-72-1	HEXACHLOROETHANE (ACGIH,DOT)
67-72-1	HEXACHLOROETHYLENE
70-30-4	HEXACHLOROFEN
30900-23-3	HEXACHLOROHEXAHYDRO-ENDO-EXO-DIMETHANONAPHTHA-LENE
115-29-7	HEXACHLOROHEXAHYDROMETHANO 2,4,3-BENZODIOX-ATHIEPIN-3-OXIDE
1335-87-1	HEXACHLORONAPHTHALENE
1335-87-1	HEXACHLORONAPHTHALENE (ACGIH)
70-30-4	HEXACHLOROPHEN
70-30-4	HEXACHLOROPHENE
16941-12-1	HEXACHLOROPLATINIC ACID
16941-12-1	HEXACHLOROPLATINIC ACID (H2PtCl6)
16941-12-1	HEXACHLOROPLATINIC(IV) ACID
116-16-5	HEXACHLOROPROPANONE
142-62-1	HEXACID 698
2646-17-5	HEXACOL OIL ORANGE SS
3761-53-3	HEXACOL PONCEAU 2R
3761-53-3	HEXACOL PONCEAU MX
5894-60-0	HEXADECYLTRICHLOROSILANE
5894-60-0	HEXADECYLTRICHLOROSILANE (DOT)
112-02-7	HEXADECYLTRIMETHYLAMMONIUM CHLORIDE
42296-74-2	HEXADIENE
42296-74-2	HEXADIENE (DOT)
72-20-8	HEXADRIN
2157-42-8	HEXAETHOXYDISILOXANE
2157-42-8	HEXAETHYL DIORTHOSILICATE
757-58-4	HEXAETHYL TETRAPHOSPHATE
757-58-4	HEXAETHYL TETRAPHOSPHATE (DOT)
757-58-4	HEXAETHYL TETRAPHOSPHATE MIXTURE, DRY, WITH > 2% HEXAETHYL TETRAPHOSPHATE
757-58-4	HEXAETHYL TETRAPHOSPHATE MIXTURE, DRY, WITH ≤ 2% HEXAETHYL TETRAPHOSPHATE
757-58-4	HEXAETHYL TETRAPHOSPHATE MIXTURE, LIQUID, WITH > 25% HEXAETHYL TETRAPHOPHATE
757-58-4	HEXAETHYL TETRAPHOSPHATE MIXTURE, LIQUID, WITH ≤ 25% HEXAETHYL TETRAPHOSPHATE
757-58-4	HEXAETHYL TETRAPHOSPHATE, LIQUID
757-58-4	HEXAETHYL TETRAPHOSPHATE, LIQUID (DOT)
757-58-4	HEXAETHYL TETRAPHOSPHATE, LIQUID, CONTAINING MORE THAN 25% HEXAETHYL TETRAPHOSPHATE (DOT)
757-58-4	HEXAETHYL TETRAPHOSPHATE, LIQUID, CONTAINING NOT MORE THAN 25% HEXAETHYL TETRAPHOSPHATE (DOT)
70-30-4	HEXAFEN
14484-64-1	HEXAFERB
428-59-1	HEXAFLUORO-1,2-EPOXYPROPANE
677-71-4	HEXAFLUORO-2,2-PROPANEDIOL
684-16-2	HEXAFLUOROACETONE
684-16-2	HEXAFLUOROACETONE (ACGIH,DOT)
677-71-4	HEXAFLUOROACETONE HYDRATE
677-71-4	HEXAFLUOROACETONE HYDRATE (DOT)
428-59-1	HEXAFLUOROEPOXYPROPANE
76-16-4	HEXAFLUOROETHANE
76-16-4	HEXAFLUOROETHANE (DOT)
16961-83-4	HEXAFLUOROKIESELSAIURE (GERMAN)
16961-83-4	HEXAFLUOROKIEZELZUUR (DUTCH)
16940-81-1	HEXAFLUOROPHOSPHORIC ACID
16940-81-1	HEXAFLUOROPHOSPHORIC ACID (DOT)
116-15-4	HEXAFLUOROPROPENE
428-59-1	HEXAFLUOROPROPENE EPOXIDE
428-59-1	HEXAFLUOROPROPENE OXIDE
116-15-4	HEXAFLUOROPROPYLENE
116-15-4	HEXAFLUOROPROPYLENE (DOT)
428-59-1	HEXAFLUOROPROPYLENE EPOXIDE
428-59-1	HEXAFLUOROPROPYLENE OXIDE
428-59-1	HEXAFLUOROPROPYLENE OXIDE (DOT)
17084-08-1	HEXAFLUOROSILICATE
17084-08-1	HEXAFLUOROSILICATE (SiF62-)
17084-08-1	HEXAFLUOROSILICATE(2-)
17084-08-1	HEXAFLUOROSILICATE(2-) ION
16961-83-4	HEXAFLUOROSILICIC ACID
2551-62-4	HEXAFLUORURE DE SOUFRE (FRENCH)
16961-83-4	HEXAFLUOSILICIC ACID
100-97-0	HEXAFORM
121-82-4	HEXAHYDRO-1,3,5-TRINITRO-1,3,5-TRIAZINE
121-82-4	HEXAHYDRO-1,3,5-TRINITRO-S-TRIAZINE
110-85-0	HEXAHYDRO-1,4-DIAZINE
111-49-9	HEXAHYDRO-1H-AZEPINE
105-60-2	HEXAHYDRO-2-AZEPINONE
105-60-2	HEXAHYDRO-2H-AZEPIN-2-ONE
105-60-2	HEXAHYDRO-2H-AZEPIN-2-ONE (9CI)
56-25-7	HEXAHYDRO-3A,7A-DIMETHYL-4,7-EPOXYISOBENZOFURAN-1,3-DIONE
108-91-8	HEXAHYDROANILINE
111-49-9	HEXAHYDROAZEPINE
110-82-7	HEXAHYDROBENZENE
25639-42-3	HEXAHYDROCRESOL
25639-42-3	HEXAHYDROMETHYLPHENOL
108-93-0	HEXAHYDROPHENOL
110-85-0	HEXAHYDROPYRAZINE
110-89-4	HEXAHYDROPYRIDINE
108-87-2	HEXAHYDROTOLUENE
58-89-9	HEXAKLOR
66-25-1	HEXALDEHYDE
66-25-1	HEXALDEHYDE (DOT)
108-93-0	HEXALIN
108-93-0	HEXALIN (ALCOHOL)
680-31-9	HEXAMETAPOL
999-97-3	HEXAMETHYL DISILAZANE
680-31-9	HEXAMETHYL PHOSPHORAMIDE
100-97-0	HEXAMETHYLENAMINE
110-82-7	HEXAMETHYLENE
124-09-4	HEXAMETHYLENE DIAMINE
124-09-4	HEXAMETHYLENE DIAMINE, SOLID (DOT)
822-06-0	HEXAMETHYLENE DIISOCYANATE
111-49-9	HEXAMETHYLENE IMINE (DOT)
100-97-0	HEXAMETHYLENE TERAMINE
822-06-0	HEXAMETHYLENE-1,6-DIISOCYANATE
100-97-0	HEXAMETHYLENEAMINE
4835-11-4	HEXAMETHYLENEDIAMINE, N,N'-DIBUTYL-
124-09-4	HEXAMETHYLENEDIAMINE, SOLID
124-09-4	HEXAMETHYLENEDIAMINE, SOLUTION
124-09-4	HEXAMETHYLENEDIAMINE, SOLUTION (DOT)
822-06-0	HEXAMETHYLENEDIISOCYANATE (DOT)
111-49-9	HEXAMETHYLENEIMINE
100-97-0	HEXAMETHYLENETETRAMINE
100-97-0	HEXAMETHYLENETETRAMINE
111-49-9	HEXAMETHYLENIMINE
105-60-2	HEXAMETHYLENIMINE, 2-OXO-
100-97-0	HEXAMETHYLENTETRAMIN (GERMAN)
100-97-0	HEXAMETHYLENTETRAMINE
131-73-7	HEXAMETHYLENTETRAMINE
680-31-9	HEXAMETHYLORTHOPHOSPHORIC TRIAMIDE
680-31-9	HEXAMETHYLPHOSPHORIC ACID TRIAMIDE
680-31-9	HEXAMETHYLPHOSPHORIC TRIAMIDE
999-97-3	HEXAMETHYLSILAZANE
100-97-0	HEXAMINE
100-97-0	HEXAMINE (1,3,5,7-TETRAAZATRICYCLO-3.3.1.13,7-DECANE)
100-97-0	HEXAMINE (DOT)
100-97-0	HEXAMINE (HETEROCYCLE)
131-73-7	HEXAMINE (POTASSIUM REAGENT)
151-21-3	HEXAMOL SLS
66-25-1	HEXANAL
123-05-7	HEXANAL, 2-ETHYL-
496-03-7	HEXANAL, 2-ETHYL-3-HYDROXY-
110-82-7	HEXANAPHTHENE
110-54-3	HEXANE
110-54-3	HEXANE (DOT)
110-54-3	HEXANE (N-HEXANE)
822-06-0	HEXANE 1,6-DIISOCYANATE
110-54-3	HEXANE AND ITS ISOMERS
112-58-3	HEXANE, 1,1'-OXYBIS-
822-06-0	HEXANE, 1,6-DIISOCYANATO-
28679-16-5	HEXANE, 1,6-DIISOCYANATOTRIMETHYL-
544-10-5	HEXANE, 1-CHLORO-
3522-94-9	HEXANE, 2,2,5-TRIMETHYL-
584-94-1	HEXANE, 2,3-DIMETHYL-

CAS No.	Chemical Name
589-43-5	HEXANE, 2,4-DIMETHYL-
78-63-7	HEXANE, 2,5-DIMETHYL-2,5-DI(T-BUTYLPEROXY)-
78-63-7	HEXANE, 2,5-DIMETHYL-2,5-DI(T-BUTYLPEROXY)-, MAXIMUM CONCENTRATION 52% WITH INERT SOLID
591-76-4	HEXANE, 2-METHYL-
3074-77-9	HEXANE, 3-ETHYL-4-METHYL-
589-34-4	HEXANE, 3-METHYL-
3074-75-7	HEXANE, 4-ETHYL-2-METHYL-
25495-90-3	HEXANE, CHLORO-
3025-88-5	HEXANE, DIMETHYL-, DIHYDROPEROXIDE
111-69-3	HEXANEDINITRILE
124-04-9	HEXANEDIOIC ACID
103-23-1	HEXANEDIOIC ACID, BIS(2-ETHYLHEXYL) ESTER
123-79-5	HEXANEDIOIC ACID, BIS(2-ETHYLHEXYL) ESTER
103-23-1	HEXANEDIOIC ACID, BIS(2-ETHYLHEXYL) ESTER (9CI)
123-79-5	HEXANEDIOIC ACID, BIS(2-ETHYLHEXYL) ESTER (9CI)
103-23-1	HEXANEDIOIC ACID, DIOCTYL ESTER
123-79-5	HEXANEDIOIC ACID, DIOCTYL ESTER
111-50-2	HEXANEDIOYL CHLORIDE
111-50-2	HEXANEDIOYL DICHLORIDE
110-54-3	HEXANEN (DUTCH)
3006-82-4	HEXANEPEROXOIC ACID, 2-ETHYL-, 1,1-DIMETHYLETHYL ESTER
686-31-7	HEXANEPEROXOIC ACID, 2-ETHYL-, 1,1-DIMETHYLPROPYL ESTER
13122-18-4	HEXANEPEROXOIC ACID, 3,5,5-TRIMETHYL-, 1,1-DIMETHYL-ETHYL ESTER
131-73-7	HEXANITRODIPHENYLAMINE
35860-31-2	HEXANITRODIPHENYLAMINE
20062-22-0	HEXANITROSTILBENE
142-62-1	HEXANOIC ACID
142-62-1	HEXANOIC ACID (DOT)
3006-82-4	HEXANOIC ACID, 2-((1,1-DIMETHYLETHYL)DIOXY)ETHYL ESTER
3006-82-4	HEXANOIC ACID, 2-((1,1-DIMETHYLETHYL)DIOXY)ETHYL ESTER, WITH AT LEAST 50% PHLEGMATIZER
149-57-5	HEXANOIC ACID, 2-ETHYL-
94-04-2	HEXANOIC ACID, 2-ETHYL-, ETHENYL ESTER
94-04-2	HEXANOIC ACID, 2-ETHYL-, VINYL ESTER
105-60-2	HEXANOIC ACID, 6-AMINO-, CYCLIC LACTAM
105-60-2	HEXANOIC ACID, 6-AMINO-, LACTAM
123-66-0	HEXANOIC ACID, ETHYL ESTER
111-27-3	HEXANOL
105-60-2	HEXANOLACTAM
108-94-1	HEXANON
105-60-2	HEXANONE ISOXIME
591-78-6	HEXANONE-2
105-60-2	HEXANONISOXIM (GERMAN)
84-74-2	HEXAPLAS M/B
100-97-0	HEXASAN
62-38-4	HEXASAN
62-38-4	HEXASAN (FUNGICIDE)
10124-56-8	HEXASODIUM HEXAMETAPHOSPHATE
10124-56-8	HEXASODIUM METAPHOSPHATE
7704-34-9	HEXASUL
786-19-6	HEXATHION
137-26-8	HEXATHIR
58-89-9	HEXATOX
58-89-9	HEXAVERM
63-25-2	HEXAVIN
110-89-4	HEXAZANE
592-41-6	HEXENE
81624-06-8	HEXENE
116-15-4	HEXFLUOROPROPYLENE
58-89-9	HEXICIDE
100-97-0	HEXILMETHYLENAAMINE
677-71-4	HEXOFLUOROACETONE DIHYDRATE
121-82-4	HEXOGEN
121-82-4	HEXOGEN (EXPLOSIVE)
121-82-4	HEXOGEN 5W
142-62-1	HEXOIC ACID
108-10-1	HEXONE
70-30-4	HEXOPHENE
70-30-4	HEXOSAN
58-89-9	HEXYCLAN
131-73-7	HEXYL
131-73-7	HEXYL (REAGENT)
108-84-9	HEXYL ACETATE
142-92-7	HEXYL ACETATE
142-92-7	HEXYL ACETATE [COMBUSTIBLE LIQUID LABEL]
142-92-7	HEXYL ACETATE [FLAMMABLE LIQUID LABEL]
111-27-3	HEXYL ALCOHOL
142-92-7	HEXYL ALCOHOL, ACETATE
544-10-5	HEXYL CHLORIDE
142-92-7	HEXYL ETHANOATE
112-58-3	HEXYL ETHER
66-25-1	HEXYLALDEHYDE
111-26-2	HEXYLAMINE
104-75-6	HEXYLAMINE, 2-ETHYL-
10137-80-1	HEXYLAMINE, 2-ETHYL-N-PHENYL-
5432-61-1	HEXYLAMINE, N-CYCLOHEXYL-2-ETHYL-
58-89-9	HEXYLAN
107-41-5	HEXYLENE GLYCOL
107-41-5	HEXYLENE GLYCOL (ACGIH)
124-09-4	HEXYLENEDIAMINE
928-65-4	HEXYLTRICHLOROSILANE
928-65-4	HEXYLTRICHLOROSILANE (DOT)
9002-84-0	HEYDEFLON
144-49-0	HFA
68334-28-1	HG 150
58-89-9	HGI
309-00-2	HHDN
56-53-1	HI-BESTROL
9005-25-8	HI-COASTAR PC 11
112-60-7	HI-DRY
93-79-8	HI-ESTER 2,4,5-T
2545-59-7	HI-ESTER 2,4,5-T
94-80-4	HI-ESTER 2,4-D
8030-30-6	HI-FLASH NAPHTHA
8002-74-2	HI-MIC 1045
1338-23-4	HI-POINT 180
7778-39-4	HI-YIELD DESICCANT H-10
999-81-5	HICO CCC
111-77-3	HICOTOL CAR
3761-53-3	HIDACID SCARLET 2R
1694-09-3	HIDACID WOOL VIOLET 5B
633-03-4	HIDACO BRILLIANT GREEN
97-56-3	HIDACO OIL YELLOW
57-41-0	HIDANTAL
16940-66-2	HIDKITEX DF
53-16-7	HIESTRONE
1401-55-4	HIFIX SL
115-32-2	HIFOL
52628-25-8	HIGH SPEED
58-89-9	HILBEECH
115-29-7	HILDAN
115-32-2	HILFOL 185 EC
2782-57-2	HILITE 60
121-75-5	HILTHION
121-75-5	HILTHION 25WDP
119-90-4	HILTONIL FAST BLUE B BASE
88-74-4	HILTONIL FAST ORANGE GR BASE
89-62-3	HILTONIL FAST RED GL BASE
95-79-4	HILTONIL FAST RED KB BASE
99-55-8	HILTONIL FAST SCARLET G BASE
99-55-8	HILTONIL FAST SCARLET G SALT
99-55-8	HILTONIL FAST SCARLET GC BASE
88-74-4	HILTOSAL FAST ORANGE GR SALT
89-62-3	HILTOSAL FAST RED GL SALT
7631-86-9	HIMESIL A
2746-19-2	HIMIC ANHYDRIDE
88-74-4	HINDASOL ORANGE GR SALT
3165-93-3	HINDASOL RED TR SALT
7722-84-1	HIOXYL
64-00-6	HIP
50-55-5	HIPOSERPIL
9002-89-5	HISELON C 300
16071-86-6	HISPALUZ BROWN BRL
1937-37-7	HISPAMIN BLACK EF
2602-46-2	HISPAMIN BLUE 2B
72-57-1	HISPAMIN BLUE 3BX
7782-42-5	HITCO HMG 50
9004-70-0	HITENOL 12
88-85-7	HIVERTOX
7631-86-9	HK 125
7631-86-9	HK 400
1420-04-8	HL 2448
62-38-4	HL-331
434-07-1	HMD
124-09-4	HMDA

CAS No.	Chemical Name
822-06-0	HMDI
999-97-3	HMDS
10124-56-8	HMP
680-31-9	HMPA
680-31-9	HMPT
680-31-9	HMPTA
100-97-0	HMT
51-75-2	HN2
55-86-7	HN2 HYDROCHLORIDE
555-77-1	HN3
97-77-8	HOCA
50-00-0	HOCH
25322-69-4	HODAG HT 11
115-29-7	HOE 2671
639-58-7	HOE 2872
24017-47-8	HOE 2960
24017-47-8	HOE 2960 OJ
13463-67-7	HOMBITAN R 101D
13463-67-7	HOMBITAN R 505
13463-67-7	HOMBITAN R 506
13463-67-7	HOMBITAN R 610D
13463-67-7	HOMBITAN R 610K
20830-75-5	HOMOLLE'S DIGITALIN
111-49-9	HOMOPIPERIDINE
58-22-0	HOMOSTERON
58-22-0	HOMOSTERONE
62-38-4	HONGNIEN
12001-29-5	HOOKER NO.1 CHRYSOTILE ASBESTOS
999-81-5	HORMOCEL-2CCC
57-83-0	HORMOFLAVEINE
53-16-7	HORMOFOLLIN
57-83-0	HORMOLUTON
93-79-8	HORMOSLYR 500T
2545-59-7	HORMOSLYR 500T
53-16-7	HORMOVARINE
11119-70-3	HORNA GL 35
13463-67-7	HORSEHEAD A 430C
13463-67-7	HORSEHEAD A 430FG
13463-67-7	HORSEHEAD R 771
58-89-9	HORTEX
111-30-8	HOSPEX
9002-84-0	HOSTAFLON SE-VP 585
9002-84-0	HOSTAFLON SE-VP 5875
9002-84-0	HOSTAFLON TF
9002-84-0	HOSTAFLON TF 2026
9002-84-0	HOSTAFLON TF 2053
9002-84-0	HOSTAFLON TF 5032
9002-84-0	HOSTAFLON TF 9205
9002-84-0	HOSTAFLON TF-VP 5034
9002-84-0	HOSTAFLON TF-VP 5444
9016-45-9	HOSTAPAL CV
9016-45-9	HOSTAPAL W
62-38-4	HOSTAQUICK
62-38-4	HOSTAQUIK
24017-47-8	HOSTATHION
24017-47-8	HOSTATION
120-82-1	HOSTETEX L-PEC
680-31-9	HPT
150-76-5	HQMME
77-47-4	HRS 1655
2385-85-5	HRS L276
9005-25-8	HRW 13
25155-30-0	HS 85S
85-44-9	HT 901
101-77-9	HT 972
112-02-7	HTAC
7778-54-3	HTH
7778-54-3	HTH (BLEACHING AGENT)
757-58-4	HTP
1314-13-2	HUBBUCK'S WHITE
62-53-3	HUILE D'ANILINE (FRENCH)
76-22-2	HUILE DE CAMPHRE (FRENCH)
8013-75-0	HUILE DE FUSEL (FRENCH)
110-02-1	HUILE H50
110-02-1	HUILE HSO
577-11-7	HUMIFEN WT 27G
57-11-4	HUMKO INDUSTRENE R
58-89-9	HUNGARIA L 7
1912-24-9	HUNGAZIN
1912-24-9	HUNGAZIN PK
1912-24-9	HUNGAZINPK
9002-89-5	HV POVAL
330-54-1	HW 920
9004-70-0	HX 3/5
12001-26-2	HX 610
112-24-3	HY 951
112-80-1	HY-PHI 1055
112-80-1	HY-PHI 1088
57-11-4	HY-PHI 1199
57-11-4	HY-PHI 1205
57-11-4	HY-PHI 1303
57-11-4	HY-PHI 1401
112-80-1	HY-PHI 2066
112-80-1	HY-PHI 2088
112-80-1	HY-PHI 2102
67-68-5	HYADUR
7681-52-9	HYCLORITE
118-52-5	HYDAN
118-52-5	HYDAN (ANTISEPTIC)
118-52-5	HYDANTOIN, 1,3-DICHLORO-5,5-DIMETHYL-
57-41-0	HYDANTOIN, 5,5-DIPHENYL-
57-57-8	HYDRACRYLIC ACID, BETA-LACTONE
109-78-4	HYDRACRYLONITRILE
106-71-8	HYDRACRYLONITRILE, ACRYLATE (ESTER)
1332-58-7	HYDRAPRINT
7774-29-0	HYDRARGYRUM BIJODATUM (GERMAN)
1332-58-7	HYDRASHEEN 90M
1305-62-0	HYDRATED LIME
80-17-1	HYDRAZIDE BSG
302-01-2	HYDRAZINE
302-01-2	HYDRAZINE (ACGIH)
302-01-2	HYDRAZINE BASE
10034-93-2	HYDRAZINE DIHYDROGEN SULFATE SALT
302-01-2	HYDRAZINE HYDRATE (DOT)
10034-93-2	HYDRAZINE HYDROGEN SULFATE
10034-93-2	HYDRAZINE MONOSULFATE
13464-97-6	HYDRAZINE NITRATE
10034-93-2	HYDRAZINE SULFATE
57-14-7	HYDRAZINE, 1,1-DIMETHYL-
1615-80-1	HYDRAZINE, 1,2-DIETHYL-
540-73-8	HYDRAZINE, 1,2-DIMETHYL-
122-66-7	HYDRAZINE, 1,2-DIPHENYL-
302-01-2	HYDRAZINE, ANHYDROUS
302-01-2	HYDRAZINE, ANHYDROUS (DOT)
302-01-2	HYDRAZINE, AQUEOUS SOLUTION (DOT)
302-01-2	HYDRAZINE, AQUEOUS SOLUTION CONTAINING MORE THAN 64% HYDRAZINE (DOT)
302-01-2	HYDRAZINE, AQUEOUS SOLUTION WITH LESS THAN 64% HYDRAZINE (DOT)
302-01-2	HYDRAZINE, HYDRATE
60-34-4	HYDRAZINE, METHYL-
100-63-0	HYDRAZINE, PHENYL-
59-88-1	HYDRAZINE, PHENYL-, HYDROCHLORIDE
59-88-1	HYDRAZINE, PHENYL-, MONOHYDROCHLORIDE
10034-93-2	HYDRAZINE, SULFATE (1:1)
10036-47-2	HYDRAZINE, TETRAFLUORO-
100-63-0	HYDRAZINE-BENZENE
2231-57-4	HYDRAZINECARBOHYDRAZONOTHIOIC ACID
79-19-6	HYDRAZINECARBOTHIOAMIDE
3570-75-0	HYDRAZINECARBOXALDEHYDE, 2-[4-(5-NITRO-2-FURANYL)-2-THIAZOLYL]-
16568-02-8	HYDRAZINECARBOXALDEHYDE, ETHYLIDENEMETHYL-
563-41-7	HYDRAZINECARBOXAMIDE MONOHYDROCHLORIDE
563-41-7	HYDRAZINECARBOXAMIDE, HYDROCHLORIDE
10034-93-2	HYDRAZINIUM SULFATE
100-63-0	HYDRAZINOBENZENE
122-66-7	HYDRAZOBENZENE
26628-22-8	HYDRAZOIC ACID, SODIUM SALT
60-34-4	HYDRAZOMETHANE
540-73-8	HYDRAZOMETHANE
10034-93-2	HYDRAZONIUM SULFATE
302-01-2	HYDRAZYNA (POLISH)
12184-88-2	HYDRIDE
12184-88-2	HYDRIDE ION
12184-88-2	HYDRIDE ION(H1-)
12184-88-2	HYDRIDE, METAL
16842-03-8	HYDRIDOCOBALT TETRACARBONYL
16842-03-8	HYDRIDOTETRACARBONYLCOBALT

CAS No.	Chemical Name
10034-85-2	HYDRIODIC ACID
10034-85-2	HYDRIODIC ACID (DOT)
10034-85-2	HYDRIODIC ACID, SOLUTION (DOT)
1332-58-7	HYDRITE UF
10035-10-6	HYDROBROMIC ACID
10035-10-6	HYDROBROMIC ACID ≥ 49% STRENGTH
12124-97-9	HYDROBROMIC ACID MONOAMMONIATE
10035-10-6	HYDROBROMIC ACID, ≤ 49% STRENGTH
10035-10-6	HYDROBROMIC ACID, ANHYDROUS (DOT)
74-96-4	HYDROBROMIC ETHER
8020-83-5	HYDROCARBON OIL
8020-83-5	HYDROCARBON OILS
8002-05-9	HYDROCARBON OILS, PETROLEUM
8002-74-2	HYDROCARBON WAXES
123-31-9	HYDROCHINON (CZECH,POLISH)
7647-01-0	HYDROCHLORIC ACID
7647-01-0	HYDROCHLORIC ACID (DOT)
7647-01-0	HYDROCHLORIC ACID GAS
7647-01-0	HYDROCHLORIC ACID, ANHYDROUS (DOT)
8007-56-5	HYDROCHLORIC ACID, MIXED WITH NITRIC ACID (3:1)
7647-01-0	HYDROCHLORIC ACID, SOLUTION (DOT)
7647-01-0	HYDROCHLORIC ACID, SOLUTION, INHIBITED (DOT)
75-00-3	HYDROCHLORIC ETHER
7647-01-0	HYDROCHLORIDE
142-04-1	HYDROCHLORIDE BENZENAMIDE
1318-59-8	HYDROCHLORITE
14998-27-7	HYDROCHLORITE
16842-03-8	HYDROCOBALT TETRACARBONYL
74-90-8	HYDROCYANIC ACID
74-90-8	HYDROCYANIC ACID (PRUSSIC), UNSTABILIZED (DOT)
74-90-8	HYDROCYANIC ACID SOLUTION, 5% HYDROCYANIC ACID
57-12-5	HYDROCYANIC ACID, ION(1-)
74-90-8	HYDROCYANIC ACID, LIQUEFIED
74-90-8	HYDROCYANIC ACID, LIQUEFIED (DOT)
151-50-8	HYDROCYANIC ACID, POTASSIUM SALT
57-12-5	HYDROCYANIC ACID, SALTS
143-33-9	HYDROCYANIC ACID, SODIUM SALT
107-12-0	HYDROCYANIC ETHER
1191-15-7	HYDRODIISOBUTYLALUMINUM
8030-30-6	HYDROFINING
8020-83-5	HYDROFINING
86290-81-5	HYDROFINING
7664-39-3	HYDROFLUORIC ACID
7664-39-3	HYDROFLUORIC ACID GAS
7664-39-3	HYDROFLUORIC ACID SOLUTION
7664-39-3	HYDROFLUORIC ACID SOLUTION (DOT)
7664-39-3	HYDROFLUORIC ACID, ANHYDROUS (DOT)
16984-48-8	HYDROFLUORIC ACID, ION(1-)
7664-39-3	HYDROFLUORIDE
16961-83-4	HYDROFLUOROSILICIC ACID
16961-83-4	HYDROFLUOROSILICIC ACID (DOT)
16961-83-4	HYDROFLUOSILICIC ACID
57-11-4	HYDROFOL 1895
57-11-4	HYDROFOL ACID 150
57-11-4	HYDROFOL ACID 1655
57-11-4	HYDROFOL ACID 1855
109-99-9	HYDROFURAN
1333-74-0	HYDROGEN
1333-74-0	HYDROGEN (DOT)
1333-74-0	HYDROGEN (H2)
7803-52-3	HYDROGEN ANTIMONIDE
7784-42-1	HYDROGEN ARSENIDE
18130-74-0	HYDROGEN BIFLUORIDE ION(1-)
298-07-7	HYDROGEN BIS(2-ETHYLHEXYL)PHOSPHATE
10035-10-6	HYDROGEN BROMIDE
10035-10-6	HYDROGEN BROMIDE (ACGIH,DOT)
10035-10-6	HYDROGEN BROMIDE (H2Br2)
10035-10-6	HYDROGEN BROMIDE (HBr)
64-18-6	HYDROGEN CARBOXYLIC ACID
7647-01-0	HYDROGEN CHLORIDE
7647-01-0	HYDROGEN CHLORIDE (ACGIH,DOT)
7647-01-0	HYDROGEN CHLORIDE (HCl)
7647-01-0	HYDROGEN CHLORIDE (LIQUEFIED GAS)
7647-01-0	HYDROGEN CHLORIDE, ANHYDROUS
7647-01-0	HYDROGEN CHLORIDE, ANHYDROUS (DOT)
7647-01-0	HYDROGEN CHLORIDE, REFRIGERATED LIQUID
7647-01-0	HYDROGEN CHLORIDE, REFRIGERATED LIQUID (DOT)
420-04-2	HYDROGEN CYANAMIDE
74-90-8	HYDROGEN CYANIDE
74-90-8	HYDROGEN CYANIDE (ACGIH)
74-90-8	HYDROGEN CYANIDE, ABSORBED
74-90-8	HYDROGEN CYANIDE, ANHYDROUS, STABILIZED (DOT)
7783-28-0	HYDROGEN DIAMMONIUM PHOSPHATE
18130-74-0	HYDROGEN DIFLUORIDE ION(1-)
7722-84-1	HYDROGEN DIOXIDE
7558-79-4	HYDROGEN DISODIUM PHOSPHATE
7664-39-3	HYDROGEN FLUORIDE
7664-39-3	HYDROGEN FLUORIDE (ACGIH,DOT)
7664-39-3	HYDROGEN FLUORIDE (HF)
18130-74-0	HYDROGEN FLUORIDE (HF2 1-)
18130-74-0	HYDROGEN FLUORIDE ION (HF2 1-)
16941-12-1	HYDROGEN HEXACHLOROPLATINATE(IV)
16940-81-1	HYDROGEN HEXAFLUOROPHOSPHATE
16940-81-1	HYDROGEN HEXAFLUOROPHOSPHATE(1-)
16961-83-4	HYDROGEN HEXAFLUOROSILICATE
10034-85-2	HYDROGEN IODIDE
10034-85-2	HYDROGEN IODIDE (HI)
10034-85-2	HYDROGEN IODIDE SOLUTION
10034-85-2	HYDROGEN IODIDE SOLUTION (DOT)
10034-85-2	HYDROGEN IODIDE, ANHYDROUS
10034-85-2	HYDROGEN IODIDE, ANHYDROUS (DOT)
1333-74-0	HYDROGEN MOLECULE
10035-10-6	HYDROGEN MONOBROMIDE
10034-85-2	HYDROGEN MONOIODIDE
7697-37-2	HYDROGEN NITRATE
3734-97-2	HYDROGEN OXALATE OF AMITON
7722-84-1	HYDROGEN PEROXIDE
7722-84-1	HYDROGEN PEROXIDE (ACGIH)
7722-84-1	HYDROGEN PEROXIDE (H2O2)
3313-92-6	HYDROGEN PEROXIDE (H2O2), COMPD WITH DISODIUM CARBONATE (3:2)
15630-89-4	HYDROGEN PEROXIDE (H2O2), COMPD WITH DISODIUM CARBONATE (3:2)
124-43-6	HYDROGEN PEROXIDE (H2O2), COMPD WITH UREA (1:1)
124-43-6	HYDROGEN PEROXIDE CARBAMIDE
7722-84-1	HYDROGEN PEROXIDE SOLUTION (40% TO 52% PEROXIDE)
7722-84-1	HYDROGEN PEROXIDE SOLUTION (8% TO 40% PEROXIDE)
7722-84-1	HYDROGEN PEROXIDE SOLUTION (8% TO 40% PEROXIDE) (DOT)
7722-84-1	HYDROGEN PEROXIDE SOLUTION, 40% TO 52% (DOT)
7722-84-1	HYDROGEN PEROXIDE, 20% TO 60%
7722-84-1	HYDROGEN PEROXIDE, 30%
7722-84-1	HYDROGEN PEROXIDE, 90%
7722-84-1	HYDROGEN PEROXIDE, SOLUTION (DOT)
7722-84-1	HYDROGEN PEROXIDE, SOLUTION (OVER 52% PEROXIDE) (DOT)
7722-84-1	HYDROGEN PEROXIDE, STABILIZED (OVER 60% PEROXIDE) (DOT)
124-43-6	HYDROGEN PEROXIDE-UREA COMPOUND (1:1)
7803-51-2	HYDROGEN PHOSPHIDE
16941-12-1	HYDROGEN PLATINUM CHLORIDE (H2PtCl6)
7789-29-9	HYDROGEN POTASSIUM DIFLUORIDE
7789-29-9	HYDROGEN POTASSIUM FLUORIDE
7789-29-9	HYDROGEN POTASSIUM FLUORIDE (HKF2)
7646-93-7	HYDROGEN POTASSIUM SULFATE
7783-07-5	HYDROGEN SELENIDE
7783-07-5	HYDROGEN SELENIDE (ACGIH,DOT)
7783-07-5	HYDROGEN SELENIDE (H2Se)
7783-07-5	HYDROGEN SELENIDE, ANHYDROUS (DOT)
7631-90-5	HYDROGEN SODIUM SULFATE
16721-80-5	HYDROGEN SODIUM SULFIDE
7664-93-9	HYDROGEN SULFATE (DOT)
7783-06-4	HYDROGEN SULFIDE
7783-06-4	HYDROGEN SULFIDE (ACGIH,DOT)
7783-06-4	HYDROGEN SULFIDE (H2S)
12136-50-4	HYDROGEN SULFIDE (H2S5), DIPOTASSIUM SALT
7631-90-5	HYDROGEN SULFITE SODIUM
7783-06-4	HYDROGEN SULFURIC ACID
7783-06-4	HYDROGEN SULPHIDE
1333-74-0	HYDROGEN, COMPRESSED (DOT)
12184-88-2	HYDROGEN, ION (H1-)
7782-39-0	HYDROGEN, ISOTOPE OF MASS 2
1333-74-0	HYDROGEN, REFRIGERATED LIQUID
1333-74-0	HYDROGEN, REFRIGERATED LIQUID (DOT)
7782-39-0	HYDROGEN-2
7782-39-0	HYDROGEN-D2
68334-28-1	HYDROGENATED MIXED VEGETABLE OILS
7783-06-4	HYDROGENE SULFURE (FRENCH)
1332-58-7	HYDROGLOSS
85-42-7	HYDROHEXAPHTHALIC ANHYDRIDE
10034-85-2	HYDROIODIC ACID

CAS No.	Chemical Name
7775-14-6	HYDROLIN
124-43-6	HYDROPERIT
124-43-6	HYDROPERITE
7722-84-1	HYDROPEROXIDE
3025-88-5	HYDROPEROXIDE, (1,1,4,4-TETRAMETHYL-1,4-BUTANEDIYL)BIS- 4,5-BIS(HYDROPEROXY)-2,5-DIMETHYLHEXANE
3025-88-5	HYDROPEROXIDE, (1,1,4,4-TETRAMETHYLTETRAMETHYLENE)DI-
5809-08-5	HYDROPEROXIDE, 1,1,3,3-TETRAMETHYLBUTYL
75-91-2	HYDROPEROXIDE, 1,1-DIMETHYLETHYL
771-29-9	HYDROPEROXIDE, 1,2,3,4-TETRAHYDRO-1-NAPHTHALENYL
771-29-9	HYDROPEROXIDE, 1,2,3,4-TETRAHYDRO-1-NAPHTHYL-
80-47-7	HYDROPEROXIDE, 1-METHYL-1-(4-METHYLCYCLOHEXYL)ETHYL
80-15-9	HYDROPEROXIDE, 1-METHYL-1-PHENYLETHYL
28324-52-9	HYDROPEROXIDE, 2,6,6-TRIMETHYLBICYCLO(311)HEPTYL-
28324-52-9	HYDROPEROXIDE, 2,6,6-TRIMETHYLBICYCLO(311)HEPTYL-, NOT OVER 45% PEROXIDE
85873-97-8	HYDROPEROXIDE, 2,6,6-TRIMETHYLBICYCLO[311]HEPT-2-YL, (1ALPHA,2ALPHA,5ALPHA)-
28324-52-9	HYDROPEROXIDE, 2,6,6-TRIMETHYLBICYCLO[311]HEPTYL
79-21-0	HYDROPEROXIDE, ACETYL
80-15-9	HYDROPEROXIDE, ALPHA,ALPHA-DIMETHYLBENZYL-
26762-93-6	HYDROPEROXIDE, BIS(1-METHYLETHYL)PHENYL
26762-93-6	HYDROPEROXIDE, DIISOPROPYLPHENYL
26762-93-6	HYDROPEROXIDE, DIISOPROPYLPHENYL-, (SOLUTION)
80-15-9	HYDROPEROXYDE DE CUMENE (FRENCH)
80-15-9	HYDROPEROXYDE DE CUMYLE (FRENCH)
771-29-9	HYDROPEROXYDE DE TETRALINE (FRENCH)
123-31-9	HYDROQUINOL
123-31-9	HYDROQUINOLE
123-31-9	HYDROQUINONE
123-31-9	HYDROQUINONE (ACGIH,DOT)
150-76-5	HYDROQUINONE METHYL ETHER
150-76-5	HYDROQUINONE MONOMETHYL ETHER
8020-83-5	HYDROREFINING
8030-30-6	HYDROREFINING
7775-14-6	HYDROS
16961-83-4	HYDROSILICOFLUORIC ACID
16961-83-4	HYDROSILICOFLUORIC ACID (DOT)
7783-06-4	HYDROSULFURIC ACID
124-68-5	HYDROXY-TERT-BUTYLAMINE
107-16-4	HYDROXYACETONITRILE
7803-49-8	HYDROXYAMINE
108-95-2	HYDROXYBENZENE
1333-39-7	HYDROXYBENZENESULFONIC ACID
9004-34-6	HYDROXYCELLULOSE
108-93-0	HYDROXYCYCLOHEXANE
1310-58-3	HYDROXYDE DE POTASSIUM (FRENCH)
75-59-2	HYDROXYDE DE TETRAMETHYLAMMONIUM (FRENCH)
75-60-5	HYDROXYDIMETHYLARSINE OXIDE
124-65-2	HYDROXYDIMETHYLARSINE OXIDE SODIUM SALT
818-61-1	HYDROXYETHYL ACRYLATE
60-24-2	HYDROXYETHYL MERCAPTAN
92-50-2	HYDROXYETHYLETHYLANILINE
10138-74-6	HYDROXYETHYLPROPYLENEDIAMINE
10039-54-0	HYDROXYL AMINE SULPHATE (DOT)
10039-54-0	HYDROXYL AMMONIUM SULFATE ((HONH3)2SO4)
25154-52-3	HYDROXYL NO 253
7803-49-8	HYDROXYLAMINE
10039-54-0	HYDROXYLAMINE NEUTRAL SULFATE
10039-54-0	HYDROXYLAMINE SULFATE
3710-84-7	HYDROXYLAMINE, N,N-DIETHYL-
135-20-6	HYDROXYLAMINE, N-NITROSO-N-PHENYL-, AMMONIUMSALT
10039-54-0	HYDROXYLAMINE, SULFATE (2:1)
10039-54-0	HYDROXYLAMINE, SULFATE (2:1) (SALT)
10039-54-0	HYDROXYLAMMONIUM SULFATE
107-16-4	HYDROXYMETHYL CYANIDE
107-16-4	HYDROXYMETHYLNITRILE
27598-85-2	HYDROXYPHENYLAMINE
25584-83-2	HYDROXYPROPYL ACRYLATE
27813-02-1	HYDROXYPROPYL METHACRYLATE
100-51-6	HYDROXYTOLUENE
1319-77-3	HYDROXYTOLUENE
1319-77-3	HYDROXYTOLUOLE (GERMAN)
13121-70-5	HYDROXYTRICYCLOHEXYLSTANNANE
127-07-1	HYDROXYUREA
563-12-2	HYLEMAX
563-12-2	HYLEMOX
101-68-8	HYLENE M50
26471-62-5	HYLENE T

CAS No.	Chemical Name
91-08-7	HYLENE TCPA
26471-62-5	HYLENE TCPA
91-08-7	HYLENE TIC
26471-62-5	HYLENE TLC
91-08-7	HYLENE TM
26471-62-5	HYLENE TM
91-08-7	HYLENE TM-65
26471-62-5	HYLENE TM-65
91-08-7	HYLENE TRF
26471-62-5	HYLENE TRF
26471-62-5	HYLENE-T
2893-78-9	HYLITE 60G
9005-25-8	HYLON
1344-28-1	HYPALOX II
80-15-9	HYPERIZ
124-43-6	HYPEROL
7631-86-9	HYPERSIL
98-86-2	HYPNON
98-86-2	HYPNONE
14380-61-1	HYPOCHLORITE
14380-61-1	HYPOCHLORITE ANION
14380-61-1	HYPOCHLORITE ION
14380-61-1	HYPOCHLORITE SOLUTION WITH ≥ 7% AVAILABLE CHLORINE BY WEIGHT
14380-61-1	HYPOCHLORITE SOLUTION WITH > 7% AVAILABLE CHLORINE BY WEIGHT
13477-10-6	HYPOCHLOROUS ACID, BARIUM SALT
7778-54-3	HYPOCHLOROUS ACID, CALCIUM SALT
7778-54-3	HYPOCHLOROUS ACID, CALCIUM SALT, DRY MIXTURE WITH 10% TO 39% AVAILABLE CHLORINE
14380-61-1	HYPOCHLOROUS ACID, ION(1-)
13840-33-0	HYPOCHLOROUS ACID, LITHIUM SALT
7778-66-7	HYPOCHLOROUS ACID, POTASSIUM SALT
7681-52-9	HYPOCHLOROUS ACID, SODIUM SALT
10022-70-5	HYPOCHLOROUS ACID, SODIUM SALT, PENTAHYDRATE
7791-21-1	HYPOCHLOROUS ANHYDRIDE
52-68-6	HYPODERMACID
10024-97-2	HYPONITROUS ACID ANHYDRIDE
11104-93-1	HYPONITROUS ACID ANHYDRIDE
25322-69-4	HYPROX DP 400
57-11-4	HYSTRENE 4516
57-11-4	HYSTRENE 5016
57-11-4	HYSTRENE 7018
57-11-4	HYSTRENE 80
57-11-4	HYSTRENE 9718
57-11-4	HYSTRENE S 97
57-11-4	HYSTRENE T 70
108-94-1	HYTROL O
314-40-9	HYVAR X
314-40-9	HYVAR X BROMACIL
314-40-9	HYVAREX
56-75-7	I 337A
123-92-2	I-AMYL ACETATE
78-81-9	I-BUTYLAMINE
3452-97-9	I-NONYL ALCOHOL
7631-86-9	IATROBEADS 6RS8060
366-70-1	IBENZMETHYZIN HYDROCHLORIDE
366-70-1	IBENZMETHYZINE HYDROCHLORIDE
139-91-3	IBIFUR
366-70-1	IBZ
21564-17-0	ICHIBAN
1954-28-5	ICI 32865
71-55-6	ICI-CF 2
13010-47-4	ICIG 1109
60-00-4	ICRF 185
21416-87-5	ICRF-159
53-86-1	IDOMETHINE
50-78-2	IDRAGIN
123-31-9	IDROCHINONE (ITALIAN)
7783-06-4	IDROGENO SOLFORATO (ITALIAN)
80-15-9	IDROPEROSSIDO DI CUMENE (ITALIAN)
80-15-9	IDROPEROSSIDO DI CUMOLO (ITALIAN)
771-29-9	IDROPEROSSIDO DI TETRALINA (ITALIAN)
206-44-0	IDRYL
30674-80-7	IEM
121-75-5	IFO 13140
7782-42-5	IG 11
9016-45-9	IGEPAL CO
9016-45-9	IGEPAL CO 210

CAS No.	Chemical Name
9016-45-9	IGEPAL CO 430
9016-45-9	IGEPAL CO 436
9016-45-9	IGEPAL CO 520
9016-45-9	IGEPAL CO 530
9016-45-9	IGEPAL CO 610
9016-45-9	IGEPAL CO 630
9016-45-9	IGEPAL CO 660
9016-45-9	IGEPAL CO 710
9016-45-9	IGEPAL CO 730
9016-45-9	IGEPAL CO 850
9016-45-9	IGEPAL CO 880
9016-45-9	IGEPAL CO 887
9016-45-9	IGEPAL CO 890
9016-45-9	IGEPAL CO 970
9016-45-9	IGEPAL CO 977
9016-45-9	IGEPAL CO 990
9016-45-9	IGEPAL CO 997
60-11-7	IKETON YELLOW EXTRA
7773-06-0	IKURIN
60-57-1	ILLOXOL
9016-45-9	IMBENTIN
9016-45-9	IMBENTIN N 52
9016-45-9	IMBENTINE
53-86-1	IMBRILON
9004-66-4	IMFERON
7440-32-6	IMI 115
7440-32-6	IMI 125
7440-32-6	IMI 130
7440-32-6	IMI 155
732-11-6	IMIDAN
732-11-6	IMIDATHION
4316-42-1	IMIDAZOLE, 1-BUTYL-
443-48-1	IMIDAZOLE-1-ETHANOL, 2-METHYL-5-NITRO-
4342-03-4	IMIDAZOLE-4(OR 5)-CARBOXAMIDE, 5(OR 4)-(3,3-DIMETHYL-1-TRIAZENO)-
4342-03-4	IMIDAZOLE-4-CARBOXAMIDE, 5-(3,3-DIMETHYL-1-TRIAZENO)-
96-45-7	IMIDAZOLIDINETHIONE
96-45-7	IMIDAZOLINE-2(3H)-THIONE
96-45-7	IMIDAZOLINE-2-THIOL
947-02-4	IMIDOCARBONIC ACID, (DIETHOXYPHOSPHINYL)DITHIO-, CYCLIC ETHYLENE ESTER
947-02-4	IMIDOCARBONIC ACID, PHOSPHONODITHIO-, CYCLIC ETHYLENE P,P-DIETHYL ESTER
21548-32-3	IMIDOCARBONIC ACID, PHOSPHONODITHIO-, CYCLIC METHYLENE P,P-DIETHYL ESTER
950-10-7	IMIDOCARBONIC ACID, PHOSPHONODITHIO-, CYCLIC PROPYLENE P,P-DIETHYL ESTER
947-02-4	IMIDOCARBONIC ACID, PHOSPHONODITHIO-, P,P-DIETHYL CYCLIC ETHYLENE ESTER
622-44-6	IMIDOCARBONYL CHLORIDE, PHENYL-
541-53-7	IMIDODICARBONIMIDOTHIOIC DIAMIDE
541-53-7	IMIDODICARBONODITHIOIC DIAMIDE
109-97-7	IMIDOLE
56-18-8	IMINO BISPROPYLAMINE (DOT)
56-18-8	IMINOBIS(PROPYLAMINE)
111-42-2	IMINODIETHANOL
56-18-8	IMINODIPROPYLAMINE
1330-78-5	IMOL S 140
12002-03-8	IMPERIAL GREEN
9005-25-8	IMPERMEX
545-55-1	IMPERON FIXER T
107-44-8	IMPF
9004-66-4	IMPOSIL
128-37-0	IMPRUVOL
7631-86-9	IMSIL 10
7631-86-9	IMSIL 1240
7631-86-9	IMSIL A 10
7631-86-9	IMSIL A 108
7631-86-9	IMSIL A 15
7631-86-9	IMSIL H
67-63-0	IMSOL A
446-86-6	IMURAN
446-86-6	IMUREK
446-86-6	IMUREL
16752-77-5	IN 1179
53-86-1	INACID
138-86-3	INACTIVE LIMONENE
1912-24-9	INAKOR
94-36-0	INCIDOL

CAS No.	Chemical Name
7758-98-7	INCRACIDE 10A
7758-98-7	INCRACIDE E 51
999-81-5	INCRECEL
53-86-1	INDACIN
95-13-6	INDEN
95-13-6	INDENE
193-39-5	INDENO (1,2,3-CD) PYRENE
124-87-8	INDIAN BERRY
2602-46-2	INDIGO BLUE 2B
7440-74-6	INDIUM
53-86-1	INDO-RECTOLMIN
53-86-1	INDO-TABLINEN
53-86-1	INDOCID
53-86-1	INDOCIN
53-86-1	INDOLE-3-ACETIC ACID, 1-(P-CHLOROBENZOYL)-5-METHOXY-2-METHYL-
53-86-1	INDOMECOL
53-86-1	INDOMED
53-86-1	INDOMEE
53-86-1	INDOMETACIN
53-86-1	INDOMETACINE
53-86-1	INDOMETHACIN
53-86-1	INDOMETHACINE
53-86-1	INDOMETICINA (SPANISH)
95-13-6	INDONAPHTHENE
53-86-1	INDOPTIC
8002-74-2	INDRAMIC 30
53-86-1	INDREN
112-80-1	INDUSTRENE 105
112-80-1	INDUSTRENE 205
112-80-1	INDUSTRENE 206
57-11-4	INDUSTRENE 5016
57-11-4	INDUSTRENE 8718
57-11-4	INDUSTRENE 9018
57-11-4	INDUSTRENE R
1336-36-3	INERTEEN
75-78-5	INERTON AW-DMCS
57-63-6	INESTRA
58-89-9	INEXIT
78-53-5	INFERNO
67-68-5	INFILTRINA
53-86-1	INFLAZON
53-86-1	INFROCIN
7722-84-1	INHIBINE
71-55-6	INHIBISOL
72-33-3	INOSTRAL
8003-34-7	INSECT POWDER
121-75-5	INSECTICIDE NO 4049
60-57-1	INSECTICIDE NO 497
121-75-5	INSECTICIDE NUMBER 4049
23135-22-0	INSECTICIDE-NEMATICIDE 1410
8003-34-7	INSECTICIDES, PYRETHRINS
62-73-7	INSECTIGAS D
122-14-5	INSECTIGAS F
60-57-1	INSECTLACK
115-29-7	INSECTOPHENE
53-86-1	INTEBAN
53-86-1	INTEBAN SP
1937-37-7	INTERCHEM DIRECT BLACK Z
7722-84-1	INTEROX
112-02-7	INTEXAN CTC 29
112-02-7	INTEXSAN CTC 29
112-02-7	INTEXSAN CTC 50
12789-03-6	INTOX
12789-03-6	INTOX (INSECTICIDE)
56-75-7	INTRAMYCETIN
640-15-3	INTRATHION
640-15-3	INTRATION
12002-48-1	INVALON TC
93-76-5	INVERTON 245
114-26-1	INVISI-GARD
7783-97-3	IODIC ACID (HIO3), SILVER(1+) SALT
7553-56-2	IODINE
7553-56-2	IODINE (127I2)
7790-99-0	IODINE CHLORIDE
7790-99-0	IODINE CHLORIDE (ICl)
7553-56-2	IODINE COLLOIDAL
7553-56-2	IODINE CRYSTALS
506-78-5	IODINE CYANIDE

CAS No.	*Chemical Name*
506-78-5	IODINE CYANIDE (I(CN))
7783-66-6	IODINE FLUORIDE
7783-66-6	IODINE FLUORIDE (IF5)
7790-99-0	IODINE MONOCHLORIDE
7790-99-0	IODINE MONOCHLORIDE (DOT)
506-78-5	IODINE MONOCYANIDE
506-78-5	IODINE MONOCYANIDE (ICN)
7783-66-6	IODINE PENTAFLUORIDE
7783-66-6	IODINE PENTAFLUORIDE (DOT)
7553-56-2	IODINE SUBLIMED
7790-99-0	IODINE(I) CHLORIDE
513-38-2	IODO METHYLPROPANE (1-IODO-2-METHYL PROPANE)
558-17-8	IODO METHYLPROPANE (2-IODO-2-METHYL PROPANE)
26914-02-3	IODO PROPANE
25267-27-0	IODOBUTANE
7790-99-0	IODOCHLORINE
506-78-5	IODOCYANIDE (ICN)
75-47-8	IODOFORM
74-88-4	IODOMETANO (ITALIAN)
74-88-4	IODOMETHANE
513-38-2	IODOMETHYLPROPANE
620-05-3	IODOPHENYLMETHANE
26914-02-3	IODOPROPANES (DOT)
15385-57-6	IODURE DE MERCURE (FRENCH)
7783-30-4	IODURE DE MERCURE (FRENCH)
74-88-4	IODURE DE METHYLE (FRENCH)
128-37-0	IONOL
128-37-0	IONOL (ANTIOXIDANT)
128-37-0	IONOL 1
128-37-0	IONOL BHT
128-37-0	IONOL CP
128-37-0	IONOLE
7778-50-9	IOPEZITE
7553-56-2	IOSAN SUPERDIP
136-60-7	IP CARRIER N 20
108-90-7	IP CARRIER T 40
121-91-5	IPA
94-75-7	IPANER
3458-22-8	IPD
4098-71-9	IPDI
114-26-1	IPMC
626-17-5	IPN
105-64-6	IPP
505-60-2	IPRIT
1333-86-4	IRA 2
7681-49-4	IRADICAV
7778-74-7	IRENAL
7778-74-7	IRENAT
10025-97-5	IRIDIUM CHLORIDE (IrCl4)
10024-97-5	IRIDIUM TETRACHLORIDE
10025-97-5	IRIDIUM TETRACHLORIDE
10025-97-5	IRIDIUM(IV) CHLORIDE
151-21-3	IRIUM
9004-66-4	IRO-JEX
10421-48-4	IRON (III) NITRATE, ANHYDROUS
8049-18-1	IRON ALLOY, BASE, (FERROCERIUM)
69523-06-4	IRON ALLOY, BASE, (FERROCERIUM)
60475-66-3	IRON ALLOY, BASE, (FERROCERIUM)
12604-58-9	IRON ALLOY, NONBASE, V,C,FE (FERROVANADIUM)
10102-50-8	IRON ARSENATE (DOT)
10102-50-8	IRON ARSENATE (Fe3(AsO4)2)
10102-49-5	IRON ARSENATE (FeAsO4)
63989-69-5	IRON ARSENITE OXIDE (Fe2(AsO3)2O3), PENTAHYDRATE
102-54-5	IRON BIS(CYCLOPENTADIENIDE)
13463-40-6	IRON CARBONYL
13463-40-6	IRON CARBONYL (DOT)
13463-40-6	IRON CARBONYL (Fe(CO)5)
13463-40-6	IRON CARBONYL (Fe(CO)5), (TB-5-11)-
12040-57-2	IRON CHLORIDE
7705-08-0	IRON CHLORIDE
7758-94-3	IRON CHLORIDE (FeCl2)
7705-08-0	IRON CHLORIDE (FeCl3)
7705-08-0	IRON CHLORIDE, SOLID (DOT)
9004-66-4	IRON DEXTRAN
9004-66-4	IRON DEXTRAN COMPLEX
9004-66-4	IRON DEXTRAN INJECTION
7758-94-3	IRON DICHLORIDE
102-54-5	IRON DICYCLOPENTADIENYL
14484-64-1	IRON DIMETHYLDITHIOCARBAMATE
7783-50-8	IRON FLUORIDE
7783-50-8	IRON FLUORIDE (FeF3)
1301-70-8	IRON GLYCEROPHOSPHATE (Fe(O6PC3H8)3)
1309-37-1	IRON MINIUM
7720-78-7	IRON MONOSULFATE
10421-48-4	IRON NITRATE
10421-48-4	IRON NITRATE (Fe(NO3)3)
1309-37-1	IRON OXIDE
1309-37-1	IRON OXIDE (Fe2O3)
1309-37-1	IRON OXIDE FUME
1309-37-1	IRON OXIDE RED
1309-37-1	IRON OXIDE RED 110M
1309-37-1	IRON OXIDE RED TRANSPARENT 288VN
13463-40-6	IRON PENTACARBONYL
13463-40-6	IRON PENTACARBONYL (ACGIH,DOT)
7705-08-0	IRON PERCHLORIDE
10028-22-5	IRON PERSULFATE
7758-94-3	IRON PROTOCHLORIDE
7720-78-7	IRON PROTOSULFATE
516-03-0	IRON PROTOXALATE
7705-08-0	IRON SESQUICHLORIDE, SOLID (DOT)
1309-37-1	IRON SESQUIOXIDE
10028-22-5	IRON SESQUISULFATE
7720-78-7	IRON SULFATE (1:1)
10028-22-5	IRON SULFATE (2:3)
10028-22-5	IRON SULFATE (Fe2(SO4)3)
7720-78-7	IRON SULFATE (FeSO4)
10028-22-5	IRON TERSULFATE
7705-08-0	IRON TRICHLORIDE
7783-50-8	IRON TRIFLUORIDE
10421-48-4	IRON TRINITRATE
1309-37-1	IRON TRIOXIDE
7720-78-7	IRON VITRIOL
7758-94-3	IRON(2+) CHLORIDE
516-03-0	IRON(2+) OXALATE
7720-78-7	IRON(2+) SULFATE
7720-78-7	IRON(2+) SULFATE (1:1)
1301-70-8	IRON(3+) GLYCEROPHOSPHATE
1309-37-1	IRON(3+) OXIDE
10028-22-5	IRON(3+) SULFATE
10102-50-8	IRON(II) ARSENATE (3:2)
7758-94-3	IRON(II) CHLORIDE
7758-94-3	IRON(II) CHLORIDE (1:2)
7758-94-3	IRON(II) CHLORIDE (FeCl2)
516-03-0	IRON(II) OXALATE
7720-78-7	IRON(II) SULFATE
7720-78-7	IRON(II) SULFATE (1:1)
10102-49-5	IRON(III) ARSENATE (1:1)
7705-08-0	IRON(III) CHLORIDE
10421-48-4	IRON(III) NITRATE
63989-69-5	IRON(III) O-ARSENITE PENTAHYDRATE
1309-37-1	IRON(III) OXIDE
10028-22-5	IRON(III) SULFATE
102-54-5	IRON, BIS(ETA 5-2,4-CYCLOPENTADIEN-1-YL)-
13463-40-6	IRON, PENTACARBONYL-
14484-64-1	IRON, TRIS(DIMETHYLCARBAMODITHIOATO-S,S')-, (OC-6-11)-
14484-64-1	IRON, TRIS(DIMETHYLDITHIOCARBAMATO)-
7720-78-7	IROSPAN
7720-78-7	IROSUL
7789-75-5	IRTRAN 3
8020-83-5	IS 45
1333-86-4	ISAF
8001-86-3	ISANO OIL
76-13-1	ISCEON 113
75-71-8	ISCEON 122
75-69-4	ISCEON 131
75-45-6	ISCEON 22
1320-37-2	ISCEON 224
74-83-9	ISCOBROME
106-93-4	ISCOBROME D
56-53-1	ISCOVESCO
56-75-7	ISMICETINA
123-51-3	ISO-AMYLALKOHOL (GERMAN)
638-11-9	ISO-PROPANOL BUTYRATE
78-59-1	ISOACETOPHORONE
123-92-2	ISOAMYL ACETATE
123-92-2	ISOAMYL ACETATE (ACGIH)
123-51-3	ISOAMYL ALCOHOL
584-02-1	ISOAMYL ALCOHOL (ACGIH)

CAS No.	Chemical Name
123-51-3	ISOAMYL ALCOHOL (ACGIH,DOT)
123-51-3	ISOAMYL ALKOHOL (CZECH)
107-82-4	ISOAMYL BROMIDE
107-84-6	ISOAMYL CHLORIDE
123-92-2	ISOAMYL ETHANOATE
110-12-3	ISOAMYL METHYL KETONE
590-86-3	ISOAMYLALDEHYDE
26760-64-5	ISOAMYLENE
123-51-3	ISOAMYLOL
70-30-4	ISOBAC 20
297-78-9	ISOBENZAN
85-44-9	ISOBENZOFURAN, 1,3-DIHYDRO-1,3-DIOXO-
9016-87-9	ISOBIND 100
78-84-2	ISOBUTALDEHYDE
78-84-2	ISOBUTANAL
75-28-5	ISOBUTANE
75-28-5	ISOBUTANE (DOT)
79-31-2	ISOBUTANOIC ACID
78-83-1	ISOBUTANOL
78-83-1	ISOBUTANOL (DOT)
109-53-5	ISOBUTANOL VINYL ETHER
124-68-5	ISOBUTANOL-2-AMINE
79-30-1	ISOBUTANOYL CHLORIDE
78-85-3	ISOBUTENAL
115-11-7	ISOBUTENE
7756-94-7	ISOBUTENE TRIMER
563-47-3	ISOBUTENYL CHLORIDE
141-79-7	ISOBUTENYL METHYL KETONE
109-53-5	ISOBUTOXYETHENE
97-86-9	ISOBUTYL 2-METHYL-2-PROPENOATE
106-63-8	ISOBUTYL 2-PROPENOATE
110-19-0	ISOBUTYL ACETATE
110-19-0	ISOBUTYL ACETATE (ACGIH,DOT)
106-63-8	ISOBUTYL ACRYLATE
106-63-8	ISOBUTYL ACRYLATE, INHIBITED (DOT)
78-83-1	ISOBUTYL ALCOHOL
78-83-1	ISOBUTYL ALCOHOL (ACGIH,DOT)
97-86-9	ISOBUTYL ALPHA-METHACRYLATE
97-86-9	ISOBUTYL ALPHA-METHYLACRYLATE
13195-76-1	ISOBUTYL BORATE ((C4H9O)3B)
78-77-3	ISOBUTYL BROMIDE
123-51-3	ISOBUTYL CARBINOL
513-36-0	ISOBUTYL CHLORIDE
542-55-2	ISOBUTYL FORMATE
542-55-2	ISOBUTYL FORMATE (DOT)
4439-24-1	ISOBUTYL GLYCOL
123-18-2	ISOBUTYL HEPTYL KETONE
19594-40-2	ISOBUTYL HEPTYL KETONE
513-38-2	ISOBUTYL IODIDE
97-85-8	ISOBUTYL ISOBUTANOATE
97-85-8	ISOBUTYL ISOBUTYRATE
1873-29-6	ISOBUTYL ISOCYANATE
1873-29-6	ISOBUTYL ISOCYANATE (DOT)
108-83-8	ISOBUTYL KETONE
97-86-9	ISOBUTYL METHACRYLATE
97-86-9	ISOBUTYL METHACRYLATE, INHIBITED (DOT)
108-10-1	ISOBUTYL METHYL KETONE
37206-20-5	ISOBUTYL METHYL KETONE PEROXIDE
37206-20-5	ISOBUTYL METHYL KETONE PEROXIDE, NO MORE THAN 62% IN SOLUTION (DOT)
19594-40-2	ISOBUTYL N-HEPTYL KETONE
540-42-1	ISOBUTYL PROPANOATE
106-63-8	ISOBUTYL PROPENOATE
540-42-1	ISOBUTYL PROPIONATE
540-42-1	ISOBUTYL PROPIONATE (DOT)
109-53-5	ISOBUTYL VINYL ETHER
78-83-1	ISOBUTYLALKOHOL (CZECH)
78-81-9	ISOBUTYLAMINE
78-81-9	ISOBUTYLAMINE (DOT)
538-93-2	ISOBUTYLBENZENE
115-11-7	ISOBUTYLENE
115-11-7	ISOBUTYLENE (DOT)
691-37-2	ISOBUTYLETHENE
97-85-8	ISOBUTYLISOBUTYRATE (DOT)
105-30-6	ISOBUTYLMETHYLCARBINOL
108-11-2	ISOBUTYLMETHYLCARBINOL
54972-97-3	ISOBUTYLMETHYLCARBINOL
108-11-2	ISOBUTYLMETHYLMETHANOL
105-30-6	ISOBUTYLMETHYLMETHANOL
54972-97-3	ISOBUTYLMETHYLMETHANOL
540-84-1	ISOBUTYLTRIMETHYLMETHANE
78-84-2	ISOBUTYRADEHYDE
78-84-2	ISOBUTYRAL
78-84-2	ISOBUTYRALDEHYDE
79-31-2	ISOBUTYRIC ACID
79-30-1	ISOBUTYRIC ACID CHLORIDE
97-62-1	ISOBUTYRIC ACID, ETHYL ESTER
97-85-8	ISOBUTYRIC ACID, ISOBUTYL ESTER
617-50-5	ISOBUTYRIC ACID, ISOPROPYL ESTER
78-84-2	ISOBUTYRIC ALDEHYDE
97-72-3	ISOBUTYRIC ANHYDRIDE
78-82-0	ISOBUTYRONITRILE
79-30-1	ISOBUTYROYL CHLORIDE
3437-84-1	ISOBUTYROYL PEROXIDE
78-84-2	ISOBUTYRYL ALDEHYDE
79-30-1	ISOBUTYRYL CHLORIDE
79-30-1	ISOBUTYRYL CHLORIDE (DOT)
3437-84-1	ISOBUTYRYL PEROXIDE
3437-84-1	ISOBUTYRYL PEROXIDE, MAXIMUM CONCENTRATION 52% IN SOLUTION (DOT)
3437-84-1	ISOBUTYRYL PEROXIDE, NOT MORE THAN 52% IN SOLUTION
11071-47-9	ISOCETENE
540-84-1	ISOCTANE
26952-21-6	ISOCTYL ALCOHOL
103-65-1	ISOCUMENE
71000-82-3	ISOCYANATE
9016-87-9	ISOCYANATE 580
624-83-9	ISOCYANATE DE METHYLE (FRENCH)
71000-82-3	ISOCYANATE ION(1-)
71000-82-3	ISOCYANATES AND SOLUTIONS, [FLAMMABLE LIQUID AND POISONOUS LIQUID LABELS]
71000-82-3	ISOCYANATES AND SOLUTIONS, [FLAMMABLE LIQUID LABEL]
71000-82-3	ISOCYANATES, BP ≥300 C
3170-57-8	ISOCYANATES, N.O.S. :OR: ISOCYANATE, SOLUTIONS, N.O.S. :- FLASHPOINT ≥23C:BOILING POINT 300C
3170-57-8	ISOCYANATES, N.O.S. :OR: ISOCYANATE, SOLUTIONS, N.O.S. FLAMMABLE
71000-82-3	ISOCYANATES, [POISON B LABEL]
103-71-9	ISOCYANATOBENZENE
71121-36-3	ISOCYANATOBENZOTRI-FLUORIDE
109-90-0	ISOCYANATOETHANE
30674-80-7	ISOCYANATOETHYL METHACRYLATE
624-83-9	ISOCYANATOMETHANE
102-36-3	ISOCYANIC ACID 3,4-DICHLOROPHENYL ESTER
30674-80-7	ISOCYANIC ACID, 2-HYDROXYETHYL ESTER METHACRYLATE (ESTER)
91-08-7	ISOCYANIC ACID, 2-METHYL-M-PHENYLENE ESTER
91-08-7	ISOCYANIC ACID, 2-METHYL-META-PHENYLENE ESTER
28479-22-3	ISOCYANIC ACID, 3-CHLORO-P-TOLYL ESTER
584-84-9	ISOCYANIC ACID, 4-METHYL-M-PHENYLENE ESTER
26471-62-5	ISOCYANIC ACID, 4-METHYL-M-PHENYLENE ESTER
111-36-4	ISOCYANIC ACID, BUTYL ESTER
822-06-0	ISOCYANIC ACID, DIESTER WITH 1,6-HEXANEDIOL
109-90-0	ISOCYANIC ACID, ETHYL ESTER
822-06-0	ISOCYANIC ACID, HEXAMETHYLENE ESTER
1873-29-6	ISOCYANIC ACID, ISOBUTYL ESTER
1795-48-8	ISOCYANIC ACID, ISOPROPYL ESTER
624-83-9	ISOCYANIC ACID, METHYL ESTER
26471-62-5	ISOCYANIC ACID, METHYL-M-PHENYLENE ESTER
4098-71-9	ISOCYANIC ACID, METHYLENE(3,5,5-TRIMETHYL-3,1-CYCLO-HEXYLENE) ESTER
5124-30-1	ISOCYANIC ACID, METHYLENEDI-4,1-CYCLOHEXYLENE ESTER
101-68-8	ISOCYANIC ACID, METHYLENEDI-P-PHENYLENE ESTER
26471-62-5	ISOCYANIC ACID, METHYLPHENYLENE ESTER
103-71-9	ISOCYANIC ACID, PHENYL ESTER
9016-87-9	ISOCYANIC ACID, POLYMETHYLENEPOLYPHENYLENEESTER
110-78-1	ISOCYANIC ACID, PROPYL ESTER
57-12-5	ISOCYANIDE
2782-57-2	ISOCYANURIC ACID, DICHLORO-
2244-21-5	ISOCYANURIC ACID, DICHLORO-, POTASSIUM SALT
87-90-1	ISOCYANURIC CHLORIDE
2782-57-2	ISOCYANURIC DICHLORIDE
3085-26-5	ISODECALDEHYDE
25339-17-7	ISODECANOL
1330-61-6	ISODECYL ACRYLATE
25339-17-7	ISODECYL ALCOHOL
1330-61-6	ISODECYL ALCOHOL, ACRYLATE
26761-40-0	ISODECYL ALCOHOL, PHTHALATE (2:1)

CAS No.	Chemical Name
26761-40-0	ISODECYL PHTHALATE
1330-61-6	ISODECYL PROPENOATE
498-15-7	ISODIPRENE
465-73-6	ISODRIN
55-91-4	ISOFLUORPHATE
26675-46-7	ISOFLURANE
78-59-1	ISOFORON
591-76-4	ISOHEPTANE
31394-54-4	ISOHEPTANE
68975-47-3	ISOHEPTENE
107-83-5	ISOHEXANE
79-29-8	ISOHEXANE (ACGIH)
27236-46-0	ISOHEXENE
105-30-6	ISOHEXYL ALCOHOL
108-11-2	ISOHEXYL ALCOHOL
67-63-0	ISOHOL
107-41-5	ISOL
119-38-0	ISOLAN
119-38-0	ISOLAN (PESTICIDE)
119-38-0	ISOLANE (FRENCH)
919-86-8	ISOMETASYSTOX
919-86-8	ISOMETHYLSYSTOX
101-68-8	ISONATE
101-68-8	ISONATE 125 MF
101-68-8	ISONATE 125M
9016-87-9	ISONATE 390P
79-46-9	ISONITROPROPANE
25322-01-4	ISONITROPROPANE
58449-37-9	ISONONANOYL PEROXIDE
540-84-1	ISOOCTANE
26635-64-3	ISOOCTANE
26952-21-6	ISOOCTANOL
27215-10-7	ISOOCTANOL, HYDROGEN PHOSPHATE
25168-26-7	ISOOCTYL 2,4-DICHLOROPHENOXYACETATE
32534-95-5	ISOOCTYL 2-(2,4,5-TRICHLOROPHENOXY)PROPIONATE
26952-21-6	ISOOCTYL ALCOHOL
25168-26-7	ISOOCTYL ALCOHOL, (2,4-DICHLOROPHENOXY)ACETATE
27215-10-7	ISOOCTYL PHOSPHATE ((C8H17O)2(HO)PO)
78-79-5	ISOPENTADIENE
590-86-3	ISOPENTALDEHYDE
590-86-3	ISOPENTANAL
78-78-4	ISOPENTANE
503-74-2	ISOPENTANOIC ACID
123-51-3	ISOPENTANOL
26760-64-5	ISOPENTENE
123-92-2	ISOPENTYL ACETATE
123-51-3	ISOPENTYL ALCOHOL
123-92-2	ISOPENTYL ALCOHOL, ACETATE
107-82-4	ISOPENTYL BROMIDE
107-84-6	ISOPENTYL CHLORIDE
123-92-2	ISOPENTYL ETHANOATE
110-12-3	ISOPENTYL METHYL KETONE
78-59-1	ISOPHORON
78-59-1	ISOPHORONE
4098-71-9	ISOPHORONE DIAMINE DIISOCYANATE
4098-71-9	ISOPHORONE DIISOCYANATE
4098-71-9	ISOPHORONE DIISOCYANATE (ACGIH,DOT)
2855-13-2	ISOPHORONEDIAMINE
3778-73-2	ISOPHOSPHAMIDE
121-91-5	ISOPHTHALATE
121-91-5	ISOPHTHALIC ACID
626-17-5	ISOPHTHALONITRILE
78-79-5	ISOPRENE
78-79-5	ISOPRENE (DOT)
78-79-5	ISOPRENE, INHIBITED (DOT)
75-29-6	ISOPRID
75-33-2	ISOPROPANETHIOL
67-63-0	ISOPROPANOL
78-96-6	ISOPROPANOLAMINE
54590-52-2	ISOPROPANOLAMINE DODECYLBENZENESULFONATE
54590-52-2	ISOPROPANOLAMINE DODECYLBENZENESULFONATE (DOT)
126-98-7	ISOPROPENE CYANIDE
98-83-9	ISOPROPENIL-BENZOLO (ITALIAN)
108-22-5	ISOPROPENYL ACETATE
108-22-5	ISOPROPENYL ACETATE (DOT)
78-80-8	ISOPROPENYL ACETYLENE
98-83-9	ISOPROPENYL BENZENE
513-42-8	ISOPROPENYL CARBINOL
557-98-2	ISOPROPENYL CHLORIDE
814-78-8	ISOPROPENYL METHYL KETONE
98-83-9	ISOPROPENYL-BENZEEN (DUTCH)
98-83-9	ISOPROPENYL-BENZOL (GERMAN)
98-83-9	ISOPROPENYLBENZENE (DOT)
126-98-7	ISOPROPENYLNITRILE
75-31-0	ISOPROPILAMINA (ITALIAN)
98-82-8	ISOPROPILBENZENE (ITALIAN)
108-21-4	ISOPROPILE (ACETATO DI) (ITALIAN)
109-59-1	ISOPROPOXYETHANOL
926-65-8	ISOPROPOXYETHENE
926-65-8	ISOPROPOXYETHYLENE
107-44-8	ISOPROPOXYMETHYLPHOSPHORYL FLUORIDE
79435-04-4	ISOPROPYL (+)-2-CHLOROPROPIONATE
94-11-1	ISOPROPYL (2,4-DICHLOROPHENOXY)ACETATE
108-21-4	ISOPROPYL (ACETATE D') (FRENCH)
94-11-1	ISOPROPYL 2,4-D ESTER
79435-04-4	ISOPROPYL 2-CHLOROPROPIONATE (DOT)
617-50-5	ISOPROPYL 2-METHYLPROPANOATE
108-21-4	ISOPROPYL ACETATE
108-21-4	ISOPROPYL ACETATE (ACGIH,DOT)
1623-24-1	ISOPROPYL ACID PHOSPHATE
1623-24-1	ISOPROPYL ACID PHOSPHATE, SOLID
1623-24-1	ISOPROPYL ACID PHOSPHATE, SOLID (DOT)
67-63-0	ISOPROPYL ALCOHOL
78-84-2	ISOPROPYL ALDEHYDE
5419-55-6	ISOPROPYL BORATE
638-11-9	ISOPROPYL BUTANOATE
638-11-9	ISOPROPYL BUTYRATE
638-11-9	ISOPROPYL BUTYRATE (DOT)
109-59-1	ISOPROPYL CELLOSOLVE
75-29-6	ISOPROPYL CHLORIDE
105-48-6	ISOPROPYL CHLOROACETATE
105-48-6	ISOPROPYL CHLOROACETATE (DOT)
108-23-6	ISOPROPYL CHLOROCARBONATE
108-23-6	ISOPROPYL CHLOROFORMATE
108-23-6	ISOPROPYL CHLOROFORMATE (DOT)
108-23-6	ISOPROPYL CHLOROMETHANOATE
79435-04-4	ISOPROPYL CHLOROPROPIONATE
78-82-0	ISOPROPYL CYANIDE
1195-42-2	ISOPROPYL CYCLOHEXYLAMINE
2275-18-5	ISOPROPYL DIETHYLDITHIOPHOSPHORYLACETAMIDE
1623-24-1	ISOPROPYL DIHYDROGEN PHOSPHATE
108-11-2	ISOPROPYL DIMETHYL CARBINOL
105-30-6	ISOPROPYL DIMETHYL CARBINOL
108-21-4	ISOPROPYL ETHANOATE
108-20-3	ISOPROPYL ETHER
108-20-3	ISOPROPYL ETHER (ACGIH)
625-55-8	ISOPROPYL FORMATE
625-55-8	ISOPROPYL FORMATE (DOT)
4016-14-2	ISOPROPYL GLYCIDYL ETHER
4016-14-2	ISOPROPYL GLYCIDYL ETHER (IGE)
1623-24-1	ISOPROPYL HYDROGEN PHOSPHATE
617-50-5	ISOPROPYL ISOBUTYRATE
617-50-5	ISOPROPYL ISOBUTYRATE (DOT)
1795-48-8	ISOPROPYL ISOCYANATE
1795-48-8	ISOPROPYL ISOCYANATE (DOT)
617-51-6	ISOPROPYL LACTATE
75-33-2	ISOPROPYL MERCAPTAN
75-33-2	ISOPROPYL MERCAPTAN (DOT)
107-44-8	ISOPROPYL METHANEFLUOROPHOSPHONATE
563-80-4	ISOPROPYL METHYL KETONE
107-44-8	ISOPROPYL METHYLFLUOROPHOSPHATE
107-44-8	ISOPROPYL METHYLFLUOROPHOSPHONATE
107-44-8	ISOPROPYL METHYLPHOSPHONOFLUORIDATE
1712-64-7	ISOPROPYL NITRATE
1712-64-7	ISOPROPYL NITRATE (DOT)
78-82-0	ISOPROPYL NITRILE
109-59-1	ISOPROPYL OXITOL
105-64-6	ISOPROPYL PERCARBONATE
105-64-6	ISOPROPYL PERCARBONATE, STABILIZED
105-64-6	ISOPROPYL PERCARBONATE, STABILIZED (DOT)
105-64-6	ISOPROPYL PERCARBONATE, UNSTABILIZED
105-64-6	ISOPROPYL PERCARBONATE, UNSTABILIZED (DOT)
105-64-6	ISOPROPYL PEROXYDICARBONATE
105-64-6	ISOPROPYL PEROXYDICARBONATE, NOT MORE THAN 52% IN SOLUTION (DOT)
105-64-6	ISOPROPYL PEROXYDICARBONATE, TECHNICALLY PURE (DOT)
1623-24-1	ISOPROPYL PHOSPHATE
1623-24-1	ISOPROPYL PHOSPHATE ((C3H7O)(HO)2PO)

CAS No.	Chemical Name
1623-24-1	ISOPROPYL PHOSPHORIC ACID, SOLID (DOT)
637-78-5	ISOPROPYL PROPANOATE
637-78-5	ISOPROPYL PROPIONATE
926-65-8	ISOPROPYL VINYL ETHER
40058-87-5	ISOPROPYL-2-CHLOROPROPIONATE
98-82-8	ISOPROPYL-BENZOL (GERMAN)
107-44-8	ISOPROPYL-METHYL-PHOSPHORYL FLUORIDE
108-21-4	ISOPROPYLACETAAT (DUTCH)
108-21-4	ISOPROPYLACETAT (GERMAN)
108-10-1	ISOPROPYLACETONE
75-31-0	ISOPROPYLAMINE
75-31-0	ISOPROPYLAMINE (ACGIH,DOT)
26264-05-1	ISOPROPYLAMINE DODECYLBENZENESULFONATE
22224-92-6	ISOPROPYLAMINO-O-ETHYL-(4-METHYLMERCAPTO-3-METHYL-PHENYL)PHOSPHATE
109-56-8	ISOPROPYLAMINOETHANOL
109-56-8	ISOPROPYLAMINOETHANOL [COMBUSTIBLE LIQUID LABEL]
109-56-8	ISOPROPYLAMINOETHANOL [FLAMMABLE LIQUID LABEL]
26264-05-1	ISOPROPYLAMMONIUM DODECYLBENZENESULFONATE
768-52-5	ISOPROPYLANILINE
643-28-7	ISOPROPYLANILINE
98-82-8	ISOPROPYLBENZEEN (DUTCH)
98-82-8	ISOPROPYLBENZENE
98-82-8	ISOPROPYLBENZENE (DOT)
80-15-9	ISOPROPYLBENZENE HYDROPEROXIDE
80-43-3	ISOPROPYLBENZENE PEROXIDE
98-82-8	ISOPROPYLBENZOL
78-83-1	ISOPROPYLCARBINOL
57-55-6	ISOPROPYLENE GLYCOL
6423-43-4	ISOPROPYLENE NITRATE
563-45-1	ISOPROPYLETHENE
563-45-1	ISOPROPYLETHYLENE
78-84-2	ISOPROPYLFORMALDEHYDE
79-31-2	ISOPROPYLFORMIC ACID
141-79-7	ISOPROPYLIDENE ACETONE
100-79-8	ISOPROPYLIDENE GLYCEROL
646-06-0	ISOPROPYLIDENE GLYCEROL
115-11-7	ISOPROPYLIDENEMETHYLENE
119-38-0	ISOPROPYLMETHYLPYRAZOLYL DIMETHYLCARBAMATE
333-41-5	ISOPROPYLMETHYLPYRIMIDYL DIETHYL THIOPHOSPHATE
75-33-2	ISOPROPYLTHIOL
51-83-2	ISOPTO CARBACHOL
120-58-1	ISOSAFROLE
9016-87-9	ISOSET CX 11
586-62-9	ISOTERPINENE
57-06-7	ISOTHIOCYANATE D'ALLYLE (FRENCH)
556-61-6	ISOTHIOCYANATE DE METHYLE (FRENCH)
556-61-6	ISOTHIOCYANATOMETHANE
57-06-7	ISOTHIOCYANIC ACID, ALLYL ESTER
556-61-6	ISOTHIOCYANIC ACID, METHYL ESTER
79-19-6	ISOTHIOSEMICARBAZIDE
62-56-6	ISOTHIOUREA
556-61-6	ISOTIOCIANATO DI METILE (ITALIAN)
58-89-9	ISOTOX
75-69-4	ISOTRON 11
75-71-8	ISOTRON 12
75-45-6	ISOTRON 22
57-13-6	ISOUREA
590-86-3	ISOVALERAL
590-86-3	ISOVALERALDEHYDE
556-24-1	ISOVALERIC ACID, METHYL ESTER
590-86-3	ISOVALERIC ALDEHYDE
108-83-8	ISOVALERONE
590-86-3	ISOVALERYLALDEHYDE
50-78-2	ISTOPIRIN
14807-96-6	IT EXTRA
105-60-2	ITAMID
105-60-2	ITAMID 250
105-60-2	ITAMIDE 25
105-60-2	ITAMIDE 250
105-60-2	ITAMIDE 250G
105-60-2	ITAMIDE 35
105-60-2	ITAMIDE 350
105-60-2	ITAMIDE S
2631-37-0	ITC
8020-83-5	ITERM 6
563-12-2	ITOPAZ
50-00-0	IVALON
9002-89-5	IVALON
109-53-5	IVE
50-29-3	IVORAN
108-95-2	IZAL
108-20-3	IZOPROPYLOWY ETER (POLISH)
2893-78-9	IZOSAN G
8002-74-2	J 1440
11138-49-1	J 242
2540-82-1	J-38
58-89-9	JACUTIN
76-22-2	JAPAN CAMPHOR
2646-17-5	JAPAN ORANGE 403
8008-51-3	JAPANESE CAMPHOR OIL
8008-51-3	JAPANESE, OIL OF CAMPHOR
7440-66-6	JASAD
8003-34-7	JASMOLIN I OR II
7778-66-7	JAVELLE WATER
7681-52-9	JAVEX
64-17-5	JAYSOL
64-17-5	JAYSOL S
101-77-9	JEFFAMINE AP-20
112-34-5	JEFFERSOL DB
109-86-4	JEFFERSOL EM
25322-69-4	JEFFOX
37286-64-9	JEFFOX OL 2700
51-83-2	JESTRYL
1309-37-1	JEWELER'S ROUGE
7439-95-4	JIS 1
7440-32-6	JIS H2151 TW 35
7439-96-5	JIS-G 1213
7429-90-5	JISC 3108
7429-90-5	JISC 3110
7440-32-6	JISTP 28
74-88-4	JOD-METHAN (GERMAN)
506-78-5	JODCYAN
8002-74-2	JOHNSONS WAX 111
13194-48-4	JOLT
74-88-4	JOODMETHAAN (DUTCH)
102-85-2	JP 304
12788-93-1	JP 504
13463-67-7	JR 701
1344-28-1	JRC-ALO 4
1344-28-1	JUBENON R
8052-42-4	JUDEAN PITCH
481-39-0	JUGLONE
118-74-1	JULIN'S CARBON CHLORIDE
5329-14-6	JUMBO
56-75-7	JUVAMYCETIN
7429-90-5	K 102
7429-90-5	K 102 (METAL)
7783-28-0	K 2
7783-28-0	K 2 (PHOSPHATE)
8020-83-5	K 315
112-80-1	K 52
1332-21-4	K 6-20
12001-29-5	K 6-30
1335-85-9	K III
534-52-1	K III
534-52-1	K IV
1335-85-9	K IV
7784-30-7	K-BOND 90
7775-14-6	K-BRITE
25231-46-3	K-CURE 1040
104-15-4	K-CURE 1040
7681-11-0	K1-N
13463-67-7	KA 10
75-94-5	KA 1003
1344-28-1	KA 101
13463-67-7	KA 15
13463-67-7	KA 35
10108-64-2	KADMIUMCHLORID (GERMAN)
2223-93-0	KADMIUMSTEARAT (GERMAN)
1306-19-0	KADMU TLENEK (POLISH)
1314-13-2	KADOX 15
1314-13-2	KADOX 72
1314-13-2	KADOX-25
7440-50-8	KAFAR COPPER
121-75-5	KAFPON
7681-11-0	KAIOD
75-71-8	KAISER CHEMICALS 12

CAS No.	Chemical Name
9016-87-9	KAISER NCO 20
95-79-4	KAKO RED TR BASE
95-69-2	KAKO RED TR BASE
10043-22-8	KALIUM OXALATE
583-52-8	KALIUM OXALATE
151-50-8	KALIUM-CYANID (GERMAN)
10124-50-2	KALIUMARSENIT (GERMAN)
3811-04-9	KALIUMCHLORAAT (DUTCH)
3811-04-9	KALIUMCHLORAT (GERMAN)
7778-50-9	KALIUMDICHROMAT (GERMAN)
7757-79-1	KALIUMNITRAT (GERMAN)
10118-76-0	KALIUMPERMANGANAT (GERMAN)
1305-62-0	KALKHYDRATE
62-44-2	KALMIN
7778-44-1	KALO
7546-30-7	KALOMEL (GERMAN)
108-01-0	KALPUR P
75-69-4	KALTRON 11
7778-44-1	KALZIUMARSENIAT (GERMAN)
57-11-4	KAM 1000
57-11-4	KAM 2000
57-11-4	KAM 3000
56-75-7	KAMAVER
95-79-4	KAMBAMINE RED TR
8001-35-2	KAMFOCHLOR
1332-58-7	KAMIG
76-22-2	KAMPFER (GERMAN)
505-60-2	KAMPSTOFF LOST
59-01-8	KANAMYCIN
1336-36-3	KANECHLOR
1336-36-3	KANECHLOR 300
1336-36-3	KANECHLOR 400
56-25-7	KANTARIDIN
56-25-7	KANTHARIDIN (GERMAN)
1332-58-7	KAO-GEL
1332-58-7	KAOBRITE
1332-58-7	KAOLIN
1332-58-7	KAOLIN COLLOIDAL
105-60-2	KAPROLIT
105-60-2	KAPROLIT B
105-60-2	KAPROLON
105-60-2	KAPROLON B
105-60-2	KAPROMINE
105-60-2	KAPRON
105-60-2	KAPRON A
105-60-2	KAPRON B
133-06-2	KAPTAN
330-54-1	KARAMEX
577-11-7	KARAWET DOSS
9000-36-6	KARAYA
14484-64-1	KARBAM BLACK
63-25-2	KARBARYL (POLISH)
63-25-2	KARBASPRAY
63-25-2	KARBATOX
63-25-2	KARBATOX 75
141-66-2	KARBICRON
121-75-5	KARBOFOS
1563-66-2	KARBOFURANU (POLISH)
63-25-2	KARBOSEP
7681-49-4	KARI-RINSE
7681-49-4	KARIDIUM
7681-49-4	KARIGEL
50-70-4	KARION
50-70-4	KARION (CARBOHYDRATE)
330-54-1	KARMEX
330-54-1	KARMEX D
330-54-1	KARMEX DIURON HERBICIDE
330-54-1	KARMEX DW
50-00-0	KARSAN
112-53-8	KARUKORU 20
7722-84-1	KASTONE
8002-03-7	KATCHUNG OIL
84-80-0	KATIV N
138-86-3	KAUTSCHIN
78-34-2	KAVADEL
614-45-9	KAYABUTYL B
3457-61-2	KAYABUTYL C
30580-75-7	KAYABUTYL C
2372-21-6	KAYACARBON BIC
80-43-3	KAYACUMYL D
109-13-7	KAYAESTER I
838-88-0	KAYAHARD MDT
78-63-7	KAYAHEXA AD
78-63-7	KAYAHEXA AD 40C
1068-27-5	KAYAHEXA YD
119-90-4	KAYAKU BLUE B BASE
2602-46-2	KAYAKU DIRECT BLUE BB
1937-37-7	KAYAKU DIRECT DEEP BLACK EX
1937-37-7	KAYAKU DIRECT DEEP BLACK GX
1937-37-7	KAYAKU DIRECT DEEP BLACK S
1937-37-7	KAYAKU DIRECT LEATHER BLACK EX
1937-37-7	KAYAKU DIRECT SPECIAL BLACK AAX
99-55-8	KAYAKU SCARLET G BASE
1338-23-4	KAYAMEK A
37206-20-5	KAYAMEK B
1338-23-4	KAYAMEK M
16071-86-6	KAYARUS SUPRA BROWN BRS
333-41-5	KAYAZINON
333-41-5	KAYAZOL
8020-83-5	KAYDOL
8020-83-5	KAYDOL WHITE MINERAL OIL
84-80-0	KAYWAN
26628-22-8	KAZOE
25155-30-0	KB
25155-30-0	KB (SURFACTANT)
154-23-4	KB 53
120-80-9	KB-53
16071-86-6	KCA LIGHT FAST BROWN BR
10103-46-5	KDV 15U
9005-25-8	KEESTAR 328
75-00-3	KELENE
115-32-2	KELTANE
115-32-2	KELTHANE
115-32-2	KELTHANE (DOT)
115-32-2	KELTHANE A
115-32-2	KELTHANE DUST BASE
115-32-2	KELTHANE, LIQUID
115-32-2	KELTHANE, SOLID
115-32-2	KELTHANETHANOL
2016-57-1	KEMAMINE P 190D
101-05-3	KEMATE
12654-97-6	KEMATE
103-23-1	KEMESTER 5652
123-79-5	KEMESTER 5652
56-75-7	KEMICETINE
732-11-6	KEMOLATE
75-99-0	KENAPON
84-80-0	KEPHTON
143-50-0	KEPONE
143-50-0	KEPONE (DOT)
143-50-0	KEPONE-2-ONE, DECACHLOROOCTAHYDRO-
128-37-0	KERABIT
69-72-7	KERALYT
23950-58-5	KERB
23950-58-5	KERB 50W
8008-20-6	KEROSENE
8008-20-6	KEROSINE
8008-20-6	KEROSINE (PETROLEUM)
7631-86-9	KESTREL 600
53-16-7	KESTRONE
463-51-4	KETENE
674-82-8	KETENE DIMER
1344-28-1	KETJEN B
1333-86-4	KETJENBLACK
120-92-3	KETOCYCLOPENTANE
53-16-7	KETODESTRIN
53-16-7	KETOHYDROXYESTRIN
67-64-1	KETONE PROPANE
591-78-6	KETONE, BUTYL METHYL
67-64-1	KETONE, DIMETHYL
119-61-9	KETONE, DIPHENYL
78-93-3	KETONE, ETHYL METHYL
110-12-3	KETONE, METHYL ISOAMYL
814-78-8	KETONE, METHYL ISOPROPENYL
110-43-0	KETONE, METHYL PENTYL
98-86-2	KETONE, METHYL PHENYL
78-94-4	KETONE, METHYL VINYL
1338-23-4	KETONOX

CAS No.	Chemical Name
120-92-3	KETOPENTAMETHYLENE
1333-86-4	KGO 250
13463-67-7	KH 360
8020-83-5	KHA
101-25-7	KHEMPOR N90
8020-83-5	KHF 22-24
8020-83-5	KHF 22S
91-63-4	KHINALDIN
10034-81-8	KHKM 300
75-69-4	KHLADON 11
76-13-1	KHLADON 113
75-63-8	KHLADON 13B1
75-45-6	KHLADON 22
8020-83-5	KHM 6
10326-21-3	KHMD 58
1344-28-1	KHP 2
16961-83-4	KIEZELFLUORWATERSTOFZUUR (DUTCH)
7758-98-7	KILCOP 53
2545-59-7	KILEX 3
93-79-8	KILEX 3
30553-04-9	KILL KANT Z
7784-46-5	KILL-ALL
2921-88-2	KILLMASTER
7778-44-1	KILMAG
88-85-7	KILOSEB
1314-84-7	KILRAT
1344-28-1	KIMAL
84-80-0	KINADION
1303-33-9	KING'S GOLD
12002-03-8	KING'S GREEN
1303-33-9	KING'S YELLOW
9004-34-6	KINGCOT
50-55-5	KITINE
3761-53-3	KITON PONCEAU 2R
3761-53-3	KITON PONCEAU R
3761-53-3	KITON SCARLET 2RC
1694-09-3	KITON VIOLET 4BNS
55-63-0	KLAVIKORDAL
52628-25-8	KLEANROL
127-95-7	KLEESALZ (GERMAN)
443-48-1	KLION
443-48-1	KLONT
55-86-7	KLORAMIN
7775-09-9	KLOREX
56-75-7	KLORITA
56-75-7	KLOROCID S
7681-52-9	KLOROCIN
1309-48-4	KM 40
7681-11-0	KNOLLIDE
333-41-5	KNOX-OUT
7440-48-4	KOBALT (GERMAN, POLISH)
7646-79-9	KOBALT CHLORID (GERMAN)
7758-98-7	KOBASIC
82-68-8	KOBUTOL
7704-34-9	KOCIDE
103-23-1	KODAFLEX DOA
123-79-5	KODAFLEX DOA
117-81-7	KODAFLEX DOP
117-84-0	KODAFLEX DOP
9004-70-0	KODAK LR 115
1332-58-7	KOG
8020-83-5	KOGASIN PROCESS
124-38-9	KOHLENDIOXYD (GERMAN)
75-15-0	KOHLENDISULFID (SCHWEFELKOHLENSTOFF) (GERMAN)
630-08-0	KOHLENMONOXID (GERMAN)
630-08-0	KOHLENOXYD (GERMAN)
124-38-9	KOHLENSAURE (GERMAN)
58-89-9	KOKOTINE
7704-34-9	KOLO 100
7704-34-9	KOLOFOG
7704-34-9	KOLOSPRAY
56-38-2	KOLPHOS
53-16-7	KOLPON
630-60-4	KOMBETIN
60-57-1	KOMBI-ALBERTAN
84-80-0	KONAKION
76-03-9	KONESTA
577-11-7	KONLAX
630-08-0	KOOLMONOXYDE (DUTCH)
75-15-0	KOOLSTOFDISULFIDE (ZWAVELKOOLSTOF) (DUTCH)
75-44-5	KOOLSTOFOXYCHLORIDE (DUTCH)
115-29-7	KOP-THIODAN
121-75-5	KOP-THION
106-93-4	KOPFUME
299-84-3	KORLAN
299-84-3	KORLANE
7782-42-5	KOROBON
309-00-2	KORTOFIN
30900-23-3	KORTOFIN
577-11-7	KOSATE
1333-86-4	KOSMAS 40
13463-67-7	KR 380
2636-26-2	KRECALVIN
137-26-8	KREGASAN
8020-83-5	KREMOL 100
8020-83-5	KREMOL 50
8020-83-5	KREMOL 90
8020-83-5	KREMOL REGULAR
1335-85-9	KRENITE (OBS)
534-52-1	KREOZAN
1335-85-9	KRESAMONE
1319-77-3	KRESOLE (GERMAN)
1319-77-3	KRESOLEN (DUTCH)
120-71-8	KREZIDIN
1319-77-3	KREZOL (POLISH)
1335-85-9	KREZOTOL 50
534-52-1	KREZOTOL 50
86-88-4	KRIPID
128-44-9	KRISTALLOSE
7704-34-9	KRISTEX
1332-58-7	KRKHS
10326-21-3	KRMD 58
1309-37-1	KROKUS
12001-28-4	KROKYDOLITH (GERMAN)
111-48-8	KROMFAX SOLVENT
3165-93-3	KROMON GREEN B
1330-78-5	KRONITEX
1330-78-5	KRONITEX R
1330-78-5	KRONITEX TCP
13463-67-7	KRONOS
13463-67-7	KRONOS 2073
13463-67-7	KRONOS CL 220
13463-67-7	KRONOS KR 380
13463-67-7	KRONOS RN 40
13463-67-7	KRONOS RN 40P
13463-67-7	KRONOS RN 56
13463-67-7	KRONOS RN 59
97-77-8	KROTENAL
2971-38-2	KROTILIN
2971-38-2	KROTILINE
314-40-9	KROVAR II
117-52-2	KRUMKIL
7439-90-9	KRYPTON
7439-90-9	KRYPTON, COMPRESSED
7439-90-9	KRYPTON, COMPRESSED (DOT)
7439-90-9	KRYPTON, LIQUID (REFRIGERATED)
7439-90-9	KRYPTON, REFRIGERATED LIQUID (DOT)
86-88-4	KRYSID
30553-04-9	KRYSID
30553-04-9	KRYSID PI
62-53-3	KRYSTALLIN
2545-59-7	KRZEWOTOKS
93-79-8	KRZEWOTOKS
93-79-8	KRZEWOTOX
2545-59-7	KRZEWOTOX
7631-86-9	KS 160
7631-86-9	KS 300
105-60-2	KS 30P
7631-86-9	KS 380
7631-86-9	KS 404
7440-32-6	KS 50
7440-32-6	KS 70
1330-20-7	KSYLEN (POLISH)
81-81-2	KUMADER
81-81-2	KUMADU
81-81-2	KUMATOX
1129-41-5	KUMIAI
7704-34-9	KUMULUS

CAS No.	Chemical Name
7704-34-9	KUMULUS FL
7758-98-7	KUPFERSULFAT (GERMAN)
9002-89-5	KURALON VP
9002-89-5	KURARAY POVAL PVA 420
9002-89-5	KURARE POVAL 120
9002-89-5	KURARE POVAL 1700
9002-89-5	KURARE PVA 205
9002-89-5	KURATE POVAL 120
93-72-1	KUROSAL G
7775-09-9	KUSA-TOHRU
7775-09-9	KUSATOL
21564-17-0	KVK 733059
64-18-6	KWAS METANIOWY (POLISH)
58-89-9	KWELL
7439-97-6	KWIK (DUTCH)
57-24-9	KWIK-KIL
62-38-4	KWIKSAN
563-12-2	KWIT
143-33-9	KYANID SODNY (CZECH)
506-64-9	KYANID STRIBRNY (CZECH)
62-53-3	KYANOL
506-61-6	KYANOSTRIBRNAN DRASELNY (CZECH)
108-77-0	KYANURCHLORID (CZECH)
1309-48-4	KYOWAMAG 100
1309-48-4	KYOWAMAG 150
1309-48-4	KYOWAMAG 30
1344-28-1	KYOWARD 200
12789-03-6	KYPCHLOR
81-81-2	KYPFARIN
121-75-5	KYPFOS
56-38-2	KYPTHION
5329-14-6	KYSELINA AMIDOSULFONOVA (CZECH)
2782-57-2	KYSELINA DICHLORISOKYANUROVA (CZECH)
110-17-8	KYSELINA FUMAROVA (CZECH)
121-91-5	KYSELINA ISOFTALOVA (CZECH)
50-21-5	KYSELINA MLECNA (CZECH)
25231-46-3	KYSELINA P-TOLUENESULFONOVA (CZECH)
104-15-4	KYSELINA P-TOLUENESULFONOVA (CZECH)
144-62-7	KYSELINA STAVELOVA (CZECH)
5329-14-6	KYSELINA SULFAMINOVA (CZECH)
87-90-1	KYSELINA TRICHLOISOKYANUROVA (CZECH)
409-21-2	KZ 3M
409-21-2	KZ 5M
409-21-2	KZ 7M
7429-90-5	L 1018
298-02-2	L 11/6
7429-90-5	L 16
9002-84-0	L 169
9002-84-0	L 169A
7440-22-4	L 3
2275-18-5	L 343
1582-09-8	L 36352
112-80-1	L'ACIDE OLEIQUE (FRENCH)
111-70-6	L'ALCOOL N-HEPTYLIQUE PRIMAIRE (FRENCH)
87-69-4	L-(+)-TARTARIC ACID
514-73-8	L-01748
65-30-5	L-1-METHYL-2-(3-PYRIDYL)-PYRROLIDINE SULFATE
54-11-5	L-3-(1-METHYL-2-PYRROLIDYL)PYRIDINE
65-30-5	L-3-(1-METHYL-2-PYRROLIDYL)PYRIDINE SULFATE
60-51-5	L-395
115-02-6	L-AZASERINE
316-42-7	L-EMETINE DIHYDROCHLORIDE
54-62-6	L-GLUTAMIC ACID, N-[4-[[(2,4-DIAMINO-6-PTERIDINYL)METHYL] AMINO]BENZOYL]-
50-70-4	L-GULITOL
54-11-5	L-NICOTINE
148-82-3	L-PAM
110-66-7	L-PENTANETHIOL
148-82-3	L-PHENYLALANINE MUSTARD
148-82-3	L-PHENYLALANINE, 4-[BIS(2-CHLOROETHYL)AMINO]-
148-82-3	L-SARCOLYSIN
148-82-3	L-SARCOLYSINE
148-82-3	L-SARKOLYSIN
115-02-6	L-SERINE, DIAZOACETATE (ESTER)
3164-29-2	L-TARTARIC ACID AMMONIUM SALT
9004-34-6	LA 01
1344-28-1	LA 6
2646-17-5	LACQUER ORANGE V
97-64-3	LACTATE D'ETHYLE (FRENCH)

CAS No.	Chemical Name
50-21-5	LACTIC ACID
138-22-7	LACTIC ACID, BUTYL ESTER
97-64-3	LACTIC ACID, ETHYL ESTER
617-51-6	LACTIC ACID, ISOPROPYL ESTER
515-98-0	LACTIC ACID, MONOAMMONIUM SALT
78-97-7	LACTONITRILE
75-86-5	LACTONITRILE, 2-METHYL-
88-85-7	LADOB
119-90-4	LAKE BLUE B BASE
89-62-3	LAKE RED G BASE
3761-53-3	LAKE SCARLET 2RBN
99-55-8	LAKE SCARLET G BASE
3761-53-3	LAKE SCARLET R
4301-50-2	LAMBROL
50-28-2	LAMDIOL
9002-89-5	LAMEPHIL OJ
1333-86-4	LAMPBLACK
79-01-6	LANADIN
8006-54-0	LANAIN
8006-54-0	LANALIN
71-63-6	LANATOXIN
13397-24-5	LANDPLASTER
8006-54-0	LANESIN
112-72-1	LANETTE K
27196-00-5	LANETTE K
112-72-1	LANETTE WAX KS
27196-00-5	LANETTE WAX KS
151-21-3	LANETTE WAX S
8006-54-0	LANICHOL
20830-75-5	LANICOR
8006-54-0	LANIOL
123-79-5	LANKROFLEX DOA
103-23-1	LANKROFLEX DOA
577-11-7	LANKROPOL KO 2
16752-77-5	LANNATE
16752-77-5	LANNATE 90SP
20830-75-5	LANOCARDIN
8006-54-0	LANOLIN
8006-54-0	LANOLIN, ANHYDROUS
8006-54-0	LANOPRODINE
20830-75-5	LANOXIN
8006-54-0	LANTROL
8006-54-0	LANUM
25322-69-4	LAPROL 1502-3-100
25322-69-4	LAPROL 1602-3-100
25322-69-4	LAPROL 2002
25322-69-4	LAPROL 2102
25322-69-4	LAPROL 503B
25322-69-4	LAPROL 702
25322-69-4	LAPROL L 1502-3-100
8016-28-2	LARD OIL
76-06-2	LARVACIDE
27176-87-0	LAS 99
88-85-7	LASEB
303-34-4	LASIOCARPINE
58-89-9	LASOCHRON
10024-97-2	LAUGHING GAS
11104-93-1	LAUGHING GAS
76-22-2	LAUREL CAMPHOR
5350-03-8	LAURIC ACID, PENTYL ESTER
112-53-8	LAURIC ALCOHOL
112-53-8	LAURINIC ALCOHOL
105-74-8	LAUROX
105-74-8	LAUROX Q
105-74-8	LAUROX W 40
105-74-8	LAUROYL PEROXIDE
105-74-8	LAUROYL PEROXIDE (DOT)
105-74-8	LAUROYL PEROXIDE, NOT MORE THAN 42%,STABLE DISPERSION, IN WATER
105-74-8	LAUROYL PEROXIDE, TECHNICALLY PURE (DOT)
105-74-8	LAURYDOL
112-53-8	LAURYL 24
112-53-8	LAURYL ALCOHOL
2235-54-3	LAURYL AMMONIUM SULFATE
112-55-0	LAURYL MERCAPTAN
151-21-3	LAURYL SODIUM SULFATE
2235-54-3	LAURYL SULFATE AMMONIUM SALT
143-00-0	LAURYL SULFATE DIETHANOLAMINE SALT
139-96-8	LAURYL SULFATE ESTER TRIETHANOLAMINE SALT

CAS No.	*Chemical Name*
151-21-3	LAURYL SULFATE SODIUM
151-21-3	LAURYL SULFATE SODIUM SALT
139-96-8	LAURYL SULFATE TRIETHANOLAMINE SALT
27176-87-0	LAURYLBENZENESULFONIC ACID
139-96-8	LAURYLSULFURIC ACID TRIETHANOLAMINE SALT
53-86-1	LAUSIT
87-86-5	LAUXTOL
87-86-5	LAUXTOL A
8007-45-2	LAVATAR
94-75-7	LAWN-KEEP
577-11-7	LAXINATE
577-11-7	LAXINATE 100
133-06-2	LE CAPTANE (FRENCH)
1335-85-9	LE DINITROCRESOL-4,6 (FRENCH)
77-81-6	LE-100
7681-49-4	LEA-COV
7439-92-1	LEAD
7758-95-4	LEAD (II) CHLORIDE
15245-44-0	LEAD 2,4,6-TRINITRORESORCINOXIDE
301-04-2	LEAD ACETATE
301-04-2	LEAD ACETATE (2+)
301-04-2	LEAD ACETATE (Pb(AC)2)
301-04-2	LEAD ACETATE (Pb(O2C2H3)2)
546-67-8	LEAD ACETATE (Pb(O2C2H3)4)
1335-32-6	LEAD ACETATE HYDROXIDE (Pb3(OAC)2(OH)4)
546-67-8	LEAD ACETATE [Pb(OAC)4]
1335-32-6	LEAD ACETATE, BASIC
7784-40-9	LEAD ACID ARSENATE
7784-40-9	LEAD ARSENATE
7645-25-2	LEAD ARSENATE
7784-40-9	LEAD ARSENATE (PbHASO4)
10031-13-7	LEAD ARSENITE
10031-13-7	LEAD ARSENITE, Pb(ASO2)2
10031-13-7	LEAD ARSENITE, SOLID
10031-13-7	LEAD ARSENITE, SOLID (DOT)
13424-46-9	LEAD AZIDE
592-87-0	LEAD BIS(THIOCYANATE)
13814-96-5	LEAD BORON FLUORIDE
7446-14-2	LEAD BOTTOMS
15739-80-7	LEAD BOTTOMS
1309-60-0	LEAD BROWN
7758-95-4	LEAD CHLORIDE
12612-47-4	LEAD CHLORIDE
7758-95-4	LEAD CHLORIDE (DOT)
7758-95-4	LEAD CHLORIDE (PbCl2)
7758-97-6	LEAD CHROMATE
18454-12-1	LEAD CHROMATE
11119-70-3	LEAD CHROMATE
18454-12-1	LEAD CHROMATE (Pb2CrO5)
7758-97-6	LEAD CHROMATE (PbCrO4)
18454-12-1	LEAD CHROMATE (VI) OXIDE
11119-70-3	LEAD CHROMATE OXIDE
18454-12-1	LEAD CHROMATE OXIDE
18454-12-1	LEAD CHROMATE OXIDE (Pb2OCrO4)
18454-12-1	LEAD CHROMATE, RED
7758-97-6	LEAD CHROMIUM OXIDE (PbCrO4)
592-05-2	LEAD CYANIDE
592-05-2	LEAD CYANIDE (DOT)
592-05-2	LEAD CYANIDE (Pb(CN)2)
301-04-2	LEAD DIACETATE
301-04-2	LEAD DIBASIC ACETATE
1344-40-7	LEAD DIBASIC PHOSPHITE
7758-95-4	LEAD DICHLORIDE
7783-46-2	LEAD DIFLUORIDE
7783-46-2	LEAD DIFLUORIDE (PbF2)
10101-63-0	LEAD DIIODIDE
10099-74-8	LEAD DINITRATE
18256-98-9	LEAD DINITRATE
1309-60-0	LEAD DIOXIDE
1309-60-0	LEAD DIOXIDE (DOT)
13637-76-8	LEAD DIPERCHLORATE
7446-27-7	LEAD DIPHOSPHATE
1072-35-1	LEAD DISTEARATE
592-87-0	LEAD DITHIOCYANATE
7439-92-1	LEAD FLAKE
13814-96-5	LEAD FLUOBORATE
13814-96-5	LEAD FLUOBORATE (DOT)
7783-46-2	LEAD FLUORIDE
7783-46-2	LEAD FLUORIDE (DOT)
7783-46-2	LEAD FLUORIDE (PbF2)
13814-96-5	LEAD FLUOROBORATE
7784-40-9	LEAD HYDROGEN ARSENATE (PbHAsO4)
10101-63-0	LEAD IODIDE
12684-19-4	LEAD IODIDE
10101-63-0	LEAD IODIDE (PbI2)
592-87-0	LEAD ISOTHIOCYANATE
1317-36-8	LEAD MONOOXIDE
7446-14-2	LEAD MONOSULFATE
15739-80-7	LEAD MONOSULFATE
1314-87-0	LEAD MONOSULFIDE
1317-36-8	LEAD MONOXIDE
18256-98-9	LEAD NITRATE
10099-74-8	LEAD NITRATE
10099-74-8	LEAD NITRATE (DOT)
18256-98-9	LEAD NITRATE (DOT)
10099-74-8	LEAD NITRATE (Pb(NO3)2)
7428-48-0	LEAD OCTADECANOATE
7446-27-7	LEAD ORTHOPHOSPHATE
7446-27-7	LEAD ORTHOPHOSPHATE (Pb3(PO4)2)
1309-60-0	LEAD OXIDE
1317-36-8	LEAD OXIDE
1317-36-8	LEAD OXIDE (PbO)
1309-60-0	LEAD OXIDE (PbO2)
1309-60-0	LEAD OXIDE BROWN
1344-40-7	LEAD OXIDE PHOSPHONATE (Pb3O2(HPO3)), HEMIHYDRATE
1344-40-7	LEAD OXIDE PHOSPHONATE, HEMIHYDRATE
1317-36-8	LEAD OXIDE YELLOW
13637-76-8	LEAD PERCHLORATE
13637-76-8	LEAD PERCHLORATE (DOT)
13637-76-8	LEAD PERCHLORATE (Pb(ClO4)2)
1309-60-0	LEAD PEROXIDE
1309-60-0	LEAD PEROXIDE (DOT)
1309-60-0	LEAD PEROXIDE (PbO2)
7446-27-7	LEAD PHOSPHATE
7446-27-7	LEAD PHOSPHATE (3:2)
7446-27-7	LEAD PHOSPHATE (Pb3(PO4)2)
1344-40-7	LEAD PHOSPHITE DIBASIC (DOT)
1344-40-7	LEAD PHOSPHITE, DIBASIC
1317-36-8	LEAD PROTOXIDE
7439-92-1	LEAD S 2
1072-35-1	LEAD STEARATE
7428-48-0	LEAD STEARATE
7428-48-0	LEAD STEARATE (DOT)
15245-44-0	LEAD STYPHNATE
1335-32-6	LEAD SUBACETATE
7446-14-2	LEAD SULFATE
15739-80-7	LEAD SULFATE
7446-14-2	LEAD SULFATE (1:1)
15739-80-7	LEAD SULFATE (1:1)
7446-14-2	LEAD SULFATE (DOT)
15739-80-7	LEAD SULFATE (DOT)
7446-14-2	LEAD SULFATE (PbSO4)
7446-14-2	LEAD SULFATE, SOLID, CONTAINING MORE THAN 3% FREE ACID (DOT)
15739-80-7	LEAD SULFATE, SOLID, CONTAINING MORE THAN 3% FREE ACID (DOT)
1314-87-0	LEAD SULFIDE
39377-56-5	LEAD SULFIDE
1314-87-0	LEAD SULFIDE (1:1)
1314-87-0	LEAD SULFIDE (DOT)
1314-87-0	LEAD SULFIDE (PbS)
592-87-0	LEAD SULFOCYANATE
7446-14-2	LEAD SULPHATE
1309-60-0	LEAD SUPEROXIDE
546-67-8	LEAD TETRAACETATE
546-67-8	LEAD TETRACETATE
13814-96-5	LEAD TETRAFLUOROBORATE
13814-96-5	LEAD TETRAFLUOROBORATE (Pb(BF4)2)
592-87-0	LEAD THIOCYANATE
592-87-0	LEAD THIOCYANATE (DOT)
592-87-0	LEAD THIOCYANATE (Pb(SCN)2)
26265-65-6	LEAD THIOSULFATE
12737-98-3	LEAD TUNGSTATE
12737-98-3	LEAD TUNGSTEN OXIDE
301-04-2	LEAD(2+) ACETATE
7758-95-4	LEAD(2+) CHLORIDE
18256-98-9	LEAD(2+) NITRATE
10099-74-8	LEAD(2+) NITRATE

CAS No.	Chemical Name
1317-36-8	LEAD(2+) OXIDE
13637-76-8	LEAD(2+) PERCHLORATE
7446-27-7	LEAD(2+) PHOSPHATE (Pb3(PO4)2)
1072-35-1	LEAD(2+) STEARATE
7446-14-2	LEAD(2+) SULFATE
15739-80-7	LEAD(2+) SULFATE
1314-87-0	LEAD(2+) SULFIDE
301-04-2	LEAD(II) ACETATE
10031-13-7	LEAD(II) ARSENITE
7758-95-4	LEAD(II) CHLORIDE
7783-46-2	LEAD(II) FLUORIDE
10101-63-0	LEAD(II) IODIDE
10099-74-8	LEAD(II) NITRATE
18256-98-9	LEAD(II) NITRATE
10099-74-8	LEAD(II) NITRATE (1:2)
18256-98-9	LEAD(II) NITRATE (1:2)
1317-36-8	LEAD(II) OXIDE
1072-35-1	LEAD(II) STEARATE
7446-14-2	LEAD(II) SULFATE
15739-80-7	LEAD(II) SULFATE
7446-14-2	LEAD(II) SULFATE (1:1)
15739-80-7	LEAD(II) SULFATE (1:1)
1314-87-0	LEAD(II) SULFIDE
592-87-0	LEAD(II) THIOCYANATE
592-05-2	LEAD(II)CYANIDE
546-67-8	LEAD(IV) ACETATE
1309-60-0	LEAD(IV) OXIDE
1335-32-6	LEAD, BIS(ACETATO)TETRAHYDROXYTRI-
1335-32-6	LEAD, BIS(ACETATO-O)TETRAHYDROXYTRI-
56189-09-4	LEAD, BIS(OCTADECANOATO)DIOXODI-
1344-40-7	LEAD, DIOXO[PHOSPHITO(2-)]TRI-, HEMIHYDRATE
7439-92-1	LEAD, INORGANIC, DUST AND FUMES
1335-32-6	LEAD, SUBACETATE
78-00-2	LEAD, TETRAETHYL-
75-74-1	LEAD, TETRAMETHYL-
928-96-1	LEAF ALCOHOL
55-38-9	LEBAYCID
112-02-7	LEBON TM 16
75-69-4	LEDON 11
75-71-8	LEDON 12
52-68-6	LEIVASOM
50-55-5	LEMISERP
7681-49-4	LEMOFLUR
9002-89-5	LEMOL
9002-89-5	LEMOL 12-88
9002-89-5	LEMOL 16-98
9002-89-5	LEMOL 24-98
9002-89-5	LEMOL 30-98
9002-89-5	LEMOL 5-88
9002-89-5	LEMOL 5-98
9002-89-5	LEMOL 51-98
9002-89-5	LEMOL 60-98
9002-89-5	LEMOL 75-98
9002-89-5	LEMOL GF 60
92-52-4	LEMONENE
58-89-9	LENDINE
55-63-0	LENITRAL
8002-74-2	LENOLENE AC
51-83-2	LENTIN
51-83-2	LENTINE (FRENCH)
58-89-9	LENTOX
57-41-0	LEPITOIN
21609-90-5	LEPTOPHOS
2971-90-6	LERBEK
9016-45-9	LEROLAT N
9016-45-9	LEROLAT N 300
56-38-2	LETHALAIRE G-54
3689-24-5	LETHALAIRE G-57
152-16-9	LETHALAIRE G-59
112-56-1	LETHANE
786-19-6	LETHOX
79-01-6	LETHURIN
51-79-6	LEUCETHANE
150-76-5	LEUCOBASAL
150-76-5	LEUCODINE B
91-22-5	LEUCOL
91-22-5	LEUCOLINE
55-98-1	LEUCOSULFAN
20830-81-3	LEUKAEMOMYCIN C
305-03-3	LEUKERAN
91-22-5	LEUKOL
56-75-7	LEUKOMYAN
56-75-7	LEUKOMYCIN
3129-91-7	LEUKORROSIN C
8070-50-6	LEUTOX
148-82-3	LEVOFALAN
56-75-7	LEVOMICETINA
56-75-7	LEVOMITSETIN
56-75-7	LEVOMYCETIN
56-75-7	LEVOVETIN
302-01-2	LEVOXINE
541-25-3	LEWISITE
541-25-3	LEWISITE (ARSENIC COMPOUND)
21087-64-9	LEXONE
62-38-4	LEYTOSAN
7646-78-8	LIBAVIUS FUMING SPIRIT
110-17-8	LICHENIC ACID
58-89-9	LIDENAL
8008-51-3	LIGHT CAMPHOR OIL
8030-30-6	LIGHT LIGROIN
8008-51-3	LIGHT OIL OF CAMPHOR
2235-25-8	LIGNASAN
2235-25-8	LIGNASAN FUNGICIDE
2235-25-8	LIGNASAN-X
8032-32-4	LIGROIN
8032-32-4	LIGROINE
999-81-5	LIHOCIN
1582-09-8	LILLY 36,352
108-62-3	LIMAX
9002-91-9	LIMAX
1305-62-0	LIMBUX
1305-78-8	LIME
7778-54-3	LIME CHLORIDE
156-62-7	LIME NITROGEN
156-62-7	LIME NITROGEN (DOT)
1305-62-0	LIME WATER
1305-78-8	LIME, BURNED
1305-78-8	LIME, UNSLAKED (DOT)
9007-13-0	LIMED ROSIN
1317-65-3	LIMESTONE
138-86-3	LIMONEN
138-86-3	LIMONENE
138-86-3	LIMONENE, INACTIVE
108-62-3	LIMOVET
9002-91-9	LIMOVET
58-89-9	LINDAFOR
58-89-9	LINDAGAM
58-89-9	LINDAGRAIN
58-89-9	LINDAGRANOX
2636-26-2	LINDAN
58-89-9	LINDANE
58-89-9	LINDANE (ACGIH,DOT)
58-89-9	LINDANE, LIQUID
58-89-9	LINDANE, SOLID
58-89-9	LINDAPOUDRE
58-89-9	LINDATOX
1344-28-1	LINDE A
58-89-9	LINDEX
1330-78-5	LINDOL
58-89-9	LINDOSEP
93-76-5	LINE RIDER
25322-69-4	LINEARTOP E
305-03-3	LINFOLIZIN
305-03-3	LINFOLYSIN
57-63-6	LINORAL
8001-26-1	LINSEED OIL
8001-26-1	LINSEED OIL, BLEACHED
58-89-9	LINTOX
58-89-9	LINVUR
9016-45-9	LIPAL 9N
534-52-1	LIPAN
1335-85-9	LIPAN
7789-75-5	LIPARITE
108-01-0	LIPARON
3691-35-8	LIPHADIONE
57-83-0	LIPO-LUTIN
8001-30-7	LIPOMUL
9016-45-9	LIPONOX NCG

CAS No.	Chemical Name
9016-45-9	LIPONOX NCH
9016-45-9	LIPONOX NCI
9016-45-9	LIPONOX NCM
81-81-2	LIQUA-TOX
106-97-8	LIQUEFIED PETROLEUM GAS
74-98-6	LIQUEFIED PETROLEUM GAS
68476-85-7	LIQUEFIED PETROLEUM GAS
75-28-5	LIQUEFIED PETROLEUM GAS (DOT)
115-11-7	LIQUEFIED PETROLEUM GAS (DOT)
68476-85-7	LIQUEFIED PETROLEUM GAS (DOT)
68476-85-7	LIQUEFIED PETROLEUM GASES
7440-06-4	LIQUID BRIGHT PLATINUM
8008-51-3	LIQUID CAMPHOR
83-79-4	LIQUID DERRIS
74-85-1	LIQUID ETHYENE
8001-58-9	LIQUID PITCH OIL
10043-52-4	LIQUIDOW
68476-85-7	LIQUIFIED PETROLEUM GAS (L.P.G)
62-38-4	LIQUIPHENE
14484-64-1	LIROMATE
94-80-4	LIRONOX
75-99-0	LIROPON
87-86-5	LIROPREM
56-38-2	LIROTHION
9016-45-9	LISSAPOL N
9016-45-9	LISSAPOL NX
9016-45-9	LISSAPOL NXP 10
9016-45-9	LISSAPOL TN 450
7428-48-0	LISTAB 28
56189-09-4	LISTAB 51
7782-89-0	LITHAMIDE
1317-36-8	LITHARGE
1317-36-8	LITHARGE PURE
1317-36-8	LITHARGE YELLOW L 28
7439-93-2	LITHIUM
39990-99-3	LITHIUM ACETYLIDE (Li(C2H)), COMPD WITH 1,2-ETHANE-DIAMINE
50475-76-8	LITHIUM ACETYLIDE ETHYLENEDIAMINE
39990-99-3	LITHIUM ACETYLIDE, COMPLEXED WITH ETHYLENEDIAMINE
39990-99-3	LITHIUM ACETYLIDE-ETHYLENE DIAMINE COMPLEX
39990-99-3	LITHIUM ACETYLIDE-ETHYLENEDIAMINE COMPLEX (DOT)
16853-85-3	LITHIUM ALANATE
68848-64-6	LITHIUM ALLOY, NONBASE, Li,Si
16853-85-3	LITHIUM ALUMINOHYDRIDE
16853-85-3	LITHIUM ALUMINUM HYDRIDE
16853-85-3	LITHIUM ALUMINUM HYDRIDE (DOT)
16853-85-3	LITHIUM ALUMINUM HYDRIDE (LiAlH4)
16853-85-3	LITHIUM ALUMINUM HYDRIDE, ETHEREAL
16853-85-3	LITHIUM ALUMINUM HYDRIDE, ETHEREAL (DOT)
16853-85-3	LITHIUM ALUMINUM TETRAHYDRIDE
7782-89-0	LITHIUM AMIDE
7782-89-0	LITHIUM AMIDE (DOT)
7782-89-0	LITHIUM AMIDE (Li(NH2))
7782-89-0	LITHIUM AMIDE, POWDERED
7782-89-0	LITHIUM AMIDE, POWDERED (DOT)
16949-15-8	LITHIUM BOROHYDRIDE
16949-15-8	LITHIUM BOROHYDRIDE (LiBH4)
554-13-2	LITHIUM CARBONATE
13840-33-0	LITHIUM CHLORIDE OXIDE (LiClO)
14307-35-8	LITHIUM CHROMATE
14307-35-8	LITHIUM CHROMATE (Li2CrO4)
14307-35-8	LITHIUM CHROMATE(VI)
7439-93-2	LITHIUM ELEMENT
70399-13-2	LITHIUM FERROSILICON
7580-67-8	LITHIUM HYDRIDE
7580-67-8	LITHIUM HYDRIDE (LiH)
1310-65-2	LITHIUM HYDROXIDE
1310-65-2	LITHIUM HYDROXIDE (Li(OH))
1310-65-2	LITHIUM HYDROXIDE (Li(OH)) (9CI)
1310-66-3	LITHIUM HYDROXIDE (Li(OH)), MONOHYDRATE
1310-66-3	LITHIUM HYDROXIDE HYDRATE
1310-66-3	LITHIUM HYDROXIDE MONOHYDRATE
1310-66-3	LITHIUM HYDROXIDE MONOHYDRATE (DOT)
1310-65-2	LITHIUM HYDROXIDE SOLUTION
1310-65-2	LITHIUM HYDROXIDE, SOLUTION (DOT)
13840-33-0	LITHIUM HYPOCHLORITE
13840-33-0	LITHIUM HYPOCHLORITE (LiClO)
13840-33-0	LITHIUM HYPOCHLORITE COMPOUND, DRY, CONTAINING MORE THAN 39% AVAILABLE CHLORINE (DOT)

CAS No.	Chemical Name
7439-93-2	LITHIUM METAL
7439-93-2	LITHIUM METAL (DOT)
7439-93-2	LITHIUM METAL, IN CARTRIDGES (DOT)
7580-67-8	LITHIUM MONOHYDRIDE
7790-69-4	LITHIUM NITRATE
7790-69-4	LITHIUM NITRATE (DOT)
26134-62-3	LITHIUM NITRIDE
26134-62-3	LITHIUM NITRIDE (DOT)
26134-62-3	LITHIUM NITRIDE (Li3N)
12031-80-0	LITHIUM OXIDE (Li2O2)
13840-33-0	LITHIUM OXYCHLORIDE
12031-80-0	LITHIUM PEROXIDE
12031-80-0	LITHIUM PEROXIDE (DOT)
12031-80-0	LITHIUM PEROXIDE (Li2(O2))
68848-64-6	LITHIUM SILICON
53095-76-4	LITHIUM SILICON
68848-64-6	LITHIUM SILICON (DOT)
16853-85-3	LITHIUM TETRAHYDRIDOALUMINATE
16853-85-3	LITHIUM TETRAHYDROALUMINATE
16853-85-3	LITHIUM TETRAHYDROALUMINATE (AlLiH4)
16853-85-3	LITHIUM TETRAHYDROALUMINATE(1-)
1317-65-3	LITHOGRAPHIC STONE
99-55-8	LITHOSOL ORANGE R BASE
89-62-3	LITHOSOL SCARLET BASE M
89-62-3	LITHOSOL SCARLET BASE MB
89-62-3	LITHOSOL SCARLET BASE MBW
89-62-3	LITHOSOL SCARLET BASE MW
8001-69-2	LIVER OILS
3691-35-8	LM 91
28772-56-7	LM-637
14807-96-6	LMR 100
1309-37-1	LN 1331
52-68-6	LOISOL
13010-47-4	LOMUSTINE
8012-74-6	LONDON PURPLE
8012-74-6	LONDON PURPLE, SOLID
8012-74-6	LONDON PURPLE, SOLID (DOT)
846-49-1	LORAZEPAM
1332-58-7	LORCO BANTAC PLUS
58-89-9	LOREXANE
112-53-8	LOROL
112-53-8	LOROL 11
111-87-5	LOROL 20
112-30-1	LOROL 22
112-53-8	LOROL 5
112-53-8	LOROL 7
56-75-7	LOROMISIN
62-38-4	LOROPHYN
94-36-0	LOROXIDE
2921-88-2	LORSBAN
2921-88-2	LORSBAN 50SL
7778-54-3	LOSANTIN
505-60-2	LOST
111-90-0	LOSUNGSMITTEL APV
50-55-5	LOWESERP
112-72-1	LOXANOL V
27196-00-5	LOXANOL V
57-11-4	LOXIOL G 20
1642-54-2	LOXURAN
16919-19-0	LPE 6
74-98-6	LPG
106-97-8	LPG
68476-85-7	LPG
75-08-1	LPG ETHYL MERCAPTAN 1010
9004-70-0	LR 115
9004-70-0	LR 115II
7440-66-6	LS 2
639-58-7	LS 4442
9002-84-0	LUBLON L 2
9002-84-0	LUBLON L 5
9016-45-9	LUBROL APN 5
9016-45-9	LUBROL L
9016-45-9	LUBROL N
9016-45-9	LUBROL N 13
1344-28-1	LUCALOX
94-36-0	LUCIDOL
94-36-0	LUCIDOL 40E
94-36-0	LUCIDOL 50P
94-36-0	LUCIDOL 98

CAS No.	Chemical Name
94-36-0	LUCIDOL B 50
94-36-0	LUCIDOL CH 50
1338-23-4	LUCIDOL DDM 9
1338-23-4	LUCIDOL DELTA X
94-36-0	LUCIDOL G 20
94-36-0	LUCIDOL KL 50
94-36-0	LUCIDOL S 50
57-83-0	LUCORTEUM SOL
7631-86-9	LUDOX
7631-86-9	LUDOX AS 30
1344-28-1	LUDOX CL
7631-86-9	LUDOX HS 30
7631-86-9	LUDOX HS 40
7631-86-9	LUDOX RS 40
7631-86-9	LUFILEN E 100
91-80-5	LULAMIN
37273-91-9	LUMACRUSK5
110-85-0	LUMBRICAL
57-11-4	LUNAC S 20
7761-88-8	LUNAR CAUSTIC
94-36-0	LUPERCO
80-43-3	LUPERCO
78-63-7	LUPERCO 101XL
1068-27-5	LUPERCO 130XL
6731-36-8	LUPERCO 231G
6731-36-8	LUPERCO 231XL
6731-36-8	LUPERCO 231XLP
3006-86-8	LUPERCO 331XL
80-43-3	LUPERCO 500-40C
80-43-3	LUPERCO 500-40KE
94-36-0	LUPERCO AA
94-36-0	LUPERCO AST
133-14-2	LUPERCO CST
80-43-3	LUPEROX
78-63-7	LUPEROX 101
2618-77-1	LUPEROX 118
1068-27-5	LUPEROX 130
3025-88-5	LUPEROX 2,5-2,5
37187-22-7	LUPEROX 224
6731-36-8	LUPEROX 231
80-43-3	LUPEROX 500
80-43-3	LUPEROX 500R
80-43-3	LUPEROX 500T
3457-61-2	LUPEROX 801
30580-75-7	LUPEROX 801
94-36-0	LUPEROX FL
105-64-6	LUPEROX IPP
614-45-9	LUPEROX P
26748-41-4	LUPERSOL 10
78-63-7	LUPERSOL 101
26748-41-4	LUPERSOL 10M75
927-07-1	LUPERSOL 11
2618-77-1	LUPERSOL 118
1068-27-5	LUPERSOL 130
26748-47-0	LUPERSOL 188
26748-47-0	LUPERSOL 188M75
3025-88-5	LUPERSOL 2,5-2,5
5809-08-5	LUPERSOL 215
2167-23-9	LUPERSOL 220
16066-38-9	LUPERSOL 221
3179-56-4	LUPERSOL 228Z
6731-36-8	LUPERSOL 231
3006-86-8	LUPERSOL 321
3006-86-8	LUPERSOL 331
3006-86-8	LUPERSOL 331-80B
80-43-3	LUPERSOL 500
29240-17-3	LUPERSOL 554
29240-17-3	LUPERSOL 554M75
686-31-7	LUPERSOL 575
107-71-1	LUPERSOL 70
109-13-7	LUPERSOL 8
3457-61-2	LUPERSOL 801
1338-23-4	LUPERSOL DDA 30
1338-23-4	LUPERSOL DDM
1338-23-4	LUPERSOL DDM 9
1338-23-4	LUPERSOL DELTA X
1338-23-4	LUPERSOL DELTA X 9
1338-23-4	LUPERSOL DNF
1338-23-4	LUPERSOL DSW

CAS No.	Chemical Name
3006-82-4	LUPERSOL PDO
29240-17-3	LUPERSOL TA 54
686-31-7	LUPERSOL TA 75
2372-21-6	LUPERSOL TBIC
2372-21-6	LUPERSOL TBIC-M 75
25322-69-4	LUPRAMOL 1010
9016-87-9	LUPRINATE M 20
79-09-4	LUPROSIL
117-52-2	LURAT
1937-37-7	LURAZOL BLACK BA
60-51-5	LURGO
7681-49-4	LURIDE
7681-49-4	LURIDE LOZI-TABS
7681-49-4	LURIDE-SF
1332-58-7	LUSTRA
109-53-5	LUTANOL LR 8500
57-83-0	LUTEAL HORMONE
57-83-0	LUTEINIQUE
9016-45-9	LUTENSOL AP 10
9016-45-9	LUTENSOL AP 20
9016-45-9	LUTENSOL AP 9
57-83-0	LUTEOCRIN NORMALE
57-83-0	LUTEODYN
57-83-0	LUTEOGAN
57-83-0	LUTEOHORMONE
57-83-0	LUTEOL
57-83-0	LUTEOPUR
57-83-0	LUTEOSAN
57-83-0	LUTEOSTAB
57-83-0	LUTEOVIS
57-83-0	LUTEX
57-83-0	LUTIDON
57-83-0	LUTIN
57-83-0	LUTOCICLINA
57-83-0	LUTOCYCLIN
57-83-0	LUTOCYCLIN M
57-83-0	LUTOCYLIN
57-83-0	LUTOFORM
57-83-0	LUTOGYL
67-63-0	LUTOSOL
57-83-0	LUTREN
57-83-0	LUTROMONE
640-15-3	LUXISTELM
1310-73-2	LYE SOLUTION
12627-52-0	LYMPHOSCAN
57-63-6	LYNESTRENOL
57-63-6	LYNORAL
105-74-8	LYP 97
105-74-8	LYP 97F
50-00-0	LYSOFORM
7440-50-8	M 1
9002-89-5	M 13/20
57-74-9	M 140
12789-03-6	M 140
67-68-5	M 176
4301-50-2	M 2060
555-84-0	M 254
7440-50-8	M 3
1332-21-4	M 3-60
13121-70-5	M 3180
7440-50-8	M 4
12789-03-6	M 410
1332-21-4	M 5-60
8001-35-2	M 5055
7446-18-6	M 7-GIFTKOERNER
10031-59-1	M 7-GIFTKOERNER
1420-04-8	M 73
298-04-4	M 74
298-04-4	M 74 (PESTICIDE)
640-15-3	M 81
9016-45-9	M 812
1066-45-1	M&T CHEMICALS 1222-45
95-76-1	M,P-DICHLOROANILINE
2454-37-7	M-(1-HYDROXYETHYL)ANILINE
98-16-8	M-(TRIFLUOROMETHYL)ANILINE
97-36-9	M-ACETOACET XYLIDIDE
94-59-7	M-ALLYLPYROCATECHIN METHYLENE ETHER
2454-37-7	M-AMINO-ALPHA-METHYLBENZYL ALCOHOL
98-16-8	M-AMINO-ALPHA,ALPHA,ALPHA-TRIFLUOROTOLUENE

CAS No.	Chemical Name
108-45-2	M-AMINOALINE
108-45-2	M-AMINOANILINE
98-16-8	M-AMINOBENZAL FLUORIDE
98-16-8	M-AMINOBENZOTRIFLUORIDE
108-44-1	M-AMINOTOLUENE
8004-09-9	M-B-C FUMIGANT
108-45-2	M-BENZENEDIAMINE
121-91-5	M-BENZENEDICARBOXYLIC ACID
626-17-5	M-BENZENEDINITRILE
108-46-3	M-BENZENEDIOL
110-69-0	M-BUTYRALDEHYDE OXIME
98-15-7	M-CHLORO-ALPHA,ALPHA,ALPHA-TRIFLUOROTOLUENE
98-15-7	M-CHLOROBENZOTRIFLUORIDE
98-15-7	M-CHLOROBENZOTRIFLUORIDE (DOT)
937-14-4	M-CHLOROBENZOYL HYDROPEROXIDE
121-73-3	M-CHLORONITROBENZENE
121-73-3	M-CHLORONITROBENZENE (DOT)
937-14-4	M-CHLOROPERBENZOIC ACID
937-14-4	M-CHLOROPEROXOBENZOIC ACID
937-14-4	M-CHLOROPEROXYBENZOIC ACID
937-14-4	M-CHLOROPEROXYBENZOIC ACID, MAXIMUM CONCENTRATION 86% (DOT)
102-50-1	M-CRESIDINE
108-39-4	M-CRESOL
108-39-4	M-CRESOL (DOT)
96-69-5	M-CRESOL, 4,4'-THIOBIS[6-TERT-BUTYL-
59-50-7	M-CRESOL, 4-CHLORO-
1321-10-4	M-CRESOL, 4-CHLORO-
122-14-5	M-CRESOL, 4-NITRO-, O-ESTER WITH O,O-DIMETHYL PHOSPHOROTHIOATE
1333-13-7	M-CRESOL, TERT-BUTYL-
108-39-4	M-CRESOLE
1129-41-5	M-CRESYL ESTER OF N-METHYLCARBAMIC ACID
1129-41-5	M-CRESYL METHYLCARBAMATE
1129-41-5	M-CRESYL N-METHYLCARBAMATE
108-39-4	M-CRESYLIC ACID
64-00-6	M-CUMENOL METHYLCARBAMATE
64-00-6	M-CUMENYL METHYLCARBAMATE
626-17-5	M-CYANOBENZONITRILE
2631-37-0	M-CYM-5-YL METHYLCARBAMATE
108-45-2	M-DIAMINOBENZENE
1477-55-0	M-DIAMINOXYLENE
121-91-5	M-DICARBOXYBENZENE
541-73-1	M-DICHLOROBENZENE
541-73-1	M-DICHLOROBENZOL
626-17-5	M-DICYANOBENZENE
141-93-5	M-DIETHYL BENZENE
1401-55-4	M-DIGALLIC ACID
108-46-3	M-DIHYDROXYBENZENE
108-38-3	M-DIMETHYLBENZENE
591-21-9	M-DIMETHYLCYCLOHEXANE
99-65-0	M-DINITROBENZENE
99-65-0	M-DINITROBENZENE (ACGIH,DOT)
108-46-3	M-DIOXYBENZENE
108-57-6	M-DIVINYLBENZENE
141-93-5	M-ETHYLETHYLBENZENE
108-45-2	M-FENYLENDIAMIN (CZECH)
1401-55-4	M-GALLOYL GALLIC ACID
108-46-3	M-HYDROQUINONE
554-84-7	M-HYDROXYNITROBENZENE
108-46-3	M-HYDROXYPHENOL
108-39-4	M-HYDROXYTOLUENE
64-00-6	M-ISOPROPYLPHENOL METHYLCARBAMATE
64-00-6	M-ISOPROPYLPHENOL N-METHYLCARBAMATE
64-00-6	M-ISOPROPYLPHENYL METHYLCARBAMATE
64-00-6	M-ISOPROPYLPHENYL N-METHYLCARBAMATE
108-39-4	M-KRESOL
108-44-1	M-METHYLANILINE
108-44-1	M-METHYLBENZENAMINE
99-08-1	M-METHYLNITROBENZENE
108-39-4	M-METHYLPHENOL
1129-41-5	M-METHYLPHENYL METHYLCARBAMATE
1129-41-5	M-METHYLPHENYL N-METHYLCARBAMATE
108-38-3	M-METHYLTOLUENE
119-32-4	M-NITRO-P-TOLUIDINE
121-73-3	M-NITROCHLOROBENZENE
554-84-7	M-NITROPHENOL
554-84-7	M-NITROPHENOL (DOT)
99-08-1	M-NITROTOLUENE
99-08-1	M-NITROTOLUENE (ACGIH,DOT)
99-51-4	M-NITROXYLENE
108-39-4	M-OXYTOLUENE
298-00-0	M-PARATHION
638-49-3	M-PENTYL FORMATE
541-73-1	M-PHENYLENE DICHLORIDE
108-45-2	M-PHENYLENEDIAMINE
108-45-2	M-PHENYLENEDIAMINE (DOT)
108-45-2	M-PHENYLENEDIAMINE OR P-PHENYLENEDIAMINE, SOLID
121-91-5	M-PHTHALIC ACID
626-17-5	M-PHTHALODINITRILE
110-78-1	M-PROPYL ISOCYANATE
872-50-4	M-PYROL
92-06-8	M-TERPHENYLS
95-80-7	M-TOLUENEDIAMINE
108-44-1	M-TOLUIDIN (CZECH)
108-44-1	M-TOLUIDINE
108-44-1	M-TOLUIDINE (ACGIH,DOT)
98-16-8	M-TOLUIDINE, ALPHA,ALPHA,ALPHA-TRIFLUORO-
102-27-2	M-TOLUIDINE, N-ETHYL-
108-39-4	M-TOLUOL
91-99-6	M-TOLYDIETHANOLAMINE
1129-41-5	M-TOLYL METHYLCARBAMATE
1129-41-5	M-TOLYL N-METHYLCARBAMATE
108-44-1	M-TOLYLAMINE
91-99-6	M-TOLYLDIETHANOLAMINE
91-08-7	M-TOLYLENE DIISOCYANATE
95-80-7	M-TOLYLENEDIAMINE
98-15-7	M-TRIFLUOROMETHYLPHENYL CHLORIDE
108-57-6	M-VINYLSTYRENE
108-38-3	M-XYLENE
108-38-3	M-XYLENE (ACGIH,DOT)
1477-55-0	M-XYLENE A,A-DIAMINE
81-15-2	M-XYLENE, 5-TERT-BUTYL-2,4,6-TRINITRO-
1477-55-0	M-XYLENE-ALPHA,ALPHA-DIAMINE
1477-55-0	M-XYLENEDIAMINE
105-67-9	M-XYLENOL
108-38-3	M-XYLOL
108-38-3	M-XYLOL (DOT)
1477-55-0	M-XYLYLENEDIAMINE
22259-30-9	M-[[(DIMETHYLAMINO)METHYLENE]AMINO]PHENYL METHYLCARBAMATE
23422-53-9	M-[[(DIMETHYLAMINO)METHYLENE]AMINO]PHENYL METHYLCARBAMATE HYDROCHLORIDE
7440-50-8	M3 (COPPER)
7440-50-8	M3R
7440-50-8	M3S
7440-50-8	M4 (COPPER)
298-00-0	M40 & 80
1333-86-4	MA 100
1333-86-4	MA 100 (CARBON)
1344-28-1	MA 11
1344-28-1	MA 11 (METAL OXIDE)
112-53-8	MA-1214
108-84-9	MAAC
55-98-1	MABLIN
532-27-4	MACE
532-27-4	MACE (LACRIMATOR)
54-11-5	MACH-NIC
7784-41-0	MACQUER'S SALT
50-28-2	MACRODIOL
107-21-1	MACROGOL 400 BPC
94-75-7	MACRONDRAY
2636-26-2	MAFU
2636-26-2	MAFU STRIP
1309-37-1	MAG 1730
1309-48-4	MAGCAL
1309-48-4	MAGLITE
1309-48-4	MAGLITE D
1309-48-4	MAGLITE DE
1309-48-4	MAGLITE K
1309-48-4	MAGLITE S
1309-48-4	MAGLITE Y
107-02-8	MAGNACIDE
107-02-8	MAGNACIDE H
1309-48-4	MAGNESA PREPRATA
1309-48-4	MAGNESIA
1309-48-4	MAGNESIA USTA
7439-95-4	MAGNESIO (ITALIAN)

CAS No.	Chemical Name
13717-00-5	MAGNESITE
13717-00-5	MAGNESITE (MG(CO3))
7439-95-4	MAGNESIUM
7803-54-5	MAGNESIUM AMIDE
7803-54-5	MAGNESIUM AMIDE (Mg(NH2)2)
10103-50-1	MAGNESIUM ARSENATE
10103-50-1	MAGNESIUM ARSENATE (DOT)
10103-50-1	MAGNESIUM ARSENATE PHOSPHOR
10103-50-1	MAGNESIUM ARSENATE, SOLID
10103-50-1	MAGNESIUM ARSENATE, SOLID (DOT)
13774-25-9	MAGNESIUM BISULFITE
13774-25-9	MAGNESIUM BISULFITE SOLUTION
7439-95-4	MAGNESIUM BORINGS (DOT)
14519-17-6	MAGNESIUM BROMATE
7789-36-8	MAGNESIUM BROMATE
14519-17-6	MAGNESIUM BROMATE (DOT)
10326-21-3	MAGNESIUM CHLORATE
10326-21-3	MAGNESIUM CHLORATE (DOT)
13423-61-5	MAGNESIUM CHROMATE
13423-61-5	MAGNESIUM CHROMATE (MgCrO4)
7439-95-4	MAGNESIUM CLIPPINGS
7439-95-4	MAGNESIUM CLIPPINGS (DOT)
7803-54-5	MAGNESIUM DIAMIDE
7803-54-5	MAGNESIUM DIAMIDE (DOT)
10326-21-3	MAGNESIUM DICHLORATE
13423-61-5	MAGNESIUM DICHROMATE
13423-61-5	MAGNESIUM DICHROMIUM TETROXIDE
10377-60-3	MAGNESIUM DINITRATE
10034-81-8	MAGNESIUM DIPERCHLORATE
555-54-4	MAGNESIUM DIPHENYL
555-54-4	MAGNESIUM DIPHENYL (DOT)
7439-95-4	MAGNESIUM ELEMENT
7439-95-4	MAGNESIUM GRANULES COATED, PARTICLE SIZE NOT LESS THAN 149 MICRONS (DOT)
60616-74-2	MAGNESIUM HYDRIDE
60616-74-2	MAGNESIUM HYDRIDE (DOT)
1309-42-8	MAGNESIUM HYDROXIDE
7439-95-4	MAGNESIUM METAL (DOT)
1309-48-4	MAGNESIUM MONOXIDE
10377-60-3	MAGNESIUM NITRATE
1309-48-4	MAGNESIUM OXIDE
1309-48-4	MAGNESIUM OXIDE (MgO)
1309-48-4	MAGNESIUM OXIDE FUME
7439-95-4	MAGNESIUM PELLETS
7439-95-4	MAGNESIUM PELLETS (DOT)
10034-81-8	MAGNESIUM PERCHLORATE
10034-81-8	MAGNESIUM PERCHLORATE (DOT)
10034-81-8	MAGNESIUM PERCHLORATE (Mg(ClO4)2)
1335-26-8	MAGNESIUM PEROXIDE
12057-74-8	MAGNESIUM PHOSPHIDE
7439-95-4	MAGNESIUM POWDER (DOT)
7439-95-4	MAGNESIUM POWDERED
7439-95-4	MAGNESIUM RIBBON (DOT)
7439-95-4	MAGNESIUM RIBBONS
7439-95-4	MAGNESIUM SCALPINGS (DOT)
7439-95-4	MAGNESIUM SCRAP (DOT)
7439-95-4	MAGNESIUM SHAVINGS (DOT)
7439-95-4	MAGNESIUM SHEET
39404-03-0	MAGNESIUM SILICIDE
16949-65-8	MAGNESIUM SILICO-FLUORIDE
13774-25-9	MAGNESIUM SULFITE (Mg(HSO3)2)
7439-95-4	MAGNESIUM TURNINGS
7439-95-4	MAGNESIUM TURNINGS (DOT)
557-18-6	MAGNESIUM, DIETHYL-
2999-74-8	MAGNESIUM, DIMETHYL-
7439-95-4	MAGNESIUM, METAL (POWDERED, PELLETS, TURNINGS, OR RIBBON)
7704-34-9	MAGNETIC 70, 90, AND 95
10377-60-3	MAGNIOSAN
1309-48-4	MAGOX
10326-21-3	MAGRON
8001-30-7	MAISE OIL
9005-25-8	MAIZENA
28772-56-7	MAKI
9016-45-9	MAKON
9016-45-9	MAKON 10
9016-45-9	MAKON 12
9016-45-9	MAKON 14
9016-45-9	MAKON 30
9016-45-9	MAKON 4
9016-45-9	MAKON 6
9016-45-9	MAKON 8
633-03-4	MALACHITE GREEN G
121-75-5	MALACIDE
121-75-5	MALAFOR
121-75-5	MALAGRAN
121-75-5	MALAKILL
121-75-5	MALAMAR
121-75-5	MALAMAR 50
121-75-5	MALAPHELE
121-75-5	MALAPHOS
121-75-5	MALASOL
121-75-5	MALASPRAY
121-75-5	MALATAF
121-75-5	MALATHION
121-75-5	MALATHION (ACGIH,DOT)
121-75-5	MALATHION E50
121-75-5	MALATHION LV CONCENTRATE
121-75-5	MALATHIOZOO
121-75-5	MALATHON
121-75-5	MALATHYL
121-75-5	MALATHYL LV CONCENTRATE & ULV CONCENTRATE
121-75-5	MALATION (POLISH)
121-75-5	MALATOL
121-75-5	MALATOX
121-75-5	MALDISON
110-16-7	MALEIC ACID
108-31-6	MALEIC ACID ANHYDRIDE
108-31-6	MALEIC ANHYDRIDE
108-31-6	MALEIC ANHYDRIDE (ACGIH,DOT)
85-43-8	MALEIC ANHYDRIDE ADDUCT OF BUTADIENE
108-31-6	MALEIC ANHYDRIDE, SOLID OR MOLTEN (DOT)
1931-62-0	MALEIC MONOPEROXY ACID, 1-TERT-BUTYL ESTER, NOT MORE THAN 55% IN SOLUTION
1931-62-0	MALEIC MONOPEROXYACID, 1-TERT-BUTYL ESTER
1931-62-0	MALEIC MONOPEROXYACID, OO-TERT-BUTYL ESTER
87-69-4	MALIC ACID, 3-HYDROXY-
133-06-2	MALIPUR
115-29-7	MALIX
136-40-3	MALLOPHENE
121-75-5	MALMED
105-56-6	MALONIC ACID ETHYL ESTER NITRILE
372-09-8	MALONIC MONONITRILE
109-77-3	MALONONITRILE
2698-41-1	MALONONITRILE, (O-CHLOROBENZYLIDENE)-
121-75-5	MALPHOS
121-75-5	MALTOX
121-75-5	MALTOX MLT
12427-38-2	MANEB
7439-96-5	MANGANESE
10377-66-9	MANGANESE (II) NITRATE
10377-66-9	MANGANESE (II) NITRATE, ANHYDROUS
7785-87-7	MANGANESE (II) SULFATE (1:1)
12079-65-1	MANGANESE CYCLOPENTADIENYL TRICARBONYL
1313-13-9	MANGANESE DIOXIDE
10377-66-9	MANGANESE NITRATE
10377-66-9	MANGANESE NITRATE (DOT)
10377-66-9	MANGANESE NITRATE (Mn(NO3)2)
9008-34-8	MANGANESE RESINATE
9008-34-8	MANGANESE RESINATE (DOT)
12108-13-3	MANGANESE TRICARBONYL METHYLCYCLOPENTADIENYL
1317-34-6	MANGANESE TRIOXIDE
10377-66-9	MANGANESE(2+) NITRATE
12079-65-1	MANGANESE, TRICARBONYL(.ETA.5-2,4-CYCLOPENTADIEN-1-YL)-
12108-13-3	MANGANESE, TRICARBONYL(METHYL-PI-CYCLOPENTADIENYL)-
12108-13-3	MANGANESE, TRICARBONYL(METHYLCYCLOPENTADIENYL)-
12079-65-1	MANGANESE, TRICARBONYL-PI-CYCLOPENTADIENYL-
12108-13-3	MANGANESE, TRICARBONYL[(1,2,3,4,5-ETA)-1-METHYL-2,4-CYCLOPENTADIEN-1-YL]-
7439-96-5	MANGANESE-55
10377-66-9	MANGANOUS DINITRATE
10377-66-9	MANGANOUS NITRATE
15825-70-4	MANNITOL HEXANITRATE
7631-86-9	MANOSIL VN 3
577-11-7	MANOXOL OP
577-11-7	MANOXOL OT
104-15-4	MANRO PTSA 65 E
25231-46-3	MANRO PTSA 65 E

CAS No.	Chemical Name
104-15-4	MANRO PTSA 65 H
25231-46-3	MANRO PTSA 65 H
104-15-4	MANRO PTSA 65 LS
25231-46-3	MANRO PTSA 65 LS
68476-26-6	MANUFACTURED GAS
68476-26-6	MANUFACTURED FUEL GASES
105-30-6	MAOH
108-11-2	MAOH
54972-97-3	MAOH
1309-37-1	MAPICO RED 347
1309-37-1	MAPICO RED R 220-3
3564-09-8	MAPLE PONCEAU 3R
59355-75-8	MAPP
59355-75-8	MAPPGAS
151-21-3	MAPROFIX 563
151-21-3	MAPROFIX LK
151-21-3	MAPROFIX NEU
2235-54-3	MAPROFIX NH
139-96-8	MAPROFIX TLS
139-96-8	MAPROFIX TLS 500
139-96-8	MAPROFIX TLS 65
151-21-3	MAPROFIX WAC
151-21-3	MAPROFIX WAC-LA
81-81-2	MAR-FRIN
72-43-5	MARALATE
25155-30-0	MARANIL
9005-25-8	MARANTA
105-60-2	MARANYL F 114
105-60-2	MARANYL F 124
105-60-2	MARANYL F 500
9016-45-9	MARCHON
129-06-6	MAREVAN
129-06-6	MAREVAN (SODIUM SALT)
1344-95-2	MARIMET 45
82-68-8	MARISAN FORTE
72-43-5	MARLATE
577-11-7	MARLINAT DF 8
25155-30-0	MARLON 375
25155-30-0	MARLON A
25155-30-0	MARLON A 350
25155-30-0	MARLON A 375
25155-30-0	MARLON A 396
27176-87-0	MARLON AS 3
27176-87-0	MARLON AS B
9016-45-9	MARLOPHEN
9016-45-9	MARLOPHEN 810
9016-45-9	MARLOPHEN 812
9016-45-9	MARLOPHEN 85
9016-45-9	MARLOPHEN 88
9016-45-9	MARLOPHEN 89
1309-48-4	MARMAG
330-54-1	MARMER
51-28-5	MAROXOL-50
74-82-8	MARSH GAS
14808-60-7	MARSHALITE
81-81-2	MARTIN'S MAR-FRIN
1344-28-1	MARTOXIN
2636-26-2	MARVEX
7631-86-9	MAS 200
52-68-6	MASOTEN
8020-83-5	MASROLAR D
1317-36-8	MASSICOT
1317-36-8	MASSICOTITE
56-75-7	MASTIPHEN
76-22-2	MATRICARIA CAMPHOR
7664-93-9	MATTING ACID (DOT)
366-70-1	MATULANE
81-81-2	MAVERAN
121-75-5	MAVIDAN
115-77-5	MAXINUTRIL
8001-30-7	MAYDOL
50-55-5	MAYSERPINE
8001-30-7	MAZOLA OIL
52-68-6	MAZOTEN
25155-30-0	MB-VR
51-75-2	MBA
55-86-7	MBA HYDROCHLORIDE
17804-35-2	MBC
8004-09-9	MBC 33

CAS No.	Chemical Name
21609-90-5	MBCP
366-70-1	MBH
101-68-8	MBI
591-78-6	MBK
115-19-5	MBY
7487-94-7	MC
24934-91-6	MC 2188
10326-21-3	MC DEFOLIANT
79-11-8	MCA
108-90-7	MCB
77-81-6	MCE
79-22-1	MCF
1333-86-4	MCF 88
1333-86-4	MCF-HS 78
1333-86-4	MCF-LS 74
7439-98-7	MCHVL
991-42-4	MCN 1025
937-14-4	MCPBA
101-77-9	MDA
1918-00-9	MDBA
105-59-9	MDEA
101-68-8	MDI
9016-87-9	MDI-CR
9016-87-9	MDI-CR 100
9016-87-9	MDI-CR 200
9016-87-9	MDI-CR 300
101-68-8	MDT
72-54-8	ME 1700
298-00-0	ME-PARATHION
141-43-5	MEA
141-43-5	MEA (ALCOHOL)
12002-03-8	MEADOW GREEN
50-78-2	MEASURIN
1332-58-7	MECA
111-77-3	MECB
150-76-5	MECHINOLUM
51-75-2	MECHLORETHAMINE
55-86-7	MECHLORETHAMINE HYDROCHLORIDE
302-70-5	MECHLORETHAMINE OXIDE HYDROCHLORIDE
137-05-3	MECRILAT
137-05-3	MECRYLATE
109-86-4	MECS
110-49-6	MECSAC
10124-56-8	MEDI-CALGON
1918-00-9	MEDIBEN
139-91-3	MEDIFURAN
50-78-2	MEDISYL
78-93-3	MEETCO
640-19-7	MEGATOX
78-93-3	MEK
1338-23-4	MEK PEROXIDE
1338-23-4	MEKP
1338-23-4	MEKPO
108-78-1	MELAMINE
151-21-3	MELANOL CL
151-21-3	MELANOL CL 30
139-96-8	MELANOL LP 20 T
56-72-4	MELDANE
56-72-4	MELDONE
88-89-1	MELINITE
3771-19-5	MELIPAN
8001-35-2	MELIPAX
9005-25-8	MELOGEL
148-82-3	MELPHALAN
9005-25-8	MELUNA
502-39-6	MEMA
25134-21-8	MEMTETRAHYDROPHTHALIC ANHYDRIDE
53-16-7	MENAGEN
63-25-2	MENAPHTAM
556-61-6	MENCS
72-20-8	MENDRIN
53-16-7	MENFORMON
7786-34-7	MENIPHOS
7786-34-7	MENITE
57-63-6	MENOLYN
56-53-1	MENOSTILBEEN
80-47-7	MENTHANE HYDROPEROXIDE, PARA
333-41-5	MEODINON
104-90-5	MEP

CAS No.	Chemical Name
122-14-5	MEP
122-14-5	MEP (PESTICIDE)
298-00-0	MEPATON
950-10-7	MEPHOSFOLAN
84-80-0	MEPHYTON
1338-23-4	MEPOX
57-53-4	MEPROBAMATE
298-00-0	MEPTOX
150-76-5	MEQUINOL
62-38-4	MERACEN
50-76-0	MERACTINOMYCIN
110-66-7	MERCAPTAN AMYLIQUE (FRENCH)
68-11-1	MERCAPTOACETATE
68-11-1	MERCAPTOACETIC ACID
108-98-5	MERCAPTOBENZENE
2032-65-7	MERCAPTODIMETHUR
2032-65-7	MERCAPTODIMETHUR (DOT)
2032-65-7	MERCAPTODIMETHUR (METHIOCARB)
75-08-1	MERCAPTOETHANE
60-24-2	MERCAPTOETHANOL
55-38-9	MERCAPTOFOS
8065-48-3	MERCAPTOFOS
96-45-7	MERCAPTOIMIDAZOLINE
74-93-1	MERCAPTOMETHANE
55-38-9	MERCAPTOPHOS
8065-48-3	MERCAPTOPHOS
79-42-5	MERCAPTOPROPIONIC ACID
121-75-5	MERCAPTOSUCCINIC ACID DIETHYL ESTER
121-75-5	MERCAPTOTHION
121-75-5	MERCAPTOTION (SPANISH)
96-45-7	MERCAZIN I
25155-30-0	MERCOL 25
25155-30-0	MERCOL 30
62-38-4	MERCRON
137-26-8	MERCURAM
2235-25-8	MERCURATE(2-), ETHYL[PHOSPHATO(3-)-O]-, DIHYDROGEN
7439-97-6	MERCURE (FRENCH)
1600-27-7	MERCURIACETATE
74-89-5	MERCURIALIN
1600-27-7	MERCURIC ACETATE
1600-27-7	MERCURIC ACETATE (DOT)
10124-48-8	MERCURIC AMIDOCHLORIDE
10124-48-8	MERCURIC AMMONIUM CHLORIDE
10124-48-8	MERCURIC AMMONIUM CHLORIDE, SOLID (DOT)
7784-37-4	MERCURIC ARSENATE
7784-37-4	MERCURIC ARSENATE (DOT)
1312-03-4	MERCURIC BASIC SULFATE
583-15-3	MERCURIC BENZOATE
583-15-3	MERCURIC BENZOATE, SOLID
583-15-3	MERCURIC BENZOATE, SOLID (DOT)
7487-94-7	MERCURIC BICHLORIDE
7789-47-1	MERCURIC BROMIDE
7789-47-1	MERCURIC BROMIDE, SOLID
7789-47-1	MERCURIC BROMIDE, SOLID (DOT)
7487-94-7	MERCURIC CHLORIDE
7487-94-7	MERCURIC CHLORIDE (DOT)
10124-48-8	MERCURIC CHLORIDE, AMMONIATED
7487-94-7	MERCURIC CHLORIDE, SOLID
7487-94-7	MERCURIC CHLORIDE, SOLID (DOT)
592-04-1	MERCURIC CYANIDE
592-04-1	MERCURIC CYANIDE (DOT)
592-04-1	MERCURIC CYANIDE, SOLID
592-04-1	MERCURIC CYANIDE, SOLID (DOT)
1600-27-7	MERCURIC DIACETATE
7789-47-1	MERCURIC DIBROMIDE
7774-29-0	MERCURIC DIIODIDE
7774-29-0	MERCURIC IODIDE
7774-29-0	MERCURIC IODIDE, RED
7774-29-0	MERCURIC IODIDE, SOLID
7774-29-0	MERCURIC IODIDE, SOLID (DOT)
7774-29-0	MERCURIC IODIDE, SOLUTION
7774-29-0	MERCURIC IODIDE, SOLUTION (DOT)
10045-94-0	MERCURIC NITRATE
10045-94-0	MERCURIC NITRATE (DOT)
1191-80-6	MERCURIC OLEATE
1191-80-6	MERCURIC OLEATE, SOLID
1191-80-6	MERCURIC OLEATE, SOLID (DOT)
21908-53-2	MERCURIC OXIDE
21908-53-2	MERCURIC OXIDE (HgO)
1335-31-5	MERCURIC OXYCYANIDE
591-89-9	MERCURIC POTASSIUM CYANIDE
5970-32-1	MERCURIC SALICYLATE
10415-73-3	MERCURIC SALICYLATE
5970-32-1	MERCURIC SALICYLATE, SOLID
10415-73-3	MERCURIC SALICYLATE, SOLID
5970-32-1	MERCURIC SALICYLATE, SOLID (DOT)
10415-73-3	MERCURIC SALICYLATE, SOLID (DOT)
1312-03-4	MERCURIC SUBSULFATE
1312-03-4	MERCURIC SUBSULFATE, SOLID
1312-03-4	MERCURIC SUBSULFATE, SOLID (DOT)
7783-35-9	MERCURIC SULFATE
7783-35-9	MERCURIC SULFATE, SOLID
7783-35-9	MERCURIC SULFATE, SOLID (DOT)
1344-48-5	MERCURIC SULFIDE
1344-48-5	MERCURIC SULFIDE RED
1344-48-5	MERCURIC SULFIDE, BLACK
592-85-8	MERCURIC SULFOCYANATE
592-85-8	MERCURIC SULFOCYANATE, SOLID
592-85-8	MERCURIC SULFOCYANATE, SOLID (DOT)
592-85-8	MERCURIC SULFOCYANIDE
7783-35-9	MERCURIC SULPHATE (DOT)
592-85-8	MERCURIC THIOCYANATE
592-85-8	MERCURIC THIOCYANATE, SOLID (DOT)
1600-27-7	MERCURIC(II) ACETATE
7439-97-6	MERCURIO (ITALIAN)
62-38-4	MERCURIPHENYL ACETATE
5970-32-1	MERCURISALICYLIC ACID
10415-73-3	MERCURISALICYLIC ACID
7546-30-7	MERCUROCHLORIDE (DUTCH)
12002-19-6	MERCUROL
12002-19-6	MERCUROL (DOT)
12002-19-6	MERCUROL, SOLID
62-38-4	MERCURON
631-60-7	MERCUROUS ACETATE
631-60-7	MERCUROUS ACETATE (DOT)
631-60-7	MERCUROUS ACETATE, SOLID
631-60-7	MERCUROUS ACETATE, SOLID (DOT)
15385-58-7	MERCUROUS BROMIDE
10031-18-2	MERCUROUS BROMIDE
10031-18-2	MERCUROUS BROMIDE (DOT)
15385-58-7	MERCUROUS BROMIDE (DOT)
15385-58-7	MERCUROUS BROMIDE (Hg2Br2)
10031-18-2	MERCUROUS BROMIDE, SOLID
15385-58-7	MERCUROUS BROMIDE, SOLID
10031-18-2	MERCUROUS BROMIDE, SOLID (DOT)
15385-58-7	MERCUROUS BROMIDE, SOLID (DOT)
7546-30-7	MERCUROUS CHLORIDE
10112-91-1	MERCUROUS CHLORIDE
7546-30-7	MERCUROUS CHLORIDE (HgCl)
63937-14-4	MERCUROUS GLUCONATE
63937-14-4	MERCUROUS GLUCONATE, SOLID
63937-14-4	MERCUROUS GLUCONATE, SOLID (DOT)
7783-30-4	MERCUROUS IODIDE
15385-57-6	MERCUROUS IODIDE
7783-30-4	MERCUROUS IODIDE, SOLID
15385-57-6	MERCUROUS IODIDE, SOLID
7783-30-4	MERCUROUS IODIDE, SOLID (DOT)
15385-57-6	MERCUROUS IODIDE, SOLID (DOT)
10415-75-5	MERCUROUS NITRATE
10415-75-5	MERCUROUS NITRATE (DOT)
7782-86-7	MERCUROUS NITRATE MONOHYDRATE
10415-75-5	MERCUROUS NITRATE, SOLID
10415-75-5	MERCUROUS NITRATE, SOLID (DOT)
15829-53-5	MERCUROUS OXIDE
15829-53-5	MERCUROUS OXIDE (Hg2O)
15829-53-5	MERCUROUS OXIDE, BLACK, SOLID
15829-53-5	MERCUROUS OXIDE, BLACK, SOLID (DOT)
7783-36-0	MERCUROUS SULFATE
7783-36-0	MERCUROUS SULFATE, SOLID
7783-36-0	MERCUROUS SULFATE, SOLID (DOT)
7439-97-6	MERCURY
7439-97-6	MERCURY (ACGIH)
62-38-4	MERCURY (II) ACETATE, PHENYL-
1600-27-7	MERCURY ACETATE
1600-27-7	MERCURY ACETATE (DOT)
1600-27-7	MERCURY ACETATE (Hg(O2C2H3)2)
631-60-7	MERCURY ACETATE (HgOAC)
11110-52-4	MERCURY ALLOY, BASE, Hg,Na

CAS No.	Chemical Name
10124-48-8	MERCURY AMIDE CHLORIDE
10124-48-8	MERCURY AMIDE CHLORIDE (Hg(NH2)Cl)
10124-48-8	MERCURY AMINE CHLORIDE
10124-48-8	MERCURY AMMONIATED
10124-48-8	MERCURY AMMONIUM CHLORIDE
10124-48-8	MERCURY AMMONIUM CHLORIDE (DOT)
583-15-3	MERCURY BENZOATE
7487-94-7	MERCURY BICHLORIDE
7774-29-0	MERCURY BINIODIDE
7783-35-9	MERCURY BISULFATE
7783-35-9	MERCURY BISULPHATE (DOT)
10031-18-2	MERCURY BROMIDE
15385-58-7	MERCURY BROMIDE
15385-58-7	MERCURY BROMIDE (Hg2Br2)
10031-18-2	MERCURY BROMIDE (HgBr)
7789-47-1	MERCURY BROMIDE (HgBr2)
10112-91-1	MERCURY CHLORIDE
7546-30-7	MERCURY CHLORIDE (HgCl)
7487-94-7	MERCURY CHLORIDE (HgCl2)
592-04-1	MERCURY CYANIDE (Hg(CN)2)
1600-27-7	MERCURY DIACETATE
7789-47-1	MERCURY DIBROMIDE
7487-94-7	MERCURY DICHLORIDE
592-04-1	MERCURY DICYANIDE
7774-29-0	MERCURY DIIODIDE
10045-94-0	MERCURY DINITRATE
592-85-8	MERCURY DITHIOCYANATE
7439-97-6	MERCURY ELEMENT
628-86-4	MERCURY FULMINATE
63937-14-4	MERCURY GLUCONATE
63937-14-4	MERCURY GLUCONATE (DOT)
15385-57-6	MERCURY IODIDE (Hg2I2)
7783-30-4	MERCURY IODIDE (HgI)
7774-29-0	MERCURY IODIDE (HgI2)
631-60-7	MERCURY MONOACETATE
10031-18-2	MERCURY MONOBROMIDE
15385-58-7	MERCURY MONOBROMIDE
7546-30-7	MERCURY MONOCHLORIDE
7783-30-4	MERCURY MONOIODIDE
1344-48-5	MERCURY MONOSULFIDE
10045-94-0	MERCURY NITRATE
7782-86-7	MERCURY NITRATE (Hg(NO3)) MONOHYDRATE
10045-94-0	MERCURY NITRATE (Hg(NO3)2)
10415-75-5	MERCURY NITRATE (Hg2(NO3)2)
10415-75-5	MERCURY NITRATE (HgNO3)
12002-19-6	MERCURY NUCLEATE
12002-19-6	MERCURY NUCLEATE, SOLID (DOT)
1191-80-6	MERCURY OLEATE
15829-53-5	MERCURY OXIDE
15829-53-5	MERCURY OXIDE (Hg2O)
21908-53-2	MERCURY OXIDE (HgO)
15829-53-5	MERCURY OXIDE BLACK
1312-03-4	MERCURY OXIDE SULFATE
1312-03-4	MERCURY OXIDE SULFATE (Hg3O2(SO4))
1312-03-4	MERCURY OXONIUM SULFATE
7487-94-7	MERCURY PERCHLORIDE
10045-94-0	MERCURY PERNITRATE
7783-35-9	MERCURY PERSULFATE
7783-33-7	MERCURY POTASSIUM IODIDE
7546-30-7	MERCURY PROTOCHLORIDE
15385-57-6	MERCURY PROTOIODIDE
7782-86-7	MERCURY PROTONITRATE
5970-32-1	MERCURY SALICYLATE
10415-73-3	MERCURY SALICYLATE
5970-32-1	MERCURY SALICYLATE (DOT)
10415-73-3	MERCURY SALICYLATE (DOT)
5970-32-1	MERCURY SUBSALICYLATE
10415-73-3	MERCURY SUBSALICYLATE
7783-36-0	MERCURY SULFATE (Hg2SO4)
7783-35-9	MERCURY SULFATE (HgSO4)
1344-48-5	MERCURY SULFIDE (HgS)
1344-48-5	MERCURY SULPHIDE
53408-91-6	MERCURY THIOCYANATE
592-85-8	MERCURY THIOCYANATE (DOT)
592-85-8	MERCURY THIOCYANATE (Hg(SCN)2)
15385-58-7	MERCURY(1)BROMIDE(1:1)
10031-18-2	MERCURY(1+) BROMIDE
7546-30-7	MERCURY(1+) CHLORIDE
7783-30-4	MERCURY(1+) IODIDE
15385-57-6	MERCURY(1+) IODIDE
1600-27-7	MERCURY(2+) ACETATE
10045-94-0	MERCURY(2+) NITRATE
1344-48-5	MERCURY(2+) SULFIDE
631-60-7	MERCURY(I) ACETATE
10031-18-2	MERCURY(I) BROMIDE (1:1)
15385-58-7	MERCURY(I) BROMIDE (1:1)
7546-30-7	MERCURY(I) CHLORIDE
63937-14-4	MERCURY(I) GLUCONATE
7783-30-4	MERCURY(I) IODIDE
10415-75-5	MERCURY(I) NITRATE
10415-75-5	MERCURY(I) NITRATE (1:1)
15829-53-5	MERCURY(I) OXIDE
7783-36-0	MERCURY(I) SULFATE
1600-27-7	MERCURY(II) ACETATE
583-15-3	MERCURY(II) BENZOATE
7789-47-1	MERCURY(II) BROMIDE
7789-47-1	MERCURY(II) BROMIDE (1:2)
7487-94-7	MERCURY(II) CHLORIDE
592-04-1	MERCURY(II) CYANIDE
1600-27-7	MERCURY(II) DIACETATE
7774-29-0	MERCURY(II) IODIDE
10045-94-0	MERCURY(II) NITRATE
10045-94-0	MERCURY(II) NITRATE (1:2)
7784-37-4	MERCURY(II) O-ARSENATE
7783-35-9	MERCURY(II) SULFATE
7783-35-9	MERCURY(II) SULFATE (1:1)
592-85-8	MERCURY(II) THIOCYANATE
502-39-6	MERCURY, (3-CYANOGUANIDINO)METHYL-
62-38-4	MERCURY, (ACETATO)PHENYL-
62-38-4	MERCURY, (ACETATO-O)PHENYL-
502-39-6	MERCURY, (CYANOGUANIDINATO) METHYL-
502-39-6	MERCURY, (CYANOGUANIDINATO-N')METHYL-
502-39-6	MERCURY, (CYANOGUANIDINE)METHYL-
63937-14-4	MERCURY, (D-GLUCONATO)-
2235-25-8	MERCURY, (DIHYDROGEN PHOSPHATO)ETHYL-
2235-25-8	MERCURY, (HYDROGEN PHOSPHATO)BIS(ETHYL-
5970-32-1	MERCURY, (SALICYLATO(2-))-
10415-73-3	MERCURY, (SALICYLATO(2-))-
62-38-4	MERCURY, ACETOXYPHENYL-
10124-48-8	MERCURY, AMMONOBASIC (HgNH2Cl)
592-85-8	MERCURY, BIS(THIOCYANATO)-
7439-97-6	MERCURY, METALLIC
7439-97-6	MERCURY, METALLIC (DOT)
5970-32-1	MERCURY, [2-HYDROXYBENZOATO(2-)-O1,O2]-
10415-73-3	MERCURY, [2-HYDROXYBENZOATO(2-)-O1,O2]-
5970-32-1	MERCURY, [SALICYLATO(2-)]-
10415-73-3	MERCURY, [SALICYLATO(2-)]-
1600-27-7	MERCURYL ACETATE
143-50-0	MEREX
62-38-4	MERGAL A 25
2163-80-6	MERGE
2163-80-6	MERGE 823
9016-45-9	MERGITAL OP 2
9016-45-9	MERITEN NF 9
13171-21-6	MERKON
35400-43-2	MERPAFOS
133-06-2	MERPAN
531-76-0	MERPHALAN
25155-30-0	MERPISAP AP 90P
9016-45-9	MERPOXEN 230
9016-45-9	MERPOXEN ON
7440-66-6	MERRILLITE
62-38-4	MERSOLITE
62-38-4	MERSOLITE 8
62-38-4	MERSOLITE D
58-22-0	MERTESTATE
577-11-7	MERVAMINE
2163-80-6	MESAMATE
2163-80-6	MESAMATE CONCENTRATE
2163-80-6	MESAMATE HC
2163-80-6	MESAMATE-400
2163-80-6	MESAMATE-600
88-05-1	MESIDIN
88-05-1	MESIDIN (CZECH)
88-05-1	MESIDINE
141-79-7	MESITYL OXIDE
141-79-7	MESITYL OXIDE (ACGIH,DOT)
88-05-1	MESITYLAMINE

CAS No.	Chemical Name
108-67-8	MESITYLENE
88-05-1	MESITYLENE, 2-AMINO-
141-79-7	MESITYLOXID (GERMAN)
141-79-7	MESITYLOXYDE (DUTCH)
16752-77-5	MESOMILE
72-33-3	MESTRANOL
2032-65-7	MESUROL
2032-65-7	MESUROL PHENOL
558-25-8	MESYL FLUORIDE
7789-75-5	MET-SPAR
108-62-3	META
9002-91-9	META
1937-37-7	META BLACK
937-14-4	META-CHLOROPERBENZOIC ACID
108-44-1	META-TOLUIDINE
91-08-7	META-TOLYLENE DIISOCYANATE
26471-62-5	META-TOLYLENE DIISOCYANATE
115-77-5	METAB-AUXIL
53-86-1	METACEN
108-62-3	METACETALDEHYDE
9002-91-9	METACETALDEHYDE
96-22-0	METACETONE
79-09-4	METACETONIC ACID
121-73-3	METACHLORONITROBENZENE
298-00-0	METACID 50
298-00-0	METACIDE
298-00-0	METACIDE (INSECTICIDE)
56-04-2	METACIL
1129-41-5	METACRATE
14464-46-1	METACRISTOBALITE
541-73-1	METADICHLOROBENZENE
86290-81-5	METAFORMING
298-00-0	METAFOS
298-00-0	METAFOS (PESTICIDE)
919-86-8	METAISOSEPTOX
919-86-8	METAISOSYSTOX
80-62-6	METAKRYLAN METYLU (POLISH)
108-62-3	METALDEHYD (GERMAN)
9002-91-9	METALDEHYD (GERMAN)
108-62-3	METALDEHYDE
9002-91-9	METALDEHYDE
108-62-3	METALDEHYDE (DOT)
9002-91-9	METALDEHYDE (DOT)
108-62-3	METALDEIDE (ITALIAN)
9002-91-9	METALDEIDE (ITALIAN)
557-05-1	METALLAC
7440-38-2	METALLIC ARSENIC
7439-97-6	METALLIC MERCURY
7440-31-5	METALLIC TIN
105-60-2	METAMID
10265-92-6	METAMIDOFOS ESTRELLA
10265-92-6	METAMIDOPHOS
7429-90-5	METANA
1333-86-4	METANEX D
67-56-1	METANOLO (ITALIAN)
108-45-2	METAPHENYLENEDIAMINE
298-00-0	METAPHOR
298-00-0	METAPHOS
7785-84-4	METAPHOSPHORIC ACID (H3P3O9), TRISODIUM SALT
10124-56-8	METAPHOSPHORIC ACID (H6P6O18), HEXASODIUM SALT
91-80-5	METAPYRILENE
60-00-4	METAQUEST A
53-86-1	METARTRIL
557-05-1	METASAP 576
62-38-4	METASOL 30
108-62-3	METASON
9002-91-9	METASON
8022-00-2	METASYSTOX
919-86-8	METASYSTOX (I)
919-86-8	METASYSTOX 55
8022-00-2	METASYSTOX FORTE
919-86-8	METASYSTOX I
919-86-8	METASYSTOX J
122-14-5	METATHION
122-14-5	METATHION E 50
122-14-5	METATHIONE
122-14-5	METATHIONINE E 50
122-14-5	METATION
122-14-5	METATION E 50
112-80-1	METAUPON
13718-26-8	METAWANADANEM SODOWYM (POLISH)
11105-06-9	METAWANADANEM SODOWYM (POLISH)
13721-39-6	METAWANADANEM SODOWYM (POLISH)
12001-29-5	METAXITE
96-22-0	METHACETONE
108-88-3	METHACIDE
56-04-2	METHACIL
78-85-3	METHACRALDEHYDE (DOT)
78-85-3	METHACROLEIN
10476-95-6	METHACROLEIN DIACETATE
513-42-8	METHACRYL ALCOHOL
78-85-3	METHACRYLALDEHYDE
78-85-3	METHACRYLALDEHYDE (DOT)
97-88-1	METHACRYLATE DE BUTYLE (FRENCH)
80-62-6	METHACRYLATE DE METHYLE (FRENCH)
79-41-4	METHACRYLIC ACID
79-41-4	METHACRYLIC ACID (ACGIH)
760-93-0	METHACRYLIC ACID ANHYDRIDE
30674-80-7	METHACRYLIC ACID BETA-ISOCYANATOETHYL ESTER
80-62-6	METHACRYLIC ACID METHYL ESTER
106-91-2	METHACRYLIC ACID, 2,3-EPOXYPROPYL ESTER
2867-47-2	METHACRYLIC ACID, 2-(DIMETHYLAMINO)ETHYL ESTER
3775-90-4	METHACRYLIC ACID, 2-(TERT-BUTYLAMINO)ETHYLESTER
30674-80-7	METHACRYLIC ACID, 2-ISOCYANATOETHYL ESTER
97-88-1	METHACRYLIC ACID, BUTYL ESTER
27813-02-1	METHACRYLIC ACID, ESTER WITH 1,2-PROPANEDIOL
97-63-2	METHACRYLIC ACID, ETHYL ESTER
79-41-4	METHACRYLIC ACID, INHIBITED (DOT)
97-86-9	METHACRYLIC ACID, ISOBUTYL ESTER
27813-02-1	METHACRYLIC ACID, MONOESTER WITH 1,2-PROPANEDIOL
78-85-3	METHACRYLIC ALDEHYDE
760-93-0	METHACRYLIC ANHYDRIDE
126-98-7	METHACRYLNITRILE
126-98-7	METHACRYLONITRILE
760-93-0	METHACRYLOYL ANHYDRIDE
920-46-7	METHACRYLOYL CHLORIDE
30674-80-7	METHACRYLOYLOXYETHYL ISOCYANATE
97-88-1	METHACRYLSAEUREBUTYLESTER (GERMAN)
80-62-6	METHACRYLSAEUREMETHYL ESTER (GERMAN)
122-14-5	METHADION
50-00-0	METHALDEHYDE
513-42-8	METHALLYL ALCOHOL
513-42-8	METHALLYL ALCOHOL (DOT)
563-47-3	METHALLYL CHLORIDE
10265-92-6	METHAMIDOPHOS
10265-92-6	METHAMIDOPHUS
100-97-0	METHAMIN
50-00-0	METHANAL
75-12-7	METHANAMIDE
74-89-5	METHANAMINE
74-89-5	METHANAMINE (9CI)
75-50-3	METHANAMINE, N,N-DIMETHYL-
6369-96-6	METHANAMINE, N,N-DIMETHYL-, (2,4,5-TRICHLOROPHENOXY)-ACETATE
124-40-3	METHANAMINE, N-METHYL-
124-40-3	METHANAMINE, N-METHYL- (9CI)
6369-97-7	METHANAMINE, N-METHYL-, (2,4,5-TRICHLOROPHENOXY)ACETATE
62-75-9	METHANAMINE, N-METHYL-N-NITROSO-
75-59-2	METHANAMINIUM, N,N,N-TRIMETHYL-, HYDROXIDE
74-82-8	METHANE
74-82-8	METHANE (DOT)
75-09-2	METHANE DICHLORIDE
558-13-4	METHANE TETRABROMIDE
56-23-5	METHANE TETRACHLORIDE
115-77-5	METHANE TETRAMETHYLOL
67-66-3	METHANE TRICHLORIDE
111-91-1	METHANE, BIS(2-CHLOROETHOXY)-
74-83-9	METHANE, BROMO-
74-83-9	METHANE, BROMO-, LIQUID, INCLUDING UP TO 2% CHLOROPICRIN
74-97-5	METHANE, BROMOCHLORO-
353-59-3	METHANE, BROMOCHLORODIFLUORO-
75-27-4	METHANE, BROMODICHLORO-
776-74-9	METHANE, BROMODIPHENYL-
75-63-8	METHANE, BROMOTRIFLUORO-
74-87-3	METHANE, CHLORO-
75-45-6	METHANE, CHLORODIFLUORO-

CAS No.	*Chemical Name*
107-30-2	METHANE, CHLOROMETHOXY-
75-72-9	METHANE, CHLOROTRIFLUORO-
74-82-8	METHANE, COMPRESSED (DOT)
75-05-8	METHANE, CYANO-
334-88-3	METHANE, DIAZO-
124-48-1	METHANE, DIBROMOCHLORO-
75-61-6	METHANE, DIBROMODIFLUORO-
75-09-2	METHANE, DICHLORO-
75-71-8	METHANE, DICHLORODIFLUORO-
75-43-4	METHANE, DICHLOROFLUORO-
462-95-3	METHANE, DIETHOXY-
109-87-5	METHANE, DIMETHOXY-
86-73-7	METHANE, DIPHENYLENE-
540-67-0	METHANE, ETHOXY-
593-53-3	METHANE, FLUORO-
75-69-4	METHANE, FLUOROTRICHLORO-
74-88-4	METHANE, IODO-
624-83-9	METHANE, ISOCYANATO-
556-61-6	METHANE, ISOTHIOCYANATO-
75-52-5	METHANE, NITRO-
115-10-6	METHANE, OXYBIS-
353-42-4	METHANE, OXYBIS-, BORON COMPLEX
353-42-4	METHANE, OXYBIS-, COMPD WITH TRIFLUOROBORANE (1:1)
542-88-1	METHANE, OXYBIS[CHLORO-
108-88-3	METHANE, PHENYL-
74-82-8	METHANE, REFRIGERATED LIQUID (DOT)
67-68-5	METHANE, SULFINYLBIS-
558-13-4	METHANE, TETRABROMO-
56-23-5	METHANE, TETRACHLORO-
75-73-0	METHANE, TETRAFLUORO-
509-14-8	METHANE, TETRANITRO-
75-18-3	METHANE, THIOBIS-
556-64-9	METHANE, THIOCYANATO-
75-25-2	METHANE, TRIBROMO-
67-66-3	METHANE, TRICHLORO-
75-69-4	METHANE, TRICHLOROFLUORO-
76-06-2	METHANE, TRICHLORONITRO-
76-06-2	METHANE, TRICHLORONITRO-, (FLAMMABLE MIXTURE)
76-06-2	METHANE, TRICHLORONITRO-, (MIXTURE)
8004-09-9	METHANE, TRICHLORONITRO-, MIXT WITH BROMOMETHANE
122-51-0	METHANE, TRIETHOXY-
75-46-7	METHANE, TRIFLUORO-
75-47-8	METHANE, TRIIODO-
2163-80-6	METHANEARSONIC ACID, MONOSODIUM SALT
2163-80-6	METHANEARSONIC ACID, SODIUM SALT
75-05-8	METHANECARBONITRILE
507-09-5	METHANECARBOTHIOLIC ACID
64-19-7	METHANECARBOXYLIC ACID
676-97-1	METHANEPHOSPHONODICHLORIDIC ACID
676-98-2	METHANEPHOSPHONOTHIOIC DICHLORIDE
676-97-1	METHANEPHOSPHONYL CHLORIDE
594-42-3	METHANESULFENYL CHLORIDE, TRICHLORO-
62-50-0	METHANESULFONIC ACID, ETHYL ESTER
66-27-3	METHANESULFONIC ACID, METHYL ESTER
558-25-8	METHANESULFONIC FLUORIDE
558-25-8	METHANESULFONYL FLUORIDE
558-25-8	METHANESULPHONYL FLUORIDE
74-93-1	METHANETHIOL
2275-14-1	METHANETHIOL, ((2,5-DICHLOROPHENYL)THIO)-, S-ESTER WITH O,O-DIETHYL PHOSPHORODITHIOATE
3735-23-7	METHANETHIOL, ((2,5-DICHLOROPHENYLTHIO)-, S-ESTER WITH O,O-DIMETHYL PHOSPHORODITHIOATE
298-02-2	METHANETHIOL, (ETHYLTHIO)-, S-ESTER WITH O,O-DIETHYL PHOSPHORODITHIOATE
13071-79-9	METHANETHIOL, (TERT-BUTYLTHIO)-, S-ESTER WITH O,O-DIETHYL PHOSPHORODITHIOATE
22259-30-9	METHANIMIDAMIDE, N,N-DIMETHYL-, N'-(3-(((METHYLAMINO)-CARBONYL)OXY)PHENYL)- (9CI)
22259-30-9	METHANIMIDAMIDE, N,N-DIMETHYL-N'-[3-[[(METHYLAMINO)-CARBONYL]OXY]PHENYL]-
23422-53-9	METHANIMIDAMIDE, N,N-DIMETHYL-N'-[3-[[(METHYLAMINO)-CARBONYL]OXY]PHENYL]-, MONOHYDROCHLORIDE
55738-54-0	METHANIMIDAMIDE, N,N-DIMETHYL-N'-[5-[2-(5-NITRO-2-FURANYL)ETHENYL]-1,3,4-OXADIAZOL-2-YL]-,
55738-54-0	METHANIMIDAMIDE, N,N-DIMETHYL-N'-[5-[2-(5-NITRO-2-FURANYL)ETHENYL]-1,3,4-OXADIAZOL-2-YL]-, (E)
64-18-6	METHANOIC ACID
109-94-4	METHANOIC ACID ETHYL ESTER
107-31-3	METHANOIC ACID METHYL ESTER

CAS No.	*Chemical Name*
67-56-1	METHANOL
67-56-1	METHANOL (DOT)
98-00-0	METHANOL, (2-FURYL)-
590-96-5	METHANOL, (METHYL-ONN-AZOXY)-
592-62-1	METHANOL, (METHYL-ONN-AZOXY)-, ACETATE (ESTER)
590-96-5	METHANOL, (METHYLAZOXY)-
592-62-1	METHANOL, (METHYLAZOXY)-, ACETATE
513-86-0	METHANOL, ACETYLMETHYL-
107-19-7	METHANOL, ETHYNYL-
98-85-1	METHANOL, METHYLPHENYL-
100-51-6	METHANOL, PHENYL-
124-41-4	METHANOL, SODIUM SALT
75-65-0	METHANOL, TRIMETHYL-
794-93-4	METHANOL, [[6-[2-(5-NITRO-2-FURANYL)ETHENYL]-1,2,4-TRIAZIN-3-YL]IMINO]BIS-
794-93-4	METHANOL, [[6-[2-(5-NITRO-2-FURYL)VINYL]-AS-TRIAZIN-3-YL] IMINO]DI-
109-78-4	METHANOLACETONITRILE
90-94-8	METHANONE, BIS[4-(DIMETHYLAMINO)PHENYL]-
119-61-9	METHANONE, DIPHENYL-
91-80-5	METHAPYRILENE
53-86-1	METHAZINE
100-97-0	METHENAMIN
100-97-0	METHENAMINE
75-25-2	METHENYL TRIBROMIDE
67-66-3	METHENYL TRICHLORIDE
56-04-2	METHIACIL
56-04-2	METHICIL
950-37-8	METHIDATHION
60-56-0	METHIMAZOLE
2032-65-7	METHIOCARB
56-04-2	METHIOCIL
3268-49-3	METHIONAL
16752-77-5	METHOMYL
16752-77-5	METHOMYL SP
59-05-2	METHOTREXATE
72-43-5	METHOXCIDE
72-43-5	METHOXO
298-81-7	METHOXSALEN WITH ULTRA-VIOLET A THERAPY
72-43-5	METHOXY-DDT
104-94-9	METHOXYANILINE
29191-52-4	METHOXYANILINE
100-66-3	METHOXYBENZENE
100-07-2	METHOXYBENZOYL CHLORIDE
79-22-1	METHOXYCARBONYL CHLORIDE
96-33-3	METHOXYCARBONYLETHYLENE
72-43-5	METHOXYCHLOR
72-43-5	METHOXYCHLOR (ACGIH,DOT)
107-30-2	METHOXYCHLOROMETHANE
111-77-3	METHOXYDIGLYCOL
540-67-0	METHOXYETHANE
109-86-4	METHOXYETHANOL
107-25-5	METHOXYETHENE
111-77-3	METHOXYETHOXYETHANOL
107-25-5	METHOXYETHYLENE
109-86-4	METHOXYETHYLENE GLYCOL
151-38-2	METHOXYETHYLMERCURIC ACETATE
76-38-0	METHOXYFLURANE
109-86-4	METHOXYHYDROXYETHANE
115-10-6	METHOXYMETHANE
107-30-2	METHOXYMETHYL CHLORIDE
6427-21-0	METHOXYMETHYL ISOCYANATE
109-87-5	METHOXYMETHYL METHYL ETHER
124-41-4	METHOXYSODIUM
1928-38-7	METHYL (2,4-DICHLOROPHENOXY)ACETATE
110-43-0	METHYL (N-AMYL)KETONE
137-05-3	METHYL ALPHA-CYANOACRYLATE
4435-53-4	METHYL 1,3-BUTYLENE GLYCOL ACETATE
17804-35-2	METHYL 1-(BUTYLCARBAMOYL)-2-BENZIMIDAZOLECARBAMATE
17804-35-2	METHYL 1-(BUTYLCARBAMOYL)-2-BENZIMIDAZOLYLCARBAMATE
23135-22-0	METHYL 1-(DIMETHYLCARBAMOYL)-N-(METHYLCARBAMOYL-OXY)THIOFORMIMIDATE
625-33-2	METHYL 1-PROPENYL KETONE
141-79-7	METHYL 2,2-DIMETHYLVINYL KETONE
23135-22-0	METHYL 2-(DIMETHYLAMINO)-N-(((METHYLAMINO)CARBONYL)-OXY)-2-OXOETHANIMIDOTHIOATE
96-32-2	METHYL 2-BROMOACETATE
80-63-7	METHYL 2-CHLORO-2-PROPENOATE
17639-93-9	METHYL 2-CHLOROPROPANOATE

CAS No.	Chemical Name
80-63-7	METHYL 2-CHLOROPROPENOATE
17639-93-9	METHYL 2-CHLOROPROPIONATE
137-05-3	METHYL 2-CYANOACRYLATE
119-36-8	METHYL 2-HYDROXYBENZOATE
141-79-7	METHYL 2-METHYL-1-PROPENYL KETONE
80-62-6	METHYL 2-METHYL-2-PROPENOATE
80-62-6	METHYL 2-METHYLPROPENOATE
7786-34-7	METHYL 3-(DIMETHOXYPHOSPHINYLOXY)CROTONATE
556-24-1	METHYL 3-METHYLBUTANOATE
556-24-1	METHYL 3-METHYLBUTYRATE
105-45-3	METHYL 3-OXOBUTANOATE
105-45-3	METHYL 3-OXOBUTYRATE
120-61-6	METHYL 4-CARBOMETHOXYBENZOATE
80-48-8	METHYL 4-METHYLBENZENESULFONATE
79-20-9	METHYL ACETATE
79-09-4	METHYL ACETIC ACID
105-45-3	METHYL ACETOACETATE
78-93-3	METHYL ACETONE
78-93-3	METHYL ACETONE (DOT)
105-45-3	METHYL ACETYLACETATE
105-45-3	METHYL ACETYLACETONATE
74-99-7	METHYL ACETYLENE
59355-75-8	METHYL ACETYLENE AND PROPADIENE MIXTURE, STABILIZED (DOT)
59355-75-8	METHYL ACETYLENE-PROPADIENE MIXTURE
59355-75-8	METHYL ACETYLENE-PROPADIENE MIXTURE (ACGIH)
59355-75-8	METHYL ACETYLENE-PROPADIENE MIXTURE (MAPP)
96-33-3	METHYL ACRYLATE
96-33-3	METHYL ACRYLATE (ACGIH)
96-33-3	METHYL ACRYLATE, INHIBITED
96-33-3	METHYL ACRYLATE, INHIBITED (DOT)
67-56-1	METHYL ALCOHOL
67-56-1	METHYL ALCOHOL (ACGIH,DOT)
50-00-0	METHYL ALDEHYDE
563-47-3	METHYL ALLYL CHLORIDE
563-47-3	METHYL ALLYL CHLORIDE (DOT)
96-32-2	METHYL ALPHA-BROMOACETATE
96-34-4	METHYL ALPHA-CHLOROACETATE
17639-93-9	METHYL ALPHA-CHLOROPROPIONATE
80-62-6	METHYL ALPHA-METHYLACRYLATE
12263-85-3	METHYL ALUMINIUM SESQUIBROMIDE (DOT)
12542-85-7	METHYL ALUMINIUM SESQUICHLORIDE (DOT)
12263-85-3	METHYL ALUMINUM SESQUIBROMIDE
12263-85-3	METHYL ALUMINUM SESQUIBROMIDE (DOT)
12542-85-7	METHYL ALUMINUM SESQUICHLORIDE
12542-85-7	METHYL ALUMINUM SESQUICHLORIDE (DOT)
108-84-9	METHYL AMYL ACETATE
108-84-9	METHYL AMYL ACETATE (DOT)
105-30-6	METHYL AMYL ALCOHOL
108-11-2	METHYL AMYL ALCOHOL
54972-97-3	METHYL AMYL ALCOHOL
110-43-0	METHYL AMYL KETONE
110-43-0	METHYL AMYL KETONE (DOT)
93-58-3	METHYL BENZENECARBOXYLATE
93-58-3	METHYL BENZOATE
121-43-7	METHYL BORATE
74-83-9	METHYL BROMIDE
8004-09-9	METHYL BROMIDE AND MORE THAN 2% CHLOROPICRIN MIXTURE, LIQUID (DOT)
74-83-9	METHYL BROMIDE, LIQUID, INCLUDING UP TO 2% CHLOROPICRIN (DOT)
74-83-9	METHYL BROMIDE, LIQUID, WITH 2% CHLOROPICRIN
96-32-2	METHYL BROMOACETATE
96-32-2	METHYL BROMOACETATE (DOT)
623-42-7	METHYL BUTANOATE
563-80-4	METHYL BUTANONE
26760-64-5	METHYL BUTENE
513-35-9	METHYL BUTENE (DOT)
26760-64-5	METHYL BUTENE (DOT)
591-78-6	METHYL BUTYL KETONE
623-42-7	METHYL BUTYRATE
623-42-7	METHYL BUTYRATE (DOT)
615-53-2	METHYL CARBAMATE
1563-66-2	METHYL CARBAMIC ACID 2,3-DIHYDRO-2,2-DIMETHYL-7-BENZOFURANYL ESTER
2032-65-7	METHYL CARBAMIC ACID 4-(METHYLTHIO)-3,5-XYLYL ESTER
85-71-2	METHYL CARBETHOXYMETHYL PHTHALATE
111-77-3	METHYL CARBITOL
616-38-6	METHYL CARBONATE

CAS No.	Chemical Name
616-38-6	METHYL CARBONATE ((MEO)2CO)
79-22-1	METHYL CARBONOCHLORIDATE
151-38-2	METHYL CELLOSOLVE
109-86-4	METHYL CELLOSOLVE
109-86-4	METHYL CELLOSOLVE (DOT)
110-49-6	METHYL CELLOSOLVE ACETATE
110-49-6	METHYL CELLOSOLVE ACETATE (DOT)
110-49-6	METHYL CELLOSOLYE ACETAAT (DUTCH)
74-87-3	METHYL CHLORIDE
74-87-3	METHYL CHLORIDE (ACGIH, DOT)
96-34-4	METHYL CHLOROACETATE
96-34-4	METHYL CHLOROACETATE (DOT)
79-22-1	METHYL CHLOROCARBONATE
79-22-1	METHYL CHLOROCARBONATE (DOT)
71-55-6	METHYL CHLOROFORM
71-55-6	METHYL CHLOROFORM (ACGIH,DOT)
79-22-1	METHYL CHLOROFORMATE
79-22-1	METHYL CHLOROFORMATE (DOT)
107-30-2	METHYL CHLOROMETHYL ETHER
107-30-2	METHYL CHLOROMETHYL ETHER, ANHYDROUS
107-30-2	METHYL CHLOROMETHYL ETHER, ANHYDROUS (DOT)
78-95-5	METHYL CHLOROMETHYL KETONE
52-68-6	METHYL CHLOROPHOS
17639-93-9	METHYL CHLOROPROPIONATE
993-00-0	METHYL CHLOROSILANE
993-00-0	METHYL CHLOROSILANE (DOT)
75-05-8	METHYL CYANIDE
75-05-8	METHYL CYANIDE (DOT)
137-05-3	METHYL CYANOACRYLATE
25639-42-3	METHYL CYCLOHEXANOL
1331-22-2	METHYL CYCLOHEXANONE
1331-22-2	METHYL CYCLOHEXANONE (DOT)
11118-65-3	METHYL CYCLOHEXANONE PEROXIDE
100-60-7	METHYL CYCLOHEXYLAMINE [CORROSIVE LABEL]
100-60-7	METHYL CYCLOHEXYLAMINE [FLAMMABLE LIQUID AND CORROSIVE LABELS]
100-60-7	METHYL CYCLOHEXYLAMINE [FLAMMABLE LIQUID LABEL]
96-37-7	METHYL CYCLOPENTANE
8022-00-2	METHYL DEMETON
919-86-8	METHYL DEMETON THIOESTER
116-54-1	METHYL DICHLOROACETATE
75-54-7	METHYL DICHLOROSILANE
75-54-7	METHYL DICHLOROSILANE (DOT)
111-77-3	METHYL DIGOL
111-77-3	METHYL DIOXITOL
624-92-0	METHYL DISULFIDE
298-00-0	METHYL E 605
79-20-9	METHYL ETHANOATE
115-10-6	METHYL ETHER
353-42-4	METHYL ETHER, COMPD WITH BF3 (1:1)
353-42-4	METHYL ETHER, COMPD WITH BORON FLUORIDE (BF3) (1:1)
109-86-4	METHYL ETHOXOL
540-67-0	METHYL ETHYL ETHER
540-67-0	METHYL ETHYL ETHER (DOT)
78-93-3	METHYL ETHYL KETONE
78-93-3	METHYL ETHYL KETONE (ACGIH,DOT)
78-93-3	METHYL ETHYL KETONE (MEK)
1338-23-4	METHYL ETHYL KETONE HYDROPEROXIDE
1338-23-4	METHYL ETHYL KETONE PEROXIDE
1338-23-4	METHYL ETHYL KETONE PEROXIDE, IN SOLUTION WITH NOT MORE THAN 9% BY WT ACTIVE OXYGEN (DOT)
1338-23-4	METHYL ETHYL KETONE PEROXIDE, WITH ≤ 50% PEROXIDE
1338-23-4	METHYL ETHYL KETONE PEROXIDE, WITH ≤ 60% PEROXIDE
104-90-5	METHYL ETHYL PYRIDINE
104-90-5	METHYL ETHYL PYRIDINE (DOT)
75-56-9	METHYL ETHYLENE OXIDE
593-53-3	METHYL FLUORIDE
593-53-3	METHYL FLUORIDE (DOT)
107-31-3	METHYL FORMATE
107-31-3	METHYL FORMATE (ACGIH,DOT)
298-00-0	METHYL FOSFERNO
109-86-4	METHYL GLYCOL
110-49-6	METHYL GLYCOL ACETATE
110-49-6	METHYL GLYCOL MONOACETATE
86-50-0	METHYL GUTHION
60-34-4	METHYL HYDRAZINE
74-82-8	METHYL HYDRIDE
67-56-1	METHYL HYDROXIDE
74-88-4	METHYL IODIDE

CAS No.	Chemical Name
74-88-4	METHYL IODIDE (ACGIH,DOT)
74-88-4	METHYL IODIDE (CH3I)
110-12-3	METHYL ISOAMYL KETONE
110-12-3	METHYL ISOAMYL KETONE (ACGIH)
141-79-7	METHYL ISOBUTENYL KETONE
105-30-6	METHYL ISOBUTYL CARBINOL
108-11-2	METHYL ISOBUTYL CARBINOL
54972-97-3	METHYL ISOBUTYL CARBINOL
105-30-6	METHYL ISOBUTYL CARBINOL (ACGIH,DOT)
108-11-2	METHYL ISOBUTYL CARBINOL (ACGIH,DOT)
54972-97-3	METHYL ISOBUTYL CARBINOL (ACGIH,DOT)
108-10-1	METHYL ISOBUTYL KETONE
37206-20-5	METHYL ISOBUTYL KETONE PEROXIDE
37206-20-5	METHYL ISOBUTYL KETONE PEROXIDE (CONCENTRATION ≥ 60%)
37206-20-5	METHYL ISOBUTYL KETONE PEROXIDE, WITH NOT MORE THAN 9% BY WEIGHT ACTIVE OXYGEN (DOT)
624-83-9	METHYL ISOCYANAT (GERMAN)
624-83-9	METHYL ISOCYANATE
624-83-9	METHYL ISOCYANATE (ACGIH,DOT)
624-83-9	METHYL ISOCYANATE SOLUTIONS (DOT)
556-24-1	METHYL ISOPENTANOATE
110-12-3	METHYL ISOPENTYL KETONE
814-78-8	METHYL ISOPROPENYL KETONE
814-78-8	METHYL ISOPROPENYL KETONE, INHIBITED
814-78-8	METHYL ISOPROPENYL KETONE, INHIBITED (DOT)
563-80-4	METHYL ISOPROPYL KETONE
563-80-4	METHYL ISOPROPYL KETONE (ACGIH)
919-86-8	METHYL ISOSYSTOX
556-61-6	METHYL ISOTHIOCYANATE
556-61-6	METHYL ISOTHIOCYANATE (DOT)
556-24-1	METHYL ISOVALERATE
67-64-1	METHYL KETONE
75-86-5	METHYL LACTONITRILE
75-16-1	METHYL MAGNESIUM BROMIDE
74-93-1	METHYL MERCAPTAN
502-39-6	METHYL MERCURIC DICYANDIAMIDE
502-39-6	METHYL MERCURY DICYANDIAMIDE
80-62-6	METHYL METHACRYLATE
80-62-6	METHYL METHACRYLATE (ACGIH)
80-62-6	METHYL METHACRYLATE MONOMER
80-62-6	METHYL METHACRYLATE MONOMER, INHIBITED (DOT)
80-62-6	METHYL METHACRYLATE MONOMER, UNINHIBITED
66-27-3	METHYL METHANESULFONATE
66-27-3	METHYL METHANESULPHONATE
107-31-3	METHYL METHANOATE
80-62-6	METHYL METHYLACRYLATE
66-27-3	METHYL METHYLSULFONATE
96-32-2	METHYL MONOBROMOACETATE
96-34-4	METHYL MONOCHLORACETATE
96-34-4	METHYL MONOCHLOROACETATE
75-18-3	METHYL MONOSULFIDE
556-61-6	METHYL MUSTARD
556-61-6	METHYL MUSTARD OIL
23135-22-0	METHYL N',N'-DIMETHYL-N-((METHYLCARBAMOYL)OXY)-1-THIOOXAMIMIDATE
110-43-0	METHYL N-AMYL KETONE
110-43-0	METHYL N-AMYL KETONE (ACGIH)
623-42-7	METHYL N-BUTANOATE
591-78-6	METHYL N-BUTYL KETONE
591-78-6	METHYL N-BUTYL KETONE (ACGIH)
110-43-0	METHYL N-PENTYL KETONE
557-17-5	METHYL N-PROPYL ETHER
16752-77-5	METHYL N-[(METHYLCARBAMOYL)OXY]THIOACETIMIDATE
298-00-0	METHYL NIRAN
624-91-9	METHYL NITRITE
624-91-9	METHYL NITRITE (DOT)
16752-77-5	METHYL O-(METHYLCARBAMOYL)THIOLACETOHYDROXAMATE
16752-77-5	METHYL O-(METHYLCARBAMYL)THIOLACETOHYDROXAMATE
119-36-8	METHYL O-HYDROXYBENZOATE
681-84-5	METHYL ORTHOSILICATE
75-56-9	METHYL OXIRANE
109-86-4	METHYL OXITOL
120-61-6	METHYL P-(METHOXYCARBONYL)BENZOATE
80-48-8	METHYL P-METHYLBENZENESULFONATE
100-17-4	METHYL P-NITROPHENYL ETHER
80-48-8	METHYL P-TOLUENESULFONATE
80-48-8	METHYL P-TOSYLATE
298-00-0	METHYL PARATHION
298-00-0	METHYL PARATHION (ACGIH)

CAS No.	Chemical Name
298-00-0	METHYL PARATHION MIXTURE, DRY
298-00-0	METHYL PARATHION MIXTURE, DRY (DOT)
298-00-0	METHYL PARATHION MIXTURE, LIQUID, WITH 25% METHYL PARATHION
298-00-0	METHYL PARATHION, LIQUID
298-00-0	METHYL PARATHION, LIQUID (DOT)
2524-03-0	METHYL PCT
43133-95-5	METHYL PENTANE
110-43-0	METHYL PENTYL KETONE
3735-23-7	METHYL PHENCAPTON
3735-23-7	METHYL PHENKAPTON
100-66-3	METHYL PHENYL ETHER
98-86-2	METHYL PHENYL KETONE
121-45-9	METHYL PHOSPHITE
676-97-1	METHYL PHOSPHONIC DICHLORIDE
676-97-1	METHYL PHOSPHONIC DICHLORIDE (DOT)
676-98-2	METHYL PHOSPHONOTHIOIC DICHLORIDE
676-98-2	METHYL PHOSPHONOTHIOIC DICHLORIDE, ANHYDROUS
676-98-2	METHYL PHOSPHONOTHIOIC DICHLORIDE, ANHYDROUS (DOT)
676-83-5	METHYL PHOSPHONOUS DICHLORIDE
676-83-5	METHYL PHOSPHONOUS DICHLORIDE (DOT)
10265-92-6	METHYL PHOSPHORAMIDOTHIOATE
2524-03-0	METHYL PHOSPHOROCHLORIDOTHIOATE ((MEO)2ClPS)
131-11-3	METHYL PHTHALATE
85-71-2	METHYL PHTHALYL ETHER
85-71-2	METHYL PHTHALYL ETHER GLYCOLATE
85-71-2	METHYL PHTHALYL ETHYL GLYCOLATE
96-33-3	METHYL PROP-2-ENOATE
554-12-1	METHYL PROPANOATE
96-33-3	METHYL PROPENATE
96-33-3	METHYL PROPENOATE
625-33-2	METHYL PROPENYL KETONE
554-12-1	METHYL PROPIONATE
554-12-1	METHYL PROPIONATE (DOT)
764-35-2	METHYL PROPYL ACETYLENE
28729-54-6	METHYL PROPYL BENZENE
557-17-5	METHYL PROPYL ETHER
557-17-5	METHYL PROPYL ETHER (DOT)
107-87-9	METHYL PROPYL KETONE
107-87-9	METHYL PROPYL KETONE (ACGIH,DOT)
554-12-1	METHYL PROPYLATE
50-55-5	METHYL RESERPATE 3,4,5-TRIMETHOXYBENZOATE (ESTER)
119-36-8	METHYL SALICYLATE
681-84-5	METHYL SILICATE
4421-95-8	METHYL SILICATE
12002-26-5	METHYL SILICATE
681-84-5	METHYL SILICATE ((CH3)4SiO4)
681-84-5	METHYL SILICATE ((MEO)4Si)
98-83-9	METHYL STYRENE
25013-15-4	METHYL STYRENE
77-78-1	METHYL SULFATE
77-78-1	METHYL SULFATE (DOT)
75-18-3	METHYL SULFIDE
75-18-3	METHYL SULFIDE (DOT)
556-64-9	METHYL SULFOCYANATE
67-68-5	METHYL SULFOXIDE
75-18-3	METHYL SULPHIDE
8022-00-2	METHYL SYSTOX
25265-68-3	METHYL TETRAHYDROFURAN
556-64-9	METHYL THIOCYANATE
556-61-6	METHYL THIOISOCYANATE
137-26-8	METHYL THIRAM
137-26-8	METHYL THIURAMDISULFIDE
1330-20-7	METHYL TOLUENE
80-48-8	METHYL TOLUENE SULFONATE
80-48-8	METHYL TOLUENE-4-SULFONATE
80-48-8	METHYL TOSYLATE
67-66-3	METHYL TRICHLORIDE
598-99-2	METHYL TRICHLOROACETATE
598-99-2	METHYL TRICHLOROACETATE (DOT)
75-79-6	METHYL TRICHLOROSILANE
75-46-7	METHYL TRIFLUORIDE
137-26-8	METHYL TUADS
615-53-2	METHYL URETHANE
123-15-9	METHYL VALERALDEHYDE
123-15-9	METHYL VALERALDEHYDE (2-METHYL PENTANAL)
107-25-5	METHYL VINYL ETHER
78-94-4	METHYL VINYL KETONE
78-94-4	METHYL VINYL KETONE (DOT)

CAS No.	Chemical Name
78-94-4	METHYL VINYL KETONE, INHIBITED
78-94-4	METHYL VINYL KETONE, INHIBITED (DOT)
1910-42-5	METHYL VIOLOGEN
1910-2-5	METHYL VIOLOGEN (REDUCED)
1910-42-5	METHYL VIOLOGEN DICHLORIDE
4685-14-7	METHYL VIOLOGEN(2+)
60-11-7	METHYL YELLOW
109-83-1	METHYL(2-HYDROXYETHYL)AMINE
109-83-1	METHYL(BETA-HYDROXYETHYL)AMINE
109-83-1	METHYL(HYDROXYETHYL)AMINE
105-30-6	METHYL-1-PENTANOL
108-11-2	METHYL-1-PENTANOL
54972-97-3	METHYL-1-PENTANOL
80-63-7	METHYL-2-CHLOROACRYLATE
17639-93-9	METHYL-2-CHLOROPROPIONATE
17639-93-9	METHYL-2-CHLOROPROPIONATE (DOT)
96-33-3	METHYL-2-PROPENOATE
96-33-3	METHYL-ACRYLAT (GERMAN)
80-63-7	METHYL-ALPHA-CHLOROACRYLATE
110-43-0	METHYL-AMYL-CETONE (FRENCH)
75-54-7	METHYL-DICHLORSILAN (CZECH)
298-00-0	METHYL-E 605
556-61-6	METHYL-ISOTHIOCYANAT (GERMAN)
26471-62-5	METHYL-M-PHENYLENE ISOCYANATE
919-86-8	METHYL-MERCAPTOFOS TEOLERY
26471-62-5	METHYL-META-PHENYLENE DIISOCYANATE
80-62-6	METHYL-METHACRYLAT (GERMAN)
623-42-7	METHYL-N-BUTYRATE
5517-17-5	METHYL-N-PROPYL ETHER
107-87-9	METHYL-N-PROPYL KETONE
107-87-9	METHYL-PROPYL-CETONE (FRENCH)
1634-04-4	METHYL-TERT-BUTYL ETHER
75-79-6	METHYL-TRICHLORSILAN (CZECH)
78-94-4	METHYL-VINYL-CETONE (FRENCH)
123-38-6	METHYLACETALDEHYDE
123-62-6	METHYLACETIC ANHYDRIDE
59355-75-8	METHYLACETYLENE-PROPADIENE, STABILIZED
59355-75-8	METHYLACETYLENE-PROPADIENE, STABILIZED (DOT)
78-85-3	METHYLACROLEIN
96-33-3	METHYLACRYLAAT (DUTCH)
78-85-3	METHYLACRYLALDEHYDE
79-41-4	METHYLACRYLIC ACID
126-98-7	METHYLACRYLONITRILE
126-98-7	METHYLACRYLONITRILE (ACGIH)
109-87-5	METHYLAL
109-87-5	METHYLAL (ACGIH,DOT)
67-56-1	METHYLALKOHOL (GERMAN)
74-89-5	METHYLAMINE
74-89-5	METHYLAMINE (ACGIH)
617-89-0	METHYLAMINE, 1-(2-FURYL)-
74-89-5	METHYLAMINE, ANHYDROUS
74-89-5	METHYLAMINE, ANHYDROUS (DOT)
74-89-5	METHYLAMINE, AQUEOUS SOLUTION
74-89-5	METHYLAMINE, AQUEOUS SOLUTION (DOT)
74-89-5	METHYLAMINEN (DUTCH)
109-83-1	METHYLAMINOETHANOL
105-30-6	METHYLAMYL ALCOHOL
108-11-2	METHYLAMYL ALCOHOL
54972-97-3	METHYLAMYL ALCOHOL
100-61-8	METHYLANILINE
2163-80-6	METHYLARSENIC ACID, SODIUM SALT
86-50-0	METHYLAZINPHOS
590-96-5	METHYLAZOXYMETHANOL
14901-08-7	METHYLAZOXYMETHANOL BETA-D-GLUCOSIDE
592-62-1	METHYLAZOXYMETHANOL ACETATE
14901-08-7	METHYLAZOXYMETHANOL GLUCOSIDE
592-62-1	METHYLAZOXYMETHYL ACETATE
108-88-3	METHYLBENZENE
104-15-4	METHYLBENZENESULFONIC ACID
25231-46-3	METHYLBENZENESULFONIC ACID
93-58-3	METHYLBENZOATE (DOT)
108-88-3	METHYLBENZOL
98-85-1	METHYLBENZYL ALCOHOL (ALPHA)
55-86-7	METHYLBIS(BETA-CHLOROETHYL)AMINE HYDROCHLORIDE
302-70-5	METHYLBIS(BETA-CHLOROETHYL)AMINE N-OXIDE HYDROCHLORIDE
51-75-2	METHYLBIS(2-CHLOROETHYL)AMINE
55-86-7	METHYLBIS(2-CHLOROETHYL)AMINE HYDROCHLORIDE
105-59-9	METHYLBIS(2-HYDROXYETHYL)AMINE
51-75-2	METHYLBIS(BETA-CHLOROETHYL)AMINE
110-68-9	METHYLBUTYLAMINE
63-25-2	METHYLCARBAMATE 1-NAPHTHALENOL
63-25-2	METHYLCARBAMATE 1-NAPHTHOL
22259-30-9	METHYLCARBAMIC ACID ESTER WITH N'-(M-HYDROXYPHENYL)-N,N-DIMETHYLFORMAMIDINE
2631-37-0	METHYLCARBAMIC ACID M-CYM-5-YL ESTER
63-25-2	METHYLCARBAMIC ACID, 1-NAPHTHYL ESTER
64-17-5	METHYLCARBINOL
79-22-1	METHYLCHLOORFORMIAAT (DUTCH)
74-87-3	METHYLCHLORID (GERMAN)
107-30-2	METHYLCHLOROMETHYL ETHER (DOT)
2971-90-6	METHYLCHLOROPINDOL
2971-90-6	METHYLCHLORPINDOL
56-49-5	METHYLCHOLANTHRENE
1331-22-2	METHYLCYCLOHEXAN-1-ONE
108-87-2	METHYLCYCLOHEXANE
108-87-2	METHYLCYCLOHEXANE (ACGIH,DOT)
25639-42-3	METHYLCYCLOHEXANOL
25639-42-3	METHYLCYCLOHEXANOL (ACGIH,DOT)
11118-65-3	METHYLCYCLOHEXANONE PEROXIDE
100-60-7	METHYLCYCLOHEXYLAMINE
12108-13-3	METHYLCYCLOPENTADIENYL MANGANESE TRICARBONYL
12108-13-3	METHYLCYMANTRENE
51-75-2	METHYLDI(2-CHLOROETHYL)AMINE
302-70-5	METHYLDI(2-CHLOROETHYL)AMINE N-OXIDE HYDROCHLORIDE
593-89-5	METHYLDICHLORARSINE
593-89-5	METHYLDICHLOROARSINE
593-89-5	METHYLDICHLOROARSINE (DOT)
676-83-5	METHYLDICHLOROPHOSPHINE
676-98-2	METHYLDICHLOROPHOSPHINE SULFIDE
105-59-9	METHYLDIETHANOLAMINE
25321-14-6	METHYLDINITROBENZENE
107-31-3	METHYLE (FORMIATE DE) (FRENCH)
77-78-1	METHYLE (SULFATE DE) (FRENCH)
563-12-2	METHYLEEN-S,S'-BIS(O,O-DIETHYL-DITHIOFOSFAAT) (DUTCH)
78-94-4	METHYLENE ACETONE
75-09-2	METHYLENE BICHLORIDE
5124-30-1	METHYLENE BIS(4-CYCLOHEXYLISOCYANATE)
101-68-8	METHYLENE BISPHENYL ISOCYANATE
101-68-8	METHYLENE BISPHENYL ISOCYANATE (ACGIH)
101-68-8	METHYLENE BISPHENYL ISOCYANATE (MDI)
74-95-3	METHYLENE BROMIDE
75-09-2	METHYLENE CHLORIDE
75-09-2	METHYLENE CHLORIDE (ACGIH,DOT)
74-97-5	METHYLENE CHLOROBROMIDE
101-68-8	METHYLENE DI(PHENYLENE ISOCYANATE) (DOT)
75-09-2	METHYLENE DICHLORIDE
109-87-5	METHYLENE DIMETHYL ETHER
50-00-0	METHYLENE GYLCOL
50-00-0	METHYLENE OXIDE
123-30-8	METHYLENE-DIPHENYLENE DI-ISOCYANATE
563-12-2	METHYLENE-S,S'-BIS(O,O-DIAETHYL-DITHIOPHOSPHAT) (GERMAN)
5124-30-1	METHYLENEBIS(1,4-CYCLOHEXYLENE) DIISOCYANATE
101-14-4	METHYLENEBIS(3-CHLORO-4-AMINOBENZENE)
5124-30-1	METHYLENEBIS(4-CYCLOHEXYL ISOCYANATE)
101-68-8	METHYLENEBIS(4-ISOCYANATOBENZENE)
5124-30-1	METHYLENEBIS(4-ISOCYANATOCYCLOHEXANE)
101-68-8	METHYLENEBIS(4-PHENYL ISOCYANATE)
101-68-8	METHYLENEBIS(4-PHENYLENE ISOCYANATE)
101-77-9	METHYLENEBIS(ANILINE)
101-68-8	METHYLENEBIS(P-PHENYL ISOCYANATE)
101-68-8	METHYLENEBIS(P-PHENYLENE ISOCYANATE)
101-68-8	METHYLENEBIS-P-PHENYLENE DIISOCYANATE
101-68-8	METHYLENEBISPHENYLENE DIISOCYANATE
5124-30-1	METHYLENEDI-1,4-CYCLOHEXYLENE ISOCYANATE
5124-30-1	METHYLENEDI-4-CYCLOHEXYLENE DIISOCYANATE
101-68-8	METHYLENEDI-P-PHENYLENE DIISOCYANATE
101-68-8	METHYLENEDI-P-PHENYLENE ISOCYANATE
101-77-9	METHYLENEDIANILINE
109-83-1	METHYLETHANOLAMINE
115-07-1	METHYLETHENE
57-55-6	METHYLETHYL GLYCOL
96-17-3	METHYLETHYLACETALDEHYDE
78-76-2	METHYLETHYLBROMOMETHANE
78-92-2	METHYLETHYLCARBINOL
115-07-1	METHYLETHYLENE
57-55-6	METHYLETHYLENE GLYCOL

CAS No.	Chemical Name
106-97-8	METHYLETHYLMETHANE
10595-95-6	METHYLETHYLNITROSAMINE
10595-95-6	METHYLETHYLNITROSOAMINE
109-83-1	METHYLETHYLOLAMINE
107-44-8	METHYLFLUOROPHOSPHORIC ACID ISOPROPYL ESTER
107-44-8	METHYLFLUORPHOSPHORSAEUREISOPROPYLESTER (GERMAN)
107-31-3	METHYLFORMIAAT (DUTCH)
107-31-3	METHYLFORMIAT (GERMAN)
27137-41-3	METHYLFURAN
27137-41-3	METHYLFURAN (DOT)
109-86-4	METHYLGLYKOL (GERMAN)
110-49-6	METHYLGLYKOLACETAT (GERMAN)
86-50-0	METHYLGUSATHION
25639-42-3	METHYLHEXALIN
110-12-3	METHYLHEXANONE
60-34-4	METHYLHYDRAZINE (ACGIH,DOT)
60-34-4	METHYLHYDRAZINE (MONO)
105-59-9	METHYLIMINODIETHANOL
108-84-9	METHYLISOAMYL ACETATE
105-30-6	METHYLISOBUTYL CARBINOL
108-11-2	METHYLISOBUTYL CARBINOL
54972-97-3	METHYLISOBUTYL CARBINOL
108-84-9	METHYLISOBUTYLCARBINOL ACETATE
108-84-9	METHYLISOBUTYLCARBINYL ACETATE
624-83-9	METHYLISOCYANAAT (DUTCH)
598-75-4	METHYLISOPROPYLCARBINOL
556-61-6	METHYLISOTHIOCYANAAT (DUTCH)
556-24-1	METHYLISOVALERATE (DOT)
74-88-4	METHYLJODID (GERMAN)
74-88-4	METHYLJODIDE (DUTCH)
8022-00-2	METHYLMERCAPTOPHOS
502-39-6	METHYLMERCURIC CYANOGUANIDINE
502-39-6	METHYLMERCURIC DICYANAMIDE
502-39-6	METHYLMERCURIC DICYANDIAMIDE
502-39-6	METHYLMERCURY DICYANDIAMIDE
80-62-6	METHYLMETHACRYLAAT (DUTCH)
74-84-0	METHYLMETHANE
109-02-4	METHYLMORPHOLINE
109-02-4	METHYLMORPHOLINE (DOT)
107-16-4	METHYLNITRILE, HYDROXY-
1321-12-6	METHYLNITROBENZENE
70-25-7	METHYLNITRONITROSOGUANIDINE
122-14-5	METHYLNITROPHOS
684-93-5	METHYLNITROSOUREA
615-53-2	METHYLNITROSOURETHANE
67-56-1	METHYLOL
71-36-3	METHYLOLPROPANE
25322-69-4	METHYLOXIRANE POLYMER
54363-49-4	METHYLPENTADIENE
73513-30-1	METHYLPENTALDEHYDE
43133-95-5	METHYLPENTANE (DOT)
1319-77-3	METHYLPHENOL
100-61-8	METHYLPHENYLAMINE
98-85-1	METHYLPHENYLCARBINOL
149-74-6	METHYLPHENYLDICHLOROSILANE
149-74-6	METHYLPHENYLDICHLOROSILANE (DOT)
26471-62-5	METHYLPHENYLENE ISOCYANATE
25376-45-8	METHYLPHENYLENEDIAMINE
30143-13-6	METHYLPHENYLENEDIAMINE
676-83-5	METHYLPHOSPHINIC DICHLORIDE
676-83-5	METHYLPHOSPHINOUS DICHLORIDE
676-97-1	METHYLPHOSPHONIC ACID DICHLORIDE
676-97-1	METHYLPHOSPHONIC DICHLORIDE
676-97-1	METHYLPHOSPHONODICHLORIDIC ACID
107-44-8	METHYLPHOSPHONOFLUORIDIC ACID 1-METHYLETHYL ESTER
107-44-8	METHYLPHOSPHONOFLUORIDIC ACID ISOPROPYL ESTER
676-98-2	METHYLPHOSPHONOTHIOIC DICHLORIDE
676-98-2	METHYLPHOSPHONOTHIOYL DICHLORIDE
676-83-5	METHYLPHOSPHONOUS DICHLORIDE
676-97-1	METHYLPHOSPHONYL CHLORIDE
676-97-1	METHYLPHOSPHONYL DICHLORIDE
676-83-5	METHYLPHOSPHORUS DICHLORIDE
109-01-3	METHYLPIPERAZINE
626-67-5	METHYLPIPERIDINE
78-84-2	METHYLPROPANAL
625-33-2	METHYLPROPENYL KETONE, INHIBITED
6032-29-7	METHYLPROPYLCARBINOL
109-08-0	METHYLPYRAZINE
1333-41-1	METHYLPYRIDINE
96-54-8	METHYLPYRROLE
27417-39-6	METHYLPYRROLE
120-94-5	METHYLPYRROLIDINE
872-50-4	METHYLPYRROLIDONE
556-64-9	METHYLRHODANID (GERMAN)
556-61-6	METHYLSENFOEL (GERMAN)
75-79-6	METHYLSILYL TRICHLORIDE
1319-73-9	METHYLSTYRENE
67-68-5	METHYLSULFINYLMETHANE
25265-68-3	METHYLTETRAHYDROFURAN (DOT)
75-18-3	METHYLTHIOMETHANE
919-86-8	METHYLTHIONODEMETON
676-98-2	METHYLTHIONOPHOSPHONIC DICHLORIDE
298-00-0	METHYLTHIOPHOS
676-98-2	METHYLTHIOPHOSPHONIC ACID DICHLORIDE
676-98-2	METHYLTHIOPHOSPHONIC DICHLORIDE
56-04-2	METHYLTHIOURACIL
71-55-6	METHYLTRICHLOROMETHANE
75-79-6	METHYLTRICHLOROSILANE
75-79-6	METHYLTRICHLOROSILANE (DOT)
108-22-5	METHYLVINYL ACETATE
78-94-4	METHYLVINYL KETONE
78-94-4	METHYLVINYLKETON (GERMAN)
4549-40-0	METHYLVINYLNITROSAMINE
1910-42-5	METHYLVIOLOGEN
1910-42-5	METHYLVIOLOGEN CHLORIDE
25551-13-7	METHYLXYLENE
2971-90-6	METICLORPINDOL
52-68-6	METIFONATE
107-31-3	METIL (FORMIATO DI) (ITALIAN)
109-86-4	METIL CELLOSOLVE (ITALIAN)
624-83-9	METIL ISOCIANATO (ITALIAN)
80-62-6	METIL METACRILATO (ITALIAN)
96-33-3	METILACRILATO (ITALIAN)
105-30-6	METILAMIL ALCOHOL (ITALIAN)
108-11-2	METILAMIL ALCOHOL (ITALIAN)
54972-97-3	METILAMIL ALCOHOL (ITALIAN)
74-89-5	METILAMINE (ITALIAN)
79-22-1	METILCLOROFORMIATO (ITALIAN)
78-93-3	METILETILCHETONE (ITALIAN)
298-00-0	METILPARATION (HUNGARIAN)
86-50-0	METILTRIA ZOTION
53-86-1	METINDOL
137-26-8	METIURAC
2032-65-7	METMERCAPTURON
72-43-5	METOKSYCHLOR (POLISH)
109-86-4	METOKSYETYLOWY ALKOHOL (POLISH)
1129-41-5	METOLCARB
557-17-5	METOPRYL
72-43-5	METOX
78-53-5	METRAMAC
78-53-5	METRAMAK
21087-64-9	METRIBUZIN
52-68-6	METRIFONATE
52-68-6	METRIPHONATE
95-79-4	METROGEN RED FORMER KB SOLN
298-00-0	METRON
298-00-0	METRON (PESTICIDE)
443-48-1	METRONIDAZOL
443-48-1	METRONIDAZOLE
6834-92-0	METSO 20
6834-92-0	METSO BEADS 2048
6834-92-0	METSO BEADS, DRYMET
6834-92-0	METSO PENTABEAD 20
109-87-5	METYLAL (POLISH)
822-06-0	METYLENO-BIS-FENYLOIZOCYJANIAN (POLISH)
75-09-2	METYLENU CHLOREK (POLISH)
74-89-5	METYLOAMINA (POLISH)
108-87-2	METYLOCYKLOHEKSAN (POLISH)
25639-42-3	METYLOCYKLOHEKSANOL (POLISH)
1331-22-2	METYLOCYKLOHEKSANON (POLISH)
78-93-3	METYLOETYLOKETON (POLISH)
60-34-4	METYLOHYDRAZYNA (POLISH)
298-00-0	METYLOPARATION (POLISH)
107-87-9	METYLOPROPYLOKETON (POLISH)
67-56-1	METYLOWY ALKOHOL (POLISH)
298-00-0	METYLPARATION (CZECH)
74-87-3	METYLU CHLOREK (POLISH)
74-88-4	METYLUJODEK (POLISH)

CAS No.	Chemical Name
7786-34-7	MEVINFOS (DUTCH)
7786-34-7	MEVINPHOS
7786-34-7	MEVINPHOS (ACGIH,DOT)
7786-34-7	MEVINPHOS MIXTURE, DRY
7786-34-7	MEVINPHOS MIXTURE, DRY (DOT)
7786-34-7	MEVINPHOS MIXTURE, WET (DOT)
315-18-4	MEXACARBATE
443-48-1	MEXIBOL
88-05-1	MEZIDINE
53-86-1	MEZOLIN
1836-75-5	MEZOTOX
72-43-5	MEZOX K
144-49-0	MFA
107-44-8	MFI
7782-42-5	MG 1
122-14-5	MGLAWIK F
58-89-9	MGLAWIK L
110-12-3	MIAK
105-30-6	MIBC
108-11-2	MIBC
54972-97-3	MIBC
141-79-7	MIBK
108-10-1	MIBK
105-30-6	MIC
108-11-2	MIC
556-61-6	MIC
624-83-9	MIC
54972-97-3	MIC
12001-26-2	MICA
12001-26-2	MICA DUST
12001-26-2	MICA-GROUP MINERALS
12001-26-2	MICA-TYPE MINERALS
1309-37-1	MICACEOUS IRON ORE
12001-26-2	MICATEX
101-61-1	MICHLER'S BASE
101-61-1	MICHLER'S HYDRIDE
90-94-8	MICHLER'S KETONE
101-61-1	MICHLER'S METHANE
56-75-7	MICLORETIN
7704-34-9	MICOWETSULF
50-78-2	MICRISTIN
1344-95-2	MICRO-CEL
1344-95-2	MICRO-CEL A
1344-95-2	MICRO-CEL B
1344-95-2	MICRO-CEL C
1344-95-2	MICRO-CEL E
1344-95-2	MICRO-CEL T
1344-95-2	MICRO-CEL T 26
1344-95-2	MICRO-CEL T 38
1344-95-2	MICRO-CEL T 41
133-06-2	MICRO-CHECK 12
1344-95-2	MICROCAL ET
56-75-7	MICROCETINA
999-81-5	MICROCIL
56-53-1	MICROEST
7704-34-9	MICROFLOTOX
57-63-6	MICROFOLLIN
1344-28-1	MICROGRIT WCA
1333-86-4	MICROLITH BLACK CA
76-06-2	MICROLYSIN
10124-56-8	MICROMET
12001-26-2	MICROMICA W 1
14807-96-6	MICRON WHITE 5000S
14807-96-6	MICRONEECE K 1
68-22-4	MICRONETT
68-22-4	MICRONOR
68-22-4	MICRONOVUM
101-25-7	MICROPOR
57-50-1	MICROSE
82-28-0	MICROSETILE ORANGE RA
7631-86-9	MICROSIL
3012-65-5	MICROSTOP
7704-34-9	MICROSULF
14807-96-6	MICROTALCO IT EXTRA
7704-34-9	MICROTHIOL
3691-35-8	MICROZUL
55-98-1	MIELOSAN
55-98-1	MIELUCIN
64-18-6	MIERENZUUR (DUTCH)

CAS No.	Chemical Name
91-20-3	MIGHTY 150
91-20-3	MIGHTY RD1
671-16-9	MIH
108-10-1	MIK
53-86-1	MIKAMETAN
101-25-7	MIKROFOR N
74-97-5	MIL-B-4394-B
20816-12-0	MILAS' REAGENT
115-32-2	MILBOL
58-89-9	MILBOL 49
50-21-5	MILCHSAURE (GERMAN)
7546-30-7	MILD MERCURY CHLORIDE
55-98-1	MILECITAN
55-98-1	MILERAN
56-53-1	MILESTROL
50-21-5	MILK ACID
1305-62-0	MILK OF LIME
7446-14-2	MILK WHITE
101-14-4	MILLIONATE M
9016-87-9	MILLIONATE MR
9016-87-9	MILLIONATE MR 100
9016-87-9	MILLIONATE MR 200
9016-87-9	MILLIONATE MR 300
9016-87-9	MILLIONATE MR 340
9016-87-9	MILLIONATE MR 400
9016-87-9	MILLIONATE MR 500
10045-94-0	MILLON'S REAGENT
7631-86-9	MILOWHITE
7681-52-9	MILTON
7631-86-9	MIN-U-SIL
2631-37-0	MINACIDE
7782-42-5	MINERAL CARBON
12002-03-8	MINERAL GREEN
8020-83-5	MINERAL OIL
8020-83-5	MINERAL OILS
8052-42-4	MINERAL PITCH
8052-42-4	MINERAL RUBBER
8030-30-6	MINERAL SPIRITS
1318-59-8	MINERALS, CHLORITE-GROUP
14998-27-7	MINERALS, CHLORITE-GROUP
12001-26-2	MINERALS, MICA-GROUP
111-76-2	MINEX BDH
68-22-4	MINI-PE
8020-83-5	MINKH 1
7631-86-9	MINUSIL 30
7631-86-9	MINUSIL 5
51-83-2	MIOSTAT
78-96-6	MIPA
54590-52-2	MIPA-DODECYLBENZENESULFONATE
371-86-8	MIPAFOX (DOT)
131-11-3	MIPAX
563-80-4	MIPK
9005-25-8	MIRA QUICK C
94-75-7	MIRACLE
105-60-2	MIRAMID H 2
105-60-2	MIRAMID WM 55
98-95-3	MIRBANE OIL
2385-85-5	MIREX
8049-20-5	MISCH METAL
8049-20-5	MISCHMETAL, POWDER
14807-96-6	MISTRON 139
14807-96-6	MISTRON 2SC
14807-96-6	MISTRON FROST P
14807-96-6	MISTRON RCS
14807-96-6	MISTRON STAR
14807-96-6	MISTRON SUPER FROST
14807-96-6	MISTRON VAPOR
51-83-2	MISTURA C
55-98-1	MISULBAN
556-61-6	MIT
50-07-7	MIT-C
9002-84-0	MITEX
115-32-2	MITIGAN
12002-03-8	MITIS GREEN
50-07-7	MITO-C
50-07-7	MITOCIN-C
302-70-5	MITOMEN
50-07-7	MITOMYCIN
50-07-7	MITOMYCIN C

CAS No.	Chemical Name
50-07-7	MITOMYCINUM
55-98-1	MITOSTAN
2465-27-2	MITSUI AURAMINE O
119-90-4	MITSUI BLUE B BASE
633-03-4	MITSUI BRILLIANT GREEN GX
1937-37-7	MITSUI DIRECT BLACK EX
1937-37-7	MITSUI DIRECT BLACK GX
2602-46-2	MITSUI DIRECT BLUE 2BN
89-62-3	MITSUI RED GL BASE
95-79-4	MITSUI RED TR BASE
95-69-2	MITSUI RED TR BASE
99-55-8	MITSUI SCARLET G BASE
8000-41-7	MIXTURE OF P-METHENOLS
7631-86-9	MIZUKASIL
7631-86-9	MIZUKASIL P 527
7631-86-9	MIZUKASIL P 801
7631-86-9	MIZUKASIL SK 7
79-11-8	MKHUK
100-37-8	MKS
121-75-5	MLT
1344-28-1	MM 21
1309-48-4	MM 469
74-89-5	MMA
80-62-6	MMA
50-07-7	MMC
502-39-6	MMD
80-62-6	MME
60-34-4	MMH
66-27-3	MMS
12108-13-3	MMT
9004-34-6	MN-CELLULOSE
591-78-6	MNBK
70-25-7	MNNG
89-62-3	MNPT
99-08-1	MNT
684-93-5	MNU
1313-27-5	MO 1202T
9016-87-9	MOBAY MRS
8002-74-2	MOBIL 150
8002-74-2	MOBIL 150-155F
13194-48-4	MOBIL V-C 9-104
53-86-1	MOBILAN
8002-74-2	MOBILCER 161
8002-74-2	MOBILCER 739
8002-74-2	MOBILCER ED 80/229
8020-83-5	MOBILSOL 30
8002-74-2	MOBILWAX 220
101-14-4	MOCA
101-14-4	MOCA (CURING AGENT)
13194-48-4	MOCAP
13194-48-4	MOCAP 10G
577-11-7	MODANE SOFT
7681-52-9	MODIFIED DAKIN'S SOLUTION
1333-86-4	MOGUL
1333-86-4	MOGUL L
10045-89-3	MOHR'S SALT
1333-86-4	MOLACCO
64-17-5	MOLASSES ALCOHOL
577-11-7	MOLATOC
577-11-7	MOLCER
57-24-9	MOLE DEATH
7782-50-5	MOLECULAR CHLORINE
1333-74-0	MOLECULAR HYDROGEN
7553-56-2	MOLECULAR IODINE
7727-37-9	MOLECULAR NITROGEN
7782-44-7	MOLECULAR OXYGEN
117-84-0	MOLLAN O
103-23-1	MOLLAN S
123-79-5	MOLLAN S
1420-04-8	MOLLUTOX
1332-58-7	MOLOCHITE
577-11-7	MOLOFAC
13106-76-8	MOLYBDATE (MoO4 2-), DIAMMONIUM, (T-4)-
13106-76-8	MOLYBDATE, HEXAAMMONIUM (9CI)
1313-27-5	MOLYBDENA
7439-98-7	MOLYBDENUM
1317-33-5	MOLYBDENUM (IV) SULFIDE
10241-05-1	MOLYBDENUM CHLORIDE (MoCl5)
1313-27-5	MOLYBDENUM OXIDE
1313-27-5	MOLYBDENUM OXIDE (MoO3)
10241-05-1	MOLYBDENUM PENTACHLORIDE
10241-05-1	MOLYBDENUM PENTACHLORIDE (DOT)
1313-27-5	MOLYBDENUM TRIOXIDE
1313-27-5	MOLYBDENUM(VI) OXIDE
1313-27-5	MOLYBDENUM(VI) TRIOXIDE
13106-76-8	MOLYBDIC ACID (H2MoO4), DIAMMONIUM SALT
1313-27-5	MOLYBDIC ACID ANHYDRIDE
13106-76-8	MOLYBDIC ACID DIAMMONIUM SALT
13106-76-8	MOLYBDIC ACID, HEXAAMMONIUM SALT
1313-27-5	MOLYBDIC ANHYDRIDE
1313-27-5	MOLYBDIC TRIOXIDE
9002-84-0	MOLYKOTE 522
1333-86-4	MONARCH 1100
1333-86-4	MONARCH 4
1333-86-4	MONARCH 700
1333-86-4	MONARCH 71
1333-86-4	MONARCH 800
1333-86-4	MONARCH 81
2163-80-6	MONATE
577-11-7	MONAWET MD 70E
577-11-7	MONAWET MO 70
577-11-7	MONAWET MO-70 RP
577-11-7	MONAWET MO-84 R2W
9016-87-9	MONDUR E 441
9016-87-9	MONDUR E 541
9016-87-9	MONDUR MR
9016-87-9	MONDUR MRS
9016-87-9	MONDUR MRS 10
103-71-9	MONDUR P
26471-62-5	MONDUR TD
26471-62-5	MONDUR TD-80
26471-62-5	MONDUR TDS
10265-92-6	MONITOR
10265-92-6	MONITOR (INSECTICIDE)
74-88-4	MONO IODURO DI METILE (ITALIAN)
9016-45-9	MONO(NONYLPHENYL)POLYETHYLENE GLYCOL
71-63-6	MONO-GLYCOCARD
84-80-0	MONO-KAY
109-73-9	MONO-N-BUTYLAMINE
111-26-2	MONO-N-HEXYLAMINE
107-10-8	MONO-N-PROPYLAMINE
18351-85-4	MONO-O-CRESYL PHOSPHATE
107-11-9	MONOALLYLAMINE
7784-30-7	MONOALUMINUM PHOSPHATE
1066-33-7	MONOAMMONIUM CARBONATE
7803-63-6	MONOAMMONIUM HYDROGEN SULFATE
7773-06-0	MONOAMMONIUM SULFAMATE
7803-63-6	MONOAMMONIUM SULFATE
10192-30-0	MONOAMMONIUM SULFITE
110-58-7	MONOAMYLAMINE
1335-32-6	MONOBASIC LEAD ACETATE
100-46-9	MONOBENZYLAMINE
790-83-0	MONOBROMESSIGSAEURE (GERMAN)
598-31-2	MONOBROMOACETONE
598-21-0	MONOBROMOACETYL BROMIDE
108-86-1	MONOBROMOBENZENE
75-27-4	MONOBROMODICHLOROMETHANE
74-96-4	MONOBROMOETHANE
74-83-9	MONOBROMOMETHANE
75-63-8	MONOBROMOTRIFLUOROMETHANE
35884-77-6	MONOBROMOXYLENE
111-76-2	MONOBUTYL GLYCOL ETHER
109-73-9	MONOBUTYLAMINE
52740-16-6	MONOCALCIUM ARSENITE
79-11-8	MONOCHLOORAZIJNZUUR (DUTCH)
108-90-7	MONOCHLOORBENZEEN (DUTCH)
79-11-8	MONOCHLORACETIC ACID
78-95-5	MONOCHLORACETONE
108-90-7	MONOCHLORBENZENE
108-90-7	MONOCHLORBENZOL (GERMAN)
79-11-8	MONOCHLORESSIGSAEURE (GERMAN)
75-00-3	MONOCHLORETHANE
96-24-2	MONOCHLORHYDRIN
107-07-3	MONOCHLORHYDRINE DU GLYCOL (FRENCH)
25497-29-4	MONOCHLORO DIFLUOROETHANE
107-20-0	MONOCHLOROACETALDEHYDE
79-11-8	MONOCHLOROACETIC ACID
96-34-4	MONOCHLOROACETIC ACID METHYL ESTER

CAS No.	Chemical Name
78-95-5	MONOCHLOROACETONE
78-95-5	MONOCHLOROACETONE, INHIBITED (DOT)
78-95-5	MONOCHLOROACETONE, STABILIZED
78-95-5	MONOCHLOROACETONE, STABILIZED (DOT)
78-95-5	MONOCHLOROACETONE, UNSTABILIZED (DOT)
107-14-2	MONOCHLOROACETONITRILE
79-04-9	MONOCHLOROACETYL CHLORIDE
108-90-7	MONOCHLOROBENZENE
542-18-7	MONOCHLOROCYCLOHEXANE
124-48-1	MONOCHLORODIBROMOMETHANE
75-45-6	MONOCHLORODIFLUORMETHANE (DOT)
75-45-6	MONOCHLORODIFLUOROMETHANE
107-30-2	MONOCHLORODIMETHYL ETHER
75-00-3	MONOCHLOROETHANE
79-11-8	MONOCHLOROETHANOIC ACID
75-01-4	MONOCHLOROETHENE
75-01-4	MONOCHLOROETHYLENE
75-01-4	MONOCHLOROETHYLENE (DOT)
96-24-2	MONOCHLOROHYDRIN
74-87-3	MONOCHLOROMETHANE
107-14-2	MONOCHLOROMETHYL CYANIDE
542-88-1	MONOCHLOROMETHYL ETHER
107-30-2	MONOCHLOROMETHYL METHYL ETHER
74-97-5	MONOCHLOROMONOBROMOMETHANE
76-15-3	MONOCHLOROPENTAFLUOROETHANE
76-15-3	MONOCHLOROPENTAFLUOROETHANE (DOT)
25167-80-0	MONOCHLOROPHENOL
28554-00-9	MONOCHLOROPROPIONIC ACID
13465-78-6	MONOCHLOROSILANE
7790-94-5	MONOCHLOROSULFONIC ACID
7790-94-5	MONOCHLOROSULFURIC ACID
63938-10-3	MONOCHLOROTETRAFLUOROETHANE
63938-10-3	MONOCHLOROTETRAFLUOROETHANE (DOT)
79-38-9	MONOCHLOROTRIFLUOROETHYLENE
75-72-9	MONOCHLOROTRIFLUOROMETHANE
75-72-9	MONOCHLOROTRIFLUOROMETHANE (DOT)
75-77-4	MONOCHLOROTRIMETHYLSILICON
13530-68-2	MONOCHROMIUM OXIDE
1333-82-0	MONOCHROMIUM TRIOXIDE
6923-22-4	MONOCIL 40
108-90-7	MONOCLOROBENZENE (ITALIAN)
1333-86-4	MONOCOL 35T
1333-86-4	MONOCOL 37T
1333-86-4	MONOCOL MX 230
6923-22-4	MONOCRON
315-22-0	MONOCROTALIN
315-22-0	MONOCROTALINE
6923-22-4	MONOCROTOPHOS
6923-22-4	MONOCROTOPHOS (ACGIH)
372-09-8	MONOCYANOACETIC ACID
2074-87-5	MONOCYANOGEN
2016-57-1	MONODECYLAMINE
84-80-0	MONODION
151-21-3	MONODODECYL SODIUM SULFATE
141-43-5	MONOETHANOLAMINE
111-41-1	MONOETHANOLETHYLENEDIAMINE
111-90-0	MONOETHYL ETHER OF DIETHYLENE GLYCOL
540-82-9	MONOETHYL SULFATE
75-04-7	MONOETHYLAMINE
75-04-7	MONOETHYLAMINE (DOT)
75-04-7	MONOETHYLAMINE, ANHYDROUS (DOT)
110-73-6	MONOETHYLAMINOETHANOL
1789-58-8	MONOETHYLDICHLOROSILANE
107-21-1	MONOETHYLENE GLYCOL
110-71-4	MONOETHYLENE GLYCOL DIMETHYL ETHER
144-49-0	MONOFLUORAZIJNZUUR (DUTCH)
144-49-0	MONOFLUORESSIGSAURE (GERMAN)
62-74-8	MONOFLUORESSIGSAURES NATRIUM (GERMAN)
640-19-7	MONOFLUOROACETAMIDE
144-49-0	MONOFLUOROACETATE
144-49-0	MONOFLUOROACETIC ACID
462-06-6	MONOFLUOROBENZENE
75-43-4	MONOFLUORODICHLOROMETHANE
353-36-6	MONOFLUOROETHANE
75-02-5	MONOFLUOROETHYLENE
13537-32-1	MONOFLUOROPHOSPHORIC ACID
13537-32-1	MONOFLUOROPHOSPHORIC ACID, ANHYDROUS
13537-32-1	MONOFLUOROPHOSPHORIC ACID, ANHYDROUS (DOT)
7789-21-1	MONOFLUOROSULFURIC ACID

CAS No.	Chemical Name
75-69-4	MONOFLUOROTRICHLOROMETHANE
151-21-3	MONOGEN Y 100
151-21-3	MONOGEN Y 500
7782-65-2	MONOGERMANE
110-71-4	MONOGLYME
7783-00-8	MONOHYDRATED SELENIUM DIOXIDE
7558-79-4	MONOHYDROGEN DISODIUM PHOSPHATE
10039-32-4	MONOHYDROGEN DISODIUM PHOSPHATE DODECAHYDRATE
108-95-2	MONOHYDROXYBENZENE
67-56-1	MONOHYDROXYMETHANE
78-81-9	MONOISOBUTYLAMINE
78-96-6	MONOISOPROPANOLAMINE
1623-24-1	MONOISOPROPYL PHOSPHATE
75-31-0	MONOISOPROPYLAMINE
109-56-8	MONOISOPROPYLAMINOETHANOL
26264-05-1	MONOISOPROPYLAMMONIUM DODECYLBENZENESULFONATE
6923-22-4	MONOKROTOFOSZ (HUNGARIAN)
25322-69-4	MONOLAN PPG 700
10415-75-5	MONOMERCURY NITRATE
1344-48-5	MONOMERCURY SULFIDE
109-86-4	MONOMETHYL ETHER OF ETHYLENE GLYCOL
60-34-4	MONOMETHYL HYDRAZINE
109-83-1	MONOMETHYL-AMINOAETHANOL (GERMAN)
74-89-5	MONOMETHYLAMINE
74-89-5	MONOMETHYLAMINE, ANHYDROUS (DOT)
74-89-5	MONOMETHYLAMINE, AQUEOUS SOLUTION (DOT)
75-04-7	MONOMETHYLAMINE, AQUEOUS SOLUTION (DOT)
109-83-1	MONOMETHYLAMINOETHANOL
100-61-8	MONOMETHYLANILINE
75-54-7	MONOMETHYLDICHLOROSILANE
109-83-1	MONOMETHYLETHANOLAMINE
109-86-4	MONOMETHYLGLYCOL
109-83-1	MONOMETHYLMONOETHANOLAMINE
1313-99-1	MONONICKEL OXIDE
27254-36-0	MONONITRONAPHTHALENE
25154-55-6	MONONITROPHENOL
1321-12-6	MONONITROTOLUENE
12167-20-3	MONONITROTOLUOL
111-86-4	MONOOCTYLAMINE
2570-26-5	MONOPENTADECYLAMINE
115-77-5	MONOPENTAERYTHRITOL
115-77-5	MONOPENTEK
110-58-7	MONOPENTYLAMINE
79-21-0	MONOPERACETIC ACID
108-95-2	MONOPHENOL
100-63-0	MONOPHENYLHYDRAZINE
64-10-8	MONOPHENYLUREA
103-23-1	MONOPLEX DOA
123-79-5	MONOPLEX DOA
9016-45-9	MONOPOL 1020
7784-41-0	MONOPOTASSIUM ARSENATE
7784-41-0	MONOPOTASSIUM DIHYDROGEN ARSENATE
127-95-7	MONOPOTASSIUM OXALATE
7646-93-7	MONOPOTASSIUM SULFATE
79-09-4	MONOPROP
107-10-8	MONOPROPYLAMINE
107-10-8	MONOPROPYLAMINE (DOT)
57-55-6	MONOPROPYLENE GLYCOL
109-97-7	MONOPYRROLE
94-75-7	MONOSAN
7803-62-5	MONOSILANE
7803-62-5	MONOSILANE (SiH4)
2163-80-6	MONOSODIUM ACID METHANEARSONATE
2163-80-6	MONOSODIUM ACID METHARSONATE
7681-38-1	MONOSODIUM HYDROGEN SULFATE
2163-80-6	MONOSODIUM METHANEARSONATE
2163-80-6	MONOSODIUM METHANEARSONIC ACID
2163-80-6	MONOSODIUM METHYL ARSONATE
70-30-4	MONOSODIUM SALT OF 2,2'-METHYLENE BIS(3,4,6-TRICHLOROPHENOL)
7681-38-1	MONOSODIUM SULFATE
7631-90-5	MONOSODIUM SULFITE
9002-89-5	MONOSOL 9000-0015-3
10545-99-0	MONOSULFUR DICHLORIDE
507-09-5	MONOTHIOACETIC ACID
60-24-2	MONOTHIOETHYLENE GLYCOL
60-24-2	MONOTHIOGLYCOL
689-97-4	MONOVINYLACETYLENE
577-11-7	MONOXOL OT

CAS No.	Chemical Name
3926-62-3	MONOXONE
122-14-5	MONSANTO CP 47114
2665-30-7	MONSANTO CP-40294
63-25-2	MONSUR
1336-36-3	MONTAR
2235-54-3	MONTOPOL LA 20
151-21-3	MONTOPOL LA PASTE
299-86-5	MONTREL
56-81-5	MOON
299-84-3	MOORMAN'S MEDICATED RID-EZY
62-73-7	MOPARI
2636-26-2	MOPARI
50-00-0	MORBICID
112-02-7	MORPAN CHA
110-91-8	MORPHOLINE
110-91-8	MORPHOLINE (ACGIH,DOT)
141-91-3	MORPHOLINE, 2,6-DIMETHYL-
2038-03-1	MORPHOLINE, 4-(2-AMINOETHYL)-
123-00-2	MORPHOLINE, 4-(3-AMINOPROPYL)-
1696-20-4	MORPHOLINE, 4-ACETYL-
123-00-2	MORPHOLINE, 4-AMINOPROPYL-
100-74-3	MORPHOLINE, 4-ETHYL-
109-02-4	MORPHOLINE, 4-METHYL-
59-89-2	MORPHOLINE, 4-NITROSO-
92-53-5	MORPHOLINE, 4-PHENYL-
110-91-8	MORPHOLINE, AQUEOUS MIXTURE [CORROSIVE LABEL]
110-91-8	MORPHOLINE, AQUEOUS MIXTURE [FLAMMABLE LIQUID LABEL]
110-91-8	MORPHOLINE, AQUEOUS, MIXTURE (DOT)
123-00-2	MORPHOLINE, N-AMINOPROPYL-
109-02-4	MORPHOLINE, N-METHYL-
92-53-5	MORPHOLINOBENZENE
2038-03-1	MORPHOLINOETHYLAMINE
502-39-6	MORSODREN
556-61-6	MORTON EP-161E
502-39-6	MORTON EP-227
2631-37-0	MORTON EP-316
502-39-6	MORTON SOIL DRENCH
502-39-6	MORTON SOIL-DRENCH-C
51-83-2	MORYL
121-75-5	MOSCARDA
12002-03-8	MOSS GREEN
94-75-7	MOTA MASKROS
91-20-3	MOTH BALLS
91-20-3	MOTH BALLS (DOT)
91-20-3	MOTH FLAKES
60-57-1	MOTH SNUB D
86290-81-5	MOTOR FUELS
8006-61-9	MOTOR SPIRIT
8001-35-2	MOTOX
25321-22-6	MOTT-EX
67-72-1	MOTTENHEXE
25321-22-6	MOTTENSCHUTZMITTEL EVAU P
1332-21-4	MOUNTAIN CORK
12002-03-8	MOUNTAIN GREEN
1332-21-4	MOUNTAIN LEATHER
1332-21-4	MOUNTAIN WOOD
1314-84-7	MOUS-CON
81-81-2	MOUSE PAK
57-24-9	MOUSE-NOTS
57-24-9	MOUSE-RID
57-24-9	MOUSE-TOX
9002-89-5	MOWIOL
9002-89-5	MOWIOL 4-88
9002-89-5	MOWIOL 4-98
9002-89-5	MOWIOL 42-99
9002-89-5	MOWIOL N 30-88
9002-89-5	MOWIOL N 50-88
9002-89-5	MOWIOL N 50-98
9002-89-5	MOWIOL N 70-98
9002-89-5	MOWIOL N 85-88
72-43-5	MOXIE
94-75-7	MOXONE
8020-83-5	MP 12
14807-96-6	MP 12-50
14807-96-6	MP 25-38
14807-96-6	MP 40-27
14807-96-6	MP 45-26
107-41-5	MPD
7782-42-5	MPG 6
107-87-9	MPK
55-38-9	MPP
55-38-9	MPP (PESTICIDE)
8020-83-5	MR 6
109-94-4	MROWCZAN ETYLU (POLISH)
114-26-1	MROWKOZOL
558-25-8	MSF
2163-80-6	MSMA
58-89-9	MSZYCOL
1333-86-4	MT
1333-86-4	MT (CARBON BLACK)
10265-92-6	MTD
1129-41-5	MTMC
56-04-2	MTU
63-25-2	MUGAN
7446-14-2	MULHOUSE WHITE
56-04-2	MURACIL
446-86-6	MURAN
56-38-2	MURFOS
10025-65-7	MURIATE OF PLATINUM
7647-01-0	MURIATIC ACID
7647-01-0	MURIATIC ACID (DOT)
75-00-3	MURIATIC ETHER
3691-35-8	MURIOL
63-25-2	MURVIN
56-72-4	MUSCATOX
2763-96-4	MUSCIMOL
56-72-4	MUSCOTOX
81-15-2	MUSK EXYLENE
81-15-2	MUSK XYLENE
81-15-2	MUSK XYLOL
14807-96-6	MUSSOLINITE
505-60-2	MUSTARD GAS
505-60-2	MUSTARD HD
57-06-7	MUSTARD OIL
8007-40-7	MUSTARD OIL
505-60-2	MUSTARD VAPOR
505-60-2	MUSTARD, SULFUR
51-75-2	MUSTARGEN
55-86-7	MUSTARGEN HYDROCHLORIDE
51-75-2	MUSTINE
55-86-7	MUSTINE HYDROCHLORIDE
302-70-5	MUSTRON
51-75-2	MUTAGEN
50-07-7	MUTAMYCIN
50-07-7	MUTAMYCIN (MITOMYCIN FOR INJECTION)
79-27-6	MUTHMANN'S LIQUID
50-29-3	MUTOXAN
1477-55-0	MXDA
56-75-7	MYCHEL
56-75-7	MYCHEL-VET
56-75-7	MYCINOL
56-75-7	MYCLOCIN
56-75-7	MYCOCHLORIN
55-98-1	MYELOSAN
55-98-1	MYLECYTAN
55-98-1	MYLERAN
107-72-2	MYLSILANE
55-63-0	MYOCON
71-63-6	MYODIGIN
9004-66-4	MYOFER 100
55-63-0	MYOGLYCERIN
112-72-1	MYRISTIC ALCOHOL
27196-00-5	MYRISTIC ALCOHOL
112-72-1	MYRISTYL ALCOHOL
27196-00-5	MYRISTYL ALCOHOL
112-72-1	MYRISTYL ALCOHOL (MIXED ISOMERS)
27196-00-5	MYRISTYL ALCOHOL (MIXED ISOMERS)
64-18-6	MYRMICYL
8002-74-2	MYSTOLENE MK 7
8002-74-2	MYSTOLENE SP 30
131-52-2	MYSTOX D
50-07-7	MYTOMYCIN
4301-50-2	MYTROL
9016-45-9	N 100
1333-86-4	N 234
944-22-9	N 2790
9002-89-5	N 300
1333-86-4	N 326

CAS No.	Chemical Name
1333-86-4	N 339
8020-83-5	N 350
1333-86-4	N 351
1333-86-4	N 358
1333-86-4	N 358 (CARBON BLACK)
1333-86-4	N 375
1333-86-4	N 472
8020-83-5	N 500
1333-86-4	N 550
1333-86-4	N 630
1333-86-4	N 630 (CARBON BLACK)
1333-86-4	N 650
1333-86-4	N 660
1333-86-4	N 660 (CARBON BLACK)
1333-86-4	N 76225
1333-86-4	N 76230
1333-86-4	N 76230 (CARBON BLACK)
1333-86-4	N 774
1333-86-4	N 787
1333-86-4	N 990
1333-86-4	N 990 (CARBON BLACK)
556-88-7	N"-NITROGUANIDINE
23135-22-0	N',N'-DIMETHYL-N-((METHYLCARBAMOYL)OXY)-1-THIO-OXAMIMIDIC ACID METHYL ESTER
330-54-1	N'-(3,4-DICHLOROPHENYL)-N,N-DIMETHYLUREA
1982-47-4	N'-4-(4-CHLOROPHENOXY)PHENYL-N,N-DIMETHYLUREA
16543-55-8	N'-NITROSONORNICOTINE
2842-38-8	N(2-HYDROXETHYL)CYCLOHEXYLAMINE
10138-74-6	N(2-HYDROXYETHYL)PROPYLENE DIAMINE
140-80-7	N(SUP 1),N(SUP 1)-DIETHYL-1,4-PENTANEDIAMINE
101-25-7	N(SUP 1),N(SUP 3)-DINITROSOPENTAMETHYLENETETRAMINE
545-55-1	N,N',N"-TRI-1,2-ETHANEDIYLPHOSPHORIC TRIAMIDE
87-90-1	N,N',N"-TRICHLOROISOCYANURIC ACID
545-55-1	N,N',N"-TRIETHYLENEPHOSPHORAMIDE
545-55-1	N,N',N"-TRIETHYLENEPHOSPHORIC TRIAMIDE
52-24-4	N,N',N"-TRIETHYLENETHIOPHOSPHORAMIDE
52-24-4	N,N',N"-TRIETHYLENETHIOPHOSPHORTRIAMIDE
137-26-8	N,N'-(DITHIODICARBONOTHIOYL)BIS(N-METHYLMETHANAMINE)
122-66-7	N,N'-BIANILINE
3081-14-9	N,N'-BIS(1,4-DIMETHYLPENTYL)-1,4-BENZENEDIAMINE
3081-14-9	N,N'-BIS(1,4-DIMETHYLPENTYL)-1,4-PHENYLENEDIAMINE
3081-14-9	N,N'-BIS(1,4-DIMETHYLPENTYL)-P-PHENYLENEDIAMINE
112-24-3	N,N'-BIS(2-AMINOETHYL)-1,2-DIAMINOETHANE
112-24-3	N,N'-BIS(2-AMINOETHYL)-1,2-ETHANEDIAMINE
112-24-3	N,N'-BIS(2-AMINOETHYL)ETHYLENEDIAMINE
78-67-1	N,N'-BIS(2-CYANO-2-PROPYL)DIAZENE
56-18-8	N,N'-BIS(TRIMETHYLENEAMINO)TRIAMINE
3081-14-9	N,N'-DI(1,4-DIMETHYLPENTYL)-P-PHENYLENEDIAMINE
101-96-2	N,N'-DI-SEC-BUTYL-P-PHENYLDIAMINE
101-96-2	N,N'-DI-SEC-BUTYL-P-PHENYLENEDIAMINE
613-35-4	N,N'-DIACETYLBENZIDINE
4835-11-4	N,N'-DIBUTYL-1,6-HEXANEDIAMINE
4835-11-4	N,N'-DIBUTYLHEXAMETHYLENEDIAMINE
118-52-5	N,N'-DICHLORO-5,5-DIMETHYLHYDANTOIN
32280-46-9	N,N'-DIETHYL-1,3-BUTANEDIAMINE
1615-80-1	N,N'-DIETHYLHYDRAZINE
371-86-8	N,N'-DIISOPROPIL-FOSFORODIAMMIDO-FLUORURO (ITALIAN)
371-86-8	N,N'-DIISOPROPYL-DIAMIDO-FOSFORZUUR-FLUORIDE (D UTCH)
371-86-8	N,N'-DIISOPROPYL-DIAMIDO-PHOSPHORSAEURE-FLUORID (GERMAN)
4685-14-7	N,N'-DIMETHYL-4,4'-BIPYRIDINIUM
4685-14-7	N,N'-DIMETHYL-4,4'-BIPYRIDINIUM DICATION
1910-42-5	N,N'-DIMETHYL-4,4'-BIPYRIDINIUM DICHLORIDE
2074-50-2	N,N'-DIMETHYL-4,4'-BIPYRIDINIUM METHOSULFATE
1910-42-5	N,N'-DIMETHYL-4,4'-DIPYRIDYLIUM DICHLORIDE
540-73-8	N,N'-DIMETHYLHYDRAZINE
101-25-7	N,N'-DINITROSOPENTAMETHYLENETETRAMINE
122-66-7	N,N'-DIPHENYLHYDRAZINE
280-57-9	N,N'-ENDO-ETHYLENEPIPERAZINE
142-59-6	N,N'-ETHYLENE BIS(DITHIOCARBAMATE DE SODIUM) (FRENCH)
62207-76-5	N,N'-ETHYLENEBIS(3-FLUOROSALICYLIDENEIMINATO)COBALT (II)
96-45-7	N,N'-ETHYLENETHIOUREA
142-59-6	N,N'-ETILEN-BIS(DITIOCARBAMMATO) DI SODIO (ITALIAN)
97-77-8	N,N,N',N'-TETRAETHYLTHIURAM DISULFIDE
110-18-9	N,N,N',N'-TETRAMETHYL-1,2-DIAMINOETHANE
110-18-9	N,N,N',N'-TETRAMETHYL-1,2-ETHANEDIAMINE
90-94-8	N,N,N',N'-TETRAMETHYL-4,4'-DIAMINOBENZOPHENONE
101-61-1	N,N,N',N'-TETRAMETHYL-4,4'-DIAMINODIPHENYLMETHANE
115-26-4	N,N,N',N'-TETRAMETHYL-DIAMIDO-FLUOR-PHOSPHIN-OXID (GERMAN)
115-26-4	N,N,N',N'-TETRAMETHYL-DIAMIDO-PHOSPHORSAEURE-FLUORID (GERMAN)
101-61-1	N,N,N',N'-TETRAMETHYL-P,P'-DIAMINODIPHENYLMETHANE
110-18-9	N,N,N',N'-TETRAMETHYLDIAMINOETHANE
110-18-9	N,N,N',N'-TETRAMETHYLETHANEDIAMINE
110-18-9	N,N,N',N'-TETRAMETHYLETHYLENEDIAMINE
115-26-4	N,N,N',N'-TETRAMETHYLPHOSPHORODIAMIDIC FLUORIDE
137-26-8	N,N,N',N'-TETRAMETHYLTHIURAM DISULFIDE
115-26-4	N,N,N',N'-TETRAMETIL-FOSFORODIAMMIDO-FLUORURO(ITAL-IAN)
111-40-0	N,N-BIS(2-AMINOETHYL)AMINE
494-03-1	N,N-BIS(2-CHLOROETHYL)-2-NAPHTHYLAMINE
50-18-0	N,N-BIS(2-CHLOROETHYL)-N',O-PROPYLENEPHOSPHORIC ACID ESTER DIAMIDE
51-75-2	N,N-BIS(2-CHLOROETHYL)METHYLAMINE
4985-85-7	N,N-BIS(2-HYDROXYETHYL)-1,3-PROPANEDIAMINE
91-99-6	N,N-BIS(2-HYDROXYETHYL)-3-METHYLANILINE
91-99-6	N,N-BIS(2-HYDROXYETHYL)-M-TOLUIDINE
111-42-2	N,N-BIS(2-HYDROXYETHYL)AMINE
102-79-4	N,N-BIS(2-HYDROXYETHYL)BUTYLAMINE
139-87-7	N,N-BIS(2-HYDROXYETHYL)ETHYLAMINE
105-59-9	N,N-BIS(2-HYDROXYETHYL)METHYLAMINE
110-96-3	N,N-BIS(2-METHYLPROPYL)AMINE
50-18-0	N,N-BIS(BETA-CHLOROETHYL)-N',O-TRIMETHYLENEPHOSPHORIC ACID ESTER DIAMIDE
91-99-6	N,N-BIS(BETA-HYDROXYETHYL)-3-METHYLANILINE
139-13-9	N,N-BIS(CARBOXYMETHYL)GLYSINE
4985-85-7	N,N-BIS(HYDROXYETHYL)-1,3-PROPANEDIAMINE
4985-85-7	N,N-BIS(HYDROXYETHYL)TRIMETHYLENEDIAMINE
3081-14-9	N,N-BIS-(1,4-METHYL PENTYL)-P-PHENYLENEDIAMINE
4985-85-7	N,N-DI(2-HYDROXYETHYL)-1,3-PROPANEDIAMINE
51-75-2	N,N-DI(CHLOROETHYL)METHYLAMINE
924-16-3	N,N-DI-N-BUTYLNITROSAMINE
1582-09-8	N,N-DI-N-PROPYL-2,6-DINITRO-4-TRIFLUOROMETHYLANILINE
101-96-2	N,N-DI-SEC-BUTYL-P-PHENYLENE-DIAMINE
613-35-4	N,N-DIACETYLBENZIDINE
102-81-8	N,N-DIBUTYL-2-HYDROXYETHYLAMINE
2109-64-0	N,N-DIBUTYL-2-HYDROXYPROPYLAMINE
102-81-8	N,N-DIBUTYL-N-(2-HYDROXYETHYL)AMINE
102-81-8	N,N-DIBUTYLAMINOETHANOL
613-29-6	N,N-DIBUTYLANILINE
613-29-6	N,N-DIBUTYLBENZENAMINE
102-81-8	N,N-DIBUTYLETHANOLAMINE
101-83-7	N,N-DICYCLOHEXYLAMINE
1212-29-9	N,N-DICYCLOHEXYLTHIOREA
111-42-2	N,N-DIETHANOLAMINE
1116-54-7	N,N-DIETHANOLNITROSAMINE
100-36-7	N,N-DIETHYL-1,2-DIAMINOETHANE
100-36-7	N,N-DIETHYL-1,2-ETHANEDIAMINE
32280-46-9	N,N-DIETHYL-1,3-BUTANEDIAMINE
104-78-9	N,N-DIETHYL-1,3-DIAMINOPROPANE
104-78-9	N,N-DIETHYL-1,3-PROPANEDIAMINE
104-78-9	N,N-DIETHYL-1,3-PROPYLENEDIAMINE
100-37-8	N,N-DIETHYL-2-AMINOETHANOL
100-37-8	N,N-DIETHYL-2-HYDROXYETHYLAMINE
93-05-0	N,N-DIETHYL-4-AMINOANILINE
1642-54-2	N,N-DIETHYL-4-METHYL-1-PIPERAZINE CARBOXAMIDE CITRATE
1642-54-2	N,N-DIETHYL-4-METHYL-1-PIPERAZINECARBOXAMIDE DIHYDRO-GEN CITRATE
100-37-8	N,N-DIETHYL-N-(BETA-HYDROXYETHYL)AMINE
93-05-0	N,N-DIETHYL-P-PHENYLENEDIAMINE
109-89-7	N,N-DIETHYLAMINE
91-66-7	N,N-DIETHYLAMINOBENZENE
2426-54-2	N,N-DIETHYLAMINOETHYL ACRYLATE
104-78-9	N,N-DIETHYLAMINOPROPYLAMINE
91-66-7	N,N-DIETHYLANILIN (CZECH)
91-66-7	N,N-DIETHYLANILINE
91-66-7	N,N-DIETHYLANILINE (DOT)
91-66-7	N,N-DIETHYLBENZENAMINE
772-54-3	N,N-DIETHYLBENZYLAMINE
88-10-8	N,N-DIETHYLCARBAMOYL CHLORIDE
110-85-0	N,N-DIETHYLENE DIAMINE (DOT)
121-44-8	N,N-DIETHYLETHANAMINE
100-37-8	N,N-DIETHYLETHANOLAMINE
100-36-7	N,N-DIETHYLETHYLENEDIAMINE
3710-84-7	N,N-DIETHYLHYDROXYAMINE
3710-84-7	N,N-DIETHYLHYDROXYLAMINE
100-37-8	N,N-DIETHYLMONOETHANOLAMINE

CAS No.	Chemical Name
55-18-5	N,N-DIETHYLNITROSOAMINE
104-78-9	N,N-DIETHYLPROPYLENEDIAMINE
105-55-5	N,N-DIETHYLTHIOUREA
104-78-9	N,N-DIETHYLTRIMETHYLENEDIAMINE
143-16-8	N,N-DIHEXYLAMINE
96-80-0	N,N-DIISOPROPYL ETHANOLAMINE (DOT)
108-18-9	N,N-DIISOPROPYLAMINE
96-80-0	N,N-DIISOPROPYLETHANOLAMINE
68-12-2	N,N-DIMETHYL FORMAMIDE
108-01-0	N,N-DIMETHYL(2-HYDROXYETHYL)AMINE
109-55-7	N,N-DIMETHYL-1,3-DIAMINOPROPANE
109-55-7	N,N-DIMETHYL-1,3-PROPANEDIAMINE
109-55-7	N,N-DIMETHYL-1,3-PROPYLENEDIAMINE
99-98-9	N,N-DIMETHYL-1,4-BENZENEDIAMINE
99-98-9	N,N-DIMETHYL-1,4-DIAMINOBENZENE
99-98-9	N,N-DIMETHYL-1,4-PHENYLENEDIAMINE
108-01-0	N,N-DIMETHYL-2-AMINOETHANOL
108-01-0	N,N-DIMETHYL-2-HYDROXYETHYLAMINE
108-16-7	N,N-DIMETHYL-2-HYDROXYPROPYLAMINE
138-89-6	N,N-DIMETHYL-4-NITROSOANILINE
2449-49-2	N,N-DIMETHYL-ALPHA-METHYLBENZYLAMINE
23135-22-0	N,N-DIMETHYL-ALPHA-METHYLCARBAMOYLOXYIMINO-ALPHA-(METHYLTHIO)ACETAMIDE
2449-49-2	N,N-DIMETHYL-ALPHA-PHENYLETHYLAMINE
108-01-0	N,N-DIMETHYL-BETA-HYDROXYETHYLAMINE
108-01-0	N,N-DIMETHYL-N-(2-HYDROXYETHYL)AMINE
109-55-7	N,N-DIMETHYL-N-(3-AMINOPROPYL)AMINE
108-01-0	N,N-DIMETHYL-N-(BETA-HYDROXYETHYL)AMINE
926-63-6	N,N-DIMETHYL-N-PROPYLAMINE
60-11-7	N,N-DIMETHYL-P-(PHENYLAZO)ANILINE
99-98-9	N,N-DIMETHYL-P-BENZENEDIAMINE
138-89-6	N,N-DIMETHYL-P-NITROSOANILINE
53324-05-3	N,N-DIMETHYL-P-NITROSOANILINE
99-98-9	N,N-DIMETHYL-P-PHENYLENEDIAMINE
127-19-5	N,N-DIMETHYLACETAMIDE
108-16-7	N,N-DIMETHYLAMINO-2-PROPANOL
121-69-7	N,N-DIMETHYLAMINOBENZENE
79-44-7	N,N-DIMETHYLAMINOCARBONYL CHLORIDE
108-01-0	N,N-DIMETHYLAMINOETHANOL
2867-47-2	N,N-DIMETHYLAMINOETHYL METHACRYLATE
121-69-7	N,N-DIMETHYLANILINE
121-69-7	N,N-DIMETHYLANILINE (DOT)
121-69-7	N,N-DIMETHYLBENZENAMINE
121-69-7	N,N-DIMETHYLBENZENEAMINE
79-44-7	N,N-DIMETHYLCARBAMIC ACID CHLORIDE
79-44-7	N,N-DIMETHYLCARBAMIDOYL CHLORIDE
79-44-7	N,N-DIMETHYLCARBAMOYL CHLORIDE
79-44-7	N,N-DIMETHYLCARBAMOYL CHLORIDE (DOT)
79-44-7	N,N-DIMETHYLCARBAMYL CHLORIDE
1467-79-4	N,N-DIMETHYLCYANAMIDE
127-19-5	N,N-DIMETHYLETHANAMIDE
108-01-0	N,N-DIMETHYLETHANOLAMINE
2867-47-2	N,N-DIMETHYLETHANOLAMINE METHACRYLATE
68-12-2	N,N-DIMETHYLFORMAMIDE
68-12-2	N,N-DIMETHYLFORMAMIDE (DOT)
57-14-7	N,N-DIMETHYLHYDRAZINE
108-16-7	N,N-DIMETHYLISOPROPANOLAMINE
68-12-2	N,N-DIMETHYLMETHANAMIDE
75-50-3	N,N-DIMETHYLMETHANAMINE
62-75-9	N,N-DIMETHYLNITROSAMINE
53324-05-3	N,N-DIMETHYLNITROSOANILINE
121-69-7	N,N-DIMETHYLPHENYLAMINE
926-63-6	N,N-DIMETHYLPROPYLAMINE
109-55-7	N,N-DIMETHYLPROPYLENEDIAMINE
109-55-7	N,N-DIMETHYLTRIMETHYLENEDIAMINE
101-25-7	N,N-DINITROSOPENTAMETHYLENETETRAMINE
86-30-6	N,N-DIPHENYL-N-NITROSOAMINE
122-39-4	N,N-DIPHENYLAMINE
603-34-9	N,N-DIPHENYLANILINE
86-30-6	N,N-DIPHENYLNITROSAMINE
1582-09-8	N,N-DIPROPYL-2,6-DINITRO-4-TRIFLUORMETHYLANILIN (GERMAN)
1582-09-8	N,N-DIPROPYL-4-TRIFLUOROMETHYL-2,6-DINITROANILINE
621-64-7	N,N-DIPROPYLNITROSAMINE
137-26-8	N,N-TETRAMETHYLTHIURAM DISULPHIDE
133-06-2	N-((TRICHLOROMETHYL)THIO)-4-CYCLOHEXENE-1,2-DICARBOXIMIDE
133-06-2	N-((TRICHLOROMETHYL)THIO)TETRAHYDROPHTHALIMIDE
2038-03-1	N-(BETA-AMINOETHYL)MORPHOLINE
122-98-5	N-(BETA-HYDROXYETHYL)ANILINE
2425-06-1	N-(1,1,2,2-TETRACHLOROETHYLTHIO)-DELTA4-TETRAHYDROPHTHALIMIDE
23950-58-5	N-(1,1-DIMETHYLPROPYNYL)-3,5-DICHLOROBENZAMIDE
108-18-9	N-(1-METHYLETHYL)-2-PROPANAMINE
768-52-5	N-(1-METHYLETHYL)BENZENAMINE
626-23-3	N-(1-METHYLPROPYL)-2-BUTANAMINE
30553-04-9	N-(1-NAPHTHYL)-2-THIOUREA
111-41-1	N-(2'-HYDROXYETHYL)ETHYLENEDIAMINE
111-40-0	N-(2-AMINOETHYL)-1,2-ETHANEDIAMINE
100-36-7	N-(2-AMINOETHYL)-N,N-DIETHYLAMINE
111-41-1	N-(2-AMINOETHYL)ETHANOLAMINE
111-40-0	N-(2-AMINOETHYL)ETHYLENEDIAMINE
2038-03-1	N-(2-AMINOETHYL)MORPHOLINE
140-31-8	N-(2-AMINOETHYL)PIPERAZINE
63-92-3	N-(2-CHLOROETHYL)-N-(1-METHYL-2-PHENOXYETHYL)BENZYLAMINE, HYDROCHLORIDE
5344-82-1	N-(2-CHLOROPHENYL)THIOUREA
702-03-4	N-(2-CYANOETHYL)-N-CYCLOHEXYLAMINE
702-03-4	N-(2-CYANOETHYL)CYCLOHEXYLAMINE
100-36-7	N-(2-DIETHYLAMINOETHYL)AMINE
5432-61-1	N-(2-ETHYLHEXYL)CYCLOHEXYLAMINE
111-41-1	N-(2-HYDROXYETHYL)-1,2-ETHANEDIAMINE
2842-38-8	N-(2-HYDROXYETHYL)-N-CYCLOHEXYLAMINE
92-50-2	N-(2-HYDROXYETHYL)-N-ETHYLANILINE
122-98-5	N-(2-HYDROXYETHYL)ANILINE
122-98-5	N-(2-HYDROXYETHYL)BENZENAMINE
100-37-8	N-(2-HYDROXYETHYL)DIETHYLAMINE
108-01-0	N-(2-HYDROXYETHYL)DIMETHYLAMINE
111-41-1	N-(2-HYDROXYETHYL)ETHYLENE DIAMINE
109-83-1	N-(2-HYDROXYETHYL)METHYLAMINE
622-40-2	N-(2-HYDROXYETHYL)MORPHOLINE
122-98-5	N-(2-HYDROXYETHYL)PHENYLAMINE
10138-74-6	N-(2-HYDROXYETHYL)PROPYLENE DIAMINE
2109-64-0	N-(2-HYDROXYPROPYL)DIBUTYLAMINE
135-88-6	N-(2-NAPHTHYL)-N-PHENYLAMINE
135-88-6	N-(2-NAPHTHYL)ANILINE
330-54-1	N-(3,4-DICHLOROPHENYL)-N',N'-DIMETHYLUREA
56-18-8	N-(3-AMINOPROPYL)-1,3-PROPANEDIAMINE
3312-60-5	N-(3-AMINOPROPYL)CYCLOHEXYLAMINE
4985-85-7	N-(3-AMINOPROPYL)DIETHANOLAMINE
123-00-2	N-(3-AMINOPROPYL)MORPHOLINE
104-78-9	N-(3-DIETHYLAMINOPROPYL)AMINE
54-62-6	N-(4-((2,4-DIAMINO-6-PTERIDINYL)METHYL)AMINO)BENZOYL)-L-GLUTAMIC ACID
531-82-8	N-(4-(5-NITRO-2-FURYL)2-THIAZOLYL)ACETAMIDE
62-44-2	N-(4-ETHOXYPHENYL)ACETAMIDE
103-89-9	N-(4-METHYLPHENYL)ACETAMIDE
53558-25-1	N-(4-NITROPHENYL)-N'-(3-PYRIDINYLMETHYL)UREA
102-01-2	N-(ACETYLACETYL)ANILINE
111-41-1	N-(AMINOETHYL)ETHANOLAMINE
111-41-1	N-(BETA-AMINOETHYL)ETHANOLAMINE
140-31-8	N-(BETA-AMINOETHYL)PIPERAZINE
111-41-1	N-(BETA-HYDROXYETHYL)ETHYLENEDIAMINE
926-64-7	N-(CYANOMETHYL)DIMETHYLAMINE
622-44-6	N-(DICHLOROMETHYLENE)ANILINE
517-16-8	N-(ETHYLMERCURIC)-P-TOLUENESULPHONANILIDE
111-41-1	N-(HYDROXYETHYL)ETHYLENEDIAMINE
732-11-6	N-(MERCAPTOMETHYL)PHTHALIMIDE S-(O,O-DIMETHYL PHOSPHORODITHIOATE)
614-78-8	N-(O-TOLYL)THIOUREA
702-03-4	N-(P-CYANOETHYL)CYCLOHEXYLAMINE
772-54-3	N-(PHENYLMETHYL)DIETHYLAMINE
2425-06-1	N-(TETRACHLOROETHYLTHIO)TETRAHYDROPHTHALIMIDE
133-06-2	N-(TRICHLOR-METHYLTHIO)-PHTHALIMID (GERMAN)
133-06-2	N-(TRICHLOROMETHYLMERCAPTO)-DELTA(SUP 4)-TETRAHYDROPHTHALIMIDE
622-40-2	N-BETA-HYDROXYETHYLMORPHOLINE
135-88-6	N-BETA-NAPHTHYL-N-PHENYLAMINE
872-05-9	N-1-DECENE
86-88-4	N-1-NAPHTHYLTHIOUREA
111-66-0	N-1-OCTENE
629-76-5	N-1-PENTADECANOL
821-95-4	N-1-UNDECENE
10137-80-1	N-2(ETHYLHEXYL)ANILINE
53-96-3	N-2-FLUORENYLACETAMIDE
56-18-8	N-3-AMINOPROPYL-1,3-DIAMINOPROPANE
53558-25-1	N-3-PYRIDYLMETHYL-N'-P-NITROPHENYLUREA
97-36-9	N-ACETOACETYL-2,4-XYLIDINE

CAS No.	Chemical Name
92-15-9	N-ACETOACETYL-O-ANISIDINE
1696-20-4	N-ACETYL MORPHOLINE
64-86-8	N-ACETYL TRIMETHYLCOLCHICINIC ACID METHYLETHER
62-44-2	N-ACETYL-4-ETHOXYANILINE
103-89-9	N-ACETYL-4-METHYLANILINE
62-44-2	N-ACETYL-P-ETHOXYANILINE
62-44-2	N-ACETYL-P-PHENETIDINE
103-89-9	N-ACETYL-P-TOLUIDINE
103-84-4	N-ACETYLANILINE
766-09-6	N-AETHYLPIPERIDIN (GERMAN)
111-41-1	N-AMINOETHYLETHANOLAMINE
2038-03-1	N-AMINOETHYLMORPHOLINE
140-31-8	N-AMINOETHYLPIPERAZINE
140-31-8	N-AMINOETHYLPIPERAZINE (DOT)
123-00-2	N-AMINOPROPYLMORPHOLINE
123-00-2	N-AMINOPROPYLMORPHOLINE (DOT)
79-19-6	N-AMINOTHIOUREA
628-63-7	N-AMYL ACETATE
628-63-7	N-AMYL ACETATE (ACGIH)
71-41-0	N-AMYL ALCOHOL
71-41-0	N-AMYL ALCOHOL (DOT)
110-53-2	N-AMYL BROMIDE
29756-38-5	N-AMYL BROMIDE
540-18-1	N-AMYL BUTYRATE
540-18-1	N-AMYL BUTYRATE (DOT)
543-59-9	N-AMYL CHLORIDE
638-49-3	N-AMYL FORMATE
5350-03-8	N-AMYL LAURATE
110-66-7	N-AMYL MERCAPTAN
110-43-0	N-AMYL METHYL KETONE
463-04-7	N-AMYL NITRITE
71-41-0	N-AMYLALKOHOL (CZECH)
110-58-7	N-AMYLAMINE
538-68-1	N-AMYLBENZENE
109-67-1	N-AMYLENE (DOT)
513-35-9	N-AMYLENE (DOT)
92-59-1	N-BENZYL-N-ETHYLANILINE
772-54-3	N-BENZYLDIETHYLAMINE
71-36-3	N-BUTAN-1-OL
123-72-8	N-BUTANAL
123-72-8	N-BUTANAL (CZECH)
106-97-8	N-BUTANE
109-74-0	N-BUTANENITRILE
109-79-5	N-BUTANETHIOL
107-92-6	N-BUTANOIC ACID
71-36-3	N-BUTANOL
106-88-7	N-BUTENE-1,2-OXIDE
109-73-9	N-BUTILAMINA (ITALIAN)
93-79-8	N-BUTYL (2,4,5-TRICHLOROPHENOXY)ACETATE
2545-59-7	N-BUTYL (2,4,5-TRICHLOROPHENOXY)ACETATE
995-33-5	N-BUTYL 4,4-DI(TERT-BUTYL PEROXY)VALERATE (MAXIMUM 52% SOLUTION)
995-33-5	N-BUTYL 4,4-DI(TERT-BUTYL PEROXY)VALERATE (STOCK PURE)
1119-49-9	N-BUTYL ACETAMIDE
123-86-4	N-BUTYL ACETATE
123-86-4	N-BUTYL ACETATE (ACGIH)
12788-93-1	N-BUTYL ACID PHOSPHATE (DOT)
141-32-2	N-BUTYL ACRYLATE
141-32-2	N-BUTYL ACRYLATE (ACGIH)
71-36-3	N-BUTYL ALCOHOL
71-36-3	N-BUTYL ALCOHOL (ACGIH)
123-72-8	N-BUTYL ALDEHYDE
136-60-7	N-BUTYL BENZOATE
85-68-7	N-BUTYL BENZYL PHTHALATE
688-74-4	N-BUTYL BORATE
109-65-9	N-BUTYL BROMIDE
109-21-7	N-BUTYL BUTANOATE
109-21-7	N-BUTYL BUTYRATE
71-41-0	N-BUTYL CARBINOL
109-69-3	N-BUTYL CHLORIDE
109-69-3	N-BUTYL CHLORIDE (DOT)
142-96-1	N-BUTYL ETHER
106-35-4	N-BUTYL ETHYL KETONE
592-84-7	N-BUTYL FORMATE
2426-08-6	N-BUTYL GLYCIDYL ETHER
111-36-4	N-BUTYL ISOCYANATE
138-22-7	N-BUTYL LACTATE
109-79-5	N-BUTYL MERCAPTAN
97-88-1	N-BUTYL METHACRYLATE
97-88-1	N-BUTYL METHACRYLATE (DOT)
591-78-6	N-BUTYL METHYL KETONE
109-21-7	N-BUTYL N-BUTYRATE
84-74-2	N-BUTYL PHTHALATE
84-74-2	N-BUTYL PHTHALATE (DOT)
590-01-2	N-BUTYL PROPANOATE
590-01-2	N-BUTYL PROPIONATE
109-79-5	N-BUTYL THIOALCOHOL
111-34-2	N-BUTYL VINYL ETHER
102-79-4	N-BUTYL-2,2'-IMINODIETHANOL
102-79-4	N-BUTYL-N,N-BIS(2-HYDROXYETHYL)AMINE
102-79-4	N-BUTYL-N,N-BIS(HYDROXYETHYL)AMINE
13360-63-9	N-BUTYL-N-ETHYLAMINE
110-68-9	N-BUTYL-N-METHYLAMINE
91-49-6	N-BUTYLACETANILIDE
109-73-9	N-BUTYLAMIN (GERMAN)
109-73-9	N-BUTYLAMINE
109-73-9	N-BUTYLAMINE (ACGIH)
1126-78-9	N-BUTYLANILINE
1126-78-9	N-BUTYLBENZENAMINE
104-51-8	N-BUTYLBENZENE
104-51-8	N-BUTYLBENZENE (DOT)
592-34-7	N-BUTYLCHLOROFORMATE (DOT)
10108-56-2	N-BUTYLCYCLOHEXYLAMINE
102-79-4	N-BUTYLDIETHANOLAMINE
93-79-8	N-BUTYLESTER KYSELINI 2,4,5-TRICHLORFENOXYOCTOVE (CZECH)
2545-59-7	N-BUTYLESTER KYSELINI 2,4,5-TRICHLORFENOXYOCTOVE (CZECH)
13360-63-9	N-BUTYLETHYLAMINE
110-68-9	N-BUTYLMETHYLAMINE
123-72-8	N-BUTYRALDEHYDE
110-69-0	N-BUTYRALDOXIME
107-92-6	N-BUTYRIC ACID
106-31-0	N-BUTYRIC ACID ANHYDRIDE
106-31-0	N-BUTYRIC ANHYDRIDE
109-74-0	N-BUTYRONITRILE
141-75-3	N-BUTYRYL CHLORIDE
66-25-1	N-CAPROALDEHYDE
142-62-1	N-CAPROIC ACID
66-25-1	N-CAPRONALDEHYDE
66-25-1	N-CAPROYLALDEHYDE
124-13-0	N-CAPRYLALDEHYDE
109-69-3	N-CHLOROBUTANE
502-39-6	N-CYANO-N'-(METHYLMERCURY)GUANIDINE
1467-79-4	N-CYANO-N-METHYLMETHANAMINE
420-04-2	N-CYANOAMINE
3312-60-5	N-CYCLOHEXYL-1,3-DIAMINOPROPANE
3312-60-5	N-CYCLOHEXYL-1,3-PROPANEDIAMINE
3312-60-5	N-CYCLOHEXYL-1,3-PROPYLENEDIAMINE
1195-42-2	N-CYCLOHEXYL-N-ISOPROPYLAMINE
2842-38-8	N-CYCLOHEXYLAMINOETHANOL
101-83-7	N-CYCLOHEXYLCYCLOHEXANAMINE
2842-38-8	N-CYCLOHEXYLETHANOLAMINE
5459-93-8	N-CYCLOHEXYLETHYLAMINE
3312-60-5	N-CYCLOHEXYLTRIMETHYLENEDIAMINE
124-18-5	N-DECANE
112-30-1	N-DECANOL
112-30-1	N-DECATYL ALCOHOL
112-30-1	N-DECYL ALCOHOL
2016-57-1	N-DECYLAMINE
104-72-3	N-DECYLBENZENE
142-96-1	N-DIBUTYL ETHER
100-37-8	N-DIETHYLAMINOETHANOL
121-69-7	N-DIMETHYL ANILINE
108-01-0	N-DIMETHYLAMINOETHANOL
117-84-0	N-DIOCTYLPHTHALATE
142-84-7	N-DIPROPYLAMINE
112-41-4	N-DODEC-1-ENE
112-53-8	N-DODECAN-1-OL
112-55-0	N-DODECANETHIOL
112-53-8	N-DODECANOL
112-53-8	N-DODECYL ALCOHOL
112-55-0	N-DODECYL MERCAPTAN
151-21-3	N-DODECYL SULFATE SODIUM
123-01-3	N-DODECYLBENZENE
27176-87-0	N-DODECYLBENZENESULFONIC ACID
122-98-5	N-ETHANOLANILINE
100-74-3	N-ETHYL MORPHOLINE

CAS No.	Chemical Name
139-87-7	N-ETHYL-2,2'-IMINODIETHANOL
110-73-6	N-ETHYL-2-AMINOETHANOL
110-73-6	N-ETHYL-2-HYDROXYETHYLAMINE
102-27-2	N-ETHYL-3-METHYLANILINE
102-27-2	N-ETHYL-3-METHYLBENZENAMINE
622-57-1	N-ETHYL-4-METHYLANILINE
622-57-1	N-ETHYL-4-TOLUIDINE
102-27-2	N-ETHYL-M-TOLUIDINE
102-27-2	N-ETHYL-M-TOLUIDINE (DOT)
110-73-6	N-ETHYL-N-(2-HYDROXYETHYL)AMINE
92-50-2	N-ETHYL-N-(2-HYDROXYETHYL)ANILINE
110-73-6	N-ETHYL-N-(BETA-HYDROXYETHYL)AMINE
92-50-2	N-ETHYL-N-(BETA-HYDROXYETHYL)ANILINE
92-50-2	N-ETHYL-N-(HYDROXYETHYL)ANILINE
92-59-1	N-ETHYL-N-BENZYLANILINE
92-59-1	N-ETHYL-N-BENZYLANILINE (DOT)
13360-63-9	N-ETHYL-N-BUTYLAMINE
55-18-5	N-ETHYL-N-NITROSOETHANAMINE
759-73-9	N-ETHYL-N-NITROSOUREA
103-69-5	N-ETHYL-N-PHENYLAMINE
92-50-2	N-ETHYL-N-PHENYLAMINOETHANOL
92-59-1	N-ETHYL-N-PHENYLBENZYLAMINE
92-50-2	N-ETHYL-N-PHENYLETHANOLAMINE
622-57-1	N-ETHYL-P-METHYLANILINE
622-57-1	N-ETHYL-P-TOLUIDINE
622-57-1	N-ETHYL-P-TOLUIDINE (DOT)
75-04-7	N-ETHYLAMINE
103-69-5	N-ETHYLAMINOBENZENE
110-73-6	N-ETHYLAMINOETHANOL
103-69-5	N-ETHYLANILINE
103-69-5	N-ETHYLANILINE (DOT)
92-50-2	N-ETHYLANILINOETHANOL
103-69-5	N-ETHYLBENZENAMINE
103-69-5	N-ETHYLBENZENAMINO
538-07-8	N-ETHYLBIS(2-CHLOROETHYL)AMINE
13360-63-9	N-ETHYLBUTANAMINE
13360-63-9	N-ETHYLBUTYLAMINE
5459-93-8	N-ETHYLCYCLOHEXANAMINE
5459-93-8	N-ETHYLCYCLOHEXYLAMINE
139-87-7	N-ETHYLDIETHANOLAMINE
110-73-6	N-ETHYLETHANOLAMINE
110-73-6	N-ETHYLMONOETHANOLAMINE
766-09-6	N-ETHYLPIPERIDINE
102-27-2	N-ETHYLTOLUIDINE
2540-82-1	N-FORMYL-N-METHYLCARBAMOYLMETHYL O,O-DIMETHYL PHOSPHORODITHIOATE
68-12-2	N-FORMYLDIMETHYLAMINE
30207-98-8	N-HENDECYLENIC ALCOHOL
592-76-7	N-HEPT-1-ENE
111-70-6	N-HEPTAN-1-OL
142-82-5	N-HEPTANE
111-70-6	N-HEPTANOL
111-70-6	N-HEPTANOL-1 (FRENCH)
592-76-7	N-HEPTENE
592-76-7	N-HEPTENE (DOT)
111-70-6	N-HEPTYL ALCOHOL
111-68-2	N-HEPTYLAMINE
112-02-7	N-HEXADECYLTRIMETHYLAMMONIUM CHLORIDE
111-27-3	N-HEXAN-1-OL
66-25-1	N-HEXANAL
110-54-3	N-HEXANE
110-54-3	N-HEXANE (ACGIH)
142-62-1	N-HEXANOIC ACID
111-27-3	N-HEXANOL
111-27-3	N-HEXANOL (DOT)
142-62-1	N-HEXOIC ACID
142-92-7	N-HEXYL ACETATE
111-27-3	N-HEXYL ALCOHOL
544-10-5	N-HEXYL CHLORIDE
112-58-3	N-HEXYL ETHER
111-26-2	N-HEXYLAMINE
143-16-8	N-HEXYLHEXANAMINE
142-62-1	N-HEXYLIC ACID
3710-84-7	N-HYDROXYDIETHYLAMINE
111-41-1	N-HYDROXYETHYL-1,2-ETHANEDIAMINE
671-16-9	N-ISOPROPYL-ALPHA-(2-METHYLHYDRAZINO)-P-TOLUAMIDE
366-70-1	N-ISOPROPYL-P-(2-METHYLHYDRAZINOMETHYL)BENZAMIDE HYDRCHLORIDE
109-56-8	N-ISOPROPYLAMINOETHANOL

CAS No.	Chemical Name
643-28-7	N-ISOPROPYLANILINE
768-52-5	N-ISOPROPYLANILINE
768-52-5	N-ISOPROPYLBENZENAMINE
1195-42-2	N-ISOPROPYLCYCLOHEXANAMINE
1195-42-2	N-ISOPROPYLCYCLOHEXYLAMINE
109-56-8	N-ISOPROPYLETHANOLAMINE
112-53-8	N-LAURYL ALCOHOL, PRIMARY
112-55-0	N-LAURYL MERCAPTAN
51-75-2	N-LOST (GERMAN)
64-00-6	N-METHYL 3-ISOPROPYLPHENYL CARBAMATE
100-61-8	N-METHYL ANILINE
64-00-6	N-METHYL M-ISOPROPYLPHENYL CARBAMATE
109-02-4	N-METHYL MORPHOLINE
872-50-4	N-METHYL PYRROLIDINONE
63-25-2	N-METHYL-1 -NAPHTHYL CARBAMATE
63-25-2	N-METHYL-1-NAFTYL-CARBAMAAT (DUTCH)
63-25-2	N-METHYL-1-NAPHTHYL-CARBAMAT (GERMAN)
51-75-2	N-METHYL-2,2'-DICHLORODIETHYLAMINE
302-70-5	N-METHYL-2,2'-DICHLORODIETHYLAMINE N-OXIDE HYDRO-CHLORIDE
105-59-9	N-METHYL-2,2'-IMINODIETHANOL
109-83-1	N-METHYL-2-AMINOETHANOL
109-83-1	N-METHYL-2-ETHANOLAMINE
109-83-1	N-METHYL-2-HYDROXYETHYLAMINE
872-50-4	N-METHYL-2-PYRROLIDINONE
872-50-4	N-METHYL-2-PYRROLIDONE
63-25-2	N-METHYL-ALPHA-NAPHTHYLCARBAMATE
63-25-2	N-METHYL-ALPHA-NAPHTHYLURETHAN
872-50-4	N-METHYL-ALPHA-PYRROLIDINONE
872-50-4	N-METHYL-ALPHA-PYRROLIDONE
51-75-2	N-METHYL-BIS(2-CHLOROETHYL)AMINE
51-75-2	N-METHYL-BIS(BETA-CHLOROETHYL)AMINE
51-75-2	N-METHYL-BIS-CHLORAETHYLAMIN (GERMAN)
872-50-4	N-METHYL-GAMMA-BUTYROLACTAM
51-75-2	N-METHYL-LOST
70-25-7	N-METHYL-N'-NITRO-N-NITROSOGUANIDINE
479-45-8	N-METHYL-N,2,4,6-N-TETRANITROANILINE
479-45-8	N-METHYL-N,2,4,6-TETRANITROANILINE
109-83-1	N-METHYL-N-(2-HYDROXYETHYL)AMINE
93-90-3	N-METHYL-N-(2-HYDROXYETHYL)ANILINE
109-83-1	N-METHYL-N-(BETA-HYDROXYETHYL)AMINE
13256-22-9	N-METHYL-N-(CARBOXYMETHYL)NITROSAMINE
110-68-9	N-METHYL-N-BUTYLAMINE
100-60-7	N-METHYL-N-CYCLOHEXYLAMINE
109-83-1	N-METHYL-N-HYDROXYETHYLAMINE
70-25-7	N-METHYL-N-NITRO-N-NITROSOGUANIDINE
70-25-7	N-METHYL-N-NITROSO-N'-NITROGUANIDINE
10595-95-6	N-METHYL-N-NITROSOETHANAMINE
10595-95-6	N-METHYL-N-NITROSOETHYLAMINE
62-75-9	N-METHYL-N-NITROSOMETHANAMINE
684-93-5	N-METHYL-N-NITROSOUREA
615-53-2	N-METHYL-N-NITROSOURETHANE
4549-40-0	N-METHYL-N-NITROSOVINYLAMINE
479-45-8	N-METHYL-N-PICRYLNITRAMINE
100-61-8	N-METHYLAMINOBENZENE
105-59-9	N-METHYLAMINODIGLYCOL
109-83-1	N-METHYLAMINOETHANOL
100-61-8	N-METHYLANILINE (ACGIH,DOT)
86-50-0	N-METHYLBENZAZIMIDE, DIMETHYLDITHIOPHOSPHORIC ACID ESTER
100-61-8	N-METHYLBENZENAMINE
51-75-2	N-METHYLBIS(2-CHLOROETHYL)AMINE
302-70-5	N-METHYLBIS(2-CHLOROETHYL)AMINE N-OXIDE HYDRO-CHLORIDE
51-75-2	N-METHYLBIS(BETA-CHLOROETHYL)AMINE
110-68-9	N-METHYLBUTANAMINE
110-68-9	N-METHYLBUTYLAMINE
63-25-2	N-METHYLCARBAMATE DE 1-NAPHTYLE (FRENCH)
100-60-7	N-METHYLCYCLOHEXANAMINE
100-60-7	N-METHYLCYCLOHEXYLAMINE
105-59-9	N-METHYLDIETHANOLAMINE
105-59-9	N-METHYLDIETHANOLIMINE
109-83-1	N-METHYLETHANOLAMINE
105-59-9	N-METHYLIMINODIETHANOL
124-40-3	N-METHYLMETHANAMINE
109-83-1	N-METHYLMONOETHANOLAMINE
100-61-8	N-METHYLPHENYLAMINE
109-01-3	N-METHYLPIPERAZINE
626-67-5	N-METHYLPIPERIDINE

CAS No.	Chemical Name
120-94-5	N-METHYLPYRROLIDINE
872-50-4	N-METHYLPYRROLIDINONE
872-50-4	N-METHYLPYRROLIDONE
120-94-5	N-METHYLTETRAHYDROPYRROLE
63-25-2	N-METIL-1-NAFTIL-CARBAMMATO (ITALIAN)
2275-18-5	N-MONOISOPROPYLAMIDE OF O,O-DIETHYLDITHIOPHOSPHORYL-ACETIC ACID
60-51-5	N-MONOMETHYLAMIDE OF O,O-DIMETHYLDITHIOPHOSPHORYL-ACETIC ACID
109-83-1	N-MONOMETHYLAMINOETHANOL
100-61-8	N-MONOMETHYLANILINE
109-83-1	N-MONOMETHYLETHANOLAMINE
120-40-1	N-N-BIS-(2-HYDROXYETHYL)DODECANAMIDE
4316-42-1	N-N-BUTYL IMIDAZOLE (DOT)
4316-42-1	N-N-BUTYL-IMIDAZOLE
1126-78-9	N-N-BUTYLANILINE (DOT)
1120-48-5	N-N-OCTYL-N-OCTYLAMINE
30553-04-9	N-NAPHTHYLTHIOUREA
70-25-7	N-NITROSO-N'-NITRO-N-METHYLGUANIDINE
55-18-5	N-NITROSO-N,N-DIETHYLAMINE
62-75-9	N-NITROSO-N,N-DIMETHYLAMINE
86-30-6	N-NITROSO-N-DIPHENYLAMINE
759-73-9	N-NITROSO-N-ETHYLUREA
70-25-7	N-NITROSO-N-METHYL-N'-NITROGUANIDINE
684-93-5	N-NITROSO-N-METHYLCARBAMIDE
10595-95-6	N-NITROSO-N-METHYLETHYLAMINE
70-25-7	N-NITROSO-N-METHYLNITROGUANIDINE
684-93-5	N-NITROSO-N-METHYLUREA
615-53-2	N-NITROSO-N-METHYLURETHANE
4549-40-0	N-NITROSO-N-METHYLVINYLAMINE
86-30-6	N-NITROSO-N-PHENYLANILINE
135-20-6	N-NITROSO-N-PHENYLHYDROXYAMINE AMMONIUM SALT
135-20-6	N-NITROSO-N-PHENYLHYDROXYLAMINE AMMONIUM SALT
924-16-3	N-NITROSODI-N-BUTYLAMINE
621-64-7	N-NITROSODI-N-PROPYLAMINE
924-16-3	N-NITROSODIBUTYLAMINE
1116-54-7	N-NITROSODIETHANOLAMINE
55-18-5	N-NITROSODIETHYLAMINE
62-75-9	N-NITROSODIMETHYLAMINE
62-75-9	N-NITROSODIMETHYLAMINE (ACGIH)
86-30-6	N-NITROSODIPHENYLAMINE
621-64-7	N-NITROSODIPROPYLAMINE
10595-95-6	N-NITROSOETHYLMETHYLAMINE
10595-95-6	N-NITROSOMETHYLETHYLAMINE
4549-40-0	N-NITROSOMETHYLVINYLAMINE
59-89-2	N-NITROSOMORPHOLINE
16543-55-8	N-NITROSONORNICOTINE
100-75-4	N-NITROSOPIPERIDINE
930-55-2	N-NITROSOPYRROLIDINE
13256-22-9	N-NITROSOSARCOSINE
99-08-1	N-NITROTOLUENE
124-11-8	N-NON-1-ENE
143-08-8	N-NONAN-1-OL
111-84-2	N-NONANE
143-08-8	N-NONYL ALCOHOL
25154-52-3	N-NONYLPHENOL
57-11-4	N-OCTADECANOIC ACID
112-04-9	N-OCTADECYLTRICHLOROSILANE
124-13-0	N-OCTALDEHYDE
111-87-5	N-OCTAN-1-OL
124-13-0	N-OCTANAL
111-65-9	N-OCTANE
111-88-6	N-OCTANETHIOL
111-87-5	N-OCTANOL
106-68-3	N-OCTANONE-3
111-64-8	N-OCTANOYL CHLORIDE
762-16-3	N-OCTANOYL PEROXIDE (DOT)
762-16-3	N-OCTANOYL PEROXIDE, TECHNICALLY PURE
762-16-3	N-OCTANOYL PEROXIDE, TECHNICALLY PURE (DOT)
112-14-1	N-OCTANYL ACETATE
112-14-1	N-OCTYL ACETATE
2499-59-4	N-OCTYL ACRYLATE
111-87-5	N-OCTYL ALCOHOL
124-13-0	N-OCTYL ALDEHYDE
111-88-6	N-OCTYL MERCAPTAN
117-84-0	N-OCTYL PHTHALATE
124-13-0	N-OCTYLAL
111-86-4	N-OCTYLAMINE
111-88-6	N-OCTYLTHIOL

CAS No.	Chemical Name
53-86-1	N-P-CHLORBENZOYL-5-METHOXY-2-METHYLINDOLE-3-ACETIC ACID
629-76-5	N-PENTADECANOL
2570-26-5	N-PENTADECYLAMINE
71-41-0	N-PENTAN-1-OL
110-62-3	N-PENTANAL
109-66-0	N-PENTANE
110-66-7	N-PENTANETHIOL
109-52-4	N-PENTANOIC ACID
71-41-0	N-PENTANOL
513-35-9	N-PENTENE
628-63-7	N-PENTYL ACETATE
71-41-0	N-PENTYL ALCOHOL
110-53-2	N-PENTYL BROMIDE
543-59-9	N-PENTYL CHLORIDE
628-63-7	N-PENTYL ETHANOATE
638-49-3	N-PENTYL FORMATE
110-66-7	N-PENTYL MERCAPTAN
110-43-0	N-PENTYL METHYL KETONE
540-18-1	N-PENTYL N-BUTYRATE
1002-16-0	N-PENTYL NITRATE
110-58-7	N-PENTYLAMINE
538-68-1	N-PENTYLBENZENE
110-66-7	N-PENTYLTHIOL
122-98-5	N-PHENYL-2-AMINOETHANOL
135-88-6	N-PHENYL-2-NAPHTHALENAMINE
135-88-6	N-PHENYL-2-NAPHTHYLAMINE
135-88-6	N-PHENYL-BETA-NAPHTHYLAMINE
92-50-2	N-PHENYL-N-ETHYL-2-AMINOETHANOL
92-50-2	N-PHENYL-N-ETHYLETHANOL AMINE
156-10-5	N-PHENYL-P-NITROSOANILINE
103-84-4	N-PHENYLACETAMIDE
102-01-2	N-PHENYLACETOACETAMIDE
122-39-4	N-PHENYLANILINE
122-39-4	N-PHENYLBENZENAMINE
622-44-6	N-PHENYLCARBIMIDE DICHLORIDE
622-44-6	N-PHENYLCARBONIMIDIC DICHLORIDE
613-29-6	N-PHENYLDIBUTYLAMINE
122-98-5	N-PHENYLETHANOLAMINE
59-88-1	N-PHENYLHYDRAZINE HYDROCHLORIDE
622-44-6	N-PHENYLIMIDOPHOSGENE
622-44-6	N-PHENYLIMINOCARBONYL DICHLORIDE
768-52-5	N-PHENYLISOPROPYLAMINE
100-61-8	N-PHENYLMETHYLAMINE
92-53-5	N-PHENYLMORPHOLINE
103-85-5	N-PHENYLTHIOUREA
64-10-8	N-PHENYLUREA
479-45-8	N-PICRYL-N-METHYLNITRAMINE
71-23-8	N-PROPAN-1-OL
67-63-0	N-PROPAN-2-OL
123-38-6	N-PROPANAL
74-98-6	N-PROPANE
107-12-0	N-PROPANENITRILE
107-03-9	N-PROPANETHIOL
71-23-8	N-PROPANOL
109-60-4	N-PROPANOL ACETATE
109-60-4	N-PROPYL ACETATE
109-60-4	N-PROPYL ACETATE (ACGIH,DOT)
71-23-8	N-PROPYL ALCOHOL
71-23-8	N-PROPYL ALCOHOL (ACGIH)
71-23-8	N-PROPYL ALKOHOL (GERMAN)
106-94-5	N-PROPYL BROMIDE
105-66-8	N-PROPYL BUTYRATE
540-54-5	N-PROPYL CHLORIDE
109-61-5	N-PROPYL CHLOROFORMATE
109-61-5	N-PROPYL CHLOROFORMATE (DOT)
111-43-3	N-PROPYL ETHER
110-74-7	N-PROPYL FORMATE
110-78-1	N-PROPYL ISOCYANATE (DOT)
107-03-9	N-PROPYL MERCAPTAN
627-13-4	N-PROPYL NITRATE
627-13-4	N-PROPYL NITRATE (ACGIH,DOT)
16066-38-9	N-PROPYL PERCARBONATE
106-36-5	N-PROPYL PROPIONATE
142-84-7	N-PROPYL-1-PROPANAMINE
107-10-8	N-PROPYLAMINE
103-65-1	N-PROPYLBENZENE
109-69-3	N-PROPYLCARBINYL CHLORIDE
107-03-9	N-PROPYLTHIOL

CAS No.	Chemical Name
28729-54-6	N-PROPYLTOLUENE
141-57-1	N-PROPYLTRICHLOROSILANE
1929-82-4	N-SERVE
3775-90-4	N-TERT-BUTYLAMINOETHYL METHACRYLATE
1120-36-1	N-TETRADEC-1-ENE
112-72-1	N-TETRADECAN-1-OL
27196-00-5	N-TETRADECAN-1-OL
112-72-1	N-TETRADECANOL
27196-00-5	N-TETRADECANOL
112-72-1	N-TETRADECANOL-1
27196-00-5	N-TETRADECANOL-1
112-72-1	N-TETRADECYL ALCOHOL
27196-00-5	N-TETRADECYL ALCOHOL
133-06-2	N-TRICHLOROMETHYLMERCAPTO-4-CYCLOHEXENE-1,2-DICARBOXIMIDE
133-06-2	N-TRICHLOROMETHYLTHIO-3A,4,7,7A-TETRAHYDROPHTHALIMIDE
133-06-2	N-TRICHLOROMETHYLTHIO-CIS-DELTA(SUP 4)-CYCLOHEXENE-1,2-DICARBOXIMIDE
133-06-2	N-TRICHLOROMETHYLTHIOCYCLOHEX-4-ENE-1,2-DICARBOXIMIDE
2437-56-1	N-TRIDEC-1-ENE
75-50-3	N-TRIMETHYLAMINE
1120-21-4	N-UNDECANE
30207-98-8	N-UNDECANOL
30207-98-8	N-UNDECYL ALCOHOL
6742-54-7	N-UNDECYLBENZENE
110-62-3	N-VALERALDEHYDE
110-62-3	N-VALERALDEHYDE (ACGIH)
109-52-4	N-VALERIC ACID
110-62-3	N-VALERIC ALDEHYDE
133-06-2	N-[(TRICHLOROMETHYL)THIO]-DELTA4-TETRAHYDROPHTHALIMIDE
133-06-2	N-[(TRICHLOROMETHYL)THIO]TETRAHYDROPHTHALIMIDE
7440-02-0	N1
140-80-7	N1,N1-DIETHYL-1,4-PENTANEDIAMINE
101-25-7	N1,N3-DINITROSOPENTAMETHYLENETETRAMINE
556-88-7	N1-NITROGUANIDINE
140-80-7	N5,N5-DIETHYL-2,5-PENTANEDIAMINE
74-90-8	NA 1051 (DOT)
10544-72-6	NA 1067 (DOT)
10102-44-0	NA 1067 (DOT)
71-36-3	NA 1120 (DOT)
55-63-0	NA 1204 (DOT)
80-62-6	NA 1247 (DOT)
124-41-4	NA 1289 (DOT)
9004-70-0	NA 1324 (DOT)
12627-52-0	NA 1325 (DOT)
9004-70-0	NA 1325 (DOT)
88-89-1	NA 1344 (DOT)
7440-70-2	NA 1401 (DOT)
7440-23-5	NA 1421 (DOT)
13530-68-2	NA 1463 (DOT)
3251-23-8	NA 1479 (DOT)
7778-50-9	NA 1479 (DOT)
10588-01-9	NA 1479 (DOT)
124-43-6	NA 1511 (DOT)
7783-56-4	NA 1549 (DOT)
7789-61-9	NA 1549 (DOT)
593-89-5	NA 1556 (DOT)
696-28-6	NA 1556 (DOT)
1303-33-9	NA 1557 (DOT)
7784-45-4	NA 1557 (DOT)
7787-49-7	NA 1566 (DOT)
7787-47-5	NA 1566 (DOT)
52740-16-6	NA 1574 (DOT)
8004-09-9	NA 1581 (DOT)
76-06-2	NA 1583 (DOT)
75-05-8	NA 1648 (DOT)
78-00-2	NA 1649 (DOT)
25168-04-1	NA 1665 (DOT)
7446-18-6	NA 1707 (DOT)
10031-59-1	NA 1707 (DOT)
25376-45-8	NA 1709 (DOT)
30143-13-6	NA 1709 (DOT)
7772-99-8	NA 1759 (DOT)
7758-94-3	NA 1759 (DOT)
75-99-0	NA 1760 (DOT)
109-52-4	NA 1760 (DOT)

CAS No.	Chemical Name
110-91-8	NA 1760 (DOT)
142-62-1	NA 1760 (DOT)
123-00-2	NA 1760 (DOT)
676-98-2	NA 1760 (DOT)
1498-51-7	NA 1760 (DOT)
929-06-6	NA 1760 (DOT)
7758-94-3	NA 1760 (DOT)
4985-85-7	NA 1760 (DOT)
7784-30-7	NA 1760 (DOT)
7697-37-2	NA 1760 (DOT)
16961-83-4	NA 1778 (DOT)
1314-56-3	NA 1807 (DOT)
7789-29-9	NA 1811 (DOT)
7783-05-3	NA 1831 (DOT)
8014-95-7	NA 1831 (DOT)
7439-95-4	NA 1869 (DOT)
298-07-7	NA 1902 (DOT)
6484-52-2	NA 1942 (DOT)
428-59-1	NA 1956 (DOT)
56-38-2	NA 1967 (DOT)
107-19-7	NA 1986 (DOT)
100-52-7	NA 1989 (DOT)
625-58-1	NA 1993 (DOT)
87-86-5	NA 2020 (DOT)
933-78-4	NA 2020 (DOT)
25167-82-2	NA 2020 (DOT)
15950-66-0	NA 2020 (DOT)
1312-03-4	NA 2025 (DOT)
110-91-8	NA 2054 (DOT)
94-36-0	NA 2085 (DOT)
762-16-3	NA 2129 (DOT)
79-21-0	NA 2131 (DOT)
105-64-6	NA 2133 (DOT)
105-64-6	NA 2134 (DOT)
96-45-7	NA 22
1314-87-0	NA 2291 (DOT)
592-87-0	NA 2291 (DOT)
7758-95-4	NA 2291 (DOT)
7446-14-2	NA 2291 (DOT)
13814-96-5	NA 2291 (DOT)
15739-80-7	NA 2291 (DOT)
41070-66-9	NA 2422 (DOT)
1113-38-8	NA 2449 (DOT)
814-91-5	NA 2449 (DOT)
2244-21-5	NA 2465 (DOT)
543-90-8	NA 2570 (DOT)
7789-42-6	NA 2570 (DOT)
10108-64-2	NA 2570 (DOT)
27176-87-0	NA 2584 (DOT)
7790-93-4	NA 2626 (DOT)
1336-21-6	NA 2672 (DOT)
7631-90-5	NA 2693 (DOT)
7681-57-4	NA 2693 (DOT)
4429-42-9	NA 2693 (DOT)
7757-74-6	NA 2693 (DOT)
10192-30-0	NA 2693 (DOT)
15512-36-4	NA 2693 (DOT)
16731-55-8	NA 2693 (DOT)
63-25-2	NA 2757 (DOT)
1563-66-2	NA 2757 (DOT)
2032-65-7	NA 2757 (DOT)
76-44-8	NA 2761 (DOT)
60-57-1	NA 2761 (DOT)
72-20-8	NA 2761 (DOT)
58-89-9	NA 2761 (DOT)
72-43-5	NA 2761 (DOT)
143-50-0	NA 2761 (DOT)
115-32-2	NA 2761 (DOT)
309-00-2	NA 2761 (DOT)
115-29-7	NA 2761 (DOT)
117-80-6	NA 2761 (DOT)
8001-35-2	NA 2761 (DOT)
30900-23-3	NA 2761 (DOT)
309-00-2	NA 2762 (DOT)
12789-03-6	NA 2762 (DOT)
93-76-5	NA 2765 (DOT)
93-72-1	NA 2765 (DOT)
94-75-7	NA 2765 (DOT)
2312-35-8	NA 2765 (DOT)

CAS No.	Chemical Name
330-54-1	NA 2767 (DOT)
1918-00-9	NA 2769 (DOT)
137-26-8	NA 2771 (DOT)
85-00-7	NA 2781 (DOT)
52-68-6	NA 2783 (DOT)
56-72-4	NA 2783 (DOT)
56-38-2	NA 2783 (DOT)
86-50-0	NA 2783 (DOT)
107-49-3	NA 2783 (DOT)
121-75-5	NA 2783 (DOT)
298-04-4	NA 2783 (DOT)
333-41-5	NA 2783 (DOT)
298-00-0	NA 2783 (DOT)
563-12-2	NA 2783 (DOT)
2636-26-2	NA 2783 (DOT)
7786-34-7	NA 2783 (DOT)
2921-88-2	NA 2783 (DOT)
2275-14-1	NA 2783 (DOT)
7439-97-6	NA 2809 (DOT)
8012-54-2	NA 2810 (DOT)
7783-46-2	NA 2811 (DOT)
7428-48-0	NA 2811 (DOT)
10101-63-0	NA 2811 (DOT)
39990-99-3	NA 2813 (DOT)
584-79-2	NA 2902 (DOT)
2524-03-0	NA 2922 (DOT)
16721-80-5	NA 2922 (DOT)
16721-80-5	NA 2923 (DOT)
76-06-2	NA 2929 (DOT)
79-92-5	NA 9011 (DOT)
131-89-5	NA 9026 (DOT)
67-72-1	NA 9037 (DOT)
631-61-8	NA 9079 (DOT)
1863-63-4	NA 9080 (DOT)
1066-33-7	NA 9081 (DOT)
1111-78-0	NA 9083 (DOT)
506-87-6	NA 9084 (DOT)
12125-02-9	NA 9085 (DOT)
7788-98-9	NA 9086 (DOT)
3012-65-5	NA 9087 (DOT)
13826-83-0	NA 9088 (DOT)
7773-06-0	NA 9089 (DOT)
10196-04-0	NA 9090 (DOT)
3164-29-2	NA 9091 (DOT)
1762-95-4	NA 9092 (DOT)
84-74-2	NA 9095 (DOT)
13765-19-0	NA 9096 (DOT)
26264-06-2	NA 9097 (DOT)
133-06-2	NA 9099 (DOT)
10101-53-8	NA 9100 (DOT)
1066-30-4	NA 9101 (DOT)
10049-05-5	NA 9102 (DOT)
7789-43-7	NA 9103 (DOT)
544-18-3	NA 9104 (DOT)
14017-41-5	NA 9105 (DOT)
142-71-2	NA 9106 (DOT)
7758-98-7	NA 9109 (DOT)
10380-29-7	NA 9110 (DOT)
815-82-7	NA 9111 (DOT)
60-00-4	NA 9117 (DOT)
1185-57-5	NA 9118 (DOT)
14221-47-7	NA 9119 (DOT)
7783-50-8	NA 9120 (DOT)
10028-22-5	NA 9121 (DOT)
10045-89-3	NA 9122 (DOT)
7720-78-7	NA 9125 (DOT)
110-17-8	NA 9126 (DOT)
54590-52-2	NA 9127 (DOT)
1338-24-5	NA 9137 (DOT)
15699-18-0	NA 9138 (DOT)
7718-54-9	NA 9139 (DOT)
12054-48-7	NA 9140 (DOT)
7786-81-4	NA 9141 (DOT)
7789-00-6	NA 9142 (DOT)
7775-11-3	NA 9145 (DOT)
25155-30-0	NA 9146 (DOT)
7558-79-4	NA 9147 (DOT)
7601-54-9	NA 9148 (DOT)
7789-06-2	NA 9149 (DOT)
27323-41-7	NA 9151 (DOT)
27774-13-6	NA 9152 (DOT)
557-34-6	NA 9153 (DOT)
52628-25-8	NA 9154 (DOT)
1332-07-6	NA 9155 (DOT)
7699-45-8	NA 9156 (DOT)
3486-35-9	NA 9157 (DOT)
7783-49-5	NA 9158 (DOT)
557-41-5	NA 9159 (DOT)
127-82-2	NA 9160 (DOT)
7733-02-0	NA 9161 (DOT)
16923-95-8	NA 9162 (DOT)
14644-61-2	NA 9163 (DOT)
8003-34-7	NA 9184 (DOT)
13446-10-1	NA 9190 (DOT)
1309-64-4	NA 9201 (DOT)
630-08-0	NA 9202 (DOT)
676-97-1	NA 9206 (DOT)
7681-49-4	NA FRINSE
76-03-9	NA TA
58933-55-4	NA URE
1333-83-1	NA-0101 T 1/8"
1313-82-2	Na2S
57-11-4	NAA 173
70-30-4	NABAC
142-59-6	NABAM
142-59-6	NABAME (FRENCH)
142-59-6	NABASAN
16940-66-2	NaBH4
9005-25-8	NABOND
63-25-2	NAC
25155-30-0	NACCANOL NR
25155-30-0	NACCANOL SW
101-68-8	NACCONATE 300
5124-30-1	NACCONATE H 12
26471-62-5	NACCONATE-100
25155-30-0	NACCONOL 35SL
25155-30-0	NACCONOL 40F
25155-30-0	NACCONOL 90F
27176-87-0	NACCONOL 98SA
108-94-1	NADONE
7681-49-4	NAFEEN
3771-19-5	NAFENOPIN
7681-49-4	NAFPAK
91-20-3	NAFTALEN (POLISH)
135-88-6	NAFTAM 2
494-03-1	NAFTICLORINA
142-59-6	NAFUN IPO
7646-69-7	NAH 80
123-30-8	NAKO BROWN R
27598-85-2	NAKO BROWN R
106-50-3	NAKO H
108-46-3	NAKO TGG
95-80-7	NAKO TMT
7631-86-9	NALCAST PLW
25322-69-4	NALCO 131S
11138-49-1	NALCO 680
2425-06-1	NALCO 7046
7631-86-9	NALCOAG 1030
7631-86-9	NALCOAG 1034A
7631-86-9	NALCOAG 1050
7631-86-9	NALCOAG 1115
300-76-5	NALED
7631-86-9	NALFLOC N 1030
7631-86-9	NALFLOC N 1050
123-01-3	NALKYLENE 500
57-83-0	NALUTRON
7681-52-9	NAMED REAGENTS AND SOLUTIONS, DAKIN'S
534-16-7	NAMED REAGENTS AND SOLUTIONS, FETIZON'S
20816-12-0	NAMED REAGENTS AND SOLUTIONS, MILAS'
10045-94-0	NAMED REAGENTS AND SOLUTIONS, MILLON'S
108-62-3	NAMEKIL
9002-91-9	NAMEKIL
299-84-3	NANCHOR
299-84-3	NANKOR
27176-87-0	NANSA 1042P
25155-30-0	NANSA 1260
25155-30-0	NANSA HF 80
25155-30-0	NANSA HS 80

CAS No.	Chemical Name
25155-30-0	NANSA HS 80S
25155-30-0	NANSA HS 85S
25155-30-0	NANSA SL
25155-30-0	NANSA SS
27176-87-0	NANSA SSA
131-52-2	NAPCLOR-G
131-52-2	NAPCP
1338-24-5	NAPHID
61789-51-3	NAPHTENATE DE COBALT (FRENCH)
8030-30-6	NAPHTHA
65996-79-4	NAPHTHA
8030-30-6	NAPHTHA (DOT)
8030-30-6	NAPHTHA DISTILLATE
8030-30-6	NAPHTHA DISTILLATE (DOT)
8030-30-6	NAPHTHA PETROLEUM (DOT)
8030-30-6	NAPHTHA VM&P
8032-32-4	NAPHTHA, LIGROINE
91-20-3	NAPHTHALENE, MOLTEN (DOT)
8030-30-6	NAPHTHA, PETROLEUM
8052-41-3	NAPHTHA, SOLVENT
8030-30-6	NAPHTHA, SOLVENT (DOT)
8030-30-6	NAPHTHA, STRAIGHT-RUN
91-20-3	NAPHTHALENE
91-20-3	NAPHTHALENE (ACGIH,DOT)
119-64-2	NAPHTHALENE 1,2,3,4-TETRAHYDRIDE
25551-28-4	NAPHTHALENE DI-ISOCYANATE
3761-53-3	NAPHTHALENE LAKE SCARLET R
8001-58-9	NAPHTHALENE OIL
3761-53-3	NAPHTHALENE SCARLET R
73090-68-3	NAPHTHALENE, (1,1-DIMETHYLETHYL)-1,2,3,4-TETRAHYDRO-
119-64-2	NAPHTHALENE, 1,2,3,4-TETRAHYDRO-
771-29-9	NAPHTHALENE, 1,2,3,4-TETRAHYDRO-, 1-HYDROPEROXIDE
90-12-0	NAPHTHALENE, 1-METHYL-
91-58-7	NAPHTHALENE, 2-CHLORO-
91-20-3	NAPHTHALENE, CRUDE OR REFINED (DOT)
91-17-8	NAPHTHALENE, DECAHYDRO-
1335-87-1	NAPHTHALENE, HEXACHLORO-
91-20-3	NAPHTHALENE, MOLTEN
27254-36-0	NAPHTHALENE, MONONITRO-
27254-36-0	NAPHTHALENE, NITRO-
2234-13-1	NAPHTHALENE, OCTACHLORO-
1321-64-8	NAPHTHALENE, PENTACHLORO-
1320-27-0	NAPHTHALENE, PENTYL-
1335-88-2	NAPHTHALENE, TETRACHLORO-
1321-65-9	NAPHTHALENE, TRICHLORO-
134-32-7	NAPHTHALIDAM
134-32-7	NAPHTHALIDINE
91-20-3	NAPHTHALIN
91-20-3	NAPHTHALIN (DOT)
91-20-3	NAPHTHALINE
2602-46-2	NAPHTHAMINE BLUE 2B
91-17-8	NAPHTHAN
119-90-4	NAPHTHANIL BLUE B BASE
89-62-3	NAPHTHANIL RED G BASE
99-55-8	NAPHTHANIL SCARLET G BASE
3761-53-3	NAPHTHAZINE SCARLET 2R
91-20-3	NAPHTHENE
1338-24-5	NAPHTHENIC ACID
1338-24-5	NAPHTHENIC ACID (DOT)
1338-02-9	NAPHTHENIC ACID COPPER SALT
61789-51-3	NAPHTHENIC ACID, COBALT SALT
8020-83-5	NAPHTHOLITE
95-79-4	NAPHTHOSOL FAST RED KB BASE
192-65-4	NAPHTHO[1,2,3,4-DEF]CHRYSENE
91-59-8	NAPHTHYLAMINE (BETA)
72-57-1	NAPHTHYLAMINE BLUE
494-03-1	NAPHTHYLAMINE MUSTARD
83-32-9	NAPHTHYLENEETHYLENE
30553-04-9	NAPHTHYLTHIOUREA
6950-84-1	NAPHTHYLUREA (1-NAPHTHALENYL UREA)
13114-62-0	NAPHTHYLUREA (2-NAPHTHALENYL UREA)
89-62-3	NAPHTOELAN FAST RED GL BASE
99-55-8	NAPHTOELAN FAST SCARLET G BASE
99-55-8	NAPHTOELAN FAST SCARLET G SALT
100-01-6	NAPHTOELAN RED GG BASE
29757-24-2	NAPHTOELAN RED GG BASE
106-49-0	NAPHTOL AS-KG
106-49-0	NAPHTOL AS-KGLL
86-88-4	NAPHTOX
30553-04-9	NAPHTOX
25322-69-4	NAPTER E 8075
66-81-9	NARAMYCIN
66-81-9	NARAMYCIN A
79-01-6	NARCOGEN
75-29-6	NARCOSOP
75-00-3	NARCOTILE
74-86-2	NARCYLEN
79-01-6	NARKOGEN
79-01-6	NARKOSOID
75-09-2	NARKOTIL
8020-83-5	NASR OIL
434-07-1	NASTENON
88-74-4	NATASOL FAST ORANGE GR SALT
3165-93-3	NATASOL FAST RED TR SALT
7681-49-4	NATRIUM FLUORIDE
7632-00-0	NATRIUM NITRIT (GERMAN)
26628-22-8	NATRIUMAZID (GERMAN)
10588-01-9	NATRIUMBICHROMAAT (DUTCH)
7775-09-9	NATRIUMCHLORAAT (DUTCH)
7775-09-9	NATRIUMCHLORAT (GERMAN)
10588-01-9	NATRIUMDICHROMAAT (DUTCH)
10588-01-9	NATRIUMDICHROMAT (GERMAN)
62-74-8	NATRIUMFLUORACETAAT (DUTCH)
62-74-8	NATRIUMFLUORACETAT (GERMAN)
26628-22-8	NATRIUMMAZIDE (DUTCH)
62-76-0	NATRIUMOXALAT (GERMAN)
7601-89-0	NATRIUMPERCHLORAAT (DUTCH)
7601-89-0	NATRIUMPERCHLORAT (GERMAN)
7558-79-4	NATRIUMPHOSPHAT (GERMAN)
540-72-7	NATRIUMRHODANID (GERMAN)
13410-01-0	NATRIUMSELENIAT (GERMAN)
10102-18-8	NATRIUMSELENIT (GERMAN)
16893-85-9	NATRIUMSILICOFLUORID (GERMAN)
366-70-1	NATULAN
366-70-1	NATULAN HYDROCHLORIDE
366-70-1	NATULANAR
1317-65-3	NATURAL CALCIUM CARBONATE
8006-14-2	NATURAL GAS
86290-81-5	NATURAL GAS CONDENSATES, GASOLINE
74-82-8	NATURAL GAS WITH A HIGH METHANE CONTENT, COMPRESSED
74-82-8	NATURAL GAS, LIQUID (REFRIGERATED) WITH A HIGH METHANE CONTENT
8006-14-2	NATURAL GAS, SWEET
1314-87-0	NATURAL LEAD SULFIDE
2312-35-8	NAUGATUCK D 014
78-34-2	NAVADEL
9005-90-7	NAVAL STORES, TURPENTINE
640-19-7	NAVRON
108-93-0	NAXOL
94-36-0	NAYPER B
94-36-0	NAYPER B AND BO
94-36-0	NAYPER BO
14797-65-0	NBT
136-40-3	NC 150
60-51-5	NC-262
12001-28-4	NCI C09007
12001-29-5	NCI C61223A
4384-82-1	NCI CO2835
30900-23-3	NCI-C00044
86-50-0	NCI-C00066
133-06-2	NCI-C00077
12789-03-6	NCI-C00099
2636-26-2	NCI-C00113
60-57-1	NCI-C00124
60-51-5	NCI-C00135
72-20-8	NCI-C00157
76-44-8	NCI-C00180
143-50-0	NCI-C00191
58-89-9	NCI-C00204
121-75-5	NCI-C00215
56-38-2	NCI-C00226
8001-35-2	NCI-C00259
78-34-2	NCI-C00395
1582-09-8	NCI-C00442
115-32-2	NCI-C00486
72-43-5	NCI-C00497
96-12-8	NCI-C00500
107-06-2	NCI-C00511

CAS No.	Chemical Name
106-93-4	NCI-C00522
76-06-2	NCI-C00533
115-29-7	NCI-C00566
138-89-6	NCI-C01821
53324-05-3	NCI-C01821
121-14-2	NCI-C01865
103-85-5	NCI-C02017
27134-26-5	NCI-C02039
7632-00-0	NCI-C02084
57-13-6	NCI-C02119
100-42-5	NCI-C02200
3165-93-3	NCI-C02368
1306-19-0	NCI-C02551
67-66-3	NCI-C02686
7772-99-8	NCI-C02722
50-00-0	NCI-C02799
933-78-4	NCI-C02904
15950-66-0	NCI-C02904
25167-82-2	NCI-C02904
156-62-7	NCI-C02937
999-81-5	NCI-C02960
298-00-0	NCI-C02971
64-17-5	NCI-C03134
85-44-9	NCI-C03601
123-91-1	NCI-C03689
62-53-3	NCI-C03736
142-04-1	NCI-C03736
7440-32-6	NCI-C04251
75-34-3	NCI-C04535
79-01-6	NCI-C04546
127-18-4	NCI-C04580
75-15-0	NCI-C04591
67-72-1	NCI-C04604
107-05-1	NCI-C04615
71-55-6	NCI-C04626
75-69-4	NCI-C04637
50-07-7	NCI-C04706
108-46-3	NCI-C05970
100-51-6	NCI-C06111
109-69-3	NCI-C06155
75-00-3	NCI-C06224
100-44-7	NCI-C06360
2385-85-5	NCI-C06428
26628-22-8	NCI-C06462
108-88-3	NCI-C07272
116-06-3	NCI-C08640
56-72-4	NCI-C08662
333-41-5	NCI-C08673
101-05-3	NCI-C08684
12654-97-6	NCI-C08684
63283-80-7	NCI-C50044
120-61-6	NCI-C50055
115-07-1	NCI-C50077
75-56-9	NCI-C50099
75-09-2	NCI-C50102
107-07-3	NCI-C50135
151-21-3	NCI-C50191
593-60-2	NCI-C50373
140-88-5	NCI-C50384
57-06-7	NCI-C50464
26471-62-5	NCI-C50533
106-99-0	NCI-C50602
25339-57-5	NCI-C50602
105-60-2	NCI-C50646
101-68-8	NCI-C50668
80-62-6	NCI-C50680
117-84-0	NCI-C52733
91-20-3	NCI-C52904
76-01-7	NCI-C53894
85-68-7	NCI-C54375
103-23-1	NCI-C54386
123-79-5	NCI-C54386
563-47-3	NCI-C54820
52-68-6	NCI-C54831
108-90-7	NCI-C54886
87-86-5	NCI-C54933
95-50-1	NCI-C54944
106-46-7	NCI-C54955
78-00-2	NCI-C54988

CAS No.	Chemical Name
532-27-4	NCI-C55107
75-25-2	NCI-C55130
1309-64-4	NCI-C55152
111-42-2	NCI-C55174
144-62-7	NCI-C55209
7681-49-4	NCI-C55221
1330-20-7	NCI-C55232
110-86-1	NCI-C55301
75-65-0	NCI-C55367
87-86-5	NCI-C55378
74-96-4	NCI-C55481
108-86-1	NCI-C55492
106-88-7	NCI-C55527
77-47-4	NCI-C55607
98-85-1	NCI-C55685
123-31-9	NCI-C55834
106-51-4	NCI-C55845
2244-16-8	NCI-C55867
509-14-8	NCI-C55947
100-02-7	NCI-C55992
540-59-0	NCI-C56031
100-52-7	NCI-C56133
53-86-1	NCI-C56144
118-96-7	NCI-C56155
98-01-1	NCI-C56177
110-00-9	NCI-C56202
98-00-0	NCI-C56224
123-73-9	NCI-C56279
4170-30-3	NCI-C56279
123-72-8	NCI-C56291
75-07-0	NCI-C56326
100-41-4	NCI-C56393
25013-15-4	NCI-C56406
10043-35-3	NCI-C56417
121-69-7	NCI-C56428
684-16-2	NCI-C56440
87-86-5	NCI-C56655
106-92-3	NCI-C56666
2163-80-6	NCI-C60071
98-95-3	NCI-C60082
13494-80-9	NCI-C60117
7487-94-7	NCI-C60173
75-38-7	NCI-C60208
75-44-5	NCI-C60219
79-11-8	NCI-C60231
7440-48-4	NCI-C60311
7786-81-4	NCI-C60344
7439-97-6	NCI-C60399
99-99-0	NCI-C60537
109-99-9	NCI-C60560
110-54-3	NCI-C60571
97-02-9	NCI-C60753
100-01-6	NCI-C60786
29757-24-2	NCI-C60786
7699-43-6	NCI-C60811
75-05-8	NCI-C60822
105-58-8	NCI-C60899
68-12-2	NCI-C60913
123-38-6	NCI-C61029
1330-78-5	NCI-C61041
127-85-5	NCI-C61176
124-09-4	NCI-C61405
9016-87-9	NCO 20
3129-91-7	NDA
62-75-9	NDMA
138-89-6	NDMA
53324-05-3	NDMA
123-91-1	NE 220
62-38-4	NEANTINA
84-66-2	NEANTINE
584-79-2	NECARBOXYLIC ACID
56-23-5	NECATORINA
56-23-5	NECATORINE
25155-30-0	NECCANOL SW
12627-52-0	NEEDLE ANTIMONY
106-93-4	NEFIS
62-73-7	NEFRAFOS
52-68-6	NEGUVON
52-68-6	NEGUVON A

CAS No.	Chemical Name
577-11-7	NEKAL WT-27
3761-53-3	NEKLACID RED RR
100-74-3	NEM
21548-32-3	NEM-A-TAK
127-18-4	NEMA
96-12-8	NEMABROM
22224-92-6	NEMACUR
22224-92-6	NEMACUR P
297-97-2	NEMAFOS
297-97-2	NEMAFOS 10 G
96-12-8	NEMAFUME
96-12-8	NEMAGON
96-12-8	NEMAGON 20
96-12-8	NEMAGON 206
96-12-8	NEMAGON 20G
96-12-8	NEMAGON 90
96-12-8	NEMAGON SOIL FUMIGANT
96-12-8	NEMAGONE
96-12-8	NEMANAX
96-12-8	NEMAPAZ
297-97-2	NEMAPHOS
96-12-8	NEMASET
21548-32-3	NEMATAK
297-97-2	NEMATOCIDE
297-97-2	NEMATOCIDE GR
96-12-8	NEMATOX
92-84-2	NEMAZENE
96-12-8	NEMAZON
26628-22-8	NEMAZYD
9016-45-9	NEMOL K 1030
9016-45-9	NEMOL K 2030
9016-45-9	NEMOL K 34
9016-45-9	NEMOL K 36
72-20-8	NENDRIN
7758-19-2	NEO SILOX D
7681-52-9	NEO-CLEANER
57-63-6	NEO-ESTRONE
57-11-4	NEO-FAT 18
57-11-4	NEO-FAT 18-53
57-11-4	NEO-FAT 18-54
57-11-4	NEO-FAT 18-55
57-11-4	NEO-FAT 18-59
57-11-4	NEO-FAT 18-61
57-11-4	NEO-FAT 18-S
112-80-1	NEO-FAT 90-04
112-80-1	NEO-FAT 92-04
56-53-1	NEO-OESTRANOL I
58-89-9	NEO-SCABICIDOL
1333-86-4	NEO-SPECTRA II
1333-86-4	NEO-SPECTRA MARK II
75-84-3	NEOAMYL ALCOHOL
1344-28-1	NEOBEAD C
80-51-3	NEOCELLBORN FE 9
80-51-3	NEOCELLBORN P 1000M
2244-21-5	NEOCHLOR 59
2893-78-9	NEOCHLOR 60P
87-90-1	NEOCHLOR 90
64093-79-4	NEOCHROMIUM
50-29-3	NEOCID
50-29-3	NEOCIDOL
333-41-5	NEOCIDOL
333-41-5	NEOCIDOL (OIL)
50-29-3	NEOCIDOL (SOLID)
26748-41-4	NEODECANEPEROXOIC ACID, 1,1-DIMETHYLETHYL ESTER
26748-47-0	NEODECANEPEROXOIC ACID, 1-METHYL-1-PHENYLETHYL ESTER
25155-30-0	NEOGEN SC
27323-41-7	NEOGEN T
75-83-2	NEOHEXANE
75-83-2	NEOHEXANE (ACGIH,DOT)
1332-58-7	NEOKAOLIN
7440-01-9	NEON
7440-01-9	NEON (DOT)
7440-01-9	NEON, COMPRESSED (DOT)
7440-01-9	NEON, REFRIGERATED LIQUID
7440-01-9	NEON, REFRIGERATED LIQUID (DOT)
25155-30-0	NEOPELEX 05
25155-30-0	NEOPELEX 25
25155-30-0	NEOPELEX 6
25155-30-0	NEOPELEX F 25
25155-30-0	NEOPELEX F 60
463-82-1	NEOPENTANE
75-98-9	NEOPENTANOIC ACID
75-84-3	NEOPENTANOL
3282-30-2	NEOPENTANOYL CHLORIDE
75-84-3	NEOPENTYL ALCOHOL
15663-27-1	NEOPLATIN
2235-54-3	NEOPON LAM
70-30-4	NEOSEPT V
7631-86-9	NEOSIL
7631-86-9	NEOSIL A
7631-86-9	NEOSIL XV
135-88-6	NEOSONE D
50-70-4	NEOSORB
50-70-4	NEOSORB 20/60DC
50-70-4	NEOSORB 70/70
7631-86-9	NEOSYL
7631-86-9	NEOSYL 186
7631-86-9	NEOSYL 224
7631-86-9	NEOSYL 81
58-22-0	NEOTESTIS
557-17-5	NEOTHYL
135-88-6	NEOZON D
135-88-6	NEOZONE
135-88-6	NEOZONE D
106-93-4	NEPHIS
3383-96-8	NEPHIS
3383-96-8	NEPHIS 1G
786-19-6	NEPHOCARP
133-06-2	NERACID
62-73-7	NERKOL
2636-26-2	NERKOL
60-00-4	NERVANAID B ACID
138-86-3	NESOL
25155-30-0	NESTAPONE
94-75-7	NETAGRONE
94-75-7	NETAGRONE 600
514-73-8	NETOCYD
7704-34-9	NETZSCHWEFEL
534-52-1	NEUDORFF DN 50
50-78-2	NEURONIKA
7788-98-9	NEUTRAL AMMONIUM CHROMATE
12125-01-8	NEUTRAL AMMONIUM FLUORIDE
7428-48-0	NEUTRAL LEAD STEARATE
7789-00-6	NEUTRAL POTASSIUM CHROMATE
7775-11-3	NEUTRAL SODIUM CHROMATE
142-71-2	NEUTRAL VERDIGRIS
7789-38-0	NEUTRALIZER K-126
7789-38-0	NEUTRALIZER K-140
7789-38-0	NEUTRALIZER K-938
151-21-3	NEUTRAZYME
9016-45-9	NEUTRONYX 640
9016-45-9	NEUTRONYX 676
3165-93-3	NEUTROSEL RED TRVA
12002-03-8	NEUWIED GREEN
577-11-7	NEVAX
999-81-5	NEW '5C' CYCOCEL
12002-03-8	NEW GREEN
2235-25-8	NEW IMPROVED CERESAN
2235-25-8	NEW IMPROVED GRANOSAN
3761-53-3	NEW PONCEAU 4R
577-11-7	NEWCOL 290M
9016-45-9	NEWCOL 504
9016-45-9	NEWCOL 520
9016-45-9	NEWCOL 560
9016-45-9	NEWCOL 561H
9016-45-9	NEWCOL 562
9016-45-9	NEWCOL 564
9016-45-9	NEWCOL 568
37286-64-9	NEWPOL LB 625
25155-30-0	NEWREX R
92-84-2	NEXARBOL
58-89-9	NEXEN FB
58-89-9	NEXIT
58-89-9	NEXIT-STARK
58-89-9	NEXOL-E
555-84-0	NF 246
139-91-3	NF 260
531-82-8	NFTA

CAS No.	Chemical Name
55-63-0	NG
9002-89-5	NH 18
7440-02-0	NI 0901-S
7440-02-0	NI 0901S (HARSHAW)
7440-02-0	NI 233
7440-02-0	NI 270
7440-02-0	NI 4303T
1563-66-2	NIA 10242
563-12-2	NIA 1240
584-79-2	NIA 249
115-29-7	NIA 5462
2778-04-3	NIA 5767
1194-65-6	NIA 5996
1563-66-2	NIAGARA 10242
563-12-2	NIAGARA 1240
115-29-7	NIAGARA 5462
2778-04-3	NIAGARA 5767
1031-47-6	NIAGARA 5943
1194-65-6	NIAGARA 5996
72-57-1	NIAGARA BLUE
2602-46-2	NIAGARA BLUE 2B
72-57-1	NIAGARA BLUE 3B
1563-66-2	NIAGARA NIA-10242
54-11-5	NIAGARA PA DUST
140-57-8	NIAGARAMITE
563-12-2	NIALATE
79-01-6	NIALK
25322-69-4	NIAX 10-25
25322-69-4	NIAX 20-25
25322-69-4	NIAX 61-58
25322-69-4	NIAX 61-582
9016-87-9	NIAX AFPI
26471-62-5	NIAX ISOCYANATE TDI
25322-69-4	NIAX POLYOL PPG 2025
25322-69-4	NIAX POLYOL PPG 4025
25322-69-4	NIAX PPG
25322-69-4	NIAX PPG 1025
25322-69-4	NIAX PPG 2025
25322-69-4	NIAX PPG 3025
25322-69-4	NIAX PPG 4025
25322-69-4	NIAX PPG 425
91-08-7	NIAX TDI
26471-62-5	NIAX TDI
91-08-7	NIAX TDI-P
26471-62-5	NIAX TDI-P
409-21-2	NICALON
7440-02-0	NICHEL (ITALIAN)
13463-39-3	NICHEL TETRACARBONILE (ITALIAN)
7440-02-0	NICKEL
7440-02-0	NICKEL (ACGIH)
7440-02-0	NICKEL (DUST)
7440-02-0	NICKEL 270
373-02-4	NICKEL ACETATE
15699-18-0	NICKEL AMMONIUM SULFATE
15699-18-0	NICKEL AMMONIUM SULFATE (DOT)
15699-18-0	NICKEL AMMONIUM SULFATE (Ni(NH4)2(SO4)2)
13138-45-9	NICKEL BIS(NITRATE)
14708-14-6	NICKEL BOROFLUORIDE
13462-88-9	NICKEL BROMIDE
13462-88-9	NICKEL BROMIDE (NiBR2)
3333-67-3	NICKEL CARBONATE
16337-84-1	NICKEL CARBONATE
17237-93-3	NICKEL CARBONATE
3333-67-3	NICKEL CARBONATE (1:1)
17237-93-3	NICKEL CARBONATE (Ni(HCO3)2)
3333-67-3	NICKEL CARBONATE (NiCO3)
13463-39-3	NICKEL CARBONYL
12612-55-4	NICKEL CARBONYL
13463-39-3	NICKEL CARBONYL (ACGIH,DOT)
13463-39-3	NICKEL CARBONYL (Ni(CO)4)
13463-39-3	NICKEL CARBONYL (Ni(CO)4), (T-4)-
13463-39-3	NICKEL CARBONYLE (FRENCH)
7440-02-0	NICKEL CATALYST
7440-02-0	NICKEL CATALYST, DRY
7440-02-0	NICKEL CATALYST, DRY :PRECIPITATED ON A CARRIER WITH A SPECIAL ACTIVATOR
7440-02-0	NICKEL CATALYST, FINELY DIVIDED, WET WITH ≥40% WATER
7718-54-9	NICKEL CHLORIDE
37211-05-5	NICKEL CHLORIDE
7718-54-9	NICKEL CHLORIDE (DOT)
7718-54-9	NICKEL CHLORIDE (NiCl2)
557-19-7	NICKEL CYANIDE
557-19-7	NICKEL CYANIDE (DOT)
557-19-7	NICKEL CYANIDE (Ni(CN)2)
557-19-7	NICKEL CYANIDE, SOLID
557-19-7	NICKEL CYANIDE, SOLID (DOT)
373-02-4	NICKEL DIACETATE
13462-88-9	NICKEL DIBROMIDE
7718-54-9	NICKEL DICHLORIDE
557-19-7	NICKEL DICYANIDE
1271-28-9	NICKEL DICYCLOPENTADIENYL
3349-06-2	NICKEL DIFORMATE
12054-48-7	NICKEL DIHYDROXIDE
13138-45-9	NICKEL DINITRATE
17861-62-0	NICKEL DINITRITE
7440-02-0	NICKEL ELEMENT
14708-14-6	NICKEL FLUOROBORATE
3349-06-2	NICKEL FORMATE
12054-48-7	NICKEL HYDROXIDE
11113-74-9	NICKEL HYDROXIDE
12054-48-7	NICKEL HYDROXIDE (DOT)
12054-48-7	NICKEL HYDROXIDE (Ni(OH)2)
1313-99-1	NICKEL MONOOXIDE
7786-81-4	NICKEL MONOSULFATE
1313-99-1	NICKEL MONOXIDE
13138-45-9	NICKEL NITRATE
13138-45-9	NICKEL NITRATE (DOT)
13138-45-9	NICKEL NITRATE (Ni(NO3)2)
17861-62-0	NICKEL NITRITE
17861-62-0	NICKEL NITRITE (DOT)
17861-62-0	NICKEL NITRITE (Ni(NO2)2)
1314-06-3	NICKEL OXIDE
1313-99-1	NICKEL OXIDE
11099-02-8	NICKEL OXIDE
1314-06-3	NICKEL OXIDE (Ni2O3)
1313-99-1	NICKEL OXIDE (NiO)
1313-99-1	NICKEL OXIDE SINTER 75
7440-02-0	NICKEL PARTICLES
1314-06-3	NICKEL PEROXIDE
1314-06-3	NICKEL SESQUIOXIDE
7440-02-0	NICKEL SPONGE
12035-72-2	NICKEL SUBSULFIDE
12035-72-2	NICKEL SUBSULFIDE (Ni3S2)
12035-72-2	NICKEL SUBSULPHIDE
7786-81-4	NICKEL SULFATE
7786-81-4	NICKEL SULFATE (1:1)
7786-81-4	NICKEL SULFATE (DOT)
7786-81-4	NICKEL SULFATE (NiSO4)
7786-81-4	NICKEL SULFATE(1:1)
12035-72-2	NICKEL SULFIDE (Ni3S2)
13463-39-3	NICKEL TETRACARBONYL
13463-39-3	NICKEL TETRACARBONYLE (FRENCH)
14708-14-6	NICKEL TETRAFLUOROBORATE
14708-14-6	NICKEL TETRAFLUOROBORATE (Ni(BF4)2)
1314-06-3	NICKEL TRIOXIDE
373-02-4	NICKEL(2+) ACETATE
13462-88-9	NICKEL(2+) BROMIDE
3333-67-3	NICKEL(2+) CARBONATE
3333-67-3	NICKEL(2+) CARBONATE (NiCO3)
7718-54-9	NICKEL(2+) CHLORIDE
3349-06-2	NICKEL(2+) FORMATE
12054-48-7	NICKEL(2+) HYDROXIDE
13138-45-9	NICKEL(2+) NITRATE
1313-99-1	NICKEL(2+) OXIDE
7786-81-4	NICKEL(2+) SULFATE
7786-81-4	NICKEL(2+) SULFATE (1:1)
14708-14-6	NICKEL(2+) TETRAFLUOROBORATE(1-)
373-02-4	NICKEL(II) ACETATE
373-02-4	NICKEL(II) ACETATE (1:2)
13462-88-9	NICKEL(II) BROMIDE
3333-67-3	NICKEL(II) CARBONATE
7718-54-9	NICKEL(II) CHLORIDE
7718-54-9	NICKEL(II) CHLORIDE (1:2)
557-19-7	NICKEL(II) CYANIDE
14708-14-6	NICKEL(II) FLUOBORATE
3349-06-2	NICKEL(II) FORMATE
12054-48-7	NICKEL(II) HYDROXIDE
13138-45-9	NICKEL(II) NITRATE

CAS No.	Chemical Name
13138-45-9	NICKEL(II) NITRATE (1:2)
1313-99-1	NICKEL(II) OXIDE
7786-81-4	NICKEL(II) SULFATE
7786-81-4	NICKEL(II) SULFATE (1:1)
14708-14-6	NICKEL(II) TETRAFLUOROBORATE
1314-06-3	NICKEL(III) OXIDE
1271-28-9	NICKEL, BIS(ETA5-2,4-CYCLOPENTADIEN-1-YL)-
1271-28-9	NICKEL, DI-PI-CYCLOPENTADIENYL-
1314-06-3	NICKELIC OXIDE
1271-28-9	NICKELOCENE
373-02-4	NICKELOUS ACETATE
17237-93-3	NICKELOUS BICARBONATE
13462-88-9	NICKELOUS BROMIDE
3333-67-3	NICKELOUS CARBONATE
7718-54-9	NICKELOUS CHLORIDE
3349-06-2	NICKELOUS FORMATE
12054-48-7	NICKELOUS HYDROXIDE
13138-45-9	NICKELOUS NITRATE
1313-99-1	NICKELOUS OXIDE
7786-81-4	NICKELOUS SULFATE
14708-14-6	NICKELOUS TETRAFLUOROBORATE
1836-75-5	NICLOFEN
1420-04-8	NICLOSAMIDE ETHANOLAMINE SALT
54-11-5	NICO-DUST
54-11-5	NICO-FUME
58-89-9	NICOCHLORAN
54-11-5	NICOCIDE
54-11-5	NICOTIN
54-11-5	NICOTINA (ITALIAN)
54-11-5	NICOTINE
54-11-5	NICOTINE (ACGIH)
54-11-5	NICOTINE (ALKALOID)
65-31-6	NICOTINE ACID TARTRATE
65-31-6	NICOTINE BITARTRATE
65-31-6	NICOTINE D-BITARTRATE
2820-51-1	NICOTINE HYDROCHLORIDE
2820-51-1	NICOTINE HYDROCHLORIDE (D,L)
2820-51-1	NICOTINE HYDROCHLORIDE (DOT)
2820-51-1	NICOTINE HYDROCHLORIDE SOLUTION
2820-51-1	NICOTINE HYDROCHLORIDE SOLUTION (DOT)
65-31-6	NICOTINE HYDROGEN TARTRATE
65-31-6	NICOTINE HYDROTARTRATE
29790-52-1	NICOTINE SALICYLATE
29790-52-1	NICOTINE SALICYLATE (DOT)
65-30-5	NICOTINE SULFATE
65-30-5	NICOTINE SULFATE, LIQUID
65-30-5	NICOTINE SULFATE, LIQUID (DOT)
65-30-5	NICOTINE SULFATE, SOLID
65-30-5	NICOTINE SULFATE, SOLID (DOT)
65-31-6	NICOTINE TARTRATE
3275-73-8	NICOTINE TARTRATE
65-31-6	NICOTINE TARTRATE (DOT)
16543-55-8	NICOTINE, 1'-DEMETHYL-1'-NITROSO-
54-11-5	NICOTINE, COMPOUNDS
54-11-5	NICOTINE, LIQUID
54-11-5	NICOTINE, LIQUID (DOT)
29790-52-1	NICOTINE, MONOSALICYLATE
54-11-5	NICOTINE, SOLID (DOT)
65-30-5	NICOTINE, SULFATE
65-30-5	NICOTINE, SULFATE (2:1)
65-31-6	NICOTINE, TARTRATE
65-31-6	NICOTINE, TARTRATE (1:1)
65-31-6	NICOTINE, TARTRATE (1:2)
65-31-6	NICOTINE-D-TARTRATE
83-79-4	NICOULINE
17702-41-9	NIDO-DECABORANE(14)
107-49-3	NIFOS
555-84-0	NIFURADEN
555-84-0	NIFURADENE
3570-75-0	NIFURTHIAZOL
3570-75-0	NIFURTHIAZOLE
55-63-0	NIGLIN
55-63-0	NIGLYCON
1333-86-4	NIGROS F
1333-86-4	NIGROS G
1333-86-4	NIGROS K
13463-39-3	NIKKELTETRACARBONYL (DUTCH)
9016-45-9	NIKKOL N P 2
9016-45-9	NIKKOL NP

CAS No.	Chemical Name
9016-45-9	NIKKOL NP 10
9016-45-9	NIKKOL NP 100
9016-45-9	NIKKOL NP 15
9016-45-9	NIKKOL NP 18TX
9016-45-9	NIKKOL NP 5
9016-45-9	NIKKOL NP 75
577-11-7	NIKKOL OTP 70
151-21-3	NIKKOL SLS
139-96-8	NIKKOL TEALS
54-11-5	NIKOTIN (GERMAN)
65-30-5	NIKOTINSULFAT (GERMAN)
54-11-5	NIKOTYNA (POLISH)
135-88-6	NILOX PBNA
26264-06-2	NINATE 401
93-58-3	NIOBE OIL
55-63-0	NIONG
1836-75-5	NIP
79-46-9	NIPAR S-20
25322-01-4	NIPAR S-20
79-46-9	NIPAR S-20 SOLVENT
25322-01-4	NIPAR S-20 SOLVENT
79-46-9	NIPAR S-30 SOLVENT
25322-01-4	NIPAR S-30 SOLVENT
100-02-7	NIPHEN
2602-46-2	NIPPON BLUE BB
1937-37-7	NIPPON DEEP BLACK
1937-37-7	NIPPON DEEP BLACK GX
333-41-5	NIPSAN
7631-86-9	NIPSIL 300A
7631-86-9	NIPSIL E 150
7631-86-9	NIPSIL E 150J
7631-86-9	NIPSIL ER
7631-86-9	NIPSIL NS
7631-86-9	NIPSIL NST
7631-86-9	NIPSIL VN 3
7631-86-9	NIPSIL VN3LP
56-38-2	NIRAN
12789-03-6	NIRAN
56-38-2	NIRAN E-4
61-57-4	NIRIDAZOLE
14797-65-0	NIRIT
112-02-7	NISSAN CATION PB 40
25155-30-0	NISSAN NEWREX R
9016-45-9	NISSAN NONION NS
9016-45-9	NISSAN NONION NS 203
9016-45-9	NISSAN NONION NS 206
9016-45-9	NISSAN NONION NS 210
9016-45-9	NISSAN NONION NS 215
9016-45-9	NISSAN NONION NS 220
9016-45-9	NISSAN NONION NS 230
577-11-7	NISSAN RAPISOL
577-11-7	NISSAN RAPISOL B 30
577-11-7	NISSAN RAPISOL B 80
151-21-3	NISSAN SINTREX L 100
25322-69-4	NISSAN UNIOL D 2000
25322-69-4	NISSAN UNIOL D 400
7697-37-2	NITAL
7631-99-4	NITER
7757-79-1	NITER
7681-38-1	NITER CAKE
1333-86-4	NITERON 55
139-94-6	NITHIAZIDE
55-63-0	NITORA
1335-85-9	NITRADOR
139-91-3	NITRALDONE
4726-14-1	NITRALIN
4726-14-1	NITRALINE
479-45-8	NITRAMINE
463-04-7	NITRAMYL
1582-09-8	NITRAN
1929-82-4	NITRAPYRIN
14797-55-8	NITRATE
1002-16-0	NITRATE D'AMYLE (FRENCH)
7761-88-8	NITRATE D'ARGENT (FRENCH)
10022-31-8	NITRATE DE BARYUM (FRENCH)
10099-74-8	NITRATE DE PLOMB (FRENCH)
18256-98-9	NITRATE DE PLOMB (FRENCH)
627-13-4	NITRATE DE PROPYLE NORMAL (FRENCH)
7631-99-4	NITRATE DE SODIUM (FRENCH)

CAS No.	Chemical Name
10042-76-9	NITRATE DE STRONTIUM (FRENCH)
7779-88-6	NITRATE DE ZINC (FRENCH)
12033-49-7	NITRATE FREE RADICAL
10544-73-7	NITRATE FREE RADICAL
14797-55-8	NITRATE ION
14797-55-8	NITRATE ION (NO3-)
14797-55-8	NITRATE ION(1-)
10415-75-5	NITRATE MERCUREUX (FRENCH)
10045-94-0	NITRATE MERCURIQUE (FRENCH)
14797-55-8	NITRATE(1-)
7631-99-4	NITRATINE
14797-55-8	NITRATO
100-01-6	NITRAZOL CF EXTRA
29757-24-2	NITRAZOL CF EXTRA
7757-79-1	NITRE
7681-38-1	NITRE CAKE
7697-37-2	NITRIC ACID
7697-37-2	NITRIC ACID (ACGIH,DOT)
6484-52-2	NITRIC ACID AMMONIUM SALT
10022-31-8	NITRIC ACID BARIUM SALT (2:1)
10124-37-5	NITRIC ACID CALCIUM SALT (2:1)
7757-79-1	NITRIC ACID POTASSIUM SALT
7757-79-1	NITRIC ACID POTASSIUM SALT (1:1)
7761-88-8	NITRIC ACID SILVER SALT (1:1)
7761-88-8	NITRIC ACID SILVER(1+) SALT
7761-88-8	NITRIC ACID SILVER(I) SALT
7631-99-4	NITRIC ACID SODIUM SALT
55-63-0	NITRIC ACID TRIESTER OF GLYCEROL
124-47-0	NITRIC ACID UREA SALT
1712-64-7	NITRIC ACID, 1-METHYLETHYL ESTER
7697-37-2	NITRIC ACID, 40% OR LESS (DOT)
10022-31-8	NITRIC ACID, BARIUM SALT
13597-99-4	NITRIC ACID, BERYLLIUM SALT
7787-55-5	NITRIC ACID, BERYLLIUM SALT, TRIHYDRATE
928-45-0	NITRIC ACID, BUTYL ESTER
10325-94-7	NITRIC ACID, CADMIUM SALT
10124-37-5	NITRIC ACID, CALCIUM SALT
7789-18-6	NITRIC ACID, CESIUM SALT
13548-38-4	NITRIC ACID, CHROMIUM SALT
13548-38-4	NITRIC ACID, CHROMIUM(3+) SALT
10141-05-6	NITRIC ACID, COBALT(2+) SALT
3251-23-8	NITRIC ACID, COPPER(2+) SALT
625-58-1	NITRIC ACID, ETHYL ESTER
7697-37-2	NITRIC ACID, FUMING
14797-55-8	NITRIC ACID, ION(1-)
10421-48-4	NITRIC ACID, IRON(3+) SALT
1712-64-7	NITRIC ACID, ISOPROPYL ESTER
10099-74-8	NITRIC ACID, LEAD SALT
18256-98-9	NITRIC ACID, LEAD SALT
10099-74-8	NITRIC ACID, LEAD(2+) SALT
18256-98-9	NITRIC ACID, LEAD(2+) SALT
7790-69-4	NITRIC ACID, LITHIUM SALT
10377-60-3	NITRIC ACID, MAGNESIUM SALT
10415-75-5	NITRIC ACID, MERCURY(1+) SALT
7782-86-7	NITRIC ACID, MERCURY(1+) SALT, MONOHYDRATE
10045-94-0	NITRIC ACID, MERCURY(2+) SALT
10415-75-5	NITRIC ACID, MERCURY(I) SALT
10045-94-0	NITRIC ACID, MERCURY(II) SALT
14216-75-2	NITRIC ACID, NICKEL SALT
13138-45-9	NITRIC ACID, NICKEL(2+) SALT
13138-45-9	NITRIC ACID, NICKEL(II) SALT
7697-37-2	NITRIC ACID, OVER 40% (DOT)
1002-16-0	NITRIC ACID, PENTYL ESTER
627-13-4	NITRIC ACID, PROPYL ESTER
10042-76-9	NITRIC ACID, STRONTIUM SALT
13746-98-0	NITRIC ACID, THALLIUM SALT
10102-45-1	NITRIC ACID, THALLIUM SALT
16901-76-1	NITRIC ACID, THALLIUM SALT
10102-45-1	NITRIC ACID, THALLIUM(1+) SALT
16901-76-1	NITRIC ACID, THALLIUM(1+) SALT
13746-98-0	NITRIC ACID, THALLIUM(3+) SALT
16901-76-1	NITRIC ACID, THALLIUM(3+) SALT
13823-29-5	NITRIC ACID, THORIUM(4+) SALT
13823-29-5	NITRIC ACID, THORIUM(4+) SALT (8CI,9CI)
15905-86-9	NITRIC ACID, URANIUM SALT
7779-88-6	NITRIC ACID, ZINC SALT
13746-89-9	NITRIC ACID, ZIRCONIUM(4+) SALT
625-58-1	NITRIC ETHER (DOT)
10102-43-9	NITRIC OXIDE
10102-43-9	NITRIC OXIDE (ACGIH,DOT)
10102-43-9	NITRIC OXIDE (NO)
10102-43-9	NITRIC OXIDE TRIMER
139-13-9	NITRILO-2,2',2"-TRIACETIC ACID
102-71-6	NITRILO-2,2',2"-TRIETHANOL
460-19-5	NITRILOACETONITRILE
139-13-9	NITRILOTRIACETIC ACID
139-13-9	NITRILOTRIACETIC ACID (NTA)
102-71-6	NITRILOTRIETHANOL
55-63-0	NITRIN
55-63-0	NITRINE
55-63-0	NITRINE-TDC
14797-65-0	NITRITE
7632-00-0	NITRITE DE SODIUM (FRENCH)
14797-65-0	NITRITE ION
14797-65-0	NITRITE ION (NO2-)
14797-65-0	NITRITE ION(1-)
10102-44-0	NITRITE RADICAL
10102-44-0	NITRITO
10102-44-0	NITRO
7782-78-7	NITRO ACID SULFITE
98-95-3	NITRO BENZOL
51-28-5	NITRO KLEENUP
55-63-0	NITRO-DUR
55-63-0	NITRO-LENT
7664-41-7	NITRO-SIL
55-63-0	NITRO-SPAN
100-01-6	NITROANILINE
29757-24-2	NITROANILINE
98-95-3	NITROBENZEEN (DUTCH)
98-95-3	NITROBENZEN (POLISH)
98-95-3	NITROBENZENE
98-95-3	NITROBENZENE (ACGIH)
98-95-3	NITROBENZENE, LIQUID
98-95-3	NITROBENZENE, LIQUID (DOT)
31212-28-9	NITROBENZENESULFONIC ACID
31212-28-9	NITROBENZENESULPHONIC ACID (DOT)
98-95-3	NITROBENZOL
98-95-3	NITROBENZOL (DOT)
98-95-3	NITROBENZOL, LIQUID (DOT)
402-54-0	NITROBENZOTRIFLUORIDE
28984-85-2	NITROBIPHENYL
586-78-7	NITROBROMOBENZENE
75-52-5	NITROCARBOL
9004-70-0	NITROCEL S
9004-70-0	NITROCELLULOSE
9004-70-0	NITROCELLULOSE (DOT)
9004-70-0	NITROCELLULOSE E 950
9004-70-0	NITROCELLULOSE, DRY (DOT)
9004-70-0	NITROCELLULOSE, IN SOLUTION IN FLAMMABLE LIQUIDS (DOT)
56-57-5	NITROCHIN
1836-75-5	NITROCHLOR
100-00-5	NITROCHLOROBENZENE
25167-93-5	NITROCHLOROBENZENE
121-73-3	NITROCHLOROBENZENE, META-, SOLID (DOT)
88-73-3	NITROCHLOROBENZENE, ORTHO, LIQUID (DOT)
100-00-5	NITROCHLOROBENZENE, PARA-, SOLID (DOT)
39974-35-1	NITROCHLOROBENZOTRI-FLUORIDE (1-NITRO-3-(TRIFLUORO-METHYL)-2-CHLO
25889-38-7	NITROCHLOROBENZOTRI-FLUORIDE (2-NITRO-1-(TRIFLUORO-METHYL)-4-CHLO
777-37-7	NITROCHLOROBENZOTRI-FLUORIDE (4-NITRO-2-(TRIFLUORO-METHYL)-1-CHLO
121-17-5	NITROCHLOROBENZOTRIFLUORIDE
76-06-2	NITROCHLOROFORM
9004-70-0	NITROCOTTON
12167-20-3	NITROCRESOL
12167-20-3	NITROCRESOLS (DOT)
1122-60-7	NITROCYCLOHEXANE
25168-04-1	NITRODIMETHYLBENZENE
79-24-3	NITROETAN (POLISH)
79-24-3	NITROETHANE
79-24-3	NITROETHANE (ACGIH,DOT)
534-52-1	NITROFAN
1836-75-5	NITROFEN
13410-72-5	NITROFEN
1836-75-5	NITROFEN (PESTICIDE)
13410-72-5	NITROFEN (PHARMACEUTICAL)
139-91-3	NITROFURMETHONE

CAS No.	Chemical Name
139-91-3	NITROFURMETON
7727-37-9	NITROGEN
7727-37-9	NITROGEN (DOT)
7727-37-9	NITROGEN (LIQUIFIED)
10102-44-0	NITROGEN DIOXIDE
10102-44-0	NITROGEN DIOXIDE (ACGIH)
10102-44-0	NITROGEN DIOXIDE (NO2)
14797-65-0	NITROGEN DIOXIDE(1-)
10544-72-6	NITROGEN DIOXIDE, DI-
14797-65-0	NITROGEN DIOXIDE, ION(1-)
10102-44-0	NITROGEN DIOXIDE, LIQUID
10102-44-0	NITROGEN DIOXIDE, LIQUID (DOT)
7783-54-2	NITROGEN FLUORIDE
10036-47-2	NITROGEN FLUORIDE
10036-47-2	NITROGEN FLUORIDE (N2F4)
7783-54-2	NITROGEN FLUORIDE (NF3)
7727-37-9	NITROGEN GAS
302-01-2	NITROGEN HYDRIDE (N2H4)
156-62-7	NITROGEN LIME
10102-43-9	NITROGEN MONOXIDE
51-75-2	NITROGEN MUSTARD
55-86-7	NITROGEN MUSTARD
55-86-7	NITROGEN MUSTARD HYDROCHLORIDE
302-70-5	NITROGEN MUSTARD N-OXIDE
126-85-2	NITROGEN MUSTARD N-OXIDE
302-70-5	NITROGEN MUSTARD N-OXIDE HYDROCHLORIDE
10024-97-2	NITROGEN OXIDE
11104-93-1	NITROGEN OXIDE
10024-97-2	NITROGEN OXIDE (N2O)
10544-73-7	NITROGEN OXIDE (N2O3)
10544-72-6	NITROGEN OXIDE (N2O4)
10102-43-9	NITROGEN OXIDE (N4O4)
10102-43-9	NITROGEN OXIDE (NO)
10102-44-0	NITROGEN OXIDE (NO2)
12033-49-7	NITROGEN OXIDE (NO3)
11104-93-1	NITROGEN OXIDE (NOX)
2696-92-6	NITROGEN OXIDE CHLORIDE (NOCl)
11104-93-1	NITROGEN OXIDES
2696-92-6	NITROGEN OXYCHLORIDE
2696-92-6	NITROGEN OXYCHLORIDE (NOCl)
10102-44-0	NITROGEN PEROXIDE
14797-65-0	NITROGEN PEROXIDE ION(1-)
10102-44-0	NITROGEN PEROXIDE, LIQUID (DOT)
10544-73-7	NITROGEN SESQUIOXIDE
12033-49-7	NITROGEN SESQUIOXIDE
10544-72-6	NITROGEN TETRAOXIDE
10544-72-6	NITROGEN TETROXIDE
10544-72-6	NITROGEN TETROXIDE (DOT)
10544-72-6	NITROGEN TETROXIDE, LIQUID
10544-72-6	NITROGEN TETROXIDE, LIQUID (DOT)
7783-54-2	NITROGEN TRIFLUORIDE
7783-54-2	NITROGEN TRIFLUORIDE (ACGIH,DOT)
12033-49-7	NITROGEN TRIOXIDE
10544-73-7	NITROGEN TRIOXIDE
10544-73-7	NITROGEN TRIOXIDE (DOT)
12033-49-7	NITROGEN TRIOXIDE (DOT)
10544-73-7	NITROGEN TRIOXIDE (N2O3)
12033-49-7	NITROGEN TRIOXIDE (NO3)
7727-37-9	NITROGEN, COMPRESSED (DOT)
7727-37-9	NITROGEN, REFRIGERATED LIQUID
7727-37-9	NITROGEN, REFRIGERATED LIQUID (DOT)
7727-37-9	NITROGEN-14
55-63-0	NITROGLICERINA (ITALIAN)
55-63-0	NITROGLICERYNA (POLISH)
55-63-0	NITROGLYCERIN
55-63-0	NITROGLYCERIN (ACGIH)
55-63-0	NITROGLYCERIN, LIQUID, DESENSITIZED (DOT)
55-63-0	NITROGLYCERIN, LIQUID, NOT DESENSITIZED (DOT)
55-63-0	NITROGLYCERIN, SPIRITS OF
55-63-0	NITROGLYCERINE
55-63-0	NITROGLYCEROL
628-96-6	NITROGLYCOL
55-63-0	NITROGLYN
55-86-7	NITROGRANULOGEN
556-88-7	NITROGUANIDINE
8007-56-5	NITROHYDROCHLORIC ACID
8007-56-5	NITROHYDROCHLORIC ACID (DOT)
8007-56-5	NITROHYDROCHLORIC ACID, DILUTED (DOT)
79-46-9	NITROISOPROPANE

CAS No.	Chemical Name
25322-01-4	NITROISOPROPANE
55-63-0	NITROL
55-63-0	NITROL (PHARMACEUTICAL)
55-63-0	NITROLAN
55-63-0	NITROLETTEN
156-62-7	NITROLIM
156-62-7	NITROLIME
55-63-0	NITROLINGUAL
55-63-0	NITROLOWE
55-63-0	NITROMEL
75-52-5	NITROMETAN (POLISH)
75-52-5	NITROMETHANE
75-52-5	NITROMETHANE (ACGIH,DOT)
302-70-5	NITROMIN
302-70-5	NITROMIN HYDROCHLORIDE
8007-56-5	NITROMURIATIC ACID
8007-56-5	NITROMURIATIC ACID (DOT)
9004-70-0	NITRON
9004-70-0	NITRON (NITROCELLULOSE)
27254-36-0	NITRONAPHTHALENE
27254-36-0	NITRONAPHTHALENE (DOT)
55-63-0	NITRONET
55-63-0	NITRONG
82-68-8	NITROPENTACHLOROBENZENE
51-28-5	NITROPHEN
1836-75-5	NITROPHEN
51-28-5	NITROPHENE
1836-75-5	NITROPHENE
25154-55-6	NITROPHENOL
25154-55-6	NITROPHENOL (DOT)
1321-12-6	NITROPHENYLMETHANE
122-14-5	NITROPHOS
88-85-7	NITROPONE C
80-17-1	NITROPORE OBSH
80-51-3	NITROPORE OBSH
25322-01-4	NITROPROPANE
55-63-0	NITRORECTAL
55-63-0	NITRORETARD
759-73-9	NITROSO-N-ETHYLUREA
684-93-5	NITROSO-N-METHYLUREA
924-16-3	NITROSODI-N-BUTYLAMINE
621-64-7	NITROSODI-N-PROPYLAMINE
924-16-3	NITROSODIBUTYLAMINE
1116-54-7	NITROSODIETHANOLAMINE
55-18-5	NITROSODIETHYLAMINE
62-75-9	NITROSODIMETHYLAMINE
139-89-6	NITROSODIMETHYLANILINE
53324-05-3	NITROSODIMETHYLANILINE
86-30-6	NITROSODIPHENYLAMINE
156-10-5	NITROSODIPHENYLAMINE
621-64-7	NITROSODIPROPYLAMINE
70-25-7	NITROSOGUANIDINE
10595-95-6	NITROSOMETHYLETHYLAMINE
684-93-5	NITROSOMETHYLUREA
615-53-2	NITROSOMETHYLURETHANE
59-89-2	NITROSOMORPHOLINE
7782-78-7	NITROSONIUM BISULFATE
2696-92-6	NITROSONIUM CHLORIDE
16543-55-8	NITROSONORNICOTINE
100-75-4	NITROSOPIPERIDINE
13256-22-9	NITROSOSARCOSINE
7782-78-7	NITROSOSULFURIC ACID
55-63-0	NITROSTABILIN
9056-38-6	NITROSTARCH
55-63-0	NITROSTAT
56-38-2	NITROSTIGMIN (GERMAN)
56-38-2	NITROSTIGMINE
56-38-2	NITROSTYGMINE
7782-78-7	NITROSULFONIC ACID
7782-78-7	NITROSYL BISULFATE
2696-92-6	NITROSYL CHLORIDE
2696-92-6	NITROSYL CHLORIDE ((NO)Cl)
2696-92-6	NITROSYL CHLORIDE (DOT)
109-95-5	NITROSYL ETHOXIDE
7782-78-7	NITROSYL HYDROGEN SULFATE
10102-43-9	NITROSYL RADICAL
7782-78-7	NITROSYL SULFATE
7782-78-7	NITROSYL SULFATE ((NO)H(SO4))
7782-78-7	NITROSYL SULFURIC ACID (NOHSO4)

CAS No.	Chemical Name
7782-78-7	NITROSYLSULFURIC ACID
7782-78-7	NITROSYLSULPHURIC ACID (DOT)
61-57-4	NITROTHIAMIDAZOLE
61-57-4	NITROTHIAZOL
61-57-4	NITROTHIAZOLE
99-08-1	NITROTOLUENE
1321-12-6	NITROTOLUENE
1321-12-6	NITROTOLUENE (DOT)
119-32-4	NITROTOLUIDINE
76-06-2	NITROTRICHLOROMETHANE
7632-00-0	NITROUS ACID SODIUM SALT (1:1)
63885-01-8	NITROUS ACID, AMMONIUM ZINC SALT (3:1:1)
109-95-5	NITROUS ACID, ETHYL ESTER
14797-65-0	NITROUS ACID, ION(1-)
624-91-9	NITROUS ACID, METHYL ESTER
17861-62-0	NITROUS ACID, NICKEL (2+) SALT
463-04-7	NITROUS ACID, PENTYL ESTER
7758-09-0	NITROUS ACID, POTASSIUM SALT
7632-00-0	NITROUS ACID, SODIUM SALT
10544-73-7	NITROUS ANHYDRIDE
12033-49-7	NITROUS ANHYDRIDE
109-95-5	NITROUS ETHER
109-95-5	NITROUS ETHER (DOT)
109-95-5	NITROUS ETHYL ETHER
11104-93-1	NITROUS OXIDE
10024-97-2	NITROUS OXIDE
11104-93-1	NITROUS OXIDE (DOT)
10024-97-2	NITROUS OXIDE (DOT)
10024-97-2	NITROUS OXIDE, COMPRESSED
10024-97-2	NITROUS OXIDE, COMPRESSED (DOT)
11104-93-1	NITROUS OXIDE, COMPRESSED (DOT)
10024-97-2	NITROUS OXIDE, REFRIGERATED LIQUID
10024-97-2	NITROUS OXIDE, REFRIGERATED LIQUID (DOT)
11104-93-1	NITROUS OXIDE, REFRIGERATED LIQUID (DOT)
298-00-0	NITROX
298-00-0	NITROX 80
88-89-1	NITROXANTHIC ACID
25168-04-1	NITROXYLENE
25168-04-1	NITROXYLOL
25168-04-1	NITROXYLOL (DOT)
7782-78-7	NITROXYLSULFURIC ACID
55-63-0	NITROZELL RETARD
7697-37-2	NITRYL HYDROXIDE
56-38-2	NIUIF 100
50-70-4	NIVITIN
9004-70-0	NIXON N/C
514-73-8	NK 136
7790-98-9	NK 270
21609-90-5	NK 711
55-63-0	NK 843
9002-89-5	NM 11
9002-89-5	NM 14
63-25-2	NMC 50
684-93-5	NMM
872-50-4	NMP
615-53-2	NMU
684-93-5	NMU
118-74-1	NO BUNT
118-74-1	NO BUNT 40
118-74-1	NO BUNT 80
118-74-1	NO BUNT LIQUID
122-39-4	NO SCALD
2636-26-2	NO-PEST
2636-26-2	NO-PEST STRIP
137-26-8	NOBECUTAN
96-45-7	NOCCELER 22
100-97-0	NOCCELER H
97-77-8	NOCCELER TET
137-26-8	NOCCELER TT
128-37-0	NOCRAC 200
96-69-5	NOCRAC 300
135-88-6	NOCRAC D
1336-36-3	NOFLAMOL
1314-13-2	NOGENOL
1314-13-2	NOGENOL BITE REGISTRATION PASTE
62-73-7	NOGOS
2636-26-2	NOGOS
62-73-7	NOGOS 50
2636-26-2	NOGOS 50
2636-26-2	NOGOS 50 EC
62-73-7	NOGOS G
2636-26-2	NOGOS G
9016-45-9	NOIGEN EA 150
9016-45-9	NOIGEN EA 50
9016-45-9	NOIGEN EA 70
9016-45-9	NOIGEN EA 80
137-26-8	NOMERSAN
9016-45-9	NONAL 208
9016-45-9	NONAL 210
143-08-8	NONALOL
143-08-8	NONAN-1-OL
3085-26-5	NONANAL, 8-METHYL-
111-84-2	NONANE
111-84-2	NONANE (ACGIH,DOT)
1941-79-3	NONANEDIPEROXOIC ACID
143-08-8	NONANOL
762-13-0	NONANOYL PEROXIDE
9016-45-9	NONARIL 910
9016-45-9	NONARIL 930
27215-95-8	NONENE
96-69-5	NONFLEX BPS
9016-45-9	NONIDET NP 40
9016-45-9	NONIDET NP 50
9016-45-9	NONIDET P 80
9016-45-9	NONIOLITE PN 4
9016-45-9	NONION NS
9016-45-9	NONION NS 203
9016-45-9	NONION NS 206
9016-45-9	NONION NS 2085
9016-45-9	NONION NS 210
9016-45-9	NONION NS 212
9016-45-9	NONION NS 215
9016-45-9	NONION NS 220
9016-45-9	NONION NS 230
9016-45-9	NONION NS 240
9016-45-9	NONIONIK NI
9016-45-9	NONIPOL
9016-45-9	NONIPOL 100
9016-45-9	NONIPOL 110
9016-45-9	NONIPOL 120
9016-45-9	NONIPOL 130
9016-45-9	NONIPOL 140
9016-45-9	NONIPOL 160
9016-45-9	NONIPOL 20
9016-45-9	NONIPOL 200
9016-45-9	NONIPOL 40
9016-45-9	NONIPOL 400
9016-45-9	NONIPOL 45
9016-45-9	NONIPOL 55
9016-45-9	NONIPOL 60
9016-45-9	NONIPOL 70
9016-45-9	NONIPOL 800
9016-45-9	NONIPOL 85
9016-45-9	NONIPOL 95
577-11-7	NONIT
135-88-6	NONOX D
135-88-6	NONOX DN
128-37-0	NONOX TBC
143-08-8	NONYL ALCOHOL
25154-52-3	NONYL PHENOL (MIXED ISOMERS)
5283-67-0	NONYL TRICHLOROSILANE (DOT)
112-30-1	NONYLCARBINOL
3452-97-9	NONYLOL
25154-52-3	NONYLPHENOL
9016-45-9	NONYLPHENOL ETHOXYLATE
9016-45-9	NONYLPHENOL POLYETHYLENE GLYCOL ETHER
9016-45-9	NONYLPHENOL POLYETHYLENE OXIDE
9016-45-9	NONYLPHENOXY POLYETHOXY ETHANOL
9016-45-9	NONYLPHENOXYPOLY(ETHYLENEOXY)ETHANOL
9016-45-9	NONYLPHENOXYPOLY(OXYETHYLENE)ETHANOL
9016-45-9	NONYLPHENYL POLYETHYLENE GLYCOL ETHER
5283-67-0	NONYLSILANE, TRICHLORO-
5283-67-0	NONYLTRICHLOROSILANE
9016-45-9	NOP 9
8002-74-2	NOPCOSIZE DS 101
7429-90-5	NORAL ALUMINIUM
7429-90-5	NORAL EXTRA FINE LINING GRADE
7429-90-5	NORAL INK GRADE ALUMINIUM

CAS No.	Chemical Name
68-22-4	NORALUTIN
8002-74-2	NORANE FH
991-42-4	NORBORMIDE
121-46-0	NORBORNADIENE
76-22-2	NORCAMPHOR, 1,7,7-TRIMETHYL-
108-01-0	NORCHOLINE
7664-93-9	NORDHAUSEN ACID (DOT)
50-28-2	NORDICOL
66-75-1	NORDOPAN
9016-45-9	NOREGAL LC 4 CONC
68-22-4	NORETHINDRONE
68-22-4	NORETHISTERON
68-22-4	NORETHISTERONE
68-22-4	NORETHYNODRONE
1982-47-4	NOREX
68-22-4	NORFOR
62-38-4	NORFORMS
68-22-4	NORGESTIN
107-21-1	NORKOOL
110-58-7	NORLEUCAMINE
68-22-4	NORLUTEN
68-22-4	NORLUTIN
68-22-4	NORLUTON
1072-35-1	NORMAL LEAD STEARATE
137-26-8	NORMERSAN
56-75-7	NORMIMYCIN V
16543-55-8	NORNICOTINE, N-NITROSO-
94-36-0	NOROX BZP-250
94-36-0	NOROX BZP-C-35
7568-93-6	NORPHEDRIN
68-22-4	NORPREGNENINOLONE
72-33-3	NORQUEN
9000-07-1	NORSK GELATAN
56-18-8	NORSPERMIDINE
577-11-7	NORVAL
109-73-9	NORVALAMINE
10124-37-5	NORWAY SALTPETER
56-38-2	NOURITHION
94-36-0	NOVADELOX
7631-86-9	NOVAKUP
19222-41-4	NOVAMMON
7723-14-0	NOVARED 120UF
57-63-6	NOVESTROL
50-78-2	NOVID
58-89-9	NOVIGAM
150-76-5	NOVO-DERMOQUINONA
140-80-7	NOVOLDIAMINE
56-75-7	NOVOMYCETIN
62-73-7	NOVOTOX
97-77-8	NOXAL
83-79-4	NOXFISH
9016-45-9	NP
9016-45-9	NP (NONIONIC SURFACTANT)
9016-45-9	NP 10
9016-45-9	NP 100
9016-45-9	NP 13
9016-45-9	NP 14
9016-45-9	NP 17
9004-70-0	NP 180
7440-02-0	NP 2
9016-45-9	NP 20
9016-45-9	NP 40
9016-45-9	NP 660
9016-45-9	NP 695
9016-45-9	NP 75
9016-45-9	NP 8
9016-45-9	NP 80
9016-45-9	NP 936
8020-83-5	NR 440
17804-35-2	NS 02
17804-35-2	NS 02 (FUNGICIDE)
1338-24-5	NS 130
1338-24-5	NS 160
9016-45-9	NS 215
15663-27-1	NSC 119875
2001-95-8	NSC 122023
23214-92-8	NSC 123127
51-28-5	NSC 1532
66-81-9	NSC 185
10265-92-6	NSC 190987
22224-92-6	NSC 195106
21923-23-9	NSC 195164
51-21-8	NSC 19893
684-93-5	NSC 23909
434-07-1	NSC 26198
50-18-0	NSC 26271
62-50-0	NSC 26805
50-07-7	NSC 26980
110-17-8	NSC 2752
615-53-2	NSC 2860
68-76-8	NSC 29215
50-76-0	NSC 3053
10124-50-2	NSC 3060
26628-22-8	NSC 3072
305-03-3	NSC 3088
127-19-5	NSC 3138
316-42-7	NSC 33669
66-75-1	NSC 34462
446-86-6	NSC 39084
87-90-1	NSC 405124
94-75-7	NSC 423
4342-03-4	NSC 45388
759-73-9	NSC 45403
68-12-2	NSC 5356
2238-07-5	NSC 54739
52-24-4	NSC 6396
2636-26-2	NSC 6738
101-25-7	NSC 73599
54-62-6	NSC 739
51-79-6	NSC 746
55-98-1	NSC 750
64-86-8	NSC 757
55-86-7	NSC 762
51-75-2	NSC 762
55-86-7	NSC 762 HYDROCHLORIDE
67-68-5	NSC 763
366-70-1	NSC 77213
13010-47-4	NSC 79037
20830-81-3	NSC 82151
18883-66-4	NSC 85998
148-82-3	NSC 8806
107-02-8	NSC 8819
70-25-7	NSC 9369
68-22-4	NSC 9564
545-55-1	NSC 9717
11099-06-2	NT 40
139-13-9	NTA
18662-53-8	NTA TRISODIUM SALT H20
55-63-0	NTG
131-11-3	NTM
35400-43-2	NTN 9306
122-20-3	NTP
9004-70-0	NTS 218
9004-70-0	NTS 222
9004-70-0	NTS 539
9004-70-0	NTS 542
9004-70-0	NTS 62
127-85-5	NUARSOL
1332-58-7	NUCAP 100
1332-58-7	NUCAP 190
333-41-5	NUCIDOL
16752-77-5	NUDRIN
7681-49-4	NUFLUOR
60-00-4	NULLAPON B ACID
60-00-4	NULLAPON BF ACID
117-84-0	NUOPLAZ DOP
8020-83-5	NUTRAL 600
7601-54-9	NUTRIFOS STP
2636-26-2	NUVA
6923-22-4	NUVACRON
6923-22-4	NUVACRON 20
62-73-7	NUVAN
2636-26-2	NUVAN
62-73-7	NUVAN 100EC
2636-26-2	NUVAN 100EC
62-73-7	NUVAN 7
122-14-5	NUVANOL
8002-74-2	NXK 501S

CAS No.	Chemical Name
7631-86-9	NYACOL 2034A
1309-64-4	NYACOL A 1510LP
1309-64-4	NYACOL A 1530
62-38-4	NYLMERATE
1333-86-4	NYLOFIL BLACK BLN
105-60-2	NYLON A1035SF
105-60-2	NYLON CM 1031
105-60-2	NYLON X 1051
82-28-0	NYLOQUINONE ORANGE JR
94-36-0	NYPER B
94-36-0	NYPER BO
55-63-0	NYSCONITRINE
14807-96-6	NYTAL 200
14807-96-6	NYTAL 400
52-68-6	O,O DIMETIL 2,2,2-TRICHLORO 1 HIDROXIETIL FOSFONATO (PORTUGESE)
133-14-2	O,O',P,P'-TETRACHLORODIBENZOYL PEROXIDE
91-94-1	O,O'-DICHLOROBENZIDINE
119-93-7	O,O'-TOLIDINE
563-12-2	O,O,O',O'-TETRAAETHYL-BIS(DITHIOPHOSPHAT) (GERMAN)
563-12-2	O,O,O',O'-TETRAETHYL S,S'-METHYLENE BIS(PHOSPORO-DITHIOATE)
563-12-2	O,O,O',O'-TETRAETHYL S,S'-METHYLENE DI(PHOSPHORO-DITHIOATE)
563-12-2	O,O,O',O'-TETRAETHYL S,S'-METHYLENEBISPHOSPHOR-DITHIOATE
3383-96-8	O,O,O',O'-TETRAMETHYL O,O'-THIODI-P-PHENYLENE PHOS-PHOROTHIOATE
3689-24-5	O,O,O,O-TETRAAETHYL-DITHIONOPYROPHOSPHAT (GERMAN)
3689-24-5	O,O,O,O-TETRAETHYL DITHIOPYROPHOSPHATE
3689-24-5	O,O,O,O-TETRAETHYL-DITHIO-DIFOSFAAT (DUTCH)
3689-24-5	O,O,O,O-TETRAETIL-DITIO-PIROFOSFATO (ITALIAN)
4104-14-7	O,O-BIS(P-CHLOROPHENYL) ACETIMIDOYLAMIDOTHIOPHOS-PHATE
4104-14-7	O,O-BIS(P-CHLOROPHENYL)ACETIMIDOYLPHOSPHORAMI-DOTHIOATE
333-41-5	O,O-DIAETHYL-O-(2-ISOPROPYL-4-METHYL)-6-PYRIMIDYL-THIONOPHOSPHAT (GERMAN)
333-41-5	O,O-DIAETHYL-O-(2-ISOPROPYL-4-METHYL-PYRIMIDIN-6-YL)-MONOTHIOPHOSPHAT (GERMAN)
297-97-2	O,O-DIAETHYL-O-(2-PYRAZINYL)-THIONOPHOSPHAT (GERMAN)
56-72-4	O,O-DIAETHYL-O-(3-CHLOR-4-METHYL-CUMARIN-7-YL)-MONOTHIOPHOSPHAT (GERMAN)
297-97-2	O,O-DIAETHYL-O-(PYRAZIN-2YL)-MONOTHIOPHOSPHAT (GER-MAN)
470-90-6	O,O-DIAETHYL-O-1-(4,5-DICHLORPHENYL)-2-CHLOR-VINYL-PHOSPHAT (GERMAN)
2921-88-2	O,O-DIAETHYL-O-3,5,6-TRICHLOR-2-PYRIDYLMONOTHIOPHOS-PHAT (GERMAN)
2275-14-1	O,O-DIAETHYL-S((2,5-DICHLOR-PHENYL-THIO)-METHYL)-DITHIOPHOSPHAT (GERMAN)
786-19-6	O,O-DIAETHYL-S-((4-CHLOR-PHENYL-THIO)-METHYL)-DITHIOPHOSPHAT (GERMAN)
2642-71-9	O,O-DIAETHYL-S-((4-OXO-3H-1,2,3-BENZOTRIAZIN-3-YL)-METHYL)-DITHIOPHOSPHAT (GERMAN)
298-04-4	O,O-DIAETHYL-S-(2-AETHYLTHIO-AETHYL)-DITHIOPHOSPHAT (GERMAN)
298-04-4	O,O-DIAETHYL-S-(3-THIA-PENTYL)-DITHIOPHOSPHAT (GERMAN)
2642-71-9	O,O-DIAETHYL-S-(4-OXOBENZOTRIAZIN-3-METHYL)-DITHIOPHOS-PHAT (GERMAN)
298-02-2	O,O-DIAETHYL-S-(AETHYLTHIO-METHYL)-DITHIOPHOSPHAT (GERMAN)
8065-48-3	O,O-DIETHYL 2-ETHYLMERCAPTOETHYL THIOPHOSPHATE
298-04-4	O,O-DIETHYL 2-ETHYLTHIOETHYL PHOSPHORODITHIOATE
333-41-5	O,O-DIETHYL 2-ISOPROPYL-4-METHYLPYRIMIDYL-6-THIOPHOS-PHATE
56-72-4	O,O-DIETHYL 3-CHLORO-4-METHYL-7-UMBELLIFERONE THIOPHOSPHATE
814-49-3	O,O-DIETHYL CHLORIDOPHOSPHATE
814-49-3	O,O-DIETHYL CHLOROPHOSPHATE
814-49-3	O,O-DIETHYL CHLOROPHOSPHONATE
786-19-6	O,O-DIETHYL DITHIOPHOSPHORIC ACID P-CHLOROPHENYL-THIOMETHYL ESTER
298-02-2	O,O-DIETHYL ETHYLTHIOMETHYL PHOSPHORODITHIOATE
8065-48-3	O,O-DIETHYL O(AND S)-2-(ETHYLTHIO)ETHYL PHOS-PHOROTHIOATE MIXTURE
297-97-2	O,O-DIETHYL O,2-PYRAZINYL PHOSPHOROTHIOATE
24017-47-8	O,O-DIETHYL O-(1-PHENYL-1H-1,2,4-TRIAZOL-3-YL)PHOS-PHOROTHIOATE
470-90-6	O,O-DIETHYL O-(2-CHLORO-1-(2',4'-DICHLOROPHENYL)VINYL) PHOSPHATE
23505-41-1	O,O-DIETHYL O-(2-DIETHYLAMINO-6-METHYL-4-PYRIMIDINYL)-PHOSPHOROTHIOATE
333-41-5	O,O-DIETHYL O-(2-ISOPROPYL-4-METHYL-6-PYRIMIDYL) THIONOPHOSPHATE
333-41-5	O,O-DIETHYL O-(2-ISOPROPYL-6-METHYL-4-PYRIMIDINYL) PHOSPHOROTHIOATE
333-41-5	O,O-DIETHYL O-(2-ISOPROPYL-6-METHYL-4-PYRIMIDIYL) PHOS-PHOROTHIOATE
56-72-4	O,O-DIETHYL O-(3 CHLORO-4-METHYL-2OXO-2H-1 BENZOPYRAN-7-YL) PHOSPHOROTHIOATE
56-72-4	O,O-DIETHYL O-(3-CHLORO-4-METHYL-2-OXO-2H-BENZOPYRAN-7-YL)PHOSPHOROTHIOATE
56-72-4	O,O-DIETHYL O-(3-CHLORO-4-METHYL-7-COUMARINYL)PHOS-PHOROTHIOATE
56-72-4	O,O-DIETHYL O-(3-CHLORO-4-METHYLCOUMARINYL-7) THIOPHOSPHATE
56-72-4	O,O-DIETHYL O-(3-CHLORO-4-METHYLUMBELLIFERYL)PHOS-PHOROTHIOATE
115-90-2	O,O-DIETHYL O-(4-METHYLSULFINYLPHENYL) MONOTHIOPHOS-PHATE
115-90-2	O,O-DIETHYL O-(P-METHYLSULFINYL PHENYL) PHOS-PHOROTHIOATE
56-38-2	O,O-DIETHYL O-(P-NITROPHENYL) PHOSPHOROTHIOATE
297-97-2	O,O-DIETHYL O-2-PYRAZINYL PHOSPHOTHIONATE
2921-88-2	O,O-DIETHYL O-3,5,6-TRICHLORO-2-PYRIDYL PHOS-PHOROTHIOATE
333-41-5	O,O-DIETHYL O-6-METHYL-2-ISOPROPYL-4-PYRIMIDINYL PHOS-PHOROTHIOATE
56-38-2	O,O-DIETHYL O-P-NITROPHENYL THIOPHOSPHATE
297-97-2	O,O-DIETHYL O-PYRAZINYL THIOPHOSPHATE
115-90-2	O,O-DIETHYL O-[4-(METHYLSULFINYL)PHENYL] PHOS-PHOROTHIOATE
786-19-6	O,O-DIETHYL P-CHLOROPHENYLMERCAPTOMETHYL DITHIOPHOSPHATE
814-49-3	O,O-DIETHYL PHOSPHOROCHLORIDATE
2642-71-9	O,O-DIETHYL PHOSPHORODITHIOATE S-ESTER WITH 3-(MERCAP-TOMETHYL)-1,2,3-BENZOTRIAZIN-4(3H)-ONE
2275-14-1	O,O-DIETHYL S-(2,5-DICHLOROPHENYLTHIOMETHYL) DITHIOPHOSPHATE
2275-14-1	O,O-DIETHYL S-(2,5-DICHLOROPHENYLTHIOMETHYL) PHOS-PHORODITHIOATE
2275-14-1	O,O-DIETHYL S-(2,5-DICHLOROPHENYLTHIOMETHYL) PHOS-PHOROTHIOLOTHIONATE
2497-07-6	O,O-DIETHYL S-(2-(ETHYLSULFINYL)ETHYL) PHOSPHORO-DITHIOATE
3734-97-2	O,O-DIETHYL S-(2-DIETHYLAMINO)ETHYL PHOSPHOROTHIOATE HYDROGEN OXALATE
78-53-5	O,O-DIETHYL S-(2-DIETHYLAMINOETHYL) THIOPHOSPHATE
298-04-4	O,O-DIETHYL S-(2-ETHTHIOETHYL) PHOSPHORODITHIOATE
298-04-4	O,O-DIETHYL S-(2-ETHTHIOETHYL) THIOTHIONOPHOSPHATE
3734-97-2	O,O-DIETHYL S-(2-ETHYL-N,N-DIETHYLAMINO)PHOS-PHOROTHIOATE HYDROGEN OXALATE
298-04-4	O,O-DIETHYL S-(2-ETHYLMERCAPTOETHYL) DITHIOPHOSPHATE
2497-07-6	O,O-DIETHYL S-(2-ETHYLTHIONYLETHYL) PHOSPHORODITHIOATE
786-19-6	O,O-DIETHYL S-(4-CHLOROPHENYLTHIOMETHYL) DITHIOPHOS-PHATE
2642-71-9	O,O-DIETHYL S-(4-OXOBENZOTRIAZINO-3-METHYL)PHOSPHORO-DITHIOATE
78-53-5	O,O-DIETHYL S-(BETA-DIETHYLAMINO)ETHYL PHOSPHOROTHIO-LATE
3734-97-2	O,O-DIETHYL S-(BETA-DIETHYLAMINO)ETHYL PHOSPHOROTHIO-LATE HYDROGEN OXALATE
24934-91-6	O,O-DIETHYL S-(CHLOROMETHYL) DITHIOPHOSPHATE
298-02-2	O,O-DIETHYL S-(ETHYLTHIO)METHYL PHOSPHORODITHIOATE
2275-18-5	O,O-DIETHYL S-(N-ISOPROPYLCARBAMOYLMETHYL) DITHIOPHOSPHATE
2275-18-5	O,O-DIETHYL S-(N-ISOPROPYLCARBAMOYLMETHYL) PHOS-PHORODITHIOATE
786-19-6	O,O-DIETHYL S-(P-CHLOROPHENYLTHIO)METHYL PHOSPHORO-DITHIOATE
13071-79-9	O,O-DIETHYL S-(TERT-BUTYLTHIO)METHYL PHOSPHORO-DITHIOATE
298-04-4	O,O-DIETHYL S-2-(ETHYLTHIO)ETHYL PHOSPHORODITHIOATE
78-53-5	O,O-DIETHYL S-2-DIETHYLAMINOETHYL PHOSPHOROTHIOATE
78-53-5	O,O-DIETHYL S-2-DIETHYLAMINOETHYL PHOSPHOROTHIOLATE
78-53-5	O,O-DIETHYL S-DIETHYLAMINOETHYL PHOSPHOROTHIOLATE
298-02-2	O,O-DIETHYL S-ETHYLMERCAPTOMETHYL DITHIOPHOSPHATE

CAS No.	Chemical Name
298-02-2	O,O-DIETHYL S-ETHYLMERCAPTOMETHYL DITHIOPHOSPHONATE
298-02-2	O,O-DIETHYL S-ETHYLTHIOMETHYL DITHIOPHOSPHATE
298-02-2	O,O-DIETHYL S-ETHYLTHIOMETHYL DITHIOPHOSPHONATE
298-02-2	O,O-DIETHYL S-ETHYLTHIOMETHYL THIOTHIONOPHOSPHATE
2275-18-5	O,O-DIETHYL S-ISOPROPYLCARBAMOYLMETHYL PHOSPHORO-DITHIOATE
786-19-6	O,O-DIETHYL S-P-CHLOROPHENYLTHIOMETHYL DITHIOPHOS-PHATE
298-02-2	O,O-DIETHYL S-[(ETHYLTHIO)METHYL] PHOSPHORODITHIOATE
298-04-4	O,O-DIETHYL S-[2-(ETHYLTHIO)ETHYL] DITHIOPHOSPHATE
13071-79-9	O,O-DIETHYL S-[[(1,1-DIMETHYLETHYL)THIO]METHYL] PHOS-PHORODITHIOATE
2524-04-1	O,O-DIETHYL THIONOPHOSPHORIC CHLORIDE
2524-04-1	O,O-DIETHYL THIONOPHOSPHOROCHLORIDATE
2524-04-1	O,O-DIETHYL THIONOPHOSPHORYL CHLORIDE
2524-04-1	O,O-DIETHYL THIOPHOSPHORIC ACID CHLORIDE
2524-04-1	O,O-DIETHYL THIOPHOSPHORYL CHLORIDE
333-41-5	O,O-DIETHYL-O-(2-ISOPROPYL-4-METHYL-6-PYRIMIDINYL)-PHOSPHOROTHIOATE
333-41-5	O,O-DIETHYL-O-(2-ISOPROPYL-4-METHYL-6-PYRIMIDYL)PHOS-PHOROTHIOATE
333-41-5	O,O-DIETHYL-O-(2-ISOPROPYL-4-METHYL-PYRIMIDIN-6-YL)-MONOTHIOFOSFAAT (DUTCH)
21923-23-9	O,O-DIETHYL-O-2,4,5-DICHLORO-(METHYLTHIO)PHENYL THIONOPHOSPHATE
2642-71-9	O,O-DIETHYL-S-((4-OXO-3H-1,2,3-BENZOTRIAZIN-3-YL)-METHYL)-DITHIOFOSFAAT (DUTCH)
2497-07-6	O,O-DIETHYL-S-((ETHYLSULFINYL)ETHYL)PHOSPHORODITHIOATE
2275-14-1	O,O-DIETHYL-S-(2,5-DICHLOROPHENYLTHIOMETHYL) DITHIOPHOSPHORAN
298-04-4	O,O-DIETHYL-S-(2-ETHYLTHIO-ETHYL)-DITHIOFOSFAAT (DUTCH)
786-19-6	O,O-DIETHYL-S-(4-CHLOOR-FENYL-THIO)-METHYL)-DITHIOFOS-FAAT (DUTCH)
2642-71-9	O,O-DIETHYL-S-(4-OXO-3H-1,2,3-BENZOTRIAZINE-3-YL)-METHYL-DITHIOPHOSPHATE
298-02-2	O,O-DIETHYL-S-(ETHYLTHIO-METHYL)-DITHIOFOSFAAT (DUTCH)
786-19-6	O,O-DIETHYL-S-P-CHLORFENYLTHIOMETHYLESTER KYSELINY DITHIOFOSFORECNE (CZECH)
2275-18-5	O,O-DIETHYLDITHIOPHOSPHORYLACETIC ACID, N-MONOISO-PROPYLAMIDE
2524-04-1	O,O-DIETHYLPHOSPHOROCHLORIDOTHIOATE
814-49-3	O,O-DIETHYLPHOSPHORYL CHLORIDE
2524-04-1	O,O-DIETHYLTHIOPHOSPHOROCHLORIDATE
333-41-5	O,O-DIETIL-O-(2-ISOPROPIL-4-METIL-PIRIMIDIN-6-IL)-MONO-TIOFOSFATO (ITALIAN)
56-72-4	O,O-DIETIL-O-(3-CLORO-4-METIL-CUMARIN-7-IL-MONOTIOFOS-FATO) (ITALIAN)
786-19-6	O,O-DIETIL-S-((4-CLORO-FENIL-TIO)-METILE)-DITIOFOSFATO (ITALIAN)
2642-71-9	O,O-DIETIL-S-((4-OXO-3H-1,2,3-BENZOTRIAZIN-3-IL)-METIL)-DITIOFOSFATO (ITALIAN)
298-04-4	O,O-DIETIL-S-(2-ETILTIO-ETIL)-DITIOFOSFATO (ITALIAN)
298-02-2	O,O-DIETIL-S-(ETILTIO-METIL)-DITIOFOSFATO (ITALIAN)
56-38-2	O,O-DIETYL-O-P-NITROFENYLTIOFOSFAT (CZECH)
52-68-6	O,O-DIMETHYL (1-HYDROXY-2,2,2-TRICHLOROETHYL)PHOSPHO-NATE
52-68-6	O,O-DIMETHYL (2,2,2-TRICHLORO-1-HYDROXYETHYL)PHOSPHO-NATE
62-73-7	O,O-DIMETHYL 2,2-DICHLOROVINYL PHOSPHATE
919-86-8	O,O-DIMETHYL 2-ETHYLMERCAPTOETHYL THIOPHOSPHATE, THIOLO ISOMER
732-11-6	O,O-DIMETHYL 5-(PHTHALIMIDOMETHYL)DITHIOPHOSPHATE
2524-03-0	O,O-DIMETHYL CHLOROTHIONOPHOSPHATE
2524-03-0	O,O-DIMETHYL CHLOROTHIOPHOSPHATE
2636-26-2	O,O-DIMETHYL DICHLOROVINYL PHOSPHATE
2540-82-1	O,O-DIMETHYL DITHIOPHOSPHORYLACETIC ACID N-METHYL-N-FORMYLAMIDE
60-51-5	O,O-DIMETHYL METHYLCARBAMOYLMETHYL PHOSPHORO-DITHIOATE
7786-34-7	O,O-DIMETHYL O-(1-CARBOMETHOXY-1-PROPEN-2-YL) PHOS-PHATE
7786-34-7	O,O-DIMETHYL O-(1-METHYL-2-CARBOXYVINYL) PHOSPHATE
299-84-3	O,O-DIMETHYL O-(2,4,5-TRICHLOROPHENYL) PHOS-PHOROTHIOATE
299-84-3	O,O-DIMETHYL O-(2,4,5-TRICHLOROPHENYL) THIOPHOSPHATE
122-14-5	O,O-DIMETHYL O-(3-METHYL) PHOSPHOROTHIOATE
122-14-5	O,O-DIMETHYL O-(3-METHYL-4-NITROPHENYL) PHOS-PHOROTHIOATE
122-14-5	O,O-DIMETHYL O-(3-METHYL-4-NITROPHENYL) THIOPHOSPHATE
2636-26-2	O,O-DIMETHYL O-(4-CYANOPHENYL) PHOSPHOROTHIOATE
3254-63-5	O,O-DIMETHYL O-(4-METHYLMERCAPTOPHENYL) PHOSPHATE
122-14-5	O,O-DIMETHYL O-(4-NITRO-3-METHYLPHENYL)THIOPHOSPHATE
141-66-2	O,O-DIMETHYL O-(N,N-DIMETHYLCARBAMOYL-1-METHYLVINYL) PHOSPHATE
2636-26-2	O,O-DIMETHYL O-(P-CYANOPHENYL) PHOSPHOROTHIOATE
298-00-0	O,O-DIMETHYL O-(P-NITROPHENYL) PHOSPHOROTHIOATE
298-00-0	O,O-DIMETHYL O-(P-NITROPHENYL) THIONOPHOSPHATE
2636-26-2	O,O-DIMETHYL O-2,2-DICHLOROVINYL PHOSPHATE
122-14-5	O,O-DIMETHYL O-4-NITRO-M-TOLYL PHOSPHOROTHIOATE
298-00-0	O,O-DIMETHYL O-P-NITROPHENYL THIOPHOSPHATE
2524-03-0	O,O-DIMETHYL PHOSPHOROCHLORIDOTHIOATE
2524-03-0	O,O-DIMETHYL PHOSPHOROCHLOROTHIOATE
756-80-9	O,O-DIMETHYL PHOSPHORODITHIOATE
2540-82-1	O,O-DIMETHYL PHOSPHORODITHIOATE N-FORMYL-2-MERCAPTO-N-METHYLACETAMIDE S-ESTER
2524-03-0	O,O-DIMETHYL PHOSPHOROTHIONOCHLORIDATE
121-75-5	O,O-DIMETHYL S-(1,2-BIS(ETHOXYCARBONYL)ETHYL)-DITHIOPHOSPHATE
121-75-5	O,O-DIMETHYL S-(1,2-DICARBETHOXYETHYL) THIOTHIONOPHOS-PHATE
121-75-5	O,O-DIMETHYL S-(1,2-DICARBETHOXYETHYL)PHOSPHORO-DITHIOATE
3735-23-7	O,O-DIMETHYL S-(2,5-DICHLOROPHENYLTHIO)METHYL PHOS-PHORODITHIOATE
640-15-3	O,O-DIMETHYL S-(2-(ETHYLTHIO)ETHYL) PHOSPHORODITHIOATE
919-86-8	O,O-DIMETHYL S-(2-(ETHYLTHIO)ETHYL)PHOSPHOROTHIOATE
60-51-5	O,O-DIMETHYL S-(2-(METHYLAMINO)-2-OXOETHYL) PHOSPHORO-DITHIOATE
640-15-3	O,O-DIMETHYL S-(2-ETHYLTHIOETHYL) DITHIOPHOSPHATE
86-50-0	O,O-DIMETHYL S-(3 ,4-DIHYDRO-4-KETO-1,2,3-BENZOTRIAZINYL-3-METHYL) DITHIOPHOSPHATE
86-50-0	O,O-DIMETHYL S-(4-OXO-1,2,3-BENZOTRIAZINO(3)-METHYL) THIOTHIONOPHOSPHATE
86-50-0	O,O-DIMETHYL S-(4-OXO-3H-1,2,3-BENZOTRIAZINE-3-METHYL)-PHOSPHORODITHIOATE
86-50-0	O,O-DIMETHYL S-(4-OXOBENZOTRIAZINO-3-METHYL)PHOS-PHORODITHIOATE
950-37-8	O,O-DIMETHYL S-(5-METHOXY-1,3,4-THIADIAZOLINYL-3-METHYL) DITHIOPHOSPHATE
2778-04-3	O,O-DIMETHYL S-(5-METHOXY-4-OXO-4H-PYRAN-2-YL)PHOS-PHOROTHIOATE
2778-04-3	O,O-DIMETHYL S-(5-METHOXYPYRONYL-2-METHYL) THIOPHOS-PHATE
2540-82-1	O,O-DIMETHYL S-(N-FORMYL-N-METHYLCARBAMOYLMETHYL) PHOSPHORODITHIOATE
2540-82-1	O,O-DIMETHYL S-(N-METHYL-N-FORMYLCARBAMOYLMETHYL)-PHOSPHORODITHIOATE
60-51-5	O,O-DIMETHYL S-(N-METHYLCARBAMOYLMETHYL) DITHIOPHOS-PHATE
60-51-5	O,O-DIMETHYL S-(N-METHYLCARBAMOYLMETHYL) PHOSPHORO-DITHIOATE
60-51-5	O,O-DIMETHYL S-(N-METHYLCARBAMYLMETHYL) THIOTHIONOPHOSPHATE
732-11-6	O,O-DIMETHYL S-(N-PHTHALIMIDOMETHYL)DITHIOPHOSPHATE
732-11-6	O,O-DIMETHYL S-(PHTHALIMIDOMETHYL) DITHIOPHOSPHATE
121-75-5	O,O-DIMETHYL S-1,2-DI(ETHOXYCARBAMYL)ETHYL PHOSPHORO-DITHIOATE
919-86-8	O,O-DIMETHYL S-2-(ETHYLTHIO)ETHYL PHOSPHOROTHIOATE
86-50-0	O,O-DIMETHYL S-4-OXO-1,2,3-BENZOTRIAZIN-3(4H)-YL METHYL PHOSPHORODITHIOATE
919-86-8	O,O-DIMETHYL S-ETHYLMERCAPTOETHYL THIOPHOSPHATE
732-11-6	O,O-DIMETHYL S-PHTHALIMIDOMETHYL PHOSPHORODITHIOATE
2524-03-0	O,O-DIMETHYL THIONOPHOSPHOROCHLORIDATE
2524-03-0	O,O-DIMETHYL THIOPHOSPHORIC ACID CHLORIDE
2524-03-0	O,O-DIMETHYL THIOPHOSPHORYL CHLORIDE
52-68-6	O,O-DIMETHYL-(1-HYDROXY-2,2,2-TRICHLORAETHYL)PHOSPHON-SAEURE ESTER (GERMAN)
52-68-6	O,O-DIMETHYL-(1-HYDROXY-2,2,2-TRICHLORATHYL)-PHOSPHAT (GERMAN)
52-68-6	O,O-DIMETHYL-(1-HYDROXY-2,2,2-TRICHLORO)ETHYL PHOS-PHATE
52-68-6	O,O-DIMETHYL-(2,2,2-TRICHLOOR-1-HYDROXY-ETHYL)-FOSFO-NAAT (DUTCH)
52-68-6	O,O-DIMETHYL-(2,2,2-TRICHLOR-1-HYDROXY-AETHYL)PHOS-PHONAT (GERMAN)
52-68-6	O,O-DIMETHYL-1-OXY-2,2,2-TRICHLOROETHYL PHOSPHONATE
52-68-6	O,O-DIMETHYL-2,2,2-TRICHLORO-1-HYDROXYETHYL PHOSPHO-NATE

CAS No.	Chemical Name
60-51-5	O,O-DIMETHYL-DITHIOPHOSPHORYLESSIGSAEURE MONOMETHYLAMID (GERMAN)
141-66-2	O,O-DIMETHYL-O-(1,4-DIMETHYL-3-OXO-4-AZA-PENT-1-ENYL)FOSFAAT (DUTCH)
141-66-2	O,O-DIMETHYL-O-(1,4-DIMETHYL-3-OXO-4-AZA-PENT-1-ENYL)-PHOSPHATE
141-66-2	O,O-DIMETHYL-O-(1-METHYL-2-N,N-DIMETHYL-CARBAMOYL)-VINYL-PHOSPHAT (GERMAN)
6923-22-4	O,O-DIMETHYL-O-(1-METHYL-2-N-METHYL-CARBAMOYL)-VINYL-PHOSPHAT (GERMAN)
2636-26-2	O,O-DIMETHYL-O-(2,2-DICHLOR-VINYL)-PHOSPHAT (GERMAN)
7786-34-7	O,O-DIMETHYL-O-(2-CARBOMETHOXY-1-METHYLVINYL) PHOSPHATE
141-66-2	O,O-DIMETHYL-O-(2-DIMETHYL-CARBAMOYL-1-METHYL-VINYL)-PHOSPHAT (GERMAN)
6923-22-4	O,O-DIMETHYL-O-(2-N-METHYLCARBAMOYL-1-METHYL)-VINYL-PHOSPHAT (GERMAN)
6923-22-4	O,O-DIMETHYL-O-(2-N-METHYLCARBAMOYL-1-METHYL-VINYL) PHOSPHATE
6923-22-4	O,O-DIMETHYL-O-(2-N-METHYLCARBAMOYL-1-METHYL-VINYL)-FOSFAAT (DUTCH)
122-14-5	O,O-DIMETHYL-O-(3-METHYL-4-NITRO-PHENYL)-MONOTHIOPHOSPHAT (GERMAN)
122-14-5	O,O-DIMETHYL-O-(3-METHYL-4-NITROFENYL)-MONOTHIOFOSFAAT (DUTCH)
122-14-5	O,O-DIMETHYL-O-(4-NITRO-5-METHYLPHENYL)-THIONOPHOSPHAT (GERMAN)
298-00-0	O,O-DIMETHYL-O-(4-NITRO-FENYL)-MONOTHIOFOSFAAT (DUTCH)
298-00-0	O,O-DIMETHYL-O-(4-NITRO-PHENYL)-MONOTHIOPHOSPHAT (GERMAN)
298-00-0	O,O-DIMETHYL-O-(4-NITROPHENYL) PHOSPHOROTHIOATE
298-00-0	O,O-DIMETHYL-O-(4-NITROPHENYL)-THIONOPHOSPHAT (GERMAN)
298-00-0	O,O-DIMETHYL-O-(P-NITROPHENYL)-THIONOPHOSPHAT (GERMAN)
7786-34-7	O,O-DIMETHYL-O-2-METHOXYCARBONYL-1-METHYL-VINYL-PHOSPHAT (GERMAN)
298-00-0	O,O-DIMETHYL-O-P-NITROFENYLESTER KYSELINY THIOFOSFORECNE (CZECH)
950-37-8	O,O-DIMETHYL-S-((2-METHOXY-1,3,4 (4H)-THIODIAZOL-5-ON-4-YL)-METHYL)-DITHIOFOSFAAT (DUTCH)
86-50-0	O,O-DIMETHYL-S-((4-OXO-3H-1,2,3-BENZOTRIAZIN-3-YL)-METHYL)-DITHIOFOSFAAT (DUTCH)
86-50-0	O,O-DIMETHYL-S-((4-OXO-3H-1,2,3-BENZOTRIAZIN-3-YL)-METHYL)-DITHIOPHOSPHAT (GERMAN)
86-50-0	O,O-DIMETHYL-S-(1,2,3-BENZOTRIAZINYL-4-KETO)METHYL PHOSPHORODITHIOATE
121-75-5	O,O-DIMETHYL-S-(1,2-DICARBETHOXYETHYL) DITHIOPHOSPHATE
640-15-3	O,O-DIMETHYL-S-(2-AETHYLTHIO-AETHYL)-DITHIO PHOSPHAT (GERMAN)
919-86-8	O,O-DIMETHYL-S-(2-AETHYLTHIO-AETHYL)-MONOTHIOPHOSPHAT (GERMAN)
640-15-3	O,O-DIMETHYL-S-(2-ETHYLMERCAPTOETHYL) DITHIOPHOSPHATE
640-15-3	O,O-DIMETHYL-S-(2-ETHYLTHIO-ETHYL)-DITHIOFOSFAAT (DUTCH)
919-86-8	O,O-DIMETHYL-S-(2-ETHYLTHIO-ETHYL)-MONOTHIOFOSFAAT (DUTCH)
950-37-8	O,O-DIMETHYL-S-(2-METHOXY-1,3,4-THIADIAZOL-5(4H)-ONYL-(4)-METHYL) PHOSPHORODITHIOATE
950-37-8	O,O-DIMETHYL-S-(2-METHOXY-1,3,4-THIADIAZOL-5-(4H)-ONYL-(4)-METHYL)-DITHIOPHOSPHAT (GERMAN)
950-37-8	O,O-DIMETHYL-S-(2-METHOXY-1,3,4-THIADIAZOL-5-ON-4-LY)-METHYL-DITHIOPHOSPHAT (GERMAN)
60-51-5	O,O-DIMETHYL-S-(2-OXO-3-AZA-BUTYL)-DITHIOPHOSPHAT (GERMAN)
2540-82-1	O,O-DIMETHYL-S-(3-METHYL-2,4-DIOXO-3-AZA-BUTYL)-DITHIOFOSFAAT (DUTCH)
2540-82-1	O,O-DIMETHYL-S-(3-METHYL-2,4-DIOXO-3-AZA-BUTYL)-DITHIOPHOSPHAT (GERMAN)
919-86-8	O,O-DIMETHYL-S-(3-THIA-PENTYL)-MONOTHIOPHOSPHAT (GERMAN)
86-50-0	O,O-DIMETHYL-S-(4-OXOBENZOTRIAZIN-3-METHYL)-DITHIOPHOSPHAT (GERMAN)
2778-04-3	O,O-DIMETHYL-S-(5-METHOXY-PYRON-2-YL)-METHYL)-THIOLPHOSPHAT (GERMAN)
86-50-0	O,O-DIMETHYL-S-(BENZAZIMINOMETHYL) DITHIOPHOSPHATE
60-51-5	O,O-DIMETHYL-S-(N-METHYL-CARBAMOYL)-METHYL-DITHIOFOSFAAT (DUTCH)
2540-82-1	O,O-DIMETHYL-S-(N-METHYL-N-FORMYL-CARBAMOYLMETHYL)-DITHIOPHOSPHAT (GERMAN)
60-51-5	O,O-DIMETHYL-S-(N-MONOMETHYL)-CARBAMYL METHYL DITHIOPHOSPHATE
121-75-5	O,O-DIMETHYL-S-1,2-(DICARBAETHOXYAETHYL)-DITHIOPHOSPHAT (GERMAN)
121-75-5	O,O-DIMETHYL-S-1,2-DIKARBETOXYLETHYLDITIOFOSFAT (CZECH)
640-15-3	O,O-DIMETHYL-S-2-ETHYLMERKAPTOETHYLESTER KYSELINY DITHIOFOSFORECNE (CZECH)
121-75-5	O,O-DIMETHYLDITHIOPHOSPHATE DIETHYLMERCAPTOSUCCINATE
60-51-5	O,O-DIMETHYLDITHIOPHOSPHORYLACETIC ACID, N-MONOMETHYLAMIDE SALT
2524-03-0	O,O-DIMETHYLESTER KYSELINY CHLORTHIOFOSFORECNE (CZECH)
2524-03-0	O,O-DIMETHYLPHOSPHOROCHLORIDOTHIOATE
2524-03-0	O,O-DIMETHYLTHIONOPHOSPHORYL CHLORIDE
52-68-6	O,O-DIMETIL-(2,2,2-TRICLORO-1-IDROSSI-ETIL)-FOSFONATO (ITALIAN)
141-66-2	O,O-DIMETIL-O-(1,4-DIMETIL-3-OXO-4-AZA-PENT-1-ENIL)-FOSFATO (ITALIAN)
6923-22-4	O,O-DIMETIL-O-(2-N-METILCARBAMOIL-1-METIL-VINIL)-FOSFATO (ITALIAN)
298-00-0	O,O-DIMETIL-O-(4-NITRO-FENIL)-MONOTIOFOSFATO (ITALIAN)
950-37-8	O,O-DIMETIL-S-((2-METOSSI-1,3,4-(4H)-TIADIAZOL-5-ON-4-IL)-METIL)-DITIFOSFATO (ITALIAN)
86-50-0	O,O-DIMETIL-S-((4-OXO-3H-1,2,3-BENZOTRIAZIN-3-IL)-METIL)-DITIOFOSFATO (ITALIAN)
919-86-8	O,O-DIMETIL-S-(2-ETILTIO-ETIL)-MONOTIOFOSFATO (ITALIAN)
640-15-3	O,O-DIMETIL-S-(ETILTIO-ETIL)-DITIOFOSFATO (ITALIAN)
2540-82-1	O,O-DIMETIL-S-(N-FORMIL-N-METIL-CARBAMOIL-METIL)-DITIOFOSFATO (ITALIAN)
60-51-5	O,O-DIMETIL-S-(N-METIL-CARBAMOIL-METIL)-DITIOFOSFATO (ITALIAN)
298-04-4	O,O-ETHYL S-2(ETHYLTHIO)ETHYL PHOSPHORODITHIOATE
10265-92-6	O,S-DIMETHYL ESTER AMIDE OF AMIDOTHIOATE
10265-92-6	O,S-DIMETHYL PHOSPHORAMIDOTHIOATE
10265-92-6	O,S-DIMETHYL THIOPHOSPHORAMIDE
2636-26-2	O-(2,2-DICHLORVINYL)-O,O-DIMETHYLPHOSPHAT (GERMAN)
21609-90-5	O-(2,5 -DICHLORO-4-BROMOPHENYL) O-METHYL PHENYLTHIOPHOSPHONATE
23505-41-1	O-(2-(DIETHYLAMINO)-6-METHYL-4-PYRIMIDINYL)O,O-DIETHYL PHOSPHOROTHIOATE
114-26-1	O-(2-ISOPROPOXYPHENYL) N-METHYLCARBAMATE
21609-90-5	O-(4-BROMO-2,5-DICHLOROPHENYL) O-METHYL PHENYLPHOSPHONOTHIOATE
2104-64-5	O-(4-NITROPHENYL) O-ETHYL PHENYL THIOPHOSPHONATE
2665-30-7	O-(4-NITROPHENYL) O-PHENYL METHYLPHOSPHONOTHIOATE
50-78-2	O-(ACETYLOXY)BENZOIC ACID
21923-23-9	O-(DICHLORO(METHYLTHIO)PHENYL) O,O-DIETHYL PHOSPHOROTHIOATE (3 ISOMERS)
2665-30-7	O-(P-NITROPHENYL) O-PHENYL METHYLPHOSPHONOTHIOATE
88-17-5	O-(TRIFLUOROMETHYL)ANILINE
88-16-4	O-(TRIFLUOROMETHYL)CHLOROBENZENE
470-90-6	O-2-CHLOOR-1-(2,4-DICHLOOR-FENYL)-VINYL-O,O-DIETHYLFOSFAAT (DUTCH)
470-90-6	O-2-CHLOR-1-(2,4-DICHLOR-PHENYL)-VINYL-O,O-DIAETHYLPHOSPHAT (GERMAN)
470-90-6	O-2-CLORO-1-(2,4-DICLORO-FENIL)-VINIL-O,O-DIETILFOSFATO (ITALIAN)
333-41-5	O-2-ISOPROPYL-4-METHYLPYRIMIDYL-O,O-DIETHYL PHOSPHOROTHIOATE
56-72-4	O-3-CHLORO-4-METHYL-7-COUMARINYL O,O-DIETHYL PHOSPHOROTHIOATE
92-15-9	O-ACETOACETANISIDIDE
50-78-2	O-ACETOXYBENZOIC ACID
50-78-2	O-ACETYLSALICYLIC ACID
327-98-0	O-AETHYL-O-(2,4,5-TRICHLORPHENYL)-AETHYLTHIONOPHOSPHONAT (GERMAN)
22224-92-6	O-AETHYL-O-(3-METHYL-4-METHYLTHIOPHENYL)-ISOPROPYLAMIDO-PHOSPHORSAEUREESTER (GERMAN)
2104-64-5	O-AETHYL-O-(4-NITRO-PHENYL)-PHENYL-MONOTHIOPHOSPHONAT (GERMAN)
944-22-9	O-AETHYL-S-PHENYL-AETHYL-DITHIOPHOSPHONAT (GERMAN)
90-04-0	O-AMINOANISOLE
29191-52-4	O-AMINOANISOLE
97-56-3	O-AMINOAZOTOLUENE
88-17-5	O-AMINOBENZOTRIFLUORIDE
90-41-5	O-AMINOBIPHENYL

CAS No.	Chemical Name
90-41-5	O-AMINODIPHENYL
578-54-1	O-AMINOETHYLBENZENE
88-74-4	O-AMINONITROBENZENE
504-29-0	O-AMINOPYRIDINE
95-53-4	O-AMINOTOLUENE
136-81-2	O-AMYL PHENOL
135-02-4	O-ANISALDEHYDE
1918-00-9	O-ANISIC ACID, 3,6-DICHLORO-
90-04-0	O-ANISIDINE
29191-52-4	O-ANISIDINE
29191-52-4	O-ANISIDINE (ACGIH,DOT)
120-71-8	O-ANISIDINE, 5-METHYL-
99-59-2	O-ANISIDINE, 5-NITRO-
134-29-2	O-ANISIDINE, HYDROCHLORIDE
29191-52-4	O-ANISYLAMINE
97-56-3	O-AT
84-66-2	O-BENZENEDICARBOXYLIC ACID DIETHYL ESTER
84-74-2	O-BENZENEDICARBOXYLIC ACID, DIBUTYL ESTER
120-80-9	O-BENZENEDIOL
81-07-2	O-BENZOIC ACID SULFIMIDE
81-07-2	O-BENZOIC SULFIMIDE
81-07-2	O-BENZOSULFIMIDE
81-07-2	O-BENZOYL SULFIMIDE
90-41-5	O-BIPHENYLAMINE
86-73-7	O-BIPHENYLENEMETHANE
95-46-5	O-BROMOMETHYLBENZENE
95-46-5	O-BROMOTOLUENE
112-34-5	O-BUTYL DIETHYLENE GLYCOL
111-76-2	O-BUTYL ETHYLENE GLYCOL
27458-20-4	O-BUTYLTOLUENE
27458-20-4	O-BUTYLTOLUENE (DOT)
69-72-7	O-CARBOXYPHENOL
50-78-2	O-CARBOXYPHENYL ACETATE
88-16-4	O-CHLORO-ALPHA,ALPHA,ALPHA-TRIFLUOROTOLUENE
92-04-6	O-CHLORO-P-PHENYLPHENOL
88-16-4	O-CHLOROBENZOTRIFLUORIDE
2698-41-1	O-CHLOROBENZYLIDENE MALO-NONITRILE (OCBM)
2698-41-1	O-CHLOROBENZYLIDENE MALONONITRILE
2698-41-1	O-CHLOROBENZYLIDENEMALONIC NITRILE
88-73-3	O-CHLORONITROBENZENE
88-73-3	O-CHLORONITROBENZENE (DOT)
95-57-8	O-CHLOROPHENOL
29154-12-9	O-CHLOROPYRIDINE
1331-28-8	O-CHLOROSTYRENE
2039-87-4	O-CHLOROSTYRENE
95-49-8	O-CHLOROTOLUENE
2039-87-4	O-CHLOROVINYLBENZENE
95-48-7	O-CRESOL
95-48-7	O-CRESOL (DOT)
534-52-1	O-CRESOL, 4,6-DINITRO-
1335-85-9	O-CRESOL, 4,6-DINITRO-
98-27-1	O-CRESOL, 4-TERT-BUTYL-
18351-85-4	O-CRESOL, DIHYDROGEN PHOSPHATE
1335-85-9	O-CRESOL, DINITRO-
29595-25-3	O-CRESOL, DINITRO-, AMMONIUM SALT
25641-53-6	O-CRESOL, DINITRO-, SODIUM SALT, WETTED WITH AT LEAST 10-15% WATER
18351-85-4	O-CRESYL DIHYDROGEN PHOSPHATE
78-30-8	O-CRESYL PHOSPHATE
95-48-7	O-CRESYLIC ACID
119-42-6	O-CYCLOHEXYLPHENOL
128-37-0	O-DI-TERT-BUTYL-P-METHYLPHENOL
119-90-4	O-DIANISIDINE
115-02-6	O-DIAZOACETYL-L-SERINE
95-50-1	O-DICHLOR BENZOL
95-50-1	O-DICHLORBENZENE
27134-27-6	O-DICHLOROANILINE
95-50-1	O-DICHLOROBENZENE
95-50-1	O-DICHLOROBENZENE (ACGIH)
95-50-1	O-DICHLOROBENZENE, LIQUID
135-01-3	O-DIETHYLBENZENE
120-80-9	O-DIHYDROXYBENZENE
95-47-6	O-DIMETHYLBENZENE
528-29-0	O-DINITROBENZENE
528-29-0	O-DINITROBENZENE (ACGIH,DOT)
573-56-8	O-DINITROPHENOL
120-80-9	O-DIOXYBENZENE
94-70-2	O-ETHOXYANILINE
62-50-0	O-ETHYL METHYLSULFONATE
35400-43-2	O-ETHYL O-(4-METHYLTHIOPHENYL) S-PROPYL DITHIOPHOSPHATE
2104-64-5	O-ETHYL O-(4-NITROPHENYL) PHENYLPHOSPHONOTHIOATE
2104-64-5	O-ETHYL O-(4-NITROPHENYL)BENZENETHIONOPHOSPHONATE
2104-64-5	O-ETHYL O-(P-NITROPHENYL) PHENYLPHOSPHONOTHIOATE
327-98-0	O-ETHYL O-2,4,5-TRICHLOROPHENYL ETHYLPHOSPHONOTHIOATE
2104-64-5	O-ETHYL O-P-NITROPHENYL BENZENETHIOPHOSPHONATE
2104-64-5	O-ETHYL O-P-NITROPHENYL PHENYLPHOSPHOROTHIOATE
2104-64-5	O-ETHYL PHENYL P-NITROPHENYL THIOPHOSPHONATE
140-89-6	O-ETHYL POTASSIUM DITHIOCARBONATE
13194-48-4	O-ETHYL S,S-DIPROPYL DITHIOPHOSPHATE
13194-48-4	O-ETHYL S,S-DIPROPYL PHOSPHORODITHIOATE
944-22-9	O-ETHYL S-PHENYL ETHYLDITHIOPHOSPHONATE
944-22-9	O-ETHYL S-PHENYL ETHYLPHOSPHONODITHIOATE
2104-64-5	O-ETHYL-O-((4-NITRO-FENYL)-FENYL)-MONOTHIOFOSFONAAT (DUTCH)
944-22-9	O-ETHYL-S-PHENYL ETHYLPHOSPHONODITHIOATE
578-54-1	O-ETHYLANILINE
111-90-0	O-ETHYLDIGOL
51-79-6	O-ETHYLURETHANE
2104-64-5	O-ETIL-O-((4-NITRO-FENIL)-FENIL)-MONOTIOFOSFONATO (ITALIAN)
90-02-8	O-FORMYLPHENOL
7440-59-7	O-HELIUM
90-02-8	O-HYDROXYBENZALDEHYDE
69-72-7	O-HYDROXYBENZOIC ACID
119-36-8	O-HYDROXYBENZOIC ACID METHYL ESTER
88-75-5	O-HYDROXYNITROBENZENE
120-80-9	O-HYDROXYPHENOL
95-48-7	O-HYDROXYTOLUENE
114-26-1	O-ISOPROPOXYPHENYL METHYLCARBAMATE
114-26-1	O-ISOPROPOXYPHENYL N-METHYLCARBAMATE
107-44-8	O-ISOPROPYL METHYLFLUOROPHOSPHONATE
107-44-8	O-ISOPROPYL METHYLPHOSPHONOFLUORIDATE
95-48-7	O-KRESOL (GERMAN)
92-15-9	O-METHOXYACETOACETANILIDE
90-04-0	O-METHOXYANILINE
29191-52-4	O-METHOXYANILINE
135-02-4	O-METHOXYBENZALDEHYDE
90-04-0	O-METHOXYPHENYLAMINE
29191-52-4	O-METHOXYPHENYLAMINE
21609-90-5	O-METHYL O-2,5-DICHLORO-4-BROMOPHENYL PHENYLTHIOPHOSPHONATE
21609-90-5	O-METHYL-O-(4-BROMO-2,5-DICHLOROPHENYL)PHENYL THIOPHOSPHONATE
95-53-4	O-METHYLANILINE
636-21-5	O-METHYLANILINE HYDROCHLORIDE
95-53-4	O-METHYLBENZENAMINE
643-58-3	O-METHYLBIPHENYL
95-46-5	O-METHYLBROMOBENZENE
583-60-8	O-METHYLCYCLOHEXANONE
88-72-2	O-METHYLNITROBENZENE
95-48-7	O-METHYLPHENOL
95-46-5	O-METHYLPHENYL BROMIDE
95-48-7	O-METHYLPHENYLOL
109-06-8	O-METHYLPYRIDINE
95-47-6	O-METHYLTOLUENE
88-74-4	O-NITRANILINE
88-74-4	O-NITROANILINE
88-73-3	O-NITROCHLOROBENZENE
88-73-3	O-NITROCHLOROBENZENE, LIQUID
88-75-5	O-NITROPHENOL
88-75-5	O-NITROPHENOL (DOT)
88-72-2	O-NITROTOLUENE
88-72-2	O-NITROTOLUENE (ACGIH,DOT)
95-48-7	O-OXYTOLUENE
136-81-2	O-PENTYLPHENOL
94-70-2	O-PHENETIDINE
90-41-5	O-PHENYLANILINE
95-54-5	O-PHENYLENEDIAMINE
95-83-0	O-PHENYLENEDIAMINE, 4-CHLORO-
120-80-9	O-PHENYLENEDIOL
193-39-5	O-PHENYLENEPYRENE
90-43-7	O-PHENYLPHENOL
109-06-8	O-PICOLINE
109-06-8	O-PICOLINE (DOT)
89-72-5	O-SEC-BUTYLPHENOL
81-07-2	O-SULFOBENZIMIDE
81-07-2	O-SULFOBENZOIC ACID IMIDE

CAS No.	Chemical Name
84-15-1	O-TERPHENYLS
119-93-7	O-TOLIDINE
148-01-6	O-TOLUAMIDE, 3,5-DINITRO-
95-53-4	O-TOLUIDIN (CZECH)
95-53-4	O-TOLUIDINE
95-53-4	O-TOLUIDINE (ACGIH,DOT)
636-21-5	O-TOLUIDINE HYDROCHLORIDE
838-88-0	O-TOLUIDINE, 4,4'-METHYLENEDI-
95-69-2	O-TOLUIDINE, 4-CHLORO-
3165-93-3	O-TOLUIDINE, 4-CHLORO-, HYDROCHLORIDE
95-79-4	O-TOLUIDINE, 5-CHLORO-
99-55-8	O-TOLUIDINE, 5-NITRO-
88-17-5	O-TOLUIDINE, ALPHA,ALPHA,ALPHA-TRIFLUORO-
636-21-5	O-TOLUIDINIUM CHLORIDE
95-53-4	O-TOLUIDYNA (POLISH)
95-48-7	O-TOLUOL
95-46-5	O-TOLYL BROMIDE
95-49-8	O-TOLYL CHLORIDE
18351-85-4	O-TOLYL PHOSPHATE
614-78-8	O-TOLYL THIOUREA
95-53-4	O-TOLYLAMINE
88-16-4	O-TRIFLUOROMETHYLPHENYL CHLORIDE
95-47-6	O-XYLENE
95-47-6	O-XYLENE (ACGIH,DOT)
87-59-2	O-XYLIDINE
87-62-7	O-XYLIDINE
95-47-6	O-XYLOL
95-47-6	O-XYLOL (DOT)
97-56-3	OAAT
999-97-3	OAP
131-74-8	OBELINE PICRATE
80-51-3	OBSH
577-11-7	OBSTON
12789-03-6	OCTA-KLOR
10210-68-1	OCTACARBONYLDICOBALT
12789-03-6	OCTACHLOR
12789-03-6	OCTACHLORO-4,7-METHANOHYDROINDANE
57-74-9	OCTACHLORO-4,7-METHANOTETRAHYDROINDANE
12789-03-6	OCTACHLORO-4,7-METHANOTETRAHYDROINDANE
297-78-9	OCTACHLORO-HEXAHYDRO-METHANOISOBENZOFURAN
8001-35-2	OCTACHLOROCAMPHENE
12789-03-6	OCTACHLORODIHYDRODICYCLOPENTADIENE
297-78-9	OCTACHLOROHEXAHYDRO-4,7-METHANOISOBENZOFURAN
2234-13-1	OCTACHLORONAPHTHALENE
57-11-4	OCTADECANOIC ACID
1002-89-7	OCTADECANOIC ACID, AMMONIUM SALT
2223-93-0	OCTADECANOIC ACID, CADMIUM SALT
56189-09-4	OCTADECANOIC ACID, LEAD COMPLEX
7428-48-0	OCTADECANOIC ACID, LEAD SALT
56189-09-4	OCTADECANOIC ACID, LEAD SALT, DIBASIC
1072-35-1	OCTADECANOIC ACID, LEAD(2+) SALT
557-05-1	OCTADECANOIC ACID, ZINC SALT
112-04-9	OCTADECYL TRICHLOROSILANE
112-04-9	OCTADECYLTRICHLOROSILANE (DOT)
63597-41-1	OCTADIENE
63597-41-1	OCTADIENE (DOT)
4521-94-2	OCTAETHOXY TRISILOXANE
360-89-4	OCTAFLUOROBUT-2-ENE
41070-66-9	OCTAFLUOROBUT-2-ENE (DOT)
41070-66-9	OCTAFLUOROBUTENE
41070-66-9	OCTAFLUOROBUTENE-2
115-25-3	OCTAFLUOROCYCLOBUTANE
115-25-3	OCTAFLUOROCYCLOBUTANE (DOT)
76-19-7	OCTAFLUOROPROPANE
76-19-7	OCTAFLUOROPROPANE (DOT)
124-13-0	OCTALDEHYDE
309-00-2	OCTALENE
60-57-1	OCTALOX
152-16-9	OCTAMETHYL
152-16-9	OCTAMETHYL PYROPHOSPHORTETRAMIDE
152-16-9	OCTAMETHYL TETRAMIDO PYROPHOSPHATE
152-16-9	OCTAMETHYL-DIFOSFORZUUR-TETRAMIDE (DUTCH)
152-16-9	OCTAMETHYL-DIPHOSPHORSAEURE-TETRAMID (GERMAN)
152-16-9	OCTAMETHYLDIPHOSPHORAMIDE
152-16-9	OCTAMETHYLPYROPHOSPHORAMIDE
152-16-9	OCTAMETHYLPYROPHOSPHORIC ACID AMIDE
152-16-9	OCTAMETHYLPYROPHOSPHORIC ACID TETRAMIDE
628-63-7	OCTAN AMYLU (POLISH)
111-76-2	OCTAN ETOKSYETYLU (POLISH)
141-78-6	OCTAN ETYLU (POLISH)
62-38-4	OCTAN FENYLRTUTNATY (CZECH)
142-71-2	OCTAN MEDNATY (CZECH)
123-86-4	OCTAN N-BUTYLU (POLISH)
109-60-4	OCTAN PROPYLU (POLISH)
108-05-4	OCTAN WINYLU (POLISH)
124-13-0	OCTANAL
124-13-0	OCTANALDEHYDE
111-86-4	OCTANAMINE
111-65-9	OCTANE
111-65-9	OCTANE (ACGIH,DOT)
3221-61-2	OCTANE, 2-METHYL-
2216-33-3	OCTANE, 3-METHYL-
2216-34-4	OCTANE, 4-METHYL-
106-32-1	OCTANOIC ACID, ETHYL ESTER
124-13-0	OCTANOIC ALDEHYDE
111-64-8	OCTANOIC CHLORIDE
111-87-5	OCTANOL
111-64-8	OCTANOYL CHLORIDE
762-16-3	OCTANOYL PEROXIDE
762-16-3	OCTANYL PEROXIDE
111-65-9	OCTENE
111-87-5	OCTILIN
117-84-0	OCTOIL
117-81-7	OCTOIL
75-07-0	OCTOWY ALDEHYD (POLISH)
108-24-7	OCTOWY BEZWODNIK (POLISH)
64-19-7	OCTOWY KWAS (POLISH)
103-09-3	OCTYL ACETATE
112-14-1	OCTYL ACETATE
103-11-7	OCTYL ACRYLATE
2499-59-4	OCTYL ACRYLATE
103-23-1	OCTYL ADIPATE
123-79-5	OCTYL ADIPATE
111-87-5	OCTYL ALCOHOL
112-14-1	OCTYL ALCOHOL ACETATE
111-87-5	OCTYL ALCOHOL, NORMAL-PRIMARY
123-05-7	OCTYL ALDEHYDE
143-08-8	OCTYL CARBINOL
111-88-6	OCTYL MERCAPTAN
27193-28-8	OCTYL PHENOL
117-84-0	OCTYL PHTHALATE
117-81-7	OCTYL PHTHALATE
5283-66-9	OCTYL TRICHLOROSILANE (DOT)
111-86-4	OCTYLAMINE
111-88-6	OCTYLTHIOL
5283-66-9	OCTYLTRICHLOROSILANE
95-50-1	ODB
95-50-1	ODCB
151-21-3	ODORIPON AL 95
8006-54-0	OESIPOS
50-28-2	OESTERGON
50-28-2	OESTRA-1,3,5(10)-TRIENE-3,17BETA-DIOL
50-28-2	OESTRADIOL
50-28-2	OESTRADIOL-17B
53-16-7	OESTRIN
53-16-7	OESTROFORM
56-53-1	OESTROGENINE
50-28-2	OESTROGLANDOL
56-53-1	OESTROMENIN
56-53-1	OESTROMENSYL
53-16-7	OESTRONE
53-16-7	OESTROPEROS
7440-50-8	OFHC COPPER
7440-50-8	OFHC CU
3165-93-3	OFNA-PERL SALT RRA
8008-51-3	OIL CAMPHOR SASSAFRASSY
8002-05-9	OIL DEPOSITS
68476-26-6	OIL GAS
8006-20-0	OIL GASES, PRODUCER GAS
8021-92-9	OIL GASES, WATER GAS
8012-95-1	OIL MIST, MINERAL
8008-51-3	OIL OF CAMPHOR
8008-51-3	OIL OF CAMPHOR RECTIFIED
8008-51-3	OIL OF CAMPHOR WHITE
8002-09-3	OIL OF FIR-SIBERIAN
98-95-3	OIL OF MIRBANE
98-95-3	OIL OF MIRBANE (DOT)
57-06-7	OIL OF MUSTARD BPC 1949

CAS No.	*Chemical Name*
57-06-7	OIL OF MUSTARD, ARTIFICIAL
8007-40-7	OIL OF MUSTARD, EXPRESSED
98-95-3	OIL OF MYRBANE
93-58-3	OIL OF NIOBE
8002-09-3	OIL OF PINE
7664-93-9	OIL OF VITRIOL
7664-93-9	OIL OF VITRIOL (DOT)
2646-17-5	OIL ORANGE 204
2646-17-5	OIL ORANGE OPEL
2646-17-5	OIL ORANGE SS
2646-17-5	OIL ORANGE TX
60-11-7	OIL YELLOW 20
97-56-3	OIL YELLOW 21
60-11-7	OIL YELLOW 2625
97-56-3	OIL YELLOW 2681
60-11-7	OIL YELLOW 2G
97-56-3	OIL YELLOW 2R
97-56-3	OIL YELLOW AT
60-11-7	OIL YELLOW BB
97-56-3	OIL YELLOW C
60-11-7	OIL YELLOW D
60-11-7	OIL YELLOW FN
60-11-7	OIL YELLOW G
60-11-7	OIL YELLOW GG
60-11-7	OIL YELLOW GR
97-56-3	OIL YELLOW I
60-11-7	OIL YELLOW II
60-11-7	OIL YELLOW N
60-11-7	OIL YELLOW PEL
60-11-7	OIL YELLOW S
8002-03-7	OILS, ARACHIS
8001-86-3	OILS, BOLEKO
8007-40-7	OILS, BRASSICA ALBA
8007-40-7	OILS, BRASSICA NIGRA
8008-51-3	OILS, CAMPHOR
8001-69-2	OILS, COD-LIVER
8001-30-7	OILS, CORN
8001-29-4	OILS, COTTONSEED
8001-26-1	OILS, FLAXSEED OR LINSEED
8013-75-0	OILS, FUSEL
8002-03-7	OILS, GROUNDNUT
8016-28-2	OILS, LARD
8007-40-7	OILS, MUSTARD
8002-03-7	OILS, PEANUT
8002-09-3	OILS, PINE
8002-09-3	OILS, PINE, SYNTHETIC
8001-22-7	OILS, SOYBEAN
68334-28-1	OILS, VEGETABLE, HYDROGENATED
7631-86-9	OK 412
2074-50-2	OK 621
1910-42-5	OK 622
9005-25-8	OK PRE-GEL
62-73-7	OKO
2636-26-2	OKO
111-65-9	OKTAN (POLISH)
111-65-9	OKTANEN (DUTCH)
72-20-8	OKTANEX
57-74-9	OKTATERR
12789-03-6	OKTATERR
141-43-5	OLAMINE
143-19-1	OLATE FLAKES
2646-17-5	OLEAL ORANGE SS
60-11-7	OLEAL YELLOW 2G
1191-80-6	OLEATE OF MERCURY
74-85-1	OLEFIANT GAS
112-80-1	OLEIC ACID
13961-86-9	OLEIC ACID DIETHANOLAMINE CONDENSATE
544-60-5	OLEIC ACID, AMMONIUM SALT
26094-13-3	OLEIC ACID, COMPD. WITH BUTYLAMINE (1:1)
1191-80-6	OLEIC ACID, MERCURY(2+) SALT
143-18-0	OLEIC ACID, POTASSIUM SALT
143-19-1	OLEIC ACID, SODIUM SALT
112-80-1	OLEINE 7503
786-19-6	OLEOAKARITHION
141-66-2	OLEOBIDRIN
333-41-5	OLEODIAZINON
2275-18-5	OLEOFAC
56-38-2	OLEOFOS 20
1912-24-9	OLEOGESAPRIM

CAS No.	*Chemical Name*
1912-24-9	OLEOGESAPRIM 200
122-14-5	OLEOMETATHION
56-38-2	OLEOPARAPHENE
56-38-2	OLEOPARATHENE
56-38-2	OLEOPARATHION
121-75-5	OLEOPHOSPHOTHION
21609-90-5	OLEOPHOSVEL
122-14-5	OLEOSUMIFENE
298-00-0	OLEOVOFOTOX
7783-05-3	OLEUM
8014-95-7	OLEUM
8014-95-7	OLEUM (DOT)
7783-05-3	OLEUM (DOT)
8002-09-3	OLEUM ABIETIS
57-06-7	OLEUM SINAPIS
57-06-7	OLEUM SINAPIS VOLATILE
8020-83-5	OLEX WT 2577
1582-09-8	OLITREF
63-25-2	OLTITOX
2312-35-8	OMAIT
88-06-2	OMAL
25167-82-2	OMAL
98-07-7	OMEGA,OMEGA,OMEGA-TRICHLOROTOLUENE
100-46-9	OMEGA-AMINOTOLUENE
70-11-1	OMEGA-BROMACETOPHENONE
70-11-1	OMEGA-BROMOACETOPHENONE
105-60-2	OMEGA-CAPROLACTAM
532-27-4	OMEGA-CHLOROACETOPHENONE
109-70-6	OMEGA-CHLOROBROMOPROPANE
100-44-7	OMEGA-CHLOROTOLUENE
140-29-4	OMEGA-CYANOTOLUENE
9016-45-9	OMEGA-HYDROXY-ALPHA-(NONYLPHENYL)POLY(OXY-1,2-ETHANEDIYL)
98-08-8	OMEGA-TRIFLUOROTOLUENE
2312-35-8	OMITE
2312-35-8	OMITE 57E
2312-35-8	OMITE 85E
514-73-8	OMNI-PASSIN
58-89-9	OMNITOX
152-16-9	OMPA
152-16-9	OMPACIDE
152-16-9	OMPATOX
152-16-9	OMPAX
299-84-3	OMS 123
470-90-6	OMS 1328

CAS No.	*Chemical Name*
8002-09-3	OILS, PINE
8002-09-3	OILS, PINE, SYNTHETIC
8001-22-7	OILS, SOYBEAN
68334-28-1	OILS, VEGETABLE, HYDROGENATED
7631-86-9	OK 412
2074-50-2	OK 621
1910-42-5	OK 622
9005-25-8	OK PRE-GEL
62-73-7	OKO
2636-26-2	OKO
111-65-9	OKTAN (POLISH)
111-65-9	OKTANEN (DUTCH)
72-20-8	OKTANEX
57-74-9	OKTATERR
12789-03-6	OKTATERR
141-43-5	OLAMINE
143-19-1	OLATE FLAKES
2646-17-5	OLEAL ORANGE SS
60-11-7	OLEAL YELLOW 2G
1191-80-6	OLEATE OF MERCURY
74-85-1	OLEFIANT GAS
112-80-1	OLEIC ACID
13961-86-9	OLEIC ACID DIETHANOLAMINE CONDENSATE
544-60-5	OLEIC ACID, AMMONIUM SALT
26094-13-3	OLEIC ACID, COMPD. WITH BUTYLAMINE (1:1)
1191-80-6	OLEIC ACID, MERCURY(2+) SALT
143-18-0	OLEIC ACID, POTASSIUM SALT
143-19-1	OLEIC ACID, SODIUM SALT
112-80-1	OLEINE 7503
786-19-6	OLEOAKARITHION
141-66-2	OLEOBIDRIN
333-41-5	OLEODIAZINON
2275-18-5	OLEOFAC
56-38-2	OLEOFOS 20

CAS No.	Chemical Name
1912-24-9	OLEOGESAPRIM
1912-24-9	OLEOGESAPRIM 200
122-14-5	OLEOMETATHION
56-38-2	OLEOPARAPHENE
56-38-2	OLEOPARATHENE
56-38-2	OLEOPARATHION
121-75-5	OLEOPHOSPHOTHION
21609-90-5	OLEOPHOSVEL
122-14-5	OLEOSUMIFENE
298-00-0	OLEOVOFOTOX
7783-05-3	OLEUM
8014-95-7	OLEUM
8014-95-7	OLEUM (DOT)
7783-05-3	OLEUM (DOT)
8002-09-3	OLEUM ABIETIS
57-06-7	OLEUM SINAPIS
57-06-7	OLEUM SINAPIS VOLATILE
8020-83-5	OLEX WT 2577
1582-09-8	OLITREF
63-25-2	OLTITOX
2312-35-8	OMAIT
88-06-2	OMAL
933-78-4	OMAL
25167-82-2	OMAL
15950-66-0	OMAL
98-07-7	OMEGA,OMEGA,OMEGA-TRICHLOROTOLUENE
100-46-9	OMEGA-AMINOTOLUENE
70-11-1	OMEGA-BROMACETOPHENONE
70-11-1	OMEGA-BROMOACETOPHENONE
105-60-2	OMEGA-CAPROLACTAM
532-27-4	OMEGA-CHLOROACETOPHENONE
109-70-6	OMEGA-CHLOROBROMOPROPANE
100-44-7	OMEGA-CHLOROTOLUENE
140-29-4	OMEGA-CYANOTOLUENE
9016-45-9	OMEGA-HYDROXY-ALPHA-(NONYLPHENYL)POLY(OXY-1,2-ETHANEDIYL)
98-08-8	OMEGA-TRIFLUOROTOLUENE
2312-35-8	OMITE
2312-35-8	OMITE 57E
2312-35-8	OMITE 85E
514-73-8	OMNI-PASSIN
58-89-9	OMNITOX
152-16-9	OMPA
152-16-9	OMPACIDE
152-16-9	OMPATOX
152-16-9	OMPAX
299-84-3	OMS 123
470-90-6	OMS 1328
21923-23-9	OMS 1342
62-73-7	OMS 14
2636-26-2	OMS 14
64-00-6	OMS 15
64-00-6	OMS 162
55-38-9	OMS 2
114-26-1	OMS 33
122-14-5	OMS 43
115-29-7	OMS 570
2631-37-0	OMS 716
116-06-3	OMS 771
1563-66-2	OMS 864
2921-88-2	OMS-0971
64-00-6	OMS-15
63-25-2	OMS-29
2032-65-7	OMS-93
2540-82-1	OMS-968
297-78-9	OMTAN
88-74-4	ONA
88-73-3	ONCB
68-76-8	ONCOREDOX
50-76-0	ONCOSTATIN K
52-24-4	ONCOTIOTEPA
2893-78-9	ONIACHLOR 60
9004-34-6	ONOZUKA P 500
88-72-2	ONT
2588-05-8	OO-DIETHYL S-ETHYLSULPHINYLMETHYL PHOSPHOROTHIOATE
2588-06-9	OO-DIETHYL S-ETHYLSULPHONYLMETHYL PHOSPHOROTHIOATE
2600-69-3	OO-DIETHYL S-ETHYLTHIOMETHYL PHOSPHOROTHIOATE
75-52-4	OO-DIETHYL S-ISOPROPYLTHIOMETHYL PHOSPHORDITHIOATE
3309-68-0	OO-DIETHYL S-PROPYLTHIOMETHYL PHOSPHORODITHIOATE
2372-21-6	OO-TERT-BUTYL O-ISOPROPYL PERCARBONATE
2372-21-6	OO-TERT-BUTYL O-ISOPROPYL PEROXYCARBONATE
25322-69-4	OOPG 1000
25322-69-4	OOPG 1002
9016-45-9	OP 2
7733-02-0	OP-THAL-ZIN
56-75-7	OPCLOR
25322-69-4	OPD 500
25322-69-4	OPD 600
101-25-7	OPEX
101-25-7	OPEX 93
56-75-7	OPHTHOCHLOR
50-00-0	OPLOSSINGEN (DUTCH)
71-23-8	OPTAL
1332-58-7	OPTIWHITE
7733-02-0	OPTRAEX
57-63-6	ORADIOL
82-66-6	ORAGULANT
88-74-4	ORANGE BASE CIBA II
88-74-4	ORANGE BASE IRGA II
88-74-4	ORANGE GRS SALT
2646-17-5	ORANGE OT
88-74-4	ORANGE SALT CIBA II
88-74-4	ORANGE SALT IRGA II
2646-17-5	ORANGE SS
1344-48-5	ORANGE VERMILION
56-04-2	ORCANON
2782-57-2	ORCED
50-21-5	ORDINARY LACTIC ACID
7440-32-6	OREMET
57-63-6	ORESTRALYN
105-60-2	ORGAMID RMNOCD
105-60-2	ORGAMIDE
110-22-5	ORGANIC PEROXIDE
14265-44-2	ORGANIC POSPHATE COMPOUND, SOLID (POISON B)
2646-17-5	ORGANOL ORANGE 2R
97-56-3	ORGANOL YELLOW 2T
60-11-7	ORGANOL YELLOW ADM
5593-70-4	ORGATICS TA 25
60-11-7	ORIENT OIL YELLOW GG
124-87-8	ORIENTAL BERRY
92-84-2	ORIMON
65-86-1	OROTIC ACID
1303-33-9	ORPIMENT
58-22-0	ORQUISTERON
106-50-3	ORSIN
2425-06-1	ORTHO 5865
10265-92-6	ORTHO 9006
7775-09-9	ORTHO C-1 DEFOLIANT & WEED KILLER
16893-85-9	ORTHO EARWIG BAIT
121-75-5	ORTHO MALATHION
10326-21-3	ORTHO MC
10265-92-6	ORTHO MONITOR
54-11-5	ORTHO N-4 AND N-5 DUSTS
54-11-5	ORTHO N-4 DUST
54-11-5	ORTHO N-5 DUST
12002-03-8	ORTHO P-G BAIT
1910-2-5	ORTHO PARAQUAT CL
16893-85-9	ORTHO WEEVIL BAIT
85-00-7	ORTHO-DIQUAT
12789-03-6	ORTHO-KLOR
140-57-8	ORTHO-MITE
95-47-6	ORTHO-XYLENE
7778-39-4	ORTHOARSENIC ACID
10043-35-3	ORTHOBORIC ACID
10043-35-3	ORTHOBORIC ACID (H3BO3)
133-06-2	ORTHOCIDE
133-06-2	ORTHOCIDE 406
133-06-2	ORTHOCIDE 50
133-06-2	ORTHOCIDE 75
133-06-2	ORTHOCIDE 83
95-48-7	ORTHOCRESOL
95-50-1	ORTHODICHLOROBENZENE
95-50-1	ORTHODICHLOROBENZOL
122-51-0	ORTHOFORMIC ACID ETHYL ESTER
122-51-0	ORTHOFORMIC ACID, TRIETHYL ESTER
1333-74-0	ORTHOHYDROGEN
69-72-7	ORTHOHYDROXYBENZOIC ACID
122-51-0	ORTHOMRAVENCAN ETHYLNATY (CZECH)

CAS No.	Chemical Name
88-74-4	ORTHONITROANILINE (DOT)
56-38-2	ORTHOPHOS
7664-38-2	ORTHOPHOSPHORIC ACID
6834-92-0	ORTHOSIL
124-43-6	ORTIZON
443-48-1	ORVAGIL
107-18-6	ORVINYLCARBINOL
151-21-3	ORVUS WA
151-21-3	ORVUS WA PASTE
96-12-8	OS 1897
7786-34-7	OS 2046
79-21-0	OSBON AC
20816-12-0	OSMIC ACID
20816-12-0	OSMIUM OXIDE (OsO4)
20816-12-0	OSMIUM OXIDE (OsO4), (T-4)-
20816-12-0	OSMIUM TETROXIDE
20816-12-0	OSMIUM TETROXIDE (ACGIH,DOT)
20816-12-0	OSMIUM(IV) OXIDE
1332-58-7	OSMO KAOLIN
56-81-5	OSMOGLYN
71-23-8	OSMOSOL EXTRA
133-06-2	OSOCIDE
7681-49-4	OSSALIN
141-79-7	OSSIDO DI MESITILE (ITALIAN)
7681-49-4	OSSIN
139-91-3	OTIFURIL
59-50-7	OTTAFACT
1321-10-4	OTTAFACT
111-65-9	OTTANI (ITALIAN)
152-16-9	OTTOMETIL-PIROFOSFORAMMIDE (ITALIAN)
630-60-4	OUABAGENIN L-RHAMNOSIDE
630-60-4	OUABAGENIN-L-RHAMNOSID (GERMAN)
630-60-4	OUABAIN
630-60-4	OUABAINE
630-60-4	OUBAIN
1314-13-2	OUTMINE
122-14-5	OVADOFOS
58-89-9	OVADZIAK
50-28-2	OVAHORMON
50-28-2	OVASTEROL
50-28-2	OVASTEVOL
72-33-3	OVASTOL
53-16-7	OVIFOLLIN
50-28-2	OVOCYCLIN
122-14-5	OWADOFOS
58-89-9	OWADZIAK
144-62-7	OXAALZUUR (DUTCH)
142-68-7	OXACYCLOHEXANE
110-00-9	OXACYCLOPENTADIENE
109-99-9	OXACYCLOPENTANE
75-21-8	OXACYCLOPROPANE
555-84-0	OXAFURADENE
107-22-2	OXAL
107-22-2	OXALALDEHYDE
144-62-7	OXALIC ACID
144-62-7	OXALIC ACID (ACGIH)
6153-56-6	OXALIC ACID DIHYDRATE
460-19-5	OXALIC ACID DINITRILE
2944-67-4	OXALIC ACID, AMMONIUM IRON(3+) SALT (3:3:1)
14258-49-2	OXALIC ACID, AMMONIUM SALT
814-91-5	OXALIC ACID, COPPER(2+) SALT (1:1)
1113-38-8	OXALIC ACID, DIAMMONIUM SALT
6009-70-7	OXALIC ACID, DIAMMONIUM SALT, MONOHYDRATE
1113-38-8	OXALIC ACID, DIAMMONIUM SALT, MONOHYDRATE
95-92-1	OXALIC ACID, DIETHYL ESTER
583-52-8	OXALIC ACID, DIPOTASSIUM SALT
10043-22-8	OXALIC ACID, DIPOTASSIUM SALT
62-76-0	OXALIC ACID, DISODIUM SALT
516-03-0	OXALIC ACID, IRON(2+) SALT (1:1)
127-95-7	OXALIC ACID, MONOPOTASSIUM SALT
583-52-8	OXALIC ACID, POTASSIUM SALT
10043-22-8	OXALIC ACID, POTASSIUM SALT
460-19-5	OXALIC NITRILE
460-19-5	OXALONITRILE
144-62-7	OXALSAEURE (GERMAN)
460-19-5	OXALYL CYANIDE
23135-22-0	OXAMIMIDIC ACID, N',N'-DIMETHYL-N-((METHYLCARBAMOYL)-OXY)-1-METHYLTHIO-
23135-22-0	OXAMIMIDIC ACID, N',N'-DIMETHYL-N-[(METHYLCARBAMOYL)-OXY]-1-THIO-, METHYL ESTER
7803-49-8	OXAMMONIUM
10039-54-0	OXAMMONIUM SULFATE
23135-22-0	OXAMYL
23135-22-0	OXAMYL (PESTICIDE)
75-21-8	OXANE
142-68-7	OXANE
604-75-1	OXAZEPAM
78-71-7	OXETANE, 3,3-BIS(CHLOROMETHYL)
93-58-3	OXIDATE LE
75-21-8	OXIDOETHANE
75-21-8	OXIRANE
26447-14-3	OXIRANE ((METHYLPHENOXY)METHYL)- (9CI)
106-92-3	OXIRANE, ((2-PROPENYLOXY)METHYL)-
3132-64-7	OXIRANE, (BROMOMETHYL)-
2426-08-6	OXIRANE, (BUTOXYMETHYL)-
106-89-8	OXIRANE, (CHLOROMETHYL)-
4016-11-9	OXIRANE, (ETHOXYMETHYL)-
4016-11-9	OXIRANE, (ETHOXYMETHYL)- (9CI)
122-60-1	OXIRANE, (PHENOXYMETHYL)-
1954-28-5	OXIRANE, 2,2'-(2,5,8,11-TETRAOXADODECANE-1,12-DIYL)BIS-
1954-28-5	OXIRANE, 2,2'-(2,5,8,11-TETRAOXADODECANE-1,12-DIYL)BIS- (9CI)
2238-07-5	OXIRANE, 2,2'-[OXYBIS(METHYLENE)]BIS-
106-89-8	OXIRANE, 2-(CHLOROMETHYL)
930-22-3	OXIRANE, ETHENYL-
106-88-7	OXIRANE, ETHYL-
75-56-9	OXIRANE, METHYL-
25322-69-4	OXIRANE, METHYL-, HOMOPOLYMER
8070-50-6	OXIRANE, MIXT WITH CARBON DIOXIDE
96-09-3	OXIRANE, PHENYL-
428-59-1	OXIRANE, TRIFLUORO(TRIFLUOROMETHYL)-
4016-14-2	OXIRANE, [(1-METHYLETHOXY)METHYL]-
26447-14-3	OXIRANE, [(METHYLPHENOXY)METHYL]-
765-34-4	OXIRANECARBOXALDEHYDE
556-52-5	OXIRANEMETHANOL
556-52-5	OXIRANYLMETHANOL
75-21-8	OXIRENE, DIHYDRO-
110-80-5	OXITOL
111-15-9	OXITOL ACETATE
109-99-9	OXOLANE
110-00-9	OXOLE
96-12-8	OXY DBCP
94-36-0	OXY WASH
94-36-0	OXY-10
94-36-0	OXY-5
108-95-2	OXYBENZENE
101-80-4	OXYBIS(4-AMINOBENZENE)
80-51-3	OXYBIS(BENZENESULFONYLHYDRAZIDE)
101-84-8	OXYBISBENZENE
542-88-1	OXYBIS[CHLOROMETHANE]
107-89-1	OXYBUTANAL
107-89-1	OXYBUTYRIC ALDEHYDE
463-58-1	OXYCARBON SULFIDE
463-58-1	OXYCARBON SULFIDE (COS)
14977-61-8	OXYCHLORURE CHROMIQUE (FRENCH)
7775-09-9	OXYCIL
106-92-3	OXYDE D'ALLYLE ET DE GLYCIDYLE (FRENCH)
60-29-7	OXYDE D'ETHYLE (FRENCH)
1304-28-5	OXYDE DE BARYUM (FRENCH)
1305-78-8	OXYDE DE CALCIUM (FRENCH)
630-08-0	OXYDE DE CARBONE (FRENCH)
111-44-4	OXYDE DE CHLORETHYLE (FRENCH)
141-79-7	OXYDE DE MESITYLE (FRENCH)
75-56-9	OXYDE DE PROPYLENE (FRENCH)
10102-43-9	OXYDE NITRIQUE (FRENCH)
101-80-4	OXYDI-P-PHENYLENEDIAMINE
7783-41-7	OXYDIFLUORIDE
101-84-8	OXYDIPHENYL
2497-07-6	OXYDISULFOTON
7722-84-1	OXYDOL
9016-45-9	OXYETHYLATED NONYLPHENOL
9016-45-9	OXYETHYLENE NONYLPHENYL ETHER
75-21-8	OXYFUME
75-21-8	OXYFUME 12
8070-50-6	OXYFUME 20
8070-50-6	OXYFUME 30
7782-44-7	OXYGEN
7782-44-7	OXYGEN (DOT)
7782-44-7	OXYGEN (LIQUID)

CAS No.	Chemical Name
7783-41-7	OXYGEN DIFLUORIDE
7783-41-7	OXYGEN DIFLUORIDE (ACGIH,DOT)
7783-41-7	OXYGEN FLUORIDE
7783-41-7	OXYGEN FLUORIDE (OF2)
7782-44-7	OXYGEN MOLECULE
7782-44-7	OXYGEN, COMPRESSED
7782-44-7	OXYGEN, COMPRESSED (DOT)
10028-15-6	OXYGEN, MOL (O3)
7782-44-7	OXYGEN, REFRIGERATED LIQUID
7782-44-7	OXYGEN, REFRIGERATED LIQUID (DOT)
8063-77-2	OXYGEN-CARBON DIOXIDE MIXTURE (DOT)
94-36-0	OXYLITE
434-07-1	OXYMETHENOLONE
434-07-1	OXYMETHOLONE
50-00-0	OXYMETHYLENE
3811-04-9	OXYMURIATE OF POTASH
4452-58-8	OXYPER
3313-92-6	OXYPER
15630-89-4	OXYPER
120-80-9	OXYPHENIC ACID
111-76-2	OXYTOL ACETATE
302-01-2	OXYTREAT 35
1314-13-2	OZIDE
1314-13-2	OZLO
10028-15-6	OZON (POLISH)
10028-15-6	OZONE
10028-15-6	OZONE (ACGIH)
5283-66-9	P 09830
2540-82-1	P 1
25322-69-4	P 1200
1333-86-4	P 1250
8002-74-2	P 127
1314-87-0	P 128
1314-87-0	P 128 (SULFIDE)
56-18-8	P 2 (HARDENER)
25322-69-4	P 2000
298-07-7	P 204
298-07-7	P 204 (ACID)
128-37-0	P 21
13463-67-7	P 25
13463-67-7	P 25 (OXIDE)
1333-86-4	P 33
1333-86-4	P 33 (CARBON BLACK)
1314-87-0	P 37
1314-87-0	P 37 (FILTER)
25322-69-4	P 400
25322-69-4	P 400 (POLYGLYCOL)
25322-69-4	P 4000
25322-69-4	P 4000 (POLYMER)
25322-69-4	P 425
105-60-2	P 6 (POLYAMIDE)
9002-89-5	P 700
12001-26-2	P 80P
151-21-3	P AND G EMULSIFIER 104
104-83-6	P,ALPHA-DICHLOROTOLUENE
92-87-5	P,P'-BIANILINE
90-94-8	P,P'-BIS(DIMETHYLAMINO)BENZOPHENONE
101-61-1	P,P'-BIS(DIMETHYLAMINO)DIPHENYLMETHANE
72-54-8	P,P'-DDD
72-55-9	P,P'-DDE
50-29-3	P,P'-DDT
92-87-5	P,P'-DIAMINOBIPHENYL
101-77-9	P,P'-DIAMINODIFENYLMETHAN (CZECH)
101-80-4	P,P'-DIAMINODIPHENYL ETHER
139-65-1	P,P'-DIAMINODIPHENYL SULFIDE
101-77-9	P,P'-DIAMINODIPHENYLMETHANE
94-17-7	P,P'-DICHLOROBENZOYL PEROXIDE
94-17-7	P,P'-DICHLORODIBENZOYL PEROXIDE
72-54-8	P,P'-DICHLORODIPHENYL-2,2-DICHLOROETHYLENE
72-54-8	P,P'-DICHLORODIPHENYLDICHLOROETHANE
72-55-9	P,P'-DICHLORODIPHENYLDICHLOROETHYLENE
50-29-3	P,P'-DICHLORODIPHENYLTRICHLOROETHANE
50-29-3	P,P'-DICHLORODIPHENYLTRICHLOROMETHYLMETHANE
72-43-5	P,P'-DIMETHOXYDIPHENYLTRICHLOROETHANE
101-68-8	P,P'-DIPHENYLMETHANE DIISOCYANATE
72-43-5	P,P'-DMDT
115-32-2	P,P'-KELTHANE
72-43-5	P,P'-METHOXYCHLOR
101-68-8	P,P'-METHYLENEBIS(PHENYL ISOCYANATE)
101-77-9	P,P'-METHYLENEDIANILINE
80-51-3	P,P'-OXYBIS(BENZENESULFOHYDRAZIDE)
80-51-3	P,P'-OXYBIS(BENZENESULFONE HYDRAZIDE)
80-51-3	P,P'-OXYBIS(BENZENESULFONYLHYDRAZINE)
80-51-3	P,P'-OXYBISBENZENE DISULFONYLHYDRAZIDE
101-80-4	P,P'-OXYBIS[ANILINE]
101-80-4	P,P'-OXYDIANILINE
72-54-8	P,P'-TDE
101-61-1	P,P'-TETRAMETHYLDIAMINODIPHENYLMETHANE
72-55-9	P,P-DDE
947-02-4	P,P-DIETHYL CYCLIC ETHYLENE ESTER OF PHOSPHONO-DITHIOIMIDOCARBONIC ACID
950-10-7	P,P-DIETHYL CYCLIC PROPYLENE ESTER OF PHOSPHONO-DITHIOIMIDOCARBONIC ACID
72-56-0	P,P-ETHYL DDD (PERTHANE)
139-65-1	P,P-THIODIANILINE
80-46-6	P-(ALPHA,ALPHA-DIMETHYLPROPYL)PHENOL
80-46-6	P-(1,1-DIMETHYLPROPYL)PHENOL
671-16-9	P-(2-METHYLHYDRAZINOMETHYL)-N-ISOPROPYLBENZAMIDE
1031-47-6	P-(5-AMINO-3-PHENYL-1H-1,2,4-TRIAZOL-1-YL)-N,N,N'-TETRA-METHYL PHOSPHONIC DIAMIDE
100-14-1	P-(CHLOROMETHYL)NITROBENZENE
93-05-0	P-(DIETHYLAMINO)ANILINE
99-98-9	P-(DIMETHYLAMINO)ANILINE
138-89-6	P-(DIMETHYLAMINO)NITROSOBENZENE
53324-05-3	P-(DIMETHYLAMINO)NITROSOBENZENE
99-98-9	P-(DIMETHYLAMINO)PHENYLAMINE
138-89-6	P-(N,N-DIMETHYLAMINO)NITROSOBENZENE
98-56-6	P-(TRIFLUOROMETHYL)CHLOROBENZENE
402-54-0	P-(TRIFLUOROMETHYL)NITROBENZENE
25155-30-0	P-1',1',4',4'-TETRAMETHYLOKTYLBENZENSULFONAN SODNY (CZECH)
13410-01-0	P-40
103-89-9	P-ACETAMIDOTOLUENE
122-82-7	P-ACETOACETOPHENETIDIDE
62-44-2	P-ACETOPHENETIDIDE
103-89-9	P-ACETOTOLUIDE
103-89-9	P-ACETOTOLUIDIDE
93-05-0	P-AMINO-N,N-DIETHYLANILINE
99-98-9	P-AMINO-N,N-DIMETHYLANILINE
106-50-3	P-AMINOANILINE
104-94-9	P-AMINOANISOLE
92-67-1	P-AMINOBIPHENYL
93-05-0	P-AMINODIETHYLANILINE
99-98-9	P-AMINODIMETHYLANILINE
92-67-1	P-AMINODIPHENYL
123-30-8	P-AMINOFENOL (CZECH)
27598-85-2	P-AMINOFENOL (CZECH)
100-01-6	P-AMINONITROBENZENE
29757-24-2	P-AMINONITROBENZENE
156-43-4	P-AMINOPHENETOLE
1321-31-9	P-AMINOPHENETOLE
123-30-8	P-AMINOPHENOL
27598-85-2	P-AMINOPHENOL
123-30-8	P-AMINOPHENOL (DOT)
27598-85-2	P-AMINOPHENOL (DOT)
101-80-4	P-AMINOPHENYL ETHER
504-24-5	P-AMINOPYRIDINE
504-24-5	P-AMINOPYRIDINE (DOT)
106-49-0	P-AMINOTOLUENE
104-94-9	P-ANISIDINE
20265-97-8	P-ANISIDINE HYDROCHLIRIDE
100-07-2	P-ANISOYL CHLORIDE
104-94-9	P-ANISYLAMINE
106-50-3	P-BENZENEDIAMINE
123-31-9	P-BENZENEDIOL
106-51-4	P-BENZOQUINONE
68-76-8	P-BENZOQUINONE, 2,3,5-TRIS(1-AZIRIDINYL)-
106-51-4	P-BENZQUINONE
92-67-1	P-BIPHENYLAMINE
92-66-0	P-BROMOBIPHENYL
92-66-0	P-BROMODIPHENYL
101-55-3	P-BROMODIPHENYL ETHER
586-78-7	P-BROMONITROBENZENE
101-55-3	P-BROMOPHENOXYBENZENE
101-55-3	P-BROMOPHENYL PHENYL ETHER
106-38-7	P-BROMOTOLUENE
28805-86-9	P-BUTYLPHENOL, LIQUID (DOT)
28805-86-9	P-BUTYLPHENOL, SOLID (DOT)

CAS No.	Chemical Name
106-51-4	P-CHINON (GERMAN)
59-50-7	P-CHLOR-M-CRESOL
1321-10-4	P-CHLOR-M-CRESOL
27134-26-5	P-CHLORANILINE
106-48-9	P-CHLORFENOL (CZECH)
25167-80-0	P-CHLORFENOL (CZECH)
98-56-6	P-CHLORO-ALPHA,ALPHA,ALPHA-TRIFLUOROTOLUENE
59-50-7	P-CHLORO-M-CRESOL
1321-10-4	P-CHLORO-M-CRESOL
95-85-2	P-CHLORO-O-AMINOPHENOL
93-50-5	P-CHLORO-O-ANISIDINE
95-69-2	P-CHLORO-O-TOLUIDINE
3165-93-3	P-CHLORO-O-TOLUIDINE HYDROCHLORIDE
106-47-8	P-CHLOROANILINE
27134-26-5	P-CHLOROANILINE
27134-26-5	P-CHLOROANILINE, LIQUID (DOT)
27134-26-5	P-CHLOROANILINE, SOLID (DOT)
104-88-1	P-CHLOROBENZALDEHYDE
104-88-1	P-CHLOROBENZENECARBOXALDEHYDE
98-56-6	P-CHLOROBENZOTRIFLUORIDE
98-56-6	P-CHLOROBENZOTRIFLUORIDE (DOT)
94-17-7	P-CHLOROBENZOYL PEROXIDE
94-17-7	P-CHLOROBENZOYL PEROXIDE (DOT)
94-17-7	P-CHLOROBENZOYL PEROXIDE, NOT MORE THAN 75% WITH WATER (DOT)
104-83-6	P-CHLOROBENZYL CHLORIDE
104-83-6	P-CHLOROBENZYL CHLORIDE (DOT)
59-50-7	P-CHLOROCRESOL
1321-10-4	P-CHLOROCRESOL
7005-72-3	P-CHLORODIPHENYL OXIDE
100-00-5	P-CHLORONITROBENZENE
100-00-5	P-CHLORONITROBENZENE (DOT)
106-48-9	P-CHLOROPHENOL
25167-80-0	P-CHLOROPHENOL
106-48-9	P-CHLOROPHENOL, LIQUID (DOT)
25167-80-0	P-CHLOROPHENOL, LIQUID (DOT)
106-48-9	P-CHLOROPHENOL, SOLID (DOT)
25167-80-0	P-CHLOROPHENOL, SOLID (DOT)
106-46-7	P-CHLOROPHENYL CHLORIDE
7005-72-3	P-CHLOROPHENYL PHENYL ETHER
104-83-6	P-CHLOROPHENYLMETHYL CHLORIDE
106-43-4	P-CHLOROTOLUENE
106-43-4	P-CHLOROTOLUENE (DOT)
98-56-6	P-CHLOROTRIFLUOROMETHYLBENZENE
120-71-8	P-CRESIDINE
106-44-5	P-CRESOL
106-44-5	P-CRESOL (DOT)
128-37-0	P-CRESOL, 2,6-DI-TERT-BUTYL-
106-44-5	P-CRESYLIC ACID
99-87-6	P-CYMENE
99-87-6	P-CYMOL
106-50-3	P-DIAMINOBENZENE
92-87-5	P-DIAMINODIPHENYL
106-46-7	P-DICHLOORBENZEEN (DUTCH)
106-46-7	P-DICHLORBENZOL (GERMAN)
106-46-7	P-DICHLOROBENZENE
106-46-7	P-DICHLOROBENZENE (ACGIH)
106-46-7	P-DICHLOROBENZENE, SOLID
106-46-7	P-DICHLOROBENZOL
106-46-7	P-DICLOROBENZENE (ITALIAN)
105-05-5	P-DIETHYL BENZENE
123-31-9	P-DIHYDROXYBENZENE
60-11-7	P-DIMETHYLAMINOAZOBENZENE
99-98-9	P-DIMETHYLAMINOPHENYLAMINE
106-42-3	P-DIMETHYLBENZENE
589-90-2	P-DIMETHYLCYCLOHEXANE
100-25-4	P-DINITROBENZENE
100-25-4	P-DINITROBENZENE (ACGIH,DOT)
123-91-1	P-DIOXAN
123-91-1	P-DIOXAN (CZECH)
123-91-1	P-DIOXANE
25136-55-4	P-DIOXANE, DIMETHYL-
78-34-2	P-DIOXANE-2,3-DITHIOL, S,S-DIESTER WITH O,O-DIETHYL PHOSPHORODITHIOATE
78-34-2	P-DIOXANE-2,3-DIYL ETHYL PHOSPHORODITHIOATE
123-91-1	P-DIOXIN, TETRAHYDRO-
123-31-9	P-DIOXOBENZENE
123-31-9	P-DIOXYBENZENE
54590-52-2	P-DODECYLBENZENESULFONIC ACID 2-HYDROXY-1-PROPYLAMINE SALT
62-44-2	P-ETHOXYACETANILIDE
156-43-4	P-ETHOXYANILINE
1321-31-9	P-ETHOXYANILINE
105-05-5	P-ETHYLETHYLBENZENE
123-07-9	P-ETHYLPHENOL
371-40-4	P-FLUOROANILINE
371-40-4	P-FLUOROPHENYLAMINE
25496-08-6	P-FLUOROTOLUENE
25496-08-6	P-FLUOROTOLUENE (DOT)
150-76-5	P-GUAIACOL
7440-59-7	P-HELIUM
123-31-9	P-HYDROQUINONE
123-30-8	P-HYDROXYANILINE
27598-85-2	P-HYDROXYANILINE
150-76-5	P-HYDROXYANISOLE
127-82-2	P-HYDROXYBENZENESULFONIC ACID ZINC SALT
150-76-5	P-HYDROXYMETHOXYBENZENE
100-02-7	P-HYDROXYNITROBENZENE
123-31-9	P-HYDROXYPHENOL
123-30-8	P-HYDROXYPHENYLAMINE
123-07-9	P-HYDROXYPHENYLETHANE
106-44-5	P-HYDROXYTOLUENE
99-87-6	P-ISOPROPYLTOLUENE
110-91-8	P-ISOXAZINE, TETRAHYDRO-
106-44-5	P-KRESOL
80-47-7	P-MENTH-8-YL HYDROPEROXIDE
586-62-9	P-MENTHA-1,4(8)-DIENE
138-86-3	P-MENTHA-1,8-DIENE
138-86-3	P-MENTHA-1,8-DIENE, DL-
2244-16-8	P-MENTHA-6,8-DIEN-2-ONE
2244-16-8	P-MENTHA-6,8-DIEN-2-ONE, (S)-(+)-
80-47-7	P-MENTHANE HYDROPEROXIDE
80-47-7	P-MENTHANE-8-HYDROPEROXIDE
104-94-9	P-METHOXYANILINE
100-07-2	P-METHOXYBENZOIC ACID CHLORIDE
100-07-2	P-METHOXYBENZOYL CHLORIDE
100-17-4	P-METHOXYNITROBENZENE
150-76-5	P-METHOXYPHENOL
104-94-9	P-METHOXYPHENYLAMINE
622-57-1	P-METHYL-N-ETHYLANILINE
103-89-9	P-METHYLACETANILIDE
106-49-0	P-METHYLANILINE
106-49-0	P-METHYLBENZENAMINE
80-48-8	P-METHYLBENZENESULFONATE METHYL ESTER
104-15-4	P-METHYLBENZENESULFONIC ACID
25231-46-3	P-METHYLBENZENESULFONIC ACID
106-38-7	P-METHYLBROMOBENZENE
99-87-6	P-METHYLCUMENE
106-44-5	P-METHYLHYDROXYBENZENE
99-87-6	P-METHYLISOPROPYLBENZENE
99-99-0	P-METHYLNITROBENZENE
106-44-5	P-METHYLPHENOL
106-38-7	P-METHYLPHENYL BROMIDE
106-49-0	P-METHYLPHENYLAMINE
104-15-4	P-METHYLPHENYLSULFONIC ACID
25231-46-3	P-METHYLPHENYLSULFONIC ACID
108-89-4	P-METHYLPYRIDINE
622-97-9	P-METHYLSTYRENE
106-42-3	P-METHYLTOLUENE
100-01-6	P-NITRANILINE
29757-24-2	P-NITRANILINE
402-54-0	P-NITRO(TRIFLUOROMETHYL)BENZENE
100-01-6	P-NITROANILINA (POLISH)
29757-24-2	P-NITROANILINA (POLISH)
100-01-6	P-NITROANILINE
29757-24-2	P-NITROANILINE
100-01-6	P-NITROANILINE (ACGIH,DOT)
29757-24-2	P-NITROANILINE (ACGIH,DOT)
100-01-6	P-NITROANILINE, SOLID
100-17-4	P-NITROANISOL
100-17-4	P-NITROANISOLE
100-17-4	P-NITROANISOLE (DOT)
402-54-0	P-NITROBENZOTRIFLUORIDE
402-54-0	P-NITROBENZOTRIFLUORIDE (DOT)
100-14-1	P-NITROBENZYL CHLORIDE
92-93-3	P-NITROBIPHENYL
586-78-7	P-NITROBROMOBENZENE
586-78-7	P-NITROBROMOBENZENE (DOT)

CAS No.	Chemical Name
100-00-5	P-NITROCHLOORBENZEEN (DUTCH)
100-00-5	P-NITROCHLOROBENZENE
100-00-5	P-NITROCHLOROBENZENE (ACGIH)
100-00-5	P-NITROCHLOROBENZOL (GERMAN)
100-00-5	P-NITROCLOROBENZENE (ITALIAN)
92-93-3	P-NITRODIPHENYL
100-17-4	P-NITROMETHOXYBENZENE
100-02-7	P-NITROPHENOL
100-02-7	P-NITROPHENOL (DOT)
586-78-7	P-NITROPHENYL BROMIDE
100-00-5	P-NITROPHENYL CHLORIDE
100-01-6	P-NITROPHENYLAMINE
29757-24-2	P-NITROPHENYLAMINE
298-00-0	P-NITROPHENYLDIMETHYLTHIONOPHOSPHATE
1124-33-0	P-NITROPYRIDINE N-OXIDE
138-89-6	P-NITROSO-N,N-DIMETHYLANILINE
53324-05-3	P-NITROSO-N,N-DIMETHYLANILINE
156-10-5	P-NITROSO-N-PHENYLANILINE
138-89-6	P-NITROSODIMETHYLANILINE
138-89-6	P-NITROSODIMETHYLANILINE (DOT)
53324-05-3	P-NITROSODIMETHYLANILINE (DOT)
156-10-5	P-NITROSODIPHENYLAMINE
99-99-0	P-NITROTOLUENE
1321-12-6	P-NITROTOLUENE
99-99-0	P-NITROTOLUENE (ACGIH,DOT)
106-44-5	P-OXYTOLUENE
156-43-4	P-PHENETIDIN
1321-31-9	P-PHENETIDIN
156-43-4	P-PHENETIDINE
1321-31-9	P-PHENETIDINE
1321-31-9	P-PHENETIDINE (DOT)
13410-72-5	P-PHENETIDINE, N-(5-NITROFURFURYLIDENE)-
101-55-3	P-PHENOXYBROMOBENZENE
101-55-3	P-PHENOXYPHENYL BROMIDE
156-10-5	P-PHENYLAMINONITROSOBENZENE
92-67-1	P-PHENYLANILINE
92-66-0	P-PHENYLBROMOBENZENE
106-50-3	P-PHENYLENE DIAMINE
3081-14-9	P-PHENYLENEDIAMINE, N,N'-BIS(1,4-DIMETHYLPENTYL)-
101-96-2	P-PHENYLENEDIAMINE, N,N'-DI-SEC-BUTYL-
93-05-0	P-PHENYLENEDIAMINE, N,N-DIETHYL-
99-98-9	P-PHENYLENEDIAMINE, N,N-DIMETHYL-
100-20-9	P-PHENYLENEDICARBONYL DICHLORIDE
100-20-9	P-PHTHALOYL CHLORIDE
100-20-9	P-PHTHALOYL DICHLORIDE
108-89-4	P-PICOLINE
106-51-4	P-QUINONE
105-11-3	P-QUINONE DIOXIME
92-94-4	P-TERPHENYLS
2049-92-5	P-TERT-AMYLANILINE
80-46-6	P-TERT-AMYLPHENOL
98-27-1	P-TERT-BUTYL-O-CRESOL
98-29-3	P-TERT-BUTYLCATECHOL
98-29-3	P-TERT-BUTYLPYROCATECHOL
98-51-1	P-TERT-BUTYLTOLUENE
2049-92-5	P-TERT-PENTYLANILINE
80-46-6	P-TERT-PENTYLPHENOL
15980-15-1	P-THIOXANE
671-16-9	P-TOLUAMIDE, N-ISOPROPYL-ALPHA-(2-METHYLHYDRAZINO)-
366-70-1	P-TOLUAMIDE, N-ISOPROPYL-ALPHA-(2-METHYLHYDRAZINO)-, MONOHYDROCHLORIDE
104-15-4	P-TOLUENESULFONIC ACID
25231-46-3	P-TOLUENESULFONIC ACID
80-48-8	P-TOLUENESULFONIC ACID, METHYL ESTER
104-15-4	P-TOLUENESULPHONIC ACID
25231-46-3	P-TOLUENESULPHONIC ACID
95-53-4	P-TOLUIDIN (CZECH)
108-44-1	P-TOLUIDIN (CZECH)
106-49-0	P-TOLUIDIN (CZECH)
106-49-0	P-TOLUIDINE
106-49-0	P-TOLUIDINE (ACGIH,DOT)
89-62-3	P-TOLUIDINE, 2-NITRO-
119-32-4	P-TOLUIDINE, 3-NITRO-
1582-09-8	P-TOLUIDINE, ALPHA,ALPHA,ALPHA-TRIFLUORO-2,6-DINITRO-N,N-DIPROPYL-
622-57-1	P-TOLUIDINE, N-ETHYL-
106-44-5	P-TOLUOL
106-44-5	P-TOLYL ALCOHOL
106-38-7	P-TOLYL BROMIDE
106-43-4	P-TOLYL CHLORIDE
106-49-0	P-TOLYLAMINE
104-15-4	P-TOLYLSULFONIC ACID
25231-46-3	P-TOLYLSULFONIC ACID
98-56-6	P-TRIFLUOROMETHYLPHENYL CHLORIDE
622-97-9	P-VINYLTOLUENE
92-67-1	P-XENYLAMINE
106-42-3	P-XYLENE
106-42-3	P-XYLENE (ACGIH,DOT)
105-05-5	P-XYLENE, ALPHA,ALPHA'-DIMETHYL-
106-42-3	P-XYLOL
106-42-3	P-XYLOL (DOT)
88-89-1	PA
9002-89-5	PA 18
9002-89-5	PA 18 (POLYOL)
9002-89-5	PA 20
105-60-2	PA 6
105-60-2	PA 6 (POLYMER)
56-53-1	PABESTROL
56-38-2	PAC
8002-74-2	PACEMAKER 37
56-38-2	PACOL
92-84-2	PADOPHENE
8032-32-4	PAINTERS' NAPHTHA
84-66-2	PALATINOL A
117-81-7	PALATINOL AH
117-84-0	PALATINOL AH
85-68-7	PALATINOL BB
84-74-2	PALATINOL C
131-11-3	PALATINOL M
26761-40-0	PALATINOL Z
56-53-1	PALESTROL
7647-10-1	PALLADIUM (II) CHLORIDE
584-79-2	PALLETHRINE
112-02-7	PALMITYLTRIMETHYLAMMONIUM CHLORIDE
57-63-6	PALONYL
62-38-4	PAMISAN
112-80-1	PAMOLYN
112-80-1	PAMOLYN 100
63-25-2	PANAM
62-73-7	PANAPLATE
8049-47-6	PANCREATIN
502-39-6	PANDRINOX
794-93-4	PANFURAN S
129-06-6	PANIVARFIN
502-39-6	PANO-DRENCH
502-39-6	PANO-DRENCH 4
502-39-6	PANODRIN A-13
502-39-6	PANOGEN
502-39-6	PANOGEN (OLD)
502-39-6	PANOGEN 15
502-39-6	PANOGEN 43
502-39-6	PANOGEN 8
502-39-6	PANOGEN PX
502-39-6	PANOGEN TURF FUNGICIDE
502-39-6	PANOGEN TURF SPRAY
62-38-4	PANOMATIC
137-26-8	PANORAM 75
60-57-1	PANORAM D-31
502-39-6	PANOSPRAY 30
94-36-0	PANOXYL
2763-96-4	PANTHERINE
56-38-2	PANTHION
56-75-7	PANTOVERNIL
129-06-6	PANWARFIN
123-30-8	PAP
27598-85-2	PAP
7429-90-5	PAP 1
9001-73-4	PAPAIN
1937-37-7	PAPER BLACK BA
1937-37-7	PAPER BLACK T
1937-37-7	PAPER DEEP BLACK C
3761-53-3	PAPER RED HRR
9016-87-9	PAPI
9016-87-9	PAPI 135
9016-87-9	PAPI 20
9016-87-9	PAPI 27
9016-87-9	PAPI 580
9016-87-9	PAPI 901

CAS No.	Chemical Name
7782-42-5	PAPYEX
106-46-7	PARA CRYSTALS
115-32-2	PARA,PARA'-KELTHANE
59-50-7	PARA-CHLORO-META-CRESOL
1321-10-4	PARA-CHLORO-META-CRESOL
1910-2-5	PARA-COL
106-44-5	PARA-CRESOL
123-63-7	PARAACETALDEHYDE
128-37-0	PARABAR 441
123-63-7	PARACETALDEHYDE
50-29-3	PARACHLOROCIDUM
106-48-9	PARACHLOROPHENOL
25167-80-0	PARACHLOROPHENOL
106-46-7	PARACIDE
8002-74-2	PARACOL 404A
8002-74-2	PARACOL 404C
8002-74-2	PARACOL 505N
83-79-4	PARADERIL
106-46-7	PARADI
106-46-7	PARADICHLORBENZOL (GERMAN)
106-46-7	PARADICHLOROBENZENE
106-46-7	PARADICHLOROBENZOL
8002-74-2	PARADIT ASP
8002-74-2	PARADIT PR
8002-74-2	PARADIT PR NEW
8002-74-2	PARADIT PR NEW A
8002-74-2	PARADIUM SS
91-80-5	PARADORMALENE
106-46-7	PARADOW
56-38-2	PARADUST
8002-74-2	PARAFFIN WAX FUME
8002-74-2	PARAFFIN WAXES AND HYDROCARBON WAXES
8002-74-2	PARAFFINS
8002-74-2	PARAFILM
8002-74-2	PARAFLINT H 1
8002-74-2	PARAFLINT HI-NI
8002-74-2	PARAFLINT HIN 3
8002-74-2	PARAFLINT RG
50-00-0	PARAFORM
30525-89-4	PARAFORM
110-88-3	PARAFORMALDEHYDE
30525-89-4	PARAFORMALDEHYDE
1332-58-7	PARAGON
1332-58-7	PARAGON (CLAY)
1333-74-0	PARAHYDROGEN
123-63-7	PARAL
123-63-7	PARALDEHYDE
56-38-2	PARAMAR
56-38-2	PARAMAR 50
52061-60-6	PARAMENTHANE HYDRO-PEROXIDE
106-44-5	PARAMETHYL PHENOL
2385-85-5	PARAMEX
1937-37-7	PARAMINE BLACK B
1937-37-7	PARAMINE BLACK E
2602-46-2	PARAMINE BLUE 2B
72-57-1	PARAMINE BLUE 3B
106-46-7	PARAMOTH
120-12-7	PARANAPHTHALENE
100-01-6	PARANITROANILINE, SOLID (DOT)
29757-24-2	PARANITROANILINE, SOLID (DOT)
100-02-7	PARANITROFENOL (DUTCH)
100-02-7	PARANITROFENOLO (ITALIAN)
100-02-7	PARANITROPHENOL (FRENCH,GERMAN)
53324-05-3	PARANITROSODIMETHYLANILIDE
123-30-8	PARANOL
27598-85-2	PARANOL
16071-86-6	PARANOL FAST BROWN BRL
106-46-7	PARANUGGETS
311-45-5	PARAOXON
298-00-0	PARAPEST M-50
56-38-2	PARAPHOS
4685-14-7	PARAQUAT
2074-50-2	PARAQUAT (ACGIH)
2074-50-2	PARAQUAT BIS(METHYL SULFATE)
1910-2-5	PARAQUAT CHLORIDE
1910-2-5	PARAQUAT Cl
4685-14-7	PARAQUAT DICATION
1910-42-5	PARAQUAT DICHLORIDE
2074-50-2	PARAQUAT DIMETHOSULFATE
2074-50-2	PARAQUAT DIMETHYL SULFATE
2074-50-2	PARAQUAT DIMETHYL SULPHATE
2074-50-2	PARAQUAT I
4685-14-7	PARAQUAT ION
2074-50-2	PARAQUAT METHOSULFATE
2074-50-2	PARAQUAT METHYLSULFATE
1910-2-5	PARAQUAT, DICHLORIDE
8002-74-2	PARASEAL
298-00-0	PARATAF
56-38-2	PARATHENE
56-38-2	PARATHION
56-38-2	PARATHION (ACGIH)
56-38-2	PARATHION AND COMPRESSED GAS MIXTURE (DOT)
56-38-2	PARATHION AND COMPRESSED GAS MIXTURES
298-00-0	PARATHION METHYL
298-00-0	PARATHION METHYL HOMOLOG
56-38-2	PARATHION MIXTURE, DRY
56-38-2	PARATHION MIXTURE, DRY (DOT)
56-38-2	PARATHION MIXTURE, LIQUID
56-38-2	PARATHION MIXTURE, LIQUID (DOT)
56-38-2	PARATHION, LIQUID
56-38-2	PARATHION, LIQUID (DOT)
56-38-2	PARATHION-AETHYL (GERMAN)
56-38-2	PARATHION-COMPRESSED GAS MIXTURE
56-38-2	PARATHION-ETHYL
298-00-0	PARATHION-METILE (ITALIAN)
298-00-0	PARATOX
298-00-0	PARATUF
56-38-2	PARAWET
56-75-7	PARAXIN
106-46-7	PARAZENE
1405-87-4	PARENTRACIN
8003-34-7	PAREXAN
12002-03-8	PARIS GREEN
12002-03-8	PARIS GREEN, SOLID (DOT)
9004-70-0	PARLODION
59-50-7	PARMETOL
1321-10-4	PARMETOL
59-50-7	PAROL
1321-10-4	PAROL
12002-03-8	PARROT GREEN
514-73-8	PARTEL
298-00-0	PARTRON M
8002-74-2	PARVEN 5250
142-59-6	PARZATE
142-59-6	PARZATE LIQUID
7440-66-6	PASCO
9005-25-8	PASSELI P
12002-03-8	PATENT GREEN
1910-2-5	PATHCLEAR
1309-64-4	PATOX L
1309-64-4	PATOX M
1309-64-4	PATOX S
112-02-7	PB 40
7439-92-1	PB-S 100
59536-65-1	PBB
135-88-6	PBNA
7789-60-8	PBR3
121-82-4	PBX(AF) 108
64-10-8	PC
1306-23-6	PC 108
9004-70-0	PC MEDIUM
1336-36-3	PCB
1336-36-3	PCB (DOT)
11097-69-1	PCB 1254
11096-82-5	PCB 1260
53469-21-9	PCB-1242 (CHLORODIPHENYL (42% Cl))
11097-69-1	PCB-1254 (CHLORODIPHENYL (54% Cl))
1336-36-3	PCBS
8001-35-2	PCC
8001-35-2	PCHK
123-63-7	PCHO
77-47-4	PCL
59-50-7	PCMC
1321-10-4	PCMC
82-68-8	PCNB
87-86-5	PCP
87-86-5	PCP (PESTICIDE)
131-52-2	PCP SODIUM SALT

CAS No.	Chemical Name
131-52-2	PCP-SODIUM
57-11-4	PD 185
7786-34-7	PD 5
106-46-7	PDB
78-87-5	PDC
106-46-7	PDCB
115-77-5	PE
115-77-5	PE 200
1333-86-4	PEACH BLACK
8002-03-7	PEANUT OIL
8002-03-7	PEANUT OIL, GROUNDNUT OIL
123-92-2	PEAR OIL
628-63-7	PEAR OIL
3811-04-9	PEARL ASH
57-11-4	PEARL STEARIC
1333-86-4	PEARLS 800
7446-70-0	PEARSALL
50-29-3	PEB1
9032-75-1	PECTINASE
7681-49-4	PEDIAFLOR
7681-49-4	PEDIDENT
131-89-5	PEDINEX (FRENCH)
58-89-9	PEDRACZAK
1937-37-7	PEERAMINE BLACK E
1937-37-7	PEERAMINE BLACK GXOO
16071-86-6	PEERAMINE FAST BROWN BRL
80-62-6	PEGALAN
60-51-5	PEI 75
10043-52-4	PELADOW
106-50-3	PELAGOL D
106-50-3	PELAGOL DR
120-80-9	PELAGOL GREY C
106-50-3	PELAGOL GREY D
95-80-7	PELAGOL GREY J
123-30-8	PELAGOL GREY P BASE
27598-85-2	PELAGOL GREY P BASE
108-46-3	PELAGOL GREY RS
95-80-7	PELAGOL J
123-30-8	PELAGOL P BASE
27598-85-2	PELAGOL P BASE
108-46-3	PELAGOL RS
143-08-8	PELARGONIC ALCOHOL
762-13-0	PELARGONOYL PEROXIDE
762-13-0	PELARGONYL PEROXIDE
762-13-0	PELARGONYL PEROXIDE, TECHNICALLY PURE
762-13-0	PELARGONYL PEROXIDE, TECHNICALLY PURE (DOT)
577-11-7	PELEX OT
577-11-7	PELEX OT-P
1333-86-4	PELLETEX
9000-07-1	PELLUGEL
25155-30-0	PELOPON A
7783-28-0	PELOR
106-50-3	PELTOL D
7778-44-1	PENCAL
87-86-5	PENCHLOROL
9000-07-1	PENCOGEL
115-77-5	PENETEK
9016-45-9	PENETRAX
9005-25-8	PENFORD GUM 380
61-33-6	PENICILLIN
7784-46-5	PENITE
1405-87-4	PENITRACIN
96-45-7	PENNAC CRA
100-37-8	PENNAD 150
94-75-7	PENNAMINE
94-75-7	PENNAMINE D
56-38-2	PENNCAP E
298-00-0	PENNCAP M
298-00-0	PENNCAP MLS
7681-49-4	PENNWHITE
8001-35-2	PENPHENE
115-90-2	PENSULFOTHION
626-38-0	PENT-2-YL ETHANOATE
628-63-7	PENT-ACETATE
628-63-7	PENT-ACETATE 28
87-86-5	PENTA
87-86-5	PENTA-KIL
19624-22-7	PENTABORANE
19624-22-7	PENTABORANE (ACGIH,DOT)
19624-22-7	PENTABORANE (B5H9)
19624-22-7	PENTABORANE(9)
7789-69-7	PENTABROMOPHOSPHORANE
7789-69-7	PENTABROMOPHOSPHORUS
13463-40-6	PENTACARBONYL IRON
76-01-7	PENTACHLOORETHAAN (DUTCH)
87-86-5	PENTACHLOORFENOL (DUTCH)
76-01-7	PENTACHLORAETHAN (GERMAN)
76-01-7	PENTACHLORETHANE (FRENCH)
50-29-3	PENTACHLORIN
7647-18-9	PENTACHLOROANTIMONY
76-01-7	PENTACHLOROETHANE
76-01-7	PENTACHLOROETHANE (DOT)
87-86-5	PENTACHLOROFENOL
10241-05-1	PENTACHLOROMOLYBDENUM
1321-64-8	PENTACHLORONAPHTHALENE
82-68-8	PENTACHLORONITROBENZENE
87-86-5	PENTACHLOROPHENATE
131-52-2	PENTACHLOROPHENATE SODIUM
87-86-5	PENTACHLOROPHENOL
87-86-5	PENTACHLOROPHENOL (ACGIH,DOT)
131-52-2	PENTACHLOROPHENOL SODIUM SALT
87-86-5	PENTACHLOROPHENOL, DOWICIDE EC-7
87-86-5	PENTACHLOROPHENOL, DP-2
87-86-5	PENTACHLOROPHENOL, TECHNICAL
131-52-2	PENTACHLOROPHENOXY SODIUM
118-74-1	PENTACHLOROPHENYL CHLORIDE
10026-13-8	PENTACHLOROPHOSPHORANE
10026-13-8	PENTACHLOROPHOSPHORUS
87-86-5	PENTACHLORPHENOL (GERMAN)
7647-18-9	PENTACHLORURE D'ANTIMOINE (FRENCH)
76-01-7	PENTACLOROETANO (ITALIAN)
87-86-5	PENTACLOROFENOLO (ITALIAN)
87-86-5	PENTACON
629-76-5	PENTADECANOL
629-76-5	PENTADECYL ALCOHOL
2570-26-5	PENTADECYLAMINE
115-77-5	PENTAERYTHRITE
78-11-5	PENTAERYTHRITE TETRANITRATE
115-77-5	PENTAERYTHRITOL
115-77-5	PENTAERYTHRITOL (ACGIH)
12772-47-3	PENTAERYTHRITOL OLEATE
112-98-1	PENTAETHER
7783-70-2	PENTAFLUOROANTIMONY
76-15-3	PENTAFLUOROCHLOROETHANE
76-15-3	PENTAFLUOROETHYL CHLORIDE
7783-66-6	PENTAFLUOROIODINE
7647-19-0	PENTAFLUOROPHOSPHORANE
7647-19-0	PENTAFLUOROPHOSPHORUS
82-68-8	PENTAGEN
76-01-7	PENTALIN
30586-18-6	PENTAMETHYL HEPTANE
287-92-3	PENTAMETHYLENE
628-76-2	PENTAMETHYLENE CHLORIDE
142-68-7	PENTAMETHYLENE OXIDE
110-89-4	PENTAMETHYLENEIMINE
101-25-7	PENTAMETHYLENETETRAMINE, DINITROSO-
110-89-4	PENTAMETHYLENIMINE
30586-18-6	PENTAMETHYLHEPTANE (DOT)
109-66-0	PENTAN (POLISH)
71-41-0	PENTAN-1-OL
584-02-1	PENTAN-3-OL
110-62-3	PENTANAL
32749-94-3	PENTANAL, 2,3-DIMETHYL-
123-15-9	PENTANAL, 2-METHYL-
73513-30-1	PENTANAL, METHYL-
109-66-0	PENTANE
109-66-0	PENTANE (ACGIH,DOT)
872-10-6	PENTANE, 1,1'-THIOBIS-
628-76-2	PENTANE, 1,5-DICHLORO-
110-53-2	PENTANE, 1-BROMO-
543-59-9	PENTANE, 1-CHLORO-
538-68-1	PENTANE, 1-PHENYL-
7154-79-2	PENTANE, 2,2,3,3-TETRAMETHYL-
1186-53-4	PENTANE, 2,2,3,4-TETRAMETHYL-
564-02-3	PENTANE, 2,2,3-TRIMETHYL-
540-84-1	PENTANE, 2,2,4-TRIMETHYL-
560-21-4	PENTANE, 2,3,3-TRIMETHYL-
565-59-3	PENTANE, 2,3-DIMETHYL-

CAS No.	Chemical Name
108-08-7	PENTANE, 2,4-DIMETHYL-
29756-38-5	PENTANE, 2-BROMO-
107-83-5	PENTANE, 2-METHYL-
1067-20-5	PENTANE, 3,3-DIETHYL-
1068-87-7	PENTANE, 3-ETHYL-2,4-DIMETHYL-
609-26-7	PENTANE, 3-ETHYL-2-METHYL-
96-14-0	PENTANE, 3-METHYL-
760-21-4	PENTANE, 3-METHYLENE-
29756-38-5	PENTANE, BROMO-
43133-95-5	PENTANE, METHYL-
123-54-6	PENTANE-2,4-DIONE
111-30-8	PENTANEDIAL
123-54-6	PENTANEDIONE
123-54-6	PENTANEDIONE-2,4
109-66-0	PENTANEN (DUTCH)
110-66-7	PENTANETHIOL
109-66-0	PENTANI (ITALIAN)
109-52-4	PENTANOIC ACID
71-41-0	PENTANOL
71-41-0	PENTANOL-1
584-02-1	PENTANOL-3
96-22-0	PENTANONE-3
638-29-9	PENTANOYL CHLORIDE
80-46-6	PENTAPHEN
131-52-2	PENTAPHENATE
1314-56-3	PENTAPHOSPHORIC ACID (H3P5O14)
131-52-2	PENTAPLASTIC
7758-29-4	PENTASODIUM TRIPHOSPHATE
7758-29-4	PENTASODIUM TRIPOLYPHOSPHATE
71-41-0	PENTASOL
87-86-5	PENTASOL
1314-80-3	PENTASULFURE DE PHOSPHORE (FRENCH)
115-77-5	PENTEK
109-67-1	PENTENE
513-35-9	PENTENE
25377-72-4	PENTENE
25167-70-8	PENTENE, 2,4,4-TRIMETHYL-
92-84-2	PENTHAZINE
50-29-3	PENTICIDUM
142-62-1	PENTIFORMIC ACID
12772-47-3	PENTOL
12772-47-3	PENTOL (EMULSIFIER)
542-92-7	PENTOLE
8066-33-9	PENTOLITE
628-63-7	PENTYL ACETATE
71-41-0	PENTYL ALCOHOL
463-04-7	PENTYL ALCOHOL, NITRITE
110-53-2	PENTYL BROMIDE
540-18-1	PENTYL BUTANOATE
540-18-1	PENTYL BUTYRATE
543-59-9	PENTYL CHLORIDE
638-49-3	PENTYL FORMATE
5350-03-8	PENTYL LAURATE
110-66-7	PENTYL MERCAPTAN
110-43-0	PENTYL METHYL KETONE
1002-16-0	PENTYL NITRATE
463-04-7	PENTYL NITRITE
872-10-6	PENTYL SULFIDE
110-58-7	PENTYLAMINE
110-58-7	PENTYLAMINE (MIXED ISOMERS)
2050-92-2	PENTYLAMINE, PENTYL-
538-68-1	PENTYLBENZENE
111-27-3	PENTYLCARBINOL
109-67-1	PENTYLENE
513-35-9	PENTYLENE
142-62-1	PENTYLFORMIC ACID
1320-27-0	PENTYLNAPHTHALENE
1322-06-1	PENTYLPHENOL
107-72-2	PENTYLSILICON TRICHLORIDE
110-66-7	PENTYLTHIOL
107-72-2	PENTYLTRICHLOROSILANE
87-86-5	PENWAR
127-18-4	PER
79-21-0	PERACETIC ACID
79-21-0	PERACETIC ACID (CONCENTRATION ≤60%)
79-21-0	PERACETIC ACID, NOT OVER 43% ACID AND NOT OVER 6% HYDROGEN PEROXIDE (DOT)
58-22-0	PERANDREN
87-86-5	PERATOX

CAS No.	Chemical Name
127-18-4	PERAWIN
39349-73-0	PERBORATE
7632-04-4	PERBORIC ACID (HBo3), SODIUM SALT
11138-47-9	PERBORIC ACID (HBo3), SODIUM SALT
39349-73-0	PERBORIC ACID, ION (NEG)
1303-96-4	PERBORIC ACID, SODIUM SALT
7632-04-4	PERBORIC ACID, SODIUM SALT
11138-47-9	PERBORIC ACID, SODIUM SALT
3457-61-2	PERBUTYL C
30580-75-7	PERBUTYL C
110-05-4	PERBUTYL D
75-91-2	PERBUTYL H
109-13-7	PERBUTYL IB
3006-82-4	PERBUTYL O
927-07-1	PERBUTYL PV
614-45-9	PERBUTYL Z
127-18-4	PERC
15520-11-3	PERCADOX 16
124-43-6	PERCARBAMID
124-43-6	PERCARBAMIDE
127-18-4	PERCHLOORETHYLEEN, PER (DUTCH)
127-18-4	PERCHLOR
127-18-4	PERCHLORAETHYLEN, PER (GERMAN)
14797-73-0	PERCHLORATE
10034-81-8	PERCHLORATE DE MAGNESIUM (FRENCH)
7601-89-0	PERCHLORATE DE SODIUM (FRENCH)
14797-73-0	PERCHLORATE ION
14797-73-0	PERCHLORATE ION(1-)
14797-73-0	PERCHLORATE(1-)
127-18-4	PERCHLORETHYLENE
127-18-4	PERCHLORETHYLENE, PER (FRENCH)
7601-90-3	PERCHLORIC ACID
13477-36-6	PERCHLORIC ACID CALCIUM SALT (2:1)
7790-98-9	PERCHLORIC ACID, AMMONIUM SALT
13465-95-7	PERCHLORIC ACID, BARIUM SALT
13477-36-6	PERCHLORIC ACID, CALCIUM SALT
14797-73-0	PERCHLORIC ACID, ION(1-)
13637-76-8	PERCHLORIC ACID, LEAD(2+) SALT
10034-81-8	PERCHLORIC ACID, MAGNESIUM SALT
7601-90-3	PERCHLORIC ACID, MORE THAN 50% BUT NOT MORE THAN 72% STRENGTH (DOT)
7601-90-3	PERCHLORIC ACID, NOT OVER 50% ACID (DOT)
7778-74-7	PERCHLORIC ACID, POTASSIUM SALT
7778-74-7	PERCHLORIC ACID, POTASSIUM SALT (1:1)
7601-89-0	PERCHLORIC ACID, SODIUM SALT
13450-97-0	PERCHLORIC ACID, STRONTIUM SALT
7487-94-7	PERCHLORIDE OF MERCURY
87-68-3	PERCHLORO-1,3-BUTADIENE
116-16-5	PERCHLOROACETONE
118-74-1	PERCHLOROBENZENE
87-68-3	PERCHLOROBUTADIENE
77-47-4	PERCHLOROCYCLOPENTADIENE
2385-85-5	PERCHLORODIHOMOCUBANE
67-72-1	PERCHLOROETHANE
127-18-4	PERCHLOROETHYLENE
127-18-4	PERCHLOROETHYLENE (ACGIH,DOT)
56-23-5	PERCHLOROMETHANE
594-42-3	PERCHLOROMETHYL MERCAPTAN
2234-13-1	PERCHLORONAPHTHALENE
2385-85-5	PERCHLOROPENTACYCLO(5210(SUP 2,6)0(SUP 3,9)0(SUP 5,8))DE-CANE
2385-85-5	PERCHLOROPENTACYCLODECANE
2385-85-5	PERCHLOROPENTACYCLO[52102,603,905,8]DECANE
10026-04-7	PERCHLOROSILANE
7616-94-6	PERCHLOROYLFLUORIDE
7647-18-9	PERCHLORURE D'ANTIMOINE (FRENCH)
7705-08-0	PERCHLORURE DE FER (FRENCH)
7616-94-6	PERCHLORYL FLUORIDE
7616-94-6	PERCHLORYL FLUORIDE (ClO3F)
127-18-4	PERCLENE
127-18-4	PERCLENE D
127-18-4	PERCLOROETILENE (ITALIAN)
732-11-6	PERCOLATE
127-18-4	PERCOSOLVE
80-43-3	PERCUMYL D
80-43-3	PERCUMYL D 40
80-15-9	PERCUMYL H
58-22-0	PERCUTACRINE ANDROGENIQUE
57-83-0	PERCUTACRINE LUTEINIQUE

CAS No.	Chemical Name
4452-58-8	PERDOX
3313-92-6	PERDOX
15630-89-4	PERDOX
60-51-5	PERFECTHION
60-51-5	PERFEKTHION
16984-48-8	PERFLUORIDE
428-59-1	PERFLUORO(METHYLOXIRANE)
116-15-4	PERFLUORO-1-PROPENE
41070-66-9	PERFLUORO-2-BUTENE (DOT)
684-16-2	PERFLUORO-2-PROPANONE
76-05-1	PERFLUOROACETIC ACID
684-16-2	PERFLUOROACETONE
7783-54-2	PERFLUOROAMMONIA
41070-66-9	PERFLUOROBUT-2-ENE
41070-66-9	PERFLUOROBUTENE
115-25-3	PERFLUOROCYCLOBUTANE
76-16-4	PERFLUOROETHANE
116-14-3	PERFLUOROETHENE
76-15-3	PERFLUOROETHYL CHLORIDE
116-14-3	PERFLUOROETHYLENE
9002-84-0	PERFLUOROETHYLENE POLYMER
10036-47-2	PERFLUOROHYDRAZINE
75-73-0	PERFLUOROMETHANE
76-19-7	PERFLUOROPROPANE
116-15-4	PERFLUOROPROPENE
116-15-4	PERFLUOROPROPYLENE
428-59-1	PERFLUOROPROPYLENE OXIDE
7783-61-1	PERFLUOROSILANE
1694-09-3	PERGACID VIOLET 2B
7681-49-4	PERGANTENE
55-63-0	PERGLOTTAL
78-63-7	PERHEXA 25B
78-63-7	PERHEXA 25B40
6731-36-8	PERHEXA 3M
78-63-7	PERHEXA 3M40
1068-27-5	PERHEXYNE 25B
1068-27-5	PERHEXYNE 25B40
124-43-6	PERHYDRIT
111-49-9	PERHYDROAZEPINE
7722-84-1	PERHYDROL
124-43-6	PERHYDROL-UREA
91-17-8	PERHYDRONAPHTHALENE
110-89-4	PERHYDROPYRIDINE
123-75-1	PERHYDROPYRROLE
83-32-9	PERI-ETHYLENENAPHTHALENE
7778-74-7	PERIODIN
59-50-7	PERITONAN
1321-10-4	PERITONAN
127-18-4	PERK
15520-11-3	PERKADOX 16
15520-11-3	PERKADOX 16W40
80-43-3	PERKADOX B
80-43-3	PERKADOX BC
80-43-3	PERKADOX BC 40
80-43-3	PERKADOX BC 9
80-43-3	PERKADOX BC 95
80-43-3	PERKADOX BC 96
80-43-3	PERKADOX SB
762-12-9	PERKADOX SE 10
762-16-3	PERKADOX SE 8
127-18-4	PERKLONE
151-21-3	PERLANDROL L
151-21-3	PERLANKROL L
53-16-7	PERLATAN
50-28-2	PERLATANOL
7446-27-7	PERLEX PASTE 500
7446-27-7	PERLEX PASTE 600A
82-28-0	PERLITON ORANGE 3R
79-01-6	PERM-A-CHLOR
79-01-6	PERM-A-CLOR
60-00-4	PERMA KLEER 50 ACID
1333-86-4	PERMABLAK 663
87-86-5	PERMACIDE
107-22-2	PERMAFRESH 114
87-86-5	PERMAGARD
8002-74-2	PERMAL
1314-13-2	PERMANENT WHITE
14333-13-2	PERMANGANATE
14333-13-2	PERMANGANATE (MnO4-) ION

CAS No.	Chemical Name
10101-50-5	PERMANGANATE DE SODIUM (FRENCH)
14333-13-2	PERMANGANATE ION
14333-13-2	PERMANGANATE ION(1-)
13446-10-1	PERMANGANIC ACID (HMnO4), AMMONIUM SALT
7787-36-2	PERMANGANIC ACID (HMnO4), BARIUM SALT
10118-76-0	PERMANGANIC ACID (HMnO4), CALCIUM SALT
14333-13-2	PERMANGANIC ACID (HMnO4), ION(1-)
7722-64-7	PERMANGANIC ACID (HMnO4), POTASSIUM SALT
10101-50-5	PERMANGANIC ACID (HMnO4), SODIUM SALT
23414-72-4	PERMANGANIC ACID (HMnO4), ZINC SALT
7722-64-7	PERMANGANIC ACID POTASSIUM SALT
87-86-5	PERMASAN
87-86-5	PERMATOX DP-2
87-86-5	PERMATOX PENTA
1338-23-4	PERMEK G
1338-23-4	PERMEK N
87-86-5	PERMITE
2234-13-1	PERNA
12033-49-7	PERNITRITE RADICAL
10544-73-7	PERNITRITE RADICAL
7778-74-7	PEROIDIN
7722-84-1	PERONE
7722-84-1	PERONE 30
7722-84-1	PERONE 35
7722-84-1	PERONE 50
94-36-0	PEROSSIDO DI BENZOILE (ITALIAN)
110-05-4	PEROSSIDO DI BUTILE TERZIARIO (ITALIAN)
7722-84-1	PEROSSIDO DI IDROGENO (ITALIAN)
57-63-6	PEROVEX
7722-84-1	PEROXAAN
7722-84-1	PEROXAN
7722-84-1	PEROXIDE
14915-07-2	PEROXIDE
14915-07-2	PEROXIDE ION
78-63-7	PEROXIDE, (1,1,4,4-TETRAMETHYL-1,4-BUTANEDIYL)BIS[(1,1-DIMETHYLETHYL)
1068-27-5	PEROXIDE, (1,1,4,4-TETRAMETHYL-2-BUTYNE-1,4-DIYL)BIS[(1,1-DIMETHYLETHYL)
1068-27-5	PEROXIDE, (1,1,4,4-TETRAMETHYL-2-BUTYNYLENE)BIS[TERT-BUTYL
78-63-7	PEROXIDE, (1,1,4,4-TETRAMETHYLTETRAMETHYLENE)BIS[TERT-BUTYL
3006-86-8	PEROXIDE, (1,1-CYCLOHEXYLIDENE)BIS(TERT-BUTYL-
3006-86-8	PEROXIDE, (1,1-CYCLOHEXYLIDENE)BIS(TERT-BUTYL-, NOT MORE THAN 77% IN SOLUTION
3006-86-8	PEROXIDE, (1,1-CYCLOHEXYLIDENE)BIS(TERT-BUTYL-, WITH AT LEAST 13% PHLEGMATIZER AND 47% INERT SOLID
3006-86-8	PEROXIDE, (1,1-CYCLOHEXYLIDENE)BIS(TERT-BUTYL-, WITH AT LEAST 50% PHLEGMATIZER
2167-23-9	PEROXIDE, (1-METHYLPROPYLIDENE)BIS(TERT-BUTYL-, NOT MORE THAN 55% IN SOLUTION
2167-23-9	PEROXIDE, (1-METHYLPROPYLIDENE)BIS[(1,1-DIMETHYLETHYL)
6731-36-8	PEROXIDE, (3,3,5-TRIMETHYLCYCLOHEXYLIDENE)BIS(TERT-BUTYL-, NOT MORE THAN 57% IN SOLUTION
6731-36-8	PEROXIDE, (3,3,5-TRIMETHYLCYCLOHEXYLIDENE)BIS[(1,1-DIMETHYLETHYL)
1068-27-5	PEROXIDE, (TETRAMETHYL-2-BUTYNYLENE)BIS[TERT-BUTYL
3457-61-2	PEROXIDE, 1,1-DIMETHYLETHYL (1-METHYLETHYL)PHENYL
30580-75-7	PEROXIDE, 1,1-DIMETHYLETHYL (1-METHYLETHYL)PHENYL
3457-61-2	PEROXIDE, 1,1-DIMETHYLETHYL 1-METHYL-1-PHENYLETHYL
30580-75-7	PEROXIDE, 1,1-DIMETHYLETHYL 1-METHYL-1-PHENYLETHYL
3457-61-2	PEROXIDE, 1,1-DIMETHYLETHYL 1-METHYL-1-PHENYLETHYL (9CI)
30580-75-7	PEROXIDE, 1,1-DIMETHYLETHYL 1-METHYL-1-PHENYLETHYL (9CI)
644-31-5	PEROXIDE, ACETYL BENZOYL
644-31-5	PEROXIDE, ACETYL BENZOYL, NOT OVER 40% PEROXIDE IN SOLUTION
3179-56-4	PEROXIDE, ACETYL CYCLOHEXYLSULFONYL
3179-56-4	PEROXIDE, ACETYL CYCLOHEXYLSULFONYL, NOT MORE THAN 32% IN SOLUTION
3179-56-4	PEROXIDE, ACETYL CYCLOHEXYLSULFONYL, NOT MORE THAN 82% WETTED WITH NOT LESS THAN 12% WATER
110-05-4	PEROXIDE, BIS(1,1-DIMETHYLETHYL)
2407-94-5	PEROXIDE, BIS(1-HYDROXYCYCLOHEXYL)
80-43-3	PEROXIDE, BIS(1-METHYL-1-PHENYLETHYL)
762-12-9	PEROXIDE, BIS(1-OXODECYL)
105-74-8	PEROXIDE, BIS(1-OXODODECYL)-
762-13-0	PEROXIDE, BIS(1-OXONONYL)

CAS No.	Chemical Name
762-16-3	PEROXIDE, BIS(1-OXOOCTYL)
762-16-3	PEROXIDE, BIS(1-OXOOCTYL) (9CI)
3248-28-0	PEROXIDE, BIS(1-OXOPROPYL)
133-14-2	PEROXIDE, BIS(2,4-DICHLOROBENZOYL)
133-14-2	PEROXIDE, BIS(2,4-DICHLOROBENZOYL)-, NOT MORE THAN 52% AS A PASTE OR IN SOLUTION
3437-84-1	PEROXIDE, BIS(2-METHYL-1-OXOPROPYL)-
123-23-9	PEROXIDE, BIS(3-CARBOXYPROPIONYL)
94-17-7	PEROXIDE, BIS(4-CHLOROBENZOYL)
80-43-3	PEROXIDE, BIS(ALPHA,ALPHA-DIMETHYLBENZYL)
94-17-7	PEROXIDE, BIS(P-CHLOROBENZOYL)-
3006-86-8	PEROXIDE, CYCLOHEXYLIDENEBIS[(1,1-DIMETHYLETHYL)
3006-86-8	PEROXIDE, CYCLOHEXYLIDENEBIS[TERT-BUTYL
110-22-5	PEROXIDE, DIACETYL
94-36-0	PEROXIDE, DIBENZOYL
628-37-5	PEROXIDE, DIETHYL
14915-07-2	PEROXIDE, INORGANIC
762-16-3	PEROXIDE, OCTANOYL
2167-23-9	PEROXIDE, SEC-BUTYLIDENEBIS[TERT-BUTYL
3457-61-2	PEROXIDE, TERT-BUTYL ALPHA,ALPHA-DIMETHYLBENZYL
30580-75-7	PEROXIDE, TERT-BUTYL ALPHA,ALPHA-DIMETHYLBENZYL
79-21-0	PEROXOACETIC ACID
3313-92-6	PEROXY SODIUM CARBONATE
4452-58-8	PEROXY SODIUM CARBONATE
15630-89-4	PEROXY SODIUM CARBONATE
79-21-0	PEROXYACETIC ACID
79-21-0	PEROXYACETIC ACID, MORE THAN 43% WITH MORE THAN 6% HYDROGEN PEROXIDE
79-21-0	PEROXYACETIC ACID, MORE THAN 43% WITH MORE THAN 6% HYDROGEN PEROXIDE (DOT)
79-21-0	PEROXYACETIC ACID, NOT MORE THAN 43% ACID AND NOT MORE THAN 6% HYDROGEN PEROXIDE (DOT)
79-21-0	PEROXYACETIC ACID, NOT OVER 43% ACID AND NOT OVER 6% HYDROGEN PEROXIDE
107-71-1	PEROXYACETIC ACID, TERT-BUTYL ESTER
1941-79-3	PEROXYAZELAIC ACID
1941-79-3	PEROXYAZELAIC ACID, MAXIMUM CONCENTRATION 27%, WITH AT LEAST 13% AZELAIC ACID AND 53%SODIUM SULFATE
2618-77-1	PEROXYBENZOIC ACID, 1,1,4,4-TETRAMETHYLTETRAMETHYLENE ESTER
937-14-4	PEROXYBENZOIC ACID, M-CHLORO-
937-14-4	PEROXYBENZOIC ACID, M-CHLORO-, MAXIMUM CONCENTRATION 86%
614-45-9	PEROXYBENZOIC ACID, TERT-BUTYL ESTER
4511-39-1	PEROXYBENZOIC ACID, TERT-PENTYL ESTER
15520-11-3	PEROXYCARBONIC ACID, BIS(4-TERT-BUTYLCYCLOHEXYL) ESTER, NOT MORE THAN 42% IN WATER
2372-21-6	PEROXYCARBONIC ACID, OO-TERT-BUTYL O-ISOPROPYL ESTER
7722-84-1	PEROXYDE D'HYDROGENE (FRENCH)
1304-29-6	PEROXYDE DE BARYUM (FRENCH)
94-36-0	PEROXYDE DE BENZOYLE (FRENCH)
110-05-4	PEROXYDE DE BUTYLE TERTIAIRE (FRENCH)
105-74-8	PEROXYDE DE LAUROYLE (FRENCH)
1309-60-0	PEROXYDE DE PLOMB (FRENCH)
105-64-6	PEROXYDICARBONATE D'ISOPROPYLE (FRENCH)
105-64-6	PEROXYDICARBONIC ACID, BIS(1-METHYLETHYL) ESTER
16111-62-9	PEROXYDICARBONIC ACID, BIS(2-ETHYLHEXYL) ESTER
15520-11-3	PEROXYDICARBONIC ACID, BIS(4-TERT-BUTYLCYCLOHEXYL) ESTER
2144-45-8	PEROXYDICARBONIC ACID, BIS(PHENYLMETHYL) ESTER
15520-11-3	PEROXYDICARBONIC ACID, BIS[4-(1,1-DIMETHYLETHYL)CYCLOHEXYL] ESTER
16111-62-9	PEROXYDICARBONIC ACID, DI(2-ETHYLHEXYL) ESTER
16111-62-9	PEROXYDICARBONIC ACID, DI(2-ETHYLHEXYL) ESTER, NOT MORE THAN 77% IN SOLUTION
2144-45-8	PEROXYDICARBONIC ACID, DIBENZYL ESTER
1561-49-5	PEROXYDICARBONIC ACID, DICYCLOHEXYL ESTER
1561-49-5	PEROXYDICARBONIC ACID, DICYCLOHEXYL ESTER, NOT MORE THAN 91% WITH WATER
14666-78-5	PEROXYDICARBONIC ACID, DIETHYL ESTER
14666-78-5	PEROXYDICARBONIC ACID, DIETHYL ESTER, NOT MORE THAN 27% IN SOLUTION
105-64-6	PEROXYDICARBONIC ACID, DIISOPROPYL ESTER
105-64-6	PEROXYDICARBONIC ACID, DIISOPROPYL ESTER, NOT MORE THAN 52% IN SOLUTION
52326-66-6	PEROXYDICARBONIC ACID, DIOCTADECYL ESTER
52326-66-6	PEROXYDICARBONIC ACID, DIOCTADECYL ESTER, NOT MORE THAN 85% WITH STEARYL ALCOHOL
16066-38-9	PEROXYDICARBONIC ACID, DIPROPYL ESTER
3313-92-6	PEROXYDICARBONIC ACID, DISODIUM SALT
4452-58-8	PEROXYDICARBONIC ACID, DISODIUM SALT
15630-89-4	PEROXYDICARBONIC ACID, DISODIUM SALT
53220-22-7	PEROXYDICARBONIC ACID, DITETRADECYL ESTER
53220-22-7	PEROXYDICARBONIC ACID, DITETRADECYL ESTER, NOT MORE THAN 22% IN WATER
123-23-9	PEROXYDISUCCINIC ACID
7727-54-0	PEROXYDISULFURIC ACID ([(HO)S(O)2]2O2), DIAMMONIUM SALT
7727-21-1	PEROXYDISULFURIC ACID ([(HO)S(O)2]2O2), DIPOTASSIUM SALT
7775-27-1	PEROXYDISULFURIC ACID ([(HO)S(O)2]2O2), DISODIUM SALT
7727-54-0	PEROXYDISULFURIC ACID, DIAMMONIUM SALT
7727-21-1	PEROXYDISULFURIC ACID, DIPOTASSIUM SALT
7775-27-1	PEROXYDISULFURIC ACID, DISODIUM SALT
3006-82-4	PEROXYHEXANOIC ACID, 2-ETHYL-, TERT-BUTYL ESTER
686-31-7	PEROXYHEXANOIC ACID, 2-ETHYL-, TERT-PENTYL ESTER
13122-18-4	PEROXYHEXANOIC ACID, 3,5,5-TRIMETHYL-, TERT-BUTYL ESTER
109-13-7	PEROXYISOBUTYRIC ACID, TERT-BUTYL ESTER
109-13-7	PEROXYISOBUTYRIC ACID, TERT-BUTYL ESTER, MORE THAN 52% BUT NOT MORE THAN 77% IN SOLUTION
109-13-7	PEROXYISOBUTYRIC ACID, TERT-BUTYL ESTER, NOT MORE THAN 52% IN SOLUTION
15593-29-0	PEROXYMONOSULFURIC ACID, DISODIUM SALT
28831-12-1	PEROXYMONOSULFURIC ACID, MONOSODIUM SALT
26748-47-0	PEROXYNEODECANOIC ACID, ALPHA,ALPHA-DIMETHYLBENZYL ESTER
26748-47-0	PEROXYNEODECANOIC ACID, DIMETHYLBENZYL ESTER, 77% IN SOLUTION
26748-41-4	PEROXYNEODECANOIC ACID, TERT-BUTYL ESTER
26748-41-4	PEROXYNEODECANOIC ACID, TERT-BUTYL ESTER, NOT MORE THAN 77% IN SOLUTION
927-07-1	PEROXYPIVALIC ACID, TERT-BUTYL ESTER
927-07-1	PEROXYPIVALIC ACID, TERT-BUTYL ESTER, NOT MORE THAN 77% IN SOLUTION
29240-17-3	PEROXYPIVALIC ACID, TERT-PENTYL ESTER
11105-06-9	PEROXYVANADIC ACID, SODIUM SALT
13718-26-8	PEROXYVANADIC ACID, SODIUM SALT
13721-39-6	PEROXYVANADIC ACID, SODIUM SALT
105-64-6	PEROYL IPP
94-36-0	PERSADOX
127-18-4	PERSEC
106-46-7	PERSIA-PERAZOL
7727-54-0	PERSULFATE D'AMMONIUM (FRENCH)
7775-27-1	PERSULFATE DE SODIUM (FRENCH)
15593-29-0	PERSULFATE DE SODIUM (FRENCH)
28831-12-1	PERSULFATE DE SODIUM (FRENCH)
62-44-2	PERTONAL
106-93-4	PESTMASTER
106-93-4	PESTMASTER EDB-85
371-86-8	PESTON XV
152-16-9	PESTOX
115-26-4	PESTOX 14
371-86-8	PESTOX 15
152-16-9	PESTOX 3
152-16-9	PESTOX III
115-26-4	PESTOX IV
56-38-2	PESTOX PLUS
115-26-4	PESTOX XIV
371-86-8	PESTOX XV
56-38-2	PETHION
67-63-0	PETROHOL
8006-61-9	PETROL
86290-81-5	PETROL
60-11-7	PETROL YELLOW WT
86290-81-5	PETROL, SYNTHETIC
8002-05-9	PETROLEUM
8030-30-6	PETROLEUM BENZIN
8002-05-9	PETROLEUM CRUDE
8030-30-6	PETROLEUM DISTILLATES
8030-30-6	PETROLEUM DISTILLATES (NAPHTHA)
8032-32-4	PETROLEUM ETHER
8030-30-6	PETROLEUM ETHER
68476-85-7	PETROLEUM GAS LIQUEFIED
68476-85-7	PETROLEUM GAS, LIQUEFIED (DOT)
8030-30-6	PETROLEUM NAPHTHA
8030-30-6	PETROLEUM NAPHTHA (DOT)
8002-05-9	PETROLEUM OIL
106-97-8	PETROLEUM PRODUCTS, LIQUEFIED GASES
68476-85-7	PETROLEUM PRODUCTS, LIQUEFIED GASES
8052-42-4	PETROLEUM REFINING RESIDUES, ASPHALTS

CAS No.	Chemical Name
8030-30-6	PETROLEUM SPIRIT
8002-74-2	PETROLITE C 400
8002-74-2	PETRONAUBA WAX C 8500
8002-74-2	PETROX P 200
79-01-6	PETZINOL
3878-19-1	PF 7402
58-89-9	PF LANZOL
1309-37-1	PFERROX 2380
57-55-6	PG 12
7782-42-5	PG 50
6423-43-4	PGDN
122-60-1	PGE
1333-86-4	PGM 33
1333-86-4	PGM 40
95-79-4	PHARMAZOID RED KB
2971-90-6	PHARMCOCCID
114-26-1	PHC
62-44-2	PHENACETIN
62-44-2	PHENACETINE
140-29-4	PHENACETONITRILE
103-80-0	PHENACETYL CHLORIDE
88-06-2	PHENACHLOR
25167-82-2	PHENACHLOR
8001-35-2	PHENACIDE
70-11-1	PHENACYL BROMIDE
70-11-1	PHENACYL BROMIDE (DOT)
532-27-4	PHENACYL CHLORIDE
92-52-4	PHENADOR-X
103-84-4	PHENALGENE
103-84-4	PHENALGIN
1937-37-7	PHENAMINE BLACK BCN-CF
1937-37-7	PHENAMINE BLACK CL
1937-37-7	PHENAMINE BLACK E
1937-37-7	PHENAMINE BLACK E 200
2602-46-2	PHENAMINE BLUE BB
22224-92-6	PHENAMIPHOS
85-01-8	PHENANTHRENE
578-94-9	PHENARSAZINE CHLORIDE
58-36-6	PHENARSAZINE OXIDE
578-94-9	PHENARSAZINE, 10-CHLORO-5,10-DIHYDRO-
1420-04-8	PHENASAL ETHANOLAMINE SALT
2275-14-1	PHENATOL
8001-35-2	PHENATOX
578-94-9	PHENAZARSINE CHLORIDE
62-44-2	PHENAZETIN
136-40-3	PHENAZODINE
94-78-0	PHENAZOPYRIDINE
136-40-3	PHENAZOPYRIDINE HYDROCHLORIDE
136-40-3	PHENAZOPYRIDINIUM CHLORIDE
2275-14-1	PHENCAPTON
2275-14-1	PHENCAPTON (DOT)
299-84-3	PHENCHLORFOS
71-43-2	PHENE
62-44-2	PHENEDINA
92-84-2	PHENEGIC
3546-10-9	PHENESTERIN
7568-93-6	PHENETHANOLAMINE
1321-31-9	PHENETHIDINE
100-42-5	PHENETHYLENE
96-09-3	PHENETHYLENE OXIDE
1321-31-9	PHENETIDINE
108-95-2	PHENIC ACID
62-44-2	PHENIDIN
62-44-2	PHENIN
122-14-5	PHENITROTHION
2275-14-1	PHENKAPTON
2275-14-1	PHENKAPTONE
62-38-4	PHENMAD
1937-37-7	PHENO BLACK EP
1937-37-7	PHENO BLACK SGN
2602-46-2	PHENO BLUE 2B
1336-36-3	PHENOCHLOR
1336-36-3	PHENOCLOR
67-72-1	PHENOHEP
108-95-2	PHENOL
122-60-1	PHENOL GLYCIDYL ETHER
139-02-6	PHENOL SODIUM
139-02-6	PHENOL SODIUM SALT
27193-28-8	PHENOL, (1,1,3,3-TETRAMETHYLBUTYL)-

CAS No.	Chemical Name
1333-13-7	PHENOL, (1,1-DIMETHYLETHYL)-3-METHYL-
70-30-4	PHENOL, 2,2'-METHYLENEBIS[3,4,6-TRICHLORO-
97-18-7	PHENOL, 2,2'-THIOBIS(4,6-DICHLORO)-
4418-66-0	PHENOL, 2,2'-THIOBIS(4-CHLORO-6-METHYL)-
62207-76-5	PHENOL, 2,2'-[1,2-ETHANEDIYLBIS(NITRILOMETHYLIDYNE)]BIS[6-FLUORO-, COBALT COMPLEX
97-18-7	PHENOL, 2,2-THIOBIS 4,6-DICHLORO
4418-66-0	PHENOL, 2,2-THIOBIS(4-CHLORO-6-METHYL)
933-78-8	PHENOL, 2,3,5-TRICHLORO-
933-75-5	PHENOL, 2,3,6-TRICHLORO-
95-95-4	PHENOL, 2,4,5-TRICHLORO-
327-98-0	PHENOL, 2,4,5-TRICHLORO-, O-ESTER WITH O-ETHYL ETHYL-PHOSPHONOTHIOATE
88-06-2	PHENOL, 2,4,6-TRICHLORO-
88-89-1	PHENOL, 2,4,6-TRINITRO-
131-74-8	PHENOL, 2,4,6-TRINITRO-, AMMONIUM SALT
131-74-8	PHENOL, 2,4,6-TRINITRO-, AMMONIUM SALT (9CI)
146-84-9	PHENOL, 2,4,6-TRINITRO-, SILVER(1+) SALT
120-83-2	PHENOL, 2,4-DICHLORO-
105-67-9	PHENOL, 2,4-DIMETHYL-
51-28-5	PHENOL, 2,4-DINITRO-
138-00-1	PHENOL, 2,4-DIPENTYL-
329-71-5	PHENOL, 2,5-DINITRO-
128-37-0	PHENOL, 2,6-BIS(1,1-DIMETHYLETHYL)-4-METHYL-
573-56-8	PHENOL, 2,6-DINITRO-
1420-07-1	PHENOL, 2-(1,1-DIMETHYLETHYL)-4,6-DINITRO-
114-26-1	PHENOL, 2-(1-METHYLETHOXY)-, METHYLCARBAMATE
89-72-5	PHENOL, 2-(1-METHYLPROPYL)-
88-85-7	PHENOL, 2-(1-METHYLPROPYL)-4,6-DINITRO-
831-52-7	PHENOL, 2-AMINO-4,6-DINITRO-, MONOSODIUM SALT
95-85-2	PHENOL, 2-AMINO-4-CHLORO-
95-57-8	PHENOL, 2-CHLORO-
42350-99-2	PHENOL, 2-CHLORO-4,6-BIS(1,1-DIMETHYLPROPYL)-
98-28-2	PHENOL, 2-CHLORO-4-(1,1-DIMETHYLETHYL)-
92-04-6	PHENOL, 2-CHLORO-4-PHENYL-
119-42-6	PHENOL, 2-CYCLOHEXYL-
131-89-5	PHENOL, 2-CYCLOHEXYL-4,6-DINITRO-
95-48-7	PHENOL, 2-METHYL-
95-48-7	PHENOL, 2-METHYL- (9CI)
29595-25-3	PHENOL, 2-METHYL-, DINITRO DERIV, AMMONIUM SALT
534-52-1	PHENOL, 2-METHYL-4,6-DINITRO-
1335-85-9	PHENOL, 2-METHYL-4,6-DINITRO- (9CI)
1335-85-9	PHENOL, 2-METHYLDINITRO-
25641-53-6	PHENOL, 2-METHYLDINITRO-, SODIUM SALT
88-75-5	PHENOL, 2-NITRO-
136-81-2	PHENOL, 2-PENTYL-
1420-07-1	PHENOL, 2-TERT-BUTYL-4,6-DINITRO-
609-19-8	PHENOL, 3,4,5-TRICHLORO-
2032-65-7	PHENOL, 3,5-DIMETHYL-4-(METHYLTHIO)-, METHYLCARBAMATE
64-00-6	PHENOL, 3-(1-METHYLETHYL)-, METHYLCARBAMATE
108-39-4	PHENOL, 3-METHYL-
108-39-4	PHENOL, 3-METHYL- (9CI)
2631-37-0	PHENOL, 3-METHYL-5-(1-METHYLETHYL)-, METHYLCARBAMATE
554-84-7	PHENOL, 3-NITRO-
56-53-1	PHENOL, 4,4'-(1,2-DIETHYL-1,2-ETHENEDIYL)BIS-, (E)-
96-69-5	PHENOL, 4,4'-THIOBIS[2-(1,1-DIMETHYLETHYL)-5-METHYL-
98-27-1	PHENOL, 4-(1,1-DIMETHYLETHYL)-2-METHYL-
80-46-6	PHENOL, 4-(1,1-DIMETHYLPROPYL)-
315-18-4	PHENOL, 4-(DIMETHYLAMINO)-3,5-DIMETHYL-, METHYLCARBAMATE (ESTER)
123-30-8	PHENOL, 4-AMINO-
106-48-9	PHENOL, 4-CHLORO-
25167-80-0	PHENOL, 4-CHLORO-
59-50-7	PHENOL, 4-CHLORO-3-METHYL-
1321-10-4	PHENOL, 4-CHLORO-3-METHYL-
59-50-7	PHENOL, 4-CHLORO-3-METHYL- (9 CI)
1321-10-4	PHENOL, 4-CHLORO-3-METHYL- (9 CI)
123-07-9	PHENOL, 4-ETHYL-
150-76-5	PHENOL, 4-METHOXY-
106-44-5	PHENOL, 4-METHYL-
106-44-5	PHENOL, 4-METHYL- (9CI)
100-02-7	PHENOL, 4-NITRO-
98-28-2	PHENOL, 4-TERT-BUTYL-2-CHLORO-
131-89-5	PHENOL, 6-CYCLOHEXYL-2,4-DINITRO-
51-28-5	PHENOL, ALPHA-DINITRO-
27598-85-2	PHENOL, AMINO-
573-56-8	PHENOL, BETA-DINITRO-
28805-86-9	PHENOL, BUTYL-
25167-80-0	PHENOL, CHLORO-

CAS No.	Chemical Name
73090-69-4	PHENOL, CHLORO-4-(1,1-DIMETHYLPROPYL)-
59-50-7	PHENOL, CHLOROMETHYL-
1321-10-4	PHENOL, CHLOROMETHYL-
1300-71-6	PHENOL, DIMETHYL-
25550-58-7	PHENOL, DINITRO-
25550-58-7	PHENOL, DINITRO-, WETTED WITH AT LEAST 15% WATER
329-71-5	PHENOL, GAMMA-DINITRO-
108-93-0	PHENOL, HEXAHYDRO-
108-46-3	PHENOL, M-HYDROXY-
64-00-6	PHENOL, M-ISOPROPYL-, METHYLCARBAMATE
554-84-7	PHENOL, M-NITRO-
22259-30-9	PHENOL, M-[[(DIMETHYLAMINO)METHYLENE]AMINO]-, METHYLCARBAMATE (ESTER)
1319-77-3	PHENOL, METHYL-
1319-77-3	PHENOL, METHYL- (9CI)
12167-20-3	PHENOL, METHYLNITRO-
25154-55-6	PHENOL, NITRO-
25154-52-3	PHENOL, NONYL-
95-57-8	PHENOL, O-CHLORO-
119-42-6	PHENOL, O-CYCLOHEXYL-
88-75-5	PHENOL, O-NITRO-
136-81-2	PHENOL, O-PENTYL-
89-72-5	PHENOL, O-SEC-BUTYL-
1420-07-1	PHENOL, O-T-BUTYL-4,6-DINITRO-
27193-28-8	PHENOL, OCTYL-
3254-63-5	PHENOL, P-(METHYLTHIO)-, DIMETHYL PHOSPHATE
123-30-8	PHENOL, P-AMINO-
27598-85-2	PHENOL, P-AMINO-
28805-86-9	PHENOL, P-BUTYL
106-48-9	PHENOL, P-CHLORO-
25167-80-0	PHENOL, P-CHLORO-
123-07-9	PHENOL, P-ETHYL-
150-76-5	PHENOL, P-METHOXY-
100-02-7	PHENOL, P-NITRO-
56-38-2	PHENOL, P-NITRO-, O-ESTER WITH O,O-DIETHYLPHOSPHOROTHIOATE
298-00-0	PHENOL, P-NITRO-, O-ESTER WITH O,O-DIMETHYLPHOSPHOROTHIOATE
2104-64-5	PHENOL, P-NITRO-, O-ESTER WITH O-ETHYL PHENYL PHOSPHONOTHIOATE
80-46-6	PHENOL, P-TERT-PENTYL-
87-86-5	PHENOL, PENTACHLORO-
131-52-2	PHENOL, PENTACHLORO-, SODIUM SALT
1322-06-1	PHENOL, PENTYL-
139-02-6	PHENOL, SODIUM SALT, (SOLID)
108-98-5	PHENOL, THIO-
933-78-4	PHENOL, TRICHLORO-
15950-66-0	PHENOL, TRICHLORO-
25167-82-2	PHENOL, TRICHLORO-
53894-28-3	PHENOL, [[(2-AMINOETHYL)AMINO]METHYL]-
69-72-7	PHENOL-2-CARBOXYLIC ACID
100-51-6	PHENOLCARBINOL
1333-39-7	PHENOLSULFONIC A
98-11-3	PHENOLSULPHONIC ACID
62-38-4	PHENOMERCURIC ACETATE
88-85-7	PHENOTAN
92-84-2	PHENOTHIAZINE
92-84-2	PHENOVERM
92-84-2	PHENOVIS
94-75-7	PHENOX
58-36-6	PHENOXARSINE, 10,10'-OXYDI-
92-84-2	PHENOXUR
2122-46-5	PHENOXY
2122-46-5	PHENOXY PESTICIDE, LIQUID [FLAMMABLE LIQUID LABEL]
2122-46-5	PHENOXY PESTICIDE, LIQUID [POISON B LABEL]
2122-46-5	PHENOXY PESTICIDE, SOLID
2122-46-5	PHENOXY RADICAL
59-96-1	PHENOXYBENZAMINE
63-92-3	PHENOXYBENZAMINE CHLORIDE
63-92-3	PHENOXYBENZAMINE HYDROCHLORIDE
101-84-8	PHENOXYBENZENE
2122-46-5	PHENOXYL
2122-46-5	PHENOXYL RADICAL
100-66-3	PHENOXYMETHANE
127-82-2	PHENOZIN
92-84-2	PHENTHIAZINE
55-38-9	PHENTHION
2275-14-1	PHENUDIN
2275-14-1	PHENUDINE
122-60-1	PHENYL 2,3-EPOXYPROPYL ETHER
140-29-4	PHENYL ACETYL NITRILE
108-95-2	PHENYL ALCOHOL
98-05-5	PHENYL ARSENIC ACID
135-88-6	PHENYL-BETA-NAPHTHYLAMINE
108-86-1	PHENYL BROMIDE
103-71-9	PHENYL CARBONIMIDE
1885-14-9	PHENYL CARBONOCHLORIDATE
108-90-7	PHENYL CHLORIDE
1885-14-9	PHENYL CHLOROCARBONATE
98-07-7	PHENYL CHLOROFORM
1885-14-9	PHENYL CHLOROFORMATE
532-27-4	PHENYL CHLOROMETHYL KETONE
100-47-0	PHENYL CYANIDE
696-28-6	PHENYL DICHLORARSINE
696-28-6	PHENYL DICHLOROARSINE (DOT)
101-84-8	PHENYL ETHER
101-84-8	PHENYL ETHER (ACGIH)
101-84-8	PHENYL ETHER VAPOR
8004-13-5	PHENYL ETHER-BIPHENYL MIXTURE VAPOR
8004-13-5	PHENYL ETHER-DIPHENYL MIXTURE
100-42-5	PHENYL ETHYLENE
462-06-6	PHENYL FLUORIDE
122-60-1	PHENYL GLYCIDYL ETHER
122-60-1	PHENYL GLYCIDYL ETHER (PGE)
108-95-2	PHENYL HYDRATE
71-43-2	PHENYL HYDRIDE
108-95-2	PHENYL HYDROXIDE
103-71-9	PHENYL ISOCYANATE
103-71-9	PHENYL ISOCYANATE (DOT)
622-44-6	PHENYL ISOCYANIDE, DICHLORIDE
622-44-6	PHENYL ISOCYANODICHLORIDE
119-61-9	PHENYL KETONE
108-98-5	PHENYL MERCAPTAN
108-98-5	PHENYL MERCAPTAN (ACGIH,DOT)
62-38-4	PHENYL MERCURIC ACETATE
100-66-3	PHENYL METHYL ETHER
98-86-2	PHENYL METHYL KETONE
101-84-8	PHENYL OXIDE
118-74-1	PHENYL PERCHLORYL
115-86-6	PHENYL PHOSPHATE ((PHO)3PO)
14684-25-4	PHENYL PHOSPHORODICHLORIDOTHIOITE
644-97-3	PHENYL PHOSPHORUS DICHLORIDE
644-97-3	PHENYL PHOSPHORUS DICHLORIDE (DOT)
98-13-5	PHENYL TRICHLOROSILANE
98-13-5	PHENYL TRICHLOROSILANE (DOT)
135-88-6	PHENYL-2-NAPHTHYLAMINE
103-80-0	PHENYLACETIC ACID CHLORIDE
140-29-4	PHENYLACETONITRILE
140-29-4	PHENYLACETONITRILE, LIQUID (DOT)
103-80-0	PHENYLACETYL CHLORIDE
103-80-0	PHENYLACETYL CHLORIDE (DOT)
148-82-3	PHENYLALANINE MUSTARD
62-53-3	PHENYLAMINE
142-04-1	PHENYLAMINE HYDROCHLORIDE
696-28-6	PHENYLARSENIC DICHLORIDE
696-28-6	PHENYLARSINEDICHLORIDE
98-05-5	PHENYLARSONIC ACID
136-40-3	PHENYLAZO TABLETS
92-52-4	PHENYLBENZENE
104-51-8	PHENYLBUTANE
64-10-8	PHENYLCARBAMIDE
103-71-9	PHENYLCARBIMIDE
100-51-6	PHENYLCARBINOL
622-44-6	PHENYLCARBONIMIDIC DICHLORIDE
65-85-0	PHENYLCARBOXYLIC ACID
622-44-6	PHENYLCARBYLAMINE CHLORIDE
622-44-6	PHENYLCARBYLAMINE CHLORIDE (DOT)
1885-14-9	PHENYLCHLOROFORMATE
1885-14-9	PHENYLCHLOROFORMATE (DOT)
827-52-1	PHENYLCYCLOHEXANE
696-28-6	PHENYLDICHLOROARSINE
644-97-3	PHENYLDICHLOROPHOSPHINE
123-01-3	PHENYLDODECAN (GERMAN)
108-45-2	PHENYLENEDIAMINE, META, SOLID (DOT)
100-41-4	PHENYLETHANE
122-98-5	PHENYLETHANOLAMINE
7568-93-6	PHENYLETHANOLAMINE
100-42-5	PHENYLETHENE

CAS No.	Chemical Name
92-59-1	PHENYLETHYLBENZYLAMINE
96-09-3	PHENYLETHYLENE OXIDE
92-50-2	PHENYLETHYLETHANOLAMINE
98-08-8	PHENYLFLUOROFORM
65-85-0	PHENYLFORMIC ACID
69-91-0	PHENYLGLYCINE ACID
100-63-0	PHENYLHYDRAZIN (GERMAN)
59-88-1	PHENYLHYDRAZIN HYDROCHLORID (GERMAN)
100-63-0	PHENYLHYDRAZINE
100-63-0	PHENYLHYDRAZINE (ACGIH,DOT)
59-88-1	PHENYLHYDRAZINE HYDROCHLORIDE
59-88-1	PHENYLHYDRAZINE MONOHYDROCHLORIDE
59-88-1	PHENYLHYDRAZINIUM CHLORIDE
108-95-2	PHENYLIC ACID
108-95-2	PHENYLIC ALCOHOL
622-44-6	PHENYLIMIDOCARBONYL CHLORIDE
622-44-6	PHENYLIMINOCARBONYL DICHLORIDE
622-44-6	PHENYLISONITRILE DICHLORIDE
555-54-4	PHENYLMAGNESIUM
62-38-4	PHENYLMERCURIACETATE
62-38-4	PHENYLMERCURIC ACETATE
62-38-4	PHENYLMERCURIC ACETATE (DOT)
100-57-2	PHENYLMERCURIC HYDROXIDE
55-68-5	PHENYLMERCURIC NITRATE
62-38-4	PHENYLMERCURY ACETATE
100-52-7	PHENYLMETHANAL
108-88-3	PHENYLMETHANE
100-53-8	PHENYLMETHANETHIOL
100-51-6	PHENYLMETHANOL
100-51-6	PHENYLMETHYL ALCOHOL
100-44-7	PHENYLMETHYL CHLORIDE
93-90-3	PHENYLMETHYL ETHANOLAMINE
100-53-8	PHENYLMETHYL MERCAPTAN
98-85-1	PHENYLMETHYLCARBINOL
149-74-6	PHENYLMETHYLDICHLOROSILANE
96-09-3	PHENYLOXIRANE
2122-46-5	PHENYLOXY
90-43-7	PHENYLPHENOL
638-21-1	PHENYLPHOSPHINE
644-97-3	PHENYLPHOSPHINE DICHLORIDE
2104-64-5	PHENYLPHOSPHONOTHIOATE, O-ETHYL-O-P-NITROPHENYL-
21609-90-5	PHENYLPHOSPHONOTHIOIC ACID O-(4-BROMO-2,5-DICHLORO-PHENYL) O-METHYL ESTER
2104-64-5	PHENYLPHOSPHONOTHIOIC ACID O-ETHYL O-P-NITROPHENYL ESTER
644-97-3	PHENYLPHOSPHONOUS ACID DICHLORIDE
644-97-3	PHENYLPHOSPHONOUS DICHLORIDE
62-38-4	PHENYLQUECKSILBERACETAT (GERMAN)
2097-19-0	PHENYLSILATRANE
98-13-5	PHENYLSILICON TRICHLORIDE
80-17-1	PHENYLSULFOHYDRAZIDE
98-09-9	PHENYLSULFONYL CHLORIDE
80-17-1	PHENYLSULFONYL HYDRAZIDE
103-85-5	PHENYLTHIOCARBAMIDE
108-98-5	PHENYLTHIOL
2104-64-5	PHENYLTHIOPHOSPHONATE DE O-ETHYLE ET O-4-NITROPHENYLE (FRENCH)
103-85-5	PHENYLTHIOUREA
98-07-7	PHENYLTRICHLOROMETHANE
98-06-6	PHENYLTRIMETHYLMETHANE
64-10-8	PHENYLUREA
64-10-8	PHENYLUREA PESTICIDE, LIQUID [FLAMMABLE LIQUID LABEL]
64-10-8	PHENYLUREA PESTICIDE, LIQUID [POISON B LABEL]
64-10-8	PHENYLUREA PESTICIDE, SOLID
57-41-0	PHENYTOIN
57-41-0	PHENYTOINE
92-84-2	PHENZEEN
7681-11-0	PHERAJOD
123-31-9	PHIAQUIN
1333-86-4	PHILBLACK
1333-86-4	PHILBLACK N 550
1333-86-4	PHILBLACK N 765
1333-86-4	PHILBLACK O
79-01-6	PHILEX
504-24-5	PHILLIPS 1861
1314-13-2	PHILOSOPHER'S WOOL
70-30-4	PHISOHEX
62-38-4	PHIX
82-68-8	PHOMASAN
298-02-2	PHORAT (GERMAN)
298-02-2	PHORATE
298-02-2	PHORATE (ACGIH)
298-02-2	PHORATE 10G
504-20-1	PHORONE
93-76-5	PHORTOX
7783-28-0	PHOS-CHEK 202A
7681-49-4	PHOS-FLUR
4104-14-7	PHOSACETIM
2310-17-0	PHOSALONE
4104-14-7	PHOSAZETIM
52-68-6	PHOSCHLOR
52-68-6	PHOSCHLOR R50
126-72-7	PHOSCON PE 60
126-72-7	PHOSCON UF-S
7786-34-7	PHOSDRIN
7786-34-7	PHOSFENE
78-30-8	PHOSFLEX 179C
115-86-6	PHOSFLEX TPP
947-02-4	PHOSFOLAN
20859-73-8	PHOSFUME
75-44-5	PHOSGEN
75-44-5	PHOSGEN (GERMAN)
75-44-5	PHOSGENE
75-44-5	PHOSGENE (ACGIH,DOT)
463-71-8	PHOSGENE, THIO-
56-38-2	PHOSKIL
12788-93-1	PHOSLEX A 4
732-11-6	PHOSMET
7784-30-7	PHOSPHALUGEL
60-51-5	PHOSPHAMID
60-51-5	PHOSPHAMIDE
13171-21-6	PHOSPHAMIDON
13171-21-6	PHOSPHAMIDONE
2778-04-3	PHOSPHATE 100
2636-26-2	PHOSPHATE DE DIMETHYLE ET DE 2,2-DICHLOROVINYLE (FRENCH)
141-66-2	PHOSPHATE DE DIMETHYLE ET DE 2-DIMETHYLCARBAMOYL 1-METHYL VINYLE (FRENCH)
7786-34-7	PHOSPHATE DE DIMETHYLE ET DE 2-METHOXYCARBONYL-1 METHYLVINYLE (FRENCH)
6923-22-4	PHOSPHATE DE DIMETHYLE ET DE 2-METHYLCARBAMOYL 1-METHYL VINYLE (FRENCH)
470-90-6	PHOSPHATE DE O,O-DIETHYLE ET DE O-2-CHLORO-1-(2,4-DI-CHLOROPHENYL) VINYLE (FRENCH)
1330-78-5	PHOSPHATE DE TRICRESYLE (FRENCH)
16940-81-1	PHOSPHATE(1-) HEXAFLUORO-, HYDROGEN
56-38-2	PHOSPHEMOL
7786-34-7	PHOSPHENE (FRENCH)
56-38-2	PHOSPHENOL
7803-51-2	PHOSPHINE
12768-82-0	PHOSPHINE
12768-82-0	PHOSPHINE (DYE)
7803-51-2	PHOSPHINE (PH3)
12768-82-0	PHOSPHINE 3R
115-26-4	PHOSPHINE OXIDE, BIS(DIMETHYLAMINO)FLUORO-
371-86-8	PHOSPHINE OXIDE, FLUOROBIS(ISOPROPYLAMINO)-
107-44-8	PHOSPHINE OXIDE, FLUOROISOPROPOXYMETHYL-
545-55-1	PHOSPHINE OXIDE, TRIS(1-AZIRIDINYL)-
12768-82-0	PHOSPHINE RN
52-24-4	PHOSPHINE SULFIDE, TRIS(1-AZIRIDINYL)-
676-83-5	PHOSPHINE, DICHLOROMETHYL-
644-97-3	PHOSPHINE, DICHLOROPHENYL-
638-21-1	PHOSPHINE, PHENYL-
7719-12-2	PHOSPHINE, TRICHLORO-
7803-65-8	PHOSPHINIC ACID, AMMONIUM SALT
1330-78-5	PHOSPHLEX 179A
947-02-4	PHOSPHOLAN
52-68-6	PHOSPHONIC ACID, (1-HYDROXY-2,2,2-TRICHLOROETHYL)-, DIMETHYL ESTER
52-68-6	PHOSPHONIC ACID, (2,2,2-TRICHLORO-1-HYDROXYETHYL)-, DIMETHYL ESTER
1809-19-4	PHOSPHONIC ACID, DIBUTYL ESTER
1031-47-6	PHOSPHONIC DIAMIDE, P-(5-AMINO-3-PHENYL-1H-1,2,4-TRIAZOL-1-YL)-N,N,N',N'-TETRAMETHYL-
676-97-1	PHOSPHONIC DICHLORIDE, METHYL-
944-22-9	PHOSPHONODITHIOIC ACID, ETHYL-, O-ETHYL S-PHENYL ESTER
107-44-8	PHOSPHONOFLUORIDIC ACID, METHYL-, 1-METHYLETHYL ESTER
107-44-8	PHOSPHONOFLUORIDIC ACID, METHYL-, ISOPROPYL ESTER

CAS No.	Chemical Name
2524-04-1	PHOSPHONOTHIOIC ACID, CHLORO-, O,O-DIETHYL ESTER
2524-03-0	PHOSPHONOTHIOIC ACID, CHLORO-, O,O-DIMETHYL ESTER
327-98-0	PHOSPHONOTHIOIC ACID, ETHYL-, O-ETHYL O-(2,4,5-TRICHLOROPHENYL) ESTER
2665-30-7	PHOSPHONOTHIOIC ACID, METHYL, O-(4-NITROPHENYL) O-PHENYL ESTER
2665-30-7	PHOSPHONOTHIOIC ACID, METHYL-, O-(P-NITROPHENYL) O-PHENYL ESTER
2703-13-1	PHOSPHONOTHIOIC ACID, METHYL-, O-ETHYL O-(4-METHYLTHIO)
50782-69-9	PHOSPHONOTHIOIC ACID, METHYL-, S-(2-(BIS(METHYL ETHYL)-AMINO)ETHYL)O-ETHYL ETHER
2104-64-5	PHOSPHONOTHIOIC ACID, PHENYL-, ETHYL P-NITROPHENYL ESTER
2104-64-5	PHOSPHONOTHIOIC ACID, PHENYL-, O-ETHYL O-(4-NITROPHENYL) ESTER
2104-64-5	PHOSPHONOTHIOIC ACID, PHENYL-, O-ETHYL O-(P-NITROPHENYL) ESTER
676-98-2	PHOSPHONOTHIOIC DICHLORIDE, METHYL-
676-98-2	PHOSPHONOTHIOIC DICHLORIDE, METHYL-, ANHYDROUS
676-83-5	PHOSPHONOUS DICHLORIDE, METHYL-
644-97-3	PHOSPHONOUS DICHLORIDE, PHENYL-
2778-04-3	PHOSPHOPYRON
2778-04-3	PHOSPHOPYRONE
545-55-1	PHOSPHORAMIDE, N,N',N"-TRIETHYLENE-
22224-92-6	PHOSPHORAMIDIC ACID, (1-METHYLETHYL)-, ETHYL (3-METHYL-4-(METHYLTHIO)PHENYL) ESTER
950-10-7	PHOSPHORAMIDIC ACID, (4-METHYL-1,3-DITHIOLAN-2-YLIDENE)-, DIETHYL ESTER
21548-32-3	PHOSPHORAMIDIC ACID, 1,3-DITHIETAN-2-YLIDENE-, DIETHYL ESTER
947-02-4	PHOSPHORAMIDIC ACID, 1,3-DITHIOLAN-2-YLIDENE-, DIETHYL ESTER
22224-92-6	PHOSPHORAMIDIC ACID, ISOPROPYL-, 4-(METHYLTHIO)-M-TOLYL ETHYL ESTER
22224-92-6	PHOSPHORAMIDIC ACID, ISOPROPYL-, ETHYL 4-(METHYLTHIO)-M-TOLYL ESTER
299-86-5	PHOSPHORAMIDIC ACID, METHYL-, 2-CHLORO-4-(1,1-DIMETHYLETHYL)PHENYL METHYL ESTER
299-86-5	PHOSPHORAMIDIC ACID, METHYL-, 4-TERT-BUTYL-2-CHLOROPHENYL METHYL ESTER
77-81-6	PHOSPHORAMIDOCYANIDIC ACID, DIMETHYL-, ETHYL ESTER
4104-14-7	PHOSPHORAMIDOTHIOIC ACID, (1-IMINOETHYL)-, O,O-BIS(4-CHLOROPHENYL) ESTER
4104-14-7	PHOSPHORAMIDOTHIOIC ACID, ACETIMIDOYL-, O,O-BIS(P-CHLOROPHENYL) ESTER
10265-92-6	PHOSPHORAMIDOTHIOIC ACID, O,S-DIMETHYL ESTER
7789-69-7	PHOSPHORANE, PENTABROMO-
10026-13-8	PHOSPHORANE, PENTACHLORO-
7647-19-0	PHOSPHORANE, PENTAFLUORO-
7723-14-0	PHOSPHORE BLANC (FRENCH)
10026-13-8	PHOSPHORE(PENTACHLORURE DE) (FRENCH)
7719-12-2	PHOSPHORE(TRICHLORURE DE) (FRENCH)
7664-38-2	PHOSPHORIC ACID
7664-38-2	PHOSPHORIC ACID (ACGIH,DOT)
141-66-2	PHOSPHORIC ACID 3-(DIMETHYLAMINO)-1-METHYL-3-OXO-1-PROPENYL DIMETHYL ESTER
7783-28-0	PHOSPHORIC ACID AMMONIUM SALT (1:2)
298-07-7	PHOSPHORIC ACID BIS(ETHYLHEXYL) ESTER
814-49-3	PHOSPHORIC ACID CHLORIDE DIETHYL ESTER
27215-10-7	PHOSPHORIC ACID DIISOOCTYL ESTER
10039-32-4	PHOSPHORIC ACID DISODIUM DODECAHYDRATE
680-31-9	PHOSPHORIC ACID HEXAMETHYLTRIAMIDE
7601-54-9	PHOSPHORIC ACID SODIUM SALT (1:3)
126-73-8	PHOSPHORIC ACID TRIBUTYL ESTER
1330-78-5	PHOSPHORIC ACID TRICRESYL ESTER
545-55-1	PHOSPHORIC ACID TRIETHYLENE IMIDE
545-55-1	PHOSPHORIC ACID TRIETHYLENEIMINE
545-55-1	PHOSPHORIC ACID TRIETHYLENEIMINE (DOT)
7786-34-7	PHOSPHORIC ACID, (1-METHOXYCARBOXYPROPEN-2-YL) DIMETHYL ESTER
300-76-5	PHOSPHORIC ACID, 1,2-DIBROMO-2,2-DICHLOROETHYL DIMETHYL ESTER
62-73-7	PHOSPHORIC ACID, 2,2-DICHLOROETHENYL DIMETHYL ESTER
2636-26-2	PHOSPHORIC ACID, 2,2-DICHLOROETHENYL DIMETHYL ESTER
62-73-7	PHOSPHORIC ACID, 2,2-DICHLOROVINYL DIMETHYL ESTER
2636-26-2	PHOSPHORIC ACID, 2,2-DICHLOROVINYL DIMETHYL ESTER
470-90-6	PHOSPHORIC ACID, 2-CHLORO-1-(2,4-DICHLOROPHENYL)-ETHENYL DIETHYL ESTER
470-90-6	PHOSPHORIC ACID, 2-CHLORO-1-(2,4-DICHLOROPHENYL)VINYL DIETHYL ESTER
13171-21-6	PHOSPHORIC ACID, 2-CHLORO-3-(DIETHYLAMINO)-1-METHYL-3-OXO-1-PROPENYL DIMETHYL ESTER
141-66-2	PHOSPHORIC ACID, 3-(DIMETHYLAMINO)-1-METHYL-3-OXO-1-PROPENYL DIMETHYL ESTER, (E)-
7784-30-7	PHOSPHORIC ACID, ALUMINUM SALT (1:1)
7784-30-7	PHOSPHORIC ACID, ALUMINUM SALT (1:1), (SOLUTION)
13598-26-0	PHOSPHORIC ACID, BERYLLIUM SALT
35089-00-0	PHOSPHORIC ACID, BERYLLIUM SALT
13598-15-7	PHOSPHORIC ACID, BERYLLIUM SALT (1:1)
13598-26-0	PHOSPHORIC ACID, BERYLLIUM SALT (2:3)
298-07-7	PHOSPHORIC ACID, BIS(2-ETHYLHEXYL) ESTER
12788-93-1	PHOSPHORIC ACID, BUTYL ESTER
10103-46-5	PHOSPHORIC ACID, CALCIUM SALT
7783-28-0	PHOSPHORIC ACID, DIAMMONIUM SALT
107-66-4	PHOSPHORIC ACID, DIBUTYL ESTER
27215-10-7	PHOSPHORIC ACID, DIISOOCTYL ESTER
6923-22-4	PHOSPHORIC ACID, DIMETHYL 1-METHYL-3-(METHYLAMINO)-3-OXO-1-PROPENYL ESTER, (E)-
141-66-2	PHOSPHORIC ACID, DIMETHYL 1-METHYL-N,N-(DIMETHYL-AMINO)-3-OXO-1-PROPENYL ESTER, (E)- (9CI)
3254-63-5	PHOSPHORIC ACID, DIMETHYL 4-(METHYLTHIO) PHENYL ESTER
6923-22-4	PHOSPHORIC ACID, DIMETHYL ESTER, ESTER WITH (E)-3-HYDROXY-N-METHYLCROTONAMIDE
13171-21-6	PHOSPHORIC ACID, DIMETHYL ESTER, ESTER WITH 2-CHLORO-N,N-DIETHYL-3-HYDROXYCROTONAMIDE
141-66-2	PHOSPHORIC ACID, DIMETHYL ESTER, ESTER WITH 3-HYDROXY-N,N-DIMETHYLCROTONAMIDE, (E)-
6923-22-4	PHOSPHORIC ACID, DIMETHYL ESTER, ESTER WITH 3-HYDROXY-N-METHYLCROTONAMIDE, (E)-
141-66-2	PHOSPHORIC ACID, DIMETHYL ESTER, ESTER WITH CIS-3-HYDROXY-N,N-DIMETHYLCROTONAMIDE
6923-22-4	PHOSPHORIC ACID, DIMETHYL ESTER, ESTER WITH CIS-3-HYDROXY-N-METHYLCROTONAMIDE
7786-34-7	PHOSPHORIC ACID, DIMETHYL ESTER, ESTER WITH METHYL 3-HYDROXYCROTONATE
3254-63-5	PHOSPHORIC ACID, DIMETHYL P-(METHYLTHIO)PHENYL ESTER
7558-79-4	PHOSPHORIC ACID, DISODIUM SALT
10039-32-4	PHOSPHORIC ACID, DISODIUM SALT, DODECAHYDRATE
10140-65-5	PHOSPHORIC ACID, DISODIUM SALT, HYDRATE
1623-24-1	PHOSPHORIC ACID, ISOPROPYL ESTER
7446-27-7	PHOSPHORIC ACID, LEAD(2+) SALT (2:3)
7664-38-2	PHOSPHORIC ACID, LIQUID (DOT)
2235-25-8	PHOSPHORIC ACID, MERCURY COMPLEX
107-44-8	PHOSPHORIC ACID, METHYLFLUORO-, ISOPROPYL ESTER
1623-24-1	PHOSPHORIC ACID, MONO(1-METHYLETHYL) ESTER
18351-85-4	PHOSPHORIC ACID, MONO(2-METHYLPHENYL) ESTER
1623-24-1	PHOSPHORIC ACID, MONOISOPROPYL ESTER
7664-38-2	PHOSPHORIC ACID, SOLID (DOT)
78-30-8	PHOSPHORIC ACID, TRI-O-TOLYL ESTER
115-86-6	PHOSPHORIC ACID, TRIPHENYL ESTER
78-30-8	PHOSPHORIC ACID, TRIS(2-METHYLPHENYL) ESTER
1330-78-5	PHOSPHORIC ACID, TRIS(METHYLPHENYL) ESTER
7601-54-9	PHOSPHORIC ACID, TRISODIUM SALT
10361-89-4	PHOSPHORIC ACID, TRISODIUM SALT, DECAHYDRATE
10101-89-0	PHOSPHORIC ACID, TRISODIUM SALT, DODECAHYDRATE
1330-78-5	PHOSPHORIC ACID, TRITOLYL ESTER
1314-56-3	PHOSPHORIC ANHYDRIDE
1314-56-3	PHOSPHORIC ANHYDRIDE (DOT)
7789-69-7	PHOSPHORIC BROMIDE
10025-87-3	PHOSPHORIC CHLORIDE
10026-13-8	PHOSPHORIC CHLORIDE
680-31-9	PHOSPHORIC HEXAMETHYLTRIAMIDE
1314-56-3	PHOSPHORIC PENTOXIDE
1314-80-3	PHOSPHORIC SULFIDE
680-31-9	PHOSPHORIC TRIAMIDE, HEXAMETHYL-
545-55-1	PHOSPHORIC TRIAMIDE, N,N',N"-TRI-1,2-ETHANEDIYL-
545-55-1	PHOSPHORIC TRIAMIDE, N,N',N"-TRIETHYLENE-
10025-87-3	PHOSPHORIC TRICHLORIDE
680-31-9	PHOSPHORIC TRIS(DIMETHYLAMIDE)
814-49-3	PHOSPHOROCHLORIDIC ACID, DIETHYL ESTER
2524-03-0	PHOSPHOROCHLORIDOTHIOIC ACID, O,O-DIMETHYL ESTER
371-86-8	PHOSPHORODI(ISOPROPYLAMIDIC) FLUORIDE
371-86-8	PHOSPHORODIAMIDIC FLUORIDE, N,N'-DIISOPROPYL-
115-26-4	PHOSPHORODIAMIDIC FLUORIDE, TETRAMETHYL-
1498-51-7	PHOSPHORODICHLORIDIC ACID, ETHYL ESTER
14684-25-4	PHOSPHORODICHLORIDOTHIOUS ACID, PHENYL ESTER
14684-25-4	PHOSPHORODICHLORIDOTHIOUS ACID, PHENYL-
13779-41-4	PHOSPHORODIFLUORIDIC ACID

CAS No.	Chemical Name
13071-79-9	PHOSPHORODITHIOIC ACID S-(((1,1-DIMETHYLETHYL)-THIO)METHYL) O,O-DIETHYL ESTER
13071-79-9	PHOSPHORODITHIOIC ACID S-((TERT-BUTYLTHIO)METHYL) O,O-DIETHYL ESTER
2540-82-1	PHOSPHORODITHIOIC ACID, 0,0-DIMETHYL ESTER, N-FORMYL-2-MERCAPTO-N-METHYLACETAMIDE S-ESTER
563-12-2	PHOSPHORODITHIOIC ACID, O,O-DIETHYL ESTER, S,S-DIESTER WITH METHANEDITHIOL
2642-71-9	PHOSPHORODITHIOIC ACID, O,O-DIETHYL ESTER, S-ESTER WITH 3-(MERCAPTOMETHYL)-1,2,3-BENZOTRIAZIN-4(3H)-
2275-18-5	PHOSPHORODITHIOIC ACID, O,O-DIETHYL ESTER, S-ESTER WITH N-ISOPROPYL-2-MERCAPTOACETAMIDE
2497-07-6	PHOSPHORODITHIOIC ACID, O,O-DIETHYL S-((ETHYLSULFINYL)-ETHYL) ESTER
298-04-4	PHOSPHORODITHIOIC ACID, O,O-DIETHYL S-(2-(ETHYLTHIO)-ETHYL) ESTER
298-02-2	PHOSPHORODITHIOIC ACID, O,O-DIETHYL S-(ETHYLTHIO)-METHYL ESTER
2275-18-5	PHOSPHORODITHIOIC ACID, O,O-DIETHYL S-(ISOPROPYLCAR-BAMOYLMETHYL) ESTER
2642-71-9	PHOSPHORODITHIOIC ACID, O,O-DIETHYL S-[(4-OXO-1,2,3-BENZOTRIAZIN-3(4H)-YL)METHYL] ESTER
2497-07-6	PHOSPHORODITHIOIC ACID, O,O-DIETHYL S-[2-(ETHYLSULFINYL)-ETHYL] ESTER
2275-18-5	PHOSPHORODITHIOIC ACID, O,O-DIETHYL S-[2-[(1-METHYL-ETHYL)AMINO]-2-OXOETHYL] ESTER
13071-79-9	PHOSPHORODITHIOIC ACID, O,O-DIETHYL-S-(((1,1-DIMETHYL-ETHYL)THIO)METHYL)-ESTER
60-51-5	PHOSPHORODITHIOIC ACID, O,O-DIMETHYL ESTER, S-ESTER WITH 2-MERCAPTO-N-METHYLACETAMIDE
950-37-8	PHOSPHORODITHIOIC ACID, O,O-DIMETHYL ESTER, S-ESTER WITH 4-(MERCAPTOMETHYL)-2-METHOXY-DELTA(SUP 2)-1
950-37-8	PHOSPHORODITHIOIC ACID, O,O-DIMETHYL ESTER, S-ESTER WITH 4-(MERCAPTOMETHYL)-2-METHOXY-DELTA2-1,3,4-T
732-11-6	PHOSPHORODITHIOIC ACID, O,O-DIMETHYL ESTER, S-ESTER WITH N-(MERCAPTOMETHYL)PHTHALIMIDE
2540-82-1	PHOSPHORODITHIOIC ACID, O,O-DIMETHYL ESTER, S-ESTER WITH N-FORMYL-2-MERCAPTO-N-METHYLACETAMIDE
640-15-3	PHOSPHORODITHIOIC ACID, O,O-DIMETHYL S-(2-ETHYL-THIO)ETHYL ESTER
60-51-5	PHOSPHORODITHIOIC ACID, O,O-DIMETHYL S-(METHYLCAR-BAMOYLMETHYL) ESTER
86-50-0	PHOSPHORODITHIOIC ACID, O,O-DIMETHYL S-[(4-OXO-1,2,3-BENZOTRIAZIN-3(4H)-YL) METHYL] ESTER
60-51-5	PHOSPHORODITHIOIC ACID, O,O-DIMETHYL S-[2-(METHYL-AMINO)-2-OXOETHYL]ESTER
60-51-5	PHOSPHORODITHIOIC ACID, O,O-DIMETHYL-S-(2-(METHYL-AMINO)-2-OXOETHYL) ESTER (9CI)
35400-43-2	PHOSPHORODITHIOIC ACID, O-ETHYL O-[4-(METHYL-THIO)PHENYL] S-PROPYL ESTER
13194-48-4	PHOSPHORODITHIOIC ACID, O-ETHYL S,S-DIPROPYL ESTER
78-34-2	PHOSPHORODITHIOIC ACID, S,S'-1,4-DIOXANE-2,3-DIYL O,O,O',O'-TETRAETHYL ESTER
78-34-2	PHOSPHORODITHIOIC ACID, S,S'-1,4-DIOXANE-2,3-DIYL O,O,O',O'-TETRAETHYL ESTER (9CI)
563-12-2	PHOSPHORODITHIOIC ACID, S,S'-METHYLENE O,O,O',O'-TETRA-ETHYL ESTER
78-34-2	PHOSPHORODITHIOIC ACID, S,S'-P-DIOXANE-2,3-DIYL O,O,O',O'-TETRAETHYL ESTER
2275-14-1	PHOSPHORODITHIOIC ACID, S-(((2,5-DICHLOROPHENYL)-THIO)METHYL) O,O-DIETHYL ESTER
3735-23-7	PHOSPHORODITHIOIC ACID, S-(((2,5-DICHLOROPHENYL)-THIO)METHYL)O,O-DIMETHYL ESTER
786-19-6	PHOSPHORODITHIOIC ACID, S-(((P-CHLOROPHENYL)-THIO)METHYL) O,O-DIETHYL ESTER
732-11-6	PHOSPHORODITHIOIC ACID, S-((1,3-DIHYDRO-1,3-DIOXO-ISOIN-DOL-2-YL)METHYL) O,O-DIMETHYL ESTER
640-15-3	PHOSPHORODITHIOIC ACID, S-(2-(ETHYLTHIO)ETHYL) O,O-DIMETHYL ESTER
2540-82-1	PHOSPHORODITHIOIC ACID, S-(2-(FORMYLMETHYLAMINO)-2-OXOETHYL) O,O-DIMETHYL ESTER
24934-91-6	PHOSPHORODITHIOIC ACID, S-(CHLOROMETHYL) O,O-DIETHYL ESTER
732-11-6	PHOSPHORODITHIOIC ACID, S-[(1,3-DIHYDRO-1,3-DIOXO-2H-ISOINDOL-2-YL)METHYL] O,O-DIMETHYL ESTER
950-37-8	PHOSPHORODITHIOIC ACID, S-[(5-METHOXY-2-OXO-1,3,4-THIADIAZOL-3(2H)-YL)METHYL] O,O-DIMETHYL
950-37-8	PHOSPHORODITHIOIC ACID, S-[(5-METHOXY-2-OXO-1,3,4-THIADIAZOL-3(2H)-YL)METHYL] O,O-DIMETHYL ESTER
2540-82-1	PHOSPHORODITHIOIC ACID, S-[2-(FORMYLMETHYLAMINO)-2-OXOETHYL] O,O-DIMETHYL ESTER
786-19-6	PHOSPHORODITHIOIC ACID, S-[[(4-CHLOROPHENYL)-THIO]METHYL] O,O-DIETHYL ESTER
298-04-4	PHOSPHORODITHIONIC ACID, S-2-(ETHYLTHIO)ETHYL-O,O-DIETHYL ESTER
13537-32-1	PHOSPHOROFLUORIDIC ACID
333-41-5	PHOSPHOROTHIOATE, O,O-DIETHYL O-6-(2-ISOPROPYL-4-METHYL-PYRIMIDYL)
52-24-4	PHOSPHOROTHIOIC ACID TRIETHYLENETRIAMIDE
55-38-9	PHOSPHOROTHIOIC ACID, DIMETHYL [4-(METHYLTHIO)-M-TOLYL] ESTER
3383-96-8	PHOSPHOROTHIOIC ACID, O,O'-(THIODI-4,1-PHENYLENE) O,O,O',O'-TETRAMETHYL ESTER
3383-96-8	PHOSPHOROTHIOIC ACID, O,O'-(THIODI-P-PHENYLENE) O,O,O',O'-TETRAMETHYL ESTER
333-41-5	PHOSPHOROTHIOIC ACID, O,O-DIETHYL 2-ISOPROPYL-6-METHYL-4-PYRIMIDINYL ESTER
21923-23-9	PHOSPHOROTHIOIC ACID, O,O-DIETHYL O-((2,5-DICHLORO-4-METHYLTHIO)PHENYL) ESTER
24017-47-8	PHOSPHOROTHIOIC ACID, O,O-DIETHYL O-(1-PHENYL-1,2,4-TRIAZOLYL) ESTER
24017-47-8	PHOSPHOROTHIOIC ACID, O,O-DIETHYL O-(1-PHENYL-1H-1,2,4-TRIAZOL-3-YL) ESTER
23505-41-1	PHOSPHOROTHIOIC ACID, O,O-DIETHYL O-(2-(DIETHYLAMINO)-6-METHYL-4-PYRIMIDINYL) ESTER
8065-48-3	PHOSPHOROTHIOIC ACID, O,O-DIETHYL O-(2-(ETHYLTHIO)ETHYL) ESTER, MIXED WITH O,O-DIETHYL S-(2-(ETHYLTH
333-41-5	PHOSPHOROTHIOIC ACID, O,O-DIETHYL O-(2-ISOPROPYL-6-METHYL-4-PYRIMIDINYL) ESTER
2921-88-2	PHOSPHOROTHIOIC ACID, O,O-DIETHYL O-(3,5,6-TRICHLORO-2-PYRIDINYL) ESTER
2921-88-2	PHOSPHOROTHIOIC ACID, O,O-DIETHYL O-(3,5,6-TRICHLORO-2-PYRIDYL) ESTER
56-38-2	PHOSPHOROTHIOIC ACID, O,O-DIETHYL O-(4-NITROPHENYL) ESTER
56-38-2	PHOSPHOROTHIOIC ACID, O,O-DIETHYL O-(P-NITROPHENYL) ESTER
56-38-2	PHOSPHOROTHIOIC ACID; O,O-DIETHYL O-(P-NITROPHENYL) ESTER (MIXTURE)
56-38-2	PHOSPHOROTHIOIC ACID, O,O-DIETHYL O-(P-NITROPHENYL)-ESTER, MIXED WITH COMPRESSED GAS
297-97-2	PHOSPHOROTHIOIC ACID, O,O-DIETHYL O-2-PYRAZINYL ESTER
297-97-2	PHOSPHOROTHIOIC ACID, O,O-DIETHYL O-PYRAZINYL ESTER
115-90-2	PHOSPHOROTHIOIC ACID, O,O-DIETHYL O-[4-(METHYLSULFINYL)PHENYL] ESTER
333-41-5	PHOSPHOROTHIOIC ACID, O,O-DIETHYL O-[6-METHYL-2-(1-METHYLETHYL)-4-PYRIMIDINYL] ESTER
115-90-2	PHOSPHOROTHIOIC ACID, O,O-DIETHYL O-[P-(METHYLSULFINYL)PHENYL] ESTER
56-72-4	PHOSPHOROTHIOIC ACID, O,O-DIETHYL ESTER, O-ESTER WITH 3-CHLORO-7-HYDROXY-4-METHYLCOUMARIN
2636-26-2	PHOSPHOROTHIOIC ACID, O,O-DIMETHYL ESTER, O-ESTER WITH P-HYDROXYBENZONITRILE
2778-04-3	PHOSPHOROTHIOIC ACID, O,O-DIMETHYL ESTER, S-ESTER WITH 2-(MERCAPTOMETHYL)-5-METHOXY-4H-PYRAN-4-ONE
299-84-3	PHOSPHOROTHIOIC ACID, O,O-DIMETHYL O-(2,4,5-TRICHLORO-PHENYL) ESTER
122-14-5	PHOSPHOROTHIOIC ACID, O,O-DIMETHYL O-(3-METHYL-4-NITROPHENYL) ESTER
122-14-5	PHOSPHOROTHIOIC ACID, O,O-DIMETHYL O-(4-NITRO-M-TOLYL) ESTER
298-00-0	PHOSPHOROTHIOIC ACID, O,O-DIMETHYL O-(4-NITROPHENYL) ESTER
298-00-0	PHOSPHOROTHIOIC ACID, O,O-DIMETHYL O-(P-NITROPHENYL) ESTER
298-00-0	PHOSPHOROTHIOIC ACID, O,O-DIMETHYL O-(P-NITROPHENYL) ESTER (DRY MIXTURE)
55-38-9	PHOSPHOROTHIOIC ACID, O,O-DIMETHYL O-[3-METHYL-4-(METHYLTHIO)PHENYL] ESTER
55-38-9	PHOSPHOROTHIOIC ACID, O,O-DIMETHYL O-[4-(METHYLTHIO)-M-TOLYL] ESTER
919-86-8	PHOSPHOROTHIOIC ACID, O,O-DIMETHYL S-(2-(ETHYL-THIO)ETHYL) ESTER
21923-23-9	PHOSPHOROTHIOIC ACID, O-(2,5-DICHLORO-4-(METHYL-THIO)PHENYL O,O-DIETHYL ESTER
56-72-4	PHOSPHOROTHIOIC ACID, O-(3-CHLORO-4-METHYL-2-OXO-2H-1-BENZOPYRAN-7-YL) O,O-DIETHYL ESTER
2636-26-2	PHOSPHOROTHIOIC ACID, O-(4-CYANOPHENYL) O,ODIMETHYL-

CAS No.	Chemical Name
	ESTER
23505-41-1	PHOSPHOROTHIOIC ACID, O-[2-(DIETHYLAMINO)-6-METHYL-4-PYRIMIDINYL] O,O-DIETHYL ESTER
8022-00-2	PHOSPHOROTHIOIC ACID, O-[2-(ETHYLTHIO)ETHYL] O,O-DIMETHYL ESTER, MIXT. WITH S-[2-(ETHYLTHIO)ETHYL] O
78-53-5	PHOSPHOROTHIOIC ACID, S-(2-(DIETHYLAMINO)ETHYL) O,O-DIETHYL ESTER
919-86-8	PHOSPHOROTHIOIC ACID, S-(2-(ETHYLTHIO)ETHYL) O,O-DIMETHYL ESTER
3734-97-2	PHOSPHOROTHIOIC ACID, S-(2-DIETHYLAMINO)ETHYL) O,O-DIETHYL ESTER, OXALATE (1:1)
2778-04-3	PHOSPHOROTHIOIC ACID, S-[(5-METHOXY-4-OXO-4H-PYRAN-2-YL)METHYL] O,O-DIMETHYL ESTER
3734-97-2	PHOSPHOROTHIOIC ACID, S-[2-(DIETHYLAMINO)ETHYL] O,O-DIETHYL ESTER, ETHANEDIOATE (1:1)
3734-97-2	PHOSPHOROTHIOIC ACID, S-[2-(DIETHYLAMINO)ETHYL] O,O-DIETHYL ESTER, HYDROGEN OXALATE
3734-97-2	PHOSPHOROTHIOIC ACID, S-[2-(DIETHYLAMINO)ETHYL] O,O-DIETHYL ESTER, OXALATE
3734-97-2	PHOSPHOROTHIOIC ACID, S-[2-(DIETHYLAMINO)ETHYL] O,O-DIETHYL ESTER, OXALATE (1:1)
52-24-4	PHOSPHOROTHIOIC TRIAMIDE, N,N',N''-TRI-1,2-ETHANEDIYL-
3982-91-0	PHOSPHOROTHIOIC TRICHLORIDE
3982-91-0	PHOSPHOROTHIONIC TRICHLORIDE
7723-14-0	PHOSPHOROUS (WHITE)
109-47-7	PHOSPHOROUS ACID, DIBUTYL ESTER
1809-19-4	PHOSPHOROUS ACID, DIBUTYL ESTER
13598-36-2	PHOSPHOROUS ACID, ORTHO
102-85-2	PHOSPHOROUS ACID, TRIBUTYL ESTER
122-52-1	PHOSPHOROUS ACID, TRIETHYL ESTER
121-45-9	PHOSPHOROUS ACID, TRIMETHYL ESTER
7789-60-8	PHOSPHOROUS BROMIDE
7789-60-8	PHOSPHOROUS BROMIDE (DOT)
7719-12-2	PHOSPHOROUS CHLORIDE
7783-55-3	PHOSPHOROUS FLUORIDE
7789-59-5	PHOSPHOROUS OXYBROMIDE
10026-13-8	PHOSPHOROUS PENTACHLORIDE
1314-56-3	PHOSPHOROUS PENTOXIDE
1314-85-8	PHOSPHOROUS SESQUISULFIDE
3982-91-0	PHOSPHOROUS SULFOCHLORIDE
3982-91-0	PHOSPHOROUS THIOCHLORIDE
7789-60-8	PHOSPHOROUS TRIBROMIDE
7719-12-2	PHOSPHOROUS TRICHLORIDE
3982-91-0	PHOSPHOROUS TRICHLORIDE SULFIDE
7783-55-3	PHOSPHOROUS TRIFLUORIDE
7723-14-0	PHOSPHOROUS YELLOW
10025-87-3	PHOSPHOROXYCHLORIDE
10025-87-3	PHOSPHOROXYTRICHLORIDE
10026-13-8	PHOSPHORPENTACHLORID (GERMAN)
7664-38-2	PHOSPHORSAEURELOESUNGEN (GERMAN)
7719-12-2	PHOSPHORTRICHLORID (GERMAN)
7723-14-0	PHOSPHORUS
7723-14-0	PHOSPHORUS (DOT)
1314-85-8	PHOSPHORUS (III) SULFIDE (IV)
7723-14-0	PHOSPHORUS (RED)
7723-14-0	PHOSPHORUS (WHITE)
7723-14-0	PHOSPHORUS (YELLOW OR WHITE)
7723-14-0	PHOSPHORUS (YELLOW)
7723-14-0	PHOSPHORUS (YELLOW) (ACGIH)
7723-14-0	PHOSPHORUS :AMORPHOUS
7789-69-7	PHOSPHORUS BROMIDE
7789-60-8	PHOSPHORUS BROMIDE (DOT)
7789-60-8	PHOSPHORUS BROMIDE (PBr3)
7789-69-7	PHOSPHORUS BROMIDE (PBr5)
7719-12-2	PHOSPHORUS CHLORIDE
7719-12-2	PHOSPHORUS CHLORIDE (Cl6P2)
7719-12-2	PHOSPHORUS CHLORIDE (DOT)
7719-12-2	PHOSPHORUS CHLORIDE (PCl3)
10026-13-8	PHOSPHORUS CHLORIDE (PCl5)
10025-87-3	PHOSPHORUS CHLORIDE OXIDE (PCl3O)
7723-14-0	PHOSPHORUS ELEMENT
7783-55-3	PHOSPHORUS FLUORIDE
7783-55-3	PHOSPHORUS FLUORIDE (PF3)
7647-19-0	PHOSPHORUS FLUORIDE (PF5)
12037-82-0	PHOSPHORUS HEPTASULFIDE
12037-82-0	PHOSPHORUS HEPTASULFIDE (DOT)
12037-82-0	PHOSPHORUS HEPTASULPHIDE, FREE FROM YELLOW OR WHITE PHOSPHORUS (DOT)
10025-87-3	PHOSPHORUS MONOXIDE TRICHLORIDE
1314-56-3	PHOSPHORUS OXIDE
1314-24-5	PHOSPHORUS OXIDE (P2O3)
1314-56-3	PHOSPHORUS OXIDE (P2O5)
10025-87-3	PHOSPHORUS OXIDE TRICHLORIDE
7789-59-5	PHOSPHORUS OXYBROMIDE
7789-59-5	PHOSPHORUS OXYBROMIDE (DOT)
7789-59-5	PHOSPHORUS OXYBROMIDE, MOLTEN
7789-59-5	PHOSPHORUS OXYBROMIDE, MOLTEN (DOT)
7789-59-5	PHOSPHORUS OXYBROMIDE, SOLID (DOT)
10025-87-3	PHOSPHORUS OXYCHLORIDE
10025-87-3	PHOSPHORUS OXYCHLORIDE (ACGIH,DOT)
7789-59-5	PHOSPHORUS OXYTRIBROMIDE
10025-87-3	PHOSPHORUS OXYTRICHLORIDE
7789-69-7	PHOSPHORUS PENTABROMIDE
7789-69-7	PHOSPHORUS PENTABROMIDE (DOT)
10026-13-8	PHOSPHORUS PENTACHLORIDE
10026-13-8	PHOSPHORUS PENTACHLORIDE (ACGIH)
10026-13-8	PHOSPHORUS PENTACHLORIDE, SOLID
10026-13-8	PHOSPHORUS PENTACHLORIDE, SOLID (DOT)
7647-19-0	PHOSPHORUS PENTAFLUORIDE
7647-19-0	PHOSPHORUS PENTAFLUORIDE (DOT)
1314-56-3	PHOSPHORUS PENTAOXIDE
1314-80-3	PHOSPHORUS PENTASULFIDE
1314-80-3	PHOSPHORUS PENTASULFIDE (ACGIH,DOT)
1314-80-3	PHOSPHORUS PENTASULFIDE (P2S5)
1314-80-3	PHOSPHORUS PENTASULFIDE (P4S10)
1314-80-3	PHOSPHORUS PENTASULPHIDE, FREE FROM YELLOW OR WHITE PHOSPHORUS (DOT)
1314-56-3	PHOSPHORUS PENTOXIDE
1314-56-3	PHOSPHORUS PENTOXIDE (DOT)
7789-69-7	PHOSPHORUS PERBROMIDE
10026-13-8	PHOSPHORUS PERCHLORIDE
1314-80-3	PHOSPHORUS PERSULFIDE
1314-85-8	PHOSPHORUS SESQUISULFIDE
1314-85-8	PHOSPHORUS SESQUISULFIDE (DOT)
1314-85-8	PHOSPHORUS SESQUISULPHIDE, FREE FROM YELLOW OR WHITE PHOSPHORUS (DOT)
1314-80-3	PHOSPHORUS SULFIDE
12165-69-4	PHOSPHORUS SULFIDE (P2S3)
1314-80-3	PHOSPHORUS SULFIDE (P2S5)
1314-85-8	PHOSPHORUS SULFIDE (P4S3)
12037-82-0	PHOSPHORUS SULFIDE (P4S7)
3982-91-0	PHOSPHORUS SULFOCHLORIDE
3982-91-0	PHOSPHORUS THIOCHLORIDE
3982-91-0	PHOSPHORUS THIOTRICHLORIDE
7789-60-8	PHOSPHORUS TRIBROMIDE
7789-60-8	PHOSPHORUS TRIBROMIDE (DOT)
7719-12-2	PHOSPHORUS TRICHLORIDE
7719-12-2	PHOSPHORUS TRICHLORIDE (ACGIH,DOT)
10025-87-3	PHOSPHORUS TRICHLORIDE OXIDE
3982-91-0	PHOSPHORUS TRICHLORIDE SULFIDE
7783-55-3	PHOSPHORUS TRIFLUORIDE
7803-51-2	PHOSPHORUS TRIHYDRIDE
1314-24-5	PHOSPHORUS TRIOXIDE
1314-24-5	PHOSPHORUS TRIOXIDE (DOT)
12165-69-4	PHOSPHORUS TRISULFIDE
12165-69-4	PHOSPHORUS TRISULFIDE (DOT)
12165-69-4	PHOSPHORUS TRISULPHIDE, FREE FROM YELLOW OR WHITE PHOSPHORUS (DOT)
7723-14-0	PHOSPHORUS WHITE, DRY (DOT)
7723-14-0	PHOSPHORUS WHITE, IN WATER (DOT)
7723-14-0	PHOSPHORUS YELLOW, DRY (DOT)
7723-14-0	PHOSPHORUS YELLOW, IN WATER (DOT)
7789-69-7	PHOSPHORUS(V) BROMIDE
10026-13-8	PHOSPHORUS(V) CHLORIDE
1314-56-3	PHOSPHORUS(V) OXIDE
7723-14-0	PHOSPHORUS, AMORPHOUS, RED
7723-14-0	PHOSPHORUS, AMORPHOUS, RED (DOT)
7723-14-0	PHOSPHORUS, WHITE OR YELLOW
7723-14-0	PHOSPHORUS, WHITE, MOLTEN
7723-14-0	PHOSPHORUS, WHITE, MOLTEN (DOT)
7723-14-0	PHOSPHORUS-31
7789-59-5	PHOSPHORYL BROMIDE
10025-87-3	PHOSPHORYL CHLORIDE
10025-87-3	PHOSPHORYL CHLORIDE (DOT)
680-31-9	PHOSPHORYL HEXAMETHYLTRIAMIDE
7789-59-5	PHOSPHORYL TRIBROMIDE
10025-87-3	PHOSPHORYL TRICHLORIDE
56-38-2	PHOSPHOSTIGMINE

CAS No.	Chemical Name
7722-88-5	PHOSPHOTEX
121-75-5	PHOSPHOTHION
563-12-2	PHOSPHOTOX
563-12-2	PHOSPHOTOX E
20770-41-6	PHOSPHURE DE POTASSIUM (FRENCH)
24167-76-8	PHOSPHURE DE SODIUM (FRENCH)
12504-13-1	PHOSPHURE DE STRONTIUM (FRENCH)
1314-84-7	PHOSPHURE DE ZINC (FRENCH)
20859-73-8	PHOSPHURES D'ALUMIUM (FRENCH)
20859-73-8	PHOSTOXIN
20859-73-8	PHOSTOXIN A
21609-90-5	PHOSVEL
1314-84-7	PHOSVIN
2636-26-2	PHOSVIT
1305-99-3	PHOTOPHOR
92-52-4	PHPH
85-44-9	PHTHALANDIONE
85-44-9	PHTHALANHYDRIDE
120-80-9	PHTHALHYDROQUINONE
88-99-3	PHTHALIC ACID
85-44-9	PHTHALIC ACID ANHYDRIDE
117-81-7	PHTHALIC ACID DIOCTYL ESTER
117-84-0	PHTHALIC ACID DIOCTYL ESTER
131-11-3	PHTHALIC ACID METHYL ESTER
85-68-7	PHTHALIC ACID, BENZYL BUTYL ESTER
117-81-7	PHTHALIC ACID, BIS(2-ETHYLHEXYL) ESTER
117-84-0	PHTHALIC ACID, BIS(2-ETHYLHEXYL) ESTER
131-17-9	PHTHALIC ACID, DIALLYL ESTER
84-74-2	PHTHALIC ACID, DIBUTYL ESTER
84-66-2	PHTHALIC ACID, DIETHYL ESTER
3648-21-3	PHTHALIC ACID, DIHEPTYL ESTER
26761-40-0	PHTHALIC ACID, DIISODECYL ESTER
131-11-3	PHTHALIC ACID, DIMETHYL ESTER
117-84-0	PHTHALIC ACID, DIOCTYL ESTER
84-72-0	PHTHALIC ACID, ETHYL ESTER, ESTER WITH ETHYL GLYCOLATE
85-71-2	PHTHALIC ACID, METHYL ESTER, ESTER WITH ETHYL GLYCOLATE
85-44-9	PHTHALIC ANHYDRIDE
85-44-9	PHTHALIC ANHYDRIDE (ACGIH)
85-43-8	PHTHALIC ANHYDRIDE, 1,2,3,6-TETRAHYDRO-
85-44-9	PHTHALIC ANHYDRIDE, SOLID OR MOLTEN (DOT)
25251-51-8	PHTHALIDE, 3-(TERT-BUTYLDIOXY)-3-PHENYL-
25251-51-8	PHTHALIDE, 3-(TERT-BUTYLPEROXY)-3-PHENYL-
732-11-6	PHTHALIMIDE, N-(MERCAPTOMETHYL)-, S-ESTER WITH O,O-DIMETHYL PHOSPHORODITHIOATE
732-11-6	PHTHALIMIDO O,O-DIMETHYL PHOSPHORODITHIOATE
732-11-6	PHTHALIMIDOMETHYL O,O-DIMETHYL PHOSPHORODITHIOATE
84-66-2	PHTHALOL
732-11-6	PHTHALOPHOS
85-44-9	PHTHALSAEUREANHYDRID (GERMAN)
131-11-3	PHTHALSAEUREDIMETHYLESTER (GERMAN)
2163-80-6	PHYBAN
2163-80-6	PHYBAN HC
117-80-6	PHYGON
117-80-6	PHYGON PASTE
117-80-6	PHYGON SEED PROTECTANT
117-80-6	PHYGON XL
84-80-0	PHYLLOCHINON (GERMAN)
84-80-0	PHYLLOQUINONE
84-80-0	PHYLLOQUINONE K1
57-47-6	PHYSOSTIGMINE
57-64-7	PHYSOSTIGMINE SALICYLATE
57-64-7	PHYSOSTIGMINE, MONOSALICYLATE
57-64-7	PHYSOSTIGMINE, SALICYLATE (1:1)
57-47-6	PHYSOSTOL
57-64-7	PHYSOSTOL SALICYLATE
75-60-5	PHYTAR
75-60-5	PHYTAR 138
59164-68-0	PHYTAR 560
7758-98-7	PHYTO-BORDEAUX
84-80-0	PHYTOMENADIONE
84-80-0	PHYTONADIONE
327-98-0	PHYTOSOL
84-80-0	PHYTYLMENADIONE
78-67-1	PIANOFOR AN
57-83-0	PIAPONON
76-06-2	PIC-CLOR
76-06-2	PICFUME
1918-02-1	PICLORAM

CAS No.	Chemical Name
109-06-8	PICOLINE
1333-41-1	PICOLINE
1333-41-1	PICOLINE (DOT)
1333-41-1	PICOLINE [COMBUSTIBLE LIQUID LABEL]
1333-41-1	PICOLINE [FLAMMABLE LIQUID LABEL]
1333-41-1	PICOLINES
1918-02-1	PICOLINIC ACID, 4-AMINO-3,5,6-TRICHLORO-
146-84-9	PICRAGOL
88-89-1	PICRAL
831-52-7	PICRAMIC ACID, SODIUM SALT
63868-82-6	PICRAMIC ACID, ZIRCONIUM SALT (WET)
131-74-8	PICRATE OF AMMONIA (DOT)
131-74-8	PICRATOL
88-89-1	PICRIC ACID
131-74-8	PICRIC ACID, AMMONIUM SALT
131-74-8	PICRIC ACID, NH4 DERIV
146-84-9	PICRIC ACID, SILVER(1+) SALT
88-89-1	PICRIC ACID, WET WITH NOT LESS THAN 10% WATER, OVER 25 POUNDS (DOT)
88-89-1	PICRIC ACID, WETTED WITH 10% TO 30% WATER
88-89-1	PICRIC ACID, WETTED WITH AT LEAST 10% WATER (DOT)
88-89-1	PICRIC ACID, WETTED WITH AT LEAST 30% WATER (DOT)
76-06-2	PICRIDE
556-88-7	PICRITE
556-88-7	PICRITE (PROPELLANT)
556-88-7	PICRITE, WET, WITH ≥20% WATER
88-89-1	PICRONITRIC ACID
124-87-8	PICROTIN, COMPD WITH PICROTOXININ (1:1)
146-84-9	PICROTOL
124-87-8	PICROTOXIN
124-87-8	PICROTOXINE
124-87-8	PICROTOXININ, COMPD WITH PICROTIN (1:1)
2217-06-3	PICRYL SULFIDE
2217-06-3	PICRYL SULFIDE, WETTED WITH AT LEAST 10% WATER
479-45-8	PICRYLMETHYLNITRAMINE
479-45-8	PICRYLNITROMETHYLAMINE
82-66-6	PID
10026-13-8	PIECIOCHLOREK FOSFORU (POLISH)
57-24-9	PIED PIPER MOUSE SEED
94-75-7	PIELIK
1309-37-1	PIGDEX 100
127-85-5	PIGLET PRO-GEN V
1333-86-4	PIGMENT BLACK 7
3761-53-3	PIGMENT PONCEAU R
1309-37-1	PIGMENT RED 101
6923-22-4	PILLARDRIN
10265-92-6	PILLARON
1910-2-5	PILLARQUAT
1910-2-5	PILLARXONE
25155-30-0	PILOT HD-90
25155-30-0	PILOT SF-40
25155-30-0	PILOT SF-40B
25155-30-0	PILOT SF-40FG
25155-30-0	PILOT SF-60
25155-30-0	PILOT SF-96
25155-30-0	PILOT SP-60
27323-41-7	PILOT TS 60
108-94-1	PIMELIC KETONE
108-94-1	PIMELIN KETONE
2104-64-5	PIN
107-41-5	PINAKON
28324-52-9	PINANE HYDROPEROXIDE
85873-97-8	PINANE HYDROPEROXIDE
85873-97-8	PINANE HYDROPEROXIDE (DOT)
85873-97-8	PINANE HYDROPEROXIDE SOLUTION, WITH >45% PEROXIDE
28324-52-9	PINANE HYDROPEROXIDE SOLUTION, WITH >45% PEROXIDE
28324-52-9	PINANE HYDROPEROXIDE, TECHNICALLY PURE (DOT)
85873-97-8	PINANYL HYDROPEROXIDE
28324-52-9	PINANYL HYDROPEROXIDE
28324-52-9	PINANYL HYDROPEROXIDE, TECHNICAL PURE (DOT)
83-26-1	PINDON (DUTCH)
83-26-1	PINDONE
83-26-1	PINDONE (ACGIH)
83-26-1	PINDONE, LIQUID (DOT)
83-26-1	PINDONE, SOLID (DOT)
9005-90-7	PINE GUM
8002-09-3	PINE OIL
8002-09-3	PINE OIL (DOT)
8002-09-3	PINE OIL, SYNTHETIC

CAS No.	Chemical Name
1330-16-1	PINENE
1330-16-1	PINENE (DOT)
78-00-2	PIOMBO TETRA-ETILE (ITALIAN)
75-74-1	PIOMBO TETRA-METILE (ITALIAN)
110-85-0	PIPERAZIDINE
110-85-0	PIPERAZIN (GERMAN)
110-85-0	PIPERAZINE
110-85-0	PIPERAZINE (DOT)
142-64-3	PIPERAZINE DIHYDROCHLORIDE
142-64-3	PIPERAZINE HYDROCHLORIDE
142-64-3	PIPERAZINE WORMER PREMIX
140-31-8	PIPERAZINE, 1-(2-AMINOETHYL)-
109-01-3	PIPERAZINE, 1-METHYL-
110-85-0	PIPERAZINE, ANHYDROUS
142-64-3	PIPERAZINE, DIHYDROCHLORIDE
110-89-4	PIPERIDIN (GERMAN)
110-89-4	PIPERIDINE
110-89-4	PIPERIDINE (DOT)
766-09-6	PIPERIDINE, 1-ETHYL-
626-67-5	PIPERIDINE, 1-METHYL-
100-75-4	PIPERIDINE, 1-NITROSO-
104-89-2	PIPERIDINE, 5-ETHYL-2-METHYL-
120-62-7	PIPERONYL SULFOXIDE
110-85-0	PIPERSOL
504-60-9	PIPERYLENE
504-60-9	PIPERYLENE CONCENTRATE
5281-13-0	PIPROTAL
136-40-3	PIRID
110-86-1	PIRIDINA (ITALIAN)
23103-98-2	PIRIMICARB
23505-41-1	PIRIMIFOS-ETHYL
23505-41-1	PIRIMIPHOS ETHYL
3689-24-5	PIROFOS
110-86-1	PIRYDYNA (POLISH)
112-53-8	PISOL
1912-24-9	PITEZIN
117-81-7	PITTSBURGH PX-138
117-84-0	PITTSBURGH PX-138
83-26-1	PIVACIN
83-26-1	PIVAL
83-26-1	PIVALDION (ITALIAN)
83-26-1	PIVALDIONE (FRENCH)
75-98-9	PIVALIC ACID
3282-30-2	PIVALIC ACID CHLORIDE
1955-45-9	PIVALOLACTONE
3282-30-2	PIVALOYL CHLORIDE
3282-30-2	PIVALOYL CHLORIDE (8CI)
3282-30-2	PIVALOYL CHLORIDE (DOT)
83-26-1	PIVALYL
3282-30-2	PIVALYL CHLORIDE
83-26-1	PIVALYL VALONE
83-26-1	PIVALYLINDAN-1,3-DIONE
83-26-1	PIVALYLINDANDIONE
83-26-1	PIVALYN
8007-45-2	PIXALBOL
105-60-2	PK 4
105-60-2	PKA
7790-98-9	PKHA
87-86-5	PKhF
131-52-2	PKHFN
84-66-2	PLACIDOL E
4726-14-1	PLANAVIN
4726-14-1	PLANAVIN 75
1929-73-3	PLANOTOX
3689-24-5	PLANT DITHIO AEROSOL
6923-22-4	PLANTDRIN
3689-24-5	PLANTFUME 103 SMOKE GENERATOR
94-75-7	PLANTGARD
4726-14-1	PLANUIN
105-60-2	PLASKIN 8200
105-60-2	PLASKON 8201
105-60-2	PLASKON 8201HS
105-60-2	PLASKON 8202C
105-60-2	PLASKON 8205
105-60-2	PLASKON 8207
105-60-2	PLASKON 8252
105-60-2	PLASKON XP 607
26499-65-0	PLASTER OF PARIS
12001-29-5	PLASTIBEST 20

CAS No.	Chemical Name
103-23-1	PLASTOMOLL DOA
123-79-5	PLASTOMOLL DOA
14808-60-7	PLASTORIT
15663-27-1	PLATIDIAM
16919-58-7	PLATINATE(2-), HEXACHLORO-, DIAMMONIUM
16919-58-7	PLATINATE(2-), HEXACHLORO-, DIAMMONIUM, (OC-6-11)-
16919-58-7	PLATINATE(2-), HEXACHLORO-, DIAMMONIUM, (OC-6-11)- (9CI)
16941-12-1	PLATINATE(2-), HEXACHLORO-, DIHYDROGEN
16941-12-1	PLATINATE(2-), HEXACHLORO-, DIHYDROGEN, (OC-6-11)-
16919-58-7	PLATINIC AMMONIUM CHLORIDE
16941-12-1	PLATINIC CHLORIDE
15663-27-1	PLATINOL
117-84-0	PLATINOL AH
117-84-0	PLATINOL DOP
10025-65-7	PLATINOUS CHLORIDE
10025-65-7	PLATINOUS DICHLORIDE
7440-06-4	PLATINUM
13454-96-1	PLATINUM (IV) CHLORIDE
7440-06-4	PLATINUM BLACK
10025-65-7	PLATINUM CHLORIDE
10025-65-7	PLATINUM CHLORIDE ($PtCl_2$)
13454-96-1	PLATINUM CHLORIDE ($PtCl_4$)
13454-96-1	PLATINUM CHLORIDE ($PtCl_4$), (SP-4-1)-
10025-65-7	PLATINUM DICHLORIDE
13454-96-1	PLATINUM TETRACHLORIDE
10025-65-7	PLATINUM(II) CHLORIDE
10025-65-7	PLATINUM(II) DICHLORIDE
13454-96-1	PLATINUM(IV) CHLORIDE
13454-96-1	PLATINUM(IV) TETRACHLORIDE
15663-27-1	PLATINUM, DIAMMINEDICHLORO-, (SP-4-2)-
15663-27-1	PLATINUM, DIAMMINEDICHLORO-, CIS-
13121-70-5	PLICTRAN
106-99-0	PLIOLITE
25339-57-5	PLIOLITE
58-89-9	PLK
7783-46-2	PLOMB FLUORURE (FRENCH)
7782-42-5	PLUMBAGO
7782-42-5	PLUMBAGO (GRAPHITE)
78-00-2	PLUMBANE, TETRAETHYL-
75-74-1	PLUMBANE, TETRAMETHYL-
546-67-8	PLUMBIC ACETATE
1309-60-0	PLUMBIC OXIDE
301-04-2	PLUMBOUS ACETATE
7758-95-4	PLUMBOUS CHLORIDE
7758-97-6	PLUMBOUS CHROMATE
7783-46-2	PLUMBOUS FLUORIDE
10101-63-0	PLUMBOUS IODIDE
10099-74-8	PLUMBOUS NITRATE
18256-98-9	PLUMBOUS NITRATE
1317-36-8	PLUMBOUS OXIDE
1314-87-0	PLUMBOUS SULFIDE
25322-69-4	PLURACOL 1010
25322-69-4	PLURACOL 2010
25322-69-4	PLURACOL P 1010
25322-69-4	PLURACOL P 2010
25322-69-4	PLURACOL P 410
25322-69-4	PLURACOL P 710
25322-69-4	PLURIOL P
25322-69-4	PLURIOL P 2000
25322-69-4	PLURIOL P 900
13121-70-5	PLYCTRAN
1333-86-4	PM 100
1333-86-4	PM 100A
1333-86-4	PM 100KRSZ
1333-86-4	PM 100V
1333-86-4	PM 105K
1333-86-4	PM 136.5
1333-86-4	PM 15
1333-86-4	PM 30V
1333-86-4	PM 50
1333-86-4	PM 50 (CARBON BLACK)
1333-86-4	PM 70
1333-86-4	PM 70 (CARBON BLACK)
1333-86-4	PM 75
1333-86-4	PM 90E
62-38-4	PMA
62-38-4	PMA (FUNGICIDE)
62-38-4	PMA 220
62-38-4	PMAC

CAS No.	Chemical Name
62-38-4	PMACETATE
62-38-4	PMAS
1333-86-4	PME 100V
1333-86-4	PME 70V
1333-86-4	PME 80V
150-76-5	PMF (ANTIOXIDANT)
1333-86-4	PMG 33
1333-86-4	PMN 130
1333-86-4	PMN 130N
1333-86-4	PMO 130
732-11-6	PMP
732-11-6	PMP (PESTICIDE)
622-97-9	PMS
1333-86-4	PMTK 90
8020-83-5	PN 6K
100-01-6	PNA
29757-24-2	PNA
100-00-5	PNCB
99-55-8	PNOT
99-99-0	PNT
7681-49-4	POINT TWO
87-86-5	POL NU
137-26-8	POL-THIURAM
52-68-6	POLFOSCHLOR
9002-84-0	POLITEF
50-78-2	POLOPIRYNA
7704-34-9	POLSULKOL EXTRA
25322-69-4	POLY(1,2-EPOXYPROPANE)
11099-06-2	POLY(ETHYL SILICATE)
9002-84-0	POLY(ETHYLENE TETRAFLUORIDE)
9016-87-9	POLY(METHYLENE PHENYLENE ISOCYANATE)
37286-64-9	POLY(OXY(METHYL-1,2-ETHANEDIYL)), ALPHA-METHYL-OMEGA-HYDROXY-
9016-45-9	POLY(OXY-1,2-ETHANEDIYL), ALPHA-(NONYLPHENYL)-OMEGA-HYDROXY-
9016-45-9	POLY(OXYETHYLENE) NONYLPHENOL ETHER
9016-45-9	POLY(OXYETHYLENE) NONYLPHENYL ETHER
9002-81-7	POLY(OXYMETHYLENE)
9015-98-9	POLY(OXYMETHYLENE) GLYCOL
9015-98-9	POLY(OXYMETHYLENE), ALPHA-HYDRO-OMEGA-HYDROXY-
37286-64-9	POLY(OXYPROPYLENE) MONOMETHYL ETHER
9016-87-9	POLY(PHENYLENEMETHYLENE ISOCYANATE)
25322-69-4	POLY(PROPYLENE OXIDE)
9002-89-5	POLY(VINYL ALCOHOL)
111-90-0	POLY- SOLV
25322-69-4	POLY-G 1020P
25322-69-4	POLY-G 20-112
25322-69-4	POLY-G 20-265
9002-81-7	POLY-S-TRIOXANE
111-96-6	POLY-SOLV
112-34-5	POLY-SOLV DB
111-90-0	POLY-SOLV DE
111-77-3	POLY-SOLV DM
111-76-2	POLY-SOLV EB
110-80-5	POLY-SOLV EE
111-15-9	POLY-SOLV EE ACETATE
109-86-4	POLY-SOLV EM
112-50-5	POLY-SOLV TE
9016-45-9	POLY-TERGENT B
9016-45-9	POLY-TERGENT B 150
9016-45-9	POLY-TERGENT B 300
9016-45-9	POLY-TERGENT B 350
78-67-1	POLY-ZOLE AZDN
108-62-3	POLYACETALDEHYDE
9002-91-9	POLYACETALDEHYDE
105-60-2	POLYAMIDE PK 4
67774-32-7	POLYBROMINATED BIPHENYLS
59536-65-1	POLYBROMINATED BIPHENYLS
59536-65-1	POLYBROMINATED BIPHENYLS (PBBs)
79620-93-2	POLYCHLOR
79620-93-2	POLYCHLOR AGRICULTURAL FUNGICIDES
8001-35-2	POLYCHLORCAMPHENE
1336-36-3	POLYCHLORINATED BIPHENYLS
1336-36-3	POLYCHLORINATED BIPHENYLS (DOT)
8001-35-2	POLYCHLORINATED CAMPHENES
1336-36-3	POLYCHLOROBIPHENYL
8001-35-2	POLYCHLOROCAMPHENE
8029-29-6	POLYCHLORODICYCLOPENTADIENE (CHLORINE CONTENT 60-62% OR 62-64%)

CAS No.	Chemical Name
84-74-2	POLYCIZER DBP
9002-89-5	POLYDESIS
9016-45-9	POLYETHOXYLATED NONYLPHENOL
9016-45-9	POLYETHYLENE GLYCOL 450 NONYL PHENYL ETHER
9016-45-9	POLYETHYLENE GLYCOL MONO(NONYLPHENOL) ETHER
9016-45-9	POLYETHYLENE GLYCOL MONO(NONYLPHENYL) ETHER
9016-45-9	POLYETHYLENE GLYCOL NONYLPHENOL ETHER
9016-45-9	POLYETHYLENE GLYCOL NONYLPHENYL ETHER
9016-45-9	POLYETHYLENEOXIDE MONO(NONYLPHENYL) ETHER
9002-84-0	POLYFENE
9004-66-4	POLYFER
9002-84-0	POLYFLON
9002-84-0	POLYFLON D 1
9002-84-0	POLYFLON EK 1108GY-A
9002-84-0	POLYFLON EK 1700
9002-84-0	POLYFLON EK 1883GB
9002-84-0	POLYFLON EK 4105GN
9002-84-0	POLYFLON EK 4108GY
9002-84-0	POLYFLON F 101
9002-84-0	POLYFLON F 103
9002-84-0	POLYFLON F 104
9002-84-0	POLYFLON M 1
9002-84-0	POLYFLON M 12
9002-84-0	POLYFLON M 12A
9002-84-0	POLYFLON M 21
9002-84-0	POLYFLON PA 10L
9002-84-0	POLYFLON PA 5L
9002-81-7	POLYFORMALDEHYDE
25322-69-4	POLYGLYCOL P 400
25322-69-4	POLYGLYCOL P 4000
25322-69-4	POLYGLYCOL TYPE P 1200
25322-69-4	POLYGLYCOL TYPE P 2000
25322-69-4	POLYGLYCOL TYPE P 250
25322-69-4	POLYGLYCOL TYPE P 3000
25322-69-4	POLYGLYCOL TYPE P 400
25322-69-4	POLYGLYCOL TYPE P 750
7758-29-4	POLYGON
25322-69-4	POLYHARDENER D 200
9002-84-0	POLYLUBE J 12
25322-69-4	POLYMER 2
9016-87-9	POLYMETHYL POLYPHENYL POLYISOCYANATE
9015-98-9	POLYMETHYLENE GLYCOL
9016-87-9	POLYMETHYLENE POLYPHENYL ISOCYANATE
9016-87-9	POLYMETHYLENE POLYPHENYL POLYISOCYANATE
9016-87-9	POLYMETHYLENE POLYPHENYLENE ISOCYANATE
9016-87-9	POLYMETHYLENE POLYPHENYLENE ISOCYANATE POLYMER
9016-87-9	POLYMETHYLENE POLYPHENYLENE POLYISOCYANATE
9002-84-0	POLYMIST F 5
9002-84-0	POLYMIST F 5A
9002-81-7	POLYOXYALKYLENES, POLYOXYMETHYLENES
25322-69-4	POLYOXYALKYLENES, POLYPROPYLENE GLYCOL
9016-45-9	POLYOXYETHYLATED NONYLPHENOL
9016-45-9	POLYOXYETHYLENE (15) NONYL PHENYL ETHER
9016-45-9	POLYOXYETHYLENE (20) NONYL PHENYL ETHER
9016-45-9	POLYOXYETHYLENE (9) NONYL PHENYL ETHER
50-00-0	POLYOXYMETHYLENE GLYCOLS
9002-84-0	POLYPERFLUOROETHYLENE
9016-87-9	POLYPHENYLENE POLYMETHYLENE POLYISOCYANATE
9016-87-9	POLYPHENYLPOLYMETHYLENE POLYISOCYANATE
25322-69-4	POLYPROPYLENE 5820
25322-69-4	POLYPROPYLENE GLYCOL
37286-64-9	POLYPROPYLENE GLYCOL METHYL ETHER
37286-64-9	POLYPROPYLENE GLYCOL MONOMETHYLETHER
25322-69-4	POLYPROPYLENGLYKOL (CZECH)
137-26-8	POLYRAM ULTRA
9002-89-5	POLYSIZER 173
9016-45-9	POLYSTEP F 6
9016-45-9	POLYSTEP F 8
14807-96-6	POLYTAL 4641
14807-96-6	POLYTAL 4725
9002-84-0	POLYTEF
9002-84-0	POLYTETRAFLUOROETHENE
9002-84-0	POLYTETRAFLUOROETHYLENE
9002-81-7	POLYTRIOXANE
9002-89-5	POLYVINOL
9002-89-5	POLYVINYL ALCOHOL
9002-89-5	POLYVIOL
9002-89-5	POLYVIOL M 05/290
9002-89-5	POLYVIOL M 13/140

CAS No.	Chemical Name
9002-89-5	POLYVIOL MO 5/140
9002-89-5	POLYVIOL V 03/20
9002-89-5	POLYVIOL W 25/140
9002-89-5	POLYVIOL W 28/20
9002-89-5	POLYVIOL W 40/140
53467-11-1	POLY[OXY(METHYL-1,2-ETHANEDIYL)], ALPHA-[(2,4-DICHLOROPHENOXY)ACETYL]-OMEGA-BUTOXY-
25322-69-4	POLY[OXY(METHYL-1,2-ETHANEDIYL)], ALPHA-HYDRO-OMEGA-HYDROXY-
37286-64-9	POLY[OXY(METHYL-1,2-ETHANEDIYL)], ALPHA-METHYL-OMEGA-HYDROXY-
137-26-8	POMARSOL
137-26-8	POMARSOL FORTE
137-26-8	POMASOL
63-25-2	POMEX
3761-53-3	PONCEAU 2R
3761-53-3	PONCEAU 2R EXTRA A EXPORT
3761-53-3	PONCEAU 2RL
3761-53-3	PONCEAU 2RX
3564-09-8	PONCEAU 3R
3564-09-8	PONCEAU 3R LAKE
3564-09-8	PONCEAU 3R SODIUM SALT
3564-09-8	PONCEAU 3RN
3761-53-3	PONCEAU BNA
3761-53-3	PONCEAU G
3761-53-3	PONCEAU MX
3761-53-3	PONCEAU PXM
3761-53-3	PONCEAU R
3761-53-3	PONCEAU RED R
3761-53-3	PONCEAU RR
3761-53-3	PONCEAU RR TYPE 8019
3761-53-3	PONCEAU RS
3761-53-3	PONCEAU XYLIDINE
1937-37-7	PONTAMINE BLACK E
1937-37-7	PONTAMINE BLACK EBN
72-57-1	PONTAMINE BLUE 3BX
2602-46-2	PONTAMINE BLUE BB
95-80-7	PONTAMINE DEVELOPER TN
16071-86-6	PONTAMINE FAST BROWN BRL
16071-86-6	PONTAMINE FAST BROWN NP
1344-28-1	PORAMINAR
7631-86-9	PORASIL
1332-58-7	PORCELAIN CLAY
9002-84-0	POREFLON FP 500
9002-84-0	POREX P 1000
78-67-1	POROFOR 57
80-17-1	POROFOR BSH
101-25-7	POROFOR CHKHC-18
80-17-1	POROFOR CHKHZ 9
101-25-7	POROFOR DNO/F
78-67-1	POROFOR N
80-17-1	POROFOR-BSH-PULVER
101-25-7	POROPHOR B
78-67-1	POROPHOR N
1332-58-7	POSALIN ACTIVE
7631-86-9	POSITIVE SOL 130M
7631-86-9	POSITIVE SOL 232
3811-04-9	POTASH CHLORATE (DOT)
3811-04-9	POTASSIO (CHLORATO DI) (ITALIAN)
7440-09-7	POTASSIUM
3811-04-9	POTASSIUM (CHLORATE DE) (FRENCH)
7784-41-0	POTASSIUM ACID ARSENATE
7789-29-9	POTASSIUM ACID FLUORIDE
127-95-7	POTASSIUM ACID OXALATE
7646-93-7	POTASSIUM ACID SULFATE
7773-03-7	POTASSIUM ACID SULFITE
11135-81-2	POTASSIUM ALLOY, NONBASE, K,NA
28300-74-5	POTASSIUM ANTIMONY TARTRATE
28300-74-5	POTASSIUM ANTIMONYL D-TARTRATE
28300-74-5	POTASSIUM ANTIMONYL TARTRATE
506-61-6	POTASSIUM ARGENTOCYANIDE (KAG(CN)2)
7784-41-0	POTASSIUM ARSENATE
7784-41-0	POTASSIUM ARSENATE (KH2AsO4)
7784-41-0	POTASSIUM ARSENATE, MONOBASIC
7784-41-0	POTASSIUM ARSENATE, SOLID
7784-41-0	POTASSIUM ARSENATE, SOLID (DOT)
10124-50-2	POTASSIUM ARSENITE
10124-50-2	POTASSIUM ARSENITE, SOLID
10124-50-2	POTASSIUM ARSENITE, SOLID (DOT)
7440-09-7	POTASSIUM ATOM
7778-50-9	POTASSIUM BICHROMATE
7789-29-9	POTASSIUM BIFLUORIDE
7789-29-9	POTASSIUM BIFLUORIDE (KHF2)
7789-29-9	POTASSIUM BIFLUORIDE, SOLID (DOT)
7646-93-7	POTASSIUM BIFLUORIDE, SOLUTION
127-95-7	POTASSIUM BINOXALATE
7646-93-7	POTASSIUM BISULFATE
7773-03-7	POTASSIUM BISULFITE
7773-03-7	POTASSIUM BISULFITE SOLUTION
7646-93-7	POTASSIUM BISULPHATE
13762-51-1	POTASSIUM BOROHYDRATE
13762-51-1	POTASSIUM BOROHYDRIDE
13762-51-1	POTASSIUM BOROHYDRIDE
13762-51-1	POTASSIUM BOROHYDRIDE (DOT)
13762-51-1	POTASSIUM BOROHYDRIDE (KBH4)
7758-01-2	POTASSIUM BROMATE
7758-01-2	POTASSIUM BROMATE (DOT)
3811-04-9	POTASSIUM CHLORATE
3811-04-9	POTASSIUM CHLORATE (DOT)
3811-04-9	POTASSIUM CHLORATE SOLUTION
3811-04-9	POTASSIUM CHLORATE, AQUEOUS SOLUTION (DOT)
7778-66-7	POTASSIUM CHLORIDE OXIDE (KClO)
7789-00-6	POTASSIUM CHROMATE
7789-00-6	POTASSIUM CHROMATE (DOT)
7789-00-6	POTASSIUM CHROMATE (K2(CrO4))
7789-00-6	POTASSIUM CHROMATE (VI)
143-18-0	POTASSIUM CIS-9-OCTADECENOIC ACID
13682-73-0	POTASSIUM COPPER(I) CYANIDE
13682-73-0	POTASSIUM CUPROCYANIDE
151-50-8	POTASSIUM CYANIDE
151-50-8	POTASSIUM CYANIDE (ACGIH)
151-50-8	POTASSIUM CYANIDE (K(CN))
151-50-8	POTASSIUM CYANIDE SOLUTION
151-50-8	POTASSIUM CYANIDE SOLUTION (DOT)
151-50-8	POTASSIUM CYANIDE, SOLID
151-50-8	POTASSIUM CYANIDE, SOLID (DOT)
23745-86-0	POTASSIUM CYMONATE
2244-21-5	POTASSIUM DICHLORO-S-TRIAZINETRIONE
2244-21-5	POTASSIUM DICHLORO-S-TRIAZINETRIONE, DRY, CONTAINING MORE THAN 39% AVAILABLE CHLORINE (DOT)
2244-21-5	POTASSIUM DICHLOROCYANURATE
2244-21-5	POTASSIUM DICHLOROISOCYANURATE
7778-50-9	POTASSIUM DICHROMATE
7778-50-9	POTASSIUM DICHROMATE (DOT)
7778-50-9	POTASSIUM DICHROMATE (K2(Cr2O7))
7778-50-9	POTASSIUM DICHROMATE (VI)
506-61-6	POTASSIUM DICYANOARGENTATE
506-61-6	POTASSIUM DICYANOARGENTATE(1-)
506-61-6	POTASSIUM DICYANOARGENTATE(I)
13682-73-0	POTASSIUM DICYANOCUPRATE(1-)
13682-73-0	POTASSIUM DICYANOCUPRATE(I)
7784-41-0	POTASSIUM DIHYDROGEN ARSENATE
7784-41-0	POTASSIUM DIHYDROGEN ARSENATE (KH2AsO4)
12030-88-5	POTASSIUM DIOXIDE
14293-73-3	POTASSIUM DITHIONITE
140-89-6	POTASSIUM ETHYLXANTHATE
140-89-6	POTASSIUM ETHYLXANTHOGENATE
7789-29-9	POTASSIUM FLUORIDE
7789-23-3	POTASSIUM FLUORIDE
7789-23-3	POTASSIUM FLUORIDE (DOT)
7789-29-9	POTASSIUM FLUORIDE (K(HF2))
7789-23-3	POTASSIUM FLUORIDE (KF)
7789-23-3	POTASSIUM FLUORIDE SOLUTION
7789-23-3	POTASSIUM FLUORIDE SOLUTION (DOT)
23745-86-0	POTASSIUM FLUOROACETATE
23745-86-0	POTASSIUM FLUOROACETATE (DOT)
16871-90-2	POTASSIUM FLUOROSILICATE
16871-90-2	POTASSIUM FLUOROSILICATE (K2SIF6)
16871-90-2	POTASSIUM FLUOROSILICATE, SOLID
16923-95-8	POTASSIUM FLUOROZIRCONATE
16923-95-8	POTASSIUM FLUOROZIRCONATE (K2ZrF6)
7789-23-3	POTASSIUM FLUORURE (FRENCH)
16871-90-2	POTASSIUM FLUOSILICATE
16923-95-8	POTASSIUM FLUOZIRCONATE
16921-30-5	POTASSIUM HEXACHLOROPLATINATE
16871-90-2	POTASSIUM HEXAFLUOROSILICATE
16871-90-2	POTASSIUM HEXAFLUOROSILICATE (K2SiF6)
16923-95-8	POTASSIUM HEXAFLUOROZIRCONATE

CAS No.	Chemical Name
16923-95-8	POTASSIUM HEXAFLUOROZIRCONATE (K2ZrF6)
16923-95-8	POTASSIUM HEXAFLUOROZIRCONATE(IV)
7784-41-0	POTASSIUM HYDROGEN ARSENATE
7784-41-0	POTASSIUM HYDROGEN ARSENATE (KH2AsO4)
7789-29-9	POTASSIUM HYDROGEN DIFLUORIDE
7789-29-9	POTASSIUM HYDROGEN FLUORIDE
7789-29-9	POTASSIUM HYDROGEN FLUORIDE (DOT)
7789-29-9	POTASSIUM HYDROGEN FLUORIDE (KHF2)
7789-29-9	POTASSIUM HYDROGEN FLUORIDE SOLUTION
7789-29-9	POTASSIUM HYDROGEN FLUORIDE, SOLUTION (DOT)
127-95-7	POTASSIUM HYDROGEN OXALATE
7646-93-7	POTASSIUM HYDROGEN SULFATE
7646-93-7	POTASSIUM HYDROGEN SULFATE, SOLID
7646-93-7	POTASSIUM HYDROGEN SULFATE, SOLID (DOT)
7773-03-7	POTASSIUM HYDROGEN SULFITE
14293-73-3	POTASSIUM HYDROSULFITE
14293-73-3	POTASSIUM HYDROSULPHITE (DOT)
1310-58-3	POTASSIUM HYDROXIDE
1310-58-3	POTASSIUM HYDROXIDE (K(OH))
7778-74-7	POTASSIUM HYPERCHLORIDE
7778-66-7	POTASSIUM HYPOCHLORITE
7778-66-7	POTASSIUM HYPOCHLORITE SOLUTION
7778-66-7	POTASSIUM HYPOCHLORITE, SOLUTION (DOT)
7681-11-0	POTASSIUM IODIDE
7681-11-0	POTASSIUM IODIDE (KI)
10124-50-2	POTASSIUM METAARSENITE
4429-42-9	POTASSIUM METABISULFITE
16731-55-8	POTASSIUM METABISULFITE
4429-42-9	POTASSIUM METABISULFITE (DOT)
16731-55-8	POTASSIUM METABISULFITE (DOT)
4429-42-9	POTASSIUM METABISULFITE (K2S2O5)
16731-55-8	POTASSIUM METABISULFITE (K2S2O5)
13769-43-2	POTASSIUM METAVANADATE
13769-43-2	POTASSIUM METAVANADATE (DOT)
13769-43-2	POTASSIUM METAVANADATE (KVO3)
7789-23-3	POTASSIUM MONOFLUORIDE
7789-29-9	POTASSIUM MONOHYDROGEN DIFLUORIDE
7681-11-0	POTASSIUM MONOIODIDE
1312-73-8	POTASSIUM MONOSULFIDE
12401-70-6	POTASSIUM MONOXIDE
583-52-8	POTASSIUM NEUTRAL OXALATE
10043-22-8	POTASSIUM NEUTRAL OXALATE
7757-79-1	POTASSIUM NITRATE
7757-79-1	POTASSIUM NITRATE (DOT)
7758-09-0	POTASSIUM NITRITE
7758-09-0	POTASSIUM NITRITE (1:1)
7758-09-0	POTASSIUM NITRITE (DOT)
140-89-6	POTASSIUM O-ETHYL DITHIOCARBONATE
143-18-0	POTASSIUM OLEATE
583-52-8	POTASSIUM OXALATE
10043-22-8	POTASSIUM OXALATE
583-52-8	POTASSIUM OXALATE (K2C2O4)
10043-22-8	POTASSIUM OXALATE (K2C2O4)
127-95-7	POTASSIUM OXALATE (KHC2O4)
12136-45-7	POTASSIUM OXIDE
12401-70-6	POTASSIUM OXIDE
12401-70-6	POTASSIUM OXIDE (DOT)
17014-71-0	POTASSIUM OXIDE (K2(O2))
12401-70-6	POTASSIUM OXIDE (KO)
3811-04-9	POTASSIUM OXYMURIATE
12401-70-6	POTASSIUM OZONIDE
12401-70-6	POTASSIUM OZONIDE (K(O3))
7778-74-7	POTASSIUM PERCHLORATE
7778-74-7	POTASSIUM PERCHLORATE (DOT)
7778-74-7	POTASSIUM PERCHLORATE (KClO4)
7722-64-7	POTASSIUM PERMANGANATE
17014-71-0	POTASSIUM PEROXIDE
17014-71-0	POTASSIUM PEROXIDE (DOT)
17014-71-0	POTASSIUM PEROXIDE (K2(O2))
7727-21-1	POTASSIUM PEROXYDISULFATE
7727-21-1	POTASSIUM PEROXYDISULFATE (K2S2O3)
7727-21-1	POTASSIUM PEROXYDISULPHATE
7727-21-1	POTASSIUM PERSULFATE
7727-21-1	POTASSIUM PERSULFATE (ACGIH,DOT)
20770-41-6	POTASSIUM PHOSPHIDE
20770-41-6	POTASSIUM PHOSPHIDE (DOT)
20770-41-6	POTASSIUM PHOSPHIDE (K3P)
37340-70-8	POTASSIUM PHOSPHOTUNGSTATE
4429-42-9	POTASSIUM PYROSULFITE
16731-55-8	POTASSIUM PYROSULFITE
4429-42-9	POTASSIUM PYROSULFITE (K2S2O5)
16731-55-8	POTASSIUM PYROSULFITE (K2S2O5)
127-95-7	POTASSIUM SALT OF SORREL
14293-86-8	POTASSIUM SELENATE
14293-86-8	POTASSIUM SELENATE (K2Se2O7)
16871-90-2	POTASSIUM SILICOFLUORIDE
16871-90-2	POTASSIUM SILICOFLUORIDE (DOT)
16871-90-2	POTASSIUM SILICOFLUORIDE (K2SiF6)
16871-90-2	POTASSIUM SILICON FLUORIDE (K2SiF6)
506-61-6	POTASSIUM SILVER CYANIDE
506-61-6	POTASSIUM SILVER CYANIDE (KAg(CN)2)
11135-81-2	POTASSIUM SODIUM ALLOY
7646-93-7	POTASSIUM SULFATE
7646-93-7	POTASSIUM SULFATE (KHSO4)
1312-73-8	POTASSIUM SULFIDE
12136-50-4	POTASSIUM SULFIDE
12136-49-1	POTASSIUM SULFIDE
37248-34-3	POTASSIUM SULFIDE
1312-73-8	POTASSIUM SULFIDE (2:1)
1312-73-8	POTASSIUM SULFIDE (2:1), HYDRATED, CONTAINING AT LEAST 30% WATER
1312-73-8	POTASSIUM SULFIDE (DOT)
12136-50-4	POTASSIUM SULFIDE (DOT)
12136-49-1	POTASSIUM SULFIDE (DOT)
1312-73-8	POTASSIUM SULFIDE (K2S)
12136-49-1	POTASSIUM SULFIDE (K2S3)
12136-49-1	POTASSIUM SULFIDE (K2S4)
12136-50-4	POTASSIUM SULFIDE (K2S5)
7773-03-7	POTASSIUM SULFITE (KHSO3)
1312-73-8	POTASSIUM SULPHIDE, ANHYDROUS OR CONTAINING LESS THAN 30% WATER (DOT)
12136-49-1	POTASSIUM SULPHIDE, ANHYDROUS OR CONTAINING LESS THAN 30% WATER (DOT)
12136-50-4	POTASSIUM SULPHIDE, ANHYDROUS OR CONTAINING LESS THAN 30% WATER (DOT)
1312-73-8	POTASSIUM SULPHIDE, HYDRATED, CONTAINING NOT LESS THAN 30% WATER (DOT)
12030-88-5	POTASSIUM SUPEROXIDE
12030-88-5	POTASSIUM SUPEROXIDE (DOT)
12030-88-5	POTASSIUM SUPEROXIDE (K(O2))
10025-99-7	POTASSIUM TETRACHLOROPLATINATE
591-89-8	POTASSIUM TETRACYANOMERCURATE (II)
13762-51-1	POTASSIUM TETRAHYDROBORATE
13762-51-1	POTASSIUM TETRAHYDROBORATE(1-)
2244-21-5	POTASSIUM TROCLOSENE
13769-43-2	POTASSIUM VANADATE (KVO3)
13769-43-2	POTASSIUM VANADATE(V) (KVO3)
13769-43-2	POTASSIUM VANADIUM TRIOXIDE
140-89-6	POTASSIUM XANTHATE
140-89-6	POTASSIUM XANTHOGENATE
37224-57-0	POTASSIUM ZINC CHROMATE
37224-57-0	POTASSIUM ZINC CHROMATE HYDROXIDE
16923-95-8	POTASSIUM ZIRCONIUM FLUORIDE
16923-95-8	POTASSIUM ZIRCONIUM FLUORIDE (K2ZrF6)
16923-95-8	POTASSIUM ZIRCONIUM HEXAFLUORIDE
7440-09-7	POTASSIUM, (LIQUID ALLOY)
7789-29-9	POTASSIUM, BIFLUORIDE, SOLUTION (DOT)
7440-09-7	POTASSIUM, METAL
7440-09-7	POTASSIUM, METAL (DOT)
7440-09-7	POTASSIUM, METAL ALLOYS (DOT)
7440-09-7	POTASSIUM, METAL LIQUID ALLOY
7440-09-7	POTASSIUM, METAL LIQUID ALLOY (DOT)
7440-09-7	POTASSIUM, METALLIC (DOT)
11135-81-2	POTASSIUM-SODIUM, ALLOY (DOT)
64-17-5	POTATO ALCOHOL
3811-04-9	POTCRATE
7681-11-0	POTIDE
9002-89-5	POVAL
9002-89-5	POVAL 105
9002-89-5	POVAL 117
9002-89-5	POVAL 120
9002-89-5	POVAL 1700
9002-89-5	POVAL 203
9002-89-5	POVAL 205
9002-89-5	POVAL 205S
9002-89-5	POVAL 217
9002-89-5	POVAL 217S
9002-89-5	POVAL 420

CAS No.	Chemical Name
9002-89-5	POVAL A
9002-89-5	POVAL B 03
9002-89-5	POVAL B 05
9002-89-5	POVAL B 17
9002-89-5	POVAL C 17
9002-89-5	POVAL HL 12
9002-89-5	POVAL K 17E
9002-89-5	POVAL K 24E
9002-89-5	POVAL PA 5
9002-89-5	POVAL UP 240G
1314-13-2	POWDER BASE 900
12002-03-8	POWDER GREEN
85-00-7	PP 100
1910-2-5	PP 148
23505-41-1	PP 211
2074-50-2	PP 910
25322-69-4	PPG
25322-69-4	PPG 15
25322-69-4	PPG 2025
25322-69-4	PPG 30
25322-69-4	PPG 400
25322-69-4	PPG DIOL 1000
25322-69-4	PPG DIOL 2000
25322-69-4	PPG DIOL 400
34590-94-8	PPG-2 METHYL ETHER
7440-06-4	PR0
51-79-6	PRACARBAMINE
7440-70-2	PRAVAL
2955-38-6	PRAZEPAM
7704-34-9	PRECIPITATED SULFUR
7704-34-9	PRECIPITATED SULPHUR
7546-30-7	PRECIPITE BLANC
7681-49-4	PREDENT
50-24-8	PREDNISOLONE
53-03-2	PREDNISONE
85-00-7	PREEGLONE
7631-86-9	PREGEL
57-83-0	PREGN-4-ENE-3,20-DIONE
88-85-7	PREMERG
88-85-7	PREMERGE
88-85-7	PREMERGE 3
115-76-4	PRENDEROL
68-76-8	PRENIMON
1836-75-5	PREPARATION 125
100-97-0	PREPARATION AF
8001-58-9	PRESERV-O-SOTE
57-13-6	PRESPERSION, 75 UREA
2235-54-3	PRESULIN
13463-67-7	PRETIOX AP 22
13463-67-7	PRETIOX RD 53
59-50-7	PREVENTOL CMK
1321-10-4	PREVENTOL CMK
95-95-4	PREVENTOL I
25167-82-2	PREVENTOL I
87-86-5	PREVENTOL P
9016-45-9	PREVOCELL W-OF 100
25154-52-3	PREVOSTSEL VON-100
87-86-5	PRILTOX
137-32-6	PRIMARY ACTIVE AMYL ALCOHOL
628-63-7	PRIMARY AMYL ACETATE
71-41-0	PRIMARY AMYL ALCOHOL
112-30-1	PRIMARY DECYL ALCOHOL
513-38-2	PRIMARY ISOBUTYL IODIDE
111-87-5	PRIMARY OCTYL ALCOHOL
1912-24-9	PRIMATOL
1912-24-9	PRIMATOL A
1912-24-9	PRIMAZE
23505-41-1	PRIMICID
119-38-0	PRIMIN
50-28-2	PRIMOFOL
57-63-6	PRIMOGYN
57-63-6	PRIMOGYN C
57-63-6	PRIMOGYN M
8020-83-5	PRIMOL 205
8020-83-5	PRIMOL 325
57-83-0	PRIMOLUT
68-22-4	PRIMOLUT N
23505-41-1	PRIMOTEC
58-22-0	PRIMOTEST

CAS No.	Chemical Name
58-22-0	PRIMOTESTON
1306-23-6	PRIMROSE 1466
23505-41-1	PRINICID
1333-86-4	PRINTEX 140
1333-86-4	PRINTEX 200
1333-86-4	PRINTEX 30
1333-86-4	PRINTEX 300
1333-86-4	PRINTEX 400
1333-86-4	PRINTEX 60
1333-86-4	PRINTEX A
1333-86-4	PRINTEX G
1333-86-4	PRINTEX U
121-75-5	PRIODERM
109-86-4	PRIST
67-63-0	PRO
127-85-5	PRO-GEN SODIUM
671-16-9	PROCARBAZINE
366-70-1	PROCARBAZINE HYDROCHLORIDE
51-52-5	PROCASIL
8006-54-0	PROCESSED LANOLIN
3383-96-8	PROCIDA
50-18-0	PROCYTOX
7784-46-5	PRODALUMNOL
7784-46-5	PRODALUMNOL DOUBLE
16893-85-9	PRODAN
16893-85-9	PRODAN (PESTICIDE)
8006-20-0	PRODUCER GAS
151-21-3	PRODUCT NO 161
151-21-3	PRODUCT NO 75
952-23-8	PROFLAVINE
50-28-2	PROFOLIOL
13194-48-4	PROFOS
76-06-2	PROFUME A
57-83-0	PROGEKAN
57-83-0	PROGESTASERT
57-83-0	PROGESTEROL
57-83-0	PROGESTERONE
57-83-0	PROGESTIN
57-83-0	PROGESTONE
57-83-0	PROGESTRON
62-38-4	PROGRAMIN
50-28-2	PROGYNON
57-63-6	PROGYNON C
50-28-2	PROGYNON-DH
534-52-1	PROKARBOL
1335-85-9	PROKARBOL
123-75-1	PROLAMINE
732-11-6	PROLATE
57-83-0	PROLETS
57-83-0	PROLIDON
9004-66-4	PROLONGAL
68-22-4	PROLUTEASI
57-83-0	PROLUTON
82-66-6	PROMAR
2631-37-0	PROMECARB
2631-37-0	PROMECARBE
5836-73-7	PROMURIT
23950-58-5	PRONAMIDE
60-29-7	PRONARCOL
107-02-8	PROP-2-EN-1-AL
51-52-5	PROPACIL
463-49-0	PROPADIENE
463-49-0	PROPADIENE, INHIBITED (DOT)
123-38-6	PROPALDEHYDE
108-01-0	PROPAMINE A
110-18-9	PROPAMINE D
123-38-6	PROPANAL
78-84-2	PROPANAL, 2-METHYL-
116-06-3	PROPANAL, 2-METHYL-2-(METHYLTHIO)-, O-((METHYLAMINO)-CARBONYL))OXIME
3268-49-3	PROPANAL, 3-(METHYLTHIO)-
3268-49-3	PROPANAL, 3-(METHYLTHIO)- (9CI)
2806-85-1	PROPANAL, 3-ETHOXY-
123-38-6	PROPANALDEHYDE
107-10-8	PROPANAMINE
74-98-6	PROPANE
68476-85-7	PROPANE MIXTURES
1120-71-4	PROPANE SULTONE
111-43-3	PROPANE, 1,1'-OXYBIS-

CAS No.	Chemical Name
111-43-3	PROPANE, 1,1'-OXYBIS- (9CI)
78-99-9	PROPANE, 1,1-DICHLORO-
595-44-8	PROPANE, 1,1-DICHLORO-1-NITRO-
96-18-4	PROPANE, 1,2,3-TRICHLORO-
96-12-8	PROPANE, 1,2-DIBROMO-3-CHLORO-
78-87-5	PROPANE, 1,2-DICHLORO-
75-56-9	PROPANE, 1,2-EPOXY-
428-59-1	PROPANE, 1,2-EPOXY-1,1,2,3,3,3-HEXAFLUORO-
26447-14-3	PROPANE, 1,2-EPOXY-3-(TOLYLOXY)-
4016-11-9	PROPANE, 1,2-EPOXY-3-ETHOXY-
4016-14-2	PROPANE, 1,2-EPOXY-3-ISOPROPOXY-
122-60-1	PROPANE, 1,2-EPOXY-3-PHENOXY-
142-28-9	PROPANE, 1,3-DICHLORO-
106-92-3	PROPANE, 1-(ALLYLOXY)-2,3-EPOXY-
109-53-5	PROPANE, 1-(ETHENYLOXY)-2-METHYL-
106-94-5	PROPANE, 1-BROMO-
3132-64-7	PROPANE, 1-BROMO-2,3-EPOXY-
109-70-6	PROPANE, 1-BROMO-3-CHLORO-
2426-08-6	PROPANE, 1-BUTOXY-2,3-EPOXY-
540-54-5	PROPANE, 1-CHLORO-
600-25-9	PROPANE, 1-CHLORO-1-NITRO-
96-12-8	PROPANE, 1-CHLORO-2,3-DIBROMO-
106-89-8	PROPANE, 1-CHLORO-2,3-EPOXY-
513-36-0	PROPANE, 1-CHLORO-2-METHYL-
628-32-0	PROPANE, 1-ETHOXY-
628-32-0	PROPANE, 1-ETHOXY- (9CI)
513-38-2	PROPANE, 1-IODO-2-METHYL-
110-78-1	PROPANE, 1-ISOCYANATO-
1873-29-6	PROPANE, 1-ISOCYANATO-2-METHYL-
1873-29-6	PROPANE, 1-ISOCYANATO-2-METHYL- (9CI)
557-17-5	PROPANE, 1-METHOXY-
557-17-5	PROPANE, 1-METHOXY- (9CI)
108-03-2	PROPANE, 1-NITRO-
108-20-3	PROPANE, 2,2'-OXYBIS-
39638-32-9	PROPANE, 2,2'-OXYBIS[2-CHLORO-
63283-80-7	PROPANE, 2,2'-OXYBIS[DICHLORO-
463-82-1	PROPANE, 2,2-DIMETHYL-
926-65-8	PROPANE, 2-(ETHENYLOXY)-
75-31-0	PROPANE, 2-AMINO-
507-19-7	PROPANE, 2-BROMO-2-METHYL-
75-29-6	PROPANE, 2-CHLORO-
507-20-0	PROPANE, 2-CHLORO-2-METHYL-
594-71-8	PROPANE, 2-CHLORO-2-NITRO-
1795-48-8	PROPANE, 2-ISOCYANATO-
1795-48-8	PROPANE, 2-ISOCYANATO- (9CI)
75-28-5	PROPANE, 2-METHYL-
79-46-9	PROPANE, 2-NITRO-
25322-01-4	PROPANE, 2-NITRO-
98-82-8	PROPANE, 2-PHENYL
3132-64-7	PROPANE, 3-BROMO-1,2-EPOXY-
75-56-9	PROPANE, EPOXY-
26914-02-3	PROPANE, IODO-
25322-01-4	PROPANE, NITRO-
76-19-7	PROPANE, OCTAFLUORO-
57-55-6	PROPANE-1,2-DIOL
107-03-9	PROPANE-1-THIOL
79869-58-2	PROPANE-1-THIOL
1712-64-7	PROPANE-2-NITRATE
2698-41-1	PROPANEDINITRILE, [(2-CHLOROPHENYL)METHYLENE]-
107-12-0	PROPANENITRILE
78-67-1	PROPANENITRILE, 2,2'-AZOBIS[2-METHYL-
78-97-7	PROPANENITRILE, 2-HYDROXY-
75-86-5	PROPANENITRILE, 2-HYDROXY-2-METHYL-
78-82-0	PROPANENITRILE, 2-METHYL-
702-03-4	PROPANENITRILE, 3-(CYCLOHEXYLAMINO)-
109-78-4	PROPANENITRILE, 3-HYDROXY-
110-67-8	PROPANENITRILE, 3-METHOXY-
927-07-1	PROPANEPEROXOIC ACID, 2,2-DIMETHYL-, 1,1-DIMETHYLETHYL ESTER
29240-17-3	PROPANEPEROXOIC ACID, 2,2-DIMETHYL-, 1,1-DIMETHYLPROPYL ESTER
109-13-7	PROPANEPEROXOIC ACID, 2-METHYL-, 1,1-DIMETHYL ETHYL ESTER
109-13-7	PROPANEPEROXOIC ACID, 2-METHYL-, 1,1-DIMETHYLETHYL ESTER
107-03-9	PROPANETHIOL
79869-58-2	PROPANETHIOL
56-81-5	PROPANETRIOL
55-63-0	PROPANETRIOL TRINITRATE

CAS No.	Chemical Name
79-09-4	PROPANOIC ACID
637-78-5	PROPANOIC ACID, 1-METHYLETHYL ESTER
75-99-0	PROPANOIC ACID, 2,2-DICHLORO-
75-98-9	PROPANOIC ACID, 2,2-DIMETHYL-
93-72-1	PROPANOIC ACID, 2-(2,4,5-TRICHLOROPHENOXY)-
32534-95-5	PROPANOIC ACID, 2-(2,4,5-TRICHLOROPHENOXY)-, ISOOCTYL ESTER
79435-04-4	PROPANOIC ACID, 2-CHLORO-, 1-METHYLETHYL ESTER, (R)-
535-13-7	PROPANOIC ACID, 2-CHLORO-, ETHYL ESTER
535-13-7	PROPANOIC ACID, 2-CHLORO-, ETHYL ESTER (9CI)
17639-93-9	PROPANOIC ACID, 2-CHLORO-, METHYL ESTER
50-21-5	PROPANOIC ACID, 2-HYDROXY-
617-51-6	PROPANOIC ACID, 2-HYDROXY-, 1-METHYLETHYL ESTER
138-22-7	PROPANOIC ACID, 2-HYDROXY-, BUTYL ESTER
16039-52-4	PROPANOIC ACID, 2-HYDROXY-, COPPER COMPLEX
97-64-3	PROPANOIC ACID, 2-HYDROXY-, ETHYL ESTER
515-98-0	PROPANOIC ACID, 2-HYDROXY-, MONOAMMONIUM SALT
79-42-5	PROPANOIC ACID, 2-MERCAPTO-
79-31-2	PROPANOIC ACID, 2-METHYL-
617-50-5	PROPANOIC ACID, 2-METHYL-, 1-METHYLETHYL ESTER
617-50-5	PROPANOIC ACID, 2-METHYL-, 1-METHYLETHYL ESTER (9CI)
97-85-8	PROPANOIC ACID, 2-METHYL-, 2-METHYLPROPYL ESTER
97-85-8	PROPANOIC ACID, 2-METHYL-, 2-METHYLPROPYL ESTER (9CI)
97-62-1	PROPANOIC ACID, 2-METHYL-, ETHYL ESTER
3771-19-5	PROPANOIC ACID, 2-METHYL-2-[4-(1,2,3,4-TETRAHYDRO-1-NAPHTHALENYL)PHENOXY]-
540-42-1	PROPANOIC ACID, 2-METHYLPROPYL ESTER
540-42-1	PROPANOIC ACID, 2-METHYLPROPYL ESTER (9CI)
10213-74-8	PROPANOIC ACID, 3-(2-ETHYLBUTOXY)-
4324-38-3	PROPANOIC ACID, 3-ETHOXY-
57-57-8	PROPANOIC ACID, 3-HYDROXY-, BETA-LACTONE
123-62-6	PROPANOIC ACID, ANHYDRIDE
590-01-2	PROPANOIC ACID, BUTYL ESTER
590-01-2	PROPANOIC ACID, BUTYL ESTER (9CI)
28554-00-9	PROPANOIC ACID, CHLORO-
28554-00-9	PROPANOIC ACID, CHLORO- (9CI)
105-38-4	PROPANOIC ACID, ETHENYL ESTER
105-37-3	PROPANOIC ACID, ETHYL ESTER
554-12-1	PROPANOIC ACID, METHYL ESTER
106-36-5	PROPANOIC ACID, PROPYL ESTER
123-62-6	PROPANOIC ANHYDRIDE
71-23-8	PROPANOL
71-23-8	PROPANOL (DOT)
34590-94-8	PROPANOL, (2-METHOXYMETHYLETHOXY)-
25265-71-8	PROPANOL, OXYBIS-
24800-44-0	PROPANOL, [(1-METHYL-1,2-ETHANEDIYL)BIS(OXY)]BIS-
71-23-8	PROPANOL-1
156-87-6	PROPANOLAMINE
71-23-8	PROPANOLE (GERMAN)
71-23-8	PROPANOLEN (DUTCH)
71-23-8	PROPANOLI (ITALIAN)
57-57-8	PROPANOLIDE
67-64-1	PROPANONE
79-03-8	PROPANOYL CHLORIDE
3282-30-2	PROPANOYL CHLORIDE, 2,2-DIMETHYL-
3282-30-2	PROPANOYL CHLORIDE, 2,2-DIMETHYL- (9CI)
79-30-1	PROPANOYL CHLORIDE, 2-METHYL-
2312-35-8	PROPARGIL
2312-35-8	PROPARGITE
2312-35-8	PROPARGITE (DOT)
107-19-7	PROPARGYL ALCOHOL
107-19-7	PROPARGYL ALCOHOL (ACGIH,DOT)
106-96-7	PROPARGYL BROMIDE
143-00-0	PROPASTE D
139-96-8	PROPASTE T
79-09-4	PROPCORN
75-69-4	PROPELLANT 11
76-15-3	PROPELLANT 115
75-71-8	PROPELLANT 12
75-45-6	PROPELLANT 22
115-25-3	PROPELLANT C318
107-18-6	PROPEN-1-OL-3
108-22-5	PROPEN-2-YL ACETATE
814-78-8	PROPEN-2-YL METHYL KETONE
107-37-9	PROPEN-3-YLTRICHLOROSILANE
107-02-8	PROPENAL
107-02-8	PROPENAL (CZECH)
3054-95-3	PROPENAL DIETHYL ACETAL
79-06-1	PROPENAMIDE

CAS No.	Chemical Name
115-07-1	PROPENE
79-10-7	PROPENE ACID
75-56-9	PROPENE OXIDE
96-19-5	PROPENE, 1,2,3-TRICHLORO-
78-87-5	PROPENE, 1,2-DICHLORO-
542-75-6	PROPENE, 1,3-DICHLORO-
590-21-6	PROPENE, 1-CHLORO-
78-88-6	PROPENE, 2,3-DICHLORO-
557-98-2	PROPENE, 2-CHLORO-
115-11-7	PROPENE, 2-METHYL-
7756-94-7	PROPENE, 2-METHYL-, TRIMER
3054-95-3	PROPENE, 3,3-DIETHOXY-
106-95-6	PROPENE, 3-BROMO-
107-05-1	PROPENE, 3-CHLORO-
563-47-3	PROPENE, 3-CHLORO-2-METHYL-
556-56-9	PROPENE, 3-IODO-
57-06-7	PROPENE, 3-ISOTHIOCYANATO-
116-15-4	PROPENE, HEXAFLUORO-
6842-15-5	PROPENE, TETRAMER
13987-01-4	PROPENE, TRIMER
107-13-1	PROPENENITRILE
79-10-7	PROPENOIC ACID
96-33-3	PROPENOIC ACID METHYL ESTER (9CI)
107-18-6	PROPENOL
814-68-6	PROPENOYL CHLORIDE
107-18-6	PROPENYL ALCOHOL
590-21-6	PROPENYL CHLORIDE
928-55-2	PROPENYL ETHYL ETHER
13194-48-4	PROPHOS
13194-48-4	PROPHOS (ESTER)
74-99-7	PROPINE
57-57-8	PROPIOLACTONE
57-57-8	PROPIOLACTONE, BETA-
123-38-6	PROPIONAL
123-38-6	PROPIONALDEHYDE
123-38-6	PROPIONALDEHYDE (DOT)
765-34-4	PROPIONALDEHYDE, 2,3-EPOXY-
116-06-3	PROPIONALDEHYDE, 2-METHYL-2-(METHYLTHIO)-, O-(METHYL-CARBAMOYL)OXIME
3268-49-3	PROPIONALDEHYDE, 3-(METHYLTHIO)-
2806-85-1	PROPIONALDEHYDE, 3-ETHOXY-
105-37-3	PROPIONATE D'ETHYLE (FRENCH)
554-12-1	PROPIONATE DE METHYLE (FRENCH)
96-22-0	PROPIONE
79-09-4	PROPIONIC ACID
79-09-4	PROPIONIC ACID (ACGIH,DOT)
123-62-6	PROPIONIC ACID ANHYDRIDE
79-03-8	PROPIONIC ACID CHLORIDE
79-09-4	PROPIONIC ACID GRAIN PRESERVER
93-72-1	PROPIONIC ACID, (2,4,5-TRICHLOROPHENOXY)-
75-99-0	PROPIONIC ACID, 2,2-DICHLORO-
75-98-9	PROPIONIC ACID, 2,2-DIMETHYL-
93-72-1	PROPIONIC ACID, 2-(2,4,5-TRICHLOROPHENOXY)-
32534-95-5	PROPIONIC ACID, 2-(2,4,5-TRICHLOROPHENOXY)-, ISOOCTYL ESTER
535-13-7	PROPIONIC ACID, 2-CHLORO-, ETHYL ESTER
79435-04-4	PROPIONIC ACID, 2-CHLORO-, ISOPROPYL ESTER
17639-93-9	PROPIONIC ACID, 2-CHLORO-, METHYL ESTER
50-21-5	PROPIONIC ACID, 2-HYDROXY-
79-42-5	PROPIONIC ACID, 2-MERCAPTO-
97-62-1	PROPIONIC ACID, 2-METHYL-, ETHYL ESTER
3771-19-5	PROPIONIC ACID, 2-METHYL-2-[P-(1,2,3,4-TETRAHYDRO-1-NAPHTHYL)PHENOXY]-
79-41-4	PROPIONIC ACID, 2-METHYLENE-
123-23-9	PROPIONIC ACID, 3,3'-(DIOXYDICARBONYL)DI-
123-23-9	PROPIONIC ACID, 3,3'-(DIOXYDICARBONYL)DI-, MAXIMUM CONCENTRATION 72%
10213-74-8	PROPIONIC ACID, 3-(2-ETHYLBUTOXY)-
4324-38-3	PROPIONIC ACID, 3-ETHOXY-
57-57-8	PROPIONIC ACID, 3-HYDROXY-, BETA-LACTONE
590-01-2	PROPIONIC ACID, BUTYL ESTER
28554-00-9	PROPIONIC ACID, CHLORO-
105-37-3	PROPIONIC ACID, ETHYL ESTER
540-42-1	PROPIONIC ACID, ISOBUTYL ESTER
554-12-1	PROPIONIC ACID, METHYL ESTER
106-36-5	PROPIONIC ACID, PROPYL ESTER
79-09-4	PROPIONIC ACID, SOLUTION
79-09-4	PROPIONIC ACID, SOLUTION (DOT)
79-09-4	PROPIONIC ACID, SOLUTION CONTAINING NOT LESS THAN 80% ACID (DOT)

CAS No.	Chemical Name
105-38-4	PROPIONIC ACID, VINYL ESTER
123-38-6	PROPIONIC ALDEHYDE
123-62-6	PROPIONIC ANHYDRIDE
123-62-6	PROPIONIC ANHYDRIDE (DOT)
79-03-8	PROPIONIC CHLORIDE
105-37-3	PROPIONIC ESTER
105-37-3	PROPIONIC ETHER
107-12-0	PROPIONIC NITRILE
107-12-0	PROPIONITRILE
107-12-0	PROPIONITRILE (DOT)
542-76-7	PROPIONITRILE 3-CHLORO
78-67-1	PROPIONITRILE, 2,2'-AZOBIS(2-METHYL-
78-97-7	PROPIONITRILE, 2-HYDROXY-
702-03-4	PROPIONITRILE, 3-(CYCLOHEXYLAMINO)-
109-78-4	PROPIONITRILE, 3-HYDROXY-
110-67-8	PROPIONITRILE, 3-METHOXY-
107-12-0	PROPIONONITRILE
79-03-8	PROPIONYL CHLORIDE
79-03-8	PROPIONYL CHLORIDE (DOT)
123-62-6	PROPIONYL OXIDE
3248-28-0	PROPIONYL PEROXIDE
3248-28-0	PROPIONYL PEROXIDE (DOT)
3248-28-0	PROPIONYL PEROXIDE, NOT MORE THAN 28% IN SOLUTION
79-09-4	PROPKORN
9002-84-0	PROPLAST
67-63-0	PROPOL
114-26-1	PROPOTOX
114-26-1	PROPOXUR
114-26-1	PROPOXYLOR
75-99-0	PROPROP
51-52-5	PROPYCIL
1928-61-6	PROPYL 2,4-DICHLOROPHENOXYACETATE
109-60-4	PROPYL ACETATE
109-60-4	PROPYL ACETATE (DOT)
71-23-8	PROPYL ALCOHOL
71-23-8	PROPYL ALCOHOL (DOT)
3087-37-4	PROPYL ALCOHOL, TITANIUM(4+) SALT
123-38-6	PROPYL ALDEHYDE
2179-59-1	PROPYL ALLYL DISULFIDE
103-65-1	PROPYL BENZENE
103-65-1	PROPYL BENZENE (DOT)
106-94-5	PROPYL BROMIDE
105-66-8	PROPYL BUTANOATE
105-66-8	PROPYL BUTYRATE
71-36-3	PROPYL CARBINOL
540-54-5	PROPYL CHLORIDE
540-54-5	PROPYL CHLORIDE (DOT)
109-61-5	PROPYL CHLOROCARBONATE
109-61-5	PROPYL CHLOROFORMATE
13889-92-4	PROPYL CHLOROTHIOFORMATE
109-74-0	PROPYL CYANIDE
109-60-4	PROPYL ETHANOATE
111-43-3	PROPYL ETHER
628-32-0	PROPYL ETHYL ETHER
110-74-7	PROPYL FORMATE
110-74-7	PROPYL FORMATE (DOT)
74-98-6	PROPYL HYDRIDE
106-97-8	PROPYL HYDRIDE
110-78-1	PROPYL ISOCYANATE
123-19-3	PROPYL KETONE
107-03-9	PROPYL MERCAPTAN
79869-58-2	PROPYL MERCAPTAN
107-03-9	PROPYL MERCAPTAN (DOT)
79869-58-2	PROPYL MERCAPTAN (DOT)
110-74-7	PROPYL METHANOATE
557-17-5	PROPYL METHYL ETHER
107-87-9	PROPYL METHYL KETONE
105-66-8	PROPYL N-BUTYRATE
627-13-4	PROPYL NITRATE
106-88-7	PROPYL OXIRANE
16066-38-9	PROPYL PEROXYDICARBONATE
106-36-5	PROPYL PROPANOATE
106-36-5	PROPYL PROPIONATE
3087-37-4	PROPYL TITANATE(IV)
141-57-1	PROPYL TRICHLOROSILANE
141-57-1	PROPYL TRICHLOROSILANE (DOT)
51-52-5	PROPYL-THIORIST
51-52-5	PROPYL-THYRACIL

CAS No.	Chemical Name
109-52-4	PROPYLACETIC ACID
627-19-0	PROPYLACETYLENE
107-10-8	PROPYLAMINE
107-10-8	PROPYLAMINE (DOT)
513-49-5	PROPYLAMINE, 1-METHYL
56-18-8	PROPYLAMINE, 3,3'-IMINOBIS-
5332-73-0	PROPYLAMINE, 3-METHOXY-
926-63-6	PROPYLAMINE, N,N-DIMETHYL-
25322-69-4	PROPYLAN 8123
25322-69-4	PROPYLAN D 1002
25322-69-4	PROPYLAN D 2002
102-69-2	PROPYLDI-N-PROPYLAMINE
926-63-6	PROPYLDIMETHYLAMINE
115-07-1	PROPYLENE
115-07-1	PROPYLENE (DOT)
123-73-9	PROPYLENE ALDEHYDE
4170-30-3	PROPYLENE ALDEHYDE
78-87-5	PROPYLENE CHLORIDE
78-89-7	PROPYLENE CHLOROHYDRIN
78-89-7	PROPYLENE CHLOROHYDRIN (DOT)
78-90-0	PROPYLENE DIAMINE (DOT)
78-87-5	PROPYLENE DICHLORIDE
6423-43-4	PROPYLENE DINITRATE
75-56-9	PROPYLENE EPOXIDE
57-55-6	PROPYLENE GLYCOL
25584-83-2	PROPYLENE GLYCOL ACRYLATE
6423-43-4	PROPYLENE GLYCOL DINITRATE
107-98-2	PROPYLENE GLYCOL METHYL ETHER
107-98-2	PROPYLENE GLYCOL MONOMETHYL ETHER
57-55-6	PROPYLENE GLYCOL USP
1331-17-5	PROPYLENE GLYCOL, ALLYL ETHER
6423-43-4	PROPYLENE NITRATE
75-56-9	PROPYLENE OXIDE
75-56-9	PROPYLENE OXIDE (ACGIH,DOT)
428-59-1	PROPYLENE OXIDE HEXAFLUORIDE
25322-69-4	PROPYLENE OXIDE HOMOPOLYMER
37286-64-9	PROPYLENE OXIDE-METHANOL ADDUCT
6842-15-5	PROPYLENE TETRAMER
6842-15-5	PROPYLENE TETRAMER (DOT)
13987-01-4	PROPYLENE TRIMER
116-15-4	PROPYLENE, HEXAFLUORO-
78-90-0	PROPYLENEDIAMINE
75-55-8	PROPYLENEIMINE
75-55-8	PROPYLENIMINE
109-67-1	PROPYLETHYLENE
107-92-6	PROPYLFORMIC ACID
71-23-8	PROPYLIC ALCOHOL
123-38-6	PROPYLIC ALDEHYDE
78-99-9	PROPYLIDENE CHLORIDE
71-36-3	PROPYLMETHANOL
107-12-0	PROPYLNITRILE
71-23-8	PROPYLOWY ALKOHOL (POLISH)
107-03-9	PROPYLTHIOL
51-52-5	PROPYLTHIORIT
51-52-5	PROPYLTHIOURACIL
28729-54-6	PROPYLTOLUENE
74-99-7	PROPYNE
106-96-7	PROPYNE, 3-BROMO-
59355-75-8	PROPYNE, MIXED WITH PROPADIENE
107-19-7	PROPYNYL ALCOHOL
23950-58-5	PROPYZAMIDE
63-25-2	PROSEVOR 85
56-04-2	PROSTRUMYL
9016-45-9	PROTACHEM 630
434-07-1	PROTANABOL
7631-86-9	PROTEK-SORB 121
51-52-5	PROTHIUCIL
51-52-5	PROTHIURONE
2275-18-5	PROTHOAT
2275-18-5	PROTHOATE
129-06-6	PROTHROMADIN
51-52-5	PROTHYCIL
51-52-5	PROTHYRAN
1333-74-0	PROTIUM
51-52-5	PROTIURAL
2275-18-5	PROTOAT (HUNGARIAN)
7790-99-0	PROTOCHLORURE D'IODE (FRENCH)
1309-37-1	PROTOHEMATITE
57-83-0	PROTORMONE
1314-13-2	PROTOX 166
1314-13-2	PROTOX 167
1314-13-2	PROTOX 168
1314-13-2	PROTOX 169
1314-13-2	PROTOX 267
1314-13-2	PROTOX 268
127-85-5	PROTOXYL
2425-06-1	PROXEL EF
79-21-0	PROXITANE 4002
52-68-6	PROXOL
79-09-4	PROZOIN
1309-37-1	PRUSSIAN RED
74-90-8	PRUSSIC ACID
74-90-8	PRUSSIC ACID (DOT)
74-90-8	PRUSSIC ACID, UNSTABILIZED
460-19-5	PRUSSITE
2275-14-1	PRZEDZIORKOFOS (POLISH)
76-06-2	PS
1344-28-1	PS 1
1344-28-1	PS 1 (ALUMINA)
9002-89-5	PS 1200
8020-83-5	PS 28
16893-85-9	PSC CO-OP WEEVIL BAIT
79-09-4	PSEUDOACETIC ACID
107-01-7	PSEUDOBUTYLENE
95-63-6	PSEUDOCUMENE
95-63-6	PSEUDOCUMOL
97-95-0	PSEUDOHEXYL ALCOHOL
62-56-6	PSEUDOTHIOUREA
57-13-6	PSEUDOUREA
62-56-6	PSEUDOUREA, 2-THIO-
95-63-6	PSI-CUMENE
21609-90-5	PSL
69-72-7	PSORIACID-S-STIFT
103-85-5	PTC
54-62-6	PTERAMINA
9002-84-0	PTFE
9002-84-0	PTFE-GM3
51-52-5	PTU
103-85-5	PTU
51-52-5	PTU (THYREOSTATIC)
101-25-7	PU 55
137-26-8	PURALIN
10025-73-7	PURATRONIC CHROMIUM CHLORIDE
13530-68-2	PURATRONIC CHROMIUM TRIOXIDE
1344-48-5	PURE ENGLISH (QUICKSILVER) VERMILION
7782-44-7	PURE OXYGEN
7681-52-9	PURIN B
446-86-6	PURINE, 6-[(1-METHYL-4-NITROIMIDAZOL-5-YL)THIO]-
71-63-6	PURODIGIN
630-60-4	PUROSTROPHAN
71-63-6	PURPURID
37273-91-9	PUZOMOR
12788-93-1	PV 1
12788-93-1	PV 1 (ANTIFOAMING AGENT)
51-83-2	PV CARBACHOL
13463-67-7	PV FAST WHITE R
9002-89-5	PVA
9002-89-5	PVA 008
9002-89-5	PVA 105
9002-89-5	PVA 110
9002-89-5	PVA 117
9002-89-5	PVA 210
9002-89-5	PVA 217
9002-89-5	PVA 420
9002-89-5	PVA-HC
9002-89-5	PVAL 45/02
9002-89-5	PVAL 55/12
9002-89-5	PVS
9002-89-5	PVS 4
1068-27-5	PX 1
84-74-2	PX 104
26761-40-0	PX 120
103-23-1	PX-238
123-79-5	PX-238
9002-89-5	PXA 105
584-79-2	PYNAMIN
584-79-2	PYNAMIN FORTE
1336-36-3	PYRALENE

CAS No.	Chemical Name
9004-70-0	PYRALIN
100-73-2	PYRAN ALDEHYDE
110-87-2	PYRAN, DIHYDRO-
3174-74-1	PYRAN, DIHYDRO-
25512-65-6	PYRAN, DIHYDRO-
1336-36-3	PYRANOL
12002-48-1	PYRANOL 1478
123-42-2	PYRANTON
123-42-2	PYRANTON A
110-85-0	PYRAZINE HEXAHYDRIDE
110-85-0	PYRAZINE, HEXAHYDRO-
109-08-0	PYRAZINE, METHYL-
297-97-2	PYRAZINOL, O-ESTER WITH O,O-DIETHYL PHOSPHOROTHIOATE
16071-86-6	PYRAZOL FAST BROWN BRL
119-38-0	PYRAZOL-5-OL, 1-ISOPROPYL-3-METHYL-, DIMETHYLCARBAMATE
644-64-4	PYRAZOLE-1-CARBOXAMIDE, 3-HYDROXY-N,N,5-TRIMETHYL-, DIMETHYLCARBAMATE (ESTER)
16071-86-6	PYRAZOLINE BROWN BRL
108-34-9	PYRAZOXON
129-00-0	PYREN (GERMAN)
129-00-0	PYRENE
584-79-2	PYRESIN
584-79-2	PYRESYN
121-29-9	PYRETHRIN
121-29-9	PYRETHRIN 2
121-21-1	PYRETHRIN I
8003-34-7	PYRETHRIN I OR II
121-29-9	PYRETHRIN II
8003-34-7	PYRETHRINS
8003-34-7	PYRETHRINS (DOT)
8003-34-7	PYRETHRINS AND PYRETHROIDS
8003-34-7	PYRETHROIDS
8003-34-7	PYRETHRUM
8003-34-7	PYRETHRUM (ACGIH)
8003-34-7	PYRETHRUM (INSECTICIDE)
136-40-3	PYRIDACIL
110-86-1	PYRIDIN (GERMAN)
26445-05-6	PYRIDINAMINE
26445-05-6	PYRIDINAMINE (9CI)
110-86-1	PYRIDINE
110-86-1	PYRIDINE (ACGIH,DOT)
61215-72-3	PYRIDINE, 1,2,3,6-TETRAHYDRO-
94-78-0	PYRIDINE, 2,6-DIAMINO-3-(PHENYLAZO)-
136-40-3	PYRIDINE, 2,6-DIAMINO-3-(PHENYLAZO)-, MONOHYDROCHLORIDE
504-29-0	PYRIDINE, 2-AMINO-
29154-12-9	PYRIDINE, 2-CHLORO-
1929-82-4	PYRIDINE, 2-CHLORO-6-(TRICHLOROMETHYL)-
5408-74-2	PYRIDINE, 2-ETHENYL-5-ETHYL-
109-06-8	PYRIDINE, 2-METHYL-
140-76-1	PYRIDINE, 2-METHYL-5-VINYL-
91-80-5	PYRIDINE, 2-[[2-(DIMETHYLAMINO)ETHYL]-2-THENYLAMINO]-
54-11-5	PYRIDINE, 3-(1-METHYL-2-PYRROLIDINYL)-
54-11-5	PYRIDINE, 3-(1-METHYL-2-PYRROLIDINYL)-, (S)-
54-11-5	PYRIDINE, 3-(1-METHYL-2-PYRROLIDINYL)-, (S)- (9CI)
29790-52-1	PYRIDINE, 3-(1-METHYL-2-PYRROLIDINYL)-, (S)-, MONO(2-HYDROXYBENZOATE)
65-30-5	PYRIDINE, 3-(1-METHYL-2-PYRROLIDINYL)-, (S)-, SULFATE
65-30-5	PYRIDINE, 3-(1-METHYL-2-PYRROLIDINYL)-, (S)-, SULFATE (2:1)
65-31-6	PYRIDINE, 3-(1-METHYL-2-PYRROLIDINYL)-, (S)-, [R-(R*,R*)]-2,3-DIHYDROXYBUTANEDIOATE
65-31-6	PYRIDINE, 3-(1-METHYL-2-PYRROLIDINYL)-, (S)-, [R-(R*,R*)]-2,3-DIHYDROXYBUTANEDIOATE (1:2)
2820-51-1	PYRIDINE, 3-(1-METHYL-2-PYRROLIDINYL)-, HYDROCHLORIDE, (S)-
16543-55-8	PYRIDINE, 3-(1-NITROSO-2-PYRROLIDINYL)-, (S)-
54-11-5	PYRIDINE, 3-(TETRAHYDRO-1-METHYLPYRROL-2-YL)
504-24-5	PYRIDINE, 4-AMINO-
108-89-4	PYRIDINE, 4-METHYL-
1124-33-0	PYRIDINE, 4-NITRO-, 1-OXIDE
140-76-1	PYRIDINE, 5-ETHENYL-2-METHYL-
104-90-5	PYRIDINE, 5-ETHYL-2-METHYL-
5408-74-2	PYRIDINE, 5-ETHYL-2-VINYL-
26445-05-6	PYRIDINE, AMINO-
29154-12-9	PYRIDINE, CHLORO-
110-89-4	PYRIDINE, HEXAHYDRO-
1333-41-1	PYRIDINE, METHYL-
61215-72-3	PYRIDINE, TETRAHYDRO-

CAS No.	Chemical Name
136-40-3	PYRIDIUM
535-89-7	PYRIMIDINE, 2-CHLORO-4-(DIMETHYLAMINO)-6-METHYL-
53558-25-1	PYRIMINIL
53558-25-1	PYRIMINYL
2921-88-2	PYRINEX
91-80-5	PYRINISTAB
91-80-5	PYRINISTOL
53558-25-1	PYRINURON
136-40-3	PYRIPYRIDIUM
7782-42-5	PYRO-CARB 406
67-64-1	PYROACETIC ACID
67-64-1	PYROACETIC ETHER
71-43-2	PYROBENZOL
71-43-2	PYROBENZOLE
1333-86-4	PYROBLACK 7007
120-80-9	PYROCATECHIN
120-80-9	PYROCATECHINE
120-80-9	PYROCATECHOL
98-29-3	PYROCATECHOL, 4-TERT-BUTYL-
9004-34-6	PYROCELLULOSE
107-49-3	PYRODUST
106-97-8	PYROFAX
68476-85-7	PYROFAX
87-66-1	PYROGALLIC ACID
87-66-1	PYROGALLOL
7782-42-5	PYROLITE
98-01-1	PYROMUCIC ALDEHYDE
542-92-7	PYROPENTYLENE
8049-18-1	PYROPHOROUS ALLOY
60475-66-3	PYROPHOROUS ALLOY
69523-06-4	PYROPHOROUS ALLOY
152-16-9	PYROPHOSPHORAMIDE, OCTAMETHYL-
152-16-9	PYROPHOSPHORIC ACID OCTAMETHYLTETRAAMIDE
107-49-3	PYROPHOSPHORIC ACID, TETRAETHYL ESTER
107-49-3	PYROPHOSPHORIC ACID, TETRAETHYL ESTER (MIXTURE)
3689-24-5	PYROPHOSPHORIC ACID, TETRAETHYLDITHIO-, (MIXTURE)
7722-88-5	PYROPHOSPHORIC ACID, TETRASODIUM SALT
3689-24-5	PYROPHOSPHORODITHIOIC ACID, O,O,O,O-TETRAETHYL ESTER
3689-24-5	PYROPHOSPHORODITHIOIC ACID, TETRAETHYL ESTER
152-16-9	PYROPHOSPHORYLTETRAKISDIMETHYLAMIDE
7783-05-3	PYROSULFURIC ACID
4429-42-9	PYROSULFUROUS ACID, DIPOTASSIUM SALT
16731-55-8	PYROSULFUROUS ACID, DIPOTASSIUM SALT
7681-57-4	PYROSULFUROUS ACID, DISODIUM SALT
7757-74-6	PYROSULFUROUS ACID, DISODIUM SALT
7681-57-4	PYROSULFUROUS ACID, DISODIUM SALT (8CI)
7757-74-6	PYROSULFUROUS ACID, DISODIUM SALT (8CI)
7791-27-7	PYROSULFURYL CHLORIDE
7791-27-7	PYROSULFURYL CHLORIDE (DOT)
7791-27-7	PYROSULFURYL CHLORIDE (S2O5Cl2)
7783-05-3	PYROSULPHURIC ACID
7791-27-7	PYROSULPHURYL CHLORIDE (DOT)
67-56-1	PYROXYLIC SPIRIT
9004-70-0	PYROXYLIN
9004-70-0	PYROXYLIN PLASTIC (DOT)
9004-70-0	PYROXYLIN PLASTIC SCRAP (DOT)
9004-70-0	PYROXYLIN RODS (DOT)
9004-70-0	PYROXYLIN ROLLS (DOT)
9004-70-0	PYROXYLIN SCRAP (DOT)
9004-70-0	PYROXYLIN SHEETS (DOT)
9004-70-0	PYROXYLIN TUBES (DOT)
109-97-7	PYRROL
109-97-7	PYRROLE
27417-39-6	PYRROLE, METHYL-
123-75-1	PYRROLE, TETRAHYDRO-
123-75-1	PYRROLIDINE
123-75-1	PYRROLIDINE (DOT)
123-75-1	PYRROLIDINE RING
120-94-5	PYRROLIDINE, 1-METHYL-
54-11-5	PYRROLIDINE, 1-METHYL-2-(3-PYRIDAL)-
65-30-5	PYRROLIDINE, 1-METHYL-2-(3-PYRIDYL)-, SULFATE
930-55-2	PYRROLIDINE, 1-NITROSO-
616-45-5	PYRROLIDINE-2-ONE
616-45-5	PYRROLIDONE
106-99-0	PYRROLYLENE
25339-57-5	PYRROLYLENE
14808-60-7	Q-CEL
1344-28-1	Q-LOID A 30
97-99-4	QO THFA

CAS No.	Chemical Name
14808-60-7	QUARTZ
14808-60-7	QUARTZ (SiO2)
1480-60-7	QUARTZ DUST
16919-58-7	QUATERNIUM-17
7439-97-6	QUECKSILBER
7439-97-6	QUECKSILBER (GERMAN)
7487-94-7	QUECKSILBER CHLORID (GERMAN)
7546-30-7	QUECKSILBER CHLORUER (GERMAN)
7546-30-7	QUECKSILBER(I)-CHLORID (GERMAN)
15829-53-5	QUECKSILBEROXID (GERMAN)
55-38-9	QUELETOX
58-89-9	QUELLADA
56-75-7	QUEMICETINA
60-00-4	QUESTEX 4H
3691-35-8	QUICK
1305-78-8	QUICKLIME
1305-78-8	QUICKLIME (DOT)
20859-73-8	QUICKPHOS
62-38-4	QUICKSAN
7439-97-6	QUICKSILVER
50-55-5	QUIESCIN
69-05-6	QUINACRINE HYDROCHLORIDE
91-63-4	QUINALDINE
123-31-9	QUINOL
91-22-5	QUINOLIN
91-22-5	QUINOLINE
91-22-5	QUINOLINE (DOT)
91-63-4	QUINOLINE, 2-METHYL-
56-57-5	QUINOLINE, 4-NITRO-, 1-OXIDE
94-36-0	QUINOLOR COMPOUND
106-51-4	QUINONE
106-51-4	QUINONE (ACGIH)
298-00-0	QUINOPHOS
82-68-8	QUINOSAN
117-80-6	QUINTAR
117-80-6	QUINTAR 540F
82-68-8	QUINTOCENE
60-57-1	QUINTOX
82-68-8	QUINTOZEN
82-68-8	QUINTOZENE
101-14-4	QUODOROLE
151-21-3	QUOLAC EX-UB
7631-86-9	QUSO 51
7631-86-9	QUSO G 30
7631-86-9	QUSO G 32
7631-86-9	QUSO WR 82
56-23-5	R 10
56-23-5	R 10 (REFRIGERANT)
75-69-4	R 11
75-69-4	R 11 (REFRIGERANT)
76-13-1	R 113
76-13-1	R 113 (HALOCARBON)
76-15-3	R 115
76-16-4	R 116
75-71-8	R 12
75-71-8	R 12 (DOT)
8020-83-5	R 12 (OIL)
75-71-8	R 12 (REFRIGERANT)
115-07-1	R 1270
353-59-3	R 12B1
75-61-6	R 12B2
75-72-9	R 13
786-19-6	R 1303
75-63-8	R 13B1
75-73-0	R 14
75-73-0	R 14 (REFRIGERANT)
25497-29-4	R 142
75-68-3	R 142B
732-11-6	R 1504
2642-71-9	R 1513
75-37-6	R 152A
25497-28-3	R 152A
86-50-0	R 1582
353-36-6	R 161
67-66-3	R 20
67-66-3	R 20 (REFRIGERANT)
75-43-4	R 21
75-43-4	R 21 (REFRIGERANT)
75-45-6	R 22

CAS No.	Chemical Name
75-45-6	R 22 (DOT)
75-46-7	R 23
75-46-7	R 23 (HALOCARBON)
74-98-6	R 290
75-09-2	R 30
74-87-3	R 40
74-83-9	R 40B1
23505-41-1	R 42211
494-03-1	R 48
74-82-8	R 50
74-82-8	R 50 (REFRIGERANT)
1309-37-1	R 5098
1309-37-1	R 5098 (OXIDE)
78-53-5	R 5158
13463-67-7	R 580
106-97-8	R 600
75-28-5	R 600A
13463-67-7	R 610D
107-31-3	R 611
75-04-7	R 631
13463-67-7	R 650
297-78-9	R 6700
13463-67-7	R 680
7664-41-7	R 717
124-38-9	R 744
13463-67-7	R 780-2
502-39-6	R 8
502-39-6	R 8 (FUNGICIDE)
115-25-3	R-C 318
542-92-7	R-PENTINE
9004-70-0	R. S. NITROCELLULOSE
50-21-5	RACEMIC LACTIC ACID
5836-29-3	RACUMIN
5836-29-3	RACUMIN 57
60-51-5	RACUSAN
124-65-2	RAD-E-CATE
124-65-2	RAD-E-CATE 16
75-60-5	RAD-E-CATE 25
124-65-2	RAD-E-CATE 25
124-65-2	RAD-E-CATE 35
75-99-0	RADAPON
1912-24-9	RADAZIN
1912-24-9	RADIZINE
137-26-8	RADOTHIRAM
137-26-8	RADOTIRAM
1335-85-9	RAFEX
534-52-1	RAFEX
534-52-1	RAFEX 35
1335-85-9	RAFEX 35
7681-49-4	RAFLUOR
133-06-2	RALLIS CAPTAF
82-66-6	RAMIK
7440-28-0	RAMOR
107-21-1	RAMP
298-02-2	RAMPART
3691-35-8	RAMUCIDE
3691-35-8	RANAC
1317-95-9	RANDANITE
1332-58-7	RANDITE
7440-02-0	RANEY ALLOY
7440-50-8	RANEY COPPER
7440-02-0	RANEY NICKEL
1335-85-9	RAPHATOX
534-52-1	RAPHATOX
577-11-7	RAPISOL
577-11-7	RAPISOL B 30
577-11-7	RAPISOL B 80
59-50-7	RASCHIT
1321-10-4	RASCHIT
59-50-7	RASCHIT K
1321-10-4	RASCHIT K
59-50-7	RASEN-ANICON
1321-10-4	RASEN-ANICON
7775-09-9	RASIKAL
1330-43-4	RASORITE 65
81-81-2	RAT & MICE BAIT
81-81-2	RAT-A-WAY
117-52-2	RAT-A-WAY
81-81-2	RAT-B-GON

CAS No.	Chemical Name
81-81-2	RAT-GARD
81-81-2	RAT-KILL
81-81-2	RAT-MIX
7723-14-0	RAT-NIP
81-81-2	RAT-O-CIDE #2
81-81-2	RAT-OLA
81-81-2	RAT-TROL
30553-04-9	RAT-TU
117-52-2	RATAFIN
62-74-8	RATBANE 1080
991-42-4	RATICATE
3691-35-8	RATICIDE
3691-35-8	RATICIDE-CAID
9002-89-5	RATIFIX F
28772-56-7	RATIMUS
82-66-6	RATINDAN
82-66-6	RATINDAN 1
3691-35-8	RATINDAN 3
3691-35-8	RATOMET
81-81-2	RATOREX
81-81-2	RATOX
10031-59-1	RATOX
7446-18-6	RATOX
81-81-2	RATOXIN
81-81-2	RATRON
81-81-2	RATRON G
81-81-2	RATS-NO-MORE
129-06-6	RATSUL SOLUBLE
10031-59-1	RATTENGIFTKONSERVE
7446-18-6	RATTENGIFTKONSERVE
86-88-4	RATTRACK
30553-04-9	RATTRACK
81-81-2	RATTUNAL
50-55-5	RAU-SED
50-55-5	RAUCAP
5836-29-3	RAUCUMIN 57
50-55-5	RAULEN
50-55-5	RAUNERVIL
50-55-5	RAUPASIL
50-55-5	RAURINE
50-55-5	RAUSEDIL
50-55-5	RAUSEDYL
50-55-5	RAUSINGLE
50-55-5	RAUWASEDIN
1333-86-4	RAVEN
1333-86-4	RAVEN 11
1333-86-4	RAVEN 30
1333-86-4	RAVEN 35
1333-86-4	RAVEN 420
1333-86-4	RAVEN 500
1333-86-4	RAVEN 520
1333-86-4	RAVEN 8000
3691-35-8	RAVIAC
63-25-2	RAVYON
81-81-2	RAX
9004-34-6	RAYOPHANE
13463-67-7	RAYOX
9004-34-6	RAYWEB Q
56-38-2	RB
1344-28-1	RC 172DBM
1120-48-5	RC 5632
117-84-0	RC PLASTICIZER DOP
7704-34-9	RC-SCHWEFEL EXTRA
7440-02-0	RCH 55/5
81-81-2	RCRA WASTE NUMBER P001
107-02-8	RCRA WASTE NUMBER P003
30900-23-3	RCRA WASTE NUMBER P004
107-18-6	RCRA WASTE NUMBER P005
20859-73-8	RCRA WASTE NUMBER P006
2763-96-4	RCRA WASTE NUMBER P007
504-24-5	RCRA WASTE NUMBER P008
131-74-8	RCRA WASTE NUMBER P009
7778-39-4	RCRA WASTE NUMBER P010
1303-28-2	RCRA WASTE NUMBER P011
1327-53-3	RCRA WASTE NUMBER P012
108-98-5	RCRA WASTE NUMBER P014
7440-41-7	RCRA WASTE NUMBER P015
598-31-2	RCRA WASTE NUMBER P017
357-57-3	RCRA WASTE NUMBER P018

CAS No.	Chemical Name
88-85-7	RCRA WASTE NUMBER P020
592-01-8	RCRA WASTE NUMBER P021
75-15-0	RCRA WASTE NUMBER P022
107-20-0	RCRA WASTE NUMBER P023
27134-26-5	RCRA WASTE NUMBER P024
5344-82-1	RCRA WASTE NUMBER P026
542-76-7	RCRA WASTE NUMBER P027
100-44-7	RCRA WASTE NUMBER P028
544-92-3	RCRA WASTE NUMBER P029
57-12-5	RCRA WASTE NUMBER P030
460-19-5	RCRA WASTE NUMBER P031
506-77-4	RCRA WASTE NUMBER P033
131-89-5	RCRA WASTE NUMBER P034
696-28-6	RCRA WASTE NUMBER P036
60-57-1	RCRA WASTE NUMBER P037
1303-33-9	RCRA WASTE NUMBER P038
298-04-4	RCRA WASTE NUMBER P039
297-97-2	RCRA WASTE NUMBER P040
60-51-5	RCRA WASTE NUMBER P044
39196-18-4	RCRA WASTE NUMBER P045
1335-85-9	RCRA WASTE NUMBER P047
51-28-5	RCRA WASTE NUMBER P048
541-53-7	RCRA WASTE NUMBER P049
115-29-7	RCRA WASTE NUMBER P050
72-20-8	RCRA WASTE NUMBER P051
151-56-4	RCRA WASTE NUMBER P054
7782-41-4	RCRA WASTE NUMBER P056
640-19-7	RCRA WASTE NUMBER P057
62-74-8	RCRA WASTE NUMBER P058
76-44-8	RCRA WASTE NUMBER P059
757-58-4	RCRA WASTE NUMBER P062
74-90-8	RCRA WASTE NUMBER P063
624-83-9	RCRA WASTE NUMBER P064
60-34-4	RCRA WASTE NUMBER P068
75-86-5	RCRA WASTE NUMBER P069
116-06-3	RCRA WASTE NUMBER P070
298-00-0	RCRA WASTE NUMBER P071
30553-04-9	RCRA WASTE NUMBER P072
13463-39-3	RCRA WASTE NUMBER P073
557-19-7	RCRA WASTE NUMBER P074
54-11-5	RCRA WASTE NUMBER P075
10102-43-9	RCRA WASTE NUMBER P076
100-01-6	RCRA WASTE NUMBER P077
29757-24-2	RCRA WASTE NUMBER P077
10102-44-0	RCRA WASTE NUMBER P078
55-63-0	RCRA WASTE NUMBER P081
62-75-9	RCRA WASTE NUMBER P082
152-16-9	RCRA WASTE NUMBER P085
20816-12-0	RCRA WASTE NUMBER P087
56-38-2	RCRA WASTE NUMBER P089
103-85-5	RCRA WASTE NUMBER P093
298-02-2	RCRA WASTE NUMBER P094
75-44-5	RCRA WASTE NUMBER P095
151-50-8	RCRA WASTE NUMBER P098
506-61-6	RCRA WASTE NUMBER P099
107-12-0	RCRA WASTE NUMBER P101
107-19-7	RCRA WASTE NUMBER P102
506-64-9	RCRA WASTE NUMBER P104
26628-22-8	RCRA WASTE NUMBER P105
143-33-9	RCRA WASTE NUMBER P106
57-24-9	RCRA WASTE NUMBER P108
3689-24-5	RCRA WASTE NUMBER P109
78-00-2	RCRA WASTE NUMBER P110
509-14-8	RCRA WASTE NUMBER P112
1314-32-5	RCRA WASTE NUMBER P113
7446-18-6	RCRA WASTE NUMBER P115
10031-59-1	RCRA WASTE NUMBER P115
79-19-6	RCRA WASTE NUMBER P116
7803-55-6	RCRA WASTE NUMBER P119
1314-62-1	RCRA WASTE NUMBER P120
557-21-1	RCRA WASTE NUMBER P121
1314-84-7	RCRA WASTE NUMBER P122
8001-35-2	RCRA WASTE NUMBER P123
75-07-0	RCRA WASTE NUMBER U001
67-64-1	RCRA WASTE NUMBER U002
75-05-8	RCRA WASTE NUMBER U003
98-86-2	RCRA WASTE NUMBER U004
75-36-5	RCRA WASTE NUMBER U006
79-06-1	RCRA WASTE NUMBER U007

CAS No.	Chemical Name
79-10-7	RCRA WASTE NUMBER U008
50-07-7	RCRA WASTE NUMBER U010
62-53-3	RCRA WASTE NUMBER U012
98-87-3	RCRA WASTE NUMBER U017
98-07-7	RCRA WASTE NUMBER U023
111-44-4	RCRA WASTE NUMBER U025
63283-80-7	RCRA WASTE NUMBER U027
117-84-0	RCRA WASTE NUMBER U028
71-36-3	RCRA WASTE NUMBER U031
13765-19-0	RCRA WASTE NUMBER U032
353-50-4	RCRA WASTE NUMBER U033
75-87-6	RCRA WASTE NUMBER U034
12789-03-6	RCRA WASTE NUMBER U036
108-90-7	RCRA WASTE NUMBER U037
59-50-7	RCRA WASTE NUMBER U039
1321-10-4	RCRA WASTE NUMBER U039
106-89-8	RCRA WASTE NUMBER U041
75-01-4	RCRA WASTE NUMBER U043
67-66-3	RCRA WASTE NUMBER U044
74-87-3	RCRA WASTE NUMBER U045
107-30-2	RCRA WASTE NUMBER U046
3165-93-3	RCRA WASTE NUMBER U049
8001-58-9	RCRA WASTE NUMBER U051
95-48-7	RCRA WASTE NUMBER U052
1319-77-3	RCRA WASTE NUMBER U052
108-39-4	RCRA WASTE NUMBER U052
106-44-5	RCRA WASTE NUMBER U052
123-73-9	RCRA WASTE NUMBER U053
4170-30-3	RCRA WASTE NUMBER U053
98-82-8	RCRA WASTE NUMBER U055
96-12-8	RCRA WASTE NUMBER U066
106-93-4	RCRA WASTE NUMBER U067
84-74-2	RCRA WASTE NUMBER U069
95-50-1	RCRA WASTE NUMBER U070
106-46-7	RCRA WASTE NUMBER U070
106-46-7	RCRA WASTE NUMBER U071
541-73-1	RCRA WASTE NUMBER U071
106-46-7	RCRA WASTE NUMBER U072
75-71-8	RCRA WASTE NUMBER U075
75-34-3	RCRA WASTE NUMBER U076
107-06-2	RCRA WASTE NUMBER U077
75-35-4	RCRA WASTE NUMBER U078
75-09-2	RCRA WASTE NUMBER U080
78-87-5	RCRA WASTE NUMBER U083
1464-53-5	RCRA WASTE NUMBER U085
124-40-3	RCRA WASTE NUMBER U092
80-15-9	RCRA WASTE NUMBER U096
79-44-7	RCRA WASTE NUMBER U097
57-14-7	RCRA WASTE NUMBER U098
540-73-8	RCRA WASTE NUMBER U099
131-11-3	RCRA WASTE NUMBER U102
77-78-1	RCRA WASTE NUMBER U103
121-14-2	RCRA WASTE NUMBER U105
606-20-2	RCRA WASTE NUMBER U106
123-91-1	RCRA WASTE NUMBER U108
142-84-7	RCRA WASTE NUMBER U110
141-78-6	RCRA WASTE NUMBER U112
140-88-5	RCRA WASTE NUMBER U113
60-29-7	RCRA WASTE NUMBER U117
97-63-2	RCRA WASTE NUMBER U118
75-69-4	RCRA WASTE NUMBER U121
50-00-0	RCRA WASTE NUMBER U122
64-18-6	RCRA WASTE NUMBER U123
110-00-9	RCRA WASTE NUMBER U124
98-01-1	RCRA WASTE NUMBER U125
765-34-4	RCRA WASTE NUMBER U126
118-74-1	RCRA WASTE NUMBER U127
87-68-3	RCRA WASTE NUMBER U128
58-89-9	RCRA WASTE NUMBER U129
77-47-4	RCRA WASTE NUMBER U130
67-72-1	RCRA WASTE NUMBER U131
302-01-2	RCRA WASTE NUMBER U133
7664-39-3	RCRA WASTE NUMBER U134
7783-06-4	RCRA WASTE NUMBER U135
75-60-5	RCRA WASTE NUMBER U136
74-88-4	RCRA WASTE NUMBER U138
78-83-1	RCRA WASTE NUMBER U140
143-50-0	RCRA WASTE NUMBER U142
108-31-6	RCRA WASTE NUMBER U147

CAS No.	Chemical Name
7439-97-6	RCRA WASTE NUMBER U151
126-98-7	RCRA WASTE NUMBER U152
67-56-1	RCRA WASTE NUMBER U154
79-22-1	RCRA WASTE NUMBER U156
78-93-3	RCRA WASTE NUMBER U159
80-62-6	RCRA WASTE NUMBER U162
91-20-3	RCRA WASTE NUMBER U165
91-59-8	RCRA WASTE NUMBER U168
98-95-3	RCRA WASTE NUMBER U169
100-02-7	RCRA WASTE NUMBER U170
79-46-9	RCRA WASTE NUMBER U171
25322-01-4	RCRA WASTE NUMBER U171
76-01-7	RCRA WASTE NUMBER U184
504-60-9	RCRA WASTE NUMBER U186
1314-80-3	RCRA WASTE NUMBER U189
85-44-9	RCRA WASTE NUMBER U190
109-06-8	RCRA WASTE NUMBER U191
107-10-8	RCRA WASTE NUMBER U194
110-86-1	RCRA WASTE NUMBER U196
106-51-4	RCRA WASTE NUMBER U197
108-46-3	RCRA WASTE NUMBER U201
12640-89-0	RCRA WASTE NUMBER U204
7783-00-8	RCRA WASTE NUMBER U204
7446-08-4	RCRA WASTE NUMBER U204
7488-56-4	RCRA WASTE NUMBER U205
127-18-4	RCRA WASTE NUMBER U210
56-23-5	RCRA WASTE NUMBER U211
109-99-9	RCRA WASTE NUMBER U213
6533-73-9	RCRA WASTE NUMBER U215
7791-12-0	RCRA WASTE NUMBER U216
13746-98-0	RCRA WASTE NUMBER U217
16901-76-1	RCRA WASTE NUMBER U217
10102-45-1	RCRA WASTE NUMBER U217
62-56-6	RCRA WASTE NUMBER U219
108-88-3	RCRA WASTE NUMBER U220
25376-45-8	RCRA WASTE NUMBER U221
30143-13-6	RCRA WASTE NUMBER U221
26471-62-5	RCRA WASTE NUMBER U223
75-25-2	RCRA WASTE NUMBER U225
71-55-6	RCRA WASTE NUMBER U226
79-01-6	RCRA WASTE NUMBER U228
933-78-4	RCRA WASTE NUMBER U231
25167-82-2	RCRA WASTE NUMBER U231
15950-66-0	RCRA WASTE NUMBER U231
93-76-5	RCRA WASTE NUMBER U232
99-35-4	RCRA WASTE NUMBER U234
1330-20-7	RCRA WASTE NUMBER U239
94-75-7	RCRA WASTE NUMBER U240
87-86-5	RCRA WASTE NUMBER U242
137-26-8	RCRA WASTE NUMBER U244
506-68-3	RCRA WASTE NUMBER U246
72-43-5	RCRA WASTE NUMBER U247
95-53-4	RCRA WASTE NUMBER U328
106-49-0	RCRA WASTE NUMBER U328
108-44-1	RCRA WASTE NUMBER U328
95-53-4	RCRA WASTE NUMBER U353
108-44-1	RCRA WASTE NUMBER U353
106-49-0	RCRA WASTE NUMBER U353
121-82-4	RDX
300-76-5	RE 4355
10265-92-6	RE 9006
60-51-5	REBELATE
1333-86-4	REBONEX H
1333-86-4	REBONEX HS
92-84-2	RECONOX
630-60-4	RECTOBAINA
100-01-6	RED 2G BASE
29757-24-2	RED 2G BASE
12044-79-0	RED ARSENIC GLASS
95-79-4	RED BASE CIBA IX
89-62-3	RED BASE CIBA VII
95-79-4	RED BASE IRGA IX
89-62-3	RED BASE IRGA VII
89-62-3	RED BASE NGL
95-79-4	RED BASE NTR
95-69-2	RED BASE NTR
1344-48-5	RED CINNABAR
89-62-3	RED G BASE
89-62-3	RED G SALT

CAS No.	Chemical Name
1309-37-1	RED IRON OXIDE
95-79-4	RED KB BASE
7774-29-0	RED MERCURIC IODIDE
1344-48-5	RED MERCURY SULPHIDE
112-80-1	RED OIL
12044-79-0	RED ORPIMENT
1309-37-1	RED OXIDE
7723-14-0	RED PHOSPHORUS
3165-93-3	RED SALT CIBA IX
89-62-3	RED SALT CIBA VII
3165-93-3	RED SALT IRGA IX
89-62-3	RED SALT IRGA VII
1314-13-2	RED SEAL 9
60-57-1	RED SHIELD
95-69-2	RED TR BASE
95-79-4	RED TR BASE
3165-93-3	RED TRS SALT
86-30-6	REDAX
93-76-5	REDDON
93-76-5	REDDOX
3691-35-8	REDENTIN
57-06-7	REDSKIN
101-61-1	REDUCED MICHLER'S KETONE
7775-14-6	REDUCTONE
8032-32-4	REFINED SOLVENT NAPHTHA
1332-58-7	REFORSIL 700
1332-58-7	REFRACTORINESS
75-69-4	REFRIGERANT 11
75-71-8	REFRIGERANT 12
75-72-9	REFRIGERANT 13
75-63-8	REFRIGERANT 13B1
75-73-0	REFRIGERANT 14
75-45-6	REFRIGERANT 22
76-13-1	REFRIGERANT R 113
75-69-4	REFRIGERANT R11
7664-41-7	REFRIGERENT R717
97-77-8	REFUSAL
1333-86-4	REGAL
1333-86-4	REGAL 300
1333-86-4	REGAL 300R
1333-86-4	REGAL 330
1333-86-4	REGAL 330R
1333-86-4	REGAL 400
1333-86-4	REGAL 400R
1333-86-4	REGAL 600
1333-86-4	REGAL 660
1333-86-4	REGAL 99
1333-86-4	REGAL SRF
85-00-7	REGLON
85-00-7	REGLONE
85-00-7	REGLOX
577-11-7	REGUTOL
105-60-2	RELON P
111-30-8	RELUGAN GT
111-30-8	RELUGAN GT 50
111-30-8	RELUGAN GTW
9016-45-9	REMCOPAL
299-84-3	REMELT
9005-25-8	REMYLINE
9005-25-8	REMYLINE AC
123-30-8	RENAL AC
27598-85-2	RENAL AC
95-80-7	RENAL MD
106-50-3	RENAL PF
9016-45-9	RENDELLS SUPPOSITORY
9016-45-9	RENEX 1000
9016-45-9	RENEX 300
9016-45-9	RENEX 647
9016-45-9	RENEX 648
9016-45-9	RENEX 650
9016-45-9	RENEX 678
9016-45-9	RENEX 679
9016-45-9	RENEX 688
9016-45-9	RENEX 690
9016-45-9	RENEX 697
9016-45-9	RENEX 698
105-60-2	RENYL MV
117-81-7	REOMOL D 79P
117-84-0	REOMOL D 79P

CAS No.	Chemical Name
103-23-1	REOMOL DOA
123-79-5	REOMOL DOA
117-84-0	REOMOL DOP
131-11-3	REPEFTAL
577-11-7	REQUTOL
7681-49-4	RESCUE SQUAD
50-55-5	RESERCAPS
50-55-5	RESERCEN
50-55-5	RESERLOR
50-55-5	RESERPAMED
50-55-5	RESERPEX
50-55-5	RESERPIC ACID METHYL ESTER 3,4,5-TRIMETHOXYBENZOATE (ESTER)
50-55-5	RESERPIL
50-55-5	RESERPINE
50-55-5	RESERPOID
9010-69-9	RESIN ACID ZINC SALT
9007-13-0	RESIN ACIDS AND ROSIN ACIDS, CALCIUM SALTS
68956-82-1	RESIN ACIDS AND ROSIN ACIDS, COBALT SALTS
9007-39-0	RESIN ACIDS AND ROSIN ACIDS, COPPER SALTS
9008-34-8	RESIN ACIDS AND ROSIN ACIDS, MANGANESE SALTS
9010-69-9	RESIN ACIDS AND ROSIN ACIDS, ZINC SALTS
50-55-5	RESINE
60-11-7	RESINOL YELLOW GR
9005-90-7	RESINS, PINE
9005-90-7	RESINS, TURPENTINE
9002-89-5	RESISTOFLEX
56-72-4	RESISTOX
56-72-4	RESITOX
108-46-3	RESO
108-46-3	RESORCIN
108-46-3	RESORCINE
1300-73-8	RESORCINE BROWN J
1300-73-8	RESORCINE BROWN R
108-46-3	RESORCINOL
108-46-3	RESORCINOL (ACGIH,DOT)
101-90-6	RESORCINOL DIGLYCIDYL ETHER
519-44-8	RESORCINOL, 2,4-DINITRO-
35860-51-6	RESORCINOL, 2,4-DINITRO-
519-44-8	RESORCINOL, DINITRO-, WETTED WITH AT LEAST 15% WATER
35860-51-6	RESORCINOL, DINITRO-, WETTED WITH AT LEAST 15% WATER
100-97-0	RESOTROPIN
50-55-5	RESPITAL
91-80-5	REST-ON
50-55-5	RESTRAN
91-80-5	RESTRYL
999-81-5	RETACEL
85-44-9	RETARDER AK
65-85-0	RETARDER BA
85-44-9	RETARDER ESEN
86-30-6	RETARDER J
85-44-9	RETARDER PD
69-72-7	RETARDER W
65-85-0	RETARDEX
9016-45-9	RETZANOL NP 100
53-86-1	REUMACIDE
577-11-7	REVAC
75-99-0	REVENGE
9016-45-9	REWOPAL HV 10
9016-45-9	REWOPAL HV 25
9016-45-9	REWOPOL HV-9
151-21-3	REWOPOL NLS 30
139-96-8	REWOPOL TLS 40
25155-30-0	REWORYL NKS 50
9004-34-6	REXCEL
137-26-8	REZIFILM
9004-70-0	RF 10
13463-67-7	RFC 5
12001-29-5	RG 600
23950-58-5	RH 315
53558-25-1	RH 787
7440-66-6	RHEINZINK
96-45-7	RHENOGRAN ETU
137-26-8	RHENOGRAN TMTD
1305-78-8	RHENOSORB C
1305-78-8	RHENOSORB F
50-78-2	RHEUMINTABLETTEN
14797-65-0	RHODANDINITROBENZOL
1762-95-4	RHODANID

CAS No.	Chemical Name
1762-95-4	RHODANIDE
96-45-7	RHODANIN S 62
94-75-7	RHODIA
78-53-5	RHODIA-6200
76-44-8	RHODIACHLOR
563-12-2	RHODIACIDE
56-38-2	RHODIASOL
56-38-2	RHODIATOX
56-38-2	RHODIATROX
137-26-8	RHODIAURAM
50-78-2	RHODINE
7440-16-6	RHODIUM
10049-07-7	RHODIUM CHLORIDE
10049-07-7	RHODIUM CHLORIDE (RhCl3)
10049-07-7	RHODIUM TRICHLORIDE
10049-07-7	RHODIUM(III) CHLORIDE
10049-07-7	RHODIUM(III) CHLORIDE (1:3)
7440-16-6	RHODIUM, METAL FUME AND DUSTS
7440-16-6	RHODIUM-103
563-12-2	RHODOCIDE
9002-89-5	RHODORICOL 4/20
9002-89-5	RHODOVIOL
9002-89-5	RHODOVIOL 16/200
9002-89-5	RHODOVIOL 4-125P
9002-89-5	RHODOVIOL 4/125
9002-89-5	RHODOVIOL 5/270P
9002-89-5	RHODOVIOL R 16/20
50-78-2	RHONAL
97-63-2	RHOPLEX AC-33 (ROHM AND HAAS)
72-54-8	RHOTHANE
94-59-7	RHYUNO OIL
9005-25-8	RICE STARCH
327-98-0	RICHLORONATE
25155-30-0	RICHONATE 1850
25155-30-0	RICHONATE 40B
25155-30-0	RICHONATE 45B
25155-30-0	RICHONATE 60B
27176-87-0	RICHONIC ACID B
151-21-3	RICHONOL A
151-21-3	RICHONOL AF
2235-54-3	RICHONOL AM
151-21-3	RICHONOL C
139-96-8	RICHONOL T
52-68-6	RICIFON
140-04-5	RICINOLEIC ACID, BUTYL ESTER, ACETATE
7439-95-4	RIEKE'S ACTIVE MAGNESIUM
85-43-8	RIKACID TH
68-76-8	RIKER 601
67-68-5	RIMSO-50
8020-83-5	RISELLA 33
50-55-5	RISERPA
70-30-4	RITOSEPT
52-68-6	RITSIFON
50-55-5	RIVASIN
7439-95-4	RMC
13463-67-7	RO 2
51-21-8	RO 2-9757
366-70-1	RO 4-6467
1309-37-1	RO 8097
81-81-2	RO-DETH
57-24-9	RO-DEX
7681-49-4	ROACH SALT
8052-42-4	ROAD ASPHALT, LIQUID
137-26-8	ROBAC TMT
434-07-1	ROBORAL
57-50-1	ROCK CANDY
14808-60-7	ROCK CRYSTAL
8002-05-9	ROCK OIL
7782-42-5	ROCOL X 7119
81-81-2	RODAFARIN
14797-65-0	RODATOX 60
5836-29-3	RODENTIN
81-81-2	RODEX
640-19-7	RODEX
81-81-2	RODEX BLOX
123-30-8	RODINAL
27598-85-2	RODINAL
563-12-2	RODOCID
563-12-2	RODOCIDE

CAS No.	Chemical Name
60-51-5	ROGODIAL
60-51-5	ROGOR
60-51-5	ROGOR 20L
60-51-5	ROGOR 40
60-51-5	ROGOR L
60-51-5	ROGOR P
25322-69-4	ROKOPOL D 2002
7758-98-7	ROMAN VITRIOL
299-84-3	RONNEL
83-79-4	RONONE
76-22-2	ROOT BARK OIL
81-81-2	ROSEX
9007-13-0	ROSIN CALCIUM SALT
8002-16-2	ROSIN OIL
25321-22-6	ROTAMOTT
83-79-4	ROTENON
83-79-4	ROTENONE
83-79-4	ROTENONE (COMMERCIAL)
83-79-4	ROTOCIDE
1309-37-1	ROUGE
1309-37-1	ROUGE (IRON OXIDE)
81-81-2	ROUGH & READY MOUSE MIX
20830-75-5	ROUGOXIN
299-84-3	ROVAN
13194-48-4	ROVOKIL
50-55-5	ROXINOID
60-51-5	ROXION
60-51-5	ROXION UA
1333-86-4	ROYAL SPECTRA
137-26-8	ROYAL TMTD
112-30-1	ROYALTAC
3691-35-8	ROZOL
20830-81-3	RP 13057
563-12-2	RP 8167
443-48-1	RP 8823
9004-70-0	RS
9004-70-0	RS 1/2
9004-70-0	RS 1/4
7439-97-6	RTEC (POLISH)
82-68-8	RTU 1010
8030-30-6	RUBBER SOLVENT
8052-42-4	RUBBER, MINERAL
67-63-0	RUBBING ALCOHOL
62-38-4	RUBERON
2235-25-8	RUBERON GRANULE
12001-26-2	RUBIDIAN MICA
7440-17-7	RUBIDIUM
1310-82-3	RUBIDIUM HYDROXIDE
1310-82-3	RUBIDIUM HYDROXIDE (Rb(OH))
1310-82-3	RUBIDIUM HYDROXIDE SOLUTION
1310-82-3	RUBIDIUM HYDROXIDE, SOLID
1310-82-3	RUBIDIUM HYDROXIDE, SOLID (DOT)
1310-82-3	RUBIDIUM HYDROXIDE, SOLUTION (DOT)
7440-17-7	RUBIDIUM METAL
7440-17-7	RUBIDIUM METAL (DOT)
7440-17-7	RUBIDIUM METAL, IN CARTRIDGES (DOT)
20830-81-3	RUBIDOMYCIN
1309-37-1	RUBIGO
101-68-8	RUBINATE 44
9016-87-9	RUBINATE M
9016-87-9	RUBINATE MF 178
26471-62-5	RUBINATE TDI
26471-62-5	RUBINATE TDI 80/20
20830-81-3	RUBOMYCIN C
12044-79-0	RUBY ARSENIC
123-79-5	RUCOFLEX PLASTICIZER DOA
103-23-1	RUCOFLEX PLASTICIZER DOA
299-86-5	RUELENE
299-86-5	RUELENE DRENCH
9002-84-0	RULON E
9002-84-0	RULON LD
56-53-1	RUMESTROL 1
56-53-1	RUMESTROL 2
1314-84-7	RUMETAN
13463-67-7	RUNA ARG
13463-67-7	RUNA ARH 20
13463-67-7	RUNA ARH 200
13463-67-7	RUNA RH 52
13463-67-7	RUNA RP

CAS No.	Chemical Name
78-34-2	RUPHOS
13463-67-7	RUTIOX CR
69-72-7	RUTRANEX
63-25-2	RYLAM
7782-42-5	S 1
7782-42-5	S 1 (GRAPHITE)
1129-41-5	S 1065
122-14-5	S 112A
112-53-8	S 1298
75-99-0	S 1315
55-38-9	S 1752
298-04-4	S 276
21923-23-9	S 2957
1333-86-4	S 300
1333-86-4	S 300 (CARBON BLACK)
1333-86-4	S 315
7429-90-5	S 40
7429-90-5	S 40 (METAL)
7758-29-4	S 400
2636-26-2	S 4084
327-98-0	S 4400
122-14-5	S 5660
7631-86-9	S 600
2540-82-1	S 6900
991-42-4	S 6999
115-90-2	S 767
75-99-0	S 95
75-99-0	S 95 (HERBICIDE)
505-60-2	S MUSTARD
8022-00-2	S(AND O)-2-(ETHYLTHIO)ETHYL O,O-DIMETHYL PHOSPHOROTHIOATE
563-12-2	S,S'-METHYLEN-BIS(O,O-DIAETHYL-DITHIOPHOSPHAT) (GERMAN)
563-12-2	S,S'-METHYLENE O,O,O',O'-TETRAETHYL PHOSPHORODITHIOATE
13194-48-4	S,S-DIPROPYL O-ETHYL PHOSPHORODITHIOATE
13071-79-9	S-(((1,1-DIMETHYLETHYL)THIO)METHYL)-O,O-DIETHYL PHOSPHORODITHIOATE
3735-23-7	S-(((2,5-DICHLOROPHENYL)THIO)METHYL) O,O-DIMETHYL PHOSPHORODITHIOATE
2778-04-3	S-(5-METHOXY-4-PYRON-2-YLMETHYL) DIMETHYL PHOSPHOROTHIOLATE
2778-04-3	S-((5-METHOXY-4H-PYRON-2-YL)-METHYL)-O,O-DIMETHYL-MONOTHIOFOSFAAT (DUTCH)
2778-04-3	S-((5-METHOXY-4H-PYRON-2-YL)-METHYL)-O,O-DIMETHYL-MONOTHIOPHOSPHAT (GERMAN)
2778-04-3	S-((5-METOSSI-4H-PIRON-2-IL)-METIL)-O,O-DIMETIL-MONOTIOFOSFATO (ITALIAN)
786-19-6	S-((P-CHLOROPHENYL THIO)METHYL) O,O-DIETHYL PHOSPHORODITHIOATE
13071-79-9	S-((TERT-BUTYLTHIO)METHYL)O,O-DIETHYLPHOSPHORODITHIOATE
54-11-5	S-(-)-NICOTINE
121-75-5	S-(1,2-BIS(AETHOXY-CARBONYL)-AETHYL)-O,O-DIMETHYL-DITHIOPHASPHAT (GERMAN)
121-75-5	S-(1,2-BIS(CARBETHOXY)ETHYL) O,O-DIMETHYL DITHIOPHOSPHATE
121-75-5	S-(1,2-BIS(ETHOXY-CARBONYL)-ETHYL)-O,O-DIMETHYL-DITHIOFOSFAAT (DUTCH)
121-75-5	S-(1,2-BIS(ETHOXYCARBONYL)ETHYL) O,O-DIMETHYL PHOSPHORODITHIOATE
121-75-5	S-(1,2-BIS(ETOSSI-CARBONIL)-ETIL)-O,O-DIMETIL-DITIOFOSFATO (ITALIAN)
121-75-5	S-(1,2-DI(ETHOXYCARBONYL)ETHYL DIMETHYL PHOSPHOROTHIOLOTHIONATE
121-75-5	S-(1,2-DICARBETHOXYETHYL) O,O-DIMETHYLDITHIOPHOSPHATE
2303-16-4	S-(2,3-DICHLOROALLYL) DIISOPROPYLTHIOCARBAMATE
950-37-8	S-(2,3-DIHYDRO-5-METHOXY-2-OXO-1,3,4-THIADIAZOL-3-METHYL) DIMETHYL PHOSPHOROTHIOLOTHIONATE
2275-14-1	S-(2,5-DICHLOROPHENYLTHIOMETHYL) DIETHYL PHOSPHOROTHIOLOTHIONATE
2275-14-1	S-(2,5-DICHLOROPHENYLTHIOMETHYL) O,O-DIETHYL PHOSPHORODITHIOATE
78-53-5	S-(2-(DIETHYLAMINO)ETHYL)PHOSPHOROTHIOIC ACID O,O-DIETHYL ESTER
919-86-8	S-(2-(ETHYLTHIO)ETHYL) DIMETHYL PHOSPHOROTHIOLATE
919-86-8	S-(2-(ETHYLTHIO)ETHYL) O,O-DIMETHYL PHOSPHOROTHIOATE
919-86-8	S-(2-(ETHYLTHIO)ETHYL) O,O-DIMETHYL THIOPHOSPHATE
640-15-3	S-(2-(ETHYLTHIO)ETHYL) O,O-DIMETHYLPHOSPHORODITHIONATE
640-15-3	S-(2-(ETHYLTHIO)ETHYL)DIMETHYL PHOSPHOROTHIOLOTHIONATE
2540-82-1	S-(2-(FORMYLMETHYLAMINO)-2-OXOETHYL) O,O-DIMETHYL-PHOSPHORODITHIOATE
3734-97-2	S-(2-DIETHYLAMINOETHYL) O,O-DIETHYL PHOSPHOROTHIOATE HYDROGEN OXALATE
3734-97-2	S-(2-DIETHYLAMINOETHYL) O,O-DIETHYLPHOSPHOROTHIOATE HYDROGENOXALATE
86-50-0	S-(3,4-DIHYDRO-4-OXO-BENZO(ALPHA)(1,2,3)TRIAZIN-3-YL-METHYL) O,O-DIMETHYL PHOSPHORODITHIOATE
2642-71-9	S-(3,4-DIHYDRO-4-OXO-1,2,3-BENZOTRIAZIN-3-YLMETHYL) O,O-DIETHYL PHOSPHORODITHIOATE
86-50-0	S-(3,4-DIHYDRO-4-OXO-1,2,3-BENZOTRIAZIN-3-YLMETHYL) O,O-DIMETHYL PHOSPHORODITHIOATE
786-19-6	S-(4-CHLOROPHENYLTHIOMETHYL)DIETHYL PHOSPHOROTHIOLOTHIONATE
24934-91-6	S-(CHLOROMETHYL) O,O-DIETHYLPHOSPHORODITHIOATE
78-53-5	S-(DIETHYLAMINOETHYL) O,O-DIETHYL PHOSPHOROTHIOATE
2540-82-1	S-(N-FORMYL-N-METHYLCARBAMOYLMETHYL) DIMETHYL PHOSPHOROTHIOLOTHIONATE
2540-82-1	S-(N-FORMYL-N-METHYLCARBAMOYLMETHYL) O,O-DIMETHYL PHOSPHORODITHIOATE
121-75-5	S-1,2-BIS(ETHOXYCARBONYL)ETHYL-O,O-DIMETHYL THIOPHOSPHATE
115-26-4	S-14
298-04-4	S-2-(ETHYLTHIO)ETHYL O,O-DIETHYL ESTER OF PHOSPHORODITHIOIC ACID
498-15-7	S-3-CARENE
2778-04-3	S-5-METHOXY-4-OXOPYRAN-2-YLMETHYL DIMETHYL PHOSPHOROTHIOATE
86-50-0	S-93
78-92-2	S-BUTANOL
94-79-1	S-BUTYL 2,4-DICHLOROPHENOXYACETATE
78-92-2	S-BUTYL ALCOHOL
24934-91-6	S-CHLOROMETHYL O,O-DIETHYL PHOSPHOROTHIOLOTHIONATE
108-83-8	S-DIISOPROPYLACETONE
121-75-5	S-ESTER WITH O,O-DIMETHYL PHOSPHOROTHIOATE
2941-64-2	S-ETHYL CARBONOCHLORIDOTHIOATE
26555-35-1	S-ETHYL CARBONOCHLORIDOTHIOATE
2941-64-2	S-ETHYL CHLOROTHIOCARBONATE
26555-35-1	S-ETHYL CHLOROTHIOCARBONATE
2941-64-2	S-ETHYL CHLOROTHIOFORMATE
26555-35-1	S-ETHYL CHLOROTHIOFORMATE
2941-64-2	S-ETHYL CHLOROTHIOLFORMATE
26555-35-1	S-ETHYL CHLOROTHIOLFORMATE
2941-64-2	S-ETHYL THIOCHLOROFORMATE
26555-35-1	S-ETHYL THIOCHLOROFORMATE
505-60-2	S-LOST
23135-22-0	S-METHYL 1-(DIMETHYLCARBAMOYL)-N-((METHYLCARBAMOYL)-OXY)THIOFORMIMIDATE
60-51-5	S-METHYLCARBAMOYLMETHYL O,O-DIMETHYL PHOSPHORODITHIOATE
14684-25-4	S-PHENYL DICHLOROTHIOPHOSPHITE
13889-92-4	S-PROPYL CHLOROTHIOFORMATE
13889-92-4	S-PROPYL THIOCHLOROFORMATE
513-49-5	S-SEC-BUTYLAMINE
79-27-6	S-TETRABROMOETHANE
79-34-5	S-TETRACHLOROETHANE
108-77-0	S-TRIAZINE TRICHLORIDE
108-77-0	S-TRIAZINE, 2,4,6-TRICHLORO-
675-14-9	S-TRIAZINE, 2,4,6-TRIFLUORO-
101-05-3	S-TRIAZINE, 2,4-DICHLORO-6-(O-CHLOROANILINO)-
12654-97-6	S-TRIAZINE, 2,4-DICHLORO-6-(O-CHLOROANILINO)-
1912-24-9	S-TRIAZINE, 2-CHLORO-4-ETHYLAMINO-6-ISOPROPYLAMINO-
121-82-4	S-TRIAZINE, HEXAHYDRO-1,3,5-TRINITRO-
87-90-1	S-TRIAZINE-2,4,6(1H,3H,5H)-TRIONE, 1,3,5-TRICHLORO-
2782-57-2	S-TRIAZINE-2,4,6(1H,3H,5H)-TRIONE, 1,3-DICHLORO-
2244-21-5	S-TRIAZINE-2,4,6(1H,3H,5H)-TRIONE, 1,3-DICHLORO-, POTASSIUM SALT
2893-78-9	S-TRIAZINE-2,4,6(1H,3H,5H)-TRIONE, 1,3-DICHLORO-, SODIUM SALT
13057-78-8	S-TRIAZINE-2,4,6(1H,3H,5H)-TRIONE, 1-CHLORO-
13057-78-8	S-TRIAZINE-2,4,6(1H,3H,5H)-TRIONE, CHLORO-
2244-21-5	S-TRIAZINE-2,4,6(1H,3H,5H)-TRIONE, DICHLORO-, POTASSIUM DERIV
87-90-1	S-TRIAZINE-2,4,6(1H,3H,5H)-TRIONE, TRICHLORO-
61-82-5	S-TRIAZOLE, 3-AMINO-
123-63-7	S-TRIMETHYLTRIOXYMETHYLENE
99-35-4	S-TRINITROBENZENE
118-96-7	S-TRINITROTOLUENE
118-96-7	S-TRINITROTOLUOL

CAS No.	Chemical Name
110-88-3	S-TRIOXANE
123-63-7	S-TRIOXANE, 2,4,6-TRIMETHYL-
505-60-2	S-YPERITE
121-75-5	S-[1,2-BIS(ETHOXYCARBONYL)ETHYL] O,O-DIMETHYL THIOPHOSPHATE
640-15-3	S-[2-(ETHYLTHIO)ETHYL] O,O-DIMETHYL PHOSPHORODITHIOATE
7439-92-1	S0
69-72-7	SA
118-74-1	SAATBEIZFUNGIZID (GERMAN)
81-07-2	SACCHARIMIDE
81-07-2	SACCHARIN
81-07-2	SACCHARIN ACID
81-07-2	SACCHARIN INSOLUBLE
128-44-9	SACCHARIN SODIUM
128-44-9	SACCHARIN SODIUM SALT
128-44-9	SACCHARIN SOLUBLE
81-07-2	SACCHARINE
81-07-2	SACCHARINOL
81-07-2	SACCHARINOSE
81-07-2	SACCHAROL
57-50-1	SACCHAROSE
57-50-1	SACCHARUM
121-75-5	SADOFOS
121-75-5	SADOFOS 30
121-75-5	SADOPHOS
137-26-8	SADOPLON
137-26-8	SADOPLON 75
53-86-1	SADOREUM
7782-41-4	SAEURE FLUORIDE (GERMAN)
1333-86-4	SAF
732-11-6	SAFIDON
94-59-7	SAFRENE
94-59-7	SAFROL
94-59-7	SAFROLE
94-59-7	SAFROLE MF
16893-85-9	SAFSAN
90-02-8	SAH
13770-96-2	SAH 22
8001-58-9	SAKRESOTE 100
12125-02-9	SAL AMMONIA
12125-02-9	SAL AMMONIAC
7646-93-7	SAL ENIXUM
50-78-2	SALACETIN
12125-02-9	SALAMMONITE
50-78-2	SALCETOGEN
14167-18-1	SALCOMINE
148-01-6	SALCOSTAT
50-78-2	SALETIN
90-02-8	SALICYLAL
135-02-4	SALICYLALDEHYDE METHYL ETHER
1420-04-8	SALICYLANILIDE, 2',5-DICHLORO-4'-NITRO-, COMPD. WITH 2-AMINOETHANOL (1:1)
69-72-7	SALICYLIC ACID
50-78-2	SALICYLIC ACID ACETATE
69-72-7	SALICYLIC ACID COLLODION
29790-52-1	SALICYLIC ACID, COMPD WITH NICOTINE (1:1)
57-64-7	SALICYLIC ACID, COMPD WITH PHYSOSTIGMINE (1:1)
5970-32-1	SALICYLIC ACID, MERCURY(2+) SALT (2:1)
10415-73-3	SALICYLIC ACID, MERCURY(2+) SALT (2:1)
119-36-8	SALICYLIC ACID, METHYL ESTER
90-02-8	SALICYLIC ALDEHYDE
12125-02-9	SALMIAC
69-72-7	SALONIL
7697-37-2	SALPETERSAURE (GERMAN)
7697-37-2	SALPETERZUUROPLOSSINGEN (DUTCH)
301-04-2	SALT OF SATURN
127-95-7	SALT OF SORREL
3811-04-9	SALT OF TARTER
7757-79-1	SALTPETER
7631-99-4	SALTPETER (CHILE)
10124-37-5	SALTPETER (NORWAY)
16893-85-9	SALUFER
75-60-5	SALVO
94-75-7	SALVO
65-85-0	SALVO LIQUID
62-38-4	SAMTOL
640-15-3	SAN 230
2540-82-1	SAN 244 I
2540-82-1	SAN 6913 I

CAS No.	Chemical Name
2540-82-1	SAN 7107 I
8002-74-2	SAN ROGUE WAX
57-24-9	SANASEED
16961-83-4	SAND ACID
16961-83-4	SAND ACID (DOT)
25155-30-0	SANDET 60
534-52-1	SANDOLIN
1335-85-9	SANDOLIN
1335-85-9	SANDOLIN A
534-52-1	SANDOLIN A
1937-37-7	SANDOPEL BLACK EX
2540-82-1	SANDOZ S-6900
50-55-5	SANDRIL
58-89-9	SANG GAMMA
106-93-4	SANHYUUM
92-04-6	SANIDRIL
62-38-4	SANITIZED SPG
62-38-4	SANMICRON
577-11-7	SANMORIN OT 70
25322-69-4	SANNIX PP 1000
25322-69-4	SANNIX PP 200
25322-69-4	SANNIX PP 400
577-11-7	SANNOL LDF 110
118-74-1	SANOCIDE
80-43-3	SANPEROX DCP
117-80-6	SANQUINON
85-68-7	SANTICIZER 160
84-72-0	SANTICIZER E-15
85-71-2	SANTICIZER M-17
87-86-5	SANTOBRITE
131-52-2	SANTOBRITE
7631-86-9	SANTOCEL
7631-86-9	SANTOCEL 54
7631-86-9	SANTOCEL 62
7631-86-9	SANTOCEL CS
7631-86-9	SANTOCEL Z
106-46-7	SANTOCHLOR
3081-14-9	SANTOFLEX 77
25155-30-0	SANTOMERSE 3
25155-30-0	SANTOMERSE ME
25155-30-0	SANTOMERSE NO 1
25155-30-0	SANTOMERSE NO 85
96-69-5	SANTONOX
96-69-5	SANTONOX BM
96-69-5	SANTONOX R
87-86-5	SANTOPHEN
87-86-5	SANTOPHEN 20
1336-36-3	SANTOTHERM
1336-36-3	SANTOTHERM FR
96-69-5	SANTOWHITE CRYSTALS
96-69-5	SANTOX
89-62-3	SANYO FAST RED GL BASE
3165-93-3	SANYO FAST RED SALT TR
95-69-2	SANYO FAST RED TR BASE
95-79-4	SANYO FAST RED TR BASE
119-38-0	SAOLAN
131-52-2	SAPCO 25
470-90-6	SAPECRON
470-90-6	SAPECRON 50EC
470-90-6	SAPRECON C
13256-22-9	SARCOSINE, N-NITROSO-
107-44-8	SARIN
107-44-8	SARIN II
333-41-5	SAROLEX
9004-70-0	SARTORIUS SM 11304
9004-70-0	SARTORIUS SM 11306
8002-74-2	SASOLWAX HI
8002-74-2	SASOLWAX M 1
1332-58-7	SATINTONE
1332-58-7	SATINTONE 5
1332-58-7	SATINTONE SPECIAL
52-68-6	SATOX 20WSC
16071-86-6	SATURN BROWN LBR
69-72-7	SAX
128-44-9	SAXIN
1314-13-2	SAZEX 2000
1314-13-2	SAZEX 4000
78-92-2	SBA
577-11-7	SBO

CAS No.	Chemical Name
68-22-4	SC 4640
72-33-3	SC 4725
122-39-4	SCALDIP
3761-53-3	SCARLET 2R
3761-53-3	SCARLET 2RB
3761-53-3	SCARLET 2RL BLUISH
99-55-8	SCARLET BASE CIBA II
99-55-8	SCARLET BASE IRGA II
99-55-8	SCARLET BASE NSP
3564-09-8	SCARLET F
99-55-8	SCARLET G BASE
3761-53-3	SCARLET R
3761-53-3	SCARLET RRA
1344-48-5	SCARLET VERMILION
2223-93-0	SCD
2631-37-0	SCH 34615
10290-12-7	SCHEELE'S MINERAL
10290-12-7	SCHEELES GREEN
2631-37-0	SCHERING 34615
152-16-9	SCHRADAN
152-16-9	SCHRADANE (FRENCH)
7782-42-5	SCHUNGITE
505-60-2	SCHWEFEL-LOST
7446-09-5	SCHWEFELDIOXYD (GERMAN)
75-15-0	SCHWEFELKOHLENSTOFF (GERMAN)
7664-93-9	SCHWEFELSAEURELOESUNGEN (GERMAN)
7783-06-4	SCHWEFELWASSERSTOFF (GERMAN)
7782-99-2	SCHWEFLIGE SAURE (GERMAN)
12002-03-8	SCHWEINFURT GREEN
12002-03-8	SCHWEINFURTERGRUN
12002-03-8	SCHWEINFURTH GREEN
106-42-3	SCINTILLAR
85-44-9	SCONOC 7
62-38-4	SCUTL
540-72-7	SCYAN
62-73-7	SD
4726-14-1	SD 11831
16752-77-5	SD 14999
62-73-7	SD 1750
96-12-8	SD 1897
309-00-2	SD 2794
60-57-1	SD 3417
72-20-8	SD 3419
869-29-4	SD 345
869-29-4	SD 345 (ESTER)
141-66-2	SD 3562
470-90-6	SD 4072
115-29-7	SD 4314
297-78-9	SD 4402
12789-03-6	SD 5532
7421-93-4	SD 7442
470-90-6	SD 7859
6923-22-4	SD 9129
2032-65-7	SD 9228
64-17-5	SD ALCOHOL 23-HYDROGEN
2636-26-2	SD-1750
25155-30-0	SDBS
540-73-8	SDMH
151-21-3	SDS
9000-07-1	SEAGEL GH
9000-07-1	SEAGEL PET
9000-07-1	SEAKEM 202
9000-07-1	SEAKEM CARRAGEENIN
1309-48-4	SEASORB
1333-86-4	SEAST 3
1333-86-4	SEAST 3H
1333-86-4	SEAST 6
1333-86-4	SEAST SO
1333-86-4	SEAST V
626-38-0	SEC-AMYL ACETATE
53496-15-4	SEC-AMYL ACETATE
53496-15-4	SEC-AMYL ACETATE (ACGIH,DOT)
584-02-1	SEC-AMYL ALCOHOL
6032-29-7	SEC-AMYL ALCOHOL
625-30-9	SEC-AMYLAMINE
41444-43-3	SEC-AMYLAMINE
513-53-1	SEC-BUTANETHIOL
78-92-2	SEC-BUTANOL
78-92-2	SEC-BUTANOL (DOT)

CAS No.	Chemical Name
94-79-1	SEC-BUTYL 2,4-DICHLOROPHENOXYACETATE
105-46-4	SEC-BUTYL ACETATE
105-46-4	SEC-BUTYL ACETATE (ACGIH,DOT)
78-92-2	SEC-BUTYL ALCOHOL
78-92-2	SEC-BUTYL ALCOHOL (ACGIH,DOT)
105-46-4	SEC-BUTYL ALCOHOL ACETATE
78-76-2	SEC-BUTYL BROMIDE
78-86-4	SEC-BUTYL CHLORIDE
25267-27-0	SEC-BUTYL IODIDE
513-53-1	SEC-BUTYL MERCAPTAN
513-53-1	SEC-BUTYL THIOALCOHOL
513-53-1	SEC-BUTYL THIOL
513-49-5	SEC-BUTYLAMINE
13952-84-6	SEC-BUTYLAMINE
513-49-5	SEC-BUTYLAMINE, (S)-
135-98-8	SEC-BUTYLBENZENE
137-32-6	SEC-BUTYLCARBINOL
2167-23-9	SEC-BUTYLIDENEBIS[TERT-BUTYL PEROXIDE]
142-92-7	SEC-HEXYL ACETATE
108-84-9	SEC-HEXYL ACETATE
5953-49-1	SEC-HEXYL ACETATE
108-84-9	SEC-HEXYL ACETATE (ACGIH)
598-75-4	SEC-ISOAMYL ALCOHOL
108-82-7	SEC-NONYL ALCOHOL
41444-43-3	SEC-PENTANAMINE
584-02-1	SEC-PENTANOL
6032-29-7	SEC-PENTANOL
53496-15-4	SEC-PENTYL ACETATE
584-02-1	SEC-PENTYL ALCOHOL
98-85-1	SEC-PHENETHYL ALCOHOL
67-63-0	SEC-PROPYL ALCOHOL
75-29-6	SEC-PROPYL CHLORIDE
75-31-0	SEC-PROPYLAMINE
127-00-4	SEC-PROPYLENE CHLOROHYDRIN
7784-44-3	SECONDARY AMMONIUM ARSENATE
7783-28-0	SECONDARY AMMONIUM PHOSPHATE
7558-79-4	SECONDARY SODIUM PHOSPHATE
577-11-7	SECOSOL DOS 70
7778-44-1	SECURITY
50-55-5	SEDARAUPIN
50-55-5	SEDSERP
136-40-3	SEDURAL
62-38-4	SEED DRESSING R
309-00-2	SEEDRIN
63-25-2	SEFFEIN
1333-86-4	SEIKA SEVEN
13410-01-0	SEL-TOX SSO2 AND SS-20
7783-07-5	SELANE
7783-08-6	SELENIC ACID
7783-08-6	SELENIC ACID (DOT)
14293-86-8	SELENIC ACID (H2Se2O7), DIPOTASSIUM SALT
7787-41-9	SELENIC ACID (H2SeO4), BARIUM SALT (1:1)
14019-91-1	SELENIC ACID (H2SeO4), CALCIUM SALT (1:1)
70027-50-8	SELENIC ACID (H2SeO4), COPPER SALT
13410-01-0	SELENIC ACID (H2SeO4), DISODIUM SALT
13597-54-1	SELENIC ACID (H2SeO4), ZINC SALT (1:1)
7787-41-9	SELENIC ACID, BARIUM SALT (1:1)
14019-91-1	SELENIC ACID, CALCIUM SALT (1:1)
70027-50-8	SELENIC ACID, COPPER SALT
14293-86-8	SELENIC ACID, DIPOTASSIUM SALT
13410-01-0	SELENIC ACID, DISODIUM SALT
7783-08-6	SELENIC ACID, LIQUID
7783-08-6	SELENIC ACID, LIQUID (DOT)
13597-54-1	SELENIC ACID, ZINC SALT (1:1)
7791-23-3	SELENINYL CHLORIDE
7791-23-3	SELENINYL DICHLORIDE
7783-00-8	SELENIOUS ACID
7783-00-8	SELENIOUS ACID (H2SeO3)
13718-59-7	SELENIOUS ACID (H2SeO3), BARIUM SALT (1:1)
10214-40-1	SELENIOUS ACID (H2SeO3), COPPER SALT
51325-42-9	SELENIOUS ACID (H2SeO3), COPPER SALT
10214-40-1	SELENIOUS ACID (H2SeO3), COPPER(2+) SALT (1:1)
51325-42-9	SELENIOUS ACID (H2SeO3), COPPER(2+) SALT (1:1)
10102-18-8	SELENIOUS ACID (H2SeO3), DISODIUM SALT
13718-59-7	SELENIOUS ACID, BARIUM SALT (1:1)
10214-40-1	SELENIOUS ACID, COPPER SALT
51325-42-9	SELENIOUS ACID, COPPER SALT
10214-40-1	SELENIOUS ACID, COPPER(2+) SALT (1:1)
51325-42-9	SELENIOUS ACID, COPPER(2+) SALT (1:1)

CAS No.	Chemical Name
10102-18-8	SELENIOUS ACID, DISODIUM SALT
7446-08-4	SELENIOUS ANHYDRIDE
12640-89-0	SELENIOUS ANHYDRIDE
7782-49-2	SELENIUM
7782-49-2	SELENIUM (COLLOIDAL)
7791-23-3	SELENIUM CHLORIDE OXIDE
7791-23-3	SELENIUM CHLORIDE OXIDE (SeCl2O)
7791-23-3	SELENIUM DICHLORIDE OXIDE
7783-07-5	SELENIUM DIHYDRIDE
7446-08-4	SELENIUM DIOXIDE
7783-00-8	SELENIUM DIOXIDE
12640-89-0	SELENIUM DIOXIDE
7488-56-4	SELENIUM DISULFIDE
7488-56-4	SELENIUM DISULPHIDE (DOT)
7782-49-2	SELENIUM ELEMENT
7783-79-1	SELENIUM FLUORIDE
7783-79-1	SELENIUM FLUORIDE (SeF6)
7783-79-1	SELENIUM FLUORIDE (SeF6), (OC-6-11)-
7783-79-1	SELENIUM HEXAFLUORIDE
7783-79-1	SELENIUM HEXAFLUORIDE (ACGIH,DOT)
7783-07-5	SELENIUM HYDRIDE
7783-07-5	SELENIUM HYDRIDE (H2Se)
7782-49-2	SELENIUM METAL, POWDER
7446-34-6	SELENIUM MONOSULFIDE
7446-08-4	SELENIUM OXIDE
12640-89-0	SELENIUM OXIDE
7446-08-4	SELENIUM OXIDE (SeO2)
13768-86-0	SELENIUM OXIDE (SeO3)
7791-23-3	SELENIUM OXYCHLORIDE
7791-23-3	SELENIUM OXYCHLORIDE (DOT)
7791-23-3	SELENIUM OXYCHLORIDE (SeOCl2)
7791-23-3	SELENIUM OXYDICHLORIDE
7488-56-4	SELENIUM SULFIDE
7446-34-6	SELENIUM SULFIDE
56093-45-9	SELENIUM SULFIDE
7446-34-6	SELENIUM SULFIDE (SeS)
7488-56-4	SELENIUM SULFIDE (SeS2)
13768-86-0	SELENIUM TRIOXIDE
13768-86-0	SELENIUM TRIOXIDE (SeO3)
7446-08-4	SELENIUM(IV) DIOXIDE (1:2)
12640-89-0	SELENIUM(IV) DIOXIDE (1:2)
7488-56-4	SELENIUM(IV) DISULFIDE (1:2)
7783-00-8	SELENOUS ACID
56-38-2	SELEPHOS
534-52-1	SELINON
1335-85-9	SELINON
8002-74-2	SELOSOL 428
7488-56-4	SELSUN
7488-56-4	SELSUN BLUE
563-41-7	SEMICARBAZIDE CHLORIDE
563-41-7	SEMICARBAZIDE HYDROCHLORIDE
79-19-6	SEMICARBAZIDE, 3-THIO-
563-41-7	SEMICARBAZIDE, MONOHYDROCHLORIDE
79-19-6	SEMICARBAZIDE, THIO-
91-80-5	SEMIKON
8020-83-5	SEMTOL 70
1309-64-4	SENARMONTITE
21087-64-9	SENCOR
21087-64-9	SENCOREX
50-18-0	SENDOXAN
114-26-1	SENDRAN
8002-05-9	SENECA OIL
57-06-7	SENF OEL (GERMAN)
505-60-2	SENFGAS
57-06-7	SENFOEL
79-09-4	SENTRY GRAIN PRESERVER
139-91-3	SEPSINOL
63-25-2	SEPTENE
56-75-7	SEPTICOL
70-30-4	SEPTISOL
70-30-4	SEPTOFEN
8065-48-3	SEPTOX
60-00-4	SEQ 100
60-00-4	SEQUESTRENE AA
60-00-4	SEQUESTRIC ACID
60-00-4	SEQUESTROL
50-55-5	SERFIN
70-30-4	SERIBAK
115-02-6	SERINE, DIAZOACETATE (ESTER), L-
82-28-0	SERISOL ORANGE YL
1937-37-7	SERISTAN BLACK B
50-55-5	SEROLFIA
50-55-5	SERP-AFD
50-55-5	SERPALOID
50-55-5	SERPANRAY
50-55-5	SERPASIL
50-55-5	SERPASIL PREMIX
50-55-5	SERPASOL
50-55-5	SERPATE
50-55-5	SERPEN
50-55-5	SERPENTINA
12001-29-5	SERPENTINE
12001-29-5	SERPENTINE CHRYSOTILE
50-55-5	SERPICON
50-55-5	SERPILOID
50-55-5	SERPINE
50-55-5	SERPINE (PHARMACEUTICAL)
50-55-5	SERPIPUR
56-53-1	SERRAL
50-55-5	SERTABS
50-55-5	SERTINA
1694-09-3	SERVA VIOLET 49
1333-86-4	SERVACARB
136-78-7	SES
136-78-7	SES-T
136-78-7	SESON
136-78-7	SESONE
12075-68-2	SESQUIETHYLALUMINUM CHLORIDE
1314-85-8	SESQUISULFURE DE PHOSPHORE (FRENCH)
119-90-4	SETACYL DIAZO NAVY R
62-38-4	SETRETE
1333-86-4	SEVACARB MT
1333-86-4	SEVALCO
63-25-2	SEVIMOL
63-25-2	SEVIN
63-25-2	SEVIN 4
56-53-1	SEXOCRETIN
108-94-1	SEXTONE
108-87-2	SEXTONE B
14808-60-7	SF 35
121-75-5	SF 60
9004-70-0	SHADOLAC MT
68308-34-9	SHALE OIL
7775-09-9	SHED-A-LEAF
7775-09-9	SHED-A-LEAF 'L'
8002-74-2	SHELL 100
470-90-6	SHELL 4072
297-78-9	SHELL 4402
1912-24-9	SHELL ATRAZINE HERBICIDE
6923-22-4	SHELL SD 9129
141-66-2	SHELL SD-3562
12789-03-6	SHELL SD-5532
7440-22-4	SHELL SILVER
107-18-6	SHELL UNKRAUTTOD A
297-78-9	SHELL WL 1650
9000-59-3	SHELLAC
8020-83-5	SHELLFLEX 273
8020-83-5	SHELLFLEX 412
111-84-2	SHELLSOL 140
8020-83-5	SHELLSOL 71
8002-74-2	SHELLWAX 100
8002-74-2	SHELLWAX 300
8002-74-2	SHELLWAX 400
94-59-7	SHIKIMOLE
94-59-7	SHIKOMOL
100-01-6	SHINNIPPON FAST RED GG BASE
29757-24-2	SHINNIPPON FAST RED GG BASE
89-62-3	SHINNIPPON FAST RED GL BASE
1333-86-4	SHOBLACK N 330
991-42-4	SHOXIN
7704-34-9	SHREESUL
7782-42-5	SHUNGITE
78-10-4	SI 42
7631-86-9	SI-O-LITE
10025-67-9	SIARKI CHLOREK (POLISH)
7446-09-5	SIARKI DWUTLENEK (POLISH)
7783-06-4	SIARKOWODOR (POLISH)
56-53-1	SIBOL

CAS No.	Chemical Name
7704-34-9	SICKOSUL
117-84-0	SICOL 150
117-81-7	SICOL 150
85-68-7	SICOL 160
26761-40-0	SICOL 184
103-23-1	SICOL 250
123-79-5	SICOL 250
14808-60-7	SIDERITE (SiO2)
1318-59-8	SIERRALITE
14998-27-7	SIERRALITE
7631-86-9	SIFLOX
9004-34-6	SIGMACELL
14808-60-7	SIKRON H 200
14808-60-7	SIKRON H 600
999-97-3	SILANAMINE, 1,1,1-TRIMETHYL-N-(TRIMETHYLSILYL)-
999-97-3	SILANAMINE, 1,1,1-TRIMETHYL-N-(TRIMETHYLSILYL)- (9CI)
7803-62-5	SILANE
7803-62-5	SILANE (DOT)
3037-72-7	SILANE, (4-AMINOBUTYL)DIETHOXYMETHYL-
107-37-9	SILANE, ALLYLTRICHLORO-
7521-80-4	SILANE, BUTYLTRICHLORO-
13465-78-6	SILANE, CHLORO-
1609-19-4	SILANE, CHLORODIETHYL-
1558-25-4	SILANE, CHLOROMETHYL(TRICHLORO)-
993-00-0	SILANE, CHLOROMETHYL-
26571-79-9	SILANE, CHLOROPHENYLTRICHLORO-
75-77-4	SILANE, CHLOROTRIMETHYL-
4109-96-0	SILANE, DICHLORO-
1719-53-5	SILANE, DICHLORODIETHYL-
75-78-5	SILANE, DICHLORODIMETHYL-
80-10-4	SILANE, DICHLORODIPHENYL-
1789-58-8	SILANE, DICHLOROETHYL-
75-54-7	SILANE, DICHLOROMETHYL-
149-74-6	SILANE, DICHLOROMETHYLPHENYL-
78-62-6	SILANE, DIETHOXYDIMETHYL-
4484-72-4	SILANE, DODECYLTRICHLORO-
115-21-9	SILANE, ETHYLTRICHLORO-
5894-60-0	SILANE, HEXADECYLTRICHLORO-
928-65-4	SILANE, HEXYLTRICHLORO-
75-79-6	SILANE, METHYLTRICHLORO-
112-04-9	SILANE, OCTADECYLTRICHLORO-
5283-66-9	SILANE, OCTYLTRICHLORO-
107-72-2	SILANE, PENTYLTRICHLORO-
98-13-5	SILANE, PHENYLTRICHLORO-
141-57-1	SILANE, PROPYLTRICHLORO-
10026-04-7	SILANE, TETRACHLORO-
78-10-4	SILANE, TETRAETHOXY-
7783-61-1	SILANE, TETRAFLUORO-
681-84-5	SILANE, TETRAMETHOXY-
75-76-3	SILANE, TETRAMETHYL-
1558-25-4	SILANE, TRICHLORO(CHLOROMETHYL)-
1558-25-4	SILANE, TRICHLORO(CHLOROMETHYL)- (9CI)
26571-79-9	SILANE, TRICHLORO(CHLOROPHENYL)-
10025-78-2	SILANE, TRICHLORO-
107-37-9	SILANE, TRICHLORO-2-PROPENYL-
107-37-9	SILANE, TRICHLOROALLYL-
98-12-4	SILANE, TRICHLOROCYCLOHEXYL-
4484-72-4	SILANE, TRICHLORODODECYL-
75-94-5	SILANE, TRICHLOROETHENYL-
115-21-9	SILANE, TRICHLOROETHYL-
5894-60-0	SILANE, TRICHLOROHEXADECYL-
928-65-4	SILANE, TRICHLOROHEXYL-
75-79-6	SILANE, TRICHLOROMETHYL-
5283-67-0	SILANE, TRICHLORONONYL-
112-04-9	SILANE, TRICHLOROOCTADECYL-
5283-66-9	SILANE, TRICHLOROOCTYL-
107-72-2	SILANE, TRICHLOROPENTYL-
98-13-5	SILANE, TRICHLOROPHENYL-
141-57-1	SILANE, TRICHLOROPROPYL-
75-94-5	SILANE, TRICHLOROVINYL-
998-30-1	SILANE, TRIETHOXY-
75-77-4	SILANE, TRIMETHYLCHLORO-
75-94-5	SILANE, VINYL TRICHLORO A-150
7631-86-9	SILANOX 101
2097-19-0	SILATRANE, PHENYL-
7761-88-8	SILBERNITRAT
1344-95-2	SILENE EF
11099-06-2	SILESTER AR
11099-06-2	SILESTER OS

CAS No.	Chemical Name
7631-86-9	SILEX
7440-22-4	SILFLAKE 135
7631-86-9	SILICA
7631-86-9	SILICA (SiO2)
7631-86-9	SILICA, AMORPHOUS, FUMED
60676-86-0	SILICA, AMORPHOUS, FUSED
14464-46-1	SILICA, CRISTOBALITE
60676-86-0	SILICA, FUSED, DUST
409-21-2	SILICA, GRAPHITE
12001-26-2	SILICA, MICA
14808-60-7	SILICA, QUARTZ
15468-32-3	SILICA, TRIDYMITE
1317-95-9	SILICA, TRIPOLI
7631-86-9	SILICAFILM
7631-86-9	SILICALITE S 115
7803-62-5	SILICANE
75-77-4	SILICANE, CHLOROTRIMETHYL-
115-21-9	SILICANE, TRICHLOROETHYL-
78-10-4	SILICATE D'ETHYLE (FRENCH)
78-10-4	SILICATE TETRAETHYLIQUE (FRENCH)
17084-08-1	SILICATE(2-), HEXAFLUORO-
16919-19-0	SILICATE(2-), HEXAFLUORO-, DIAMMONIUM
16961-83-4	SILICATE(2-), HEXAFLUORO-, DIHYDROGEN
16871-90-2	SILICATE(2-), HEXAFLUORO-, DIPOTASSIUM
16893-85-9	SILICATE(2-), HEXAFLUORO-, DISODIUM
16871-71-9	SILICATE(2-), HEXAFLUORO-, ZINC
16871-71-9	SILICATE(2-), HEXAFLUORO-, ZINC (1:1)
65997-15-1	SILICATE, PORTLAND CEMENT
10025-78-2	SILICI-CHLOROFORME (FRENCH)
6834-92-0	SILICIC ACID (H2SiO3), DISODIUM SALT
15191-85-2	SILICIC ACID (H4SiO4), BERYLLIUM SALT (1:2)
78-10-4	SILICIC ACID (H4SiO4), TETRAETHYL ESTER
681-84-5	SILICIC ACID (H4SiO4), TETRAMETHYL ESTER
75364-04-4	SILICIC ACID (H6Si2O7), COBALT(2+) MAGNESIUM SALT (1:2:1)
2157-42-8	SILICIC ACID (H6Si2O7), HEXAETHYL ESTER
4521-94-2	SILICIC ACID (H8Si3O10), OCTAETHYL ESTER
4421-95-8	SILICIC ACID (H8Si3O10), OCTAMETHYL ESTER
58500-38-2	SILICIC ACID, BERYLLIUM SALT
39413-47-3	SILICIC ACID, BERYLLIUM ZINC SALT
1344-95-2	SILICIC ACID, CALCIUM SALT
11099-06-2	SILICIC ACID, ETHYL ESTER
12002-26-5	SILICIC ACID, METHYL ESTER
78-10-4	SILICIC ACID, TETRAETHYL ESTER
10026-04-7	SILICIO(TETRACLORURO DI)
10026-04-7	SILICIUM(TETRACHLORURE DE) (FRENCH)
10025-78-2	SILICIUMCHLOROFORM (GERMAN)
10026-04-7	SILICIUMTETRACHLORID (GERMAN)
10026-04-7	SILICIUMTETRACHLORIDE (DUTCH)
10025-78-2	SILICOCHLOROFORM
16961-83-4	SILICOFLUORIC ACID
39630-75-6	SILICOFLUORIC ACID
16961-83-4	SILICOFLUORIC ACID (DOT)
17084-08-1	SILICOFLUORIDE
17084-08-1	SILICOFLUORIDE, SOLID
7440-21-3	SILICON
7440-21-3	SILICON (ACGIH)
68848-64-6	SILICON ALLOY, NONBASE, Li,Si
409-21-2	SILICON CARBIDE
12327-32-1	SILICON CARBIDE
12327-32-1	SILICON CARBIDE (Si2C3)
409-21-2	SILICON CARBIDE (SiC)
10026-04-7	SILICON CHLORIDE
10026-04-7	SILICON CHLORIDE (DOT)
10026-04-7	SILICON CHLORIDE (SiCl4)
10025-78-2	SILICON CHLORIDE HYDRIDE (SiHCl3)
7631-86-9	SILICON DIOXIDE
7631-86-9	SILICON DIOXIDE (SiO2)
7440-21-3	SILICON ELEMENT
78-10-4	SILICON ETHOXIDE
7783-61-1	SILICON FLUORIDE
7783-61-1	SILICON FLUORIDE (SiF4)
16961-83-4	SILICON HEXAFLUORIDE DIHYDRIDE
17084-08-1	SILICON HEXAFLUORIDE ION
7803-62-5	SILICON HYDRIDE (SiH4)
12002-26-5	SILICON METHYLATE
409-21-2	SILICON MONOCARBIDE
7631-86-9	SILICON OXIDE (SiO2)
98-13-5	SILICON PHENYL TRICHLORIDE
7440-21-3	SILICON POWDER, AMORPHOUS

CAS No.	Chemical Name
7440-21-3	SILICON POWDER, AMORPHOUS (DOT)
16893-85-9	SILICON SODIUM FLUORIDE
16893-85-9	SILICON SODIUM FLUORIDE (Na2SiF6)
10026-04-7	SILICON TETRACHLORIDE
10026-04-7	SILICON TETRACHLORIDE (DOT)
78-10-4	SILICON TETRAETHOXIDE
78-10-4	SILICON TETRAETHOXIDE (Si(OET)4)
7783-61-1	SILICON TETRAFLUORIDE
7783-61-1	SILICON TETRAFLUORIDE (DOT)
7803-62-5	SILICON TETRAHYDRIDE
7803-62-5	SILICON TETRAHYDRIDE (ACGIH)
681-84-5	SILICON TETRAMETHOXIDE
16871-71-9	SILICON ZINC FLUORIDE
16871-71-9	SILICON ZINC FLUORIDE (ZnSiF6)
78-10-4	SILIKAN L
7631-86-9	SILIKIL
7631-86-9	SILIPUR
7631-86-9	SILLIKOLLOID
7631-86-9	SILMOS
1344-95-2	SILMOS T
133-14-2	SILOPREN-VERNETZER CL 40
60-11-7	SILOTRAS YELLOW T 2G
7631-86-9	SILOXID
7440-22-4	SILPOWDER 130
7631-86-9	SILTON A
7631-86-9	SILTON A 2
7631-86-9	SILTON R 2
409-21-2	SILUNDUM
534-22-5	SILVAN
27137-41-3	SILVAN (CZECH)
58-89-9	SILVANOL
25168-26-7	SILVAPRON D
7440-22-4	SILVER
563-63-3	SILVER ACETATE
563-63-3	SILVER ACETATE (AgOAC)
7784-08-9	SILVER ARSENITE
7784-08-9	SILVER ARSENITE (DOT)
7440-22-4	SILVER ATOM
534-16-7	SILVER CARBONATE
534-16-7	SILVER CARBONATE (2:1)
506-64-9	SILVER CYANIDE
506-64-9	SILVER CYANIDE (Ag(CN))
506-64-9	SILVER CYANIDE (Ag2(CN)2)
506-64-9	SILVER CYANIDE (DOT)
7783-95-1	SILVER DIFLUORIDE
7783-95-1	SILVER FLUORIDE
7783-95-1	SILVER FLUORIDE (AgF2)
7782-42-5	SILVER GRAPHITE
7783-97-3	SILVER IODATE
7783-97-3	SILVER IODATE (AgIO3)
7440-31-5	SILVER MATT POWDER
7440-22-4	SILVER METAL
563-63-3	SILVER MONOACETATE
7761-88-8	SILVER MONONITRATE
7761-88-8	SILVER NITRATE
7761-88-8	SILVER NITRATE (AgNO3)
7761-88-8	SILVER NITRATE (DOT)
20667-12-3	SILVER OXIDE
20667-12-3	SILVER OXIDE (Ag2O)
146-84-9	SILVER PICRATE
146-84-9	SILVER PICRATE, DRY (DOT)
146-84-9	SILVER PICRATE, WETTED WITH AT LEAST 30% WATER (DOT)
506-61-6	SILVER POTASSIUM CYANIDE
506-61-6	SILVER POTASSIUM CYANIDE (AgK(CN)2)
10294-26-5	SILVER SULFATE
10294-26-5	SILVER SULFATE (Ag2SO4)
563-63-3	SILVER(1+) ACETATE
506-64-9	SILVER(1+) CYANIDE
7783-97-3	SILVER(1+) IODATE
7761-88-8	SILVER(1+) NITRATE
20667-12-3	SILVER(1+) OXIDE
563-63-3	SILVER(I) ACETATE
534-16-7	SILVER(I) CARBONATE
7761-88-8	SILVER(I) NITRATE (1:1)
20667-12-3	SILVER(I) OXIDE
7783-95-1	SILVER(II) FLUORIDE
146-84-9	SILVER, (PICRYLOXY)-
7440-22-4	SILVEST TCG 1
93-72-1	SILVEX
32534-95-5	SILVEX ISOOCTYL ESTER
93-72-1	SILVI-RHAP
124-65-2	SILVISAR
75-60-5	SILVISAR 510
2163-80-6	SILVISAR 550
13465-78-6	SILYL CHLORIDE
6834-92-0	SIMET A
732-11-6	SIMIDAN
2893-78-9	SIMPLA
298-00-0	SINAFID M 48
87-86-5	SINITUHO
151-21-3	SINNOPON LS 100
151-21-3	SINNOPON LS 95
25155-30-0	SINNOZON
26264-06-2	SINNOZON NCX 70
27323-41-7	SINNOZON NT 45
2235-54-3	SINOPON
60-51-5	SINORATOX
1335-85-9	SINOX
534-52-1	SINOX
88-85-7	SINOX GENERAL
151-21-3	SINTAPON L
56-75-7	SINTOMICETIN
56-75-7	SINTOMICETINA
56-75-7	SINTOMICETINE R
50-70-4	SIONIT
50-70-4	SIONITE
50-70-4	SIONON
7631-86-9	SIONOX
50-70-4	SIOSAN
105-60-2	SIPAS 60
7631-86-9	SIPERNAT 22
151-21-3	SIPEX OP
151-21-3	SIPEX SB
151-21-3	SIPEX SD
151-21-3	SIPEX SP
151-21-3	SIPEX UB
112-30-1	SIPOL L 10
112-53-8	SIPOL L 12
111-87-5	SIPOL L8
2235-54-3	SIPON L 22
2235-54-3	SIPON LA 30
143-00-0	SIPON LD
151-21-3	SIPON LS
151-21-3	SIPON LS 100
151-21-3	SIPON LSB
139-96-8	SIPON LT
139-96-8	SIPON LT 40
139-96-8	SIPON LT 6
151-21-3	SIPON PD
151-21-3	SIPON WD
25155-30-0	SIPONATE DS 10
25155-30-0	SIPONATE DS 4
25155-30-0	SIPONATE LDS 10
112-53-8	SIPONOL 25
112-53-8	SIPONOL L 2
112-53-8	SIPONOL L 5
2235-54-3	SIPROL L 22
121-75-5	SIPTOX I
7631-86-9	SIPUR 1500
16071-86-6	SIRIUS SUPRA BROWN BR
16071-86-6	SIRIUS SUPRA BROWN BRL
16071-86-6	SIRIUS SUPRA BROWN BRS
57-55-6	SIRLENE
1314-80-3	SIRNIK FOSFORECNY (CZECH)
60-51-5	SISTEMIN
55-63-0	SK-106N
545-55-1	SK-3818
20830-75-5	SK-DIGOXIN
106-89-8	SKEKHG
109-66-0	SKELLYSOLVE A
110-54-3	SKELLYSOLVE B
142-82-5	SKELLYSOLVE C
8032-32-4	SKELLYSOLVE F
8032-32-4	SKELLYSOLVE G
110-69-0	SKINO #1
9005-90-7	SKIPIDAR
7782-42-5	SKLN 1
9004-70-0	SL 1

CAS No.	Chemical Name
25322-69-4	SL 1000
8002-74-2	SLACKWAX 11
8002-74-2	SLACKWAX 30
1305-62-0	SLAKED LIME
91-80-5	SLEEPWELL
107-02-8	SLIMICIDE
1309-48-4	SLO 369
1309-48-4	SLO 469
8002-74-2	SLOPVOX
7720-78-7	SLOW-FE
151-21-3	SLS
64742-24-1	SLUDGE ACID
108-62-3	SLUG-TOX
9002-91-9	SLUG-TOX
108-62-3	SLUGIT
9002-91-9	SLUGIT
7428-48-0	SLW
3926-62-3	SMA
3926-62-3	SMA (HERBICIDE)
3926-62-3	SMCA
86-88-4	SMEESANA
7440-32-6	SMELLOFF-CUTTER TITANIUM
62-74-8	SMFA
732-11-6	SMIDAN
26628-22-8	SMITE
8002-74-2	SMITHWAX 117
12001-26-2	SMM 160
82-28-0	SMOKE ORANGE LK 6044
118-74-1	SMUT-GO
2631-37-0	SN 34615
131-89-5	SN 46
37273-91-9	SNAIL-KIL
712-48-1	SNEEZING GAS
55-63-0	SNG
118-74-1	SNIECIOTOX
644-64-4	SNIP
644-64-4	SNIP FLY
644-64-4	SNIP FLY BANDS
10043-52-4	SNOMELT
1314-13-2	SNOW WHITE
9005-25-8	SNOWFLAKE 12615
9005-25-8	SNOWFLAKE 30091
7631-86-9	SNOWTEX
7631-86-9	SNOWTEX 20
7631-86-9	SNOWTEX 30
7631-86-9	SNOWTEX C
7631-86-9	SNOWTEX N
7631-86-9	SNOWTEX O
7631-86-9	SNOWTEX OL
56-38-2	SNP
8020-83-5	SNPKH 7R2
7681-49-4	SO-FLO
127-85-5	SOAMIN
577-11-7	SOBITAL
8020-83-5	SOCAL 226
8020-83-5	SOCAL NO. 226
7775-09-9	SODA CHLORATE
7775-09-9	SODA CHLORATE (DOT)
8006-28-8	SODA LIME
8006-28-8	SODA LIME (DOT)
8006-28-8	SODA LIME, SOLID
8006-28-8	SODA LIME, SOLID (DOT)
7631-99-4	SODA NITER
7558-79-4	SODA PHOSPHATE
1310-73-2	SODA, CAUSTIC
7782-92-5	SODAMIDE
7784-46-5	SODANIT
57-41-0	SODANTON
7775-09-9	SODIO (CLORATO DI) (ITALIAN)
10588-01-9	SODIO (DICROMATO DI) (ITALIAN)
7601-89-0	SODIO (PERCLORATO DI) (ITALIAN)
62-74-8	SODIO, FLUORACETATO DI (ITALIAN)
7440-23-5	SODIUM
11135-81-2	SODIUM POTASSIUM ALLOY
13560-99-1	SODIUM (2,4,5-TRICHLOROPHENOXY)ACETATE
7601-89-0	SODIUM (PERCHLORATE DE) (FRENCH)
577-11-7	SODIUM 1,2-BIS(2-ETHYLHEXYLOXYCARBONYL)-1-ETHANE-SULFONATE
577-11-7	SODIUM 1,4-BIS(2-ETHYLHEXYL) SULFOSUCCINATE
18811-72-8	SODIUM 2,1-DIAZONAPHTHOL-5-SULFONATE
136-78-7	SODIUM 2,4-DICHLOROPHENOXYETHYL SULPHATE
136-78-7	SODIUM 2-(2,4-DICHLOROPHENOXY)ETHYL SULFATE
18811-72-8	SODIUM 2-DIAZO-1-NAPHTHOL-5-SULFONATE
577-11-7	SODIUM 2-ETHYLHEXYLSULFOSUCCINATE
831-52-7	SODIUM 6-AMINO-2,4-DINITROPHENOLATE
2163-80-6	SODIUM ACID METHANEARSONATE
7681-38-1	SODIUM ACID SULFATE
7681-38-1	SODIUM ACID SULFATE, SOLID (DOT)
7681-38-1	SODIUM ACID SULFATE, SOLUTION (DOT)
7631-90-5	SODIUM ACID SULFITE
11110-52-4	SODIUM ALLOY, NONBASE, Hg,Na
11135-81-2	SODIUM ALLOY, NONBASE, K,Na
11138-49-1	SODIUM ALUMINATE
11138-49-1	SODIUM ALUMINATE SOLUTION
11138-49-1	SODIUM ALUMINATE, SOLID
11138-49-1	SODIUM ALUMINATE, SOLID (DOT)
11138-49-1	SODIUM ALUMINATE, SOLUTION (DOT)
15096-52-3	SODIUM ALUMINUM FLUORIDE
13770-96-2	SODIUM ALUMINUM HYDRIDE
13770-96-2	SODIUM ALUMINUM HYDRIDE (DOT)
13770-96-2	SODIUM ALUMINUM HYDRIDE (NaAlH4)
11138-49-1	SODIUM ALUMINUM OXIDE
13770-96-2	SODIUM ALUMINUM TETRAHYDRIDE
11110-52-4	SODIUM AMALGAM
11110-52-4	SODIUM AMALGAM (DOT)
7782-92-5	SODIUM AMIDE
7782-92-5	SODIUM AMIDE (DOT)
7782-92-5	SODIUM AMIDE (Na(NH2))
127-85-5	SODIUM AMINARSONATE
127-85-5	SODIUM AMINOPHENOL ARSONATE
13718-26-8	SODIUM AMMONIUM VANADATE
127-85-5	SODIUM ANILARSONATE
128-56-3	SODIUM ANTHRAQUINONE-1-SULFONATE
127-85-5	SODIUM ARSANILATE
127-85-5	SODIUM ARSANILATE (DOT)
7631-89-2	SODIUM ARSENATE
7784-46-5	SODIUM ARSENIC OXIDE (NaAsO2)
7784-46-5	SODIUM ARSENITE
13464-37-4	SODIUM ARSENITE
13464-37-4	SODIUM ARSENITE (Na3AsO3)
7784-46-5	SODIUM ARSENITE (NaAsO2)
7784-46-5	SODIUM ARSENITE SOLUTION
7784-46-5	SODIUM ARSENITE, LIQUID (SOLUTION) (DOT)
7784-46-5	SODIUM ARSENITE, SOLID
7784-46-5	SODIUM ARSENITE, SOLID (DOT)
127-85-5	SODIUM ARSONILATE
7440-23-5	SODIUM ATOM
26628-22-8	SODIUM AZIDE
26628-22-8	SODIUM AZIDE (ACGIH,DOT)
26628-22-8	SODIUM AZIDE (Na(N3))
11138-49-1	SODIUM BETA ALUMINA
1330-43-4	SODIUM BIBORATE
1303-96-4	SODIUM BIBORATE DECAHYDRATE
10588-01-9	SODIUM BICHROMATE
1333-83-1	SODIUM BIFLUORIDE
1333-83-1	SODIUM BIFLUORIDE, SOLID
1333-83-1	SODIUM BIFLUORIDE, SOLID (DOT)
1333-83-1	SODIUM BIFLUORIDE, SOLUTION
1333-83-1	SODIUM BIFLUORIDE, SOLUTION (DOT)
577-11-7	SODIUM BIS(2-ETHYLHEXYL) SULFOSUCCINATE
7681-38-1	SODIUM BISULFATE
7681-38-1	SODIUM BISULFATE, FUSED
7681-38-1	SODIUM BISULFATE, SOLID (DOT)
7681-38-1	SODIUM BISULFATE, SOLUTION (DOT)
16721-80-5	SODIUM BISULFIDE
7631-90-5	SODIUM BISULFITE
7631-90-5	SODIUM BISULFITE (1:1)
7631-90-5	SODIUM BISULFITE (ACGIH)
7631-90-5	SODIUM BISULFITE (NaHSO3)
7631-90-5	SODIUM BISULFITE, (SOLID)
7631-90-5	SODIUM BISULFITE, SOLID (DOT)
7631-90-5	SODIUM BISULFITE, SOLUTION (DOT)
7681-38-1	SODIUM BISULPHATE
7631-90-5	SODIUM BISULPHITE
1303-96-4	SODIUM BORATE
7632-04-4	SODIUM BORATE
11138-47-9	SODIUM BORATE
1303-96-4	SODIUM BORATE (Na2B4O7), DECAHYDRATE

CAS No.	Chemical Name
7632-04-4	SODIUM BORATE (NaBO3)
11138-47-9	SODIUM BORATE (NaBO3)
1303-96-4	SODIUM BORATES
16940-66-2	SODIUM BOROHYDRATE
16940-66-2	SODIUM BOROHYDRIDE
16940-66-2	SODIUM BOROHYDRIDE (DOT)
16940-66-2	SODIUM BOROHYDRIDE (Na(BH4))
7789-38-0	SODIUM BROMATE
7789-38-0	SODIUM BROMATE (DOT)
7789-38-0	SODIUM BROMATE (NaBrO3)
124-65-2	SODIUM CACODYLATE
124-65-2	SODIUM CACODYLATE (DOT)
139-02-6	SODIUM CARBOLATE
3313-92-6	SODIUM CARBONATE
3313-92-6	SODIUM CARBONATE (Na2C2O6)
3313-92-6	SODIUM CARBONATE PEROXIDE
4452-58-8	SODIUM CARBONATE PEROXIDE
15630-89-4	SODIUM CARBONATE PEROXIDE
4452-58-8	SODIUM CARBONATE-HYDROGEN PEROXIDE (2:3)
3313-92-6	SODIUM CARBONATE-HYDROGEN PEROXIDE (2:3)
15630-89-4	SODIUM CARBONATE-HYDROGEN PEROXIDE (2:3)
4452-58-8	SODIUM CARBONATE-HYDROGEN PEROXIDE ADDUCT
3313-92-6	SODIUM CARBONATE-HYDROGEN PEROXIDE ADDUCT
15630-89-4	SODIUM CARBONATE-HYDROGEN PEROXIDE ADDUCT
7775-09-9	SODIUM CHLORATE
7775-09-9	SODIUM CHLORATE
7775-09-9	SODIUM CHLORATE (DOT)
7775-09-9	SODIUM CHLORATE (NaClO3)
7775-09-9	SODIUM CHLORATE SOLUTION
7775-09-9	SODIUM CHLORATE, AQUEOUS SOLUTION (DOT)
7758-19-2	SODIUM CHLORITE
7758-19-2	SODIUM CHLORITE (DOT)
7758-19-2	SODIUM CHLORITE, NOT MORE THAN 42% (DOT)
7758-19-2	SODIUM CHLORITE, SOLUTION CONTAINING MORE THAN 5% AVAILABLE CHLORINE (DOT)
3926-62-3	SODIUM CHLOROACETATE
3926-62-3	SODIUM CHLOROACETATE (DOT)
51580-86-0	SODIUM CHLOROCYANURATE DIHYDRATE
1307-82-0	SODIUM CHLOROPLATINATE
7775-11-3	SODIUM CHROMATE
12680-48-7	SODIUM CHROMATE
10588-01-9	SODIUM CHROMATE
7775-11-3	SODIUM CHROMATE (DOT)
10588-01-9	SODIUM CHROMATE (Na2Cr2O7)
7775-11-3	SODIUM CHROMATE (Na2CrO4)
7775-11-3	SODIUM CHROMATE SOLUTION
7775-11-3	SODIUM CHROMATE(VI)
12680-48-7	SODIUM CHROMITE
129-06-6	SODIUM COUMADIN
3564-09-8	SODIUM CUMENEAZO-BETA-NAPHTHOL DISULFONATE
14264-31-4	SODIUM CUPROCYANIDE SOLUTION
14264-31-4	SODIUM CUPROCYANIDE SOLUTION (DOT)
14264-31-4	SODIUM CUPROCYANIDE, SOLID
14264-31-4	SODIUM CUPROCYANIDE, SOLID (DOT)
143-33-9	SODIUM CYANIDE
143-33-9	SODIUM CYANIDE (ACGIH)
143-33-9	SODIUM CYANIDE (Na(CN))
143-33-9	SODIUM CYANIDE SOLUTION
143-33-9	SODIUM CYANIDE, SOLID
143-33-9	SODIUM CYANIDE, SOLID (DOT)
143-33-9	SODIUM CYANIDE, SOLUTION (DOT)
4384-82-1	SODIUM DEDT
577-11-7	SODIUM DI(2-ETHYLHEXYL) SULFOSUCCINATE
2893-78-9	SODIUM DICHLORO-ISOCYANATE
2893-78-9	SODIUM DICHLORO-S-TRIAZINETRONE
2893-78-9	SODIUM DICHLOROCYANURATE
2893-78-9	SODIUM DICHLOROISOCYANURATE
51580-86-0	SODIUM DICHLOROISOCYANURATE DIHYDRATE
10588-01-9	SODIUM DICHROMATE
10588-01-9	SODIUM DICHROMATE (DOT)
10588-01-9	SODIUM DICHROMATE (Na2(Cr2O7))
10588-01-9	SODIUM DICHROMATE(VI)
4384-82-1	SODIUM DIETHYLDITHIOCARBAMATE
124-65-2	SODIUM DIMETHYLARSINATE
124-65-2	SODIUM DIMETHYLARSONATE
25641-53-6	SODIUM DINITRO-O-CRESOLATE, WETTED WITH AT LEAST 10% WATER (DOT)
25641-53-6	SODIUM DINITRO-O-CRESOLATE, WETTED WITH AT LEAST 15% WATER (DOT)

CAS No.	Chemical Name
25641-53-6	SODIUM DINITRO-O-CRESYLATE
2312-76-7	SODIUM DINITRO-ORTHO-CRESOLATE
577-11-7	SODIUM DIOCTYL SULFOSUCCINATE
577-11-7	SODIUM DIOCTYL SULPHOSUCCINATE
1313-60-6	SODIUM DIOXIDE
7722-88-5	SODIUM DIPHOSPHATE
7722-88-5	SODIUM DIPHOSPHATE (Na4P2O7)
7681-57-4	SODIUM DISULFITE
7757-74-6	SODIUM DISULFITE
7775-14-6	SODIUM DITHIONITE
7775-14-6	SODIUM DITHIONITE (DOT)
7775-14-6	SODIUM DITHIONITE (Na2S2O4)
151-21-3	SODIUM DODECYL SULFATE
151-21-3	SODIUM DODECYL SULPHATE
25155-30-0	SODIUM DODECYLBENZENE SULFONATE
25155-30-0	SODIUM DODECYLBENZENESULFONATE (DOT)
25155-30-0	SODIUM DODECYLBENZENESULFONATE, DRY
25155-30-0	SODIUM DODECYLBENZENESULPHONATE
25155-30-0	SODIUM DODECYLPHENYLSULFONATE
13601-19-9	SODIUM FERROCYANIDE
13601-19-9	SODIUM FERROCYANIDE (Na4[Fe(CN)6])
62-74-8	SODIUM FLUOACETATE
62-74-8	SODIUM FLUOACETIC ACID
62-74-8	SODIUM FLUORACETATE
1333-83-1	SODIUM FLUORIDE
7681-49-4	SODIUM FLUORIDE
1333-83-1	SODIUM FLUORIDE (Na(HF2))
7681-49-4	SODIUM FLUORIDE (NaF)
7681-49-4	SODIUM FLUORIDE CYCLIC DIMER
7681-49-4	SODIUM FLUORIDE, SOLID
7681-49-4	SODIUM FLUORIDE, SOLID (DOT)
7681-49-4	SODIUM FLUORIDE, SOLUTION
7681-49-4	SODIUM FLUORIDE, SOLUTION (DOT)
62-74-8	SODIUM FLUOROACETATE
62-74-8	SODIUM FLUOROACETATE (ACGIH,DOT)
62-74-8	SODIUM FLUOROACETATE DE (FRENCH)
16893-85-9	SODIUM FLUOROSILICATE
16893-85-9	SODIUM FLUOROSILICATE (Na2SiF6)
7681-49-4	SODIUM FLUORURE (FRENCH)
16893-85-9	SODIUM FLUOSILICATE
16893-85-9	SODIUM FLUOSILICATE (Na2(SiF6))
1307-82-0	SODIUM HEXACHLOROPLATINATE (VI)
13601-19-9	SODIUM HEXACYANOFERRATE (II)
16893-85-9	SODIUM HEXAFLUOROSILICATE
16893-85-9	SODIUM HEXAFLUOROSILICATE (Na2SiF6)
16893-85-9	SODIUM HEXAFLUOSILICATE
10124-56-8	SODIUM HEXAMETAPHOSPHATE
10124-56-8	SODIUM HEXAMETAPHOSPHATE (Na6P6O18)
7646-69-7	SODIUM HYDRIDE
7646-69-7	SODIUM HYDRIDE (DOT)
7646-69-7	SODIUM HYDRIDE (NaH)
16940-66-2	SODIUM HYDROBORATE
7681-49-4	SODIUM HYDROFLUORIDE
1333-83-1	SODIUM HYDROGEN DIFLUORIDE
1333-83-1	SODIUM HYDROGEN FLUORIDE (DOT)
7558-79-4	SODIUM HYDROGEN PHOSPHATE
7681-38-1	SODIUM HYDROGEN SULFATE
7681-38-1	SODIUM HYDROGEN SULFATE (NaHSO4)
7681-38-1	SODIUM HYDROGEN SULFATE SOLUTION
7681-38-1	SODIUM HYDROGEN SULFATE, SOLID
7681-38-1	SODIUM HYDROGEN SULFATE, SOLID (DOT)
7681-38-1	SODIUM HYDROGEN SULFATE, SOLUTION (DOT)
16721-80-5	SODIUM HYDROGEN SULFIDE
16721-80-5	SODIUM HYDROGEN SULFIDE (NaHS)
7631-90-5	SODIUM HYDROGEN SULFITE
7631-90-5	SODIUM HYDROGEN SULFITE, SOLID
7631-90-5	SODIUM HYDROGEN SULFITE, SOLID (DOT)
7631-90-5	SODIUM HYDROGEN SULFITE, SOLUTION
7631-90-5	SODIUM HYDROGEN SULFITE, SOLUTION (DOT)
7681-38-1	SODIUM HYDROSULFATE
16721-80-5	SODIUM HYDROSULFIDE
16721-80-5	SODIUM HYDROSULFIDE (Na(HS))
16721-80-5	SODIUM HYDROSULFIDE, SOLID, WITH 25% WATER OF CRYSTAL-LIZATION
16721-80-5	SODIUM HYDROSULFIDE, SOLUTION (DOT)
7775-14-6	SODIUM HYDROSULFITE
7775-14-6	SODIUM HYDROSULFITE (DOT)
7775-14-6	SODIUM HYDROSULFITE (Na2S2O4)
16721-80-5	SODIUM HYDROSULPHIDE, SOLID (DOT)

CAS No.	Chemical Name
16721-80-5	SODIUM HYDROSULPHIDE, WITH LESS THAN 25% WATER OF CRYSTALLIZATION (DOT)
7775-14-6	SODIUM HYDROSULPHITE
1310-73-2	SODIUM HYDROXIDE
1310-73-2	SODIUM HYDROXIDE (Na(OH))
8006-28-8	SODIUM HYDROXIDE (Na(OH)), MIXT WITH LIME
1310-73-2	SODIUM HYDROXIDE (SOLUTION)
7681-52-9	SODIUM HYPOCHLORITE
10022-70-5	SODIUM HYPOCHLORITE PENTAHYDRATE
7775-14-6	SODIUM HYPOSULFITE
540-72-7	SODIUM ISOTHIOCYANATE
151-21-3	SODIUM LAURYL SULFATE
151-21-3	SODIUM LAURYL SULPHATE
25155-30-0	SODIUM LAURYLBENZENESULFONATE
7681-57-4	SODIUM M-BISULFITE
10101-50-5	SODIUM MANGANATE
16721-80-5	SODIUM MERCAPTAN
16721-80-5	SODIUM MERCAPTIDE
7784-46-5	SODIUM METAARSENITE
7681-57-4	SODIUM METABISULFITE
7757-74-6	SODIUM METABISULFITE
7681-57-4	SODIUM METABISULFITE (ACGIH,DOT)
7757-74-6	SODIUM METABISULFITE (ACGIH,DOT)
7757-74-6	SODIUM METABISULFITE (Na2S2O5)
7681-57-4	SODIUM METABISULFITE (Na2S2O5)
7681-57-4	SODIUM METABISULPHITE
7757-74-6	SODIUM METABISULPHITE
7440-23-5	SODIUM METAL
7785-84-4	SODIUM METAPHOSPHATE
7785-84-4	SODIUM METAPHOSPHATE ((NaPO3)3)
10124-56-8	SODIUM METAPHOSPHATE (Na6P6O18)
6834-92-0	SODIUM METASILICATE
6834-92-0	SODIUM METASILICATE (Na2SiO3)
6834-92-0	SODIUM METASILICATE, ANHYDROUS
13718-26-8	SODIUM METAVANADATE
11105-06-9	SODIUM METAVANADATE
11105-06-9	SODIUM METAVANADATE (NaVO3)
2163-80-6	SODIUM METHANEARSONATE
124-41-4	SODIUM METHANOLATE
124-41-4	SODIUM METHOXIDE
124-41-4	SODIUM METHYLATE
124-41-4	SODIUM METHYLATE (ALCOHOL MIXTURE)
124-41-4	SODIUM METHYLATE (DOT)
124-41-4	SODIUM METHYLATE, ALCOHOL MIXTURE (DOT)
124-41-4	SODIUM METHYLATE, DRY
124-41-4	SODIUM METHYLATE, DRY (DOT)
124-41-4	SODIUM METHYLATE-ALCOHOL MIXTURE [COMBUSTIBLE LIQUID LABEL]
124-41-4	SODIUM METHYLATE-ALCOHOL MIXTURE [CORROSIVE MATER-IAL LABEL]
124-41-4	SODIUM METHYLATE-ALCOHOL MIXTURE [FLAMMABLE LIQUID LABEL]
3926-62-3	SODIUM MONOCHLORACETATE
3926-62-3	SODIUM MONOCHLOROACETATE
151-21-3	SODIUM MONODODECYL SULFATE
7681-49-4	SODIUM MONOFLUORIDE
62-74-8	SODIUM MONOFLUOROACETATE
7646-69-7	SODIUM MONOHYDRIDE
7558-79-4	SODIUM MONOHYDROGEN PHOSPHATE
7558-79-4	SODIUM MONOHYDROGEN PHOSPHATE (2:1:1)
151-21-3	SODIUM MONOLAURYL SULFATE
15593-29-0	SODIUM MONOPERSULFATE
28831-12-1	SODIUM MONOPERSULFATE
1313-82-2	SODIUM MONOSULFIDE
12401-86-4	SODIUM MONOXIDE
12401-86-4	SODIUM MONOXIDE (DOT)
12401-86-4	SODIUM MONOXIDE, SOLID
12401-86-4	SODIUM MONOXIDE, SOLID (DOT)
4384-82-1	SODIUM N,N-DIETHYLDITHIOCARBAMATE
151-21-3	SODIUM N-DODECYL SULFATE
7631-99-4	SODIUM NITRATE
7631-99-4	SODIUM NITRATE (DOT)
7632-00-0	SODIUM NITRITE
7632-00-0	SODIUM NITRITE (DOT)
7632-00-0	SODIUM NITRITE (NaNO2)
128-44-9	SODIUM O-BENZOSULFIMIDE
143-19-1	SODIUM OLEATE
7631-89-2	SODIUM ORTHOARSENATE
10101-89-0	SODIUM ORTHOPHOSPHATE DODECAHYDRATE
13721-39-6	SODIUM ORTHOVANADATE
62-76-0	SODIUM OXALATE
12401-86-4	SODIUM OXIDE
1313-60-6	SODIUM OXIDE (Na2O2)
12401-86-4	SODIUM OXIDE (NaO)
127-85-5	SODIUM P-AMINOBENZENEARSONATE
127-85-5	SODIUM P-AMINOPHENYLARSONATE
127-85-5	SODIUM P-ARSANILATE
131-52-2	SODIUM PCP
131-52-2	SODIUM PENTACHLOROPHENATE
131-52-2	SODIUM PENTACHLOROPHENATE (DOT)
131-52-2	SODIUM PENTACHLOROPHENOL
131-52-2	SODIUM PENTACHLOROPHENOLATE
131-52-2	SODIUM PENTACHLOROPHENOXIDE
131-52-2	SODIUM PENTACHLORPHENATE
1303-96-4	SODIUM PERBORATE
7632-04-4	SODIUM PERBORATE
11138-47-9	SODIUM PERBORATE
7632-04-4	SODIUM PERBORATE (NaBO3)
11138-47-9	SODIUM PERBORATE (NaBO3)
3313-92-6	SODIUM PERCARBONATE
4452-58-8	SODIUM PERCARBONATE
15630-89-4	SODIUM PERCARBONATE
15630-89-4	SODIUM PERCARBONATE (CARBONIC ACID DISODIUM SALT, COMPOUND WITH WATER)
4452-58-8	SODIUM PERCARBONATE (CARBONOPEROXOIC ACID, DISODIUM SALT)
3313-92-6	SODIUM PERCARBONATE (DOT)
4452-58-8	SODIUM PERCARBONATE (DOT)
15630-89-4	SODIUM PERCARBONATE (DOT)
3313-92-6	SODIUM PERCARBONATE (Na2C2O6)
15630-89-4	SODIUM PERCARBONATE (Na2C2O6)
15630-89-4	SODIUM PERCARBONATE (Na2CO315H2O2)
3313-92-6	SODIUM PERCARBONATE (PEROXYDICARBONIC ACID, DISODIUM SALT)
7601-89-0	SODIUM PERCHLORATE
7601-89-0	SODIUM PERCHLORATE (DOT)
10101-50-5	SODIUM PERMANGANATE
10101-50-5	SODIUM PERMANGANATE (DOT)
10101-50-5	SODIUM PERMANGANATE (NaMnO4)
1313-60-6	SODIUM PEROXIDE
1313-60-6	SODIUM PEROXIDE (DOT)
1313-60-6	SODIUM PEROXIDE (Na2(O2))
1303-96-4	SODIUM PEROXOBORATE
7632-04-4	SODIUM PEROXOBORATE
11138-47-9	SODIUM PEROXOBORATE
4452-58-8	SODIUM PEROXYCARBONATE
3313-92-6	SODIUM PEROXYCARBONATE
15630-89-4	SODIUM PEROXYCARBONATE
4452-58-8	SODIUM PEROXYCARBONATE (Na2CO4)
3313-92-6	SODIUM PEROXYCARBONATE (Na2CO4)
15630-89-4	SODIUM PEROXYCARBONATE (Na2CO4)
3313-92-6	SODIUM PEROXYDICARBONATE
4452-58-8	SODIUM PEROXYDICARBONATE
15630-89-4	SODIUM PEROXYDICARBONATE
3313-92-6	SODIUM PEROXYDICARBONATE (Na2C2O6)
15630-89-4	SODIUM PEROXYDICARBONATE (Na2C2O6)
7775-27-1	SODIUM PEROXYDISULFATE
28831-12-1	SODIUM PEROXYDISULFATE
7775-27-1	SODIUM PEROXYDISULFATE (Na2S2O8)
4452-58-8	SODIUM PEROXYMONOCARBONATE
15593-29-0	SODIUM PEROXYMONOSULFATE (Na2(SO5))
11105-06-9	SODIUM PEROXYVANADATE
13718-26-8	SODIUM PEROXYVANADATE
13721-39-6	SODIUM PEROXYVANADATE
7775-27-1	SODIUM PERSULFATE
28831-12-1	SODIUM PERSULFATE
15593-29-0	SODIUM PERSULFATE
7775-27-1	SODIUM PERSULFATE (ACGIH,DOT)
15593-29-0	SODIUM PERSULFATE (ACGIH,DOT)
28831-12-1	SODIUM PERSULFATE (ACGIH,DOT)
7775-27-1	SODIUM PERSULFATE (Na2S2O8)
7775-27-1	SODIUM PERSULFATE (PEROXYDISULFURIC ACID, DISODIUM SALT)
139-02-6	SODIUM PHENATE
139-02-6	SODIUM PHENOLATE
139-02-6	SODIUM PHENOLATE, SOLID
139-02-6	SODIUM PHENOLATE, SOLID (DOT)
139-02-6	SODIUM PHENOXIDE

CAS No.	Chemical Name
139-02-6	SODIUM PHENYLATE
7601-54-9	SODIUM PHOSPHATE
7785-84-4	SODIUM PHOSPHATE ((NaPO3)3)
7558-79-4	SODIUM PHOSPHATE (Na2HPO4)
10039-32-4	SODIUM PHOSPHATE (Na2HPO4) DODECAHYDRATE
7601-54-9	SODIUM PHOSPHATE (Na3PO4)
10101-89-0	SODIUM PHOSPHATE (Na3PO4) DODECAHYDRATE
7722-88-5	SODIUM PHOSPHATE (Na4P2O7)
7758-29-4	SODIUM PHOSPHATE (Na5P3O10)
10124-56-8	SODIUM PHOSPHATE (Na6P6O18)
7601-54-9	SODIUM PHOSPHATE, ANHYDROUS
7558-79-4	SODIUM PHOSPHATE, DIBASIC
7558-79-4	SODIUM PHOSPHATE, DIBASIC (DOT)
7601-54-9	SODIUM PHOSPHATE, TRIBASIC
7601-54-9	SODIUM PHOSPHATE, TRIBASIC (DOT)
12058-85-4	SODIUM PHOSPHIDE
24167-76-8	SODIUM PHOSPHIDE
24167-76-8	SODIUM PHOSPHIDE (DOT)
24167-76-8	SODIUM PHOSPHIDE (Na(H2P))
831-52-7	SODIUM PICRAMATE
831-52-7	SODIUM PICRAMATE, WET (WITH AT LEAST 20% WATER) (DOT)
11138-49-1	SODIUM POLYALUMINATE
11135-81-2	SODIUM POTASSIUM ALLOY (LIQUID)
11135-81-2	SODIUM POTASSIUM ALLOY (SOLID)
11135-81-2	SODIUM POTASSIUM ALLOY, LIQUID (DOT)
11135-81-2	SODIUM POTASSIUM ALLOY, SOLID (DOT)
1135-81-2	SODIUM POTASSIUM ALLOYS
1303-96-4	SODIUM PYROBORATE
1303-96-4	SODIUM PYROBORATE DECAHYDRATE
7722-88-5	SODIUM PYROPHOSPHATE (Na4P2O7)
7681-38-1	SODIUM PYROSULFATE
7757-74-6	SODIUM PYROSULFITE
7681-57-4	SODIUM PYROSULFITE
7681-57-4	SODIUM PYROSULFITE (Na2S2O5)
7757-74-6	SODIUM PYROSULFITE (Na2S2O5)
540-72-7	SODIUM RHODANATE
540-72-7	SODIUM RHODANIDE
128-44-9	SODIUM SACCHARIDE
128-44-9	SODIUM SACCHARIN
128-44-9	SODIUM SACCHARINATE
128-44-9	SODIUM SACCHARINE
124-65-2	SODIUM SALT OF CACODYLIC ACID
4384-82-1	SODIUM SALT OF N,N-DIETHYLDITHIOCARBAMIC ACID
630-93-3	SODIUM SALT OF PHENYTOIN
13410-01-0	SODIUM SELENATE
13410-01-0	SODIUM SELENATE (Na2SeO4)
10102-18-8	SODIUM SELENITE
10102-18-8	SODIUM SELENITE (DOT)
10102-18-8	SODIUM SELENIUM OXIDE (Na2SeO3)
13410-01-0	SODIUM SELENIUM OXIDE (Na2SeO4)
6834-92-0	SODIUM SILICATE
6834-92-0	SODIUM SILICATE (Na2SiO3)
16893-85-9	SODIUM SILICOFLUORIDE
16893-85-9	SODIUM SILICOFLUORIDE (DOT)
16893-85-9	SODIUM SILICOFLUORIDE (Na2SiF6)
16893-85-9	SODIUM SILICOFLUORIDE, SOLID
16893-85-9	SODIUM SILICON FLUORIDE
16893-85-9	SODIUM SILICON FLUORIDE (Na2SiF6)
7757-82-6	SODIUM SULFATE (SOLUTION)
16721-80-5	SODIUM SULFHYDRATE
1313-82-2	SODIUM SULFIDE
1313-82-2	SODIUM SULFIDE (ANHYDROUS)
16721-80-5	SODIUM SULFIDE (Na(HS))
1313-82-2	SODIUM SULFIDE (Na2S)
1313-82-2	SODIUM SULFIDE, ANHYDROUS
1313-82-2	SODIUM SULFIDE, HYDRATED, WITH AT LEAST 30% WATER
7631-90-5	SODIUM SULFITE (NaHSO3)
540-72-7	SODIUM SULFOCYANATE
540-72-7	SODIUM SULFOCYANIDE
577-11-7	SODIUM SULFODI-(2-ETHYLHEXYL)-SULFOSUCCINATE
7775-14-6	SODIUM SULFOXYLATE
7631-90-5	SODIUM SULHYDRATE
1313-82-2	SODIUM SULPHIDE
1313-82-2	SODIUM SULPHIDE, ANHYDROUS OR CONTAINING LESS THAN 30% WATER (DOT)
1313-82-2	SODIUM SULPHIDE, HYDRATED, WITH AT LEAST 30% WATER (DOT)
12034-12-7	SODIUM SUPEROXIDE
12034-12-7	SODIUM SUPEROXIDE (DOT)
12034-12-7	SODIUM SUPEROXIDE (Na(O2))
76-03-9	SODIUM TCA SOLUTION
10102-20-2	SODIUM TELLURATE (Na2TeO3)
10102-20-2	SODIUM TELLURATE(IV)
10102-20-2	SODIUM TELLURATE(IV) (Na2TeO3)
10102-20-2	SODIUM TELLURITE
10102-20-2	SODIUM TELLURITE (Na2TeO3)
10102-20-2	SODIUM TELLURIUM OXIDE (Na2TeO3)
7601-54-9	SODIUM TERTIARY PHOSPHATE
1330-43-4	SODIUM TETRABORATE
1303-96-4	SODIUM TETRABORATE
1303-96-4	SODIUM TETRABORATE DECAHYDRATE
13770-96-2	SODIUM TETRAHYDROALUMINATE
13770-96-2	SODIUM TETRAHYDROALUMINATE (NaAlH4)
13770-96-2	SODIUM TETRAHYDROALUMINATE(1-)
16940-66-2	SODIUM TETRAHYDROBORATE
16940-66-2	SODIUM TETRAHYDROBORATE(1-)
540-72-7	SODIUM THIOCYANATE
540-72-7	SODIUM THIOCYANIDE
14264-31-4	SODIUM TRICYANOCUPRATE(I)
7785-84-4	SODIUM TRIMETAPHOSPHATE
7785-84-4	SODIUM TRIMETAPHOSPHATE (Na3P3O9)
7758-29-4	SODIUM TRIPHOSPHATE (Na5P3O10)
7758-29-4	SODIUM TRIPOLYPHOSPHATE
7758-29-4	SODIUM TRIPOLYPHOSPHATE (Na5P3O10)
13718-26-8	SODIUM VANADATE
11105-06-9	SODIUM VANADATE
13721-39-6	SODIUM VANADATE
13721-39-6	SODIUM VANADATE (Na3VO4)
13718-26-8	SODIUM VANADATE (NaVO3)
11105-06-9	SODIUM VANADATE (NaVO3)
13718-26-8	SODIUM VANADATE(V) (NaVO3)
11105-06-9	SODIUM VANADATE(V) (NaVO3)
11105-06-9	SODIUM VANADIUM OXIDE
13718-26-8	SODIUM VANADIUM OXIDE
13721-39-6	SODIUM VANADIUM OXIDE
13721-39-6	SODIUM VANADIUM OXIDE (Na3VO4)
11105-06-9	SODIUM VANADIUM OXIDE (NaVO3)
13718-26-8	SODIUM VANADIUM OXIDE (NaVO3)
129-06-6	SODIUM WARFARIN
7775-09-9	SODIUM(CHLORATE DE) (FRENCH)
10588-01-9	SODIUM(DICHROMATE DE) (FRENCH)
7631-99-4	SODIUM(I) NITRATE (1:1)
129-06-6	SODIUM, ((3-(ALPHA-ACETONYLBENZYL)-2-OXO-2H-1-BENZOPYRAN-4-YL)OXY)-
831-52-7	SODIUM, (2-AMINO-4,6-DINITROPHENOXY)-
7440-23-5	SODIUM, (LIQUID ALLOY)
131-52-2	SODIUM, (PENTACHLOROPHENOXY)-
26628-22-8	SODIUM, AZOTURE DE (FRENCH)
26628-22-8	SODIUM, AZOTURO DI (ITALIAN)
7440-23-5	SODIUM, METAL LIQUID ALLOY
7440-23-5	SODIUM, METAL LIQUID ALLOY (DOT)
124-65-2	SODIUM, [(DIMETHYLARSINO)OXY]-, AS-OXIDE
25641-53-6	SODIUM, [(DINITRO-O-TOLYL)OXY]-
129-06-6	SODIUM, [[2-OXO-3-(3-OXO-1-PHENYLBUTYL)-2H-1-BENZOPYRAN-4-YL]OXY]-
7440-23-5	SODIUM-23
127-85-5	SODIUM-ANILINE ARSENATE
11135-81-2	SODIUM-POTASSIUM ALLOY (DOT)
7704-34-9	SOFRIL
577-11-7	SOFTIL
1317-65-3	SOHNHOFEN STONE
106-93-4	SOILBROM
106-93-4	SOILBROM-100
106-93-4	SOILBROM-40
106-93-4	SOILBROM-85
106-93-4	SOILBROM-90
106-93-4	SOILBROM-90EC
106-93-4	SOILBROME-85
106-93-4	SOILFUME
63-25-2	SOK
25155-30-0	SOL SODOWA KWASU LAURYLOBENZENOSULFONOWEGO (POLISH)
97-64-3	SOLACTOL
75-09-2	SOLAESTHIN
16071-86-6	SOLANTINE BROWN BRL
25155-30-0	SOLAR 40
25155-30-0	SOLAR 90
16071-86-6	SOLAR BROWN PL
57-55-6	SOLAR WINTER BAN

CAS No.	Chemical Name
57-55-6	SOLARGARD P
52-68-6	SOLDEP
1344-95-2	SOLEX
16071-86-6	SOLEX BROWN R
81-81-2	SOLFARIN
51-28-5	SOLFO BLACK 2B SUPRA
51-28-5	SOLFO BLACK B
51-28-5	SOLFO BLACK BB
51-28-5	SOLFO BLACK G
51-28-5	SOLFO BLACK SB
75-15-0	SOLFURO DI CARBONIO (ITALIAN)
23505-41-1	SOLGARD
3724-65-0	SOLID CROTONIC ACID
633-03-4	SOLID GREEN
16071-86-6	SOLIUS LIGHT BROWN BRLL
16071-86-6	SOLIUS LIGHT BROWN BRS
577-11-7	SOLIWAX
9004-34-6	SOLKA-FIL
9004-34-6	SOLKA-FLOC
9004-34-6	SOLKA-FLOC BW
9004-34-6	SOLKA-FLOC BW 100
9004-34-6	SOLKA-FLOC BW 20
9004-34-6	SOLKA-FLOC BW 200
9004-34-6	SOLKA-FLOC BW 2030
100-79-8	SOLKETAL
646-06-0	SOLKETAL
53-16-7	SOLLICULIN
75-09-2	SOLMETHINE
577-11-7	SOLOVET
1313-60-6	SOLOZONE
151-21-3	SOLSOL NEEDLES
9004-70-0	SOLUBLE GUN COTTON
128-44-9	SOLUBLE SACCHARIN
1303-96-4	SOLUBOR
630-60-4	SOLUFANTINA
577-11-7	SOLUSOL
577-11-7	SOLUSOL-100%
577-11-7	SOLUSOL-75%
55-63-0	SOLUTION GLYCERYL TRINITRATE
82-66-6	SOLVAN
84-66-2	SOLVANOL
131-11-3	SOLVANOM
9002-89-5	SOLVAR
131-11-3	SOLVARONE
71-55-6	SOLVENT 111
8020-83-5	SOLVENT C-IX
60-29-7	SOLVENT ETHER
8052-41-3	SOLVENT NAPHTHA, STODDARD
8030-30-6	SOLVENTS, NAPHTHAS
8052-41-3	SOLVENTS, NAPHTHAS
298-04-4	SOLVIREX
65-85-0	SOLVO POWDER
111-90-0	SOLVOLSOL
7778-54-3	SOLVOX KS
60-11-7	SOMALIA YELLOW A
97-56-3	SOMALIA YELLOW R
67-68-5	SOMIPRONT
7664-38-2	SONAC
111-30-8	SONACIDE
127-85-5	SONATE
56-38-2	SOPRATHION
563-12-2	SOPRATHION
26264-06-2	SOPROFOR S 70
50-70-4	SORBEX M
50-70-4	SORBEX R
50-70-4	SORBEX RP
50-70-4	SORBEX S
50-70-4	SORBEX X
50-70-4	SORBICOLAN
50-70-4	SORBILANDE
50-70-4	SORBIT
50-70-4	SORBITE
50-70-4	SORBITOL
50-70-4	SORBITOL SYRUP C
9005-25-8	SORBITOSE C 5
50-70-4	SORBO
50-70-4	SORBOL
50-70-4	SORBOSTYL
7631-86-9	SORBSIL MSG

CAS No.	Chemical Name
9002-84-0	SOREFLON
9002-84-0	SOREFLON 5A
9002-84-0	SOREFLON 604
9002-84-0	SOREFLON 7
9002-84-0	SOREFLON 7A20
9002-84-0	SOREFLON 8G
9005-25-8	SORGHUM GUM
127-95-7	SORREL SALT
50-70-4	SORVILANDE
52-68-6	SOTIPOX
55-63-0	SOUP
1336-36-3	SOVOL
12002-03-8	SOWBUG & CUTWORM BAIT
12002-03-8	SOWBUG CUTWORM CONTROL
96-45-7	SOXINOL 22
8001-22-7	SOYBEAN OIL
8002-74-2	SP 1030
50-78-2	SP 189
1338-24-5	SP 230
57-63-6	SPANESTRIN
88-85-7	SPARIC
9004-34-6	SPARTOSE OM-22
1333-86-4	SPECIAL BLACK 15
1333-86-4	SPECIAL BLACK 4
1333-86-4	SPECIAL BLACK 5
95-50-1	SPECIAL TERMITE FLUID
333-41-5	SPECTRACIDE
333-41-5	SPECTRACIDE 25EC
95-79-4	SPECTROLENE RED KB
1309-37-1	SPECULAR IRON
105-60-2	SPENCER 401
105-60-2	SPENCER 601
2540-82-1	SPENCER S-6900
7664-93-9	SPENT SULFURIC ACID (DOT)
7704-34-9	SPERLOX-S
7704-34-9	SPERSUL
7704-34-9	SPERSUL THIOVIT
1332-58-7	SPESWHITE
1333-86-4	SPF 35
1333-86-4	SPHERON 6
1333-86-4	SPHERON 9
577-11-7	SPILON 8
8025-81-8	SPIRAMYCIN
76-22-2	SPIRIT OF CAMPHOR
109-95-5	SPIRIT OF ETHYL NITRITE
55-63-0	SPIRIT OF GLONOIN
55-63-0	SPIRIT OF GLYCERYL TRINITRATE
7664-41-7	SPIRIT OF HARTSHORN
55-63-0	SPIRIT OF TRINITROGLYCERIN
55-63-0	SPIRITS OF NITROGLYCERIN, (1 TO 10%)
55-63-0	SPIRITS OF NITROGLYCERIN, (1 TO 10%) (DOT)
7647-01-0	SPIRITS OF SALT
7647-01-0	SPIRITS OF SALT (DOT)
64-17-5	SPIRITS OF WINE
64-17-5	SPIRT
93-76-5	SPONTOX
111-30-8	SPORICIDIN
75-87-6	SPOROTAL 100
7429-90-5	SPOTA MOBIL 801
137-26-8	SPOTRETE
137-26-8	SPOTRETE-F
55-38-9	SPOTTON
7778-44-1	SPRA-CAL
81-81-2	SPRAY-TROL BRAND RODEN-TROL
8020-83-5	SPRAYTEX
142-59-6	SPRING-BAK
94-75-7	SPRITZ-HORMIN/2,4-D
94-75-7	SPRITZ-HORMIT/2,4-D
58-89-9	SPRITZ-RAPIDIN
58-89-9	SPRITZLINDANE
62-38-4	SPRUCE SEAL
58-89-9	SPRUEHPFLANZOL
88-85-7	SPURGE
137-26-8	SQ 1489
67-68-5	SQ 9453
13463-67-7	SR 1
13463-67-7	SR 1 (OXIDE)
133-06-2	SR 406
1420-04-8	SR 73

CAS No.	Chemical Name
7440-22-4	SR 999
22224-92-6	SRA 3886
10265-92-6	SRA 5172
9004-70-0	SS
7631-86-9	SS 10
7429-90-5	SS 3666
7631-86-9	SSA 1
7631-86-9	SSK 5
9005-25-8	ST 1500
93-72-1	STA-FAST
9005-25-8	STA-RX 1500
999-81-5	STABILAN
2223-93-0	STABILISATOR SCD
64-10-8	STABILISATOR VH
135-88-6	STABILIZATOR AR
56-38-2	STABILIZED ETHYL PARATHION
135-88-6	STABILIZER AR
2223-93-0	STABILIZER SCD
64-10-8	STABILIZER VH
8020-83-5	STABILOIL 18
8020-83-5	STABILOIL 62
1072-35-1	STABINEX NC18
95-79-4	STABLE RED KB BASE
84-74-2	STAFLEX DBP
103-23-1	STAFLEX DOA
123-79-5	STAFLEX DOA
117-84-0	STAFLEX DOP
117-81-7	STAFLEX DOP
7646-78-8	STAGNO (TETRACLORURO DI) (ITALIAN)
151-21-3	STANDAPOL 112 CONC
139-96-8	STANDAPOL T
139-96-8	STANDAPOL TLS 40
151-21-3	STANDAPOL WA-AC
151-21-3	STANDAPOL WAQ
151-21-3	STANDAPOL WAQ SPECIAL
151-21-3	STANDAPOL WAS 100
1066-45-1	STANNANE, CHLOROTRIMETHYL-
639-58-7	STANNANE, CHLOROTRIPHENYL-
7646-78-8	STANNANE, TETRACHLORO-
10026-06-9	STANNANE, TETRACHLORO-, PENTAHYDRATE
597-64-8	STANNANE, TETRAETHYL-
594-27-4	STANNANE, TETRAMETHYL-
595-90-4	STANNANE, TETRAPHENYL-
13121-70-5	STANNANE, TRICYCLOHEXYLHYDROXY-
7646-78-8	STANNIC CHLORIDE
10026-06-9	STANNIC CHLORIDE PENTAHYDRATE
7646-78-8	STANNIC CHLORIDE, ANHYDROUS
7646-78-8	STANNIC CHLORIDE, ANHYDROUS (DOT)
10026-06-9	STANNIC CHLORIDE, HYDRATED
25324-56-5	STANNIC PHOSPHIDE
25324-56-5	STANNIC PHOSPHIDE (DOT)
7646-78-8	STANNIC TETRACHLORIDE
10026-06-9	STANNIC TETRACHLORIDE PENTAHYDRATE
7772-99-8	STANNOCHLOR
7772-99-8	STANNOUS CHLORIDE
7772-99-8	STANNOUS CHLORIDE, SOLID
7772-99-8	STANNOUS CHLORIDE, SOLID (DOT)
7772-99-8	STANNOUS DICHLORIDE
7783-47-3	STANNOUS FLUORIDE
7783-47-3	STANNOUS FLUORIDE (SnF2)
56-75-7	STANOMYCETIN
56-81-5	STAR
9005-25-8	STARAMIC 747
9005-25-8	STARCH
12789-03-6	STARCHLOR
1344-95-2	STARLEX L
1333-86-4	STATEX B
1333-86-4	STATEX B 12
1333-86-4	STATEX M 70
1333-86-4	STATEX N 550
56-38-2	STATHION
70-11-1	STAUFFER 4644
133-06-2	STAUFFER CAPTAN
14484-64-1	STAUFFER FERBAM
944-22-9	STAUFFER N 2790
786-19-6	STAUFFER R 1303
732-11-6	STAUFFER R 1504
327-98-0	STAUFFER N 3049
557-05-1	STAVINOR ZN-E

CAS No.	Chemical Name
128-37-0	STAVOX
7681-49-4	STAY-FLO
9005-25-8	STAYCO S
60-11-7	STEAR YELLOW JB
106-51-4	STEARA PBQ
57-11-4	STEAREX BEADS
57-11-4	STEARIC ACID
1002-89-7	STEARIC ACID, AMMONIUM SALT
2223-93-0	STEARIC ACID, CADMIUM SALT
7428-48-0	STEARIC ACID, LEAD SALT
1072-35-1	STEARIC ACID, LEAD(2+) SALT
557-05-1	STEARIC ACID, ZINC SALT
57-11-4	STEAROPHANIC ACID
14807-96-6	STEAWHITE
105-60-2	STEELON
151-21-3	STEINAPOL NLS 90
139-96-8	STEINAPOL TLS 40
25155-30-0	STEINARYL NKS 100
25155-30-0	STEINARYL NKS 50
78-10-4	STEINFESTIGER OH
470-90-6	STELADONE
25155-30-0	STEPAN DS 60
143-00-0	STEPANOL DEA
151-21-3	STEPANOL ME
151-21-3	STEPANOL ME DRY
151-21-3	STEPANOL ME DRY AW
151-21-3	STEPANOL METHYL
151-21-3	STEPANOL METHYL DRY AW
151-21-3	STEPANOL T 28
151-21-3	STEPANOL WA
151-21-3	STEPANOL WA 100
151-21-3	STEPANOL WA PASTE
151-21-3	STEPANOL WAC
151-21-3	STEPANOL WAQ
139-96-8	STEPANOL WAT
70-30-4	STERAL
70-30-4	STERASKIN
10048-13-2	STERIGMATOCYSTIN
111-30-8	STERIHYDE
1333-86-4	STERLING 142
2235-54-3	STERLING AM
1333-86-4	STERLING FT
1333-86-4	STERLING MT
1333-86-4	STERLING MTG
1333-86-4	STERLING N 765
1333-86-4	STERLING NS
1333-86-4	STERLING R
1333-86-4	STERLING SO
1333-86-4	STERLING SO 1
1333-86-4	STERLING V
1333-86-4	STERLING VH
1333-86-4	STERLING VL
151-21-3	STERLING WA PASTE
151-21-3	STERLING WAQ-CH
151-21-3	STERLING WAQ-COSMETIC
139-96-8	STERLING WAT
102-71-6	STEROLAMIDE
7803-52-3	STIBINE
7803-52-3	STIBINE (ACGIH,DOT)
7789-61-9	STIBINE, TRIBROMO-
10025-91-9	STIBINE, TRICHLORO-
7783-56-4	STIBINE, TRIFLUORO-
7783-56-4	STIBINE, TRIFLUORO- (9CI)
1309-64-4	STIBIOX MS
7440-36-0	STIBIUM
10102-43-9	STICKMONOXYD (GERMAN)
10102-44-0	STICKSTOFFDIOXID (GERMAN)
10102-44-0	STIKSTOFDIOXYDE (DUTCH)
56-53-1	STIL
56-53-1	STIL-ROL
56-53-1	STILBESTROL
56-53-1	STILBETIN
56-53-1	STILBOEFRAL
56-53-1	STILBOESTROFORM
56-53-1	STILBOESTROL
56-53-1	STILKAP
105-60-2	STILON
102-71-6	STING-KILL
7783-06-4	STINK DAMP

CAS No.	Chemical Name
100-42-5	STIROLO (ITALIAN)
1420-07-1	STIRPAN FORTE
8052-41-3	STODDARD SOLVENT
97-77-8	STOPETYL
10043-52-4	STOPIT
7782-42-5	STOVE BLACK
7758-29-4	STPP
56-38-2	STRATHION
1912-24-9	STRAZINE
57-92-1	STREPTOMYCIN
18883-66-4	STREPTOZOCIN
18883-66-4	STREPTOZOTICIN
18883-66-4	STREPTOZOTOCIN
58-89-9	STREUNEX
57-24-9	STRICNINA (ITALIAN)
71-55-6	STROBANE
8001-35-2	STROBANE T
8001-35-2	STROBANE T-90
630-60-4	STRODIVAL
7440-24-6	STRONTIUM
91724-16-2	STRONTIUM ARSENITE
15195-06-9	STRONTIUM ARSENITE
91724-16-2	STRONTIUM ARSENITE (DOT)
91724-16-2	STRONTIUM ARSENITE (Sr(As2O4))
91724-16-2	STRONTIUM ARSENITE, SOLID
91724-16-2	STRONTIUM ARSENITE, SOLID (DOT)
1635-05-2	STRONTIUM CARBONATE
7791-10-8	STRONTIUM CHLORATE
7791-10-8	STRONTIUM CHLORATE (DOT)
7791-10-8	STRONTIUM CHLORATE (Sr(ClO3)2)
7791-10-8	STRONTIUM CHLORATE, WET (DOT)
7789-06-2	STRONTIUM CHROMATE
7789-06-2	STRONTIUM CHROMATE (1:1)
7789-06-2	STRONTIUM CHROMATE (DOT)
7789-06-2	STRONTIUM CHROMATE (SrCrO4)
7789-06-2	STRONTIUM CHROMATE (VI)
7789-06-2	STRONTIUM CHROMATE 12170
7789-06-2	STRONTIUM CHROMATE A
7789-06-2	STRONTIUM CHROMATE X 2396
10042-76-9	STRONTIUM DINITRATE
1314-18-7	STRONTIUM DIOXIDE
13450-97-0	STRONTIUM DIPERCHLORATE
10042-76-9	STRONTIUM NITRATE
10042-76-9	STRONTIUM NITRATE (DOT)
10042-76-9	STRONTIUM NITRATE (Sr(NO3)2)
13450-97-0	STRONTIUM PERCHLORATE
13450-97-0	STRONTIUM PERCHLORATE (DOT)
13450-97-0	STRONTIUM PERCHLORATE (Sr(ClO4)2)
1314-18-7	STRONTIUM PEROXIDE
1314-18-7	STRONTIUM PEROXIDE (DOT)
1314-18-7	STRONTIUM PEROXIDE (Sr(O2))
12504-13-1	STRONTIUM PHOSPHIDE
12504-13-1	STRONTIUM PHOSPHIDE (DOT)
12504-13-1	STRONTIUM PHOSPHIDE (SrP)
7789-06-2	STRONTIUM YELLOW
10042-76-9	STRONTIUM(II) NITRATE (1:2)
630-60-4	STROPHALEN
630-60-4	STROPHANTHIN G
630-60-4	STROPHOPERM
630-60-4	STROPHOSAN
56-04-2	STRUMACIL
60-41-3	STRYCHININE SULFATE
57-24-9	STRYCHNIDIN-10-ONE
357-57-3	STRYCHNIDIN-10-ONE, 2,3-DIMETHOXY-
357-57-3	STRYCHNIDIN-10-ONE, 2,3-DIMETHOXY- (9CI)
60-41-3	STRYCHNIDIN-10-ONE, SULFATE (2:1)
57-24-9	STRYCHNIN
57-24-9	STRYCHNIN (GERMAN)
57-24-9	STRYCHNINE
57-24-9	STRYCHNINE (ACGIH)
57-24-9	STRYCHNINE AND ITS SALTS
60-41-3	STRYCHNINE SULFATE
357-57-3	STRYCHNINE, 2,3-DIMETHOXY-
57-24-9	STRYCHNINE, LIQUID (DOT)
57-24-9	STRYCHNINE, SOLID
57-24-9	STRYCHNINE, SOLID (DOT)
60-41-3	STRYCHNINE, SULFATE (2:1)
60-41-3	STRYCHNINIUM SULFATE
57-24-9	STRYCHNOS

CAS No.	Chemical Name
18883-66-4	STRZ
7681-49-4	STUDAFLUOR
1314-84-7	STUTOX
1314-84-7	STUTOX I
105-60-2	STYLON
82-71-3	STYPHNIC ACID
98-85-1	STYRALLYL ALCOHOL
98-85-1	STYRALYL ALCOHOL
100-42-5	STYREEN (DUTCH)
100-42-5	STYREN (CZECH)
100-42-5	STYRENE
96-09-3	STYRENE 7,8-OXIDE
96-09-3	STYRENE EPOXIDE
100-42-5	STYRENE MONOMER
100-42-5	STYRENE MONOMER, INHIBITED
100-42-5	STYRENE MONOMER, INHIBITED (DOT)
96-09-3	STYRENE OXIDE
6607-45-0	STYRENE, ALPHA,BETA-DICHLORO-
98-83-9	STYRENE, ALPHA-METHYL-
30030-25-2	STYRENE, AR-(CHLOROMETHYL)-
25013-15-4	STYRENE, AR-METHYL-
1319-73-9	STYRENE, METHYL-
25013-15-4	STYRENE, METHYL-
100-42-5	STYRENE, MONOMER (ACGIH)
2039-87-4	STYRENE, O-CHLORO-
622-97-9	STYRENE, P-METHYL-
100-42-5	STYROL
100-42-5	STYROL (GERMAN)
100-42-5	STYROLE
100-42-5	STYROLENE
100-42-5	STYRON
100-42-5	STYROPOL
100-42-5	STYROPOL SO
100-42-5	STYROPOR
96-09-3	STYRYL OXIDE
8020-83-5	SU (OIL)
3771-19-5	SU 13437
1582-09-8	SU SEGURO CARPIDOR
7546-30-7	SUBCHLORIDE OF MERCURY
88-85-7	SUBITEX
7487-94-7	SUBLIMAT (CZECH)
7487-94-7	SUBLIMATE
7704-34-9	SUBLIMED SULFUR
7704-34-9	SUBLIMED SULPHUR
1395-21-7	SUBTILISINS (PROTEOLYTIC ENZ AS 100% PURE CRYST ENZYME)
577-11-7	SUCCINATE STD
123-23-9	SUCCINIC ACID PEROXIDE
123-23-9	SUCCINIC ACID PEROXIDE (DOT)
123-23-9	SUCCINIC ACID PEROXIDE, >72%, IN WATER
123-23-9	SUCCINIC ACID PEROXIDE, TECHNICALLY PURE (DOT)
87-69-4	SUCCINIC ACID, 2,3-DIHYDROXY-
121-75-5	SUCCINIC ACID, MERCAPTO-, DIETHYL ESTER, S-ESTER WITH O,O-DIMETHYL PHOSPHORODITHIOATE
577-11-7	SUCCINIC ACID, SULFO-, 1,4-BIS(2-ETHYLHEXYL) ESTER, SODIUM SALT
123-23-9	SUCCINIC MONOPEROXYANHYDRIDE
123-23-9	SUCCINIC PEROXIDE
110-61-2	SUCCINONITRILE
3333-52-6	SUCCINONITRILE, TETRAMETHYL-
123-23-9	SUCCINOYL PEROXIDE
123-23-9	SUCCINYL PEROXIDE
110-63-4	SUCOL B
57-50-1	SUCROSE
60-11-7	SUDAN YELLOW GG
60-11-7	SUDAN YELLOW GGA
1300-71-6	SUDOL
7704-34-9	SUFRAN
7704-34-9	SUFRAN D
99-55-8	SUGAI FAST SCARLET G BASE
57-50-1	SUGAR
301-04-2	SUGAR OF LEAD
142-04-1	SUL ANILINOVA (CZECH)
7487-94-7	SULEM
7487-94-7	SULEMA (RUSSIAN)
55-98-1	SULFABUTIN
95-06-7	SULFALLATE
126-33-0	SULFALONE
7773-06-0	SULFAMATE
723-46-6	SULFAMETHOXAZOLE

CAS No.	Chemical Name
5329-14-6	SULFAMIC ACID
14017-41-5	SULFAMIC ACID, COBALT SALT
14017-41-5	SULFAMIC ACID, COBALT(2+) SALT (2:1)
7773-06-0	SULFAMIC ACID, MONOAMMONIUM SALT
5329-14-6	SULFAMIDIC ACID
5329-14-6	SULFAMINIC ACID
7773-06-0	SULFAMINSAURE (GERMAN)
7446-11-9	SULFAN
25155-30-0	SULFAPOL
25155-30-0	SULFAPOLU (POLISH)
25155-30-0	SULFARIL PASTE
7758-98-7	SULFATE DE CUIVRE (FRENCH)
77-78-1	SULFATE DE METHYLE (FRENCH)
65-30-5	SULFATE DE NICOTINE (FRENCH)
7446-14-2	SULFATE DE PLOMB (FRENCH)
7733-02-0	SULFATE DE ZINC (FRENCH)
77-78-1	SULFATE DIMETHYLIQUE (FRENCH)
7783-35-9	SULFATE MERCURIQUE (FRENCH)
3689-24-5	SULFATEP
72-14-0	SULFATHIAZOLE
10043-01-3	SULFATODIALUMINUM DISULFATE
7720-78-7	SULFERROUS
139-96-8	SULFETAL KT 400
151-21-3	SULFETAL L 95
139-96-8	SULFETAL LT
540-82-9	SULFETHYLIC ACID
7704-34-9	SULFEX
68-11-1	SULFHYDRYLACETIC ACID
7704-34-9	SULFIDAL
505-60-2	SULFIDE, BIS(2-CHLOROETHYL)
55-38-9	SULFIDOPHOS
577-11-7	SULFIMEL DOS
7719-09-7	SULFINYL CHLORIDE
67-68-5	SULFINYLBIS(METHANE)
9004-34-6	SULFITE CELLULOSE
126-33-0	SULFOLAN
126-33-0	SULFOLANE
25103-58-6	SULFOLE 120
7790-94-5	SULFONIC ACID, MONOCHLORIDE
7791-25-5	SULFONYL CHLORIDE
7791-25-5	SULFONYL DICHLORIDE
2699-79-8	SULFONYL FLUORIDE
151-21-3	SULFOPON WA 1
151-21-3	SULFOPON WA 1 SPECIAL
151-21-3	SULFOPON WA 2
151-21-3	SULFOPON WA 3
7704-34-9	SULFORON
577-11-7	SULFOSUCCINIC ACID BIS(2-ETHYLHEXYL)ESTER SODIUM SALT
577-11-7	SULFOSUCCINIC ACID DI-2-ETHYLHEXYL ESTER SODIUM SALT
3689-24-5	SULFOTEP
3689-24-5	SULFOTEP (ACGIH)
3689-24-5	SULFOTEPP
151-21-3	SULFOTEX WA
151-21-3	SULFOTEX WALA
540-82-9	SULFOVINIC ACID
3569-57-1	SULFOXIDE, 3-CHLOROPROPYL OCTYL
94-36-0	SULFOXYL
25155-30-0	SULFRAMIN 1238 SLURRY
25155-30-0	SULFRAMIN 1240
25155-30-0	SULFRAMIN 1250 SLURRY
25155-30-0	SULFRAMIN 40
25155-30-0	SULFRAMIN 40 FLAKES
25155-30-0	SULFRAMIN 40 GRANULAR
25155-30-0	SULFRAMIN 40RA
25155-30-0	SULFRAMIN 85
25155-30-0	SULFRAMIN 90 FLAKES
27176-87-0	SULFRAMIN ACID 1298
7704-34-9	SULFUR
7704-34-9	SULFUR ATOM
12771-08-3	SULFUR CHLORIDE
10025-67-9	SULFUR CHLORIDE
10545-99-0	SULFUR CHLORIDE
10545-99-0	SULFUR CHLORIDE (8CI,9CI)
10545-99-0	SULFUR CHLORIDE (DI)
10025-67-9	SULFUR CHLORIDE (MONO)
10025-67-9	SULFUR CHLORIDE (S2Cl2)
10545-99-0	SULFUR CHLORIDE (SCl2)
7719-09-7	SULFUR CHLORIDE OXIDE
7719-09-7	SULFUR CHLORIDE OXIDE (Cl2SO)
7791-25-5	SULFUR CHLORIDE OXIDE (SO2Cl2)
10025-67-9	SULFUR CHLORIDE(DI) (DOT)
10545-99-0	SULFUR CHLORIDE(MONO) (DOT)
10545-99-0	SULFUR DICHLORIDE
10545-99-0	SULFUR DICHLORIDE (SCl2)
2699-79-8	SULFUR DIFLUORIDE DIOXIDE
7783-06-4	SULFUR DIHYDRIDE
7446-09-5	SULFUR DIOXIDE
7446-09-5	SULFUR DIOXIDE (ACGIH,DOT)
7446-09-5	SULFUR DIOXIDE (SO2)
2699-79-8	SULFUR DIOXIDE DIFLUORIDE
7782-99-2	SULFUR DIOXIDE SOLUTION
7704-34-9	SULFUR ELEMENT
7704-34-9	SULFUR FLOWER (DOT)
2551-62-4	SULFUR FLUORIDE
7783-60-0	SULFUR FLUORIDE (SF4)
7783-60-0	SULFUR FLUORIDE (SF4), (T-4)-
10546-01-7	SULFUR FLUORIDE (SF5)
2551-62-4	SULFUR FLUORIDE (SF6)
2551-62-4	SULFUR FLUORIDE (SF6), (OC-6-11)-
2551-62-4	SULFUR HEXAFLUORIDE
2551-62-4	SULFUR HEXAFLUORIDE (ACGIH,DOT)
7783-06-4	SULFUR HYDRIDE
10025-67-9	SULFUR MONOCHLORIDE
10025-67-9	SULFUR MONOCHLORIDE (ACGIH)
10025-67-9	SULFUR MONOCHLORIDE (S2Cl2)
505-60-2	SULFUR MUSTARD
505-60-2	SULFUR MUSTARD GAS
7704-34-9	SULFUR OINTMENT
7446-09-5	SULFUR OXIDE
7446-09-5	SULFUR OXIDE (SO2)
7446-11-9	SULFUR OXIDE (SO3)
7791-25-5	SULFUR OXYCHLORIDE (SO2Cl2)
7719-09-7	SULFUR OXYCHLORIDE (SOCl2)
5714-22-7	SULFUR PENTAFLUORIDE
10546-01-7	SULFUR PENTAFLUORIDE
7791-27-7	SULFUR PENTOXYDICHLORIDE
1314-80-3	SULFUR PHOSPHIDE
7446-34-6	SULFUR SELENIDE (SSe)
10025-67-9	SULFUR SUBCHLORIDE
7783-60-0	SULFUR TETRAFLUORIDE
7783-60-0	SULFUR TETRAFLUORIDE (ACGIH,DOT)
7446-11-9	SULFUR TRIOXIDE
7446-11-9	SULFUR TRIOXIDE (DOT)
7446-11-9	SULFUR TRIOXIDE, INHIBITED
7446-11-9	SULFUR TRIOXIDE, STABILIZED (DOT)
7704-34-9	SULFUR, MOLTEN
7704-34-9	SULFUR, SOLID
7704-34-9	SULFUR, SOLID (DOT)
75-18-3	SULFURE DE METHYLE (FRENCH)
7783-06-4	SULFURETED HYDROGEN
7664-93-9	SULFURIC ACID
7664-93-9	SULFURIC ACID (ACGIH,DOT)
10043-01-3	SULFURIC ACID ALUMINUM(3+) SALT (3:2)
7783-05-3	SULFURIC ACID ANHYDRIDE
7790-94-5	SULFURIC ACID CHLOROHYDRIN
7758-98-7	SULFURIC ACID COPPER(2+) SALT (1:1)
7783-20-2	SULFURIC ACID DIAMMONIUM SALT
10294-26-5	SULFURIC ACID DISILVER SALT
151-21-3	SULFURIC ACID DODECYL ESTER SODIUM SALT
8014-95-7	SULFURIC ACID FUMING
7720-78-7	SULFURIC ACID IRON SALT (1:1)
151-21-3	SULFURIC ACID MONODODECYL ESTER SODIUM SALT
139-96-8	SULFURIC ACID MONODODECYL ESTER TRIETHANOLAMINE SALT
139-96-8	SULFURIC ACID MONOLAURYL ESTER TRIETHANOLAMINE SALT
7786-81-4	SULFURIC ACID NICKEL(2+) SALT
7646-93-7	SULFURIC ACID POTASSIUM SALT (1:1)
10294-26-5	SULFURIC ACID SILVER SALT (1:2)
7681-38-1	SULFURIC ACID SODIUM SALT (1:1)
7446-18-6	SULFURIC ACID THALLIUM(1+) SALT (1:2)
7733-02-0	SULFURIC ACID ZINC SALT
10045-89-3	SULFURIC ACID, AMMONIUM IRON(2+) SALT (2:2:1)
15699-18-0	SULFURIC ACID, AMMONIUM NICKEL(2+) SALT (2:2:1)
13510-49-1	SULFURIC ACID, BERYLLIUM SALT (1:1)
10124-36-4	SULFURIC ACID, CADMIUM SALT (1:1)
10124-36-4	SULFURIC ACID, CADMIUM(2+) SALT
14489-25-9	SULFURIC ACID, CHROMIUM SALT
10101-53-8	SULFURIC ACID, CHROMIUM(3+) SALT (3:2)

CAS No.	Chemical Name
10124-43-3	SULFURIC ACID, COBALT(2+) SALT (1:1)
10380-29-7	SULFURIC ACID, COPPER(2+) SALT, AMMONIATED
64-67-5	SULFURIC ACID, DIETHYL ESTER
7783-36-0	SULFURIC ACID, DIMERCURY(1+) SALT
77-78-1	SULFURIC ACID, DIMETHYL ESTER
10294-26-5	SULFURIC ACID, DISILVER(1+) SALT
7446-18-6	SULFURIC ACID, DITHALLIUM(1+) SALT
7446-18-6	SULFURIC ACID, DITHALLIUM(1+) SALT (8CI,9CI)
139-96-8	SULFURIC ACID, DODECYL ESTER, TRIETHANOLAMINE SALT
7783-05-3	SULFURIC ACID, FUMING
7720-78-7	SULFURIC ACID, IRON(2+) SALT (1:1)
10028-22-5	SULFURIC ACID, IRON(3+) SALT (3:2)
2235-54-3	SULFURIC ACID, LAURYL ESTER, AMMONIUM SALT
7446-14-2	SULFURIC ACID, LEAD SALT
7446-14-2	SULFURIC ACID, LEAD(2+) SALT (1:1)
7783-35-9	SULFURIC ACID, MERCURY(2+) SALT (1:1)
8014-95-7	SULFURIC ACID, MIXT WITH SULFUR TRIOXIDE
7803-63-6	SULFURIC ACID, MONOAMMONIUM SALT
7782-78-7	SULFURIC ACID, MONOANHYDRIDE WITH NITROUS ACID
7782-78-7	SULFURIC ACID, MONOANHYDRIDE WITH NITROUS ACID (9CI)
2235-54-3	SULFURIC ACID, MONODODECYL ESTER, AMMONIUM SALT
139-96-8	SULFURIC ACID, MONODODECYL ESTER, COMPD WITH 2,2',2"-NITRILOTRIETHANOL (1:1)
139-96-8	SULFURIC ACID, MONODODECYL ESTER, COMPD WITH 2,2',2"-NITRILOTRIS(ETHANOL)
143-00-0	SULFURIC ACID, MONODODECYL ESTER, COMPD WITH 2,2'-IMINOBIS[ETHANOL] (1:1)
143-00-0	SULFURIC ACID, MONODODECYL ESTER, COMPD WITH 2,2'-IMINODIETHANOL (1:1)
540-82-9	SULFURIC ACID, MONOETHYL ESTER
7646-93-7	SULFURIC ACID, MONOPOTASSIUM SALT
7681-38-1	SULFURIC ACID, MONOSODIUM SALT
7786-81-4	SULFURIC ACID, NICKEL(2+) SALT (1:1)
7664-93-9	SULFURIC ACID, SPENT
7664-93-9	SULFURIC ACID, SPENT (DOT)
7446-18-6	SULFURIC ACID, THALLIUM SALT
10031-59-1	SULFURIC ACID, THALLIUM SALT
13693-11-3	SULFURIC ACID, TITANIUM SALT
7733-02-0	SULFURIC ACID, ZINC SALT (1:1)
14644-61-2	SULFURIC ACID, ZIRCONIUM(4+) SALT (2:1)
7446-11-9	SULFURIC ANHYDRIDE
7446-11-9	SULFURIC ANHYDRIDE (DOT)
7790-94-5	SULFURIC CHLOROHYDRIN
7791-25-5	SULFURIC DICHLORIDE
60-29-7	SULFURIC ETHER
7446-11-9	SULFURIC OXIDE
7791-25-5	SULFURIC OXYCHLORIDE
2699-79-8	SULFURIC OXYFLUORIDE
25155-30-0	SULFURIL 50
7782-99-2	SULFUROUS ACID
7782-99-2	SULFUROUS ACID (DOT)
7782-99-2	SULFUROUS ACID (H2SO3)
7446-09-5	SULFUROUS ACID ANHYDRIDE
7782-99-2	SULFUROUS ACID SOLUTION
2312-35-8	SULFUROUS ACID, 2-(4-(1,1-DIMETHYLETHYL)PHENOXY)CYCLOHEXYL 2-PROPYNYL ESTER
140-57-8	SULFUROUS ACID, 2-(P-TERT-BUTYLPHENOXY)-1-METHYLETHYL 2-CHLOROETHYL ESTER
2312-35-8	SULFUROUS ACID, 2-(P-TERT-BUTYLPHENOXY)CYCLOHEXYL-2-PROPYNYL ESTER
140-57-8	SULFUROUS ACID, 2-CHLOROETHYL 2-[4-(1,1-DIMETHYLETHYL)-PHENOXY]-1-METHYLETHYL ESTER
115-29-7	SULFUROUS ACID, CYCLIC ESTER WITH 1,4,5,6,7,7-HEXACHLORO-5-NORBORNENE-2,3-DIMETHANOL
10196-04-0	SULFUROUS ACID, DIAMMONIUM SALT
13774-25-9	SULFUROUS ACID, MAGNESIUM SALT (2:1)
10192-30-0	SULFUROUS ACID, MONOAMMONIUM SALT
7773-03-7	SULFUROUS ACID, MONOPOTASSIUM SALT
7631-90-5	SULFUROUS ACID, MONOSODIUM SALT
7446-09-5	SULFUROUS ANHYDRIDE
7719-09-7	SULFUROUS DICHLORIDE
7446-09-5	SULFUROUS OXIDE
7719-09-7	SULFUROUS OXYCHLORIDE
7791-25-5	SULFURYL CHLORIDE
7791-25-5	SULFURYL CHLORIDE (DOT)
7791-25-5	SULFURYL CHLORIDE (SO2Cl2)
7791-25-5	SULFURYL DICHLORIDE
2699-79-8	SULFURYL FLUORIDE
2699-79-8	SULFURYL FLUORIDE (ACGIH,DOT)
7704-34-9	SULIKOL
7704-34-9	SULIKOL K
7704-34-9	SULKOL
62-56-6	SULOUREA
5329-14-6	SULPHAMIC ACID
5329-14-6	SULPHAMIC ACID (DOT)
75-15-0	SULPHOCARBONIC ANHYDRIDE
126-33-0	SULPHOLANE
56-38-2	SULPHOS
126-33-0	SULPHOXALINE
7704-34-9	SULPHUR
7704-34-9	SULPHUR (DOT)
10025-67-9	SULPHUR CHLORIDE (S2Cl2)
10545-99-0	SULPHUR DICHLORIDE
7446-09-5	SULPHUR DIOXIDE
7446-09-5	SULPHUR DIOXIDE, LIQUEFIED (DOT)
2551-62-4	SULPHUR HEXAFLUORIDE
505-60-2	SULPHUR MUSTARD
505-60-2	SULPHUR MUSTARD GAS
7704-34-9	SULPHUR, LUMP OR POWDER (DOT)
7704-34-9	SULPHUR, MOLTEN (DOT)
7664-93-9	SULPHURIC ACID
10124-36-4	SULPHURIC ACID, CADMIUM SALT (1:1)
7782-99-2	SULPHUROUS ACID (DOT)
7791-25-5	SULPHURYL CHLORIDE
35400-43-2	SULPROFOS
7704-34-9	SULSOL
7704-34-9	SULTAF
101-77-9	SUMICURE M
9016-87-9	SUMIDUR 44V10
9016-87-9	SUMIDUR 44V20
122-14-5	SUMIFENE
16071-86-6	SUMILIGHT SUPRA BROWN BRS
128-37-0	SUMILIZER BHT
96-69-5	SUMILIZER WX
96-69-5	SUMILIZER WX-R
9002-89-5	SUMITEX H 10
122-14-5	SUMITHION
122-14-5	SUMITHION 20F
2636-26-2	SUMITOMO S 4084
121-75-5	SUMITOX
1338-24-5	SUNAPTIC ACID B
1338-24-5	SUNAPTIC ACID C
1338-24-5	SUNAPTIC B
114-26-1	SUNCIDE
13171-21-6	SUNDARAM 1975
8020-83-5	SUNISO 4GS
1401-55-4	SUNLIFE TN
8002-74-2	SUNNOC
8002-74-2	SUNNOC B
8002-74-2	SUNNOC N
8002-74-2	SUNOCO 4415
8002-74-2	SUNOCO 4417
8002-74-2	SUNOCO P 116
8020-83-5	SUNSPRAY
8020-83-5	SUNSPRAY (HYDROCARBON)
8020-83-5	SUNTEMP
56-72-4	SUNTOL
8020-83-5	SUNVIS 31
8002-74-2	SUNWAX 5512
28772-56-7	SUP'OPERATS
330-54-1	SUP'R FLO
7440-48-4	SUPER COBALT
7704-34-9	SUPER COSAN
94-75-7	SUPER D WEEDONE
93-76-5	SUPER D WEEDONE
16893-85-9	SUPER PRODAN
56-38-2	SUPER RODIATOX
8030-30-6	SUPER VMP
28772-56-7	SUPER-CAID
7631-86-9	SUPER-CEL
7681-49-4	SUPER-DENT
28772-56-7	SUPER-ROZOL
1333-86-4	SUPERCARBOVAR
57-13-6	SUPERCEL 3000
21564-17-0	SUPERDAVLOXAN
10043-52-4	SUPERFLAKE ANHYDROUS
7631-86-9	SUPERFLOSS
50-00-0	SUPERLYSOFORM

CAS No.	Chemical Name
56-81-5	SUPEROL
94-75-7	SUPERORMONE CONCENTRE
94-36-0	SUPEROX
94-36-0	SUPEROX 744
7722-84-1	SUPEROXOL
470-90-6	SUPONA
470-90-6	SUPONE
82-28-0	SUPRACET ORANGE R
950-37-8	SUPRACID
950-37-8	SUPRACIDE
9016-87-9	SUPRASEC 1042
9016-87-9	SUPRASEC DC
16071-86-6	SUPRAZO BROWN BRL
14807-96-6	SUPREME
14807-96-6	SUPREME DENSE
1332-58-7	SUPREX CLAY
16071-86-6	SUPREXCEL BROWN BRL
2893-78-9	SURCHLOR GR 60
112-02-7	SURFROYAL CTAC
70-30-4	SURGI-CEN
70-30-4	SUROFENE
2921-88-2	SUSCON
128-37-0	SUSTANE BHT
58-22-0	SUSTANON
58-22-0	SUSTANONE
6923-22-4	SUSVIN
12001-26-2	SUZORITE
12001-26-2	SUZORITE 60S
577-11-7	SV 10 EX-WET 1001
7704-34-9	SVOVEL
7704-34-9	SVOVL
128-37-0	SWANOX BHT
151-21-3	SWASCOL 1P
151-21-3	SWASCOL 3L
151-21-3	SWASCOL 4L
122-10-1	SWAT
7782-42-5	SWEDISH BLACK LEAD
12002-03-8	SWEDISH GREEN
10290-12-7	SWEDISH GREEN
1910-2-5	SWEEP
8006-14-2	SWEET NATURAL GAS
109-95-5	SWEET SPIRIT OF NITER
128-44-9	SWEETA
128-44-9	SYKOSE
12001-29-5	SYLODEX
534-22-5	SYLVAN
27137-41-3	SYLVAN
75-60-5	SYLVICOR
463-49-0	SYM-ALLYLENE
106-93-4	SYM-DIBROMOETHANE
534-07-6	SYM-DICHLOROACETONE
107-06-2	SYM-DICHLOROETHANE
111-44-4	SYM-DICHLOROETHYL ETHER
540-59-0	SYM-DICHLOROETHYLENE
542-88-1	SYM-DICHLOROMETHYL ETHER
108-83-8	SYM-DIISOPROPYLACETONE
504-20-1	SYM-DIISOPROPYLIDENE ACETONE
540-73-8	SYM-DIMETHYLHYDRAZINE
56-18-8	SYM-NORSPERMIDINE
76-12-0	SYM-TETRACHLORODIFLUOROETHANE
79-34-5	SYM-TETRACHLOROETHANE
108-77-0	SYM-TRICHLOROTRIAZINE
108-67-8	SYM-TRIMETHYLBENZENE
99-35-4	SYM-TRINITROBENZENE
129-66-8	SYM-TRINITROBENZOIC ACID
118-96-7	SYM-TRINITROTOLUENE
118-96-7	SYM-TRINITROTOLUOL
110-88-3	SYM-TRIOXANE
87-90-1	SYMCLOSEN
87-90-1	SYMCLOSENE
540-73-8	SYMETRYCZNA DWUMETYLOHYDRAZYNA (POLISH)
99-35-4	SYMMETRIC TRINITROBENZENE
99-55-8	SYMULON SCARLET G BASE
1330-78-5	SYN-O-AD 8484
108-77-0	SYN-TRICHLOTRIAZIN (CZECH)
50-28-2	SYNDIOL
56-53-1	SYNESTRIN
7632-00-0	SYNFAT 1004
10124-37-5	SYNFAT 1006

CAS No.	Chemical Name
1582-09-8	SYNFLORAN
86290-81-5	SYNFUELS
57-83-0	SYNGESTRETS
12789-03-6	SYNKLOR
9004-70-0	SYNPOR
557-05-1	SYNPRO ABG
557-05-1	SYNPRO STEARATE
151-21-3	SYNTAPON L
151-21-3	SYNTAPON L PASTA (CZECH)
67-68-5	SYNTEXAN
8001-35-2	SYNTHETIC 3956
86290-81-5	SYNTHETIC GASOLINE
56-81-5	SYNTHETIC GLYCERIN
57-06-7	SYNTHETIC MUSTARD OIL
584-79-2	SYNTHETIC PYRETHRINS
84-80-0	SYNTHEX P
56-53-1	SYNTHOESTRIN
56-53-1	SYNTHOFOLIN
56-75-7	SYNTHOMYCETIN
56-53-1	SYNTOFOLIN
57-83-0	SYNTOLUTAN
152-16-9	SYSTAM
9016-87-9	SYSTANAT MR
9016-87-9	SYSTANATE MR
60-51-5	SYSTEMIN
8065-48-3	SYSTEMOX
60-51-5	SYSTOATE
25322-69-4	SYSTOL T 121
152-16-9	SYSTOPHOS
8065-48-3	SYSTOX
152-16-9	SYTAM
7631-86-9	SYTON
7631-86-9	SYTON 2X
7631-86-9	SYTON FM
7631-86-9	SYTON W 15
7631-86-9	SYTON W 3
7631-86-9	SYTON X 30
822-06-0	SZESCIOMETYLENODWUIZOCYJANIAN (POLISH)
62-73-7	SZKLARNIAK
2636-26-2	SZKLARNIAK
26471-62-5	T 100
112-30-1	T 148
7440-32-6	T 160
126-72-7	T 23P
9002-84-0	T 30
9002-84-0	T 30 (FLUOROPOLYMER)
25322-69-4	T 32/75
7440-32-6	T 40
9002-84-0	T 5B
7440-32-6	T 60
7440-32-6	T 60 (METAL)
9002-84-0	T 8A
107-44-8	T-144
115-26-4	T-2002
77-81-6	T-2104
107-44-8	T-2106
2631-37-0	T-32
56-38-2	T-47
75-85-4	T-AMYL ALCOHOL
75-65-0	T-BUTANOL
540-88-5	T-BUTYL ACETATE
75-65-0	T-BUTYL HYDROXIDE
110-05-4	T-BUTYL PEROXIDE
75-64-9	T-BUTYLAMINE
98-06-6	T-BUTYLBENZENE
3282-30-2	T-BUTYLCARBONYL CHLORIDE
75-66-1	T-BUTYLMERCAPTAN
7778-54-3	T-EUSOL
7681-49-4	T-FLUORIDE
75-21-8	T-GAS
50-55-5	T-SERP
7722-84-1	T-STUFF
121-82-4	T4
6369-97-7	TA 20
13463-67-7	TA 400
62-55-5	TAA
77-81-6	TABOON A
77-81-6	TABUN
7705-07-9	TAC 121

CAS No.	Chemical Name
7705-07-9	TAC 131
7705-07-9	TAC 132
71-55-6	TAFCLEAN
50-29-3	TAFIDEX
62-38-4	TAG
62-38-4	TAG 331
62-38-4	TAG FUNGICIDE
62-38-4	TAG HL 331
10265-92-6	TAHMABON
121-75-5	TAK
121-75-5	TAK (PESTICIDE)
9002-81-7	TAKAFEST
25322-69-4	TAKELAC P 21
9016-87-9	TAKENATE 300C
443-48-1	TAKIMETOL
67-63-0	TAKINEOCOL
14807-96-6	TALC
14807-96-6	TALC (FIBROUS) DUST
14807-96-6	TALC (Mg3H2(SiO3)4)
14807-96-6	TALCAN PK-P
14807-96-6	TALCRON CP 44-31
557-05-1	TALCULIN Z
14807-96-6	TALCUM
55-38-9	TALODEX
10265-92-6	TAMARON
1401-55-4	TANAPHEN P 500
53-86-1	TANNEX
1401-55-4	TANNIC ACID
1401-55-4	TANNIN ANAL
1401-55-4	TANNINS
1401-55-4	TANNINS, GALLO-
7440-25-7	TANTALUM
7440-25-7	TANTALUM-181
58-89-9	TAP 85
62-73-7	TAP 9VP
2636-26-2	TAP 9VP
9005-25-8	TAPIOCA STARCH
545-55-1	TAPO
9005-25-8	TAPON
8007-45-2	TAR
91-20-3	TAR CAMPHOR
8001-58-9	TAR OIL
8007-45-2	TAR, COAL
8007-45-2	TAR, COAL, COLORLESS PURIFIED
65996-89-6	TAR, COAL, HIGH-TEMP
65996-90-9	TAR, COAL, LOW-TEMP
8007-45-2	TAR, COAL, PURIFIED COLORLESS
60-51-5	TARA 909
7631-86-9	TARANOX 500
7789-00-6	TARAPACAITE
151-21-3	TARAPON K 12
8007-45-2	TARCRONE 180
71-63-6	TARDIGAL
470-90-6	TARENE
9002-84-0	TARFLEN
2163-80-6	TARGET MSMA
105-60-2	TARLON X-A
105-60-2	TARLON XB
105-60-2	TARNAMID T
105-60-2	TARNAMID T 2
105-60-2	TARNAMID T 27
28300-74-5	TARTAR EMETIC
87-69-4	TARTARIC ACID
28300-74-5	TARTARIC ACID, ANTIMONY POTASSIUM SALT
815-82-7	TARTARIC ACID, COPPER(2+) SALT
815-82-7	TARTARIC ACID, COPPER(2+) SALT (1:1), (+)-
3164-29-2	TARTARIC ACID, DIAMMONIUM SALT
28300-74-5	TARTARIZED ANTIMONY
28300-74-5	TARTOX
28300-74-5	TARTRATE ANTIMONIO-POTASSIQUE (FRENCH)
65-31-6	TARTRATE DE NICOTINE (FRENCH)
28300-74-5	TARTRATED ANTIMONY
7705-07-9	TAS 101
62-73-7	TASK
2636-26-2	TASK
2636-26-2	TASK TABS
57-74-9	TAT CHLOR 4
12789-03-6	TAT CHLOR 4
309-00-2	TATUZINHO
79-27-6	TBE
126-73-8	TBP
76-03-9	TCA
12002-48-1	TCB
1746-01-6	TCDBD
1746-01-6	TCDD
79-01-6	TCE
7440-22-4	TCG 7R
2231-57-4	TCH
67-66-3	TCM
21564-17-0	TCMTB
95-95-4	TCP
1330-78-5	TCP
25167-82-2	TCP
13463-67-7	TCR 10
7440-50-8	TCUP1
95-80-7	TDA
126-72-7	TDBPP
72-54-8	TDE
91-08-7	TDI
26471-62-5	TDI
26471-62-5	TDI 80-20
26471-62-5	TDI-80
9002-84-0	TE 30
102-71-6	TEA
97-93-8	TEA
121-44-8	TEA
139-96-8	TEA LAURYL SULFATE
7704-34-9	TECHNETIUM TC 99M SULFUR COLLOID
608-73-1	TECHNICAL HCH
123-31-9	TECQUINOL
64-17-5	TECSOL
64-17-5	TECSOL C
112-24-3	TECZA
280-57-9	TED
280-57-9	TEDA
111-48-8	TEDEGYL
9016-87-9	TEDIMON 31
3689-24-5	TEDP
3689-24-5	TEDTP
545-55-1	TEF
9002-84-0	TEFLON
9002-84-0	TEFLON 110
9002-84-0	TEFLON 30
9002-84-0	TEFLON 30J
9002-84-0	TEFLON 327
9002-84-0	TEFLON 5
9002-84-0	TEFLON 6
9002-84-0	TEFLON 6C
9002-84-0	TEFLON 6C-J
9002-84-0	TEFLON 6J
9002-84-0	TEFLON 7A
9002-84-0	TEFLON 7J
9002-84-0	TEFLON 851-204
9002-84-0	TEFLON K
9002-84-0	TEFLON T 30
9002-84-0	TEFLON T 5
9002-84-0	TEFLON T 6
9002-84-0	TEFLON T 8A
9002-84-0	TEFLON TFE
112-27-6	TEG
112-27-6	TEG (GLYCOL)
1333-86-4	TEG 10
112-80-1	TEGO-OLEIC 130
57-11-4	TEGOSTEARIC 254
57-11-4	TEGOSTEARIC 255
57-11-4	TEGOSTEARIC 272
68-76-8	TEIB
1319-77-3	TEKRESOL
298-00-0	TEKWAISA
78-00-2	TEL
2275-18-5	TELEFOS
13494-80-9	TELLOY
13494-80-9	TELLUR (POLISH)
10102-20-2	TELLURIC ACID (H2TeO3), DISODIUM SALT
10102-20-2	TELLURIC ACID, DISODIUM SALT
13494-80-9	TELLURIUM
13494-80-9	TELLURIUM (ACGIH)
13494-80-9	TELLURIUM ELEMENT

CAS No.	Chemical Name
7783-80-4	TELLURIUM FLUORIDE (TeF6)
7783-80-4	TELLURIUM FLUORIDE (TeF6), (OC-6-11)-
7783-80-4	TELLURIUM HEXAFLUORIDE
7783-80-4	TELLURIUM HEXAFLUORIDE (ACGIH,DOT)
10102-20-2	TELLUROUS ACID, DISODIUM SALT
514-73-8	TELMICID
514-73-8	TELMID
514-73-8	TELMIDE
297-78-9	TELODRIN
1937-37-7	TELON FAST BLACK E
542-75-6	TELONE
542-75-6	TELONE II
330-54-1	TELVAR
330-54-1	TELVAR DIURON WEED KILLER
846-50-4	TEMAZEPAM
110-18-9	TEMED
3383-96-8	TEMEFOS
3383-96-8	TEMEPHOS
116-06-3	TEMIC
116-06-3	TEMIK
116-06-3	TEMIK 10 G
116-06-3	TEMIK G 10
50-78-2	TEMPERAL
55-63-0	TEMPONITRIN
50-55-5	TEMPOSERPINE
28772-56-7	TEMUS
81-81-2	TEMUS W
121-44-8	TEN
2636-26-2	TENAC
91-80-5	TENALIN
101-96-2	TENAMENE 2
128-37-0	TENAMENE 3
3081-14-9	TENAMENE 4
114-26-1	TENDEX
54-11-5	TENDUST
65-85-0	TENN-PLAS
1982-47-4	TENORAN
128-37-0	TENOX BHT
123-31-9	TENOX HQ
79-09-4	TENOX P GRAIN PRESERVATIVE
1309-37-1	TENYO 501
78-10-4	TEOS
107-49-3	TEP
112-57-2	TEP
545-55-1	TEPA
107-49-3	TEPP
123-31-9	TEQUINOL
74-83-9	TERABOL
137-26-8	TERAMETHYL THIURAM DISULFIDE
13071-79-9	TERBUFOS
63-25-2	TERCYL
100-21-0	TEREPHTHALIC ACID
100-20-9	TEREPHTHALIC ACID CHLORIDE
100-20-9	TEREPHTHALIC ACID DICHLORIDE
120-61-6	TEREPHTHALIC ACID METHYL ESTER
120-61-6	TEREPHTHALIC ACID, DIMETHYL ESTER
100-20-9	TEREPHTHALIC DICHLORIDE
100-20-9	TEREPHTHALOYL CHLORIDE
100-20-9	TEREPHTHALOYL DICHLORIDE
79-20-9	TERETON
9016-45-9	TERGITOL TP-9 (NONIONIC)
87-86-5	TERM-I-TROL
12789-03-6	TERMEX
95-50-1	TERMITKIL
60-57-1	TERMITOX
8020-83-5	TERMOL 190
68956-56-9	TERPENE HYDROCARBONS N.O.S.
8001-50-1	TERPENE POLYCHLORINATES (65 OR 66% CHLORINE)
8002-09-3	TERPENTINOEL (GERMAN)
92-94-4	TERPHENYLS
8000-41-7	TERPINEOL
8000-41-7	TERPINEOL 318
8000-41-7	TERPINEOLS
586-62-9	TERPINOLEN
586-62-9	TERPINOLENE
586-62-9	TERPINOLENE (DOT)
8004-09-9	TERR-O-GAS
8004-09-9	TERR-O-GEL
115-26-4	TERRA-SYTAM
82-68-8	TERRACHLOR
82-68-8	TERRACLOR
82-68-8	TERRACLOR 30 G
115-90-2	TERRACUR P
82-68-8	TERRAFUN
1910-2-5	TERRAKLENE
115-26-4	TERRASYTUM
2425-06-1	TERRAZOL
137-26-8	TERSAN
17804-35-2	TERSAN 1991
137-26-8	TERSAN 75
70-30-4	TERSASEPTIC
686-31-7	TERT-AMYL 2-ETHYLPEROXYHEXANOATE
625-16-1	TERT-AMYL ACETATE
75-85-4	TERT-AMYL ALCOHOL
75-85-4	TERT-AMYL ALCOHOL (DOT)
594-36-5	TERT-AMYL CHLORIDE
4511-39-1	TERT-AMYL PERBENZOATE
686-31-7	TERT-AMYL PEROXY-2-ETHYLHEXANOATE, TECHNICAL PURE (DOT)
4511-39-1	TERT-AMYL PEROXYBENZOATE
29240-17-3	TERT-AMYL PEROXYPIVALATE
29240-17-3	TERT-AMYL PERPIVALATE
26760-64-5	TERT-AMYLENE
75-66-1	TERT-BUTANETHIOL
75-65-0	TERT-BUTANOL
75-65-0	TERT-BUTANOL (DOT)
3457-61-2	TERT-BUTYL 1-METHYL-1-PHENYLETHYL PEROXIDE
30580-75-7	TERT-BUTYL 1-METHYL-1-PHENYLETHYL PEROXIDE
927-07-1	TERT-BUTYL 2,2-DIMETHYLPEROXYPROPIONATE
927-07-1	TERT-BUTYL 2,2-DIMETHYLPROPANEPEROXOATE
3006-82-4	TERT-BUTYL 2-ETHYLHEXANEPEROXOATE
3006-82-4	TERT-BUTYL 2-ETHYLPERHEXANOATE
3006-82-4	TERT-BUTYL 2-ETHYLPEROXYHEXANOATE
109-13-7	TERT-BUTYL 2-METHYLPROPANEPEROXOATE
13122-18-4	TERT-BUTYL 3,5,5-TRIMETHYLPEROXYHEXANOATE
540-88-5	TERT-BUTYL ACETATE
540-88-5	TERT-BUTYL ACETATE (ACGIH,DOT)
75-65-0	TERT-BUTYL ALCOHOL
75-65-0	TERT-BUTYL ALCOHOL (ACGIH,DOT)
507-19-7	TERT-BUTYL BROMIDE
507-20-0	TERT-BUTYL CHLORIDE
1189-85-1	TERT-BUTYL CHROMATE
1189-85-1	TERT-BUTYL CHROMATE(VI)
3457-61-2	TERT-BUTYL CUMENE PEROXIDE, TECHNICAL PURE (DOT)
30580-75-7	TERT-BUTYL CUMENE PEROXIDE, TECHNICAL PURE (DOT)
3457-61-2	TERT-BUTYL CUMYL PEROXIDE
30580-75-7	TERT-BUTYL CUMYL PEROXIDE
3457-61-2	TERT-BUTYL CUMYL PEROXIDE, TECHNICAL PURE (DOT)
30580-75-7	TERT-BUTYL CUMYL PEROXIDE, TECHNICAL PURE (DOT)
3457-61-2	TERT-BUTYL CUMYL PEROXIDE, TECHNICALLY PURE
30580-75-7	TERT-BUTYL CUMYL PEROXIDE, TECHNICALLY PURE
75-91-2	TERT-BUTYL HYDROPEROXIDE
1609-86-5	TERT-BUTYL ISOCYANATE
30026-92-7	TERT-BUTYL ISOPRPYL BENZENE HYDROPEROXIDE
75-66-1	TERT-BUTYL MERCAPTAN
1931-62-0	TERT-BUTYL MONOPERMALEATE
1931-62-0	TERT-BUTYL MONOPEROXYMALEATE
1931-62-0	TERT-BUTYL MONOPEROXYMALEATE, MAXIMUM CONCENTRATION 55% AS A PASTE (DOT)
1931-62-0	TERT-BUTYL MONOPEROXYMALEATE, MAXIMUM CONCENTRATION 55% IN SOLUTION (DOT)
13122-18-4	TERT-BUTYL PER-3,5,5-TRIMETHYLHEXANOATE
107-71-1	TERT-BUTYL PERACETATE
614-45-9	TERT-BUTYL PERBENZOATE
614-46-9	TERT-BUTYL PERBENZOATE
109-13-7	TERT-BUTYL PERISOBUTYRATE
26748-41-4	TERT-BUTYL PERNEODECANOATE
110-05-4	TERT-BUTYL PEROXIDE
110-05-4	TERT-BUTYL PEROXIDE (DOT)
13122-18-4	TERT-BUTYL PEROXY 3,5,5-TRIMETHYLHEXANOATE
3006-82-4	TERT-BUTYL PEROXY-2-ETHYLHEXANOATE
3006-82-4	TERT-BUTYL PEROXY-2-ETHYLHEXANOATE, >30% WITH 2,2-DI-(TERT-BUTYLPEROXY)BUTANE, WITH ≤35%
3006-82-4	TERT-BUTYL PEROXY-2-ETHYLHEXANOATE, TECHNICAL PURE (DOT)
3006-82-4	TERT-BUTYL PEROXY-2-ETHYLHEXANOATE, TECHNICALLY PURE
3006-82-4	TERT-BUTYL PEROXY-2-ETHYLHEXANOATE, WITH AT LEAST 50% PHLEGMATIZER (DOT)

CAS No.	*Chemical Name*
13122-18-4	TERT-BUTYL PEROXY-3,5,5-TRIMETHYL HEXANOATE, TECHNICAL PURE (DOT)
13122-18-4	TERT-BUTYL PEROXY-3,5,5-TRIMETHYLHEXANOATE, TECHNICALLY PURE
107-71-1	TERT-BUTYL PEROXYACETATE
614-45-9	TERT-BUTYL PEROXYBENZOATE
23474-91-1	TERT-BUTYL PEROXYCROTONATE
2550-33-6	TERT-BUTYL PEROXYDIETHYLACETATE, TECHNICAL PURE (DOT)
2550-33-6	TERT-BUTYL PEROXYDIETHYLACETATE, TECHNICALLY PURE
3006-82-4	TERT-BUTYL PEROXYETHYLHEXANOATE
109-13-7	TERT-BUTYL PEROXYISOBUTYRATE
109-13-7	TERT-BUTYL PEROXYISOBUTYRATE, ≤52% IN SOLUTION
109-13-7	TERT-BUTYL PEROXYISOBUTYRATE, MORE THAN 52% BUT NOT MORE THAN 77% IN SOLUTION (DOT)
109-13-7	TERT-BUTYL PEROXYISOBUTYRATE, NOT MORE THAN 52% IN SOLUTION (DOT)
13122-18-4	TERT-BUTYL PEROXYISONONANOATE, TECHNICAL PURE (DOT)
2372-21-6	TERT-BUTYL PEROXYISOPROPYL CARBONATE (CONCENTRATION ≤80%)
2372-21-6	TERT-BUTYL PEROXYISOPROPYL CARBONATE, TECHNICAL PURE (DOT)
2372-21-6	TERT-BUTYL PEROXYISOPROPYL CARBONATE, TECHNICALLY PURE
1931-62-0	TERT-BUTYL PEROXYMALEATE
1931-62-0	TERT-BUTYL PEROXYMALEATE (CONCENTRATION ≤80%)
1931-62-0	TERT-BUTYL PEROXYMALEATE, = 55% AS A PASTE
1931-62-0	TERT-BUTYL PEROXYMALEATE, MAXIMUM CONCENTRATION 55% AS A PASTE (DOT)
1931-62-0	TERT-BUTYL PEROXYMALEATE, NOT MORE THAN 55% IN SOLUTION (DOT)
1931-62-0	TERT-BUTYL PEROXYMALEIC ACID
26748-41-4	TERT-BUTYL PEROXYNEODECANOATE
26748-41-4	TERT-BUTYL PEROXYNEODECANOATE, NOT MORE THAN 77% IN SOLUTION (DOT)
26748-41-4	TERT-BUTYL PEROXYNEODECANOATE, TECHNICALLY PURE
927-07-1	TERT-BUTYL PEROXYPIVALATE
927-07-1	TERT-BUTYL PEROXYPIVALATE, NOT MORE THAN 77% IN SOLUTION (DOT)
927-07-1	TERT-BUTYL PERPIVALATE
927-07-1	TERT-BUTYL TERT-VALERYL PEROXIDE
73090-68-3	TERT-BUTYL TETRALIN
927-07-1	TERT-BUTYL TRIMETHYLPEROXYACETATE
1333-13-7	TERT-BUTYL-M-CRESOL
75-64-9	TERT-BUTYLAMINE
3775-90-4	TERT-BUTYLAMINOETHYL METHACRYLATE
98-06-6	TERT-BUTYLBENZENE
75-84-3	TERT-BUTYLCARBINOL
70042-58-9	TERT-BUTYLCYCLOHEXYL CHLOROFORMATE
70042-58-9	TERT-BUTYLCYCLOHEXYLCHLOROFORMATE (DOT)
2160-93-2	TERT-BUTYLDIETHANOLAMINE
2372-21-6	TERT-BUTYLPEROXY ISOPROPYL CARBONATE
75-66-1	TERT-BUTYLTHIOL
30174-58-4	TERT-DECANETHIOL
30174-58-4	TERT-DECYL MERCAPTAN
110-05-4	TERT-DIBUTYL PEROXIDE
25103-58-6	TERT-DODECANETHIOL
25103-58-6	TERT-DODECYL MERCAPTAN
25103-58-6	TERT-DODECYLTHIOL
25360-10-5	TERT-NONANETHIOL
25360-10-5	TERT-NONYL MERCAPTAN
5809-08-5	TERT-OCTYL HYDROPEROXIDE
141-59-3	TERT-OCTYL MERCAPTAN
107-45-9	TERT-OCTYLAMINE
27193-28-8	TERT-OCTYLPHENOL
141-59-3	TERT-OCTYLTHIOL
463-82-1	TERT-PENTANE
463-82-1	TERT-PENTANE (DOT)
75-98-9	TERT-PENTANOIC ACID
75-85-4	TERT-PENTANOL
625-16-1	TERT-PENTYL ACETATE
75-85-4	TERT-PENTYL ALCOHOL
686-31-7	TERT-PENTYL ALCOHOL, 2-ETHYLPEROXYHEXANOATE
625-16-1	TERT-PENTYL ALCOHOL, ACETATE
29240-17-3	TERT-PENTYL ALCOHOL, PEROXYPIVALATE
594-36-5	TERT-PENTYL CHLORIDE
29240-17-3	TERT-PENTYL PEROXYPIVALATE
28983-37-1	TERT-TETRADECANETHIOL
28983-37-1	TERT-TETRADECYL MERCAPTAN
1694-09-3	TERTRACID BRILLIANT VIOLET 6B

CAS No.	*Chemical Name*
3761-53-3	TERTRACID PONCEAU 2R
106-50-3	TERTRAL D
123-30-8	TERTRAL P BASE
27598-85-2	TERTRAL P BASE
1937-37-7	TERTRODIRECT BLACK E
1937-37-7	TERTRODIRECT BLACK EFD
2602-46-2	TERTRODIRECT BLUE 2B
16071-86-6	TERTRODIRECT FAST BROWN BR
633-03-4	TERTROPHENE BRILLIANT GREEN G
51-28-5	TERTROSULPHUR BLACK PB
51-28-5	TERTROSULPHUR PBR
78-10-4	TES 28
11099-06-2	TES 40
107-21-1	TESCOL
58-22-0	TESLEN
52-24-4	TESPAMIN
52-24-4	TESPAMINE
58-22-0	TESTANDRONE
58-22-0	TESTICULOSTERONE
58-22-0	TESTOBASE
58-22-0	TESTOPROPON
58-22-0	TESTOSTEROID
58-22-0	TESTOSTERON
58-22-0	TESTOSTERONE
58-22-0	TESTOVIRON SCHERING
58-22-0	TESTOVIRON T
58-22-0	TESTRONE
58-22-0	TESTRYL
7704-34-9	TESULOID
597-64-8	TET
112-24-3	TETA
97-77-8	TETD
127-18-4	TETLEN
120-12-7	TETRA OLIVE N2G
115-26-4	TETRA SYTAM
115-77-5	TETRA(HYDROXYMETHYL)METHANE
681-84-5	TETRA-METHYL ORTHOSILICATE
3087-37-4	TETRA-N-PROPYL TITANATE
1299-86-1	TETRAALUMINUM TRICARBIDE
12656-43-8	TETRAALUMINUM TRICARBIDE
10380-29-7	TETRAAMINECOPPER SULFATE MONOHYDRATE
10380-29-7	TETRAAMMINECOPPER SULFATE, HYDRATE
10380-29-7	TETRAAMMINECOPPER(2+) SULFATE (1:1) MONOHYDRATE
10380-29-7	TETRAAMMINECOPPER(II) SULFATE MONOHYDRATE
101-61-1	TETRABASE
79-27-6	TETRABROMOACETYLENE
79-27-6	TETRABROMOETHANE
25167-20-8	TETRABROMOETHANE
558-13-4	TETRABROMOMETHANE
5593-70-4	TETRABUTOXYTITANIUM
5593-70-4	TETRABUTYL ORTHOTITANATE
5593-70-4	TETRABUTYL TITANATE
5593-70-4	TETRABUTYLTITANATE (CZECH)
127-18-4	TETRACAP
16842-03-8	TETRACARBONYLHYDRIDOCOBALT
16842-03-8	TETRACARBONYLHYDROCOBALT
13463-39-3	TETRACARBONYLNICKEL
127-18-4	TETRACHLOORETHEEN (DUTCH)
56-23-5	TETRACHLOORKOOLSTOF (DUTCH)
56-23-5	TETRACHLOORMETAAN
127-18-4	TETRACHLORAETHEN (GERMAN)
127-18-4	TETRACHLORETHYLENE
56-23-5	TETRACHLORKOHLENSTOFF, TETRA (GERMAN)
56-23-5	TETRACHLORMETHAN (GERMAN)
76-12-0	TETRACHLORO-1,2-DIFLUOROETHANE
56-23-5	TETRACHLOROCARBON
1746-01-6	TETRACHLORODIBENZO-P-DIOXIN
79-34-5	TETRACHLOROETHANE
25322-20-7	TETRACHLOROETHANE
25322-20-7	TETRACHLOROETHANE (DOT)
127-18-4	TETRACHLOROETHENE
127-18-4	TETRACHLOROETHYLENE
127-18-4	TETRACHLOROETHYLENE
127-18-4	TETRACHLOROETHYLENE (DOT)
2425-06-1	TETRACHLOROETHYLTHIOTETRAHYDROPHTHALIMIDE
56-23-5	TETRACHLOROMETHANE
1335-88-2	TETRACHLORONAPHTHALENE
117-08-8	TETRACHLOROPHTHALIC ANHYDRIDE
10026-04-7	TETRACHLOROSILANE

CAS No.	Chemical Name
10026-04-7	TETRACHLOROSILICON
7646-78-8	TETRACHLOROSTANNANE
10026-06-9	TETRACHLOROSTANNANE PENTAHYDRATE
6012-97-1	TETRACHLOROTHIOPHENE
7646-78-8	TETRACHLOROTIN
7550-45-0	TETRACHLOROTITANIUM
7632-51-1	TETRACHLOROVANADIUM
961-11-5	TETRACHLOROVINPHOS
10026-11-6	TETRACHLOROZIRCONIUM
56-23-5	TETRACHLORURE DE CARBONE (FRENCH)
10026-04-7	TETRACHLORURE DE SILICIUM (FRENCH)
7550-45-0	TETRACHLORURE DE TITANE (FRENCH)
961-11-5	TETRACHLORVINPHOS
127-18-4	TETRACLOROETENE (ITALIAN)
56-23-5	TETRACLOROMETANO (ITALIAN)
56-23-5	TETRACLORURO DI CARBONIO (ITALIAN)
60-54-8	TETRACYCLINE
17702-41-9	TETRADECAHYDRODECABORANE
1459-10-5	TETRADECANE, 1-PHENYL-
112-72-1	TETRADECANOL
27196-00-5	TETRADECANOL
27196-00-5	TETRADECANOL, MIXED ISOMERS
112-72-1	TETRADECYL ALCOHOL
27196-00-5	TETRADECYL ALCOHOL
1459-10-5	TETRADECYLBENZENE
97-77-8	TETRADIN
97-77-8	TETRADINE
78-10-4	TETRAETHOXYSILANE
78-10-4	TETRAETHOXYSILICON
107-49-3	TETRAETHYL DIPHOSPHATE
3689-24-5	TETRAETHYL DITHIONOPYROPHOSPHATE
3689-24-5	TETRAETHYL DITHIOPYROPHOSPHATE
3689-24-5	TETRAETHYL DITHIOPYROPHOSPHATE MIXTURE, DRY
3689-24-5	TETRAETHYL DITHIOPYROPHOSPHATE MIXTURE, DRY (DOT)
3689-24-5	TETRAETHYL DITHIOPYROPHOSPHATE MIXTURE, LIQUID
3689-24-5	TETRAETHYL DITHIOPYROPHOSPHATE MIXTURE, LIQUID (DOT)
3689-24-5	TETRAETHYL DITHIOPYROPHOSPHATE, LIQUID
78-00-2	TETRAETHYL LEAD
78-00-2	TETRAETHYL LEAD (ACGHI,DOT)
78-00-2	TETRAETHYL LEAD, LIQUID
78-00-2	TETRAETHYL LEAD, LIQUID (INCLUDING FLASHPOINT FOR EXPORT SHIPMENT BY WATER) (DOT)
78-10-4	TETRAETHYL ORTHOSILICATE
78-10-4	TETRAETHYL ORTHOSILICATE (DOT)
107-49-3	TETRAETHYL PYROPHOSPHATE
107-49-3	TETRAETHYL PYROPHOSPHATE MIXTURE, DRY
107-49-3	TETRAETHYL PYROPHOSPHATE MIXTURE, DRY (DOT)
107-49-3	TETRAETHYL PYROPHOSPHATE MIXTURE, LIQUID
107-49-3	TETRAETHYL PYROPHOSPHATE MIXTURE, LIQUID (DOT)
107-49-3	TETRAETHYL PYROPHOSPHATE, LIQUID [FLAMMABLE LIQUID LABEL]
107-49-3	TETRAETHYL PYROPHOSPHATE, LIQUID [POISON B LABEL]
563-12-2	TETRAETHYL S,S'-METHYLENE BIS(PHOSPHOROTHIOLOTHIONATE)
78-10-4	TETRAETHYL SILICATE
78-10-4	TETRAETHYL SILICATE (DOT)
597-64-8	TETRAETHYL TIN
112-60-7	TETRAETHYLENE GLYCOL
112-98-1	TETRAETHYLENE GLYCOL DIBUTYL ETHER
112-57-2	TETRAETHYLENE PENTAMINE
112-57-2	TETRAETHYLENEPENTAMINE (DOT)
1067-20-5	TETRAETHYLMETHANE
78-10-4	TETRAETHYLOXYSILANE
78-00-2	TETRAETHYLPLUMBANE
597-64-8	TETRAETHYLSTANNANE
97-77-8	TETRAETHYLTHIOPEROXYDICARBONIC DIAMIDE
97-77-8	TETRAETHYLTHIRAM DISULFIDE
97-77-8	TETRAETHYLTHIURAM DISULFIDE
97-77-8	TETRAETHYLTHIURAM SULFIDE
97-77-8	TETRAETIL
56-23-5	TETRAFINOL
116-14-3	TETRAFLUORETHYLENE
75-73-0	TETRAFLUOROCARBON
63938-10-3	TETRAFLUOROCHLOROETHANE
1320-37-2	TETRAFLUORODICHLOROETHANE
116-14-3	TETRAFLUOROETHENE
116-14-3	TETRAFLUOROETHENE (9CI)
9002-84-0	TETRAFLUOROETHENE HOMOPOLYMER
9002-84-0	TETRAFLUOROETHENE POLYMER
116-14-3	TETRAFLUOROETHYLENE
9002-84-0	TETRAFLUOROETHYLENE HOMOPOLYMER
116-14-3	TETRAFLUOROETHYLENE MONOMER
9002-84-0	TETRAFLUOROETHYLENE POLYMER
9002-84-0	TETRAFLUOROETHYLENE RESIN
116-14-3	TETRAFLUOROETHYLENE, INHIBITED
116-14-3	TETRAFLUOROETHYLENE, INHIBITED (DOT)
10036-47-2	TETRAFLUOROHYDRAZINE
10036-47-2	TETRAFLUOROHYDRAZINE (DOT)
75-73-0	TETRAFLUOROMETHANE
75-73-0	TETRAFLUOROMETHANE (DOT)
63938-10-3	TETRAFLUOROMONOCHLOROETHANE
7783-61-1	TETRAFLUOROSILANE
7783-60-0	TETRAFLUOROSULFURANE
56-23-5	TETRAFORM
7723-14-0	TETRAFOSFOR (DUTCH)
127-18-4	TETRAGUER
123-91-1	TETRAHYDRO-1,4-DIOXIN
110-91-8	TETRAHYDRO-1,4-ISOXAZINE
110-91-8	TETRAHYDRO-1,4-OXAZINE
97-99-4	TETRAHYDRO-2-FURANCARBINOL
97-99-4	TETRAHYDRO-2-FURANMETHANOL
97-99-4	TETRAHYDRO-2-FURANYLMETHANOL
97-99-4	TETRAHYDRO-2-FURYLMETHANOL
96-47-9	TETRAHYDRO-2-METHYLFURAN
110-91-8	TETRAHYDRO-2H-1,4-OXAZINE
96-45-7	TETRAHYDRO-2H-IMIDAZOLE-2-THIONE
142-68-7	TETRAHYDRO-2H-PYRAN
123-91-1	TETRAHYDRO-P-DIOXIN
110-91-8	TETRAHYDRO-P-OXAZINE
1321-16-0	TETRAHYDROBENZALDEHYDE
110-83-8	TETRAHYDROBENZENE
85-43-8	TETRAHYDROFTALANHYDRID (CZECH)
109-99-9	TETRAHYDROFURAAN (DUTCH)
109-99-9	TETRAHYDROFURAN
109-99-9	TETRAHYDROFURAN (ACGIH,DOT)
109-99-9	TETRAHYDROFURANNE (FRENCH)
97-99-4	TETRAHYDROFURFURYL ALCOHOL
4795-29-3	TETRAHYDROFURFURYLAMINE
4795-29-3	TETRAHYDROFURFURYLAMINE (DOT)
97-99-4	TETRAHYDROFURYLALKOHOL (CZECH)
7782-65-2	TETRAHYDROGERMANE
119-64-2	TETRAHYDRONAPHTHALENE
54-11-5	TETRAHYDRONICOTYRINE, DL-
85-43-8	TETRAHYDROPHTHALIC ACID ANHYDRIDE
85-43-8	TETRAHYDROPHTHALIC ANHYDRIDE
85-43-8	TETRAHYDROPHTHALIC ANHYDRIDE (DOT)
142-68-7	TETRAHYDROPYRAN
142-68-7	TETRAHYDROPYRANE
61215-72-3	TETRAHYDROPYRIDINE
123-75-1	TETRAHYDROPYRROLE
96-47-9	TETRAHYDROSYLVAN
110-01-0	TETRAHYDROTHIOPHEN
110-01-0	TETRAHYDROTHIOPHENE
110-01-0	TETRAHYDROTHIOPHENE (DOT)
126-33-0	TETRAHYDROTHIOPHENE 1,1-DIOXIDE
126-33-0	TETRAHYDROTHIOPHENE DIOXIDE
115-77-5	TETRAHYDROXYMETHYLMETHANE
109-99-9	TETRAIDROFURANO (ITALIAN)
115-77-5	TETRAKIS(HYDROXYMETHYL)METHANE
152-16-9	TETRAKISDIMETHYLAMINOPHOSPHONOUS ANHYDRIDE
127-18-4	TETRALENO
127-18-4	TETRALEX
119-61-2	TETRALIN
771-29-9	TETRALIN 1-HYDROPEROXIDE
771-29-9	TETRALIN HYDROPEROXIDE
771-29-9	TETRALIN HYDROPEROXIDE, TECHNICALLY PURE
771-29-9	TETRALIN HYDROPEROXIDE, TECHNICALLY PURE (DOT)
71-29-9	TETRALIN PEROXIDE
119-64-2	TETRALINA (POLISH)
119-64-2	TETRALINE
771-29-9	TETRALINIHYDROPEROXYDE (DUTCH)
771-29-9	TETRALINIHYDROPEROXIDE (GERMAN)
479-45-8	TETRALIT
479-45-8	TETRALITE
771-29-9	TETRALYL HYDROPEROXIDE
3734-97-2	TETRAM
3734-97-2	TETRAM 75
3734-97-2	TETRAM MONOOXALATE

CAS No.	Chemical Name
3734-97-2	TETRAM, ACID OXALATE
110-18-9	TETRAMEEN
681-84-5	TETRAMETHOXYSILANE
75-59-2	TETRAMETHYL AMMONIUM HYDROXIDE, LIQUID (DOT)
110-18-9	TETRAMETHYL ETHYLENE DIAMINE
75-74-1	TETRAMETHYL LEAD
75-74-1	TETRAMETHYL LEAD (ACGIH)
51-80-9	TETRAMETHYL METHYLENEDIAMINE
75-76-3	TETRAMETHYL SILANE
681-84-5	TETRAMETHYL SILICATE
3333-52-6	TETRAMETHYL SUCCINONITRILE
137-26-8	TETRAMETHYLTHIURANE DISULFIDE
137-26-8	TETRAMETHYL THIURANE DISULPHIDE
594-27-4	TETRAMETHYL TIN
137-26-8	TETRAMETHYL-THIRAM DISULFID (GERMAN)
75-59-2	TETRAMETHYLAMMONIUM HYDROXIDE
75-59-2	TETRAMETHYLAMMONIUM HYDROXIDE (DOT)
75-59-2	TETRAMETHYLAMMONIUM HYDROXIDE, LIQUID
115-26-4	TETRAMETHYLDIAMIDOPHOSPHORIC FLUORIDE
90-94-8	TETRAMETHYLDIAMINOBENZOPHENE
101-61-1	TETRAMETHYLDIAMINODIPHENYLMETHANE
137-26-8	TETRAMETHYLDIURANE SULPHITE
287-23-0	TETRAMETHYLENE
110-63-4	TETRAMETHYLENE 1,4-DIOL
55-98-1	TETRAMETHYLENE BIS[METHANESULFONATE]
109-99-9	TETRAMETHYLENE OXIDE
110-01-0	TETRAMETHYLENE SULFIDE
126-33-0	TETRAMETHYLENE SULFONE
80-12-6	TETRAMETHYLENEDISULPHOTETRAMINE
137-26-8	TETRAMETHYLENETHIURAM DISULPHIDE
123-75-1	TETRAMETHYLENIMINE
563-79-1	TETRAMETHYLETHENE
563-79-1	TETRAMETHYLETHYLENE
463-82-1	TETRAMETHYLMETHANE
115-77-5	TETRAMETHYLOLMETHANE
115-26-4	TETRAMETHYLPHOSPHORODIAMIDIC FLUORIDE
75-74-1	TETRAMETHYLPLUMBANE
75-76-3	TETRAMETHYLSILANE (DOT)
594-27-4	TETRAMETHYLSTANNANE
3333-52-6	TETRAMETHYLSUCCINIC ACID DINITRILE
3333-52-6	TETRAMETHYLSUCCINODINITRILE
3333-52-6	TETRAMETHYLSUCCINONITRILE
137-26-8	TETRAMETHYLTHIOCARBAMOYLDISULPHIDE
137-26-8	TETRAMETHYLTHIOPEROXYDICARBONIC DIAMIDE
137-26-8	TETRAMETHYLTHIORAMDISULFIDE (DUTCH)
137-26-8	TETRAMETHYLTHIURAM
137-26-8	TETRAMETHYLTHIURAM BISULFIDE
137-26-8	TETRAMETHYLTHIURAM BISULPHIDE
137-26-8	TETRAMETHYLTHIURAM DISULFIDE
137-26-8	TETRAMETHYLTHIURAM DISULPHIDE
137-26-8	TETRAMETHYLTHIURUM DISULFIDE
137-26-8	TETRAMETHYLTHIURUM DISULPHIDE
16071-86-6	TETRAMINE FAST BROWN BRDN EXTRA
16071-86-6	TETRAMINE FAST BROWN BRP
16071-86-6	TETRAMINE FAST BROWN BRS
9002-84-0	TETRAN 30
9002-84-0	TETRAN PTFE
119-64-2	TETRANAP
13746-89-9	TETRANITRATOZIRCONIUM
53014-37-2	TETRANITRO-ANILINE
509-14-8	TETRANITROMETHANE
509-14-8	TETRANITROMETHANE (ACGIH,DOT)
50-00-0	TETRAOXYMETHYLENE
56-55-3	TETRAPHENE
595-90-4	TETRAPHENYL TIN
595-90-4	TETRAPHENYLSTANNANE
757-58-4	TETRAPHOSPHATE HEXAETHYLIQUE (FRENCH)
7723-14-0	TETRAPHOSPHOR (GERMAN)
757-58-4	TETRAPHOSPHORIC ACID, HEXAETHYL ESTER
757-58-4	TETRAPHOSPHORIC ACID, HEXAETHYL ESTER, LIQUID
1314-80-3	TETRAPHOSPHORUS DECASULFIDE
12037-82-0	TETRAPHOSPHORUS HEPTASULFIDE
1314-85-8	TETRAPHOSPHORUS TRISULFIDE
137-26-8	TETRAPOM
3087-37-4	TETRAPROPOXY TITANATE
3087-37-4	TETRAPROPYL ORTHOTITANATE
3087-37-4	TETRAPROPYL TITANATE
3087-37-4	TETRAPROPYL-O-TITANATE
6842-15-5	TETRAPROPYLENE
3087-37-4	TETRAPROPYLORTHOTINATE (DOT)
137-26-8	TETRASIPTON
7722-88-5	TETRASODIUM DIPHOSPATE
7722-88-5	TETRASODIUM DIPHOSPHATE (Na4P207)
13601-19-1	TETRASODIUM FERROCYANIDE
13601-19-9	TETRASODIUM HEXACYANOFERRATE
13601-19-9	TETRASODIUM HEXACYANOFERRATE(4-)
7722-88-5	TETRASIUM PYROPHOSPHATE
7722-88-5	TETRASODIUM PYROPHOSPHATE (Na4P2O7)
56-23-5	TETRASOL
107-49-3	TETRASTIGMINE
137-26-8	TETRATHIURAM DISULFIDE
137-26-8	TETRATHIURAM DISULPHIDE
127-18-4	TETRAVEC
2636-26-2	TETRAVOS
1937-37-7	TETRAZO DEEP BLACK G
21732-17-2	TETRAZOL-1-ACETIC ACID
112-57-2	TETREN
479-45-8	TETRIL
60-00-4	TETRINE ACID
127-18-4	TETROGUER
110-00-9	TETROLE
107-49-3	TETRON
7722-88-5	TETRON
7722-88-5	TETRON (DISPERSANT)
107-49-3	TETRON (PESTICIDE)
107-49-3	TETRON 100
127-18-4	TETROPIL
92-52-4	TETROSIN LY
108-90-7	TETROSIN SP
479-45-8	TETRYL
479-45-8	TETRYL (ACGIH,DOT)
542-55-2	TETRYL FORMATE
97-77-8	TETURAM
56-75-7	TEVCOCIN
540-88-5	TEXACO LEAD APPRECIATOR
2235-54-3	TEXAPON A 400
2235-54-3	TEXAPON ALS
151-21-3	TEXAPON DL CONC
143-00-0	TEXAPON DLS
151-21-3	TEXAPON K 12
151-21-3	TEXAPON K 1296
151-21-3	TEXAPON L 100
2235-54-3	TEXAPON SPECIAL
139-96-8	TEXAPON T 35
139-96-8	TEXAPON T 42
139-96-8	TEXAPON TH
151-21-3	TEXAPON V HC
151-21-3	TEXAPON V HC POWDER
151-21-3	TEXAPON Z HIGH CONC NEEDLES
151-21-3	TEXAPON ZHC
7758-19-2	TEXTILE
7758-19-2	TEXTONE
116-14-3	TFE
1333-86-4	TG 10
7440-32-6	TG-TV
14808-60-7	TGL 16319
85-44-9	TGL 6525
7705-07-9	TGY 24
4301-50-2	TH 367-1
8002-74-2	TH 44
13746-98-0	THALLIC NITRATE
10102-45-1	THALLIC NITRATE
16901-76-1	THALLIC NITRATE
1314-32-5	THALLIC OXIDE
7440-28-0	THALLIUM
6533-73-9	THALLIUM CARBONATE
6533-73-9	THALLIUM CARBONATE (Tl2CO3)
13453-30-0	THALLIUM CHLORATE
13453-30-0	THALLIUM CHLORATE (DOT)
13453-30-0	THALLIUM CHLORATE (TlClO3)
7791-12-0	THALLIUM CHLORIDE
7791-12-0	THALLIUM CHLORIDE (TlCl)
7791-12-0	THALLIUM MONOCHLORIDE
10102-45-1	THALLIUM MONONITRATE
16901-76-1	THALLIUM MONONITRATE
13746-98-0	THALLIUM NITRATE
10102-45-1	THALLIUM NITRATE
16901-76-1	THALLIUM NITRATE

CAS No.	Chemical Name
10102-45-1	THALLIUM NITRATE (DOT)
13746-98-0	THALLIUM NITRATE (DOT)
16901-76-1	THALLIUM NITRATE (DOT)
10102-45-1	THALLIUM NITRATE (Tl(NO3))
13746-98-0	THALLIUM NITRATE (Tl(NO3)3)
1314-32-5	THALLIUM OXIDE
1314-32-5	THALLIUM OXIDE (2:3)
1314-32-5	THALLIUM OXIDE (8CI,9CI)
1314-32-5	THALLIUM OXIDE (Tl2O3)
1314-32-5	THALLIUM PEROXIDE
1314-32-5	THALLIUM SESQUIOXIDE
7446-18-6	THALLIUM SULFATE
10031-59-1	THALLIUM SULFATE
7446-18-6	THALLIUM SULFATE (Tl2(SO4))
10031-59-1	THALLIUM SULFATE (Tl2(SO4))
7446-18-6	THALLIUM SULFATE, SOLID
10031-59-1	THALLIUM SULFATE, SOLID
7446-18-6	THALLIUM SULFATE, SOLID (DOT)
10031-59-1	THALLIUM SULFATE, SOLID (DOT)
13746-98-0	THALLIUM TRINITRATE
1314-32-5	THALLIUM TRIOXIDE
7446-18-6	THALLIUM(1) SULFATE
10031-59-1	THALLIUM(1) SULFATE
7791-12-0	THALLIUM(1+) CHLORIDE
10102-45-1	THALLIUM(1+) NITRATE
16901-76-1	THALLIUM(1+) NITRATE
1314-32-5	THALLIUM(111) OXIDE
1314-32-5	THALLIUM(3+) OXIDE
6533-73-9	THALLIUM(I) CARBONATE (2:1)
7791-12-0	THALLIUM(I) CHLORIDE
10102-45-1	THALLIUM(I) NITRATE
16901-76-1	THALLIUM(I) NITRATE
10102-45-1	THALLIUM(I) NITRATE (1:1)
16901-76-1	THALLIUM(I) NITRATE (1:1)
7446-18-6	THALLIUM(I) SULFATE
10031-59-1	THALLIUM(I) SULFATE
7446-18-6	THALLIUM(I) SULFATE (2:1)
13746-98-0	THALLIUM(III) NITRATE
1314-32-5	THALLIUM(III) OXIDE
6533-73-9	THALLOUS CARBONATE
7791-12-0	THALLOUS CHLORIDE
2757-18-8	THALLOUS MALONATE
10102-45-1	THALLOUS NITRATE
7446-18-6	THALLOUS SULFATE
10031-59-1	THALLOUS SULFATE
9016-87-9	THANATE P 210
9016-87-9	THANATE P 220
9016-87-9	THANATE P 270
280-57-9	THANCAT TD 33
94-36-0	THE RADERM
53-16-7	THEELIN
50-28-2	THEELIN, DIHYDRO-
53-16-7	THELYKININ
124-43-6	THENARDOL
91-80-5	THENYLENE
91-80-5	THENYLPYRAMINE
7681-49-4	THERA FLUR
7681-49-4	THERA-FLUR-N
1304-56-9	THERMALOX
1304-56-9	THERMALOX 995
1333-86-4	THERMAX
1336-36-3	THERMINOL FR-1
1309-64-4	THERMOGUARD B
1309-64-4	THERMOGUARD L
1309-64-4	THERMOGUARD S
7758-29-4	THERMPHOS
7758-29-4	THERMPHOS L 50
7758-29-4	THERMPHOS N
7758-29-4	THERMPHOS SPR
109-99-9	THF
97-99-4	THFA
62-55-5	THIACETAMIDE
507-09-5	THIACETIC ACID
110-02-1	THIACYCLOPENTADIENE
110-01-0	THIACYCLOPENTANE
126-33-0	THIACYCLOPENTANE DIOXIDE
712-68-5	THIAFUR
638-17-5	THIALDINE
640-15-3	THIAMETON
3268-49-3	THIAPENTANAL
110-02-1	THIAPHENE
79-19-6	THIASEMICARBAZIDE
531-82-8	THIAZOLE, 2-ACETAMIDO-4-(5-NITRO-2-FURYL)-
115-29-7	THIFOR
110-01-0	THILANE
137-26-8	THILLATE
56-04-2	THIMECIL
137-26-8	THIMER
298-02-2	THIMET
298-02-2	THIMET 10G
298-02-2	THIMET G
115-29-7	THIMUL
52-24-4	THIO-TEPA
62-55-5	THIOACETAMIDE
507-09-5	THIOACETIC ACID
507-09-5	THIOACETIC ACID (DOT)
96-69-5	THIOALKOFEN BM
96-69-5	THIOALKOFEN BM 4
96-69-5	THIOALKOFEN BMCH
96-69-5	THIOALKOFEN MBCH
95-06-7	THIOALLATE
139-65-1	THIOANILINE
100-53-8	THIOBENZYL ALCOHOL
75-18-3	THIOBIS(METHANE)
109-79-5	THIOBUTYL ALCOHOL
4384-82-1	THIOCARB
62-56-6	THIOCARBAMIDE
79-19-6	THIOCARBAMOYLHYDRAZINE
79-19-6	THIOCARBAMYLHYDRAZINE
2231-57-4	THIOCARBAZIDE
2231-57-4	THIOCARBOHYDRAZIDE
463-71-8	THIOCARBONIC DICHLORIDE
2231-57-4	THIOCARBONIC DIHYDRAZIDE
2231-57-4	THIOCARBONOHYDRAZIDE
463-71-8	THIOCARBONYL CHLORIDE
463-71-8	THIOCARBONYL DICHLORIDE
463-71-8	THIOCARBONYL-CHLORIDE (DOT)
6533-73-9	THIOCHROMAN-4-ONE, OXIME
540-72-7	THIOCYANATE SODIUM
21564-17-0	THIOCYANIC ACID, (2-BENZOTHIAZOLYLTHIO)METHYL ESTER
14797-65-0	THIOCYANIC ACID, 2,4-DINITROPHENYL ESTER
21564-17-0	THIOCYANIC ACID, 2-(BENZOTHIAZOLYLTHIO)METHYL ESTER, 60%
1762-95-4	THIOCYANIC ACID, AMMONIUM SALT
542-90-5	THIOCYANIC ACID, ETHYL ESTER
592-87-0	THIOCYANIC ACID, LEAD(2+) SALT
592-85-8	THIOCYANIC ACID, MERCURY(2+) SALT
592-85-8	THIOCYANIC ACID, MERCURY(II) SALT
556-64-9	THIOCYANIC ACID, METHYL ESTER
540-72-7	THIOCYANIC ACID, SODIUM SALT
126-33-0	THIOCYCLOPENTANE-1,1-DIOXIDE
115-29-7	THIODAN
115-29-7	THIODAN 35
115-29-7	THIODAN 35EC
1031-07-8	THIODAN SULFATE
298-04-4	THIODEMETON
298-04-4	THIODEMETRON
139-65-1	THIODI-P-PHENYLENEDIAMINE
111-48-8	THIODIETHANOL
111-48-8	THIODIETHYLENE GLYCOL
111-48-8	THIODIGLYCOL
92-84-2	THIODIPHENYLAMINE
3689-24-5	THIODIPHOSPHORIC ACID ([(HO)2P(S)]2O), TETRAETHYL ESTER
3689-24-5	THIODIPHOSPHORIC ACID TETRAETHYL ESTER
75-08-1	THIOETHANOL
75-08-1	THIOETHYL ALCOHOL
60-24-2	THIOETHYLENE GLYCOL
141-43-5	THIOFACO M 50
102-71-6	THIOFACO T-35
39196-18-4	THIOFANOX
39196-18-4	THIOFAX
115-29-7	THIOFOR
463-71-8	THIOFOSGEN (CZECH)
52-24-4	THIOFOZIL
110-02-1	THIOFURAN
110-02-1	THIOFURFURAN
60-24-2	THIOGLYCOL
60-24-2	THIOGLYCOL (DOT)

CAS No.	Chemical Name
68-11-1	THIOGLYCOLIC ACID
68-11-1	THIOGLYCOLIC ACID (ACGIH,DOT)
2941-64-2	THIOHYPOCHLOROUS ACID, ANHYDROSULFIDE WITH O-ETHYL THIOCARBONATE
26555-35-1	THIOHYPOCHLOROUS ACID, ANHYDROSULFIDE WITH O-ETHYL THIOCARBONATE
541-53-7	THIOIMIDODICARBONIC DIAMIDE ([(H2N)C(S)]2NH)
463-71-8	THIOKARBONYLCHLORID (CZECH)
507-09-5	THIOLACETIC ACID
79-42-5	THIOLACTIC ACID
79-42-5	THIOLACTIC ACID (DOT)
110-01-0	THIOLANE
126-33-0	THIOLANE-1,1-DIOXIDE
110-02-1	THIOLE
7704-34-9	THIOLUX
640-15-3	THIOMETON
56-38-2	THIOMEX
60-24-2	THIOMONOGLYCOL
115-29-7	THIOMUL
297-97-2	THIONAZIN
297-97-2	THIONAZINE
115-29-7	THIONEX
507-09-5	THIONOACETIC ACID
7719-09-7	THIONYL CHLORIDE
7719-09-7	THIONYL CHLORIDE (ACGIH,DOT)
7719-09-7	THIONYL DICHLORIDE
91-80-5	THIONYLAN
97-77-8	THIOPEROXYDICARBONIC DIAMIDE ([(H2N)C(S)]2S2), TETRAETHYL-
137-26-8	THIOPEROXYDICARBONIC DIAMIDE ([(H2N)C(S)]2S2), TETRAMETHYL-
126-33-0	THIOPHAN SULFONE
110-01-0	THIOPHANE
126-33-0	THIOPHANE 1,1-DIOXIDE
126-33-0	THIOPHANE DIOXIDE
110-02-1	THIOPHEN
110-02-1	THIOPHENE
110-02-1	THIOPHENE (DOT)
110-01-0	THIOPHENE, TETRAHYDRO-
126-33-0	THIOPHENE, TETRAHYDRO-, 1,1-DIOXIDE
298-00-0	THIOPHENIT
108-98-5	THIOPHENOL
108-98-5	THIOPHENOL (DOT)
56-38-2	THIOPHOS
56-38-2	THIOPHOS 3422
463-71-8	THIOPHOSGENE
463-71-8	THIOPHOSGENE (DOT)
52-24-4	THIOPHOSPHAMIDE
56-72-4	THIOPHOSPHATE DE O,O-DIETHYLE ET DE O-(3-CHLORO-4-METHYL-7-COUMARINYLE) (FRENCH)
56-38-2	THIOPHOSPHATE DE O,O-DIETHYLE ET DE O-(4-NITROPHENYLE) (FRENCH)
333-41-5	THIOPHOSPHATE DE O,O-DIETHYLE ET DE O-2-ISOPROPYL-4-METHYL-6-PYRIMIDYLE (FRENCH)
298-00-0	THIOPHOSPHATE DE O,O-DIMETHYLE ET DE O-(4-NITROPHENYLE) (FRENCH)
2778-04-3	THIOPHOSPHATE DE O,O-DIMETHYLE ET DE S-((5-METHOXY-4-PYRONYL)-METHYLE) (FRENCH)
919-86-8	THIOPHOSPHATE DE O,O-DIMETHYLE ET DE S-2-ETHYLTHIO-ETHYLE (FRENCH)
52-24-4	THIOPHOSPHORAMIDE, N,N',N''-TRI-1,2-ETHANEDIYL-
1314-80-3	THIOPHOSPHORIC ANHYDRIDE
3982-91-0	THIOPHOSPHORIC TRICHLORIDE
10265-92-6	THIOPHOSPHORSAEURE-O,S-DIMETHYLESTERAMID (GERMAN)
3982-91-0	THIOPHOSPHORYL CHLORIDE
3982-91-0	THIOPHOSPHORYL CHLORIDE (DOT)
3982-91-0	THIOPHOSPHORYL CHLORIDE (PSCl3)
3982-91-0	THIOPHOSPHORYL TRICHLORIDE
3689-24-5	THIOPYROPHOSPHORIC ACID ([(HO)2PS]2O), TETRAETHYL ESTER
3689-24-5	THIOPYROPHOSPHORIC ACID, TETRAETHYL ESTER
137-26-8	THIOSAN
137-26-8	THIOSCABIN
79-19-6	THIOSEMICARBAZIDE
7704-34-9	THIOSOL
115-29-7	THIOSULFAN
7783-18-8	THIOSULFURIC ACID (H2S2O3), DIAMMONIUM SALT
26265-65-6	THIOSULFURIC ACID (H2S2O3), LEAD SALT
7783-18-8	THIOSULFURIC ACID, DIAMMONIUM SALT
26265-65-6	THIOSULFURIC ACID, LEAD SALT
10025-67-9	THIOSULFUROUS DICHLORIDE
52-24-4	THIOTEF
52-24-4	THIOTEPA
3689-24-5	THIOTEPP
110-02-1	THIOTETROLE
137-26-8	THIOTEX
56-04-2	THIOTHYMIN
137-26-8	THIOTOX
115-29-7	THIOTOX
137-26-8	THIOTOX (FUNGICIDE)
115-29-7	THIOTOX (INSECTICIDE)
52-24-4	THIOTRIETHYLENEPHOSPHORAMIDE
62-56-6	THIOUREA
62-56-6	THIOUREA (DOT)
5344-82-1	THIOUREA, (2-CHLOROPHENYL)-
614-78-8	THIOUREA, (2-METHYLPHENYL)-
86-88-4	THIOUREA, 1-NAPHTHALENYL-
30553-04-9	THIOUREA, 1-NAPHTHALENYL-
96-45-7	THIOUREA, N,N'-(1,2-ETHANEDIYL)-
30553-04-9	THIOUREA, NAPHTHALENYL-
103-85-5	THIOUREA, PHENYL-
68-11-1	THIOVANIC ACID
7704-34-9	THIOVIT
23135-22-0	THIOXAMYL
7704-34-9	THIOZOL
137-26-8	THIRAM
137-26-8	THIRAM (ACGIH,DOT)
137-26-8	THIRAM 75
137-26-8	THIRAM 80
137-26-8	THIRAM B
137-26-8	THIRAMAD
137-26-8	THIRAME (FRENCH)
137-26-8	THIRASAN
137-26-8	THIRIDE
137-26-8	THIULIN
137-26-8	THIULIX
137-26-8	THIURAD
51-52-5	THIURAGYL
137-26-8	THIURAM
137-26-8	THIURAM D
137-26-8	THIURAM DISULFIDE, TETRAMETHYL-
97-77-8	THIURAM E
137-26-8	THIURAM M
137-26-8	THIURAM M RUBBER ACCELERATOR
137-26-8	THIURAM TMTD
137-26-8	THIURAMIN
137-26-8	THIURAMYL
97-77-8	THIURANIDE
115-77-5	THME
87-86-5	THOMPSON'S WOOD FIX
1314-20-1	THORIA
7440-29-1	THORIUM
1314-20-1	THORIUM DIOXIDE
7440-29-1	THORIUM METAL, PYROPHORIC
7440-29-1	THORIUM METAL, PYROPHORIC (DOT)
13823-29-5	THORIUM NITRATE
13823-29-5	THORIUM NITRATE ((Th(NO3)4))
13823-29-5	THORIUM NITRATE (DOT)
1314-20-1	THORIUM OXIDE
1314-20-1	THORIUM OXIDE (ThO2)
13823-29-5	THORIUM TETRANITRATE
13823-29-5	THORIUM(4+) NITRATE
13823-29-5	THORIUM(IV) NITRATE
1314-20-1	THORIUM(IV) OXIDE
7440-29-1	THORIUM-232
1314-20-1	THOROTRAST
1314-20-1	THORTRAST
142-68-7	THP
85-43-8	THPA
78-96-6	THREAMINE
87-69-4	THREARIC ACID
10043-35-3	THREE ELEPHANT
1464-53-5	THREITOL, 1,2:3,4-DIANHYDRO-
299-75-2	THREITOL, 1,4-DIMETHANESULFONATE, (2S,3S)-
299-75-2	THREOSULPHAN
79-01-6	THRETHYLEN
79-01-6	THRETHYLENE
62-56-6	THU
137-26-8	THYLATE

CAS No.	Chemical Name
298-00-0	THYLPAR M-50
53-16-7	THYNESTRON
56-04-2	THYREONORM
56-04-2	THYREOSTAT
56-04-2	THYREOSTAT I
51-52-5	THYREOSTAT II
7440-32-6	TI 160
7440-32-6	TI 35A
7440-32-6	TI 40
7440-32-6	TI 40A
7440-32-6	TI 50A
7440-32-6	TI 75A
7440-32-6	TI-A55
13463-67-7	TI-PURE
13463-67-7	TI-PURE R 100
13463-67-7	TI-PURE R 101
13463-67-7	TI-PURE R 900
13463-67-7	TI-PURE R 901
13463-67-7	TI-PURE R 915
712-68-5	TIAFUR
56-75-7	TIFOMYCINE
137-26-8	TIGAM
14808-60-7	TIGER-EYE
55-38-9	TIGUVON
82-68-8	TILCAREX
51-21-8	TIMAZIN
298-02-2	TIMET
1309-64-4	TIMONOX
7440-31-5	TIN
25324-56-5	TIN (IV) PHOSPHIDE
7783-47-3	TIN BIFLUORIDE
7647-18-9	TIN CHLORIDE
7772-99-8	TIN CHLORIDE
7772-99-8	TIN CHLORIDE (SnCl2)
7646-78-8	TIN CHLORIDE (SnCl4)
10026-06-9	TIN CHLORIDE (SnCl4), PENTAHYDRATE
7646-78-8	TIN CHLORIDE, FUMING (DOT)
7772-99-8	TIN DICHLORIDE
7772-99-8	TIN DICHLORIDE (SnCl2)
7783-47-3	TIN DIFLUORIDE
7440-31-5	TIN FLAKE
7783-47-3	TIN FLUORIDE
7783-47-3	TIN FLUORIDE (SnF2)
25324-56-5	TIN MONOPHOSPHIDE
7646-78-8	TIN PERCHLORIDE
7646-78-8	TIN PERCHLORIDE (DOT)
25324-56-5	TIN PHOSPHIDE (SnP)
7440-31-5	TIN POWDER
7772-99-8	TIN PROTOCHLORIDE
7646-78-8	TIN TETRACHLORIDE
10026-06-9	TIN TETRACHLORIDE PENTAHYDRATE
7646-78-8	TIN TETRACHLORIDE, ANHYDROUS
7646-78-8	TIN TETRACHLORIDE, ANHYDROUS (DOT)
10026-06-9	TIN TETRACHLORIDE, HYDRATED
7772-99-8	TIN(II) CHLORIDE
7772-99-8	TIN(II) CHLORIDE (1:2)
7646-78-8	TIN(IV) CHLORIDE
7646-78-8	TIN(IV) CHLORIDE (1:4)
7646-78-8	TIN(IV) TETRACHLORIDE
597-64-8	TIN, TETRAETHYL-
13121-70-5	TIN, TRICYCLOHEXYLHYDROXY-
60-24-2	TINCTURES, MEDICINAL
639-58-7	TINMATE
7646-85-7	TINNING FLUX (DOT)
7646-78-8	TINTETRACHLORIDE (DUTCH)
129-06-6	TINTORANE
52-24-4	TIO-TEF
56-38-2	TIOFOS
52-24-4	TIOFOSFAMID
52-24-4	TIOFOSYL
52-24-4	TIOFOZIL
56-04-2	TIOMERACIL
13463-67-7	TIONA 113
13463-67-7	TIONA VB
115-29-7	TIONEX
56-04-2	TIORALE M
115-29-7	TIOVEL
30553-04-9	TIOX 30
13463-67-7	TIOXIDE AD-M
13463-67-7	TIOXIDE R-CR
13463-67-7	TIOXIDE R-CR 3
13463-67-7	TIOXIDE R-SM
13463-67-7	TIOXIDE R-TC 2
13463-67-7	TIOXIDE R-TC 60
13463-67-7	TIOXIDE R.XL
13463-67-7	TIOXIDE RHD
7704-34-9	TIOZOL 80
30207-98-8	TIP-NIP
122-20-3	TIPA
13463-67-7	TIPAQUE CR 50
13463-67-7	TIPAQUE CR 90
13463-67-7	TIPAQUE CR 95
13463-67-7	TIPAQUE R 820
93-76-5	TIPPON
137-26-8	TIRAMPA
26419-73-8	TIRPATE
1332-58-7	TISYN
7550-45-0	TITAANTETRACHLORIDE (DUTCH)
13463-67-7	TITAN A
13463-67-7	TITAN RC-K 20
7440-32-6	TITAN VT 1-1
7550-45-0	TITANE (TETRACHLORURE DE) (FRENCH)
13463-67-7	TITANIA
13463-67-7	TITANIA (TiO2)
5593-70-4	TITANIC ACID, TETRABUTYL ESTER
7550-45-0	TITANIO (TETRACLORURO DI) (ITALIAN)
7440-32-6	TITANIUM
7440-32-6	TITANIUM 50A
7440-32-6	TITANIUM A 40
7440-32-6	TITANIUM ALLOY
5593-70-4	TITANIUM BUTOXIDE (Ti(OBU)4)
7550-45-0	TITANIUM CHLORIDE
7705-07-9	TITANIUM CHLORIDE
7705-07-9	TITANIUM CHLORIDE (TiCl3)
7550-45-0	TITANIUM CHLORIDE (TiCl4)
7550-45-0	TITANIUM CHLORIDE (TiCl4) (T-4)-
7705-07-9	TITANIUM CHLORIDE, MIXTURES, NON-PYROPHORIC
7704-98-5	TITANIUM DIHYDRIDE
11140-68-4	TITANIUM DIHYDRIDE
7704-98-5	TITANIUM DIHYDRIDE (TiH2)
11140-68-4	TITANIUM DIHYDRIDE (TiH2)
13463-67-7	TITANIUM DIOXIDE
13463-67-7	TITANIUM DIOXIDE (TiO2)
13693-11-3	TITANIUM DISULFATE
7440-32-6	TITANIUM ELEMENT
7704-98-5	TITANIUM HYDRIDE
11140-68-4	TITANIUM HYDRIDE
7704-98-5	TITANIUM HYDRIDE (DOT)
11140-68-4	TITANIUM HYDRIDE (DOT)
7704-98-5	TITANIUM HYDRIDE (TiH2)
11140-68-4	TITANIUM HYDRIDE (TiH2)
7440-32-6	TITANIUM METAL POWDER, DRY
7440-32-6	TITANIUM METAL POWDER, DRY OR WET WITH 20% WATER
7440-32-6	TITANIUM METAL POWDER, WET WITH NOT LESS THAN 25% WATER (DOT)
7440-32-6	TITANIUM METAL, POWDER, DRY (DOT)
13463-67-7	TITANIUM OXIDE
13463-67-7	TITANIUM OXIDE (TiO2)
13463-67-7	TITANIUM PEROXIDE (TiO2)
3087-37-4	TITANIUM PROPOXIDE (Ti(OPR)4)
7440-32-6	TITANIUM SPONGE GRANULES (DOT)
7440-32-6	TITANIUM SPONGE POWDERS (DOT)
7440-32-6	TITANIUM SPONGE, GRANULES OR POWDER
13693-11-3	TITANIUM SULFATE
13825-74-6	TITANIUM SULFATE
13693-11-3	TITANIUM SULFATE (1:2)
13693-11-3	TITANIUM SULFATE (TI(SO4)2)
13693-11-3	TITANIUM SULFATE SOLUTION WITH ≤ 45% SULFURIC ACID
5593-70-4	TITANIUM TETRABUTOXIDE
5593-70-4	TITANIUM TETRABUTYLATE
7550-45-0	TITANIUM TETRACHLORIDE
7550-45-0	TITANIUM TETRACHLORIDE (DOT)
3087-37-4	TITANIUM TETRAPROPOXIDE
3087-37-4	TITANIUM TETRAPROPYLATE
7705-07-9	TITANIUM TRICHLORIDE
7705-07-9	TITANIUM TRICHLORIDE MIXTURE
7705-07-9	TITANIUM TRICHLORIDE MIXTURES, NON-PYROPHORIC (DOT)
7705-07-9	TITANIUM TRICHLORIDE, PYROPHORIC

CAS No.	Chemical Name
7705-07-9	TITANIUM TRICHLORIDE, PYROPHORIC (DOT)
7440-32-6	TITANIUM VT 1
5593-70-4	TITANIUM(4+) BUTOXIDE
13693-11-3	TITANIUM(4+) SULFATE
7705-07-9	TITANIUM(III) CHLORIDE
7550-45-0	TITANIUM(IV) CHLORIDE
13463-67-7	TITANIUM(IV) OXIDE
13693-11-3	TITANIUM(IV) SULFATE
7440-32-6	TITANIUM, POWDER WET WITH 20% OR MORE WATER (DOT)
5593-70-4	TITANIUM, TETRABUTOXY-
3087-37-4	TITANIUM, TETRAPROPOXY-
7440-32-6	TITANIUM, WET, WITH LESS THAN 20% WATER (DOT)
7440-32-6	TITANIUM-125
7705-07-9	TITANOUS CHLORIDE
13463-67-7	TITANOX
13463-67-7	TITANOX 2010
13463-67-7	TITANOX RANC
7550-45-0	TITANTETRACHLORID (GERMAN)
13463-67-7	TITONE R 5N
60-00-4	TITRIPLEX
139-13-9	TITRIPLEX I
137-26-8	TIURAM (POLISH)
137-26-8	TIURAMYL
7631-86-9	TIX-O-SIL 33J
7631-86-9	TIX-O-SIL 38A
7631-86-9	TK 900
577-11-7	TKB 20
76-03-9	TKHU
76-03-9	TKHUK
9002-84-0	TL 102
9002-84-0	TL 102 (POLYMER)
14486-19-2	TL 1026
9002-84-0	TL 103F
538-07-8	TL 1149
9002-84-0	TL 115
75-77-4	TL 1163
9002-84-0	TL 125
9002-84-0	TL 126
57-64-7	TL 1380
555-77-1	TL 145
624-83-9	TL 1450
51-75-2	TL 146
77-81-6	TL 1578
107-44-8	TL 1618
627-63-4	TL 189
627-11-2	TL 207
598-14-1	TL 214
3982-91-0	TL 262
593-89-5	TL 294
538-07-8	TL 329
151-56-4	TL 337
79-44-7	TL 389
541-41-3	TL 423
79-22-1	TL 438
51-83-2	TL 457
359-06-8	TL 670
696-28-6	TL 69
371-62-0	TL 741
7783-55-3	TL 75
822-06-0	TL 78
115-26-4	TL 792
506-68-3	TL 822
62-74-8	TL 869
7487-94-7	TL 898
9002-84-0	TL-R
9002-84-0	TL-V
111-46-6	TL4N
540-88-5	TLA
9002-84-0	TLP 10
9002-84-0	TLP 10F
75-59-2	TM
1333-86-4	TM 30
121-75-5	TM 4049
1333-86-4	TM 70
75-50-3	TMA
108-67-8	TMB
56-93-9	TMBAC
109-76-2	TMEDA
75-74-1	TML
137-26-8	TMT
137-26-8	TMTD
137-26-8	TMTDS
99-35-4	TNB
7758-98-7	TNCS 53
55-63-0	TNG
105-60-2	TNK 2G5
509-14-8	TNM
118-96-7	TNT
118-96-7	TNT-TOLITE (FRENCH)
78-30-8	TOCP
8020-83-5	TOGASTAN
1836-75-5	TOK
1836-75-5	TOK E
1836-75-5	TOK E 25
1836-75-5	TOK E 40
78-30-8	TOKF
53-16-7	TOKOKIN
7631-86-9	TOKUSIL GU
7631-86-9	TOKUSIL GU-N
7631-86-9	TOKUSIL GV-N
7631-86-9	TOKUSIL N
7631-86-9	TOKUSIL P
7631-86-9	TOKUSIL TPLM
7631-86-9	TOKUSIL U
633-03-4	TOKYO ANILINE BRILLIANT GREEN
118-96-7	TOLIT
118-96-7	TOLITE
298-00-0	TOLL
108-88-3	TOLU-SOL
97-56-3	TOLUAZOTOLUIDINE
108-88-3	TOLUEEN (DUTCH)
26471-62-5	TOLUEEN-DIISOCYANAAT (DUTCH)
108-88-3	TOLUEN (CZECH)
26471-62-5	TOLUEN-DISOCIANATO (ITALIAN)
108-88-3	TOLUENE
108-88-3	TOLUENE (ACGIH,DOT)
91-08-7	TOLUENE DIISOCYANATE
26471-62-5	TOLUENE DIISOCYANATE
26471-62-5	TOLUENE DIISOCYANATE (DOT)
108-87-2	TOLUENE HEXAHYDRIDE
25231-46-3	TOLUENE SULFONIC ACID
104-15-4	TOLUENE SULFONIC ACID, LIQUID
25231-46-3	TOLUENE SULFONIC ACID, LIQUID
104-15-4	TOLUENE SULFONIC ACID, LIQUID (DOT)
25231-46-3	TOLUENE SULFONIC ACID, LIQUID (DOT)
104-15-4	TOLUENE SULFONIC ACID, SOLID
25231-46-3	TOLUENE SULFONIC ACID, SOLID
104-15-4	TOLUENE SULPHONIC ACID, LIQUID, CONTAINING MORE THAN 5% FREE SULPHURIC ACID (DOT)
25231-46-3	TOLUENE SULPHONIC ACID, LIQUID, CONTAINING MORE THAN 5% FREE SULPHURIC ACID (DOT)
104-15-4	TOLUENE SULPHONIC ACID, LIQUID, CONTAINING NOT MORE THAN 5% FREE SULPHURIC ACID (DOT)
25231-46-3	TOLUENE SULPHONIC ACID, LIQUID, CONTAINING NOT MORE THAN 5% FREE SULPHURIC ACID (DOT)
104-15-4	TOLUENE SULPHONIC ACID, SOLID, CONTAINING MORE THAN 5% FREE SULPHURIC ACID (DOT)
25231-46-3	TOLUENE SULPHONIC ACID, SOLID, CONTAINING MORE THAN 5% FREE SULPHURIC ACID (DOT)
104-15-4	TOLUENE SULPHONIC ACID, SOLID, CONTAINING NOT MORE THAN 5% FREE SULPHURIC ACID (DOT)
25231-46-3	TOLUENE SULPHONIC ACID, SOLID, CONTAINING NOT MORE THAN 5% FREE SULPHURIC ACID (DOT)
98-07-7	TOLUENE TRICHLORIDE
118-96-7	TOLUENE, 2,4,6-TRINITRO-
118-96-7	TOLUENE, 2,4,6-TRINITRO- (WET)
121-14-2	TOLUENE, 2,4-DINITRO-
606-20-2	TOLUENE, 2,6-DINITRO-
610-39-9	TOLUENE, 3,4-DINITRO-
98-16-8	TOLUENE, 3-AMINO-ALPHA,ALPHA,ALPHA-TRIFLUORO-
777-37-7	TOLUENE, 2-CHLORO-ALPHA,ALPHA,ALPHA-TRIFLUORO-5-NITRO-
121-17-5	TOLUENE, 4-CHLORO-3-NITRO-ALPHA,ALPHA,ALPHA-TRIFLUORO-
121-17-5	TOLUENE, 4-CHLORO-ALPHA,ALPHA,ALPHA-TRIFLUORO-3-NITRO-
402-54-0	TOLUENE, 4-NITRO-ALPHA,ALPHA,ALPHA-TRIFLUORO-
98-07-7	TOLUENE, ALPHA,ALPHA,ALPHA-TRICHLORO-
98-08-8	TOLUENE, ALPHA,ALPHA,ALPHA-TRIFLUORO-
402-54-0	TOLUENE, ALPHA,ALPHA,ALPHA-TRIFLUORO-P-NITRO-
98-87-3	TOLUENE, ALPHA,ALPHA-DICHLORO-

CAS No.	Chemical Name
100-44-7	TOLUENE, ALPHA-CHLORO-
100-14-1	TOLUENE, ALPHA-CHLORO-P-NITRO-
30030-25-2	TOLUENE, ALPHA-CHLOROVINYL-
140-29-4	TOLUENE, ALPHA-CYANO-
620-05-3	TOLUENE, ALPHA-IODO-
25321-14-6	TOLUENE, AR,AR-DINITRO-
25168-05-2	TOLUENE, AR-CHLORO-
25496-08-6	TOLUENE, AR-FLUORO-
1321-12-6	TOLUENE, AR-NITRO-
27458-20-4	TOLUENE, BUTYL-
25567-68-4	TOLUENE, CHLORO-O-NITRO-
61878-61-3	TOLUENE, CHLORO-O-NITRO-
25567-68-4	TOLUENE, CHLORONITRO-
61878-61-3	TOLUENE, CHLORONITRO-
25321-14-6	TOLUENE, DINITRO-
108-87-2	TOLUENE, HEXAHYDRO-
98-15-7	TOLUENE, M-CHLORO-ALPHA,ALPHA,ALPHA-TRIFLUORO-
99-08-1	TOLUENE, M-NITRO-
95-46-5	TOLUENE, O-BROMO-
27458-20-4	TOLUENE, O-BUTYL-
95-49-8	TOLUENE, O-CHLORO-
88-16-4	TOLUENE, O-CHLORO-ALPHA,ALPHA,ALPHA-TRIFLUORO-
88-72-2	TOLUENE, O-NITRO-
104-83-6	TOLUENE, P,ALPHA-DICHLORO-
106-38-7	TOLUENE, P-BROMO-
106-43-4	TOLUENE, P-CHLORO-
98-56-6	TOLUENE, P-CHLORO-ALPHA,ALPHA,ALPHA-TRIFLUORO-
25496-08-6	TOLUENE, P-FLUORO-
99-99-0	TOLUENE, P-NITRO-
98-51-1	TOLUENE, P-TERT-BUTYL-
1320-01-0	TOLUENE, PENTYL-
28729-54-6	TOLUENE, PROPYL-
25013-15-4	TOLUENE, VINYL- (MIXED ISOMERS)
95-80-7	TOLUENE-2,4-DIAMINE
584-84-9	TOLUENE-2,4-DIISOCYANATE
26471-62-5	TOLUENE-2,4-DIISOCYANATE
26471-62-5	TOLUENE-2,4-DIISOCYANATE (ACGIH)
584-84-9	TOLUENE-2,4-DIISOCYANATE (TDI)
91-08-7	TOLUENE-2,6-DIISOCYANATE
2646-17-5	TOLUENE-2-AZONAPHTHOL-2
30143-13-6	TOLUENE-AR,AR-DIAMINE
25376-45-8	TOLUENE-AR,AR-DIAMINE
25376-45-8	TOLUENEDIAMINE
30143-13-6	TOLUENEDIAMINE
30143-13-6	TOLUENEDIAMINE (DOT)
25376-45-8	TOLUENEDIAMINE (DOT)
104-15-4	TOLUENESULFONIC ACID
25231-46-3	TOLUENESULFONIC ACID
26471-62-5	TOLUILENODWUIZOCYJANIAN (POLISH)
108-88-3	TOLUOL
108-88-3	TOLUOL (DOT)
108-88-3	TOLUOLO (ITALIAN)
26471-62-5	TOLUYLENE-2,4-DIISOCYANATE
100-44-7	TOLYL CHLORIDE
26447-14-3	TOLYL GLYCIDYL ETHER
108-44-1	TOLYLAMINE
106-49-0	TOLYLAMINE
95-53-4	TOLYLAMINE
91-08-7	TOLYLENE 2,6-DIISOCYANATE
26471-62-5	TOLYLENE DIISOCYANATE
26471-62-5	TOLYLENE ISOCYANATE
26471-62-5	TOLYLENE-2,4-DIISOCYANATE
30143-13-6	TOLYLENEDIAMINE
25376-45-8	TOLYLENEDIAMINE
117-52-2	TOMARIN
9004-34-6	TOMOFAN
101-77-9	TONOX
123-73-9	TOPANEL
4170-30-3	TOPANEL
128-37-0	TOPANOL
128-37-0	TOPANOL BHT
101-96-2	TOPANOL M
128-37-0	TOPANOL O
128-37-0	TOPANOL OC
94-36-0	TOPEX
12789-03-6	TOPICHLOR 20
12789-03-6	TOPICLOR
12789-03-6	TOPICLOR 20
3691-35-8	TOPITOX

CAS No.	Chemical Name
1405-87-4	TOPITRACIN
1405-87-4	TOPITRASIN
2540-82-1	TOPROSE
67-68-5	TOPSYM
105-60-2	TORAYCA N 6
1918-02-1	TORDON
93-79-8	TORMONA
104-15-4	TOSIC ACID
25231-46-3	TOSIC ACID
7631-86-9	TOSIL
7631-86-9	TOSIL P
1910-2-5	TOTACOL
78-30-8	TOTP
56-38-2	TOX 47
81-81-2	TOX-HID
8001-35-2	TOXADUST
8001-35-2	TOXAFEEN (DUTCH)
8001-35-2	TOXAKIL
63-25-2	TOXAN
8001-35-2	TOXAPHEN
8001-35-2	TOXAPHEN (GERMAN)
8001-35-2	TOXAPHENE
8001-35-2	TOXAPHENE (DOT)
1910-2-5	TOXER TOTAL
57-74-9	TOXICHLOR
12789-03-6	TOXICHLOR
110-16-7	TOXILIC ACID
108-31-6	TOXILIC ANHYDRIDE
1402-68-2	TOXINS, AFLA
8001-35-2	TOXON 63
8001-35-2	TOXPHENE
8001-35-2	TOXYPHEN
89-62-3	TOYO FAST RED GL BASE
60-11-7	TOYO OIL YELLOW G
115-86-6	TP
7440-32-6	TP 28
7440-32-6	TP 28C
3771-19-5	TPIA
639-58-7	TPTC
13463-67-7	TR 840
9000-65-1	TRAGACANTH GUM
137-26-8	TRAMETAN
15271-41-7	TRANID
111-90-0	TRANS CUTOL
154-23-4	TRANS-(+)-3,3',4',5,7-FLAVANPENTOL
156-60-5	TRANS-1,2-DICHLOROETHENE
156-60-5	TRANS-1,2-DICHLOROETHYLENE
624-64-6	TRANS-1,2-DIMETHYLETHYLENE
110-17-8	TRANS-1,2-ETHYLENEDICARBOXYLIC ACID
10061-02-6	TRANS-1,3-DICHLOROPROPENE
110-57-6	TRANS-1,4-DICHLOROBUTENE
2207-04-7	TRANS-1,4-DIMETHYLCYCLOHEXANE
55738-54-0	TRANS-2((DIMETHYLAMINO) METHYLIMINO)-5-(2-(5-NITRO-2-FURYL)VINYL)-1,3,4-OXADIAZOLE
123-73-9	TRANS-2-BUTENAL
4170-30-3	TRANS-2-BUTENAL
624-64-6	TRANS-2-BUTENE
10544-63-5	TRANS-2-BUTENOIC ACID ETHYL ESTER
14686-13-6	TRANS-2-HEPTENE
646-04-8	TRANS-2-PENTENE
56-53-1	TRANS-4,4'-DIHYDROXY-ALPHA,BETA-DIETHYLSTILBENE
107-02-8	TRANS-ACROLEIN
1762-95-4	TRANS-AID
584-79-2	TRANS-ALLETHRIN
56-53-1	TRANS-ALPHA,ALPHA-DIETHYL-4,4'-STILBENEDIOL
646-04-8	TRANS-BETA-AMYLENE
141-66-2	TRANS-BIDRIN
624-64-6	TRANS-BUTENE
624-64-6	TRANS-BUTENE-2
110-17-8	TRANS-BUTENEDIOIC ACID
56-53-1	TRANS-DIETHYLSTILBESTEROL
56-53-1	TRANS-DIETHYLSTILBESTROL
84-80-0	TRANS-PHYLLOQUINONE
2163-80-6	TRANS-VERT
94-75-7	TRANSAMINE
93-76-5	TRANSAMINE
556-61-6	TRAPEX
556-61-6	TRAPEXIDE
7775-09-9	TRAVEX

CAS No.	Chemical Name
1582-09-8	TREFANOCIDE
1582-09-8	TREFICON
1582-09-8	TREFLAN
1582-09-8	TREFLANOCIDE ELANCOLAN
1332-21-4	TREMOLITE
14567-73-8	TREMOLITE
143-18-0	TRENAMINE D 200
143-18-0	TRENAMINE D 201
68-76-8	TRENIMON
68-76-8	TRENINON
56-75-7	TREOMICETINA
299-75-2	TREOSULFAN
299-75-2	TREOSULPHAN
151-21-3	TREPENOL WA
25155-30-0	TREPOLATE F 40
25155-30-0	TREPOLATE F 95
299-75-2	TRESULFAN
79-01-6	TRETHYLENE
79-01-6	TRI
545-55-1	TRI(1-AZIRIDINYL)PHOSPHINE OXIDE
545-55-1	TRI(AZIRIDINYL)PHOSPHINE OXIDE
52-24-4	TRI(ETHYLENEIMINO)THIOPHOSPHORAMIDE
102-71-6	TRI(HYDROXYETHYL)AMINE
545-55-1	TRI-1-AZIRIDINYLPHOSPHINE OXIDE
52-24-4	TRI-1-AZIRIDINYLPHOSPHINE SULFIDE
555-77-1	TRI-2-CHLOROETHYLAMINE
122-20-3	TRI-2-PROPANOLAMINE
58-89-9	TRI-6
83-26-1	TRI-BAN
79-01-6	TRI-CLENE
76-06-2	TRI-CLOR
71-63-6	TRI-DIGITOXOSIDE (GERMAN)
71-55-6	TRI-ETHANE
688-74-4	TRI-N-BUTOXYBORANE
688-74-4	TRI-N-BUTYL BORATE
126-73-8	TRI-N-BUTYL PHOSPHATE
102-85-2	TRI-N-BUTYL PHOSPHITE
1116-70-7	TRI-N-BUTYLALUMINUM
102-82-9	TRI-N-BUTYLAMINE
102-69-2	TRI-N-PROPYLAMINE
78-30-8	TRI-O-CRESYL PHOSPHATE
78-30-8	TRI-O-TOLYL PHOSPHATE
79-01-6	TRI-PLUS
79-01-6	TRI-PLUS M
14797-65-0	TRI-RODAZENE
79-01-6	TRIAD
102-71-6	TRIAETHANOLAMIN-NG
121-44-8	TRIAETHYLAMIN (GERMAN)
545-55-1	TRIAETHYLENPHOSPHORSAEUREAMID (GERMAN)
712-68-5	TRIAFUR
79-01-6	TRIAL
1693-71-6	TRIALLYL BORATE
1693-71-6	TRIALLYL BORATE (DOT)
102-70-5	TRIALLYLAMINE
102-70-5	TRIALLYLAMINE (DOT)
1693-71-6	TRIALLYLOXYBORANE
1031-47-6	TRIAMIFOS (GERMAN, DUTCH, ITALIAN)
1031-47-6	TRIAMIPHOS
14639-98-6	TRIAMMONIUM PENTACHLOROZINCATE
14221-47-7	TRIAMMONIUM TRIS(OXALATO)FERRATE(3-)
14221-47-7	TRIAMMONIUM TRIS-(ETHANEDIOATO(2-)-O,O')FERRATE(3-1)
1031-47-6	TRIAMPHOS
621-77-2	TRIAMYLAMINE
16071-86-6	TRIANTINE BROWN BRS
16071-86-6	TRIANTINE FAST BROWN OG
16071-86-6	TRIANTINE FAST BROWN OR
16071-86-6	TRIANTINE LIGHT BROWN BRS
16071-86-6	TRIANTINE LIGHT BROWN OG
79-01-6	TRIASOL
101-05-3	TRIASYN
12654-97-6	TRIASYN
10028-15-6	TRIATOMIC OXYGEN
101-05-3	TRIAZIN
12654-97-6	TRIAZIN
101-05-3	TRIAZINE
12654-97-6	TRIAZINE
101-05-3	TRIAZINE (PESTICIDE)
12654-97-6	TRIAZINE (PESTICIDE)
1912-24-9	TRIAZINE A 1294
101-05-3	TRIAZINE PESTICIDE, LIQUID [FLAMMABLE LIQUID LABEL]
12654-97-6	TRIAZINE PESTICIDE, LIQUID [FLAMMABLE LIQUID LABEL]
101-05-3	TRIAZINE PESTICIDE, LIQUID [POISON B LABEL]
12654-97-6	TRIAZINE PESTICIDE, LIQUID [POISON B LABEL]
101-05-3	TRIAZINE PESTICIDE, SOLID
12654-97-6	TRIAZINE PESTICIDE, SOLID
68-76-8	TRIAZIQUINONE
68-76-8	TRIAZIQUINONUM
68-76-8	TRIAZIQUON
68-76-8	TRIAZIQUONE
545-55-1	TRIAZIRIDINOPHOSPHINE OXIDE
545-55-1	TRIAZIRIDINYLPHOSPHINE OXIDE
52-24-4	TRIAZIRIDINYLPHOSPHINE SULFIDE
24017-47-8	TRIAZOFOS
24017-47-8	TRIAZOFOSZ (HUNGARIAN)
28911-01-5	TRIAZOLAM
24017-47-8	TRIAZOPHOS
2642-71-9	TRIAZOTION (RUSSIAN)
7601-54-9	TRIBASIC SODIUM ORTHOPHOSPHATE
7601-54-9	TRIBASIC SODIUM PHOSPHATE
75-25-2	TRIBROMMETHAAN (DUTCH)
75-25-2	TRIBROMMETHAN (GERMAN)
7727-15-3	TRIBROMOALUMINUM
64973-06-4	TRIBROMOARSINE
10294-33-4	TRIBROMOBORANE
10294-33-4	TRIBROMOBORON
75-25-2	TRIBROMOMETAN (ITALIAN)
75-25-2	TRIBROMOMETHANE
7789-60-8	TRIBROMOPHOSPHINE
7789-59-5	TRIBROMOPHOSPHORUS OXIDE
7789-61-9	TRIBROMOSTIBINE
12263-85-3	TRIBROMOTRIMETHYLDIALUMINUM
94-75-7	TRIBUTON
93-76-5	TRIBUTON
688-74-4	TRIBUTOXYBORANE
688-74-4	TRIBUTOXYBORON
102-85-2	TRIBUTOXYPHOSPHINE
126-73-8	TRIBUTOXYPHOSPHINE OXIDE
1116-70-7	TRIBUTYL ALUMINUM
688-74-4	TRIBUTYL BORATE
688-74-4	TRIBUTYL ORTHOBORATE
126-73-8	TRIBUTYL PHOSPHATE
102-85-2	TRIBUTYL PHOSPHITE
102-82-9	TRIBUTYLAMINE
56-36-0	TRIBUTYLTIN ACETATE
7778-44-1	TRICALCIUM ARSENATE
7778-44-1	TRICALCIUMARSENAT (GERMAN)
12079-65-1	TRICARBONYL(ETA5-CYCLOPENTADIENYL)MANGANESE
12108-13-3	TRICARBONYL(ETA5-METHYLCYCLOPENTADIENYL)MANGANESE
12108-13-3	TRICARBONYL(2-METHYLCYCLOPENTADIENYL)MANGANESE
12108-13-3	TRICARBONYL(METHYL-PI-CYCLOPENTADIENYL)MANGANESE
12108-13-3	TRICARBONYL(METHYLCYCLOPENTADIENYL)MANGANESE
12079-65-1	TRICARBONYL-PI-CYCLOPENTADIENYLMANGANESE
12079-65-1	TRICARBONYLCYCLOPENTADIENYLMANGANESE
63-25-2	TRICARNAM
76-03-9	TRICHLOORAZIJNZUUR (DUTCH)
79-01-6	TRICHLOORETHEEN (DUTCH)
79-01-6	TRICHLOORETHYLEEN, TRI (DUTCH)
52-68-6	TRICHLOORFON (DUTCH)
67-66-3	TRICHLOORMETHAAN (DUTCH)
98-07-7	TRICHLOORMETHYLBENZEEN (DUTCH)
76-06-2	TRICHLOORNITROMETHAAN (DUTCH)
10025-78-2	TRICHLOORSILAAN (DUTCH)
76-03-9	TRICHLORACETIC ACID
79-01-6	TRICHLORAETHEN (GERMAN)
79-01-6	TRICHLORAETHYLEN, TRI (GERMAN)
79-01-6	TRICHLORAN
51023-22-4	TRICHLORBUTYLENE
79-01-6	TRICHLOREN
76-03-9	TRICHLORESSIGSAEURE (GERMAN)
79-01-6	TRICHLORETHENE (FRENCH)
79-01-6	TRICHLORETHYLENE
79-01-6	TRICHLORETHYLENE, TRI (FRENCH)
52-68-6	TRICHLORFON
52-68-6	TRICHLORFON (DOT)
87-90-1	TRICHLORINATED ISOCYANURIC ACID
299-84-3	TRICHLORMETAPHOS
67-66-3	TRICHLORMETHAN (CZECH)
555-77-1	TRICHLORMETHINE

CAS No.	Chemical Name
98-07-7	TRICHLORMETHYLBENZOL (GERMAN)
76-06-2	TRICHLORNITROMETHAN (GERMAN)
1558-25-4	TRICHLORO(CHLOROMETHYL)SILANE
26571-79-9	TRICHLORO(CHLOROPHENYL)SILANE
27137-85-5	TRICHLORO(DICHLOROPHENYL)SILANE
71-55-6	TRICHLORO-1,1,1-ETHANE (FRENCH)
141-57-1	TRICHLORO-N-PROPYLSILANE
108-77-0	TRICHLORO-S-TRIAZINE
87-90-1	TRICHLORO-S-TRIAZINE-2,4,6(1H,3H,5H)-TRIONE
87-90-1	TRICHLORO-S-TRIAZINETRIONE
87-90-1	TRICHLORO-S-TRIAZINETRIONE, DRY, CONTAINING OVER 39% AVAILABLE CHLORINE (DOT)
75-87-6	TRICHLOROACETALDEHYDE
76-03-9	TRICHLOROACETIC ACID
76-03-9	TRICHLOROACETIC ACID (ACGIH)
76-02-8	TRICHLOROACETIC ACID CHLORIDE
76-03-9	TRICHLOROACETIC ACID SOLUTION
76-03-9	TRICHLOROACETIC ACID SOLUTION (DOT)
76-03-9	TRICHLOROACETIC ACID, SOLID
76-03-9	TRICHLOROACETIC ACID, SOLID (DOT)
76-02-8	TRICHLOROACETOCHLORIDE
76-02-8	TRICHLOROACETYL CHLORIDE
76-02-8	TRICHLOROACETYL CHLORIDE (DOT)
7446-70-0	TRICHLOROALUMINUM
107-72-2	TRICHLOROAMYLSILANE
10025-91-9	TRICHLOROANTIMONY
7784-34-1	TRICHLOROARSINE
12002-48-1	TRICHLOROBENZENE
12002-48-1	TRICHLOROBENZENE, LIQUID
12002-48-1	TRICHLOROBENZENE, LIQUID (DOT)
12002-48-1	TRICHLOROBENZENES
50-29-3	TRICHLOROBIS(4'-CHLOROPHENYL)ETHANE
51023-22-4	TRICHLOROBUTENE
51023-22-4	TRICHLOROBUTENE (DOT)
7521-80-4	TRICHLOROBUTYLSILANE
10025-73-7	TRICHLOROCHROMIUM
108-77-0	TRICHLOROCYANIDINE
87-90-1	TRICHLOROCYANURIC ACID
98-12-4	TRICHLOROCYCLOHEXYLSILANE
4484-72-4	TRICHLORODODECYLSILANE
75-87-6	TRICHLOROETHANAL
71-55-6	TRICHLOROETHANE
76-03-9	TRICHLOROETHANOIC ACID
79-01-6	TRICHLOROETHENE
79-01-6	TRICHLOROETHYLENE
79-01-6	TRICHLOROETHYLENE (ACGIH,DOT)
115-21-9	TRICHLOROETHYLSILANE
115-21-9	TRICHLOROETHYLSILICANE
115-21-9	TRICHLOROETHYLSILICON
75-69-4	TRICHLOROFLUOROCARBON
75-69-4	TRICHLOROFLUOROMETHANE
75-69-4	TRICHLOROFLUOROMETHANE (ACGIH)
52-68-6	TRICHLOROFON
67-66-3	TRICHLOROFORM
13450-90-3	TRICHLOROGALLIUM
928-65-4	TRICHLOROHEXYLSILANE
96-18-4	TRICHLOROHYDRIN
87-90-1	TRICHLOROISOCYANIC ACID
87-90-1	TRICHLOROISOCYANURIC ACID
87-90-1	TRICHLOROISOCYANURIC ACID, DRY (DOT)
299-84-3	TRICHLOROMETAPHOS
67-66-3	TRICHLOROMETHANE
594-42-3	TRICHLOROMETHANESULFENYL CHLORIDE
594-42-3	TRICHLOROMETHANESULPHENYL CHLORIDE
98-07-7	TRICHLOROMETHYLBENZENE
71-55-6	TRICHLOROMETHYLMETHANE
75-79-6	TRICHLOROMETHYLSILANE
75-79-6	TRICHLOROMETHYLSILICON
594-42-3	TRICHLOROMETHYLSULPHENYL CHLORIDE
133-06-2	TRICHLOROMETHYLTHIO-1,2,5,6-TETRAHYDROPHTHALAMIDE
75-69-4	TRICHLOROMONOFLUOROMETHANE
10025-78-2	TRICHLOROMONOSILANE
1321-65-9	TRICHLORONAPHTHALENE
327-98-0	TRICHLORONAT
327-98-0	TRICHLORONATE
76-06-2	TRICHLORONITROMETHANE
5283-67-0	TRICHLORONONYLSILANE
112-04-9	TRICHLOROOCTADECYLSILANE
5283-66-9	TRICHLOROOCTYLSILANE
7727-18-6	TRICHLOROOXOVANADIUM
107-72-2	TRICHLOROPENTYLSILANE
933-78-4	TRICHLOROPHENOL
15950-66-0	TRICHLOROPHENOL
25167-82-2	TRICHLOROPHENOL
933-78-4	TRICHLOROPHENOL (DOT)
25167-82-2	TRICHLOROPHENOL (DOT)
15950-66-0	TRICHLOROPHENOL (DOT)
93-72-1	TRICHLOROPHENOXYPROPIONIC ACID
6047-17-2	TRICHLOROPHENOXYPROPIONIC ACID ESTER
73826-29-6	TRICHLOROPHENOXYPROPIONIC ACID ESTER
98-13-5	TRICHLOROPHENYL ETHER
98-07-7	TRICHLOROPHENYLMETHANE
98-13-5	TRICHLOROPHENYLSILANE
52-68-6	TRICHLOROPHON
7719-12-2	TRICHLOROPHOSPHINE
10025-87-3	TRICHLOROPHOSPHINE OXIDE
3982-91-0	TRICHLOROPHOSPHINE SULFIDE
10025-87-3	TRICHLOROPHOSPHORUS OXIDE
141-57-1	TRICHLOROPROPYLSILANE
10049-07-7	TRICHLORORHODIUM
10025-78-2	TRICHLOROSILANE
10025-78-2	TRICHLOROSILANE (DOT)
10025-78-2	TRICHLOROSILANE (HSiCl3)
10025-91-9	TRICHLOROSTIBINE
7705-07-9	TRICHLOROTITANIUM
12075-68-2	TRICHLOROTRIETHYLDIALUMINIUM
12075-68-2	TRICHLOROTRIETHYLDIALUMINUM
12542-85-7	TRICHLOROTRIMETHYLDIALUMINUM
7727-18-6	TRICHLOROVANADIUM OXIDE
75-94-5	TRICHLOROVINYL SILICANE
75-94-5	TRICHLOROVINYLSILANE
75-94-5	TRICHLOROVINYLSILICON
52-68-6	TRICHLORPHENE
52-68-6	TRICHLORPHON
52-68-6	TRICHLORPHON FN
10025-78-2	TRICHLORSILAN (GERMAN)
10025-91-9	TRICHLORURE D'ANTIMOINE (FRENCH)
7784-34-1	TRICHLORURE D'ARSENIC (FRENCH)
443-48-1	TRICHOMOL
443-48-1	TRICHOPAL
79-01-6	TRICLENE
79-01-6	TRICLORETENE (ITALIAN)
76-06-2	TRICLORO-NITRO-METANO (ITALIAN)
79-01-6	TRICLOROETILENE (ITALIAN)
67-66-3	TRICLOROMETANO (ITALIAN)
98-07-7	TRICLOROMETILBENZENE (ITALIAN)
10025-78-2	TRICLOROSILANO (ITALIAN)
98-07-7	TRICLOROTOLUENE (ITALIAN)
60-00-4	TRICON BW
1330-78-5	TRICRESILFOSFATI (ITALIAN)
1319-77-3	TRICRESOL
1330-78-5	TRICRESOL PHOSPHATE
1330-78-5	TRICRESYL PHOSPHATE
1330-78-5	TRICRESYLFOSFATEN (DUTCH)
1330-78-5	TRICRESYLPHOSPHATE, WITH MORE THAN 3% ORTHO ISOMER (DOT)
108-77-0	TRICYANOGEN CHLORIDE
13121-70-5	TRICYCLOHEXYLHYDROXYSTANNANE
13121-70-5	TRICYCLOHEXYLHYDROXYTIN
13121-70-5	TRICYCLOHEXYLSTANNANOL
13121-70-5	TRICYCLOHEXYLSTANNYL HYDROXIDE
13121-70-5	TRICYCLOHEXYLTIN HYDROXIDE
77-73-6	TRICYCLO[52102,6]DECA-3,8-DIENE
26248-42-0	TRIDECANOL
137-26-8	TRIDIPAM
15468-32-3	TRIDYMITE
15468-32-3	TRIDYMITE (SiO2)
15468-32-3	TRIDYMITE DUST
79-01-6	TRIELENE
79-01-6	TRIELIN
79-01-6	TRIELINA (ITALIAN)
79-01-6	TRIELINE
112-24-3	TRIEN
112-24-3	TRIENTINE
8003-34-7	TRIESTE FLOWERS
102-71-6	TRIETHANOLAMIN
102-71-6	TRIETHANOLAMINE
139-96-8	TRIETHANOLAMINE DODECYL SULFATE

CAS No.	Chemical Name
27323-41-7	TRIETHANOLAMINE DODECYLBENZENESULFONATE
27323-41-7	TRIETHANOLAMINE DODECYLBENZENESULFONATE (DOT)
27323-41-7	TRIETHANOLAMINE DODECYLBENZENESULFONATE (SALT)
27323-41-7	TRIETHANOLAMINE DODECYLBENZENESULFONIC ACID SALT
27323-41-7	TRIETHANOLAMINE DODECYLBENZOSULFONATE
139-96-8	TRIETHANOLAMINE LAURYL SULFATE
139-96-8	TRIETHANOLAMINE MONODODECYL SULFATE
139-96-8	TRIETHANOLAMMONIUM DODECYL SULFATE
27323-41-7	TRIETHANOLAMMONIUM DODECYLBENZENESULFONATE
139-96-8	TRIETHANOLAMMONIUM LAURYL SULFATE
122-51-0	TRIETHOXYMETHANE
122-52-1	TRIETHOXYPHOSPHINE
998-30-1	TRIETHOXYSILANE
122-51-0	TRIETHYL ORTHOFORMATE
122-52-1	TRIETHYL PHOSPHITE
122-52-1	TRIETHYL PHOSPHITE (DOT)
97-93-8	TRIETHYLALANE
97-93-8	TRIETHYLALUMINIUM
97-93-8	TRIETHYLALUMINIUM (DOT)
97-93-8	TRIETHYLALUMINUM
12075-68-2	TRIETHYLALUMINUM SESQUICHLORIDE
121-44-8	TRIETHYLAMINE
121-44-8	TRIETHYLAMINE (ACGIH,DOT)
555-77-1	TRIETHYLAMINE, 2,2',2''-TRICHLORO-
102-71-6	TRIETHYLAMINE, 2,2',2''-TRIHYDROXY-
538-07-8	TRIETHYLAMINE, 2,2'-DICHLORO-
25340-18-5	TRIETHYLBENZENE
97-94-9	TRIETHYLBORANE
97-94-9	TRIETHYLBORON
12075-68-2	TRIETHYLDIALUMINUM TRICHLORIDE
280-57-9	TRIETHYLENE DIAMINE
112-27-6	TRIETHYLENE GLYCOL
112-26-5	TRIETHYLENE GLYCOL DICHLORIDE
1954-28-5	TRIETHYLENE GLYCOL DIGLYCIDYL ETHER
112-50-5	TRIETHYLENE GLYCOL ETHYL ETHER
112-50-5	TRIETHYLENE GLYCOL MONOETHYL ETHER
68-76-8	TRIETHYLENEIMINOBENZOQUINONE
51-18-3	TRIETHYLENEMELAMINE
545-55-1	TRIETHYLENEPHOSPHORAMIDE
545-55-1	TRIETHYLENEPHOSPHORIC TRIAMIDE
545-55-1	TRIETHYLENEPHOSPHOROTRIAMIDE
112-24-3	TRIETHYLENETETRAMINE
112-24-3	TRIETHYLENETETRAMINE (DOT)
52-24-4	TRIETHYLENETHIOPHOSPHORAMIDE
52-24-4	TRIETHYLENETHIOPHOSPHOROTRIAMIDE
68-76-8	TRIETHYLENIMINOBENZOQUINONE
102-71-6	TRIETHYLOLAMINE
994-31-0	TRIETHYLTIN CHLORIDE
12075-68-2	TRIETHYLTRICHLORODIALUMINUM
121-44-8	TRIETILAMINA (ITALIAN)
1582-09-8	TRIFLORAN
76-05-1	TRIFLUORACETIC ACID
1582-09-8	TRIFLUORALIN
428-59-1	TRIFLUORO(TRIFLUOROMETHYL)OXIRANE
675-14-9	TRIFLUORO-1,3,5-TRIAZINE
675-14-9	TRIFLUORO-S-TRIAZINE
76-05-1	TRIFLUOROACETIC ACID
76-05-1	TRIFLUOROACETIC ACID (DOT)
7783-54-2	TRIFLUOROAMINE
7783-54-2	TRIFLUOROAMMONIA
7783-56-4	TRIFLUOROANTIMONY
7637-07-2	TRIFLUOROBORANE
109-63-7	TRIFLUOROBORANE DIETHYL ETHERATE
109-63-7	TRIFLUOROBORANE-1,1'-OXYBIS[ETHANE] (1:1)
7637-07-2	TRIFLUOROBORON
109-63-7	TRIFLUOROBORON ETHERATE
109-63-7	TRIFLUOROBORON-DIETHYL ETHER COMPLEX
598-73-2	TRIFLUOROBROMOETHYLENE
75-63-8	TRIFLUOROBROMOMETHANE
75-63-8	TRIFLUOROBROMOMETHANE (ACGIH,DOT)
1330-45-6	TRIFLUOROCHLOROETHANE
1330-45-6	TRIFLUOROCHLOROETHANE (DOT)
79-38-9	TRIFLUOROCHLOROETHYLENE
79-38-9	TRIFLUOROCHLOROETHYLENE (DOT)
79-38-9	TRIFLUOROCHLOROETHYLENE, INHIBITED (DOT)
75-72-9	TRIFLUOROCHLOROMETHANE
75-72-9	TRIFLUOROCHLOROMETHANE (DOT)
27987-06-0	TRIFLUOROETHANE
27987-06-0	TRIFLUOROETHANE (DOT)
76-05-1	TRIFLUOROETHANOIC ACID
75-46-7	TRIFLUOROMETHANE
75-46-7	TRIFLUOROMETHANE (DOT)
75-63-8	TRIFLUOROMETHYL BROMIDE
75-72-9	TRIFLUOROMETHYL CHLORIDE
75-63-8	TRIFLUOROMONOBROMOMETHANE
75-72-9	TRIFLUOROMONOCHLOROCARBON
79-38-9	TRIFLUOROMONOCHLOROETHYLENE
7783-55-3	TRIFLUOROPHOSPHINE
7783-56-4	TRIFLUOROSTIBINE
675-14-9	TRIFLUOROTRIAZINE
598-73-2	TRIFLUOROVINYL BROMIDE
79-38-9	TRIFLUOROVINYL CHLORIDE
7790-91-2	TRIFLUORURE DE CHLORE (FRENCH)
1582-09-8	TRIFLURALIN
1582-09-8	TRIFLURALINE
1335-85-9	TRIFOCIDE
110-88-3	TRIFORMOL
1335-85-9	TRIFRINA
1582-09-8	TRIFUREX
112-27-6	TRIGEN
139-13-9	TRIGLYCINE
112-27-6	TRIGLYCOL
112-26-5	TRIGLYCOL DICHLORIDE
112-50-5	TRIGLYCOL MONOETHYL ETHER
139-13-9	TRIGLYCOLLAMIC ACID
112-27-6	TRIGOL
78-63-7	TRIGONOX 101
78-63-7	TRIGONOX 101-101/45
3006-86-8	TRIGONOX 22B50
3006-86-8	TRIGONOX 22B75
927-07-1	TRIGONOX 25/75
6731-36-8	TRIGONOX 29
6731-36-8	TRIGONOX 29/40
6731-36-8	TRIGONOX 29B50
6731-36-8	TRIGONOX 29B75
11118-65-3	TRIGONOX 38
37187-22-7	TRIGONOX 40
13122-18-4	TRIGONOX 42
37187-22-7	TRIGONOX 44B
110-05-4	TRIGONOX B
614-45-9	TRIGONOX C
2167-23-9	TRIGONOX DB 50
2167-23-9	TRIGONOX DM 50
37206-20-5	TRIGONOX HM 80
80-15-9	TRIGONOX K 80
1338-23-4	TRIGONOX M 50
3457-61-2	TRIGONOX T
30580-75-7	TRIGONOX T
3006-82-4	TRIGONOX T 21S
3457-61-2	TRIGONOX T 40
30580-75-7	TRIGONOX T 40
5809-08-5	TRIGONOX TMPH
78-63-7	TRIGONOX XQ 8
62-38-4	TRIGOSAN
10043-35-3	TRIHYDROXYBORANE
56-81-5	TRIHYDROXYPROPANE
102-71-6	TRIHYDROXYTRIETHYLAMINE
7784-45-4	TRIIODOARSINE
75-47-8	TRIIODOMETHANE
100-99-2	TRIISOBUTYL ALUMINUM
13195-76-1	TRIISOBUTYL BORATE
13195-76-1	TRIISOBUTYL ORTHOBORATE
100-99-2	TRIISOBUTYLALANE
100-99-2	TRIISOBUTYLALUMINIUM
100-99-2	TRIISOBUTYLALUMINIUM (DOT)
7756-94-7	TRIISOBUTYLENE
7756-94-7	TRIISOBUTYLENE (DOT)
7756-94-7	TRIISOCYANATOISOCYANURATE OF ISOPHORONEDIISOCYANATE, SOLUTION
122-20-3	TRIISOPROPANOLAMINE
5419-55-6	TRIISOPROPOXYBORANE
5419-55-6	TRIISOPROPOXYBORON
5419-55-6	TRIISOPROPYL BORATE
5419-55-6	TRIISOPROPYL BORATE (DOT)
5419-55-6	TRIISOPROPYL ORTHOBORATE
1582-09-8	TRIKEPIN
79-01-6	TRIKLONE
1330-78-5	TRIKRESYLPHOSPHATE (GERMAN)

CAS No.	Chemical Name
7446-27-7	TRILEAD PHOSPHATE
79-01-6	TRILEN
79-01-6	TRILENE
79-01-6	TRILINE
26134-62-3	TRILITHIUM NITRIDE
26134-62-3	TRILITHIUM NITRIDE (Li3N)
77-81-6	TRILON 83
60-00-4	TRILON B
60-00-4	TRILON BW
107-44-8	TRILONE 46
1582-09-8	TRIM
79-01-6	TRIMAR
133-06-2	TRIMEGOL
552-30-7	TRIMELLITIC ACID 1,2-ANHYDRIDE
552-30-7	TRIMELLITIC ACID ANHYDRIDE
552-30-7	TRIMELLITIC ACID CYCLIC 1,2-ANHYDRIDE
552-30-7	TRIMELLITIC ANHYDRIDE
7785-84-4	TRIMETAPHOSPHORIC ACID (H3P3O9), TRISODIUM SALT
127-48-0	TRIMETHADIONE
2275-18-5	TRIMETHOATE
738-70-5	TRIMETHOPRIM
121-43-7	TRIMETHOXYBORANE
121-43-7	TRIMETHOXYBORINE
121-43-7	TRIMETHOXYBORON
121-45-9	TRIMETHOXYPHOSPHINE
3282-30-2	TRIMETHYL ACETYL CHLORIDE (DOT)
25551-13-7	TRIMETHYL BENZENE
95-63-6	TRIMETHYL BENZENE (ACGIH)
108-67-8	TRIMETHYL BENZENE (ACGIH)
25551-13-7	TRIMETHYL BENZENE (ACGIH)
121-43-7	TRIMETHYL BORATE
121-43-7	TRIMETHYL BORATE (DOT)
75-77-4	TRIMETHYL CHLOROSILANE
57-55-6	TRIMETHYL GLYCOL
28324-52-9	TRIMETHYL NORPINANYL HYDROPEROXIDE
121-45-9	TRIMETHYL PHOSPHITE
121-45-9	TRIMETHYL PHOSPHITE (ACGIH,DOT)
25620-58-0	TRIMETHYL-1,6-HEXANEDIAMINE
999-81-5	TRIMETHYL-BETA-CHLORETHYLAMMONIUMCHLORID (CZECH)
999-81-5	TRIMETHYL-BETA-CHLOROETHYLAMMONIUM CHLORIDE
75-98-9	TRIMETHYLACETIC ACID
3282-30-2	TRIMETHYLACETYL CHLORIDE
75-24-1	TRIMETHYLALANE
75-24-1	TRIMETHYLALUMINIUM
75-24-1	TRIMETHYLALUMINIUM (DOT)
75-24-1	TRIMETHYLALUMINUM
75-50-3	TRIMETHYLAMINE
75-50-3	TRIMETHYLAMINE (ACGIH)
6369-96-6	TRIMETHYLAMINE, (2,4,5-TRICHLOROPHENOXY)ACETATE
75-50-3	TRIMETHYLAMINE, ANHYDROUS
75-50-3	TRIMETHYLAMINE, ANHYDROUS (DOT)
75-50-3	TRIMETHYLAMINE, AQUEOUS SOLUTION
75-50-3	TRIMETHYLAMINE, AQUEOUS SOLUTION (DOT)
75-50-3	TRIMETHYLAMINE, AQUEOUS SOLUTIONS CONTAINING NOT MORE THAN 30% OF TRIMETHYLAMINE (DOT)
75-64-9	TRIMETHYLAMINOMETHANE
108-67-8	TRIMETHYLBENZOL
56-93-9	TRIMETHYLBENZYLAMMONIUM CHLORIDE
507-19-7	TRIMETHYLBROMOMETHANE
75-65-0	TRIMETHYLCARBINOL
112-02-7	TRIMETHYLCETYLAMMONIUM CHLORIDE
507-20-0	TRIMETHYLCHLOROMETHANE
75-77-4	TRIMETHYLCHLOROSILANE (DOT)
1066-45-1	TRIMETHYLCHLOROSTANNANE
1066-45-1	TRIMETHYLCHLOROTIN
1321-60-4	TRIMETHYLCYCLOHEXANOL
34216-34-7	TRIMETHYLCYCLOHEXYLAMINE
34216-34-7	TRIMETHYLCYCLOHEXYLAMINE (DOT)
75-19-4	TRIMETHYLENE
75-19-4	TRIMETHYLENE (CYCLIC)
109-70-6	TRIMETHYLENE BROMIDE CHLORIDE
109-70-6	TRIMETHYLENE CHLOROBROMIDE
627-30-5	TRIMETHYLENE CHLOROHYDRIN
142-28-9	TRIMETHYLENE DICHLORIDE
109-76-2	TRIMETHYLENEDIAMINE
121-82-4	TRIMETHYLENETRINITRAMINE
513-35-9	TRIMETHYLETHYLENE
26760-64-5	TRIMETHYLETHYLENE
112-02-7	TRIMETHYLHEXADECYLAMMONIUM CHLORIDE
25620-58-0	TRIMETHYLHEXAMETHYLENE DIAMINE (DOT)
25620-58-0	TRIMETHYLHEXAMETHYLENE DIAMINES
28679-16-5	TRIMETHYLHEXAMETHYLENE DIISOCYANATE
28679-16-5	TRIMETHYLHEXAMETHYLENE DIISOCYANATE (DOT)
28679-16-5	TRIMETHYLHEXAMETHYLENE ISOCYANATE
25620-58-0	TRIMETHYLHEXAMETHYLENEDIAMINE
75-28-5	TRIMETHYLMETHANE
75-65-0	TRIMETHYLMETHANOL
824-11-3	TRIMETHYLOLPROPANE PHOSPHITE
98-06-6	TRIMETHYLPHENYLMETHANE
512-56-1	TRIMETHYLPHOSPHATE
75-77-4	TRIMETHYLSILYL CHLORIDE
1066-45-1	TRIMETHYLSTANNYL CHLORIDE
2489-77-2	TRIMETHYLTHIOUREA
1066-45-1	TRIMETHYLTIN CHLORIDE
60-51-5	TRIMETION
7758-98-7	TRINAGLE
55-63-0	TRINALGON
7601-54-9	TRINATRIUMPHOSPHAT (GERMAN)
52-68-6	TRINEX
12035-72-2	TRINICKEL DISULFIDE
55-63-0	TRINITRIN
602-99-3	TRINITRO-M-CRESOL
26952-42-1	TRINITROANILINE
606-35-9	TRINITROANISOLE
99-35-4	TRINITROBENZENE
25377-32-6	TRINITROBENZENE
99-35-4	TRINITROBENZENE, WET, AT LEAST 10% WATER, OVER 16 OZS IN ONE OUTSIDE PACKAGING (DOT)
99-35-4	TRINITROBENZENE, WET, CONTAINING AT LEAST 10% WATER (DOT)
99-35-4	TRINITROBENZENE, WET, CONTAINING LESS THAN 30% WATER (DOT)
99-35-4	TRINITROBENZENE, WETTED WITH NOT LESS THAN 30% WATER (DOT)
2508-19-2	TRINITROBENZENESULFONIC ACID
129-66-8	TRINITROBENZOIC ACID
35860-50-5	TRINITROBENZOIC ACID
129-66-8	TRINITROBENZOIC ACID, WET, CONTAINING AT LEAST 10% WATER (DOT)
129-66-8	TRINITROBENZOIC ACID, WET, CONTAINING LESS THAN 30% WATER (DOT)
28260-61-9	TRINITROCHLOROBENZENE
28905-71-7	TRINITROCRESOL
25322-14-9	TRINITROFLUORENONE
55-63-0	TRINITROGLYCERIN
55-63-0	TRINITROGLYCEROL
55-63-0	TRINITROL
55810-17-8	TRINITRONAPHTHALENE
4732-14-3	TRINITROPHENETOLE
88-89-1	TRINITROPHENOL
88-89-1	TRINITROPHENOL, WETTED WITH AT LEAST 10% WATER (DOT)
88-89-1	TRINITROPHENOL, WETTED WITH AT LEAST 30% WATER (DOT)
479-45-8	TRINITROPHENYLMETHYLNITRAMINE
479-45-8	TRINITROPHENYLMETHYLNITRAMINE (DOT)
82-71-3	TRINITRORESORCINOL
118-96-7	TRINITROTOLUENE
118-96-7	TRINITROTOLUENE, DRY OR CONTAINING, BY WEIGHT, LESS THAN 30% WATER (DOT)
118-96-7	TRINITROTOLUENE, DRY OR WET, WITH 10% WATER
118-96-7	TRINITROTOLUENE, WET CONTAINING AT LEAST 10% WATER (DOT)
118-96-7	TRINITROTOLUENE, WET CONTAINING AT LEAST 10% WATER, OVER 16 OZS IN ONE OUTSIDE PACKAGING
118-96-7	TRINITROTOLUENE, WETTED WITH NOT LESS THAN 30% WATER (DOT)
94-75-7	TRINOXOL
93-76-5	TRINOXOL
2545-59-7	TRINOXOL
79-01-6	TRIOL
8020-83-5	TRIONA
8020-83-5	TRIONA B
78-30-8	TRIORTHOCRESYL PHOSPHATE
110-88-3	TRIOXAN
110-88-3	TRIOXANE
93-76-5	TRIOXON
93-79-8	TRIOXONE
7616-94-6	TRIOXYCHLOROFLUORIDE
110-88-3	TRIOXYMETHYLENE

CAS No.	Chemical Name
621-77-2	TRIPENTYLAMINE
115-86-6	TRIPHENOXYPHOSPHINE OXIDE
603-34-9	TRIPHENYL AMINE
115-86-6	TRIPHENYL PHOSPHATE
639-58-7	TRIPHENYLCHLOROSTANNANE
639-58-7	TRIPHENYLCHLOROTIN
639-58-7	TRIPHENYLTIN CHLORIDE
76-87-9	TRIPHENYLTIN HYDROXIDE
7758-29-4	TRIPHOSPHORIC ACID, PENTASODIUM SALT
50-78-2	TRIPLE-SAL
1317-95-9	TRIPOLI
1317-95-9	TRIPOLI DUST
137-26-8	TRIPOMOL
75-99-0	TRIPON
20770-41-6	TRIPOTASSIOPHOSPHINE
102-67-0	TRIPROPYL ALUMINUM
102-67-0	TRIPROPYLALUMINUM (DOT)
102-69-2	TRIPROPYLAMINE
102-69-2	TRIPROPYLAMINE (8CI)
102-69-2	TRIPROPYLAMINE (DOT)
102-69-2	TRIPROPYLAMINE [COMBUSTIBLE LIQUID LABEL]
102-69-2	TRIPROPYLAMINE [FLAMMABLE LIQUID LABEL]
13987-01-4	TRIPROPYLENE
13987-01-4	TRIPROPYLENE (DOT)
24800-44-0	TRIPROPYLENE GLYCOL
464-06-2	TRIPTAN
464-06-2	TRIPTANE
594-56-9	TRIPTENE
14797-65-0	TRIRODAZEEN
126-72-7	TRIS
126-72-7	TRIS (FLAME RETARDANT)
545-55-1	TRIS(1-AZIRIDINE)PHOSPHINE OXIDE
68-76-8	TRIS(1-AZIRIDINYL)-P-BENZOQUINONE
545-55-1	TRIS(1-AZIRIDINYL)PHOSPHINE OXIDE
52-24-4	TRIS(1-AZIRIDINYL)PHOSPHINE SULFIDE
126-72-7	TRIS(2,3-DIBROMOPROPYL)PHOSPHATE
555-77-1	TRIS(2-CHLOROETHYL)AMINE
122-20-3	TRIS(2-HYDROXY-1-PROPYL)AMINE
102-71-6	TRIS(2-HYDROXYETHYL)AMINE
122-20-3	TRIS(2-HYDROXYPROPYL)AMINE
122-20-3	TRIS(2-PROPANOL)AMINE
102-70-5	TRIS(2-PROPENYL)AMINE
1693-71-6	TRIS(ALLYLOXY)BORANE
68-76-8	TRIS(AZIRIDINYL)-P-BENZOQUINONE
545-55-1	TRIS(AZIRIDINYL)PHOSPHINE OXIDE
52-24-4	TRIS(AZIRIDINYL)PHOSPHINE SULFIDE
555-77-1	TRIS(BETA-CHLOROETHYL)AMINE
102-71-6	TRIS(BETA-HYDROXYETHYL)AMINE
688-74-4	TRIS(BUTOXY)BORANE
680-31-9	TRIS(DIMETHYLAMINO)PHOSPHINE OXIDE
14484-64-1	TRIS(DIMETHYLDITHIOCARBAMATO)IRON
68-76-8	TRIS(ETHYLENEIMINO)BENZOQUINONE
100-99-2	TRIS(ISO-BUTYL)ALUMINUM
100-99-2	TRIS(ISOBUTYL)ALUMINUM(III)
14484-64-1	TRIS(N,N-DIMETHYLDITHIOCARBAMATO)IRON
14484-64-1	TRIS(N,N-DIMETHYLDITHIOCARBAMATO)IRON(III)
545-55-1	TRIS(N-ETHYLENE)PHOSPHOROTRIAMIDATE
78-30-8	TRIS(O-CRESYL) PHOSPHATE
78-30-8	TRIS(O-METHYLPHENYL) PHOSPHATE
78-30-8	TRIS(O-TOLYL) PHOSPHATE
1330-78-5	TRIS(TOLYLOXY)PHOSPHINE OXIDE
545-55-1	TRIS-(1-AZIRIDINYL)PHOSPHINE OXIDE (DOT)
545-55-1	TRIS-(1-AZIRIDINYL)PHOSPHINE OXIDE, SOLUTION (DOT)
102-82-9	TRIS-N-BUTYLAMINE
50-55-5	TRISERPIN
68-76-8	TRISETHYLENEIMINOQUINONE
4521-94-2	TRISILOXANE, OCTAETHOXY-
4421-95-8	TRISILOXANE, OCTAMETHOXY-
5419-55-6	TRISISOPROPOXYBORANE
7631-89-2	TRISODIUM ARSENATE
13464-37-4	TRISODIUM ARSENITE
7785-84-4	TRISODIUM CYCLOTRIPHOSPHATE
7785-84-4	TRISODIUM METAPHOSPHATE
7601-54-9	TRISODIUM ORTHOPHOSPHATE
10101-89-0	TRISODIUM ORTHOPHOSPHATE DODECAHYDRATE
13721-39-6	TRISODIUM ORTHOVANADATE
7601-54-9	TRISODIUM PHOSPHATE
10101-89-0	TRISODIUM PHOSPHATE (Na3PO4.12H2O)
10361-89-4	TRISODIUM PHOSPHATE DECAHYDRATE
10101-89-0	TRISODIUM PHOSPHATE DODECAHYDRATE
10101-89-0	TRISODIUM PHOSPHATE DODECAHYDRATE (Na3PO4.12H2O)
7681-49-4	TRISODIUM TRIFLUORIDE
7785-84-4	TRISODIUM TRIMETAPHOSPHATE
7785-84-4	TRISODIUM TRIMETAPHOSPHATE (Na3P3O9)
13721-39-6	TRISODIUM VANADATE
1314-85-8	TRISULFURATED PHOSPHORUS
79-38-9	TRITHENE
786-19-6	TRITHION
786-19-6	TRITHION MITICIDE
82-68-8	TRITISAN
118-96-7	TRITOL
1330-78-5	TRITOLYL PHOSPHATE
118-96-7	TRITON
577-11-7	TRITON GR 5
577-11-7	TRITON GR 5M
577-11-7	TRITON GR 7
577-11-7	TRITON GR 7M
443-48-1	TRIVAZOL
1836-75-5	TRIZILIN
1314-84-7	TRIZINC DIPHOSPHIDE
2782-57-2	TROCLOSENE
2244-21-5	TROCLOSENE POTASSIUM
9005-25-8	TROGUM
7719-12-2	TROJCHLOREK FOSFORU (POLISH)
118-96-7	TROJNITROTOLUEN (POLISH)
102-71-6	TROLAMINE
299-84-3	TROLEN
299-84-3	TROLENE
7601-54-9	TROMETE
13463-67-7	TRONEX CR 800
556-61-6	TROPEX
118-96-7	TROTYL
118-96-7	TROTYL OIL
75-01-4	TROVIDUR
110-69-0	TROYKYD ANTI-SKIN BTO
62-38-4	TROYSAN 30
62-38-4	TROYSAN PMA 30
62450-06-0	TRP-P 1
62450-07-1	TRP-P 2
12135-76-1	TRUE AMMONIUM SULFIDE
103-23-1	TRUFLEX DOA
123-79-5	TRUFLEX DOA
117-81-7	TRUFLEX DOP
117-84-0	TRUFLEX DOP
72-57-1	TRYPAN BLUE
72-57-1	TRYPAN BLUE BPC
72-57-1	TRYPAN BLUE, COMMERCIAL GRADE
72-57-1	TRYPANE BLUE
127-85-5	TRYPOXYL
9002-07-7	TRYPSIN
62450-06-0	TRYPTOPHAN P 1
62450-07-1	TRYPTOPHAN P 2
555-77-1	TS 160
104-15-4	TSA-HP
25231-46-3	TSA-HP
104-15-4	TSA-MH
25231-46-3	TSA-MH
420-04-2	TSAKS
9004-70-0	TSAPOLAK 964
79-19-6	TSC
62-56-6	TSIZP 34
7439-98-7	TSM1
7601-54-9	TSP
52-24-4	TSPA
7722-88-5	TSPP
1129-41-5	TSUMACIDE
1129-41-5	TSUMAUNKA
79-19-6	TSZ
97-77-8	TTD
137-26-8	TTD
97-77-8	TTS
137-26-8	TUADS
83-79-4	TUBATOXIN
544-19-4	TUBERCUPROSE
137-26-8	TUEX
7429-90-5	TUFF MIC
8020-83-5	TUFFLO 6204
52-68-6	TUGON

CAS No.	Chemical Name
52-68-6	TUGON FLY BAIT
52-68-6	TUGON STABLE SPRAY
89-62-3	TULABASE FAST RED GL
95-79-4	TULABASE FAST RED TR
137-26-8	TULISAN
7631-86-9	TULLANOX TM 500
108-88-3	TULUOL
26471-62-5	TULUYLENDIISOCYANAT (GERMAN)
7775-09-9	TUMBLEAF
7440-33-7	TUNGSTEN
12070-12-1	TUNGSTEN CARBIDE
7783-82-6	TUNGSTEN FLUORIDE
7783-82-6	TUNGSTEN FLUORIDE (WF6)
7783-82-6	TUNGSTEN FLUORIDE (WF6), (OC-6-11)-
7783-82-6	TUNGSTEN HEXAFLUORIDE
7783-82-6	TUNGSTEN HEXAFLUORIDE (DOT)
7783-82-6	TUNGSTEN HEXAFLUORIDE (WF6)
12737-98-3	TUNGSTIC ACID, LEAD
9004-34-6	TUNICIN
999-81-5	TUR
7778-44-1	TURF-CAL
1309-37-1	TURKEY RED
9005-90-7	TURPENTINE
9005-90-7	TURPENTINE GUM
1312-03-4	TURPETH MINERAL
137-26-8	TUTAN
513-49-5	TUTANE
13952-84-6	TUTANE
151-21-3	TV M 474
8006-54-0	TW 30
81-81-2	TWIN LIGHT RAT AWAY
139-96-8	TYLOROL LT 50
137-26-8	TYRADIN
123-42-2	TYRANTON
13463-67-7	TYTANPOL R 323
5593-70-4	TYZOR TBT
66-81-9	TZA
82-66-6	U 1363
7631-86-9	U 333
66-81-9	U 4527
94-75-7	U 46
93-76-5	U 46
94-75-7	U 46DP
86-88-4	U 5227
96-24-2	U 5897
103-85-5	U 6324
110-17-8	U-1149
26628-22-8	U-3886
68-12-2	U-4224
94-75-7	U-5043
30553-04-9	U-5227
127-19-5	U-5954
66-75-1	U-8344
51-21-8	U-8953
57-14-7	U-DIMETHYLHYDRAZINE
409-21-2	UA 1
409-21-2	UA 2
409-21-2	UA 3
409-21-2	UA 4
630-60-4	UABAINA
630-60-4	UABANIN
8002-74-2	UBATOL EXP 21
105-60-2	UBE 1022B
64-00-6	UC 10854
15271-41-7	UC 20047
15271-41-7	UC 20047A
116-06-3	UC 21149
15271-41-7	UC 26089
63-25-2	UC 7744
2631-37-0	UC 9880
107-21-1	UCAR 17
57-55-6	UCAR 35
7782-42-5	UCAR 38
109-59-1	UCAR AC
111-30-8	UCARCIDE 250
76-12-0	UCON 112
75-71-8	UCON 12
75-71-8	UCON 12/HALOCARBON 12
75-45-6	UCON 22
75-45-6	UCON 22/HALOCARBON 22
75-69-4	UCON FLUOROCARBON 11
37286-64-9	UCON LB 1715
75-69-4	UCON REFRIGERANT 11
57-14-7	UDMH
57-14-7	UDMH (DOT)
2636-26-2	UDVF
138-89-6	ULTRA BRILLIANT BLUE P
53324-05-3	ULTRA BRILLIANT BLUE P
1332-58-7	ULTRA COTE
151-21-3	ULTRA SULFATE SL-1
7704-34-9	ULTRA SULFUR
1332-58-7	ULTRA WHITE 90
950-37-8	ULTRACID
950-37-8	ULTRACID 40
8002-74-2	ULTRAFLEX
8002-74-2	ULTRAFLEX (WAX)
139-91-3	ULTRAFUR
105-60-2	ULTRAMID B 3
105-60-2	ULTRAMID B 4
105-60-2	ULTRAMID B 5
105-60-2	ULTRAMID BMK
8020-83-5	ULTRASENE
7631-86-9	ULTRASIL VH 3
7631-86-9	ULTRASIL VN 2
7631-86-9	ULTRASIL VN 3
25155-30-0	ULTRAWET 1T
25155-30-0	ULTRAWET 60K
27323-41-7	ULTRAWET 60L
25155-30-0	ULTRAWET 99LS
25155-30-0	ULTRAWET K
25155-30-0	ULTRAWET KX
25155-30-0	ULTRAWET SK
51-21-8	ULUP
8065-48-3	ULV
6923-22-4	ULVAIR
8020-83-5	ULVAPRON
56-72-4	UMBETHION
1314-20-1	UMBRATHOR
131-74-8	UN 0004 (DOT)
25550-58-7	UN 0076 (DOT)
55-63-0	UN 0143 (DOT)
55-63-0	UN 0144 (DOT)
479-45-8	UN 0208 (DOT)
118-96-7	UN 0209 (DOT)
99-35-4	UN 0214 (DOT)
129-66-8	UN 0215 (DOT)
6484-52-2	UN 0222 (DOT)
9004-70-0	UN 0340 (DOT)
7790-98-9	UN 0402 (DOT)
74-86-2	UN 1001 (DOT)
7664-41-7	UN 1005 (DOT)
7440-37-1	UN 1006 (DOT)
75-63-8	UN 1009 (DOT)
106-99-0	UN 1010 (DOT)
25339-57-5	UN 1010 (DOT)
106-97-8	UN 1011 (DOT)
124-38-9	UN 1013 (DOT)
8063-77-2	UN 1014 (DOT)
53569-62-3	UN 1015 (DOT)
630-08-0	UN 1016 (DOT)
7782-50-5	UN 1017 (DOT)
75-45-6	UN 1018 (DOT)
76-15-3	UN 1020 (DOT)
63938-10-3	UN 1021 (DOT)
75-72-9	UN 1022 (DOT)
460-19-5	UN 1026 (DOT)
75-71-8	UN 1028 (DOT)
75-43-4	UN 1029 (DOT)
75-37-6	UN 1030 (DOT)
25497-28-3	UN 1030 (DOT)
124-40-3	UN 1032 (DOT)
115-10-6	UN 1033 (DOT)
74-84-0	UN 1035 (DOT)
75-04-7	UN 1036 (DOT)
75-00-3	UN 1037 (DOT)
74-85-1	UN 1038 (DOT)
540-67-0	UN 1039 (DOT)
8070-50-6	UN 1041 (DOT)

CAS No.	Chemical Name
7782-41-4	UN 1045 (DOT)
7440-59-7	UN 1046 (DOT)
10035-10-6	UN 1048 (DOT)
1333-74-0	UN 1049 (DOT)
7647-01-0	UN 1050 (DOT)
74-90-8	UN 1051 (DOT)
7664-39-3	UN 1052 (DOT)
7783-06-4	UN 1053 (DOT)
115-11-7	UN 1055 (DOT)
7439-90-9	UN 1056 (DOT)
59355-75-8	UN 1060 (DOT)
74-89-5	UN 1061 (DOT)
74-83-9	UN 1062 (DOT)
74-87-3	UN 1063 (DOT)
7440-01-9	UN 1065 (DOT)
7727-37-9	UN 1066 (DOT)
10102-44-0	UN 1067 (DOT)
10544-72-6	UN 1067 (DOT)
2696-92-6	UN 1069 (DOT)
11104-93-1	UN 1070 (DOT)
10024-97-2	UN 1070 (DOT)
7782-44-7	UN 1072 (DOT)
7782-44-7	UN 1073 (DOT)
75-28-5	UN 1075 (DOT)
115-11-7	UN 1075 (DOT)
115-07-1	UN 1075 (DOT)
68476-85-7	UN 1075 (DOT)
75-44-5	UN 1076 (DOT)
115-07-1	UN 1077 (DOT)
7446-09-5	UN 1079 (DOT)
2551-62-4	UN 1080 (DOT)
116-14-3	UN 1081 (DOT)
79-38-9	UN 1082 (DOT)
75-50-3	UN 1083 (DOT)
593-60-2	UN 1085 (DOT)
75-01-4	UN 1086 (DOT)
107-25-5	UN 1087 (DOT)
105-57-7	UN 1088 (DOT)
75-07-0	UN 1089 (DOT)
67-64-1	UN 1090 (DOT)
107-02-8	UN 1092 (DOT)
107-18-6	UN 1098 (DOT)
106-95-6	UN 1099 (DOT)
107-05-1	UN 1100 (DOT)
96-10-6	UN 1101 (DOT)
97-93-8	UN 1102 (DOT)
75-24-1	UN 1103 (DOT)
628-63-7	UN 1104 (DOT)
53496-15-4	UN 1104 (DOT)
75-85-4	UN 1105 (DOT)
71-41-0	UN 1105 (DOT)
123-51-3	UN 1105 (DOT)
110-58-7	UN 1106 (DOT)
543-59-9	UN 1107 (DOT)
513-35-9	UN 1108 (DOT)
109-67-1	UN 1108 (DOT)
638-49-3	UN 1109 (DOT)
110-43-0	UN 1110 (DOT)
110-66-7	UN 1111 (DOT)
463-04-7	UN 1113 (DOT)
71-36-3	UN 1120 (DOT)
75-65-0	UN 1120 (DOT)
78-92-2	UN 1120 (DOT)
105-46-4	UN 1123 (DOT)
110-19-0	UN 1123 (DOT)
123-86-4	UN 1123 (DOT)
540-88-5	UN 1123 (DOT)
109-73-9	UN 1125 (DOT)
109-65-9	UN 1126 (DOT)
109-69-3	UN 1127 (DOT)
123-72-8	UN 1129 (DOT)
8008-51-3	UN 1130 (DOT)
75-15-0	UN 1131 (DOT)
108-90-7	UN 1134 (DOT)
107-07-3	UN 1135 (DOT)
8001-58-9	UN 1136 (DOT)
123-73-9	UN 1143 (DOT)
4170-30-3	UN 1143 (DOT)
503-17-3	UN 1144 (DOT)

CAS No.	Chemical Name
287-92-3	UN 1146 (DOT)
123-42-2	UN 1148 (DOT)
142-96-1	UN 1149 (DOT)
109-89-7	UN 1154 (DOT)
60-29-7	UN 1155 (DOT)
96-22-0	UN 1156 (DOT)
108-83-8	UN 1157 (DOT)
108-18-9	UN 1158 (DOT)
108-20-3	UN 1159 (DOT)
124-40-3	UN 1160 (DOT)
616-38-6	UN 1161 (DOT)
75-78-5	UN 1162 (DOT)
57-14-7	UN 1163 (DOT)
75-18-3	UN 1164 (DOT)
123-91-1	UN 1165 (DOT)
100-79-8	UN 1166 (DOT)
646-06-0	UN 1166 (DOT)
109-93-3	UN 1167 (DOT)
64-17-5	UN 1170 (DOT)
111-76-2	UN 1172 (DOT)
141-78-6	UN 1173 (DOT)
100-41-4	UN 1175 (DOT)
51845-86-4	UN 1176 (DOT)
97-96-1	UN 1178 (DOT)
97-95-0	UN 1179 (DOT)
105-54-4	UN 1180 (DOT)
105-39-5	UN 1181 (DOT)
541-41-3	UN 1182 (DOT)
1789-58-8	UN 1183 (DOT)
107-06-2	UN 1184 (DOT)
151-56-4	UN 1185 (DOT)
109-86-4	UN 1188 (DOT)
110-49-6	UN 1189 (DOT)
109-94-4	UN 1190 (DOT)
123-05-7	UN 1191 (DOT)
97-64-3	UN 1192 (DOT)
78-93-3	UN 1193 (DOT)
109-95-5	UN 1194 (DOT)
105-37-3	UN 1195 (DOT)
115-21-9	UN 1196 (DOT)
50-00-0	UN 1198 (DOT)
98-01-1	UN 1199 (DOT)
8013-75-0	UN 1201 (DOT)
55-63-0	UN 1204 (DOT)
9000-32-2	UN 1205 (DOT)
142-82-5	UN 1206 (DOT)
66-25-1	UN 1207 (DOT)
75-83-2	UN 1208 (DOT)
110-54-3	UN 1208 (DOT)
78-83-1	UN 1212 (DOT)
110-19-0	UN 1213 (DOT)
78-81-9	UN 1214 (DOT)
78-79-5	UN 1218 (DOT)
108-21-4	UN 1220 (DOT)
75-31-0	UN 1221 (DOT)
1712-64-7	UN 1222 (DOT)
141-79-7	UN 1229 (DOT)
67-56-1	UN 1230 (DOT)
78-93-3	UN 1232 (DOT)
108-84-9	UN 1233 (DOT)
109-87-5	UN 1234 (DOT)
74-89-5	UN 1235 (DOT)
623-42-7	UN 1237 (DOT)
79-22-1	UN 1238 (DOT)
107-30-2	UN 1239 (DOT)
75-54-7	UN 1242 (DOT)
107-31-3	UN 1243 (DOT)
60-34-4	UN 1244 (DOT)
814-78-8	UN 1246 (DOT)
80-62-6	UN 1247 (DOT)
554-12-1	UN 1248 (DOT)
107-87-9	UN 1249 (DOT)
75-79-6	UN 1250 (DOT)
78-94-4	UN 1251 (DOT)
8030-30-6	UN 1255 (DOT)
8030-30-6	UN 1256 (DOT)
13463-39-3	UN 1259 (DOT)
75-52-5	UN 1261 (DOT)
111-65-9	UN 1262 (DOT)

CAS No.	Chemical Name
109-66-0	UN 1265 (DOT)
463-82-1	UN 1265 (DOT)
8002-09-3	UN 1272 (DOT)
71-23-8	UN 1274 (DOT)
123-38-6	UN 1275 (DOT)
109-60-4	UN 1276 (DOT)
107-10-8	UN 1277 (DOT)
540-54-5	UN 1278 (DOT)
75-56-9	UN 1280 (DOT)
110-74-7	UN 1281 (DOT)
625-55-8	UN 1281 (DOT)
110-86-1	UN 1282 (DOT)
124-41-4	UN 1289 (DOT)
78-10-4	UN 1292 (DOT)
108-88-3	UN 1294 (DOT)
10025-78-2	UN 1295 (DOT)
121-44-8	UN 1296 (DOT)
75-50-3	UN 1297 (DOT)
75-77-4	UN 1298 (DOT)
108-05-4	UN 1301 (DOT)
109-92-2	UN 1302 (DOT)
75-35-4	UN 1303 (DOT)
109-53-5	UN 1304 (DOT)
75-94-5	UN 1305 (DOT)
95-47-6	UN 1307 (DOT)
106-42-3	UN 1307 (DOT)
108-38-3	UN 1307 (DOT)
1330-20-7	UN 1307 (DOT)
7440-67-7	UN 1308 (DOT)
131-74-8	UN 1310 (DOT)
9007-13-0	UN 1313 (DOT)
9007-13-0	UN 1314 (DOT)
68956-82-1	UN 1318 (DOT)
25550-58-7	UN 1320 (DOT)
519-44-8	UN 1322 (DOT)
35860-51-6	UN 1322 (DOT)
8049-18-1	UN 1323 (DOT)
60475-66-3	UN 1323 (DOT)
69523-06-4	UN 1323 (DOT)
9004-70-0	UN 1324 (DOT)
7440-58-6	UN 1326 (DOT)
100-97-0	UN 1328 (DOT)
131-73-7	UN 1328 (DOT)
9008-34-8	UN 1330 (DOT)
108-62-3	UN 1332 (DOT)
9002-91-9	UN 1332 (DOT)
37273-91-9	UN 1332 (DOT)
7440-45-1	UN 1333 (DOT)
91-20-3	UN 1334 (DOT)
7723-14-0	UN 1338 (DOT)
12037-82-0	UN 1339 (DOT)
1314-80-3	UN 1340 (DOT)
1314-85-8	UN 1341 (DOT)
12165-69-4	UN 1343 (DOT)
88-89-1	UN 1344 (DOT)
7440-21-3	UN 1346 (DOT)
146-84-9	UN 1347 (DOT)
25641-53-6	UN 1348 (DOT)
831-52-7	UN 1349 (DOT)
7704-34-9	UN 1350 (DOT)
7440-32-6	UN 1352 (DOT)
99-35-4	UN 1354 (DOT)
129-66-8	UN 1355 (DOT)
118-96-7	UN 1356 (DOT)
124-47-0	UN 1357 (DOT)
7440-67-7	UN 1358 (DOT)
1305-99-3	UN 1360 (DOT)
557-20-0	UN 1366 (DOT)
557-18-6	UN 1367 (DOT)
2999-74-8	UN 1368 (DOT)
138-89-6	UN 1369 (DOT)
53324-05-3	UN 1369 (DOT)
544-97-8	UN 1370 (DOT)
19624-22-7	UN 1380 (DOT)
7723-14-0	UN 1381 (DOT)
1312-73-8	UN 1382 (DOT)
12136-49-1	UN 1382 (DOT)
12136-50-4	UN 1382 (DOT)
7440-46-2	UN 1383 (DOT)

CAS No.	Chemical Name
7440-66-6	UN 1383 (DOT)
7775-14-6	UN 1384 (DOT)
1313-82-2	UN 1385 (DOT)
1299-86-1	UN 1394 (DOT)
12656-43-8	UN 1394 (DOT)
20859-73-8	UN 1397 (DOT)
7440-39-3	UN 1399 (DOT)
7440-39-3	UN 1400 (DOT)
7440-70-2	UN 1401 (DOT)
75-20-7	UN 1402 (DOT)
156-62-7	UN 1403 (DOT)
7789-78-8	UN 1404 (DOT)
57308-10-8	UN 1404 (DOT)
12737-18-7	UN 1405 (DOT)
12737-18-7	UN 1406 (DOT)
7440-46-2	UN 1407 (DOT)
8049-17-0	UN 1408 (DOT)
16853-85-3	UN 1410 (DOT)
16853-85-3	UN 1411 (DOT)
7782-89-0	UN 1412 (DOT)
7439-93-2	UN 1415 (DOT)
68848-64-6	UN 1417 (DOT)
7439-95-4	UN 1418 (DOT)
7440-09-7	UN 1420 (DOT)
11135-81-2	UN 1422 (DOT)
7440-17-7	UN 1423 (DOT)
11110-52-4	UN 1424 (DOT)
7782-92-5	UN 1425 (DOT)
16940-66-2	UN 1426 (DOT)
124-41-4	UN 1431 (DOT)
24167-76-8	UN 1432 (DOT)
25324-56-5	UN 1433 (DOT)
7440-66-6	UN 1436 (DOT)
7704-99-6	UN 1437 (DOT)
11105-16-1	UN 1437 (DOT)
7789-09-5	UN 1439 (DOT)
7790-98-9	UN 1442 (DOT)
7727-54-0	UN 1444 (DOT)
13477-00-4	UN 1445 (DOT)
10022-31-8	UN 1446 (DOT)
13465-95-7	UN 1447 (DOT)
7787-36-2	UN 1448 (DOT)
1304-29-6	UN 1449 (DOT)
7789-18-6	UN 1451 (DOT)
10137-74-3	UN 1452 (DOT)
14674-72-7	UN 1453 (DOT)
10124-37-5	UN 1454 (DOT)
13477-36-6	UN 1455 (DOT)
10118-76-0	UN 1456 (DOT)
1305-79-9	UN 1457 (DOT)
13530-68-2	UN 1463 (DOT)
10421-48-4	UN 1466 (DOT)
506-93-4	UN 1467 (DOT)
52470-25-4	UN 1467 (DOT)
18256-98-9	UN 1469 (DOT)
10099-74-8	UN 1469 (DOT)
13637-76-8	UN 1470 (DOT)
13840-33-0	UN 1471 (DOT)
12031-80-0	UN 1472 (DOT)
14519-17-6	UN 1473 (DOT)
10034-81-8	UN 1475 (DOT)
7758-01-2	UN 1484 (DOT)
3811-04-9	UN 1485 (DOT)
7757-79-1	UN 1486 (DOT)
7758-09-0	UN 1488 (DOT)
7778-74-7	UN 1489 (DOT)
17014-71-0	UN 1491 (DOT)
7727-21-1	UN 1492 (DOT)
7761-88-8	UN 1493 (DOT)
7789-38-0	UN 1494 (DOT)
7775-09-9	UN 1495 (DOT)
7758-19-2	UN 1496 (DOT)
7631-99-4	UN 1498 (DOT)
7632-00-0	UN 1500 (DOT)
7601-89-0	UN 1502 (DOT)
10101-50-5	UN 1503 (DOT)
1313-60-6	UN 1504 (DOT)
15593-29-0	UN 1505 (DOT)
7775-27-1	UN 1505 (DOT)

CAS No.	Chemical Name
28831-12-1	UN 1505 (DOT)
7791-10-8	UN 1506 (DOT)
10042-76-9	UN 1507 (DOT)
13450-97-0	UN 1508 (DOT)
1314-18-7	UN 1509 (DOT)
509-14-8	UN 1510 (DOT)
124-43-6	UN 1511 (DOT)
63885-01-8	UN 1512 (DOT)
10361-95-2	UN 1513 (DOT)
7779-88-6	UN 1514 (DOT)
23414-72-4	UN 1515 (DOT)
1314-22-3	UN 1516 (DOT)
63868-82-6	UN 1517 (DOT)
75-86-5	UN 1541 (DOT)
57-06-7	UN 1545 (DOT)
7784-44-3	UN 1546 (DOT)
62-53-3	UN 1547 (DOT)
142-04-1	UN 1548 (DOT)
28300-74-5	UN 1551 (DOT)
7778-39-4	UN 1553 (DOT)
7778-39-4	UN 1554 (DOT)
64973-06-4	UN 1555 (DOT)
1303-33-9	UN 1557 (DOT)
7440-38-2	UN 1558 (DOT)
1303-28-2	UN 1559 (DOT)
7784-34-1	UN 1560 (DOT)
1327-53-3	UN 1561 (DOT)
8028-73-7	UN 1562 (DOT)
7440-41-7	UN 1567 (DOT)
598-31-2	UN 1569 (DOT)
357-57-3	UN 1570 (DOT)
18810-58-7	UN 1571 (DOT)
75-60-5	UN 1572 (DOT)
7778-44-1	UN 1573 (DOT)
592-01-8	UN 1575 (DOT)
97-00-7	UN 1577 (DOT)
25567-67-3	UN 1577 (DOT)
88-73-3	UN 1578 (DOT)
100-00-5	UN 1578 (DOT)
121-73-3	UN 1578 (DOT)
3165-93-3	UN 1579 (DOT)
76-06-2	UN 1580 (DOT)
8004-09-9	UN 1581 (DOT)
76-06-2	UN 1583 (DOT)
124-87-8	UN 1584 (DOT)
12002-03-8	UN 1585 (DOT)
10290-12-7	UN 1586 (DOT)
57-12-5	UN 1588 (DOT)
506-77-4	UN 1589 (DOT)
27134-27-6	UN 1590 (DOT)
95-50-1	UN 1591 (DOT)
106-46-7	UN 1592 (DOT)
75-09-2	UN 1593 (DOT)
77-78-1	UN 1595 (DOT)
26471-56-7	UN 1596 (DOT)
99-65-0	UN 1597 (DOT)
100-25-4	UN 1597 (DOT)
528-29-0	UN 1597 (DOT)
25154-54-5	UN 1597 (DOT)
25550-58-7	UN 1599 (DOT)
25321-14-6	UN 1600 (DOT)
105-36-2	UN 1603 (DOT)
106-93-4	UN 1605 (DOT)
10102-49-5	UN 1606 (DOT)
63989-69-5	UN 1607 (DOT)
10102-50-8	UN 1608 (DOT)
757-58-4	UN 1611 (DOT)
74-90-8	UN 1614 (DOT)
10031-13-7	UN 1618 (DOT)
592-05-2	UN 1620 (DOT)
8012-74-6	UN 1621 (DOT)
10103-50-1	UN 1622 (DOT)
7784-37-4	UN 1623 (DOT)
7487-94-7	UN 1624 (DOT)
10045-94-0	UN 1625 (DOT)
10415-75-5	UN 1627 (DOT)
7783-36-0	UN 1628 (DOT)
631-60-7	UN 1629 (DOT)
1600-27-7	UN 1629 (DOT)

CAS No.	Chemical Name
10124-48-8	UN 1630 (DOT)
583-15-3	UN 1631 (DOT)
7783-35-9	UN 1633 (DOT)
7789-47-1	UN 1634 (DOT)
15385-58-7	UN 1634 (DOT)
10031-18-2	UN 1634 (DOT)
592-04-1	UN 1636 (DOT)
63937-14-4	UN 1637 (DOT)
15385-57-6	UN 1638 (DOT)
7783-30-4	UN 1638 (DOT)
7774-29-0	UN 1638 (DOT)
12002-19-6	UN 1639 (DOT)
1191-80-6	UN 1640 (DOT)
15829-53-5	UN 1641 (DOT)
5970-32-1	UN 1644 (DOT)
10415-73-3	UN 1644 (DOT)
7783-35-9	UN 1645 (DOT)
592-85-8	UN 1646 (DOT)
75-05-8	UN 1648 (DOT)
78-00-2	UN 1649 (DOT)
91-59-8	UN 1650 (DOT)
30553-04-9	UN 1651 (DOT)
557-19-7	UN 1653 (DOT)
54-11-5	UN 1654 (DOT)
2820-51-1	UN 1656 (DOT)
29790-52-1	UN 1657 (DOT)
65-30-5	UN 1658 (DOT)
6505-86-8	UN 1658 (DOT)
65-31-6	UN 1659 (DOT)
10102-43-9	UN 1660 (DOT)
100-01-6	UN 1661 (DOT)
29757-24-2	UN 1661 (DOT)
98-95-3	UN 1662 (DOT)
88-75-5	UN 1663 (DOT)
100-02-7	UN 1663 (DOT)
554-84-7	UN 1663 (DOT)
25154-55-6	UN 1663 (DOT)
99-99-0	UN 1664 (DOT)
99-08-1	UN 1664 (DOT)
88-72-2	UN 1664 (DOT)
1321-12-6	UN 1664 (DOT)
76-01-7	UN 1669 (DOT)
622-44-6	UN 1672 (DOT)
108-45-2	UN 1673 (DOT)
7784-41-0	UN 1677 (DOT)
10124-50-2	UN 1678 (DOT)
13682-73-0	UN 1679 (DOT)
151-50-8	UN 1680 (DOT)
7784-08-9	UN 1683 (DOT)
506-64-9	UN 1684 (DOT)
7784-46-5	UN 1686 (DOT)
26628-22-8	UN 1687 (DOT)
124-65-2	UN 1688 (DOT)
143-33-9	UN 1689 (DOT)
7681-49-4	UN 1690 (DOT)
91724-16-2	UN 1691 (DOT)
57-24-9	UN 1692 (DOT)
5798-79-8	UN 1694 (DOT)
78-95-5	UN 1695 (DOT)
532-27-4	UN 1697 (DOT)
578-94-9	UN 1698 (DOT)
35884-77-6	UN 1701 (DOT)
25322-20-7	UN 1702 (DOT)
3689-24-5	UN 1704 (DOT)
95-53-4	UN 1708 (DOT)
106-49-0	UN 1708 (DOT)
108-44-1	UN 1708 (DOT)
79-01-6	UN 1710 (DOT)
1300-73-8	UN 1711 (DOT)
13464-44-3	UN 1712 (DOT)
10326-24-6	UN 1712 (DOT)
13464-33-0	UN 1712 (DOT)
28838-01-9	UN 1712 (DOT)
557-21-1	UN 1713 (DOT)
1314-84-7	UN 1714 (DOT)
108-24-7	UN 1715 (DOT)
506-96-7	UN 1716 (DOT)
75-36-5	UN 1717 (DOT)
12788-93-1	UN 1718 (DOT)

CAS No.	Chemical Name
2937-50-0	UN 1722 (DOT)
556-56-9	UN 1723 (DOT)
107-37-9	UN 1724 (DOT)
7727-15-3	UN 1725 (DOT)
7446-70-0	UN 1726 (DOT)
1341-49-7	UN 1727 (DOT)
107-72-2	UN 1728 (DOT)
100-07-2	UN 1729 (DOT)
7647-18-9	UN 1730 (DOT)
7647-18-9	UN 1731 (DOT)
7783-70-2	UN 1732 (DOT)
10025-91-9	UN 1733 (DOT)
98-88-4	UN 1736 (DOT)
100-44-7	UN 1738 (DOT)
7726-95-6	UN 1744 (DOT)
7521-80-4	UN 1747 (DOT)
7778-54-3	UN 1748 (DOT)
7790-91-2	UN 1749 (DOT)
79-11-8	UN 1750 (DOT)
79-11-8	UN 1751 (DOT)
79-04-9	UN 1752 (DOT)
26571-79-9	UN 1753 (DOT)
7790-94-5	UN 1754 (DOT)
13530-68-2	UN 1755 (DOT)
7788-97-8	UN 1756 (DOT)
7788-97-8	UN 1757 (DOT)
14977-61-8	UN 1758 (DOT)
13426-91-0	UN 1761 (DOT)
79-43-6	UN 1764 (DOT)
79-36-7	UN 1765 (DOT)
1719-53-5	UN 1767 (DOT)
13779-41-4	UN 1768 (DOT)
80-10-4	UN 1769 (DOT)
776-74-9	UN 1770 (DOT)
4484-72-4	UN 1771 (DOT)
7705-08-0	UN 1773 (DOT)
13537-32-1	UN 1776 (DOT)
7789-21-1	UN 1777 (DOT)
16961-83-4	UN 1778 (DOT)
64-18-6	UN 1779 (DOT)
627-63-4	UN 1780 (DOT)
5894-60-0	UN 1781 (DOT)
16940-81-1	UN 1782 (DOT)
124-09-4	UN 1783 (DOT)
928-65-4	UN 1784 (DOT)
10034-85-2	UN 1787 (DOT)
10035-10-6	UN 1788 (DOT)
7647-01-0	UN 1789 (DOT)
7664-39-3	UN 1790 (DOT)
7778-66-7	UN 1791 (DOT)
7790-99-0	UN 1792 (DOT)
1623-24-1	UN 1793 (DOT)
7446-14-2	UN 1794 (DOT)
15739-80-7	UN 1794 (DOT)
8007-56-5	UN 1798 (DOT)
5283-67-0	UN 1799 (DOT)
112-04-9	UN 1800 (DOT)
5283-66-9	UN 1801 (DOT)
7601-90-3	UN 1802 (DOT)
98-13-5	UN 1804 (DOT)
7664-38-2	UN 1805 (DOT)
10026-13-8	UN 1806 (DOT)
1314-56-3	UN 1807 (DOT)
7789-60-8	UN 1808 (DOT)
7719-12-2	UN 1809 (DOT)
10025-87-3	UN 1810 (DOT)
7789-29-9	UN 1811 (DOT)
7789-23-3	UN 1812 (DOT)
79-03-8	UN 1815 (DOT)
141-57-1	UN 1816 (DOT)
7791-27-7	UN 1817 (DOT)
10026-04-7	UN 1818 (DOT)
11138-49-1	UN 1819 (DOT)
7681-38-1	UN 1821 (DOT)
12401-86-4	UN 1825 (DOT)
7646-78-8	UN 1827 (DOT)
10025-67-9	UN 1828 (DOT)
10545-99-0	UN 1828 (DOT)
7446-11-9	UN 1829 (DOT)
7664-93-9	UN 1830 (DOT)
7783-05-3	UN 1831 (DOT)
8014-95-7	UN 1831 (DOT)
7664-93-9	UN 1832 (DOT)
7782-99-2	UN 1833 (DOT)
7791-25-5	UN 1834 (DOT)
75-59-2	UN 1835 (DOT)
7719-09-7	UN 1836 (DOT)
3982-91-0	UN 1837 (DOT)
7550-45-0	UN 1838 (DOT)
76-03-9	UN 1839 (DOT)
7646-85-7	UN 1840 (DOT)
75-39-8	UN 1841 (DOT)
29595-25-3	UN 1843 (DOT)
124-38-9	UN 1845 (DOT)
56-23-5	UN 1846 (DOT)
1312-73-8	UN 1847 (DOT)
79-09-4	UN 1848 (DOT)
1313-82-2	UN 1849 (DOT)
7440-39-3	UN 1854 (DOT)
7440-70-2	UN 1855 (DOT)
116-15-4	UN 1858 (DOT)
7783-61-1	UN 1859 (DOT)
75-02-5	UN 1860 (DOT)
10544-63-5	UN 1862 (DOT)
627-13-4	UN 1865 (DOT)
7439-95-4	UN 1869 (DOT)
13762-51-1	UN 1870 (DOT)
7704-98-5	UN 1871 (DOT)
11140-68-4	UN 1871 (DOT)
1309-60-0	UN 1872 (DOT)
7601-90-3	UN 1873 (DOT)
1304-28-5	UN 1884 (DOT)
98-87-3	UN 1886 (DOT)
74-97-5	UN 1887 (DOT)
67-66-3	UN 1888 (DOT)
506-68-3	UN 1889 (DOT)
74-96-4	UN 1891 (DOT)
598-14-1	UN 1892 (DOT)
127-18-4	UN 1897 (DOT)
507-02-8	UN 1898 (DOT)
27215-10-7	UN 1902 (DOT)
7783-08-6	UN 1905 (DOT)
8006-28-8	UN 1907 (DOT)
7758-19-2	UN 1908 (DOT)
1305-78-8	UN 1910 (DOT)
19287-45-7	UN 1911 (DOT)
7440-01-9	UN 1913 (DOT)
590-01-2	UN 1914 (DOT)
111-44-4	UN 1916 (DOT)
140-88-5	UN 1917 (DOT)
98-82-8	UN 1918 (DOT)
96-33-3	UN 1919 (DOT)
111-84-2	UN 1920 (DOT)
123-75-1	UN 1922 (DOT)
15512-36-4	UN 1923 (DOT)
563-43-9	UN 1924 (DOT)
12075-68-2	UN 1925 (DOT)
12263-85-3	UN 1926 (DOT)
12542-85-7	UN 1927 (DOT)
14293-73-3	UN 1929 (DOT)
100-99-2	UN 1930 (DOT)
7779-86-4	UN 1931 (DOT)
57-12-5	UN 1935 (DOT)
790-83-0	UN 1938 (DOT)
7789-59-5	UN 1939 (DOT)
68-11-1	UN 1940 (DOT)
6484-52-2	UN 1942 (DOT)
7440-37-1	UN 1951 (DOT)
10036-47-2	UN 1955 (DOT)
7782-39-0	UN 1957 (DOT)
1320-37-2	UN 1958 (DOT)
75-38-7	UN 1959 (DOT)
74-84-0	UN 1961 (DOT)
74-85-1	UN 1962 (DOT)
7440-59-7	UN 1963 (DOT)
1333-74-0	UN 1966 (DOT)
75-28-5	UN 1969 (DOT)
7439-90-9	UN 1970 (DOT)

CAS No.	Chemical Name
74-82-8	UN 1971 (DOT)
74-82-8	UN 1972 (DOT)
353-59-3	UN 1974 (DOT)
115-25-3	UN 1976 (DOT)
7727-37-9	UN 1977 (DOT)
75-73-0	UN 1982 (DOT)
1330-45-6	UN 1983 (DOT)
75-46-7	UN 1984 (DOT)
13463-40-6	UN 1994 (DOT)
61789-51-3	UN 2001 (DOT)
7803-54-5	UN 2004 (DOT)
555-54-4	UN 2005 (DOT)
7440-67-7	UN 2008 (DOT)
7440-67-7	UN 2009 (DOT)
60616-74-2	UN 2010 (DOT)
20770-41-6	UN 2012 (DOT)
12504-13-1	UN 2013 (DOT)
7722-84-1	UN 2014 (DOT)
7722-84-1	UN 2015 (DOT)
27134-26-5	UN 2018 (DOT)
27134-26-5	UN 2019 (DOT)
106-48-9	UN 2020 (DOT)
25167-80-0	UN 2020 (DOT)
106-48-9	UN 2021 (DOT)
25167-80-0	UN 2021 (DOT)
1319-77-3	UN 2022 (DOT)
106-89-8	UN 2023 (DOT)
7784-46-5	UN 2027 (DOT)
302-01-2	UN 2029 (DOT)
302-01-2	UN 2030 (DOT)
7697-37-2	UN 2031 (DOT)
12401-70-6	UN 2033 (DOT)
27987-06-0	UN 2035 (DOT)
7440-63-3	UN 2036 (DOT)
25321-14-6	UN 2038 (DOT)
463-82-1	UN 2044 (DOT)
77-73-6	UN 2048 (DOT)
25340-17-4	UN 2049 (DOT)
25167-70-8	UN 2050 (DOT)
108-01-0	UN 2051 (DOT)
138-86-3	UN 2052 (DOT)
105-30-6	UN 2053 (DOT)
108-11-2	UN 2053 (DOT)
54972-97-3	UN 2053 (DOT)
110-91-8	UN 2054 (DOT)
100-42-5	UN 2055 (DOT)
109-99-9	UN 2056 (DOT)
13987-01-4	UN 2057 (DOT)
110-62-3	UN 2058 (DOT)
9004-70-0	UN 2059 (DOT)
9004-70-0	UN 2060 (DOT)
57608-40-9	UN 2070 (DOT)
7664-41-7	UN 2073 (DOT)
79-06-1	UN 2074 (DOT)
75-87-6	UN 2075 (DOT)
95-48-7	UN 2076 (DOT)
106-44-5	UN 2076 (DOT)
108-39-4	UN 2076 (DOT)
1319-77-3	UN 2076 (DOT)
26471-62-5	UN 2078 (DOT)
111-40-0	UN 2079 (DOT)
37187-22-7	UN 2080 (DOT)
644-31-5	UN 2081 (DOT)
3179-56-4	UN 2082 (DOT)
3179-56-4	UN 2083 (DOT)
110-22-5	UN 2084 (DOT)
94-36-0	UN 2085 (DOT)
94-36-0	UN 2086 (DOT)
94-36-0	UN 2088 (DOT)
94-36-0	UN 2089 (DOT)
3457-61-2	UN 2091 (DOT)
30580-75-7	UN 2091 (DOT)
1931-62-0	UN 2100 (DOT)
1931-62-0	UN 2101 (DOT)
110-05-4	UN 2102 (DOT)
2372-21-6	UN 2103 (DOT)
13122-18-4	UN 2104 (DOT)
927-07-1	UN 2110 (DOT)
2167-23-9	UN 2111 (DOT)
94-17-7	UN 2113 (DOT)
80-15-9	UN 2116 (DOT)
762-12-9	UN 2120 (DOT)
80-43-3	UN 2121 (DOT)
16111-62-9	UN 2122 (DOT)
16111-62-9	UN 2123 (DOT)
105-74-8	UN 2124 (DOT)
37206-20-5	UN 2126 (DOT)
1338-23-4	UN 2127 (DOT)
762-16-3	UN 2129 (DOT)
762-13-0	UN 2130 (DOT)
79-21-0	UN 2131 (DOT)
3248-28-0	UN 2132 (DOT)
105-64-6	UN 2133 (DOT)
105-64-6	UN 2134 (DOT)
123-23-9	UN 2135 (DOT)
771-29-9	UN 2136 (DOT)
133-14-2	UN 2138 (DOT)
133-14-2	UN 2139 (DOT)
109-13-7	UN 2142 (DOT)
3006-82-4	UN 2143 (DOT)
2550-33-6	UN 2144 (DOT)
6731-36-8	UN 2146 (DOT)
2407-94-5	UN 2148 (DOT)
2144-45-8	UN 2149 (DOT)
1561-49-5	UN 2152 (DOT)
1561-49-5	UN 2153 (DOT)
15520-11-3	UN 2154 (DOT)
78-63-7	UN 2155 (DOT)
78-63-7	UN 2156 (DOT)
1068-27-5	UN 2158 (DOT)
1068-27-5	UN 2159 (DOT)
5809-08-5	UN 2160 (DOT)
28324-52-9	UN 2162 (DOT)
85873-97-8	UN 2162 (DOT)
54693-46-8	UN 2163 (DOT)
26762-93-6	UN 2171 (DOT)
2618-77-1	UN 2172 (DOT)
2618-77-1	UN 2173 (DOT)
3025-88-5	UN 2174 (DOT)
14666-78-5	UN 2175 (DOT)
16066-38-9	UN 2176 (DOT)
26748-41-4	UN 2177 (DOT)
3006-86-8	UN 2179 (DOT)
3006-86-8	UN 2180 (DOT)
3437-84-1	UN 2182 (DOT)
7647-01-0	UN 2186 (DOT)
124-38-9	UN 2187 (DOT)
7784-42-1	UN 2188 (DOT)
4109-96-0	UN 2189 (DOT)
7783-41-7	UN 2190 (DOT)
2699-79-8	UN 2191 (DOT)
7782-65-2	UN 2192 (DOT)
76-16-4	UN 2193 (DOT)
7783-79-1	UN 2194 (DOT)
7783-80-4	UN 2195 (DOT)
7783-82-6	UN 2196 (DOT)
10034-85-2	UN 2197 (DOT)
7647-19-0	UN 2198 (DOT)
463-49-0	UN 2200 (DOT)
10024-97-2	UN 2201 (DOT)
11104-93-1	UN 2201 (DOT)
7783-07-5	UN 2202 (DOT)
7803-62-5	UN 2203 (DOT)
463-58-1	UN 2204 (DOT)
7778-54-3	UN 2208 (DOT)
50-00-0	UN 2209 (DOT)
12001-28-4	UN 2212 (DOT)
85-44-9	UN 2214 (DOT)
108-31-6	UN 2215 (DOT)
79-10-7	UN 2218 (DOT)
106-92-3	UN 2219 (DOT)
100-66-3	UN 2222 (DOT)
100-47-0	UN 2224 (DOT)
98-07-7	UN 2226 (DOT)
97-88-1	UN 2227 (DOT)
28805-86-9	UN 2228 (DOT)
28805-86-9	UN 2229 (DOT)
107-20-0	UN 2232 (DOT)

CAS No.	Chemical Name
98-56-6	UN 2234 (DOT)
98-15-7	UN 2234 (DOT)
104-83-6	UN 2235 (DOT)
28479-22-3	UN 2236 (DOT)
41587-36-4	UN 2237 (DOT)
106-43-4	UN 2238 (DOT)
14489-25-9	UN 2240 (DOT)
110-71-4	UN 2252 (DOT)
121-69-7	UN 2253 (DOT)
7440-09-7	UN 2257 (DOT)
78-90-0	UN 2258 (DOT)
112-24-3	UN 2259 (DOT)
102-69-2	UN 2260 (DOT)
1300-71-6	UN 2261 (DOT)
79-44-7	UN 2262 (DOT)
589-90-2	UN 2263 (DOT)
68-12-2	UN 2265 (DOT)
926-63-6	UN 2266 (DOT)
56-18-8	UN 2269 (DOT)
75-04-7	UN 2270 (DOT)
106-68-3	UN 2271 (DOT)
103-69-5	UN 2272 (DOT)
578-54-1	UN 2273 (DOT)
92-59-1	UN 2274 (DOT)
104-75-6	UN 2276 (DOT)
97-63-2	UN 2277 (DOT)
592-76-7	UN 2278 (DOT)
87-68-3	UN 2279 (DOT)
124-09-4	UN 2280 (DOT)
822-06-0	UN 2281 (DOT)
111-27-3	UN 2282 (DOT)
97-86-9	UN 2283 (DOT)
30586-18-6	UN 2286 (DOT)
4098-71-9	UN 2290 (DOT)
100-61-8	UN 2294 (DOT)
96-34-4	UN 2295 (DOT)
108-87-2	UN 2296 (DOT)
1331-22-2	UN 2297 (DOT)
104-90-5	UN 2300 (DOT)
27137-41-3	UN 2301 (DOT)
110-12-3	UN 2302 (DOT)
98-83-9	UN 2303 (DOT)
91-20-3	UN 2304 (DOT)
31212-28-9	UN 2305 (DOT)
402-54-0	UN 2306 (DOT)
121-17-5	UN 2307 (DOT)
7782-78-7	UN 2308 (DOT)
63597-41-1	UN 2309 (DOT)
123-54-6	UN 2310 (DOT)
1321-31-9	UN 2311 (DOT)
109-06-8	UN 2313 (DOT)
1333-41-1	UN 2313 (DOT)
1336-36-3	UN 2315 (DOT)
14264-31-4	UN 2316 (DOT)
14264-31-4	UN 2317 (DOT)
16721-80-5	UN 2318 (DOT)
112-57-2	UN 2320 (DOT)
12002-48-1	UN 2321 (DOT)
51023-22-4	UN 2322 (DOT)
122-52-1	UN 2323 (DOT)
7756-94-7	UN 2324 (DOT)
108-67-8	UN 2325 (DOT)
34216-34-7	UN 2326 (DOT)
25620-58-0	UN 2327 (DOT)
28679-16-5	UN 2328 (DOT)
121-45-9	UN 2329 (DOT)
1120-21-4	UN 2330 (DOT)
7646-85-7	UN 2331 (DOT)
107-29-9	UN 2332 (DOT)
591-87-7	UN 2333 (DOT)
107-11-9	UN 2334 (DOT)
557-31-3	UN 2335 (DOT)
1838-59-1	UN 2336 (DOT)
108-98-5	UN 2337 (DOT)
98-08-8	UN 2338 (DOT)
78-76-2	UN 2339 (DOT)
592-55-2	UN 2340 (DOT)
107-82-4	UN 2341 (DOT)
507-19-7	UN 2342 (DOT)

CAS No.	Chemical Name
29756-38-5	UN 2343 (DOT)
106-94-5	UN 2344 (DOT)
106-96-7	UN 2345 (DOT)
141-32-2	UN 2348 (DOT)
111-34-2	UN 2352 (DOT)
141-75-3	UN 2353 (DOT)
3188-13-4	UN 2354 (DOT)
629-20-9	UN 2358 (DOT)
110-96-3	UN 2361 (DOT)
75-34-3	UN 2362 (DOT)
75-08-1	UN 2363 (DOT)
103-65-1	UN 2364 (DOT)
105-58-8	UN 2366 (DOT)
123-15-9	UN 2367 (DOT)
1330-16-1	UN 2368 (DOT)
592-41-6	UN 2370 (DOT)
110-18-9	UN 2372 (DOT)
462-95-3	UN 2373 (DOT)
3054-95-3	UN 2374 (DOT)
926-64-7	UN 2378 (DOT)
108-09-8	UN 2379 (DOT)
78-62-6	UN 2380 (DOT)
624-92-0	UN 2381 (DOT)
540-73-8	UN 2382 (DOT)
142-84-7	UN 2383 (DOT)
111-43-3	UN 2384 (DOT)
97-62-1	UN 2385 (DOT)
766-09-6	UN 2386 (DOT)
462-06-6	UN 2387 (DOT)
25496-08-6	UN 2388 (DOT)
110-00-9	UN 2389 (DOT)
25267-27-0	UN 2390 (DOT)
26914-02-3	UN 2392 (DOT)
542-55-2	UN 2393 (DOT)
540-42-1	UN 2394 (DOT)
79-30-1	UN 2395 (DOT)
78-85-3	UN 2396 (DOT)
563-80-4	UN 2397 (DOT)
626-67-5	UN 2399 (DOT)
556-24-1	UN 2400 (DOT)
110-89-4	UN 2401 (DOT)
75-33-2	UN 2402 (DOT)
107-03-9	UN 2402 (DOT)
79869-58-2	UN 2402 (DOT)
108-22-5	UN 2403 (DOT)
107-12-0	UN 2404 (DOT)
638-11-9	UN 2405 (DOT)
617-50-5	UN 2406 (DOT)
108-23-6	UN 2407 (DOT)
61215-72-3	UN 2410 (DOT)
109-74-0	UN 2411 (DOT)
110-01-0	UN 2412 (DOT)
3087-37-4	UN 2413 (DOT)
110-02-1	UN 2414 (DOT)
121-43-7	UN 2416 (DOT)
353-50-4	UN 2417 (DOT)
7783-60-0	UN 2418 (DOT)
598-73-2	UN 2419 (DOT)
684-16-2	UN 2420 (DOT)
10544-73-7	UN 2421 (DOT)
12033-49-7	UN 2421 (DOT)
41070-66-9	UN 2422 (DOT)
76-19-7	UN 2424 (DOT)
6484-52-2	UN 2426 (DOT)
3811-04-9	UN 2427 (DOT)
7775-09-9	UN 2428 (DOT)
10137-74-3	UN 2429 (DOT)
29191-52-4	UN 2431 (DOT)
91-66-7	UN 2432 (DOT)
61878-61-3	UN 2433 (DOT)
25567-68-4	UN 2433 (DOT)
507-09-5	UN 2436 (DOT)
149-74-6	UN 2437 (DOT)
3282-30-2	UN 2438 (DOT)
1333-83-1	UN 2439 (DOT)
7705-07-9	UN 2441 (DOT)
76-02-8	UN 2442 (DOT)
7727-18-6	UN 2443 (DOT)
7632-51-1	UN 2444 (DOT)

CAS No.	Chemical Name
12167-20-3	UN 2446 (DOT)
7723-14-0	UN 2447 (DOT)
7704-34-9	UN 2448 (DOT)
7783-54-2	UN 2451 (DOT)
107-00-6	UN 2452 (DOT)
353-36-6	UN 2453 (DOT)
593-53-3	UN 2454 (DOT)
557-98-2	UN 2456 (DOT)
79-29-8	UN 2457 (DOT)
42296-74-2	UN 2458 (DOT)
513-35-9	UN 2460 (DOT)
26760-64-5	UN 2460 (DOT)
43133-95-5	UN 2462 (DOT)
7784-21-6	UN 2463 (DOT)
13597-99-4	UN 2464 (DOT)
2782-57-2	UN 2465 (DOT)
2244-21-5	UN 2465 (DOT)
12030-88-5	UN 2466 (DOT)
3313-92-6	UN 2467 (DOT)
4452-58-8	UN 2467 (DOT)
15630-89-4	UN 2467 (DOT)
87-90-1	UN 2468 (DOT)
14519-07-4	UN 2469 (DOT)
140-29-4	UN 2470 (DOT)
20816-12-0	UN 2471 (DOT)
83-26-1	UN 2472 (DOT)
127-85-5	UN 2473 (DOT)
463-71-8	UN 2474 (DOT)
7718-98-1	UN 2475 (DOT)
556-61-6	UN 2477 (DOT)
624-83-9	UN 2480 (DOT)
109-90-0	UN 2481 (DOT)
110-78-1	UN 2482 (DOT)
1795-48-8	UN 2483 (DOT)
1873-29-6	UN 2486 (DOT)
103-71-9	UN 2487 (DOT)
101-68-8	UN 2489 (DOT)
63283-80-7	UN 2490 (DOT)
111-49-9	UN 2493 (DOT)
7783-66-6	UN 2495 (DOT)
123-62-6	UN 2496 (DOT)
139-02-6	UN 2497 (DOT)
100-50-5	UN 2498 (DOT)
545-55-1	UN 2501 (DOT)
638-29-9	UN 2502 (DOT)
10026-11-6	UN 2503 (DOT)
79-27-6	UN 2504 (DOT)
12125-01-8	UN 2505 (DOT)
7803-63-6	UN 2506 (DOT)
10241-05-1	UN 2508 (DOT)
7646-93-7	UN 2509 (DOT)
28554-00-9	UN 2511 (DOT)
123-30-8	UN 2512 (DOT)
27598-85-2	UN 2512 (DOT)
598-21-0	UN 2513 (DOT)
108-86-1	UN 2514 (DOT)
75-25-2	UN 2515 (DOT)
558-13-4	UN 2516 (DOT)
25497-29-4	UN 2517 (DOT)
29965-97-7	UN 2520 (DOT)
2867-47-2	UN 2522 (DOT)
122-51-0	UN 2524 (DOT)
95-92-1	UN 2525 (DOT)
617-89-0	UN 2526 (DOT)
106-63-8	UN 2527 (DOT)
97-85-8	UN 2528 (DOT)
79-41-4	UN 2531 (DOT)
598-99-2	UN 2533 (DOT)
993-00-0	UN 2534 (DOT)
109-02-4	UN 2535 (DOT)
25265-68-3	UN 2536 (DOT)
27254-36-0	UN 2538 (DOT)
586-62-9	UN 2541 (DOT)
7440-58-6	UN 2545 (DOT)
7440-32-6	UN 2546 (DOT)
12034-12-7	UN 2547 (DOT)
13637-63-3	UN 2548 (DOT)
1338-23-4	UN 2550 (DOT)
677-71-4	UN 2552 (DOT)

CAS No.	Chemical Name
8030-30-6	UN 2553 (DOT)
563-47-3	UN 2554 (DOT)
3132-64-7	UN 2558 (DOT)
109-13-7	UN 2562 (DOT)
76-03-9	UN 2564 (DOT)
131-52-2	UN 2567 (DOT)
540-82-9	UN 2571 (DOT)
100-63-0	UN 2572 (DOT)
13453-30-0	UN 2573 (DOT)
1330-78-5	UN 2574 (DOT)
7789-59-5	UN 2576 (DOT)
103-80-0	UN 2577 (DOT)
1314-24-5	UN 2578 (DOT)
110-85-0	UN 2579 (DOT)
7727-15-3	UN 2580 (DOT)
7446-70-0	UN 2581 (DOT)
7705-08-0	UN 2582 (DOT)
104-15-4	UN 2583 (DOT)
25231-46-3	UN 2583 (DOT)
104-15-4	UN 2584 (DOT)
25231-46-3	UN 2584 (DOT)
104-15-4	UN 2585 (DOT)
25231-46-3	UN 2585 (DOT)
104-15-4	UN 2586 (DOT)
25231-46-3	UN 2586 (DOT)
106-51-4	UN 2587 (DOT)
2549-51-1	UN 2589 (DOT)
12001-29-5	UN 2590 (DOT)
7440-63-3	UN 2591 (DOT)
52326-66-6	UN 2592 (DOT)
53220-22-7	UN 2595 (DOT)
25251-51-8	UN 2596 (DOT)
287-23-0	UN 2601 (DOT)
100-73-2	UN 2607 (DOT)
79-46-9	UN 2608 (DOT)
25322-01-4	UN 2608 (DOT)
1693-71-6	UN 2609 (DOT)
102-70-5	UN 2610 (DOT)
78-89-7	UN 2611 (DOT)
557-17-5	UN 2612 (DOT)
513-42-8	UN 2614 (DOT)
628-32-0	UN 2615 (DOT)
5419-55-6	UN 2616 (DOT)
25639-42-3	UN 2617 (DOT)
25013-15-4	UN 2618 (DOT)
540-18-1	UN 2620 (DOT)
513-86-0	UN 2621 (DOT)
765-34-4	UN 2622 (DOT)
7790-93-4	UN 2626 (DOT)
23745-86-0	UN 2628 (DOT)
62-74-8	UN 2629 (DOT)
10102-18-8	UN 2630 (DOT)
144-49-0	UN 2642 (DOT)
96-32-2	UN 2643 (DOT)
74-88-4	UN 2644 (DOT)
70-11-1	UN 2645 (DOT)
77-47-4	UN 2646 (DOT)
534-07-6	UN 2649 (DOT)
101-77-9	UN 2651 (DOT)
620-05-3	UN 2653 (DOT)
16871-90-2	UN 2655 (DOT)
91-22-5	UN 2656 (DOT)
7488-56-4	UN 2657 (DOT)
3926-62-3	UN 2659 (DOT)
116-16-5	UN 2661 (DOT)
123-31-9	UN 2662 (DOT)
105-56-6	UN 2666 (DOT)
27458-20-4	UN 2667 (DOT)
107-14-2	UN 2668 (DOT)
108-77-0	UN 2670 (DOT)
504-24-5	UN 2671 (DOT)
95-85-2	UN 2673 (DOT)
16893-85-9	UN 2674 (DOT)
7803-52-3	UN 2676 (DOT)
1310-82-3	UN 2677 (DOT)
1310-82-3	UN 2678 (DOT)
1310-65-2	UN 2679 (DOT)
1310-66-3	UN 2680 (DOT)
21351-79-1	UN 2681 (DOT)

CAS No.	Chemical Name
21351-79-1	UN 2682 (DOT)
12135-76-1	UN 2683 (DOT)
104-78-9	UN 2684 (DOT)
110-85-0	UN 2685 (DOT)
100-37-8	UN 2686 (DOT)
109-70-6	UN 2688 (DOT)
96-24-2	UN 2689 (DOT)
4316-42-1	UN 2690 (DOT)
7789-69-7	UN 2691 (DOT)
7631-90-5	UN 2693 (DOT)
85-43-8	UN 2698 (DOT)
76-05-1	UN 2699 (DOT)
75-33-2	UN 2703 (DOT)
107-03-9	UN 2704 (DOT)
79869-58-2	UN 2704 (DOT)
12772-47-3	UN 2705 (DOT)
584-02-1	UN 2706 (DOT)
25136-55-4	UN 2707 (DOT)
104-51-8	UN 2709 (DOT)
123-19-3	UN 2710 (DOT)
26249-12-7	UN 2711 (DOT)
260-94-6	UN 2713 (DOT)
9010-69-9	UN 2714 (DOT)
110-65-6	UN 2716 (DOT)
10070-67-0	UN 2716 (DOT)
76-22-2	UN 2717 (DOT)
102-67-0	UN 2718 (DOT)
13967-90-3	UN 2719 (DOT)
26506-47-8	UN 2721 (DOT)
7790-69-4	UN 2722 (DOT)
10326-21-3	UN 2723 (DOT)
10377-66-9	UN 2724 (DOT)
13138-45-9	UN 2725 (DOT)
17861-62-0	UN 2726 (DOT)
16901-76-1	UN 2727 (DOT)
13746-98-0	UN 2727 (DOT)
10102-45-1	UN 2727 (DOT)
13746-89-9	UN 2728 (DOT)
118-74-1	UN 2729 (DOT)
100-17-4	UN 2730 (DOT)
586-78-7	UN 2732 (DOT)
1126-78-9	UN 2738 (DOT)
106-31-0	UN 2739 (DOT)
109-61-5	UN 2740 (DOT)
13477-10-6	UN 2741 (DOT)
592-34-7	UN 2743 (DOT)
81228-87-7	UN 2744 (DOT)
22128-62-7	UN 2745 (DOT)
1885-14-9	UN 2746 (DOT)
70042-58-9	UN 2747 (DOT)
24468-13-1	UN 2748 (DOT)
75-76-3	UN 2749 (DOT)
2524-04-1	UN 2751 (DOT)
4016-11-9	UN 2752 (DOT)
102-27-2	UN 2754 (DOT)
622-57-1	UN 2754 (DOT)
937-14-4	UN 2755 (DOT)
371-86-8	UN 2783 (DOT)
757-58-4	UN 2783 (DOT)
3268-49-3	UN 2785 (DOT)
64-19-7	UN 2789 (DOT)
64-19-7	UN 2790 (DOT)
644-97-3	UN 2798 (DOT)
14684-25-4	UN 2799 (DOT)
7440-55-3	UN 2803 (DOT)
26134-62-3	UN 2806 (DOT)
7439-97-6	UN 2809 (DOT)
11138-49-1	UN 2812 (DOT)
140-31-8	UN 2815 (DOT)
1341-49-7	UN 2817 (DOT)
9080-17-5	UN 2818 (DOT)
107-92-6	UN 2820 (DOT)
29154-12-9	UN 2822 (DOT)
3724-65-0	UN 2823 (DOT)
96-80-0	UN 2825 (DOT)
2941-64-2	UN 2826 (DOT)
26555-35-1	UN 2826 (DOT)
71-55-6	UN 2831 (DOT)
13770-96-2	UN 2835 (DOT)

CAS No.	Chemical Name
7681-38-1	UN 2837 (DOT)
123-20-6	UN 2838 (DOT)
107-89-1	UN 2839 (DOT)
110-69-0	UN 2840 (DOT)
2050-92-2	UN 2841 (DOT)
79-24-3	UN 2842 (DOT)
557-20-0	UN 2845 (DOT)
676-83-5	UN 2845 (DOT)
627-30-5	UN 2849 (DOT)
6842-15-5	UN 2850 (DOT)
13319-75-0	UN 2851 (DOT)
2217-06-3	UN 2852 (DOT)
16919-19-0	UN 2854 (DOT)
16871-71-9	UN 2855 (DOT)
7440-67-7	UN 2858 (DOT)
7803-55-6	UN 2859 (DOT)
1314-34-7	UN 2860 (DOT)
1314-62-1	UN 2862 (DOT)
13769-43-2	UN 2864 (DOT)
10039-54-0	UN 2865 (DOT)
7705-07-9	UN 2869 (DOT)
16962-07-5	UN 2870 (DOT)
7440-36-0	UN 2871 (DOT)
96-12-8	UN 2872 (DOT)
98-00-0	UN 2874 (DOT)
108-46-3	UN 2876 (DOT)
62-56-6	UN 2877 (DOT)
7440-32-6	UN 2878 (DOT)
7791-23-3	UN 2879 (DOT)
3006-86-8	UN 2885 (DOT)
3006-82-4	UN 2888 (DOT)
53220-22-7	UN 2892 (DOT)
15520-11-3	UN 2894 (DOT)
686-31-7	UN 2898 (DOT)
27774-13-6	UN 2931 (DOT)
17639-93-9	UN 2933 (DOT)
79435-04-4	UN 2934 (DOT)
535-13-7	UN 2935 (DOT)
79-42-5	UN 2936 (DOT)
98-85-1	UN 2937 (DOT)
93-58-3	UN 2938 (DOT)
88-17-5	UN 2942 (DOT)
4795-29-3	UN 2943 (DOT)
371-40-4	UN 2944 (DOT)
140-80-7	UN 2946 (DOT)
105-48-6	UN 2947 (DOT)
98-16-8	UN 2948 (DOT)
7439-95-4	UN 2950 (DOT)
78-67-1	UN 2952 (DOT)
1941-79-3	UN 2958 (DOT)
16111-62-9	UN 2960 (DOT)
123-23-9	UN 2962 (DOT)
26748-47-0	UN 2963 (DOT)
353-42-4	UN 2965 (DOT)
60-24-2	UN 2966 (DOT)
5329-14-6	UN 2967 (DOT)
7440-29-1	UN 2975 (DOT)
13823-29-5	UN 2976 (DOT)
1344-40-7	UN 2989 (DOT)
123-30-8	UNAL
27598-85-2	UNAL
1120-21-4	UNDECANE
1120-21-4	UNDECANE (DOT)
6742-54-7	UNDECANE, 1-PHENYL-
30207-98-8	UNDECANOL
30207-98-8	UNDECYL ALCOHOL
6742-54-7	UNDECYLBENZENE
114-26-1	UNDEN
114-26-1	UNDEN (PESTICIDE)
53-16-7	UNDEN (PHARMACEUTICAL)
12789-03-6	UNEXAN-KOEDER
101-25-7	UNICEL 100
101-25-7	UNICEL ND
101-25-7	UNICEL NDX
88-85-7	UNICROP DNBP
71-63-6	UNIDIGIN
330-54-1	UNIDRON
103-23-1	UNIFLEX DOA
123-79-5	UNIFLEX DOA

CAS No.	Chemical Name
62-73-7	UNIFOS
2636-26-2	UNIFOS
62-73-7	UNIFOS (PESTICIDE)
2636-26-2	UNIFOS 50 EC
106-93-4	UNIFUME
139-91-3	UNIFUR
85-68-7	UNIMOLL BB
84-66-2	UNIMOLL DA
84-74-2	UNIMOLL DB
131-11-3	UNIMOLL DM
56-75-7	UNIMYCETIN
1937-37-7	UNION BLACK EM
64-00-6	UNION CARBIDE 10854
116-06-3	UNION CARBIDE 21149
63-25-2	UNION CARBIDE 7744
75-94-5	UNION CARBIDE A-150
15271-41-7	UNION CARBIDE UC 20047
64-00-6	UNION CARBIDE UC-10854
116-06-3	UNION CARBIDE UC-21149
2631-37-0	UNION CARBIDE UC-9880
8002-09-3	UNIPINE
75-99-0	UNIPON
117-80-6	UNIROYAL
2312-35-8	UNIROYAL D014
7772-99-8	UNISTON CR-HT 200
13463-67-7	UNITANE
13463-67-7	UNITANE OR 450
13463-67-7	UNITANE OR 572
13463-67-7	UNITANE OR 650
1333-86-4	UNITED 3017
7775-09-9	UNITED CHEMICAL DEFOLIANT NO 1
1333-86-4	UNITED SL 90
62-73-7	UNITOX
470-90-6	UNITOX
56-23-5	UNIVERM
9002-84-0	UNON P
9002-84-0	UNON P 300
57-14-7	UNS-DIMETHYLHYDRAZINE
57-14-7	UNSYM-DIMETHYLHYDRAZINE
120-82-1	UNSYM-TRICHLOROBENZENE
57-14-7	UNSYMMETRICAL DIMETHYLHYDRAZINE
101-96-2	UOP 5
3081-14-9	UOP 788
66-75-1	URACIL MUSTARD
66-75-1	URACIL, 5-(BIS(2-CHLOROETHYL)AMINO)-
314-40-9	URACIL, 5-BROMO-3-SEC-BUTYL-6-METHYL-
51-21-8	URACIL, 5-FLUORO-
66-75-1	URACIL, 5-[BIS(2-CHLOROETHYL)AMINO]-
56-04-2	URACIL, 6-METHYL-2-THIO-
51-52-5	URACIL, 6-PROPYL-2-THIO-
66-75-1	URACILLOST
100-97-0	URAMIN
10043-52-4	URAMINE MC
66-75-1	URAMUSTIN
66-75-1	URAMUSTINE
7440-61-1	URANIUM
541-09-3	URANIUM DIACETATE DIOXIDE
7783-81-5	URANIUM HEXAFLUORIDE
7440-61-1	URANIUM I (238U)
15905-86-9	URANIUM NITRATE
10102-06-4	URANIUM NITRATE OXIDE (UO2(NO3)2)
541-09-3	URANIUM OXYACETATE
10026-10-5	URANIUM TETRACHLORIDE
10049-14-6	URANIUM TETRAFLUORIDE
10025-93-1	URANIUM TRICHLORIDE
1344-58-7	URANIUM TRIOXIDE
7440-61-1	URANIUM(NATURAL)
541-09-3	URANIUM, BIS(ACETATO)DIOXO-
541-09-3	URANIUM, BIS(ACETATO-O)DIOXO-
10102-06-4	URANIUM, BIS(NITRATO-O)DIOXO-, (T-4)-
36478-76-9	URANIUM, BIS(NITRATO-O,O')DIOXO-, (OC-6-11)-
10102-06-4	URANIUM, DINITRATODIOXO-
7440-61-1	URANIUM-238
541-09-3	URANYL ACETATE
541-09-3	URANYL ACETATE (UO2(OAC)2)
7791-26-6	URANYL CHLORIDE
541-09-3	URANYL DIACETATE
10102-06-4	URANYL DINITRATE
1344-57-6	URANYL DIOXIDE
10102-06-4	URANYL NITRATE
36478-76-9	URANYL NITRATE
10102-06-4	URANYL NITRATE (UO2(NO3)2)
13520-83-7	URANYL NITRATE HEXAHYDRATE
19525-15-6	URANYL PEROXIDE
18433-48-2	URANYL PHOSPHATE
1314-64-3	URANYL SULFATE
541-09-3	URANYL(2+) ACETATE
100-97-0	URATRINE
57-13-6	UREA
124-43-6	UREA DIOXIDE
124-43-6	UREA HYDROGEN PEROXIDE
124-43-6	UREA HYDROGEN PEROXIDE (DOT)
124-43-6	UREA HYDROGEN PEROXIDE SALT
124-43-6	UREA HYDROPEROXIDE
124-47-0	UREA NITRATE
124-47-0	UREA NITRATE (1:1)
124-47-0	UREA NITRATE (WET)
124-47-0	UREA NITRATE, WET WITH 10% OR MORE WATER (DOT)
124-47-0	UREA NITRATE, WET WITH 10% OR MORE WATER, OVER 25 LBS IN ONE OUTSIDE PACKAGING (DOT)
124-47-0	UREA NITRATE, WETTED WITH NOT LESS THAN 20% WATER (DOT)
124-43-6	UREA PEROXIDE
124-43-6	UREA PEROXIDE (DOT)
86-88-4	UREA, 1-(1-NAPHTHYL)-2-THIO-
30553-04-9	UREA, 1-(1-NAPHTHYL)-2-THIO-
13010-47-4	UREA, 1-(2-CHLOROETHYL)-3-CYCLOHEXYL-1-NITROSO-
30553-04-9	UREA, 1-(NAPHTHYL)-2-THIO-
5344-82-1	UREA, 1-(O-CHLOROPHENYL)-2-THIO-
759-73-9	UREA, 1-ETHYL-1-NITROSO-
684-93-5	UREA, 1-METHYL-1-NITROSO-
53558-25-1	UREA, 1-NITROPHENYL-3-(3-PYRIDYLMETHYL)-
103-85-5	UREA, 1-PHENYL-2-THIO-
62-56-6	UREA, 2-THIO-
541-53-7	UREA, 2-THIO-1-(THIOCARBAMOYL)-
614-78-8	UREA, 2-THIO-1-O-TOLYL-
330-54-1	UREA, 3-(3,4-DICHLOROPHENYL)-1,1-DIMETHYL-
1982-47-4	UREA, 3-(P-(P-CHLOROPHENOXY)PHENYL)-1,1-DIMETHYL-
124-43-6	UREA, COMPD WITH H2O2
124-43-6	UREA, COMPD WITH HYDROGEN PEROXIDE (1:1)
124-43-6	UREA, COMPD WITH HYDROGEN PEROXIDE (H2O2) (1:1)
124-47-0	UREA, MONONITRATE
124-47-0	UREA, MONONITRATE (8CI,9CI)
330-54-1	UREA, N'-(3,4-DICHLOROPHENYL)-N,N-DIMETHYL-
1982-47-4	UREA, N'-[4-(4-CHLOROPHENOXY)PHENYL]-N,N-DIMETHYL-
13010-47-4	UREA, N-(2-CHLOROETHYL)-N'-CYCLOHEXYL-N-NITROSO-
53558-25-1	UREA, N-(4-NITROPHENYL)-N'-(3-PYRIDINYLMETHYL)-
759-73-9	UREA, N-ETHYL-N-NITROSO-
684-93-5	UREA, N-METHYL-N-NITROSO-
124-47-0	UREA, NITRATE
64-10-8	UREA, PHENYL-
62-56-6	UREA, THIO-
62-56-6	UREA, THIO- (8CI)
57-13-6	UREAPHIL
57-13-6	UREOPHIL
57-13-6	UREPEARL
51-79-6	URETHAN
51-79-6	URETHANE
57-13-6	UREVERT
136-40-3	URIDINAL
100-97-0	URITONE
131-73-7	URITONE
79-43-6	URNER'S LIQUID
100-97-0	URODEINE
136-40-3	URODINE
100-97-0	UROTROPIN
100-97-0	UROTROPINE
330-54-1	UROX D
106-50-3	URSOL D
95-83-0	URSOL OLIVE 6G
123-30-8	URSOL P
27598-85-2	URSOL P
123-30-8	URSOL P BASE
27598-85-2	URSOL P BASE
117-80-6	US RUBBER 604
2312-35-8	US RUBBER D-014
3564-09-8	USACERT RED NO. 1
107-16-4	USAF A-8565
542-76-7	USAF A-8798

CAS No.	Chemical Name
107-29-9	USAF AM-5
110-69-0	USAF AM-6
137-26-8	USAF B-30
541-53-7	USAF B-44
91-59-8	USAF CB-22
68-11-1	USAF CB-35
1405-87-4	USAF CB-7
156-62-7	USAF CY-2
105-57-7	USAF DO-45
140-31-8	USAF DO-46
109-83-1	USAF DO-50
105-59-9	USAF DO-52
79-19-6	USAF EK-1275
103-85-5	USAF EK-1569
110-02-1	USAF EK-1860
137-26-8	USAF EK-2089
91-22-5	USAF EK-218
4384-82-1	USAF EK-2596
123-31-9	USAF EK-356
60-24-2	USAF EK-4196
142-04-1	USAF EK-442
75-05-8	USAF EK-488
98-86-2	USAF EK-496
62-56-6	USAF EK-497
2231-57-4	USAF EK-7372
1762-95-4	USAF EK-P-433
110-17-8	USAF EK-P-583
30553-04-9	USAF EK-P-5976
541-53-7	USAF EK-P-6281
507-09-5	USAF EK-P-737
540-72-7	USAF EK-T-434
372-09-8	USAF KF-17
140-29-4	USAF KF-21
105-56-6	USAF KF-25
107-14-2	USAF KF-5
98-08-8	USAF MA-16
98-16-8	USAF MA-4
106-51-4	USAF P-220
137-26-8	USAF P-5
330-54-1	USAF P-7
617-89-0	USAF Q-1
4795-29-3	USAF Q-2
2867-47-2	USAF RH-3
27193-28-8	USAF RH-6
109-78-4	USAF RH-7
75-86-5	USAF RH-8
126-98-7	USAF ST-40
108-98-5	USAF XR-19
330-54-1	USAF XR-42
3006-86-8	USP 400
2618-77-1	USP 711
117-80-6	USR 604
110-85-0	UVILON
1332-58-7	UW 90
17804-35-2	UZGEN
7440-22-4	V 9
7775-14-6	V-BRITE
7775-14-6	V-BRITE B
13194-48-4	V-C 9-104
13194-48-4	V-C CHEMICAL V-C 9-104
8032-32-4	V.M. AND P. NAPHTHA
8032-32-4	V.M.&P. NAPHTHA
7440-33-7	VA
7440-33-7	VA (TUNGSTEN)
108-05-4	VAC
53558-25-1	VACOR
443-48-1	VAGIMID
7775-09-9	VAL-DROP
78-81-9	VALAMINE
1309-64-4	VALENTINITE
110-62-3	VALERAL
110-62-3	VALERALDEHYDE
110-62-3	VALERALDEHYDE (DOT)
32749-94-3	VALERALDEHYDE, 2,3-DIMETHYL-
123-15-9	VALERALDEHYDE, 2-METHYL-
109-52-4	VALERIANIC ACID
110-62-3	VALERIANIC ALDEHYDE
109-52-4	VALERIC ACID
109-52-4	VALERIC ACID (DOT)

CAS No.	Chemical Name
110-62-3	VALERIC ACID ALDEHYDE
110-62-3	VALERIC ALDEHYDE
108-83-8	VALERONE
638-29-9	VALEROYL CHLORIDE
638-29-9	VALERYL CHLORIDE
638-29-9	VALERYL CHLORIDE (DOT)
110-62-3	VALERYLALDEHYDE
9002-84-0	VALFLON
9002-84-0	VALFLON 7000
9002-84-0	VALFLON 7790
9002-84-0	VALFLON 7990
2001-95-8	VALINOMICIN
2001-95-8	VALINOMYCIN
139-91-3	VALSYN
8020-83-5	VALVATA 85
81-81-2	VAMPIRINIP II
81-81-2	VAMPIRINIP III
60-41-3	VAMPIROL
7803-55-6	VANADATE (VO3 1-), AMMONIUM
13769-43-2	VANADATE (VO3 1-), POTASSIUM
13721-39-4	VANADATE (VO3 1-), SODIUM
11105-06-9	VANADATE (VO3 1-), SODIUM
13718-26-8	VANADATE (VO3 1-), SODIUM
13721-39-6	VANADATE (VO4 3-), TRISODIUM, (T-4)-
13721-39-4	VANADIC ACID (H3VO4), TRISODIUM SALT
11105-06-9	VANADIC ACID (H3VO4), TRISODIUM SALT
7803-55-6	VANADIC ACID (HVO3), AMMONIUM SALT
13769-43-2	VANADIC ACID (HVO3), POTASSIUM SALT
11105-06-9	VANADIC ACID (HVO3), SODIUM SALT
13718-26-8	VANADIC ACID (HVO3), SODIUM SALT
7803-55-6	VANADIC ACID, AMMONIUM SALT
13718-26-8	VANADIC ACID, MONOSODIUM SALT
11105-06-9	VANADIC ACID, MONOSODIUM SALT
13769-43-2	VANADIC ACID, POTASSIUM SALT
13718-26-8	VANADIC ACID, SODIUM SALT
11105-06-9	VANADIC ACID, SODIUM SALT
13721-39-6	VANADIC ACID, SODIUM SALT
1314-62-1	VANADIC ANHYDRIDE
1314-34-7	VANADIC OXIDE
13721-39-6	VANADIC(II) ACID, TRISODIUM SALT
1314-62-1	VANADIO, PENTOSSIDO DI (ITALIAN)
7440-62-2	VANADIUM
7440-62-2	VANADIUM (FUME OR DUST)
12604-58-9	VANADIUM ALLOY, BASE, V,C,FE (FERROVANADIUM)
11130-21-5	VANADIUM CARBIDE
7632-51-1	VANADIUM CHLORIDE
7718-98-1	VANADIUM CHLORIDE (VCl3)
7632-51-1	VANADIUM CHLORIDE (VCl4)
7632-51-1	VANADIUM CHLORIDE (VCl4), (T-4)-
7727-18-6	VANADIUM CHLORIDE OXIDE (VCl3O)
1314-62-1	VANADIUM DUST AND FUME (ACGIH)
7727-18-6	VANADIUM MONOXIDE TRICHLORIDE
1314-34-7	VANADIUM OXIDE
1314-34-7	VANADIUM OXIDE (V2O3)
1314-62-1	VANADIUM OXIDE (V2O5)
27774-13-6	VANADIUM OXIDE SULFATE (VO(SO4))
7727-18-6	VANADIUM OXIDE TRICHLORIDE
27774-13-6	VANADIUM OXOSULFATE
7727-18-6	VANADIUM OXYCHLORIDE
27774-13-6	VANADIUM OXYSULFATE (VOSO4)
7727-18-6	VANADIUM OXYTRICHLORIDE
7727-18-6	VANADIUM OXYTRICHLORIDE (DOT)
1314-62-1	VANADIUM PENTAOXIDE
1314-62-1	VANADIUM PENTOXIDE
1314-62-1	VANADIUM PENTOXIDE (DOT)
1314-62-1	VANADIUM PENTOXIDE, DUST AND FUME
1314-62-1	VANADIUM PENTOXIDE, NON-FUSED FORM (DOT)
1314-34-7	VANADIUM SESQUIOXIDE
7632-51-1	VANADIUM TETRACHLORIDE
7632-51-1	VANADIUM TETRACHLORIDE (DOT)
7718-98-1	VANADIUM TRICHLORIDE
7718-98-1	VANADIUM TRICHLORIDE (DOT)
7727-18-6	VANADIUM TRICHLORIDE MONOOXIDE
7727-18-6	VANADIUM TRICHLORIDE MONOXIDE
7727-18-6	VANADIUM TRICHLORIDE OXIDE
1314-34-7	VANADIUM TRIOXIDE
1314-34-7	VANADIUM TRIOXIDE, NON-FUSED FORM (DOT)
7718-98-1	VANADIUM(3+) CHLORIDE
1314-34-7	VANADIUM(3+) OXIDE

CAS No.	Chemical Name
7718-98-1	VANADIUM(III) CHLORIDE
7632-51-1	VANADIUM(IV) CHLORIDE
7727-18-6	VANADIUM(V) OXYTRICHLORIDE
27774-13-6	VANADIUM, OXOSULFATO-
27774-13-6	VANADIUM, OXO[SULFATO(2-)-O]-
27774-13-6	VANADIUM, OXYSULFATO-
1314-62-1	VANADIUM, PENTOXYDE DE (FRENCH)
7727-18-6	VANADIUM, TRICHLOROOXO-
7440-62-2	VANADIUM-51
1314-62-1	VANADIUMPENTOXID (GERMAN)
1314-62-1	VANADIUMPENTOXYDE (DUTCH)
7727-18-6	VANADYL CHLORIDE
7727-18-6	VANADYL CHLORIDE (VOCl3)
27774-13-6	VANADYL SULFATE
27774-13-6	VANADYL SULFATE (DOT)
27774-13-6	VANADYL SULFATE (VO(SO4))
7727-18-6	VANADYL TRICHLORIDE
137-26-8	VANCIDA TM-95
133-06-2	VANCIDE 89
133-06-2	VANCIDE 89RE
133-06-2	VANCIDE P-75
137-26-8	VANCIDE TM
1314-13-2	VANDEM VAC
1314-13-2	VANDEM VOC
1314-13-2	VANDEM VPC
133-06-2	VANGARD K
133-06-2	VANGUARD K
133-06-2	VANICIDE
57-11-4	VANICOL
128-37-0	VANLUBE PC
128-37-0	VANLUBE PCX
94-36-0	VANOXIDE
62-73-7	VAPONA
2636-26-2	VAPONA
62-73-7	VAPONA INSECTICIDE
62-73-7	VAPONITE
2636-26-2	VAPONITE
56-38-2	VAPOPHOS
8020-83-5	VAPOR 52
2636-26-2	VAPORA II
107-49-3	VAPOTONE
129-06-6	VARFINE
6484-52-2	VARIOFORM I
57-13-6	VARIOFORM II
112-02-7	VARIQUAT E 228
76-03-9	VARITOX
8032-32-4	VARNISH MAKERS' AND PAINTERS' NAPHTHA
8032-32-4	VARNISH MAKERS' NAPHTHA
8032-32-4	VARNISH MAKERS' NAPHTHA AND PAINTERS' NAPHTHA
8052-41-3	VARNOLINE
78-63-7	VAROX
6731-36-8	VAROX 231XL
78-63-7	VAROX 50
80-43-3	VAROX DCP-R
80-43-3	VAROX DCP-T
8052-41-3	VARSOL
55-63-0	VASOGLYN
51-83-2	VASOPERIF
7775-14-6	VATROLITE
577-11-7	VATSOL OT
78-67-1	VAZO
78-67-1	VAZO 64
75-01-4	VC
75-01-4	VCM
107-13-1	VCN
21609-90-5	VCS
21609-90-5	VCS 506
75-35-4	VDC
75-38-7	VDF
8002-74-2	VEBAFINE FT 300
8002-74-2	VEBAWAX SH 105
1912-24-9	VECTAL
1912-24-9	VECTAL SC
95-06-7	VEGADEX
95-06-7	VEGADEX SUPER
298-02-2	VEGFRU
563-12-2	VEGFRU FOSMITE
121-75-5	VEGFRU MALATOX
2540-82-1	VEL 4284

CAS No.	Chemical Name
577-11-7	VELMOL
8020-83-5	VELOSIT
8020-83-5	VELOSITE
76-44-8	VELSICOL 104
12789-03-6	VELSICOL 1068
21609-90-5	VELSICOL 506
1024-57-3	VELSICOL 53CS17
1918-00-9	VELSICOL 58-CS-11
1918-00-9	VELSICOL COMPOUND R
76-44-8	VELSICOL HEPTACHLOR
21609-90-5	VELSICOL VCS 506
640-15-3	VELTIN
1309-37-1	VENETIAN RED
107-13-1	VENTOX
93-76-5	VEON
93-76-5	VEON 245
1420-07-1	VERALINE CREME
7733-02-0	VERAZINC
514-73-8	VERCIDON
62-38-4	VERDASAN
2636-26-2	VERDICAN
2636-26-2	VERDIPOR
94-75-7	VERGEMASTER
298-02-2	VERGFRU FORATOX
58-89-9	VERINDAL ULTRA
110-85-0	VERMEX
52-68-6	VERMICIDE BAYER 2349
1318-00-9	VERMICULITE
1344-48-5	VERMILION
56-23-5	VERMOESTRICID
60-00-4	VERSENE
60-00-4	VERSENE ACID
139-13-9	VERSENE NTA ACID
121-69-7	VERSNELLER NL 63/10
8001-35-2	VERTAC 90%
88-85-7	VERTAC DINITRO WEED KILLER
88-85-7	VERTAC GENERAL WEED KILLER
298-00-0	VERTAC METHYL PARATHION TECHNISCH 80%
88-85-7	VERTAC SELECTIVE WEED KILLER
8001-35-2	VERTAC TOXAPHENE 90
8004-09-9	VERTAFUME
122-14-5	VERTHION
7631-86-9	VERTICURINE
94-75-7	VERTON
94-75-7	VERTON 2D
93-76-5	VERTON 2T
94-75-7	VERTON D
94-75-7	VERTRON 2D
117-84-0	VESTINOL AH
117-81-7	VESTINOL AH
26761-40-0	VESTINOL DZ
103-23-1	VESTINOL OA
123-79-5	VESTINOL OA
79-01-6	VESTROL
121-75-5	VETIOL
64-10-8	VH
7429-90-5	VI 5
10108-64-2	VI-CAD
128-37-0	VIANOL
9002-89-5	VIBATEX S
7757-79-1	VICKNITE
7722-88-5	VICTOR TSPP
8002-74-2	VICTORY
8003-19-8	VIDDEN D
105-60-2	VIDLON
94-75-7	VIDON 638
12002-03-8	VIENNA GREEN
2699-79-8	VIKANE
2699-79-8	VIKANE FUMIGANT
7681-49-4	VILLIAUMITE
9002-89-5	VINACOL MH
9002-89-5	VINALAK
109-92-2	VINAMAR
9002-89-5	VINAROL
9002-89-5	VINAROL DT
9002-89-5	VINAROL ST
9002-89-5	VINAROL SVH
9002-89-5	VINAROLE
9002-89-5	VINAVILOL 2-98

CAS No.	Chemical Name
64-19-7	VINEGAR ACID
141-78-6	VINEGAR NAPHTHA
109-93-3	VINESTHENE
109-93-3	VINESTHESIN
109-93-3	VINETHEN
109-93-3	VINETHENE
109-93-3	VINETHER
117-84-0	VINICIZER 80
117-81-7	VINICIZER 80
117-84-0	VINICIZER 85
109-93-3	VINIDYL
108-05-4	VINILE (ACETATO DI) (ITALIAN)
593-60-2	VINILE (BROMURO DI) (ITALIAN)
75-01-4	VINILE (CLORURO DI) (ITALIAN)
60-00-4	VINKEIL 100
9002-89-5	VINNAROL
109-53-5	VINOFLEX MO 400
9002-89-5	VINOL
9002-89-5	VINOL 107
9002-89-5	VINOL 125
9002-89-5	VINOL 165
9002-89-5	VINOL 205
9002-89-5	VINOL 205S
9002-89-5	VINOL 325
9002-89-5	VINOL 350
9002-89-5	VINOL 351
9002-89-5	VINOL 523
9002-89-5	VINOL UNISIZE
9002-89-5	VINOL WS 42
109-93-3	VINYDAN
14861-06-4	VINYL 2-BUTENOATE
94-04-2	VINYL 2-ETHYLHEXANOATE
108-05-4	VINYL A MONOMER
108-05-4	VINYL ACETATE
108-05-4	VINYL ACETATE (ACGIH,DOT)
108-05-4	VINYL ACETATE HQ
108-05-4	VINYL ACETATE MONOMER
108-05-4	VINYL ACETATE, INHIBITED (DOT)
689-97-4	VINYL ACETYLENE
9002-89-5	VINYL ALCOHOL POLYMER
2636-26-2	VINYL ALCOHOL, 2,2-DICHLORO-, DIMETHYL PHOSPHATE
3917-15-5	VINYL ALLYL ETHER
79-06-1	VINYL AMIDE
110-75-8	VINYL BETA-CHLOROETHYL ETHER
593-60-2	VINYL BROMIDE
593-60-2	VINYL BROMIDE (ACGIH)
593-60-2	VINYL BROMIDE, INHIBITED (DOT)
123-20-6	VINYL BUTANOATE
111-34-2	VINYL BUTYL ETHER
123-20-6	VINYL BUTYRATE
123-20-6	VINYL BUTYRATE, INHIBITED (DOT)
75-01-4	VINYL C MONOMER
75-01-4	VINYL CHLORIDE
75-01-4	VINYL CHLORIDE (ACGIH,DOT)
75-01-4	VINYL CHLORIDE MONOMER
2549-51-1	VINYL CHLOROACETATE
2549-51-1	VINYL CHLOROACETATE (DOT)
14861-06-4	VINYL CROTONATE
107-13-1	VINYL CYANIDE
930-22-3	VINYL EPOXIDE
108-05-4	VINYL ETHANOATE
109-93-3	VINYL ETHER
109-92-2	VINYL ETHYL ETHER
109-92-2	VINYL ETHYL ETHER, INHIBITED
109-92-2	VINYL ETHYL ETHER, INHIBITED (DOT)
75-02-5	VINYL FLUORIDE
75-02-5	VINYL FLUORIDE, INHIBITED
75-02-5	VINYL FLUORIDE, INHIBITED (DOT)
109-53-5	VINYL ISOBUTYL ETHER
109-53-5	VINYL ISOBUTYL ETHER (DOT)
109-53-5	VINYL ISOBUTYL ETHER, INHIBITED (DOT)
926-65-8	VINYL ISOPROPYL ETHER
107-25-5	VINYL METHYL ETHER
107-25-5	VINYL METHYL ETHER (DOT)
107-25-5	VINYL METHYL ETHER, INHIBITED (DOT)
78-94-4	VINYL METHYL KETONE
2549-51-1	VINYL MONOCHLOROACETATE
111-34-2	VINYL N-BUTYL ETHER
105-38-4	VINYL PROPANOATE
105-38-4	VINYL PROPIONATE
25013-15-4	VINYL TOLUENE
25013-15-4	VINYL TOLUENE (ACGIH)
25013-15-4	VINYL TOLUENES (MIXED ISOMERS), INHIBITED (DOT)
75-94-5	VINYL TRICHLOROSILANE
75-94-5	VINYL TRICHLOROSILANE (DOT)
75-94-5	VINYL TRICHLOROSILANE, INHIBITED (DOT)
110-75-8	VINYL-2-CHLOROETHYL ETHER
94-04-2	VINYL-2-ETHYLHEXOATE
103-44-6	VINYL-2-ETHYLHEXYL ETHER
1663-35-0	VINYL-2-METHOXYETHYL ETHER
108-05-4	VINYLACETAAT (DUTCH)
108-05-4	VINYLACETAT (GERMAN)
593-67-9	VINYLAMINE
4549-40-0	VINYLAMINE, N-METHYL-N-NITROSO-
100-42-5	VINYLBENZEN (CZECH)
100-42-5	VINYLBENZENE
100-42-5	VINYLBENZOL
30030-25-2	VINYLBENZYL CHLORIDE
593-60-2	VINYLBROMID (GERMAN)
107-18-6	VINYLCARBINOL
75-01-4	VINYLCHLORID (GERMAN)
106-87-6	VINYLCYCLOHEXENE DIEPOXIDE
106-87-6	VINYLCYCLOHEXENE DIOXIDE
108-05-4	VINYLE (ACETATE DE) (FRENCH)
593-60-2	VINYLE (BROMURE DE) (FRENCH)
75-01-4	VINYLE(CHLORURE DE) (FRENCH)
74-86-2	VINYLENE
25339-57-5	VINYLETHYLENE
106-99-0	VINYLETHYLENE
79-10-7	VINYLFORMIC ACID
75-35-4	VINYLIDENE CHLORIDE
75-35-4	VINYLIDENE CHLORIDE, INHIBITED
75-35-4	VINYLIDENE CHLORIDE, INHIBITED (DOT)
75-38-7	VINYLIDENE DIFLUORIDE
75-38-7	VINYLIDENE FLUORIDE
3048-64-4	VINYLNORBORNENE
62-73-7	VINYLOFOS
2636-26-2	VINYLOFOS
9002-89-5	VINYLON FILM 2000
9002-89-5	VINYLON FILM 3000
9002-89-5	VINYLON FILM VF-A 2500
2636-26-2	VINYLOPHOS
62-73-7	VINYLOPHOS
930-22-3	VINYLOXIRANE
470-90-6	VINYLPHATE
75-94-5	VINYLSILICON TRICHLORIDE
1321-74-0	VINYLSTYRENE
79-00-5	VINYLTRICHLORIDE
75-94-5	VINYLTRICHLOROSILANE
470-90-6	VINYPHATE
50-55-5	VIO-SERPINE
1330-20-7	VIOLET 3
1694-09-3	VIOLET 6B
1694-09-3	VIOLET NO. 1
1910-2-5	VIOLOGEN, METHYL-
63-25-2	VIOXAN
299-84-3	VIOZENE
7775-14-6	VIRCHEM
13194-48-4	VIRGINIA-CAROLINA VC 9-104
58-22-0	VIRORMONE
1397-94-0	VIROSIN
58-22-0	VIROSTERONE
7775-14-6	VIRTEX CC
7775-14-6	VIRTEX D
7775-14-6	VIRTEX L
7775-14-6	VIRTEX RD
93-76-5	VISKO RHAP LOW VOLATILE ESTER
94-75-7	VISKO-RHAP
94-75-7	VISKO-RHAP LOW DRIFT HERBICIDES
94-75-7	VISKO-RHAP LOW VOLATILE 4L
84-80-0	VITAMIN K1
84-80-0	VITAMIN K1(20)
57-83-0	VITARRINE
7631-86-9	VITASIL 1500
7631-86-9	VITASIL 1600
7631-86-9	VITASIL 220
58-89-9	VITON
79-01-6	VITRAN

CAS No.	Chemical Name
12627-52-0	VITREOUS ANTIMONY
56-38-2	VITREX
7664-93-9	VITRIOL BROWN OIL
7664-93-9	VITRIOL, OIL OF (DOT)
8020-83-5	VM 4
8030-30-6	VM&P NAPHTHA
504-24-5	VMI 10-3
298-00-0	VOFATOX
57-06-7	VOLATILE MUSTARD OIL
57-06-7	VOLATILE OIL OF MUSTARD
52-68-6	VOLFARTOL
62-38-4	VOLPAR
1937-37-7	VONDACEL BLACK N
2602-46-2	VONDACEL BLUE 2B
133-06-2	VONDCAPTAN
330-54-1	VONDURON
112-80-1	VOPCOLENE 27
25322-69-4	VORANOL P 1010
25322-69-4	VORANOL P 2000
25322-69-4	VORANOL P 4000
556-61-6	VORLEX
3878-19-1	VORONIT
3878-19-1	VORONITE
556-61-6	VORTEX
52-68-6	VOTEXIT
2167-23-9	VP 1200
9002-89-5	VPB 105-2
7440-32-6	VT 1
7440-32-6	VT 1-0
7440-32-6	VT 1-1
7440-32-6	VT 1-2
8001-22-7	VT 18
8001-22-7	VT 18 (OIL)
7440-32-6	VT 1D
7440-32-6	VT 1L
75-94-5	VTCS
7440-32-6	VTL 0
298-04-4	VUAGT 1-4
115-90-2	VUAGT 108
298-02-2	VUAGT 182
298-04-4	VUAGT 1964
115-90-2	VUAGT 96
137-26-8	VUAGT-I-4
101-25-7	VULCACEL B 40
101-25-7	VULCACEL BN
101-25-7	VULCACEL BN 94
137-26-8	VULCAFOR TMT
137-26-8	VULCAFOR TMTD
1333-86-4	VULCAN
1333-86-4	VULCAN C
1333-86-4	VULCAN XC 72R
86-30-6	VULCATARD A
100-97-0	VULKACIT H 30
5459-93-8	VULKACIT HX
137-26-8	VULKACIT MTIC
96-45-7	VULKACIT NPV/C
137-26-8	VULKACIT TH
137-26-8	VULKACIT THIURAM
137-26-8	VULKACIT THIURAM/C
86-30-6	VULKALENT A
85-44-9	VULKALENT B/C
128-37-0	VULKANOX KB
135-88-6	VULKANOX PBN
7631-86-9	VULKASIL C
7631-86-9	VULKASIL S
1863-63-4	VULNOC AB
8002-74-2	VULTEX 8
86-30-6	VULTROL
7782-42-5	VVP 66-95
108-05-4	VYAC
23135-22-0	VYDATE
23135-22-0	VYDATE L
23135-22-0	VYDATE L INSECTICIDE/NEMATICIDE
23135-22-0	VYDATE L OXAMYL INSECTICIDE/NEMATOCIDE
23135-22-0	VYDATE-G
14464-46-1	W 006
9005-25-8	W 13 STABILIZER
136-40-3	W 1655
25322-69-4	W 166
535-89-7	W 491
3878-19-1	W VII/117
9005-25-8	W-GUM
115-26-4	WACKER S 14/10
1314-62-1	WANADU PIECIOTLENEK (POLISH)
7440-31-5	WANG
1305-78-8	WAPNIOWY TLENEK (POLISH)
151-21-3	WAQE
129-06-6	WARAN
9002-89-5	WARCOPOLYMER A 20
129-06-6	WARCOUMIN
577-11-7	WARCOWET 060
96-45-7	WARECURE C
81-81-2	WARF 42
81-81-2	WARF COMPOUND 42
81-81-2	WARFARAT
81-81-2	WARFARIN
81-81-2	WARFARIN (ACGIH)
81-81-2	WARFARIN PLUS
81-81-2	WARFARIN Q
129-06-6	WARFARIN SODIUM
129-06-6	WARFARIN SODIUM SALT
129-06-6	WARFARIN, SODIUM DERIV
81-81-2	WARFARINE (FRENCH)
81-81-2	WARFICIDE
129-06-6	WARFILONE
60-00-4	WARKEELATE ACID
8001-58-9	WASH OIL
7722-84-1	WASSERSTOFFPEROXID (GERMAN)
8021-92-9	WATER GAS
6834-92-0	WATER GLASS
7722-84-1	WATERSTOFPEROXYDE (DUTCH)
8002-74-2	WAXES AND WAXY SUBSTANCES, HYDROCARBON
60-11-7	WAXOLINE YELLOW ADS
577-11-7	WAXSOL
7681-38-1	WC 00
7681-38-1	WC-KLOSETTREINIGER
7681-38-1	WC-PERFECT
7664-38-2	WC-REINIGER
7681-38-1	WC-SUPER
52-68-6	WEC 50
112-80-1	WECOLINE OO
2163-80-6	WEED 108
107-18-6	WEED DRENCH
94-75-7	WEED TOX
94-75-7	WEED-AG-BAR
94-75-7	WEED-B-GON
2163-80-6	WEED-E-RAD
2163-80-6	WEED-HOE
94-75-7	WEED-RHAP
93-76-5	WEEDAR
94-75-7	WEEDAR
94-75-7	WEEDAR-64
94-75-7	WEEDATUL
61-82-5	WEEDAZOL
1762-95-4	WEEDAZOL TL
131-52-2	WEEDBEADS
1912-24-9	WEEDEX A
94-75-7	WEEDEZ WONDER BAR
1910-2-5	WEEDOL
87-86-5	WEEDONE
94-75-7	WEEDONE
93-76-5	WEEDONE
94-11-1	WEEDONE 128
93-76-5	WEEDONE 2,4,5-T
1929-73-3	WEEDONE LV 4
94-75-7	WEEDONE LV4
85-00-7	WEEDTRINE-D
94-75-7	WEEDTROL
75-15-0	WEEVILTOX
75-15-0	WEGLA DWUSIARCZEK (POLISH)
630-08-0	WEGLA TLENEK (POLISH)
7723-14-0	WEISS PHOSPHOR (GERMAN)
1309-64-4	WEISSPIESSGLANZ (GERMAN)
1031-47-6	WEPSIN
1031-47-6	WEPSYN
1031-47-6	WEPSYN 155
7631-86-9	WESSALON S
79-01-6	WESTROSOL

CAS No.	Chemical Name
577-11-7	WETAID SR
7704-34-9	WETSULF
7704-34-9	WETTASUL
9004-34-6	WHATMAN CC-31
9002-84-0	WHITCON 5TFE
9002-84-0	WHITCON 7
1327-53-3	WHITE ARSENIC
12001-29-5	WHITE ASBESTOS
8008-51-3	WHITE CAMPHOR OIL
7631-86-9	WHITE CARBON
1310-73-2	WHITE CAUSTIC
7733-02-0	WHITE COPPERAS
10124-48-8	WHITE MERCURIC PRECIPITATE
10124-48-8	WHITE MERCURY PRECIPITATED
8020-83-5	WHITE MINERAL OIL
8008-51-3	WHITE OIL OF CAMPHOR
7723-14-0	WHITE PHOSPHORUS
7723-14-0	WHITE PHOSPHORUS, DRY (DOT)
7723-14-0	WHITE PHOSPHORUS, WET (DOT)
10124-48-8	WHITE PRECIPITATE
1314-13-2	WHITE SEAL 7
80052-41-3	WHITE SPIRITS
57-50-1	WHITE SUGAR
91-20-3	WHITE TAR
7733-02-0	WHITE VITRIOL
1314-13-2	WHITE ZINC
1332-58-7	WHITETEX 2
123-79-5	WICKENOL 158
103-23-1	WICKENOL 158
105-60-2	WIDLON
7790-99-0	WIJS' CHLORIDE
128-44-9	WILLOSETTEN
534-52-1	WINTERWASH
1335-85-9	WINTERWASH
62-73-7	WINYLOPHOS
75-01-4	WINYLU CHLOREK (POLISH)
327-98-0	WIRKSTOFF 37289
103-23-1	WITAMOL 320
123-79-5	WITAMOL 320
84-74-2	WITCIZER 300
117-81-7	WITCIZER 312
117-84-0	WITCIZER 312
1333-86-4	WITCOBLAK NO. 100
25155-30-0	WITCONATE 1238
8052-42-4	WITCURB 22L
1338-02-9	WITTOX C
297-78-9	WL 1650
16752-77-5	WL 18236
556-61-6	WN 12
68476-26-6	WOBBE INDEX
112-80-1	WOCHEM NO 320
298-00-0	WOFATOX
7440-33-7	WOLFRAM
1912-24-9	WONUK
67-56-1	WOOD ALCOHOL
67-56-1	WOOD ALCOHOL (DOT)
115-10-6	WOOD ETHER
67-56-1	WOOD NAPHTHA
67-56-1	WOOD SPIRIT
87-86-5	WOODTREAT A
8006-54-0	WOOL FAT, PURIFIED
1694-09-3	WOOL VIOLET 4BN
1694-09-3	WOOL VIOLET 5BN
110-85-0	WORM-A-TON
8002-74-2	WOSTEN 65
52-68-6	WOTEXIT
1031-47-6	WP 155
2971-90-6	WR 61112
999-81-5	WR 62
88-85-7	WSX-8365
12002-03-8	WUERZBERG GREEN
110-85-0	WURMIRAZIN
53-16-7	WYNESTRON
1333-86-4	X 1303
1333-86-4	X 1341
1333-86-4	X 1341 (CARBON BLACK)
25155-30-0	X 2073
7779-88-6	X 4
7779-88-6	X 4 (NITRATE)

CAS No.	Chemical Name
8002-74-2	X 7905
100-97-0	XAMETRIN
50-78-2	XAXA
1333-86-4	XC 3017L
1333-86-4	XC 72
92-52-4	XENENE
7440-63-3	XENON
7440-63-3	XENON (DOT)
7440-63-3	XENON ATOM
7440-63-3	XENON, LIQUID
7440-63-3	XENON, REFRIGERATED LIQUID (DOT)
92-67-1	XENYLAMINE
94-36-0	XERAC
1300-71-6	XILENOLI (ITALIAN)
1300-73-8	XILIDINE (ITALIAN)
1330-20-7	XILOLI (ITALIAN)
8002-74-2	XL 165
54-11-5	XL ALL INSECTICIDE
1314-13-2	XX 203
1314-13-2	XX 601
1314-13-2	XX 78
1330-20-7	XYLENE
1330-20-7	XYLENE (ACGIH,DOT)
1330-20-7	XYLENE (MIXED ISOMERS)
81-15-2	XYLENE MUSK
81-15-2	XYLENE MUSK
827-21-4	XYLENE SULFONIC ACID, SODIUM SALT
28347-13-9	XYLENE, ALPHA,ALPHA'-DICHLORO-
25168-04-1	XYLENE, AR-NITRO-
35884-77-6	XYLENE, BROMO-
25168-04-1	XYLENE, NITRO-
1330-20-7	XYLENEN (DUTCH)
1330-20-7	XYLENES
1300-71-6	XYLENOL
1300-71-6	XYLENOL (DOT)
1300-71-6	XYLENOLEN (DUTCH)
1300-73-8	XYLIDINE
1300-73-8	XYLIDINE (ACGIH,DOT)
3761-53-3	XYLIDINE PONCEAU
3761-53-3	XYLIDINE PONCEAU 2R
3761-53-3	XYLIDINE RED
9004-70-0	XYLOIDIN
1330-20-7	XYLOL
1330-20-7	XYLOL (DOT)
1330-20-7	XYLOLE (GERMAN)
28258-59-5	XYLYL BROMIDE
35884-77-6	XYLYL BROMIDE
35884-77-6	XYLYL BROMIDE (DOT)
28347-13-9	XYLYLENE CHLORIDE
28347-13-9	XYLYLENE DICHLORIDE
3037-72-7	Y 1902
1563-66-2	YALTOX
640-19-7	YANOCK
8002-09-3	YARMOR
8002-09-3	YARMOR PINE OIL
62-74-8	YASOKNOCK
50-78-2	YASTA
409-21-2	YE 5626
505-60-2	YELLOW CROSS LIQUID
60-11-7	YELLOW G SOLUBLE IN GREASE
1317-36-8	YELLOW LEAD OCHER
15385-57-6	YELLOW MERCURY IODIDE
7783-30-4	YELLOW MERCURY IODIDE
7723-14-0	YELLOW PHOSPHORUS
7723-14-0	YELLOW PHOSPHORUS, DRY (DOT)
7723-14-0	YELLOW PHOSPHORUS, WET (DOT)
13765-19-0	YELLOW ULTRAMARINE
1309-37-1	YLO 2288B
96-69-5	YOSHINOX S
96-69-5	YOSHINOX SR
505-60-2	YPERITE
7440-65-5	YTTRIUM
7440-65-5	YTTRIUM-89
8020-83-5	Z 26
140-89-6	Z 3
140-89-6	Z 3 (PESTICIDE)
928-96-1	Z-3-HEXENOL
112-80-1	Z-9-OCTADECENOIC ACID
112-24-3	Z1

CAS No.	Chemical Name
52628-25-8	ZACLON
74-90-8	ZACLONDISCOIDS
315-18-4	ZACTRAN
1335-85-9	ZAHLREICHE BEZEICHNUNGEN (GERMAN)
9004-70-0	ZAPON
470-90-6	ZAPRAWA ENOLOFOS
502-39-6	ZAPRAWA NASIENNA PLYNNA
62-38-4	ZAPRAWA NASIENNA R
118-74-1	ZAPRAWA NASIENNA SNECIOTOX
137-26-8	ZAPRAWA NASIENNA T
999-81-5	ZAR
1332-07-6	ZB 112
1332-07-6	ZB 237
1912-24-9	ZEAPOS
1912-24-9	ZEAZIN
1912-24-9	ZEAZINE
315-18-4	ZECTANE
315-18-4	ZECTRAN
9002-84-0	ZEDEFLON
9002-84-0	ZEFLUOR
10031-59-1	ZELIO
7446-18-6	ZELIO
57-41-0	ZENTROPIL
7631-86-9	ZEO 49
7631-86-9	ZEODENT 113
7631-86-9	ZEOFREE 80
1318-02-1	ZEOLITE
1912-24-9	ZEOPOS
7631-86-9	ZEOSYL 200
7631-86-9	ZEOSYL 2000
7631-86-9	ZEOTHIX 265
9002-84-0	ZEPEL C-SF
50-29-3	ZERDANE
107-21-1	ZEREX
108-05-4	ZESET T
8007-45-2	ZETAR
126-72-7	ZETOFEX ZN
315-18-4	ZEXTRAN
8020-83-5	ZHF 12-18
62-38-4	ZIARNIK
7440-66-6	ZINC
7646-85-7	ZINC (CHLORURE DE) (FRENCH)
7440-66-6	ZINC (FUME OR DUST)
557-34-6	ZINC ACETATE
557-34-6	ZINC ACETATE (DOT)
52628-25-8	ZINC AMMONIUM CHLORIDE
52628-25-8	ZINC AMMONIUM CHLORIDE (DOT)
63885-01-8	ZINC AMMONIUM NITRITE
63885-01-8	ZINC AMMONIUM NITRITE (DOT)
13464-44-3	ZINC ARSENATE
13464-33-0	ZINC ARSENATE
28838-01-9	ZINC ARSENATE
1303-39-5	ZINC ARSENATE
13464-33-0	ZINC ARSENATE (DOT)
28838-01-9	ZINC ARSENATE (DOT)
13464-44-3	ZINC ARSENATE (DOT)
13464-44-3	ZINC ARSENATE (Zn3(AsO4)2)
13464-33-0	ZINC ARSENATE, BASIC
13464-44-3	ZINC ARSENATE, BASIC
28838-01-9	ZINC ARSENATE, BASIC
13464-44-3	ZINC ARSENATE, SOLID (DOT)
28838-01-9	ZINC ARSENATE, SOLID (DOT)
13464-33-0	ZINC ARSENATE, SOLID (DOT)
10326-24-6	ZINC ARSENITE
10326-24-6	ZINC ARSENITE, Zn(AsO2)2
10326-24-6	ZINC ARSENITE, SOLID
10326-24-6	ZINC ARSENITE, SOLID (DOT)
39413-47-3	ZINC BERYLLIUM SILICATE
14018-95-2	ZINC BICHROMATE
13826-88-5	ZINC BIS(TETRAFLUOROBORATE)
1332-07-6	ZINC BORATE
1332-07-6	ZINC BORATE (DOT)
13826-88-5	ZINC BOROFLUORIDE
13826-88-5	ZINC BOROFLUORIDE [Zn(BF4)2]
14519-07-4	ZINC BROMATE
14519-07-4	ZINC BROMATE (DOT)
7699-45-8	ZINC BROMIDE
7699-45-8	ZINC BROMIDE (DOT)
7699-45-8	ZINC BROMIDE (ZnBr2)

CAS No.	Chemical Name
7646-85-7	ZINC BUTTER
3486-35-9	ZINC CARBONATE
3486-35-9	ZINC CARBONATE (1:1)
3486-35-9	ZINC CARBONATE (DOT)
3486-35-9	ZINC CARBONATE (ZnCO3)
10361-95-2	ZINC CHLORATE
10361-95-2	ZINC CHLORATE
10361-95-2	ZINC CHLORATE (DOT)
7646-85-7	ZINC CHLORIDE
7646-85-7	ZINC CHLORIDE (ACGIH)
7646-85-7	ZINC CHLORIDE (ZnCl2)
7646-85-7	ZINC CHLORIDE FUME
7646-85-7	ZINC CHLORIDE SOLUTION
7646-85-7	ZINC CHLORIDE, ANHYDROUS (DOT)
7646-85-7	ZINC CHLORIDE, SOLID
7646-85-7	ZINC CHLORIDE, SOLID (DOT)
7646-85-7	ZINC CHLORIDE, SOLUTION (DOT)
14639-97-5	ZINC CHLORIDE-AMMONIUM CHLORIDE COMPLEX (1:2)
13530-65-9	ZINC CHROMATE
14018-95-2	ZINC CHROMATE
14018-95-2	ZINC CHROMATE (ACGIH)
13530-65-9	ZINC CHROMATE (ACGIH)
14018-95-2	ZINC CHROMATE (ZnCr2O7)
13530-65-9	ZINC CHROMATE (ZnCrO4)
37224-57-0	ZINC CHROME
13530-65-9	ZINC CHROMIUM OXIDE
14018-95-2	ZINC CHROMIUM OXIDE
14018-95-2	ZINC CHROMIUM OXIDE (ZnCr2O7)
13530-65-9	ZINC CHROMIUM OXIDE (ZnCrO4)
557-21-1	ZINC CYANIDE
557-21-1	ZINC CYANIDE (DOT)
557-21-1	ZINC CYANIDE (Zn(CN)2)
557-34-6	ZINC DIACETATE
7699-45-8	ZINC DIBROMIDE
7646-85-7	ZINC DICHLORIDE
14018-95-2	ZINC DICHROMATE
13530-65-9	ZINC DICHROMATE
14018-95-2	ZINC DICHROMATE (VI)
13530-65-9	ZINC DICHROMATE (VI)
14018-95-2	ZINC DICHROMATE (ZnCr2O7)
14018-95-2	ZINC DICHROMATE(VI)
557-21-1	ZINC DICYANIDE
7783-49-5	ZINC DIFLUORIDE
557-41-5	ZINC DIFORMATE
7779-88-6	ZINC DINITRATE
1314-22-3	ZINC DIOXIDE
557-05-1	ZINC DISTEARATE
7779-86-4	ZINC DITHIONITE
7779-86-4	ZINC DITHIONITE (ZnS2O4)
7440-66-6	ZINC DUST
7440-66-6	ZINC ELEMENT
557-20-0	ZINC ETHIDE
557-20-0	ZINC ETHYL (DOT)
13826-88-5	ZINC FLUOBORATE
7783-49-5	ZINC FLUORIDE
7783-49-5	ZINC FLUORIDE (DOT)
7783-49-5	ZINC FLUORIDE (ZnF2)
13826-88-5	ZINC FLUOROBORATE
16871-71-9	ZINC FLUOROSILICATE
7783-49-5	ZINC FLUORURE (FRENCH)
16871-71-9	ZINC FLUOSILICATE
557-41-5	ZINC FORMATE
557-41-5	ZINC FORMATE (DOT)
1314-13-2	ZINC GELATIN
16871-71-9	ZINC HEXAFLUOROSILICATE
16871-71-9	ZINC HEXAFLUOROSILICATE (ZnSiF6)
16871-71-9	ZINC HEXAFLUOROSILICATE(2-)
7779-86-4	ZINC HYDROSULFITE
7779-86-4	ZINC HYDROSULFITE (DOT)
7779-86-4	ZINC HYDROSULPHITE
10326-24-6	ZINC METAARSENITE
7440-66-6	ZINC METAL, POWDER OR DUST
10326-24-6	ZINC METHARSENITE
3486-35-9	ZINC MONOCARBONATE
1314-13-2	ZINC MONOXIDE
7646-85-7	ZINC MURIATE, SOLUTION (DOT)
7779-88-6	ZINC NITRATE
7779-88-6	ZINC NITRATE (DOT)
7779-88-6	ZINC NITRATE (Zn(NO3)2)

CAS No.	Chemical Name
557-05-1	ZINC OCTADECANOATE
1314-13-2	ZINC OXIDE
1314-13-2	ZINC OXIDE (ZnO)
1314-22-3	ZINC OXIDE (ZnO2)
23414-72-4	ZINC OXIDE FUME
1314-13-2	ZINC OXIDE FUME
127-82-2	ZINC P-HYDROXYBENZENESULFONATE
127-82-2	ZINC P-PHENOL SULFONATE
23414-72-4	ZINC PERMANGANATE
23414-72-4	ZINC PERMANGANATE (DOT)
1314-22-3	ZINC PEROXIDE
1314-22-3	ZINC PEROXIDE (DOT)
1314-22-3	ZINC PEROXIDE (Zn(O2))
127-82-2	ZINC PHENOLSULFONATE
127-82-2	ZINC PHENOLSULFONATE (DOT)
1314-84-7	ZINC PHOSPHIDE
51810-70-9	ZINC PHOSPHIDE
1314-84-7	ZINC PHOSPHIDE (DOT)
1314-84-7	ZINC PHOSPHIDE (Zn3P2)
37224-57-0	ZINC POTASSIUM CHROMATE.
7440-66-6	ZINC POWDER
7440-66-6	ZINC POWDER OR DUST, PYROPHORIC
9010-69-9	ZINC RESINATE
9010-69-9	ZINC RESINATE (DOT)
13597-54-1	ZINC SELENATE
13597-54-1	ZINC SELENATE (ZnSeO4)
54497-43-7	ZINC SELENITE
16871-71-9	ZINC SILICOFLUORIDE
16871-71-9	ZINC SILICOFLUORIDE (DOT)
557-05-1	ZINC STEARATE
7733-02-0	ZINC SULFATE
7733-02-0	ZINC SULFATE (1:1)
7733-02-0	ZINC SULFATE (DOT)
7733-02-0	ZINC SULFATE (ZnSO4)
127-82-2	ZINC SULFOCARBOLATE
127-82-2	ZINC SULFOPHENATE
7733-02-0	ZINC SULPHATE
1314-22-3	ZINC SUPEROXIDE
13826-88-5	ZINC TETRAFLUOROBORATE
13530-65-9	ZINC TETRAOXYCHROMATE
13530-65-9	ZINC TETROXYCHROMATE
7733-02-0	ZINC VITRIOL
1314-13-2	ZINC WHITE
37224-57-0	ZINC YELLOW
13826-88-5	ZINC(2+) BIS[TETRAFLUOROBORATE(1-)]
557-34-6	ZINC(II) ACETATE
7646-85-7	ZINC(II) CHLORIDE
13826-88-5	ZINC(II) FLUOBORATE
1314-84-7	ZINC(PHOSPHURE DE) (FRENCH)
58270-08-9	ZINC, DICHLORO[4,4-DIMETHYL-5-(((METHYLAMINO)CARBONYL)-OXY)IMINO)PENTANE NITRILE]
557-20-0	ZINC, DIETHYL-
544-97-8	ZINC, DIMETHYL-
7440-66-6	ZINC, POWDER OR DUST, NON-PYROPHORIC (DOT)
7440-66-6	ZINC, POWDER OR DUST, PYROPHORIC (DOT)
1314-84-7	ZINC-TOX
1314-13-2	ZINCA 20
14639-97-5	ZINCATE(2-), TETRACHLORO-, DIAMMONIUM
14639-97-5	ZINCATE(2-), TETRACHLORO-, DIAMMONIUM, (T-4)-
14639-98-6	ZINCATE(3-), PENTACHLORO-, TRIAMMONIUM
1318-59-8	ZINCIAN CHLORITE
7646-85-7	ZINCO (CLORURO DI) (ITALIAN)
1314-84-7	ZINCO(FOSFURO DI) (ITALIAN)
1314-13-2	ZINCOID
7733-02-0	ZINCOMED
12122-67-7	ZINEB
7646-85-7	ZINKCHLORID (GERMAN)
7646-85-7	ZINKCHLORIDE (DUTCH)
1314-84-7	ZINKFOSFIDE (DUTCH)
7733-02-0	ZINKOSITE
1314-84-7	ZINKPHOSPHID (GERMAN)
7646-78-8	ZINNTETRACHLORID (GERMAN)
12654-97-6	ZINOCHLOR
101-05-3	ZINOCHLOR
297-97-2	ZINOPHOS
7631-86-9	ZIPAX
7440-67-7	ZIRCAT
16923-95-8	ZIRCONATE(2-), HEXAFLUORO-, DIPOTASSIUM
16923-95-8	ZIRCONATE(2-), HEXAFLUORO-, DIPOTASSIUM, (OC-6-11)-

CAS No.	Chemical Name
7440-67-7	ZIRCONIUM
7440-67-7	ZIRCONIUM (ACGIH)
4229-34-9	ZIRCONIUM ACETATE
3227-63-2	ZIRCONIUM ACETATE
5153-24-2	ZIRCONIUM ACETATE
3227-63-2	ZIRCONIUM ACETATE (WATERPROOFING AGENT)
5153-24-2	ZIRCONIUM ACETATE (WATERPROOFING AGENT)
4229-34-9	ZIRCONIUM ACETATE (WATERPROOFING AGENT)
4229-34-9	ZIRCONIUM ACETATE [Zr(OAC)4]
10026-11-6	ZIRCONIUM CHLORIDE
10026-11-6	ZIRCONIUM CHLORIDE (Zr2Cl8)
10026-11-6	ZIRCONIUM CHLORIDE (ZrCl4)
10026-11-6	ZIRCONIUM CHLORIDE (ZrCl4), (T-4)-
7699-43-6	ZIRCONIUM CHLORIDE OXIDE (ZrCl2O)
7699-43-6	ZIRCONIUM CHLORIDE, BASIC
5153-24-2	ZIRCONIUM DIACETATE
3227-63-2	ZIRCONIUM DIACETATE
7699-43-6	ZIRCONIUM DICHLORIDE MONOXIDE
7699-43-6	ZIRCONIUM DICHLORIDE OXIDE
7704-99-6	ZIRCONIUM DIHYDRIDE
14644-61-2	ZIRCONIUM DISULFATE
7440-67-7	ZIRCONIUM ELEMENT
7704-99-6	ZIRCONIUM HYDRIDE
11105-16-1	ZIRCONIUM HYDRIDE
7704-99-6	ZIRCONIUM HYDRIDE (DOT)
11105-16-1	ZIRCONIUM HYDRIDE (DOT)
7704-99-6	ZIRCONIUM HYDRIDE (ZrH2)
7440-67-7	ZIRCONIUM METAL POWDER, WETTED WITH NOT LESS THAN 25% WATER (DOT)
7440-67-7	ZIRCONIUM METAL, DRY, CHEMICALLY PRODUCED, FINER THAN 20 MESH PARTICLE SIZE
7440-67-7	ZIRCONIUM METAL, DRY, CHEMICALLY PRODUCED, FINER THAN 270 MESH PARTICLE SIZE (DOT)
7440-67-7	ZIRCONIUM METAL, DRY, MECHANICALLY PRODUCED, FINER THAN 270 MESH PARTICLE SIZE (DOT)
7440-67-7	ZIRCONIUM METAL, LIQUID SUSPENSIONS (DOT)
7440-67-7	ZIRCONIUM METAL, WET, CHEMICALLY PRODUCED, FINER THAN 20 MESH PARTICLE SIZE
7440-67-7	ZIRCONIUM METAL, WET, CHEMICALLY PRODUCED, FINER THAN 270 MESH PARTICLE SIZE (DOT)
7440-67-7	ZIRCONIUM METAL, WET, MECHANICALLY PRODUCED, FINER THAN 270 MESH PARTICLE SIZE (DOT)
13746-89-9	ZIRCONIUM NITRATE
13746-89-9	ZIRCONIUM NITRATE (Zr(NO3)4)
14644-61-2	ZIRCONIUM ORTHOSULFATE
7699-43-6	ZIRCONIUM OXIDE CHLORIDE (ZrOCl2)
7699-43-6	ZIRCONIUM OXYCHLORIDE
7699-43-6	ZIRCONIUM OXYDICHLORIDE
63868-82-6	ZIRCONIUM PICRAMATE
63868-82-6	ZIRCONIUM PICRAMATE, WET (WITH AT LEAST 20% WATER) (DOT)
16923-95-8	ZIRCONIUM POTASSIUM FLUORIDE
16923-95-8	ZIRCONIUM POTASSIUM FLUORIDE (DOT)
7440-67-7	ZIRCONIUM SHAVINGS (DOT)
7440-67-7	ZIRCONIUM SHEETS (DOT)
14644-61-2	ZIRCONIUM SULFATE
14644-61-2	ZIRCONIUM SULFATE (1:2)
14644-61-2	ZIRCONIUM SULFATE (DOT)
14644-61-2	ZIRCONIUM SULPHATE
4229-34-9	ZIRCONIUM TETRAACETATE
10026-11-6	ZIRCONIUM TETRACHLORIDE
10026-11-6	ZIRCONIUM TETRACHLORIDE (DOT)
10026-11-6	ZIRCONIUM TETRACHLORIDE, SOLID
10026-11-6	ZIRCONIUM TETRACHLORIDE, SOLID (DOT)
13746-89-9	ZIRCONIUM TETRANITRATE
7440-67-7	ZIRCONIUM TURNINGS
10026-11-6	ZIRCONIUM(IV) CHLORIDE (1:4)
14644-61-2	ZIRCONIUM(IV) SULFATE (1:2)
5153-24-2	ZIRCONIUM, BIS(ACETATO)DIHYDROXY-
5153-24-2	ZIRCONIUM, BIS(ACETATO)OXO-
5153-24-2	ZIRCONIUM, BIS(ACETATO-O)DIHYDROXY-
5153-24-2	ZIRCONIUM, BIS(ACETATO-O)OXO-
7699-43-6	ZIRCONIUM, DICHLOROOXO-
5153-24-2	ZIRCONIUM, DIHYDROXYBIS(ACETATO)-
7440-67-7	ZIRCONIUM, METAL, DRY, COILED WIRE, FINISHED METAL SHEETS THINNER THAN 18 MICRONS (DOT)
7440-67-7	ZIRCONIUM, METAL, LIQUID, SUSPENSIONS
7699-43-6	ZIRCONYL CHLORIDE
7699-43-6	ZIRCONYL CHLORIDE (ZrOCl2)
5153-24-2	ZIRCONYL DIACETATE

CAS No.	Chemical Name
7699-43-6	ZIRCONYL DICHLORIDE
14644-61-2	ZIRCONYL SULFATE
9002-84-0	ZITEX H 662-124
9002-84-0	ZITEX K 233-122
121-75-5	ZITHIOL
10326-24-6	ZMA
1314-13-2	ZN 0701T
1332-07-6	ZN 100
148-01-6	ZOALENE
148-01-6	ZOAMIX
106-50-3	ZOBA BLACK D
123-30-8	ZOBA BROWN P BASE
27598-85-2	ZOBA BROWN P BASE
95-80-7	ZOBA GKE
81-81-2	ZOOCOUMARIN
81-81-2	ZOOCOUMARIN (RUSSIAN)
129-06-6	ZOOCOUMARIN SODIUM SALT

CAS No.	Chemical Name
13463-67-7	ZOPAQUE LDC
7631-86-9	ZORBAX SIL
7631-86-9	ZORBAX SILICA
7778-39-4	ZOTOX
1303-28-2	ZOTOX
7778-39-4	ZOTOX CRAB GRASS KILLER
1314-84-7	ZP
1314-22-3	ZPO
137-26-8	ZUPA S 80
1405-87-4	ZUTRACIN
7783-06-4	ZWAVELWATERSTOF (DUTCH)
7664-93-9	ZWAVELZUUROPLOSSINGEN (DUTCH)
12002-03-8	ZWICKAU GREEN
7681-49-4	ZYMAFLUOR
297-97-2	ZYNOPHOS
105-60-2	ZYTEL 211

CHAPTER 3

CHEMICAL BY CAS NUMBER

CAS No.	IN	Chemical Name
50-00-0		
		ALDEHYDE FORMIQUE (FRENCH)
		ALDEIDE FORMICA (ITALIAN)
		BFV
		FA
		FANNOFORM
		FORMALDEHYD (CZECH, POLISH)
	*	FORMALDEHYDE
		FORMALDEHYDE (ACGIH)
		FORMALDEHYDE (CONCENTRATION ≤90%)
		FORMALDEHYDE (GAS)
		FORMALDEHYDE SOLUTION
		FORMALDEHYDE SOLUTION (DOT)
		FORMALDEHYDE SOLUTION (FLASH POINT >141 DEG F; IN CONTAINERS > 110 GALLONS
		FORMALDEHYDE, AS FORMALIN SOLUTION (DOT)
		FORMALIN
		FORMALIN (DOT)
		FORMALIN 40
		FORMALIN-LOESUNGEN (GERMAN)
		FORMALINA (ITALIAN)
		FORMALINE (GERMAN)
		FORMALITH
		FORMIC ALDEHYDE
		FORMOL
		FYDE
		HOCH
		IVALON
		KARSAN
		LYSOFORM
		METHALDEHYDE
		METHANAL
		METHYL ALDEHYDE
		METHYLENE GYLCOL
		METHYLENE OXIDE
		MORBICID
		NCI-C02799
		OPLOSSINGEN (DUTCH)
		OXYMETHYLENE
		PARAFORM
		POLYOXYMETHYLENE GLYCOLS
		RCRA WASTE NUMBER U122
		SUPERLYSOFORM
		TETRAOXYMETHYLENE
		TRIOXANE
		UN 1198 (DOT)
		UN 2209 (DOT)
50-07-7		
		7-AMINO-9-ALPHA-METHOXYMITOSANE
		AMETYCIN
		AMETYCINE
		MIT-C
		MITO-C
		MITOCIN-C
		MITOMYCIN
		MITOMYCIN C
		MITOMYCINUM
		MMC
		MUTAMYCIN
		MUTAMYCIN (MITOMYCIN FOR INJECTION)
		MYTOMYCIN
		NCI-C04706
		NSC 26980
		RCRA WASTE NUMBER U010
50-14-6		
		9,10-SECOERGOSTA-5,7,10(19),22-TETRAEN-3-OL, (3BETA,-5Z,7E,22E)-
		BUCO-D
		CALCIFEROL
		CONDOC
		D-ARTHIN
		ERGOCALCIFEROL
50-18-0		
		2H-1,3,2-OXAZAPHOSPHORIN-2-AMINE, N,N-BIS(2-CHLOROETHYL)TETRAHYDRO-, 2-OXIDE
		2H-1,3,2-OXAZAPHOSPHORINE, 2-[BIS(2-CHLOROETHYL)-AMINO]TETRAHYDRO-, 2-OXIDE
		ASTA B 518
		B 518
		BIS(2-CHLOROETHYL)PHOSPHORAMIDE CYCLIC PROPANOL-AMIDE ESTER
		CLAFEN
		CLAPHENE
		CYCLOPHOSPHAMID
		CYCLOPHOSPHAMIDE
		CYCLOPHOSPHAN
		CYCLOPHOSPHANE
		CYTOPHOSPHAN
		CYTOXAN
		ENDOXAN
		GENOXAL
		N,N-BIS(BETA-CHLOROETHYL)-N',O-TRIMETHYLENEPHOSPHORIC ACID ESTER DIAMIDE
		N,N-BIS(2-CHLOROETHYL)-N',O-PROPYLENEPHOSPHORIC ACID ESTER DIAMIDE
		NSC 26271
		PROCYTOX
		SENDOXAN
50-21-5		
		1-HYDROXYETHANECARBOXYLIC ACID
		2-HYDROXYPROPANOIC ACID
		2-HYDROXYPROPIONIC ACID
		ACETONIC ACID
		ALPHA-HYDROXYPROPANOIC ACID
		ALPHA-HYDROXYPROPIONIC ACID
		CHEM-CAST
		DL-LACTIC ACID
		ETHYLIDENELACTIC ACID
		KYSELINA MLECNA (CZECH)
		LACTIC ACID
		MILCHSAURE (GERMAN)
		MILK ACID
		ORDINARY LACTIC ACID
		PROPANOIC ACID, 2-HYDROXY-
		PROPIONIC ACID, 2-HYDROXY-
		RACEMIC LACTIC ACID
50-24-8		
		PREDNISOLONE
50-28-2		
		17BETA-ESTRADIOL
		17BETA-OESTRADIOL
		3,17-EPIDIHYDROXYESTRATRIENE
		3,17BETA-DIHYDROXYESTRA-1,3,5(10)-TRIENE
		3,17BETA-ESTRADIOL
		ALTRAD
		AQUADIOL
		BARDIOL
		BETA-ESTRADIOL
		COMPUDOSE
		CORPAGEN
		D-3,17BETA-ESTRADIOL
		DIHYDROFOLLICULAR HORMONE
		DIHYDROFOLLICULIN
		DIHYDROMENFORMON
		DIHYDROTHEELIN
		DIHYDROXYESTRIN
		DIMENFORMON
		DIOGYN
		DIOGYNETS
		ESTRA-1,3,5(10)-TRIENE-3,17-DIOL (17.BETA.)-
		ESTRA-1,3,5(10)-TRIENE-3,17BETA-DIOL
		ESTRACE
		ESTRADIOL
		ESTRALDINE
		ESTROGENS (NOT CONJUGATED): ESTRADIOL 17BETA
		ESTROVITE
		FEMESTRAL
		FEMOGEN
		FOLLICYCLIN
		GINOSEDOL
		GYNERGON
		GYNOESTRYL

CAS No.	IN	Chemical Name
		LAMDIOL
		MACRODIOL
		NORDICOL
		OESTERGON
		OESTRA-1,3,5(10)-TRIENE-3,17.BETA.-DIOL
		OESTRADIOL
		OESTRADIOL-17B
		OESTROGLANDOL
		OVAHORMON
		OVASTEROL
		OVASTEVOL
		OVOCYCLIN
		PERLATANOL
		PRIMOFOL
		PROFOLIOL
		PROGYNON
		PROGYNON-DH
		SYNDIOL
		THEELIN, DIHYDRO-
50-29-3		
		1,1,1-TRICHLORO-2,2-BIS(P-CHLOROPHENYL)ETHANE
		1,1-BIS(4-CHLOROPHENYL)-2,2,2-TRICHLOROETHANE
		1,1-BIS(P-CHLOROPHENYL)-2,2,2-TRICHLOROETHANE
		2,2,2-TRICHLORO-1,1-BIS(4-CHLOROPHENYL)ETHANE
		2,2-BIS(P-CHLOROPHENYL)-1,1,1-TRICHLOROETHANE
		4,4'-DDT
		4,4'-DICHLORODIPHENYLTRICHLOROETHANE
		AAVERO-EXTRA
		AGRITAN
		ALPHA,ALPHA-BIS(P-CHLOROPHENYL)-BETA,BETA,BETA-TRICHLORETHANE
		ARKOTINE
		AZOTOX M-33
		BENZENE, 1,1'-(2,2,2-TRICHLOROETHYLIDENE)BIS[4-CHLORO-
		BOSAN SUPRA
		BOVIDERMOL
		CHLOROPHENOTHANE
		CHLORPHENOTHAN
		CHLORPHENOTOXUM
		CITOX
		CLOFENOTAN
		CLOFENOTANE
	*	DDT
		DEOVAL
		DETOX
		DETOXAN
		DIBOVIN
		DICHLORO DIPHENYL TRICHLOROETHANE
		DICOPHANE
		DODAT
		DYKOL
		ENT-1506
		ESTONATE
		ETHANE, 1,1,1-TRICHLORO-2,2-BIS(4-CHLOROPHENYL)-
		ETHANE, 1,1,1-TRICHLORO-2,2-BIS(P-CHLOROPHENYL)-
		GESAFID
		GESAROL
		IVORAN
		MUTOXAN
		NEOCID
		NEOCIDOL
		NEOCIDOL (SOLID)
		P,P'-DDT
		P,P'-DICHLORODIPHENYLTRICHLOROETHANE
		P,P'-DICHLORODIPHENYLTRICHLOROMETHYLMETHANE
		PARACHLOROCIDUM
		PEB1
		PENTACHLORIN
		PENTICIDUM
		TAFIDEX
		TRICHLOROBIS(4'-CHLOROPHENYL)ETHANE
		ZERDANE
50-31-7		
		2,3,6-TBA
50-32-8		
		3,4-BENZOPYRENE
		3,4-BENZPYRENE
		3,4-BENZ[A]PYRENE
		4,5-BENZPYRENE
		6,7-BENZOPYRENE
		BENZO(A)PYRENE
		BENZOPYRENE
		BENZO[D,E,F]CHRYSENE
		BENZ[A]PYRENE
		BP
50-41-9		
		CLOMIPHENE CITRATE
50-55-5		
		3,4,5-TRIMETHOXYBENZOYL METHYL RESERPATE
		ANQUIL
		APOPLON
		BANASIL
		BENZ[G]INDOLO[2,3-A]QUINOLIZINE, YOHIMBAN-16-CARBOXYLIC ACID DERIV.
		BIOSERPINE
		CARPACIL
		CRYSTOSERPINE
		ELSERPINE
		ENT 50146
		ESERPINE
		ESKASERP
		HIPOSERPIL
		KITINE
		LEMISERP
		LOWESERP
		MAYSERPINE
		METHYL RESERPATE 3,4,5-TRIMETHOXYBENZOATE (ESTER)
		QUIESCIN
		RAU-SED
		RAUCAP
		RAULEN
		RAUNERVIL
		RAUPASIL
		RAURINE
		RAUSEDIL
		RAUSEDYL
		RAUSINGLE
		RAUWASEDIN
		RESERCAPS
		RESERCEN
		RESERLOR
		RESERPAMED
		RESERPEX
		RESERPIC ACID METHYL ESTER 3,4,5-TRIMETHOXYBENZOATE (ESTER)
		RESERPIL
	*	RESERPINE
		RESERPOID
		RESINE
		RESPITAL
		RESTRAN
		RISERPA
		RIVASIN
		ROXINOID
		SANDRIL
		SEDARAUPIN
		SEDSERP
		SERFIN
		SEROLFIA
		SERP-AFD
		SERPALOID
		SERPANRAY
		SERPASIL
		SERPASIL PREMIX
		SERPASOL
		SERPATE
		SERPEN
		SERPENTINA
		SERPICON
		SERPILOID
		SERPINE

CAS No.	IN	Chemical Name
		SERPINE (PHARMACEUTICAL)
		SERPIPUR
		SERTABS
		SERTINA
		T-SERP
		TEMPOSERPINE
		TRISERPIN
		VIO-SERPINE
50-70-4		
		CHOLAXINE
		D-(-)-SORBITOL
		D-GLUCITOL
		D-SORBITOL
		D-SORBOL
		DIAKARMON
		ESASORB
		GLUCITOL
		GLUCITOL, D-
		GULITOL
		KARION
		KARION (CARBOHYDRATE)
		L-GULITOL
		NEOSORB
		NEOSORB 20/60DC
		NEOSORB 70/70
		NIVITIN
		SIONIT
		SIONITE
		SIONON
		SIOSAN
		SORBEX M
		SORBEX R
		SORBEX RP
		SORBEX S
		SORBEX X
		SORBICOLAN
		SORBILANDE
		SORBIT
		SORBITE
		SORBITOL
		SORBITOL SYRUP C
		SORBO
		SORBOL
		SORBOSTYL
		SORVILANDE
50-76-0		
		1H-PYRROLO[2,1-I][1,4,7,10,13]OXATETRAAZACYCLOHEXA-DECINE, CYCLIC PEPTIDE DERIV.
		3H-PHENOXAZINE, ACTINOMYCIN D DERIV.
		ACTINOMYCIN AIV
		ACTINOMYCIN C1
		ACTINOMYCIN D
		ACTINOMYCIN I1
		ACTINOMYCIN IV
		ACTINOMYCINDIOIC D ACID, DILACTONE
		AD
		COSMEGEN
		DACTINOMYCIN
		DACTINOMYCIN D
		MERACTINOMYCIN
		NSC 3053
		ONCOSTATIN K
50-78-2		
		2-(ACETYLOXY)BENZOIC ACID
		2-ACETOXYBENZOIC ACID
		2-CARBOXYPHENYL ACETATE
		A.S.A. EMPIRIN
		AC 5230
		ACENTERINE
		ACESAL
		ACETARD
		ACETICYL
		ACETILUM ACIDULATUM
		ACETISAL
		ACETOL
		ACETOPHEN
		ACETOSAL
		ACETOSALIC ACID
		ACETOSALIN
		ACETYLIN
		ACETYLSAL
	*	ACETYLSALICYLIC ACID
		ACETYSAL
		ACIDUM ACETYLSALICYLICUM
		ACISAL
		ACYLPYRIN
		ASA
		ASAGRAN
		ASPIRIN
		ASPRO
		ASTERIC
		BENASPIR
		BENZOIC ACID, 2-(ACETYLOXY)-
		BIALPIRINA
		CAPRIN
		COLFARIT
		DOLEAN PH 8
		DURAMAX
		ECM
		ECOTRIN
		EMPIRIN
		ENDOSPRIN
		ENDYDOL
		ENTEROSAREIN
		ENTEROSARINE
		ENTROPHEN
		GLOBENTYL
		GLOBOID
		HELICON
		IDRAGIN
		ISTOPIRIN
		MEASURIN
		MEDISYL
		MICRISTIN
		NEURONIKA
		NOVID
		O-(ACETYLOXY)BENZOIC ACID
		O-ACETOXYBENZOIC ACID
		O-ACETYLSALICYLIC ACID
		O-CARBOXYPHENYL ACETATE
		POLOPIRYNA
		RHEUMINTABLETTEN
		RHODINE
		RHONAL
		SALACETIN
		SALCETOGEN
		SALETIN
		SALICYLIC ACID ACETATE
		SP 189
		TEMPERAL
		TRIPLE-SAL
		XAXA
		YASTA
51-18-3		
		TRIETHYLENEMELAMINE
51-21-8		
		2,4(1H,3H)-PYRIMIDINEDIONE, 5-FLUORO-
		5-FLUORACIL (GERMAN)
		5-FLUORO-2,4(1H,3H)-PYRIMIDINEDIONE
		5-FLUORO-2,4-PYRIMIDINEDIONE
		5-FLUOROPYRIMIDINE-2,4-DIONE
		5-FLUOROURACIL
		5-FLUOROURACIL 2,4(1H,3H)-PYRIMIDINEDIONE 5-FLUORO
		5-FLUORURACIL (GERMAN)
		5-FU
		ADRUCIL
		ARUMEL
		CARZONAL
		EFFLUDERM (FREE BASE)
		EFUDEX
		EFUDIX
		FLUOROBLASTIN
		FLUOROPLEX

CAS No.	IN	Chemical Name
		FLUOROURACIL
		FLURACIL
		FLURI
		FLURIL
		FU
		NSC 19893
		RO 2-9757
		TIMAZIN
		U-8953
		ULUP
		URACIL, 5-FLUORO-
51-28-5		
		1-HYDROXY-2,4-DINITROBENZENE
		2,4-DINITROFENOL (DUTCH)
		2,4-DINITROPHENOL
		2,4-DNP
		ALDIFEN
		ALPHA-DINITROPHENOL
		CHEMOX PE
		DINITROFENOLO (ITALIAN)
		DINITROPHENOL, WETTED, WITH, BY WEIGHT, AT LEAST 15% WATER
		DINOFAN
		FENOXYL CARBON N
		MAROXOL-50
		NITRO KLEENUP
		NITROPHEN
		NITROPHENE
		NSC 1532
		PHENOL, 2,4-DINITRO-
		PHENOL, ALPHA-DINITRO-
		RCRA WASTE NUMBER P048
		SOLFO BLACK 2B SUPRA
		SOLFO BLACK B
		SOLFO BLACK BB
		SOLFO BLACK G
		SOLFO BLACK SB
		TERTROSULPHUR BLACK PB
		TERTROSULPHUR PBR
51-52-5		
		4(1H)-PYRIMIDINONE, 2,3-DIHYDRO-6-PROPYL-2-THIOXO-
		6-N-PROPYL-2-THIOURACIL
		6-N-PROPYLTHIOURACIL
		6-PROPYL-2-THIO-2,4(1H,3H)PYRIMIDINEDIONE
		6-PROPYL-2-THIOURACIL
		6-PROPYLTHIOURACIL
		PROCASIL
		PROPACIL
		PROPYCIL
		PROPYL-THIORIST
		PROPYL-THYRACIL
		PROPYLTHIORIT
		PROPYLTHIOURACIL
		PROTHIUCIL
		PROTHIURONE
		PROTHYCIL
		PROTHYRAN
		PROTIURAL
		PTU
		PTU (THYREOSTATIC)
		THIURAGYL
		THYREOSTAT II
		URACIL, 6-PROPYL-2-THIO-
51-75-2		
		2,2'-DICHLORO-N-METHYLDIETHYLAMINE
		BETA,BETA'-DICHLORODIETHYL-N-METHYLAMINE
		BIS(2-CHLOROETHYL)METHYLAMINE
		BIS(BETA-CHLOROETHYL)METHYLAMINE
		CARYOLYSIN
		CHLORETHAZINE
		CHLORMETHINE
		CLORAMIN
		DI(2-CHLOROETHYL)METHYLAMINE
		DICHLORAMINE
		DICHLOREN (GERMAN)
		DIETHYLAMINE, 2,2'-DICHLORO-N-METHYL-
		EMBICHIN
		ENT-25294
		ETHANAMINE, 2-CHLORO-N-(2-CHLOROETHYL)-N-METHYL-
		HN2
		MBA
		MECHLORETHAMINE
		METHYLBIS(2-CHLOROETHYL)AMINE
		METHYLBIS(BETA-CHLOROETHYL)AMINE
		METHYLDI(2-CHLOROETHYL)AMINE
		MUSTARGEN
		MUSTINE
		MUTAGEN
		N,N-BIS(2-CHLOROETHYL)METHYLAMINE
		N,N-DI(CHLOROETHYL)METHYLAMINE
		N-LOST (GERMAN)
		N-METHYL-2,2'-DICHLORODIETHYLAMINE
		N-METHYL-BIS(2-CHLOROETHYL)AMINE
		N-METHYL-BIS(BETA-CHLOROETHYL)AMINE
		N-METHYL-BIS-CHLORAETHYLAMIN (GERMAN)
		N-METHYL-LOST
		N-METHYLBIS(2-CHLOROETHYL)AMINE
		N-METHYLBIS(BETA-CHLOROETHYL)AMINE
		NITROGEN MUSTARD
		NSC 762
		TL 146
51-79-6		
		CARBAMIC ACID, ETHYL ESTER
		ETHYL CARBAMATE
		ETHYL URETHANE
		LEUCETHANE
		NSC 746
		O-ETHYLURETHANE
		PRACARBAMINE
		URETHAN
	*	URETHANE
51-80-9		
		TETRAMETHYL METHYLENEDIAMINE
51-83-2		
		(2-CARBAMOYLOXYETHYL)TRIMETHYLAMMONIUM CHLORIDE
		(2-HYDROXYETHYL)TRIMETHYL AMMONIUM CHLORIDE CARBAMATE
		2-((AMINOCARBONYL)OXY)-N,N,N-TRIMETHYLETHANAMINIUM CHLORIDE
		CARBACHOL
		CARBACHOL CHLORIDE
		CARBACHOLIN
		CARBACHOLINE
		CARBACHOLINE CHLORIDE
		CARBACOLINA
		CARBAMIC ACID, ESTER WITH CHOLINE CHLORIDE
		CARBAMINOCHOLINE CHLORIDE
		CARBAMINOYLCHOLINE CHLORIDE
		CARBAMIOTIN
		CARBAMOYLCHOLINE CHLORIDE
		CARBAMOYLCHOLINE-HYDROCHLORIDE
		CARBAMYLCHOLINE CHLORIDE
		CARBOCHOL
		CARBOCHOLIN
		CARBOCHOLINE
		CARBYL
		CARCHOLIN
		CHOLINE CARBAMATE CHLORIDE
		CHOLINE CHLORIDE, CARBAMOYL-
		CHOLINE CHLORINE CARBAMATE
		CHOLINE, CHLORIDE CARBAMATE(ESTER)
		CHOLINE, CHLORIDE, CARBAMATE
		COLETYL
		DORYL
		DORYL (PHARMACEUTICAL)
		ETHANAMINIUM, 2-(AMINOCARBONYL)OXY-N,N,N-TRIMETHYL-, CHLORIDE
		GAMMA-CARBAMOYL CHOLINE CHLORIDE
		ISOPTO CARBACHOL
		JESTRYL
		LENTIN

CAS No.	IN	Chemical Name
		LENTINE (FRENCH)
		MIOSTAT
		MISTURA C
		MORYL
		PV CARBACHOL
		TL 457
		VASOPERIF
52-24-4		
		1,1',1''-PHOSPHINOTHIOYLIDYNETRISAZIRIDINE
		AI 3-24916
		AZIRIDINE, 1,1',1''-PHOSPHINOTHIOYLIDYNETRIS-
		GIROSTAN
		N,N',N''-TRIETHYLENETHIOPHOSPHORAMIDE
		N,N',N''-TRIETHYLENETHIOPHOSPHORTRIAMIDE
		NSC 6396
		ONCOTIOTEPA
		PHOSPHINE SULFIDE, TRIS(1-AZIRIDINYL)-
		PHOSPHOROTHIOIC ACID TRIETHYLENETRIAMIDE
		PHOSPHOROTHIOIC TRIAMIDE, N,N',N''-TRI-1,2-ETHANEDIYL-
		TESPAMIN
		TESPAMINE
		THIO-TEPA
		THIOFOZIL
		THIOPHOSPHAMIDE
		THIOPHOSPHORAMIDE, N,N',N''-TRI-1,2-ETHANEDIYL-
		THIOTEF
		THIOTEPA
		THIOTRIETHYLENEPHOSPHORAMIDE
		TIO-TEF
		TIOFOSFAMID
		TIOFOSYL
		TIOFOZIL
		TRI(ETHYLENEIMINO)THIOPHOSPHORAMIDE
		TRI-1-AZIRIDINYLPHOSPHINE SULFIDE
		TRIAZIRIDINYLPHOSPHINE SULFIDE
		TRIETHYLENETHIOPHOSPHORAMIDE
		TRIETHYLENETHIOPHOSPHOROTRIAMIDE
		TRIS(1-AZIRIDINYL)PHOSPHINE SULFIDE
		TRIS(AZIRIDINYL)PHOSPHINE SULFIDE
		TSPA
52-68-6		
		((2,2,2-TRICHLORO-1-HYDROXYETHYL) DIMETHYLPHOSPHONATE)
		(2,2,2-TRICHLORO-1-HYDROXYETHYL)PHOSPHONIC ACID DIMETHYL ESTER
		1-HYDROXY-2,2,2-TRICHLORO-ETHYLE PHOSPHONATE DE DIMETHYLE (FRENCH)
		1-HYDROXY-2,2,2-TRICHLOROETHYLPHOSPHONIC ACID DIMETHYL ESTER
		2,2,2-TRICHLORO-1-HYDROXYETHYL-PHOSPHONATE, DIMETHYL ESTER
		AEROL 1
		AEROL 1 (PESTICIDE)
		AGROFOROTOX
		ANTHON
		BAY 15922
		BAYER 15922
		BAYER L 13/59
		BILARCIL
		BOVINOX
		BRITON
		BRITTEN
		CEKUFON
		CHLORAK
		CHLORFOS
		CHLOROFOS
		CHLOROFTALM
		CHLOROPHOS
		CHLOROPHOSE
		CHLOROPHTHALM
		CHLOROXYPHOS
		CICLOSOM
		CLOROFOS (RUSSIAN)
		COMBOT
		COMBOT EQUINE
		DANEX
		DEP
		DEP (PESTICIDE)
		DEPTHON
		DETF
		DIMETHOXY-2,2,2-TRICHLORO-1-HYDROXY-ETHYL-PHOSPHINE OXIDE
		DIMETHYL (2,2,2-TRICHLORO-1-HYDROXYETHYL)PHOSPHONATE
		DIMETHYL 1-HYDROXY-2,2,2-TRICHLOROETHYL PHOSPHONATE
		DIMETHYLTRICHLOROHYDROXYETHYL PHOSPHONATE
		DIMETOX
		DIOXAPHOS
		DIPTERAX
		DIPTEREX
		DIPTEREX 50
		DIPTEVUR
		DITRIFON
		DYLOX
		DYLOX-METASYSTOX-R
		DYREX
		DYVON
		ENT 19,763
		EQUINO-ACID
		EQUINO-AID
		FLIBOL E
		FLIEGENTELLER
		FOROTOX
		FOSCHLOR
		FOSCHLOR 25
		FOSCHLOR R
		FOSCHLOR R-50
		FOSCHLOREM (POLISH)
		HYPODERMACID
		LEIVASOM
		LOISOL
		MASOTEN
		MAZOTEN
		METHYL CHLOROPHOS
		METIFONATE
		METRIFONATE
		METRIPHONATE
		NA 2783 (DOT)
		NCI-C54831
		NEGUVON
		NEGUVON A
		O,O DIMETIL 2,2,2-TRICHLORO 1 HIDROXIETIL FOSFONATO (PORTUGESE)
		O,O-DIMETHYL (1-HYDROXY-2,2,2-TRICHLOROETHYL)-PHOSPHONATE
		O,O-DIMETHYL (2,2,2-TRICHLORO-1-HYDROXYETHYL)-PHOSPHONATE
		O,O-DIMETHYL-(1-HYDROXY-2,2,2-TRICHLORAETHYL)-PHOSPHONSAEURE ESTER (GERMAN)
		O,O-DIMETHYL-(1-HYDROXY-2,2,2-TRICHLORATHYL)-PHOSPHAT (GERMAN)
		O,O-DIMETHYL-(1-HYDROXY-2,2,2-TRICHLORO)ETHYL PHOSPHATE
		O,O-DIMETHYL-(2,2,2-TRICHLOOR-1-HYDROXY-ETHYL)-FOSFONAAT (DUTCH)
		O,O-DIMETHYL-(2,2,2-TRICHLOR-1-HYDROXY-AETHYL)-PHOSPHONAT (GERMAN)
		O,O-DIMETHYL-1-OXY-2,2,2-TRICHLOROETHYL PHOSPHONATE
		O,O-DIMETIL-(2,2,2-TRICLORO-1-IDROSSI-ETIL)-FOSFONATO (ITALIAN)
		PHOSCHLOR
		PHOSCHLOR R50
		PHOSPHONIC ACID, (1-HYDROXY-2,2,2-TRICHLOROETHYL)-, DIMETHYL ESTER
		PHOSPHONIC ACID, (2,2,2-TRICHLORO-1-HYDROXYETHYL)-, DIMETHYL ESTER
		POLFOSCHLOR
		PROXOL
		RICIFON
		RITSIFON
		SATOX 20WSC
		SOLDEP
		SOTIPOX

CAS No.	IN	Chemical Name
		TRICHLOORFON (DUTCH)
		TRICHLORFON
		TRICHLORFON (DOT)
		TRICHLOROFON
		TRICHLOROPHON
		TRICHLORPHENE
		TRICHLORPHON
		TRICHLORPHON FN
		TRINEX
		TUGON
		TUGON FLY BAIT
		TUGON STABLE SPRAY
		VERMICIDE BAYER 2349
		VOLFARTOL
		VOTEXIT
		WEC 50
		WOTEXIT
52-85-7		
		FAMPHUR
53-03-2		
		PREDNISONE
53-16-7		
		AQUACRINE
		CRINOVARYL
		CRISTALLOVAR
		CRYSTOGEN
		DESTRONE
		DISYNFORMON
		ENDOFOLLICULINA
		ESTRA-1,3,5(10)-TRIEN-17-ONE, 3-HYDROXY-
		ESTROGENS (NOT CONJUGATED): ESTRONE
		ESTRON
		ESTRONE
		ESTROVARIN
		ESTRUGENONE
		ESTRUSOL
		FEMESTRONE INJECTION
		FEMIDYN
		FOLIKRIN
		FOLIPEX
		FOLISAN
		FOLLESTRINE
		FOLLESTROL
		FOLLICULAR HORMONE
		FOLLICULIN
		FOLLICUNODIS
		FOLLIDRIN
		GLANDUBOLIN
		HIESTRONE
		HORMOFOLLIN
		HORMOVARINE
		KESTRONE
		KETODESTRIN
		KETOHYDROXYESTRIN
		KOLPON
		MENAGEN
		MENFORMON
		OESTRIN
		OESTROFORM
		OESTRONE
		OESTROPEROS
		OVIFOLLIN
		PERLATAN
		SOLLICULIN
		THEELIN
		THELYKININ
		THYNESTRON
		TOKOKIN
		UNDEN (PHARMACEUTICAL)
		WYNESTRON
53-70-3		
		1,2:5,6-BENZANTHRACENE
		1,2:5,6-DIBENZANTHRACENE
		1,2:5,6-DIBENZOANTHRACENE
		1,2:5,6-DIBENZ[A]ANTHRACENE
		DBA
		DIBENZ(A,H)ANTHRACENE
		DIBENZO[A,H]ANTHRACENE
53-86-1		
		(1-P-CHLOROBENZOYL-5-METHOXY-2-METHYLINDOL-3-YL)ACETIC ACID
		1-(P-CHLOROBENZOYL)-2-METHYL-5-METHOXY-3-INDOLE-ACETIC ACID
		1-(P-CHLOROBENZOYL)-2-METHYL-5-METHOXYINDOLE-3-ACETIC ACID
		1-(P-CHLOROBENZOYL)-5-METHOXY-2-METHYLINDOLE-3-ACETIC ACID
		1-P-CLORO-BENZOIL-5-METOXI-2-METILINDOL-3-ACIDO ACETICO (SPANISH)
		1H-INDOLE-3-ACETIC ACID, 1-(4-CHLOROBENZOYL)-5-METHOXY-2-METHYL-
		ALPHA-(1-(P-CHLOROBENZOYL)-2-METHYL-5-METHOXY-3-INDOLYL)ACETIC ACID
		AMUNO
		ARTRACIN
		ARTRINOVO
		ARTRIVIA
		CONFORTID
		DOLOVIN
		IDOMETHINE
		IMBRILON
		INACID
		INDACIN
		INDO-RECTOLMIN
		INDO-TABLINEN
		INDOCID
		INDOCIN
		INDOLE-3-ACETIC ACID, 1-(P-CHLOROBENZOYL)-5-METHOXY-2-METHYL-
		INDOMECOL
		INDOMED
		INDOMEE
		INDOMETACIN
		INDOMETACINE
		INDOMETHACIN
		INDOMETHACINE
		INDOMETHAZINE
		INDOMETICINA (SPANISH)
		INDOPTIC
		INDREN
		INFLAZON
		INFROCIN
		INTEBAN
		INTEBAN SP
		LAUSIT
		METACEN
		METARTRIL
		METHAZINE
		METINDOL
		MEZOLIN
		MIKAMETAN
		MOBILAN
		N-P-CHLORBENZOYL-5-METHOXY-2-METHYLINDOLE-3-ACETIC ACID
		NCI-C56144
		REUMACIDE
		SADOREUM
		TANNEX
53-96-3		
		2-AAF
		2-ACETAMIDOFLUORENE
	*	2-ACETYLAMINOFLUORENE
		2-FAA
		AAF
		ACETAMIDE, N-9H-FLUOREN-2-YL-
		ACETAMIDE, N-FLUOREN-2-YL-
		FAA
		N-2-FLUORENYLACETAMIDE
54-05-7		
		CHLOROQUINE

CAS No.	IN	Chemical Name
54-11-5		
		(-)-3-(1-METHYL-2-PYRROLIDYL)PYRIDINE
		(-)-3-(N-METHYLPYRROLIDINO)PYRIDINE
		(-)-NICOTINE
		(S)-NICOTINE
		1-METHYL-2-(3-PYRIDYL)PYRROLIDINE
		3-(N-METHYLPYROLLIDINO)PYRIDINE
		3-(N-METHYLPYRROLIDINO)PYRIDINE
		BETA-PYRIDYL-ALPHA-N-METHYLPYRROLIDINE
		BLACK LEAF
		BLACK LEAF 40
		DESTRUXOL ORCHID SPRAY
		EMO-NIK
		ENT 3,424
		FLUX MAAG
		FUMETOBAC
		L-3-(1-METHYL-2-PYRROLIDYL)PYRIDINE
		L-NICOTINE
		MACH-NIC
		NIAGARA PA DUST
		NICO-DUST
		NICO-FUME
		NICOCIDE
		NICOTIN
		NICOTINA (ITALIAN)
	*	NICOTINE
		NICOTINE (ACGIH)
		NICOTINE (ALKALOID)
		NICOTINE, COMPOUNDS
		NICOTINE, LIQUID
		NICOTINE, LIQUID (DOT)
		NICOTINE, SOLID (DOT)
		NIKOTIN (GERMAN)
		NIKOTYNA (POLISH)
		ORTHO N-4 AND N-5 DUSTS
		ORTHO N-4 DUST
		ORTHO N-5 DUST
		PYRIDINE, 3-(1-METHYL-2-PYRROLIDINYL)-
		PYRIDINE, 3-(1-METHYL-2-PYRROLIDINYL)-, (S)-
		PYRIDINE, 3-(1-METHYL-2-PYRROLIDINYL)-, (S)- (9CI)
		PYRIDINE, 3-(TETRAHYDRO-1-METHYLPYRROL-2-YL)
		PYRROLIDINE, 1-METHYL-2-(3-PYRIDAL)-
		RCRA WASTE NUMBER P075
		S-(-)-NICOTINE
		TENDUST
		TETRAHYDRONICOTYRINE, DL-
		UN 1654 (DOT)
		XL ALL INSECTICIDE
54-62-6		
		4-AMINO-4-DEOXYPTEROYLGLUTAMATE
		4-AMINO-PGA
		4-AMINOFOLIC ACID
		4-AMINOPTEROYLGLUTAMIC ACID
		A-NINOPTERIN
		AMINOPTERIDINE
		AMINOPTERIN
		AMINOPTERINE
		APGA
		ENT-26079
		FOLIC ACID, 4-AMINO-
		GLUTAMIC ACID, N-(P-(((2,4-DIAMINO-6-PTERIDINYL)-METHYL)AMINO)BENZOYL)-, L-
		L-GLUTAMIC ACID, N-[4-[[(2,4-DIAMINO-6-PTERIDINYL)-METHYL]AMINO]BENZOYL]-
		N-(4-((2,4-DIAMINO-6-PTERIDINYL)METHYL)AMINO)BEN-ZOYL)-L-GLUTAMIC ACID
		NSC 739
		PTERAMINA
55-18-5		
		DEN
		DENA
		DIETHYLAMINE, N-NITROSO-
		DIETHYLNITROSAMIDE
		DIETHYLNITROSAMINE
		DIETHYLNITROSOAMINE
		ETHANAMINE, N-ETHYL-N-NITROSO-
		N,N-DIETHYLNITROSOAMINE

CAS No.	IN	Chemical Name
		N-ETHYL-N-NITROSOETHANAMINE
		N-NITROSO-N,N-DIETHYLAMINE
		N-NITROSODIETHYLAMINE
		NITROSODIETHYLAMINE
55-21-0		
		BENZAMIDE
55-38-9		
		BAY 29493
		BAYCID
		BAYER 9007
		BAYTEX
		ENT 25540
		ENTEX
		FENTHION
		FENTHION-METHYL
		LEBAYCID
		MERCAPTOFOS
		MERCAPTOPHOS
		MPP
		MPP (PESTICIDE)
		OMS 2
		PHENTHION
		PHOSPHOROTHIOIC ACID, DIMETHYL [4-(METHYLTHIO)-M-TOLYL] ESTER
		PHOSPHOROTHIOIC ACID, O,O-DIMETHYL O-[3-METHYL-4-(METHYLTHIO)PHENYL] ESTER
		PHOSPHOROTHIOIC ACID, O,O-DIMETHYL O-[4-(METHYL-THIO)-M-TOLYL] ESTER
		QUELETOX
		S 1752
		SPOTTON
		SULFIDOPHOS
		TALODEX
		TIGUVON
55-63-0		
		1,2,3-PROPANETRIOL, TRINITRATE
		1,2,3-PROPANETRIYL NITRATE
		ANGIBID
		ANGININE
		ANGIOLINGUAL
		ANGORIN
		BLASTING GELATIN
		BLASTING GELATIN (DOT)
		BLASTING OIL
		CARDAMIST
		GILUCOR NITRO
		GLONOIN
		GLYCERIN TRINITRATE
		GLYCERINTRINITRATE (CZECH)
		GLYCEROL TRINITRATE
		GLYCEROL(TRINITRATE DE) (FRENCH)
		GLYCEROL, NITRIC ACID TRIESTER
		GLYCEROLTRINITRAAT (DUTCH)
		GLYCERYL NITRATE
		GLYCERYL TRINITRATE
		GLYCERYL TRINITRATE SOLUTION
		GTN
		KLAVIKORDAL
		LENITRAL
		MYOCON
		MYOGLYCERIN
		NA 1204 (DOT)
		NG
		NIGLIN
		NIGLYCON
		NIONG
		NITORA
		NITRIC ACID TRIESTER OF GLYCEROL
		NITRIN
		NITRINE
		NITRINE-TDC
		NITRO-DUR
		NITRO-LENT
		NITRO-SPAN
		NITROGLICERINA (ITALIAN)
		NITROGLICERYNA (POLISH)

CAS No.	IN	Chemical Name
		NITROGLYCERIN
		NITROGLYCERIN (ACGIH)
		NITROGLYCERIN, LIQUID, DESENSITIZED (DOT)
		NITROGLYCERIN, LIQUID, NOT DESENSITIZED (DOT)
		NITROGLYCERIN, SPIRITS OF
		NITROGLYCERINE
		NITROGLYCEROL
		NITROGLYN
		NITROL
		NITROL (PHARMACEUTICAL)
		NITROLAN
		NITROLETTEN
		NITROLINGUAL
		NITROLOWE
		NITROMEL
		NITRONET
		NITRONG
		NITRORECTAL
		NITRORETARD
		NITROSTABILIN
		NITROSTAT
		NITROZELL RETARD
		NK 843
		NTG
		NYSCONITRINE
		PERGLOTTAL
		PROPANETRIOL TRINITRATE
		RCRA WASTE NUMBER P081
		SK-106N
		SNG
		SOLUTION GLYCERYL TRINITRATE
		SOUP
		SPIRIT OF GLONOIN
		SPIRIT OF GLYCERYL TRINITRATE
		SPIRIT OF TRINITROGLYCERIN
		SPIRITS OF NITROGLYCERIN, (1 TO 10%)
		SPIRITS OF NITROGLYCERIN, (1 TO 10%) (DOT)
		TEMPONITRIN
		TNG
		TRINALGON
		TRINITRIN
		TRINITROGLYCERIN
		TRINITROGLYCEROL
		TRINITROL
		UN 0143 (DOT)
		UN 0144 (DOT)
		UN 1204 (DOT)
		VASOGLYN
55-68-5		
		PHENYLMERCURIC NITRATE
55-86-7		
		1,5-DICHLORO-3-METHYL-3-AZAPENTANE HYDROCHLORIDE
		2,2'-DICHLORO-N-METHYLDIETHYLAMINE HYDROCHLORIDE
		ANTIMIT
		BIS(2-CHLOROETHYL)METHYLAMINE HYDROCHLORIDE
		CARYOLYSINE
		CHLORAMIN
		CHLORAMINE
		CHLORMETHINUM
		DICHLOREN
		DIETHYLAMINE, 2,2'-DICHLORO-N-METHYL-, HYDROCHLORIDE
		DIMITAN
		EMBICHIN
		ERASOL
		ETHANAMINE, 2-CHLORO-N-(2-CHLOROETHYL)-N-METHYL, HYDROCHLORIDE
		HN2 HYDROCHLORIDE
		KLORAMIN
		MBA HYDROCHLORIDE
		MECHLORETHAMINE HYDROCHLORIDE
		METHYLBIS(BETA-CHLOROETHYL)AMINE HYDROCHLORIDE
		METHYLBIS(2-CHLOROETHYL)AMINE HYDROCHLORIDE
		MUSTARGEN HYDROCHLORIDE
		MUSTINE HYDROCHLORIDE
		NITROGEN MUSTARD
		NITROGEN MUSTARD HYDROCHLORIDE
		NITROGRANULOGEN
		NSC 762
		NSC-762 HYDROCHLORIDE
55-91-4		
		ISOFLUORPHATE
55-98-1		
		1,4-BIS(METHANESULFONYLOXY)BUTANE
		1,4-BIS[METHANESULFONOXY]BUTANE
		1,4-BUTANEDIOL DIMESYLATE
		1,4-BUTANEDIOL DIMETHANESULPHONATE
		1,4-BUTANEDIOL DIMETHYLSULFONATE
		1,4-BUTANEDIOL, DIMETHANESULFONATE
		1,4-DIMETHANESULFONOXYBUTANE
		1,4-DIMETHYLSULFONYLOXYBUTANE
	*	BUSULFAN
		BUSULPHAN
		CB 2041
		GT 41
		LEUCOSULFAN
		MABLIN
		MIELOSAN
		MIELUCIN
		MILECITAN
		MILERAN
		MISULBAN
		MITOSTAN
		MYELOSAN
		MYLECYTAN
		MYLERAN
		NSC 750
		SULFABUTIN
		TETRAMETHYLENE BIS[METHANESULFONATE]
56-04-2		
		2-THIO-6-METHYLURACIL
		4(1H)-PYRIMIDINONE, 2,3-DIHYDRO-6-METHYL-2-THIOXO-
		4-METHYL-2-THIOURACIL
		6-METHYL-2-MERCAPTOURACIL
		6-METHYL-2-THIOURACIL
		6-THIO-4-METHYLURACIL
		ALKIRON
		ANTIBASON
		BASECIL
		METACIL
		METHACIL
		METHIACIL
		METHICIL
		METHIOCIL
		METHYLTHIOURACIL
		MTU
		MURACIL
		ORCANON
		PROSTRUMYL
		STRUMACIL
		THIMECIL
		THIOTHYMIN
		THYREONORM
		THYREOSTAT
		THYREOSTAT I
		TIOMERACIL
		TIORALE M
		URACIL, 6-METHYL-2-THIO-
56-18-8		
		1,2-PROPANEDIAMINE, N-(AMINOMETHYLETHYL)-
		1,3-PROPANEDIAMINE, N-(3-AMINOPROPYL)-
		1,5,9-TRIAZANONANE
		1,7-DIAMINO-4-AZAHEPTANE
		1-PROPANAMINE, 3,3'-IMINOBIS-
		2,2'-DIMETHYLBIS(2-AMINOETHYL)AMINE
		2-DIPROPYLENETRIAMINE
		3,3'-DIAMINODIPROPYLAMINE
		3,3'-IMINOBISPROPYLAMINE
		3,3'-IMINOBISPROPYLAMINE (DOT)

CAS No.	IN	Chemical Name
		3,3'-IMINODIPROPYLAMINE
		3,3-DIAMINODIPROPYLAMINE
		4-AZAHEPTAMETHYLENEDIAMINE
		4-AZAHEPTANE-1,7-DIAMINE
		AMINOBIS(PROPYLAMINE)
		BIS(2-AMINOPROPYL)AMINE
		BIS(3-AMINOPROPYL)AMINE
		BIS(AMINOPROPYL)AMINE
		CALDINE
		DI-1,2-PROPANETRIAMINE
		DIETHYLENETRIAMINE, 2,6-DIMETHYL-
		DIETHYLENETRIAMINE, DIMETHYL-
		DIPROPYLAMINE, 3,3'-DIAMINO-
		DIPROPYLENE TRIAMINE
		DIPROPYLENTRIAMIN (GERMAN)
		IMINO BISPROPYLAMINE (DOT)
		IMINOBIS(PROPYLAMINE)
		IMINODIPROPYLAMINE
		N,N'-BIS(TRIMETHYLENEAMINO)TRIAMINE
		N-(3-AMINOPROPYL)-1,3-PROPANEDIAMINE
		N-3-AMINOPROPYL-1,3-DIAMINOPROPANE
		NORSPERMIDINE
		P 2 (HARDENER)
		PROPYLAMINE, 3,3'-IMINOBIS-
		SYM-NORSPERMIDINE
		UN 2269 (DOT)
56-23-5		
		BENZINOFORM
		CARBON CHLORIDE
		CARBON CHLORIDE (CCl4)
		CARBON TET
	*	CARBON TETRACHLORIDE
		CARBON TETRACHLORIDE (ACGIH,DOT)
		CARBONA
		CZTEROCHLOREK WEGLA (POLISH)
		ENT 4,705
		FASCIOLIN
		FLUKOIDS
		METHANE TETRACHLORIDE
		METHANE, TETRACHLORO-
		NECATORINA
		NECATORINE
		PERCHLOROMETHANE
		R 10
		R 10 (REFRIGERANT)
		RCRA WASTE NUMBER U211
		TETRACHLOORKOOLSTOF (DUTCH)
		TETRACHLOORMETAAN
		TETRACHLORKOHLENSTOFF, TETRA (GERMAN)
		TETRACHLORMETHAN (GERMAN)
		TETRACHLOROCARBON
		TETRACHLOROMETHANE
		TETRACHLORURE DE CARBONE (FRENCH)
		TETRACLOROMETANO (ITALIAN)
		TETRACLORURO DI CARBONIO (ITALIAN)
		TETRAFINOL
		TETRAFORM
		TETRASOL
		UN 1846 (DOT)
		UNIVERM
		VERMOESTRICID
56-25-7		
		1,2-DIMETHYL-3,6-EPOXYPERHYDROPHTHALIC ANHYDRIDE
		2,3-DIMETHYL-7-OXABICYCLO(221)HEPTANE-2,3-DICARBOXYLIC ANHYDRIDE
		4,7-EPOXYISOBENZOFURAN-1,3-DIONE, HEXAHYDRO-3A,7A-DIMETHYL-, (3AALPHA,4BETA,7BETA,7AALPHA)-
		7-OXABICYCLO(221)HEPTANE-2,3-DICARBOXYLIC ANHYDRIDE, 2,3-DIMETHYL-
		CAN
		CANTHARIDES CAMPHOR
		CANTHARIDIN
		CANTHARIDINE
		CANTHARONE
		EXO-1,2-CIS-DIMETHYL-3,6-EPOXYHEXAHYDROPHTHALIC ANHYDRIDE
		HEXAHYDRO-3A,7A-DIMETHYL-4,7-EPOXYISOBENZOFURAN-1,3-DIONE
		KANTARIDIN
		KANTHARIDIN (GERMAN)
56-36-0		
		TRIBUTYLTIN ACETATE
56-38-2		
		AAT
		AATP
		ALKRON
		ALLERON
		AMERICAN CYANAMID 3422
		APHAMITE
		ARALO
		BAYER E-605
		BLADAN F
		DIETHYL 4-NITROPHENYL PHOSPHOROTHIONATE
		DIETHYL P-NITROPHENYL PHOSPHOROTHIONATE
		DIETHYL P-NITROPHENYL THIONOPHOSPHATE
		DIETHYL PARATHION
		DNTP
		DPP
		DREXEL PARATHION 8E
		E 605
		E 605 F
		E 605 FORTE
		ECATOX
		EKATIN WF & WF ULV
		EKATOX
		ENT 15,108
		ETHLON
		ETHYL PARATHION
		ETILON
		ETYLPARATION (CZECH)
		FOLI
		FOLIDOL
		FOLIDOL E
		FOLIDOL E & E 605
		FOLIDOL E605
		FOLIDOL OIL
		FOSFERNO
		FOSFEX
		FOSFIVE
		FOSOVA
		FOSTERN
		FOSTOX
		GEARPHOS
		GENITHION
		KOLPHOS
		KYPTHION
		LETHALAIRE G-54
		LIROTHION
		MURFOS
		NA 1967 (DOT)
		NA 2783 (DOT)
		NCI-C00226
		NIRAN
		NIRAN E-4
		NITROSTIGMIN (GERMAN)
		NITROSTIGMINE
		NITROSTYGMINE
		NIUIF 100
		NOURITHION
		O,O-DIETHYL O-(P-NITROPHENYL) PHOSPHOROTHIOATE
		O,O-DIETHYL O-P-NITROPHENYL THIOPHOSPHATE
		O,O-DIETYL-O-P-NITROFENYLTIOFOSFAT (CZECH)
		OLEOFOS 20
		OLEOPARAPHENE
		OLEOPARATHENE
		OLEOPARATHION
		ORTHOPHOS
		PAC
		PACOL
		PANTHION
		PARADUST
		PARAMAR
		PARAMAR 50

CAS No.	IN	Chemical Name
		PARAPHOS
		PARATHENE
		PARATHION
		PARATHION (ACGIH)
		PARATHION AND COMPRESSED GAS MIXTURE (DOT)
		PARATHION AND COMPRESSED GAS MIXTURES
		PARATHION MIXTURE, DRY
		PARATHION MIXTURE, DRY (DOT)
		PARATHION MIXTURE, LIQUID
		PARATHION MIXTURE, LIQUID (DOT)
		PARATHION, LIQUID
		PARATHION, LIQUID (DOT)
		PARATHION-AETHYL (GERMAN)
		PARATHION-COMPRESSED GAS MIXTURE
		PARATHION-ETHYL
		PARAWET
		PENNCAP E
		PESTOX PLUS
		PETHION
		PHENOL, P-NITRO-, O-ESTER WITH O,O-DIETHYLPHOSPHOROTHIOATE
		PHOSKIL
		PHOSPHEMOL
		PHOSPHENOL
		PHOSPHOROTHIOIC ACID, O,O-DIETHYL O-(4-NITROPHENYL) ESTER
		PHOSPHOROTHIOIC ACID, O,O-DIETHYL O-(P-NITROPHENYL) ESTER
		PHOSPHOROTHIOIC ACID, O,O-DIETHYL O-(P-NITROPHENYL) ESTER (MIXTURE)
		PHOSPHOROTHIOIC ACID, O,O-DIETHYL O-(P-NITROPHENYL)ESTER, MIXED WITH COMPRESSED GAS
		PHOSPHOSTIGMINE
		RB
		RCRA WASTE NUMBER P089
		RHODIASOL
		RHODIATOX
		RHODIATROX
		SELEPHOS
		SNP
		SOPRATHION
		STABILIZED ETHYL PARATHION
		STATHION
		STRATHION
		SULPHOS
		SUPER RODIATOX
		T-47
		THIOMEX
		THIOPHOS
		THIOPHOS 3422
		THIOPHOSPHATE DE O,O-DIETHYLE ET DE O-(4-NITROPHENYLE) (FRENCH)
		TIOFOS
		TOX 47
		VAPOPHOS
		VITREX
56-49-5		
		20-MC
		20-METHYLCHOLANTHRENE
		3-MC
		3-METHYLCHOLANTHRENE
		BENZ[J]ACEANTHRYLENE, 1,2-DIHYDRO-3-METHYL-
		CHOLANTHRENE, 3-METHYL-
		METHYLCHOLANTHRENE
56-53-1		
		ALPHA,ALPHA'-DIETHYL-4,4'-STILBENEDIOL
		ALPHA,ALPHA'-DIETHYLSTILBENEDIOL
		4,4'-DIHYDROXY-ALPHA,BETA-DIETHYLSTILBENE
		4,4'-DIHYDROXYDIETHYLSTILBENE
		4,4'-STILBENEDIOL, ALPHA,ALPHA'-DIETHYL-, (E)-
		AGOSTILBEN
		ANTIGESTIL
		BIO-DES
		BUFON
		COMESTROL
		CYREN
		CYREN A
		DAWE'S DESTROL
		DEB
		DES
		DES (SYNTHETIC ESTROGEN)
		DI-ESTRYL
		DIBESTROL 2 PREMIX
		DIETHYLSTILBESTROL
		DISTILBENE
		DOMESTROL
		ESTILBIN MCO
		ESTROBENE
		ESTROMENIN
		ESTROSYN
		FONATOL
		GRAFESTROL
		HI-BESTROL
		ISCOVESCO
56-53-1		
		MENOSTILBEEN
		MICROEST
		MILESTROL
		NEO-OESTRANOL I
		OESTROGENINE
		OESTROMENIN
		OESTROMENSYL
		PABESTROL
		PALESTROL
		PHENOL, 4,4'-(1,2-DIETHYL-1,2-ETHENEDIYL)BIS-, (E)-
		RUMESTROL 1
		RUMESTROL 2
		SERRAL
		SEXOCRETIN
		SIBOL
		STIL
		STIL-ROL
		STILBESTROL
		STILBETIN
		STILBOEFRAL
		STILBOESTROFORM
		STILBOESTROL
		STILKAP
		SYNESTRIN
		SYNTHOESTRIN
		SYNTHOFOLIN
		SYNTOFOLIN
		TRANS-ALPHA,ALPHA'-DIETHYL-4,4'-STILBENEDIOL
		TRANS-4,4'-DIHYDROXY-ALPHA,BETA-DIETHYLSTILBENE
		TRANS-DIETHYLSTILBESTEROL
		TRANS-DIETHYLSTILBESTROL
		E-3,4-BIS(4-HYDROXYPHENYL)-3-HEXENE
56-55-3		
		1,2-BENZANTHRACENE
		1,2-BENZANTHRENE
		1,2-BENZOANTHRACENE
		1,2-BENZ[A]ANTHRACENE
		2,3-BENZOPHENANTHRENE
		BENZ(A)ANTHRACENE
		BENZANTHRACENE
		BENZANTHRENE
		BENZOANTHRACENE
		BENZO[A]ANTHRACENE
		BENZO[B]PHENANTHRENE
		TETRAPHENE
56-57-5		
		4-NITROQUINOLINE 1-OXIDE
		4-NITROQUINOLINE N-OXIDE
		4-NITROQUINOLINE OXIDE
		4-NQO
		NITROCHIN
	*	QUINOLINE, 4-NITRO-, 1-OXIDE
56-72-4		
		3-CHLORO-4-METHYL-7-COUMARINYL DIETHYL PHOSPHOROTHIOATE
		3-CHLORO-4-METHYL-7-HYDROXYCOUMARIN DIETHYL THIOPHOSPHORIC ACID ESTER

CAS No. *IN* *Chemical Name*

3-CHLORO-4-METHYLUMBELLIFERONE O-ESTER WITH O,O-DIETHYL PHOSPHOROTHIOATE
3-CHLORO-7-HYDROXY-4-METHYL-COUMARIN O,O-DI-ETHYL PHOSPHOROTHIOATE
3-CHLORO-7-HYDROXY-4-METHYL-COUMARIN O-ESTER WITH O,O-DIETHYL PHOSPHOROTHIOATE
AGRIDIP
ASUNTHOL
ASUNTOL
AZUNTHOL
BAY 21/199
BAYER 21/199
BAYMIX
BAYMIX 50
CO-RAL
COUMAFOS
COUMAPHOS
COUMAPHOS (DOT)
COUMARIN, 3-CHLORO-7-HYDROXY-4-METHYL-, O-ESTER WITH O,O-DIETHYL PHOSPHOROTHIOATE
CUMAFOS (DUTCH)
DIETHYL 3-CHLORO-4-METHYLUMBELLIFERYL THIONOPHOSPHATE
DIETHYL THIOPHOSPHORIC ACID ESTER OF 3-CHLORO-4-METHYL-7-HYDROXYCOUMARIN
DIOLICE
ENT 17,957
ENT 7,957
MELDANE
MELDONE
MUSCATOX
MUSCOTOX
NA 2783 (DOT)
NCI-C08662
O,O-DIAETHYL-O-(3-CHLOR-4-METHYL-CUMARIN-7-YL)-MONOTHIOPHOSPHAT (GERMAN)
O,O-DIETHYL 3-CHLORO-4-METHYL-7-UMBELLIFERONE THIOPHOSPHATE
O,O-DIETHYL O-(3 CHLORO-4-METHYL-2OXO-2H-1 BENZO-PYRAN-7-YL) PHOSPHOROTHIOATE
O,O-DIETHYL O-(3-CHLORO-4-METHYL-2-OXO-2H-BENZO-PYRAN-7-YL)PHOSPHOROTHIOATE
O,O-DIETHYL O-(3-CHLORO-4-METHYL-7-COUMARINYL)-PHOSPHOROTHIOATE
O,O-DIETHYL O-(3-CHLORO-4-METHYLCOUMARINYL-7) THIOPHOSPHATE
O,O-DIETHYL O-(3-CHLORO-4-METHYLUMBELLIFERYL)-PHOSPHOROTHIOATE
O,O-DIETIL-O-(3-CLORO-4-METIL-CUMARIN-7-IL-MONO-TIOFOSFATO) (ITALIAN)
O-3-CHLORO-4-METHYL-7-COUMARINYL O,O-DIETHYL PHOSPHOROTHIOATE
PHOSPHOROTHIOIC ACID, O,O-DIETHYL ESTER, O-ESTER WITH 3-CHLORO-7-HYDROXY-4-METHYLCOUMARIN
PHOSPHOROTHIOIC ACID, O-(3-CHLORO-4-METHYL-2-OXO-2H-1-BENZOPYRAN-7-YL) O,O-DIETHYL ESTER
RESISTOX
RESITOX
SUNTOL
THIOPHOSPHATE DE O,O-DIETHYLE ET DE O-(3-CHLORO-4-METHYL-7-COUMARINYLE) (FRENCH)
UMBETHION

56-75-7

ACETAMIDE, 2,2-DICHLORO-N-[2-HYDROXY-1-(HYDROXY-METHYL)-2-(4-NITROPHENYL)ETHYL]-, [R-(R*,R*)]-
ALFICETYN
AMPHENICOL
AMPHICOL
AMSECLOR
AUSTRACOL
CAF
CAF (PHARMACEUTICAL)
CAM
CAP
CATILAN
CHEMICETIN
CHEMICETINA
CHLOMYCOL
CHLORAMEX
CHLORAMPHENICOL
CHLORAMSAAR
CHLOROCAPS
CHLOROCID
CHLOROCIDE
CHLOROCIDIN C
CHLOROCIDIN C TETRAN
CHLOROMYCETIN
CHLORONITRIN
CHLOROPTIC
CIPLAMYCETIN
CLORAMFICIN
CLORAMICOL
CLOROCYN
CLOROMISAN
CPH
CYLPHENICOL
D-(-)-CHLORAMPHENICOL
D-(-)-THREO-CHLORAMPHENICOL
D-CHLORAMPHENICOL
D-THREO-CHLORAMPHENICOL
DETREOMYCIN
EMBACETIN
ENICOL
ENTEROMYCETIN
FARMICETINA
FENICOL
GLOBENICOL
GLOVETICOL
I 337A
INTRAMYCETIN
ISMICETINA
JUVAMYCETIN
KAMAVER
KEMICETINE
KLORITA
KLOROCID S
LEUKOMYAN
LEUKOMYCIN
LEVOMICETINA
LEVOMITSETIN
LEVOMYCETIN
LEVOVETIN
LOROMISIN
MASTIPHEN
MICLORETIN
MICROCETINA
MYCHEL
MYCHEL-VET
MYCINOL
MYCLOCIN
MYCOCHLORIN
NORMIMYCIN V
NOVOMYCETIN
OPCLOR
OPHTHOCHLOR
PANTOVERNIL
PARAXIN
QUEMICETINA
SEPTICOL
SINTOMICETIN
SINTOMICETINA
SINTOMICETINE R
STANOMYCETIN
SYNTHOMYCETIN
TEVCOCIN
TIFOMYCINE
TREOMICETINA
UNIMYCETIN

56-81-5

1,2,3-PROPANETRIOL
1,2,3-TRIHYDROXYPROPANE
90 TECHNICAL GLYCERINE
GLYCERIN
GLYCERIN (ACGIH)
GLYCERIN, ANHYDROUS
GLYCERIN, SYNTHETIC

CAS No.	IN	Chemical Name
		GLYCERINE
		GLYCERINE, CRUDE, CONCENTRATED
		GLYCERITOL
	*	GLYCEROL
		GLYCYL ALCOHOL
		GLYROL
		GLYSANIN
		GROCOLENE
		MOON
		OSMOGLYN
		PROPANETRIOL
		STAR
		SUPEROL
		SYNTHETIC GLYCERIN
		TRIHYDROXYPROPANE
56-93-9		
		AMMONIUM, BENZYLTRIMETHYL-, CHLORIDE
		BENZENEMETHANAMINIUM, N,N,N-TRIMETHYL-, CHLORIDE
		BENZENEMETHANAMINIUM, N,N,N-TRIMETHYL-, CHLORIDE (9CI)
		BENZYLTRIMETHYLAMMONIUM CHLORIDE
		BTM
		TMBAC
		TRIMETHYLBENZYLAMMONIUM CHLORIDE
57-06-7		
		1-PROPENE, 3-ISOTHIOCYANATO-
		2-PROPENYL ISOTHIOCYANATE
		3-ISOTHIOCYANATO-1-PROPENE
		AITC
		AITK
		ALLYL ISORHODANIDE
		ALLYL ISOSULFOCYANATE
		ALLYL ISOSULPHOCYANATE
	*	ALLYL ISOTHIOCYANATE
		ALLYL ISOTHIOCYANATE, INHIBITED
		ALLYL ISOTHIOCYANATE, STABILIZED (DOT)
		ALLYL MUSTARD OIL
		ALLYL SEVENOLUM
		ALLYL THIOCARBONIMIDE
		ALLYLSENEVOL
		ALLYLSENFOEL
		ALLYLSENFOEL (GERMAN)
		ALLYLSEVENOLUM
		ARTIFICIAL MUSTARD OIL
		ARTIFICIAL OIL OF MUSTARD
		CARBOSPOL
		FEMA NO 2034
		ISOTHIOCYANATE D'ALLYLE (FRENCH)
		ISOTHIOCYANIC ACID, ALLYL ESTER
		MUSTARD OIL
		NCI-C50464
		OIL OF MUSTARD BPC 1949
		OIL OF MUSTARD, ARTIFICIAL
		OLEUM SINAPIS
		OLEUM SINAPIS VOLATILE
		PROPENE, 3-ISOTHIOCYANATO-
		REDSKIN
		SENF OEL (GERMAN)
		SENFOEL
		SYNTHETIC MUSTARD OIL
		UN 1545 (DOT)
		VOLATILE MUSTARD OIL
		VOLATILE OIL OF MUSTARD
57-11-4		
		1-HEPTADECANECARBOXYLIC ACID
		BAROLUB FTA
		CENTURY 1210
		CENTURY 1220
		CENTURY 1230
		CENTURY 1240
		DAR-CHEM 14
		EMERSOL 120
		EMERSOL 132
		EMERSOL 150
		EMERSOL 153
		EMERSOL 6349
		FORMULA 300
		GLYCON DP
		GLYCON S-70
		GLYCON S-80
		GLYCON S-90
		GLYCON TP
		GROCO 54
		GROCO 55
		GROCO 55L
		GROCO 58
		GROCO 59
		HUMKO INDUSTRENE R
		HY-PHI 1199
		HY-PHI 1205
		HY-PHI 1303
		HY-PHI 1401
		HYDROFOL 1895
		HYDROFOL ACID 150
		HYDROFOL ACID 1655
		HYDROFOL ACID 1855
		HYSTRENE 4516
		HYSTRENE 5016
		HYSTRENE 7018
		HYSTRENE 80
		HYSTRENE 9718
		HYSTRENE S 97
		HYSTRENE T 70
		INDUSTRENE 5016
		INDUSTRENE 8718
		INDUSTRENE 9018
		INDUSTRENE R
		KAM 1000
		KAM 2000
		KAM 3000
		LOXIOL G 20
		LUNAC S 20
		N-OCTADECANOIC ACID
		NAA 173
		NEO-FAT 18
		NEO-FAT 18-53
		NEO-FAT 18-54
		NEO-FAT 18-55
		NEO-FAT 18-59
		NEO-FAT 18-61
		NEO-FAT 18-S
		OCTADECANOIC ACID
		PD 185
		PEARL STEARIC
		STEAREX BEADS
		STEARIC ACID
		STEAROPHANIC ACID
		TEGOSTEARIC 254
		TEGOSTEARIC 255
		TEGOSTEARIC 272
		VANICOL
57-12-5		
		CARBON NITRIDE ION (CN(SUP 1-))
		CARBON NITRIDE ION (CN1-)
	*	CYANIDE
		CYANIDE (CN(SUP 1-))
		CYANIDE (CN1-)
		CYANIDE ANION
		CYANIDE COMPOUNDS
		CYANIDE ION
		CYANIDE OR CYANIDE MIXTURE, DRY (DOT)
		CYANIDE SOLUTION
		CYANIDE(1-)
		CYANIDE(1-) ION
		CYANIDE, DRY
		CYANIDE, SOLUTION (DOT)
		CYANIDES
		CYANURE (FRENCH)
		HYDROCYANIC ACID, ION(1-)
		HYDROCYANIC ACID, SALTS
		ISOCYANIDE
		RCRA WASTE NUMBER P030
		UN 1588 (DOT)

CAS No.	IN	Chemical Name
		UN 1935 (DOT)
57-13-6		
		B-I-K
		BENURAL 70
		CARBAMIDE
		CARBAMIDE RESIN
		CARBAMIMIDIC ACID
		CARBONYL DIAMIDE
		CARBONYLDIAMINE
		ISOUREA
		NCI-C02119
		PRESPERSION, 75 UREA
		PSEUDOUREA
		SUPERCEL 3000
		UREA
		UREAPHIL
		UREOPHIL
		UREPEARL
		UREVERT
		VARIOFORM II
57-14-7		
		1,1-DIMETHYL HYDRAZINE
		1,1-DIMETHYLHYDRAZIN (GERMAN)
		1,1-DIMETHYLHYDRAZINE (ACGIH)
		AS-DIMETHYLHYDRAZINE
		ASYMMETRIC DIMETHYLHYDRAZINE
		DIMAZIN
		DIMAZINE
		DIMETHYLHYDRAZINE
		DIMETHYLHYDRAZINE UNSYMMETRICAL (DOT)
		DIMETHYLHYDRAZINE, UNSYMMETRICAL
		DMH
		HYDRAZINE, 1,1-DIMETHYL-
		N,N-DIMETHYLHYDRAZINE
		RCRA WASTE NUMBER U098
		U-DIMETHYLHYDRAZINE
		UDMH
		UDMH (DOT)
		UN 1163 (DOT)
		UNS-DIMETHYLHYDRAZINE
		UNSYM-DIMETHYLHYDRAZINE
		UNSYMMETRICAL DIMETHYLHYDRAZINE
57-24-9		
		4,6-METHANO-6H,14H-INDOLO[3,2,1-IJ]OXEPINO[2,3,4-DE]PYRROLO[2,3-H]QUINOLINE, STRYCHNIDIN-1
		CERTOX
		DOLCO MOUSE CEREAL
		KWIK-KIL
		MOLE DEATH
		MOUSE-NOTS
		MOUSE-RID
		MOUSE-TOX
		PIED PIPER MOUSE SEED
		RCRA WASTE NUMBER P108
		RO-DEX
		SANASEED
		STRICNINA (ITALIAN)
		STRYCHNIDIN-10-ONE
		STRYCHNIN
		STRYCHNIN (GERMAN)
		STRYCHNINE
		STRYCHNINE (ACGIH)
		STRYCHNINE AND ITS SALTS
		STRYCHNINE, LIQUID (DOT)
		STRYCHNINE, SOLID
		STRYCHNINE, SOLID (DOT)
		STRYCHNOS UN 1692 (DOT)
57-41-0		
		2,4-IMIDAZOLIDINEDIONE, 5,5-DIPHENYL-
		5,5-DIPHENYL-2,4-IMIDAZOLIDINEDIONE
		5,5-DIPHENYLHYDANTOIN
		ALEVIATIN
		DENYL
		DI-HYDAN
		DI-LAN
		DIHYCON
		DILABID
		DINTOINA
		DIPHANTOIN
		DIPHEDAN
		DIPHENYLHYDANTOIN
		DPH
		HIDANTAL
		HYDANTOIN, 5,5-DIPHENYL-
		LEPITOIN
		PHENYTOIN
		PHENYTOINE
		SODANTON
		ZENTROPIL
57-47-6		
		(-)-PHYSOSTIGMINE
		CARBAMIC ACID, METHYL-, ESTER WITH ESEROLINE
		CS 58525
		ERSERINE
		ESERINE
		ESEROLEIN, METHYLCARBAMATE (ESTER)
		ESROMIOTIN
		EZERIN
		PHYSOSTIGMINE
		PHYSOSTOL
57-50-1		
		ALPHA-D-GLUCOPYRANOSIDE, BETA-D-FRUCTOFURANOSYL
		BETA-D-FRUCTOFURANOSYL ALPHA-D-GLUCOPYRANOSIDE
		AMERFOND
		BEET SUGAR
		CANE SUGAR
		CONFECTIONER'S SUGAR
		D-(+)-SACCHAROSE
		D-(+)-SUCROSE
		D-SUCROSE
		GRANULATED SUGAR
		MICROSE
		ROCK CANDY
		SACCHAROSE
		SACCHARUM
	*	SUCROSE
		SUGAR
		WHITE SUGAR
57-53-4		
		MEPROBAMATE
57-55-6		
		1,2-DIHYDROXYPROPANE
		1,2-PROPANEDIOL
		1,2-PROPYLENE GLYCOL
		2,3-PROPANEDIOL
		2-HYDROXYPROPANOL
		ALPHA-PROPYLENE GLYCOL
		DOWFROST
		ISOPROPYLENE GLYCOL
		METHYL GLYCOL
		METHYLETHYL GLYCOL
		METHYLETHYLENE GLYCOL
		MONOPROPYLENE GLYCOL
		PG 12
		PROPANE-1,2-DIOL
	*	PROPYLENE GLYCOL
		PROPYLENE GLYCOL USP
		SIRLENE
		SOLAR WINTER BAN
		SOLARGARD P
		TRIMETHYL GLYCOL
		UCAR 35
57-57-8		
		1,3-PROPIOLACTONE
		2-OXETANONE
		3-HYDROXYPROPIONIC ACID LACTONE
		3-PROPANOLIDE
		3-PROPIOLACTONE

CAS No.	IN	Chemical Name
		BETA-PROPIOLACTONE
		BETA-PROPIOLACTONE (ACGIH)
		BETA-PROPIONOLACTONE
		BETAPRONE
		BPL
		HYDRACRYLIC ACID BETA-LACTONE
		PROPANOIC ACID, 3-HYDROXY-, BETA-LACTONE
		PROPANOLIDE
		PROPIOLACTONE
		PROPIOLACTONE, BETA-
		PROPIONIC ACID, 3-HYDROXY-, BETA-LACTONE
57-63-6		
		17-ETHINYL-3,17-ESTRADIOL
		17-ETHINYLESTRADIOL
		17-ETHYNYLESTRADIOL
		17ALPHA-ETHINYL-17BETA-ESTRADIOL
		17ALPHA-ETHINYLESTRADIOL
		17ALPHA-ETHYNYLESTRADIOL
		19-NOR-17ALPHA-PREGNA-1,3,5(10)-TRIEN-20-YNE-3,17-DIOL
		19-NORPREGNA-1,3,5(10)-TRIEN-20-YNE-3,17-DIOL, (17ALPHA)-
		AMENORON
		CHEE-O-GEN
		CHEE-O-GENF
		DIOGYN-E
		DYLOFORM
		ESTEED
		ESTIGYN
		ESTINYL
		ESTON-E
		ESTORAL
		ESTORALS
		ESTRADIOL, 17-ETHYNYL-
		ESTROGENS (NOT CONJUGATED): ETHINYLOESTRADIOL
		ETHIDOL
		ETHINORAL
		ETHINYLESTRADIOL
		ETHINYLESTRIOL
		ETHINYLOESTRADIOL
		ETHYNYLESTRADIOL
		ETICYCLIN
		ETICYCLOL
		ETINESTROL
		ETINESTRYL
		ETINOESTRYL
		ETISTRADIOL
		FOLLICORAL
		GINESTRENE
		INESTRA
		LINORAL
		LYNESTRENOL
		LYNORAL
		MENOLYN
		MICROFOLLIN
		NEO-ESTRONE
		NOVESTROL
		ORADIOL
		ORESTRALYN
		PALONYL
		PEROVEX
		PRIMOGYN
		PRIMOGYN C
		PRIMOGYN M
		PROGYNON C
		SPANESTRIN
57-64-7		
		AR-44
		ESERINE SALICYLATE
		PHYSOSTIGMINE SALICYLATE
		PHYSOSTIGMINE, MONOSALICYLATE
		PHYSOSTIGMINE, SALICYLATE (1:1)
		PHYSOSTOL SALICYLATE
		SALICYLIC ACID, COMPD WITH PHYSOSTIGMINE (1:1)
		TL-1380
57-74-9		
		4,7-METHANO-1H-INDENE, 1,2,4,5,6,7,8,8-OCTACHLORO-2,3,3A,4,7,7A-HEXAHYDRO-
		4,7-METHANOINDAN, 1,2,4,5,6,7,8,8-OCTACHLORO-3A,4,7,7A-TETRAHYDRO-
		CD 68
	*	CHLORDANE
		CHLORINDAN
		CORTILAN-NEU
		DOWCHLOR
		ENT 9932
		HCS 3260
		M 140
		OCTACHLORO-4,7-METHANOTETRAHYDROINDANE
		OKTATERR
		TAT CHLOR 4
		TOXICHLOR
57-83-0		
		AGOLUTIN
		BIO-LUTON
		CORLUTIN
		CORLUTINA
		CORLUVITE
		CORPORIN
		CORPUS LUTEUM HORMONE
		DELTA4-PREGNENE-3,20-DIONE
		FLAVOLUTAN
		FOLOGENON
		GESTEROL
		GESTONE
		GESTORMONE
		GESTRON
		GLANDUCORPIN
		GYNLUTIN
		GYNOLUTONE
		HORMOFLAVEINE
		HORMOLUTON
		LIPO-LUTIN
		LUCORTEUM SOL
		LUTEAL HORMONE
		LUTEINIQUE
		LUTEOCRIN NORMALE
		LUTEODYN
		LUTEOGAN
		LUTEOHORMONE
		LUTEOL
		LUTEOPUR
		LUTEOSAN
		LUTEOSTAB
		LUTEOVIS
		LUTEX
		LUTIDON
		LUTIN
		LUTOCICLINA
		LUTOCYCLIN
		LUTOCYCLIN M
		LUTOCYLIN
		LUTOFORM
		LUTOGYL
		LUTREN
		LUTROMONE
		NALUTRON
		PERCUTACRINE LUTEINIQUE
		PIAPONON
		PREGN-4-ENE-3,20-DIONE
		PRIMOLUT
		PROGEKAN
		PROGERSTERONE
		PROGESTASERT
		PROGESTEROL
	*	PROGESTERONE
		PROGESTIN
		PROGESTONE
		PROGESTRON
		PROLETS
		PROLIDON
		PROLUTON
		PROTORMONE
		SYNGESTRETS
		SYNTOLUTAN
		VITARRINE

CAS No.	IN	Chemical Name
57-92-1		
	*	STREPTOMYCIN
57-97-6		
		7,12-DIMETHYL-1,2-BENZANTHRACENE
		7,12-DIMETHYLBENZ(A)ANTHRACENE
		7,12-DIMETHYLBENZANTHRACENE
		7,12-DIMETHYLBENZO[A]ANTHRACENE
		7,12-DMBA
		9,10-DIMETHYL-1,2-BENZANTHRACENE
		9,10-DIMETHYLBENZ[A]ANTHRACENE
		BENZ(A)ANTHRACENE, 7,12-DIMETHYL-
		DIMETHYLBENZANTHRACENE
		DIMETHYLBENZ[A]ANTHRACENE
		DMBA
58-22-0		
		ANDROLIN
		ANDRONAQ
		ANDROST-4-EN-3-ONE, 17-HYDROXY-, (17.BETA.)-
		ANDRUSOL
		CRISTERONA T
		DELTA4-ANDROSTEN-17BETA-OL-3-ONE
		GENO-CRISTAUX GREMY
		HOMOSTERON
		HOMOSTERONE
		MERTESTATE
		NEOTESTIS
		ORQUISTERON
		PERANDREN
		PERCUTACRINE ANDROGENIQUE
		PRIMOTEST
		PRIMOTESTON
		SUSTANON
		SUSTANONE
		TESLEN
		TESTANDRONE
		TESTICULOSTERONE
		TESTOBASE
		TESTOPROPON
		TESTOSTEROID
		TESTOSTERON
	*	TESTOSTERONE
		TESTOVIRON SCHERING
		TESTOVIRON T
		TESTRONE
		TESTRYL
		VIRORMONE
		VIROSTERONE
58-22-0		
		CHLORDIAZEPOXIDE
58-36-6		
		PHENARSAZINE OXIDE
		PHENOXARSINE, 10,10'-OXYDI-
58-89-9		
		1,2,3,4,5,6-HEXACHLOROCYCLOHEXANE
		1,2,3,4,5,6-HEXACHLOROCYCLOHEXANE, GAMMA-ISOMER
		1-ALPHA,2-ALPHA,3-BETA,4-ALPHA,5-ALPHA,6-BETA-HEXACHLOROCYCLOHEXANE
		AALINDAN
		AFICIDE
		AGRISOL G-20
		AGROCIDE
		AGROCIDE 2
		AGROCIDE 6G
		AGROCIDE 7
		AGROCIDE III
		AGROCIDE WP
		AGRONEXIT
		AMEISENATOD
		AMEISENMITTEL MERCK
		AMEISENTOD
		APARASIN
		APHTIRIA
		APLIDAL
		ARBITEX
		BBH
		BEN-HEX
		BENHEXOL
		BENTOX 10
		BENZENE HEXACHLORIDE
		BENZENE HEXACHLORIDE-GAMMA-ISOMER
		BEXOL
		BHC
		CELANEX
		CHLORESENE
		CODECHINE
		COMPOUND-666
		CYCLOHEXANE, 1,2,3,4,5,6-HEXACHLORO-
		CYCLOHEXANE, 1,2,3,4,5,6-HEXACHLORO-, (1ALPHA,2ALPHA,3BETA,4ALPHA,5ALPHA,6BETA)-
		CYCLOHEXANE, 1,2,3,4,5,6-HEXACHLORO-, GAMMA-
		CYCLOHEXANE, 1,2,3,4,5,6-HEXACHLORO-, GAMMA-ISOMER
		DBH
		DETMOL-EXTRAKT
		DETOX 25
		DEVORAN
		DOL GRANULE
		DRILL TOX-SPEZIAL AGLUKON
		ENT 7,796
		ENT 8,601
		ENTOMOXAN
		EXAGAMA
		FENOFORM FORTE
		FORLIN
		FORST-NEXEN
		GALLOGAMA
		GAMACARBATOX
		GAMACID
		GAMAPHEX
		GAMENE
		GAMISO
		GAMMA BENZENE HEXACHLORIDE
		GAMMA ISOMER OF BENZENE HEXACHLORIDE
		GAMMA-1,2,3,4,5,6-HEXACHLOROCYCLOHEXANE
		GAMMA-666
		GAMMA-BENZENE HEXACHLORIDE
		GAMMA-BENZOHEXACHLORIDE
		GAMMA-BHC
		GAMMA-COL
		GAMMA-HCH
		GAMMA-HEXACHLOR
		GAMMA-HEXACHLORAN
		GAMMA-HEXACHLORANE
		GAMMA-HEXACHLOROBENZENE
		GAMMA-HEXACHLOROCYCLOHEXANE
		GAMMA-LINDANE
		GAMMAHEXA
		GAMMAHEXANE
		GAMMALIN
		GAMMALIN 20
		GAMMATERR
		GAMMEX
		GAMMEXANE
		GAMMOPAZ
		GEOBILAN
		GEOLIN G 3
		GEXANE
		HCC
		HCCH
		HCH
		HECLOTOX
		HEXA
		HEXACHLOR
		HEXACHLORAN
		HEXACHLORANE
		HEXACHLOROCYCLOHEXANE
		HEXACHLOROCYCLOHEXANE, GAMMA-ISOMER
		HEXAKLOR
		HEXATOX
		HEXAVERM
		HEXICIDE
		HEXYCLAN
		HEXYLAN

CAS No.	IN	Chemical Name
		HGI
		HILBEECH
		HORTEX
		HUNGARIA L 7
		INEXIT
		ISOTOX
		JACUTIN
		KOKOTINE
		KWELL
		LASOCHRON
		LENDINE
		LENTOX
		LIDENAL
		LINDAFOR
		LINDAGAM
		LINDAGRAIN
		LINDAGRANOX
	*	LINDANE
		LINDANE (ACGIH,DOT)
		LINDANE, LIQUID
		LINDANE, SOLID
		LINDAPOUDRE
		LINDATOX
		LINDEX
		LINDOSEP
		LINTOX
		LINVUR
		LOREXANE
		MGLAWIK L
		MILBOL 49
		MSZYCOL
		NA 2761 (DOT)
		NCI-C00204
		NEO-SCABICIDOL
		NEXEN FB
		NEXIT
		NEXIT-STARK
		NEXOL-E
		NICOCHLORAN
		NOVIGAM
		OMNITOX
		OVADZIAK
		OWADZIAK
		PEDRACZAK
		PF LANZOL
		PLK
		QUELLADA
		RCRA WASTE NUMBER U129
		SANG GAMMA
		SILVANOL
		SPRITZ-RAPIDIN
		SPRITZLINDANE
		SPRUEHPFLANZOL
		STREUNEX
		TAP 85
		TRI-6
		VERINDAL ULTRA
		VITON
59-01-8		
		KANAMYCIN
59-05-2		
		METHOTREXATE
59-50-7		
		1-CHLORO-2-METHYL-4-HYDROXYBENZENE
		2-CHLORO-5-HYDROXYTOLUENE
		2-CHLORO-HYDROXYTOLUENE
		3-METHYL-4-CHLOROPHENOL
		4-CHLORO-3-CRESOL
		4-CHLORO-3-METHYLPHENOL
		4-CHLORO-5-METHYLPHENOL
		4-CHLORO-M-CRESOL
		6-CHLORO-3-HYDROXYTOLUENE
		6-CHLORO-M-CRESOL
		APTAL
		BAKTOL
		BAKTOLAN

CAS No.	IN	Chemical Name
		CANDASEPTIC
		CHLOROCRESOL
		CRESOL, CHLORO-
		M-CRESOL, 4-CHLORO-
		OTTAFACT
		P-CHLOR-M-CRESOL
		P-CHLORO-M-CRESOL
		P-CHLOROCRESOL
		PARA-CHLORO-META-CRESOL
		PARMETOL
		PAROL
		PCMC
		PERITONAN
		PHENOL, 4-CHLORO-3-METHYL-
		PHENOL, 4-CHLORO-3-METHYL- (9 CI)
		PHENOL, CHLOROMETHYL-
		PREVENTOL CMK
		RASCHIT
		RASCHIT K
		RASEN-ANICON
		RCRA WASTE NUMBER U039
59-88-1		
		HYDRAZINE, PHENYL-, HYDROCHLORIDE
		HYDRAZINE, PHENYL-, MONOHYDROCHLORIDE
		N-PHENYLHYDRAZINE HYDROCHLORIDE
		PHENYLHYDRAZIN HYDROCHLORID (GERMAN)
		PHENYLHYDRAZINE HYDROCHLORIDE
		PHENYLHYDRAZINE MONOHYDROCHLORIDE
		PHENYLHYDRAZINIUM CHLORIDE
59-89-2		
		4-NITROSOMORPHOLINE
		MORPHOLINE, 4-NITROSO-
		N-NITROSOMORPHOLINE
		NITROSOMORPHOLINE
59-96-1		
		BENSYLYT
		BENZENEMETHANAMINE, N-(2-CHLOROETHYL)-N-(1-METHYL-2-PHENOXYETHYL)-
		BENZYLAMINE, N-(2-CHLOROETHYL)-N-(1-METHYL-2-PHENOXYETHYL)-
		BENZYLYT
		DIBENYLIN
		DIBENYLINE
		DIBENZYLINE
		PHENOXYBENZAMINE
60-00-4		
		3,6-DIAZAOCTANEDIOIC ACID, 3,6-BIS(CARBOXYMETHYL)-
		ACETIC ACID, (ETHYLENEDINITRILO)TETRA-
		ACETIC ACID, 2,2',2",2'"-(1,2-ETHANEDIYLDINITRILO)TETRAKIS-
		ACIDE ETHYLENEDIAMINETETRACETIQUE (FRENCH)
		CELON A
		CELON ATH
		CHEELOX
		CHEELOX BF ACID
		CHEMCOLOX 340
		CLEWAT TAA
		COMPLEXON II
		EDATHAMIL
		EDETIC
		EDETIC ACID
		EDTA
		EDTA (CHELATING AGENT)
		EDTA (DOT)
		EDTA ACID
		ENDRATE
		ETHYLENEDIAMINE TETRA-ACETIC ACID
		ETHYLENEDIAMINE-N,N,N',N'-TETRAACETIC ACID
		ETHYLENEDIAMINETETRAACETATE
	*	ETHYLENEDIAMINETETRAACETIC ACID
		ETHYLENEDIAMINETETRAACETIC ACID (DOT)
		ETHYLENEDIAMINOTETRAACETIC ACID
		ETHYLENEDINITRILOTETRAACETIC ACID
		GLYCINE, N,N'-1,2-ETHANEDIYLBIS(N-(CARBOXYMETHYL)-(9CI)

CAS No.	IN	Chemical Name
		GLYCINE, N,N'-1,2-ETHANEDIYLBIS[N-(CARBOXYMETHYL)-
		HAMP-ENE ACID
		HAVIDOTE
		ICRF 185
		METAQUEST A
		NA 9117 (DOT)
		NERVANAID B ACID
		NULLAPON B ACID
		NULLAPON BF ACID
		PERMA KLEER 50 ACID
		QUESTEX 4H
		SEQ 100
		SEQUESTRENE AA
		SEQUESTRIC ACID
		SEQUESTROL
		TETRINE ACID
		TITRIPLEX
		TRICON BW
		TRILON B
		TRILON BW
		VERSENE
		VERSENE ACID
		VINKEIL 100
		WARKEELATE ACID
60-09-3		
		4-AMINOAZOBENZENE
		C.I. SOLVENT YELLOW 1
60-11-7		
		4-(N,N-DIMETHYLAMINO)AZOBENZENE
		4-(PHENYLAZO)-N,N-DIMETHYLANILINE
	*	4-DIMETHYLAMINOAZOBENZENE
		BENZENAMINE, N,N-DIMETHYL-4-(PHENYLAZO)-
		BRILLIANT FAST OIL YELLOW
		BRILLIANT FAST SPIRIT YELLOW
		BRILLIANT OIL YELLOW
		BUTTER YELLOW
		C.I. 11020
		C.I. SOLVENT YELLOW 2
		CERASINE YELLOW GG
		DAB
		DAB (CARCINOGEN)
		DIMETHYL YELLOW
		DMAB
		ENIAL YELLOW 2G
		FAST OIL YELLOW B
		FAT YELLOW
		FAT YELLOW A
		FAT YELLOW AD OO
		FAT YELLOW ES
		FAT YELLOW ES EXTRA
		FAT YELLOW EXTRA CONC
		FAT YELLOW R
		GRASAL BRILLIANT YELLOW
		IKETON YELLOW EXTRA
		METHYL YELLOW
		N,N-DIMETHYL-P-(PHENYLAZO)ANILINE
		OIL YELLOW 20
		OIL YELLOW 2625
		OIL YELLOW 2G
		OIL YELLOW BB
		OIL YELLOW D
		OIL YELLOW FN
		OIL YELLOW G
		OIL YELLOW GG
		OIL YELLOW GR
		OIL YELLOW II
		OIL YELLOW N
		OIL YELLOW PEL
		OIL YELLOW S
		OLEAL YELLOW 2G
		ORGANOL YELLOW ADM
		ORIENT OIL YELLOW GG
		P-DIMETHYLAMINOAZOBENZENE
		PETROL YELLOW WT
		RESINOL YELLOW GR
		SILOTRAS YELLOW T 2G
		SOMALIA YELLOW A
		STEAR YELLOW JB
		SUDAN YELLOW GG
		SUDAN YELLOW GGA
		TOYO OIL YELLOW G
		WAXOLINE YELLOW ADS
		YELLOW G SOLUBLE IN GREASE
60-24-2		
		1-ETHANOL-2-THIOL
		1-HYDROXY-2-MERCAPTOETHANE
		1-MERCAPTO-2-HYDROXYETHANE
		2-HYDROXY-1-ETHANETHIOL
		2-HYDROXYETHANETHIOL
		2-HYDROXYETHYL MERCAPTAN
		2-ME
		2-MERCAPTO-1-ETHANOL
		2-MERCAPTOETHANOL
		2-MERCAPTOETHYL ALCOHOL
		2-THIOETHANOL
		BETA-HYDROXYETHANETHIOL
		BETA-HYDROXYETHYLMERCAPTAN
		BETA-MERCAPTOETHANOL
		EMERY 5791
		ETHANOL, 2-MERCAPTO-
		ETHYLENE GLYCOL, MONOTHIO-
		HYDROXYETHYL MERCAPTAN
		MERCAPTOETHANOL
		MONOTHIOETHYLENE GLYCOL
		MONOTHIOETHYLENEGLYCOL
		MONOTHIOGLYCOL
		THIOETHYLENE GLYCOL
		THIOGLYCOL
		THIOGLYCOL (DOT)
		THIOMONOGLYCOL
		TINCTURES, MEDICINAL
		UN 2966 (DOT)
		USAF EK-4196
60-29-7		
		3-OXAPENTANE
		AETHER
		ANAESTHETIC ETHER
		ANESTHESIA ETHER
		ANESTHETIC ETHER
		DIAETHYLAETHER (GERMAN)
	*	DIETHYL ETHER
		DIETHYL ETHER (DOT)
		DIETHYL OXIDE
		DWUETYLOWY ETER (POLISH)
		ETERE ETILICO (ITALIAN)
		ETHANE, 1,1'-OXYBIS-
		ETHER
		ETHER ETHYLIQUE (FRENCH)
		ETHOXYETHANE
		ETHYL ETHER
		ETHYL ETHER (ACGIH,DOT)
		OXYDE D'ETHYLE (FRENCH)
		PRONARCOL
		RCRA WASTE NUMBER U117
		SOLVENT ETHER
		SULFURIC ETHER
		UN 1155 (DOT)
60-34-4		
		1-METHYLHYDRAZINE
		HYDRAZINE, METHYL-
		HYDRAZOMETHANE
		METHYL HYDRAZINE
		METHYLHYDRAZINE (ACGIH,DOT)
		METHYLHYDRAZINE (MONO)
		METYLOHYDRAZYNA (POLISH)
		MMH
		MONOMETHYL HYDRAZINE
		RCRA WASTE NUMBER P068
		UN 1244 (DOT)
60-35-5		
	*	ACETAMIDE

CAS No.	IN	Chemical Name
60-41-3		
		ANTIVAMPIRE
		STRYCHININE SULFATE
		STRYCHNIDIN-10-ONE, SULFATE (2:1)
		STRYCHNINE SULFATE
		STRYCHNINE, SULFATE (2:1)
		STRYCHNINIUM SULFATE
		VAMPIROL
60-51-5		
		(O,O-DIMETHYL-S-(N-METHYL-CARBAMOYL-METHYL)-DITHIOPHOSPHAT) (GERMAN)
		8014 BIS HC
		AADIMETHOAL
		AC-12880
		AC-18682
		ACETIC ACID, O,O-DIMETHYLDITHIOPHOSPHORYL-, N-MONOMETHYLAMIDE SALT
		AMERICAN CYANAMID 12,880
		BI 58
		BI 58 EC
		CEKUTHOATE
		CL 12880
		CYGON
		CYGON 2E
		CYGON 4E
		CYGON INSECTICIDE
		DAPHENE
		DE-FEND
		DEMOS-L40
		DEVIGON
		DIMATE 267
		DIMETATE
		DIMETHOAAT (DUTCH)
		DIMETHOAT (GERMAN)
		DIMETHOAT TECHNISCH 95%
		DIMETHOATE
		DIMETHOATE-267
		DIMETHOGEN
		DIMETON
		DIMEVUR
		DITHIOPHOSPHATE DE O,O-DIMETHYLE ET DE S(-N-METHYLCARBAMOYL-METHYLE) (FRENCH)
		EI-12880
		ENT 24,650
		EXPERIMENTAL INSECTICIDE 12,880
		FERKETHION
		FIP
		FORTION NM
		FOSFAMID
		FOSFATOX R
		FOSFOTOX
		FOSFOTOX R
		FOSFOTOX R 35
		FOSTION MM
		L-395
		LURGO
		N-MONOMETHYLAMIDE OF O,O-DIMETHYLDITHIOPHOSPHORYLACETIC ACID
		NC-262
		NCI-C00135
		O,O-DIMETHYL METHYLCARBAMOYLMETHYL PHOSPHORODITHIOATE
		O,O-DIMETHYL S-(2-(METHYLAMINO)-2-OXOETHYL) PHOSPHORODITHIOATE
		O,O-DIMETHYL S-(N-METHYLCARBAMOYLMETHYL) DITHIOPHOSPHATE
		O,O-DIMETHYL S-(N-METHYLCARBAMOYLMETHYL) PHOSPHORODITHIOATE
		O,O-DIMETHYL S-(N-METHYLCARBAMYLMETHYL) THIOTHIONOPHOSPHATE
		O,O-DIMETHYL-DITHIOPHOSPHORYLESSIGSAEURE MONOMETHYLAMID (GERMAN)
		O,O-DIMETHYL-S-(2-OXO-3-AZA-BUTYL)-DITHIOPHOSPHAT (GERMAN)
		O,O-DIMETHYL-S-(N-METHYL-CARBAMOYL)-METHYL-DITHIOFOSFAAT (DUTCH)
		O,O-DIMETHYL-S-(N-MONOMETHYL)-CARBAMYL METHYL DITHIOPHOSPHATE
		O,O-DIMETHYLDITHIOPHOSPHORYLACETIC ACID, N-MONOMETHYLAMIDE SALT
		O,O-DIMETIL-S-(N-METIL-CARBAMOIL-METIL)-DITIOFOSFATO (ITALIAN)
		PEI 75
		PERFECTHION
		PERFEKTHION
		PERFEKTION
		PHOSPHAMID
		PHOSPHAMIDE
		PHOSPHORODITHIOIC ACID, O,O-DIMETHYL ESTER, S-ESTER WITH 2-MERCAPTO-N-METHYLACETAMIDE
		PHOSPHORODITHIOIC ACID, O,O-DIMETHYL S-(METHYLCARBAMOYLMETHYL) ESTER
		PHOSPHORODITHIOIC ACID, O,O-DIMETHYL S-[2-(METHYLAMINO)-2-OXOETHYL]ESTER
		PHOSPHORODITHIOIC ACID, O,O-DIMETHYL-S-(2-(METHYLAMINO)-2-OXOETHYL) ESTER (9CI)
		RACUSAN
		RCRA WASTE NUMBER P044
		REBELATE
		ROGODIAL
		ROGOR
		ROGOR 20L
		ROGOR 40
		ROGOR L
		ROGOR P
		ROXION
		ROXION UA
		S-METHYLCARBAMOYLMETHYL O,O-DIMETHYL PHOSPHORODITHIOATE
		SINORATOX
		SISTEMIN
		SYSTEMIN
		SYSTOATE
		TARA 909
		TRIMETION
60-54-8		
		TETRACYCLINE
60-56-0		
		METHIMAZOLE
60-57-1		
		3,4,5,6,9,9-HEXACHLORO-1A,2,2A,3,6,6A,7,7A-OCTAHYDRO-2,7:3,6-DIMETHANONAPHTH(2,3-B)OXIRENE
		ALDRIN EPOXIDE
		ALVIT
		ALVIT 55
		COMPOUND 497
		DIELDREX
	*	DIELDRIN
		DIELDRIN (ACGIH,DOT)
		DIELDRINE (FRENCH)
		DIELDRITE
		DIELMOTH
		DORYTOX
		ENT 16,225
		EXO-DIELDRIN
		HEOD
		HEXACHLOROEPOXYOCTAHYDRO-ENDO,EXO-DIMETHANONAPHTHALENE
		ILLOXOL
		INSECTICIDE NO 497
		INSECTLACK
		KOMBI-ALBERTAN
		MOTH SNUB D
		NA 2761 (DOT)
		NCI-C00124
		OCTALOX
		PANORAM D-31
		QUINTOX
		RCRA WASTE NUMBER P037
		RED SHIELD
		SD 3417
		TERMITOX

CAS No.	IN	Chemical Name
61-33-6		
		BENZYLPENICILLIN
		PENICILLIN
61-57-4		
		2-IMIDAZOLIDINONE, 1-(5-NITRO-2-THIAZOLYL)-
		AMBILHAR
		BA 32644
		CIBA 32644
		CIBA 32644-BA
		NIRIDAZOLE
		NITROTHIAMIDAZOLE
		NITROTHIAZOL
		NITROTHIAZOLE
61-82-5		
		1,2,4-TRIAZOLE-3-AMINE
		1H-1,2,4-TRIAZOL-3-AMINE
		2-AMINO-1,3,4-TRIAZOLE
		2-AMINOTRIAZOLE
		3,A-T
		3-AMINO-1,2,4-TRIAZOLE
		3-AMINO-1H-1,2,4-TRIAZOLE
		3-AMINO-A-TRIAZOLE
		3-AMINO-S-TRIAZOLE
		3-AMINOTRIAZOLE
		5-AMINO-1,2,4-TRIAZOLE
		5-AMINO-1H-1,2,4-TRIAZOLE
		AMINOTRIAZOLE
		AMINOTRIAZOLE (PLANT REGULATOR)
		AMITROL
		AMITROL 90
		AMITROLE
		AMIZOL
		AT
		ATA
		AZAPLANT
		CYTROL
		CYTROLE
		ENT 25445
		HERBIDAL TOTAL
		S-TRIAZOLE, 3-AMINO-
		WEEDAZOL
62-38-4		
		(ACETATO)PHENYLMERCURY
		(ACETATO-O)PHENYLMERCURY
		(ACETOXYMERCURI)BENZENE
		ACETATE PHENYLMERCURIQUE (FRENCH)
		ACETATOPHENYLMERCURY
		ACETIC ACID, PHENYLMERCURY DERIV
		ACETOXYPHENYLMERCURY
		AGROSAN
		AGROSAN D
		AGROSAN GN 5
		ALGIMYCIN
		ALGIMYCIN 200
		ANTICON
		ANTIMUCIN WBR
		ANTIMUCIN WDR
		BENZENE, (ACETOXYMERCURI)-
		BENZENE, (ACETOXYMERCURIO)-
		BUFEN
		BUFEN 30
		CEKUSIL
		CELMER
		CERESAN
		CERESAN UNIVERSAL
		CERESOL
		CONTRA CREME
		DYANACIDE
		FEMMA
		FENYLMERCURIACETAT (CZECH)
		FMA
		FUNGICIDE R
		FUNGITOX OR
		GALLOTOX
		HEXASAN
		HEXASAN (FUNGICIDE)
		HL-331
		HONGNIEN
		HOSTAQUICK
		HOSTAQUIK
		KWIKSAN
		LEYTOSAN
		LIQUIPHENE
		LOROPHYN
		MERACEN
		MERCRON
		MERCURIPHENYL ACETATE
		MERCURON
		MERCURY (II) ACETATE, PHENYL-
		MERCURY, (ACETATO)PHENYL-
		MERCURY, (ACETATO-O)PHENYL-
		MERCURY, ACETOXYPHENYL-
		MERGAL A 25
		MERSOLITE
		MERSOLITE 8
		MERSOLITE D
		METASOL 30
		NEANTINA
		NORFORMS
		NYLMERATE
		OCTAN FENYLRTUTNATY (CZECH)
		PAMISAN
		PANOMATIC
		PHENMAD
		PHENOMERCURIC ACETATE
		PHENYL MERCURIC ACETATE
		PHENYLMERCURIACETATE
	*	PHENYLMERCURIC ACETATE
		PHENYLMERCURIC ACETATE (DOT)
		PHENYLMERCURY ACETATE
		PHENYLQUECKSILBERACETAT (GERMAN)
		PHIX
		PMA
		PMA (FUNGICIDE)
		PMA 220
		PMAC
		PMACETATE
		PMAS
		PROGRAMIN
		QUICKSAN
		RUBERON
		SAMTOL
		SANITIZED SPG
		SANMICRON
		SCUTL
		SEED DRESSING R
		SETRETE
		SPRUCE SEAL
		TAG
		TAG 331
		TAG FUNGICIDE
		TAG HL 331
		TRIGOSAN
		TROYSAN 30
		TROYSAN PMA 30
		VERDASAN
		VOLPAR
		ZAPRAWA NASIENNA R
		ZIARNIK
62-44-2		
		4'-ETHOXYACETANILIDE
		4-ETHOXYACETANILIDE
		ACETAMIDE, N-(4-ETHOXYPHENYL)-
		ACETO-4-PHENETIDINE
		ACETOPHENETIDIN
		ACETOPHENETIDINE
		ACETOPHENETIN
		FENIDINA
		FENINA
		KALMIN
		N-(4-ETHOXYPHENYL)ACETAMIDE
		N-ACETYL-4-ETHOXYANILINE
		N-ACETYL-P-ETHOXYANILINE
		N-ACETYL-P-PHENETIDINE

CAS No.	IN	Chemical Name
		P-ACETOPHENETIDIDE
		P-ETHOXYACETANILIDE
		PERTONAL
	*	PHENACETIN
		PHENACETINE
		PHENAZETIN
		PHENEDINA
		PHENIDIN
		PHENIN
62-50-0		
		EMS
		ETHYL MESYLATE
		ETHYL METHANESULFONATE
		ETHYL METHANESULPHONATE
		METHANESULFONIC ACID, ETHYL ESTER
		NSC 26805
		O-ETHYL METHYLSULFONATE
62-53-3		
		AMINOBENZENE
		AMINOPHEN
		ANILIN (CZECH)
		ANILINA (ITALIAN, POLISH)
	*	ANILINE
		ANILINE (ACGIH,DOT)
		ANILINE OIL
		ANILINE OIL, LIQUID
		ANILINE OIL, LIQUID (DOT)
		ANYVIM
		BENZENAMINE
		BENZENE, AMINO-
		BENZIDAM
		BLUE OIL
		C.I. 76000
		CI OXIDATION BASE 1
		CYA NOL
		HUILE D'ANILINE (FRENCH)
		KRYSTALLIN
		KYANOL
		NCI-C03736
		PHENYLAMINE
		RCRA WASTE NUMBER U012
		UN 1547 (DOT)
62-55-5		
		ACETAMIDE, THIO-
		ACETIMIDIC ACID, THIO-
		ACETOTHIOAMIDE
		ETHANETHIOAMIDE
		TAA
		THIACETAMIDE
	*	THIOACETAMIDE
62-56-6		
		2-THIOUREA
		BETA-THIOPSEUDOUREA
		ISOTHIOUREA
		PSEUDOTHIOUREA
		PSEUDOUREA, 2-THIO-
		RCRA WASTE NUMBER U219
		SULOUREA
		THIOCARBAMIDE
	*	THIOUREA
		THIOUREA (DOT)
		THU
		TSIZP 34
		UN 2877 (DOT)
		UREA, 2-THIO-
		UREA, THIO-
		UREA, THIO- (8CI)
		USAF EK-497
62-73-7		
		2,2-DICHLOROETHENYL DIMETHYL PHOSPHATE
		2,2-DICHLOROVINYL DIMETHYL PHOSPHATE
		ATGARD
		ATGARD V
		BIBESOL

CAS No.	IN	Chemical Name
		BREVINYL
		BREVINYL E 50
		CANOGARD
		CHLORVINPHOS
		DDVP
		DDVP (INSECTICIDE)
		DEDEVAP
		DES
		DICHLOFOS
		DICHLORMAN
		DICHLOROVOS
		DICHLORVOS
		DICHLORVOS MIXTURE, DRY
		DIMETHYL 2,2-DICHLOROETHENYL PHOSPHATE
		DIMETHYL 2,2-DICHLOROVINYL PHOSPHATE
		DIMETHYL DICHLOROVINYL PHOSPHATE
		DIMETHYL-O-O-DICHLOROVINYL-2-2-PHOSPHATE (TECHNICAL)
		DIVIPAN
		ENT 20738
		EQUIGARD
		EQUIGEL
		ESTROSEL
		ESTROSOL
		ETHENOL, 2,2-DICHLORO-, DIMETHYL PHOSPHATE
		FECAMA
		FEKAMA
		INSECTIGAS D
		MOPARI
		NEFRAFOS
		NERKOL
		NOGOS
		NOGOS 50
		NOGOS G
		NOVOTOX
		NUVAN
		NUVAN 100EC
		NUVAN 7
		O,O-DIMETHYL 2,2-DICHLOROVINYL PHOSPHATE
		OKO
		OMS 14
		PANAPLATE
		PHOSPHORIC ACID, 2,2-DICHLOROETHENYL DIMETHYL ESTER
		PHOSPHORIC ACID, 2,2-DICHLOROVINYL DIMETHYL ESTER
		SD
		SD 1750
		SZKLARNIAK
		TAP 9VP
		TASK
		UNIFOS
		UNIFOS (PESTICIDE)
		UNITOX
		VAPONA
		VAPONA INSECTICIDE
		VAPONITE
		VINYLOFOS
		VINYLOPHOS
		WINYLOPHOS
62-74-8		
		ACETIC ACID, FLUORO-, SODIUM SALT
		COMPOUND NO 1080
		FLUORESSIGAEURE (GERMAN)
		FLUOROACETIC ACID SODIUM SALT
		FRATOL
		FURATOL
		MONOFLUORESSIGSAURES NATRIUM (GERMAN)
		NATRIUMFLUORACETAAT (DUTCH)
		NATRIUMFLUORACETAT (GERMAN)
		RATBANE 1080
		RCRA WASTE NUMBER P058
		SMFA
		SODIO, FLUORACETATO DI (ITALIAN)
		SODIUM FLUOACETATE
		SODIUM FLUOACETIC ACID
		SODIUM FLUORACETATE
		SODIUM FLUOROACETATE
		SODIUM FLUOROACETATE (ACGIH,DOT)

CAS No.	IN	Chemical Name
		SODIUM FLUOROACETATE DE (FRENCH)
		SODIUM MONOFLUOROACETATE
		TL 869
		UN 2629 (DOT)
		YASOKNOCK
62-75-9		
		DIMETHYLAMINE, N-NITROSO-
		DIMETHYLNITROSAMIN (GERMAN)
		DIMETHYLNITROSAMINE
		DIMETHYLNITROSOAMINE
		DMN
		DMNA
		METHANAMINE, N-METHYL-N-NITROSO-
		N,N-DIMETHYLNITROSAMINE
		N-METHYL-N-NITROSOMETHANAMINE
		N-NITROSO-N,N-DIMETHYLAMINE
		N-NITROSODIMETHYLAMINE
		N-NITROSODIMETHYLAMINE (ACGIH)
		NDMA
		NITROSODIMETHYLAMINE
		RCRA WASTE NUMBER P082
62-76-0		
		DISODIUM OXALATE
		ETHANEDIOIC ACID, DISODIUM SALT
		NATRIUMOXALAT (GERMAN)
		OXALIC ACID, DISODIUM SALT
	*	SODIUM OXALATE
63-25-2		
		1-NAPHTHALENOL, METHYLCARBAMATE
		1-NAPHTHOL N-METHYLCARBAMATE
		1-NAPHTHYL METHYLCARBAMATE
		1-NAPHTHYL N-METHYLCARBAMATE
		1-NAPHTHYL-N-METHYL-KARBAMAT (GERMAN)
		ALPHA-NAFTYL-N-METHYLKARBAMAT (CZECH)
		ALPHA-NAPHTHALENYL METHYLCARBAMATE
		ALPHA-NAPHTHYL METHYLCARBAMATE
		ALPHA-NAPHTHYL- N-METHYLCARBAMATE
		ARILAT
		ARILATE
		ARYLAM
		ATOXAN
		BERCEMA NMC50
		CAPROLIN
		CARBAMIC ACID, METHYL-, 1-NAPHTHYL ESTER
		CARBAMINE
		CARBARIL (ITALIAN)
	*	CARBARYL
		CARBARYL (ACGIH,DOT)
		CARBATOX
		CARBATOX 60
		CARBATOX 75
		CARBAVUR
		CARBOMATE
		CARPOLIN
		CARYLDERM
		CEKUBARYL
		COMPOUND 7744
		CRAG SEVIN
		DENAPON
		DEVICARB
		DICARBAM
		DYNA-CARBYL
		ENT 23,969
		EXPERIMENTAL INSECTICIDE 7744
		GAMONIL
		GERMAIN'S
		HEXAVIN
		KARBARYL (POLISH)
		KARBASPRAY
		KARBATOX
		KARBATOX 75
		KARBOSEP
		MENAPHTAM
		METHYLCARBAMATE 1-NAPHTHALENOL
		METHYLCARBAMATE 1-NAPHTHOL
		METHYLCARBAMIC ACID, 1-NAPHTHYL ESTER
		MONSUR
		MUGAN
		MURVIN
		N-METHYL-1 -NAPHTHYL CARBAMATE
		N-METHYL-1-NAFTYL-CARBAMAAT (DUTCH)
		N-METHYL-1-NAPHTHYL-CARBAMAT (GERMAN)
		N-METHYL-ALPHA-NAPHTHYLCARBAMATE
		N-METHYL-ALPHA-NAPHTHYLURETHAN
		N-METHYLCARBAMATE DE 1-NAPHTYLE (FRENCH)
		N-METIL-1-NAFTIL-CARBAMMATO (ITALIAN)
		NA 2757 (DOT)
		NAC
		NMC 50
		OLTITOX
		OMS-29
		PANAM
		POMEX
		PROSEVOR 85
		RAVYON
		RYLAM
		SEFFEIN
		SEPTENE
		SEVIMOL
		SEVIN
		SEVIN 4
		SEWIN
		SOK
		TERCYL
		TOXAN
		TRICARNAM
		UC 7744
		UNION CARBIDE 7,744
		VIOXAN
63-92-3		
		BENZENEMETHANAMINE, N-(2-CHLOROETHYL)-N-(1-METHYL-2-PHENOXYETHYL)-, HYDROCHLORIDE
		BENZYLAMINE, N-(2-CHLOROETHYL)-N-(1-METHYL-2-PHENOXYETHYL)-, HYDROCHLORIDE
		DIBENZYLINE CHLORIDE
		DIBENZYLINE HYDROCHLORIDE
		N-(2-CHLOROETHYL)-N-(1-METHYL-2-PHENOXYETHYL)BENZYLAMINE, HYDROCHLORIDE
		PHENOXYBENZAMINE CHLORIDE
		PHENOXYBENZAMINE HYDROCHLORIDE
64-00-6		
		1PC
		3-ISOPROPYLPHENYL METHYLCARBAMATE
		3-ISOPROPYLPHENYL N-METHYLCARBAMATE
		AC 5727
		CARBAMIC ACID, METHYL-, 3-(1-METHYLETHYL)PHENYL ESTER
		CARBAMIC ACID, METHYL-, M-CUMENYL ESTER
		CARBAMIC ACID, N-METHYL-, 3-ISOPROPYLPHENYL ESTER
		COMPOUND 10854
		ENT 25,500
		ENT 25,543
		H 5727
		H 8757
		HER 5727
		HERCULES 5727
		HERCULES AC 5727
		HIP
		M-CUMENOL METHYLCARBAMATE
		M-CUMENYL METHYLCARBAMATE
		M-ISOPROPYLPHENOL METHYLCARBAMATE
		M-ISOPROPYLPHENOL N-METHYLCARBAMATE
		M-ISOPROPYLPHENYL METHYLCARBAMATE
		M-ISOPROPYLPHENYL N-METHYLCARBAMATE
		N-METHYL 3-ISOPROPYLPHENYL CARBAMATE
		N-METHYL M-ISOPROPYLPHENYL CARBAMATE
		OMS 15
		OMS 162
		PHENOL, 3-(1-METHYLETHYL)-, METHYLCARBAMATE
		PHENOL, M-ISOPROPYL-, METHYLCARBAMATE
		UC 10854
		UNION CARBIDE 10854
		UNION CARBIDE UC-10,854

CAS No.	IN	Chemical Name
64-10-8		
		1-PHENYLUREA
		MONOPHENYLUREA
		N-PHENYLUREA
		PC
		PHENYLCARBAMIDE
		PHENYLUREA
		PHENYLUREA PESTICIDE, LIQUID [FLAMMABLE LIQUID LABEL]
		PHENYLUREA PESTICIDE, LIQUID [POISON B LABEL]
		PHENYLUREA PESTICIDE, SOLID
		STABILISATOR VH
		STABILIZER VH
		UREA, PHENYL-
		VH
64-17-5		
		ABSOLUTE ETHANOL
		AETHANOL (GERMAN)
		AETHYLALKOHOL (GERMAN)
		ALCARE HAND DEGERMER
		ALCOHOL
		ALCOHOL (ETHYL)
		ALCOHOL ANHYDROUS
		ALCOHOL DEHYDRATED
		ALCOOL ETHYLIQUE (FRENCH)
		ALCOOL ETILICO (ITALIAN)
		ALGRAIN
		ALKOHOL (GERMAN)
		ALKOHOLU ETYLOWEGO (POLISH)
		ANHYDROL
		COLOGNE SPIRIT
		COLOGNE SPIRITS
		COLOGNE SPIRITS (ALCOHOL) (DOT)
		DENATURATED ALCOHOL
		DENATURED ALCOHOL
		ETANOLO (ITALIAN)
		ETHANOL
		ETHANOL (DOT)
		ETHANOL 200 PROOF
		ETHANOL SOLUTION (DOT)
	*	ETHYL ALCOHOL
		ETHYL ALCOHOL (ACGIH,DOT)
		ETHYL ALCOHOL ANHYDROUS
		ETHYL HYDRATE
		ETHYL HYDROXIDE
		ETHYLALCOHOL (DUTCH)
		ETYLOWY ALKOHOL (POLISH)
		FERMENTATION ALCOHOL
		GRAIN ALCOHOL
		JAYSOL
		JAYSOL S
		METHYLCARBINOL
		MOLASSES ALCOHOL
		NCI-C03134
		POTATO ALCOHOL
		SD ALCOHOL 23-HYDROGEN
		SPIRITS OF WINE
		SPIRT
		TECSOL
		TECSOL C
		UN 1170 (DOT)
64-18-6		
		ACIDE FORMIQUE (FRENCH)
		ACIDO FORMICO (ITALIAN)
		ADD-F
		AMEISENSAEURE (GERMAN)
		AMINIC ACID
		BILORIN
		COLLO-BUEGLATT
		COLLO-DIDAX
	*	FORMIC ACID
		FORMIC ACID (ACGIH,DOT)
		FORMIC ACID SOLUTION
		FORMIC ACID, SOLUTION (DOT)
		FORMIRA
		FORMISOTON
		FORMYLIC ACID
		HYDROGEN CARBOXYLIC ACID
		KWAS METANIOWY (POLISH)
		METHANOIC ACID
		MIERENZUUR (DUTCH)
		MYRMICYL
		RCRA WASTE NUMBER U123
		UN 1779 (DOT)
64-19-7		
	*	ACETIC ACID
		ACETIC ACID (ACGIH)
		ACETIC ACID (AQUEOUS SOLUTION)
		ACETIC ACID (AQUEOUS SOLUTION) (DOT)
		ACETIC ACID SOLUTION, MORE THAN 25% BUT NOT MORE THAN 80% ACID, BY WEIGHT (DOT)
		ACETIC ACID, GLACIAL
		ACETIC ACID, GLACIAL (DOT)
		ACETIC ACID, GLACIAL OR ACETIC ACID SOLUTION, MORE THAN 80% ACID
		ACETIC ACID, MORE THAN 25% BUT NOT MORE THAN 80% ACID, BY WEIGHT
		ACIDE ACETIQUE (FRENCH)
		ACIDO ACETICO (ITALIAN)
		AZIJNZUUR (DUTCH)
		ESSIGSAEURE (GERMAN)
		ETHANOIC ACID
		ETHYLIC ACID
		GLACIAL ACETIC ACID
		METHANECARBOXYLIC ACID
		OCTOWY KWAS (POLISH)
		UN 2789 (DOT)
		UN 2790 (DOT)
		VINEGAR ACID
64-67-5		
		DES
	*	DIETHYL SULFATE
		DIETHYL SULPHATE
		ETHYL SULFATE
		SULFURIC ACID, DIETHYL ESTER
64-86-8		
		7-ALPHA-H-COLCHICINE
		ACETAMIDE, N-(5,6,7,9-TETRAHYDRO-1,2,3,10-TETRA-METHOXY-9-OXOBENZO(ALPHA)HEPTALEN-7-YL)-
		ACETAMIDE, N-(5,6,7,9-TETRAHYDRO-1,2,3,10-TETRA-METHOXY-9-OXOBENZO[A]HEPTALEN-7-YL)-, (S)-
		BENZO(A)HEPTALEN-9(5H)-ONE, 7-ACETAMIDO-6,7-DIHY-DRO-1,2,3,10-TETRAMETHOXY-
		BENZO[A]HEPTALENE, ACETAMIDE DERIV
		COLCHICIN (GERMAN)
		COLCHICINA (ITALIAN)
	*	COLCHICINE
		COLCHINEOS
		COLCHISOL
		COLCIN
		COLSALOID
		CONDYLON
		N-ACETYL TRIMETHYLCOLCHICINIC ACID METHYLETHER
		NSC 757
65-30-5		
		(S)-3-(1-METHYL-2-PYRROLIDINYL)PYRIDINE SULFATE (2:1)
		BLACK LEAF 40
		ENT 2,435
		L-1-METHYL-2-(3-PYRIDYL)-PYRROLIDINE SULFATE
		L-3-(1-METHYL-2-PYRROLIDYL)PYRIDINE SULFATE
	*	NICOTINE SULFATE
		NICOTINE SULFATE, LIQUID
		NICOTINE SULFATE, LIQUID (DOT)
		NICOTINE SULFATE, SOLID
		NICOTINE SULFATE, SOLID (DOT)
		NICOTINE, SULFATE
		NICOTINE, SULFATE (2:1)
		NIKOTINSULFAT (GERMAN)
		PYRIDINE, 3-(1-METHYL-2-PYRROLIDINYL)-, (S)-, SULFATE
		PYRIDINE, 3-(1-METHYL-2-PYRROLIDINYL)-, (S)-, SULFATE (2:1)
		PYRROLIDINE, 1-METHYL-2-(3-PYRIDYL)-, SULFATE

CAS No.	IN	Chemical Name
		SULFATE DE NICOTINE (FRENCH)
		UN 1658 (DOT)
65-31-6		
		(-)-NICOTINE BITARTRATE
		(-)-NICOTINE HYDROGEN TARTRATE
		4921453
		NICOTINE ACID TARTRATE
		NICOTINE BITARTRATE
		NICOTINE D-BITARTRATE
		NICOTINE HYDROGEN TARTRATE
		NICOTINE HYDROTARTRATE
		NICOTINE TARTRATE
		NICOTINE TARTRATE (DOT)
		NICOTINE, TARTRATE
		NICOTINE, TARTRATE (1:1)
		NICOTINE, TARTRATE (1:2)
		NICOTINE-D-TARTRATE
		PYRIDINE, 3-(1-METHYL-2-PYRROLIDINYL)-, (S)-, [R-(R*,R*)]- 2,3-DIHYDROXYBUTANEDIOATE
		PYRIDINE, 3-(1-METHYL-2-PYRROLIDINYL)-, (S)-, [R-(R*,R*)]- 2,3-DIHYDROXYBUTANEDIOATE (1:1)
		PYRIDINE, 3-(1-METHYL-2-PYRROLIDINYL)-, (S)-, [R-(R*,R*)]- 2,3-DIHYDROXYBUTANEDIOATE (1:2)
		PYRIDINE, 3-(1-METHYL-2-PYRROLIDINYL-, (S)-, (R-(R,R))- 2,3-DIHYDROXYBUTANEDIOATE (1:2)
		QT0350000
		TARTRATE DE NICOTINE (FRENCH)
		UN 1659 (DOT)
65-85-0		
		BENZENECARBOXYLIC ACID
		BENZENEFORMIC ACID
		BENZENEMETHANOIC ACID
	*	BENZOIC ACID
		CARBOXYBENZENE
		DRACYLIC ACID
		HA 1
		PHENYLCARBOXYLIC ACID
		PHENYLFORMIC ACID
		RETARDER BA
		RETARDEX
		SALVO LIQUID
		SOLVO POWDER
		TENN-PLAS
65-86-1		
		OROTIC ACID
66-25-1		
		1-HEXANAL
		ALDEHYDE C-6
		CAPROALDEHYDE
		CAPROIC ALDEHYDE
		CAPRONALDEHYDE
		HEXALDEHYDE
		HEXALDEHYDE (DOT)
		HEXANAL
		HEXYLALDEHYDE
		N-CAPROALDEHYDE
		N-CAPRONALDEHYDE
		N-CAPROYLALDEHYDE
		N-HEXANAL
		UN 1207 (DOT)
66-27-3		
		METHANESULFONIC ACID, METHYL ESTER
		METHYL METHANESLFONATE
		METHYL METHANESULFONATE
		METHYL METHANESULPHONATE
		METHYL METHYLSULFONATE
		MMS
66-75-1		
		2,4(1H,3H)-PYRIMIDINEDIONE, 5-[BIS(2-CHLOROETHY- L)AMINO]-
		2,6-DIHYDROXY-5-BIS[2-CHLOROETHYL]AMINOPYRIMIDINE
		5-(BIS(2-CHLOROETHYL)AMINO)URACIL 2,4(1H,3H) PYRIMI- DINEDIONE, 5-BIS(2-CHLOROETHYL)AMINO
		5-N,N-BIS(2-CHLOROETHYL)AMINOURACIL
		5-[BIS(2-CHLOROETHYL)AMINO]URACIL
		5-[DI(.BETA.-CHLOROETHYL)AMINO]URACIL
		AMINOURACIL MUSTARD
		CHLORETHAMINACIL
		DEMETHYLDOPAN
		DESMETHYLDOPAN
		ENT 50439
		NORDOPAN
		NSC-34462
		U-8344
		URACIL MUSTARD
		URACIL, 5-(BIS(2-CHLOROETHYL)AMINO)-
		URACILLOST
		URAMUSTIN
		URAMUSTINE
66-81-9		
		2,6-PIPERIDINEDIONE, 4-[2-(3,5-DIMETHYL-2-OXOCYCLO- HEXYL)-2-HYDROXYETHYL]-, [1S-[1ALPHA(S*),3ALPHA
		ACTI-AID
		ACTI-DIONE
		ACTI-DIONE BR
		ACTI-DIONE PM
		ACTI-DIONE TGF
		ACTIDION
		CYCLOHEXIMIDE
		GLUTARIMIDE, 3-[2-(3,5-DIMETHYL-2-OXOCYCLOHEXYL)-2- HYDROXYETHYL]-
		NARAMYCIN
		NARAMYCIN A
		NSC 185
		TZA
		U 4527
67-56-1		
		ALCOOL METHYLIQUE (FRENCH)
		ALCOOL METILICO (ITALIAN)
		CARBINOL
		COLONIAL SPIRIT
		COLUMBIAN SPIRIT
		COLUMBIAN SPIRITS
		COLUMBIAN SPIRITS (DOT)
		METANOLO (ITALIAN)
		METHANOL
		METHANOL (DOT)
	*	METHYL ALCOHOL
		METHYL ALCOHOL (ACGIH,DOT)
		METHYL HYDROXIDE
		METHYLALKOHOL (GERMAN)
		METHYLOL
		METYLOWY ALKOHOL (POLISH)
		MONOHYDROXYMETHANE
		PYROXYLIC SPIRIT
		RCRA WASTE NUMBER U154
		UN 1230 (DOT)
		WOOD ALCOHOL
		WOOD ALCOHOL (DOT)
		WOOD NAPHTHA
		WOOD SPIRIT
67-63-0		
		2-PROPANOL
		ALCOJEL
		ALCOSOLVE 2
		AVANTIN
		AVANTINE
		COMBI-SCHUTZ
		DIMETHYLCARBINOL
		HARTOSOL
		IMSOL A
		ISOHOL
		ISOPROPANOL
	*	ISOPROPYL ALCOHOL
		LUTOSOL
		N-PROPAN-2-OL
		PETROHOL
		PRO

CAS No.	IN	Chemical Name
		PROPOL
		RUBBING ALCOHOL
		SEC-PROPYL ALCOHOL
		TAKINEOCOL
67-64-1		
		2-PROPANONE
		ACETON (GERMAN, DUTCH, POLISH)
	*	ACETONE
		ACETONE (ACGIH,DOT)
		BETA-KETOPROPANE
		CHEVRON ACETONE
		DIMETHYL KETONE
		DIMETHYLFORMALDEHYDE
		DIMETHYLKETAL
		KETONE PROPANE
		KETONE, DIMETHYL
		METHYL KETONE
		PROPANONE
		PYROACETIC ACID
		PYROACETIC ETHER
		RCRA WASTE NUMBER U002
		UN 1090 (DOT)
67-66-3		
	*	CHLOROFORM
		CHLOROFORM (ACGIH,DOT)
		CHLOROFORME (FRENCH)
		CLOROFORMIO (ITALIAN)
		FORMYL TRICHLORIDE
		METHANE TRICHLORIDE
		METHANE, TRICHLORO-
		METHENYL TRICHLORIDE
		METHYL TRICHLORIDE
		NCI-C02686
		R 20
		R 20 (REFRIGERANT)
		RCRA WASTE NUMBER U044
		TCM
		TRICHLOORMETHAAN (DUTCH)
		TRICHLORMETHAN (CZECH)
		TRICHLOROFORM
		TRICHLOROMETHANE
		TRICLOROMETANO (ITALIAN)
		UN 1888 (DOT)
67-68-5		
		A 10846
		DELTAN
		DEMASORB
		DEMAVET
		DEMESO
		DEMSODROX
		DERMASORB
		DIMETHYL SULFOXIDE
		DIMETHYL SULPHOXIDE
		DIMEXIDE
		DIPIRARTRIL-TROPICO
		DMS 70
		DMS 90
		DMSO
		DOLICUR
		DOLIGUR
		DOMOSO
		DROMISOL
		DURASORB
		GAMASOL 90
		HYADUR
		INFILTRINA
		M 176
		METHANE, SULFINYLBIS-
		METHYL SULFOXIDE
		METHYLSULFINYLMETHANE
		NSC-763
		RIMSO-50
		SOMIPRONT
		SQ 9453
		SULFINYLBIS(METHANE)
		SULFINYLBISMETHANE

CAS No.	IN	Chemical Name
		SYNTEXAN
		TOPSYM
67-72-1		
		1,1,1,2,2,2-HEXACHLOROETHANE
		AVLOTHANE
		CARBON HEXACHLORIDE
		DISTOKAL
		DISTOPAN
		DISTOPIN
		EGITOL
		ETHANE HEXACHLORIDE
		ETHANE, HEXACHLORO-
		ETHYLENE HEXACHLORIDE
		FALKITOL
		FASCIOLIN
		HEXACHLOR-AETHAN (GERMAN)
		HEXACHLORETHANE
	*	HEXACHLOROETHANE
		HEXACHLOROETHANE (ACGIH,DOT)
		HEXACHLOROETHYLENE
		MOTTENHEXE
		NA 9037 (DOT)
		NCI-C04604
		PERCHLOROETHANE
		PHENOHEP
		RCRA WASTE NUMBER U131
68-11-1		
		2-MERCAPTOACETIC ACID
		2-MERCAPTOETHANOIC ACID
		2-THIOGLYCOLIC ACID
		ACETIC ACID, MERCAPTO-
		ACIDE THIOGLYCOLIQUE (FRENCH)
		ALPHA-MERCAPTOACETIC ACID
		GLYCOLIC ACID, 2-THIO-
		GLYCOLIC ACID, THIO-
		MERCAPTOACETATE
		MERCAPTOACETIC ACID
		SULFHYDRYLACETIC ACID
	*	THIOGLYCOLIC ACID
		THIOGLYCOLIC ACID (ACGIH,DOT)
		THIOGLYCOLLIC ACID
		THIOVANIC ACID
		UN 1940 (DOT)
		USAF CB-35
68-12-2		
	*	DIMETHYL FORMAMIDE
		DIMETHYLFORMAMID (GERMAN)
		DIMETHYLFORMAMIDE (ACGIH)
		DIMETILFORMAMIDE (ITALIAN)
		DIMETYLFORMAMIDU (CZECH)
		DMF
		DMF (AMIDE)
		DMFA
		DWUMETYLOFORMAMID (POLISH)
		FORMAMIDE, N,N-DIMETHYL-
		N,N-DIMETHYL FORMAMIDE
		N,N-DIMETHYLFORMAMIDE (DOT)
		N,N-DIMETHYLMETHANAMIDE
		N-FORMYLDIMETHYLAMINE
		NCI-C60913
		NSC 5356
		U-4224
		UN 2265 (DOT)
68-22-4		
		17ALPHA-ETHINYL-19-NORTESTOSTERONE
		17ALPHA-ETHYNYL-19-NORTESTOSTERONE
		19-NOR-17ALPHA-ETHYNYLTESTOSTERONE
		19-NOR-17ALPHA-PREGN-4-EN-20-YN-3-ONE, 17-HYDROXY-
		19-NORPREGN-4-EN-20-YN-3-ONE, 17-HYDROXY-, (17ALPHA)-
		19-NORTESTOSTERONE, 17-ETHYNYL-
		ANOVULE
		CONLUDAF
		ETHINYLNORTESTOSTERONE
		ETHYNYLNORTESTOSTERONE
		GESTEST

CAS No.	IN	Chemical Name
		MICRONETT
		MICRONOR
		MICRONOVUM
		MINI-PE
		NORALUTIN
		NORETHINDRONE
		NORETHISTERON
		NORETHISTERONE
		NORETHYNODRONE
		NORFOR
		NORGESTIN
		NORLUTEN
		NORLUTIN
		NORLUTON
		NORPREGNENINOLONE
		NSC-9564
		PRIMOLUT N
		PROLUTEASI
		SC 4640
68-76-8		
		10257RP
		2,3,5-ETHYLENIMINE-1,4-BENZOQUINONE
		2,3,5-TRIETHYLENEIMINO-P-BENZOQUINONE
		2,3,5-TRIS(AZIRIDINYL)-1,4-BENZOQUINONE
		2,3,5-TRIS(ETHYLENIMINO)-1,4-BENZOQUINONE
		2,3,5-TRIS(ETHYLENIMINO)-P-BENZOQUINONE
		2,3,5-TRISETHYLENEIMINOBENZOQUINONE
		2,5-CYCLOHEXADIENE-1,4-DIONE, 2,3,5-TRIS(1-AZIRIDINYL)-
		BAY 3231
		NSC-29215
		ONCOREDOX
		P-BENZOQUINONE, 2,3,5-TRIS(1-AZIRIDINYL)-
		PRENIMON
		RIKER 601
		TEIB
		TRENIMON
		TRENINON
		TRIAZIQUINONE
		TRIAZIQUINONUM
		TRIAZIQUON
		TRIAZIQUONE
		TRIETHYLENEIMINOBENZOQUINONE
		TRIETHYLENIMINOBENZOQUINONE
		TRIS(1-AZIRIDINYL)-P-BENZOQUINONE
		TRIS(AZIRIDINYL)-P-BENZOQUINONE
		TRIS(ETHYLENEIMINO)BENZOQUINONE
		TRISETHYLENEIMINOQUINONE
69-05-6		
	*	QUINACRINE HYDROCHLORIDE
69-53-4		
		AMPICILLIN
69-72-7		
		2-CARBOXYPHENOL
		2-HYDROXYBENZENECARBOXYLIC ACID
		2-HYDROXYBENZOIC ACID
		ACIDO SALICILICO (ITALIAN)
		BENZOIC ACID, 2-HYDROXY-
		KERALYT
		O-CARBOXYPHENOL
		O-HYDROXYBENZOIC ACID
		ORTHOHYDROXYBENZOIC ACID
		PHENOL-2-CARBOXYLIC ACID
		PSORIACID-S-STIFT
		RETARDER W
		RUTRANEX
		SA
		SALICYLIC ACID
		SALICYLIC ACID COLLODION
		SALONIL
		SAX
69-91-0		
		PHENYLGLYCINE ACID
70-11-1		
		2-BROMO-1-PHENYLETHANONE
		2-BROMOACETOPHENONE
		ACETOPHENONE, 2-BROMO-
		ALPHA-BROMOACETOPHENONE
		BENZOYLMETHYL BROMIDE
		BROMOMETHYL PHENYL KETONE
		ETHANONE, 2-BROMO-1-PHENYL-
		ETHANONE, 2-BROMO-1-PHENYL- (9CI)
		OMEGA-BROMACETOPHENONE
		OMEGA-BROMOACETOPHENONE
		PHENACYL BROMIDE
		PHENACYL BROMIDE (DOT)
		STAUFFER 4644
		UN 2645 (DOT)
70-25-7		
		1-METHYL 3-NITRO 1-NITROSO-GUANIDINE
		1-METHYL-1-NITROSO-3-NITROGUANIDINE
		1-NITROSO-3-NITRO-1-METHYLGUANIDINE
		GUANIDINE, 1-METHYL-3-NITRO-1-NITROSO-
		GUANIDINE, N-METHYL-N'-NITRO-N-NITROSO-
		METHYLNITRONITROSOGUANIDINE
		MNNG
		N-METHYL-N'-NITRO-N-NITROSOGUANIDINE
		N-METHYL-N-NITRO-N-NITROSOGUANIDINE
		N-METHYL-N-NITROSO-N'-NITROGUANIDINE
		N-NITROSO-N'-NITRO-N-METHYLGUANIDINE
		N-NITROSO-N-METHYL-N'-NITROGUANIDINE
		N-NITROSO-N-METHYLNITROGUANIDINE
		NITROSOGUANIDINE
		NSC 9369
70-30-4		
		2,2',3,3',5,5'-HEXACHLORO-6,6'-DIHYDROXYDIPHENYLMETHANE
		2,2'-DIHYDROXY-3,3',5,5',6,6'-HEXACHLORODIPHENYLMETHANE
		2,2'-DIHYDROXY-3,5,6,3',5',6'-HEXACHLORODIPHENYLMETHANE
		2,2'-METHYLENEBIS[3,4,6-TRICHLOROPHENOL]
		ACIGENA
		ALMEDERM
		AT 7
		B 32
		BIS(2-HYDROXY-3,5,6-TRICHLOROPHENYL)METHANE
		BIS(3,5,6-TRICHLORO-2-HYDROXYPHENYL)METHANE
		COTOFILM
		DISTODIN
		EXOFENE
		FESIA-SIN
		FOSTRIL
		G 11
		GAMOPHEN
		GAMOPHENE
		GERMA-MEDICA
		HEXABALM
		HEXACHLOROFEN
		HEXACHLOROPHEN
	*	HEXACHLOROPHENE
		HEXAFEN
		HEXOPHENE
		HEXOSAN
		ISOBAC 20
		MONOSODIUM SALT OF 2,2'-METHYLENE BIS(3,4,6-TRICHLOROPHENOL)
		NABAC
		NEOSEPT V
		PHENOL, 2,2'-METHYLENEBIS(3,4,6-TRICHLORO-, SODIUM SALT
		PHENOL, 2,2'-METHYLENEBIS[3,4,6-TRICHLORO-
		PHISOHEX
		RITOSEPT
		SEPTISOL
		SEPTOFEN
		SERIBAK
		STERAL
		STERASKIN
		SURGI-CEN

CAS No.	IN	Chemical Name
		SUROFENE
		TERSASEPTIC
71-23-8		
		1-HYDROXYPROPANE
		1-PROPANOL
		1-PROPYL ALCOHOL
		ALCOOL PROPILICO (ITALIAN)
		ALCOOL PROPYLIQUE (FRENCH)
		ETHYL CARBINOL
		N-PROPAN-1-OL
		N-PROPANOL
		N-PROPYL ALCOHOL
		N-PROPYL ALCOHOL (ACGIH)
		N-PROPYL ALKOHOL (GERMAN)
		OPTAL
		OSMOSOL EXTRA
		PROPANOL
		PROPANOL (DOT)
		PROPANOL-1
		PROPANOLE (GERMAN)
		PROPANOLEN (DUTCH)
		PROPANOLI (ITALIAN)
	*	PROPYL ALCOHOL
		PROPYL ALCOHOL (DOT)
		PROPYLIC ALCOHOL
		PROPYLOWY ALKOHOL (POLISH)
		UN 1274 (DOT)
71-36-3		
		1-BUTANOL
		1-BUTYL ALCOHOL
		1-HYDROXYBUTANE
		ALCOOL BUTYLIQUE (FRENCH)
		BUTAN-1-OL
		BUTANOL
		BUTANOL (DOT)
		BUTANOL (FRENCH)
		BUTANOLEN (DUTCH)
		BUTANOLO (ITALIAN)
		BUTYL ALCOHOL
		BUTYL ALCOHOL (DOT)
		BUTYL HYDROXIDE
		BUTYLOWY ALKOHOL (POLISH)
		BUTYRIC OR NORMAL PRIMARY BUTYL ALCOHOL
		CCS 203
		HEMOSTYP
		METHYLOLPROPANE
		N-BUTAN-1-OL
		N-BUTANOL
	*	N-BUTYL ALCOHOL
		N-BUTYL ALCOHOL (ACGIH)
		NA 1120 (DOT)
		PROPYL CARBINOL
		PROPYLMETHANOL
		RCRA WASTE NUMBER U031
		UN 1120 (DOT)
71-41-0		
		1-PENTANOL
		1-PENTYL ALCOHOL
		ALCOOL AMYLIQUE (FRENCH)
	*	AMYL ALCOHOL
		AMYL ALCOHOL, NORMAL
		AMYLOL
		N-AMYL ALCOHOL
		N-AMYL ALCOHOL (DOT)
		N-AMYLALKOHOL (CZECH)
		N-BUTYL CARBINOL
		N-PENTAN-1-OL
		N-PENTANOL
		N-PENTYL ALCOHOL
		PENTAN-1-OL
		PENTANOL
		PENTANOL-1
		PENTASOL
		PENTYL ALCOHOL
		PRIMARY AMYL ALCOHOL
		UN 1105 (DOT)

CAS No.	IN	Chemical Name
71-43-2		
	*	BENZENE
		BENZOL
		BENZOLE
		COAL NAPHTHA
		CYCLOHEXATRIENE
		PHENE
		PHENYL HYDRIDE
		PYROBENZOL
		PYROBENZOLE
		[6]ANNULENE
71-48-7		
		ACETIC ACID, COBALT(2+) SALT
		BIS(ACETATO)COBALT
		COBALT ACETATE
		COBALT ACETATE (Co(OAC)2)
		COBALT DIACETATE
		COBALT(2+) ACETATE
		COBALT(II) ACETATE
		COBALTOUS ACETATE
		COBALTOUS DIACETATE
71-55-6		
		1,1,1-TCE
		1,1,1-TRICHLOORETHAAN (DUTCH)
		1,1,1-TRICHLORAETHAN (GERMAN)
		1,1,1-TRICHLORETHANE
		1,1,1-TRICHLOROETHANE
		1,1,1-TRICHLOROETHANE (DOT)
		1,1,1-TRICLOROETANO (ITALIAN)
		AEROTHENE TT
		ALPHA-T
		ALPHA-TRICHLOROETHANE
		CF 2
		CHLOROETHENE
		CHLOROETHENE NU
		CHLOROFORM, METHYL-
		CHLOROTENE
		CHLOROTHANE NU
		CHLOROTHENE
		CHLOROTHENE NU
		CHLOROTHENE SM
		CHLOROTHENE VG
		CHLOROTHENE(INHIBITED)
		CHLORTEN
		ETHANA NU
		ETHANE, 1,1,1-TRICHLORO-
		ICI-CF 2
		INHIBISOL
	*	METHYL CHLOROFORM
		METHYL CHLOROFORM (ACGIH,DOT)
		METHYLTRICHLOROMETHANE
		NCI-C04626
		RCRA WASTE NUMBER U226
		SOLVENT 111
		STROBANE
		TAFCLEAN
		TRI-ETHANE
		TRICHLORO-1,1,1-ETHANE (FRENCH)
		TRICHLOROETHANE
		TRICHLOROMETHYLMETHANE
		UN 2831 (DOT)
71-63-6		
		ACEDOXIN
		ASTHENTHILO
		CARDIDIGIN
		CARDIGIN
		CARDITOXIN
		CRISTAPURAT
		CRYSTALLINE
		CRYSTALLINE DIGITALIN
		CRYSTODIGIN
		DIGILONG
		DIGIMED
		DIGIMERCK
		DIGISIDIN
		DIGITALIN

CAS No.	IN	Chemical Name
		DIGITALINE (FRENCH)
		DIGITALINE CRISTALLISEE
		DIGITALINE NATIVELLE
		DIGITALINUM VERUM
		DIGITOPHYLLIN
		DIGITOXIGENIN-TRIDIGITOXOSID (GERMAN)
		DIGITOXIGENIN TRIDIGITOXOSIDE
		DIGITOXIN
		DIGITOXOSIDE
		DIGITRIN
		DITAVEN
		GLUCODIGIN
		LANATOXIN
		MONO-GLYCOCARD
		MYODIGIN
		PURODIGIN
		PURPURID
		TARDIGAL
		TRI-DIGITOXOSIDE (GERMAN)
		UNIDIGIN
72-14-0		
	*	SULFATHIAZOLE
72-20-8		
		3,4,5,6,9,9-HEXACHLORO-1A,2,2A,3,6,6A,7,7A-OCTAHYDRO-2,7:3,6-DIMETHANONAPHTH(2,3-B)OXIRENE
		COMPOUND 269
		EN 57
		ENDREX
		ENDRICOL
	*	ENDRIN
		ENDRIN (ACGIH,DOT)
		ENDRINE (FRENCH)
		ENT 17,251
		EXPERIMENTAL INSECTICIDE 269
		HEXACHLOROEPOXYOCTAHYDRO-ENDO,ENDO-DIMETHANONAPHTHALENE
		HEXADRIN
		MENDRIN
		NA 2761 (DOT)
		NCI-C00157
		NENDRIN
		OKTANEX
		RCRA WASTE NUMBER P051
		SD 3419
72-33-3		
		17ALPHA-ETHINYLESTRADIOL 3-METHYL ETHER
		17ALPHA-ETHYNYLESTRADIOL 3-METHYL ETHER
		17ALPHA-ETHYNYLESTRADIOL METHYL ETHER
		17-ETHYNYLESTRADIOL 3-METHYL ETHER
		19-NOR-17ALPHA-PREGNA-1,3,5(10)-TRIEN-20-YN-17-OL, 3-METHOXY-
		19-NORPREGNA-1,3,5(10)-TRIEN-20-YN-17-OL, 3-METHOXY-, (17ALPHA)-
		3-METHOXY-17ALPHA-ETHINYLESTRADIOL
		3-METHOXY-17ALPHA-ETHYNYLESTRADIOL
		3-METHOXYETHYNYLESTRADIOL
		3-METHYLETHYNYLESTRADIOL
		3-O-METHYLETHYNYLESTRADIOL
		8027 C. B.
		DELTA-MVE
		DEVOCIN
		EE 3ME
		ESTROGENS (NOT CONJUGATED): MESTRANOL
		ETHINYLESTRADIOL 3-METHYL ETHER
		ETHYNYLESTRADIOL 3-METHYL ETHER
		ETHYNYLESTRADIOL METHYL ETHER
		INOSTRAL
		MESTRANOL
		NORQUEN
		OVASTOL
		SC 4725
72-43-5		
		1,1'-(2,2,2-TRICHLOROETHYLIDENE)BIS(4-METHOXYBENZENE)
		1,1,1-TRICHLOR-2,2-BIS(4-METHOXY-PHENYL)-AETHAN (GERMAN)
		1,1,1-TRICHLORO-2,2-BIS(4-METHOXYPHENYL)ETHANE
		1,1,1-TRICHLORO-2,2-BIS(P-ANISYL)ETHANE
		1,1,1-TRICHLORO-2,2-BIS(P-METHOXYPHENOL)ETHANOL
		1,1,1-TRICHLORO-2,2-BIS(P-METHOXYPHENYL)ETHANE
		1,1,1-TRICHLORO-2,2-DI(4-METHOXYPHENYL)ETHANE
		1,1-BIS(P-METHOXYPHENYL)-2,2,2-TRICHLOROETHANE
		2,2,2-TRICHLORO-1,1-BIS(4-METHOXYPHENYL)ETHANE
		2,2-BIS(P-ANISYL)-1,1,1-TRICHLOROETHANE
		2,2-BIS(P-METHOXYPHENYL)-1,1,1-TRICHLOROETHANE
		2,2-DI-(P-METHOXYPHENYL)-1,1,1-TRICHLOROETHANE
		2,2-DI-P-ANISYL-1,1,1-TRICHLOROETHANE
		4,4'-(2,2,2-TRICHLOROETHYLIDENE)DIANISOLE
		BENZENE, 1,1'-(2,2,2-TRICHLOROETHYLIDENE)BIS[4-METHOXY-
		CHEMFORM
		DI(P-METHOXYPHENYL)-TRICHLOROMETHYL METHANE
		DIANISYLTRICHLORETHANE
		DIMETHOXY-DDT
		DMDT
		ENT 1,716
		ETHANE, 1,1,1-TRICHLORO-2,2-BIS(P-METHOXYPHENYL)-
		ETHANE, 2,2-BIS(P-ANISYL)-1,1,1-TRICHLORO-
		FLO PRO MCSEED PROTECTANT
		MARALATE
		MARLATE
		METHOXCIDE
		METHOXO
		METHOXY-DDT
	*	METHOXYCHLOR
		METHOXYCHLOR (ACGIH,DOT)
		METOKSYCHLOR (POLISH)
		METOX
		MEZOX K
		MOXIE
		NA 2761 (DOT)
		NCI-C00497
		P,P'-DIMETHOXYDIPHENYLTRICHLOROETHANE
		P,P'-DMDT
		P,P'-METHOXYCHLOR
		RCRA WASTE NUMBER U247
72-54-8		
		1,1-BIS(4-CHLOROPHENYL)-2,2-DICHLOROETHANE
		1,1-BIS(P-CHLOROPHENYL)-2,2-DICHLOROETHANE
		1,1-DICHLORO-2,2-BIS(4-CHLOROPHENYL)ETHANE
	*	1,1-DICHLORO-2,2-BIS(P-CHLOROPHENYL)ETHANE
		2,2-BIS(4-CHLOROPHENYL)-1,1-DICHLOROETHANE
		2,2-BIS(P-CHLOROPHENYL)-1,1-DICHLOROETHANE
		4,4'-DDD
		BENZENE, 1,1'-(2,2-DICHLOROETHYLIDENE)BIS[4-CHLORO-
		DDD
		DICHLORODIPHENYL DICHLOROETHANE
		DILENE
		ETHANE, 1,1-DICHLORO-2,2-BIS(P-CHLOROPHENYL)-
		ME 1700
		P,P'-DDD
		P,P'-DICHLORODIPHENYL-2,2-DICHLOROETHYLENE
		P,P'-DICHLORODIPHENYLDICHLOROETHANE
		P,P'-TDE
		RHOTHANE
		TDE
72-55-9		
		1,1-BIS(P-CHLOROPHENYL)-2,2-DICHLOROETHYLENE
		1,1-DICHLORO-2,2-BIS(P-CHLOROPHENYL)ETHYLENE
		1,1-DICHLORO-2,2-DI(P-CHLOROPHENYL)ETHYLENE
		2,2-BIS(4-CHLOROPHENYL)-1,1-DICHLOROETHYLENE
		2,2-DICHLORO-1,1-BIS(4-CHLOROPHENYL)ETHYLENE
		4,4'-DDE
		BENZENE, 1,1'-(DICHLOROETHENYLIDENE)BIS[4-CHLORO-
	*	DDE
		ETHYLENE, 1,1-DICHLORO-2,2-BIS(P-CHLOROPHENYL)-
		P,P'-DDE
		P,P'-DICHLORODIPHENYLDICHLOROETHYLENE
		P,P-DDE

CAS No.	IN	Chemical Name
72-56-0		
		P,P-ETHYL DDD (PERTHANE)
72-57-1		
		AMANIL SKY BLUE R
		BENCIDAL BLUE 3B
		BENZAMINE BLUE
		BENZANIL BLUE 3BN
		BENZO BLUE
		BENZO BLUE 3BS
		BLUE 3B
		BLUE EMB
		BRASILAMINA BLUE 3B
		C.I. 23850
		C.I. DIRECT BLUE 14
	*	C.I. DIRECT BLUE 14, TETRASODIUM SALT
		CHLORAMINE BLUE 3B
		CHROME LEATHER BLUE 3B
		CONGO BLUE
		CRESOTINE BLUE 3B
		DIAMINE BLUE 3B
		DIAMINEBLUE
		DIAPHTAMINE BLUE TH
		DIAZINE BLUE 3B
		DIAZOL BLUE 3B
		DIPHENYL BLUE 3B
		DIRECT BLUE 14
		DIRECT BLUE 3B
		DIRECT BLUE M3B
		HISPAMIN BLUE 3BX
		NAPHTHYLAMINE BLUE
		NIAGARA BLUE
		NIAGARA BLUE 3B
		PARAMINE BLUE 3B
		PONTAMINE BLUE 3BX
		TRYPAN BLUE
		TRYPAN BLUE BPC
		TRYPAN BLUE, COMMERCIAL GRADE
		TRYPANE BLUE
74-82-8		
		BIOGAS
		FIRE DAMP
		MARSH GAS
	*	METHANE
		METHANE (DOT)
		METHANE, COMPRESSED (DOT)
		METHANE, REFRIGERATED LIQUID (DOT)
		METHYL HYDRIDE
		NATURAL GAS WITH A HIGH METHANE CONTENT, COMPRESSED
		NATURAL GAS, LIQUID (REFRIGERATED) WITH A HIGH METHANE CONTENT
		R 50
		R 50 (REFRIGERANT)
		UN 1971 (DOT)
		UN 1972 (DOT)
74-83-9		
		BROMOMETHANE
		CURAFUME
		EMBAFUME
		HALON 1001
		HALTOX
		ISCOBROME
		METHANE, BROMO-
		METHANE, BROMO-, LIQUID, INCLUDING UP TO 2% CHLOROPICRIN
		METHYL BROMIDE
		METHYL BROMIDE, LIQUID, INCLUDING UP TO 2% CHLOROPICRIN (DOT)
		METHYL BROMIDE, LIQUID, WITH 2% CHLOROPICRIN
		MONOBROMOMETHANE
		R 40B1
		TERABOL
		UN 1062 (DOT)
74-84-0		
		BIMETHYL
		DIMETHYL
	*	ETHANE
		ETHANE, REFRIGERATED
		ETHANE, REFRIGERATED LIQUID (DOT)
		ETHYL HYDRIDE
		METHYLMETHANE
		UN 1035 (DOT)
		UN 1961 (DOT)
74-85-1		
		ACETENE
		ATHYLEN (GERMAN)
		BICARBURRETTED HYDROGEN
		ELAYL
		ETHENE
	*	ETHYLENE
		ETHYLENE (DOT)
		ETHYLENE, COMPRESSED (DOT)
		ETHYLENE, REFRIGERATED LIQUID
		ETHYLENE, REFRIGERATED LIQUID (DOT)
		LIQUID ETHYENE
		OLEFIANT GAS
		UN 1038 (DOT)
		UN 1962 (DOT)
74-86-2		
		ACETYLEN
	*	ACETYLENE
		ACETYLENE (DOT)
		ACETYLENE (LIQUID)
		ACETYLENE, DISSOLVED (DOT)
		ETHINE
		ETHYNE
		NARCYLEN
		UN 1001 (DOT)
		VINYLENE
74-87-3		
		ARTIC
		CHLOOR-METHAAN (DUTCH)
		CHLOR-METHAN (GERMAN)
		CHLOROMETHANE
		CHLORURE DE METHYLE (FRENCH)
		CLOROMETANO (ITALIAN)
		CLORURO DI METILE (ITALIAN)
		METHANE, CHLORO-
	*	METHYL CHLORIDE
		METHYL CHLORIDE (ACGIH, DOT)
		METHYLCHLORID (GERMAN)
		METYLU CHLOREK (POLISH)
		MONOCHLOROMETHANE
		R 40
		RCRA WASTE NUMBER U045
		UN 1063 (DOT)
74-88-4		
		HALON 10001
		IODOMETANO (ITALIAN)
		IODOMETHANE
		IODURE DE METHYLE (FRENCH)
		JOD-METHAN (GERMAN)
		JOODMETHAAN (DUTCH)
		METHANE, IODO-
	*	METHYL IODIDE
		METHYL IODIDE (ACGIH,DOT)
		METHYL IODIDE (CH3I)
		METHYLJODID (GERMAN)
		METHYLJODIDE (DUTCH)
		METYLUJODEK (POLISH)
		MONO IODURO DI METILE (ITALIAN)
		RCRA WASTE NUMBER U138
		UN 2644 (DOT)
74-89-5		
		AMINOMETHANE
		CARBINAMINE
		MERCURIALIN

CAS No.	IN	Chemical Name
		METHANAMINE
		METHANAMINE (9CI)
	*	METHYLAMINE
		METHYLAMINE (ACGIH)
		METHYLAMINE, ANHYDROUS
		METHYLAMINE, ANHYDROUS (DOT)
		METHYLAMINE, AQUEOUS SOLUTION
		METHYLAMINE, AQUEOUS SOLUTION (DOT)
		METHYLAMINEN (DUTCH)
		METILAMINE (ITALIAN)
		METYLOAMINA (POLISH)
		MMA
		MONOMETHYLAMINE
		MONOMETHYLAMINE, ANHYDROUS (DOT)
		MONOMETHYLAMINE, AQUEOUS SOLUTION (DOT)
		UN 1061 (DOT)
		UN 1235 (DOT)
74-90-8		
		ACIDE CYANHYDRIQUE (FRENCH)
		ACIDO CIANIDRICO (ITALIAN)
		AERO LIQUID HCN
		BLAUSAEURE (GERMAN)
		BLAUWZUUR (DUTCH)
		CARBON HYDRIDE NITRIDE (CHN)
		CYAANWATERSTOF (DUTCH)
		CYANWASSERSTOFF (GERMAN)
		CYCLON
		CYCLONE B
		CYJANOWODOR (POLISH)
		EVERCYN
		FLUOHYDRIC ACID GAS
		FORMIC ANAMMONIDE
		FORMONITRILE
		HCN
		HYDROCYANIC ACID
		HYDROCYANIC ACID (PRUSSIC), UNSTABILIZED (DOT)
		HYDROCYANIC ACID SOLUTION, 5% HYDROCYANIC ACID
		HYDROCYANIC ACID, LIQUEFIED
		HYDROCYANIC ACID, LIQUEFIED (DOT)
		HYDROFLUORIC ACID GAS
	*	HYDROGEN CYANIDE
		HYDROGEN CYANIDE (ACGIH)
		HYDROGEN CYANIDE, ABSORBED
		HYDROGEN CYANIDE, ANHYDROUS, STABILIZED (DOT)
		NA 1051 (DOT)
		PRUSSIC ACID
		PRUSSIC ACID (DOT)
		PRUSSIC ACID, UNSTABILIZED
		RCRA WASTE NUMBER P063
		UN 1051 (DOT)
		UN 1614 (DOT)
		ZACLONDISCOIDS
74-93-1		
		MERCAPTOMETHANE
		METHANETHIOL
		METHYL MERCAPTAN
74-95-3		
	*	METHYLENE BROMIDE
74-96-4		
		BROMIC ETHER
		BROMOETHANE
		BROMURE D'ETHYLE
		ETHANE, BROMO-
	*	ETHYL BROMIDE
		ETHYL BROMIDE (ACGIH,DOT)
		ETYLU BROMEK (POLISH)
		HALON 2001
		HYDROBROMIC ETHER
		MONOBROMOETHANE
		NCI-C55481
		UN 1891 (DOT)
74-97-5		
		BROMOCHLOROMETHANE
		BROMOCHLOROMETHANE (DOT)
		CHLOROBROMOMETHANE
		CHLOROBROMOMETHANE (ACGIH)
		FLUOROCARBON 1011
		HALON 1011
		METHANE, BROMOCHLORO-
		METHYLENE CHLOROBROMIDE
		MIL-B-4394-B
		MONOCHLOROMONOBROMOMETHANE
		UN 1887 (DOT)
74-98-6		
		DIMETHYLMETHANE
		LIQUEFIED PETROLEUM GAS
		LPG
		N-PROPANE
	*	PROPANE
		PROPYL HYDRIDE
		R 290
74-99-7		
		1-PROPYNE
		ALLYLENE
	*	METHYL ACETYLENE
		PROPINE
		PROPYNE
75-00-3		
		AETHYLCHLORID (GERMAN)
		AETHYLIS
		AETHYLIS CHLORIDUM
		ANODYNON
		CHELEN
		CHLOORETHAAN (DUTCH)
		CHLORENE
		CHLORETHYL
		CHLORIDUM
		CHLOROAETHAN (GERMAN)
		CHLOROETHANE
		CHLORURE D'ETHYLE (FRENCH)
		CHLORYL
		CHLORYL ANESTHETIC
		CLORETILO
		CLOROETANO (ITALIAN)
		CLORURO DI ETILE (ITALIAN)
		DUBLOFIX
		ETHANE, CHLORO-
		ETHER CHLORATUS
		ETHER HYDROCHLORIC
		ETHER MURIATIC
	*	ETHYL CHLORIDE
		ETHYL CHLORIDE (ACGIH,DOT)
		ETYLU CHLOREK (POLISH)
		HYDROCHLORIC ETHER
		KELENE
		MONOCHLORETHANE
		MONOCHLOROETHANE
		MURIATIC ETHER
		NARCOTILE
		NCI-C06224
		UN 1037 (DOT)
75-01-4		
		CHLORETHENE
		CHLORETHYLENE
		CHLOROETHENE
		CHLOROETHYLENE
		CHLORURE DE VINYLE (FRENCH)
		CLORURO DI VINILE (ITALIAN)
		ETHENE, CHLORO-
		ETHYLENE MONOCHLORIDE
		ETHYLENE, CHLORO-
		MONOCHLOROETHENE
		MONOCHLOROETHYLENE
		MONOCHLOROETHYLENE (DOT)
		RCRA WASTE NUMBER U043
		TROVIDUR
		UN 1086 (DOT)
		VC
		VCM

CAS No.	IN	Chemical Name
		VINILE (CLORURO DI) (ITALIAN)
		VINYL C MONOMER
	*	VINYL CHLORIDE
		VINYL CHLORIDE (ACGIH,DOT)
		VINYL CHLORIDE MONOMER
		VINYLCHLORID (GERMAN)
		VINYLE(CHLORURE DE) (FRENCH)
		WINYLU CHLOREK (POLISH)
75-02-5		
		ETHENE, FLUORO-
		ETHYLENE, FLUORO-
		ETHYLENE, FLUORO- (8CI)
		FLUOROETHENE
		FLUOROETHYLENE
		MONOFLUOROETHYLENE
		UN 1860 (DOT)
		VINYL FLUORIDE
		VINYL FLUORIDE, INHIBITED
		VINYL FLUORIDE, INHIBITED (DOT)
75-04-7		
		1-AMINOETHANE
		AETHYLAMINE (GERMAN)
		AMINOETHANE
		ETHANAMINE
		ETHANAMINE, (AQUEOUS SOLUTION)
	*	ETHYLAMINE
		ETHYLAMINE (ACGIH,DOT)
		ETHYLAMINE SOLUTION
		ETHYLAMINE SOLUTION, IN WATER, CONCENTRATIONS UP TO 70% (DOT)
		ETILAMINA (ITALIAN)
		ETYLOAMINA (POLISH)
		MONOETHYLAMINE
		MONOETHYLAMINE (DOT)
		MONOETHYLAMINE, ANHYDROUS (DOT)
		MONOMETHYLAMINE, AQUEOUS SOLUTION (DOT)
		N-ETHYLAMINE
		R 631
		UN 1036 (DOT)
		UN 2270 (DOT)
75-05-8		
		ACETONITRIL (GERMAN, DUTCH)
	*	ACETONITRILE
		ACETONITRILE (ACGIH,DOT)
		CYANOMETHANE
		CYANURE DE METHYL (FRENCH)
		ETHANENITRILE
		ETHYL NITRILE
		METHANE, CYANO-
		METHANECARBONITRILE
		METHYL CYANIDE
		METHYL CYANIDE (DOT)
		NA 1648 (DOT)
		NCI-C60822
		RCRA WASTE NUMBER U0 03
		UN 1648 (DOT)
		USAF EK-488
75-07-0		
		ACETALDEHYD (GERMAN)
	*	ACETALDEHYDE
		ACETALDEHYDE (ACGIH,DOT)
		ACETIC ALDEHYDE
		ACETYLALDEHYDE
		ALDEHYDE ACETIQUE (FRENCH)
		ALDEIDE ACETICA (ITALIAN)
		ETHANAL
		ETHYL ALDEHYDE
		ETHYL ALDEHYDE (DOT)
		NCI-C56326
		OCTOWY ALDEHYD (POLISH)
		RCRA WASTE NUMBER U001
		UN 1089 (DOT)
75-08-1		
		AETHANETHIOL (GERMAN)
		AETHYLMERCAPTAN (GERMAN)
		ETANTIOLO (ITALIAN)
		ETHAANTHIOL (DUTCH)
		ETHANETHIOL
		ETHYL HYDROSULFIDE
	*	ETHYL MERCAPTAN
		ETHYL MERCAPTAN (ACGIH,DOT)
		ETHYL SULFHYDRATE
		ETHYL THIOALCOHOL
		ETHYLMERCAPTAAN (DUTCH)
		ETHYLMERKAPTAN (CZECH)
		ETILMERCAPTANO (ITALIAN)
		LPG ETHYL MERCAPTAN 1010
		MERCAPTOETHANE
		THIOETHANOL
		THIOETHYL ALCOHOL
		UN 2363 (DOT)
75-09-2		
		AEROTHENE MM
		CHLORURE DE METHYLENE (FRENCH)
		DCM
		DICHLOROMETHANE
		DICHLOROMETHANE (DOT)
		FREON 30
		METHANE DICHLORIDE
		METHANE, DICHLORO-
		METHYLENE BICHLORIDE
	*	METHYLENE CHLORIDE
		METHYLENE CHLORIDE (ACGIH,DOT)
		METHYLENE DICHLORIDE
		METYLENU CHLOREK (POLISH)
		NARKOTIL
		NCI-C50102
		R 30
		RCRA WASTE NUMBER U080
		SOLAESTHIN
		SOLMETHINE
		UN 1593 (DOT)
75-11-6		
		DIIODOMETHANE
75-12-7		
		CARBAMALDEHYDE
	*	FORMAMIDE
		FORMIMIDIC ACID
		METHANAMIDE
75-15-0		
		CARBON BISULFIDE
		CARBON BISULFIDE (DOT)
		CARBON BISULPHIDE
	*	CARBON DISULFIDE
		CARBON DISULFIDE (ACGIH,DOT)
		CARBON DISULPHIDE
		CARBON SULFIDE
		CARBON SULFIDE (CS2)
		CARBON SULPHIDE (DOT)
		CARBONE (SUFURE DE) (FRENCH)
		CARBONIO (SOLFURO DI) (ITALIAN)
		DITHIOCARBONIC ANHYDRIDE
		KOHLENDISULFID (SCHWEFELKOHLENSTOFF) (GERMAN)
		KOOLSTOFDISULFIDE (ZWAVELKOOLSTOF) (DUTCH)
		NCI-C04591
		RCRA WASTE NUMBER P022
		SCHWEFELKOHLENSTOFF (GERMAN)
		SOLFURO DI CARBONIO (ITALIAN)
		SULPHOCARBONIC ANHYDRIDE
		UN 1131 (DOT)
		WEEVILTOX
		WEGLA DWUSIARCZEK (POLISH)
75-16-1		
		METHYL MAGNESIUM BROMIDE

CAS No.	IN	Chemical Name
75-18-3		
		2-THIAPROPANE
		2-THIOPROPANE
		DIMETHYL MONOSULFIDE
		DIMETHYL SULFIDE
		DIMETHYL SULFIDE (DOT)
		DIMETHYL SULPHIDE
		DIMETHYL THIOETHER
		DIMETHYLSULFID (CZECH)
		DMS
		EXACT-S
		METHANE, THIOBIS-
		METHYL MONOSULFIDE
		METHYL SULFIDE
		METHYL SULFIDE (DOT)
		METHYL SULPHIDE
		METHYLTHIOMETHANE
		SULFURE DE METHYLE (FRENCH)
		THIOBIS(METHANE)
		UN 1164 (DOT)
75-19-4		
	*	CYCLOPROPANE
		TRIMETHYLENE
		TRIMETHYLENE (CYCLIC)
75-20-7		
		ACETYLENOGEN
		CALCIUM ACETYLIDE
		CALCIUM ACETYLIDE (Ca(C2))
	*	CALCIUM CARBIDE
		CALCIUM CARBIDE (CaC2)
		CALCIUM CARBIDE (DOT)
		CALCIUM DICARBIDE
		ETHYNE, CALCIUM DERIV
		UN 1402 (DOT)
75-21-8		
		1,2-EPOXYETHANE
		DIHYDROOXIRENE
		DIMETHYLENE OXIDE
		EPOXYETHANE
		ETHENE OXIDE
	*	ETHYLENE OXIDE
		ETO
		OXACYCLOPROPANE
		OXANE
		OXIDOETHANE
		OXIRANE
		OXIRENE, DIHYDRO-
		OXYFUME
		OXYFUME 12
		T-GAS
75-24-1		
		ALANE, TRIMETHYL-
		ALUMINUM TRIMETHYL
		ALUMINUM TRIMETHYL (DOT)
		TRIMETHYLALANE
		TRIMETHYLALUMINIUM
		TRIMETHYLALUMINIUM (DOT)
		TRIMETHYLALUMINUM
		UN 1103 (DOT)
75-25-2		
	*	BROMOFORM
		BROMOFORM (ACGIH,DOT)
		BROMOFORME (FRENCH)
		BROMOFORMIO (ITALIAN)
		METHANE, TRIBROMO-
		METHENYL TRIBROMIDE
		NCI-C55130
		RCRA WASTE NUMBER U225
		TRIBROMMETHAAN (DUTCH)
		TRIBROMMETHAN (GERMAN)
		TRIBROMOMETAN (ITALIAN)
		TRIBROMOMETHANE
		UN 2515 (DOT)

CAS No.	IN	Chemical Name
75-27-4		
		BROMODICHLOROMETHANE
		DICHLOROBROMOMETHANE
		DICHLOROMONOBROMOMETHANE
		METHANE, BROMODICHLORO-
		MONOBROMODICHLOROMETHANE
75-28-5		
		1,1-DIMETHYLETHANE
		2-METHYLPROPANE
		A 31
		A 31 (HYDROCARBON)
	*	ISOBUTANE
		ISOBUTANE (DOT)
		LIQUEFIED PETROLEUM GAS (DOT)
		PROPANE, 2-METHYL-
		R 600A
		TRIMETHYLMETHANE
		UN 1075 (DOT)
		UN 1969 (DOT)
75-29-6		
		2-CHLOROPROPANE
		2-PROPYL CHLORIDE
		CHLORODIMETHYLMETHANE
		ISOPRID
		ISOPROPYL CHLORIDE
		NARCOSOP
		PROPANE, 2-CHLORO-
		SEC-PROPYL CHLORIDE
75-31-0		
		1-METHYLETHYLAMINE
		2-AMINO-PROPAAN (DUTCH)
		2-AMINO-PROPANO (ITALIAN)
		2-AMINOPROPAN (GERMAN)
		2-AMINOPROPANE
		2-PROPANAMINE
		2-PROPYLAMINE
		ISOPROPILAMINA (ITALIAN)
		ISOPROPYLAMINE
		ISOPROPYLAMINE (ACGIH,DOT)
		MONOISOPROPYLAMINE
		PROPANE, 2-AMINO-
		SEC-PROPYLAMINE
		UN 1221 (DOT)
75-33-2		
		1-METHYLETHANETHIOL
		2-MERCAPTOPROPANE
		2-PROPANETHIOL
		2-PROPANETHIOL (DOT)
		2-PROPYLMERCAPTAN
		ISOPROPANETHIOL
		ISOPROPYL MERCAPTAN
		ISOPROPYL MERCAPTAN (DOT)
		ISOPROPYLTHIOL
		UN 2402 (DOT)
		UN 2703 (DOT)
75-34-3		
		1,1-DICHLOORETHAAN (DUTCH)
		1,1-DICHLORAETHAN (GERMAN)
		1,1-DICHLORETHANE
	*	1,1-DICHLOROETHANE
		1,1-DICHLOROETHANE (ACGIH,DOT)
		1,1-DICLOROETANO (ITALIAN)
		1,1-ETHYLIDENE DICHLORIDE
		AETHYLIDENCHLORID (GERMAN)
		CHLORINATED HYDROCHLORIC ETHER
		CHLORURE D'ETHYLIDENE (FRENCH)
		CLORURO DI ETILIDENE (ITALIAN)
		DICHLOROMETHYLMETHANE
		ETHANE, 1,1-DICHLORO-
		ETHYLIDENE CHLORIDE
		ETHYLIDENE DICHLORIDE
		NCI-C04535
		RCRA WASTE NUMBER U076
		UN 2362 (DOT)

CAS No.	IN	Chemical Name
75-34-4		1,1-DICHLOROETHANE
75-35-4		1,1-DICHLOROETHENE
		1,1-DICHLOROETHYLENE
		ETHENE, 1,1-DICHLORO-
		ETHYLENE, 1,1-DICHLORO-
		ETHYLENE, 1,1-DICHLORO- (8CI)
		ETHYLENE, 1,1-DICHLORO-, INHIBITED
		RCRA WASTE NUMBER U078
		UN 1303 (DOT)
		VDC
	*	VINYLIDENE CHLORIDE
		VINYLIDENE CHLORIDE, INHIBITED
		VINYLIDENE CHLORIDE, INHIBITED (DOT)
75-36-5		ACETIC ACID, CHLORIDE
	*	ACETIC CHLORIDE
		ACETYL CHLORIDE
		ACETYL CHLORIDE (DOT)
		ETHANOYL CHLORIDE
		RCRA WASTE NUMBER U006
		UN 1717 (DOT)
75-37-6		1,1-DIFLUOROETHANE
		1,1-DIFLUOROETHANE (DOT)
		ALGOFRENE TYPE 67
		DIFLUOROETHANE
		DYMEL 152
		ETHANE, 1,1-DIFLUORO-
		ETHANE, DIFLUORO-
		ETHYLENE FLUORIDE
		ETHYLIDENE DIFLUORIDE
		ETHYLIDENE FLUORIDE
		FC 152A
		FREON 152
		GENETRON 100
		GENETRON 152A
		HALOCARBON 152A
		R 152A
		UN 1030 (DOT)
75-38-7		1,1-DIFLUOROETHENE
		1,1-DIFLUOROETHYLENE
		1,1-DIFLUOROETHYLENE (DOT)
		ETHENE, 1,1-DIFLUORO-
		ETHYLENE, 1,1-DIFLUORO-
		GENETRON 1132A
		HALOCARBON 1132A
		NCI-C60208
		UN 1959 (DOT)
		VDF
		VINYLIDENE DIFLUORIDE
		VINYLIDENE FLUORIDE
75-39-8		1-AMINOETHANOL
		ACETALDEHYDE AMMONIA
		ACETALDEHYDE AMMONIA (DOT)
		ACETALDEHYDE, AMINE SALT
		ALDEHYDE AMMONIA
		ALPHA-AMINOETHYL ALCOHOL
		ETHANOL, 1-AMINO-
		ETHANOL, 1-AMINO- (8CI,9CI)
		UN 1841 (DOT)
75-43-4		ALGOFRENE TYPE 5
		ARCTON 7
		CFC 21
		DICHLOROFLUOROMETHANE
		DICHLOROFLUOROMETHANE (ACGIH)
		DICHLOROMONOFLUOROMETHANE
		DICHLOROMONOFLUOROMETHANE (DOT)
		DWUCHLOROFLUOROMETAN (POLISH)
		F 21
		FC 21
		FLUORODICHLOROMETHANE
		FREON 21
		FREON F 21
		GENETRON 21
		METHANE, DICHLOROFLUORO-
		MONOFLUORODICHLOROMETHANE
		R 21
		R 21 (REFRIGERANT)
		UN 1029 (DOT)
75-44-5		CARBON DICHLORIDE OXIDE
		CARBON OXYCHLORIDE
		CARBONE (OXYCHLORURE DE) (FRENCH)
		CARBONIC DICHLORIDE
		CARBONIO (OSSICLORURO DI) (ITALIAN)
		CARBONYL CHLORIDE
		CARBONYL CHLORIDE (DOT)
		CARBONYL DICHLORIDE
		CARBONYLCHLORID (GERMAN)
		CG
		CHLOROFORMYL CHLORIDE
		FOSGEEN (DUTCH)
		FOSGEN (POLISH)
		FOSGENE (ITALIAN)
		KOOLSTOFOXYCHLORIDE (DUTCH)
		NCI-C60219
		PHOSGEN
		PHOSGEN (GERMAN)
	*	PHOSGENE
		PHOSGENE (ACGIH,DOT)
		RCRA WASTE NUMBER P095
		UN 1076 (DOT)
75-45-6		ALGEON 22
		ALGOFRENE 22
		ALGOFRENE 6
		ALGOFRENE TYPE 6
		ARCTON 22
		ARCTON 4
		CFC 22
	*	CHLORODIFLUOROMETHANE
		CHLORODIFLUOROMETHANE (ACGIH,DOT)
		DAIFLON 22
		DIFLUOROCHLOROMETHANE
		DIFLUOROMONOCHLOROMETHANE
		DYMEL 22
		ELECTRO-CF 22
		ESKIMON 22
		F 22
		FC 22
		FLON 22
		FLUGENE 22
		FLUOROCARBON 22
		FLUOROCARBON-22
		FORANE 22
		FREON
		FREON 22
		FRIGEN
		FRIGEN 22
		GENETRON 22
		HALTRON 22
		ISCEON 22
		ISOTRON 22
		KHLADON 22
		METHANE, CHLORODIFLUORO-
		MONOCHLORODIFLUORMETHANE (DOT)
		MONOCHLORODIFLUOROMETHANE
		PROPELLANT 22
		R 22
		R 22 (DOT)
		REFRIGERANT 22
		UCON 22
		UCON 22/HALOCARBON 22
		UN 1018 (DOT)

CAS No.	IN	Chemical Name
75-46-7		
		ARCTON
		ARCTON 1
		CARBON TRIFLUORIDE
		FLUOROFORM
		FLUORYL
		FREON 23
		FREON F-23
		GENETRON 23
		HALOCARBON 23
		METHANE, TRIFLUORO-
		METHYL TRIFLUORIDE
		R 23
		R 23 (HALOCARBON)
		TRIFLUOROMETHANE
		TRIFLUOROMETHANE (DOT)
		UN 1984 (DOT)
75-47-8		
		CARBON TRIIODIDE
	*	IODOFORM
		METHANE, TRIIODO-
		TRIIODOMETHANE
75-50-3		
		METHANAMINE, N,N-DIMETHYL-
		N,N-DIMETHYLMETHANAMINE
		N-TRIMETHYLAMINE
		TMA
	*	TRIMETHYLAMINE
		TRIMETHYLAMINE (ACGIH)
		TRIMETHYLAMINE, ANHYDROUS
		TRIMETHYLAMINE, ANHYDROUS (DOT)
		TRIMETHYLAMINE, AQUEOUS SOLUTION
		TRIMETHYLAMINE, AQUEOUS SOLUTION (DOT)
		TRIMETHYLAMINE, AQUEOUS SOLUTIONS CONTAINING NOT MORE THAN 30% OF TRIMETHYLAMINE (DOT)
		UN 1083 (DOT)
		UN 1297 (DOT)
75-52-4		
		OO-DIETHYL S-ISOPROPYLTHIOMETHYL PHOSPHORDITHIOATE
75-52-5		
		METHANE, NITRO-
		NITROCARBOL
		NITROMETAN (POLISH)
	*	NITROMETHANE
		NITROMETHANE (ACGIH,DOT)
		UN 1261 (DOT)
75-54-7		
		DICHLOROHYDRIDOMETHYLSILICON
		DICHLOROMETHYLSILANE
		METHYL DICHLOROSILANE
		METHYL DICHLOROSILANE (DOT)
		METHYL-DICHLORSILAN (CZECH)
		MONOMETHYLDICHLOROSILANE
		SILANE, DICHLOROMETHYL-
		UN 1242 (DOT)
75-55-8		
		1,2-PROPYLENIMINE
		2-METHYLAZIRIDINE
		2-METHYLETHYLENIMINE
		AZIRIDINE, 2-METHYL-
		PROPYLENEIMINE
		PROPYLENIMINE
75-56-9		
		1,2-EPOXYPROPANE
		1,2-PROPYLENE OXIDE
		2,3-EPOXYPROPANE
		AD 6
		AD 6 (SUSPENDING AGENT)
		EPOXYPROPANE
		ETHYLENE OXIDE, METHYL-
		METHYL ETHYLENE OXIDE

CAS No.	IN	Chemical Name
		METHYL OXIRANE
		NCI-C50099
		OXIRANE, METHYL-
		OXYDE DE PROPYLENE (FRENCH)
		PROPANE, 1,2-EPOXY-
		PROPANE, EPOXY-
		PROPENE OXIDE
		PROPYLENE EPOXIDE
	*	PROPYLENE OXIDE
		PROPYLENE OXIDE (ACGIH,DOT)
		UN 1280 (DOT)
75-59-2		
		AMMONIUM, TETRAMETHYL-, HYDROXIDE
		HYDROXYDE DE TETRAMETHYLAMMONIUM (FRENCH)
		METHANAMINIUM, N,N,N-TRIMETHYL-, HYDROXIDE
		TETRAMETHYL AMMONIUM HYDROXIDE, LIQUID (DOT)
		TETRAMETHYLAMMONIUM HYDROXIDE
		TETRAMETHYLAMMONIUM HYDROXIDE (DOT)
		TETRAMETHYLAMMONIUM HYDROXIDE, LIQUID
		TM
		UN 1835 (DOT)
75-59-8		
		ALDEHYDE AMMONIA
75-60-5		
		ACIDE CACODYLIQUE (FRENCH)
		ACIDE DIMETHYLARSINIQUE (FRENCH)
		AGENT BLUE
		ANSAR
		ANSAR 138
		ARSAN
		ARSINE OXIDE, DIMETHYLHYDROXY-
		ARSINE OXIDE, HYDROXYDIMETHYL-
		ARSINIC ACID, DIMETHYL-
		ARSINIC ACID, DIMETHYL- (9CI)
		BOLLS-EYE
	*	CACODYLIC ACID
		CACODYLIC ACID (DOT)
		CHEXMATE
		DILIC
		DIMETHYLARSENIC ACID
		DIMETHYLARSINIC ACID
		DMAA
		ERASE
		HYDROXYDIMETHYLARSINE OXIDE
		PHYTAR
		PHYTAR 138
		PHYTAR 560
		RAD-E-CATE 25
		RCRA WASTE NUMBER U136
		SALVO
		SILVISAR 510
		SYLVICOR
		UN 1572 (DOT)
75-61-6		
		DIBROMODIFLUOROMETHANE
		DIFLUORODIBROMOMETHANE
		FREON 1282
		FREON 12B2
		FREON 21
		HALON 1202
		METHANE, DIBROMODIFLUORO-
		R 12B2
75-63-8		
		BROMOFLUOROFORM
		BROMOTRIFLUOROMETHANE
		BROMOTRIFLUOROMETHANE (DOT)
		CARBON MONOBROMIDE TRIFLUORIDE
		F 13B1
		FC 13B1
		FLUGEX 13B1
		FLUOROCARBON 1301
		FREON 13B1
		HALON 1301
		KHLADON 13B1

CAS No.	IN	Chemical Name
		METHANE, BROMOTRIFLUORO-
		MONOBROMOTRIFLUOROMETHANE
		R 13B1
		REFRIGERANT 13B1
		TRIFLUOROBROMOMETHANE
		TRIFLUOROBROMOMETHANE (ACGIH,DOT)
		TRIFLUOROMETHYL BROMIDE
		TRIFLUOROMONOBROMOMETHANE
		UN 1009 (DOT)
75-64-9		
		1,1-DIMETHYLETHYLAMINE
		2-AMINO-2-METHYLPROPANE
		2-AMINOISOBUTANE
		2-METHYL-2-AMINOPROPANE
		2-METHYL-2-PROPANAMINE
		2-PROPANAMINE, 2-METHYL-
		BUTYLAMINE, TERTIARY
		T-BUTYLAMINE
		TERT-BUTYLAMINE
		TRIMETHYLAMINOMETHANE
75-65-0		
		1,1-DIMETHYLETHANOL
		2-METHYL-2-PROPANOL
		2-PROPANOL, 2-METHYL-
		ALCOOL BUTYLIQUE TERTIAIRE (FRENCH)
		BUTANOL TERTIAIRE (FRENCH)
		METHANOL, TRIMETHYL-
		NCI-C55367
		T-BUTANOL
		T-BUTYL HYDROXIDE
		TERT-BUTANOL
		TERT-BUTANOL (DOT)
	*	TERT-BUTYL ALCOHOL
		TERT-BUTYL ALCOHOL (ACGIH,DOT)
		TRIMETHYLCARBINOL
		TRIMETHYLMETHANOL
		UN 1120 (DOT)
75-66-1		
		1,1-DIMETHYLETHANETHIOL
		2-ISOBUTANETHIOL
		2-METHYL-2-PROPANETHIOL
		2-PROPANETHIOL, 2-METHYL-
		T-BUTYLMERCAPTAN
		TERT-BUTANETHIOL
		TERT-BUTYL MERCAPTAN
		TERT-BUTYLTHIOL
75-68-3		
		1,1-DIFLUORO-1-CHLOROETHANE
		1-CHLORO-1,1-DIFLUOROETHANE
		ALPHA-CHLOROETHYLIDENE FLUORIDE
		CHLORODIFLUOROETHANE
		DIFLUORO-1-CHLOROETHANE
		DYMEL 142
		ETHANE, 1-CHLORO-1,1-DIFLUORO-
		FC 142B
		GENETRON 101
		GENETRON 142B
		R 142B
75-69-4		
		ALGOFRENE TYPE 1
		ARCTON 11
		ARCTON 9
		CFC 11
		DAIFLON 11
		DAIFLON S 1
		ELECTRO CF 11
		ESKIMON 11
		F 11
		F 11 (HALOCARBON)
		FC 11
		FC 11 (HALOCARBON)
		FKW 11
		FLUOROCARBON 11
		FLUOROCARBON NO 11
		FLUOROCHLOROFORM
		FLUOROTRICHLOROMETHANE
		FLUOROTROJCHLOROMETAN (POLISH)
		FREON 11
		FREON 11A
		FREON 11B
		FREON HE
		FREON MF
		FRIGEN 11
		FRIGEN 11A
		FRIGEN S 11
		GENETRON 11
		HALOCARBON 11
		HALON 11
		ISCEON 131
		ISOTRON 11
		KALTRON 11
		KHLADON 11
		LEDON 11
		METHANE, FLUOROTRICHLORO-
		METHANE, TRICHLOROFLUORO-
		MONOFLUOROTRICHLOROMETHANE
		NCI-C04637
		PROPELLANT 11
		R 11
		R 11 (REFRIGERANT)
		RCRA WASTE NUMBER U121
		REFRIGERANT 11
		REFRIGERANT R11
		TRICHLOROFLUOROCARBON
	*	TRICHLOROFLUOROMETHANE
		TRICHLOROFLUOROMETHANE (ACGIH)
		TRICHLOROMONOFLUOROMETHANE
		UCON FLUOROCARBON 11
		UCON REFRIGERANT 11
75-71-8		
		ALGOFRENE TYPE 2
		ARCTON 12
		ARCTON 6
		CF 12
		CFC 12
		CHLOROFLUOROCARBON 12
	*	DICHLORODIFLUOROMETHANE
		DICHLORODIFLUOROMETHANE (ACGIH,DOT)
		DIFLUORODICHLOROMETHANE
		DWUCHLORODWUFLUOROMETAN (POLISH)
		ELECTRO-CF 12
		ESKIMON 12
		F 12
		FC 12
		FLUOROCARBON 12
		FORANE 12
		FREON 12
		FREON F-12
		FRIGEN 12
		GENETRON 12
		HALON
		ISCEON 122
		ISOTRON 12
		KAISER CHEMICALS 12
		LEDON 12
		METHANE, DICHLORODIFLUORO-
		PROPELLANT 12
		R 12
		R 12 (DOT)
		R 12 (REFRIGERANT)
		RCRA WASTE NUMBER U075
		REFRIGERANT 12
		UCON 12
		UCON 12/HALOCARBON 12
		UN 1028 (DOT)
75-72-9		
		ARCTON 3
		CHLOROTRIFLUOROMETHANE
		CHLOROTRIFLUOROMETHANE (DOT)
		F 13
		FC 13

CAS No.	IN	Chemical Name
		FREON 13
		FRIGEN 13
		GENETRON 13
		HALOCARBON 13/UCON 13
		METHANE, CHLOROTRIFLUORO-
		MONOCHLOROTRIFLUOROMETHANE
		MONOCHLOROTRIFLUOROMETHANE (DOT)
		R 13
		REFRIGERANT 13
		TRIFLUOROCHLOROMETHANE
		TRIFLUOROCHLOROMETHANE (DOT)
		TRIFLUOROMETHYL CHLORIDE
		TRIFLUOROMONOCHLOROCARBON
		UN 1022 (DOT)
75-73-0		
		ARCTON 0
		CARBON FLUORIDE
		CARBON FLUORIDE (CF4)
		CARBON TETRAFLUORIDE
		F 14
		FC 14
		FREON 14
		HALOCARBON 14
		HALON 14
		METHANE, TETRAFLUORO-
		PERFLUOROMETHANE
		R 14
		R 14 (REFRIGERANT)
		REFRIGERANT 14
		TETRAFLUOROCARBON
		TETRAFLUOROMETHANE
		TETRAFLUOROMETHANE (DOT)
		UN 1982 (DOT)
75-74-1		
		LEAD, TETRAMETHYL-
		PIOMBO TETRA-METILE (ITALIAN)
		PLUMBANE, TETRAMETHYL-
		TETRAMETHYL LEAD
		TETRAMETHYL LEAD (ACGIH)
		TETRAMETHYLPLUMBANE
		TML
75-75-2		
		ALKANE SULFONIC ACID
75-76-3		
		SILANE, TETRAMETHYL-
	*	TETRAMETHYL SILANE
		TETRAMETHYLSILANE (DOT)
		UN 2749 (DOT)
75-77-4		
		CHLOROTRIMETHYLSILANE
		MONOCHLOROTRIMETHYLSILICON
		SILANE, CHLOROTRIMETHYL-
		SILANE, TRIMETHYLCHLORO-
		SILICANE, CHLOROTRIMETHYL-
		TL 1163
	*	TRIMETHYL CHLOROSILANE
		TRIMETHYLCHLOROSILANE (DOT)
		TRIMETHYLSILYL CHLORIDE
		UN 1298 (DOT)
75-78-5		
		DICHLORODIMETHYLSILANE
		DICHLORODIMETHYLSILICON
		DIMETHYL-DICHLORSILAN (CZECH)
		DIMETHYLDICHLOROSILANE
		DIMETHYLDICHLOROSILANE (DOT)
		INERTON AW-DMCS
		SILANE, DICHLORODIMETHYL-
		UN 1162 (DOT)
75-79-6		
		METHYL TRICHLOROSILANE
		METHYL-TRICHLORSILAN (CZECH)
		METHYLSILYL TRICHLORIDE
		METHYLTRICHLOROSILANE
		METHYLTRICHLOROSILANE (DOT)
		SILANE, METHYLTRICHLORO-
		SILANE, TRICHLOROMETHYL-
		TRICHLOROMETHYLSILANE
		TRICHLOROMETHYLSILICON
		UN 1250 (DOT)
75-83-2		
		2,2-DIMETHYLBUTANE
		BUTANE, 2,2-DIMETHYL-
		NEOHEXANE
		NEOHEXANE (ACGIH,DOT)
		UN 1208 (DOT)
75-84-3		
		1-PROPANOL, 2,2-DIMETHYL-
		2,2-DIMETHYL-1-PROPANOL
		2,2-DIMETHYLPROPYL ALCOHOL
		NEOAMYL ALCOHOL
		NEOPENTANOL
		NEOPENTYL ALCOHOL
		TERT-BUTYLCARBINOL
75-85-4		
		1,1-DIMETHYL-1-PROPANOL
		2-BUTANOL, 2-METHYL-
		2-ETHYL-2-PROPANOL
		2-METHYL BUTANOL-2
		2-METHYL-2-BUTANOL
		2-METHYL-2-HYDROXYBUTANE
		3-METHYLBUTAN-3-OL
		AMYLENE HYDRATE
		DIMETHYLETHYLCARBINOL
		ETHYLDIMETHYLCARBINOL
		T-AMYL ALCOHOL
	*	TERT-AMYL ALCOHOL
		TERT-AMYL ALCOHOL (DOT)
		TERT-PENTANOL
		TERT-PENTYL ALCOHOL
		UN 1105 (DOT)
75-86-5		
		2-CYANO-2-HYDROXYPROPANE
		2-CYANO-2-PROPANOL
		2-CYANOPROPAN-2-OL
		2-HYDROXY-2-CYANOPROPANE
		2-HYDROXY-2-METHYLPROPANENITRILE
		2-HYDROXY-2-METHYLPROPIONITRILE
		2-HYDROXYISOBUTYRONITRILE
		2-METHYL LACTONITRILE
		2-PROPANONE, CYANOHYDRIN
		ACETONCYAANHYDRINE (DUTCH)
		ACETONCIANHIDRINEI (ROMANIAN)
		ACETONCIANIDRINA (ITALIAN)
		ACETONCYANHYDRIN (GERMAN)
		ACETONE CYANOHYDRIN
		ACETONE CYANOHYDRIN (DOT)
		ACETONE CYANOHYDRIN, STABILIZED (DOT)
		ACETONECYANHYDRINE (FRENCH)
		ACETONKYANHYDRIN (CZECH)
		ALPHA-HYDROXYISOBUTYRONITRILE
		CYANHYDRINE D'ACETONE (FRENCH)
		LACTONITRILE, 2-METHYL-
		METHYL LACTONITRILE
		PROPANENITRILE, 2-HYDROXY-2-METHYL-
		RCRA WASTE NUMBER P069
		UN 1541 (DOT)
		USAF RH-8
75-87-6		
		2,2,2-TRICHLOROACETALDEHYDE
		2,2,2-TRICHLOROETHANAL
		ACETALDEHYDE, TRICHLORO-
		ACETALDEHYDE, TRICHLORO- (9CI)
		ANHYDROUS CHLORAL
	*	CHLORAL
		CHLORAL, ANHYDROUS, INHIBITED
		CHLORAL, ANHYDROUS, INHIBITED (DOT)

CAS No.	IN	Chemical Name
		CLORALIO (ITALIAN)
		GRASEX
		RCRA WASTE NUMBER U034
		SPOROTAL 100
		TRICHLOROACETALDEHYDE
		TRICHLOROETHANAL
		UN 2075 (DOT)
75-91-2		1,1-DIMETHYLETHYL HYDROPEROXIDE
		2-HYDROPEROXY-2-METHYLPROPANE
		CADOX TBH
		HYDROPEROXIDE, 1,1-DIMETHYLETHYL
		PERBUTYL H
		TERT-BUTYL HYDROPEROXIDE
75-94-5		A 150
		A 150 (SILANE)
		KA 1003
		SILANE, TRICHLOROETHENYL-
		SILANE, TRICHLOROVINYL-
		SILANE, VINYL TRICHLORO A-150
		TRICHLOROVINYL SILICANE
		TRICHLOROVINYLSILANE
		TRICHLOROVINYLSILICON
		UN 1305 (DOT)
		UNION CARBIDE A-150
		VINYL TRICHLOROSILANE
		VINYL TRICHLOROSILANE (DOT)
		VINYL TRICHLOROSILANE, INHIBITED (DOT)
		VINYLSILICON TRICHLORIDE
		VINYLTRICHLOROSILANE
		VTCS
75-98-9		2,2-DIMETHYLPROPANOIC ACID
		2,2-DIMETHYLPROPIONIC ACID
		ACETIC ACID, TRIMETHYL-
		ALPHA,ALPHA-DIMETHYLPROPIONIC ACID
		NEOPENTANOIC ACID
		PIVALIC ACID
		PROPANOIC ACID
		PROPANOIC ACID, 2,2-DIMETHYL-
		PROPIONIC ACID, 2,2-DIMETHYL-
		TERT-PENTANOIC ACID
		TRIMETHYLACETIC ACID
75-99-0		2,2-DICHLOROPROPANOIC ACID
		2,2-DICHLOROPROPIONIC ACID
		2,2-DICHLOROPROPIONIC ACID (ACGIH,DOT)
		2,2-DPA
		ALATEX
		ALPHA,ALPHA-DICHLOROPROPIONIC ACID
		ALPHA,ALPHA-DICHLOROPROPIONIC ACID
		BASFAPON
		BASFAPON B
		BASFAPON/BASFAPON N
		BASINEX
		BASINEX P
		BH DALAPON
		CRISAPON
		DALAPON
		DALAPON 85
		DED-WEED
		DEVIPON
		DOWPON
		DOWPON M
		DPA
		GRAMEVIN
		KENAPON
		LIROPON
		NA 1760 (DOT)
		PROPANOIC ACID, 2,2-DICHLORO-
		PROPIONIC ACID, 2,2-DICHLORO-
		PROPROP
		RADAPON
		REVENGE
		S 1315
		S 95
		S 95 (HERBICIDE)
		TRIPON
		UNIPON
76-01-7		ETHANE PENTACHLORIDE
		ETHANE, PENTACHLORO-
		NCI-C53894
		PENTACHLOORETHAAN (DUTCH)
		PENTACHLORAETHAN (GERMAN)
		PENTACHLORETHANE (FRENCH)
		PENTACHLOROETHANE
		PENTACHLOROETHANE (DOT)
		PENTACLOROETANO (ITALIAN)
		PENTALIN
		RCRA WASTE NUMBER U184
		UN 1669 (DOT)
76-02-8		ACETYL CHLORIDE, TRICHLORO-
		TRICHLOROACETIC ACID CHLORIDE
		TRICHLOROACETOCHLORIDE
		TRICHLOROACETYL CHLORIDE
		TRICHLOROACETYL CHLORIDE (DOT)
		UN 2442 (DOT)
76-03-9		ACETIC ACID, TRICHLORO-
		ACETO-CAUSTIN
		ACIDE TRICHLORACETIQUE (FRENCH)
		ACIDO TRICLOROACETICO (ITALIAN)
		AMCHEM GRASS KILLER
		DOW SODIUM TCA INHIBITED
		KONESTA
		NA TA
		SODIUM TCA SOLUTION
		TCA
		TKHU
		TKHUK
		TRICHLOORAZIJNZUUR (DUTCH)
		TRICHLORACETIC ACID
		TRICHLORESSIGSAEURE (GERMAN)
	*	TRICHLOROACETIC ACID
		TRICHLOROACETIC ACID (ACGIH)
		TRICHLOROACETIC ACID SOLUTION
		TRICHLOROACETIC ACID SOLUTION (DOT)
		TRICHLOROACETIC ACID, SOLID
		TRICHLOROACETIC ACID, SOLID (DOT)
		TRICHLOROETHANOIC ACID
		UN 1839 (DOT)
		UN 2564 (DOT)
		VARITOX
76-05-1		2,2,2-TRIFLUOROACETIC ACID
		ACETIC ACID, TRIFLUORO-
		PERFLUOROACETIC ACID
		TRIFLUORACETIC ACID
	*	TRIFLUOROACETIC ACID
		TRIFLUOROACETIC ACID (DOT)
		TRIFLUOROETHANOIC ACID
		UN 2699 (DOT)
76-06-2		ACQUINITE
		CHLOORPIKRINE (DUTCH)
		CHLOR-O-PIC
		CHLOROFORM, NITRO-
		CHLOROPICRIN
		CHLOROPICRIN (ACGIH,DOT)
		CHLOROPICRIN MIXTURE (DOT)
		CHLOROPICRIN MIXTURE, FLAMMABLE (DOT)
		CHLOROPICRIN MIXTURE, FLAMMABLE (PRESSURE = 147 PSIA; FLASH POINT 100 DEG F)
		CHLOROPICRIN MIXTURE, WITH NO COMPRESSED GAS OR POISON A LIQUID
		CHLOROPICRIN, ABSORBED

CAS No.	IN	Chemical Name
		CHLOROPICRIN, ABSORBED (DOT)
		CHLOROPICRIN, LIQUID
		CHLOROPICRIN, LIQUID (DOT)
		CHLOROPICRINE (FRENCH)
		CHLORPIKRIN (GERMAN)
		CLOROPICRINA (ITALIAN)
		DOJYOPICRIN
		DOLOCHLOR
		G 25
		LARVACIDE
		METHANE, TRICHLORONITRO-
		METHANE, TRICHLORONITRO-, (FLAMMABLE MIXTURE)
		METHANE, TRICHLORONITRO-, (MIXTURE)
		MICROLYSIN
		NA 1583 (DOT)
		NA 2929 (DOT)
		NCI-C00533
		NITROCHLOROFORM
		NITROTRICHLOROMETHANE
		PIC-CLOR
		PICFUME
		PICRIDE
		PROFUME A
		PS
		S 1
		TRI-CLOR
		TRICHLOORNITROMETHAAN (DUTCH)
		TRICHLORNITROMETHAN (GERMAN)
		TRICHLORONITROMETHANE
		TRICLORO-NITRO-METANO (ITALIAN)
		UN 1580 (DOT)
		UN 1583 (DOT)
76-11-9		1,1,1,2-TETRACHLORO-2,2-DIFLUOROETHANE
		1,1-DIFLUOROPERCHLOROETHANE
		2,2-DIFLUORO-1,1,1,2-TETRACHLOROETHANE
		ETHANE, 1,1,1,2-TETRACHLORO-2,2-DIFLUORO-
		FLUOROCARBON 112A
76-12-0		1,1,2,2-TETRACHLORO-1,2-DIFLUOROETHANE
		1,2-DIFLUORO-1,1,2,2-TETRACHLOROETHANE
		ETHANE, 1,1,2,2-TETRACHLORO-1,2-DIFLUORO-
		F 112
		FC 112
		FLUOROCARBON 112
		FREON 112
		FREON R 112
		GENETRON 112
		SYM-TETRACHLORODIFLUOROETHANE
		TETRACHLORO-1,2-DIFLUOROETHANE
		UCON 112
76-13-1	*	1,1,2-TRICHLORO-1,2,2-TRIFLUOROETHANE
		1,1,2-TRICHLOROTRIFLUOROETHANE
		1,1,2-TRIFLUORO-1,2,2-TRICHLOROETHANE
		1,1,2-TRIFLUOROTRICHLOROETHANE
		1,2,2-TRICHLOROTRIFLUOROETHANE
		ARCTON 63
		ARKLONE P
		CHLORINATED FLUOROCARBON
		DAIFLON S 3
		ETHANE, 1,1,2-TRICHLORO-1,2,2-TRIFLUORO-
		F 113
		FC 113
		FLUOROCARBON 113
		FORANE 113
		FREON 113
		FREON 113 TR-T
		FREON F-113
		FREON TF
		FRIGEN 113
		FRIGEN 113A
		FRIGEN 113TR
		FRIGEN 113TR-N
		FRIGEN 113TR-T
		GENETRON 113
		ISCEON 113
		KHLADON 113
		R 113
		R 113 (HALOCARBON)
		REFRIGERANT R 113
76-14-2		DICHLOROTETRAFLUOROETHANE
		FLUOROCARBON 114
76-15-3		1-CHLORO-1,1,2,2,2-PENTAFLUOROETHANE
		CHLOROPENTAFLUOROETHANE
		CHLOROPENTAFLUOROETHANE (ACGIH,DOT)
		CHLOROPERFLUOROETHANE
		ETHANE, CHLOROPENTAFLUORO-
		F 115
		FC 115
		FLUOROCARBON 115
		FREON 115
		GENETRON 115
		HALOCARBON 115
		MONOCHLOROPENTAFLUOROETHANE
		MONOCHLOROPENTAFLUOROETHANE (DOT)
		PENTAFLUOROCHLOROETHANE
		PENTAFLUOROETHYL CHLORIDE
		PERFLUOROETHYL CHLORIDE
		PROPELLANT 115
		R 115
		UN 1020 (DOT)
76-16-4		ETHANE, HEXAFLUORO-
		F 116
		FREON 116
		HEXAFLUOROETHANE
		HEXAFLUOROETHANE (DOT)
		PERFLUOROETHANE
		R 116
		UN 2193 (DOT)
76-19-7		FREON 218
		GENETRON 218
		OCTAFLUOROPROPANE
		OCTAFLUOROPROPANE (DOT)
		PERFLUOROPROPANE
		PROPANE, OCTAFLUORO-
		UN 2424 (DOT)
76-22-2		1,7,7-TRIMETHYLBICYCLO(221)-2-HEPTANONE
		1,7,7-TRIMETHYLNORCAMPHOR
		2-BORNANONE
		2-CAMPHANONE
		2-KETO-1,7,7-TRIMETHYLNORCAMPHANE
		BICYCLO(221)HEPTAN-2-ONE, 1,7,7-TRIMETHYL-
		BORNANE, 2-OXO-
	*	CAMPHOR
		CAMPHOR, SYNTHETIC
		CAMPHOR, SYNTHETIC (ACGIH,DOT)
		CAMPHOR-NATURAL
		FORMOSA CAMPHOR
		GUM CAMPHOR
		HUILE DE CAMPHRE (FRENCH)
		JAPAN CAMPHOR
		KAMPFER (GERMAN)
		LAUREL CAMPHOR
		MATRICARIA CAMPHOR
		NORCAMPHOR, 1,7,7-TRIMETHYL-
		ROOT BARK OIL
		SPIRIT OF CAMPHOR
		UN 2717 (DOT)
76-36-3		BUTYL ALCOHOL

CAS No.	IN	Chemical Name
76-38-0		
		METHOXYFLURANE
76-44-8		
		1(3A),4,5,6,7,8,8-HEPTACHLORO-3A(1),4,7,7A-TETRAHYDRO-4,7-METHANOINDENE
		1,4,5,6,7,10,10-HEPTACHLORO-4,7,8,9-TETRAHYDRO-4,7-ENDOMETHYLENEINDENE
		1,4,5,6,7,10,10-HEPTACHLORO-4,7,8,9-TETRAHYDRO-4,7-METHYLENEINDENE
		1,4,5,6,7,8,8-EPTACLORO-3A,4,7,7A-TETRAIDRO-4,7-ENDO-METANO-INDENE (ITALIAN)
		1,4,5,6,7,8,8-HEPTACHLOOR-3A,4,7,7A-TETRAHYDRO-4,7-ENDO-METHANO-INDEEN (DUTCH)
		1,4,5,6,7,8,8-HEPTACHLOR-3A,4,7,7,7A-TETRAHYDRO-4,7-ENDO-METHANO-INDEN (GERMAN)
		1,4,5,6,7,8,8-HEPTACHLORO-3A,4,7,7,7A-TETRAHYDRO-4,7-METHYLENE INDENE
		1,4,5,6,7,8,8-HEPTACHLORO-3A,4,7,7A-TETRAHYDRO-4,7-ENDOMETHANOINDENE
		1,4,5,6,7,8,8-HEPTACHLORO-3A,4,7,7A-TETRAHYDRO-4,7-METHANOINDENE
		1,4,5,6,7,8,8-HEPTACHLORO-3A,4,7,7A-TETRAHYDRO-4,7-METHANOL-1H-INDENE
		1,4,5,6,7,8,8A-HEPTACHLORO-3A,4,7,7A-TETRAHYDRO-4,7-METHANOINDANE
		3,4,5,6,7,8,8-HEPTACHLORODICYCLOPENTADIENE
		3,4,5,6,7,8,8A-HEPTACHLORODICYCLOPENTADIENE
		3-CHLOROCHLORDENE
		4,7-METHANO-1H-INDENE, 1,4,5,6,7,8,8-HEPTACHLORO-3A,4,7,7A-TETRAHYDRO-
		4,7-METHANOINDENE, 1,4,5,6,7,8,8-HEPTACHLORO-3A,4,7,7A-TETRAHYDRO-
		AAHEPTA
		AGROCERES
		ARBINEX 30TN
		DICYCLOPENTADIENE, 3,4,5,6,7,8,8A-HEPTACHLORO-
		DRINOX
		DRINOX H-34
		E 3314
		ENT 15,152
		EPTACLORO (ITALIAN)
		GPKH
		H
		H-34
		HEPTA
		HEPTACHLOOR (DUTCH)
	*	HEPTACHLOR
		HEPTACHLOR (ACGIH,DOT)
		HEPTACHLORANE
		HEPTACHLORE(FRENCH)
		HEPTAGRAN
		HEPTAMUL
		NA 2761 (DOT)
		NCI-C00180
		RCRA WASTE NUMBER P059
		RHODIACHLOR
		VELSICOL 104
		VELSICOL HEPTACHLOR
76-87-9		
		TRIPHENYLTIN HYDROXIDE
77-47-4		
		1,2,3,4,5,5-HEXACHLORO-1,3-CYCLOPENTADIENE
		1,3-CYCLOPENTADIENE, 1,2,3,4,5,5-HEXACHLORO-
		C 56
		GRAPHLOX
		HCCPD
		HEXACHLORCYKLOPENTADIEN (CZECH)
		HEXACHLORO-1,3-CYCLOPENTADIENE
	*	HEXACHLOROCYCLOPENTADIENE
		HEXACHLOROCYCLOPENTADIENE (ACGIH,DOT)
		HRS 1655
		NCI-C55607
		PCL
		PERCHLOROCYCLOPENTADIENE
		RCRA WASTE NUMBER U130
		UN 2646 (DOT)
77-73-6		
		1,3-CYCLOPENTADIENE, DIMER
		3A,4,7,7A-TETRAHYDRO-4,7-METHANOINDENE
		4,7-METHANO-1H-INDENE, 3A,4,7,7A-TETRAHYDRO-
		4,7-METHANOINDENE, 3A,4,7,7A-TETRAHYDRO-
		BICYCLOPENTADIENE
		BISCYCLOPENTADIENE
		CYCLOPENTADIENE DIMER
	*	DICYCLOPENTADIENE
		DICYCLOPENTADIENE (ACGIH,DOT)
		DICYKLOPENTADIEN (CZECH)
		DIMER CYKLOPENTADIENU (CZECH)
		TRICYCLO[52102,6]DECA-3,8-DIENE
		UN 2048 (DOT)
77-78-1		
		DIMETHYL MONOSULFATE
		DIMETHYL SULFATE
		DIMETHYL SULFATE (ACGIH,DOT)
		DIMETHYL SULPHATE
		DIMETHYLESTER KYSELINY SIROVE (CZECH)
		DIMETHYLSULFAAT (DUTCH)
		DIMETHYLSULFAT (CZECH)
		DIMETILSOLFATO (ITALIAN)
		DMS
		DWUMETYLOWY SIARCZAN (POLISH)
		METHYL SULFATE
		METHYL SULFATE (DOT)
		METHYLE (SULFATE DE) (FRENCH)
		RCRA WASTE NUMBER U103
		SULFATE DE METHYLE (FRENCH)
		SULFATE DIMETHYLIQUE (FRENCH)
		SULFURIC ACID, DIMETHYL ESTER
		UN 1595 (DOT)
77-81-6		
		DIMETHYLAMIDOETHOXYPHOSPHORYL CYANIDE
		DIMETHYLAMINOCYANPHOSPHORSAEUREAETHYLESTER (GERMAN)
		DIMETHYLPHOSPHORAMIDOCYANIDIC ACID, ETHYL ESTER
		EA 1205
		ETHYL DIMETHYLAMIDOCYANOPHOSPHATE
		ETHYL DIMETHYLPHOSPHORAMIDOCYANIDATE
		ETHYL N,N-DIMETHYLAMINO CYANOPHOSPHATE
		ETHYL N,N-DIMETHYLPHOSPHORAMIDOCYANIDATE
		GA
		GA (CHEMICAL WARFARE AGENT)
		GELAN I
		LE-100
		MCE
		PHOSPHORAMIDOCYANIDIC ACID, DIMETHYL-, ETHYL ESTER
		T-2104
		TABOON A
		TABUN
		TL 1578
		TRILON 83
78-00-2		
		CZTEROETYLEK OLOWIU (POLISH)
		LEAD, TETRAETHYL-
		NA 1649 (DOT)
		NCI-C54988
		PIOMBO TETRA-ETILE (ITALIAN)
		PLUMBANE, TETRAETHYL-
		RCRA WASTE NUMBER P110
		TEL
		TETRAETHYL LEAD
		TETRAETHYL LEAD (ACGIH,DOT)
		TETRAETHYL LEAD, LIQUID
		TETRAETHYL LEAD, LIQUID (INCLUDING FLASHPOINT FOR EXPORT SHIPMENT BY WATER) (DOT)
		TETRAETHYLPLUMBANE
		UN 1649 (DOT)
78-10-4		
		DYNASIL A
		ES 100
		ES 28

CAS No.	IN	Chemical Name
		ES 28 (ESTER)
		ETHYL ORTHOSILICATE
	*	ETHYL SILICATE
		ETHYL SILICATE ((ETO)4Si)
		ETHYL SILICATE (ACGIH,DOT)
		ETYLU KRZEMIAN (POLISH)
		EXTREMA
		SI 42
		SILANE, TETRAETHOXY-
		SILICATE D'ETHYLE (FRENCH)
		SILICATE TETRAETHYLIQUE (FRENCH)
		SILICIC ACID (H4SiO4), TETRAETHYL ESTER
		SILICIC ACID, TETRAETHYL ESTER
		SILICON ETHOXIDE
		SILICON TETRAETHOXIDE
		SILICON TETRAETHOXIDE (Si(OET)4)
		SILIKAN L
		STEINFESTIGER OH
		TEOS
		TES 28
		TETRAETHOXYSILANE
		TETRAETHOXYSILICON
		TETRAETHYL ORTHOSILICATE
		TETRAETHYL ORTHOSILICATE (DOT)
		TETRAETHYL SILICATE
		TETRAETHYL SILICATE (DOT)
		TETRAETHYLOXYSILANE
		UN 1292 (DOT)
78-11-5		
		PENTAERYTHRITE TETRANITRATE
78-30-8		
		O-CRESYL PHOSPHATE
		PHOSFLEX 179C
		PHOSPHORIC ACID, TRI-O-TOLYL ESTER
		PHOSPHORIC ACID, TRIS(2-METHYLPHENYL) ESTER
		TOCP
		TOKF
		TOTP
		TRI-O-CRESYL PHOSPHATE
		TRI-O-TOLYL PHOSPHATE
		TRIORTHOCRESYL PHOSPHATE
		TRIS(O-CRESYL) PHOSPHATE
		TRIS(O-METHYLPHENYL) PHOSPHATE
		TRIS(O-TOLYL) PHOSPHATE
78-34-2		
		1,4-DIOSSAN-2,3-DIYL-BIS(O,O-DIETIL-DITIOFOSFATO) (ITALIAN)
		1,4-DIOXAAN-2,3-DIYL-BIS(O,O-DIETHYL-DITHIOFOSFAAT) (DUTCH)
		1,4-DIOXAN-2,3-DIYL BIS(O,O-DIETHYL PHOSPHOROTHIOLOTHIONATE)
		1,4-DIOXAN-2,3-DIYL BIS(O,O-DIETHYLPHOSPHOROTHIOLOTHIONATE)
		1,4-DIOXAN-2,3-DIYL O,O,O',O'-TETRAETHYL DI(PHOSPHOROMITHIOATE)
		1,4-DIOXAN-2,3-DIYL-BIS(O,O-DIAETHYL-DITHIOPHOSPHAT) (GERMAN)
		2,3-BIS(DIETHOXYPHOSPHINOTHIOYLTHIO)-1,4-DIOXANE
		2,3-P-DIOXAN-S,S'-BIS(O,O-DIAETHYLDITHIOPHOSPHAT) (GERMAN)
		2,3-P-DIOXANDITHIOL S,S-BIS(O,O-DIETHYL PHOSPHORODITHIOATE)
		2,3-P-DIOXANE S,S-BIS(O,O-DIETHYLPHOSPHOROITHIOATE)
		2,3-P-DIOXANEDITHIOL S,S-BIS(O,O-DIETHYL PHOSPHORODITHIOATE)
		AC 528
		BIS(DITHIOPHOSPHATE DE O,O-DIETHYLE) DE S,S'-(1,4-DIOXANNE-2,3-DIYLE) (FRENCH)
		DELANOV
		DELNATEX
		DELNAV
		DIOXATHION
		DIOXATHION (ACGIH)
		DIOXATION
		DIOXOTHION
		ENT 22,897
		HERCULES 528
		HERCULES AC528
		KAVADEL
		NAVADEL
		NCI-C00395
		P-DIOXANE-2,3-DITHIOL, S,S-DIESTER WITH O,O-DIETHYL PHOSPHORODITHIOATE
		P-DIOXANE-2,3-DIYL ETHYL PHOSPHORODITHIOATE
		PHOSPHORODITHIOIC ACID, S,S'-1,4-DIOXANE-2,3-DIYL O,O,O',O'-TETRAETHYL ESTER
		PHOSPHORODITHIOIC ACID, S,S'-1,4-DIOXANE-2,3-DIYL O,O,O',O'-TETRAETHYL ESTER (9CI)
		PHOSPHORODITHIOIC ACID, S,S'-P-DIOXANE-2,3-DIYL O,O,O',O'-TETRAETHYL ESTER
		RUPHOS
78-53-5		
		(2-DIETHYLAMINO)ETHYLPHOSPHOROTHIOIC ACID O,O-DIETHYL ESTER
		AMITON
		CHIPMAN 6200
		CITRAM
		DIETHYL S-2-DIETHYLAMINOETHYL PHOSPHOROTHIOATE
		DSDP
		ENT 24,980-X
		INFERNO
		METRAMAC
		METRAMAK
		O,O-DIETHYL S-(2-DIETHYLAMINOETHYL) THIOPHOSPHATE
		O,O-DIETHYL S-(BETA-DIETHYLAMINO)ETHYL PHOSPHOROTHIOLATE
		O,O-DIETHYL S-2-DIETHYLAMINOETHYL PHOSPHOROTHIOATE
		O,O-DIETHYL S-2-DIETHYLAMINOETHYL PHOSPHOROTHIOLATE
		O,O-DIETHYL S-DIETHYLAMINOETHYL PHOSPHOROTHIOLATE
		PHOSPHOROTHIOIC ACID, S-(2-(DIETHYLAMINO)ETHYL) O,O-DIETHYL ESTER
		R 5158
		RHODIA-6200
		S-(2-(DIETHYLAMINO)ETHYL)PHOSPHOROTHIOIC ACID O,O-DIETHYL ESTER
		S-(DIETHYLAMINOETHYL) O,O-DIETHYL PHOSPHOROTHIOATE
		TETRAM
78-59-1		
		1,5,5-TRIMETHYL-3-OXOCYCLOHEXENE
		2-CYCLOHEXEN-1-ONE, 3,5,5-TRIMETHYL-
		3,5,5-TRIMETHYL-2-CYCLOHEXENONE
		ALPHA-ISOPHORON
		ALPHA-ISOPHORONE
		ISOACETOPHORONE
		ISOFORON
		ISOPHORON
		ISOPHORONE
78-62-6		
		DIETHOXYDIMETHYL SILANE
		DIMETHYL-DIETHOXYSILAN (CZECH)
		DIMETHYLDIETHOXYSILANE
		DIMETHYLDIETHOXYSILANE (DOT)
		SILANE, DIETHOXYDIMETHYL-
		UN 2380 (DOT)
78-63-7		
		(1,1,4,4-TETRAMETHYLTETRAMETHYLENE)BIS(TERT-BUTYL PEROXIDE)
		2,5-BIS(TERT-BUTYLDIOXY)-2,5-DIMETHYLHEXANE
		2,5-BIS(TERT-BUTYLPEROXY)-2,5-DIMETHYLHEXANE
		2,5-DIMETHYL-2,5-BIS(TERT-BUTYLDIOXY)HEXANE
		2,5-DIMETHYL-2,5-BIS(TERT-BUTYLPEROXY)HEXANE
		2,5-DIMETHYL-2,5-BIS-(TERT-BUTYLPEROXY)HEXANE, MAX CONCENTRATION 52% WITH INERT SOLID (DOT)
		2,5-DIMETHYL-2,5-DI(T-BUTYLPEROXY)HEXANE
		2,5-DIMETHYL-2,5-DI(TERT-BUTYLPEROXY)HEXANE
		2,5-DIMETHYL-2,5-DI-(TERT-BUTYLPEROXY)HEXANE, TECHNICALLY PURE

CAS No.	IN	Chemical Name
		2,5-DIMETHYL-2,5-DI-(TERT-BUTYLPEROXY)HEXANE, TECHNICALLY PURE (DOT)
		HEXANE, 2,5-DIMETHYL-2,5-DI(T-BUTYLPEROXY)-
		HEXANE, 2,5-DIMETHYL-2,5-DI(T-BUTYLPEROXY)-, MAXIMUM CONCENTRATION 52% WITH INERT SOLID
		KAYAHEXA AD
		KAYAHEXA AD 40C
		LUPERCO 101XL
		LUPEROX 101
		LUPERSOL 101
		PERHEXA 25B
		PERHEXA 25B40
		PERHEXA 3M40
		PEROXIDE, (1,1,4,4-TETRAMETHYL-1,4-BUTANEDIYL)-BIS[(1,1-DIMETHYLETHYL)
		PEROXIDE, (1,1,4,4-TETRAMETHYLTETRAMETHYLENE)BIS-[TERT-BUTYL
		TRIGONOX 101
		TRIGONOX 101-101/45
		TRIGONOX XQ 8
		UN 2155 (DOT)
		UN 2156 (DOT)
		VAROX
		VAROX 50
78-67-1		
		2,2'-AZOBIS(2-CYANOBUTANE)
		2,2'-AZOBIS(2-METHYLPROPIONITRILE)
		2,2'-AZOBIS(ISOBUTYRONITRILE)
		2,2'-AZOBIS[2-METHYLPROPANENITRILE]
		2,2'-AZODIISOBUTYRONITRILE
		2,2'-DICYANO-2,2'-AZOPROPANE
		2,2'-DIMETHYL-2,2'-AZODIPROPIONITRILE
		ACETO AZIB
		AIBN
		AIVN
		ALPHA,ALPHA'-AZOBIS(ISOBUTYRONITRILE)
		ALPHA,ALPHA'-AZOBISISOBUTYLONITRILE
		ALPHA,ALPHA'-AZODIISOBUTYRIC ACID DINITRILE
		ALPHA,ALPHA'-AZODIISOBUTYRONITRILE
		AZDH
		AZDN
		AZOBISISOBUTYLONITRILE
		AZOBISISOBUTYRONITRILE
		AZODIISOBUTYRONITRILE
		AZODIISOBUTYRONITRILE (DOT)
		CHKHZ 57
		GENITRON
		GENITRON AZDN
		GENITRON AZDN-FF
		N,N'-BIS(2-CYANO-2-PROPYL)DIAZENE
		PIANOFOR AN
		POLY-ZOLE AZDN
		POROFOR 57
		POROFOR N
		POROPHOR N
		PROPANENITRILE, 2,2'-AZOBIS[2-METHYL-
		PROPIONITRILE, 2,2'-AZOBIS(2-METHYL-
		UN 2952 (DOT)
		VAZO
		VAZO 64
78-71-7		
		OXETANE, 3,3-BIS(CHLOROMETHYL)
78-76-2		
	*	2-BROMOBUTANE
		2-BROMOBUTANE (DOT)
		2-BUTYL BROMIDE
		BUTANE, 2-BROMO-
		METHYLETHYLBROMOMETHANE
		SEC-BUTYL BROMIDE
		UN 2339 (DOT)
78-77-3		
		1-BROMO-2-METHYLPROPANE
		ISOBUTYL BROMIDE
78-78-4		
		1,1,2-TRIMETHYLETHANE
		2-METHYLBUTANE
		BUTANE, 2-METHYL-
		ISOPENTANE
78-79-5		
		1,3-BUTADIENE, 2-METHYL-
		2-METHYL-1,3-BUTADIENE
		2-METHYL-1,3-BUTADIENE (DOT)
		2-METHYLBUTADIENE
		3-METHYL-1,3-BUTADIENE
		BETA-METHYLBIVINYL
		ISOPENTADIENE
	*	ISOPRENE
		ISOPRENE (DOT)
		ISOPRENE, INHIBITED (DOT)
		UN 1218 (DOT)
78-80-8		
		1-BUTEN-3-YNE, 2-METHYL-
		2-METHYL-1-BUTEN-3-YNE
		2-METHYL-1-BUTENYNE
		2-METHYLBUTENYNE
		3-METHYL-3-BUTEN-1-YNE
		ISOPROPENYL ACETYLENE
78-81-9		
		1-AMINO-2-METHYLPROPANE
		1-PROPANAMINE, 2-METHYL-
		2-METHYL-1-PROPANAMINE
		2-METHYLPROPYLAMINE
		3-METHYL-2-PROPYLAMINE
		I-BUTYLAMINE
		ISOBUTYLAMINE
		ISOBUTYLAMINE (DOT)
		MONOISOBUTYLAMINE
		UN 1214 (DOT)
		VALAMINE
78-82-0		
		1-CYANO-1-METHYLETHANE
		2-CYANOPROPANE
		2-METHYLPROPANENITRILE
		2-METHYLPROPIONITRILE
		ALPHA-METHYLPROPANENITRILE
		DIMETHYLACETONITRILE
		ISOBUTYRONITRILE
		ISOPROPYL CYANIDE
		ISOPROPYL NITRILE
		PROPANENITRILE, 2-METHYL-
78-83-1		
		1-HYDROXYMETHYLPROPANE
		1-PROPANOL, 2-METHYL-
		2-METHYL PROPANOL
		2-METHYL-1-PROPANOL
		2-METHYLPROPAN-1-OL
		2-METHYLPROPYL ALCOHOL
		ALCOOL ISOBUTYLIQUE (FRENCH)
		FERMENTATION BUTYL ALCOHOL
		ISO-BUTYL ALCOHOL
		ISOBUTANOL
		ISOBUTANOL (DOT)
	*	ISOBUTYL ALCOHOL
		ISOBUTYL ALCOHOL (ACGIH,DOT)
		ISOBUTYLALKOHOL (CZECH)
		ISOPROPYLCARBINOL
		RCRA WASTE NUMBER U140
		UN 1212 (DOT)
78-84-2		
		2-METHYL-1-PROPANAL
		2-METHYLPROPANAL
		2-METHYLPROPIONALDEHYDE
		ALPHA-METHYLPROPIONALDEHYDE
		ISOBUTALDEHYDE
		ISOBUTANAL

CAS No.	IN	Chemical Name
		ISOBUTYRAL
	*	ISOBUTYRALDEHYD E
		ISOBUTYRALDEHYDE
		ISOBUTYRIC ALDEHYDE
		ISOBUTYRYL ALDEHYDE
		ISOPROPYL ALDEHYDE
		ISOPROPYLFORMALDEHYDE
		METHYLPROPANAL
		PROPANAL, 2-METHYL-
78-85-3		
		2-METHYL-2-PROPENAL
		2-METHYLACROLEIN
		2-METHYLENEPROPANAL
		2-METHYLPROPENAL
		2-METHYLPROPENAL (CZECH)
		2-PROPENAL, 2-METHYL-
		ACROLEIN, 2-METHYL-
		ALPHA-METHACROLEIN
		ALPHA-METHYLACROLEIN
		ALPHA-METHYLACRYLALDEHYDE
		ISOBUTENAL
		METHACRALDEHYDE (DOT)
		METHACROLEIN
		METHACRYLALDEHYDE
		METHACRYLALDEHYDE (DOT)
		METHACRYLIC ALDEHYDE
		METHYLACROLEIN
		METHYLACRYLALDEHYDE
		UN 2396 (DOT)
78-86-4		
		1-METHYLPROPYL CHLORIDE
	*	2-CHLOROBUTANE
		BUTANE, 2-CHLORO-
		SEC-BUTYL CHLORIDE
78-87-5		
	*	1,2-DICHLOROPROPANE
		1,2-DICHLOROPROPYLENE
		DICHLOR
		DICHLORPROPEN-GEMISCH (GERMAN)
		PDC
		PROPANE, 1,2-DICHLORO-
		PROPENE, 1,2-DICHLORO-
		PROPYLENE CHLORIDE
		PROPYLENE DICHLORIDE
		RCRA WASTE NUMBER U083
78-88-6		
		1-PROPENE, 2,3-DICHLORO-
		2,3-DICHLORO-1-PROPENE
		2,3-DICHLOROPROPENE
		2,3-DICHLOROPROPYLENE
		2-CHLOROALLYL CHLORIDE
		PROPENE, 2,3-DICHLORO-
78-89-7		
		1-HYDROXY-2-CHLOROPROPANE
		1-PROPANOL, 2-CHLORO-
		2-CHLORO-1-PROPANOL
		2-CHLOROPROPANOL
		2-CHLOROPROPYL ALCOHOL
		PROPYLENE CHLOROHYDRIN
		PROPYLENE CHLOROHYDRIN (DOT)
		UN 2611 (DOT)
78-90-0		
		1,2-DIAMINOPROPANE
		1,2-PROPANEDIAMINE
		1,2-PROPYLENEDIAMINE
		PROPYLENE DIAMINE (DOT)
		PROPYLENEDIAMINE
		UN 2258 (DOT)
78-92-2		
		1-METHYL PROPANOL
		1-METHYL-1-PROPANOL
		1-METHYLPROPYL ALCOHOL

CAS No.	IN	Chemical Name
		2-BUTANOL
		2-BUTYL ALCOHOL
		2-HYDROXYBUTANE
		ALCOOL BUTYLIQUE SECONDAIRE (FRENCH)
		BUTAN-2-OL
		BUTANOL SECONDAIRE (FRENCH)
		BUTANOL-2
		BUTYLENE HYDRATE
		CCS 301
		ETHYLMETHYL CARBINOL
		METHYLETHYLCARBINOL
		S-BUTANOL
		S-BUTYL ALCOHOL
		SBA
		SEC-BUTANOL
		SEC-BUTANOL (DOT)
	*	SEC-BUTYL ALCOHOL
		SEC-BUTYL ALCOHOL (ACGIH,DOT)
		UN 1120 (DOT)
78-93-3		
		2-BUTANONE
		3-BUTANONE
		ACETONE, METHYL-
		AETHYLMETHYLKETON (GERMAN)
		BUTANONE
		BUTANONE 2 (FRENCH)
		ETHYL METHYL CETONE (FRENCH)
		ETHYL METHYL KETONE
		ETHYL METHYL KETONE (DOT)
		ETHYLMETHYLKETON (DUTCH)
		KETONE, ETHYL METHYL
		MEETCO
		MEK
		METHYL ACETONE
		METHYL ACETONE (DOT)
	*	METHYL ETHYL KETONE
		METHYL ETHYL KETONE (ACGIH,DOT)
		METHYL ETHYL KETONE (MEK)
		METILETILCHETONE (ITALIAN)
		METYLOETYLOKETON (POLISH)
		RCRA WASTE NUMBER U159
		UN 1193 (DOT)
		UN 1232 (DOT)
78-94-4		
		1-BUTEN-3-ONE
		2-BUTENONE
		3-BUTEN-2-ONE
		3-BUTENE-2-ONE
		3-OXOBUTENE
		ACETONE, METHYLENE-
		ACETYL ETHYLENE
		BUTENONE
		DELTA(SUP 3)-2-BUTENONE
		GAMMA-OXO-ALPHA-BUTYLENE
		KETONE, METHYL VINYL
		METHYL VINYL KETONE
		METHYL VINYL KETONE (DOT)
		METHYL VINYL KETONE, INHIBITED
		METHYL VINYL KETONE, INHIBITED (DOT)
		METHYL-VINYL-CETONE (FRENCH)
		METHYLENE ACETONE
		METHYLVINYL KETONE
		METHYLVINYLKETON (GERMAN)
		UN 1251 (DOT)
		VINYL METHYL KETONE
78-95-5		
		1-CHLORO-2-PROPANONE
		1-CHLOROACETONE
		2-PROPANONE, 1-CHLORO-
		ACETONE, CHLORO-
		ACETONYL CHLORIDE
		ALPHA-CHLOROACETONE
		CHLORACETONE (FRENCH)
		CHLORO-2-PROPANONE
	*	CHLOROACETONE
		CHLOROACETONE, STABILIZED (DOT)

CAS No.	IN	Chemical Name
		CHLOROMETHYL METHYL KETONE
		CHLOROPROPANONE
		METHYL CHLOROMETHYL KETONE
		MONOCHLORACETONE
		MONOCHLOROACETONE
		MONOCHLOROACETONE, INHIBITED (DOT)
		MONOCHLOROACETONE, STABILIZED
		MONOCHLOROACETONE, STABILIZED (DOT)
		MONOCHLOROACETONE, UNSTABILIZED (DOT)
		UN 1695 (DOT)
78-96-6		
		1-AMINO-2-HYDROXYPROPANE
		1-AMINO-2-PROPANOL
		1-AMINOPROPAN-2-OL
		1-METHYL-2-AMINOETHANOL
		2-AMINO-1-METHYLETHANOL
		2-HYDROXY-1-PROPANAMINE
		2-HYDROXY-1-PROPYLAMINE
		2-HYDROXYPROPANAMINE
		2-HYDROXYPROPYLAMINE
		2-PROPANOL, 1-AMINO-
		ALPHA-AMINOISOPROPYL ALCOHOL
		BETA-AMINOISOPROPANOL
		ISOPROPANOLAMINE
		MIPA
		MONOISOPROPANOLAMINE
		THREAMINE
78-97-7		
		2-HYDROXYPROPIONITRILE
		ACETALDEHYDE, CYANOHYDRIN
		ACETOCYANOHYDRIN
		ALPHA-HYDROXYPROPIONITRILE
		LACTONITRILE
		PROPANENITRILE, 2-HYDROXY-
		PROPIONITRILE, 2-HYDROXY-
78-98-3		
		MEK
78-99-9		
		1,1-DICHLOROPROPANE
		PROPANE, 1,1-DICHLORO-
		PROPYLIDENE CHLORIDE
79-00-5		
	*	1,1,2-TRICHLOROETHANE
		BETA-T
		BETA-TRICHLOROETHANE
		ETHANE, 1,1,2-TRICHLORO-
		VINYLTRICHLORIDE
79-01-6		
		1,1,2-TRICHLOROETHYLENE
		1,1-DICHLORO-2-CHLOROETHYLENE
		1,2,2-TRICHLOROETHYLENE
		1-CHLORO-2,2-DICHLOROETHYLENE
		ACETYLENE TRICHLORIDE
		ALGYLEN
		ANAMENTH
		BENZINOL
		BLACOSOLV
		BLANCOSOLV
		CECOLENE
		CHLORILEN
		CHLORYLEA
		CHLORYLEN
		CHORYLEN
		CIRCOSOLV
		CRAWHASPOL
		DENSINFLUAT
		DOW-TRI
		DUKERON
		ETHENE, TRICHLORO-
		ETHINYL TRICHLORIDE
		ETHYLENE TRICHLORIDE
		ETHYLENE, TRICHLORO-
		FLECK-FLIP
		FLOCK FLIP
		FLUATE
		GEMALGENE
		GERMALGENE
		LANADIN
		LETHURIN
		NARCOGEN
		NARKOGEN
		NARKOSOID
		NCI-C04546
		NIALK
		PERM-A-CHLOR
		PERM-A-CLOR
		PETZINOL
		PHILEX
		RCRA WASTE NUMBER U228
		TCE
		THRETHYLEN
		THRETHYLENE
		TRETHYLENE
		TRI
		TRI-CLENE
		TRI-PLUS
		TRI-PLUS M
		TRIAD
		TRIAL
		TRIASOL
		TRICHLOORETHEEN (DUTCH)
		TRICHLOORETHYLEEN, TRI (DUTCH)
		TRICHLORAETHEN (GERMAN)
		TRICHLORAETHYLEN, TRI (GERMAN)
		TRICHLORAN
		TRICHLOREN
		TRICHLORETHENE (FRENCH)
		TRICHLORETHYLENE
		TRICHLORETHYLENE, TRI (FRENCH)
		TRICHLOROETHENE
	*	TRICHLOROETHYLENE
		TRICHLOROETHYLENE (ACGIH,DOT)
		TRICLENE
		TRICLORETENE (ITALIAN)
		TRICLOROETILENE (ITALIAN)
		TRIELENE
		TRIELIN
		TRIELINA (ITALIAN)
		TRIELINE
		TRIKLONE
		TRILEN
		TRILENE
		TRILINE
		TRIMAR
		TRIOL
		UN 1710 (DOT)
		VESTROL
		VITRAN
		WESTROSOL
79-03-8		
		PROPANOYL CHLORIDE
		PROPIONIC ACID CHLORIDE
		PROPIONIC CHLORIDE
		PROPIONYL CHLORIDE
		PROPIONYL CHLORIDE (DOT)
		UN 1815 (DOT)
79-04-9		
		ACETYL CHLORIDE, CHLORO-
		CHLORACETYL CHLORIDE
		CHLOROACETIC ACID CHLORIDE
		CHLOROACETIC CHLORIDE
		CHLOROACETYL CHLORIDE
		CHLOROACETYL CHLORIDE (ACGIH,DOT)
		CHLORURE DE CHLORACETYLE (FRENCH)
		MONOCHLOROACETYL CHLORIDE
		UN 1752 (DOT)

CAS No.	IN	Chemical Name
79-06-1		
		2-PROPENAMIDE
		2-PROPENAMIDE (9CI)
	*	ACRYLAMIDE
		ACRYLAMIDE (ACGIH,DOT)
		ACRYLAMIDE SOLUTION [COMBUSTIBLE LIQUID LABEL]
		ACRYLAMIDE SOLUTION [FLAMMABLE LIQUID LABEL]
		ACRYLIC AMIDE
		AKRYLAMID (CZECH)
		ETHYLENECARBOXAMIDE
		PROPENAMIDE
		RCRA WASTE NUMBER U007
		UN 2074 (DOT)
		VINYL AMIDE
79-08-3		
		BROMOACETIC ACID, SOLID OR SOLUTION
79-09-4		
		ACIDE PROPIONIQUE (FRENCH)
		ANTISCHIM B
		CARBOXYETHANE
		ETHANECARBOXYLIC ACID
		ETHYLFORMIC ACID
		LUPROSIL
		METACETONIC ACID
		METHYL ACETIC ACID
		MONOPROP
		PROPANOIC ACID
		PROPCORN
	*	PROPIONIC ACID
		PROPIONIC ACID (ACGIH,DOT)
		PROPIONIC ACID GRAIN PRESERVER
		PROPIONIC ACID, SOLUTION
		PROPIONIC ACID, SOLUTION (DOT)
		PROPIONIC ACID, SOLUTION CONTAINING NOT LESS THAN 80% ACID (DOT)
		PROPKORN
		PROZOIN
		PSEUDOACETIC ACID
		SENTRY GRAIN PRESERVER
		TENOX P GRAIN PRESERVATIVE
		UN 1848 (DOT)
79-10-7		
		2-PROPENOIC ACID
		2-PROPENOIC ACID (9CI)
		ACROLEIC ACID
	*	ACRYLIC ACID
		ACRYLIC ACID (ACGIH,DOT)
		ACRYLIC ACID (GLACIAL)
		ACRYLIC ACID, INHIBITED (DOT)
		ETHYLENECARBOXYLIC ACID
		PROPENE ACID
		PROPENOIC ACID
		RCRA WASTE NUMBER U008
		UN 2218 (DOT)
		VINYLFORMIC ACID
79-11-8		
		ACETIC ACID, CHLORO-
		ACIDE CHLORACETIQUE (FRENCH)
		ACIDE MONOCHLORACETIQUE (FRENCH)
		ACIDOMONOCLOROACETICO (ITALIAN)
		ALPHA-CHLOROACETIC ACID
		CHLORACETIC ACID
	*	CHLOROACETIC ACID
		CHLOROACETIC ACID, LIQUID
		CHLOROACETIC ACID, LIQUID (DOT)
		CHLOROACETIC ACID, SOLID
		CHLOROACETIC ACID, SOLID (DOT)
		CHLOROACETIC ACID, SOLUTION (DOT)
		CHLOROETHANOIC ACID
		MCA
		MKHUK
		MONOCHLOORAZIJNZUUR (DUTCH)
		MONOCHLORACETIC ACID
		MONOCHLORESSIGSAEURE (GERMAN)
		MONOCHLOROACETIC ACID
		MONOCHLOROETHANOIC ACID
		NCI-C60231
		UN 1750 (DOT)
		UN 1751 (DOT)
79-19-6		
		1-AMINO-2-THIOUREA
		1-AMINOTHIOUREA
		2-THIOSEMICARBAZIDE
		HYDRAZINECARBOTHIOAMIDE
		ISOTHIOSEMICARBAZIDE
		N-AMINOTHIOUREA
		RCRA WASTE NUMBER P116
		SEMICARBAZIDE, 3-THIO-
		SEMICARBAZIDE, THIO-
		THIASEMICARBAZIDE
		THIOCARBAMOYLHYDRAZINE
		THIOCARBAMYLHYDRAZINE
		THIOSEMICARBAZIDE
		TSC
		TSZ
		USAF EK-1275
79-20-9		
		ACETIC ACID, METHYL ESTER
		DEVOTON
	*	METHYL ACETATE
		METHYL ETHANOATE
		TERETON
79-21-0		
		ACETIC PEROXIDE
		ACETYL HYDROPEROXIDE
		DESOXON 1
		ESTOSTERIL
		ETHANE PEROXOIC ACID
		HYDROPEROXIDE, ACETYL
		MONOPERACETIC ACID
		NA 2131 (DOT)
		OSBON AC
		PERACETIC ACID
		PERACETIC ACID (CONCENTRATION ≤60%)
		PERACETIC ACID, NOT OVER 43% ACID AND NOT OVER 6% HYDROGEN PEROXIDE (DOT)
		PEROXOACETIC ACID
		PEROXYACETIC ACID
		PEROXYACETIC ACID, MORE THAN 43% WITH MORE THAN 6% HYDROGEN PEROXIDE
		PEROXYACETIC ACID, MORE THAN 43% WITH MORE THAN 6% HYDROGEN PEROXIDE (DOT)
		PEROXYACETIC ACID, NOT MORE THAN 43% ACID AND NOT MORE THAN 6% HYDROGEN PEROXIDE (DOT)
		PEROXYACETIC ACID, NOT OVER 43% ACID AND NOT OVER 6% HYDROGEN PEROXIDE
		PROXITANE 4002
		UN 2131 (DOT)
79-22-1		
		CARBONOCHLORIDIC ACID, METHYL ESTER
		CHLORAMEISENSAEURE METHYLESTER (GERMAN)
		CHLOROCARBONATE DE METHYLE (FRENCH)
		CHLOROCARBONIC ACID METHYL ESTER
		CHLOROFORMIATE DE METHYLE (FRENCH)
		CHLOROFORMIC ACID METHYL ESTER
		FORMIC ACID, CHLORO-, METHYL ESTER
		MCF
		METHOXYCARBONYL CHLORIDE
		METHYL CARBONOCHLORIDATE
		METHYL CHLOROCARBONATE
		METHYL CHLOROCARBONATE (DOT)
		METHYL CHLOROFORMATE
		METHYL CHLOROFORMATE (DOT)
		METHYLCHLOORFORMIAAT (DUTCH)
		METILCLOROFORMIATO (ITALIAN)
		RCRA WASTE NUMBER U156
		TL 438
		UN 1238 (DOT)

CAS No.	IN	Chemical Name
79-24-3		
		ETHANE, NITRO-
		NITROETAN (POLISH)
	*	NITROETHANE
		NITROETHANE (ACGIH,DOT)
		UN 2842 (DOT)
79-27-6		
		1,1,2,2-TETRABROMAETHAN (GERMAN)
		1,1,2,2-TETRABROMOETANO (ITALIAN)
		1,1,2,2-TETRABROMOETHANE
		1,1,2,2-TETRABROMOETHYLENE
		1,1,2,2-TETRABROOMETHAAN (DUTCH)
	*	ACETYLENE TETRABROMIDE
		ACETYLENE TETRABROMIDE (ACGIH,DOT)
		ETHANE, 1,1,2,2-TETRABROMO-
		MUTHMANN'S LIQUID
		S-TETRABROMOETHANE
		TBE
		TETRABROMOACETYLENE
		TETRABROMOETHANE
		UN 2504 (DOT)
79-29-8		
		1,1,2,2-TETRAMETHYLETHANE
		2,3-DIMETHYLBUTANE
		2,3-DIMETHYLBUTANE (DOT)
		BIISOPROPYL
		BUTANE, 2,3-DIMETHYL-
		DIISOPROPYL
		ISOHEXANE (ACGIH)
		UN 2457 (DOT)
79-30-1		
		2-METHYLPROPANOYL CHLORIDE
		2-METHYLPROPIONYL CHLORIDE
		ALPHA-METHYLPROPIONYL CHLORIDE
		ISOBUTANOYL CHLORIDE
		ISOBUTYRIC ACID CHLORIDE
		ISOBUTYROYL CHLORIDE
		ISOBUTYRYL CHLORIDE
		ISOBUTYRYL CHLORIDE (DOT)
		PROPANOYL CHLORIDE, 2-METHYL-
		UN 2395 (DOT)
79-31-2		
		2-METHYLPROPANOIC ACID
		2-METHYLPROPIONIC ACID
		ALPHA-METHYLPROPANOIC ACID
		ALPHA-METHYLPROPIONIC ACID
		DIMETHYLACETIC ACID
		ISOBUTANOIC ACID
	*	ISOBUTYRIC ACID
		ISOPROPYLFORMIC ACID
		PROPANOIC ACID, 2-METHYL-
79-34-5		
		1,1,2,2-TETRACHLORETHANE
	*	1,1,2,2-TETRACHLOROETHANE
		ACETYLENE TETRACHLORIDE
		BONOFORM
		CELLON
		ETHANE, 1,1,2,2-TETRACHLORO-
		S-TETRACHLOROETHANE
		SYM-TETRACHLOROETHANE
		TETRACHLOROETHANE
79-36-7		
		2,2-DICHLOROACETYL CHLORIDE
		ACETYL CHLORIDE, DICHLORO-
		ALPHA,ALPHA-DICHLOROACETYL CHLORIDE
		CHLORURE DE DICHLORACETYLE (FRENCH)
	*	DICHLORACETYL CHLORIDE
		DICHLORO ACETYL CHLORIDE (DOT)
		DICHLOROACETIC ACID CHLORIDE
		DICHLOROACETYL CHLORIDE
		DICHLOROETHANOYL CHLORIDE
		UN 1765 (DOT)

CAS No.	IN	Chemical Name
79-38-9		
		1,1,2-TRIFLUORO-2-CHLOROETHYLENE
		1-CHLORO-1,2,2-TRIFLUOROETHENE
		1-CHLORO-1,2,2-TRIFLUOROETHYLENE
		2-CHLORO-1,1,2-TRIFLUOROETHYLENE
		CHLORO TRIFLUOROETHYLENE (DOT)
		CHLOROTRIFLUOROETHENE
		CHLOROTRIFLUOROETHYLENE
		CHLORTRIFLUORAETHYLEN (GERMAN)
		CTFE
		DAIFLON
		ETHENE, CHLOROTRIFLUORO-
		ETHYLENE, CHLOROTRIFLUORO-
		ETHYLENE, TRIFLUOROCHLORO-
		FLUOROPLAST 3
		GENETRON 1113
		MONOCHLOROTRIFLUOROETHYLENE
		TRIFLUOROCHLOROETHYLENE
		TRIFLUOROCHLOROETHYLENE (DOT)
		TRIFLUOROCHLOROETHYLENE, INHIBITED (DOT)
		TRIFLUOROMONOCHLOROETHYLENE
		TRIFLUOROVINYL CHLORIDE
		TRITHENE
		UN 1082 (DOT)
79-41-4		
		2-METHYL-2-PROPENOIC ACID
		2-METHYLACRYLIC ACID
		2-METHYLPROPENOIC ACID
		2-PROPENOIC ACID, 2-METHYL-
		ACRYLIC ACID, 2-METHYL-
		ALPHA-METHACRYLIC ACID
		ALPHA-METHYLACRYLIC ACID
		METHACRYLIC ACID
		METHACRYLIC ACID (ACGIH)
		METHACRYLIC ACID, INHIBITED (DOT)
		METHYLACRYLIC ACID
		PROPIONIC ACID, 2-METHYLENE-
		UN 2531 (DOT)
79-42-5		
		2-MERCAPTOPROPANOIC ACID
		2-MERCAPTOPROPIONIC ACID
		2-THIOLACTIC ACID
		ALPHA-MERCAPTOPROPANOIC ACID
		ALPHA-MERCAPTOPROPIONIC ACID
		MERCAPTOPROPIONIC ACID
		PROPANOIC ACID, 2-MERCAPTO-
		PROPIONIC ACID, 2-MERCAPTO-
	*	THIOLACTIC ACID
		THIOLACTIC ACID (DOT)
		UN 2936 (DOT)
79-43-6		
		2,2-DICHLOROACETIC ACID
		ACETIC ACID, DICHLORO-
		BICHLORACETIC ACID
		DCA
		DCA (ACID)
	*	DICHLORACETIC ACID
		DICHLORETHANOIC ACID
		DICHLOROACETIC ACID
		DICHLOROACETIC ACID (DOT)
		DICHLOROETHANOIC ACID
		DKHUK
		UN 1764 (DOT)
		URNER'S LIQUID
79-44-7		
		(DIMETHYLAMINO)CARBONYL CHLORIDE
		CARBAMIC CHLORIDE, DIMETHYL-
		CARBAMOYL CHLORIDE, DIMETHYL-
		CARBAMYL CHLORIDE, N,N-DIMETHYL-
		CHLOROFORMIC ACID DIMETHYLAMIDE
		DDC
		DIMETHOXYL CARBAMYL CHLORIDE
		DIMETHYL CARBAMOYL CHLORIDE (ACGIH)
		DIMETHYLCARBAMIC ACID CHLORIDE
		DIMETHYLCARBAMIC CHLORIDE

CAS No.	IN	Chemical Name
		DIMETHYLCARBAMIDOYL CHLORIDE
		DIMETHYLCARBAMOYL CHLORIDE
		DIMETHYLCARBAMYL CHLORIDE
		DIMETHYLCHLOROFORMAMIDE
		DMCC
		N,N-DIMETHYLAMINOCARBONYL CHLORIDE
		N,N-DIMETHYLCARBAMIC ACID CHLORIDE
		N,N-DIMETHYLCARBAMIDOYL CHLORIDE
		N,N-DIMETHYLCARBAMOYL CHLORIDE
		N,N-DIMETHYLCARBAMOYL CHLORIDE (DOT)
		N,N-DIMETHYLCARBAMYL CHLORIDE
		RCRA WASTE NUMBER U097
		TL 389
		UN 2262 (DOT)
79-46-9		
	*	2-NITROPROPANE
		2-NITROPROPANE (ACGIH,DOT)
		2-NP
		BETA-NITROPROPANE
		DIMETHYLNITROMETHANE
		ISONITROPROPANE
		NIPAR S-20
		NIPAR S-20 SOLVENT
		NIPAR S-30 SOLVENT
		NITROISOPROPANE
		PROPANE, 2-NITRO-
		RCRA WASTE NUMBER U171
		UN 2608 (DOT)
79-88-1		
		BUTYL METHACRYLATE
79-92-5		
		2,2-DIMETHYL-3-METHYLENEBICYCLO[221]HEPTANE
		2,2-DIMETHYL-3-METHYLENENORBORNANE
		3,3-DIMETHYL-2-METHYLENENORBORNANE
		3,3-DIMETHYL-2-METHYLENENORCAMPHANE
		BICYCLO[221]HEPTANE, 2,2-DIMETHYL-3-METHYLENE-
		CAMPHENE
		CAMPHENE (DOT)
		NA 9011 (DOT)
80-05-7		
		4,4-ISOPROPYLIDENDIPHENOL
		4,4-ISOPROPYLIDENEDIPHENOL
80-08-0		
		DAPSONE
80-10-4		
		DICHLORODIPHENYLSILANE
		DIPHENYL DICHLOROSILANE
		DIPHENYL DICHLOROSILANE (DOT)
		DIPHENYLSILICON DICHLORIDE
		DIPHENYLSILYL DICHLORIDE
		SILANE, DICHLORODIPHENYL-
		UN 1769 (DOT)
80-12-6		
		TETRAMETHYLENEDISULPHOTETRAMINE
80-15-9		
		1-METHYL-1-PHENYLETHYL HYDROPEROXIDE
		2-PHENYL-2-PROPYL HYDROPEROXIDE
		7-CUMYL HYDROPEROXIDE
		ALPHA,ALPHA-DIMETHYLBENZYL HYDROPEROXIDE
		ALPHA-CUMENE HYDROPEROXIDE
		ALPHA-CUMYL HYDROPEROXIDE
		CUMEENHYDROPEROXYDE (DUTCH)
		CUMENE HYDROPEROXIDE
		CUMENE HYDROPEROXIDE (DOT)
		CUMENE HYDROPEROXIDE, TECHNICALLY PURE (DOT)
		CUMENYL HYDROPEROXIDE
		CUMOLHYDROPEROXID (GERMAN)
		CUMYL HYDROPEROXIDE
		CUMYL HYDROPEROXIDE, TECHNICAL PURE (DOT)
		HYDROPEROXIDE, 1-METHYL-1-PHENYLETHYL
		HYDROPEROXIDE, ALPHA,ALPHA-DIMETHYLBENZYL-
		HYDROPEROXYDE DE CUMENE (FRENCH)
		HYDROPEROXYDE DE CUMYLE (FRENCH)
		HYPERIZ
		IDROPEROSSIDO DI CUMENE (ITALIAN)
		IDROPEROSSIDO DI CUMOLO (ITALIAN)
		ISOPROPYLBENZENE HYDROPEROXIDE
		PERCUMYL H
		RCRA WASTE NUMBER U096
		TRIGONOX K 80
		UN 2116 (DOT)
80-17-1		
		BENZENESULFOHYDRAZIDE
		BENZENESULFONIC ACID, HYDRAZIDE
		BENZENESULFONIC HYDRAZIDE
		BENZENESULFONOHYDRAZIDE
		BENZENESULFONYLHYDRAZINE
		CELOGEN BSH
		CHKHZ 9
		GENITRON BSH
		HYDRAZIDE BSG
		NITROPORE OBSH
		PHENYLSULFOHYDRAZIDE
		PHENYLSULFONYL HYDRAZIDE
		POROFOR BSH
		POROFOR CHKHZ 9
		POROFOR-BSH-PULVER
80-43-3		
		ACTIVE DICUMYL PEROXIDE
		ALPHA,ALPHA'-DICUMYL PEROXIDE
		ALPHA,ALPHA-DIMETHYLBENZYL PEROXIDE
		ALPHA-CUMYL PEROXIDE
		BIS(1-METHYL-1-PHENYLETHYL) PEROXIDE
		BIS(2-PHENYL-2-PROPYL) PEROXIDE
		BIS(ALPHA,ALPHA-DIMETHYLBENZYL)PEROXIDE
		CUMENE PEROXIDE
		CUMYL PEROXIDE
		DCP
		DI-ALPHA-CUMYL PEROXIDE
		DI-CUP
		DI-CUP 40C
		DI-CUP 40HAF
		DI-CUP 40KE
		DI-CUP R
		DI-CUP T
		DICUMENE HYDROPEROXIDE
		DICUMENYL PEROXIDE
		DICUMYL PEROXIDE
		DICUMYL PEROXIDE, DRY
		DICUMYL PEROXIDE, DRY (DOT)
		DICUMYL PEROXIDE, TECHNICAL PURE OR WITH INERT SOLID (DOT)
		DICUP 40
		DIISOPROPYLBENZENE PEROXIDE
		ISOPROPYLBENZENE PEROXIDE
		KAYACUMYL D
		LUPERCO
		LUPERCO 500-40C
		LUPERCO 500-40KE
		LUPEROX
		LUPEROX 500
		LUPEROX 500R
		LUPEROX 500T
		LUPERSOL 500
		PERCUMYL D
		PERCUMYL D 40
		PERKADOX B
		PERKADOX BC
		PERKADOX BC 40
		PERKADOX BC 9
		PERKADOX BC 95
		PERKADOX BC 96
		PERKADOX SB
		PEROXIDE, BIS(1-METHYL-1-PHENYLETHYL)
		PEROXIDE, BIS(ALPHA,ALPHA-DIMETHYLBENZYL)
		SANPEROX DCP
		UN 2121 (DOT)
		VAROX DCP-R

CAS No.	IN	Chemical Name
		VAROX DCP-T
80-46-6		
		4-(1,1-DIMETHYLPROPYL)PHENOL
		4-T-AMYLPHENOL
		4-TERT-AMYLPHENOL
		4-TERT-PENTYLPHENOL
		AMILFENOL
		P-(ALPHA,ALPHA-DIMETHYLPROPYL)PHENOL
		P-(1,1-DIMETHYLPROPYL)PHENOL
		P-TERT-AMYLPHENOL
		P-TERT-PENTYLPHENOL
		PENTAPHEN
		PHENOL, 4-(1,1-DIMETHYLPROPYL)-
		PHENOL, P-TERT-PENTYL-
80-47-7		
		HYDROPEROXIDE, 1-METHYL-1-(4-METHYLCYCLOHEXYL)ETHYL
		MENTHANE HYDROPEROXIDE, PARA
		P-MENTH-8-YL HYDROPEROXIDE
		P-MENTHANE HYDROPEROXIDE
		P-MENTHANE-8-HYDROPEROXIDE
80-48-8		
		BENZENESULFONIC ACID, 4-METHYL-, METHYL ESTER
		METHYL 4-METHYLBENZENESULFONATE
		METHYL P-METHYLBENZENESULFONATE
		METHYL P-TOLUENESULFONATE
		METHYL P-TOSYLATE
		METHYL TOLUENE SULFONATE
		METHYL TOLUENE-4-SULFONATE
		METHYL TOSYLATE
		P-METHYLBENZENESULFONATE METHYL ESTER
		P-TOLUENESULFONIC ACID, METHYL ESTER
80-51-3		
		4,4'-BIS(HYDRAZINOSULFONYL)DIPHENYL ETHER
		4,4'-OXYBIS(BENZENESULFONIC ACID HYDRAZIDE)
		4,4'-OXYBIS(BENZENESULFONIC ACID) DIHYDRAZIDE
		4,4'-OXYBIS(BENZENESULFONYL HYDRAZIDE)
		4,4'-OXYDI(BENZENESULFONIC ACID HYDRAZIDE)
		4,4'-OXYDIBENZENESULFONIC ACID DIHYDRAZIDE
		BENZENESULFONIC ACID, 4,4'-OXYBIS-, DIHYDRAZIDE
		BENZENESULFONIC ACID, 4,4'-OXYDI-, DIHYDRAZIDE
		BENZENESULFONIC ACID, OXYBIS-, DIHYDRAZIDE (9CI)
		CELLMIC S
		CELOGEN OT
		CENITRON OB
		DIHYDRAZIDE SDO
		DIPHENYL ETHER 4,4'-DISULFOHYDRAZIDE
		DIPHENYL OXIDE 4,4'-DISULFOHYDRAZIDE
		FE 9
		GENITRON OB
		NEOCELLBORN FE 9
		NEOCELLBORN P 1000M
		NITROPORE OBSH
		OBSH
		OXYBIS(BENZENESULFONYLHYDRAZIDE)
		P,P'-OXYBIS(BENZENESULFOHYDRAZIDE)
		P,P'-OXYBIS(BENZENESULFONE HYDRAZIDE)
		P,P'-OXYBIS(BENZENESULFONYLHYDRAZINE)
		P,P'-OXYBISBENZENE DISULFONYLHYDRAZIDE
80-56-8		
		2,6,6-TRIMETHYLBICYCLO[3.1.1]HEPT-2-ENE
		2-PINENE
		ALPHA-PINENE
		ALPHA-(+)-PINENE
		BICYCLO[3.1.1]HEPT-2-ENE, 2,6,6-TRIMETHYL-
80-62-6		
		2-METHYL-2-PROPENOIC ACID METHYL ESTER
		2-PROPENOIC ACID, 2-METHYL-, METHYL ESTER
		ACRYLIC ACID, 2-METHYL-, METHYL ESTER
		DIAKON
		METAKRYLAN METYLU (POLISH)
		METHACRYLATE DE METHYLE (FRENCH)

CAS No.	IN	Chemical Name
		METHACRYLIC ACID METHYL ESTER
		METHACRYLSAEUREMETHYL ESTER (GERMAN)
		METHYL 2-METHYL-2-PROPENOATE
		METHYL 2-METHYLPROPENOATE
		METHYL ALPHA-METHYLACRYLATE
	*	METHYL METHACRYLATE
		METHYL METHACRYLATE (ACGIH)
		METHYL METHACRYLATE MONOMER
		METHYL METHACRYLATE MONOMER, INHIBITED (DOT)
		METHYL METHACRYLATE MONOMER, UNINHIBITED
		METHYL METHYLACRYLATE
		METHYL-METHACRYLAT (GERMAN)
		METHYLMETHACRYLAAT (DUTCH)
		METIL METACRILATO (ITALIAN)
		MMA
		MME
		'MONOCITE' METHACRYLATE MONOMER
		NA 1247 (DOT)
		NCI-C50680
		PEGALAN
		RCRA WASTE NUMBER U162
		UN 1247 (DOT)
80-63-7		
		2-CHLOROACRYLIC ACID METHYL ESTER
		2-PROPENOIC ACID, 2-CHLORO-, METHYL ESTER
		2-PROPENOIC ACID, 2-CHLORO-, METHYL ESTER (9CI)
		ACRYLIC ACID, 2-CHLORO-, METHYL ESTER
		METHYL 2-CHLORO-2-PROPENOATE
		METHYL 2-CHLOROPROPENOATE
		METHYL-2-CHLOROACRYLATE
		METHYL-ALPHA-CHLOROACRYLATE
81-07-2		
		1,2-BENZISOTHIAZOL-3(2H)-ONE, 1,1-DIOXIDE
		1,2-BENZISOTHIAZOLIN-3-ONE, 1,1-DIOXIDE
		1,2-DIHYDRO-2-KETOBENZISOSULFONAZOLE
		2,3-DIHYDRO-3-OXOBENZISOSULFONAZOLE
		3-BENZISOTHIAZOLINONE 1,1-DIOXIDE
		3-HYDROXYBENZISOTHIAZOLE-S,S-DIOXIDE
		550 SACCHARINE
		ANHYDRO-O-SULFAMINEBENZOIC ACID
		BENZOIC SULFIMIDE
		BENZOIC SULPHINIDE
		BENZOSULFINIDE
		GARANTOSE
		GLUCID
		GLUSIDE
		O-BENZOIC ACID SULFIMIDE
		O-BENZOIC SULFIMIDE
		O-BENZOSULFIMIDE
		O-BENZOYL SULFIMIDE
		O-SULFOBENZIMIDE
		O-SULFOBENZOIC ACID IMIDE
		SACCHARIMIDE
	*	SACCHARIN
		SACCHARIN ACID
		SACCHARIN INSOLUBLE
		SACCHARINE
		SACCHARINOL
		SACCHARINOSE
		SACCHAROL
81-11-8		
		4,4'-DIAMINO-2,2'-STILBENEDISULFONIC ACID
81-15-2		
		2,4,6-TRINITRO-1,3-DIMETHYL-5-TERT-BUTYLBENZENE
		2,4,6-TRINITRO-3,5-DIMETHYL-TERT-BUTYLBENZENE
		BENZENE, 1-(1,1-DIMETHYLETHYL)-3,5-DIMETHYL-2,4,6-TRINITRO-
		BENZENE, 1-TERT-BUTYL-3,5-DIMETHYL-2,4,6-TRINITRO-
		M-XYLENE, 5-TERT-BUTYL-2,4,6-TRINITRO-
		MUSK EXYLENE
		MUSK XYLENE
		MUSK XYLOL
		XYLENE MUSK

CAS No.	IN	Chemical Name
81-81-2		
		(PHENYL-1 ACETYL-2 ETHYL) 3-HYDROXY-4 COUMARINE (FRENCH)
		1-(4'-HYDROXY-3'-COUMARINYL)-1-PHENYL-3-BUTANONE
		2H-1-BENZOPYRAN-2-ONE, 4-HYDROXY-3-(3-OXO-1-PHENYL-BUTYL)-
		2H-1-BENZOPYRAN-2-ONE, 4-HYDROXY-3-(3-OXO-1-PHENYL-BUTYL)- (9CI)
		3-(1'-PHENYL-2'-ACETYLETHYL)-4-HYDROXYCOUMARIN
		3-(ACETONYLBENZYL)-4-HYDROXYCOUMARIN
		3-(ALPHA-ACETONYLBENZYL)-4-HYDROXYCOUMARIN
		3-(ALPHA-PHENYL-BETA-ACETYLAETHYL)-4-HYDROXYCU-MARIN (GERMAN)
		3-(ALPHA-PHENYL-BETA-ACETYLETHYL)-4-HYDROXY-COUMARIN
		4-HYDROXY-3-(3-OXO-1-FENYL-BUTYL) CUMARINE (DUTCH)
		4-HYDROXY-3-(3-OXO-1-PHENYL-BUTYL)-CUMARIN (GER-MAN)
		4-HYDROXY-3-(3-OXO-1-PHENYLBUTYL)-2H-1-BENZOPYRAN-2-ONE
		4-IDROSSI-3-(3-OXO-)-FENIL-BUTIL)-CUMARINE (ITALIAN)
		ARAB RAT DETH
		ATHROMBIN-K
		ATHROMBINE-K
		BRUMOLIN
		CO-RAX
		COMPOUND 42
		COUMADIN
		COUMAFEN
		COUMAFENE
		COUMAPHEN
		COUMARIN
		COUMARIN, 3-(ALPHA-ACETONYLBENZYL)-4-HYDROXY-
		COUMEFENE
		COV-R-TOX
		D-CON
		DETHMOR
		DETHNEL
		EASTERN STATES DUOCIDE
		FASCO FASCRAT POWDER
		KUMADER
		KUMADU
		KUMATOX
		KYPFARIN
		LIQUA-TOX
		MAR-FRIN
		MARTIN'S MAR-FRIN
		MAVERAN
		MOUSE PAK
		PROTHROMADIN
		RAT & MICE BAIT
		RAT-A-WAY
		RAT-B-GON
		RAT-GARD
		RAT-KILL
		RAT-MIX
		RAT-O-CIDE #2
		RAT-OLA
		RAT-TROL
		RATOREX
		RATOX
		RATOXIN
		RATRON
		RATRON G
		RATS-NO-MORE
		RATTUNAL
		RAX
		RCRA WASTE NUMBER P001
		RO-DETH
		RODAFARIN
		RODEX
		RODEX BLOX
		ROSEX
		ROUGH & READY MOUSE MIX
		SOLFARIN
		SPRAY-TROL BRAND RODEN-TROL
		TEMUS W
		TOX-HID
		TWIN LIGHT RAT AWAY
		VAMPIRINIP II
		VAMPIRINIP III
		WARAN
		WARF 42
		WARF COMPOUND 42
		WARFARAT
	*	WARFARIN
		WARFARIN (ACGIH)
		WARFARIN PLUS
		WARFARIN Q
		WARFARINE (FRENCH)
		WARFICIDE
		ZOOCOUMARIN
		ZOOCOUMARIN (RUSSIAN)
81-88-9		
		C.I. FOOD RED 15
82-28-0		
		1-AMINO-2-METHYLANTHRAQUINONE
		2-METHYL-1-ANTHRAQUINONYLAMINE
		9,10-ANTHRACENEDIONE, 1-AMINO-2-METHYL-
		ACETATE FAST ORANGE R
		ACETOQUINONE LIGHT ORANGE JL
		ANTHRAQUINONE, 1-AMINO-2-METHYL-
		ARTISIL ORANGE 3RP
		C.I. 60700
		C.I. DISPERSE ORANGE 11
		C.I. SOLVENT ORANGE 35
		CELLITON ORANGE R
		CILLA ORANGE R
		DISPERSE ORANGE
		DISPERSE ORANGE (ANTHRAQUINONE DYE)
		DURANOL ORANGE G
		MICROSETILE ORANGE RA
		NYLOQUINONE ORANGE JR
		PERLITON ORANGE 3R
		SERISOL ORANGE YL
		SMOKE ORANGE LK 6044
		SUPRACET ORANGE R
82-66-6		
		1,3-INDANDIONE, 2-DIPHENYLACETYL-
		1H-INDENE-1,3(2H)-DIONE, 2-(DIPHENYLACETYL)-
		2-(DIPHENYLACETYL)-1H-INDENE-1,3(2H)-DIONE
		2-(DIPHENYLACETYL)INDAN-1,3-DIONE
		2-DIPHENYLACETYL-1,3-DIKETOHYDRINDENE
		2-DIPHENYLACETYL-1,3-INDANDIONE
		DIDANDIN
		DIDION
		DIPAXIN
		DIPHACIN
		DIPHACINON
	*	DIPHACINONE
		DIPHENACIN
		DIPHENADION
		DIPHENADIONE
		ORAGULANT
		PID
		PROMAR
		RAMIK
		RATINDAN
		RATINDAN 1
		SOLVAN
		U 1363
82-68-8		
		AVICOL (PESTICIDE)
		BATRILEX
		BENZENE, PENTACHLORONITRO-
		BOTRILEX
		BRASSICOL
		BRASSICOL 75
		BRASSICOL SUPER
		CHINOZAN
		FARTOX
		FOLOSAN
		FOMAC 2
		GC 3944-3-4

CAS No.	IN	Chemical Name
		KOBUTOL
		MARISAN FORTE
		NITROPENTACHLOROBENZENE
		PCNB
		PENTACHLORONITROBENZENE
		PENTAGEN
		PHOMASAN
		QUINOSAN
		QUINTOCENE
		QUINTOZEN
		QUINTOZENE
		RTU 1010
		TERRACHLOR
		TERRACLOR
		TERRACLOR 30 G
		TERRAFUN
		TILCAREX
		TRITISAN
82-71-3		
		2,4,6-TRINITRORESORCINOL
		STYPHNIC ACID
		TRINITRORESORCINOL
83-26-1		
		1,3-INDANDIONE, 2-PIVALOYL-
		1,3-INDANDIONE, 2-PIVALYL-
		1H-INDENE-1,3(2H)-DIONE, 2-(2,2-DIMETHYL-1-OXOPROPYL)-
		2-(2,2-DIMETHYL-1-OXOPROPYL)-1H-INDENE-1,3(2H)-DIONE
		2-(TRIMETHYLACETYL)-1,3-INDANDIONE
		2-(TRIMETIL-ACETIL)-INDAN-1,3-DIONE (ITALIAN)
		2-PIVALOYL-1,3-INDANDIONE
		2-PIVALOYL-INDAAN-1,3-DION (DUTCH)
		2-PIVALOYL-INDAN-1,3-DION (GERMAN)
		2-PIVALOYLINDANE-1,3-DIONE
		2-PIVALYL-1,3-INDANDIONE
		CHEMRAT
		PINDON (DUTCH)
		PINDONE
		PINDONE (ACGIH)
		PINDONE, LIQUID (DOT)
		PINDONE, SOLID (DOT)
		PIVACIN
		PIVAL
		PIVALDION (ITALIAN)
		PIVALDIONE (FRENCH)
		PIVALYL
		PIVALYL VALONE
		PIVALYLINDAN-1,3-DIONE
		PIVALYLINDANDIONE
		PIVALYN
		TRI-BAN
		UN 2472 (DOT)
83-32-9		
		1,2-DIHYDROACENAPHTHYLENE
		1,8-ETHYLENENAPHTHALENE
		ACENAPHTHENE
		ACENAPHTHYLENE, 1,2-DIHYDRO-
		NAPHTHYLENEETHYLENE
		PERI-ETHYLENENAPHTHALENE
83-79-4		
		(-)-ROTENONE
		5'BETA-ROTENONE
		CUBE-PULVER
		DACTINOL
		DERIL
		DERRIN
		DERRIS
		DERRIS (INSECTICIDE)
		DRI-KIL
		LIQUID DERRIS
		NICOULINE
		NOXFISH
		PARADERIL
		RONONE
		ROTENON
		ROTENONE
		ROTENONE (COMMERCIAL)
		ROTOCIDE
		TUBATOXIN
84-15-1		
		O-TERPHENYLS
84-17-3		
		DIENOESTROL
84-65-1		
	*	ANTHRAQUINONE
84-66-2		
		1,2-BENZENEDICARBOXYLIC ACID, DIETHYL ESTER
		ANOZOL
		DIETHYL 1,2-BENZENEDICARBOXYLATE
	*	DIETHYL PHTHALATE
		DIETHYLESTER PHTHALIC ACID
		ETHYL PHTHALATE
		NEANTINE
		O-BENZENEDICARBOXYLIC ACID DIETHYL ESTER
		PALATINOL A
		PHTHALIC ACID, DIETHYL ESTER
		PHTHALOL
		PLACIDOL E
		SOLVANOL
		UNIMOLL DA
84-72-0		
		1,2-BENZENEDICARBOXYLIC ACID, 2-ETHOXY-2-OXOETHYL ETHYL ESTER
		CARBETHOXYMETHYL ETHYL PHTHALATE
		DIETHYL O-CARBOXYBENZOYLOXYACETATE
		ETHOXYCARBONYLMETHYL ETHYL PHTHALATE
		ETHYL CARBETHOXYMETHYL PHTHALATE
		ETHYL PHTHALYL ETHYL GLYCOLATE
		GLYCOLIC ACID, ETHYL ESTER, ETHYL PHTHALATE
		PHTHALIC ACID, ETHYL ESTER, ESTER WITH ETHYL GLYCOLATE
		SANTICIZER E-15
84-74-2		
		1,2-BENZENEDICARBOXYLIC ACID, DIBUTYL ESTER
		BENZENE-O-DICARBOXYLIC ACID DI-N-BUTYL ESTER
		BIS-N-BUTYL PHTHALATE
		BUTYL PHTHALATE
		CELLUFLEX DPB
		DBP
		DBP (ESTER)
		DI(N-BUTYL) 1,2-BENZENEDICARBOXYLATE
	*	DI-N-BUTYL PHTHALATE
		DIBUTYL 1,2-BENZENEDICARBOXYLATE
		DIBUTYL ESTER PHTHALIC ACID
		DIBUTYL O-PHTHALATE
		DIBUTYL PHTHALATE
		DIBUTYL PHTHALATE (ACGIH)
		ELAOL
		ERGOPLAST FDB
		GENOPLAST B
		HEXAPLAS M/B
		N-BUTYL PHTHALATE
		N-BUTYL PHTHALATE (DOT)
		NA 9095 (DOT)
		O-BENZENEDICARBOXYLIC ACID, DIBUTYL ESTER
		PALATINOL C
		PHTHALIC ACID, DIBUTYL ESTER
		POLYCIZER DBP
		PX 104
		RCRA WASTE NUMBER U069
		STAFLEX DBP
		UNIMOLL DB
		WITCIZER 300
84-80-0		
		1,4-NAPHTHALENEDIONE, 2-METHYL-3-(3,7,11,15-TETRAMETHYL-2-HEXADECENYL)-
		1,4-NAPHTHALENEDIONE, 2-METHYL-3-(3,7,11,15-TETRA-

CAS No.	IN	Chemical Name
		METHYL-2-HEXADECENYL)-, [R-[R*,R*-(E)]]-
		1,4-NAPHTHOQUINONE, 2-METHYL-3-PHYTYL-
		2',3'-TRANS-VITAMIN K1
		2-METHYL-3-(3,7,11,15-TETRAMETHYL-2-HEXADECENYL)-1,4-NAPHTHALENEDIONE
		2-METHYL-3-PHYTYL-1,4-NAPHTHOCHINON (GERMAN)
		ALPHA-PHYLLOQUINONE
		ANTIHEMORRHAGIC VITAMIN
		AQUAMEPHYTON
		COMBINAL K1
		KATIV N
		KAYWAN
		KEPHTON
		KINADION
		KONAKION
		MEPHYTON
		MONO-KAY
		MONODION
		PHYLLOCHINON (GERMAN)
		PHYLLOQUINONE
		PHYLLOQUINONE K1
		PHYTOMENADIONE
		PHYTONADIONE
		PHYTYLMENADIONE
		SYNTHEX P
		TRANS-PHYLLOQUINONE
		VITAMIN K1
		VITAMIN K1(20)
85-00-7		
		1,1'-AETHYLEN-2,2'-BIPYRIDINIUM-DIBROMID (GERMAN)
		1,1'-ETHYLENE-2,2'-BIPYRIDINIUM DIBROMIDE
		1,1'-ETHYLENE-2,2'-BIPYRIDYLIUM DIBROMIDE
		1,1'-ETHYLENE-2,2'-DIPYRIDYLIUM DIBROMIDE
		1,1-ETHYLENE 2,2-DIPYRIDYLIUM DIBROMIDE
		5,6-DIHYDRO-DIPYRIDO(1,2A;2,1C)PYRAZINIUM DIBROMIDE
		6,7-DIHYDROPYRIDO(1,2-A;2',1'-C)PYRAZINEDIUM DIBROMIDE
		9,10-DIHYDRO-8A,10,-DIAZONIAPHENANTHRENE DIBROMIDE
		9,10-DIHYDRO-8A,10A-DIAZONIAPHENANTHRENE DIBROMIDE
		9,10-DIHYDRO-8A,10A-DIAZONIAPHENANTHRENE(1,1'-ETHYLENE-2,2'-BIPYRIDYLIUM)DIBROMIDE
		AQUACIDE
		DEIQUAT
		DEXTRONE
		DIPYRIDO(1,2-A;2',1'-C)PYRAZINEDIIUM, 6,7-DIHYDRO-, DIBROMIDE
		DIQUAT
		DIQUAT (ACGIH,DOT)
		DIQUAT DIBROMIDE
		ETHYLENE DIPYRIDYLIUM DIBROMIDE
		FB/2
		NA 2781 (DOT)
		ORTHO-DIQUAT
		PP 100
		PREEGLONE
		REGLON
		REGLONE
		REGLOX
		WEEDTRINE-D
85-01-8		
	*	PHENANTHRENE
85-42-7		
		HYDROHEXAPHTHALIC ANHYDRIDE
85-43-8		
		1,2,3,6-TETRAHYDROPHTHALIC ACID ANHYDRIDE
		1,2,3,6-TETRAHYDROPHTHALIC ANHYDRIDE
		1,3-ISOBENZOFURANDIONE, 3A,4,7,7A-TETRAHYDRO-
		4-CYCLOHEXENE-1,2-DICARBOXYLIC ACID ANHYDRIDE
		4-CYCLOHEXENE-1,2-DICARBOXYLIC ANHYDRIDE
		ANHYDRID KYSELINY TETRAHYDROFTALOVE (CZECH)
		BUTADIENE-MALEIC ANHYDRIDE ADDUCT
		DELTA(SUP 4)-TETRAHYDROPHTHALIC ANHYDRIDE
		DELTA4-TETRAHYDROPHTHALIC ANHYDRIDE
		MALEIC ANHYDRIDE ADDUCT OF BUTADIENE
		PHTHALIC ANHYDRIDE, 1,2,3,6-TETRAHYDRO-
		RIKACID TH
		TETRAHYDROFTALANHYDRID (CZECH)
		TETRAHYDROPHTHALIC ACID ANHYDRIDE
		TETRAHYDROPHTHALIC ANHYDRIDE
		TETRAHYDROPHTHALIC ANHYDRIDE (DOT)
		THPA
		UN 2698 (DOT)
85-44-9		
		1,2-BENZENEDICARBOXYLIC ACID ANHYDRIDE
		1,2-BENZENEDICARBOXYLIC ANHYDRIDE
		1,3-DIOXOPHTHALAN
		1,3-ISOBENZOFURANDIONE
		1,3-PHTHALANDIONE
		ANHYDRIDE PHTALIQUE (FRENCH)
		ANIDRIDE FTALICA (ITALIAN)
		ARALDITE HT 901
		ESEN
		FTAALZUURANHYDRIDE (DUTCH)
		FTALOWY BEZWODNIK (POLISH)
		HT 901
		ISOBENZOFURAN, 1,3-DIHYDRO-1,3-DIOXO-
		NCI-C03601
		PHTHALANDIONE
		PHTHALANHYDRIDE
		PHTHALIC ACID ANHYDRIDE
	*	PHTHALIC ANHYDRIDE
		PHTHALIC ANHYDRIDE (ACGIH)
		PHTHALIC ANHYDRIDE, SOLID OR MOLTEN (DOT)
		PHTHALSAEUREANHYDRID (GERMAN)
		RCRA WASTE NUMBER U190
		RETARDER AK
		RETARDER ESEN
		RETARDER PD
		SCONOC 7
		TGL 6525
		UN 2214 (DOT)
		VULKALENT B/C
85-68-7		
		1,2-BENZENEDICARBOXYLIC ACID, BUTYL PHENYLMETHYL ESTER
		BBP
		BENZYL BUTYL ESTER PHTHALIC ACID
		BENZYL BUTYL PHTHALATE
		BENZYL N-BUTYL PHTHALATE
	*	BUTYL BENZYL PHTHALATE
		BUTYL PHENYLMETHYL 1,2-BENZENEDICARBOXYLATE
		N-BUTYL BENZYL PHTHALATE
		NCI-C54375
		PALATINOL BB
		PHTHALIC ACID, BENZYL BUTYL ESTER
		SANTICIZER 160
		SICOL 160
		UNIMOLL BB
85-71-2		
		1,2-BENZENEDICARBOXYLIC ACID, 2-ETHOXY-2-OXOETHYL METHYL ESTER
		ETHOXYCARBONYLMETHYL METHYL PHTHALATE
		ETHYL O-[O-(METHOXYCARBONYL)BENZOYL]GLYCOLATE
		GLYCOLIC ACID, ETHYL ESTER, METHYL PHTHALATE
		METHYL CARBETHOXYMETHYL PHTHALATE
		METHYL PHTHALYL ETHER
		METHYL PHTHALYL ETHER GLYCOLATE
		METHYL PHTHALYL ETHYL GLYCOLATE
		PHTHALIC ACID, METHYL ESTER, ESTER WITH ETHYL GLYCOLATE
		SANTICIZER M-17
86-30-6		
		BENZENAMINE, N-NITROSO-N-PHENYL-
		DIPHENYLAMINE, N-NITROSO-
		DIPHENYLNITROSAMINE
		N,N-DIPHENYL-N-NITROSOAMINE
		N,N-DIPHENYLNITROSAMINE
		N-NITROSO-N-DIPHENYLAMINE

CAS No.	IN	Chemical Name
		N-NITROSO-N-PHENYLANILINE
		N-NITROSODIPHENYLAMINE
		NITROSODIPHENYLAMINE
		REDAX
		RETARDER J
		VULCATARD A
		VULKALENT A
		VULTROL
86-50-0		
		1,2,3-BENZOTRIAZIN-4(3H)-ONE, 3-(MERCAPTOMETHYL)-, O,O-DIMETHYL PHOSPHORODITHIOATE
		1,2,3-BENZOTRIAZINE, PHOSPHORODITHIOIC ACID DERIV
		3-(MERCAPTOMETHYL)-1,2,3-BENZOTRIAZIN-4(3H)-ONE O,O-DIMETHYL PHOSPHORODITHIOATE S-ESTER
		AZINFOS-METHYL
		AZINFOS-METHYL (DUTCH)
		AZINOPHOS-METHYL
		AZINPHOS
		AZINPHOS METHYL, LIQUID
		AZINPHOS METHYL, LIQUID (DOT)
		AZINPHOS-METHYL
		AZINPHOS-METHYL (ACGIH,DOT)
		AZINPHOS-METILE (ITALIAN)
		BAY 17147
		BAY 9027
		BAYER 17147
		BAYER 9027
		BENZOTRIAZINE DERIVATIVE OF A METHY L DITHIOPHOS-PHATE
		BENZOTRIAZINEDITHIOPHOSPHORIC ACID DIMETHOXY ESTER
		CARFENE
		COTNEON
		COTNION
		COTNION METHYL
		CRYSTHION 2L
		CRYSTHYON
		DBD
		DIMETHYLDITHIOPHOSPHORIC ACID N-METHYLBENZA-ZIMIDE ESTER
		ENT 23,233
		GOTHNION
		GUSATHION
		GUSATHION 25
		GUSATHION K
		GUSATHION M
		GUSATHION METHYL
		GUSATHION-20
		GUTHION
		GUTHION (DOT)
		GUTHION, LIQUID (DOT)
		METHYL GUTHION
		METHYLAZINPHOS
		METHYLGUSATHION
		METILTRIA ZOTION
		N-METHYLBENZAZIMIDE, DIMETHYLDITHIOPHOSPHORIC ACID ESTER
		NA 2783 (DOT)
		NCI-C00066
		O,O-DIMETHYL S-(3 ,4-DIHYDRO-4-KETO-1,2,3-BENZO-TRIAZINYL-3-METHYL) DITHIOPHOSPHATE
		O,O-DIMETHYL S-(4-OXO-1,2,3-BENZOTRIAZINO(3)-METHYL) THIOTHIONOPHOSPHATE
		O,O-DIMETHYL S-(4-OXO-3H-1,2,3-BENZOTRIAZINE-3-METHYL)PHOSPHORODITHIOATE
		O,O-DIMETHYL S-(4-OXOBENZOTRIAZINO-3-METHYL)PHOS-PHORODITHIOATE
		O,O-DIMETHYL S-4-OXO-1,2,3-BENZOTRIAZIN-3(4H)-YL METHYL PHOSPHORODITHIOATE
		O,O-DIMETHYL-S-((4-OXO-3H-1,2,3-BENZOTRIAZIN-3-YL)-METHYL)-DITHIOFOSFAAT (DUTCH)
		O,O-DIMETHYL-S-((4-OXO-3H-1,2,3-BENZOTRIAZIN-3-YL)-METHYL)-DITHIOPHOSPHAT (GERMAN)
		O,O-DIMETHYL-S-(1,2,3-BENZOTRIAZINYL-4-KETO)METHYL PHOSPHORODITHIOATE
		O,O-DIMETHYL-S-(4-OXOBENZOTRIAZIN-3-METHYL)-DITHIOPHOSPHAT (GERMAN)
		O,O-DIMETHYL-S-(BENZAZIMINOMETHYL) DITHIOPHOS-PHATE
		O,O-DIMETIL-S-((4-OXO-3H-1,2,3-BENZOTRIAZIN-3-IL)-METIL)-DITIOFOSFATO (ITALIAN)
		PHOSPHORODITHIOIC ACID, O,O-DIMETHYL S-[(4-OXO-1,2,3-BENZOTRIAZIN-3(4H)-YL)METHYL] ESTER
		R 1582
		S-(3,4-DIHYDRO-4-OXO-BENZO(ALPHA)(1,2,3)TRIAZIN-3-YL METHYL) O,O-DIMETHYL PHOSPHORODITHIOATE
		S-(3,4-DIHYDRO-4-OXO-1,2,3-BENZOTRIAZIN-3-YLMETHYL) O,O-DIMETHYL PHOSPHORODITHIOATE
		S-93
86-73-7		
		2,2'-METHYLENEBIPHENYL
		9H-FLUORENE
		DIPHENYLENEMETHANE
	*	FLUORENE
		METHANE, DIPHENYLENE-
		O-BIPHENYLENEMETHANE
		1-(1-NAPHTHYL)-2-THIOUREA
		1-(1-NAPHTHYL)THIOUREA
		1-NAPHTHYLTHIOUREA
		ALPHA-NAPHTHYLTHIOCARBAMIDE
		ALPHA-NAPHTHYLTHIOUREA
		ALPHA-NAPTHYL THIOUREA (ANTU)
		ALRATO
86-88-4		
		ANTU
		ANTURAT
		BANTU
		CHEMICAL 109
		DIRAX
		KRIPID
		KRYSID
		N-1-NAPHTHYLTHIOUREA
		NAPHTOX
		RATTRACK
		SMEESANA
		THIOUREA, 1-NAPHTHALENYL-
		U 5227
		UREA, 1-(1-NAPHTHYL)-2-THIO-
87-31-0		
		DIAZODINITROPHENOL
87-59-2		
		2,3-DIMETHYLANILINE
		2,3-DIMETHYLBENZENAMINE
		2,3-DIMETHYLPHENYLAMINE
		2,3-XYLIDINE
		2,3-XYLYLAMINE
		3-AMINO-O-XYLENE
		BENZENAMINE, 2,3-DIMETHYL-
		O-XYLIDINE
87-60-5		
		3-CHLORO-2-METHYL ANILINE (CHLOROTOLUIDINE)
87-61-6		
	*	1,2,3-TRICHLOROBENZENE
87-62-7		
		1-AMINO-2,6-DIMETHYLBENZENE
		2,6-DIMETHYLANILINE
		2,6-DIMETHYLBENZENAMINE
		2,6-XYLIDENE
		2,6-XYLIDINE
		2,6-XYLYLAMINE
		2-AMINO-1,3-DIMETHYLBENZENE
		2-AMINO-1,3-XYLENE
		2-AMINO-M-XYLENE
		BENZENAMINE, 2,6-DIMETHYL-
		O-XYLIDINE
87-63-8		
		6-CHLORO-2-METHYL ANILINE

CAS No.	IN	Chemical Name
87-66-1		
		1,2,3-BENZENETRIOL
		1,2,3-TRIHYDROXYBENZEN (CZECH)
		1,2,3-TRIHYDROXYBENZENE
		2,3-DIHYDROXYPHENOL
		BENZENE, 1,2,3-TRIHYDROXY-
		CI 76515
		CI OXIDATION BASE 32
		FOURAMINE BROWN AP
		FOURRINE 85
		FOURRINE PG
		PYROGALLIC ACID
		PYROGALLOL
87-68-3		
		1,1,2,3,4,4-HEXACHLORO-1,3-BUTADIENE
		1,3-BUTADIENE, 1,1,2,3,4,4-HEXACHLORO-
		1,3-BUTADIENE, HEXACHLORO-
		1,3-HEXACHLOROBUTADIENE
		BUTADIENE, HEXACHLORO-
		C 46
		DOLEN-PUR
		GP-40-66:120
		HCBD
		HEXACHLOR-1,3-BUTADIEN (CZECH)
	*	HEXACHLORBUTADIENE
		HEXACHLORO-1,3-BUTADIENE
		HEXACHLOROBUTADIENE
		HEXACHLOROBUTADIENE (ACGIH,DOT)
		PERCHLORO-1,3-BUTADIENE
		PERCHLOROBUTADIENE
		RCRA WASTE NUMBER U128
		UN 2279 (DOT)
87-69-4		
		2,3-DIHYDROSUCCINIC ACID
		BUTANEDIOIC ACID, 2,3-DIHYROXY-
		L-(+)-TARTARIC ACID
		MALIC ACID, 3-HYDROXY-
		SUCCINIC ACID, 2,3-DIHYDROXY-
		TARTARIC ACID
		THREARIC ACID
87-86-5		
		1-HYDROXYPENTACHLOROBENZENE
		2,3,4,5,6-PENTACHLOROPHENOL
		CHEM-TOL
		CHLON
		CHLOROPHEN
		CRYPTOGIL OL
		DOW PENTACHLOROPHENOL DP-2 ANTIMICROBIAL
		DOWICIDE 7
		DOWICIDE EC-7
		DOWICIDE G
		DURA TREET II
		DUROTOX
		EP 30
		FUNGIFEN
		GLAZD PENTA
		GRUNDIER ARBEZOL
		LAUXTOL
		LAUXTOL A
		LIROPREM
		NA 2020 (DOT)
		NCI-C54933
		NCI-C55378
		NCI-C56655
		PCP
		PCP (PESTICIDE)
		PENCHLOROL
		PENTA
		PENTA-KIL
		PENTACHLOORFENOL (DUTCH)
		PENTACHLOROFENOL
		PENTACHLOROPHENATE
	*	PENTACHLOROPHENOL
		PENTACHLOROPHENOL (ACGIH,DOT)
		PENTACHLOROPHENOL, DOWICIDE EC-7
		PENTACHLOROPHENOL, DP-2
		PENTACHLOROPHENOL, TECHNICAL
		PENTACHLORPHENOL (GERMAN)
		PENTACLOROFENOLO (ITALIAN)
		PENTACON
		PENTASOL
		PENWAR
		PERATOX
		PERMACIDE
		PERMAGARD
		PERMASAN
		PERMATOX DP-2
		PERMATOX PENTA
		PERMITE
		PHENOL, PENTACHLORO-
		PKHF
		POL NU
		PREVENTOL P
		PRILTOX
		RCRA WASTE NUMBER U242
		SANTOBRITE
		SANTOPHEN
		SANTOPHEN 20
		SINITUHO
		TERM-I-TROL
		THOMPSON'S WOOD FIX
		WEEDONE
		WOODTREAT A
87-90-1		
		1,3,5-TRIAZINE-2,4,6(1H,3H,5H)-TRIONE, 1,3,5-TRICHLORO-
		1,3,5-TRICHLORO-2,4,6-TRIOXOHEXAHYDRO-S-TRIAZINE
		1,3,5-TRICHLORO-S-TRIAZINE-2,4,6(1H,3H,5H)-TRIONE
		1,3,5-TRICHLOROISOCYANURIC ACID
		ACL 85
		CBD 90
		CDB 90
		CHLOREAL
		FICHLOR 91
		ISOCYANURIC CHLORIDE
		KYSELINA TRICHLOISOKYANUROVA (CZECH)
		N,N',N''-TRICHLOROISOCYANURIC ACID
		NEOCHLOR 90
		NSC-405124
		S-TRIAZINE-2,4,6(1H,3H,5H)-TRIONE, 1,3,5-TRICHLORO-
		S-TRIAZINE-2,4,6(1H,3H,5H)-TRIONE, TRICHLORO-
		SYMCLOSEN
		SYMCLOSENE
		TRICHLORINATED ISOCYANURIC ACID
		TRICHLORO-S-TRIAZINE-2,4,6(1H,3H,5H)-TRIONE
		TRICHLORO-S-TRIAZINETRIONE
		TRICHLORO-S-TRIAZINETRIONE, DRY, CONTAINING OVER 39% AVAILABLE CHLORINE (DOT)
		TRICHLOROCYANURIC ACID
		TRICHLOROISOCYANIC ACID
	*	TRICHLOROISOCYANURIC ACID
		TRICHLOROISOCYANURIC ACID, DRY (DOT)
		UN 2468 (DOT)
88-05-1		
		1-AMINO-2,4,6-TRIMETHYL BENZENE
		1-AMINO-2,4,6-TRIMETHYLBENZEN (CZECH)
		2,4,6-TRIMETHYL ANILINE
		2,4,6-TRIMETHYLPHENYLAMINE
		2-AMINO-1,3,5-TRIMETHYLBENZENE
		2-AMINOMESITYLENE
		AMINOMESITYLENE
		ANILINE, 2,4,6-TRIMETHYL-
		BENZENAMINE, 2,4,6-TRIMETHYL-
		BENZENAMINE, 2,4,6-TRIMETHYL- (9CI)
		MESIDIN
		MESIDIN (CZECH)
		MESIDINE
		MESITYLAMINE
		MESITYLENE, 2-AMINO-
		MEZIDINE
88-06-2		
	*	2,4,6-TRICHLOROPHENOL
		DOWICIDE 2S

CAS No.	IN	Chemical Name
		OMAL
		PHENACHLOR
		PHENOL, 2,4,6-TRICHLORO-
88-09-5		
		2-ETHYL-N-BUTYRIC ACID
		2-ETHYLBUTANOIC ACID
		2-ETHYLBUTYRIC ACID
		3-PENTANECARBOXYLIC ACID
		ACETIC ACID, DIETHYL-
		ALPHA-ETHYLBUTYRIC ACID
		BUTANOIC ACID, 2-ETHYL-
		BUTYRIC ACID, 2-ETHYL-
		DIETHYLACETIC ACID
88-10-8		
		CARBAMIC CHLORIDE, DIETHYL-
		CARBAMOYL CHLORIDE, DIETHYL-
		DIETHYL CARBAMYL CHLORIDE
		DIETHYLCARBAMOYL CHLORIDE
		N,N-DIETHYLCARBAMOYL CHLORIDE
88-16-4		
		1-CHLORO-2-(TRIFLUOROMETHYL)BENZENE
		2-CHLORO(TRIFLUOROMETHYL)BENZENE
		2-CHLORO-ALPHA,ALPHA,ALPHA-TRIFLUOROTOLUENE
		2-CHLOROBENZOTRIFLUORIDE
		BENZENE, 1-CHLORO-2-(TRIFLUOROMETHYL)-
		O-(TRIFLUOROMETHYL)CHLOROBENZENE
		O-CHLORO-ALPHA,ALPHA,ALPHA-TRIFLUOROTOLUENE
		O-CHLOROBENZOTRIFLUORIDE
		O-TRIFLUOROMETHYLPHENYL CHLORIDE
		TOLUENE, O-CHLORO-ALPHA,ALPHA,ALPHA-TRIFLUORO-
88-17-5		
		2-(TRIFLUOROMETHYL)ANILINE
		2-(TRIFLUOROMETHYL)BENZENAMINE
		2-AMINOBENZOTRIFLUORIDE
		2-TRIFLUOROMETHYL ANILINE
		2-TRIFLUOROMETHYL ANILINE (DOT)
		ALPHA,ALPHA,ALPHA-TRIFLUORO-O-TOLUIDINE
		BENZENAMINE, 2-(TRIFLUOROMETHYL)-
		BENZENAMINE, 2-(TRIFLUOROMETHYL)- (9CI)
		O-(TRIFLUOROMETHYL)ANILINE
		O-AMINOBENZOTRIFLUORIDE
		O-TOLUIDINE, ALPHA,ALPHA,ALPHA-TRIFLUORO-
		UN 2942 (DOT)
88-72-2		
		1-METHYL-2-NITROBENZENE
		2-METHYL-1-NITROBENZENE
		2-METHYLNITROBENZENE
		2-NITROTOLUENE
		BENZENE, 1-METHYL-2-NITRO-
		O-METHYLNITROBENZENE
		O-NITROTOLUENE
		O-NITROTOLUENE (ACGIH,DOT)
		ONT
		TOLUENE, O-NITRO-
		UN 1664 (DOT)
88-73-3		
		1-CHLORO-2-NITROBENZENE
		1-NITRO-2-CHLOROBENZENE
		2-CHLORO-1-NITROBENZENE
		2-CHLORONITROBENZENE
		2-NITROCHLOROBENZENE
		BENZENE, 1-CHLORO-2-NITRO-
		CHLORO-O-NITROBENZENE
		NITROCHLOROBENZENE, ORTHO, LIQUID (DOT)
		O-CHLORONITROBENZENE
		O-CHLORONITROBENZENE (DOT)
		O-NITROCHLOROBENZENE
		O-NITROCHLOROBENZENE, LIQUID
		ONCB
		UN 1578 (DOT)

CAS No.	IN	Chemical Name
88-74-4		
		1-AMINO-2-NITROBENZENE
		1-NITRO-2-AMINOBENZENE
		2-AMINONITROBENZENE
		2-NITROANILINE
		2-NITROBENZENAMINE
		ANILINE, O-NITRO-
		AZOENE FAST ORANGE GR BASE
		AZOENE FAST ORANGE GR SALT
		AZOFIX ORANGE GR SALT
		AZOGENE FAST ORANGE GR
		AZOIC DIAZO COMPONENT 6
		BENZENAMINE, 2-NITRO-
		BRENTAMINE FAST ORANGE GR BASE
		BRENTAMINE FAST ORANGE GR SALT
		CI 37025
		CI AZOIC DIAZO COMPONENT 6
		DEVOL ORANGE B
		DEVOL ORANGE SALT B
		DIAZO FAST ORANGE GR
		FAST ORANGE BASE GR
		FAST ORANGE BASE JR
		FAST ORANGE GR BASE
		FAST ORANGE GR SALT
		FAST ORANGE O BASE
		FAST ORANGE O SALT
		FAST ORANGE SALT GR
		FAST ORANGE SALT JR
		HILTONIL FAST ORANGE GR BASE
		HILTOSAL FAST ORANGE GR SALT
		HINDASOL ORANGE GR SALT
		NATASOL FAST ORANGE GR SALT
		O-AMINONITROBENZENE
		O-NITRANILINE
	*	O-NITROANILINE
		ONA
		ORANGE BASE CIBA II
		ORANGE BASE IRGA II
		ORANGE GRS SALT
		ORANGE SALT CIBA II
		ORANGE SALT IRGA II
		ORTHONITROANILINE (DOT)
88-75-5		
		2-HYDROXYNITROBENZENE
	*	2-NITROPHENOL
		O-HYDROXYNITROBENZENE
		O-NITROPHENOL
		O-NITROPHENOL (DOT)
		PHENOL, 2-NITRO-
		PHENOL, O-NITRO-
		UN 1663 (DOT)
88-85-7		
		2,4-DINITRO-6-(1-METHYL-PROPYL)PHENOL (FRENCH)
		2,4-DINITRO-6-(1-METHYLPROPYL)PHENOL
		2,4-DINITRO-6-SEC-BUTYLPHENOL
		2-(1-METHYLPROPYL)-4,6-DINITROPHENOL
		2-SEC-BUTYL-4,6-DINITROPHENOL
		4,6-DINITRO-2-(1-METHYL-N-PROPYL)PHENOL
		4,6-DINITRO-2-SEC-BUTYLPHENOL
		4,6-DINITRO-2-SECBUTYLFENOL (CZECH)
		4,6-DINITRO-O-SEC-BUTYLPHENOL
		6-(1-METHYL-PROPYL)-2,4-DINITROFENOL (DUTCH)
		6-(1-METIL-PROPIL)-2,4-DINITRO-FENOLO (ITALIAN)
		6-SEC-BUTYL-2,4-DINITROPHENOL
		AATOX
		ARETIT
		BASANITE
		BNP 20
		BNP 30
		BUTAPHEN
		BUTAPHENE
		CALDON
		CHEMOX GENERAL
		CHEMOX PE
		DBNF
		DESICOIL
		DIBUTOX

CAS No.	IN	Chemical Name
		DINITRO
		DINITRO-3
		DINITROBUTYLPHENOL
		DINOSEB
		DINOSEBE (FRENCH)
		DN 289
		DNBP
		DNOSBP
		DNSBP
		DOW GENERAL
		DOW GENERAL WEED KILLER
		DOW SELECTIVE WEED KILLER
		DYTOP
		ELGETOL
		ELGETOL 318
		ENT 1,122
		GEBUTOX
		HEL-FIRE
		HIVERTOX
		KILOSEB
		LADOB
		LASEB
		NITROPONE C
		PHENOL, 2-(1-METHYLPROPYL)-4,6-DINITRO-
		PHENOTAN
		PREMERG
		PREMERGE
		PREMERGE 3
		RCRA WASTE NUMBER P020
		SINOX GENERAL
		SPARIC
		SPURGE
		SUBITEX
		UNICROP DNBP
		VERTAC DINITRO WEED KILLER
		VERTAC GENERAL WEED KILLER
		VERTAC SELECTIVE WEED KILLER
		WSX-8365
88-89-1		
	*	2,4,6-TRINITROPHENOL
		C.I. 10305
		CARBAZOTIC ACID
		MELINITE
		NA 1344 (DOT)
		NITROXANTHIC ACID
		PA
		PHENOL, 2,4,6-TRINITRO-
		PICRAL
		PICRIC ACID
		PICRIC ACID, WET WITH NOT LESS THAN 10% WATER, OVER 25 POUNDS (DOT)
		PICRIC ACID, WETTED WITH 10% TO 30% WATER
		PICRIC ACID, WETTED WITH AT LEAST 10% WATER (DOT)
		PICRIC ACID, WETTED WITH AT LEAST 30% WATER (DOT)
		PICRONITRIC ACID
		TRINITROPHENOL
		TRINITROPHENOL, WETTED WITH AT LEAST 10% WATER (DOT)
		TRINITROPHENOL, WETTED WITH AT LEAST 30% WATER (DOT)
		UN 1344 (DOT)
88-99-3		
		PHTHALIC ACID
89-62-3		
		1-AMINO-2-NITRO-4-METHYLBENZENE
		2-NITRO-4-METHYLANILINE
		2-NITRO-P-TOLUIDINE
		3-NITRO-4-AMINOTOLUENE
		4-AMINO-3-NITROTOLUENE
		4-METHYL-2-NITROANILINE
		4-METHYL-2-NITROBENZENAMINE
		4-METHYL-6-NITROANILINE
		4-METHYL-O-NITROANILINE
		AMARTHOL FAST RED GL BASE
		AMARTHOL FAST RED GL SALT
		AZOAMINE RED A
		AZOBASE NAT
		AZOENE FAST RED RED GL SALT
		AZOFIX RED GL SALT
		AZOIC DIAZO COMPONENT 8
		BENZENAMINE, 4-METHYL-2-NITRO-
		C.I. 37110
		C.I. AZOIC DIAZO COMPONENT 8
		DEVOL RED G
		DEVOL RED SALT G
		DIAZO FAST RED GL
		FAST RED 3NT BASE
		FAST RED 3NT SALT
		FAST RED BASE GL
		FAST RED BASE JL
		FAST RED G BASE
		FAST RED GL
		FAST RED GL BASE
		FAST RED MGL BASE
		HD FAST RED GL BASE
		HILTONIL FAST RED GL BASE
		HILTOSAL FAST RED GL SALT
		LAKE RED G BASE
		LITHOSOL SCARLET BASE M
		LITHOSOL SCARLET BASE MB
		LITHOSOL SCARLET BASE MBW
		LITHOSOL SCARLET BASE MW
		MITSUI RED GL BASE
		MNPT
		NAPHTHANIL RED G BASE
		NAPHTOELAN FAST RED GL BASE
		P-TOLUIDINE, 2-NITRO-
		RED BASE CIBA VII
		RED BASE IRGA VII
		RED BASE NGL
		RED G BASE
		RED G SALT
		RED SALT CIBA VII
		RED SALT IRGA VII
		SANYO FAST RED GL BASE
		SHINNIPPON FAST RED GL BASE
		TOYO FAST RED GL BASE
		TULABASE FAST RED GL
89-72-5		
		2-(1-METHYLPROPYL)PHENOL
		2-SEC-BUTYLPHENOL
		O-SEC-BUTYLPHENOL
		PHENOL, 2-(1-METHYLPROPYL)-
		PHENOL, O-SEC-BUTYL-
90-02-8		
		2-FORMYLPHENOL
		2-HYDROXYBENZALDEHYDE
		BENZALDEHYDE, 2-HYDROXY-
		BENZALDEHYDE, O-HYDROXY-
		O-FORMYLPHENOL
		O-HYDROXYBENZALDEHYDE
		SAH
		SALICYLAL
		SALICYLALDEHYDE
		SALICYLIC ALDEHYDE
90-04-0		
		1-AMINO-2-METHOXYBENZENE
		2-AMINOANISOLE
		2-ANISIDINE
		2-METHOXY-1-AMINOBENZENE
		2-METHOXYANILINE
		2-METHOXYBENZENAMINE
		BENZENAMINE, 2-METHOXY-
		O-AMINOANISOLE
		O-ANISIDINE
		O-METHOXYANILINE
		O-METHOXYPHENYLAMINE
90-12-0		
	*	1-METHYLNAPHTHALENE
		ALPHA-METHYLNAPHTHALENE
		NAPHTHALENE, 1-METHYL-

CAS No.	IN	Chemical Name
90-41-5		
		[1,1'-BIPHENYL]-2-AMINE
		2-AMINOBIPHENYL
		2-AMINODIPHENYL
		2-BIPHENYLAMINE
		2-PHENYLANILINE
		2-PHENYLBENZENAMINE
		O-AMINOBIPHENYL
		O-AMINODIPHENYL
		O-BIPHENYLAMINE
		O-PHENYLANILINE
90-43-7		
		2-PHENYLPHENOL
	*	O-PHENYLPHENOL
		PHENYLPHENOL
90-94-8		
		4,4'-BIS(DIMETHYLAMINO)BENZOPHENONE
		BENZOPHENONE, 4,4'-BIS(DIMETHYLAMINO)-
		BIS(4-DIMETHYLAMINOPHENYL) KETONE
		BIS[P-(N,N-DIMETHYLAMINO)PHENYL] KETONE
		METHANONE, BIS[4-(DIMETHYLAMINO)PHENYL]-
		MICHLERS KETONE
		N,N,N',N'-TETRAMETHYL-4,4'-DIAMINOBENZOPHENONE
		P,P'-BIS(DIMETHYLAMINO)BENZOPHENONE
		TETRAMETHYLDIAMINOBENZOPHENE
91-08-7		
		2,6-DIISOCYANATO-1-METHYLBENZENE
		2,6-DIISOCYANATOTOLUENE
		2,6-TDI
		2,6-TOLUENE DIISOCYANATE
		2-METHYL-M-PHENYLENE ISOCYANATE
		2-METHYL-META-PHENYLENE DIISOCYANATE
		2-METHYL-META-PHENYLENE ISOCYANATE
		BENZENE, 1,3-DIISOCYANATO-2-METHYL-
		BENZENE, 2,6-DIISOCYANATO-1-METHYL-
		HYLENE TCPA
		HYLENE TIC
		HYLENE TM
		HYLENE TM-65
		HYLENE TRF
		ISOCYANIC ACID, 2-METHYL-M-PHENYLENE ESTER
		ISOCYANIC ACID, 2-METHYL-META-PHENYLENE ESTER
		M-TOLYLENE DIISOCYANATE
		META-TOLYLENE DIISOCYANATE
		NIAX TDI
		NIAX TDI-P
		TDI
		TOLUENE DIISOCYANATE
		TOLUENE-2,6-DIISOCYANATE
		TOLYLENE 2,6-DIISOCYANATE
91-17-8		
		BICYCLO[4.4.0]DECANE
		DEC
	*	DECAHYDRONAPHTHALENE
		DECALIN
		DEKALIN
		NAPHTHALENE, DECAHYDRO-
		NAPHTHAN
		PERHYDRONAPHTHALENE
91-20-3		
		ALBOCARBON
		CAMPHOR TAR
		DEZODORATOR
		MIGHTY 150
		MIGHTY RD1
		MOTH BALLS
		MOTH BALLS (DOT)
		MOTH FLAKES
		NAFTALEN (POLISH)
	*	NAPHTHALENE
		NAPHTHALENE (ACGIH,DOT)
		NAPHTHALENE, CRUDE OR REFINED (DOT)
		NAPHTHALENE, MOLTEN
		NAPHTHALIN
		NAPHTHALIN (DOT)
		NAPHTHALINE
		NAPHTHENE
		NAPTHALENE, MOLTEN (DOT)
		NCI-C52904
		RCRA WASTE NUMBER U165
		TAR CAMPHOR
		UN 1334 (DOT)
		UN 2304 (DOT)
		WHITE TAR
91-22-5		
		1-AZANAPHTHALENE
		1-BENZAZINE
		1-BENZINE
		B 500
		BENZO(B)PYRIDINE
		BENZOPYRIDINE
		CHINOLEINE
		CHINOLIN (CZECH)
		CHINOLINE
		LEUCOL
		LEUCOLINE
		LEUKOL
		QUINOLIN
	*	QUINOLINE
		QUINOLINE (DOT)
		UN 2656 (DOT)
		USAF EK-218
91-49-6		
		ACETAMIDE, N-BUTYL-N-PHENYL-
		ACETANILIDE, N-BUTYL-
		BAA
		BUTYLACETANILIDE
		N-BUTYLACETANILIDE
91-58-7		
		2-CHLORONAPHTHALENE
	*	BETA-CHLORONAPHTHALENE
		NAPHTHALENE, 2-CHLORO-
91-59-8		
		2-AMINONAPHTHALENE
		2-NAPHTHALENAMINE
		2-NAPHTHYLAMIN (GERMAN)
		2-NAPHTHYLAMINE
		2-NAPHTHYLAMINE MUSTARD
		6-NAPHTHYLAMINE
		A-NAPHTHYLAMIN (GERMAN)
		BETA-NAPHTHYLAMINE
		BETA-NAPHTHYLAMINE (ACGIH,DOT)
		C.I. 37270
		FAST SCARLET BASE B
		NAPHTHYLAMINE (BETA)
		RCRA WASTE NUMBER U168
		UN 1650 (DOT)
		USAF CB-22
91-63-4		
		2-METHYLQUINOLINE
		CHINALDINE
		KHINALDIN
		QUINALDINE
		QUINOLINE, 2-METHYL-
91-66-7		
		(DIETHYLAMINO)BENZENE
		ANILINE, N,N-DIETHYL-
		BENZENAMINE, N,N-DIETHYL-
		BENZENAMINE, N,N-DIETHYL- (9CI)
		DEA
		DIAETHYLANILIN (GERMAN)
	*	DIETHYL ANILINE
		DIETHYLPHENYLAMINE
		N,N-DIETHYLAMINOBENZENE
		N,N-DIETHYLANILIN (CZECH)
		N,N-DIETHYLANILINE

CAS No.	IN	Chemical Name
		N,N-DIETHYLANILINE (DOT)
		N,N-DIETHYLBENZENAMINE
		UN 2432 (DOT)
91-80-5		
		1,2-ETHANEDIAMINE, N,N-DIMETHYL-N'-2-PYRIDINYL-N'-(2-THIENYLMETHYL)-
		2-[[2-(DIMETHYLAMINO)ETHYL]-2-THENYLAMINO]PYRIDINE
		A 3322
		AH-42
		LULAMIN
		METAPYRILENE
		METHAPYRILENE
		PARADORMALENE
		PYRIDINE, 2-[[2-(DIMETHYLAMINO)ETHYL]-2-THENYLAMINO]-
		PYRINISTAB
		PYRINISTOL
		REST-ON
		RESTRYL
		SEMIKON
		SLEEPWELL
		TENALIN
		THENYLENE
		THENYLPYRAMINE
		THIONYLAN
91-92-9		
		C.I. AZOIC COUPLING COMPONENT
91-93-0		
		3,3-DIMETHOXYBENZIDINE-4,4-DIISOCYANATE
91-94-1		
		[1,1'-BIPHENYL]-4,4'-DIAMINE, 3,3'-DICHLORO-
		3,3'-DICHLORO-4,4'-DIAMINOBIPHENYL
		3,3'-DICHLORO-4,4'-DIAMINODIPHENYL
	*	3,3'-DICHLOROBENZIDINE
		3,3'-DICHLOROBIPHENYL-4,4'-DIAMINE
		3,3-DICHLOROBENZIDENE
		4,4'-DIAMINO-3,3'-DICHLOROBIPHENYL
		4,4'-DIAMINO-3,3'-DICHLORODIPHENYL
		BENZIDINE, 3,3'-DICHLORO-
		C.I. 23060
		CURITHANE C 126
		DICHLOROBENZIDINE
		O,O'-DICHLOROBENZIDINE
91-96-3		
		C.I. AZOIC COUPLING COMPONENT 5
91-99-6		
		2,2'-(M-TOLYLIMINO)DIETHANOL
		DIETHANOL-M-TOLUIDINE
		ETHANOL, 2,2'-(M-TOLYLIMINO)DI-
		ETHANOL, 2,2'-[(3-METHYLPHENYL)IMINO]BIS-
		M-TOLYDIETHANOLAMINE
		M-TOLYLDIETHANOLAMINE
		N,N-BIS(2-HYDROXYETHYL)-3-METHYLANILINE
		N,N-BIS(2-HYDROXYETHYL)-M-TOLUIDINE
		N,N-BIS(BETA-HYDROXYETHYL)-3-METHYLANILINE
92-04-6		
		[1,1'-BIPHENYL]-4-OL, 3-CHLORO-
		2-CHLORO-4-PHENYLPHENOL
		2-CHLORO-P-PHENYLPHENOL
		3-CHLORO-4-BIPHENYLOL
		4-BIPHENYLOL, 3-CHLORO-
		4-HYDROXY-3-CHLOROBIPHENYL
		4-PHENYL-2-CHLOROPHENOL
		DOWICIDE 4
		O-CHLORO-P-PHENYLPHENOL
		PHENOL, 2-CHLORO-4-PHENYL-
		SANIDRIL
92-06-8		
		M-TERPHENYLS
92-15-9		
		1-ACETOACETYLAMINO-2-METHOXYBENZENE
		2'-METHOXYACETOACETANILIDE
		2-ACETOACETYLAMINOANISOLE
		2-METHOXYACETOACETANILIDE
		ACETOACETIC ACID O-ANISIDIDE
		ACETOACETYL-O-ANISIDIDE
		ACETOACETYL-O-ANISIDINE
		BUTANAMIDE, N-(2-METHOXYPHENYL)-3-OXO-
		N-ACETOACETYL-O-ANISIDINE
		O-ACETOACETANISDIDE
		O-ACETOACETANISIDIDE
		O-METHOXYACETOACETANILIDE
92-50-2		
		2-(ETHYLPHENYLAMINO)ETHANOL
		2-(N-ETHYL-N-PHENYLAMINO)ETHANOL
		2-(N-ETHYLANILINO)ETHANOL
		BETA-ETHYLANILINOETHYL ALCOHOL
		ETHANOL, 2-(ETHYLPHENYLAMINO)-
		ETHANOL, 2-(N-ETHYLANILINO)-
		ETHYL(BETA-HYDROXYETHYL)ANILINE
		ETHYLPHENYLETHANOLAMINE
		HYDROXYETHYLETHYLANILINE
		N-(2-HYDROXYETHYL)-N-ETHYLANILINE
		N-ETHYL(BETA-HYDROXYETHYL)ANILINE
		N-ETHYL-N-(BETA-HYDROXYETHYL)ANILINE
		N-ETHYL-N-(2-HYDROXYETHYL)ANILINE
		N-ETHYL-N-(HYDROXYETHYL)ANILINE
		N-ETHYL-N-PHENYLAMINOETHANOL
		N-ETHYL-N-PHENYLETHANOLAMINE
		N-ETHYLANILINOETHANOL
		N-PHENYL-N-ETHYL-2-AMINOETHANOL
		N-PHENYL-N-ETHYLETHANOL AMINE
	*	PHENYLETHYLETHANOLAMINE
92-52-4		
		1,1'-BIPHENYL
		1,1'-DIPHENYL
		1,1-BIPHENYL
		BIBENZENE
		BIPHENYL
		BIPHENYL (ACGIH)
		CAROLIDAL
	*	DIPHENYL
		LEMONENE
		PHENADOR-X
		PHENYLBENZENE
		PHPH
		TETROSIN LY
		XENENE
92-53-5		
		4-PHENYLMORPHOLINE
		MORPHOLINE, 4-PHENYL-
		MORPHOLINOBENZENE
		N-PHENYLMORPHOLINE
92-59-1		
		BENZENEMETHANAMINE, N-ETHYL-N-PHENYL-
		BENZYLAMINE, N-ETHYL-N-PHENYL-
		BENZYLETHYLPHENYLAMINE
		ETHYLBENZYLANILINE
		N-BENZYL-N-ETHYLANILINE
		N-ETHYL-N-BENZYLANILINE
		N-ETHYL-N-BENZYLANILINE (DOT)
		N-ETHYL-N-PHENYLBENZYLAMINE
		PHENYLETHYLBENZYLAMINE
		UN 2274 (DOT)
92-66-0		
		1,1'-BIPHENYL, 4-BROMO-
		4-BIPHENYL BROMIDE
		4-BROMO-1,1'-BIPHENYL
		4-BROMOBIPHENYL
		4-BROMODIPHENYL
		BIPHENYL, 4-BROMO-
		P-BROMOBIPHENYL
		P-BROMODIPHENYL

CAS No.	IN	Chemical Name
		P-PHENYLBROMOBENZENE
92-67-1		
		[1,1'-BIPHENYL]-4-AMINE
		4-AMINOBIPHENYL
		4-AMINODIPHENYL
		4-BIPHENYLAMINE
		4-BIPHENYLYLAMINE
		4-PHENYLANILINE
		P-AMINOBIPHENYL
		P-AMINODIPHENYL
		P-BIPHENYLAMINE
		P-PHENYLANILINE
		P-XENYLAMINE
		XENYLAMINE
92-84-2		
		10H-PHENOTHIAZINE
		CONTAVERM
		DANIKOROPA
		DIBENZO-1,4-THIAZINE
		DIBENZOTHIAZINE
		EARLY BIRD WORMER
		ENT 38
		FEENO
		FENOVERM
		NEMAZENE
		NEXARBOL
		ORIMON
		PADOPHENE
		PENTHAZINE
		PHENEGIC
		PHENOTHIAZINE
		PHENOVERM
		PHENOVIS
		PHENOXUR
		PHENTHIAZINE
		PHENZEEN
		RECONOX
		THIODIPHENYLAMINE
92-87-5		
		[1,1'-BIPHENYL]-4,4'-DIAMINE
		4,4'-BIANILINE
		4,4'-BIPHENYLDIAMINE
		4,4'-DIAMINO-1,1'-BIPHENYL
		4,4'-DIAMINOBIPHENYL
		4,4'-DIAMINODIPHENYL
		4,4'-DIPHENYLENEDIAMINE
	*	BENZIDINE
		C.I. AZOIC DIAZO COMPONENT 112
		FAST CORINTH BASE B
		P,P'-BIANILINE
		P,P'-DIAMINOBIPHENYL
		P-DIAMINODIPHENYL
92-93-3		
		1,1'-BIPHENYL, 4-NITRO-
		1-NITRO-4-PHENYLBENZENE
	*	4-NITROBIPHENYL
		4-NITRODIPHENYL
		BA 2794
		BIPHENYL, 4-NITRO-
		P-NITROBIPHENYL
		P-NITRODIPHENYL
92-94-4		
		P-TERPHENYLS
		TERPHENYLS
93-05-0		
		1,4-BENZENEDIAMINE, N,N-DIETHYL-
		4- AMINO-N,N-DIETHYLANILINE
		4-(DIETHYLAMINO)ANILINE
		DIETHYL-P-PHENYLENEDIAMINE
		DIETHYL-PARA-PHENYLENEDIAMINE
		DPD
		N,N-DIETHYL-4-AMINOANILINE
		N,N-DIETHYL-P-PHENYLENEDIAMINE
		P-(DIETHYLAMINO)ANILINE
		P-AMINO-N,N-DIETHYLANILINE
		P-AMINODIETHYLANILINE
		P-PHENYLENEDIAMINE, N,N-DIETHYL-
93-50-5		
		P-CHLORO-O-ANISIDINE
93-58-3		
		BENZOIC ACID, METHYL ESTER
		CLORIUS
		ESSENCE OF NIOBE
		METHYL BENZENECARBOXYLATE
	*	METHYLBENZOATE
		METHYLBENZOATE (DOT)
		NIOBE OIL
		OIL OF NIOBE
		OXIDATE LE
		UN 2938 (DOT)
93-68-5		
		ACETOACET-O-TOLUIDIDE
93-72-1		
		2,4,5-TP
		2,4,5-TP (DOT)
		2,4,5-TRICHLOROPHENOXYPROPIONIC ACID
		2,4,5-TRICHLOROPHENOXYPROPIONIC ACID (DOT)
		2-(2,4,5-TRICHLOROPHENOXY)PROPIONIC ACID
		ALPHA-(2,4,5-TRICHLOROPHENOXY)PROPIONIC ACID
		COLOR-SET
		FENOPROP
		FENORMONE
		FRUITONE T
		KUROSAL G
		NA 2765 (DOT)
		PROPANOIC ACID, 2-(2,4,5-TRICHLOROPHENOXY)-
		PROPIONIC ACID, (2,4,5-TRICHLOROPHENOXY)-
		PROPIONIC ACID, 2-(2,4,5-TRICHLOROPHENOXY)-
		SILVEX
		SILVI-RHAP
		STA-FAST
		TRICHLOROPHENOXYPROPIONIC ACID
93-76-5		
		(2,4,5-TRICHLOOR-FENOXY)-AZIJNZUUR (DUTCH)
		(2,4,5-TRICHLOR-PHENOXY)-ESSIGSAEURE (GERMAN)
		2,4,5-T
		2,4,5-T (ACGIH,DOT)
	*	2,4,5-TRICHLOROPHENOXY ACETIC ACID
		2,4,5-TRICHLOROPHENOXYACETIC ACID (DOT)
		ACETIC ACID, (2,4,5-TRICHLOROPHENOXY)-
		ACIDE 2,4,5-TRICHLORO PHENOXYACETIQUE (FRENCH)
		ACIDO (2,4,5-TRICLORO-FENOSSI)-ACETICO (ITALIAN)
		AMINE 2,4,5-T FOR RICE
		BCF-BUSHKILLER
		BRUSH RHAP
		BRUSH-OFF 445 LOW VOLATILE BRUSH KILLER
		BRUSHTOX
		DACAMINE
		DEBROUSSAILLANT CONCENTRE
		DEBROUSSAILLANT SUPER CONCENTRE
		DECAMINE 4T
		DED-WEED BRUSH KILLER
		DED-WEED LV-6 BRUSH KIL AND T-5 BRUSH KIL
		DINOXOL
		ENVERT-T
		ESTERCIDE T-2 AND T-245
		ESTERON
		ESTERON 245
		ESTERON 245 BE
		ESTERON BRUSH KILLER
		FARMCO FENCE RIDER
		FENCE RIDER
		FORRON
		FORST U 46
		FORTEX
		FRUITONE A

CAS No.	IN	Chemical Name
		INVERTON 245
		LINE RIDER
		NA 2765 (DOT)
		PHORTOX
		RCRA WASTE NUMBER U232
		REDDON
		REDDOX
		SPONTOX
		SUPER D WEEDONE
		TIPPON
		TORMONA
		TRANSAMINE
		TRIBUTON
		TRINOXOL
		TRIOXON
		TRIOXONE
		U 46
		VEON
		VEON 245
		VERTON 2T
		VISKO RHAP LOW VOLATILE ESTER
		WEEDAR
		WEEDONE
		WEEDONE 2,4,5-T
93-78-7		
		2,4,5-T ESTER [ISOPROPYL ESTER]
93-79-8		
		(2,4,5-TRICHLOROPHENOXY)ACETIC ACID 2-BUTOXYETHYL ESTER
		(2,4,5-TRICHLOROPHENOXY)ACETIC ACID BUTYL ESTER
		2,4,5-T BUTOXYETHANOL ESTER
		2,4,5-T BUTOXYETHYL ESTER
		2,4,5-T BUTYL ESTER
		2,4,5-T ESTER [BUTYL ESTER]
		2,4,5-T N-BUTYL ESTER
		2,4,5-TRICHLOROPHENOXYACETIC ACID AMINE, ESTER, OR SALT
		2,4,5-TRICHLOROPHENOXYACETIC ACID, BUTYL ESTER
		ACETIC ACID, (2,4,5-TRICHLOROPHENOXY)-, 2-BUTOXY-ETHYL ESTER
		ACETIC ACID, (2,4,5-TRICHLOROPHENOXY)-, BUTYL ESTER
		ARBORICID
		BLADEX H
		BUTOXYETHYL 2,4,5-T
		BUTYL 2,4,5-T
		BUTYL 2,4,5-TRICHLOROPHENOXYACETATE
		BUTYLATE 2,4,5-T
		FLOMORE
		HI-ESTER 2,4,5-T
		HORMOSLYR 500T
		KILEX 3
		KRZEWOTOKS
		KRZEWOTOX
		N-BUTYL (2,4,5-TRICHLOROPHENOXY)ACETATE
		N-BUTYLESTER KYSELINI 2,4,5-TRICHLORFENOXYOCTOVE (CZECH)
		TORMONA
		TRINOXOL
		TRIOXONE
		U 46KW
		U 46T
		U46KW
93-89-0		
		BENZOIC ACID, ETHYL ESTER
		BENZOYL ETHYL ETHER
		ETHYL BENZENECARBOXYLATE
	*	ETHYL BENZOATE
93-90-3		
		2-(METHYLPHENYLAMINO)ETHANOL
		2-(N-METHYL-N-PHENYLAMINO)ETHANOL
		2-(N-METHYLANILINO)ETHANOL
		ETHANOL, 2-(METHYLPHENYLAMINO)-
		ETHANOL, 2-(N-METHYLANILINO)-
		N-METHYL-N-(2-HYDROXYETHYL)ANILINE
		PHENYLMETHYL ETHANOLAMINE
93-96-9		
		1,1'-DIPHENYLDIETHYL ETHER
		ALPHA-METHYLBENZYL ETHER
		BENZENE, 1,1'-(OXYDIETHYLIDENE)BIS-
		BIS(ALPHA-METHYLBENZYL) ETHER
		BIS(ALPHA-PHENYLETHYL) ETHER
		ETHER, BIS(ALPHA-METHYLBENZYL)
94-04-2		
		2-ETHYLHEXANOIC ACID, VINYL ESTER
		HEXANOIC ACID, 2-ETHYL-, ETHENYL ESTER
		HEXANOIC ACID, 2-ETHYL-, VINYL ESTER
		VINYL 2-ETHYLHEXANOATE
		VINYL-2-ETHYLHEXOATE
94-11-1		
		2,4-D ISOPROPYL ESTER
		2,4-DICHLOROPHENOXYACETIC ACID ISOPROPYL ESTER
		ACETIC ACID, (2,4-DICHLOROPHENOXY)-, 1-METHYLETHYL ESTER
		ACETIC ACID, (2,4-DICHLOROPHENOXY)-, ISOPROPYL ESTER
		ESTERON 44
		ISOPROPYL (2,4-DICHLOROPHENOXY)ACETATE
		ISOPROPYL 2,4-D ESTER
		WEEDONE 128
94-17-7		
		4-CHLOROBENZOYL PEROXIDE
		BIS(4-CHLOROBENZOYL) PEROXIDE
		BIS(P-CHLOROBENZOYL) PEROXIDE
		CADOX PS
		DI-(4-CHLOROBENZOYL) PEROXIDE, NOT MORE THAN 75% WITH WATER (DOT)
		P,P'-DICHLOROBENZOYL PEROXIDE
		P,P'-DICHLORODIBENZOYL PEROXIDE
		P-CHLOROBENZOYL PEROXIDE
		P-CHLOROBENZOYL PEROXIDE (DOT)
		P-CHLOROBENZOYL PEROXIDE, NOT MORE THAN 75% WITH WATER (DOT)
		PEROXIDE, BIS(4-CHLOROBENZOYL)
		PEROXIDE, BIS(P-CHLOROBENZOYL)-
		UN 2113 (DOT)
94-36-0		
		ACETOXYL
		ACNEGEL
		AZTEC BPO
		BENOXYL
		BENZAC
		BENZAKNEW
		BENZOIC ACID, PEROXIDE
		BENZOPEROXIDE
		BENZOYL
	*	BENZOYL PEROXIDE
		BENZOYL PEROXIDE (ACGIH,DOT)
		BENZOYL PEROXIDE (DOT)
		BENZOYL PEROXIDE, MORE THAN 72% BUT LESS THAN 95%
		BENZOYL PEROXIDE, MORE THAN 72% BUT LESS THAN 95% AS A PASTE (DOT)
		BENZOYL PEROXIDE, MORE THAN 77% BUT LESS THAN 95% WITH WATER (DOT)
		BENZOYL PEROXIDE, NOT LESS THAN 30% BUT NOT MORE THAN 52% WITH INERT SOLID
		BENZOYL PEROXIDE, TECHNICAL PURE (DOT)
		BENZOYL SUPEROXIDE
		BENZOYL-PEROXIDE, MORE THAN 52% WITH INERT SOLID (DOT)
		BENZOYLPEROXID (GERMAN)
		BENZOYLPEROXYDE (DUTCH)
		BZF-60
		CADET
		CADOX
		CADOX 40E
		CADOX B 50 P
		CADOX BS
		CLEARASIL BENZOYL PEROXIDE LOTION
		CLEARASIL BP ACNE TREATMENT
		CUTICURA ACNE CREAM

CAS No.	IN	Chemical Name
		DEBROXIDE
		DIBENZOYLPEROXID (GERMAN)
		DIBENZOYL PEROXIDE
		DIBENZOYLPEROXYDE (DUTCH)
		DIPHENYLGLYOXAL PEROXIDE
		DRY AND CLEAR
		EPI-CLEAR
		FOSTEX
		G 20
		GAROX
		INCIDOL
		LOROXIDE
		LUCIDOL
		LUCIDOL 40E
		LUCIDOL 50P
		LUCIDOL 98
		LUCIDOL B 50
		LUCIDOL CH 50
		LUCIDOL G 20
		LUCIDOL KL 50
		LUCIDOL S 50
		LUPERCO
		LUPERCO AA
		LUPERCO AST
		LUPEROX FL
		NA 2085 (DOT)
		NAYPER B
		NAYPER B AND BO
		NAYPER BO
		NOROX BZP-250
		NOROX BZP-C-35
		NOVADELOX
		NYPER B
		NYPER BO
		OXY WASH
		OXY-10
		OXY-5
		OXYLITE
		PANOXYL
		PEROSSIDO DI BENZOILE (ITALIAN)
		PEROXIDE, DIBENZOYL
		PEROXYDE DE BENZOYLE (FRENCH)
		PERSADOX
		QUINOLOR COMPOUND
		SULFOXYL
		SUPEROX
		SUPEROX 744
		THE RADERM
		TOPEX
		UN 2085 (DOT)
		UN 2086 (DOT)
		UN 2088 (DOT)
		UN 2089 (DOT)
		VANOXIDE
		XERAC
94-52-0		
		6-NITROBENZIMIDAZOLE
94-58-6		
	*	1,2-(METHYLENEDIOXY)-4-PROPYLBENZENE
		1,3-BENZODIOXOLE, 5-PROPYL-
		2',3'-DIHYDROSAFROLE
		5-PROPYL-1,3-BENZODIOXOLE
		BENZENE, 1,2-(METHYLENEDIOXY)-4-PROPYL-
		DIHYDROSAFROLE
94-59-7		
		1,3-BENZODIOXOLE, 5-(2-PROPENYL)-
		1-ALLYL-3,4-METHYLENEDIOXYBENZENE
		3,4-METHYLENEDIOXY-ALLYLBENZENE
		3-(3,4-METHYLENEDIOXYPHENYL)PROP-1-ENE
		4-ALLYL-1,2-(METHYLENEDIOXY)BENZENE
		5-ALLYL-1,3-BENZODIOXOLE
		ALLYLCATECHOL METHYLENE ETHER
		ALLYLDIOXYBENZENE METHYLENE ETHER
		ALLYLPYROCATECHOL METHYLENE ETHER
		BENZENE, 4-ALLYL-1,2-(METHYLENEDIOXY)-
		M-ALLYLPYROCATECHIN METHYLENE ETHER
		RHYUNO OIL
		SAFRENE
		SAFROL
	*	SAFROLE
		SAFROLE MF
		SHIKIMOLE
		SHIKOMOL
94-70-2		
		2-AMINOPHENETOLE
		2-ETHOXYANILINE
		BENZENAMINE, 2-ETHOXY-
		O-ETHOXYANILINE
		O-PHENETIDINE
94-75-7		
		(2,4-DICHLOOR-FENOXY)-AZIJNZUUR (DUTCH)
		(2,4-DICHLOR-PHENOXY)-ESSIGSAEURE (GERMAN)
		(2,4-DICHLOROPHENOXY)ACETIC ACID
	*	2,4-D
		2,4-D (ACGIH,DOT)
		2,4-D ACID
		2,4-DICHLOROPHENOXYACETIC ACID
		2,4-DICHLOROPHENOXYACETIC ACID (DOT)
		2,4-DICHLOROPHENOXYACETIC ACID (HIGH VOLATILE ESTERS)
		2,4-DICHLOROPHENOXYETHANOIC ACID
		2,4-DICHLORPHENOXYACETIC ACID
		2,4-DWUCHLOROFENOKSYOCTOWY KWAS (POLISH)
		2,4-PA
		ACETIC ACID, (2,4-DICHLOROPHENOXY)-
		ACIDE 2,4-DICHLORO PHENOXYACETIQUE (FRENCH)
		ACIDO(2,4-DICLORO-FENOSSI)-ACETICO (ITALIAN)
		AGROTECT
		AMIDOX
		AMOXONE
		AQUA-KLEEN
		B-SELEKTONON
		BH 2,4-D
		BRUSH-RHAP
		CHLOROXONE
		CROP RIDER
		CROTILIN
		D 50
		DACAMINE
		DEBROUSSAILLANT 600
		DECAMINE
		DED-WEED
		DED-WEED LV-69
		DESORMONE
		DICHLOROPHENOXYACETIC ACID
		DICLORDON
		DICOPUR
		DICOTOX
		DINOXOL
		DMA-4
		DORMONE
		EMULSAMINE BK
		EMULSAMINE E-3
		ENT 8,538
		ENVERT 171
		ENVERT DT
		ESTERON
		ESTERON 44 WEED KILLER
		ESTERON 76 BE
		ESTERON 99
		ESTERON 99 CONCENTRATE
		ESTERON BRUSH KILLER
		ESTERONE FOUR
		ESTONE
		FARMCO
		FERNESTA
		FERNIMINE
		FERNOXONE
		FERXONE
		FOREDEX 75
		FORMULA 40
		HEDONAL
		HEDONAL (HERBICIDE)

CAS No.	IN	Chemical Name
		HERBIDAL
		IPANER
		KROTILINE
		LAWN-KEEP
		MACRONDRAY
		MIRACLE
		MONOSAN
		MOTA MASKROS
		MOXONE
		NA 2765 (DOT)
		NETAGRONE
		NETAGRONE 600
		NSC 423
		PENNAMINE
		PENNAMINE D
		PHENOX
		PIELIK
		PLANOTOX
		PLANTGARD
		RCRA WASTE NUMBER U240
		RHODIA
		SALVO
		SPRITZ-HORMIN/2,4-D
		SPRITZ-HORMIT/2,4-D
		SUPER D WEEDONE
		SUPERORMONE CONCENTRE
		TRANSAMINE
		TRIBUTON
		TRINOXOL
		U 46
		U 46DP
		U-5043
		VERGEMASTER
		VERTON
		VERTON 2D
		VERTON D
		VERTRON 2D
		VIDON 638
		VISKO-RHAP
		VISKO-RHAP LOW DRIFT HERBICIDES
		VISKO-RHAP LOW VOLATILE 4L
		WEED TOX
		WEED-AG-BAR
		WEED-B-GON
		WEED-RHAP
		WEEDAR
		WEEDAR-64
		WEEDATUL
		WEEDEZ WONDER BAR
		WEEDONE
		WEEDONE LV4
		WEEDTROL
94-78-0		
		2,6-DIAMINO-3-PHENYLAZOPYRIDINE
		2,6-PYRIDINEDIAMINE, 3-(PHENYLAZO)-
		GASTRACID
		PHENAZOPYRIDINE
		PYRIDINE, 2,6-DIAMINO-3-(PHENYLAZO)-
94-79-1		
		(2,4-DICHLOROPHENOXY)ACETIC ACID, SEC-BUTYL ESTER
		ACETIC ACID, (2,4-DICHLOROPHENOXY)-, 1-METHYLPROPYL ESTER
		ACETIC ACID, (2,4-DICHLOROPHENOXY)-, SEC-BUTYL ESTER
		S-BUTYL 2,4-DICHLOROPHENOXYACETATE
		SEC-BUTYL 2,4-DICHLOROPHENOXYACETATE
94-80-4		
		2,4-D BUTYL ESTER
		2,4-DBE
		2,4-DICHLOROPHENOXYACETIC ACID BUTYL ESTER
		2,4-DICHLOROPHENOXYACETIC ACID N-BUTYL ESTER
		ACETIC ACID, (2,4-DICHLOROPHENOXY)-, BUTYL ESTER
		BUTAPON
		BUTYL (2,4-DICHLOROPHENOXY)ACETATE
		BUTYL 2,4-D
		BUTYL DICHLOROPHENOXYACETATE
		ESSO HERBICIDE 10
		FERNESTA
		HI-ESTER 2,4-D
		LIRONOX
95-06-7		
		2-CHLORALLYL DIETHYLDITHIOCARBAMATE
		2-CHLOROALLYL DIETHYLDITHIOCARBAMATE
		2-CHLOROALLYL N,N-DIETHYLDITHIOCARBAMATE
		CARBAMIC ACID, DIETHYLDITHIO-, 2-CHLOROALLYL ESTER
		CARBAMODITHIOIC ACID, DIETHYL-, 2-CHLORO-2-PROPENYL ESTER
		CDEC
		CHLORALLYL DIETHYLDITHIOCARBAMATE
		CP 4,742
		DIETHYLDITHIOCARBAMIC ACID 2-CHLOROALLYL ESTER
		SULFALLATE
		THIOALLATE
		VEGADEX
		VEGADEX SUPER
95-13-6		
		1H-INDENE
		INDEN
		INDENE
		INDONAPHTHENE
95-14-7		
		1H-BENZOTRIAZOLE
95-46-5		
		1-BROMO-2-METHYLBENZENE
	*	2-BROMOTOLUENE
		2-METHYLBROMOBENZENE
		2-TOLYL BROMIDE
		BENZENE, 1-BROMO-2-METHYL-
		O-BROMOMETHYLBENZENE
		O-BROMOTOLUENE
		O-METHYLBROMOBENZENE
		O-METHYLPHENYL BROMIDE
		O-TOLYL BROMIDE
		TOLUENE, O-BROMO-
95-47-6		
		1,2-DIMETHYLBENZENE
		1,2-XYLENE
		BENZENE, 1,2-DIMETHYL-
		O-DIMETHYLBENZENE
		O-METHYLTOLUENE
		O-XYLENE (ACGIH,DOT)
		O-XYLOL
		O-XYLOL (DOT)
		ORTHO-XYLENE
		UN 1307 (DOT)
95-48-7		
		1-HYDROXY-2-METHYLBENZENE
		2-CRESOL
		2-HYDROXYTOLUENE
		2-METHYLPHENOL
	*	O-CRESOL
		O-CRESOL (DOT)
		O-CRESYLIC ACID
		O-HYDROXYTOLUENE
		O-KRESOL (GERMAN)
		O-METHYLPHENOL
		O-METHYLPHENYLOL
		O-OXYTOLUENE
		O-TOLUOL
		ORTHOCRESOL
		PHENOL, 2-METHYL-
		PHENOL, 2-METHYL- (9CI)
		RCRA WASTE NUMBER U052
		UN 2076 (DOT)
95-49-8		
		1-CHLORO-2-METHYLBENZENE
		1-METHYL-2-CHLOROBENZENE
		2-CHLORO-1-METHYLBENZENE

CAS No.	IN	Chemical Name
		2-CHLOROTOLUENE
		2-METHYLCHLOROBENZENE
		BENZENE, 1-CHLORO-2-METHYL-
	*	O-CHLOROTOLUENE
		O-TOLYL CHLORIDE
		TOLUENE, O-CHLORO-
95-50-1		
	*	1,2-DICHLOROBENZENE
		BENZENE, 1,2-DICHLORO-
		BENZENE, O-DICHLORO-
		CHLOROBEN
		CHLORODEN
		CLOROBEN
		DCB
		DICHLOROBENZENE, ORTHO, LIQUID
		DICHLOROBENZENE, ORTHO, LIQUID (DOT)
		DILANTIN DB
		DILATIN DB
		DIZENE
		DOWTHERM E
		NCI-C54944
		O-DICHLOR BENZOL
		O-DICHLORBENZENE
		O-DICHLOROBENZENE
		O-DICHLOROBENZENE (ACGIH)
		O-DICHLOROBENZENE, LIQUID
		ODB
		ODCB
		ORTHODICHLOROBENZENE
		ORTHODICHLOROBENZOL
		RCRA WASTE NUMBER U070
		SPECIAL TERMITE FLUID
		TERMITKIL
		UN 1591 (DOT)
95-53-4		
		1-AMINO-2-METHYLBENZENE
		1-METHYL-2-AMINOBENZENE
		2-AMINO-1-METHYLBENZENE
		2-AMINOTOLUENE
		2-METHYL-1-AMINOBENZENE
		2-METHYLANILINE
		2-METHYLBENZENAMINE
		2-METHYLPHENYLAMINE
		2-TOLUIDINE
		ANILINE, 2-METHYL-
		ANILINE, P-METHYL-
		BENZENAMINE, 2-METHYL-
		BENZENAMINE, 2-METHYL- (9CI)
		CI 37077
		CI 37107
		CI AZOIC COUPLING COMPONENT 107
		NAPHTOL AS-KG
		NAPHTOL AS-KGLL
		O-AMINOTOLUENE
		O-METHYLANILINE
		O-METHYLBENZENAMINE
		O-TOLUIDIN (CZECH)
	*	O-TOLUIDINE
		O-TOLUIDINE (ACGIH,DOT)
		O-TOLUIDYNA (POLISH)
		O-TOLYLAMINE
		RCRA WASTE NUMBER U328
		RCRA WASTE NUMBER U353
		TOLYLAMINE
		UN 1708 (DOT)
95-54-5		
		O-PHENYLENEDIAMINE
95-57-8		
	*	2-CHLOROPHENOL
		2-HYDROXYCHLOROBENZENE
		O-CHLOROPHENOL
		PHENOL, 2-CHLORO-
		PHENOL, O-CHLORO-

CAS No.	IN	Chemical Name
95-63-6		
		1,2,4-TRIMETHYL BENZENE
		1,2,5-TRIMETHYLBENZENE
		1,3,4-TRIMETHYLBENZENE
		AS-TRIMETHYLBENZENE
		ASYMMETRICAL TRIMETHYLBENZENE
		BENZENE, 1,2,4-TRIMETHYL-
		BENZENE, 1,2,5-TRIMETHYL-
		PSEUDOCUMENE
		PSEUDOCUMOL
		PSI-CUMENE
		TRIMETHYL BENZENE (ACGIH)
95-69-2		
		2-AMINO-5-CHLOROTOLUENE
		2-METHYL-4-CHLOROANILINE
		4-CHLORO-2-METHYLANILINE
		4-CHLORO-2-METHYLBENZENAMINE
		4-CHLORO-2-TOLUIDINE
		4-CHLORO-6-METHYLANILINE
		4-CHLORO-O-TOLUIDINE
		5-CHLORO-2-AMINOTOLUENE
		BENZENAMINE, 4-CHLORO-2-METHYL-
		DAITO RED BASE TR
		FAST RED 5CT BASE
		FAST RED BASE TR
		FAST RED TR BASE
		FAST RED TR-T BASE
		FAST RED TRO BASE
		KAKO RED TR BASE
		MITSUI RED TR BASE
		O-TOLUIDINE, 4-CHLORO-
		P-CHLORO-O-TOLUIDINE
		RED BASE NTR
		RED TR BASE
		SANYO FAST RED TR BASE
95-74-9		
		3-CHLORO-4-METHYL ANILINE (CHLOROTOLUIDINE)
95-76-1		
		3,4-DCA
		3,4-DICHLORANILINE
	*	3,4-DICHLOROANILINE
		3,4-DICHLOROBENZENAMINE
		4,5-DICHLOROANILINE
		ANILINE, 3,4-DICHLORO-
		BENZENAMINE, 3,4-DICHLORO-
		DCA
		M,P-DICHLOROANILINE
95-79-4		
		2-AMINO-4-CHLOROTOLUENE
		2-AMINO-5-CHLOROTOLUENE
		2-METHYL-4-CHLOROANILINE
		2-METHYL-5-CHLOROANILINE
		4-CHLORO-2-AMINOTOLUENE
		4-CHLORO-2-METHYLANILINE
		4-CHLORO-2-METHYLBENZENEAMINE
		4-CHLORO-2-TOLUIDINE
		4-CHLORO-6-METHYLANILINE
		4-CHLORO-O-TOLUIDINE
		5-CHLORO-2-AMINOTOLUENE
		5-CHLORO-2-METHYL ANILINE
	*	5-CHLORO-2-METHYLANILINE
		5-CHLORO-2-METHYLBENZENAMINE
		5-CHLORO-2-TOLUIDINE
		5-CHLORO-O-TOLUIDINE
		ACCO FAST RED KB BASE
		AMARTHOL FAST RED TR BASE
		ANSIBASE RED KB
		AZOENE FAST RED KB BASE
		AZOENE FAST RED TR BASE
		AZOGENE FAST RED TR
		AZOIC DIAZO COMPONENT 11, BASE
		BENZENAMINE, 5-CHLORO-2-METHYL-
		BRENTAMINE FAST RED TR BASE
		DAITO RED BASE TR
		DEVAL RED K

CAS No.	IN	Chemical Name
		DEVAL RED TR
		DIAZO FAST RED TRA
		FAST RED 5CT BASE
		FAST RED BASE TR
		FAST RED KB AMINE
		FAST RED KB BASE
		FAST RED KB SALT
		FAST RED KB SALT SUPRA
		FAST RED KB-T BASE
		FAST RED KBS SALT
		FAST RED TR
		FAST RED TR BASE
		FAST RED TR11
		FAST RED TRO BASE
		GENAZO RED KB SOLN
		HILTONIL FAST RED KB BASE
		KAKO RED TR BASE
		KAMBAMINE RED TR
		METROGEN RED FORMER KB SOLN
		MITSUI RED TR BASE
		NAPHTHOSOL FAST RED KB BASE
		O-TOLUIDINE, 4-CHLORO-
		O-TOLUIDINE, 5-CHLORO-
		PHARMAZOID RED KB
		RED BASE CIBA IX
		RED BASE IRGA IX
		RED BASE NTR
		RED KB BASE
		RED TR BASE
		SANYO FAST RED TR BASE
		SPECTROLENE RED KB
		STABLE RED KB BASE
		TULABASE FAST RED TR
95-80-7		
		1,3-BENZENEDIAMINE, 4-METHYL-
		1,3-DIAMINO-4-METHYLBENZENE
		2,4-DIAMINO-1-METHYLBENZENE
		2,4-DIAMINOTOLUENE
		2,4-TOLUENEDIAMINE
		2,4-TOLYLENEDIAMINE
		4-METHYL-1,3-BENZENEDIAMINE
		4-METHYL-1,3-PHENYLENEDIAMINE
		4-METHYL-M-PHENYLENEDIAMINE
		C.I. OXIDATION BASE
		C.I. OXIDATION BASE 35
		EUCANINE GB
		FOURAMINE J
		FOURRINE 94
		FOURRINE M
		M-TOLUENEDIAMINE
		M-TOLYLENEDIAMINE
		NAKO TMT
		PELAGOL GREY J
		PELAGOL J
		PONTAMINE DEVELOPER TN
		RENAL MD
		TDA
		TOLUENE-2,4-DIAMINE
		ZOBA GKE
95-81-8		
		2-CHLORO-5-METHYL ANILINE (CHLOROTOLUIDINE)
95-83-0		
		1,2-BENZENEDIAMINE, 4-CHLORO-
		1,2-DIAMINO-4-CHLOROBENZENE
		2-AMINO-4-CHLOROANILINE
		3,4-DIAMINO-1-CHLOROBENZENE
		3,4-DIAMINOCHLOROBENZENE
		4-CHLORO-1,2-BENZENEDIAMINE
		4-CHLORO-1,2-DIAMINOBENZENE
		4-CHLORO-1,2-PHENYLENEDIAMINE
		4-CHLORO-O-PHENYLENEDIAMINE
		C.I. 76015
		O-PHENYLENEDIAMINE, 4-CHLORO-
		URSOL OLIVE 6G
95-85-2		
		2-AMINO-4-CHLOROPHENOL
		2-AMINO-4-CHLOROPHENOL (DOT)
		2-HYDROXY-5-CHLOROANILINE
		4-CHLORO-2-AMINOPHENOL
		5-CHLORO-2-HYDROXYANILINE
		AMINOCHLOROPHENOL
		CI 76525
		CI OXIDATION BASE 18
		FOURAMINE PY
		P-CHLORO-O-AMINOPHENOL
		PHENOL, 2-AMINO-4-CHLORO-
		UN 2673 (DOT)
95-92-1		
		DIETHYL ETHANEDIOATE
		DIETHYL OXALATE
		ETHANEDIOIC ACID, DIETHYL ESTER
	*	ETHYL OXALATE
		ETHYL OXALATE (DOT)
		OXALIC ACID, DIETHYL ESTER
		UN 2525 (DOT)
95-95-4		
		2,4,5-TRICHLOROPHENOL
		DOWICIDE 2
		PHENOL, 2,4,5-TRICHLORO-
		PREVENTOL I
		TCP
96-09-3		
		1,2-EPOXYETHYLBENZENE
		1-PHENYL-1,2-EPOXYETHANE
		2-PHENYLOXIRANE
		BENZENE, (EPOXYETHYL)-
		EPOXYSTYRENE
		ETHANE, 1,2-EPOXY-1-PHENYL-
		OXIRANE, PHENYL-
		PHENETHYLENE OXIDE
		PHENYLETHYLENE OXIDE
		PHENYLOXIRANE
		STYRENE 7,8-OXIDE
		STYRENE EPOXIDE
		STYRENE OXIDE
		STYRYL OXIDE
		ALUMINUM, CHLORODIETHYL-
		ALUMINUM, DICHLOROTETRAETHYLDI-
		CHLORODIETHYLALUMINUM
		DEAK
		DIETHYLALUMINIUM CHLORIDE
		DIETHYLALUMINIUM CHLORIDE (DOT)
		DIETHYLALUMINUM CHLORIDE
		DIETHYLALUMINUM MONOCHLORIDE
		DIETHYLCHLOROALUMINUM
96-10-6		
		UN 1101 (DOT)
96-12-8		
		1,2-DIBROM-3-CHLOR-PROPAN (GERMAN)
		1,2-DIBROMO-3-CHLOROPROPANE
		1,2-DIBROMO-3-CHLOROPROPANE (DOT)
		1,2-DIBROMO-3-CLORO-PROPANO (ITALIAN)
		1,2-DIBROOM-3-CHLOORPROPAAN (DUTCH)
		1-CHLORO-2,3-DIBROMOPROPANE
		3-CHLORO-1,2-DIBROMOPROPANE
		BBC 12
		DBCP
		DIBROMCHLORPROPAN (GERMAN)
		DIBROMOCHLOROPROPANE
		FUMAGON
		FUMAZONE
		FUMAZONE 86
		FUMAZONE 86E
		NCI-C00500
		NEMABROM
		NEMAFUME
		NEMAGON
		NEMAGON 20

CAS No.	IN	Chemical Name
		NEMAGON 206
		NEMAGON 20G
		NEMAGON 90
		NEMAGON SOIL FUMIGANT
		NEMAGONE
		NEMANAX
		NEMAPAZ
		NEMASET
		NEMATOCIDE
		NEMATOX
		NEMAZON
		OS 1897
		OXY DBCP
		PROPANE, 1,2-DIBROMO-3-CHLORO-
		PROPANE, 1-CHLORO-2,3-DIBROMO-
		RCRA WASTE NUMBER U066
		SD 1897
		UN 2872 (DOT)
96-14-0		
		3-METHYLPENTANE
		PENTANE, 3-METHYL-
96-17-3		
		2-FORMYLBUTANE
		2-METHYLBUTANAL
		2-METHYLBUTYRALDEHYDE
		2-METHYLBUTYRIC ALDEHYDE
		ALPHA-METHYL-N-BUTANAL
		ALPHA-METHYLBUTANAL
		ALPHA-METHYLBUTYRALDEHYDE
		ALPHA-METHYLBUTYRIC ALDEHYDE
		BUTANAL, 2-METHYL-
		BUTYRALDEHYDE, 2-METHYL-
		METHYLETHYLACETALDEHYDE
96-18-4		
		1,2,3-TRICHLOROPROPANE
		ALLYL TRICHLORIDE
		GLYCEROL TRICHLOROHYDRIN
		GLYCERYL TRICHLOROHYDRIN
		PROPANE, 1,2,3-TRICHLORO-
		TRICHLOROHYDRIN
96-19-5		
		1,2,3-TRICHLORO-1-PROPENE
		1,2,3-TRICHLOROPROPENE
		1-PROPENE, 1,2,3-TRICHLORO-
		2,3-DICHLOROALLYL CHLORIDE
		PROPENE, 1,2,3-TRICHLORO-
96-20-8		
		1-(HYDROXYMETHYL)PROPYLAMINE
		1-BUTANOL, 2-AMINO-
		1-HYDROXY-2-BUTYLAMINE
		1-HYDROXY-SEC-BUTYLAMINE
		2-AMINO-1-BUTANOL
		2-AMINO-1-HYDROXYBUTANE
		2-AMINOBUTYL ALCOHOL
96-22-0		
		3-PENTANONE
		DEK
	*	DIETHYL KETONE
		DIETHYL KETONE (ACGIH,DOT)
		DIETHYLCETONE (FRENCH)
		DIMETHYLACETONE
		ETHYL KETONE
		METACETONE
		METHACETONE
		PENTANONE-3
		PROPIONE
		UN 1156 (DOT)
96-23-1		
		1,3-DICHLORO-2-PROPANOL
		1,3-DICHLOROPROPANOL
		1,3-DICHLOROPROPANOL-2

CAS No.	IN	Chemical Name
96-24-2		
		1,2-DIHYDROXY-3-CHLOROPROPANE
		1,2-PROPANEDIOL, 3-CHLORO-
		1-CHLORO-1-DEOXYGLYCEROL
		1-CHLORO-2,3-DIHYDROXYPROPANE
		1-CHLORO-2,3-PROPANEDIOL
		1-CHLOROPROPANE-2,3-DIOL
		2,3-DIHYDROXYPROPYL CHLORIDE
		3-CHLORO-1,2-DIHYDROXYPROPANE
		3-CHLORO-1,2-PROPANEDIOL
		3-CHLORO-1,2-PROPYLENE GLYCOL
		3-CHLOROPROPANE-1,2-DIOL
		3-CHLOROPROPANEDIOL
		3-CHLOROPROPYLENE GLYCOL
		ALPHA-CHLORHYDRIN
		ALPHA-CHLOROHYDRIN
		ALPHA-MONOCHLOROHYDRIN
		BETA,BETA'-DIHYDROXYISOPROPYL CHLORIDE
		CHLORHYDRIN
		CHLORODEOXYGLYCEROL
		CHLOROHYDRIN
		CHLOROPROPANEDIOL
		EPIBLOC
		GLYCERIN ALPHA-MONOCHLORHYDRIN
		GLYCERIN EPICHLOROHYDRIN
		GLYCEROL 3-CHLOROHYDRIN
		GLYCEROL ALPHA-CHLOROHYDRIN
		GLYCEROL ALPHA-MONOCHLOROHYDRIN
		GLYCEROL CHLOROHYDRIN
		GLYCEROL-ALPHA-MONOCHLOROHYDRIN
		GLYCEROL-ALPHA-MONOCHLOROHYDRIN (DOT)
		GLYCERYL CHLORIDE
		GLYCERYL-ALPHA-CHLOROHYDRIN
		MONOCHLORHYDRIN
		MONOCHLOROHYDRIN
		U 5897
		UN 2689 (DOT)
96-32-2		
		ACETIC ACID, BROMO-, METHYL ESTER
		BROMOACETIC ACID METHYL ESTER
		METHYL 2-BROMOACETATE
		METHYL ALPHA-BROMOACETATE
		METHYL BROMOACETATE
		METHYL BROMOACETATE (DOT)
		METHYL MONOBROMOACETATE
		UN 2643 (DOT)
96-33-3		
		2-PROPENOIC ACID METHYL ESTER
		2-PROPENOIC ACID, METHYL ESTER (9CI)
		ACRYLATE DE METHYLE (FRENCH)
		ACRYLIC ACID METHYL ESTER
		ACRYLSAEUREM ETHYLESTER (GERMAN)
		CURITHANE 103
		METHOXYCARBONYLETHYLENE
	*	METHYL ACRYLATE
		METHYL ACRYLATE (ACGIH)
		METHYL ACRYLATE, INHIBITED
		METHYL ACRYLATE, INHIBITED (DOT)
		METHYL PROP-2-ENOATE
		METHYL PROPENATE
		METHYL PROPENOATE
		METHYL-2-PROPENOATE
		METHYL-ACRYLAT (GERMAN)
		METHYLACRYLAAT (DUTCH)
		METILACRILATO (ITALIAN)
		PROPENOIC ACID METHYL ESTER (9CI)
		UN 1919 (DOT)
96-34-4		
		ACETIC ACID, CHLORO-, METHYL ESTER
		METHYL ALPHA-CHLOROACETATE
		METHYL CHLOROACETATE
		METHYL CHLOROACETATE (DOT)
		METHYL MONOCHLORACETATE
		METHYL MONOCHLOROACETATE
		MONOCHLOROACETIC ACID METHYL ESTER
		UN 2295 (DOT)

CAS No.	IN	Chemical Name
96-37-7		CYCLOPENTANE, METHYL-
	*	METHYL CYCLOPENTANE
96-41-3	*	CYCLOPENTANOL
96-45-7		1,3-ETHYLENETHIOUREA
		2-IMIDAZOLIDINETHIONE
		2-IMIDAZOLINE-2-THIOL
		2-MERCAPTO-2-IMIDAZOLINE
		2-MERCAPTOIMIDAZOLINE
		2-THIOIMIDAZOLIDINE
		2-THIONOIMIDAZOLIDINE
		4,5-DIHYDRO-2-MERCAPTOIMIDAZOLE
	*	ETHYLENE THIOUREA
		ETU
		IMIDAZOLIDINETHIONE
		IMIDAZOLINE-2(3H)-THIONE
		IMIDAZOLINE-2-THIOL
		MERCAPTOIMIDAZOLINE
		MERCAZIN I
		N,N'-ETHYLENETHIOUREA
		NA 22
		NOCCELER 22
		PENNAC CRA
		RHENOGRAN ETU
		RHODANIN S 62
		SOXINOL 22
		TETRAHYDRO-2H-IMIDAZOLE-2-THIONE
		THIOUREA, N,N'-(1,2-ETHANEDIYL)-
		VULKACIT NPV/C
		WARECURE C
96-47-9		2-METHYLTETRAHYDROFURAN
		FURAN, TETRAHYDRO-2-METHYL-
		TETRAHYDRO-2-METHYLFURAN
		TETRAHYDROSYLVAN
96-49-1		1,3-DIOXOLAN-2-ONE
		CARBONIC ACID, CYCLIC ETHYLENE ESTER
		CYCLIC ETHYLENE CARBONATE
		ETHYLENE CARBONATE
		ETHYLENE GLYCOL CARBONATE
		GLYCOL CARBONATE
96-54-8		METHYLPYRROLE
96-69-5		2,2'-DI-TERT-BUTYL-5,5'-DIMETHYL-4,4'-THIODIPHENOL
		4,4'-THIOBIS(3-METHYL-6-TERT-BUTYLPHENOL)
		4,4'-THIOBIS[2-TERT-BUTYL-5-METHYLPHENOL]
		4,4'-THIOBIS[6-TERT-BUTYL-3-METHYLPHENOL]
		4,4'-THIOBIS[6-TERT-BUTYL-M-CRESOL]
		AO 4
		BIS(2-METHYL-4-HYDROXY-5-TERT-BUTYLPHENYL) SULFIDE
		BIS(2-METHYL-5-TERT-BUTYL-4-HYDROXYPHENYL) SULFIDE
		BIS(4-HYDROXY-5-TERT-BUTYL-2-METHYLPHENYL) SULFIDE
		DISPERSE MB 61
		M-CRESOL, 4,4'-THIOBIS[6-TERT-BUTYL-
		NOCRAC 300
		NONFLEX BPS
		PHENOL, 4,4'-THIOBIS[2-(1,1-DIMETHYLETHYL)-5-METHYL-
		SANTONOX
		SANTONOX BM
		SANTONOX R
		SANTOWHITE CRYSTALS
		SANTOX
		SUMILIZER WX
		SUMILIZER WX-R
		THIOALKOFEN BM
		THIOALKOFEN BM 4
		THIOALKOFEN BMCH
		THIOALKOFEN MBCH
		YOSHINOX S
		YOSHINOX SR
96-80-0		(N,N-DIISOPROPYLAMINO)ETHANOL
		2-(DIISOPROPYLAMINO)ETHYL ALCOHOL
		2-DIISOPROPYLAMINOETHANOL
		DIISOPROPYL ETHANOLAMINE
		ETHANOL, 2-(DIISOPROPYLAMINO)-
		ETHANOL, 2-[BIS(1-METHYLETHYL)AMINO]-
		N,N-DIISOPROPYL ETHANOLAMINE (DOT)
		N,N-DIISOPROPYLETHANOLAMINE
		UN 2825 (DOT)
97-00-7		1,3-DINITRO-4-CHLOROBENZENE
		1-CHLOOR-2,4-DINITROBENZEEN (DUTCH)
		1-CHLOR-2,4-DINITROBENZENE
	*	1-CHLORO-2,4-DINITROBENZENE
		1-CHLORO-2,4-DINITROBENZENE 4-CHLORO-1,3-DINITROBENZENE
		1-CHLORO-2,4-DINITROBENZOL (GERMAN)
		1-CLORO-2,4-DINITROBENZENE (ITALIAN)
		2,4-DINITRO-1-CHLOROBENZENE
		2,4-DINITROCHLOROBENZENE
		2,4-DINITROPHENYL CHLORIDE
		4-CHLORO-1,3-DINITROBENZENE
		6-CHLORO-1,3-DINITROBENZENE
		BENZENE, 1-CHLORO-2,4-DINITRO-
		BENZENE, CHLORODINITRO-
		BENZENE, CHLORODINITRO- (MIXED ISOMERS)
		CDNB
		CHLORODINITROBENZENE
		CHLORODINITROBENZENE (DOT)
		DINITROCHLOROBENZENE
		DINITROCHLOROBENZENE (DOT)
		DINITROCHLOROBENZOL
		DINITROCHLOROBENZOL (DOT)
		DNCB
		UN 1577 (DOT)
97-02-9		1-AMINO-2,4-DINITROBENZENE
		2,4-DINITRANILINE
		2,4-DINITROANILIN (GERMAN)
		2,4-DINITROANILINA (ITALIAN)
		2,4-DINITROANILINE
		ANILINE, 2,4-DINITRO-
		BENZENAMINE, 2,4-DINITRO-
		BENZENAMINE, 2,4-DINITRO- (9CI)
		DNA
		NCI-C60753
97-18-7		2,2-THIOBIS(4,6-DICHLORO-PHENOL
		PHENOL, 2,2'-THIOBIS(4,6-DICHLORO)-
		PHENOL, 2,2-THIOBIS 4,6-DICHLORO
97-36-9		1-ACETOACETYLAMINO-2,4-DIMETHYLBENZENE
		2',4'-ACETOACETOXYLIDIDE
		2',4'-DIMETHYLACETOACETANILIDE
		2,4-ACETOACETOXYLIDIDE
		ACETOACETIC ACID M-XYLIDIDE
		ACETOACETO-M-XYLIDIDE
		ACETOACETYL-M-XYLIDIDE
		BUTANAMIDE, N-(2,4-DIMETHYLPHENYL)-3-OXO-
		M-ACETOACET XYLIDIDE
		N-ACETOACETYL-2,4-XYLIDINE
97-56-3		2',3-DIMETHYL-4-AMINOAZOBENZENE
		2-METHYL-4-[(O-TOLYL)AZO]ANILINE
		4-(O-TOLYLAZO)-O-TOLUIDINE
		4-AMINO-2',3-DIMETHYLAZOBENZENE

CAS No.	IN	Chemical Name
		BENZENAMINE, 2-METHYL-4-[(2-METHYLPHENYL)AZO]-
		BRASILAZINA OIL YELLOW R
		C.I. 11160
		C.I. 11160B
		C.I. SOLVENT YELLOW 3
		FAST YELLOW AT
		FAT YELLOW B
		HIDACO OIL YELLOW
		O-AMINOAZOTOLUENE
		O-AT
		OAAT
		OIL YELLOW 21
		OIL YELLOW 2681
		OIL YELLOW 2R
		OIL YELLOW AT
		OIL YELLOW C
		OIL YELLOW I
		ORGANOL YELLOW 2T
		SOMALIA YELLOW R
		TOLUAZOTOLUIDINE
97-62-1		
		ETHYL 2-METHYLPROPANOATE
		ETHYL 2-METHYLPROPIONATE
		ETHYL ISOBUTANOATE
		ETHYL ISOBUTYRATE
		ETHYLISOBUTYRATE (DOT)
		ISOBUTYRIC ACID, ETHYL ESTER
		PROPANOIC ACID, 2-METHYL-, ETHYL ESTER
		PROPIONIC ACID, 2-METHYL-, ETHYL ESTER
		UN 2385 (DOT)
97-63-2		
		2-PROPENOIC ACID, 2-METHYL-, ETHYL ESTER
		ETHYL 2-METHYL-2-PROPENOATE
		ETHYL 2-METHYLACRYLATE
		ETHYL ALPHA-METHYL ACRYLATE
	*	ETHYL METHACRYLATE
		ETHYL METHACRYLATE, INHIBITED (DOT)
		METHACRYLIC ACID, ETHYL ESTER
		RCRA WASTE NUMBER U118
		RHOPLEX AC-33 (ROHM AND HAAS)
		UN 2277 (DOT)
97-64-3		
		ACTYLOL
		ACYTOL
		ETHYL 2-HYDROXYPROPANOATE
		ETHYL 2-HYDROXYPROPIONATE
		ETHYL ALPHA-HYDROXYPROPIONATE
		ETHYL LACTATE
		ETHYL LACTATE (DOT)
		LACTATE D'ETHYLE (FRENCH)
		LACTIC ACID, ETHYL ESTER
		PROPANOIC ACID, 2-HYDROXY-, ETHYL ESTER
		SOLACTOL
		UN 1192 (DOT)
97-72-3		
		ISOBUTYRIC ANHYDRIDE
97-77-8		
		ABSTENSIL
		ABSTINIL
		ABSTINYL
		ACCEL TET
		ALCOPHOBIN
		ANTABUS
		ANTABUSE
		ANTADIX
		ANTAETHYL
		ANTALCOL
		ANTETHAN
		ANTETIL
		ANTICOL
		ANTIETANOL
		ANTIETIL
		ANTIKOL
		ANTIVITIUM

CAS No.	IN	Chemical Name
		AVERSAN
		AVERZAN
		BIS(DIETHYLTHIOCARBAMOYL) DISULFIDE
		BIS(N,N-DIETHYLTHIOCARBAMOYL) DISULFIDE
		CONTRALIN
		CRONETAL
		DICUPRAL
		DISULFIDE, BIS(DIETHYLTHIOCARBAMOYL)
		DISULFIRAM
		DISULFURAM
		EKAGOM DTET
		EKAGOM TEDS
		EKAGOM TETDS
		ESPENAL
		ESPERAL
		ETABUS
		ETHYL THIRAM
		ETHYL THIURAD
		ETHYL TUADS
		ETHYL TUEX
		EXHORRAN
		HOCA
		KROTENAL
		N,N,N',N'-TETRAETHYLTHIURAM DISULFIDE
		NOCCELER TET
		NOXAL
		REFUSAL
		STOPETYL
		TETD
		TETRADIN
		TETRADINE
		TETRAETHYLTHIOPEROXYDICARBONIC DIAMIDE
		TETRAETHYLTHIRAM DISULFIDE
		TETRAETHYLTHIURAM DISULFIDE
		TETRAETHYLTHIURAM SULFIDE
		TETRAETIL
		TETURAM
		THIOPEROXYDICARBONIC DIAMIDE ([(H2N)C(S)]2S2), TETRAETHYL-
		THIURAM E
		THIURANIDE
		TTD
		TTS
97-85-8		
		2-METHYLPROPIONIC ACID ISOBUTYL ESTER
		2-METHYLPROPYL 2-METHYLPROPANOATE
		2-METHYLPROPYL 2-METHYLPROPIONATE
		2-METHYLPROPYL ISOBUTYRATE
		ISOBUTYL ISOBUTANOATE
	*	ISOBUTYL ISOBUTYRATE
		ISOBUTYLISOBUTYRATE (DOT)
		ISOBUTYRIC ACID, ISOBUTYL ESTER
		PROPANOIC ACID, 2-METHYL-, 2-METHYLPROPYL ESTER
		PROPANOIC ACID, 2-METHYL-, 2-METHYLPROPYL ESTER (9CI)
		UN 2528 (DOT)
97-86-9		
		2-METHYLPROPYL METHACRYLATE
		2-PROPENOIC ACID, 2-METHYL-, 2-METHYLPROPYL ESTER
		ISOBUTYL 2-METHYL-2-PROPENOATE
		ISOBUTYL ALPHA-METHACRYLATE
		ISOBUTYL ALPHA-METHYLACRYLATE
		ISOBUTYL METHACRYLATE
		ISOBUTYL METHACRYLATE, INHIBITED (DOT)
		METHACRYLIC ACID, ISOBUTYL ESTER
		UN 2283 (DOT)
97-88-1		
		2-METHYL-BUTYLACRYLAAT (DUTCH)
		2-METHYL-BUTYLACRYLAT (GERMAN)
		2-METHYL-BUTYLACRYLATE
		2-PROPENOIC ACID, 2-METHYL-, BUTYL ESTER
		BUTIL METACRILATO (ITALIAN)
		BUTYL 2-METHACRYLATE
		BUTYL 2-METHYL-2-PROPENOATE
		BUTYL METHACRYLATE
		BUTYLMETHACRYLAAT (DUTCH)

CAS No.	IN	Chemical Name
		METHACRYLATE DE BUTYLE (FRENCH)
		METHACRYLIC ACID, BUTYL ESTER
		METHACRYLSAEUREBUTYLESTER (GERMAN)
		N-BUTYL METHACRYLATE
		N-BUTYL METHACRYLATE (DOT)
		UN 2227 (DOT)
97-93-8		
		ALUMINUM, TRIETHYL-
		TEA
		TRIETHYLALANE
		TRIETHYLALUMINIUM
		TRIETHYLALUMINIUM (DOT)
		TRIETHYLALUMINUM
		UN 1102 (DOT)
97-94-9		
		BORANE, TRIETHYL-
		TRIETHYLBORANE
		TRIETHYLBORON
97-95-0		
		1-BUTANOL, 2-ETHYL-
		2-ETHYL-1-BUTANOL
		2-ETHYLBUTANOL
		2-ETHYLBUTYL ALCOHOL
		3-METHYLOLPENTANE
		ETHER ETHYLBUTYLIQUE (FRENCH)
		ETHYL BUTYL ETHER (DOT)
		ETHYL N-BUTYL ETHER
		ETHYLBUTANOL
		PSEUDOHEXYL ALCOHOL
		UN 1179 (DOT)
97-96-1		
		2-ETHYLBUTANAL
		2-ETHYLBUTYRALDEHYDE
		2-ETHYLBUTYRALDEHYDE (DOT)
		2-ETHYLBUTYRIC ALDEHYDE
		3-FORMYLPENTANE
		ALDEHYDE 2-ETHYLBUTYRIQUE (FRENCH)
		ALPHA-ETHYLBUTYRALDEHYDE
		BUTANAL, 2-ETHYL-
		BUTYRALDEHYDE, 2-ETHYL-
		DIETHYL ACETALDEHYDE
	*	ETHYL BUTYRALDEHYDE
		ETHYL BUTYRALDEHYDE (DOT)
		UN 1178 (DOT)
97-97-2		
		1,1-DIMETHOXY-2-CHLOROETHANE
		1-CHLORO-2,2-DIMETHOXYETHANE
		2-CHLORO-1,1-DIMETHOXYETHANE
		2-CHLOROACETALDEHYDE DIMETHYL ACETAL
		ACETALDEHYDE, CHLORO-, DIMETHYL ACETAL
		CHLOROACETALDEHYDE DIMETHYL ACETAL
		DIMETHYL CHLORACETAL
		ETHANE, 2-CHLORO-1,1-DIMETHOXY-
97-99-4		
		2-FURANMETHANOL, TETRAHYDRO-
		FURFURYL ALCOHOL, TETRAHYDRO-
		QO THFA
		TETRAHYDRO-2-FURANCARBINOL
		TETRAHYDRO-2-FURANMETHANOL
		TETRAHYDRO-2-FURANYLMETHANOL
		TETRAHYDRO-2-FURYLMETHANOL
		TETRAHYDROFURFURYL ALCOHOL
		TETRAHYDROFURYLALKOHOL (CZECH)
		THFA
98-00-0		
		2-FURANCARBINOL
		2-FURANMETHANOL
		2-FURANYLMETHANOL
		2-FURFUROL
		2-FURFURYL ALCOHOL
		2-FURFURYLALKOHOL (CZECH)
		2-FURYLCARBINOL
		2-FURYLMETHANOL
		2-HYDROXYMETHYLFURAN
		5-HYDROXYMETHYLFURAN
		ALPHA-FURFURYL ALCOHOL
		ALPHA-FURYLCARBINOL
		CHEM-REZ 200
		FURFURAL ALCOHOL
		FURFURALCOHOL
	*	FURFURYL ALCOHOL
		FURFURYL ALCOHOL (ACGIH,DOT)
		FURYL ALCOHOL
		FURYLCARBINOL
		METHANOL, (2-FURYL)-
		NCI-C56224
		UN 2874 (DOT)
98-01-1		
		2-FORMYLFURAN
		2-FURALDEHYDE
		2-FURANALDEHYDE
		2-FURANCARBONAL
		2-FURANCARBOXALDEHYDE
		2-FURFURAL
		2-FURFURALDEHYDE
		2-FURIL-METANALE (ITALIAN)
		2-FURYL-METHANAL
		2-FURYLALDEHYDE
		2-FURYLCARBOXALDEHYDE
		ALPHA-FUROLE
		ARTIFICIAL ANT OIL
		ARTIFICIAL OIL OF ANTS
		FURAL
		FURALDEHYDE
		FURALE
		FURANCARBONAL
	*	FURFURAL
		FURFURAL (ACGIH,DOT)
		FURFURALDEHYDE
		FURFURALE (ITALIAN)
		FURFURALU (POLISH)
		FURFUROL
		FURFUROLE
		FURFURYLALDEHYDE
		FUROLE
		NCI-C56177
		PYROMUCIC ALDEHYDE
		RCRA WASTE NUMBER U125
		UN 1199 (DOT)
98-05-5		
		ARSONIC ACID, PHENYL-
		BENZENEARSONIC ACID
		PHENYL ARSENIC ACID
		PHENYLARSONIC ACID
98-06-6		
		2-METHYL-2-PHENYLPROPANE
		BENZENE, (1,1-DIMETHYLETHYL)-
		BENZENE, TERT-BUTYL-
		DIMETHYLETHYLBENZENE
		PHENYLTRIMETHYLMETHANE
		T-BUTYLBENZENE
		TERT-BUTYLBENZENE
		TRIMETHYLPHENYLMETHANE
98-07-7		
		(TRICHLOROMETHYL)BENZENE
		1-(TRICHLOROMETHYL)BENZENE
		ALPHA,ALPHA,ALPHA-TRICHLOROTOLUENE
		BENZENE, (TRICHLOROMETHYL)-
		BENZENYL CHLORIDE
		BENZENYL TRICHLORIDE
		BENZOIC TRICHLORIDE
		BENZOTRICHLORIDE
		BENZOTRICHLORIDE (DOT)
		BENZYL TRICHLORIDE
		BENZYLIDYNE CHLORIDE
		CHLORURE DE BENZENYLE (FRENCH)
		OMEGA,OMEGA,OMEGA-TRICHLOROTOLUENE

CAS No.	IN	Chemical Name
		PHENYL CHLOROFORM
		PHENYLTRICHLOROMETHANE
		RCRA WASTE NUMBER U023
		TOLUENE TRICHLORIDE
		TOLUENE, ALPHA,ALPHA,ALPHA-TRICHLORO-
		TRICHLOORMETHYLBENZEEN (DUTCH)
		TRICHLORMETHYLBENZOL (GERMAN)
		TRICHLOROMETHYLBENZENE
		TRICHLOROPHENYLMETHANE
		TRICLOROMETILBENZENE (ITALIAN)
		TRICLOROTOLUENE (ITALIAN)
		UN 2226 (DOT)
98-08-8		
		(TRIFLUOROMETHYL)BENZENE
		ALPHA,ALPHA,ALPHA-TRIFLUOROTOLUENE
		BENZENE, (TRIFLUOROMETHYL)-
		BENZENYL FLUORIDE
		BENZOTRIFLUORIDE
		BENZOTRIFLUORIDE (DOT)
		BENZYLIDYNE FLUORIDE
		OMEGA-TRIFLUOROTOLUENE
		PHENYLFLUOROFORM
		TOLUENE, ALPHA,ALPHA,ALPHA-TRIFLUORO-
		UN 2338 (DOT)
		USAF MA-16
98-09-9		
		BENZENE SULFOCHLORIDE
	*	BENZENE SULFONYL CHLORIDE
		BENZENESULFONIC CHLORIDE
		PHENYLSULFONYL CHLORIDE
98-11-3		
		PHENOLSULPHONIC ACID
98-12-4		
		CYCLOHEXYL TRICHLOROSILANE
		SILANE, TRICHLOROCYCLOHEXYL-
		TRICHLOROCYCLOHEXYLSILANE
98-13-5		
		PHENYL TRICHLOROSILANE
		PHENYL TRICHLOROSILANE (DOT)
		PHENYLSILICON TRICHLORIDE
		SILANE, PHENYLTRICHLORO-
		SILANE, TRICHLOROPHENYL-
		SILICON PHENYL TRICHLORIDE
		TRICHLOROPHENYL ETHER
		TRICHLOROPHENYLSILANE
		UN 1804 (DOT)
98-15-7		
		1-CHLORO-3-(TRIFLUOROMETHYL)BENZENE
		3-CHLORO(TRIFLUOROMETHYL)BENZENE
		3-CHLORO-ALPHA,ALPHA,ALPHA-TRIFLUOROTOLUENE
		3-CHLOROBENZOTRIFLUORIDE
		BENZENE, 1-CHLORO-3-(TRIFLUOROMETHYL)-
		BENZENE, 1-CHLORO-3-(TRIFLUOROMETHYL)- (9CI)
		CHLOROBENZOTRIFLUORIDE
		M-CHLORO-ALPHA,ALPHA,ALPHA-TRIFLUOROTOLUENE
		M-CHLOROBENZOTRIFLUORIDE
		M-CHLOROBENZOTRIFLUORIDE (DOT)
		M-TRIFLUOROMETHYLPHENYL CHLORIDE
		TOLUENE, M-CHLORO-ALPHA,ALPHA,ALPHA-TRIFLUORO-
		UN 2234 (DOT)
98-16-8		
		1-AMINO-3-(TRIFLUOROMETHYL)BENZENE
		3-(TRIFLUOROMETHYL)BENZENAMINE
		3-AMINOBENZOTRIFLUORIDE
		3-TRIFLUOROMETHYL ANILINE
		3-TRIFLUOROMETHYL ANILINE (DOT)
		ALPHA,ALPHA,ALPHA-TRIFLUORO-M-TOLUIDINE
		BENZENAMINE, 3-(TRIFLUOROMETHYL)-
		M-(TRIFLUOROMETHYL)ANILINE
		M-AMINO-ALPHA,ALPHA,ALPHA-TRIFLUOROTOLUENE
		M-AMINOBENZAL FLUORIDE
		M-AMINOBENZOTRIFLUORIDE

CAS No.	IN	Chemical Name
		M-TOLUIDINE, ALPHA,ALPHA,ALPHA-TRIFLUORO-
		TOLUENE, 3-AMINO-ALPHA,ALPHA,ALPHA-TRIFLUORO-
		UN 2948 (DOT)
		USAF MA-4
98-27-1		
		2-METHYL-4-TERT-BUTYLPHENOL
		4-TERT-BUTYL-2-METHYLPHENOL
		4-TERT-BUTYL-O-CRESOL
		O-CRESOL, 4-TERT-BUTYL-
		P-TERT-BUTYL-O-CRESOL
		PHENOL, 4-(1,1-DIMETHYLETHYL)-2-METHYL-
98-28-2		
		2-CHLORO-4-TERT-BUTYLPHENOL
		4-TERT-BUTYL-2-CHLOROPHENOL
		PHENOL, 2-CHLORO-4-(1,1-DIMETHYLETHYL)-
		PHENOL, 4-TERT-BUTYL-2-CHLORO-
98-29-3		
		1,2-BENZENEDIOL, 4-(1,1-DIMETHYLETHYL)-
		1,2-DIHYDROXY-4-TERT-BUTYLBENZENE
		4-(1,1-DIMETHYLETHYL)BENZENE-1,2-DIOL
		4-(1,1-DIMETHYLETHYL)CATECHOL
		4-T-BUTYLCATECHOL
		4-TBC
		4-TERT-BUTYL CATECHOL
		4-TERT-BUTYL-1,2-BENZENEDIOL
		4-TERT-BUTYL-1,2-DIHYDROXYBENZENE
		4-TERT-BUTYLCATECHIN
		4-TERT-BUTYLCATECHOL
		4-TERT-BUTYLPYROCATECHOL
		P-TERT-BUTYLCATECHOL
		P-TERT-BUTYLPYROCATECHOL
		PYROCATECHOL, 4-TERT-BUTYL-
98-46-4		
		3-NITROBENZOTRIFLUORIDE
98-51-1		
		1-METHYL-4-TERT-BUTYLBENZENE
		1-TERT-BUTYL-4-METHYLBENZENE
		4-METHYL-TERT-BUTYLBENZENE
		4-TERT-BUTYL-1-METHYLBENZENE
		4-TERT-BUTYLTOLUENE
		BENZENE, 1-(1,1-DIMETHYLETHYL)-4-METHYL-
		P-TERT-BUTYLTOLUENE
		TOLUENE, P-TERT-BUTYL-
98-56-6		
		(P-CHLOROPHENYL)TRIFLUOROMETHANE
		1-CHLORO-4-(TRIFLUOROMETHYL)BENZENE
		4-(TRIFLUOROMETHYL)CHLOROBENZENE
		4-CHLORO-ALPHA,ALPHA,ALPHA-TRIFLUOROTOLUENE
		4-CHLOROBENZOTRIFLUORIDE
		ALPHA,ALPHA,ALPHA-TRIFLUORO-4-CHLOROTOLUENE
		BENZENE, 1-CHLORO-4-(TRIFLUOROMETHYL)-
		BENZENE, 1-CHLORO-4-(TRIMETHYL)- (9CI)
		P-(TRIFLUOROMETHYL)CHLOROBENZENE
		P-CHLORO-ALPHA,ALPHA,ALPHA-TRIFLUOROTOLUENE
		P-CHLOROBENZOTRIFLUORIDE
		P-CHLOROBENZOTRIFLUORIDE (DOT)
		P-CHLOROTRIFLUOROMETHYLBENZENE
		P-TRIFLUOROMETHYLPHENYL CHLORIDE
		TOLUENE, P-CHLORO-ALPHA,ALPHA,ALPHA-TRIFLUORO-
		UN 2234 (DOT)
98-82-8		
		(1-METHYLETHYL)BENZENE
		2-FENILPROPANO (ITALIAN)
		2-FENYL-PROPAAN (DUTCH)
		2-PHENYLPROPANE
		BENZENE, (1-METHYLETHYL)-
		BENZENE, (1-METHYLETHYL)- (9CI)
		BENZENE, ISOPROPYL
		CUMEEN (DUTCH)
	*	CUMENE
		CUMENE (ACGIH)
		CUMOL

CAS No.	IN	Chemical Name
		ISOPROPILBENZENE (ITALIAN)
		ISOPROPYL-BENZOL (GERMAN)
		ISOPROPYLBENZEEN (DUTCH)
		ISOPROPYLBENZENE
		ISOPROPYLBENZENE (DOT)
		ISOPROPYLBENZOL
		PROPANE, 2-PHENYL
		RCRA WASTE NUMBER U055
		UN 1918 (DOT)
98-83-9		
		(1-METHYLETHENYL)BENZENE
		1-METHYL-1-PHENYLETHYLENE
		1-PHENYL-1-METHYLETHYLENE
		1-PROPENE, 2-PHENYL-
		2-PHENYL-1-PROPENE
		2-PHENYL-2-PROPENE
		2-PHENYLPROPENE
		2-PHENYLPROPYLENE
		ALPHA-METHYL STYRENE
		ALPHA-METHYL STYRENE (ACGIH)
		ALPHA-METHYL-STYROL (GERMAN)
		ALPHA-METHYLSTYREEN (DUTCH)
		ALPHA-METHYLSTYROL
		ALPHA-METIL-STIROLO (ITALIAN)
		AS-METHYLPHENYLETHYLENE
		BENZENE, (1-METHYLETHENYL)-
		BENZENE, ISOPROPENYL-
		BETA-PHENYLPROPENE
		BETA-PHENYLPROPYLENE
		ISOPROPENIL-BENZOLO (ITALIAN)
		ISOPROPENYL-BENZEEN (DUTCH)
		ISOPROPENYL-BENZOL (GERMAN)
		ISOPROPENYLBENZENE
		ISOPROPENYLBENZENE (DOT)
		METHYL STYRENE
		STYRENE, ALPHA-METHYL-
		UN 2303 (DOT)
98-84-0		
		(1-AMINOETHYL)BENZENE
		1-AMINO-1-PHENYLETHANE
		1-PHENYL-1-ETHANAMINE
		1-PHENYLETHANAMINE
		1-PHENYLETHYLAMINE
		ALPHA-METHYLBENZYLAMINE
		ALPHA-PHENETHYLAMINE
		ALPHA-PHENYLETHYLAMINE
		BENZENEMETHANAMINE, ALPHA-METHYL-
		BENZYLAMINE, ALPHA-METHYL-
		ETHANAMINE, 1-PHENYL-
98-85-1		
		(1-HYDROXYETHYL)BENZENE
		1-PHENETHYL ALCOHOL
		1-PHENYL-1-HYDROXYETHANE
		1-PHENYLETHANOL
		1-PHENYLETHYL ALCOHOL
		ALPHA-HYDROXYETHYLBENZENE
		ALPHA-METHYLBENZENEMETHANOL
		ALPHA-METHYLBENZYL ALCOHOL
		ALPHA-METHYLBENZYL ALCOHOL (DOT)
		ALPHA-PHENETHYL ALCOHOL
		ALPHA-PHENYLETHANOL
		ALPHA-PHENYLETHYL ALCOHOL
		BENZENEMETHANOL, ALPHA-METHYL-
		BENZYL ALCOHOL, ALPHA-METHYL-
		ETHANOL, 1-PHENYL-
		METHANOL, METHYLPHENYL-
		METHYLBENZYL ALCOHOL (ALPHA)
		METHYLPHENYLCARBINOL
		NCI-C55685
		PHENYLMETHYLCARBINOL
		SEC-PHENETHYL ALCOHOL
		STYRALLYL ALCOHOL
		STYRALYL ALCOHOL
		UN 2937 (DOT)

CAS No.	IN	Chemical Name
98-86-2		
		1-PHENYL-1-ETHANONE
		1-PHENYLETHANONE
		ACETOPHENON
	*	ACETOPHENONE
		ACETYLBENZENE
		BENZENE, ACETYL-
		BENZOYL METHIDE
		DYMEX
		ETHANONE, 1-PHENYL-
		ETHANONE, 1-PHENYL- (9CI)
		HYPNON
		HYPNONE
		KETONE, METHYL PHENYL
		METHYL PHENYL KETONE
		PHENYL METHYL KETONE
		RCRA WASTE NUMBER U004
		USAF EK-496
98-87-3		
		(DICHLOROMETHYL)BENZENE
		ALPHA,ALPHA-DICHLOROTOLUENE
	*	BENZAL CHLORIDE
		BENZENE, (DICHLOROMETHYL)-
		BENZYL DICHLORIDE
		BENZYLENE CHLORIDE
		BENZYLIDENE CHLORIDE
		BENZYLIDENE CHLORIDE (DOT)
		CHLOROBENZAL
		CHLORURE DE BENZYLIDENE (FRENCH)
		DICHLOROPHENYLMETHANE
		RCRA WASTE NUMBER U017
		TOLUENE, ALPHA,ALPHA-DICHLORO-
		UN 1886 (DOT)
98-88-4		
		ALPHA-CHLOROBENZALDEHYDE
		BENZALDEHYDE, ALPHA-CHLORO-
		BENZENECARBONYL CHLORIDE
		BENZOIC ACID, CHLORIDE
	*	BENZOYL CHLORIDE
		BENZOYL CHLORIDE (DOT)
		UN 1736 (DOT)
98-94-2		
		DIMETHYLCYCLOHEXYL AMINE
98-95-3		
		BENZENE, NITRO-
		ESSENCE OF MIRBANE
		ESSENCE OF MYRBANE
		MIRBANE OIL
		NCI-C60082
		NITRO BENZOL
		NITROBENZEEN (DUTCH)
		NITROBENZEN (POLISH)
	*	NITROBENZENE
		NITROBENZENE (ACGIH)
		NITROBENZENE, LIQUID
		NITROBENZENE, LIQUID (DOT)
		NITROBENZOL
		NITROBENZOL (DOT)
		NITROBENZOL, LIQUID (DOT)
		OIL OF MIRBANE
		OIL OF MIRBANE (DOT)
		OIL OF MYRBANE
		RCRA WASTE NUMBER U169
		UN 1662 (DOT)
99-08-1		
		1-METHYL-3-NITROBENZENE
		3-METHYLNITROBENZENE
		3-NITROTOLUENE
		3-NITROTOLUOL
		BENZENE, 1-METHYL-3-NITRO-
		M-METHYLNITROBENZENE
	*	M-NITROTOLUENE
		M-NITROTOLUENE (ACGIH,DOT)
		MNT

CAS No.	IN	Chemical Name
		N-NITROTOLUENE
		NITROTOLUENE
		TOLUENE, M-NITRO-
		UN 1664 (DOT)
99-35-4		
		1,3,5-TRINITROBENZENE
		BENZENE, 1,3,5-TRINITRO-
		BENZENE, TRINITRO- (10% TO 30% WATER)
		RCRA WASTE NUMBER U234
		S-TRINITROBENZENE
		SYM-TRINITROBENZENE
		SYMMETRIC TRINITROBENZENE
		TNB
		TRINITROBENZENE
		TRINITROBENZENE, WET, AT LEAST 10% WATER, OVER 16 OZS IN ONE OUTSIDE PACKAGING (DOT)
		TRINITROBENZENE, WET, CONTAINING AT LEAST 10% WATER (DOT)
		TRINITROBENZENE, WET, CONTAINING LESS THAN 30% WATER (DOT)
		TRINITROBENZENE, WETTED WITH NOT LESS THAN 30% WATER (DOT)
		UN 0214 (DOT)
		UN 1354 (DOT)
99-51-4		
		M-NITROXYLENE
99-53-4		
		TRINITROBENZENE
99-55-8		
		2-AMINO-4-NITROTOLUENE
		2-METHYL-5-NITROANILINE
		4-NITRO-2-AMINOTOLUENE
		5-NITRO-2-METHYLANILINE
		5-NITRO-2-TOLUIDINE
	*	5-NITRO-O-TOLUIDINE
		6-METHYL-3-NITROANILINE
		AMARTHOL FAST SCARLET G BASE
		AMARTHOL FAST SCARLET G SALT
		AZOENE FAST SCARLET GC BASE
		AZOENE FAST SCARLET GC SALT
		AZOFIX SCARLET G SALT
		AZOGENE FAST SCARLET G
		BENZENAMINE, 2-METHYL-5-NITRO-
		C.I. 37105
		C.I. AZOIC DIAZO COMPONENT 12
		DAINICHI FAST SCARLET G BASE
		DAITO SCARLET BASE G
		DEVOL SCARLET B
		DEVOL SCARLET G SALT
		DIABASE SCARLET G
		DIAZO FAST SCARLET G
		FAST RED SG BASE
		FAST SCARLET BASE J
		FAST SCARLET G
		FAST SCARLET G BASE
		FAST SCARLET G SALT
		FAST SCARLET GC BASE
		FAST SCARLET J SALT
		FAST SCARLET M 4NT BASE
		FAST SCARLET T BASE
		HILTONIL FAST SCARLET G BASE
		HILTONIL FAST SCARLET G SALT
		HILTONIL FAST SCARLET GC BASE
		KAYAKU SCARLET G BASE
		LAKE SCARLET G BASE
		LITHOSOL ORANGE R BASE
		MITSUI SCARLET G BASE
		NAPHTHANIL SCARLET G BASE
		NAPHTOELAN FAST SCARLET G BASE
		NAPHTOELAN FAST SCARLET G SALT
		O-TOLUIDINE, 5-NITRO-
		PNOT
		SCARLET BASE CIBA II
		SCARLET BASE IRGA II
		SCARLET BASE NSP
		SCARLET G BASE
		SUGAI FAST SCARLET G BASE
		SYMULON SCARLET G BASE
99-59-2		
		1-AMINO-2-METHOXY-5-NITROBENZENE
		1-METHOXY-2-AMINO-4-NITROBENZENE
		2-AMINO-1-METHOXY-4-NITROBENZENE
		2-AMINO-4-NITROANISOLE
		2-METHOXY-5-NITROANILINE
		2-METHOXY-5-NITROBENZENAMINE
		3-NITRO-6-METHOXYANILINE
		5-NITRO-2-METHOXYANILINE
		5-NITRO-O-ANISIDINE
		AZOAMINE SCARLET K
		BENZENAMINE, 2-METHOXY-5-NITRO-
		O-ANISIDINE, 5-NITRO-
99-65-0		
		1,3-DINITROBENZENE
		1,3-DINITROBENZOL
		BENZENE, 1,3-DINITRO-
		BENZENE, M-DINITRO-
		BINITROBENZENE 2,4-DINITROBENZENE
		DWUNITROBENZEN (POLISH)
		M-DINITROBENZENE
		M-DINITROBENZENE (ACGIH,DOT)
		UN 1597 (DOT)
99-87-6		
		1-ISOPROPYL-4-METHYLBENZENE
		1-METHYL-4-(1-METHYLETHYL)BENZENE
		1-METHYL-4-ISOPROPYLBENZENE
		2-P-TOLYLPROPANE
		4-ISOPROPYL-1-METHYLBENZENE
		4-ISOPROPYLTOLUENE
		4-METHYLISOPROPYLBENZENE
		BENZENE, 1-METHYL-4-(1-METHYLETHYL)-
		CAMPHOGEN
		DOLCYMENE
		P-CYMENE
		P-CYMOL
		P-ISOPROPYLTOLUENE
		P-METHYLCUMENE
		P-METHYLISOPROPYLBENZENE
99-98-9		
		1,4-BENZENEDIAMINE, N,N-DIMETHYL-
		1,4-BENZENEDIAMINE, N,N-DIMETHYL- (9CI)
		1-AMINO-4-(DIMETHYLAMINO)BENZENE
		4-(DIMETHYLAMINO)ANILINE
		4-(DIMETHYLAMINO)BENZENAMINE
		4-AMINO-N,N-DIMETHYLANILINE
		CI 76075
	*	DIMETHYL-P-PHENYLENEDIAMINE
		DMPD
		N,N-DIMETHYL-1,4-BENZENEDIAMINE
		N,N-DIMETHYL-1,4-DIAMINOBENZENE
		N,N-DIMETHYL-1,4-PHENYLENEDIAMINE
		N,N-DIMETHYL-P-BENZENEDIAMINE
		N,N-DIMETHYL-P-PHENYLENEDIAMINE
		P-(DIMETHYLAMINO)ANILINE
		P-(DIMETHYLAMINO)PHENYLAMINE
		P-AMINO-N,N-DIMETHYLANILINE
		P-AMINODIMETHYLANILINE
		P-DIMETHYLAMINOPHENYLAMINE
		P-PHENYLENEDIAMINE, N,N'-DIMETHYL-
		P-PHENYLENEDIAMINE, N,N-DIMETHYL-
99-99-0		
		1-METHYL-4-NITROBENZENE
		4-METHYLNITROBENZENE
		4-NITROTOLUENE
		4-NITROTOLUOL
		BENZENE, 1-METHYL-4-NITRO-
		NCI-C60537
		P-METHYLNITROBENZENE
	*	P-NITROTOLUENE
		P-NITROTOLUENE (ACGIH,DOT)

CAS No.	IN	Chemical Name
		PNT
		TOLUENE, P-NITRO-
		UN 1664 (DOT)
100-00-5		
		1-CHLOOR-4-NITROBENZEEN (DUTCH)
		1-CHLOR-4-NITROBENZOL (GERMAN)
	*	1-CHLORO-4-NITROBENZENE
		1-CLORO-4-NITROBENZENE (ITALIAN)
		1-NITRO-4-CHLOROBENZENE
		4-CHLORO-1-NITROBENZENE
		4-CHLORONITROBENZENE
		4-NITRO-1-CHLOROBENZENE
		4-NITROCHLOROBENZENE
		BENZENE, 1-CHLORO-4-NITRO-
		NITROCHLOROBENZENE
		NITROCHLOROBENZENE, PARA-, SOLID (DOT)
		P-CHLORONITROBENZENE
		P-CHLORONITROBENZENE (DOT)
		P-NITROCHLOORBENZEEN (DUTCH)
		P-NITROCHLOROBENZENE
		P-NITROCHLOROBENZENE (ACGIH)
		P-NITROCHLOROBENZOL (GERMAN)
		P-NITROCLOROBENZENE (ITALIAN)
		P-NITROPHENYL CHLORIDE
		PNCB
		UN 1578 (DOT)
100-01-6		
		1-AMINO-4-NITROBENZENE
		4-AMINONITROBENZENE
		4-NITRANILINE
		4-NITROANILINE
		4-NITROBENZENAMINE
		ANILINE, 4-NITRO-
		ANILINE, P-NITRO-
		AZOAMINE RED ZH
		AZOFIX RED GG SALT
		AZOIC DIAZO COMPONENT 37
		BENZENAMINE, 4-NITRO-
		BENZENAMINE, 4-NITRO- (9CI)
		C.I. 37035
		C.I. AZOIC DIAZO COMPONENT 37
		C.I. DEVELOPER 17
		DEVELOPER P
		DEVOL RED GG
		DIAZO FAST RED GG
		FAST RED 2G BASE
		FAST RED 2G SALT
		FAST RED BASE 2J
		FAST RED BASE GG
		FAST RED GG BASE
		FAST RED GG SALT
		FAST RED MP BASE
		FAST RED P BASE
		FAST RED P SALT
		FAST RED SALT 2J
		FAST RED SALT GG
		NAPHTOELAN RED GG BASE
		NCI-C60786
		NITRAZOL CF EXTRA
		NITROANILINE
		P-AMINONITROBENZENE
		P-NITRANILINE
		P-NITROANILINA (POLISH)
	*	P-NITROANILINE
		P-NITROANILINE (ACGIH,DOT)
		P-NITROANILINE, SOLID
		P-NITROPHENYLAMINE
		PARANITROANILINE, SOLID (DOT)
		PNA
		RCRA WASTE NUMBER P077
		RED 2G BASE
		SHINNIPPON FAST RED GG BASE
		UN 1661 (DOT)
100-02-7		
		1-HYDROXY-4-NITROBENZENE
		4-HYDROXYNITROBENZENE
		4-NITROFENOL (DUTCH)
	*	4-NITROPHENOL
		NCI-C55992
		NIPHEN
		P-HYDROXYNITROBENZENE
		P-NITROPHENOL
		P-NITROPHENOL (DOT)
		PARANITROFENOL (DUTCH)
		PARANITROFENOLO (ITALIAN)
		PARANITROPHENOL (FRENCH,GERMAN)
		PHENOL, 4-NITRO-
		PHENOL, P-NITRO-
		RCRA WASTE NUMBER U170
		UN 1663 (DOT)
100-07-2		
		4-ANISOYL CHLORIDE
		4-METHOXYBENZOIC ACID CHLORIDE
		4-METHOXYBENZOYL CHLORIDE
		ANISOYL CHLORIDE
		ANISOYL CHLORIDE (DOT)
		BENZOYL CHLORIDE, 4-METHOXY-
		BENZOYL CHLORIDE, METHOXY-
		BENZOYL CHLORIDE, METHOXY- (9CI)
		METHOXYBENZOYL CHLORIDE
		P-ANISOYL CHLORIDE
		P-METHOXYBENZOIC ACID CHLORIDE
		P-METHOXYBENZOYL CHLORIDE
		UN 1729 (DOT)
100-14-1		
		1-(CHLOROMETHYL)-4-NITROBENZENE
		4-(CHLOROMETHYL)NITROBENZENE
		4-NITROBENZYLCHLORIDE
		ALPHA-CHLORO-4-NITROTOLUENE
		ALPHA-CHLORO-P-NITROTOLUENE
		BENZENE, 1-(CHLOROMETHYL)-4-NITRO-
		P-(CHLOROMETHYL)NITROBENZENE
		P-NITROBENZYL CHLORIDE
		TOLUENE, ALPHA-CHLORO-P-NITRO-
100-17-4		
		1-METHOXY-4-NITROBENZENE
		4-METHOXY-1-NITROBENZENE
		4-METHOXYNITROBENZENE
		4-NITROANISOLE
		4-NITROPHENYL METHYL ETHER
		ANISOLE, P-NITRO-
		BENZENE, 1-METHOXY-4-NITRO-
		BENZENE, 1-METHOXY-4-NITRO- (9CI)
		METHYL P-NITROPHENYL ETHER
		P-METHOXYNITROBENZENE
		P-NITROANISOL
		P-NITROANISOLE
		P-NITROANISOLE (DOT)
		P-NITROMETHOXYBENZENE
		UN 2730 (DOT)
100-20-9		
		1,4-BENZENEDICARBONYL CHLORIDE
		1,4-BENZENEDICARBONYL DICHLORIDE
		P-PHENYLENEDICARBONYL DICHLORIDE
		P-PHTHALOYL CHLORIDE
		P-PHTHALOYL DICHLORIDE
		TEREPHTHALIC ACID CHLORIDE
		TEREPHTHALIC ACID DICHLORIDE
		TEREPHTHALIC DICHLORIDE
		TEREPHTHALOYL CHLORIDE
		TEREPHTHALOYL DICHLORIDE
100-21-0		
		TEREPHTHALIC ACID
100-25-4		
		1,4-DINITROBENZENE
		BENZENE, 1,4-DINITRO-
		BENZENE, P-DINITRO-
		DITHANE A-4

CAS No.	IN	Chemical Name
		P-DINITROBENZENE
		P-DINITROBENZENE (ACGIH,DOT)
		UN 1597 (DOT)
100-36-7		
		1,2-ETHANEDIAMINE, N,N-DIETHYL-
		1-AMINO-2-DIETHYLAMINOETHANE
		2-(DIETHYLAMINO)ETHYLAMINE
		2-(N,N-DIETHYLAMINO)ETHYLAMINE
		BETA-DIETHYLAMINOETHYLAMINE
		DIETHYLETHYLENE DIAMINE
		ETHYLENEDIAMINE, N,N-DIETHYL-
		N,N-DIETHYL-1,2-DIAMINOETHANE
		N,N-DIETHYL-1,2-ETHANEDIAMINE
		N,N-DIETHYLETHYLENEDIAMINE
		N-(2-AMINOETHYL)-N,N-DIETHYLAMINE
		N-(2-DIETHYLAMINOETHYL)AMINE
100-37-8		
		(2-HYDROXYETHYL)DIETHYLAMINE
		(DIETHYLAMINO)ETHANOL
		2-(DIETHYLAMINO)ETHANOL
		2-(DIETHYLAMINO)ETHYL ALCOHOL
		2-(N,N-DIETHYLAMINO)ETHANOL
		2-DIETHYLAMINOETHANOL (ACGIH)
		2-HYDROXYTRIETHYLAMINE
		2-N-DIETHYLAMINOETHANOL
		BETA-DIETHYLAMINOETHANOL
		BETA-DIETHYLAMINOETHYL ALCOHOL
		DEAE
		DIAETHYLAMINOAETHANOL (GERMAN)
		DIETHYL(2-HYDROXYETHYL)AMINE
	*	DIETHYLAMINOETHANOL
		DIETHYLAMINOETHANOL (DOT)
		DIETHYLETHANOLAMINE
		DIETHYLMONOETHANOLAMINE
		ETHANOL, 2-(DIETHYLAMINO)-
		MKS
		N,N-DIETHYL-2-AMINOETHANOL
		N,N-DIETHYL-2-HYDROXYETHYLAMINE
		N,N-DIETHYL-N-(BETA-HYDROXYETHYL)AMINE
		N,N-DIETHYLETHANOLAMINE
		N,N-DIETHYLMONOETHANOLAMINE
		N-(2-HYDROXYETHYL)DIETHYLAMINE
		N-DIETHYLAMINOETHANOL
		PENNAD 150
		UN 2686 (DOT)
100-39-0		
		(BROMOMETHYL)BENZENE
		BENZENE, (BROMOMETHYL)-
	*	BENZYL BROMIDE
		BROMOPHENYLMETHANE
100-40-3		
		1-VINYL-3-CYCLOHEXENE
		4-VINYL-1-CYCLOHEXENE
		4-VINYLCYCLOHEXENE
		CYCLOHEXENE, 4-ETHENYL-
		CYCLOHEXENE, 4-VINYL-
100-41-4		
		AETHYLBENZOL (GERMAN)
		ALPHA-METHYLTOLUENE
		BENZENE, ETHYL-
		EB
	*	ETHYL BENZENE
		ETHYL BENZENE (ACGIH,DOT)
		ETHYLBENZEEN (DUTCH)
		ETHYLBENZOL
		ETILBENZENE (ITALIAN)
		ETYLOBENZEN (POLISH)
		NCI-C56393
		PHENYLETHANE
		UN 1175 (DOT)
100-42-5		
		BENZENE, ETHENYL-
		BENZENE, VINYL-
		CINNAMENE
		CINNAMENOL
		CINNAMOL
		DIAREX HF 77
		ETHENYLBENZENE
		ETHYLENE, PHENYL-
		NCI-C02200
		PHENETHYLENE
		PHENYL ETHYLENE
		PHENYLETHENE
		STIROLO (ITALIAN)
		STYREEN (DUTCH)
		STYREN (CZECH)
		STYRENE
	*	STYRENE MONOMER
		STYRENE MONOMER, INHIBITED
		STYRENE MONOMER, INHIBITED (DOT)
		STYRENE, MONOMER (ACGIH)
		STYROL
		STYROL (GERMAN)
		STYROLE
		STYROLENE
		STYRON
		STYROPOL
		STYROPOL SO
		STYROPOR
		UN 2055 (DOT)
		VINYLBENZEN (CZECH)
		VINYLBENZENE
		VINYLBENZOL
100-44-7		
		ALPHA-CHLOROTOLUENE
		ALPHA-CHLORTOLUOL (GERMAN)
		BENZENE, (CHLOROMETHYL)-
		BENZILE (CLORURO DI) (ITALIAN)
	*	BENZYL CHLORIDE
		BENZYL CHLORIDE (ACGIH,DOT)
		BENZYLCHLORID (GERMAN)
		BENZYLE (CHLORURE DE) (FRENCH)
		CHLOROMETHYLBENZENE
		CHLOROPHENYLMETHANE
		CHLOROTOLUENE
		CHLORURE DE BENZYLE (FRENCH)
		NCI-C06360
		OMEGA-CHLOROTOLUENE
		PHENYLMETHYL CHLORIDE
		RCRA WASTE NUMBER P028
		TOLUENE, ALPHA-CHLORO-
		TOLYL CHLORIDE
		UN 1738 (DOT)
100-46-9		
		(AMINOMETHYL)BENZENE
		(PHENYLMETHYL)AMINE
		ALPHA-AMINOTOLUENE
		BENZENEMETHANAMINE
		BENZYLAMINE
		MONOBENZYLAMINE
		OMEGA-AMINOTOLUENE
100-47-0		
		BENZENE, CYANO-
		BENZENENITRILE
		BENZOIC ACID NITRILE
		BENZONITRIL
	*	BENZONITRILE
		BENZONITRILE (DOT)
		CYANOBENZENE
		PHENYL CYANIDE
		UN 2224 (DOT)
100-50-5		
		1,2,3,6-TETRAHYDROBENZALDEHYDE
		1,2,3,6-TETRAHYDROBENZALDEHYDE (DOT)
		1,2,5,6-TETRAHYDROBENZALDEHYDE
		1-CYCLOHEXENE-4-CARBOXALDEHYDE
		1-FORMYL-3-CYCLOHEXENE
		3-CYCLOHEXEN-1-ALDEHYDE

CAS No. IN Chemical Name

3-CYCLOHEXENE-1-CARBOXALDEHYDE
4-CYCLOHEXENE-1-CARBOXALDEHYDE
4-FORMYLCYCLOHEXENE
CYCLOHEXENE-4-CARBOXALDEHYDE
UN 2498 (DOT)

100-51-6
(HYDROXYMETHYL)BENZENE
ALPHA-HYDROXYTOLUENE
ALPHA-TOLUENOL
BENZAL ALCOHOL
BENZENECARBINOL
BENZENEMETHANOL
BENZOYL ALCOHOL
* BENZYL ALCOHOL
HYDROXYTOLUENE
METHANOL, PHENYL-
NCI-C06111
PHENOLCARBINOL
PHENYLCARBINOL
PHENYLMETHANOL
PHENYLMETHYL ALCOHOL

100-52-7
ALMOND ARTIFICIAL ESSENTIAL OIL
ARTIFICIAL ALMOND OIL
ARTIFICIAL ESSENTIAL OIL OF ALMOND
* BENZALDEHYDE
BENZALDEHYDE (DOT)
BENZALDEHYDE FFC
BENZENE CARBALDEHYDE
BENZENECARBONAL
BENZENECARBOXALDEHYDE
BENZOIC ALDEHYDE
NA 1989 (DOT)
NCI-C5 6133
PHENYLMETHANAL

100-53-8
(MERCAPTOMETHYL)BENZENE
ALPHA-MERCAPTOTOLUENE
ALPHA-TOLUENETHIOL
ALPHA-TOLUOLTHIOL
ALPHA-TOLYL MERCAPTAN
BENZENEMETHANETHIOL
BENZYL MERCAPTAN
BENZYLTHIOL
PHENYLMETHANETHIOL
PHENYLMETHYL MERCAPTAN
THIOBENZYL ALCOHOL

100-57-2
PHENYLMERCURIC HYDROXIDE

100-60-7
CYCLOHEXANAMINE, N-METHYL-
CYCLOHEXANAMINE, N-METHYL- (9CI)
CYCLOHEXYLAMINE, N-METHYL-
CYCLOHEXYLMETHYLAMINE
METHYL CYCLOHEXYLAMINE [CORROSIVE LABEL]
METHYL CYCLOHEXYLAMINE [FLAMMABLE LIQUID AND CORROSIVE LABELS]
METHYL CYCLOHEXYLAMINE [FLAMMABLE LIQUID LABEL]
METHYLCYCLOHEXYLAMINE
N-METHYL-N-CYCLOHEXYLAMINE
N-METHYLCYCLOHEXANAMINE
N-METHYLCYCLOHEXYLAMINE

100-61-8
(METHYLAMINO)BENZENE
ANILINE, N-METHYL-
ANILINOMETHANE
BENZENAMINE, N-METHYL-
BENZENENAMINE, N-METHYL- (9CI)
* METHYLANILINE
METHYLPHENYLAMINE
MONOMETHYLANILINE
N-METHYL ANILINE
N-METHYLAMINOBENZENE
N-METHYLANILINE (ACGIH,DOT)
N-METHYLBENZENAMINE
N-METHYLPHENYLAMINE
N-MONOMETHYLANILINE
N-PHENYLMETHYLAMINE
UN 2294 (DOT)

100-63-0
FENILIDRAZINA (ITALIAN)
FENYLHYDRAZINE (DUTCH)
HYDRAZINE, PHENYL-
HYDRAZINE-BENZENE
HYDRAZINOBENZENE
MONOPHENYLHYDRAZINE
PHENYLHYDRAZIN (GERMAN)
* PHENYLHYDRAZINE
PHENYLHYDRAZINE (ACGIH,DOT)
UN 2572 (DOT)

100-66-3
ANISOL
* ANISOLE
ANISOLE (DOT)
BENZENE, METHOXY-
ETHER, METHYL PHENYL
METHOXYBENZENE
METHYL PHENYL ETHER
PHENOXYMETHANE
PHENYL METHYL ETHER
UN 2222 (DOT)

100-73-2
2-FORMYL-3,4-DIHYDRO-2H-PYRAN
2-PROPENAL, DIMER
2H-PYRAN-2-CARBOXALDEHYDE, 3,4-DIHYDRO-
3,4-DIHYDRO-2-FORMYL-2H-PYRAN
3,4-DIHYDRO-2H-PYRAN-2-CARBOXALDEHYDE
5-HEXENAL, 2,6-EPOXY-
ACROLEIN DIMER
ACROLEIN DIMER, STABILIZED
ACROLEIN DIMER, STABILIZED (DOT)
PYRAN ALDEHYDE
UN 2607 (DOT)

100-74-3
4-ETHYLMORPHOLINE
ETHYLMORPHOLINE
MORPHOLINE, 4-ETHYL-
N-ETHYL MORPHOLINE
NEM

100-75-4
1-NITROSOPIPERIDINE
N-NITROSOPIPERIDINE
NITROSOPIPERIDINE
PIPERIDINE, 1-NITROSO-

100-79-8
1,2-ISOPROPYLIDENEGLYCERIN
1,2-ISOPROPYLIDENEGLYCEROL
1,2-O,O-ISOPROPYLIDENEGLYCERIN
1,2-O-ISOPROPYLIDENEGLYCEROL
1,3-DIOXACYCLOPENTANE
1,3-DIOXOLAN
1,3-DIOXOLANE
1,3-DIOXOLANE-4-METHANOL, 2,2-DIMETHYL-
1,3-DIOXOLE, DIHYDRO-
2,2-DIMETHYL-1,3-DIOXOLAN-4-YLMETHANOL
2,2-DIMETHYL-1,3-DIOXOLANE-4-METHANOL
2,2-DIMETHYL-4-HYDROXYMETHYLDIOXOLANE
2,2-DIMETHYL-4-OXYMETHYL-1,3-DIOXOLANE
2,2-DIMETHYL-5-HYDROXYMETHYL-1,3-DIOXOLANE
2,3-(ISOPROPYLIDENEDIOXY)PROPANOL
2,3-ISOPROPYLIDENEGLYCEROL
2,3-O-ISOPROPYLIDENEGLYCEROL
4-HYDROXYMETHYL-2,2-DIMETHYL-1,3-DIOXOLANE

CAS No.	IN	Chemical Name
		ACETONE GLYCERIN KETAL
		ACETONE MONOGLYCEROL KETAL
		ACETONE, CYCLIC (HYDROXYMETHYL)ETHYLENE ACETAL
		ALPHA,BETA-ISOPROPYLIDENEGLYCEROL
		DIOXOLAN
		DIOXOLANE
		DIOXOLANE (DOT)
		ETHYLENE GLYCOL FORMAL
		FORMAL GLYCOL
		GIE
		GLYCERIN ISOPROPYLIDENE ETHER
		GLYCEROL ACETONIDE
		GLYCEROL ALPHA,BETA-ISOPROPYLIDENE ETHER
		GLYCEROL DIMETHYLKETAL
		GLYCEROL, 1,2-O-ISOPROPYLIDENE
		GLYCEROLACETONE
		GLYCOL FORMAL
		ISOPROPYLIDENE GLYCEROL
		SOLKETAL
		UN 1166 (DOT)
100-97-0		
		1,3,5,7-TETRAAZAADAMANTANE
		1,3,5,7-TETRAAZATRICYCLO[33113,7]DECANE
		2,2',4,4',6,6'-HEXANITRODIPHENYLAMINE
		2,4,6,2',4',6'-HEXANITRODIPHENYLAMINE
		2,4,6-TRINITRO-N-(2,4,6-TRINITROPHENYL)BENZENAMINE
		ACETO HMT
		AMINOFORM
		AMINOFORMALDEHYDE
		AMMOFORM
		AMMONIOFORMALDEHYDE
		ANTIHYDRAL
		BENZENAMINE, 2,4,6-TRINITRO-N-(2,4,6-TRINITROPHENYL)-
		BIS(2,4,6-TRINITROPHENYL)AMINE
		CYSTAMIN
		CYSTOGEN
		DIPHENYLAMINE, 2,2',4,4',6,6'-HEXANITRO-
		DIPICRYLAMINE
		DUIREXOL
		EKAGOM H
		ESAMETILENTETRAMINA (ITALIAN)
		FORMAMINE
		FORMIN
		FORMIN (HETEROCYCLE)
		HERAX UTS
		HETERIN
		HEXA
		HEXA (VULCANIZATION ACCELERATOR)
		HEXA-FLO-PULVER
		HEXAFORM
		HEXAMETHYLENAMINE
		HEXAMETHYLENE TERAMINE
		HEXAMETHYLENEAMINE
		HEXAMETHYLENETETRAAMINE
		HEXAMETHYLENETETRAMINE
		HEXAMETHYLENTETRAMIN (GERMAN)
		HEXAMETHYLENTETRAMINE
	*	HEXAMINE
		HEXAMINE (1,3,5,7-TETRAAZATRICYCLO-3.3.1.13,7-DECANE)
		HEXAMINE (DOT)
		HEXAMINE (HETEROCYCLE)
		HEXAMINE (POTASSIUM REAGENT)
		HEXANITRODIPHENYLAMINE
		HEXASAN
		HEXILMETHYLENAMINE
		HEXYL
		HEXYL (REAGENT)
		HMT
		METHAMIN
		METHENAMIN
		METHENAMINE
		NOCCELER H
		PREPARATION AF
		RESOTROPIN
		UN 1328 (DOT)
		URAMIN
		URATRINE
		URITONE
		URODEINE
		UROTROPIN
		UROTROPINE
		VULKACIT H 30
		XAMETRIN
100-99-2		
		ALUMINUM, TRIISOBUTYL-
		ALUMINUM, TRIS(2-METHYLPROPYL)-
		ALUMINUM, TRIS(2-METHYLPROPYL)- (9CI)
		TRIISOBUTYL ALUMINUM
		TRIISOBUTYLALANE
		TRIISOBUTYLALUMINIUM
		TRIISOBUTYLALUMINIUM (DOT)
		TRIS(ISO-BUTYL)ALUMINUM
		TRIS(ISOBUTYL)ALUMINUM(III)
		UN 1930 (DOT)
101-05-3		
		(O-CHLOROANILINO)DICHLOROTRIAZINE
		1,3,5-TRIAZIN-2-AMINE, 4,6-DICHLORO-N-(2-CHLOROPHENYL)-
		2,4-DICHLORO-6-(2-CHLOROANILINO)-1,3,5-TRIAZINE
		2,4-DICHLORO-6-(O-CHLOROANILINO)-S-TRIAZINE
		2,4-DICHLORO-6-O-CHLORANILINO-S-TRIAZINE
		2-(2-CHLORANILIN)-4,6-DICHLOR-1,3,5-TRIAZIN (GERMAN)
		4,6-DICHLORO-N-(2-CHLOROPHENYL)-1,3,5-TRIAZIN-2-AMINE
		ANILAZIN
		ANILAZINE
		ANIYALINE
		B-622
		BORTRYSAN
		DIREZ
		DYRENE
		DYRENE 50W
		ENT 26,058
		KEMATE
		NCI-C08684
		S-TRIAZINE, 2,4-DICHLORO-6-(O-CHLOROANILINO)-
		TRIASYN
		TRIAZIN
		TRIAZINE
		TRIAZINE (PESTICIDE)
		TRIAZINE PESTICIDE, LIQUID [FLAMMABLE LIQUID LABEL]
		TRIAZINE PESTICIDE, LIQUID [POISON B LABEL]
		TRIAZINE PESTICIDE, SOLID
		ZINOCHLOR
101-14-4		
		3,3'-DICHLORO-4,4'-DIAMINODIPHENYLMETHANE
		4,4'-METHYLENE-BIS(2-CHLOROANILINE)
		4,4'-METHYLENEBIS(2-CHLORANILINE)
		4,4'-METHYLENEBIS(O-CHLOROANILINE)
		4,4'-METHYLENEBIS-(2-CHLOROANILINE)
		ANILINE, 4,4'-METHYLENEBIS[2-CHLORO-
		BENZENAMINE, 4,4'-METHYLENEBIS[2-CHLORO-
		BIS(3-CHLORO-4-AMINOPHENYL)METHANE
		BIS(4-AMINO-3-CHLOROPHENYL)METHANE
		DIAMET KH
		METHYLENEBIS(3-CHLORO-4-AMINOBENZENE)
		MILLIONATE M
		MOCA
		MOCA (CURING AGENT)
		QUODOROLE
101-25-7		
		1,3,5,7-TETRAAZABICYCLO(331)NONANE, 3,7-DINITROSO-
		1,5-ENDOMETHYLENE-3,7-DINITROSO-1,3,5,7-TETRAAZACYCLOOCTANE
		1,5-METHYLENE-3,7-DINITROSO-1,3,5,7-TETRAAZACYCLOOCTANE
		3,4-DI-N-NITROSOPENTAMETHYLENETETRAMINE
		3,7-DI-N-NITROSOPENTAMETHYLENETETRAMINE
		3,7-DINITROSO-1,3,5,7-TETRAAZABICYCLO-(3,3,1)-NONANE
		ACETO DNPT 100
		ACETO DNPT 40
		ACETO DNPT 80

CAS No. IN Chemical Name

CELLMIC A
CELLMIC A 80
CHEMPOR N 90
CHEMPOR PC 65
CHKHZ 18
DI-N-NITROSOPENTAMETHYLENETETRAMINE
DINITROSOPENTAMETHENETETRAMINE
DINITROSOPENTAMETHYLENE TETRAMINE
DIPENTAX
DNPMT
DNPT
KHEMPOR N90
MICROPOR
MIKROFOR N
N(SUP 1),N(SUP 3)-DINITROSOPENTAMETHYLENE-
TETRAMINE
N,N'-DINITROSOPENTAMETHYLENETETRAMINE
N,N-DINITROSOPENTAMETHYLENETETRAMINE
N1,N3-DINITROSOPENTAMETHYLENETETRAMINE
NSC 73599
OPEX
OPEX 93
PENTAMETHYLENETETRAMINE, DINITROSO-
POROFOR CHKHC-18
POROFOR DNO/F
POROPHOR B
PU 55
UNICEL 100
UNICEL ND
UNICEL NDX
VULCACEL B 40
VULCACEL BN
VULCACEL BN 94

101-55-3

1-BROMO-4-PHENOXYBENZENE
4-BROMODIPHENYL ETHER
4-BROMOPHENOXYBENZENE
4-BROMOPHENYL PHENYL ETHER
BENZENE, 1-BROMO-4-PHENOXY-
ETHER, P-BROMOPHENYL PHENYL
P-BROMODIPHENYL ETHER
P-BROMOPHENOXYBENZENE
P-BROMOPHENYL PHENYL ETHER
P-PHENOXYBROMOBENZENE
P-PHENOXYPHENYL BROMIDE

101-61-1

4,4'-BIS(DIMETHYLAMINO)DIPHENYLMETHANE
4,4'-BIS(DIMETHYLAMINOPHENYL)METHANE
4,4'-METHYLENEBIS[N,N-DIMETHYLANILINE]
4,4'-TETRAMETHYLDIAMINODIPHENYLMETHANE
4,4-METHYLENEBIS(N-N-DIMETHYL)BENZENEAMINE
ANILINE, 4,4'-METHYLENEBIS[N,N-DIMETHYL-
ARNOLD'S BASE
BENZENAMINE, 4,4'-METHYLENEBIS[N,N-DIMETHYL-
BIS[4-(DIMETHYLAMINO)PHENYL]METHANE
BIS[4-(N,N-DIMETHYLAMINO)PHENYL]METHANE
BIS[P-(DIMETHYLAMINO)PHENYL]METHANE
BIS[P-(N,N-DIMETHYLAMINO)PHENYL]METHANE
MICHLER'S BASE
MICHLER'S HYDRIDE
MICHLER'S METHANE
N,N,N',N'-TETRAMETHYL-4,4'-DIAMINODIPHENYL-
METHANE
N,N,N',N'-TETRAMETHYL-P,P'-DIAMINODIPHENYL-
METHANE
P,P'-BIS(DIMETHYLAMINO)DIPHENYLMETHANE
P,P'-TETRAMETHYLDIAMINODIPHENYLMETHANE
REDUCED MICHLER'S KETONE
TETRABASE
TETRAMETHYLDIAMINODIPHENYLMETHANE

101-68-8

1,1'-METHYLENEBIS[4-ISOCYANATOBENZENE]
1,1-METHYLENEBIS(4-ISOCYANATOBENZENE)
4,4'-DIISOCYANATODIPHENYLMETHANE
4,4'-DIPHENYLMETHANE DIISOCYANATE
4,4'-DIPHENYLMETHANE ISOCYANATE
4,4'-METHYLENEBIS(ISOCYANATOBENZENE)
4,4'-METHYLENEBIS(PHENYL ISOCYANATE)
4,4'-METHYLENEDI-P-PHENYLENE DIISOCYANATE
4,4'-METHYLENEDIPHENYL DIISOCYANATE
4,4'-METHYLENEDIPHENYL ISOCYANATE
4,4'-METHYLENEDIPHENYLENE ISOCYANATE
4,4-DIISOCYANATE DIPHENYL METHANE
4-4'-DIISOCYANATE DE DIPHENYLMETHANE (FRENCH)
BENZENE, 1,1'-METHYLENEBIS(4-ISOCYANATO- (9CI)
BENZENE, 1,1'-METHYLENEBIS[4-ISOCYANATO-
BIS(1,4-ISOCYANATOPHENYL)METHANE
BIS(4-ISOCYANATOPHENYL)METHANE
BIS(P-ISOCYANATOPHENYL)METHANE
CARADATE 30
DESMODUR 44
DI(4-ISOCYANATOPHENYL)METHANE
DIFENIL-METAN-DIISOCIANATO (ITALIAN)
DIFENYLMETHAAN-DISSOCYANAAT (DUTCH)
DIPHENYL METHANE DIISOCYANATE
DIPHENYLMETHAN-4,4'-DIISOCYANAT (GERMAN)
DIPHENYLMETHANE 4,4'-DIISOCYANATE
DIPHENYLMETHANE DIISOCYANATE
DIPHENYLMETHANE-4,4'-DIISOCYANATE (DOT)
HYLENE M50
ISOCYANIC ACID, METHYLENEDI-P-PHENYLENE ESTER
ISONATE
ISONATE 125 MF
ISONATE 125M
MBI
MDI
MDT
METHYLENE BISPHENYL ISOCYANATE
METHYLENE BISPHENYL ISOCYANATE (ACGIH)
METHYLENE BISPHENYL ISOCYANATE (MDI)
METHYLENE DI(PHENYLENE ISOCYANATE) (DOT)
METHYLENEBIS(4-ISOCYANATOBENZENE)
METHYLENEBIS(4-PHENYL ISOCYANATE)
METHYLENEBIS(4-PHENYLENE ISOCYANATE)
METHYLENEBIS(P-PHENYL ISOCYANATE)
METHYLENEBIS(P-PHENYLENE ISOCYANATE)
METHYLENEBIS-P-PHENYLENE DIISOCYANATE
METHYLENEBISPHENYLENE DIISOCYANATE
METHYLENEDI-P-PHENYLENE DIISOCYANATE
METHYLENEDI-P-PHENYLENE ISOCYANATE
NACCONATE 300
NCI-C50668
P,P'-DIPHENYLMETHANE DIISOCYANATE
P,P'-METHYLENEBIS(PHENYL ISOCYANATE)
RUBINATE 44
UN 2489 (DOT)

101-77-9

4,4'-DIAMINODIPHENYLMETHAN (GERMAN)
4,4'-DIAMINODIPHENYLMETHANE
4,4'-DIAMINODIPHENYLMETHANE (DOT)
4,4'-DIPHENYLMETHANEDIAMINE
4,4'-METHYLENE DIANILINE
4,4'-METHYLENEBIS(ANILINE)
4,4'-METHYLENEBIS(BENZENEAMINE)
4,4'-METHYLENEBISANILINE
4,4'-METHYLENEDIANILINE
4,4'-METHYLENEDIBENZENAMINE
4,4-METHYLENE DIANILINE
4,4-METHYLENEDIANILINE (ACGIH)
4-(4-AMINOBENZYL)ANILINE
ANCAMINE TL
ANILINE, 4,4'-METHYLENEDI-
ARALDITE HARDENER 972
BENZENAMINE, 4,4'-METHYLENEBIS-
BIS(4-AMINOPHENYL)METHANE
BIS(AMINOPHENYL)METHANE
BIS(P-AMINOPHENYL)METHANE
BIS-P-AMINOFENYLMETHAN (CZECH)
CURITHANE
DADPM
DAPM
DDM
DI-(4-AMINOPHENYL)METHANE
DIAMINODIPHENYL METHANE

CAS No.	IN	Chemical Name
		DIANILINEMETHANE
		DIANILINO METHANE
		EPICURE DDM
		EPIKURE DDM
		HT 972
		JEFFAMINE AP-20
		MDA
		METHYLENEBIS(ANILINE)
		METHYLENEDIANILINE
		P,P'-DIAMINODIFENYLMETHAN (CZECH)
		P,P'-DIAMINODIPHENYLMETHANE
		P,P'-METHYLENEDIANILINE
		SUMICURE M
		TONOX
		UN 2651 (DOT)
101-80-4		
		4,4'-DIAMINODIPHENYL ETHER
		4,4'-DIAMINODIPHENYL OXIDE
		4,4'-DIAMINOPHENYL ETHER
		4,4'-OXYBIS(ANILINE)
		4,4'-OXYDIANILINE
		4,4'-OXYDIPHENYLAMINE
		4-AMINOPHENYL ETHER
		ANILINE, 4,4'-OXYDI-
		BENZENAMINE, 4,4'-OXYBIS-
		BIS(4-AMINOPHENYL) ETHER
		BIS(P-AMINOPHENYL) ETHER
		DIAMINODIPHENYL ETHER
		OXYBIS(4-AMINOBENZENE)
		OXYDI-P-PHENYLENEDIAMINE
		P,P'-DIAMINODIPHENYL ETHER
		P,P'-OXYBIS[ANILINE]
		P,P'-OXYDIANILINE
		P-AMINOPHENYL ETHER
101-83-7		
		AMINODICYCLOHEXANE
		CYCLOHEXANAMINE, N-CYCLOHEXYL-
	*	DICYCLOHEXYLAMINE
		DODECAHYDRODIPHENYLAMINE
		N,N-DICYCLOHEXYLAMINE
		N-CYCLOHEXYLCYCLOHEXANAMINE
101-84-8		
		BENZENE, 1,1'-OXYBIS-
		BENZENE, PHENOXY-
		BIPHENYL OXIDE
		CHEMCRYL JK-EB
		DIPHENYL ETHER
		DIPHENYL OXIDE
		ETHER, DIPHENYL
		GERANIUM CRYSTALS
		OXYBISBENZENE
		OXYDIPHENYL
		PHENOXYBENZENE
		PHENYL ETHER
		PHENYL ETHER (ACGIH)
		PHENYL ETHER VAPOR
		PHENYL OXIDE
101-90-6		
		RESORCINOL DIGLYCIDYL ETHER
101-96-2		
		1,4-BENZENEDIAMINE, N,N'-BIS(1-METHYLPROPYL)-
		ANTIOXIDANT 22
		DU PONT GASOLINE ANTIOXIDANT NO. 22
		N,N'-DI-SEC-BUTYL-P-PHENYLDIAMINE
		N,N'-DI-SEC-BUTYL-P-PHENYLENEDIAMINE
		N,N-DI-SEC-BUTYL-P-PHENYLENE-DIAMINE
		P-PHENYLENEDIAMINE, N,N'-DI-SEC-BUTYL-
		TENAMENE 2
		TOPANOL M
		UOP 5
102-01-2		
		1-(PHENYLCARBAMOYL)-2-PROPANONE
		3-OXO-N-PHENYLBUTANAMIDE
		ACETOACETAMIDOBENZENE
		ACETOACETANILIDE
		ACETOACETIC ACID ANILIDE
		ACETOACETIC ANILIDE
		ACETOACETYLANILINE
		ALPHA-ACETYL-N-PHENYLACETAMIDE
		ALPHA-ACETYLACETANILIDE
		BETA-KETOBUTYRANILIDE
		BUTANAMIDE, 3-OXO-N-PHENYL-
		N-(ACETYLACETYL)ANILINE
		N-PHENYLACETOACETAMIDE
		[(ACETOACETYL)AMINO]BENZENE
102-27-2		
		3-METHYL-N-ETHYLANILINE
		BENZENAMINE, N-ETHYL-3-METHYL-
		BENZENAMINE, N-ETHYL-3-METHYL- (9CI)
		M-TOLUIDINE, N-ETHYL-
		N-ETHYL-3-METHYLANILINE
		N-ETHYL-3-METHYLBENZENAMINE
		N-ETHYL-M-TOLUIDINE
		N-ETHYL-M-TOLUIDINE (DOT)
		N-ETHYLTOLUIDINE
		UN 2754 (DOT)
102-36-3		
		3,4-DICHLOROPHENYL ISOCYANATE
		BENZENE, 1,2-DICHLORO-4-ISOCYANATO-
		DICHLOROPHENYL-ISOCYANATE (1,2-DICHLORO-4-ISO-CYANATO-BENZENE)
		ISOCYANIC ACID 3,4-DICHLOROPHENYL ESTER
102-50-1		
		M-CRESIDINE
102-54-5		
		BIS(CYCLOPENTADIENYL)IRON
		BIS(ETA-CYCLOPENTADIENYL)IRON
		DI-2,4-CYCLOPENTADIEN-1-YLIRON
		DI-PI-CYCLOPENTADIENYL IRON
	*	DICYCLOPENTADIENYL IRON
		FERROCENE
		FERROTSEN
		IRON BIS(CYCLOPENTADIENIDE)
		IRON DICYCLOPENTADIENYL
		IRON, BIS(ETA5-2,4-CYCLOPENTADIEN-1-YL)-
102-56-7		
		1-AMINO-2,5-DIMETHOXYBENZENE
		2,5-DIMETHOXYANILINE
		2,5-DIMETHOXYBENZENAMINE
		AMINOHYDROQUINONE DIMETHYL ETHER
		ANILINE, 2,5-DIMETHOXY-
		BENZENAMINE, 2,5-DIMETHOXY-
		C.I. 35811
102-67-0		
		ALUMINUM, TRIPROPYL-
	*	TRIPROPYLALUMINUM
		TRIPROPYLALUMINUM (DOT)
		UN 2718 (DOT)
102-69-2		
		1-PROPANAMINE, N,N-DIPROPYL-
		PROPYLDI-N-PROPYLAMINE
		TRI-N-PROPYLAMINE
	*	TRIPROPYLAMINE
		TRIPROPYLAMINE (8CI)
		TRIPROPYLAMINE (DOT)
		TRIPROPYLAMINE [COMBUSTIBLE LIQUID LABEL]
		TRIPROPYLAMINE [FLAMMABLE LIQUID LABEL]
		UN 2260 (DOT)
102-70-5		
		2-PROPEN-1-AMINE, N,N-DI-2-PROPENYL-
		2-PROPEN-1-AMINE, N-N-DI-2-PROPENYL- (9CI)
		TRIALLYLAMINE
		TRIALLYLAMINE (DOT)
		TRIS(2-PROPENYL)AMINE

CAS No.	IN	Chemical Name
		UN 2610 (DOT)
102-71-6		
		2,2',2"-NITRILOTRIETHANOL
		2,2',2"-NITRILOTRIS[ETHANOL]
		ALKANOLAMINE 244
		DALTOGEN
		ETHANOL, 2,2',2"-NITRILOTRI-
		ETHANOL, 2,2',2"-NITRILOTRIS-
		NITRILO-2,2',2"-TRIETHANOL
		NITRILOTRIETHANOL
		STEROLAMIDE
		STING-KILL
		TEA
		THIOFACO T-35
		TRI(HYDROXYETHYL)AMINE
		TRIAETHANOLAMIN-NG
		TRIETHANOLAMIN
	*	TRIETHANOLAMINE
		TRIETHYLAMINE, 2,2',2"-TRIHYDROXY-
		TRIETHYLOLAMINE
		TRIHYDROXYTRIETHYLAMINE
		TRIS(2-HYDROXYETHYL)AMINE
		TRIS(BETA-HYDROXYETHYL)AMINE
		TROLAMINE
102-79-4		
		2,2'-(BUTYLIMINO)DIETHANOL
		BIDE
		BIS(BETA-HYDROXYETHYL)BUTYLAMINE
		BUTYLBIS(2-HYDROXYETHYL)AMINE
		BUTYLDIETHANOLAMINE
		ETHANOL, 2,2'-(BUTYLIMINO)BIS-
		ETHANOL, 2,2'-(BUTYLIMINO)DI-
		N,N-BIS(2-HYDROXYETHYL)BUTYLAMINE
		N-BUTYL-2,2'-IMINODIETHANOL
		N-BUTYL-N,N-BIS(2-HYDROXYETHYL)AMINE
		N-BUTYL-N,N-BIS(HYDROXYETHYL)AMINE
		N-BUTYLDIETHANOLAMINE
102-81-8		
		2-(DIBUTYLAMINO)ETHANOL
		2-N-DIBUTYLAMINOETHANOL
		DIBUTYLAMINOETHANOL
		DIBUTYLETHANOLAMINE
		ETHANOL, 2-(DIBUTYLAMINO)-
		N,N-DIBUTYL-2-HYDROXYETHYLAMINE
		N,N-DIBUTYL-N-(2-HYDROXYETHYL)AMINE
		N,N-DIBUTYLAMINOETHANOL
		N,N-DIBUTYLETHANOLAMINE
102-82-9		
		1-BUTANAMINE, N,N-DIBUTYL-
		TRI-N-BUTYLAMINE
	*	TRIBUTYLAMINE
		TRIS-N-BUTYLAMINE
102-85-2		
		JP 304
		PHOSPHOROUS ACID, TRIBUTYL ESTER
		TRI-N-BUTYL PHOSPHITE
		TRIBUTOXYPHOSPHINE
		TRIBUTYL PHOSPHITE
103-09-3		
		1-HEXANOL, 2-ETHYL-, ACETATE
		2-ETHYL-1-HEXANOL ACETATE
		2-ETHYL-1-HEXYL ACETATE
		2-ETHYLHEXANYL ACETATE
		2-ETHYLHEXYL ACETATE
		2-ETHYLHEXYL ETHANOATE
		ACETIC ACID ALPHA-ETHYLHEXYL ESTER
		ACETIC ACID, 2-ETHYLHEXYL ESTER
		ACETIC ACID, BIS(2-ETHYLHEXYL) ESTER
		BETA-ETHYLHEXYL ACETATE
		ETHYLHEXYL ACETATE
		OCTYL ACETATE
103-11-7		
		1-HEXANOL, 2-ETHYL-, ACRYLATE
		2-ETHYLHEXYL 2-PROPENOATE
		2-ETHYLHEXYL ACRYLATE
		2-PROPENOIC ACID, 2-ETHYLHEXYL ESTER
		2-PROPENOIC ACID, 2-ETHYLHEXYL ESTER (9CI)
		ACRYLIC ACID, 2-ETHYLHEXYL ESTER
		OCTYL ACRYLATE
103-23-1		
		ADIMOLL DO
		ADIPIC ACID, BIS(2-ETHYLHEXYL) ESTER
		ADIPIC ACID, DIOCTYL ESTER
		ADIPOL 2EH
		BEHA
		BIS(2-ETHYLHEXYL)ADIPATE
		BISOFLEX DOA
		DEHA
		DI-2-ETHYLHEXYL ADIPATE
		DI-N-OCTYL ADIPATE
		DIETHYLHEXYL ADIPATE
		DIOCTYL ADIPATE
		DOA
		EFFEMOLL DOA
		EFFOMOLL DOA
		ERGOPLAST ADDO
		FLEXOL A 26
		FLEXOL PLASTICIZER 10-A
		FLEXOL PLASTICIZER A-26
		HEXANEDIOIC ACID, BIS(2-ETHYLHEXYL) ESTER
		HEXANEDIOIC ACID, BIS(2-ETHYLHEXYL) ESTER (9CI)
		HEXANEDIOIC ACID, DIOCTYL ESTER
		KEMESTER 5652
		KODAFLEX DOA
		LANKROFLEX DOA
		MOLLAN S
		MONOPLEX DOA
		NCI-C54386
		OCTYL ADIPATE
		PLASTOMOLL DOA
		PX-238
		REOMOL DOA
		RUCOFLEX PLASTICIZER DOA
		SICOL 250
		STAFLEX DOA
		TRUFLEX DOA
		UNIFLEX DOA
		VESTINOL OA
		WICKENOL 158
		WITAMOL 320
103-33-3		
		AZOBENZENE
103-44-6		
		1-ETHENOXY-2-ETHYLHEXANE
		2-ETHYLHEXYL VINYL ETHER
		ETHER, 2-ETHYLHEXYL VINYL
		HEPTANE, 3-[(ETHENYLOXY)METHYL]-
		VINYL-2-ETHYLHEXYL ETHER
103-65-1		
		1-PHENYLPROPANE
		1-PROPYLBENZENE
		BENZENE, PROPYL-
		ISOCUMENE
		N-PROPYLBENZENE
		PROPYL BENZENE
		PROPYL BENZENE (DOT)
		UN 2364 (DOT)
103-69-5		
		AETHYLANILIN (GERMAN)
		ANILINE, N-ETHYL-
		ANILINOETHANE
		BENZENAMINE, N-ETHYL-
		BENZENAMINE, N-ETHYL- (9CI)
		ETHYLANILINE
		ETHYLPHENYLAMINE

CAS No.	IN	Chemical Name
		N-ETHYL-N-PHENYLAMINE
		N-ETHYLAMINOBENZENE
		N-ETHYLANILINE
		N-ETHYLANILINE (DOT)
		N-ETHYLBENZENAMINE
		N-ETHYLBENZENAMINO
		UN 2272 (DOT)
103-71-9		
		BENZENE, ISOCYANATO-
		CARBANIL
		ISOCYANATOBENZENE
		ISOCYANIC ACID, PHENYL ESTER
		MONDUR P
		PHENYL CARBONIMIDE
	*	PHENYL ISOCYANATE
		PHENYL ISOCYANATE (DOT)
		PHENYLCARBIMIDE
		UN 2487 (DOT)
103-75-3		
		2-ETHOXY-2,3-DIHYDRO-4H-PYRAN
		2-ETHOXY-2,3-DIHYDRO-GAMMA-PYRAN
		2-ETHOXY-3,4-DIHYDRO-1,2-PYRAN
		2-ETHOXY-3,4-DIHYDRO-2H-PYRAN
		2-ETHOXY-3,4-DIHYDROPYRAN
		2-ETHOXYDIHYDROPYRAN, IN PREGNANCY DIAGNOSIS
		2H-PYRAN, 2-ETHOXY-3,4-DIHYDRO-
		3,4-DIHYDRO-2-ETHOXY-2H-PYRAN
		3,4-DIHYDRO-2-ETHOXYPYRAN
		ETHOXYDIHYDROPYRAN
103-80-0		
		ACETYL CHLORIDE, PHENYL-
		ALPHA-PHENYLACETYL CHLORIDE
		BENZENEACETYL CHLORIDE
		BENZENEACETYL CHLORIDE (9CI)
		PHENACETYL CHLORIDE
		PHENYLACETIC ACID CHLORIDE
		PHENYLACETYL CHLORIDE
		PHENYLACETYL CHLORIDE (DOT)
		UN 2577 (DOT)
103-83-3		
		BENZYL DIMETHYLAMINE
103-84-4		
		ACETAMIDE, N-PHENYL-
		ACETAMIDOBENZENE
		ACETANIL
	*	ACETANILIDE
		ACETOANILIDE
		ACETYLANILINE
		ANTIFEBRIN
		BENZENAMINE, N-ACETYL-
		N-ACETYLANILINE
		N-PHENYLACETAMIDE
		PHENALGENE
		PHENALGIN
103-85-5		
		1-PHENYL-2-THIOUREA
		1-PHENYLTHIOUREA
		ALPHA-PHENYLTHIOUREA
		N-PHENYLTHIOUREA
		NCI-C02017
		PHENYLTHIOCARBAMIDE
	*	PHENYLTHIOUREA
		PTC
		PTU
		RCRA WASTE NUMBER P093
		THIOUREA, PHENYL-
		U 6324
		UREA, 1-PHENYL-2-THIO-
		USAF EK-1569
103-89-9		
		1-ACETAMIDO-4-METHYLBENZENE
		4'-METHYLACETANILIDE
		4-(ACETYLAMINO)TOLUENE
		4-ACETOTOLUIDE
		4-METHYLACETANILIDE
		ACETAMIDE, N-(4-METHYLPHENYL)-
		ACETYL-P-TOLUIDINE
		N-(4-METHYLPHENYL)ACETAMIDE
		N-ACETYL-4-METHYLANILINE
		N-ACETYL-P-TOLUIDINE
		P-ACETAMIDOTOLUENE
		P-ACETOTOLUIDE
		P-ACETOTOLUIDIDE
		P-METHYLACETANILIDE
103-95-3		
		2-ETHOXY-3-4-DIHYDRO-2-PYRAN
104-15-4		
		4-METHYLBENZENESULFONIC ACID
		4-TOLUENESULFONIC ACID
		AR-TOLUENESULFONIC ACID
		BENZENESULFONIC ACID, 4-METHYL-
		BENZENESULFONIC ACID, METHYL-
		CYZAC 4040
		K-CURE 1040
		KYSELINA P-TOLUENESULFONOVA (CZECH)
		MANRO PTSA 65 E
		MANRO PTSA 65 H
		MANRO PTSA 65 LS
		METHYLBENZENESULFONIC ACID
		P-METHYLBENZENESULFONIC ACID
		P-METHYLPHENYLSULFONIC ACID
	*	P-TOLUENESULFONIC ACID
		P-TOLUENESULPHONIC ACID
		P-TOLYLSULFONIC ACID
		TOLUENE SULFONIC ACID, LIQUID
		TOLUENE SULFONIC ACID, LIQUID (DOT)
	*	TOLUENE SULFONIC ACID, SOLID
		TOLUENE SULPHONIC ACID, LIQUID, CONTAINING MORE THAN 5% FREE SULPHURIC ACID (DOT)
		TOLUENE SULPHONIC ACID, LIQUID, CONTAINING NOT MORE THAN 5% FREE SULPHURIC ACID (DOT)
		TOLUENE SULPHONIC ACID, SOLID, CONTAINING MORE THAN 5% FREE SULPHURIC ACID (DOT)
		TOLUENE SULPHONIC ACID, SOLID, CONTAINING NOT MORE THAN 5% FREE SULPHURIC ACID (DOT)
		TOLUENESULFONIC ACID
		TOSIC ACID
		TSA-HP
		TSA-MH
		UN 2583 (DOT)
		UN 2584 (DOT)
		UN 2585 (DOT)
		UN 2586 (DOT)
104-51-8		
		1-BUTYLBENZENE
		1-PHENYLBUTANE
		BENZENE, BUTYL-
		BUTYL BENZENE
		N-BUTYLBENZENE
		N-BUTYLBENZENE (DOT)
		PHENYLBUTANE
		UN 2709 (DOT)
104-72-3		
		1-PHENYLDECANE
		BENZENE, DECYL-
		DECANE, 1-PHENYL-
		DECYLBENZENE
		N-DECYLBENZENE
104-75-6		
		1-AMINO-2-ETHYLHEXAN (CZECH)
		1-AMINO-2-ETHYLHEXANE
		1-HEXANAMINE, 2-ETHYL-
		2-ETHYL HEXYLAMINE
		2-ETHYL-1-HEXANAMINE
		2-ETHYL-1-HEXYLAMINE
		2-ETHYLHEXANAMINE

CAS No.	IN	Chemical Name
		2-ETHYLHEXYLAMINE
		2-ETHYLHEXYLAMINE (DOT)
		BETA-ETHYLHEXYLAMINE
		ETHYL HEXYLAMINE
		HEXYLAMINE, 2-ETHYL-
		UN 2276 (DOT)
104-76-7		
		1-HEXANOL, 2-ETHYL-
		2-AETHYLHEXANOL (GERMAN)
		2-ETHYL HEXANOL
	*	2-ETHYL-1-HEXANOL
		2-ETHYLHEXYL ALCOHOL
		ETHYLHEXANOL
104-78-9		
		1,3-PROPANEDIAMINE, N,N-DIETHYL-
		1-AMINO-3-(DIETHYLAMINO)PROPANE
		3-(DIETHYLAMINO) PROPYLAMINE
		3-(DIETHYLAMINO)-N-PROPYLAMINE
		3-(DIETHYLAMINO)PROPYLAMINE
		3-(DIETHYLAMINO)PROPYLAMINE (DOT)
		3-(N,N-DIETHYLAMINO)-1-PROPYLAMINE
		3-DIETHYLAMINO-1-PROPYLAMINE
		DIETHYLAMINOPROPYLAMINE
		DIETHYLAMINOTRIMETHYLENAMINE
		GAMMA-DIETHYLAMINOPROPYLAMINE
		N,N-DIETHYL-1,3-DIAMINOPROPANE
		N,N-DIETHYL-1,3-PROPANEDIAMINE
		N,N-DIETHYL-1,3-PROPYLENEDIAMINE
		N,N-DIETHYLAMINOPROPYLAMINE
		N,N-DIETHYLPROPYLENEDIAMINE
		N,N-DIETHYLTRIMETHYLENEDIAMINE
		N-(3-DIETHYLAMINOPROPYL)AMINE
		UN 2684 (DOT)
104-83-6		
		(4-CHLOROPHENYL)METHYL CHLORIDE
		1-CHLORO-4-CHLOROMETHYLBENZENE
		4-CHLOROBENZYL CHLORIDE
		ALPHA,4-DICHLOROTOLUENE
		ALPHA,P-DICHLOROTOLUENE
		BENZENE, 1-CHLORO-4-(CHLOROMETHYL)-
		CHLOROBENZYL CHLORIDE
		P,ALPHA-DICHLOROTOLUENE
		P-CHLOROBENZYL CHLORIDE
		P-CHLOROBENZYL CHLORIDE (DOT)
		P-CHLOROPHENYLMETHYL CHLORIDE
		TOLUENE, P,ALPHA-DICHLORO-
		UN 2235 (DOT)
104-88-1		
		4-CHLOROBENZALDEHYDE
		BENZALDEHYDE, 4-CHLORO-
		BENZALDEHYDE, P-CHLORO-
	*	P-CHLOROBENZALDEHYDE
		P-CHLOROBENZENECARBOXALDEHYDE
104-89-2		
		2-METHYL-5-ETHYLPIPERIDINE
		5-ETHYL-2-METHYLPIPERIDINE
		5-ETHYL-2-PIPECOLINE
		COPELLIDINE
		PIPERIDINE, 5-ETHYL-2-METHYL-
104-90-5		
		2-METHYL-5-ETHYLPYRIDINE
		2-METHYL-5-ETHYLPYRIDINE (DOT)
		2-PICOLINE, 5-ETHYL-
		3-ETHYL-6-METHYLPYRIDINE
		5-ETHYL-2-METHYLPYRIDINE
		5-ETHYL-2-PICOLINE
		5-ETHYL-ALPHA-PICOLINE
		6-METHYL-3-ETHYLPYRIDINE
		ALDEHYDECOLLIDINE
		ALDEHYDINE
		COLLIDINE, ALDEHYDECOLLIDINE
		ETHYL MORPHINE

CAS No.	IN	Chemical Name
		MEP
		METHYL ETHYL PYRIDINE
		METHYL ETHYL PYRIDINE (DOT)
		PYRIDINE, 5-ETHYL-2-METHYL-
		UN 2300 (DOT)
104-94-9		
		1-AMINO-4-METHOXYBENZENE
		4-AMINOANISOLE
		4-ANISIDINE
		4-METHOXYANILINE
		4-METHOXYBENZENAMINE
		BENZENAMINE, 4-METHOXY-
		METHOXYANILINE
		P-AMINOANISOLE
	*	P-ANISIDINE
		P-ANISYLAMINE
		P-METHOXYANILINE
		P-METHOXYPHENYLAMINE
105-05-5		
		1,4-DIETHYLBENZENE
		BENZENE, 1,4-DIETHYL-
		BENZENE, P-DIETHYL-
		P-DIETHYL BENZENE
		P-ETHYLETHYLBENZENE
		P-XYLENE, ALPHA,ALPHA'-DIMETHYL-
105-11-3		
		P-QUINONE DIOXIME
105-30-6		
		1,3-DIMETHYL BUTANOL
		1,3-DIMETHYL-1-BUTANOL
		1-PENTANOL, 2-METHYL-
		1-PENTANOL, METHYL-
		2-METHYL-1-PENTANOL
		2-METHYL-2-PROPYLETHANOL
		2-METHYL-4-PENTANOL
		2-METHYLPENTANOL-1
		2-PENTANOL, 4-METHYL-
		3-MIC
		4-METHYL-2-PENTANOL
		4-METHYL-2-PENTYL ALCOHOL
		4-METHYLPENTANOL-2
		4-METILPENTAN-2-OLO (ITALIAN)
		ALCOOL METHYL AMYLIQUE (FRENCH)
		AMYL METHYL ALCOHOL
		ISOBUTYLMETHYLCARBINOL
		ISOBUTYLMETHYLMETHANOL
		ISOHEXYL ALCOHOL
		ISOPROPYL DIMETHYL CARBINOL
		MAOH
		METHYL AMYL ALCOHOL
		METHYL ISOBUTYL CARBINOL
		METHYL ISOBUTYL CARBINOL (ACGIH,DOT)
		METHYL-1-PENTANOL
		METHYLAMYL ALCOHOL
		METHYLISOBUTYL CARBINOL
		METILAMIL ALCOHOL (ITALIAN)
		MIBC
		MIC
		UN 2053 (DOT)
105-36-2		
		ACETIC ACID, BROMO-, ETHYL ESTER
		ANTOL
		BROMOACETIC ACID ETHYL ESTER
		ETHOXYCARBONYLMETHYL BROMIDE
		ETHYL 2-BROMOACETATE
		ETHYL ALPHA-BROMOACETATE
		ETHYL BROMACETATE
		ETHYL BROMOACETATE
		ETHYL BROMOACETATE (DOT)
		ETHYL MONOBROMOACETATE
		UN 1603 (DOT)

CAS No.	IN	Chemical Name
105-37-3		ETHYL PROPANOATE
	*	ETHYL PROPIONATE
		ETHYL PROPIONATE (DOT)
		PROPANOIC ACID, ETHYL ESTER
		PROPIONATE D'ETHYLE (FRENCH)
		PROPIONIC ACID, ETHYL ESTER
		PROPIONIC ESTER
		PROPIONIC ETHER
		UN 1195 (DOT)
105-38-4		PROPANOIC ACID, ETHENYL ESTER
		PROPIONIC ACID, VINYL ESTER
		VINYL PROPANOATE
		VINYL PROPIONATE
105-39-5		ACETIC ACID, CHLORO-, ETHYL ESTER
		CHLOROACETIC ACID ETHYL ESTER
		ETHYL 2-CHLOROACETATE
		ETHYL ALPHA-CHLOROACETATE
		ETHYL CHLORACETATE
	*	ETHYL CHLOROACETATE
		ETHYL CHLOROACETATE (DOT)
		ETHYL CHLOROETHANOATE
		ETHYL MONOCHLORACETATE
		ETHYL MONOCHLOROACETATE
		UN 1181 (DOT)
105-45-3		1-METHOXYBUTANE-1,3-DIONE
		3-OXOBUTANOIC ACID METHYL ESTER
		ACETOACETIC ACID, METHYL ESTER
		ACETOACETIC METHYL ESTER
		BUTANOIC ACID, 3-OXO-, METHYL ESTER
		BUTANOIC ACID, 3-OXO-, METHYL ESTER (9CI)
		METHYL 3-OXOBUTANOATE
		METHYL 3-OXOBUTYRATE
		METHYL ACETOACETATE
		METHYL ACETYLACETATE
		METHYL ACETYLACETONATE
105-46-4		1-METHYLPROPYL ACETATE
		2-ACETOXYBUTANE
		2-BUTANOL ACETATE
		2-BUTYL ACETATE
		ACETATE DE BUTYLE SECONDAIRE (FRENCH)
		ACETIC ACID, 1-METHYLPROPYL ESTER (9CI)
		ACETIC ACID, 1-METHYLPROPYL ESTER
		ACETIC ACID, 2-BUTOXY ESTER
		ACETIC ACID, SEC-BUTYL ESTER
	*	SEC-BUTYL ACETATE
		SEC-BUTYL ACETATE (ACGIH,DOT)
		SEC-BUTYL ALCOHOL ACETATE
		UN 1123 (DOT)
105-48-6		1-METHYLETHYL MONOCHLOROACETATE
		ACETIC ACID, CHLORO-, 1-METHYLETHYL ESTER
		ACETIC ACID, CHLORO-, ISOPROPYL ESTER
		CHLOROACETIC ACID ISOPROPYL ESTER
		ISOPROPYL CHLOROACETATE
		ISOPROPYL CHLOROACETATE (DOT)
		UN 2947 (DOT)
105-54-4		BUTANOIC ACID ETHYL ESTER
		BUTYRIC ACID, ETHYL ESTER
		BUTYRIC ETHER
		ETHYL BUTANOATE
	*	ETHYL BUTYRATE
		ETHYL BUTYRATE (DOT)
		ETHYL N-BUTYRATE
		UN 1180 (DOT)
105-55-5		N,N-DIETHYLTHIOUREA
105-56-6		ACETIC ACID, CYANO-, ETHYL ESTER
		CYANACETATE ETHYLE (GERMAN)
		CYANOACETIC ACID ETHYL ESTER
		CYANOACETIC ESTER
		ETHYL 2-CYANOACETATE
		ETHYL CYANACETATE
	*	ETHYL CYANOACETATE
		ETHYL CYANOACETATE (DOT)
		ETHYL CYANOETHANOATE
		MALONIC ACID ETHYL ESTER NITRILE
		UN 2666 (DOT)
		USAF KF-25
105-57-7		1,1-DIAETHOXY-AETHAN (GERMAN)
		1,1-DIETHOXY-ETHAAN (DUTCH)
		1,1-DIETHOXYETHANE
		1,1-DIETOSSIETANO (ITALIAN)
		ACETAAL (DUTCH)
		ACETAL
		ACETAL (DOT)
		ACETAL DIETHYLIQUE (FRENCH)
		ACETALDEHYDE ETHYL ACETAL
		ACETALDEHYDE, DIETHYL ACETAL
		ACETALE (ITALIAN)
		ACETOL
		DIAETHYLACETAL (GERMAN)
		DIETHYL ACETAL
		ETHANE, 1,1-DIETHOXY-
		ETHYLIDENE DIETHYL ETHER
		UN 1088 (DOT)
		USAF DO-45
105-58-8		CARBONIC ACID, DIETHYL ESTER
		DEC
		DIAETHYLCARBONAT (GERMAN)
		DIATOL
		DIETHYL CARBONATE
		DIETHYL CARBONATE (DOT)
		DIETHYLESTER CARBONIC ACID
		ETHOXYFORMIC ANHYDRIDE
		ETHYL CARBONATE
		ETHYL CARBONATE ((ETO)2CO)
		EUFIN
		NCI-C60899
		UN 2366 (DOT)
105-59-9		2,2'-(METHYLIMINO)DIETHANOL
		BIS(2-HYDROXYETHYL) METHYL AMINE
		DIETHANOLMETHYLAMINE
		ETHANOL, 2,2'-(METHYLIMINO)BIS-
		ETHANOL, 2,2'-(METHYLIMINO)DI-
		MDEA
		METHYLBIS(2-HYDROXYETHYL)AMINE
		METHYLDIETHANOLAMINE
		METHYLIMINODIETHANOL
		N,N-BIS(2-HYDROXYETHYL)METHYLAMINE
		N-METHYL-2,2'-IMINODIETHANOL
		N-METHYLAMINODIGLYCOL
		N-METHYLDIETHANOLAMINE
		N-METHYLDIETHANOLIMINE
		N-METHYLIMINODIETHANOL
		USAF DO-52
105-60-2		1,6-HEXOLACTAM
		1-AZA-2-CYCLOHEPTANONE
		2-AZACYCLOHEPTANONE
		2-KETOHEXAMETHYLENEIMINE
		2-KETOHEXAMETHYLENIMINE
		2-OXOHEXAMETHYLENEIMINE
		2-OXOHEXAMETHYLENIMINE
		2-PERHYDROAZEPINONE

CAS No.	IN	Chemical Name
		2H-AZEPIN-2-ONE, HEXAHYDRO-
		2H-AZEPIN-7-ONE, HEXAHYDRO-
		6-AMINOCAPROIC ACID LACTAM
		6-AMINOHEXANOIC ACID CYCLIC LACTAM
		6-CAPROLACTAM
		6-HEXANELACTAM
		A1030
		A1030N0
		AKULON
		AKULON M 2W
		ALKAMID
		AMILAN CM 1001
		AMILAN CM 1001C
		AMILAN CM 1001G
		AMILAN CM 1011
		AMINOCAPROIC LACTAM
		ATM 2(NYLON)
		BONAMID
		CAPRAN 77C
		CAPRAN 80
	*	CAPROLACTAM
		CAPROLACTAM MONOMER
		CAPROLATTAME (FRENCH)
		CAPROLON B
		CAPROLON V
		CAPRON
		CAPRON 8250
		CAPRON 8252
		CAPRON 8253
		CAPRON 8256
		CAPRON 8257
		CAPRON B
		CAPRON GR 8256
		CAPRON GR 8258
		CAPRON PK4
		CHEMLON
		CM 1001
		CM 1011
		CM 1031
		CM 1041
		CYCLOHEXANONE ISO-OXIME
		DANAMID
		DULL 704
		DURETHAN BK
		DURETHAN BK 30S
		DURETHAN BKV 30H
		DURETHAN BKV 55H
		E-KAPROLAKTAM (CZECH)
		EPSILON-CAPROLACTAM
		EPSILON-CAPROLACTAM (ACGIH)
		EPSYLON KAPROLAKTAM (POLISH)
		ERTALON 6SA
		EXTROM 6N
		GRILON
		HEXAHYDRO-2-AZEPINONE
		HEXAHYDRO-2H-AZEPIN-2-ONE
		HEXAHYDRO-2H-AZEPIN-2-ONE (9CI)
		HEXAMETHYLENIMINE, 2-OXO-
		HEXANOIC ACID, 6-AMINO-, CYCLIC LACTAM
		HEXANOIC ACID, 6-AMINO-, LACTAM
		HEXANOLACTAM
		HEXANONE ISOXIME
		HEXANONISOXIM (GERMAN)
		ITAMID
		ITAMID 250
		ITAMIDE 25
		ITAMIDE 250
		ITAMIDE 250G
		ITAMIDE 35
		ITAMIDE 350
		ITAMIDE S
		KAPROLIT
		KAPROLIT B
		KAPROLON
		KAPROLON B
		KAPROMINE
		KAPRON
		KAPRON A
		KAPRON B
		KS 30P
		MARANYL F 114
		MARANYL F 124
		MARANYL F 500
		METAMID
		MIRAMID H 2
		MIRAMID WM 55
		NCI-C50646
		NYLON A1035SF
		NYLON CM 1031
		NYLON X 1051
		OMEGA-CAPROLACTAM
		ORGAMID RMNOCD
		ORGAMIDE
		P 6 (POLYAMIDE)
		PA 6
		PA 6 (POLYMER)
		PK 4
		PKA
		PLASKIN 8200
		PLASKON 8201
		PLASKON 8201HS
		PLASKON 8202C
		PLASKON 8205
		PLASKON 8207
		PLASKON 8252
		PLASKON XP 607
		POLYAMIDE PK 4
		RELON P
		RENYL MV
		SIPAS 60
		SPENCER 401
		SPENCER 601
		STEELON
		STILON
		STYLON
		TARLON X-A
		TARLON XB
		TARNAMID T
		TARNAMID T 2
		TARNAMID T 27
		TNK 2G5
		TORAYCA N 6
		UBE 1022B
		ULTRAMID B 3
		ULTRAMID B 4
		ULTRAMID B 5
		ULTRAMID BMK
		VIDLON
		WIDLON
		ZYTEL 211
105-64-6		DIISOPROPYL PERCARBONATE
		DIISOPROPYL PERDICARBONATE
		DIISOPROPYL PEROXYDICARBONATE
		DIISOPROPYL PEROXYDICARBONATE, MAXIMUM CONCENTRATION 52% IN SOLUTION (DOT)
		DIISOPROPYL PEROXYDICARBONATE, TECHNICAL PURE (DOT)
		DIISOPROPYL PEROXYDIFORMATE
		IPP
		ISOPROPYL PERCARBONATE
		ISOPROPYL PERCARBONATE, STABILIZED
		ISOPROPYL PERCARBONATE, STABILIZED (DOT)
		ISOPROPYL PERCARBONATE, UNSTABILIZED
		ISOPROPYL PERCARBONATE, UNSTABILIZED (DOT)
		ISOPROPYL PEROXYDICARBONATE
		ISOPROPYL PEROXYDICARBONATE, NOT MORE THAN 52% IN SOLUTION (DOT)
		ISOPROPYL PEROXYDICARBONATE, TECHNICALLY PURE (DOT)
		LUPEROX IPP
		NA 2133 (DOT)
		NA 2134 (DOT)
		PEROXYDICARBONATE D'ISOPROPYLE (FRENCH)
		PEROXYDICARBONIC ACID, BIS(1-METHYLETHYL) ESTER
		PEROXYDICARBONIC ACID, DIISOPROPYL ESTER
		PEROXYDICARBONIC ACID, DIISOPROPYL ESTER, NOT

CAS No.	IN	Chemical Name
		MORE THAN 52% IN SOLUTION
		PEROYL IPP
		UN 2133 (DOT)
		UN 2134 (DOT)
105-66-8		
		1-PROPYL BUTYRATE
		BUTANOIC ACID, PROPYL ESTER
		BUTYRIC ACID, PROPYL ESTER
		N-PROPYL BUTYRATE
		PROPYL BUTANOATE
		PROPYL BUTYRATE
		PROPYL N-BUTYRATE
105-67-9		
		1-HYDROXY-2,4-DIMETHYLBENZENE
	*	2,4-DIMETHYLPHENOL
		2,4-XYLENOL
		4,6-DIMETHYLPHENOL
		4-HYDROXY-1,3-DIMETHYLBENZENE
		M-XYLENOL
		PHENOL, 2,4-DIMETHYL-
105-74-8		
		ALPEROX C
		ALPEROX F
		DIDODECANOYL PEROXIDE
		DILAUROYL PEROXIDE
		DILAUROYL PEROXIDE, TECHNICAL PURE (DOT)
		DILAURYL PEROXIDE
		DODECANOYL PEROXIDE
		DYP-97F
		LAUROX
		LAUROX Q
		LAUROX W 40
		LAUROYL PEROXIDE
		LAUROYL PEROXIDE (DOT)
		LAUROYL PEROXIDE, NOT MORE THAN 42%,STABLE DISPERSION, IN WATER
		LAUROYL PEROXIDE, TECHNICALLY PURE (DOT)
		LAURYDOL
		LYP 97
		LYP 97F
		PEROXIDE, BIS(1-OXODODECYL)-
		PEROXYDE DE LAUROYLE (FRENCH)
		UN 2124 (DOT)
106-20-7		
		1-HEXANAMINE, 2-ETHYL-N-(2-ETHYLHEXYL)-
		2,2'-DIETHYLDIHEXYLAMINE
		BIS(2-ETHYLHEXYL)AMINE
		DI(2-ETHYLHEXYL)AMINE
		DIHEXYLAMINE, 2,2'-DIETHYL-
106-31-0		
		BUTANOIC ACID, ANHYDRIDE
		BUTANOIC ACID, ANHYDRIDE (9CI)
		BUTANOIC ANHYDRIDE
		BUTYRIC ACID ANHYDRIDE
	*	BUTYRIC ANHYDRIDE
		BUTYRIC ANHYDRIDE (DOT)
		BUTYRYL OXIDE
		N-BUTYRIC ACID ANHYDRIDE
		N-BUTYRIC ANHYDRIDE
		UN 2739 (DOT)
106-32-1		
		ETHYL CAPRYLATE
		ETHYL N-OCTANOATE
		ETHYL OCTANOATE
		ETHYL OCTOATE
		OCTANOIC ACID, ETHYL ESTER
106-35-4		
		3-HEPTANONE
		BUTYL ETHYL KETONE
	*	ETHYL BUTYL KETONE
		N-BUTYL ETHYL KETONE

CAS No.	IN	Chemical Name
106-36-5		
		N-PROPYL PROPIONATE
		PROPANOIC ACID, PROPYL ESTER
		PROPIONIC ACID, PROPYL ESTER
		PROPYL PROPANOATE
		PROPYL PROPIONATE
106-38-7		
		1-BROMO-4-METHYLBENZENE
		4-BROMO-1-METHYLBENZENE
	*	4-BROMOTOLUENE
		4-METHYL-1-BROMOBENZENE
		4-METHYLBROMOBENZENE
		4-METHYLPHENYL BROMIDE
		BENZENE, 1-BROMO-4-METHYL-
		P-BROMOTOLUENE
		P-METHYLBROMOBENZENE
		P-METHYLPHENYL BROMIDE
		P-TOLYL BROMIDE
		TOLUENE, P-BROMO-
106-42-3		
		1,4-DIMETHYLBENZENE
		1,4-XYLENE
		4-METHYLTOLUENE
		BENZENE, 1,4-DIMETHYL-
		CHROMAR
		P-DIMETHYLBENZENE
		P-METHYLTOLUENE
	*	P-XYLENE
		P-XYLENE (ACGIH,DOT)
		P-XYLOL
		P-XYLOL (DOT)
		SCINTILLAR
		UN 1307 (DOT)
106-43-4		
		1-CHLORO-4-METHYLBENZENE
		4-CHLORO-1-METHYLBENZENE
		4-CHLOROTOLUENE
		BENZENE, 1-CHLORO-4-METHYL-
		CHLOROTOLUENE
		P-CHLOROTOLUENE
		P-CHLOROTOLUENE (DOT)
		P-TOLYL CHLORIDE
		TOLUENE, P-CHLORO-
		UN 2238 (DOT)
106-44-5		
		1-HYDROXY-4-METHYLBENZENE
		1-METHYL-4-HYDROXYBENZENE
		4-CRESOL
		4-HYDROXYTOLUENE
		4-METHYLPHENOL
	*	P-CRESOL
		P-CRESOL (DOT)
		P-CRESYLIC ACID
		P-HYDROXYTOLUENE
		P-KRESOL
		P-METHYLHYDROXYBENZENE
		P-METHYLPHENOL
		P-OXYTOLUENE
		P-TOLUOL
		P-TOLYL ALCOHOL
		PARA-CRESOL
		PARAMETHYL PHENOL
		PHENOL, 4-METHYL-
		PHENOL, 4-METHYL- (9CI)
		RCRA WASTE NUMBER U052
		UN 2076 (DOT)
106-46-7		
		1,4-DICHLOORBENZEEN (DUTCH)
		1,4-DICHLOR-BENZOL (GERMAN)
	*	1,4-DICHLOROBENZENE
		1,4-DICLOROBENZENE (ITALIAN)
		BENZENE, 1,4-DICHLORO-
		BENZENE, P-DICHLORO-
		DI-CHLORICIDE

CAS No.	IN	Chemical Name
		DICHLOROBENZENE, PARA, SOLID
		DICHLOROBENZENE, PARA, SOLID (DOT)
		EVOLA
		NCI-C54955
		P-CHLOROPHENYL CHLORIDE
		P-DICHLOORBENZEEN (DUTCH)
		P-DICHLORBENZOL (GERMAN)
		P-DICHLOROBENZENE
		P-DICHLOROBENZENE (ACGIH)
		P-DICHLOROBENZENE, SOLID
		P-DICHLOROBENZOL
		P-DICLOROBENZENE (ITALIAN)
		PARA CRYSTALS
		PARACIDE
		PARADI
		PARADICHLORBENZOL (GERMAN)
		PARADICHLOROBENZENE
		PARADICHLOROBENZOL
		PARADOW
		PARAMOTH
		PARANUGGETS
		PARAZENE
		PDB
		PDCB
		PERSIA-PERAZOL
		RCRA WASTE NUMBER U070
		RCRA WASTE NUMBER U071
		RCRA WASTE NUMBER U072
		SANTOCHLOR
		UN 1592 (DOT)
106-47-8		
		P-CHLOROANILINE
106-48-9		
	*	4-CHLOROPHENOL
		4-CHLOROPHENOL (CHLOROPHENATE)
		4-HYDROXYCHLOROBENZENE
		APPLIED 3-78
		CHLOROPHENOL, SOLID
		P-CHLORFENOL (CZECH)
		P-CHLOROPHENOL
		P-CHLOROPHENOL, LIQUID (DOT)
		P-CHLOROPHENOL, SOLID (DOT)
		PARACHLOROPHENOL
		PHENOL, 4-CHLORO-
		PHENOL, P-CHLORO-
		UN 2020 (DOT)
		UN 2021 (DOT)
106-49-0		
		1-AMINO-4-METHYLBENZENE
		4-AMINO-1-METHYLBENZENE
		4-AMINOTOLUEN (CZECH)
		4-AMINOTOLUENE
		4-METHYLANILINE
		4-METHYLBENZENAMINE
		4-METHYLPHENYLAMINE
		4-TOLUIDINE
		ANILINE, P-METHYL-
		BENZENAMINE, 4-METHYL-
		C.I. 37107
		C.I. AZOIC COUPLING COMPONENT 107
		CI 37077
		CI 37107
		CI AZOIC COUPLING COMPONENT 107
		NAPHTOL AS-KG
		NAPHTOL AS-KGLL
		P-AMINOTOLUENE
		P-METHYLANILINE
		P-METHYLBENZENAMINE
		P-METHYLPHENYLAMINE
		P-TOLUIDIN (CZECH)
	*	P-TOLUIDINE
		P-TOLUIDINE (ACGIH,DOT)
		P-TOLYLAMINE
		RCRA WASTE NUMBER U328
		RCRA WASTE NUMBER U353
		TOLYLAMINE

CAS No.	IN	Chemical Name
		UN 1708 (DOT)
106-50-3		
		1,4-BENZENEDIAMINE
		1,4-DIAMINOBENZENE
		1,4-DIAMINOBENZOL
		1,4-PHENYLENEDIAMINE
		4-AMINOANILINE
		BASF URSOL D
		BENZOFUR D
		C.I. 76060
		C.I. DEVELOPER 13
		C.I. OXIDATION BASE 10
		DEVELOPER PF
		DURAFUR BLACK R
		FOURAMINE D
		FOURRINE 1
		FOURRINE D
		FUR BLACK 41867
		FUR BROWN 41866
		FUR YELLOW
		FURRO D
		FUTRAMINE D
		NAKO H
		ORSIN
		P-AMINOANILINE
		P-BENZENEDIAMINE
		P-DIAMINOBENZENE
		P-PHENYLENE DIAMINE
		P-PHENYLENEDIAMINE
		PELAGOL D
		PELAGOL DR
		PELAGOL GREY D
		PELTOL D
		RENAL PF
		TERTRAL D
		URSOL D
		ZOBA BLACK D
106-51-4		
		1,4-BENZOQUINE
		1,4-BENZOQUINONE
		1,4-CYCLOHEXADIENE DIOXIDE
		1,4-CYCLOHEXADIENEDIONE
		1,4-DIOSSIBENZENE (ITALIAN)
		1,4-DIOXY-BENZOL (GERMAN)
		1,4-DIOXYBENZENE
		2,5-CYCLOHEXADIENE-1,4-DIONE
		BENZO-CHINON (GERMAN)
		BENZOQUINONE
		BENZOQUINONE (DOT)
		CHINON (DUTCH, GERMAN)
		CHINONE
		CYCLOHEXADIENEDIONE
		NCI-C55845
		P-BENZOQUINONE
		P-BENZQUINONE
		P-CHINON (GERMAN)
		P-QUINONE
		QUINONE
		QUINONE (ACGIH)
		RCRA WASTE NUMBER U197
		STEARA PBQ
		UN 2587 (DOT)
		USAF P-220
106-63-8		
		2-METHYLPROPYL ACRYLATE
		2-PROPENOIC ACID, 2-METHYLPROPYL ESTER
		2-PROPENOIC ACID, 2-METHYLPROPYL ESTER (9CI)
		ACRYLIC ACID, ISOBUTYL ESTER
		ISOBUTYL 2-PROPENOATE
		ISOBUTYL ACRYLATE
		ISOBUTYL ACRYLATE, INHIBITED (DOT)
		ISOBUTYL PROPENOATE
		UN 2527 (DOT)
		3-OCTANONE
		5-METHYL 3-HEPTATONE
		AMYL ETHYL KETONE

CAS No.	IN	Chemical Name
		EAK
		ETHYL AMYL KETONE
		ETHYL AMYL KETONE (DOT)
		ETHYL N-AMYL KETONE
		ETHYL PENTYL KETONE
		N-OCTANONE-3
		UN 2271 (DOT)
106-71-8		
		2-CYANOETHYL ACRYLATE
		2-CYANOETHYL PROPENOATE
		2-PROPENOIC ACID, 2-CYANOETHYL ESTER
		ACRYLIC ACID, ESTER WITH HYDRACRYLONITRILE
		HYDRACRYLONITRILE, ACRYLATE (ESTER)
106-87-6		
		1,2-EPOXY-4-(EPOXYETHYL)CYCLOHEXANE
		1-(EPOXYETHYL)-3,4-EPOXYCYCLOHEXANE
		1-VINYL-3-CYCLOHEXENE DIOXIDE
		3-(1,2-EPOXYETHYL)-7-OXABICYCLO[4.1.0]HEPTANE
		3-(EPOXYETHYL)-7-OXABICYCLO[4.1.0]HEPTANE
		4-VINYL-1,2-CYCLOHEXENE DIEPOXIDE
		4-VINYL-1-CYCLOHEXENE DIEPOXIDE
		4-VINYL-1-CYCLOHEXENE DIOXIDE
		4-VINYLCYCLOHEXENE DIEPOXIDE
		4-VINYLCYCLOHEXENE DIOXIDE
		7-OXABICYCLO[4.1.0]HEPTANE, 3-(EPOXYETHYL)-
		7-OXABICYCLO[4.1.0]HEPTANE, 3-OXIRANYL-
		CHISSONOX 206 MONOMER
		VINYLCYCLOHEXENE DIEPOXIDE
		VINYLCYCLOHEXENE DIOXIDE
106-88-7		
		1,2-BUTENE OXIDE
		1,2-BUTYLENE EPOXIDE
		1,2-BUTYLENE OXIDE
		1,2-EPOXYBUTANE
		1-BUTENE OXIDE
		1-BUTYLENE OXIDE
		2-ETHYLOXIRANE
		ALPHA-BUTYLENE OXIDE
		BUTANE, 1,2-EPOXY-
		BUTYLENE OXIDE
		EPOXYBUTANE
		ETHYLENE OXIDE, ETHYL-
		ETHYLETHYLENE OXIDE
		ETHYLOXIRANE
		N-BUTENE-1,2-OXIDE
		NCI-C55527
		OXIRANE, ETHYL-
		PROPYL OXIRANE
106-89-8		
		(CHLOROMETHYL)ETHYLENE OXIDE
		(DL)-ALPHA-EPICHLOROHYDRIN
		1,2-EPOXY-3-CHLOROPROPANE
		1-CHLOOR-2,3-EPOXY-PROPAAN (DUTCH)
		1-CHLOR-2,3-EPOXY-PROPAN (GERMAN)
		1-CHLORO-2,3-EPOXYPROPANE
		1-CLORO-2,3-EPOSSIPROPANO (ITALIAN)
		2,3-EPOXYPROPYL CHLORIDE
		2-(CHLOROMETHYL)OXIRANE
		3-CHLORO-1,2-EPOXYPROPANE
		3-CHLORO-1,2-PROPYLENE OXIDE
		3-CHLOROPROPENE-1,2-OXIDE
		3-CHLOROPROPYLENE OXIDE
		ALPHA-EPICHLOROHYDRIN
		CHLOROHYDRINS
		CHLOROMETHYLOXIRANE
		CHLOROPROPYLENE OXIDE
		ECH
		EPICHLOORHYDRINE (DUTCH)
		EPICHLORHYDRIN (GERMAN)
		EPICHLORHYDRINE (FRENCH)
		EPICHLOROHYDRIN
		EPICHLOROHYDRIN (ACGIH,DOT)
		EPICHLOROHYDRYNA (POLISH)
		EPICLORIDRINA (ITALIAN)
		GAMMA-CHLOROPROPYLENE OXIDE
		GLYCEROL EPICHLORHYDRIN
		GLYCEROL EPICHLOROHYDRIN
		GLYCIDYL CHLORIDE
		OXIRANE, (CHLOROMETHYL)-
		OXIRANE, 2-(CHLOROMETHYL)
		PROPANE, 1-CHLORO-2,3-EPOXY-
		RCRA WASTE NUMBER U041
		SKEKHG
		UN 2023 (DOT)
106-91-2		
		1-PROPANOL, 2,3-EPOXY-, METHACRYLATE
		2,3-EPOXYPROPYL METHACRYLATE
		2-PROPENOIC ACID, 2-METHYL-, OXIRANYLMETHYL ESTER
		CP 105
		GLYCIDOL METHACRYLATE
		GLYCIDYL ALPHA-METHYL ACRYLATE
		GLYCIDYL METHACRYLATE
		METHACRYLIC ACID, 2,3-EPOXYPROPYL ESTER
106-92-3		
		1,2-EPOXY-3-ALLYLOXYPROPANE
		1-ALLILOSSI-2,3 EPOSSIPROPANO (ITALIAN)
		1-ALLYLOXY-2,3-EPOXY-PROPAAN (DUTCH)
		1-ALLYLOXY-2,3-EPOXYPROPAN (GERMAN)
		1-ALLYLOXY-2,3-EPOXYPROPANE
		AGE
		ALLIL-GLICIDIL-ETERE (ITALIAN)
		ALLYL 2,3-EPOXYPROPYL ETHER
		ALLYL GLYCIDYL ETHER
		ALLYL GLYCIDYL ETHER (ACGIH,DOT)
		ALLYLGLYCIDAETHER (GERMAN)
		ETHER, ALLYL 2,3-EPOXYPROPYL
		GLYCIDYL 2-PROPENYL ETHER
		GLYCIDYL ALLYL ETHER
		NCI-C56666
		OXIRANE, ((2-PROPENYLOXY)METHYL)-
		OXYDE D'ALLYLE ET DE GLYCIDYLE (FRENCH)
		PROPANE, 1-(ALLYLOXY)-2,3-EPOXY-
		UN 2219 (DOT)
		[(ALLYLOXY)METHYL]OXIRANE
106-93-4		
		1,2-DIBROMAETHAN (GERMAN)
		1,2-DIBROMOETANO (ITALIAN)
		1,2-DIBROMOETHANE
		1,2-DIBROMOETHANE (DOT)
		1,2-DIBROOMETHAAN (DUTCH)
		1,2-ETHYLENE DIBROMIDE
		AADIBROOM
		AETHYLENBROMID (GERMAN)
		ALPHA,BETA-DIBROMOETHANE
		BROMOFUME
		BROMURO DI ETILE (ITALIAN)
		CELMIDE
		DBE
		DIBROMOETHANE
		DIBROMURE D' ETHYLENE (FRENCH)
		DOWFUME 40
		DOWFUME EDB
		DOWFUME W-100
		DOWFUME W-8
		DOWFUME W-85
		DOWFUME W-90
		DWUBROMOETAN (POLISH)
		E-D-BEE
		EDB
		EDB-85
		ENT 15,349
		ETHANE, 1,2-DIBROMO-
		ETHYLENE BROMIDE
	*	ETHYLENE DIBROMIDE
		ETHYLENE DIBROMIDE (ACGIH,DOT)
		FUMO-GAS
		GLYCOL BROMIDE
		GLYCOL DIBROMIDE
		ISCOBROME D
		KOPFUME
		NCI-C00522

CAS No.	IN	Chemical Name
		NEFIS
		NEPHIS
		PESTMASTER
		PESTMASTER EDB-85
		RCRA WASTE NUMBER U067
		SANHYUUM
		SOILBROM
		SOILBROM-100
		SOILBROM-40
		SOILBROM-85
		SOILBROM-90
		SOILBROM-90EC
		SOILBROME-85
		SOILFUME
		SYM-DIBROMOETHANE
		UN 1605 (DOT)
		UNIFUME
106-94-5		
		1-BROMOPROPANE
		1-BROMOPROPANE (DOT)
		BROMOPROPANE
		N-PROPYL BROMIDE
		PROPANE, 1-BROMO-
		PROPYL BROMIDE
		UN 2344 (DOT)
106-95-6		
		1-BROMO-2-PROPENE
		1-PROPENE, 3-BROMO-
		2-PROPENYL BROMIDE
		3-BROMO-1-PROPENE
		3-BROMOPROPENE
		3-BROMOPROPYLENE
		ALLYL BROMIDE
		ALLYL BROMIDE (DOT)
		BROMALLYLENE
		PROPENE, 3-BROMO-
		UN 1099 (DOT)
106-96-7		
		1-BROMO-2-PROPYNE
		1-PROPYNE, 3-BROMO-
		2-PROPYNYL BROMIDE
		3-BROMO-1-PROPYNE
		3-BROMOPROPYNE
		3-BROMOPROPYNE (DOT)
		BROMOPROPYNE
		GAMMA-BROMOALLYLENE
		PROPARGYL BROMIDE
		PROPYNE, 3-BROMO-
		UN 2345 (DOT)
106-97-8		
		BURSHANE
	*	BUTANE
		BUTANE (ACGIH,DOT)
		BUTANEN (DUTCH)
		BUTANI (ITALIAN)
		DIETHYL
		DIMETHYLMETHANE
		FUELS, LIQUEFIED PETROLEUM GAS
		LIQUEFIED PETROLEUM GAS
		LPG
		METHYLETHYLMETHANE
		N-BUTANE
		N-PROPANE
		PETROLEUM PRODUCTS, LIQUEFIED GASES
		PROPYL HYDRIDE
		PYROFAX
		R 290
		R 600
		UN 1011 (DOT)
106-98-9		
		1-BUTENE
		1-BUTYLENE
		ALPHA-BUTENE
		ALPHA-BUTYLENE
		BUTENE-1
		ETHYLETHYLENE
106-99-0		
		1,3-BUTADIENE
		1,3-BUTADIENE (ACGIH)
		ALPHA,GAMMA-BUTADIENE
		BIETHYLENE
		BIVINYL
		BUTA-1,3-DIEEN (DUTCH)
		BUTA-1,3-DIEN (GERMAN)
		BUTA-1,3-DIENE
		BUTADIEEN (DUTCH)
		BUTADIEN (POLISH)
		BUTADIENE
		BUTADIENE, INHIBITED
		BUTADIENE, INHIBITED (DOT)
		BUTYNE
		DIVINYL
		ERYTHRENE
		NCI-C50602
		PLIOLITE
		PYRROLYLENE
		UN 1010 (DOT)
		VINYLETHYLENE
107-00-6		
		1-BUTYNE
		ETHYL ACETYLENE
		ETHYL ACETYLENE, INHIBITED
		ETHYL ACETYLENE, INHIBITED (DOT)
		ETHYLETHYNE
		UN 2452 (DOT)
107-01-7		
		2-BUTENE
		BETA-BUTENE
		BETA-BUTYLENE
		PSEUDOBUTYLENE
107-02-8		
		2-PROPEN-1-ONE
		2-PROPENAL
		ACQUINITE
		ACRALDEHYDE
		ACRALDEHYDEACROLEINA (ITALIAN)
		ACROLEIN
		ACROLEIN (ACGIH)
		ACROLEIN, INHIBITED
		ACROLEIN, INHIBITED (DOT)
		ACROLEINA (ITALIAN)
		ACROLEINE (DUTCH, FRENCH)
		ACRYLALDEHYD (GERMAN)
		ACRYLALDEHYDE
		ACRYLIC ALDEHYDE
		AKROLEIN (CZECH)
		AKROLEINA (POLISH)
		ALDEHYDE ACRYLIQUE (FRENCH)
		ALDEIDE ACRILICA (ITALIAN)
		ALLYL ALDEHYDE
		AQUALIN
		AQUALINE
		BIOCIDE
		CROLEAN
		ETHYLENE ALDEHYDE
		MAGNACIDE
		MAGNACIDE H
		NSC 8819
		PROP-2-EN-1-AL
		PROPENAL
		PROPENAL (CZECH)
		PROPYLENE ALDEHYDE
		RCRA WASTE NUMBER P003
		SLIMICIDE
		TRANS-ACROLEIN
		UN 1092 (DOT)

CAS No.	IN	Chemical Name
107-03-9		
		1-PROPANETHIOL
		1-PROPANETHIOL (DOT)
		1-PROPYL MERCAPTAN
		3-MERCAPTOPROPANOL
		N-PROPANETHIOL
		N-PROPYL MERCAPTAN
		N-PROPYLTHIOL
		PROPANE-1-THIOL
		PROPANETHIOL
	*	PROPYL MERCAPTAN
		PROPYL MERCAPTAN (DOT)
		PROPYLTHIOL
		UN 2402 (DOT)
		UN 2704 (DOT)
107-05-1		
		1-CHLOROPROPENE-2
		1-CHLORO-2-PROPENE
		1-PROPENE, 3-CHLORO-
		2-PROPENYL CHLORIDE
		3-CHLORO-1-PROPENE
		3-CHLORO-1-PROPYLENE
		3-CHLOROPRENE
		3-CHLOROPROPENE
		3-CHLOROPROPENE-1
		3-CHLOROPROPYLENE
		3-CHLORPROPEN (GERMAN)
		ALLILE (CLORURO DI) (ITALIAN)
	*	ALLYL CHLORIDE
		ALLYL CHLORIDE (ACGIH,DOT)
		ALLYLCHLORID (GERMAN)
		ALLYLE (CHLORURE D') (FRENCH)
		ALPHA-CHLOROPROPYLENE
		CHLORALLYLENE
		CHLOROALLYLENE
		NCI-C04615
		PROPENE, 3-CHLORO-
		UN 1100 (DOT)
107-06-2		
		1,2-BICHLOROETHANE
		1,2-DCE
		1,2-DICHLOORETHAAN (DUTCH)
		1,2-DICHLOR-AETHAN (GERMAN)
		1,2-DICHLORETHANE
	*	1,2-DICHLOROETHANE
		1,2-DICLOROETANO (ITALIAN)
		1,2-ETHYLENE DICHLORIDE
		1,2-ETHYLIDENE DICHLORIDE
		AETHYLENCHLORID (GERMAN)
		ALPHA,BETA-DICHLOROETHANE
		BICHLORURE D'ETHYLENE (FRENCH)
		BORER SOL
		BROCIDE
		CHLORURE D'ETHYLENE (FRENCH)
		CLORURO DI ETHENE (ITALIAN)
		DESTRUXOL BORER-SOL
		DICHLOR-MULSION
		DICHLOREMULSION
		DICHLORO-1,2-ETHANE (FRENCH)
		DICHLOROETHYLENE
		DUTCH LIQUID
		DUTCH OIL
		EDC
		ENT 1,656
		ETHANE DICHLORIDE
		ETHANE, 1,2-DICHLORO-
		ETHYLEENDICHLORIDE (DUTCH)
		ETHYLENE CHLORIDE
		ETHYLENE DICHLORIDE
		ETHYLENE DICHLORIDE (ACGIH,DOT)
		GLYCOL DICHLORIDE
		NCI-C00511
		RCRA WASTE NUMBER U077
		SYM-DICHLOROETHANE
		UN 1184 (DOT)

CAS No.	IN	Chemical Name
107-07-3		
		2-CHLOORETHANOL (DUTCH)
		2-CHLORAETHANOL (GERMAN)
		2-CHLORETHANOL (GERMAN)
		2-CHLORO-1-ETHANOL
		2-CHLOROETHANOL
		2-CHLOROETHYL ALCOHOL
		2-CLOROETANOLO (ITALIAN)
		2-HYDROXYETHYL CHLORIDE
		2-MONOCHLOROETHANOL
		AETHYLENECHLORHYDRIN (GERMAN)
		BETA-CHLOROETHANOL
		BETA-CHLOROETHYL ALCOHOL
		BETA-HYDROXYETHYL CHLORIDE
		CHLOROETHANOL
		CHLOROETHYLOWY ALKOHOL (POLISH)
		DELTA-CHLOROETHANOL
		ETHANOL, 2-CHLORO-
		ETHENE CHLOROHYDRIN
		ETHYLCHLOROHYDRIN
		ETHYLEEN-CHLOORHYDRINE (DUTCH)
		ETHYLENE CHLORHYDRIN
	*	ETHYLENE CHLOROHYDRIN
		ETHYLENE CHLOROHYDRIN (ACGIH,DOT)
		ETHYLENE GLYCOL, CHLOROHYDRIN
		GLICOL MONOCLORIDRINA (ITALIAN)
		GLYCOL CHLOROHYDRIN
		GLYCOL MONOCHLOROHYDRIN
		GLYCOLMONOCHLOORHYDRINE (DUTCH)
		GLYCOMONOCHLORHYDRIN
		MONOCHLORHYDRINE DU GLYCOL (FRENCH)
		NCI-C50135
		UN 1135 (DOT)
107-10-8		
		1-AMINOPROPANE
		1-PROPANAMINE
		1-PROPYLAMINE
		MONO-N-PROPYLAMINE
		MONOPROPYLAMINE
		MONOPROPYLAMINE (DOT)
		N-PROPYLAMINE
		PROPANAMINE
	*	PROPYLAMINE
		PROPYLAMINE (DOT)
		RCRA WASTE NUMBER U194
		UN 1277 (DOT)
107-11-9		
		1-AMINOPROP-2-ENE
		2-PROPEN-1-AMINE
		2-PROPEN-1-AMINE (9CI)
		2-PROPENAMINE
		2-PROPENYLAMINE
		3-AMINO-1-PROPENE
		3-AMINOPROPENE
		3-AMINOPROPYLENE
	*	ALLYLAMINE
		ALLYLAMINE (DOT)
		MONOALLYLAMINE
		UN 2334 (DOT)
107-12-0		
		CYANOETHANE
		ETHER CYANATUS
		ETHYL CYANIDE
		HYDROCYANIC ETHER
		N-PROPANENITRILE
		PROPANENITRILE
		PROPIONIC NITRILE
	*	PROPIONITRILE
		PROPIONITRILE (DOT)
		PROPIONONITRILE
		PROPYLNITRILE
		RCRA WASTE NUMBER P101
		UN 2404 (DOT)

CAS No.	IN	Chemical Name
107-13-1		
		2-PROPENENITRILE
		ACRYLON
	*	ACRYLONITRILE
		CARBACRYL
		CYANOETHENE
		CYANOETHYLENE
		FUMIGRAIN
		PROPENENITRILE
		VCN
		VENTOX
		VINYL CYANIDE
107-14-2		
		2-CHLORO ACETONITRILE
		ACETONITRILE, CHLORO-
		ALPHA-CHLOROACETONITRILE
		CHLORACETONITRILE
		CHLOROACETONITRILE
		CHLOROACETONITRILE (DOT)
		CHLOROMETHYL CYANIDE
		MONOCHLOROACETONITRILE
		MONOCHLOROMETHYL CYANIDE
		UN 2668 (DOT)
		USAF KF-5
107-15-3		
		1,2-DIAMINOETHANE
		1,2-ETHANEDIAMINE
		1,2-ETHYLENEDIAMINE
		BETA-AMINOETHYLAMINE
		DIAMINOETHANE
		DIMETHYLENEDIAMINE
	*	ETHYLENEDIAMINE
107-16-4		
		2-HYDROXYACETONITRILE
		CYANOMETHANOL
		FORMALDEHYDE CYANOHYDRIN
		GLYCOLIC NITRILE
		GLYCOLONITRILE
		GLYCOLONITRILE (8CI)
		GLYCONITRILE
		HYDROXYACETONITRILE
		HYDROXYMETHYL CYANIDE
		HYDROXYMETHYLNITRILE
		METHYLNITRILE, HYDROXY-
		USAF A-8565
107-18-6		
		1-PROPEN-3-OL
		1-PROPENOL-3
		2-PROPEN-1-OL
		2-PROPENE-1-OL
		2-PROPENOL
		2-PROPENYL ALCOHOL
		3-HYDROXY-1-PROPENE
		3-HYDROXYPROPENE
		AA
		ALCOOL ALLILCO (ITALIAN)
		ALCOOL ALLYLIQUE (FRENCH)
		ALLILOWY ALKOHOL (POLISH)
		ALLYL AL
		ALLYL ALCOHOL
		ALLYL ALCOHOL (ACGIH,DOT)
		ALLYLALKOHOL (GERMAN)
		ALLYLIC ALCOHOL
		ORVINYLCARBINOL
		PROPEN-1-OL-3
		PROPENOL
		PROPENYL ALCOHOL
		RCRA WASTE NUMBER P005
		SHELL UNKRAUTTOD A
		UN 1098 (DOT)
		VINYLCARBINOL
		WEED DRENCH

CAS No.	IN	Chemical Name
107-19-7		
		1-HYDROXY-2-PROPYNE
		1-PROPYN-3-OL
		1-PROPYN-3-YL ALCOHOL
		1-PROPYNE-3-OL
		2-PROPYN-1-OL
		2-PROPYNOL
		2-PROPYNYL ALCOHOL
		3-HYDROXY-1-PROPYNE
		3-PROPYNOL
		ETHYNYLCARBINOL
		METHANOL, ETHYNYL-
		NA 1986 (DOT)
		PROPARGYL ALCOHOL
		PROPARGYL ALCOHOL (ACGIH,DOT)
		PROPYNYL ALCOHOL
		RCRA WASTE NUMBER P102
107-20-0		
		2-CHLORO-1-ETHANAL
		2-CHLOROACETALDEHYDE
		2-CHLOROETHANAL
		ACETALDEHYDE, CHLORO-
		ALPHA-CHLOROACETALDEHYDE
	*	CHLOROACETALDEHYDE
		CHLOROACETALDEHYDE (ACGIH,DOT)
		CHLOROACETALDEHYDE MONOMER
		MONOCHLOROACETALDEHYDE
		RCRA WASTE NUMBER P023
		UN 2232 (DOT)
107-21-1		
		1,2-DIHYDROXYETHANE
		1,2-ETHANEDIOL
		2-HYDROXYETHANOL
		DOWTHERM SR 1
		ETHYLENE ALCOHOL
		ETHYLENE DIHYDRATE
	*	ETHYLENE GLYCOL
		ETHYLENE GLYCOL PARTICULATE AND VAPOR
		FRIDEX
		GLYCOL
		GLYCOL ALCOHOL
		MACROGOL 400 BPC
		MONOETHYLENE GLYCOL
		NORKOOL
		RAMP
		TESCOL
		UCAR 17
		ZEREX
107-22-2		
		1,2-ETHANEDIONE
		BIFORMAL
		BIFORMYL
		DIFORMAL
		DIFORMYL
		ETHANDIAL
		ETHANEDIAL
		ETHANEDIONE
		GLYOXAL
		GLYOXAL ALDEHYDE
		GOHSEZAL P
		OXAL
		OXALALDEHYDE
		PERMAFRESH 114
107-25-5		
		1-METHOXYETHYLENE
		ETHENE, METHOXY-
		ETHER, METHYL VINYL
		METHOXYETHENE
		METHOXYETHYLENE
		METHYL VINYL ETHER
		UN 1087 (DOT)
		VINYL METHYL ETHER
		VINYL METHYL ETHER (DOT)
		VINYL METHYL ETHER, INHIBITED (DOT)

CAS No.	IN	Chemical Name
107-27-7		
		ETHYLMERCURIC CHLORIDE
107-29-9		
		ACETALDEHYDE OXIME
		ACETALDEHYDE OXIME (DOT)
		ACETALDOXIME
		ALDOXIME
		ETHANAL OXIME
		ETHYLIDENEHYDROXYLAMINE
		UN 2332 (DOT)
		USAF AM-5
107-30-2		
		ALPHA,ALPHA-DICHLORODIMETHYL ETHER
		CHLORDIMETHYLETHER (CZECH)
		CHLORODIMETHYL ETHER
		CHLOROMETHOXYMETHANE
		CHLOROMETHYL METHYL ETHER
		CHLOROMETHYL METHYL ETHER (TECHNICAL GRADE)
		CMME
		DIMETHYLCHLOROETHER
		ETHER METHYLIQUE MONOCHLORE (FRENCH)
		ETHER, CHLOROMETHYL METHYL
		ETHER, DIMETHYL CHLORO
		METHANE, CHLOROMETHOXY-
		METHOXYCHLOROMETHANE
		METHOXYMETHYL CHLORIDE
		METHYL CHLOROMETHYL ETHER
		METHYL CHLOROMETHYL ETHER, ANHYDROUS
		METHYL CHLOROMETHYL ETHER, ANHYDROUS (DOT)
		METHYLCHLOROMETHYL ETHER (DOT)
		MONOCHLORODIMETHYL ETHER
		MONOCHLOROMETHYL METHYL ETHER
		RCRA WASTE NUMBER U046
		UN 1239 (DOT)
107-31-3		
		FORMIATE DE METHYLE (FRENCH)
		FORMIC ACID, METHYL ESTER
		METHANOIC ACID METHYL ESTER
	*	METHYL FORMATE
		METHYL FORMATE (ACGIH,DOT)
		METHYL METHANOATE
		METHYLE (FORMIATE DE) (FRENCH)
		METHYLFORMIAAT (DUTCH)
		METHYLFORMIAT (GERMAN)
		METIL (FORMIATO DI) (ITALIAN)
		R 611
		UN 1243 (DOT)
107-37-9		
		ALLYL TRICHLOROSILANE
		ALLYL TRICHLOROSILANE (DOT)
		ALLYL TRICHLOROSILANE, STABILIZED (DOT)
		PROPEN-3-YLTRICHLOROSILANE
		SILANE, ALLYLTRICHLORO-
		SILANE, TRICHLORO-2-PROPENYL-
		SILANE, TRICHLOROALLYL-
		UN 1724 (DOT)
107-39-1		
		1-PENTENE, 2,4,4-TRIMETHYL-
		2,2,4-TRIMETHYL-4-PENTENE
		2,4,4-TRIMETHYL-1-PENTENE
107-40-4		
		2,2,4-TRIMETHYL-3-PENTENE
		2,4,4-TRIMETHYL-2-PENTENE
		2-PENTENE, 2,4,4-TRIMETHYL-
107-41-5		
		1,1,3-TRIMETHYLTRIMETHYLENEDIOL
		1,2-HEXANEDIOL
		2,4-DIHYDROXY-2-METHYLPENTANE
		2,4-PENTANEDIOL, 2-METHYL-
		2-METHYL PENTANE-2,4-DIOL
		2-METHYL-2,4-PENTANDIOL
		2-METHYL-2,4-PENTANEDIOL
		4-METHYL-2,4-PENTANEDIOL
		ALPHA,ALPHA,ALPHA'-TRIMETHYLTRIMETHYLENE GLYCOL
		DIOLANE
	*	HEXYLENE GLYCOL
		HEXYLENE GLYCOL (ACGIH)
		ISOL
		MPD
		PINAKON
107-44-8		
		EA 1208
		GB
		IMPF
		ISOPROPOXYMETHYLPHOSPHORYL FLUORIDE
		ISOPROPYL METHANEFLUOROPHOSPHONATE
		ISOPROPYL METHYLFLUOROPHOSPHATE
		ISOPROPYL METHYLFLUOROPHOSPHONATE
		ISOPROPYL METHYLPHOSPHONOFLUORIDATE
		ISOPROPYL-METHYL-PHOSPHORYL FLUORIDE
		METHYLFLUOROPHOSPHORIC ACID ISOPROPYL ESTER
		METHYLFLUORPHOSPHORSAEUREISOPROPYLESTER (GERMAN)
		METHYLPHOSPHONOFLUORIDIC ACID 1-METHYLETHYL ESTER
		METHYLPHOSPHONOFLUORIDIC ACID ISOPROPYL ESTER
		MFI
		O-ISOPROPYL METHYLFLUOROPHOSPHONATE
		O-ISOPROPYL METHYLPHOSPHONOFLUORIDATE
		PHOSPHINE OXIDE, FLUOROISOPROPOXYMETHYL-
		PHOSPHONOFLUORIDIC ACID, METHYL-, 1-METHYLETHYL ESTER
		PHOSPHONOFLUORIDIC ACID, METHYL-, ISOPROPYL ESTER
		PHOSPHORIC ACID, METHYLFLUORO-, ISOPROPYL ESTER
		SARIN
		SARIN II
		T-144
		T-2106
		TL 1618
		TRILONE 46
107-45-9		
		1,1,3,3-TETRAMETHYLBUTANAMINE
		1,1,3,3-TETRAMETHYLBUTYLAMINE
		2,4,4-TRIMETHYL-2-PENTYLAMINE
		2-PENTANAMINE, 2,4,4-TRIMETHYL-
		BUTYLAMINE, 1,1,3,3-TETRAMETHYL-
		TERT-OCTYLAMINE
107-49-3		
		BIS-O,O-DIETHYLPHOSPHORIC ANHYDRIDE
		BLADAN
		DIPHOSPHORIC ACID, TETRAETHYL ESTER
		ETHYL PYROPHOSPHATE (ET4P2O7)
		NA 2783 (DOT)
		NIFOS
		PYRODUST
		PYROPHOSPHORIC ACID, TETRAETHYL ESTER
		PYROPHOSPHORIC ACID, TETRAETHYL ESTER (MIXTURE)
		TEP
		TEPP
		TETRAETHYL DIPHOSPHATE
		TETRAETHYL PYROPHOSPHATE
		TETRAETHYL PYROPHOSPHATE MIXTURE, DRY
		TETRAETHYL PYROPHOSPHATE MIXTURE, DRY (DOT)
		TETRAETHYL PYROPHOSPHATE MIXTURE, LIQUID
		TETRAETHYL PYROPHOSPHATE MIXTURE, LIQUID (DOT)
		TETRAETHYL PYROPHOSPHATE, LIQUID [FLAMMABLE LIQUID LABEL]
		TETRAETHYL PYROPHOSPHATE, LIQUID [POISON B LABEL]
		TETRASTIGMINE
		TETRON
		TETRON (PESTICIDE)
		TETRON 100
		VAPOTONE
107-66-4		
		DI-N-BUTYL PHOSPHATE
		DIBUTYL ACID PHOSPHATE
		DIBUTYL HYDROGEN PHOSPHATE

CAS No.	IN	Chemical Name
	*	DIBUTYL PHOSPHATE
		PHOSPHORIC ACID, DIBUTYL ESTER
107-70-0		
	*	4-METHOXY-4-METHYLPENTAN-2-ONE
		GUAIACOL
107-71-1		
		ACETYL TERT-BUTYL PEROXIDE
		ETHANEPEROXOIC ACID, 1,1-DIMETHYLETHYL ESTER
		LUPERSOL 70
		PEROXYACETIC ACID, TERT-BUTYL ESTER
		TERT-BUTYL PERACETATE
		TERT-BUTYL PEROXYACETATE
107-72-2		
		AMYL TRICHLOROSILANE
		AMYL TRICHLOROSILANE (DOT)
		MYLSILANE
		PENTYLSILICON TRICHLORIDE
		PENTYLTRICHLOROSILANE
		SILANE, PENTYLTRICHLORO-
		SILANE, TRICHLOROPENTYL-
		TRICHLOROA
		TRICHLOROAMYLSILANE
		TRICHLOROPENTYLSILANE
		UN 1728 (DOT)
107-82-4		
		1-BROMO-3-METHYLBUTANE
		1-BROMO-3-METHYLBUTANE (DOT)
		3-METHYL-1-BROMOBUTANE
		3-METHYLBUTYL BROMIDE
		4-BROMO-2-METHYLBUTANE
	*	BROMOMETHYLBUTANE
		BUTANE, 1-BROMO-3-METHYL-
		ISOAMYL BROMIDE
		ISOPENTYL BROMIDE
		UN 2341 (DOT)
107-83-5		
		2-METHYLPENTANE
		ISOHEXANE
		PENTANE, 2-METHYL-
107-84-6		
		1-CHLORO-3,3-DIMETHYLPROPANE
		1-CHLORO-3-METHYLBUTANE
		3-METHYLBUTYL CHLORIDE
		4-CHLORO-2-METHYLBUTANE
		BUTANE, 1-CHLORO-3-METHYL-
		ISOAMYL CHLORIDE
		ISOPENTYL CHLORIDE
107-87-9		
		2-PENTANONE
		ETHYL ACETONE
	*	METHYL PROPYL KETONE
		METHYL PROPYL KETONE (ACGIH,DOT)
		METHYL-N-PROPYL KETONE
		METHYL-PROPYL-CETONE (FRENCH)
		METYLOPROPYLOKETON (POLISH)
		MPK
		PROPYL METHYL KETONE
		UN 1249 (DOT)
107-89-1		
		3-BUTANOLAL
		3-HYDROXYBUTANAL
		3-HYDROXYBUTYRALDEHYDE
		ACETALDOL
	*	ALDOL
		ALDOL (DOT)
		BETA-HYDROXYBUTYRALDEHYDE
		BUTANAL, 3-HYDROXY-
		BUTYRALDEHYDE, 3-HYDROXY-
		OXYBUTANAL
		OXYBUTYRIC ALDEHYDE
		UN 2839 (DOT)
107-92-6		
		1-PROPANECARBOXYLIC ACID
		BUTANIC ACID
		BUTANOIC ACID
		BUTRIC ACID
		BUTTERSAEURE (GERMAN)
		BUTYRATE
	*	BUTYRIC ACID
		BUTYRIC ACID (DOT)
		ETHYLACETIC ACID
		N-BUTANOIC ACID
		N-BUTYRIC ACID
		PROPYLFORMIC ACID
		UN 2820 (DOT)
107-98-2		
		1-METHOXY-2-HYDROXYPROPANE
		1-METHOXY-2-PROPANOL
		2-METHOXY-1-METHYLETHANOL
		2-PROPANOL, 1-METHOXY-
		DOWANOL-33B
		DOWTHERM 209
		PROPYLENE GLYCOL METHYL ETHER
	*	PROPYLENE GLYCOL MONOMETHYL ETHER
108-01-0		
		(2-HYDROXYETHYL)DIMETHYLAMINE
		(DIMETHYLAMINO)ETHANOL
		2-(DIMETHYLAMINO)-1-ETHANOL
		2-(N,N-DIMETHYLAMINO)ETHANOL
		2-DIMETHYLAMINOETHANOL
		AMIETOL M 21
		BETA-(DIMETHYLAMINO)ETHANOL
		BETA-DIMETHYLAMINOETHANOL
		BETA-DIMETHYLAMINOETHYL ALCOHOL
		BETA-HYDROXYETHYLDIMETHYLAMINE
		BIMANOL
		DEANOL
		DIMETHYL(2-HYDROXYETHYL)AMINE
		DIMETHYL(HYDROXYETHYL)AMINE
		DIMETHYLAETHANOLAMIN (GERMAN)
		DIMETHYLAMINOAETHANOL (GERMAN)
		DIMETHYLAMINOETHANOL
		DIMETHYLETHANOLAMINE
		DIMETHYLETHANOLAMINE (DOT)
		DIMETHYLMONOETHANOLAMINE
		DMAE
		ETHANOL, 2-DIMETHYLAMINO-
		KALPUR P
		LIPARON
		N,N-DIMETHYL(2-HYDROXYETHYL)AMINE
		N,N-DIMETHYL-2-AMINOETHANOL
		N,N-DIMETHYL-2-HYDROXYETHYLAMINE
		N,N-DIMETHYL-BETA-HYDROXYETHYLAMINE
		N,N-DIMETHYL-N-(2-HYDROXYETHYL)AMINE
		N,N-DIMETHYL-N-(BETA-HYDROXYETHYL)AMINE
		N,N-DIMETHYLAMINOETHANOL
		N,N-DIMETHYLETHANOLAMINE
		N-(2-HYDROXYETHYL)DIMETHYLAMINE
		N-DIMETHYLAMINOETHANOL
		NORCHOLINE
		PROPAMINE A
		UN 2051 (DOT)
108-03-2		
	*	1-NITROPROPANE
		PROPANE, 1-NITRO-
108-05-4		
		1-ACETOXYETHYLENE
		ACETATE DE VINYLE (FRENCH)
		ACETIC ACID ETHENYL ESTER
		ACETIC ACID VINYL ESTER
		ACETIC ACID, ETHYLENE ETHER
		ACETOXYETHYLENE
		ETHANOIC ACID, ETHENYL ESTER
		ETHENYL ACETATE
		ETHENYL ETHANOATE
		OCTAN WINYLU (POLISH)

CAS No.	IN	Chemical Name
		UN 1301 (DOT)
		VAC
		VINILE (ACETATO DI) (ITALIAN)
		VINYL A MONOMER
	*	VINYL ACETATE
		VINYL ACETATE (ACGIH,DOT)
		VINYL ACETATE HQ
		VINYL ACETATE MONOMER
		VINYL ACETATE, INHIBITED (DOT)
		VINYL ETHANOATE
		VINYLACETAAT (DUTCH)
		VINYLACETAT (GERMAN)
		VINYLE (ACETATE DE) (FRENCH)
		VYAC
		ZESET T
108-08-7		
		2,4-DIMETHYLPENTANE
		PENTANE, 2,4-DIMETHYL-
108-09-8		
		1,3-DIMETHYLBUTANAMINE
		1,3-DIMETHYLBUTYLAMINE
		1,3-DIMETHYLBUTYLAMINE (DOT)
		2-AMINO-4-METHYLPENTANE
		2-PENTANAMINE, 4-METHYL-
		4-METHYL-2-AMINOPENTANE
		BUTYLAMINE, 1,3-DIMETHYL-
		UN 2379 (DOT)
108-10-1		
		2-METHYL-4-PENTANONE
		2-METHYLPROPYL METHYL KETONE
		2-PENTANONE, 4-METHYL-
		4-METHYL-2-OXOPENTANE
		4-METHYL-2-PENTANONE
		HEXONE
		ISOBUTYL METHYL KETONE
		ISOPROPYLACETONE
	*	METHYL ISOBUTYL KETONE
		MIBK
		MIK
108-11-2		
		1,3-DIMETHYL BUTANOL
		1,3-DIMETHYL-1-BUTANOL
		1-PENTANOL, 2-METHYL-
		1-PENTANOL, METHYL-
		2-METHYL-1-PENTANOL
		2-METHYL-2-PROPYLETHANOL
		2-METHYL-4-PENTANOL
		2-METHYLPENTANOL-1
		2-PENTANOL, 4-METHYL-
		3-MIC
		4-METHYL-2-PENTANOL
		4-METHYL-2-PENTYL ALCOHOL
		4-METHYLPENTANOL-2
		4-METILPENTAN-2-OLO (ITALIAN)
		ALCOOL METHYL AMYLIQUE (FRENCH)
		AMYL METHYL ALCOHOL
		ISOBUTYLMETHYLCARBINOL
		ISOBUTYLMETHYLMETHANOL
		ISOHEXYL ALCOHOL
		ISOPROPYL DIMETHYL CARBINOL
		MAOH
		METHYL AMYL ALCOHOL
		METHYL ISOBUTYL CARBINOL
		METHYL ISOBUTYL CARBINOL (ACGIH,DOT)
		METHYL-1-PENTANOL
	*	METHYLAMYL ALCOHOL
		METHYLISOBUTYL CARBINOL
		METILAMIL ALCOHOL (ITALIAN)
		MIBC
		MIC
		UN 2053 (DOT)
108-16-7		
		1-(DIMETHYLAMINO)-2-PROPANOL
		2-PROPANOL, 1-(DIMETHYLAMINO)-
		DIMETHYL(2-HYDROXYPROPYL)AMINE
		DIMETHYLISOPROPANOLAMINE
		N,N-DIMETHYL-2-HYDROXYPROPYLAMINE
		N,N-DIMETHYLAMINO-2-PROPANOL
		N,N-DIMETHYLISOPROPANOLAMINE
108-18-9		
		2-PROPANAMINE, N-(1-METHYLETHYL)-
		DIISOPROPYLAMINE
		DIISOPROPYLAMINE (ACGIH,DOT)
		DIPA
		N,N-DIISOPROPYLAMINE
		N-(1-METHYLETHYL)-2-PROPANAMINE
		UN 1158 (DOT)
108-20-3		
		2-ISOPROPOXYPROPANE
		BIS(ISOPROPYL) ETHER
		DIISOPROPYL ETHER
		DIISOPROPYL OXIDE
		DIISOPROPYLETHER (DOT)
		ETHER ISOPROPYLIQUE (FRENCH)
		ETHER, ISOPROPYL
		ISOPROPYL ETHER
		ISOPROPYL ETHER (ACGIH)
		IZOPROPYLOWY ETER (POLISH)
		PROPANE, 2,2'-OXYBIS-
		UN 1159 (DOT)
108-21-4		
		1-METHYLETHYL ACETATE
		2-ACETOXYPROPANE
		2-PROPYL ACETATE
		ACETATE D'ISOPROPYLE (FRENCH)
		ACETIC ACID, 1-METHYLETHYL ESTER
		ACETIC ACID, 1-METHYLETHYL ESTER (9CI)
		ACETIC ACID, ISOPROPYL ESTER
		ISOPROPILE (ACETATO DI) (ITALIAN)
		ISOPROPYL (ACETATE D') (FRENCH)
		ISOPROPYL ACETATE
		ISOPROPYL ACETATE (ACGIH,DOT)
		ISOPROPYL ETHANOATE
		ISOPROPYLACETAAT (DUTCH)
		ISOPROPYLACETAT (GERMAN)
		UN 1220 (DOT)
108-22-5		
		1-ACETOXY-1-METHYLETHYLENE
		1-METHYLVINYL ACETATE
		1-PROPEN-2-OL, ACETATE
		1-PROPEN-2-YL ACETATE
		2-ACETOXYPROPENE
		2-ACETOXYPROPYLENE
		ACETIC ACID ISOPROPENYL ESTER
		ISOPROPENYL ACETATE
		ISOPROPENYL ACETATE (DOT)
		METHYLVINYL ACETATE
		PROPEN-2-YL ACETATE
		UN 2403 (DOT)
108-23-6		
		CARBONOCHLORIDE ACID, 1-METHYLETHYL ESTER
		CARBONOCHLORIDIC ACID, 1-METHYLETHYL ESTER
		CHLOROFORMIC ACID ISOPROPYL ESTER
		FORMIC ACID, CHLORO-, ISOPROPYL ESTER
		ISOPROPYL CHLOROCARBONATE
	*	ISOPROPYL CHLOROFORMATE
		ISOPROPYL CHLOROFORMATE (DOT)
		ISOPROPYL CHLOROMETHANOATE
		UN 2407 (DOT)
108-24-7		
		ACETIC ACID, ANHYDRIDE
		ACETIC ACID, ANHYDRIDE (9CI)
	*	ACETIC ANHYDRIDE
		ACETIC ANHYDRIDE (ACGIH,DOT)
		ACETIC OXIDE
		ACETYL ACETATE

CAS No.	IN	Chemical Name
		ACETYL ANHYDRIDE
		ACETYL ETHER
		ACETYL OXIDE
		ANHYDRIDE ACETIQUE (FRENCH)
		ANIDRIDE ACETICA (ITALIAN)
		AZIJNZUURANHYDRIDE (DUTCH)
		ESSIGSAEUREANHYDRID (GERMAN)
		ETHANOIC ANHYDRATE
		ETHANOIC ANHYDRIDE
		OCTOWY BEZWODNIK (POLISH)
		UN 1715 (DOT)
108-31-6		
		2,5-FURANDIONE
		BM 10
		CIS-BUTENEDIOIC ANHYDRIDE
		DIHYDRO-2,5-DIOXOFURAN
		MALEIC ACID ANHYDRIDE
	*	MALEIC ANHYDRIDE
		MALEIC ANHYDRIDE (ACGIH,DOT)
		MALEIC ANHYDRIDE, SOLID OR MOLTEN (DOT)
		RCRA WASTE NUMBER U147
		TOXILIC ANHYDRIDE
		UN 2215 (DOT)
108-34-9		
		PYRAZOXON
108-38-3		
		1,3-DIMETHYLBENZENE
		1,3-XYLENE
		BENZENE, 1,3-DIMETHYL-
		M-DIMETHYLBENZENE
		M-METHYLTOLUENE
	*	M-XYLENE
		M-XYLENE (ACGIH,DOT)
		M-XYLOL
		M-XYLOL (DOT)
		UN 1307 (DOT)
108-39-4		
		1-HYDROXY-3-METHYLBENZENE
		3-CRESOL
		3-HYDROXYTOLUENE
		3-METHYLPHENOL
	*	M-CRESOL
		M-CRESOL (DOT)
		M-CRESOLE
		M-CRESYLIC ACID
		M-HYDROXYTOLUENE
		M-KRESOL
		M-METHYLPHENOL
		M-OXYTOLUENE
		M-TOLUOL
		PHENOL, 3-METHYL-
		PHENOL, 3-METHYL- (9CI)
		RCRA WASTE NUMBER U052
		UN 2076 (DOT)
108-43-0		
		3-CHLOROPHENOL (CHLOROPHENATE)
108-44-1		
		1-AMINO-3-METHYLBENZENE
		3-AMINO-1-METHYLBENZENE
		3-AMINOPHENYLMETHANE
		3-AMINOTOLUEN (CZECH)
		3-AMINOTOLUENE
		3-METHYLANILINE
		3-METHYLBENZENAMINE
		3-METHYLPHENYLAMINE
		3-TOLUIDINE
		ANILINE, 3-METHYL-
		BENZENAMINE, 3-METHYL-
		CI 37077
		CI 37107
		CI AZOIC COUPLING COMPONENT 107
		M-AMINOTOLUENE
		M-METHYLANILINE
		M-METHYLBENZENAMINE
		M-TOLUIDIN (CZECH)
		M-TOLUIDINE
		M-TOLUIDINE (ACGIH,DOT)
		M-TOLYLAMINE
		META-TOLUIDINE
		NAPHTOL AS-KG
		NAPHTOL AS-KGLL
		RCRA WASTE NUMBER U328
		RCRA WASTE NUMBER U353
		TOLYLAMINE
		UN 1708 (DOT)
108-45-2		
		1,3-BENZENEDIAMINE
		1,3-DIAMINOBENZENE
		1,3-PHENYLENEDIAMINE
		3-AMINOANILINE
		APCO 2330
		CI 76025
		CI DEVELOPER 11
		DEVELOPER 11
		DEVELOPER C
		DEVELOPER H
		DEVELOPER M
		DIRECT BROWN BR
		DIRECT BROWN GG
		M-AMINOALINE
		M-AMINOANILINE
		M-BENZENEDIAMINE
		M-DIAMINOBENZENE
		M-FENYLENDIAMIN (CZECH)
	*	M-PHENYLENEDIAMINE
		M-PHENYLENEDIAMINE (DOT)
		METAPHENYLENEDIAMINE
		PHENYLENEDIAMINE, META, SOLID (DOT)
		UN 1673 (DOT)
108-46-3		
		1,3-BENZENEDIOL
		1,3-DIHYDROXYBENZENE
		3-HYDROXYCYCLOHEXADIEN-1-ONE
		3-HYDROXYPHENOL
		BENZENE, M-DIHYDROXY-
		C.I. 76505
		C.I. DEVELOPER 4
		C.I. OXIDATION BASE 31
		DEVELOPER O
		DEVELOPER R
		DEVELOPER RS
		DURAFUR DEVELOPER G
		FOURAMINE RS
		FOURRINE 79
		FOURRINE EW
		M-BENZENEDIOL
		M-DIHYDROXYBENZENE
		M-DIOXYBENZENE
		M-HYDROQUINONE
		M-HYDROXYPHENOL
		NAKO TGG
		NCI-C05970
		PELAGOL GREY RS
		PELAGOL RS
		PHENOL, M-HYDROXY-
		RCRA WASTE NUMBER U201
		RESO
		RESORCIN
		RESORCINE
	*	RESORCINOL
		RESORCINOL (ACGIH,DOT)
		UN 2876 (DOT)
108-57-6		
		1,3-DIVINYLBENZENE
		BENZENE, 1,3-DIETHENYL-
		BENZENE, M-DIVINYL-
	*	DIVINYL BENZENE
		M-DIVINYLBENZENE

CAS No.	IN	Chemical Name
		M-VINYLSTYRENE
108-57-7		
		DIVINYLBENZENE
108-60-1		
		2,2-DICHLORO ISOPROPYL ETHER
		BIS(2-CHLORO-1-METHYLETHYL)ETHER
108-62-3		
		1,3,5,7-TETROXOCANE, 2,4,6,8-TETRAMETHYL-
		ACETALDEHYDE, HOMOPOLYMER
		ACETALDEHYDE, POLYMERS
		ACETALDEHYDE, TETRAMER
		ANTIMILACE
		ARIOTOX
		CEKUMETA
		LIMAX
		LIMOVET
		META
		METACETALDEHYDE
		METALDEHYD (GERMAN)
		METALDEHYDE
		METALDEHYDE (DOT)
		METALDEIDE (ITALIAN)
		METASON
		NAMEKIL
		POLYACETALDEHYDE
		SLUG-TOX
		SLUGIT
		UN 1332 (DOT)
108-67-8		
	*	1,3,5-TRIMETHYLBENZENE
		1,3,5-TRIMETHYLBENZENE (DOT)
		3,5-DIMETHYLTOLUENE
		BENZENE, 1,3,5-TRIMETHYL-
		FLEET-X
		MESITYLENE
		SYM-TRIMETHYLBENZENE
		TMB
		TRIMETHYL BENZENE (ACGIH)
		TRIMETHYLBENZOL
		UN 2325 (DOT)
108-77-0		
		1,3,5-TRIAZINE, 2,4,6-TRICHLORO-
		1,3,5-TRICHLOROTRIAZINE
		2,4,6-TRICHLORO-1,3,5-TRIAZINE
		2,4,6-TRICHLORO-S-TRIAZINE
		2,4,6-TRICHLORO-SYM-TRIAZINE
		2,4,6-TRICHLOROTRIAZINE
		CHLOROTRIAZINE
		CYANUR CHLORIDE
		CYANURCHLORIDE
		CYANURIC ACID CHLORIDE
		CYANURIC CHLORIDE
		CYANURIC CHLORIDE (DOT)
		CYANURIC TRICHLORIDE
		CYANURIC TRICHLORIDE (DOT)
		CYANURYL CHLORIDE
		KYANURCHLORID (CZECH)
		S-TRIAZINE TRICHLORIDE
		S-TRIAZINE, 2,4,6-TRICHLORO-
		SYM-TRICHLOROTRIAZINE
		SYN-TRICHLOTRIAZIN (CZECH)
		TRICHLORO-S-TRIAZINE
		TRICHLOROCYANIDINE
		TRICYANOGEN CHLORIDE
		UN 2670 (DOT)
108-78-1		
		MELAMINE
108-82-7		
		2,6-DIMETHYL HEPTANOL-4
		2,6-DIMETHYL-4-HEPTANOL
		4-HEPTANOL, 2,6-DIMETHYL-
		DIISOBUTYL CARBINOL

CAS No.	IN	Chemical Name
		SEC-NONYL ALCOHOL
108-83-8		
	*	2,6-DIMETHYL-4-HEPTANONE
		2,6-DIMETHYL-HEPTAN-4-ON (DUTCH, GERMAN)
		2,6-DIMETHYLHEPTAN-4-ONE
		2,6-DIMETHYLHEPTANONE
		2,6-DIMETIL-EPTAN-4-ONE (ITALIAN)
		4-HEPTANONE, 2,6-DIMETHYL-
		DI-ISOBUTYLCETONE (FRENCH)
		DIBC
		DIBK
		DIISOBUTILCHETONE (ITALIAN)
		DIISOBUTYL KETONE
		DIISOBUTYL KETONE (ACGIH,DOT)
		DIISOBUTYLKETON (DUTCH, GERMAN)
		ISOBUTYL KETONE
		ISOVALERONE
		S-DIISOPROPYLACETONE
		SYM-DIISOPROPYLACETONE
		UN 1157 (DOT)
		VALERONE
108-84-9		
		1,3-DIMETHYLBUTYL ACETATE
		2-PENTANOL, 4-METHYL-, ACETATE
		4-METHYL-2-PENTANOL, ACETATE
		4-METHYL-2-PENTYL ACETATE
		4-METHYLPENT-2-YL ETHANOATE
		ACETIC ACID, 1,3-DIMETHYLBUTYL ESTER
		HEXYL ACETATE
		MAAC
	*	METHYL AMYL ACETATE
		METHYL AMYL ACETATE (DOT)
		METHYLISOAMYL ACETATE
		METHYLISOBUTYLCARBINOL ACETATE
		METHYLISOBUTYLCARBINYL ACETATE
		SEC-HEXYL ACETATE
		SEC-HEXYL ACETATE (ACGIH)
		UN 1233 (DOT)
108-86-1		
		BENZENE, BROMO-
	*	BROMOBENZENE
		BROMOBENZENE (DOT)
		MONOBROMOBENZENE
		NCI-C55492
		PHENYL BROMIDE
		UN 2514 (DOT)
108-87-2		
		1-METHYLCYCLOHEXANE
		CYCLOHEXANE, METHYL-
		CYCLOHEXYLMETHANE
		HEXAHYDROTOLUENE
	*	METHYLCYCLOHEXANE
		METHYLCYCLOHEXANE (ACGIH,DOT)
		METYLOCYKLOHEKSAN (POLISH)
		SEXTONE B
		TOLUENE HEXAHYDRIDE
		TOLUENE, HEXAHYDRO-
		UN 2296 (DOT)
108-88-3		
		ANTISAL 1A
		BENZENE, METHYL-
		CP 25
		METHACIDE
		METHANE, PHENYL-
		METHYLBENZENE
		METHYLBENZOL
		NCI-C07272
		PHENYLMETHANE
		RCRA WASTE NUMBER U220
		TOLU-SOL
		TOLUEEN (DUTCH)
		TOLUEN (CZECH)
	*	TOLUENE
		TOLUENE (ACGIH,DOT)

CAS No.	IN	Chemical Name
		TOLUOL
		TOLUOL (DOT)
		TOLUOLO (ITALIAN)
		TULUOL
		UN 1294 (DOT)
108-89-4		
		4-METHYLPYRIDINE
	*	4-PICOLINE
		BA 35846
		GAMMA-METHYLPYRIDINE
		GAMMA-PICOLINE
		P-METHYLPYRIDINE
		P-PICOLINE
		PYRIDINE, 4-METHYL-
108-90-7		
		BENZENE CHLORIDE
		BENZENE, CHLORO-
		CHLOORBENZEEN (DUTCH)
		CHLORBENZENE
		CHLORBENZOL
		CHLOROBENZEN (POLISH)
	*	CHLOROBENZENE
		CHLOROBENZENE (ACGIH,DOT)
		CHLOROBENZENU (CZECH)
		CHLOROBENZOL (DOT)
		CLOROBENZENE (ITALIAN)
		CP 27
		IP CARRIER T 40
		MCB
		MONOCHLOORBENZEEN (DUTCH)
		MONOCHLORBENZENE
		MONOCHLORBENZOL (GERMAN)
		MONOCHLOROBENZENE
		MONOCLOROBENZENE (ITALIAN)
		NCI-C54886
		PHENYL CHLORIDE
		RCRA WASTE NUMBER U037
		TETROSIN SP
		UN 1134 (DOT)
108-91-8		
		1-AMINOCYCLOHEXANE
		AMINOCYCLOHEXANE
		AMINOHEXAHYDROBENZENE
		BENZENAMINE, HEXAHYDRO-
		CYCLOHEXANAMINE
	*	CYCLOHEXYLAMINE
		HEXAHYDROANILINE
108-93-0		
		1-CYCLOHEXANOL
		ADRONAL
		ADRONOL
		ANOL
	*	CYCLOHEXANOL
		CYCLOHEXYL ALCOHOL
		HEXAHYDROPHENOL
		HEXALIN
		HEXALIN (ALCOHOL)
		HYDROXYCYCLOHEXANE
		NAXOL
		PHENOL, HEXAHYDRO-
108-94-1		
		ANON
		ANONE
	*	CYCLOHEXANONE
		CYCLOHEXENONE
		HEXANON
		HYTROL O
		NADONE
		PIMELIC KETONE
		PIMELIN KETONE
		SEXTONE

CAS No.	IN	Chemical Name
108-95-2		
		BENZENOL
		CARBOLIC ACID
		HYDROXYBENZENE
		IZAL
		MONOHYDROXYBENZENE
		MONOPHENOL
		OXYBENZENE
		PHENIC ACID
	*	PHENOL
		PHENYL ALCOHOL
		PHENYL HYDRATE
		PHENYL HYDROXIDE
		PHENYLIC ACID
		PHENYLIC ALCOHOL
108-98-5		
		BENZENETHIOL
		BENZENETHIOL (DOT)
		MERCAPTOBENZENE
		PHENOL, THIO-
		PHENYL MERCAPTAN
		PHENYL MERCAPTAN (ACGIH,DOT)
		PHENYLTHIOL
		RCRA WASTE NUMBER P014
		THIOPHENOL
		THIOPHENOL (DOT)
		UN 2337 (DOT)
		USAF XR-19
109-01-3		
		1-METHYL PIPERAZINE
		METHYLPIPERAZINE
		N-METHYLPIPERAZINE
		PIPERAZINE, 1-METHYL-
109-02-4		
		1-METHYLMORPHOLINE
		4-METHYLMORPHOLINE
		METHYLMORPHOLINE
		METHYLMORPHOLINE (DOT)
		MORPHOLINE, 4-METHYL-
		MORPHOLINE, N-METHYL-
		N-METHYL MORPHOLINE
		UN 2535 (DOT)
109-06-8		
		2-METHYLPYRIDINE
	*	2-PICOLINE
		ALPHA-METHYLPYRIDINE
		ALPHA-PICOLINE
		O-METHYLPYRIDINE
		O-PICOLINE
		O-PICOLINE (DOT)
		PICOLINE
		PYRIDINE, 2-METHYL-
		RCRA WASTE NUMBER U191
		UN 2313 (DOT)
109-08-0		
		2-METHYL-1,4-DIAZINE
		2-METHYLPYRAZINE
		METHYLPYRAZINE
		PYRAZINE, METHYL-
109-09-1		
		2-CHLOROPYRIDINE
109-13-7		
		ESPEROX 24M
		KAYAESTER I
		LUPERSOL 8
		PERBUTYL IB
		PEROXYISOBUTYRIC ACID, TERT-BUTYL ESTER
		PEROXYISOBUTYRIC ACID, TERT-BUTYL ESTER, MORE THAN 52% BUT NOT MORE THAN 77% IN SOLUTION
		PEROXYISOBUTYRIC ACID, TERT-BUTYL ESTER, NOT MORE THAN 52% IN SOLUTION
		PROPANEPEROXOIC ACID, 2-METHYL-, 1,1-DIMETHYL

CAS No.	IN	Chemical Name
		ETHYL ESTER
		PROPANEPEROXOIC ACID, 2-METHYL-, 1,1-DIMETHYL-ETHYL ESTER
		TERT-BUTYL 2-METHYLPROPANEPEROXOATE
		TERT-BUTYL PERISOBUTYRATE
		TERT-BUTYL PEROXYISOBUTYRATE
		TERT-BUTYL PEROXYISOBUTYRATE, MORE THAN 52% BUT NOT MORE THAN 77% IN SOLUTION (DOT)
		TERT-BUTYL PEROXYISOBUTYRATE, NOT MORE THAN 52% IN SOLUTION (DOT)
		UN 2142 (DOT)
		UN 2562 (DOT)
109-19-2		
		BUTYL ISOVALERATE
109-21-7		
		1-BUTYL BUTYRATE
		BUTANOIC ACID, BUTYL ESTER
		BUTYL BUTANOATE
		BUTYL BUTYLATE
	*	BUTYL BUTYRATE
		BUTYRIC ACID, BUTYL ESTER
		N-BUTYL BUTANOATE
		N-BUTYL BUTYRATE
		N-BUTYL N-BUTYRATE
109-27-3		
		1-GUANYL-4-NITROAMINOGUANYL-1-TETRAZENE
109-47-7		
		BUTYL ALCOHOL, HYDROGEN PHOSPHITE
		DIBUTYL HYDROGEN PHOSPHITE
		DIBUTYL PHOSPHITE
		PHOSPHOROUS ACID, DIBUTYL ESTER
109-52-4		
		1-BUTANECARBOXYLIC ACID
		BUTANECARBOXYLIC ACID
		N-PENTANOIC ACID
		N-VALERIC ACID
		NA 1760 (DOT)
		PENTANOIC ACID
		PROPYLACETIC ACID
		VALERIANIC ACID
		VALERIC ACID
		VALERIC ACID (DOT)
109-53-5		
		ETHER, ISOBUTYL VINYL
		ISOBUTANOL VINYL ETHER
		ISOBUTOXYETHENE
		ISOBUTYL VINYL ETHER
		IVE
		LUTANOL LR 8500
		PROPANE, 1-(ETHENYLOXY)-2-METHYL-
		UN 1304 (DOT)
		VINOFLEX MO 400
		VINYL ISOBUTYL ETHER
		VINYL ISOBUTYL ETHER (DOT)
		VINYL ISOBUTYL ETHER, INHIBITED (DOT)
109-55-7		
		(3-AMINOPROPYL)DIMETHYLAMINE
		1,3-PROPANEDIAMINE, N,N-DIMETHYL-
		1-AMINO-3-(DIMETHYLAMINO)PROPANE
		1-DIMETHYLAMINO-3-AMINOPROPANE
		3-(DIMETHYLAMINO)-1-AMINOPROPANE
		3-(DIMETHYLAMINO)-1-PROPANAMINE
		3-(DIMETHYLAMINO)-1-PROPYLAMINE
		3-(DIMETHYLAMINO)PROPANAMINE
		3-(DIMETHYLAMINO)PROPYLAMINE
		3-AMINO-1-(DIMETHYLAMINO)PROPANE
		GAMMA-DIMETHYLAMINOPROPYLAMINE
		N,N-DIMETHYL-1,3-DIAMINOPROPANE
		N,N-DIMETHYL-1,3-PROPANEDIAMINE
		N,N-DIMETHYL-1,3-PROPYLENEDIAMINE
		N,N-DIMETHYL-N-(3-AMINOPROPYL)AMINE
		N,N-DIMETHYLPROPYLENEDIAMINE
		N,N-DIMETHYLTRIMETHYLENEDIAMINE
109-56-8		
		(N-HYDROXYETHYL)ISOPROPYLAMINE
		2-ISOPROPYLAMINOETHANOL
		ETHANOL, 2-((1-METHYLETHYL)AMINO)- (9CI)
		ETHANOL, 2-(ISOPROPYLAMINO)-
		ETHANOL, 2-[(1-METHYLETHYL)AMINO]-
		ETHANOLISOPROPYLAMINE
		ISOPROPYLAMINOETHANOL
		ISOPROPYLAMINOETHANOL [COMBUSTIBLE LIQUID LABEL]
		ISOPROPYLAMINOETHANOL [FLAMMABLE LIQUID LABEL]
		MONOISOPROPYLAMINOETHANOL
		N-ISOPROPYLAMINOETHANOL
		N-ISOPROPYLETHANOLAMINE
109-59-1		
		2-ISOPROPOXYETHANOL
		BETA-HYDROXYETHYL ISOPROPYL ETHER
		DOWANAL EIPAT
		ETHANOL, 2-(1-METHYLETHOXY)-
		ETHANOL, 2-ISOPROPOXY-
		ETHYLENE GLYCOL ISOPROPYL ETHER
		ETHYLENE GLYCOL MONISOPROPYL ETHER
		ETHYLENE GLYCOL MONOISOPROPYL ETHER
		ISOPROPOXYETHANOL
		ISOPROPYL CELLOSOLVE
		ISOPROPYL OXITOL
		UCAR AC
109-60-4		
		1-ACETOXYPROPANE
		1-PROPYL ACETATE
		ACETATE DE PROPYLE NORMAL (FRENCH)
		ACETIC ACID N-PROPYL ESTER
		ACETIC ACID, PROPYL ESTER
		N-PROPANOL ACETATE
	*	N-PROPYL ACETATE
		N-PROPYL ACETATE (ACGIH,DOT)
		OCTAN PROPYLU (POLISH)
		PROPYL ACETATE
		PROPYL ACETATE (DOT)
		PROPYL ETHANOATE
		UN 1276 (DOT)
109-61-5		
		CARBONOCHLORIDIC ACID, PROPYL ESTER
		FORMIC ACID, CHLORO-, PROPYL ESTER
		N-PROPYL CHLOROFORMATE
		N-PROPYL CHLOROFORMATE (DOT)
		PROPYL CHLOROCARBONATE
		PROPYL CHLOROFORMATE
		UN 2740 (DOT)
109-63-7		
		BORANE, TRIFLUORO-, COMPD. WITH 1,1'-OXYBIS[ETHANE] (1:1)
		BORON FLUORIDE (BF3), COMPD. WITH ETHYL ETHER (1:1)
		BORON FLUORIDE DIETHYL ETHER COMPLEX
		BORON FLUORIDE DIETHYL ETHERATE
		BORON FLUORIDE ETHERATE
		BORON FLUORIDE MONOETHERATE
		BORON FLUORIDE-DIETHYL ETHER COMPOUND
		BORON FLUORIDE-ETHYL ETHER COMPLEX
		BORON FLUORIDE-ETHYL ETHERATE
	*	BORON TRIFLUORIDE DIETHYL ETHERATE
		BORON TRIFLUORIDE ETHERATE
		BORON TRIFLUORIDE ETHYL ETHERATE (1:1)
		BORON TRIFLUORIDE-DIETHYL ETHER 1:1 COMPLEX
		BORON TRIFLUORIDE-DIETHYL ETHER COMPLEX
		BORON TRIFLUORIDE-DIETHYL ETHER COMPLEX (1:1)
		BORON TRIFLUORIDE-ETHER COMPLEX
		BORON TRIFLUORIDE-ETHYL ETHER
		BORON TRIFLUORIDE-ETHYL ETHER (1:1)
		BORON TRIFLUORIDE-ETHYL ETHER COMPLEX
		BORON TRIFLUORIDE-ETHYL ETHERATE
		BORON, TRIFLUORO[1,1'-OXYBIS[ETHANE]]-, (T-4)-
		DIETHYL ETHER COMPOUND WITH BORON TRIFLUORIDE
		DIETHYL ETHER TRIFLUOROBORANE COMPLEX

CAS No.	IN	Chemical Name
		ETHANE, 1,1'-OXYBIS-, COMPD. WITH TRIFLUOROBORANE (1:1)
		ETHYL ETHER, COMPD. WITH BORON FLUORIDE (BF3) (1:1)
		ETHYL ETHER-BORON TRIFLUORIDE COMPLEX
		TRIFLUOROBORANE DIETHYL ETHERATE
		TRIFLUOROBORANE-1,1'-OXYBIS[ETHANE] (1:1)
		TRIFLUOROBORON ETHERATE
		TRIFLUOROBORON-DIETHYL ETHER COMPLEX
109-65-9		
		1-BROMOBUTANE
		BUTANE, 1-BROMO-
	*	BUTYL BROMIDE
		BUTYL BROMIDE (DOT)
		BUTYL BROMIDE, NORMAL
		BUTYL BROMIDE, NORMAL (DOT)
		N-BUTYL BROMIDE
		UN 1126 (DOT)
109-66-0		
		AMYL HYDRIDE (DOT)
		N-PENTANE
		PENTAN (POLISH)
	*	PENTANE
		PENTANE (ACGIH,DOT)
		PENTANEN (DUTCH)
		PENTANI (ITALIAN)
		SKELLYSOLVE A
		UN 1265 (DOT)
109-67-1		
	*	1-PENTENE
		ALPHA-N-AMYLENE
		AMYLENE
		AMYLENE, NORMAL (DOT)
		N-AMYLENE (DOT)
		PENTENE
		PENTYLENE
		PROPYLETHYLENE
		UN 1108 (DOT)
109-69-3		
		1-CHLOROBUTANE
		1-CHLOROBUTANE (DOT)
		BUTANE, 1-CHLORO-
	*	BUTYL CHLORIDE
		BUTYL CHLORIDE (DOT)
		CHLOROBUTANE
		CHLORURE DE BUTYLE (FRENCH)
		N-BUTYL CHLORIDE
		N-BUTYL CHLORIDE (DOT)
		N-CHLOROBUTANE
		N-PROPYLCARBINYL CHLORIDE
		NCI-C06155
		UN 1127 (DOT)
109-70-6		
		1,3-CHBP
		1-BROMO-3-CHLOROPROPANE
		1-CHLORO-3-BROMOPROPANE
		1-CHLORO-3-BROMOPROPANE (DOT)
		3-BROMO-1-CHLOROPROPANE
		3-BROMOPROPYL CHLORIDE
		3-CHLORO-1-BROMOPROPANE
		3-CHLOROPROPYL BROMIDE
		CHLOROBROMOPROPANE
		OMEGA-CHLOROBROMOPROPANE
		PROPANE, 1-BROMO-3-CHLORO-
		TRIMETHYLENE BROMIDE CHLORIDE
		TRIMETHYLENE CHLOROBROMIDE
		UN 2688 (DOT)
109-73-9		
		1-AMINO-BUTAAN (DUTCH)
		1-AMINOBUTAN (GERMAN)
		1-AMINOBUTANE
		1-BUTANAMINE
		1-BUTYLAMINE

CAS No.	IN	Chemical Name
	*	BUTYLAMINE
		BUTYLAMINE (DOT)
		MONO-N-BUTYLAMINE
		MONOBUTYLAMINE
		N-BUTILAMINA (ITALIAN)
		N-BUTYLAMIN (GERMAN)
		N-BUTYLAMINE
		N-BUTYLAMINE (ACGIH)
		NORVALAMINE
		UN 1125 (DOT)
109-74-0		
		1-CYANOPROPANE
		BUTANENITRILE
		BUTYRIC ACID NITRILE
	*	BUTYRONITRILE
		BUTYRONITRILE (DOT)
		BUTYRYLONITRILE
		N-BUTANENITRILE
		N-BUTYRONITRILE
		PROPYL CYANIDE
		UN 2411 (DOT)
109-76-2		
		1,3-DIAMINOPROPANE
		1,3-PROPANEDIAMINE
		1,3-PROPYLENEDIAMINE
		1,3-TRIMETHYLENEDIAMINE
		TMEDA
		TRIMETHYLENEDIAMINE
109-77-3		
		MALONONITRILE
109-78-4		
		2-CYANO-1-ETHANOL
		2-CYANOETHANOL
		2-CYANOETHYL ALCOHOL
		2-HYDROXYCYANOETHANE
		2-HYDROXYETHYL CYANIDE
		3-HYDROXYPROPANENITRILE
		3-HYDROXYPROPIONITRILE
		BETA-CYANOETHANOL
		BETA-HPN
		BETA-HYDROXYPROPIONITRILE
		ETHYLENE CYANOHYDRIN
		GLYCOL CYANOHYDRIN
		HYDRACRYLONITRILE
		METHANOLACETONITRILE
		PROPANENITRILE, 3-HYDROXY-
		PROPIONITRILE, 3-HYDROXY-
		USAF RH-7
109-79-5		
		1-BUTANETHIOL
		1-BUTYL MERCAPTAN
		BUTANETHIOL
		BUTYL MERCAPTAN
		N-BUTANETHIOL
		N-BUTYL MERCAPTAN
		N-BUTYL THIOALCOHOL
		THIOBUTYL ALCOHOL
109-83-1		
		(2-HYDROXYETHYL)METHYLAMINE
		(HYDROXYETHYL)METHYLAMINE
		2-(METHYLAMINO)ETHANOL
		2-(N-METHYLAMINO)ETHANOL
		2-HYDROXY-N-METHYLETHYLAMINE
		2-HYDROXYETHYL-N-METHYLAMINE
		2-METHYLAMINO-1-ETHANOL
		2-METHYLAMINOETHANOL
		2-N-MONOMETHYLAMINOETHANOL
		AMIETOL M 11
		BETA-(METHYLAMINO)ETHANOL
		ETHANOL, 2-(METHYLAMINO)-
		METHYL(2-HYDROXYETHYL)AMINE
		METHYL(BETA-HYDROXYETHYL)AMINE

CAS No.	IN	Chemical Name
		METHYL(HYDROXYETHYL)AMINE
		METHYLAMINOETHANOL
		METHYLETHANOLAMINE
		METHYLETHYLOLAMINE
		MONOMETHYL-AMINOAETHANOL (GERMAN)
		MONOMETHYLAMINOETHANOL
		MONOMETHYLETHANOLAMINE
		MONOMETHYLMONOETHANOLAMINE
		N-(2-HYDROXYETHYL)METHYLAMINE
		N-METHYL-2-AMINOETHANOL
		N-METHYL-2-ETHANOLAMINE
		N-METHYL-2-HYDROXYETHYLAMINE
		N-METHYL-N-(2-HYDROXYETHYL)AMINE
		N-METHYL-N-(BETA-HYDROXYETHYL)AMINE
		N-METHYL-N-HYDROXYETHYLAMINE
		N-METHYLAMINOETHANOL
		N-METHYLETHANOLAMINE
		N-METHYLMONOETHANOLAMINE
		N-MONOMETHYLAMINOETHANOL
		N-MONOMETHYLETHANOLAMINE
		USAF DO-50
109-86-4		
		1-HYDROXY-2-METHOXYETHANE
		2-METHOXY-1-ETHANOL
		2-METHOXY-AETHANOL (GERMAN)
	*	2-METHOXYETHANOL
		2-METHOXYETHANOL (ACGIH)
		2-METHOXYETHYL ALCOHOL
		2-METOSSIETANOLO (ITALIAN)
		AETHYLENGLYKOL-MONOMETHYLAETHER (GERMAN)
		BETA-METHOXYETHANOL
		DOWANOL EM
		EGM
		EGME
		EKTASOLVE EM
		ETHANOL, 2-METHOXY-
		ETHER MONOMETHYLIQUE DE L'ETHYLENE-GLYCOL (FRENCH)
		ETHYLENE GLYCOL METHYL ETHER
		ETHYLENE GLYCOL MONOMETHYL ETHER
		ETHYLENE GLYCOL MONOMETHYL ETHER (DOT)
		ETHYLENE GLYCOL, MONOMETHYL ETHER
		GLYCOL ETHER EM
		GLYCOL METHYL ETHER
		GLYCOL MONOMETHYL ETHER
		JEFFERSOL EM
		MECS
		METHOXYETHANOL
		METHOXYETHYLENE GLYCOL
		METHOXYHYDROXYETHANE
		METHYL CELLOSOLVE
		METHYL CELLOSOLVE (DOT)
		METHYL ETHOXOL
		METHYL GLYCOL
		METHYL OXITOL
		METHYLGLYKOL (GERMAN)
		METIL CELLOSOLVE (ITALIAN)
		METOKSYETYLOWY ALKOHOL (POLISH)
		MONOMETHYL ETHER OF ETHYLENE GLYCOL
		MONOMETHYLGLYCOL
		POLY-SOLV EM
		PRIST
		UN 1188 (DOT)
109-87-5		
		2,4-DIOXAPENTANE
		ANESTHENYL
		DIMETHOXYMETHANE
		DIMETHYL FORMAL
		FORMAL
		FORMALDEHYDE DIMETHYL ACETAL
		FORMALDEHYDE METHYL KETAL
		METHANE, DIMETHOXY-
		METHOXYMETHYL METHYL ETHER
	*	METHYLAL
		METHYLAL (ACGIH,DOT)
		METHYLENE DIMETHYL ETHER
		METYLAL (POLISH)
		UN 1234 (DOT)
109-89-7		
		DEA
		DEN
		DIAETHYLAMIN (GERMAN)
		DIETHAMINE
	*	DIETHYLAMINE
		DIETHYLAMINE (ACGIH,DOT)
		DIETILAMINA (ITALIAN)
		DWUETYLOAMINA (POLISH)
		ETHANAMINE, N-ETHYL-
		N,N-DIETHYLAMINE
		UN 1154 (DOT)
109-90-0		
		ETHANE, ISOCYANATO-
		ETHYL ISOCYANATE
		ETHYL ISOCYANATE (DOT)
		ISOCYANATOETHANE
		ISOCYANIC ACID, ETHYL ESTER
		UN 2481 (DOT)
109-92-2		
		1-ETHOXYETHENE
		1-ETHOXYETHYLENE
		ETHENE, ETHOXY-
		ETHER, ETHYL VINYL
		ETHER, VINYL ETHYL
		ETHOXYETHENE
		ETHOXYETHYLENE
		ETHYL VINYL ETHER
		ETHYLOXYETHENE
		EVE
		UN 1302 (DOT)
		VINAMAR
		VINYL ETHYL ETHER
		VINYL ETHYL ETHER, INHIBITED
		VINYL ETHYL ETHER, INHIBITED (DOT)
109-93-3		
		1,1'-OXYBISETHENE
		DIVINYL ETHER
		DIVINYL ETHER (DOT)
		DIVINYL ETHER, INHIBITED (DOT)
		DIVINYL OXIDE
		DIVYNYL OXIDE
		ETHENE, 1,1'-OXYBIS-
		ETHENYLOXYETHENE
		ETHER, DIVINYL
		UN 1167 (DOT)
		VINESTHENE
		VINESTHESIN
		VINETHEN
		VINETHENE
		VINETHER
		VINIDYL
		VINYDAN
		VINYL ETHER
109-94-4		
		AETHYLFORMIAT (GERMAN)
		AREGINAL
	*	ETHYL FORMATE
		ETHYL FORMATE (ACGIH,DOT)
		ETHYL FORMIC ESTER
		ETHYL METHANOATE
		ETHYLE (FORMIATE D') (FRENCH)
		ETHYLFORMIAAT (DUTCH)
		ETILE (FORMIATO DI) (ITALIAN)
		FORMIC ACID, ETHYL ESTER
		FORMIC ETHER
		METHANOIC ACID ETHYL ESTER
		MROWCZAN ETYLU (POLISH)
		UN 1190 (DOT)
109-95-5		
	*	ETHYL NITRITE
		ETHYL NITRITE (DOT)

CAS No.	IN	Chemical Name
		ETHYL NITRITE, SOLUTION (DOT)
		NITROSYL ETHOXIDE
		NITROUS ACID, ETHYL ESTER
		NITROUS ETHER
		NITROUS ETHER (DOT)
		NITROUS ETHYL ETHER
		SPIRIT OF ETHYL NITRITE
		SWEET SPIRIT OF NITER
		UN 1194 (DOT)
109-97-7		
		1-AZA-2,4-CYCLOPENTADIENE
		1H-PYRROLE
		AZOLE
		DIVINYLENIMINE
		IMIDOLE
		MONOPYRROLE
		PYRROL
	*	PYRROLE
109-99-9		
		BUTANE ALPHA,DELTA-OXIDE
		BUTANE, 1,4-EPOXY-
		BUTYLENE OXIDE
		CYCLOTETRAMETHYLENE OXIDE
		DIETHYLENE OXIDE
		FURAN, TETRAHYDRO-
		FURANIDINE
		HYDROFURAN
		NCI-C60560
		OXACYCLOPENTANE
		OXOLANE
		RCRA WASTE NUMBER U213
		TETRAHYDROFURAAN (DUTCH)
	*	TETRAHYDROFURAN
		TETRAHYDROFURAN (ACGIH,DOT)
		TETRAHYDROFURANNE (FRENCH)
		TETRAIDROFURANO (ITALIAN)
		TETRAMETHYLENE OXIDE
		THF
		UN 2056 (DOT)
110-00-9		
		DIVINYLENE OXIDE
	*	FURAN
		FURAN (DOT)
		FURFURAN
		NCI-C56202
		OXACYCLOPENTADIENE
		OXOLE
		RCRA WASTE NUMBER U124
		TETROLE
		UN 2389 (DOT)
110-01-0		
		TETRAHYDROTHIOPHEN
	*	TETRAHYDROTHIOPHENE
		TETRAHYDROTHIOPHENE (DOT)
		TETRAMETHYLENE SULFIDE
		THIACYCLOPENTANE
		THILANE
		THIOLANE
		THIOPHANE
		THIOPHENE, TETRAHYDRO-
		UN 2412 (DOT)
110-02-1		
		CP 34
		DIVINYLENE SULFIDE
		FURAN, THIO-
		HUILE H50
		HUILE HSO
		THIACYCLOPENTADIENE
		THIAPHENE
		THIOFURAN
		THIOFURFURAN
		THIOLE
		THIOPHEN
	*	THIOPHENE
		THIOPHENE (DOT)
		THIOTETROLE
		UN 2414 (DOT)
		USAF EK-1860
110-05-4		
		(TRIBUTYL)PEROXIDE
		BIS(1,1-DIMETHYLETHYL) PEROXIDE
		BIS(TERT-BUTYL) PEROXIDE
		CADOX
		CADOX TBP
		DI-T-BUTYL PEROXIDE
		DI-TERT-BUTYL PEROXIDE
		DI-TERT-BUTYL PEROXIDE, TECHNICALLY PURE
		DI-TERT-BUTYL PEROXIDE, TECHNICALLY PURE (DOT)
		DI-TERT-BUTYL PEROXYDE (DUTCH)
		DI-TERT-BUTYLPEROXID (GERMAN)
		DTBP
		PERBUTYL D
		PEROSSIDO DI BUTILE TERZIARIO (ITALIAN)
		PEROXIDE, BIS(1,1-DIMETHYLETHYL)
		PEROXYDE DE BUTYLE TERTIAIRE (FRENCH)
		T-BUTYL PEROXIDE
		TERT-BUTYL PEROXIDE
		TERT-BUTYL PEROXIDE (DOT)
		TERT-DIBUTYL PEROXIDE
		TRIGONOX B
		UN 2102 (DOT)
110-12-3		
		2-HEXANONE, 5-METHYL-
		2-METHYL-5-HEXANONE
		5-METHYL-2-HEXANONE
		5-METHYLHEXAN-2-ONE
		5-METHYLHEXAN-2-ONE (DOT)
		ISOAMYL METHYL KETONE
		ISOPENTYL METHYL KETONE
		KETONE, METHYL ISOAMYL
	*	METHYL ISOAMYL KETONE
		METHYL ISOAMYL KETONE (ACGIH)
		METHYL ISOPENTYL KETONE
		METHYLHEXANONE
		MIAK
		UN 2302 (DOT)
110-16-7		
		2-BUTENEDIOIC ACID, (Z)-
		CIS-1,2-ETHYLENEDICARBOXYLIC ACID
		CIS-BUTENEDIOIC ACID
	*	MALEIC ACID
		TOXILIC ACID
110-17-8		
		1,2-ETHENEDICARBOXYLIC ACID, TRANS-
		1,2-ETHYLENEDICARBOXYLIC ACID, (E)
		2-BUTENEDIOIC ACID, (E)-
		ALLOMALEIC ACID
		BOLETIC ACID
		BUTENEDIOIC ACID, (E)-
	*	FUMARIC ACID
		FUMARIC ACID (DOT)
		KYSELINA FUMAROVA (CZECH)
		LICHENIC ACID
		NA 9126 (DOT)
		NSC-2752
		TRANS-1,2-ETHYLENEDICARBOXYLIC ACID
		TRANS-BUTENEDIOIC ACID
		U-1149
		USAF EK-P-583
110-18-9		
		1,2-BIS(DIMETHYLAMINO)ETHANE
		1,2-BIS(DIMETHYLAMINO)ETHANE (DOT)
		1,2-DI(DIMETHYLAMINO)ETHANE
		1,2-DI(DIMETHYLAMINO)ETHANE (DOT)
		1,2-ETHANEDIAMINE, N,N,N',N'-TETRAMETHYL-
		1,2-ETHANEDIAMINE, N,N,N',N'-TETRAMETHYL- (9CI)
		2,5-DIMETHYL-2,5-DIAZAHEXANE
		DIMETHYL[2-(DIMETHYLAMINO)ETHYL]AMINE

CAS No.	IN	Chemical Name
		ETHYLENEDIAMINE, N,N,N',N'-TETRAMETHYL-
		N,N,N',N'-TETRAMETHYL-1,2-DIAMINOETHANE
		N,N,N',N'-TETRAMETHYL-1,2-ETHANEDIAMINE
		N,N,N',N'-TETRAMETHYLDIAMINOETHANE
		N,N,N',N'-TETRAMETHYLETHANEDIAMINE
		N,N,N',N'-TETRAMETHYLETHYLENEDIAMINE
		PROPAMINE D
		TEMED
		TETRAMEEN
		TETRAMETHYL ETHYLENE DIAMINE
		TMEDA
		UN 2372 (DOT)
110-19-0		
		2-METHYL PROPYL ACETATE
		2-METHYL-1-PROPYL ACETATE
		ACETATE D'ISOBUTYLE (FRENCH)
		ACETIC ACID, 2-METHYLPROPYL ESTER
		ACETIC ACID, ISOBUTYL ESTER
		BETA-METHYLPROPYL ETHANOATE
	*	ISOBUTYL ACETATE
		ISOBUTYL ACETATE (ACGIH,DOT)
		UN 1123 (DOT)
		UN 1213 (DOT)
110-22-5		
		ACETYL PEROXIDE
		ACETYL PEROXIDE (SOLUTION)
		ACETYL PEROXIDE SOLUTION, NOT OVER 25% PEROXIDE (DOT)
		DIACETYL PEROXIDE
		DIACETYL PEROXIDE (SOLUTION)
		ORGANIC PEROXIDE
		PEROXIDE, DIACETYL
		UN 2084 (DOT)
110-43-0		
		1-METHYLHEXANAL
		2-HEPTANONE
		AMYL METHYL KETONE
		AMYL METHYL KETONE (DOT)
		AMYL-METHYL-CETONE (FRENCH)
		BUTYLACETONE
		KETONE, METHYL PENTYL
		METHYL (N-AMYL)KETONE
		METHYL AMYL KETONE
		METHYL AMYL KETONE (DOT)
	*	METHYL N-AMYL KETONE
		METHYL N-AMYL KETONE (ACGIH)
		METHYL N-PENTYL KETONE
		METHYL PENTYL KETONE
		METHYL-AMYL-CETONE (FRENCH)
		N-AMYL METHYL KETONE
		N-PENTYL METHYL KETONE
		PENTYL METHYL KETONE
		UN 1110 (DOT)
110-46-3		
		AMYL NITRITE
110-49-6		
		2-METHOXY-ETHYL ACETAAT (DUTCH)
		2-METHOXYAETHYLACETAT (GERMAN)
		2-METHOXYETHANOL ACETATE
		2-METHOXYETHYL ACETATE
		2-METHOXYETHYL ACETATE (ACGIH)
		2-METHOXYETHYLE, ACETATE DE (FRENCH)
		2-METOSSIETILACETATO (ITALIAN)
		ACETATE DE L'ETHER MONOMETHYLIQUE DE L'ETHYLENE-GLYCOL (FRENCH)
		ACETATE DE METHYLE GLYCOL (FRENCH)
		ACETATO DI METIL CELLOSOLVE (ITALIAN)
		ACETYL METHYL CELLOSOLVE
		AETHYLENGLYKOLMETHYLAETHERACETAT (GERMAN)
		BETA-METHOXYETHYL ACETATE
		ETHANOL, 2-METHOXY-, ACETATE
		ETHYLENE GLYCOL ACETATE MONOMETHYL ETHER
		ETHYLENE GLYCOL METHYL ETHER ACETATE
		ETHYLENE GLYCOL MONOMETHYL ETHER ACETATE
		ETHYLENE GLYCOL MONOMETHYL ETHER ACETATE (DOT)
		GLYCOL ETHER EM ACETATE
		GLYCOL MONOMETHYL ETHER ACETATE
		MECSAC
		METHYL CELLOSOLVE ACETATE
		METHYL CELLOSOLVE ACETATE (DOT)
		METHYL CELLOSOLYE ACETAAT (DUTCH)
		METHYL GLYCOL ACETATE
		METHYL GLYCOL MONOACETATE
		METHYLGLYKOLACETAT (GERMAN)
		UN 1189 (DOT)
110-53-2		
	*	1-BROMOPENTANE
		1-PENTYL BROMIDE
		AMYL BROMIDE
		N-AMYL BROMIDE
		N-PENTYL BROMIDE
		PENTANE, 1-BROMO-
		PENTYL BROMIDE
110-54-3		
		ESANI (ITALIAN)
		GETTYSOLVE-B
		HEKSAN (POLISH)
		HEXANE
		HEXANE (DOT)
		HEXANE (N-HEXANE)
		HEXANE AND ITS ISOMERS
		HEXANEN (DUTCH)
	*	N-HEXANE
		N-HEXANE (ACGIH)
		NCI-C60571
		SKELLYSOLVE B
		UN 1208 (DOT)
110-56-5		
		1,4-DICHLOROBUTANE
		BUTANE, 1,4-DICHLORO-
110-57-6		
		TRANS-1,4-DICHLOROBUTENE
110-58-7		
		1-AMINOPENTANE
		1-PENTANAMINE
		1-PENTYLAMINE
		AMYLAMINE
		AMYLAMINE (DOT)
		AMYLAMINE (MIXED ISOMERS)
		MONOAMYLAMINE
		MONOPENTYLAMINE
		N-AMYLAMINE
		N-PENTYLAMINE
		NORLEUCAMINE
		PENTYLAMINE
		PENTYLAMINE (MIXED ISOMERS)
		UN 1106 (DOT)
110-61-2		
	*	SUCCINONITRILE
110-62-3		
		AMYL ALDEHYDE
		BUTYL FORMAL
		N-PENTANAL
		N-VALERALDEHYDE
		N-VALERALDEHYDE (ACGIH)
		N-VALERIC ALDEHYDE
		PENTANAL
		UN 2058 (DOT)
		VALERAL
	*	VALERALDEHYDE
		VALERALDEHYDE (DOT)
		VALERIANIC ALDEHYDE
		VALERIC ACID ALDEHYDE
		VALERIC ALDEHYDE
		VALERYLALDEHYDE

CAS No.	*IN*	*Chemical Name*
110-63-4		
		1,4-BUTANEDIOL
		1,4-BUTYLENE GLYCOL
		1,4-DIHYDROXYBUTANE
		DIOL 14B
		SUCOL B
		TETRAMETHYLENE 1,4-DIOL
110-65-6		
		1,4-BUTYNEDIOL
		1,4-BUTYNEDIOL (DOT)
		1,4-DIHYDROXY-2-BUTYNE
		2-BUTYNE-1,4-DIOL
		2-BUTYNEDIOL
		BIS(HYDROXYMETHYL)ACETYLENE
		BUTYNEDIOL
		UN 2716 (DOT)
110-66-7		
		1-MERCAPTOPENTANE
		1-PENTANETHIOL
		2-METHYL 2-BUTANETHIOL
		AMYL HYDROSULFIDE
		AMYL MERCAPTAN
		AMYL MERCAPTAN (DOT)
		AMYL SULFHYDRATE
		AMYL THIOALCOHOL
		L-PENTANETHIOL
		MERCAPTAN AMYLIQUE (FRENCH)
		N-AMYL MERCAPTAN
		N-PENTANETHIOL
		N-PENTYL MERCAPTAN
		N-PENTYLTHIOL
		PENTANETHIOL
		PENTYL MERCAPTAN
		PENTYLTHIOL
		UN 1111 (DOT)
110-67-8		
		1-CYANO-2-METHOXYETHANE
		2-CYANOETHYL METHYL ETHER
		3-METHOXYPROPANENITRILE
		3-METHOXYPROPIONITRILE
		3-METHOXYPROPYLNITRILE
		BETA-METHOXYPROPIONITRILE
		BETA-METHYLOXYPROPIONITRILE
		PROPANENITRILE, 3-METHOXY-
		PROPIONITRILE, 3-METHOXY-
110-68-9		
		1-BUTANAMINE, N-METHYL-
		BUTYLAMINE, N-METHYL-
		BUTYLMETHYLAMINE
		METHYLBUTYLAMINE
		N-BUTYL-N-METHYLAMINE
		N-BUTYLMETHYLAMINE
		N-METHYL-N-BUTYLAMINE
		N-METHYLBUTANAMINE
		N-METHYLBUTYLAMINE
110-69-0		
		BUTANAL OXIME
		BUTYL ALDEHYDE, OXIME
		BUTYLALDOXIME
		BUTYRALDEHYDE, OXIME
		BUTYRALDOXIME
		BUTYRALDOXIME (DOT)
		M-BUTYRALDEHYDE OXIME
		N-BUTYRALDOXIME
		SKINO #1
		TROYKYD ANTI-SKIN BTO
		UN 2840 (DOT)
		USAF AM-6
110-71-4		
		1,2-DIMETHOXYETHANE
		1,2-DIMETHOXYETHANE (DOT)
		1,2-ETHANEDIOL, DIMETHYL ETHER
		2,5-DIOXAHEXANE
		ALPHA,BETA-DIMETHOXYETHANE
		DIMETHOXYETHANE
		DIMETHYL CELLOSOLVE
		DME
		EGDME
		ETHANE, 1,2-DIMETHOXY-
		ETHYLENE DIMETHYL ETHER
		ETHYLENE GLYCOL DIMETHYL ETHER
		GLYCOL DIMETHYL ETHER
		GLYME
		MONOETHYLENE GLYCOL DIMETHYL ETHER
		MONOGLYME
		UN 2252 (DOT)
110-73-6		
		(2-HYDROXYETHYL)ETHYLAMINE
		2-ETHYLAMINO-1-ETHANOL
		2-ETHYLAMINOETHANOL
		2-N-MONOETHYLAMINOETHANOL
		ETHANOL, 2-(ETHYLAMINO)-
		ETHYLAMINOETHANOL
		ETHYLAMINOETHANOL [COMBUSTIBLE LIQUID LABEL]
		ETHYLAMINOETHANOL [FLAMMABLE LIQUID LABEL]
		ETHYLETHANOLAMINE
		MONOETHYLAMINOETHANOL
		N-ETHYL-2-AMINOETHANOL
		N-ETHYL-2-HYDROXYETHYLAMINE
		N-ETHYL-N-(2-HYDROXYETHYL)AMINE
		N-ETHYL-N-(BETA-HYDROXYETHYL)AMINE
		N-ETHYLAMINOETHANOL
		N-ETHYLETHANOLAMINE
		N-ETHYLMONOETHANOLAMINE
110-74-7		
		FORMIATE DE PROPYLE (FRENCH)
		FORMIC ACID, PROPYL ESTER
		N-PROPYL FORMATE
		PROPYL FORMATE
		PROPYL FORMATE (DOT)
		PROPYL METHANOATE
		UN 1281 (DOT)
110-75-8		
		(2-CHLOROETHOXY)ETHENE
	*	2-CHLOROETHYL VINYL ETHER
		2-VINYLOXYETHYL CHLORIDE
		BETA-CHLOROETHYL VINYL ETHER
		ETHENE, (2-CHLOROETHOXY)-
		ETHER, 2-CHLOROETHYL VINYL
		VINYL-2-CHLOROETHYL ETHER
		VINYL BETA-CHLOROETHYL ETHER
110-78-1		
		1-ISOCYANATOPROPANE
		1-PROPYL ISOCYANATE
		ISOCYANIC ACID, PROPYL ESTER
		M-PROPYL ISOCYANATE
		N-PROPYL ISOCYANATE (DOT)
		PROPANE, 1-ISOCYANATO-
		PROPYL ISOCYANATE
		UN 2482 (DOT)
110-80-5		
	*	2-ETHOXYETHANOL
		2-ETHOXYETHYL ALCOHOL
		BETA-ETHOXYETHANOL
		CELLOSOLVE
		DOWANOL EE
		EKTASOLVE EE
		EMKANOL
		ETHANOL, 2-ETHOXY-
		ETHYL CELLOSOLVE
		ETHYL GLYCOL
		ETHYLENE GLYCOL ETHYL ETHER
		ETHYLENE GLYCOL MONOETHYL ETHER
		GLYCOL MONOETHYL ETHER
		OXITOL
		POLY-SOLV EE

CAS No.	IN	Chemical Name
110-82-7		
		BENZENE, HEXAHYDRO-
	*	CYCLOHEXANE
		HEXAHYDROBENZENE
		HEXAMETHYLENE
		HEXANAPHTHENE
110-83-8		
		1,2,3,4-TETRAHYDROBENZENE
		BENZENE TETRAHYDRIDE
		BENZENE, TETRAHYDRO-
		CYCLOHEX-1-ENE
	*	CYCLOHEXENE
		TETRAHYDROBENZENE
110-85-0		
		1,4-DIAZACYCLOHEXANE
		1,4-DIETHYLENEDIAMINE
		1,4-PIPERAZINE
		ANTIREN
		DIETHYLENEDIAMINE
		DIETHYLENEIMINE
		DISPERMINE
		ERAVERM
		HEXAHYDRO-1,4-DIAZINE
		HEXAHYDROPYRAZINE
		LUMBRICAL
		N,N-DIETHYLENE DIAMINE (DOT)
		PIPERAZIDINE
		PIPERAZIN (GERMAN)
		PIPERAZINE
		PIPERAZINE (DOT)
		PIPERAZINE, ANHYDROUS
		PIPERSOL
		PYRAZINE HEXAHYDRIDE
		PYRAZINE, HEXAHYDRO-
		UN 2579 (DOT)
		UN 2685 (DOT)
		UVILON
		VERMEX
		WORM-A-TON
		WURMIRAZIN
110-86-1		
		AZABENZENE
		AZINE
		CP 32
		NCI-C55301
		PIRIDINA (ITALIAN)
		PIRYDYNA (POLISH)
		PYRIDIN (GERMAN)
	*	PYRIDINE
		PYRIDINE (ACGIH,DOT)
		RCRA WASTE NUMBER U196
		UN 1282 (DOT)
110-87-2		
		2,3-DIHYDRO-4H-PYRAN
		2,3-DIHYDROPYRAN
		2H-3,4-DIHYDROPYRAN
		2H-PYRAN, 3,4-DIHYDRO-
		2H-PYRAN, 3,6-DIHYDRO-
		2H-PYRAN, DIHYDRO-
		3,4-DIHYDRO-2H-PYRAN
		3,4-DIHYDROPYRAN
		3,6-DIHYDRO-2H-PYRAN
		5,6-DIHYDRO-2H-PYRAN
		5,6-DIHYDRO-4H-PYRAN
		DELTA(SUP 2)-DIHYDROPYRAN
		DELTA2-DIHYDROPYRAN
		DIHYDROPYRAN
		DIHYDROPYRAN (DOT)
		PYRAN, DIHYDRO-
110-88-3		
		1,3,5-TRIOXANE
		FORMALDEHYDE, TRIMER
	*	PARAFORMALDEHYDE
		S-TRIOXANE
		SYM-TRIOXANE
		TRIFORMOL
		TRIOXAN
		TRIOXANE
		TRIOXYMETHYLENE
110-89-4		
		AZACYCLOHEXANE
		CYCLOPENTIMINE
		CYPENTIL
		HEXAHYDROPYRIDINE
		HEXAZANE
		PENTAMETHYLENEIMINE
		PENTAMETHYLENIMINE
		PERHYDROPYRIDINE
		PIPERIDIN (GERMAN)
	*	PIPERIDINE
		PIPERIDINE (DOT)
		PYRIDINE, HEXAHYDRO-
		UN 2401 (DOT)
110-90-0		
		2-ETHOXY ETHANOL
		ETHYLENE GLYCOL MONOETHYL ETHER
110-91-8		
		1-OXA-4-AZACYCLOHEXANE
		2H-1,4-OXAZINE, TETRAHYDRO-
		4H-1,4-OXAZINE, TETRAHYDRO-
		BASF 238
		DIETHYLENE IMIDOXIDE
		DIETHYLENE OXIMIDE
		DIETHYLENEIMIDE OXIDE
		DIETHYLENIMIDE OXIDE
		DREWAMINE
	*	MORPHOLINE
		MORPHOLINE (ACGIH,DOT)
		MORPHOLINE, AQUEOUS MIXTURE [CORROSIVE LABEL]
		MORPHOLINE, AQUEOUS MIXTURE [FLAMMABLE LIQUID LABEL]
		MORPHOLINE, AQUEOUS, MIXTURE (DOT)
		NA 1760 (DOT)
		NA 2054 (DOT)
		P-ISOXAZINE, TETRAHYDRO-
		TETRAHYDRO-1,4-ISOXAZINE
		TETRAHYDRO-1,4-OXAZINE
		TETRAHYDRO-2H-1,4-OXAZINE
		TETRAHYDRO-P-OXAZINE
		UN 2054 (DOT)
110-96-3		
		1-PROPANAMINE, 2-METHYL-N-(2-METHYLPROPYL)-
		DIISOBUTYLAMINE
		DIISOBUTYLAMINE (DOT)
		N,N-BIS(2-METHYLPROPYL)AMINE
		UN 2361 (DOT)
110-97-4		
		1,1'-IMINOBIS[2-PROPANOL]
		1,1'-IMINODI-2-PROPANOL
		2-PROPANOL, 1,1'-IMINOBIS-
		2-PROPANOL, 1,1'-IMINOBIS- (9CI)
		2-PROPANOL, 1,1'-IMINODI-
		BIS(2-HYDROXYPROPYL)AMINE
		BIS(2-PROPANOL)AMINE
		DIISOPROPANOLAMINE
		DIPA
		DIPROPYL-2,2'-DIHYDROXY-AMINE
111-15-9		
		1-ACETOXY-2-ETHOXYETHANE
		2-ETHOXYETHANOL ACETATE
	*	2-ETHOXYETHYL ACETATE
		BETA-ETHOXYETHYL ACETATE
		CELLOSOLVE ACETATE
		ETHANOL, 2-ETHOXY-, ACETATE
		ETHYL CELLOSOLVE ACETATE
		ETHYL GLYCOL ACETATE

CAS No.	IN	Chemical Name
		ETHYLENE GLYCOL ETHYL ETHER ACETATE
		ETHYLENE GLYCOL MONOETHYL ETHER ACETATE
		GLYCOL MONOETHYL ETHER ACETATE
		OXITOL ACETATE
		POLY-SOLV EE ACETATE
111-26-2		
		1-AMINOHEXANE
		1-HEXANAMINE
		1-HEXYLAMINE
	*	HEXYLAMINE
		MONO-N-HEXYLAMINE
		N-HEXYLAMINE
111-27-3		
		1-HEXANOL
		1-HEXYL ALCOHOL
		1-HYDROXYHEXANE
		AMYLCARBINOL
		CAPROYL ALCOHOL
		EPAL 6
		HEXANOL
		HEXYL ALCOHOL
		N-HEXAN-1-OL
	*	N-HEXANOL
		N-HEXANOL (DOT)
		N-HEXYL ALCOHOL
		PENTYLCARBINOL
		UN 2282 (DOT)
111-30-8		
		1,5-PENTANEDIAL
		1,5-PENTANEDIONE
		ALDESAN
		CIDEX 7
		GLUTACLEAN
		GLUTARAL
	*	GLUTARALDEHYDE
		GLUTARALDEHYDE SOLUTION
		GLUTARDIALDEHYDE
		GLUTAREX 28
		GLUTARIC ACID DIALDEHYDE
		GLUTARIC DIALDEHYDE
		HOSPEX
		PENTANEDIAL
		RELUGAN GT
		RELUGAN GT 50
		RELUGAN GTW
		SONACIDE
		SPORICIDIN
		STERIHYDE
		UCARCIDE 250
111-31-9		
		1-HEXANETHIOL
111-34-2		
		BUTANE, 1-(ETHENYLOXY)-
		BUTOXYETHENE
		BUTOXYETHYLENE
		BUTYL VINYL ETHER
		BUTYL VINYL ETHER, INHIBITED (DOT)
		BUTYLOXYETHENE
		BVE
		ETHER, BUTYL VINYL
		N-BUTYL VINYL ETHER
		UN 2352 (DOT)
		VINYL BUTYL ETHER
		VINYL N-BUTYL ETHER
111-36-4		
		1-ISOCYANATOBUTANE
		BUTANE, 1-ISOCYANATO-
		BUTYL ISOCYANATE
		ISOCYANIC ACID, BUTYL ESTER
		N-BUTYL ISOCYANATE
111-40-0		
		1,2-ETHANEDIAMINE, N-(2-AMINOETHYL)-
		1,4,7-TRIAZAHEPTANE
		1,5-DIAMINO-3-AZAPENTANE
		2, 2'-DIAMINODIETHYLAMINE
		2,2'-DIAMINODIETHYLAMINE
		2,2'-IMINOBIS(ETHANAMINE)
		3-AZAPENTANE-1,5-DIAMINE
		AMINOETHYLETHANDIAMINE
		BIS(2-AMINOETHYL)AMINE
		BIS(BETA-AMINOETHYL)AMINE
		CHS-P 1
		DEH 20
		DETA
		DIETHYLAMINE, 2,2'-DIAMINO-
	*	DIETHYLENETRIAMINE
		DIETHYLENETRIAMINE (ACGIH,DOT)
		ETHYLAMINE, 2,2'-IMINOBIS-
		ETHYLENEDIAMINE, N-(2-AMINOETHYL)-
		N,N-BIS(2-AMINOETHYL)AMINE
		N-(2-AMINOETHYL)-1,2-ETHANEDIAMINE
		N-(2-AMINOETHYL)ETHYLENEDIAMINE
		UN 2079 (DOT)
111-41-1		
		(2-AMINOETHYL)ETHANOLAMINE
		(2-HYDROXYETHYL)ETHYLENEDIAMINE
		(BETA-HYDROXYETHYL)ETHYLENEDIAMINE
		1-(2-HYDROXYETHYLAMINO)-2-AMINOETHANE
		2-(2-HYDROXYETHYLAMINO)ETHYLAMINE
		2-[(2-AMINOETHYL)AMINO]ETHANOL
		AMINOETHYL ETHANOLAMINE
		ETHANOL, 2-((2-AMINOETHYL)AMINO)-
		ETHANOLETHYLENE DIAMINE
		MONOETHANOLETHYLENEDIAMINE
		N-(2'-HYDROXYETHYL)ETHYLENEDIAMINE
		N-(2-AMINOETHYL)ETHANOLAMINE
		N-(2-HYDROXYETHYL)-1,2-ETHANEDIAMINE
		N-(2-HYDROXYETHYL)ETHYLENE DIAMINE
		N-(AMINOETHYL)ETHANOLAMINE
		N-(BETA-AMINOETHYL)ETHANOLAMINE
		N-(BETA-HYDROXYETHYL)ETHYLENEDIAMINE
		N-(HYDROXYETHYL)ETHYLENEDIAMINE
		N-AMINOETHYLETHANOLAMINE
		N-HYDROXYETHYL-1,2-ETHANEDIAMINE
111-42-2		
		2,2'-DIHYDROXYDIETHYLAMINE
		2,2'-IMINOBISETHANOL
		2,2'-IMINODI-1-ETHANOL
		2,2'-IMINODIETHANOL
		2,2-IMINODIETHANOL
		2-[(2-HYDROXYETHYL)AMINO]ETHANOL
		BIS(2-HYDROXYETHYL)AMINE
		BIS(HYDROXYETHYL)AMINE
		DEA
		DI(2-HYDROXYETHYL)AMINE
		DIAETHANOLAMIN (GERMAN)
		DIETHANOLAMIN (CZECH)
	*	DIETHANOLAMINE
		DIETHANOLAMINE (ACGIH)
		DIETHYLAMINE, 2,2'-DIHYDROXY-
		DIETHYLOLAMINE
		DIOLAMINE
		ETHANOL, 2,2'-IMINOBIS-
		ETHANOL, 2,2'-IMINODI-
		IMINODIETHANOL
		N,N-BIS(2-HYDROXYETHYL)AMINE
		N,N-DIETHANOLAMINE
		NCI-C55174
111-43-3		
		1,1'-OXYBIS[PROPANE]
		4-OXAHEPTANE
		DI-N-PROPYL ETHER
		DIPROPYL ETHER
		DIPROPYL ETHER (DOT)
		DIPROPYL OXIDE
		ETHER, DI-N-PROPYL-

CAS No.	IN	Chemical Name
		N-PROPYL ETHER
		PROPANE, 1,1'-OXYBIS-
		PROPANE, 1,1'-OXYBIS- (9CI)
		PROPYL ETHER
		UN 2384 (DOT)
111-44-4		1,1'-OXYBIS(2-CHLORO)ETHANE
		1,5-DICHLORO-3-OXAPENTANE
		1-CHLORO-2-(BETA-CHLOROETHOXY)ETHANE
		2,2'-DICHLOORETHYLETHER (DUTCH)
		2,2'-DICHLOR-DIAETHYLAETHER (GERMAN)
		2,2'-DICHLORETHYL ETHER
		2,2'-DICHLORODIETHYL ETHER
		2,2'-DICHLOROETHYL ETHER
		2,2'-DICLOROETILETERE (ITALIAN)
		2-CHLOROETHYL ETHER
		BETA,BETA'-DICHLORODIETHYL ETHER
		BETA,BETA'-DICHLOROETHYL ETHER
		BETA,BETA-DICHLORODIETHYL ETHER
		BIS(2-CHLOROETHYL) ETHER
		BIS(BETA-CHLOROETHYL) ETHER
		BIS(CHLORO-2-ETHYL) OXIDE
		CHLOREX
		CHLOROETHYL ETHER
		CLOREX
		DCEE
		DI(2-CHLOROETHYL) ETHER
		DI(BETA-CHLOROETHYL) ETHER
		DICHLOROETHER
		DICHLOROETHYL ETHER
		DICHLOROETHYL ETHER (ACGIH,DOT)
		DICHLOROETHYL OXIDE
		DIETHYLENE GLYCOL DICHLORIDE
		DWUCHLORODWUETYLOWY ETER (POLISH)
		ENT 4,504
		ETHANE, 1,1'-OXYBIS(2-CHLORO-
		ETHER DICHLORE (FRENCH)
		ETHER, BIS(2-CHLOROETHYL)
		ETHER, BIS(CHLOROETHYL)
		OXYDE DE CHLORETHYLE (FRENCH)
		RCRA WASTE NUMBER U025
		SYM-DICHLOROETHYL ETHER
		UN 1916 (DOT)
111-46-6		2,2'-DIHYDROXYETHYL ETHER
		2,2'-OXYBISETHANOL
		2,2'-OXYDIETHANOL
		2,2'-OXYETHANOL
		2-(2-HYDROXYETHOXY)ETHANOL
		3-OXA-1,5-PENTANEDIOL
		3-OXAPENTAMETHYLENE-1,5-DIOL
		3-OXAPENTANE-1,5-DIOL
		BETA,BETA'-DIHYDROXYDIETHYL ETHER
		BIS(2-HYDROXYETHYL) ETHER
		BIS(BETA-HYDROXYETHYL) ETHER
		BRECOLANE NDG
		CARBITOL
		DEACTIVATOR E
		DEACTIVATOR H
		DEG
		DICOL
	*	DIETHYLENE GLYCOL
		DIGENOS
		DIGLYCOL
		DIGOL
		DIHYDROXYDIETHYL ETHER
		DISSOLVANT APV
		ETHANOL, 2,2'-OXYBIS-
		ETHANOL, 2,2'-OXYDI-
		ETHYLENE DIGLYCOL
		GLYCOL ETHER
		GLYCOL ETHYL ETHER
		TL4N
111-48-8		2,2'-THIOBISETHANOL
		2,2'-THIODIETHANOL
		2,2'-THIODIGLYCOL
		BETA,BETA'-DIHYDROXYDIETHYL SULFIDE
		BETA,BETA'-DIHYDROXYETHYL SULFIDE
		BETA-HYDROXYETHYL SULFIDE
		BETA-THIODIGLYCOL
		BIS(BETA-HYDROXYETHYL) SULFIDE
		BIS(2-HYDROXYETHYL) SULFIDE
		BIS(2-HYDROXYETHYL) THIOETHER
		DI(2-HYDROXYETHYL) SULFIDE
		DIETHANOL SULFIDE
		ETHANOL, 2,2'-THIOBIS-
		ETHANOL, 2,2'-THIODI-
		KROMFAX SOLVENT
		TEDEGYL
		THIODIETHANOL
		THIODIETHYLENE GLYCOL
		THIODIGLYCOL
111-49-9		1-AZACYCLOHEPTANE
		1H-AZEPINE, HEXAHYDRO-
		AZACYCLOHEPTANE
		CYCLOHEXAMETHYLENIMINE
		G 0
		G 0 (AMINE)
		HEXAHYDRO-1H-AZEPINE
		HEXAHYDROAZEPINE
		HEXAMETHYLENE IMINE (DOT)
		HEXAMETHYLENEIMINE
		HEXAMETHYLENIMINE
		HOMOPIPERIDINE
		PERHYDROAZEPINE
		UN 2493 (DOT)
111-50-2		1,4-BIS(CHLOROCARBONYL)BUTANE
		ADIPIC ACID DICHLORIDE
		ADIPIC DICHLORIDE
		ADIPOYL CHLORIDE
		ADIPOYL DICHLORIDE
		ADIPYL CHLORIDE
		HEXANEDIOYL CHLORIDE
		HEXANEDIOYL DICHLORIDE
111-55-7		1,2-DIACETOXYETHANE
		1,2-ETHANEDIOL DIACETATE
		ETHANEDIOL DIACETATE
		ETHYLENE ACETATE
		ETHYLENE DIACETATE
		ETHYLENE DIETHANOATE
		ETHYLENE GLYCOL ACETATE
		ETHYLENE GLYCOL DIACETATE
		GLYCOL DIACETATE
111-64-8		CAPRYLIC ACID CHLORIDE
		CAPRYLOYL CHLORIDE
		CAPRYLYL CHLORIDE
		N-OCTANOYL CHLORIDE
		OCTANOIC CHLORIDE
		OCTANOYL CHLORIDE
111-65-9		N-OCTANE
	*	OCTANE
		OCTANE (ACGIH,DOT)
		OCTENE
		OKTAN (POLISH)
		OKTANEN (DUTCH)
		OTTANI (ITALIAN)
		UN 1262 (DOT)
111-66-0		1-OCTANE
		1-OCTENE
		ALPHA-OCTENE
		ALPHA-OCTYLENE

CAS No.	IN	Chemical Name
		CAPRYLENE
		GULFTENE 8
		N-1-OCTENE
111-68-2		1-AMINOHEPTANE
		1-HEPTANAMINE
		1-HEPTYLAMINE
	*	HEPTYLAMINE
		N-HEPTYLAMINE
111-69-3		1,4-DICYANOBUTANE
		ADIPIC ACID DINITRILE
		ADIPIC ACID NITRILE
		ADIPODINITRILE
		ADIPONITRILE
		HEXANEDINITRILE
111-70-6		1-HEPTANOL
		1-HYDROXYHEPTANE
		ENANTHIC ALCOHOL
		GENTANOL
		HEPTANOL
		HEPTYL ALCOHOL
		L'ALCOOL N-HEPTYLIQUE PRIMAIRE (FRENCH)
		N-HEPTAN-1-OL
		N-HEPTANOL
		N-HEPTANOL-1 (FRENCH)
		N-HEPTYL ALCOHOL
111-76-2		2-AETHOXY-AETHYLACETAT (GERMAN)
	*	2-BUTOXY ETHANOL
		2-BUTOXY-1-ETHANOL
		2-ETHOXY-ETHYLACETAAT (DUTCH)
		2-ETHOXYETHANOL ACETATE
		2-ETHOXYETHANOL, ESTER WITH ACETIC ACID
		2-ETHOXYETHYL ACETATE
		2-ETHOXYETHYL ACETATE (ACGIH)
		2-ETHOXYETHYLE, ACETATE DE (FRENCH)
		2-ETOS SIETIL-ACETATO (ITALIAN)
		3-OXA-1-HEPTANOL
		ACETATE D'ETHYLGLYCOL (FRENCH)
		ACETATE DE CELLOSOLVE (FRENCH)
		ACETATE DE L'ETHER MONOETHYLIQUE DE L'ETHYLENE-GLYCOL (FRENCH)
		ACETATO DI CELLOSOLVE (ITALIAN)
		ACETIC ACID, 2-ETHOXYETHYL ESTER
		AETHYLENGLYKOLA ETHERACETAT (GERMAN)
		BETA-BUTOXYETHANOL
		BETA-ETHOXYETHYL ACETATE
		BUTYL CELLOSOLVE
		BUTYL CELLU-SOL
		BUTYL GLYCOL
		BUTYL MONOETHER GLYCOL
		BUTYL OXITOL
		CELLOSOLVE ACETATE (DOT)
		CHIMEC NR
		CSAC
		DOWANOL EB
		EKTASOLVE EB
		EKTASOLVE EE ACETATE SOLVENT
		ETHANOL, 2-BUTOXY-
		ETHOXY ACETATE
		ETHOXYETHYL ACETATE
		ETHYL CELLOSOLVE ACETAAT (DUTCH)
		ETHYL CELLOSOLVE ACETATE
		ETHYL GLYCOL ACETATE
		ETHYLENE GLYCOL BUTYL ETHER
		ETHYLENE GLYCOL ETHYL ETHER ACETATE
		ETHYLENE GLYCOL MONO-N-BUTYL ETHER
		ETHYLENE GLYCOL MONOBUTYL ETHER
		ETHYLENE GLYCOL MONOETHYL ETHER ACETATE
		ETHYLENE GLYCOL MONOETHYL ETHER ACETATE (DOT)
		ETHYLENE GLYCOL N-BUTYL ETHER
		ETHYLGLYCOL ACETATE
		ETHYLGLYKOLACETAT (GERMAN)
		GAFCOL EB
		GLYCOL BUTYL ETHER
		GLYCOL ETHER EE ACETATE
		GLYCOL MONOBUTYL ETHER
		GLYCOL MONOBUTYL ETHER ACETATE
		GLYCOL MONOETHYL ETHER ACETATE
		MINEX BDH
		MONOBUTYL GLYCOL ETHER
		NG
		O-BUTYL ETHYLENE GLYCOL
		OCTAN ETOKSYETYLU (POLISH)
		OXITOL ACETATE
		OXYTOL ACETATE
		POLY-SOLV EB
		POLY-SOLV EE ACETATE
		UN 1172 (DOT)
111-77-3		2-(2-METHOXYETHOXY)ETHANOL
		3,6-DIOXA-1-HEPTANOL
		BETA-METHOXY-BETA'-HYDROXYDIETHYL ETHER
		DIETHYLENE GLYCOL METHYL ETHER
	*	DIETHYLENE GLYCOL MONOMETHYL ETHER
		DIGLYCOL MONOMETHYL ETHER
		DOWANOL DM
		EKTASOLVE DM
		ETHANOL, 2,2'-OXYBIS-, MONOMETHYL ETHER
		ETHANOL, 2-(2-METHOXYETHOXY)-
		ETHYLENE DIGLYCOL MONOMETHYL ETHER
		HICOTOL CAR
		MECB
		METHOXYDIGLYCOL
		METHOXYETHOXYETHANOL
		METHYL CARBITOL
		METHYL DIGOL
		METHYL DIOXITOL
		POLY-SOLV DM
111-78-4	*	1,5-CYCLOOCTADIENE
111-84-2		N-NONANE
	*	NONANE
		NONANE (ACGIH,DOT)
		SHELLSOL 140
		UN 1920 (DOT)
111-86-4		1-AMINOOCTANE
		1-OCTANAMINE
		1-OCTYLAMINE
		ARMEEN 8
		ARMEEN 8D
		CAPRYLAMINE
		CAPRYLYLAMINE
		MONOOCTYLAMINE
		N-OCTYLAMINE
		OCTANAMINE
		OCTYLAMINE
111-87-5		1-HYDROXYOCTANE
		1-OCTANOL
		ALCOHOL C-8
		ALFOL 8
		CAPRYLIC ALCOHOL
		DYTOL M-83
		EPAL 8
		HEPTYL CARBINOL
		LOROL 20
		N-OCTAN-1-OL
		N-OCTANOL
		N-OCTYL ALCOHOL
		OCTANOL
		OCTILIN
	*	OCTYL ALCOHOL
		OCTYL ALCOHOL, NORMAL-PRIMARY
		PRIMARY OCTYL ALCOHOL

CAS No.	IN	Chemical Name
		SIPOL L8
111-88-6		
		1-MERCAPTOOCTANE
		1-OCTANETHIOL
		1-OCTYL MERCAPTAN
		1-OCTYL THIOL
		N-OCTANETHIOL
		N-OCTYL MERCAPTAN
		N-OCTYLTHIOL
		OCTYL MERCAPTAN
		OCTYLTHIOL
111-90-0		
		1-HYDROXY-3,6-DIOXAOCTANE
		2-(2-ETHOXYETHOXY)ETHANOL
		APV
		CARBITOL
		CARBITOL CELLOSOLVE
		CARBITOL SOLVENT
		DIETHYLENE GLYCOL ETHYL ETHER
		DIETHYLENE GLYCOL MONOETHYL ESTER
	*	DIETHYLENE GLYCOL MONOETHYL ETHER
		DIGLYCOL MONOETHYL ETHER
		DIOXITOL
		DOWANOL
		DOWANOL DE
		EKTASOLVE DE
		ETHANOL, 2,2'-OXYBIS-, MONOETHYL ETHER
		ETHANOL, 2-(2-ETHOXYETHOXY)-
		ETHOXY DIGLYCOL
		ETHYL CARBITOL
		ETHYL DIETHYLENE GLYCOL
		ETHYL DIGOL
		ETHYLENE DIGLYCOL MONOETHYL ETHER
		LOSUNGSMITTEL APV
		MONOETHYL ETHER OF DIETHYLENE GLYCOL
		O-ETHYLDIGOL
		POLY- SOLV
		POLY-SOLV DE
		SOLVOLSOL
		TRANS CUTOL
111-91-1		
		BIS(2-CHLOROETHOXY)METHANE
		BIS(2-CHLOROETHYL) FORMAL
		BIS(BETA-CHLOROETHYL) FORMAL
		DI(2-CHLOROETHOXY)METHANE
		DI-2-CHLOROETHYL FORMAL
		ETHANE, 1,1'-[METHYLENEBIS(OXY)]BIS[2-CHLORO-
		FORMALDEHYDE BIS(2-CHLOROETHYL) ACETAL
		FORMALDEHYDE BIS(BETA-CHLOROETHYL) ACETAL
		METHANE, BIS(2-CHLOROETHOXY)-
111-92-2		
		1-BUTANAMINE, N-BUTYL-
		DI-N-BUTYLAMINE
	*	DIBUTYLAMINE
111-96-6		
		(2-METHOXYETHYL) ETHER
		2,5,8-TRIOXANONANE
		2-(2-METHOXYETHOXY)-1-METHOXYETHANE
		BIS(2-METHOXYETHYL)ETHER
		DIETHYL GLYCOL DIMETHYL ETHER
		DIETHYLENE GLYCOL DIMETHYL ETHER
		DIGLYME
		DIMETHYL CARBITOL
		ETHANE, 1,1'-OXYBIS(2-METHOXY- (9CI)
		ETHANE, 1,1'-OXYBIS[2-METHOXY-
		ETHANOL, 2,2'-OXYBIS-, DIMETHYL ETHER
		ETHER, BIS(2-METHOXYETHYL)
		GLYME-2
		POLY-SOLV
112-02-7		
		1-HEXADECANAMINIUM, N,N,N-TRIMETHYL-, CHLORIDE
		ADOGEN 444
		ALIQUAT 6
		AMMONIUM, HEXADECYLTRIMETHYL-, CHLORIDE
		ARQUAD 16-29
		ARQUAD 16-50
		BARQUAT CT 29
		CATION PB 40
		CETAC
		CETRIMONIUM CHLORIDE
		CETYLTRIMETHYLAMMONIUM CHLORIDE
		CTAC
		CTMA
		DEHYQUART A
		DODIGEN 1383
		GENAMIN CTAC
	*	HEXADECYLTRIMETHYLAMMONIUM CHLORIDE
		HTAC
		INTEXAN CTC 29
		INTEXSAN CTC 29
		INTEXSAN CTC 50
		LEBON TM 16
		MORPAN CHA
		N-HEXADECYLTRIMETHYLAMMONIUM CHLORIDE
		NISSAN CATION PB 40
		PALMITYLTRIMETHYLAMMONIUM CHLORIDE
		PB 40
		SURFROYAL CTAC
		TRIMETHYLCETYLAMMONIUM CHLORIDE
		TRIMETHYLHEXADECYLAMMONIUM CHLORIDE
		VARIQUAT E 228
112-04-9		
		N-OCTADECYLTRICHLOROSILANE
		OCTADECYL TRICHLOROSILANE
		OCTADECYLTRICHLOROSILANE (DOT)
		SILANE, OCTADECYLTRICHLORO-
		SILANE, TRICHLOROOCTADECYL-
		TRICHLOROOCTADECYLSILANE
		UN 1800 (DOT)
112-14-1		
		1-OCTANOL ACETATE
	*	1-OCTYL ACETATE
		ACETATE C-8
		ACETIC ACID, OCTYL ESTER
		CAPRYLYL ACETATE
		N-OCTANYL ACETATE
		N-OCTYL ACETATE
		OCTYL ACETATE
		OCTYL ALCOHOL ACETATE
112-24-3		
		1,2-ETHANEDIAMINE, N,N'-BIS(2-AMINOETHYL)-
		1,4,7,10-TETRAAZADECANE
		1,8-DIAMINO-3,6-DIAZAOCTANE
		3,6-DIAZAOCTANE-1,8-DIAMINE
		ARALDITE HARDENER HY 951
		ARALDITE HY 951
		DEH 24
		ETHYLENEDIAMINE, N,N'-BIS(2-AMINOETHYL)-
		HY 951
		N,N'-BIS(2-AMINOETHYL)-1,2-DIAMINOETHANE
		N,N'-BIS(2-AMINOETHYL)-1,2-ETHANEDIAMINE
		N,N'-BIS(2-AMINOETHYL)ETHYLENEDIAMINE
		TECZA
		TETA
		TRIEN
		TRIENTINE
	*	TRIETHYLENETETRAMINE
		TRIETHYLENETETRAMINE (DOT)
		UN 2259 (DOT)
		Z1
112-26-5		
		1,2-BIS(2-CHLOROETHOXY)ETHANE
	*	1,2-BIS(CHLOROETHOXY)ETHANE
		1,8-DICHLORO-3,6-DIOXAOCTANE
		2-(2-CHLOROETHOXY)ETHYL 2'-CHLOROETHYL ETHER
		BIS(2-CHLOROETHOXY)ETHANE
		ETHANE, 1,2-BIS(2-CHLOROETHOXY)-
		TRIETHYLENE GLYCOL DICHLORIDE

CAS No.	IN	Chemical Name
		TRIGLYCOL DICHLORIDE
112-27-6		
		1,2-BIS(2-HYDROXYETHOXY)ETHANE
		2,2'-(1,2-ETHANEDIYLBIS(OXY))BISETHANOL
		2,2'-ETHYLENEDIOXYBIS(ETHANOL)
		2,2'-ETHYLENEDIOXYDIETHANOL
		2,2'-ETHYLENEDIOXYETHANOL
		3,6-DIOXAOCTANE-1,8-DIOL
		DI-BETA-HYDROXYETHOXYETHANE
		ETHANOL, 2,2'-(ETHYLENEDIOXY)DI-
		ETHANOL, 2,2'-[1,2-ETHANEDIYLBIS(OXY)]BIS-
		ETHYLENE GLYCOL DIHYDROXYDIETHYL ETHER
		ETHYLENE GLYCOL-BIS-(2-HYDROXYETHYL ETHER)
		GLYCOL BIS(HYDROXYETHYL) ETHER
		TEG
		TEG (GLYCOL)
		TRIETHYLENE GLYCOL
		TRIGEN
		TRIGLYCOL
		TRIGOL
112-30-1		
		1-DECANOL
		AGENT 504
		ALCOHOL C-10
		ALFOL 10
		ANTAK
		C 10 ALCOHOL
		CAPRIC ALCOHOL
		CAPRINIC ALCOHOL
		DECANOL
	*	DECYL ALCOHOL
		DECYLIC ALCOHOL
		DYTOL S-91
		EPAL 10
		LOROL 22
		N-DECANOL
		N-DECATYL ALCOHOL
		N-DECYL ALCOHOL
		NONYLCARBINOL
		PRIMARY DECYL ALCOHOL
		ROYALTAC
		SIPOL L 10
		T 148
112-34-5		
		2-(2-BUTOXYETHOXY) ETHANOL
		3,6-DIOXA-1-DECANOL
		BUCB
		BUTADIGOL
		BUTOXYDIETHYLENE GLYCOL
		BUTOXYDIGLYCOL
		BUTOXYETHOXYETHANOL
		BUTYL CARBITOL
		BUTYL DIGLYCOL
		BUTYL DIGOL
		BUTYL DIOXITOL
		BUTYL OXITOL GLYCOL ETHER
		DIETHYLENE GLYCOL BUTYL ETHER
		DIETHYLENE GLYCOL MONO-N-BUTYL ETHER
		DIETHYLENE GLYCOL MONOBUTYL ETHER
		DIETHYLENE GLYCOL N-BUTYL ETHER
		DIGLYCOL MONOBUTYL ETHER
		DOWANOL DB
		EKTA SOLVE DB
		ETHANOL, 2,2'-OXYBIS-, MONOBUTYL ETHER
		ETHANOL, 2-(2-BUTOXYETHOXY)-
		GLYCOL ETHER DB
		JEFFERSOL DB
		O-BUTYL DIETHYLENE GLYCOL
		POLY-SOLV DB
112-41-4		
		1-DODECENE
		ADACENE 12
		ALPHA-DODECENE
		N-DODEC-1-ENE
112-50-5		
		2-(2-(2-ETHOXYETHOXY)ETHOXY)ETHANOL
		3,6,9-TRIOXAUNDECAN-1-OL
		DOWANOL TE
		ETHANOL, 2-(2-(2-ETHOXYETHOXY)ETHOXY)-
		ETHOXY TRIGLYCOL
		ETHOXYTRIETHYLENE GLYCOL
		ETHYLTRIGLYCOL
		POLY-SOLV TE
		TRIETHYLENE GLYCOL ETHYL ETHER
		TRIETHYLENE GLYCOL MONOETHYL ETHER
		TRIGLYCOL MONOETHYL ETHER
112-53-8		
		1-DODECANOL
		1-DODECYL ALCOHOL
		1-HYDROXYDODECANE
		ALCOHOL C-12
		ALFOL 12
		CACHALOT L-50
		CACHALOT L-90
		CO 12
		CO-1214
		CO-1214N
		CO-1214S
		DODECANOL
		DODECYL ALCOHOL
		DYTOL J-68
		EPAL 12
		KARUKORU 20
		LAURIC ALCOHOL
		LAURINIC ALCOHOL
		LAURYL 24
		LAURYL ALCOHOL
		LOROL
		LOROL 11
		LOROL 5
		LOROL 7
		MA-1214
		N-DODECAN-1-OL
		N-DODECANOL
		N-DODECYL ALCOHOL
		N-LAURYL ALCOHOL, PRIMARY
		PISOL
		S 1298
		SIPOL L 12
		SIPONOL 25
		SIPONOL L 2
		SIPONOL L 5
112-55-0		
		1-DODECANETHIOL
		1-DODECYL MERCAPTAN
		1-MERCAPTODODECANE
		DODECYL MERCAPTAN
		LAURYL MERCAPTAN
		N-DODECANETHIOL
		N-DODECYL MERCAPTAN
		N-LAURYL MERCAPTAN
112-56-1		
		LETHANE
112-57-2		
		1,11-DIAMINO-3,6,9-TRIAZAUNDECANE
		1,2-ETHANEDIAMINE, N-(2-AMINOETHYL)-N'-(2-((2-AMINOETHYL)AMINO)ETHYL)-
		1,4,7,10,13-PENTAAZATRIDECANE
		3,6,9-TRIAZAUNDECANE-1,11-DIAMINE
		DEH 26
		TEP
		TETRAETHYLENEPENTAMINE
		TETRAETHYLENEPENTAMINE (DOT)
		TETREN
		UN 2320 (DOT)
112-58-3		
		BIS(1-HEXYL) ETHER
		DI-N-HEXYL ETHER

CAS No.	IN	Chemical Name
		DIHEXYL ETHER
		HEXANE, 1,1'-OXYBIS-
		HEXYL ETHER
		N-HEXYL ETHER
112-60-7		
		2,2'-[OXYBIS(2,1-ETHANEDIYLOXY)]BISETHANOL
		3,6,9-TRIOXAUNDECANE-1,11-DIOL
		ETHANOL, 2,2'-(OXYBIS(ETHYLENEOXY))DI-
		ETHANOL, 2,2'-[OXYBIS(2,1-ETHANEDIYLOXY)]BIS-
		HI-DRY
		TETRAETHYLENE GLYCOL
112-72-1		
		1-TETRADECANOL
		ALFOL 14
		DYTOL R-52
		LANETTE K
		LANETTE WAX KS
		LOXANOL V
		MYRISTIC ALCOHOL
		MYRISTYL ALCOHOL
		MYRISTYL ALCOHOL (MIXED ISOMERS)
		N-TETRADECAN-1-OL
		N-TETRADECANOL
		N-TETRADECANOL-1
		N-TETRADECYL ALCOHOL
		TETRADECANOL
		TETRADECANOL, MIXED ISOMERS
		TETRADECYL ALCOHOL
112-80-1		
		9,10-OCTADECENOIC ACID
		9-CIS-OCTADECENOIC ACID
		9-OCTADECENOIC ACID (Z)-
		9-OCTADECENOIC ACID, CIS-
		CENTURY CD FATTY ACID
		CIS-9-OCTADECENOIC ACID
		CIS-DELTA(SUP 9)-OCTADECENOIC ACID
		CIS-DELTA9-OCTADECENOIC ACID
		CIS-OCTADEC-9-ENOIC ACID
		CIS-OLEIC ACID
		DELTA9-CIS-OCTADECENOIC ACID
		DELTA9-CIS-OLEIC ACID
		EMERSOL 210
		EMERSOL 211
		EMERSOL 213
		EMERSOL 220 WHITE OLEIC ACID
		EMERSOL 221
		EMERSOL 221 LOW TITER WHITE OLEIC ACID
		EMERSOL 233LL
		EMERSOL 6321
		GLYCON RO
		GLYCON WO
		GROCO 2
		GROCO 4
		GROCO 5L
		GROCO 6
		HY-PHI 1055
		HY-PHI 1088
		HY-PHI 2066
		HY-PHI 2088
		HY-PHI 2102
		INDUSTRENE 105
		INDUSTRENE 205
		INDUSTRENE 206
		K 52
		L'ACIDE OLEIQUE (FRENCH)
		METAUPON
		NEO-FAT 90-04
		NEO-FAT 92-04
	*	OLEIC ACID
		OLEINE 7503
		PAMOLYN
		PAMOLYN 100
		RED OIL
		TEGO-OLEIC 130
		VOPCOLENE 27
		WECOLINE OO
		WOCHEM NO 320
		Z-9-OCTADECENOIC ACID
112-80-1		
		5,8,11,14,17-PENTAOXAHENEICOSANE
		DIBUTOXY TETRAGLYCOL
		ETHER, BIS[2-(2-BUTOXYETHOXY)ETHYL]
		PENTAETHER
		TETRAETHYLENE GLYCOL DIBUTYL ETHER
114-26-1		
		2-(1-METHYLETHOXY)PHENYL N-METHYLCARBAMATE
		2-ISOPROPOXYPHENYL METHYLCARBAMATE
		2-ISOPROPOXYPHENYL N-METHYLCARBAMATE
		ARPROCARB
		BAY 39007
		BAY 5122
		BAYER 39007
		BAYER B 5122
		BAYGON
		BLATTANEX
		BLATTOSEP
		BOLFO
		BORUHO
		BORUHO 50
		BRYGOU
		CARBAMIC ACID, METHYL-, O-ISOPROPOXYPHENYL ESTER
		DALF DUST
		DDVP
		DDVP (PROPOXUR)
		ENT 25,671
		INVISI-GARD
		IPMC
		MROWKOZOL
		O-(2-ISOPROPOXYPHENYL) N-METHYLCARBAMATE
		O-ISOPROPOXYPHENYL METHYLCARBAMATE
		O-ISOPROPOXYPHENYL N-METHYLCARBAMATE
		OMS 33
		PHC
		PHENOL, 2-(1-METHYLETHOXY)-, METHYLCARBAMATE
		PROPOTOX
	*	PROPOXUR
		PROPOXYLOR
		SENDRAN
		SUNCIDE
		TENDEX
		UNDEN
		UNDEN (PESTICIDE)
115-02-6		
		AZASERIN
		AZASERINE
		C.I. 337
		CN 15757
		L-AZASERINE
		L-SERINE, DIAZOACETATE (ESTER)
		O-DIAZOACETYL-L-SERINE
		SERINE, DIAZOACETATE (ESTER), L-
115-07-1		
		1-PROPENE
		1-PROPENE (9CI)
		1-PROPYLENE
		METHYLETHENE
		METHYLETHYLENE
		NCI-C50077
		PROPENE
		PROPYLENE
		PROPYLENE (DOT)
		R 1270
		UN 1075 (DOT)
		UN 1077 (DOT)
115-10-6		
	*	DIMETHYL ETHER
		DIMETHYL ETHER (DOT)
		DIMETHYL OXIDE
		DYMEL A
		ETHER, DIMETHYL

CAS No.	IN	Chemical Name
		ETHER, METHYL
		METHANE, OXYBIS-
		METHOXYMETHANE
		METHYL ETHER
		UN 1033 (DOT)
		WOOD ETHER
115-11-7		
		1,1-DIMETHYLETHYLENE
		1-PROPENE, 2-METHYL-
		2-METHYL-1-PROPENE
		2-METHYLPROPENE
		GAMMA-BUTYLENE
		ISOBUTENE
		ISOBUTYLENE
		ISOBUTYLENE (DOT)
		ISOPROPYLIDENEMETHYLENE
		LIQUEFIED PETROLEUM GAS (DOT)
		PROPENE, 2-METHYL-
		UN 1055 (DOT)
		UN 1075 (DOT)
115-19-5		
		1,1-DIMETHYL-2-PROPYN-1-OL
		1,1-DIMETHYL-2-PROPYNOL
		1,1-DIMETHYLPROPARGYL ALCOHOL
		1,1-DIMETHYLPROPYNOL
		2-HYDROXY-2-METHYL-3-BUTYNE
		2-METHYL-2-BUTYNOL
		2-METHYL-2-HYDROXY-3-BUTYNE
	*	2-METHYL-3-BUTYN-2-OL
		2-PROPYN-1-OL, 1,1-DIMETHYL-
		3-BUTYN-2-OL, 2-METHYL-
		3-HYDROXY-3-METHYL-1-BUTYNE
		3-METHYL BUTYNOL
		3-METHYL-1-BUTYN-3-OL
		3-METHYL-BUTIN-(1)-OL-(3) (GERMAN)
		3-METHYLBUTYN-3-OL
		AB 32
		ALPHA,ALPHA-DIMETHYLPROPARGYL ALCOHOL
		DIMETHYLACETYLENECARBINOL
		DIMETHYLACETYLENYLCARBINOL
		DIMETHYLETHYNYLCARBINOL
		DIMETHYLETHYNYLMETHANOL
		ETHYNYLDIMETHYLCARBINOL
		MBY
115-21-9		
		ETHYL SILICON TRICHLORIDE
		ETHYL TRICHLOROSILANE
		ETHYL TRICHLOROSILANE (DOT)
		SILANE, ETHYLTRICHLORO-
		SILANE, TRICHLOROETHYL-
		SILICANE, TRICHLOROETHYL-
		TRICHLOROETHYLSILANE
		TRICHLOROETHYLSILICANE
		TRICHLOROETHYLSILICON
		UN 1196 (DOT)
115-25-3		
		CYCLOBUTANE, OCTAFLUORO-
		CYCLOOCTAFLUOROBUTANE
		FC-C 318
		FREON C 318
		HALOCARBON C-138
		OCTAFLUOROCYCLOBUTANE
		OCTAFLUOROCYCLOBUTANE (DOT)
		PERFLUOROCYCLOBUTANE
		PROPELLANT C318
		R-C 318
		UN 1976 (DOT)
115-26-4		
		BFP
		BFPO
		BIS(DIMETHYLAMIDO)FLUOROPHOSPHATE
		BIS(DIMETHYLAMIDO)FLUOROPHOSPHINE OXIDE
		BIS(DIMETHYLAMIDO)PHOSPHORYL FLUORIDE
		BIS(DIMETHYLAMINO)FLUOROPHOSPHATE
		BISDIMETHYLAMINOFLUOROPHOSPHINE OXIDE
		CR 409
		DIFO
		DIMEFOX
		DMF
		ENT 19,109
		FLUOPHOSPHORIC ACID DI(DIMETHYLAMIDE)
		FLUORURE DE N,N,N',N'-TETRAMETHYLE PHOSPHORODIAMIDE (FRENCH)
		HANANE
		N,N,N',N'-TETRAMETHYL-DIAMIDO-FLUOR-PHOSPHINOXID (GERMAN)
		N,N,N',N'-TETRAMETHYL-DIAMIDO-PHOSPHORSAEURE-FLUORID (GERMAN)
		N,N,N',N'-TETRAMETHYLPHOSPHORODIAMIDIC FLUORIDE
		N,N,N',N'-TETRAMETIL-FOSFORODIAMMIDO-FLUORURO (ITALIAN)
		PESTOX 14
		PESTOX IV
		PESTOX XIV
		PHOSPHINE OXIDE, BIS(DIMETHYLAMINO)FLUORO-
		PHOSPHORODIAMIDIC FLUORIDE, TETRAMETHYL-
		S-14
		T-2002
		TERRA-SYTAM
		TERRASYTUM
		TETRA SYTAM
		TETRAMETHYLDIAMIDOPHOSPHORIC FLUORIDE
		TETRAMETHYLPHOSPHORODIAMIDIC FLUORIDE
		TL 792
		WACKER S 14/10
115-29-7		
		1,2,3,4,7,7-HEXACHLOROBICYCLO(221)HEPTEN-5,6-BIOXYMETHYLENESULFITE
		1,4,5,6,7,7-HEXACHLORO-5-NORBORNENE-2,3-DIMETHANOL CYCLIC SULFITE
		5-NORBORNENE-2,3-DIMETHANOL, 1,4,5,6,7,7-HEXACHLORO-, CYCLIC SULFITE
		6,7,8,9,10,10-HEXACHLORO-1,5,5A,6,9,9A-HEXAHYDRO-6,9-METHANO-2,4,3-BENZODIOXATHIEPIN-3-OXIDE
		ALPHA,BETA-1,2,3,4,7,7-HEXACHLOROBICYCLO(221)-2-HEPTENE-5,6-BISOXYMETHYLENE SULFITE
		BENZOEPIN
		BEOSIT
		BIO 5,462
		CHLORTHIEPIN
		CRISULFAN
		CYCLODAN
		DEVISULPHAN
		ENDOCEL
		ENDOSOL
		ENDOSULFAN
		ENDOSULFAN (ACGIH,DOT)
		ENDOSULFAN 35EC
		ENDOSULPHAN
		ENSURE
		ENT 23,979
		FMC 5462
		HEXACHLOROHEXAHYDROMETHANO 2,4,3-BENZODIOXATHIEPIN-3-OXIDE
		HILDAN
		HOE 2,671
		INSECTOPHENE
		KOP-THIODAN
		MALIX
		NA 2761 (DOT)
		NCI-C00566
		NIA 5462
		NIAGARA 5462
		OMS 570
		RCRA WASTE NUMBER P050
		SD 4314
		SULFUROUS ACID, CYCLIC ESTER WITH 1,4,5,6,7,7-HEXACHLORO-5-NORBORNENE-2,3-DIMETHANOL
		THIFOR
		THIMUL
		THIODAN
		THIODAN 35

CAS No.	IN	Chemical Name
		THIODAN 35EC
		THIOFOR
		THIOMUL
		THIONEX
		THIOSULFAN
		THIOSULFAN TIONEL
		THIOTOX
		THIOTOX (INSECTICIDE)
		TIONEX
		TIOVEL
115-32-2		
		1,1,1-TRICHLOR-2,2-BIS(4-CHLORPHENYL)-AETHANOL (GERMAN)
		1,1-BIS(4-CHLOROPHENYL)-2,2,2-TRICHLOROETHANOL
		1,1-BIS(CHLOROPHENYL)-2,2,2-TRICHLOROETHANOL
		1,1-BIS(P-CHLOROPHENYL)-2,2,2-TRICHLOROETHANOL
		2,2,2-TRICHLOOR-1,1-BIS(4-CHLOORFENYL)-ETHANOL (DUTCH)
		2,2,2-TRICHLOR-1,1-BIS(4-CHLOR-PHENYL)-AETHANOL (GERMAN)
		2,2,2-TRICHLORO-1,1-BIS(4-CHLOROPHENYL)-ETHANOL (FRENCH)
		2,2,2-TRICHLORO-1,1-BIS(4-CHLOROPHENYL)ETHANOL
		2,2,2-TRICHLORO-1,1-BIS(4-CLORO-FENIL)-ETANOLO (ITALIAN)
		2,2,2-TRICHLORO-1,1-BIS(P-CHLOROPHENYL)ETHANOL
		2,2,2-TRICHLORO-1,1-DI(4-CHLOROPHENYL)ETHANOL
		4,4'-DICHLORO-ALPHA-(TRICHLOROMETHYL)BENZHYDROL
		4-CHLORO-ALPHA-(4-CHLOROPHENYL)-ALPHA-(TRI-CHLOROMETHYL)BENZENEMETHANOL
		ACARIN
		BENZENEMETHANOL, 4-CHLORO-ALPHA-(4-CHLORO-PHENYL)-ALPHA-(TRICHLOROMETHYL)-
		BENZHYDROL, 4,4'-DICHLORO-ALPHA-(TRICHLORO-METHYL)-
		CARBAX
		CEKUDIFOL
		CPCA
		DECOFOL
		DI-(P-CHLOROPHENYL)TRICHLOROMETHYLCARBINOL
		DICHLOROKELTHANE
		DICOFOL
		DTMC
		ENT 23,648
		ETHANOL, 2,2,2-TRICHLORO-1,1-BIS(4-CHLOROPHENYL)-
		ETHANOL, 2,2,2-TRICHLORO-1,1-BIS(P-CHLOROPHENYL)-
		FW 293
		HIFOL
		HILFOL 185 EC
		KELTANE
		KELTHANE
		KELTHANE (DOT)
		KELTHANE A
		KELTHANE DUST BASE
		KELTHANE, LIQUID
		KELTHANE, SOLID
		KELTHANETHANOL
		MILBOL
		MITIGAN
		NA 2761 (DOT)
		NCI-C00486
		P,P'-KELTHANE
		PARA,PARA'-KELTHANE
115-76-4		
		1,3-PROPANEDIOL, 2,2-DIETHYL-
		2,2-DIETHYL-1,3-PROPANEDIOL
		DEP
		PRENDEROL
115-77-5		
		1,1,1-TRIS(HYDROXYMETHYL)ETHANOL
		1,3-PROPANEDIOL, 2,2-BIS(HYDROXYMETHYL)-
		2,2-BIS(HYDROXYMETHYL)-1,3-PROPANEDIOL
		AUXINUTRIL
		HERCULES P6
		MAXINUTRIL
		METAB-AUXIL
		METHANE TETRAMETHYLOL
		MONOPENTAERYTHRITOL
		MONOPENTEK
		PE
		PE 200
		PENETEK
		PENTAERYTHRITE
		PENTAERYTHRITOL
		PENTAERYTHRITOL (ACGIH)
		PENTEK
		TETRA(HYDROXYMETHYL)METHANE
		TETRAHYDROXYMETHYLMETHANE
		TETRAKIS(HYDROXYMETHYL)METHANE
		TETRAMETHYLOLMETHANE
		THME
115-84-4		
		1,3-PROPANEDIOL, 2-BUTYL-2-ETHYL-
		2-ETHYL-2-BUTYL-1-3 PROPANEDIOL
		2-ETHYL-2-BUTYLPROPANEDIOL-1,3
		3,3-BIS(HYDROXYMETHYL)HEPTANE
		BEP
115-86-6		
		CELLUFLEX TPP
		DISFLAMOLL TP
		PHENYL PHOSPHATE ((PHO)3PO)
		PHOSFLEX TPP
		PHOSPHORIC ACID, TRIPHENYL ESTER
		TP
		TRIPHENOXYPHOSPHINE OXIDE
		TRIPHENYL PHOSPHATE
115-90-2		
		AGRICUR
		B 25141
		BAY 25141
		CHEMAGRO 25141
		DACONIT
		DASANIT
		DMSP
		ENT 24,945
		FENSULFOTHION
		O,O-DIETHYL O-(4-METHYLSULFINYLPHENYL) MONOTHIOPHOSPHATE
		O,O-DIETHYL O-(P-METHYLSULFINYL PHENYL) PHOS-PHOROTHIOATE
		O,O-DIETHYL O-[4-(METHYLSULFINYL)PHENYL] PHOS-PHOROTHIOATE
		PENSULFOTHION
		PHOSPHOROTHIOIC ACID, O,O-DIETHYL O-[4-(METHYL-SULFINYL)PHENYL] ESTER
		PHOSPHOROTHIOIC ACID, O,O-DIETHYL O-[P-(METHYL-SULFINYL)PHENYL] ESTER
		S 767
		TERRACUR P
		VUAGT 108
		VUAGT 96
116-02-9		
		3,3,5-TRIMETHYL-1-CYCLOHEXANOL
		3,3,5-TRIMETHYLCYCLOHEXANOL
		CYCLOHEXANOL, 3,3,5-TRIMETHYL-
		CYCLONOL
116-06-3		
		2-METHYL-2-(METHYLTHIO)PROPANAL, O-((METHYL-AMINO)CARBONYL)OXIME
		2-METHYL-2-(METHYLTHIO)PROPIONALDEHYDE O-(METHYLCARBAMOYL)OXIME
		2-METHYL-2-METHYLTHIO-PROPIONALDEHYD-O-(N-METHYL-CARBAMOYL)-OXIM (GERMAN)
		2-METIL-2-TIOMETIL-PROPIONALDEID-O-(N-METIL-CARBA-MOIL)-OSSIMA (ITALIAN)
		ALDECARB
		ALDICARB
		ALDICARBE (FRENCH)
		AMBUSH
		CARBAMIC ACID, METHYL-, 0-((2-METHYL-2-(METHYL-

CAS No.	IN	Chemical Name
		THIO)PROPYLIDENE)AMINO) DERIV
		CARBANOLATE
		ENT 27,093
		NCI-C08640
		OMS 771
		PROPANAL, 2-METHYL-2-(METHYLTHIO)-, O-((METHYLAMINO)CARBONYL))OXIME
		PROPIONALDEHYDE, 2-METHYL-2-(METHYLTHIO)-, O-(METHYLCARBAMOYL)OXIME
		RCRA WASTE NUMBER P070
		TEMIC
		TEMIK
		TEMIK 10 G
		TEMIK G 10
		UC 21149
		UNION CARBIDE 21149
		UNION CARBIDE UC-21149
116-14-3		
		1,1,2,2-TETRAFLUOROETHYLENE
		ETHENE, TETRAFLUORO-
		ETHYLENE TETRAFLUORIDE
		ETHYLENE, TETRAFLUORO-
		FLUOROPLAST 4
		PERFLUOROETHENE
		PERFLUOROETHYLENE
		TETRAFLUORETHYLENE
		TETRAFLUOROETHENE
		TETRAFLUOROETHENE (9CI)
	*	TETRAFLUOROETHYLENE
		TETRAFLUOROETHYLENE MONOMER
		TETRAFLUOROETHYLENE, INHIBITED
		TETRAFLUOROETHYLENE, INHIBITED (DOT)
		TFE
		UN 1081 (DOT)
116-15-4		
		1,1,2,3,3,3-HEXAFLUORO-1-PROPENE
		1-PROPENE, 1,1,2,3,3,3-HEXAFLUORO-
		HEXAFLUOROPROPENE
		HEXAFLUOROPROPYLENE
		HEXAFLUOROPROPYLENE (DOT)
		HEXFLUOROPROPYLENE
		PERFLUORO-1-PROPENE
		PERFLUOROPROPENE
		PERFLUOROPROPYLENE
		PROPENE, HEXAFLUORO-
		PROPYLENE, HEXAFLUORO-
		UN 1858 (DOT)
116-16-5		
		1,1,1,3,3,3-HEXACHLORO-2-PROPANONE
		1,1,1,3,3,3-HEXACHLOROPROPANONE
		2-PROPANONE, 1,1,1,3,3,3-HEXACHLORO-
		2-PROPANONE, HEXACHLORO-
		ACETONE, HEXACHLORO-
		BIS(TRICHLOROMETHYL) KETONE
		GC-1106
		HCA
		HEXACHLORO-2-PROPANONE
		HEXACHLOROACETONE
		HEXACHLOROACETONE (DOT)
		HEXACHLOROPROPANONE
		PERCHLOROACETONE
		UN 2661 (DOT)
116-54-1		
		METHYL DICHLOROACETATE
117-08-8		
		TETRACHLOROPHTHALIC ANHYDRIDE
117-52-2		
		2H-1-BENZOPYRAN-2-ONE, 3-(1-(2-FURANYL)-3-OXOBUTYL)-4-HYDROXY- (9CI)
		2H-1-BENZOPYRAN-2-ONE, 3-[1-(2-FURANYL)-3-OXOBUTYL]-4-HYDROXY-
		3-(1-FURYL-3-ACETYLETHYL)-4-HYDROXYCOUMARIN
		3-(ALPHA-ACETONYLFURFURYL)-4-HYDROXYCOUMARIN
		3-(ALPHA-FURYL-BETA-ACETYLAETHYL)-4-HYDROXYCUMARIN (GERMAN)
		COUMAFURYL
		COUMARIN, 3-(ALPHA-ACETONYLFURFURYL)-4-HYDROXY-
		CUMAFURYL (GERMAN)
		FOUMARIN
		FUMARIN
		FUMASOL
		FURMARIN
		KRUMKIL
		LURAT
		RAT-A-WAY
		RATAFIN
		TOMARIN
117-79-3		
		2-AMINO-9,10-ANTHRAQUINONE
		2-AMINOANTHRAQUINONE
		9,10-ANTHRACENEDIONE, 2-AMINO-
		ANTHRAQUINONE, 2-AMINO-
		BETA-AMINOANTHRAQUINONE
117-80-6		
		1,4-NAPHTHALENEDIONE, 2,3-DICHLORO-
		1,4-NAPHTHOQUINONE, 2,3-DICHLORO-
		2,3-DICHLORO-1,4-NAPHTHALENEDIONE
		2,3-DICHLORO-1,4-NAPHTHOQUINONE
		2,3-DICHLORONAPHTHOQUINONE-1,4
		ALGISTAT
		COMPOUND 604
		DICHLONE
		DICLONE
		ENT 3,776
		NA 2761 (DOT)
		PHYGON
		PHYGON PASTE
		PHYGON SEED PROTECTANT
		PHYGON XL
		QUINTAR
		QUINTAR 540F
		SANQUINON
		UNIROYAL
		US RUBBER 604
		USR 604
117-81-7		
		1,2-BENZENEDICARBOXYLIC ACID, BIS(2-ETHYLHEXYL) ESTER
		2-ETHYLHEXYL PHTHALATE
		BIS(2-ETHYLHEXYL) 1,2-BENZENEDICARBOXYLATE
		BIS(2-ETHYLHEXYL) O-PHTHALATE
	*	BIS(2-ETHYLHEXYL)PHTHALATE
		BISOFLEX 81
		BISOFLEX DOP
		COMPOUND 889
		DEHP
		DI(2-ETHYLHEXYL)PHTHALATE
		DI(ETHYLHEXYL) PHTHALATE
		DIOCTYL PHTHALATE
		DOP
		ERGOPLAST FDO
		ERGOPLAST FDO-S
		ETHYLHEXYL PHTHALATE
		EVIPLAST 80
		EVIPLAST 81
		FLEXIMEL
		FLEXOL DOP
		GOOD-RITE GP 264
		KODAFLEX DOP
		OCTOIL
		OCTYL PHTHALATE
		PALATINOL AH
		PHTHALIC ACID DIOCTYL ESTER
		PHTHALIC ACID, BIS(2-ETHYLHEXYL) ESTER
		PITTSBURGH PX-138
		REOMOL D 79P
		SICOL 150
		STAFLEX DOP

CAS No.	IN	Chemical Name
		TRUFLEX DOP
		VESTINOL AH
		VINICIZER 80
		WITCIZER 312
117-84-0		
		1,2-BENZENEDICARBOXYLIC ACID, BIS(2-ETHYLHEXYL) ESTER
		1,2-BENZENEDICARBOXYLIC ACID, DIOCTYL ESTER
		2-ETHYLHEXYL PHTHALATE
		BEHP
		BIS(2-ETHYLHEXYL)-1,2-BENZENEDICARBOXYLATE
		BIS(2-ETHYLHEXYL)PHTHALATE
		BISOFLEX 81
		BISOFLEX DOP
		COMPOUND 889
		DAF 68
		DEHP
		DI(2-ETHYLHEXYL)ORTHOPHTHALATE
		DI(2-ETHYLHEXYL)PHTHALATE
	*	DI-N-OCTYL PHTHALATE
		DI-SEC-OCTYL PHTHALATE
		DI-SEC-OCTYL PHTHALATE (ACGIH)
		DINOPOL NOP
		DIOCTYL O-PHTHALATE
		DIOCTYL PHTHALATE
		DNOP
		DOP
		ERGOPLAST FDO
		ETHYLHEXYL PHTHALATE
		EVIPLAST 80
		EVIPLAST 81
		FLEXIMEL
		FLEXOL DOP
		FLEXOL PLASTICIZER DOP
		GOOD-RITE GP 264
		HATCOL DOP
		HERCOFLEX 260
		KODAFLEX DOP
		MOLLAN O
		N-DIOCTYLPHTHALATE
		N-OCTYL PHTHALATE
		NCI-C52733
		NUOPLAZ DOP
		OCTOIL
		OCTYL PHTHALATE
		PALATINOL AH
		PHTHALIC ACID DIOCTYL ESTER
		PHTHALIC ACID, BIS(2-ETHYLHEXYL) ESTER
		PHTHALIC ACID, DIOCTYL ESTER
		PITTSBURGH PX-138
		PLATINOL AH
		PLATINOL DOP
		RC PLASTICIZER DOP
		RCRA WASTE NUMBER U028
		REOMOL D 79P
		REOMOL DOP
		SICOL 150
		STAFLEX DOP
		TRUFLEX DOP
		VESTINOL AH
		VINICIZER 80
		VINICIZER 85
		WITCIZER 312
118-52-5		
		1,3-DICHLORO-5,5-DIMETHYL-2,4-IMIDAZOLIDINEDIONE
	*	1,3-DICHLORO-5,5-DIMETHYLHYDANTOIN
		2,4-IMIDAZOLIDINEDIONE, 1,3-DICHLORO-5,5-DIMETHYL-
		DACTIN
		DAKTIN
		DANTOIN
		DICHLORANTIN
		HALANE
		HYDAN
		HYDAN (ANTISEPTIC)
		HYDANTOIN, 1,3-DICHLORO-5,5-DIMETHYL-
		N,N'-DICHLORO-5,5-DIMETHYLHYDANTOIN
118-74-1		
		AMATIN
		ANTICARIE
		BENZENE, HEXACHLORO-
		BUNT-CURE
		BUNT-NO-MORE
		CO-OP HEXA
		ESACLOROBENZENE (ITALIAN)
		GRANOX NM
		HCB
		HEXA CB
		HEXACHLORBENZOL (GERMAN)
	*	HEXACHLOROBENZENE
		HEXACHLOROBENZENE (DOT)
		JULIN'S CARBON CHLORIDE
		NO BUNT
		NO BUNT 40
		NO BUNT 80
		NO BUNT LIQUID
		PENTACHLOROPHENYL CHLORIDE
		PERCHLOROBENZENE
		PHENYL PERCHLORYL
		RCRA WASTE NUMBER U127
		SAATBEIZFUNGIZID (GERMAN)
		SANOCIDE
		SMUT-GO
		SNIECIOTOX
		UN 2729 (DOT)
		ZAPRAWA NASIENNA SNECIOTOX
118-83-2		
		BENZENE, 4-CHLORO, 1-NITRO, 4-(TRIFLUOROMETHYL)
118-96-7		
		1-METHYL-2,4,6-TRINITROBENZENE
		2,4,6-TRINITROTOLUEEN (DUTCH)
		2,4,6-TRINITROTOLUENE
		2,4,6-TRINITROTOLUENE (ACGIH)
		2,4,6-TRINITROTOLUENE (TNT)
		2,4,6-TRINITROTOLUOL (GERMAN)
		2-METHYL-1,3,5-TRINITROBENZENE
		ALPHA-TNT
		BENZENE, 2-METHYL-1,3,5-TRINITRO-
		ENTSUFON
		NCI-C56155
		S-TRINITROTOLUENE
		S-TRINITROTOLUOL
		SYM-TRINITROTOLUENE
		SYM-TRINITROTOLUOL
		TNT
		TNT-TOLITE (FRENCH)
		TOLIT
		TOLITE
		TOLUENE, 2,4,6-TRINITRO-
		TOLUENE, 2,4,6-TRINITRO- (WET)
		TRINITROTOLUENE
		TRINITROTOLUENE, DRY OR CONTAINING, BY WEIGHT, LESS THAN 30% WATER (DOT)
		TRINITROTOLUENE, DRY OR WET, WITH 10% WATER
		TRINITROTOLUENE, WET CONTAINING AT LEAST 10% WATER (DOT)
		TRINITROTOLUENE, WET CONTAINING AT LEAST 10% WATER, OVER 16 OZS IN ONE OUTSIDE PACKAGING
		TRINITROTOLUENE, WETTED WITH NOT LESS THAN 30% WATER (DOT)
		TRITOL
		TRITON
		TROJNITROTOLUEN (POLISH)
		TROTYL
		TROTYL OIL
		UN 0209 (DOT)
		UN 1356 (DOT)
119-32-4		
		1-AMINO-3-NITRO-4-METHYLBENZENE
		2-NITRO-4-AMINOTOLUENE
		3-NITRO-4-METHYLANILINE
		3-NITRO-4-TOLUIDIN (CZECH)
		3-NITRO-4-TOLUIDINE

CAS No.	IN	Chemical Name
		3-NITRO-P-TOLUIDINE
		4-AMINO-2-NITROTOLUENE
		4-METHYL-3-NITROANILINE
		4-METHYL-3-NITROBENZENAMINE
		5-NITRO-4-TOLUIDINE
		BENZENAMINE, 4-METHYL-3-NITRO-
		M-NITRO-P-TOLUIDINE
		NITROTOLUIDINE
		P-TOLUIDINE, 3-NITRO-
119-34-6		
		4-AMINO-2-NITROPHENOL
119-36-8		
		2-(METHOXYCARBONYL)PHENOL
		2-HYDROXYBENZOIC ACID METHYL ESTER
		ANALGIT
		ANTHRAPOLE ND
		BENZOIC ACID, 2-HYDROXY-, METHYL ESTER
		EXAGIEN
		FLUCARMIT
		METHYL 2-HYDROXYBENZOATE
		METHYL O-HYDROXYBENZOATE
	*	METHYL SALICYLATE
		O-HYDROXYBENZOIC ACID METHYL ESTER
		SALICYLIC ACID, METHYL ESTER
119-38-0		
		(1-ISOPROPIL-3-METIL-1H-PIRAZOL-5-IL)-N,N-DIMETIL-CARBAMMATO (ITALIAN)
		(1-ISOPROPYL-3-METHYL-1H-PYRAZOL-5-YL)-N,N-DI-METHYL-CARBAMAT (GERMAN)
		(1-ISOPROPYL-3-METHYL-1H-PYRAZOL-5-YL)-N,N-DI-METHYLCARBAMAAT (DUTCH)
		1-ISOPROPYL-3-METHYL-5-PYRAZOLYL DIMETHYLCARBA-MATE
		1-ISOPROPYL-3-METHYLPYRAZOLYL-(5)-DIMETHYLCARBA-MATE
		5-METHYL-2-ISOPROPYL-3-PYRAZOLYL DIMETHYLCARBA-MATE
		CARBAMIC ACID, DIMETHYL-, 1-ISOPROPYL-3-METHYLPY-RAZOL-5-YL ESTER
		CARBAMIC ACID, DIMETHYL-, 3-METHYL-1-(1-METHYL-ETHYL)-1H-PYRAZOL-5-YL ESTER
		DIMETHYL-5-(L-ISOPROPYL-3-METHYL-PYRAZOLYL)-CARBAMATE
		DIMETHYLCARBAMATE D'L-ISOPROPYL 3-METHYL 5-PYRAZOLYLE (FRENCH)
		DIMETHYLCARBAMIC ACID 3-METHYL-1-(1-METHYL-ETHYL)-1H-PYRAZOL-5-YL ESTER
		ENT 19,060
		G 23611
		GEIGY G-23611
		ISOLAN
		ISOLAN (PESTICIDE)
		ISOLANE (FRENCH)
		ISOPROPYLMETHYLPYRAZOLYL DIMETHYLCARBAMATE
		PRIMIN
		PYRAZOL-5-OL, 1-ISOPROPYL-3-METHYL-, DIMETHYLCAR-BAMATE
		SAOLAN
119-42-6		
		2-CYCLOHEXYLPHENOL
		O-CYCLOHEXYLPHENOL
		PHENOL, 2-CYCLOHEXYL-
		PHENOL, O-CYCLOHEXYL-
119-53-9		
		BENZOIN
119-61-9		
		ALPHA-OXODIPHENYLMETHANE
		ALPHA-OXODITANE
		BENZENE, BENZOYL-
		BENZOPHENONE
		BENZOYLBENZENE
		DIPHENYL KETONE
		DIPHENYLMETHANONE
		KETONE, DIPHENYL
		METHANONE, DIPHENYL-
		PHENYL KETONE
119-64-2		
		1,2,3,4-TETRAHYDRONAPHTHALENE
		BENZOCYCLOHEXANE
		DELTA(SUP 5,7,9)-NAPHTHANTRIENE
		NAPHTHALENE 1,2,3,4-TETRAHYDRIDE
		NAPHTHALENE, 1,2,3,4-TETRAHYDRO-
		TETRAHYDRONAPHTHALENE
		TETRALIN
		TETRALINA (POLISH)
		TETRALINE
		TETRANAP
119-90-4		
		3,3'-DIMETHOXY-4,4'-DIAMINODIPHENYL
	*	3,3'-DIMETHOXYBENZIDINE
		3,3-DIMETHOXYBENZIDINE
		4,4'-BI-O-ANISIDINE
		4,4'-DIAMINO-3,3'-DIMETHOXYBIPHENYL
		4,4'-DIAMINO-3,3'-DIMETHOXYDIPHENYL
		AMACEL DEVELOPED NAVY SD
		AZOGENE FAST BLUE B
		BENZIDINE, 3,3'-DIMETHOXY-
		BLUE BASE IRGA B
		BLUE BASE NB
		BLUE BN BASE
		C.I. DISPERSE BLACK 6
		CELLITAZOL B
		CIBACETE DIAZO NAVY BLUE 2B
		DIACEL NAVY DC
		DIANISIDINE
		FAST BLUE B BASE
		FAST BLUE BASE B
		FAST BLUE DSC BASE
		HILTONIL FAST BLUE B BASE
		KAYAKU BLUE B BASE
		LAKE BLUE B BASE
		MITSUI BLUE B BASE
		NAPHTHANIL BLUE B BASE
		O-DIANISIDINE
		SETACYL DIAZO NAVY R
		[1,1'-BIPHENYL]-4,4'-DIAMINE, 3,3'-DIMETHOXY-
119-93-7		
		3,3'-DIMETHYL-4,4'-BIPHENYLDIAMINE
		3,3'-DIMETHYL-4,4'-DIAMINOBIPHENYL
	*	3,3'-DIMETHYLBENZIDINE
		3,3'-TOLIDINE
		3,3-DIMETHYLBENZIDINE
		4,4'-DIAMINO-3,3'-DIMETHYLBIPHENYL
		BENZIDINE, 3,3'-DIMETHYL-
		C.I. 37230
		C.I. AZOIC DIAZO COMPONENT 113
		FAST DARK BLUE BASE R
		O,O'-TOLIDINE
		O-TOLIDINE
		[1,1'-BIPHENYL]-4,4'-DIAMINE, 3,3'-DIMETHYL-
119-94-8		
		ETHYLBENZYLTOLUIDINE
120-12-7		
		ANTHRACEN (GERMAN)
	*	ANTHRACENE
		ANTHRACIN
		GREEN OIL
		PARANAPHTHALENE
		TETRA OLIVE N2G
120-22-9		
		DIETHYL-P-NITROSOANILINE
120-40-1		
		N-N-BIS-(2-HYDROXYETHYL)DODECANAMIDE

CAS No.	IN	Chemical Name
120-58-1		1,2-(METHYLENEDIOXY)-4-PROPENYLBENZENE
		1,3-BENZODIOXOLE, 5-(1-PROPENYL)-
		3,4-(METHYLENEDIOXY)-1-PROPENYLBENZENE
		6-(1-PROPENYL)-1,3-BENZODIOXOLE
		BENZENE, 1,2-(METHYLENEDIOXY)-4-PROPENYL-
		ISOSAFROLE
120-61-6		1,4-BENZENEDICARBOXYLIC ACID, DIMETHYL ESTER
		1,4-BENZENEDICARBOXYLIC ACID, DIMETHYL ESTER (9CI)
		DIMETHYL 1,4-BENZENEDICARBOXYLATE
		DIMETHYL ESTER TERAPHTHALIC ACID
		DIMETHYL P-BENZENEDICARBOXYLATE
		DIMETHYL P-PHTHALATE
		DIMETHYL TEREPHTHALATE
		DIMETHYLESTER KYSELINY TEREFTALOVE (CZECH)
		DMT
		METHYL 4-CARBOMETHOXYBENZOATE
		METHYL P-(METHOXYCARBONYL)BENZOATE
		NCI-C50055
		TEREPHTHALIC ACID METHYL ESTER
		TEREPHTHALIC ACID, DIMETHYL ESTER
120-62-7		PIPERONYL SULFOXIDE
120-71-8		1-AMINO-2-METHOXY-5-METHYLBENZENE
		2-METHOXY-5-METHYLANILINE
		3-AMINO-4-METHOXYTOLUENE
		4-METHYL-2-AMINOANISOLE
		5-METHYL-O-ANISIDINE
		BENZENAMINE, 2-METHOXY-5-METHYL-
		CRESIDINE
		KREZIDIN
		O-ANISIDINE, 5-METHYL-
		P-CRESIDINE
120-80-9		(+)-CATECHIN
		(+)-CATECHOL
		1,2-BENZENEDIOL
	*	1,2-DIHYDROXYBENZENE
		2-HYDROXYPHENOL
		2H-1-BENZOPYRAN-3,5,7-TRIOL, 2-(3,4-DIHYDROXY-PHENYL)-3,4-DIHYDRO-, (2R-TRANS)-
		CATECHIN
		CATECHIN (FLAVAN)
		CATECHIN (PHENOL)
		CATECHINIC ACID
		CATECHOL
		CATECHOL (FLAVAN)
		CATECHOL (PHENOL)
		CATECHUIC ACID
		CATERGEN
		CI 76500
		CI OXIDATION BASE 26
		CIANIDANOL
		D-(+)-CATECHIN
		D-CATECHIN
		D-CATECHOL
		DURAFUR DEVELOPER C
		FOURAMINE PCH
		FOURRINE 68
		KB-53
		O-BENZENEDIOL
		O-DIHYDROXYBENZENE
		O-DIOXYBENZENE
		O-HYDROXYPHENOL
		O-PHENYLENEDIOL
		OXYPHENIC ACID
		PELAGOL GREY C
		PHTHALHYDROQUINONE
		PYROCATECHIN
		PYROCATECHINE
		PYROCATECHOL
120-82-1	*	1,2,4-TRICHLOROBENZENE
		1,2,4-TRICHLOROBENZOL
		1,2,5-TRICHLOROBENZENE
		1,3,4-TRICHLOROBENZENE
		BENZENE, 1,2,4-TRICHLORO-
		HOSTETEX L-PEC
		UNSYM-TRICHLOROBENZENE
120-83-2	*	2,4-DICHLOROPHENOL
		4,6-DICHLOROPHENOL
		DCP
		PHENOL, 2,4-DICHLORO-
120-92-3		ADIPIC KETONE
		ADIPIN KETON
	*	CYCLOPENTANONE
		DUMASIN
		KETOCYCLOPENTANE
		KETOPENTAMETHYLENE
120-94-5		1-METHYLPYRROLIDINE
		METHYLPYRROLIDINE
	*	N-METHYLPYRROLIDINE
		N-METHYLTETRAHYDROPYRROLE
		PYRROLIDINE, 1-METHYL-
121-14-2		1-METHYL-2,4-DINITROBENZENE
	*	2,4-DINITROTOLUENE
		2,4-DINITROTOLUOL
		2,4-DNT
		BENZENE, 1-METHYL-2,4-DINITRO-
		DINITROTOLUENE
		DINITROTOLUENE (ACGIH)
		DNT
		NCI-C01865
		RCRA WASTE NUMBER U105
		TOLUENE, 2,4-DINITRO-
121-17-5		(3-NITRO-4-CHLOROPHENYL)TRIFLUOROMETHANE
		2-CHLORO-5-(TRIFLUOROMETHYL)NITROBENZENE
		2-NITRO-4-(TRIFLUOROMETHYL)-1-CHLOROBENZENE
		2-NITRO-4-TRIFLUOROMETHYLCHLOROBENZENE
		3-NITRO-4-CHLORO-ALPHA,ALPHA,ALPHA-TRIFLUORO-TOLUENE
		3-NITRO-4-CHLOROBENZOTRIFLUORIDE (DOT)
		3-NITRO-4-CHLOROTRIFLUOROMETHYLBENZENE
		4-CHLORO-3-NITRO-1-(TRIFLUOROMETHYL)BENZENE
		4-CHLORO-3-NITRO-ALPHA,ALPHA,ALPHA-TRIFLUORO-TOLUENE
		4-CHLORO-3-NITROBENZOTRIFLUORIDE
		4-CHLORO-3-NITROBENZYLIDYNE FLUORIDE
		4-CHLORO-ALPHA,ALPHA,ALPHA-TRIFLUORO-3-NITRO-TOLUENE
		BENZENE, 1-CHLORO-2-NITRO-4-(TRIFLUOROMETHYL)-
		BENZOTRIFLUORIDE, 4-CHLORO-3-NITRO-
		NITROCHLOROBENZOTRIFLUORIDE
		TOLUENE, 4-CHLORO-3-NITRO-ALPHA,ALPHA,ALPHA-TRIFLUORO-
		TOLUENE, 4-CHLORO-ALPHA,ALPHA,ALPHA-TRIFLUORO-3-NITRO-
		UN 2307 (DOT)
121-21-1		(+)-PYRETHRONYL (+)-TRANS-CHRYSANTHEMATE
		CHRYSANTHEMUMMONOCARBOXYLIC ACID PYRETHRO-LONE ESTER
		PYRETHRIN I
121-29-9		(+)-PYRETHRONYL (+)-PYRETHRATE
		PYRETHRIN
		PYRETHRIN 2
		PYRETHRIN II

CAS No.	IN	Chemical Name
121-43-7		BORESTER O
		BORIC ACID (H3BO3), TRIMETHYL ESTER
		BORIC ACID TRIMETHYL ESTER
		METHYL BORATE
		TRIMETHOXYBORANE
		TRIMETHOXYBORINE
		TRIMETHOXYBORON
		TRIMETHYL BORATE
		TRIMETHYL BORATE (DOT)
		UN 2416 (DOT)
121-44-8		(DIETHYLAMINO)ETHANE
		ETHANAMINE, N,N-DIETHYL-
		N,N-DIETHYLETHANAMINE
		TEA
		TEN
		TRIAETHYLAMIN (GERMAN)
	*	TRIETHYLAMINE
		TRIETHYLAMINE (ACGIH,DOT)
		TRIETILAMINA (ITALIAN)
		UN 1296 (DOT)
121-45-9		FOSFORYN TROJMETYLOWY (CZECH)
		METHYL PHOSPHITE
		PHOSPHOROUS ACID, TRIMETHYL ESTER
		TRIMETHOXYPHOSPHINE
		TRIMETHYL PHOSPHITE
		TRIMETHYL PHOSPHITE (ACGIH,DOT)
		UN 2329 (DOT)
121-46-0		2,5-NORBORNADIENE
		3,6-METHANO-1,4-CYCLOHEXADIENE
		BICYCLO[2.2.1]HEPTA-2,5-DIENE
		BICYCLO[2.2.1]HEPTADIENE
		NORBORNADIENE
121-66-4		2-AMINO-5-NITROTHIAZOLE
121-69-7		(DIMETHYLAMINO)BENZENE
		ANILINE, N,N-DIMETHYL-
		BENZENAMINE, N,N,-DIMETHYL- (9CI)
		BENZENAMINE, N,N-DIMETHYL-
		DIMETHLYANILINE (ACGIH)
	*	DIMETHYLANILINE
		DIMETHYLPHENYLAMINE
		DWUMETYLOANILINA (POLISH)
		N,N-DIMETHYLAMINOBENZENE
		N,N-DIMETHYLANILINE
		N,N-DIMETHYLANILINE (DOT)
		N,N-DIMETHYLBENZENAMINE
		N,N-DIMETHYLBENZENEAMINE
		N,N-DIMETHYLPHENYLAMINE
		N-DIMETHYL ANILINE
		NCI-C56428
		UN 2253 (DOT)
		VERSNELLER NL 63/10
121-73-3		1-CHLORO-3-NITROBENZENE
		3-CHLORO-1-NITROBENZENE
		3-CHLORONITROBENZENE
		3-NITROCHLOROBENZENE
		BENZENE, 1-CHLORO-3-NITRO-
		CHLORO-M-NITROBENZENE
		M-CHLORONITROBENZENE
		M-CHLORONITROBENZENE (DOT)
	*	M-NITROCHLOROBENZENE
		METACHLORONITROBENZENE
		NITROCHLOROBENZENE, META-, SOLID (DOT)
		UN 1578 (DOT)
121-75-5		((DIMETHOXYPHOSPHINOTHIOYL)THIO)BUTANEDIOIC ACID DIETHYL ESTER
		1,2-DI(ETHOXYCARBONYL)ETHYL O,O-DIMETHYL PHOSPHORODITHIOATE
		8059HC
		AMERICAN CYANAMID 4,049
		BUTANEDIOIC ACID, [(DIMETHOXYPHOSPHINOTHIOYL)-THIO]-, DIETHYL ESTER
		CALMATHION
		CARBETHOXY MALATHION
		CARBETOVUR
		CARBETOX
		CARBOFOS
		CARBOPHOS
		CELTHION
		CHEMATHION
		CIMEXAN
		COMPOUND 4049
		CYTHION
		DETMOL MA
		DETMOL MA 96%
		DICARBOETHOXYETHYL O,O-DIMETHYL PHOSPHORODITHIOATE
		DIETHYL (DIMETHOXYPHOSPHINOTHIOYLTHIO) BUTANEDIOATE
		DIETHYL (DIMETHOXYPHOSPHINOTHIOYLTHIO)SUCCINATE
		DIETHYL MERCAPTOSUCCINATE S-ESTER WITH O,O-DIMETHYL PHOSPHORODITHIOATE
		DIETHYL MERCAPTOSUCCINATE, O,O-DIMETHYL DITHIOPHOSPHATE, S-ESTER
		DIETHYL MERCAPTOSUCCINATE, O,O-DIMETHYL PHOSPHORODITHIOATE
		DIETHYL MERCAPTOSUCCINATE, O,O-DIMETHYL THIOPHOSPHATE
		DIETHYL MERCAPTOSUCCINIC ACID O,O-DIMETHYL PHOSPHORODITHIOATE
		DITHIOPHOSPHATE DE O,O-DIMETHYLE ET DE S-(1,2-DICARBOETHOXYETHYLE) (FRENCH)
		EL 4049
		EMMATOS
		EMMATOS EXTRA
		ENT 17,034
		ETHIOLACAR
		ETIOL
		EXPERIMENTAL INSECTICIDE 4049
		EXTERMATHION
		FOG 3
		FORMAL
		FORTHION
		FOSFOTHION
		FOSFOTION
		FYFANON
		HILTHION
		HILTHION 25WDP
		IFO 13140
		INSECTICIDE NO 4049
		INSECTICIDE NUMBER 4049
		KAFPON
		KARBOFOS
		KOP-THION
		KYPFOS
		MALACIDE
		MALAFOR
		MALAGRAN
		MALAKILL
		MALAMAR
		MALAMAR 50
		MALAPHELE
		MALAPHOS
		MALASOL
		MALASPRAY
		MALATAF
	*	MALATHION
		MALATHION (ACGIH,DOT)
		MALATHION E50
		MALATHION LV CONCENTRATE
		MALATHION ULV CONCENTRATE
		MALATHIOZOO

CAS No.	IN	Chemical Name
		MALATHON
		MALATHYL
		MALATHYL LV CONCENTRATE & ULV CONCENTRATE
		MALATION (POLISH)
		MALATOL
		MALATOX
		MALDISON
		MALMED
		MALPHOS
		MALTOX
		MALTOX MLT
		MAVIDAN
		MERCAPTOSUCCINIC ACID DIETHYL ESTER
		MERCAPTOTHION
		MERCAPTOTION (SPANISH)
		MLT
		MOSCARDA
		NA 2783 (DOT)
		NCI-C00215
		O,O-DIMETHYL S-(1,2-BIS(ETHOXYCARBONYL)ETHYL)-DITHIOPHOSPHATE
		O,O-DIMETHYL S-(1,2-DICARBETHOXYETHYL) THIOTHIONOPHOSPHATE
		O,O-DIMETHYL S-(1,2-DICARBETHOXYETHYL)PHOSPHORO-DITHIOATE
		O,O-DIMETHYL S-1,2-DI(ETHOXYCARBAMYL)ETHYL PHOSPHORODITHIOATE
		O,O-DIMETHYL-S-(1,2-DICARBETHOXYETHYL) DITHIOPHOS-PHATE
		O,O-DIMETHYL-S-1,2-(DICARBAETHOXYAETHYL)-DITHIOPHOSPHAT (GERMAN)
		O,O-DIMETHYL-S-1,2-DIKARBETOXYLETHYLDITIOFOSFAT (CZECH)
		O,O-DIMETHYLDITHIOPHOSPHATE DIETHYLMERCAPTO-SUCCINATE
		OLEOPHOSPHOTHION
		ORTHO MALATHION
		PHOSPHOTHION
		PRIODERM
		S-(1,2-BIS(AETHOXY-CARBONYL)-AETHYL)-O,O-DIMETHYL-DITHIOPHASPHAT (GERMAN)
		S-(1,2-BIS(CARBETHOXY)ETHYL) O,O-DIMETHYL DITHIOPHOSPHATE
		S-(1,2-BIS(ETHOXY-CARBONYL)-ETHYL)-O,O-DIMETHYL-DITHIOFOSFAAT (DUTCH)
		S-(1,2-BIS(ETHOXYCARBONYL)ETHYL) O,O-DIMETHYL PHOSPHORODITHIOATE
		S-(1,2-BIS(ETOSSI-CARBONIL)-ETIL)-O,O-DIMETIL-DITIOFOS-FATO (ITALIAN)
		S-(1,2-DI(ETHOXYCARBONYL)ETHYL DIMETHYL PHOS-PHOROTHIOLOTHIONATE
		S-(1,2-DICARBETHOXYETHYL) O,O-DIMETHYLDITHIOPHOS-PHATE
		S-1,2-BIS(ETHOXYCARBONYL)ETHYL-O,O-DIMETHYL THIOPHOSPHATE
		S-ESTER WITH O,O-DIMETHYL PHOSPHOROTHIOATE
		S-[1,2-BIS(ETHOXYCARBONYL)ETHYL] O,O-DIMETHYL THIOPHOSPHATE
		SADOFOS
		SADOFOS 30
		SADOPHOS
		SF 60
		SIPTOX I
		SUCCINIC ACID, MERCAPTO-, DIETHYL ESTER, S-ESTER WITH O,O-DIMETHYL PHOSPHORODITHIOATE
		SUMITOX
		TAK
		TAK (PESTICIDE)
		TM 4049
		VEGFRU MALATOX
		VETIOL
		ZITHIOL
121-82-4		
		1,3,5-TRIAZA-1,3,5-TRINITROCYCLOHEXANE
		1,3,5-TRIAZINE, HEXAHYDRO-1,3,5-TRINITRO-
		1,3,5-TRINITRO-1,3,5-TRIAZACYCLOHEXANE
		1,3,5-TRINITROHEXAHYDRO-1,3,5-TRIAZINE
		1,3,5-TRINITROHEXAHYDRO-S-TRIAZINE
		1,3,5-TRINITROPERHYDRO-1,3,5-TRIAZINE
		CYCLONITE
		CYCLOTRIMETHYLENENITRAMINE
		CYCLOTRIMETHYLENETRINITRAMINE
		HEXAHYDRO-1,3,5-TRINITRO-1,3,5-TRIAZINE
		HEXAHYDRO-1,3,5-TRINITRO-S-TRIAZINE
		HEXOGEN
		HEXOGEN (EXPLOSIVE)
		HEXOGEN 5W
		PBX(AF) 108
		RDX
		S-TRIAZINE, HEXAHYDRO-1,3,5-TRINITRO-
		T4
		TRIMETHYLENETRINITRAMINE
121-91-5		
		1,3-BENZENEDICARBOXYLIC ACID
		ACIDE ISOPHTALIQUE (FRENCH)
		BENZENE-1,3-DICARBOXYLIC ACID
		IPA
		ISOPHTHALATE
		ISOPHTHALIC ACID
		KYSELINA ISOFTALOVA (CZECH)
		M-BENZENEDICARBOXYLIC ACID
		M-DICARBOXYBENZENE
		M-PHTHALIC ACID
122-10-1		
		DIMETHYL 3-HYDROXYGLUTACONATE DIMETHYL PHOS-PHATE
		SWAT
122-14-5		
		009
		8057HC
		AC-47300
		ACCOTHION
		ACEOTHION
		AGRIA 1050
		AGRIYA 1050
		AGROTHION
		AMERICAN CYANAMID CL-47,300
		ARBOGAL
		BAY 41831
		BAY S 5660
		BAYER 41831
		BAYER S 5660
		CEKUTROTHION
		CL 47300
		CP 47114
		CYFEN
		CYTEL
		CYTEN
		DIMETHYL 3-METHYL-4-NITROPHENYL PHOSPHOROTHIO-NATE
		DIMETHYL 4-NITRO-M-TOLYL PHOSPHOROTHIONATE
		EI 47300
		ENT 25,715
		FENITION
		FENITROTHION
		FOLITHION
		FOLITHION EC 50
		INSECTIGAS F
		M-CRESOL, 4-NITRO-, O-ESTER WITH O,O-DIMETHYL PHOSPHOROTHIOATE
		MEP
		MEP (PESTICIDE)
		METATHION
		METATHION E 50
		METATHIONE
		METATHIONINE E 50
		METATION
		METATION E 50
		METHADION
		METHYLNITROPHOS
		MGLAWIK F
		MONSANTO CP 47114
		NITROPHOS
		NUVANOL

CAS No.	IN	Chemical Name
		O,O-DIMETHYL O-(3-METHYL) PHOSPHOROTHIOATE
		O,O-DIMETHYL O-(3-METHYL-4-NITROPHENYL) PHOSPHOROTHIOATE
		O,O-DIMETHYL O-(3-METHYL-4-NITROPHENYL) THIOPHOSPHATE
		O,O-DIMETHYL O-(4-NITRO-3-METHYLPHENYL)THIOPHOSPHATE
		O,O-DIMETHYL O-4-NITRO-M-TOLYL PHOSPHOROTHIOATE
		O,O-DIMETHYL-O-(3-METHYL-4-NITRO-PHENYL)-MONOTHIOPHOSPHAT (GERMAN)
		O,O-DIMETHYL-O-(3-METHYL-4-NITROFENYL)-MONOTHIOFOSFAAT (DUTCH)
		O,O-DIMETHYL-O-(4-NITRO-5-METHYLPHENYL)-THIONOPHOSPHAT (GERMAN)
		OLEOMETATHION
		OLEOSUMIFENE
		OMS 43
		OVADOFOS
		OWADOFOS
		PHENITROTHION
		PHOSPHOROTHIOIC ACID, O,O-DIMETHYL O-(3-METHYL-4-NITROPHENYL) ESTER
		PHOSPHOROTHIOIC ACID, O,O-DIMETHYL O-(4-NITRO-M-TOLYL) ESTER
		S 112A
		S 5660
		SUMIFENE
		SUMITHION
		SUMITHION 20F
		VERTHION
122-20-3		
		1,1',1''-NITRILOTRI-2-PROPANOL
		1,1',1''-NITRILOTRIS(2-PROPANOL)
		2-PROPANOL, 1,1',1''-NITRILOTRI-
		2-PROPANOL, 1,1',1''-NITRILOTRIS-
		NTP
		TIPA
		TRI-2-PROPANOLAMINE
		TRIISOPROPANOLAMINE
		TRIS(2-HYDROXY-1-PROPYL)AMINE
		TRIS(2-HYDROXYPROPYL)AMINE
		TRIS(2-PROPANOL)AMINE
122-39-4		
		ANILINE, N-PHENYL-
		ANILINOBENZENE
		BENZENAMINE, N-PHENYL-
		BENZENAMINE, N-PHENYL- (9CI)
		BENZENE, (PHENYLAMINO)-
		BENZENE, ANILINO-
		BIG DIPPER
		CI 10355
		DBA
		DFA
	*	DIPHENYLAMINE
		DIPHENYLAMINE (ACGIH)
		DPA
		N,N-DIPHENYLAMINE
		N-PHENYLANILINE
		N-PHENYLBENZENAMINE
		NO SCALD
		PHENARAZINE CHLORIDE
		SCALDIP
122-51-0		
		1,1',1'-(METHYLIDYNETRIS(OXY))TRIS(ETHANE)
		AETHON
		ETHANE, 1,1',1''-[METHYLIDYNETRIS(OXY)]TRIS-ETHONE
		ETHYL FORMATE (ORTHO)
		ETHYL ORTHOFORMATE
		ETHYL ORTHOFORMATE (DOT)
		ETHYLESTER KYSELINY ORTHOMRAVENCI (CZECH)
		METHANE, TRIETHOXY-
		ORTHOFORMIC ACID ETHYL ESTER
		ORTHOFORMIC ACID, TRIETHYL ESTER
		ORTHOMRAVENCAN ETHYLNATY (CZECH)
		TRIETHOXYMETHANE
		TRIETHYL ORTHOFORMATE
		UN 2524 (DOT)
122-52-1		
		FOSFORYN TROJETYLOWY (CZECH)
		PHOSPHOROUS ACID, TRIETHYL ESTER
		TRIETHOXYPHOSPHINE
		TRIETHYL PHOSPHITE
		TRIETHYL PHOSPHITE (DOT)
		UN 2323 (DOT)
122-60-1		
		(PHENOXYMETHYL)OXIRANE
		1,2-EPOXY-3-PHENOXYPROPANE
		1-PHENOXY-2,3-EPOXYPROPANE
		2,3-EPOXY-1-PHENOXYPROPANE
		2,3-EPOXYPROPOXYBENZENE
		2,3-EPOXYPROPYL PHENYL ETHER
		3-(PHENYLOXY)-1,2-EPOXYPROPANE
		3-PHENOXY-1,2-EPOXYPROPANE
		3-PHENOXY-1,2-PROPYLENE OXIDE
		GAMMA-PHENOXYPROPYLENE OXIDE
		GLYCIDOL PHENYL ETHER
		GLYCIDYL PHENYL ETHER
		OXIRANE, (PHENOXYMETHYL)-
		PGE
		PHENOL GLYCIDYL ETHER
		PHENYL 2,3-EPOXYPROPYL ETHER
		PHENYL GLYCIDYL ETHER
		PHENYL GLYCIDYL ETHER (PGE)
		PROPANE, 1,2-EPOXY-3-PHENOXY-
122-66-7		
		1,2-DIPHENYLHYDRAZINE
		BENZENE, 1,1'-HYDRAZOBIS-
		HYDRAZINE, 1,2-DIPHENYL-
		HYDRAZOBENZENE
		N,N'-BIANILINE
		N,N'-DIPHENYLHYDRAZINE
122-82-7		
		4'-ETHOXYACETOACETANILIDE
		ACETOACET-P-PHENETIDIDE
		ACETOACETIC ACID P-PHENETIDIDE
		BUTANAMIDE, N-(4-ETHOXYPHENYL)-3-OXO-
	*	P-ACETOACETOPHENETIDIDE
122-98-5		
		2-(PHENYLAMINO)ETHANOL
		2-ANILINOETHANOL
		BENZENAMINE, N-(2-HYDROXYETHYL)-
		BETA-ANILINOETHANOL
		ETHANOL, 2-(PHENYLAMINO)-
		ETHANOL, 2-ANILINO-
		N-(2-HYDROXYETHYL)ANILINE
		N-(2-HYDROXYETHYL)BENZENAMINE
		N-(2-HYDROXYETHYL)PHENYLAMINE
		N-(BETA-HYDROXYETHYL)ANILINE
		N-ETHANOLANILINE
		N-PHENYL-2-AMINOETHANOL
		N-PHENYLETHANOLAMINE
		PHENYLETHANOLAMINE
123-00-2		
		1-AMINO-3-MORPHOLINOPROPANE
		3-(4-MORPHOLINYL)PROPYLAMINE
		3-(N-MORPHOLINO)-1-AMINOPROPANE
		3-MORPHOLINO-1-PROPYLAMINE
		3-MORPHOLINOPROPANAMINE
		3-MORPHOLINOPROPYLAMINE
		4-(3-AMINOPROPYL)MORPHOLINE
		4-(GAMMA-AMINOPROPYL)MORPHOLINE
		4-AMINOPROPYLMORPHOLINE
		4-MORPHOLINEPROPANAMINE
		4-MORPHOLINEPROPYLAMINE
		AMINOPROPYLMORPHOLINE
		GAMMA-MORPHOLINOPROPYLAMINE
		MORPHOLINE, 4-(3-AMINOPROPYL)-

CAS No.	IN	Chemical Name
		MORPHOLINE, 4-AMINOPROPYL-
		MORPHOLINE, N-AMINOPROPYL-
		N-(3-AMINOPROPYL)MORPHOLINE
		N-AMINOPROPYLMORPHOLINE
		N-AMINOPROPYLMORPHOLINE (DOT)
		NA 1760 (DOT)
123-01-3		
		1-PHENYLDODECANE
		ALKYLATE P 1
		BENZENE, DODECYL-
		DETERGENT ALKYLATE
		DETERGENT ALKYLATE NO 2
		DODECANE, 1-PHENYL-
		DODECYLBENZENE
		N-DODECYLBENZENE
		NALKYLENE 500
		PHENYLDODECAN (GERMAN)
123-04-6		
		1-CHLORO-2-ETHYLHEXANE
		2-ETHYLHEXYL CHLORIDE
		3-(CHLOROMETHYL)HEPTANE
		HEPTANE, 3-(CHLOROMETHYL)-
123-05-7		
		2-ETHYLHEXALDEHYDE
		2-ETHYLHEXANAL
		2-ETHYLHEXYLALDEHYDE
		3-FORMYLHEPTANE
		ALPHA-ETHYLCAPROALDEHYDE
		ALPHA-ETHYLHEXANAL
		BETA-PROPYL-ALPHA-ETHYLACROLEIN
		BUTYL ETHYL ACETALDEHYDE
		ETHYL HEXALDEHYDE (2-ETHYL HEXANAL)
		ETHYLBUTYLACETALDEHYDE
	*	ETHYLHEXALDEHYDE
		ETHYLHEXALDEHYDE (DOT)
		HEXANAL, 2-ETHYL-
		OCTYL ALDEHYDE
		UN 1191 (DOT)
123-07-9		
		1-ETHYL-4-HYDROXYBENZENE
		4-ETHYLPHENOL
		4-HYDROXYPHENYLETHANE
		P-ETHYLPHENOL
		P-HYDROXYPHENYLETHANE
		PHENOL, 4-ETHYL-
		PHENOL, P-ETHYL-
123-15-9		
		2-FORMYLPENTANE
		2-METHYLPENTALDEHYDE
		2-METHYLPENTANAL
		2-METHYLVALERALDEHYDE
		ALPHA-METHYL VALERALDEHYDE
		ALPHA-METHYL VALERALDEHYDE (DOT)
		ALPHA-METHYLPENTENAL
	*	METHYL VALERALDEHYDE
		METHYL VALERALDEHYDE (2-METHYL PENTANAL)
		PENTANAL, 2-METHYL-
		UN 2367 (DOT)
		VALERALDEHYDE, 2-METHYL-
123-17-1		
		2,6,8-TRIMETHYL-4-NONANOL
		4-NONANOL, 2,6,8-TRIMETHYL-
123-18-2		
		2,6,8-TRIMETHYL-4-NONANONE
		4-NONANONE, 2,6,8-TRIMETHYL-
		ISOBUTYL HEPTYL KETONE
123-19-3		
		4-HEPTANONE
		BUTYRONE
		BUTYRONE (DOT)
		DI-N-PROPYL KETONE
	*	DIPROPYL KETONE
		DIPROPYL KETONE (ACGIH,DOT)
		GBL
		HEPTAN-4-ONE
		PROPYL KETONE
		UN 2710 (DOT)
123-20-6		
		BUTANOIC ACID, ETHENYL ESTER
		BUTYRIC ACID, VINYL ESTER
		UN 2838 (DOT)
		VINYL BUTANOATE
		VINYL BUTYRATE
		VINYL BUTYRATE, INHIBITED (DOT)
123-23-9		
		ALFOZONO
		ALPHOZONE
		BIS(3-CARBOXYPROPIONYL) PEROXIDE
		BUTANOIC ACID, 4,4'-DIOXYBIS[4-OXO-
		DISUCCINIC ACID PEROXIDE
		DISUCCINIC ACID PEROXIDE, MAXIMUM CONCENTRATION 72% (DOT)
		DISUCCINIC ACID PEROXIDE, TECHNICAL PURE (DOT)
		DISUCCINOYL PEROXIDE
		PEROXIDE, BIS(3-CARBOXYPROPIONYL)
		PEROXYDISUCCINIC ACID
		PROPIONIC ACID, 3,3'-(DIOXYDICARBONYL)DI-
		PROPIONIC ACID, 3,3'-(DIOXYDICARBONYL)DI-, MAXIMUM CONCENTRATION 72%
		SUCCINIC ACID PEROXIDE
		SUCCINIC ACID PEROXIDE (DOT)
		SUCCINIC ACID PEROXIDE, ≤72%, IN WATER
		SUCCINIC ACID PEROXIDE, TECHNICALLY PURE (DOT)
		SUCCINIC MONOPEROXYANHYDRIDE
		SUCCINIC PEROXIDE
		SUCCINOYL PEROXIDE
		SUCCINYL PEROXIDE
		UN 2135 (DOT)
		UN 2962 (DOT)
123-30-8		
		1-AMINO-4-HYDROXYBENZENE
		4-AMINO-1-HYDROXYBENZENE
		4-AMINOPHENOL
		4-HYDROXYANILINE
		4-HYDROXYBENZENAMINE
		ACTIVOL
		AZOL
		BASF URSOL P BASE
		BENZOFUR P
		CERTINAL
		CI 76550
		CI OXIDATION BASE 6
		CI OXIDATION BASE 6A
		CITOL
		DURAFUR BROWN RB
		FOURAMINE P
		FOURRINE 84
		FOURRINE P BASE
		FURRO P BASE
		METHYLENE-DIPHENYLENE DI-ISOCYANATE
		NAKO BROWN R
		P-AMINOFENOL (CZECH)
		P-AMINOPHENOL
		P-AMINOPHENOL (DOT)
		P-HYDROXYANILINE
		P-HYDROXYPHENYLAMINE
		PAP
		PARANOL
		PELAGOL GREY P BASE
		PELAGOL P BASE
		PHENOL, 4-AMINO-
		PHENOL, P-AMINO-
		RENAL AC
		RODINAL
		TERTRAL P BASE
		UN 2512 (DOT)
		UNAL

CAS No.	IN	Chemical Name
		URSOL P
		URSOL P BASE
		ZOBA BROWN P BASE
123-31-9		
		1,4-BENZENEDIOL
		1,4-DIHYDROXY-BENZEEN (DUTCH)
		1,4-DIHYDROXY-BENZOL (GERMAN)
		1,4-DIHYDROXYBENZEN (CZECH)
		1,4-DIHYDROXYBENZENE
		1,4-DIIDROBENZENE (ITALIAN)
		4-HYDROXYPHENOL
		ALPHA-HYDROQUINONE
		ARCTUVIN
		BENZENE, P-DIHYDROXY-
		BENZOHYDROQUINONE
		BENZOQUINOL
		BETA-QUINOL
		BLACK AND WHITE BLEACHING CREAM
		DIAK 5
		DIHYDROQUINONE
		DIHYDROXYBENZENE
		ELDOPAQUE
		ELDOQUIN
		HE 5
		HYDROCHINON (CZECH,POLISH)
		HYDROQUINOL
		HYDROQUINOLE
	*	HYDROQUINONE
		HYDROQUINONE (ACGIH,DOT)
		IDROCHINONE (ITALIAN)
		NCI-C55834
		P-BENZENEDIOL
		P-DIHYDROXYBENZENE
		P-DIOXOBENZENE
		P-DIOXYBENZENE
		P-HYDROQUINONE
		P-HYDROXYPHENOL
		PHIAQUIN
		QUINOL
		TECQUINOL
		TENOX HQ
		TEQUINOL
		UN 2662 (DOT)
		USAF EK-356
123-38-6		
		1-PROPANAL
		1-PROPANONE
		ALDEHYDE PROPIONIQUE (FRENCH)
		METHYLACETALDEHYDE
		N-PROPANAL
		NCI-C61029
		PROPALDEHYDE
		PROPANAL
		PROPANALDEHYDE
		PROPIONAL
	*	PROPIONALDEHYDE
		PROPIONALDEHYDE (DOT)
		PROPIONIC ALDEHYDE
		PROPYL ALDEHYDE
		PROPYLIC ALDEHYDE
		UN 1275 (DOT)
123-42-2		
		2-METHYL-2-PENTANOL-4-ONE
		2-PENTANONE, 4-HYDROXY-4-METHYL-
		4-HYDROXY-2-KETO-4-METHYLPENTANE
		4-HYDROXY-4-METHYL PENTAN-2-ONE
		4-HYDROXY-4-METHYL-2-PENTANONE
		4-HYDROXY-4-METHYL-PENTAN-2-ON (GERMAN, DUTCH)
		4-HYDROXY-4-METHYLPENTANONE-2
		4-IDROSSI-4-METIL-PENTAN-2-ONE (ITALIAN)
		4-METHYL-4-HYDROXY-2-PENTANONE
		ACETONYLDIMETHYLCARBINOL
		DIACETONALCOHOL (DUTCH)
		DIACETONALCOOL (ITALIAN)
		DIACETONALKOHOL (GERMAN)
		DIACETONE
	*	DIACETONE ALCOHOL
		DIACETONE ALCOHOL (ACGIH,DOT)
		DIACETONE ALCOHOL [COMBUSTIBLE LIQUID LABEL]
		DIACETONE ALCOHOL [FLAMMABLE LIQUID LABEL]
		DIACETONE-ALCOHOL (FRENCH)
		DIKETONE ALCOHOL
		PYRANTON
		PYRANTON A
		TYRANTON
		UN 1148 (DOT)
123-51-3		
		1-BUTANOL, 3-METHYL-
		2-METHYL-4-BUTANOL
		3-METHYL BUTANOL
		3-METHYL-1-BUTANOL
		3-METHYL-1-BUTANOL (CZECH)
		3-METHYLBUTAN-1-OL
		3-METIL-BUTANOLO (ITALIAN)
		ALCOOL AMILICO (ITALIAN)
		ALCOOL ISOAMYLIQUE (FRENCH)
		AMYLOWY ALKOHOL (POLISH)
		FERMENTATION AMYL ALCOHOL
		ISO-AMYLALKOHOL (GERMAN)
	*	ISOAMYL ALCOHOL
		ISOAMYL ALCOHOL (ACGIH,DOT)
		ISOAMYL ALKOHOL (CZECH)
		ISOAMYLOL
		ISOBUTYL CARBINOL
		ISOPENTANOL
		ISOPENTYL ALCOHOL
		UN 1105 (DOT)
123-54-6		
		2,4-DIOXOPENTANE
		2,4-PENTADIONE
		2,4-PENTANEDIONE
		2,4-PENTANEDIONE (DOT)
		2-PROPANONE, ACETYL-
		ACAC
		ACETOACETONE
		ACETONE, ACETYL-
		ACETYLACETONE
		DIACETYLMETHANE
	*	PENTANE-2,4-DIONE
		PENTANEDIONE
		PENTANEDIONE-2,4
		UN 2310 (DOT)
123-62-6		
		METHYLACETIC ANHYDRIDE
		PROPANOIC ACID, ANHYDRIDE
		PROPANOIC ANHYDRIDE
		PROPIONIC ACID ANHYDRIDE
	*	PROPIONIC ANHYDRIDE
		PROPIONIC ANHYDRIDE (DOT)
		PROPIONYL OXIDE
		UN 2496 (DOT)
123-63-7		
		1,3,5-TRIMETHYL-2,4,6-TRIOXANE
		1,3,5-TRIOXANE, 2,4,6-TRIMETHYL-
		2,4,6-TRIMETHYL-1,3,5-TRIOXACYCLOHEXANE
		2,4,6-TRIMETHYL-1,3,5-TRIOXANE
		2,4,6-TRIMETHYL-S-TRIOXANE
		ACETALDEHYDE, TRIMER
		ELALDEHYDE
		PARAACETALDEHYDE
		PARACETALDEHYDE
		PARAL
	*	PARALDEHYDE
		PCHO
		S-TRIMETHYLTRIOXYMETHYLENE
		S-TRIOXANE, 2,4,6-TRIMETHYL-
123-66-0		
	*	ETHYL CAPROATE
		ETHYL HEXANOATE
		HEXANOIC ACID, ETHYL ESTER

CAS No.	IN	Chemical Name
123-72-8		
		ALDEHYDE BUTYRIQUE (FRENCH)
		ALDEIDE BUTIRRICA (ITALIAN)
		BUTAL
		BUTALDEHYDE
		BUTALYDE
		BUTANAL
		BUTANALDEHYDE
		BUTYL ALDEHYDE
		BUTYRAL
	*	BUTYRALDEHYDE
		BUTYRALDEHYDE (DOT)
		BUTYRIC ALDEHYDE
		BUTYRYLALDEHYDE
		N-BUTANAL
		N-BUTANAL (CZECH)
		N-BUTYL ALDEHYDE
		N-BUTYRALDEHYDE
		NCI-C56291
		UN 1129 (DOT)
123-73-9		
		1,2-ETHANEDIOL, DIPROPANOATE (9CI)
		2-BUTENAL
		2-BUTENAL (9CI)
		2-BUTENAL, (E)-
		ALDEHYDE CROTONIQUE (FRENCH)
		BETA-METHYL ACROLEIN (DOT)
		BETA-METHYLACROLEIN
		CROTENALDEHYDE
		CROTONAL
	*	CROTONALDEHYDE
		CROTONALDEHYDE (ACGIH)
		CROTONALDEHYDE (DOT)
		CROTONALDEHYDE, (E)-
		CROTONALDEHYDE, INHIBITED (DOT)
		CROTONIC ALDEHYDE
		CROTYLALDEHYDE
		ETHYLENE DIPROPIONATE
		ETHYLENE GLYCOL, DIPROPIONATE (8CI)
		ETHYLENE PROPIONATE
		NCI-C56279
		PROPYLENE ALDEHYDE
		RCRA WASTE NUMBER U053
		TOPANEL
		TRANS-2-BUTENAL
		UN 1143 (DOT)
123-75-1		
		AZACYCLOPENTANE
		AZOLIDINE
		BUTYLENIMINE
		PERHYDROPYRROLE
		PROLAMINE
		PYRROLE, TETRAHYDRO-
	*	PYRROLIDINE
		PYRROLIDINE (DOT)
		PYRROLIDINE RING
		TETRAHYDROPYRROLE
		TETRAMETHYLENIMINE
		UN 1922 (DOT)
123-79-5		
		ADIMOLL DO
		ADIPIC ACID, BIS(2-ETHYLHEXYL) ESTER
		ADIPIC ACID, DIOCTYL ESTER
		ADIPOL 2EH
		BEHA
		BIS(2-ETHYLHEXYL) ADIPATE
		BISOFLEX DOA
		DEHA
		DI-2-ETHYLHEXYL ADIPATE
		DI-N-OCTYL ADIPATE
		DIETHYLHEXYL ADIPATE
		DIOCTYL ADIPATE
		DOA
		EFFEMOLL DOA
		EFFOMOLL DOA
		ERGOPLAST ADDO
		FLEXOL A 26
		FLEXOL PLASTICIZER 10-A
		FLEXOL PLASTICIZER A-26
		HEXANEDIOIC ACID, BIS(2-ETHYLHEXYL) ESTER
		HEXANEDIOIC ACID, BIS(2-ETHYLHEXYL) ESTER (9CI)
		HEXANEDIOIC ACID, DIOCTYL ESTER
		KEMESTER 5652
		KODAFLEX DOA
		LANKROFLEX DOA
		MOLLAN S
		MONOPLEX DOA
		NCI-C54386
		OCTYL ADIPATE
		PLASTOMOLL DOA
		PX-238
		REOMOL DOA
		RUCOFLEX PLASTICIZER DOA
		SICOL 250
		STAFLEX DOA
		TRUFLEX DOA
		UNIFLEX DOA
		VESTINOL OA
		WICKENOL 158
		WITAMOL 320
123-81-9		
		ACETIC ACID, MERCAPTO-, 1,2-ETHANEDIYL ESTER
		ACETIC ACID, MERCAPTO-, ETHYLENE ESTER
		ETHYLENE BIS(MERCAPTOACETATE)
		ETHYLENE GLYCOL BIS(MERCAPTOACETATE)
		ETHYLENE GLYCOL BIS(THIOGLYCOLATE)
		ETHYLENE GLYCOL BIS(THIOGLYCOLIC ESTER)
		ETHYLENE MERCAPTOACETATE
		ETHYLENEBIS(THIOGLYCOLATE)
		GLYCOL BIS(MERCAPTOACETATE)
		GLYCOL DIMERCAPTOACETATE
123-86-4		
		1-BUTYL ACETATE
		ACETATE DE BUTYLE (FRENCH)
		ACETIC ACID N-BUTYL ESTER
		ACETIC ACID, (P-METHOXYPHENOXY), BUTYL ESTER
		ACETIC ACID, BUTYL ESTER
		BUTILE (ACETATI DI) (ITALIAN)
		BUTYL ACETATE
		BUTYL ACETATE (DOT)
		BUTYL ETHANOATE
		BUTYLACETA TEN (DUTCH)
		BUTYLACETAT (GERMAN)
		BUTYLE (ACETATE DE) (FRENCH)
	*	N-BUTYL ACETATE
		N-BUTYL ACETATE (ACGIH)
		OCTAN N-BUTYLU (POLISH)
		UN 1123 (DOT)
123-91-1		
		1,4-DIETHYLENE DIOXIDE
		1,4-DIOXACYCLOHEXANE
		1,4-DIOXAN
	*	1,4-DIOXANE
		1,4-DIOXIN, TETRAHYDRO-
		DIETHYLENE DIOXIDE
		DIETHYLENE ETHER
		DIETHYLENE OXIDE
		DIOKAN
		DIOKSAN (POLISH)
		DIOSSANO-1,4 (ITALIAN)
		DIOXAAN-1,4 (DUTCH)
		DIOXAN
		DIOXAN-1,4 (GERMAN)
		DIOXANE
		DIOXANE (ACGIH,DOT)
		DIOXANE, TECH. GRADE
		DIOXANE-1,4
		DIOXANNE (FRENCH)
		DIOXYETHYLENE ETHER
		GLYCOL ETHYLENE ETHER
		NCI-C03689
		NE 220

CAS No.	IN	Chemical Name
		P-DIOXAN
		P-DIOXAN (CZECH)
		P-DIOXANE
		P-DIOXIN, TETRAHYDRO-
		RCRA WASTE NUMBER U108
		TETRAHYDRO-1,4-DIOXIN
		TETRAHYDRO-P-DIOXIN
		UN 1165 (DOT)
123-92-2		
		1-BUTANOL, 3-METHYL-, ACETATE
		3-METHYL-1-BUTYL ACETATE
		3-METHYLBUTYL ACETATE
		3-METHYLBUTYL ETHANOATE
		ACETIC ACID 3-METHYLBUTYL ESTER
		ACETIC ACID, ISOPENTYL ESTER
		BANANA OIL
		I-AMYL ACETATE
	*	ISOAMYL ACETATE
		ISOAMYL ACETATE (ACGIH)
		ISOAMYL ETHANOATE
		ISOPENTYL ACETATE
		ISOPENTYL ALCOHOL, ACETATE
		ISOPENTYL ETHANOATE
		PEAR OIL
124-02-7		
		DIALLYLAMINE
124-04-9		
		1,4-BUTANEDICARBOXYLIC ACID
		1,6-HEXANEDIOIC ACID
		ACIFLOCTIN
		ACINETTEN
		ADILACTETTEN
	*	ADIPIC ACID
		HEXANEDIOIC ACID
124-09-4		
		1,6-DIAMINO-N-HEXANE
		1,6-DIAMINOHEXANE
		1,6-HEXAMETHYLENEDIAMINE
		1,6-HEXANEDIAMINE
		1,6-HEXYLENEDIAMINE
	*	HEXAMETHYLENE DIAMINE
		HEXAMETHYLENE DIAMINE, SOLID (DOT)
		HEXAMETHYLENEDIAMINE, SOLID
		HEXAMETHYLENEDIAMINE, SOLUTION
		HEXAMETHYLENEDIAMINE, SOLUTION (DOT)
		HEXYLENEDIAMINE
		HMDA
		NCI-C61405
		UN 1783 (DOT)
		UN 2280 (DOT)
124-11-8		
		1-N-NONENE
		1-NONENE
		ALPHA-NONENE
		N-NON-1-ENE
124-13-0		
		1-OCTANAL
		ALDEHYDE C-8
		ANTIFOAM LF
		C-8 ALDEHYDE
		CAPRYLALDEHYDE
		CAPRYLIC ALDEHYDE
		N-CAPRYLALDEHYDE
		N-OCTALDEHYDE
		N-OCTANAL
		N-OCTYL ALDEHYDE
		N-OCTYLAL
		OCTALDEHYDE
		OCTANAL
		OCTANALDEHYDE
		OCTANOIC ALDEHYDE
		OCTYLALDEHYDE

CAS No.	IN	Chemical Name
124-16-3		
		1-(2-BUTOXYETHOXY)-2-PROPANOL
		1-(BUTOXYETHOXY)-2-PROPANOL
		2-BUTOXY-1-(2-HYDROXYPROPOXY)ETHANE
		2-PROPANOL, 1-(2-BUTOXYETHOXY)-
		4,7-DIOXAUNDECAN-2-OL
124-17-4		
		2-(2-BUTOXYETHOXY)ETHANOL ACETATE
		2-(2-BUTOXYETHOXY)ETHYL ACETATE
		ACEATATE
		BUTOXYETHOXYETHYL ACETATE
		BUTYL CARBITOL ACETATE
		BUTYL DIETHYLENE GLYCOL ACETATE
		DIETHYLENE GLYCOL BUTYL ETHER ACETATE
		DIETHYLENE GLYCOL MONOBUTYL ETHER ACETATE
		DIGLYCOL MONOBUTYL ETHER ACETATE
		EKTASOLVE DB ACETATE
		ETHANOL, 2-(2-BUTOXYETHOXY)-, ACETATE
124-18-5		
	*	DECANE
		N-DECANE
124-38-9		
		ANHYDRIDE CARBONIQUE (FRENCH)
	*	CARBON DIOXIDE
		CARBON DIOXIDE (ACGIH,DOT)
		CARBON DIOXIDE, LIQUEFIED (DOT)
		CARBON DIOXIDE, REFRIGERATED LIQUID
		CARBON DIOXIDE, REFRIGERATED LIQUID (DOT)
		CARBON DIOXIDE, SOLID
		CARBON DIOXIDE, SOLID (DOT)
		CARBON ICE (DOT)
		CARBON OXIDE (CO2)
		CARBONIC ACID GAS
		CARBONIC ANHYDRIDE
		DRY ICE
		DRY ICE (DOT)
		KOHLENDIOXYD (GERMAN)
		KOHLENSAURE (GERMAN)
		R 744
		UN 1013 (DOT)
		UN 1845 (DOT)
		UN 2187 (DOT)
124-40-3		
	*	DIMETHYLAMINE
		DIMETHYLAMINE (ACGIH)
		DIMETHYLAMINE, ANHYDROUS
		DIMETHYLAMINE, ANHYDROUS (DOT)
		DIMETHYLAMINE, AQUEOUS SOLUTION
		DIMETHYLAMINE, AQUEOUS SOLUTION (DOT)
		DIMETHYLAMINE, SOLUTION (DOT)
		DMA
		METHANAMINE, N-METHYL-
		METHANAMINE, N-METHYL- (9CI)
		N-METHYLMETHANAMINE
		RCRA WASTE NUMBER U092
		UN 1032 (DOT)
		UN 1160 (DOT)
124-41-4		
		FELDALAT NM
		METHANOL, SODIUM SALT
		METHOXYSODIUM
		NA 1289 (DOT)
		SODIUM METHANOLATE
		SODIUM METHOXIDE
	*	SODIUM METHYLATE
		SODIUM METHYLATE (ALCOHOL MIXTURE)
		SODIUM METHYLATE (DOT)
		SODIUM METHYLATE, ALCOHOL MIXTURE (DOT)
		SODIUM METHYLATE, DRY
		SODIUM METHYLATE, DRY (DOT)
		SODIUM METHYLATE-ALCOHOL MIXTURE [COMBUSTIBLE LIQUID LABEL]
		SODIUM METHYLATE-ALCOHOL MIXTURE [CORROSIVE MATERIAL LABEL]

CAS No. *IN* *Chemical Name*

SODIUM METHYLATE-ALCOHOL MIXTURE [FLAMMABLE LIQUID LABEL]
UN 1289 (DOT)
UN 1431 (DOT)

124-43-6

CARBAMIDE PEROXIDE
GLY-OXIDE
HYDROGEN PEROXIDE (H2O2), COMPD WITH UREA (1:1)
HYDROGEN PEROXIDE CARBAMIDE
HYDROGEN PEROXIDE, COMPD WITH UREA (1:1)
HYDROGEN PEROXIDE-UREA COMPOUND (1:1)
HYDROPERIT
HYDROPERITE
HYPEROL
NA 1511 (DOT)
ORTIZON
PERCARBAMID
PERCARBAMIDE
PERHYDRIT
PERHYDROL-UREA
THENARDOL
UN 1511 (DOT)
UREA DIOXIDE
UREA HYDROGEN PEROXIDE
UREA HYDROGEN PEROXIDE (DOT)
UREA HYDROGEN PEROXIDE SALT
UREA HYDROPEROXIDE
UREA PEROXIDE
UREA PEROXIDE (DOT)
UREA, COMPD WITH H2O2
UREA, COMPD WITH HYDROGEN PEROXIDE (1:1)
UREA, COMPD WITH HYDROGEN PEROXIDE (H2O2) (1:1)

124-47-0

ACIDOGEN NITRATE
NITRIC ACID UREA SALT
UN 1357 (DOT)
UREA NITRATE
UREA NITRATE (1:1)
UREA NITRATE (WET)
UREA NITRATE, WET WITH 10% OR MORE WATER (DOT)
UREA NITRATE, WET WITH 10% OR MORE WATER, OVER 25 LBS IN ONE OUTSIDE PACKAGING (DOT)
UREA NITRATE, WETTED WITH NOT LESS THAN 20% WATER (DOT)
UREA, MONONITRATE
UREA, MONONITRATE (8CI,9CI)
UREA, NITRATE

124-48-1

* CHLORODIBROMOMETHANE
DIBROMOCHLOROMETHANE
DIBROMOMONOCHLOROMETHANE
METHANE, DIBROMOCHLORO-
MONOCHLORODIBROMOMETHANE

124-65-2

((DIMETHYLARSINO)OXY)SODIUM-AS-OXIDE
ALKARSODYL
ANSAR 160
ANSAR 560
ARSECODILE
ARSICODILE
ARSINE OXIDE, DIMETHYLHYDROXY-, SODIUM SALT
ARSINE OXIDE, HYDROXYDIMETHYL-, SODIUM SALT
ARSINIC ACID, DIMETHYL-, SODIUM SALT
ARSINIC ACID, DIMETHYL-, SODIUM SALT (9CI)
ARSYCODILE
BOLLS-EYE
CACODYLATE DE SODIUM (FRENCH)
CACODYLIC ACID SODIUM SALT
CHEMAID
DUTCH-TREAT
HYDROXYDIMETHYLARSINE OXIDE SODIUM SALT
PHYTAR 560
RAD-E-CATE
RAD-E-CATE 16
RAD-E-CATE 25
RAD-E-CATE 35
SILVISAR
SODIUM CACODYLATE
SODIUM CACODYLATE (DOT)
SODIUM DIMETHYLARSINATE
SODIUM DIMETHYLARSONATE
SODIUM SALT OF CACODYLIC ACID
SODIUM, [(DIMETHYLARSINO)OXY]-, AS-OXIDE
UN 1688 (DOT)

124-68-5

1,1-DIMETHYL-2-HYDROXYETHYLAMINE
1-PROPANOL, 2-AMINO-2-METHYL-
2-AMINO-1-HYDROXY-2-METHYLPROPANE
2-AMINO-2,2-DIMETHYLETHANOL
2-AMINO-2-METHYL-1-PROPANOL
2-AMINO-2-METHYLPROPANOL
2-AMINOISOBUTANOL
2-HYDROXYMETHYL-2-PROPYLAMINE
2-METHYL-2-AMINO-1-PROPANOL
2-METHYL-2-AMINOPROPANOL
AMP 95
AMP REGULAR
BETA-AMINOISOBUTANOL
HYDROXY-TERT-BUTYLAMINE
ISOBUTANOL-2-AMINE

124-87-8

COCCULIN
COCCULUS
COCCULUS, SOLID
COCCULUS, SOLID (FISHBERRY) (DOT)
COQUES DU LEVANT (FRENCH)
FISH BERRY
INDIAN BERRY
ORIENTAL BERRY
PICROTIN, COMPD WITH PICROTOXININ (1:1)
PICROTOXIN
PICROTOXINE
PICROTOXININ, COMPD WITH PICROTIN (1:1)
UN 1584 (DOT)

126-33-0

1,1-DIOXIDETETRAHYDROTHIOFURAN
1,1-DIOXIDETETRAHYDROTHIOPHENE
1,1-DIOXOTHIOLAN
2,3,4,5-TETRAHYDROTHIOPHENE-1,1-DIOXIDE
BONDELANE A
BONDOLANE A
CYCLIC TETRAMETHYLENE SULFONE
CYCLOTETRAMETHYLENE SULFONE
DIHYDROBUTADIENE SULPHONE
DIOXOTHIOLAN
SULFALONE
SULFOLAN
SULFOLANE
SULPHOLANE
SULPHOXALINE
TETRAHYDROTHIOPHENE 1,1-DIOXIDE
TETRAHYDROTHIOPHENE DIOXIDE
TETRAMETHYLENE SULFONE
THIACYCLOPENTANE DIOXIDE
THIOCYCLOPENTANE-1,1-DIOXIDE
THIOLANE-1,1-DIOXIDE
THIOPHAN SULFONE
THIOPHANE 1,1-DIOXIDE
THIOPHANE DIOXIDE
THIOPHENE, TETRAHYDRO-, 1,1-DIOXIDE

126-39-6

1,3-DIOXOLANE, 2-ETHYL-2-METHYL-
2-BUTANONE ETHYLENE KETAL
2-BUTANONE, CYCLIC 1,2-ETHANEDIYL ACETAL
2-ETHYL-2-METHYL-1,3-DIOXOLANE
2-ETHYL-2-METHYLDIOXOLANE
2-METHYL-2-ETHYL-1,3-DIOXOLANE
2-METHYL-2-ETHYLDIOXOLANE

CAS No.	IN	Chemical Name
126-72-7		
		1-PROPANOL, 2,3-DIBROMO-, PHOSPHATE (3:1)
		3PBR
		ANFRAM 3PB
		APEX 462-5
		BROMKAL P 67-6HP
		ES 685
		FIREMASTER LV-T 23P
		FIREMASTER T 23
		FIREMASTER T 23P
		FLACAVON R
		FLAMMEX AP
		FLAMMEX LV-T 23P
		FLAMMEX T 23P
		FYROL HB 32
		PHOSCON PE 60
		PHOSCON UF-S
		T 23P
		TDBPP
		TRIS
		TRIS (FLAME RETARDANT)
		TRIS(2,3-DIBROMOPROPYL)PHOSPHATE
		ZETOFEX ZN
126-73-8		
		BUTYL PHOSPHATE
		CELLUPHOS 4
		DISFLAMOLL TB
		PHOSPHORIC ACID TRIBUTYL ESTER
		TBP
		TRI-N-BUTYL PHOSPHATE
		TRIBUTOXYPHOSPHINE OXIDE
	*	TRIBUTYL PHOSPHATE
126-85-2		
		NITROGEN MUSTARD N-OXIDE
126-98-7		
		2-CYANO-1-PROPENE
		2-CYANOPROPENE
		2-CYANOPROPENE-1
		2-METHYL-2-PROPENENITRILE
		2-METHYLACRYLONITRILE
		2-METHYLPROPENENITRILE
		2-PROPENENITRILE, 2-METHYL-
		ALPHA-METHACRYLONITRILE
		ALPHA-METHYLACRYLONITRILE
		ISOPROPENE CYANIDE
		ISOPROPENYLNITRILE
		METHACRYLNITRILE
		METHACRYLONITRILE
		METHYLACRYLONITRILE
		METHYLACRYLONITRILE (ACGIH)
		RCRA WASTE NUMBER U152
		USAF ST-40
126-99-8		
		1,3-BUTADIENE, 2-CHLORO-
		2-CHLORO-1,3-BUTADIENE
		2-CHLOROBUTADIENE
		BETA-CHLOROPRENE
		CHLOROPRENE
127-00-4		
		1-CHLORO-2-HYDROXYPROPANE
		1-CHLORO-2-PROPANOL
		1-CHLOROISOPROPYL ALCOHOL
		2-PROPANOL, 1-CHLORO-
		ALPHA-PROPYLENE CHLOROHYDRIN
		SEC-PROPYLENE CHLOROHYDRIN
127-07-1		
		HYDROXYUREA
127-18-4		
		1,1,2,2-TETRACHLOROETHENE
		1,1,2,2-TETRACHLOROETHYLENE
		ANKILOSTIN
		ANTISAL 1
		ANTISOL 1
		CARBON BICHLORIDE
		CARBON DICHLORIDE
		CZTEROCHLOROETYLEN (POLISH)
		DIDAKENE
		DILATIN PT
		DOW-PER
		ENT 1,860
		ETHENE, TETRACHLORO-
		ETHYLENE TETRACHLORIDE
		ETHYLENE, TETRACHLORO-
		FEDAL-UN
		FREON 1110
		NCI-C04580
		NEMA
		PER
		PERAWIN
		PERC
		PERCHLOORETHYLEEN, PER (DUTCH)
		PERCHLOR
		PERCHLORAETHYLEN, PER (GERMAN)
		PERCHLORETHYLENE
		PERCHLORETHYLENE, PER (FRENCH)
		PERCHLOROETHYLENE
		PERCHLOROETHYLENE (ACGIH,DOT)
		PERCLENE
		PERCLENE D
		PERCLOROETILENE (ITALIAN)
		PERCOSOLVE
		PERK
		PERKLONE
		PERSEC
		RCRA WASTE NUMBER U210
		TETLEN
		TETRACAP
		TETRACHLOORETHEEN (DUTCH)
		TETRACHLORAETHEN (GERMAN)
		TETRACHLORETHYLENE
		TETRACHLOROETHENE
	*	TETRACHLOROETHYLENE
		TETRACHLOROETHYLENE (DOT)
		TETRACLOROETENE (ITALIAN)
		TETRAGUER
		TETRALENO
		TETRALEX
		TETRAVEC
		TETROGUER
		TETROPIL
		UN 1897 (DOT)
127-19-5		
		ACETAMIDE, N,N-DIMETHYL-
		ACETDIMETHYLAMIDE
		ACETIC ACID, DIMETHYLAMIDE
		DIMETHOXYL ACETAMIDE
		DIMETHYL ACETAMIDE
		DIMETHYL ACETAMIDE (ACGIH)
		DIMETHYLACETONE AMIDE
		DIMETHYLAMIDE ACETATE
		DMA
		DMAC
		N,N-DIMETHYLACETAMIDE
		N,N-DIMETHYLETHANAMIDE
		NSC 3138
		U-5954
127-48-0		
		TRIMETHADIONE
127-82-2		
		1-PHENOL-4-SULFONIC ACID ZINC SALT
		BENZENESULFONIC ACID, 4-HYDROXY-, ZINC SALT (2:1)
		BENZENESULFONIC ACID, P-HYDROXY-, ZINC SALT (2:1)
		NA 9160 (DOT)
		P-HYDROXYBENZENESULFONIC ACID ZINC SALT
		PHENOZIN
		ZINC P-HYDROXYBENZENESULFONATE
		ZINC P-PHENOL SULFONATE
		ZINC PHENOLSULFONATE

CAS No.	IN	Chemical Name
		ZINC PHENOLSULFONATE (DOT)
		ZINC SULFOCARBOLATE
		ZINC SULFOPHENATE
127-85-5		
		(4-AMINOPHENYL)ARSONIC ACID SODIUM SALT
		ANHYDROUS SODIUM ARSANILATE
		ARSAMIN
		ARSANILIC ACID SODIUM SALT
		ARSANILIC ACID, MONOSODIUM SALT
		ARSINOSOLVIN
		ARSONIC ACID, (4-AMINOPHENYL)-, MONOSODIUM SALT
		ARSONIC ACID, (4-AMINOPHENYL)-, MONOSODIUM SALT (9CI)
		ATOXYL
		NCI-C61176
		NUARSOL
		PIGLET PRO-GEN V
		PRO-GEN SODIUM
		PROTOXYL
		SOAMIN
		SODIUM AMINARSONATE
		SODIUM AMINOPHENOL ARSONATE
		SODIUM ANILARSONATE
		SODIUM ARSANILATE
		SODIUM ARSANILATE (DOT)
		SODIUM ARSONILATE
		SODIUM P-AMINOBENZENEARSONATE
		SODIUM P-AMINOPHENYLARSONATE
		SODIUM P-ARSANILATE
		SODIUM-ANALINE ARSONATE
		SONATE
		TRYPOXYL
		UN 2473 (DOT)
127-95-7		
		ETHANEDIOIC ACID, MONOPOTASSIUM SALT
		KLEESALZ (GERMAN)
		MONOPOTASSIUM OXALATE
		OXALIC ACID, MONOPOTASSIUM SALT
		POTASSIUM ACID OXALATE
		POTASSIUM BINOXALATE
		POTASSIUM HYDROGEN OXALATE
		POTASSIUM OXALATE (KHC2O4)
		POTASSIUM SALT OF SORREL
		SALT OF SORREL
		SORREL SALT
128-37-0		
		2,6-BIS(1,1-DIMETHYLETHYL)-4-METHYLPHENOL
		2,6-DI-TERT-BUTYL-4-HYDROXYTOLUENE
		2,6-DI-TERT-BUTYL-4-METHYLPHENOL
	*	2,6-DI-TERT-BUTYL-P-CRESOL
		2,6-DI-TERT-BUTYL-P-METHYLPHENOL
		2,6-DI-TERT-BUTYLCRESOL
		2,6-DI-TERT-BUTYLMETHYLPHENOL
		3,5-DI-TERT-BUTYL-4-HYDROXYTOLUENE
		4-HYDROXY-3,5-DI-TERT-BUTYLTOLUENE
		4-METHYL-2,6-DI-TERT-BUTYLPHENOL
		ADVASTAB 401
		AGIDOL
		AGIDOL 1
		ALKOFEN BP
		ANTIOXIDANT 264
		ANTIOXIDANT 29
		ANTIOXIDANT 30
		ANTIOXIDANT 4
		ANTIOXIDANT 4K
		ANTIOXIDANT DBPC
		ANTIOXIDANT KB
		AO 29
		AO 4
		AO 4K
		AOX 4
		AOX 4K
		BHT
		BUKS
		BUTYLATED HYDROXYTOLUENE
		CAO 1
		CAO 3
		CATALIN CAO-3
		CHEMANOX 11
		DALPAC
		DBPC
		DEENAX
		DI-TERT-BUTYL-4-METHYLPHENOL
		DI-TERT-BUTYL-P-CRESOL
		DI-TERT-BUTYLCRESOL
		DIBUNOL
		DIBUTYLATED HYDROXYTOLUENE
		IMPRUVOL
		IONOL
		IONOL (ANTIOXIDANT)
		IONOL 1
		IONOL BHT
		IONOL CP
		IONOLE
		KERABIT
		NOCRAC 200
		NONOX TBC
		O-DI-TERT-BUTYL-P-METHYLPHENOL
		P 21
		P-CRESOL, 2,6-DI-TERT-BUTYL-
		PARABAR 441
		PHENOL, 2,6-BIS(1,1-DIMETHYLETHYL)-4-METHYL-
		STAVOX
		SUMILIZER BHT
		SUSTANE BHT
		SWANOX BHT
		TENAMENE 3
		TENOX BHT
		TOPANOL
		TOPANOL BHT
		TOPANOL O
		TOPANOL OC
		VANLUBE PC
		VANLUBE PCX
		VIANOL
		VULKANOX KB
128-44-9		
		1,2-BENZISOTHIAZOL-3(2H)-ONE, 1,1-DIOXIDE, SODIUM SALT
		1,2-BENZISOTHIAZOLIN-3-ONE, 1,1-DIOXIDE, SODIUM DERIV.
		1,2-BENZISOTHIAZOLIN-3-ONE, 1,1-DIOXIDE, SODIUM SALT
		1,2-BENZOTHIAZOL-3(2H)-ONE 1,1-DIOXIDE SODIUM SALT
		CRISTALLOSE
		CRYSTALLOSE
		KRISTALLOSE
		SACCHARIN SODIUM
		SACCHARIN SODIUM SALT
		SACCHARIN SOLUBLE
		SAXIN
		SODIUM O-BENZOSULFIMIDE
		SODIUM SACCHARIDE
		SODIUM SACCHARIN
		SODIUM SACCHARINATE
		SODIUM SACCHARINE
		SOLUBLE SACCHARIN
		SWEETA
		SYKOSE
		WILLOSETTEN
128-46-1		
		DIHYDROSTREPTOMYCIN
128-56-3		
		SODIUM ANTHRAQUINONE-1-SULFONATE
128-66-5		
		C.I. VAT YELLOW 4
129-00-0		
		BENZO(DEF)PHENANTHRENE
		BETA-PYRENE
		PYREN (GERMAN)
	*	PYRENE

CAS No.	IN	Chemical Name
129-06-6		2H-1-BENZOPYRAN-2-ONE, 4-HYDROXY-3-(3-OXO-1-PHENYLBUTYL)-, SODIUM SALT
		2H-1-BENZOPYRAN-2-ONE, 4-HYDROXY-3-(3-OXO-1-PHENYLBUTYL)-, SODIUM SALT (9CI)
		3-(ALPHA-ACETONYLBENZYL)-4-HYDROXY-COUMARIN SODIUM SALT
		3-(ALPHA-ACETONYLBENZYL)-4-HYDROXYCOUMARIN SODIUM
		ATHROMBIN
		COUMADIN
		COUMADIN SODIUM
		COUMAFENE SODIUM
		COUMARIN, 3-(ALPHA-ACETONYLBENZYL)-4-HYDROXY-, SODIUM SALT
		CUMADIN
		MAREVAN
		MAREVAN (SODIUM SALT)
		PANIVARFIN
		PANWARFIN
		PROTHROMADIN
		RATSUL SOLUBLE
		SODIUM COUMADIN
		SODIUM WARFARIN
		SODIUM, ((3-(ALPHA-ACETONYLBENZYL)-2-OXO-2H-1-BENZOPYRAN-4-YL)OXY)-
		SODIUM, [[2-OXO-3-(3-OXO-1-PHENYLBUTYL)-2H-1-BENZOPYRAN-4-YL]OXY]-
		TINTORANE
		VARFINE
		WARAN
		WARCOUMIN
		WARFARIN SODIUM
		WARFARIN SODIUM SALT
		WARFARIN, SODIUM DERIV
		WARFILONE
		ZOOCOUMARIN SODIUM SALT
129-15-7		1-NITRO-2-METHYLANTHRAQUINONE
		2-METHYL-1-NITROANTHRAQUINONE
		9,10-ANTHRACENEDIONE, 2-METHYL-1-NITRO-
		ANTHRAQUINONE, 2-METHYL-1-NITRO-
129-66-8		2,4,6-TRINITROBENZOIC ACID
		BENZOIC ACID, 2,4,6-TRINITRO-
		BENZOIC ACID, TRINITRO-
		BENZOIC ACID, TRINITRO- (10% TO 30% WATER)
		SYM-TRINITROBENZOIC ACID
	*	TRINITROBENZOIC ACID
		TRINITROBENZOIC ACID, WET, CONTAINING AT LEAST 10% WATER (DOT)
		TRINITROBENZOIC ACID, WET, CONTAINING LESS THAN 30% WATER (DOT)
		UN 0215 (DOT)
		UN 1355 (DOT)
129-67-9		ENDOTHALL
129-99-8		2-CHLORO-1,3-BUTADIENE
131-11-3		1,2-BENZENEDICARBOXYLIC ACID, DIMETHYL ESTER
		AVOLIN
		DIMETHYL 1,2-BENZENEDICARBOXYLATE
		DIMETHYL BENZENEORTHODICARBOXYLATE
		DIMETHYL ESTER PHTHALIC ACID
		DIMETHYL O-PHTHALATE
	*	DIMETHYL PHTHALATE
		DIMETHYL PHTHALATE (ACGIH)
		DMF (INSECT REPELLENT)
		DMP
		ENT 262
		FERMINE
		METHYL PHTHALATE
		MIPAX
		NTM
		PALATINOL M
		PHTHALIC ACID METHYL ESTER
		PHTHALIC ACID, DIMETHYL ESTER
		PHTHALSAEUREDIMETHYLESTER (GERMAN)
		RCRA WASTE NUMBER U102
		REPEFTAL
		SOLVANOM
		SOLVARONE
		UNIMOLL DM
131-17-9		1,2-BENZENEDICARBOXYLIC ACID, DI-2-PROPENYL ESTER
		ALLYL PHTHALATE
		DAPON R
		DAPPU
		DIALLYL PHTHALATE
		PHTHALIC ACID, DIALLYL ESTER
131-24-8		CHLOROACETOPHENONE
131-52-2		DOW DORMANT FUNGICIDE
		DOWICIDE G
		DOWICIDE G-ST
		GR 48-11PS
		GR 48-32S
		MYSTOX D
		NAPCLOR-G
		NAPCP
		PCP SODIUM SALT
		PCP-SODIUM
		PENTACHLOROPHENATE SODIUM
		PENTACHLOROPHENOL SODIUM SALT
		PENTACHLOROPHENOXY SODIUM
		PENTAPHENATE
		PENTAPLASTIC
		PHENOL, PENTACHLORO-, SODIUM SALT
		PKHFN
		SANTOBRITE
		SAPCO 25
		SODIUM PCP
		SODIUM PENTACHLOROPHENATE
		SODIUM PENTACHLOROPHENATE (DOT)
		SODIUM PENTACHLOROPHENOL
		SODIUM PENTACHLOROPHENOLATE
		SODIUM PENTACHLOROPHENOXIDE
		SODIUM PENTACHLORPHENATE
		SODIUM, (PENTACHLOROPHENOXY)-
		UN 2567 (DOT)
		WEEDBEADS
131-73-7		1,3,5,7-TETRAAZAADAMANTANE
		1,3,5,7-TETRAAZATRICYCLO[33113,7]DECANE
		2,2',4,4',6,6'-HEXANITRODIPHENYLAMINE
		2,4,6,2',4',6'-HEXANITRODIPHENYLAMINE
		2,4,6-TRINITRO-N-(2,4,6-TRINITROPHENYL)BENZENAMINE
		ACETO HMT
		AMINOFORM
		AMINOFORMALDEHYDE
		AMMOFORM
		AMMONIOFORMALDEHYDE
		ANTIHYDRAL
		BENZENAMINE, 2,4,6-TRINITRO-N-(2,4,6-TRINITROPHENYL)-
		BIS(2,4,6-TRINITROPHENYL)AMINE
		CYSTAMIN
		CYSTOGEN
		DIPHENYLAMINE, 2,2',4,4',6,6'-HEXANITRO-
		DIPICRYLAMINE
		DUIREXOL
		EKAGOM H
		ESAMETILENTETRAMINA (ITALIAN)
		FORMAMINE
		FORMIN
		FORMIN (HETEROCYCLE)
		HERAX UTS
		HETERIN

CAS No.	IN	Chemical Name
		HEXA
		HEXA (VULCANIZATION ACCELERATOR)
		HEXA-FLO-PULVER
		HEXAFORM
		HEXAMETHYLENAMINE
		HEXAMETHYLENEAMINE
		HEXAMETHYLENETETRAAMINE
		HEXAMETHYLENETETRAMINE
		HEXAMETHYLENTETRAMIN (GERMAN)
		HEXAMETHYLENTETRAMINE
		HEXAMINE
		HEXAMINE (DOT)
		HEXAMINE (HETEROCYCLE)
		HEXAMINE (POTASSIUM REAGENT)
		HEXANITRODIPHENYLAMINE
		HEXASAN
		HEXILMETHYLENAMINE
		HEXYL
		HEXYL (REAGENT)
		HMT
		METHAMIN
		METHENAMIN
		METHENAMINE
		NOCCELER H
		PREPARATION AF
		RESOTROPIN
		UN 1328 (DOT)
		URAMIN
		URATRINE
		URITONE
		URODEINE
		UROTROPIN
		UROTROPINE
		VULKACIT H 30
		XAMETRIN
131-74-8		
		2,4,6-TRINITROPHENOL AMMONIUM SALT
		AMMONIUM CARBAZOATE
		AMMONIUM PICRATE
		AMMONIUM PICRATE (WET)
		AMMONIUM PICRATE, DRY
		AMMONIUM PICRATE, DRY OR CONTAINING, BY WEIGHT, LESS THAN 10% WATER (DOT)
		AMMONIUM PICRATE, WET WITH 10% OR MORE WATER (DOT)
		AMMONIUM PICRONITRATE
		EXPLOSIVE D
		FLAMMABLE SOLID
		OBELINE PICRATE
		PHENOL, 2,4,6-TRINITRO-, AMMONIUM SALT
		PHENOL, 2,4,6-TRINITRO-, AMMONIUM SALT (9CI)
		PICRATE OF AMMONIA (DOT)
		PICRATOL
		PICRIC ACID, AMMONIUM SALT
		PICRIC ACID, NH4 DERIV
		RCRA WASTE NUMBER P009
		UN 0004 (DOT)
		UN 1310 (DOT)
131-89-5		
		2,4-DINITRO-6-CYCLOHEXYLPHENOL
		2-CYCLOHEXYL-4,6-DINITROFENOL (DUTCH)
		2-CYCLOHEXYL-4,6-DINITROPHENOL
		4,6-DINITRO-O-CYCLOHEXYLPHENOL
		6-CICLOESIL-2,4-DINITR-FENOLO (ITALIAN)
		6-CYCLOHEXYL-2,4-DINITROPHENOL
		DINEX
		DINITRO-O-CYCLOHEXYLPHENOL
		DINITROCYCLOHEXYLPHENOL
		DINITROCYCLOHEXYLPHENOL (DOT)
		DN
		DN (PESTICIDE)
		DN 1
		DN DRY MIX NO 1
		DN DUST NO 12
		DNOCHP
		DOWSPRAY 17
		DRY MIX NO 1
		ENT 157
		NA 9026 (DOT)
		PEDINEX (FRENCH)
		PHENOL, 2-CYCLOHEXYL-4,6-DINITRO-
		PHENOL, 6-CYCLOHEXYL-2,4-DINITRO-
		RCRA WASTE NUMBER P034
		SN 46
132-32-1		
		3-AMINO-9-ETHYLCARBAZOLE, HYDROCHLORIDE
132-64-9		
		DIBENZOFURAN
133-06-2		
		1,2,3,6-TETRAHYDRO-N-(TRICHLOROMETHYLTHIO)-PHTHALIMIDE
		1H-ISOINDOLE-1,3(2H)-DIONE, 3A,4,7,7A-TETRAHYDRO-2-((TRICHLOROMETHYL)THIO)-
		3A,4,7,7A-TETRAHYDRO-N-(TRICHLOROMETHANE-SULPHENYL)PHTHALIMIDE
		4-CYCLOHEXENE-1,2-DICARBOXIMIDE, N-(TRICHLORO-METHYL)THIO-
		AACAPTAN
		AGROSOL S
		AGROX 2-WAY AND 3-WAY
		AMERCIDE
		BANGTON
		BEAN SEED PROTECTANT
		CAPTAF
		CAPTAF 85W
		CAPTAN
		CAPTAN (ACGIH,DOT)
		CAPTAN 50W
		CAPTAN-STREPTOMYCIN 75-01 POTATO SEED PIECE PROTECTANT
		CAPTANCAPTENEET 26,538
		CAPTANE
		CAPTEX
		ENT 26,538
		ESSO FUNGICIDE 406
		FLIT 406
		FUNGUS BAN TYPE II
		GLYODEX 3722
		GRANOX PFM
		GUSTAFSON CAPTAN 30-DD
		HEXACAP
		KAPTAN
		LE CAPTANE (FRENCH)
		MALIPUR
		MERPAN
		MICRO-CHECK 12
		N-((TRICHLOROMETHYL)THIO)-4-CYCLOHEXENE-1,2-DICARBOXIMIDE
		N-((TRICHLOROMETHYL)THIO)TETRAHYDROPHTHALIMIDE
		N-(TRICHLOR-METHYLTHIO)-PHTHALIMID (GERMAN)
		N-(TRICHLOROMETHYLMERCAPTO)-DELTA(SUP 4)-TETRA-HYDROPHTHALIMIDE
		N-TRICHLOROMETHYLMERCAPTO-4-CYCLOHEXENE-1,2-DICARBOXIMIDE
		N-TRICHLOROMETHYLTHIO-3A,4,7,7A-TETRAHYDRO-PHTHALIMIDE
		N-TRICHLOROMETHYLTHIO-CIS-DELTA(SUP 4)-CYCLOHEX-ENE-1,2-DICARBOXIMIDE
		N-TRICHLOROMETHYLTHIOCYCLOHEX-4-ENE-1,2-DICAR-BOXIMIDE
		N-[(TRICHLOROMETHYL)THIO]-DELTA4-TETRAHYDROPH-THALIMIDE
		N-[(TRICHLOROMETHYL)THIO]TETRAHYDROPHTHALIMIDE
		NA 9099 (DOT)
		NCI-C00077
		NERACID
		ORTHOCIDE
		ORTHOCIDE 406
		ORTHOCIDE 50
		ORTHOCIDE 75
		ORTHOCIDE 83
		OSOCIDE
		RALLIS CAPTAF

CAS No.	IN	Chemical Name
		SR 406
		STAUFFER CAPTAN
		TRICHLOROMETHYLTHIO-1,2,5,6-TETRAHYDROPHTHALAMIDE
		TRIMEGOL
		VANCIDE 89
		VANCIDE 89RE
		VANCIDE P-75
		VANGARD K
		VANGUARD K
		VANICIDE
		VONDCAPTAN
133-14-2		
		2,4-DICHLOROBENZOYL PEROXIDE
		2,4-DICHLOROBENZOYL PEROXIDE, NOT MORE THAN 52% AS A PASTE (DOT)
		2,4-DICHLOROBENZOYL PEROXIDE, NOT MORE THAN 52% IN SOLUTION (DOT)
		BIS(2,4-DICHLOROBENZOYL) PEROXIDE
		CADOX TDP
		CADOX TS
		CADOX TS 40,50
		DCBP
		DI-2,4-DICHLOROBENZOYL PEROXIDE, MAXIMUM CONCENTRATION 52% AS A PASTE OR IN SOLUTION (DOT)
		LUPERCO CST
		O,O',P,P'-TETRACHLORODIBENZOYL PEROXIDE
		PEROXIDE, BIS(2,4-DICHLOROBENZOYL)
		PEROXIDE, BIS(2,4-DICHLOROBENZOYL)-, NOT MORE THAN 52% AS A PASTE OR IN SOLUTION
		SILOPREN-VERNETZER CL 40
		UN 2138 (DOT)
		UN 2139 (DOT)
133-90-4		
		CHLORAMBEN
134-29-2		
		2-METHOXYANILINE HYDROCHLORIDE
		BENZENAMINE, 2-METHOXY-, HYDROCHLORIDE
		C.I. 37115
		FAST RED BB BASE
		O-ANISIDINE HYDROCHLORIDE
134-32-7		
		1-AMINONAPHTHALENE
		1-NAPHTHALAMINE
		1-NAPHTHALENAMINE
	*	1-NAPHTHYLAMINE
		1-NAPHTHYLAMINE, TECHNICAL GRADE
		ALPHA-AMINONAPHTHALENE
		ALPHA-NAPHTHYLAMINE
		C.I. 37265
		C.I. AZOIC DIAZO COMPONENT 114
		FAST GARNET BASE B
		NAPHTHALIDAM
		NAPHTHALIDINE
135-01-3		
		1,2-DIETHYLBENZENE
		BENZENE, 1,2-DIETHYL-
		BENZENE, O-DIETHYL-
		O-DIETHYL BENZENE
135-02-4		
		2-ANISALDEHYDE
		2-METHOXYBENZALDEHYDE
		2-METHOXYBENZENECARBOXALDEHYDE
		6-METHOXYBENZALDEHYDE
		BENZALDEHYDE, 2-METHOXY-
		O-ANISALDEHYDE
		O-METHOXYBENZALDEHYDE
		SALICYLALDEHYDE METHYL ETHER
135-20-6		
		BENZENAMINE, N-HYDROXY-N-NITROSO-, AMMONIUM SALT
	*	CUPFERRON
		HYDROXYLAMINE, N-NITROSO-N-PHENYL-, AMMONIUM SALT
		N-NITROSO-N-PHENYLHYDROXYAMINE AMMONIUM SALT
		N-NITROSO-N-PHENYLHYDROXYLAMINE AMMONIUM SALT
135-88-6		
		2-(PHENYLAMINO)NAPHTHALENE
		2-ANILINONAPHTHALENE
		2-NAPHTHALENAMINE, N-PHENYL-
		2-NAPHTHYLAMINE, N-PHENYL-
		2-NAPHTHYLPHENYLAMINE
		ACETO PBN
		AGERITE POWDER
		ANTIOXIDANT 116
		ANTIOXIDANT D
		ANTIOXIDANT PBN
		BETA-NAPHTHYLPHENYLAMINE
		N-(2-NAPHTHYL)-N-PHENYLAMINE
		N-(2-NAPHTHYL)ANILINE
		N-BETA-NAPHTHYL-N-PHENYLAMINE
		N-PHENYL-2-NAPHTHALENAMINE
		N-PHENYL-2-NAPHTHYLAMINE
	*	N-PHENYL-BETA-NAPHTHYLAMINE
		NAFTAM 2
		NEOSONE D
		NEOZON D
		NEOZONE
		NEOZONE D
		NILOX PBNA
		NOCRAC D
		NONOX D
		NONOX DN
		PBNA
		PHENYL-BETA-NAPHTHYLAMINE
		PHENYL-2-NAPHTHYLAMINE
		STABILIZATOR AR
		STABILIZER AR
		VULKANOX PBN
135-98-8		
		(1-METHYLPROPYL)BENZENE
		2-PHENYLBUTANE
		BENZENE, (1-METHYLPROPYL)-
		BENZENE, SEC-BUTYL-
		SEC-BUTYLBENZENE
136-40-3		
		BETA-PHENYLAZO-ALPHA,ALPHA'-DIAMINOPYRIDINE HYDROCHLORIDE
		2,6-DIAMINO-3-PHENYLAZOPYRIDINE HYDROCHLORIDE
		2,6-DIAMINO-3-PHENYLAZOPYRIDINE MONOHYDROCHLORIDE
		2,6-PYRIDINEDIAMINE, 3-(PHENYLAZO)-, MONOHYDROCHLORIDE
		AZODYNE
		BISTERIL
		DIRIDONE
		MALLOPHENE
		NC 150
		PHENAZODINE
		PHENAZOPYRIDINE HYDROCHLORIDE
		PHENAZOPYRIDINIUM CHLORIDE
		PHENYLAZO TABLETS
		PIRID
		PYRIDACIL
		PYRIDINE, 2,6-DIAMINO-3-(PHENYLAZO)-, MONOHYDROCHLORIDE
		PYRIDIUM
		PYRIPYRIDIUM
		SEDURAL
		URIDINAL
		URODINE
		W 1655
136-60-7		
		BENZOIC ACID, BUTYL ESTER
		BUTYL BENZOATE
		CHEMCRYL C 101N
		IP CARRIER N 20

CAS No.	IN	Chemical Name
		N-BUTYL BENZOATE
136-78-7		
		2,4-DES SODIUM
		CRAG HERBICIDE
		CRAG HERBICIDE 1
		CRAG SESONE
		DISUL-SODIUM
		ETHANOL, 2-(2,4-DICHLOROPHENOXY)-, HYDROGEN SULFATE SODIUM SALT
		ETHANOL, 2-(2,4-DICHLOROPHENOXY)-, HYDROGEN SULFATE, SODIUM SALT
		EXPERIMENTAL HERBICIDE 1
		SES
		SES-T
		SESON
		SESONE
		SODIUM 2,4-DICHLOROPHENOXYETHYL SULPHATE
		SODIUM 2-(2,4-DICHLOROPHENOXY)ETHYL SULFATE
136-81-2		
		2-PENTYLPHENOL
		O-AMYL PHENOL
		O-PENTYLPHENOL
		PHENOL, 2-PENTYL-
		PHENOL, O-PENTYL-
137-05-3		
		2-CYANOACRYLIC ACID METHYL ESTER
		2-PROPENOIC ACID, 2-CYANO-, METHYL ESTER
		ACRYLIC ACID, 2-CYANO-, METHYL ESTER
		ADHERE
		ALPHA-CYANOACRYLIC ACID METHYL ESTER
		CEMEDINE 3000
		CEMEDINE 3000 TYPE-II
		COAPT.
		CYANOBOND SS
		CYANOLIT
		EASTMAN 910
		MECRILAT
		MECRYLATE
		METHYL ALPHA-CYANOACRYLATE
		METHYL 2-CYANOACRYLATE
		METHYL CYANOACRYLATE
137-17-7		
		2,4,5-TRIMETHYLANILINE
137-26-8		
		AAPIROL
		AATACK
		AATIRAM
		ACCEL TMT
		ACCELERANT T
		ACCELERATOR T
		ACCELERATOR THIURAM
		ACETO TETD
		ALPHA,ALPHA'-DITHIOBIS(DIMETHYLTHIO)FORMAMIDE
		ANLES
		ARASAN
		ARASAN 42S
		ARASAN 50 RED
		ARASAN 70
		ARASAN 70-S RED
		ARASAN 75
		ARASAN M
		ARASAN-SF
		ARASAN-SF-X
		ATIRAM
		AULES
		BETOXIN
		BIS((DIMETHYLAMINO)CARBONOTHIOYL) DISULPHIDE
		BIS(DIMETHYL-THIOCARBAMOYL)-DISULFID (GERMAN)
		BIS(DIMETHYLTHIOCARBAMOYL) DISULFIDE
		BIS(DIMETHYLTHIOCARBAMOYL) DISULPHIDE
		BIS(DIMETHYLTHIOCARBAMYL) DISULFIDE
		CHIPCO THIRAM 75
		CYURAM DS
		DISOLFURO DI TETRAMETILTIOURAME (ITALIAN)
		DISULFIDE, BIS(DIMETHYLTHIOCARBAMOYL)
		DISULFURE DE TETRAMETHYLTHIOURAME (FRENCH)
		EKAGOM TB
		FALITIRAM
		FERMIDE
		FERNACOL
		FERNASAN
		FERNASAN A
		FERNIDE
		FLO PRO T SEED PROTECTANT
		FORMALSOL
		FORMAMIDE, 1,1'-DITHIOBIS(N,N-DIMETHYLTHIO-
		HERMAL
		HERMAT TMT
		HERYL
		HEXATHIR
		KREGASAN
		MERCURAM
		METHYL THIRAM
		METHYL THIURAMDISULFIDE
		METHYL TUADS
		METIURAC
		N,N'-(DITHIODICARBONOTHIOYL)BIS(N-METHYL-METHANAMINE)
		N,N,N',N'-TETRAMETHYLTHIURAM DISULFIDE
		N,N-TETRAMETHYLTHIURAM DISULPHIDE
		NA 2771 (DOT)
		NOBECUTAN
		NOCCELER TT
		NOMERSAN
		NORMERSAN
		PANORAM 75
		POL-THIURAM
		POLYRAM ULTRA
		POMARSOL
		POMARSOL FORTE
		POMASOL
		PURALIN
		RADOTHIRAM
		RADOTIRAM
		RCRA WASTE NUMBER U244
		REZIFILM
		RHENOGRAN TMTD
		RHODIAURAM
		ROBAC TMT
		ROYAL TMTD
		SADOPLON
		SADOPLON 75
		SPOTRETE
		SPOTRETE-F
		SQ 1489
		TERAMETHYL THIURAM DISULFIDE
		TERSAN
		TERSAN 75
		TETRAMETHYL THIURANE DISULFIDE
		TETRAMETHYL THIURANE DISULPHIDE
		TETRAMETHYL-THIRAM DISULFID (GERMAN)
		TETRAMETHYLDIURANE SULPHITE
		TETRAMETHYLENETHIURAM DISULPHIDE
		TETRAMETHYLTHIOCARBAMOYLDISULPHIDE
		TETRAMETHYLTHIOPEROXYDICARBONIC DIAMIDE
		TETRAMETHYLTHIORAMDISULFIDE (DUTCH)
		TETRAMETHYLTHIURAM
		TETRAMETHYLTHIURAM BISULFIDE
		TETRAMETHYLTHIURAM BISULPHIDE
		TETRAMETHYLTHIURAM DISULFIDE
		TETRAMETHYLTHIURAM DISULPHIDE
		TETRAMETHYLTHIURUM DISULFIDE
		TETRAMETHYLTHIURUM DISULPHIDE
		TETRAPOM
		TETRASIPTON
		TETRATHIURAM DISULFIDE
		TETRATHIURAM DISULPHIDE
		THILLATE
		THIMER
		THIOPEROXYDICARBONIC DIAMIDE ([(H2N)C(S)]2S2), TETRAMETHYL-
		THIOSAN
		THIOSCABIN

CAS No.	IN	Chemical Name
		THIOTEX
		THIOTOX
		THIOTOX (FUNGICIDE)
		THIRAM
		THIRAM (ACGIH,DOT)
		THIRAM 75
		THIRAM 80
		THIRAM B
		THIRAMAD
		THIRAME (FRENCH)
		THIRASAN
		THIRIDE
		THIULIN
		THIULIX
		THIURAD
		THIURAM
		THIURAM D
		THIURAM DISULFIDE, TETRAMETHYL-
		THIURAM M
		THIURAM M RUBBER ACCELERATOR
		THIURAM TMTD
		THIURAMIN
		THIURAMYL
		THYLATE
		TIGAM
		TIRAMPA
		TIURAM (POLISH)
		TIURAMYL
		TMT
		TMTD
		TMTDS
		TRAMETAN
		TRIDIPAM
		TRIPOMOL
		TTD
		TUADS
		TUEX
		TULISAN
		TUTAN
		TYRADIN
		USAF B-30
		USAF EK-2089
		USAF P-5
		VANCIDA TM-95
		VANCIDE TM
		VUAGT-I-4
		VULCAFOR TMT
		VULCAFOR TMTD
		VULKACIT MTIC
		VULKACIT TH
		VULKACIT THIURAM
		VULKACIT THIURAM/C
		ZAPRAWA NASIENNA T
		ZUPA S 80
137-32-6		
		1-BUTANOL, 2-METHYL-
	*	2-METHYL-1-BUTANOL
		2-METHYL-N-BUTANOL
		2-METHYLBUTYL ALCOHOL
		ACTIVE AMYL ALCOHOL
		ACTIVE PRIMARY AMYL ALCOHOL
		PRIMARY ACTIVE AMYL ALCOHOL
		SEC-BUTYLCARBINOL
137-88-2		
		AMPROLIUM HYDROCHLORIDE
138-00-1		
		2,4-DI-N-PENTYLPHENOL
		2,4-DIAMYLPHENOL
		2,4-DIPENTYLPHENOL
		PHENOL, 2,4-DIPENTYL-
138-22-7		
		2-HYDROXYPROPANOIC ACID BUTYL ESTER
		BUTYL ALPHA-HYDROXYPROPIONATE
		BUTYL LACETATE
		BUTYL LACTATE
		LACTIC ACID, BUTYL ESTER
	*	N-BUTYL LACTATE
		PROPANOIC ACID, 2-HYDROXY-, BUTYL ESTER
138-86-3		
		1,8(9)-P-MENTHADIENE
		1,8-P-MENTHADIENE
		1-METHYL-4-(1-METHYLETHENYL)CYCLOHEXENE
		1-METHYL-4-ISOPROPENYL-1-CYCLOHEXENE
		1-METHYL-4-ISOPROPENYLCYCLOHEXENE
		4-ISOPROPENYL-1-METHYL-1-CYCLOHEXENE
		4-ISOPROPENYL-1-METHYLCYCLOHEXENE
		ACINTENE DP
		ACINTENE DP DIPENTENE
		ALPHA-LIMONENE
		CAJEPUTEN
		CAJEPUTENE
		CINEN
		CINENE
		CYCLOHEXENE, 1-METHYL-4-(1-METHYLETHENYL)-
		DELTA-1,8-TERPODIENE
		DIPANOL
		DIPENTEN
	*	DIPENTENE
		DIPENTENE (DOT)
		DL-LIMONENE
		EULIMEN
		INACTIVE LIMONENE
		KAUTSCHIN
		LIMONEN
		LIMONENE
		LIMONENE, INACTIVE
		NESOL
		P-MENTHA-1,8-DIENE
		P-MENTHA-1,8-DIENE, DL-
		UN 2052 (DOT)
138-89-6		
		1-(DIMETHYLAMINO)-4-NITROSOBENZENE
		4-(DIMETHYLAMINO)NITROSOBENZENE
		4-NITROSO-N,N-DIMETHYLANILINE
		4-NITROSODIMETHYLANILINE
		ACCELERINE
		ANILINE, N,N-DIMETHYL-P-NITROSO-
		BENZENAMINE, N,N-DIMETHYL-4-NITROSO-
		BENZENAMINE, N,N-DIMETHYL-4-NITROSO- (9CI)
		DIMETHYL(P-NITROSOPHENYL)AMINE
		DIMETHYL-P-NITROSOANILINE
		DIMETHYL-P-NITROSOANILINE (DOT)
		N,N-DIMETHYL-4-NITROSOANILINE
		N,N-DIMETHYL-P-NITROSOANILINE
		NCI-C01821
		NDMA
		P-(DIMETHYLAMINO)NITROSOBENZENE
		P-(N,N-DIMETHYLAMINO)NITROSOBENZENE
		P-NITROSO-N,N-DIMETHYLANILINE
		P-NITROSODIMETHYLANILINE
		P-NITROSODIMETHYLANILINE (DOT)
		ULTRA BRILLIANT BLUE P
		UN 1369 (DOT)
139-02-6		
		PHENOL SODIUM
		PHENOL SODIUM SALT
		PHENOL, SODIUM SALT, (SOLID)
		SODIUM CARBOLATE
		SODIUM PHENATE
		SODIUM PHENOLATE
		SODIUM PHENOLATE, SOLID
		SODIUM PHENOLATE, SOLID (DOT)
		SODIUM PHENOXIDE
		SODIUM PHENYLATE
		UN 2497 (DOT)
139-13-9		
		ACETIC ACID, NITRILOTRI-
		ALPHA,ALPHA',ALPHA''-TRIMETHYLAMINETRICARBOXYLIC ACID
		AMINOTRIACETIC ACID

CAS No.	IN	Chemical Name
		CHEL 300
		COMPLEXON I
		GLYCINE, N,N-BIS(CARBOXYMETHYL)-
		GLYCINE, N,N-BIS(CARBOXYMETHYL)- (9CI)
		HAMPSHIRE NTA ACID
		N,N-BIS(CARBOXYMETHYL)GLYSINE
		NITRILO-2,2',2"-TRIACETIC ACID
	*	NITRILOTRIACETIC ACID
		NITRILOTRIACETIC ACID (NTA)
		NTA
		TITRIPLEX I
		TRIGLYCINE
		TRIGLYCOLLAMIC ACID
		VERSENE NTA ACID
139-65-1		
		4,4'-DIAMINODIPHENYL SULFIDE
		4,4'-DIAMINOPHENYL SULFIDE
		4,4'-THIOBIS[ANILINE]
		4,4'-THIODIANILINE
		4,4-THIODIANILINE
		ANILINE, 4,4'-THIODI-
		BENZENAMINE, 4,4'-THIOBIS-
		BIS(4-AMINOPHENYL) SULFIDE
		DI(P-AMINOPHENYL) SULFIDE
		P,P'-DIAMINODIPHENYL SULFIDE
		P,P-THIODIANILINE
		THIOANILINE
		THIODI-P-PHENYLENEDIAMINE
139-87-7		
		2,2'-(ETHYLIMINO)DIETHANOL
		DIETHANOLETHYLAMINE
		ETHANOL, 2,2'-(ETHYLIMINO)BIS-
		ETHANOL, 2,2'-(ETHYLIMINO)DI-
		ETHYLBIS(2-HYDROXYETHYL)AMINE
		ETHYLDIETHANOLAMINE
		N,N-BIS(2-HYDROXYETHYL)ETHYLAMINE
		N-ETHYL-2,2'-IMINODIETHANOL
		N-ETHYLDIETHANOLAMINE
139-89-6		
		NITROSODIMETHYLANILINE
139-91-3		
		2-OXAZOLIDINONE, 5-(4-MORPHOLINYLMETHYL)-3-[[(5-NITRO-2-FURANYL)METHYLENE]AMINO]-
		2-OXAZOLIDINONE, 5-(MORPHOLINOMETHYL)-3-[(5-NITROFURFURYLIDENE)AMINO]-
		5-MORPHOLINOMETHYL-3-(5-NITROFURFURYLIDENE-AMINO)OXAZOLIDONE
		ALTABACTINA
		ALTAFUR
		F 150
		FURALTADONE
		FURAZOLIN
		FURAZOLINE
		FURMETHANOL
		FURMETHONOL
		FURMETONOL
		IBIFUR
		MEDIFURAN
		NF 260
		NITRALDONE
		NITROFURMETHONE
		NITROFURMETON
		OTIFURIL
		SEPSINOL
		ULTRAFUR
		UNIFUR
		VALSYN
139-94-6		
		NITHIAZIDE
139-96-8		
		AKYPOSAL TLS
		CYCLORYL TAWF
		CYCLORYL WAT

CAS No.	IN	Chemical Name
		DODECYL SULFATE TRIETHANOLAMINE SALT
		DRENE
		ELFAN 4240 T
		EMAL 20T
		EMAL T
		EMAL TD
		EMERSAL 6434
		ETHANOL, 2,2',2"-NITRILOTRI-, DODECYL SULFATE (SALT)
		LAURYL SULFATE ESTER TRIETHANOLAMINE SALT
		LAURYL SULFATE TRIETHANOLAMINE SALT
		LAURYLSULFURIC ACID TRIETHANOLAMINE SALT
		MAPROFIX TLS
		MAPROFIX TLS 500
		MAPROFIX TLS 65
		MELANOL LP 20 T
		NIKKOL TEALS
		PROPASTE T
		REWOPOL TLS 40
		RICHONOL T
		SIPON LT
		SIPON LT 40
		SIPON LT 6
		STANDAPOL T
		STANDAPOL TLS 40
		STEINAPOL TLS 40
		STEPANOL WAT
		STERLING WAT
		SULFETAL KT 400
		SULFETAL LT
		SULFURIC ACID MONODODECYL ESTER TRIETHANOL-AMINE SALT
		SULFURIC ACID MONOLAURYL ESTER TRIETHANOLAMINE SALT
		SULFURIC ACID, DODECYL ESTER, TRIETHANOLAMINE SALT
		SULFURIC ACID, MONODODECYL ESTER, COMPD WITH 2,2',2"-NITRILOTRIETHANOL (1:1)
		SULFURIC ACID, MONODODECYL ESTER, COMPD WITH 2,2',2"-NITRILOTRIS(ETHANOL)
		SULFURIC ACID, MONODODECYL ESTER, COMPD WITH 2,2',2"-NITRILOTRIS[ETHANOL] (1:1)
		TEA LAURYL SULFATE
		TEXAPON T 35
		TEXAPON T 42
		TEXAPON TH
		TRIETHANOLAMINE DODECYL SULFATE
		TRIETHANOLAMINE LAURYL SULFATE
		TRIETHANOLAMINE MONODODECYL SULFATE
		TRIETHANOLAMMONIUM DODECYL SULFATE
		TRIETHANOLAMMONIUM LAURYL SULFATE
		TYLOROL LT 50
140-04-5		
		9-OCTADECENOIC ACID, 12-(ACETYLOXY)-, BUTYL ESTER, [R-(Z)]-
		BAKERS P 6
		BARYL
		BUTYL ACETYL RICINOLEATE
		FLEXRICIN P 6
		RICINOLEIC ACID, BUTYL ESTER, ACETATE
140-29-4		
		(CYANOMETHYL)BENZENE
		2-PHENYLACETONITRILE
		ACETONITRILE, PHENYL-
		ALPHA-CYANOTOLUENE
		ALPHA-TOLUNITRILE
		BENZENEACETONITRILE
		BENZENEACETONITRILE (9CI)
		BENZENEDIACETONITRILE
		BENZYL CYANIDE
		BENZYL NITRILE
		OMEGA-CYANOTOLUENE
		PHENACETONITRILE
		PHENYL ACETYL NITRILE
	*	PHENYLACETONITRILE
		PHENYLACETONITRILE, LIQUID (DOT)
		TOLUENE, ALPHA-CYANO-
		UN 2470 (DOT)

CAS No.	IN	Chemical Name
		USAF KF-21
140-31-8		
		1-(2-AMINOETHYL)PIPERAZINE
		1-AMINOETHYLPIPERAZINE
		1-PIPERAZINEETHANAMINE
		1-PIPERAZINEETHYLAMINE
		2-PIPERAZINYLETHYLAMINE
		AMINOETHYLPIPERAZINE
		N-(2-AMINOETHYL)PIPERAZINE
		N-(BETA-AMINOETHYL)PIPERAZINE
		N-AMINOETHYLPIPERAZINE
		N-AMINOETHYLPIPERAZINE (DOT)
		PIPERAZINE, 1-(2-AMINOETHYL)-
		UN 2815 (DOT)
		USAF DO-46
140-56-7		
		FENAMINOSULF
		LESAN
140-57-8		
		2-(P-TERT-BUTYLPHENOXY)ISOPROPYL 2-CHLOROETHYL SULFITE
		ARAMIT
	*	ARAMITE
		CES
		ETHANOL, 2-CHLORO-, 2-(P-TERT-BUTYLPHENOXY)-1-METHYLETHYL SULFITE
		NIAGARAMITE
		ORTHO-MITE
		SULFUROUS ACID, 2-(P-TERT-BUTYLPHENOXY)-1-METHYL-ETHYL 2-CHLOROETHYL ESTER
		SULFUROUS ACID, 2-CHLOROETHYL 2-[4-(1,1-DIMETHYL-ETHYL)PHENOXY]-1-METHYLETHYL ESTER
140-76-1		
		2-METHYL-5-VINYLPYRIDINE
		2-PICOLINE, 5-VINYL-
		5-VINYL-2-PICOLINE
		PYRIDINE, 2-METHYL-5-VINYL-
		PYRIDINE, 5-ETHENYL-2-METHYL-
140-80-7		
		1,4-PENTANEDIAMINE, N(SUP 1),N(SUP 1)-DIETHYL-
		1,4-PENTANEDIAMINE, N1,N1-DIETHYL-
		1-(DIETHYLAMINO)-4-AMINOPENTANE
		1-METHYL-4-(DIETHYLAMINO)BUTYLAMINE
		2-AMINO-5-DIETHYLAMINOPENTANE
		2-AMINO-5-DIETHYLAMINOPENTANE (DOT)
		4-(DIETHYLAMINO)-1-METHYLBUTYLAMINE
		4-AMINO-1-(DIETHYLAMINO)PENTANE
		5-(DIETHYLAMINO)-2-PENTYLAMINE
		DELTA-(DIETHYLAMINO)-ALPHA-METHYLBUTYLAMINE
		N(SUP 1),N(SUP 1)-DIETHYL-1,4-PENTANEDIAMINE
		N1,N1-DIETHYL-1,4-PENTANEDIAMINE
		N5,N5-DIETHYL-2,5-PENTANEDIAMINE
		NOVOLDIAMINE
		UN 2946 (DOT)
140-82-9		
		2-BETA-DIETHYLAMINOETHOXYETHANOL
		2-[2-(DIETHYLAMINO)ETHOXY]ETHANOL
		DIETHYLAMINOETHOXYETHANOL
		DIETHYLAMINOETHOXYETHANOL [COMBUSTIBLE LIQUID LABEL]
		DIETHYLAMINOETHOXYETHANOL [FLAMMABLE LIQUID LABEL]
		ETHANOL, 2-[2-(DIETHYLAMINO)ETHOXY]-
140-88-5		
		2-PROPENOIC ACID ETHYL ESTER
		ACRYLATE D'ETHYLE (FRENCH)
		ACRYLIC ACID ETHYL ESTER
		ACRYLSAEUREAETHYLESTER (GERMAN)
		AETHYLACRYLAT (GERMAN)
		AKRYLANEM ETYLU (POLISH)
		CARBOSET 511
		ETHOXYCARBONYLETHYLENE
		ETHYL 2-PROPENOATE
	*	ETHYL ACRYLATE
		ETHYL ACRYLATE (ACGIH)
		ETHYL ACRYLATE, INHIBITED
		ETHYL ACRYLATE, INHIBITED (DOT)
		ETHYL ESTER ACRYLIC ACID
		ETHYL PROPENOATE
		ETHYLACRYLAAT (DUTCH)
		ETHYLAKRYLAT (CZECH)
		ETIL ACRILATO (ITALIAN)
		ETILACRILATULUI (ROMANIAN)
		NCI-C50384
		RCRA WASTE NUMBER U113
		UN 1917 (DOT)
140-89-6		
		(O-ETHYL DITHIOCARBONATO)POTASSIUM
		CARBONIC ACID, DITHIO-, O-ETHYL ESTER, POTASSIUM SALT
		CARBONODITHIOIC ACID, O-ETHYL ESTER, POTASSIUM SALT
		ETHYL POTASSIUM XANTHATE
		ETHYL POTASSIUM XANTHOGENATE
		ETHYLXANTHIC ACID POTASSIUM SALT
		O-ETHYL POTASSIUM DITHIOCARBONATE
		POTASSIUM ETHYLXANTHATE
		POTASSIUM ETHYLXANTHOGENATE
		POTASSIUM O-ETHYL DITHIOCARBONATE
	*	POTASSIUM XANTHATE
		POTASSIUM XANTHOGENATE
		Z 3
		Z 3 (PESTICIDE)
141-32-2		
		2-PROPENOIC ACID BUTYL ESTER
		ACRYLIC ACID BUTYL ESTER
		ACRYLIC ACID N-BUTYL ESTER
		BUTYL 2-PROPENOATE
		BUTYL ACRYLATE
		BUTYLACRYLATE, INHIBITED (DOT)
		N-BUTYL ACRYLATE
		N-BUTYL ACRYLATE (ACGIH)
		UN 2348 (DOT)
141-43-5		
		1-AMINO-2-HYDROXYETHANE
		2-AMINO-1-ETHANOL
		2-AMINOETHANOL
		2-ETHANOLAMINE
		2-HYDROXYETHANAMINE
		2-HYDROXYETHYLAMINE
		AMINOETHANOL
		BETA-AMINOETHANOL
		BETA-AMINOETHYL ALCOHOL
		BETA-ETHANOLAMINE
		BETA-HYDROXYETHYLAMINE
		COLAMINE
		ETHANOL, 2-AMINO-
	*	ETHANOLAMINE
		ETHYLOLAMINE
		GLYCINOL
		MEA
		MEA (ALCOHOL)
		MONOETHANOLAMINE
		OLAMINE
		THIOFACO M 50
141-57-1		
		N-PROPYLTRICHLOROSILANE
		PROPYL TRICHLOROSILANE
		PROPYL TRICHLOROSILANE (DOT)
		SILANE, PROPYLTRICHLORO-
		SILANE, TRICHLOROPROPYL-
		TRICHLORO-N-PROPYLSILANE
		TRICHLOROPROPYLSILANE
		UN 1816 (DOT)

CAS No.	IN	Chemical Name
141-59-3		
		2,4,4-TRIMETHYL-2-PENTANETHIOL
		2-PENTANETHIOL, 2,4,4-TRIMETHYL-
		TERT-OCTYL MERCAPTAN
		TERT-OCTYLTHIOL
141-66-2		
		2-DIMETHYL CIS-2-DIMETHYL-CARBAMOYL-1-METHYL-VINYL PHOSPHATE
		3-(DIMETHOXYPHOSPHINYLOXY)-N,N DIMETHYLISO-CROTONAMIDE
		3-(DIMETHOXYPHOSPHINYLOXY)-N,N-DIMETHYL-CIS-CROTONAMIDE
		3-(DIMETHYLAMINO)-1-METHYL-3-OXO-1-PROPENYL DIMETHYL PHOSPHATE
		3-HYDROXY-N,N-DIMETHYL-CIS-CROTONAMIDE DIMETHYL PHOSPHATE
		3-HYDROXYDIMETHYL CROTONAMIDE DIMETHYL PHOSPHATE
		BIDRIN
		C 709
		CARBICRON
		CARBOMICRON
		CIBA 709
		CIS-2-DIMETHYLCARBAMOYL-1-METHYLVINYL DIMETHYL-PHOSPHATE
		CROTONAMIDE, 3-HYDROXY-N,N-DIMETHYL-, CIS-, DI-METHYL PHOSPHATE
		CROTONAMIDE, 3-HYDROXY-N-N-DIMETHYL-, DIMETHYL PHOSPHATE, (E)-
		CROTONAMIDE, 3-HYDROXY-N-N-DIMETHYL-, DIMETHYL PHOSPHATE, CIS-
		DIAPADRIN
		DICROTOFOS (DUTCH)
		DICROTOPHOS
		DICROTOPHOS (ACGIH)
		DIMETHYL 2-DIMETHYLCARBAMOYL-1-METHYLVINYL PHOSPHATE
		DIMETHYL PHOSPHATE ESTER WITH 3-HYDROXY-N,N-DIMETHYL-CIS-CROTONAMIDE
		EKTAFOS
		ENT 24,482
		KARBICRON
		O,O-DIMETHYL O-(N,N-DIMETHYLCARBAMOYL-1-METHYL-VINYL) PHOSPHATE
		O,O-DIMETHYL-O-(1,4-DIMETHYL-3-OXO-4-AZA-PENT-1-ENYL)FOSFAAT (DUTCH)
		O,O-DIMETHYL-O-(1,4-DIMETHYL-3-OXO-4-AZA-PENT-1-ENYL)PHOSPHATE
		O,O-DIMETHYL-O-(1-METHYL-2-N,N-DIMETHYL-CARBAMOYL)-VINYL-PHOSPHAT (GERMAN)
		O,O-DIMETHYL-O-(2-DIMETHYL-CARBAMOYL-1-METHYL-VINYL)PHOSPHAT (GERMAN)
		O,O-DIMETIL-O-(1,4-DIMETIL-3-OXO-4-AZA-PENT-1-ENIL)-FOSFATO (ITALIAN)
		OLEOBIDRIN
		PHOSPHATE DE DIMETHYLE ET DE 2-DIMETHYLCARBAMOYL 1-METHYL VINYLE (FRENCH)
		PHOSPHORIC ACID 3-(DIMETHYLAMINO)-1-METHYL-3-OXO-1-PROPENYL DIMETHYL ESTER
		PHOSPHORIC ACID, 3-(DIMETHYLAMINO)-1-METHYL-3-OXO-1-PROPENYL DIMETHYL ESTER, (E)-
		PHOSPHORIC ACID, DIMETHYL 1-METHYL-N,N-(DIMETHYL-AMINO)-3-OXO-1-PROPENYL ESTER, (E)- (9CI)
		PHOSPHORIC ACID, DIMETHYL ESTER, ESTER WITH 3-HYDROXY-N,N-DIMETHYLCROTONAMIDE, (E)-
		PHOSPHORIC ACID, DIMETHYL ESTER, ESTER WITH CIS-3-HYDROXY-N,N-DIMETHYLCROTONAMIDE
		SD 3562
		SHELL SD-3562
		TRANS-BIDRIN
141-75-3		
		BUTANOYL CHLORIDE
		BUTYRIC ACID CHLORIDE
		BUTYRIC CHLORIDE
	*	BUTYRYL CHLORIDE
		BUTYRYL CHLORIDE (DOT)
		N-BUTYRYL CHLORIDE
		UN 2353 (DOT)
141-78-6		
		ACETIC ACID ETHYL ESTER
		ACETIC ETHER
		ACETIDIN
		ACETOXYETHANE
		AETHYLACETAT (GERMAN)
		ESSIGESTER (GERMAN)
	*	ETHYL ACETATE
		ETHYL ACETATE (ACGIH,DOT)
		ETHYL ACETIC ESTER
		ETHYL ETHANOATE
		ETHYLACETAAT (DUTCH)
		ETHYLE (ACETATE D') (FRENCH)
		ETILE (ACETATO DI) (ITALIAN)
		OCTAN ETYLU (POLISH)
		RCRA WASTE NUMBER U112
		UN 1173 (DOT)
		VINEGAR NAPHTHA
141-79-7		
		1-ISOBUTENYL METHYL KETONE
		2-METHYL-2-PENTEN-4-ONE
		2-METHYL-2-PENTENONE-4
		2-METHYL-4-OXO-2-PENTENE
		3-PENTEN-2-ONE, 4-METHYL-
		4-METHYL-3-PENTEN-2-ON (DUTCH, GERMAN)
		4-METHYL-3-PENTEN-2-ONE
		4-METHYL-3-PENTENE-2-ONE
		4-METIL-3-PENTEN-2-ONE (ITALIAN)
		ACETONE, ISOPROPYLIDENE-
		ISOBUTENYL METHYL KETONE
		ISOPROPYLIDENE ACETONE
		MESITYL OXIDE
		MESITYL OXIDE (ACGIH,DOT)
		MESITYLOXID (GERMAN)
		MESITYLOXYDE (DUTCH)
		METHYL 2,2-DIMETHYLVINYL KETONE
		METHYL 2-METHYL-1-PROPENYL KETONE
		METHYL ISOBUTENYL KETONE
		MIBK
		OSSIDO DI MESITILE (ITALIAN)
		OXYDE DE MESITYLE (FRENCH)
		UN 1229 (DOT)
141-91-3		
		2,6-DIMETHYLMORPHOLINE
		MORPHOLINE, 2,6-DIMETHYL-
141-93-5		
		1,3-DIETHYLBENZENE
		BENZENE, 1,3-DIETHYL-
		BENZENE, M-DIETHYL-
		M-DIETHYLBENZENE
		M-ETHYLETHYLBENZENE
141-97-9		
		1-ETHOXYBUTANE-1,3-DIONE
		3-OXOBUTANOIC ACID ETHYL ESTER
		ACETOACETIC ACID, ETHYL ESTER
		ACETOACETIC ESTER
		ACTIVE ACETYL ACETATE
		BUTANOIC ACID, 3-OXO-, ETHYL ESTER
		DIACETIC ETHER
		EAA
		ETHYL 3-OXOBUTANOATE
		ETHYL 3-OXOBUTYRATE
	*	ETHYL ACETOACETATE
		ETHYL ACETONECARBOXYLATE
		ETHYL ACETYL ACETATE
		ETHYL ACETYLACETONATE
142-04-1		
		ANILINE CHLORIDE
	*	ANILINE HYDROCHLORIDE
		ANILINE HYDROCHLORIDE (DOT)
		BENZENAMINE, HYDROCHLORIDE

CAS No.	IN	Chemical Name
		C.I. 76001
		CHLORHYDRATE D'ANILINE (FRENCH)
		CHLORID ANILINU (CZECH)
		CI 76001
		HYDROCHLORIDE BENZENAMIDE
		NCI-C03736
		PHENYLAMINE HYDROCHLORIDE
		SUL ANILINOVA (CZECH)
		UN 1548 (DOT)
		USAF EK-442
142-28-9		
		1,3-DICHLOROPROPANE
		PROPANE, 1,3-DICHLORO-
		TRIMETHYLENE DICHLORIDE
142-29-0		
		CYCLOPENTENE
142-46-1		
		2,5-DITHIOBIUREA
142-59-6		
		1,2-ETHANEDIYLBISCARBAMODITHIOIC ACID DISODIUM SALT
		CARBAMIC ACID, ETHYLENEBIS(DITHIO-, DISODIUM SALT
		CARBAMODITHIOIC ACID, 1,2-ETHANEDIYLBIS-, DISODIUM SALT
		CARBON D
		CHEMBAM
		DI-NATRIUM-AETHYLENBISDITHIOCARBAMAT (GERMAN)
		DINATRIUM-(N,N'-AETHYLEN-BIS(DITHIOCARBAMAT)) (GERMAN)
		DINATRIUM-(N,N'-ETHYLEEN-BIS(DITHIOCARBAMAAT)) (DUTCH)
		DISODIUM ETHYLENE-1,2-BISDITHIOCARBAMATE
		DISODIUM ETHYLENEBIS(DITHIOCARBAMATE)
		DITHANE A 40
		DITHANE D 14
		DSE
		ETHYLENE BIS DITHIOCARBAMATE
		ETHYLENEBIS(DITHIOCARBAMATE), DISODIUM SALT
		ETHYLENEBIS(DITHIOCARBAMIC ACID) DISODIUM SALT
		N,N'-ETHYLENE BIS(DITHIOCARBAMATE DE SODIUM) (FRENCH)
		N,N'-ETILEN-BIS(DITIOCARBAMMATO) DI SODIO (ITALIAN)
		NABAM
		NABAME (FRENCH)
		NABASAN
		NAFUN IPO
		PARZATE
		PARZATE LIQUID
		SPRING-BAK
142-62-1		
		1-HEXANOIC ACID
		1-PENTANECARBOXYLIC ACID
		BUTYLACETIC ACID
	*	CAPROIC ACID
		CAPRONIC ACID
		HEXACID 698
		HEXANOIC ACID
		HEXANOIC ACID (DOT)
		HEXOIC ACID
		N-CAPROIC ACID
		N-HEXANOIC ACID
		N-HEXOIC ACID
		N-HEXYLIC ACID
		NA 1760 (DOT)
		PENTIFORMIC ACID
		PENTYLFORMIC ACID
142-64-3		
		DIETHYLENEDIAMINE DIHYDROCHLORIDE
		DIHYDRO PIP WORMER
		DOWZENE DHC
		PIPERAZINE DIHYDROCHLORIDE
		PIPERAZINE HYDROCHLORIDE
		PIPERAZINE WORMER PREMIX
		PIPERAZINE, DIHYDROCHLORIDE
142-68-7		
		2H-PYRAN, TETRAHYDRO-
		OXACYCLOHEXANE
		OXANE
		PENTAMETHYLENE OXIDE
		TETRAHYDRO-2H-PYRAN
		TETRAHYDROPYRAN
		TETRAHYDROPYRANE
		THP
142-71-2		
		ACETATE DE CUIVRE (FRENCH)
		ACETIC ACID CUPRIC SALT
		ACETIC ACID, COPPER(2+) SALT
		COPPER ACETATE
		COPPER ACETATE (Cu(C2H3O2)2)
		COPPER DIACETATE
		COPPER(2+) ACETATE
		COPPER(2+) DIACETATE
		COPPER(II) ACETATE
		CRYSTALLIZED VERDIGRIS
		CRYSTALS OF VENUS
		CUPRIC ACEATE (DOT)
	*	CUPRIC ACETATE
		CUPRIC DIACETATE
		NA 9106 (DOT)
		NEUTRAL VERDIGRIS
		OCTAN MEDNATY (CZECH)
142-82-5		
		DIPROPYL METHANE
		EPTANI (ITALIAN)
		GETTYSOLVE-C
		HEPTAN (POLISH)
		HEPTANE
		HEPTANE (ACGIH,DOT)
		HEPTANE :AND ITS ISOMERS:
		HEPTANE(N-HEPTANE)
		HEPTANEN (DUTCH)
		HEPTYL HYDRIDE
	*	N-HEPTANE
		SKELLYSOLVE C
		UN 1206 (DOT)
142-83-6		
		2,4-HEXADIENAL
142-84-7		
		1-PROPANAMINE, N-PROPYL-
		DI-N-PROPYLAMINE
		DIPROPYLAMINE
		DIPROPYLAMINE (DOT)
		N-DIPROPYLAMINE
		N-PROPYL-1-PROPANAMINE
		RCRA WASTE NUMBER U110
		UN 2383 (DOT)
142-92-7		
		1-HEXYL ACETATE
		ACETIC ACID, HEXYL ESTER
		HEXYL ACETATE
		HEXYL ACETATE [COMBUSTIBLE LIQUID LABEL]
		HEXYL ACETATE [FLAMMABLE LIQUID LABEL]
		HEXYL ALCOHOL, ACETATE
		HEXYL ETHANOATE
		N-HEXYL ACETATE
		SEC-HEXYL ACETATE
142-96-1		
		1,1'-OXYBISBUTANE
		1-BUTOXYBUTANE
		BUTANE, 1,1'-OXYBIS-
	*	BUTYL ETHER
		BUTYL ETHER (DOT)
		BUTYL OXIDE
		DI-N-BUTYL ETHER

CAS No.	IN	Chemical Name
		DI-N-BUTYL ETHER (DOT)
		DIBUTYL ETHER
		DIBUTYL OXIDE
		ETHER BUTYLIQUE (FRENCH)
		N-BUTYL ETHER
		N-DIBUTYL ETHER
		UN 1149 (DOT)
143-00-0		
		BIS(2-HYDROXYETHYL)AMMONIUM LAURYL SULFATE
		CONDANOL DLS
		DEA-LAURYL SULFATE
		DIETHANOLAMINE LAURYL SULFATE
		DODECYL SULFATE DIETHANOLAMINE SALT
		DODECYL SULFATE, COMPD WITH 2,2'-IMINODIETHANOL
		ETHANOL, 2,2'-IMINOBIS-, DODECYL SULFATE (SALT)
		ETHANOL, 2,2'-IMINODI-, DODECYL SULFATE (SALT)
		ETHANOL, 2,2'-IMINODI-, COMPD WITH DODECYL SULFATE
		LAURYL SULFATE DIETHANOLAMINE SALT
		PROPASTE D
		SIPON LD
		STEPANOL DEA
		SULFURIC ACID, MONODODECYL ESTER, COMPD WITH 2,2'-IMINOBIS[ETHANOL] (1:1)
		SULFURIC ACID, MONODODECYL ESTER, COMPD WITH 2,2'-IMINODIETHANOL (1:1)
		TEXAPON DLS
143-08-8		
		1-NONANOL
		ALCOHOL C-9
		N-NONAN-1-OL
		N-NONYL ALCOHOL
		NONALOL
		NONAN-1-OL
		NONANOL
		NONYL ALCOHOL
		OCTYL CARBINOL
		PELARGONIC ALCOHOL
143-10-2		
		1-DECANETHIOL
143-16-8		
		1-HEXANAMINE, N-HEXYL-
		BIS(1-HEXYL)AMINE
		DI-N-HEXYLAMINE
		DIHEXYLAMINE
		N,N-DIHEXYLAMINE
		N-HEXYLHEXANAMINE
143-18-0		
		9-OCTADECENOIC ACID (Z)-, POTASSIUM SALT
		FR 14
		OLEIC ACID, POTASSIUM SALT
		POTASSIUM CIS-9-OCTADECENOIC ACID
		POTASSIUM OLEATE
		TRENAMINE D 200
		TRENAMINE D 201
143-19-1		
		9-OCTADECENOIC ACID (Z)-, SODIUM SALT
		EUNATROL
		OLATE FLAKES
		OLEIC ACID, SODIUM SALT
		SODIUM OLEATE
143-33-9		
		CIANURO DI SODIO (ITALIAN)
		CYANIDE OF SODIUM
		CYANOBRIK
		CYANOGRAN
		CYANURE DE SODIUM (FRENCH)
		CYMAG
		HYDROCYANIC ACID, SODIUM SALT
		KYANID SODNY (CZECH)
		RCRA WASTE NUMBER P106
	*	SODIUM CYANIDE
		SODIUM CYANIDE (ACGIH)
		SODIUM CYANIDE (Na(CN))
		SODIUM CYANIDE SOLUTION
		SODIUM CYANIDE, SOLID
		SODIUM CYANIDE, SOLID (DOT)
		SODIUM CYANIDE, SOLUTION (DOT)
		UN 1689 (DOT)
143-50-0		
		1,1A,3,3A,4,5,5,5A,5B,6-DECACHLOROOCTAHYDRO-1,3,4-METHENO-2H-CYCLOBUTA(CD)PENTALEN-2-ONE
		1,2,3,5,6,7,8,9,10,10-DECACHLORO(5210(SUP 2,6)0(SUP 3,9)0(SUP 5,8))DECANO-4-ONE
		1,3,4-METHENO-2H-CYCLOBUTA(CD)PENTALEN-2-ONE, 1,1A,3,3A,4,5,5,5A,5B,6-DECACHLOROCTAHYDRO-
		CHLORDECONE
		CIBA 8514
		CLORDECONE
		COMPOUND 1189
		DECACHLORO-1,3,4-METHENO-2H-CYCLOBUTA(CD)PENTALEN-2-ONE
		DECACHLOROKETONE
		DECACHLOROOCTAHYDRO-1,3,4-METHENO-2H-CYCLOBUTA(CD)PENTALEN-2-ONE
		DECACHLOROPENTACYCLO(5210(SUP 2,6)0(SUP 3,9)0(SUP 5,8))DECAN-4-ONE
		DECACHLOROPENTACYCLO(5300(SUP 2,6)0(SUP 4,10)0(SUP 5,9)DECAN-3-ONE
		DECACHLOROPENTACYCLO[52102,603,905,8]DECAN-4-ONE
		DECACHLOROTETRACYCLODECANONE
		DECACHLOROTETRAHYDRO-4,7-METHANOINDENEONE
		ENT 16,391
		GC 1189
		GENERAL CHEMICALS 1189
		KEPONE
		KEPONE (DOT)
		KEPONE-2-ONE, DECACHLOROOCTAHYDRO-
		MEREX
		NA 2761 (DOT)
		NCI-C00191
		RCRA WASTE NUMBER U142
144-49-0		
		2-FLUOROACETIC ACID
		ACETIC ACID, FLUORO-
		ACIDE-MONOFLUORACETIQUE (FRENCH)
		ACIDO MONOFLUOROACETIO (ITALIAN)
		ALPHA-FLUOROACETIC ACID
		CYMONIC ACID
		FAA
		FLUORO ACETIC ACID (DOT)
		FLUOROACETATE
		FLUOROACETIC ACID
		FLUOROETHANOIC ACID
		GIFBLAAR POISON
		HFA
		MFA
		MONOFLUORAZIJNZUUR (DUTCH)
		MONOFLUORESSIGSAURE (GERMAN)
		MONOFLUOROACETATE
		MONOFLUOROACETIC ACID
		UN 2642 (DOT)
144-62-7		
		ACIDE OXALIQUE (FRENCH)
		ACIDO OSSALICO (ITALIAN)
		AKTISAL
		AQUISAL
		ETHANE DIONIC ACID
		ETHANEDIOIC ACID
		KYSELINA STAVELOVA (CZECH)
		NCI-C55209
		OXAALZUUR (DUTCH)
	*	OXALIC ACID
		OXALIC ACID (ACGIH)
		OXALSAEURE (GERMAN)
146-84-9		
		PHENOL, 2,4,6-TRINITRO-, SILVER(1+) SALT

CAS No.	IN	Chemical Name
		PICRAGOL
		PICRIC ACID, SILVER(1+) SALT
		PICROTOL
		SILVER PICRATE
		SILVER PICRATE, DRY (DOT)
		SILVER PICRATE, WETTED WITH AT LEAST 30% WATER (DOT)
		SILVER, (PICRYLOXY)-
		UN 1347 (DOT)
147-47-7		
		1,2-DIHYDRO-2,2,4-TRIMETHYLQUINOLINE
147-94-4		
		CYSTOSINE ARABINOSIDE
148-01-6		
		2-METHYL-3,5-DINITROBENZAMIDE
		3,5-DINITRO-2-METHYLBENZAMIDE
		3,5-DINITRO-O-TOLUAMIDE
		BENZAMIDE, 2-METHYL-3,5-DINITRO-
		COCCIDINE A
		COCCIDOT
		DINITOLMIDE
		O-TOLUAMIDE, 3,5-DINITRO-
		SALCOSTAT
		ZOALENE
		ZOAMIX
148-82-3		
		3025 C. B.
		ALANINE, 3-[P-[BIS(2-CHLOROETHYL)AMINO]PHENYL]-, L-
		ALKERAN
		CB 3025
		L-PAM
		L-PHENYLALANINE MUSTARD
		L-PHENYLALANINE, 4-[BIS(2-CHLOROETHYL)AMINO]-
		L-SARCOLYSIN
		L-SARCOLYSINE
		L-SARKOLYSIN
		LEVOFALAN
		MELPHALAN
		NSC-8806
		PHENYLALANINE MUSTARD
149-30-4		
		2-BENZOTHIAZOLETHIOL
149-31-5		
		1,3-PENTANEDIOL, 2-METHYL-
		2-METHYL-1,3-PENTANEDIOL
149-57-5		
		2-BUTYLBUTANOIC ACID
		2-ETHYL-1-HEXANOIC ACID
		2-ETHYLCAPROIC ACID
		2-ETHYLHEXANOIC ACID
		2-ETHYLHEXOIC ACID
		3-HEPTANECARBOXYLIC ACID
		ALPHA-ETHYLCAPROIC ACID
		ALPHA-ETHYLHEXANOIC ACID
		BUTYLETHYLACETIC ACID
		ETHYLHEXANOIC ACID
		ETHYLHEXOIC ACID
		HEXANOIC ACID, 2-ETHYL-
149-74-6		
		DICHLOROMETHYLPHENYL SILANE
		METHYLPHENYLDICHLOROSILANE
		METHYLPHENYLDICHLOROSILANE (DOT)
		PHENYLMETHYLDICHLOROSILANE
		SILANE, DICHLOROMETHYLPHENYL-
		UN 2437 (DOT)
149-91-7		
		3,4,5-TRIHYDROXYBENZOIC ACID
		BENZOIC ACID, 3,4,5-TRIHYDROXY-
		GALLIC ACID
150-76-5		
		1-HYDROXY-4-METHOXYBENZENE
		4-HYDROXYANISOLE
	*	4-METHOXYPHENOL
		HQMME
		HYDROQUINONE METHYL ETHER
		HYDROQUINONE MONOMETHYL ETHER
		LEUCOBASAL
		LEUCODINE B
		MECHINOLUM
		MEQUINOL
		NOVO-DERMOQUINONA
		P-GUAIACOL
		P-HYDROXYANISOLE
		P-HYDROXYMETHOXYBENZENE
		P-METHOXYPHENOL
		PHENOL, 4-METHOXY-
		PHENOL, P-METHOXY-
		PMF (ANTIOXIDANT)
151-21-3		
		AKYPOSAL SDS
		AQUAREX ME
		AQUAREX METHYL
		AVIROL 101
		AVIROL 118 CONC
		BEROL 452
		CARSONOL SLS
		CARSONOL SLS PASTE B
		CARSONOL SLS SPECIAL
		CONCO SULFATE WA
		CONCO SULFATE WA-12 45
		CONCO SULFATE WA-1200
		CONCO SULFATE WAG
		CONCO SULFATE WAN
		CONCO SULFATE WAS
		CONCO SULFATE WN
		CYCLORYL 21
		CYCLORYL 31
		CYCLORYL 580
		CYCLORYL 585N
		DEHYDAG SULFATE GL EMULSION
		DEHYDAG SULPHATE GL EMULSION
		DETERGENT 66
		DODECYL ALCOHOL, HYDROGEN SULFATE, SODIUM SALT
		DODECYL HYDROGEN SULFATE, SODIUM SALT
		DODECYL SODIUM SULFATE
		DODECYL SULFATE SODIUM
		DODECYL SULFATE SODIUM SALT
		DREFT
		DUPONAL
		DUPONAL WAQE
		DUPONOL
		DUPONOL C
		DUPONOL ME
		DUPONOL METHYL
		DUPONOL QC
		DUPONOL QX
		DUPONOL WA
		DUPONOL WA DRY
		DUPONOL WAQ
		DUPONOL WAQA
		DUPONOL WAQE
		DUPONOL WAQM
		EMAL 0
		EMAL 10
		EMAL O
		EMERSAL 6400
		EMPICOL LPZ
		EMPICOL LS 30
		EMPICOL LX 28
		EMULSIFIER NO 10 4
		EQUEX S
		FINASOL OSR(SUB 2)
		FINASOL OSR2
		HEXAMOL SLS
		IRIUM
		LANETTE WAX S
		LAURYL SODIUM SULFATE

CAS No.	IN	Chemical Name
		LAURYL SULFATE SODIUM
		LAURYL SULFATE SODIUM SALT
		MAPROFIX 563
		MAPROFIX LK
		MAPROFIX NEU
		MAPROFIX WAC
		MAPROFIX WAC-LA
		MELANOL CL
		MELANOL CL 30
		MONODODECYL SODIUM SULFATE
		MONOGEN Y 100
		MONOGEN Y 500
		MONTOPOL LA PASTE
		N-DODECYL SULFATE SODIUM
		NCI-C50191
		NEUTRAZYME
		NIKKOL SLS
		NISSAN SINTREX L 100
		ODORIPON AL 95
		ORVUS WA
		ORVUS WA PASTE
		P AND G EMULSIFIER 104
		PERLANDROL L
		PERLANKROL L
		PRODUCT NO 161
		PRODUCT NO 75
		QUOLAC EX-UB
		REWOPOL NLS 30
		RICHONOL A
		RICHONOL AF
		RICHONOL C
		SDS
		SINNOPON LS 100
		SINNOPON LS 95
		SINTAPON L
		SIPEX OP
		SIPEX SB
		SIPEX SD
		SIPEX SP
		SIPEX UB
		SIPON LS
		SIPON LS 100
		SIPON LSB
		SIPON PD
		SIPON WD
		SLS
		SODIUM DODECYL SULFATE
		SODIUM DODECYL SULPHATE
		SODIUM LAURYL SULFATE
		SODIUM LAURYL SULPHATE
		SODIUM MONODODECYL SULFATE
		SODIUM MONOLAURYL SULFATE
		SODIUM N-DODECYL SULFATE
		SOLSOL NEEDLES
		STANDAPOL 112 CONC
		STANDAPOL WA-AC
		STANDAPOL WAQ
		STANDAPOL WAQ SPECIAL
		STANDAPOL WAS 100
		STEINAPOL NLS 90
		STEPANOL ME
		STEPANOL ME DRY
		STEPANOL ME DRY AW
		STEPANOL METHYL
		STEPANOL METHYL DRY AW
		STEPANOL T 28
		STEPANOL WA
		STEPANOL WA 100
		STEPANOL WA PASTE
		STEPANOL WAC
		STEPANOL WAQ
		STERLING WA PASTE
		STERLING WAQ-CH
		STERLING WAQ-COSMETIC
		SULFETAL L 95
		SULFOPON WA 1
		SULFOPON WA 1 SPECIAL
		SULFOPON WA 2
		SULFOPON WA 3
		SULFOTEX WA
		SULFOTEX WALA
		SULFURIC ACID DODECYL ESTER SODIUM SALT
		SULFURIC ACID MONODODECYL ESTER SODIUM SALT
		SWASCOL 1P
		SWASCOL 3L
		SWASCOL 4L
		SYNTAPON L
		SYNTAPON L PASTA (CZECH)
		TARAPON K 12
		TEXAPON DL CONC
		TEXAPON K 12
		TEXAPON K 1296
		TEXAPON K12
		TEXAPON L 100
		TEXAPON V HC
		TEXAPON V HC POWDER
		TEXAPON Z HIGH CONC NEEDLES
		TEXAPON ZHC
		TREPENOL WA
		TV M 474
		ULTRA SULFATE SL-1
		WAQE
151-38-2		
		METHOXYETHYLMERCURIC ACETATE
		METHYL CELLOSOLVE
151-50-8		
		CYANIDE OF POTASSIUM
		CYANURE DE POTASSIUM (FRENCH)
		HYDROCYANIC ACID, POTASSIUM SALT
		KALIUM-CYANID (GERMAN)
	*	POTASSIUM CYANIDE
		POTASSIUM CYANIDE (ACGIH)
		POTASSIUM CYANIDE (K(CN))
		POTASSIUM CYANIDE SOLUTION
		POTASSIUM CYANIDE SOLUTION (DOT)
		POTASSIUM CYANIDE, SOLID
		POTASSIUM CYANIDE, SOLID (DOT)
		RCRA WASTE NUMBER P098
		UN 1680 (DOT)
151-56-4		
		1H-AZIRINE, DIHYDRO-
		AETHYLENIMIN (GERMAN)
		AMINOETHYLENE
		AZACYCLOPROPANE
		AZIRAN
		AZIRANE
		AZIRIDIN (GERMAN)
		AZIRIDINE
		DIHYDRO-1H-AZIRINE
		DIHYDROAZIRENE
		DIMETHYLENEIMINE
		DIMETHYLENIMINE
		EI
		ENT-50324
		ETHYLEENIMINE (DUTCH)
		ETHYLENE IMINE , INHIBITED
		ETHYLENE IMINE, INHIBITED (DOT)
	*	ETHYLENEIMINE
		ETHYLENEIMINE (ACGIH)
		ETHYLENIMINE
		ETHYLIMINE
		ETILENIMINA (ITALIAN)
		RCRA WASTE NUMBER P054
		TL 337
		UN 1185 (DOT)
151-67-7		
		HALOTHANE
152-16-9		
		BIS(BISDIMETHYLAMINOPHOSPHONOUS)ANHYDRIDE
		BIS(DIMETHYLAMINO)PHOSPHONOUS ANHYDRIDE
		BIS(DIMETHYLAMINO)PHOSPHORIC ANHYDRIDE
		BIS-N,N,N',N'-TETRAMETHYLPHOSPHORODIAMIDIC ANHYDRIDE

CAS No.	IN	Chemical Name
		DIPHOSPHORAMIDE, OCTAMETHYL-
		DIPHOSPHORAMIDE, OCTAMETHYL- (9CI)
		ENT 17,291
		LETHALAIRE G-59
		OCTAMETHYL
		OCTAMETHYL PYROPHOSPHORTETRAMIDE
		OCTAMETHYL TETRAMIDO PYROPHOSPHATE
		OCTAMETHYL-DIFOSFORZUUR-TETRAMIDE (DUTCH)
		OCTAMETHYL-DIPHOSPHORSAEURE-TETRAMID (GERMAN)
		OCTAMETHYLDIPHOSPHORAMIDE
		OCTAMETHYLPYROPHOSPHORAMIDE
		OCTAMETHYLPYROPHOSPHORIC ACID AMIDE
		OCTAMETHYLPYROPHOSPHORIC ACID TETRAMIDE
		OMPA
		OMPACIDE
		OMPATOX
		OMPAX
		OTTOMETIL-PIROFOSFORAMMIDE (ITALIAN)
		PESTOX
		PESTOX 3
		PESTOX III
		PYROPHOSPHORAMIDE, OCTAMETHYL-
		PYROPHOSPHORIC ACID OCTAMETHYLTETRAAMIDE
		PYROPHOSPHORYLTETRAKISDIMETHYLAMIDE
		RCRA WASTE NUMBER P085
		SCHRADAN
		SCHRADANE (FRENCH)
		SYSTAM
		SYSTOPHOS
		SYTAM
		TETRAKISDIMETHYLAMINOPHOSPHONOUS ANHYDRIDE
154-23-4		
		(+)-3',4',5,7-TETRAHYDROXY-2,3-TRANS-FLAVAN-3-OL
		(+)-(2R:3S)-5,7,3',4'-TETRAHYDROXYFLAVAN-3-OL
		(+)-CATECHIN
		(+)-CATECHOL
		(+)-CYANIDAN-3-OL
		(+)-CYANIDANOL-3
		2H-1-BENZOPYRAN-3,5,7-TRIOL, 2-(3,4-DIHYDROXY-PHENYL)-3,4-DIHYDRO-, (2R-TRANS)-
		3-CYANIDANOL, (+)-
		CATECHIN
		CATECHIN (FLAVAN)
		CATECHINIC ACID
		CATECHOL
		CATECHOL (FLAVAN)
		CATECHUIC ACID
		CATERGEN
		CIANIDANOL
		CYANIDANOL
		D-(+)-CATECHIN
		D-CATECHIN
		D-CATECHOL
		KB 53
		TRANS-(+)-3,3',4',5,7-FLAVANPENTOL
154-93-8		
		BISCHLOROETHYL NITROSOUREA
		CARMUSTINE
156-10-5		
		4-NITROSO-N-PHENYLANILINE
		4-NITROSODIPHENYLAMINE
		BENZENAMINE, 4-NITROSO-N-PHENYL-
		DIPHENYLAMINE, 4-NITROSO-
		N-PHENYL-P-NITROSOANILINE
		NITROSODIPHENYLAMINE
		P-NITROSO-N-PHENYLANILINE
		P-NITROSODIPHENYLAMINE
		P-PHENYLAMINONITROSOBENZENE
156-43-4		
		4-AMINOPHENETOLE
		4-ETHOXYANILINE
		4-ETHOXYBENZENAMINE
		4-PHENETIDINE
		BENZENAMINE, 4-ETHOXY-
		P-AMINOPHENETOLE
		P-ETHOXYANILINE
		P-PHENETIDIN
		P-PHENETIDINE
156-59-2		
		(Z)-1,2-DICHLOROETHENE
		(Z)-1,2-DICHLOROETHYLENE
		1,2-CIS-DICHLOROETHYLENE
		CIS-1,2-DICHLORETHYLENE
		CIS-1,2-DICHLOROETHENE
		CIS-1,2-DICHLOROETHYLENE
		CIS-DICHLOROETHYLENE
		DICHLOROETHYLENE-CIS
		ETHENE, 1,2-DICHLORO-, (Z)-
		ETHYLENE, 1,2-DICHLORO-, (Z)-
156-60-5		
		(E)-1,2-DICHLOROETHENE
		(E)-1,2-DICHLOROETHYLENE
		1,2-TRANS-DICHLOROETHENE
		1,2-TRANS-DICHLOROETHYLENE
		DICHLOROETHYLENE-TRANS
		ETHENE, 1,2-DICHLORO-, (E)-
		ETHYLENE, 1,2-DICHLORO-, (E)-
		TRANS-1,2-DICHLOROETHENE
		TRANS-1,2-DICHLOROETHYLENE
156-62-7		
		AERO CYANAMID GRANULAR
		AERO CYANAMID SPECIAL GRADE
		AERO-CYANAMID
		ALZODEF
		CALCIUM CARBIMIDE
		CALCIUM CYANAMID
	*	CALCIUM CYANAMIDE
		CALCIUM CYANAMIDE (ACGIH)
		CALCIUM CYANAMIDE (CaCN2)
		CALCIUM CYANAMIDE, NOT HYDRATED (CONTAINING MORE THAN 0.1% CALCIUM CARBIDE) (DOT)
		CCC
		CY-L 500
		CYANAMID
		CYANAMID GRANULAR
		CYANAMID SPECIAL GRADE
		CYANAMIDE
		CYANAMIDE CALCIQUE (FRENCH)
		CYANAMIDE, CALCIUM SALT (1:1)
		LIME NITROGEN
		LIME NITROGEN (DOT)
		NCI-C02937
		NITROGEN LIME
		NITROLIM
		NITROLIME
		UN 1403 (DOT)
		USAF CY-2
156-87-6		
		1,3-PROPANOLAMINE
		1-AMINO-3-HYDROXYPROPANE
		1-AMINO-3-PROPANOL
		1-PROPANOL, 3-AMINO-
		3-AMINO-1-PROPANOL
		3-AMINOPROPANOL
		3-AMINOPROPYL ALCOHOL
		3-HYDROXY-1-PROPYLAMINE
		3-HYDROXYPROPYLAMINE
		3-PROPANOLAMINE
		BETA-ALANINOL
		GAMMA-AMINOPROPANOL
		GAMMA-HYDROXY-1-PROPYLAMINE
		PROPANOLAMINE
160-51-4		
	*	BENZOQUINONE
189-55-9		
		1,2:7,8-DIBENZPYRENE
		3,4:9,10-DIBENZOPYRENE
		BENZO[RST]PENTAPHENE

CAS No.	IN	Chemical Name
		DIBENZO(A,I)PYRENE
		DIBENZO[B,H]PYRENE
189-64-0		
		3,4:8,9-DIBENZOPYRENE
		DIBENZO(A,H)PYRENE
		DIBENZO[B,DEF]CHRYSENE
191-24-2		
		1,12-BENZOPERYLENE
		1,12-BENZPERYLENE
		BENZO[GHI]PERYLENE
192-65-4		
		1,2:4,5-DIBENZOPYRENE
		DIBENZO(A,E)PYRENE
		NAPHTHO[1,2,3,4-DEF]CHRYSENE
193-39-5		
		1,10-(1,2-PHENYLENE)PYRENE
		1,10-(O-PHENYLENE)PYRENE
		INDENO (1,2,3-CD) PYRENE
		O-PHENYLENEPYRENE
		PYRENE
194-59-2		
		3,4,5,6-DIBENZOCARBAZOLE
		7-AZA-7H-DIBENZO[C,G]FLUORENE
		7H-DIBENZO[C,G]CARBAZOLE
205-82-3		
		10,11-BENZOFLUORANTHENE
		7,8-BENZFLUORANTHENE
		BENZO-12,13-FLUORANTHENE
		BENZO[J]FLUORANTHENE
		BENZO[L]FLUORANTHENE
		DIBENZO[A,JK]FLUORENE
205-99-2		
		2,3-BENZFLUORANTHENE
		3,4-BENZFLUORANTHENE
		3,4-BENZOFLUORANTHENE
		3,4-BENZ[E]ACEPHENANTHRYLENE
		BENZO(B)FLUORANTHENE
		BENZO[E]FLUORANTHENE
		BENZ[E]ACEPHENANTHRYLENE
206-44-0		
		1,2-(1,8-NAPHTHYLENE)BENZENE
		BENZENE, 1,2-(1,8-NAPHTHALENEDIYL)-
		BENZO[JK]FLUORENE
		FLUORANTHENE
		IDRYL
207-08-9		
		11,12-BENZOFLUORANTHENE
		2,3,1',8'-BINAPHTHYLENE
		8,9-BENZFLUORANTHENE
		8,9-BENZOFLUORANTHENE
		BENZO[K]FLUORANTHENE
		DIBENZO[B,JK]FLUORENE
208-96-8		
		ACENAPHTHYLENE
		CYCLOPENTA[DE]NAPHTHALENE
218-01-9		
		1,2-BENZOPHENANTHRENE
		1,2-BENZPHENANTHRENE
		BENZO[A]PHENANTHRENE
	*	CHRYSENE
224-42-0		
		1,2,7,8-DIBENZACRIDINE
		7-AZADIBENZ[A,J]ANTHRACENE
		DIBENZ(A,J)ACRIDINE
		DIBENZ[A,F]ACRIDINE

CAS No.	IN	Chemical Name
226-36-8		
		1,2,5,6-DIBENZACRIDINE
		1,2,5,6-DIBENZOACRIDINE
		7-AZADIBENZ[A,H]ANTHRACENE
		DIBENZ(A,H)ACRIDINE
		DIBENZ[A,D]ACRIDINE
260-94-6		
		10-AZAANTHRACENE
		2,3-BENZOQUINOLINE
		9-AZAANTHRACENE
	*	ACRIDINE
		ACRIDINE (DOT)
		BENZO(B)QUINOLINE
		DIBENZO(B,E)PYRIDINE
		UN 2713 (DOT)
280-57-9		
		1,4-DIAZABICYCLO(2,2,2)OCTANE
		1,4-DIAZABICYCLOOCTANE
		1,4-ETHYLENEPIPERAZINE
		BICYCLO(2,2,2)-1,4-DIAZAOCTANE
		D 33LV
		DABCO
		DABCO 33LV
		DABCO CRYSTAL
		DABCO EG
		DABCO R-8020
		DABCO S-25
		DIAZABICYCLOOCTANE
		N,N'-ENDO-ETHYLENEPIPERAZINE
		TED
		TEDA
		THANCAT TD 33
		TRIETHYLENE DIAMINE
287-23-0		
		CYCLOBUTANE
		CYCLOBUTANE (DOT)
		TETRAMETHYLENE
		UN 2601 (DOT)
287-92-3		
	*	CYCLOPENTANE
		CYCLOPENTANE (ACGIH,DOT)
		PENTAMETHYLENE
		UN 1146 (DOT)
291-64-5		
		CYCLOHEPTANE
297-78-9		
		1,3,4,5,6,7,10,10-OCTACHLORO-4,7-ENDO-METHYLENE-4,7,8,9-TETRAHYDROPHTHALAN
		1,3,4,5,6,7,8,8-OCTACHLORO-2-OXA-3A,4,7,7A-TETRAHYDRO-4,7-METHANOINDENE
		1,3,4,5,6,8,8-OCTACHLORO-1,3,3A,4,7,7A-HEXAHYDRO-4,7-METHANOISOBENZOFURAN
		4,7-METHANOISOBENZOFURAN, 1,3,4,5,6,7,8,8-OCTA-CHLORO-1,3,3A,4,7,7A-HEXAHYDRO-
		4,7-METHANOISOBENZOFURAN, 1,3,4,5,6,7,8,8-OCTA-CHLORO-3A,4,7,7A-TETRAHYDRO-
		CP 14,957
		ENT 25,545
		ENT 25,545-X
		ISOBENZAN
		OCTACHLORO-HEXAHYDRO-METHANOISOBENZOFURAN
		OCTACHLOROHEXAHYDRO-4,7-METHANOISOBENZOFURAN
		OMTAN
		R 6700
		SD 4402
		SHELL 4402
		SHELL WL 1650
		TELODRIN
		WL 1650
297-97-2		
		AC 18133
		ACC 18133

CAS No.	IN	Chemical Name
		AMERICAN CYANAMID 18133
		CYNEM
		CYNOPHOS
		DIETHYL O-2-PYRAZINYL PHOSPHOROTHIONATE
		EN 18133
		ENT 25,580
		ETHYL PYRAZINYL PHOSPHOROTHIOATE
		EXPERIMENTAL NEMATOCIDE 18,133
		NEMAFOS
		NEMAFOS 10 G
		NEMAPHOS
		NEMATOCIDE
		NEMATOCIDE GR
		O,O-DIAETHYL-O-(2-PYRAZINYL)-THIONOPHOSPHAT (GERMAN)
		O,O-DIAETHYL-O-(PYRAZIN-2YL)-MONOTHIOPHOSPHAT (GERMAN)
		O,O-DIETHYL O,2-PYRAZINYL PHOSPHOROTHIOATE
		O,O-DIETHYL O-2-PYRAZINYL PHOSPHOTHIONATE
		O,O-DIETHYL O-PYRAZINYL THIOPHOSPHATE
		PHOSPHOROTHIOIC ACID, O,O-DIETHYL O-2-PYRAZINYL ESTER
		PHOSPHOROTHIOIC ACID, O,O-DIETHYL O-PYRAZINYL ESTER
		PYRAZINOL, O-ESTER WITH O,O-DIETHYL PHOSPHOROTHIOATE
		RCRA WASTE NUMBER P040
		THIONAZIN
		THIONAZINE
		ZINOPHOS
		ZYNOPHOS
298-00-0		
		8056HC
		A-GRO
		AZOFOS
		AZOPHOS
		BAY 11405
		BAY E-601
		BLADAN-M
		CEKUMETHION
		DALF
		DEMETHYLFENITROTHION
		DEVITHION
		DIMETHYL 4-NITROPHENYL PHOSPHOROTHIONATE
		DIMETHYL P-NITROPHENYL MONOTHIOPHOSPHATE
		DIMETHYL P-NITROPHENYL PHOSPHOROTHIONATE
		DIMETHYL P-NITROPHENYL THIOPHOSPHATE
		DIMETHYL PARATHION
		DIMETHYL-P-NITROPHENYL THIONPHOSPHATE
		DREXEL METHYL PARATHION 4E
		E 601
		ENT 17,292
		FOLIDOL M
		FOLIDOL M 40
		FOLIDOL-80
		FOSFERNO M 50
		GEARPHOS
		M-PARATHION
		M40 & 80
		ME-PARATHION
		MEPATON
		MEPTOX
		METACID 50
		METACIDE
		METACIDE (INSECTICIDE)
		METAFOS
		METAFOS (PESTICIDE)
		METAPHOR
		METAPHOS
		METHYL E 605
		METHYL FOSFERNO
		METHYL NIRAN
		METHYL PARATHION
		METHYL PARATHION (ACGIH)
		METHYL PARATHION MIXTURE, DRY
		METHYL PARATHION MIXTURE, DRY (DOT)
		METHYL PARATHION MIXTURE, LIQUID, WITH 25% METHYL PARATHION
		METHYL PARATHION, LIQUID
		METHYL PARATHION, LIQUID (DOT)
		METHYL-E 605
		METHYLTHIOPHOS
		METILPARATION (HUNGARIAN)
		METRON
		METRON (PESTICIDE)
		METYLOPARATION (POLISH)
		METYLPARATION (CZECH)
		NA 2783 (DOT)
		NCI-C02971
		NITROX
		NITROX 80
		O,O-DIMETHYL O-(P-NITROPHENYL) PHOSPHOROTHIOATE
		O,O-DIMETHYL O-(P-NITROPHENYL) THIONOPHOSPHATE
		O,O-DIMETHYL O-P-NITROPHENYL THIOPHOSPHATE
		O,O-DIMETHYL-O-(4-NITRO-FENYL)-MONOTHIOFOSFAAT (DUTCH)
		O,O-DIMETHYL-O-(4-NITRO-PHENYL)-MONOTHIOPHOSPHAT (GERMAN)
		O,O-DIMETHYL-O-(4-NITROPHENYL) PHOSPHOROTHIOATE
		O,O-DIMETHYL-O-(4-NITROPHENYL)-THIONOPHOSPHAT (GERMAN)
		O,O-DIMETHYL-O-(P-NITROPHENYL)-THIONOPHOSPHAT (GERMAN)
		O,O-DIMETHYL-O-P-NITROFENYLESTER KYSELINY THIOFOSFORECNE (CZECH)
		O,O-DIMETIL-O-(4-NITRO-FENIL)-MONOTIOFOSFATO (ITALIAN)
		OLEOVOFOTOX
		P-NITROPHENYLDIMETHYLTHIONOPHOSPHATE
		PARAPEST M-50
		PARATAF
		PARATHION METHYL
		PARATHION METHYL HOMOLOG
		PARATHION-METILE (ITALIAN)
		PARATOX
		PARATUF
		PARTON-M
		PARTRON M
		PENNCAP M
		PENNCAP MLS
		PHENOL, P-NITRO-, O-ESTER WITH O,O-DIMETHYLPHOSPHOROTHIOATE
		PHOSPHOROTHIOIC ACID, O,O-DIMETHYL O-(4-NITROPHENYL) ESTER
		PHOSPHOROTHIOIC ACID, O,O-DIMETHYL O-(P-NITROPHENYL) ESTER
		PHOSPHOROTHIOIC ACID, O,O-DIMETHYL O-(P-NITROPHENYL) ESTER (DRY MIXTURE)
		QUINOPHOS
		RCRA WASTE NUMBER P071
		SINAFID M 48
		SIXTY-THREE SPECIAL EC INSECTICIDE
		TEKWAISA
		THIOPHENIT
		THIOPHOSPHATE DE O,O-DIMETHYLE ET DE O-(4-NITROPHENYLE) (FRENCH)
		THYLPAR M-50
		TOLL
		VERTAC METHYL PARATHION TECHNISCH 80%
		VOFATOX
		WOFATOS
		WOFATOX
		WOFOTOX
298-02-2		
		AC 3911
		AMERICAN CYANAMID 3,911
		DITHIOPHOSPHATE DE O,O-DIETHYLE ET D'ETHYLTHIOMETHYLE (FRENCH)
		EI3911
		ENT 24,042
		EXPERIMENTAL INSECTICIDE 3911
		FORAAT (DUTCH)
		GRANUTOX
		L 11/6
		METHANETHIOL, (ETHYLTHIO)-, S-ESTER WITH O,O-DIETHYL PHOSPHORODITHIOATE

CAS No.	IN	Chemical Name
		O,O-DIAETHYL-S-(AETHYLTHIO-METHYL)-DITHIOPHOSPHAT (GERMAN)
		O,O-DIETHYL ETHYLTHIOMETHYL PHOSPHORODITHIOATE
		O,O-DIETHYL S-(ETHYLTHIO)METHYL PHOSPHORODITHIOATE
		O,O-DIETHYL S-ETHYLMERCAPTOMETHYL DITHIOPHOSPHATE
		O,O-DIETHYL S-ETHYLMERCAPTOMETHYL DITHIOPHOSPHONATE
		O,O-DIETHYL S-ETHYLTHIOMETHYL DITHIOPHOSPHATE
		O,O-DIETHYL S-ETHYLTHIOMETHYL DITHIOPHOSPHONATE
		O,O-DIETHYL S-ETHYLTHIOMETHYL THIOTHIONOPHOSPHATE
		O,O-DIETHYL S-[(ETHYLTHIO)METHYL] PHOSPHORODITHIOATE
		O,O-DIETHYL-S-(ETHYLTHIO-METHYL)-DITHIOFOSFAAT (DUTCH)
		O,O-DIETIL-S-(ETILTIO-METIL)-DITIOFOSFATO (ITALIAN)
		PHORAT (GERMAN)
		PHORATE
		PHORATE (ACGIH)
		PHORATE 10G
		PHOSPHORODITHIOIC ACID, O,O-DIETHYL S-(ETHYLTHIO)METHYL ESTER
		RAMPART
		RCRA WASTE NUMBER P094
		THIMET
		THIMET 10G
		THIMET G
		TIMET
		VEGFRU
		VERGFRU FORATOX
		VUAGT 182
298-04-4		
		BAY 19639
		BAYER 19639
		DI-SYSTON
		DI-SYSTON G
		DIMAZ
		DISULFATON
		DISULFOTON
		DISULFOTON (ACGIH,DOT)
		DISULFOTON MIXTURE, DRY
		DISULFOTON MIXTURE, LIQUID
		DISYSTON
		DISYSTOX
		DITHIODEMETON
		DITHIOPHOSPHATE DE O,O-DIETHYLE ET DE S-(2-ETHYLTHIO-ETHYLE) (FRENCH)
		DITHIOSYSTOX
		DUTION
		EKATIN TD
		ENT 23,437
		ETHYL THIOMETON
		ETHYLTHIOMETON B
		FRUMIN
		FRUMIN AL
		FRUMIN G
		GLEBOFOS
		M 74
		M 74 (PESTICIDE)
		NA 2783 (DOT)
		O,O-DIAETHYL-S-(2-AETHYLTHIO-AETHYL)-DITHIOPHOSPHAT (GERMAN)
		O,O-DIAETHYL-S-(3-THIA-PENTYL)-DITHIOPHOSPHAT (GERMAN)
		O,O-DIETHYL 2-ETHYLTHIOETHYL PHOSPHORODITHIOATE
		O,O-DIETHYL S-(2-ETHTHIOETHYL) PHOSPHORODITHIOATE
		O,O-DIETHYL S-(2-ETHTHIOETHYL) THIOTHIONOPHOSPHATE
		O,O-DIETHYL S-(2-ETHYLMERCAPTOETHYL) DITHIOPHOSPHATE
		O,O-DIETHYL S-2-(ETHYLTHIO)ETHYL PHOSPHORODITHIOATE
		O,O-DIETHYL S-[2-(ETHYLTHIO)ETHYL] DITHIOPHOSPHATE
		O,O-DIETHYL-S-(2-ETHYLTHIO-ETHYL)-DITHIOFOSFAAT (DUTCH)
		O,O-DIETIL-S-(2-ETILTIO-ETIL)-DITIOFOSFATO (ITALIAN)
		O,O-ETHYL S-2(ETHYLTHIO)ETHYL PHOSPHORODITHIOATE
		PHOSPHORODITHIOIC ACID, O,O-DIETHYL S-(2-(ETHYLTHIO)ETHYL) ESTER
		PHOSPHORODITHIONIC ACID, S-2-(ETHYLTHIO)ETHYL-O,O-DIETHYL ESTER
		RCRA WASTE NUMBER P039
		S 276
		S-2-(ETHYLTHIO)ETHYL O,O-DIETHYL ESTER OF PHOSPHORODITHIOIC ACID
		SOLVIREX
		THIODEMETON
		THIODEMETRON
		VUAGT 1-4
		VUAGT 1964
298-07-7		
		1-HEXANOL, 2-ETHYL-, HYDROGEN PHOSPHATE
		2-ETHYL-1-HEXANOL HYDROGEN PHOSPHATE
		BIS(2-ETHYLHEXYL)HYDROGEN PHOSPHATE
		BIS(2-ETHYLHEXYL)ORTHOPHOSPHORIC ACID
		BIS(2-ETHYLHEXYL)PHOSPHATE
		BIS(2-ETHYLHEXYL)PHOSPHORIC ACID
		D 2EHPA
		DEHPA EXTRACTANT
		DI(2-ETHYLHEXYL)ORTHOPHOSPHORIC ACID
		DI(2-ETHYLHEXYL)PHOSPHATE
		DI(2-ETHYLHEXYL)PHOSPHORIC ACID
		DI(2-ETHYLHEXYL)PHOSPHORIC ACID (DOT)
		DI-2-ETHYLHEXYL HYDROGEN PHOSPHATE
		DP 8R
		ESCAID 100
		HDEHP
		HYDROGEN BIS(2-ETHYLHEXYL)PHOSPHATE
		NA 1902 (DOT)
		P 204
		P 204 (ACID)
		PHOSPHORIC ACID BIS(ETHYLHEXYL) ESTER
		PHOSPHORIC ACID, BIS(2-ETHYLHEXYL) ESTER
298-81-7		
		METHOXSALEN WITH ULTRA-VIOLET A THERAPY
299-75-2		
		(2S,3S)-THREITOL 1,4-BISMETHANESULFONATE
		1,2,3,4-BUTANETETROL, 1,4-DIMETHANESULFONATE, [S-(R*,R*)]-
		THREITOL, 1,4-DIMETHANESULFONATE, (2S,3S)-
		THREOSULPHAN
		TREOSULFAN
		TREOSULPHAN
		TRESULFAN
299-84-3		
		BLITEX
		DERMAFOS
		DOW ET 14
		DOW ET 57
		ECTORAL
		ENT 23,284
		ET 14
		ET 57
		ETROLENE
		FENCHLORFOS
		FENCHLORPHOS
		FENCLOFOS
		FENCLORVUR
		GESEKTIN K
		KORLAN
		KORLANE
		MOORMAN'S MEDICATED RID-EZY
		NANCHOR
		NANKOR
		O,O-DIMETHYL O-(2,4,5-TRICHLOROPHENYL) PHOSPHOROTHIOATE
		O,O-DIMETHYL O-(2,4,5-TRICHLOROPHENYL) THIOPHOSPHATE
		OMS 123
		PHENCHLORFOS
		PHOSPHOROTHIOIC ACID, O,O-DIMETHYL O-(2,4,5-TRI-

CAS No.	IN	Chemical Name
		CHLOROPHENYL) ESTER
		REMELT
		RONNEL
		ROVAN
		TRICHLORMETAPHOS
		TRICHLOROMETAPHOS
		TROLEN
		TROLENE
		VIOZENE
299-86-5		
		4-TERT-BUTYL-2-CHLOROPHENYL METHYL METHYLPHOSPHORAMIDATE
		AMIDOFOS
		AMIDOPHOS
		CRUFOMAT
		CRUFOMATE
		DOWCO 132
		MONTREL
		PHOSPHORAMIDIC ACID, METHYL-, 2-CHLORO-4-(1,1-DIMETHYLETHYL)PHENYL METHYL ESTER
		PHOSPHORAMIDIC ACID, METHYL-, 4-TERT-BUTYL-2-CHLOROPHENYL METHYL ESTER
		RUELENE
		RUELENE DRENCH
300-62-9		
		AMPHETAMINE
300-76-5		
		1,2-DIBROMO-2,2-DICHLOROETHYL DIMETHYL PHOSPHATE
		ALVORA
		BROMCHLOPHOS
		BROMEX
		BROMEX (INSECTICIDE)
		BROMEX 50
		BRP
		DIBROM
		DIBROMFOS
		DIMETHYL 1,2-DIBROMO-2,2-DICHLOROETHYL PHOSPHATE
		DIMETHYL 1,2-DIBROMO-2,2-DICHLOROETHYL PHOSPHATE (NALED)
		DIMETHYL-1,2-DIBROMO-2-DICHLOROETHYL PHOSPHATE
		ENT 24988
		FOSBROM
		NALED
		PHOSPHORIC ACID, 1,2-DIBROMO-2,2-DICHLOROETHYL DIMETHYL ESTER
		RE 4355
301-04-2		
		ACETIC ACID LEAD SALT (2:1)
		ACETIC ACID, LEAD(2+) SALT
		DIBASIC LEAD ACETATE
		LEAD ACETATE
	*	LEAD ACETATE (2+)
		LEAD ACETATE (Pb(AC)2)
		LEAD ACETATE (Pb(O2C2H3)2)
		LEAD DIACETATE
		LEAD DIBASIC ACETATE
		LEAD(2+) ACETATE
		LEAD(II) ACETATE
		PLUMBOUS ACETATE
		SALT OF SATURN
		SUGAR OF LEAD
302-01-2		
		ANHYDROUS HYDRAZINE (DOT)
		DIAMIDE
		DIAMINE
	*	HYDRAZINE
		HYDRAZINE (ACGIH)
		HYDRAZINE BASE
		HYDRAZINE HYDRATE (DOT)
		HYDRAZINE, ANHYDROUS
		HYDRAZINE, ANHYDROUS (DOT)
		HYDRAZINE, AQUEOUS SOLUTION (DOT)
		HYDRAZINE, AQUEOUS SOLUTION CONTAINING MORE THAN 64% HYDRAZINE (DOT)
		HYDRAZINE, AQUEOUS SOLUTION WITH LESS THAN 64% HYDRAZINE (DOT)
		HYDRAZINE, HYDRATE
		HYDRAZYNA (POLISH)
		LEVOXINE
		NITROGEN HYDRIDE (N2H4)
		OXYTREAT 35
		RCRA WASTE NUMBER U133
		UN 2029 (DOT)
		UN 2030 (DOT)
302-70-5		
		2,2'-DICHLORO-N-METHYLDIETHYLAMINE N-OXIDE HYDROCHLORIDE
		DIETHYLAMINE, 2,2'-DICHLORO-N-METHYL-, N-OXIDE, COMPD. WITH HYDROCHLORIC ACID
		DIETHYLAMINE, 2,2'-DICHLORO-N-METHYL-, N-OXIDE, HYDROCHLORIDE
		ETHANAMINE, 2-CHLORO-N-(2-CHLOROETHYL)-N-METHYL, N-OXIDE, HYDROCHLORIDE
		MECHLORETHAMINE OXIDE HYDROCHLORIDE
		METHYLBIS(BETA-CHLOROETHYL)AMINE N-OXIDE HYDROCHLORIDE
		METHYLDI(2-CHLOROETHYL)AMINE N-OXIDE HYDROCHLORIDE
		MITOMEN
		MUSTRON
		N-METHYL-2,2'-DICHLORODIETHYLAMINE N-OXIDE HYDROCHLORIDE
		N-METHYLBIS(2-CHLOROETHYL)AMINE N-OXIDE HYDROCHLORIDE
		NITROGEN MUSTARD N-OXIDE
		NITROGEN MUSTARD N-OXIDE HYDROCHLORIDE
		NITROMIN
		NITROMIN HYDROCHLORIDE
303-34-4		
		2-BUTENOIC ACID, 2-METHYL-, 7-[[2,3-DIHYDROXY-2-(1-METHOXYETHYL)-3-METHYL-1-OXYBUTOXY]METHYL]-2,3,5,
		LASIOCARPINE
305-03-3		
		4-[BIS(2-CHLOROETHYL)AMINO]PHENYLBUTYRIC ACID
		4-[P-[BIS(2-CHLOROETHYL)AMINO]PHENYL]BUTYRIC ACID
		AMBOCHLORIN
		AMBOCLORIN
		BENZENEBUTANOIC ACID, 4-[BIS(2-CHLOROETHYL)AMINO]-
		BUTYRIC ACID, 4-[P-[BIS(2-CHLOROETHYL)AMINO]-PHENYL]-
		CB 1348
		CHLORAMBUCIL
		CHLORAMINOPHENE
		CHLORBUTIN
		CHLOROBUTINE
		ECLORIL
		GAMMA-[P-DI(2-CHLOROETHYL)AMINOPHENYL]BUTYRIC ACID
		LEUKERAN
		LINFOLIZIN
		LINFOLYSIN
		NSC 3088
309-00-2		
		1,4:5,8-DIMETHANONAPHTHALENE, 1,2,3,4,10,10-HEXACHLORO-1,4,4A,5, 8,8A-HEXAHYDRO-, (1ALPHA,4ALPHA,4AB
		1,4:5,8-DIMETHANONAPHTHALENE, 1,4,4A,5,8,8A-HEXAHYDRO- 1,2,3,4,10,10-HEXACHLORO-, ENDO, EXOMIXTURE (
		ALDOCIT
		ALDREX 40
	*	ALDRIN
		ALDRIN MIXTURE, DRY (WITH 65% OR LESS ALDRIN) (DOT)
		ALDRIN MIXTURE, DRY (WITH MORE THAN 65% ALDRIN) (DOT)
		ALDRIN MIXTURE, LIQUID (WITH 60% OR LESS ALDRIN) (DOT)
		ALDRIN MIXTURE, LIQUID (WITH MORE T HAN 60% ALDRIN) (DOT)
		ALDRIN MIXTURE, LIQUID (WITH MORE THAN 60% AL-

CAS No.	IN	Chemical Name
		DRIN) (DOT)
		COMPOUND 118
		ENT 15,949
		HHDN
		KORTOFIN
		NA 2761 (DOT)
		NA 2762 (DOT)
		OCTALENE
		SD 2794
		SEEDRIN
		TATUZINHO
311-45-5		
		DIETHYL 4-NITROPHENYL PHOSPHATE
		PARAOXON
314-40-9		
		2,4(1H,3H)-PYRIMIDINEDIONE, 5-BROMO-6-METHYL-3-(1-METHYLPROPYL)-
	*	BROMACIL
		BROMAZIL
		HERBICIDE 976
		HYVAR X
		HYVAR X BROMACIL
		HYVAREX
		KROVAR II
		URACIL, 5-BROMO-3-SEC-BUTYL-6-METHYL-
315-18-4		
		4-DIMETHYLAMINO-3,5-DIMETHYLPHENYL N-METHYLCARBAMATE
		4-DIMETHYLAMINO-3,5-XYLYL METHYLCARBAMATE
		4-DIMETHYLAMINO-3,5-XYLYL N-METHYLCARBAMATE
		CARBAMIC ACID, METHYL-, 4-(DIMETHYLAMINO)-3,5-XYLYL ESTER
		DOWCO 139
		ENT 25,766
		MEXACARBATE
		PHENOL, 4-(DIMETHYLAMINO)-3,5-DIMETHYL-, METHYLCARBAMATE (ESTER)
		ZACTRAN
		ZECTANE
		ZECTRAN
		ZEXTRAN
315-22-0		
		20-NORCROTALANAN-11,15-DIONE, 14,19-DIHYDRO-12,13-DIHYDROXY-, (13.ALPHA.,14.ALPHA.)-
		2H-[1,6]DIOXACYCLOUNDECINO[2,3,4-GH]PYRROLIZINE, 20-NORCROTALANAN-11,15-DIONE DERIV.
		MONOCROTALIN
		MONOCROTALINE
316-42-7		
		(-)-EMETINE DIHYDROCHLORIDE
		2H-BENZO[A]QUINOLIZINE, EMETAN DERIV
		AMEBICIDE
		EMETAN, 6',7',10,11-TETRAMETHOXY-, DIHYDROCHLORIDE
		EMETINE HYDROCHLORIDE
		EMETINE, DIHYDROCHLORIDE
		L-EMETINE DIHYDROCHLORIDE
		NSC-33669
319-84-6		
		ALPHA-1,2,3,4,5,6-HEXACHLORCYCLOHEXANE
		ALPHA-1,2,3,4,5,6-HEXACHLOROCYCLOHEXANE
		ALPHA-BENZENE HEXACHLORIDE
		ALPHA-BHC
		ALPHA-HEXACHLORAN
		ALPHA-HEXACHLORANE
	*	ALPHA-HEXACHLORCYCLOHEXANE
		ALPHA-HCH
		ALPHA-HEXACHLOROCYCLOHEXANE
		ALPHA-LINDANE
		CYCLOHEXANE, 1,2,3,4,5,6-HEXACHLORO-, (1ALPHA,2ALPHA,3BETA,4ALPHA,5BETA,6BETA)-
319-85-7		
	*	BETA-1,2,3,4,5,6-HEXACHLOROCYCLOHEXANE
		BETA-BENZENE HEXACHLORIDE
		BETA-BHC
		BETA-HEXACHLORAN
		BETA-HEXACHLOROBENZENE
		BETA -HCH
		BETA-HEXACHLOROCYCLOHEXANE
		BETA-LINDANE
		CYCLOHEXANE, 1,2,3,4,5,6-HEXACHLORO-, (1ALPHA,2BETA,3ALPHA,4BETA,5ALPHA,6BETA)-
		CYCLOHEXANE, 1,2,3,4,5,6-HEXACHLORO-, BETA-ISOMER
319-86-8		
		DELTA-(AEEEE)-1,2,3,4,5,6-HEXACHLOROCYCLOHEXANE
		DELTA-1,2,3,4,5,6-HEXACHLOROCYCLOHEXANE
		DELTA-BENZENE HEXACHLORIDE
		DELTA-BHC
		DELTA-HCH
	*	DELTA-HEXACHLOROCYCLOHEXANE
		DELTA-LINDANE
		CYCLOHEXANE, 1,2,3,4,5,6-HEXACHLORO-, (1ALPHA,2ALPHA,3ALPHA,4BETA,5ALPHA,6BETA)-
320-67-2		
		5-AZACYTIDINE
327-98-0		
		5082A
		AGRISIL
		AGRITOX
		BAY 37289
		BAYER 37289
		BAYER 5081
		BAYER S 4400
		CHEMAGRO 37289
		ENT 25,712
		ETHYL TRICHLOROPHENYLETHYLPHOSPHONOTHIOATE
		FENOPHOSPHON
		FITOSOL
		O-AETHYL-O-(2,4,5-TRICHLORPHENYL)-AETHYLTHIONOPHOSPHONAT (GERMAN)
		O-ETHYL O-2,4,5-TRICHLOROPHENYL ETHYLPHOSPHONOTHIOATE
		PHENOL, 2,4,5-TRICHLORO-, O-ESTER WITH O-ETHYL ETHYLPHOSPHONOTHIOATE
		PHOSPHONOTHIOIC ACID, ETHYL-, O-ETHYL O-(2,4,5-TRICHLOROPHENYL) ESTER
		PHYTOSOL
		RICHLORONATE
		S 4400
		STAUFFER N-3049
		TRICHLORONAT
		TRICHLORONATE
		WIRKSTOFF 37289
329-71-5		
		2,5-DINITROPHENOL
		2,5-DNP
		GAMMA-DINITROPHENOL
		PHENOL, 2,5-DINITRO-
		PHENOL, GAMMA-DINITRO-
330-54-1		
		1,1-DIMETHYL-3-(3,4-DICHLOROPHENYL)UREA
		1-(3,4-DICHLOROPHENYL)-3,3-DIMETHYLUREA
		1-(3,4-DICHLOROPHENYL)-3,3-DIMETHYLUREE (FRENCH)
		3-(3,4-DICHLOOR-FENYL)-1,1-DIMETHYLUREUM (DUTCH)
		3-(3,4-DICHLOR-PHENYL)-1,1-DIMETHYL-HARNSTOFF (GERMAN)
		3-(3,4-DICHLOROPHENOL)-1,1-DIMETHYLUREA
		3-(3,4-DICHLOROPHENYL)-1,1-DIMETHYLUREA
		3-(3,4-DICLORO-FENYL)-1,1-DIMETIL-UREA (ITALIAN)
		AF 101
		CEKIURON
		CRISURON
		DAILON
		DCMU
		DI-ON

CAS No.	IN	Chemical Name
		DIATER
		DICHLORFENIDIM
		DIREX 4L
		DIUREX
		DIUROL
		DIURON
		DIURON (ACGIH,DOT)
		DIURON 4L
		DMU
		DREXEL
		DREXEL DIURON 4L
		DURAN
		DYNEX
		FARMCO DIURON
		HERBATOX
		HW 920
		KARAMEX
		KARMEX
		KARMEX D
		KARMEX DIURON HERBICIDE
		KARMEX DW
		MARMER
		N'-(3,4-DICHLOROPHENYL)-N,N-DIMETHYLUREA
		N-(3,4-DICHLOROPHENYL)-N',N'-DIMETHYLUREA
		NA 2767 (DOT)
		SUP'R FLO
		TELVAR
		TELVAR DIURON WEED KILLER
		UNIDRON
		UREA, 3-(3,4-DICHLOROPHENYL)-1,1-DIMETHYL-
		UREA, N'-(3,4-DICHLOROPHENYL)-N,N-DIMETHYL-
		UROX D
		USAF P-7
		USAF XR-42
		VONDURON
333-41-5		
		4-PYRIMIDINOL, 2-ISOPROPYL-6-METHYL-, O-ESTER WITH O,O-DIETHYL PHOSPHOROTHIOATE
		ALFA-TOX
		ANTIGAL
		BASSADINON
		BASUDIN
		BASUDIN 10G
		BASUDIN 5G
		BAZINON
		BAZUDEN
		CIAZINON
		DACUTOX
		DASSITOX
		DAZZEL
		DESAPON
		DIANON
		DIATERR-FOS
		DIAZAJET
		DIAZATOL
		DIAZIDE
	*	DIAZINON
		DIAZINON (ACGIH,DOT)
		DIAZINON AG 500
		DIAZINONE
		DIAZITOL
		DIAZOL
		DICID
		DIETHYL 2-ISOPROPYL-4-METHYL-6-PYRIMIDINYL PHOSPHOROTHIONATE
		DIETHYL 4-(2-ISOPROPYL-6-METHYLPYRIMIDINYL)PHOSPHOROTHIONATE
		DIMPYLAT
		DIMPYLATE
		DIPOFENE
		DISONEX
		DIZINON
		DYZOL
		ENT 19,507
		EXODIN
		FLYTROL
		G 24480
		G 301
		GALESAN
		GARDEN TOX
		GEIGY 24480
		ISOPROPYLMETHYLPYRIMIDYL DIETHYL THIOPHOSPHATE
		KAYAZINON
		KAYAZOL
		KNOX-OUT
		MEODINON
		NA 2783 (DOT)
		NCI-C08673
		NEOCIDOL
		NEOCIDOL (OIL)
		NIPSAN
		NUCIDOL
		O,O-DIAETHYL-O-(2-ISOPROPYL-4-METHYL)-6-PYRIMIDYL-THIONOPHOSPHAT (GERMAN)
		O,O-DIAETHYL-O-(2-ISOPROPYL-4-METHYL-PYRIMIDIN-6-YL)-MONOTHIOPHOSPHAT (GERMAN)
		O,O-DIETHYL 2-ISOPROPYL-4-METHYLPYRIMIDYL-6-THIOPHOSPHATE
		O,O-DIETHYL O-(2-ISOPROPYL-4-METHYL-6-PYRIMIDINYL) PHOSPHOROTHIOATE
		O,O-DIETHYL O-(2-ISOPROPYL-4-METHYL-6-PYRIMIDYL) THIONOPHOSPHATE
		O,O-DIETHYL O-(2-ISOPROPYL-6-METHYL-4-PYRIMIDINYL) PHOSPHOROTHIOATE
		O,O-DIETHYL O-(2-ISOPROPYL-6-METHYL-4-PYRIMIDIYL) PHOSPHOROTHIOATE
		O,O-DIETHYL O-6-METHYL-2-ISOPROPYL-4-PYRIMIDINYL PHOSPHOROTHIOATE
		O,O-DIETHYL-O-(2-ISOPROPYL-4-METHYL-6-PYRIMIDYL)-PHOSPHOROTHIOATE
		O,O-DIETHYL-O-(2-ISOPROPYL-4-METHYL-PYRIMIDIN-6-YL)-MONOTHIOFOSFAAT (DUTCH)
		O,O-DIETIL-O-(2-ISOPROPIL-4-METIL-PIRIMIDIN-6-IL)-MONOTIOFOSFATO (ITALIAN)
		O-2-ISOPROPYL-4-METHYLPYRIMIDYL-O,O-DIETHYL PHOSPHOROTHIOATE
		OLEODIAZINON
		PHOSPHOROTHIOATE, O,O-DIETHYL O-6-(2-ISOPROPYL-4-METHYLPYRIMIDYL)
		PHOSPHOROTHIOIC ACID, O,O-DIETHYL 2-ISOPROPYL-6-METHYL-4-PYRIMIDINYL ESTER
		PHOSPHOROTHIOIC ACID, O,O-DIETHYL O-(2-ISOPROPYL-6-METHYL-4-PYRIMIDINYL) ESTER
		PHOSPHOROTHIOIC ACID, O,O-DIETHYL O-[6-METHYL-2-(1-METHYLETHYL)-4-PYRIMIDINYL] ESTER
		SAROLEX
		SPECTRACIDE
		SPECTRACIDE 25EC
		THIOPHOSPHATE DE O,O-DIETHYLE ET DE O-2-ISOPROPYL-4-METHYL-6-PYRIMIDYLE (FRENCH)
334-88-3		
		AZIMETHYLENE
		DIAZIRINE
		DIAZOMETHANE
		DIAZONIUM METHYLIDE
		METHANE, DIAZO-
335-42-4		
		BORON TRIFLUORIDE COMPOUND WITH METHYL ETHER (1:1)
352-93-2		
		DIETHYL SULFIDE
		ETHYL SULFIDE
353-36-6		
		ETHANE, FLUORO-
		ETHYL FLUORIDE
		ETHYL FLUORIDE (DOT)
		FLUOROETHANE
		MONOFLUOROETHANE
		R 161
		UN 2453 (DOT)

CAS No.	IN	Chemical Name
353-42-4		
		BORANE, TRIFLUORO-, COMPD WITH OXYBIS[METHANE] (1:1)
		BORON FLUORIDE (BF3), COMPD WITH METHYL ETHER (1:1)
		BORON FLUORIDE COMPLEX WITH DIMETHYL ETHER
		BORON TRIFLUORIDE COMPD WITH METHYL ETHER
		BORON TRIFLUORIDE COMPOUND WITH METHYL ETHER (1:1)
		BORON TRIFLUORIDE DIMETHYL ETHERATE
		BORON TRIFLUORIDE DIMETHYL ETHERATE (DOT)
		BORON TRIFLUORIDE-DIMETHYL ETHER
		BORON TRIFLUORIDE-DIMETHYL ETHER COMPLEX
		BORON, TRIFLUORO[OXYBIS[METHANE]]-, (T-4)-
		METHANE, OXYBIS-, BORON COMPLEX
		METHANE, OXYBIS-, COMPD WITH TRIFLUOROBORANE (1:1)
		METHYL ETHER, COMPD WITH BF3 (1:1)
		METHYL ETHER, COMPD WITH BORON FLUORIDE (BF3) (1:1)
		UN 2965 (DOT)
353-50-4		
		CARBON DIFLUORIDE OXIDE
		CARBON FLUORIDE OXIDE
		CARBON FLUORIDE OXIDE (COF2)
		CARBON OXYFLUORIDE
		CARBON OXYFLUORIDE (COF2)
		CARBONIC DIFLUORIDE
		CARBONIC DIFLUORIDE (9CI)
		CARBONYL DIFLUORIDE
		CARBONYL DIFLUORIDE (COF2)
		CARBONYL FLUORIDE
		CARBONYL FLUORIDE (ACGIH,DOT)
		CARBONYL FLUORIDE (COF2)
		DIFLUOROFORMALDEHYDE
		DIFLUOROOXOMETHANE
		DIFLUOROPHOSGENE
		FLUOPHOSGENE
		FLUOROFORMYL FLUORIDE
		FLUOROPHOSGENE
		RCRA WASTE NUMBER U033
		UN 2417 (DOT)
353-59-3		
		BCF
		BROMOCHLORODIFLUOROMETHANE
		CHLOROBROMODIFLUOROMETHANE
		CHLORODIFLUOROBROMOMETHANE
		CHLORODIFLUOROBROMOMETHANE (DOT)
		CHLORODIFLUOROMONOBROMOMETHANE
		DIFLUOROCHLOROBROMOMETHANE
		FLUGEX 12B1
		FLUOROCARBON 1211
		FREON 12B1
		HALON 1211
		METHANE, BROMOCHLORODIFLUORO-
		R 12B1
		UN 1974 (DOT)
357-57-3		
		(-)-BRUCINE
		10,11-DIMETHYSTRYCHNINE
		2,3-DIMETHOXYSTRYCHNINE
		4,6-METHANO-6H,14H-INDOLO[3,2,1-IJ]OXEPINO[2,3,4-DE]-PYRROLO[2,3-H]QUINOLINE, STRYCHNIDIN-10-ONE DERI
		BRUCIN (GERMAN)
		BRUCINA (ITALIAN)
	*	BRUCINE
		BRUCINE (DOT)
		BRUCINE ALKALOID
		BRUCINE, SOLID
		BRUCINE, SOLID (DOT)
		DIMETHOXY STRYCHNINE
		DIMETHOXY STRYCHNINE (DOT)
		RCRA WASTE NUMBER P018
		STRYCHNIDIN-10-ONE, 2,3-DIMETHOXY-
		STRYCHNIDIN-10-ONE, 2,3-DIMETHOXY- (9CI)
		STRYCHNINE, 2,3-DIMETHOXY-
		UN 1570 (DOT)
359-06-8		
		ACETYL CHLORIDE, FLUORO-
		FLUOROACETYL CHLORIDE
		TL 670
360-89-4		
		OCTAFLUOROBUT-2-ENE
366-70-1		
		1-METHYL-2-P-(ISOPROPYLCARBAMOYL)BENZYLHYDRAZINE HYDROCHLORIDE
		2-[P-(ISOPROPYLCARBAMOYL)BENZYL]-1-METHYLHYDRAZINE HYDROCHLORIDE
		BENZAMIDE, N-(1-METHYLETHYL)-4-[(2-METHYLHYDRAZINO)METHYL]-, MONOHYDROCHLORIDE
		IBENZMETHYZIN HYDROCHLORIDE
		IBENZMETHYZINE HYDROCHLORIDE
		IBZ
		MATULANE
		MBH
		N-ISOPROPYL-P-(2-METHYLHYDRAZINOMETHYL)BENZAMIDE HYDRCHLORIDE
		NATULAN
		NATULAN HYDROCHLORIDE
		NATULANAR
		NSC 77213
		P-TOLUAMIDE, N-ISOPROPYL-.ALPHA.-(2-METHYLHYDRAZINO)-, MONOHYDROCHLORIDE
		PROCARBAZINE HYDROCHLORIDE
		RO 4-6467
371-40-4		
		1-AMINO-4-FLUOROBENZENE
		4-FLUORANILIN (CZECH)
		4-FLUOROANILINE
		4-FLUOROANILINE (DOT)
		4-FLUOROBENZENAMINE
		ANILINE, 4-FLUORO-
		ANILINE, P-FLUORO-
		BENZENAMINE, 4-FLUORO-
		FLUOROANILINE
		P-FLUOROANILINE
		P-FLUOROPHENYLAMINE
		UN 2944 (DOT)
371-62-0		
		2-FLUOROETHANOL
		BETA-FLUOROETHANOL
		ETHANOL, 2-FLUORO-
		ETHYLENE FLUOROHYDRIN
		FLUTRITEX 2
		TL 741
371-86-8		
		BIS(ISOPROPYLAMIDO) FLUOROPHOSPHATE
		BIS(MONOISOPROPYLAMINO)FLUOROPHOSPHATE
		BISISOPROPYLAMINOFLUOROPHOSPHINE OXIDE
		FLUOROBISISOPROPYLAMINOPHOSPHINE OXIDE
		FLUORURE DE N,N'-DIISOPROPYLE PHOSPHORODIAMIDE (FRENCH)
		MIPAFOX
		MIPAFOX (DOT)
		N,N'-DIISOPROPIL-FOSFORODIAMMIDO-FLUORURO (ITALIAN)
		N,N'-DIISOPROPYL-DIAMIDO-FOSFORZUUR-FLUORIDE (DUTCH)
		N,N'-DIISOPROPYL-DIAMIDO-PHOSPHORSAEURE-FLUORID (GERMAN)
		PESTON XV
		PESTOX 15
		PESTOX XV
		PHOSPHINE OXIDE, FLUOROBIS(ISOPROPYLAMINO)-
		PHOSPHORODI(ISOPROPYLAMIDIC) FLUORIDE
		PHOSPHORODIAMIDIC FLUORIDE, N,N'-DIISOPROPYL-
		UN 2783 (DOT)

CAS No.	IN	Chemical Name
372-09-8		
		2-CYANOACETIC ACID
		ACETIC ACID, CYANO-
		ACIDE CYANACETIQUE (FRENCH)
		CAA
		CYANESSIGSAEURE (GERMAN)
		CYANOACETIC ACID
		MALONIC MONONITRILE
		MONOCYANOACETIC ACID
		USAF KF-17
373-02-4		
		ACETIC ACID, NICKEL(2+) SALT
		NICKEL ACETATE
		NICKEL DIACETATE
		NICKEL(2+) ACETATE
		NICKEL(II) ACETATE
		NICKEL(II) ACETATE (1:2)
		NICKELOUS ACETATE
379-79-3		
		ERGOTAMINE TARTRATE
402-54-0		
		1-NITRO-4-(TRIFLUOROMETHYL)BENZENE
		4-(TRIFLUOROMETHYL)NITROBENZENE
		ALPHA,ALPHA,ALPHA-TRIFLUORO-P-NITROTOLUENE
		BENZENE, 1-NITRO-4-(TRIFLUOROMETHYL)-
		BENZENE, 1-NITRO-4-(TRIFLUOROMETHYL)- (9CI)
		NITROBENZOTRIFLUORIDE
		P-(TRIFLUOROMETHYL)NITROBENZENE
		P-NITRO(TRIFLUOROMETHYL)BENZENE
		P-NITROBENZOTRIFLUORIDE
		P-NITROBENZOTRIFLUORIDE (DOT)
		TOLUENE, 4-NITRO-ALPHA,ALPHA,ALPHA-TRIFLUORO-
		TOLUENE, ALPHA,ALPHA,ALPHA-TRIFLUORO-P-NITRO-
		UN 2306 (DOT)
409-21-2		
		CARBOFRAX M
		CARBON SILICIDE
		CARBORUNDUM
		CRYSTOLON B
		KZ 3M
		KZ 5M
		KZ 7M
		NICALON
	*	SILICA, GRAPHITE
		SILICON CARBIDE
		SILICON CARBIDE (SiC)
		SILICON MONOCARBIDE
		SILUNDUM
		UA 1
		UA 2
		UA 3
		UA 4
		YE 5626
420-04-2		
		ALZOGUR
		AMIDOCYANOGEN
		CARBAMONITRILE
		CARBIMIDE
	*	CYANAMIDE
		CYANOAMINE
		CYANOGEN NITRIDE
		CYANOGENAMIDE
		HYDROGEN CYANAMIDE
		N-CYANOAMINE
		TSAKS
428-59-1		
		(TRIFLUOROMETHYL)TRIFLUOROOXIRANE
		2,2,3-TRIFLUORO-3-(TRIFLUOROMETHYL)OXIRANE
		HEXAFLUORO-1,2-EPOXYPROPANE
		HEXAFLUOROEPOXYPROPANE
		HEXAFLUOROPROPENE EPOXIDE
		HEXAFLUOROPROPENE OXIDE
		HEXAFLUOROPROPYLENE EPOXIDE
		HEXAFLUOROPROPYLENE OXIDE
		HEXAFLUOROPROPYLENE OXIDE (DOT)
		NA 1956 (DOT)
		OXIRANE, TRIFLUORO(TRIFLUOROMETHYL)-
		PERFLUORO(METHYLOXIRANE)
		PERFLUOROPROPYLENE OXIDE
		PROPANE, 1,2-EPOXY-1,1,2,3,3,3-HEXAFLUORO-
		PROPYLENE OXIDE HEXAFLUORIDE
		TRIFLUORO(TRIFLUOROMETHYL)OXIRANE
431-03-8		
		2,3-BUTADIONE
		2,3-BUTANEDIONE
		2,3-DIKETOBUTANE
		2,3-DIOXOBUTANE
		BIACETYL
	*	BUTANEDIONE
		DIACETYL
		DIMETHYL DIKETONE
		DIMETHYLGLYOXAL
		ERYTHRITOL ANHYDRIDE
434-07-1		
		2-(HYDROXYMETHYLENE)-17-METHYLDIHYDROTESTOS-TERONE
		5ALPHA-ANDROSTAN-3-ONE, 17BETA-HYDROXY-2-(HY-DROXYMETHYLENE)-17-METHYL-
		ADROYD
		ANADROL
		ANAPOLON
		ANASTERON
		ANASTERONAL
		ANASTERONE
		ANDROSTAN-3-ONE, 17-HYDROXY-2-(HYDROXYMETHYL-ENE)-17-METHYL-, (5ALPHA,17BETA)-
		BECOREL
		C.I. 406
		HMD
		NASTENON
		NSC-26198
		OXYMETHENOLONE
		OXYMETHOLONE
		PROTANABOL
		ROBORAL
438-41-5		
		CHLORDIAZEPOXIDE HYDROCHLORIDE
439-14-5		
		DIAZEPAM
443-48-1		
		1H-IMIDAZOLE-1-ETHANOL, 2-METHYL-5-NITRO-
		ANAGIARDIL
		BAYER 5360
		CLONT
		DEFLAMON
		ENTIZOL
		FLAGESOL
		FLAGIL
		FLAGYL
		GINEFLAVIR
		IMIDAZOLE-1-ETHANOL, 2-METHYL-5-NITRO-
		KLION
		KLONT
		METRONIDAZOL
		METRONIDAZOLE
		MEXIBOL
		ORVAGIL
		RP 8823
		TAKIMETOL
		TRICHOMOL
		TRICHOPAL
		TRIVAZOL
		VAGIMID

CAS No.	*IN*	*Chemical Name*
446-86-6		
		1H-PURINE, 6-[(1-METHYL-4-NITRO-1H-IMIDAZOL-5-YL)THIO]-
		6-(1-METHYL-4-NITROIMIDAZOL-5-YL)THIOPURINE
		AZATHIOPRIN
		AZATHIOPRINE
		AZOTHIOPRINE
		BW 57-322
		IMURAN
		IMUREK
		IMUREL
		MURAN
		NSC 39084
		PURINE, 6-[(1-METHYL-4-NITROIMIDAZOL-5-YL)THIO]-
460-19-5		
		CARBON NITRIDE
		CARBON NITRIDE (C2N2)
		CYANOGEN
		CYANOGEN (ACGIH,DOT)
		CYANOGEN (C2N2)
		CYANOGEN GAS
		CYANOGEN GAS (DOT)
		CYANOGENE (FRENCH)
		DICYAN
		DICYANOGEN
		ETHANEDINITRILE
		NITRILOACETONITRILE
		OXALIC ACID DINITRILE
		OXALIC NITRILE
		OXALONITRILE
		OXALYL CYANIDE
		PRUSSITE
		RCRA WASTE NUMBER P031
		UN 1026 (DOT)
462-06-6		
		BENZENE, FLUORO-
		FLUOROBENZENE
		FLUOROBENZENE (DOT)
		MONOFLUOROBENZENE
		PHENYL FLUORIDE
		UN 2387 (DOT)
462-08-8		
		3-AMINO PYRIDINE
462-23-7		
		4-FLUOROBUTYRIC ACID
462-95-3		
		1,1-DIETHOXYMETHANE
	*	DIETHOXYMETHANE
		DIETHOXYMETHANE (DOT)
		DIETHYLFORMAL
		ETHANE, 1,1'-[METHYLENEBIS(OXY)]BIS-
		ETHOXYMETHYL ETHYL ETHER
		ETHYLAL
		FORMALDEHYDE DIETHYL ACETAL
		METHANE, DIETHOXY-
		UN 2373 (DOT)
463-04-7		
		1-NITROPENTANE
		1-PENTYL NITRITE
		AMYL NITRITE
		AMYL NITRITE (DOT)
		N-AMYL NITRITE
		NITRAMYL
		NITROUS ACID, PENTYL ESTER
		PENTYL ALCOHOL, NITRITE
		PENTYL NITRITE
		UN 1113 (DOT)
463-49-0		
		1,2-PROPADIENE
		1,2-PROPADIENE (9CI)
		ALLENE
		DIMETHYLENEMETHANE
	*	PROPADIENE
		PROPADIENE, INHIBITED (DOT)
		SYM-ALLYLENE
		UN 2200 (DOT)
463-51-4		
		CARBOMETHENE
		ETHENONE
		KETENE
463-58-1		
		CARBON MONOXIDE MONOSULFIDE
		CARBON OXIDE SULFIDE
		CARBON OXIDE SULFIDE (9CI)
		CARBON OXIDE SULFIDE (COS)
		CARBON OXYSULFIDE
		CARBON OXYSULFIDE (COS)
		CARBONYL SULFIDE
		CARBONYL SULFIDE (COS)
		CARBONYL SULFIDE (DOT)
		CARBONYL SULFIDE-(SUP 32)S
		CARBONYL SULFIDE-32S
		OXYCARBON SULFIDE
		OXYCARBON SULFIDE (COS)
		UN 2204 (DOT)
463-71-8		
		CARBON CHLOROSULFIDE
		CARBONIC DICHLORIDE, THIO-
		CARBONOTHIOIC DICHLORIDE
		CARBONOTHIOIC DICHLORIDE (9CI)
		CARBONYL CHLORIDE, THIO-
		CARBONYL SULFIDE DICHLORIDE
		DICHLOROTHIOCARBONYL
		DICHLOROTHIOFORMALDEHYDE
		PHOSGENE, THIO-
		THIOCARBONIC DICHLORIDE
		THIOCARBONYL CHLORIDE
		THIOCARBONYL DICHLORIDE
		THIOCARBONYL-CHLORIDE (DOT)
		THIOFOSGEN (CZECH)
		THIOKARBONYLCHLORID (CZECH)
		THIOPHOSGENE
		THIOPHOSGENE (DOT)
		UN 2474 (DOT)
463-82-1		
		1,1,1-TRIMETHYLETHANE
		2,2-DIMETHYLPROPANE
		2,2-DIMETHYLPROPANE (DOT)
		DIMETHYLPROPANE
		NEOPENTAENE
		NEOPENTANE
		PROPANE, 2,2-DIMETHYL-
		TERT-PENTANE
		TERT-PENTANE (DOT)
		TETRAMETHYLMETHANE
		UN 1265 (DOT)
		UN 2044 (DOT)
464-06-2		
		2,2,3-TRIMETHYLBUTANE
		BUTANE, 2,2,3-TRIMETHYL-
		TRIPTAN
		TRIPTANE
465-73-6		
		ISODRIN
470-90-6		
		2,4-DICHLORO-ALPHA-(CHLOROMETHYLENE)BENZYL-DIETHYL PHOSPHATE
		2-CHLORO-1-(2,4-DICHLOROPHENYL)VINYL DIETHYL PHOSPHATE
		APACHLOR
		BENZYL ALCOHOL, 2,4-DICHLORO-ALPHA-(CHLOROMETHYLENE)-, DIETHYL PHOSPHATE
		BETA-2-CHLORO-1-(2',4'-DICHLOROPHENYL) VINYL DIETHYLPHOSPHATE

CAS No.	IN	Chemical Name
		BIRLAN
		BIRLANE
		BIRLANE 10G
		BIRLANE 24
		C-10015
		C8949
		CFV
		CGA 26351
		CHLORFENVINFOS
		CHLORFENVINPHOS
		CHLORFENWINFOSEM (POLISH)
		CHLOROFENVINPHOS
		CHLORPHENVINFOS
		CHLORPHENVINPHOS
		CLOFENVINFOS
		COMPOUND 4072
		CVP
		CVP (PESTICIDE)
		DERMATON
		DIETHYL 1-(2,4-DICHLOROPHENYL)-2-CHLOROVINYL PHOSPHATE
		DIETHYL 2-CHLORO-1-(2,4-DICHLOROPHENYL)VINYL PHOSPHATE
		ENOLOFOS
		ENT 24969
		GC 4072
		HAPTASOL
		O,O-DIAETHYL-O-1-(4,5-DICHLORPHENYL)-2-CHLOR-VINYL-PHOSPHAT (GERMAN)
		O,O-DIETHYL O-(2-CHLORO-1-(2',4'-DICHLOROPHENYL)-VINYL) PHOSPHATE
		O-2-CHLOOR-1-(2,4-DICHLOOR-FENYL)-VINYL-O,O-DIETHYLFOSFAAT (DUTCH)
		O-2-CHLOR-1-(2,4-DICHLOR-PHENYL)-VINYL-O,O-DIAETHYLPHOSPHAT (GERMAN)
		O-2-CLORO-1-(2,4-DICLORO-FENIL)-VINIL-O,O-DIETILFOS-FATO (ITALIAN)
		OMS 1328
		PHOSPHATE DE O,O-DIETHYLE ET DE O-2-CHLORO-1-(2,4-DICHLOROPHENYL) VINYLE (FRENCH)
		PHOSPHORIC ACID, 2-CHLORO-1-(2,4-DICHLOROPHENYL)-ETHENYL DIETHYL ESTER
		PHOSPHORIC ACID, 2-CHLORO-1-(2,4-DICHLOROPHENYL)-VINYL DIETHYL ESTER
		SAPECRON
		SAPECRON 50EC
		SAPRECON C
		SD 4072
		SD 7859
		SHELL 4072
		STELADONE
		SUPONA
		SUPONE
		TARENE
		UNITOX
		VINYLPHATE
		VINYPHATE
		ZAPRAWA ENOLOFOS
479-45-8		
		2,4,6-TETRYL
		2,4,6-TRINITROPHENYL METHYLNITRAMINE
		2,4,6-TRINITROPHENYL-N-METHYLNITRAMINE
		2,4,6-TRINITROPHENYLMETHYLNITROAMINE
		ANILINE, N-METHYL-N,2,4,6-TETRANITRO-
		BENZENAMINE, N-METHYL-N,2,4,6-TETRANITRO-
		BENZENAMINE, N-METHYL-N,2,4,6-TETRANITRO- (9CI)
		CE
		N-METHYL-N,2,4,6-N-TETRANITROANILINE
		N-METHYL-N,2,4,6-TETRANITROANILINE
		N-METHYL-N-PICRYLNITRAMINE
		N-PICRYL-N-METHYLNITRAMINE
		NITRAMINE
		PICRYLMETHYLNITRAMINE
		PICRYLNITROMETHYLAMINE
		TETRALIT
		TETRALITE
		TETRIL
		TETRYL
		TETRYL (ACGIH,DOT)
		TRINITROPHENYLMETHYLNITRAMINE
		TRINITROPHENYLMETHYLNITRAMINE (DOT)
		UN 0208 (DOT)
481-39-0		
		5-HYDROXYNAPHTHALENE-1,4-DIONE
		JUGLONE
492-80-8		
		AURAMINE
494-03-1		
		2-NAPHTHALENAMINE, N,N-BIS(2-CHLOROETHYL)-
		2-NAPHTHYLAMINE, N,N-BIS(2-CHLOROETHYL)-
		ALEUKON
		BETA-NAPHTHYLBIS(BETA-CHLOROETHYL)AMINE
		BETA-NAPHTHYLDI(2-CHLOROETHYL)AMINE
		CB 1048
		CHLORNAPHAZINE
		CHLORONAPHTHINA
		CHLORONAPHTHINE
		CLORONAFTINA
		DI(2-CHLOROETHYL)-BETA-NAPHTHYLAMINE
		ERYSAN
		N,N-BIS(2-CHLOROETHYL)-2-NAPHTHYLAMINE
		NAFTICLORINA
		NAPHTHYLAMINE MUSTARD
		R 48
494-52-0		
		ANABASINE
496-03-7		
		2-ETHYL-3-HYDROXYHEXANAL
		3-HYDROXY-2-ETHYLHEXANAL
		BUTYRALDOL
		HEXANAL, 2-ETHYL-3-HYDROXY-
497-80-8		
		AURAMINE
		AURAMINE(TECHNICAL GRADE)
498-15-7		
		(+)-3-CARENE
		(+)-CARENE-3
		3,7,7-TRIMETHYLBICYCLO(410)-3-HEPTENE
		3-CARENE
		3-CARENE, (1S,6R)-(+)-
		3-NORCARENE, 3,7,7-TRIMETHYL-
		4,7,7-TRIMETHYL-3-NORCARENE
		BICYCLO(410)HEPT-3-ENE, 3,7,7-TRIMETHYL- (9CI)
		BICYCLO[410]HEPT-3-ENE, 3,7,7-TRIMETHYL-, (1S)-
		CARENE
		DELTA(SUP 3)-CARENE
		ISODIPRENE
		S-3-CARENE
501-53-1		
		BENZYL CHLOROFORMATE
		CARBOBENZOXY CHLORIDE
502-39-6		
		(CYANOGUANIDINO)METHYLMERCURY
		AGROSOL
		CYANO(METHYLMERCURI)GUANIDINE
		GUANIDINE, CYANO(METHYLMERCURIO)-
		GUANIDINE, CYANO-, MERCURY COMPLEX
		GUANIDINE, CYANO-, METHYLMERCURY DERIV
		MEMA
		MERCURY, (3-CYANOGUANIDINO)METHYL-
		MERCURY, (CYANOGUANIDINATO) METHYL-
		MERCURY, (CYANOGUANIDINATO-N')METHYL-
		MERCURY, (CYANOGUANIDINE)METHYL-
		METHYL MERCURIC DICYANDIAMIDE
		METHYL MERCURY DICYANDIAMIDE
		METHYLMERCURIC CYANOGUANIDINE
		METHYLMERCURIC DICYANAMIDE
		METHYLMERCURIC DICYANDIAMIDE

CAS No.	IN	Chemical Name
		METHYLMERCURY DICYANDIAMIDE
		MMD
		MORSODREN
		MORTON EP-227
		MORTON SOIL DRENCH
		MORTON SOIL-DRENCH-C
		N-CYANO-N'-(METHYLMERCURY)GUANIDINE
		PANDRINOX
		PANO-DRENCH
		PANO-DRENCH 4
		PANODRIN A-13
		PANOGEN
		PANOGEN (OLD)
		PANOGEN 15
		PANOGEN 43
		PANOGEN 8
		PANOGEN PX
		PANOGEN TURF FUNGICIDE
		PANOGEN TURF SPRAY
		PANOSPRAY 30
		R 8
		R 8 (FUNGICIDE)
		ZAPRAWA NASIENNA PLYNNA
503-17-3		
		2-BUTYNE
		2-BUTYNE (8CI,9CI)
		CROTONYLENE
		CROTONYLENE (DOT)
		DIMETHYLACETYLENE
		UN 1144 (DOT)
503-74-2		
		ISOPENTANOIC ACID
504-20-1		
		2,5-HEPTADIEN-4-ONE, 2,6-DIMETHYL-
		DIISOBUTENYL KETONE
		DIISOPROPYLIDENE ACETONE
		FORON
		PHORONE
		SYM-DIISOPROPYLIDENE ACETONE
504-24-5		
		4-AMINOPYRIDIN
		4-AMINOPYRIDINE
		4-AMINOPYRIDINE (DOT)
		4-AP
		4-PYRIDINAMINE
		4-PYRIDYLAMINE
		AMINO-4-PYRIDINE
		AVITROL
		GAMMA-AMINOPYRIDINE
		P-AMINOPYRIDINE
		P-AMINOPYRIDINE (DOT)
		PHILLIPS 1861
		PYRIDINE, 4-AMINO-
		RCRA WASTE NUMBER P008
		UN 2671 (DOT)
		VMI 10-3
504-29-0		
		1,2-DIHYDRO-2-IMINOPYRIDINE
	*	2-AMINOPYRIDINE
		2-PYRIDINAMINE
		2-PYRIDYLAMINE
		ALPHA-AMINOPYRIDINE
		ALPHA-PYRIDINAMINE
		ALPHA-PYRIDYLAMINE
		O-AMINOPYRIDINE
		PYRIDINE, 2-AMINO-
504-60-9		
		1,3-PENTADIENE
		1,3-PENTADIENE(CIS & TRANS-MIXED)
		1-METHYL-1,3-BUTADIENE
		1-METHYLBUTADIENE
		PIPERYLENE
		PIPERYLENE CONCENTRATE
		RCRA WASTE NUMBER U186
504-88-1		
		3-NITROPROPIONIC ACID
505-60-2		
		1,1'-THIOBIS(2-CHLOROETHANE)
		1-CHLORO-2-(BETA-CHLOROETHYLTHIO)ETHANE
		2,2'-DICHLORODIETHYL SULFIDE
		2,2'-DICHLOROETHYL SULFIDE
		BETA,BETA'-DICHLORODIETHYL SULFIDE
		BETA,BETA'-DICHLOROETHYL SULFIDE
		BETA,BETA-DICHLOR-ETHYL-SULPHIDE
		BIS(2-CHLOROETHYL)SULFIDE
		BIS(2-CHLOROETHYL)SULPHIDE
		BIS(BETA-CHLOROETHYL)SULFIDE
		DI-2-CHLOROETHYL SULFIDE
		DISTILLED MUSTARD
		ETHANE, 1,1'-THIOBIS[2-CHLORO-
		H
		HD
		IPRIT
		KAMPSTOFF LOST
		LOST
	*	MUSTARD GAS
		MUSTARD HD
		MUSTARD VAPOR
		MUSTARD, SULFUR
		S MUSTARD
		S-LOST
		S-YPERITE
		SCHWEFEL-LOST
		SENFGAS
		SULFIDE, BIS(2-CHLOROETHYL)
		SULFUR MUSTARD
		SULFUR MUSTARD GAS
		SULPHUR MUSTARD
		SULPHUR MUSTARD GAS
		YELLOW CROSS LIQUID
		YPERITE
506-61-6		
		ARGENTATE(1-), BIS(CYANO-C)-, POTASSIUM
		ARGENTATE(1-), DICYANO-, POTASSIUM
		KYANOSTRIBRNAN DRASELNY (CZECH)
		POTASSIUM ARGENTOCYANIDE (KAg(CN)2)
		POTASSIUM DICYANOARGENTATE
		POTASSIUM DICYANOARGENTATE(1-)
		POTASSIUM DICYANOARGENTATE(I)
		POTASSIUM SILVER CYANIDE
		POTASSIUM SILVER CYANIDE (KAg(CN)2)
		RCRA WASTE NUMBER P099
		SILVER POTASSIUM CYANIDE
		SILVER POTASSIUM CYANIDE (AgK(CN)2)
506-64-9		
		CYANURE D'ARGENT (FRENCH)
		KYANID STRIBRNY (CZECH)
		RCRA WASTE NUMBER P104
		SILVER CYANIDE
		SILVER CYANIDE (Ag(CN))
		SILVER CYANIDE (Ag2(CN)2)
		SILVER CYANIDE (DOT)
		SILVER(1+) CYANIDE
		UN 1684 (DOT)
506-68-3		
		BROMINE CYANIDE
		BROMINE CYANIDE (BrCN)
		BROMINE MONOCYANIDE
		BROMOCYAN
		BROMOCYANIDE
		BROMOCYANIDE (BRCN)
		BROMOCYANOGEN
		BROMURE DE CYANOGEN (FRENCH)
		CAMPILIT
		CYANOBROMIDE
		CYANOGEN BROMIDE

CAS No.	IN	Chemical Name
		CYANOGEN BROMIDE ((CN)Br)
		CYANOGEN BROMIDE (BrCN)
		CYANOGEN BROMIDE (DOT)
		CYANOGEN MONOBROMIDE
		RCRA WASTE NUMBER U246
		TL 822
		UN 1889 (DOT)
506-77-4		
		CHLORCYAN
		CHLORINE CYANIDE
		CHLORINE CYANIDE (ClCN)
		CHLOROCYAN
		CHLOROCYANIDE
		CHLOROCYANIDE (ClCN)
		CHLOROCYANOGEN
		CHLORURE DE CYANOGENE (FRENCH)
		CYANOCHLORIDE (CNCl)
		CYANOGEN CHLORIDE
		CYANOGEN CHLORIDE ((CN)Cl)
		CYANOGEN CHLORIDE (ACGIH)
		CYANOGEN CHLORIDE (ClCN)
		CYANOGEN CHLORIDE, CONTAINING LESS THAN 0.9% WATER (DOT)
		CYANOGEN CHLORIDE, INHIBITED (DOT)
		CYANOGEN CHLORIDE, WITH 0.9% WATER
		RCRA WASTE NUMBER P033
		UN 1589 (DOT)
506-78-5		
		CYANOGEN IODIDE
		CYANOGEN MONOIODIDE
		IODINE CYANIDE
		IODINE CYANIDE (I(CN))
		IODINE MONOCYANIDE
		IODINE MONOCYANIDE (ICN)
		IODOCYANIDE (ICN)
		JODCYAN
506-87-6		
	*	AMMONIUM CARBONATE
		AMMONIUM CARBONATE ((NH4)2CO3)
		AMMONIUM CARBONATE (DOT)
		AMMONIUMCARBONAT (GERMAN)
		BIS(AMMONIUM) CARBONATE
		CARBONIC ACID, AMMONIUM SALT
		CARBONIC ACID, DIAMMONIUM SALT
		CARBONIC ACID, DIAMMONIUM SALT (8CI,9CI)
		DIAMMONIUM CARBONATE
		DICARBONIC ACID, DIAMMONIUM SALT
		NA 9084 (DOT)
506-93-4		
	*	GUANIDINE NITRATE
		GUANIDINE NITRATE (1:1)
		GUANIDINE NITRATE (DOT)
		GUANIDINE, MONONITRATE
		GUANIDINIUM NITRATE
		UN 1467 (DOT)
506-96-7		
		ACETYL BROMIDE
		ACETYL BROMIDE (DOT)
		UN 1716 (DOT)
507-02-8		
		ACETYL IODIDE
		ACETYL IODIDE (DOT)
		UN 1898 (DOT)
507-09-5		
		ACETIC ACID, THIO-
		ACETYL MERCAPTAN
		ETHANETHIOIC ACID
		ETHANETHIOLIC ACID
		METHANECARBOTHIOLIC ACID
		MONOTHIOACETIC ACID
		THIACETIC ACID
	*	THIOACETIC ACID
		THIOACETIC ACID (DOT)
		THIOLACETIC ACID
		THIONOACETIC ACID
		UN 2436 (DOT)
		USAF EK-P-737
507-19-7		
		1,1-DIMETHYLETHYL BROMIDE
		2-BROMO-2-METHYLPROPANE
		2-BROMO-2-METHYLPROPANE (DOT)
		2-BROMOISOBUTANE
		2-METHYL-2-BROMOPROPANE
		BROMOMETHYLPROPANE
		BROMOTRIMETHYLMETHANE
		PROPANE, 2-BROMO-2-METHYL-
		TERT-BUTYL BROMIDE
		TRIMETHYLBROMOMETHANE
		UN 2342 (DOT)
507-20-0		
		1,1-DIMETHYLETHYL CHLORIDE
		2-CHLORO-2-METHYLPROPANE
		2-CHLOROISOBUTANE
		2-METHYL-2-CHLOROPROPANE
		2-METHYL-2-PROPYL CHLORIDE
		CHLOROTRIMETHYLMETHANE
		PROPANE, 2-CHLORO-2-METHYL-
	*	TERT-BUTYL CHLORIDE
		TRIMETHYLCHLOROMETHANE
507-70-0		
		BICYCLO[2.2.1]HEPTAN-2-OL, 1,7,7-TRIMETHYL-, ENDO-
	*	BORNEOL
		CAMPHOL
		ENDO-2-HYDROXY-1,7,7-TRIMETHYLNORBORNANE
		ENDO-BORNEOL
509-14-8		
		METHANE, TETRANITRO-
		NCI-C55947
		RCRA WASTE NUMBER P112
		TETRANITROMETHANE
		TETRANITROMETHANE (ACGIH,DOT)
		TNM
		UN 1510 (DOT)
510-15-6		
		4,4'-DICHLOROBENZILIC ACID ETHYL ESTER
		ACAR
		ACARABEN
		AKAR
		AKAR 338
		BENZENEACETIC ACID, 4-CHLORO-ALPHA-(4-CHLOROPHENYL)-ALPHA-HYDROXY-, ETHYL ESTER
		BENZILAN
		BENZILIC ACID, 4,4'-DICHLORO-, ETHYL ESTER
		CHLORBENZILAT
		CHLORBENZYLATE
		CHLOROBENZILATE
		ECB
	*	ETHYL 4,4'-DICHLOROBENZILATE
		FOLBEX
		G 23992
		G 338
		GEIGY 338
512-56-1		
		TRIMETHYLPHOSPHATE
513-35-9		
		1,1,2-TRIMETHYLETHYLENE
		2-BUTENE, 2-METHYL-
		2-METHYL-2-BUTENE
		2-METHYL-2-BUTENE (DOT)
		3-METHYL-2-BUTENE
		AMYLENE
		AMYLENE (DOT)
	*	AMYLENE, NORMAL
		AMYLENE, NORMAL (DOT)

CAS No.	IN	Chemical Name
		BETA-ISOAMYLENE
		ETHYLENE, TRIMETHYL-
		METHYL BUTENE (DOT)
		N-AMYLENE (DOT)
		N-PENTENE
		PENTENE
		PENTYLENE
		TRIMETHYLETHYLENE
		UN 1108 (DOT)
		UN 2460 (DOT)
513-36-0		
		1-CHLORO-2-METHYLPROPANE
		2-METHYL-1-CHLOROPROPANE
		2-METHYLPROPYL CHLORIDE
		ISOBUTYL CHLORIDE
		PROPANE, 1-CHLORO-2-METHYL-
513-38-2		
		1-IODO-2-METHYLPROPANE
		IODO METHYLPROPANE (1-IODO-2-METHYL PROPANE)
		IODOMETHYLPROPANE
		ISOBUTYL IODIDE
		PRIMARY ISOBUTYL IODIDE
		PROPANE, 1-IODO-2-METHYL-
513-42-8		
		2-METHALLYL ALCOHOL
		2-METHYL-2-PROPEN-1-OL
		2-METHYL-2-PROPENOL
		2-METHYLALLYL ALCOHOL
		2-METHYLPROP-1-EN-3-OL
		2-PROPEN-1-OL, 2-METHYL-
		3-HYDROXY-2-METHYLPROPENE
		BETA-METHALLYL ALCOHOL
		BETA-METHYLALLYL ALCOHOL
		ISOPROPENYL CARBINOL
		METHACRYL ALCOHOL
		METHALLYL ALCOHOL
		METHALLYL ALCOHOL (DOT)
		UN 2614 (DOT)
513-48-4		
	*	2-IODOBUTANE
513-49-5		
		(+)-2-BUTYLAMINE
		(+)-SEC-BUTYLAMINE
		(S)-2-BUTANAMINE
		2-AB
		2-AMINOBUTANE
		2-AMINOBUTANE BASE
		2-BUTANAMINE
		2-BUTANAMINE, (S)-
		BUTAFUME
		FRUCOTE
		PROPYLAMINE, 1-METHYL
		S-SEC-BUTYLAMINE
		SEC-BUTYLAMINE
		SEC-BUTYLAMINE, (S)-
		TUTANE
513-53-1		
		1-METHYL-1-PROPANETHIOL
		2-BUTANETHIOL
		2-BUTYL MERCAPTAN
		2-MERCAPTOBUTANE
		SEC-BUTANETHIOL
		SEC-BUTYL MERCAPTAN
		SEC-BUTYL THIOALCOHOL
		SEC-BUTYL THIOL
513-77-9		
		BARIUM CARBONATE
		BARIUM CARBONATE (1:1)
		BARIUM CARBONATE (BaCO3)
		BARIUM MONOCARBONATE
		CARBONIC ACID, BARIUM SALT (1:1)
		CI 77099

CAS No.	IN	Chemical Name
		CI PIGMENT WHITE 10
513-86-0		
		1-HYDROXYETHYL METHYL KETONE
		2,3-BUTANOLONE
		2-BUTANOL-3-ONE
		2-BUTANONE, 3-HYDROXY-
		2-HYDROXY-3-BUTANONE
		3-HYDROXY-2-BUTANONE
		ACETOIN
		ACETYL METHYL CARBINOL
		ACETYL METHYLCARBINOL (DOT)
		DIMETHYLKETOL
		GAMMA-HYDROXY-BETA-OXOBUTANE
		METHANOL, ACETYLMETHYL-
		UN 2621 (DOT)
514-73-8		
		3,3'-DIETHYL-2,2'-THIADICARBOCYANINE IODIDE
		3,3'-DIETHYLDITHIACARBODICYANINE IODIDE
		3,3'-DIETHYLPENTAMETHINETHIACYANINE IODIDE
		3,3'-DIETHYLTHIADICARBOCYANINE IODIDE
		3-ETHYL-2-(5-(3-ETHYL-2-BENZOTHIAZOLINYLIDENE)-1,3-PENTADIENYL)BENZOTHIAZOLIUM IODIDE
		[2-BIS(3-ETHYLBENZOTHIAZOLYL)] PENTAMETHINE CYANINE IODIDE
		ABMINTHIC
		ANELMID
		ANGUIFUGAN
		BENZOTHIAZOLIUM, 3-ETHYL-2-(5-(3-ETHYL-2-BENZOTHIAZOLINYLIDENE)-1,3-PENTADIENYL)-, IODIDE
		BENZOTHIAZOLIUM, 3-ETHYL-2-[5-(3-ETHYL-2(3H)-BENZOTHIAZOLYLIDENE)-1,3-PENTADIENYL]-, IODIDE
		COMPOUND 01748
		DEJO
		DELVEX
		DIETHYLTHIADICARBOCYANINE IODIDE
		DILOMBRIN
		DILOMBRINE
		DITHIAZANIN IODIDE
		DITHIAZANINE IODIDE
		DITHIAZINE
		DITHIAZINE (DYE)
		DITHIAZININE
		DTDC
		EASTMAN 7663
		L-01748
		NETOCYD
		NK 136
		OMNI-PASSIN
		PARTEL
		TELMICID
		TELMID
		TELMIDE
		VERCIDON
515-98-0		
		AMMONIUM LACTATE
		LACTIC ACID, MONOAMMONIUM SALT
		PROPANOIC ACID, 2-HYDROXY-, MONOAMMONIUM SALT
516-03-0		
		ETHANEDIOIC ACID, IRON(2+) SALT (1:1)
		FERROUS OXALATE
		FERROUS OXALATE (1:1)
		FERROX
		IRON(2+) OXALATE
		IRON PROTOXALATE
		IRON(II) OXALATE
		OXALIC ACID, IRON(2+) SALT (1:1)
517-16-8		
		(ETHYLMERCURIC)-P-TOLUENESULPHONANILIDE
		N-(ETHYLMERCURIC)-P-TOLUENESULPHONANILIDE
519-44-8		
		1,3-BENZENEDIOL, 2,4-DINITRO-
		1,3-BENZENEDIOL, DINITRO-
		2,4-DINITRORESORCINOL

CAS No.	IN	Chemical Name
		3-HYDROXY-2,4-DINITROPHENOL
		DINITRORESORCINOL
		DINITRORESORCINOL, WETTED WITH, BY WEIGHT, AT LEAST 15% WATER (DOT)
		RESORCINOL, 2,4-DINITRO-
		RESORCINOL, DINITRO-, WETTED WITH AT LEAST 15% WATER
		UN 1322 (DOT)
528-29-0		
		1,2-DINITROBENZENE
		1,2-DINITROBENZOL
		BENZENE, 1,2-DINITRO-
		BENZENE, O-DINITRO-
		DINITROBENZENE
	*	O-DINITROBENZENE
		O-DINITROBENZENE (ACGIH,DOT)
		UN 1597 (DOT)
531-76-0		
		ALANINE, 3-[P-[BIS(2-CHLOROETHYL)AMINO]PHENYL]-, DL-
		DL-PHENYLALANINE MUSTARD
		DL-PHENYLALANINE, 4-[BIS(2-CHLOROETHYL)AMINO]-
		DL-SARCOLYSIN
		DL-SARCOLYSINE
		MERPHALAN
531-82-8		
		ACETAMIDE, N-[4-(5-NITRO-2-FURANYL)-2-THIAZOLYL]-
		ACETAMIDE, N-[4-(5-NITRO-2-FURYL)-2-THIAZOLYL]-
		FURATHIAZOLE
		FURIUM
		FUROTHIAZOLE
		N-(4-(5-NITRO-2-FURYL)2-THIAZOLYL)ACETAMIDE
		NFTA
		THIAZOLE, 2-ACETAMIDO-4-(5-NITRO-2-FURYL)-
532-27-4		
		1-CHLOROACETOPHENONE
		2-CHLORO-1-PHENYLETHANONE
		2-CHLOROACETOPHENONE
		ACETOPHENONE, 2-CHLORO-
	*	ALPHA-CHLOROACETOPHENONE
		ALPHA-CHLOROACETOPHENONE (ACGIH)
		CAF
		CAP
		CHEMICAL MACE
		CHLORACETOPHENONE
		CHLOROACETOPHENONE
		CHLOROACETOPHENONE (DOT)
		CHLOROACETOPHENONE, GAS, LIQUID, OR SOLID (DOT)
		CHLOROMETHYL PHENYL KETONE
		CN
		ETHANONE, 2-CHLORO-1-PHENYL-
		MACE
		MACE (LACRIMATOR)
		NCI-C55107
		OMEGA-CHLOROACETOPHENONE
		PHENACYL CHLORIDE
		PHENYLCHLOROMETHYLKETONE
		UN 1697 (DOT)
534-07-6		
		1,3-DICHLORO-2-PROPANONE
		1,3-DICHLOROACETONE
		1,3-DICHLOROACETONE (DOT)
		2-PROPANONE, 1,3-DICHLORO-
		ALPHA,ALPHA'-DICHLOROACETONE
		ALPHA,GAMMA-DICHLOROACETONE
		BIS(CHLOROMETHYL)KETONE
		DICHLOROPROPANE
		SYM-DICHLOROACETONE
		UN 2649 (DOT)
534-15-6		
	*	1,1-DIMETHOXYETHANE
		DIMETHYLACETAL
534-16-7		
		CARBONIC ACID, DISILVER(1+) SALT
		DISILVER CARBONATE
		FETIZON'S REAGENT
		NAMED REAGENTS AND SOLUTIONS, FETIZON'S
		SILVER CARBONATE
		SILVER CARBONATE (2:1)
		SILVER(I) CARBONATE
534-22-5		
		2-METHYLFURAN
		5-METHYLFURAN
		ALPHA-METHYLFURAN
		FURAN, 2-METHYL-
		SILVAN
		SYLVAN
534-52-1		
		2,4-DINITRO-6-METHYLPHENOL
		2-METHYL-4,6-DINITROPHENOL
		3,5-DINITRO-2-HYDROXYTOLUENE
		4,6-DINITRO-2-METHYLPHENOL
	*	4,6-DINITRO-O-CRESOL
		6-METHYL-2,4-DINITROPHENOL
		ANTINONIN
		ANTINONNIN
		ARBOROL
		DEGRASSAN
		DEKRYSIL
		DETAL
		DILLEX
		DINITRO
		DINITRO-O-CRESOL
		DINITROCRESOL
		DINITRODENDTROXAL
		DINITROL
		DINOC
		DINURANIA
		DITROSOL
		DNOC
		EFFUSAN
		EFFUSAN 3436
		ELGETOL
		ELGETOL 30
		ELIPOL
		EXTRAR
		FLAVIN-SANDOZ
		HEDOLIT
		HEDOLITE
		K III
		K IV
		KREOZAN
		KREZOTOL 50
		LIPAN
		NEUDORFF DN 50
		NITROFAN
		O-CRESOL, 4,6-DINITRO-
		PHENOL, 2-METHYL-4,6-DINITRO-
		PROKARBOL
		RAFEX
		RAFEX 35
		RAPHATOX
		SANDOLIN
		SANDOLIN A
		SELINON
		SINOX
		WINTERWASH
535-13-7		
		ETHYL ALPHA-CHLOROPROPIONATE
		ETHYL CHLOROPROPIONATE
		ETHYL-2-CHLOROPROPIONATE
		ETHYL-2-CHLOROPROPIONATE (DOT)
		PROPANOIC ACID, 2-CHLORO-, ETHYL ESTER
		PROPANOIC ACID, 2-CHLORO-, ETHYL ESTER (9CI)
		PROPIONIC ACID, 2-CHLORO-, ETHYL ESTER
		UN 2935 (DOT)

CAS No.	IN	Chemical Name
535-89-7		
		2-CHLOOR-4-DIMETHYLAMINO-6-METHYL-PYRIMIDINE (DUTCH)
		2-CHLOR-4-DIMETHYLAMINO-6-METHYLPYRIMIDIN (GERMAN)
		2-CHLORO-4-(DIMETHYLAMINO)-6-METHYLPYRIMIDINE
		2-CHLORO-4-METHYL-6-(DIMETHYLAMINO)PYRIMIDINE
		2-CHLORO-N,N-6-TRIMETHYL-4-PYRIMIDINAMINE
		2-CLORO-4-DIMETILAMINO-6-METIL-PIRIMIDINA (ITALIAN)
		4-PYRIMIDINAMINE, 2-CHLORO-N,N,6-TRIMETHYL-
		CASTRIX
		CRIMIDIN (GERMAN)
		CRIMIDINA (ITALIAN)
		CRIMIDINE
		CRIMITOX
		PYRIMIDINE, 2-CHLORO-4-(DIMETHYLAMINO)-6-METHYL-
		W 491
536-33-4		
	*	ETHIONAMIDE
538-07-8		
		2,2'-DICHLOROTRIETHYLAMINE
		BIS(2-CHLOROETHYL)ETHYLAMINE
		ETHANAMINE, 2-CHLORO-N-(2-CHLOROETHYL)-N-ETHYL-
		ETHYL-S
		ETHYLBIS(2-CHLOROETHYL)AMINE
		ETHYLBIS(BETA-CHLOROETHYL)AMINE
		H N1
		N-ETHYLBIS(2-CHLOROETHYL)AMINE
		TL 1149
		TL 329
		TRIETHYLAMINE, 2,2'-DICHLORO-
538-68-1		
		1-PHENYL-N-PENTANE
		1-PHENYLPENTANE
		AMYLBENZENE
		BENZENE, PENTYL-
		N-AMYLBENZENE
		N-PENTYLBENZENE
		PENTANE, 1-PHENYL-
		PENTYLBENZENE
538-93-2		
		(2-METHYLPROPYL)BENZENE
		2-METHYL-1-PHENYLPROPANE
		BENZENE, (2-METHYLPROPYL)-
		BENZENE, ISOBUTYL-
		ISOBUTYLBENZENE
540-18-1		
		1-PENTYL BUTYRATE
		AMYL BUTYRATE
		BUTANOIC ACID PENTYL ESTER
		BUTYRIC ACID, PENTYL ESTER
		N-AMYL BUTYRATE
		N-AMYL BUTYRATE (DOT)
		N-PENTYL N-BUTYRATE
		PENTYL BUTANOATE
		PENTYL BUTYRATE
		UN 2620 (DOT)
540-42-1		
		2-METHYLPROPYL PROPANOATE
		2-METHYLPROPYL PROPIONATE
		ISOBUTYL PROPANOATE
		ISOBUTYL PROPIONATE
		ISOBUTYL PROPIONATE (DOT)
		PROPANOIC ACID, 2-METHYLPROPYL ESTER
		PROPANOIC ACID, 2-METHYLPROPYL ESTER (9CI)
		PROPIONIC ACID, ISOBUTYL ESTER
		UN 2394 (DOT)
540-54-5		
		1-CHLOROPROPANE
		N-PROPYL CHLORIDE
		PROPANE, 1-CHLORO-
		PROPYL CHLORIDE

CAS No.	IN	Chemical Name
		PROPYL CHLORIDE (DOT)
		UN 1278 (DOT)
540-59-0		
		1,2-DICHLOR-AETHEN (GERMAN)
		1,2-DICHLOROETHENE
	*	1,2-DICHLOROETHYLENE
		1,2-DICHLOROETHYLENE (ACGIH)
		ACETYLENE DICHLORIDE
		CIS-DICHLOROETHYLENE
		DICHLORO-1,2-ETHYLENE (FRENCH)
		DIOFORM
		ETHENE, 1,2-DICHLORO-
		ETHYLENE, 1,2-DICHLORO-
		NCI-C56031
		SYM-DICHLOROETHYLENE
540-67-0		
		ETHANE, METHOXY-
		ETHER, ETHYL METHYL
		ETHYL METHYL ETHER
		ETHYL METHYL ETHER (DOT)
		METHANE, ETHOXY-
		METHOXYETHANE
	*	METHYL ETHYL ETHER
		METHYL ETHYL ETHER (DOT)
		UN 1039 (DOT)
540-69-2		
		AMMONIUM FORMATE
		FORMIC ACID AMMONIUM SALT
540-72-7		
		HAIMASED
		NATRIUMRHODANID (GERMAN)
		SCYAN
		SODIUM ISOTHIOCYANATE
		SODIUM RHODANATE
		SODIUM RHODANIDE
		SODIUM SULFOCYANATE
		SODIUM SULFOCYANIDE
		SODIUM THIOCYANATE
		SODIUM THIOCYANIDE
		THIOCYANATE SODIUM
		THIOCYANIC ACID, SODIUM SALT
		USAF EK-T-434
540-73-8		
		1,2-DIMETHYLHYDRAZIN (GERMAN)
		1,2-DIMETHYLHYDRAZINE
		DIMETHYLHYDRAZINE, SYMMETRICAL
		DIMETHYLHYDRAZINE, SYMMETRICAL (DOT)
		DMH
		HYDRAZINE, 1,2-DIMETHYL-
		HYDRAZOMETHANE
		N,N'-DIMETHYLHYDRAZINE
		RCRA WASTE NUMBER U099
		SDMH
		SYM-DIMETHYLHYDRAZINE
		SYMETRYCZNA DWUMETYLOHYDRAZYNA (POLISH)
		UN 2382 (DOT)
540-82-9		
		DES
		DIETHYL SULFATE
		DIETHYL SULPHATE
		ETHYL HYDROGEN SULFATE
		ETHYL SULFATE
		ETHYLSULFURIC ACID
		ETHYLSULPHURIC ACID (DOT)
		MONOETHYL SULFATE
		SULFETHYLIC ACID
		SULFOVINIC ACID
		SULFURIC ACID, MONOETHYL ESTER
		UN 2571 (DOT)
540-84-1		
		2,2,4-TRIMETHYLPENTANE
		ISOBUTYLTRIMETHYLMETHANE

CAS No.	IN	Chemical Name
		ISOCTANE
	*	ISOOCTANE
		PENTANE, 2,2,4-TRIMETHYL-
540-88-5		
		1,1-DIMETHYLETHYL ACETATE
		ACETIC ACID, 1,1-DIMETHYLETHYL ESTER
		ACETIC ACID, 1,1-DIMETHYLETHYL ESTER (9CI)
		ACETIC ACID, TERT-BUTYL ESTER
		T-BUTYL ACETATE
		TERT-BUTYL ACETATE
		TERT-BUTYL ACETATE (ACGIH,DOT)
		TEXACO LEAD APPRECIATOR
		TLA
		UN 1123 (DOT)
541-09-3		
		BIS(ACETATO)DIOXOURANIUM
		URANIUM DIACETATE DIOXIDE
		URANIUM OXYACETATE
		URANIUM, BIS(ACETATO)DIOXO-
		URANIUM, BIS(ACETATO-O)DIOXO-
	*	URANYL ACETATE
		URANYL ACETATE (UO2(OAC)2)
		URANYL DIACETATE
		URANYL(2+) ACETATE
541-25-3		
		2-CHLOROVINYLDICHLOROARSINE
		ARSINE, (2-CHLOROVINYL)DICHLORO-
		ARSINE, DICHLORO(2-CHLOROVINYL)-
		ARSONOUS DICHLORIDE, (2-CHLOROETHENYL)-
		ARSONOUS DICHLORIDE, (2-CHLOROETHENYL)- (9CI)
		BETA-CHLOROVINYLBICHLOROARSINE
		CHLOROVINYLARSINE DICHLORIDE
		DICHLORO(2-CHLOROVINYL)ARSINE
		LEWISITE
		LEWISITE (ARSENIC COMPOUND)
541-41-3		
		CARBONOCHLORIDIC ACID, ETHYL ESTER
		CATHYL CHLORIDE
		CHLORAMEISENSAEUREAETHYLESTER (GERMAN)
		CHLOROCARBONATE D'ETHYLE (FRENCH)
		CHLOROCARBONIC ACID ETHYL ESTER
		CHLOROFORMIC ACID ETHYL ESTER
		ECF
		ETHOXYCARBONYL CHLORIDE
		ETHYL CARBONOCHLORIDATE
		ETHYL CHLOROCARBONATE
		ETHYL CHLOROCARBONATE (DOT)
		ETHYL CHLOROFORMATE
		ETHYL CHLOROFORMATE (DOT)
		ETHYLCHLOORFORMIAAT (DUTCH)
		ETHYLE, CHLOROFORMIAT D' (FRENCH)
		ETIL CLOROCARBONATO (ITALIAN)
		ETIL CLOROFORMIATO (ITALIAN)
		FORMIC ACID, CHLORO-, ETHYL ESTER
		TL 423
		UN 1182 (DOT)
541-53-7		
		2,4-DITHIOBIURET
		ALLOPHANIMIDIC ACID, DITHIO-
		BIURET, 2,4-DITHIO-
		BIURET, DITHIO-
		DITHIOBIURET
		DTB
		IMIDODICARBONIMIDOTHIOIC DIAMIDE
		IMIDODICARBONODITHIOIC DIAMIDE
		RCRA WASTE NUMBER P049
		THIOIMIDODICARBONIC DIAMIDE ([(H2N)C(S)]2NH)
		UREA, 2-THIO-1-(THIOCARBAMOYL)-
		USAF B-44
		USAF EK-P-6281
541-73-1		
	*	1,3-DICHLOROBENZENE
		BENZENE, 1,3-DICHLORO-
		BENZENE, 1,3-DICHLORO- (9CI)
		BENZENE, M-DICHLORO-
		M-DICHLOROBENZENE
		M-DICHLOROBENZOL
		M-PHENYLENE DICHLORIDE
		METADICHLOROBENZENE
		RCRA WASTE NUMBER U071
541-85-5		
		3-HEPTANONE, 5-METHYL-
		3-METHYL-5-HEPTANONE
		5-METHYL-3-HEPTANONE
		ETHYL 2-METHYLBUTYL KETONE
		ETHYL AMYL KETONE
		ETHYL SEC-AMYL KETONE
542-18-7		
		CHLOROCYCLOHEXANE
		CYCLOHEXANE, CHLORO-
	*	CYCLOHEXYL CHLORIDE
		MONOCHLOROCYCLOHEXANE
542-53-5		
		2-CHLOROETHYL ACETATE
542-55-2		
		2-METHYLPROPYL FORMATE
		FORMIC ACID, 2-METHYLPROPYL ESTER
		FORMIC ACID, ISOBUTYL ESTER
	*	ISOBUTYL FORMATE
		ISOBUTYL FORMATE (DOT)
		TETRYL FORMATE
		UN 2393 (DOT)
542-58-5		
		1-ACETOXY-2-CHLOROETHANE
		2-ACETOXY-1-CHLOROETHANE
		2-ACETOXYETHYL CHLORIDE
		2-CHLOROETHYL ACETATE
		BETA-CHLOROETHYL ACETATE
		ETHANOL, 2-CHLORO-, ACETATE
542-62-1		
		BARIUM CYANIDE
		BARIUM CYANIDE (Ba(CN)2)
		BARIUM DICYANIDE
542-75-6		
		1,3-D
		1,3-DICHLORO-1-PROPENE
		1,3-DICHLORO-2-PROPENE
	*	1,3-DICHLOROPROPENE
		1,3-DICHLOROPROPYLENE
		1-PROPENE, 1,3-DICHLORO-
		3-CHLOROALLYL CHLORIDE
		3-CHLOROPROPENYL CHLORIDE
		ALPHA,GAMMA-DICHLOROPROPYLENE
		DICHLOROPROPENE
		GAMMA-CHLOROALLYL CHLORIDE
		PROPENE, 1,3-DICHLORO-
		TELONE
		TELONE II
542-76-7		
		1-CHLORO-2-CYANOETHANE
		3-CHLOROPROPANENITRILE
		3-CHLOROPROPANONITRILE
	*	3-CHLOROPROPIONITRILE
		BETA-CHLOROPROPIONITRILE
		PROPIONITRILE 3-CHLORO
		RCRA WASTE NUMBER P027
		USAF A-8798
542-88-1		
		ALPHA,ALPHA'-DICHLORODIMETHYL ETHER
		BIS(2-CHLOROMETHYL)ETHER
		BIS(CHLOROMETHYL)ETHER
		CHLOROMETHYL ETHER
		DICHLORODIMETHYL ETHER

CAS No.	IN	Chemical Name
		DICHLOROMETHYL ETHER
		ETHER, BIS(CHLOROMETHYL)
		METHANE, OXYBIS[CHLORO-
		MONOCHLOROMETHYL ETHER
		OXYBIS[CHLOROMETHANE]
		SYM-DICHLOROMETHYL ETHER
542-90-5		
		AETHYLRHODANID (GERMAN)
		ETHANE, THIOCYANATO-
		ETHYL RHODANATE
		ETHYL SULFOCYANATE
		ETHYL THIOCYANATE
		THIOCYANIC ACID, ETHYL ESTER
542-92-7		
		1,3-CYCLOPENTADIENE
		CYCLOPENTADIENE
		PENTOLE
		PYROPENTYLENE
		R-PENTINE
543-59-9		
		1-CHLOROPENTANE
		AMYL CHLORIDE
		AMYL CHLORIDE (DOT)
		N-AMYL CHLORIDE
		N-PENTYL CHLORIDE
		PENTANE, 1-CHLORO-
		PENTYL CHLORIDE
		UN 1107 (DOT)
543-90-8		
		ACETIC ACID, CADMIUM SALT
		BIS(ACETOXY)CADMIUM
		C.I. 77185
	*	CADMIUM ACETATE
		CADMIUM ACETATE (DOT)
		CADMIUM DIACETATE
		CADMIUM(II) ACETATE
		NA 2570 (DOT)
544-10-5		
		1-CHLOROHEXANE
		1-HEXYL CHLORIDE
		CHLOROHEXANE
		HEXANE, 1-CHLORO-
		HEXYL CHLORIDE
		N-HEXYL CHLORIDE
544-16-1		
		BUTYL NITRITE
544-18-3		
		COBALT DIFORMATE
		COBALT FORMATE
		COBALT FORMATE (Co(O2CH)2)
		COBALT(2+) FORMATE
		COBALTOUS FORMATE
		COBALTOUS FORMATE (DOT)
		FORMIC ACID, COBALT(2+) SALT
		NA 9104 (DOT)
544-19-4		
		COPPER DIFORMATE
		COPPER FORMATE
		COPPER(2+) FORMATE
		COPPER(II) FORMATE
		CUPRIC DIFORMATE
		CUPRIC FORMATE
		FORMIC ACID, COPPER(2+) SALT
		FORMIC ACID, COPPER(2+) SALT (1:1)
		TUBERCUPROSE
544-25-2		
		CYCLOHEPTATRIENE

CAS No.	IN	Chemical Name
544-60-5		
		9-OCTADECENOIC ACID (Z)-, AMMONIUM SALT
		AMMONIA SOAP
		AMMONIUM OLEATE
		OLEIC ACID, AMMONIUM SALT
544-92-3		
		COPPER CYANIDE
		COPPER CYANIDE (Cu(CN))
		COPPER CYANIDE (DOT)
		COPPER(I) CYANIDE
		CUPRICIN
		CUPROUS CYANIDE
		RCRA WASTE NUMBER P029
544-97-8		
		DIMETHYLZINC
		DIMETHYLZINC (DOT)
		UN 1370 (DOT)
		ZINC, DIMETHYL-
545-55-1		
		1,1',1"-PHOSPHINYLIDYNETRISAZIRIDINE
		1-AZIRIDINYL PHOSPHINE OXIDE (TRIS)
		1-AZIRIDINYL PHOSPHINE OXIDE (TRIS) (DOT)
		A 6366
		APHOXIDE
		APO
		AZIRIDINE, 1,1',1"-PHOSPHINYLIDYNETRIS-
		CBC 906288
		ENT 24915
		IMPERON FIXER T
		N,N',N"-TRI-1,2-ETHANEDIYLPHOSPHORIC TRIAMIDE
		N,N',N"-TRIETHYLENEPHOSPHORAMIDE
		N,N',N"-TRIETHYLENEPHOSPHORIC TRIAMIDE
		NSC 9717
		PHOSPHINE OXIDE, TRIS(1-AZIRIDINYL)-
		PHOSPHORAMIDE, N,N',N"-TRIETHYLENE-
		PHOSPHORIC ACID TRIETHYLENE IMIDE
		PHOSPHORIC ACID TRIETHYLENEIMINE
		PHOSPHORIC ACID TRIETHYLENEIMINE (DOT)
		PHOSPHORIC TRIAMIDE, N,N',N"-TRI-1,2-ETHANEDIYL-
		PHOSPHORIC TRIAMIDE, N,N',N"-TRIETHYLENE-
		SK-3818
		TAPO
		TEF
		TEPA
		TRI(1-AZIRIDINYL)PHOSPHINE OXIDE
		TRI(AZIRIDINYL)PHOSPHINE OXIDE
		TRI-1-AZIRIDINYLPHOSPHINE OXIDE
		TRIAETHYLENPHOSPHORSAEUREAMID (GERMAN)
		TRIAZIRIDINOPHOSPHINE OXIDE
		TRIAZIRIDINYLPHOSPHINE OXIDE
		TRIETHYLENEPHOSPHORAMIDE
		TRIETHYLENEPHOSPHORIC TRIAMIDE
		TRIETHYLENEPHOSPHOROTRIAMIDE
		TRIS(1-AZIRIDINE)PHOSPHINE OXIDE
		TRIS(1-AZIRIDINYL)PHOSPHINE OXIDE
		TRIS(AZIRIDINYL)PHOSPHINE OXIDE
		TRIS(N-ETHYLENE)PHOSPHOROTRIAMIDATE
		TRIS-(1-AZIRIDINYL)PHOSPHINE OXIDE (DOT)
		TRIS-(1-AZIRIDINYL)PHOSPHINE OXIDE, SOLUTION (DOT)
		UN 2501 (DOT)
546-67-8		
		ACETIC ACID, LEAD(4+) SALT
		LEAD ACETATE (Pb(O2C2H3)4)
		LEAD ACETATE [Pb(OAC)4]
		LEAD TETRAACETATE
		LEAD TETRACETATE
		LEAD(IV) ACETATE
		PLUMBIC ACETATE
551-16-6		
		6-AMINO-PENICILLANIC ACID
552-30-7		
		1,2,4-BENZENETRICARBOXYLIC ACID, ANHYDRIDE
		1,2,4-BENZENETRICARBOXYLIC ACID, CYCLIC 1,2-ANHY-

CAS No.	IN	Chemical Name
		DRIDE
		1,2,4-BENZENETRICARBOXYLIC ANHYDRIDE
		1,3-DIOXO-5-PHTHALANCARBOXYLIC ACID
		4-CARBOXYPHTHALIC ANHYDRIDE
		5-ISOBENZOFURANCARBOXYLIC ACID, 1,3-DIHYDRO-1,3-DIOXO-
		ANHYDROTRIMELLITIC ACID
		TRIMELLITIC ACID 1,2-ANHYDRIDE
		TRIMELLITIC ACID ANHYDRIDE
		TRIMELLITIC ACID CYCLIC 1,2-ANHYDRIDE
		TRIMELLITIC ANHYDRIDE
554-12-1		
		METHYL PROPANOATE
	*	METHYL PROPIONATE
		METHYL PROPIONATE (DOT)
		METHYL PROPYLATE
		PROPANOIC ACID, METHYL ESTER
		PROPIONATE DE METHYLE (FRENCH)
		PROPIONIC ACID, METHYL ESTER
		UN 1248 (DOT)
554-13-2		
	*	LITHIUM CARBONATE
554-84-7		
		1-HYDROXY-3-NITROBENZENE
		3-HYDROXYNITROBENZENE
	*	3-NITROPHENOL
		M-HYDROXYNITROBENZENE
		M-NITROPHENOL
		M-NITROPHENOL (DOT)
		PHENOL, 3-NITRO-
		PHENOL, M-NITRO-
		UN 1663 (DOT)
555-30-6		
		ALDOMET
		ALPHA-METHYLDOPA
555-54-4		
		DIPHENYLMAGNESIUM
		MAGNESIUM DIPHENYL
		MAGNESIUM DIPHENYL (DOT)
		PHENYLMAGNESIUM
		UN 2005 (DOT)
555-77-1		
		2,2’,2”-TRICHLOROTRIETHYLAMINE
		ETHANAMINE, 2-CHLORO-N,N-BIS(2-CHLOROETHYL)-
		HN3
		TL 145
		TRI-2-CHLOROETHYLAMINE
		TRICHLORMETHINE
		TRIETHYLAMINE, 2,2’,2”-TRICHLORO-
		TRIS(2-CHLOROETHYL)AMINE
		TRIS(BETA-CHLOROETHYL)AMINE
		TS 160
555-84-0		
		1-(5-NITROFURFURYLIDENE)AMINO)-2-IMIDAZOLIDINONE
		1-(5-NITROFURFURYLIDENEAMINO)IMIDAZOLIDIN-2-ONE
		2-IMIDAZOLIDINONE, 1-[(5-NITROFURFURYLIDENE)AMINO]-
		2-IMIDAZOLIDINONE, 1-[[(5-NITRO-2-FURANYL)METHYLENE]AMINO]-
		M 254
		NF 246
		NIFURADEN
		NIFURADENE
		OXAFURADENE
556-24-1		
		BUTANOIC ACID, 3-METHYL-, METHYL ESTER
		BUTANOIC ACID, 3-METHYL-, METHYL ESTER (9CI)
		ISOVALERIC ACID, METHYL ESTER
		METHYL 3-METHYLBUTANOATE
		METHYL 3-METHYLBUTYRATE
		METHYL ISOPENTANOATE
		METHYL ISOVALERATE
		METHYLISOVALERATE (DOT)
		UN 2400 (DOT)
556-52-5		
		1,2-EPOXY-3-HYDROXYPROPANE
		1-HYDROXY-2,3-EPOXYPROPANE
		1-PROPANOL, 2,3-EPOXY-
		2,3-EPOXY-1-PROPANOL
		2-(HYDROXYMETHYL)OXIRANE
		3-HYDROXY-1,2-EPOXYPROPANE
		3-HYDROXYPROPYLENE OXIDE
		ALLYL ALCOHOL OXIDE
		EPIHYDRIN ALCOHOL
		GLYCIDE
		GLYCIDOL
		GLYCIDYL ALCOHOL
		OXIRANEMETHANOL
		OXIRANYLMETHANOL
556-56-9		
		1-PROPENE, 3-IODO-
		1-PROPENE, 3-IODO- (9CI)
		3-IODO-1-PROPENE
		3-IODOPROPENE
		3-IODOPROPYLENE
		ALLYL IODIDE
		ALLYL IODIDE (DOT)
		PROPENE, 3-IODO-
		UN 1723 (DOT)
556-61-6		
		EP-161E
		ISOTHIOCYANATE DE METHYLE (FRENCH)
		ISOTHIOCYANATOMETHANE
		ISOTHIOCYANIC ACID, METHYL ESTER
		ISOTIOCIANATO DI METILE (ITALIAN)
		MENCS
		METHANE, ISOTHIOCYANATO-
		METHYL ISOTHIOCYANATE
		METHYL ISOTHIOCYANATE (DOT)
		METHYL MUSTARD
		METHYL MUSTARD OIL
		METHYL THIOISOCYANATE
		METHYL-ISOTHIOCYANAT (GERMAN)
		METHYLISOTHIOCYANAAT (DUTCH)
		METHYLSENFOEL (GERMAN)
		MIC
		MIT
		MITC
		MORTON EP-161E
		TRAPEX
		TRAPEXIDE
		TROPEX
		UN 2477 (DOT)
		VORLEX
		VORTEX
		WN 12
556-64-9		
		METHANE, THIOCYANATO-
		METHYL SULFOCYANATE
		METHYL THIOCYANATE
		METHYLRHODANID (GERMAN)
		THIOCYANIC ACID, METHYL ESTER
556-88-7		
		1-NITROGUANIDINE
		ALPHA-NITROGUANIDINE
		BETA-NITROGUANIDINE
		GUANIDINE, NITRO-
		N”-NITROGUANIDINE
		N1-NITROGUANIDINE
		NITROGUANIDINE
		PICRITE
		PICRITE (PROPELLANT)
		PICRITE, WET, WITH ≥20% WATER

CAS No.	IN	Chemical Name
557-05-1		DERMARONE
		DIBASIC ZINC STEARATE
		METALLAC
		METASAP 576
		OCTADECANOIC ACID, ZINC SALT
		STAVINOR ZN-E
		STEARIC ACID, ZINC SALT
		SYNPRO ABG
		SYNPRO STEARATE
		TALCULIN Z
		ZINC DISTEARATE
		ZINC OCTADECANOATE
		ZINC STEARATE
557-17-5		1-METHOXYPROPANE
		ALPHA-METHOXYPROPANE
		ETHER, METHYL PROPYL
		ETHYL ACETONE
		METHYL N-PROPYL ETHER
		METHYL PROPYL ETHER
		METHYL PROPYL ETHER (DOT)
		METOPRYL
		NEOTHYL
		PROPANE, 1-METHOXY-
		PROPANE, 1-METHOXY- (9CI)
		PROPYL METHYL ETHER
		UN 2612 (DOT)
557-18-6		DIETHYLMAGNESIUM
		DIETHYLMAGNESIUM (DOT)
		MAGNESIUM, DIETHYL-
		UN 1367 (DOT)
557-19-7		DICYANONICKEL
		NICKEL CYANIDE
		NICKEL CYANIDE (DOT)
		NICKEL CYANIDE (Ni(CN)2)
		NICKEL CYANIDE, SOLID
		NICKEL CYANIDE, SOLID (DOT)
		NICKEL DICYANIDE
		NICKEL(II) CYANIDE
		RCRA WASTE NUMBER P074
		UN 1653 (DOT)
557-20-0		DIETHYL ZINC
		DIETHYL ZINC (DOT)
		UN 1366 (DOT)
		UN 2845 (DOT)
		ZINC ETHIDE
		ZINC ETHYL (DOT)
		ZINC, DIETHYL-
557-21-1		CYANURE DE ZINC (FRENCH)
		RCRA WASTE NUMBER P121
		UN 1713 (DOT)
	*	ZINC CYANIDE
		ZINC CYANIDE (DOT)
		ZINC CYANIDE (Zn(CN)2)
		ZINC DICYANIDE
557-31-3		1-PROPENE, 3-ETHOXY-
		1-PROPENE, 3-ETHOXY- (9CI)
		3-ETHOXY-1-PROPENE
		ALLYL ETHYL ETHER
		ALLYL ETHYL ETHER (DOT)
		ETHER, ALLYL ETHYL
		ETHYL 2-PROPENYL ETHER
		ETHYL ALLYL ETHER
		UN 2335 (DOT)
557-34-6		ACETIC ACID, ZINC SALT
		ACETIC ACID, ZINC SALT (8CI,9CI)
		ACETIC ACID, ZINC(II) SALT
		DICARBOMETHOXYZINC
		NA 9153 (DOT)
	*	ZINC ACETATE
		ZINC ACETATE (DOT)
		ZINC DIACETATE
		ZINC(II) ACETATE
557-40-4		1-PROPENE, 3,3'-OXYBIS-
		ALLYL ETHER
		DIALLYL ETHER
557-41-5		FORMIC ACID, ZINC SALT
		NA 9159 (DOT)
		ZINC DIFORMATE
		ZINC FORMATE
		ZINC FORMATE (DOT)
557-98-2		1-PROPENE, 2-CHLORO-
		2-CHLORO-1-PROPENE
		2-CHLOROPROPENE
		2-CHLOROPROPENE (DOT)
		2-CHLOROPROPYLENE
		BETA-CHLOROPROPENE
		BETA-CHLOROPROPYLENE
		ISOPROPENYL CHLORIDE
		PROPENE, 2-CHLORO-
		UN 2456 (DOT)
558-13-4		CARBON BROMIDE
		CARBON BROMIDE (CBr4)
	*	CARBON TETRABROMIDE
		CARBON TETRABROMIDE (ACGIH,DOT)
		CBr4
		METHANE TETRABROMIDE
		METHANE, TETRABROMO-
		TETRABROMOMETHANE
		UN 2516 (DOT)
558-17-8		IODO METHYLPROPANE (2-IODO-2-METHYL PROPANE)
558-25-8		FLUOROMETHYL SULFONE
		FUMETTE
		MESYL FLUORIDE
		METHANESULFONIC FLUORIDE
		METHANESULFONYL FLUORIDE
		METHANESULPHONYL FLUORIDE
		MSF
560-21-4		2,3,3-TRIMETHYLPENTANE
		PENTANE, 2,3,3-TRIMETHYL-
563-12-2		AC 3422
		BIS(S-(DIETHOXYPHOSPHINOTHIOYL)MERCAPTO)METHANE
		BLADAN
		DIETHION
		EMBATHION
		ENT 24,105
		ETHANOX
		ETHIOL
		ETHIOL 100
		ETHION
		ETHION (ACGIH,DOT)
		ETHODAN
		ETHOPAZ
		ETHYL METHYLENE PHOSPHORODITHIOATE
		ETHYL METHYLENE PHOSPHORODITHIOATE ([(ETO)2P(S)S]2CH2)

CAS No.	IN	Chemical Name
		FMC-1240
		FOSFATOX E
		FOSFONO 50
		HYLEMAX
		HYLEMOX
		ITOPAZ
		KWIT
		METHYLEEN-S,S'-BIS(O,O-DIETHYL-DITHIOFOSFAAT) (DUTCH)
		METHYLENE-S,S'-BIS(O,O-DIAETHYL-DITHIOPHOSPHAT) (GERMAN)
		NA 2783 (DOT)
		NIA 1240
		NIAGARA 1240
		NIALATE
		O,O,O',O'-TETRAAETHYL-BIS(DITHIOPHOSPHAT) (GERMAN)
		O,O,O',O'-TETRAETHYL S,S'-METHYLENE BIS(PHOSPHORO-DITHIOATE)
		O,O,O',O'-TETRAETHYL S,S'-METHYLENE DI(PHOSPHORO-DITHIOATE)
		O,O,O',O'-TETRAETHYL S,S'-METHYLENEBISPHOSPHOR-DITHIOATE
		PHOSPHORODITHIOIC ACID, O,O-DIETHYL ESTER, S,S-DIESTER WITH METHANEDITHIOL
		PHOSPHORODITHIOIC ACID, S,S'-METHYLENE O,O,O',O'-TETRAETHYL ESTER
		PHOSPHOTOX
		PHOSPHOTOX E
		RHODIACIDE
		RHODOCIDE
		RODOCID
		RODOCIDE
		RP 8167
		S,S'-METHYLEN-BIS(O,O-DIAETHYL-DITHIOPHOSPHAT) (GERMAN)
		S,S'-METHYLENE O,O,O',O'-TETRAETHYL PHOSPHORO-DITHIOATE
		SOPRATHION
		TETRAETHYL S,S'-METHYLENE BIS(PHOS-PHOROTHIOLOTHIONATE)
		VEGFRU FOSMITE
563-41-7		
		AMIDOUREA HYDROCHLORIDE
		AMINOUREA HYDROCHLORIDE
		CARBAMYLHYDRAZINE HYDROCHLORIDE
		CH
		HYDRAZINECARBOXAMIDE MONOHYDROCHLORIDE
		HYDRAZINECARBOXAMIDE, HYDROCHLORIDE
		SEMICARBAZIDE CHLORIDE
		SEMICARBAZIDE HYDROCHLORIDE
		SEMICARBAZIDE, MONOHYDROCHLORIDE
563-43-9		
		ALUMINUM, DICHLOROETHYL-
		DICHLOROETHYLALUMINUM
		DICHLOROMONOETHYLALUMINUM
		ETHYALUMINUM DICHLORIDE
		ETHYL ALUMINUM DICHLORIDE
		ETHYL ALUMINUM DICHLORIDE (DOT)
		ETHYLALUMINIUM DICHLORIDE
		ETHYLDICHLOROALUMINUM
		UN 1924 (DOT)
563-45-1		
		1-BUTENE, 3-METHYL-
		2-METHYL-3-BUTENE
		3-METHYL-1-BUTENE
		ALPHA-ISOAMYLENE
		ISOPROPYLETHENE
		ISOPROPYLETHYLENE
563-46-2		
		1-BUTENE, 2-METHYL-
		1-ISOAMYLENE
		2-METHYL-1-BUTENE
		2-METHYL-1-BUTENE (TECHNICAL)
		GAMMA-ISOAMYLENE

CAS No.	IN	Chemical Name
563-47-3		
		1-PROPENE, 3-CHLORO-2-METHYL-
		2-METHALLYL CHLORIDE
		2-METHYL-2-PROPENYL CHLORIDE
		2-METHYL-3-CHLOROPROPENE
		2-METHYL-ALLYLCHLORID (GERMAN)
		2-METHYLALLYL CHLORIDE
		3-CHLOR-2-METHYL-PROP-1-EN (GERMAN)
		3-CHLORO-2-METHYL-1-PROPENE
		3-CHLORO-2-METHYLPROPENE
		3-CLORO-2-METIL-PROP-1-ENE (ITALIAN)
		BETA-METHALLYL CHLORIDE
		BETA-METHYLALLYL CHLORIDE
		CHLORURE DE METHALLYLE (FRENCH)
		CLORURO DI METALLILE (ITALIAN)
		GAMMA-CHLOROISOBUTYLENE
		ISOBUTENYL CHLORIDE
		METHALLYL CHLORIDE
	*	METHYL ALLYL CHLORIDE
		METHYL ALLYL CHLORIDE (DOT)
		NCI-C54820
		PROPENE, 3-CHLORO-2-METHYL-
		UN 2554 (DOT)
563-63-3		
		ACETIC ACID, SILVER(1+) SALT
		SILVER ACETATE
		SILVER ACETATE (AgOAC)
		SILVER MONOACETATE
		SILVER(1+) ACETATE
		SILVER(I) ACETATE
563-78-0		
		1-BUTENE, 2,3-DIMETHYL-
		2,3-DIMETHYL-1-BUTENE
563-79-1		
		1,1,2,2-TETRAMETHYLETHYLENE
		2,3-DIMETHYL-2-BUTENE
		2-BUTENE, 2,3-DIMETHYL-
		TETRAMETHYLETHENE
		TETRAMETHYLETHYLENE
563-80-4		
		2-ACETYLPROPANE
		2-BUTANONE, 3-METHYL-
		2-METHYLBUTAN-3-ONE
		3-METHYL BUTAN-2-ONE (DOT)
		3-METHYL-2-BUTANONE
		3-METHYLBUTANONE
		ISOPROPYL METHYL KETONE
		METHYL BUTANONE
	*	METHYL ISOPROPYL KETONE
		METHYL ISOPROPYL KETONE (ACGIH)
		MIPK
		UN 2397 (DOT)
564-02-3		
		2,2,3-TRIMETHYLPENTANE
		2-TERT-BUTYLBUTANE
		PENTANE, 2,2,3-TRIMETHYL-
565-59-3		
		2,3-DIMETHYLPENTANE
		3,4-DIMETHYLPENTANE
		PENTANE, 2,3-DIMETHYL-
565-76-4		
		1-PENTENE, 2,3,4-TRIMETHYL-
		2,3,4-TRIMETHYL-1-PENTENE
569-64-2		
	*	C.I. BASIC GREEN 4
573-56-8		
		2,6-DINITROPHENOL
		BETA-DINITROPHENOL
		O-DINITROPHENOL
		PHENOL, 2,6-DINITRO-

CAS No.	IN	Chemical Name
		PHENOL, BETA-DINITRO-
573-58-0		
		C.I. DIRECT RED 28
577-11-7		
		1,4-BIS(2-ETHYLHEXYL) SODIUM SULFOSUCCINATE
		2-ETHYLHEXYL SULFOSUCCINATE SODIUM
		AEROSOL GPG
		AEROSOL OT
		AEROSOL OT 100
		AEROSOL OT 75
		AEROSOL OT-B
		AEROSOL OT-S
		ALCOPOL O
		ALPHASOL OT
		AOT
		AOT I
		BEROL 478
		BIS(2-ETHYLHEXYL) S-SODIUM SULFOSUCCINATE
		BIS(2-ETHYLHEXYL) SODIOSULFOSUCCINATE
		BIS(2-ETHYLHEXYL) SULFOSUCCINATE SODIUM SALT
		BIS(2-ETHYLHEXYL)SODIUM SULFOSUCCINATE
		BIS(ETHYLHEXYL) ESTER OF SODIUM SULFOSUCCINIC ACID
		BUTANEDIOIC ACID, SULFO-, 1,4-BIS(2-ETHYLHEXYL) ESTER, SODIUM SALT
		BUTANEDIOIC ACID, SULFO-, 1,4-BIS(2-ETHYLHEXYL) ESTER, SODIUM SALT (9CI)
		CELANOL DOS 65
		CELANOL DOS 75
		CLESTOL
		COLACE
		COMPLEMIX
		CONSTONATE
		COPROL
		D-S-S
		DEFILIN
		DI(2-ETHYLHEXYL)SULFOSUCCINATE SODIUM SALT
		DI-(2-ETHYLHEXYL) SODIUM SULFOSUCCINATE
		DIOCTLYN
		DIOCTYL ESTER OF SODIUM SULFOSUCCINATE
		DIOCTYL ESTER OF SODIUM SULFOSUCCINIC ACID
		DIOCTYL SODIUM SULFOSUCCINATE
		DIOCTYL SULFOSUCCINATE SODIUM
		DIOCTYL SULFOSUCCINATE SODIUM SALT
		DIOCTYL-MEDO FORTE
		DIOCTYLAL
		DIOMEDICONE
		DIOSUCCIN
		DIOTILAN
		DIOVAC
		DIOX
		DISONATE
		DOCUSATE SODIUM
		DOSS
		DOSS 70
		DOXINATE
		DOXOL
		DREWFAX 007
		DSS
		DULSIVAC
		DUOSOL
		HUMIFEN WT 27G
		KARAWET DOSS
		KONLAX
		KOSATE
		LANKROPOL KO 2
		LAXINATE
		LAXINATE 100
		MANOXOL OP
		MANOXOL OT
		MARLINAT DF 8
		MERVAMINE
		MODANE SOFT
		MOLATOC
		MOLCER
		MOLOFAC
		MONAWET MD 70E
		MONAWET MO 70
		MONAWET MO-70 RP
		MONAWET MO-84 R2W
		MONOXOL OT
		NEKAL WT-27
		NEVAX
		NEWCOL 290M
		NIKKOL OTP 70
		NISSAN RAPISOL
		NISSAN RAPISOL B 30
		NISSAN RAPISOL B 80
		NONIT
		NORVAL
		OBSTON
		PELEX OT
		PELEX OT-P
		RAPISOL
		RAPISOL B 30
		RAPISOL B 80
		REGUTOL
		REQUTOL
		REVAC
		SANMORIN OT 70
		SANNOL LDF 110
		SBO
		SECOSOL DOS 70
		SOBITAL
		SODIUM 1,2-BIS(2-ETHYLHEXYLOXYCARBONYL)-1-ETHANE-SULFONATE
		SODIUM 1,4-BIS(2-ETHYLHEXYL) SULFOSUCCINATE
		SODIUM 2-ETHYLHEXYLSULFOSUCCINATE
		SODIUM BIS(2-ETHYLHEXYL) SULFOSUCCINATE
		SODIUM DI(2-ETHYLHEXYL) SULFOSUCCINATE
		SODIUM DIOCTYL SULFOSUCCINATE
		SODIUM DIOCTYL SULPHOSUCCINATE
		SODIUM SULFODI-(2-ETHYLHEXYL)-SULFOSUCCINATE
		SOFTIL
		SOLIWAX
		SOLOVET
		SOLUSOL
		SOLUSOL-100%
		SOLUSOL-75%
		SPILON 8
		SUCCINATE STD
		SUCCINIC ACID, SULFO-, 1,4-BIS(2-ETHYLHEXYL) ESTER, SODIUM SALT
		SULFIMEL DOS
		SULFOSUCCINIC ACID BIS(2-ETHYLHEXYL)ESTER SODIUM SALT
		SULFOSUCCINIC ACID DI-2-ETHYLHEXYL ESTER SODIUM SALT
		SV 10 EX-WET 1001
		TKB 20
		TRITON GR 5
		TRITON GR 5M
		TRITON GR 7
		TRITON GR 7M
		VATSOL OT
		VELMOL
		WARCOWET 060
		WAXSOL
		WETAID SR
578-54-1		
		2-ETHYL BENZENAMINE
		2-ETHYLANILINE
		2-ETHYLANILINE (DOT)
		2-ETHYLPHENYLAMINE
		ANILINE, 2-ETHYL-
		ANILINE, O-ETHYL-
		ANILINE, O-ETHYL- (8CI)
		BENZENAMINE, 2-ETHYL-
		BENZENAMINE, 2-ETHYL- (9CI)
		O-AMINOETHYLBENZENE
		O-ETHYLANILINE
		UN 2273 (DOT)

CAS No.	IN	Chemical Name
578-94-9		
		10-CHLORO-5,10-DIHYDROARSACRIDINE
		10-CHLORO-5,10-DIHYDROPHENARSAZINE
		5-AZA-10-ARSENAANTHRACENE CHLORIDE
		5-CHLORO-5,10-DIHYDROPHENARSAZINE
		ADAMSIT
		ADAMSITE
		DIPHENYLAMINECHLORARSINE
		DIPHENYLAMINECHLOROARSINE
		DIPHENYLAMINECHLOROARSINE (DOT)
		DM
		DM (ARSENIC COMPOUND)
		PHENARSAZINE CHLORIDE
		PHENARSAZINE, 10-CHLORO-5,10-DIHYDRO-
		PHENAZARSINE CHLORIDE
		UN 1698 (DOT)
583-15-3		
		BENZOIC ACID, MERCURY(2+) SALT
		BETA-MERCURIBENZOATE
		MERCURIC BENZOATE
		MERCURIC BENZOATE, SOLID
		MERCURIC BENZOATE, SOLID (DOT)
		MERCURY BENZOATE
		MERCURY(II) BENZOATE
		UN 1631 (DOT)
583-52-8		
		DIPOTASSIUM OXALATE
		ETHANEDIOIC ACID, DIPOTASSIUM SALT
		ETHANEDIOIC ACID, DIPOTASSIUM SALT (9CI)
		ETHANEDIOIC ACID, POTASSIUM SALT
		KALIUM OXALATE
		OXALIC ACID, DIPOTASSIUM SALT
		OXALIC ACID, POTASSIUM SALT
		POTASSIUM NEUTRAL OXALATE
		POTASSIUM OXALATE
		POTASSIUM OXALATE (K2C2O4)
583-60-8		
		2-METHYL-1-CYCLOHEXANONE
		2-METHYLCYCLOHEXANONE
		ALPHA-METHYLCYCLOHEXANONE
		CYCLOHEXANONE, 2-METHYL-
		O-METHYLCYCLOHEXANONE
584-02-1		
		1-ETHYL-1-PROPANOL
		3-PENTANOL
		3-PENTYL ALCOHOL
		DIETHYLCARBINOL
		DIETHYLCARBINOL (DOT)
		ISOAMYL ALCOHOL (ACGIH)
		PENTAN-3-OL
		PENTANOL-3
		SEC-AMYL ALCOHOL
		SEC-PENTANOL
		SEC-PENTYL ALCOHOL
		UN 2706 (DOT)
584-79-2		
		(+)-ALLELRETHONYL (+)-CIS,TRANS-CHRYSANTHEMATE
		2-CYCLOPENTEN-1-ONE, 2-ALLYL-4-HYDROXY-3-METHYL-, 2,2-DIMETHYL-3-(2-METHYLPROPENYL)CYCLOPROPANE-CARBO
		3-ALLYL-2-METHYL-4-OXO-2-CYCLOPENTEN-1-YL CHRYSANTHEMATE
		3-ALLYL-4-KETO-2-METHYLCYCLOPENTENYL CHRYSANTHEMUMMONOCARBOXYLATE
		ALLETHRIN
		ALLETHRIN (DOT)
		ALLETHRIN I
		ALLETHRINE
		ALLYL CINERIN
		ALLYL CINERIN I
		ALLYL HOMOLOG OF CINERIN I
		ALLYLRETHRONYL DL-CIS-TRANS-CHRYSANTHEMATE
		BINAMIN FORTE
		BIOALLETHRIN
		CINERIN I ALLYL HOMOLOG
		CYCLOPROPANECARBOXYLIC ACID, 2,2-DIMETHYL-3-(2-METHYL-1-PROPENYL)-,2-METHYL-4-OXO-3-(2-PROPENYL)-2-C
		CYCLOPROPANECARBOXYLIC ACID, 2,2-DIMETHYL-3-(2-METHYLPROPENYL)-, ESTER WITH 2-ALLYL-4-HYDROXY-3-METH
		D,L-2-ALLYL-4-HYDROXY-3-METHYL-2-CYCLOPENTEN-1-ONE-D,L-CHRYSANTHEMUM MONOCARBOXYLATE
		D-ALLETHRIN
		DL-3-ALLYL-2-METHYL-4-OXOCYCLOPENT-2-ENYL DL-CIS TRANS CHRYSANTHEMATE
		ENT 17,510
		EXTHRIN
		FDA 1446
		FMC 249
		NA 2902 (DOT)
		NECARBOXYLIC ACID
		NIA 249
		PALLETHRINE
		PYNAMIN
		PYNAMIN FORTE
		PYRESIN
		PYRESYN
		SYNTHETIC PYRETHRINS
		TRANS-ALLETHRIN
584-84-9		
		2,4-DIISOCYANATO-1-METHYLBENZENE
		2,4-DIISOCYANATOTOLUENE
		2,4-TDI
		2,4-TOLUENE DIISOCYANATE
		2,4-TOLYLENE DIISOCYANATE
		4-METHYL-M-PHENYLENE DIISOCYANATE
		4-METHYL-M-PHENYLENE ISOCYANATE
		BENZENE, 2,4-DIISOCYANATO-1-METHYL-
		ISOCYANIC ACID, 4-METHYL-M-PHENYLENE ESTER
	*	TOLUENE-2,4-DIISOCYANATE
		TOLUENE-2,4-DIISOCYANATE(TDI)
584-94-1		
		2,3-DIMETHYLHEXANE
		HEXANE, 2,3-DIMETHYL-
586-62-9		
		4-ISOPROPYLIDENE-1-METHYLCYCLOHEXENE
		CYCLOHEXENE, 1-METHYL-4-(1-METHYLETHYLIDENE)-
		ISOTERPINENE
		P-MENTHA-1,4(8)-DIENE
		TERPINOLEN
		TERPINOLENE
		TERPINOLENE (DOT)
		UN 2541 (DOT)
586-78-7		
		1-BROMO-4-NITROBENZENE
		4-BROMONITROBENZENE
		4-NITROBROMOBENZENE
		4-NITROPHENYL BROMIDE
		BENZENE, 1-BROMO-4-NITRO-
		NITROBROMOBENZENE
		P-BROMONITROBENZENE
		P-NITROBROMOBENZENE
		P-NITROBROMOBENZENE (DOT)
		P-NITROPHENYL BROMIDE
		UN 2732 (DOT)
589-34-4		
		2-ETHYLPENTANE
		3-METHYLHEXANE
		HEXANE, 3-METHYL-
589-38-8		
	*	3-HEXANONE
		ETHYL PROPYL KETONE

CAS No.	IN	Chemical Name
589-43-5		
		2,4-DIMETHYLHEXANE
		HEXANE, 2,4-DIMETHYL-
589-90-2		
	*	1,4-DIMETHYLCYCLOHEXANE
		1,4-DIMETHYLCYCLOHEXANE (DOT)
		CYCLOHEXANE, 1,4-DIMETHYL-
		P-DIMETHYLCYCLOHEXANE
		UN 2263 (DOT)
590-01-2		
		BUTYL PROPANOATE
		BUTYL PROPIONATE
		BUTYL PROPIONATE (DOT)
		N-BUTYL PROPANOATE
		N-BUTYL PROPIONATE
		PROPANOIC ACID, BUTYL ESTER
		PROPANOIC ACID, BUTYL ESTER (9CI)
		PROPIONIC ACID, BUTYL ESTER
		UN 1914 (DOT)
590-18-1		
		(Z)-2-BUTENE
		2-BUTENE, (Z)-
		2-BUTENE-CIS
		BETA-CIS-BUTYLENE
		CIS-1,2-DIMETHYLETHYLENE
		CIS-2-BUTENE
		CIS-2-BUTYLENE
		CIS-BUTENE
		CIS-BUTENE-2
		CIS-BUTYLENE
590-21-6		
		1-CHLORO-1-PROPENE
		1-CHLOROPROPENE
		1-CHLOROPROPYLENE
		1-PROPENE, 1-CHLORO-
		PROPENE, 1-CHLORO-
		PROPENYL CHLORIDE
590-86-3		
		1-BUTANAL, 3-METHYL-
		3-METHYL-1-BUTANAL
		3-METHYLBUTANAL
		3-METHYLBUTYRALDEHYDE
		BETA-METHYLBUTANAL
		BUTANAL, 3-METHYL-
		BUTYRALDEHYDE, 3-METHYL-
		ISOAMYLALDEHYDE
		ISOPENTALDEHYDE
		ISOPENTANAL
		ISOVALERAL
	*	ISOVALERALDEHYDE
		ISOVALERIC ALDEHYDE
		ISOVALERYLALDEHYDE
590-88-5		
		1,3-BUTANEDIAMINE
		1,3-DIAMINOBUTANE
590-96-5		
		METHANOL, (METHYL-ONN-AZOXY)-
		METHANOL, (METHYLAZOXY)-
		METHYLAZOXYMETHANOL
591-21-9		
	*	1,3-DIMETHYL CYCLOHEXANE
		CYCLOHEXANE, 1,3-DIMETHYL-
		M-DIMETHYLCYCLOHEXANE
591-47-9		
		4-METHYL-1-CYCLOHEXENE
		4-METHYLCYCLOHEXENE
		CYCLOHEXENE, 4-METHYL-
591-76-4		
		2-METHYLHEXANE
		HEXANE, 2-METHYL-
		ISOHEPTANE
591-78-6		
		2-HEXANONE
		2-OXOHEXANE
		BUTYL METHYL KETONE
		HEXANONE-2
		KETONE, BUTYL METHYL
		MBK
		METHYL BUTYL KETONE
	*	METHYL N-BUTYL KETONE
		METHYL N-BUTYL KETONE (ACGIH)
		MNBK
		N-BUTYL METHYL KETONE
591-87-7		
		3-ACETOXY-1-PROPENE
		3-ACETOXYPROPENE
		ACETIC ACID, 2-PROPENYL ESTER
		ACETIC ACID, 2-PROPENYL ESTER (9CI)
		ACETIC ACID, ALLYL ESTER
		ALLYL ACETATE
		ALLYL ACETATE (DOT)
		UN 2333 (DOT)
591-89-8		
		POTASSIUM TETRACYANOMERCURATE (II)
		MERCURIC POTASSIUM CYANIDE
591-97-9		
		1-CHLORO-2-BUTENE
		1-CROTYL CHLORIDE
		2-BUTENE, 1-CHLORO-
		2-BUTENYL CHLORIDE
		ALPHA-CHLORO-BETA-BUTYLENE
	*	CROTYL CHLORIDE
		GAMMA-METHALLYL CHLORIDE
		GAMMA-METHYLALLYL CHLORIDE
592-01-8		
		CALCID
	*	CALCIUM CYANIDE
		CALCIUM CYANIDE (Ca(CN)2)
		CALCIUM CYANIDE, SOLID
		CALCIUM CYANIDE, SOLID (DOT)
		CALCYAN
		CALCYANIDE
		CYANOGAS
		CYANURE DE CALCIUM (FRENCH)
		RCRA WASTE NUMBER P021
		UN 1575 (DOT)
592-04-1		
		CIANURINA
		CYANURE DE MERCURE (FRENCH)
		DICYANOMERCURY
		MERCURIC CYANIDE
		MERCURIC CYANIDE (DOT)
		MERCURIC CYANIDE, SOLID
		MERCURIC CYANIDE, SOLID (DOT)
		MERCURY CYANIDE (Hg(CN)2)
		MERCURY DICYANIDE
		MERCURY(II) CYANIDE
		UN 1636 (DOT)
592-05-2		
		CI 77610
		CI PIGMENT YELLOW 48
		CYANURE DE PLOMB (FRENCH)
		LEAD CYANIDE
		LEAD CYANIDE (DOT)
		LEAD CYANIDE (Pb(CN)2)
		LEAD(II)CYANIDE
		UN 1620 (DOT)

CAS No.	IN	Chemical Name
592-34-7		
		BUTOXYCARBONYL CHLORIDE
		BUTYL CHLOROCARBONATE
		BUTYLCHLOROFORMATE
		CARBONOCHLORIDIC ACID, BUTYL ESTER
		CARBONOCHLORIDIC ACID, BUTYL ESTER (9CI)
		CHLOROFORMIC ACID BUTYL ESTER
		FORMIC ACID, CHLORO-, BUTYL ESTER
		N-BUTYLCHLOROFORMATE (DOT)
		UN 2743 (DOT)
592-41-6		
		1-HEXENE
		1-HEXENE (DOT)
		1-N-HEXENE
		HEX-1-ENE
		HEX-1-ENE (DOT)
	*	HEXENE
		UN 2370 (DOT)
592-43-8		
		2-HEXENE(MIXED CIS & TRANS)
592-45-0		
		1,4-HEXADIENE
592-55-2		
		1-BROMO-2-ETHOXYETHANE
		1-BROMO-2-ETHOXYETHYLENE
		1-ETHOXY-2-BROMOETHANE
		2-BROMOETHOXYETHANE
		2-BROMOETHYL ETHYL ETHER
		2-BROMOETHYL ETHYL ETHER (DOT)
		2-ETHOXYETHYL BROMIDE
		BROMOETHYL ETHYL ETHER
		ETHANE, 1-BROMO-2-ETHOXY-
		ETHANE, 1-BROMO-2-ETHOXY- (9CI)
		ETHER, 2-BROMOETHYL ETHYL
		UN 2340 (DOT)
592-62-1		
		METHANOL, (METHYL-ONN-AZOXY)-, ACETATE (ESTER)
		METHANOL, (METHYLAZOXY)-, ACETATE
		METHYLAZOXYMETHANOL ACETATE
		METHYLAZOXYMETHYL ACETATE
592-76-7		
		1-HEPTENE
		1-N-HEPTENE
		HEPTENE
		N-HEPT-1-ENE
		N-HEPTENE
		N-HEPTENE (DOT)
		UN 2278 (DOT)
592-84-7		
		BUTYL FORMATE
		FORMIC ACID, BUTYL ESTER
		N-BUTYL FORMATE
592-85-8		
		MERCURIC SULFOCYANATE
		MERCURIC SULFOCYANATE, SOLID
		MERCURIC SULFOCYANATE, SOLID (DOT)
		MERCURIC SULFOCYANIDE
		MERCURIC THIOCYANATE
		MERCURIC THIOCYANATE, SOLID (DOT)
		MERCURY DITHIOCYANATE
		MERCURY THIOCYANATE (DOT)
		MERCURY THIOCYANATE (Hg(SCN)2)
	*	MERCURY(II) THIOCYANATE
		MERCURY, BIS(THIOCYANATO)-
		THIOCYANIC ACID, MERCURY(2+) SALT
		THIOCYANIC ACID, MERCURY(II) SALT
		UN 1646 (DOT)
592-87-0		
		FERROUS ISOTHIOCYANATE
		LEAD BIS(THIOCYANATE)
		LEAD DITHIOCYANATE
		LEAD ISOTHIOCYANATE
		LEAD SULFOCYANATE
		LEAD THIOCYANATE
		LEAD THIOCYANATE (DOT)
		LEAD THIOCYANATE (Pb(SCN)2)
		LEAD(II) THIOCYANATE
		NA 2291 (DOT)
		THIOCYANIC ACID, LEAD(2+) SALT
593-53-3		
		FLUOROMETHANE
		FLUOROMETHANE (CH3F)
		FREON 41
		METHANE, FLUORO-
		METHYL FLUORIDE
		METHYL FLUORIDE (DOT)
		UN 2454 (DOT)
593-60-2		
		BROMOETHENE
		BROMOETHENE (9CI)
		BROMOETHYLENE
		BROMURE DE VINYLE (FRENCH)
		ETHENE, BROMO-
		ETHYLENE, BROMO-
		NCI-C50373
		UN 1085 (DOT)
		VINILE (BROMURO DI) (ITALIAN)
		VINYL BROMIDE
		VINYL BROMIDE (ACGIH)
		VINYL BROMIDE, INHIBITED (DOT)
		VINYLBROMID (GERMAN)
		VINYLE (BROMURE DE) (FRENCH)
593-67-9		
		AMINOETHYLENE
		ETHENAMINE
		ETHENAMINE (9CI)
		ETHYLENAMINE
		ETHYLENE AMINES [COMBUSTIBLE LIQUID LABEL]
		ETHYLENE AMINES [CORROSIVE LABEL]
		ETHYLENE AMINES [FLAMMABLE LIQUID LABEL]
		ETHYLENE AMINES [FLAMMABLE LIQUID, CORROSIVE LABELS]
		ETHYLENEAMINE
		VINYLAMINE
593-74-8		
		DIMETHYL MERCURY
593-89-5		
		ARSONOUS DICHLORIDE, METHYL- (9CI)
		METHYLDICHLORARSINE
		METHYLDICHLOROARSINE
		METHYLDICHLOROARSINE (DOT)
		NA 1556 (DOT)
		TL 294
594-27-4		
		STANNANE, TETRAMETHYL-
		TETRAMETHYL TIN
		TETRAMETHYLSTANNANE
594-36-5		
		1,1-DIMETHYLPROPYL CHLORIDE
		2-CHLORO-2-METHYLBUTANE
		2-METHYL-2-CHLOROBUTANE
		BUTANE, 2-CHLORO-2-METHYL-
		TERT-AMYL CHLORIDE
		TERT-PENTYL CHLORIDE
594-42-3		
		METHANESULFENYL CHLORIDE, TRICHLORO-
		PERCHLOROMETHYL MERCAPTAN
		TRICHLOROMETHANESULFENYL CHLORIDE
		TRICHLOROMETHANESULPHENYL CHLORIDE
		TRICHLOROMETHYLSULPHENYL CHLORIDE

CAS No.	IN	Chemical Name
594-56-9		1-BUTENE, 2,3,3-TRIMETHYL-
		1-METHYL-1-TERT-BUTYLETHYLENE
		2,3,3-TRIMETHYL-1-BUTENE
		2,3,3-TRIMETHYLBUTENE
		TRIPTENE
594-71-8		2-CHLORO-2-NITROPROPANE
		PROPANE, 2-CHLORO-2-NITRO-
594-72-9		1,1-DICHLORO-1-NITROETHANE
		ETHANE, 1,1-DICHLORO-1-NITRO-
		ETHIDE
595-44-8		1,1-DICHLORO-1-NITROPROPANE
		PROPANE, 1,1-DICHLORO-1-NITRO-
595-90-4		STANNANE, TETRAPHENYL-
		TETRAPHENYL TIN
		TETRAPHENYLSTANNANE
597-64-8		STANNANE, TETRAETHYL-
		TET
		TETRAETHYL TIN
		TETRAETHYLSTANNANE
		TIN, TETRAETHYL-
598-14-1		ARSONOUS DICHLORIDE, ETHYL- (9CI)
		DICK (GERMAN)
		ED
		ETHYLDICHLORARSINE
		ETHYLDICHLOROARSINE (DOT)
		TL 214
		UN 1892 (DOT)
598-21-0		2-BROMOACETYL BROMIDE
		ACETYL BROMIDE, BROMO-
		BROMOACETYL BROMIDE
		BROMOACETYL BROMIDE (DOT)
		MONOBROMOACETYL BROMIDE
		UN 2513 (DOT)
598-31-2		1-BROMO-2-PROPANONE
		2-PROPANONE, 1-BROMO-
		2-PROPANONE, 1-BROMO- (9CI)
		2-PROPANONE, BROMO-
		ACETONYL BROMIDE
		ACETYL METHYL BROMIDE
		ALPHA-BROMOACETONE
		ALPHA-BROMOPROPANONE
		BROMO-2-PROPANONE
		BROMOACETONE
		BROMOACETONE (DOT)
		BROMOACETONE, LIQUID
		BROMOACETONE, LIQUID (DOT)
		BROMOMETHYL METHYL KETONE
		MONOBROMOACETONE
		RCRA WASTE NUMBER P017
		UN 1569 (DOT)
598-73-2		1-BROMO-1,2,2-TRIFLUOROETHYLENE
		BROMOTRIFLUOROETHENE
		BROMOTRIFLUOROETHYLENE
		BROMOTRIFLUOROETHYLENE (DOT)
		ETHENE, BROMOTRIFLUORO-
		ETHENE, BROMOTRIFLUORO- (9CI)
		ETHYLENE, BROMOTRIFLUORO-
		TRIFLUOROBROMOETHYLENE
		TRIFLUOROVINYL BROMIDE
		UN 2419 (DOT)

CAS No.	IN	Chemical Name
598-75-4		1,2-DIMETHYLPROPANOL
		2-BUTANOL, 3-METHYL-
		2-METHYL-3-BUTANOL
		3-METHYL-2-BUTANOL
		METHYLISOPROPYLCARBINOL
		SEC-ISOAMYL ALCOHOL
598-92-5		1-CHLORO-1-NITROETHANE
		1-CHLORONITROETHANE
		ETHANE, 1-CHLORO-1-NITRO-
598-96-9		2-PENTENE, 3,4,4-TRIMETHYL-
		3,4,4-TRIMETHYL-2-PENTENE
598-99-2		ACETIC ACID, TRICHLORO-, METHYL ESTER
		METHYL TRICHLOROACETATE
		METHYL TRICHLOROACETATE (DOT)
		UN 2533 (DOT)
600-25-9		1-CHLORO-1-NITROPROPANE
		PROPANE, 1-CHLORO-1-NITRO-
602-87-9		5-NITROACENAPHTHENE
		ACENAPHTHENE, 5-NITRO-
		ACENAPHTHYLENE, 1,2-DIHYDRO-5-NITRO-
602-99-3		TRINITRO-M-CRESOL
603-34-9		BENZENAMINE, N,N-DIPHENYL-
		N,N-DIPHENYLANILINE
		TRIPHENYL AMINE
604-75-1		OXAZEPAM
606-20-2		1-METHYL-2,6-DINITROBENZENE
	*	2,6-DINITROTOLUENE
		2,6-DNT
		2-METHYL-1,3-DINITROBENZENE
		BENZENE, 2-METHYL-1,3-DINITRO-
		BENZENE, 2-METHYL-1,3-DINITRO- (9CI)
		RCRA WASTE NUMBER U106
		TOLUENE, 2,6-DINITRO-
606-35-9		2,4,6-TRINITROANISOLE
		TRINITROANISOLE
608-73-1		1,2,3,4,5,6-HEXACHLOROCYCLOHEXANE
		CYCLOHEXANE, 1,2,3,4,5,6-HEXACHLORO-
		TECHNICAL HCH
609-19-8		3,4,5-TRICHLOROPHENOL
		PHENOL, 3,4,5-TRICHLORO-
609-26-7		2-METHYL-3-ETHYLPENTANE
		3-ETHYL-2-METHYLPENTANE
		PENTANE, 3-ETHYL-2-METHYL-
610-39-9		1-METHYL-3,4-DINITROBENZENE
		3,4-DINITROTOLUENE
		3,4-DNT
		4-METHYL-1,2-DINITROBENZENE

CAS No.	IN	Chemical Name
		BENZENE, 4-METHYL-1,2-DINITRO-
		BENZENE, 4-METHYL-1,2-DINITRO- (9CI)
		TOLUENE, 3,4-DINITRO-
613-29-6		ANILINE, N,N-DIBUTYL-
		BENZENAMINE, N,N-DIBUTYL-
		DIBUTYLANILINE
		N,N-DIBUTYLANILINE
		N,N-DIBUTYLBENZENAMINE
		N-PHENYLDIBUTYLAMINE
613-35-4		4',4'''-BIACETANILIDE
		4,4'-DIACETAMIDOBIPHENYL
		4,4'-DIACETYLAMINOBIPHENYL
		4,4'-DIACETYLBENZIDINE
		ACETAMIDE, N,N'-[1,1'-BIPHENYL]-4,4'-DIYLBIS-
		DIACETYLBENZIDINE
		N,N'-DIACETYLBENZIDINE
		N,N-DIACETYLBENZIDINE
614-45-9		BENZENECARBOPEROXOIC ACID, 1,1-DIMETHYLETHYL ESTER
		BENZOYL TERT-BUTYL PEROXIDE
		BUTYL PEROXYBENZOATE
		CHALOXYD TBPB
		ESPEROX 10
		KAYABUTYL B
		LUPEROX P
		PERBUTYL Z
		PEROXYBENZOIC ACID, TERT-BUTYL ESTER
		TERT-BUTYL PERBENZOATE
		TERT-BUTYL PEROXYBENZOATE
		TRIGONOX C
614-46-9		TERT-BUTYL PERBENZOATE
614-78-8		1-O-TOLYL-2-THIOUREA
		2-METHYLPHENYLTHIOUREA
		N-(O-TOLYL)THIOUREA
		O-TOLYL THIOUREA
		THIOUREA, (2-METHYLPHENYL)-
		UREA, 2-THIO-1-O-TOLYL-
615-05-4		2,4-DIAMINOANISOLE
615-53-2		CARBAMIC ACID, METHYLNITROSO-, ETHYL ESTER
		ETHYL N-METHYLNITROSOCARBAMATE
		METHYL CARBAMATE
		METHYL URETHANE
		METHYLNITROSOURETHANE
		N-METHYL-N-NITROSOURETHANE
		N-NITROSO-N-METHYLURETHANE
		NITROSOMETHYLURETHANE
		NMU
		NSC 2860
615-65-6		2-CHLORO-4-METHYL ANILINE (CHLOROTOLUIDINE)
616-21-7		1,2-DICHLOROBUTANE
		3,4-DICHLOROBUTANE
		BUTANE, 1,2-DICHLORO-
616-29-5		1,3-DIAMINO-2-HYDROXYPROPANE
		1,3-DIAMINO-2-PROPANOL
		2-HYDROXY-1,3-DIAMINOPROPANE
		2-HYDROXY-1,3-PROPANEDIAMINE
		2-PROPANOL, 1,3-DIAMINO-
616-38-6		CARBONIC ACID, DIMETHYL ESTER
	*	DIMETHYL CARBONATE
		DIMETHYL CARBONATE (DOT)
		METHYL CARBONATE
		METHYL CARBONATE ((MEO)2CO)
		UN 1161 (DOT)
616-45-5		2-OXOPYRROLIDINE
		2-PYROL
		2-PYRROLIDINONE
		2-PYRROLIDONE
		4-AMINOBUTYRIC ACID LACTAM
		ALPHA-PYRROLIDINONE
		ALPHA-PYRROLIDONE
		BUTANOIC ACID, 4-AMINO-, LACTAM
		BUTYROLACTAM
		GAMMA-AMINOBUTYRIC ACID LACTAM
		GAMMA-AMINOBUTYRIC LACTAM
		GAMMA-AMINOBUTYROLACTAM
		GAMMA-BUTYROLACTAM
		PYRROLIDINE-2-ONE
	*	PYRROLIDONE
617-50-5		ISOBUTYRIC ACID, ISOPROPYL ESTER
		ISOPROPYL 2-METHYLPROPANOATE
		ISOPROPYL ISOBUTYRATE
		ISOPROPYL ISOBUTYRATE (DOT)
		PROPANOIC ACID, 2-METHYL-, 1-METHYLETHYL ESTER
		PROPANOIC ACID, 2-METHYL-, 1-METHYLETHYL ESTER (9CI)
		UN 2406 (DOT)
617-51-6		ISOPROPYL LACTATE
		LACTIC ACID, ISOPROPYL ESTER
		PROPANOIC ACID, 2-HYDROXY-, 1-METHYLETHYL ESTER
617-89-0		2-(AMINOMETHYL)FURAN
		2-FURANMETHANAMINE
		2-FURANMETHYLAMINE
		2-FURANYLMETHYLAMINE
		2-FURFURYLAMINE
		2-FURYLMETHYLAMINE
		ALPHA-FURFURYLAMINE
		FURFURYLAMINE
		FURFURYLAMINE (DOT)
		METHYLAMINE, 1-(2-FURYL)-
		UN 2526 (DOT)
		USAF Q-1
620-05-3		(IODOMETHYL)BENZENE
		ALPHA-IODOTOLUENE
		BENZENE, (IODOMETHYL)-
		BENZENE, (IODOMETHYL)- (9CI)
		BENZYL IODIDE
		BENZYL IODIDE (DOT)
		IODOPHENYLMETHANE
		TOLUENE, ALPHA-IODO-
		UN 2653 (DOT)
621-64-7		1-PROPANAMINE, N-NITROSO-N-PROPYL-
		DIPROPYLAMINE, N-NITROSO-
		DIPROPYLNITROSAMINE
		N,N-DIPROPYLNITROSAMINE
		N-NITROSODI-N-PROPYLAMINE
		N-NITROSODIPROPYLAMINE
		NITROSODI-N-PROPYLAMINE
		NITROSODIPROPYLAMINE
621-77-2		1-PENTANAMINE, N,N-DIPENTYL-
		TRIAMYLAMINE
		TRIPENTYLAMINE

CAS No.	IN	Chemical Name
622-08-2		2-(BENZYLOXY)ETHANOL
		ETHANOL, 2-(BENZYLOXY)-
		ETHANOL, 2-(PHENYLMETHOXY)-
		ETHYLENE GLYCOL MONOBENZYL ETHER
		GLYCOL MONOBENZYL ETHER
622-40-2		2-(4-MORPHOLINYL)ETHANOL
		2-MORPHOLINOETHANOL
		4(2-HYDROXYETHYL)MORPHOLINE
		4-MORPHOLINEETHANOL
		BÈTA-MORPHOLINOETHANOL
		N-(2-HYDROXYETHYL)MORPHOLINE
		N-BETA-HYDROXYETHYLMORPHOLINE
622-44-6		1,1-DICHLORO-N-PHENYLMETHANIMINE
		BENZENAMINE, N-(DICHLOROMETHYLENE)-
		CARBONIC DICHLORIDE, (PHENYLIMINO)-
		CARBONIMIDIC DICHLORIDE, PHENYL-
		CARBONIMIDIC DICHLORIDE, PHENYL- (9CI)
		DICHLORO(PHENYLIMINO)METHANE
		IMIDOCARBONYL CHLORIDE, PHENYL-
		N-(DICHLOROMETHYLENE)ANILINE
		N-PHENYLCARBIMIDE DICHLORIDE
		N-PHENYLCARBONIMIDIC DICHLORIDE
		N-PHENYLIMIDOPHOSGENE
		N-PHENYLIMINOCARBONYL DICHLORIDE
		PHENYL ISOCYANIDE, DICHLORIDE
		PHENYL ISOCYANODICHLORIDE
		PHENYLCARBONIMIDIC DICHLORIDE
		PHENYLCARBYLAMINE CHLORIDE
		PHENYLCARBYLAMINE CHLORIDE (DOT)
		PHENYLIMIDOCARBONYL CHLORIDE
		PHENYLIMINOCARBONYL DICHLORIDE
		PHENYLISONITRILE DICHLORIDE
		UN 1672 (DOT)
622-45-7		CYCLOHEXYL ACETATE
622-57-1		BENZENAMINE, N-ETHYL-4-METHYL-
		ETHYL TOLUIDINE
		N-ETHYL-4-METHYLANILINE
		N-ETHYL-4-TOLUIDINE
		N-ETHYL-P-METHYLANILINE
		N-ETHYL-P-TOLUIDINE
		N-ETHYL-P-TOLUIDINE (DOT)
		P-METHYL-N-ETHYLANILINE
		P-TOLUIDINE, N-ETHYL-
		UN 2754 (DOT)
622-97-9		1-METHYL-4-VINYLBENZENE
		1-P-TOLYLETHENE
		4-ETHENYLMETHYLBENZENE
		4-METHYLSTYRENE
		4-VINYLTOLUENE
		BENZENE, 1-ETHENYL-4-METHYL-
		BENZENE, 1-ETHENYL-4-METHYL- (9CI)
		P-METHYLSTYRENE
		P-VINYLTOLUENE
		PMS
		STYRENE, P-METHYL-
623-42-7		BUTANOIC ACID, METHYL ESTER
		BUTYRIC ACID, METHYL ESTER
		METHYL BUTANOATE
	*	METHYL BUTYRATE
		METHYL BUTYRATE (DOT)
		METHYL N-BUTANOATE
		METHYL-N-BUTYRATE
		UN 1237 (DOT)
623-70-1		ETHYL CROTONATE
624-29-3		1,4-DIMETHYLCYCLOHEXANE-CIS
		CIS-1,4-DIMETHYLCYCLOHEXANE
		CYCLOHEXANE, 1,4-DIMETHYL-, CIS-
624-64-6		(E)-2-BUTENE
		2-BUTENE, (E)-
		2-BUTENE-TRANS
		2-TRANS-BUTENE
		BETA-TRANS-BUTYLENE
		TRANS-1,2-DIMETHYLETHYLENE
		TRANS-2-BUTENE
		TRANS-BUTENE
		TRANS-BUTENE-2
624-83-9		ISOCYANATE DE METHYLE (FRENCH)
		ISOCYANATOMETHANE
		ISOCYANIC ACID, METHYL ESTER
		METHANE, ISOCYANATO-
		METHYL ISOCYANAT (GERMAN)
		METHYL ISOCYANATE
		METHYL ISOCYANATE (ACGIH,DOT)
		METHYL ISOCYANATE SOLUTIONS (DOT)
		METHYLISOCYANAAT (DUTCH)
		METIL ISOCIANATO (ITALIAN)
		MIC
		RCRA WASTE NUMBER P064
		TL 1450
		UN 2480 (DOT)
624-91-9		METHYL NITRITE
		METHYL NITRITE (DOT)
		NITROUS ACID, METHYL ESTER
624-92-0		(METHYLDITHIO)METHANE
		2,3-DITHIABUTANE
		DIMETHYL DISULFIDE
		DIMETHYL DISULFIDE (DOT)
		DIMETHYL DISULPHIDE
		DISULFIDE, DIMETHYL
		METHYL DISULFIDE
		UN 2381 (DOT)
625-16-1		1,1-DIMETHYLPROPYL ACETATE
		2-BUTANOL, 2-METHYL-, ACETATE
		2-METHYL-2-BUTYL ACETATE
	*	TERT-AMYL ACETATE
		TERT-PENTYL ACETATE
		TERT-PENTYL ALCOHOL, ACETATE
625-27-4		2-ETHYL-1,1-DIMETHYLETHYLENE
		2-METHYL-2-PENTENE
		2-PENTENE, 2-METHYL-
		4-METHYL-3-PENTENE
625-30-9		1-METHYL-N-BUTYLAMINE
		1-METHYLBUTYLAMINE
		2-AMINOPENTANE
		2-PENTANAMINE
		2-PENTYLAMINE
		BUTYLAMINE, 1-METHYL-
		SEC-AMYLAMINE
625-33-2		2-OXO-3-PENTENE
		2-PENTEN-4-ONE
		3-PENTEN-2-ONE
		ETHYLIDENEACETONE
		METHYL 1-PROPENYL KETONE

CAS No.	IN	Chemical Name
		METHYL PROPENYL KETONE
		METHYLPROPENYL KETONE, INHIBITED
625-55-8		
		FORMIC ACID, 1-METHYLETHYL ESTER
		FORMIC ACID, ISOPROPYL ESTER
		ISOPROPYL FORMATE
		ISOPROPYL FORMATE (DOT)
		UN 1281 (DOT)
625-58-1		
		ETHYL NITRATE
		ETHYL NITRATE (DOT)
		NA 1993 (DOT)
		NITRIC ACID, ETHYL ESTER
		NITRIC ETHER (DOT)
625-86-5		
		2,5-DIMETHYLFURAN
		FURAN, 2,5-DIMETHYL-
626-17-5		
		1,3-BENZENEDICARBONITRILE
		1,3-BENZODINITRILE
		1,3-DICYANOBENZENE
		3-CYANOBENZONITRILE
		IPN
		ISOPHTHALONITRILE
		M-BENZENEDINITRILE
		M-CYANOBENZONITRILE
		M-DICYANOBENZENE
		M-PHTHALODINITRILE
626-23-3		
		2-BUTANAMINE, N-(1-METHYLPROPYL)-
		BIS(1-METHYLPROPYL)AMINE
		DI-2-BUTYLAMINE
		DI-SEC-BUTYLAMINE
		N-(1-METHYLPROPYL)-2-BUTANAMINE
626-38-0		
		1-METHYLBUTYL ACETATE
		2-ACETOXYPENTANE
		2-PENTANOL, ACETATE
		2-PENTYL ACETATE
		PENT-2-YL ETHANOATE
	*	SEC-AMYL ACETATE
626-67-5		
		1-METHYLPIPERIDINE
		1-METHYLPIPERIDINE (DOT)
		METHYLPIPERIDINE
		N-METHYLPIPERIDINE
		PIPERIDINE, 1-METHYL-
		UN 2399 (DOT)
627-11-2		
		(2-CHLOROETHOXY)CARBONYL CHLORIDE
		2-CHLOROETHYL CHLOROCARBONATE
		2-CHLOROETHYL CHLOROFORMATE
		BETA-CHLOROETHYL CHLOROFORMATE
		CARBONOCHLORIDIC ACID, 2-CHLOROETHYL ESTER
		CHLOROETHYL CHLOROFORMATE
		CHLOROFORMIC ACID, 2-CHLOROETHYL ESTER
		FORMIC ACID, CHLORO-, 2-CHLOROETHYL ESTER
		TL 207
627-13-4		
		N-PROPYL NITRATE
		N-PROPYL NITRATE (ACGIH,DOT)
		NITRATE DE PROPYLE NORMAL (FRENCH)
		NITRIC ACID, PROPYL ESTER
		PROPYL NITRATE
		UN 1865 (DOT)
627-19-0		
		1-PENTYNE
		PROPYLACETYLENE
627-20-3		
		(Z)-2-PENTENE
		2-PENTENE, (Z)-
		BETA-AMYLENE-CIS
		CIS-2-PENTENE
		CIS-BETA-AMYLENE
		CIS-PENTENE
627-30-5		
		1-CHLORO-3-HYDROXYPROPANE
		1-CHLORO-3-PROPANOL
		1-PROPANOL, 3-CHLORO-
		3-CHLORO-1-HYDROXYPROPANE
		3-CHLORO-1-PROPANOL
		3-CHLOROPROPANOL
		3-CHLOROPROPANOL-1
		3-CHLOROPROPANOL-1 (DOT)
		3-CHLORPROPAN-1-OL (GERMAN)
		TRIMETHYLENE CHLOROHYDRIN
		UN 2849 (DOT)
627-53-2		
		DIETHYL SELENIDE
		DIETHYL SELENIUM
		DIETHYLMONOSELENIDE
		ETHANE, 1,1'-SELENOBIS-
		ETHYL SELENIDE
627-63-4		
		2-BUTENEDIOYL DICHLORIDE, (E)-
		CHLORURE DE FUMARYLE (FRENCH)
		DICHLORID KYSELINY FUMAROVE (CZECH)
		FUMARIC ACID CHLORIDE
		FUMARIC ACID DICHLORIDE
		FUMARIC DICHLORIDE
		FUMAROYL CHLORIDE
		FUMAROYL DICHLORIDE
		FUMARYL CHLORIDE
		FUMARYL CHLORIDE (DOT)
		FUMARYLCHLORID (CZECH)
		TL 189
		UN 1780 (DOT)
628-20-6		
		4-CHLOROBUTANENITRILE
		4-CHLOROBUTYRONITRILE
		BUTANENITRILE, 4-CHLORO-
		BUTANENITRILE, 4-CHLORO- (9CI)
		BUTYRONITRILE, 4-CHLORO-
		GAMMA-CHLOROBUTYRONITRILE
628-32-0		
		1-ETHOXYPROPANE
		ETHER, ETHYL PROPYL
		ETHYL N-PROPYL ETHER
		ETHYL PROPYL ETHER
		ETHYL PROPYL ETHER (DOT)
		PROPANE, 1-ETHOXY-
		PROPANE, 1-ETHOXY- (9CI)
		PROPYL ETHYL ETHER
		UN 2615 (DOT)
628-37-5		
		DIETHYL DIOXIDE
		DIETHYL PEROXIDE
		ETHYL PEROXIDE
		PEROXIDE, DIETHYL
628-63-7		
		1-PENTANOL ACETATE
		1-PENTYL ACETATE
		ACETATE D'AMYLE (FRENCH)
		ACETIC ACID, AMYL ESTER
		ACETIC ACID, PENTYL ESTER
		AMYL ACETATE
		AMYL ACETATE (DOT)
		AMYL ACETIC ESTER
		AMYL ACETIC ETHER
		AMYL AZETAT (GERMAN)

CAS No.	IN	Chemical Name
		BIRNENOEL
	*	N-AMYL ACETATE
		N-AMYL ACETATE (ACGIH)
		N-PENTYL ACETATE
		N-PENTYL ETHANOATE
		OCTAN AMYLU (POLISH)
		PEAR OIL
		PENT-ACETATE
		PENT-ACETATE 28
		PENTYL ACETATE
		PRIMARY AMYL ACETATE
		UN 1104 (DOT)
628-76-2		
		1,5-DICHLOROPENTANE
		PENTAMETHYLENE CHLORIDE
		PENTANE, 1,5-DICHLORO-
628-81-9		
		1-ETHOXYBUTANE
		BUTANE, 1-ETHOXY-
		BUTYL ETHYL ETHER
		ETHER, BUTYL ETHYL
		ETHYL BUTYL ETHER
		ETHYL N-BUTYL ETHER
628-86-4		
		MERCURY FULMINATE
628-92-2		
		CYCLOHEPTENE
628-96-6		
		1,2-ETHANEDIOL, DINITRATE
		EGDN
		ETHYLENE DINITRATE
		ETHYLENE GLYCOL DINITRATE
		ETHYLENE NITRATE
		GLYCOL DINITRATE
		NITROGLYCOL
629-14-1		
		1,2-DIAETHOXY-AETHEN (GERMAN)
		1,2-DIETHOXY ETHENE
		1,2-DIETHOXY ETHYLENE
		1,2-DIETHOXYETHANE
		2-ETHOXYETHYL ETHYL ETHER
		3,6-DIOXAOCTANE
		DIETHOXYLETHANE
		DIETHYL CELLOSOLVE
		ETHANE, 1,2-DIETHOXY-
		ETHER DIETHYLIQUE DE L'ETHYLENE-GLYCOL (FRENCH)
		ETHYLENE GLYCOL DIETHYL ETHER
		ETHYLENE, 1,2-DIETHOXY-
		GLYCOL DIETHYL ETHER
629-20-9		
		(8)ANNULENE
		1,3,5,7-CYCLOOCTATETRAENE
		CYCLOOCTATETRAENE
		CYCLOOCTATETRAENE (DOT)
		UN 2358 (DOT)
629-76-5		
		1-PENTADECANOL
		N-1-PENTADECANOL
		N-PENTADECANOL
		PENTADECANOL
		PENTADECYL ALCOHOL
630-08-0		
	*	CARBON MONOXIDE
		CARBON MONOXIDE (ACGIH,DOT)
		CARBON MONOXIDE, CRYOGENIC
		CARBON MONOXIDE, CRYOGENIC LIQUID (DOT)
		CARBON OXIDE (CO)
		CARBONE (OXYDE DE) (FRENCH)
		CARBONIC OXIDE
		CARBONIO (OSSIDO DI) (ITALIAN)
		EXHAUST GAS
		FLUE GAS
		KOHLENMONOXID (GERMAN)
		KOHLENOXYD (GERMAN)
		KOOLMONOXYDE (DUTCH)
		NA 9202 (DOT)
		OXYDE DE CARBONE (FRENCH)
		UN 1016 (DOT)
		WEGLA TLENEK (POLISH)
630-60-4		
		ACOCANTHERIN
		ASTROBAIN
		CARD-20(22)-ENOLIDE, 3-[(6-DEOXY-ALPHA-L-MANNO-PYRANOSYL)OXY]-1,5,11,14,19-PENTAHYDROXY-, (1BETA,-3BET
		G-STROPHANTHIN
		G-STROPHICOR
		GRATIBAIN
		GRATUS STROPHANTHIN
		KOMBETIN
		OUABAGENIN L-RHAMNOSIDE
		OUABAGENIN-L-RHAMNOSID (GERMAN)
		OUABAIN
		OUABAINE
		OUBAIN
		PUROSTROPHAN
		RECTOBAINA
		SOLUFANTINA
		STRODIVAL
		STROPHALEN
		STROPHANTHIN G
		STROPHOPERM
		STROPHOSAN
		UABAINA
		UABANIN
630-93-3		
		SODIUM SALT OF PHENYTOIN
631-60-7		
		ACETIC ACID, MERCURY(1+) SALT
		MERCUROUS ACETATE
		MERCUROUS ACETATE (DOT)
		MERCUROUS ACETATE, SOLID
		MERCUROUS ACETATE, SOLID (DOT)
		MERCURY ACETATE (HgOAC)
		MERCURY MONOACETATE
		MERCURY(I) ACETATE
		UN 1629 (DOT)
631-61-8		
		ACETIC ACID, AMMONIUM SALT
	*	AMMONIUM ACETATE
		AMMONIUM ACETATE (DOT)
		NA 9079 (DOT)
633-03-4		
		12415 GREEN
		ADC BRILLIANT GREEN CRYSTALS
		AIZEN DIAMOND GREEN GH
		AIZEN MALACHITE GREEN GH
		AMMONIUM, (4-(P-(DIETHYLAMINO)-ALPHA-PHENYLBEN-ZYLIDENE)-2,5-CYCLOHEXADIEN-1-YLIDENE)-DIETHYL-, SULFA
		ANILINE GREEN
		ASTRA DIAMOND GREEN GX
		ASTRAZON GREEN D
		AVON GREEN A 4379
		AZIEN MALACHITE GREEN GH
		BASIC BRIGHT GREEN
		BASIC BRIGHT GREEN SULFATE
		BASIC BRILLIANT GREEN
		BASIC GREEN 1
		BASIC GREEN V
		BENZALDEHYDE GREEN
		BRILLANT-GRUN (GERMAN)
		BRILLIANT GREEN
		BRILLIANT GREEN ASEPTIC

CAS No.	IN	Chemical Name
		BRILLIANT GREEN B
		BRILLIANT GREEN BP
		BRILLIANT GREEN BP CRYSTALS
		BRILLIANT GREEN BPC
		BRILLIANT GREEN CRYSTALS
		BRILLIANT GREEN CRYSTALS H
		BRILLIANT GREEN DSC
		BRILLIANT GREEN G
		BRILLIANT GREEN GX
		BRILLIANT GREEN LAKE
		BRILLIANT GREEN P
		BRILLIANT GREEN SPECIAL
		BRILLIANT GREEN SULFATE
		BRILLIANT GREEN WP CRYSTALS
		BRILLIANT GREEN Y
		BRILLIANT GREEN YN
		BRILLIANT GREEN YNS
		BRILLIANT LAKE GREEN Y
		BRILLIANT TUNGSTATE GREEN TONER GT 288
		C.I. BASIC GREEN 1
		CALCOZINE BRILLIANT GREEN G
		CI 42040
		CI BASIC GREEN 1, SULFATE (1:1)
		DEORLENE GREEN JJO
		DIAMOND GREEN G
		EMERALD GREEN
		ETHANAMINIUM, N-[4-[[4-(DIETHYLAMINO)PHENYL]-PHENYLMETHYLENE]-2,5-CYCLOHEXADIEN-1-YLIDENE]-N-ETHYL-,
		ETHYL GREEN
		FAST GREEN J
		FAST GREEN JJO
		GREEN EN
		HIDACO BRILLIANT GREEN
		MALACHITE GREEN G
		MITSUI BRILLIANT GREEN GX
		SOLID GREEN
		TERTROPHENE BRILLIANT GREEN G
		TOKYO ANILINE BRILLIANT GREEN
636-21-5		
		2-METHYLANILINE HYDROCHLORIDE
		BENZENAMINE, 2-METHYL-, HYDROCHLORIDE
		O-METHYLANILINE HYDROCHLORIDE
		O-TOLUIDINE HYDROCHLORIDE
		O-TOLUIDINIUM CHLORIDE
637-78-5		
		ISOPROPYL PROPANOATE
		ISOPROPYL PROPIONATE
		PROPANOIC ACID, 1-METHYLETHYL ESTER
638-11-9		
		1-METHYLETHYL BUTANOATE
		BUTANOIC ACID, 1-METHYLETHYL ESTER
		BUTYRIC ACID, ISOPROPYL ESTER
		ISO-PROPANOL BUTYRATE
		ISOPROPYL BUTANOATE
		ISOPROPYL BUTYRATE
		ISOPROPYL BUTYRATE (DOT)
		UN 2405 (DOT)
638-17-5		
		2,4,6-TRIMETHYLDIHYDRO-1,3,5-DITHIAZINE
		2,4,6-TRIMETHYLPERHYDRO-1,3-DITHIAZINE
		4H-1,3,5-DITHIAZINE, DIHYDRO-2,4,6-TRIMETHYL-, (2ALPHA,4ALPHA,6ALPHA)-
		THIALDINE
638-21-1		
		PHENYLPHOSPHINE
		PHOSPHINE, PHENYL-
638-29-9		
		PENTANOYL CHLORIDE
		UN 2502 (DOT)
		VALEROYL CHLORIDE
		VALERYL CHLORIDE
		VALERYL CHLORIDE (DOT)
638-49-3		
		AMYL FORMATE
		AMYL FORMATE (DOT)
		FORMIC ACID, PENTYL ESTER
		M-PENTYL FORMATE
		N-AMYL FORMATE
		N-PENTYL FORMATE
		PENTYL FORMATE
		UN 1109 (DOT)
638-56-2		
		BIS(2-(2-CHLOROETHOXY)ETHYL)ETHER
638-63-7		
		AMYL ACETATE
639-58-7		
		AQUATIN
		BRESTANOL
		CHLOROTRIPHENYLSTANNANE
		CHLOROTRIPHENYLTIN
		FENTIN CHLORIDE
		GC 8993
		GENERAL CHEMICALS 8993
		HOE 2872
		LS 4442
		STANNANE, CHLOROTRIPHENYL-
		TINMATE
		TPTC
		TRIPHENYLCHLOROSTANNANE
		TRIPHENYLCHLOROTIN
		TRIPHENYLTIN CHLORIDE
640-15-3		
		2-ETHYLTHIOETHYL O,O-DIMETHYL PHOSPHORODITHIOATE
		BAY 23129
		COMPOUND M-81
		DITHIOMETHON
		DITHIOMETON (FRENCH)
		DITHIOPHOSPHATE DE O,O-DIMETHYLE ET DE S-(2-ETHYLTHIO-ETHYLE) (FRENCH)
		EKATIN
		EKATIN AEROSOL
		EKATIN ULV
		EKATINE-25
		ETHANETHIOL, 2-(ETHYLTHIO)-, S-ESTER WITH O,O-DIMETHYL PHOSPHORODITHIOATE
		INTRATHION
		INTRATION
		LUXISTELM
		M 81
		O,O-DIMETHYL S-(2-(ETHYLTHIO)ETHYL) PHOSPHORODITHIOATE
		O,O-DIMETHYL S-(2-ETHYLTHIOETHYL) DITHIOPHOSPHATE
		O,O-DIMETHYL-S-(2-AETHYLTHIO-AETHYL)-DITHIO PHOSPHAT (GERMAN)
		O,O-DIMETHYL-S-(2-ETHYLMERCAPTOETHYL) DITHIOPHOSPHATE
		O,O-DIMETHYL-S-(2-ETHYLTHIO-ETHYL)-DITHIOFOSFAAT (DUTCH)
		O,O-DIMETHYL-S-2-ETHYLMERKAPTOETHYLESTER KYSELINY DITHIOFOSFORECNE (CZECH)
		O,O-DIMETIL-S-(ETILTIO-ETIL)-DITIOFOSFATO (ITALIAN)
		PHOSPHORODITHIOIC ACID, O,O-DIMETHYL S-(2-ETHYLTHIO)ETHYL ESTER
		PHOSPHORODITHIOIC ACID, S-(2-(ETHYLTHIO)ETHYL) O,O-DIMETHYL ESTER
		S-(2-(ETHYLTHIO)ETHYL) O,O-DIMETHYLPHOSPHORODITHIONATE
		S-(2-(ETHYLTHIO)ETHYL)DIMETHYL PHOSPHOROTHIOLOTHIONATE
		S-[2-(ETHYLTHIO)ETHYL] O,O-DIMETHYL PHOSPHORODITHIOATE
		SAN 230
		THIAMETON
		THIOMETON

CAS No.	IN	Chemical Name
		VELTIN
640-19-7		
		1081
		2-FLUOROACETAMIDE
		ACETAMIDE, 2-FLUORO-
		AFL 1081
		COMPOUND 1081
		FAA
		FLUORAKIL 100
		FLUOROACETAMIDE
		FLUOROACETAMIDE/1081
		FLUOROACETIC ACID AMIDE
		FLUTRITEX 1
		FUSSOL
		MEGATOX
		MONOFLUOROACETAMIDE
		NAVRON
		RCRA WASTE NUMBER P057
		RODEX
		YANOCK
643-28-7		
		ISOPROPYLANILINE
		N-ISOPROPYLANILINE
643-58-3		
		1,1'-BIPHENYL, 2-METHYL-
		2-METHYL-1,1'-BIPHENYL
		2-METHYLBIPHENYL
		BIPHENYL, 2-METHYL-
		O-METHYLBIPHENYL
644-31-5		
		ACETOZONE
		ACETYL BENZOYL PEROXIDE
		ACETYL BENZOYL PEROXIDE SOLUTION, NOT OVER 40% PEROXIDE (DOT)
		BENZOYL ACETYL PEROXIDE
		BENZOZONE
		PEROXIDE, ACETYL BENZOYL
		PEROXIDE, ACETYL BENZOYL, NOT OVER 40% PEROXIDE IN SOLUTION
		UN 2081 (DOT)
644-64-4		
		1-DIMETHYLCARBAMOYL-5-METHYL-3-PYRAZOLYL DIMETHYLCARBAMATE
		2-(N,N-DIMETHYLCARBAMYL)-3-METHYLPYRAZOLYL-5 N,N-DIMETHYLCARBAMATE
		2-DIMETHYLCARBAMOYL-3-METHYL-5-PYRAZOLYL DIMETHYLCARBAMATE
		2-DIMETHYLCARBAMOYL-3-METHYLPYRAZOLYL-(5)-N,N-DIMETHYLCARBAMAT (GERMAN)
		5-METHYL-1H-PYRAZOL-3-YL DIMETHYLCARBAMATE
		CARBAMIC ACID, DIMETHYL-, 1-((DIMETHYLAMINO)CARBONYL)-5-METHYL-1H-PYRAZOL-3-YL ESTER
		CARBAMIC ACID, DIMETHYL-, 1-DIMETHYLCARBAMOYL-5-METHYLPYRAZOL-3-YL ESTER
		CARBAMIC ACID, DIMETHYL-, 1-[(DIMETHYLAMINO)CARBONYL]-5-METHYL-1H-PYRAZOL-3-YL ESTER
		CARBAMIC ACID, DIMETHYL-, 5-METHYL-1H-PYRAZOL-3-YL ESTER
		CARBAMIC ACID, DIMETHYL-, ESTER WITH 3-HYDROXY-N,N,5-TRIMETHYLPYRAZOLE-1-CARBOXAMIDE
		DIMETHYL 2-CARBAMYL-3-METHYLPYRAZOLYLDIMETHYLCARBAMATE
		DIMETHYLCARBAMIC ACID 1-((DIMETHYLAMINO)CARBONYL)-5-METHYL-1H-PYRAZOL-3-YL ESTER
		DIMETHYLCARBAMIC ACID ESTER WITH 3-HYDROXY-N,N,5-TRIMETHYLPYRAZOLE-1-CARBOXAMIDE
		DIMETILAN
		DIMETILANE
		ENT 25,922
		ENT 25595-X
		G-22870
		GEIGY 22870
		GEIGY GS-13332
		GS-13332
		PYRAZOLE-1-CARBOXAMIDE, 3-HYDROXY-N,N,5-TRIMETHYL-, DIMETHYLCARBAMATE (ESTER)
		SNIP
		SNIP FLY
		SNIP FLY BANDS
644-97-3		
		BENZENE PHOSPHOROUS DICHLORIDE
		BENZENE PHOSPHORUS DICHLORIDE
		BENZENE PHOSPHORUS DICHLORIDE (DOT)
		DICHLOROPHENYLPHOSPHINE
		PHENYL PHOSPHORUS DICHLORIDE
		PHENYL PHOSPHORUS DICHLORIDE (DOT)
		PHENYLDICHLOROPHOSPHINE
		PHENYLPHOSPHINE DICHLORIDE
		PHENYLPHOSPHONOUS ACID DICHLORIDE
		PHENYLPHOSPHONOUS DICHLORIDE
		PHOSPHINE, DICHLOROPHENYL-
		PHOSPHONOUS DICHLORIDE, PHENYL-
		UN 2798 (DOT)
645-62-5		
		2-ETHYL-2-HEXEN-1-AL
		2-ETHYL-2-HEXENAL
		2-ETHYL-3-PROPYLACROLEIN
		2-ETHYLHEXENAL
		2-HEXENAL, 2-ETHYL-
		3-FORMYLHEPT-3-ENE
		ACROLEIN, 2-ETHYL-3-PROPYL-
		ALPHA-ETHYL-2-HEXENAL
		ALPHA-ETHYL-BETA-N-PROPYLACROLEIN
		ALPHA-ETHYL-BETA-PROPYLACROLEIN
646-04-8		
		(E)-2-PENTENE
		2-PENTENE, (E)-
		2-TRANS-PENTENE
		BETA-AMYLENE-TRANS
		TRANS-BETA-AMYLENE
		TRANS-2-PENTENE
646-06-0		
		1,2-ISOPROPYLIDENEGLYCERIN
		1,2-ISOPROPYLIDENEGLYCEROL
		1,2-O,O-ISOPROPYLIDENEGLYCERIN
		1,2-O-ISOPROPYLIDENEGLYCEROL
		1,3-DIOXACYCLOPENTANE
		1,3-DIOXOLAN
		1,3-DIOXOLANE
		1,3-DIOXOLANE-4-METHANOL, 2,2-DIMETHYL-
		1,3-DIOXOLE, DIHYDRO-
		2,2-DIMETHYL-1,3-DIOXOLAN-4-YLMETHANOL
		2,2-DIMETHYL-1,3-DIOXOLANE-4-METHANOL
		2,2-DIMETHYL-4-HYDROXYMETHYLDIOXOLANE
		2,2-DIMETHYL-4-OXYMETHYL-1,3-DIOXOLANE
		2,2-DIMETHYL-5-HYDROXYMETHYL-1,3-DIOXOLANE
		2,3-(ISOPROPYLIDENEDIOXY)PROPANOL
		2,3-ISOPROPYLIDENEGLYCEROL
		2,3-O-ISOPROPYLIDENEGLYCEROL
		4-HYDROXYMETHYL-2,2-DIMETHYL-1,3-DIOXOLANE
		ACETONE GLYCERIN KETAL
		ACETONE MONOGLYCEROL KETAL
		ACETONE, CYCLIC (HYDROXYMETHYL)ETHYLENE ACETAL
		ALPHA,BETA-ISOPROPYLIDENEGLYCEROL
		DIOXOLAN
		DIOXOLANE
		DIOXOLANE (DOT)
		ETHYLENE GLYCOL FORMAL
		FORMAL GLYCOL
		GIE
		GLYCERIN ISOPROPYLIDENE ETHER
		GLYCEROL ACETONIDE
		GLYCEROL ALPHA,BETA-ISOPROPYLIDENE ETHER
		GLYCEROL DIMETHYLKETAL
		GLYCEROL, 1,2-O-ISOPROPYLIDENE
		GLYCEROLACETONE
		GLYCOL FORMAL
		ISOPROPYLIDENE GLYCEROL
		SOLKETAL

CAS No. | IN | Chemical Name

UN 1166 (DOT)

671-16-9
1-METHYL-2-[P-(ISOPROPYLCARBAMOYL)BENZYL]HYDRAZINE
4-[(2-METHYLHYDRAZINO)METHYL]-N-ISOPROPYLBENZAMIDE
BENZAMIDE, N-(1-METHYLETHYL)-4-[(2-METHYLHYDRAZINO)METHYL]-
CB 400-497
MIH
N-ISOPROPYL-ALPHA-(2-METHYLHYDRAZINO)-P-TOLUAMIDE
P-(2-METHYLHYDRAZINOMETHYL)-N-ISOPROPYLBENZAMIDE
P-TOLUAMIDE, N-ISOPROPYL-ALPHA-(2-METHYLHYDRAZINO)-
PROCARBAZINE

674-82-8
2-OXETANONE, 4-METHYLENE-
3-BUTENO-BETA-LACTONE
3-BUTENOIC ACID, 3-HYDROXY-, BETA-LACTONE
4-METHYLENE-2-OXETANONE
DIKETENE
DIKETENE, INHIBITED (DOT)
ETHENONE, DIMER
KETENE DIMER

675-14-9
1,3,5-TRIAZINE, 2,4,6-TRIFLUORO-
2,4,6-TRIFLUORO-1,3,5-TRIAZINE
2,4,6-TRIFLUORO-S-TRIAZINE
2,4,6-TRIFLUOROTRIAZINE
CYANURIC FLUORIDE
CYANURIC TRIFLUORIDE
S-TRIAZINE, 2,4,6-TRIFLUORO-
TRIFLUORO-1,3,5-TRIAZINE
TRIFLUORO-S-TRIAZINE
TRIFLUOROTRIAZINE

676-83-5
DICHLOROMETHYLPHOSPHINE
METHYL PHOSPHONOUS DICHLORIDE
METHYL PHOSPHONOUS DICHLORIDE (DOT)
METHYLDICHLOROPHOSPHINE
METHYLPHOSPHINIC DICHLORIDE
METHYLPHOSPHINOUS DICHLORIDE
METHYLPHOSPHONOUS DICHLORIDE
METHYLPHOSPHORUS DICHLORIDE
PHOSPHINE, DICHLOROMETHYL-
PHOSPHONOUS DICHLORIDE, METHYL-
UN 2845 (DOT)

676-97-1
DICHLOROMETHYLPHOSPHINE OXIDE
METHANEPHOSPHONODICHLORIDIC ACID
METHANEPHOSPHONYL CHLORIDE
METHYL PHOSPHONIC DICHLORIDE
METHYL PHOSPHONIC DICHLORIDE (DOT)
METHYLPHOSPHONIC ACID DICHLORIDE
METHYLPHOSPHONIC DICHLORIDE
METHYLPHOSPHONODICHLORIDIC ACID
METHYLPHOSPHONYL CHLORIDE
METHYLPHOSPHONYL DICHLORIDE
NA 9206 (DOT)
PHOSPHONIC DICHLORIDE, METHYL-

676-98-2
DICHLOROMETHYLPHOSPHINE SULFIDE
METHANEPHOSPHONOTHIOIC DICHLORIDE
METHYL PHOSPHONOTHIOIC DICHLORIDE
METHYL PHOSPHONOTHIOIC DICHLORIDE, ANHYDROUS
METHYL PHOSPHONOTHIOIC DICHLORIDE, ANHYDROUS (DOT)
METHYLDICHLOROPHOSPHINE SULFIDE
METHYLPHOSPHONOTHIOIC DICHLORIDE
METHYLPHOSPHONOTHIOYL DICHLORIDE
METHYLTHIONOPHOSPHONIC DICHLORIDE
METHYLTHIOPHOSPHONIC ACID DICHLORIDE
METHYLTHIOPHOSPHONIC DICHLORIDE
NA 1760 (DOT)
PHOSPHONOTHIOIC DICHLORIDE, METHYL-
PHOSPHONOTHIOIC DICHLORIDE, METHYL-, ANHYDROUS

677-71-4
2,2-PROPANEDIOL, 1,1,1,3,3,3-HEXAFLUORO-
2,2-PROPANEDIOL, HEXAFLUORO-
2-PROPANONE, HEXAFLUORO-, HYDRATE
ACETONE, HEXAFLUORO-, HYDRATE
HEXAFLUORO-2,2-PROPANEDIOL
HEXAFLUOROACETONE HYDRATE
HEXAFLUOROACETONE HYDRATE (DOT)
HEXOFLUOROACETONE DIHYDRATE
UN 2552 (DOT)

680-31-9
EASTMAN INHIBITOR HPT
ENT 50,882
HEMPA
HEXAMETAPOL
* HEXAMETHYL PHOSPHORAMIDE
HEXAMETHYLORTHOPHOSPHORIC TRIAMIDE
HEXAMETHYLPHOSPHORIC ACID TRIAMIDE
HEXAMETHYLPHOSPHORIC TRIAMIDE
HMPA
HMPT
HMPTA
HPT
PHOSPHORIC ACID HEXAMETHYLTRIAMIDE
PHOSPHORIC HEXAMETHYLTRIAMIDE
PHOSPHORIC TRIAMIDE, HEXAMETHYL-
PHOSPHORIC TRIS(DIMETHYLAMIDE)
PHOSPHORYL HEXAMETHYLTRIAMIDE
TRIS(DIMETHYLAMINO)PHOSPHINE OXIDE

681-84-5
METHYL ORTHOSILICATE
METHYL SILICATE
METHYL SILICATE ((CH3)4SiO4)
METHYL SILICATE ((MEO)4Si)
SILANE, TETRAMETHOXY-
SILICIC ACID (H4SiO4), TETRAMETHYL ESTER
SILICON TETRAMETHOXIDE
TETRA-METHYL ORTHOSILICATE
TETRAMETHOXYSILANE
TETRAMETHYL SILICATE

684-16-2
1,1,1,3,3,3-HEXAFLUORO-2-PROPANONE
2-PROPANONE, 1,1,1,3,3,3-HEXAFLUORO-
2-PROPANONE, HEXAFLUORO-
6FK
ACETONE, HEXAFLUORO-
GC 7887
HEXAFLUOROACETONE
HEXAFLUOROACETONE (ACGIH,DOT)
NCI-C56440
PERFLUORO-2-PROPANONE
PERFLUOROACETONE
UN 2420 (DOT)

684-93-5
1-METHYL-1-NITROSOUREA
1-NITROSO-1-METHYLUREA
METHYLNITROSOUREA
MNU
N-METHYL-N-NITROSOUREA
N-NITROSO-N-METHYLCARBAMIDE
N-NITROSO-N-METHYLUREA
NITROSO-N-METHYLUREA
NITROSOMETHYLUREA
NMM
NMU
NSC 23909
UREA, 1-METHYL-1-NITROSO-
UREA, N-METHYL-N-NITROSO-

CAS No.	IN	Chemical Name
686-31-7		
		2-ETHYLPEROXYHEXANOIC ACID TERT-PENTYL ESTER
		HEXANEPEROXOIC ACID, 2-ETHYL-, 1,1-DIMETHYLPROPYL ESTER
		LUPERSOL 575
		LUPERSOL TA 75
		PEROXYHEXANOIC ACID, 2-ETHYL-, TERT-PENTYL ESTER
		TERT-AMYL 2-ETHYLPEROXYHEXANOATE
		TERT-AMYL PEROXY-2-ETHYLHEXANOATE, TECHNICAL PURE (DOT)
		TERT-PENTYL ALCOHOL, 2-ETHYLPEROXYHEXANOATE
		UN 2898 (DOT)
688-74-4		
		BORESTER 2
		BORIC ACID (H3BO3), TRIBUTYL ESTER
		BORON TRIBUTOXIDE
		BUTYL BORATE
		N-BUTYL BORATE
		TRI-N-BUTOXYBORANE
		TRI-N-BUTYL BORATE
		TRIBUTOXYBORANE
		TRIBUTOXYBORON
		TRIBUTYL BORATE
		TRIBUTYL ORTHOBORATE
		TRIS(BUTOXY)BORANE
689-97-4		
		1-BUTEN-3-YNE
		1-BUTENYNE
		1-BUTYN-3-ENE
		3-BUTEN-1-YNE
		BUTENYNE
		ETHENE, ETHYNYL-
		MONOVINYLACETYLENE
		VINYL ACETYLENE
691-37-2		
		1-PENTENE, 4-METHYL-
	*	4-METHYL-1-PENTENE
		ISOBUTYLETHENE
693-21-0		
		DIETHYLENE GLYCOL DINITRATE
694-05-3		
		1,2,3,6-TETRAHYDROPYRIDINE
696-28-6		
		ARSINE, DICHLOROPHENYL-
		ARSONOUS DICHLORIDE, PHENYL-
		ARSONOUS DICHLORIDE, PHENYL- (9CI)
		DICHLOROPHENYLARSINE
		FDA
		FENILDICLOROARSINA (ITALIAN)
		NA 1556 (DOT)
		PHENYL DICHLORARSINE
		PHENYL DICHLOROARSINE (DOT)
		PHENYLARSENIC DICHLORIDE
		PHENYLARSINEDICHLORIDE
		PHENYLDICHLOROARSINE
		RCRA WASTE NUMBER P036
		TL 69
702-03-4		
		3-(CYCLOHEXYLAMINO)PROPIONITRILE
		CYCLOHEXYLAMINE, CYANOETHYL-
		N-(2-CYANOETHYL)-N-CYCLOHEXYLAMINE
		N-(2-CYANOETHYL)CYCLOHEXYLAMINE
		N-(P-CYANOETHYL)CYCLOHEXYLAMINE
		PROPANENITRILE, 3-(CYCLOHEXYLAMINO)-
		PROPIONITRILE, 3-(CYCLOHEXYLAMINO)-
712-48-1		
		ARSINE, CHLORODIPHENYL-
		ARSINOUS CHLORIDE, DIPHENYL-
		ARSINOUS CHLORIDE, DIPHENYL- (9CI)
		BLUE CROSS
		CHLORODIPHENYLARSINE
		CLARK I
		DA
		DIPHENYLCHLOORARSINE (DUTCH)
		DIPHENYLCHLOROARSINE
		DIPHENYLCHLOROARSINE (DOT)
		SNEEZING GAS
712-68-5		
		1,3,4-THIADIAZOL-2-AMINE, 5-(5-NITRO-2-FURANYL)-
		1,3,4-THIADIAZOLE, 2-AMINO-5-(5-NITRO-2-FURYL)-
		2-AMINO-5(5-NITRO-2-FURYL)-1,3,4-THIADIZOLE
		FURIDIAZINA
		THIAFUR
		TIAFUR
		TRIAFUR
723-46-6		
		SULFAMETHOXAZOLE
732-11-6		
		(O,O-DIMETHYL-PHTHALIMIDIOMETHYL-DITHIOPHOSPHATE)
		APPA
		DECEMTHION
		DECEMTHION P-6
		ENT 25,705
		FTALOPHOS
		IMIDAN
		IMIDATHION
		KEMOLATE
		N-(MERCAPTOMETHYL)PHTHALIMIDE S-(O,O-DIMETHYL PHOSPHORODITHIOATE)
		O,O-DIMETHYL 5-(PHTHALIMIDOMETHYL)DITHIOPHOSPHATE
		O,O-DIMETHYL S-(N-PHTHALIMIDOMETHYL)DITHIOPHOSPHATE
		O,O-DIMETHYL S-(PHTHALIMIDOMETHYL) DITHIOPHOSPHATE
		O,O-DIMETHYL S-PHTHALIMIDOMETHYL PHOSPHORODITHIOATE
		PERCOLATE
		PHOSMET
		PHOSPHORODITHIOIC ACID, O,O-DIMETHYL ESTER, S-ESTER WITH N-(MERCAPTOMETHYL)PHTHALIMIDE
		PHOSPHORODITHIOIC ACID, S-((1,3-DIHYDRO-1,3-DIOXO-ISOINDOL-2-YL)METHYL) O,O-DIMETHYL ESTER
		PHOSPHORODITHIOIC ACID, S-[(1,3-DIHYDRO-1,3-DIOXO-2H-ISOINDOL-2-YL)METHYL] O,O-DIMETHYL ESTER
		PHTHALIMIDE, N-(MERCAPTOMETHYL)-, S-ESTER WITH O,O-DIMETHYL PHOSPHORODITHIOATE
		PHTHALIMIDO O,O-DIMETHYL PHOSPHORODITHIOATE
		PHTHALIMIDOMETHYL O,O-DIMETHYL PHOSPHORODITHIOATE
		PHTHALOPHOS
		PMP
		PMP (PESTICIDE)
		PROLATE
		R 1504
		SAFIDON
		SIMIDAN
		SMIDAN
		STAUFFER R 1504
738-70-5		
		TRIMETHOPRIM
753-53-7		
		BORON TRIFLUORIDE ACETIC ACID COMPLEX
756-80-9		
		O,O-DIMETHYL PHOSPHORODITHIOATE
757-58-4		
		BLADAN
		BLADAN BASE
		ETHYL TETRAPHOSPHATE
		ETHYL TETRAPHOSPHATE, HEXA-
		HET
		HETP

CAS No.	IN	Chemical Name
		HEXAETHYL TETRAPHOSPHATE
		HEXAETHYL TETRAPHOSPHATE (DOT)
		HEXAETHYL TETRAPHOSPHATE MIXTURE, DRY, WITH >2% HEXAETHYL TETRAPHOSPHATE
		HEXAETHYL TETRAPHOSPHATE MIXTURE, DRY, WITH ≤2% HEXAETHYL TETRAPHOSPHATE
		HEXAETHYL TETRAPHOSPHATE MIXTURE, LIQUID, WITH >25% HEXAETHYL TETRAPHOPHATE
		HEXAETHYL TETRAPHOSPHATE MIXTURE, LIQUID, WITH ≤25% HEXAETHYL TETRAPHOSPHATE
		HEXAETHYL TETRAPHOSPHATE, LIQUID
		HEXAETHYL TETRAPHOSPHATE, LIQUID (DOT)
		HEXAETHYL TETRAPHOSPHATE, LIQUID, CONTAINING MORE THAN 25% HEXAETHYL TETRAPHOSPHATE (DOT)
		HEXAETHYL TETRAPHOSPHATE, LIQUID, CONTAINING NOT MORE THAN 25% HEXAETHYL TETRAPHOSPHATE (DOT)
		HTP
		RCRA WASTE NUMBER P062
		TETRAPHOSPHATE HEXAETHYLIQUE (FRENCH)
		TETRAPHOSPHORIC ACID, HEXAETHYL ESTER
		TETRAPHOSPHORIC ACID, HEXAETHYL ESTER, LIQUID
		UN 1611 (DOT)
		UN 2783 (DOT)
759-73-9		
		1-ETHYL-1-NITROSOUREA
		ENU
		N-ETHYL-N-NITROSOUREA
		N-NITROSO-N-ETHYLUREA
		NITROSO-N-ETHYLUREA
		NSC 45403
		UREA, 1-ETHYL-1-NITROSO-
		UREA, N-ETHYL-N-NITROSO-
760-21-4		
		1,1-DIETHYLETHENE
		1-BUTENE, 2-ETHYL-
		2-ETHYL-1-BUTENE
		3-METHYLENEPENTANE
		PENTANE, 3-METHYLENE-
760-93-0		
		2-METHYL-2-PROPENOIC ACID ANHYDRIDE
		2-PROPENOIC ACID, 2-METHYL-, ANHYDRIDE
		2-PROPENOIC ACID, 2-METHYL-, ANHYDRIDE (9CI)
		METHACRYLIC ACID ANHYDRIDE
		METHACRYLIC ANHYDRIDE
		METHACRYLOYL ANHYDRIDE
762-12-9		
		DECANOX
		DECANOX F
		DECANOYL PEROXIDE
		DECANOYL PEROXIDE (DOT)
		DECANOYL PEROXIDE, TECHNICALLY PURE
		DIDECANOYL PEROXIDE
		DIDECANOYL PEROXIDE, TECHNICAL PURE (DOT)
		PERKADOX SE 10
		PEROXIDE, BIS(1-OXODECYL)
		UN 2120 (DOT)
762-13-0		
		DI-N-NONANOYL PEROXIDE
		DI-N-NONANOYL PEROXIDE, TECHNICALLY PURE (DOT)
		DIPELARGONYL PEROXIDE
		NONANOYL PEROXIDE
		PELARGONOYL PEROXIDE
		PELARGONYL PEROXIDE
		PELARGONYL PEROXIDE, TECHNICALLY PURE
		PELARGONYL PEROXIDE, TECHNICALLY PURE (DOT)
		PEROXIDE, BIS(1-OXONONYL)
		UN 2130 (DOT)
762-16-3		
		CAPROLYL PEROXIDE
		CAPRYL PEROXIDE
		CAPRYLOYL PEROXIDE (DOT)
		CAPRYLYL PEROXIDE
		CAPRYLYL PEROXIDE SOLUTION
		CAPRYLYL PEROXIDE SOLUTION (DOT)
		DI-N-OCTANOYL PEROXIDE
		DI-N-OCTANOYL PEROXIDE, TECHNICAL PURE (DOT)
		DICAPRYLYL PEROXIDE
		DIOCTANOYL PEROXIDE
		N-OCTANOYL PEROXIDE (DOT)
		N-OCTANOYL PEROXIDE, TECHNICALLY PURE
		N-OCTANOYL PEROXIDE, TECHNICALLY PURE (DOT)
		NA 2129 (DOT)
		OCTANOYL PEROXIDE
		OCTANYL PEROXIDE
		PERKADOX SE 8
		PEROXIDE, BIS(1-OXOOCTYL)
		PEROXIDE, BIS(1-OXOOCTYL) (9CI)
		PEROXIDE, OCTANOYL
		UN 2129 (DOT)
763-29-1		
		1-PENTENE, 2-METHYL-
	*	2-METHYL-1-PENTENE
		2-METHYL-PENTENE-1
		2-METHYLPENTENE
		4-METHYL-4-PENTENE
764-35-2		
		1-METHYL-2-PROPYLACETYLENE
		2-HEXYNE
		METHYL PROPYL ACETYLENE
765-34-4		
		2,3-EPOXY-1-PROPANAL
		2,3-EPOXYPROPANAL
		2,3-EPOXYPROPIONALDEHYDE
		EPIHYDRINALDEHYDE
		EPIHYDRINE ALDEHYDE
		EPOXYPROPANAL
		GLYCIDAL
		GLYCIDALDEHYDE
		GLYCIDALDEHYDE (DOT)
		GLYCIDYLALDEHYDE
		OXIRANECARBOXALDEHYDE
		PROPIONALDEHYDE, 2,3-EPOXY-
		RCRA WASTE NUMBER U126
		UN 2622 (DOT)
766-09-6		
		1-ETHYL PIPERIDINE
		1-ETHYL PIPERIDINE (DOT)
		ETHYL PIPERIDINE
		N-AETHYLPIPERIDIN (GERMAN)
		N-ETHYLPIPERIDINE
		PIPERIDINE, 1-ETHYL-
		UN 2386 (DOT)
768-52-5		
		ANILINE, N-ISOPROPYL-
		BENZENAMINE, N-(1-METHYLETHYL)-
		ISOPROPYLANILINE
		N-(1-METHYLETHYL)BENZENAMINE
		N-ISOPROPYLANILINE
		N-ISOPROPYLBENZENAMINE
		N-PHENYLISOPROPYLAMINE
771-29-9		
		1,2,3,4-TETRAHYDRO-1-NAPHTHYL HYDROPEROXIDE
		1-HYDROPEROXYTETRALIN
		ALPHA-TETRALIN HYDROPEROXIDE
		HYDROPEROXIDE, 1,2,3,4-TETRAHYDRO-1-NAPHTHALENYL
		HYDROPEROXIDE, 1,2,3,4-TETRAHYDRO-1-NAPHTHYL-
		HYDROPEROXYDE DE TETRALINE (FRENCH)
		IDROPEROSSIDO DI TETRALINA (ITALIAN)
		NAPHTHALENE, 1,2,3,4-TETRAHYDRO-, 1-HYDROPEROXIDE
		TETRALIN 1-HYDROPEROXIDE
		TETRALIN HYDROPEROXIDE
		TETRALIN HYDROPEROXIDE, TECHNICALLY PURE
		TETRALIN HYDROPEROXIDE, TECHNICALLY PURE (DOT)
		TETRALIN PEROXIDE

CAS No.	IN	Chemical Name
		TETRALINEHYDROPEROXYDE (DUTCH)
		TETRALINHYDROPEROXID (GERMAN)
		TETRALYL HYDROPEROXIDE
		UN 2136 (DOT)
772-54-3		
		BENZENEMETHANAMINE, N,N-DIETHYL-
		BENZYLAMINE, N,N-DIETHYL-
		BENZYLDIETHYLAMINE
		DIETHYLBENZYLAMINE
		N,N-DIETHYLBENZYLAMINE
		N-(PHENYLMETHYL)DIETHYLAMINE
		N-BENZYLDIETHYLAMINE
776-74-9		
		ALPHA-BROMODIPHENYLMETHANE
		BENZENE, 1,1'-(BROMOMETHYLENE)BIS-
		BENZHYDRYL BROMIDE
		BROMODIPHENYLMETHANE
		DIPHENYL METHYL BROMIDE SOLUTION
		DIPHENYL METHYL BROMIDE, SOLID
		DIPHENYL METHYL BROMIDE, SOLID (DOT)
		DIPHENYL METHYL BROMIDE, SOLUTION (DOT)
		DIPHENYLBROMOMETHANE
		DIPHENYLMETHYL BROMIDE
		DIPHENYLMETHYL BROMIDE (DOT)
		METHANE, BROMODIPHENYL-
		UN 1770 (DOT)
777-37-7		
		1-CHLORO-4-NITRO-2-(TRIFLUOROMETHYL)BENZENE
		2-(TRIFLUOROMETHYL)-4-NITROCHLOROBENZENE
		2-CHLORO-ALPHA,ALPHA,ALPHA-TRIFLUORO-5-NITROTOLUENE
		2-CHLORO-5-NITRO-1-TRIFLUOROMETHYLBENZENE
		2-CHLORO-5-NITROBENZOTRIFLUORIDE
		3-(TRIFLUOROMETHYL)-4-CHLORONITROBENZENE
		4-CHLORO-3-(TRIFLUOROMETHYL)NITROBENZENE
		BENZENE, 1-CHLORO-4-NITRO-2-(TRIFLUOROMETHYL)-
		NITROCHLOROBENZOTRI-FLUORIDE (4-NITRO-2-(TRIFLUOROMETHYL)-1-CHLO
		TOLUENE, 2-CHLORO-ALPHA,ALPHA,ALPHA-TRIFLUORO5-NITRO-
778-74-7		
		POTASSIUM PERCHLORATE
786-19-6		
		ACARITHION
		AKARITHION
		CARBOFENOTHION
		CARBOFENOTHION (DUTCH)
		CARBOFENTHION
		CARBOPHENOTHION
		DAGADIP
		DITHIOPHOSPHATE DE O,O-DIETHYLE ET DE (4-CHLOROPHENYL) THIOMETHYLE (FRENCH)
		ENDYL
		ENT 23,708
		ETHYL CARBOPHENOTHION
		GARRATHION
		HEXATHION
		LETHOX
		NEPHOCARP
		O,O-DIAETHYL-S-((4-CHLOR-PHENYL-THIO)-METHYL)-DITHIOPHOSPHAT (GERMAN)
		O,O-DIETHYL DITHIOPHOSPHORIC ACID P-CHLOROPHENYLTHIOMETHYL ESTER
		O,O-DIETHYL P-CHLOROPHENYLMERCAPTOMETHYL DITHIOPHOSPHATE
		O,O-DIETHYL S-(4-CHLOROPHENYLTHIOMETHYL) DITHIOPHOSPHATE
		O,O-DIETHYL S-(P-CHLOROPHENYLTHIO)METHYL PHOSPHORODITHIOATE
		O,O-DIETHYL S-P-CHLOROPHENYLTHIOMETHYL DITHIOPHOSPHATE
		O,O-DIETHYL-S-(4-CHLOOR-FENYL-THIO)-METHYL)-DITHIOFOSFAAT (DUTCH)
		O,O-DIETHYL-S-P-CHLORFENYLTHIOMETHYLESTER
		KYSELINY DITHIOFOSFORECNE (CZECH)
		O,O-DIETIL-S-((4-CLORO-FENIL-TIO)-METILE)-DITIOFOSFATO (ITALIAN)
		OLEOAKARITHION
		PHOSPHORODITHIOIC ACID, S-(((P-CHLOROPHENYL)-THIO)METHYL) O,O-DIETHYL ESTER
		PHOSPHORODITHIOIC ACID, S-[[(4-CHLOROPHENYL)-THIO]METHYL] O,O-DIETHYL ESTER
		R 1303
		S-((P-CHLOROPHENYL THIO)METHYL) O,O-DIETHYL PHOSPHORODITHIOATE
		S-(4-CHLOROPHENYLTHIOMETHYL)DIETHYL PHOSPHOROTHIOLOTHIONATE
		STAUFFER R 1303
		TRITHION
		TRITHION MITICIDE
789-61-7		
		B-TGDR
790-83-0		
		ALPHA-BROMOETHANOIC ACID
		BROMOACETIC ACID, SOLUTION (DOT)
		BROMOACETIC ACID SOLUTION
		MONOBROMESSIGSAEURE (GERMAN)
		UN 1938 (DOT)
794-93-4		
		BIS(HYDROXYMETHYL)FURATRIZINE
		DIHYDROXYMETHYLFURATRIZINE
		FURATONE
		FURATRIZINE, BIS(HYDROXYMETHYL)-
		METHANOL, [[6-[2-(5-NITRO-2-FURANYL)ETHENYL]-1,2,4-TRIAZIN-3-YL]IMINO]BIS-
		METHANOL, [[6-[2-(5-NITRO-2-FURYL)VINYL]-AS-TRIAZIN-3-YL]IMINO]DI-
		PANFURAN S
814-49-3		
		CHLORODIETHOXYPHOSPHINE OXIDE
		CHLOROPHOSPHORIC ACID, DIETHYL ESTER
		DIETHOXYPHOSPHORUS OXYCHLORIDE
		DIETHOXYPHOSPHORYL CHLORIDE
		DIETHYL CHLOROPHOSPHATE
		DIETHYL CHLOROPHOSPHONATE
		DIETHYL PHOSPHOROCHLORIDATE
		DIETHYL PHOSPHOROCHLORIDE
		DIETHYLPHOSPHORIC ACID CHLORIDE
		ETHYL PHOSPHOROCHLORIDATE (Cl(ETO)2PO)
		ETHYL PHOSPHOROCHLORIDATE (ETO)2ClPO
		O,O-DIETHYL CHLORIDOPHOSPHATE
		O,O-DIETHYL CHLOROPHOSPHATE
		O,O-DIETHYL CHLOROPHOSPHONATE
		O,O-DIETHYL PHOSPHOROCHLORIDATE
		O,O-DIETHYLPHOSPHORYL CHLORIDE
		PHOSPHORIC ACID CHLORIDE DIETHYL ESTER
		PHOSPHOROCHLORIDIC ACID, DIETHYL ESTER
814-68-6		
		2-PROPENOYL CHLORIDE
		2-PROPENOYL CHLORIDE (9CI)
		ACRYLIC ACID CHLORIDE
		ACRYLOYL CHLORIDE
		ACRYLYL CHLORIDE
		PROPENOYL CHLORIDE
814-78-8		
		2-METHYL-1-BUTEN-3-ONE
		3-BUTEN-2-ONE, 3-METHYL-
		3-METHYL-3-BUTEN-2-ON (GERMAN)
		3-METHYLENE-2-BUTANONE
		ISOPROPENYL METHYL KETONE
		KETONE, METHYL ISOPROPENYL
		METHYL ISOPROPENYL KETONE
		METHYL ISOPROPENYL KETONE, INHIBITED
		METHYL ISOPROPENYL KETONE, INHIBITED (DOT)
		PROPEN-2-YL METHYL KETONE
		UN 1246 (DOT)

CAS No.	IN	Chemical Name
814-91-5		
		COPPER OXALATE
		COPPER OXALATE (CuC2O4)
		COPPER(2+) OXALATE
		COPPER(II) OXALATE
	*	CUPRIC OXALATE
		CUPRIC OXALATE (1:1)
		CUPRIC OXALATE (DOT)
		ETHANEDIOIC ACID, COPPER(2+) SALT (1:1)
		NA 2449 (DOT)
		OXALIC ACID, COPPER(2+) SALT (1:1)
815-82-7		
		BUTANEDIOIC ACID, 2,3-DIHYDROXY-, (R-(R*,R*)-, COPPER (2+) SALT
		CUPRIC TARTRATE
		CUPRIC TARTRATE (DOT)
		NA 9111 (DOT)
		TARTARIC ACID, COPPER(2+) SALT
		TARTARIC ACID, COPPER(2+) SALT (1:1), (+)-
818-61-1		
		2-(ACRYLOYLOXY)ETHANOL
		2-HYDROXYETHYL ACRYLATE
		2-PROPENOIC ACID, 2-HYDROXYETHYL ESTER
		2-PROPENOIC ACID, 2-HYDROXYETHYL ESTER (9CI)
		ACRYLIC ACID, 2-HYDROXYETHYL ESTER
		BETA-HYDROXYETHYL ACRYLATE
		BISOMER 2HEA
		ETHYLENE GLYCOL MONOACRYLATE
		ETHYLENE GLYCOL, ACRYLATE
		HYDROXYETHYL ACRYLATE
821-08-9		
		1,5-HEXADIEN-3-YNE
		DIVINYL ACETYLENE
821-95-4		
		1-UNDECENE
		ALPHA-UNDECENE
		N-1-UNDECENE
822-06-0		
		1,6-DIISOCYANATOHEXANE
		1,6-HEXAMETHYLENE DIISOCYANATE
		1,6-HEXANEDIOL DIISOCYANATE
		1,6-HEXYLENE DIISOCYANATE
		HDI
	*	HEXAMETHYLENE DIISOCYANATE
		HEXAMETHYLENE-1,6-DIISOCYANATE
		HEXAMETHYLENEDIISOCYANATE (DOT)
		HEXANE 1,6-DIISOCYANATE
		HEXANE, 1,6-DIISOCYANATO-
		HMDI
		ISOCYANIC ACID, DIESTER WITH 1,6-HEXANEDIOL
		ISOCYANIC ACID, HEXAMETHYLENE ESTER
		METYLENO-BIS-FENYLOIZOCYJANIAN (POLISH)
		SZESCIOMETYLENODWUIZOCYJANIAN (POLISH)
		TL 78
		UN 2281 (DOT)
824-11-3		
		TRIMETHYLOLPROPANE PHOSPHITE
827-21-4		
		XYLENE SULFONIC ACID, SODIUM SALT
827-52-1		
		1,1'-BIPHENYL, 1,2,3,4,5,6-HEXAHYDRO-
		4-CYCLOHEXYLBENZENE
		BENZENE, CYCLOHEXYL-
		CYCLOHEXANE, PHENYL-
		CYCLOHEXYLBENZENE
		PHENYLCYCLOHEXANE
		PHENOL, 2-AMINO-4,6-DINITRO-, MONOSODIUM SALT
		PICRAMIC ACID, SODIUM SALT
		SODIUM 6-AMINO-2,4-DINITROPHENOLATE
		SODIUM PICRAMATE
		SODIUM PICRAMATE, WET (WITH AT LEAST 20% WATER) (DOT)
		SODIUM PICRAMATE, WETTED :WITH, BY WEIGHT, AT LEAST 20% WATER:
		SODIUM, (2-AMINO-4,6-DINITROPHENOXY)-
		UN 1349 (DOT)
838-88-0		
		2,2'-DIMETHYL-4,4'-METHYLENEDIANILINE
		3,3'-DIMETHYL-4,4'-DIAMINODIPHENYLMETHANE
		4,4'-DIAMINO-3,3'-DIMETHYLDIPHENYLMETHANE
		4,4'-METHYLENEBIS(O-TOLUIDINE)
		4,4'-METHYLENEBIS[2-METHYLANILINE]
		4,4'-METHYLENEDI-O-TOLUIDINE
		BENZENAMINE, 4,4'-METHYLENEBIS[2-METHYL-
		BIS(3-METHYL-4-AMINOPHENYL)METHANE
		KAYAHARD MDT
		O-TOLUIDINE, 4,4'-METHYLENEDI-
842-07-9		
		C.I. SOLVENT YELLOW 14
846-49-1		
		LORAZEPAM
846-50-4		
		TEMAZEPAM
869-29-4		
		2-PROPENAL, MONOHYDRATE, DIACETATE
		2-PROPENE-1,1-DIOL, DIACETATE
		ACROLEIN DIACETATE
		ALLYLIDENE ACETATE
		ALLYLIDENE DIACETATE
		DAP
		DAP (PESTICIDE)
		SD 345
		SD 345 (ESTER)
871-27-2		
		ALUMINUM, DIETHYLHYDRO-
		DIETHYLALANE
		DIETHYLALUMINUM HYDRIDE
		DIETHYLALUMINUM MONOHYDRIDE
		DIETHYLHYDROALUMINUM
872-05-9		
		1-DECENE
		1-N-DECENE
		ALPHA-DECENE
		GULFTENE 10
		N-1-DECENE
872-10-6		
		6-THIAUNDECANE
		AMYL SULFIDE
		DI-N-PENTYL SULFIDE
		DIAMYL SULFIDE
		DIPENTYL SULFIDE
		PENTANE, 1,1'-THIOBIS-
		PENTYL SULFIDE
872-50-4		
	*	1-METHYL-2-PYRROLIDINONE
		1-METHYL-2-PYRROLIDONE
		1-METHYL-5-PYRROLIDINONE
		1-METHYLAZACYCLOPENTAN-2-ONE
		1-METHYLPYRROLIDINONE
		1-METHYLPYRROLIDONE
		2-PYRROLIDINONE, 1-METHYL-
		M-PYROL
		METHYLPYRROLIDONE
		N-METHYL PYRROLIDINONE
		N-METHYL-2-PYRROLIDINONE
		N-METHYL-2-PYRROLIDONE
		N-METHYL-ALPHA-PYRROLIDINONE
		N-METHYL-ALPHA-PYRROLIDONE
		N-METHYL-GAMMA-BUTYROLACTAM
		N-METHYLPYRROLIDINONE
		N-METHYLPYRROLIDONE

CAS No.	IN	Chemical Name
		NMP
919-86-8		
		BAY 18436
		BAYER 25/154
		DEMENTON-S-METHYL
		DEMETON-S-METHYL
		DEMETON-S-METILE (ITALIAN)
		DIMETHYL S-(2-ETHTHIOETHYL)THIOPHOSPHATE
		DURATOX
		ETHANETHIOL, 2-(ETHYLTHIO)-, S-ESTER WITH O,O-DIMETHYL PHOSPHOROTHIOATE
		ISOMETASYSTOX
		ISOMETHYLSYSTOX
		METAISOSEPTOX
		METAISOSYSTOX
		METASYSTOX (I)
		METASYSTOX 55
		METASYSTOX FORTE
		METASYSTOX I
		METASYSTOX J
		METHYL DEMETON THIOESTER
		METHYL ISOSYSTOX
		METHYL-MERCAPTOFOS TEOLERY
		METHYLTHIONODEMETON
		O,O-DIMETHYL 2-ETHYLMERCAPTOETHYL THIOPHOSPHATE, THIOLO ISOMER
		O,O-DIMETHYL S-(2-(ETHYLTHIO)ETHYL)PHOSPHOROTHIOATE
		O,O-DIMETHYL S-(2-ETHTHIOETHYL)PHOSPHOROTHIOATE
		O,O-DIMETHYL S-2-(ETHYLTHIO)ETHYL PHOSPHOROTHIOATE
		O,O-DIMETHYL S-ETHYLMERCAPTOETHYL THIOPHOSPHATE
		O,O-DIMETHYL-S-(2-AETHYLTHIO-AETHYL)-MONOTHIOPHOSPHAT (GERMAN)
		O,O-DIMETHYL-S-(2-ETHYLTHIO-ETHYL)-MONOTHIOFOSFAAT (DUTCH)
		O,O-DIMETHYL-S-(3-THIA-PENTYL)-MONOTHIOPHOSPHAT (GERMAN)
		O,O-DIMETIL-S-(2-ETILTIO-ETIL)-MONOTIOFOSFATO (ITALIAN)
		PHOSPHOROTHIOIC ACID, O,O-DIMETHYL S-(2-(ETHYLTHIO)ETHYL) ESTER
		PHOSPHOROTHIOIC ACID, S-(2-(ETHYLTHIO)ETHYL) O,O-DIMETHYL ESTER
		S-(2-(ETHYLTHIO)ETHYL) DIMETHYL PHOSPHOROTHIOLATE
		S-(2-(ETHYLTHIO)ETHYL) O,O-DIMETHYL PHOSPHOROTHIOATE
		S-(2-(ETHYLTHIO)ETHYL) O,O-DIMETHYL THIOPHOSPHATE
		THIOPHOSPHATE DE O,O-DIMETHYLE ET DE S-2-ETHYLTHIO-ETHYLE (FRENCH)
920-46-7		
		METHACRYLOYL CHLORIDE
924-16-3		
		1-BUTANAMINE, N-BUTYL-N-NITROSO-
		DIBUTYLAMINE, N-NITROSO-
		DIBUTYLNITROSAMINE
		N,N-DI-N-BUTYLNITROSAMINE
		N-NITROSODI-N-BUTYLAMINE
		N-NITROSODIBUTYLAMINE
		NITROSODI-N-BUTYLAMINE
		NITROSODIBUTYLAMINE
926-56-7		
		1,1-DIMETHYLBUTADIENE
		1,3-PENTADIENE, 4-METHYL-
		4-METHYL-1,3-PENTADIENE
926-57-8		
		1,3-DICHLORO-2-BUTENE
		1,3-DICHLOROBUTENE-2
		2-BUTENE, 1,3-DICHLORO-
926-63-6		
		1-PROPANAMINE, N,N-DIMETHYL-
		DIMETHYL-N-PROPYLAMINE

CAS No.	IN	Chemical Name
		DIMETHYL-N-PROPYLAMINE (DOT)
		DIMETHYLPROPYLAMINE
		N,N-DIMETHYL-N-PROPYLAMINE
		N,N-DIMETHYLPROPYLAMINE
		PROPYLAMINE, N,N-DIMETHYL-
		PROPYLDIMETHYLAMINE
		UN 2266 (DOT)
926-64-7		
		2-DIMETHYLAMINOACETO-NITRILE
		2-DIMETHYLAMINOACETONITRILE (DOT)
		ACETONITRILE, (DIMETHYLAMINO)-
		DIMETHYLAMINOACETONITRILE
		DIMETHYLCYANOMETHYLAMINE
		GLYCINONITRILE, N,N-DIMETHYL-
		N-(CYANOMETHYL)DIMETHYLAMINE
		UN 2378 (DOT)
926-65-8		
		ETHER, ISOPROPYL VINYL
		ISOPROPOXYETHENE
		ISOPROPOXYETHYLENE
		ISOPROPYL VINYL ETHER
		PROPANE, 2-(ETHENYLOXY)-
		VINYL ISOPROPYL ETHER
927-07-1		
		ESPEROX 31M
		LUPERSOL 11
		PERBUTYL PV
		PEROXYPIVALIC ACID, TERT-BUTYL ESTER
		PEROXYPIVALIC ACID, TERT-BUTYL ESTER, NOT MORE THAN 77% IN SOLUTION
		PROPANEPEROXOIC ACID, 2,2-DIMETHYL-, 1,1-DIMETHYLETHYL ESTER
		TERT-BUTYL 2,2-DIMETHYLPEROXYPROPIONATE
		TERT-BUTYL 2,2-DIMETHYLPROPANEPEROXOATE
		TERT-BUTYL PEROXYPIVALATE
		TERT-BUTYL PEROXYPIVALATE, NOT MORE THAN 77% IN SOLUTION (DOT)
		TERT-BUTYL PERPIVALATE
		TERT-BUTYL TERT-VALERYL PEROXIDE
		TERT-BUTYL TRIMETHYLPEROXYACETATE
		TRIGONOX 25/75
		UN 2110 (DOT)
927-80-0		
		ETHER, ETHYL ETHYNYL
		ETHOXYACETYLENE
		ETHOXYETHYNE
		ETHYL ETHYNYL ETHER
		ETHYNE, ETHOXY-
928-45-0		
		BUTYL NITRATE
		NITRIC ACID, BUTYL ESTER
928-55-2		
		1-ETHOXY-1-PROPENE
		1-ETHOXYPROPENE
		1-PROPENE, 1-ETHOXY-
		ETHER, ETHYL PROPENYL
		ETHYL 1-PROPENYL ETHER
		ETHYL PROPENYL ETHER
		PROPENYL ETHYL ETHER
928-65-4		
		HEXYLTRICHLOROSILANE
		HEXYLTRICHLOROSILANE (DOT)
		SILANE, HEXYLTRICHLORO-
		SILANE, TRICHLOROHEXYL-
		TRICHLOROHEXYLSILANE
		UN 1784 (DOT)
928-96-1		
		(Z)-HEX-3-EN-1-OL
		3-(Z)-HEXENOL
		3-HEXEN-1-OL, (Z)-
		3-Z-HEXEN-1-OL

CAS No.	IN	Chemical Name
		BLATTERALKOHOL
		CIS-3-HEXEN-1-OL
		CIS-3-HEXENOL
		LEAF ALCOHOL
		Z-3-HEXENOL
929-06-6		
		1-AMINO-2-(2-HYDROXYETHOXY)ETHANE
		2-(2-AMINOETHOXY)ETHANOL
		2-(2-AMINOETHOXY)ETHANOL (DOT)
		2-(2-HYDROXYETHOXY)ETHYLAMINE
		2-(HYDROXYETHOXY)ETHYLAMINE
		2-AMINO-2'-HYDROXYDIETHYL ETHER
		2-AMINOETHOXYETHANOL
		2-AMINOETHYL 2-HYDROXYETHYL ETHER
		2-HYDROXYETHYLOXYETHYLAMINE
		BETA-(BETA-HYDROXYETHOXY)ETHYLAMINE
		BETA-HYDROXY-BETA'-AMINOETHYL ETHER
		DIETHYLENE GLYCOL AMINE
		DIETHYLENE GLYCOL MONOAMINE
		DIGLYCOLAMINE
		ETHANOL, 2-(2-AMINOETHOXY)-
		NA 1760 (DOT)
930-22-3		
		1,2-EPOXY-3-BUTENE
		1,2-EPOXY-3-BUTYLENE
		1,2-OXIDO-3-BUTENE
		1,3-BUTADIENE MONOEPOXIDE
		1,3-BUTADIENE OXIDE
		1-BUTENE, 3,4-EPOXY-
		3,4-EPOXY-1-BUTENE
		3,4-EPOXYBUTENE
		BUTADIENE EPOXIDE
		BUTADIENE MONOEPOXIDE
		BUTADIENE MONOOXIDE
		BUTADIENE MONOXIDE
		BUTADIENE OXIDE
		ETHENYLOXIRANE
		OXIRANE, ETHENYL-
		VINYL EPOXIDE
		VINYLOXIRANE
930-55-2		
		1-NITROSOPYRROLIDINE
		N-NITROSOPYRROLIDINE
		PYRROLIDINE, 1-NITROSO-
933-75-5		
		2,3,6-TRICHLOROPHENOL
		PHENOL, 2,3,6-TRICHLORO-
933-78-4		
		2,4,5-TRICHLOROPHENOL
		2,4,6-TRICHLORFENOL (CZECH)
		2,4,6-TRICHLOROPHENOL
		DOWICIDE 2
		DOWICIDE 2S
		NA 2020 (DOT)
		NCI-C02904
		OMAL
		PHENACHLOR
		PHENOL, 2,4,5-TRICHLORO-
		PHENOL, 2,4,6-TRICHLORO-
		PHENOL, TRICHLORO-
		PREVENTOL I
		RCRA WASTE NUMBER U231
		TCP
		TRICHLOROPHENOL
		TRICHLOROPHENOL (DOT)
933-78-8		
		2,3,5-TRICHLOROPHENOL
		PHENOL, 2,3,5-TRICHLORO-
937-14-4		
		3-CHLOROPERBENZOIC ACID
		3-CHLOROPEROXYBENZOIC ACID
		3-CHLOROPEROXYBENZOIC ACID, ≤86% WITH 3-CHLORO- BENZOIC ACID
		3-CHLOROPEROXYBENZOIC ACID, MAXIMUM CONCENTRATION 86% (DOT)
		BENZENECARBOPEROXOIC ACID, 3-CHLORO-
		M-CHLOROBENZOYL HYDROPEROXIDE
		M-CHLOROPERBENZOIC ACID
		M-CHLOROPEROXOBENZOIC ACID
		M-CHLOROPEROXYBENZOIC ACID
		M-CHLOROPEROXYBENZOIC ACID, MAXIMUM CONCENTRATION 86% (DOT)
		MCPBA
		META-CHLOROPERBENZOIC ACID
		PEROXYBENZOIC ACID, M-CHLORO-
		PEROXYBENZOIC ACID, M-CHLORO-, MAXIMUM CONCENTRATION 86%
		UN 2755 (DOT)
944-22-9		
		DIFONATE
		DIFONATUL
		DYFONAT
		DYFONATE
		DYFONATE 10G
		DYPHONATE
		ENT 25,796
		FONOFOS
		FONOFOS (ACGIH)
		FONOPHOS
		N 2790
		O-AETHYL-S-PHENYL-AETHYL-DITHIOPHOSPHONAT (GERMAN)
		O-ETHYL S-PHENYL ETHYLDITHIOPHOSPHONATE
		O-ETHYL S-PHENYL ETHYLPHOSPHONODITHIOATE
		O-ETHYL-S-PHENYL ETHYLPHOSPHONODITHIOATE
		PHOSPHONODITHIOIC ACID, ETHYL-, O-ETHYL S-PHENYL ESTER
		STAUFFER N 2790
947-02-4		
		(DIETHOXYPHOSPHINYL)DITHIOIMIDOCARBONIC ACID CYCLIC ETHYLENE ESTER
		1,2-ETHANEDITHIOL, CYCLIC ESTER WITH P,P-DIETHYL PHOSPHONODITHIOIMIDOCARBONATE
		1,2-ETHANEDITHIOL, CYCLIC S,S-ESTER WITH PHOSPHONODITHIOIMIDOCARBONIC ACID P,P-DIETHYL ESTER
		2-(DIETHOXYPHOSPHINYLIMINO)-1,3-DITHIOLANE
		AC 47031
		AMERICAN CYANAMID 47031
		AMERICAN CYANAMID AC 47,031
		AMERICAN CYANAMID CL-47031
		AMERICAN CYANAMIDE 47031
		CI-47031
		CL 47031
		CYCLIC ETHYLENE P,P-DIETHYL PHOSPHONODITHIOIMIDOCARBONATE
		CYCLIC ETHYLENE(DIETHOXYPHOSPHINOTHIOYL)-DITHIOIMIDOCARBONATE
		CYLAN
		CYOLANE
		CYOLANE CYLAN
		CYOLANE INSECTICIDE
		DIETHYL 1,3-DITHIOLAN-2-YLIDENEPHOSPHORAMIDATE
		EI 47031
		ENT 25,830
		IMIDOCARBONIC ACID, (DIETHOXYPHOSPHINYL)DITHIO-, CYCLIC ETHYLENE ESTER
		IMIDOCARBONIC ACID, PHOSPHONODITHIO-, CYCLIC ETHYLENE P,P-DIETHYL ESTER
		IMIDOCARBONIC ACID, PHOSPHONODITHIO-, P,P-DIETHYL CYCLIC ETHYLENE ESTER
		P,P-DIETHYL CYCLIC ETHYLENE ESTER OF PHOSPHONODITHIOIMIDOCARBONIC ACID
		PHOSFOLAN
		PHOSPHOLAN
		PHOSPHORAMIDIC ACID, 1,3-DITHIOLAN-2-YLIDENE-, DIETHYL ESTER

CAS No.	IN	Chemical Name
950-10-7		
		(DIETHOXYPHOSPHINYL)DITHIOIMIDOCARBONIC ACID CYCLIC PROPYLENE ESTER
		1,2-PROPANEDITHIOL, CYCLIC ESTER WITH P,P-DIETHYL PHOSPHONODITHIOIMIDOCARBONATE
		1,3-DITHIOLANE, 2-(DIETHOXYPHOSPHINYLIMINO)-4-METHYL-
		2-(DIETHOXYPHOSPHINYLIMINO)-4-METHYL-1,3-DITHIOLANE
		AC 47470
		AMERICAN CYANAMID CL-47470
		CL 47470
		CYCLIC PROPYLENE (DIETHOXYPHOSPHINYL)DITHIO-IMIDOCARBONATE
		CYTROLANE
		DIETHYL (4-METHYL-1,3-DITHIOLAN-2-YLIDENE)PHOSPHO-ROAMIDATE
		EI 47470
		ENT-25,991
		IMIDOCARBONIC ACID, PHOSPHONODITHIO-, CYCLIC PROPYLENE P,P-DIETHYL ESTER
		MEPHOSFOLAN
		P,P-DIETHYL CYCLIC PROPYLENE ESTER OF PHOSPHONO-DITHIOIMIDOCARBONIC ACID
		PHOSPHORAMIDIC ACID, (4-METHYL-1,3-DITHIOLAN-2-YLIDENE)-, DIETHYL ESTER
950-37-8		
		(O,O-DIMETHYL)-S-(-2-METHOXY-DELTA(SUP 2)-1,3,4-THIADIAZOLIN-5-ON-4-YLMETHYL)DITHIOPHOSPHATE
		CIBA-GEIGY GS 13005
		DMTP
		DMTP(JAPAN)
		GEIGY 13005
		GEIGY GS 13005
		GS 13005
		METHIDATHION
		O,O-DIMETHYL S-(5-METHOXY-1,3,4-THIADIAZOLINYL-3-METHYL) DITHIOPHOSPHATE
		O,O-DIMETHYL-S-((2-METHOXY-1,3,4 (4H)-THIODIAZOL-5-ON-4-YL)-METHYL)-DITHIOFOSFAAT (DUTCH)
		O,O-DIMETHYL-S-(2-METHOXY-1,3,4-THIADIAZOL-5(4H)-ONYL-(4)-METHYL) PHOSPHORODITHIOATE
		O,O-DIMETHYL-S-(2-METHOXY-1,3,4-THIADIAZOL-5-(4H)-ONYL-(4)-METHYL)-DITHIOPHOSPHAT (GERMAN)
		O,O-DIMETHYL-S-(2-METHOXY-1,3,4-THIADIAZOL-5-ON-4-LY)-METHYL-DITHIOPHOSPHAT (GERMAN)
		O,O-DIMETIL-S-((2-METOSSI-1,3,4-(4H)-TIADIAZOL-5-ON-4-IL)-METIL)-DITIFOSFATO (ITALIAN)
		PHOSPHORODITHIOIC ACID, O,O-DIMETHYL ESTER, S-ESTER WITH 4-(MERCAPTOMETHYL)-2-METHOXY-DELTA-(SUP 2)-1
		PHOSPHORODITHIOIC ACID, O,O-DIMETHYL ESTER, S-ESTER WITH 4-(MERCAPTOMETHYL)-2-METHOXY-DELTA2-1,3,4-T
		PHOSPHORODITHIOIC ACID, S-[(5-METHOXY-2-OXO-1,3,4-THIADIAZOL-3(2H)-YL)METHYL] O,O-DIMETHYL
		PHOSPHORODITHIOIC ACID, S-[(5-METHOXY-2-OXO-1,3,4-THIADIAZOL-3(2H)-YL)METHYL] O,O-DIMETHYL ESTER
		S-(2,3-DIHYDRO-5-METHOXY-2-OXO-1,3,4-THIADIAZOL-3-METHYL) DIMETHYL PHOSPHOROTHIOLOTHIONATE
		SUPRACID
		SUPRACIDE
		ULTRACID
		ULTRACID 40
952-23-8		
		PROFLAVINE
959-98-8		
		5-NORBORNENE-2,3-DIMETHANOL, 1,4,5,6,7,7-HEXA-CHLORO-, CYCLIC SULFITE, ENDO-
		6,9-METHANO-2,4,3-BENZODIOXATHIEPIN, 6,7,8,9,10,10-HEXACHLORO-1,5,5A,6,9,9A-HEXAHYDRO-, 3-OXIDE, (3.
		ALPHA-BENZOEPIN
	*	ALPHA-ENDOSULFAN
		ALPHA-THIODAN
		BETA-THIONEX
		ENDOSULFAN 1
		ENDOSULFAN A
		ENDOSULFAN I
961-11-5		
		TETRACHLOROVINPHOS
		TETRACHLORVINPHOS
968-81-0		
		ACETOHEXAMIDE
989-38-8		
		C.I. BASIC RED 1
991-42-4		
		4,7-METHANO-1H-ISOINDOLE-1,3(2H)-DIONE, 3A,4,7,7A-TETRAHYDRO-5-(HYDROXYPHENYL-2-PYRIDINYL-METHYL)-8-(
		5-NORBORNENE-2,3-DICARBOXIMIDE, 5-(ALPHA-HYDROXY-ALPHA-2-PYRIDYLBENZYL)-7-(ALPHA-2-PYRIDYLBENZYL-IDEN
		COMPOUND S-6,999
991-42-4		
		ENT 51,762
		MCN 1025
		NORBORMIDE
		RATICATE
		S 6999
		SHOXIN
992-59-6		
		C.I. DIRECT RED 2, DISODIUM SALT
993-00-0		
		CHLOROMETHYLSILANE
		METHYL CHLOROSILANE
		METHYL CHLOROSILANE (DOT)
		SILANE, CHLOROMETHYL-
		UN 25 34 (DOT)
994-31-0		
		TRIETHYLTIN CHLORIDE
995-33-5		
		N-BUTYL 4,4-DI(TERT-BUTYL PEROXY)VALERATE (MAXI-MUM 52% SOLUTION)
		N-BUTYL 4,4-DI(TERT-BUTYL PEROXY)VALERATE (STOCK PURE)
995-58-8		
		5-NITRO-O-TOLUIDINE
998-30-1		
		SILANE, TRIETHOXY-
		TRIETHOXYSILANE
999-61-1		
		1,2-PROPANEDIOL, 1-ACRYLATE
		2-HYDROXYPROPYL ACRYLATE
		2-PROPENOIC ACID, 2-HYDROXYPROPYL ESTER
		ACRYLIC ACID, 2-HYDROXYPROPYL ESTER
		BETA-HYDROXYPROPYL ACRYLATE
999-81-5		
		(2-CHLOROETHYL)TRIMETHYLAMMONIUM CHLORIDE
		(BETA-CHLOROETHYL)TRIMETHYLAMMONIUM CHLORIDE
		2-CHLORAETHYL-TRIMETHYLAMMONIUMCHLORID (GERMAN)
		2-CHLORO-N,N,N-TRIMETHYLETHANAMINIUM CHLORIDE
		60-CS-16
		AC 38555
		AMMONIUM, (2-CHLOROETHYL)TRIMETHYL-, CHLORIDE
		ANTYWYLEGACZ
		BARLEYQUAT B
		BETA-CHLOROETHYLTRIMETHYLAMMONIUM CHLORIDE
		CCC
		CCC PLANT GROWTH REGULANT
		CE CE CE
		CHLORCHLOINCHLORIDE

CAS No.	IN	Chemical Name
		CHLORCHOLINCHLORID (CZECH,GERMAN)
		CHLORCHOLINE CHLORIDE
		CHLORMEQUAT
		CHLORMEQUAT CHLORIDE
		CHLOROCHOLINE CHLORIDE
		CHOLINE DICHLORIDE
		CYCLOCEL
		CYCOCEL
		CYCOCEL-EXTRA
		CYCOGAN
		CYCOGAN EXTRA
		CYOCEL
		EI 38,555
		ETHANAMINIUM, 2-CHLORO-N,N,N-TRIMETHYL-, CHLORIDE
		ETHANAMINIUM, 2-CHLORO-N,N,N-TRIMETHYL-, CHLORIDE (9CI)
		HALLOWEEN
		HICO CCC
		HORMOCEL-2CCC
		INCRECEL
		LIHOCIN
		MICROCIL
		NCI-C02960
		NEW '5C' CYCOCEL
		RETACEL
		STABILAN
		TRIMETHYL-BETA-CHLORETHYLAMMONIUMCHLORID (CZECH)
		TRIMETHYL-BETA-CHLOROETHYLAMMONIUM CHLORIDE
		TUR
		WR 62
		ZAR
999-97-3		
		1,1,1,3,3,3-HEXAMETHYLDISILAZANE
		BIS(TRIMETHYLSILYL)AMINE
		DISILAZANE, 1,1,1,3,3,3-HEXAMETHYL-
		HEXAMETHYL DISILAZANE
		HEXAMETHYLSILAZANE
		HMDS
		OAP
		SILANAMINE, 1,1,1-TRIMETHYL-N-(TRIMETHYLSILYL)-
		SILANAMINE, 1,1,1-TRIMETHYL-N-(TRIMETHYLSILYL)- (9CI)
1002-16-0		
		1-PENTYL NITRATE
		AMYL NITRATE
		AMYL NITRATE (DOT)
		N-PENTYL NITRATE
		NITRATE D'AMYLE (FRENCH)
		NITRIC ACID, PENTYL ESTER
		PENTYL NITRATE
1002-89-7		
		AMMONIUM STEARATE
		OCTADECANOIC ACID, AMMONIUM SALT
		STEARIC ACID, AMMONIUM SALT
1024-57-3		
		2,5-METHANO-2H-INDENO[1,2-B]OXIRENE, 2,3,4,5,6,7,7-HEPTACHLORO-1A,1B,5,5A,6,6A-HEXAHYDRO-, (1A.ALPHA
		4,7-METHANOINDAN, 1,4,5,6,7,8,8-HEPTACHLORO-2,3-EPOXY-3A,4,7,7A-TETRAHYDRO-
		BETA-HEPTACHLOREPOXIDE
		ENT 25584
		EPOXYHEPTACHLOR
		GPKH EPOXIDE
		HCE
		HEPTACHLOR CIS-OXIDE
	*	HEPTACHLOR EPOXIDE
		VELSICOL 53CS17
1031-07-8		
		5-NORBORNENE-2,3-DIMETHANOL, 1,4,5,6,7,7-HEXACHLORO-, CYCLIC SULFATE
		6,9-METHANO-2,4,3-BENZODIOXATHIEPIN, 6,7,8,9,10,10-HEXACHLORO-1,5,5A,6,9,9A-HEXAHYDRO-, 3,3-DIOXIDE
		BENZOEPIN SULFATE
		ENDOSULFAN SULFATE
		THIODAN SULFATE
1031-47-6		
		3-PHENYL-5-AMINO-1,2,4-TRIAZOLYL-(1)-(N,N'-TETRAMETHYL) DIAMIDOPHOSPHONATE
		5-AMINO-1-(BIS(DIMETHYLAMINO)PHOSPHINYL)-3-PHENYL-1,2,4-TRIAZOLE
		5-AMINO-1-BIS(DIMETHYLAMIDE)PHOSPHORYL-3-PHENYL-1,2,4-TRIAZOLE
		5-AMINO-1-BIS(DIMETHYLAMIDO)PHOSPHORYL-3-PHENYL-1,2,4-TRIAZOLE
		5-AMINO-3-FENIL-1-BIS(-DIMETILAMINO)-FOSFORIL-1,2,4-TRIAZOLO (ITALIAN)
		5-AMINO-3-FENYL-1-BIS(DIMETHYL-AMINO)-FOSFORYL-1,2,4-TRIAZOOL (DUTCH)
		5-AMINO-3-PHENYL-1,2,4-TRIAZOLE-1-YL-N,N,N',N'-TETRAMETHYLPHOSPHODIAMIDE
		5-AMINO-3-PHENYL-1,2,4-TRIAZOLYL-1-BIS(DIMETHYL-AMIDO)PHEOSPHATE
		5-AMINO-3-PHENYL-1,2,4-TRIAZOLYL-NNN'N'-TETRAMETHYL-PHOSPHONAMIDE
		5-AMINO-3-PHENYL-1,2,4-TRIAZOLYLBIS(DIMETHYL-AMINO)-PHOSPHINOXID (GERMAN)
		5-AMINO-3-PHENYL-1-BIS (DIMETHYL-AMINO)-PHOSPHORYLE-1,2,4-TRIAZOLE (FRENCH)
		5-AMINO-3-PHENYL-1-BIS(DIMETHYLAMINO)-PHOSPHORYL-1H-1,2,4-TRIAZOL (GERMAN)
		BIS(DIMETHYLAMINO)-3-AMINO-5-PHENYLTRIAZOLYL PHOSPHINE OXIDE
		NIAGARA 5943
		P-(5-AMINO-3-PHENYL-1H-1,2,4-TRIAZOL-1-YL)-N,N,N'-TETRAMETHYL PHOSPHONIC DIAMIDE
		PHOSPHONIC DIAMIDE, P-(5-AMINO-3-PHENYL-1H-1,2,4-TRIAZOL-1-YL)-N,N,N',N'-TETRAMETHYL-
		TRIAMIFOS (GERMAN, DUTCH, ITALIAN)
		TRIAMIPHOS
		TRIAMPHOS
		WEPSIN
		WEPSYN
		WEPSYN 155
		WP 155
1036-95-2		
		ZINC CHLORATE
1066-30-4		
		ACETIC ACID, CHROMIUM(3+) SALT
	*	CHROMIC ACETATE
		CHROMIC ACETATE (DOT)
		CHROMIC ACETATE (III)
		CHROMIUM ACETATE
		CHROMIUM TRIACETATE
		CHROMIUM(III) ACETATE
		NA 9101 (DOT)
1066-33-7		
		ACID AMMONIUM CARBONATE
	*	AMMONIUM BICARBONATE
		AMMONIUM BICARBONATE (1:1)
		AMMONIUM BICARBONATE (DOT)
		AMMONIUM CARBONATE
		AMMONIUM HYDROGEN CARBONATE
		CARBONIC ACID, MONOAMMONIUM SALT
		MONOAMMONIUM CARBONATE
		NA 9081 (DOT)
1066-45-1		
		CHLOROTRIMETHYLSTANNANE
		CHLOROTRIMETHYLTIN
		M&T CHEMICALS 1222-45
		STANNANE, CHLOROTRIMETHYL-
		TRIMETHYLCHLOROSTANNANE
		TRIMETHYLCHLOROTIN
		TRIMETHYLSTANNYL CHLORIDE
		TRIMETHYLTIN CHLORIDE

CAS No.	IN	Chemical Name
1067-20-5		3,3-DIETHYLPENTANE
		PENTANE, 3,3-DIETHYL-
		TETRAETHYLMETHANE
1067-33-0		DIBUTYLTIN DIACETATE
1068-27-5		2,5-BIS(TERT-BUTYLPEROXY)-2,5-DIMETHYL-3-HEXYNE
		2,5-DIMETHYL-2,5-BIS-(TERT-BUTYLPEROXY)HEXYNE-3, MAX CONCENT 52% WITH INERT SOLID (DOT)
		2,5-DIMETHYL-2,5-DI(T-BUTYLPEROXY)HEXYNE-3
		2,5-DIMETHYL-2,5-DI(TERT-BUTYLPEROXY)-3-HEXYNE
		2,5-DIMETHYL-2,5-DI-(TERT-BUTYLPEROXY)HEXYNE-3, MAXIMUM 52% PEROXIDE IN INERT SOLID
		2,5-DIMETHYL-2,5-DI-(TERT-BUTYLPEROXY)HEXYNE-3, TECHNICALLY PURE
		2,5-DIMETHYL-2,5-DI-(TERT-BUTYLPEROXY)HEXYNE-3, TECHNICALLY PURE (DOT)
		3,4,9,10-TETRAOXADODEC-6-YNE, 2,2,5,5,8,8,11,11-OCTA-METHYL-
		3-HEXYNE, 2,5-DIMETHYL-2,5-DI(T-BUTYLPEROXY)-
		3-HEXYNE, 2,5-DIMETHYL-2,5-DI(T-BUTYLPEROXY)-, MAXIMUM CONCENTRATION 52% WITH INERT SOLID
		CAB-O-CURE 2P
		KAYAHEXA YD
		LUPERCO 130XL
		LUPEROX 130
		LUPERSOL 130
		PERHEXYNE 25B
		PERHEXYNE 25B40
		PEROXIDE, (1,1,4,4-TETRAMETHYL-2-BUTYNE-1,4-DIYL)-BIS[(1,1-DIMETHYLETHYL)
		PEROXIDE, (1,1,4,4-TETRAMETHYL-2-BUTYNYLENE)BIS-[TERT-BUTYL
		PEROXIDE, (TETRAMETHYL-2-BUTYNYLENE)BIS[TERT-BUTYL
		PX 1
		UN 2158 (DOT)
		UN 2159 (DOT)
1068-87-7		2,4-DIMETHYL-3-ETHYLPENTANE
		3-ETHYL-2,4-DIMETHYLPENTANE
		PENTANE, 3-ETHYL-2,4-DIMETHYL-
1072-35-1		LEAD DISTEARATE
		LEAD STEARATE
		LEAD(2+) STEARATE
		LEAD(II) STEARATE
		NORMAL LEAD STEARATE
		OCTADECANOIC ACID, LEAD(2+) SALT
		STABINEX NC18
		STEARIC ACID, LEAD(2+) SALT
1111-78-0		AMMONIUM AMINOFORMATE
	*	AMMONIUM CARBAMATE
		AMMONIUM CARBAMATE (DOT)
		CARBAMIC ACID, AMMONIUM SALT
		CARBAMIC ACID, MONOAMMONIUM SALT
		NA 9083 (DOT)
1113-38-8	*	AMMONIUM OXALATE
		AMMONIUM OXALATE ((NH4)2C2O4) MONOHYDRATE
		AMMONIUM OXALATE (DOT)
		AMMONIUM OXALATE MONOHYDRATE
		DIAMMONIUM OXALATE
		DIAMMONIUM OXALATE MONOHYDRATE
		ETHANEDIOIC ACID DIAMMONIUM SALT
		ETHANEDIOIC ACID, DIAMMONIUM SALT, MONOHYDRATE
		NA 2449 (DOT)
		OXALIC ACID, DIAMMONIUM SALT
		OXALIC ACID, DIAMMONIUM SALT, MONOHYDRATE
1116-54-7		DIETHANOLNITROSAMINE
		ETHANOL, 2,2'-(NITROSOIMINO)BIS-
		ETHANOL, 2,2'-NITROSIMINODI-
		N,N-DIETHANOLNITROSAMINE
		N-NITROSODIETHANOLAMINE
		NITROSODIETHANOLAMINE
1116-70-7		ALUMINUM TRIBUTYL
		TRI-N-BUTYLALUMINUM
		TRIBUTYL ALUMINUM
1118-58-7		1,3-DIMETHYL-1,3-BUTADIENE
		1,3-PENTADIENE, 2-METHYL-
		2-METHYL-1,3-PENTADIENE
1119-49-9		ACETAMIDE, N-BUTYL-
		BUTYLACETAMIDE
		N-BUTYL ACETAMIDE
1120-21-4		HENDECANE
		N-UNDECANE
		UN 2330 (DOT)
		UNDECANE
		UNDECANE (DOT)
1120-23-6		2,BETA-BUTOXYETHOXYETHYL CHLORIDE
1120-36-1		1-TETRADECENE
		ALPHA-TETRADECENE
		N-TETRADEC-1-ENE
1120-48-5		1-OCTANAMINE, N-OCTYL-
		DI-N-OCTYLAMINE
		DIOCTYLAMINE
		N-N-OCTYL-N-OCTYLAMINE
		RC 5632
1120-71-4		1,2-OXATHIOLANE 2,2-DIOXIDE
		1,3-PROPANESULTONE
		1-PROPANESULFONIC ACID, 3-HYDROXY-, GAMMA-SUL-TONE
		3-HYDROXY-1-PROPANESULFONIC ACID SULTONE
		GAMMA-PROPANE SULTONE
		PROPANE SULTONE
1122-60-7		CYCLOHEXANE, NITRO-
		NITROCYCLOHEXANE
1124-33-0		4-NITROPYRIDINE 1-OXIDE
		4-NITROPYRIDINE OXIDE
		4-NITROPYRIDINE-N-OXIDE
		P-NITROPYRIDINE N-OXIDE
		PYRIDINE, 4-NITRO, 1-OXIDE
1125-27-5		ETHYL PHENYL DICHLORO-SILANE
1126-78-9		4-(PHENYLAMINO)BUTANE
		ANILINE, N-BUTYL-
		BENZENAMINE, N-BUTYL-
		BENZENAMINE, N-BUTYL- (9CI)
	*	BUTYLANILINE
		N-BUTYLANILINE
		N-BUTYLBENZENAMINE
		N-N-BUTYLANILINE (DOT)
		UN 2738 (DOT)

CAS No.	IN	Chemical Name
1129-41-5		
		3-METHYLPHENYL METHYLCARBAMATE
		3-METHYLPHENYL N-METHYLCARBAMATE
		3-TOLYL METHYLCARBAMATE
		3-TOLYL-N-METHYLCARBAMATE
		CARBAMIC ACID, METHYL-, 3-METHYLPHENYL ESTER
		CARBAMIC ACID, METHYL-, 3-TOLYL ESTER
		CARBAMIC ACID, METHYL-, M-TOLYL ESTER
		DICRESYL
		DRC 3341
		KUMIAI
		M-CRESYL ESTER OF N-METHYLCARBAMIC ACID
		M-CRESYL METHYLCARBAMATE
		M-CRESYL N-METHYLCARBAMATE
		M-METHYLPHENYL METHYLCARBAMATE
		M-METHYLPHENYL N-METHYLCARBAMATE
		M-TOLYL METHYLCARBAMATE
		M-TOLYL N-METHYLCARBAMATE
		METACRATE
		METOLCARB
		MTMC
		S 1065
		TSUMACIDE
		TSUMAUNKA
1135-81-2		
		SODIUM POTASSIUM ALLOYS
1163-19-5		
		DECABROMODIPHENYL ETHER
1165-39-5		
		AFLATOXIN G1
1185-57-5		
		1,2,3-PROPANETRICARBOXYLIC ACID, 2-HYDROXY-, AMMONIUM IRON(3+) SALT
		1,2,3-PROPANETRICARBOXYLIC ACID, 2-HYDROXY-, AMMONIUM IRON(III) SALT
		AMMONIUM FERRIC CITRATE
		AMMONIUM IRON(III) CITRATE
		CITRIC ACID, AMMONIUM IRON(3+) SALT
		FAC
	*	FERRIC AMMONIUM CITRATE
		FERRIC AMMONIUM CITRATE (DOT)
		NA 9118 (DOT)
1186-53-4		
		2,2,3,4-TETRAMETHYL PENTANE
		PENTANE, 2,2,3,4-TETRAMETHYL-
1189-85-1		
		CHROMIC ACID (H2CrO4), BIS(1,1-DIMETHYLETHYL) ESTER
		CHROMIC ACID (H2CrO4), DI-TERT-BUTYL ESTER
		CHROMIC ACID, BIS(1,1-DIMETHYLETHYL) ESTER
		DI-TERT-BUTOXYCHROMYL
		DI-TERT-BUTYL CHROMATE
		TERT-BUTYL CHROMATE
		TERT-BUTYL CHROMATE(VI)
1191-15-7		
		ALUMINUM, HYDROBIS(2-METHYLPROPYL)-
		ALUMINUM, HYDRODIISOBUTYL-
		BIS(ISO-BUTYL)ALUMINUM HYDRIDE
		BIS(ISOBUTYL)HYDROALUMINUM
		DIBAL
		DIISOBUTYLALUMINUM
		DIISOBUTYLALUMINUM HYDRIDE
		DIISOBUTYLHYDROALUMINUM
		HYDRODIISOBUTYLALUMINUM
1191-80-6		
		9-OCTADECENOIC ACID (Z)-, MERCURY(2+) SALT
		MERCURIC OLEATE
		MERCURIC OLEATE, SOLID
		MERCURIC OLEATE, SOLID (DOT)
		MERCURY OLEATE
		OLEATE OF MERCURY
		OLEIC ACID, MERCURY(2+) SALT
		UN 1640 (DOT)
1194-65-6		
		2,6-DICHLOROBENZONITRILE
		2,6-DICHLOROCYANOBENZENE
		BENZONITRILE, 2,6-DICHLORO-
		CASORON
		CASORON 133
		DBN
		DBN (PESTICIDE)
		DICHLOBENIL
		H 133
		NIA 5996
		NIAGARA 5996
1195-42-2		
		CYCLOHEXANAMINE, N-(1-METHYLETHYL)-
		CYCLOHEXYLAMINE, N-ISOPROPYL-
		CYCLOHEXYLISOPROPYLAMINE
		ISOPROPYL CYCLOHEXYLAMINE
		N-CYCLOHEXYL-N-ISOPROPYLAMINE
		N-ISOPROPYLCYCLOHEXANAMINE
		N-ISOPROPYLCYCLOHEXYLAMINE
1212-29-9		
		N,N-DICYCLOHEXYLTHIOREA
1271-28-9		
		BIS(ETA5-2,4-CYCLOPENTADIEN-1-YL)NICKEL
		DI-PI-CYCLOPENTADIENYLNICKEL
		DICYCLOPENTADIENYLNICKEL
		NICKEL DICYCLOPENTADIENYL
		NICKEL, BIS(ETA5-2,4-CYCLOPENTADIEN-1-YL)-
		NICKEL, DI-PI-CYCLOPENTADIENYL-
		NICKELOCENE
1299-86-1		
		ALUMINUM CARBIDE
		ALUMINUM CARBIDE (Al4C3)
		ALUMINUM CARBIDE (DOT)
		TETRAALUMINUM TRICARBIDE
		UN 1394 (DOT)
1300-64-7		
		ANISOYL CHLORIDE
1300-71-6		
		2,4-XYLENOL
		DIMETHYLPHENOL
		PHENOL, DIMETHYL-
		SUDOL
		UN 2261 (DOT)
		XILENOLI (ITALIAN)
	*	XYLENOL
		XYLENOL (DOT)
		XYLENOLEN (DUTCH)
1300-73-8		
		11460 BROWN
		ACID LEATHER BROWN 2G
		ACID ORANGE 24
		AMINODIMETHYLBENZENE
		BENZENAMINE, AR,AR-DIMETHYL-
		DIMETHYLAMINOBENZENE
		DIMETHYLANILINE
		DIMETHYLPHENYLAMINE
		RESORCINE BROWN J
		RESORCINE BROWN R
		UN 1711 (DOT)
		XILIDINE (ITALIAN)
		XYLIDINE
		XYLIDINE (ACGIH,DOT)
1301-70-8		
		1,2,3-PROPANETRIOL, MONO(DIHYDROGEN PHOSPHATE), IRON(3+) SALT (3:2)
		FERRIC GLYCEROPHOSPHATE
		GLYCEROL, MONO(DIHYDROGEN PHOSPHATE), IRON(3+) SALT (3:2)

CAS No.	IN	Chemical Name
		IRON GLYCEROPHOSPHATE (Fe(O6PC3H8)3)
		IRON(3+) GLYCEROPHOSPHATE
1302-52-9		
		BERYL
		BERYL (Al2Be3(SiO3)6)
		BERYL ORE
1303-28-2		
		ANHYDRIDE ARSENIQUE (FRENCH)
		ARSENIC (V) OXIDE
		ARSENIC ACID
		ARSENIC ACID ANHYDRIDE
		ARSENIC ANHYDRIDE
		ARSENIC OXIDE
		ARSENIC OXIDE (As2O5)
		ARSENIC PENTAOXIDE
	*	ARSENIC PENTOXIDE
		ARSENIC PENTOXIDE (DOT)
		ARSENIC PENTOXIDE, SOLID
		ARSENIC PENTOXIDE, SOLID (DOT)
		DIARSENIC PENTOXIDE
		DIARSONIC PENTOXIDE
		RCRA WASTE NUMBER P011
		UN 1559 (DOT)
		ZOTOX
1303-33-9		
		ARSENIC SESQUISULFIDE
		ARSENIC SESQUISULFIDE (As2S3)
		ARSENIC SESQUISULPHIDE
		ARSENIC SULFIDE
		ARSENIC SULFIDE (As2S3)
		ARSENIC SULFIDE YELLOW
		ARSENIC SULFIDE, SOLID (DOT)
		ARSENIC SULPHIDE
		ARSENIC TERSULPHIDE
		ARSENIC TRISULFIDE
		ARSENIC TRISULFIDE (As2S3)
		ARSENIC TRISULFIDE (DOT)
		ARSENIC YELLOW
		ARSENIC(III) SULFIDE
		ARSENIOUS SULFIDE
		ARSENIOUS SULPHIDE
		ARSENOUS SULFIDE
		AURIPIGMENT
		C.I. 77086
		C.I. PIGMENT YELLOW 39
		CI PIGMENT YELLOW
		CI PIGMENT YELLOW 39
		DIARSENIC TRISULFIDE
		DIARSENIC TRISULPHIDE
		KING'S GOLD
		KING'S YELLOW
		NA 1557 (DOT)
		ORPIMENT
		RCRA WASTE NUMBER P038
		UN 1557 (DOT)
1303-39-5		
		ZINC ARSENATE
1303-86-2		
		BORIC ACID (HBO2), ANHYDRIDE
		BORIC ANHYDRIDE
		BORIC OXIDE
		BORIC OXIDE (B2O3)
	*	BORON OXIDE
		BORON OXIDE (B2O3)
		BORON SESQUIOXIDE
		BORON TRIOXIDE
		DIBORON TRIOXIDE
		FUSED BORIC ACID
1303-96-4		
		BORASCU
		BORATE, TETRASODIUM SALT
		BORAX
		BORAX (B4Na2O7.10H2O)
		BORAX DECAHYDRATE
		BORIC ACID (H2B4O7), DISODIUM SALT, DECAHYDRATE
		BORICIN
		BORON SODIUM OXIDE (B4Na2O7), DECAHYDRATE
		BURA
		DISODIUM TETRABORATE DECAHYDRATE
		GERSTLEY BORATE
		PERBORIC ACID, SODIUM SALT
		SODIUM BIBORATE DECAHYDRATE
		SODIUM BORATE
		SODIUM BORATE (Na2B4O7), DECAHYDRATE
	*	SODIUM BORATES
		SODIUM PERBORATE
		SODIUM PEROXOBORATE
		SODIUM PYROBORATE
		SODIUM PYROBORATE DECAHYDRATE
		SODIUM TETRABORATE
		SODIUM TETRABORATE DECAHYDRATE
		SOLUBOR
1304-28-5		
		BARIUM MONOOXIDE
		BARIUM MONOXIDE
	*	BARIUM OXIDE
		BARIUM OXIDE (BaO)
		BARIUM OXIDE (DOT)
		BARIUM PROTOXIDE
		BARYTA
		CALCINED BARYTA
		OXYDE DE BARYUM (FRENCH)
		UN 1884 (DOT)
1304-29-6		
		BARIO (PEROSSIDO DI) (ITALIAN)
		BARIUM BINOXIDE
		BARIUM DIOXIDE
		BARIUM OXIDE (BaO2)
	*	BARIUM PEROXIDE
		BARIUM PEROXIDE (Ba(O2))
		BARIUM PEROXIDE (DOT)
		BARIUM SUPEROXIDE
		BARIUMPEROXID (GERMAN)
		BARIUMPEROXYDE (DUTCH)
		DIOXYDE DE BARYUM (FRENCH)
		PEROXYDE DE BARYUM (FRENCH)
		UN 1449 (DOT)
1304-56-9		
		BERYLLIA
		BERYLLIUM MONOXIDE
		BERYLLIUM OXIDE
		BERYLLIUM OXIDE (BeO)
		THERMALOX
		THERMALOX 995
1304-82-1		
		BISMUTH SESQUITELLURIDE
		BISMUTH TELLURIDE
		BISMUTH TELLURIDE (Bi2Te3)
		BISMUTH(3+) TELLURIDE
		DIBISMUTH TRITELLURIDE
1305-62-0		
		BELL MINE
		BIOCALC
		CALCIUM DIHYDROXIDE
		CALCIUM HYDRATE
	*	CALCIUM HYDROXIDE
		CALCIUM HYDROXIDE (ACGIH)
		CALCIUM HYDROXIDE (Ca(OH)2)
		CALVIT
		CALVITAL
		CARBOXIDE
		HYDRATED LIME
		KALKHYDRATE
		LIMBUX
		LIME WATER
		MILK OF LIME

CAS No.	IN	Chemical Name
		SLAKED LIME
1305-78-8		
		BURNT LIME
		CALCIA
		CALCIUM MONOXIDE
	*	CALCIUM OXIDE
		CALCIUM OXIDE (ACGIH,DOT)
		CALCIUM OXIDE (CaO)
		CALOXOL CP 2
		CALOXOL W 3
		CALX
		CALXYL
		DESICAL P
		LIME
		LIME, BURNED
		LIME, UNSLAKED (DOT)
		OXYDE DE CALCIUM (FRENCH)
		QUICKLIME
		QUICKLIME (DOT)
		RHENOSORB C
		RHENOSORB F
		UN 1910 (DOT)
		WAPNIOWY TLENEK (POLISH)
1305-79-9		
		CALCIUM DIOXIDE
		CALCIUM OXIDE (CaO2)
		CALCIUM PEROXIDE
		CALCIUM PEROXIDE (Ca(O2))
		CALCIUM PEROXIDE (DOT)
		CALPER
		CALPER G
		UN 1457 (DOT)
1305-99-3		
		CALCIUM PHOSPHIDE
		CALCIUM PHOSPHIDE (Ca3P2)
		CALCIUM PHOSPHIDE (DOT)
		CALCIUM PHOTOPHOR
		PHOTOPHOR
		UN 1360 (DOT)
1306-19-0		
		CADMIUM MONOXIDE
	*	CADMIUM OXIDE
		CADMIUM OXIDE (ACGIH)
		CADMIUM OXIDE (CdO)
		KADMU TLENEK (POLISH)
		NCI-C02551
1306-23-6		
		C. P. GOLDEN YELLOW 55
		CADMIUM GOLDEN 366
		CADMIUM LEMON YELLOW 527
		CADMIUM MONOSULFIDE
		CADMIUM PRIMROSE 819
		CADMIUM SULFIDE
		CADMIUM SULFIDE (CdS)
		CADMIUM SULFIDE YELLOW
		CADMIUM SULPHIDE
		CADMIUM YELLOW
		CADMIUM YELLOW 000
		CADMIUM YELLOW 10G CONC
		CADMIUM YELLOW 892
		CADMIUM YELLOW CONC. DEEP
		CADMIUM YELLOW CONC. GOLDEN
		CADMIUM YELLOW CONC. LEMON
		CADMIUM YELLOW CONC. PRIMROSE
		CADMIUM YELLOW OZ DARK
		CADMIUM YELLOW PRIMROSE 47-4100
		CADMOPUR GOLDEN YELLOW N
		CADMOPUR YELLOW
		CAPSEBON
		FERRO LEMON YELLOW
		FERRO ORANGE YELLOW
		FERRO YELLOW
		PC 108
		PRIMROSE 1466

CAS No.	IN	Chemical Name
1307-82-0		
		SODIUM CHLOROPLATINATE
		SODIUM HEXACHLOROPLATINATE (VI)
1308-38-9		
	*	CHROMIUM(III) OXIDE(2:3)
1309-32-6		
		AMMONIUM FLUOSILICATE
1309-37-1		
		ALPHA-FERRIC OXIDE
		ALPHA-IRON OXIDE
		BAUXITE RESIDUE
		BAYER S11
		BAYFERROX 110M
		BAYFERROX 130M
		C.I. 77491
		C.I. PIGMENT RED 101
		CAPUT MORTUUM
		CAPUT MORTUUM LIGHT
		CERVEN H
		COLCOTHAR
		COLLIRON
		COLLOIDAL FERRIC OXIDE
		CROCUS (IRON OXIDE)
		DEANOX
		DIIRON TRIOXIDE
		ENGLISH IRON OXIDE RED
		FELAC
		FERRIC OXIDE
		FERRUGO
		GAMMA-FERRIC OXIDE
		GAMMA-IRON OXIDE (Fe2O3)
		GAMMA-MYD
		IRON MINIUM
		IRON OXIDE
		IRON OXIDE (Fe2O3)
	*	IRON OXIDE FUME
		IRON OXIDE RED
		IRON OXIDE RED 110M
		IRON OXIDE RED TRANSPARENT 288VN
		IRON SESQUIOXIDE
		IRON TRIOXIDE
		IRON(3+) OXIDE
		IRON(III) OXIDE
		JEWELER'S ROUGE
		KROKUS
		LN 1331
		MAG 1730
		MAPICO RED 347
		MAPICO RED R 220-3
		MICACEOUS IRON ORE
		PFERROX 2380
		PIGDEX 100
		PIGMENT RED 101
		PROTOHEMATITE
		PRUSSIAN RED
		R 5098
		R 5098 (OXIDE)
		RED IRON OXIDE
		RED OXIDE
		RO 8097
		ROUGE
		ROUGE (IRON OXIDE)
		RUBIGO
		SPECULAR IRON
		TENYO 501
		TURKEY RED
		VENETIAN RED
		YLO 2288B
1309-42-8		
		MAGNESIUM HYDROXIDE
1309-48-4		
		ANIMAG
		ANSCOR P
		CALCINED MAGNESIA

CAS No.	IN	Chemical Name
		CAUSTIC MAGNESITE
		ELASTOMAG 100
		ELASTOMAG 170
		KM 40
		KYOWAMAG 100
		KYOWAMAG 150
		KYOWAMAG 30
		MAGCAL
		MAGLITE
		MAGLITE D
		MAGLITE DE
		MAGLITE K
		MAGLITE S
		MAGLITE Y
		MAGNESA PREPRATA
		MAGNESIA
		MAGNESIA USTA
		MAGNESIUM MONOXIDE
	*	MAGNESIUM OXIDE
		MAGNESIUM OXIDE (MgO)
		MAGNESIUM OXIDE FUME
		MAGOX
		MARMAG
		MM 469
		SEASORB
		SLO 369
		SLO 469
1309-60-0		
		BIOXYDE DE PLOMB (FRENCH)
		CI 77580
		LEAD BROWN
	*	LEAD DIOXIDE
		LEAD DIOXIDE (DOT)
		LEAD OXIDE
		LEAD OXIDE (PbO2)
		LEAD OXIDE BROWN
		LEAD PEROXIDE
		LEAD PEROXIDE (DOT)
		LEAD PEROXIDE (PbO2)
		LEAD SUPEROXIDE
		LEAD(IV) OXIDE
		PEROXYDE DE PLOMB (FRENCH)
		PLUMBIC OXIDE
		UN 1872 (DOT)
1309-64-4		
		A 1530
		A 1582
		A 1588LP
		ANTIMONIOUS OXIDE
		ANTIMONY OXIDE
		ANTIMONY OXIDE (Sb2O3)
		ANTIMONY PEROXIDE
		ANTIMONY SESQUIOXIDE
	*	ANTIMONY TRIOXIDE
		ANTIMONY TRIOXIDE (ACGIH,DOT)
		ANTIMONY TRIOXIDE (Sb2O3)
		ANTIMONY WHITE
		ANTIMONY(3+) OXIDE
		ANTOX
		AP 50
		ATOX S
		C.I. 77052
		C.I. PIGMENT WHITE 11
		CHEMETRON FIRE SHIELD
		DECHLORANE A-O
		DIANTIMONY TRIOXIDE
		EXITELITE
		EXTREMA
		FLOWERS OF ANTIMONY
		NA 9201 (DOT)
		NCI-C55152
		NYACOL A 1510LP
		NYACOL A 1530
		PATOX L
		PATOX M
		PATOX S
		SENARMONTITE
		STIBIOX MS
		THERMOGUARD B
		THERMOGUARD L
		THERMOGUARD S
		TIMONOX
		VALENTINITE
		WEISSPIESSGLANZ (GERMAN)
1310-58-3		
		CAUSTIC POTASH
		CAUSTIC POTASH, DRY, SOLID, FLAKE, BEAD, OR GRANULAR (DOT)
		CAUSTIC POTASH, LIQUID OR SOLUTION (DOT)
		HYDROXYDE DE POTASSIUM (FRENCH)
	*	POTASSIUM HYDROXIDE
		POTASSIUM HYDROXIDE (K(OH))
1310-65-2		
		LITHIUM HYDROXIDE
		LITHIUM HYDROXIDE (Li(OH))
		LITHIUM HYDROXIDE (Li(OH)) (9CI)
		LITHIUM HYDROXIDE SOLUTION
		LITHIUM HYDROXIDE, SOLUTION (DOT)
		UN 2679 (DOT)
1310-66-3		
		LITHIUM HYDROXIDE (Li(OH)), MONOHYDRATE
		LITHIUM HYDROXIDE HYDRATE
	*	LITHIUM HYDROXIDE MONOHYDRATE
		LITHIUM HYDROXIDE MONOHYDRATE (DOT)
		UN 2680 (DOT)
1310-73-2		
		AETZNATRON
		ASCARITE
		CAUSTIC SODA
		CAUSTIC SODA, DRY, SOLID
		CAUSTIC SODA, SOLUTION
		COLLO-GRILLREIN
		COLLO-TAPETTA
		LYE SOLUTION
		SODA, CAUSTIC
	*	SODIUM HYDROXIDE
		SODIUM HYDROXIDE (NA(OH))
		SODIUM HYDROXIDE (SOLUTION)
		WHITE CAUSTIC
1310-82-3		
		RUBIDIUM HYDROXIDE
		RUBIDIUM HYDROXIDE (Rb(OH))
		RUBIDIUM HYDROXIDE SOLUTION
		RUBIDIUM HYDROXIDE, SOLID
		RUBIDIUM HYDROXIDE, SOLID (DOT)
		RUBIDIUM HYDROXIDE, SOLUTION (DOT)
		UN 2677 (DOT)
		UN 2678 (DOT)
1312-03-4		
		BASIC MERCURIC SULFATE
		DIOXOTRIMERCURY SULFATE
		MERCURIC BASIC SULFATE
		MERCURIC SUBSULFATE
		MERCURIC SUBSULFATE, SOLID
		MERCURIC SUBSULFATE, SOLID (DOT)
		MERCURY OXIDE SULFATE
		MERCURY OXIDE SULFATE (Hg3O2(SO4))
		MERCURY OXONIUM SULFATE
		NA 2025 (DOT)
		TURPETH MINERAL
1312-73-8		
		DIPOTASSIUM MONOSULFIDE
		DIPOTASSIUM MONOSULFIDE (K2S)
		DIPOTASSIUM SULFIDE
		POTASSIUM MONOSULFIDE
	*	POTASSIUM SULFIDE
		POTASSIUM SULFIDE (2:1)
		POTASSIUM SULFIDE (2:1), HYDRATED, CONTAINING AT LEAST 30% WATER

CAS No.	IN	Chemical Name
		POTASSIUM SULFIDE (DOT)
		POTASSIUM SULFIDE (K2S)
		POTASSIUM SULPHIDE, ANHYDROUS OR CONTAINING LESS THAN 30% WATER (DOT)
		POTASSIUM SULPHIDE, HYDRATED, CONTAINING NOT LESS THAN 30% WATER (DOT)
		UN 1382 (DOT)
		UN 1847 (DOT)
1313-13-9		
	*	MANGANESE DIOXIDE
1313-27-5		
		MO 1202T
		MOLYBDENA
		MOLYBDENUM OXIDE
		MOLYBDENUM OXIDE (MoO3)
	*	MOLYBDENUM TRIOXIDE
		MOLYBDENUM(VI) OXIDE
		MOLYBDENUM(VI) TRIOXIDE
		MOLYBDIC ACID ANHYDRIDE
		MOLYBDIC ANHYDRIDE
		MOLYBDIC TRIOXIDE
1313-60-6		
		DISODIUM DIOXIDE
		DISODIUM PEROXIDE
		FLOCOOL 180
		SODIUM DIOXIDE
		SODIUM OXIDE (Na2O2)
	*	SODIUM PEROXIDE
		SODIUM PEROXIDE (DOT)
		SODIUM PEROXIDE (Na2(O2))
		SOLOZONE
		UN 1504 (DOT)
1313-82-2		
		DISODIUM MONOSULFIDE
		DISODIUM SULFIDE
		Na2S
		SODIUM MONOSULFIDE
	*	SODIUM SULFIDE
		SODIUM SULFIDE (ANHYDROUS)
		SODIUM SULFIDE (Na2S)
		SODIUM SULFIDE, ANHYDROUS
		SODIUM SULFIDE, HYDRATED, WITH AT LEAST 30% WATER
		SODIUM SULPHIDE
		SODIUM SULPHIDE, ANHYDROUS OR CONTAINING LESS THAN 30% WATER (DOT)
		SODIUM SULPHIDE, HYDRATED, WITH AT LEAST 30% WATER (DOT)
		UN 1385 (DOT)
		UN 1849 (DOT)
1313-99-1		
		MONONICKEL OXIDE
		NICKEL MONOOXIDE
		NICKEL MONOXIDE
		NICKEL OXIDE
		NICKEL OXIDE (NiO)
		NICKEL OXIDE SINTER 75
		NICKEL(2+) OXIDE
		NICKEL(II) OXIDE
		NICKELOUS OXIDE
1314-06-3		
		DINICKEL TRIOXIDE
		NICKEL OXIDE
		NICKEL OXIDE (Ni2O3)
		NICKEL PEROXIDE
		NICKEL SESQUIOXIDE
		NICKEL TRIOXIDE
		NICKEL(III) OXIDE
		NICKELIC OXIDE
1314-13-2		
		ACTOX 14
		ACTOX 16
		ACTOX 216
		AMALOX
		AZO 22
		AZODOX
		CADOX XX 78
		ELECTROX 2500
		EMAR
		FLOWERS OF ZINC
		GIAP 10
		GREEN SEAL 8
		HUBBUCK'S WHITE
		KADOX 15
		KADOX 72
		KADOX-25
		NOGENOL
		NOGENOL BITE REGISTRATION PASTE
		OUTMINE
		OZIDE
		OZLO
		PERMANENT WHITE
		PHILOSOPHER'S WOOL
		POWDER BASE 900
		PROTOX 166
		PROTOX 167
		PROTOX 168
		PROTOX 169
		PROTOX 267
		PROTOX 268
		RED SEAL 9
		SAZEX 2000
		SAZEX 4000
		SNOW WHITE
		VANDEM VAC
		VANDEM VOC
		VANDEM VPC
		WHITE SEAL 7
		WHITE ZINC
		XX 203
		XX 601
		XX 78
		ZINC GELATIN
		ZINC MONOXIDE
	*	ZINC OXIDE
		ZINC OXIDE (ZnO)
		ZINC OXIDE FUME
		ZINC WHITE
		ZINCA 20
		ZINCOID
		ZN 0701T
1314-18-7		
		STRONTIUM DIOXIDE
		STRONTIUM PEROXIDE
		STRONTIUM PEROXIDE (DOT)
		STRONTIUM PEROXIDE (Sr(O2))
		UN 1509 (DOT)
1314-20-1		
		THORIA
	*	THORIUM DIOXIDE
		THORIUM OXIDE
		THORIUM OXIDE (ThO2)
		THORIUM(IV) OXIDE
		THOROTRAST
		THORTRAST
		UMBRATHOR
1314-22-3		
		UN 1516 (DOT)
		ZINC DIOXIDE
		ZINC OXIDE (ZnO2)
	*	ZINC PEROXIDE
		ZINC PEROXIDE (DOT)
		ZINC PEROXIDE (Zn(O2))
		ZINC SUPEROXIDE
		ZPO
1314-24-5		
		DIPHOSPHORUS TRIOXIDE
		PHOSPHORUS OXIDE (P2O3)

CAS No.	IN	Chemical Name
		PHOSPHORUS TRIOXIDE
		PHOSPHORUS TRIOXIDE (DOT)
		UN 2578 (DOT)
1314-32-5		
		DITHALLIUM TRIOXIDE
		RCRA WASTE NUMBER P113
		THALLIC OXIDE
		THALLIUM OXIDE
		THALLIUM OXIDE (2:3)
		THALLIUM OXIDE (8CI,9CI)
		THALLIUM OXIDE (Tl2O3)
		THALLIUM PEROXIDE
		THALLIUM SESQUIOXIDE
		THALLIUM TRIOXIDE
		THALLIUM(111) OXIDE
		THALLIUM(3+) OXIDE
		THALLIUM(III) OXIDE
1314-34-7		
		DIVANADIUM TRIOXIDE
		UN 2860 (DOT)
		VANADIC OXIDE
		VANADIUM OXIDE
		VANADIUM OXIDE (V2O3)
		VANADIUM SESQUIOXIDE
		VANADIUM TRIOXIDE
		VANADIUM TRIOXIDE, NON-FUSED FORM (DOT)
		VANADIUM(3+) OXIDE
1314-56-3		
		DIPHOSPHORUS PENTAOXIDE
		DIPHOSPHORUS PENTOXIDE
		NA 1807 (DOT)
		PENTAPHOSPHORIC ACID (H3P5O14)
	*	PHOSPHORIC ANHYDRIDE
		PHOSPHORIC ANHYDRIDE (DOT)
		PHOSPHORIC PENTOXIDE
		PHOSPHOROUS PENTOXIDE
		PHOSPHORUS OXIDE
		PHOSPHORUS OXIDE (P2O5)
		PHOSPHORUS PENTAOXIDE
		PHOSPHORUS PENTOXIDE
		PHOSPHORUS PENTOXIDE (DOT)
		PHOSPHORUS(V) OXIDE
		UN 1807 (DOT)
1314-62-1		
		ANHYDRIDE VANADIQUE (FRENCH)
		C.I. 77938
		DIVANADIUM PENTAOXIDE
		DIVANADIUM PENTOXIDE
		RCRA WASTE NUMBER P120
		UN 2862 (DOT)
		VANADIC ANHYDRIDE
		VANADIO, PENTOSSIDO DI (ITALIAN)
		VANADIUM DUST AND FUME (ACGIH)
		VANADIUM OXIDE (V2O5)
		VANADIUM PENTAOXIDE
	*	VANADIUM PENTOXIDE
		VANADIUM PENTOXIDE (DOT)
		VANADIUM PENTOXIDE, DUST AND FUME
		VANADIUM PENTOXIDE, NON-FUSED FORM (DOT)
		VANADIUM, PENTOXYDE DE (FRENCH)
		VANADIUMPENTOXID (GERMAN)
		VANADIUMPENTOXYDE (DUTCH)
		WANADU PIECIOTLENEK (POLISH)
1314-64-3		
		URANYL SULFATE
1314-77-3		
		CRESOL
1314-80-3		
		DIPHOPSHORUS PENTASULFIDE
		PENTASULFURE DE PHOSPHORE (FRENCH)
		PHOSPHORIC SULFIDE
	*	PHOSPHORUS PENTASULFIDE
		PHOSPHORUS PENTASULFIDE (ACGIH,DOT)
		PHOSPHORUS PENTASULFIDE (P2S5)
		PHOSPHORUS PENTASULFIDE (P4S10)
		PHOSPHORUS PENTASULPHIDE, FREE FROM YELLOW OR WHITE PHOSPHORUS (DOT)
		PHOSPHORUS PERSULFIDE
		PHOSPHORUS SULFIDE
		PHOSPHORUS SULFIDE (P2S5)
		RCRA WASTE NUMBER U189
		SIRNIK FOSFORECNY (CZECH)
		SULFUR PHOSPHIDE
		TETRAPHOSPHORUS DECASULFIDE
		THIOPHOSPHORIC ANHYDRIDE
		UN 1340 (DOT)
1314-84-7		
		ARREX E
		BLUE-OX
		DELUSAL
		KILRAT
		MOUS-CON
		PHOSPHURE DE ZINC (FRENCH)
		PHOSVIN
		RCRA WASTE NUMBER P122
		RUMETAN
		STUTOX
		STUTOX I
		TRIZINC DIPHOSPHIDE
		UN 1714 (DOT)
	*	ZINC PHOSPHIDE
		ZINC PHOSPHIDE (DOT)
		ZINC PHOSPHIDE (Zn3P2)
		ZINC(PHOSPHURE DE) (FRENCH)
		ZINC-TOX
		ZINCO(FOSFURO DI) (ITALIAN)
		ZINKFOSFIDE (DUTCH)
		ZINKPHOSPHID (GERMAN)
		ZP
1314-85-8		
		PHOSPHOROUS SESQUISULFIDE
		PHOSPHORUS (III) SULFIDE (IV)
		PHOSPHORUS SESQUISULFIDE
		PHOSPHORUS SESQUISULFIDE (DOT)
		PHOSPHORUS SESQUISULPHIDE, FREE FROM YELLOW OR WHITE PHOSPHORUS (DOT)
		PHOSPHORUS SULFIDE (P4S3)
		SESQUISULFURE DE PHOSPHORE (FRENCH)
		TETRAPHOSPHORUS TRISULFIDE
		TRISULFURATED PHOSPHORUS
		UN 1341 (DOT)
1314-87-0		
		C.I. 77640
		GALENA
		LEAD MONOSULFIDE
	*	LEAD SULFIDE
		LEAD SULFIDE (1:1)
		LEAD SULFIDE (DOT)
		LEAD SULFIDE (PbS)
		LEAD(2+) SULFIDE
		LEAD(II) SULFIDE
		NA 2291 (DOT)
		NATURAL LEAD SULFIDE
		P 128
		P 128 (SULFIDE)
		P 37
		P 37 (FILTER)
		PLUMBOUS SULFIDE
1315-04-4		
		ANTIMONIAL SAFFRON
		ANTIMONIC SULFIDE
	*	ANTIMONY PENTASULFIDE
		ANTIMONY RED
		ANTIMONY SULFIDE (Sb2S5)
		ANTIMONY SULFIDE GOLDEN
		C.I. 77061
		GOLDEN ANTIMONY SULFIDE

CAS No.	IN	Chemical Name
1317-33-5		
	*	MOLYBDENUM (IV) SULFIDE
1317-34-6		
		MANGANESE TRIOXIDE
1317-36-8		
		CI 77577
		CI PIGMENT YELLOW 46
		LEAD MONOOXIDE
		LEAD MONOXIDE
		LEAD OXIDE
		LEAD OXIDE (PbO)
		LEAD OXIDE YELLOW
		LEAD PROTOXIDE
		LEAD(2+) OXIDE
	*	LEAD(II) OXIDE
		LITHARGE
		LITHARGE PURE
		LITHARGE YELLOW L 28
		MASSICOT
		MASSICOTITE
		PLUMBOUS OXIDE
		YELLOW LEAD OCHER
1317-65-3		
		AGRICULTURAL LIMESTONE
		AGSTONE
		CALMOT AD
	*	LIMESTONE
		LITHOGRAPHIC STONE
		NATURAL CALCIUM CARBONATE
		SOHNHOFEN STONE
1317-95-9		
		RANDANITE
		SILICA, TRIPOLI
		TRIPOLI
		TRIPOLI DUST
1318-00-9		
		VERMICULITE
1318-02-1		
		ZEOLITE
1318-59-8		
		CHLORINE DIOXIDE ION(1-)
		CHLORITE
		CHLORITE (MINERAL CLASS)
		CHLORITE ION
		CHLORITE, INORGANIC
		CHLORITE-GROUP MINERALS
		CHLORITE-TYPE MINERALS
		CHLOROUS ACID, ION(1-)
		GREENITE
		HYDROCHLORITE
		MINERALS, CHLORITE-GROUP
		SIERRALITE
		ZINCIAN CHLORITE
1319-72-8		
		2,4,5-TRICHLOROPHENOXYACETIC ACID, ISOPROPANOL AMINE
		2-PROPANOL, 1-AMINO-, (2,4,5-TRICHLOROPHENOXY)ACETATE (SALT)
		2-PROPANOL, 1-AMINO-, COMPD. WITH (2,4,5-TRICHLOROPHENOXY)ACETIC ACID (1:1)
		ACETIC ACID, (2,4,5-TRICHLOROPHENOXY)-, COMPD. WITH 1-AMINO-2-PROPANOL (1:1)
1319-73-9		
		BENZENE, ETHENYL-, MONOMETHYL DERIV.
		METHYLSTYRENE
		STYRENE, METHYL-
1319-77-3		
		ACEDE CRESYLIQUE (FRENCH)
		AR-TOLUENOL
		BACILLOL
		CRESOL
		CRESOL (ACGIH,DOT)
		CRESOL (MIXED ISOMERS)
		CRESOLI (ITALIAN)
	*	CRESYLIC ACID
		CRESYLIC ACID (DOT)
		HYDROXYTOLUENE
		HYDROXYTOLUOLE (GERMAN)
		KRESOLE (GERMAN)
		KRESOLEN (DUTCH)
		KREZOL (POLISH)
		METHYLPHENOL
		PHENOL, METHYL-
		PHENOL, METHYL- (9CI)
		RCRA WASTE NUMBER U052
		TEKRESOL
		TRICRESOL
		UN 2022 (DOT)
		UN 2076 (DOT)
1320-01-0		
		AMYL TOLUENE
		BENZENE, METHYLPENTYL-
		TOLUENE, PENTYL-
1320-18-9		
		2,4-D PROPYLENE GLYCOL BUTYL ETHER ESTER
		2,4-DICHLOROPHENOXYACETIC ACID, PROPYLENE GLYCOL BUTYL ETHER ESTER
		ACETIC ACID, (2,4-DICHLOROPHENOXY)-, 2-BUTOXYMETHYLETHYL ESTER
		ACETIC ACID, (2,4-DICHLOROPHENOXY)-, BUTOXY PROPYLENE DERIV
1320-21-4		
		AMYL XYLYL ETHER
		BENZENE, DIMETHYL(PENTYLOXY)-
		ETHER, PENTYL XYLYL
1320-27-0		
		AMYLNAPHTHALENE
		NAPHTHALENE, PENTYL-
		PENTYLNAPHTHALENE
1320-37-2		
	*	DICHLOROTETRAFLUOROETHANE
		DICHLOROTETRAFLUOROETHANE (DOT)
		DWUCHLOROCZTEROFLUOROETAN (POLISH)
		ETHANE, DICHLOROTETRAFLUORO-
		ISCEON 224
		TETRAFLUORODICHLOROETHANE
		UN 1958 (DOT)
1321-10-4		
		1-CHLORO-2-METHYL-4-HYDROXYBENZENE
		2-CHLORO-5-HYDROXYTOLUENE
		2-CHLORO-HYDROXYTOLUENE
		3-METHYL-4-CHLOROPHENOL
		4-CHLORO-3-CRESOL
		4-CHLORO-3-METHYLPHENOL
		4-CHLORO-5-METHYLPHENOL
		4-CHLORO-M-CRESOL
		6-CHLORO-3-HYDROXYTOLUENE
		6-CHLORO-M-CRESOL
		APTAL
		BAKTOL
		BAKTOLAN
		CANDASEPTIC
		CHLOROCRESOLS
		CRESOL, CHLORO-
		M-CRESOL, 4-CHLORO-
		OTTAFACT
		P-CHLOR-M-CRESOL
		P-CHLORO-M-CRESOL

CAS No.	IN	Chemical Name
		P-CHLOROCRESOL
		PARA-CHLORO-META-CRESOL
		PARMETOL
		PAROL
		PCMC
		PERITONAN
		PHENOL, 4-CHLORO-3-METHYL-
		PHENOL, 4-CHLORO-3-METHYL- (9 CI)
		PHENOL, CHLOROMETHYL-
		PREVENTOL CMK
		RASCHIT
		RASCHIT K
		RASEN-ANICON
		RCRA WASTE NUMBER U039
1321-12-6		
		BENZENE, METHYLNITRO-
		METHYLNITROBENZENE
		MONONITROTOLUENE
		NITROPHENYLMETHANE
	*	NITROTOLUENE
		NITROTOLUENE (DOT)
		P-NITROTOLUENE
		TOLUENE, AR-NITRO-
		UN 1664 (DOT)
1321-16-0		
		1,2,3,6-TETRAHYDROBENZALDEHYDE
		CYCLOHEXENECARBOXALDEHYDE
		TETRAHYDROBENZALDEHYDE
1321-31-9		
		4-AMINOPHENETOLE
		4-ETHOXYANILINE
		ANILINE, P-ETHOXY-
		BENZENAMINE, 4-ETHOXY- (9CI)
		BENZENAMINE, AR-ETHOXY-
		ETHOXYANILINE
		P-AMINOPHENETOLE
		P-ETHOXYANILINE
		P-PHENETIDIN
		P-PHENETIDINE
		P-PHENETIDINE (DOT)
		PHENETHIDINE
		PHENETIDINE
		UN 2311 (DOT)
1321-60-4		
		CYCLOHEXANOL, TRIMETHYL-
		TRIMETHYLCYCLOHEXANOL
1321-64-8		
		NAPHTHALENE, PENTACHLORO-
		PENTACHLORONAPHTHALENE
1321-65-9		
		HALOWAX
		NAPHTHALENE, TRICHLORO-
		TRICHLORONAPHTHALENE
1321-74-0		
		BENZENE, DIETHENYL-
		BENZENE, DIVINYL-
		DIETHENYLBENZENE
		DIVINYL BENZENE
		VINYLSTYRENE
1322-06-1		
		AMYL PHENOL
		AMYLHYDROXYBENZENE
		PENTYLPHENOL
		PHENOL, PENTYL-
1327-53-3		
		ACIDE ARSENIEUX (FRENCH)
		ANHYDRIDE ARSENIEUX (FRENCH)
		ARSENIC (III) OXIDE
		ARSENIC BLANC (FRENCH)
		ARSENIC OXIDE
		ARSENIC OXIDE (As2O3)
		ARSENIC SESQUIOXIDE
	*	ARSENIC TRIOXIDE
		ARSENIC TRIOXIDE (ACGIH)
		ARSENIC TRIOXIDE, SOLID
		ARSENIC TRIOXIDE, SOLID (DOT)
		ARSENIC, WHITE, SOLID (DOT)
		ARSENICUM ALBUM
		ARSENIGEN SAURE (GERMAN)
		ARSENIOUS ACID
		ARSENIOUS ACID, SOLID (DOT)
		ARSENIOUS OXIDE
		ARSENIOUS TRIOXIDE
		ARSENITE
		ARSENOLITE
		ARSENOUS ACID
		ARSENOUS ACID ANHYDRIDE
		ARSENOUS ANHYDRIDE
		ARSENOUS OXIDE
		ARSENOUS OXIDE ANHYDRIDE
		ARSENTRIOXIDE
		ARSODENT
		CLAUDELITE
		CLAUDETITE
		CRUDE ARSENIC
		DIARSENIC TRIOXIDE
		RCRA WASTE NUMBER P012
		UN 1561 (DOT)
		WHITE ARSENIC
1330-16-1		
		BICYCLO(311)HEPTANE, 2,6,6-TRIMETHYL-, DIDEHYDRO DERIV
	*	PINENE
		PINENE (DOT)
		UN 2368 (DOT)
1330-20-7		
		BENZENE, DIMETHYL-
		DILAN
		DIMETHYLBENZENE
		KSYLEN (POLISH)
		METHYL TOLUENE
		NCI-C55232
		RCRA WASTE NUMBER U239
		UN 1307 (DOT)
		VIOLET 3
		XILOLI (ITALIAN)
	*	XYLENE
		XYLENE (ACGIH,DOT)
		XYLENE (MIXED ISOMERS)
		XYLENEN (DUTCH)
		XYLOL
		XYLOL (DOT)
		XYLOLE (GERMAN)
1330-43-4		
		ANHYDROUS BORAX
		BORAX GLASS
		BORIC ACID (H2B4O7), DISODIUM SALT
		BORON SODIUM OXIDE (B4Na2O7)
		DISODIUM TETRABORATE
		FR 28
		FUSED BORAX
		RASORITE 65
		SODIUM BIBORATE
		SODIUM TETRABORATE
1330-45-6		
		ARCTON 50
		CHLOROTRIFLUOROETHANE
		CHLOROTRIFLUOROETHANE (DOT)
		ETHANE, CHLOROTRIFLUORO-
		TRIFLUOROCHLOROETHANE
		TRIFLUOROCHLOROETHANE (DOT)
		UN 1983 (DOT)

CAS No.	IN	Chemical Name
1330-61-6		
		2-PROPENOIC ACID, ISODECYL ESTER
		2-PROPENOIC ACID, ISODECYL ESTER (9CI)
		ACRYLIC ACID, ISODECYL ESTER
		AGEFLEX FA-10
		ISODECYL ACRYLATE
		ISODECYL ALCOHOL, ACRYLATE
		ISODECYL PROPENOATE
1330-78-5		
		CELLUFLEX 179C
		CRESYL PHOSPHATE
		DISFLAMOLL TKP
		DURAD
		FLEXOL PLASTICIZER TCP
		FYRQUEL 150
		IMOL S 140
		KRONITEX
		KRONITEX R
		KRONITEX TCP
		LINDOL
		NCI-C61041
		PHOSPHATE DE TRICRESYLE (FRENCH)
		PHOSPHLEX 179A
		PHOSPHORIC ACID TRICRESYL ESTER
		PHOSPHORIC ACID, TRIS(METHYLPHENYL) ESTER
		PHOSPHORIC ACID, TRITOLYL ESTER
		SYN-O-AD 8484
		TCP
		TRICRESILFOSFATI (ITALIAN)
		TRICRESOL PHOSPHATE
		TRICRESYL PHOSPHATE
		TRICRESYLFOSFATEN (DUTCH)
		TRICRESYLPHOSPHATE, WITH MORE THAN 3% ORTHO ISOMER (DOT)
		TRIKRESYLPHOSPHATE (GERMAN)
		TRIS(TOLYLOXY)PHOSPHINE OXIDE
		TRITOLYL PHOSPHATE
		UN 2574 (DOT)
1331-11-9		
		3-ETHOXYPROPIONIC ACID
1331-17-5		
	*	PROPYLENE GLYCOL, ALLYL ETHER
1331-22-2		
		CYCLOHEXANONE, METHYL-
		METHYL CYCLOHEXANONE
		METHYL CYCLOHEXANONE (DOT)
		METHYLCYCLOHEXAN-1-ONE
		METYLOCYKLOHEKSANON (POLISH)
		UN 2297 (DOT)
1331-28-8		
		O-CHLOROSTYRENE
1331-43-7		
		CYCLOHEXANE, DIETHYL-
		DIETHYLCYCLOHEXANE
1332-07-6		
		BORAX 2335
		BORIC ACID, ZINC SALT
		NA 9155 (DOT)
		ZB 112
		ZB 237
		ZINC BORATE
		ZINC BORATE (DOT)
		ZN 100
1332-21-4		
		4T04
		7N05
		7RF10
		AMOSITE
		ANTHOPHYLLITE
	*	ASBESTOS
		ASBESTOS (FRIABLE)
		ASBESTOS DUST
		ASBESTOS, AMPHIBOLE
		ASBESTOS, FIBERS
		AT 7-1
		CALIDREA HPP
		CALIDRIA R-G 244
		CHLOROBESTOS 25
		FAPM 410-120
		FERODO C3C
		K 6-20
		M 3-60
		M 5-60
		MOUNTAIN CORK
		MOUNTAIN LEATHER
		MOUNTAIN WOOD
		TREMOLITE
1332-58-7		
		AIRFLO V 8
		ALPHACOTE
		AMAZON 88
		APSILEX
		ARGIFLEX
		ARGILLA
		ARGIREC KN 15
		ARGIREK B 22
		ASP 170
		BOLUS ALBA
		CHINA CLAY
		CLAY 347
		CLAYS, CHINA
		DEVOLITE
		DINKIE A
		ELECTROS
		HYDRAPRINT
		HYDRASHEEN 90M
		HYDRITE UF
		HYDROGLOSS
		KAMIG
		KAO-GEL
		KAOBRITE
		KAOLIN
		KAOLIN COLLOIDAL
		KOG
		KRKHS
		LORCO BANTAC PLUS
		LUSTRA
		MECA
		MOLOCHITE
		NEOKAOLIN
		NUCAP 100
		NUCAP 190
		OPTIWHITE
		OSMO KAOLIN
		PARAGON
		PARAGON (CLAY)
		PORCELAIN CLAY
		POSALIN ACTIVE
		RANDITE
		REFORSIL 700
		REFRACTORINESS
		SATINTONE
		SATINTONE 5
		SATINTONE SPECIAL
		SPESWHITE
		SUPREX CLAY
		TISYN
		ULTRA COTE
		ULTRA WHITE 90
		UW 90
		WHITETEX 2
1333-13-7		
		M-CRESOL, TERT-BUTYL-
		PHENOL, (1,1-DIMETHYLETHYL)-3-METHYL-
		TERT-BUTYL-M-CRESOL

CAS No.	IN	Chemical Name
1333-39-7		
		BENZENESULFONIC ACID, HYDROXY-
		HYDROXYBENZENESULFONIC ACID
		PHENOLSULFONIC A
1333-41-1		
		METHYLPYRIDINE
		PICOLINE
		PICOLINE (DOT)
		PICOLINE [COMBUSTIBLE LIQUID LABEL]
		PICOLINE [FLAMMABLE LIQUID LABEL]
		PICOLINES
		PYRIDINE, METHYL-
		UN 2313 (DOT)
1333-74-0		
		DIHYDROGEN
	*	HYDROGEN
		HYDROGEN (DOT)
		HYDROGEN (H2)
		HYDROGEN MOLECULE
		HYDROGEN, COMPRESSED (DOT)
		HYDROGEN, REFRIGERATED LIQUID
		HYDROGEN, REFRIGERATED LIQUID (DOT)
		MOLECULAR HYDROGEN
		ORTHOHYDROGEN
		PARAHYDROGEN
		PROTIUM
		UN 1049 (DOT)
		UN 1966 (DOT)
1333-82-0		
		CHROMIC ANHYDRIDE
		CHROMIC TRIOXIDE
		CHROMIUM ANHYDRIDE
		CHROMIUM OXIDE (Cr4O12)
		CHROMIUM OXIDE (CrO3)
		CHROMIUM TRIOXIDE
		CHROMIUM(VI) OXIDE
	*	CHROMIUM(VI) OXIDE(1:3)
		MONOCHROMIUM TRIOXIDE
1333-83-1		
		Na-0101 T 1/8"
		SODIUM BIFLUORIDE
		SODIUM BIFLUORIDE, SOLID
		SODIUM BIFLUORIDE, SOLID (DOT)
		SODIUM BIFLUORIDE, SOLUTION
		SODIUM BIFLUORIDE, SOLUTION (DOT)
		SODIUM FLUORIDE
		SODIUM FLUORIDE (Na(HF2))
		SODIUM HYDROGEN DIFLUORIDE
		SODIUM HYDROGEN FLUORIDE
		SODIUM HYDROGEN FLUORIDE (DOT)
		UN 2439 (DOT)
1333-86-4		
		ACETYLENE BLACK
		ACTICARBON AC 35
		AD 200
		AM BLACK
		ASAHITHERMAL
		ATG 60
		ATG 70
		AUSTIN BLACK
		AX 3023
		BLACK FW
		BLACK PEARLS
		C.I. 77266
		C.I. PIGMENT BLACK 7
		CABOT 607
		CARBALAC 2
		CARBOLAC 1
	*	CARBON BLACK
		CARBON BLACK MONARCH 81
		CC 40-220
		CHANNEL BLACK
		CHESACARB K 2
		CK 4
		CK 4 (CARBON BLACK)
		COLUMBIA CARBON
		CONDUCTEX 40-220
		CONDUCTEX 900
		CONDUCTEX 950
		CONDUCTEX 975
		CONDUCTEX CC 40-220
		CONDUCTEX N 472
		CONDUCTEX SC
		CONTINEX N 356
		CORAX 3HS
		CORAX A
		CORAX L
		CORAX L 6
		CORAX P
		CSX 147
		CSX 150A2
		CSX 174
		CSX 200A
		DEGUSSA BLACK FW
		DENKABLACK
		DERMMAPOL BLACK G
		DG 100
		DIABLACK G
		DMG 105A
		DUREX O
		EDO
		EDO (CARBON BLACK)
		ELF 78
		ELF-O
		ELFTEX 150
		ELFTEX 5
		ELFTEX 8
		EPC
		EPC (CARBON BLACK)
		EXP
		EXP (CARBON BLACK)
		EXP 1
		EXP 2
		F 122
		FARBRUSS FW 1
		FARBRUSS S 160
		FLAMMRUSS 101
		FURNACE BLACK
		FURNAL 500
		FURNEX
		FURNEX N 765
		FW 200
		FW 200 (CARBON)
		G 2
		G 2 (CARBON BLACK)
		GAS BLACK
		GRAPHON C
		GRAPHTOL BLACK BLN
		IRA 2
		ISAF
		KETJENBLACK
		KGO 250
		KOSMAS 40
		LAMPBLACK
		MA 100
		MA 100 (CARBON)
		MCF 88
		MCF-HS 78
		MCF-LS 74
		METANEX D
		MICROLITH BLACK CA
		MOGUL
		MOGUL L
		MOLACCO
		MONARCH 1100
		MONARCH 4
		MONARCH 700
		MONARCH 71
		MONARCH 800
		MONARCH 81
		MONOCOL 35T
		MONOCOL 37T
		MONOCOL MX 230

CAS No.	IN	Chemical Name
		MT
		MT (CARBON BLACK)
		N 234
		N 326
		N 339
		N 351
		N 358
		N 358 (CARBON BLACK)
		N 375
		N 472
		N 550
		N 630
		N 630 (CARBON BLACK)
		N 650
		N 660
		N 660 (CARBON BLACK)
		N 76225
		N 76230
		N 76230 (CARBON BLACK)
		N 774
		N 787
		N 990
		N 990 (CARBON BLACK)
		NEO-SPECTRA II
		NEO-SPECTRA MARK II
		NIGROS F
		NIGROS G
		NIGROS K
		NITERON 55
		NYLOFIL BLACK BLN
		P 1250
		P 33
		P 33 (CARBON BLACK)
		PEACH BLACK
		PEARLS 800
		PELLETEX
		PERMABLAK 663
		PGM 33
		PGM 40
		PHILBLACK
		PHILBLACK N 550
		PHILBLACK N 765
		PHILBLACK O
		PIGMENT BLACK 7
		PM 100
		PM 100A
		PM 100KRSZ
		PM 100V
		PM 105K
		PM 136.5
		PM 15
		PM 30V
		PM 50
		PM 50 (CARBON BLACK)
		PM 70
		PM 70 (CARBON BLACK)
		PM 75
		PM 90E
		PME 100V
		PME 70V
		PME 80V
		PMG 33
		PMN 130
		PMN 130N
		PMO 130
		PMTK 90
		PRINTEX 140
		PRINTEX 200
		PRINTEX 30
		PRINTEX 300
		PRINTEX 400
		PRINTEX 60
		PRINTEX A
		PRINTEX G
		PRINTEX U
		PYROBLACK 7007
		RAVEN
		RAVEN 11
		RAVEN 30
		RAVEN 35
		RAVEN 420
		RAVEN 500
		RAVEN 520
		RAVEN 8000
		REBONEX H
		REBONEX HS
		REGAL
		REGAL 300
		REGAL 300R
		REGAL 330
		REGAL 330R
		REGAL 400
		REGAL 400R
		REGAL 600
		REGAL 660
		REGAL 99
		REGAL SRF
		ROYAL SPECTRA
		S 300
		S 300 (CARBON BLACK)
		S 315
		SAF
		SEAST 3
		SEAST 3H
		SEAST 6
		SEAST SO
		SEAST V
		SEIKA SEVEN
		SERVACARB
		SEVACARB MT
		SEVALCO
		SHOBLACK N 330
		SPECIAL BLACK 15
		SPECIAL BLACK 4
		SPECIAL BLACK 5
		SPF 35
		SPHERON 6
		SPHERON 9
		STATEX B
		STATEX B 12
		STATEX M 70
		STATEX N 550
		STERLING 142
		STERLING FT
		STERLING MT
		STERLING MTG
		STERLING N 765
		STERLING NS
		STERLING R
		STERLING SO
		STERLING SO 1
		STERLING V
		STERLING VH
		STERLING VL
		SUPERCARBOVAR
		TEG 10
		TG 10
		THERMAX
		TM 30
		TM 70
		UNITED 3017
		UNITED SL 90
		VULCAN
		VULCAN C
		VULCAN XC 72R
		WITCOBLAK NO. 100
		X 1303
		X 1341
		X 1341 (CARBON BLACK)
		XC 3017L
		XC 72
1335-23-5		
		COPPER IODIDE
1335-26-8		
		MAGNESIUM PEROXIDE

CAS No.	IN	Chemical Name
1335-31-5		
		MERCURIC OXYCYANIDE
1335-32-6		
		BIS(ACETATO)DIHYDROXYTRILEAD
		LEAD ACETATE HYDROXIDE (Pb3(OAC)2(OH)4)
		LEAD ACETATE, BASIC
		LEAD SUBACETATE
		LEAD, BIS(ACETATO)TETRAHYDROXYTRI-
		LEAD, BIS(ACETATO-O)TETRAHYDROXYTRI-
		LEAD, SUBACETATE
		MONOBASIC LEAD ACETATE
1335-85-9		
		2,4-DINITRO-6-METHYLPHENOL
		2,4-DINITRO-O-CRESOL
		3,5-DINITRO-2-HYDROXYTOLUENE
		4,6-DINITRO-O-CRESOL
		4,6-DINITRO-O-CRESOLO (ITALIAN)
		4,6-DINITRO-O-KRESOL (CZECH)
		4,6-DINITROKRESOL (DUTCH)
		ANTINONIN
		ANTINONNIN
		ARBOROL
		CAPSINE
		CHEMSECT DNOC
		DEGRASSAN
		DEKRYSIL
		DETAL
		DINITRO-O-CRESOL
		DINITRO-O-CRESOL (ACGIH)
		DINITROCRESOL
		DINITRODENDTROXAL
		DINITROL
		DINITROMETHYL CYCLOHEXYLTRIENOL
		DINOC
		DINURANIA
		DITROSOL
		DN
		DN-DRY MIX NO 2
		DNC
		DNOC
		DNOK (CZECH)
		DWUNITRO-O-KREZOL (POLISH)
		EFFUSAN
		EFFUSAN 3436
		ELGETOL
		ELGETOL 30
		ELIPOL
		ENT 154
		EXTRAR
		HEDOLIT
		HEDOLITE
		K III
		K IV
		KRENITE (OBS)
		KRESAMONE
		KREZOTOL 50
		LE DINITROCRESOL-4,6 (FRENCH)
		LIPAN
		NITRADOR
		NITROFAN
		O-CRESOL, 4,6-DINITRO-
		O-CRESOL, DINITRO-
		PHENOL, 2-METHYL-4,6-DINITRO- (9CI)
		PHENOL, 2-METHYLDINITRO-
		PROKARBOL
		RAFEX
		RAFEX 35
		RAPHATOX
		RCRA WASTE NUMBER P047
		SANDOLIN
		SANDOLIN A
		SELINON
		SINOX
		TRIFOCIDE
		TRIFRINA
		WINTERWASH
		ZAHLREICHE BEZEICHNUNGEN (GERMAN)
1335-87-1		
		HALOWAX 1014
		HEXACHLORONAPHTHALENE
		HEXACHLORONAPHTHALENE (ACGIH)
		NAPHTHALENE, HEXACHLORO-
1335-88-2		
		NAPHTHALENE, TETRACHLORO-
		TETRACHLORONAPHTHALENE
1336-21-6		
		AMMMONIUM HYDROXIDE ((NH4)(OH))
		AMMONIA AQUEOUS
		AMMONIA SOLUTION, CONTAINING 44% OR LESS AMMONIA (DOT)
		AMMONIA WATER
		AMMONIA WATER 29%
		AMMONIA, AQUA
		AMMONIA, MONOHYDRATE
	*	AMMONIUM HYDROXIDE
		AMMONIUM HYDROXIDE ((NH4)(OH))
		AMMONIUM HYDROXIDE (DOT)
		AMMONIUM HYDROXIDE, CONTAINING LESS THAN 12% AMMONIA
		AMMONIUM HYDROXIDE, CONTAINING LESS THAN 12% AMMONIA (DOT)
		AMMONIUM HYDROXIDE, CONTAINING NOT LESS THAN 12% BUT NOT MORE THAN 44% AMMONIA (DOT)
		AMMONIUM HYDROXIDE, WITH 12% AMMONIA
		AQUA AMMONIA
		AQUA AMMONIA, SOLUTION (DOT)
		NA 2672 (DOT)
1336-36-3		
		1,1'-BIPHENYL, CHLORO DERIVS.
		AROCLOR
		AROCLOR 1221
		AROCLOR 1232
		AROCLOR 1242
		AROCLOR 1248
		AROCLOR 1254
		AROCLOR 1260
		AROCLOR 1262
		AROCLOR 1268
		AROCLOR 2565
		AROCLOR 4465
		AROCLOR 5442
		BIPHENYL, CHLORINATED
		BIPHENYL, POLYCHLORO-
		CHLOPHEN
		CHLOREXTOL
		CHLORINATED BIPHENYL
		CHLORINATED DIPHENYL
		CHLORINATED DIPHENYLENE
		CHLORO 1,1-BIPHENYL
		CHLORO BIPHENYL
		CLOPHEN
		DIPHENYL, CHLORINATED
		DYKANOL
		FENCLOR
		INERTEEN
		KANECHLOR
		KANECHLOR 300
		KANECHLOR 400
		MONTAR
		NOFLAMOL
		PCB
		PCB (DOT)
		PCBS
		PHENOCHLOR
		PHENOCLOR
	*	POLYCHLORINATED BIPHENYLS
		POLYCHLORINATED BIPHENYLS (DOT)
		POLYCHLOROBIPHENYL
		PYRALENE
		PYRANOL
		SANTOTHERM
		SANTOTHERM FR
		SOVOL

CAS No.	IN	Chemical Name
		THERMINOL FR-1
		UN 2315 (DOT)
1338-02-9		
		COPPER NAPHTHANATE SOLUTION, NOT MORE THAN 8% COPPER NAPHTHANATE
		COPPER NAPHTHENATE
		COPPER UVERSOL
		CUPRINOL
		NAPHTHENIC ACID COPPER SALT
		WITTOX C
1338-23-4		
		2-BUTANONE PEROXIDE, MAXIMUM CONCENTRATION 60%
		2-BUTANONE PEROXIDE, WITH NOT MORE THAN 9% BY WEIGHT ACTIVE OXYGEN
		2-BUTANONE, PEROXIDE
		BUTANOX LPT
		BUTANOX M 105
		BUTANOX M 50
		CHALOXYD MEKP-HA 1
		CHALOXYD MEKP-LA 1
		ETHYL METHYL KETONE PEROXIDE
		ETHYL METHYL KETONE PEROXIDE, MAXIMUM CONCENTRATION 50% (DOT)
		ETHYL METHYL KETONE PEROXIDE, MAXIMUM CONCENTRATION 60% (DOT)
		FR 222
		HI-POINT 180
		KAYAMEK A
		KAYAMEK M
		KETONOX
		LUCIDOL DDM 9
		LUCIDOL DELTA X
		LUPERSOL DDA 30
		LUPERSOL DDM
		LUPERSOL DDM 9
		LUPERSOL DELTA X
		LUPERSOL DELTA X 9
		LUPERSOL DNF
		LUPERSOL DSW
		MEK PEROXIDE
		MEKP
		MEKPO
		MEPOX
		METHYL ETHYL KETONE HYDROPEROXIDE
	*	METHYL ETHYL KETONE PEROXIDE
		METHYL ETHYL KETONE PEROXIDE, IN SOLUTION WITH NOT MORE THAN 9% BY WT ACTIVE OXYGEN (DOT)
		METHYL ETHYL KETONE PEROXIDE, WITH NOT MORE THAN 50% PEROXIDE
		METHYL ETHYL KETONE PEROXIDE, WITH NOT MORE THAN 60% PEROXIDE
		PERMEK G
		PERMEK N
		TRIGONOX M 50
		UN 2127 (DOT)
		UN 2550 (DOT)
1338-24-5		
		ACIDOL (PETROLEUM BY-PRODUCT)
		ACIDS, CARBOXYLIC
		ACIDS, NAPHTHENIC
		AGENAP
		CARBOXYLIC ACIDS, NAPHTHENIC
		NA 9137 (DOT)
		NAPHID
	*	NAPHTHENIC ACID
		NAPHTHENIC ACID (DOT)
		NS 130
		NS 160
		SP 230
		SUNAPTIC ACID B
		SUNAPTIC ACID C
		SUNAPTIC B
1341-24-8		
		ACETOPHENONE, CHLORO-
		CHLOROACETOPHENONE
		ETHANONE, 1-PHENYL-, MONOCHLORO DERIV.
1341-49-7		
		ACID AMMONIUM FLUORIDE
	*	AMMONIUM BIFLUORIDE
		AMMONIUM BIFLUORIDE (NH4HF2)
		AMMONIUM BIFLUORIDE (NH5F2)
		AMMONIUM BIFLUORIDE, SOLID (DOT)
		AMMONIUM BIFLUORIDE, SOLUTION (DOT)
		AMMONIUM DIFLUORIDE
		AMMONIUM DIFLUORIDE (NH4HF2)
		AMMONIUM FLUORIDE ((NH4)(HF2))
		AMMONIUM FLUORIDE COMP WITH HYDROGEN FLUORIDE (1:1)
		AMMONIUM FLUORIDE COMPD WITH HYDROGEN FLUORIDE (1:1)
		AMMONIUM FLUORIDE COMPD. WITH HYDROGEN FLUORIDE (1:1)
		AMMONIUM HYDROFLUORIDE
		AMMONIUM HYDROFLUORIDE (NH4HF2)
		AMMONIUM HYDROFLUORIDE (NH5F2)
		AMMONIUM HYDROGEN BIFLUORIDE
		AMMONIUM HYDROGEN DIFLUORIDE
		AMMONIUM HYDROGEN DIFLUORIDE (NH4(HF2))
		AMMONIUM HYDROGEN FLUORIDE
		AMMONIUM HYDROGEN FLUORIDE ((NH4)(HF2))
		AMMONIUM HYDROGEN FLUORIDE SOLUTION
		AMMONIUM HYDROGEN FLUORIDE SOLUTION (DOT)
		AMMONIUM HYDROGEN FLUORIDE, SOLID
		AMMONIUM HYDROGEN FLUORIDE, SOLID (DOT)
		AMMONIUM MONOHYDROGEN DIFLUORIDE
		FLAMMON
		UN 1727 (DOT)
		UN 2817 (DOT)
1344-28-1		
		A 1
		A 1 (SORBENT)
		AERO 100
		AL 13
		AL 13 (OXIDE)
		ALCOA F 1
		ALMITE
		ALON
		ALON C
		ALOXITE
		ALPHA-ALUMINA
		ALPHA-ALUMINUM OXIDE
		ALUMINA
		ALUMINA CGAMMA
		ALUMINASOL 100
		ALUMINITE 37
	*	ALUMINUM OXIDE
		ALUMINUM OXIDE (Al2O3)
		ALUMINUM OXIDE (BROCKMANN)
		ALUMINUM SESQUIOXIDE
		ALUMINUM TRIOXIDE
		ALUMITE
		ALUMITE (OXIDE)
		ALUMOGEL A 1
		ALUNDUM
		ALUNDUM 600
		AQ 10
		AQ 25
		AQ 50
		BETA-ALUMINUM OXIDE
		CAB-O-GRIP
		CATAPAL S
		CKA
		COMPALOX
		CONOPAL
		DELTA-ALUMINUM OXIDE
		DIALUMINUM TRIOXIDE
		DISPAL
		DISPAL M
		DOTMENT 324
		DOTMENT 358
		ETA-ALUMINA
		EXOLON XW 60

CAS No.	IN	Chemical Name
		F 360
		F 360 (ALUMINA)
		FASERTON
		FASERTONERDE
		G 0
		G 0 (OXIDE)
		G 2
		G 2 (OXIDE)
		GAMMA-ALUMINA
		GAMMA-ALUMINUM OXIDE
		GK
		GK (OXIDE)
		HYPALOX II
		JRC-ALO 4
		JUBENON R
		KA 101
		KETJEN B
		KHP 2
		KIMAL
		KYOWARD 200
		LA 6
		LINDE A
		LUCALOX
		LUDOX CL
		MA 11
		MA 11 (METAL OXIDE)
		MARTOXIN
		MICROGRIT WCA
		MM 21
		NEOBEAD C
		PORAMINAR
		PS 1
		PS 1 (ALUMINA)
		Q-LOID A 30
		RC 172DBM
1344-40-7		
		CI 77620
		DIBASIC LEAD METAPHOSPHATE
		DIBASIC LEAD PHOSPHITE
		LEAD DIBASIC PHOSPHITE
		LEAD OXIDE PHOSPHONATE (Pb3O2(HPO3)), HEMIHYDRATE
		LEAD OXIDE PHOSPHONATE, HEMIHYDRATE
		LEAD PHOSPHITE DIBASIC (DOT)
		LEAD PHOSPHITE, DIBASIC
		LEAD, DIOXO[PHOSPHITO(2-)]TRI-, HEMIHYDRATE
		UN 2989 (DOT)
1344-48-5		
		BETA-MERCURIC SULFIDE
		CHINESE VERMILION
		CI 77766
		CI PIGMENT RED 106
		ETHIOPS MINERAL
		MERCURIC SULFIDE
		MERCURIC SULFIDE RED
		MERCURIC SULFIDE, BLACK
		MERCURY MONOSULFIDE
		MERCURY SULFIDE (HgS)
		MERCURY SULPHIDE
		MERCURY(2+) SULFIDE
		MONOMERCURY SULFIDE
		ORANGE VERMILION
		PURE ENGLISH (QUICKSILVER) VERMILION
		RED CINNABAR
		RED MERCURY SULPHIDE
		SCARLET VERMILION
		VERMILION
1344-57-6		
		URANYL DIOXIDE
1344-58-7		
		URANIUM TRIOXIDE
1344-67-8		
	*	COPPER CHLORIDE
		CUPRIC CHLORIDE
1344-95-2		
		CALCIUM HYDROSILICATE
		CALCIUM MONOSILICATE
		CALCIUM POLYSILICATE
		CALCIUM SILICATE
		CALFLO E
		CALSIL
		CS LAFARGE
		FLOLITE R
		MARIMET 45
		MICRO-CEL
		MICRO-CEL A
		MICRO-CEL B
		MICRO-CEL C
		MICRO-CEL E
		MICRO-CEL T
		MICRO-CEL T 26
		MICRO-CEL T 38
		MICRO-CEL T 41
		MICROCAL ET
		SILENE EF
		SILICIC ACID, CALCIUM SALT
		SILMOS T
		SOLEX
		STARLEX L
1395-21-7		
		SUBTILISINS (PROTEOLYTIC ENZ AS 100% PURE CRYST ENZYME)
1397-94-0		
		ANTIMYCIN
		ANTIMYCIN A
		ANTIPIRICULLIN
		VIROSIN
1401-55-4		
		3,4-DIHYDROXY-5-((3,4,5-TRIHYDROXYBENZOYL)OXY)BENZOIC ACID
		3-GALLOYL GALLIC ACID
		ACIDE TANNIQUE (FRENCH)
		BENZOIC ACID, 3,4-DIHYDROXY-5-((3,4,5-TRIHYDROXYBENZOYL)OXY)- (9CI)
		CATECHINS
		DIGALLIC ACID
		ELLAGI-TANNIN ANALYSIS
		GALLIC ACID 3-MONOGALLATE
		GALLIC ACID, 3-GALLATE
		GALLO-TANNIN ANALYSIS
		GALLOTANNIC ACIDS
		GALLOTANNINS
		HIFIX SL
		M-DIGALLIC ACID
		M-GALLOYL GALLIC ACID
		SUNLIFE TN
		TANAPHEN P 500
		TANNIC ACID
		TANNIN ANAL
		TANNINS
		TANNINS, GALLO-
1402-68-2		
		AFLATOXINS
		FLAVATOXINS
		TOXINS, AFLA
1405-87-4		
		AYFIVIN
		BACI-JEL
		BACIGUENT
		BACILIQU IN
		BACITEK OINTMENT
		BACITRACIN
		FORTRACIN
		PARENTRACIN
		PENITRACIN
		TOPITRACIN
		TOPITRASIN
		USAF CB-7

CAS No.	IN	Chemical Name
		ZUTRACIN
1420-04-8		
		2',5-DICHLORO-4'-NITROSALICYLANILIDE, 2-AMINO-ETHANOL SALT
		BAY 6076
		BAY 73
		BAYER 6076
		BAYER 73
		BAYLUCIT
		BAYLUSCIDE
		BENZAMIDE, 5-CHLORO-N-(2-CHLORO-4-NITROPHENYL)-2-HYDROXY-, COMPD. WITH 2-AMINOETHANOL (1:1)
		CLONITRALID
		CLONITRALIDE
		ETHANOL, 2-AMINO-, COMPD. WITH 2',5-DICHLORO-4'-NITROSALICYLANILIDE (1:1)
		HL 2448
		M 73
		MOLLUTOX
		NICLOSAMIDE ETHANOLAMINE SALT
		PHENASAL ETHANOLAMINE SALT
		SALICYLANILIDE, 2',5-DICHLORO-4'-NITRO-, COMPD. WITH 2-AMINOETHANOL (1:1)
		SR 73
1420-07-1		
		2,4-DINITRO-6-TERT-BUTYLPHENOL
		2-(1,1-DIMETHYLETHYL)-4,6-DINITROPHENOL
		2-TERT-BUTYL-4,6-DINITROPHENOL
		DINOTERB
		DINOTERBE
		DNTBP
		HERBOGIL
		PHENOL, 2-(1,1-DIMETHYLETHYL)-4,6-DINITRO-
		PHENOL, 2-TERT-BUTYL-4,6-DINITRO-
		PHENOL, O-T-BUTYL-4,6-DINITRO-
		STIRPAN FORTE
		VERALINE CREME
1455-21-6		
		1-NONANETHIOL
1459-10-5		
		1-PHENYLTETRADECANE
		BENZENE, TETRADECYL-
		TETRADECANE, 1-PHENYL-
		TETRADECYLBENZENE
1464-53-5		
		1,1'-BI(ETHYLENE OXIDE)
		1,2,3,4-DIEPOXYBUTANE
		1,3-BUTADIENE DIEPOXIDE
		2,2'-BIOXIRANE
		2,4-DIEPOXYBUTANE
		BIOXIRAN
		BIOXIRANE
		BUTADIENDIOXYD (GERMAN)
		BUTADIENE DIEPOXIDE
		BUTADIENE DIOXIDE
		BUTANE DIEPOXIDE
		BUTANE, 1,2:3,4-DIEPOXY-
		DEB
		DIEPOXYBUTANE
		DIOXYBUTADIENE
		ENT-26592
		ERYTHRITOL ANHYDRIDE
		RCRA WASTE NUMBER U085
		THREITOL, 1,2:3,4-DIANHYDRO-
1467-79-4		
		CYANAMIDE, DIMETHYL-
		CYANODIMETHYLAMINE
		DIMETHYLCYANAMIDE
		N,N-DIMETHYLCYANAMIDE
		N-CYANO-N-METHYLMETHANAMINE
1477-55-0		
		1,3-BENZENEDIMETHANAMINE
		1,3-BIS(AMINOMETHYL)BENZENE
		1,3-XYLYLENEDIAMINE
		3-(AMINOMETHYL)BENZYLAMINE
		ALPHA,ALPHA'-DIAMINO-M-XYLENE
		ALPHA,ALPHA'-M-XYLENEDIAMINE
		M-DIAMINOXYLENE
		M-XYLENE A,A-DIAMINE
		M-XYLENE-ALPHA,ALPHA-DIAMINE
		M-XYLENEDIAMINE
		M-XYLYLENEDIAMINE
		MXDA
1480-60-7		
		QUARTZ DUST
1498-40-4		
		ETHYL PHOSPHONOUS DICHLORIDE
1498-51-7		
		DICHLOROPHOSPHORIC ACID, ETHYL ESTER
		ETHYL DICHLOROPHOSPHATE
		ETHYL PHOSPHORODICHLORIDATE
		ETHYL PHOSPHORODICHLORIDATE (DOT)
		ETHYLPHOSPHORIC ACID DICHLORIDE
		NA 1760 (DOT)
		PHOSPHORODICHLORIDIC ACID, ETHYL ESTER
1552-12-1		
		1,5-CYCLOOCTADIENE
1558-25-4		
		(CHLOROMETHYL)TRICHLOROSILANE
		SILANE, CHLOROMETHYL(TRICHLORO)-
		SILANE, TRICHLORO(CHLOROMETHYL)-
		SILANE, TRICHLORO(CHLOROMETHYL)- (9CI)
		TRICHLORO(CHLOROMETHYL)SILANE
1561-49-5		
		CHPC
		DICYCLOHEXYL PEROXIDE CARBONATE
		DICYCLOHEXYL PEROXYDICARBONATE
		DICYCLOHEXYL PEROXYDICARBONATE (DOT)
		DICYCLOHEXYL PEROXYDICARBONATE, NOT MORE THAN 91% WITH WATER
		DICYCLOHEXYL PEROXYDICARBONATE, NOT MORE THAN 91% WITH WATER (DOT)
		DICYCLOHEXYL PEROXYDICARBONATE, TECHNICALLY PURE
		PEROXYDICARBONIC ACID, DICYCLOHEXYL ESTER
		PEROXYDICARBONIC ACID, DICYCLOHEXYL ESTER, NOT MORE THAN 91% WITH WATER
		UN 2152 (DOT)
		UN 2153 (DOT)
1563-66-2		
		2,2-DIMETHYL-2,3-DIHYDRO-7-BENZOFURANYL N-METHYL-CARBAMATE
		2,2-DIMETHYL-7-COUMARANYL N-METHYLCARBAMATE
		2,3-DIHYDRO-2,2-DIMETHYL-7-BENZOFURANYL METHYL-CARBAMATE
		2,3-DIHYDRO-2,2-DIMETHYLBENZOFURANYL-7-N-METHYL-CARBAMATE
		7-BENZOFURANOL, 2,3-DIHYDRO-2,2-DIMETHYL-, METHYL-CARBAMATE
		BAY 70143
		CARBAMIC ACID, METHYL-, 2,2-DIMETHYL-2,3-DIHYDRO-BENZOFURAN-7-YL ESTER
		CARBAMIC ACID, METHYL-, 2,3-DIHYDRO-2,2-DIMETHYL-7-BENZOFURANYL ESTER
		CARBOFURAN
		CARBOFURAN (ACGIH,DOT)
		CARBOFURAN MIXTURE, LIQUID
		CHINUFUR
		CURATERR
		D 1221
		ENT 27,164
		FMC 10242

CAS No.	IN	Chemical Name
		FURADAN
		FURADAN 3G
		FURADAN 75 WP
		FURODAN
		KARBOFURANU (POLISH)
		METHYL CARBAMIC ACID 2,3-DIHYDRO-2,2-DIMETHYL-7-BENZOFURANYL ESTER
		NA 2757 (DOT)
		NIA 10242
		NIAGARA 10242
		NIAGARA NIA-10242
		OMS 864
		YALTOX
1569-69-3		
		CYCLOHEXANETHIOL
1582-09-8		
		2,6-DINITRO-4-TRIFLUORMETHYL-N,N-DIPROPYLANILIN (GERMAN)
		2,6-DINITRO-N,N-DI-N-PROPYL-ALPHA,ALPHA,ALPHA-TRIFLUORO-P-TOLUIDINE
		2,6-DINITRO-N,N-DIPROPYL-4-(TRIFLUOROMETHYL)BENZENAMINE
		4-(DI-N-PROPYLAMINO)-3,5-DINITRO-1-TRIFLUOROMETHYLBENZENE
		AGREFLAN
		AGRIFLAN 24
		ALPHA,ALPHA,ALPHA-TRIFLUORO-2,6-DINITRO-N,N-DIPROPYL-P-TOLUIDINE
		BENZENAMINE, 2,6-DINITRO-N,N-DIPROPYL-4-(TRIFLUOROMETHYL)-
		BENZENAMINE, 2,6-DINITRO-N,N-DIPROPYL-4-(TRIFLUOROMETHYL)- (9CI)
		CRISALIN
		DIGERMIN
		ELANCOLAN
		L 36352
		LILLY 36,352
		N,N-DI-N-PROPYL-2,6-DINITRO-4-TRIFLUOROMETHYLANILINE
		N,N-DIPROPYL-2,6-DINITRO-4-TRIFLUORMETHYLANILIN (GERMAN)
		N,N-DIPROPYL-4-TRIFLUOROMETHYL-2,6-DINITROANILINE
		NCI-C00442
		NITRAN
		OLITREF
		P-TOLUIDINE, ALPHA,ALPHA,ALPHA-TRIFLUORO-2,6-DINITRO-N,N-DIPROPYL-
		SU SEGURO CARPIDOR
		SYNFLORAN
		TREFANOCIDE
		TREFICON
		TREFLAN
		TREFLANOCIDE ELANCOLAN
		TRIFLORAN
		TRIFLUORALIN
	*	TRIFLURALIN
		TRIFLURALINE
		TRIFUREX
		TRIKEPIN
		TRIM
1596-84-5		
		DIAMINOZIDE
1600-27-7		
		ACETIC ACID, MERCURIDI-
		ACETIC ACID, MERCURY(2+) SALT
		BIS(ACETYLOXY)MERCURY
		DIACETOXYMERCURY
		MERCURIACETATE
	*	MERCURIC ACETATE
		MERCURIC ACETATE (DOT)
		MERCURIC DIACETATE
		MERCURIC(II) ACETATE
		MERCURY ACETATE
		MERCURY ACETATE (DOT)
		MERCURY ACETATE (Hg(O2C2H3)2)
		MERCURY DIACETATE
		MERCURY(2+) ACETATE
		MERCURY(II) ACETATE
		MERCURY(II) DIACETATE
		MERCURYL ACETATE
		UN 1629 (DOT)
1609-19-4		
		CHLORODIETHYLSILANE
		DIETHYLCHLOROSILANE
		SILANE, CHLORODIETHYL-
1609-86-5		
		TERT-BUTYL ISOCYANATE
1615-80-1		
		1,2-DIETHYL HYDRAZINE
		HYDRAZINE, 1,2-DIETHYL-
		N,N'-DIETHYLHYDRAZINE
1622-32-8		
		ETHANESULFONYL CHLORIDE, 2-CHLORO
1623-24-1		
		DIHYDROGEN ISOPROPYL PHOSPHATE
		ISOPROPYL ACID PHOSPHATE
		ISOPROPYL ACID PHOSPHATE, SOLID
		ISOPROPYL ACID PHOSPHATE, SOLID (DOT)
		ISOPROPYL DIHYDROGEN PHOSPHATE
		ISOPROPYL HYDROGEN PHOSPHATE
		ISOPROPYL PHOSPHATE
		ISOPROPYL PHOSPHATE ((C3H7O)(HO)2PO)
		ISOPROPYL PHOSPHORIC ACID, SOLID (DOT)
		MONOISOPROPYL PHOSPHATE
		PHOSPHORIC ACID, ISOPROPYL ESTER
		PHOSPHORIC ACID, MONO(1-METHYLETHYL) ESTER
		PHOSPHORIC ACID, MONOISOPROPYL ESTER
		UN 1793 (DOT)
1634-04-4		
		METHYL-TERT-BUTYL ETHER
1635-05-2		
		STRONTIUM CARBONATE
1639-09-4		
		1-HEPTANETHIOL
1640-89-7		
		CYCLOPENTANE, ETHYL-
		ETHYL CYCLOPENTANE
1642-54-2		
		1-DIETHYLCARBAMOYL-4-METHYLPIPERAZINE DIHYDROGEN CITRATE
		1-METHYL-4-DIETHYLCARBAMOYLPIPERAZINE CITRATE
		1-PIPERAZINECARBOXAMIDE, N,N-DIETHYL-4-METHYL-, 2-HYDROXY-1,2,3-PROPANETRICARBOXYLATE (1:1)
		1-PIPERAZINECARBOXAMIDE, N,N-DIETHYL-4-METHYL-, CITRATE (1:1)
		BANOCIDE
		CARICIDE
		CARITROL
		DICAROCIDE
		DIETHYLCARBAMAZANE CITRATE
		DIETHYLCARBAMAZINE ACID CITRATE
		DIETHYLCARBAMAZINE CITRATE
		DIETHYLCARBAMAZINE HYDROGEN CITRATE
		DIROCIDE
		DITRAZIN
		DITRAZIN CITRATE
		DITRAZINE
		DITRAZINE CITRATE
		ETHODRYL CITRATE
		ETHYLAMINOAZINE CITRATE
		FILAZINE
		FRANOCIDE

CAS No.	IN	Chemical Name
		FRANOZAN
		HETRAZAN
		LOXURAN
		N,N-DIETHYL-4-METHYL-1-PIPERAZINE CARBOXAMIDE CITRATE
		N,N-DIETHYL-4-METHYL-1-PIPERAZINECARBOXAMIDE DIHYDROGEN CITRATE
1653-19-6		
		1,3-BUTADIENE, 2,3-DICHLORO-
		2,3-DICHLORO-1,3-BUTADIENE
		2,3-DICHLOROBUTA-1,3-DIYNE
		2,3-DICHLOROBUTADIENE
		2,3-DICHLOROBUTADIENE-1,3
		DICHLOROBUTADIENE
1663-35-0		
		1-METHOXY-2-(VINYLOXY)ETHANE
		2-METHOXYETHYL ETHENYL ETHER
		2-METHOXYETHYL VINYL ETHER
		ETHANE, 1-METHOXY-2-(VINYLOXY)-
		ETHENE, (2-METHOXYETHOXY)-
		VINYL-2-METHOXYETHYL ETHER
1675-54-3		
	*	DIGLYCIDYL ETHER OF BISPHENOL A
1678-91-7		
		CYCLOHEXANE, ETHYL-
		ETHYL CYCLOHEXANE
1693-71-6		
		ALLYL BORATE
		ALLYL BORATE ((C3H5O)3B)
		BORIC ACID (H3BO3), TRI-2-PROPENYL ESTER
		BORIC ACID (H3BO3), TRIALLYL ESTER
		BORIC ACID, TRI-2-PROPENYL ESTER
		BORIC ACID, TRIALLYL ESTER
		TRIALLYL BORATE
		TRIALLYL BORATE (DOT)
		TRIALLYLOXYBORANE
		TRIS(ALLYLOXY)BORANE
		UN 2609 (DOT)
1694-09-3		
		11386 VIOLET
		A.F. VIOLET NO. 1
		ACID FAST VIOLET 5BN
		ACID VIOLET
		ACID VIOLET 49
		ACID VIOLET 4BNP
		ACID VIOLET 4BNS
		ACID VIOLET 5B
		ACID VIOLET 5BN
		ACID VIOLET 6B
		ACID VIOLET S
		ACILAN VIOLET S 4BN
		AIZEN ACID VIOLET 5BH
		ATLANTIC ACID VIOLET 4BNS
		BENZYL VIOLET
		BENZYL VIOLET 3B
		BENZYL VIOLET 4B
		C.I. 42640
		C.I. ACID VIOLET 49
		C.I. ACID VIOLET 49, SODIUM SALT
		C.I. FOOD VIOLET 2
		COOMASSIE VIOLET
		D AND C VIOLET NO. 1
		FAST ACID VIOLET 5BN
		FD AND C VIOLET 1
		FD AND C VIOLET NO. 1
		FOOD VIOLET 2
		HIDACID WOOL VIOLET 5B
		KITON VIOLET 4BNS
		PERGACID VIOLET 2B
		SERVA VIOLET 49
		TERTRACID BRILLIANT VIOLET 6B
		VIOLET 6B
		VIOLET NO. 1
		WOOL VIOLET 4BN
		WOOL VIOLET 5BN
1696-20-4		
		4-ACETYLMORPHOLINE
		MORPHOLINE, 4-ACETYL-
		N-ACETYL MORPHOLINE
1712-64-7		
		ISOPROPYL NITRATE
		ISOPROPYL NITRATE (DOT)
		NITRIC ACID, 1-METHYLETHYL ESTER
		NITRIC ACID, ISOPROPYL ESTER
		PROPANE-2-NITRATE
		UN 1222 (DOT)
1719-53-5		
		DICHLORODIETHYLSILANE
		DIETHYL DICHLOROSILANE
		DIETHYLDICHLOROSILANE (DOT)
		DIETHYLDICHLOROSILICON
		SILANE, DICHLORODIETHYL-
		UN 1767 (DOT)
1738-25-6		
		BETA-DIMETHYLAMINOPROPIONITRILE
1746-01-6		
		2,3,7,8-TCDD
		2,3,7,8-TETRACHLORODIBENZO-1,4-DIOXIN
		2,3,7,8-TETRACHLORODIBENZO-P-DIOXIN
		DIBENZO-P-DIOXIN, 2,3,7,8-TETRACHLORO-
		DIBENZO[B,E][1,4]DIOXIN, 2,3,7,8-TETRACHLORO-
		DIOXIN
		DIOXIN (HERBICIDE CONTAMINANT)
		TCDBD
		TCDD
		TETRACHLORODIBENZO-P-DIOXIN
1752-30-3		
		ACETONE THIOSEMICARBAZIDE
1758-61-8		
		DI(1-HYDROXY CYCLOHEXYL)PEROXIDE
1762-95-4		
		AMMONIUM ISOTHIOCYANATE
		AMMONIUM RHODANATE
		AMMONIUM RHODANIDE
		AMMONIUM SULFOCYANATE
		AMMONIUM SULFOCYANIDE
	*	AMMONIUM THIOCYANATE
		AMMONIUM THIOCYANATE (DOT)
		AMTHIO
		NA 9092 (DOT)
		RHODANID
		RHODANIDE
		THIOCYANIC ACID, AMMONIUM SALT
		TRANS-AID
		USAF EK-P-433
		WEEDAZOL TL
1777-84-0		
		3-NITRO-P-ACETOPHENETIDE
1789-58-8		
		DICHLOROETHYLSILANE
		ETHYL DICHLOROSILANE
		ETHYL DICHLOROSILANE (DOT)
		MONOETHYLDICHLOROSILANE
		SILANE, DICHLOROETHYL-
		UN 1183 (DOT)
1795-48-8		
		2-ISOCYANATOPROPANE
		ISOCYANIC ACID, ISOPROPYL ESTER
	*	ISOPROPYL ISOCYANATE
		ISOPROPYL ISOCYANATE (DOT)

CAS No.	IN	Chemical Name
		PROPANE, 2-ISOCYANATO-
		PROPANE, 2-ISOCYANATO- (9CI)
		UN 2483 (DOT)
1809-19-4		
		BUTYL ALCOHOL, HYDROGEN PHOSPHITE
		BUTYL PHOSPHITE
		BUTYL PHOSPHONATE ((BUO)2HPO)
		DI-N-BUTYL HYDROGEN PHOSPHITE
		DIBUTOXYPHOSPHINE OXIDE
		DIBUTYL HYDROGEN PHOSPHITE
		DIBUTYL PHOSPHATE
		DIBUTYL PHOSPHITE
		DIBUTYL PHOSPHONATE
		PHOSPHONIC ACID, DIBUTYL ESTER
		PHOSPHOROUS ACID, DIBUTYL ESTER
1836-75-5		
		2,4-DICHLORO-4'-NITRODIPHENYL ETHER
		2,4-DICHLOROPHENYL 4-NITROPHENYL ETHER
		2,4-DICHLOROPHENYL P-NITROPHENYL ETHER
		4'-NITRO-2,4-DICHLORODIPHENYL ETHER
		4-(2,4-DICHLOROPHENOXY)NITROBENZENE
		4-NITRO-2',4'-DICHLOROPHENYL ETHER
		BENZENE, 2,4-DICHLORO-1-(4-NITROPHENOXY)-
		ETHER, 2,4-DICHLOROPHENYL P-NITROPHENYL
		FW 925
		MEZOTOX
		NICLOFEN
		NIP
		NITROCHLOR
		NITROFEN
		NITROFEN (PESTICIDE)
		NITROPHEN
		NITROPHENE
		PREPARATION 125
		TOK
		TOK E
		TOK E 25
		TOK E 40
		TRIZILIN
1838-59-1		
		ALLYL FORMATE
		ALLYL FORMATE (DOT)
		FORMIC ACID, 2-PROPENYL ESTER
		FORMIC ACID, ALLYL ESTER
		UN 2336 (DOT)
1863-63-4		
		AMMONIUM BENZOATE
		AMMONIUM BENZOATE (DOT)
		BENZOIC ACID, AMMONIUM SALT
		NA 9080 (DOT)
		VULNOC AB
1873-29-6		
		ISOBUTYL ISOCYANATE
		ISOBUTYL ISOCYANATE (DOT)
		ISOCYANIC ACID, ISOBUTYL ESTER
		PROPANE, 1-ISOCYANATO-2-METHYL-
		PROPANE, 1-ISOCYANATO-2-METHYL- (9CI)
		UN 2486 (DOT)
1885-14-9		
		CARBONOCHLORIDIC ACID, PHENYL ESTER
		CHLOROFORMIC ACID PHENYL ESTER
		FENYLESTER KYSELINY CHLORMRAVENCI (CZECH)
		FORMIC ACID, CHLORO-, PHENYL ESTER
		PHENYL CARBONOCHLORIDATE
		PHENYL CHLOROCARBONATE
		PHENYL CHLOROFORMATE
		PHENYLCHLOROFORMATE
		PHENYLCHLOROFORMATE (DOT)
		UN 27 46 (DOT)
1897-45-6		
		CHLOROTHALONIL

CAS No.	IN	Chemical Name
1910-2-5		
		1,1'-DIMETHYL-4,4'-DIPYRIDINIUM-DICHLORID (GERMAN)
		1,1'-DIMETHYL-4,4'-DIPYRIDYLIUM CHLORIDE
		4,4'-BIPYRIDINIUM, 1,1'-DIMETHYL-, DICHLORIDE
		4,4'-DIMETHYLDIPYRIDYL DICHLORIDE
		ESGRAM
		GRAMONOL
		GRAMOXON
		GRAMURON
		HERBAXON
		HERBOXONE
		METHYL VIOLOGEN (REDUCED)
		METHYLVIOLOGEN
		ORTHO PARAQUAT CL
		PARA-COL
		PARAQUAT (ACGIH)
		PARAQUAT CHLORIDE
		PARAQUAT CL
		PARAQUAT, DICHLORIDE
		PATHCLEAR
		PILLARQUAT
		PILLARXONE
		PP148
		SWEEP
		TERRAKLENE
		TOTACOL
		TOXER TOTAL
		VIOLOGEN, METHYL-
		WEEDOL
1910-42-5		
		1,1'-DIMETHYL-4,4'-BIPYRIDINIUM DICHLORIDE
		1,1'-DIMETHYL-4,4'-DIPYRIDYLIUM DICHLORIDE
		4,4'-BIPYRIDINIUM, 1,1'-DIMETHYL-, DICHLORIDE
		AH 501
		DIMETHYL VIOLOGEN CHLORIDE
		GRAMIXEL
		GRAMOXONE
		GRAMOXONE D
		GRAMOXONE DICHLORIDE
		GRAMOXONE S
		GRAMOXONE W
		METHYL VIOLOGEN
		METHYL VIOLOGEN DICHLORIDE
		METHYLVIOLOGEN CHLORIDE
		N,N'-DIMETHYL-4,4'-BIPYRIDINIUM DICHLORIDE
		N,N'-DIMETHYL-4,4'-DIPYRIDYLIUM DICHLORIDE
		OK 622
		PARAQUAT
		PARAQUAT DICHLORIDE
		PARAQUAT DITCHLORIDE
1912-24-9		
		1,3,5-TRIAZINE-2,4-DIAMINE, 6-CHLORO-N-ETHYL-N'-(1-METHYLETHYL)-
		1,3,5-TRIAZINE-2,4-DIAMINE, 6-CHLORO-N-ETHYL-N'-(1-METHYLETHYL)- (9CI)
		1-CHLORO-3-ETHYLAMINO-5-ISOPROPYLAMINO-2,4,6-TRIAZINE
		1-CHLORO-3-ETHYLAMINO-5-ISOPROPYLAMINO-S-TRIAZINE
		2 -AETHYLAMINO-4-ISOPROPYLAMINO-6-CHLOR-1,3,5-TRIAZIN (GERMAN)
		2-AETHYLAMINO-4-CHLOR-6-ISOPROPYLAMINO-1,3,5-TRIAZIN (GERMAN)
		2-CHLORO-4- ETHYLAMINEISOPROPYLAMINE-S-TRIAZINE
		2-CHLORO-4-(2-PROPYLAMINO)-6-ETHYLAMINO-S-TRIAZINE
		2-CHLORO-4-(ETHYLAMINO)-6-(ISOPROPYLAMINO)-TRIAZINE
		2-CHLORO-4-ETHYLAMINO-6-ISOPROPYLAMI NO-S-TRIAZINE
		2-CHLORO-4-ETHYLAMINO-6-ISOPROPYLAMINO-1,3,5-TRIAZINE
		2-CHLORO-4-ETHYLAMINO-6-ISOPROPYLAMINO-S-TRIAZINE
		6-CHLORO-N-ETHY L-N'-(1-METHYLETHYL)-1,3,5-TRIAZINE-2,4-DIAMINE
		A 361
		AATREX

CAS No.	IN	Chemical Name
		AATREX 4L
		AATREX 80W
		AATREX NINE-O
		ACTINITE PK
		AKTICON
		AKTIKON
		AKTIKON PK
		AKTINIT A
		AKTINIT PK
		ARGEZIN
		ATAZINAX
		ATRANEX
		ATRASINE
		ATRATAF
		ATRATOL A
		ATRAZIN
		ATRAZINE
		ATRAZINE (ACGIH)
		ATRED
		ATREX
		ATZ
		CANDEX
		CEKUZINA-T
		CET
		CHROMOZIN
		CRISATRINA
		CRISAZINE
		CYAZIN
		FARMCO ATRAZINE
		FE NATROL
		FENAMIN
		FENAMINE
		G 30027
		GEIGY 30,027
		GESAPRIM
		GESAPRIM 50
		GESAPRIM 500
		GESOPRIM
		GRIFFEX
		HERBATOXOL
		HUNGAZIN
		HUNGAZIN PK
		INAKOR
		OLEOGESAPRIM
		OLEOGESAPRIM 200
		PITEZIN
		PRIMATOL
		PRIMATOL A
		PRIMAZE
		RADAZIN
		RADIZINE
		S-TRIAZINE, 2-CHLORO-4-ETHYLAMINO-6-ISOPROPYL-AMINO-
		SHELL ATRAZINE HERBICIDE
		STRAZINE
		TRIAZINE A 1294
		VECTAL
		VECTAL SC
		WEEDEX A
		WONUK
		ZEAPOS
		ZEAZIN
		ZEAZINE
		ZEOPOS
1918-00-9		
		2,5-DICHLORO-6-METHOXYBENZOIC ACID
		2-METHOXY-3,6-DICHLOROBENZOIC ACID
		3,6-DICHLOOR-2-METHOXY-BENZOEIZUUR (DUTCH)
		3,6-DICHLOR-3-METHOXY-BENZOESAEURE (GERMAN)
		3,6-DICHLORO-2-METHOXYBENZOIC ACID
		3,6-DICHLORO-O-ANISIC ACID
		ACIDO (3,6-DICLORO-2-METOSSI)-BENZOICO (ITALIAN)
		BANEX
		BANLEN
		BANVEL
		BANVEL 4S
		BANVEL 4WS
		BANVEL CST
		BANVEL D
		BANVEL HERBICIDE
		BANVEL II HERBICIDE
		BENZOIC ACID, 3,6-DICHLORO-2-METHOXY-
		BRUSH BUSTER
		COMPOUND B DICAMBA
		DIANAT
		DIANAT (RUSSIAN)
		DIANATE
		DICAMBA
		DICAMBA (DOT)
		MDBA
		MEDIBEN
		NA 2769 (DOT)
		O-ANISIC ACID, 3,6-DICHLORO-
		VELSICOL 58-CS-11
		VELSICOL COMPOUND R
1918-02-1		
		2-PYRIDINECARBOXYLIC ACID, 4-AMINO-3,5,6-TRICHLORO-
		3,5,6-TRICHLORO-4-AMINOPICOLINIC ACID
		4-AMINO-3,5,6-TRICHLOROPICOLINIC ACID
		4-AMINOTRICHLOROPICOLINIC ACID
		ATCP
		PICLORAM
		PICOLINIC ACID, 4-AMINO-3,5,6-TRICHLORO-
		TORDON
1928-38-7		
		2,4-D METHYL ESTER
		2,4-DICHLOROPHENOXYACETIC ACID METHYL ESTER
		ACETIC ACID, (2,4-DICHLOROPHENOXY)-, METHYL ESTER
		METHYL (2,4-DICHLOROPHENOXY)ACETATE
1928-61-6		
		ACETIC ACID, (2,4-DICHLOROPHENOXY)-, PROPYL ESTER
		PROPYL 2,4-DICHLOROPHENOXYACETATE
1929-73-3		
		2,4-D 2-BUTOXYETHYL ESTER
		2,4-D BUTOXYETHANOL ESTER
		2,4-D BUTOXYETHYL ESTER
		2,4-DBEE
		2,4-DICHLOROPHENOXYACETIC ACID BUTOXYETHANOL ESTER
		2,4-DICHLOROPHENOXYACETIC ACID BUTOXYETHYL ESTER
		2-BUTOXYETHYL 2,4-DICHLOROPHENOXYACETATE
		ACETIC ACID, (2,4-DICHLOROPHENOXY)-, 2-BUTOXYETHYL ESTER
		AQUA-KLEEN
		BLADEX B
		BRUSH KILLER 64
		BUTOXYETHANOL ESTER OF 2,4-D
		BUTOXYETHYL (2,4-DICHLOROPHENOXY)ACETATE
		ETHANOL, 2-BUTOXY-, (2,4-DICHLOROPHENOXY)ACETATE
		PLANOTOX
		WEEDONE LV 4
1929-82-4		
		2-CHLORO-6-(TRICHLOROMETHYL)PYRIDINE
		6-CHLORO-2-(TRICHLOROMETHYL)PYRIDINE
		CP
		N-SERVE
		NITRAPYRIN
		PYRIDINE, 2-CHLORO-6-(TRICHLOROMETHYL)-
1931-62-0		
		2-PROPENEPEROXOIC ACID, 3-CARBOXY-, 1-(1,1-DIMETHYL-ETHYL) ESTER, (Z)-
		MALEIC MONOPEROXY ACID, 1-TERT-BUTYL ESTER, NOT MORE THAN 55% IN SOLUTION
		MALEIC MONOPEROXYACID, 1-TERT-BUTYL ESTER
		MALEIC MONOPEROXYACID, OO-TERT-BUTYL ESTER
		TERT-BUTYL MONOPERMALEATE
		TERT-BUTYL MONOPEROXYMALEATE

CAS No.	IN	Chemical Name
		TERT-BUTYL MONOPEROXYMALEATE, MAXIMUM CONCENTRATION 55% AS A PASTE (DOT)
		TERT-BUTYL MONOPEROXYMALEATE, MAXIMUM CONCENTRATION 55% IN SOLUTION (DOT)
		TERT-BUTYL PEROXYMALEATE
		TERT-BUTYL PEROXYMALEATE (CONCENTRATION NOT MORE THAN 80%)
		TERT-BUTYL PEROXYMALEATE, MAXIMUM CONCENTRATION 55% AS A PASTE (DOT)
		TERT-BUTYL PEROXYMALEATE, NOT MORE THAN 55% IN SOLUTION (DOT)
		TERT-BUTYL PEROXYMALEIC ACID
		UN 2100 (DOT)
		UN 2101 (DOT)
1937-35-5		
		C.I. DIRECT RED 13, DISODIUM SALT
1937-37-7		
		2,7-NAPHTHALENEDISULFONIC ACID, 4-AMINO-3-[[4'-[(2,4-DIAMINOPHENYL)AZO][1,1'-BIPHENYL]-4-YL]AZO]-5-H
		AHCO DIRECT BLACK GX
		AIREDALE BLACK ED
		AIZEN DIRECT DEEP BLACK EH
		AIZEN DIRECT DEEP BLACK GH
		AIZEN DIRECT DEEP BLACK RH
		AMANIL BLACK GL
		AMANIL BLACK WD
		APOMINE BLACK GX
		ATLANTIC BLACK BD
		ATLANTIC BLACK C
		ATLANTIC BLACK E
		ATLANTIC BLACK EA
		ATLANTIC BLACK GAC
		ATLANTIC BLACK GG
		ATLANTIC BLACK GXCW
		ATLANTIC BLACK GXOO
		ATLANTIC BLACK SD
		ATUL DIRECT BLACK E
		AZINE DEEP BLACK EW
		AZOCARD BLACK EW
		AZOMINE BLACK EWO
		BELAMINE BLACK GX
		BENCIDAL BLACK E
		BENZANIL BLACK E
		BENZO DEEP BLACK E
		BENZO LEATHER BLACK E
		BENZOFORM BLACK BCN-CF
		BLACK 2EMBL
		BLACK 4EMBL
		BRASILAMINA BLACK GN
		BRILLIANT CHROME LEATHER BLACK H
		C.I. 30235
		C.I. DIRECT BLACK 38
		C.I. DIRECT BLACK 38, DISODIUM SALT
		CALCOMINE BLACK
		CALCOMINE BLACK EXL
		CARBIDE BLACK E
		CHLORAMINE BLACK C
		CHLORAMINE BLACK EC
		CHLORAMINE BLACK ERT
		CHLORAMINE BLACK EX
		CHLORAMINE BLACK EXR
		CHLORAMINE BLACK XO
		CHLORAMINE CARBON BLACK S
		CHLORAMINE CARBON BLACK SJ
		CHLORAMINE CARBON BLACK SN
		CHLORAZOL BLACK E
		CHLORAZOL BLACK EA
		CHLORAZOL BLACK EN
		CHLORAZOL BURL BLACK E
		CHLORAZOL LEATHER BLACK ENP
		CHLORAZOL SILK BLACK G
		CHROME LEATHER BLACK E
		CHROME LEATHER BLACK EC
		CHROME LEATHER BLACK EM
		CHROME LEATHER BLACK G
		CHROME LEATHER BRILLIANT BLACK ER
		COIR DEEP BLACK C
		COLUMBIA BLACK EP
		CORANIL DIRECT BLACK F
		DIACOTTON DEEP BLACK
		DIACOTTON DEEP BLACK RX
		DIAMINE DEEP BLACK EC
		DIAMINE DIRECT BLACK E
		DIAPHTAMINE BLACK V
		DIAZINE BLACK E
		DIAZINE DIRECT BLACK E
		DIAZINE DIRECT BLACK G
		DIAZOL BLACK 2V
		DIPHENYL DEEP BLACK G
		DIRECT BLACK 38
		DIRECT BLACK 38(TECHNICAL) GRADE
		DIRECT BLACK A
		DIRECT BLACK BRN
		DIRECT BLACK CX
		DIRECT BLACK CXR
		DIRECT BLACK E
		DIRECT BLACK EW
		DIRECT BLACK EX
		DIRECT BLACK FR
		DIRECT BLACK GAC
		DIRECT BLACK GW
		DIRECT BLACK GX
		DIRECT BLACK GXR
		DIRECT BLACK JET
		DIRECT BLACK META
		DIRECT BLACK METHYL
		DIRECT BLACK N
		DIRECT BLACK RX
		DIRECT BLACK SD
		DIRECT BLACK WS
		DIRECT BLACK Z
		DIRECT BLACK ZSH
		DIRECT DEEP BLACK E
		DIRECT DEEP BLACK E EXTRA
		DIRECT DEEP BLACK E-EX
		DIRECT DEEP BLACK EA-CF
		DIRECT DEEP BLACK EAC
		DIRECT DEEP BLACK EW
		DIRECT DEEP BLACK EX
		DIRECT DEEP BLACK WX
		ENIANIL BLACK CN
		ERIE BLACK B
		ERIE BLACK BF
		ERIE BLACK GAC
		ERIE BLACK GXOO
		ERIE BLACK JET
		ERIE BLACK NUG
		ERIE BLACK RXOO
		ERIE BRILLIANT BLACK S
		ERIE FIBRE BLACK VP
		FENAMIN BLACK E
		FIBRE BLACK VF
		FIXANOL BLACK E
		FORMALINE BLACK C
		FORMIC BLACK C
		FORMIC BLACK CW
		FORMIC BLACK EA
		FORMIC BLACK MTG
		FORMIC BLACK TG
		HISPAMIN BLACK EF
		INTERCHEM DIRECT BLACK Z
		KAYAKU DIRECT DEEP BLACK EX
		KAYAKU DIRECT DEEP BLACK GX
		KAYAKU DIRECT DEEP BLACK S
		KAYAKU DIRECT LEATHER BLACK EX
		KAYAKU DIRECT SPECIAL BLACK AAX
		LURAZOL BLACK BA
		META BLACK
		MITSUI DIRECT BLACK EX
		MITSUI DIRECT BLACK GX
		NIPPON DEEP BLACK
		NIPPON DEEP BLACK GX
		PAPER BLACK BA
		PAPER BLACK T
		PAPER DEEP BLACK C
		PARAMINE BLACK B

CAS No.	IN	Chemical Name
		PARAMINE BLACK E
		PEERAMINE BLACK E
		PEERAMINE BLACK GXOO
		PHENAMINE BLACK BCN-CF
		PHENAMINE BLACK CL
		PHENAMINE BLACK E
		PHENAMINE BLACK E 200
		PHENO BLACK EP
		PHENO BLACK SGN
		PONTAMINE BLACK E
		PONTAMINE BLACK EBN
		SANDOPEL BLACK EX
		SERISTAN BLACK B
		TELON FAST BLACK E
		TERTRODIRECT BLACK E
		TERTRODIRECT BLACK EFD
		TETRAZO DEEP BLACK G
		UNION BLACK EM
		VONDACEL BLACK N
1940-66-2		
		SODIUM BOROHYDRIDE
1941-79-3		
		AZELAIC DIPERACID
		DIPERAZELAIC ACID
		DIPEROXYAZELAIC ACID
		DIPEROXYAZELAIC ACID, MAXIMUM CONCENTRATION 27% (DOT)
		NONANEDIPEROXOIC ACID
		PEROXYAZELAIC ACID
		PEROXYAZELAIC ACID, MAXIMUM CONCENTRATION 27%, WITH AT LEAST 13% AZELAIC ACID AND 53%SODIUM SULFATE
		UN 2958 (DOT)
1954-28-5		
		1,2-BIS(2-(2,3-EPOXYPROPOXY)ETHOXY)ETHANE
		1,2:15,16-DIEPOXY-4,7,10,13-TETRAOXAHEXADECANE
		2,2'-(2,5,8,11-TETRAOXA-1,12-DODECANEDIYL)BISOXIRANE
		4,7,10,13-TETRAOXAHEXADECANE, 1,2:15,16-DIEPOXY-
		4,7,10,13-TETRAOXAHEXADECANE, 1,2:15,16-DIEPOXY- (8CI)
		AYERST 62013
		DIGLYCIDYLTRIETHYLENE GLYCOL
		EPODYL
		ETHANE, 1,2-BIS[2-(2,3-EPOXYPROPOXY)ETHOXY]-
		ETHOGLUCID
		ETHOGLUCIDE
		ETOGLUCID
		ETOGLUCIDE
		ICI 32865
		OXIRANE, 2,2'-(2,5,8,11-TETRAOXADODECANE-1,12-DIYL)-BIS-
		OXIRANE, 2,2'-(2,5,8,11-TETRAOXADODECANE-1,12-DIYL)-BIS- (9CI)
		TDE
		TDE (1,1-DICHLORO-2,2-BIS(P-CHLOROPHENYL) ETHANE)
		TRIETHYLENE GLYCOL DIGLYCIDYL ETHER
1955-45-9		
		PIVALOLACTONE
1982-47-4		
		1-(4-(4-CHLORO-PHENOXY)PHENYL)-3,3-D'METHYLUREE (FRENCH)
		3-(4-(4-CHLOOR-FENOXY)-FENOXY)-FENYL)-1,1-DIMETHYLUREUM (DUTCH)
		3-(4-(4-CHLOR-PHENOXY)-PHENYL)-1,1-DIMETHYLHARNSTOFF (GERMAN)
		3-(4-(4-CHLORO-FENOSSIL)-1,1-DIMETIL-UREA (ITALIAN)
		3-(P-(P-CHLOROPHENOXY)PHENYL)-1,1-DIMETHYLUREA
		C 1983
		CHLOROXIFENIDIM
		CHLOROXURON
		CIBA 1983
		N'-4-(4-CHLOROPHENOXY)PHENYL-N,N-DIMETHYLUREA
		NOREX
		TENORAN
		UREA, 3-(P-(P-CHLOROPHENOXY)PHENYL)-1,1-DIMETHYL-
		UREA, N'-[4-(4-CHLOROPHENOXY)PHENYL]-N,N-DIMETHYL-
2001-95-8		
		1,7,13,19,25,31-HEXAOXA-4,10,16,22,28,34-HEXAAZACYCLOHEXATRIACONTANE, CYCLIC PEPTIDE DERIV
		ANTIBIOTIC N-329 B
		NSC 122023
		VALINOMICIN
		VALINOMYCIN
2016-57-1		
		1-AMINODECANE
		1-DECANAMINE
		1-DECYLAMINE
		DECANAMINE
		DECYLAMINE
		KEMAMINE P 190D
		MONODECYLAMINE
		N-DECYLAMINE
2025-56-1		
		ETHYL
		ETHYL FLUID
		ETHYL RADICAL
2032-59-9		
		4-(DIMETHYLAMINO)-M-TOLYL METHYLCARBAMATE
2032-65-7		
		3,5-DIMETHYL-4-(METHYLTHIO)PHENOL METHYLCARBAMATE
		3,5-DIMETHYL-4-(METHYLTHIO)PHENYL METHYLCARBAMATE
		3,5-DIMETHYL-4-METHYL-THIOPHENYL-N-CARBAMAT (GERMAN)
		3,5-DIMETHYL-4-METHYLTHIOPHENYL N-METHYLCARBAMATE
		3,5-XYLENOL, 4-(METHYLTHIO)-, METHYLCARBAMATE
		4-(METHYLTHIO)-3,5-XYLYL METHYLCARBAMATE
		4-METHYLMERCAPTO-3,5-DIMETHYLPHENYL N-METHYLCARBAMATE
		4-METHYLMERCAPTO-3,5-XYLYL METHYLCARBAMATE
		4-METHYLTHIO-3,5-DIMETHYLPHENYL METHYLCARBAMATE
		B 37344
		BAY 37344
		BAY 5024
		BAY 9026
		BAYER 37344
		CARBAMIC ACID, METHYL-, 3,5-DIMETHYL-4-(METHYLTHIO)PHENYL ESTER
		CARBAMIC ACID, METHYL-, 4-(METHYLTHIO)-3,5-XYLYL ESTER
		CARBAMIC ACID, N-METHYL-, 4-(METHYLTHIO)-3,5-XYLYL ESTER
		DCR 736
		DRAZA
		ENT 25,726
		H 321
		MERCAPTODIMETHUR
		MERCAPTODIMETHUR (DOT)
		MERCAPTODIMETHUR (METHIOCARB)
		MESUROL
		MESUROL PHENOL
		METHIOCARB
		METHYL CARBAMIC ACID 4-(METHYLTHIO)-3,5-XYLYL ESTER
		METMERCAPTURON
		NA 2757 (DOT)
		OMS-93
		PHENOL, 3,5-DIMETHYL-4-(METHYLTHIO)-, METHYLCARBAMATE
		SD 9228
2036-15-9		
		ALUMINUM, HYDRODIPROPYL-
		DIPROPYLALUMINUM HYDRIDE

CAS No.	IN	Chemical Name
2038-03-1		
		1-AMINO-2-MORPHOLINOETHANE
		2-(4-MORPHOLINYL)ETHYLAMINE
		2-MORPHOLINOETHANAMINE
		2-MORPHOLINOETHYLAMINE
		4-(2-AMINOETHYL)MORPHOLINE
		4-MORPHOLINEETHANAMINE
		BETA-MORPHOLINOETHYLAMINE
		MORPHOLINE, 4-(2-AMINOETHYL)-
		MORPHOLINOETHYLAMINE
		N-(2-AMINOETHYL)MORPHOLINE
		N-(BETA-AMINOETHYL)MORPHOLINE
		N-AMINOETHYLMORPHOLINE
2039-87-4		
		2-CHLOROSTYRENE
		BENZENE, 1-CHLORO-2-ETHENYL-
		O-CHLOROSTYRENE
		O-CHLOROVINYLBENZENE
		STYRENE, O-CHLORO-
2049-92-5		
		4-TERT-AMYLANILINE
		BENZENAMINE, 4-(1,1-DIMETHYLPROPYL)-
		P-TERT-AMYLANILINE
		P-TERT-PENTYLANILINE
2050-92-2		
		1-PENTANAMINE, N-PENTYL-
		DI-N-AMYLAMINE
		DI-N-AMYLAMINE (DOT)
		DI-N-PENTYLAMINE
		DIAMYL AMINE
		DIPENTYLAMINE
		PENTYLAMINE, PENTYL-
		UN 2841 (DOT)
2074-50-2		
		1,1'-DIMETHYL-4,4'-BIPYRIDINIUM BIS(METHYL SULFATE)
		1,1'-DIMETHYL-4,4'-BIPYRIDINIUM DIMETHOSULFATE
		1,1'-DIMETHYL-4,4'-BIPYRIDYLIUM DIMETHYLSULFATE
		1,1'-DIMETHYL-4,4'-BIPYRIDYNIUM DIMETHYLSULFATE
		1,1'-DIMETHYL-4,4'-DIPYRIDINIUM DI(METHYL SULFATE)
		1,1'-DIMETHYL-4,4'-DIPYRIDINIUM DIMETHOSULFATE
		4,4'-BIPYRIDINIUM, 1,1'-DIMETHYL-, BIS(METHYL SULFATE)
		GRAMOXONE METHYL SULFATE
		N,N'-DIMETHYL-4,4'-BIPYRIDINIUM METHOSULFATE
		OK 621
		PARAQUAT (ACGIH)
		PARAQUAT BIS(METHYL SULFATE)
		PARAQUAT DIMETHOSULFATE
		PARAQUAT DIMETHYL SULFATE
		PARAQUAT DIMETHYL SULPHATE
		PARAQUAT I
		PARAQUAT METHOSULFATE
		PARAQUAT METHYLSULFATE
		PP 910
2074-87-5		
		CYANIDE RADICAL
		CYANO RADICAL
		CYANO RADICAL (CN)
		CYANOGEN
		CYANOGEN (CN)
		CYANOGEN RADICAL
		MONOCYANOGEN
2084-18-6		
		2-BUTANETHIOL, 3-METHYL-
		3-METHYL-2-BUTANETHIOL
2097-19-0		
		1-PHENYLSILATRANE
		2,8,9-TRIOXA-5-AZA-1-SILABICYCLO(333)UNDECANE, PHENYL-
		2,8,9-TRIOXA-5-AZA-1-SILABICYCLO[333]UNDECANE, 1-PHENYL-
		FENYLSILATRAN (CZECH)
		PHENYLSILATRANE
		SILATRANE, PHENYL-
2100-42-7		
		1-CHLORO-2,5-DIMETHOXYBENZENE
		2,5-DIMETHOXYCHLOROBENZENE
		2-CHLORO-1,4-DIMETHOXYBENZENE
		BENZENE, 2-CHLORO-1,4-DIMETHOXY-
		CHLORO-1,4-DIMETHOXYBENZENE
		CHLOROHYDROQUINONE DIMETHYL ETHER
2104-64-5		
		BENZENEPHOSPHONIC ACID, THIONO-, ETHYL-P-NITROPHENYL ESTER
		ENT 17,798
		EPN
		EPN (ACGIH)
		EPN 300
		ETHOXY-4-NITROPHENOXYPHENYLPHOSPHINE SULFIDE
		ETHYL P-NITROPHENYL BENZENETHIONOPHOSPHONATE
		ETHYL P-NITROPHENYL BENZENETHIOPHOSPHATE
		ETHYL P-NITROPHENYL BENZENETHIOPHOSPHONATE
		ETHYL P-NITROPHENYL PHENYLPHOSPHONOTHIOATE
		ETHYL P-NITROPHENYL THIONOBENZENEPHOSPHATE
		ETHYL P-NITROPHENYL THIONOBENZENEPHOSPHONATE
		O-(4-NITROPHENYL) O-ETHYL PHENYL THIOPHOSPHONATE
		O-AETHYL-O-(4-NITRO-PHENYL)-PHENYL-MONOTHIOPHOSPHONAT (GERMAN)
		O-ETHYL O-(4-NITROPHENYL) PHENYLPHOSPHONOTHIOATE
		O-ETHYL O-(4-NITROPHENYL)BENZENETHIONOPHOSPHONATE
		O-ETHYL O-(P-NITROPHENYL) PHENYLPHOSPHONOTHIOATE
		O-ETHYL O-P-NITROPHENYL BENZENETHIOPHOSPHONATE
		O-ETHYL O-P-NITROPHENYL PHENYLPHOSPHOROTHIOATE
		O-ETHYL PHENYL P-NITROPHENYL THIOPHOSPHONATE
		O-ETHYL-O-((4-NITRO-FENYL)-FENYL)-MONOTHIOFOSFONAAT (DUTCH)
		O-ETIL-O-((4-NITRO-FENIL)-FENIL)-MONOTIOFOSFONATO (ITALIAN)
		PHENOL, P-NITRO-, O-ESTER WITH O-ETHYL PHENYL PHOSPHONOTHIOATE
		PHENYLPHOSPHONOTHIOATE, O-ETHYL-O-P-NITROPHENYL-
		PHENYLPHOSPHONOTHIOIC ACID O-ETHYL O-P-NITROPHENYL ESTER
		PHENYLTHIOPHOSPHONATE DE O-ETHYLE ET O-4-NITROPHENYLE (FRENCH)
		PHOSPHONOTHIOIC ACID, PHENYL-, ETHYL P-NITROPHENYL ESTER
		PHOSPHONOTHIOIC ACID, PHENYL-, O-ETHYL O-(4-NITROPHENYL) ESTER
		PHOSPHONOTHIOIC ACID, PHENYL-, O-ETHYL O-(P-NITROPHENYL) ESTER
		PIN
		SANTOX
2109-64-0		
		2-PROPANOL, 1-(DIBUTYLAMINO)-
		DI-N-BUTYLAMINO-2-PROPANOL
		DIBUTYLISOPROPANOLAMINE
		N,N-DIBUTYL-2-HYDROXYPROPYLAMINE
		N-(2-HYDROXYPROPYL)DIBUTYLAMINE
2122-46-5		
		PHENOXY
		PHENOXY PESTICIDE, LIQUID [FLAMMABLE LIQUID LABEL]
		PHENOXY PESTICIDE, LIQUID [POISON B LABEL]
		PHENOXY PESTICIDE, SOLID
		PHENOXY RADICAL
		PHENOXYL
		PHENOXYL RADICAL
		PHENYLOXY
2144-45-8		
		DIBENZYL PEROXYDICARBONATE
		DIBENZYL PEROXYDICARBONATE (CONCENTRATION ≤90%)
		DIBENZYL PEROXYDICARBONATE (DOT)
		DIBENZYL PEROXYDICARBONATE, MAXIMUM CONCEN-

CAS No.	IN	Chemical Name
		TRATION 87% WITH WATER (DOT)
		PEROXYDICARBONIC ACID, BIS(PHENYLMETHYL) ESTER
		PEROXYDICARBONIC ACID, DIBENZYL ESTER
		UN 2149 (DOT)
2150-54-1		
		C.I. DIRECT BLUE 25, TETRASODIUM SALT
2155-71-7		
		DI-TERT-BUTYLPEROXYPHTHALATE, MAXIMUM 55% IN PASTE
		DI-TERT-BUTYLPEROXYPHTHALATE, MAXIMUM 55% IN SOLUTION
		DI-TERT-BUTYLPEROXYPHTHALATE, TECHNICAL PURE
2156-96-9		
		2-PROPENOIC ACID, DECYL ESTER
		ACRYLIC ACID, DECYL ESTER
		DECYL ACRYLATE
2157-42-8		
		DISILOXANE, HEXAETHOXY-
		ETHYL SILICATE
		HEXAETHOXYDISILOXANE
		HEXAETHYL DIORTHOSILICATE
		SILICIC ACID (H6Si2O7), HEXAETHYL ESTER
2160-93-2		
		2,2'-(TERT-BUTYLIMINO)DIETHANOL
		ETHANOL, 2,2'-(TERT-BUTYLIMINO)DI-
		ETHANOL, 2,2'-[(1,1-DIMETHYLETHYL)IMINO]BIS-
		TERT-BUTYLDIETHANOLAMINE
2163-80-6		
		ANSAR 170
		ANSAR 170 HC
		ANSAR 170L
		ANSAR 529
		ANSAR 529 HC
		ARSONATE LIQUID
		ARSONIC ACID, METHYL-, MONOSODIUM SALT
		ASAZOL
		BUENO
		BUENO 6
		DACONATE
		DACONATE 6
		DAL-E-RAD
		DAL-E-RAD 120
		GEPIRON
		HERB-ALL
		HERBAN M
		MERGE
		MERGE 823
		MESAMATE
		MESAMATE CONCENTRATE
		MESAMATE HC
		MESAMATE-400
		MESAMATE-600
		METHANEARSONIC ACID, MONOSODIUM SALT
		METHANEARSONIC ACID, SODIUM SALT
		METHYLARSENIC ACID, SODIUM SALT
		MONATE
		MONOSODIUM ACID METHANEARSONATE
		MONOSODIUM ACID METHARSONATE
		MONOSODIUM METHANEARSONATE
		MONOSODIUM METHANEARSONIC ACID
		MONOSODIUM METHYL ARSONATE
		MSMA
		NCI-C60071
		PHYBAN
		PHYBAN HC
		SILVISAR 550
		SODIUM ACID METHANEARSONATE
		SODIUM METHANEARSONATE
		TARGET MSMA
		TRANS-VERT
		WEED 108
		WEED-E-RAD
		WEED-HOE
2164-17-2		
		FLUOMETURON
2167-23-9		
		2,2-BIS(TERT-BUTYLDIOXY)BUTANE
		2,2-BIS(TERT-BUTYLPEROXY)BUTANE
		2,2-BIS-(TERT-BUTYLPEROXY)-BUTANE (CONCENTRATION ≤70%)
		2,2-DI(TERT-BUTYLPEROXY)BUTANE
		2,2-DI-(TERT-BUTYLPEROXY)BUTANE, NOT MORE THAN 55% IN SOLUTION (DOT)
		3,4,6,7-TETRAOXANONANE, 5-ETHYL-2,2,5,8,8-PENTA-METHYL-
		CHALOXYD P 1200AL
		CHALOXYD P 1293AL
		LUPERSOL 220
		PEROXIDE, (1-METHYLPROPYLIDENE)BIS(TERT-BUTYL-, NOT MORE THAN 55% IN SOLUTION
		PEROXIDE, (1-METHYLPROPYLIDENE)BIS[(1,1-DIMETHYL-ETHYL)
		PEROXIDE, SEC-BUTYLIDENEBIS[TERT-BUTYL
		SEC-BUTYLIDENEBIS[TERT-BUTYL PEROXIDE]
		TRIGONOX DB 50
		TRIGONOX DM 50
		UN 2111 (DOT)
		VP 1200
2179-59-1		
		4,5-DITHIA-1-OCTENE
		ALLYL PROPYL DISULFIDE
		DISULFIDE, 2-PROPENYL PROPYL
		DISULFIDE, ALLYL PROPYL
		PROPYL ALLYL DISULFIDE
2207-04-7		
		1,4-DIMETHYLCYCLOHEXANE-TRANS
		CYCLOHEXANE, 1,4-DIMETHYL-, TRANS-
		TRANS-1,4-DIMETHYLCYCLOHEXANE
2216-33-3		
		3-METHYLOCTANE
		OCTANE, 3-METHYL-
2216-34-4		
		4-METHYLOCTANE
		OCTANE, 4-METHYL-
2217-06-3		
		2,2',4,4',6,6'-HEXANITRODIPHENYL SULFIDE
		2,4,6,2',4',6'-HEXANITRODIPHENYL SULFIDE
		2,4,6-TRINITROPHENYL SULFIDE
		BENZENE, 1,1'-THIOBIS[2,4,6-TRINITRO-
		BIS(2,4,6-TRINITROPHENYL)SULFIDE
		DIPICRYL SULFIDE
		DIPICRYL SULPHIDE, WETTED WITH, BY WEIGHT, AT LEAST 10% WATER (DOT)
		PICRYL SULFIDE
		PICRYL SULFIDE, WETTED WITH AT LEAST 10% WATER
		UN 2852 (DOT)
2223-93-0		
		ALAIXOL 11
		CADMIUM DISTEARATE
		CADMIUM OCTADECANOATE
		CADMIUM STEARATE
		KADMIUMSTEARAT (GERMAN)
		OCTADECANOIC ACID, CADMIUM SALT
		SCD
		STABILISATOR SCD
		STABILIZER SCD
		STEARIC ACID, CADMIUM SALT
2231-57-4		
		1,3-DIAMINO-2-THIOUREA
		CARBOHYDRAZIDE, THIO-
		CARBONOTHIOIC DIHYDRAZIDE
		CARBONOTHIOIC DIHYDRAZIDE (9CI)

CAS No.	IN	Chemical Name
		HYDRAZINECARBOHYDRAZONOTHIOIC ACID
		TCH
		THIOCARBAZIDE
		THIOCARBOHYDRAZIDE
		THIOCARBONIC DIHYDRAZIDE
		THIOCARBONOHYDRAZIDE
		USAF EK-7372
2234-13-1		
		1,2,3,4,5,6,7,8-OCTACHLORONAPHTHALENE
		HALOWAX 1051
		NAPHTHALENE, OCTACHLORO-
		OCTACHLORONAPHTHALENE
		PERCHLORONAPHTHALENE
		PERNA
2235-25-8		
		BIS(ETHYLMERCURI)PHOSPHATE
		CERESAN NI
		EMP
		ETHYL MERCURY PHOSPHATE
		ETHYLMERCURIC PHOSPHATE
		FUSARIOL UNIVERSAL
		GRANOSAN M
		LIGNASAN
		LIGNASAN FUNGICIDE
		LIGNASAN-X
		MERCURATE(2-), ETHYL[PHOSPHATO(3-)-O]-, DIHYDROGEN
		MERCURY, (DIHYDROGEN PHOSPHATO)ETHYL-
		MERCURY, (HYDROGEN PHOSPHATO)BIS(ETHYL-
		NEW IMPROVED CERESAN
		NEW IMPROVED GRANOSAN
		PHOSPHORIC ACID, MERCURY COMPLEX
		RUBERON GRANULE
2235-54-3		
		AKYPOSAL ALS 33
		AMMONIUM DODECYL SULFATE
		AMMONIUM LAURYL SULFATE
		AMMONIUM N-DODECYL SULFATE
		AVIROL 200
		CONCO SULFATE A
		DODECYL AMMONIUM SULFATE
		EMAL A
		EMAL AD
		EMERSOL 6430
		LAURYL AMMONIUM SULFATE
		LAURYL SULFATE AMMONIUM SALT
		MAPROFIX NH
		MONTOPOL LA 20
		NEOPON LAM
		PRESULIN
		RICHONOL AM
		SINOPON
		SIPON L 22
		SIPON LA 30
		SIPROL L 22
		STERLING AM
		SULFURIC ACID, LAURYL ESTER, AMMONIUM SALT
		SULFURIC ACID, MONODODECYL ESTER, AMMONIUM SALT
		TEXAPON A 400
		TEXAPON ALS
		TEXAPON SPECIAL
2238-07-5		
		BIS(2,3-EPOXYPROPYL)ETHER
		DGE
		DGE (DIGLYCIDYL ETHER)
		DI(2,3-EPOXY)PROPYL ETHER
		DIGLYCIDYL ETHER
		DIGLYCIDYL ETHER (ACGIH)
		ETHER, BIS(2,3-EPOXYPROPYL)
		ETHER, DIGLYCIDYL
		GLYCIDYL ETHER
		NSC 54739
		OXIRANE, 2,2'-[OXYBIS(METHYLENE)]BIS-

CAS No.	IN	Chemical Name
2243-62-1		
		1-5-NAPHTHALENEDIAMINE
2244-16-8		
		(+)-CARVONE
		(S)-(+)-CARVONE
		(S)-CARVONE
		2-CYCLOHEXEN-1-ONE, 2-METHYL-5-(1-METHYLETHENYL)-, (S)-
		6,8(9)-P-MENTHADIEN-2-ONE
		CARVOL
		CARVONE
		CARVONE, (+)-
		D-(+)-CARVONE
		D-CARVONE
		DELTA(SUP 6,8)-(9)-TERPADIENONE-2
		DELTA-1-METHYL-4-ISOPROPENYL-6-CYCLOHEXEN-2-ONE
		NCI-C55867
		P-MENTHA-6,8-DIEN-2-ONE
		P-MENTHA-6,8-DIEN-2-ONE, (S)-(+)-
2244-21-5		
		1,3,5-TRIAZINE-2,4,6(1H,3H,5H)-TRIONE, 1,3-DICHLORO-, POTASSIUM SALT
		1,3-DICHLORO-S-TRIAZINE-2,4,6(1H,3H,5H)TRIONE POTASSIUM SALT
		ACL 59
		DICHLOR-S-TRIAZIN-2,4,6(1H,3H,5H)TRIONE POTASSIUM
		DICHLORO-S-TRIAZINE-2,4,6(1H,3H,5H)-TRIONE POTASSIUM
		DICHLORO-S-TRIAZINE-2,4,6(1H,3H,5H)-TRIONE POTASSIUM DERIV
		DICHLOROISOCYANURIC ACID POTASSIUM SALT
		DICHLOROISOCYANURIC ACID POTASSIUM SALT (DOT)
		ISOCYANURIC ACID, DICHLORO-, POTASSIUM SALT
		NA 2465 (DOT)
		NEOCHLOR 59
		POTASSIUM DICHLORO-ISOCYANURATE
		POTASSIUM DICHLORO-S-TRIAZINETRIONE
		POTASSIUM DICHLORO-S-TRIAZINETRIONE, DRY, CONTAINING MORE THAN 39% AVAILABLE CHLORINE (DOT)
		POTASSIUM DICHLOROCYANURATE
		POTASSIUM DICHLOROISOCYANURATE
		POTASSIUM TROCLOSENE
		S-TRIAZINE-2,4,6(1H,3H,5H)-TRIONE, 1,3-DICHLORO-, POTASSIUM SALT
		S-TRIAZINE-2,4,6(1H,3H,5H)-TRIONE, DICHLORO-, POTASSIUM DERIV.
		TROCLOSENE POTASSIUM
		UN 2465 (DOT)
2275-14-1		
		2,5-DICHLOROPHENYLTHIOMETHYL O,O-DIETHYL PHOSPHORODITHIOATE
		CMP
		CMP (PESTICIDE)
		DITHIOPHOSPHATE DE O,O-DIETHYLE ET DE S(2,5-DICHLOROPHENYL) THIOMETHYLE (FRENCH)
		ENT 25,585
		FENKAPTON (DUTCH)
		G 28029
		GEIGY 28029
		GEIGY G 28029
		METHANETHIOL, ((2,5-DICHLOROPHENYL)THIO)-, S-ESTER WITH O,O-DIETHYL PHOSPHORODITHIOATE
		NA 2783 (DOT)
		O,O-DIAETHYL-S((2,5-DICHLOR-PHENYL-THIO)-METHYL)-DITHIOPHOSPHAT (GERMAN)
		O,O-DIETHYL S-(2,5-DICHLOROPHENYLTHIOMETHYL) DITHIOPHOSPHATE
		O,O-DIETHYL S-(2,5-DICHLOROPHENYLTHIOMETHYL) PHOSPHORODITHIOATE
		O,O-DIETHYL S-(2,5-DICHLOROPHENYLTHIOMETHYL) PHOSPHOROTHIOLOTHIONATE
		O,O-DIETHYL-S-(2,5-DICHLOROPHENYLTHIOMETHYL) DITHIOPHOSPHORAN
		PHENATOL
		PHENCAPTON
		PHENCAPTON (DOT)
		PHENKAPTON

CAS No.	IN	Chemical Name
		PHENKAPTONE
		PHENUDIN
		PHENUDINE
		PHOSPHORODITHIOIC ACID, S-(((2,5-DICHLOROPHENYL)-THIO)METHYL) O,O-DIETHYL ESTER
		PRZEDZIORKOFOS (POLISH)
		S-(2,5-DICHLOROPHENYLTHIOMETHYL) DIETHYL PHOSPHOROTHIOLOTHIONATE
		S-(2,5-DICHLOROPHENYLTHIOMETHYL) O,O-DIETHYL PHOSPHORODITHIOATE
2275-18-5		
		AC 18682
		ACETAMIDE, N-ISOPROPYL-2-MERCAPTO-, S-ESTER WITH O,O-DIETHYL PHOSPHORODITHIOATE
		AMERICAN CYANAMID 18,682
		ENT 24,652
		FAC
		FAC 20
		FAK-40
		FOSTION
		ISOPROPYL DIETHYLDITHIOPHOSPHORYLACETAMIDE
		L 343
		N-MONOISOPROPYLAMIDE OF O,O-DIETHYLDITHIOPHOSPHORYLACETIC ACID
		O,O-DIETHYL S-(N-ISOPROPYLCARBAMOYLMETHYL) DITHIOPHOSPHATE
		O,O-DIETHYL S-(N-ISOPROPYLCARBAMOYLMETHYL) PHOSPHORODITHIOATE
		O,O-DIETHYL S-ISOPROPYLCARBAMOYLMETHYL PHOSPHORODITHIOATE
		O,O-DIETHYLDITHIOPHOSPHORYLACETIC ACID, N-MONOISOPROPYLAMIDE
		OLEOFAC
		PHOSPHORODITHIOIC ACID, O,O-DIETHYL ESTER, S-ESTER WITH N-ISOPROPYL-2-MERCAPTOACETAMIDE
		PHOSPHORODITHIOIC ACID, O,O-DIETHYL S-(ISOPROPYLCARBAMOYLMETHYL) ESTER
		PHOSPHORODITHIOIC ACID, O,O-DIETHYL S-[2-[(1-METHYLETHYL)AMINO]-2-OXOETHYL] ESTER
		PROTHOAT
		PROTHOATE
		PROTOAT (HUNGARIAN)
		TELEFOS
		TRIMETHOATE
2303-16-4		
		2,3-DCDT
		2,3-DICHLOROALLYL N,N-DIISOPROPYLTHIOLCARBAMATE
		AVADEX
		CARBAMIC ACID, DIISOPROPYLTHIO-, S-(2,3-DICHLOROALLYL) ESTER
		CARBAMOTHIOIC ACID, BIS(1-METHYLETHYL)-, S-(2,3-DICHLORO-2-PROPENYL) ESTER
		CP 15336
		DATC
		DI-ALLATE
		DIALLATE
		S-(2,3-DICHLOROALLYL) DIISOPROPYLTHIOCARBAMATE
2310-17-0		
		PHOSALONE
2312-35-8		
		2-(4-(1,1-DIMETHYLETHYL)PHENOXY)CYCLOHEXYL 2-PROPYNYL SULFITE
		2-(P-T-BUTYLPHENOXY)CYCLOHEXYL PROPARGYL SULFITE
		2-(P-TERT-BUTYLPHENOXY)CYCLOHEXYL 2-PROPYNYL SULFITE
		2-(P-TERT-BUTYLPHENOXY)CYCLOHEXYL PROPARGYL SULFITE
		BPPS
		COMITE
		CYCLOSULFYNE
		D 014
		ENT 27226
		NA 2765 (DOT)
		NAUGATUCK D 014
		OMAIT
		OMITE
		OMITE 57E
		OMITE 85E
		PROPARGIL
		PROPARGITE
		PROPARGITE (DOT)
		SULFUROUS ACID, 2-(4-(1,1-DIMETHYLETHYL)PHENOXY)-CYCLOHEXYL 2-PROPYNYL ESTER
		SULFUROUS ACID, 2-(P-TERT-BUTYLPHENOXY)CYCLOHEXYL-2-PROPYNYL ESTER
		UNIROYAL D014
		US RUBBER D-014
2312-76-7		
		SODIUM DINITRO-ORTHO-CRESOLATE
2338-12-7		
		5-NITROBENZOTRIAZOL
2372-21-6		
		CARBONOPEROXOIC ACID, OO-(1,1-DIMETHYLETHYL) O-(1-METHYLETHYL) ESTER
		KAYACARBON BIC
		LUPERSOL TBIC
		LUPERSOL TBIC-M 75
		OO-TERT-BUTYL O-ISOPROPYL PERCARBONATE
		OO-TERT-BUTYL O-ISOPROPYL PEROXYCARBONATE
		PEROXYCARBONIC ACID, OO-TERT-BUTYL O-ISOPROPYL ESTER
		TERT-BUTYL PEROXYISOPROPYL CARBONATE (CONCENTRATION NOT MORE THAN 80%)
		TERT-BUTYL PEROXYISOPROPYL CARBONATE, TECHNICAL PURE (DOT)
		TERT-BUTYL PEROXYISOPROPYL CARBONATE, TECHNICALLY PURE
		TERT-BUTYLPEROXY ISOPROPYL CARBONATE
		UN 2103 (DOT)
2385-85-5		
		1,1A,2,2,3,3A,4,5,5,5A,5B,6-DODECACHLOROOCTAHYDRO-1,3,4-METHENO-1H-CYCLOBUTA(CD)PENTALENE
		1,2,3,4,5,5-HEXACHLORO-1,3-CYCLOPENTADIENE DIMER
		1,3,4-METHENO-1H-CYCLOBUTA(CD)PENTALENE, 1,1A,2,2,3,3A,4,5,5,5A,5B,6-DODECACHLOROOCTAHYDRO-
		1,3,4-METHENO-1H-CYCLOBUTA(CD)PENTALENE, DODECACHLOROOCTAHYDRO-
		1,3-CYCLOPENTADIENE, 1,2,3,4,5,5-HEXACHLORO-, DIMER
		BICHLORENDO
		CG-1283
		CYCLOPENTADIENE, HEXACHLORO-, DIMER
		DECANE,PERCHLOROPENTACYCLO-
		DECHLORANE
		DECHLORANE 4070
		DODECACHLOROOCTAHYDRO-1,3,4-METHENO-2H-CYCLOBUTA(C,D)PENTALENE
		DODECACHLOROPENTACYCLO(3220(SUP 2,6),0(SUP 3,9),0(SUP 5,10))DECANE
		DODECACHLOROPENTACYCLODECANE
		ENT 25,719
		FERRIAMICIDE
		GC 1283
		HEXACHLOROCYCLOPENTADIENE DIMER
		HRS L276
		MIREX
		NCI-C06428
		PARAMEX
		PERCHLORODIHOMOCUBANE
		PERCHLOROPENTACYCLO(5210(SUP 2,6)0(SUP 3,9)0(SUP 5,8))DECANE
		PERCHLOROPENTACYCLODECANE
		PERCHLOROPENTACYCLO[52102,603,905,8]DECANE
2407-94-5		
		1,1'-DIHYDROXYDICYCLOHEXYL PEROXIDE
		1,1'-PEROXYDICYCLOHEXANOL
		BIS(1-HYDROXYCYCLOHEXYL)PEROXIDE
		BIS-(1-HYDROXYCYCLOHEXYL)PEROXIDE, TECHNICAL PURE (DOT)
		CYCLOHEXANOL, 1,1'-DIOXYBIS-

CAS No.	IN	Chemical Name
		CYCLOHEXANOL, 1,1'-DIOXYBIS- (9CI)
		CYCLOHEXANOL, 1,1'-DIOXYDI-
		DI-(1-HYDROXYCYCLOHEXYL) PEROXIDE, TECHNICALLY PURE
		DI-(1-HYDROXYCYCLOHEXYL)PEROXIDE, TECHNICAL PURE (DOT)
		PEROXIDE, BIS(1-HYDROXYCYCLOHEXYL)
		UN 2148 (DOT)
2425-06-1		
		1H-ISOINDOLE-1,3(2H)-DIONE, 3A,4,7,7A-TETRAHYDRO-2-[(1,1,2,2-TETRACHLOROETHYL)THIO]-
		4-CYCLOHEXENE-1,2-DICARBOXIMIDE, N-[(1,1,2,2-TETRA-CHLOROETHYL)THIO]-
		ALFLOC 7020
		ALFLOC 7046
		ARBORSEAL
		CAPTAFOL
		CS 5623
		DIFOLATAN
		DIFOLATAN 4F
		DIFOLATAN 4F1
		DIFOLATAN BOW
		FOLCID
		HAIPEN 50
		N-(1,1,2,2-TETRACHLOROETHYLTHIO)-DELTA4-TETRAHY-DROPHTHALIMIDE
		N-(TETRACHLOROETHYLTHIO)TETRAHYDROPHTHALIMIDE
		NALCO 7046
		ORTHO 5865
		PROXEL EF
		TERRAZOL
		TETRACHLOROETHYLTHIOTETRAHYDROPHTHALIMIDE
2426-08-6		
		(BUTOXYMETHYL)OXIRANE
		1-BUTOXY-2,3-EPOXYPROPANE
		2,3-EPOXYPROPYL BUTYL ETHER
		3-BUTOXY-1,2-EPOXYPROPANE
		BUTYL GLYCIDYL ETHER
		ERL 0810
		GLYCIDYL BUTYL ETHER
		GLYCIDYL N-BUTYL ETHER
		N-BUTYL GLYCIDYL ETHER
		OXIRANE, (BUTOXYMETHYL)-
		PROPANE, 1-BUTOXY-2,3-EPOXY-
2426-54-2		
		2-(DIETHYLAMINO)ETHYL ACRYLATE
		2-PROPENOIC ACID, 2-(DIETHYLAMINO)ETHYL ESTER
		ACRYLIC ACID, 2-(DIETHYLAMINO)ETHYL ESTER
		BETA-(DIETHYLAMINO)ETHYL ACRYLATE
		DIETHYLAMINOETHYL ACRYLATE
		N,N-DIETHYLAMINOETHYL ACRYLATE
2429-70-1		
		C.I. DIRECT RED 10, DISODIUM SALT
2429-71-2		
		C.I. DIRECT BLUE 8, DISODIUM SALT
2429-73-4		
		C.I. DIRECT BLUE 2, TRISODIUM SALT
2429-74-5		
		C.I. DIRECT BLUE 15
2429-80-3		
		C.I. ACID ORANGE 45
2429-81-4		
		C.I. DIRECT BROWN 31, TETRASODIUM SALT
2429-82-5		
		C.I. DIRECT BROWN 2, DISODIUM SALT
2429-83-6		
		C.I. DIRECT BLACK 4, DISODIUM SALT
2429-84-7		
		C.I. DIRECT RED 1, DISODIUM SALT
2437-56-1		
		1-TRIDECENE
		ALPHA-TRIDECENE
		N-TRIDEC-1-ENE
2449-49-2		
		ALPHA-METHYLBENZYL DIMETHYL AMINE
		ALPHA,N,N-TRIMETHYLBENZYLAMINE
		BENZENEMETHANAMINE, N,N,ALPHA-TRIMETHYL-
		BENZYLAMINE, N,N,ALPHA-TRIMETHYL-
		N,N-DIMETHYL-ALPHA-METHYLBENZYLAMINE
		N,N-DIMETHYL-ALPHA-PHENYLETHYLAMINE
2454-37-7		
		(M-AMINOPHENYL)METHYL CARBINOL
		3-AMINO-ALPHA-METHYLBENZYL ALCOHOL
		BENZENEMETHANOL, 3-AMINO-ALPHA-METHYL-
		BENZYL ALCOHOL, M-AMINO-ALPHA-METHYL-
		M-(1-HYDROXYETHYL)ANILINE
		M-AMINO-ALPHA-METHYLBENZYL ALCOHOL
2465-27-2		
		ADC AURAMINE O
		AIZEN AURAMINE
		AIZEN AURAMINE CONC.SFA
		AIZEN AURAMINE OH
		AURAMIN
		AURAMINE
		AURAMINE 0-100
		AURAMINE A1
		AURAMINE CHLORIDE
		AURAMINE EXTRA
		AURAMINE EXTRA CONC. A
		AURAMINE FA
		AURAMINE FWA
		AURAMINE HYDROCHLORIDE
		AURAMINE II
		AURAMINE LAKE YELLOW O
		AURAMINE N
		AURAMINE O
		AURAMINE O EXTRA CONC. A EXPORT
		AURAMINE ON
		AURAMINE OO
		AURAMINE OOO
		AURAMINE OS
		AURAMINE PURE
		AURAMINE SP
		AURAMINE YELLOW
		BASIC YELLOW 2
		BENZENAMINE, 4,4'-CARBONIMIDOYLBIS[N,N-DIMETHYL-, MONOHYDROCHLORIDE
		C.I. 41000
		C.I. BASIC YELLOW 2
	*	C.I. BASIC YELLOW 2, MONOHYDROCHLORIDE
		CALCOZINE YELLOW OX
		MITSUI AURAMINE O
2489-77-2		
		TRIMETHYLTHIOUREA
2497-07-6		
		BAY 23323
		DEPD
		DISULFOTON DISULIDE
		DISULFOTON SULFOXIDE
		DISYSTON S
		DISYSTON SULFOXIDE
		DISYSTON SULPHOXIDE
		ETHYLTHIOMETON SULFOXIDE
		O,O-DIETHYL S-(2-(ETHYLSULFINYL)ETHYL) PHOSPHORO-DITHIOATE
		O,O-DIETHYL S-(2-ETHYLTHIONYLETHYL) PHOSPHORO-DITHIOATE
		O,O-DIETHYL-S-((ETHYLSULFINYL)ETHYL)PHOSPHORO-DITHIOATE

CAS No.	IN	Chemical Name
		OXYDISULFOTON
		PHOSPHORODITHIOIC ACID, O,O-DIETHYL S-((ETHYL-SULFINYL)ETHYL) ESTER
		PHOSPHORODITHIOIC ACID, O,O-DIETHYL S-[2-(ETHYL-SULFINYL)ETHYL] ESTER
2499-59-4		
		2-PROPENOIC ACID, OCTYL ESTER
		ACRYLIC ACID, OCTYL ESTER
		ENT 3827
		N-OCTYL ACRYLATE
		OCTYL ACRYLATE
2508-19-2		
		TRINITROBENZENESULFONIC ACID
2524-03-0		
		CHLORODIMETHOXYPHOSPHINE SULFIDE
		DIMETHOXY THIOPHOSPHONYL CHLORIDE
		DIMETHYL CHLOROTHIONOPHOSPHATE
		DIMETHYL CHLOROTHIOPHOSPHATE
		DIMETHYL CHLOROTHIOPHOSPHATE (DOT)
		DIMETHYL PHOSPHOROCHLORIDOTHIOATE
		DIMETHYL PHOSPHOROCHLORIDOTHIOATE (DOT)
		DIMETHYL THIONOPHOSPHOROCHLORIDATE
		DIMETHYL THIOPHOSPHOROCHLORIDATE
		DIMETHYL THIOPHOSPHORYL CHLORIDE
		DIMETHYLCHLORTHIOFOSFAT (CZECH)
		METHYL PCT
		METHYL PHOSPHOROCHLORIDOTHIOATE ((MEO)2ClPS)
		NA 2922 (DOT)
		O,O-DIMETHYL CHLOROTHIONOPHOSPHATE
		O,O-DIMETHYL CHLOROTHIOPHOSPHATE
		O,O-DIMETHYL PHOSPHOROCHLORIDOTHIOATE
		O,O-DIMETHYL PHOSPHOROCHLOROTHIOATE
		O,O-DIMETHYL PHOSPHOROTHIONOCHLORIDATE
		O,O-DIMETHYL THIONOPHOSPHOROCHLORIDATE
		O,O-DIMETHYL THIOPHOSPHORIC ACID CHLORIDE
		O,O-DIMETHYL THIOPHOSPHORYL CHLORIDE
		O,O-DIMETHYLESTER KYSELINY CHLORTHIOFOSFORECNE (CZECH)
		O,O-DIMETHYLPHOSPHOROCHLORIDOTHIOATE
		O,O-DIMETHYLTHIONOPHOSPHORYL CHLORIDE
		PHOSPHONOTHIOIC ACID, CHLORO-, O,O-DIMETHYL ESTER
		PHOSPHOROCHLORIDOTHIOIC ACID, O,O-DIMETHYL ESTER
2524-04-1		
		DIETHYL THIOPHOSPHORIC CHLORIDE
		DIETHYL THIOPHOSPHORYL CHLORIDE
		DIETHYLCHLOROTHIOPHOSPHATE
		DIETHYLCHLORTHIOFOSFAT (CZECH)
		DIETHYLTHIOPHOSPHORYL CHLORIDE
		DIETHYLTHIOPHOSPHORYL CHLORIDE (DOT)
		O,O-DIETHYL THIONOPHOSPHORIC CHLORIDE
		O,O-DIETHYL THIONOPHOSPHOROCHLORIDATE
		O,O-DIETHYL THIONOPHOSPHORYL CHLORIDE
		O,O-DIETHYL THIOPHOSPHORIC ACID CHLORIDE
		O,O-DIETHYL THIOPHOSPHORYL CHLORIDE
		O,O-DIETHYLPHOSPHOROCHLORIDOTHIOATE
		O,O-DIETHYLTHIOPHOSPHOROCHLORIDATE
		PHOSPHONOTHIOIC ACID, CHLORO-, O,O-DIETHYL ESTER
		UN 2751 (DOT)
2540-82-1		
		ACETAMIDE, N-FORMYL-2-MERCAPTO-N-METHYL-, S-ESTER WITH O,O-DIMETHYL PHOSPHORODITHIOATE
		AFLIX
		ANTHIO
		ANTHIO 25
		ANTIO
		CP 53926
		ENT 27,257
		FORMOTHION
		J-38
		N-FORMYL-N-METHYLCARBAMOYLMETHYL O,O-DI-METHYL PHOSPHORODITHIOATE
		O,O-DIMETHYL DITHIOPHOSPHORYLACETIC ACID N-METHYL-N-FORMYLAMIDE
		O,O-DIMETHYL PHOSPHORODITHIOATE N-FORMYL-2-MERCAPTO-N-METHYLACETAMIDE S-ESTER
		O,O-DIMETHYL S-(N-FORMYL-N-METHYLCARBAMOYL-METHYL) PHOSPHORODITHIOATE
		O,O-DIMETHYL S-(N-METHYL-N-FORMYLCARBAMOYL-METHYL)PHOSPHORODITHIOATE
		O,O-DIMETHYL-S-(3-METHYL-2,4-DIOXO-3-AZA-BUTYL)-DITHIOFOSFAAT (DUTCH)
		O,O-DIMETHYL-S-(3-METHYL-2,4-DIOXO-3-AZA-BUTYL)-DITHIOPHOSPHAT (GERMAN)
		O,O-DIMETHYL-S-(N-METHYL-N-FORMYL-CARBAMOYL-METHYL)-DITHIOPHOSPHAT (GERMAN)
		O,O-DIMETIL-S-(N-FORMIL-N-METIL-CARBAMOIL-METIL)-DITIOFOSFATO (ITALIAN)
		OMS-968
		P 1
		PHOSPHORODITHIOIC ACID, 0,0-DIMETHYL ESTER, N-FORMYL-2-MERCAPTO-N-METHYLACETAMIDE S-ESTER
		PHOSPHORODITHIOIC ACID, O,O-DIMETHYL ESTER, S-ESTER WITH N-FORMYL-2-MERCAPTO-N-METHYL-ACETAMIDE
		PHOSPHORODITHIOIC ACID, S-(2-(FORMYLMETHYLAMINO)-2-OXOETHYL) O,O-DIMETHYL ESTER
		S 6900
		S-(2-(FORMYLMETHYLAMINO)-2-OXOETHYL) O,O-DI-METHYLPHOSPHORODITHIOATE
		S-(N-FORMYL-N-METHYLCARBAMOYLMETHYL) DIMETHYL PHOSPHOROTHIOLOTHIONATE
		S-(N-FORMYL-N-METHYLCARBAMOYLMETHYL) O,O-DIMETHYL PHOSPHORODITHIOATE
		SAN 244 I
		SAN 6913 I
		SAN 7107 I
		SANDOZ S-6900
		SPENCER S-6900
		TOPROSE
		VEL 4284
2545-59-7		
		(2,4,5-TRICHLOROPHENOXY)ACETIC ACID 2-BUTOXYETHYL ESTER
		(2,4,5-TRICHLOROPHENOXY)ACETIC ACID BUTOXY-ETHANOL ESTER
		(2,4,5-TRICHLOROPHENOXY)ACETIC ACID BUTYL ESTER
		2,4,5-T BUTOXYETHANOL ESTER
		2,4,5-T BUTOXYETHYL ESTER
		2,4,5-T BUTYL ESTER
		2,4,5-T ISOOCTYL ESTER
		2,4,5-T N-BUTYL ESTER
		2,4,5-TRICHLOROPHENOXYACETIC ACID AMINE, ESTER, OR SALT
		2,4,5-TRICHLOROPHENOXYACETIC ACID BUTOXYETHYL ESTER
		2,4,5-TRICHLOROPHENOXYACETIC ACID, BUTYL ESTER
		ACETIC ACID, (2,4,5-TRICHLOROPHENOXY)-, 2-BUTOXY-ETHYL ESTER
		ACETIC ACID, (2,4,5-TRICHLOROPHENOXY)-, BUTYL ESTER
		ACETIC ACID, 2,4,5-TRICHLOROPHENOXY-, ISOOCTYL ESTER
		ARBORICID
		BLADEX H
		BUTOXYETHYL 2,4,5-T
		BUTOXYETHYL 2,4,5-TRICHLOROPHENOXYACETATE
		BUTYL (2,4,5-TRICHLOROPHENOXY)ACETATE
		BUTYL 2,4,5-T
		BUTYL 2,4,5-TRICHLOROPHENOXYACETATE
		BUTYLATE 2,4,5-T
		ETHANOL, 2-BUTOXY-, (2,4,5-TRICHLOROPHENOXY)ACE-TATE
		FLOMORE
		HI-ESTER 2,4,5-T
		HORMOSLYR 500T
		KILEX 3
		KRZEWOTOKS
		KRZEWOTOX
		N-BUTYL (2,4,5-TRICHLOROPHENOXY)ACETATE
		N-BUTYLESTER KYSELINI 2,4,5-TRICHLORFENOXYOCTOVE (CZECH)
		TORMONA
		TRINOXOL

CAS No.	IN	Chemical Name
		TRIOXONE
		WEEDONE 2,4,5-T
2549-51-1		
		ACETIC ACID, CHLORO-, ETHENYL ESTER
		ACETIC ACID, CHLORO-, VINYL ESTER
		UN 2589 (DOT)
		VINYL CHLOROACETATE
		VINYL CHLOROACETATE (DOT)
		VINYL MONOCHLOROACETATE
2550-33-6		
		TERT-BUTYL PEROXYDIETHYLACETATE, TECHNICAL PURE (DOT)
		TERT-BUTYL PEROXYDIETHYLACETATE, TECHNICALLY PURE
		UN 2144 (DOT)
2551-62-4		
		ELEGAS
		HEXAFLUORURE DE SOUFRE (FRENCH)
		SULFUR FLUORIDE
		SULFUR FLUORIDE (SF6)
		SULFUR FLUORIDE (SF6), (OC-6-11)-
		SULFUR HEXAFLUORIDE
		SULFUR HEXAFLUORIDE (ACGIH,DOT)
		SULPHUR HEXAFLUORIDE
		UN 1080 (DOT)
2570-26-5		
		1-AMINOPENTADECANE
		1-PENTADECANAMINE
		1-PENTADECYLAMINE
		MONOPENTADECYLAMINE
		N-PENTADECYLAMINE
		PENTADECYLAMINE
2586-57-4		
		C.I. DIRECT BLUE 22, DISODIUM SALT
2586-60-9		
		C.I. DIRECT VIOLET 1, DISODIUM SALT
2588-05-8		
		OO-DIETHYL S-ETHYLSULPHINYLMETHYL PHOSPHOROTHIOATE
2588-06-9		
		OO-DIETHYL S-ETHYLSULPHONYLMETHYL PHOSPHOROTHIOATE
2600-69-3		
		OO-DIETHYL S-ETHYLTHIOMETHYL PHOSPHOROTHIOATE
2602-46-2		
		2,7-NAPHTHALENEDISULFONIC ACID, 3,3'-[[1,1'-BIPHENYL]-4,4'-DIYLBIS(AZO)]BISP5-AMINO-4-HYDROXY-, TETR
		AIREDALE BLUE 2BD
		AIZEN DIRECT BLUE 2BH
		AMANIL BLUE 2BX
		ATLANTIC BLUE 2B
		ATUL DIRECT BLUE 2B
		AZOCARD BLUE 2B
		AZOMINE BLUE 2B
		BELAMINE BLUE 2B
		BENCIDAL BLUE 2B
		BENZANIL BLUE 2B
		BENZO BLUE 2B
		BENZO BLUE BBA-CF
		BENZO BLUE BBN-CF
		BENZO BLUE GS
		BLUE 2B
		BLUE 2B SALT
		BRASILAMINA BLUE 2B
		C.I. 22610
		C.I. DIRECT BLUE 6
		C.I. DIRECT BLUE 6, TETRASODIUM SALT
		CALCOMINE BLUE 2B
		CHLORAMINE BLUE 2B
		CHLORAZOL BLUE B
		CHLORAZOL BLUE BP
		CHROME LEATHER BLUE 2B
		CRESOTINE BLUE 2B
		DIACOTTON BLUE BB
		DIAMINE BLUE 2B
		DIAMINE BLUE BB
		DIAPHTAMINE BLUE BB
		DIAZINE BLUE 2B
		DIAZOL BLUE 2B
		DIPHENYL BLUE 2B
		DIPHENYL BLUE KF
		DIPHENYL BLUE M 2B
		DIRECT BLUE 6
		DIRECT BLUE 2B
		DIRECT BLUE 6
		DIRECT BLUE 6 (TECHNICAL GRADE)
		DIRECT BLUE A
		DIRECT BLUE BB
		DIRECT BLUE GS
		DIRECT BLUE K
		DIRECT BLUE M 2B
		ENIANIL BLUE 2BN
		FENAMIN BLUE 2B
		FIXANOL BLUE 2B
		HISPAMIN BLUE 2B
		INDIGO BLUE 2B
		KAYAKU DIRECT BLUE BB
		MITSUI DIRECT BLUE 2BN
		NAPHTHAMINE BLUE 2B
		NIAGARA BLUE 2B
		NIPPON BLUE BB
		PARAMINE BLUE 2B
		PHENAMINE BLUE BB
		PHENO BLUE 2B
		PONTAMINE BLUE BB
		TERTRODIRECT BLUE 2B
		VONDACEL BLUE 2B
2610-05-1		
		C.I. DIRECT BLUE 1
2612-57-9		
		DICHLOROPHENYL-ISOCYANATE (2,4-DICHLORO-1-ISOCYANATO-BENZENE)
2614-76-8		
		2,2-DIHYDROPEROXY PROPANE
		2,2-DIHYDROPEROXYPROPANE (CONCENTRATION NOT MORE THAN 30%)
2618-77-1		
		2,5-BIS(BENZOYLDIOXY)-2,5-DIMETHYLHEXANE
		2,5-BIS(BENZOYLPEROXY)-2,5-DIMETHYLHEXANE
		2,5-DIBENZOYLPEROXY-2,5-DIMETHYLHEXANE
		2,5-DIMETHYL-2,5-BIS(BENZOYLPEROXY)HEXANE
		2,5-DIMETHYL-2,5-DI(BENZOYLPEROXY)HEXANE
		2,5-DIMETHYL-2,5-DI-(BENZOYLPEROXY)HEXANE TECHNICALLY PURE (DOT)
		2,5-DIMETHYL-2,5-DI-(BENZOYLPEROXY)HEXANE, NOT MORE THAN 82% WITH INERT SOLID (DOT)
		2,5-DIMETHYL-2,5-DI-(BENZOYLPEROXY)HEXANE, TECHNICALLY PURE
		2,5-DIMETHYL-2,5-HEXANEDIHYDROPEROXIDE DIBENZOATE
		2,5-DIMETHYL-2,5-HEXANEDIOL BIS(PEROXYBENZOATE)
		2,5-DIMETHYL-2,5-HEXANEDIOL DIPEROXYBENZOATE
		2,5-DIMETHYL-2,5-HEXANEDIYL BIS(PEROXYBENZOATE)
		2,5-DIMETHYL-2,5-HEXANEDIYL PEROXYBENZOATE
		2,5-DIMETHYLHEXANE-2,5-DIPEROXYBENZOATE
		2,5-DIMETHYLHEXANE-2,5-DIYL DIPERBENZOATE
		2,5-HEXANEDIOL, 2,5-DIMETHYL-, BIS(PEROXYBENZOATE)
		BENZENECARBOPEROXOIC ACID, 1,1,4,4-TETRAMETHYL-1,4-BUTANEDIYL ESTER
		LUPEROX 118
		LUPERSOL 118
		PEROXYBENZOIC ACID, 1,1,4,4-TETRAMETHYLTETRA-

CAS No.	IN	Chemical Name
		METHYLENE ESTER
		UN 2172 (DOT)
		UN 2173 (DOT)
		USP 711
2631-37-0		(3-METHYL-5-ISOPROPYLPHENYL)-N-METHYLCARBAMAT (GERMAN)
		3-ISOPROPYL-5-METHYLPHENYL METHYLCARBAMATE
		3-ISOPROPYL-5-METHYLPHENYL N-METHYLCARBAMATE
		3-METHYL-5-(1-METHYLETHYL)PHENOLMETHYLCARBAMATE
		3-METHYL-5-ISOPROPYL N-METHYLCARBAMATE
		3-METHYL-5-ISOPROPYLPHENYL METHYLCARBAMATE
		3-METHYL-5-ISOPROPYLPHENYL-N-METHYLCARBAMATE
		CARBAMIC ACID, METHYL-, 3-METHYL-5-(1-METHYLETHYL)PHENYL ESTER
		CARBAMIC ACID, METHYL-, M-CYM-5-YL ESTER
		CARBAMIC ACID, N-METHYL-, 3-METHYL-5-ISOPROPYLPHENYL ESTER
		CARBAMULT
		ENT 27,300-A
		ENT 27300
		EP 316
		ITC
		M-CYM-5-YL METHYLCARBAMATE
		METHYLCARBAMIC ACID M-CYM-5-YL ESTER
		MINACIDE
		MORTON EP-316
		OMS 716
		PHENOL, 3-METHYL-5-(1-METHYLETHYL)-, METHYLCARBAMATE
		PROMECARB
		PROMECARBE
		SCH 34615
		SCHERING 34615
		SN 34615
		T-32
		UC 9880
		UNION CARBIDE UC-9880
2636-26-2		(2,2-DICHLOOR-VINYL)-DIMETHYL-FOSFAAT (DUTCH)
		(2,2-DICHLOR-VINYL)-DIMETHYL-PHOSPHAT (GERMAN)
		(2,2-DICHLORO-VINIL)DIMETIL-FOSFATO (ITALIAN)
		2,2-DICHLOROETHENYL DIMETHYL PHOSPHATE
		2,2-DICHLOROETHENYL PHOSPHORIC ACID DIMETHYL ESTER
		2,2-DICHLOROVINYL DIMETHYL PHOSPHATE
		2,2-DICHLOROVINYL DIMETHYL PHOSPHORIC ACID ESTER
		APAVAP
		ASTROBOT
		ATGARD
		ATGARD C
		ATGARD V
		BAY 34727
		BAY-19149
		BENFOS
		BIBESOL
		BREVINYL
		BREVINYL E50
		CANOGARD
		CEKUSAN
		CHLORVINPHOS
		CIAFOS
		CYANOPHOS
		CYANOX
		CYAP
		CYPONA
		DDVF
		DDVP
		DEDEVAP
		DERIBAN
		DERRIBANTE
		DEVIKOL
		DICHLOORVO (DUTCH)
		DICHLORFOS (POLISH)
		DICHLORMAN
		DICHLOROPHOS
		DICHLOROVOS
		DICHLORPHOS
		DICHLORVOS
		DICHLORVOS (ACGIH,DOT)
		DIMETHYL 2,2-DICHLOROETHENYL PHOSPHATE
		DIMETHYL 2,2-DICHLOROVINYL PHOSPHATE
		DIMETHYL DICHLOROVINYL PHOSPHATE
		DIVIPAN
		DUO-KILL
		DURAVOS
		ENT 20738
		EQUIGARD
		EQUIGEL
		ETHENOL, 2,2-DICHLORO-, DIMETHYL PHOSPHATE
		FECAMA
		FLY FIGHTER
		FLY-DIE
		HERKAL
		HERKOL
		KRECALVIN
		LINDAN
		MAFU
		MAFU STRIP
		MARVEX
		MOPARI
		NA 2783 (DOT)
		NCI-C00113
		NERKOL
		NO-PEST
		NO-PEST STRIP
		NOGOS
		NOGOS 50
		NOGOS 50 EC
		NOGOS G
		NSC-6738
		NUVA
		NUVAN
		NUVAN 100EC
		O,O-DIMETHYL DICHLOROVINYL PHOSPHATE
		O,O-DIMETHYL O-(4-CYANOPHENYL) PHOSPHOROTHIOATE
		O,O-DIMETHYL O-(P-CYANOPHENYL) PHOSPHOROTHIOATE
		O,O-DIMETHYL O-2,2-DICHLOROVINYL PHOSPHATE
		O,O-DIMETHYL-O-(2,2-DICHLOR-VINYL)-PHOSPHAT (GERMAN)
		O-(2,2-DICHLORVINYL)-O,O-DIMETHYLPHOSPHAT (GERMAN)
		OKO
		OMS 14
		PHOSPHATE DE DIMETHYLE ET DE 2,2-DICHLOROVINYLE (FRENCH)
		PHOSPHORIC ACID, 2,2-DICHLOROETHENYL DIMETHYL ESTER
		PHOSPHORIC ACID, 2,2-DICHLOROVINYL DIMETHYL ESTER
		PHOSPHOROTHIOIC ACID, O,O-DIMETHYL ESTER, O-ESTER WITH P-HYDROXYBENZONITRILE
		PHOSPHOROTHIOIC ACID, O-(4-CYANOPHENYL) O,O-DIMETHYL ESTER
		PHOSVIT
		S 4084
		SD-1750
		SUMITOMO S 4084
		SZKLARNIAK
		TAP 9VP
		TASK
		TASK TABS
		TENAC
		TETRAVOS
		UDVF
		UNIFOS
		UNIFOS 50 EC
		VAPONA
		VAPONITE
		VAPORA II
		VERDICAN
		VERDIPOR
		VINYL ALCOHOL, 2,2-DICHLORO-, DIMETHYL PHOSPHATE
		VINYLOFOS
		VINYLOPHOS

CAS No.	IN	Chemical Name
2642-71-9		1,2,3-BENZOTRIAZINE, PHOSPHORODITHIOIC ACID DERIV
		3,4-DIHYDRO-4-OXO-3-BENZOTRIAZINYLMETHYL O,O-DIETHYL PHOSPHORODITHIOATE
		ATHYL-GUSATHION
		AZINFOS-ETHYL (DUTCH)
		AZINOS
		AZINPHOS ETHYL
		AZINPHOS-AETHYL (GERMAN)
		AZINPHOS-ETILE (ITALIAN)
		BAY 16255
		BAYER 16259
		BENZOTRIAZINE DERIVATIVE OF AN ETHYL DITHIOPHOSPHATE
		COTNION-ETHYL
		CRYSTHION
		ENT 22,014
		ETHYL AZINPHOS
		ETHYL GUSATHION
		ETHYL GUTHION
		GUSATHION A
		GUSATHION ETHYL
		GUSATHION H AND K
		GUTHION ETHYL
		O,O-DIAETHYL-S-((4-OXO-3H-1,2,3-BENZOTRIAZIN-3-YL)-METHYL)-DITHIOPHOSPHAT (GERMAN)
		O,O-DIAETHYL-S-(4-OXOBENZOTRIAZIN-3-METHYL)-DITHIOPHOSPHAT (GERMAN)
		O,O-DIETHYL PHOSPHORODITHIOATE S-ESTER WITH 3-(MERCAPTOMETHYL)-1,2,3-BENZOTRIAZIN-4(3H)-ONE
		O,O-DIETHYL S-(4-OXOBENZOTRIAZINO-3-METHYL)PHOSPHORODITHIOATE
		O,O-DIETHYL-S-((4-OXO-3H-1,2,3-BENZOTRIAZIN-3-YL)-METHYL)-DITHIOFOSFAAT (DUTCH)
		O,O-DIETHYL-S-(4-OXO-3H-1,2,3-BENZOTRIAZINE-3-YL)-METHYL-DITHIOPHOSPHATE
		O,O-DIETIL-S-((4-OXO-3H-1,2,3-BENZOTRIAZIN-3-IL)-METIL)-DITIOFOSFATO (ITALIAN)
		PHOSPHORODITHIOIC ACID, O,O-DIETHYL ESTER, S-ESTER WITH 3-(MERCAPTOMETHYL)-1,2,3-BENZOTRIAZIN-4(3H)-
		PHOSPHORODITHIOIC ACID, O,O-DIETHYL S-[(4-OXO-1,2,3-BENZOTRIAZIN-3(4H)-YL)METHYL] ESTER
		R 1513
		S-(3,4-DIHYDRO-4-OXO-1,2,3-BENZOTRIAZIN-3-YLMETHYL) O,O-DIETHYL PHOSPHORODITHIOATE
		TRIAZOTION (RUSSIAN)
2646-17-5		1-O-TOLUENEAZO-2-NAPHTHOL
		1-O-TOLYLAZO-2-NAPHTHOL
		1-[(2-METHYLPHENYL)AZO]-2-NAPHTHOL
		2-NAPHTHALENOL, 1-[(2-METHYLPHENYL)AZO]-
		A.F. ORANGE NO. 2
		C.I. 12100
		C.I. SOLVENT ORANGE 2
		DOLKWAL ORANGE SS
		EXT D AND C ORANGE NO. 4
		FAT ORANGE II
		FAT ORANGE RR
		FD AND C ORANGE NO. 2
		HEXACOL OIL ORANGE SS
		JAPAN ORANGE 403
		LACQUER ORANGE V
		OIL ORANGE 204
		OIL ORANGE OPEL
		OIL ORANGE SS
		OIL ORANGE TX
		OLEAL ORANGE SS
		ORANGE OT
		ORANGE SS
		ORGANOL ORANGE 2R
		TOLUENE-2-AZONAPHTHOL-2
2650-18-2		C.I. ACID BLUE 9, DIAMMONIUM SALT
2665-30-7		COLEP
		CP 40294
		ENT 25,787
		MONSANTO CP-40294
		O-(4-NITROPHENYL) O-PHENYLMETHYLPHOSPHONOTHIOATE
		O-(P-NITROPHENYL) O-PHENYL METHYLPHOSPHONOTHIOATE
		PHOSPHONOTHIOIC ACID, METHYL, O-(4-NITROPHENYL) O-PHENYL ESTER
		PHOSPHONOTHIOIC ACID, METHYL-, O-(P-NITROPHENYL) O-PHENYL ESTER
2691-41-0		CYCLOTETRAMETHYLENETETRANITRAMINE
2696-92-6		NITROGEN OXIDE CHLORIDE (NOCl)
		NITROGEN OXYCHLORIDE
		NITROGEN OXYCHLORIDE (NOCl)
		NITROSONIUM CHLORIDE
		NITROSYL CHLORIDE
		NITROSYL CHLORIDE ((NO)Cl)
		NITROSYL CHLORIDE (DOT)
		UN 1069 (DOT)
2698-41-1		(O-CHLOROBENZAL)MALONONITRILE
		(O-CHLOROBENZYLIDENE)MALONONITRILE
		2-CHLOROBENZALMALONONITRILE
		2-CHLOROBENZYLIDENEMALONINITRILE
		2-CHLOROBENZYLIDENEMALONONITRILE
		BETA,BETA-DICYANO-O-CHLOROSTYRENE
		CS
		CS (LACRIMATOR)
		MALONONITRILE, (O-CHLOROBENZYLIDENE)-
		O-CHLOROBENZYLIDENE MALO-NONITRILE (OCBM)
		O-CHLOROBENZYLIDENE MALONONITRILE
		O-CHLOROBENZYLIDENEMALONIC NITRILE
		PROPANEDINITRILE, [(2-CHLOROPHENYL)METHYLENE]-
2699-79-8		FLUORURE DE SULFURYLE (FRENCH)
		SULFONYL FLUORIDE
		SULFUR DIFLUORIDE DIOXIDE
		SULFUR DIOXIDE DIFLUORIDE
		SULFURIC OXYFLUORIDE
		SULFURYL FLUORIDE
		SULFURYL FLUORIDE (ACGIH,DOT)
		UN 2191 (DOT)
		VIKANE
		VIKANE FUMIGANT
2703-13-1		PHOSPHONOTHIOIC ACID, METHYL-, O-ETHYL O-(4-METHYLTHIO)
2746-19-2		HIMIC ANHYDRIDE
2757-18-8		THALLOUS MALONATE
2763-96-4		3(2H)-ISOXAZOLONE, 5-(AMINOMETHYL)-
		3-HYDROXY-5-AMINOMETHYLISOXAZOLE
		3-HYDROXY-5-AMINOMETHYLISOXAZOLE-AGARIN
		3-ISOXAZOLOL, 5-(AMINOMETHYL)-
		4-ISOXAZOLIN-3-ONE, 5-(AMINOMETHYL)-
		5-(AMINOMETHYL)-3(2H)-ISOXAZOLONE
		5-(AMINOMETHYL)-3-ISOXAZOLOL
		5-AMINOMETHYL-3-HYDROXYISOXAZOLE
		5-AMINOMETHYL-3-ISOXYZOLE
		AGARIN
		AGARINE
		MUSCIMOL
		PANTHERINE
		RCRA WASTE NUMBER P007

CAS No.	IN	Chemical Name
2764-72-9		1,1'-ETHYLENE-2,2'-BIPYRIDINIUM
		2,2'-BIPYRIDINIUM, 1,1'-(1,2-ETHANEDIYL)-
		9,10-DIHYDRO-8A,10A-DIAZONIAPHENANTHRENE
		DIPYRIDO[1,2-A:2',1'-C]PYRAZINEDIIUM, 6,7-DIHYDRO-
		DIQUAT
		DIQUAT DICATION
2778-04-3		5-METHOXY-2-(DIMETHOXYPHOSPHINYLTHIOMETHYL)PYRONE-4
		AC-18,737
		ENDOCID
		ENDOCIDE
		ENDOTHION
		ENT 24,653
		EXOTHION
		NIA-5767
		NIAGARA 5767
		O,O-DIMETHYL S-(5-METHOXY-4-OXO-4H-PYRAN-2-YL)-PHOSPHOROTHIOATE
		O,O-DIMETHYL S-(5-METHOXYPYRONYL-2-METHYL) THIOPHOSPHATE
		O,O-DIMETHYL-S-(5-METHOXY-PYRON-2-YL)-METHYL)-THIOLPHOSPHAT (GERMAN)
		PHOSPHATE 100
		PHOSPHOPYRON
		PHOSPHOPYRONE
		PHOSPHOROTHIOIC ACID, O,O-DIMETHYL ESTER, S-ESTER WITH 2-(MERCAPTOMETHYL)-5-METHOXY-4H-PYRAN-4-ONE
		PHOSPHOROTHIOIC ACID, S-[(5-METHOXY-4-OXO-4H-PYRAN-2-YL)METHYL] O,O-DIMETHYL ESTER
		S-((5-METHOXY-4H-PYRON-2-YL)-METHYL)-O,O-DIMETHYL-MONOTHIOFOSFAAT (DUTCH)
		S-((5-METHOXY-4H-PYRON-2-YL)-METHYL)-O,O-DIMETHYL-MONOTHIOPHOSPHAT (GERMAN)
		S-((5-METOSSI-4H-PIRON-2-IL)-METIL)-O,O-DIMETIL-MONO-TIOFOSFATO (ITALIAN)
		S-(5-M ETHOXY-4-PYRON-2-YLMETHYL) DIMETHYL PHOSPHOROTHIOLATE
		S-5-METHOXY-4-OXOPYRAN-2-YLMETHYL DIMETHYL PHOSPHOROTHIOATE
		THIOPHOSPHATE DE O,O-DIMETHYLE ET DE S-((5-METHOXY-4-PYRONYL)-METHYLE) (FRENCH)
2782-57-2		1,3,5-TRIAZINE-2,4,6(1H,3H,5H)-TRIONE, 1,3-DICHLORO-
		1,3-DICHLORO-S-TRIAZINE-2,4,6-TRIONE
		ACL 70
		CDB 60
		DICHLORO-S-TRIAZINETRIONE
		DICHLOROCYANURIC ACID
		DICHLOROISOCYANURATE
		DICHLOROISOCYANURIC ACID
		DICHLOROISOCYANURIC ACID, DRY
		DICHLOROISOCYANURIC ACID, DRY (DOT)
		FICLOR 71
		HILITE 60
		ISOCYANURIC ACID, DICHLORO-
		ISOCYANURIC DICHLORIDE
		KYSELINA DICHLORISOKYANUROVA (CZECH)
		ORCED
		S-TRIAZINE-2,4,6(1H,3H,5H)-TRIONE, 1,3-DICHLORO-
		TROCLOSENE
		UN 2465 (DOT)
2806-85-1		3-ETHOXYPROPANAL
		3-ETHOXYPROPIONALDEHYDE
		BETA-ETHOXYPROPIONALDEHYDE
		PROPANAL, 3-ETHOXY-
		PROPIONALDEHYDE, 3-ETHOXY-
2812-73-9		ETHYL CHLOROTHIOFORMATE (CARBONOCHLORIDOTHIOC ACID, S-ETHYL ESTE
2820-51-1		CHLORHYDRATE DE NICOTINE (FRENCH)
		NICOTINE HYDROCHLORIDE
		NICOTINE HYDROCHLORIDE (D,L)
		NICOTINE HYDROCHLORIDE (DOT)
		NICOTINE HYDROCHLORIDE SOLUTION
		NICOTINE HYDROCHLORIDE SOLUTION (DOT)
		PYRIDINE, 3-(1-METHYL-2-PYRROLIDINYL)-, HYDROCHLORIDE, (S)-
		UN 1656 (DOT)
2832-40-8		C.I. DISPERSE YELLOW 3
2842-38-8		2-(CYCLOHEXYLAMINO)ETHANOL
		ETHANOL, 2-(CYCLOHEXYLAMINO)-
		N(2-HYDROXETHYL)CYCLOHEXYLAMINE
		N-(2-HYDROXYETHYL)-N-CYCLOHEXYLAMINE
		N-CYCLOHEXYLAMINOETHANOL
		N-CYCLOHEXYLETHANOLAMINE
2855-13-2		ISOPHORONEDIAMINE
2867-47-2		2-(N,N-DIMETHYLAMINO)ETHYL METHACRYLATE
		2-DIMETHYLAMINOETHYL METHACRYLATE
		2-PROPENOIC ACID, 2-METHYL-, 2-(DIMETHYLAMINO)-ETHYL ESTER
		AGEFLEX FM-1
		BETA-(N,N-DIMETHYLAMINO)ETHYL METHACRYLATE
		BETA-DIMETHYLAMINOETHYL METHACRYLATE
		DIMETHYLAMINOETHYL METHACRYLATE
		DIMETHYLAMINOETHYL METHACRYLATE (DOT)
		ETHANOL, 2-(DIMETHYLAMINO)-, METHACRYLATE
		METHACRYLIC ACID, 2-(DIMETHYLAMINO)ETHYL ESTER
		N,N-DIMETHYLAMINOETHYL METHACRYLATE
		N,N-DIMETHYLETHANOLAMINE METHACRYLATE
		UN 2522 (DOT)
		USAF RH-3
2885-00-9		1-OCTADECANETHIOL
2893-78-9		1,3,5-TRIAZINE-2,4,6(1H,3H,5H)-TRIONE, 1,3-DICHLORO-, SODIUM SALT
		ACL 60
		CDB 63
		CDB CLEARON
		CLEARON
		DICHLOROISOCYANURIC ACID SODIUM SALT
		DIKONIT
		FI CLOR 60S
		FI CLOR CLEARON
		HYLITE 60G
		IZOSAN G
		NEOCHLOR 60P
		ONIACHLOR 60
		S-TRIAZINE-2,4,6(1H,3H,5H)-TRIONE, 1,3-DICHLORO-, SODIUM SALT
		SIMPLA
		SODIUM DICHLORISOCYANURATE
		SODIUM DICHLORO-ISOCYANATE
		SODIUM DICHLORO-S-TRIAZINETRONE
		SODIUM DICHLOROCYANURATE
	*	SODIUM DICHLOROISOCYANURATE
		SURCHLOR GR 60
2893-80-3		C.I. DIRECT BROWN 6, DISODIUM SALT
2917-26-2		1-HEXADECANETHIOL

CAS No.	IN	Chemical Name
2921-88-2		2-PYRIDINOL, 3,5,6-TRICHLORO-, O-ESTER WITH O,O-DIETHYL PHOSPHOROTHIOATE
		BRODAN
		CHLOROPYRIFOS
		CHLOROPYRIPHOS
		CHLORPYRIFOS
		CHLORPYRIFOS (ACGIH,DOT)
		CHLORPYRIFOS-ETHYL
		CHLORPYRIPHOS
		COROBAN
		DETMOL UA
		DOWCO 179
		DURSBAN
		DURSBAN 10CR
		DURSBAN 4E
		DURSBAN F
		ENT 27,311
		ERADEX
		ETHION, DRY
		KILLMASTER
		LORSBAN
		LORSBAN 50SL
		NA 2783 (DOT)
		O,O-DIAETHYL-O-3,5,6-TRICHLOR-2-PYRIDYLMONO-THIOPHOSPHAT (GERMAN)
		O,O-DIETHYL O-3,5,6-TRICHLORO-2-PYRIDYL PHOSPHOROTHIOATE
		OMS-0971
		PHOSPHOROTHIOIC ACID, O,O-DIETHYL O-(3,5,6-TRICHLORO-2-PYRIDINYL) ESTER
		PHOSPHOROTHIOIC ACID, O,O-DIETHYL O-(3,5,6-TRICHLORO-2-PYRIDYL) ESTER
		PYRINEX
		SUSCON
2935-44-6		2,5-DIHYDROXYHEXANE
		2,5-HEXANEDIOL
		DIISOPROPANOL
2937-50-0		ALLYL CHLOROCARBONATE
		ALLYL CHLOROCARBONATE (DOT)
		ALLYL CHLOROFORMATE
		ALLYL CHLOROFORMATE (DOT)
		ALLYLOXYCARBONYL CHLORIDE
		CARBONOCHLORIDIC ACID, 2-PROPENYL ESTER
		FORMIC ACID, CHLORO-, ALLYL ESTER
		UN 1722 (DOT)
2941-64-2		CARBONIC ACID, THIO-, ANHYDROSULFIDE WITH THIOHYPOCHLOROUS ACID, ETHYL ESTER
		CARBONOCHLORIDOTHIOIC ACID, S-ETHYL ESTER
		CARBONOTHIOIC ACID, ANHYDROSULFIDE WITH THIOHYPOCHLOROUS ACID, ETHYL ESTER
		ETHOXYCARBONYLSULFENYL CHLORIDE
		ETHYL (CHLOROSULFENYL)FORMATE
		ETHYL CHLOROTHIOFORMATE
		ETHYL CHLOROTHIOLFORMATE
		ETHYL CHLOROTHIOLOFORMATE
		ETHYL THIOCHLOROFORMATE
		FORMIC ACID, CHLOROTHIO-, ETHYL ESTER
		FORMIC ACID, CHLOROTHIO-, S-ETHYL ESTER
		S-ETHYL CARBONOCHLORIDOTHIOATE
		S-ETHYL CHLOROTHIOCARBONATE
		S-ETHYL CHLOROTHIOFORMATE
		S-ETHYL CHLOROTHIOLFORMATE
		S-ETHYL THIOCHLOROFORMATE
		THIOHYPOCHLOROUS ACID, ANHYDROSULFIDE WITH O-ETHYL THIOCARBONATE
		UN 2826 (DOT)
2944-67-4		ETHANEDIOIC ACID, AMMONIUM IRON(3+) SALT (3:3:1)
		OXALIC ACID, AMMONIUM IRON(3+) SALT (3:3:1)

CAS No.	IN	Chemical Name
2955-38-6		PRAZEPAM
2971-38-2		2,4-D CHLOROCROTYL ESTER
		ACETIC ACID, (2,4-DICHLOROPHENOXY)-, 4-CHLORO-2-BUTENYL ESTER
		CROTILIN
		CROTILINE
		CROTYLIN
		KROTILIN
		KROTILINE
2971-90-6		2,6-DIMETHYL-3,5-DICHLORO-4-PYRIDINOL
		3,5-DICHLORO-2,6-DIMETHYL-4-HYDROXYPYRIDINE
		3,5-DICHLORO-2,6-DIMETHYL-4-PYRIDINOL
		3,5-DICHLORO-2,6-DIMETHYLPYRIDINOL
		3,5-DICHLORO-4-HYDROXY-2,6-DIMETHYLPYRIDINE
		4-PYRIDINOL, 3,5-DICHLORO-2,6-DIMETHYL-
		CLOPIDOL
		COCCIDIOSTAT C
		COYDEN
		COYDEN 25
		FARMCOCCID
		LERBEK
		METHYLCHLOROPINDOL
		METHYLCHLORPINDOL
		METICLORPINDOL
		PHARMCOCCID
		WR 61112
2980-64-5		AMMONIUM DINITRO-O-CRESOLATE
2999-74-8		DIMETHYLMAGNESIUM
		DIMETHYLMAGNESIUM (DOT)
		MAGNESIUM, DIMETHYL-
		UN 1368 (DOT)
3006-82-4		CHALOXYD P 1310
		CHALOXYD P 1327
		ESPEROX 28
		HEXANEPEROXOIC ACID, 2-ETHYL-, 1,1-DIMETHYLETHYL ESTER
		HEXANOIC ACID, 2-((1,1-DIMETHYLETHYL)DIOXY)ETHYL ESTER
		HEXANOIC ACID, 2-((1,1-DIMETHYLETHYL)DIOXY)ETHYL ESTER, WITH AT LEAST 50% PHLEGMATIZER
		LUPERSOL PDO
		PERBUTYL O
		PEROXYHEXANOIC ACID, 2-ETHYL-, TERT-BUTYL ESTER
		TERT-BUTYL 2-ETHYLHEXANEPEROXOATE
		TERT-BUTYL 2-ETHYLPERHEXANOATE
		TERT-BUTYL 2-ETHYLPEROXYHEXANOATE
		TERT-BUTYL PEROXY-2-ETHYLHEXANOATE
		TERT-BUTYL PEROXY-2-ETHYLHEXANOATE, MORE THAN 30% WITH 2,2-DI-(TERT-BUTYLPEROXY)BUTANE, WITH NOT MORE THAN 35%
		TERT-BUTYL PEROXY-2-ETHYLHEXANOATE, TECHNICAL PURE (DOT)
		TERT-BUTYL PEROXY-2-ETHYLHEXANOATE, TECHNICALLY PURE
		TERT-BUTYL PEROXY-2-ETHYLHEXANOATE, WITH AT LEAST 50% PHLEGMATIZER (DOT)
		TERT-BUTYL PEROXYETHYLHEXANOATE
		TRIGONOX T 21S
		UN 2143 (DOT)
		UN 2888 (DOT)
3006-86-8		1,1- DI-(TERT-BUTYLPEROXY)CYCLOHEXANE, AT LEAST 40% WITH INERT INORGANIC SOLID, WITH AT LEAST 13% PHLEGMATIZER
		1,1-BIS(TERT-BUTYLDIOXY)CYCLOHEXANE
		1,1-BIS(TERT-BUTYLPEROXY)CYCLOHEXANE

CAS No.	IN	Chemical Name
		1,1-DI(TERT-BUTYLPEROXY)CYCLOHEXANE
		1,1-DI-(TERT-BUTYLPEROXY)CYCLOHEXANE (DOT)
		1,1-DI-(TERT-BUTYLPEROXY)CYCLOHEXANE, NOT MORE THAN 77% IN SOLUTION (DOT)
		1,1-DI-(TERT-BUTYLPEROXY)CYCLOHEXANE, TECHNICALLY PURE
		CHALOXYD P 1250AL
		LUPERCO 331XL
		LUPERSOL 321
		LUPERSOL 331
		LUPERSOL 331-80B
		PEROXIDE, (1,1-CYCLOHEXYLIDENE)BIS(TERT-BUTYL-
		PEROXIDE, (1,1-CYCLOHEXYLIDENE)BIS(TERT-BUTYL-, NOT MORE THAN 77% IN SOLUTION
		PEROXIDE, (1,1-CYCLOHEXYLIDENE)BIS(TERT-BUTYL-, WITH AT LEAST 13% PHLEGMATIZER AND 47% INERT SOLID
		PEROXIDE, (1,1-CYCLOHEXYLIDENE)BIS(TERT-BUTYL-, WITH AT LEAST 50% PHLEGMATIZER
		PEROXIDE, CYCLOHEXYLIDENEBIS[(1,1-DIMETHYLETHYL)
		PEROXIDE, CYCLOHEXYLIDENEBIS[TERT-BUTYL
		TRIGONOX 22B50
		TRIGONOX 22B75
		UN 2179 (DOT)
		UN 2180 (DOT)
		UN 2885 (DOT)
		USP 400
3012-65-5		1,2,3-PROPANETRICARBOXYLIC ACID, 2-HYDROXY-, DIAMMONIUM SALT
		AMMONIUM CITRATE
	*	AMMONIUM CITRATE DIBASIC
		AMMONIUM CITRATE DIBASIC (DOT)
		AMMONIUM MONOHYDROGEN CITRATE
		CITRIC ACID, DIAMMONIUM SALT
		DIAMMONIUM CITRATE
		DIAMMONIUM HYDROGEN CITRATE
		DIBASIC AMMONIUM CITRATE
		MICROSTOP
		NA 9087 (DOT)
3025-88-5		1,1,4,4-TETRAMETHYLTETRAMETHYLENE DIHYDROPEROXIDE
		2,5-BIS(HYDROPEROXY)-2,5-DIMETHYLHEXANE
		2,5-DIHYDROPEROXY-2,5-DIMETHYLHEXANE
		2,5-DIMETHYL-2,5-BIS(HYDROPEROXY)HEXANE
		2,5-DIMETHYL-2,5-DIHYDROPEROXYHEXANE
		2,5-DIMETHYL-2,5-HEXYLENE DIHYDROPEROXIDE
		2,5-DIMETHYLHEXANE 2,5-BIS(HYDROPEROXIDE)
		2,5-DIMETHYLHEXANE-2,5-DIHYDROPEROXIDE
		2,5-DIMETHYLHEXYL-2,5-DIHYDROPEROXIDE
		DIMETHYLHEXANE DIHYDROPEROXIDE, (WITH 18% OR MORE WATER) (DOT)
		DIMETHYLHEXANE DIHYDROPEROXIDE, DRY (DOT)
		HEXANE, DIMETHYL-, DIHYDROPEROXIDE
		HYDROPEROXIDE, (1,1,4,4-TETRAMETHYL-1,4-BUTANEDIYL)BIS-4,5-BIS(HYDROPEROXY)-2,5-DIMETHYLHEXANE
		HYDROPEROXIDE, (1,1,4,4-TETRAMETHYLTETRAMETHYLENE)DI-
		LUPEROX 2,5-2,5
		LUPERSOL 2,5-2,5
		UN 2174 (DOT)
3034-79-5		DI(2-METHYLBENZOYL)PEROXIDE
3037-72-7		(4-AMINOBUTYL)DIETHOXYMETHYLSILANE
		(DELTA-AMINOBUTYL)METHYLDIETHOXYSILANE
		1-BUTANAMINE, 4-(DIETHOXYMETHYLSILYL)-
		4-(DIETHOXYMETHYLSILYL)BUTYLAMINE
		4-AMINOBUTYL(METHYL)DIETHOXYSILANE
		BUTYLAMINE, 4-(DIETHOXYMETHYLSILYL)-
		DELTA-AMINOBUTYLMETHYLDIETHOXYSILANE
		SILANE, (4-AMINOBUTYL)DIETHOXYMETHYL-
		Y 1902
3048-64-4		VINYLNORBORNENE
3054-95-3		1,1-DIETHOXY-2-PROPENE
		1-PROPENE, 3,3-DIETHOXY-
		3,3-DIETHOXY-1-PROPENE
		3,3-DIETHOXYPROPENE
		3,3-DIETHOXYPROPENE (DOT)
		ACROLEIN ACETAL
		ACROLEIN, DIETHYL ACETAL
		ACRYLALDEHYDE DIETHYL ACETAL
		DIETHOXYPROPENE
		PROPENAL DIETHYL ACETAL
		PROPENE, 3,3-DIETHOXY-
		UN 2374 (DOT)
3058-38-6		1,3,5-TRIAMINO-2,4,6-TRINITROBENZENE
3068-88-0		2-OXETANONE, 4-METHYL-
		3-HYDROXYBUTYRIC ACID LACTONE
		4-METHYL-2-OXETANONE
		BETA-BUTYROLACTONE
		BETA-METHYL-BETA-PROPIOLACTONE
		BETA-METHYLPROPIOLACTONE
		BUTANOIC ACID, 3-HYDROXY-, BETA-LACTONE
3074-75-7		2-METHYL-4-ETHYLHEXANE
		4-ETHYL-2-METHYLHEXANE
		HEXANE, 4-ETHYL-2-METHYL-
3074-77-9		3-ETHYL-4-METHYLHEXANE
		3-METHYL-4-ETHYLHEXANE
		4-ETHYL-3-METHYLHEXANE
		HEXANE, 3-ETHYL-4-METHYL-
3081-14-9		1,4-BENZENEDIAMINE, N,N'-BIS(1,4-DIMETHYLPENTYL)-
		ANTIOXIDANT 4030
		EASTOZONE 33
		FLEXZONE 4L
		N,N'-BIS(1,4-DIMETHYLPENTYL)-1,4-BENZENEDIAMINE
		N,N'-BIS(1,4-DIMETHYLPENTYL)-1,4-PHENYLENEDIAMINE
		N,N'-BIS(1,4-DIMETHYLPENTYL)-P-PHENYLENEDIAMINE
		N,N'-DI(1,4-DIMETHYLPENTYL)-P-PHENYLENEDIAMINE
		N,N-BIS-(1,4-METHYL PENTYL)-P-PHENYLENEDIAMINE
		P-PHENYLENEDIAMINE, N,N'-BIS(1,4-DIMETHYLPENTYL)-
		SANTOFLEX 77
		TENAMENE 4
		UOP 788
3085-26-5		8-METHYLNONANAL
		ISODECALDEHYDE
		NONANAL, 8-METHYL-
3087-37-4		1-PROPANOL, TITANIUM(4+) SALT
		1-PROPANOL, TITANIUM(4+) SALT (9CI)
		PROPYL ALCOHOL, TITANIUM(4+) SALT
		PROPYL TITANATE(IV)
		TETRA-N-PROPYL TITANATE
		TETRAPROPOXY TITANATE
		TETRAPROPOXYTITANIUM
		TETRAPROPYL ORTHOTITANATE
		TETRAPROPYL TITANATE
		TETRAPROPYL-O-TITANATE
		TETRAPROPYLORTHOTITANATE (DOT)
		TITANIUM PROPOXIDE (Ti(OPR)4)
		TITANIUM TETRAPROPOXIDE
		TITANIUM TETRAPROPYLATE
		TITANIUM, TETRAPROPOXY-
		UN 2413 (DOT)

CAS No.	IN	Chemical Name

3118-97-6

C.I. SOLVENT ORANGE 7

3129-91-7

CYCLOHEXANAMINE, N-CYCLOHEXYL-, NITRITE
DECHAN
DIANA
DICHAN
DICHAN (CZECH)
DICYCLOHEXYLAMINE, NITRITE
DICYCLOHEXYLAMINONITRITE
DICYCLOHEXYLAMMONIUM NITRITE
DICYKLOHEXYLAMIN NITRIT (CZECH)
DICYNIT (CZECH)
DODECAHYDROPHENYLAMINE NITRITE
DUSITAN DICYKLOHEXYLAMINU (CZECH)
LEUKORROSIN C
NDA

3132-64-7

(BROMOMETHYL)ETHYLENE OXIDE
(BROMOMETHYL)OXIRANE
1,2-EPOXY-3-BROMOPROPANE
1-BROMO-2,3-EPOXYPROPANE
2-(BROMOMETHYL)OXIRANE
3-BROMO-1,2-EPOXYPROPANE
BROMOHYDRIN
EPIBROMHYDRIN
EPIBROMOHYDRIN
EPIBROMOHYDRIN (DOT)
EPIBROMOHYDRINE
OXIRANE, (BROMOMETHYL)-
PROPANE, 1-BROMO-2,3-EPOXY-
PROPANE, 3-BROMO-1,2-EPOXY-
UN 2558 (DOT)

3164-29-2

2,3-DIHYDROXY-BUTANEDIOIC ACID, DIAMMONIUM SALT (9CI)
AMMONIUM D-TARTRATE
* AMMONIUM TARTRATE
AMMONIUM TARTRATE (DOT)
BUTANEDIOIC ACID, 2,3-DIHYDROXY- [R-(R*,R*)]-, DIAMMONIUM SALT
DIAMMONIUM D-TARTRATE
DIAMMONIUM L-(+)-TARTRATE
DIAMMONIUM TARTRATE
L-TARTARIC ACID AMMONIUM SALT
NA 9091 (DOT)
TARTARIC ACID, DIAMMONIUM SALT

3165-93-3

2-AMINO-5-CHLOROTOLUENE HYDROCHLORIDE
2-METHYL-4-CHLOROANILINE HYDROCHLORIDE
4-CHLORO-2-METHYLANILINE HYDROCHLORIDE
4-CHLORO-2-METHYLBENZENAMINE HYDROCHLORIDE
4-CHLORO-2-TOLUIDINE HYDROCHLORIDE
4-CHLORO-6-METHYLANILINE HYDROCHLORIDE
4-CHLORO-O-TOLUIDINE HYDROCHLORIDE
4-CHLORO-O-TOLUIDINE HYDROCHLORIDE (DOT)
5-CHLORO-2-AMINOTOLUENE HYDROCHLORIDE
AMARTHOL FAST RED TR BASE
AMARTHOL FAST RED TR SALT
AZANIL RED SALT TRD
AZOENE FAST RED TR SALT
AZOGENE FAST RED TR
AZOIC DIAZO COMPONENT 11 BASE
BENZENAMINE, 4-CHLORO-2-METHYL-, HYDROCHLORIDE
BRENTAMINE FAST RED TR SALT
C I 37085
CHLORHYDRATE DE 4-CHLOROORTHOTOLUIDINE (FRENCH)
CI AZOIC DIAZO COMPONENT 11
DAITO RED SALT TR
DEVOL RED K
DEVOL RED TA SALT
DEVOL RED TR
DIAZO FAST RED TR
DIAZO FAST RED TRA
FAST RED 5CT SALT
FAST RED SALT TR
FAST RED SALT TRA
FAST RED SALT TRN
FAST RED TR SALT
HINDASOL RED TR SALT
KROMON GREEN B
NATASOL FAST RED TR SALT
NCI-C02368
NEUTROSEL RED TRVA
O-TOLUIDINE, 4-CHLORO-, HYDROCHLORIDE
OFNA-PERL SALT RRA
P-CHLORO-O-TOLUIDINE HYDROCHLORIDE
RCRA WASTE NUMBER U049
RED BASE CIBA IX
RED BASE IRGA IX
RED SALT CIBA IX
RED SALT IRGA IX
RED TRS SALT
SANYO FAST RED SALT TR
UN 1579 (DOT)

3170-57-8

* ISOCYANATES, N.O.S. :OR: ISOCYANATE, SOLUTIONS, N.O.S. :FLASHPOINT ≥23C:BOILING POINT 300C
ISOCYANATES, N.O.S. :OR: ISOCYANATE, SOLUTIONS, N.O.S. FLAMMABLE

3173-53-3

CYCLOHEXYL ISOCYANATE

3174-74-1

2,3-DIHYDRO-4H-PYRAN
2H-3,4-DIHYDROPYRAN
2H-PYRAN, 3,4-DIHYDRO-
2H-PYRAN, 3,6-DIHYDRO-
2H-PYRAN, DIHYDRO-
3,4-DIHYDRO-2H-PYRAN
3,4-DIHYDROPYRAN
3,6-DIHYDRO-2H-PYRAN
5,6-DIHYDRO-2H-PYRAN
5,6-DIHYDRO-4H-PYRAN
DELTA(SUP 2)-DIHYDROPYRAN
DELTA2-DIHYDROPYRAN
DIHYDROPYRAN
DIHYDROPYRAN (DOT)
PYRAN, DIHYDRO-

3179-56-4

ACETYL CYCLOHEXANE SULFONYL PEROXIDE
ACETYL CYCLOHEXANE SULPHONYL PEROXIDE :MAXIMUM CONCENTRATION 32%
ACETYL CYCLOHEXANE SULPHONYL PEROXIDE :MAXIMUM CONCENTRATION 82%
ACETYL CYCLOHEXANEPERSULFONATE
ACETYL CYCLOHEXANESULFONYL PEROXIDE
ACETYL CYCLOHEXANESULFONYL PEROXIDE (DOT)
ACETYL CYCLOHEXANESULFONYL PEROXIDE, ≤82%, WETTED WITH ≥12% WATER
ACETYL CYCLOHEXANESULFONYL PEROXIDE, NOT MORE THAN 32% IN SOLUTION (DOT)
ACETYL CYCLOHEXYLSULFONYL PEROXIDE
LUPERSOL 228Z
PEROXIDE, ACETYL CYCLOHEXYLSULFONYL
PEROXIDE, ACETYL CYCLOHEXYLSULFONYL, NOT MORE THAN 32% IN SOLUTION
PEROXIDE, ACETYL CYCLOHEXYLSULFONYL, NOT MORE THAN 82% WETTED WITH NOT LESS THAN 12% WATER
UN 2082 (DOT)
UN 2083 (DOT)

3188-13-4

CHLOROMETHOXYETHANE
CHLOROMETHYL ETHYL ETHER
CHLOROMETHYL ETHYL ETHER (DOT)
ETHANE, (CHLOROMETHOXY)-
ETHANE, (CHLOROMETHOXY)- (9CI)
ETHER, CHLOROMETHYL ETHYL
ETHOXYCHLOROMETHANE

CAS No.	IN	Chemical Name
		ETHOXYMETHYL CHLORIDE
		ETHYL CHLOROMETHYL ETHER
		UN 2354 (DOT)
3221-61-2		
		2-METHYLOCTANE
		OCTANE, 2-METHYL-
3227-63-2		
		ACETIC ACID, ZIRCONIUM SALT
		ACETIC ACID, ZIRCONIUM(2+) SALT
		BIS(ACETATO)OXOZIRCONIUM
		ZIRCONIUM ACETATE
		ZIRCONIUM ACETATE (WATERPROOFING AGENT)
		ZIRCONIUM DIACETATE
3248-28-0		
		DIPROPIONYL PEROXIDE
		DIPROPIONYL PEROXIDE (DOT)
		PEROXIDE, BIS(1-OXOPROPYL)
		PROPIONYL PEROXIDE
		PROPIONYL PEROXIDE (DOT)
		PROPIONYL PEROXIDE, NOT MORE THAN 28% IN SOLUTION
		UN 2132 (DOT)
3251-23-8		
		CLAYCOP
		COPPER (II) NITRATE
		COPPER DINITRATE
		COPPER NITRATE (Cu(NO3)2)
		COPPER(2+) NITRATE
		COPPER(II) NITRATE
		CUPRIC DINITRATE
	*	CUPRIC NITRATE
		CUPRIC NITRATE (DOT)
		CUPRIC NITRATE SOLUTION
		NA 1479 (DOT)
		NITRIC ACID, COPPER(2+) SALT
3254-63-5		
		4-METHYLTHIOPHENYLDIMETHYL PHOSPHATE
		ALLIED GC 6506
		DIMETHYL P-(METHYLTHIO)PHENYL PHOSPHATE
		ENT 25734
		GC 6506
		HA-1200
		O,O-DIMETHYL O-(4-METHYLMERCAPTOPHENYL) PHOSPHATE
		PHENOL, P-(METHYLTHIO)-, DIMETHYL PHOSPHATE
		PHOSPHORIC ACID, DIMETHYL 4-(METHYLTHIO)PHENYL ESTER
		PHOSPHORIC ACID, DIMETHYL P-(METHYLTHIO)PHENYL ESTER
3268-49-3		
		3-(METHYLMERCAPTO)PROPIONALDEHYDE
		3-(METHYLTHIO)PROPANAL
		3-(METHYLTHIO)PROPIONALDEHYDE
		4-THIAPENTANAL
		4-THIAPENTANAL (DOT)
		BETA-(METHYLMERCAPTO)PROPIONALDEHYDE
		BETA-(METHYLTHIO)PROPIONALDEHYDE
		METHIONAL
		PROPANAL, 3-(METHYLTHIO)-
		PROPANAL, 3-(METHYLTHIO)- (9CI)
		PROPIONALDEHYDE, 3-(METHYLTHIO)-
		THIAPENTANAL
		UN 2785 (DOT)
3275-73-8		
		NICOTINE TARTRATE
3282-30-2		
		2,2-DIMETHYLPROPANOYL CHLORIDE
		2,2-DIMETHYLPROPIONIC ACID CHLORIDE
		2,2-DIMETHYLPROPIONYL CHLORIDE
		ACETYL CHLORIDE, TRIMETHYL-
		NEOPENTANOYL CHLORIDE
		PIVALIC ACID CHLORIDE
		PIVALOLYL CHLORIDE
		PIVALOYL CHLORIDE
		PIVALOYL CHLORIDE (8CI)
		PIVALOYL CHLORIDE (DOT)
		PIVALYL CHLORIDE
		PROPANOYL CHLORIDE, 2,2-DIMETHYL-
		PROPANOYL CHLORIDE, 2,2-DIMETHYL- (9CI)
		T-BUTYLCARBONYL CHLORIDE
		TRIMETHYL ACETYL CHLORIDE (DOT)
		TRIMETHYLACETYL CHLORIDE
		UN 2438 (DOT)
3309-68-0		
		OO-DIETHYL S-PROPYLTHIOMETHYL PHOSPHORODITHIOATE
3312-60-5		
		1,3-PROPANEDIAMINE, N-CYCLOHEXYL-
		1-(CYCLOHEXYLAMINO)-3-AMINOPROPANE
		3-(CYCLOHEXYLAMINO)-1-PROPYLAMINE
		3-AMINO-1-(CYCLOHEXYLAMINO)PROPANE
		CYCLOHEXYL-1,3-PROPANEDIAMINE
		N-(3-AMINOPROPYL)CYCLOHEXYLAMINE
		N-CYCLOHEXYL-1,3-DIAMINOPROPANE
		N-CYCLOHEXYL-1,3-PROPANEDIAMINE
		N-CYCLOHEXYL-1,3-PROPYLENEDIAMINE
		N-CYCLOHEXYLTRIMETHYLENEDIAMINE
3313-92-6		
		CARBONIC ACID DISODIUM SALT, COMPD WITH HYDROGEN PEROXIDE (H2O2) (2:3)
		CARBONOPEROXOIC ACID, DISODIUM SALT
		CARBONOPEROXOIC ACID, DISODIUM SALT (9CI)
		DISODIUM CARBONATE COMPOUND WITH HYDROGEN PEROXIDE (Na2CO3CNTDOT15 H2O2)
		DISODIUM PEROXYDICARBONATE
		FB SODIUM PERCARBONATE
		HYDROGEN PEROXIDE (H2O2), COMPD WITH DISODIUM CARBONATE (3:2)
		OXYPER
		PERDOX
		PEROXY SODIUM CARBONATE
		PEROXYDICARBONIC ACID, DISODIUM SALT
		SODIUM CARBONATE
		SODIUM CARBONATE (Na2C2O6)
		SODIUM CARBONATE PEROXIDE
		SODIUM CARBONATE-HYDROGEN PEROXIDE (2:3)
		SODIUM CARBONATE-HYDROGEN PEROXIDE ADDUCT
		SODIUM PERCARBONATE
		SODIUM PERCARBONATE (DOT)
		SODIUM PERCARBONATE (Na2C2O6)
		SODIUM PERCARBONATE (PEROXYDICARBONIC ACID, DISODIUM SALT)
		SODIUM PEROXYCARBONATE
		SODIUM PEROXYCARBONATE (Na2CO4)
		SODIUM PEROXYDICARBONATE
		SODIUM PEROXYDICARBONATE (Na2C2O6)
		SODIUM PEROXYMONOCARBONATE
		UN 2467 (DOT)
3333-52-6		
		2,3-DICYANO-2,3-DIMETHYLBUTANE
		BUTANEDINITRILE, TETRAMETHYL-
		SUCCINONITRILE, TETRAMETHYL-
		TETRAMETHYL SUCCINONITRILE
		TETRAMETHYLSUCCINIC ACID DINITRILE
		TETRAMETHYLSUCCINODINITRILE
		TETRAMETHYLSUCCINONITRILE
3333-67-3		
		CARBONIC ACID, NICKEL(2+) SALT (1:1)
		NICKEL CARBONATE
		NICKEL CARBONATE (1:1)
		NICKEL CARBONATE (NiCO3)
		NICKEL(2+) CARBONATE
		NICKEL(2+) CARBONATE (NiCO3)
		NICKEL(II) CARBONATE
		NICKELOUS CARBONATE

CAS No. *IN* *Chemical Name*

3349-06-2
FORMIC ACID, NICKEL(2+) SALT
NICKEL DIFORMATE
NICKEL FORMATE
NICKEL(2+) FORMATE
NICKEL(II) FORMATE
NICKELOUS FORMATE

3383-96-8
ABATE
ABATHION
AC 52160
BIOTHION
DIFENPHOS
DIFOS
DIPHOS
DIPHOS (PESTICIDE)
ECOPRO 1707
EI 52160
EXPERIMENTAL INSECTICIDE 52,160
NEPHIS
NEPHIS 1G
O,O,O',O'-TETRAMETHYL O,O'-THIODI-P-PHENYLENE PHOSPHOROTHIOATE
PHOSPHOROTHIOIC ACID, O,O'-(THIODI-4,1-PHENYLENE) O,O,O',O'-TETRAMETHYL ESTER
PHOSPHOROTHIOIC ACID, O,O'-(THIODI-P-PHENYLENE) O,O,O',O'-TETRAMETHYL ESTER
PROCIDA
TEMEFOS
TEMEPHOS

3437-84-1
DIISOBUTYRYL PEROXIDE
DIISOBUTYRYL PEROXIDE (DOT)
DIISOBUTYRYL PEROXIDE, NOT MORE THAN 52% IN SOLUTION
ISOBUTYROYL PEROXIDE
ISOBUTYRYL PEROXIDE
ISOBUTYRYL PEROXIDE, MAXIMUM CONCENTRATION 52% IN SOLUTION (DOT)
ISOBUTYRYL PEROXIDE, NOT MORE THAN 52% IN SOLUTION
PEROXIDE, BIS(2-METHYL-1-OXOPROPYL)-
UN 2182 (DOT)

3452-97-9
1-HEXANOL, 3,5,5-TRIMETHYL-
3,5,5-TRIMETHYL-1-HEXANOL
3,5,5-TRIMETHYLHEXANOL
3,5,5-TRIMETHYLHEXYL ALCOHOL
I-NONYL ALCOHOL
NONYLOL

3457-61-2
7-(TERT-BUTYLPEROXY)CUMENE
BCP
CUMYL TERT-BUTYL PEROXIDE
KAYABUTYL C
LUPEROX 801
LUPERSOL 801
PERBUTYL C
PEROXIDE, 1,1-DIMETHYLETHYL (1-METHYLETHYL)-PHENYL
PEROXIDE, 1,1-DIMETHYLETHYL 1-METHYL-1-PHENYL-ETHYL
PEROXIDE, 1,1-DIMETHYLETHYL 1-METHYL-1-PHENYL-ETHYL (9CI)
PEROXIDE, TERT-BUTYL ALPHA,ALPHA-DIMETHYLBENZYL
TERT-BUTYL 1-METHYL-1-PHENYLETHYL PEROXIDE
TERT-BUTYL CUMENE PEROXIDE
TERT-BUTYL CUMENE PEROXIDE, TECHNICAL PURE (DOT)
TERT-BUTYL CUMYL PEROXIDE
TERT-BUTYL CUMYL PEROXIDE, TECHNICAL PURE (DOT)
TERT-BUTYL CUMYL PEROXIDE, TECHNICALLY PURE
TRIGONOX T
TRIGONOX T 40
UN 2091 (DOT)

CAS No. *IN* *Chemical Name*

3458-22-8
IPD

3476-90-2
C.I. DIRECT BROWN 59, DISODIUM SALT

3479-86-5
DIBROMOBUTANONE

3486-35-9
C.I. 77950
CARBONIC ACID, ZINC SALT (1:1)
CI 77950
NA 9157 (DOT)
* ZINC CARBONATE
ZINC CARBONATE (1:1)
ZINC CARBONATE (DOT)
ZINC CARBONATE ($ZnCO_3$)
ZINC MONOCARBONATE

3522-94-9
2,2,5-TRIMETHYLHEXANE
HEXANE, 2,2,5-TRIMETHYL-

3530-19-6
C.I. DIRECT RED 37

3546-10-9
PHENESTERIN

3564-09-8
2,7-NAPHTHALENEDISULFONIC ACID, 3-HYDROXY-4-[(2,4,5-TRIMETHYLPHENYL)AZO]-, DISODIUM SALT
C.I. 16155
D AND C RED 15
DOLKWAL PONCEAU 3R
EXT D AND C RED NO. 15
FD AND C RED NO. 1
FDC RED 1
MAPLE PONCEAU 3R
PONCEAU 3R
PONCEAU 3R LAKE
PONCEAU 3R SODIUM SALT
PONCEAU 3RN
SCARLET F
SODIUM CUMENEAZO-BETA-NAPHTHOL DISULFONATE
USACERT RED NO. 1

3565-26-2
4-NITROQUINOLINE-1-OXIDE

3567-65-5
C.I. ACID RED 85

3569-57-1
SULFOXIDE, 3-CHLOROPROPYL OCTYL

3570-75-0
2(2-FORMYLHYDRAZINO)4-(5-NITRO-2-FURYL)THIAZOLE
AS 17665
FNT
FORMIC ACID, 2-[4-(5-NITRO-2-FURYL)-2-THIAZOLYL]HYDRAZIDE
HYDRAZINECARBOXALDEHYDE, 2-[4-(5-NITRO-2-FURANYL)-2-THIAZOLYL]-
NIFURTHIAZOL
NIFURTHIAZOLE

3626-28-6
C.I. DIRECT GREEN 1

3648-21-3
1,2-BENZENEDICARBOXYLIC ACID, DIHEPTYL ESTER
DI-N-HEPTYL PHTHALATE
DIHEPTYL PHTHALATE
HEPTYL PHTHALATE
PHTHALIC ACID, DIHEPTYL ESTER

CAS No.	IN	Chemical Name
3688-53-7		2-FURANACETAMIDE, ALPHA-[(5-NITRO-2-FURANYL)-METHYLENE]-
		2-FURANACRYLAMIDE, ALPHA-2-FURYL-5-NITRO-
		AF 2
		AF 2 (PRESERVATIVE)
		FURYLAMIDE
		FURYLFURAMIDE
3689-24-5		ASP 47
		BAY-E-393
		BAYER-E 393
		BIS-O,O-DIETHYLPHOSPHOROTHIONIC ANHYDRIDE
		BLADAFUM
		BLADAFUME
		DI(THIOPHOSPHORIC) ACID, TETRAETHYL ESTER
		DITHIODIPHOSPHORIC ACID, TETRAETHYL ESTER
		DITHIOFOS
		DITHION
		DITHIONE
		DITHIOPHOS
		DITHIOPYROPHOSPHATE DE TETRAETHYLE (FRENCH)
		DITHIOTEP
		E393
		ENT 16,273
		ETHYL THIOPYROPHOSPHATE
		ETHYL THIOPYROPHOSPHATE ([(ETO)2PS]2O)
		LETHALAIRE G-57
		O,O,O,O-TETRAAETHYL-DITHIONOPYROPHOSPHAT (GERMAN)
		O,O,O,O-TETRAETHYL DITHIOPYROPHOSPHATE
		O,O,O,O-TETRAETHYL-DITHIO-DIFOSFAAT (DUTCH)
		O,O,O,O-TETRAETIL-DITIO-PIROFOSFATO (ITALIAN)
		PIROFOS
		PLANT DITHIO AEROSOL
		PLANTFUME 103 SMOKE GENERATOR
		PYROPHOSPHORIC ACID, TETRAETHYLDITHIO-, (MIXTURE)
		PYROPHOSPHORODITHIOIC ACID, O,O,O,O-TETRAETHYL ESTER
		PYROPHOSPHORODITHIOIC ACID, TETRAETHYL ESTER
		RCRA WASTE NUMBER P109
		SULFATEP
		SULFOTEP
		SULFOTEP (ACGIH)
		SULFOTEPP
		TEDP
		TEDTP
		TETRAETHYL DITHIONOPYROPHOSPHATE
		TETRAETHYL DITHIOPYROPHOSPHATE
		TETRAETHYL DITHIOPYROPHOSPHATE MIXTURE, DRY
		TETRAETHYL DITHIOPYROPHOSPHATE MIXTURE, DRY (DOT)
		TETRAETHYL DITHIOPYROPHOSPHATE MIXTURE, LIQUID
		TETRAETHYL DITHIOPYROPHOSPHATE MIXTURE, LIQUID (DOT)
		TETRAETHYL DITHIOPYROPHOSPHATE, LIQUID
		THIODIPHOSPHORIC ACID ([(HO)2P(S)]2O), TETRAETHYL ESTER
		THIODIPHOSPHORIC ACID TETRAETHYL ESTER
		THIOPYROPHOSPHORIC ACID ([(HO)2PS]2O), TETRAETHYL ESTER
		THIOPYROPHOSPHORIC ACID, TETRAETHYL ESTER
		THIOTEPP
		UN 1704 (DOT)
3691-35-8		((4-CHLORPHENYL)-1-PHENYL)-ACETYL-1,3-INDANDION (GERMAN)
		1,3-INDANDIONE, 2-((P-CHLOROPHENYL)PHENYLACETYL)-
		1-(4-CHLORPHENYL)-1-PHENYL-ACETYL-INDAN-1,3-DION (GERMAN)
		1H-INDENE-1,3(2H)-DIONE, 2-[(4-CHLOROPHENYL)PHENYL-ACETYL]-
		2(2-(4-CHLOOR-FENYL-2-FENYL)-ACETYL)-INDAAN-1,3-DION (DUTCH)
		2(2-(4-CHLOR-PHENYL-2-PHENYL)ACETYL)INDAN-1,3-DION (GERMAN)
		2(2-(4-CHLOROPHENYL)-2-PHENYLACETYL)INDAN-1,3-DIONE
		2(2-(4-CLORO-FENIL-2FENIL)-ACETIL)INDAN-1,3-DIONE (ITALIAN)
		2-((4-CHLOROPHENYL)PHENYLACETYL)-1H-INDENE-1,3(2H)-DIONE
		2-((P-CHLOROPHENYL)PHENYLACETYL)-1,3-INDANDIONE
		2-(2-PHENYL-2-(4-CHLOROPHENYL)ACETYL)-1,3-INDANDIONE
		2-(ALPHA-P-CHLOROPHENYLACETYL)INDANE-1,3-DIONE
		AFNOR
		CAID
		CHLOORFACINON (DUTCH)
		CHLORFACINON (GERMAN)
		CHLORODIPHACINONE
		CHLOROPHACINON
		CHLOROPHACINONE
		CHLOROPHACINONE (ROZOL)
		CHLORPHACINON (ITALIAN)
		CHLORPHACINONE
		CHLORPHENACONE
		DELTA
		DRAT
		LIPHADIONE
		LM 91
		MICROZUL
		MURIOL
		QUICK
		RAMUCIDE
		RANAC
		RATICIDE
		RATICIDE-CAID
		RATINDAN 3
		RATOMET
		RAVIAC
		REDENTIN
		ROZOL
		TOPITOX
3697-24-3		5-METHYLCHRYSENE
		CHRYSENE, 5-METHYL-
3710-84-7		DEHA
		DIETHYLHYDROXYLAMINE
		ETHANAMINE, N-ETHYL-N-HYDROXY-
		HYDROXYLAMINE, N,N-DIETHYL-
		N,N-DIETHYLHYDROXYAMINE
		N,N-DIETHYLHYDROXYLAMINE
		N-HYDROXYDIETHYLAMINE
3724-65-0		2-BUTENOIC ACID
		2-BUTENOIC ACID (9CI)
		3-METHYLACRYLIC ACID
		ACRYLIC ACID, 3-METHYL-
		ALPHA-BUTENOIC ACID
		ALPHA-CROTONIC ACID
		BETA-METHYLACRYLIC ACID
		CROTONIC ACID
		CROTONIC ACID (DOT)
		SOLID CROTONIC ACID
		UN 2823 (DOT)
3734-95-0		CYANTHOATE
3734-97-2		AMITON OXALATE
		CHIPMAN 6199
		CHIPMAN R-6,199
		CITRAM
		ENT 20,993
		HYDROGEN OXALATE OF AMITON
		O,O-DIETHYL S-(2-DIETHYLAMINO)ETHYL PHOSPHOROTHIOATE HYDROGEN OXALATE
		O,O-DIETHYL S-(2-ETHYL-N,N-DIETHYLAMINO)PHOSPHOROTHIOATE HYDROGEN OXALATE

CAS No.	IN	Chemical Name
		O,O-DIETHYL S-(BETA-DIETHYLAMINO)ETHYL PHOSPHOROTHIOLATE HYDROGEN OXALATE
		PHOSPHOROTHIOIC ACID, S-(2-DIETHYLAMINO)ETHYL) O,O-DIETHYL ESTER, OXALATE (1:1)
		PHOSPHOROTHIOIC ACID, S-[2-(DIETHYLAMINO)ETHYL] O,O-DIETHYL ESTER, ETHANEDIOATE (1:1)
		PHOSPHOROTHIOIC ACID, S-[2-(DIETHYLAMINO)ETHYL] O,O-DIETHYL ESTER, HYDROGEN OXALATE
		PHOSPHOROTHIOIC ACID, S-[2-(DIETHYLAMINO)ETHYL] O,O-DIETHYL ESTER, OXALATE
		PHOSPHOROTHIOIC ACID, S-[2-(DIETHYLAMINO)ETHYL] O,O-DIETHYL ESTER, OXALATE (1:1)
		S-(2-DIETHYLAMINOETHYL) O,O-DIETHYL PHOSPHOROTHIOATE HYDROGEN OXALATE
		S-(2-DIETHYLAMINOETHYL) O,O-DIETHYLPHOSPHOROTHIOATE HYDROGENOXALATE
		TETRAM
		TETRAM 75
		TETRAM MONOOXALATE
		TETRAM, ACID OXALATE
3735-23-7		
		ENT 25,554-X
		G 30494
		GEIGY 30494
		GEIGY G-30494
		METHANETHIOL, ((2,5-DICHLOROPHENYLTHIO)-, S-ESTER WITH O,O-DIMETHYL PHOSPHORODITHIOATE
		METHYL PHENCAPTON
		METHYL PHENKAPTON
		O,O-DIMETHYL S-(2,5-DICHLOROPHENYLTHIO)METHYL PHOSPHORODITHIOATE
		PHOSPHORODITHIOIC ACID, S-(((2,5-DICHLOROPHENYL)-THIO)METHYL)O,O-DIMETHYL ESTER
		S-(((2,5-DICHLOROPHENYL)THIO)METHYL) O,O-DIMETHYL PHOSPHORODITHIOATE
3761-53-3		
		1695 RED
		2,7-NAPHTHALENEDISULFONIC ACID, 4-[(2,4-DIMETHYLPHENYL)AZO]-3-HYDROXY-, DISODIUM SALT
		ACID LEATHER RED KPR
		ACID LEATHER RED P 2R
		ACID LEATHER SCARLET IRW
		ACID PONCEAU 2RL
		ACID PONCEAU R
		ACID PONCEAU SPECIAL
		ACID RED 26
		ACID SCARLET
		ACID SCARLET 2B
		ACID SCARLET 2R
		ACID SCARLET 2RL
		ACIDAL PONCEAU G
		AHCOCID FAST SCARLET R
		AIZEN PONCEAU RH
		AMACID LAKE SCARLET 2R
		C.I. 16150
		C.I. ACID RED 26
		C.I. ACID RED 26, DISODIUM SALT
		C.I. FOOD RED 5
		CALCOCID SCARLET 2R
		CALCOLAKE SCARLET 2R
		CERTICOL PONCEAU MXS
		COLACID PONCEAU SPECIAL
		D AND C RED NO. 5
		EDICOL SUPRA PONCEAU R
		FENAZO SCARLET 2R
		FOOD RED 5
		HEXACOL PONCEAU 2R
		HEXACOL PONCEAU MX
		HIDACID SCARLET 2R
		KITON PONCEAU 2R
		KITON PONCEAU R
		KITON SCARLET 2RC
		LAKE SCARLET 2RBN
		LAKE SCARLET R
		NAPHTHALENE LAKE SCARLET R
		NAPHTHALENE SCARLET R
		NAPHTHAZINE SCARLET 2R
		NEKLACID RED RR
		NEW PONCEAU 4R
		PAPER RED HRR
		PIGMENT PONCEAU R
		PONCEAU 2R
		PONCEAU 2R EXTRA A EXPORT
		PONCEAU 2RL
		PONCEAU 2RX
		PONCEAU BNA
		PONCEAU G
		PONCEAU MX
		PONCEAU PXM
		PONCEAU R
		PONCEAU RED R
		PONCEAU RR
		PONCEAU RR TYPE 8019
		PONCEAU RS
		PONCEAU XYLIDINE
		SCARLET 2R
		SCARLET 2RB
		SCARLET 2RL BLUISH
		SCARLET R
		SCARLET RRA
		TERTRACID PONCEAU 2R
		XYLIDINE PONCEAU
		XYLIDINE PONCEAU 2R
		XYLIDINE RED
3771-19-5		
		C 13437SU
		CH 13-437
		CIBA 13437SU
		MELIPAN
		NAFENOPIN
		PROPANOIC ACID, 2-METHYL-2-[4-(1,2,3,4-TETRAHYDRO-1-NAPHTHALENYL)PHENOXY]-
		PROPIONIC ACID, 2-METHYL-2-[P-(1,2,3,4-TETRAHYDRO-1-NAPHTHYL)PHENOXY]-
		SU 13437
		TPIA
3775-90-4		
		2-(TERT-BUTYLAMINO)ETHYL METHACRYLATE
		2-PROPENOIC ACID, 2-METHYL-, 2-[(1,1-DIMETHYLETHYL)-AMINO]ETHYL ESTER
		ETHANOL, 2-(TERT-BUTYLAMINO)-, METHACRYLATE (ESTER)
		METHACRYLIC ACID, 2-(TERT-BUTYLAMINO)ETHYL ESTER
		N-TERT-BUTYLAMINOETHYL METHACRYLATE
		TERT-BUTYLAMINOETHYL METHACRYLATE
3778-73-2		
		ISOPHOSPHAMIDE
3803-51-2		
		PHOSPHINE
3811-04-9		
		BERTHOLLET'S SALT
		CHLORATE DE POTASSIUM (FRENCH)
		CHLORATE OF POTASH
		CHLORATE OF POTASH (DOT)
		CHLORIC ACID, POTASSIUM SALT
		FEKABIT
		KALIUMCHLORAAT (DUTCH)
		KALIUMCHLORAT (GERMAN)
		OXYMURIATE OF POTASH
		PEARL ASH
		POTASH CHLORATE (DOT)
		POTASSIO (CHLORATO DI) (ITALIAN)
		POTASSIUM (CHLORATE DE) (FRENCH)
	*	POTASSIUM CHLORATE
		POTASSIUM CHLORATE (DOT)
		POTASSIUM CHLORATE SOLUTION
		POTASSIUM CHLORATE, AQUEOUS SOLUTION (DOT)
		POTASSIUM OXYMURIATE
		POTCRATE
		SALT OF TARTER
		UN 1485 (DOT)

CAS No.	IN	Chemical Name
		UN 2427 (DOT)
3811-71-0		
		C.I. DIRECT BROWN 1
3813-14-7		
		2,4,5-T TRIETHANOLAMINE SALT
		2,4,5-TRICHLOROPHENOXYACETIC ACID TRIETHANOLAMINE SALT
		ACETIC ACID, (2,4,5-TRICHLOROPHENOXY)-, COMPD. WITH 2,2',2"-NITRILOTRIETHANOL (1:1)
		ACETIC ACID, (2,4,5-TRICHLOROPHENOXY)-, COMPD. WITH 2,2',2"-NITRILOTRIS[ETHANOL] (1:1)
		ETHANOL, 2,2',2"-NITRILOTRI-, (2,4,5-TRICHLOROPHENOXY)ACETATE (SALT)
		ETHANOL, 2,2',2"-NITRILOTRIS-, (2,4,5-TRICHLOROPHENOXY)ACETATE (1:1) (SALT)
3844-45-9		
		C.I. ACID BLUE 9, DISODIUM SALT
3878-19-1		
		1H-BENZIMIDAZOLE, 2-(2-FURANYL)-
		2-(2'-FURYL)BENZIMIDAZOLE
		2-(2-FURYL)BENZIMIDAZOLE
		B-33172
		BAY 33172
		BAYER 33172
		BENZIMIDAZOLE, 2-(2-FURYL)-
		FUBERIDATOL
		FUBERIDAZOL
		FUBERIDAZOLE
		FUBERISAZOL
		FUBRIDAZOLE
		FURIDAZOL
		FURIDAZOLE
		PF 7402
		VORONIT
		VORONITE
		W VII/117
3882-06-2		
		DICYCLOHEXYLAMMONIUM NITRATE
3917-15-5		
		1-PROPENE, 3-(ETHENYLOXY)-
		ALLYL VINYL ETHER
		ETHER, ALLYL VINYL
		VINYL ALLYL ETHER
3926-62-3		
		ACETIC ACID, CHLORO-, SODIUM SALT
		CHLOROACETIC ACID SODIUM SALT
		CHLOROCTAN SODNY (CZECH)
		DOW DEFOLIANT
		MONOXONE
		SMA
		SMA (HERBICIDE)
		SMCA
		SODIUM CHLOROACETATE
		SODIUM CHLOROACETATE (DOT)
		SODIUM MONOCHLORACETATE
		SODIUM MONOCHLOROACETATE
		UN 2659 (DOT)
3953-10-4		
		2-ETHYLBUTYL ACRYLATE
		2-PROPENOIC ACID, 2-ETHYLBUTYL ESTER
		ACRYLIC ACID, 2-ETHYLBUTYL ESTER
3982-91-0		
		PHOSPHOROTHIOIC TRICHLORIDE
		PHOSPHOROTHIONIC TRICHLORIDE
		PHOSPHOROUS SULFOCHLORIDE
		PHOSPHOROUS THIOCHLORIDE
		PHOSPHOROUS TRICHLORIDE SULFIDE
		PHOSPHORUS SULFOCHLORIDE
		PHOSPHORUS THIOCHLORIDE
		PHOSPHORUS THIOTRICHLORIDE

CAS No.	IN	Chemical Name
		PHOSPHORUS TRICHLORIDE SULFIDE
		THIOPHOSPHORIC TRICHLORIDE
		THIOPHOSPHORYL CHLORIDE
		THIOPHOSPHORYL CHLORIDE (DOT)
		THIOPHOSPHORYL CHLORIDE (PSCl3)
		THIOPHOSPHORYL TRICHLORIDE
		TL 262
		TRICHLOROPHOSPHINE SULFIDE
		UN 1837 (DOT)
4016-11-9		
		(ETHOXYMETHYL)OXIRANE
		1,2-EPOXY-3-ETHOXYPROPANE
		1,2-EPOXY-3-ETHYLOXY PROPANE (DOT)
		1,2-EPOXY-3-ETHYLOXYPROPANE
		3-ETHOXY-1,2-EPOXYPROPANE
		EPOXY ETHYLOXY PROPANE
		ETHYL GLYCIDYL ETHER
		OXIRANE, (ETHOXYMETHYL)-
		OXIRANE, (ETHOXYMETHYL)- (9CI)
		PROPANE, 1,2-EPOXY-3-ETHOXY-
		UN 2752 (DOT)
4016-14-2		
		(ISOPROPOXYMETHYL)OXIRANE
		1,2-EPOXY-3-ISOPROPOXYPROPANE
		3-ISOPROPOXY-1,2-EPOXYPROPANE
		3-ISOPROPYLOXYPROPYLENE OXIDE
		GLYCIDYL ISOPROPYL ETHER
		ISOPROPYL GLYCIDYL ETHER
		ISOPROPYL GLYCIDYL ETHER (IGE)
		OXIRANE, [(1-METHYLETHOXY)METHYL]-
		PROPANE, 1,2-EPOXY-3-ISOPROPOXY-
4032-86-4		
		3,3-DIMETHYLHEPTANE
		HEPTANE, 3,3-DIMETHYL-
4044-65-9		
		BITOSCANATE
4098-71-9		
		1-ISOCYANATO-3,3,5-TRIMETHYL-5-ISOCYANATOMETHYL-CYCLOHEXANE
		1-ISOCYANATO-3-ISOCYANATOMETHYL-3,5,5-TRIMETHYL-CYCLOHEXANE
		1-ISOCYANATO-5-(ISOCYANATOMETHYL)-3,3,5-TRIMETHYL-CYCLOHEXANE
		3,3,5-TRIMETHYL-5-(ISOCYANATOMETHYL)CYCLOHEXYL ISOCYANATE
		3-ISOCYANATOMETHYL-3,5,5-TRIMETHYLCYCLOHEXYLISOCYANATE
		5-ISOCYANATO-1-ISOCYANATOMETHYL-1,3,3-TRIMETHYL-CYCLOHEXANE
		CYCLOHEXANE, 5-ISOCYANATO-1-(ISOCYANATOMETHYL)-1,3,3-TRIMETHYL-
		CYCLOHEXANE, 5-ISOCYANATO-1-(ISOCYANATOMETHYL)-1,3,3-TRIMETHYL- (9CI)
		IPDI
		ISOCYANIC ACID, METHYLENE(3,5,5-TRIMETHYL-3,1-CYCLOHEXYLENE) ESTER
		ISOPHORONE DIAMINE DIISOCYANATE
		ISOPHORONE DIISOCYANATE
		ISOPHORONE DIISOCYANATE (ACGIH,DOT)
		UN 2290 (DOT)
4104-14-7		
		BAY 38819
		BAYER 38819
		DRC 714
		GOPHACIDE
		O,O-BIS(P-CHLOROPHENYL) ACETIMIDOYLAMIDOTHIOPHOSPHATE
		O,O-BIS(P-CHLOROPHENYL) ACETIMIDOYLPHOSPHORAMIDOTHIOATE
		PHOSACETIM
		PHOSAZETIM
		PHOSPHORAMIDOTHIOIC ACID, (1-IMINOETHYL)-, O,O-BIS(4-CHLOROPHENYL) ESTER

CAS No.	IN	Chemical Name
		PHOSPHORAMIDOTHIOIC ACID, ACETIMIDOYL-, O,O-BIS(P-CHLOROPHENYL) ESTER
4109-96-0		
		CHLOROSILANE (SiH2Cl2)
		DICHLOROSILANE
		DICHLOROSILANE (DOT)
		SILANE, DICHLORO-
		UN 2189 (DOT)
4170-30-3		
		1,2-ETHANEDIOL, DIPROPANOATE (9CI)
		2-BUTENAL
		2-BUTENAL (9CI)
		2-BUTENAL, (E)-
		ALDEHYDE CROTONIQUE (FRENCH)
		BETA-METHYL ACROLEIN
		BETA-METHYL ACROLEIN (DOT)
		CROTENALDEHYDE
		CROTONAL
		CROTONALDEHYDE
		CROTONALDEHYDE (ACGIH)
		CROTONALDEHYDE (DOT)
		CROTONALDEHYDE, (E)-
		CROTONALDEHYDE, INHIBITED (DOT)
		CROTONIC ALDEHYDE
		CROTYLALDEHYDE
		ETHYLENE DIPROPIONATE
		ETHYLENE GLYCOL, DIPROPIONATE (8CI)
		ETHYLENE PROPIONATE
		NCI-C56279
		PROPYLENE ALDEHYDE
		RCRA WASTE NUMBER U053
		TOPANEL
		TRANS-2-BUTENAL
		UN 1143 (DOT)
4229-34-9		
		ACETIC ACID, ZIRCONIUM SALT
		ACETIC ACID, ZIRCONIUM(4+) SALT
		BIS(ACETATO)OXOZIRCONIUM
		ZIRCONIUM ACETATE
		ZIRCONIUM ACETATE (WATERPROOFING AGENT)
		ZIRCONIUM ACETATE [Zr(OAC)4]
		ZIRCONIUM TETRAACETATE
4301-50-2		
		(1,1'-BIPHENYL)-4-ACETIC ACID, 2-FLUOROETHYL ESTER
		4-BIPHENYLACETIC ACID, 2-FLUOROETHYL ESTER
		ACETIC ACID, 4-BIPHENYLYL-, 2-FLUOROETHYL ESTER
		BETA-FLUOROETHYL 4-BIPHENYLACETATE
		ETHANOL, 2-FLUORO-, 4-BIPHENYLACETATE
		FLUENETHYL
		FLUENETIL
		FLUENYL
		LAMBROL
		M 2060
		MYTROL
		TH 367-1
4316-42-1		
		1H-IMIDAZOLE, 1-BUTYL- (9CI)
		1H-IMIDAZOLE, BUTYL-
		BUTYL IMIDAZOLE
		IMIDAZOLE, 1-BUTYL-
		N-N-BUTYL IMIDAZOLE (DOT)
		N-N-BUTYL-IMIDAZOLE
		UN 2690 (DOT)
4324-38-3		
		3-ETHOXYPROPANOIC ACID
		3-ETHOXYPROPIONIC ACID
		PROPANOIC ACID, 3-ETHOXY-
		PROPIONIC ACID, 3-ETHOXY-
4335-09-5		
		C.I. DIRECT GREEN 6, DISODIUM SALT
4342-03-4		
		(DIMETHYLTRIAZENO)IMIDAZOLECARBOXAMIDE
		1H-IMIDAZOLE-4-CARBOXAMIDE, 5-(3,3-DIMETHYL-1-TRIAZENYL)-
		4-(DIMETHYLTRIAZENO)IMIDAZOLE-5-CARBOXAMIDE
		5-(3,3-DIMETHYLTRIAZENO)IMIDAZOLE-4-CARBOXAMIDE
		5-(DIMETHYLTRIAZENO)IMIDAZOLE-4-CARBOXAMIDE
		BIOCARBAZINE R
		DACARBAZINE
		DTIC
		IMIDAZOLE-4(OR 5)-CARBOXAMIDE, 5(OR 4)-(3,3-DI-METHYL-1-TRIAZENO)-
		IMIDAZOLE-4-CARBOXAMIDE, 5-(3,3-DIMETHYL-1-TRIAZENO)-
		NSC 45388
4384-82-1		
		CARBAMIC ACID, DIETHYLDITHIO-, SODIUM SALT
		CARBAMIC ACID, DITHIO-, ION(1-)
		CARBAMODITHIOIC ACID, DIETHYL-, SODIUM SALT (9CI)
		CARBAMODITHIOIC ACID, ION(1-)
		CUPRAL
		DDC
		DEDC
		DEDK
		DIETHYL SODIUM DITHIOCARBAMATE
		DIETHYLCARBAMODITHIOIC ACID, SODIUM SALT
		DIETHYLDITHIOCARBAMATE SODIUM
		DIETHYLDITHIOCARBAMIC ACID SODIUM
		DIETHYLDITHIOCARBAMIC ACID, SODIUM SALT
		DITHIOCARB
		DITHIOCARBAMATE
		DITHIOCARBAMATE ANION
		DITHIOCARBAMATE PESTICIDE, LIQUID [FLAMMABLE LIQUID LABEL]
		DITHIOCARBAMATE PESTICIDE, LIQUID [POISON B LABEL]
		DITHIOCARBAMATE PESTICIDE, SOLID
		NCI CO2835
		SODIUM DEDT
		SODIUM DIETHYLDITHIOCARBAMATE
		SODIUM N,N-DIETHYLDITHIOCARBAMATE
		SODIUM SALT OF N,N-DIETHYLDITHIOCARBAMIC ACID
		THIOCARB
		USAF EK-2596
4418-66-0		
		2,2-THIOBIS(4-CHLORO-6-METHYL)PHENOL
		PHENOL, 2,2'-THIOBIS(4-CHLORO-6-METHYL)-
		PHENOL, 2,2-THIOBIS(4-CHLORO-6-METHYL)
4421-95-8		
		METHYL SILICATE
		SILICIC ACID (H8Si3O10), OCTAMETHYL ESTER
		TRISILOXANE, OCTAMETHOXY-
4429-42-9		
		DIPOTASSIUM DISULFITE
		DIPOTASSIUM METABISULFITE
		DIPOTASSIUM PYROSULFITE
		DISULFUROUS ACID, DIPOTASSIUM SALT
		NA 2693 (DOT)
		POTASSIUM METABISULFITE
		POTASSIUM METABISULFITE (DOT)
		POTASSIUM METABISULFITE (K2S2O5)
		POTASSIUM PYROSULFITE
		POTASSIUM PYROSULFITE (K2S2O5)
		PYROSULFUROUS ACID, DIPOTASSIUM SALT
4435-53-4		
		BUTOXYL
		METHYL 1,3-BUTYLENE GLYCOL ACETATE
4439-24-1		
		2-ISOBUTOXYETHANOL
		ETHANOL, 2-(2-METHYLPROPOXY)-
		ETHANOL, 2-ISOBUTOXY-
		ETHYLENE GLYCOL MONOISOBUTYL ETHER
		ISOBUTYL GLYCOL

CAS No.	IN	Chemical Name
4452-58-8		CARBONIC ACID DISODIUM SALT, COMPD WITH HYDROGEN PEROXIDE (H2O2) (2:3) CARBONOPEROXOIC ACID, DISODIUM SALT CARBONOPEROXOIC ACID, DISODIUM SALT (9CI) DISODIUM CARBONATE COMPOUND WITH HYDROGEN PEROXIDE (2:3) DISODIUM PEROXYDICARBONATE FB SODIUM PERCARBONATE HYDROGEN PEROXIDE, COMPD WITH DISODIUM CARBONATE (3:2) OXYPER PERDOX PEROXY SODIUM CARBONATE PEROXYDICARBONIC ACID, DISODIUM SALT SODIUM CARBONATE SODIUM CARBONATE (Na2C2O6) SODIUM CARBONATE PEROXIDE SODIUM CARBONATE-HYDROGEN PEROXIDE (2:3) SODIUM CARBONATE-HYDROGEN PEROXIDE ADDUCT SODIUM PERCARBONATE SODIUM PERCARBONATE (CARBONOPEROXOIC ACID, DISODIUM SALT) SODIUM PERCARBONATE (DOT) SODIUM PERCARBONATE (Na2C2O6) SODIUM PEROXYCARBONATE SODIUM PEROXYCARBONATE (Na2CO4) SODIUM PEROXYDICARBONATE SODIUM PEROXYMONOCARBONATE UN 2467 (DOT)
4461-41-0		2-BUTENE, 2-CHLORO- 2-CHLORO-2-BUTENE 2-CHLOROBUTENE-2 3-CHLORO-2-BUTENE
4461-48-7		1,1-DIMETHYL-2-BUTENE 2-METHYL-3-PENTENE 2-PENTENE, 4-METHYL-
	*	4-METHYL-2-PENTENE
4484-72-4		DODECYL TRICHLOROSILANE DODECYL TRICHLOROSILANE (DOT) SILANE, DODECYLTRICHLORO- SILANE, TRICHLORODODECYL- TRICHLORODODECYLSILANE UN 1771 (DOT)
4511-39-1		BENZENECARBOPEROXOIC ACID, 1,1-DIMETHYLPROPYL ESTER PEROXYBENZOIC ACID, TERT-PENTYL ESTER TERT-AMYL PERBENZOATE TERT-AMYL PEROXYBENZOATE
4521-94-2		ETHYL SILICATE OCTAETHOXY TRISILOXANE SILICIC ACID (H8Si3O10), OCTAETHYL ESTER TRISILOXANE, OCTAETHOXY-
4549-40-0		ETHENAMINE, N-METHYL-N-NITROSO- METHYLVINYLNITROSAMINE N-METHYL-N-NITROSOVINYLAMINE N-NITROSO-N-METHYLVINYLAMINE N-NITROSOMETHYLVINYLAMINE VINYLAMINE, N-METHYL-N-NITROSO-
4680-78-8		C.I. ACID GREEN 3
4685-14-7		1,1'-DIMETHYL-4,4'-BIPYRIDINIUM 1,1'-DIMETHYL-4,4'-BIPYRIDINIUM CATION 4,4'-BIPYRIDINIUM, 1,1'-DIMETHYL- DIMETHYL VIOLOGEN METHYL VIOLOGEN(2+) N,N'-DIMETHYL-4,4'-BIPYRIDINIUM N,N'-DIMETHYL-4,4'-BIPYRIDINIUM DICATION PARAQUAT PARAQUAT DICATION PARAQUAT ION
4726-14-1		2,6-DINITRO-4-METHYLSULFONYL-N,N-DIPROPYLANILINE 4-(METHYLSULFONYL)-2,6-DINITRO-N,N-DIPROPYLANILINE 4-(METHYLSULFONYL)-2,6-DINITRO-N,N-DIPROPYLBENZENEAMINE ANILINE, 2,6-DINITRO-N,N-DIPROPYL-4-(METHYLSULFONYL)- ANILINE, 4-(METHYLSULFONYL)-2,6-DINITRO-N,N-DIPROPYL- BENZENAMINE, 4-(METHYLSULFONYL)-2,6-DINITRO-N,N-DIPROPYL- BENZENAMINE, 4-(METHYLSULFONYL)-2,6-DINITRO-N,N-DIPROPYL- (9CI) NITRALIN NITRALINE PLANAVIN PLANAVIN 75 PLANUIN SD 11831
4732-14-3		2,4,6-TRINITROPHENETOLE TRINITROPHENETOLE
4784-77-4		1-BROMO-2-BUTENE 1-CROTYL BROMIDE 2-BUTENE, 1-BROMO- 2-BUTENYL BROMIDE CROTONYL BROMIDE
	*	CROTYL BROMIDE
4795-29-3		2-AMINOMETHYLTETRAHYDROFURAN 2-FURANMETHANAMINE, TETRAHYDRO- FURFURYLAMINE, TETRAHYDRO- TETRAHYDROFURFURYLAMINE TETRAHYDROFURFURYLAMINE (DOT) UN 2943 (DOT) USAF Q-2
4806-61-5		CYCLOBUTANE, ETHYL- ETHYL CYCLOBUTANE
4835-11-4		1,6-HEXANEDIAMINE, N,N'-DIBUTYL- 1,6-N,N'-DIBUTYLHEXANEDIAMINE DBHMD DIBUTYLHEXAMETHYLENEDIAMINE HEXAMETHYLENEDIAMINE, N,N'-DIBUTYL- N,N'-DIBUTYL-1,6-HEXANEDIAMINE N,N'-DIBUTYLHEXAMETHYLENEDIAMINE
4904-61-4		1,5,9-CYCLODODECATRIENE
4985-85-7		2,2'-[(3-AMINOPROPYL)IMINO]DIETHANOL 3-(AMINOPROPYL)DIETHANOLAMINE 3-[BIS(2-HYDROXYETHYL)AMINO]PROPYLAMINE AMINOPROPYLDIETHANOLAMINE AMINOPROPYLDIETHANOLAMINE (DOT) ETHANOL, 2,2'-(AMINOPROPYLIMINO)- ETHANOL, 2,2'-[(3-AMINOPROPYL)IMINO]BIS- ETHANOL, 2,2'-[(3-AMINOPROPYL)IMINO]DI- N,N-BIS(2-HYDROXYETHYL)-1,3-PROPANEDIAMINE N,N-BIS(HYDROXYETHYL)-1,3-PROPANEDIAMINE N,N-BIS(HYDROXYETHYL)TRIMETHYLENEDIAMINE N,N-DI(2-HYDROXYETHYL)-1,3-PROPANEDIAMINE N-(3-AMINOPROPYL)DIETHANOLAMINE

CAS No.	IN	Chemical Name
		NA 1760 (DOT)
5124-30-1		4,4'-DIISOCYANATODICYCLOHEXYLMETHANE
		4,4'-METHYLENEBIS(CYCLOHEXYL ISOCYANATE)
		BIS(4-ISOCYANATOCYCLOHEXYL)METHANE
		CYCLOHEXANE, 1,1'-METHYLENEBIS[4-ISOCYANATO-
		DICYCLOHEXYLMETHANE 4,4'-DIISOCYANATE
		ISOCYANIC ACID, METHYLENEDI-4,1-CYCLOHEXYLENE ESTER
		METHYLENE BIS(4-CYCLOHEXYLISOCYANATE)
		METHYLENEBIS(1,4-CYCLOHEXYLENE) DIISOCYANATE
		METHYLENEBIS(4-CYCLOHEXYL ISOCYANATE)
		METHYLENEBIS(4-ISOCYANATOCYCLOHEXANE)
		METHYLENEDI-1,4-CYCLOHEXYLENE ISOCYANATE
		METHYLENEDI-4-CYCLOHEXYLENE DIISOCYANATE
		NACCONATE H 12
5131-60-2		4-CHLORO-M-PHENYLENEDIAMINE
5153-24-2		ACETIC ACID, ZIRCONIUM SALT
		BIS(ACETATO)OXOZIRCONIUM
		ZIRCONIUM ACETATE
		ZIRCONIUM ACETATE (WATERPROOFING AGENT)
		ZIRCONIUM DIACETATE
		ZIRCONIUM, BIS(ACETATO)DIHYDROXY-
		ZIRCONIUM, BIS(ACETATO)OXO-
		ZIRCONIUM, BIS(ACETATO-O)DIHYDROXY-
		ZIRCONIUM, BIS(ACETATO-O)OXO-
		ZIRCONIUM, DIHYDROXYBIS(ACETATO)-
		ZIRCONYL DIACETATE
5281-13-0		PIPROTAL
5283-66-9		OCTYL TRICHLOROSILANE (DOT)
		OCTYLTRICHLOROSILANE
		P 09830
		SILANE, OCTYLTRICHLORO-
		SILANE, TRICHLOROOCTYL-
		TRICHLOROOCTYLSILANE
		UN 1801 (DOT)
5283-67-0		NONYL TRICHLOROSILANE (DOT)
		NONYLSILANE, TRICHLORO-
		NONYLTRICHLOROSILANE
		SILANE, TRICHLORONONYL-
		TRICHLORONONYLSILANE
		UN 1799 (DOT)
5307-14-2		2-NITRO-P-PHENYLENEDIAMINE
5309-52-4		2-ETHYL-2-HEXENOIC ACID
		2-ETHYL-3-PROPYLACRYLIC ACID
		2-HEXENOIC ACID, 2-ETHYL-
5329-14-6		AMIDOSULFONIC ACID
		AMIDOSULFURIC ACID
		AMINESULFONIC ACID
		AMINOSULFONIC ACID
		AMINOSULFURIC ACID
		JUMBO
		KYSELINA AMIDOSULFONOVA (CZECH)
		KYSELINA SULFAMINOVA (CZECH)
		SULFAMIC ACID
		SULFAMIDIC ACID
		SULFAMINIC ACID
	*	SULPHAMIC ACID
		SULPHAMIC ACID (DOT)
		UN 2967 (DOT)
5332-52-5		1-UNDECANETHIOL
5332-73-0		1-AMINO-3-METHOXYPROPANE
		1-PROPANAMINE, 3-METHOXY-
		3-AMINOPROPYL METHYL ETHER
		3-METHOXY-1-PROPANAMINE
		3-METHOXY-N-PROPYLAMINE
		3-METHOXYPROPYLAMINE
		GAMMA-METHOXYPROPYLAMINE
		PROPYLAMINE, 3-METHOXY-
5344-82-1		(O-CHLOROPHENYL)THIOUREA
		1-(2-CHLOROPHENYL)-2-THIOUREA
		1-(2-CHLOROPHENYL)THIOUREA
		1-(O-CHLOROPHENYL)THIOUREA
		2-CHLOROPHENYL THIOUREA
		N-(2-CHLOROPHENYL)THIOUREA
		RCRA WASTE NUMBER P026
		THIOUREA, (2-CHLOROPHENYL)-
		UREA, 1-(O-CHLOROPHENYL)-2-THIO-
5350-03-8		AMYL LAURATE
		DODECANOIC ACID, PENTYL ESTER
		LAURIC ACID, PENTYL ESTER
		N-AMYL LAURATE
		PENTYL LAURATE
5392-82-5		DICHLOROPHENYL-ISOCYANATE (1,4-DICHLORO-2-ISO-CYANATO-BENZENE)
5408-74-2		2-VINYL-5-ETHYLPYRIDINE
		5-ETHYL-2-VINYLPYRIDINE
		PYRIDINE, 2-ETHENYL-5-ETHYL-
		PYRIDINE, 5-ETHYL-2-VINYL-
5419-55-6		BORIC ACID, TRIISOPROPYL ESTER
		BORIC ACID, TRIS(1-METHYLETHYL) ESTER
		BORON ISOPROPOXIDE
		BORON TRIISOPROPOXIDE
		ISOPROPYL BORATE
		TRIISOPROPOXYBORANE
		TRIISOPROPOXYBORON
		TRIISOPROPYL BORATE
		TRIISOPROPYL BORATE (DOT)
		TRIISOPROPYL ORTHOBORATE
		TRISISOPROPOXYBORANE
		UN 2616 (DOT)
5422-17-3		C.I. DIRECT GREEN 8, TRISODIUM SALT
5432-61-1		CYCLOHEXANAMINE, N-(2-ETHYLHEXYL)-
		HEXYLAMINE, N-CYCLOHEXYL-2-ETHYL-
		N-(2-ETHYLHEXYL)CYCLOHEXYLAMINE
5459-93-8		ACCELERATOR HX
		CYCLOHEXANAMINE, N-ETHYL-
		CYCLOHEXYLAMINE, N-ETHYL-
		ETHYL-N-CYCLOHEXYLAMINE
		N-CYCLOHEXYLETHYLAMINE
		N-ETHYLCYCLOHEXANAMINE
		N-ETHYLCYCLOHEXYLAMINE
		VULKACIT HX
5517-17-5		METHYL-N-PROPYL ETHER

CAS No.	IN	Chemical Name
5593-70-4		
		1-BUTANOL, TITANIUM(4+) SALT
		B 1
		B 1 (TITANATE)
		BUTYL ALCOHOL, TITANIUM(4+) SALT
		BUTYL ORTHOTITANATE
		BUTYL TITANATE
		BUTYL TITANATE(IV)
		BUTYL TITANATE(IV) ((BUO)4Ti)
		ORGATICS TA 25
		TETRABUTOXYTITANIUM
		TETRABUTYL ORTHOTITANATE
		TETRABUTYL TITANATE
		TETRABUTYLTITANATE (CZECH)
		TITANIC ACID, TETRABUTYL ESTER
		TITANIUM BUTOXIDE (Ti(OBU)4)
		TITANIUM TETRABUTOXIDE
		TITANIUM TETRABUTYLATE
		TITANIUM(4+) BUTOXIDE
		TITANIUM, TETRABUTOXY-
		TYZOR TBT
5714-22-7		
		SULFUR PENTAFLUORIDE
5798-79-8		
		ACETIC ACID, BROMOPHENYL-, NITRILE
		ACETONITRILE, BROMOPHENYL-
		ALPHA-BROMO-ALPHA-TOLUNITRILE
		ALPHA-BROMOBENZENEACETONITRILE
		ALPHA-BROMOBENZYL CYANIDE
		ALPHA-BROMOBENZYLNITRILE
		ALPHA-BROMOPHENYLACETONITRILE
		BBC
		BBN
		BENZENEACETONITRILE, ALPHA-BROMO-
		BENZENEACETONITRILE, ALPHA-BROMO- (9CI)
		BROMBENZYL CYANIDE
		BROMOBENZYL CYANIDE
		BROMOBENZYL CYANIDE (DOT)
		BROMOBENZYLNITRILE
		CA
		CAMITE
		UN 1694 (DOT)
5809-08-5		
		1,1,3,3-TETRAMETHYLBUTYL HYDROPEROXIDE
		1,1,3,3-TETRAMETHYLBUTYL HYDROPEROXIDE, TECHNICALLY PURE
		1,1,3,3-TETRAMETHYLBUTYL HYDROPEROXIDE, TECHNICALLY PURE (DOT)
		2,4,4-TRIMETHYL-2-HYDROPEROXYPENTANE
		2,4,4-TRIMETHYL-2-PENTYL HYDROPEROXIDE
		2-NEOPENTYL-2-PROPYL HYDROPEROXIDE
		BUTYL HYDROPEROXIDE, 1,1,3,3-TETRAMETHYL-
		HYDROPEROXIDE, 1,1,3,3-TETRAMETHYLBUTYL
		LUPERSOL 215
		TERT-OCTYL HYDROPEROXIDE
		TRIGONOX TMPH
		UN 2160 (DOT)
5836-29-3		
		2H-1-BENZOPYRAN-2-ONE, 4-HYDROXY-3-(1,2,3,4-TETRAHYDRO-1-NAPHTHALENYL)-
		2H-1-BENZOPYRAN-2-ONE, 4-HYDROXY-3-(1,2,3,4-TETRAHYDRO-1-NAPHTHALENYL)- (9CI)
		3-(1,2,3,4-TETRAHYDRO-1-NAPHTHYL)-4-HYDROXYCUMARIN (GERMAN)
		3-(1,2,3,4-TETRAHYDRO-1-NAPHTYL)-4-HYDROXY-COUMARINE (FRENCH)
		3-(ALPHA-TETRAL)-4-OXYCOUMARIN
		3-(ALPHA-TETRALINYL)-4-HYDROXYCOUMARIN
		3-(ALPHA-TETRALYL)-4-HYDROXYCOUMARIN
		3-(D-TETRALYL)-4-HYDROXYCOUMARIN
		4-HYDROXY-3-(1,2,3,4-TETRAHYDRO-1-NAFTYL)-CUMARINE (DUTCH)
		4-HYDROXY-3-(1,2,3,4-TETRAHYDRO-1-NAPHTHYL)-CUMARIN
		4-IDROSSI-3-(1,2,3,4-TETRAIDRO-1-NAFTIL)-CUMARINA (ITALIAN)
		BAY 25634
		BAY ENE 11183 B
		BAYER 25 634
		COUMARIN, 4-HYDROXY-3-(1,2,3,4-TETRAHYDRO-1-NAPHTHYL)-
		COUMATETRALYL
		CUMATETRALYL (GERMAN, DUTCH)
		ENDOX
		ENDROCID
		ENDROCIDE
		ENE 11183 B
		RACUMIN
		RACUMIN 57
		RAUCUMIN 57
		RODENTIN
5836-73-7		
		1-(3,4-DICHLOROPHENYL)-3-TRIAZENETHIO-CARBOXAMIDE
		PROMURIT
5894-60-0		
		HEXADECYLTRICHLOROSILANE
		HEXADECYLTRICHLOROSILANE (DOT)
		SILANE, HEXADECYLTRICHLORO-
		SILANE, TRICHLOROHEXADECYL-
		UN 1781 (DOT)
5953-49-1		
		SEC-HEXYL ACETATE
5970-32-1		
		MERCURIC SALICYLATE
		MERCURIC SALICYLATE, SOLID
		MERCURIC SALICYLATE, SOLID (DOT)
		MERCURISALICYLIC ACID
		MERCURY SALICYLATE
		MERCURY SALICYLATE (DOT)
		MERCURY SUBSALICYLATE
		MERCURY, (SALICYLATO(2-))-
		MERCURY, [2-HYDROXYBENZOATO(2-)-O1,O2]-
		MERCURY, [SALICYLATO(2-)]-
		SALICYLIC ACID, MERCURY(2+) SALT (2:1)
		UN 1644 (DOT)
5972-73-6		
		AMMONIUM BINOXALATE MONOHYDRATE
		AMMONIUM OXALATE, NH4C2HO4, HYDRATE
		ETHANEDIOIC ACID, MONOAMMONIUM SALT, MONOHYDRATE
6009-70-7		
		AMMONIUM OXALATE ((NH4)2C2O4) MONOHYDRATE
	*	AMMONIUM OXALATE MONOHYDRATE
		DIAMMONIUM OXALATE MONOHYDRATE
		ETHANEDIOIC ACID, DIAMMONIUM SALT, MONOHYDRATE
		OXALIC ACID, DIAMMONIUM SALT, MONOHYDRATE
6012-97-1		
		TETRACHLOROTHIOPHENE
6032-29-7		
		1-METHYL-1-BUTANOL
		1-METHYLBUTANOL
	*	2-PENTANOL
		2-PENTYL ALCOHOL
		METHYLPROPYLCARBINOL
		SEC-AMYL ALCOHOL
		SEC-PENTANOL
6047-17-2		
		TRICHLOROPHENOXYPROPIONIC ACID ESTER
6094-40-2		
		PIPERAZINE HYDROCHLORIDE

CAS No.	IN	Chemical Name
6108-11-8		CYCLOHEXANE, 1,2,3,4,5,6-HEXACHLORO-, ALPHA-ISOMER
6117-91-5		1-HYDROXY-2-BUTENE
		2-BUTEN-1-OL
		2-BUTENYL ALCOHOL
		3-METHYLALLYL ALCOHOL
		CROTONOL
		CROTONYL ALCOHOL
		CROTYL ALCOHOL
6153-56-6	*	ETHANEDIOIC ACID, DIHYDRATE
		OXALIC ACID DIHYDRATE
6247-51-4		C.I. DIRECT BROWN 59
6341-97-5		DICHLORPHENOXYACETIC ACID ESTER (2,4-DICHLOROACETATE PHENOL)
6358-29-8		C.I. DIRECT RED 39, DISODIUM SALT
6358-53-8		1-(2,5-DIMETHOXYPHENYLAZO)-2-NAPHTHOL
		2,5-DIMETHOXYBENZENEAZO-BETA-NAPHTHOL
		2-NAPHTHALENOL, 1-[(2,5-DIMETHOXYPHENYL)AZO]-
		C.I. 12156
		C.I. SOLVENT RED 80
	*	CITRUS RED 2
		CITRUS RED NO. 2
6360-54-9		C.I. DIRECT BROWN 154
6369-96-6		2,4,5-T TRIMETHYLAMINE SALT
		ACETIC ACID, (2,4,5-TRICHLOROPHENOXY)-, COMPD. WITH N,N-DIMETHYLMETHANAMINE (1:1)
		ACETIC ACID, (2,4,5-TRICHLOROPHENOXY)-, COMPD. WITH TRIMETHYLAMINE (1:1)
		METHANAMINE, N,N-DIMETHYL-, (2,4,5-TRICHLOROPHENOXY)ACETATE
		TRIMETHYLAMINE, (2,4,5-TRICHLOROPHENOXY)ACETATE
6369-97-7		2,4,5-T DIMETHYLAMINE SALT
		2,4,5-TRICHLOROPHENOXYACETIC ACID DIMETHYL AMINE SALT
		ACETIC ACID, (2,4,5-TRICHLOROPHENOXY)-, COMPD. WITH DIMETHYLAMINE (1:1)
		ACETIC ACID, (2,4,5-TRICHLOROPHENOXY)-, COMPD. WITH N-METHYLMETHANAMINE (1:1)
		DIMETHYLAMINE, (2,4,5-TRICHLOROPHENOXY)ACETATE
		FARMCO TA 20
		METHANAMINE, N-METHYL-, (2,4,5-TRICHLOROPHENOXY)-ACETATE
		TA 20
6423-43-4		1,2-PROPANEDIOL, DINITRATE
		1,2-PROPYLENE GLYCOL DINITRATE
		ISOPROPYLENE NITRATE
		PGDN
		PROPYLENE DINITRATE
		PROPYLENE GLYCOL DINITRATE
		PROPYLENE NITRATE
6426-62-6		C.I. DIRECT YELLOW 20
6426-67-1		C.I. DIRECT VIOLET 22, TRISODIUM SALT
6427-21-0		METHOXYMETHYL ISOCYANATE
6459-94-5		C.I. ACID RED 114, DISODIUM SALT
6484-52-2	*	AMMONIUM NITRATE
		AMMONIUM NITRATE (DOT)
		AMMONIUM NITRATE (NO ORGANIC COATING)
		AMMONIUM NITRATE (ORGANIC COATING)
		AMMONIUM NITRATE (SOLUTION)
		AMMONIUM NITRATE, SOLUTION (CONTAINING NOT LESS THAN 15% WATER) (DOT)
		AMMONIUM NITRATE, WITH MORE THAN 0.2% COMBUSTIBLE SUBSTANCES (DOT)
		AMMONIUM NITRATE, WITH NOT MORE THAN 0.2% COMBUSTIBLE SUBSTANCES (DOT)
		AMMONIUM(I) NITRATE (1:1)
		HERCO PRILLS
		NA 1942 (DOT)
		NITRIC ACID AMMONIUM SALT
		UN 0222 (DOT)
		UN 1942 (DOT)
		UN 2426 (DOT)
		VARIOFORM I
6505-86-8		(S)-3-(1-METHYL-2-PYRROLIDINYL)PYRIDINE SULFATE (2:1)
		BLACK LEAF 40
		ENT 2,435
		L-1-METHYL-2-(3-PYRIDYL)-PYRROLIDINE SULFATE
		L-3-(1-METHYL-2-PYRROLIDYL)PYRIDINE SULFATE
		NICOTINE SULFATE
		NICOTINE SULFATE, LIQUID
		NICOTINE SULFATE, LIQUID (DOT)
		NICOTINE SULFATE, SOLID
		NICOTINE SULFATE, SOLID (DOT)
		NICOTINE, SULFATE
		NICOTINE, SULFATE (2:1)
		NIKOTINSULFAT (GERMAN)
		PYRIDINE, 3-(1-METHYL-2-PYRROLIDINYL)-, (S)-, SULFATE
		PYRIDINE, 3-(1-METHYL-2-PYRROLIDINYL)-, (S)-, SULFATE (2:1)
		PYRROLIDINE, 1-METHYL-2-(3-PYRIDYL)-, SULFATE
		SULFATE DE NICOTINE (FRENCH)
		UN 1658 (DOT)
6533-73-9		CARBONIC ACID, DITHALLIUM(1+) SALT
		DITHALLIUM CARBONATE
		RCRA WASTE NUMBER U215
		THALLIUM CARBONATE
		THALLIUM CARBONATE (Tl2CO3)
		THALLIUM(I) CARBONATE (2:1)
		THALLOUS CARBONATE
		THIOCHROMAN-4-ONE, OXIME
6607-45-0		ALPHA, BETA-DICHLOROSTYRENE
		BENZENE, (1,2-DICHLOROETHENYL)-
		STYRENE, ALPHA,BETA-DICHLORO-
6637-88-3		C.I. DIRECT ORANGE 6, DISODIUM SALT
6731-36-8		1,1-BIS(TERT-BUTYLDIOXY)-3,3,5-TRIMETHYLCYCLOHEXANE
		1,1-BIS(TERT-BUTYLPEROXY)-3,3,5-TRIMETHYLCYCLOHEXANE
		1,1-DI-(TERT-BUTYLPEROXY)-3,3,5-TRIMETHYL CYCLOHEXANE
		1,1-DI-(TERT-BUTYLPEROXY)-3,3,5-TRIMETHYL CYCLOHEXANE, NOT MORE THAN 57% IN SOLUTION
		1,1-DI-(TERT-BUTYLPEROXY)-3,3,5-TRIMETHYLCYCLOHEXANE (DOT)
		3,3,5-TRIMETHYL-1,1-BIS(TERT-BUTYLPEROXY)CYCLOHEXANE

CAS No.	IN	Chemical Name
		GEM-BIS(TERT-BUTYLPEROXY)-3,3,5-TRIMETHYLCYCLO-HEXANE
		GEM-DI-TERT-BUTYLPEROXY-3,3,5-TRIMETHYLCYCLOHEX-ANE
		LUPERCO 231G
		LUPERCO 231XL
		LUPERCO 231XLP
		LUPEROX 231
		LUPERSOL 231
		PERHEXA 3M
		PEROXIDE, (3,3,5-TRIMETHYLCYCLOHEXYLIDENE)BIS-(TERT-BUTYL-, NOT MORE THAN 57% IN SOLUTION
		PEROXIDE, (3,3,5-TRIMETHYLCYCLOHEXYLIDENE)BIS[(1,1-DIMETHYLETHYL)
		TRIGONOX 29
		TRIGONOX 29/40
		TRIGONOX 29B50
		TRIGONOX 29B75
		UN 2146 (DOT)
		VAROX 231XL
6739-62-4		
		C.I. DIRECT BLACK 91, TRISODIUM SALT
6742-54-7		
		BENZENE, UNDECYL-
		N-UNDECYLBENZENE
		UNDECANE, 1-PHENYL-
		UNDECYLBENZENE
6806-86-6		
		CHLOROMETHYL
6834-92-0		
		B-W
		CRYSTAMET
		DISODIUM METASILICATE
		DISODIUM MONOSILICATE
		DISODIUM SILICATE
		METSO 20
		METSO BEADS 2048
		METSO BEADS, DRYMET
		METSO PENTABEAD 20
		ORTHOSIL
		SILICIC ACID (H2SiO3), DISODIUM SALT
		SIMET A
		SODIUM METASILICATE
		SODIUM METASILICATE (Na2SiO3)
		SODIUM METASILICATE, ANHYDROUS
		SODIUM SILICATE
		SODIUM SILICATE (Na2SiO3)
		WATER GLASS
6842-15-5		
		1-PROPENE, TETRAMER
		1-PROPENE, TETRAMER (9CI)
		AMSCO TETRAMER
		DODECENE
		DODECYLENE
		PROPENE, TETRAMER
		PROPYLENE TETRAMER
		PROPYLENE TETRAMER (DOT)
		TETRAPROPYLENE
		UN 2850 (DOT)
6923-22-4		
		(E)-DIMETHYL 1-METHYL-3-(METHYLAMINO)-3-OXO-1-PROPENYL PHOSPHATE
		3-(DIMETHOXYPHOSPHINYLOXY)N-METHYL-CIS-CRO-TONAMIDE
		3-HYDROXY-N-METHYL-CIS-CROTONAMIDE DIMETHYL PHOSPHATE
		3-HYDROXY-N-METHYLCROTONAMIDE DIMETHYL PHOS-PHATE
		APADRIN
		AZODRIN
		AZODRIN INSECTICIDE
		BILOBORN
		BILOBRAN
		C 1414
		CIBA 1414
		CIS-1-METHYL-2-METHYL CARBAMOYL VINYL PHOSPHATE
		CRISODIN
		CRISODRIN
		CROTONAMIDE, 3-HYDROXY-N-METHYL-, DIMETHYLPHOS-PHATE, (E)-
		CROTONAMIDE, 3-HYDROXY-N-METHYL-, DIMETHYLPHOS-PHATE, CIS-
		DIMETHYL 1-METHYL-2-(METHYLCARBAMOYL)VINYL PHOSPHATE, CIS
		DIMETHYL PHOSPHATE ESTER OF 3-HYDROXY-N-METHYL-CIS-CROTONAMIDE
		DIMETHYL PHOSPHATE ESTER WITH (E)-3-HYDROXY-N-METHYLCROTONAMIDE
		DIMETHYL PHOSPHATE OF 3-HYDROXY-N-METHYL-CIS-CROTONAMINE
		E-MONOCROTOPHOS
		ENT 27,129
		HAZODRIN
		MONOCIL 40
		MONOCRON
		MONOCROTOPHOS
		MONOCROTOPHOS (ACGIH)
		MONOKROTOFOSZ (HUNGARIAN)
		NUVACRON
		NUVACRON 20
		O,O-DIMETHYL-O-(1-METHYL-2-N-METHYL-CARBAMOYL)-VINYL-PHOSPHAT (GERMAN)
		O,O-DIMETHYL-O-(2-N-METHYLCARBAMOYL-1-METHYL)-VINYL-PHOSPHAT (GERMAN)
		O,O-DIMETHYL-O-(2-N-METHYLCARBAMOYL-1-METHYL-VINYL) PHOSPHATE
		O,O-DIMETHYL-O-(2-N-METHYLCARBAMOYL-1-METHYL-VINYL)-FOSFAAT (DUTCH)
		O,O-DIMETIL-O-(2-N-METILCARBAMOIL-1-METIL-VINIL)-FOSFATO (ITALIAN)
		PHOSPHATE DE DIMETHYLE ET DE 2-METHYLCARBAMOYL 1-METHYL VINYLE (FRENCH)
		PHOSPHORIC ACID, DIMETHYL 1-METHYL-3-(METHYL-AMINO)-3-OXO-1-PROPENYL ESTER, (E)-
		PHOSPHORIC ACID, DIMETHYL ESTER, ESTER WITH (E)-3-HYDROXY-N-METHYLCROTONAMIDE
		PHOSPHORIC ACID, DIMETHYL ESTER, ESTER WITH 3-HYDROXY-N-METHYLCROTONAMIDE, (E)-
		PHOSPHORIC ACID, DIMETHYL ESTER, ESTER WITH CIS-3-HYDROXY-N-METHYLCROTONAMIDE
		PILLARDRIN
		PLANTDRIN
		SD 9129
		SHELL SD 9129
		SUSVIN
		ULVAIR
6950-84-1		
		NAPHTHYLUREA (1-NAPHTHALENYL UREA)
6959-48-4		
		3-(CHLOROMETHYL)PYRIDINE HYDROCHLORIDE
7005-72-3		
		1-CHLORO-4-PHENOXYBENZENE
		4-CHLORODIPHENYL ETHER
		4-CHLOROPHENYL PHENYL ETHER
		BENZENE, 1-CHLORO-4-PHENOXY-
		ETHER, P-CHLOROPHENYL PHENYL
		P-CHLORODIPHENYL OXIDE
		P-CHLOROPHENYL PHENYL ETHER
7008-42-6		
		ACRONYCINE
7008-81-3		
		DIAZODINITROPHENOL
7149-75-9		
		4-CHLORO-3-METHYL ANILINE (CHLOROTOLUIDINE)

CAS No.	IN	Chemical Name
7154-79-2		
		2,2,3,3-TETRAMETHYL PENTANE
		PENTANE, 2,2,3,3-TETRAMETHYL-
7209-38-3		
		BIS(AMINOPROPYL)PIPERAZINE
7220-81-7		
		AFLATOXIN B2
7415-31-8		
		1,3-DICHLORO-2-BUTENE
7421-93-4		
		1,2,4-METHENOCYCLOPENTA[CD]PENTALENE-5-CARBOXALDEHYDE, 2,2A,3,3,4,7-HEXACHLORODECAHYDRO-, (1ALPHA,
		ENDRIN ALDEHYDE
		SD 7442
7428-48-0		
		AUSTROSTAB 110E
		BLEISTEARAT (GERMAN)
		HAL-LUB-N
		LEAD OCTADECANOATE
		LEAD STEARATE
		LEAD STEARATE (DOT)
		LISTAB 28
		NA 2811 (DOT)
		NEUTRAL LEAD STEARATE
		OCTADECANOIC ACID, LEAD SALT
		SLW
		STEARIC ACID, LEAD SALT
7429-90-5		
		A 00
		A 1200P
		A 97
		A 99
		A 99 (METAL)
		A 99N
		A999
		A999V
		AA 1099
		AA 1193
		AA 1199
		AE
		ALPASTE 0230T
		ALPASTE 1500MA
		ALPASTE 240T
	*	ALUMINIUM
		ALUMINIUM FLAKE
		ALUMINUM
		ALUMINUM (FUME OR DUST)
		ALUMINUM A 00
		ALUMINUM DEHYDRATED
		ALUMINUM POWDER
		ALUMINUM, METAL AND OXIDE AND WELDING FUMES
		ALUMINUM-27
		AR 2
		AV 000
		AV00
		C.I. 77000
		JISC 3108
		JISC 3110
		K 102
		K 102 (METAL)
		L 1018
		L 16
		METANA
		NORAL ALUMINIUM
		NORAL EXTRA FINE LINING GRADE
		NORAL INK GRADE ALUMINIUM
		PAP 1
		S 40
		S 40 (METAL)
		SPOTA MOBIL 801
		SS 3666
		TUFF MIC
		VI 5
7439-90-9		
		KRYPTON
		KRYPTON, COMPRESSED
		KRYPTON, COMPRESSED (DOT)
		KRYPTON, LIQUID (REFRIGERATED)
		KRYPTON, REFRIGERATED LIQUID (DOT)
		UN 1056 (DOT)
		UN 1970 (DOT)
7439-92-1		
		C.I. 77575
		C.I. PIGMENT METAL 4
	*	LEAD
		LEAD FLAKE
		LEAD S 2
		LEAD, INORGANIC, DUST AND FUMES
		PB-S 100
		SO
7439-93-2		
	*	LITHIUM
		LITHIUM ELEMENT
		LITHIUM METAL
		LITHIUM METAL (DOT)
		LITHIUM METAL, IN CARTRIDGES (DOT)
		UN 1415 (DOT)
7439-95-4		
		JIS 1
		MAGNESIO (ITALIAN)
	*	MAGNESIUM
		MAGNESIUM BORINGS (DOT)
		MAGNESIUM CLIPPINGS
		MAGNESIUM CLIPPINGS (DOT)
		MAGNESIUM ELEMENT
		MAGNESIUM GRANULES COATED, PARTICLE SIZE NOT LESS THAN 149 MICRONS (DOT)
		MAGNESIUM METAL (DOT)
		MAGNESIUM PELLETS
		MAGNESIUM PELLETS (DOT)
		MAGNESIUM POWDER (DOT)
		MAGNESIUM POWDERED
		MAGNESIUM RIBBON (DOT)
		MAGNESIUM RIBBONS
		MAGNESIUM SCALPINGS (DOT)
		MAGNESIUM SCRAP (DOT)
		MAGNESIUM SHAVINGS (DOT)
		MAGNESIUM SHEET
		MAGNESIUM TURNINGS
		MAGNESIUM TURNINGS (DOT)
		MAGNESIUM, METAL (POWDERED, PELLETS, TURNINGS, OR RIBBON)
		NA 1869 (DOT)
		RIEKE'S ACTIVE MAGNESIUM
		RMC
		UN 1418 (DOT)
		UN 1869 (DOT)
		UN 2950 (DOT)
7439-96-5		
		COLLOIDAL MANGANESE
		CUTAVAL
		JIS-G 1213
	*	MANGANESE
		MANGANESE-55
7439-97-6		
		COLLOIDAL MERCURY
		KWIK (DUTCH)
		MERCURE (FRENCH)
		MERCURIO (ITALIAN)
	*	MERCURY
		MERCURY (ACGIH)
		MERCURY ELEMENT
		MERCURY, METALLIC

CAS No.	IN	Chemical Name
		MERCURY, METALLIC (DOT)
		METALLIC MERCURY
		NA 2809 (DOT)
		NCI-C60399
		QUECKSILBER
		QUECKSILBER (GERMAN)
		QUICKSILVER
		RCRA WASTE NUMBER U151
		RTEC (POLISH)
		UN 2809 (DOT)
7439-98-7		
		MCHVL
	*	MOLYBDENUM
		TSM1
7440-01-9		
		NEON
		NEON (DOT)
		NEON, COMPRESSED (DOT)
		NEON, REFRIGERATED LIQUID
		NEON, REFRIGERATED LIQUID (DOT)
		UN 1065 (DOT)
		UN 1913 (DOT)
7440-02-0		
		C.I. 77775
		FM 1208
		HCA 1
		N1
		NI 0901-S
		NI 0901S (HARSHAW)
		NI 233
		NI 270
		NI 4303T
		NICHEL (ITALIAN)
	*	NICKEL
		NICKEL (ACGIH)
		NICKEL (DUST)
		NICKEL 270
		NICKEL CATALYST
		NICKEL CATALYST, DRY
		NICKEL CATALYST, DRY :PRECIPITATED ON A CARRIER WITH A SPECIAL ACTIVATOR
		NICKEL CATALYST, FINELY DIVIDED, WET WITH ≥40% WATER
		NICKEL ELEMENT
		NICKEL PARTICLES
		NICKEL SPONGE
		NP 2
		NP-2
		RANEY ALLOY
		RANEY NICKEL
		RCH 55/5
7440-06-4		
		C.I. 77795
		LIQUID BRIGHT PLATINUM
	*	PLATINUM
		PLATINUM BLACK
		PRO
7440-09-7		
	*	POTASSIUM
		POTASSIUM ATOM
		POTASSIUM, (LIQUID ALLOY)
		POTASSIUM, METAL
		POTASSIUM, METAL (DOT)
		POTASSIUM, METAL ALLOYS (DOT)
		POTASSIUM, METAL LIQUID ALLOY
		POTASSIUM, METAL LIQUID ALLOY (DOT)
		POTASSIUM, METALLIC (DOT)
		UN 1420 (DOT)
		UN 2257 (DOT)
7440-16-6		
		RHODIUM
		RHODIUM, METAL FUME AND DUSTS
		RHODIUM-103
7440-17-7		
		RUBIDIUM
		RUBIDIUM METAL
		RUBIDIUM METAL (DOT)
		RUBIDIUM METAL, IN CARTRIDGES (DOT)
		UN 1423 (DOT)
7440-21-3		
		DEFOAMER S-10
	*	SILICON
		SILICON (ACGIH)
		SILICON ELEMENT
		SILICON POWDER, AMORPHOUS
		SILICON POWDER, AMORPHOUS (DOT)
		UN 1346 (DOT)
7440-22-4		
		ALGAEDYN
		ARGENTUM
		C.I. 77820
		E 20
		L 3
		SHELL SILVER
		SILFLAKE 135
		SILPOWDER 130
	*	SILVER
		SILVER ATOM
		SILVER METAL
		SILVEST TCG 1
		SR 999
		TCG 7R
		V 9
7440-23-5		
		NA 1421 (DOT)
	*	SODIUM
		SODIUM ATOM
		SODIUM METAL
		SODIUM, (LIQUID ALLOY)
		SODIUM, METAL LIQUID ALLOY
		SODIUM, METAL LIQUID ALLOY (DOT)
		SODIUM-23
7440-24-6		
	*	STRONTIUM
7440-25-7		
	*	TANTALUM
		TANTALUM-181
7440-28-0		
		RAMOR
		THALLIUM
7440-29-1		
	*	THORIUM
		THORIUM METAL, PYROPHORIC
		THORIUM METAL, PYROPHORIC (DOT)
		THORIUM-232
		UN 2975 (DOT)
7440-31-5		
		C.I. 77860
		C.I. PIGMENT METAL 5
		METALLIC TIN
		SILVER MATT POWDER
	*	TIN
		TIN FLAKE
		TIN POWDER
		WANG
7440-32-6		
		ALPHA-VT 1-0
		ASTM B348 GR 2
		ATI 24
		BS 2TA6
		CONTIMET 30
		CONTIMET 55
		CP TITANIUM

CAS No.	IN	Chemical Name
		EMO 140
		IMI 115
		IMI 125
		IMI 130
		IMI 155
		JIS H2151 TW 35
		JISTP 28
		KS 50
		KS 70
		NCI-C04251
		OREMET
		SMELLOFF-CUTTER TITANIUM
		T 160
		T 40
		T 60
		T 60 (METAL)
		TG-TV
		TI 160
		TI 35A
		TI 40
		TI 40A
		TI 50A
		TI 75A
		TI-A55
		TITAN VT 1-1
	*	TITANIUM
		TITANIUM 50A
		TITANIUM A 40
		TITANIUM ALLOY
		TITANIUM ELEMENT
		TITANIUM METAL POWDER, DRY
		TITANIUM METAL POWDER, DRY OR WET WITH 20% WATER
		TITANIUM METAL POWDER, WET WITH NOT LESS THAN 25% WATER (DOT)
		TITANIUM METAL, POWDER, DRY (DOT)
		TITANIUM SPONGE GRANULES (DOT)
		TITANIUM SPONGE POWDERS (DOT)
		TITANIUM SPONGE, GRANULES OR POWDER
		TITANIUM VT 1
		TITANIUM, POWDER WET WITH 20% OR MORE WATER (DOT)
		TITANIUM, WET, WITH LESS THAN 20% WATER (DOT)
		TITANIUM-125
		TP 28
		TP 28C
		UN 1352 (DOT)
		UN 2546 (DOT)
		UN 2878 (DOT)
		VT 1
		VT 1-0
		VT 1-1
		VT 1-2
		VT 1D
		VT 1L
		VTL 0
7440-33-7		
	*	TUNGSTEN
		VA
		VA (TUNGSTEN)
		WOLFRAM
7440-36-0		
	*	ANTIMONY
		ANTIMONY (ACGIH)
		ANTIMONY BLACK
		ANTIMONY ELEMENT
		ANTIMONY POWDER
		ANTIMONY POWDER (DOT)
		ANTIMONY, REGULUS
		ANTYMON (POLISH)
		C.I. 77050
		STIBIUM
		UN 2871 (DOT)
7440-37-1		
	*	ARGON
		ARGON, ISOTOPE OF MASS 40
		ARGON, LIQUID PRESSURIZED (DOT)
		ARGON, REFRIGERATED LIQUID
		ARGON, REFRIGERATED LIQUID (DOT)
		ARGON-40
		UN 1006 (DOT)
		UN 1951 (DOT)
7440-38-2		
		ARSEN (GERMAN,POLISH)
	*	ARSENIC
		ARSENIC (ACGIH)
		ARSENIC BLACK
		ARSENIC ELEMENT
		ARSENIC, METALLIC (DOT)
		ARSENIC, SOLID
		ARSENIC, SOLID (DOT)
		ARSENIC-75
		ARSENICALS
		COLLOIDAL ARSENIC
		GREY ARSENIC
		METALLIC ARSENIC
		UN 1558 (DOT)
7440-39-3		
	*	BARIUM
		BARIUM (ACGIH)
		BARIUM ALLOY
		BARIUM ALLOY, PYROPHORIC
		BARIUM ELEMENT
		BARIUM, ALLOYS, NON-PYROPHORIC (DOT)
		BARIUM, ALLOYS, PYROPHORIC (DOT)
		BARIUM, METAL, NON-PYROPHORIC (DOT)
		UN 1399 (DOT)
		UN 1400 (DOT)
		UN 1854 (DOT)
7440-41-7		
		BERYLLIUM
		BERYLLIUM (ACGIH)
		BERYLLIUM ELEMENT
		BERYLLIUM, METAL POWDER (DOT)
		BERYLLIUM, POWDER
		BERYLLIUM-9
		GLUCINIUM
		GLUCINUM
		RCRA WASTE NUMBER P015
		UN 1567 (DOT)
7440-43-9		
		C.I. 77180
	*	CADMIUM
7440-45-1		
	*	CERIUM
		CERIUM, CRUDE
		CERIUM, CRUDE, POWDER (DOT)
		CERIUM, CRUDE, SLABS OR INGOTS (DOT)
		UN 1333 (DOT)
7440-46-2		
	*	CESIUM
		CESIUM METAL
		CESIUM METAL (DOT)
		CESIUM, POWDERED (DOT)
		CESIUM-133
		UN 1383 (DOT)
		UN 1407 (DOT)
7440-47-3		
		CHROME
	*	CHROMIUM
		CHROMIUM (METAL)
7440-48-4		
		AQUACAT
		C.I. 77320
	*	COBALT
		COBALT (ACGIH)
		COBALT ELEMENT
		COBALT METAL, DUST, AND FUME
		COBALT-59

CAS No.	IN	Chemical Name
		KOBALT (GERMAN, POLISH)
		NCI-C60311
		SUPER COBALT
7440-50-8		
		1721 GOLD
		ALLBRI NATURAL COPPER
		ANAC 110
		ARWOOD COPPER
		C 10200
		C 12200
		C.I. 77400
		C.I. PIGMENT METAL 2
		CA 122
		CDA 101
		CDA 102
		CDA 110
		CDA 122
		CE 1110
	*	COPPER
		COPPER M 1
		COPPER POWDER
		CU M2
		CU M3
		CUEP
		CUEPP
		DCUP1
		E 115
		E 115 (METAL)
		E-COPPER
		E-CU57
		GE 1110
		KAFAR COPPER
		M 1
		M 3
		M 4
		M3 (COPPER)
		M3R
		M3S
		M4 (COPPER)
		OFHC COPPER
		OFHC CU
		RANEY COPPER
		TCUP1
7440-55-3		
	*	GALLIUM
		GALLIUM ELEMENT
		GALLIUM METAL, LIQUID
		GALLIUM METAL, LIQUID (DOT)
		GALLIUM METAL, SOLID
		GALLIUM METAL, SOLID (DOT)
		UN 2803 (DOT)
7440-58-6		
		CELTIUM
		HAFNIUM
		HAFNIUM (ACGIH)
		HAFNIUM ELEMENT
		HAFNIUM METAL, DRY
		HAFNIUM METAL, DRY (DOT)
		HAFNIUM METAL, WET
		HAFNIUM METAL, WET (DOT)
		HAFNIUM, WET WITH NOT LESS THAN 25% WATER (DOT)
		UN 1326 (DOT)
		UN 2545 (DOT)
7440-59-7		
		ATOMIC HELIUM
	*	HELIUM
		HELIUM (DOT)
		HELIUM, COMPRESSED (DOT)
		HELIUM, REFRIGERATED LIQUID
		HELIUM, REFRIGERATED LIQUID (DOT)
		HELIUM-4
		O-HELIUM
		P-HELIUM
		UN 1046 (DOT)
		UN 1963 (DOT)
7440-61-1		
	*	URANIUM
		URANIUM I (238U)
		URANIUM(NATURAL)
		URANIUM-238
7440-62-2		
	*	VANADIUM
		VANADIUM (FUME OR DUST)
		VANADIUM-51
7440-63-3		
		UN 2036 (DOT)
		UN 2591 (DOT)
		XENON
		XENON (DOT)
		XENON ATOM
		XENON, LIQUID
		XENON, REFRIGERATED LIQUID
		XENON, REFRIGERATED LIQUID (DOT)
7440-65-5		
		YTTRIUM
		YTTRIUM-89
7440-66-6		
		ASARCO L 15
		BLUE POWDER
		CI 77945
		CI PIGMENT BLACK 16
		CI PIGMENT METAL 6
		EMANAY ZINC DUST
		GRANULAR ZINC
		JASAD
		LS 2
		MERRILLITE
		PASCO
		RHEINZINK
		UN 1383 (DOT)
		UN 1436 (DOT)
	*	ZINC
		ZINC (FUME OR DUST)
		ZINC DUST
		ZINC ELEMENT
		ZINC METAL, POWDER OR DUST
		ZINC POWDER
		ZINC POWDER OR DUST, PYROPHORIC
		ZINC, POWDER OR DUST, NON-PYROPHORIC (DOT)
		ZINC, POWDER OR DUST, PYROPHORIC (DOT)
7440-67-7		
		UN 1308 (DOT)
		UN 1358 (DOT)
		UN 2008 (DOT)
		UN 2009 (DOT)
		UN 2858 (DOT)
		ZIRCAT
		ZIRCONIUM
		ZIRCONIUM (ACGIH)
		ZIRCONIUM ELEMENT
		ZIRCONIUM METAL POWDER, WETTED WITH NOT LESS THAN 25% WATER (DOT)
		ZIRCONIUM METAL, DRY, CHEMICALLY PRODUCED, FINER THAN 20 MESH PARTICLE SIZE
		ZIRCONIUM METAL, DRY, CHEMICALLY PRODUCED, FINER THAN 270 MESH PARTICLE SIZE (DOT)
		ZIRCONIUM METAL, DRY, MECHANICALLY PRODUCED, FINER THAN 270 MESH PARTICLE SIZE (DOT)
		ZIRCONIUM METAL, LIQUID SUSPENSIONS (DOT)
		ZIRCONIUM METAL, WET, CHEMICALLY PRODUCED, FINER THAN 20 MESH PARTICLE SIZE
		ZIRCONIUM METAL, WET, CHEMICALLY PRODUCED, FINER THAN 270 MESH PARTICLE SIZE (DOT)
		ZIRCONIUM METAL, WET, MECHANICALLY PRODUCED, FINER THAN 270 MESH PARTICLE SIZE (DOT)
		ZIRCONIUM SHAVINGS (DOT)
		ZIRCONIUM SHEETS (DOT)
		ZIRCONIUM TURNINGS

CAS No.	IN	Chemical Name
		ZIRCONIUM, METAL, DRY, COILED WIRE, FINISHED METAL SHEETS THINNER THAN 18 MICRONS (DOT)
		ZIRCONIUM, METAL, LIQUID, SUSPENSIONS
7440-70-2		ATOMIC CALCIUM
		BLOOD-COAGULATION FACTOR IV
		CALCICAT
	*	CALCIUM
		CALCIUM ATOM
		CALCIUM ELEMENT
		CALCIUM, METAL
		CALCIUM, METAL (DOT)
		CALCIUM, METAL, CRYSTALLINE
		CALCIUM, METAL, CRYSTALLINE (DOT)
		CALCIUM, METAL, PYROPHORIC
		CALCIUM, NON-PYROPHORIC (DOT)
		CALCIUM, PYROPHORIC (DOT)
		DIETARY CALCIUM
		NA 1401 (DOT)
		PRAVAL
		UN 1401 (DOT)
		UN 1855 (DOT)
7440-74-6	*	INDIUM
7446-08-4		RCRA WASTE NUMBER U204
		SELENIOUS ANHYDRIDE
		SELENIUM DIOXIDE
	*	SELENIUM OXIDE
		SELENIUM OXIDE (SeO2)
		SELENIUM(IV) DIOXIDE (1:2)
7446-09-5		BISULFITE
		FERMENICIDE LIQUID
		FERMENICIDE POWDER
		FERMENTICIDE LIQUID
		SCHWEFELDIOXYD (GERMAN)
		SIARKI DWUTLENEK (POLISH)
	*	SULFUR DIOXIDE
		SULFUR DIOXIDE (ACGIH,DOT)
		SULFUR DIOXIDE (SO2)
		SULFUR OXIDE
		SULFUR OXIDE (SO2)
		SULFUROUS ACID ANHYDRIDE
		SULFUROUS ANHYDRIDE
		SULFUROUS OXIDE
		SULPHUR DIOXIDE
		SULPHUR DIOXIDE, LIQUEFIED (DOT)
		UN 1079 (DOT)
7446-11-9		SULFAN
		SULFUR OXIDE (SO3)
		SULFUR TRIOXIDE
		SULFUR TRIOXIDE (DOT)
		SULFUR TRIOXIDE, INHIBITED
		SULFUR TRIOXIDE, STABILIZED (DOT)
		SULFURIC ANHYDRIDE
		SULFURIC ANHYDRIDE (DOT)
		SULFURIC OXIDE
		UN 1829 (DOT)
7446-14-2		ANGLISLITE
		BLEISULFAT (GERMAN)
		C.I. 77630
		C.I. PIGMENT WHITE 3
		FAST WHITE
		FREEMANS WHITE LEAD
		LEAD BOTTOMS
		LEAD DROSS (DOT)
		LEAD MONOSULFATE
		LEAD SULFATE
		LEAD SULFATE (1:1)
		LEAD SULFATE (DOT)
		LEAD SULFATE (PbSO4)
		LEAD SULFATE, SOLID, CONTAINING MORE THAN 3% FREE ACID (DOT)
	*	LEAD SULPHATE
		LEAD(2+) SULFATE
		LEAD(II) SULFATE
		LEAD(II) SULFATE (1:1)
		MILK WHITE
		MULHOUSE WHITE
		NA 2291 (DOT)
		SULFATE DE PLOMB (FRENCH)
		SULFURIC ACID, LEAD SALT
		SULFURIC ACID, LEAD(2+) SALT (1:1)
		UN 1794 (DOT)
7446-18-6		CFS
		CSF-GIFTWEIZEN
		DITHALLIUM SULFATE
		DITHALLIUM(1+) SULFATE
		ECCOTHAL
		M7-GIFTKOERNER
		NA 1707 (DOT)
		RATOX
		RATTENGIFTKONSERVE
		RCRA WASTE NUMBER P115
		SULFURIC ACID THALLIUM(1+) SALT (1:2)
		SULFURIC ACID, DITHALLIUM(1+) SALT
		SULFURIC ACID, DITHALLIUM(1+) SALT (8CI,9CI)
		SULFURIC ACID, THALLIUM SALT
		THALLIUM SULFATE
		THALLIUM SULFATE (Tl2SO4)
		THALLIUM SULFATE, SOLID
		THALLIUM SULFATE, SOLID (DOT)
		THALLIUM(1) SULFATE
		THALLIUM(I) SULFATE
		THALLIUM(I) SULFATE (2:1)
		THALLOUS SULFATE
		ZELIO
7446-27-7		C.I. 77622
		LEAD DIPHOSPHATE
		LEAD ORTHOPHOSPHATE
		LEAD ORTHOPHOSPHATE (Pb3(PO4)2)
		LEAD PHOSPHATE
		LEAD PHOSPHATE (3:2)
		LEAD PHOSPHATE (Pb3(PO4)2)
		LEAD(2+) PHOSPHATE (Pb3(PO4)2)
		PERLEX PASTE 500
		PERLEX PASTE 600A
		PHOSPHORIC ACID, LEAD(2+) SALT (2:3)
		TRILEAD PHOSPHATE
7446-34-6		SELENIUM MONOSULFIDE
		SELENIUM SULFIDE
		SELENIUM SULFIDE (SeS)
		SULFUR SELENIDE (SSe)
7446-70-0		ALLUMINIO(CLORURO DI) (ITALIAN)
		ALUMINIUM CHLORIDE
		ALUMINIUMCHLORID (GERMAN)
	*	ALUMINUM CHLORIDE
		ALUMINUM CHLORIDE (1:3)
		ALUMINUM CHLORIDE (AlCl3)
		ALUMINUM CHLORIDE SOLUTION (DOT)
		ALUMINUM CHLORIDE, ANHYDROUS
		ALUMINUM CHLORIDE, ANHYDROUS (DOT)
		ALUMINUM CHLORIDE, SOLUTION
		ALUMINUM TRICHLORIDE
		ALUMINUM TRICHLORIDE (AlCl3)
		CHLORURE D'ALUMINIUM (FRENCH)
		PEARSALL
		TRICHLOROALUMINUM
		UN 1726 (DOT)
		UN 2581 (DOT)

CAS No.	IN	Chemical Name
7447-39-4		CHLORID MEDNY (CZECH)
		COPPER BICHLORIDE
	*	COPPER CHLORIDE
		COPPER CHLORIDE (CuCl2)
		COPPER DICHLORIDE
		COPPER(2+) CHLORIDE
		COPPER(II) CHLORIDE
		CUPRIC CHLORIDE
		CUPRIC DICHLORIDE
		CUPROUS CHLORIDE
		CUPROUS DICHLORIDE
7487-94-7		BICHLORIDE OF MERCURY
		BICHLORURE DE MERCURE (FRENCH)
		CALOCHLOR
		CHLORID RTUTNATY (CZECH)
		CHLORURE MERCURIQUE (FRENCH)
		CLORURO DI MERCURIO (ITALIAN)
		CORROSIVE MERCURY CHLORIDE
		CORROSIVE SUBLIMATE
		CRC
		DICHLOROMERCURY
		EMISAN 6
		FUNGCHEX
		MC
		MERCURIC BICHLORIDE
	*	MERCURIC CHLORIDE
		MERCURIC CHLORIDE (DOT)
		MERCURIC CHLORIDE, SOLID
		MERCURIC CHLORIDE, SOLID (DOT)
		MERCURY BICHLORIDE
		MERCURY CHLORIDE (HgCl2)
		MERCURY DICHLORIDE
		MERCURY PERCHLORIDE
		MERCURY(II) CHLORIDE
		NCI-C60173
		PERCHLORIDE OF MERCURY
		QUECKSILBER CHLORID (GERMAN)
		SUBLIMAT (CZECH)
		SUBLIMATE
		SULEM
		SULEMA (RUSSIAN)
		TL 898
		UN 1624 (DOT)
7488-56-4		EXSEL
		RCRA WASTE NUMBER U205
		SELENIUM DISULFIDE
		SELENIUM DISULPHIDE (DOT)
		SELENIUM SULFIDE
		SELENIUM SULFIDE (SeS2)
		SELENIUM(IV) DISULFIDE (1:2)
		SELSUN
		SELSUN BLUE
		UN 2657 (DOT)
7521-80-4		BUTYL TRICHLOROSILANE
		BUTYL TRICHLOROSILANE (DOT)
		SILANE, BUTYLTRICHLORO-
		TRICHLOROBUTYLSILANE
		UN 1747 (DOT)
7530-07-6		CAPRYLYL PEROXIDE (N-OCTANOYL PEROXIDE)
7546-30-7		CALOGREEN
		CALOMEL
		CALOMELANO (ITALIAN)
		CALOSAN
		CHLORURE MERCUREUX (FRENCH)
		CI 77764
		CLORURO DI MERCURIO (ITALIAN)
		CLORURO MERCUROSO (ITALIAN)
		CYCLOSAN
		KALOMEL (GERMAN)
		MERCUROCHLORIDE (DUTCH)
	*	MERCUROUS CHLORIDE
		MERCUROUS CHLORIDE (HgCl)
		MERCURY CHLORIDE (HgCl)
		MERCURY MONOCHLORIDE
		MERCURY PROTOCHLORIDE
		MERCURY(1+) CHLORIDE
		MERCURY(I) CHLORIDE
		MILD MERCURY CHLORIDE
		PRECIPITE BLANC
		QUECKSILBER CHLORUER (GERMAN)
		QUECKSILBER(I)-CHLORID (GERMAN)
		SUBCHLORIDE OF MERCURY
7550-45-0		TETRACHLOROTITANIUM
		TETRACHLORURE DE TITANE (FRENCH)
		TITAANTETRACHLORIDE (DUTCH)
		TITANE (TETRACHLORURE DE) (FRENCH)
		TITANIO (TETRACLORURO DI) (ITALIAN)
		TITANIUM CHLORIDE
		TITANIUM CHLORIDE (TiCl4)
		TITANIUM CHLORIDE (TiCl4) (T-4)-
	*	TITANIUM TETRACHLORIDE
		TITANIUM TETRACHLORIDE (DOT)
		TITANIUM(IV) CHLORIDE
		TITANTETRACHLORID (GERMAN)
		UN 1838 (DOT)
7553-56-2		ACTOMAR
		DIIODINE
		ERANOL
	*	IODINE
		IODINE (127I2)
		IODINE COLLOIDAL
		IODINE CRYSTALS
		IODINE SUBLIMED
		IOSAN SUPERDIP
		MOLECULAR IODINE
7558-79-4		ACETEST
		ANHYDROUS SODIUM ACID PHOSPHATE
		DIBASIC SODIUM PHOSPHATE
		DISODIUM ACID ORTHOPHOSPHATE
		DISODIUM ACID PHOSPHATE
		DISODIUM HYDROGEN PHOSPHATE
		DISODIUM HYDROPHOSPHATE
		DISODIUM MONOHYDROGEN PHOSPHATE
		DISODIUM ORTHOPHOSPHATE
		DISODIUM PHOSPHATE
		DISODIUM PHOSPHATE (Na2HPO4)
		DISODIUM PHOSPHORIC ACID
		DSP
		EXSICCATED SODIUM PHOSPHATE
		HYDROGEN DISODIUM PHOSPHATE
		MONOHYDROGEN DISODIUM PHOSPHATE
		NA 9147 (DOT)
		NATRIUMPHOSPHAT (GERMAN)
		PHOSPHORIC ACID, DISODIUM SALT
		SECONDARY SODIUM PHOSPHATE
		SODA PHOSPHATE
		SODIUM HYDROGEN PHOSPHATE
		SODIUM MONOHYDROGEN PHOSPHATE
		SODIUM MONOHYDROGEN PHOSPHATE (2:1:1)
		SODIUM PHOSPHATE (Na2HPO4)
	*	SODIUM PHOSPHATE, DIBASIC
		SODIUM PHOSPHATE, DIBASIC (DOT)
7568-93-6		1-PHENYL-2-AMINOETHANOL
		2-AMINO-1-PHENYLETHANOL
		2-HYDROXY-2-PHENYLETHYLAMINE
		2-PHENYL-2-HYDROXYETHYLAMINE
		ALPHA-(AMINOMETHYL)BENZYL ALCOHOL
		APOPHEDRIN
		BENZENEETHANAMINE, BETA-HYDROXY-

CAS No.	IN	Chemical Name
		BENZENEMETHANOL, ALPHA-(AMINOMETHYL)-
		BENZYL ALCOHOL, ALPHA-(AMINOMETHYL)-
		BETA-HYDROXY-BETA-PHENYLETHYLAMINE
		BETA-HYDROXYPHENETHYLAMINE
		BETA-HYDROXYPHENYLETHYLAMINE
		BETA-PHENETHANOLAMINE
		BETA-PHENYLETHANOLAMINE
		BISNOREPHEDRINE
		NORPHEDRIN
		PHENETHANOLAMINE
		PHENYLETHANOLAMINE
7572-29-4		
		ACETYLENE, DICHLORO-
		DICHLOROACETYLENE
		DICHLOROETHYNE
		ETHYNE, DICHLORO-
7580-67-8		
	*	LITHIUM HYDRIDE
		LITHIUM HYDRIDE (LiH)
		LITHIUM MONOHYDRIDE
7581-97-7		
		2,3-DICHLOROBUTANE
		BUTANE, 2,3-DICHLORO-
7601-54-9		
		ANTISAL 4
		DRI-TRI
		EMULSIPHOS 440/660
		NA 9148 (DOT)
		NUTRIFOS STP
		PHOSPHORIC ACID SODIUM SALT (1:3)
		PHOSPHORIC ACID, TRISODIUM SALT
		SODIUM PHOSPHATE
		SODIUM PHOSPHATE (Na3PO4)
		SODIUM PHOSPHATE, ANHYDROUS
	*	SODIUM PHOSPHATE, TRIBASIC
		SODIUM PHOSPHATE, TRIBASIC (DOT)
		SODIUM TERTIARY PHOSPHATE
		TRIBASIC SODIUM ORTHOPHOSPHATE
		TRIBASIC SODIUM PHOSPHATE
		TRINATRIUMPHOSPHAT (GERMAN)
		TRISODIUM ORTHOPHOSPHATE
		TRISODIUM PHOSPHATE
		TROMETE
		TSP
7601-89-0		
		NATRIUMPERCHLORAAT (DUTCH)
		NATRIUMPERCHLORAT (GERMAN)
		PERCHLORATE DE SODIUM (FRENCH)
		PERCHLORIC ACID, SODIUM SALT
		SODIO (PERCLORATO DI) (ITALIAN)
		SODIUM (PERCHLORATE DE) (FRENCH)
	*	SODIUM PERCHLORATE
		SODIUM PERCHLORATE (DOT)
		UN 1502 (DOT)
7601-90-3		
	*	PERCHLORIC ACID
		PERCHLORIC ACID, MORE THAN 50% BUT NOT MORE THAN 72% STRENGTH (DOT)
		PERCHLORIC ACID, NOT OVER 50% ACID (DOT)
		UN 1802 (DOT)
		UN 1873 (DOT)
7616-94-6		
		CHLORINE FLUORIDE OXIDE (ClO3F)
		CHLORINE OXYFLUORIDE (ClO3F)
		PERCHLOROYLFLUORIDE
		PERCHLORYL FLUORIDE
		PERCHLORYL FLUORIDE (ClO3F)
		TRIOXYCHLOROFLUORIDE
7631-86-9		
		ACTICEL
		ADELITE 30

CAS No.	IN	Chemical Name
		ADELITE A
		AEROGEL 200
		AEROSIL
		AEROSIL 130
		AEROSIL 130V
		AEROSIL 175
		AEROSIL 200
		AEROSIL 200V
		AEROSIL 300
		AEROSIL 308
		AEROSIL 380
		AEROSIL A 130
		AEROSIL A 175
		AEROSIL A 300
		AEROSIL A 380
		AEROSIL BS 50
		AEROSIL E 300
		AEROSIL K 7
		AEROSIL M 300
		AEROSIL OX 50
		AEROSIL PST
		AEROSIL TT 600
		AEROSIL-DEGUSSA
		AMORPHOUS SILICA
		APASIL
		AQUAFIL
		AROGEN 500
		ARSIL
		BS 30
		BS 30 (FILLER)
		BS 50
		BS 50 (SILICA)
		CAB-O-SIL
		CAB-O-SIL H 5
		CAB-O-SIL L 5
		CAB-O-SIL M 5
		CAB-O-SIL MS 7
		CABOSIL N 5
		CABOSIL ST 1
		CARPLEX
		CARPLEX 1120
		CARPLEX 30
		CARPLEX 67
		CARPLEX 80
		CARPLEX FPS 1
		CARPLEX FPS 3
		CATALOID
		CATALOID HS 40
		CATALOID S 30H
		CATALOID S 30L
		CATALOID SI 350
		CELITE SUPERFLOSS
		COLLOIDAL SILICA
		COLLOIDAL SILICON DIOXIDE
		CORASIL II
		CRYSTALITE A 1
		DAVISON 951
		DRI-DIE
		ECCOSPHERES SI
		EXTRUSIL
		F 307
		FINESIL B
		FK 160
		FLOLITE S 700
		FOSSIL FLOUR
		FRANSIL 251
		GAROSIL GB
		GAROSIL N
		GP 71
		HDK-N 20
		HDK-S 15
		HDK-V 15
		HIMESIL A
		HK 125
		HK 400
		HYPERSIL
		IATROBEADS 6RS8060
		IMSIL 10
		IMSIL 1240

CAS No.	IN	Chemical Name
		IMSIL A 10
		IMSIL A 108
		IMSIL A 15
		IMSIL H
		KESTREL 600
		KS 160
		KS 300
		KS 380
		KS 404
		LUDOX
		LUDOX AS 30
		LUDOX HS 30
		LUDOX HS 40
		LUDOX RS 40
		LUFILEN E 100
		MANOSIL VN 3
		MAS 200
		MICROSIL
		MILOWHITE
		MIN-U-SIL
		MINUSIL 30
		MINUSIL 5
		MIZUKASIL
		MIZUKASIL P 527
		MIZUKASIL P 801
		MIZUKASIL SK 7
		NALCAST PLW
		NALCOAG 1030
		NALCOAG 1034A
		NALCOAG 1050
		NALCOAG 1115
		NALFLOC N 1030
		NALFLOC N 1050
		NEOSIL
		NEOSIL A
		NEOSIL XV
		NEOSYL
		NEOSYL 186
		NEOSYL 224
		NEOSYL 81
		NIPSIL 300A
		NIPSIL E 150
		NIPSIL E 150J
		NIPSIL ER
		NIPSIL NS
		NIPSIL NST
		NIPSIL VN 3
		NIPSIL VN3LP
		NOVAKUP
		NYACOL 2034A
		OK 412
		PORASIL
		POSITIVE SOL 130M
		POSITIVE SOL 232
		PREGEL
		PROTEK-SORB 121
		QUSO 51
		QUSO G 30
		QUSO G 32
		QUSO WR 82
		S 600
		SANTOCEL
		SANTOCEL 54
		SANTOCEL 62
		SANTOCEL CS
		SANTOCEL Z
		SI-O-LITE
		SIFLOX
		SILANOX 101
		SILEX
		SILICA
		SILICA (SiO2)
	*	SILICA, AMORPHOUS, FUMED
		SILICAFILM
		SILICALITE S 115
		SILICON DIOXIDE
		SILICON DIOXIDE (SiO2)
		SILICON OXIDE (SiO2)
		SILIKIL
		SILIPUR
		SILLIKOLLOID
		SILMOS
		SILOXID
		SILTON A
		SILTON A 2
		SILTON R 2
		SIONOX
		SIPERNAT 22
		SIPUR 1500
		SNOWTEX
		SNOWTEX 20
		SNOWTEX 30
		SNOWTEX C
		SNOWTEX N
		SNOWTEX O
		SNOWTEX OL
		SORBSIL MSG
		SS 10
		SSA 1
		SSK 5
		SUPER-CEL
		SUPERFLOSS
		SYTON
		SYTON 2X
		SYTON FM
		SYTON W 15
		SYTON W 3
		SYTON X 30
		TARANOX 500
		TIX-O-SIL 33J
		TIX-O-SIL 38A
		TK 900
		TOKUSIL GU
		TOKUSIL GU-N
		TOKUSIL GV-N
		TOKUSIL N
		TOKUSIL P
		TOKUSIL TPLM
		TOKUSIL U
		TOSIL
		TOSIL P
		TULLANOX TM 500
		U 333
		ULTRASIL VH 3
		ULTRASIL VN 2
		ULTRASIL VN 3
		VERTICURINE
		VITASIL 1500
		VITASIL 1600
		VITASIL 220
		VULKASIL C
		VULKASIL S
		WESSALON S
		WHITE CARBON
		ZEO 49
		ZEODENT 113
		ZEOFREE 80
		ZEOSYL 200
		ZEOSYL 2000
		ZEOTHIX 265
		ZIPAX
		ZORBAX SIL
		ZORBAX SILICA
7631-89-2		
		ARSENIC ACID (H3AsO4), SODIUM SALT
		ARSENIC ACID, TRISODIUM SALT
	*	SODIUM ARSENATE
		SODIUM ORTHOARSENATE
		TRISODIUM ARSENATE
7631-90-5		
		BISULFITE DE SODIUM (FRENCH)
		FR-62
		HYDROGEN SODIUM SULFATE
		HYDROGEN SULFITE SODIUM
		MONOSODIUM SULFITE
		NA 2693 (DOT)

CAS No.	IN	Chemical Name
		SODIUM ACID SULFITE
	*	SODIUM BISULFITE
		SODIUM BISULFITE (1:1)
		SODIUM BISULFITE (ACGIH)
		SODIUM BISULFITE (NaHSO3)
		SODIUM BISULFITE, (SOLID)
		SODIUM BISULFITE, SOLID (DOT)
		SODIUM BISULFITE, SOLUTION (DOT)
		SODIUM BISULPHITE
		SODIUM HYDROGEN SULFITE
		SODIUM HYDROGEN SULFITE, SOLID
		SODIUM HYDROGEN SULFITE, SOLID (DOT)
		SODIUM HYDROGEN SULFITE, SOLUTION
		SODIUM HYDROGEN SULFITE, SOLUTION (DOT)
		SODIUM SULFITE (NaHSO3)
		SODIUM SULHYDRATE
7631-90-5		
		SULFUROUS ACID, MONOSODIUM SALT
		UN 2693 (DOT)
7631-99-4		
		CHILE SALTPETER
		CUBIC NITER
		NITER
		NITRATE DE SODIUM (FRENCH)
		NITRATINE
		NITRIC ACID SODIUM SALT
		SALTPETER (CHILE)
		SODA NITER
		SODIUM NITRATE
		SODIUM NITRATE (DOT)
		SODIUM(I) NITRATE (1:1)
		UN 1498 (DOT)
7632-00-0		
		ANTI-RUST
		DIAZOTIZING SALTS
		DUSITAN SODNY (CZECH)
		ERINITRIT
		FILMERINE
		NATRIUM NITRIT (GERMAN)
		NCI-C02084
		NITRITE DE SODIUM (FRENCH)
		NITROUS ACID SODIUM SALT (1:1)
		NITROUS ACID, SODIUM SALT
	*	SODIUM NITRITE
		SODIUM NITRITE (DOT)
		SODIUM NITRITE (NaNO2)
		SYNFAT 1004
		UN 1500 (DOT)
7632-04-4		
		PERBORIC ACID (HBO3), SODIUM SALT
		PERBORIC ACID, SODIUM SALT
		SODIUM BORATE
		SODIUM BORATE (NaBO3)
		SODIUM PERBORATE
		SODIUM PERBORATE (NaBO3)
		SODIUM PEROXOBORATE
7632-51-1		
		TETRACHLOROVANADIUM
		UN 2444 (DOT)
		VANADIUM CHLORIDE
		VANADIUM CHLORIDE (VCl4)
		VANADIUM CHLORIDE (VCl4), (T-4)-
		VANADIUM TETRACHLORIDE
		VANADIUM TETRACHLORIDE (DOT)
		VANADIUM(IV) CHLORIDE
7637-07-2		
		ANCA 1040
		BORANE, TRIFLUORO-
		BORON FLUORIDE
		BORON FLUORIDE (BF3)
	*	BORON TRIFLUORIDE
		TRIFLUOROBORANE
		TRIFLUOROBORON
7645-25-2		
		ARSENIC ACID (H3AsO4), LEAD SALT
		LEAD ARSENATE
7646-69-7		
		NAH 80
	*	SODIUM HYDRIDE
		SODIUM HYDRIDE (DOT)
		SODIUM HYDRIDE (NaH)
		SODIUM MONOHYDRIDE
7646-78-8		
		ETAIN (TETRACHLORURE D') (FRENCH)
		LIBAVIUS FUMING SPIRIT
		STAGNO (TETRACLORURO DI) (ITALIAN)
		STANNANE, TETRACHLORO-
		STANNIC CHLORIDE
		STANNIC CHLORIDE, ANHYDROUS
		STANNIC CHLORIDE, ANHYDROUS (DOT)
		STANNIC TETRACHLORIDE
		TETRACHLOROSTANNANE
		TETRACHLOROTIN
		TIN CHLORIDE (SnCl4)
		TIN CHLORIDE, FUMING (DOT)
		TIN PERCHLORIDE
		TIN PERCHLORIDE (DOT)
	*	TIN TETRACHLORIDE
		TIN TETRACHLORIDE, ANHYDROUS
		TIN TETRACHLORIDE, ANHYDROUS (DOT)
		TIN(IV) CHLORIDE
		TIN(IV) CHLORIDE (1:4)
		TIN(IV) TETRACHLORIDE
		TINTETRACHLORIDE (DUTCH)
		UN 1827 (DOT)
		ZINNTETRACHLORID (GERMAN)
7646-79-9		
		COBALT CHLORIDE
		COBALT CHLORIDE (CoCl2)
		COBALT DICHLORIDE
		COBALT DICHLORIDE (CoCl2)
		COBALT MURIATE
		COBALT(2+) CHLORIDE
		COBALT(II) CHLORIDE
		COBALTOUS CHLORIDE
		COBALTOUS DICHLORIDE
		KOBALT CHLORID (GERMAN)
7646-85-7		
		BUTTER OF ZINC
		CHLORURE DE ZINC (FRENCH)
		TINNING FLUX (DOT)
		UN 1840 (DOT)
		UN 2331 (DOT)
		ZINC (CHLORURE DE) (FRENCH)
		ZINC BUTTER
	*	ZINC CHLORIDE
		ZINC CHLORIDE (ACGIH)
		ZINC CHLORIDE (ZnCl2)
		ZINC CHLORIDE FUME
		ZINC CHLORIDE SOLUTION
		ZINC CHLORIDE, ANHYDROUS (DOT)
		ZINC CHLORIDE, SOLID
		ZINC CHLORIDE, SOLID (DOT)
		ZINC CHLORIDE, SOLUTION (DOT)
		ZINC DICHLORIDE
		ZINC MURIATE, SOLUTION (DOT)
		ZINC(II) CHLORIDE
		ZINCO (CLORURO DI) (ITALIAN)
		ZINKCHLORID (GERMAN)
		ZINKCHLORIDE (DUTCH)
7646-93-7		
		ACID POTASSIUM SULFATE
		HYDROGEN POTASSIUM SULFATE
		MONOPOTASSIUM SULFATE

CAS No.	IN	Chemical Name
		POTASSIUM ACID SULFATE
		POTASSIUM BIFLUORIDE, SOLUTION
		POTASSIUM BISULFATE
		POTASSIUM BISULPHATE
	*	POTASSIUM HYDROGEN SULFATE
		POTASSIUM HYDROGEN SULFATE, SOLID
		POTASSIUM HYDROGEN SULFATE, SOLID (DOT)
		POTASSIUM SULFATE
		POTASSIUM SULFATE (KHSO4)
		SAL ENIXUM
		SULFURIC ACID POTASSIUM SALT (1:1)
		SULFURIC ACID, MONOPOTASSIUM SALT
		UN 2509 (DOT)
7647-01-0		
		ACIDE CHLORHYDRIQUE (FRENCH)
		ACIDO CLORIDRICO (ITALIAN)
		ANHYDROUS HYDROCHLORIC ACID
		CHLOORWATERSTOF (DUTCH)
		CHLOROHYDRIC ACID
		CHLOROWODOR (POLISH)
		CHLORWASSERSTOFF (GERMAN)
		DILUTE HYDROCHLORIC ACID
		HCL
		HYDROCHLORIC ACID
		HYDROCHLORIC ACID (DOT)
		HYDROCHLORIC ACID GAS
		HYDROCHLORIC ACID, ANHYDROUS (DOT)
		HYDROCHLORIC ACID, SOLUTION (DOT)
		HYDROCHLORIC ACID, SOLUTION, INHIBITED (DOT)
		HYDROCHLORIDE
	*	HYDROGEN CHLORIDE
		HYDROGEN CHLORIDE (ACGIH,DOT)
		HYDROGEN CHLORIDE (HCl)
		HYDROGEN CHLORIDE (LIQUEFIED GAS)
		HYDROGEN CHLORIDE, ANHYDROUS
		HYDROGEN CHLORIDE, ANHYDROUS (DOT)
		HYDROGEN CHLORIDE, REFRIGERATED LIQUID
		HYDROGEN CHLORIDE, REFRIGERATED LIQUID (DOT)
		MURIATIC ACID
		MURIATIC ACID (DOT)
		SPIRITS OF SALT
		SPIRITS OF SALT (DOT)
		UN 1050 (DOT)
		UN 1789 (DOT)
		UN 2186 (DOT)
7647-01-1		
		HYDROGEN CHLORIDE
7647-10-1		
		PALLADIUM (II) CHLORIDE
7647-18-9		
		ANTIMOINE (PENTACHLORURE D') (FRENCH)
		ANTIMONIO (PENTACLORURO DI) (ITALIA N)
		ANTIMONIO (PENTACLORURO DI) (ITALIAN)
		ANTIMONPENTACHLORID (GERMAN)
		ANTIMONY (V) CHLORIDE
		ANTIMONY CHLORIDE (SbCl5)
	*	ANTIMONY PENTACHLORIDE
		ANTIMONY PENTACHLORIDE (DOT)
		ANTIMONY PENTACHLORIDE (SbCl5)
		ANTIMONY PENTACHLORIDE SOLUTION
		ANTIMONY PENTACHLORIDE SOLUTION (DOT)
		ANTIMONY PENTACHLORIDE, LIQUID (DOT)
		ANTIMONY PERCHLORIDE
		ANTIMONY(V) CHLORIDE
		ANTIMOONPENTACHLORIDE (DUTCH)
		BUTTER OF ANTIMONY
		PENTACHLOROANTIMONY
		PENTACHLORURE D'ANTIMOINE (FRENCH)
		PERCHLORURE D'ANTIMOINE (FRENCH)
		TIN CHLORIDE
		UN 1730 (DOT)
		UN 1731 (DOT)
7647-19-0		
		PENTAFLUOROPHOSPHORANE
		PENTAFLUOROPHOSPHORUS
		PHOSPHORANE, PENTAFLUORO-
		PHOSPHORUS FLUORIDE (PF5)
		PHOSPHORUS PENTAFLUORIDE
		PHOSPHORUS PENTAFLUORIDE (DOT)
		UN 2198 (DOT)
7664-38-2		
		ACIDE PHOSPHORIQUE (FRENCH)
		ACIDO FOSFORICO (ITALIAN)
		DECON 4512
		EVITS
		FOSFORZUUROPLOSSINGEN (DUTCH)
		ORTHOPHOSPHORIC ACID
	*	PHOSPHORIC ACID
		PHOSPHORIC ACID (ACGIH,DOT)
		PHOSPHORIC ACID, LIQUID (DOT)
		PHOSPHORIC ACID, SOLID (DOT)
		PHOSPHORSAEURELOESUNGEN (GERMAN)
		SONAC
		UN 1805 (DOT)
		WC-REINIGER
7664-39-3		
		ACIDE FLUORHYDRIQUE (FRENCH)
		ACIDO FLUORIDRICO (ITALIAN)
		ANHYDROUS HYDROFLUORIC ACID
		ANHYDROUS HYDROFLUORIC ACID (DOT)
		ANTISAL 2B
		FLUORHYDRIC ACID
		FLUOROWODOR (POLISH)
		FLUORWASSERSTOFF (GERMAN)
		FLUORWATERSTOF (DUTCH)
		HYDROFLUORIC ACID
		HYDROFLUORIC ACID GAS
		HYDROFLUORIC ACID SOLUTION
		HYDROFLUORIC ACID SOLUTION (DOT)
		HYDROFLUORIC ACID, ANHYDROUS (DOT)
		HYDROFLUORIDE
	*	HYDROGEN FLUORIDE
		HYDROGEN FLUORIDE (ACGIH,DOT)
		HYDROGEN FLUORIDE (HF)
		RCRA WASTE NUMBER U134
		UN 1052 (DOT)
		UN 1790 (DOT)
7664-41-7		
		AM-FOL
	*	AMMONIA
		AMMONIA (ACGIH)
		AMMONIA ANHYDROUS
		AMMONIA GAS
		AMMONIA SOLUTION, CONTAINING MORE THAN 44% AMMONIA (DOT)
		AMMONIA SOLUTION, CONTAINING MORE THAN 50% AMMONIA (DOT)
		AMMONIA, ANHYDROUS
		AMMONIA, ANHYDROUS (DOT)
		AMMONIA-14N
		AMMONIAC (FRENCH)
		AMMONIACA (ITALIAN)
		AMMONIAK (GERMAN)
		AMMONIUM AMIDE
		AMONIAK (POLISH)
		ANHYDROUS AMMONIA (DOT)
		NITRO-SIL
		R 717
		REFRIGERENT R717
		SPIRIT OF HARTSHORN
		UN 1005 (DOT)
		UN 2073 (DOT)
7664-93-9		
		ACIDE SULFURIQUE (FRENCH)
		ACIDO SOLFORICO (ITALIAN)
		BOV
		DIHYDROGEN SULFATE

CAS No.	IN	Chemical Name
		DIPPING ACID
		HYDROGEN SULFATE (DOT)
		MATTING ACID (DOT)
		NORDHAUSEN ACID (DOT)
		OIL OF VITRIOL
		OIL OF VITRIOL (DOT)
		SCHWEFELSAEURELOESUNGEN (GERMAN)
		SPENT SULFURIC ACID (DOT)
	*	SULFURIC ACID
		SULFURIC ACID (ACGIH,DOT)
		SULFURIC ACID, SPENT
		SULFURIC ACID, SPENT (DOT)
		SULPHURIC ACID
		UN 1830 (DOT)
		UN 1832 (DOT)
		VITRIOL BROWN OIL
		VITRIOL, OIL OF (DOT)
		ZWAVELZUUROPLOSSINGEN (DUTCH)
7681-11-0		
		ASMOFUG E
		K1-N
		KAIOD
		KNOLLIDE
		PHERAJOD
		POTASSIUM IODIDE
		POTASSIUM IODIDE (KI)
		POTASSIUM MONOIODIDE
		POTIDE
7681-38-1		
		BIF
		BISULFATE OF SODA
		FANAL
		GBS
		MONOSODIUM HYDROGEN SULFATE
		MONOSODIUM SULFATE
		NITER CAKE
		NITRE CAKE
		SODIUM ACID SULFATE
		SODIUM ACID SULFATE, SOLID (DOT)
		SODIUM ACID SULFATE, SOLUTION (DOT)
		SODIUM BISULFATE
		SODIUM BISULFATE, FUSED
		SODIUM BISULFATE, SOLID (DOT)
		SODIUM BISULFATE, SOLUTION (DOT)
		SODIUM BISULPHATE
	*	SODIUM HYDROGEN SULFATE
		SODIUM HYDROGEN SULFATE (NaHSO4)
		SODIUM HYDROGEN SULFATE SOLUTION
		SODIUM HYDROGEN SULFATE, SOLID
		SODIUM HYDROGEN SULFATE, SOLID (DOT)
		SODIUM HYDROGEN SULFATE, SOLUTION (DOT)
		SODIUM HYDROSULFATE
		SODIUM PYROSULFATE
		SULFURIC ACID SODIUM SALT (1:1)
		SULFURIC ACID, MONOSODIUM SALT
		UN 1821 (DOT)
		UN 2837 (DOT)
		WC 00
		WC-KLOSETTREINIGER
		WC-PERFECT
		WC-SUPER
7681-49-4		
		ACT
		ALCOA SODIUM FLUORIDE
		ANTIBULIT
		CAVI-TROL
		CHEMIFLUOR
		CREDO
		DISODIUM DIFLUORIDE
		DURAPHAT
		F1-TABS
		FDA 0101
		FLORIDINE
		FLOROCID
		FLOZENGES
		FLUDENT
		FLUOR-O-KOTE
		FLUORADAY
		FLUORAL
		FLUORID SODNY (CZECH)
		FLUORIDE, SODIUM
		FLUORIDENT
		FLUORIGARD
		FLUORINEED
		FLUORINSE
		FLUORITAB
		FLUOROCID
	*	FLUOROL
		FLUORURE DE SODIUM (FRENCH)
		FLURA
		FLURA DROPS
		FLURA-GEL
		FLURA-LOZ
		FLURCARE
		FLURSOL
		FUNGOL B
		GEL II
		GELUTION
		GLEEM
		IRADICAV
		KARI-RINSE
		KARIDIUM
		KARIGEL
		LEA-COV
		LEMOFLUR
		LURIDE
		LURIDE LOZI-TABS
		LURIDE-SF
		NA FRINSE
		NAFEEN
		NAFPAK
		NATRIUM FLUORIDE
		NCI-C55221
		NUFLUOR
		OSSALIN
		OSSIN
		PEDIAFLOR
		PEDIDENT
		PENNWHITE
		PERGANTENE
		PHOS-FLUR
		POINT TWO
		PREDENT
		RAFLUOR
		RESCUE SQUAD
		ROACH SALT
		SO-FLO
	*	SODIUM FLUORIDE
		SODIUM FLUORIDE (NaF)
		SODIUM FLUORIDE CYCLIC DIMER
		SODIUM FLUORIDE, SOLID
		SODIUM FLUORIDE, SOLID (DOT)
		SODIUM FLUORIDE, SOLUTION
		SODIUM FLUORIDE, SOLUTION (DOT)
		SODIUM FLUORURE (FRENCH)
		SODIUM HYDROFLUORIDE
		SODIUM MONOFLUORIDE
		STAY-FLO
		STUDAFLUOR
		SUPER-DENT
		T-FLUORIDE
		THERA FLUR
		THERA-FLUR-N
		TRISODIUM TRIFLUORIDE
		UN 1690 (DOT)
		VILLIAUMITE
		ZYMAFLUOR
7681-52-9		
		ANTIFORMIN
		B-K LIQUID
		CARREL-DAKIN SOLUTION
		CHLOROS
		CLOROX
		DAKIN'S SOLUTION

CAS No.	IN	Chemical Name
		DEOSAN
		HYCLORITE
		HYPOCHLOROUS ACID, SODIUM SALT
		JAVEX
		KLOROCIN
		MILTON
		MODIFIED DAKIN'S SOLUTION
		NAMED REAGENTS AND SOLUTIONS, DAKIN'S
		NEO-CLEANER
		PURIN B
	*	SODIUM HYPOCHLORITE
7681-57-4		
		DISODIUM DISULFITE
		DISODIUM PYROSULFITE
		DISULFUROUS ACID, DISODIUM SALT
		FERTISILO
		NA 2693 (DOT)
		PYROSULFUROUS ACID, DISODIUM SALT
		PYROSULFUROUS ACID, DISODIUM SALT (8CI)
		SODIUM DISULFITE
		SODIUM M-BISULFITE
	*	SODIUM METABISULFITE
		SODIUM METABISULFITE (ACGIH,DOT)
		SODIUM METABISULFITE (Na2S2O5)
		SODIUM METABISULPHITE
		SODIUM PYROSULFITE
		SODIUM PYROSULFITE (Na2S2O5)
7688-21-3		
		2-HEXENE-CIS
7697-37-2		
		ACIDE NITRIQUE (FRENCH)
		ACIDO NITRICO (ITALIAN)
		AQUA FORTIS
		AZOTIC ACID
		AZOTOWY KWAS (POLISH)
		HYDROGEN NITRATE
		NA 1760 (DOT)
		NITAL
	*	NITRIC ACID
		NITRIC ACID (ACGIH,DOT)
		NITRIC ACID, 40% OR LESS (DOT)
		NITRIC ACID, FUMING
		NITRIC ACID, OVER 40% (DOT)
		NITRYL HYDROXIDE
		SALPETERSAURE (GERMAN)
		SALPETERZUUROPLOSSINGEN (DUTCH)
		UN 2031 (DOT)
7699-43-6		
		BASIC ZIRCONIUM CHLORIDE
		CHLOROZIRCONYL
		DICHLOROOXOZIRCONIUM
		NCI-C60811
		ZIRCONIUM CHLORIDE OXIDE (ZrCl2O)
		ZIRCONIUM CHLORIDE, BASIC
		ZIRCONIUM DICHLORIDE MONOXIDE
		ZIRCONIUM DICHLORIDE OXIDE
		ZIRCONIUM OXIDE CHLORIDE (ZrOCl2)
		ZIRCONIUM OXYCHLORIDE
		ZIRCONIUM OXYDICHLORIDE
		ZIRCONIUM, DICHLOROOXO-
		ZIRCONYL CHLORIDE
		ZIRCONYL CHLORIDE (ZrOCl2)
		ZIRCONYL DICHLORIDE
7699-45-8		
		NA 9156 (DOT)
	*	ZINC BROMIDE
		ZINC BROMIDE (DOT)
		ZINC BROMIDE (ZnBr2)
		ZINC DIBROMIDE
7704-34-9		
		ASULFA-SUPRA
		ATOMIC SULFUR
		BENSULFOID
		BRIMSTONE
		COLLOIDAL SULFUR
		COLLOIDAL-S
		COLLOKIT
		COLSUL
		COROSUL D AND S
		COSAN
		COSAN 80
		CRYSTEX
		DEVISULPHUR
		ELOSAL
		FLOUR SULPHUR
		FLOWERS OF SULFUR (DOT)
		FLOWERS OF SULPHUR
		GROUND VOCLE SULPHUR
		HEXASUL
		KOCIDE
		KOLO 100
		KOLOFOG
		KOLOSPRAY
		KRISTEX
		KUMULUS
		KUMULUS FL
		MAGNETIC 70, 90, AND 95
		MICOWETSULF
		MICROFLOTOX
		MICROSULF
		MICROTHIOL
		NETZSCHWEFEL
		POLSULKOL EXTRA
		PRECIPITATED SULFUR
		PRECIPITATED SULPHUR
		RC-SCHWEFEL EXTRA
		SHREESUL
		SICKOSUL
		SOFRIL
		SPERLOX-S
		SPERSUL
		SPERSUL THIOVIT
		SUBLIMED SULFUR
		SUBLIMED SULPHUR
		SUFRAN
		SUFRAN D
		SULFEX
		SULFIDAL
		SULFORON
	*	SULFUR
		SULFUR ATOM
		SULFUR ELEMENT
		SULFUR FLOWER (DOT)
		SULFUR OINTMENT
		SULFUR, MOLTEN
		SULFUR, SOLID
		SULFUR, SOLID (DOT)
		SULIKOL
		SULIKOL K
		SULKOL
		SULPHUR
		SULPHUR (DOT)
		SULPHUR, LUMP OR POWDER (DOT)
		SULPHUR, MOLTEN (DOT)
		SULSOL
		SULTAF
		SUPER COSAN
		SVOVEL
		SVOVL
		TECHNETIUM TC 99M SULFUR COLLOID
		TESULOID
		THIOLUX
		THIOSOL
		THIOVIT
		THIOZOL
		TIOZOL 80
		ULTRA SULFUR
		UN 1350 (DOT)
		UN 2448 (DOT)
		WETSULF
		WETTASUL

CAS No.	*IN*	*Chemical Name*
7704-98-5		
		TITANIUM DIHYDRIDE
		TITANIUM DIHYDRIDE (TiH2)
		TITANIUM HYDRIDE
		TITANIUM HYDRIDE (DOT)
		TITANIUM HYDRIDE (TiH2)
		UN 1871 (DOT)
7704-99-6		
		UN 1437 (DOT)
		ZIRCONIUM DIHYDRIDE
		ZIRCONIUM HYDRIDE
		ZIRCONIUM HYDRIDE (DOT)
		ZIRCONIUM HYDRIDE (ZrH2)
7705-07-9		
		TAC 121
		TAC 131
		TAC 132
		TAS 101
		TGY 24
		TITANIUM CHLORIDE
		TITANIUM CHLORIDE (TiCl3)
		TITANIUM CHLORIDE, MIXTURES, NON-PYROPHORIC
		TITANIUM TRICHLORIDE
		TITANIUM TRICHLORIDE MIXTURE
		TITANIUM TRICHLORIDE MIXTURES, NON-PYROPHORIC (DOT)
		TITANIUM TRICHLORIDE, PYROPHORIC
		TITANIUM TRICHLORIDE, PYROPHORIC (DOT)
		TITANIUM(III) CHLORIDE
		TITANOUS CHLORIDE
		TRICHLOROTITANIUM
		UN 2441 (DOT)
		UN 2869 (DOT)
7705-08-0		
		CHLORURE PERRIQUE (FRENCH)
		FERRIC CHLORIDE
		FERRIC CHLORIDE SOLUTION
		FERRIC CHLORIDE, SOLID (DOT)
		FERRIC CHLORIDE, SOLID, ANHYDROUS
		FERRIC CHLORIDE, SOLID, ANHYDROUS (DOT)
		FERRIC CHLORIDE, SOLUTION (DOT)
		FERRIC TRICHLORIDE
		FLORES MARTIS
	*	IRON CHLORIDE
		IRON CHLORIDE (FeCl3)
		IRON CHLORIDE, SOLID (DOT)
		IRON PERCHLORIDE
		IRON SESQUICHLORIDE, SOLID (DOT)
		IRON TRICHLORIDE
		IRON(III) CHLORIDE
		PERCHLORURE DE FER (FRENCH)
		UN 1773 (DOT)
		UN 2582 (DOT)
7718-54-9		
		NA 9139 (DOT)
	*	NICKEL CHLORIDE
		NICKEL CHLORIDE (DOT)
		NICKEL CHLORIDE (NiCl2)
		NICKEL DICHLORIDE
		NICKEL(2+) CHLORIDE
		NICKEL(II) CHLORIDE
		NICKEL(II) CHLORIDE (1:2)
		NICKELOUS CHLORIDE
7718-98-1		
		UN 2475 (DOT)
		VANADIUM CHLORIDE (VCl3)
		VANADIUM TRICHLORIDE
		VANADIUM TRICHLORIDE (DOT)
		VANADIUM(3+) CHLORIDE
		VANADIUM(III) CHLORIDE
7719-09-7		
		SULFINYL CHLORIDE
		SULFUR CHLORIDE OXIDE
		SULFUR CHLORIDE OXIDE (Cl2SO)
		SULFUR OXYCHLORIDE (SOCl2)
		SULFUROUS DICHLORIDE
		SULFUROUS OXYCHLORIDE
	*	THIONYL CHLORIDE
		THIONYL CHLORIDE (ACGIH,DOT)
		THIONYL DICHLORIDE
		UN 1836 (DOT)
7719-12-2		
		CHLORIDE OF PHOSPHORUS (DOT)
		FOSFORO(TRICLORURO DI) (ITALIAN)
		FOSFORTRICHLORIDE (DUTCH)
		PHOSPHINE, TRICHLORO-
		PHOSPHORE(TRICHLORURE DE) (FRENCH)
		PHOSPHOROUS CHLORIDE
		PHOSPHOROUS TRICHLORIDE
		PHOSPHORTRICHLORID (GERMAN)
		PHOSPHORUS CHLORIDE
		PHOSPHORUS CHLORIDE (Cl6P2)
		PHOSPHORUS CHLORIDE (DOT)
		PHOSPHORUS CHLORIDE (PCl3)
	*	PHOSPHORUS TRICHLORIDE
		PHOSPHORUS TRICHLORIDE (ACGIH,DOT)
		TRICHLOROPHOSPHINE
		TROJCHLOREK FOSFORU (POLISH)
		UN 1809 (DOT)
7720-78-7		
		COPPERAS
		DURETTER
		DUROFERON
		EXSICCATED FERROUS SULFATE
		EXSICCATED FERROUS SULPHATE
		FEOSOL
		FEOSPAN
		FER-IN-SOL
		FERO-GRADUMET
		FERRALYN
		FERRO-GRADUMET
		FERRO-THERON
		FERROSULFAT (GERMAN)
		FERROSULFATE
	*	FERROUS SULFATE
		FERROUS SULFATE (1:1)
		FERROUS SULFATE (DOT)
		FERROUS SULPHATE
		FERSOLATE
		GREEN VITRIOL
		IRON MONOSULFATE
		IRON PROTOSULFATE
		IRON SULFATE (1:1)
		IRON SULFATE (FeSO4)
		IRON VITRIOL
		IRON(2+) SULFATE
		IRON(2+) SULFATE (1:1)
		IRON(II) SULFATE
		IRON(II) SULFATE (1:1)
		IROSPAN
		IROSUL
		NA 9125 (DOT)
		SLOW-FE
		SULFERROUS
		SULFURIC ACID IRON SALT (1:1)
		SULFURIC ACID, IRON(2+) SALT (1:1)
7722-64-7		
		C.I. 77755
		CAIROX
		CHAMELEON MINERAL
		CONDY'S CRYSTALS
		PERMANGANIC ACID (HMnO4), POTASSIUM SALT
		PERMANGANIC ACID POTASSIUM SALT
	*	POTASSIUM PERMANGANATE
7722-84-1		
		ALBONE
		ALBONE 35
		ALBONE 35CG

CAS No.	IN	Chemical Name
		ALBONE 50
		ALBONE 50CG
		ALBONE 70
		ALBONE 70CG
		ALBONE DS
		DIHYDROGEN DIOXIDE
		HIOXYL
		HYDROGEN DIOXIDE
	*	HYDROGEN PEROXIDE
		HYDROGEN PEROXIDE (ACGIH)
		HYDROGEN PEROXIDE (H2O2)
		HYDROGEN PEROXIDE SOLUTION (40% TO 52% PEROXIDE)
		HYDROGEN PEROXIDE SOLUTION (8% TO 40% PEROXIDE)
		HYDROGEN PEROXIDE SOLUTION (8% TO 40% PEROXIDE) (DOT)
		HYDROGEN PEROXIDE SOLUTION, 40% TO 52% (DOT)
		HYDROGEN PEROXIDE, 20% TO 60%
		HYDROGEN PEROXIDE, 30%
		HYDROGEN PEROXIDE, 90%
		HYDROGEN PEROXIDE, SOLUTION (DOT)
		HYDROGEN PEROXIDE, SOLUTION (OVER 52% PEROXIDE) (DOT)
		HYDROGEN PEROXIDE, STABILIZED (OVER 60% PEROXIDE) (DOT)
		HYDROPEROXIDE
		INHIBINE
		INTEROX
		KASTONE
		OXYDOL
		PERHYDROL
		PERONE
		PERONE 30
		PERONE 35
		PERONE 50
		PEROSSIDO DI IDROGENO (ITALIAN)
		PEROXAAN
		PEROXAN
		PEROXIDE
		PEROXYDE D'HYDROGENE (FRENCH)
		SUPEROXOL
		T-STUFF
		UN 2014 (DOT)
		UN 2015 (DOT)
		WASSERSTOFFPEROXID (GERMAN)
		WATERSTOFPEROXYDE (DUTCH)
7722-88-5		
		ANHYDROUS TETRASODIUM PYROPHOSPHATE
		DIPHOSPHORIC ACID, TETRASODIUM SALT
		PHOSPHOTEX
		PYROPHOSPHORIC ACID, TETRASODIUM SALT
		SODIUM DIPHOSPHATE
		SODIUM DIPHOSPHATE (Na4P2O7)
		SODIUM PHOSPHATE (Na4P2O7)
		SODIUM PYROPHOSPHATE (Na4P2O7)
		TETRASODIUM DIPHOSPHATE
		TETRASODIUM DIPHOSPHATE (Na4P2O7)
	*	TETRASODIUM PYROPHOSPHATE
		TETRASODIUM PYROPHOSPHATE (Na4P2O7)
		TETRON
		TETRON (DISPERSANT)
		TSPP
		VICTOR TSPP
7723-14-0		
		1 PALLET 120UF
		BONIDE BLUE DEATH RAT KILLER
		COMMON SENSE COCKROACH AND RAT PREPARATIONS
		EXOLIT 405
		EXOLIT LPKN 275
		EXOLIT VPK-N 361
		EXOLITE 405
		FOSFORO BIANCO (ITALIAN)
		GELBER PHOSPHOR (GERMAN)
		NOVARED 120UF
		PHOSPHORE BLANC (FRENCH)
		PHOSPHOROUS (WHITE)
		PHOSPHOROUS YELLOW
		PHOSPHORUS
		PHOSPHORUS (RED)
		PHOSPHORUS (WHITE)
	*	PHOSPHORUS (YELLOW OR WHITE)
		PHOSPHORUS (YELLOW)
		PHOSPHORUS (YELLOW) (ACGIH)
		PHOSPHORUS :AMORPHOUS
		PHOSPHORUS ELEMENT
		PHOSPHORUS WHITE, DRY (DOT)
		PHOSPHORUS WHITE, IN WATER (DOT)
		PHOSPHORUS YELLOW, DRY (DOT)
		PHOSPHORUS YELLOW, IN WATER (DOT)
	*	PHOSPHORUS, AMORPHOUS, RED
		PHOSPHORUS, AMORPHOUS, RED (DOT)
		PHOSPHORUS, WHITE OR YELLOW
		PHOSPHORUS, WHITE, MOLTEN
		PHOSPHORUS, WHITE, MOLTEN (DOT)
		PHOSPHORUS-31
		RAT-NIP
		RED PHOSPHORUS
		TETRAFOSFOR (DUTCH)
		TETRAPHOSPHOR (GERMAN)
		UN 1338 (DOT)
		UN 1381 (DOT)
		UN 2447 (DOT)
		WEISS PHOSPHOR (GERMAN)
		WHITE PHOSPHORUS
		WHITE PHOSPHORUS, DRY (DOT)
		WHITE PHOSPHORUS, WET (DOT)
		YELLOW PHOSPHORUS
		YELLOW PHOSPHORUS, DRY (DOT)
		YELLOW PHOSPHORUS, WET (DOT)
7726-95-6		
		BROM (GERMAN)
		BROME (FRENCH)
	*	BROMINE
		BROMINE (ACGIH,DOT)
		BROMINE ELEMENT
		BROMINE MOLECULE (Br2)
		BROMINE SOLUTION
		BROMINE SOLUTION (DOT)
		BROMO (ITALIAN)
		BROOM (DUTCH)
		DIATOMIC BROMINE
		DIBROMINE
		UN 1744 (DOT)
7727-15-3		
		ALUMINUM BROMIDE
		ALUMINUM BROMIDE (AlBr3)
		ALUMINUM BROMIDE (ANHYDROUS)
		ALUMINUM BROMIDE SOLUTION
		ALUMINUM BROMIDE SOLUTION (DOT)
		ALUMINUM BROMIDE, ANHYDROUS
		ALUMINUM BROMIDE, ANHYDROUS (DOT)
		ALUMINUM TRIBROMIDE
		ALUMINUM TRIBROMIDE (AlBr3)
		TRIBROMOALUMINUM
		UN 1725 (DOT)
		UN 2580 (DOT)
7727-18-6		
		TRICHLOROOXOVANADIUM
		TRICHLOROVANADIUM OXIDE
		UN 2443 (DOT)
		VANADIUM CHLORIDE OXIDE (VCl3O)
		VANADIUM MONOXIDE TRICHLORIDE
		VANADIUM OXIDE TRICHLORIDE
		VANADIUM OXYCHLORIDE
	*	VANADIUM OXYTRICHLORIDE
		VANADIUM OXYTRICHLORIDE (DOT)
		VANADIUM TRICHLORIDE MONOOXIDE
		VANADIUM TRICHLORIDE MONOXIDE
		VANADIUM TRICHLORIDE OXIDE
		VANADIUM(V) OXYTRICHLORIDE
		VANADIUM, TRICHLOROOXO-
		VANADYL CHLORIDE
		VANADYL CHLORIDE (VOCl3)
		VANADYL TRICHLORIDE

CAS No.	IN	Chemical Name
7727-21-1		
		ANTHION
		DIPOTASSIUM PEROXODISULFATE
		DIPOTASSIUM PEROXYDISULFATE
		DIPOTASSIUM PERSULFATE
		PEROXYDISULFURIC ACID ([(HO)S(O)2]2O2), DIPOTASSIUM SALT
		PEROXYDISULFURIC ACID, DIPOTASSIUM SALT
		POTASSIUM PEROXYDISULFATE
		POTASSIUM PEROXYDISULFATE (K2S2O3)
		POTASSIUM PEROXYDISULPHATE
	*	POTASSIUM PERSULFATE
		POTASSIUM PERSULFATE (ACGIH,DOT)
		UN 1492 (DOT)
7727-37-9		
		DIATOMIC NITROGEN
		DINITROGEN
		MOLECULAR NITROGEN
	*	NITROGEN
		NITROGEN (DOT)
		NITROGEN (LIQUIFIED)
		NITROGEN GAS
		NITROGEN, COMPRESSED (DOT)
		NITROGEN, REFRIGERATED LIQUID
		NITROGEN, REFRIGERATED LIQUID (DOT)
		NITROGEN-14
		UN 1066 (DOT)
		UN 1977 (DOT)
7727-54-0		
		AMMONIUM PEROXIDODISULFATE
		AMMONIUM PEROXYDISULFATE
		AMMONIUM PEROXYDISULFATE ((NH4)2S2O8)
		AMMONIUM PEROXYSULFATE
	*	AMMONIUM PERSULFATE
		AMMONIUM PERSULFATE (ACGIH,DOT)
		DIAMMONIUM PEROXYDISULFATE
		DIAMMONIUM PEROXYDISULPHATE
		DIAMMONIUM PERSULFATE
		PEROXYDISULFURIC ACID ([(HO)S(O)2]2O2), DIAMMONIUM SALT
		PEROXYDISULFURIC ACID, DIAMMONIUM SALT
		PERSULFATE D'AMMONIUM (FRENCH)
		UN 1444 (DOT)
7733-02-0		
		BONAZEN
		BUFOPTO ZINC SULFATE
		NA 9161 (DOT)
		OP-THAL-ZIN
		OPTRAEX
		SULFATE DE ZINC (FRENCH)
		SULFURIC ACID ZINC SALT
		SULFURIC ACID, ZINC SALT (1:1)
		VERAZINC
		WHITE COPPERAS
		WHITE VITRIOL
	*	ZINC SULFATE
		ZINC SULFATE (1:1)
		ZINC SULFATE (DOT)
		ZINC SULFATE (ZnSO4)
		ZINC SULPHATE
		ZINC VITRIOL
		ZINCOMED
		ZINKOSITE
7738-94-5		
	*	CHROMIC ACID
		CHROMIC ACID, SOLUTION
7756-94-7		
		1-PROPENE, 2-METHYL-, TRIMER
		1-PROPENE, 2-METHYL-, TRIMER (9CI)
		ISOBUTENE TRIMER
		PROPENE, 2-METHYL-, TRIMER
		TRIISOBUTYLENE
		TRIISOBUTYLENE (DOT)
		TRIISOCYANATOISOCYANURATE OF ISOPHORONEDIISOCYANATE, SOLUTION
		UN 2324 (DOT)
7757-74-6		
		DISODIUM DISULFITE
		DISODIUM PYROSULFITE
		DISULFUROUS ACID, DISODIUM SALT
		FERTISILO
		NA 2693 (DOT)
		PYROSULFUROUS ACID, DISODIUM SALT
		PYROSULFUROUS ACID, DISODIUM SALT (8CI)
		SODIUM DISULFITE
		SODIUM METABISULFITE
		SODIUM METABISULFITE (ACGIH,DOT)
		SODIUM METABISULFITE (Na2S2O5)
		SODIUM METABISULPHITE
		SODIUM PYROSULFITE
		SODIUM PYROSULFITE (Na2S2O5)
7757-79-1		
		COLLO-BO
		KALIUMNITRAT (GERMAN)
		NITER
		NITRE
		NITRIC ACID POTASSIUM SALT
		NITRIC ACID POTASSIUM SALT (1:1)
	*	POTASSIUM NITRATE
		POTASSIUM NITRATE (DOT)
		SALTPETER
		UN 1486 (DOT)
		VICKNITE
7757-82-6		
	*	SODIUM SULFATE (SOLUTION)
7758-01-2		
		BROMIC ACID, POTASSIUM SALT
	*	POTASSIUM BROMATE
		POTASSIUM BROMATE (DOT)
		UN 1484 (DOT)
7758-09-0		
		NITROUS ACID, POTASSIUM SALT
	*	POTASSIUM NITRITE
		POTASSIUM NITRITE (1:1)
		POTASSIUM NITRITE (DOT)
		UN 1488 (DOT)
7758-19-2		
		CHLOROUS ACID, SODIUM SALT
		NEO SILOX D
	*	SODIUM CHLORITE
		SODIUM CHLORITE (DOT)
		SODIUM CHLORITE, NOT MORE THAN 42% (DOT)
		SODIUM CHLORITE, SOLUTION CONTAINING MORE THAN 5% AVAILABLE CHLORINE (DOT)
		TEXTILE
		TEXTONE
		UN 1496 (DOT)
		UN 1908 (DOT)
7758-29-4		
		ARMOFOS
		PENTASODIUM TRIPHOSPHATE
		PENTASODIUM TRIPOLYPHOSPHATE
		POLYGON
		S 400
		SODIUM PHOSPHATE (Na5P3O10)
		SODIUM TRIPHOSPHATE (Na5P3O10)
	*	SODIUM TRIPOLYPHOSPHATE
		SODIUM TRIPOLYPHOSPHATE (Na5P3O10)
		STPP
		THERMPHOS
		THERMPHOS L 50
		THERMPHOS N
		THERMPHOS SPR
		TRIPHOSPHORIC ACID, PENTASODIUM SALT

CAS No.	IN	Chemical Name
7758-94-3		
		FERRO 66
	*	FERROUS CHLORIDE
		FERROUS CHLORIDE, SOLID
		FERROUS CHLORIDE, SOLID (DOT)
		FERROUS CHLORIDE, SOLUTION
		FERROUS CHLORIDE, SOLUTION (DOT)
		FERROUS DICHLORIDE
		IRON CHLORIDE (FeCl2)
		IRON DICHLORIDE
		IRON PROTOCHLORIDE
		IRON(2+) CHLORIDE
		IRON(II) CHLORIDE
		IRON(II) CHLORIDE (1:2)
		IRON(II) CHLORIDE (FeCl2)
		NA 1759 (DOT)
		NA 1760 (DOT)
7758-95-4		
		LEAD (II) CHLORIDE
	*	LEAD CHLORIDE
		LEAD CHLORIDE (DOT)
		LEAD CHLORIDE (PbCl2)
		LEAD DICHLORIDE
		LEAD(2+) CHLORIDE
		LEAD(II) CHLORIDE
		NA 2291 (DOT)
		PLUMBOUS CHLORIDE
7758-97-6		
		CHROMIC ACID (H2CrO4), LEAD(2+) SALT (1:1)
	*	LEAD CHROMATE
		LEAD CHROMATE (PbCrO4)
		LEAD CHROMIUM OXIDE (PbCrO4)
		PLUMBOUS CHROMATE
7758-98-7		
		BCS COPPER FUNGICIDE
		BLUE COPPER
		BLUE STONE
		BLUE VITRIOL
		COPPER (II) SULFATE (1:1)
		COPPER MONOSULFATE
		COPPER SULFATE
		COPPER SULFATE (1:1)
		COPPER SULFATE (CuSO4)
		COPPER SULFATE BASIC
		COPPER(2+) SULFATE
		COPPER(2+) SULFATE (1:1)
		COPPER(II) SULFATE
		CP BASIC SULFATE
	*	CUPRIC SULFATE
		CUPRIC SULFATE (DOT)
		CUPRIC SULFATE ANHYDROUS
		CUPRIC SULPHATE
		GRIFFIN SUPER CU
		INCRACIDE 10A
		INCRACIDE E 51
		KILCOP 53
		KOBASIC
		KUPFERSULFAT (GERMAN)
		NA 9109 (DOT)
		PHYTO-BORDEAUX
		ROMAN VITRIOL
		SULFATE DE CUIVRE (FRENCH)
		SULFURIC ACID COPPER(2+) SALT (1:1)
		TNCS 53
		TRINAGLE
7761-88-8		
		LUNAR CAUSTIC
		NITRATE D'ARGENT (FRENCH)
		NITRIC ACID SILVER SALT (1:1)
		NITRIC ACID SILVER(1+) SALT
		NITRIC ACID SILVER(I) SALT
		SILBERNITRAT
		SILVER MONONITRATE
	*	SILVER NITRATE
		SILVER NITRATE (AgNO3)
		SILVER NITRATE (DOT)
		SILVER(1+) NITRATE
		SILVER(I) NITRATE (1:1)
		UN 1493 (DOT)
7772-99-8		
		CI 77864
		DICHLOROTIN
		NA 1759 (DOT)
		NCI-C02722
		STANNOCHLOR
	*	STANNOUS CHLORIDE
		STANNOUS CHLORIDE, SOLID
		STANNOUS CHLORIDE, SOLID (DOT)
		STANNOUS DICHLORIDE
		TIN CHLORIDE
		TIN CHLORIDE (SnCl2)
		TIN DICHLORIDE
		TIN DICHLORIDE (SnCl2)
		TIN PROTOCHLORIDE
		TIN(II) CHLORIDE
		TIN(II) CHLORIDE (1:2)
		UNISTON CR-HT 200
7773-03-7		
		POTASSIUM ACID SULFITE
		POTASSIUM BISULFITE
		POTASSIUM BISULFITE SOLUTION
		POTASSIUM HYDROGEN SULFITE
		POTASSIUM SULFITE (KHSO3)
		SULFUROUS ACID, MONOPOTASSIUM SALT
7773-06-0		
		AMCIDE
		AMICIDE
		AMMAT
		AMMATE
		AMMATE X
		AMMONIUM AMIDOSULFATE
		AMMONIUM AMIDOSULFONATE
		AMMONIUM AMIDOSULPHATE
		AMMONIUM AMINOSULFONATE
		AMMONIUM SULFAMATE
		AMMONIUM SULFAMATE (ACGIH,DOT)
		AMMONIUM SULPHAMATE
		AMMONIUMSALZ DER AMIDOSULFONSAURE (GERMAN)
		AMS
		AMS (SALT)
		FELIDERM K
		FYRAN 206K
		IKURIN
		MONOAMMONIUM SULFAMATE
		NA 9089 (DOT)
		SULFAMATE
		SULFAMIC ACID, MONOAMMONIUM SALT
		SULFAMINSAURE (GERMAN)
7774-29-0		
		DIIODOMERCURY
		HYDRARGYRUM BIJODATUM (GERMAN)
		MERCURIC DIIODIDE
	*	MERCURIC IODIDE
		MERCURIC IODIDE, RED
		MERCURIC IODIDE, SOLID
		MERCURIC IODIDE, SOLID (DOT)
		MERCURIC IODIDE, SOLUTION
		MERCURIC IODIDE, SOLUTION (DOT)
		MERCURY BINIODIDE
		MERCURY DIIODIDE
		MERCURY IODIDE (HgI2)
		MERCURY(II) IODIDE
		RED MERCURIC IODIDE
		UN 1638 (DOT)
7775-09-9		
		AGROSAN
		ASEX
		ATLACIDE
		ATRATOL

CAS No.	IN	Chemical Name
		B-HERBATOX
		CHLORATE OF SODA
		CHLORATE OF SODA (DOT)
		CHLORATE SALT OF SODIUM
		CHLORAX
		CHLORIC ACID, SODIUM SALT
		CHLORSAURE (GERMAN)
		DE-FOL-ATE
		DESOLET
		DREXEL DEFOL
		DROP LEAF
		EVAU-SUPER
		FALL
		GRAIN SORGHUM HARVEST-AID
		GRANEX O
		HARVEST-AID
		KLOREX
		KUSA-TOHRU
		KUSATOL
		NATRIUMCHLORAAT (DUTCH)
		NATRIUMCHLORAT (GERMAN)
		ORTHO C-1 DEFOLIANT & WEED KILLER
		OXYCIL
		RASIKAL
		SHED-A-LEAF
		SHED-A-LEAF 'L'
		SODA CHLORATE
		SODA CHLORATE (DOT)
		SODIO (CLORATO DI) (ITALIAN)
	*	SODIUM CHLORATE
		SODIUM CHLORATE (DOT)
		SODIUM CHLORATE (NaClO3)
		SODIUM CHLORATE SOLUTION
		SODIUM CHLORATE, AQUEOUS SOLUTION (DOT)
		SODIUM(CHLORATE DE) (FRENCH)
		TRAVEX
		TUMBLEAF
		UN 1495 (DOT)
		UN 2428 (DOT)
		UNITED CHEMICAL DEFOLIANT NO 1
		VAL-DROP
7775-11-3		
		CHROMATE OF SODA
		CHROMIC ACID (H2CrO4), DISODIUM SALT
		CHROMIC ACID, DISODIUM SALT
		CHROMIUM DISODIUM OXIDE
		CHROMIUM SODIUM OXIDE
		CHROMIUM SODIUM OXIDE (CrNa2O4)
		DISODIUM CHROMATE
		DISODIUM CHROMATE (Na2CrO4)
		NA 9145 (DOT)
		NEUTRAL SODIUM CHROMATE
	*	SODIUM CHROMATE
		SODIUM CHROMATE (DOT)
		SODIUM CHROMATE (Na2CrO4)
		SODIUM CHROMATE SOLUTION
		SODIUM CHROMATE(VI)
7775-14-6		
		BLANKIT
		BLANKIT IN
		BURMOL
		D-OX
		DISODIUM DITHIONITE
		DISODIUM HYDROSULFITE
		DITHIONOUS ACID, DISODIUM SALT
		HYDROLIN
		HYDROS
		K-BRITE
		REDUCTONE
	*	SODIUM DITHIONITE
		SODIUM DITHIONITE (DOT)
		SODIUM DITHIONITE (Na2S2O4)
		SODIUM HYDROSULFITE
		SODIUM HYDROSULFITE (DOT)
		SODIUM HYDROSULFITE (Na2S2O4)
		SODIUM HYDROSULPHITE
		SODIUM HYPOSULFITE
		SODIUM SULFOXYLATE
		UN 1384 (DOT)
		V-BRITE
		V-BRITE B
		VATROLITE
		VIRCHEM
		VIRTEX CC
		VIRTEX D
		VIRTEX L
		VIRTEX RD
7775-27-1		
		DISODIUM PEROXODISULFATE
		DISODIUM PEROXYDISULFATE
		DISODIUM PERSULFATE
		PEROXYDISULFURIC ACID ([(HO)S(O)2]2O2), DISODIUM SALT
		PEROXYDISULFURIC ACID, DISODIUM SALT
		PERSULFATE DE SODIUM (FRENCH)
	*	SODIUM PEROXYDISULFATE
		SODIUM PEROXYDISULFATE (Na2S2O8)
		SODIUM PERSULFATE
		SODIUM PERSULFATE (ACGIH,DOT)
		SODIUM PERSULFATE (Na2S2O8)
		SODIUM PERSULFATE (PEROXYDISULFURIC ACID, DISODIUM SALT)
		UN 1505 (DOT)
7778-39-4		
		ACIDE ARSENIQUE LIQUIDE (FRENCH)
		ARSENATE
		ARSENIC (V) ACID
	*	ARSENIC ACID
		ARSENIC ACID (H3AsO4)
		ARSENIC ACID SOLUTION
		ARSENIC ACID SOLUTION (DOT)
		ARSENIC ACID, (SOLUTION)
		ARSENIC ACID, LIQUID
		ARSENIC ACID, LIQUID (DOT)
		ARSENIC ACID, SOLID
		ARSENIC ACID, SOLID (DOT)
		DESICCANT L-10
		HI-YIELD DESICCANT H-10
		ORTHOARSENIC ACID
		RCRA WASTE NUMBER P010
		UN 1553 (DOT)
		UN 1554 (DOT)
		ZOTOX
		ZOTOX CRAB GRASS KILLER
7778-44-1		
		ARSENIATE DE CALCIUM (FRENCH)
		ARSENIC ACID (H3AsO4), CALCIUM SALT (2:3)
		ARSENIC ACID, CALCIUM SALT(2:3)
	*	CALCIUM ARSENATE
		CALCIUM ARSENATE (Ca3(AsO4)2)
		CALCIUM ARSENATE (DOT)
		CALCIUM ARSENATE, SOLID
		CALCIUM ARSENATE, SOLID (DOT)
		CALCIUM ORTHOARSENATE
		CALCIUMARSENAT
		CHIP-CAL
		CHIP-CAL GRANULAR
		CUCUMBER DUST
		FENCAL
		FLAC
		KALO
		KALZIUMARSENIAT (GERMAN)
		KILMAG
		PENCAL
		SECURITY
		SPRA-CAL
		TRICALCIUM ARSENATE
		TRICALCIUMARSENAT (GERMAN)
		TURF-CAL
		UN 1573 (DOT)
7778-50-9		
		BICHROMATE OF POTASH
		CHROMIC ACID (H2Cr2O7), DIPOTASSIUM SALT

CAS No.	IN	Chemical Name
		CHROMIC ACID, DIPOTASSIUM SALT
		DICHROMIC ACID (H2Cr2O7), DIPOTASSIUM SALT
		DICHROMIC ACID DIPOTASSIUM SALT
		DIPOTASSIUM BICHROMATE
		DIPOTASSIUM BICHROMATE (K2Cr2O7)
		DIPOTASSIUM DICHROMATE
		IOPEZITE
		KALIUMDICHROMAT (GERMAN)
		NA 1479 (DOT)
		POTASSIUM BICHROMATE
	*	POTASSIUM DICHROMATE
		POTASSIUM DICHROMATE (DOT)
		POTASSIUM DICHROMATE (K2Cr2O7)
		POTASSIUM DICHROMATE (VI)
7778-54-3		
		B-K POWDER
		BLEACHING POWDER
	*	CALCIUM HYPOCHLORITE
		CALCIUM HYPOCHLORITE MIXTURE, DRY (CONTAINING MORE THAN 39% AVAILABLE CHLORINE) (DOT)
		CALCIUM HYPOCHLORITE MIXTURE, DRY, WITH >10% BUT ≤39% AVAILABLE CHLORINE
		CALCIUM HYPOCHLORITE MIXTURE, DRY, WITH >39% AVAILABLE CHLORINE
		CALCIUM HYPOCHLORITE, HYDRATED >55% BUT <10% WATER, AND CONTAINING >39% AVAILABLE CHLORINE)
		CALCIUM OXYCHLORIDE
		CAPORIT
		CHEMICHLON G
		CHLORIDE OF LIME
		CHLORINATED LIME
		CHLORO LIME CHEMICAL
		EUSOL BPC
		HTH
		HTH (BLEACHING AGENT)
		HYPOCHLOROUS ACID, CALCIUM SALT
		HYPOCHLOROUS ACID, CALCIUM SALT, DRY MIXTURE WITH 10% TO 39% AVAILABLE CHLORINE
		LIME CHLORIDE
		LOSANTIN
		SOLVOX KS
		T-EUSOL
		UN 1748 (DOT)
		UN 2208 (DOT)
7778-66-7		
		HYPOCHLOROUS ACID, POTASSIUM SALT
		JAVELLE WATER
		POTASSIUM CHLORIDE OXIDE (KClO)
	*	POTASSIUM HYPOCHLORITE
		POTASSIUM HYPOCHLORITE SOLUTION
		POTASSIUM HYPOCHLORITE, SOLUTION (DOT)
		UN 1791 (DOT)
7778-74-7		
		ASTRUMAL
		IRENAL
		IRENAT
		PERCHLORIC ACID, POTASSIUM SALT
		PERCHLORIC ACID, POTASSIUM SALT (1:1)
		PERIODIN
		PEROIDIN
		POTASSIUM HYPERCHLORIDE
	*	POTASSIUM PERCHLORATE
		POTASSIUM PERCHLORATE (DOT)
		POTASSIUM PERCHLORATE (KClO4)
		UN 1489 (DOT)
7779-86-4		
		DITHIONOUS ACID, ZINC SALT (1:1)
		UN 1931 (DOT)
		ZINC DITHIONITE
		ZINC DITHIONITE (ZnS2O4)
		ZINC HYDROSULFITE
		ZINC HYDROSULFITE (DOT)
		ZINC HYDROSULPHITE

CAS No.	IN	Chemical Name
7779-88-6		
		AEROTEX ACCELERATOR NUMBER 5
		CELLOXAN
		NITRATE DE ZINC (FRENCH)
		NITRIC ACID, ZINC SALT
		UN 1514 (DOT)
		X 4
		X 4 (NITRATE)
		ZINC DINITRATE
	*	ZINC NITRATE
		ZINC NITRATE (DOT)
		ZINC NITRATE (Zn(NO3)2)
7782-39-0		
	*	DEUTERIUM
		DEUTERIUM (D2)
		DEUTERIUM (DOT)
		DEUTERIUM MOLECULE
		DIPLOGEN
		HYDROGEN, ISOTOPE OF MASS 2
		HYDROGEN-2
		HYDROGEN-D2
		UN 1957 (DOT)
7782-41-4		
	*	FLUORINE
		FLUORINE, CRYOGENIC LIQUID
		FLUORINE-19
		RCRA WASTE NUMBER P056
		SAEURE FLUORIDE (GERMAN)
		UN 1045 (DOT)
7782-42-5		
		AERODAG G
		AG 1500
		AQUADAG
		AS 1
		AT 20
		ATJ-S
		ATJ-S GRAPHITE
		BLACK LEAD
		C.I. 77265
		C.I. PIGMENT BLACK 10
		CANLUB
		CB 50
		CEYLON BLACK LEAD
		CPB 5000
		DC 2
		EG 0
		ELECTROGRAPHITE
		EXP-F
		FORTAFIL 5Y
		GK 2
		GK 3
		GP 60
		GP 60S
		GP 63
		GRAFOIL
		GRAFOIL GTA
	*	GRAPHITE
		GRAPHITE (NATURAL) DUST
		GRAPHNOL N 3M
		GS 2
		GY 70
		H 451
		HITCO HMG 50
		IG 11
		KOROBON
		MG 1
		MINERAL CARBON
		MPG 6
		PAPYEX
		PG 50
		PLUMBAGO
		PLUMBAGO (GRAPHITE)
		PYRO-CARB 406
		PYROLITE
		ROCOL X 7119
		S 1

CAS No.	IN	Chemical Name
		S 1 (GRAPHITE)
		SCHUNGITE
		SHUNGITE
		SILVER GRAPHITE
		SKLN 1
		STOVE BLACK
		SWEDISH BLACK LEAD
		UCAR 38
		VVP 66-95
7782-44-7		
		DIOXYGEN
		MOLECULAR OXYGEN
	*	OXYGEN
		OXYGEN (DOT)
		OXYGEN (LIQUID)
		OXYGEN MOLECULE
		OXYGEN, COMPRESSED
		OXYGEN, COMPRESSED (DOT)
		OXYGEN, REFRIGERATED LIQUID
		OXYGEN, REFRIGERATED LIQUID (DOT)
		PURE OXYGEN
		UN 1072 (DOT)
		UN 1073 (DOT)
7782-49-2		
		C.I. 77805
	*	SELENIUM
		SELENIUM (COLLOIDAL)
		SELENIUM ELEMENT
		SELENIUM METAL, POWDER
7782-50-5		
		BERTHOLITE
		CHLOOR (DUTCH)
		CHLOR (GERMAN)
		CHLORE (FRENCH)
	*	CHLORINE
		CHLORINE (ACGIH,DOT)
		CHLORINE MOLECULE (Cl2)
		CLORO (ITALIAN)
		DIATOMIC CHLORINE
		DICHLORINE
		MOLECULAR CHLORINE
		UN 1017 (DOT)
7782-65-2		
		GERMANE
		GERMANE (DOT)
		GERMANIUM HYDRIDE
		GERMANIUM HYDRIDE (GeH4)
		GERMANIUM TETRAHYDRIDE
		GERMANIUM TETRAHYDRIDE (ACGIH)
		MONOGERMANE
		TETRAHYDROGERMANE
		UN 2192 (DOT)
7782-78-7		
		NITRO ACID SULFITE
		NITROSONIUM BISULFATE
		NITROSOSULFURIC ACID
		NITROSULFONIC ACID
		NITROSYL BISULFATE
		NITROSYL HYDROGEN SULFATE
		NITROSYL SULFATE
		NITROSYL SULFATE ((NO)H(SO4))
		NITROSYL SULFURIC ACID (NOHSO4)
		NITROSYLSULFURIC ACID
		NITROSYLSULPHURIC ACID (DOT)
		NITROXYLSULFURIC ACID
		SULFURIC ACID, MONOANHYDRIDE WITH NITROUS ACID
		SULFURIC ACID, MONOANHYDRIDE WITH NITROUS ACID (9CI)
		UN 2308 (DOT)
7782-86-7		
		MERCUROUS NITRATE MONOHYDRATE
		MERCURY NITRATE (Hg(NO3)) MONOHYDRATE
		MERCURY PROTONITRATE
		NITRIC ACID, MERCURY(1+) SALT, MONOHYDRATE
7782-89-0		
		LITHAMIDE
		LITHIUM AMIDE
		LITHIUM AMIDE (DOT)
		LITHIUM AMIDE (Li(NH2))
		LITHIUM AMIDE, POWDERED
		LITHIUM AMIDE, POWDERED (DOT)
		UN 1412 (DOT)
7782-92-5		
		SODAMIDE
	*	SODIUM AMIDE
		SODIUM AMIDE (DOT)
		SODIUM AMIDE (Na(NH2))
		UN 1425 (DOT)
7782-99-2		
		SCHWEFLIGE SAURE (GERMAN)
		SULFUR DIOXIDE SOLUTION
		SULFUROUS ACID
		SULFUROUS ACID (DOT)
		SULFUROUS ACID (H2SO3)
	*	SULFUROUS ACID SOLUTION
		SULPHUROUS ACID (DOT)
		UN 1833 (DOT)
7783-00-8		
		MONOHYDRATED SELENIUM DIOXIDE
		RCRA WASTE NUMBER U204
		SELENIOUS ACID
		SELENIOUS ACID (H2SeO3)
		SELENIUM DIOXIDE
		SELENOUS ACID
7783-05-3		
		DISULFURIC ACID
		DISULPHURIC ACID
		DITHIONIC ACID
		FUMING SULFURIC ACID
		FUMING SULFURIC ACID (DOT)
		NA 1831 (DOT)
		OLEUM
		OLEUM (DOT)
		PYROSULFURIC ACID
		PYROSULPHURIC ACID
		SULFURIC ACID ANHYDRIDE
		SULFURIC ACID, FUMING
		UN 1831 (DOT)
7783-06-4		
		ACIDE SULFHYDRIQUE (FRENCH)
		DIHYDROGEN MONOSULFIDE
		DIHYDROGEN SULFIDE
	*	HYDROGEN SULFIDE
		HYDROGEN SULFIDE (ACGIH,DOT)
		HYDROGEN SULFIDE (H2S)
		HYDROGEN SULFURIC ACID
		HYDROGEN SULPHIDE
		HYDROGENE SULFURE (FRENCH)
		HYDROSULFURIC ACID
		IDROGENO SOLFORATO (ITALIAN)
		RCRA WASTE NUMBER U135
		SCHWEFELWASSERSTOFF (GERMAN)
		SIARKOWODOR (POLISH)
		STINK DAMP
		SULFUR DIHYDRIDE
		SULFUR HYDRIDE
		SULFURETED HYDROGEN
		UN 1053 (DOT)
		ZWAVELWATERSTOF (DUTCH)
7783-07-5		
		DIHYDROGEN SELENIDE
		ELECTRONIC E-2
		HYDROGEN SELENIDE
		HYDROGEN SELENIDE (ACGIH,DOT)
		HYDROGEN SELENIDE (H2Se)

CAS No.	IN	Chemical Name
		HYDROGEN SELENIDE, ANHYDROUS (DOT)
		SELANE
		SELENIUM DIHYDRIDE
		SELENIUM HYDRIDE
		SELENIUM HYDRIDE (H2Se)
		UN 2202 (DOT)
7783-08-6		
		SELENIC ACID
		SELENIC ACID (DOT)
		SELENIC ACID, LIQUID
		SELENIC ACID, LIQUID (DOT)
		UN 1905 (DOT)
7783-18-8		
		AMMONIUM HYPOSULFITE
	*	AMMONIUM THIOSULFATE
		AMMONIUM THIOSULFATE ((NH4)2S2O3)
		DIAMMONIUM THIOSULFATE
		THIOSULFURIC ACID (H2S2O3), DIAMMONIUM SALT
		THIOSULFURIC ACID, DIAMMONIUM SALT
7783-20-2		
	*	AMMONIUM SULFATE
		AMMONIUM SULFATE ((NH4)2SO4)
		AMMONIUM SULFATE (SOLUTION)
		AMMONIUM SULPHATE
		COALTROL LPA 40
		DIAMMONIUM SULFATE
		DIAMMONIUM SULPHATE
		DOLAMIN
		SULFURIC ACID DIAMMONIUM SALT
7783-28-0		
		AKOUSTAN A
		AMMONIUM DIBASIC PHOSPHATE
		AMMONIUM HYDROGEN PHOSPHATE
		AMMONIUM HYDROGEN PHOSPHATE ((NH4)2HPO4)
		AMMONIUM MONOHYDROGEN ORTHOPHOSPHATE
		AMMONIUM PHOSPHATE
		AMMONIUM PHOSPHATE ((NH4)2HPO4)
		AMMONIUM PHOSPHATE DIBASIC
		COALTROL LPA 445
		DIAMMONIUM ACID PHOSPHATE
		DIAMMONIUM HYDROGEN ORTHOPHOSPHATE
		DIAMMONIUM HYDROGEN PHOSPHATE
		DIAMMONIUM MONOHYDROGEN PHOSPHATE
		DIAMMONIUM ORTHOPHOSPHATE
		DIAMMONIUM PHOSPHATE
		DIBASIC AMMONIUM PHOSPHATE
		HYDROGEN DIAMMONIUM PHOSPHATE
		K 2
		K 2 (PHOSPHATE)
		PELOR
		PHOS-CHEK 202A
		PHOSPHORIC ACID AMMONIUM SALT (1:2)
		PHOSPHORIC ACID, DIAMMONIUM SALT
		SECONDARY AMMONIUM PHOSPHATE
7783-30-4		
		IODURE DE MERCURE (FRENCH)
		MERCUROUS IODIDE
		MERCUROUS IODIDE, SOLID
		MERCUROUS IODIDE, SOLID (DOT)
		MERCURY IODIDE (HgI)
		MERCURY MONOIODIDE
		MERCURY(1+) IODIDE
	*	MERCURY(I) IODIDE
		UN 1638 (DOT)
		YELLOW MERCURY IODIDE
7783-33-7		
		MERCURY POTASSIUM IODIDE
7783-35-9		
	*	MERCURIC SULFATE
		MERCURIC SULFATE, SOLID
		MERCURIC SULFATE, SOLID (DOT)
		MERCURIC SULPHATE (DOT)
		MERCURY BISULFATE
		MERCURY BISULPHATE (DOT)
		MERCURY PERSULFATE
		MERCURY SULFATE (HgSO4)
		MERCURY(II) SULFATE
		MERCURY(II) SULFATE (1:1)
		SULFATE MERCURIQUE (FRENCH)
		SULFURIC ACID, MERCURY(2+) SALT (1:1)
		UN 1633 (DOT)
		UN 1645 (DOT)
7783-36-0		
		MERCUROUS SULFATE
		MERCUROUS SULFATE, SOLID
		MERCUROUS SULFATE, SOLID (DOT)
		MERCURY SULFATE (Hg2SO4)
		MERCURY(I) SULFATE
		SULFURIC ACID, DIMERCURY(1+) SALT
		UN 1628 (DOT)
7783-41-7		
		DIFLUORINE MONOOXIDE
		DIFLUORINE MONOXIDE
		DIFLUORINE OXIDE
		FLUORINE MONOXIDE
		FLUORINE MONOXIDE (F2O)
		FLUORINE OXIDE
		FLUORINE OXIDE (F2O)
		OXYDIFLUORIDE
		OXYGEN DIFLUORIDE
		OXYGEN DIFLUORIDE (ACGIH,DOT)
		OXYGEN FLUORIDE
		OXYGEN FLUORIDE (OF2)
		UN 2190 (DOT)
7783-46-2		
		LEAD DIFLUORIDE
		LEAD DIFLUORIDE (PbF2)
	*	LEAD FLUORIDE
		LEAD FLUORIDE (DOT)
		LEAD FLUORIDE (PbF2)
		LEAD(II) FLUORIDE
		NA 2811 (DOT)
		PLOMB FLUORURE (FRENCH)
		PLUMBOUS FLUORIDE
7783-47-3		
		FLUORISTAN
		GEL-TIN
	*	STANNOUS FLUORIDE
		STANNOUS FLUORIDE (SnF2)
		TIN BIFLUORIDE
		TIN DIFLUORIDE
		TIN FLUORIDE
		TIN FLUORIDE (SnF2)
7783-49-5		
		NA 9158 (DOT)
		ZINC DIFLUORIDE
	*	ZINC FLUORIDE
		ZINC FLUORIDE (DOT)
		ZINC FLUORIDE (ZnF2)
		ZINC FLUORURE (FRENCH)
7783-50-8		
		FERRIC FLUORIDE
		FERRIC FLUORIDE (DOT)
		FERRIC FLUORIDE (FeF3)
		FERRIC TRIFLUORIDE
		IRON FLUORIDE
		IRON FLUORIDE (FeF3)
		IRON TRIFLUORIDE
		NA 9120 (DOT)
7783-54-2		
		NITROGEN FLUORIDE
		NITROGEN FLUORIDE (NF3)
		NITROGEN TRIFLUORIDE
		NITROGEN TRIFLUORIDE (ACGIH,DOT)

CAS No.	IN	Chemical Name
		PERFLUOROAMMONIA
		TRIFLUOROAMINE
		TRIFLUOROAMMONIA
		UN 2451 (DOT)
7783-55-3		PHOSPHOROUS FLUORIDE
		PHOSPHOROUS TRIFLUORIDE
		PHOSPHORUS FLUORIDE
		PHOSPHORUS FLUORIDE (PF3)
		PHOSPHORUS TRIFLUORIDE
		TL 75
		TRIFLUOROPHOSPHINE
7783-56-4		ANTIMOINE FLUORURE (FRENCH)
		ANTIMONOUS FLUORIDE
		ANTIMONY FLUORIDE (SbF3)
		ANTIMONY TRIFLUORIDE
		ANTIMONY TRIFLUORIDE SOLUTION
		ANTIMONY TRIFLUORIDE, SOLID
		ANTIMONY TRIFLUORIDE, SOLID (DOT)
		ANTIMONY(III) FLUORIDE (1:3)
		NA 1549 (DOT)
		STIBINE, TRIFLUORO-
		STIBINE, TRIFLUORO- (9CI)
		TRIFLUOROANTIMONY
		TRIFLUOROSTIBINE
7783-60-0		SULFUR FLUORIDE (SF4)
		SULFUR FLUORIDE (SF4), (T-4)-
		SULFUR TETRAFLUORIDE
		SULFUR TETRAFLUORIDE (ACGIH,DOT)
		TETRAFLUOROSULFURANE
		UN 2418 (DOT)
7783-61-1		PERFLUOROSILANE
		SILANE, TETRAFLUORO-
		SILICON FLUORIDE
		SILICON FLUORIDE (SiF4)
		SILICON TETRAFLUORIDE
		SILICON TETRAFLUORIDE (DOT)
		TETRAFLUOROSILANE
		UN 1859 (DOT)
7783-66-6		IODINE FLUORIDE
		IODINE FLUORIDE (IF5)
		IODINE PENTAFLUORIDE
		IODINE PENTAFLUORIDE (DOT)
		PENTAFLUOROIODINE
		UN 2495 (DOT)
7783-70-2		ANTIMONY (V) FLUORIDE
		ANTIMONY FLUORIDE
		ANTIMONY FLUORIDE (SbF5)
	*	ANTIMONY PENTAFLUORIDE
		ANTIMONY PENTAFLUORIDE (DOT)
		ANTIMONY(V) FLUORIDE
		ANTIMONY(V) PENTAFLUORIDE
		PENTAFLUOROANTIMONY
		UN 1732 (DOT)
7783-79-1		SELENIUM FLUORIDE
		SELENIUM FLUORIDE (SeF6)
		SELENIUM FLUORIDE (SeF6), (OC-6-11)-
		SELENIUM HEXAFLUORIDE
		SELENIUM HEXAFLUORIDE (ACGIH,DOT)
		UN 2194 (DOT)
7783-80-4		TELLURIUM FLUORIDE (TeF6)
		TELLURIUM FLUORIDE (TeF6), (OC-6-11)-
		TELLURIUM HEXAFLUORIDE
		TELLURIUM HEXAFLUORIDE (ACGIH,DOT)
		UN 2195 (DOT)
7783-81-5		URANIUM HEXAFLUORIDE
7783-82-6		TUNGSTEN FLUORIDE
		TUNGSTEN FLUORIDE (WF6)
		TUNGSTEN FLUORIDE (WF6), (OC-6-11)-
	*	TUNGSTEN HEXAFLUORIDE
		TUNGSTEN HEXAFLUORIDE (DOT)
		TUNGSTEN HEXAFLUORIDE (WF6)
		UN 2196 (DOT)
7783-95-1		ARGENT FLUORURE (FRENCH)
		ARGENTIC FLUORIDE
		SILVER DIFLUORIDE
		SILVER FLUORIDE
		SILVER FLUORIDE (AgF2)
		SILVER(II) FLUORIDE
7783-97-3		IODIC ACID (HIO3), SILVER(1+) SALT
		SILVER IODATE
		SILVER IODATE (AgIO3)
		SILVER(1+) IODATE
7784-08-9		ARSENIOUS ACID (H3AsO3), TRISILVER(1+) SALT
		ARSENIOUS ACID, TRISILVER(1+) SALT
		ARSENOUS ACID, TRISILVER(1+) SALT
		ARSENOUS ACID, TRISILVER(1+) SALT (9CI)
	*	SILVER ARSENITE
		SILVER ARSENITE (DOT)
		UN 1683 (DOT)
7784-18-1	*	ALUMINUM FLUORIDE
7784-21-6		ALANE
		ALPHA-ALUMINUM TRIHYDRIDE
	*	ALUMINUM HYDRIDE
		ALUMINUM HYDRIDE (AlH3)
		ALUMINUM HYDRIDE (DOT)
		ALUMINUM TRIHYDRIDE
		UN 2463 (DOT)
7784-30-7		ALUMINOPHOSPHORIC ACID
		ALUMINUM ACID PHOSPHATE
		ALUMINUM MONOPHOSPHATE
		ALUMINUM ORTHOPHOSPHATE
	*	ALUMINUM PHOSPHATE
		ALUMINUM PHOSPHATE (1:1)
		ALUMINUM PHOSPHATE (Al(PO4))
		ALUMINUM PHOSPHATE SOLUTION
		ALUMINUM PHOSPHATE SOLUTION (DOT)
		ALUPHOS
		FB 67
		FFB 32
		K-BOND 90
		MONOALUMINUM PHOSPHATE
		NA 1760 (DOT)
		PHOSPHALUGEL
		PHOSPHORIC ACID, ALUMINUM SALT (1:1)
		PHOSPHORIC ACID, ALUMINUM SALT (1:1), (SOLUTION)
7784-33-0		ARSENIC BROMIDE
7784-34-1		ARSENIC BUTTER
		ARSENIC CHLORIDE
		ARSENIC CHLORIDE (AsCl3)
		ARSENIC CHLORIDE (DOT)

CAS No.	IN	Chemical Name
		ARSENIC CHLORIDE, LIQUID (DOT)
	*	ARSENIC TRICHLORIDE
		ARSENIC TRICHLORIDE (ARSENOUS)
		ARSENIC TRICHLORIDE (DOT)
		ARSENIC TRICHLORIDE, LIQUID
		ARSENIC TRICHLORIDE, LIQUID (DOT)
		ARSENIC(III) CHLORIDE
		ARSENIOUS CHLORIDE
		ARSENOUS CHLORIDE
		ARSENOUS TRICHLORIDE
		ARSENOUS TRICHLORIDE (9CI)
		BUTTER OF ARSENIC
		CHLORURE ARSENIEUX (FRENCH)
		CHLORURE D'ARSENIC (FRENCH)
		FUMING LIQUID ARSENIC
		TRICHLOROARSINE
		TRICHLORURE D'ARSENIC (FRENCH)
		UN 1560 (DOT)
7784-37-4		ARSENIC ACID (H3AsO4), MERCURY(2+) SALT (1:1)
		CI 77762
		MERCURIC ARSENATE
		MERCURIC ARSENATE (DOT)
		MERCURY(II) O-ARSENATE
		UN 1623 (DOT)
7784-40-9		ACID LEAD ARSENATE
		ARSENIC ACID (H3AsO4), LEAD(2+) SALT (1:1)
		ARSENIC ACID LEAD SALT
		LEAD ACID ARSENATE
	*	LEAD ARSENATE
		LEAD ARSENATE (PbHAsO4)
		LEAD HYDROGEN ARSENATE (PbHAsO4)
7784-41-0		ARSENIC ACID (H3AsO4), MONOPOTASSIUM SALT
		ARSENIC ACID, MONOPOTASSIUM SALT
		MACQUER'S SALT
		MONOPOTASSIUM ARSENATE
		MONOPOTASSIUM DIHYDROGEN ARSENATE
		POTASSIUM ACID ARSENATE
	*	POTASSIUM ARSENATE
		POTASSIUM ARSENATE (KH2AsO4)
		POTASSIUM ARSENATE, MONOBASIC
		POTASSIUM ARSENATE, SOLID
		POTASSIUM ARSENATE, SOLID (DOT)
		POTASSIUM DIHYDROGEN ARSENATE
		POTASSIUM DIHYDROGEN ARSENATE (KH2AsO4)
		POTASSIUM HYDROGEN ARSENATE
		POTASSIUM HYDROGEN ARSENATE (KH2AsO4)
		UN 1677 (DOT)
7784-42-1		ARSENIC HYDRID
		ARSENIC HYDRIDE
		ARSENIC HYDRIDE (AsH3)
		ARSENIC TRIHYDRIDE
		ARSENIURETTED HYDROGEN
		ARSENOUS HYDRIDE
		ARSENOWODOR (POLISH)
		ARSENWASSERSTOFF (GERMAN)
		ARSINE
		ARSINE (ACGIH,DOT)
		HYDROGEN ARSENIDE
		UN 2188 (DOT)
7784-44-3		AMMONIUM ACID ARSENATE
		AMMONIUM ARSENATE
		AMMONIUM ARSENATE ((NH4)2HAsO4)
		AMMONIUM ARSENATE (DOT)
		AMMONIUM ARSENATE, SOLID
		AMMONIUM ARSENATE, SOLID (DOT)
		ARSENIC ACID (H3AsO4), DIAMMONIUM SALT
		ARSENIC ACID, DIAMMONIUM SALT
		DIAMMONIUM ARSENATE
		DIAMMONIUM HYDROGEN ARSENATE
		DIAMMONIUM MONOHYDROGEN ARSENATE
		DIBASIC AMMONIUM ARSENATE
		SECONDARY AMMONIUM ARSENATE
		UN 1546 (DOT)
7784-45-4		ARSENIC IODIDE
		ARSENIC IODIDE (AsI3)
		ARSENIC IODIDE, SOLID
		ARSENIC IODIDE, SOLID (DOT)
		ARSENIC TRIIODIDE
		ARSENOUS IODIDE
		ARSENOUS TRIIODIDE
		ARSENOUS TRIIODIDE (9CI)
		NA 1557 (DOT)
		TRIIODOARSINE
7784-46-5		ARSENENOUS ACID, SODIUM SALT
		ARSENENOUS ACID, SODIUM SALT (9CI)
		ARSENIOUS ACID, MONOSODIUM SALT
		ARSENIOUS ACID, SODIUM SALT
		ARSENITE DE SODIUM (FRENCH)
		ATLAS 'A'
		CHEM PELS C
		CHEM-SEN 56
		KILL-ALL
		PENITE
		PRODALUMNOL
		PRODALUMNOL DOUBLE
		SODANIT
		SODIUM ARSENIC OXIDE (NaAsO2)
	*	SODIUM ARSENITE
		SODIUM ARSENITE (NaAsO2)
		SODIUM ARSENITE SOLUTION
		SODIUM ARSENITE, LIQUID (SOLUTION) (DOT)
		SODIUM ARSENITE, SOLID
		SODIUM ARSENITE, SOLID (DOT)
		SODIUM METAARSENITE
		UN 1686 (DOT)
		UN 2027 (DOT)
7785-84-4		CYCLIC SODIUM TRIMETAPHOSPHATE
		METAPHOSPHORIC ACID (H3P3O9), TRISODIUM SALT
		SODIUM METAPHOSPHATE
		SODIUM METAPHOSPHATE (Na3P3O9)
		SODIUM PHOSPHATE ((NaPO3)3)
		SODIUM TRIMETAPHOSPHATE
		SODIUM TRIMETAPHOSPHATE (Na3P3O9)
		TRIMETAPHOSPHORIC ACID (H3P3O9), TRISODIUM SALT
		TRISODIUM CYCLOTRIPHOSPHATE
		TRISODIUM METAPHOSPHATE
		TRISODIUM TRIMETAPHOSPHATE
		TRISODIUM TRIMETAPHOSPHATE (Na3P3O9)
7785-87-7		MANGANESE (II) SULFATE (1:1)
7786-34-7		(2-METHOXYCARBONYL-1-METHYL-VINYL)-DIMETHYL-PHOSPHAT (GERMAN)
		(2-METHOXYCARBONYL-1-METHYL-VINYL)-DIMETHYL-FOSFAAT (DUTCH)
		(2-METHOXYCARBONYL-1-METHYL-VINYL)-DIMETHYL-PHOSPHAT (GERMAN)
		(2-METOSSICARBONIL-1-METIL-VINIL)-DIMETIL-FOSFATO (ITALIAN)
		1-METHOXYCARBONYL-1-PROPEN-2-YL DIMETHYL PHOSPHATE
		2-BUTENOIC ACID, 3-((DIMETHOXYPHOSPHINYL)OXY)-, METHYL ESTER (9CI)
		2-BUTENOIC ACID, 3-[(DIMETHOXYPHOSPHINYL)OXY]-, METHYL ESTER
		2-CARBOMETHOXY-1-METHYLVINYL DIMETHYL PHOSPHATE
		2-CARBOMETHOXY-1-PROPEN-2-YL DIMETHYL PHOSPHATE
		2-METHOXYCARBONYL-1-METHYLVINYL DIMETHYL PHOSPHATE

CAS No.	IN	Chemical Name
		3-((DIMETHOXYPHOSPHINYL)OXY)-2-BUTENOIC ACID METHYL ESTER
		3-HYDROXYCROTONIC ACID METHYL ESTER DIMETHYL PHOSPHATE
		ALPHA-2-CARBOMETHOXY-1-METHYLVINYL DIMETHYL PHOSPHATE
		APAVINPHOS
		CMDP
		COMPOUND 2046
		CROTONIC ACID, 3-HYDROXY-, METHYL ESTER, DIMETHYL PHOSPHATE
		DIMETHYL (1-METHOXYCARBOXYPROPEN-2-YL)PHOSPHATE
		DIMETHYL 2-METHOXYCARBONYL-1-METHYLVINYL PHOSPHATE
		DIMETHYL METHOXYCARBONYLPROPENYL PHOSPHATE
		DIMETHYL-1-CARBOMETHOXY-1-PROPEN-2-YL PHOSPHATE
		DURAPHOS
		ENT 22,374
		FOSDRIN
		GESFID
		GESTID
		MENIPHOS
		MENITE
		METHYL 3-(DIMETHOXYPHOSPHINYLOXY)CROTONATE
		MEVINFOS (DUTCH)
		MEVINPHOS
		MEVINPHOS (ACGIH,DOT)
		MEVINPHOS MIXTURE, DRY
		MEVINPHOS MIXTURE, DRY (DOT)
		MEVINPHOS MIXTURE, WET (DOT)
		NA 2783 (DOT)
		O,O-DIMETHYL O-(1-CARBOMETHOXY-1-PROPEN-2-YL) PHOSPHATE
		O,O-DIMETHYL O-(1-METHYL-2-CARBOXYVINYL) PHOSPHATE
		O,O-DIMETHYL-O-(2-CARBOMETHOXY-1-METHYLVINYL) PHOSPHATE
		O,O-DIMETHYL-O-2-METHOXYCARBONYL-1-METHYL-VINYL-PHOSPHAT (GERMAN)
		OS 2046
		PD 5
		PHOSDRIN
		PHOSFENE
		PHOSPHATE DE DIMETHYLE ET DE 2-METHOXYCARBONYL-1 METHYLVINYLE (FRENCH)
		PHOSPHENE (FRENCH)
		PHOSPHORIC ACID, (1-METHOXYCARBOXYPROPEN-2-YL) DIMETHYL ESTER
		PHOSPHORIC ACID, DIMETHYL ESTER, ESTER WITH METHYL 3-HYDROXYCROTONATE
7786-81-4		
		NA 9141 (DOT)
		NCI-C60344
		NICKEL MONOSULFATE
	*	NICKEL SULFATE
		NICKEL SULFATE (1:1)
		NICKEL SULFATE (DOT)
		NICKEL SULFATE (NiSO4)
		NICKEL SULFATE(1:1)
		NICKEL(2+) SULFATE
		NICKEL(2+) SULFATE (1:1)
		NICKEL(II) SULFATE
		NICKEL(II) SULFATE (1:1)
		NICKELOUS SULFATE
		SULFURIC ACID NICKEL(2+) SALT
		SULFURIC ACID, NICKEL(2+) SALT (1:1)
7787-36-2		
		BARIUM PERMANGANATE
		BARIUM PERMANGANATE (DOT)
		PERMANGANIC ACID (HMnO4), BARIUM SALT
		PERMANGANIC ACID, BARIUM SALT
		UN 1448 (DOT)
7787-41-9		
		BARIUM SELENATE
		SELENIC ACID (H2SeO4), BARIUM SALT (1:1)
		SELENIC ACID, BARIUM SALT (1:1)
7787-47-5		
		BERYLLIUM CHLORIDE
		BERYLLIUM CHLORIDE (BeCl2)
		BERYLLIUM CHLORIDE (DOT)
		BERYLLIUM DICHLORIDE
		NA 1566 (DOT)
7787-49-7		
		BERYLLIUM DIFLUORIDE
		BERYLLIUM FLUORIDE
		BERYLLIUM FLUORIDE (BeF2)
		BERYLLIUM FLUORIDE (DOT)
		NA 1566 (DOT)
7787-55-5		
		BERYLLIUM NITRATE TRIHYDRATE
		NITRIC ACID, BERYLLIUM SALT, TRIHYDRATE
7787-70-4		
		COPPER BROMIDE (Cu4Br4)
		COPPER BROMIDE (CuBr)
		COPPER BROMIDE (OUS)
		COPPER MONOBROMIDE
		COPPER(1+) BROMIDE
		COPPER(1+) BROMIDE TETRAMER
		COPPER(I) BROMIDE
		CUPROUS BROMIDE
		CUPROUS BROMIDE (CuBr)
7787-71-5		
		BROMINE FLUORIDE (BrF3)
		BROMINE TRIFLUORIDE
7788-97-8		
		CHROME FLUORURE (FRENCH)
	*	CHROMIC FLUORIDE
		CHROMIC FLUORIDE SOLUTION
		CHROMIC FLUORIDE, SOLID
		CHROMIC FLUORIDE, SOLID (DOT)
		CHROMIC FLUORIDE, SOLUTION (DOT)
		CHROMIC TRIFLUORIDE
		CHROMIUM FLUORIDE (CrF3)
		CHROMIUM TRIFLUORIDE
		CHROMIUM(III) FLUORIDE
		UN 1756 (DOT)
		UN 1757 (DOT)
7788-98-9		
	*	AMMONIUM CHROMATE
		AMMONIUM CHROMATE ((NH4) 2CrO4)
		AMMONIUM CHROMATE (DOT)
		AMMONIUM CHROMATE(VI)
		CHROMIC ACID (H2CrO4), DIAMMONIUM SALT
		CHROMIC ACID AMMONIUM SALT
		CHROMIC ACID, DIAMMONIUM SALT
		DIAMMONIUM CHROMATE
		DIAMMONIUM CHROMATE ((NH4)2CrO4)
		NA 9086 (DOT)
		NEUTRAL AMMONIUM CHROMATE
7789-00-6		
		BIPOTASSIUM CHROMATE
		CHROMATE OF POTASSIUM
		CHROMIC ACID (H2CrO4), DIPOTASSIUM SALT
		CHROMIC ACID, DIPOTASSIUM SALT
		DIPOTASSIUM CHROMATE
		DIPOTASSIUM CHROMATE (K2CrO4)
		DIPOTASSIUM MONOCHROMATE
		NA 9142 (DOT)
		NEUTRAL POTASSIUM CHROMATE
	*	POTASSIUM CHROMATE
		POTASSIUM CHROMATE (DOT)
		POTASSIUM CHROMATE (K2(CrO4))
		POTASSIUM CHROMATE (VI)
		POTASSIUM CHROMATE(VI)
		TARAPACAITE

CAS No.	IN	Chemical Name
7789-06-2		
		C.I. PIGMENT YELLOW 32
		CHROMIC ACID (H2CrO4), STRONTIUM SALT (1:1)
		CHROMIC ACID, STRONTIUM SALT (1:1)
		DEEP LEMON YELLOW
		NA 9149 (DOT)
	*	STRONTIUM CHROMATE
		STRONTIUM CHROMATE (1:1)
		STRONTIUM CHROMATE (DOT)
		STRONTIUM CHROMATE (SrCrO4)
		STRONTIUM CHROMATE (VI)
		STRONTIUM CHROMATE 12170
		STRONTIUM CHROMATE A
		STRONTIUM CHROMATE X 2396
		STRONTIUM CHROMATE(VI)
		STRONTIUM YELLOW
7789-09-5		
		AMMONIO (BICROMATO DI) (ITALIAN)
		AMMONIO (DICROMATO DI) (ITALIAN)
		AMMONIUM (DICHROMATE D') (FRENCH)
		AMMONIUM BICHROMATE
		AMMONIUM BICHROMATE (DOT)
		AMMONIUM CHROMATE ((NH4)2Cr2O7)
	*	AMMONIUM DICHROMATE
		AMMONIUM DICHROMATE (DOT)
		AMMONIUM DICHROMATE (VI)
		AMMONIUMBICHROMAAT (DUTCH)
		AMMONIUMDICHROMAAT (DUTCH)
		AMMONIUMDICHROMAT (GERMAN)
		BICHROMATE D'AMMONIUM (FRENCH)
		CHROMIC ACID (H2Cr2O7), DIAMMONIUM SALT
		DIAMMONIUM DICHROMATE
		DICHROMIC ACID (H2Cr2O7), DIAMMONIUM SALT
		DICHROMIC ACID, DIAMMONIUM SALT
		UN 1439 (DOT)
7789-18-6		
	*	CESIUM NITRATE
		CESIUM NITRATE (DOT)
		CESIUM(I) NITRATE (1:1)
		NITRIC ACID, CESIUM SALT
		UN 1451 (DOT)
7789-21-1		
		FLUOROSULFONIC ACID
		FLUOROSULFONIC ACID (DOT)
		FLUOROSULFURIC ACID
		FLUOROSULFURIC ACID (HSO3F)
		FLUOROSULPHONIC ACID
		FLUOSULFONIC ACID
		FLUOSULFONIC ACID (DOT)
		FLUOSULFURIC ACID
		MONOFLUOROSULFURIC ACID
		UN 1777 (DOT)
7789-23-3		
		FLUORURE DE POTASSIUM (FRENCH)
	*	POTASSIUM FLUORIDE
		POTASSIUM FLUORIDE (DOT)
		POTASSIUM FLUORIDE (KF)
		POTASSIUM FLUORIDE SOLUTION
		POTASSIUM FLUORIDE SOLUTION (DOT)
		POTASSIUM FLUORURE (FRENCH)
		POTASSIUM MONOFLUORIDE
		UN 1812 (DOT)
7789-29-9		
		BIFLUORURE DE POTASSIUM (FRENCH)
		HYDROGEN POTASSIUM DIFLUORIDE
		HYDROGEN POTASSIUM FLUORIDE
		HYDROGEN POTASSIUM FLUORIDE (HKF2)
		NA 1811 (DOT)
		POTASSIUM ACID FLUORIDE
		POTASSIUM BIFLUORIDE
		POTASSIUM BIFLUORIDE (KHF2)
		POTASSIUM BIFLUORIDE, SOLID (DOT)
		POTASSIUM FLUORIDE
		POTASSIUM FLUORIDE (K(HF2))
		POTASSIUM HYDROGEN DIFLUORIDE
	*	POTASSIUM HYDROGEN FLUORIDE
		POTASSIUM HYDROGEN FLUORIDE (DOT)
		POTASSIUM HYDROGEN FLUORIDE (KHF2)
		POTASSIUM HYDROGEN FLUORIDE SOLUTION
		POTASSIUM HYDROGEN FLUORIDE, SOLUTION (DOT)
		POTASSIUM MONOHYDROGEN DIFLUORIDE
		POTASSIUM, BIFLUORIDE, SOLUTION (DOT)
		UN 1811 (DOT)
7789-30-2		
		BROMINE FLUORIDE (BrF5)
		BROMINE PENTAFLUORIDE
7789-36-8		
		MAGNESIUM BROMATE
7789-38-0		
		BROMATE DE SODIUM (FRENCH)
		BROMIC ACID, SODIUM SALT
		DYETONE
		NEUTRALIZER K-126
		NEUTRALIZER K-140
		NEUTRALIZER K-938
	*	SODIUM BROMATE
		SODIUM BROMATE (DOT)
		SODIUM BROMATE (NaBrO3)
		UN 1494 (DOT)
7789-42-6		
	*	CADMIUM BROMIDE
		CADMIUM BROMIDE (CdBr2)
		CADMIUM BROMIDE (DOT)
		CADMIUM DIBROMIDE
		NA 2570 (DOT)
7789-43-7		
		COBALT BROMIDE (CoBr2)
		COBALT DIBROMIDE
		COBALT(II) BROMIDE
		COBALTOUS BROMIDE
		COBALTOUS BROMIDE (DOT)
		NA 9103 (DOT)
7789-47-1		
		DIBROMOMERCURY
	*	MERCURIC BROMIDE
		MERCURIC BROMIDE, SOLID
		MERCURIC BROMIDE, SOLID (DOT)
		MERCURIC DIBROMIDE
		MERCURY BROMIDE (HgBr2)
		MERCURY DIBROMIDE
		MERCURY(II) BROMIDE
		MERCURY(II) BROMIDE (1:2)
		UN 1634 (DOT)
7789-59-5		
		PHOSPHOROUS OXYBROMIDE
		PHOSPHORUS OXYBROMIDE
		PHOSPHORUS OXYBROMIDE (DOT)
		PHOSPHORUS OXYBROMIDE, MOLTEN
		PHOSPHORUS OXYBROMIDE, MOLTEN (DOT)
		PHOSPHORUS OXYBROMIDE, SOLID (DOT)
		PHOSPHORUS OXYTRIBROMIDE
		PHOSPHORYL BROMIDE
		PHOSPHORYL TRIBROMIDE
		TRIBROMOPHOSPHORUS OXIDE
		UN 1939 (DOT)
		UN 2576 (DOT)
7789-60-8		
		EXTREMA
		PBr3
		PHOSPHOROUS BROMIDE
		PHOSPHOROUS BROMIDE (DOT)
		PHOSPHOROUS TRIBROMIDE
		PHOSPHORUS BROMIDE (DOT)
		PHOSPHORUS BROMIDE (PBr3)
		PHOSPHORUS TRIBROMIDE

CAS No.	IN	Chemical Name
		PHOSPHORUS TRIBROMIDE (DOT)
		TRIBROMOPHOSPHINE
		UN 1808 (DOT)
7789-61-9		
		ANTIMONY BROMIDE
		ANTIMONY BROMIDE (SbBr3)
		ANTIMONY TRIBROMIDE
		ANTIMONY TRIBROMIDE SOLUTION
		ANTIMONY TRIBROMIDE, SOLID
		ANTIMONY TRIBROMIDE, SOLID (DOT)
		ANTIMONY TRIBROMIDE, SOLUTION (DOT)
		NA 1549 (DOT)
		STIBINE, TRIBROMO-
		TRIBROMOSTIBINE
7789-69-7		
		PENTABROMOPHOSPHORANE
		PENTABROMOPHOSPHORUS
		PHOSPHORANE, PENTABROMO-
		PHOSPHORIC BROMIDE
		PHOSPHORUS BROMIDE
		PHOSPHORUS BROMIDE (PBr5)
		PHOSPHORUS PENTABROMIDE
		PHOSPHORUS PENTABROMIDE (DOT)
		PHOSPHORUS PERBROMIDE
		PHOSPHORUS(V) BROMIDE
		UN 2691 (DOT)
7789-75-5		
		ACID-SPAR
		CALCIUM DIFLUORIDE
	*	CALCIUM FLUORIDE
		CALCIUM FLUORIDE (CaF2)
		FLUORITE (9CI)
		FLUORSPAR
		IRTRAN 3
		LIPARITE
		MET-SPAR
7789-78-8		
		CALCIUM DIHYDRIDE
		CALCIUM HYDRIDE
		CALCIUM HYDRIDE (CaH2)
		CALCIUM HYDRIDE (DOT)
		UN 1404 (DOT)
7790-69-4		
	*	LITHIUM NITRATE
		LITHIUM NITRATE (DOT)
		NITRIC ACID, LITHIUM SALT
		UN 2722 (DOT)
7790-91-2		
		CHLORINE FLUORIDE
		CHLORINE FLUORIDE (Cl2F6)
		CHLORINE FLUORIDE (ClF3)
		CHLORINE TRIFLUORIDE
		CHLORINE TRIFLUORIDE (ACGIH,DOT)
		CHLORINE TRIFLUORIDE (ClF3)
		CHLOROTRIFLUORIDE
		TRIFLUORURE DE CHLORE (FRENCH)
		UN 1749 (DOT)
7790-93-4		
		CHLORIC ACID
		CHLORIC ACID (DOT)
		CHLORIC ACID SOLUTION, CONTAINING NOT MORE THAN 10% ACID (DOT)
		CHLORINE DIOXIDE HYDRATE, FROZEN
		NA 2626 (DOT)
		UN 2626 (DOT)
7790-94-5		
	*	CHLOROSULFONIC ACID
		CHLOROSULFONIC ACID (DOT)
		CHLOROSULFURIC ACID
		CHLOROSULPHONIC ACID
		MONOCHLOROSULFONIC ACID
		MONOCHLOROSULFURIC ACID
		SULFONIC ACID, MONOCHLORIDE
		SULFURIC ACID CHLOROHYDRIN
		SULFURIC CHLOROHYDRIN
		UN 1754 (DOT)
7790-98-9		
		AMMONIUM PERCHLORATE
		AMMONIUM PERCHLORATE (DOT)
		AMMONIUM PERCHLORATE (NH4ClO4)
		AMMONIUM PERCHLORATE, AVERAGE PARTICLE SIZE LESS THAN 45 MICRONS (DOT)
		NK 270
		PERCHLORIC ACID, AMMONIUM SALT
		PKHA
		UN 0402 (DOT)
		UN 1442 (DOT)
7790-99-0		
		CHLORINE IODIDE
		CHLORINE IODIDE (ClI)
		CHLORINE MONOIODIDE
		IODINE CHLORIDE
		IODINE CHLORIDE (ICl)
	*	IODINE MONOCHLORIDE
		IODINE MONOCHLORIDE (DOT)
		IODINE(I) CHLORIDE
		IODOCHLORINE
		PROTOCHLORURE D'IODE (FRENCH)
		UN 1792 (DOT)
		WIJS' CHLORIDE
7791-10-8		
		CHLORIC ACID, STRONTIUM SALT
	*	STRONTIUM CHLORATE
		STRONTIUM CHLORATE (DOT)
		STRONTIUM CHLORATE (Sr(ClO3)2)
		STRONTIUM CHLORATE, WET (DOT)
		UN 1506 (DOT)
7791-12-0		
		RCRA WASTE NUMBER U216
		THALLIUM CHLORIDE
		THALLIUM CHLORIDE (TlCl)
		THALLIUM MONOCHLORIDE
		THALLIUM(1+) CHLORIDE
		THALLIUM(I) CHLORIDE
		THALLOUS CHLORIDE
7791-21-1		
		CHLORINE MONOOXIDE
		CHLORINE MONOXIDE
		CHLORINE MONOXIDE (Cl2O)
		CHLORINE OXIDE
		CHLORINE OXIDE (Cl2O)
		DICHLORINE MONOXIDE
		DICHLORINE OXIDE
		DICHLOROMONOXIDE
		DICHLOROXIDE
		HYPOCHLOROUS ANHYDRIDE
7791-23-3		
		SELENINYL CHLORIDE
		SELENINYL DICHLORIDE
		SELENIUM CHLORIDE OXIDE
		SELENIUM CHLORIDE OXIDE (SeCl2O)
		SELENIUM DICHLORIDE OXIDE
		SELENIUM OXYCHLORIDE
		SELENIUM OXYCHLORIDE (DOT)
		SELENIUM OXYCHLORIDE (SeOCl2)
		SELENIUM OXYDICHLORIDE
		UN 2879 (DOT)
7791-25-5		
		SULFONYL CHLORIDE
		SULFONYL DICHLORIDE
		SULFUR CHLORIDE OXIDE (SO2Cl2)
		SULFUR OXYCHLORIDE (SO2Cl2)
		SULFURIC DICHLORIDE

CAS No.	IN	Chemical Name
		SULFURIC OXYCHLORIDE
		SULFURYL CHLORIDE
		SULFURYL CHLORIDE (DOT)
		SULFURYL CHLORIDE (SO2Cl2)
		SULFURYL DICHLORIDE
		SULPHURYL CHLORIDE
		UN 1834 (DOT)
7791-26-6		
		URANYL CHLORIDE
7791-27-7		
		CHLOROSULFONIC ANHYDRIDE
		DISULFUR PENTOXYDICHLORIDE
		DISULFURYL CHLORIDE
		PYROSULFURYL CHLORIDE
		PYROSULFURYL CHLORIDE (DOT)
		PYROSULFURYL CHLORIDE (S2O5Cl2)
		PYROSULPHURYL CHLORIDE
		PYROSULPHURYL CHLORIDE (DOT)
		SULFUR PENTOXYDICHLORIDE
		UN 1817 (DOT)
7803-49-8		
		HYDROXYAMINE
		HYDROXYLAMINE
		OXAMMONIUM
7803-51-2		
		HYDROGEN PHOSPHIDE
		PHOSPHINE
		PHOSPHINE (PH3)
		PHOSPHORUS TRIHYDRIDE
7803-52-3		
		ANTIMONWASSERSTOFFES (GERMAN)
		ANTIMONY HYDRIDE
		ANTIMONY HYDRIDE (SbH3)
		ANTIMONY TRIHYDRIDE
		ANTYMONOWODOR (POLISH)
		HYDROGEN ANTIMONIDE
		STIBINE
		STIBINE (ACGIH,DOT)
		UN 2676 (DOT)
7803-54-5		
		MAGNESIUM AMIDE
		MAGNESIUM AMIDE (Mg(NH2)2)
		MAGNESIUM DIAMIDE
		MAGNESIUM DIAMIDE (DOT)
		UN 2004 (DOT)
7803-55-6		
	*	AMMONIUM METAVANADATE
		AMMONIUM METAVANADATE (DOT)
		AMMONIUM METAVANADATE (NH4VO3)
		AMMONIUM MONOVANADATE
		AMMONIUM VANADATE
		AMMONIUM VANADATE (NH4VO3)
		AMMONIUM VANADATE(V) ((NH4)VO3)
		AMMONIUM VANADIUM OXIDE (NH4VO3)
		AMMONIUM VANADIUM TRIOXIDE
		RCRA WASTE NUMBER P119
		UN 2859 (DOT)
		VANADATE (VO3 1-), AMMONIUM
		VANADIC ACID (HVO3), AMMONIUM SALT
		VANADIC ACID, AMMONIUM SALT
7803-62-5		
		FLOTS 100SCO
		MONOSILANE
		MONOSILANE (SiH4)
		SILANE
		SILANE (DOT)
		SILICANE
		SILICON HYDRIDE (SiH4)
		SILICON TETRAHYDRIDE
		SILICON TETRAHYDRIDE (ACGIH)
		UN 2203 (DOT)
7803-63-6		
		ACID AMMONIUM SULFATE
		AMMONIUM ACID SULFATE
		AMMONIUM BISULFATE
		AMMONIUM HYDROGEN SULFATE
		AMMONIUM HYDROGEN SULFATE (DOT)
		AMMONIUM HYDROGEN SULFATE (NH4HSO4)
		AMMONIUM MONOHYDROGEN SULFATE
		AMMONIUM SULFATE ((NH4)HSO4)
		MONOAMMONIUM HYDROGEN SULFATE
		MONOAMMONIUM SULFATE
		SULFURIC ACID, MONOAMMONIUM SALT
		UN 2506 (DOT)
7803-65-8		
		AMMONIUM HYPOPHOSPHITE
		PHOSPHINIC ACID, AMMONIUM SALT
7864-93-9		
		SULFURIC ACID
8000-41-7		
		MIXTURE OF P-METHENOLS
		TERPINEOL
		TERPINEOL 318
		TERPINEOLS
8001-22-7		
		DEGUMMED SOYBEAN OIL
		OILS, SOYBEAN
		SOYBEAN OIL
		VT 18
		VT 18 (OIL)
8001-26-1		
		FLAXSEED OIL
	*	LINSEED OIL
		LINSEED OIL, BLEACHED
		OILS, FLAXSEED OR LINSEED
8001-29-4		
	*	COTTONSEED OIL
		FMC 710
		OILS, COTTONSEED
8001-30-7		
		CORN OIL
		LIPOMUL
		MAISE OIL
		MAYDOL
		MAZOLA OIL
		OILS, CORN
8001-31-8		
	*	COPRA
8001-35-2		
		AGRICIDE MAGGOT KILLER (F)
		ALLTEX
		ALLTOX
		ANATOX
		ATTAC 4-2
		ATTAC 4-4
		ATTAC 6
		ATTAC 6-3
		ATTAC 8
		CAMPECHLOR
		CAMPHECHLOR
		CAMPHOCHLOR
		CAMPHOCLOR
		CAMPHOFENE HUILEUX
		CANFECLOR
		CHEM-PHENE
		CHLORINATED CAMPHENE
		CHLORINATED CAMPHENE 60% (ACGIH)
		CHLOROCAMPHENE
		CLOR CHEM T-590
		COMPOUND 3956
		CRESTOXO

CAS No.	IN	Chemical Name
		CRISTOXO
		CRISTOXO 90
		ENT 9,735
		ESTONOX
		FASCO-TERPENE
		GENIPHENE
		GY-PHENE
		HERCULES 3956
		HERCULES TOXAPHENE
		KAMFOCHLOR
		M 5055
		MELIPAX
		MOTOX
		NA 2761 (DOT)
		NCI-C00259
		OCTACHLOROCAMPHENE
		PCC
		PCHK
		PENPHENE
		PHENACIDE
		PHENATOX
		PKHF
		POLYCHLORCAMPHENE
		POLYCHLORINATED CAMPHENES
		POLYCHLOROCAMPHENE
		RCRA WASTE NUMBER P123
		STROBANE T
		STROBANE T-90
		SYNTHETIC 3956
		TOXADUST
		TOXAFEEN (DUTCH)
		TOXAKIL
		TOXAPHEN
		TOXAPHEN (GERMAN)
	*	TOXAPHENE
		TOXAPHENE (DOT)
		TOXON 63
		TOXPHENE
		TOXYPHEN
		VERTAC 90%
		VERTAC TOXAPHENE 90
8001-50-1		
		TERPENE POLYCHLORINATES (65 OR 66 % CHLORINE)
8001-58-9		
		AWPA #1
		BRICK OIL
		COAL CREOSOTE
	*	COAL TAR CREOSOTE
		COAL TAR OIL
		COAL TAR OIL (DOT)
		CREOSOTE
		CREOSOTE OIL
		CREOSOTE P1
		CREOSOTE, COAL TAR
		CREOSOTE, FROM COAL TAR
		CREOSOTUM
		CRESOTE
		CRESOTE OIL
		CRESYLIC CREOSOTE
		HEAVY OIL
		LIQUID PITCH OIL
		NAPHTHALENE OIL
		PRESERV-O-SOTE
		RCRA WASTE NUMBER U051
		SAKRESOTE 100
		TAR OIL
		UN 1136 (DOT)
		WASH OIL
8001-69-2		
		COD LIVER OIL DISTILLATE
		COD OIL
		COD-LIVER OIL
		FISH-LIVER OILS
		GADISTOL
		GADUOL
		LIVER OILS
		OILS, COD-LIVER
8001-86-3		
		BOLAKO OIL
		BOLEKO OIL
		ISANO OIL
		OILS, BOLEKO
8002-03-7		
		ARACHIS OIL
		EARTHNUT OIL
		GROUNDNUT OIL
		KATCHUNG OIL
		OILS, ARACHIS
		OILS, GROUNDNUT
		OILS, PEANUT
		PEANUT OIL
		PEANUT OIL, GROUNDNUT OIL
8002-05-9		
		COAL OIL
		CONTAINERS, OIL TANKS
		CRUDE OIL
		HYDROCARBON OILS, PETROLEUM
		MINERAL OILS
		OIL DEPOSITS
		PETROLEUM
		PETROLEUM CRUDE
		PETROLEUM OIL
		ROCK OIL
		SENECA OIL
8002-09-3		
		ARIZOLE
		OIL OF FIR-SIBERIAN
		OIL OF PINE
		OILS, PINE
		OILS, PINE, SYNTHETIC
		OLEUM ABIETIS
	*	PINE OIL
		PINE OIL (DOT)
		PINE OIL, SYNTHETIC
		TERPENTINOEL (GERMAN)
		UN 1272 (DOT)
		UNIPINE
		YARMOR
		YARMOR PINE OIL
8002-16-2		
		ROSIN OIL
8002-74-2		
		ADVAWAX 165
		APIEZON M
		APIEZON N
		APIEZON W
		ARCOWAX 1150G
		ARCOWAX 2143G
		ARCOWAX 4154G
		ARCOWAX 4158G
		ARISTOWAX
		BARECO 170/175
		CERATAK
		CERETAL 165
		CROLENE LC
		DP 652
		DUROWAX FT 300
		ESKAR R 25
		ESSO 3150
		EVORAL PL
		EVORAL SP
		FISCHER-TROPSCH-WAX
		FLEXOWAX C
		FREEMAN 155/160
		FT 150
		FT 300
		GATCH
		GULFWAX

CAS No.	IN	Chemical Name
		H 1N3
		HARD PARAFFIN
		HAROWAX L 1
		HAROWAX L 2
		HI-MIC 1045
		HYDROCARBON WAXES
		INDRAMIC 30
		J 1440
		JOHNSONS WAX 111
		LENOLENE AC
		MOBIL 150
		MOBIL 150-155F
		MOBILCER 161
		MOBILCER 739
		MOBILCER ED 80/229
		MOBILWAX 220
		MYSTOLENE MK 7
		MYSTOLENE SP 30
		NOPCOSIZE DS 101
		NORANE FH
		NXK 501S
		P 127
		PACEMAKER 37
		PARACOL 404A
		PARACOL 404C
		PARACOL 505N
		PARADIT ASP
		PARADIT PR
		PARADIT PR NEW
		PARADIT PR NEW A
		PARADIUM SS
		PARAFFIN WAX FUME
		PARAFFIN WAXES AND HYDROCARBON WAXES
		PARAFFINS
		PARAFILM
		PARAFLINT H 1
		PARAFLINT HI-NI
		PARAFLINT HIN 3
		PARAFLINT RG
		PARASEAL
		PARVEN 5250
		PERMAL
		PETROLITE C 400
		PETRONAUBA WAX C 8500
		PETROX P 200
		SAN ROGUE WAX
		SASOLWAX HI
		SASOLWAX M 1
		SELOSOL 428
		SHELL 100
		SHELLWAX 100
		SHELLWAX 300
		SHELLWAX 400
		SLACKWAX 11
		SLACKWAX 30
		SLOPVOX
		SMITHWAX 117
		SP 1030
		SUNNOC
		SUNNOC B
		SUNNOC N
		SUNOCO 4415
		SUNOCO 4417
		SUNOCO P 116
		SUNWAX 5512
		TH 44
		UBATOL EXP 21
		ULTRAFLEX
		ULTRAFLEX (WAX)
		VEBAFINE FT 300
		VEBAWAX SH 105
		VICTORY
		VULTEX 8
		WAXES AND WAXY SUBSTANCES, HYDROCARBON
		WOSTEN 65
		X 7905
		XL 165
8003-19-8		
		1,2-DICHLOROPROPANE-1,3-DICHLOROPROPENE MIXT.
		1-PROPENE, 1,3-DICHLORO-, MIXT. WITH 1,2-DICHLOROPROPANE
		D-D
		D-D (PESTICIDE)
		D-D SOIL FUMIGANT
		DD NEMATOCIDE
		DOWFUME N
		VIDDEN D
8003-34-7		
		BUHACH
		CHRYSANTHEMUM CINERAREAEFOLIUM
		CINERIN I OR II
		DALMATION INSECT FLOWERS
		FIRMOTOX
		INSECT POWDER
		INSECTICIDES, PYRETHRINS
		JASMOLIN I OR II
		NA 9184 (DOT)
		PAREXAN
		PYRETHRIN I OR II
		PYRETHRINS
		PYRETHRINS (DOT)
		PYRETHRINS AND PYRETHROIDS
		PYRETHROIDS
	*	PYRETHRUM
		PYRETHRUM (ACGIH)
		PYRETHRUM (INSECTICIDE)
		TRIESTE FLOWERS
8004-09-9		
		AGEL TG 37
		AGEL TG 67
		BROM-O-GAS
		BROZONE
		CHLOROPICRIN AND METHYL BROMIDE, MIXTURE (DOT)
		CHLOROPICRIN-METHYL BROMIDE MIXT
		DOWFUME MC 2
		DOWFUME MC 33
		M-B-C FUMIGANT
		MBC 33
		METHANE, TRICHLORONITRO-, MIXT WITH BROMOMETHANE
		METHYL BROMIDE AND MORE THAN 2% CHLOROPICRIN MIXTURE, LIQUID
		METHYL BROMIDE AND MORE THAN 2% CHLOROPICRIN MIXTURE, LIQUID (DOT)
		NA 1581 (DOT)
		TERR-O-GAS
		TERR-O-GEL
		UN 1581 (DOT)
		VERTAFUME
8004-13-5		
		1,1'-BIPHENYL, MIXT WITH 1,1'-OXYBIS[BENZENE]
		BENZENE, 1,1'-OXYBIS-, MIXT CONTG
		BIPHENYL, MIXED WITH BIPHENYL OXIDE (3:7)
		BIPHENYL-DIPHENYL ETHER MIXTURE
		BIPHENYL-PHENYL ETHER MIXTURE
		DINIL
		DINYL
		DIPHENYL MIXED WITH DIPHENYL OXIDE
		DIPHYL
		DOWTHERM
		DOWTHERM A
		PHENYL ETHER-BIPHENYL MIXTURE VAPOR
		PHENYL ETHER-DIPHENYL MIXTURE
8006-14-2		
		FUELS, GAS
		GAS NATURAL
		GASES, NATURAL
		NATURAL GAS
		NATURAL GAS, SWEET
		SWEET NATURAL GAS

CAS No.	IN	Chemical Name
8006-20-0		
		BLOW GAS
		FUEL GASES, PRODUCER GAS
		GAS PRODUCER
		OIL GASES, PRODUCER GAS
		PRODUCER GAS
8006-28-8		
	*	SODA LIME
		SODA LIME (DOT)
		SODA LIME, SOLID
		SODA LIME, SOLID (DOT)
		SODIUM HYDROXIDE (Na(OH)), MIXT WITH LIME
		SODIUM HYDROXIDE, MIXT WITH LIME
		UN 1907 (DOT)
8006-54-0		
		ADEPS LANE
		AGNOLIN
		AGNOLIN NO. 1
		ALAPURIN
		AMBER LANOLIN
		ANHYDROUS LANOLIN
		ANHYDROUS LANUM
		ARGOWAX
		COSMELAN
		CRODAPUR
		FATS, LANOLIN
		FATS, WOOL
		LANAIN
		LANALIN
		LANESIN
		LANICHOL
		LANIOL
	*	LANOLIN
		LANOLIN, ANHYDROUS
		LANOPRODINE
		LANTROL
		LANUM
		OESIPOS
		PROCESSED LANOLIN
		TW 30
		WOOL FAT, PURIFIED
8006-61-9		
	*	GASOLINE
		MOTOR SPIRIT
		PETROL
8006-64-2		
	*	TURPENTINE
8007-40-7		
		MUSTARD OIL
		OIL OF MUSTARD, EXPRESSED
		OILS, BRASSICA ALBA
		OILS, BRASSICA NIGRA
		OILS, MUSTARD
8007-45-2		
		COAL TAR
		COAL TAR (COAL TAR PITCH)
		COAL TAR EXTRACT
		COAL TAR OINTMENT
	*	COAL TAR PITCH VOLATILES
		ESTAR
		ESTAR (SKIN TREATMENT)
		LAVATAR
		PIXALBOL
		TAR
		TAR, COAL
		TAR, COAL, COLORLESS PURIFIED
		TAR, COAL, PURIFIED COLORLESS
		TARCRONE 180
		ZETAR
8007-56-5		
		AQUA REGIA
		CHLOROAZOTIC ACID
		CHLORONITROUS ACID
		HYDROCHLORIC ACID, MIXED WITH NITRIC ACID (3:1)
		NITROHYDROCHLORIC ACID
		NITROHYDROCHLORIC ACID (DOT)
		NITROHYDROCHLORIC ACID, DILUTED (DOT)
		NITROMURIATIC ACID
		NITROMURIATIC ACID (DOT)
		UN 1798 (DOT)
8008-15-3		
		CAMPHOR OIL (LIGHT)
8008-20-6		
		FUEL OIL #1
	*	KEROSENE
		KEROSINE
		KEROSINE (PETROLEUM)
8008-51-3		
		CAMPHOR OIL
		CAMPHOR OIL (DOT)
		CAMPHOR OIL WHITE
		CAMPHOR OIL YELLOW
		CAMPHOR OIL, RECTIFIED
		FORMOSA CAMPHOR OIL
		FORMOSE OIL OF CAMPHOR
		JAPANESE CAMPHOR OIL
		JAPANESE, OIL OF CAMPHOR
		LIGHT CAMPHOR OIL
		LIGHT OIL OF CAMPHOR
		LIQUID CAMPHOR
		OIL CAMPHOR SASSAFRASSY
		OIL OF CAMPHOR
		OIL OF CAMPHOR RECTIFIED
		OIL OF CAMPHOR WHITE
		OILS, CAMPHOR
		UN 1130 (DOT)
		WHITE CAMPHOR OIL
		WHITE OIL OF CAMPHOR
8012-54-2		
		ARSENIC TRIIODIDE MIXED WITH MERCURIC IODIDE
		ARSENIOUS AND MERCURIC IODIDE SOLUTION
		ARSENIOUS AND MERCURIC IODIDE SOLUTION (DOT)
		DONOVAN'S SOLUTION
		NA 2810 (DOT)
8012-74-6		
		LONDON PURPLE
		LONDON PURPLE, SOLID
		LONDON PURPLE, SOLID (DOT)
		UN 1621 (DOT)
8012-95-1		
	*	MINERAL OIL
	*	OIL MIST, MINERAL
8013-75-0		
		ALCOHOLS, FUSEL
		FUSEL OIL
		FUSEL OIL (DOT)
		FUSELOEL (GERMAN)
		HUILE DE FUSEL (FRENCH)
		OILS, FUSEL
		UN 1201 (DOT)
8014-91-3		
		C.I. DIRECT BROWN 74
8014-95-7		
		DISULPHURIC ACID
		DITHIONIC ACID
		FUMING SULFURIC ACID
		FUMING SULFURIC ACID (DOT)
		NA 1831 (DOT)
		OLEUM
		OLEUM (DOT)
		PYROSULPHURIC ACID
	*	SULFURIC ACID FUMING

CAS No.	IN	Chemical Name
		SULFURIC ACID, MIXT WITH SULFUR TRIOXIDE
		UN 1831 (DOT)
8016-28-2		
		LARD OIL
		OILS, LARD
8020-83-5		
		AAR 1
		ABOLIUM
		ACTIPRON
		AMSCO OMS
		AROMAX 3
		ARTOL 10
		AS 6
		AVIOL
		AVTOL 10
		AW 409
		AWK 1
		BIPHAGITTOL
		BLANDOL WHITE MINERAL OIL
		BUKOMKLEEN
		CARNEA 21
		CARNEA OIL 31
		CERTREX 39
		CHEMKLEEN
		CITOL OIL
		DEOBASE
		DP 11
		DRAKEOL
		DRAKEOL 10
		DRAKEOL 13
		DRAKEOL 15
		DRAKEOL 19
		DRAKEOL 21
		DRAKEOL 32
		DRAKEOL 33
		DRAKEOL 9
		DUTEREX
		ERVOL
		ERVOL WHITE MINERAL OIL
		EUPHYTAN EXTRA
		FLAVEX 937
		FLEXON 791
		FM 5.6AP
		GLORIA
		GLORIA WHITE MINERAL OIL
		HYDROCARBON OIL
		HYDROCARBON OILS
		HYDROFINING
		HYDROREFINING
		IS 45
		ITERM 6
		K 315
		KAYDOL
		KAYDOL WHITE MINERAL OIL
		KHA
		KHF 22-24
		KHF 22S
		KHM 6
		KOGASIN PROCESS
		KREMOL 100
		KREMOL 50
		KREMOL 90
		KREMOL REGULAR
		MASROLAR D
		MINERAL OIL
		MINERAL OILS
		MINKH 1
		MOBILSOL 30
		MP 12
		MR 6
		N 350
		N 500
		NAPHTHOLITE
		NASR OIL
		NR 440
		NUTRAL 600
		OLEX WT 2577
		PN 6K
		PRIMOL 205
		PRIMOL 325
		PS 28
		R 12 (OIL)
		RISELLA 33
		SEMTOL 70
		SHELLFLEX 273
		SHELLFLEX 412
		SHELLSOL 71
		SNPKH 7R2
		SOCAL 226
		SOCAL NO. 226
		SOLVENT C-IX
		SPRAYTEX
		STABILOIL 18
		STABILOIL 62
		SU (OIL)
		SUNISO 4GS
		SUNSPRAY
		SUNSPRAY (HYDROCARBON)
		SUNTEMP
		SUNVIS 31
		TERMOL 190
		TOGASTAN
		TRIONA
		TRIONA B
		TUFFLO 6204
		ULTRASENE
		ULVAPRON
		VALVATA 85
		VAPOR 52
		VELOSIT
		VELOSITE
		VM 4
		WHITE MINERAL OIL
		Z 26
		ZHF 12-18
8021-92-9		
		BLUE GAS
		FUEL GASES, WATER GAS
		OIL GASES, WATER GAS
		WATER GAS
8022-00-2		
		BAYER 21/116
		DEMETON METHYL
		METASYSTOX
		METASYSTOX FORTE
		METHYL DEMETON
		METHYL SYSTOX
		METHYLMERCAPTOPHOS
		PHOSPHOROTHIOIC ACID, O-[2-(ETHYLTHIO)ETHYL] O,O-DIMETHYL ESTER, MIXT. WITH S-[2-(ETHYLTHIO)ETHYL] O
		S(AND O)-2-(ETHYLTHIO)ETHYL O,O-DIMETHYL PHOSPHOROTHIOATE
8023-53-8		
		DICHLOROBENZALKONIUM CHLORIDE
8025-81-8		
		SPIRAMYCIN
8028-73-7		
		ARSENICAL DUST
		ARSENICAL DUST (DOT)
		ARSENICAL FLUE DUST
		FLUE DUST, ARSENIC-CONTG
		UN 1562 (DOT)
8029-29-6		
		POLYCHLORODICYCLOPENTADIENE (CHLORINE CONTENT 60-62% OR 62-64%)

CAS No.	IN	Chemical Name
8030-30-6		
		160 DEGREE BENZOL
		AMSCO H-J
		AMSCO H-SB
		BENZIN
		BENZIN B70
		BENZINE
		COAL TAR NAPHTHA (DOT)
		HI-FLASH NAPHTHA
		HYDROFINING
		HYDROREFINING
		LIGHT LIGROIN
		MINERAL SPIRITS
		NAPHTHA (DOT)
		NAPHTHA
		NAPHTHA DISTILLATE
		NAPHTHA DISTILLATE (DOT)
		NAPHTHA PETROLEUM (DOT)
		NAPHTHA VM&P
		NAPHTHA, PETROLEUM
		NAPHTHA, SOLVENT (DOT)
		NAPHTHA, STRAIGHT-RUN
		PETROLEUM BENZIN
		PETROLEUM DISTILLATES
		PETROLEUM DISTILLATES (NAPHTHA)
		PETROLEUM ETHER
		PETROLEUM NAPHTHA
		PETROLEUM NAPHTHA (DOT)
		PETROLEUM SPIRIT
		RUBBER SOLVENT
		SOLVENTS, NAPHTHAS
		SUPER VMP
		UN 1255 (DOT)
		UN 1256 (DOT)
		UN 2553 (DOT)
		VM&P NAPHTHA
8032-32-4		
		BENZINE (LIGHT PETROLEUM DISTILLATE)
		BENZOLINE
		CANADOL
		LIGROIN
		LIGROINE
		NAPHTHA, LIGROINE
		PAINTERS' NAPHTHA
	*	PETROLEUM ETHER
		REFINED SOLVENT NAPHTHA
		SKELLYSOLVE F
		SKELLYSOLVE G
		V.M. AND P. NAPHTHA
		V.M.& P. NAPHTHA
		VARNISH MAKERS' AND PAINTERS' NAPHTHA
		VARNISH MAKERS' NAPHTHA
		VARNISH MAKERS' NAPHTHA AND PAINTERS' NAPHTHA
8049-17-0		
		FERROSILICON
		FERROSILICON (DOT)
		FERROSILICON, CONTAINING MORE THAN 30% BUT LESS THAN 90% SILICON (DOT)
		FERROSILICON, WITH >30% BUT <70% SILICON
		UN 1408 (DOT)
8049-18-1		
		AUER METAL
		CERIUM ALLOY, BASE, (FERROCERIUM)
		FERROCERIUM
		FERROCERIUM (DOT)
		IRON ALLOY, BASE, (FERROCERIUM)
		PYROPHOROUS ALLOY
		SPARK METAL
		UN 1323 (DOT)
8049-20-5		
		CERIUM MISCH METAL
		MISCH METAL
		MISCHMETAL, POWDER
8049-47-6		
	*	PANCREATIN
8052-41-3		
		NAPHTHA, SOLVENT
		SOLVENT NAPHTHA, STODDARD
		SOLVENTS, NAPHTHAS
	*	STODDARD SOLVENT
		VARNOLINE
		VARSOL
8052-42-4		
		AC 20
		AC 8 (ASPHALT)
		ASPHALT
		ASPHALT CUTBACK
	*	ASPHALT FUMES
		ASPHALT(LIQUID RAPID-CURING)
		ASPHALTUM
		AZ-IP 90
		BITUMEN
		BITUMENS, ASPHALT
		BITUMINOUS MATERIALS, ASPHALT
		BITUSIZE B
		DACHOLEUM
		JUDEAN PITCH
		MINERAL PITCH
		MINERAL RUBBER
		PETROLEUM REFINING RESIDUES, ASPHALTS
		ROAD ASPHALT, LIQUID
		RUBBER, MINERAL
		WITCURB 22L
8063-77-2		
		CARBOGEN
		CARBOGEN (8CI)
		CARBON DIOXIDE MIXED WITH OXYGEN
		CARBON DIOXIDE-OXYGEN MIXTURE
		CARBON DIOXIDE-OXYGEN MIXTURE (DOT)
		OXYGEN-CARBON DIOXIDE MIXTURE (DOT)
		UN 1014 (DOT)
8065-48-3		
		BAY 10756
		BAYER 8169
		DEMETON
		DEMETON (ACGIH)
		DEMETON-O + DEMETON-S
		DEMOX
		DIETHOXY THIOPHOSPHORIC ACID ESTER OF 2-ETHYLMERCAPTOETHANOL
		E 1059
		ENT 17,295
		ETHYL SYSTOX
		MERCAPTOFOS
		MERCAPTOPHOS
		O,O-DIETHYL 2-ETHYLMERCAPTOETHYL THIOPHOSPHATE
		O,O-DIETHYL O(AND S)-2-(ETHYLTHIO)ETHYL PHOSPHOROTHIOATE MIXTURE
		PHOSPHOROTHIOIC ACID, O,O-DIETHYL O-(2-(ETHYLTHIO) ETHYL) ESTER, MIXED WITH O,O-DIETHYL S-(2-(ETHYLTH
		SEPTOX
		SYSTEMOX
		SYSTOX
		ULV
8066-33-9		
		PENTOLITE
8070-50-6		
		ANHYDRIDE CARBONIQUE ET OXYDE D'ETHYLENE MELANGES (FRENCH)
		CARBON DIOXIDE AND ETHYLENE OXIDE MIXTURES, WITH MORE THAN 6% ETHYLENE OXIDE (DOT)
		CARBON DIOXIDE-ETHYLENE OXIDE MIXTURE, WITH >6% ETHYLENE OXIDE
		CARBON DIOXIDE-ETHYLENE OXIDE MIXTURE, WITH ≤6% ETHYLENE OXIDE

CAS No.	IN	Chemical Name
		CARBOXID
		CARBOXIDE
		CARBOXIDE (PESTICIDE)
		CARTOX
		ETHYLENE OXIDE AND CARBON DIOXIDE MIXTURES (DOT)
		ETHYLENE OXIDE, MIXED WITH CARBON DIOXIDE
		ETHYLENE OXIDE-CARBON DIOXIDE MIXTURE
		ETOX
		FUMIGEN 10
		LEUTOX
		OXIRANE, MIXT WITH CARBON DIOXIDE
		OXYFUME 20
		OXYFUME 30
		UN 1041 (DOT)
9000-01-5		
		ACACIA
	*	GUM ARABIC
9000-07-1		
		AUBYGUM X 2
		CARRAGEENAN
		CARRAGEENAN GH
		CARRAGEENAN GUM
		CARRAGEENIN
		CARRAGHEEN
		CARRAGHEENAN
		GELCARIN HWG
		GELLOID J
		GELOZONE
		GENUGOL RLV
		GENUVISCO J
		GUM CARRAGEENAN
		GUM CHON 2
		GUM CHOND
		NORSK GELATAN
		PELLUGEL
		PENCOGEL
		SEAGEL GH
		SEAGEL PET
		SEAKEM 202
		SEAKEM CARRAGEENIN
		KAPPA LAMBDA-CARRAGEENAN
9000-32-2		
		GUTTA PERCHA, SOLUTION (DOT)
		GUTTA-PERCHA
		GUTTA-PERCHA SOLUTION
		UN 1205 (DOT)
9000-36-6		
		KARAYA
9000-59-3		
	*	SHELLAC
9000-65-1		
		TRAGACANTH GUM
9000-92-4		
	*	DIASTASE
9001-37-0		
		GLUCOSE OXIDASE
9001-73-4		
		PAPAIN
9002-07-7		
	*	TRYPSIN
9002-81-7		
		1,3,5-TRIOXANE, HOMOPOLYMER
		ALPHA-POLY(OXYMETHYLENE)
		FORMALDEHYDE POLYMER
		FORMALDEHYDE, HOMOPOLYMER
		POLY(OXYMETHYLENE)
		POLY-S-TRIOXANE
		POLYFORMALDEHYDE
		POLYOXYALKYLENES, POLYOXYMETHYLENES
		POLYTRIOXANE
		TAKAFEST
9002-84-0		
		AFG 80VS
		AFLON G 8
		AFLON G 80
		ALGOFLON
		ALGOFLON SV
		AMIP 15M
		ARMALON XT 2663
		AVCOAT 8029-1
		AVCOAT 8029-2
		AVCOAT 8029-3
		BDH 29-801
		CHROMOSORB T
		DF 100
		DIXON 164
		DLX 6000
		DLX 7000
		DUROID 5650
		DUROID 5650M
		DUROID 5813
		DUROID 5870
		DUROID 5870M
		DUROID X
		DUROID X 026
		EK 1108GY-A
		ETHENE, TETRAFLUORO-, HOMOPOLYMER
		ETHICON PTFE
		ETHYLENE, TETRAFLUORO-, POLYMERS
		F 103
		F 4DP
		F 4K20
		F 4ZH20
		FBF 74D
		FG 15
		FLOROLON 4M
		FLUO-KEM
		FLUON
		FLUON 169
		FLUON AD 704
		FLUON CD 023
		FLUON CD 042
		FLUON CD 1
		FLUON G 163
		FLUON G 201
		FLUON G 308
		FLUON G 4
		FLUON GPI
		FLUON L 169
		FLUON L 169A
		FLUON L 169B
		FLUON L 170
		FLUON L 171
		FLUON VP 25
		FLUON VX 2
		FLUON VXI
		FLUOROFLEX
		FLUOROLON 4
		FLUOROLON 4D
		FLUORON AD 2
		FLUOROPAK 80
		FLUOROPLAST 4
		FLUOROPLAST 4B
		FLUOROPLAST 4D
		FLUOROPLAST 4M
		FLUOROPORE F 045
		FLUOROPORE FP 022
		FLUOROPORE FP 045
		FLUOROPORE FP 120
		FLUOROPORE FP 200
		FLUOROPORE FP 500
		FN 3
		FP 4
		FT 4
		FTORLON 4
		FTORLON 4D

CAS No.	IN	Chemical Name
		FTORLON 4DP
		FTORLON 4K20
		FTORLON 4M
		FTOROLON 4
		FTOROPLAST 4
		FTOROPLAST 4B
		FTOROPLAST 4D
		FTOROPLAST 4DP
		FTOROPLAST 4G10
		FTOROPLAST 4K15M5L-EA
		FTOROPLAST 4K20
		FTOROPLAST 4M
		FTOROPLAST AMIP 15M
		FTOROPLAST F 4
		FTOROPLAST FBF 74D
		FTOROPLAST FP 4D
		G 163
		GLASROCK POREX P 1000
		GORE-TEX
		HALON G 183
		HALON G 700
		HALON G 80
		HALON TFE
		HALON TFEG 180
		HEYDEFLON
		HOSTAFLON SE-VP 585
		HOSTAFLON SE-VP 5875
		HOSTAFLON TF
		HOSTAFLON TF 2026
		HOSTAFLON TF 2053
		HOSTAFLON TF 5032
		HOSTAFLON TF 9205
		HOSTAFLON TF-VP 5034
		HOSTAFLON TF-VP 5444
		L 169
		L 169A
		LUBLON L 2
		LUBLON L 5
		MITEX
		MOLYKOTE 522
		PERFLUOROETHYLENE POLYMER
		POLITEF
		POLY(ETHYLENE TETRAFLUORIDE)
		POLYFENE
		POLYFLON
		POLYFLON D 1
		POLYFLON EK 1108GY-A
		POLYFLON EK 1700
		POLYFLON EK 1883GB
		POLYFLON EK 4105GN
		POLYFLON EK 4108GY
		POLYFLON F 101
		POLYFLON F 103
		POLYFLON F 104
		POLYFLON M 1
		POLYFLON M 12
		POLYFLON M 12A
		POLYFLON M 21
		POLYFLON PA 10L
		POLYFLON PA 5L
		POLYLUBE J 12
		POLYMIST F 5
		POLYMIST F 5A
		POLYPERFLUOROETHYLENE
		POLYTEF
		POLYTETRAFLUOROETHENE
		POLYTETRAFLUOROETHYLENE
		POREFLON FP 500
		POREX P 1000
		PROPLAST
		PTFE
		PTFE-GM3
		RULON E
		RULON LD
		SOREFLON
		SOREFLON 5A
		SOREFLON 604
		SOREFLON 7
		SOREFLON 7A20
		SOREFLON 8G
		T 30
		T 30 (FLUOROPOLYMER)
		T 5B
		T 8A
		TARFLEN
		TE 30
	*	TEFLON
		TEFLON 110
		TEFLON 30
		TEFLON 30J
		TEFLON 327
		TEFLON 5
		TEFLON 6
		TEFLON 6C
		TEFLON 6C-J
		TEFLON 6J
		TEFLON 7A
		TEFLON 7J
		TEFLON 851-204
		TEFLON K
		TEFLON T 30
		TEFLON T 5
		TEFLON T 6
		TEFLON T 8A
		TEFLON TFE
		TETRAFLUOROETHENE HOMOPOLYMER
		TETRAFLUOROETHENE POLYMER
		TETRAFLUOROETHYLENE HOMOPOLYMER
		TETRAFLUOROETHYLENE POLYMER
		TETRAFLUOROETHYLENE RESIN
		TETRAN 30
		TETRAN PTFE
		TL 102
		TL 102 (POLYMER)
		TL 103F
		TL 115
		TL 125
		TL 126
		TL-R
		TL-V
		TLP 10
		TLP 10F
		UNON P
		UNON P 300
		VALFLON
		VALFLON 7000
		VALFLON 7790
		VALFLON 7990
		WHITCON 5TFE
		WHITCON 7
		ZEDEFLON
		ZEFLUOR
		ZEPEL C-SF
		ZITEX H 662-124
		ZITEX K 233-122
9002-89-5		
		ALCOTEX 17F-H
		ALCOTEX 725L
		ALCOTEX 75L
		ALCOTEX 88/05
		ALCOTEX 88/10
		ALCOTEX 99/10
		ALKOTEX
		ALVYL
		ARACET APV
		ARACET APV 50/88
		ATACTIC POLY(VINYL ALCOHOL)
		BOVLON
		C 17
		CHEMTREND 39
		CIPOVIOL W 72
		COVAL 9700
		COVOL
		COVOL 971
		DENKA POVAL B 17
		DENKA POVAL G 05
		EG 40

CAS No.	IN	Chemical Name
		ELVANOL
		ELVANOL 50-42
		ELVANOL 51-05G
		ELVANOL 5105
		ELVANOL 52-22
		ELVANOL 52-22G
		ELVANOL 522-22
		ELVANOL 70-05
		ELVANOL 71-30
		ELVANOL 73125G
		ELVANOL 90-50
		ELVANOL T 25
		ENBRA OV
		EP 160
		ETHENOL HOMOPOLYMER (9CI)
		ETHENOL, HOMOPOLYMER
		GALVATOL 1-60
		GELUTOL
		GELVATOL
		GELVATOL 1-30
		GELVATOL 1-60
		GELVATOL 1-90
		GELVATOL 20-30
		GELVATOL 2060
		GELVATOL 209
		GELVATOL 3-60
		GELVATOL 3-91
		GH 20
		GL 02
		GL 03
		GL 05
		GLO 5
		GM 14
		GOHSEFIMER L 7514
		GOHSENOL
		GOHSENOL AH 22
		GOHSENOL EG 40
		GOHSENOL GH
		GOHSENOL GH 17
		GOHSENOL GH 20
		GOHSENOL GH 23
		GOHSENOL GL 02
		GOHSENOL GL 03
		GOHSENOL GL 05
		GOHSENOL GL 08
		GOHSENOL GM 14
		GOHSENOL GM 14L
		GOHSENOL GM 94
		GOHSENOL KH 17
		GOHSENOL KH 20
		GOHSENOL KL 05
		GOHSENOL KP 06
		GOHSENOL KP 08
		GOHSENOL L 5307
		GOHSENOL MG 14
		GOHSENOL N 300
		GOHSENOL NH 05
		GOHSENOL NH 14
		GOHSENOL NH 17
		GOHSENOL NH 18
		GOHSENOL NH 20
		GOHSENOL NH 26
		GOHSENOL NK 114
		GOHSENOL NL 05
		GOHSENOL NM 11
		GOHSENOL NM 114
		GOHSENOL NM 14
		GOHSENOL NM 300
		GOHSENOL T
		GOHSENOL T 330
		HISELON C 300
		HV POVAL
		IVALON
		KURALON VP
		KURARAY POVAL PVA 420
		KURARE POVAL 120
		KURARE POVAL 1700
		KURARE PVA 205
		KURATE POVAL 120

CAS No.	IN	Chemical Name
		LAMEPHIL OJ
		LEMOL
		LEMOL 12-88
		LEMOL 16-98
		LEMOL 24-98
		LEMOL 30-98
		LEMOL 5-88
		LEMOL 5-98
		LEMOL 51-98
		LEMOL 60-98
		LEMOL 75-98
		LEMOL GF 60
		M 13/20
		MONOSOL 9000-0015-3
		MOWIOL
		MOWIOL 4-88
		MOWIOL 4-98
		MOWIOL 42-99
		MOWIOL N 30-88
		MOWIOL N 50-88
		MOWIOL N 50-98
		MOWIOL N 70-98
		MOWIOL N 85-88
		N 300
		NH 18
		NM 11
		NM 14
		P 700
		PA 18
		PA 18 (POLYOL)
		PA 20
		POLY(VINYL ALCOHOL)
		POLYDESIS
		POLYSIZER 173
		POLYVINOL
		POLYVINYL ALCOHOL
		POLYVIOL
		POLYVIOL M 05/290
		POLYVIOL M 13/140
		POLYVIOL MO 5/140
		POLYVIOL V 03/20
		POLYVIOL W 25/140
		POLYVIOL W 28/20
		POLYVIOL W 40/140
		POVAL
		POVAL 105
		POVAL 117
		POVAL 120
		POVAL 1700
		POVAL 203
		POVAL 205
		POVAL 205S
		POVAL 217
		POVAL 217S
		POVAL 420
		POVAL A
		POVAL B 03
		POVAL B 05
		POVAL B 17
		POVAL C 17
		POVAL HL 12
		POVAL K 17E
		POVAL K 24E
		POVAL PA 5
		POVAL UP 240G
		PS 1200
		PVA
		PVA 008
		PVA 105
		PVA 110
		PVA 117
		PVA 210
		PVA 217
		PVA 420
		PVA-HC
		PVAL 45/02
		PVAL 55/12
		PVS
		PVS 4

CAS No.	IN	Chemical Name
		PXA 105
		RATIFIX F
		RESISTOFLEX
		RHODORICOL 4/20
		RHODOVIOL
		RHODOVIOL 16/200
		RHODOVIOL 4-125P
		RHODOVIOL 4/125
		RHODOVIOL 5/270P
		RHODOVIOL R 16/20
		SOLVAR
		SUMITEX H 10
		VIBATEX S
		VINACOL MH
		VINALAK
		VINAROL
		VINAROL DT
		VINAROL ST
		VINAROL SVH
		VINAROLE
		VINAVILOL 2-98
		VINNAROL
		VINOL
		VINOL 107
		VINOL 125
		VINOL 165
		VINOL 205
		VINOL 205S
		VINOL 325
		VINOL 350
		VINOL 351
		VINOL 523
		VINOL UNISIZE
		VINOL WS 42
		VINYL ALCOHOL POLYMER
		VINYLON FILM 2000
		VINYLON FILM 3000
		VINYLON FILM VF-A 2500
		VPB 105-2
		WARCOPOLYMER A 20
9002-91-9		
		1,3,5,7-TETROXOCANE, 2,4,6,8-TETRAMETHYL-
		ACETALDEHYDE POLYMER
		ACETALDEHYDE, HOMOPOLYMER
		ACETALDEHYDE, TETRAMER
		ARIOTOX
		CEKUMETA
		LIMAX
		LIMOVET
		META
		METACETALDEHYDE
		METALDEHYD (GERMAN)
		METALDEHYDE
		METALDEHYDE (DOT)
		METALDEIDE (ITALIAN)
		METASON
		NAMEKIL
		POLYACETALDEHYDE
		SLUG-TOX
		SLUGIT
		UN 1332 (DOT)
9004-34-6		
		ABICEL
		ALPHA-CELLULOSE
		ARBOCEL
		ARBOCEL BC 200
		ARBOCELL B 600/30
		AVICEL
		AVICEL 101
		AVICEL 102
		AVICEL PH 101
		AVICEL PH 105
		BETA-AMYLOSE
		CELLEX MX
		CELLULOSE
		CELLULOSE 248
		CELLULOSE CRYSTALLINE
		CELUFI
		CEPO
		CEPO CFM
		CEPO S 20
		CEPO S 40
		CHROMEDIA CC 31
		CHROMEDIA CF 11
		CUPRICELLULOSE
		ELCEMA F 150
		ELCEMA G 250
		ELCEMA P 050
		ELCEMA P 100
		FRESENIUS D 6
		HEWETEN 10
		HYDROXYCELLULOSE
		KINGCOT
		LA 01
		MN-CELLULOSE
		ONOZUKA P 500
		PYROCELLULOSE
		RAYOPHANE
		RAYWEB Q
		REXCEL
		SIGMACELL
		SOLKA-FIL
		SOLKA-FLOC
		SOLKA-FLOC BW
		SOLKA-FLOC BW 100
		SOLKA-FLOC BW 20
		SOLKA-FLOC BW 200
		SOLKA-FLOC BW 2030
		SPARTOSE OM-22
		SULFITE CELLULOSE
		TOMOFAN
		TUNICIN
		WHATMAN CC-31
9004-66-4		
		A 100
		A 100 (PHARMACEUTICAL)
		B 75
		CHINOFER
		DEXTROFER 100
		DEXTROFER 75
		FE-DEXTRAN
		FERDEX 100
		FERRIC DEXTRAN
		FERRIDEXTRAN
		FERRODEXTRAN
		FERROGLUCIN
		FERROGLUKIN
		FERROGLUKIN 75
		IMFERON
		IMPOSIL
		IRO-JEX
		IRON DEXTRAN
		IRON DEXTRAN COMPLEX
		IRON DEXTRAN INJECTION
		MYOFER 100
		POLYFER
		PROLONGAL
9004-70-0		
		A 5021
		AS
		BA 85
		BK2-W
		BK2-Z
		BOX TOE GUM
		C 2018
		CA 80
		CA 80-15
		CELEX
		CELLINE 200
		CELLOIDIN
		CELLULOSE NITRATE
		CELLULOSE NITRATE (CONTAINING NOT MORE THAN 12.6% NITROGEN)
		CELLULOSE TETRANITRATE

CAS No.	IN	Chemical Name
		CELLULOSE, NITRATE (9CI)
		CELNOVA BTH 1/2
		CN 85
		CN 88
		COLLODION
		COLLODION (DOT)
		COLLODION COTTON
		COLLODION WOOL
		COLLOXYLIN
		COLLOXYLIN VNV
		CORIAL EM FINISH F
		DAICEL RS 1
		DAICEL RS 1/2H
		DAICEL RS 7
		DHX 30/50
		E 1440
		E 375
		FILM (DOT)
		FLEXIBLE COLLODION
		FM-NTS
		GUNCOTTON
		GUNCOTTON (DOT)
		H 1/2
		HE 2000
		HITENOL 12
		HX 3/5
		KODAK LR 115
		LR 115
		LR 115II
		NA 1324 (DOT)
		NA 1325 (DOT)
		NITROCEL S
	*	NITROCELLULOSE
		NITROCELLULOSE (DOT)
		NITROCELLULOSE E 950
		NITROCELLULOSE, DRY (DOT)
		NITROCELLULOSE, IN SOLUTION IN FLAMMABLE LIQUIDS (DOT)
		NITROCOTTON
		NITRON
		NITRON (NITROCELLULOSE)
		NIXON N/C
		NP 180
		NTS 218
		NTS 222
		NTS 539
		NTS 542
		NTS 62
		PARLODION
		PC MEDIUM
		PYRALIN
		PYROXYLIN
		PYROXYLIN PLASTIC (DOT)
		PYROXYLIN PLASTIC SCRAP (DOT)
		PYROXYLIN RODS (DOT)
		PYROXYLIN ROLLS (DOT)
		PYROXYLIN SCRAP (DOT)
		PYROXYLIN SHEETS (DOT)
		PYROXYLIN TUBES (DOT)
		R. S. NITROCELLULOSE
		RF 10
		RS
		RS 1/2
		RS 1/4
		SARTORIUS SM 11304
		SARTORIUS SM 11306
		SHADOLAC MT
		SL 1
		SOLUBLE GUN COTTON
		SS
		SYNPOR
		TSAPOLAK 964
		UN 0340 (DOT)
		UN 1324 (DOT)
		UN 2059 (DOT)
		UN 2060 (DOT)
		XYLOIDIN
		ZAPON
9005-25-8		
		ALPHA-STARCH
		AMAIZO 310
		AMAIZO W 13
		AMICOL 1B
		AMICOL C
		AMIGEL 12014
		AMIGEL 30076
		AMYLOMAIZE VII
		AMYLOSE, MIXT. WITH AMYLOPECTIN
		AMYLUM
		AMYSIL K
		ARROWROOT STARCH
		CALOREEN
		CLARO 5591
		CLEARJEL
		CORN STARCH
		CPC 3005
		CPC 6448
		EMJEL 200
		EMJEL 300
		FARINEX 100
		FARINEX TSC
		HI-COASTAR PC 11
		HRW 13
		HYLON
		IMPERMEX
		KEESTAR 328
		MAIZENA
		MARANTA
		MELOGEL
		MELUNA
		MIRA QUICK C
		NABOND
		OK PRE-GEL
		PASSELI P
		PENFORD GUM 380
		REMYLINE
		REMYLINE AC
		RICE STARCH
		SNOWFLAKE 12615
		SNOWFLAKE 30091
		SORBITOSE C 5
		SORGHUM GUM
		ST 1500
		STA-RX 1500
		STARAMIC 747
	*	STARCH
		STAYCO S
		TAPIOCA STARCH
		TAPON
		TROGUM
		W 13 STABILIZER
		W-GUM
9005-90-7		
		GALIPOT
		GUM THUS
		GUM TURPENTINE
		NAVAL STORES, TURPENTINE
		PINE GUM
		RESINS, PINE
		RESINS, TURPENTINE
		SKIPIDAR
		TURPENTINE
		TURPENTINE GUM
9007-13-0		
		CALCIUM RESINATE
		CALCIUM RESINATE (DOT)
		CALCIUM RESINATE, FUSED
		CALCIUM RESINATE, FUSED (DOT)
		CALCIUM RESINATE, TECHNICALLY PURE (DOT)
		LIMED ROSIN
		RESIN ACIDS AND ROSIN ACIDS, CALCIUM SALTS
		ROSIN CALCIUM SALT
		UN 1313 (DOT)
		UN 1314 (DOT)

CAS No.	IN	Chemical Name
9007-39-0		
		COPPER RESINATE
		RESIN ACIDS AND ROSIN ACIDS, COPPER SALTS
9008-34-8		
		MANGANESE RESINATE
		MANGANESE RESINATE (DOT)
		RESIN ACIDS AND ROSIN ACIDS, MANGANESE SALTS
		UN 1330 (DOT)
9010-69-9		
		RESIN ACID ZINC SALT
		RESIN ACIDS AND ROSIN ACIDS, ZINC SALTS
		UN 2714 (DOT)
	*	ZINC RESINATE
		ZINC RESINATE (DOT)
9015-68-3		
		ASPARAGINASE
9015-98-9		
		ALPHA-HYDRO-OMEGA-HYDROXYPOLY(OXYMETHYLENE)
		POLY(OXYMETHYLENE) GLYCOL
		POLY(OXYMETHYLENE), ALPHA-HYDRO-OMEGA-HYDROXY-
		POLYMETHYLENE GLYCOL
9016-45-9		
		(NONYLPHENOXY)POLYETHYLENE OXIDE
		A 730
		A 730 (SURFACTANT)
		ADEKATOL NP 700
		AGRAL
		AGRAL 90
		AGRAL LN
		AGRAL R
		AKYPO NP 70
		AKYPOROX NP 105
		AKYPOROX NP 95
		ALFENOL
		ALFENOL 10
		ALFENOL 18
		ALFENOL 22
		ALFENOL 28
		ALFENOL 710
		ALFENOL 8
		ALKASURF NP
		ALKASURF NP 11
		ALKASURF NP 8
		ALPHA-(NONYLPHENYL)-OMEGA-HYDROXYPOLY(OXY-1,2-ETHANEDIYL)
		ALPHA-(NONYLPHENYL)-OMEGA-HYDROXYPOLYOXYETHYLENE
		ANTAROX CO
		ANTAROX CO 430
		ANTAROX CO 530
		ANTAROX CO 630
		ANTAROX CO 730
		ANTAROX CO 850
		ANTAROX CO 880
		ANTAROX CO 970
		ARCOPAL N 100
		ARKOPAL N
		ARKOPAL N 040
		ARKOPAL N 060
		ARKOPAL N 080
		ARKOPAL N 090
		ARKOPAL N 100
		ARKOPAL N 110
		ARKOPAL N 150
		ARKOPAL N 300
		AUXIPON NP
		B 350
		BEROL 02
		BEROL 09
		BEROL 259
		BEROL 26
		BEROL 267
		BEROL 296
		BLM
		BO
		BURTEMUL N
		CCC JELLY
		CEMULSOL NP 10
		CEMULSOL NP 8
		CEMULSOL NP 9
		CEMULSOL NP-EO 6
		CHIMIPAL WN 6
		CO 630
		CONCO NI
		CONCO NI 190
		DEHSCOXID 781
		DISPERGATOR BO
		DME
		DOWFAX 9N9
		DS 3195
		EA 120
		EA 80
		ELFAPUR N 70
		EMALEX NP 15
		EMMON 15332
		EMPILAN NP 9
		EMU 02
		EMU 09
		EMULGATOR NP 10
		EMULGEN 900
		EMULGEN 903
		EMULGEN 905
		EMULGEN 906
		EMULGEN 909
		EMULGEN 910
		EMULGEN 911
		EMULGEN 913
		EMULGEN 920
		EMULGEN 930
		EMULGEN 931
		EMULGEN 935
		EMULGEN 950
		EMULGEN 985
		EMULGEN PI 20T
		EMULSON 20B
		EMULSON 9B
		ETHER
		ETHOXYLATED NONYLPHENOL
		ETHYLAN 20
		ETHYLAN 44
		ETHYLAN 55
		ETHYLAN BCP
		ETHYLAN HA
		ETHYLAN N
		ETHYLAN N 55
		ETHYLAN TU
		ETHYLENE OXIDE-NONYLPHENOL CONDENSATE
		ETHYLENE OXIDE-NONYLPHENOL POLYMER
		ETOLAT 914
		FENOPAL
		GAFAC CO 99 0
		GLYCOLS, POLYETHYLENE, MONO(NONYLPHENYL) ETHER
		HOSTAPAL CV
		HOSTAPAL W
		IGEPAL CO
		IGEPAL CO 210
		IGEPAL CO 430
		IGEPAL CO 436
		IGEPAL CO 520
		IGEPAL CO 530
		IGEPAL CO 610
		IGEPAL CO 630
		IGEPAL CO 660
		IGEPAL CO 710
		IGEPAL CO 730
		IGEPAL CO 850
		IGEPAL CO 880
		IGEPAL CO 887
		IGEPAL CO 890
		IGEPAL CO 970
		IGEPAL CO 977
		IGEPAL CO 990

CAS No.	IN	Chemical Name
		IGEPAL CO 997
		IMBENTIN
		IMBENTIN N 52
		IMBENTINE
		LEROLAT N
		LEROLAT N 300
		LIPAL 9N
		LIPONOX NCG
		LIPONOX NCH
		LIPONOX NCI
		LIPONOX NCM
		LISSAPOL N
		LISSAPOL NX
		LISSAPOL NXP 10
		LISSAPOL TN 450
		LUBROL APN 5
		LUBROL L
		LUBROL N
		LUBROL N 13
		LUTENSOL AP 10
		LUTENSOL AP 20
		LUTENSOL AP 9
		M 812
		MAKON
		MAKON 10
		MAKON 12
		MAKON 14
		MAKON 30
		MAKON 4
		MAKON 6
		MAKON 8
		MARCHON
		MARLOPHEN
		MARLOPHEN 810
		MARLOPHEN 812
		MARLOPHEN 85
		MARLOPHEN 88
		MARLOPHEN 89
		MERGITAL OP 2
		MERITEN NF 9
		MERPOXEN 230
		MERPOXEN ON
		MONO(NONYLPHENYL)POLYETHYLENE GLYCOL
		MONOPOL 1020
		N 100
		NEMOL K 1030
		NEMOL K 2030
		NEMOL K 34
		NEMOL K 36
		NEUTRONYX 640
		NEUTRONYX 676
		NEWCOL 504
		NEWCOL 520
		NEWCOL 560
		NEWCOL 561H
		NEWCOL 562
		NEWCOL 564
		NEWCOL 568
		NIKKOL N P 2
		NIKKOL NP
		NIKKOL NP 10
		NIKKOL NP 100
		NIKKOL NP 15
		NIKKOL NP 18TX
		NIKKOL NP 5
		NIKKOL NP 75
		NISSAN NONION NS
		NISSAN NONION NS 203
		NISSAN NONION NS 206
		NISSAN NONION NS 210
		NISSAN NONION NS 215
		NISSAN NONION NS 220
		NISSAN NONION NS 230
		NOIGEN EA 150
		NOIGEN EA 50
		NOIGEN EA 70
		NOIGEN EA 80
		NONAL 208
		NONAL 210

CAS No.	IN	Chemical Name
		NONARIL 910
		NONARIL 930
		NONIDET NP 40
		NONIDET NP 50
		NONIDET P 80
		NONIOLITE PN 4
		NONION NS
		NONION NS 203
		NONION NS 206
		NONION NS 2085
		NONION NS 210
		NONION NS 212
		NONION NS 215
		NONION NS 220
		NONION NS 230
		NONION NS 240
		NONIONIK NI
		NONIPOL
		NONIPOL 100
		NONIPOL 110
		NONIPOL 120
		NONIPOL 130
		NONIPOL 140
		NONIPOL 160
		NONIPOL 20
		NONIPOL 200
		NONIPOL 40
		NONIPOL 400
		NONIPOL 45
		NONIPOL 55
		NONIPOL 60
		NONIPOL 70
		NONIPOL 800
		NONIPOL 85
		NONIPOL 95
		NONYLPHENOL ETHOXYLATE
		NONYLPHENOL POLYETHYLENE GLYCOL ETHER
		NONYLPHENOL POLYETHYLENE OXIDE
		NONYLPHENOXY POLYETHOXY ETHANOL
		NONYLPHENOXYPOLY(ETHYLENEOXY)ETHANOL
		NONYLPHENOXYPOLY(OXYETHYLENE)ETHANOL
		NONYLPHENYL POLYETHYLENE GLYCOL ETHER
		NOP 9
		NOREGAL LC 4 CONC
		NP
		NP (NONIONIC SURFACTANT)
		NP 10
		NP 100
		NP 13
		NP 14
		NP 17
		NP 20
		NP 40
		NP 660
		NP 695
		NP 75
		NP 8
		NP 80
		NP 936
		NS 215
		OMEGA-HYDROXY-ALPHA-(NONYLPHENYL)POLY(OXY-1,2-ETHANEDIYL)
		OP 2
		OXYETHYLATED NONYLPHENOL
		OXYETHYLENE NONYLPHENYL ETHER
		PENETRAX
		POLY(OXY-1,2-ETHANEDIYL), ALPHA-(NONYLPHENYL)-OMEGA-HYDROXY-
		POLY(OXYETHYLENE) NONYLPHENOL ETHER
		POLY(OXYETHYLENE) NONYLPHENYL ETHER
		POLY-TERGENT B
		POLY-TERGENT B 150
		POLY-TERGENT B 300
		POLY-TERGENT B 350
		POLYETHOXYLATED NONYLPHENOL
		POLYETHYLENE GLYCOL 450 NONYL PHENYL ETHER
		POLYETHYLENE GLYCOL MONO(NONYLPHENOL) ETHER
		POLYETHYLENE GLYCOL MONO(NONYLPHENYL) ETHER
		POLYETHYLENE GLYCOL NONYLPHENOL ETHER

CAS No.	IN	Chemical Name
		POLYETHYLENE GLYCOL NONYLPHENYL ETHER
		POLYETHYLENEOXIDE MONO(NONYLPHENYL) ETHER
		POLYOXYETHYLATED NONYLPHENOL
		POLYOXYETHYLENE (15) NONYL PHENYL ETHER
		POLYOXYETHYLENE (20) NONYL PHENYL ETHER
		POLYOXYETHYLENE (9) NONYL PHENYL ETHER
		POLYSTEP F 6
		POLYSTEP F 8
		PREVOCELL W-OF 100
		PROTACHEM 630
		REMCOPAL
		RENDELLS SUPPOSITORY
		RENEX 1000
		RENEX 300
		RENEX 647
		RENEX 648
		RENEX 650
		RENEX 678
		RENEX 679
		RENEX 688
		RENEX 690
		RENEX 697
		RENEX 698
		RETZANOL NP 100
		REWOPAL HV 10
		REWOPAL HV 25
		REWOPOL HV-9
		TERGITOL TP-9 (NONIONIC)
9016-87-9		
		CORONATE MR 200
		DESMODUR PU 1520A20
		E 534
		ISOBIND 100
		ISOCYANATE 580
		ISOCYANIC ACID, POLYMETHYLENEPOLYPHENYLENE ESTER
		ISONATE 390P
		ISOSET CX 11
		KAISER NCO 20
		LUPRINATE M 20
		MDI-CR
		MDI-CR 100
		MDI-CR 200
		MDI-CR 300
		MILLIONATE MR
		MILLIONATE MR 100
		MILLIONATE MR 200
		MILLIONATE MR 300
		MILLIONATE MR 340
		MILLIONATE MR 400
		MILLIONATE MR 500
		MOBAY MRS
		MONDUR E 441
		MONDUR E 541
		MONDUR MR
		MONDUR MRS
		MONDUR MRS 10
		NCO 20
		NIAX AFPI
		PAPI
		PAPI 135
		PAPI 20
		PAPI 27
		PAPI 580
		PAPI 901
		POLY(METHYLENE PHENYLENE ISOCYANATE)
		POLY(PHENYLENEMETHYLENE ISOCYANATE)
		POLYMETHYL POLYPHENYL POLYISOCYANATE
		POLYMETHYLENE POLYPHENYL ISOCYANATE
		POLYMETHYLENE POLYPHENYL POLYISOCYANATE
		POLYMETHYLENE POLYPHENYLENE ISOCYANATE
		POLYMETHYLENEPOLYPHENYLENE ISOCYANATE POLYMER
		POLYMETHYLENEPOLYPHENYLENE POLYISOCYANATE
		POLYPHENYLENE POLYMETHYLENE POLYISOCYANATE
		POLYPHENYLPOLYMETHYLENE POLYISOCYANATE
		RUBINATE M
		RUBINATE MF 178
		SUMIDUR 44V10
		SUMIDUR 44V20
		SUPRASEC 1042
		SUPRASEC DC
		SYSTANAT MR
		SYSTANATE MR
		TAKENATE 300C
		TEDIMON 31
		THANATE P 210
		THANATE P 220
		THANATE P 270
9032-75-1		
	*	PECTINASE
9056-38-6		
		NITROSTARCH
9080-17-5		
		AMMONIA POLYSULFIDE
		AMMONIUM POLYSULFIDE
		AMMONIUM POLYSULFIDE SOLUTION
		AMMONIUM POLYSULFIDE SOLUTION (DOT)
		AMMONIUM SULFIDE ((NH4)2(SX))
		AMMONIUM SULFIDE (POLY-)
		AMMONIUM SULFIDE SOLUTION, RED
		AMMONIUM TRISULFIDE
		AP-S
		DIAMMONIUM POLYSULFIDE
		DIAMMONIUM TRISULFIDE
		UN 2818 (DOT)
10022-31-8		
		BARIUM DINITRATE
	*	BARIUM NITRATE
		BARIUM NITRATE (Ba(NO3)2)
		BARIUM NITRATE (DOT)
		BARIUM(II) NITRATE (1:2)
		DUSICNAN BARNATY (CZECH)
		NITRATE DE BARYUM (FRENCH)
		NITRIC ACID BARIUM SALT (2:1)
		NITRIC ACID, BARIUM SALT
		UN 1446 (DOT)
10022-70-5		
		HYPOCHLOROUS ACID, SODIUM SALT, PENTAHYDRATE
		SODIUM HYPOCHLORITE PENTAHYDRATE
10024-97-2		
		DINITROGEN MONOXIDE
		DINITROGEN OXIDE
		FACTITIOUS AIR
		HYPONITROUS ACID ANHYDRIDE
		LAUGHING GAS
		NITROGEN OXIDE
		NITROGEN OXIDE (N2O)
	*	NITROUS OXIDE
		NITROUS OXIDE (DOT)
		NITROUS OXIDE, COMPRESSED
		NITROUS OXIDE, COMPRESSED (DOT)
		NITROUS OXIDE, REFRIGERATED LIQUID
		NITROUS OXIDE, REFRIGERATED LIQUID (DOT)
		UN 1070 (DOT)
		UN 2201 (DOT)
10024-97-5		
		IRIDIUM TETRACHLORIDE
10025-65-7		
		MURIATE OF PLATINUM
		PLATINOUS CHLORIDE
		PLATINOUS DICHLORIDE
		PLATINUM CHLORIDE
		PLATINUM CHLORIDE (PtCl2)
		PLATINUM DICHLORIDE
		PLATINUM(II) CHLORIDE
		PLATINUM(II) DICHLORIDE

CAS No.	IN	Chemical Name
10025-67-9		
		CHLORIDE OF SULFUR (DOT)
		CHLOROSULFANE
		DICHLORODISULFANE
		DISULFUR DICHLORIDE
		SIARKI CHLOREK (POLISH)
		SULFUR CHLORIDE
		SULFUR CHLORIDE (MONO)
		SULFUR CHLORIDE (S2Cl2)
		SULFUR CHLORIDE(DI) (DOT)
		SULFUR MONOCHLORIDE
		SULFUR MONOCHLORIDE (ACGIH)
		SULFUR MONOCHLORIDE (S2Cl2)
		SULFUR SUBCHLORIDE
		SULPHUR CHLORIDE (S2Cl2)
		THIOSULFUROUS DICHLORIDE
		UN 1828 (DOT)
10025-73-7		
	*	CHROMIC CHLORIDE
		CHROMIC CHLORIDE (CrCl3)
		CHROMIUM CHLORIDE
		CHROMIUM CHLORIDE (CrCl3)
		CHROMIUM CHLORIDE, ANHYDROUS
		CHROMIUM TRICHLORIDE
		CHROMIUM TRICHLORIDE (CrCl3)
		CHROMIUM(III) CHLORIDE
		CHROMIUM(III) CHLORIDE (1:3)
		CI 77295
		PURATRONIC CHROMIUM CHLORIDE
		TRICHLOROCHROMIUM
10025-78-2		
		SILANE, TRICHLORO-
		SILICI-CHLOROFORME (FRENCH)
		SILICIUMCHLOROFORM (GERMAN)
		SILICOCHLOROFORM
		SILICON CHLORIDE HYDRIDE (SIHCL3)
		TRICHLOORSILAAN (DUTCH)
		TRICHLOROMONOSILANE
		TRICHLOROSILANE
		TRICHLOROSILANE (DOT)
		TRICHLOROSILANE (HSiCl3)
		TRICHLORSILAN (GERMAN)
		TRICLOROSILANO (ITALIAN)
		UN 1295 (DOT)
10025-87-3		
		PHOSPHORIC CHLORIDE
		PHOSPHORIC TRICHLORIDE
		PHOSPHOROXYCHLORIDE
		PHOSPHOROXYTRICHLORIDE
		PHOSPHORUS CHLORIDE OXIDE (PCl3O)
		PHOSPHORUS MONOXIDE TRICHLORIDE
		PHOSPHORUS OXIDE TRICHLORIDE
		PHOSPHORUS OXYCHLORIDE
		PHOSPHORUS OXYCHLORIDE (ACGIH,DOT)
		PHOSPHORUS OXYTRICHLORIDE
		PHOSPHORUS TRICHLORIDE OXIDE
		PHOSPHORYL CHLORIDE
		PHOSPHORYL CHLORIDE (DOT)
		PHOSPHORYL TRICHLORIDE
		TRICHLOROPHOSPHINE OXIDE
		TRICHLOROPHOSPHORUS OXIDE
		UN 1810 (DOT)
10025-91-9		
		ANTIMOINE (TRICHLORURE D') (FRENCH)
		ANTIMONIO (TRICLORURO DI) (ITALIAN)
		ANTIMONOUS CHLORIDE
		ANTIMONOUS CHLORIDE (DOT)
		ANTIMONTRICHLORID (GERMAN)
		ANTIMONY (III) CHLORIDE
		ANTIMONY BUTTER
		ANTIMONY CHLORIDE
		ANTIMONY CHLORIDE (DOT)
		ANTIMONY CHLORIDE (Sb2Cl6)
		ANTIMONY CHLORIDE (SbCl3)
	*	ANTIMONY TRICHLORIDE
		ANTIMONY TRICHLORIDE SOLUTION
		ANTIMONY TRICHLORIDE SOLUTION (DOT)
		ANTIMONY TRICHLORIDE, LIQUID (DOT)
		ANTIMONY TRICHLORIDE, SOLID
		ANTIMONY TRICHLORIDE, SOLID (DOT)
		ANTIMONY(III) CHLORIDE
		ANTIMOONTRICHLRIDE (DUTCH)
		BUTTER OF ANTIMONY
		C.I. 77056
		CHLORID ANTIMONITY (CZECH)
		CHLORURE ANTIMONIEUX (FRENCH)
		STIBINE, TRICHLORO-
		TRICHLOROANTIMONY
		TRICHLOROSTIBINE
		TRICHLORURE D'ANTIMOINE (FRENCH)
		UN 1733 (DOT)
10025-93-1		
		URANIUM TRICHLORIDE
10025-97-5		
		IRIDIUM CHLORIDE (IrCl4)
		IRIDIUM TETRACHLORIDE
		IRIDIUM(IV) CHLORIDE
10025-99-7		
		POTASSIUM TETRACHLOROPLATINATE
10026-04-7		
		CHLORID KREMICITY (CZECH)
		EXTREMA
		PERCHLOROSILANE
		SILANE, TETRACHLORO-
		SILICIO(TETRACLORURO DI)
		SILICIUM(TETRACHLORURE DE) (FRENCH)
		SILICIUMTETRACHLORID (GERMAN)
		SILICIUMTETRACHLORIDE (DUTCH)
		SILICON CHLORIDE
		SILICON CHLORIDE (DOT)
		SILICON CHLORIDE (SiCl4)
	*	SILICON TETRACHLORIDE
		SILICON TETRACHLORIDE (DOT)
		TETRACHLOROSILANE
		TETRACHLOROSILICON
		TETRACHLORURE DE SILICIUM (FRENCH)
		UN 1818 (DOT)
10026-06-9		
		STANNANE, TETRACHLORO-, PENTAHYDRATE
		STANNIC CHLORIDE PENTAHYDRATE
	*	STANNIC CHLORIDE, HYDRATED
		STANNIC TETRACHLORIDE PENTAHYDRATE
		TETRACHLOROSTANNANE PENTAHYDRATE
		TIN CHLORIDE (SnCl4), PENTAHYDRATE
		TIN TETRACHLORIDE PENTAHYDRATE
		TIN TETRACHLORIDE, HYDRATED
10026-10-5		
		URANIUM TETRACHLORIDE
10026-11-6		
		TETRACHLOROZIRCONIUM
		UN 2503 (DOT)
		ZIRCONIUM CHLORIDE
		ZIRCONIUM CHLORIDE (Zr2Cl8)
		ZIRCONIUM CHLORIDE (ZrCl4)
		ZIRCONIUM CHLORIDE (ZrCl4), (T-4)-
		ZIRCONIUM TETRACHLORIDE
		ZIRCONIUM TETRACHLORIDE (DOT)
		ZIRCONIUM TETRACHLORIDE, SOLID
		ZIRCONIUM TETRACHLORIDE, SOLID (DOT)
		ZIRCONIUM(IV) CHLORIDE (1:4)
10026-13-8		
		FOSFORO(PENTACLORURO DI) (ITALIAN)
		FOSFORPENTACHLORIDE (DUTCH)
		PENTACHLOROPHOSPHORANE
		PENTACHLOROPHOSPHORUS
		PHOSPHORANE, PENTACHLORO-

CAS No.	IN	Chemical Name
		PHOSPHORE(PENTACHLORURE DE) (FRENCH)
		PHOSPHORIC CHLORIDE
		PHOSPHOROUS PENTACHLORIDE
		PHOSPHORPENTACHLORID (GERMAN)
		PHOSPHORUS CHLORIDE (PCl5)
	*	PHOSPHORUS PENTACHLORIDE
		PHOSPHORUS PENTACHLORIDE (ACGIH)
		PHOSPHORUS PENTACHLORIDE, SOLID
		PHOSPHORUS PENTACHLORIDE, SOLID (DOT)
		PHOSPHORUS PERCHLORIDE
		PHOSPHORUS(V) CHLORIDE
		PIECIOCHLOREK FOSFORU (POLISH)
		UN 1806 (DOT)
10026-17-2		
		COBALT DIFLUORIDE
		COBALT FLUORIDE
		COBALT FLUORIDE (CoF2)
		COBALT(II) FLUORIDE
		COBALTOUS FLUORIDE
10028-15-6		
		OXYGEN, MOL. (O3)
		OZON (POLISH)
		OZONE
		OZONE (ACGIH)
		TRIATOMIC OXYGEN
10028-22-5		
		DIIRON TRISULFATE
	*	FERRIC SULFATE
		FERRIC SULFATE (DOT)
		IRON PERSULFATE
		IRON SESQUISULFATE
		IRON SULFATE (2:3)
		IRON SULFATE (Fe2(SO4)3)
		IRON TERSULFATE
		IRON(3+) SULFATE
		IRON(III) SULFATE
		NA 9121 (DOT)
		SULFURIC ACID, IRON(3+) SALT (3:2)
10031-13-7		
		ARSENENOUS ACID, LEAD(2+) SALT
		ARSENIOUS ACID (HAsO2), LEAD(2+) SALT
		LEAD ARSENITE
		LEAD ARSENITE, Pb(AsO2)2
		LEAD ARSENITE, SOLID
		LEAD ARSENITE, SOLID (DOT)
		LEAD(II) ARSENITE
		UN 1618 (DOT)
10031-18-2		
		MERCUROUS BROMIDE
		MERCUROUS BROMIDE (DOT)
		MERCUROUS BROMIDE, SOLID
		MERCUROUS BROMIDE, SOLID (DOT)
		MERCURY BROMIDE
		MERCURY BROMIDE (HgBr)
		MERCURY MONOBROMIDE
		MERCURY(1+) BROMIDE
		MERCURY(I) BROMIDE (1:1)
		UN 1634 (DOT)
10031-59-1		
		CFS
		CSF-GIFTWEIZEN
		DITHALLIUM SULFATE
		DITHALLIUM(1+) SULFATE
		ECCOTHAL
		M7-GIFTKOERNER
		NA 1707 (DOT)
		RATOX
		RATTENGIFTKONSERVE
		RCRA WASTE NUMBER P115
		SULFURIC ACID, DITHALLIUM(1+) SALT
		SULFURIC ACID, DITHALLIUM(1+) SALT (8CI,9CI)
		SULFURIC ACID, THALLIUM SALT
	*	THALLIUM SULFATE
		THALLIUM SULFATE (Tl2SO4)
		THALLIUM SULFATE, SOLID
		THALLIUM SULFATE, SOLID (DOT)
		THALLIUM(1) SULFATE
		THALLIUM(I) SULFATE
		THALLOUS SULFATE
		ZELIO
10031-87-5		
		2-ETHYLBUTYL ACETATE
		ACETIC ACID, 2-ETHYLBUTYL ESTER
10034-81-8		
		ANHYDRONE
		ANHYDROUS MAGNESIUM PERCHLORATE
		DEHYDRITE
		KHKM 300
		MAGNESIUM DIPERCHLORATE
	*	MAGNESIUM PERCHLORATE
		MAGNESIUM PERCHLORATE (DOT)
		MAGNESIUM PERCHLORATE (Mg(ClO4)2)
		PERCHLORATE DE MAGNESIUM (FRENCH)
		PERCHLORIC ACID, MAGNESIUM SALT
		UN 1475 (DOT)
10034-85-2		
		ANHYDROUS HYDRIODIC ACID
	*	HYDRIODIC ACID
		HYDRIODIC ACID (DOT)
		HYDRIODIC ACID, SOLUTION (DOT)
		HYDROGEN IODIDE
		HYDROGEN IODIDE (HI)
		HYDROGEN IODIDE SOLUTION
		HYDROGEN IODIDE SOLUTION (DOT)
		HYDROGEN IODIDE, ANHYDROUS
		HYDROGEN IODIDE, ANHYDROUS (DOT)
		HYDROGEN MONOIODIDE
		HYDROIODIC ACID
		UN 1787 (DOT)
		UN 2197 (DOT)
10034-93-2		
		HYDRAZINE DIHYDROGEN SULFATE SALT
		HYDRAZINE HYDROGEN SULFATE
		HYDRAZINE MONOSULFATE
	*	HYDRAZINE SULFATE
		HYDRAZINE, SULFATE (1:1)
		HYDRAZINIUM SULFATE
		HYDRAZONIUM SULFATE
10035-10-6		
		ACIDE BROMHYDRIQUE (FRENCH)
		ACIDO BROMIDRICO (ITALIAN)
		ANHYDROUS HYDROBROMIC ACID
		BROMOWODOR (POLISH)
		BROMWASSERSTOFF (GERMAN)
		BROOMWATERSTOF (DUTCH)
		HYDROBROMIC ACID
		HYDROBROMIC ACID MORE THAN 49% STRENGTH
		HYDROBROMIC ACID, NOT MORE THAN 49% STRENGTH
		HYDROBROMIC ACID, ANHYDROUS (DOT)
	*	HYDROGEN BROMIDE
		HYDROGEN BROMIDE (ACGIH,DOT)
		HYDROGEN BROMIDE (H2Br2)
		HYDROGEN BROMIDE (HBr)
		HYDROGEN MONOBROMIDE
		UN 1048 (DOT)
		UN 1788 (DOT)
10036-47-2		
		DINITROGEN TETRAFLUORIDE
		HYDRAZINE, TETRAFLUORO-
		NITROGEN FLUORIDE
		NITROGEN FLUORIDE (N2F4)
		PERFLUOROHYDRAZINE
		TETRAFLUOROHYDRAZINE
		TETRAFLUOROHYDRAZINE (DOT)

CAS No.	IN	Chemical Name
		UN 1955 (DOT)
10039-32-4		DISODIUM HYDROGEN ORTHOPHOSPHATE DODECAHYDRATE
		DISODIUM HYDROGEN PHOSPHATE DODECAHYDRATE
		DISODIUM MONOHYDROGEN PHOSPHATE DODECAHYDRATE
		DISODIUM ORTHOPHOSPHATE DODECAHYDRATE
		DISODIUM PHOSPHATE (Na2HPO4) DODECAHYDRATE
		DISODIUM PHOSPHATE (Na2HPO4.12H2O)
		DISODIUM PHOSPHATE DODECAHYDRATE
		DISODIUM PHOSPHATE DODECAHYDRATE (Na2HPO4.12H2O)
		MONOHYDROGEN DISODIUM PHOSPHATE DODECAHYDRATE
		PHOSPHORIC ACID DISODIUM DODECAHYDRATE
		PHOSPHORIC ACID, DISODIUM SALT, DODECAHYDRATE
		SODIUM PHOSPHATE (Na2HPO4) DODECAHYDRATE
10039-54-0		BIS(HYDROXYLAMINE) SULFATE
		HYDROXYL AMINE SULPHATE (DOT)
		HYDROXYL AMMONIUM SULFATE ((HONH3)2SO4)
		HYDROXYLAMINE NEUTRAL SULFATE
	*	HYDROXYLAMINE SULFATE
		HYDROXYLAMINE, SULFATE (2:1)
		HYDROXYLAMINE, SULFATE (2:1) (SALT)
		HYDROXYLAMMONIUM SULFATE
		OXAMMONIUM SULFATE
		UN 2865 (DOT)
10042-76-9		NITRATE DE STRONTIUM (FRENCH)
		NITRIC ACID, STRONTIUM SALT
		STRONTIUM DINITRATE
	*	STRONTIUM NITRATE
		STRONTIUM NITRATE (DOT)
		STRONTIUM NITRATE (Sr(NO3)2)
		STRONTIUM(II) NITRATE (1:2)
		UN 1507 (DOT)
10043-01-3		ALUM
		ALUMINUM ALUM
		ALUMINUM SESQUISULFATE
	*	ALUMINUM SULFATE
		ALUMINUM SULFATE (2:3)
		ALUMINUM SULFATE (Al2(SO4)3)
		ALUMINUM SULPHATE
		ALUMINUM TRISULFATE
		ALUMINUM(III) SULFATE
		CAKE ALUM
		DIALUMINUM SULFATE
		DIALUMINUM TRISULFATE
		SULFATODIALUMINUM DISULFATE
		SULFURIC ACID ALUMINUM(3+) SALT (3:2)
		SULFURIC ACID, ALUMINUM SALT (3:2)
10043-22-8		DIPOTASSIUM OXALATE
		ETHANEDIOIC ACID, DIPOTASSIUM SALT
		ETHANEDIOIC ACID, DIPOTASSIUM SALT (9CI)
		ETHANEDIOIC ACID, POTASSIUM SALT
		KALIUM OXALATE
		OXALIC ACID, DIPOTASSIUM SALT
		OXALIC ACID, POTASSIUM SALT
		POTASSIUM NEUTRAL OXALATE
		POTASSIUM OXALATE
		POTASSIUM OXALATE (K2C2O4)
10043-35-3		BORACIC ACID
		BORIC ACID
		BORIC ACID (H3BO3)
		BOROFAX
		BORON TRIHYDROXIDE
		BORSAURE (GERMAN)
		NCI-C56417
		ORTHOBORIC ACID
		ORTHOBORIC ACID (H3BO3)
		THREE ELEPHANT
		TRIHYDROXYBORANE
10043-52-4		CALCIUM CHLORIDE
		CALCIUM CHLORIDE (CaCl2)
		CALCIUM DICHLORIDE
		CALCOSAN
		CALPLUS
		CALTAC
		DOWFLAKE
		LIQUIDOW
		PELADOW
		SNOMELT
		STOPIT
		SUPERFLAKE ANHYDROUS
		URAMINE MC
10045-89-3		AMMONIUM FERROUS SULFATE
		AMMONIUM IRON SULFATE
		AMMONIUM IRON SULFATE (2:2:1)
		DIAMMONIUM FERROUS DISULFATE
		DIAMMONIUM IRON DISULFATE
	*	FERROUS AMMONIUM SULFATE
		FERROUS AMMONIUM SULFATE (DOT)
		FERROUS AMMONIUM SULFATE (Fe(NH4)2(SO4)2)
		FERROUS DIAMMONIUM DISULFATE
		MOHR'S SALT
		NA 9122 (DOT)
		SULFURIC ACID, AMMONIUM IRON(2+) SALT(2:2:1)
10045-94-0		CITRINE OINTMENT
	*	MERCURIC NITRATE
		MERCURIC NITRATE (DOT)
		MERCURY DINITRATE
		MERCURY NITRATE
		MERCURY NITRATE (Hg(NO3)2)
		MERCURY PERNITRATE
		MERCURY(2+) NITRATE
		MERCURY(II) NITRATE
		MERCURY(II) NITRATE (1:2)
		MILLON'S REAGENT
		NAMED REAGENTS AND SOLUTIONS, MILLON'S
		NITRATE MERCURIQUE (FRENCH)
		NITRIC ACID, MERCURY(2+) SALT
		NITRIC ACID, MERCURY(II) SALT
		UN 1625 (DOT)
10048-13-2		7H-FURO[3',2':4,5]FURO[2,3-C]XANTHEN-7-ONE, 3A,12C-DIHYDRO-8-HYDROXY-6-METHOXY-, (3AR-CIS)-
		STERIGMATOCYSTIN
10049-04-4		ALCIDE
		ANTHIUM DIOXCIDE
	*	CHLORINE DIOXIDE
		CHLORINE OXIDE
		CHLORINE OXIDE (ClO2)
		CHLORINE PEROXIDE
		CHLORINE(IV) OXIDE
		CHLOROPEROXYL
		CHLORYL RADICAL
		DOXCIDE 50
10049-05-5		CHROMIUM CHLORIDE
		CHROMIUM CHLORIDE (CrCl2)
		CHROMIUM DICHLORIDE
		CHROMIUM(II) CHLORIDE
		CHROMIUM(II) CHLORIDE (1:2)
	*	CHROMOUS CHLORIDE
		CHROMOUS CHLORIDE (DOT)
		NA 9102 (DOT)

CAS No.	IN	Chemical Name
10049-07-7		RHODIUM CHLORIDE
		RHODIUM CHLORIDE (RhCl3)
		RHODIUM TRICHLORIDE
		RHODIUM(III) CHLORIDE
		RHODIUM(III) CHLORIDE (1:3)
		TRICHLORORHODIUM
10049-14-6		URANIUM TETRAFLUORIDE
10061-01-5		CIS-1,3-DICHLOROPROPENE
10061-02-6		TRANS-1,3-DICHLOROPROPENE
10070-67-0		1,4-BUTYNEDIOL
		1,4-BUTYNEDIOL (DOT)
		1,4-DIHYDROXY-2-BUTYNE
		2-BUTYNE-1,4-DIOL
		2-BUTYNEDIOL
		BIS(HYDROXYMETHYL)ACETYLENE
		BUTYNEDIOL
		UN 2716 (DOT)
10099-74-8		LEAD DINITRATE
	*	LEAD NITRATE
		LEAD NITRATE (DOT)
		LEAD NITRATE (Pb(NO3)2)
		LEAD(2+) NITRATE
		LEAD(II) NITRATE
		LEAD(II) NITRATE (1:2)
		NITRATE DE PLOMB (FRENCH)
		NITRIC ACID, LEAD SALT
		NITRIC ACID, LEAD(2+) SALT
		PLUMBOUS NITRATE
		UN 1469 (DOT)
10101-50-5		PERMANGANATE DE SODIUM (FRENCH)
		PERMANGANIC ACID (HMnO4), SODIUM SALT
		PERMANGANIC ACID, SODIUM SALT
		SODIUM MANGANATE
	*	SODIUM PERMANGANATE
		SODIUM PERMANGANATE (DOT)
		SODIUM PERMANGANATE (NaMnO4)
		UN 1503 (DOT)
10101-53-8		BAYCHROM A
		BAYCHROM F
		C.I. 77305
	*	CHROMIC SULFATE
		CHROMIC SULFATE (Cr2(SO4)3)
		CHROMIC SULFATE (DOT)
		CHROMIC SULPHATE
		CHROMITAN B
		CHROMITAN MS
		CHROMITAN NA
		CHROMIUM (III) SULFATE (2:3)
		CHROMIUM III SULFATE
		CHROMIUM SULFATE
		CHROMIUM SULFATE (2:3)
		CHROMIUM SULFATE (Cr2(SO4)3)
		CHROMIUM SULPHATE
		CHROMIUM SULPHATE (2:3)
		DICHROMIUM SULFATE
		DICHROMIUM SULPHATE
		DICHROMIUM TRIS(SULFATE)
		DICHROMIUM TRISULFATE
		DICHROMIUM TRISULPHATE
		NA 9100 (DOT)
		SULFURIC ACID, CHROMIUM(3+) SALT (3:2)
10101-63-0		C.I. 77613
		LEAD DIIODIDE
	*	LEAD IODIDE
		LEAD IODIDE (PbI2)
		LEAD(II) IODIDE
		NA 2811 (DOT)
		PLUMBOUS IODIDE
10101-89-0		PHOSPHORIC ACID, TRISODIUM SALT, DODECAHYDRATE
		SODIUM ORTHOPHOSPHATE DODECAHYDRATE
		SODIUM PHOSPHATE (Na3PO4) DODECAHYDRATE
		TRISODIUM ORTHOPHOSPHATE DODECAHYDRATE
		TRISODIUM PHOSPHATE (Na3PO4.12H2O)
		TRISODIUM PHOSPHATE DODECAHYDRATE
		TRISODIUM PHOSPHATE DODECAHYDRATE (Na3PO4.12H2O)
10102-06-4		DINITRATODIOXOURANIUM
		URANIUM NITRATE OXIDE (UO2(NO3)2)
		URANIUM, BIS(NITRATO-O)DIOXO-, (T-4)-
		URANIUM, DINITRATODIOXO-
		URANYL DINITRATE
	*	URANYL NITRATE
		URANYL NITRATE (UO2(NO3)2)
10102-18-8		DISODIUM SELENITE
		DISODIUM SELENIUM TRIOXIDE
		NATRIUMSELENIT (GERMAN)
		SELENIOUS ACID (H2SeO3), DISODIUM SALT
		SELENIOUS ACID, DISODIUM SALT
	*	SODIUM SELENITE
		SODIUM SELENITE (DOT)
		SODIUM SELENIUM OXIDE (Na2SeO3)
		UN 2630 (DOT)
10102-20-2		DISODIUM TELLURITE
		SODIUM TELLURATE (Na2TeO3)
		SODIUM TELLURATE(IV)
		SODIUM TELLURATE(IV) (Na2TeO3)
		SODIUM TELLURITE
		SODIUM TELLURITE (Na2TeO3)
		SODIUM TELLURIUM OXIDE (Na2TeO3)
		TELLURIC ACID (H2TeO3), DISODIUM SALT
		TELLURIC ACID, DISODIUM SALT
		TELLUROUS ACID, DISODIUM SALT
10102-43-9		AMIDOGEN, OXO-
		BIOXYDE D'AZOTE (FRENCH)
	*	NITRIC OXIDE
		NITRIC OXIDE (ACGIH,DOT)
		NITRIC OXIDE (NO)
		NITRIC OXIDE TRIMER
		NITROGEN MONOXIDE
		NITROGEN OXIDE (N4O4)
		NITROGEN OXIDE (NO)
		NITROSYL RADICAL
		OXYDE NITRIQUE (FRENCH)
		RCRA WASTE NUMBER P076
		STICKMONOXYD (GERMAN)
		UN 1660 (DOT)
10102-44-0		AZOTE (FRENCH)
		AZOTO (ITALIAN)
		NA 1067 (DOT)
		NITRITE RADICAL
		NITRITO
		NITRO
	*	NITROGEN DIOXIDE
		NITROGEN DIOXIDE (ACGIH)
		NITROGEN DIOXIDE (NO2)
		NITROGEN DIOXIDE, LIQUID
		NITROGEN DIOXIDE, LIQUID (DOT)

CAS No.	IN	Chemical Name
		NITROGEN OXIDE (NO2)
		NITROGEN PEROXIDE
		NITROGEN PEROXIDE, LIQUID (DOT)
		RCRA WASTE NUMBER P078
		STICKSTOFFDIOXID (GERMAN)
		STIKSTOFDIOXYDE (DUTCH)
		UN 1067 (DOT)
10102-45-1		
		NITRIC ACID, THALLIUM SALT
		NITRIC ACID, THALLIUM(1+) SALT
10102-49-5		
		ARSENATE OF IRON, FERRIC
		ARSENIC ACID (H3AsO4), IRON(3+) SALT (1:1)
		FERRIC ARSENATE
		FERRIC ARSENATE, SOLID
		FERRIC ARSENATE, SOLID (DOT)
		IRON ARSENATE (FeAsO4)
		IRON(III) ARSENATE (1:1)
		UN 1606 (DOT)
10102-50-8		
		ARSENATE OF IRON, FERROUS
		ARSENIC ACID (H3AsO4), IRON(2+) SALT (2:3)
		FERROUS ARSENATE
		FERROUS ARSENATE (DOT)
		FERROUS ARSENATE, SOLID
		FERROUS ARSENATE, SOLID (DOT)
		IRON ARSENATE (DOT)
		IRON ARSENATE (Fe3(AsO4)2)
		IRON(II) ARSENATE (3:2)
		UN 1608 (DOT)
10103-46-5		
		CALCIUM PHOSPHATE
		DIKAL 21
		DYNAFOS
		KDV 15U
		PHOSPHORIC ACID, CALCIUM SALT
10103-50-1		
		ARSENIATE DE MAGNESIUM (FRENCH)
		ARSENIC ACID (H3AsO4), MAGNESIUM SALT
		ARSENIC ACID MAGNESIUM SALT
		MAGNESIUM ARSENATE
		MAGNESIUM ARSENATE (DOT)
		MAGNESIUM ARSENATE PHOSPHOR
		MAGNESIUM ARSENATE, SOLID
		MAGNESIUM ARSENATE, SOLID (DOT)
		UN 1622 (DOT)
10108-56-2		
		(BUTYLAMINO)CYCLOHEXANE
		BUTYLCYCLOHEXYLAMINE
		CYCLOHEXANAMINE, N-BUTYL-
		CYCLOHEXYL-N-BUTYLAMINE
		CYCLOHEXYLAMINE, N-BUTYL-
		N-BUTYLCYCLOHEXYLAMINE
10108-64-2		
		CADDY
	*	CADMIUM CHLORIDE
		CADMIUM CHLORIDE (CdCl2)
		CADMIUM CHLORIDE (DOT)
		CADMIUM DICHLORIDE
		DICHLOROCADMIUM
		KADMIUMCHLORID (GERMAN)
		NA 2570 (DOT)
		VI-CAD
10112-91-1		
	*	MERCUROUS CHLORIDE
		MERCURY CHLORIDE
10118-72-6		
		1-PROPEN-2-CHLORO-1,3-DIOL-DIACETATE
10118-76-0		
		ACERDOL
		CALCIUM PERMANGANATE
		CALCIUM PERMANGANATE (DOT)
		CALCIUM PERMANGANATE, Ca(MnO4)2
		KALIUMPERMANGANAT (GERMAN)
		PERMANGANIC ACID (HMnO4), CALCIUM SALT
		UN 1456 (DOT)
10124-36-4		
		CADMIUM MONOSULFATE
	*	CADMIUM SULFATE
		CADMIUM SULFATE (1:1)
		CADMIUM SULPHATE
		SULFURIC ACID, CADMIUM SALT (1:1)
		SULFURIC ACID, CADMIUM(2+) SALT
		SULPHURIC ACID, CADMIUM SALT (1:1)
10124-37-5		
		CALCIUM DINITRATE
	*	CALCIUM NITRATE
		CALCIUM NITRATE (Ca(NO3)2)
		CALCIUM NITRATE (DOT)
		CALCIUM(II) NITRATE (1:2)
		NITRIC ACID CALCIUM SALT (2:1)
		NITRIC ACID, CALCIUM SALT
		NORWAY SALTPETER
		SALTPETER (NORWAY)
		SYNFAT 1006
		UN 1454 (DOT)
10124-43-3		
		COBALT (2+) SULFATE
		COBALT (II) SULFATE (1:1)
		COBALT SULFATE
		COBALT SULFATE (1:1)
		COBALT SULFATE (CoSO4)
		COBALT(2+) SULFATE
		COBALT(II) SULFATE
		COBALT(II) SULPHATE
		COBALTOUS SULFATE
		SULFURIC ACID, COBALT(2+) SALT (1:1)
10124-48-8		
		AMINOMERCURIC CHLORIDE
		AMMONIATED MERCURIC CHLORIDE
		AMMONIATED MERCURY
		MERCURIC AMIDOCHLORIDE
		MERCURIC AMMONIUM CHLORIDE
		MERCURIC AMMONIUM CHLORIDE, SOLID (DOT)
		MERCURIC CHLORIDE, AMMONIATED
		MERCURY AMIDE CHLORIDE
		MERCURY AMIDE CHLORIDE (Hg(NH2)Cl)
		MERCURY AMINE CHLORIDE
		MERCURY AMMONIATED
		MERCURY AMMONIUM CHLORIDE
		MERCURY AMMONIUM CHLORIDE (DOT)
		MERCURY, AMMONOBASIC (HgNH2Cl)
		UN 1630 (DOT)
		WHITE MERCURIC PRECIPITATE
		WHITE MERCURY PRECIPITATED
		WHITE PRECIPITATE
10124-50-2		
		ARSENENOUS ACID, POTASSIUM SALT
		ARSENIOUS ACID (H3AsO3), POTASSIUM SALT
		ARSENIOUS ACID, POTASSIUM SALT
		ARSENITE DE POTASSIUM (FRENCH)
		ARSENOUS ACID, POTASSIUM SALT
		ARSONIC ACID, POTASSIUM SALT
		KALIUMARSENIT (GERMAN)
		NSC 3060
	*	POTASSIUM ARSENITE
		POTASSIUM ARSENITE, SOLID
		POTASSIUM ARSENITE, SOLID (DOT)
		POTASSIUM METAARSENITE
		UN 1678 (DOT)

CAS No.	IN	Chemical Name
10124-56-8		
		CALGON
		CALGON (OLD)
		CALGON S
		CHEMI-CHARL
		GILTEX
		HAGAN PHOSPHATE
		HEXASODIUM HEXAMETAPHOSPHATE
		HEXASODIUM METAPHOSPHATE
		HMP
		MEDI-CALGON
		METAPHOSPHORIC ACID (H6P6O18), HEXASODIUM SALT
		MICROMET
	*	SODIUM HEXAMETAPHOSPHATE
		SODIUM HEXAMETAPHOSPHATE (Na6P6O18)
		SODIUM METAPHOSPHATE (Na6P6O18)
		SODIUM PHOSPHATE (Na6P6O18)
10137-69-6		
		CYCLOHEXENYL TRICHLOROSILANE
10137-74-3		
	*	CALCIUM CHLORATE
		CALCIUM CHLORATE (DOT)
		CALCIUM CHLORATE :AQUEOUS: SOLUTION
		CALCIUM CHLORATE SOLUTION
		CALCIUM CHLORATE, AQUEOUS SOLUTION (DOT)
		CHLORATE DE CALCIUM (FRENCH)
		CHLORIC ACID, CALCIUM SALT
		UN 1452 (DOT)
		UN 2429 (DOT)
10137-80-1		
		BENZENAMINE, N-(2-ETHYLHEXYL)-
		HEXYLAMINE, 2-ETHYL-N-PHENYL-
		N-2(ETHYLHEXYL)ANILINE
10138-74-6		
		ETHANOL, 2-[(2-AMINO-1-METHYLETHYL)AMINO]-
		HYDROXYETHYLPROPYLENEDIAMINE
		N-(2-HYDROXYETHYL)PROPYLENE DIAMINE
10140-65-5		
		PHOSPHORIC ACID, DISODIUM SALT, HYDRATE
10140-87-1		
		1,2-DICHLOROETHYL ACETATE
		ETHANOL :OR: ETHANOL SOLUTIONS :INCLUDING: ALCOHOLIC BEVERAGES
		ETHANOL, 1,2-DICHLORO-, ACETATE
10141-05-6		
		COBALT BIS(NITRATE)
		COBALT DINITRATE
	*	COBALT NITRATE
		COBALT NITRATE (Co(NO3)2)
		COBALT(2+) NITRATE
		COBALT(II) NITRATE
		COBALT(II) NITRATE (1:2)
		COBALTOUS NITRATE
		NITRIC ACID, COBALT(2+) SALT
10192-30-0		
		AMMONIUM ACID SULFITE
		AMMONIUM BISULFITE
		AMMONIUM BISULFITE (NH4HSO3)
		AMMONIUM BISULFITE SOLUTION
		AMMONIUM BISULFITE, SOLID
		AMMONIUM BISULFITE, SOLID (DOT)
		AMMONIUM BISULFITE, SOLID (DOT)
		AMMONIUM BISULFITE, SOLUTION (DOT)
		AMMONIUM HYDROGEN SULFITE
		AMMONIUM MONOSULFITE
		AMMONIUM SULFITE (NH4HSO3)
		MONOAMMONIUM SULFITE
		NA 2693 (DOT)
		SULFUROUS ACID, MONOAMMONIUM SALT

CAS No.	IN	Chemical Name
10196-04-0		
	*	AMMONIUM SULFITE
		AMMONIUM SULFITE (DOT)
		DIAMMONIUM SULFITE
		NA 9090 (DOT)
		SULFUROUS ACID, DIAMMONIUM SALT
10201-68-1		
		COBALT CARBONYL
		COBALT CARBONYL
		COBALT CARBONYL (ACGIH)
		COBALT CARBONYL (Co2(CO)8)
		COBALT OCTACARBONYL
		COBALT TETRACARBONYL
		COBALT TETRACARBONYL DIMER
		COBALT, DI-MU-CARBONYLHEXACARBONYLDI-, (CO-CO)
		DI-MU-CARBONYLHEXACARBONYLDICOBALT
		DICOBALT CARBONYL
		DICOBALT CARBONYL (Co2(CO)8)
		DICOBALT OCTACARBONYL
		OCTACARBONYLDICOBALT
10213-74-8		
		3-(2-ETHYLBUTOXY)PROPIONIC ACID
		PROPANOIC ACID, 3-(2-ETHYLBUTOXY)-
		PROPIONIC ACID, 3-(2-ETHYLBUTOXY)-
10214-40-1		
		COPPER SELENITE
		COPPER SELENITE (Cu(SEO3))
		COPPER(II) SELENITE
		CUPRIC SELENITE
		SELENIOUS ACID (H2SeO3), COPPER SALT
		SELENIOUS ACID (H2SeO3), COPPER(2+) SALT (1:1)
		SELENIOUS ACID, COPPER SALT
		SELENIOUS ACID, COPPER(2+) SALT (1:1)
10241-05-1		
		MOLYBDENUM CHLORIDE (MoCl5)
		MOLYBDENUM PENTACHLORIDE
		MOLYBDENUM PENTACHLORIDE (DOT)
		PENTACHLOROMOLYBDENUM
		UN 2508 (DOT)
10265-92-6		
		ACEPHATE-MET
		BAY 71628
		BAYER 5546
		BAYER 71628
		CHEVRON 9006
		CHEVRON ORTHO 9006
		CKB 1220
		ENT 27,396
		FILITOX
		HAMIDOP
		METAMIDOFOS ESTRELLA
		METAMIDOPHOS
		METHAMIDOPHOS
		METHAMIDOPHUS
		METHYL PHOSPHORAMIDOTHIOATE
		MONITOR
		MONITOR (INSECTICIDE)
		MTD
		NSC 190987
		O,S-DIMETHYL ESTER AMIDE OF AMIDOTHIOATE
		O,S-DIMETHYL PHOSPHORAMIDOTHIOATE
		O,S-DIMETHYL THIOPHOSPHORAMIDE
		ORTHO 9006
		ORTHO MONITOR
		PHOSPHORAMIDOTHIOIC ACID, O,S-DIMETHYL ESTER
		PILLARON
		RE 9006
		SRA 5172
		TAHMABON
		TAMARON
		THIOPHOSPHORSAEURE-O,S-DIMETHYLESTERAMID (GERMAN)

CAS No.	IN	Chemical Name
10290-12-7		
		ACID COPPER ARSENITE
		AIR-FLO GREEN
		ARSENIOUS ACID (H3AsO3), COPPER(2+) SALT (1:1)
		ARSENIOUS ACID, COPPER(II) SALT (1:1)
		ARSONIC ACID, COPPER(2+) SALT (1:1)
		ARSONIC ACID, COPPER(2+) SALT (1:1) (9CI)
		COPPER ARSENITE
		COPPER ARSENITE (CuHAsO3)
		COPPER ARSENITE (DOT)
		COPPER ARSENITE, SOLID
		COPPER ARSENITE, SOLID (DOT)
		COPPER ORTHOARSENITE
		CUPRIC ARSENITE
		CUPRIC GREEN
		SCHEELE'S MINERAL
		SCHEELES GREEN
		SWEDISH GREEN
		UN 1586 (DOT)
10294-26-5		
		DISILVER MONOSULFATE
		DISILVER SULFATE
		DISILVER(1+) SULFATE
		SILVER SULFATE
		SILVER SULFATE (Ag2SO4)
		SULFURIC ACID DISILVER SALT
		SULFURIC ACID SILVER SALT (1:2)
		SULFURIC ACID, DISILVER(1+) SALT
10294-33-4		
		BORANE, TRIBROMO-
		BORON BROMIDE
		BORON BROMIDE (BBr3)
	*	BORON TRIBROMIDE
		TRIBROMOBORANE
		TRIBROMOBORON
10294-34-5		
	*	BORON TRICHLORIDE
10311-84-9		
		DIALIFOS
10325-94-7		
		CADMIUM DINITRATE
		CADMIUM NITRATE
		NITRIC ACID, CADMIUM SALT
10326-21-3		
		CHLORATE SALT OF MAGNESIUM
		CHLORIC ACID, MAGNESIUM SALT
		DE-FOL-ATE
		E-Z-OFF
		KHMD 58
		KRMD 58
	*	MAGNESIUM CHLORATE
		MAGNESIUM CHLORATE (DOT)
		MAGNESIUM DICHLORATE
		MAGRON
		MC DEFOLIANT
		ORTHO MC
		UN 2723 (DOT)
10326-24-6		
		ARSENENOUS ACID, ZINC SALT
		ARSENENOUS ACID, ZINC SALT (9CI)
		ARSENIOUS ACID (HAsO2), ZINC SALT
		ARSENIOUS ACID, ZINC SALT
		UN 1712 (DOT)
		ZINC ARSENITE
		ZINC ARSENITE, Zn(AsO2)2
		ZINC ARSENITE, SOLID
		ZINC ARSENITE, SOLID (DOT)
		ZINC METAARSENITE
		ZINC METHARSENITE
		ZMA
10361-29-2		
		AMMONIUM CARBONATE
		CARBONIC ACID, AMMONIUM SALT
10361-89-4		
		PHOSPHORIC ACID, TRISODIUM SALT, DECAHYDRATE
		TRISODIUM PHOSPHATE DECAHYDRATE
10361-95-2		
		CHLORIC ACID, ZINC SALT
		UN 1513 (DOT)
		ZINC CHLORATE
		ZINC CHLORATE (DOT)
10377-60-3		
		MAGNESIUM DINITRATE
	*	MAGNESIUM NITRATE
		MAGNIOSAN
		NITRIC ACID, MAGNESIUM SALT
10377-66-9		
		MANGANESE (II) NITRATE
		MANGANESE (II) NITRATE, ANHYDROUS
	*	MANGANESE NITRATE
		MANGANESE NITRATE (DOT)
		MANGANESE NITRATE (Mn(NO3)2)
		MANGANESE(2+) NITRATE
		MANGANOUS DINITRATE
		MANGANOUS NITRATE
		UN 2724 (DOT)
10380-29-7		
		(TETRAAMMINE)COPPER SULFATE HYDRATE
		AMMONIATED CUPRIC SULFATE MONOHYDRATE
		COPPER TETRAAMMINE SULFATE MONOHYDRATE
		COPPER(2+), TETRAAMMINE-, SULFATE (1:1), MONOHYDRATE
		CUPRIC SULFATE, AMMONIATED
		CUPRIC SULFATE, AMMONIATED (DOT)
		NA 9110 (DOT)
		SULFURIC ACID, COPPER(2+) SALT, AMMONIATED
		TETRAAMMINECOPPER SULFATE MONOHYDRATE
		TETRAAMMINECOPPER SULFATE, HYDRATE
		TETRAAMMINECOPPER(2+) SULFATE (1:1) MONOHYDRATE
		TETRAAMMINECOPPER(II) SULFATE MONOHYDRATE
10415-73-3		
		MERCURIC SALICYLATE
		MERCURIC SALICYLATE, SOLID
		MERCURIC SALICYLATE, SOLID (DOT)
		MERCURISALICYLIC ACID
		MERCURY SALICYLATE
		MERCURY SALICYLATE (DOT)
		MERCURY SUBSALICYLATE
		MERCURY, (SALICYLATO(2-))-
		MERCURY, [2-HYDROXYBENZOATO(2-)-O1,O2]-
		MERCURY, [SALICYLATO(2-)]-
		SALICYLIC ACID, MERCURY(2+) SALT (2:1)
		UN 1644 (DOT)
10415-75-5		
	*	MERCUROUS NITRATE
		MERCUROUS NITRATE (DOT)
		MERCUROUS NITRATE, SOLID
		MERCUROUS NITRATE, SOLID (DOT)
		MERCURY NITRATE (Hg2(NO3)2)
		MERCURY NITRATE (HgNO3)
		MERCURY(I) NITRATE
		MERCURY(I) NITRATE (1:1)
		MONOMERCURY NITRATE
		NITRATE MERCUREUX (FRENCH)
		NITRIC ACID, MERCURY(1+) SALT
		NITRIC ACID, MERCURY(I) SALT
		UN 1627 (DOT)
10421-48-4		
		CLAYFEN
	*	FERRIC NITRATE
		FERRIC NITRATE (DOT)

CAS No.	IN	Chemical Name
		IRON (III) NITRATE, ANHYDROUS
		IRON NITRATE
		IRON NITRATE (Fe(NO3)3)
		IRON TRINITRATE
		IRON(III) NITRATE
		NITRIC ACID, IRON(3+) SALT
		UN 1466 (DOT)
10476-95-6		
		METHACROLEIN DIACETATE
10544-63-5		
		2-BUTENOIC ACID, ETHYL ESTER
		2-BUTENOIC ACID, ETHYL ESTER, (E)- (9CI)
		ALPHA-CROTONIC ACID ETHYL ESTER
		CROTONATE D'ETHYLE (FRENCH)
		CROTONIC ACID, ETHYL ESTER
		CROTONIC ACID, ETHYL ESTER, (E)-
		ETHYL (E)-CROTONATE
		ETHYL 2-BUTENOATE
		ETHYL CROTONATE
		ETHYL CROTONATE (DOT)
		ETHYL TRANS-CROTONATE
		ETHYLCROTONATE
		TRANS-2-BUTENOIC ACID ETHYL ESTER
		UN 1862 (DOT)
10544-72-6		
		DINITROGEN TETRAOXIDE
		DINITROGEN TETROXIDE
		DINITROGEN TETROXIDE (DOT)
		NA 1067 (DOT)
		NITROGEN DIOXIDE, DI-
		NITROGEN OXIDE (N2O4)
		NITROGEN TETRAOXIDE
		NITROGEN TETROXIDE
		NITROGEN TETROXIDE (DOT)
		NITROGEN TETROXIDE, LIQUID
		NITROGEN TETROXIDE, LIQUID (DOT)
		UN 1067 (DOT)
10544-73-7		
		DINITROGEN TRIOXIDE
		NITRATE FREE RADICAL
		NITROGEN OXIDE (N2O3)
		NITROGEN SESQUIOXIDE
		NITROGEN TRIOXIDE
		NITROGEN TRIOXIDE (DOT)
		NITROGEN TRIOXIDE (N2O3)
		NITROUS ANHYDRIDE
		PERNITRITE RADICAL
		UN 2421 (DOT)
10545-99-0		
		CHLORIDE OF SULFUR (DOT)
		CHLORINE SULFIDE
		CHLORINE SULFIDE (Cl2S)
		DICHLOROSULFANE
		MONOSULFUR DICHLORIDE
		SULFUR CHLORIDE
		SULFUR CHLORIDE (8CI,9CI)
		SULFUR CHLORIDE (DI)
		SULFUR CHLORIDE (SCl2)
		SULFUR CHLORIDE(MONO) (DOT)
		SULFUR DICHLORIDE
		SULFUR DICHLORIDE (SCl2)
		UN 1828 (DOT)
10546-01-7		
		SULFUR FLUORIDE (SF5)
		SULFUR PENTAFLUORIDE
10588-01-9		
		BICHROMATE DE SODIUM (FRENCH)
		BICHROMATE OF SODA
		CHROMIC ACID (H2Cr2O7), DISODIUM SALT
		CHROMIC ACID, DISODIUM SALT
		CHROMIUM SODIUM OXIDE
		CHROMIUM SODIUM OXIDE (Cr3Na2O7)
		DICHROMIC ACID (H2Cr2O7), DISODIUM SALT
		DICHROMIC ACID, DISODIUM SALT
		DISODIUM DICHROMATE
		NA 1479 (DOT)
		NATRIUMBICHROMAAT (DUTCH)
		NATRIUMDICHROMAAT (DUTCH)
		NATRIUMDICHROMAT (GERMAN)
		SODIO (DICROMATO DI) (ITALIAN)
		SODIUM BICHROMATE
		SODIUM CHROMATE
		SODIUM CHROMATE (Na2Cr2O7)
	*	SODIUM DICHROMATE
		SODIUM DICHROMATE (DOT)
		SODIUM DICHROMATE (Na2(Cr2O7))
		SODIUM DICHROMATE(VI)
		SODIUM(DICHROMATE DE) (FRENCH)
10595-95-6		
		ETHANAMINE, N-METHYL-N-NITROSO-
		ETHYLAMINE, N-METHYL-N-NITROSO-
		ETHYLMETHYLNITROSAMINE
		METHYLETHYLNITROSAMINE
		METHYLETHYLNITROSOAMINE
		N-METHYL-N-NITROSOETHANAMINE
		N-METHYL-N-NITROSOETHYLAMINE
		N-NITROSO-N-METHYLETHYLAMINE
		N-NITROSOETHYLMETHYLAMINE
		N-NITROSOMETHYLETHYLAMINE
		NITROSOMETHYLETHYLAMINE
10599-90-3		
		CHLORAMINE
11069-19-5		
		DICHLOROBUTENE
11071-47-9		
		ISOCETENE
11096-82-5		
		AROCHLOR 1260
		AROCLOR 1260
		PCB 1260
11097-69-1		
		AROCHLOR 1254
		AROCLOR 1254
		CHLORODIPHENYL (54% CHLORINE)
		PCB 1254
		PCB-1254 (CHLORODIPHENYL (54% Cl))
11099-02-8		
		C 11-2S
		C 11-9
		NICKEL OXIDE
11099-06-2		
		ES 32
		ES 40
		ES 40 (SILICATE)
		ETHYL POLYSILICATE
		ETHYL SILICATE
		ETHYL SILICATE 32
		ETHYL SILICATE 40
		ETHYL SILICATE 50
		ETS 32
		ETS 40
		NT 40
		POLY(ETHYL SILICATE)
		SILESTER AR
		SILESTER OS
		SILICIC ACID, ETHYL ESTER
		TES 40
11104-28-2		
		AROCHLOR 1221
		AROCLOR 1221

CAS No.	IN	Chemical Name
11104-93-1		
		DINITROGEN MONOXIDE
		FACTITIOUS AIR
		HYPONITROUS ACID ANHYDRIDE
		LAUGHING GAS
		NITROGEN OXIDE
		NITROGEN OXIDE (NOX)
		NITROGEN OXIDES
		NITROUS OXIDE
		NITROUS OXIDE (DOT)
		NITROUS OXIDE, COMPRESSED (DOT)
		NITROUS OXIDE, REFRIGERATED LIQUID (DOT)
		UN 1070 (DOT)
		UN 2201 (DOT)
11105-06-9		
		METAWANADANEM SODOWYM (POLISH)
		PEROXYVANADIC ACID, SODIUM SALT
		SODIUM METAVANADATE
		SODIUM METAVANADATE (NaVO3)
		SODIUM PEROXYVANADATE
		SODIUM VANADATE
		SODIUM VANADATE (NaVO3)
		SODIUM VANADATE(V) (NaVO3)
		SODIUM VANADIUM OXIDE
		SODIUM VANADIUM OXIDE (NaVO3)
		TRISODIUM VANADATE
		VANADATE (VO31-), SODIUM
		VANADIC ACID (HVO3), SODIUM SALT
		VANADIC ACID, MONOSODIUM SALT
		VANADIC ACID, SODIUM SALT
11105-16-1		
		UN 1437 (DOT)
		ZIRCONIUM DIHYDRIDE
		ZIRCONIUM HYDRIDE
		ZIRCONIUM HYDRIDE (DOT)
		ZIRCONIUM HYDRIDE (ZrH2)
11110-52-4		
		MERCURY ALLOY, BASE, Hg,Na
		SODIUM ALLOY, NONBASE, Hg,Na
	*	SODIUM AMALGAM
		SODIUM AMALGAM (DOT)
		UN 1424 (DOT)
11113-57-8		
		COBALT DIFLUORIDE
		COBALT FLUORIDE
		COBALT FLUORIDE (CoF2)
		COBALT(II) FLUORIDE
		COBALTOUS FLUORIDE
11113-74-9		
		NICKEL HYDROXIDE
11118-65-3		
		CYCLOHEXANONE, METHYL-, PEROXIDE
		METHYLCYCLOHEXANONE PEROXIDE
		TRIGONOX 38
11119-70-3		
		BASIC LEAD CHROMATE
		CHROME ORANGE
		CHROMIC ACID, LEAD SALT
		CHROMIC ACID, LEAD SALT, BASIC
		CHROMIUM LEAD OXIDE
		HORNA GL 35
		LEAD CHROMATE
		LEAD CHROMATE OXIDE
11129-27-4		
		COPPER BROMIDE
11130-21-5		
		VANADIUM CARBIDE
11135-81-2		
		POTASSIUM ALLOY, NONBASE, K,Na
		POTASSIUM SODIUM ALLOY
		POTASSIUM-SODIUM, ALLOY (DOT)
		SODIUM POTASSIUM ALLOY
		SODIUM ALLOY, NONBASE, K,Na
		SODIUM POTASSIUM ALLOY (LIQUID)
		SODIUM POTASSIUM ALLOY (SOLID)
		SODIUM POTASSIUM ALLOY, LIQUID (DOT)
		SODIUM POTASSIUM ALLOY, SOLID (DOT)
		UN 1422 (DOT)
11138-47-9		
		PERBORIC ACID (HBo(O2)), SODIUM SALT
		PERBORIC ACID (HBo3), SODIUM SALT
		PERBORIC ACID, SODIUM SALT
		SODIUM BORATE
		SODIUM BORATE (NaBo3)
		SODIUM PERBORATE
		SODIUM PERBORATE (NaBo3)
		SODIUM PEROXOBORATE
11138-49-1		
		ALUMINUM SODIUM OXIDE
		BETA"-ALUMINA
		BETA-ALUMINA
		J 242
		NALCO 680
		SODIUM ALUMINATE
		SODIUM ALUMINATE SOLUTION
		SODIUM ALUMINATE, SOLID
		SODIUM ALUMINATE, SOLID (DOT)
		SODIUM ALUMINATE, SOLUTION (DOT)
		SODIUM ALUMINUM OXIDE
		SODIUM BETA" ALUMINA
		SODIUM BETA-ALUMINA
		SODIUM POLYALUMINATE
		UN 1819 (DOT)
		UN 2812 (DOT)
11140-68-4		
		TITANIUM DIHYDRIDE (TiH2)
		TITANIUM HYDRIDE
		TITANIUM HYDRIDE (DOT)
		TITANIUM HYDRIDE (TiH2)
		UN 1871 (DOT)
11141-16-5		
		AROCLOR 1232
12001-26-2		
		A 21 (MINERAL)
		A 41 (MINERAL)
		ABHRAK
		C 1000
		DAVENITE P 12
		DIMONITE
		DIMONITE DM(NA-TS)
		DR 1
		FLUORMICA
		HX 610
		MICA
		MICA DUST
		MICA-GROUP MINERALS
		MICA-TYPE MINERALS
		MICATEX
		MICROMICA W 1
		MINERALS, MICA-GROUP
		P 80P
		RUBIDIAN MICA
	*	SILICA, MICA
		SMM 160
		SUZORITE
		SUZORITE 60S

CAS No.	IN	Chemical Name
12001-28-4		
		AMORPHOUS CROCIDOLITE ASBESTOS
		ASBESTOS (ACGIH)
		ASBESTOS, BLUE
		ASBESTOS, CROCIDOLITE
		BLUE ASBESTOS
		BLUE ASBESTOS (DOT)
		CROCIDOLITE
		CROCIDOLITE (DOT)
		CROCIDOLITE ASBESTOS
		CROCIDOLITE DUST
		FIBROUS CROCIDOLITE ASBESTOS
		KROKYDOLITH (GERMAN)
		NCI C09007
		UN 2212 (DOT)
12001-29-5		
		5R04
		5RO4
		7-45 ASBESTOS
		ASBESTOS (ACGIH)
	*	ASBESTOS, CHRYSOTILE
		ASBESTOS, WHITE
		ASBESTOS, WHITE (DOT)
		AVIBEST C
		CALIDRIA RG 100
		CALIDRIA RG 144
		CALIDRIA RG 600
		CASSIAR AK
		CHRYSOTILE
		CHRYSOTILE (DOT)
		CHRYSOTILE (H4Mg3(Si2O9))
		CHRYSOTILE (Mg3H2(SiO4)2.H2O)
		CHRYSOTILE ASBESTOS
		CHRYSOTILE DUST
		HOOKER NO.1 CHRYSOTILE ASBESTOS
		K 6-30
		METAXITE
		NCI C61223A
		PLASTIBEST 20
		RG 600
		SERPENTINE
		SERPENTINE CHRYSOTILE
		SYLODEX
		UN 2590 (DOT)
		WHITE ASBESTOS
12002-03-8		
		(ACETATO-O)(TRIMETAARSENITO)DICOPPER
		ACETOARSENITE DE CUIVRE (FRENCH)
		BASLE GREEN
		CI 77410
		CI PIGMENT GREEN 21
		CI PIGMENT GREEN 21 (9CI)
		COPPER ACETATE ARSENITE
	*	COPPER ACETOARSENITE
		COPPER ACETOARSENITE (DOT)
		COPPER ACETOARSENITE, SOLID
		COPPER ACETOARSENITE, SOLID (DOT)
		COPPER, BIS(ACETATO)HEXAMETAARSENITOTETRA-
		CUPRIC ACETOARSENITE
		EMERALD GREEN
		ENT 884
		FRENCH GREEN
		GENUINE PARIS GREEN
		IMPERIAL GREEN
		KING'S GREEN
		MEADOW GREEN
		MINERAL GREEN
		MITIS GREEN
		MOSS GREEN
		MOUNTAIN GREEN
		NEUWIED GREEN
		NEW GREEN
		ORTHO P-G BAIT
		PARIS GREEN
		PARIS GREEN, SOLID (DOT)
		PARROT GREEN
		PATENT GREEN
		POWDER GREEN
		SCHWEINFURT GREEN
		SCHWEINFURTERGRUN
		SCHWEINFURTH GREEN
		SOWBUG & CUTWORM BAIT
		SOWBUG CUTWORM CONTROL
		SWEDISH GREEN
		UN 1585 (DOT)
		VIENNA GREEN
		WUERZBERG GREEN
		ZWICKAU GREEN
12002-19-6		
		MERCUROL
		MERCUROL (DOT)
		MERCUROL, SOLID
		MERCURY NUCLEATE
		MERCURY NUCLEATE, SOLID (DOT)
		UN 1639 (DOT)
12002-25-4		
		DIPROPYLENE GLYCOL METHYL ETHER
12002-26-5		
		METHYL SILICATE
		SILICIC ACID, METHYL ESTER
		SILICON METHYLATE
12002-48-1		
		BENZENE, TRICHLORO-
		INVALON TC
		PYRANOL 1478
		TCB
		TRICHLOROBENZENE
		TRICHLOROBENZENE, LIQUID
		TRICHLOROBENZENE, LIQUID (DOT)
		TRICHLOROBENZENES
		UN 2321 (DOT)
12003-41-7		
		ALUMINUM FERROSILICON
12007-89-5		
		AMMONIUM BORATE (NH4B5O8)
		AMMONIUM BORON OXIDE ((NH4)B5O8)
		AMMONIUM PENTABORATE
		AMMONIUM PENTABORATE ((NH4)B5O8)
		BORIC ACID (HB5O8), AMMONIUM SALT
12013-56-8		
		CALCIUM SILICON
12027-06-4		
		AMMONIUM IODIDE
		AMMONIUM IODIDE ((NH4)I)
12030-88-5		
		POTASSIUM DIOXIDE
		POTASSIUM SUPEROXIDE
		POTASSIUM SUPEROXIDE (DOT)
		POTASSIUM SUPEROXIDE (K(O2))
		UN 2466 (DOT)
12031-80-0		
		DILITHIUM PEROXIDE
		LITHIUM OXIDE (Li2O2)
		LITHIUM PEROXIDE
		LITHIUM PEROXIDE (DOT)
		LITHIUM PEROXIDE (Li2(O2))
		UN 1472 (DOT)

CAS No.	IN	Chemical Name
12033-49-7		
		DINITROGEN TRIOXIDE
		NITRATE FREE RADICAL
		NITROGEN OXIDE (NO3)
		NITROGEN SESQUIOXIDE
		NITROGEN TRIOXIDE
		NITROGEN TRIOXIDE (DOT)
		NITROGEN TRIOXIDE (NO3)
		NITROUS ANHYDRIDE
		PERNITRITE RADICAL
		UN 2421 (DOT)
12034-12-7		
		SODIUM SUPEROXIDE
		SODIUM SUPEROXIDE (DOT)
		SODIUM SUPEROXIDE (Na(O2))
		UN 2547 (DOT)
12035-72-2		
		NICKEL SUBSULFIDE
		NICKEL SUBSULPHIDE
		NICKEL SUBSULFIDE (Ni3S2)
		NICKEL SULFIDE (Ni3S2)
		TRINICKEL DISULFIDE
12037-82-0		
		PHOSPHORUS HEPTASULFIDE
		PHOSPHORUS HEPTASULFIDE (DOT)
		PHOSPHORUS HEPTASULPHIDE, FREE FROM YELLOW OR WHITE PHOSPHORUS (DOT)
		PHOSPHORUS SULFIDE (P4S7)
		TETRAPHOSPHORUS HEPTASULFIDE
		UN 1339 (DOT)
12040-57-2		
		IRON CHLORIDE
12042-55-6		
		ALUMINUM SILICON (AlSi)
12044-79-0		
		ARSENIC MONOSULFIDE
		ARSENIC SULFIDE (As2S2)
		ARSENIC SULFIDE (As3S3)
		ARSENIC SULFIDE (AsS)
		ARSENIC SULFIDE RED
		ARSINO, THIOXO-
		C.I. 77085
		C.I. PIGMENT YELLOW 39
		RED ARSENIC GLASS
		RED ORPIMENT
		RUBY ARSENIC
12054-48-7		
		NA 9140 (DOT)
		NICKEL DIHYDROXIDE
		NICKEL HYDROXIDE
		NICKEL HYDROXIDE (DOT)
		NICKEL HYDROXIDE (Ni(OH)2)
		NICKEL(2+) HYDROXIDE
		NICKEL(II) HYDROXIDE
		NICKELOUS HYDROXIDE
12057-74-8		
		MAGNESIUM PHOSPHIDE
12058-85-4		
		SODIUM PHOSPHIDE
12070-12-1		
		TUNGSTEN CARBIDE
12075-68-2		
		ALUMINUM, TRICHLOROTRIETHYLDI-
		ETHYL ALUMINUM SESQUICHLORIDE
		ETHYL ALUMINUM SESQUICHLORIDE (DOT)
		SESQUIETHYLALUMINUM CHLORIDE
		TRICHLOROTRIETHYLDIALUMINIUM
		TRICHLOROTRIETHYLDIALUMINUM
		TRIETHYLALUMINUM SESQUICHLORIDE
		TRIETHYLDIALUMINUM TRICHLORIDE
		TRIETHYLTRICHLORODIALUMINUM
		UN 1925 (DOT)
12079-65-1		
		CYCLOPENTADIENYLMANGANESE TRICARBONYL
		CYCLOPENTADIENYLTRICARBONYLMANGANESE
		CYMANTRENE
		MANGANESE CYCLOPENTADIENYL TRICARBONYL
		MANGANESE, TRICARBONYL(ETA5-2,4-CYCLOPENTADIEN-1-YL)-
		MANGANESE, TRICARBONYL-PI-CYCLOPENTADIENYL-
		PI-CYCLOPENTADIENYLMANGANESE TRICARBONYL
		TRICARBONYL(ETA5-CYCLOPENTADIENYL)MANGANESE
		TRICARBONYL-PI-CYCLOPENTADIENYLMANGANESE
		TRICARBONYLCYCLOPENTADIENYLMANGANESE
12108-13-3		
		(METHYLCYCLOPENTADIENYL)TRICARBONYLMANGANESE
		2-METHYLCYCLOPENTADIENYLMANGANESE TRICARBONYL
		MANGANESE TRICARBONYL METHYLCYCLOPENTADIENYL
		MANGANESE, TRICARBONYL(METHYL-PI-CYCLOPENTADIENYL)-
		MANGANESE, TRICARBONYL(METHYLCYCLOPENTADIENYL)-
		MANGANESE, TRICARBONYL[(1,2,3,4,5-ETA)-1-METHYL-2,4-CYCLOPENTADIEN-1-YL]-
		METHYLCYCLOPENTADIENYL MANGANESE TRICARBONYL
		METHYLCYMANTRENE
		MMT
		PI-METHYLCYCLOPENTADIENYLMANGANESE TRICARBONYL
		TRICARBONYL(2-METHYLCYCLOPENTADIENYL)-MANGANESE
		TRICARBONYL(ETA5-METHYLCYCLOPENTADIENYL)-MANGANESE
		TRICARBONYL(METHYL-PI-CYCLOPENTADIENYL)-MANGANESE
		TRICARBONYL(METHYLCYCLOPENTADIENYL)MANGANESE
12122-67-7		
		ZINEB
12124-97-9		
	*	AMMONIUM BROMIDE
		AMMONIUM BROMIDE ((NH4)Br)
		HYDROBROMIC ACID MONOAMMONIATE
12124-99-1		
		AMMONIUM HYDROSULFIDE SOLUTION
12125-01-8		
	*	AMMONIUM FLUORIDE
		AMMONIUM FLUORIDE (DOT)
		AMMONIUM FLUORIDE (NH4F)
		AMMONIUM FLUORURE (FRENCH)
		NEUTRAL AMMONIUM FLUORIDE
		UN 2505 (DOT)
12125-02-9		
		AMCHLOR
		AMMONERIC
	*	AMMONIUM CHLORIDE
		AMMONIUM CHLORIDE ((NH4)Cl)
		AMMONIUM CHLORIDE (ACGIH,DOT)
		AMMONIUM CHLORIDE FUME
		AMMONIUM MURIATE
		AMMONIUMCHLORID (GERMAN)
		CHLORID AMONNY (CZECH)
		DARAMMON
		NA 9085 (DOT)
		SAL AMMONIA
		SAL AMMONIAC
		SALAMMONITE
		SALMIAC

CAS No.	IN	Chemical Name
12135-76-1		
		AMMONIUM MONOSULFIDE
	*	AMMONIUM SULFIDE
		AMMONIUM SULFIDE ((NH4)2S)
		AMMONIUM SULFIDE SOLUTION
		AMMONIUM SULFIDE SOLUTION (DOT)
		AMMONIUM SULPHIDE, SOLUTION (DOT)
		DIAMMONIUM SULFIDE
		TRUE AMMONIUM SULFIDE
		UN 2683 (DOT)
12136-45-7		
	*	POTASSIUM OXIDE
12136-49-1		
		DIPOTASSIUM SULFIDE
		DIPOTASSIUM TETRASULFIDE
		POTASSIUM SULFIDE
		POTASSIUM SULFIDE (DOT)
		POTASSIUM SULFIDE (K2S3)
		POTASSIUM SULPHIDE, ANHYDROUS OR CONTAINING LESS THAN 30% WATER (DOT)
		UN 1382 (DOT)
12136-50-4		
		DIPOTASSIUM SULFIDE
		DIPOTASSIUM TETRASULFIDE
		HYDROGEN SULFIDE (H2S5), DIPOTASSIUM SALT
		POTASSIUM SULFIDE
		POTASSIUM SULFIDE (DOT)
		POTASSIUM SULFIDE (K2S3)
		POTASSIUM SULPHIDE, ANHYDROUS OR CONTAINING LESS THAN 30% WATER (DOT)
		UN 1382 (DOT)
12165-69-4		
		DIPHOSPHORUS TRISULFIDE
		PHOSPHORUS SULFIDE (P2S3)
		PHOSPHORUS TRISULFIDE
		PHOSPHORUS TRISULFIDE (DOT)
		PHOSPHORUS TRISULPHIDE, FREE FROM YELLOW OR WHITE PHOSPHORUS (DOT)
		UN 1343 (DOT)
12167-20-3		
		CRESOL, NITRO-
		MONONITROTOLUOL
		NITROCRESOL
		NITROCRESOLS (DOT)
		PHENOL, METHYLNITRO-
		UN 2446 (DOT)
12172-73-5		
		AMOSITE
		AMOSITE ASBESTOS
		AMOSITE, GRUNERITE
	*	ASBESTOS, AMOSITE
		ASBESTOS, FERROGEDRITE
		ASBESTOS, GRUNERITE
		FERROGEDRITE ASBESTOS
		GRUNERITE ASBESTOS
12184-88-2		
		HYDRIDE
		HYDRIDE ION
		HYDRIDE ION(H1-)
		HYDRIDE, METAL
		HYDROGEN, ION (H1-)
12222-20-7		
		C.I. DIRECT BROWN 111
12259-92-6		
	*	AMMONIUM POLYSULFIDE
12262-58-7		
	*	CYCLOHEXANONE PEROXIDE
12263-85-3		
		ALUMINUM, TRIBROMOTRIMETHYLDI-
		METHYL ALUMINIUM SESQUIBROMIDE (DOT)
		METHYL ALUMINUM SESQUIBROMIDE
		METHYL ALUMINUM SESQUIBROMIDE (DOT)
		TRIBROMOTRIMETHYLDIALUMINUM
		UN 1926 (DOT)
12327-32-1		
		SILICON CARBIDE
		SILICON CARBIDE (Si2C3)
12385-08-9		
		BENZENEDIOL
		DIHYDROXYBENZENE
12401-70-6		
		POTASSIUM MONOXIDE
		POTASSIUM OXIDE
		POTASSIUM OXIDE (DOT)
		POTASSIUM OXIDE (KO)
		POTASSIUM OZONIDE
		POTASSIUM OZONIDE (K(O3))
		UN 2033 (DOT)
12401-86-4		
		CALCINED SODA
		DISODIUM MONOXIDE
		DISODIUM OXIDE
	*	SODIUM MONOXIDE
		SODIUM MONOXIDE (DOT)
		SODIUM MONOXIDE, SOLID
		SODIUM MONOXIDE, SOLID (DOT)
		SODIUM OXIDE
		SODIUM OXIDE (NaO)
		UN 1825 (DOT)
12415-34-8		
		EMERY
12427-38-2		
		MANEB
12504-13-1		
		PHOSPHURE DE STRONTIUM (FRENCH)
		STRONTIUM PHOSPHIDE
		STRONTIUM PHOSPHIDE (DOT)
		STRONTIUM PHOSPHIDE (SrP)
		UN 2013 (DOT)
12542-85-7		
		ALUMINUM, TRICHLOROTRIMETHYLDI-
		METHYL ALUMINIUM SESQUICHLORIDE (DOT)
		METHYL ALUMINUM SESQUICHLORIDE
		METHYL ALUMINUM SESQUICHLORIDE (DOT)
		TRICHLOROTRIMETHYLDIALUMINUM
		UN 1927 (DOT)
12604-53-4		
		FERROMANGANESE
12604-58-9		
		CARBON ALLOY, NONBASE, V,C,Fe (FERROVANADIUM)
	*	FERROVANADIUM
		FERROVANADIUM DUST
		IRON ALLOY, NONBASE, V,C,Fe (FERROVANADIUM)
		VANADIUM ALLOY, BASE, V,C,Fe (FERROVANADIUM)
12612-47-4		
		LEAD CHLORIDE
12612-55-4		
		NICKEL CARBONYL
12627-52-0		
		ANTIMONIAL GLASS
		ANTIMONOUS SULFIDE
		ANTIMONY GLANCE
		ANTIMONY ORANGE

CAS No.	IN	Chemical Name
		ANTIMONY SULFIDE
		ANTIMONY SULFIDE, SOLID
		ANTIMONY SULFIDE, SOLID (DOT)
		ANTIMONY TRISULFIDE
		ANTIMONY TRISULFIDE COLLOID
		CI 77060
		CI PIGMENT RED 107
		CRIMSON ANTIMONY
		LYMPHOSCAN
		NA 1325 (DOT)
		NEEDLE ANTIMONY
		VITREOUS ANTIMONY
12640-89-0		
		RCRA WASTE NUMBER U204
		SELENIOUS ANHYDRIDE
		SELENIUM DIOXIDE
		SELENIUM OXIDE
		SELENIUM(IV) DIOXIDE (1:2)
12654-97-6		
		(O-CHLOROANILINO)DICHLOROTRIAZINE
		1,3,5-TRIAZIN-2-AMINE, 4,6-DICHLORO-N-(2-CHLOROPHENYL)-
		2,4-DICHLORO-6-(2-CHLOROANILINO)-1,3,5-TRIAZINE
		2,4-DICHLORO-6-O-CHLORANILINO-S-TRIAZINE
		2-(2-CHLORANILIN)-4,6-DICHLOR-1,3,5-TRIAZIN (GERMAN)
		4,6-DICHLORO-N-(2-CHLOROPHENYL)-1,3,5-TRIAZIN-2-AMINE
		ANILAZIN
		ANILAZINE
		ANIYALINE
		B-622
		BORTRYSAN
		DIREZ
		DYRENE
		DYRENE 50W
		ENT 26,058
		KEMATE
		NCI-C08684
		S-TRIAZINE, 2,4-DICHLORO-6-(O-CHLOROANILINO)-
		TRIASYN
		TRIAZIN
		TRIAZINE
		TRIAZINE (PESTICIDE)
		TRIAZINE PESTICIDE, LIQUID [FLAMMABLE LIQUID LABEL]
		TRIAZINE PESTICIDE, LIQUID [POISON B LABEL]
		TRIAZINE PESTICIDE, SOLID
		ZINOCHLOR
12656-43-8		
	*	ALUMINUM CARBIDE
		ALUMINUM CARBIDE (Al4C3)
		ALUMINUM CARBIDE (DOT)
		TETRAALUMINUM TRICARBIDE
		UN 1394 (DOT)
12672-29-6		
		AROCLOR 1248
12674-11-2		
		AROCLOR 1016
12680-48-7		
		CHROMIC ACID, SODIUM SALT
		CHROMIUM SODIUM OXIDE
		SODIUM CHROMATE
		SODIUM CHROMITE
12684-19-4		
		LEAD IODIDE
12737-18-7		
		CALCIUM SILICIDE
		CALCIUM SILICIDE (DOT)
		CALCIUM SILICON (DOT)
		CALCIUM SILICON (POWDER)
		CALCIUM SILICON, POWDER (DOT)
		UN 1405 (DOT)
		UN 1406 (DOT)
12737-98-3		
		LEAD TUNGSTATE
		LEAD TUNGSTEN OXIDE
		TUNGSTIC ACID, LEAD
12768-82-0		
		C.I. BASIC ORANGE 15
		PHOSPHINE
		PHOSPHINE (DYE)
		PHOSPHINE 3R
		PHOSPHINE RN
12771-08-3		
		SULFUR CHLORIDE
12772-47-3		
		1-PENTOL
		1-PENTOL (DOT)
		2-PENTEN-4-YN-1-OL, 3-METHYL-
		3-METHYL-2-PENTEN-4-YN-1-OL
		9-OCTADECENOIC ACID (Z)-, ESTER WITH 2,2-BIS(HYDROXYMETHYL)-1,3-PROPANEDIOL
		PENTAERYTHRITOL OLEATE
		PENTOL
		PENTOL (EMULSIFIER)
		UN 2705 (DOT)
12788-93-1		
		ACID BUTYL PHOSPHATE
		ACID BUTYL PHOSPHATE (DOT)
		BAP
		BAP (ESTER)
		BUTYL ACID PHOSPHATE
		BUTYL HYDROGEN PHOSPHATE
		BUTYL PHOSPHORIC ACID
		BUTYL PHOSPHORIC ACID (DOT)
		JP 504
		N-BUTYL ACID PHOSPHATE (DOT)
		PHOSLEX A 4
		PHOSPHORIC ACID, BUTYL ESTER
		PV 1
		PV 1 (ANTIFOAMING AGENT)
		UN 1718 (DOT)
12789-03-6		
		1,2,4,5,6,7,10,10-OCTACHLORO-4,7,8,9-TETRAHYDRO-4,7-METHYLENEINDANE
		1,2,4,5,6,7,8,8-OCTACHLOR-2,3,3A,4,7,7A-HEXAHYDRO-4,7-METHANOINDANE
		1,2,4,5,6,7,8,8-OCTACHLORO-2,3,3A,4,7,7A-HEXAHYDRO-4,7-METHANO-1H-INDENE
		1,2,4,5,6,7,8,8-OCTACHLORO-2,3,3A,4,7,7A-HEXAHYDRO-4,7-METHANOINDENE
		1,2,4,5,6,7,8,8-OCTACHLORO-3A,4,7,7A-HEXAHYDRO-4,7-METHYLENE INDANE
		1,2,4,5,6,7,8,8-OCTACHLORO-3A,4,7,7A-TETRAHYDRO-4,7-METHANOINDAN
		1,2,4,5,6,7,8,8-OCTACHLORO-3A,4,7,7A-TETRAHYDRO-4,7-METHANOINDANE
		1,2,4,5,6,7,8,8-OCTACHLORO-4,7-METHANO-3A,4,7,7A-TETRAHYDROINDANE
		4,7-METHANO-1H-INDENE, 1,2,4,5,6,7,8,8-OCTACHLORO-2,3,3A,4,7,7A-HEXAHYDRO-
		4,7-METHANOINDAN, 1,2,4,5,6,7,8,8-OCTACHLORO-3A,4,7,7A-TETRAHYDRO-
		4,7-METHANOINDEN, 1,2,4,5,6,7,8,9-OCTACHLORO-3A,4,7,7A-TETRAHYDRO-
		ASPON-CHLORDANE
		BELT
		CD 68
		CHLOORDAAN (DUTCH)
		CHLOR KIL
		CHLORDAN
		CHLORDANE
		CHLORDANE (ACGIH)
		CHLORDANE [FLAMMABLE LIQUID LABEL]

CAS No.	IN	Chemical Name
		CHLORDANE, TECHNICAL
		CHLORDANE,LIQUID (DOT)
		CHLORINDAN
		CHLORODANE
		CHLORTOX
		CLORDAN (ITALIAN)
		CORODANE
		CORTILAN-NEU
		DICHLOROCHLORDENE
		DOWCHLOR
		ENT 25,552-X
		ENT 9,932
		GAMMA-CHLORDAN
		HCS 3260
		INTOX
		INTOX (INSECTICIDE)
		KYPCHLOR
		M 140
		M 410
		NA 2762 (DOT)
		NCI-C00099
		NIRAN
		OCTA-KLOR
		OCTACHLOR
		OCTACHLORO-4,7-METHANOHYDROINDANE
		OCTACHLORO-4,7-METHANOTETRAHYDROINDANE
		OCTACHLORODIHYDRODICYCLOPENTADIENE
		OKTATERR
		ORTHO-KLOR
		RCRA WASTE NUMBER U036
		SD 5532
		SHELL SD-55 32
		STARCHLOR
		SYNKLOR
		TAT CHLOR 4
		TERMEX
		TOPICHLOR 20
		TOPICLOR
		TOPICLOR 20
		TOXICHLOR
		UNEXAN-KOEDER
		VELSICOL 1068
12789-46-7		
		AMYL ACID PHOSPHATE
13010-47-4		
		1-(2-CHLOROETHYL)-3-CYCLO-HEXYL-1-NITROSOUREA
		1-(2-CHLOROETHYL)-3-CYCLOHEXYL-1-NITROSOUREA
		1-(2-CHLOROETHYL)-3-CYCLOHEXYLNITROSOUREA
		BELUSTINE
		CCNU
		CHLOROETHYLCYCLOHEXYLNITROSOUREA
		ICIG 1109
		LOMUSTINE
		NSC 79037
		UREA, 1-(2-CHLOROETHYL)-3-CYCLOHEXYL-1-NITROSO-
		UREA, N-(2-CHLOROETHYL)-N'-CYCLOHEXYL-N-NITROSO-
13057-78-8		
		1,3,5-TRIAZINE-2,4,6(1H,3H,5H)-TRIONE, 1-CHLORO-
		CHLOROISOCYANURIC ACID
		S-TRIAZINE-2,4,6(1H,3H,5H)-TRIONE, 1-CHLORO-
		S-TRIAZINE-2,4,6(1H,3H,5H)-TRIONE, CHLORO-
13071-79-9		
		AC 92100
		COUNTER
		COUNTER 15G
		COUNTER 15G SOIL INSECTICIDE
		COUNTER 15G SOIL INSECTICIDE-NEMATICIDE
		METHANETHIOL, (TERT-BUTYLTHIO)-, S-ESTER WITH O,O-DIETHYL PHOSPHORODITHIOATE
		O,O-DIETHYL S-(TERT-BUTYLTHIO)METHYL PHOSPHORODITHIOATE
		O,O-DIETHYL S-[[(1,1-DIMETHYLETHYL)THIO]METHYL] PHOSPHORODITHIOATE
		PHOSPHORODITHIOIC ACID S-(((1,1-DIMETHYLETHYL)-THIO)METHYL) O,O-DIETHYL ESTER
		PHOSPHORODITHIOIC ACID S-((TERT-BUTYLTHIO)METHYL) O,O-DIETHYL ESTER
		PHOSPHORODITHIOIC ACID, O,O-DIETHYL-S-(((1,1-DI-METHYLETHYL)THIO)METHYL)-ESTER
		S-(((1,1-DIMETHYLETHYL)THIO)METHYL)-O,O-DIETHYL PHOSPHORODITHIOATE
		S-((TERT-BUTYLTHIO)METHYL)O,O-DIETHYLPHOSPHORO-DITHIOATE
		TERBUFOS
13106-47-3		
		BERYLLIUM CARBONATE
		BERYLLIUM CARBONATE ($BeCO_3$)
		CARBONIC ACID, BERYLLIUM SALT (1:1)
13106-76-8		
		AMMONIUM HEPTAMOLYBDATE
	*	AMMONIUM MOLYBDATE
		AMMONIUM MOLYBDATE [$(NH_4)_2MoO_4$]
		AMMONIUM PARAMOLYBDATE
		DIAMMONIUM MOLYBDATE
		DIAMMONIUM MOLYBDATE ($(NH_4)_2MoO_4$)
		MOLYBDATE (MoO_4^{2-}), DIAMMONIUM, (T-4)-
		MOLYBDATE, HEXAAMMONIUM (9CI)
		MOLYBDIC ACID (H_2MoO_4), DIAMMONIUM SALT
		MOLYBDIC ACID DIAMMONIUM SALT
		MOLYBDIC ACID, HEXAAMMONIUM SALT
13114-62-0		
		NAPHTHYLUREA (2-NAPHTHALENYL UREA)
13121-70-5		
		CYHEXATIN
		DOWCO 213
		HYDROXYTRICYCLOHEXYLSTANNANE
		M 3180
		PLICTRAN
		PLYCTRAN
		STANNANE, TRICYCLOHEXYLHYDROXY-
		TIN, TRICYCLOHEXYLHYDROXY-
		TRICYCLOHEXYLHYDROXYSTANNANE
		TRICYCLOHEXYLHYDROXYTIN
		TRICYCLOHEXYLSTANNANOL
		TRICYCLOHEXYLSTANNYL HYDROXIDE
		TRICYCLOHEXYLTIN HYDROXIDE
13122-18-4		
		HEXANEPEROXOIC ACID, 3,5,5-TRIMETHYL-, 1,1-DIMETHYLETHYL ESTER
		PEROXYHEXANOIC ACID, 3,5,5-TRIMETHYL-, TERT-BUTYL ESTER
		TERT-BUTYL 3,5,5-TRIMETHYLPEROXYHEXANOATE
		TERT-BUTYL PER-3,5,5-TRIMETHYLHEXANOATE
		TERT-BUTYL PEROXY 3,5,5-TRIMETHYLHEXANOATE
		TERT-BUTYL PEROXY-3,5,5-TRIMETHYL HEXANOATE, TECHNICAL PURE (DOT)
		TERT-BUTYL PEROXY-3,5,5-TRIMETHYLHEXANOATE, TECHNICALLY PURE
		TERT-BUTYL PEROXYISONONANOATE, TECHNICAL PURE (DOT)
		TRIGONOX 42
		UN 2104 (DOT)
13138-45-9		
		NICKEL BIS(NITRATE)
		NICKEL DINITRATE
	*	NICKEL NITRATE
		NICKEL NITRATE (DOT)
		NICKEL NITRATE ($Ni(NO_3)_2$)
		NICKEL(2+) NITRATE
		NICKEL(II) NITRATE
		NICKEL(II) NITRATE (1:2)
		NICKELOUS NITRATE
		NITRIC ACID, NICKEL(2+) SALT
		NITRIC ACID, NICKEL(II) SALT
		UN 2725 (DOT)

CAS No.	IN	Chemical Name
13171-21-6		
		DIMECRON
		DIMECRON 100
		DIMECRON 50
		DIMECRON-20
		MERKON
		PHOSPHAMIDON
		PHOSPHAMIDONE
		PHOSPHORIC ACID, 2-CHLORO-3-(DIETHYLAMINO)-1-METHYL-3-OXO-1-PROPENYL DIMETHYL ESTER
		PHOSPHORIC ACID, DIMETHYL ESTER, ESTER WITH 2-CHLORO-N,N-DIETHYL-3-HYDROXYCROTONAMIDE
		SUNDARAM 1975
13194-48-4		
		ENT 27,318
		ETHOPROP
		ETHOPROPHOS
		ETHYL PROPYL PHOSPHORODITHIOATE ((ETO)(PRS)2PO)
		JOLT
		MOBIL V-C 9-104
		MOCAP
		MOCAP 10G
		O-ETHYL S,S-DIPROPYL DITHIOPHOSPHATE
		O-ETHYL S,S-DIPROPYL PHOSPHORODITHIOATE
		PHOSPHORODITHIOIC ACID, O-ETHYL S,S-DIPROPYL ESTER
		PROFOS
		PROPHOS
		PROPHOS (ESTER)
		ROVOKIL
		S,S-DIPROPYL O-ETHYL PHOSPHORODITHIOATE
		V-C 9-104
		V-C CHEMICAL V-C 9-104
		VIRGINIA-CAROLINA VC 9-104
13195-76-1		
		BORIC ACID (H3BO3), TRIISOBUTYL ESTER
		BORIC ACID (H3BO3), TRIS(2-METHYLPROPYL) ESTER
		ISOBUTYL BORATE ((C4H9O)3B)
		TRIISOBUTYL BORATE
		TRIISOBUTYL ORTHOBORATE
13256-22-9		
		GLYCINE, N-METHYL-N-NITROSO-
		N-METHYL-N-(CARBOXYMETHYL)NITROSAMINE
		N-NITROSOSARCOSINE
		NITROSOSARCOSINE
		SARCOSINE, N-NITROSO-
13319-75-0		
		BORANE, TRIFLUORO-, DIHYDRATE
		BORON FLUORIDE (BF3) DIHYDRATE
		BORON FLUORIDE DIHYDRATE
		BORON TRIFLUORIDE DIHYDRATE
		BORON TRIFLUORIDE DIHYDRATE (DOT)
		UN 2851 (DOT)
13327-32-7		
		BERYLLIUM DIHYDROXIDE
		BERYLLIUM HYDRATE
		BERYLLIUM HYDROXIDE
		BERYLLIUM HYDROXIDE (Be(OH)2)
13360-63-9		
		1-BUTANAMINE, N-ETHYL-
		BUTYLAMINE, N-ETHYL-
		BUTYLETHYLAMINE
		ETHYLBUTYLAMINE
		N-BUTYL-N-ETHYLAMINE
		N-BUTYLETHYLAMINE
		N-ETHYL-N-BUTYLAMINE
		N-ETHYLBUTANAMINE
		N-ETHYLBUTYLAMINE
13387-02-0		
		9-PHOSPHABICYCLONONANE [...3,3,1-NONANE]

CAS No.	IN	Chemical Name
13396-80-0		
		9-PHOSPHABICYCLONONANE [...4,2,1-NONANE]
13397-24-5		
		GIPS
		GYPSITE
	*	GYPSUM
		GYPSUM (Ca(SO4).2H2O)
		LANDPLASTER
13410-01-0		
		DISODIUM SELENATE
		NATRIUMSELENIAT (GERMAN)
		P-40
		SEL-TOX SSO2 AND SS-20
		SELENIC ACID (H2SeO4), DISODIUM SALT
		SELENIC ACID, DISODIUM SALT
		SODIUM SELENATE
		SODIUM SELENATE (Na2SeO4)
		SODIUM SELENIUM OXIDE (Na2SeO4)
13410-72-5		
		BENZENAMINE, 4-ETHOXY-N-[(5-NITRO-2-FURANYL)-METHYLENE]-
		NITROFEN
		NITROFEN (PHARMACEUTICAL)
		P-PHENETIDINE, N-(5-NITROFURFURYLIDENE)-
13423-61-5		
		CHROMIC ACID (H2CrO4), MAGNESIUM SALT (1:1)
		CHROMIUM MAGNESIUM OXIDE (MgCrO4)
		MAGNESIUM CHROMATE
		MAGNESIUM CHROMATE (MgCrO4)
		MAGNESIUM DICHROMATE
		MAGNESIUM DICHROMIUM TETROXIDE
13424-46-9		
		LEAD AZIDE
13426-91-0		
		BIS(ETHYLENEDIAMINE)COPPER ION
		BIS(ETHYLENEDIAMINE)COPPER(2+)
		BIS(ETHYLENEDIAMINE)COPPER(2+) ION
		BIS(ETHYLENEDIAMINE)COPPER(II)
		COPPER(2+), BIS(1,2-ETHANEDIAMINE-N,N')-
		COPPER(2+), BIS(ETHYLENEDIAMINE)-, ION
		COPPER-ETHYLENEDIAMINE COMPLEX
		CUPRIETHYLENE DIAMINE
		CUPRIETHYLENE-DIAMINE SOLUTION
		CUPRIETHYLENE-DIAMINE SOLUTION (DOT)
		ETHANE, 1,2-DIAMINO-, COPPER COMPLEX
		UN 1761 (DOT)
13446-10-1		
		AMMONIUM PERMANGANATE
		AMMONIUM PERMANGANATE (DOT)
		NA 9190 (DOT)
		PERMANGANIC ACID (HMnO4), AMMONIUM SALT
		PERMANGANIC ACID, AMMONIUM SALT
13450-90-3		
		GALLIUM CHLORIDE
		GALLIUM CHLORIDE (GaCl3)
		GALLIUM TRICHLORIDE
		GALLIUM(3+) CHLORIDE
		TRICHLOROGALLIUM
13450-97-0		
		PERCHLORIC ACID, STRONTIUM SALT
		STRONTIUM DIPERCHLORATE
		STRONTIUM PERCHLORATE
		STRONTIUM PERCHLORATE (DOT)
		STRONTIUM PERCHLORATE (Sr(ClO4)2)
		UN 1508 (DOT)
13453-30-0		
		CHLORIC ACID, THALLIUM(1+) SALT
		THALLIUM CHLORATE
		THALLIUM CHLORATE (DOT)

CAS No.	IN	Chemical Name
		THALLIUM CHLORATE (TlClO3)
		UN 2573 (DOT)
13454-96-1		
		PLATINUM (IV) CHLORIDE
		PLATINUM CHLORIDE (PtCl4)
		PLATINUM CHLORIDE (PtCl4), (SP-4-1)-
		PLATINUM TETRACHLORIDE
		PLATINUM(IV) CHLORIDE
		PLATINUM(IV) TETRACHLORIDE
13462-88-9		
		NICKEL BROMIDE
		NICKEL BROMIDE (NiBr2)
		NICKEL DIBROMIDE
		NICKEL(2+) BROMIDE
		NICKEL(II) BROMIDE
		NICKELOUS BROMIDE
13463-39-3		
		NICHEL TETRACARBONILE (ITALIAN)
	*	NICKEL CARBONYL
		NICKEL CARBONYL (ACGIH,DOT)
		NICKEL CARBONYL (Ni(CO)4)
		NICKEL CARBONYL (Ni(CO)4), (T-4)-
		NICKEL CARBONYLE (FRENCH)
		NICKEL TETRACARBONYL
		NICKEL TETRACARBONYLE (FRENCH)
		NIKKELTETRACARBONYL (DUTCH)
		RCRA WASTE NUMBER P073
		TETRACARBONYLNICKEL
		UN 1259 (DOT)
13463-40-6		
		FER PENTACARBONYLE (FRENCH)
		IRON CARBONYL
		IRON CARBONYL (DOT)
		IRON CARBONYL (Fe(CO)5)
		IRON CARBONYL (Fe(CO)5), (TB-5-11)-
		IRON PENTACARBONYL
		IRON PENTACARBONYL (ACGIH,DOT)
		IRON, PENTACARBONYL-
		PENTACARBONYL IRON
		UN 1994 (DOT)
13463-67-7		
		A-FIL CREAM
		AUSTIOX R-CR 3
		BAYERTITAN A
		BAYERTITAN R-FD 1
		BAYERTITAN R-U-F
		C.I. 77891
		C.I. PIGMENT WHITE 6
		CAB-O-TI
		CRS 31
		FINNTITAN
		FINNTITAN RF 2
		FLAMENCO
		HOMBITAN R 101D
		HOMBITAN R 505
		HOMBITAN R 506
		HOMBITAN R 610D
		HOMBITAN R 610K
		HORSEHEAD A 430C
		HORSEHEAD A 430FG
		HORSEHEAD R 771
		JR 701
		KA 10
		KA 15
		KA 35
		KH 360
		KR 380
		KRONOS
		KRONOS 2073
		KRONOS CL 220
		KRONOS KR 380
		KRONOS RN 40
		KRONOS RN 40P
		KRONOS RN 56
		KRONOS RN 59
		P 25
		P 25 (OXIDE)
		PRETIOX AP 22
		PRETIOX RD 53
		PV FAST WHITE R
		R 580
		R 610D
		R 650
		R 680
		R 780-2
		RAYOX
		RFC 5
		RO 2
		RUNA ARG
		RUNA ARH 20
		RUNA ARH 200
		RUNA RH 52
		RUNA RP
		RUTIOX CR
		SR 1
		SR 1 (OXIDE)
		TA 400
		TCR 10
		TI-PURE
		TI-PURE R 100
		TI-PURE R 101
		TI-PURE R 900
		TI-PURE R 901
		TI-PURE R 915
		TIONA 113
		TIONA VB
		TIOXIDE AD-M
		TIOXIDE R-CR
		TIOXIDE R-CR 3
		TIOXIDE R-SM
		TIOXIDE R-TC 2
		TIOXIDE R-TC 60
		TIOXIDE R.XL
		TIOXIDE RHD
		TIPAQUE CR 50
		TIPAQUE CR 90
		TIPAQUE CR 95
		TIPAQUE R 820
		TITAN A
		TITAN RC-K 20
		TITANIA
		TITANIA (TiO2)
	*	TITANIUM DIOXIDE
		TITANIUM DIOXIDE (TiO2)
		TITANIUM OXIDE
		TITANIUM OXIDE (TiO2)
		TITANIUM PEROXIDE (TiO2)
		TITANIUM(IV) OXIDE
		TITANOX
		TITANOX 2010
		TITANOX RANC
		TITONE R 5N
		TR 840
		TRONEX CR 800
		TYTANPOL R 323
		UNITANE
		UNITANE OR 450
		UNITANE OR 572
		UNITANE OR 650
		ZOPAQUE LDC
13464-33-0		
		ARSENENIC ACID, ZINC SALT
		ARSENIC ACID (H3AsO4) ZINC SALT
		ARSENIC ACID (H3AsO4), ZINC SALT (1:1)
		ARSENIC ACID (HASO3), ZINC SALT
		ARSENIC ACID, ZINC SALT
		UN 1712 (DOT)
		ZINC ARSENATE
		ZINC ARSENATE (DOT)
		ZINC ARSENATE, BASIC
		ZINC ARSENATE, SOLID (DOT)

CAS No.	IN	Chemical Name
13464-37-4		
		ARSENIOUS ACID (H3AsO3), TRISODIUM SALT
		ARSENIOUS ACID SODIUM SALT (Na3AsO3)
		ARSENOUS ACID, TRISODIUM SALT
		SODIUM ARSENITE
		SODIUM ARSENITE (Na3AsO3)
		TRISODIUM ARSENITE
13464-44-3		
		ARSENENIC ACID, ZINC SALT
		ARSENIC ACID (H3AsO4) ZINC SALT
		ARSENIC ACID (H3AsO4), ZINC SALT (2:3)
		ARSENIC ACID (HAsO3), ZINC SALT
		ARSENIC ACID, ZINC SALT
		UN 1712 (DOT)
		ZINC ARSENATE
		ZINC ARSENATE (DOT)
		ZINC ARSENATE (Zn3(AsO4)2)
		ZINC ARSENATE, BASIC
		ZINC ARSENATE, SOLID (DOT)
13464-97-6		
		HYDRAZINE NITRATE
13465-78-6		
		CHLOROSILANE
		CHLOROSILANE (ClSiH3)
		CHLOROSILANE, [CORROSIVE LABEL]
		CHLOROSILANE, [EMITS FLAMMABLE GAS WHEN WET, CORROSIVE LABELS]
		CHLOROSILANE, [FLAMMABLE, CORROSIVE LABELS]
		MONOCHLOROSILANE
		SILANE, CHLORO-
		SILYL CHLORIDE
13465-95-7		
		BARIUM DIPERCHLORATE
		BARIUM PERCHLORATE
		BARIUM PERCHLORATE (DOT)
		PERCHLORIC ACID, BARIUM SALT
		UN 1447 (DOT)
13473-90-0		
	*	ALUMINUM NITRATE
13477-00-4		
		BARIUM CHLORATE
		BARIUM CHLORATE (Ba(ClO3)2)
		BARIUM CHLORATE (DOT)
		CHLORIC ACID, BARIUM SALT
		UN 1445 (DOT)
13477-10-6		
	*	BARIUM HYPOCHLORITE
		BARIUM HYPOCHLORITE, CONTAINING MORE THAN 22% AVAILABLE CHLORINE (DOT)
		HYPOCHLOROUS ACID, BARIUM SALT
		UN 2741 (DOT)
13477-36-6		
		CALCIUM DIPERCHLORATE
		CALCIUM PERCHLORATE
		CALCIUM PERCHLORATE (DOT)
		PERCHLORIC ACID CALCIUM SALT (2:1)
		PERCHLORIC ACID, CALCIUM SALT
		UN 1455 (DOT)
13479-54-4		
		BIS(AMINOACETATO)COPPER
		BIS(GLYCINATO)COPPER
		BIS(GLYCINATO)COPPER(II)
		COPPER AMINOACETATE
		COPPER DIGLYCINATE
		COPPER GLYCINATE
		COPPER(GLYCINE)2
		COPPER(II) GLYCINATE
		COPPER, BIS(GLYCINATO)-
		COPPER, BIS(GLYCINATO-N,O)-
		CUPRIC AMINOACETATE
		CUPRIC GLYCINATE
		GLYCINE, COPPER(2+) SALT (2:1)
13494-80-9		
		NCI-C60117
		TELLOY
		TELLUR (POLISH)
		TELLURIUM
		TELLURIUM (ACGIH)
		TELLURIUM ELEMENT
13494-80-9		
		BERYLLIUM SULFATE
		BERYLLIUM SULFATE (1:1)
		BERYLLIUM SULFATE (BeSO4)
		BERYLLIUM SULPHATE
		SULFURIC ACID, BERYLLIUM SALT (1:1)
13520-83-7		
		URANYL NITRATE HEXAHYDRATE
13530-65-9		
		BUTTERCUP YELLOW
		CHROMIC ACID (H2CrO4), ZINC SALT (1:1)
		CHROMIC ACID, ZINC SALT (ZnCrO4)
		CHROMIUM ZINC OXIDE (ZnCrO4)
		DICHROMIC ACID, ZINC SALT (1:1)
		ZINC BICHROMATE
	*	ZINC CHROMATE
		ZINC CHROMATE (ACGIH)
		ZINC CHROMATE (ZnCrO4)
		ZINC CHROMIUM OXIDE
		ZINC CHROMIUM OXIDE (ZnCrO4)
		ZINC DICHROMATE
		ZINC DICHROMATE (VI)
		ZINC TETRAOXYCHROMATE
		ZINC TETROXYCHROMATE
13530-68-2		
		ANHYDRIDE CHROMIQUE (FRENCH)
		ANIDRIDE CROMICA (ITALIAN)
		CHROME (TRIOXYDE DE) (FRENCH)
		CHROMIC (VI) ACID
		CHROMIC ACID
		CHROMIC ACID (H2Cr2O7)
		CHROMIC ACID (MIXTURE)
		CHROMIC ACID MIXTURE, DRY
		CHROMIC ACID MIXTURE, DRY (DOT)
		CHROMIC ACID SOLUTION
		CHROMIC ACID, SOLID
		CHROMIC ACID, SOLID (DOT)
		CHROMIC ACID, SOLUTION (DOT)
		CHROMIC ANHYDRIDE
		CHROMIC ANHYDRIDE (DOT)
		CHROMIC TRIOXIDE
		CHROMIC TRIOXIDE (DOT)
		CHROMIUM (VI) OXIDE
		CHROMIUM OXIDE
		CHROMIUM TRIOXIDE
		CHROMIUM TRIOXIDE, ANHYDROUS (DOT)
		CHROMIUM(6+) TRIOXIDE
		CHROMIUM(VI) OXIDE (1:3)
		CHROMSAEUREANHYDRID (GERMAN)
		CHROMTRIOXID (GERMAN)
		CHROOMTRIOXYDE (DUTCH)
		CHROOMZUURANHYDRIDE (DUTCH)
		CROMO(TRIOSSIDO DI) (ITALIAN)
		DICHROMIC ACID
		DICHROMIC ACID (H2Cr2O7)
		DICHROMIC(VI) ACID
		MONOCHROMIUM OXIDE
		MONOCHROMIUM TRIOXIDE
		NA 1463 (DOT)
		PURATRONIC CHROMIUM TRIOXIDE
		UN 1463 (DOT)
		UN 1755 (DOT)

CAS No.	*IN*	*Chemical Name*
13537-32-1		
		FLUOROPHOSPHONIC ACID (F(HO)2PO)
	*	FLUOROPHOSPHORIC ACID
		FLUOROPHOSPHORIC ACID, ANHYDROUS
		FLUOROPHOSPHORIC ACID, ANHYDROUS (DOT)
		MONOFLUOROPHOSPHORIC ACID
		MONOFLUOROPHOSPHORIC ACID, ANHYDROUS
		MONOFLUOROPHOSPHORIC ACID, ANHYDROUS (DOT)
		PHOSPHOROFLUORIDIC ACID
		UN 1776 (DOT)
13548-38-4		
	*	CHROMIC NITRATE
		CHROMIUM NITRATE
		CHROMIUM NITRATE (Cr(NO3)3)
		CHROMIUM TRINITRATE
		CHROMIUM(3+) NITRATE
		CHROMIUM(III) NITRATE
		NITRIC ACID, CHROMIUM SALT
		NITRIC ACID, CHROMIUM(3+) SALT
13552-44-8		
		4,4'-METHYLENDIANILINE DIHYDROCHLORIDE
13560-99-1		
		2,4,5-T SODIUM SALT
		2,4,5-TRICHLOROPHENOXYACETIC ACID, SODIUM SALT
		ACETIC ACID, (2,4,5-TRICHLOROPHENOXY)-, SODIUM SALT
		SODIUM (2,4,5-TRICHLOROPHENOXY)ACETATE
13597-54-1		
		SELENIC ACID (H2SeO4), ZINC SALT (1:1)
		SELENIC ACID, ZINC SALT (1:1)
		ZINC SELENATE
		ZINC SELENATE (ZnSeO4)
13597-99-4		
		BERYLLIUM DINITRATE
	*	BERYLLIUM NITRATE
		BERYLLIUM NITRATE (Be(NO3)2)
		BERYLLIUM NITRATE (DOT)
		NITRIC ACID, BERYLLIUM SALT
		UN 2464 (DOT)
13598-15-7		
		BERYLLIUM PHOSPHATE
		BERYLLIUM PHOSPHATE (BeHPO4)
		PHOSPHORIC ACID, BERYLLIUM SALT (1:1)
13598-26-0		
		BERYLLIUM PHOSPHATE
		PHOSPHORIC ACID, BERYLLIUM SALT
		PHOSPHORIC ACID, BERYLLIUM SALT (2:3)
13598-36-2		
		PHOSPHOROUS ACID, ORTHO
13601-19-9		
		FERRATE(4-), HEXACYANO-, TETRASODIUM
		FERRATE(4-), HEXAKIS(CYANO-C)-, TETRASODIUM, (OC-6-11)-
		SODIUM FERROCYANIDE
		SODIUM FERROCYANIDE (Na4[Fe(CN)6])
		SODIUM HEXACYANOFERRATE (II)
		TETRASODIUM FERROCYANIDE
		TETRASODIUM HEXACYANOFERRATE
		TETRASODIUM HEXACYANOFERRATE(4-)
13637-63-3		
		CHLORINE FLUORIDE (ClF5)
		CHLORINE PENTAFLUORIDE
		CHLORINE PENTAFLUORIDE (DOT)
		UN 2548 (DOT)
13637-76-8		
		LEAD DIPERCHLORATE
		LEAD PERCHLORATE
		LEAD PERCHLORATE (DOT)
		LEAD PERCHLORATE (Pb(ClO4)2)
		LEAD(2+) PERCHLORATE
		PERCHLORIC ACID, LEAD(2+) SALT
		UN 1470 (DOT)
13682-73-0		
		COPPER(I) POTASSIUM CYANIDE
		CUPRATE(1-), BIS(CYANO-C)-, POTASSIUM
		CUPRATE(1-), DICYANO-, POTASSIUM
		CUPROUS POTASSIUM CYANIDE
		POTASSIUM COPPER(I) CYANIDE
		POTASSIUM CUPROCYANIDE
		POTASSIUM DICYANOCUPRATE(1-)
		POTASSIUM DICYANOCUPRATE(I)
		UN 1679 (DOT)
13693-11-3		
		SULFURIC ACID, TITANIUM SALT
		SULFURIC ACID, TITANIUM(4+) SALT (2:1)
		TITANIUM DISULFATE
		TITANIUM SULFATE
		TITANIUM SULFATE (1:2)
		TITANIUM SULFATE (Ti(SO4)2)
		TITANIUM SULFATE SOLUTION WITH NO MORE THAN 45% SULFURIC ACID
		TITANIUM(4+) SULFATE
		TITANIUM(IV) SULFATE
13717-00-5		
		AUSTRALIAN MAGNESITE
		MAGNESITE
		MAGNESITE (Mg(CO3))
13718-26-8		
		METAWANADANEM SODOWYM (POLISH)
		PEROXYVANADIC ACID, SODIUM SALT
		SODIUM AMMONIUM VANADATE
		SODIUM PEROXYVANADATE
		SODIUM VANADATE
		SODIUM VANADATE (NaVO3)
		SODIUM VANADATE(V) (NaVO3)
		SODIUM VANADIUM OXIDE
		SODIUM VANADIUM OXIDE (NaVO3)
		TRISODIUM VANADATE
		VANADATE (VO31-), SODIUM
		VANADIC ACID (HVO3), SODIUM SALT
		VANADIC ACID, MONOSODIUM SALT
		VANADIC ACID, SODIUM SALT
13718-59-7		
		BARIUM SELENITE
		SELENIOUS ACID (H2SeO3), BARIUM SALT (1:1)
		SELENIOUS ACID, BARIUM SALT (1:1)
13721-39-6		
		METAWANADANEM SODOWYM (POLISH)
		PEROXYVANADIC ACID, SODIUM SALT
		SODIUM ORTHOVANADATE
		SODIUM PEROXYVANADATE
		SODIUM VANADATE
		SODIUM VANADATE(V) (NaVO3)
		SODIUM VANADIUM OXIDE
		SODIUM VANADIUM OXIDE (Na3VO4)
		TRISODIUM ORTHOVANADATE
		TRISODIUM VANADATE
		VANADATE (VO43-), TRISODIUM, (T-4)-
		VANADIC ACID (H3VO4), TRISODIUM SALT
		VANADIC ACID, SODIUM SALT
		VANADIC(II) ACID, TRISODIUM SALT
13746-89-9		
		DUSICNAN ZIRKONICITY (CZECH)
		NITRIC ACID, ZIRCONIUM(4+) SALT
		TETRANITRATOZIRCONIUM
		UN 2728 (DOT)
	*	ZIRCONIUM NITRATE
		ZIRCONIUM NITRATE (Zr(NO3)4)
		ZIRCONIUM TETRANITRATE

CAS No.	IN	Chemical Name
13746-98-0		
		NITRIC ACID, THALLIUM SALT
		NITRIC ACID, THALLIUM(3+) SALT
		RCRA WASTE NUMBER U217
		THALLIC NITRATE
		THALLIUM NITRATE
		THALLIUM NITRATE (DOT)
		THALLIUM NITRATE (Tl(NO3)3)
		THALLIUM TRINITRATE
		THALLIUM(III) NITRATE
		UN 2727 (DOT)
13762-51-1		
		BORATE(1-), TETRAHYDRO-, POTASSIUM
		BORATE(1-), TETRAHYDRO-, POTASSIUM (8CI,9CI)
		BOROHYDRURE DE POTASSIUM (FRENCH)
		POTASSIUM BOROHYDRATE
		POTASSIUM BOROHYDRIDE
		POTASSIUM BOROHYDRIDE (DOT)
		POTASSIUM BOROHYDRIDE (KBH4)
		POTASSIUM TETRAHYDROBORATE
		POTASSIUM TETRAHYDROBORATE(1-)
		UN 1870 (DOT)
13765-19-0		
		CALCIUM CHROMATE
		CALCIUM CHROMATE (CaCrO4)
		CALCIUM CHROMATE (DOT)
		CALCIUM CHROMATE (VI)
		CALCIUM CHROMATE(VI)
		CALCIUM CHROME YELLOW
		CALCIUM CHROMIUM OXIDE (CaCrO4)
		CALCIUM MONOCHROMATE
		CHROMIC ACID (H2CrO4), CALCIUM SALT (1:1)
		CHROMIC ACID, CALCIUM SALT (1:1)
		CI 77223
		CI PIGMENT YELLOW 33
		GELBIN
		NA 9096 (DOT)
		RCRA WASTE NUMBER U032
		YELLOW ULTRAMARINE
13768-86-0		
		SELENIUM OXIDE (SeO3)
		SELENIUM TRIOXIDE
		SELENIUM TRIOXIDE (SeO3)
13769-43-2		
		POTASSIUM METAVANADATE
		POTASSIUM METAVANADATE (DOT)
		POTASSIUM METAVANADATE (KVO3)
		POTASSIUM VANADATE (KVO3)
		POTASSIUM VANADATE(V) (KVO3)
		POTASSIUM VANADIUM TRIOXIDE
		UN 2864 (DOT)
		VANADATE (VO31-), POTASSIUM
		VANADIC ACID (HVO3), POTASSIUM SALT
		VANADIC ACID, POTASSIUM SALT
13770-96-2		
		ALUMINATE (1-), TETRAHYDRO-, SODIUM
		ALUMINATE(1-), TETRAHYDRO-, SODIUM, (T-4)-
		ALUMINATE(1-), TETRAHYDRO-, SODIUM, (T-4)- (9CI)
		ALUMINUM SODIUM HYDRIDE
		ALUMINUM SODIUM HYDRIDE (AlNaH4)
		SAH 22
		SODIUM ALUMINUM HYDRIDE
		SODIUM ALUMINUM HYDRIDE (DOT)
		SODIUM ALUMINUM HYDRIDE (NaAlH4)
		SODIUM ALUMINUM TETRAHYDRIDE
		SODIUM TETRAHYDROALUMINATE
		SODIUM TETRAHYDROALUMINATE (NaAlH4)
		SODIUM TETRAHYDROALUMINATE(1-)
		UN 2835 (DOT)
13774-25-9		
		MAGNESIUM BISULFITE
		MAGNESIUM BISULFITE SOLUTION
		MAGNESIUM SULFITE (Mg(HSO3)2)
		SULFUROUS ACID, MAGNESIUM SALT (2:1)
13779-41-4		
		DIFLUOROPHOSPHORIC ACID
		DIFLUOROPHOSPHORIC ACID, ANHYDROUS
		DIFLUOROPHOSPHORIC ACID, ANHYDROUS (DOT)
		FLUOPHOSPHORIC ACID (F2(HO)PO)
		PHOSPHORODIFLUORIDIC ACID
		UN 1768 (DOT)
13780-03-5		
		CALCIUM BISULFIDE
	*	CALCIUM HYDROGEN SULFITE
13814-96-5		
		BORATE(1-), TETRAFLUORO-, LEAD (2+)
		BORATE(1-), TETRAFLUORO-, LEAD(2+) (2:1)
		LEAD BORON FLUORIDE
		LEAD FLUOBORATE
		LEAD FLUOBORATE (DOT)
		LEAD FLUOROBORATE
		LEAD TETRAFLUOROBORATE
		LEAD TETRAFLUOROBORATE (Pb(BF4)2)
		NA 2291 (DOT)
13820-41-2		
		AMMONIUM PLATINIC CHLORIDE
		AMMONIUM TETRACHLOROPLATINATE
13823-29-5		
		NITRIC ACID, THORIUM(4+) SALT
		NITRIC ACID, THORIUM(4+) SALT (8CI,9CI)
	*	THORIUM NITRATE
		THORIUM NITRATE ((Th(NO3)4))
		THORIUM NITRATE (DOT)
		THORIUM TETRANITRATE
		THORIUM(4+) NITRATE
		THORIUM(IV) NITRATE
		UN 2976 (DOT)
13825-74-6		
		TITANIUM SULFATE
13826-83-0		
		AMMONIUM BOROFLUORIDE
		AMMONIUM BOROFLUORIDE (NH4BF4)
		AMMONIUM FLUOBORATE
		AMMONIUM FLUOBORATE (DOT)
		AMMONIUM FLUOROBORATE
		AMMONIUM FLUOROBORATE (NH4BF4)
		AMMONIUM TETRAFLUOROBORATE
		AMMONIUM TETRAFLUOROBORATE (NH4BF4)
		AMMONIUM TETRAFLUOROBORATE(1-)
		BORATE(1-), TETRAFLUORO-, AMMONIUM
		NA 9088 (DOT)
13826-88-5		
		BORATE(1-), TETRAFLUORO-, ZINC
		BORATE(1-), TETRAFLUORO-, ZINC (2:1)
		ZINC BIS(TETRAFLUOROBORATE)
		ZINC BOROFLUORIDE
		ZINC BOROFLUORIDE [Zn(BF4)2]
		ZINC FLUOBORATE
		ZINC FLUOROBORATE
		ZINC TETRAFLUOROBORATE
		ZINC(2+) BIS[TETRAFLUOROBORATE(1-)]
		ZINC(II) FLUOBORATE
13838-16-9		
		ENFLURANE
13840-33-0		
		HYPOCHLOROUS ACID, LITHIUM SALT
		LITHIUM CHLORIDE OXIDE (LiClO)
	*	LITHIUM HYPOCHLORITE
		LITHIUM HYPOCHLORITE (LiClO)
		LITHIUM HYPOCHLORITE COMPOUND, DRY, CONTAINING

CAS No.	IN	Chemical Name
		MORE THAN 39% AVAILABLE CHLORINE (DOT)
		LITHIUM OXYCHLORIDE
		UN 1471 (DOT)
13863-41-7		
		BROMINE CHLORIDE
13889-92-4		
		CARBONOCHLORIDOTHIOIC ACID, S-PROPYL ESTER
		FORMIC ACID, CHLOROTHIO-, S-PROPYL ESTER
		PROPYL CHLOROTHIOFORMATE
		S-PROPYL CHLOROTHIOFORMATE
		S-PROPYL THIOCHLOROFORMATE
13952-84-6		
		1-METHYLPROPANAMINE
		1-METHYLPROPYLAMINE
		2-AB
		2-AMINOBUTANE
		2-BUTANAMINE
		2-BUTYLAMINE
		BUTAFUME
		SEC-BUTYLAMINE
		DECCOTANE
		TUTANE
13961-86-9		
		OLEIC ACID DIETHANOLAMINE CONDENSATE
13967-90-3		
		BARIUM BROMATE
		BARIUM BROMATE (DOT)
		BROMIC ACID, BARIUM SALT
		UN 2719 (DOT)
13987-01-4		
		1-PROPENE, TRIMER
		PROPENE, TRIMER
		PROPYLENE TRIMER
		TRIPROPYLENE
		TRIPROPYLENE (DOT)
		UN 2057 (DOT)
14017-41-5		
		COBALT SULFAMATE
		COBALTOUS SULFAMATE
		COBALTOUS SULFAMATE (DOT)
		NA 9105 (DOT)
		SULFAMIC ACID, COBALT SALT
		SULFAMIC ACID, COBALT(2+) SALT (2:1)
14018-95-2		
		CHROMIC ACID (H2Cr2O7), ZINC SALT (1:1)
		DICHROMIC ACID (H2Cr2O7), ZINC SALT (1:1)
		DICHROMIC ACID, ZINC SALT (1:1)
		ZINC BICHROMATE
		ZINC CHROMATE
		ZINC CHROMATE (ACGIH)
		ZINC CHROMATE (ZnCr2O7)
		ZINC CHROMIUM OXIDE
		ZINC CHROMIUM OXIDE (ZnCr2O7)
		ZINC DICHROMATE
		ZINC DICHROMATE (VI)
		ZINC DICHROMATE (ZnCr2O7)
		ZINC DICHROMATE(VI)
14019-91-1		
		CALCIUM SELENATE
		CALCIUM SELENATE (CaSeO4)
		SELENIC ACID (H2SeO4), CALCIUM SALT (1:1)
		SELENIC ACID, CALCIUM SALT (1:1)
14167-18-1		
		SALCOMINE
14216-75-2		
		NITRIC ACID, NICKEL SALT
14221-47-7		
		AMMONIUM FERRIC OXALATE
		AMMONIUM FERRIOXALATE
		AMMONIUM TRIOXALATOFERRATE(III)
		AMMONIUM TRIS(OXALATO)FERRATE(III)
		FERRATE(3-), TRIS(ETHANEDIOATO(2-)-O,O')-, TRIAMMONIUM, (OC-6-11)- (9CI)
		FERRATE(3-), TRIS(OXALATO)-, TRIAMMONIUM
		FERRATE(3-), TRIS[ETHANEDIOATO(2-)-O,O']-, TRIAMMONIUM, (OC-6-11)-
		FERRIC AMMONIUM OXALATE
		FERRIC AMMONIUM OXALATE (DOT)
		NA 9119 (DOT)
		TRIAMMONIUM TRIS(OXALATO)FERRATE(3-)
		TRIAMMONIUM TRIS-(ETHANEDIOATO(2-)-O,O')FERRATE(3-1)
14258-49-2		
		ETHANEDIOIC ACID, AMMONIUM SALT
		OXALIC ACID, AMMONIUM SALT
14264-31-4		
		COPPER SODIUM CYANIDE
		CUPRATE(2-), TRIS(CYANO-C)-, DISODIUM
		SODIUM CUPROCYANIDE SOLUTION
		SODIUM CUPROCYANIDE SOLUTION (DOT)
		SODIUM CUPROCYANIDE, SOLID
		SODIUM CUPROCYANIDE, SOLID (DOT)
		SODIUM TRICYANOCUPRATE(I)
		UN 2316 (DOT)
		UN 2317 (DOT)
14265-44-2		
		ORGANIC POSPHATE COMPOUND, SOLID (POISON B)
14293-73-3		
		DITHIONOUS ACID, DIPOTASSIUM SALT
		POTASSIUM DITHIONITE
		POTASSIUM HYDROSULFITE
		POTASSIUM HYDROSULPHITE (DOT)
		UN 1929 (DOT)
14293-86-8		
		POTASSIUM SELENATE
		POTASSIUM SELENATE (K2Se2O7)
		SELENIC ACID (H2Se2O7), DIPOTASSIUM SALT
		SELENIC ACID, DIPOTASSIUM SALT
14307-33-6		
		CALCIUM BICHROMATE
		CALCIUM CHROMATE (CaCr2O7)
		CALCIUM DICHROMATE
		CALCIUM DICHROMATE (CaCr2O7)
		CHROMIC ACID (H2Cr2O7), CALCIUM SALT (1:1)
		DICHROMIC ACID (H2Cr2O7), CALCIUM SALT (1:1)
		DICHROMIC(VI) ACID, CALCIUM SALT (1:1)
14307-35-8		
		CHROMIC ACID (H2CrO4), DILITHIUM SALT
		CHROMIUM LITHIUM OXIDE (CrLi2O4)
		DILITHIUM CHROMATE (Li2CrO4)
		LITHIUM CHROMATE
		LITHIUM CHROMATE (Li2CrO4)
		LITHIUM CHROMATE(VI)
14333-13-2		
		PERMANGANATE
		PERMANGANATE (MnO4-) ION
		PERMANGANATE (MNO4 1-)
		PERMANGANATE ION
		PERMANGANATE ION(1-)
		PERMANGANIC ACID (HMnO4), ION(1-)
14380-61-1		
		CHLORINE OXIDE (ClO 1-)
		HYPOCHLORITE
		HYPOCHLORITE ANION
		HYPOCHLORITE ION
		HYPOCHLORITE SOLUTION WITH MORE THAN 7% AVAILA-

CAS No.	IN	Chemical Name
		BLE CHLORINE BY WEIGHT
		HYPOCHLORITE SOLUTION WITH NOT MORE THAN 7% AVAILABLE CHLORINE BY WEIGHT
		HYPOCHLOROUS ACID, ION(1-)
14464-46-1		
		ALPHA-CRISTOBALITE
		ALPHA-CRYSTOBALITE
		CRISTOBALITE
		CRISTOBALITE (SiO2)
		CRISTOBALITE DUST
		CRISTOBALLITE
		CRYSTOBALITE
		CRYSTOBALLITE
		METACRISTOBALITE
		SILICA, CRISTOBALITE
		W 006
14484-64-1		
		AAFERTIS
		BERCEMA FERTAM 50
		CARBAMODITHIOIC ACID, DIMETHYL-, IRON COMPLEX
		DIMETHYLDITHIOCARBAMIC ACID IRON(3+) SALT
		FERBAM
		FERBAM 50
		FERBERK
		FERMATE
		FERRADOW
		FERRIC DIMETHYLDITHIOCARBAMATE
		FERRIC N,N-DIMETHYLDITHIOCARBAMATE
		FUKLASIN ULTRA
		FUKLAZIN
		HEXAFERB
		IRON DIMETHYLDITHIOCARBAMATE
		IRON, TRIS(DIMETHYLCARBAMODITHIOATO-S,S')-, (OC-6-11)-
		IRON, TRIS(DIMETHYLDITHIOCARBAMATO)-
		KARBAM BLACK
		LIROMATE
		STAUFFER FERBAM
		TRIS(DIMETHYLDITHIOCARBAMATO)IRON
		TRIS(N,N-DIMETHYLDITHIOCARBAMATO)IRON
		TRIS(N,N-DIMETHYLDITHIOCARBAMATO)IRON(III)
14486-19-2		
		BORATE(1-), TETRAFLUORO-, CADMIUM
		BORATE(1-), TETRAFLUORO-, CADMIUM (2:1)
		CADMIUM FLUOBORATE
		CADMIUM FLUORBORATE
		CADMIUM FLUOROBORATE
		CADMIUM TETRAFLUOROBORATE
		TL 1026
14489-25-9		
		CHROMIUM SULFATE
		CHROMOSULFURIC ACID
		CHROMOSULPHURIC ACID (DOT)
		CHRONISULFAT (GERMAN)
		SULFURIC ACID, CHROMIUM SALT
		UN 2240 (DOT)
14519-07-4		
		BROMIC ACID, ZINC SALT
		UN 2469 (DOT)
		ZINC BROMATE
		ZINC BROMATE (DOT)
14519-17-6		
		BROMIC ACID, MAGNESIUM SALT
		MAGNESIUM BROMATE
		MAGNESIUM BROMATE (DOT)
		UN 1473 (DOT)
14567-73-8		
		TREMOLITE
14639-97-5		
		AMMONIUM TETRACHLOROZINCATE
		AMMONIUM ZINC CHLORIDE (2NH4Cl.ZnCl2)
		DIAMMONIUM TETRACHLOROZINCATE
		DIAMMONIUM TETRACHLOROZINCATE(2-)
		ZINC CHLORIDE-AMMONIUM CHLORIDE COMPLEX (1:2)
		ZINCATE(2-), TETRACHLORO-, DIAMMONIUM
		ZINCATE(2-), TETRACHLORO-, DIAMMONIUM, (T-4)-
14639-98-6		
		AMMONIUM PENTACHLOROZINCATE
		AMMONIUM ZINC CHLORIDE ((NH4)3ZnCl5)
		TRIAMMONIUM PENTACHLOROZINCATE
		ZINCATE(3-), PENTACHLORO-, TRIAMMONIUM
14644-61-2		
		DISULFATOZIRCONIC ACID
		NA 9163 (DOT)
		SULFURIC ACID, ZIRCONIUM(4+) SALT (2:1)
		ZIRCONIUM DISULFATE
		ZIRCONIUM ORTHOSULFATE
		ZIRCONIUM SULFATE
		ZIRCONIUM SULFATE (1:2)
		ZIRCONIUM SULFATE (DOT)
		ZIRCONIUM SULPHATE
		ZIRCONIUM(IV) SULFATE (1:2)
		ZIRCONYL SULFATE
14666-78-5		
		DIETHYL PEROXYDICARBONATE
		DIETHYL PEROXYDICARBONATE (CONCENTRATION NOT MORE THAN 30%)
		DIETHYL PEROXYDICARBONATE, NOT MORE THAN 27% IN SOLUTION (DOT)
		DIETHYL PEROXYDIFORMATE
		ETHYL PEROXYCARBONATE
		PEROXYDICARBONIC ACID, DIETHYL ESTER
		PEROXYDICARBONIC ACID, DIETHYL ESTER, NOT MORE THAN 27% IN SOLUTION
		UN 2175 (DOT)
14674-72-7		
	*	CALCIUM CHLORITE
		CALCIUM CHLORITE (DOT)
		CHLOROUS ACID, CALCIUM SALT
		UN 1453 (DOT)
14684-25-4		
		BENZENE PHOSPHOROUS THIODICHLORIDE
		BENZENE PHOSPHORUS THIODICHLORIDE
		BENZENE PHOSPHORUS THIODICHLORIDE (DOT)
		DICHLORO(PHENYLTHIO)PHOSPHINE
		PHENYL PHOSPHORODICHLORIDOTHIOITE
		PHOSPHORODICHLORIDOTHIOUS ACID, PHENYL ESTER
		PHOSPHORODICHLORIDOTHIOUS ACID, PHENYL-
		S-PHENYL DICHLOROTHIOPHOSPHITE
		UN 2799 (DOT)
14686-13-6		
		(E)-2-HEPTENE
		2-HEPTENE, (E)-
		HEPTYLENE-2-TRANS
		TRANS-2-HEPTENE
14708-14-6		
		BORATE(1-), TETRAFLUORO-, NICKEL(2+)
		BORATE(1-), TETRAFLUORO-, NICKEL(2+) (2:1)
		BORATE(1-), TETRAFLUORO-, NICKEL(2+) (2:1) (9CI)
		NICKEL BOROFLUORIDE
		NICKEL FLUOROBORATE
		NICKEL TETRAFLUOROBORATE
		NICKEL TETRAFLUOROBORATE (Ni(BF4)2)
		NICKEL(2+) TETRAFLUOROBORATE(1-)
		NICKEL(II) FLUOBORATE
		NICKEL(II) TETRAFLUOROBORATE
		NICKELOUS TETRAFLUOROBORATE

CAS No.	IN	Chemical Name
14763-77-0		
	*	COPPER CYANIDE
14797-55-8		
		NITRATE
		NITRATE ION
		NITRATE ION (NO3-)
		NITRATE ION (NO3 1-)
		NITRATE ION(1-)
		NITRATE(1-)
		NITRATO
		NITRIC ACID, ION(1-)
14797-65-0		
		2,4-DINITRO-1-THIOCYANOBENZENE
		2,4-DINITRO-RHODANBENZOL (GERMAN)
		2,4-DINITROPHENYL THIOCYANATE
		2,4-DINITROTHIOCYANATOBENZENE
		2,4-DINITROTHIOCYANOBENZENE
		DNRB
		DNTB
		DRB
		GRYZBOL
		GRZYBOL
		NBT
		NIRIT
		NITRITE
		NITRITE ION
		NITRITE ION (NO2-)
		NITRITE ION(1-)
		NITROGEN DIOXIDE(1-)
		NITROGEN DIOXIDE, ION(1-)
		NITROGEN PEROXIDE ION(1-)
		NITROUS ACID, ION(1-)
		RHODANDINITROBENZOL
		RODATOX 60
		THIOCYANIC ACID, 2,4-DINITROPHENYL ESTER
		TRI-RODAZENE
		TRIRODAZEEN
14797-73-0		
		PERCHLORATE
		PERCHLORATE ION
		PERCHLORATE ION (ClO4 1-)
		PERCHLORATE ION(1-)
		PERCHLORATE(1-)
		PERCHLORIC ACID, ION(1-)
14807-96-6		
		AGALITE
		ASBESTINE
		BEAVER WHITE 200
		CP 10-40
		CP 38-33
		CRYSTALITE CRS 6002
		DESERTALC 57
		EMTAL 500
		EMTAL 549
		EMTAL 596
		EMTAL 599
		EX-II
		FIBRENE C 400
		FW-XO
		IT EXTRA
		LMR 100
		MICRON WHITE 5000S
		MICRONEECE K 1
		MICROTALCO IT EXTRA
		MISTRON 139
		MISTRON 2SC
		MISTRON FROST P
		MISTRON RCS
		MISTRON STAR
		MISTRON SUPER FROST
		MISTRON VAPOR
		MP 12-50
		MP 25-38
		MP 40-27
		MP 45-26
		MUSSOLINITE
		NYTAL 200
		NYTAL 400
		POLYTAL 4641
		POLYTAL 4725
		STEAWHITE
		SUPREME
		SUPREME DENSE
	*	TALC
		TALC (FIBROUS) DUST
		TALC (Mg3H2(SiO3)4)
		TALCAN PK-P
		TALCRON CP 44-31
		TALCUM
14808-60-7		
		ALPHA-QUARTZ
		AVENTURINE
		AVENTURINE (QUARTZ)
		BETA-QUARTZ
		CRYSTALITE AA
		DQ 12
		MARSHALITE
		PLASTORIT
		Q-CEL
		QUARTZ
		QUARTZ (SiO2)
		ROCK CRYSTAL
		SF 35
		SIDERITE (SiO2)
		SIKRON H 200
		SIKRON H 600
		SILICA, QUARTZ
		TGL 16319
		TIGER-EYE
14861-06-4		
		2-BUTENOIC ACID, ETHENYL ESTER
		CROTONIC ACID, VINYL ESTER
		VINYL 2-BUTENOATE
		VINYL CROTONATE
14866-68-3		
		CHLORATE
		CHLORATE ION
		CHLORATE ION (ClO3-)
		CHLORATE(1-)
		CHLORIC ACID, ION(1-)
		CHLORINE OXIDE (ClO3 1-)
14901-08-7		
		BETA-D-GLUCOPYRANOSIDE, (METHYL-ONN-AZOXY)-METHYL
		CYCASIN
		METHYLAZOXYMETHANOL BETA-D-GLUCOSIDE
		METHYLAZOXYMETHANOL GLUCOSIDE
14915-07-2		
		DIOXYGEN ION(2-)
		PEROXIDE
		PEROXIDE (O2 2-)
		PEROXIDE ION
		PEROXIDE, INORGANIC
14977-61-8		
		CHLORURE DE CHROMYLE (FRENCH)
		CHROMIC OXYCHLORIDE
		CHROMIUM (VI) DIOXYCHLORIDE
		CHROMIUM CHLORIDE OXIDE
		CHROMIUM CHLORIDE OXIDE (CrCl2O2)
		CHROMIUM DICHLORIDE DIOXIDE
		CHROMIUM DIOXIDE DICHLORIDE
		CHROMIUM OXYCHLORIDE
		CHROMIUM OXYCHLORIDE (CrO2Cl2)
		CHROMIUM OXYCHLORIDE (DOT)
		CHROMIUM, DICHLORODIOXO-
		CHROMIUM, DICHLORODIOXO-, (T-4)-
		CHROMOXYCHLORID (GERMAN)

CAS No.	IN	Chemical Name
		CHROMYL CHLORIDE CHROMYL CHLORIDE (ACGIH,DOT) CHROMYL CHLORIDE (CrO2Cl2) CHROMYLCHLORID (GERMAN) CHROOMOXYCHLORIDE (DUTCH) CROMILE, CLORURO DI (ITALIAN) CROMO, OSSICLORURO DI (ITALIAN) DICHLORODIOXOCHROMIUM DIOXODICHLOROCHROMIUM OXYCHLORURE CHROMIQUE (FRENCH) UN 1758 (DOT)
14998-27-7		CHLORINE DIOXIDE ION(1-) CHLORITE CHLORITE (MINERAL CLASS) CHLORITE ION CHLORITE, INORGANIC CHLORITE-GROUP MINERALS CHLORITE-TYPE MINERALS CHLOROUS ACID, ION(1-) GREENITE HYDROCHLORITE MINERALS, CHLORITE-GROUP SIERRALITE ZINCIAN CHLORITE
15096-52-3		CRYOLITE SODIUM ALUMINUM FLUORIDE
15159-40-7		4-(CHLOROFORMYL) MORPHOLINE
15191-85-2		BERYLLIUM ORTHOSILICATE BERYLLIUM ORTHOSILICATE (Be2SiO4) BERYLLIUM SILICATE BERYLLIUM SILICATE (Be2SiO4) BERYLLIUM SILICON OXIDE (Be2SiO4) DIBERYLLIUM MONOSILICATE SILICIC ACID (H4SiO4), BERYLLIUM SALT (1:2)
15195-06-9		STRONTIUM ARSENITE
15245-44-0		LEAD 2,4,6-TRINITRORESORCINOXIDE LEAD STYPHNATE
15271-41-7		2-EXO-CHLORO-6-ENDO-CYANO-2-NORBORNANONE-O-(METHYLCARBAMOYL)OXIME 2-NORBORNANECARBONITRILE, 5-CHLORO-6-OXO-, O-(METHYLCARBAMOYL) OXIME, (E)-ENDO-2,EXO-5- 2-NORBORNANONE, ENDO-3-CHLORO-EXO-6-CYANO-, O-(METHYLCARBAMOYL)OXIME 3-CHLORO-6-CYANO-2-NORBORNANONE-O-(METHYLCARBAMOYL)OXIME 3-CHLORO-6-CYANONORBORNANONE-2 OXIME O,N-METHYLCARBAMATE 5-CHLORO-6-((((METHYLAMINO)CARBONYL)OXY)IMINO)BICYCLO(221)HEPTANE-2-CARBONITRILE BICYCLOHEPTANE-2-CARBONITRILE, 5-CHLORO-6-((((METHYLAMINO)CARBONYL)OXY)IMINO)- BICYCLO[221]HEPTANE-2-CARBONITRILE, 5-CHLORO-6-[[[(METHYLAMINO)CARBONYL]OXY]IMINO]-, [1S-(1ALPHA,2BE COMPOUND UC-20047 A ENDO-3-CHLORO-EXO-6-CYANO-2-NORBORNANONE O-(METHYLCARBAMOYL)OXIME ENT 25,962 EXO-5-CHLORO-6-OXO-ENDO-2-NORBORNANECARBONITRILE O-(METHYLCARBAMOYL)OXIME TRANID UC 20,047A UC 20047 UC 26089 UNION CARBIDE UC 20047
15385-57-6		DIMERCURY DIIODIDE IODURE DE MERCURE (FRENCH) MERCUROUS IODIDE MERCUROUS IODIDE, SOLID MERCUROUS IODIDE, SOLID (DOT) MERCURY IODIDE (Hg2I2) MERCURY PROTOIODIDE UN 1638 (DOT) YELLOW MERCURY IODIDE
15385-58-7		DIMERCURY DIBROMIDE MERCUROUS BROMIDE MERCUROUS BROMIDE (DOT) MERCUROUS BROMIDE (Hg2Br2) MERCUROUS BROMIDE, SOLID MERCUROUS BROMIDE, SOLID (DOT) MERCURY BROMIDE MERCURY BROMIDE (Hg2Br2) MERCURY MONOBROMIDE MERCURY(1)BROMIDE(1:1) MERCURY(1+) BROMIDE MERCURY(I) BROMIDE (1:1) UN 1634 (DOT)
15468-32-3		ALPHA-TRIDYMITE CHRISTENSENITE SILICA, TRIDYMITE TRIDYMITE TRIDYMITE (SiO2) TRIDYMITE DUST
15512-36-4		CALCIUM BISULFITE CALCIUM BISULFITE SOLUTION (DOT) CALCIUM DITHIONITE CALCIUM HYDROGEN SULFITE SOLUTION CALCIUM HYDROGEN SULFITE SOLUTION (DOT) CALCIUM HYDROSULFITE CALCIUM HYDROSULPHITE (DOT) DITHIONOUS ACID, CALCIUM SALT (1:1) NA 2693 (DOT) UN 1923 (DOT)
15520-11-3		BIS(4-TERT-BUTYLCYCLOHEXYL) PEROXYDICARBONATE CYCLOHEXANOL, 4-TERT-BUTYL-, PEROXYDICARBONATE (2:1) DI(4-TERT -BUTYLCYCLOHEXYL) PEROXYDICARBONATE DI(4-TERT-BUTYLCYCLOHEXYL) PEROXYDICARBONATE DI-(4-TERT-BUTYLCYCLOHEXYL)PEROXYDICARBONATE, NOT MORE THAN 42% IN WATER (DOT) DI-(4-TERT-BUTYLCYCLOHEXYL)PEROXYDICARBONATE, TECHNICAL PURE (DOT) DI-(4-TERT-BUTYLCYCLOHEXYL)PEROXYDICARBONATE, TECHNICALLY PURE DI-TERT-BUTYLDICYCLOHEXYL PEROXYDICARBONATE PERCADOX 16 PERKADOX 16 PERKADOX 16W40 PEROXYCARBONIC ACID, BIS(4-TERT-BUTYLCYCLOHEXYL) ESTER, NOT MORE THAN 42% IN WATER PEROXYDICARBONIC ACID, BIS(4-TERT-BUTYLCYCLOHEXYL) ESTER PEROXYDICARBONIC ACID, BIS[4-(1,1-DIMETHYLETHYL)-CYCLOHEXYL] ESTER UN 2154 (DOT) UN 2894 (DOT)
15545-97-8		2,2'-AZODI-(2,4-DIMETHYL-4-METHOXYVALERONITRILE)
15593-29-0	*	DISODIUM PERSULFATE PEROXYMONOSULFURIC ACID, DISODIUM SALT PERSULFATE DE SODIUM (FRENCH)

CAS No.	IN	Chemical Name
		SODIUM MONOPERSULFATE
		SODIUM PEROXYMONOSULFATE (Na2(SO5))
		SODIUM PEROXYMONOSULFATE (NaHSO5)
		SODIUM PERSULFATE
		SODIUM PERSULFATE (ACGIH,DOT)
		UN 1505 (DOT)
15630-89-4		
		CARBONIC ACID DISODIUM SALT, COMPD WITH HYDROGEN PEROXIDE (2:3)
		CARBONIC ACID DISODIUM SALT, COMPD WITH HYDROGEN PEROXIDE (H2O2) (2:3)
		CARBONOPEROXOIC ACID, DISODIUM SALT
		CARBONOPEROXOIC ACID, DISODIUM SALT (9CI)
		DISODIUM CARBONATE COMPOUND WITH HYDROGEN PEROXIDE (2:3)
		DISODIUM PEROXYDICARBONATE
		FB SODIUM PERCARBONATE
		HYDROGEN PEROXIDE (H2O2), COMPD WITH DISODIUM CARBONATE (3:2)
		HYDROGEN PEROXIDE, COMPD WITH DISODIUM CARBON-ATE (3:2)
		OXYPER
		PERDOX
		PEROXY SODIUM CARBONATE
		PEROXYDICARBONIC ACID, DISODIUM SALT
		SODIUM CARBONATE
		SODIUM CARBONATE (Na2C2O6)
		SODIUM CARBONATE PEROXIDE
		SODIUM CARBONATE-HYDROGEN PEROXIDE (2:3)
		SODIUM CARBONATE-HYDROGEN PEROXIDE ADDUCT
		SODIUM PERCARBONATE
		SODIUM PERCARBONATE (CARBONIC ACID DISODIUM SALT, COMPD. W/H2O2)
		SODIUM PERCARBONATE (DOT)
		SODIUM PERCARBONATE (Na2C2O6)
		SODIUM PERCARBONATE (Na2CO3 15H2O2)
		SODIUM PEROXYCARBONATE
		SODIUM PEROXYCARBONATE (Na2CO4)
		SODIUM PEROXYDICARBONATE
		SODIUM PEROXYDICARBONATE (Na2C2O6)
		SODIUM PEROXYMONOCARBONATE
		UN 2467 (DOT)
15663-27-1		
		BIOCISPLATINUM
		CIS-DDP
		CIS-DIAMINODICHLOROPLATINUM(II)
		CIS-DIAMMINEDICHLOROPLATINUM
		CIS-DIAMMINEDICHLOROPLATINUM(II)
		CIS-DICHLORODIAMMINEPLATINUM
		CIS-DICHLORODIAMMINEPLATINUM(II)
		CIS-DPP
		CIS-PLATINE
		CIS-PLATINOUS DIAMMINODICHLORIDE
		CIS-PLATINUM
		CIS-PLATINUM DIAMINODICHLORIDE
		CIS-PLATINUM II
		CIS-PLATINUM(II) DIAMINODICHLORIDE
		CIS-PLATINUM(II) DIAMMINEDICHLORIDE
		CIS-PLATINUMDIAMINE DICHLORIDE
		CIS-PLATINUMDIAMMINE DICHLORIDE
		CISPLATIN
		CISPLATINUM
		CPDD
		NEOPLATIN
		NSC 119875
		PLATIDIAM
		PLATINOL
		PLATINUM, DIAMMINEDICHLORO-, (SP-4-2)-
		PLATINUM, DIAMMINEDICHLORO-, CIS-
15699-18-0		
		AMMONIUM DISULFATONICKELATE(II)
		AMMONIUM NICKEL SULFATE
		AMMONIUM NICKEL SULFATE ((NH4)2Ni(SO4)2)
		NA 9138 (DOT)
	*	NICKEL AMMONIUM SULFATE
		NICKEL AMMONIUM SULFATE (DOT)
		NICKEL AMMONIUM SULFATE (Ni(NH4)2(SO4)2)
		SULFURIC ACID, AMMONIUM NICKEL(2+) SALT (2:2:1)
15739-80-7		
		ANGLISLITE
		BLEISULFAT (GERMAN)
		CI 77630
		CI PIGMENT WHITE 3
		FAST WHITE
		FREEMANS WHITE LEAD
		LEAD BOTTOMS
		LEAD DROSS (DOT)
		LEAD MONOSULFATE
		LEAD SULFATE
		LEAD SULFATE (1:1)
		LEAD SULFATE (DOT)
		LEAD SULFATE (PbSO4)
		LEAD SULFATE, SOLID, CONTAINING MORE THAN 3% FREE ACID (DOT)
		LEAD(2+) SULFATE
		LEAD(II) SULFATE
		LEAD(II) SULFATE (1:1)
		MILK WHITE
		MULHOUSE WHITE
		NA 2291 (DOT)
		SULFATE DE PLOMB (FRENCH)
		SULFURIC ACID, LEAD SALT
		SULFURIC ACID, LEAD(2+) SALT (1:1)
		UN 1794 (DOT)
15825-70-4		
		MANNITOL HEXANITRATE
15829-53-5		
		MERCUROUS OXIDE
		MERCUROUS OXIDE (Hg2O)
		MERCUROUS OXIDE, BLACK, SOLID
		MERCUROUS OXIDE, BLACK, SOLID (DOT)
		MERCURY OXIDE
		MERCURY OXIDE (Hg2O)
		MERCURY OXIDE BLACK
		MERCURY(I) OXIDE
		QUECKSILBEROXID (GERMAN)
		UN 1641 (DOT)
15845-52-0		
		LEAD PHOSPHITE, DIBASIC
15905-86-9		
		NITRIC ACID, URANIUM SALT
	*	URANIUM NITRATE
15950-66-0		
		DOWICIDE 2
		DOWICIDE 2S
		NA 2020 (DOT)
		NCI-C02904
		OMAL
		PHENACHLOR
		PHENOL, TRICHLORO-
		PREVENTOL I
		RCRA WASTE NUMBER U231
		TCP
		TRICHLOROPHENOL
		TRICHLOROPHENOL (DOT)
15980-15-1		
		1,4-OXATHIANE
		1,4-OXATHIIN, 2,3,5,6-TETRAHYDRO-
		1,4-THIOXANE
		1-OXA-4-THIACYCLOHEXANE
		P-THIOXANE
16039-52-4		
		COPPER LACTATE
		COPPER, BIS(2-HYDROXYPROPANOATO-O1,O2)-
		COPPER, BIS(LACTATO)-
		PROPANOIC ACID, 2-HYDROXY-, COPPER COMPLEX

CAS No.	IN	Chemical Name
16065-83-1		CHROMIC ION
		CHROMIUM ($Cr3+$)
		CHROMIUM ION ($Cr3+$)
		CHROMIUM ION(3+)
		CHROMIUM(3+)
		CHROMIUM(III)
		CHROMIUM(III) CATION
		CHROMIUM(III) ION
		CHROMIUM, ION ($Cr3+$)
16066-38-9		DI-N-PROPYL PEROXYDICARBONATE
		DI-N-PROPYL PEROXYDICARBONATE (CONCENTRATION NOT MORE THAN 80%)
		DI-N-PROPYL PEROXYDICARBONATE, TECHNICALLY PURE
		DI-N-PROPYL PEROXYDICARBONATE, TECHNICALLY PURE (DOT)
		DIPROPYL PEROXYDICARBONATE
		LUPERSOL 221
		N-PROPYL PERCARBONATE
		PEROXYDICARBONIC ACID, DIPROPYL ESTER
		PROPYL PEROXYDICARBONATE
		UN 2176 (DOT)
16071-86-6		AIZEN PRIMULA BROWN BRLH
		AIZEN PRIMULA BROWN PLH
		AMANIL FAST BROWN BRL
		AMANIL SUPRA BROWN LBL
		ATLANTIC FAST BROWN BRL
		ATLANTIC RESIN FAST BROWN BRL
		BELAMINE FAST BROWN BRLL
		BENZANIL SUPRA BROWN BRLL
		BENZANIL SUPRA BROWN BRLN
		BROWN 4EMBL
		C.I. 30145
		C.I. DIRECT BROWN 95
		CALCODUR BROWN BRL
		CHLORAMINE FAST BROWN BRL
		CHLORAMINE FAST CUTCH BROWN PL
		CHLORANTINE FAST BROWN BRLL
		CHROME LEATHER BROWN BRLL
		CHROME LEATHER BROWN BRSL
		CUPROFIX BROWN GL
		DERMA FAST BROWN W-GL
		DERMAFIX BROWN PL
		DIALUMINOUS BROWN BRS
		DIAPHTAMINE LIGHT BROWN BRLL
		DIAZINE FAST BROWN RSL
		DIAZOL LIGHT BROWN BRN
		DICOREL BROWN LMR
		DIPHENYL FAST BROWN BRL
		DIRECT BROWN 95
		DIRECT BROWN 95 (TECHNICAL GRADE)
		DIRECT BROWN BRL
		DIRECT FAST BROWN BRL
		DIRECT FAST BROWN LMR
		DIRECT LIGHT BROWN BRS
		DIRECT LIGHTFAST BROWN M
		DIRECT SUPRA LIGHT BROWN ML
		DURAZOL BROWN BR
		DUROFAST BROWN BRL
		ELIAMINA LIGHT BROWN BRL
		ENIANIL LIGHT BROWN BRL
		FASTOLITE BROWN BRL
		FASTUSOL BROWN LBRSA
		FASTUSOL BROWN LBRSN
		FENALUZ BROWN BRL
		HELION BROWN BRSL
		HISPALUZ BROWN BRL
		KAYARUS SUPRA BROWN BRS
		KCA LIGHT FAST BROWN BR
		PARANOL FAST BROWN BRL
		PEERAMINE FAST BROWN BRL
		PONTAMINE FAST BROWN BRL
		PONTAMINE FAST BROWN NP
		PYRAZOL FAST BROWN BRL
		PYRAZOLINE BROWN BRL
		SATURN BROWN LBR
		SIRIUS SUPRA BROWN BR
		SIRIUS SUPRA BROWN BRL
		SIRIUS SUPRA BROWN BRS
		SOLANTINE BROWN BRL
		SOLAR BROWN PL
		SOLEX BROWN R
		SOLIUS LIGHT BROWN BRLL
		SOLIUS LIGHT BROWN BRS
		SUMILIGHT SUPRA BROWN BRS
		SUPRAZO BROWN BRL
		SUPREXCEL BROWN BRL
		TERTRODIRECT FAST BROWN BR
		TETRAMINE FAST BROWN BRDN EXTRA
		TETRAMINE FAST BROWN BRP
		TETRAMINE FAST BROWN BRS
		TRIANTINE BROWN BRS
		TRIANTINE FAST BROWN OG
		TRIANTINE FAST BROWN OR
		TRIANTINE LIGHT BROWN BRS
		TRIANTINE LIGHT BROWN OG
16111-62-9		BIS(2-ETHYLHEXYL) PERDICARBONATE
		BIS(2-ETHYLHEXYL) PEROXYDICARBONATE
		BIS(ETHYLHEXYL) PEROXYDICARBONATE
		DI(2-ETHYLHEXYL) PEROXYDICARBONATE
		DI(2-ETHYLHEXYL)PEROXYDICARBONATE
		DI(2-ETHYLHEXYL)PEROXYDICARBONATE, 77% IN SOLUTION (DOT)
		DI-(2-ETHYLHEXYL) PEROXYDICARBONATE, TECHNICALLY PURE
		DI-(2-ETHYLHEXYL)PEROXYDICARBONATE, MAXIMUM CONCENTRATION 32% (DOT)
		DI-(2-ETHYLHEXYL)PEROXYDICARBONATE, TECHNICAL PURE (DOT)
		DI-2-ETHYLHEXYL PEROXYDICARBONATE
		ESPERCARB 840M
		PEROXYDICARBONIC ACID, BIS(2-ETHYLHEXYL) ESTER
		PEROXYDICARBONIC ACID, DI(2-ETHYLHEXYL) ESTER
		PEROXYDICARBONIC ACID, DI(2-ETHYLHEXYL) ESTER, NOT MORE THAN 77% IN SOLUTION
		UN 2122 (DOT)
		UN 2123 (DOT)
		UN 2960 (DOT)
16215-49-9		BUTYL PEROXYDICARBONATE
16219-75-3		2-NORBORNENE, 5-ETHYLIDENE-
		5-ETHYLIDENE-2-NORBORNENE
		5-ETHYLIDENEBICYCLO(221)HEPT-2-ENE
		BICYCLO[221]HEPT-2-ENE, 5-ETHYLIDENE-
		ETHYLIDENE NORBORNENE
		ETHYLIDENE NORBORNENE (ACGIH)
16337-84-1		CARBONIC ACID, NICKEL SALT
		NICKEL CARBONATE
16543-55-8		1'-NITROSONORNICOTINE
		N'-NITROSONORNICOTINE
		N-NITROSONORNICOTINE
		NICOTINE, 1'-DEMETHYL-1'-NITROSO-
		NITROSONORNICOTINE
		NORNICOTINE, N-NITROSO-
		PYRIDINE, 3-(1-NITROSO-2-PYRROLIDINYL)-, (S)-
16568-02-8		ACETALDEHYDE METHYLFORMYLHYDRAZONE
		ACETALDEHYDE N-FORMYL-N-METHYLHYDRAZONE
		ACETALDEHYDE-N-METHYL-N-FORMYLHYDRAZONE
		ETHYLIDENE GYROMITRIN
		FORMIC ACID, ETHYLIDENEMETHYLHYDRAZIDE
		GYROMITRIN
		HYDRAZINECARBOXALDEHYDE, ETHYLIDENEMETHYL-

CAS No.	IN	Chemical Name
16721-80-5		
		HYDROGEN SODIUM SULFIDE
		NA 2922 (DOT)
		NA 2923 (DOT)
		SODIUM BISULFIDE
		SODIUM HYDROGEN SULFIDE
		SODIUM HYDROGEN SULFIDE (NaHS)
	*	SODIUM HYDROSULFIDE
		SODIUM HYDROSULFIDE (Na(HS))
		SODIUM HYDROSULFIDE, SOLID, WITH 25% WATER OF CRYSTALLIZATION
		SODIUM HYDROSULFIDE, SOLUTION (DOT)
		SODIUM HYDROSULPHIDE, SOLID (DOT)
		SODIUM HYDROSULPHIDE, WITH LESS THAN 25% WATER OF CRYSTALLIZATION (DOT)
		SODIUM MERCAPTAN
		SODIUM MERCAPTIDE
		SODIUM SULFHYDRATE
		SODIUM SULFIDE
		SODIUM SULFIDE (NA(SH))
		UN 2318 (DOT)
16731-55-8		
		DIPOTASSIUM DISULFITE
		DIPOTASSIUM METABISULFITE
		DIPOTASSIUM PYROSULFITE
		DISULFUROUS ACID, DIPOTASSIUM SALT
		NA 2693 (DOT)
		POTASSIUM METABISULFITE
		POTASSIUM METABISULFITE (DOT)
		POTASSIUM METABISULFITE (K2S2O5)
		POTASSIUM PYROSULFITE
		POTASSIUM PYROSULFITE (K2S2O5)
		PYROSULFUROUS ACID, DIPOTASSIUM SALT
16752-77-5		
		DU PONT 1179
		DU PONT INSECTICIDE 1179
		ETHANIMIDOTHIOIC ACID, N-[[(METHYLAMINO)CARBONYL]OXY]-, METHYL ESTER
		IN 1179
		LANNATE
		LANNATE 90SP
		MESOMILE
	*	METHOMYL
		METHOMYL SP
		METHYL N-[(METHYLCARBAMOYL)OXY]THIOACETIMIDATE
		METHYL O-(METHYLCARBAMOYL)THIOLACETOHYDROXAMATE
		METHYL O-(METHYLCARBAMYL)THIOLACETOHYDROXAMATE
		NUDRIN
		SD 14999
		WL 18236
16842-03-8		
		COBALT HYDROCARBONYL
		COBALT HYDROCARBONYL [CoH(CO)4]
		COBALT TETRACARBONYL HYDRIDE
		COBALT, TETRACARBONYLHYDRO-
		HYDRIDOCOBALT TETRACARBONYL
		HYDRIDOTETRACARBONYLCOBALT
		HYDROCOBALT TETRACARBONYL
		TETRACARBONYLHYDRIDOCOBALT
		TETRACARBONYLHYDROCOBALT
16853-85-3		
		ALUMINATE (1-), TETRAHYDRO-, LITHIUM
		ALUMINATE(1-), TETRAHYDRO-, LITHIUM, (T-4)-
		ALUMINATE(1-), TETRAHYDRO-, LITHIUM, (T-4)- (9CI)
		ALUMINUM LITHIUM HYDRIDE
		ALUMINUM LITHIUM HYDRIDE (LiAlH4)
		ALUMINUM LITHIUM TETRAHYDRIDE
		LITHIUM ALANATE
		LITHIUM ALUMINOHYDRIDE
	*	LITHIUM ALUMINUM HYDRIDE
		LITHIUM ALUMINUM HYDRIDE (DOT)
		LITHIUM ALUMINUM HYDRIDE (LiAlH4)
		LITHIUM ALUMINUM HYDRIDE, ETHEREAL
		LITHIUM ALUMINUM HYDRIDE, ETHEREAL (DOT)
		LITHIUM ALUMINUM TETRAHYDRIDE
		LITHIUM TETRAHYDRIDOALUMINATE
		LITHIUM TETRAHYDROALUMINATE
		LITHIUM TETRAHYDROALUMINATE (AlLiH4)
		LITHIUM TETRAHYDROALUMINATE(1-)
		UN 1410 (DOT)
		UN 1411 (DOT)
16871-71-9		
		FLUAT-VOGEL
		FUNGOL
		FUNGONIT GF 2
		SILICATE(2-), HEXAFLUORO-, ZINC
		SILICATE(2-), HEXAFLUORO-, ZINC (1:1)
		SILICON ZINC FLUORIDE
		SILICON ZINC FLUORIDE (ZnSiF6)
		UN 2855 (DOT)
		ZINC FLUOROSILICATE
		ZINC FLUOSILICATE
		ZINC HEXAFLUOROSILICATE
		ZINC HEXAFLUOROSILICATE (ZnSiF6)
		ZINC HEXAFLUOROSILICATE(2-)
		ZINC SILICOFLUORIDE
		ZINC SILICOFLUORIDE (DOT)
16871-90-2		
		DIPOTASSIUM HEXAFLUOROSILICATE
		DIPOTASSIUM HEXAFLUOROSILICATE(2-)
		POTASSIUM FLUOROSILICATE
		POTASSIUM FLUOROSILICATE (K2SiF6)
		POTASSIUM FLUOROSILICATE, SOLID
		POTASSIUM FLUOSILICATE
		POTASSIUM HEXAFLUOROSILICATE
		POTASSIUM HEXAFLUOROSILICATE (K2SiF6)
	*	POTASSIUM SILICOFLUORIDE
		POTASSIUM SILICOFLUORIDE (DOT)
		POTASSIUM SILICOFLUORIDE (K2SiF6)
		POTASSIUM SILICON FLUORIDE (K2SiF6)
		SILICATE(2-), HEXAFLUORO-, DIPOTASSIUM
		UN 2655 (DOT)
16872-11-0		
	*	FLUOBORIC ACID
16893-85-9		
		DESTRUXOL APPLEX
		DISODIUM HEXAFLUOROSILICATE
		DISODIUM HEXAFLUOROSILICATE (2-)
		DISODIUM HEXAFLUOROSILICATE (Na2SiF6)
		DISODIUM SILICOFLUORIDE
		ENS-ZEM WEEVIL BAIT
		ENT 1,501
		FLUOSILICATE DE SODIUM
		NATRIUMSILICOFLUORID (GERMAN)
		ORTHO EARWIG BAIT
		ORTHO WEEVIL BAIT
		PRODAN
		PRODAN (PESTICIDE)
		PSC CO-OP WEEVIL BAIT
		SAFSAN
		SALUFER
		SILICATE(2-), HEXAFLUORO-, DISODIUM
		SILICON SODIUM FLUORIDE
		SILICON SODIUM FLUORIDE (Na2SiF6)
	*	SODIUM FLUOROSILICATE
		SODIUM FLUOROSILICATE (Na2SiF6)
		SODIUM FLUOSILICATE
		SODIUM FLUOSILICATE (Na2(SiF6))
		SODIUM HEXAFLUOROSILICATE
		SODIUM HEXAFLUOROSILICATE (Na2SiF6)
		SODIUM HEXAFLUOSILICATE
		SODIUM SILICOFLUORIDE
		SODIUM SILICOFLUORIDE (DOT)
		SODIUM SILICOFLUORIDE (Na2SiF6)
		SODIUM SILICOFLUORIDE, SOLID
		SODIUM SILICON FLUORIDE
		SODIUM SILICON FLUORIDE (Na2SiF6)
		SUPER PRODAN

CAS No.	IN	Chemical Name
		UN 2674 (DOT)
16901-76-1		
		NITRIC ACID, THALLIUM SALT
		NITRIC ACID, THALLIUM(1+) SALT
		RCRA WASTE NUMBER U217
		THALLIC NITRATE
		THALLIUM MONONITRATE
		THALLIUM NITRATE
		THALLIUM NITRATE (DOT)
		THALLIUM NITRATE (Tl(NO3))
		THALLIUM(1+) NITRATE
		THALLIUM(I) NITRATE
		THALLIUM(I) NITRATE (1:1)
		UN 2727 (DOT)
16919-19-0		
		AMMONIUM FLUOROSILICATE
		AMMONIUM FLUOROSILICATE ((NH4)2SiF6)
		AMMONIUM FLUOSILICATE
		AMMONIUM HEXAFLUOROSILICATE
		AMMONIUM SILICOFLUORIDE
		AMMONIUM SILICOFLUORIDE (DOT)
		AMMONIUM SILICON FLUORIDE ((NH4)2SiF6)
		CRYPTOHALITE
		DIAMMONIUM FLUOSILICATE ((NH4)2SiF6)
		DIAMMONIUM HEXAFLUOROSILICATE
		DIAMMONIUM HEXAFLUOROSILICATE(2-)
		DIAMMONIUM SILICON HEXAFLUORIDE
		FLUOSILICATE DE AMMONIUM (FRENCH)
		LPE 6
		SILICATE(2-), HEXAFLUORO-, DIAMMONIUM
		UN 2854 (DOT)
16919-58-7		
		1-HEXADECANAMINIUM, N-ETHYL-N,N-DIMETHYL-, BROMIDE (9CI)
		AMMONIUM CHLOROPLATINATE
		AMMONIUM HEXACHLOROPLATINATE
		AMMONIUM HEXACHLOROPLATINATE(IV)
		AMMONIUM PLATINIC CHLORIDE
		DIAMMONIUM HEXACHLOROPLATINATE
		DIAMMONIUM HEXACHLOROPLATINATE(2-)
		DIAMMONIUM PLATINUM HEXACHLORIDE
		PLATINATE(2-), HEXACHLORO-, DIAMMONIUM
		PLATINATE(2-), HEXACHLORO-, DIAMMONIUM, (OC-6-11)-
		PLATINATE(2-), HEXACHLORO-, DIAMMONIUM, (OC-6-11)- (9CI)
		PLATINIC AMMONIUM CHLORIDE
		QUATERNIUM-17
16921-30-5		
	*	POTASSIUM HEXACHLOROPLATINATE
16923-95-8		
		DIPOTASSIUM HEXAFLUOROZIRCONATE
		DIPOTASSIUM HEXAFLUOROZIRCONATE(2-)
		DIPOTASSIUM ZIRCONIUM HEXAFLUORIDE
		NA 9162 (DOT)
		POTASSIUM FLUOROZIRCONATE
		POTASSIUM FLUOROZIRCONATE (K2ZrF6)
		POTASSIUM FLUOZIRCONATE
		POTASSIUM HEXAFLUOROZIRCONATE
		POTASSIUM HEXAFLUOROZIRCONATE (K2ZrF6)
		POTASSIUM HEXAFLUOROZIRCONATE(IV)
		POTASSIUM ZIRCONIUM FLUORIDE
		POTASSIUM ZIRCONIUM FLUORIDE (K2ZrF6)
		POTASSIUM ZIRCONIUM HEXAFLUORIDE
		ZIRCONATE(2-), HEXAFLUORO-, DIPOTASSIUM
		ZIRCONATE(2-), HEXAFLUORO-, DIPOTASSIUM, (OC-6-11)-
		ZIRCONIUM POTASSIUM FLUORIDE
		ZIRCONIUM POTASSIUM FLUORIDE (DOT)
16940-66-2		
		BORATE(1-), TETRAHYDRO-, SODIUM
		BOROHYDRURE DE SODIUM (FRENCH)
		BOROL
		HIDKITEX DF
		NABH4
		SODIUM BOROHYDRATE
	*	SODIUM BOROHYDRIDE
		SODIUM BOROHYDRIDE (DOT)
		SODIUM BOROHYDRIDE (Na(BH4))
		SODIUM HYDROBORATE
		SODIUM TETRAHYDROBORATE
		SODIUM TETRAHYDROBORATE(1-)
		UN 1426 (DOT)
16940-81-1		
		HEXAFLUOROPHOSPHORIC ACID
		HEXAFLUOROPHOSPHORIC ACID (DOT)
		HYDROGEN HEXAFLUOROPHOSPHATE
		HYDROGEN HEXAFLUOROPHOSPHATE(1-)
		PHOSPHATE(1-) HEXAFLUORO-, HYDROGEN
		UN 1782 (DOT)
16941-12-1		
		CHLOROPLATINIC ACID
		CHLOROPLATINIC(IV) ACID
		CHLOROPLATINIC(IV) ACID (H2PtCl6)
		DIHYDROGEN HEXACHLOROPLATINATE
		DIHYDROGEN HEXACHLOROPLATINATE(2-)
		HEXACHLOROPLATINIC ACID
		HEXACHLOROPLATINIC ACID (H2PtCl6)
		HEXACHLOROPLATINIC(IV) ACID
		HYDROGEN HEXACHLOROPLATINATE(IV)
		HYDROGEN PLATINUM CHLORIDE (H2PtCl6)
		PLATINATE(2-), HEXACHLORO-, DIHYDROGEN
		PLATINATE(2-), HEXACHLORO-, DIHYDROGEN, (OC-6-11)-
		PLATINIC CHLORIDE
16949-15-8		
		BORATE(1-), TETRAHYDRO-, LITHIUM
	*	LITHIUM BOROHYDRIDE
		LITHIUM BOROHYDRIDE (LiBH4)
16949-65-8		
		MAGNESIUM SILICO-FLUORIDE
16961-83-4		
		ACIDE FLUOROSILICIQUE (FRENCH)
		ACIDE FLUOSILICIQUE (FRENCH)
		ACIDO FLUOSILICICO (ITALIAN)
		DIHYDROGEN HEXAFLUOROSILCATE(2-)
		DIHYDROGEN HEXAFLUOROSILICATE
		FKS
		FLUOROSILICIC ACID
		FLUOROSILICIC ACID (DOT)
		FLUOROSILICIC ACID (H2SiF6)
		FLUOSILICIC ACID
		FLUOSILICIC ACID (DOT)
		HEXAFLUOROKIESELSAIURE (GERMAN)
		HEXAFLUOROKIEZELZUUR (DUTCH)
		HEXAFLUOROSILICIC ACID
		HEXAFLUOSILICIC ACID
		HYDROFLUOROSILICIC ACID
		HYDROFLUOROSILICIC ACID (DOT)
		HYDROFLUOSILICIC ACID
		HYDROGEN HEXAFLUOROSILICATE
		HYDROSILICOFLUORIC ACID
		HYDROSILICOFLUORIC ACID (DOT)
		KIEZELFLUORWATERSTOFZUUR (DUTCH)
		NA 1778 (DOT)
		SAND ACID
		SAND ACID (DOT)
		SILICATE(2-), HEXAFLUORO-, DIHYDROGEN
		SILICOFLUORIC ACID
		SILICOFLUORIC ACID (DOT)
		SILICON HEXAFLUORIDE DIHYDRIDE
		UN 1778 (DOT)
16962-07-5		
		ALUMINUM BOROHYDRIDE
		ALUMINUM BOROHYDRIDE (DOT)
		ALUMINUM HYDROBORATE
		ALUMINUM HYDROBORATE (Al[BH4]3)

CAS No. IN Chemical Name

ALUMINUM TETRAHYDROBORATE
BORATE(1-), TETRAHYDRO-, ALUMINUM
BORATE(1-), TETRAHYDRO-, ALUMINUM (3:1)
BORATE(1-), TETRAHYDRO-, ALUMINUM (3:1) (9CI)
BORATE(1-), TETRAHYDRO-, ALUMINUM SALT
UN 2870 (DOT)

16984-48-8
FLUORIDE
FLUORIDE ION
FLUORIDE ION (F-)
FLUORIDE(1-)
FLUORINE ION(1-)
FLUORINE ION(F1-)
FLUORINE, ION
HYDROFLUORIC ACID, ION(1-)
PERFLUORIDE

17014-71-0
DIPOTASSIUM PEROXIDE
POTASSIUM OXIDE (K2(O2))
POTASSIUM PEROXIDE
POTASSIUM PEROXIDE (DOT)
POTASSIUM PEROXIDE (K2(O2))
UN 1491 (DOT)

17026-81-2
3-AMINO-4-ETHOXYACETANILIDE

17084-08-1
FLUOROSILICATE
FLUOROSILICATE (SiF6 2-)
FLUOSILICATE
HEXAFLUOROSILICATE
HEXAFLUOROSILICATE (SiF6 2-)
HEXAFLUOROSILICATE(2-)
HEXAFLUOROSILICATE(2-) ION
SILICATE(2-), HEXAFLUORO-
SILICOFLUORIDE
SILICOFLUORIDE, SOLID
SILICON HEXAFLUORIDE ION

17237-93-3
CARBONIC ACID, NICKEL(2+) SALT (2:1)
NICKEL CARBONATE
NICKEL CARBONATE (Ni(HCO3)2)
NICKELOUS BICARBONATE

17617-23-1
FLURAZEPAM

17639-93-9
2-CHLOROPROPANOIC ACID METHYL ESTER
2-CHLOROPROPIONIC ACID METHYL ESTER
ALPHA-CHLOROPROPIONIC ACID METHYL ESTER
METHYL 2-CHLOROPROPANOATE
METHYL 2-CHLOROPROPIONATE
METHYL ALPHA-CHLOROPROPIONATE
METHYL CHLOROPROPIONATE
METHYL-2-CHLOROPROPIONATE
METHYL-2-CHLOROPROPIONATE (DOT)
PROPANOIC ACID, 2-CHLORO-, METHYL ESTER
PROPIONIC ACID, 2-CHLORO-, METHYL ESTER
UN 2933 (DOT)

17702-41-9
BORON HYDRIDE (B10H14)
DECABORANE
DECABORANE (B10H14)
DECABORANE(14)
NIDO-DECABORANE(14)
TETRADECAHYDRODECABORANE

17702-57-7
FORMPARANATE

CAS No. IN Chemical Name

17804-35-2
2-BENZIMIDAZOLECARBAMIC ACID, 1-(BUTYLCARBAMOYL)-, METHYL ESTER
AGROCIT
BBC
BC 6597
BENLATE
BENLATE 50
BENLATE 50W
BENOMYL
BENOMYL-IMEX
CARBAMIC ACID, [1-[(BUTYLAMINO)CARBONYL]-1H-BENZIMIDAZOL-2-YL]-, METHYL ESTER
DU PONT 1991
FUNDAZOL
FUNGICIDE D-1991
FUNGOCHROM
MBC
METHYL 1-(BUTYLCARBAMOYL)-2-BENZIMIDAZOLECARBAMATE
METHYL 1-(BUTYLCARBAMOYL)-2-BENZIMIDAZOLYLCARBAMATE
NS 02
NS 02 (FUNGICIDE)
TERSAN 1991
UZGEN

17861-62-0
NICKEL DINITRITE
NICKEL NITRITE
NICKEL NITRITE (DOT)
NICKEL NITRITE (Ni(NO2)2)
NITROUS ACID, NICKEL(2+) SALT
UN 2726 (DOT)

18130-44-4
SULFURIC ACID, TITANIUM SALT
SULFURIC ACID, TITANIUM(4+) SALT (2:1)
TITANIUM DISULFATE
TITANIUM SULFATE
TITANIUM SULFATE (1:2)
TITANIUM SULFATE (Ti(SO4)2)
TITANIUM SULFATE SOLUTION WITH NO MORE THAN 45% SULFURIC ACID
TITANIUM(4+) SULFATE
TITANIUM(IV) SULFATE

18130-74-0
BIFLUORIDE
BIFLUORIDE (HF2-)
BIFLUORIDE ANION (HF2 1-)
BIFLUORIDE, N.O.S.
FLUORIDE (HF2-)
FLUORIDE (HF2 1-)
HYDROGEN BIFLUORIDE ION(1-)
HYDROGEN DIFLUORIDE ION(1-)
HYDROGEN FLUORIDE (HF2 1-)
HYDROGEN FLUORIDE ION (HF2 1-)

18256-98-9
LEAD DINITRATE
LEAD NITRATE
LEAD NITRATE (DOT)
LEAD(2+) NITRATE
LEAD(II) NITRATE
LEAD(II) NITRATE (1:2)
NITRATE DE PLOMB (FRENCH)
NITRIC ACID, LEAD SALT
NITRIC ACID, LEAD(2+) SALT
PLUMBOUS NITRATE
UN 1469 (DOT)

18351-85-4
MONO-O-CRESYL PHOSPHATE
O-CRESOL, DIHYDROGEN PHOSPHATE
O-CRESYL DIHYDROGEN PHOSPHATE
O-TOLYL PHOSPHATE
PHOSPHORIC ACID, MONO(2-METHYLPHENYL) ESTER

CAS No.	IN	Chemical Name
18414-36-3		DIBENZYLDICHLOROSILANE
18433-48-2		URANYL PHOSPHATE
18454-12-1		AUSTRIAN CINNABAR BASIC CHROMIUM LEAD OXIDE (CrPb2O5) CHROMIC ACID (H4CrO5), LEAD(2+) SALT (1:2) CHROMIUM DILEAD PENTAOXIDE CHROMIUM LEAD OXIDE (CrPb2O5) LEAD CHROMATE LEAD CHROMATE (Pb2rRO5) LEAD CHROMATE (VI) OXIDE LEAD CHROMATE OXIDE LEAD CHROMATE OXIDE (Pb2OCrO4) LEAD CHROMATE, RED
18540-29-9		CHROMIUM (CR6+) CHROMIUM HEXAVALENT ION CHROMIUM(6+) CHROMIUM(6+) ION CHROMIUM(VI) CHROMIUM, ION (CR6+)
18662-53-8		NTA TRISODIUM SALT H20
18810-58-7		BARIUM AZIDE BARIUM AZIDE (Ba(N3)2) BARIUM AZIDE, (WET) BARIUM AZIDE, :DRY OR CONTAINING, BY WEIGHT, LESS THAN 50% WATER BARIUM AZIDE, WET , 50% OR MORE WATER (DOT) BARIUM AZIDE, WETTED :WITH NOT LESS THAN 50% WATER, BY WEIGHT: UN 1571 (DOT)
18811-72-8		2-NAPHTHALENEDIAZONIUM, 1-HYDROXY-5-SULFO-, HYDROXIDE, INNER SALT, SODIUM SALT SODIUM 2,1-DIAZONAPHTHOL-5-SULFONATE SODIUM 2-DIAZO-1-NAPHTHOL-5-SULFONATE
18883-66-4		D-GLUCOSE, 2-DEOXY-2-[[(METHYLNITROSOAMINO)CARBONYL]AMINO]- GLUCOPYRANOSE, 2-DEOXY-2-(3-METHYL-3-NITROSOUREIDO)-, D- NSC 85998 STREPTOZOCIN STREPTOZOTICIN STREPTOZOTOCIN STRZ
19222-41-4		AMMONIUM GLUCONATE D-GLUCONIC ACID, MONOAMMONIUM SALT GLUCAMONIX GLUCONIC ACID, MONOAMMONIUM SALT, D- NOVAMMON
19287-45-7		BORANE (B2H6) BOROETHANE BORON HYDRIDE BORON HYDRIDE (B2H6) DIBORANE DIBORANE (ACGIH,DOT) DIBORANE (B2H6) DIBORANE(6) DIBORON HEXAHYDRIDE UN 1911 (DOT)
19408-74-3		1,2,3,7,8,9-HEXACHLORODIBENZO-P-DIOXIN
19525-15-6		URANYL PEROXIDE
19594-40-2		2-METHYL-4-UNDECANONE 4-UNDECANONE, 2-METHYL- HEPTYL ISOBUTYL KETONE ISOBUTYL HEPTYL KETONE ISOBUTYL N-HEPTYL KETONE
19624-22-7		PENTABORANE PENTABORANE (ACGIH,DOT) PENTABORANE (B5H9) PENTABORANE(9) UN 1380 (DOT)
19910-65-7		DI-SEC-BUTYL PEROXYDICARBONATE DI-SEC-BUTYL PEROXYDICARBONATE (CONCENTRATION NOT MORE THAN 80%) DI-SEC-BUTYL PEROXYDICARBONATE :MAXIMUM CONCENTRATION 52% IN SOLUTION
20062-22-0		2,2',4,4',6,6'-HEXANITROSTILBENE HEXANITROSTILBENE
20265-97-8		P-ANISIDINE HYDROCHLIRIDE
20667-12-3		ARGENTOUS OXIDE DISILVER MONOXIDE DISILVER OXIDE SILVER OXIDE SILVER OXIDE (Ag2O) SILVER(1+) OXIDE SILVER(I) OXIDE
20770-41-6		PHOSPHURE DE POTASSIUM (FRENCH) POTASSIUM PHOSPHIDE POTASSIUM PHOSPHIDE (DOT) POTASSIUM PHOSPHIDE (K3P) TRIPOTASSIOPHOSPHINE UN 2012 (DOT)
20816-12-0		MILAS' REAGENT NAMED REAGENTS AND SOLUTIONS, MILAS' OSMIC ACID OSMIUM OXIDE (OsO4) OSMIUM OXIDE (OsO4), (T-4)- OSMIUM TETROXIDE OSMIUM TETROXIDE (ACGIH,DOT) OSMIUM(IV) OXIDE RCRA WASTE NUMBER P087 UN 2471 (DOT)
20830-75-5		12BETA-HYDROXYDIGITOXIN CHLOROFORMIC DIGITALIN CORDIOXIL DAVOXIN DIGACIN DIGITALIS GLYCOSIDE DIGOSIN DIGOXIGENIN-TRIDIGITOXOSID (GERMAN) DIGOXIN DIGOXINE DILANACIN DIXINA HOMOLLE'S DIGITALIN

CAS No.	IN	Chemical Name
		LANICOR
		LANOCARDIN
		LANOXIN
		ROUGOXIN
		SK-DIGOXIN
20830-81-3		
		(+)-DAUNOMYCIN
		5,12-NAPHTHACENEDIONE, 8-ACETYL-10-[(3-AMINO-2,3,6-TRIDEOXY-ALPHA-L-LYXO-HEXOPYRANOSYL)OXY]-7,8,9,
		ACETYLADRIAMYCIN
		CERUBIDIN
		DAUNOMYCIN
		DAUNORUBICIN
		DAUNORUBICINE
		LEUKAEMOMYCIN C
		NSC 82151
		RP 13057
		RUBIDOMYCIN
		RUBOMYCIN C
20859-73-8		
		AIP
		AL-PHOS
		ALUMINIUM FOSFIDE (DUTCH)
		ALUMINIUM PHOSPHIDE
		ALUMINIUM PHOSPHIDE (AlP)
		ALUMINUM MONOPHOSPHIDE
		ALUMINUM PHOSPHIDE
		ALUMINUM PHOSPHIDE (AlP)
		ALUMINUM PHOSPHIDE (DOT)
		CELPHIDE
		CELPHINE
		CELPHOS
		DELICIA
		DELICIA GASTOXIN
		DETIA
		DETIA GAS EX-B
		DETIA-EX-B
		FOSFURI DI ALLUMINIO (ITALIAN)
		FUMITOXIN
		GASTION
		PHOSFUME
		PHOSPHURES D'ALUMIUM (FRENCH)
		PHOSTOXIN
		PHOSTOXIN A
		PHOSTOXIN-A
		QUICKPHOS
		RCRA WASTE NUMBER P006
		UN 1397 (DOT)
21087-64-9		
		1,2,4-TRIAZIN-5(4H)-ONE, 4-AMINO-6-(1,1-DIMETHYL-ETHYL)-3-(METHYLTHIO)-
		AS-TRIAZIN-5(4H)-ONE, 4-AMINO-6-TERT-BUTYL-3-(METHY-LTHIO)-
		BAY 6159
		BAY 61597
		BAY 6159H
		BAY 94337
		BAYER 6159
		LEXONE
		METRIBUZIN
		SENCOR
		SENCOREX
21351-79-1		
		CESIUM HYDRATE
		CESIUM HYDROXIDE
		CESIUM HYDROXIDE (ACGIH)
		CESIUM HYDROXIDE (Cs(OH))
		CESIUM HYDROXIDE DIMER
		CESIUM HYDROXIDE SOLUTION
		CESIUM HYDROXIDE, SOLID (DOT)
		CESIUM HYDROXIDE, SOLUTION (DOT)
		UN 2681 (DOT)
		UN 2682 (DOT)
21416-87-5		
		ICRF-159
21548-32-3		
		(DIETHOXYPHOSPHINYLIMINO)-1,3-DITHIETANE
		1,3-DITHIETANE, PHOSPHORAMIDIC ACID DERIV
		AC 64475
		ACCONEM
		CL 64475
		DIETHOXYPHOSPHINYLIMINO-2 DITHIETANNE-1,3 (FRENCH)
		FOSTHIETAN
		GEOFOS
		IMIDOCARBONIC ACID, PHOSPHONODITHIO-, CYCLIC METHYLENE P,P-DIETHYL ESTER
		NEM-A-TAK
		NEMATAK
		PHOSPHORAMIDIC ACID, 1,3-DITHIETAN-2-YLIDENE-, DIETHYL ESTER
21564-17-0		
		2-(THIOCYANOMETHYLTHIO)BENZOTHIAZOLE
		2-(THIOCYANOMETHYLTHIO)BENZOTHIAZOLE, 60%
		2-[(THIOCYANATOMETHYL)THIO]BENZOTHIAZOLE
		ALENTISAN
		BENTHIAZOLE
		BENZOTHIAZOLE, 2-[(THIOCYANATOMETHYL)THIO]-
		BUSAN
		BUSAN 15
		BUSAN 30
		BUSAN 30-1
		BUSAN 30A
		BUSAN 30I
		BUSAN 70
		BUSAN 71
		BUSAN 72
		BUSAN 72A
		ICHIBAN
		KVK 733059
		SUPERDAVLOXAN
		TCMTB
		THIOCYANIC ACID, (2-BENZOTHIAZOLYLTHIO)METHYL ESTER
		THIOCYANIC ACID, 2-(BENZOTHIAZOLYLTHIO)METHYL ESTER, 60%
21609-90-5		
		LEPTOPHOS
		MBCP
		NK 711
		O-(2,5 -DICHLORO-4-BROMOPHENYL) O-METHYL PHENYL-THIOPHOSPHONATE
		O-(4-BROMO-2,5-DICHLOROPHENYL) O-METHYL PHENYL-PHOSPHONOTHIOATE
		O-METHYL O-2,5-DICHLORO-4-BROMOPHENYL PHENYL-THIOPHOSPHONATE
		O-METHYL-O-(4-BROMO-2,5-DICHLOROPHENYL)PHENYL THIOPHOSPHONATE
		OLEOPHOSVEL
		PHENYLPHOSPHONOTHIOIC ACID O-(4-BROMO-2,5-DI-CHLOROPHENYL) O-METHYL ESTER
		PHOSVEL
		PSL
		VCS
		VCS 506
		VELSICOL 506
		VELSICOL VCS 506
21725-46-2		
		BLADEX
21732-17-2		
		TETRAZOL-1-ACETIC ACID
21908-53-2		
		CI 77760
	*	MERCURIC OXIDE
		MERCURIC OXIDE (HgO)

CAS No.	IN	Chemical Name
		MERCURY OXIDE (HgO)
21923-23-9		CELA S 2957
		CELAMERCK S 2957
		CELATHION
		CELATHION (PESTICIDE)
		CHLORTHIOPHOS
		CHLORTHIOPHOS I
		CM S 2957
		ENT 27635
		NSC 195164
		O,O-DIETHYL-O-2,4,5-DICHLORO-(METHYLTHIO)PHENYL THIONOPHOSPHATE
		O-(DICHLORO(METHYLTHIO)PHENYL) O,O-DIETHYL PHOSPHOROTHIOATE (3 ISOMERS)
		OMS 1342
		PHOSPHOROTHIOIC ACID, O,O-DIETHYL O-((2,5-DICHLORO-4-METHYLTHIO)PHENYL) ESTER
		PHOSPHOROTHIOIC ACID, O-(2,5-DICHLORO-4-(METHYL-THIO)PHENYL O,O-DIETHYL ESTER
		S 2957
22128-62-7		CARBONOCHLORIDIC ACID, CHLOROMETHYL ESTER
		CARBONOCHLORIDIC ACID, CHLOROMETHYL ESTER (9CI)
		CHLOROMETHOXYCARBONYL CHLORIDE
		CHLOROMETHYL CHLOROFORMATE
		CHLOROMETHYLCHLOROFORMATE (DOT)
		FORMIC ACID, CHLORO-, CHLOROMETHYL ESTER
		UN 2745 (DOT)
22224-92-6		1-(METHYLETHYL)-ETHYL 3-METHYL-4-(METHYL-THIO)PHENYL PHOSPHORAMIDATE
		B 68138
		BAY 68138
		BAYER 68138
		ENT 27572
		ETHYL 3-METHYL-4-(METHYLTHIO)PHENYL (1-METHYL-ETHYL)PHOSPHORAMIDATE
		ETHYL 4-(METHYLTHIO)-M-TOLYL ISOPROPYLPHOS-PHORAMIDATE
		FENAMIPHOS
		FENAMIPHOS (ACGIH)
		ISOPROPYLAMINO-O-ETHYL-(4-METHYLMERCAPTO-3-METHYLPHENYL)PHOSPHATE
		NEMACUR
		NEMACUR P
		NSC 195106
		O-AETHYL-O-(3-METHYL-4-METHYLTHIOPHENYL)-ISOPRO-PYLAMIDO-PHOSPHORSAEUREESTER (GERMAN)
		PHENAMIPHOS
		PHOSPHORAMIDIC ACID, (1-METHYLETHYL)-, ETHYL (3-METHYL-4-(METHYLTHIO)PHENYL) ESTER
		PHOSPHORAMIDIC ACID, ISOPROPYL-, 4-(METHYLTHIO)-M-TOLYL ETHYL ESTER
		PHOSPHORAMIDIC ACID, ISOPROPYL-, ETHYL 4-(METHYL-THIO)-M-TOLYL ESTER
		SRA 3886
22259-30-9		3-[(DIMETHYLAMINO)METHYLENIMINO]PHENYL N-METHYLCARBAMATE
		CARBAMIC ACID, METHYL-, ESTER WITH N'-(M-HYDROXY-PHENYL)-N,N-DIMETHYLFORMAMIDINE
		CARBAMIC ACID, METHYL-, M-(((DIMETHYLAMINO)-METHYLENE)AMINO)PHENYL ESTER
		FORMAMIDINE, N'-(M-HYDROXYPHENYL)-N,N-DIMETHYL-, METHYLCARBAMATE (ESTER)
		FORMETANAT
		FORMETANATE
		M-[[(DIMETHYLAMINO)METHYLENE]AMINO]PHENYL METHYLCARBAMATE
		METHANIMIDAMIDE, N,N-DIMETHYL-, N'-(3-(((METHYL-AMINO)CARBONYL)OXY)PHENYL)- (9CI)
		METHYLCARBAMIC ACID ESTER WITH N'-(M-HYDROXY-PHENYL)-N,N-DIMETHYLFORMAMIDINE
		PHENOL, M-[[(DIMETHYLAMINO)METHYLENE]AMINO]-, METHYLCARBAMATE (ESTER)
22288-43-3		1,1,3,3-TETRAMETHYL BUTYL PEROXY-2-ETHYL HEX-ANOATE, *TECHNICAL PURE
22326-55-2		BARIUM HYDROXIDE (Ba(OH)2), MONOHYDRATE
		BARIUM HYDROXIDE MONOHYDRATE
22397-33-7		3,3,6,6,9,9-HEXAMETHYL-1,2,4,5-TETRAOXOCY-CLONONANE, TECHNICAL PURE
		3,3,6,6,9,9-HEXAMETHYL-1,2,4,5-TETROXACYCLONONANE (CONC. NOT MORE THAN 75%)
22541-79-3		CHROMIUM (II)
		CHROMIUM(2+)
		CHROMIUM(2+) ION
		CHROMIUM(II)
		CHROMIUM(II) ION
		CHROMIUM, ION (Cr2+)
		CHROMOUS ION
22781-23-3		BENDIOCARB
22966-79-6		ESTRADIOL MUSTARD
23092-17-3		HALAZEPAM
23103-98-2		PIRIMICARB
23135-22-0		2-(DIMETHYLAMINO)-N-(((METHYLAMINO)CARBONYL)-OXY)-2-OXOETHANIMIDOTHIOIC ACID METHYL ESTER
		2-DIMETHYLAMINO-1-(METHYLTHIO)GLYOXAL O-METHYL-CARB AMOYLMONOXIME
		CARBAMIC ACID, METHYL-, O-[[[(DIMETHYLCARBAMOYL)-METHYLTHIO]METHYLENE]AMINO] DERIV
		D-1410
		DPX 1410
		DPX 1410L
		DU PONT 1410
		ETHANIMIDOTHIOIC ACID, 2-(DIMETHYLAMINO)-N-[[(METHYLAMINO)CARBONYL]OXY]-2-OXO-, METHYL ESTER
		INSECTICIDE-NEMATICIDE 1410
		METHYL 1-(DIMETHYLCARBAMOYL)-N-(METHYLCARBA-MOYLOXY)THIOFORMIMIDATE
		METHYL 2-(DIMETHYLAMINO)-N-(((METHYLAMINO)CAR-BONYL)OXY)-2-OXOETHANIMIDOTHIOATE
		METHYL N',N'-DIMETHYL-N-((METHYLCARBAMOYL)OXY)-1-THIOOXAMIMIDATE
		N',N'-DIMETHYL-N-((METHYLCARBAMOYL)OXY)-1-THIO-OXAMIMIDIC ACID METHYL ESTER
		N,N-DIMETHYL-ALPHA-METHYLCARBAMOYLOXYIMINO-ALPHA-(METHYLTHIO)ACETAMIDE
		OXAMIMIDIC ACID, N',N'-DIMETHYL-N-((METHYLCARBA-MOYL)OXY)-1-METHYLTHIO-
		OXAMIMIDIC ACID, N',N'-DIMETHYL-N-[(METHYLCARBA-MOYL)OXY]-1-THIO-, METHYL ESTER
		OXAMYL
		OXAMYL (PESTICIDE)
		S-METHYL 1-(DIMETHYLCARBAMOYL)-N-((METHYLCARBA-MOYL)OXY)THIOFORMIMIDATE
		THIOXAMYL
		VYDATE
		VYDATE L
		VYDATE L INSECTICIDE/NEMATICIDE
		VYDATE L OXAMYL INSECTICIDE/NEMATOCIDE
		VYDATE-G

CAS No.	IN	Chemical Name
23214-92-8		
		14-HYDROXYDAUNOMYCIN
		5,12-NAPHTHACENEDIONE, 10-[(3-AMINO-2,3,6-TRIDEOXY-ALPHA-L-LYXO-HEXOPYRANOSYL)OXY]-7,8,9,10-TETRAH
		ADRIAMYCIN
		ADRIBLASTIN
		DOXORUBICIN
		F.I. 106
		NSC 123127
23414-72-4		
		PERMANGANIC ACID (HMnO4), ZINC SALT
		UN 1515 (DOT)
		ZINC OXIDE FUME
		ZINC PERMANGANATE
		ZINC PERMANGANATE (DOT)
23422-53-9		
		CARBAMIC ACID, METHYL-, ESTER WITH N'-(M-HYDROXY-PHENYL)-N,N-DIMETHYLFORMAMIDINE, MONOHY-DROCHLORIDE
		CARZOL
		CARZOL SP
		EP 332
		FORMETANATE HYDROCHLORIDE
		FORMETANATE MONOHYDROCHLORIDE
		M-[[(DIMETHYLAMINO)METHYLENE]AMINO]PHENYL METHYLCARBAMATE HYDROCHLORIDE
		METHANIMIDAMIDE, N,N-DIMETHYL-N'-[3-[[(METHYL-AMINO)CARBONYL]OXY]PHENYL]-, MONOHYDRO-CHLORIDE
23474-91-1		
		BUTYL PEROXYCROTONATE
		TERT-BUTYL PEROXYCROTONATE
23505-41-1		
		2-DIETHYLAMINO-6-METHYLPYRIMIDIN-4-YL DIETHYL-PHOSPHOROTHIONATE
		ETHYL PIRIMIPHOS
		FERNEX
		O,O-DIETHYL O-(2-DIETHYLAMINO-6-METHYL-4-PYRIMI-DINYL)PHOSPHOROTHIOATE
		O-(2-(DIETHYLAMINO)-6-METHYL-4-PYRIMIDINYL)O,O-DIETHYL PHOSPHOROTHIOATE
		PHOSPHOROTHIOIC ACID, O,O-DIETHYL O-(2-(DIETHYL-AMINO)-6-METHYL-4-PYRIMIDINYL) ESTER
		PHOSPHOROTHIOIC ACID, O-[2-(DIETHYLAMINO)-6-METHYL-4-PYRIMIDINYL] O,O-DIETHYL ESTER
		PIRIMIFOS-ETHYL
		PIRIMIPHOS ETHYL
		PIRIMPHOS-ETHYL
		PP 211
		PRIMICID
		PRIMOTEC
		PRINICID
		R 42211
		SOLGARD
23745-86-0		
		ACETIC ACID, FLUORO-, POTASSIUM SALT
		DICHAPETULUM CYMOSUM (HOOK) ENGL
		GIFBLAAR
		POTASSIUM CYMONATE
		POTASSIUM FLUOROACETATE
		POTASSIUM FLUOROACETATE (DOT)
		UN 2628 (DOT)
23950-58-5		
		3,5-DICHLORO-N-(1,1-DIMETHYL-2-PROPYNYL)BENZAMIDE
		3,5-DICHLORO-N-(1,1-DIMETHYLPROPYNYL)BENZAMIDE
		BENZAMIDE, 3,5-DICHLORO-N-(1,1-DIMETHYL-2-PRO-PYNYL)-
		KERB
		KERB 50W
		N-(1,1-DIMETHYLPROPYNYL)-3,5-DICHLOROBENZAMIDE
		PRONAMIDE
		PROPYZAMIDE
		RH 315
24017-47-8		
		1-PHENYL-1,2,4-TRIAZOLYL-3-(O,O-DIETHYLTHIONOPHOS-PHATE)
		1-PHENYL-3-(O,O-DIETHYL-THIONOPHOSPHORYL)-1,2,4-TRIAZOLE
		1H-1,2,4-TRIAZOL-3-OL, 1-PHENYL-, O-ESTER WITH O,O-DIETHYL PHOSPHOROTHIOATE
		HOE 2960
		HOE 2960 OJ
		HOSTATHION
		HOSTATION
		O,O-DIETHYL O-(1-PHENYL-1H-1,2,4-TRIAZOL-3-YL)PHOS-PHOROTHIOATE
		PHOSPHOROTHIOIC ACID, O,O-DIETHYL O-(1-PHENYL-1,2,4-TRIAZOLYL) ESTER
		PHOSPHOROTHIOIC ACID, O,O-DIETHYL O-(1-PHENYL-1H-1,2,4-TRIAZOL-3-YL) ESTER
		TRIAZOFOS
		TRIAZOFOSZ (HUNGARIAN)
		TRIAZOPHOS
24167-76-8		
		PHOSPHURE DE SODIUM (FRENCH)
		SODIUM PHOSPHIDE
		SODIUM PHOSPHIDE (DOT)
		SODIUM PHOSPHIDE (Na(H2P))
		UN 1432 (DOT)
24468-13-1		
		2-ETHYLHEXYL CHLOROFORMATE
		2-ETHYLHEXYL CHLOROFORMATE (DOT)
		CARBONOCHLORIDIC ACID, 2-ETHYLHEXYL ESTER
		CHLOROFORMIC ACID 2-ETHYLHEXYL ESTER
		ETHYL HEXYLCHLOROFORMATE
		FORMIC ACID, CHLORO-, 2-ETHYLHEXYL ESTER
		UN 2748 (DOT)
24800-44-0		
		2-(2-(2-HYDROXYPROPOXY)PROPOXY-1-PROPANOL
		PROPANOL, [(1-METHYL-1,2-ETHANEDIYL)BIS(OXY)]BIS-
		TRIPROPYLENE GLYCOL
24934-91-6		
		CHLORMEPHOS
		CHLORMETHYLFOS
		DOTAN
		MC 2188
		O,O-DIETHYL S-(CHLOROMETHYL) DITHIOPHOSPHATE
		PHOSPHORODITHIOIC ACID, S-(CHLOROMETHYL) O,O-DIETHYL ESTER
		S-(CHLOROMETHYL) O,O-DIETHYLPHOSPHORODITHIOATE
		S-CHLOROMETHYL O,O-DIETHYL PHOSPHOROTHIOLOTHIO-NATE
25013-15-4		
		BENZENE, ETHENYLMETHYL-
		METHYL STYRENE
		NCI-C56406
		STYRENE, AR-METHYL-
		STYRENE, METHYL-
		TOLUENE, VINYL- (MIXED ISOMERS)
		UN 2618 (DOT)
	*	VINYL TOLUENE
		VINYL TOLUENE (ACGIH)
		VINYL TOLUENES (MIXED ISOMERS), INHIBITED (DOT)
25103-58-6		
		SULFOLE 120
		TERT-DODECANETHIOL
		TERT-DODECYL MERCAPTAN
		TERT-DODECYLTHIOL
25134-21-8		
		MEMTETRAHYFROPHTHALIC ANHYDRIDE

CAS No.	IN	Chemical Name
25136-55-4		
		1,4-DIOXANE, DIMETHYL-
		DIMETHYL DIOXANE
		DIMETHYL-P-DIOXANE (DOT)
		DIMETHYLDIOXANE
		P-DIOXANE, DIMETHYL-
		UN 2707 (DOT)
25154-42-1		
		1-CHLOROBUTANE
	*	CHLOROBUTANE
		N-BUTYL CHLORIDE
25154-52-3		
		HYDROXYL NO 253
		N-NONYLPHENOL
		NONYL PHENOL (MIXED ISOMERS)
		NONYLPHENOL
		PHENOL, NONYL-
		PREVOSTSEL VON-100
25154-54-5		
		BENZENE, DINITRO-
	*	DINITROBENZENE
		DINITROBENZENE (ACGIH)
		DINITROBENZENE SOLID (DOT)
		DINITROBENZENE SOLUTION
		DINITROBENZENE, SOLID
		DINITROBENZENE, SOLUTION (DOT)
		DINITROBENZOL SOLID (DOT)
		UN 1597 (DOT)
25154-55-6		
		MONONITROPHENOL
	*	NITROPHENOL
		NITROPHENOL (DOT)
		PHENOL, NITRO-
		UN 1663 (DOT)
25155-15-1		
		CYMENE
25155-30-0		
		A 1-1575
		AA-10
		AA-9
		ABESON NAM
		ALKANATE DC
		ARYLAN SBC
		BENZENESULFONIC ACID, DODECYL-, SODIUM SALT
		BIO-SOFT D 40
		BIO-SOFT D-35X
		BIO-SOFT D-40
		BIO-SOFT D-60
		BIO-SOFT D-62
		C 550
		CALSOFT F 90
		CALSOFT L 40
		CALSOFT L-60
		CONCO AAS 35
		CONCO AAS 35H
		CONCO AAS 45S
		CONCO AAS-35
		CONCO AAS-40
		CONCO AAS-65
		CONCO AAS-90
		CONOCO C 550
		CONOCO C-50
		CONOCO C-60
		CONOCO SD 40
		DETERGENT HD-90
		DETERLON
		DODECYL BENZENE SODIUM SULFONATE
		DODECYLBENZENESULFONIC ACID SODIUM SALT
		DODECYLBENZENESULPHONATE, SODIUM SALT
		DODECYLBENZENSULFONAN SODNY (CZECH)
		DS 60
		ELFAN WA 35
		ELFAN WA 50
		ELFAN WA POWDER
		EMULIN B 22
		F 90
		HS 85S
		KB
		KB (SURFACTANT)
		MARANIL
		MARLON 375
		MARLON A
		MARLON A 350
		MARLON A 375
		MARLON A 396
		MB-VR
		MERCOL 25
		MERCOL 30
		MERPISAP AP 90P
		NA 9146 (DOT)
		NACCANOL NR
		NACCANOL SW
		NACCONOL 35SL
		NACCONOL 40F
		NACCONOL 90F
		NANSA 1260
		NANSA HF 80
		NANSA HS 80
		NANSA HS 80S
		NANSA HS 85S
		NANSA SL
		NANSA SS
		NECCANOL SW
		NEOGEN SC
		NEOPELEX 05
		NEOPELEX 25
		NEOPELEX 6
		NEOPELEX F 25
		NEOPELEX F 60
		NESTAPONE
		NEWREX R
		NISSAN NEWREX R
		P-1',1',4',4'-TETRAMETHYLOKTYLBENZENSULFONAN SODNY (CZECH)
		PELOPON A
		PILOT HD-90
		PILOT SF-40
		PILOT SF-40B
		PILOT SF-40FG
		PILOT SF-60
		PILOT SF-96
		PILOT SP-60
		REWORYL NKS 50
		RICHONATE 1850
		RICHONATE 40B
		RICHONATE 45B
		RICHONATE 60B
		SANDET 60
		SANTOMERSE 3
		SANTOMERSE ME
		SANTOMERSE NO 1
		SANTOMERSE NO 85
		SDBS
		SINNOZON
		SIPONATE DS 10
		SIPONATE DS 4
		SIPONATE LDS 10
	*	SODIUM DODECYLBENZENE SULFONATE
		SODIUM DODECYLBENZENESULFONATE (DOT)
		SODIUM DODECYLBENZENESULFONATE, DRY
		SODIUM DODECYLBENZENESULPHONATE
		SODIUM DODECYLPHENYLSULFONATE
		SODIUM LAURYLBENZENESULFONATE
		SOL SODOWA KWASU LAURYLOBENZENOSULFONOWEGO (POLISH)
		SOLAR 40
		SOLAR 90
		STEINARYL NKS 100
		STEINARYL NKS 50
		STEPAN DS 60
		SULFAPOL

CAS No.	IN	Chemical Name
		SULFAPOLU (POLISH)
		SULFARIL PASTE
		SULFRAMIN 1238 SLURRY
		SULFRAMIN 1240
		SULFRAMIN 1250 SLURRY
		SULFRAMIN 40
		SULFRAMIN 40 FLAKES
		SULFRAMIN 40 GRANULAR
		SULFRAMIN 40RA
		SULFRAMIN 85
		SULFRAMIN 90 FLAKES
		SULFURIL 50
		TREPOLATE F 40
		TREPOLATE F 95
		ULTRAWET 1T
		ULTRAWET 60K
		ULTRAWET 99LS
		ULTRAWET K
		ULTRAWET KX
		ULTRAWET SK
		WITCONATE 1238
		X 2073
25156-49-4		
		C.I. DIRECT BLACK 4
25167-20-8		
		TETRABROMOETHANE
25167-67-3		
		BUTYLENE
25167-70-8		
		2,4,4-TRIMETHYL PENTENE
	*	DI-ISOBUTYLENE
		DIISOBUTENE
		DIISOBUTYLENE
		DIISOBUTYLENE (DOT)
		PENTENE, 2,4,4-TRIMETHYL-
		UN 2050 (DOT)
25167-80-0		
		4-CHLOROPHENOL
		CHLOROPHENOL
		CHLOROPHENOL, LIQUID
		MONOCHLOROPHENOL
		P-CHLORFENOL (CZECH)
		P-CHLOROPHENOL
		P-CHLOROPHENOL, LIQUID (DOT)
		P-CHLOROPHENOL, SOLID (DOT)
		PARACHLOROPHENOL
		PHENOL, 4-CHLORO-
		PHENOL, CHLORO-
		PHENOL, P-CHLORO-
		UN 2020 (DOT)
		UN 2021 (DOT)
25167-82-2		
		DOWICIDE 2
		DOWICIDE 2S
		NA 2020 (DOT)
		NCI-C02904
		OMAL
		PHENACHLOR
		PHENOL, TRICHLORO-
		PREVENTOL I
		RCRA WASTE NUMBER U231
		TCP
		TRICHLOROPHENOL
		TRICHLOROPHENOL (DOT)
25167-83-3		
		TRICHLOROPHENOL
25167-93-5		
		ALTRITAN
		BENZENE, CHLORONITRO-
		CHLORONITROBENZENE
		CHLORONITROBENZENES
		NITROCHLOROBENZENE
25168-04-1		
		BENZENE, DIMETHYLNITRO-
		NA 1665 (DOT)
		NITRODIMETHYLBENZENE
		NITROXYLENE
		NITROXYLOL
		NITROXYLOL (DOT)
		XYLENE, AR-NITRO-
		XYLENE, NITRO-
25168-05-2		
		AR-CHLOROTOLUENE
		BC
		BENZENE, CHLOROMETHYL-
		CHLOROMETHYLBENZENE
	*	CHLOROTOLUENE
		TOLUENE, AR-CHLORO-
25168-15-4		
		2,4,5-T ESTER [ISOOCTYL ESTER]
		2,4,5-T ISOOCTYL ESTER
		ACETIC ACID, 2,4,5-TRICHLOROPHENOXY-, ISOOCTYL ESTER
		N-BUTYLESTER KYSELINI 2,4,5-TRICHLORFENOXYOCTOVE (CZECH)
25168-26-7		
		2,4-D ISOOCTYL ESTER
		2,4-DICHLOROPHENOXYACETIC ACID ISOOCTYL ESTER
		ACETIC ACID, (2,4-DICHLOROPHENOXY)-, ISOOCTYL ESTER
		ISOOCTYL 2,4-DICHLOROPHENOXYACETATE
		ISOOCTYL ALCOHOL, (2,4-DICHLOROPHENOXY)ACETATE
		SILVAPRON D
25180-19-2		
		C.I. DIRECT BLUE 2
25180-27-2		
		C.I. DIRECT BLUE 25
25180-39-6		
		C.I. DIRECT BROWN 6
25180-41-0		
		C.I. DIRECT BROWN 31
25180-45-4		
		C.I. DIRECT GREEN 1
25180-46-5		
		C.I. DIRECT GREEN 6, DISODIUM SALT
25180-47-6		
		C.I. DIRECT GREEN 8
25188-24-3		
		C.I. DIRECT RED 1
25188-29-8		
		C.I. DIRECT RED 10
25188-30-1		
		C.I. DIRECT RED 13
25188-44-7		
		C.I. DIRECT VIOLET 1
25231-46-3		
		4-METHYLBENZENESULFONIC ACID
		4-TOLUENESULFONIC ACID
		AR-TOLUENESULFONIC ACID
		BENZENESULFONIC ACID, 4-METHYL-
		BENZENESULFONIC ACID, METHYL-
		CYZAC 4040
		K-CURE 1040
		KYSELINA P-TOLUENESULFONOVA (CZECH)
		MANRO PTSA 65 E

CAS No.	IN	Chemical Name
		MANRO PTSA 65 H
		MANRO PTSA 65 LS
		METHYLBENZENESULFONIC ACID
		P-METHYLBENZENESULFONIC ACID
		P-METHYLPHENYLSULFONIC ACID
		P-TOLUENESULFONIC ACID
		P-TOLUENESULPHONIC ACID
		P-TOLYLSULFONIC ACID
		TOLUENE SULFONIC ACID
		TOLUENE SULFONIC ACID, LIQUID
		TOLUENE SULFONIC ACID, LIQUID (DOT)
		TOLUENE SULFONIC ACID, SOLID
		TOLUENE SULPHONIC ACID, LIQUID, CONTAINING MORE THAN 5% FREE SULPHURIC ACID (DOT)
		TOLUENE SULPHONIC ACID, LIQUID, CONTAINING NOT MORE THAN 5% FREE SULPHURIC ACID (DOT)
		TOLUENE SULPHONIC ACID, SOLID, CONTAINING MORE THAN 5% FREE SULPHURIC ACID (DOT)
		TOLUENE SULPHONIC ACID, SOLID, CONTAINING NOT MORE THAN 5% FREE SULPHURIC ACID (DOT)
		TOLUENESULFONIC ACID
		TOLUENESULFONIC ACID, CONTAINING MORE THAN 5% FREE SULPHURIC ACID
		TOLUENESULFONIC ACID, CONTAINING NOT MORE THAN 5% FREE SULPHURIC ACID
		TOSIC ACID
		TSA-HP
		TSA-MH
		UN 2583 (DOT)
		UN 2584 (DOT)
		UN 2585 (DOT)
		UN 2586 (DOT)
25251-51-8		
		1(3H)-ISOBENZOFURANONE, 3-[(1,1-DIMETHYLETHYL)-DIOXY]-3-PHENYL-
		3-PHENYL-3-(TERT-BUTYLPEROXY)PHTHALIDE
		3-TERT-BUTYL PEROXY-3-PHENYLPHTHALIDE, TECHNICALLY PURE
		3-TERT-BUTYLPEROXY-3-PHENYLPHTHALIDE
		3-TERT-BUTYLPEROXY-3-PHENYLPHTHALIDE, TECHNICAL PURE (DOT)
		PHTHALIDE, 3-(TERT-BUTYLDIOXY)-3-PHENYL-
		PHTHALIDE, 3-(TERT-BUTYLPEROXY)-3-PHENYL-
		UN 2596 (DOT)
25255-06-5		
		C.I. DIRECT BROWN 2
25265-68-3		
		FURAN, TETRAHYDROMETHYL-
		METHYL TETRAHYDROFURAN
		METHYLTETRAHYDROFURAN (DOT)
		UN 2536 (DOT)
25265-71-8		
		1,1'-OXYDI-2-PROPANOL
		2,2'-DIHYDROXYDIPROPYL ETHER
		2,2'-DIHYDROXYISOPROPYL ETHER
		2-PROPANOL, 1,1'-OXYDI-
		DIPROPYLENE GLYCOL
		PROPANOL, OXYBIS-
25267-27-0		
		2-IODOBUTANE
		2-IODOBUTANE (DOT)
		BUTANE, 2-IODO-
		BUTANE, IODO-
		IODOBUTANE
		SEC-BUTYL IODIDE
		UN 2390 (DOT)
25316-40-9		
		5,12-NAPHTHACENEDIONE, 10-[(3-AMINO-2,3,6-TRIDEOXY-ALPHA-L-LYXO-HEXOPYRANOSYL)OXY]-7,8,9,10-TETRAH
		ADRIAMYCIN, HYDROCHLORIDE
		DOXORUBICIN HYDROCHLORIDE
25321-14-6		
		BENZENE, METHYLDINITRO-
		DINITROPHENYLMETHANE
		DINITROTOLUENE
		DINITROTOLUENE, LIQUID
		DINITROTOLUENE, LIQUID (DOT)
		DINITROTOLUENE, MOLTEN (DOT)
		DINITROTOLUENE, SOLID
		DINITROTOLUENE, SOLID (DOT)
		DNT
		METHYLDINITROBENZENE
		TOLUENE, AR,AR-DINITRO-
		TOLUENE, DINITRO-
		UN 1600 (DOT)
		UN 2038 (DOT)
25321-22-6		
		AMISIA-MOTTENSCHUTZ
		BENZENE, DICHLORO-
		DCB
		DICHLOROBENZENE
		DICHLOROBENZENE (MIXED ISOMERS)
		MOTT-EX
		MOTTENSCHUTZMITTEL EVAU P
		ROTAMOTT
25322-01-4		
		2-NITROPROPANE
		2-NITROPROPANE (ACGIH,DOT)
		2-NP
		BETA-NITROPROPANE
		DIMETHYLNITROMETHANE
		ISONITROPROPANE
		NIPAR S-20
		NIPAR S-20 SOLVENT
		NIPAR S-30 SOLVENT
		NITROISOPROPANE
		NITROPROPANE
		PROPANE, 2-NITRO-
		PROPANE, NITRO-
		RCRA WASTE NUMBER U171
		UN 2608 (DOT)
25322-14-9		
		TRINITROFLUORENONE
25322-20-7		
		ETHANE, TETRACHLORO-
		TETRACHLOROETHANE
		TETRACHLOROETHANE (DOT)
		UN 1702 (DOT)
25322-69-4		
		1,2-EPOXYPROPANE POLYMER
		ACTOCOL 51-530
		ADEKA P 1000
		ADEKA P 3000
		ADEKA P 700
		ALKAPOL PPG-1200
		ALKAPOL PPG-2000
		ALKAPOL PPG-4000
		ALPHA-HYDRO-OMEGA-HYDROXYPOLY(OXYPROPYLENE)
		DESMOPHEN 360C
		DIELECTROL VI
		DIOL 1000
		DIOL 2000
		DIOL 400
		EMKAPYL
		EP 240
		EXCENOL 1020
		FRA 1173
		GLYCOLS, POLYPROPYLENE
		HODAG HT 11
		HYPROX DP 400
		JEFFOX
		LAPROL 1502-3-100
		LAPROL 1602-3-100
		LAPROL 2002
		LAPROL 2102

CAS No.	IN	Chemical Name
		LAPROL 503B
		LAPROL 702
		LAPROL L 1502-3-100
		LINEARTOP E
		LUPRAMOL 1010
		METHYLOXIRANE POLYMER
		MONOLAN PPG 700
		NALCO 131S
		NAPTER E 8075
		NIAX 10-25
		NIAX 20-25
		NIAX 61-58
		NIAX 61-582
		NIAX POLYOL PPG 2025
		NIAX POLYOL PPG 4025
		NIAX PPG
		NIAX PPG 1025
		NIAX PPG 2025
		NIAX PPG 3025
		NIAX PPG 4025
		NIAX PPG 425
		NISSAN UNIOL D 2000
		NISSAN UNIOL D 400
		OOPG 1000
		OOPG 1002
		OPD 500
		OPD 600
		OXIRANE, METHYL-, HOMOPOLYMER
		P 1200
		P 2000
		P 400
		P 400 (POLYGLYCOL)
		P 4000
		P 4000 (POLYMER)
		P 425
		PLURACOL 1010
		PLURACOL 2010
		PLURACOL P 1010
		PLURACOL P 2010
		PLURACOL P 410
		PLURACOL P 710
		PLURIOL P
		PLURIOL P 2000
		PLURIOL P 900
		POLY(1,2-EPOXYPROPANE)
		POLY(PROPYLENE OXIDE)
		POLY-G 1020P
		POLY-G 20-112
		POLY-G 20-265
		POLY-G 20-56
		POLYGLYCOL P 400
		POLYGLYCOL P 4000
		POLYGLYCOL TYPE P 1200
		POLYGLYCOL TYPE P 2000
		POLYGLYCOL TYPE P 250
		POLYGLYCOL TYPE P 3000
		POLYGLYCOL TYPE P 400
		POLYGLYCOL TYPE P 750
		POLYHARDENER D 200
		POLYMER 2
		POLYOXYALKYLENES, POLYPROPYLENE GLYCOL
		POLYPROPYLENE 5820
		POLYPROPYLENE GLYCOL
		POLYPROPYLENGLYKOL (CZECH)
		POLY[OXY(METHYL-1,2-ETHANEDIYL)], ALPHA-HYDRO-OMEGA-HYDROXY-
		PPG
		PPG 15
		PPG 2025
		PPG 30
		PPG 400
		PPG DIOL 1000
		PPG DIOL 2000
		PPG DIOL 400
		PROPYLAN 8123
		PROPYLAN D 1002
		PROPYLAN D 2002
		PROPYLENE OXIDE HOMOPOLYMER
		ROKOPOL D 2002
		SANNIX PP 1000
		SANNIX PP 200
		SANNIX PP 400
		SL 1000
		SYSTOL T 121
		T 32/75
		TAKELAC P 21
		VORANOL P 1010
		VORANOL P 2000
		VORANOL P 4000
		W 166
25324-56-5		
		STANNIC PHOSPHIDE
		STANNIC PHOSPHIDE (DOT)
		TIN (IV) PHOSPHIDE
		TIN MONOPHOSPHIDE
		TIN PHOSPHIDE (SnP)
		UN 1433 (DOT)
25329-82-2		
		C.I. DIRECT VIOLET 22
25339-17-7		
		ISODECANOL
		ISODECYL ALCOHOL
25339-56-4		
		HEPTENE
		HEPTYLENE
25339-57-5		
		1,3-BUTADIENE
		1,3-BUTADIENE (ACGIH)
		ALPHA,GAMMA-BUTADIENE
		BIETHYLENE
		BIVINYL
		BUTA-1,3-DIEEN (DUTCH)
		BUTA-1,3-DIEN (GERMAN)
		BUTA-1,3-DIENE
		BUTADIEEN (DUTCH)
		BUTADIEN (POLISH)
		BUTADIENE
		BUTADIENE, INHIBITED
		BUTADIENE, INHIBITED (DOT)
		BUTYNE
		DIVINYL
		ERYTHRENE
		NCI-C50602
		PLIOLITE
		PYRROLYLENE
		UN 1010 (DOT)
		VINYLETHYLENE
25340-17-4		
		BENZENE, DIETHYL-
		DIETHYL BENZENE
		DIETHYLBENZENE (DOT)
		UN 2049 (DOT)
25340-18-5		
		BENZENE, TRIETHYL-
		BENZENE, TRIETHYL- (MIXED ISOMERS)
		TRIETHYLBENZENE
25360-10-5		
		TERT-NONANETHIOL
		TERT-NONYL MERCAPTAN
25376-45-8		
		2,4-TOLUYLENEDIAMINE
		BENZENE, METHYL-, DIAMINO DERIV
		BENZENEDIAMINE, AR-METHYL-
		DIAMINOTOLUENE
		DIAMINOTOLUENE (MIXED ISOMERS)
		METHYLPHENYLENEDIAMINE
		NA 1709 (DOT)
		RCRA WASTE NUMBER U221
		TOLUENE-AR,AR-DIAMINE

CAS No.	IN	Chemical Name
		TOLUENEDIAMINE
		TOLUENEDIAMINE (DOT)
		TOLYLENEDIAMINE
25377-32-6		
		TRINITROBENZENE
25377-72-4		
		AMYLENE, NORMAL
		PENTENE
25495-90-3		
		CHLOROHEXANE
		HEXANE, CHLORO-
25496-08-6		
		BENZENE, FLUOROMETHYL-
		FLUOROTOLUENE
		P-FLUOROTOLUENE
		P-FLUOROTOLUENE (DOT)
		TOLUENE, AR-FLUORO-
		TOLUENE, P-FLUORO-
		UN 2388 (DOT)
25497-28-3		
		1,1-DIFLUOROETHANE
		1,1-DIFLUOROETHANE (DOT)
		ALGOFRENE TYPE 67
		DIFLUOROETHANE
		DYMEL 152
		ETHANE, 1,1-DIFLUORO-
		ETHANE, DIFLUORO-
		ETHYLENE FLUORIDE
		ETHYLIDENE DIFLUORIDE
		ETHYLIDENE FLUORIDE
		FC 152A
		FREON 152
		GENETRON 100
		GENETRON 152A
		HALOCARBON 152A
		R 152A
		UN 1030 (DOT)
25497-29-4		
		1,1-DIFLUORO-1-CHLOROETHANE
		1-CHLORO-1,1-DIFLUOROETHANE
		1-CHLORO-1,1-DIFLUOROETHANE (DOT)
		2517
		ALPHA-CHLOROETHYLIDENE FLUORIDE
		CHLORODIFLUOROETHANE
		CHLORODIFLUOROETHANE (DOT)
		CHLORODIFLUOROETHANES
		CHLOROETHYLIDENE FLUORIDE
		DIFLUOROMONOCHLOROETHANE
		DIFLUOROMONOCHLOROETHANE (DOT)
		ETHANE, 1-CHLORO-1,1-DIFLUORO-
		ETHANE, CHLORODIFLUORO-
		FC142B
		FLUOROCARBON FC142B
		FREON 142
		FREON 142B
		GENETRON 101
		GENETRON 142B
		GENTRON 142B
		MONOCHLORO DIFLUOROETHANE
		R 142
		UN 2517 (DOT)
25512-65-6		
		2H-PYRAN, DIHYDRO-
		DELTA(SUP 2)-DIHYDROPYRAN
		DELTA2-DIHYDROPYRAN
		DIHYDROPYRAN
		DIHYDROPYRAN (DOT)
		PYRAN, DIHYDRO-
25550-55-4		
		DINITROSOBENZENE
25550-58-7		
	*	DINITROPHENOL
		DINITROPHENOL SOLUTION
		DINITROPHENOL SOLUTION (DOT)
		DINITROPHENOL, DRY OR CONTAINING, BY WEIGHT, LESS THAN 15% WATER (DOT)
		DINITROPHENOL, WETTED WITH, BY WEIGHT, AT LEAST 15% WATER (DOT)
		PHENOL, DINITRO-
		PHENOL, DINITRO-, WETTED WITH AT LEAST 15% WATER
		UN 0076 (DOT)
		UN 1320 (DOT)
		UN 1599 (DOT)
25551-13-7		
		BENZENE, TRIMETHYL-
		BENZENE, TRIMETHYL- (MIXED ISOMERS)
		METHYLXYLENE
	*	TRIMETHYL BENZENE
		TRIMETHYL BENZENE (ACGIH)
25551-14-8		
		AZODI-(1,1'-HEXAHYDROBENZONITRILE)
25551-28-4		
		NAPHTHALENE DI-ISOCYANATE
25567-67-3		
		1,3-DINITRO-4-CHLOROBENZENE
		1-CHLOOR-2,4-DINITROBENZEEN (DUTCH)
		1-CHLOR-2,4-DINITROBENZENE
		1-CHLORO-2,4-DINITROBENZENE
		1-CHLORO-2,4-DINITROBENZOL (GERMAN)
		1-CLORO-2,4-DINITROBENZENE (ITALIAN)
		2,4-DINITRO-1-CHLOROBENZENE
		2,4-DINITROCHLOROBENZENE
		2,4-DINITROPHENYL CHLORIDE
		4-CHLORO-1,3-DINITROBENZENE
		6-CHLORO-1,3-DINITROBENZENE
		BENZENE, 1-CHLORO-2,4-DINITRO-
		BENZENE, CHLORODINITRO-
		BENZENE, CHLORODINITRO- (MIXED ISOMERS)
		CDNB
		CHLORODINITROBENZENE
		CHLORODINITROBENZENE (DOT)
		DINITROCHLOROBENZENE
		DINITROCHLOROBENZENE (DOT)
		DINITROCHLOROBENZOL
		DINITROCHLOROBENZOL (DOT)
		DNCB
		UN 1577 (DOT)
25567-68-4		
		BENZENE, CHLOROMETHYLNITRO-
		BENZENE, METHYL-, MONOCHLORO MONONITRO DERIV
		CHLORO-O-NITROTOLUENE
		CHLORO-O-NITROTOLUENE (DOT)
		CHLORONITROTOLUENE
		TOLUENE, CHLORO-O-NITRO-
		TOLUENE, CHLORONITRO-
		UN 2433 (DOT)
25584-83-2		
		1,2-PROPANEDIOL, MONOACRYLATE
		2-PROPENOIC ACID, MONOESTER WITH 1,2-PROPANEDIOL
		ACRYLIC ACID, MONOESTER WITH 1,2-PROPANEDIOL
		HYDROXYPROPYL ACRYLATE
		PROPYLENE GLYCOL ACRYLATE
25620-58-0		
		1,6-HEXANEDIAMINE, TRIMETHYL-
		TRIMETHYL-1,6-HEXANEDIAMINE
		TRIMETHYLHEXAMETHYLENE DIAMINE (DOT)
		TRIMETHYLHEXAMETHYLENEDIAMINE
		UN 2327 (DOT)
25639-42-3		
		CYCLOHEXANOL, METHYL-
		HEXAHYDROCRESOL

CAS No.	IN	Chemical Name
		HEXAHYDROMETHYLPHENOL
	*	METHYL CYCLOHEXANOL
		METHYLCYCLOHEXANOL (ACGIH,DOT)
		METHYLHEXALIN
		METYLOCYKLOHEKSANOL (POLISH)
		UN 2617 (DOT)
25641-53-6		
		O-CRESOL, DINITRO-, SODIUM SALT, WETTED WITH AT LEAST 10-15% WATER
		PHENOL, 2-METHYLDINITRO-, SODIUM SALT
		SODIUM DINITRO-O-CRESOLATE, WETTED WITH AT LEAST 10% WATER (DOT)
		SODIUM DINITRO-O-CRESOLATE, WETTED WITH AT LEAST 15% WATER (DOT)
		SODIUM DINITRO-O-CRESYLATE
		SODIUM, [(DINITRO-O-TOLYL)OXY]-
		UN 1348 (DOT)
25889-38-7		
		NITROCHLOROBENZOTRI-FLUORIDE (2-NITRO-1-(TRI-FLUOROMETHYL)-4-CHLO
26094-13-3		
		1-BUTANAMINE, (Z)-9-OCTADECENOATE
		9-OCTADECENOIC ACID (Z)-, COMPD. WITH 1-BUTANAMINE (1:1)
		BUTYLAMINE OLEATE
		BUTYLAMMONIUM OLEATE
		OLEIC ACID, COMPD. WITH BUTYLAMINE (1:1)
26134-62-3		
	*	LITHIUM NITRIDE
		LITHIUM NITRIDE (DOT)
		LITHIUM NITRIDE (Li3N)
		TRILITHIUM NITRIDE
		TRILITHIUM NITRIDE (Li3N)
		UN 2806 (DOT)
26248-42-0		
		TRIDECANOL
26249-12-7		
		BENZENE, DIBROMO-
		DIBROMOBENZENE
		DIBROMOBENZENE (DOT)
		UN 2711 (DOT)
26264-05-1		
		2-PROPANAMINE, DODECYLBENZENESULFONATE
		ARYLAN PWS
		ATLAS G 3300
		ATLAS G 711
		BENZENESULFONIC ACID, DODECYL-, COMPD WITH 2-PROPANAMINE (1:1)
		BENZENESULFONIC ACID, DODECYL-, COMPD WITH ISOPROPYLAMINE
		BENZENESULFONIC ACID, DODECYL-, COMPD WITH ISOPROPYLAMINE (1:1)
		DODECYLBENZENESULFONATE ISOPROPYLAMINE SALT
		DODECYLBENZENESULFONIC ACID ISOPROPYLAMINE SALT
		DODECYLBENZENESULFONIC ACID, ISOPROPYLAMINE SALT
		G 3300
		G 711
		ISOPROPYLAMINE DODECYLBENZENESULFONATE
		ISOPROPYLAMMONIUM DODECYLBENZENESULFONATE
		MONOISOPROPYLAMMONIUM DODECYLBENZENESULFO-NATE
26264-06-2		
		BENZENESULFONIC ACID, DODECYL-, CALCIUM SALT
		CALCIUM DODECYLBENZENE SULFONATE
		CALCIUM DODECYLBENZENESULFONATE (DOT)
		CALCIUM N-DODECYLBENZENESULFONATE
		CASUL 70HF
		DODECYLBENZENESULFONIC ACID CALCIUM SALT
		NA 9097 (DOT)

CAS No.	IN	Chemical Name
		NINATE 401
		SINNOZON NCX 70
		SOPROFOR S 70
26265-65-6		
		LEAD THIOSULFATE
		THIOSULFURIC ACID (H2S2O3), LEAD SALT
		THIOSULFURIC ACID, LEAD SALT
26322-14-5		
		DICETYL PEROXYDICARBONATE
26419-73-8		
		2,4-DIMETHYL-1,3-DITHIOLANE-2-CARBOXALDEHYDE O-METHYCARBAMOYLOXIME
		TIRPATE
26445-05-6		
		AMINOPYRIDINE
		AMINOPYRINE
		PYRIDINAMINE
		PYRIDINAMINE (9CI)
		PYRIDINE, AMINO-
26446-77-5		
		BROMOPROPANE
26447-14-3		
		((METHYLPHENOXY)METHYL)OXIRANE
		1,2-EPOXY-3-(TOLYLOXY)PROPANE
		CRESOL GLYCIDYL ETHER
		CRESYL GLYCIDYL ETHER
		CRESYLGLYCIDE ETHER
		GLYCIDYL METHYLPHENYL ETHER
		GLYCIDYL TOLYL ETHER
		OXIRANE ((METHYLPHENOXY)METHYL)- (9CI)
		OXIRANE, [(METHYLPHENOXY)METHYL]-
		PROPANE, 1,2-EPOXY-3-(TOLYLOXY)-
		TOLYL GLYCIDYL ETHER
26471-56-7		
		ANILINE, DINITRO-
		ANILINE, DINITRO- (MIXED ISOMERS)
		BENZENAMINE, AR,AR-DINITRO-
		BENZENAMINE, AR,AR-DINITRO- (9CI)
		DINITROANILINE
		DINITROANILINES
		DINITROANILINES (DOT)
		UN 1596 (DOT)
26471-62-5		
		2,4-DIISOCYANATO-1-METHYLBENZENE
		2,4-DIISOCYANATO-1-METHYLBENZENE (9CI)
		2,4-DIISOCYANATOTOLUENE
		2,4-TDI
		2,4-TOLUENE DIISOCYANATE
		2,4-TOLYLENE DIISOCYANATE
		4-METHYL-M-PHENYLENE DIISOCYANATE
		4-METHYL-M-PHENYLENE ISOCYANATE
		4-METHYL-PHENYLENE DIISOCYANATE
		4-METHYL-PHENYLENE ISOCYANATE
		BENZENE, 2,4-DIISOCYANATO-1-METHYL-
		CRESORCINOL DIISOCYANATE
		DESMODUR T100
		DESMODUR T80
		DI-ISO-CYANATOLUENE
		DI-ISOCYANATE DE TOLUYLENE (FRENCH)
		DIISOCYANAT-TOLUOL (GERMAN)
		DIISOCYANATOMETHYLBENZENE
		DIISOCYANATOTOLUENE
		HYLENE T
		HYLENE TCPA
		HYLENE TLC
		HYLENE TM
		HYLENE TM-65
		HYLENE TRF
		HYLENE-T
		ISOCYANIC ACID, 4-METHYL-M-PHENYLENE ESTER
		ISOCYANIC ACID, METHYL-M-PHENYLENE ESTER

CAS No.	IN	Chemical Name
		ISOCYANIC ACID, METHYLPHENYLENE ESTER
		META-TOLYLENE DIISOCYANATE
		METHYL-M-PHENYLENE ISOCYANATE
		METHYL-META-PHENYLENE DIISOCYANATE
		METHYLPHENYLENE ISOCYANATE
		MONDUR TD
		MONDUR TD-80
		MONDUR TDS
		MONDUR-TD
		MONDUR-TD-80
		NACCONATE-100
		NCI-C50533
		NIAX ISOCYANATE TDI
		NIAX TDI
		NIAX TDI-P
		RCRA WASTE NUMBER U223
		RUBINATE TDI
		RUBINATE TDI 80/20
		T 100
		TDI
		TDI 80-20
		TDI-80
		TOLUEEN-DIISOCYANAAT (DUTCH)
		TOLUEN-DISOCIANATO (ITALIAN)
		TOLUENE 2,4-DIISOCYANATE
		TOLUENE DIISOCYANATE
		TOLUENE DIISOCYANATE (DOT)
		TOLUENE-2,4-DIISOCYANATE
		TOLUENE-2,4-DIISOCYANATE (ACGIH)
		TOLUILENODWUIZOCYJANIAN (POLISH)
		TOLUYLENE-2,4-DIISOCYANATE
		TOLYENE 2,4-DIISOCYANATE
		TOLYLENE DIISOCYANATE
		TOLYLENE ISOCYANATE
		TOLYLENE-2,4-DIISOCYANATE
		TULUYLENDIISOCYANAT (GERMAN)
		UN 2078 (DOT)
26499-65-0		
		GYPSUM HEMIHYDRATE
		HEMIHYDRATE GYPSUM
	*	PLASTER OF PARIS
26506-47-8		
		CHLORIC ACID, COPPER SALT
		COPPER CHLORATE
		COPPER CHLORATE (DOT)
		CUPRIC CHLORATE
		UN 2721 (DOT)
26555-35-1		
		CARBONIC ACID, THIO-, ANHYDROSULFIDE WITH THIO-HYPOCHLOROUS ACID, ETHYL ESTER
		CARBONOCHLORIDOTHIOIC ACID, S-ETHYL ESTER
		CARBONOTHIOIC ACID, ANHYDROSULFIDE WITH THIO-HYPOCHLOROUS ACID, ETHYL ESTER
		ETHOXYCARBONYLSULFENYL CHLORIDE
		ETHYL (CHLOROSULFENYL)FORMATE
		ETHYL CHLOROTHIOFORMATE
		ETHYL CHLOROTHIOLFORMATE
		ETHYL CHLOROTHIOLOFORMATE
		ETHYL THIOCHLOROFORMATE
		FORMIC ACID, CHLOROTHIO-, ETHYL ESTER
		FORMIC ACID, CHLOROTHIO-, S-ETHYL ESTER
		S-ETHYL CARBONOCHLORIDOTHIOATE
		S-ETHYL CHLOROTHIOCARBONATE
		S-ETHYL CHLOROTHIOFORMATE
		S-ETHYL CHLOROTHIOLFORMATE
		S-ETHYL THIOCHLOROFORMATE
		THIOHYPOCHLOROUS ACID, ANHYDROSULFIDE WITH O-ETHYL THIOCARBONATE
		UN 2826 (DOT)
26571-79-9		
		CHLOROPHENYL TRICHLOROSILANE
		CHLOROPHENYLTRICHLOROSILANE (DOT)
		SILANE, CHLOROPHENYLTRICHLORO-
		SILANE, TRICHLORO(CHLOROPHENYL)-
		TRICHLORO(CHLOROPHENYL)SILANE
		UN 1753 (DOT)
26628-22-8		
		AZIDE
		AZIUM
		AZOTURE DE SODIUM (FRENCH)
		HYDRAZOIC ACID, SODIUM SALT
		KAZOE
		NATRIUMAZID (GERMAN)
		NATRIUMMAZIDE (DUTCH)
		NCI-C06462
		NEMAZYD
		NSC 3072
		RCRA WASTE NUMBER P105
		SMITE
	*	SODIUM AZIDE
		SODIUM AZIDE (ACGIH,DOT)
		SODIUM AZIDE (Na(N3))
		SODIUM, AZOTURE DE (FRENCH)
		SODIUM, AZOTURO DI (ITALIAN)
		U-3886
		UN 1687 (DOT)
26635-64-3		
		ISOOCTANE
26675-46-7		
		ISOFLURANE
26748-41-4		
		ESPEROX 33M
		LUPERSOL 10
		LUPERSOL 10M75
		NEODECANEPEROXOIC ACID, 1,1-DIMETHYLETHYL ESTER
		PEROXYNEODECANOIC ACID, TERT-BUTYL ESTER
		PEROXYNEODECANOIC ACID, TERT-BUTYL ESTER, NOT MORE THAN 77% IN SOLUTION
		TERT-BUTYL PERNEODECANOATE
		TERT-BUTYL PEROXYNEODECANOATE
		TERT-BUTYL PEROXYNEODECANOATE, ≤ 77% IN SOLU-TION
		TERT-BUTYL PEROXYNEODECANOATE, NOT MORE THAN 77% IN SOLUTION (DOT)
		TERT-BUTYL PEROXYNEODECANOATE, TECHNICALLY PURE
		UN 2177 (DOT)
26748-47-0		
		ALPHA-CUMYL PEROXYNEODECANOATE
		BENZYL ALCOHOL, ALPHA,ALPHA-DIMETHYL-, PEROXY-NEODECANOATE
		CUMYL PERNEODECANOATE
		CUMYL PEROXYNEODECANOATE
		CUMYL PEROXYNEODECANOATE, MAXIMUM CONCENTRA-TION 77% IN SOLUTION (DOT)
		ESPEROX 939M
		LUPERSOL 188
		LUPERSOL 188M75
		NEODECANEPEROXOIC ACID, 1-METHYL-1-PHENYLETHYL ESTER
		PEROXYNEODECANOIC ACID, ALPHA,ALPHA-DIMETHYL-BENZYL ESTER
		PEROXYNEODECANOIC ACID, DIMETHYLBENZYL ESTER, 77% IN SOLUTION
		UN 2963 (DOT)
26760-64-5		
		2-BUTENE, 2-METHYL-
		2-METHYL-2-BUTENE
		2-METHYL-2-BUTENE (DOT)
		2-METHYLBUTENE
		AMYLENE
		AMYLENE (DOT)
		BETA-ISO-AMYLENE
		BUTENE, 2-METHYL-
		ETHYLENE, TRIMETHYL-
		ISOAMYLENE
		ISOPENTENE

CAS No.	IN	Chemical Name
		METHYL BUTENE
		METHYL BUTENE (DOT)
		TERT-AMYLENE
		TRIMETHYLETHYLENE
		UN 2460 (DOT)
26761-40-0		
		1,2-BENZENEDICARBOXYLIC ACID, DIISODECYL ESTER
		1,2-BENZENEDICARBOXYLIC ACID, DIISODECYL ESTER (9CI)
		BIS(ISODECYL)PHTHALATE
		DIDP
		DIDP (PLASTICIZER)
		DIISODECYL PHTHALATE
		ISODECYL ALCOHOL, PHTHALATE (2:1)
		ISODECYL PHTHALATE
		PALATINOL Z
		PHTHALIC ACID, DIISODECYL ESTER
		PX 120
		SICOL 184
		VESTINOL DZ
26762-93-6		
		DIISOPROPYLBENZENE HYDROPEROXIDE
		DIISOPROPYLBENZENE HYDROPEROXIDE SOLUTION , NOT OVER 72% PEROXIDE (DOT)
		DIISOPROPYLBENZENE HYDROPEROXIDE SOLUTION, NOT OVER 72% PEROXIDE (DOT)
		DIISOPROPYLBENZENE HYDROPEROXIDE, NOT MORE THAN 72% IN SOLUTION (DOT)
		HYDROPEROXIDE, BIS(1-METHYLETHYL)PHENYL
		HYDROPEROXIDE, DIISOPROPYLPHENYL
		HYDROPEROXIDE, DIISOPROPYLPHENYL-, (SOLUTION)
		UN 2171 (DOT)
26914-02-3		
		IODO PROPANE
		IODOPROPANES (DOT)
		PROPANE, IODO-
		UN 2392 (DOT)
26952-21-6		
		ISOCTYL ALCOHOL
		ISOOCTANOL
		ISOOCTYL ALCOHOL
26952-23-8		
		1-PROPENE, DICHLORO-
		DICHLOROPROPENE
		DICHLOROPROPYLENE
26952-42-1		
		TRINITRO-ANILINE
		TRINITROANILINE
27134-26-5		
		1-AMINO-4-CHLORO BENZENE
		4-CHLORANILIN (CZECH)
		4-CHLOROANILINE
		4-CHLOROBENZENAMINE
		4-CHLOROPHENYLAMINE
		ANILINE, 4-CHLORO-
		ANILINE, CHLORO-
		ANILINE, P-CHLORO-
		BENZENAMINE, CHLORO-
		BENZENEAMINE, 4-CHLORO
		CHLOROANILINE
		CHLOROANILINE, LIQUID
		CHLOROANILINE, SOLID
		NCI-C02039
		P-CHLORANILINE
		P-CHLOROANILINE
		P-CHLOROANILINE, LIQUID (DOT)
		P-CHLOROANILINE, SOLID (DOT)
		RCRA WASTE NUMBER P024
		UN 2018 (DOT)
		UN 2019 (DOT)
27134-27-6		
		ANILINE, DICHLORO-
		ANILINE, DICHLORO- (MIXED ISOMERS)
		BENZENAMINE, AR,AR-DICHLORO-
		BENZENAMINE, AR,AR-DICHLORO- (9CI)
		DICHLOROANILINE
		DICHLOROANILINES
		DICHLOROANILINES (DOT)
		O-DICHLOROANILINE
		UN 1590 (DOT)
27137-41-3		
		2-METHYLFURAN (CZECH)
		2-METHYLFURAN (DOT)
		FURAN, 2-METHYL-
		FURAN, METHYL-
		METHYLFURAN
		METHYLFURAN (DOT)
		SILVAN (CZECH)
		SYLVAN
		UN 2301 (DOT)
27137-85-5		
		DICHLOROPHENYL TRI-CHLOROSILANE
		TRICHLORO(DICHLOROPHENYL)SILANE
27156-03-2		
		DICHLORODIFLUORO-ETHYLENE
27176-87-0		
		BENZENESULFONIC ACID, DODECYL-
		BIO-SOFT S 100
		CALSOFT LAS 99
		DOBANIC ACID 83
		DOBANIC ACID JN
	*	DODECYLBENZENESULFONIC ACID
		DODECYLBENZENESULFONIC ACID (DOT)
		DODECYLBENZENESULPHONIC ACID
		E 7256
		ELFAN WA SULPHONIC ACID
		LAS 99
		LAURYLBENZENESULFONIC ACID
		MARLON AS 3
		MARLON AS B
		N-DODECYLBENZENESULFONIC ACID
		NA 2584 (DOT)
		NACCONOL 98SA
		NANSA 1042P
		NANSA SSA
		RICHONIC ACID B
		SULFRAMIN ACID 1298
27193-28-8		
		OCTYL PHENOL
		PHENOL, (1,1,3,3-TETRAMETHYLBUTYL)-
		PHENOL, OCTYL-
		TERT-OCTYLPHENOL
		USAF RH-6
27196-00-5		
		1-TETRADECANOL
		ALFOL 14
		DYTOL R-52
		LANETTE K
		LANETTE WAX KS
		LOXANOL V
		MYRISTIC ALCOHOL
		MYRISTYL ALCOHOL
		MYRISTYL ALCOHOL (MIXED ISOMERS)
		N-TETRADECAN-1-OL
		N-TETRADECANOL
		N-TETRADECANOL-1
		N-TETRADECYL ALCOHOL
		TETRADECANOL
		TETRADECANOL, MIXED ISOMERS
		TETRADECYL ALCOHOL

CAS No.	IN	Chemical Name
27215-10-7		DIISOOCTYL ACID PHOSPHATE
		DIISOOCTYL ACID PHOSPHATE (DOT)
		DIISOOCTYL PHOSPHATE
		ISOOCTANOL, HYDROGE N PHOSPHATE
		ISOOCTYL PHOSPHATE ((C8H17O)2(HO)PO)
		PHOSPHORIC ACID DIISOOCTYL ESTER
		PHOSPHORIC ACID, DIISOOCTYL ESTER
		UN 1902 (DOT)
27215-95-8		NONENE
27236-46-0		ISOHEXENE
27254-36-0		MONONITRONAPHTHALENE
		NAPHTHALENE, MONONITRO-
		NAPHTHALENE, NITRO-
		NITRONAPHTHALENE
		NITRONAPHTHALENE (DOT)
		UN 2538 (DOT)
27323-18-8		AROCHLOR-1254
27323-41-7		BENZENESULFONIC ACID, DODECYL-, COMPD WITH 2,2',2"-NITRILOTRIETHANOL (1:1)
		BENZENESULFONIC ACID, DODECYL-, COMPD WITH 2,2',2"-NITRILOTRIS[ETHANOL] (1:1)
		CALSOFT T 60
		CONCO AAS SPECIAL
		DECOL T 70
		DODECYLBENZENESULFONIC ACID TRIETHANOLAMINE SALT
		ELFAN WAT
		ETHANOL, 2,2',2"-NITRILOTRI-, DODECYLBENZENESULFONATE (SALT)
		ETHANOL, 2,2',2"-NITRILOTRIS-, DODECYLBENZENE-SULFONATE (SALT)
		NA 9151 (DOT)
		NEOGEN T
		PILOT TS 60
		SINNOZON NT 45
		TRIETHANOLAMINE DODECYLBENZENESULFONATE
		TRIETHANOLAMINE DODECYLBENZENESULFONATE (DOT)
		TRIETHANOLAMINE DODECYLBENZENESULFONATE (SALT)
		TRIETHANOLAMINE DODECYLBENZENESULFONIC ACID SALT
		TRIETHANOLAMINE DODECYLBENZOSULFONATE
		TRIETHANOLAMMONIUM DODECYLBENZENESULFONATE
		ULTRAWET 60L
27417-39-6		1H-PYRROLE, METHYL-
		METHYLPYRROLE
		PYRROLE, METHYL-
27458-20-4		BENZENE, 1-BUTYL-2-METHYL- (9CI)
		BENZENE, BUTYLMETHYL-
		BUTYL TOLUENE
		O-BUTYLTOLUENE
		O-BUTYLTOLUENE (DOT)
		TOLUENE, BUTYL-
		TOLUENE, O-BUTYL-
		UN 2667 (DOT)
27598-85-2		4-AMINO-1-HYDROXYBENZENE
		4-AMINOPHENOL
		4-HYDROXYANILINE
		ACTIVOL
		AMINOPHENOL
		AMINOPHENOLS
		AMINOPHENOLS (:O-, M-, P-:)
		AZOL
		BASF URSOL P BASE
		BENZOFUR P
		CERTINAL
		CI OXIDATION BASE 6A
		CITOL
		DURAFUR BROWN RB
		FOURAMINE P
		FOURRINE 84
		FOURRINE P BASE
		FURRO P BASE
		HYDROXYPHENYLAMINE
		NAKO BROWN R
		P-AMINOFENOL (CZECH)
		P-AMINOPHENOL
		P-AMINOPHENOL (DOT)
		P-HYDROXYANILINE
		PAP
		PARANOL
		PELAGOL GREY P BASE
		PELAGOL P BASE
		PHENOL, AMINO-
		PHENOL, P-AMINO-
		RENAL AC
		RODINAL
		TERTRAL P BASE
		UN 2512 (DOT)
		UNAL
		URSOL P
		URSOL P BASE
		ZOBA BROWN P BASE
27774-13-6		C.I. 77940
		NA 9152 (DOT)
		UN 2931 (DOT)
		VANADIUM OXIDE SULFATE (VO(SO4))
		VANADIUM OXOSULFATE
		VANADIUM OXYSULFATE (VOSO4)
		VANADIUM, OXOSULFATO-
		VANADIUM, OXO[SULFATO(2-)-O]-
		VANADIUM, OXYSULFATO-
		VANADYL SULFATE
		VANADYL SULFATE (DOT)
		VANADYL SULFATE (VO(SO4))
27813-02-1		1,2-PROPANEDIOL, MONOMETHACRYLATE
		2-PROPENOIC ACID, 2-METHYL-, MONOESTER WITH 1,2-PROPANEDIOL
		HYDROXYPROPYL METHACRYLATE
		METHACRYLIC ACID, ESTER WITH 1,2-PROPANEDIOL
		METHACRYLIC ACID, MONOESTER WITH 1,2-PROPANEDIOL
27987-06-0		ETHANE, TRIFLUORO-
		TRIFLUOROETHANE
		TRIFLUOROETHANE (DOT)
		UN 2035 (DOT)
28165-71-1		DICHLORPHENOXYACETIC ACID ESTER (2,6-DICHLOROACETATE PHENOL)
28258-59-5		XYLYL BROMIDE
28260-61-9		CHLOROTRINITROBENZENE
		TRINITROCHLOROBENZENE
28300-74-5		ANTIMONATE(1-), AQUA[TARTRATO(4-)]-, POTASSIUM, HEMIHYDRATE, DIMER
		ANTIMONATE(1-), OXO(TARTRATO)-, POTASSIUM HEMIHYDRATE, DIMER
		ANTIMONATE(2)-, BIS(MU-TARTRATO(4-))DI-, DIPOTASSIUM, TRIHYDRATE
	*	ANTIMONY POTASSIUM TARTRATE
		ANTIMONY POTASSIUM TARTRATE (DOT)

CAS No.	IN	Chemical Name
		ANTIMONY POTASSIUM TARTRATE SOLID (DOT)
		ANTIMONY POTASSIUM TARTRATE, SOLID
		ANTIMONYL POTASSIUM TARTRATE
		ANTIMONYL POTASSIUM TARTRATE, HEMIHYDRATE
		BUTANEDIOIC ACID, 2,3-DIHYDROXY- [R-(R*,R*)]-, ANTIMONY COMPLEX
		EMETIQUE (FRENCH)
		ENT 50,434
		POTASSIUM ANTIMONY TARTRATE
		POTASSIUM ANTIMONYL D-TARTRATE
		POTASSIUM ANTIMONYL TARTRATE
		TARTAR EMETIC
		TARTARIC ACID, ANTIMONY POTASSIUM SALT
		TARTARIZED ANTIMONY
		TARTOX
		TARTRATE ANTIMONIO-POTASSIQUE (FRENCH)
		TARTRATED ANTIMONY
		UN 1551 (DOT)
28314-03-6		
		1-ACETYLAMINOFLUORENE
		ACETAMIDE, N-9H-FLUOREN-1-YL-
		ACETAMIDE, N-FLUOREN-1-YL-
		ALPHA-ACETAMIDOFLUORENE
28324-52-9		
		2,6,6-TRIMETHYL NORPINANYL HYDROPEROXIDE, TECHNICALLY PURE (DOT)
		BORNYL CHLORIDE
		HYDROPEROXIDE, 2,6,6-TRIMETHYLBICYCLO(311)HEPTYL-
		HYDROPEROXIDE, 2,6,6-TRIMETHYLBICYCLO(311)HEPTYL-, NOT OVER 45% PEROXIDE
		HYDROPEROXIDE, 2,6,6-TRIMETHYLBICYCLO[311]HEPT-2-YL, (1ALPHA,2ALPHA,5ALPHA)-
		HYDROPEROXIDE, 2,6,6-TRIMETHYLBICYCLO[311]HEPTYL
		PINANE HYDROPEROXIDE
		PINANE HYDROPEROXIDE (DOT)
		PINANE HYDROPEROXIDE SOLUTION, WITH NOT MORE THAN 45% PEROXIDE
		PINANE HYDROPEROXIDE, TECHNICALLY PURE (DOT)
		PINANYL HYDROPEROXIDE
		PINANYL HYDROPEROXIDE, TECHNICAL PURE (DOT)
		TRIMETHYL NORPINANYL HYDROPEROXIDE
		UN 2162 (DOT)
28347-13-9		
		ALPHA,ALPHA'-DICHLOROXYLENE
		BENZENE, BIS(CHLOROMETHYL)-
		BENZENE, BIS(CHLOROMETHYL)- (9CI)
		BIS(CHLOROMETHYL)BENZENE
		DICHLOROXYLYLENE
		XYLENE, ALPHA,ALPHA'-DICHLORO-
		XYLYLENE CHLORIDE
		XYLYLENE DICHLORIDE
28434-86-8		
		3,3'-DICHLORO-4,4'-DIAMINODIPHENYL ETHER
		3,3'-DICHLORO-4-4'-DIAMINODIPHENYL ETHER
		4,4'-OXYBIS(2-CHLOROANILINE)
		ANILINE, 4,4'-OXYBIS[2-CHLORO-
		BENZENAMINE, 4,4'-OXYBIS[2-CHLORO-
28479-22-3		
		3-CHLORO-4-METHYLPHENYL ISOCYANATE
		3-CHLORO-4-METHYLPHENYL ISOCYANATE (DOT)
		3-CHLORO-P-TOLYL ISOCYANATE
		BENZENE, 2-CHLORO-4-ISOCYANATO-1-METHYL-
		CHLOROMETHYLPHENYLISOCYANATE
		ISOCYANIC ACID, 3-CHLORO-P-TOLYL ESTER
		UN 2236 (DOT)
28554-00-9		
		CHLOROPROPIONIC ACID
		CHLOROPROPIONIC ACID (DOT)
		MONOCHLOROPROPIONIC ACID
		PROPANOIC ACID, CHLORO-
		PROPANOIC ACID, CHLORO- (9CI)
		PROPIONIC ACID, CHLORO-
		UN 2511 (DOT)
28604-91-3		
		2,2'-AZODI-(2,4-DIMETHYLVALERONITRILE)
28679-16-5		
		HEXANE, 1,6-DIISOCYANATOTRIMETHYL-
		TRIMETHYLHEXAMETHYLENE DIISOCYANATE
		TRIMETHYLHEXAMETHYLENE DIISOCYANATE (DOT)
		TRIMETHYLHEXAMETHYLENE ISOCYANATE
		UN 2328 (DOT)
28729-54-6		
		BENZENE, METHYLPROPYL-
		METHYL PROPYL BENZENE
		N-PROPYLTOLUENE
		PROPYLTOLUENE
		TOLUENE, PROPYL-
28772-56-7		
		(HYDROXY-4 COUMARINYL 3)-3 PHENYL-3 (BROMO-4 BIPHENYLYL-4)-1 PROPANOL-1 (FRENCH)
		2H-1-BENZOPYRAN-2-ONE, 3-[3-(4'-BROMO[1,1'-BIPHENYL]-4-YL)-3-HYDROXY-1-PHENYLPROPYL]-4-HYDROXY-
		3-(3-(4'-BROMO(1,1'-BIPHENYL)-4-YL)3-HYDROXY-1-PHENYLPROPYL)-4-HYDROXY-2H-1-BENZOPYRAN-2-ONE
		3-(ALPHA-(P-(P-BROMOPHENYL)-BETA-HYDROXYPHENETHYL)BENZYL)-4-HYDROXYCOUMARIN
		BROMADIALONE
		BROMADIOLON
		BROMADIOLONE
		BROMONE
		CANADIEN 2000
		CONTRAC
		COUMARIN, 3-(3-(4'-BROMO-1,1'-BIPHENYL-4-YL)-3-HYDROXY-1-PHENYLPROPYL)-4-HYDROXY-
		COUMARIN, 3-(ALPHA-(P-(P-BROMOPHENYL)-BETA-HYDROXYPHENETHYL)BENZYL)-4-HYDROXY- (8CI)
		COUMARIN, 3-[ALPHA-[P-(P-BROMOPHENYL)-BETA-HYDROXYPHENETHYL]BENZYL]-4-HYDROXY-
		LM-637
		MAKI
		RATIMUS
		SUP'OPERATS
		SUPER-CAID
		SUPER-ROZOL
		TEMUS
28805-86-9		
		4-N-BUTYLPHENOL
		BUTYL PHENOL
		BUTYL PHENOL, LIQUID
		P-BUTYLPHENOL, LIQUID (DOT)
		P-BUTYLPHENOL, SOLID (DOT)
		PHENOL, BUTYL-
		PHENOL, P-BUTYL
		UN 2228 (DOT)
		UN 2229 (DOT)
28831-12-1		
		PEROXYMONOSULFURIC ACID, MONOSODIUM SALT
		PERSULFATE DE SODIUM (FRENCH)
		SODIUM MONOPERSULFATE
		SODIUM PEROXYDISULFATE
		SODIUM PERSULFATE
		SODIUM PERSULFATE (ACGIH,DOT)
		UN 1505 (DOT)
28838-02-0		
		ARSENENIC ACID, ZINC SALT
		ARSENIC ACID (H4As2O7), ZINC SALT (1:2)
		ARSENIC ACID, ZINC SALT
		DIARSENIC ACID, ZINC SALT (1:2)
		UN 1712 (DOT)
		ZINC ARSENATE
		ZINC ARSENATE (DOT)
		ZINC ARSENATE, BASIC
		ZINC ARSENATE, SOLID (DOT)

CAS No.	IN	Chemical Name
28905-71-7		TRINITROCRESOL
28911-01-5		TRIAZOLAM
28981-97-7		ALPRAZOLAM
28983-37-1		TERT-TETRADECANETHIOL
		TERT-TETRADECYL MERCAPTAN
28984-85-2		1,1'-BIPHENYL, NITRO-
		BIPHENYL, NITRO-
		NITROBIPHENYL
29027-17-6		2-CHLORO-3-METHYL ANILINE (CHLOROTOLUIDINE)
29154-12-9		2-CHLOROPYRIDINE
		2-CHLOROPYRIDINE (DOT)
		ALPHA-CHLOROPYRIDINE
		CHLOROPYRIDINE
		O-CHLOROPYRIDINE
		PYRIDINE, 2-CHLORO-
		PYRIDINE, CHLORO-
		UN 2822 (DOT)
29191-52-4		(METHOXYPHENYL)AMINE
		1-AMINO-2-METHOXYBENZENE
		2-AMINOANISOLE
		2-ANISIDINE
		2-METHOXY-1-AMINOBENZENE
		2-METHOXYANILINE
		2-METHOXYBENZENAMINE
		ANISIDINE
		ANISIDINE, ISOMERS
		BENZENAMINE, 2-METHOXY- (9CI)
		BENZENAMINE, AR-METHOXY-
		METHOXYANILINE
		O-AMINOANISOLE
		O-ANISIDINE
		O-ANISIDINE (ACGIH,DOT)
		O-ANISYLAMINE
		O-METHOXYANILINE
		O-METHOXYPHENYLAMINE
		UN 2431 (DOT)
29240-17-3		ESPEROX 551M
		LUPERSOL 554
		LUPERSOL 554M75
		LUPERSOL TA 54
		PEROXYPIVALIC ACID, TERT-PENTYL ESTER
		PROPANEPEROXOIC ACID, 2,2-DIMETHYL-, 1,1-DIMETHYLPROPYL ESTER
		TERT-AMYL PEROXYPIVALATE
		TERT-AMYL PERPIVALATE
		TERT-PENTYL ALCOHOL, PEROXYPIVALATE
		TERT-PENTYL PEROXYPIVALATE
29595-25-3		AMMONIUM DINITRO-O-CRESOLATE
		AMMONIUM DINITRO-O-CRESOLATE (DOT)
		O-CRESOL, DINITRO-, AMMONIUM SALT
		PHENOL, 2-METHYL-, DINITRO DERIV, AMMONIUM SALT
		UN 1843 (DOT)
29756-38-5		2-BROMOPENTANE
		2-BROMOPENTANE (DOT)
		BROMOPENTANE
		N-AMYL BROMIDE
		PENTANE, 2-BROMO-
		PENTANE, BROMO-
		UN 2343 (DOT)
29757-24-2		1-AMINO-4-NITROBENZENE
		4-NITRANILINE
		4-NITROANILINE
		4-NITROBENZENAMINE
		AMINONITROBENZENE
		ANILINE, 4-NITRO-
		ANILINE, NITRO-
		ANILINE, P-NITRO-
		AZOAMINE RED ZH
		AZOFIX RED GG SALT
		AZOIC DIAZO COMPONENT 37
		BENZENAMINE, 4-NITRO- (9CI)
		BENZENAMINE, AR-NITRO-
		CI 37035
		CI AZOIC DIAZO COMPONENT 37
		CI DEVELOPER 17
		DEVELOPER P
		DEVOL RED GG
		DIAZO FAST RED GG
		FAST RED 2G BASE
		FAST RED 2G SALT
		FAST RED BASE 2J
		FAST RED BASE GG
		FAST RED GG BASE
		FAST RED GG SALT
		FAST RED MP BASE
		FAST RED P BASE
		FAST RED P SALT
		FAST RED SALT 2J
		FAST RED SALT GG
		NAPHTOELAN RED GG BASE
		NCI-C60786
		NITRAZOL CF EXTRA
		NITROANILINE
		P-AMINONITROBENZENE
		P-NITRANILINE
		P-NITROANILINA (POLISH)
		P-NITROANILINE
		P-NITROANILINE (ACGIH,DOT)
		P-NITROPHENYLAMINE
		PARANITROANILINE, SOLID (DOT)
		PNA
		RCRA WASTE NUMBER P077
		RED 2G BASE
		SHINNIPPON FAST RED GG BASE
		UN 1661 (DOT)
29790-52-1		1-METHYL-2-(3-PYRIDYL)PYRROLIDINE SALICYLATE
		BENZOIC ACID, 2-HYDROXY-, COMPD WITH (S)-3-(1-METHYL-2-PYRROLIDINYL)PYRIDINE (1:1)
		EUDERMOL
		NICOTINE SALICYLATE
		NICOTINE SALICYLATE (DOT)
		NICOTINE, MONOSALICYLATE
		PYRIDINE, 3-(1-METHYL-2-PYRROLIDINYL)-, (S)-, MONO(2-HYDROXYBENZOATE)
		SALICYLIC ACID, COMPD WITH NICOTINE (1:1)
		UN 1657 (DOT)
29965-97-7		CYCLOOCTADIENE
		CYCLOOCTADIENES (DOT)
		UN 2520 (DOT)
30026-92-7		TERT-BUTYL ISOPRPYL BENZENE HYDROPEROXIDE
30030-25-2		AR-(CHLOROMETHYL)STYRENE
		AR-VINYLBENZYL CHLORIDE
		BENZENE, (CHLOROMETHYL)ETHENYL-
		STYRENE, AR-(CHLOROMETHYL)-
		TOLUENE, ALPHA-CHLOROVINYL-

CAS No.	IN	Chemical Name
		VINYLBENZYL CHLORIDE
30143-13-6		BENZENE, METHYL-, DIAMINO DERIV
		BENZENEDIAMINE, AR-METHYL-
		DIAMINOTOLUENE
		METHYLPHENYLENEDIAMINE
		NA 1709 (DOT)
		RCRA WASTE NUMBER U221
		TOLUENE-AR,AR-DIAMINE
		TOLUENEDIAMINE
		TOLUENEDIAMINE (DOT)
		TOLYLENEDIAMINE
30145-38-1		ETHYL TELLURAC
30174-58-4		TERT-DECANETHIOL
		TERT-DECYLMERCAPTAN
30207-98-8		1-HENDECANOL
		ALCOHOL C-11
		HENDECANOIC ALCOHOL
		HENDECYL ALCOHOL
		N-HENDECYLENIC ALCOHOL
		N-UNDECANOL
		N-UNDECYL ALCOHOL
		TIP-NIP
		UNDECANOL
		UNDECYL ALCOHOL
30525-89-4		ALDACIDE
		FLO-MOR
		PARAFORM
		PARAFORMALDEHYDE
30525-9-4		PARAFORMALDEHYDE
		1-(1-NAPHTHYL)-2-THIOUREA
		1-NAFTIL-TIOUREA (ITALIAN)
		1-NAFTYLTHIOUREUM (DUTCH)
		1-NAPHTHALENYLTHIOUREA
		1-NAPHTHYL-THIOHARNSTOFF (GERMAN)
		1-NAPHTHYL-THIOUREE (FRENCH)
		ALPHA-NAPHTHALTHIOHARNSTOFF (GERMAN)
		ALPHA-NAPHTHOTHIOUREA
		ALPHA-NAPHTHYLTHIOCARBAMIDE
		ALPHA-NAPHTHYLTHIOUREA
		ALPHA-NAPHTHYLTHIOUREA (DOT)
		ALPHANAPHTYL THIOUREE (FRENCH)
		ALRATO
		ANTU
		ANTU (ACGIH)
		ANTURAT
		BANTU
		CHEMICAL 109
		DIRAX
		KILL KANT Z
		KRYSID
		KRYSID PI
		N-(1-NAPHTHYL)-2-THIOUREA
		N-NAPHTHYLTHIOUREA
		NAPHTHYLTHIOUREA
		NAPHTOX
		RAT-TU
		RATTRACK
		RCRA WASTE NUMBER P072
		SMEESANA
		THIOUREA, 1-NAPHTHALENYL-
		THIOUREA, NAPHTHALENYL-
		TIOX 30
		U-5227
		UN 1651 (DOT)
		UREA, 1-(1-NAPHTHYL)-2-THIO-
		UREA, 1-(NAPHTHYL)-2-THIO-

CAS No.	IN	Chemical Name
		USAF EK-P-5976
30580-75-7		7-(TERT-BUTYLPEROXY)CUMENE
		BCP
		CUMYL TERT-BUTYL PEROXIDE
		KAYABUTYL C
		LUPEROX 801
		LUPERSOL 801
		PERBUTYL C
		PEROXIDE, 1,1-DIMETHYLETHYL (1-METHYLETHYL)-PHENYL
		PEROXIDE, 1,1-DIMETHYLETHYL 1-METHYL-1-PHENYL-ETHYL
		PEROXIDE, 1,1-DIMETHYLETHYL 1-METHYL-1-PHENYL-ETHYL (9CI)
		PEROXIDE, TERT-BUTYL ALPHA,ALPHA-DIMETHYLBENZYL
		TERT-BUTYL 1-METHYL-1-PHENYLETHYL PEROXIDE
		TERT-BUTYL CUMENE PEROXIDE, TECHNICAL PURE (DOT)
		TERT-BUTYL CUMYL PEROXIDE
		TERT-BUTYL CUMYL PEROXIDE, TECHNICAL PURE (DOT)
		TERT-BUTYL CUMYL PEROXIDE, TECHNICALLY PURE
		TRIGONOX T
		TRIGONOX T 40
		UN 2091 (DOT)
30586-10-8		DICHLOROPENTANE
30586-18-6		HEPTANE, PENTAMETHYL-
		PENTAMETHYL HEPTANE
		PENTAMETHYLHEPTANE (DOT)
		UN 2286 (DOT)
30674-80-7		2-ISOCYANATOETHYL METHACRYLATE
		2-PROPENOIC ACID, 2-METHYL-, 2-ISOCYANATOETHYL ESTER
		BETA-ISOCYANATOETHYL METHACRYLATE
		IEM
		ISOCYANATOETHYL METHACRYLATE
		ISOCYANIC ACID, 2-HYDROXYETHYL ESTER METHACRYLATE (ESTER)
		METHACRYLIC ACID BETA-ISOCYANATOETHYL ESTER
		METHACRYLIC ACID, 2-ISOCYANATOETHYL ESTER
		METHACRYLOYLOXYETHYL ISOCYANATE
		METHARCYLIC ACID, 2-ISOCYANATOETHYL ESTER
30714-78-4		CARBONIC ACID, BUTYL ETHYL ESTER
		ETHYL BUTYL CARBONATE
30900-23-3		1,2,3,4,10,10-HEXACHLORO-1,4,4A,5,8,8A-HEXAHYDRO-EXO-1,4-ENDO-5,8-DIMETHANONAPHTHALENE
		1,2,3,4,10,10-HEXACHLORO-1,4,4A,5,8,8A-HEXAHYDRO-1,4,5,8-DIMETHANONAPHTHALENE
		1,2,3,4,10,10-HEXACHLORO-1,4,4A,5,8,8A-HEXAHYDRO-1,4-ENDO-EXO-5,8-DIMETHANONA-PHTHALENE
		1,4:5,8-DIMETHANONAPHTHALENE, 1,2,3,4,10,10-HEXA-CHLORO-1,4,4A,5,8,8A-HEXAHYDRO-, ENDO,EXO-
		ALDOCIT
		ALDREX
		ALDREX 30
		ALDREX 40
		ALDRIN
		ALDRIN (ACGIH,DOT)
		ALDRIN, CAST SOLID (DOT)
		ALDRINE (FRENCH)
		ALDRITE
		ALDROSOL
		ALTOX
		COMPOUND 118
		DRINOX
		ENT 15,949
		HEXACHLOROHEXAHYDRO-ENDO-EXO-DIMETHANONAPH-THALENE

CAS No.	IN	Chemical Name
		HHDN
		KORTOFIN
		NA 2761 (DOT)
		NCI-C00044
		OCTALENE
		RCRA WASTE NUMBER P004
		SD 2794
		SEEDRIN
		TATUZINHO
31212-28-9		
		BENZENESULFONIC ACID, NITRO-
		NITROBENZENESULFONIC ACID
		NITROBENZENESULPHONIC ACID (DOT)
		UN 2305 (DOT)
31394-54-4		
		ISOHEPTANE
32280-46-9		
		1,3-BUTANEDIAMINE, N,N'-DIETHYL-
		ETHYLAMINE, N,N'-(1-METHYLTRIMETHYLENE)BIS-
		N,N'-DIETHYL-1,3-BUTANEDIAMINE
		N,N-DIETHYL-1,3-BUTANEDIAMINE
32534-95-5		
		ALPHA-(2,4,5-TRICHLOROPHENOXY)PROPIONIC ACID ISOOCTYL ESTER
		2,4,5-TRICHLOROPHENOXYPROPIONIC ACID ESTER
		ISOOCTYL 2-(2,4,5-TRICHLOROPHENOXY)PROPIONATE
		PROPANOIC ACID, 2-(2,4,5-TRICHLOROPHENOXY)-, ISOOCTYL ESTER
		PROPIONIC ACID, 2-(2,4,5-TRICHLOROPHENOXY)-, ISOOCTYL ESTER
		SILVEX ISOOCTYL ESTER
32749-94-3		
		2,3-DIMETHYL PENTALDEHYDE
		2,3-DIMETHYLPENTANAL
		2,3-DIMETHYLVALERALDEHYDE
		PENTANAL, 2,3-DIMETHYL-
		VALERALDEHYDE, 2,3-DIMETHYL-
33213-65-9		
		ALPHA-THIONEX
		BETA-BENZOEPIN
		BETA-ENDOSULFAN
		BETA-THIODAN
		5-NORBORNENE-2,3-DIMETHANOL, 1,4,5,6,7,7-HEXACHLORO-, CYCLIC SULFITE, EXO-
		6,9-METHANO-2,4,3-BENZODIOXATHIEPIN, 6,7,8,9,10,10-HEXACHLORO-1,5,5A,6,9,9A-HEXAHYDRO-, 3-OXIDE, (3
		ENDOSULFAN 2
		ENDOSULFAN B
		ENDOSULFAN II
		GENERAL WEED KILLER
33382-64-8		
		COPPER ARSENITE
33857-26-0		
		2,7-DICHLORODIBENZO-P-DIOXIN
34099-73-5		
		ETHYL BORATE
34216-34-7		
		CYCLOHEXYLAMINE, TRIMETHYL-
		TRIMETHYLCYCLOHEXYLAMINE
		TRIMETHYLCYCLOHEXYLAMINE (DOT)
		UN 2326 (DOT)
34590-94-8		
		DIPROPYLENE GLYCOL METHYL ETHER
		DIPROPYLENE GLYCOL MONOMETHYL ETHER
		DOWANOL DPM
		PPG-2 METHYL ETHER
		PROPANOL, (2-METHOXYMETHYLETHOXY)-
34731-32-3		
		CARBAMODITHIOIC ACID, 1,2-ETHANEDIYL ESTER
		ETHYLENE BISDITHIOCARBAMATE
34893-92-0		
		DICHLOROPHENYL-ISOCYANATE (1,3-DICHLORO-5-ISOCYANATO-BENZENE)
35089-00-0		
		BERYLLIUM PHOSPHATE
		PHOSPHORIC ACID, BERYLLIUM SALT
35400-43-2		
		BAY-NTN 9306
		BAYER NTN 9306
		BOLSTAR
		HELOTHION
		MERPAFOS
		NTN 9306
		O-ETHYL O-(4-METHYLTHIOPHENYL) S-PROPYL DITHIOPHOSPHATE
		PHOSPHORODITHIOIC ACID, O-ETHYL O-[4-(METHYLTHIO)PHENYL] S-PROPYL ESTER
		SULPROFOS
35860-31-2		
		HEXANITRODIPHENYLAMINE
35860-50-5		
		TRINITROBENZOIC ACID
35860-51-6		
		1,3-BENZENEDIOL, 2,4-DINITRO-
		1,3-BENZENEDIOL, DINITRO-
		2,4-DINITRORESORCINOL
		3-HYDROXY-2,4-DINITROPHENOL
		DINITRORESORCINOL
		DINITRORESORCINOL, WET, WITH AT LEAST 15% WATER
		DINITRORESORCINOL, WETTED WITH, BY WEIGHT, AT LEAST 15% WATER (DOT)
		RESORCINOL, 2,4-DINITRO-
		RESORCINOL, DINITRO-, WETTED WITH AT LEAST 15% WATER
		UN 1322 (DOT)
35884-77-6		
		BENZENE, BROMODIMETHYL-
		BROMURE DE XYLYLE (FRENCH)
		MONOBROMOXYLENE
		UN 1701 (DOT)
		XYLENE, BROMO-
		XYLYL BROMIDE
		XYLYL BROMIDE (DOT)
36478-76-9		
		URANIUM, BIS(NITRATO-O,O')DIOXO-, (OC-6-11)-
	*	URANYL NITRATE
37187-22-7		
		2,4-PENTANEDIONE PEROXIDE, WITH NO MORE THAN 9% BY WEIGHT ACTIVE OXYGEN
		2,4-PENTANEDIONE, PEROXIDE
		ACETYL ACETONE PEROXIDE
		ACETYL ACETONE PEROXIDE SOLUTION WITH NO MORE THAN 9% BY WEIGHT ACTIVE OXYGEN
		ACETYLACETONE PEROXIDE
		ACETYLACETONE PEROXIDE (DOT)
		CHALOXYD AAP-NA 1
		LUPEROX 224
		TRIGONOX 40
		TRIGONOX 44B
		UN 2080 (DOT)
37206-20-5		
		2-PENTANONE, 4-METHYL-, PEROXIDE
		2-PENTANONE, 4-METHYL-, PEROXIDE, WITH NOT MORE THAN 9% BY WEIGHT ACTIVE OXYGEN

CAS No.	IN	Chemical Name
		ISOBUTYL METHYL KETONE PEROXIDE
		ISOBUTYL METHYL KETONE PEROXIDE, NO MORE THAN 62% IN SOLUTION (DOT)
		KAYAMEK B
		METHYL ISOBUTYL KETONE PEROXIDE
		METHYL ISOBUTYL KETONE PEROXIDE (CONCENTRATION NO MORE THAN 60%)
		METHYL ISOBUTYL KETONE PEROXIDE, IN SOLUTION WITH NO MORE THAN 9% BY WEIGHT ACTIVE OXYGEN
		METHYL ISOBUTYL KETONE PEROXIDE, WITH NOT MORE THAN 9% BY WEIGHT ACTIVE OXYGEN (DOT)
		TRIGONOX HM 80
		UN 2126 (DOT)
37211-05-5		
		NICKEL CHLORIDE
37224-57-0		
		BUTTERCUP YELLOW
		CHROMATE(1-), HYDROXYOCTAOXODIZINCATEDI-, POTASSIUM
		CHROMIC ACID, POTASSIUM ZINC SALT (2:2:1)
		CHROMIUM POTASSIUM ZINC OXIDE
		CITRON YELLOW
		POTASSIUM ZINC CHROMATE
		POTASSIUM ZINC CHROMATE HYDROXIDE
		ZINC CHROME
		ZINC POTASSIUM CHROMATE
		ZINC YELLOW
37226-49-6		
		ARSENIC CHLORIDE
37248-34-3		
		POTASSIUM SULFIDE
37264-96-3		
		COBALT CARBONYL
37273-91-9		
		1,3,5,7-TETROXOCANE, 2,4,6,8-TETRAMETHYL-
		ACETALDEHYDE POLYMER
		ACETALDEHYDE, HOMOPOLYMER
		ACETALDEHYDE, TETRAMER
		AGRIMORT
		ARIOTOX
		CEKUMETA
		LUMACRUSK5
		METASON
		PUZOMOR
		UN 1332 (DOT)
37286-64-9		
		DOWFROTH 250
		JEFFOX OL 2700
		NEWPOL LB 625
		POLY(OXY(METHYL-1,2-ETHANEDIYL)), ALPHA-METHYL-OMEGA-HYDROXY-
		POLY(OXYPROPYLENE) MONOMETHYL ETHER
		POLYPROPYLENE GLYCOL METHYL ETHER
		POLYPROPYLENE GLYCOL MONOMETHYL ETHER
		POLY[OXY(METHYL-1,2-ETHANEDIYL)], ALPHA-METHYL-OMEGA-HYDROXY-
		PROPYLENE OXIDE-METHANOL ADDUCT
		UCON LB 1715
37293-14-4		
		BISMUTH TELLURIDE
37340-70-8		
		POTASSIUM PHOSPHOTUNGSTATE
37759-72-1		
		4-FLUOROCROTONIC ACID
39156-41-7		
		1,3-BENZENEDIAMINE, 4-METHOXY-, SULFATE (1:1)
		2,4-DIAMINOANISOLE SULFATE
		2,4-DIAMINOANISOLE SULPHATE
		4-METHOXY-M-PHENYLENEDIAMINE SULFATE
		C.I. 76051
39196-18-4		
		2-BUTANONE, 3,3-DIMETHYL-1-(METHYLTHIO)-, O-((METHYLAMINO)CARBONYL)OXIME
		3,3-DIMETHYL-1-(METHYLTHIO)-2-BUTANONE-O-((METHYLAMINO)CARBONYL)OXIME
		DACAMOX
		DIAMOND SHAMROCK DS-15647
		DS 15647
		ENT 27851
		RCRA WASTE NUMBER P045
		THIOFANOX
		THIOFAX
39349-73-0		
		PERBORATE
		PERBORIC ACID, ION (NEG)
39377-56-5		
		LEAD SULFIDE
39404-03-0		
		MAGNESIUM SILICIDE
39413-47-3		
		SILICIC ACID, BERYLLIUM ZINC SALT
		ZINC BERYLLIUM SILICATE
39630-75-6		
		SILICOFLUORIC ACID
39638-32-9		
		BIS(2-CHLOROISOPROPYL) ETHER
		PROPANE, 2,2'-OXYBIS[2-CHLORO-
39920-37-1		
		DICHLOROPHENYL-ISOCYANATE (1,3-DICHLORO-2-ISOCYANATO-BENZENE)
39974-35-1		
		NITROCHLOROBENZOTRI-FLUORIDE (1-NITRO-3-(TRIFLUOROMETHYL)-2-CHLO
39990-99-3		
		1,2-ETHANEDIAMINE, COMPD WITH LITHIUM ACETYLIDE (Li(C2H))
		LITHIUM ACETYLIDE (Li(C2H)), COMPD WITH 1,2-ETHANEDIAMINE
		LITHIUM ACETYLIDE, COMPLEXED WITH ETHYLENEDIAMINE
		LITHIUM ACETYLIDE-ETHYLENE DIAMINE COMPLEX
		LITHIUM ACETYLIDE-ETHYLENEDIAMINE COMPLEX (DOT)
		NA 2813 (DOT)
40058-87-5		
		ISOPROPYL-2-CHLOROPROPIONATE
40780-64-1		
		ETHYLBUTYL ACETATE
41070-66-9		
		2-BUTENE, 1,1,1,2,3,4,4,4-OCTAFLUORO-
		BUTENE, OCTAFLUORO-
		FC-1318
		NA 2422 (DOT)
		OCTAFLUOROBUT-2-ENE (DOT)
		OCTAFLUOROBUTENE
		OCTAFLUOROBUTENE-2
		PERFLUORO-2-BUTENE (DOT)
		PERFLUOROBUT-2-ENE
		PERFLUOROBUTENE
		UN 2422 (DOT)
41083-11-8		
		1-TRI(CYCLOHEXYL) STANNYL-1H-1,2,4-TRIAZOLE

CAS No.	IN	Chemical Name
41444-43-3		
		SEC-AMYLAMINE
		SEC-PENTANAMINE
41587-36-4		
		ANILINE, CHLORONITRO-
		BENZENAMINE, CHLORONITRO-
		CHLORONITROANILINE
		CHLORONITROANILINES (DOT)
		UN 2237 (DOT)
42195-92-6		
		2,3-DIMETHYLCYCLOHEXYLAMINE
		CYCLOHEXANAMINE, 2,3-DIMETHYL-
		CYCLOHEXYLAMINE, 2,3-DIMETHYL-
42296-74-2		
		HEXADIENE
		HEXADIENE (DOT)
		UN 2458 (DOT)
42350-99-2		
		2-CHLORO-4,6-DI-TERT-AMYL-PHENOL
		PHENOL, 2-CHLORO-4,6-BIS(1,1-DIMETHYLPROPYL)-
42589-07-1		
		2,4,5-T AMINE
43133-95-5		
		METHYL PENTANE
		METHYLPENTANE (DOT)
		PENTANE, METHYL-
		UN 2462 (DOT)
50475-76-8		
	*	LITHIUM ACETYLIDE ETHYLENEDIAMINE
50782-69-9		
		PHOSPHONOTHIOIC ACID, METHYL-, S-(2-(BIS(METHYL ETHYL)AMINO)ETHYL)O-ETHYL ETHER
50810-25-8		
		ALUMINUM SILICON (AlSi5)
51004-61-6		
		AF 2
		AF 2 (FOAMING AGENT)
51023-22-4		
		BUTENE, TRICHLORO-
		TRICHLORBUTYLENE
	*	TRICHLOROBUTENE
		TRICHLOROBUTENE (DOT)
		UN 2322 (DOT)
51289-10-2		
		CHLORINATED DIPHENYL OXIDE
51325-42-9		
		COPPER SELENITE
		COPPER SELENITE (Cu(SeO3))
		COPPER(II) SELENITE
		CUPRIC SELENITE
		SELENIOUS ACID (H2SeO3), COPPER SALT
		SELENIOUS ACID (H2SeO3), COPPER(2+) SALT (1:1)
		SELENIOUS ACID, COPPER SALT
		SELENIOUS ACID, COPPER(2+) SALT (1:1)
51580-86-0		
		1,3,5-TRIAZINE-2,4,6(1H,3H,5H)-TRIONE, 1,3-DICHLORO-, SODIUM SALT, DIHYDRATE
		SODIUM CHLOROCYANURATE DIHYDRATE
		SODIUM DICHLOROISOCYANURATE DIHYDRATE
51787-44-1		
		5,7,8,9-TETRAMETHYL-3,4-BENZOACRIDINE
		BENZ[C]ACRIDINE, 7,8,9,11-TETRAMETHYL-
51810-70-9		
		ZINC PHOSPHIDE
51845-86-4		
		BORIC ACID, ETHYL ESTER
		ETHYL BORATE
		ETHYL BORATE (DOT)
		UN 1176 (DOT)
52061-60-6		
		PARAMENTHANE HYDRO-PEROXIDE
52181-51-8		
		CHLOROBENZOTRIFLUORIDE
52326-66-6		
		DISTEARYL PEROXYDICARBONATE
		DISTEARYL PEROXYDICARBONATE, NOT MORE THAN 85% WITH STEARYL ALCOHOL
		DISTEARYL PEROXYDICARBONATE, NOT MORE THAN 85% WITH STEARYL ALCOHOL (DOT)
		PEROXYDICARBONIC ACID, DIOCTADECYL ESTER
		PEROXYDICARBONIC ACID, DIOCTADECYL ESTER, NOT MORE THAN 85% WITH STEARYL ALCOHOL
		UN 2592 (DOT)
52470-25-4		
		GUANIDINE NITRATE
		GUANIDINE NITRATE (1:1)
		GUANIDINE NITRATE (DOT)
		GUANIDINE, MONONITRATE
		GUANIDINIUM NITRATE
		UN 1467 (DOT)
52628-25-8		
		AMMONIUM ZINC CHLORIDE
		HIGH SPEED
		KLEANROL
		NA 9154 (DOT)
		ZACLON
		ZINC AMMONIUM CHLORIDE
		ZINC AMMONIUM CHLORIDE (DOT)
52740-16-6		
		ARSENIOUS ACID, CALCIUM SALT
		ARSENOUS ACID, CALCIUM SALT
		ARSONIC ACID, CALCIUM SALT (1:1)
		CALCIUM ARSENITE
		CALCIUM ARSENITE (CaHAsO3)
		CALCIUM ARSENITE, SOLID
		CALCIUM ARSENITE, SOLID (DOT)
		MONOCALCIUM ARSENITE
		NA 1574 (DOT)
53014-37-2		
		TETRANITRO-ANILINE
53095-76-4		
		LITHIUM SILICON
53220-22-7		
		DIMYRISTYL PEROXYDICARBONATE
		DIMYRISTYL PEROXYDICARBONATE, NOT MORE THAN 22%, STABLE DISPERSION, IN WATER
		DIMYRISTYL PEROXYDICARBONATE, NOT MORE THAN 42%, IN WATER
		DIMYRISTYL PEROXYDICARBONATE, NOT MORE THAN 22% IN WATER (DOT)
		DIMYRISTYL PEROXYDICARBONATE, TECHNICALLY PURE (DOT)
		DITETRADECYL PEROXYDICARBONATE
		PEROXYDICARBONIC ACID, DITETRADECYL ESTER
		PEROXYDICARBONIC ACID, DITETRADECYL ESTER, NOT MORE THAN 22% IN WATER
		UN 2595 (DOT)
		UN 2892 (DOT)

CAS No. *IN* *Chemical Name*

53324-05-3

4-(DIMETHYLAMINO)NITROSOBENZENE
4-NITROSODIMETHYLANILINE
ACCELERINE
ANILINE, N,N-DIMETHYL-P-NITROSO-
ANILINE, N,N-DIMETHYLNITROSO-
BENZENAMINE, N,N-DIMETHYL-4-NITROSO- (9CI)
BENZENAMINE, N,N-DIMETHYLNITROSO-
DIMETHYL(P-NITROSOPHENYL)AMINE
DIMETHYL-P-NITROSOANILINE (DOT)
N,N-DIMETHYL-P-NITROSOANILINE
N,N-DIMETHYLNITROSOANILINE
NCI-C01821
NDMA
NITROSODIMETHYLANILINE
P-(DIMETHYLAMINO)NITROSOBENZENE
P-NITROSO-N,N-DIMETHYLANILINE
P-NITROSODIMETHYLANILINE (DOT)
PARANITROSODIMETHYLANILIDE
ULTRA BRILLIANT BLUE P
UN 1369 (DOT)

53408-91-6

MERCURY THIOCYANATE

53449-21-9

CHLORODIPHENYL (42% CHLORINE)

53467-11-1

POLY[OXY(METHYL-1,2-ETHANEDIYL)], ALPHA-[(2,4-DICHLOROPHENOXY)ACETYL]-OMEGA-BUTOXY-

53469-21-9

AROCLOR 1242
PCB-1242 (CHLORODIPHENYL (42% Cl))

53496-15-4

1-METHYLBUTYL ACETATE
2-ACETOXYPENTANE
2-PENTANOL, ACETATE (8CI, 9CI)
2-PENTYL ACETATE
ACETIC ACID, 2-PENTYL ESTER
ACETIC ACID, SEC-PENTYL ESTER
SEC-AMYL ACETATE
SEC-AMYL ACETATE (ACGIH,DOT)
SEC-PENTYL ACETATE
UN 1104 (DOT)

53558-25-1

1-(3-PYRIDYLMETHYL)-3-(4-NITROPHENYL)UREA
DLP 787
DLP-87
N-(4-NITROPHENYL)-N'-(3-PYRIDINYLMETHYL)UREA
N-3-PYRIDYLMETHYL-N'-P-NITROPHENYLUREA
PYRIMINIL
PYRIMINYL
PYRINURON
RH 787
UREA, 1-NITROPHENYL-3-(3-PYRIDYLMETHYL)-
UREA, N-(4-NITROPHENYL)-N'-(3-PYRIDINYLMETHYL)-
VACOR

53569-62-3

CARBON DIOXIDE MIXED WITH NITROUS OXIDE
CARBON DIOXIDE, MIXT WITH NITROGEN OXIDE (N2O)
CARBON DIOXIDE, MIXTURE WITH NITROGEN OXIDE (N2O) (9CI)
CARBON DIOXIDE-NITROGEN OXIDE (N2O) MIXTURE
CARBON DIOXIDE-NITROUS OXIDE MIXTURE
CARBON DIOXIDE-NITROUS OXIDE MIXTURE (DOT)
NITROGEN OXIDE (N2O), MIXT CONTG
UN 1015 (DOT)

53799-46-5

CROCIDOLITE
CROCIDOLITE (Fe5Na2(SiO3)8)

CAS No. *IN* *Chemical Name*

53894-28-3

AF 2
AF 2 (CROSS-LINKING AGENT)
PHENOL, [[(2-AMINOETHYL)AMINO]METHYL]-

54363-49-4

METHYLPENTADIENE

54497-43-7

DISELENIOUS ACID, ZINC SALT (1:1)
ZINC SELENITE

54566-73-3

BORON OXIDE

54579-28-1

C.I. DIRECT ORANGE 1

54590-52-2

2-PROPANOL, 1-AMINO-, 4-DODECYLBENZENESULFONATE (SALT)
BENZENESULFONIC ACID, 4-DODECYL-, COMPD WITH 1-AMINO-2-PROPANOL (1:1)
CONCO AAS SPECIAL 3
ISOPROPANOLAMINE DODECYLBENZENESULFONATE
ISOPROPANOLAMINE DODECYLBENZENESULFONATE (DOT)
MIPA-DODECYLBENZENESULFONATE
NA 9127 (DOT)
P-DODECYLBENZENESULFONIC ACID 2-HYDROXY-1-PROPYLAMINE SALT

54693-46-8

2-PENTANONE, 4-HYDROXY-4-METHYL-, PEROXIDE, MORE THAN 57% IN SOLUTION
2-PENTANONE, 4-HYDROXY-4-METHYL-, PEROXIDE
2-PENTANONE, 4-HYDROXY-4-METHYL-, PEROXIDE, NOT MORE THAN 57% IN SOLUTION
DIACETONE ALCOHOL PEROXIDE
DIACETONE ALCOHOL PEROXIDE, MORE THAN 57% IN SOLUTION WITH NOT LESS THAN 9% HYDROGEN PEROXIDE, LESS THAN 26% DIACETONE ALCOHOL
DIACETONE ALCOHOL PEROXIDE, MORE THAN 57% IN SOLUTION (DOT)
DIACETONE ALCOHOL PEROXIDE, NOT MORE THAN 57% IN SOLUTION (DOT)
UN 2163 (DOT)

54972-97-3

1,3-DIMETHYL BUTANOL
1,3-DIMETHYL-1-BUTANOL
1-PENTANOL, 2-METHYL-
1-PENTANOL, METHYL-
2-METHYL-1-PENTANOL
2-METHYL-2-PROPYLETHANOL
2-METHYL-4-PENTANOL
2-METHYLPENTANOL-1
2-PENTANOL, 4-METHYL-
3-MIC
4-METHYL-2-PENTANOL
4-METHYL-2-PENTYL ALCOHOL
4-METHYLPENTANOL-2
4-METILPENTAN-2-OLO (ITALIAN)
ALCOOL METHYL AMYLIQUE (FRENCH)
AMYL METHYL ALCOHOL
ISOBUTYLMETHYLCARBINOL
ISOBUTYLMETHYLMETHANOL
ISOHEXYL ALCOHOL
ISOPROPYL DIMETHYL CARBINOL
MAOH
METHYL AMYL ALCOHOL
METHYL ISOBUTYL CARBINOL
METHYL ISOBUTYL CARBINOL (ACGIH,DOT)
METHYL-1-PENTANOL
METHYLAMYL ALCOHOL
METHYLISOBUTYL CARBINOL
METILAMIL ALCOHOL (ITALIAN)
MIBC
MIC
UN 2053 (DOT)

CAS No.	IN	Chemical Name
55720-99-5		
		CHLORINATED DIPHENYL OXIDE
		CHLORODIPHENYL OXIDE
		HEXACHLORO DIPHENYL OXIDE
55738-54-0		
		METHANIMIDAMIDE, N,N-DIMETHYL-N'-[5-[2-(5-NITRO-2-FURANYL)ETHENYL]-1,3,4-OXADIAZOL-2-YL]-,
		METHANIMIDAMIDE, N,N-DIMETHYL-N'-[5-[2-(5-NITRO-2-FURANYL)ETHENYL]-1,3,4-OXADIAZOL-2-YL]-, (E)
		TRANS-2((DIMETHYLAMINO) METHYLIMINO)-5-(2-(5-NITRO-2-FURYL)VINYL)-1,3,4-OXADIAZOLE
55810-17-8		
		TRINITRONAPHTHALENE
56093-45-9		
		SELENIUM SULFIDE
56189-09-4		
		DIBASIC LEAD STEARATE
		LEAD, BIS(OCTADECANOATO)DIOXODI-
		LISTAB 51
		OCTADECANOIC ACID, LEAD COMPLEX
		OCTADECANOIC ACID, LEAD SALT, DIBASIC
56320-22-0		
	*	ARSENIC DISULFIDE
57109-90-7		
		CLORAZEPATE DIPOTASSIUM
57308-10-8		
		CALCIUM DIHYDRIDE
	*	CALCIUM HYDRIDE
		CALCIUM HYDRIDE (CaH2)
		CALCIUM HYDRIDE (DOT)
		UN 1404 (DOT)
57485-31-1		
		ALUMINUM SILICON (AlSi2)
57608-40-9		
		AMMONIUM NITRATE MIXED WITH AMMONIUM PHOSPHATE
		AMMONIUM NITRATE PHOSPHATE
		AMMONIUM NITRATE-PHOSPHATE (DOT)
		AMMONIUM PHOSPHATE, MIXED WITH AMMONIUM NITRATE
		UN 2070 (DOT)
58164-88-8		
		ANTIMONY LACTATE
58270-08-9		
		ZINC, DICHLORO[4,4-DIMETHYL-5-(((METHYLAMINO)CARBONYL)OXY)IMINO)PENTANE NITRILE]
58449-37-9		
		ISONONANOYL PEROXIDE
58500-38-2		
		BERYLLIUM SILICATE
		SILICIC ACID, BERYLLIUM SALT
58933-55-4		
		HELIUM-OXYGEN (MIXTURE)
		HELIUM-OXYGEN MIXTURE (DOT)
		NA URE
59355-75-8		
		MAPP
	*	MAPPGAS
		METHYL ACETYLENE AND PROPADIENE MIXTURE, STABILIZED (DOT)
		METHYL ACETYLENE-PROPADIENE MIXTURE
		METHYL ACETYLENE-PROPADIENE MIXTURE (ACGIH)
		METHYL ACETYLENE-PROPADIENE MIXTURE (MAPP)
		METHYLACETYLENE-PROPADIENE, STABILIZED
		METHYLACETYLENE-PROPADIENE, STABILIZED (DOT)
		PROPYNE, MIXED WITH PROPADIENE
		UN 1060 (DOT)
59536-65-1		
		FIREMASTER BP 6
		PBB
		POLYBROMINATED BIPHENYLS
		POLYBROMINATED BIPHENYLS (PBBs)
60475-66-3		
		AUER METAL
		CERIUM ALLOY, BASE, (FERROCERIUM)
		FERROCERIUM
		FERROCERIUM (DOT)
		IRON ALLOY, BASE, (FERROCERIUM)
		PYROPHOROUS ALLOY
		SPARK METAL
		UN 1323 (DOT)
60616-74-2		
		MAGNESIUM HYDRIDE
		MAGNESIUM HYDRIDE (DOT)
		UN 2010 (DOT)
60646-36-8		
		ARSENIC TRICHLORIDE (ARSENIC)
60676-86-0		
	*	SILICA, AMORPHOUS, FUSED
		SILICA, FUSED, DUST
61105-31-5		
		CROCIDOLITE
		CROCIDOLITE (Fe2Mg3Na2(SiO3)8)
61215-72-3		
		1,2,3,6-TETRAHYDROPYRIDINE
		1,2,3,6-TETRAHYDROPYRIDINE (DOT)
		1,2,5,6-TETRAHYDROPYRIDINE
		DELTA(SUP 3)-PIPERIDINE
		PYRIDINE, 1,2,3,6-TETRAHYDRO-
		PYRIDINE, TETRAHYDRO-
		TETRAHYDROPYRIDINE
		UN 2410 (DOT)
61702-44-1		
		2-CHLORO-P-PHENYLENEDIAMINE SULFATE
61703-05-7		
		C.I. DIRECT BLACK 114
61789-51-3		
	*	COBALT NAPHTHENATE
		COBALT NAPHTHENATE, POWDER
		COBALT NAPHTHENATE, POWDER (DOT)
		COBALT NAPHTHENATES
		NAPHTENATE DE COBALT (FRENCH)
		NAPHTHENIC ACID, COBALT SALT
		NAPHTHENIC ACIDS, COBALT SALTS
		UN 2001 (DOT)
61789-65-9		
		ALUMINUM RESINATE
61790-53-2		
	*	DIATOMACEOUS EARTH
61878-61-3		
		BENZENE, CHLOROMETHYLNITRO-
		BENZENE, METHYL-, MONOCHLORO MONONITRO DERIV
		CHLORO-O-NITROTOLUENE (DOT)
		CHLORONITROTOLUENE
		TOLUENE, CHLORO-O-NITRO-
		TOLUENE, CHLORONITRO-
		UN 2433 (DOT)

CAS No.	IN	Chemical Name
62207-76-5		
		BIS(3-FLUOROSALICYLAL)ETHYLENEDIAMINECOBALT(II)
		COBALT(II), N,N'-ETHYLENEBIS(3-FLUOROSALIĊYLIDE-NEIMINATO)-
		COBALT, ((2,2'-(1,2-ETHANEDIYLBIS(NITRILOMETHYLI-DENE)BIS(PHENALOTO))(2-)-N,N',O,O')-
		COBALT, BIS(3-FLUOROSALICYLALDEHYDE)-ETHYLENE-DIIMINE-
		COBALT, [[2,2'-[1,2-ETHANEDIYLBIS(NITRILOMETHYLI-DYNE)]BIS[6-FLUOROPHENOLATO]](2-)-N,N',O,O']-, (SP-4
		FLUOMIN
		FLUOMINE
		FLUOMINE DUST
		N,N'-ETHYLENEBIS(3-FLUOROSALICYLIDENEIMINATO)CO-BALT (II)
		PHENOL, 2,2'-[1,2-ETHANEDIYLBIS(NITRILOMETHYLI-DYNE)]BIS[6-FLUORO-, COBALT COMPLEX
62450-06-0		
		3-AMINO-1,4-DIMETHYL-5H-PYRIDO[4,3-B]INDOLE
		5H-PYRIDO[4,3-B]INDOL-3-AMINE, 1,4-DIMETHYL-
		TRP-P 1
		TRYPTOPHAN P 1
62450-07-1		
		1-METHYL-3-AMINO-5H-PYRIDO[4,3-B]INDOLE
		3-AMINO-1-METHYL-5H-PYRIDO[4,3-B]INDOLE
		5H-PYRIDO[4,3-B]INDOL-3-AMINE, 1-METHYL-
		TRP-P 2
		TRYPTOPHAN P 2
63283-80-7		
		(2-CHLORO-1-METHYLETHYL) ETHER
		2,2'-DICHLOROISOPROPYL ETHER
		BIS(2-CHLORO-1-METHYLETHYL) ETHER
		BIS(2-CHLOROISOPROPYL) ETHER
		DICHLORODIISOPROPYL ETHER
		DICHLOROISOPROPYL ETHER
		DICHLOROISOPROPYL ETHER (DOT)
		ETHER, BIS(2-CHLORO-1- METHYLETHYL)
		NCI-C50044
		PROPANE, 2,2'-OXYBIS[DICHLORO-
		RCRA WASTE NUMBER U027
		UN 2490 (DOT)
63597-41-1		
		OCTADIENE
		OCTADIENE (DOT)
		UN 2309 (DOT)
63868-82-6		
		PICRAMIC ACID, ZIRCONIUM SALT (WET)
		UN 1517 (DOT)
		ZIRCONIUM PICRAMATE
		ZIRCONIUM PICRAMATE, WET (WITH AT LEAST 20% WATER) (DOT)
		ZIRCONIUM PICRAMATE, WET, WITH AT LEAST 20% OF WATER
63885-01-8		
		NITROUS ACID, AMMONIUM ZINC SALT (3:1:1)
		UN 1512 (DOT)
		ZINC AMMONIUM NITRITE
		ZINC AMMONIUM NITRITE (DOT)
63917-41-9		
		DIMETHYL PHOSPHORAMIDOCYANIDIC ACID
63937-14-4		
		D-GLUCONIC ACID, MERCURY COMPLEX
		MERCUROUS GLUCONATE
		MERCUROUS GLUCONATE, SOLID
		MERCUROUS GLUCONATE, SOLID (DOT)
		MERCURY GLUCONATE
		MERCURY GLUCONATE (DOT)
		MERCURY(I) GLUCONATE
		MERCURY, (D-GLUCONATO)-
		UN 1637 (DOT)
63938-10-3		
		CHLOROTETRAFLUOROETHANE
		CHLOROTETRAFLUOROETHANE (DOT)
		ETHANE, CHLOROTETRAFLUORO-
		MONOCHLOROTETRAFLUOROETHANE
		MONOCHLOROTETRAFLUOROETHANE (DOT)
		TETRAFLUOROCHLOROETHANE
		TETRAFLUOROMONOCHLOROETHANE
		UN 1021 (DOT)
63989-69-5		
		FERRIC ARSENITE
		FERRIC ARSENITE, BASIC
		FERRIC ARSENITE, SOLID
		FERRIC ARSENITE, SOLID (DOT)
		IRON ARSENITE OXIDE (Fe2(AsO3)2O3), PENTAHYDRATE
		IRON(III) O-ARSENITE PENTAHYDRATE
		UN 1607 (DOT)
64037-54-3		
		1-BUTENE, 3,4-DICHLORO-, (.+-.)-
		3,4-DICHLOROBUTENE-1
64083-59-6		
		C.I. DIRECT ORANGE 8
64093-79-4		
		NEOCHROMIUM
64741-44-2		
		GAS OIL
64742-24-1		
		ACID SLUDGE
		SLUDGE ACID
64973-06-4		
		ARSENIC BROMIDE
		ARSENIC BROMIDE (DOT)
		ARSENIC BROMIDE, SOLID
		ARSENIC BROMIDE, SOLID (DOT)
		ARSENIC TRIBROMIDE
		ARSENIC(II) BROMIDE
		ARSENOUS BROMIDE
		ARSENOUS TRIBROMIDE
		TRIBROMOARSINE
		UN 1555 (DOT)
65996-68-1		
		BLAST FURNACE GAS (FERROUS METAL)
		FLUE GASES, FERROUS METAL, BLAST FURNACE
		GAS BLAST FURNACE
65996-79-4		
		COAL TAR NAPHTHA
		NAPHTHA
65996-89-6		
		TAR, COAL, HIGH-TEMP
65996-90-9		
		TAR, COAL, LOW -TEMP
65996-91-0		
		COAL TAR LIGHT OIL
		COAL TAR UPPER DISTILLATE
		DISTILLATES (COAL TAR), UPPER
65996-92-1		
		COAL TAR DISTILLATE, COMBUSTIBLE, LIQUID
		COAL TAR DISTILLATE, FLAMMABLE, LIQUID
65997-15-1		
		CEMENT, PORTLAND, CHEMICALS
		SILICATE, PORTLAND CEMENT
67711-90-4		
		FLUE DUST, POISONOUS

CAS No.	IN	Chemical Name
67774-32-7		POLYBROMINATED BIPHENYLS
68308-34-9		SHALE OIL
68334-28-1		HG 150
		HYDROGENATED MIXED VEGETABLE OILS
		OILS, VEGETABLE, HYDROGENATED
68425-31-0		GASOLINE (NATURAL GAS), NATURAL
68476-26-6		FUEL GAS, GENERATOR
		FUEL GASES
		FUEL GASES, GENERATOR
		FUEL GASES, MANUFD.
		FUELS, GAS
		GASES, FUEL
		GENERATOR GAS
		MANUFACTURED GAS
		MANUFD. FUEL GASES
		METHANE NUMBER
		OIL GAS
		WOBBE INDEX
68476-85-7		BURSHANE
		FUELS, LIQUEFIED PETROLEUM GAS
	*	LIQUEFIED PETROLEUM GAS
		LIQUEFIED PETROLEUM GAS (DOT)
		LIQUEFIED PETROLEUM GASES
		LIQUIFIED PETROLEUM GAS (L.P.G)
		LPG
		PETROLEUM GAS LIQUEFIED
		PETROLEUM GAS, LIQUEFIED (DOT)
		PETROLEUM GASES, LIQUEFIED
		PETROLEUM PRODUCTS, LIQUEFIED GASES
		PROPANE MIXTURES
		PYROFAX
		UN 1075 (DOT)
68848-64-6		LITHIUM ALLOY, NONBASE, Li,Si
		LITHIUM SILICON
		LITHIUM SILICON (DOT)
		SILICON ALLOY, NONBASE, Li,Si
		UN 1417 (DOT)
68956-56-9		TERPENE HYDROCARBONS N.O.S.
68956-82-1		COBALT RESINATE, PRECIPITATED
		COBALT RESINATE, PRECIPITATED (DOT)
		RESIN ACIDS AND ROSIN ACIDS, COBALT SALTS
		UN 1318 (DOT)
68975-47-3		ISOHEPTENE
69523-06-4		AUER METAL
		CERIUM ALLOY, BASE, (FERROCERIUM)
		FERROCERIUM
		FERROCERIUM (DOT)
		IRON ALLOY, BASE, (FERROCERIUM)
		PYROPHOROUS ALLOY
		SPARK METAL
		UN 1323 (DOT)
70027-50-8		COPPER SELENATE
		SELENIC ACID (H2SeO4), COPPER SALT
		SELENIC ACID, COPPER SALT
70042-58-9		BUTYLCYCLOHEXYCHLOROFORMATE
		CARBONOCHLORIDIC ACID, (1,1-DIMETHYLETHYL)CYCLOHEXYL ESTER
		TERT-BUTYLCYCLOHEXYL CHLOROFORMATE
		TERT-BUTYLCYCLOHEXYLCHLOROFORMATE (DOT)
		UN 2747 (DOT)
70399-13-2		LITHIUM FERROSILICON
71000-82-3		ISOCYANATE
		ISOCYANATE ION(1-)
		ISOCYANATES AND SOLUTIONS, [FLAMMABLE LIQUID AND POISONOUS LIQUID LABELS]
		ISOCYANATES AND SOLUTIONS, [FLAMMABLE LIQUID LABEL]
		ISOCYANATES, BP MORE THAN 300 C
		ISOCYANATES, [POISON B LABEL]
71121-36-3		ISOCYANATOBENZOTRI-FLUORIDE
73090-68-3		NAPHTHALENE, (1,1-DIMETHYLETHYL)-1,2,3,4-TETRAHYDRO-
		TERT-BUTYL TETRALIN
73090-69-4		CHLORO-4-TERT-AMYLPHENOL
		PHENOL, CHLORO-4-(1,1-DIMETHYLPROPYL)-
73513-30-1		METHYLPENTALDEHYDE
		PENTANAL, METHYL-
73826-29-6		TRICHLOROPHENOXYPROPIONIC ACID ESTER
75364-04-4		CHRYSOTILE
		SILICIC ACID (H6Si2O7), COBALT(2+) MAGNESIUM SALT (1:2:1)
77049-67-7		ZIRCONIUM HYDRIDE
77536-67-5	*	ASBESTOS, ANTHOPHYLLITE
79435-04-4		2-CHLOROPROPIONIC ACID ISOPROPYL ESTER
		ISOPROPYL (+)-2-CHLOROPROPIONATE
		ISOPROPYL 2-CHLOROPROPIONATE (DOT)
		ISOPROPYL CHLOROPROPIONATE
		PROPANOIC ACID, 2-CHLORO-, 1-METHYLETHYL ESTER, (R)-
		PROPIONIC ACID, 2-CHLORO-, ISOPROPYL ESTER
		UN 2934 (DOT)
79620-93-2		POLYCHLOR
		POLYCHLOR AGRICULTURAL FUNGICIDES
79869-58-2		1-PROPANETHIOL (DOT)
		3-MERCAPTOPROPANOL
		N-PROPYL MERCAPTAN
		PROPANE-1-THIOL
		PROPANETHIOL
		PROPYL MERCAPTAN
		PROPYL MERCAPTAN (DOT)
		UN 2402 (DOT)
		UN 2704 (DOT)
80052-41-3		WHITE SPIRITS

CAS No.	IN	Chemical Name
80466-34-8		
		2,4-HEXADIENAL
81228-87-7		
		CARBONOCHLORIDIC ACID, CYCLOBUTYL ESTER
		CYCLOBUTYL CHLOROFORMATE
		CYCLOBUTYLCHLOROFORMATE (DOT)
		UN 2744 (DOT)
81624-04-6		
	*	HEPTENE
81624-06-8		
		HEXENE
85873-97-8		
		HYDROPEROXIDE, 2,6,6-TRIMETHYLBICYCLO(311)HEPTYL-, NOT OVER 45% PEROXIDE
		HYDROPEROXIDE, 2,6,6-TRIMETHYLBICYCLO[311]HEPT-2-YL, (1ALPHA,2ALPHA,5ALPHA)-
		HYDROPEROXIDE, 2,6,6-TRIMETHYLBICYCLO[311]HEPTYL
		PINANE HYDROPEROXIDE
		PINANE HYDROPEROXIDE (DOT)
		PINANE HYDROPEROXIDE SOLUTION, WITH NOT MORE THAN 45% PEROXIDE
		PINANYL HYDROPEROXIDE
		UN 2162 (DOT)
86290-81-5		
		BENZINE (MOTOR FUEL)
		FUELS, GASOLINE
		GASOLINE
		GASOLINE, SYNTHETIC
		HERBICIDE ES
		HYDROFINING
		METAFORMING
		MOTOR FUELS
		NATURAL GAS CONDENSATES, GASOLINE
		PETROL
		PETROL, SYNTHETIC
		SYNFUELS
		SYNTHETIC GASOLINE
91681-63-9		
		4H-1-BENZOPYRAN-4-ONE, 6-(2,3-DIHYDROXY-3-METHYL-BUTYL)-3-(2,4-DIHYDROXYPHENYL)-7-HYDROXY-
		AF 2
		BC 3
91724-16-2		
		ARSENIOUS ACID, STRONTIUM SALT
		STRONTIUM ARSENITE
		STRONTIUM ARSENITE (DOT)
		STRONTIUM ARSENITE (Sr(As2O4))
		STRONTIUM ARSENITE, SOLID
		STRONTIUM ARSENITE, SOLID (DOT)
		UN 1691 (DOT)

SERIES DESCRIPTION

CHEMICAL AND ENVIRONMENTAL SAFETY AND HEALTH IN SCHOOLS AND COLLEGES

Comprehensive Hazard Communication and Right-to-Know Compliance Publications

Now being compiled and written by The Forum for Scientific Excellence, Inc., this series is designed to facilitate school and college compliance with federal and state hazardous chemical regulations and to provide required employee training and understanding. The information contained in the series is the result of the author's extensive and ongoing consulting experience in this field.

These publications cover essentially every aspect of the OSHA Standard and individual state Right-to-Know laws, and encompass the following:

1. Policies, procedures, and administration
2. Employee training and communication
3. Hazardous chemical and product information

Policies, Procedures and Administrative References

Compact School and College Administrator's Guide for Compliance with Federal and State Right-to-Know Regulations

Translates the "legalese" and technical language of current hazardous chemical standards and regulations into meaningful, easily understood terminology. Provides a simplified, step-by-step program to assure that an institution or system fulfills all compliance requirements. Valuable introduction for all concerned with developing a workable, effective program.

Written Hazard Communication Program for Schools and Colleges

Identifies and describes the objectives an educational institution must achieve to assure that its hazardous chemical compliance program is appropriate and adequate. Includes specific examples of the written program requirements, including the MSDS format and hazard assessment criteria, as well as sample letters. Essential for the administrator assigned responsibility for the compliance program. Practical ring binding allows individualized inclusions of additional local information specific to the site.

Employee Training and Communication Manuals—Five-Year Program

The following texts are designed for quantity distribution and use in fulfilling the ongoing training requirement in the OSHA Standard and various state regulations. They are useful for formal in-house training programs and seminars and also are effective individual self-study tools. As a five-year program the order of presentation suggested may be changed to fit local needs, although initial use of *Volume 1: Basic Principles* is strongly recommended. Additionally, since the workforce rarely is static, concurrent use of different texts with employee subgroups is a natural outcome. All five texts fulfill federal and state ongoing training requirements. Special discounts are available on quantity orders.

Concise Manuals of Chemical and Environmental Safety in Schools and Colleges

Volume 1: Basic Principles

A comprehensive primer that provides individual employees the basic definitions, concepts, and methods for dealing with hazardous materials safely. Reviews federal and state standards and regulations and discusses MSDSs, various definitions, labeling, occupational medicine and industrial hygiene, protective equipment, and proper work practices and procedures.

Volume 2: Hazardous Chemical Classes

A second-year training and communication resource for individual employees. Expands hazardous chemical information into classes of chemical hazards, including routes of absorption, flammables, corrosives, poisons, oxidizers and reactives/explosives, radioactive materials, carcinogens, mutagens, and teratogens, as well as general human and environmental safety. Gives emergency treatment procedures for groups of hazardous chemicals within each class.

Volume 3: Chemical Interactions

Third-year training and communication text module. Describes serious, "hidden" hazards and incidents that can arise when two or more chemicals interact inappropriately. Outlines mechanisms for prevention of such unnecessary hazards.

Volume 4: Safe Chemical Storage

An essential, ready reference that identifies procedures for safely storing chemicals and products in a variety of educational areas, including laboratories, art and graphics areas, industrial and vocational shops, and custodial and storage areas. Describes severe dangers created by storing chemicals in alphabetical order. Lists chemicals that should not be present and those that become more hazardous with age.

Volume 5: Safe Chemical Disposal

A detailed description of proper procedures for safely disposing of various classes of hazardous chemicals and products in educational institutions. Identifies safe interim storage procedures and potentially hazardous chemical interactions that must be avoided. Describes mechanisms to minimize costs and assure proper regulatory compliance. Fulfills OSHA, EPA, and many state requirements

Pocket Guides to Chemical and Environmental Safety in Schools and Colleges

Five condensed, portable "field" guides with the same volume numbers and titles as the *Concise Manuals* above. Each includes essential information and checklists found in the *Concise Manuals* but in abbreviated form with less theory and examples. Designed for quantity distribution and use; handy pocket size; quantity discounts also are available.

Hazardous Chemical and Product Information

Handbook of Chemical and Environmental Safety in Schools and Colleges

A comprehensive master reference containing all of the information presented in the five *Concise Manuals* and *Pocket Guides* described above, as well as additional regulatory data and checklists. Includes basic chemical safety and health principles, hazardous chemical classes, hazardous interactions, and safe storage and disposal procedures. A "total" sourcebook for anyone wanting substantial treatment of principles, techniques, and legal requirements.

Compendium of Hazardous Chemicals in Schools and Colleges

Encyclopedic coverage of more than 900 hazardous chemicals commonly found in schools and colleges. Includes all data necessary for properly identifying and defining the acute and chronic health hazards of each chemical, as well as fire, explosion, environmental, and special risks such as radiation or oxidation for each of the substances. Also includes actual hazard assessments derived from published hazard assessment criteria, as well as labeling data. The information fulfills all aspects of federal and state data requirements and also includes additional nonrequired information that has been found useful in managing effective programs. Provides a basis for evaluating the accuracy of supplier MSDS data, regulations, and employee concerns.

Index of Hazardous Contents of Commercial Products in Schools and Colleges

Lists hazardous components found in nearly 10,000 commercial products used in educational facilities, and includes specific component chemical identities and, whenever possible, the percentage of the total content. When used in conjunction with the *Compendium* above and the *Cross-Reference Index of Hazardous Chemicals, Synonyms, and CAS Registry Numbers,* this work provides the data needed for safe practices and compliance with the OSHA Standard and state Right-to-Know laws.

List of Lists of Worldwide Hazardous Chemicals and Pollutants

The most extensive compilation of regulated hazardous chemicals and environmental pollutants available. Includes separate lists of substances regulated by more than 25 state, federal, and international agencies, along with many of their selection criteria. Also includes a master list in both alphabetical and Chemical Abstract Service (CAS) Registry Number sequence. The first place to look to see if a substance is regulated in any major area or country.

Cross-Reference Index of Hazardous Chemicals, Synonyms, and CAS Registry Numbers

An extensive cross-reference listing of more than 40,000 synonyms for the hazardous chemicals and environmental pollutants identified in the worldwide *List of Lists* above. Represents the most comprehensive source available for properly identifying common names, chemical names, and trade names associated with those regulated chemicals. Provides a ready reference for identifying the CAS numbers for common or product names that "hide" their true chemical identity.

For ordering information, prices, and delivery please contact:

J.B. Lippincott Company
East Washington Square
Philadelphia, PA 19105

To discuss or suggest editorial content please contact:

The Forum for Scientific Excellence, Inc.
200 Woodport Road
Sparta, NJ 07871